A Dictionary of Natural Products

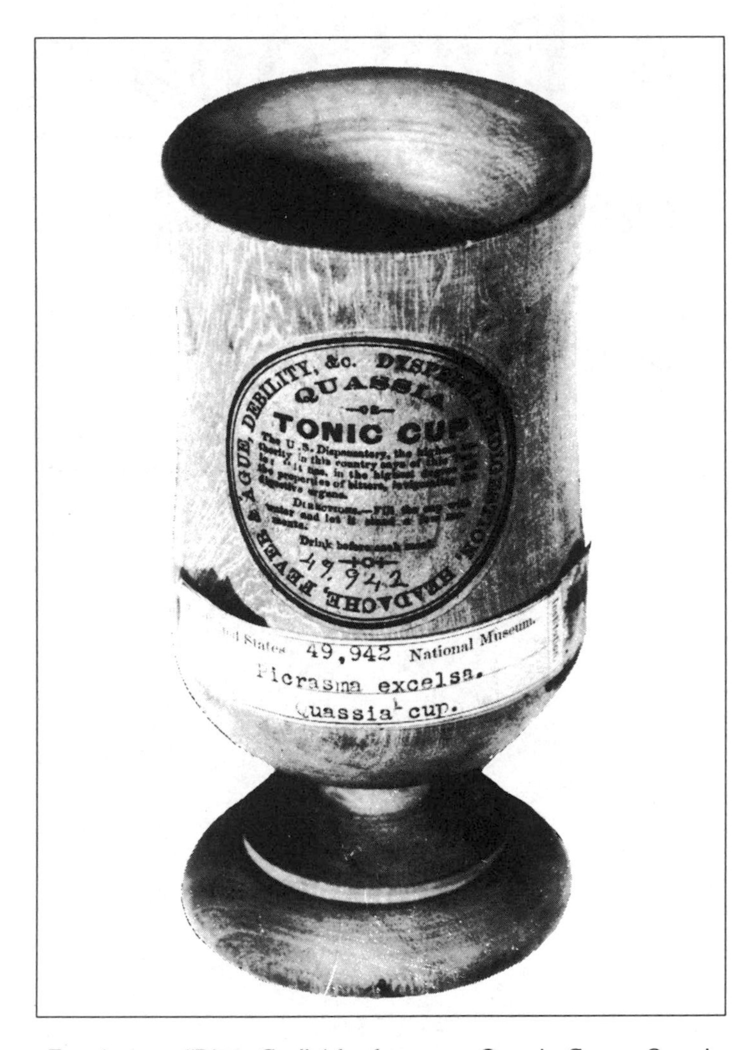

Frontispiece: "Bitter Cup" (also known as Quassia Cup or Quassia Goblet) made by turning quassia billets on a lathe (see text, page 114). (Courtesy of Division of Medicine and Public Health, United States National Museum, Washington, DC.)

A Dictionary of Natural Products

Terms in the Field of

Pharmacognosy

Relating to Natural Medicinal and Pharmaceutical Materials and the Plants, Animals, and Minerals from Which They Are Derived

By

George Macdonald Hocking, Ph.D.

Professor Emeritus of Pharmacognosy, Auburn University (Auburn, AL); former Chairman, Subcommittee on Pharmacognosy, Committee on National Formulary; Member, Commissão de Estudos de Plantas Brasileiras, Medicinais e Toxicas, Universidade de São Paulo; formerly Consultant on Medicinal Plants to the government of Pakistan (FAO-ETAP); formerly Member, Subcommittee on Medicinal Plants, Pacific Science Association

Printed in the United States of America.

Library of Congress Cataloging-in-Publication Data

Hocking, George Macdonald.
A dictionary of natural products : terms in the field of pharmacognosy relating to natural medicinal and pharmaceutical materials and the plants, animals, and minerals from which they are derived / George Macdonald Hocking.
p. cm.
Includes bibliographical references and index.
ISBN 0-937548-31-6 (hardbound)
1. Pharmacognosy—Dictionaries. 2. Natural products—Dictionaries. 3. Medicinal plants—Dictionaries. 4. Plants—Dictionaries. 5. Animals—Dictionaries. I. Title.
RS160.H55 1997
615'.321'03—dc21 97-321
CIP

Price: $139.50 (hardbound)

This is a second edition of a reference work formerly titled *Dictionary of Terms in Pharmacognosy and Other Divisions of Economic Botany* (1955).

Editor: Lauree Padgett
Cover Design: Jeanne Wachter

Cover photographs (left to right): *Aconitum napellus, Artemisia argyi, Allium sativum* (Steven Foster)

CONTENTS

To My Parents

"O, mickle is the powerful grace that lies
In herbs, plants, stones, and their true qualities:
For naught so vile that on the earth doth live
But to the earth some special good doth give."
Shakespeare, *Romeo and Juliet*
1595

"There is a remedy for everything
could men find it."
George Herbert, *Jecula Prudentum*
1651

"Building a dictionary is exceedingly
interesting work, but tough."
Mark Twain, *Papers of the Adam Family*
(published 1939)

"If you know not the names, the knowledge
of things is too wasted."
Carolus Linnaeus

"God made the forest but man made the desert."
Anonymous

"One of the advantages of being disorderly
is that one is constantly making
exciting discoveries."
A. A. Milne

"He only deserves to be remembered by
posterity who treasures up and preserves
the history of his ancestors."
Edmund Burke

PREFACE

"The writer of dictionaries . . . is . . . exposed to censure without hope of praise, . . . disgraced by miscarriage or punished for neglect, where success would have been without applause, and diligence without reward."

Dr. Samuel Johnson

To compile a dictionary, one must surely have extraordinarily good reasons for engaging at all in such an arduous labor, since this requires unending industry, meticulous care, and expenditure of many hours of precious time. Such a project entails incessant visiting of libraries at home and abroad, an endless page-turning and consultation, a constant groping for words and phrases, a perpetual writing by hand and machine, interminable correcting, improving, filing, and recopying of annotated pages which stretch on endlessly. About the only justification can be in the realization by the compiler that he has accomplished a wearisome but concrete task, and one that he may reasonably hope will prove useful enough to merit the time and labor expended.

The present Dictionary compilation had its genesis in the endless difficulties and delays encountered by the writer in his continuing search over a period of many years for texts denoting and describing crude drugs and related materials, their originating organisms, and their derivatives. No single work had been found at all adequate in this complex and sometimes even bizarre field of pharmacognosy and related branches of economic botany and zoology. Various scattered works published both in our own country and in foreign lands (many of them out of print and generally accessible only in a large public library, very often in fact rare) were found of considerable utility, although these reference books were often out of date in this ever-developing field, occasionally in error, and nearly always "spotty" in coverage.

It became obvious that a consultant in this field would have to record and compile his information and file it in accessible form. Hence, much of the data going into this compilation was initially collected and arranged to suit the author's own convenience. In the course of time, the very real need in this field for a dictionary—a guide to furnish a pathway through the maze of innumerable Latin titles, biological names, vernacular terms, and other specialized terminology —came to be appreciated.

The need for a dictionary of pharmacognosy and economic botany might in the opinion of some people call for justification. Hence, a statement concerning this field of interest might profitably be included.

Pharmacognosy is the science which treats in detail those medicinal, pharmaceutical, and related products of crude or primary type obtained from the vegetable, animal, and mineral kingdoms as they appear on the market as raw materials, as they occur in the drug store and in folk medical practice, and as they are used in chemical manufacturing processes.

"Herbal" remedies in the form of the drug itself or of the simplest preparations (as infusion, decoction, tincture, etc.), are in recent years becoming of greater and greater interest. At the same time it appears likely that a larger proportion than ever of the medicinal agents in present use are derived from plant and animal substances, if we take into account the antibiotics, the endocrine products, the vaccines and sera, and such active isolates as alkaloids and glycosides. There is currently in evidence an increasing trend to restudy and reintroduce into medical practice vegetable drugs and simple preparations which had appeared to have gone into a decline in medical acceptance (ex., in recent years aconite, Veratrum viride, Podophyllum, Serenoa, etc.). The tendency in modern medicine is to isolate and use active purified principles from plant and animal drugs (as such products from *Ammi visnaga, Veratrum album, Taxus* species, and *Rauvolfia serpentina*). Needless to say, the crude material remains important. A great many crude drugs or derivatives are also still employed in proprietary medicines.

Many pharmacologists have evidenced a disdain amounting almost to contempt for the crude drug materia medica. These same individuals, however, applaud the success of the organic chemist in his synthesis of naturally occurring alkaloids and other phytochemicals. Might not this attitude be interpreted as somewhat illogical? The frantic efforts to accomplish the synthesis of many natural compounds might be interpreted as a tribute to the therapeutic qualities of those crude materials from which they originate. Such syntheses when they have been accomplished stand as academically interesting enterprises but as a general rule have little value as practical sources of the compound. While they demonstrate Man's genius, they fail to satisfy Man's necessities. For the complex compounds at any rate, the laboratories of Nature have generally been found to provide the most economical production.

In most countries of the world, perhaps among 80 percent of the world's peoples, crude drugs still constitute the basis of prevailing medical practices. The failure of "occidental" medicine to make use of many potent drugs of the more primitive areas of the globe stems perhaps from ignorance of these materials and their sources more than from any innate lack of therapeutic value.

This Dictionary is designed to meet the special needs of students and practitioners in the various health professions (pharmacy, medicine, dentistry, nursing, veterinary medicine, etc.); of people in trade and industry (wholesalers, jobbers, dealers, manufacturers, importers, and other suppliers); and of others including laymen with an amateur interest in the useful plants and animals of our world.

A thoroughly successful dictionary would be one comprehensive, accurate, lucid, compact, handy to use, and not uninteresting in content. These have been held up as objectives by the compiler and author, and he would be happy at any measure of success in their attainment.

In view of the dearth of compilations comparable to the present work, it is confidently felt that the latter will fill a gap in the convenient reference literature and will be of genuine utility for those seeking an understanding of economic plants and animals.

Information helpful to using the Dictionary effectively is presented in the Explanatory Foreword which follows this preface. It is hoped that mistakes are few and far between, but then the reader will, it is trusted, recall that "To err is human," and "As for pledges, the gods give none!"

To all those many kind people who for years past have bestowed so unstintingly of their special knowledge in many fields, much of which is incorporated in this Dictionary, the compiler desires to express his most sincere thanks.

George M. Hocking
Auburn, Alabama
August 1997

EXPLANATORY FOREWORD

SCOPE

The present Dictionary is devoted primarily to an arrangement and explanation of terms (words and expressions) which relate to crude drugs from the three kingdoms (vegetable, animal, and mineral). There is also briefer mention of other economic plants, animals, and minerals, particularly those which furnish alimentary products. Explicitly, it was endeavored to include data on representative crude (that is, natural and non-artificial) materials having the following classes of usage besides those strictly medicinal and pharmaceutical: foodstuffs, condiments (spices), beverages, saccharine or sweetening products (sugars, etc.), masticatories, fumitories, aromatics (perfumes), tinctorial products (dyestuffs), tans (agents in tanning), toxic (poisonous and noxious) plants and animals, addictive narcotics, hallucinogens, detergent (or "soap") plants, organisms furnishing materials to the chemical and other technical industries (so-called chemurgic products, including oleaginous [yield fatty oils], waxes, gums, volatile oils, resins, oleoresins, gum-resins, oleo-gum-resins, plastics, etc.), rubber-bearing and related plants, and miscellaneous plants (ex. fish baits, pesticidal and pest-repelling agents, desert water sources, etc.).

No attempt has been made in any systematic way to cover what might for convenience be termed the mechanical protective and decorative plants (that is, the many fiber and textile plants, timber trees, ornamentals, shade-trees, etc.), weeds, and other classes of plants. These groups have been so thoroughly treated in many technical and lay works that it was thought better not to include them regularly here, unless they are mentioned in the text incidental to other uses.

COVERAGE

A dictionary by its very nature must be basically a work of compilation; yet in a work of genuine quality there must be a considerable degree of what one might term originality, if we take into account the inclusion of newer or further pertinent facts in the definition (such as new sources), the elucidation or broadening of definitions, the variations in mode of presentation, and most importantly the inclusion of terms not previously included in a dictionary. In all of these ways, a dictionary may be considered a kind of creation—inasmuch as it represents an original presentation of information.

Thus, a serious effort has been made in compiling this Dictionary to include as often as possible terms not listed in the reference works available, and hence to make this a work of more than ordinary interest and value to those engaged in pharmacognosy and other divisions of economic botany. Frustration from time consumption and failure in dictionary searches is an especially common experience to people in these fields.

Even with the resources available from some of our larger libraries in Washington, New York, and elsewhere, a considerable number of the terms in regular use could not be discovered in any of the numerous compilations and other reference works examined. A number of these terms have here been defined—perhaps for the first time—by a careful direct study of the materials, etc., as presented. Hundreds of the entries (words or phrases) were found only in some single obscure dictionary or manual, long out of print and now scarcely known; some were gotten from manuscripts, correspondence, drug museum and herbarium labels, samples and other specimens, personal conversations, and verbal inquiries (such as with representatives of commercial firms). Other defined words were drawn from scattered references in books, journals, pamphlets, and even newspapers. Only a limited proportion of the terms may be found in the larger unabridged dictionaries (such as *Webster's Unabridged, Funk and Wagnall's,* the *Century Dictionary, The New English* [or *Oxford*] *Dictionary,* etc.).

While the present book will help to fill in where such general works as those cited are mostly deficient, it is not intended that the present compilation should pretend to substitute for the valuable and detailed dictionaries of such sciences as chemistry, medicine, agriculture, biology, botany, zoology, etc. While these fields are more or less closely related to or even overlapping the field of pharmacognosy to a certain extent, space limitations prevented inclusion of many words in these fields which are clearly and fully defined in the many specialized works available.

The coverage even for medicinal plants (especially for those used in folk medicines all over the world) is admittedly incomplete. This deficiency is practically unavoidable and arises from the fact that the number of such agents used worldwide is so very great, and also from the necessity of limiting book size to a practical dimension in this day of high printing costs. The need naturally arises of selecting the most important entries. Despite this effort to economize space, relatively more emphasis is often given to the minor drugs; this was done because of the relatively greater difficulty of securing facts on these items. The descriptive matter throughout has been cut to a minimum.

It may be objected after a study of the book that not all the plant and animal species indicated under common names are represented and defined under their scientific names, where the chief discussion (or most complete definition) is inserted. Such omission is an indication that as far as could be determined the organism has not been and is not currently of interest in medicine or pharmacy. Why then has the common name often been admitted and defined? The answer is that this has been done because of the relatively common occurrence or distinctiveness of the plant; its experimental interest in such fields as chemistry, pharmacology, medicine, or biology; its use in the extraction of substances with medicinal or related interest; or for other good reasons.

David Fairchild has said: "A new idea...may make necessary for the civilization of the great temperate region some as yet unsuspected resource of the jungle, some tree perhaps known only in the dry tomes of a systematic flora..." (*Exploring for Plants*, p. 338).

One cannot predict as to what uses may eventually be made of materials previously of no material interest. It was here felt that a definition if only of common name would present some kind of a minimum treatment and thus broaden the scope of usefulness of the work. The plant and animal kingdoms represent such valuable and productive fields for medical-chemical research and endeavor, with increasingly great potentialities of new discoveries in the future, that it is now often useless to draw a hard and fast line between plants of medicinal value and those not now so classified. What were for instance yesteryear's harmful or useless molds and bacteria are today the source of our most valuable and life-saving medicines, the antibiotics.

In a world of such great variety and complexity as that of the living plants and animals on the one hand, and of the manifold cultures of the human race on the other, it is not surprising that a book of this size could include anything but a portion of those words relating to crude drugs and so on from a few of the more important languages (mostly German [Ge], French [Fr], Spanish [Sp], Portuguese [Por], Dutch, and Italian [It]). A selection of such words has been included here, but it is manifest that there may be many important omissions. The chief general criterion used has been the personal contact of the author with these words in the course of his work as a professional and teaching pharmacognosist. It is hoped that the listing as references of several foreign language dictionaries and other reference texts applicable to definite geographical areas (in Appendix B) will eke out the present work and provide real help in those directions where the present Dictionary may be lacking.

DEFINITIONS

The definitions in the body of this Dictionary compilation fall into the following categories:

1. Botanical (or zoological) generic and specific names: hereunder are concentrated most of the data on plant and animal species and their products.
2. Families (plant and animal) and sometimes other classification divisions: a brief description with geographic range and giving a few typical genera.
3. Latin titles (official or formerly so) as used here or abroad.
4. English official titles (US, GB, etc.).
5. Vernacular or common names of plants and drugs in several languages.
6. Active plant principles and other phytochemicals; brief definitions for generic and specific terms used in phytochemistry.
7. Special technical terms used in pharmacognosy proper.
8. Botanic terms: a limited number of important or specialized words.
9. Pharmaceutical preparations containing crude drug materials or derivatives (official, officinal, proprietary); sometimes primarily of historic interest (limited treatment), and adjuvants.
10. Animal products—such as hormones, sera, vaccines, toxins, toxoids, antitoxins, blood fractions, etc.
11. Minerals used in pharmacy/medicine (limited).
12. Agricultural and processing industry terms (ex. fertilizers) (limited).
13. Commercial and trade terms (limited).
14. Biographical notes on pharmacognosists (very limited coverage).

15. Miscellaneous, including terms relating to pharmacy, pharmaceutical chemistry, pharmacology, and related fields; foods, liqueurs, etc., associated with plants materials; slang expressions for narcotic drugs; common incorrect spellings, etc.

NOMENCLATURE

A. Latin Names

Many Latin titles have been included, particularly when representing official pharmacopeial names (USP, NF, HP, BP, BPC, etc.). When sufficiently important, these titles occur in the alphabetical sequence of the text and are thus readily found; but in many cases, especially when primarily used abroad, they occur in the text in drug plant definitions and are not separately cross-indexed (ex. *"Radix Musteltoe,"* Latin title applied in Europe to the root of *Rauwolfia serpentina).*

Although in the United States, Latin in the medical-pharmaceutical field has declined in importance since now of relatively restricted usage, yet the language of classical science still continues to be very widely employed by professional people in Europe and elsewhere; consequently, the need of a guide to Latin terms for drugs is still present. Sometimes the Latin title is the only identification on the label. This ancient language is also of the greatest value in understanding many technical words in the Romance languages and other European languages including English.

B. English Pharmacopeial Titles

Many pharmacopeial English titles now represent names taken directly from the Latin without pretense of any anglicization process.[1] Thus, Myristica, originally strictly a Latin word, has become recognized as the formal English title for nutmeg by action of the revisers of official compendia. This transformation of words used in medicine and pharmacy appears to be part of a continuing process whereby physicians and pharmacists are being weaned away from the language traditionally associated with prescriptions and drug orders.

C. Vernacular Names

An effort has been made to present at least one good, that is, widely or commonly used, vernacular, common, or vulgar name for each species discussed. It is hoped that this will oftentimes provide a more convenient "handle" than the scientific name, especially for the layman's use. The layman is often confused and even repelled by the often elaborate and invariably strange names represented in scientific nomenclature.

Synthetic common names, that is, names invented by a botanist or even by a committee of plant experts, and representing often only a translation of the scientific name, have largely been avoided in the present compilation. While it is true that such names are often more accurate in describing some characteristic of the plant important from the standpoint of scientific classification, it was felt that more good would be done in the present work by employing names long used by natives in those areas where the plant grows. Sometimes such names are grotesque and meritless because non-descriptive, but more often they express an outstanding characteristic of the species' habit or structure by which it may be usefully and appropriately tagged.

Preference has been given in the chief definition for each plant or animal to an English (American) name where available; where not, a name from some foreign language or even sometimes from a dialect may be introduced. The objective has been to have, if possible, a common or vernacular name for each described species. For economy's sake, only the most important of such common names so listed are cross-indexed in alphabetical sequence.

Vernacular names are useful because of the importance they hold in communications with laymen, businessmen, collectors, and others. Even in the scientific literature of many fields, such as chemistry, agriculture, forestry, the technical industries, etc., authors often do not trouble themselves to use scientific names. Botanical works may employ unusual or even original botanical names which may be unfamiliar even to the technical consultant, and sometimes here the common name proves useful in pinning down the identity of the plant under discussion.

An effort has been made in some cases to explain the special meaning, significance, or origin of a name in the vernacular, in Latin, etc. It was felt that developing this logical relationship might often be of assistance in retaining the term in memory.

The reader should emphatically be warned not to depend too much upon a vernacular name, because of the frequent differences of usage in various areas, which is often reflected in a completely different interpretation by the compilers of reference works. This fact underlines the value, even necessity, of scientific names.

Many common names are included; however, it would often have been impossible to cover more than a small fraction of such names within the confines of a single volume. To illustrate, it may be pointed out that in Gerth van Wijk's *Dictionary of Plant Names* (one of the most complete vernacular listings ever published) a total of 235 German and 68 Dutch names have been pre-

[1] It is also true that in the case of synthetic chemical drugs the opposite process of manufacturing Latin titles from the modern scientific name has commonly been resorted to.

sented for *Juniperus communis* and 62 German and 68 Dutch names for its chief pharmaceutical product, Juniper Berries. It is said that the Marshallese employ more than 60 words to describe the coconut tree and fruit and the respective parts of these. Such are typical, not exceptional, instances of the multiplicity of terms applying to economic plants and their products.

The frequent unreliability of common names is exemplified by such names as "hogweed," a term which has been applied to eight or more genera of plants. It should be kept in mind that there are many variations in spelling, especially in transliterations, and it would be practically impossible to list all of these in the present Dictionary. The peruser is therefore warned not to discontinue searching for a vernacular word too easily but to glance under various possible spellings. As an example, "y" is often rendered as "i" (ex. aiapana for ayapana).

The vernacular names besides English chiefly represented are German, French, Spanish, Portuguese, Dutch, and Italian, since these are currently of chief importance from the scientific and commercial standpoints.

German. The modern simplified spelling is used. As far as possible, the individual words are represented, and these can be combined into the long words so characteristic of the language. For instance, Sternanisoel is composed of the three words Stern (star), Anis (anise), and Oel (oil); Siebenfingerkraut is made up of Sieben (seven), Finger (finger), and Kraut (herb); Schwarzkiefer is from schwarz (black) and Kiefer (pine or fir).

In German, C and K, and C and Z are often mutually interchangeable, hence a search should be made under these various letters in the alphabetical sequence (where one or the other represents the initial letter) as well as throughout the word, where one or the other occurs. The same applies to PH and F.

French. Although of lesser importance in the scientific field, French is highly essential as constituting with English a leading language for purposes of international exchange of knowledge. Many non-French areas of the world predominantly use French (ex. Egypt, much of north and central Africa, etc.).

Spanish. Like French, this language is of much importance as a means of communication in the Western Hemisphere; it is also of prime commercial significance. Hence, the many definitions of Spanish terms included.

Both the languages used in Spain (Castilian and Catalan) and the various American dialects (Spanish Americanisms) are when widely used referred to by the abbreviation "Sp." Where, however, a Spanish language term is used only locally, this is indicated by an abbreviation referring to the place, as for instance "Pe" for Peru.

PRONUNCIATION AND ACCENTUATION

Pronunciation of words is indicated only where special difficulties arise, as where the word is irregular or where variations from the usual exist, in the case of new words, ambiguities, etc. General suggestions for pronunciation follow:

English. When a vowel is followed in the syllable by a consonant, it has the short sound, as the *o* in *pop´-py* (*o* to rhyme with *hop*). A vowel not followed by a consonant in the same syllable is given the long sound, as the *o* in *ro´-sa* (*o* to rhyme with *foe).*

Latin. Where two syllables constitute the word, the stress or accent is placed on the pentultimate (or next to the last) syllable. Thus, *dur´-us* (hard). When there are three or more syllables, the third from the last is stressed (ex. *bar´-bar-us* [foreign]); however, when the pentultimate syllable is or contains a long vowel, then the stress is on this pentultimate syllable (ex. *Cer-a´-sus* [cherry]; *lu-te´-um* [yellow]). All vowels are short which are followed by two or more consonants (ex. *flamma* [flame]).

BOTANICAL NOMENCLATURE

As far as practicable, the botanical names used are those generally accepted in standardized works (modern or recently revised) such as Gray's *Manual of Botany* (8th ed.), *Bailey's Hortus II* and *Manual of Cultivated Plants (*ed. 2), Rehder's *Manual of Cultivated Trees and Shrubs (*2nd ed.) (these last two for cultivated species), Kartesz's *Synonymized Checklist of the Vascular Flora of the US, Canada, and Greenland,* ed. 2, 2 vol., 1994, *Mabberley's Plant-Book (1989),* and other similar authoritative works[2]. With the taxonomic field divided between the "lumpers" and "splitters" it has not always been possible to select names for which there is universal approval, but most difficulties were avoided by using, as far as they could be employed, the nomenclatural works available. Beyond this, an effort was made wherever possible to use the names accepted in monographic studies of selected plant groups (ex. Verbenaceae by H. N. Moldenke).

The *nomina conservanda* of genera and families, as far as they have been published in such works as the *International Rules of Botanical Nomenclature*

[2]Many other works were used, such as Schinz and Keller's *Flora der Schweiz* (ed. 3, 4); Kearney and Peebles' *Flowering Plants and Ferns of Arizona,* Standley's *Trees and Shrubs of Mexico,* etc. The works of J. K. Small and P. A. Rydberg (among others) while valuable are encumbered with a scientific nomenclature (based on the American Code of Botanical Nomenclature) which has largely now been superseded by the older more conservative names, as recognized by the International Code of Botanical Nomenclature.

(compiled by W. H. Camp et al., [unofficial] special ed., 1948; Chronica Botanica) have been used to the exclusion of others[3].

Where the genus consists of only a single species, this is noted by insertion of "(monotypic)" or "(1)" immediately following the species name. Ex. *Cocos nucifera* L. (monotypic) (or [1]).

Initial Letters of Specific Names

The initial letters of all specific names have been rendered in lower case in conformity with the practice of the U.S. Department of Agriculture, etc.[4] Many authors in the past preferred to use capital initial letters for specific names in the following instances: 1) when derived from names of individuals (ex. *Hypericum Richeri*, the hypericum of Richer) 2) where the specific name was at some previous time a generic name (ex. *Polygonum Persicaria*) 3) where the specific name represents a vernacular name (ex. *Schinus Molle*; "Molle" is a Peruvian vernacular word)[5].

Scientific Synonyms are names precisely equivalent to but less acceptable than the currently approved scientific name. Literally millions of scientific synonyms for plant species have been published, and there is only one publication presumably complete in listing these, the valuable *Index Kewensis*. In the present work, it was obviously possible to record only a relatively small proportion representing but a few of the more important or confusing synonyms. These synonyms have been represented in either of two ways: 1) following after the accepted scientific name and enclosed within parentheses; ex. *Eriogonum alatum* Torr. (E. triste S. Wats.); while identical with the entity represented by the first name is usually less acceptable in scientific circles 2) preceding a colon or equals sign which is followed by the currently accepted scientific name; ex. *Eriogonum triste* S. Wats.: *E. alatum* Torr.[6] (The same species has been utilized in both examples in order to show relationships more clearly.)

Family Names are indicated in parentheses following each genus or first species name. Family characteristics are extremely important and often provide an excellent clue to the chemical constituents, pharmacological, and therapeutic properties, etc., of the included species. Several plant and animal families have been briefly defined, so as to present the equivalent common (or lay) name(s), a minimal botanical description (usually no more than a brief note of some striking feature of the group), sometimes a word about the chemical components found in common among the members, and then regularly to finish, a few examples, either by common or genus name.

In the case of the Cryptogams and of Animalia, the classification of genus, species, and family has often been amplified by introducing the names of orders, classes, and even sometimes sub-phyla and phyla, where this was thought advantageous to the user.

GEOGRAPHICAL DISTRIBUTION

The range of a species, genus, family, etc., has been expressed in the simplest way but with the best accuracy conformable with brevity. Thus, where the range is given as North America (NA), it will be understood that the plant occurs over a major portion of that continent; thus, it might be expected to be found in parts of Canada, the United States, Mexico, the West Indies, and Central America. However, the range within one or more of these geographic divisions may be restricted by such factors as elevation, temperature, insolation, rainfall, humidity, soils, etc. As far as possible, the most inclusive terms have been used: thus, instead of stating North and South America (NA, SA) as the range, it is preferred to render it as Western Hemisphere (wHemisph) or Old World (OW); the range of spruce pine would be given as the southeastern United States (seUS) since it is found in the tier of states running from South Carolina to Louisiana. In the case of continents, countries, larger states, etc., dependence is largely placed on the use of compass directions, as southwestern US (swUS), northern Florida (nFla), eastern Asia (eAs).

If not qualified, the geographic location will indicate place of origin of the species, or at least where believed native or indigenous; in some cases, as will be noted, areas of naturalization and cultivation are additionally given. It is assumed that many species have been introduced into some portion of the Western Hemisphere, even though not generally so stated in the definition.

[3]Efforts to set up a list of Nomina Specifica Conservanda have so far been unsuccessful due to the wide diversity of opinion of specialists and the generally fluid nature of taxonomy. A special effort to prepare such listing for the relatively small group of economic species has also been attempted. Much help is now available for North American plants by using Kartesz's *Synonymized Checklist of the Vascular Flora of the US, Canada and Greenland*, 2 volumes, 2nd ed. 1994; some local floras have also been very helpful, such as *The Jepson Manual* (California) (1993).

[4]The 7th International Congress adopted a rule that all scientific epithets below genus rank are to be written with a lower case initial letter.

[5]Under the American Code, geographical specific names were also capitalized (ex. *Urtica Californica* Greene), but this is not done under the International Code (ex. *Phoebe mexicana* Meissner).

[6]Boldface was used for the titles of all entries being defined; when botanical or zoological titles serve as entry headings, it will be understood that the boldface type here used supersedes the italics only for purposes of emphasis, uniformity of style, and guidance.

PART(S) USED

The proper structural portions as used in medicine/pharmacy are accurately given in each chief definition. Thus, white pine bark is noted as the inner bark; anise seed is actually the entire fruit; and so on. The condition of the drug is understood as dried unless otherwise qualified; thus, as fresh, preserved, candied, etc.

CONSTITUENTS

Where known, the chief components of a plant or animal material are generally named, either as a class of constituents (such as alkaloids) or by specific name (such as aconitine, an alkaloid). In some instances, the approximate percentage present in the drug may also be indicated. Where a constituent is separated out and used in practice as a preferred article with properties differing in a significant manner from the mother product, this component is discussed at greater length under the name of such component rather than under the definition for the plant species (etc.). Thus, *Theobroma cacao* (chocolate plant) and its important products, cacao butter (cocoa butter) and theobromine, are each discussed separately.

APPLICATIONS AND USES

The many materials defined in this Dictionary show a considerable variety of usage. Among the medicinal products, there is a broad range extending from those materials which are compounded largely of tradition, superstition, and empiricism, as found in a relatively unsophisticated society, to the products having well-established and entirely rational uses among regular physicians. An effort has been made in some cases to separate as it were the chaff from the grain, but many purported uses are here reported (especially for products from non-Occidental countries), which have not been confirmed as to effectiveness and safety by experimental pharmacological or clinical procedures. Such indiscriminate treatment is a result of a policy of including as many crude drug materials as possible, but is also a reflection on the pitiably small percentage of the medical materials in use in the world today which have been given adequate scientific study. In some instances where such studies have been conducted, a brief resume of applicable results, positive or negative, many be given.

It should be emphasized that many of the drugs for which certain medicinal uses are cited are actually probably worthless, at least in the manner or for the purposes for which ordinarily applied. It should be clearly understood that the objective in the Dictionary is in a majority of instances to indicate usage not usefulness and that there is intended no implication of support or warranty of effectiveness in the therapeutic applications presented.

The same disclaimer would not be required with regards to technical applications of materials, since it is generally quite obvious from a simple test to know whether a certain material will serve for instance as a dyestuff or not. Besides that, nonmedicinal applications are usually of trivial importance when compared with those affecting the most precious possessions of the individual—his health and his life.

Precautions regarding usage, handling, storage, incompatibilities, etc., are occasionally given.

PRODUCTION AREAS

Where thought to be helpful, data on production areas for the wild-growing or cultivated plant species have been included. It should be understood, however, that often this information changes radically with the development or loss of markets and altered economic and political factors otherwise. This is particularly true during times of war or of other political-economic stresses.

APPENDICES

The appendices at the end of the volume are primarily intended to furnish the interested person with further sources of information (Appendices A, B, C), and to explain some of the technical terms occurring in the Dictionary (Appendices D, E, F, G).

Some definitions or parts of definitions (uses, etc.), usually of obscure or relatively recent words, etc., have been established or fortified by citing documentary references. Such citations are indicated in the body of the text by Arabic numerals enclosed in parentheses, which are keyed to match a table of numbered reference citations appearing as the Bibliography (Appendix A).

Appendix D presents an alphabetical listing and definitions of therapeutic terms. Appendix E furnishes convenient memoranda on terms for botanical structures. Appendix F presents the plant and animal classification lists. Appendix G lists plants reported as a source of rubber.

STYLE AND TECHNIQUE OF DEFINITIONS

Principally as a means of saving space, several devices have been employed in the text, as follows:

1. Condensation of Definition. Definitions are presented in the form of brief, didactic, and practical abstracts, with many abbreviated terms, and from which all superfluous words have been deleted. The data have been concentrated chiefly into the definitions coming under the scientific name (for plants and animals) or under the chief English title (for inorganic materials, etc.). Thus, "felwort" is defined simply as *Gentiana lutea,* whereas under this latter name in its proper sequence are given the classification, constituents, uses, geographic areas, and commerical sources. Of all the common names by which plants or animals are known, generally one or two are presented under the chief discussion. Many common names scattered in alphabetical order throughout the text have *not* been repeated usually under the chief definition, neither have all of the names given under the chief definition been cross-indexed in the text. Thus, "Hungarian gentian," the common name for *Gentiana pannonica,* is not indexed under "**gentian, Hungarian**," although inserted after the botanical name, while on the other hand "**gentian, blue**," while listed under the common name, is not included under the definition for *Gentiana saponaria.*

2. Abbreviations. Many abbreviations and signs have been used throughout this work to save space, and reference should be made when necessary to the tables of abbreviations and symbols which follow this Foreword and precede the text. Some self-explanatory and readily recognizable abbreviations not listed in the key have been used (ex. many plant families, as Irid, for Iridaceae).

3. Condensation of Expressions Having a Common Word or Words. A device has been adopted here which is in frequent use in dictionary texts as a space-saver. This is to assemble into one paragraph terms and phrases having one word in common, which later is rendered as the simplest possible abbreviation—ordinarily the initial letter followed by a period. Thus, under **"fever"** in the text have been collected nine terms, including **fever bark, f.** (for fever) **bush, California f. b.** (= California fever bush), etc. The associated qualifying term or terms may either follow or precede the abbreviated common term: the terms are arranged in strict alphabetical sequence.

4. Sharing of Definitions. Repetition is avoided as far as possible. Thus, to avoid repeating a definition (unless it be a very brief one), the reader is constantly referred to other entries, usually nearby in the text. Thus, for example, haricot, kidney, dwarf, pole, snap, and some other beans are all referred to common bean, *Phaseolus vulgaris.*

When such items are adjacent, the sequence used may be illustrated with the following: **fig juice: f. latex: f. milk:** (the definition follows the last term)... **strangling f.: wild f.:** (the definition follows the last term). In this sequence, the first three items are equivalent one to the other, in the last two items, strangling fig is shown as the same as wild fig. The groups are defined as units, at a saving of space, without (it is hoped) any loss in clarity of meaning.

In this connection, it should be pointed out that there are no definite rules for hyphenating common words and the practice is so variable that no attention need be paid to this punctuation mark. Thus, while garden-sage is customarily hyphenated, in the present text the words are separated and hence the term appears under "sage" rather than under "garden," where otherwise it would have to go.

5. Avoidance of Entry Duplication. A common name made up of two or more words is not ordinarily repeated in the text; that is, articles are listed only under generic and not specific names. Thus, "flat bean" occurs only under "bean," and thus under the letter "F," there is no entry for "flat bean." However, in some instances, compound entries (i.e., of two or more words) are put under the uncommon term (thus, marble flower appears under its first name).

6. Deletion of Commas and Other Punctuation Marks. In recording a series of constituents, uses, etc., the abbreviations used are separated by periods or semi-colons, commas being used only where necessary to avoid ambiguity of meaning or error; punctuation marks may be omitted where the meaning is clear.

7. Parentheses. These have been used freely to indicate alternative wordings. Thus, **"(i)lozane"** means that both ilozane and lozane are correct renderings of the word; **"(yellow) chestnut oak"** means that yellow chestnut oak is equivalent to chestnut oak properly applied. When parentheses are placed around the comma separating words this then indicates that the word order may be reversed: thus, *Primula veris* might be indicated as "primrose (,) cowslip"; which would mean that it is properly called either "cowslip primrose" or "primrose cowslip."

8. Process Formulas. The figures used will be understood as representing parts by weight (unless other wise qualified).

9. Examples have been cited wherever possible in each definition; especially for class terms, processes, botanical adjectives, etc.

10. Direct Alphabetical Sequence of entries has been employed, except as this may be interfered with by the paragraph condensation procedure mentioned in (3) above. Thus, for instance, the series **"gall, gall berry,..."** is presented in a paragraph preceding **"Galla,"** which by direct alphabetical sequence otherwise would fall between "gall" and "gall berry." A hyphenated word is treated as a fused, unhyphenated word in alphabetical sequence. Scientific and common names, constituents, and so on have been presented wherever possible in alphabetized sequence. Emphasis has been secured through the use of bold-face type; thus, for instance, under **Andira** are given eight species all in proper order from *a* to *v*.

11. Diacritical Marks. To accord with established American dictionary practice, words are given in strict alphabetical sequence without reference to the diacritical marks. Thus, *ä* is treated as though it were *a* and not *ae*; *ñ* as though simply *n* and not *ny*; and so on.

12. Common Names of Related Species. Often the vernacular name or names given for one species applies (or apply) also to one or more or perhaps all other species of the same genus (or other taxon); to save space, these names are mostly not repeated. Hence, it should be understood that in many cases, certainly where the species show considerable resemblance, some or even all of the common names may often be used in common for the various members of the group, often with additional qualifying terms to denote individual characteristics.

13. Transliteration of Names. Names transliterated (as from Urdu into Roman characters) may often be represented by various spellings according to the national language or scholarship of the author; an effort has been made here to use the best phonetic spelling or the commonest recognized spelling for such words.

14. Diphthongs. Many diphthongs (except in scientific botanical and zoological names) have been Americanized: thus, *ae* is mostly rendered as *e, oe* as *e*, etc.

15. Identical or Closely Similar Foreign Language Names. These are generally not included when closely approaching the English word; ex. *Sandarakharz* (German for Sandarac Resin); *gentiane* (French for gentian); *Tolubalsam* (German for Tolu Balsam); *eucalipto* (Spanish, etc, for eucalyptus). Such words may be readily recognized and hence are not repeated in the text.

16. Order of Technical Terms. Scientific names and common names of identical spelling are not ordinarily admixed in a paragraph; the scientific name precedes the common name or series of common names. Thus, *Cassia* species come together in a paragraph which precedes the paragraph of items containing "cassia" as the common name.

ABBREVIATIONS

Many of the following abbreviations are used in the text of this Dictionary. Others listed below for ready reference will be found occurring in publications, commercial correspondence, labels, and so on. Where two or more letter abbreviations occur together, the period has been eliminated. Thus, "Chocolate Coated Tablets" is abbreviated CCT rather than C.C.T.

a.	are; metric unit of area
AACP	American Association of Colleges of Pharmacy
AB	1) abortion 2) antiobiotic
ABC	Atomic—Biological—Chemical (Warfare); abscess
ABC Lin.	Compound Liniment of Aconite, Belladonna, and Chloroform
abd.	abdomen; abdominal
abs.	absolute
absorb.	absorbent
absorpt.	absorption
ACA	American College of Apothecaries
acc.	according to
ACE	Adrenal Cortex Extract; acetylcholinesterase
ACPE	American Council on Pharmaceutical Education
ACS	American Chemical Society; American Cancer Society; Antireticular Cytotoxic Serum
act.	acting, active, action
ACTH	Adrenocorticotrophic Hormone
AD	addict(ion)
ad.	adulterant(s); adulteration
ADD	Attention Deficit Disorder (hyperactivity, mostly in children)
addn.	addition
ADH	antidiuretic hormone
adj.	adjacent (org. chem.), adjective
adulter.	adulterant(s), adulterate, adulteration
AGFA	Aktien Gesellschaft fuer Anilinfabrikation (Ge) (Joint Stock Company for Anilin Manufacture)
agt.	agent
A&H	Arm & Hammer™: a brand of sodas
AHA	American Heart Association
AIDS	Acquired Immunodeficiency Syndrome
AIHP	American Institute of the History of Pharmacy
alc.	alcohol (ethanol unless qualified)
ALH	anterior lobe of hypophysis
alim.	alimentary
alk(al)	alkaloid; alkali(ne)
ALS	amylotropic lateral sclerosis (Lou Gehrig's disease)
alt(er)	alterative
AMA	American Medical Association
Am Indians	American Indians
amm.	ammonia; ammonium
amorph.	amorphous
anal.	analysis, analytical
analg.	analgesic
ann.	annual
anod.	anodyne
antidiarrh.	antidiarrheal
antiepil.	antiepileptic

antiphlog.	antiphlogistic
antipyr.	antipyretic
antirh.	antirheumatic
antis(ep).	antiseptic
AFP	Animal Protein Factor
APhA	American Pharmaceutical Association
aphr(od).	aphrodisiac
APL	Anterior Pituitary-Like
appln.	application
aq.	*aqua* (L.): water, aqueous
AS&B	Aloin, Strychnine, & Belladonna (Pills) (NF)
ASB&I	Aloin, Strychnine, Belladonna, & Ipecac (Pills) (NF)
AT10	Dihydrotachysterol; prep. u. in treating parathyroid tetany
ATP	Adenosine Triphosphate
ÅU	Ångstrom Unit(s)
Bals.	Balsam(um)
BB	Blue Book (American Druggist), directory to proprietaries
BCG	Bacillus Calmette-Guerin
BD	Belladonna
BHC	Benzene Hexachloride, 666 (insecticide)
BM	bowel movement
BMA	British Medical Association
BMR	Basal Metabolic Rate
BO	Boiled (Linseed) Oil; Botanical (or Biological) Origin (genus, species, family)
BOD	Biochemical Oxygen Demand
BP	British Pharmacopoeia
BP	boiling point; blood pressure
BPC	British Pharmaceutical Codex
br.	branch
Brit. Phar.	British Pharmacopoeia
bronch.	bronchial; bronchitis
BW	biological warfare
c.	contain(s), containing, constituents; with
c. (or ca.)	*circa* (L.): about, approx.
C	Centigrade; Celsius degree(s); Central; Gallon (US.) (*Congius*, L.); Gallon (Imperial); Cocaine; carbon
CA, Ca.	Cancer; Carcinoma; Chemical Abstracts; cathecol tannins
ca. (or c.)	*circa* (L.): about, approx.
CaOx	calcium oxalate
caps.	capsule
card.	cardiac
carm(in).	carminative
cath.	cathartic
CBC	complete blood count
cc.	cubic centimeter(s)
CC	column chromatography
CC	Compound Pills of Mild Mercurous Chloride ("Compound Cathartic Pills")
CCD	counter current distribution
CCT	coated compressed tablet(s); chocolate coated tablet(s)
CD	Communicable Disease(s); Civil Defense
C & D	Chemist & Druggist
CDC	Communicable Disease Center (now Centers for Disease Control and Prevention)
cf.	compare (conficere, L.); cultivated form
CF	Crude Fiber; Change in Formula; citrovorum factor
CFC	citrus flavonoid complex
C & F	cost & freight
CG	chorionic gonadotropin
char(act.)	characteristic(s)
CHD	congestive heart disease
chem.	chemistry; chemical
CHO/CH_2O	carbohydrate(s)
chol(ag).	cholagog
Cia. (Sp):	Cie. (Fr): Company
CIF	cost, insurance, & freight
cl.	clone
cm.	centimeter(s)
CNS	Central Nervous System
Co.	Compagnie (Ge): Company
CO	Castor Oil
coll.	collected; collection; collector
combd.	combined
Comm.	Commercial (Information); communicated
Comp.	compound
conc.	cut up (*concissus*, L.); concentrated; concentrate; concentration
condim.	condiment (spice)
Cong.	gallon (*Congius*, L.)
const.	constituent(s)
constip.	constipation
cont.	contain(s)
corresp.	corresponding
cosmop.	cosmopolitan
CP	chemically pure
CPE	cytopathogenic effect
CRC	calomel, rhubarb, colocynth (cathartic) (army pills)
cryst.	crystal; crystallization
CS	cross-section(s)
CSF	cerebrospinal fluid
CT	Compressed Tablet(s); Coated T.; Calcitonin

CTC	chlortetracycline
CTS	carpal tunnel syndrome
Cu	copper
cult.	cultivated
cv.	cultivar (cultivated variety)
CV(S)	Cardio-Vascular System
CVP	citrus vitamin P
d.	penny, pence (GB) (dinar); density
D	Dose(s); dextro-
DA	dopamine
D/B/A	doing business as
DBED	Dibenzyl-Ethylene-Diamine (Penicillin)
DC	Duennschichtchromatographie (Ge): thin layer chromatography
D&C	dilatation & curettage (of cervix uteri)
DD (GB)	Dangerous Drugs (Regulation of 1928) (Great Britain)
DDT	Dichloro-Diphenyl Trichloro-Ethane (insecticide)
decoc.	decoction
dehydr.	dehydrated
dem.	demulcent
depil.	depilatory
deriv.	derivative; derived from
Descr.	Description
dest., dist.	*destillatus* (L.): distilled
detd.	determined
DFP	diisopropyl fluorophosphate
DHQ	dihydroquercetin
diab.	diabetes
Diag.	Diagnosis; Diagnostic Features
diaph.	diaphoretic
diar(rh).	diarrhea
dicot.	dicotyledon(ous)
dil.	dilute(d); dilution
Dim.	Dimension(s)
dis.	disease(s); disorder(s)
disinf.	disinfection; disinfectant
Disp.	Dispensing (*cf.* DP)
diss.	dissolve; dissolution
dist(n).	distilled (distillation)
distr.	distribution
diur.	diuretic
DNA	deoxyribonucleic acid
DNAase	deoxyribonuclease
DNP	dinitrophenol
DOCA	Desoxycorticosterone Acetate
DP	Dispensing Precaution(s); Dementia Praecox
DPN	diphosphopyridine nucleotide
dr.	dried; dra(ch)m.
DT	Delirium Tremens
Dx.	diagnosis
Dy.	deputy
dys.	dysentery
dysm(en).	dysmenorrhea
dyspep.	dyspepsia
DZ	Doppelzentner (Ge) (100 kg. [ca. 220.5 lbs.])
EAA	essential amino acids
EAL	effective (biological) activity load
EAT cells	Ehrlich ascites tumor cells
EBA	effective biological activity
ECP	Estradiol Cyclopentyl Propionate
EDTA	ethylene diamino tetra acetate
EEG	electroencephalograph(y)
e.g.	*exempli gratia* (L.): for the sake of example; for (an) instance
EKG	encephalocardiogram
EM	Elutionsmittel (Ge): eluting agent (chromatography)
EMC	encephalomyocarditis
emet.	emetic
emm.	emmenagog
emuls.	emulsion
ENT	ear, nose, & throat
eq.	equal
esp.	especially
ess.	essential
et al.	*et allii* (L.); and other persons
etc.	*et cetera* (L.): and other things or items; *et alia.*
ev.	eventuell (Ge): possibly, perhaps
EV	Eingang Vorbehalten (Ge): rights reserved; copyrighted; eingetragene Verein (Ge): registered society
evap.	evaporate(d); evaporation
EW	Eingetragenes Warenzeichen (Ge): registered trademark
ex.	(for) example
ex p.	*ex parte* (L.): in part
exp(ect).	expectorant
expr.	expressed
ext.	extract(ed); extraction; *Extractum* (L.)
exter.	external(ly)
exud.	exudation
f.	feminine; female
F	Fahrenheit degree(s); Fracture; Fluid
Fa	Firma (Ge): firm, company
FA	fatty acid(s); folic acid; fluorescent antibody
FACA	Fellow of the ACA
FAD	flavine adenine dinucleotide
Fam.	Family, Families
FAO	Food and Agricultural Organization (of the United Nations Organization)
FD	Froschdosen (Ge); frog doses

FDA — Food and Drug Administration (of the U.S. Department of Health, Education, and Welfare)
FE — Fluid Extract
febr(if). — febrifuge
ferment. — fermentation
fi. — for (an) instance
fl. — *flos, flores* (L.): flower(s); flowering (also flg.); *floruit* (L.): flourished (biog.); fluid
fl. oz. — fluid-ounce
flt. — floret
FO — Fatty (or Fixed) Oil(s)
foll. — following (item)
fr. — from
FRS — Fellow of the Royal Society (London, Eng)
frt. — fruit
FSH — follicle stimulating hormone
FU — Farmacopeia Ufficiale (Italian Pharmacopeia)
GA — gibberellic acid
GABA — gamma amino butyric acid
gal. — gallon
gb. — gallbladder
GC — Gonococcus; gonorrhea; gonorrheal infection; gas chromatography
GCT — gelatin coated tablets
GD — Geographic Distribution: area in which organism is found growing
GG — Gamma Globulin (u. in anterior poliomyelitis, measles, etc.)
GI(T), GI(S) — Gastro-Intestinal (Tract) (or System)
G(m)/g(m) — gram
GmbH — Gesellschaft mit beschränkter Haftung (Ge): limited company
gr. — grain (apothecary measure) (in some countries, may stand for Gram)
grd. — ground
grn. — grain
GU — Genito-Urinary Tract (or System)
Gum. — *Gummi* (L.): gum.
gyn. — gynecology
H — Hypoderm(at)ic (Injection); Heroin; hour
Hab. — Habitat (in pharmacognosy, equivalent to GD)
hb. — herb
Hb. — hemoglobin; herbarium
HC — hydrocarbon
HCH — hexachlorocyclohexane
HDL-C — high density lipoprotein cholesterol
hemost(at). — hemostat(ic)
Herb. — Herbarium
HGB — hemoglobin
HGC — hard gelatin capsules
HGF — hyperglycemic glycogenolytic factor (glucagon)
HIV — human immunodeficiency (retro) virus
HLA — human leukocyte antigen
HM — His (Her) Majesty's
HMB — Homatropine Methyl Bromide
HMC — Tablets of Hyoscine, Morphine, & Cactine (US); (US addicts): mixture of Heroin, morphine, & cocaine
Hno(s). — Hermano(s) (Sp): brother(s)
HOH — water
Homeop. — Hom(o)eopathic Pharmacopoeia (of the US)
hort. — horticultural name
HP — Homoeopathic Pharmacopoeia (of the US)
hpf. — high power field
hr(s). — hour(s)
H & T — Hooker and Thellung
HTH — Chlorinated Soda Soln.™
HTLV — human T-cell lymphoma virus
HV — Hochvacuum (Ge): high vacuum
HVC — Hayden's Vibunum Compound™, prep. fr. *V. opulus*
hydrag. — hydragog
hypn. — hypnotic
IAPT — International Association for Plant Taxonomy (Netherlands)
IAT — Ionenaustauschen (Ge): ion exchange
ib(id) — ibidem: same place (i.e., same reference)
I/C — in charge of
i.d. — inside diameter
IDDS — implantable drug delivery system
idem — in the same reference
i.e. — *id est* (L.): that is
IGFA — Interessen Gemeinschaft der Farbenindustrie Aktiengesellschaft (Ge): Common Interests of the Dye Industry Joint Stock Companies (dye trust)
IM — Intramuscular (injection); im Mittel (Ge): on an average
Imp. meas. — imperial measure (British)
incl. — including; inclusive of
incr. — increase(d)
indig. — indigenous
indiges. — indigestion
indy. — industry
inf. — infused; infusion
infec. — infection(s)
infl. — inflorescence
inflamm(n). — inflammation
inj. — injection
insol. — insoluble

int. internal(ly)
intest. intestine; intestinal
IPA isopropyl alcohol
IPSF International Pharmacy Students' Federation
I Q & S Elixir of Iron, Quinine, & Strychnine
irrit. irritate; irritation
IS Intra-Spinal (Injection)
IU International Unit(s)
IV Intravenous (Injection)
JIC(P) *v.* NJIC
KE Kendall's Compound E (Cortisone)
Kg. (or kg.) Kilo(gram)
KS keto-steroids
L. Limnaeus; Linné; Latin; liter (also l.); levo-
LA local anesthetic; long acting
lax. laxative
lb. *libra* (L.): pound (avoirdupois); pound (apothecary)
LD lethal dose
LDL-C low density lipoprotein-cholesterol
LE lupus erythematosus; light eruptions
lf. leaf
L.f. Linnaeus filius (Linnacus junior)
lft. leaflet
lge. large
LH luteinizing hormone
linim. liniment
liq. liquid; Liquor (L.)
lithon. lithontryptic
LS Longitudinal Section
LSD lysergic acid diethylamide (from Lysergsaeure Diäthylamid [Ge])
LTH Luteotrophic Hormone
lvs. leaves
m. male, masculine, meta-, minim
M *Misce* (L.): mix; Morphine
m.a. matière active (Fr): active principle
Mac. Macroscopic Pharmacognosy
mal. malaria
manuf. manufacture; manufacturing
MAO monoamine oxidase
max. maximum
M & C (US addicts): Morphine & Cocaine Mixture
MCH mean corpuscular hemoglobin
MCHC mean corpuscular hemoglobin concentration
MCT multiple coated tablets
MCV mean corpuscular volume
MD muscular dystrophy
MDE Modern Drug Encyclopedia (Gutman: Howard)
MDR Minimum Daily Requirements
med(ic). medicine; medic(in)al; medicated
menstr. menstruation; menstruate
mfr. manufacture(r); manufacturing
mg(m). milligram
MHV mice with hepatic virus
Mic. Microscopic Pharmacognosy
microg. microgram (gamma); 0.001 mg.
min. minute(s); minimum; minim
Mis(c). Miscellaneous; mix
Mist. *Mistura* (L.): mixture
mixt. mixture(s)
mk. mikron
mkg. meter-kilogram
ml. milliliter(s)
mm. millimeter
mo. microorganism
molec. molecule; molecular
MP melting point
MP Member of Parliament (GB)
MPS Member of the Pharmaceutical Society (GB)
MPZ Modified Protamine Zinc (Insulin)
MRC Medical Research Council (GB)
MRD Medical Research Division
MS manuscript
MSG Mono-Sodium Glutamate
muc. *mucilago* (L.): mucilage; mucous; mucin
mydr. mydriatic
n. normal; neuter; neutral
natd. naturalized
NBP Neutral Bitter Principle
NDGA Nor-Dihydro-Guaiaretic Acid
nerv. nervous; nervine; nerve
neur. neuron; neurology; &c.
neut. neutral
NF National Formulary (US)
NFIP National Foundation for Infantile Paralysis
NHI National Health Insurance (GB)
NIH National Institute(s) of Health (US)
NJIC National Joint Industrial Council (for Pharmacy) (GB)
NLT Not Less Than
nm. nanometer
NMR Neutral Magnetic Resonance
NMT Not More Than
NND New and Non-Official Drugs
NNR New & Non-Official Remedies (AMA); Nebennierenrinde (Ge): adrenal cortex
no. number
NO Non-Official
NPH Neutral Protamine Hagedorn

	(Insulin): insulin having intermediate persistence of action (one injection per day)
NPN	Non-Protein Nitrogen
nr.	*Non Repetatur* (L.): not to be repeated
NRC	National Research Council
NTA	National Tuberculosis Association
nutr.	nutrient, nutrition
O	oral(ly) (medication by mouth); pint (fr. *octarius*, L.); oxide or ether (ex. in et_2O: ethyl ether)
o-	ortho-
OB	Obstetrical
obt.	obtain(ed)
OC	order card
O/C	over counter
occ.	occur(s); occurrence
occas.	occasional(ly)
OET	Official English Title
off.	official(ly)
OH	a hydroxide or alcohol; ex. buOH: butyl alcohol
oint(m).	ointment
OJ	Orange Juice (abbreviation u. in hospital)
Ol.	*Oleum* (L.): oil
OL	Old Latin
OLT	Official Latin Title
o.O.	ohne Ort (Ge.): without place (of publication, etc.)
o.ö. (P.):	ordentlicher oeffentlicher Professor (Ge): Professor in Ordinary in a public institution of higher learning
OR	Oleoresin
ord.	order; ordinary
ORD	Optical Rotatory Dispersion
OS	Over the State (re-distribution)
oxid.	Oxidation
p.	page (*pagina*, L.); para-
p.a.	per annum
PA	pernicious anemia; pro analys. (L.); analytical grade (chem.); pantothenic acid; plasminogen activator
PAA	Proprietary Association of America
PAB(A)	para-amino-benzoic acid
par.	parenteral
PAS	para-aminosalicylic acid
PBI	protein-bound iodine
pc.	percutaneous
PC	Purchase Card; Pharmaceutical Corps (after company name); paper chromatography; personal computer
Pcog.	Pharmacognosy
pd.	powder(ed)
PDR	Physician's Desk Reference
PE	Powdered Extract; paper electrophoresis
Pg.	Pharmacognosy
Ph.	Phylum
Phar.	Pharmacop(o)eia
Phcog.	Pharmacognosy
per.	perennial
percut.	percutaneous
petr(ol).	petroleum
PF	point de fusion (Fr): melting point
PGA	pteroyl glutamic acid
Ph(ar).	Pharmacopeia
Phcol.	Pharmacology
Phm.	Pharmacist (sometimes used before personal name like "Mister")
Phy.	Pharmacy
PI	Protocol Internationale (Fr): *Prae scriptio* (or *Protocol*) *Internation alis* (L.): Official International Formulary (superseded by the International Pharmacopoeia) (Pharmacopoeia Internationalis) (also Ph.I.)
PID	Pelvic Inflammatory Disease (acute salpingitis, &c.)
Pil.	*Pilula(e)* (L.): pill(s)
pl.	plant
plur.	plural
PM	Post Mortem
PMN	Phenyl Mercuric Nitrate; polymorphonuclear leukocytes
PMS	premenstrual syndrome; Pregnant Mare's Serum
p.o.	per ora; orally administered
POB	Peanut Oil & Beeswax (formerly u. as penicillin base)
powd. (or pd.)	powder(ed)
PP	Parkinson's Paralysis (or disease)
pp.	pages; in part (in bot.), indicating a partial synonym shared by two botanical names; *praepter propter* (L.): approximately (also p. pt.)
ppt.	precipitate(d); precipitation
pr.	*v.* prep.
prec.	preceding
prep.	preparation(s); prepare(d)
pres.	preserved; preservation
princ.	principle
prn.	pro re nata (as desired or required)
prob.	probable; probably
prod.	produce(d); production; product(s)
prop.	property; properties
PSS	Plant Science Seminar (former in formal organization of pharmacognosists)

pt.	part
pto.	puerto (Sp): port
PU	Part(s) Used
pulm(on).	pulmonary
purg.	purgative
PVP	Polyvinyl Pyrrolidone; u. as blood plasma substitute
Py.	Pharmacy
q.	*quaque* (L.): every, each: (Fr) quintal (50 kg.): square (Carró)
Q (fever)	Queensland fever; a tick-transmitted dis., first found in farmers of Queensland, Australia ("query fever")
QS	Quantum Sufficat (L.): a sufficient quantity (or amount)
quant.	quantity; quantitative
qv.	*quod vide* (L.): which see; cross-reference to the word, etc., which it follows
R	Réaumur degree(s); Rio (Sp) River; alkyl gp. (chem.)
Rad.	*Radix* (L.): root (often collectively for rhizome, corm, tuber, etc.)
RB	Red Book (Drug Topics)
RBC	Red Blood Count; Red Blood Corpuscles (or cells) (also rbc)
R/C	Rubber Capped
Rect.	*Rectificatus* (L.): rectified
reduc.	reduced; reducing
reduct.	reduction
refrig.	refrigerate; refrigeration
reg.	region(s)
rel.	relative(ly)
rem.	remedy
res.	residue
Res.	*Resina* (L.): resin(s)
RES	reticuloendothelial system
Resp.	Respiratory (System) or Respiration
rheum.	rheumatism (often for arthritis)
rhiz.	rhizome
RNA	ribonucleic acid
R. P(h).	Registered Pharmacist; abbrev. often u. after surname
R & S	Roemer and Schultes
rt.	root
rubef.	rubefacient
Rx	prescription; prescribing
s.	shilling (GB, etc.); *seu, sive* (L.): or
S.	Section; Seite (Ger): page
SA	Société Anonyme (Fr): Soc. Anon.: Sociedad Anónima (Sp): Corporation
SAB	Society of American Bacteriologists
satd.	saturated
SBT	Sodium Bitartrate
SC	So-called because...; Subcutaneous (Injection)
SCT	Sugar Coated Tablet(s)
sd.	seed: sine dato (without date)
SD	Streptodornase
SE	Solid Extract(s)
sec.	second
secrn.	secretion
sed.	sedative; sediment
Sem.	*Semen, Semina* (L.): seed(s)
sep.	separate; separation
SG	Specific Gravity
sic	that is so!
SID(S)	sudden infant dealth (syndrome)
sing.	singular; single
SI System	Système Internationale
SK	Streptokinase
sl.	slight(ly)
s.l./s.	lat. sensu lato (L.): in the broad sense
sine numero	undated
Soc.	Society
sol(n).	solution; solvent; soluble in...
sp.	species (spp., plural)
sp. g(r).	specific gravity
specif.	specific(ally)
sphalm	(L.): in error
Spir.	*Spiritus* (L.): spirit
Sr.	Señor (Sp): Signor (It): Mister
SR	sustained release
SRS	slow reacting substance
SS	Sterile Solution; Standardized Soln.; less commonly, Saturated Soln.
s.st(r).	*sensu stricto* (L.): in the strictest (or narrowest) sense
st.	stem; stet (L.): let stand
St.	Standard
stat.	*statim* (L.): immediately, refers to initial dose given (u. on graphs, kymograms, prescriptions, etc.)
STD	sexually transmitted disease(s); VD (incl. some 25 separate dis. chiefly syph. GC; AIDS)
sternut(at).	sternutatory
stet	(L.): leave to remain the same
STH	somatotropic hormone
stim.	stimulate; stimulant
STM	Streptomycin
Str.	*Streptomyces*
styp.	styptic
sublm.	sublimation
subst.	substance(s)
substit.	substitute(s)
Suc./Succs.	Successors (to a business)
suppos(it).	suppository

susp.	suspension
SV	*v*SVR
SVI	Spiritus Vini Industrialis (L.): industrial alcohol
SV Meth.	Methanol
SVR	Spiritus Vini Rectificatus (L.): (rectified) alcohol
Syn.	Synonym(s); Synonymous
syn(th).	synthetic; synthesize(d)
syph.	syphilis; syphilitic
syr.	Syrupus (L.): syrup(y)
T	tabula (L.): plate; tome: volume
T_3	triiodothyronine
T_4	thyroxin
TA	Total Alkaloids
TAB	typhoid-paratlyphoid A&B vaccine
TAC	Total Alkaloids of Cinchona
TAT	Tetanus Antitoxin; toxin-antitoxin
TB/Tbc	Tuberculosis; tubercle
TB-1	Tibione; drug u. in treating TB
TBV	Tubercle Bacillus Vaccine
TCP	Tri-Chloro-Phenoxy-Acetic Acid (u. as weed killer); Tri-Cresyl Phosphate (solvent, etc.); Testos terone ß-cyclopentylpropionate
techn.	technical; technics
temp.	temperature
tenif.	tenifuge
tert.	tertiary
Tex.	Texture
THC	tetrahydro-cannabinol
therap.	therapy; therapeutics; therapeutic(al[ly]); etc.
tid	*ter in diem* (L.): three times a day
Tinct.	*Tinctura* (L.): tincture
TJ	Tomato Juice (abbrev. u. in hospitals)
TLC	thin layer chromatography
TM	Trademark(ed) (Registered)
TMV	tobacco mosaic virus
ton.	tonic
TOFA	tall oil fatty acid
tox.	toxic(ity)
TPR	Temperature, Pulse, & Respiration
Tr.	*v*. Tinct.
tr(op).	tropics, tropical
TS	Transverse (or Cross) Section
TSH	thyroid stimulating hormone
TT	Tablet Triturate(s)
TV	*Trichomonas vaginalis* (vaginitis)
u.	used; unit; use(s)
Ud.	Usted(es) (Sp): you
UNO	United Nations Organization
URI	upper respiratory infection(s)
urin.	urine; urinary
USP	United States Pharmacop(o)eia
uter.	uterus; uterine
UTI	urinary tract infection(s)
UV(L)	ultraviolet (light)
v.	*vide* (L.): see; *varietas* (L.): variety; *vel* (L.): volts, voltage
VA	Veteran's Administration; Vincent's Angina
va.	vel ante (or before)
var.	variety; variable; various
Vd.	Ud.
VD	Venereal Disease(s)
VEM	Vaso-excitor Material, formed in kidney; has hypertensive effect
vermif.	vermifuge
Veter.	Veterinary; veterinarian
vit.	vitamin
viz.	videlicet (L): namely
VO	Volatile Oil(s)
vol.	volatile; volume(tric)
vuln(er).	vulnerary
WBC	White Blood Cells (or Corpuscles); white blood count (also wbc)
wd.	wood
wh.	which (when, where, who)
WHO	World Health Organization (of the UNO)
W/P	Water Proof
ws.	water-soluble
wt.	weight
xal(l).	crystal(line); crystallized; etc.
XS	cross-section; excess
Z	(Ge) Zentner (50 kg)
ZO	Zoological Origin
zT	zum Teil (Ger): partly
ZZ	Zingiber (formerly stood for Myrrh)

The symbols used for elements and some radicals are those commonly used in chemistry texts, *Chemical Abstracts*, etc.

GEOGRAPHICAL ABBREVIATIONS

Af	Africa
Afgh	Afghanistan
Afrasia	Africa & Asia
Ala	Alabama (US)
Alas	Alaska (US)
Alg	Algeria
Ang	Angola
Alta	Alberta (Canada)
Am	America(n)
Arab	Arabic, Arabian
Arg	Argentina
Ariz	Arizona (US)
Ark	Arkansas (US)
As	Asia
Austr	Austria
Austral	Australia, Australasia
BanglaD	Bangladesh
BC	British Columbia (Canada)
Bo	Bolivia
Br	Brazil, Brasil
Bulg	Bulgaria
Bur	Burma
C	Central
c	central (region)
CA	Central America
Calif	California (US)
Cana	Canada
Cey	Ceylon (now Sri Lanka)
Ch	Chile
Chin	China, Chinese
Co(l)	Columbia
Colo	Colorado (US)
Conn	Connecticut (US)
CR	Costa Rica
CU	Cuba
Czech	Czechoslavakia
DC	District of Columbia (US)
Del(a)	Delaware (US)
Den	Denmark
Dom/DR	Dominican Republic
Du	Dutch (language)
e	eastern (region)
E	East
Ec	Ecuador
Eg(y)	Egypt
EI	East Indies; East Indian
Eth	Ethiopia
Eu	Europe
Euras	Eurasia
Fla	Florida (US)
Fr	France; French
Ga	Georgia (US)
GB	Great Britain
Ge(r)	Germany, German
Gr	Greece; Greek
Guat	Guatemala
Haw	Hawaii (US)
Haw Is	Hawaiian Islands (US)
Hind	Hindi; Hindustani
Hond	Honduras
Hung	Hungary
Ia	Iowa (US)
Icel	Iceland
Ida	Idaho (US)
Ill	Illinois (US)
Ind	India; Indian (as native of India)
Indones	Indonesia
Ir(e)	Ireland, Irish
Is	Island(s)
It	Italy; Italian
Jam	Jamaica
Jap	Japan(ese)
Kans	Kansas (US)
Kor	Korea
Ky	Kentucky (US)
La	Louisiana (US)
Leb	Lebanon
Lib	Liberia

Mad(ag)	Madagascar
Mal	Malay(a); Malaysia
Man	Manitoba (Canada)
Mass	Massachusetts (US)
Md	Maryland (US)
Me	Maine (US)
Med	Mediterranean
Mex	México; Mexican
Mich	Michigan (US)
Minn	Minnesota (US)
Miss	Mississippi (US)
Mo	Missouri (US)
Mont	Montana (US)
Mor	Morocco
Moz	Mozambique
n	north(ern)
N	North
NA	North America(n)
Nam	Namibia
NB	New Brunswick (Canada)
NCar	North Carolina (US)
NDak	North Dakota (US)
NE	New England (neUS)
Neb	Nebraska (US)
Neth	Netherlands
Nev	Nevada (US)
NewCaled	New Caledonia
Newf	Newfoundland (Canada)
NG	New Guinea
NH	New Hampshire (US)
nHemisph	Northern Hemisphere
NJ	New Jersey (US)
NM	New Mexico (US)
Nor	Norway
NS	Nova Scotia (Canada)
NW	New World
NWI	Netherland West Indies
NY	New York (US)
NZ	New Zealand
Oc	Ocean (as in Pac Oc)
Oh	Ohio (US)
Okla	Oklahoma (US)
Ont	Ontario (Canada)
Or(e)	Oregon (US)
OW	Old World
Pa	Pennsylvania (US)
Pac	Pacific
Pak	Pakistan
PapNG/PNG	Papua New Guinea
Par	Paraguay
Pe	Peru
PEI	Prince Edward Island (Canada)
PI	Philipine Islands, Philipines
Pol	Poland, Polish
Por	Portugal; Portuguese
PR	Puerto Rico
Qu	Québec (Canada)
RI	Rhode Island (US)
Rocky Mts.	Rocky Mountain(s)
Rom	Ro(u)mania
Russ	Russia(n)
s	southern (region)
S	South
SA	South America
SAf	South Africa
Salv	Salvador
SArab	Saudi Arabia
Sask	Saskatchewan (Canada)
SCar	South Carolina (US)
Scot	Scotland; Scottish
SDak	South Dakota (US)
Sen	Senegal (wAf)
Sib	Siberia
So(m)	Somalia (eAf)
Sp	Spain, Spanish
SriL	Sri Lanka (formerly Ceylon)
Sud	Sudan (eAf)
Sw(ed)	Sweden
Switz	Switzerland
Tanz	Tanzania
Tenn	Tennessee (US)
Tex	Texas (US)
Tob	Tobago (WI)
Trin	Trinidad
trop	tropics; tropical
Tun	Tunisia
Turk	Turkey; Turkish
Ur	Uruguay
US/USA	United States of America
USSR	Union of Soviet Socialist Republics (now dismembered)
Ut	Utah (US)
Va	Virginia (US)
Ve	Venezuela
VI/VanIs	Vancouver Island (BC, Canada)
VirIs	Virgin Islands
Vt	Vermont (US)
w	western (region)
W	West
Wash/Wa	Washington (US)
WI	West Indies
Wisc	Wisconsin (US)
WV(a)	West Virginia (US)
Wy	Wyoming (US)
Yem	Yemen
Yuc	Yucatán (México)
Yug	Yugoslavia
Zim	Zimbabwe

SYMBOLS USED IN PHARMACOGNOSY AND BOTANY

*	original (or unpublished) data (in this text); date of birth
!	personally verified; positively correct
=	is equal to; equals
≎ ≠ ~	is equivalent to; approx.
⌢	resembles
√	satisfactory; approved; checked; OK
√ or √√	doubly checked
?	doubtful; questioned; to be verified
#	number; space
X	in error; wrong; hybrid
0	worth nothing; zero; absent
⊖	search conducted but nothing found
Ø	diameter
∞	indefinite number, numerous; infinite; more than 20 (stamens)
†	died; deceased; toxic
>	greater than; derived from; over
<	less than; source of or productive of; under
♂	male; staminate; iron; Mars, Tuesday
♀	female; pistillate (carpellate); copper; Venus; mirror, Friday
⚥	hermaphrodite
♂♀	diecious
♃	perennial
Δ	evergreen; heat; delta (double bond position
①	annual
→	produces; is source of
②	biennial
§	section (or subgenus)
℞	prescribed; prescription; to be dispensed
☠	poisonous; toxic; lethal; dangerous: take care!
⊙	exempt Federal narcotic; gold; sun; perfection; monocarpic (flowers once only); annual plant
•	under Federal Narcotic Act limitations
⊖	salt (alchemical sign)
ø	pharmacop(o)eia, phenyl
℈	scruple
ʒ	dra(ch)m
℥	apothecary ounce
♏	minim— minims
γ	microgram; gamma; 0.000001 Gm.
γγ	micro-microgram; 0.000000001 Gm.
μ	micron; micro-millimeter: 0.001mm.
μμ	micro-micron; 0.000001 mm.
O_2	both eyes (ocular prescriptions)
£/s/d	pound Sterling: shilling; pence (respectively) (UK)
606	arsphenamine (antiluetic)
666	benzene hexachloride (insecticide)
914	neoarsphenamine (antiluetic)
°/oo	per mille: per thousand
°/ooo	per ten thousand

THE GREEK ALPHABET

Greek Letter			Name	English Letter(s) Equivalent
Α	α		Alpha	a
Β	β		Beta	b
Γ	γ		Gamma	g (*as in "go"*)
Δ	δ		Delta	d
Ε	ϵ		Epsilon	ĕ (*short as in "met"*)
Ζ	ζ		Zeta	z
Η	η		Eta	ē (*long as in "be"*)
Θ	ϑ	θ	Theta	th
Ι	ι		Iōta	i (*as in "pine"*)
Κ	κ		Kappa	k
Λ	λ		Lambda	l
Μ	μ		Mu	m
Ν	ν		Nu	n
Ξ	ξ		Xi	x
Ο	ο		Omicron	ŏ (*short as in "hog"*)
Π	π		Pi	p
Ρ	ρ		Rho	r
Σ	σ	ς	Sigma	s
Τ	τ		Tau	t
Υ	υ		Upsilon	u or y
Φ	φ	ϕ	Phi	ph
Χ	χ		Chi	ch (*as in "chemist"*)
Ψ	ψ		Psi	ps (*as in "psychic"—silent "p"*)
Ω	ω		Omega	ō (*long as in "note"*)

PLAN OF DEFINITIONS

The sequence of data is as follows[1]:

1. **When listed under SCIENTIFIC NAME:**

Botanical (or zoological) origin, i.e., **genus, species, subspecies, variety,** etc. (Family) (sometimes Division [Phylum–Class–Order]); (geographical distribution, usually where indigenous, sometimes also where naturalized); common name or names; type of plant (optional); part or parts used; leading constituents of same; properties (optional); uses in medicine, pharmacy, food industries, technology, etc.; chief areas of cultivation or production (optional); origin (source) of name, etc. (optional); other miscellaneous data (optional).

2. **When listed under COMMON NAME:**

After item entry (area where or language of name employed)[2]; scientific name or common name or both.

NB—Look under the botanical (or zoological) name entry for the most complete definition of the organism and its drug or other product; also under other definitions where cross-reference is made to them. For inorganic materials, most animals, etc., the chief definition occurs under the leading English title.

The asterisk (*) after an item or statement indicates original data, that is, data not so far as known previously published. (These notations are not complete.)

FAMILY NAMES

In recording family names, the common ending "-aceae" applied to most such names has been omitted, so that for instance Solan. should be read as "Solanaceae." The exceptions are a few large families which (in most systems) do not have the common ending and appear as follows:

Comp. + Compositae (Asteraceae)
Cruc. = Cruciferae (Brassicaceae)
Gram. = Gramineae (Poaceae)
Gutt. = Guttiferae (Clusiaceae)
Lab. = Libiatae (Menthaceae)
Leg. - Leguminosae (Fabaceae)
Palm. = Palmae (Arecaceae)
Umb. = Umbelliferae (Apiaceae)

The names shown in parentheses are family name preferred by some modern authors in order to make uniform the ending "-aceae" for all plant families.

[1]In the sequences of both types shown, parentheses are used to match those actually used in the definitions.

[2]Where no such qualification is used, one may assume the English (or American) language (or name as used in the US) is intended (or it may be that the term occurs in several languages or where the language may not be known [in a few instances]).

Aa

Aalbaum (Ge): honeysuckle; *Stapelia* spp.
Aalbeere (Ge): *Ribes nigrum.*
aalkirsch(e): 1) *Rhamnus frangula* 2) *Prunus padus.*
aalklim (Ind): *Bauhinia* sp.
Aalkraut (Ge): taxa of *Rumex, Teucrium,* & *Satureia.*
aamilis (Arg): *Rhamnus alaternus.*
aardappel (lit. earth apple) (Dutch): potato (tuber).
aardpeeren (Dutch): *Helianthus tuberosus.*
Aar von Arabien (Ge): *Acacia.*
Aaron (Ge): **Aaronstab** (Ge): **Aaronwurzel** (Ge, Dutch): *Arum maculatum* & rt.
Aaron's rod: 1) *Thermopsis caroliniana* 2) *Verbascum thapsus* 3) *Sedum purpureum.*
Aaspflanze (Ge): *Stapelia* spp.
abacá: manila hemp; inner fibers of *Musa textilis* - **abacate** (Br): avocado - **abacateiro** (Br): avocado tree - **abacaxi** (Br): pineapple.
abadejo (Sp): codfish.
abalone: *Haliotis* spp. (Haliotidae - Gastropoda - Mollusca): snail-like shellfish, prized seafood of Pac (Calif [US]) & Atl Oc areas, with fine flavor (*cf.* scallop); meat popular food world wide; *H. rufescens* Swains. chief comm. source; shells u. as currency (Indians); & u. to make costume jewelry; very hard shell has stimulated protein structural research (materials res.: ex. outer surface of automobiles).
abamectin: generic name for avermectins.
abaremo (Br): *Acacia cochleocarpos* Gomes.
abawayno (Calif [US]): *Micromeria chamissonis* (prob. corruption of "yerba santa").
Abbuiss (scabiose) (Ge): *Scabiosa succisa.*
abedul (Sp): 1) *Betula pendula* 2) (Mex) *Alnus acuminata.*
abeille (Fr-f.): bee.
abela (seUS): trumpet weed rt. fr. *Eupatorium purpureum*; u. aphrod. (1982).
abele: *Populus alba.*
abelemosc(h) (Fr): **abelmosco**: 1) (Mex): okra 2) (Sp, It): musk seed.
Abelmoschus crinitus Wall. (Malv.) (Ind, Pak): kamraj; rts. & lvs. as decoc. u. in blood "purificn." (174) - **A. esculentus** (L) Moench (Hibiscus e. L.) (eHemisph, trop; cult sEu, sUS): okra; gombo; lady's fingers; bhindi (Ind); unripe pods smooth or hairy c. mucilage, pectin, starch; u. green as vegetab. (ex. to thicken soups); emoll. dem. in gastric ulcer, diur. in beverages; lvs. u. emoll. poultice; roasted seeds as good coffee substit. (gumbo coffee); (wAf, CR &c), rts. as substit. for marshmallow rt.; mucilage as plasma replacement, also claimed to reduce friction of blood flow & of promise in atherosclerosis (939); sds. c. FO (okra oil) (940) - **A. ficulneus** (L) W. & A. (tropAs, Austral): native rosella; ripe frts. u. like okra (*A. esculentus*); cult. - **A. manihot** Medik. (Hibiscus m. L.) (eAs, orig. fr Korea, Jap): "hibiscus root" u. as poultice for wounds, burns, swellings; to treat GC (Korea); in enteritis, for cough (Chin); med. (Jap); young lvs. & buds u. as veget. (Tonga) (86); subsp. *tetraphyllus* (PI): pl. esp. rt. u. in gastric cancer - **A. moschatus** Medik. (H. abelmoschus L.) (tropics worldwide, incl. Ind, WI, Java, Eg): musk-mallow; musk-okra; wild o.; sds. (Hibiscus, L.) (ambrette [or musk] seed; vegetable m. seed) with musk-like odor; u. in aromatics, ton. stim. diur. antispasm.; antipyr., (Mariana Is.); expressed prepn. of ambrette sd. = oil (VO+FO) (prod. WI, Java, Ind); in smoking

mixts., for insect sting; pl. u. med., as dye (Tonga) (86); lvs. u. for boils (nBorneo).

abeto (Sp): *Abies alba*, etc.

Abfuehrbeeren (Ge): (lit. laxative berries): *Rhamnus cathartica* frts.

Abies Miller (Pin): the firs; handsome conifers which generally prefer relatively cool & humid climates; most significant in the lumber industry - **A. alba** Mill. (A. nobilis A. Dietr.) (c&sEu): (common or European) silver fir; tree to 50 m; exudation from uninjured bk. (blisters) constitutes Strassburg (or Alsace) turpentine; u. like other turpentines (formerly); branch tips u. in sprays for resp. dis., in bath tubs.; disinfectant in perfume & soap industries; in homeopath. med. (2088); VO fr. wd. & lvs. by distn. u. for bruises & sprains; green cones u. to make templin oil (u. in varnishes); chief Xmas tree of Eu; tallest tree of Eu; as lumber source; now mostly replaced by *A. grandis* - **A. balsamea** (L) Mill. (Cana, neUS; Lab.-Va): balsam fir (tree); fir balsam; oleoresin forms in bk. in blisters & is collected as Canada balsam, u. in folk med. of both Am Indians (various antiseptic ointments) & whites (ex. cough drops); in cements, mountant on microslides, in varnishes, &c.; VO c. pinene, juvabione, &c.; bk. u. as decoc. in kidney dis. (Cana); rt. u. for sores in mouth (Am Indians); lf. VO called Canadian fir needle oil (or balsam fir needle oil) obt. by steam distn. fr. lvs. (common names confusing); u. in perfumes, air fresheners, disinfectants, &c. - **A. canadensis** Mx. = *Tsuga c.* - **A. cilicica** (Ant. & Kotschy) Carr. (AsMin, nSyr): Cilician fir; resin u. for sprains, backache, fractures, rheum., bone setting, &c.; mixed with 20% pitch, melted, poured on cloth, & wrapped around ailing site for 6 days, then removed & site rubbed with vaseline (2469) - **A. communis** Hort. = **A. excelsa** (Lam.) DC = *Picea abies* Karst.- **A. firma** Sieb. & Zucc. (A. momi Sieb.) (Jap): Japanese (silver) fir; momi; wd. valuable, u. in joinery, buildings, paper pulp - **A. fraseri** (Pursh) Poir. (Va, NCar [US]): southern (balsam) fir; Fraser's (balsam) fir; she-balsam; oleoresin sometimes u. in place of Canada balsam; lvs. (needles, tips) u. in medl. pillows, bedding for campers (fragrant), &c.; cult. for use as Christmas trees - **A. grandis** Lindl. (wNA; VancIs [Cana] to Calif [US]): grand (or giant or white) fir; large tree (-100 m); valuable wd. u. for packing cases, paper pulp; blisters in bk. c. oleoresin u. like Canada balsam as vulner., cement - **A. lasiocarpa** (Hook.) Nutt. (wNA): Rocky Mountain (or white) fir; alpine f.; gum oleoresin (Oregon balsam) fr. wd. u. as antisep. in wounds; lvs. & wd. u. as incense (Am Indian) - **A. momi** Siebold: *A. firma* - **A. nordmanniana** Spach (cAs, Crimea, AsMin): Nordmann (or Caucasian) fir; oleoresin employed; bk. c. 10% tannin; young trees u. as Christmas trees; wd. useful, as pulpwood (941) - **A. pectinata** DC = **A. picea** Lindl. = *A. alba* - **A. pindrow** (Royle) Spach (cAs, Afgh, Nepal) (943): (west) Himalayan fir; Pakistan silver f.; Himalayan s. f.; wd. much u. for home building, &c. - **A. procera** Rehd. (A. nobilis Lindl.) (Wash-Calif [US]): noble (red) fir; r. f.; white f.; excellent lumber "larch", u. for sashes, doors, airplane mfr., &c.; VO of interest (942); wood for paper pulp - **A. religiosa** (HBK) Schlecht. & Cham. (sMex, CA): sacred fir (tree); oyamel (Mex); trunk source of oleoresin (turpentine) ("balsam") u. like regular balsams in med. (Mex, Hond); distn. furnishes VO, "aceite de palo," a turpentine oil u. in paints & med.; wd. for general building, paper pulp; branches u. for decoration of churches & homes (hence name) - **A rubra** Poiret = *Picea rubens* Sarg. - **A. sibirica** Ledeb. (nRuss to neAs): "Siberian fir"; fresh lvs. yield needle VO u. in med. (expect., antisep. in chronic rheum.) & perfume ("Siberian fir needle oil," confusing comm. name since this sp. is not a fir) - **A. taxifolia** Desf. = *A. alba*.

abietic acid: sylvic a.: a very abundant & the most inexpensive organic acid; derived fr. rosin by heating alone or with acids; u. mfr. metallic resinates, plastics, soaps, &c. - **a. anhydride**: the normal constituent of rosin fr. which a. acid is obtained.

abietin: coniferin.

abilo (Zagala): tree producing gum elemi.

abim (Singhal.): **abin atta** (Singhal.): opium poppy.

abinca (Pe): squash.

abir (Hind): scented pwd. u. in worship; c. *Hedychium spicatum* tubers + *Andropogon muricatus/-squarrosus* rt. + sandalwood.

abkari: v. Opium.

ablab (Eg): *Dolichos lablab*.

abobora (Pe, Br): pumpkin, squash, gourd; sd. of *Curcurbita pepo* (Br).

Abolboda poorchon Seub. (Abolbod./Xyrid. - Monocots) (Br): "marirca" (Br); rt. u. as lax.; fresh pressed juice of rt. u. as substit. for rhubarb syrup.

abordagem (Por): screening.

abouchi: almaciga (gum).

abrasin (Fr, Jap): mu (tung) oil; fr. *Aleurites cordata*.

abrico (do Pará) (Br): *Mammea americana* - **abricot** (Fr.-m): **abridor** (Co., etc.): apricot.

abrin: potent toxalbumin found in *Abrus precatorius*; u. in cancer research; *cf.* ricin.

Abri Semen (L.): **A. Semina** (plur.): sd. of *Abrus precatorius*.

abrojo (rojo) (Mex): *Tribulus cistoides* L.

Abroma (Ambroma) augusta (L.) L.f. (*A. fastuosa* Jacq.) (Sterculi.) (Ind to PI; cult. Af): devil's cotton; cotton abroma; olutkombul (Ind); rt. bk. c. mucil., resins, phlobaphene; fresh or dr. rt. bk. or juice of fresh bk. u. as emmen. & uter. ton.; rt. used to treat itch (Indones); stim. diur.; in female dis. (dysmenorrh.); bk. with strong fiber u. for cordage, clothes lines (cf. jute).

Abronia Juss. (Nyctagin.) (wNA): sand verbena; attractive herbaceous pl.; ca. 35 spp.; u. med. for lacteal stim.; ornam. - **A. fragrans** Nutt. & Hook. (wNA [Neb, NM, Tex+]): sweet sand verbena; ground rts. mixed with maize & u. for food (Am Indians) - **A. latifolia** Esch. (BC [Cana] to coastal Calif [US]): yellow sand verbena; y. abronia; rts. food of Chinook Indians (formerly).

Abrotanum (L): *Artemisia abrotanum*, southernwood.

abrotine: alk. fr. *Abrotanum*.

Abrus fruticosus Wall. (A. pulchellus Wall.) (Leg.) (Ind to Austral): rt. u. as licorice substit. - **A. precatorius** L. (trop&subtrop, esp. Ind; also cult. Fla [US], WI, CA, SA, Mex, HawIs): jequirity (bean); precatory beans, love b., lucky b., rosary b.; lick weed; abrus plant; love pea; prayer beads; ripe sds. c. abrine, agglutinins, abrin (jequiritin, toxalbumin), abric acid, resin; extern. u. dem. for chronic eye dis., in lupus & other skin dis.; abort.; sd. contracep. very toxic, criminal poison (Ind) (one sd. may be lethal by parenteral route) but edible if boiled; (cakes eaten); standard wt. for druggists, goldsmiths (Ind), jewelers (SriL, &c.) (231 grains = 150 mg.) (2 sds. = 1 carat); attractive ornam. novelty in pins, necklaces (Haiti), decorative table pads, &c.; sds. u. in maraca (Br) (rattles made of dr. hollowed gourd & beans) u. in orchestras, religious ceremonies, &c. as percussion instruments; lvs. c. glycyrrhizin, u. in eye & mouth dis.; rt. ("Indian licorice") u. int. as substit. for licorice, as dem. antipyr., aphrod., in snakebite (1467); exter. in leucoderma; folk drug in trachoma; lf. u. for diarrh., stomachache in babies (Mayas) (6); lf. inf. in hemorrhoids; cough (Congo) - **A. schimperi** Hochst. (Tanzania): fresh rt. chewed as aphrod.; rt. decoc. for menstr. irregul.; macerated lf. as poultice for scorpion sting; st. u. as toothbrush.

abscess root: *Polemonium reptans*.

abscisic acid = abscisin = plant hormone wh. accelerates detachment of deciduous lvs. (abscission) at time of lf. fall; pl. growth regulator.

absint (SpAm): **absinto** (Sp): absinth.

absinth: wormwood - **absinthe** (Fr): a liqueur in which the very bitter principles fr. *Artemisia absinthium* prevail; the cause of serious & dangerous addiction; hence now beverage outlawed in civilized countries (often made in combn. with anise, peppermint, hyssop & sometimes fennel & coriander); alc. (2089) **- a. aluine** (Fr): **a. aluyne: a. grande** (Fr): *Art. absinthium* - **a. petite: a. pontique: a romaine**: *Art. pontica* - **a. suisse** = genepi blanc = *A. mutellina* Vell.

absinthi(i)n: chief bitter princ.; glycoside; occ. in *Artemisia absinthium* (lvs. & tops), hence name; *Art. arborescens* L., *Ambrosia artemisiifolia*.

absinthism: chronic poisoning by *Artemisia absinthium* manifested by restlessness, emesis, vertigo, convulsions & insanity (psychoses); formerly quite frequent in Fr (habituation).

Absinthium (L): common wormwood, *Artemisia absinthium* - **A. maritimum** (L) = *Art. maritima* - **A. officinale** Lam. = **A. vulgare** (L) = *Art. absinthium* - **A. ponticum** (L) = *Art. pontica* (Roman wormwood) - **A. Santonicum Judaicum** (L) = *Art. judaica* - **A. Sant. Alexandrinum** (L) = *Art. santonicum= A. cina*.

absint(i)o (Por): 1) *Artemisia absinthium* 2) alc. liquor contg. same **- a. pontico** (Sp, Por): *Art. mutellina* Vill. (or *A. pontica*) - **a. romano**: Roman or Latin absinthe, Herba A. romani, fr. *Art. pontica*.

absinto gentil (Por): Herba Absinthii Gentil.

absolute: florescence; in perfumery, odoriferous principles of fls. in most conc. & alcohol-soluble form.

absorbance: absorbancy: extinction: log of reciprocal of transmittance of light.

abstract (US): kind of dry ext. made with milk sugar & alc. ext. of drug.

Absud (Ge): decoction; ext.

abucá: *Musa textilis*.

abuova (Ind): *Trichosanthes bracteata*.

abuta (raiz) (Br): pareira; *Chondodendron platyphyllum*.

Abuta amara Aublet = *Aristolochia glaucescens* (*qv.*) - **A. candollei** Tr. & Planch. (Menisperm.) (Guianas): rts. source of "White Pareira" (not Pareira Brava); u. as ton. stom.; added to curare - **A. grandifolia** Sandw. (Co): u. similarly to *A. candollei*; lf. inf. u. to treat fevers (Siona Indians, Co) - **A. imene** (Mart.) Eichl. (Co): rt. bk. u. in prepn. curare - **A. rufescens** Aubl. (Guyana, Br): white pareira (brava); rt. u. in prepn. curare (arrow poison, along Rio Caqueta); rt. u. diur.; bitter ton.; stom. - **A. yaupesensis** Krukoff & Barneby (Co), likewise.

Abutilon Mill. (Malv) (trop & subtr worldwide): pls. esp. fls. c. mucil.; bk. u. for fiber - **A. abutiloides** Garcke (Mex to Co; WI): bushy abutilon; mizbil; lf. decoc. in diarrh.; lvs. as poultice for tumors (Haiti); emoll. (Yuc); fl. inf. diur. - **A. angulatum** Mast. (Af pantrop): sts. yield fiber; lvs. (raw or boiled) remedy in hiccup, fever - **A. asiaticum** G. Don (trop): st. gives fiber u. for sacks; lvs. u. as remedy in GC, ulcers, diur. in bladder stone - **A. bidentatum** Hochst. ex. Rich. (Pak, Ind, sUSSR): pl. u. med., to make ropes; sds. furnish FO (Russian oil, similar to cottonseed oil) - **A. graveolens** W. & A. = *A. hirtum* - **A. hilderbrandtii** Bak. f. (trop eAf): rt. u. as anthelm. - **A. hirtum** Sweet (tropInd, Pak; Russ): sds. rich in FO, u. much like cottonseed or soy oils; sds. u. as lax. - **A. hypoleucum** A. Gray (Mex, sTex [US]): whiteleaf abutilon; u. med. (Mex) - **A. inclusum** Urb. (Cu, Hispaniola): petit mahot; inf. of fls. &c. u. as diur.; lf. inf. in asthma & as pectoral (bathing & poulticing); lf. & fl. inf. u. expect.; decoc. to cleanse wounds; st. furnishes coarse fiber (Ind) - **A. indicum** (L) Sweet (pantrop): lf. inf. antipyr. (Indochin, Mal): emoll. in poultices & fomentations; lvs. u. as adulter. of Mallow Lvs.; lf. & fl. u. int. as expect.; decoc. to cleanse wounds; st. furnishes strong fiber (Ind) - **A. mauritianum** Sweet (w tropAf): lf. inf. u. with spices in VD; pl. juice appl. as emoll. to relieve soreness of nates in young children (sNigeria) - **A. permolle** (Willd.) Sweet (sFla Keys [US], WI, CA, sMex): velvet leaf (Bah); butter weed; butter print; (common) velvet leaf; Indian mallow; heated lvs. appl. to draw boils (Bah); lf. decoc. u. to bathe sores - **A. pictum** Walp. (A. striatum Dickson) (sBr, Ur, Arg): pl. has colored variegated lvs.; ornam. - **A. sonneratianum** Sweet (Af): u. as "stim." for bulls in the spring - **A. theophrasti** Medik. (A. avicennae Gaertn.) (subtrop incl. seUS, sAs): butterweed; butterprint; velvet leaf; Indian mallow; pl. & esp. fls. u. as mucil. (*cf.* marshmallow) (sAs, esp. Ind); sds. u. diaph. antipyr. lax. dem.; antisep. in dys. (Chin); source of semi-drying fixed oil; frt. u. to imprint butter; fiber u. like jute ("China [or Chinese] jute") (cult.) & to caulk boats; generally considered a weed - **A. trisulcatum** Urban (Mex to Nic; WI): tronador(a); amantillo; lvs. crushed & rubbed on canker sores in mouth (Yuc); st. fibers u. to make ropes, hammocks & nets; u. int. for asthma in children (6) - **A. venosum** Paxt. (A. striatum Dicks.) (Mex): monacillo (amarillo); u. med. (Mex) - other *Abutilon* spp. u. in folk med. include *A. bidentatum* Hochst. and *A. muticum* Sweet.

abutua (raiz) (Br): Pareira.

Abyssinian tea: *Catha edulis*.

acacatechin: characteristic catechin of cutch (black catechu).

Acachmena H. P. Fuchs (Cruc) (Euras: Balkans; Black Sea Basin): c. digitaloid glycosides, hence a potential cardioactive drug.

Acacia Mill.: 1) genus (Leg.) of ca. 1200 trop & subtr. spp., mostly trees; most spp. are xerophytes & many bear thorns; source of many important economic products, such as gums, tannins (wattles; u. in tanning); fr. wood 2) (specif.): (L): gum arabic, acacia gum; gummy exudation fr. *A. senegal* & other Af *Acacia* spp. (produced in Sudan, sArabia) 3) (Brit horticul.): *Robinia* spp. 4) (sUS): mimosa tree - **A. albida** Del. (s&swAf): winter (or white) thorn; bk. decoc. u. in diarrh.; gum, bk. & lf. u. for colds, hemorrhage (wAf) - **A. angustissima** Ktze (sUS, Med, CA to CR): prairie acacia; guajillo; guajito; u. for enlarged spleen (Yuc); antitumor activity; bk. u. as tan.; pods as food (Mex, Ind); var. *hirta* B. L. Robins. (seUS); "acacia"; fern a.; whiteball a.; ornam. tree; pollen often produces hay fever - **A. anthelmintica** Baill. = *Albizia a.* - **A. arabica** Willd. (A. adansonii Guill. & Pers.) (tropAs&Af; indig. to Sind [Pak]): babool (tree); Indian gum arabic tree; very beautiful tree; pods c. superior tanning material (42% tannins with gallic & ellagic acids); gummy exudn. resembles crude gum arabic & u. much like off. Acacia in diarrh., dys., diabetes; dem.; acacia substit.; adhesive; prod. Babul Bark. u. astr. in diarrh., dys., sore throat, leukorrhea, stomatitis, tan; lvs. c. tannin.; lf. juice u. in conjunctivitis; frt. (Gambia pods) edible (when young); also u. as tan (esp. for sheep leather; —> glacé leather); sds. roasted & eaten; grd. bk. a famine food; (mature) wd. hard, u. for railroad ties, tool handles, &c. - **A. aroma** Gill. (Arg): aromo negro (Arg); tusca; frt. u. as "antiveneral," ext. antisep., lf. u. also - **A. caesia**: Willd. (Ind): aila; rt. u. in pleurisy; bk. in measles, bronchitis; smallpox (formerly) - **A. capensis** Burch. = *A. Karoo* - **A. catechu** Willd. (Ind, sChin, EI, Borneo): catechu (or khair) tree; heartwood source of black catechu by boiling water extn.; (gum) cutch, terra japonica; c. catechins & catechin tannins, esp. epi-catechin; u. as astr. in dys., hemorrhages, mouth washes; wd. chips boiled to furnish brown dye for textiles (khaki dye); tan; exhausted wd. u. for fuel; exudn. fr. trunk wounds is a gum called khair gum, c. aldobionic acids (glucuronosido-1: 1-galactose); u. mucil. dem. - **A. concinna** DC (Ind, Chin+): sik(k)akai, sikekai, kochai, &c. (Ind); frt. a pod, may c. saponin, u. as detergent for washing hair (shampoos), textiles &c.; bk. c. alkaloid, u.

against leprosy; sds. & resin u. med.; young lvs. edible - **A. constricta** Benth. (sUS, nMex): Mescat (or whitethorn) acacia; pods eaten by livestock; lf. & sd. in med.; ornam. - **A. cornigera** (L.) Willd. (Mex [Veracruz+] CA): zubin; cornezuelo; young shoots of shrub/small tree cooked & eaten; lvs. u. for insect bites; bk. & rt. med.; cultd. - **A. decurrens** Willd. (Austral, sAf): (Sydney) black wattle; green w.; bk. decoc. astr. in dys. (extreme cases); source of tannin (tan bark); trunk furnishes wattle gum, rich in bassorin; sds. toxic to stock - var. *dealbata* F.v.M. (A. d. Link) (Austral, Tasmania; cult. sAf): gum like acacia (Australian g.), u. in bronchial dis.; fl. VO u. in perfumes - **A. detinens** Burch. = *A. mellifera* Benth. var. d. Brenan - **A. ehrenbergiana** Hayne (tropAf): amleh; trunk source of one var. of gum arabic - **A. farnesiana** (L) Willd. (believed to have orig. in WI or nSA, now widely distr. in trops & sUS): sweet acacia; "stinking bean" (Fla [US]); cassie; huisache (Mex); fls. u. in perfumery (mostly as VO), antispasm. in cramps; gummy exudn. fr. trunk u. as emoll.; frt. u. in diarrh., dys., gynecol. dis. (Mex); fls. mixed with grease & rubbed on head for headache (Mex); bk. u. in typhoid fever (Co); st. decoc. u. in fevers, edema (Laos); gum fr. pod strong adhesive; forage; honey pl. - **A. Germanica** (L): juice (dried) of *Prunus spinosa* (German article) - **A. giraffae** Willd. (sAf): giraffe acacia; Kameldorn (=camel thorn); bk. u. tan., pd. pod as flour & fodder; gum exudation of trunk = Cape gum, an inferior form of gum arabic; sds. u. as coffee substitute - **A. glauca** Moench (WI): mata di galina; rt. chewed for sore throat; boiled in vinegar & liquid u. as gargle to loosen phlegm in throat; similarly decoc. of branch & of lvs.; rts. sap u. as styptic; rt. decoc. u. as gargle to prevent diphtheria (epidemic, Curaçao, 1875) - **A. greggii** Gray (swUS, nMex): cat's claw (acacia); cat claw; devil's claw; semi-desert shrub or small tree; pod u. —> meal, food of Am Indians; fodder; bee flower; stem produces gum *cf.* acacia, u. locally; wd. strong, for lumber, fuel; strong prickles on stem hazardous to clothes - **A. hindsii** Benth. (wMex to Pan): iscanal (negro); bk. decoc. appl. to scorpion stings - **A. hockii** De Wild. (sChad-Af): med. u. in Guinea - **A. holosericea** Cunn. (Austral): rt. inf. drunk by aborigines for laryngitis (2263) - **A. (h)omalophylla** Cunn. (Austral): (curley) yarran; wd. valuable u. in turning; stem furnishes an Austral kind of gum arabic - **A. horrida** Willd. = *A. karoo* - **A. julibrissin** Willd. = *Albizzia j.* - **A. jurema** Mart. (Br): "Barbatimão" (Por Pharm.): bk. introd. 1919 into Europe as Cortex adstringens brasiliensis; may be narcotic (Caminhoa) - **A. karoo** Hayne (A. horrida Willd.) (s& tropAf): karoo (or mimosa) thorn; "gum arabic tree"; sour (or sweet) thorn; "thorn tree"; bk. u. in diarrh., dys. (Cape), inf. of crushed rt. in colic; gummy exudate (lf. & bk.) u. for colds, hemorrhage; source of Cape (or South African) (acacia) gum; bk. decoc. emetic (Zulu); legumes & lvs. fodder (no HCN); sd. coffee substit.; bk. tan., in cordage.; wd. u. to make articles, fuel; tree u. to form hedges - **A. koa** Gray (Hawaii): koa (acacia): timber u. for furniture, implements; sd. u. necklaces - **A. leucophloea** Willd. (Ind, Burma, seAs): arinj; young pods & sds. edible; pd. bk. famine food; with bajra (*Setaria italica*), to flavor alc. beverages; tan bk.; bk. fibers u. to make nets & ropes; lvs. black dye; st. produces gum u. to adulterate off. Acacia & other gums; wd. u. to make agr. implements, fuel - **A. melanoxylon** R. Br. (Austral) blackwood; pl. has phyllodes (no true lvs.): wd. u. in cabinet making, &c. - **A. mellifera** Benth. var. *detinens* Brenan (sAf): hookthorn; pod and lf. edible; trunk yields gum, u. food, earth floor hardener, cohesive in arrow poisons; bk. chewed for cough (Tanz); wd. u. fences, fuel - **A. milleriana** Standl. (sMex): lvs. u. in coughs & colds (Yuc) (Mayas) (6); wd. for construction - **A. modesta** Wall. (Ind, Pak): phula(h)i; yields small amt. of gum ("Amritsar gum") (u. as "restorative"); lvs. & fls. fodder of cattle, camels, &c. - **A. mollissima** Willd. (Austral): black wattle; u. as tan bark (with ca. 25% tannin); toxic to stock; close to *A. decurrens* - **A. nilotica** (L.) Del. = *A. arabica* (*A. n.* may possibly have priority) - **A. Nostra** (**A. Nostrates**) (L): gum of *Prunus spinosa* (*qv.*) - **A. pennata** (L.) Willd. (A. pinnata Dalz. & Gibs.) (As, Af): arar; rugeus; bk. & frt. pulp u. as fish poison, tan; lvs. u. for fever (Indones); bk. in snakebite; for dandruff; bk. fibers u. as cordage; rts. boiled to give poultice for rheum. (Mal); twigs & rts. u. tooth brushes; rts. u. for aroma (Laos) - **A. p.** subsp. *insuavis* (Lace) Nielsen (seAs); birmanie; nau (Laotian); rts. u. for aroma (Laos) - **A. pennatula** (Schl. & Cham.) Benth. (Mex, CA): algarrobo; palo garobo; shrub or small tree; bk. u. in indigestion, as dye; wd. to make charcoal; VO prepd. fr. fls.; sds. eaten with pod when food scarce - **A. planifrons** Wight & Arn. (Ind & EI): umbrella thorn; wd. u. to make agr. implements; fibers u. to make mats; to prepare alpha-cellulose - **A. polyacantha** Willd. subsp. *campylacantha* Brenan (s&tropAf): African catechu tree; gum adhesive, agreeable to palate & u. in confectionery; gum or piece of bk. sucked for sore throat; leprosy remedy; in pneumonia; rt. repellant to snakes & crocodiles; astr., in dyeing

& tanning; wd. u. for mine shafting - **A. pycnantha** Benth. (Austral): golden wattle; bk. c. 18-50% tannin, inf. in diarrhea, eye dis., ozena, typhoid fever, hemorrhage; in tanning; sts. & branches furnish Australian (or wattle) gum (*cf.* Acacia) - **A. rigidula** Benth. (Mex): "gavia"; bk. of shrub medicinal - **A. senegal** Willd. (A. senegalensis Roberty) (tropAf): fissures in bk. produces gum (Acacia; gum arabic), chief source of the best sorts (Kordofan gum); this water-diffusible (w. "soluble") gum u. dem. emulsifier, taste modifier; mucilage for postage stamps, labels; in making inks, matches; in textile printing, sizes; in photo. paper, paper & dye industries; c. arabic acid, complex of galactose, arabinose, rhamnose, glucuronic acid, &c., as salts (Ca, Mg, K); traces of tannin, enzymes (mostly oxydases & peroxidases); prod. mostly in Sudan & Senegal - **A. seyal** Del. (Eg, Alg, Mor): shittah tree; thirsty thorn; wd. u. as fumigant for rheum. pain; lvs. & bk. in gastric ulcer; lvs. u. for fever; bk. chewed to quench thirst (Tanz); gum (Sp Phar.) edible, u. for rheum., respir. inflammn.; pods livestock fodder - **A. sieberiana** DC (tropAf) st. yields gum like gum arabic, u. emoll.; ext. rts. rich in tannins, u. tenifuge - **A. stuhlmannii** A. Taub (Tanz): mgunga (Swahili); rts. pounded & eaten with food for infertility; believed onc source of gum acacia - **A. subcoerulea** Lindl. (wAustral): silvery or blue-leaved acacia; source of a yellow dye - **A. suma** Kurz (Ind, SriL, Burma, eAf, natd. WI): khair tree; one source of Catechu (cutch) (*qv.*); rt. c. tannin, decoc. u. gonorrhea, sore throat, stomachache (eAf); yields gum - **A. sundra** Willd. (Ind): source of a gum - **A. tortilis** Hayne (nAf, wAs): talha; source of gomme rouge (red gum) (resin); u. in indigenous medicine; important forage plant - **A. tortuosa** Willd. (Prosopis microphylla Kth.) (WI, sFla, sTex, [US], Ve, Co, Ec): (Rio Grande) acacia bush; "sweet briar"; huisachillo; lf. decoc. cold remedy; to wash eruptions; rt. decoc. in GC; oil exudn. appl. to freckles, rubbed on head for headache; gum u. as chewing gum - **A. vera** Willd. = *A. arabica* - **A. verek** Guill. & Perr. = *A. senegal* - **A. victoriae** Benth. (A. sentis F. v. Mueller) (eAustralia): prickly acacia; p. wattle; gundabluie; elegant wattle; source of Sydney Gum (kind of acacia gum) - **A. xanthophloea** Benth. (Moz): fever tree (Kenya); bk. emet.; mal.; "prophylactic"(!); gum u. also.

acacia: 1) (sUS): *Albizzia julibrissin* 2) **false a.** (nAustral): *Robinia pseudoacacia* - **a. bark**: fr. *A. arabica* & *A. decurrens* - **a. blossom oil**: VO of *Acacia farnesiana* - **cat's claw a.**: *Acacia greggii* - **coast a.**: *Acacia cibaria* F. v. Muell. - **Farnesian a**: **a. flowers**: 1) (Ind, Turk): *A. farnesiana* fl. = Cassie (flowers) 2) (Ger, It): sloe fls. *Prunus spinosa* - **rose a.**: *Robinia pseudoacacia* **sweet a.**: *Acacia farnesiana* - **true Egyptian a.**: *A. arabica* - **white a.**: galam gum.

Acaciae Gummi (L): **A. Nostras**: acacia gum.

Acaena affinis Hook. (Ros.) (Kerguelen Is): Kerguelen tea; u. by whalers as febrifuge - **A. anserinifolia** Druce (A. sanguisorbae Vahl; A. novaezelanicae Kirk) (Austral, NZ): biddy biddy; lf. inf. u. to treat wounds, contusions, painful urination; rheum., kidney, bladder & gastric dis.; general ton. (NZ) - **A. argentea** Ruiz et Pav. (Chile): amores secos; astr. refriger. diur. u. vulnerary, in VD; antitumor activity against human naso-pharyngeal carcinoma - **A. cylindrostachya** R. & P. (Ve, Co): entire pl. boiled with other rts. producing a gel u. at the menstrual period - **A. eupatoria** Bitter (Ur.): u. to control fertility - **A. pinnatifida** R. & P. (Chile, Co): hb. inf. astr. in diarrh., ton. vulner. diur., for amenorrhea - **A. splendens** R. & P. (Chile): cepa caballo; hb. u. diur. emoll., refriger., in liver and kidney dis., rheum. gout; u. exter. to wash wounds.

açafrão (Por): saffron.

acahual (Mex): sunflower.

acajou (Br, Fr): **acaju**: 1) cachou or cashew frt. fr. *Anacardium occidentale* 2) a. wine - **a. balsam**; "raw cardol"; u. epispastic; mfr. indelible ink - **a. cica** (Br): **a. gum**: fr. wd. of *Anacardium occidentale* - **a. resin**: fr. frt. of *A. occidentale.*

acalifa (It): *Acalypha indica.*

Acalypha alopecuroides Jacq. (Euphorbi.) (sMex, WI, CA, nSA): (hierba) meuna; rt. decoc. diur. (Cu, Ven): for GC & bladder stone (2504) - **A. amentacea** Roxb. var. *trukensis* Fosb. (A. t. Pax & Hoffm.) (Caroline Is): fl. (& lf.) juice in water u. for eyes - **A. arvensis** Poepp. & Endl. (sMex to Pe & Bo; Martin): guasanito; pl. decoc. u. for skin dis., vener. sores, snake bite (Guat) - **A. boehmerioides** Miq. (A. fallax Muell.-Arg.) (Ind & Indochin): couppemenysakalaty (Tamil): similar in properties to *A. indica*; medication for headache (Indochin) - **A. cardiophylla** Merr. (PI): lvs. appl. to head for headache; bk. appl. to "deep" ulcers - **A. evrardii** Gagnep. (Vietnam): tra ring; fls. & lvs. u. in Vietnamese med. (diur. beverage) - **A. fruticosa** Forsk. (warm As&Af): sinnicedy (Tamil); lvs. u. as stom. (in dyspeps.) & diaph. (Ind); sts. u. as poultice to cure sores, scabies; rt. decoc. for pains in left side; lvs. pounded & placed on irritating sores (Tanz); lf. juice u. for ophthalmia (in drops); rt. inf. drunk for chancre (Paré); GC; antipyr.; lvs. u. for GIT dis, cholera (Ind) - **A. guatemalensis** Pax & Hoffman (Guat,

Hond): hierba del cancer; pl. decoc. drunk as ton. & diur.; appl. to sores & wounds - **A. hispida** Burmann f. (EI, New Guinea): chenille plant; red-hot cat-tail; rts. & fls. u. in hemoptysis; for dys. (2448); lvs. for thrush, boils (Indones); ornam. - **A. indica** L. (Ind, sChin, natd. Af): Indian nettle (or acalypha); cat's n.; Brennkraut (Ge); 3-seeded mercury; hb. u. as expect. (in bronchitis, phthisis), emet.; lax. (in supposit. in obstipation); anthelm. (vermif.); GIT irrit.; exter. in skin dis. as eczema; ipecac substit.; rt. inf. decoc. purg.; hb. c. acalyphine, VO, HCN glycoside - **A. lindheimeri** Muell.-Arg. (wTex-sAriz [US], Mex): Lindheimer's copperleaf; yerba del cancer (Mex); u. purg.; u. as wash for sore gums, loose teeth; appl. to ulcers (2230) - **A. monostachya** Cav. (A. hederacea Torr.) (Mex) "hierba del cáncer"; hb. u. in med. - **A. novoguineensis** Warb. (Solomon Is): lvs. rubbed into wounds; lvs. put in boiling water & sore eyes exposed to steam; lvs. heated & appl. to swellings - **A. ornata** Hochst. (c&eAf): rt. u. in leprosy & magic (Tanz); lvs. soaked in water u. to wash scabies (Tanz); cooked lf. u. in relief of postpartum pain; rt. decoc. lax. (cAf, Moz) - **A. peduncularia** Meissn. (sAf): rt. inf. expect. ton.; emet. - **A. petiolaris** Hochst. (tropAf): u. as vulner., contraceptive, abortif. (Zimbabwe) - **A. phleoides** Cav. (sMex): yerba del cáncer; lvs. u. to cure ulcers, pimples, inflamm.; cataplasms for snakebite; rt. decoc. u. for sore jaw & loose teeth; cancer; foul ulcers, &c. - **A. punctata** Meissn. (sAf): inf. of decorticated rt. u. as emet.; in chest dis.; vulner., rt. decoc. u. as contracept. abort. - **A. rhomboidea** Raf. (e&cNA): rhomboid copperleaf; pl. reputed medicinal - **A. sinensis** Klotzsch. (Af): in rheum. & neuralgia - **A. virginica** L. (e&cUS): Virginia copperleaf; mercury weed; pl. u. diur. & expect. - **A. volkense** Pax (Tanz): whole pl. u. for scabies; rts. u. in earache; vapor of crushed lvs. for bad cough; rt. decoc. with milk in GC - **A. wilkesiana** Muell.-Arg. (WI+, Pacif) (A. amentacea subsp. w. [Muell.-Arg.] Fosberg): beefsteak plant; copperleaf; Jacob's coat; "carpet" (Trin); lf. u. as poultice for swellings, headache (Trin); pulped shoot appl. to lacerations, open sores (2263); cult. orn.

Acampe papillosa Lindl. (Orchid.) (Ind): rt. u. as bitter ton. & in rheum. - **A. praemorsa** Blatter & McCann (Ind): similar to prec. in usage.

acanta (Sp): *Acanthus mollis*.

Acanthea virilis: pharm. name; some think may refer to Lignum Muira Puama (*Ptychopetalum olacoides* Benth. & *P. uncinatum* Anselmino [Rebourgeon] or *Liriosma ovata* Miers).

Acanthocephalus chinensis (Lam.) Rich. (Comp); (eAs, PR): laran tree; bk. u. tonic, antipyr.; lvs. u. to gargle; tree very rapid growing.

Acanthocereus (Berger) Britt. & Rose = *Cereus*.

Acantholippia Griseb. (Verben.) (SA): some spp. stim. because of VO content - **A. deserticola** (R.A. Phil.) Mold. (Lippia d. R.A. Phil.; L. salsoloides Benth.) (Arg, Bo, Ch): rica-rica; pl. u. as stom. abortient (490) - **A. hastulata** Griseb. (Bo, Arg): "lippia leaves"; "salvia sija" (?); arom. pl. considered med. (inf. for stomachache); VO u. in prepn. of artificial essences of salvia & vermouth - **A. seriphioides** (A. Gray) Mold. (Arg): tomillo (del campo); u. in fragrant teas & as condim.; suggested thyme substit.; VO c. thymol, carvacrol, citral, p-cymene - **A. trifida** (C. Gay) Mold. (Lippia t. Remy): (Ch): suggested as thyme substit. (high thymol content).

Acanthopanax Miq. (Arali.) (Chin, PI, Mal): tsu-wu-cha (Chin): some spp. u. to prevent & treat dental caries - **A. gracilistylus** W. W. Smith (Chin, Mal): rt. bk. u. to make medl. wine popular in Chin for rheum. general debility, impotency, lumbago; st. in flatulence - **A. henryi** Harms (Chin): rt. bk. u. in rheum. syph. respir. dis.; in flatulence (Mal); scabies, edema (Chin) - **A. senticosus** (Rupr. & Maxim.) Harms = *Eleutherococcus s.* Maxim. - **A. sessiliflora** Seem. (Kor): bk. of st. & branches u. by people in their dotage for arthritis, rheum. hypertension (Kor Ph.) - **A. spinosus** Miq. = *A. gracilistylus* - **A. trifoliatus** Merr. (eAs, PI): st. in wash to treat leprosy (Chin); lf. decoc. a ton. for general weakness; bk. inf. for young children late in walking. as tinct. to correct nerve & bone dis. (Indochin) (—> *Eleutherococcus*).

Acanthopeltis japonica Okamura (Gelidi. - Rhodophyceae - Algae) (Jap & Kor coasts) toriashi; dorakusa; source of a good grade of agar ("toriashi agar"); thallus u. as food.

Acanthophyllum glandulosum Bge., **A. gypsophiloides** Regel (Caryophyll.) (cAs-steppes): rt. (Turkestan [or Levant] soap root) u. as expect. & emet.; vulner. for horses; in glanders; detergent.

Acanthosicyos horridus Welw. ex. Hook. f. (Cucurbit.) (swAf, sand dunes): narras; sd. FO with fine flavor like almond oil; dr. sds. ("butter pits") eaten as nuts & in pastries; frt. edible - **A. naudiniana** (Sonder) C. Jeffrey (sAf): important as source of food & water (Kalahari Desert).

Acanthospermum australe Ktze (Comp.) (Par, Ur, Co): tapeque; u. for sores on feet; contracep. (Ur); sold in markets (Villarrica, Par) - **A. hispidum** DC (Br to Arg, natd. Af): hb. u. as bitter, arom. diur. diaph. (SA); in rheum (Af); bad weed (sAf).

Acanthosphaera ulei Warb. (Mor.) (Br esp. Amazon reg.): balsamo; bitter latex u. locally as antipyr.

Acanthosyris falcata Griseb. (Santal.) (Arg, Bo, Par): saucillo; frts. sweet, edible; u. prep. brandy (liqueur); wd. u. make furniture.

Acanthus ebracteatus Vahl (Acanth.) (tropAs): sea holly; salt bush; sds. in combn. for cough; anthelm, in children (Mal); rts. in combn. for shingles (Mal); lf. decoc. as cough med. - **A. ilicifolius** L. (Ind, Pak, seAs): harcuch kanta (Hind); lvs. u. in rheum. neuralgia, asthma, dyspepsia, snakebite; lvs. & sts. u. as purg. (Indones); (mangrove plant) - **A. mollis** L. (sEu): bear's breech; cultivated acanthus; hb. & rts. rich in mucil. u. int. (in diarrh.) & exter. in emoll. plasters & cataplasms; dr. lvs. comml. drug (Eu); fresh flowering pl. off. in Ge. Homeop. Ph. - **A. pubescens** Engl. (eAf): rt. u. in jaundice (Haya, Tanzania) - **A. spinosus** L. (sEu): (spiny) bear's breech; wild acanthus; thistle-like perenn. hb. common in It & Gr; hb. u. astr. diur.; extensively u. in art designs of ancient times (944); model for Corinthian (column) capitals (ancient Rome) (2638) - **A. syriacus** Boiss. (Syr, Saudi Arab): lvs. u. as model for scroll decorations; similar to *A. spinosus*.

Acapulco: *Diospyros ebenum* - **A. gold** (US slang): *Cannabis*.

acaricoba (Br): *Hydrocotyle*.

acarien (Fr): mite, acarus.

acaroid resin, red: fr. *Xanthorrhoea australis* & several other *X.* spp. - **yellow a. r.**: fr. *X. hastilis*; u. mostly in technology.

acarology: science dealing with mites.

acataia (Por): *Polygonum hydropiper.*

acaucil (Arg): artichoke.

accroides gum (Austral): acaroid resin; yacca gum (*qv.*).

Accutane: v. retinoids.

ACE: adrenal cortex extract; angiotensin converting enzyme.

acebo (Sp): *Ilex aquifolium.*

acedeka (Mex): *Oxalis tetraphylla.*

acedra (CR): *Rumex vesicarius.*

aceite (Sp): oil - **a. animal** (Sp, Arg): neatsfoot oil - **a. de arrellanos** (Mex): mixt. cottonseed & lavender oils - **a. de Castilla** (Sp): **a. de comar** (Sp) = **a. de oliva(s):** olive oil - **a. de palo** (Mex, laity): fir oil - **a. de patas** (Arg): neatsfoot oil - **a. lubricante** (Arg): lubricating oil - **a. mineral** (Arg): mineral oil - **a. pulga** (NM [US], Sp): castor oil - **a. secante** (Sp): drying oil - **a. volatil** (Sp): volatile oil.

aceituna (Sp): olive.

acelga (Sp): Swiss chard; beet (root).

Acer L. (Aceraceae): the maples; genus of c. 150 spp. native to all continents exc. Austral & SA; important econom. as source of timber, charcoal, maple sugar & as ornam. & shade plants (929) - **A. barbatum** Mx. (seUS, Fla): (southern) sugar maple; Florida m.; hammock m.; occasional source of m. syrup & sugar - **A. circinatum** Pursh (Pac area, NA, BC to Calif [US]): vine maple; wd. useful; shade & ornam. tree - **A. ginnala** Maxim. (Kor, Jap, Manchur, Siber): young lvs. u. as tea (Jap) - **A. macrophyllum** Pursh. (Pac area, NA): bigleaf (or Oregon) maple; largest native tree lf. in BC [Cana] (-30 cm); wd. useful, for tool & ax handles; good shade ornam. tree - **A. mono** Max. (A. pictum Thunb.) (eAs, Kor, Chin, Jap, Manchuria): lvs. irrit., bk. astr.; sap u. to make alc. beverage (Ainu); wd. useful - **A. negundo** L. (Negundo aceroides Moench.) (e&cNA s to Guat): box elder; ash-leaved maple; distinctive fr. most other *A.* spp. in having compd. lvs. with 3-7 leaflets; good shade tree (hardy, grows rapidly); wd. u. for paper pulp; woodware; several vars. recognized; bk. u. astr.; fresh bk. u. by Homeopaths (930) - **A. nigrum** Mx. f. (eUS, sCana): black maple; important m. syr. & sugar source; wd. valuable - **A. pensylvanicum** L. (A. striatum Lindl.) (eCana, neUS): striped maple; moosewood; ornam. tree (-12 m high); lvs. appl. topically for mastitis (breast inflammn.); inner bk. of branches put in cloth & boiled; the expressed liquid u. as emetic; bk. u. in poultices for swelling of limbs; also in GC, kidney dis. & hemoptysis - **A. platanoides** L. (Euras): Norway maple; bk. astr. u. in tan.; source of maple manna (manne d'érable); wd. valuable, u. in mfr. rifle stocks - **A. pseudoplatanus** L. (Euras, natd. US): sycamore maple; Scottish m.; juice u. as antisep. in urin. dis.; bk. astr.; excellent wd. u. for shoe lasts; violins, floors; large ornam. trees (-34 m) - **A. rubrum** L. (e&cNA): red (or soft) maple; bk. u. astr., to make black ink (Colonial Am) (2090); wd. much u. in mfr. furniture; sap obt. fr. trunk is like maple syrup & u. in beverages - **A. saccharinum** L. (e&cNA): silver (leaf) (or soft) maple; inner bk. expect., ton. (inf.), in arthritis, &c.; trunk source of m. sugar & syr. (fr. sap) rel. low yields (coll. in spring; chief prodn. Vt [US]); tree banned on Atlanta (Ga [US]) streets because extensive rt. system blocked water lines - **A. saccharum** Marsh. (sCana, n&cUS): sugar (or rock) maple; silver (leaf) m.; tree u. as syr. & sugar source (neUS, eCana); wd. ("hard maple") much esteemed for lumber, u. in turnery; said most valuable hardwood tree in the US (931) - **A. spicatum** Lam. (e&cNA, s to Ga): mountain (or moose) maple; shrub or small tree; bk. astr.; was widely u. as substit. for Viburnum Opulus - **A. striatum** Lindl. = *A. pensylvanicum* - **A. tataricum** L. (seEu, AsMin): Tatarian maple; u. diur. (Turk).

Aceraceae: Maple Fam.; shrubs or trees; lvs. opposite, petiolate, exstipulate; frt. double samara separating when ripe; 2 genera: *Acer, Dipteronia.*

acerin: ext. of dr. frts. of *Acer platanoides*; with virucidal, antivirotic, and antibacterial properties; low toxicity.

Acerola: *Malpighia glabra.*

Acetannin™: acetyltannic acid.

acetates: salts or esters of **acetic acid** (CH_3COOH); occ. widely in nature; much u. in pharm./med.; acidulant; preservative; solvent; manuf.

acetic acid: CH_3COOH; Essigsaeure (Ge); the chief fatty acid; widely distr. in plants (ex. frts. of *Ginkgo biloba*); usually as salts or esters (latter in VO); known & u. for centuries; obt. by fermentation of wines, &c.; dil. acid u. as acidulant, solvent, preservative, in prepn. org. compds. - **a. ether**: ethyl acetate - **glacial a. a.:** (99.5%): u. as caustic to remove warts.

aceto (It-m.): vinegar.

Acetosa Mill. = *Rumex.*

Acetosella acid: oxalic acid - **acetoselle** (Fr): *Oxalis acetosella.*

Acetosella Ktze: former generic name; now distributed in *Oxalis, Rumex.*

Acetum (L): vinegar.

acetylcholine (chloride): Ac(h); biogenic amine found in ergot ext. & animal tissues; released by cholinergic nerves, serving as neurotransmitter, to activate smooth & voluntary muscles, glands, &c.; esp. active for the autonomic nervous systems; parasympathomimetic; important vasodilator; sometimes u. by direct injection into brain in Alzheimer's disease (1984); representing a natural cholinergic (parasympathomimetic) subst.; acts to slow heart; incr. intest. movements; incr. secretion of lachrymal, salivary, bronchial glands; off. in several pharmacopeias; because of its fugitive action, synth. compds. are preferred such as methacholine, carbachol, &c.

acetyldigitoxins: represents digitoxigenin (aglycone) + 3 digitose (one with AC gp) (glycone); extd. fr. lvs. of *Digitalis ferruginea* L.; u. cardioton.

acetylsalicylic acid: Aspirin; ASA; synthetic compd. but closely related to natural substance, salicyclic acid; u. popular analg, antipy.; recently as a prophylactic against stroke & systemic embolism (in small dosage) to reduce risk of cardiac arrest; u. in inflamm., rheum.

ache (Fr): parsley - **a. céleri**: celery frt. ("c. seed") - **a. des marais** (Fr): wild celery, *Apium graveolens.*

achecchari (SA Indians): Quassia (945).

achee: akee.

achene: akene: hard 1-seeded frt., with frt. wall separated fr. sd. wall, the sd. lying free, attached to the frt. wall (placenta) only by the funicle (sd. stalk); ex. sunflower "seed."

achicoria (Sp): chicory - **achilea** (It): **Achillea** (L.): *Achillea millefolium*, milfoil.

Achillea L. (Comp.) (mostly OW, nTemp): yarrow; perennial hbs. ca. 80 spp.; fl. spring to fall - **A. ageratum** L. (sEu): sweet yarrow; flg. hb. u. fr. ancient times in Med reg. as stom. (also fls. & frts.) &c. - **A. atrata** L. (Eu-Alps): "black milfoil"; hb. (Herba Genipi nigri seu veri) u. like *A. moschata* in prepn. Iva liqueurs, u. as stom. ton. vulnerary - **A. clavenae** L. (Alpine parts of Austria & the Balkans): Steinraute; hb. u. in domestic med. - **A. filipendulina** Lam. (wEuras, cult. US): fernleaf yarrow; yellow flowered y.; fresh pl. u. in homeopathy; compass pl. - **A. fragrantissima** Sch. Bip. (Eg, Syr): u. as replacement for chamomile; stom. - **A. millefolium** L. (Euras, natd. NA): (common) yarrow; milfoil; nosebleed; sneezewort; olive-gray arom. pl. characterized by flat-topped clusters of small grayish-white fls. (2603) (sometimes pinkish or blueish), finely dissected lvs. (932); lvs with VO c. achilleine (= betonicine); azulene (formed during distill.); u. alter. bitter ton.; pd. u. as snuff (—> sneezing); chewed for toothache; inf. appl. locally to bruises & bumps; u. to prepare beer (Sweden; Eng [Cornwall]) - var. *occidentalis* DC (var. lanulosa Piper; A. l. Nutt.) (wNA): yarrow; mashed lvs. u. as poultice; rt. u. in toothache; chewed for colds & gas pains (Nev [US] Indians) - **A. moschata** Wulfen (A. erbarota All.) (cEu - alpine regions of Austria, &c.): black milfoil; musk yarrow; Iva (Ge); génépi blanc (Fr): genipp; hb. c. VO with cineole; u. ton. arom. bitter, stom. stim. diaph. (home rem.); exter. for wounds; u. in It & Switz for prepn. Esprit d'Iva (Ivas liqueur; "Iva bitters"); med. in Brandt's pills - **A. nana** L. = *A. moschata* - **A. nobilis** L. (s&cEu): noble yarrow; u. as arom. similar to *A. millefolium* but finer in odor and flavor; in stomachache (Denmark) - **A. ptarmica** L. (Euras, natd. eNA): sneezewort; sneezeweed; powd. rts. formerly u. as snuff (to invoke sneezing); ton. astr.; fl. heads adulter. for Roman chamomile - **A. santolina** L. (Euras, e&nAf): zawal (Ind); hb. u. in respir. disorders (Iran); carm., anthelm., stom.; insecticide; fl. heads u. in refrig. bever. & syrups; VO u. like that of *Santolina chamaecyparissus* - **A. sibirica** Ledeb. (As): perenn. hb. u. as ton. in dyspep., constip., for boils (broth) (Chin) - **A. tomentosa** L. (sEu, ex. It - on higher areas); felzige Schafgarbe (Ge); fls. yellow; u. as stim. ton. antipyr.; in chlorosis.

achilleine: betonicine: alk. of *Achillea millefolium*; pyridine type; with blood coagulation properties.

Achimenes erecta H. P. Fuchs (A. coccinea Pers.) (Gesneri) (Mex): hb. u. med.; cult. ornam. - **A. mexicana** (Seem.) B. & H. f. (NA): "storm cloud" (US); u. med. (Mex).

achiote (Por, Sp): **achiotl** (Mex); **achiotta** (Por): coloring matter fr. *Bixa orellana*.

Achlys triphylla DC (Podophyll./Berberid.) (wNA, BC [Cana]-Calif [US]): deer-foot; vanilla leaf; "elk weed"; "sweet after death"; "may apple"; with vanilla odor (dead lvs.); c. coumarin; lf. inf. u. in TB; lvs. u. in hairwashes, to repel flies.

Achras sapota L. = **A. zapota** L. = *Manilkara z.*

Achromycin™: tetracycline; the original name (Lederle).

achroodextrin: that one of the series of dextrins formed in hydration of starch in which I_2 first produces no color; converted into maltodextrin.

achromatic: 1) lens corrected for chromatic aberration (with rainbow colors at field margin) 2) structure staining poorly with standard dyes 3) a point in starch test with I_2.

achupalla (Pe): pineapple.

Achyranthes alba Eckl. & Zeyh. (Amaranth.) (seAf, As): hb. u. for uterine effects (Chin) (imported) - **A. aspera** L. (A. indica [L.] P. Mill.) (pantropics) (As, esp. Ind, PI, Af): "man-better-man" (Trin); latjira (Hind): pl. u. astr. diur. antispasm, hb. inf. u. in fever, influenza; pl. decoc. in VD (St. Lucia); ton. for anemic people (VirginIs); antipyr. (Indoch); antirheum.; rt. inf. emet. (Zulus): abortif. (Ind); flg. tops & sds. u. in rabies (!); rt. decoc. in menstrual pains & pain on urin. (Tanz); fresh lvs. crushed in fingers & appl. to scorpion stings; alkaline ash appl. to chancres; to aid delivery (Chin); mixed with tobacco to improve flavor (Tanz); bad weed - **A. bidentata** Bl. (tropAs): bankhat (Assam); sds. c. oleanolic acid (aglycone of saponin); pl. u. diur. astr. (Jap Pharm.); in rheum., kidney dis.; uterine inertia (Chin) - **A. calea** Ibantz = *Iresine diffusa* - **A. fauriei** Laveille & Vaniot (As): rts. (Radix Achyranthis, Jap Phar.) c. inokosterone & iso-inokosterone, latter identical with ecdysterone, with moulting hormone activity; u. leprosy - **A. indica** (L) P. Mill. (pantropics); "man-better-man" (Trin): hb. inf. u. in fever, influenza; pl. decoc. in VD (St. Lucia); ton. for anemic people (Virgin Is) - **A. japonica** Nakai (Jap, Kor): subterr. st. u. as diur. - **A. obtusifolia** Lam. (Ind, PI): hangor (Tagalog); diur. in edema; in snake bite & scorpion sting - **A. philoxeroides** Standl. = *Alternanthera p.* Griseb. - **A. repens** L. = *Alternanthera pungens* HBK.

Achyrocline flaccida DC (Comp.) (nArg, sBr): marcela hembra; shrub c. flavonoids; u. antispasm., in epilepsy; emmen.; anthelm. - **A. satureioides** DC (Br, Co, Arg, Par+): macel(l)a (do campo) (Br. Phar); marcella; fls. ("alecrim da parede") c. isognaphalin (flavone); VO c. limonene, cymol, citronellol; pl. inf. agreeable, tasty; u. ton. bitter ton. (prep Bitter Species); diaph. antidiarrh.; in kidney dis.; physicians u. in intest. dis. esp. appendicitis; contracep. (Ur).

Achyropsis leptostachya Hook. f. (Amaranth.) (Af): pl. u. emet. expect. antipyr.; rt. decoc. in rheum. (2495).

acibar (Sp): Aloe.

"acid" (US slang): LSD - **a. wood** (NCar [US]): chemical wood (*qv.*).

acids: see under individual specific names.

Acidum Ascorbicum (L): vitamin C - **A. Borussicum**: hydrocyanic acid - **A. Carbazoticum**: picric acid - **A. Catholicum**: sulfuric acid - **A. Caeruleum**: hydrocyanic acid - **A. Halleri**: sulfuric acid mixture - **A. Saccharum**: oxalic acid - **A. scytodepticum**: tannic acid.

aciminier (Fr): *Asimina triloba*.

acintro (Por): *Artemisia absinthium*.

Acioa Aublet (Chrysobalan.) (tropAm): sds. edible; rich in FO.

Acipenser huso L. (Acipenseridae - Chondrostei - Ganoidei - Pisces) (Black, Adriatic & Caspian Seas): beluga; great (white) sturgeon; Hausen (Ge); European sturgeon; large fish (-9 m); dr. swimming bladder u. in mfr. isinglass (Ichthyocolla; fish glue); c. 80% collagen; grades: Saliansky (natural & bleached); Osetrova, Servuga; u. exter. in hemorrhages; in plasters, cements, &c.; clearing agt. for wines & beers; fish source of true caviar (roe; food delicacy) - **A. stellata** Pallas (Caspian & Black Seas): sturgeon; sevruga; Scherg (Ge); meat u. fresh, frozen, smoked, canned, semi-processed; roe is caviar, food delicacy.

Acipenseridae: family of ganoid fishes incl. most sturgeons.

Ackermuenze (Ge): *Mentha arvensis* (lit. field mint).

Ackersalat (Ge): *Valerianella clitoria* - **Ackerspergel** (Ge): *Spergula arvensis*.

aclacinomycins: series of antibiotics prod. by *Streptomyces galilaeus* Ettlinger *et al.*; u. as antineoplastic.

Acleisanthes longiflora Gray (Nyctagin.) (sUS, Mex): angel trumpets; yerba de la rabia; macerated lvs. appl. to wounds made by angry animals.

Acnida L. = *Amaranthus* L.

Acnistus arborescens (L) Schl. (l) (Solan) (WI): (tree) wild tobacco (Trin); batard sirio (Dominica); lf. u. in poultices for mumps, migraine (Trin); colds, fevers (Dom.) neuralgia; substit. for belladonna; in cancer (proven value).

acoita-cavallo (Br): *Luehea* spp.

Acokanthera G. Don f. (Toxicophloea Lindl.) (Apocyn.) (tropAf): some 10 spp. of shrubs or small trees; several spp. are u. for arrow poisons; thus, the following: **A. deflersii** Schweinf.; **A. friesiorum** Markg., **A. longiflora** Stapf, &c. - **A. oppositifolia** (Lam) L. E. Codd (A. venenata G. Don) (trop eTanz & sAf): bushman's poison (bush); wd. latex, rt. c. ouabain (- 1.1%); u. prepn. Bushman, Kaffir, & Zulu arrow poisons; lf. decoc. u. int. for snake bite, stomachache, toothache, colds; also appl. locally in snake & spider bite; wd. ext. u. as ordeal poison as anthelm. for tapeworm - **A. ouabaïo** Cathel. (Ethiopia, Somalia): sds. are one source of ouabain (*qv.*) (glycoside); acoshimperosides & derivs.; wd. (& rt.) u. to make arrow poison; no indications of therap. usage by aborigines - **A. schimperi** Schweinf. (A. abyssinica Benth. & Hook.) (Ethiopia): lf. fl. frt. bk. wd. u. to prep. arrow poisons; ouabain as chief glycoside, acovenoside; wd. richest in same; poison rapidly lethal when introduced parenterally - **A. spectabilis** Hook. f. (A. oblongifolia (Lam.) L.E. Codd) (sAf): bushman's poison bush; Hottentot poison tree; winter sweet; bk. & wd. u. as suicidal & homicidal poisons, rapidly lethal (also thru accident); rt. scrapings in ointment for itch; wd. c. ouabain; ext. u. as arrow poison; u. in snakebite.

aconine: amorphous alk. of *Aconitum napellus* &c.; cardiac stim.

aconite (root; leaves): fr. *Aconitum napellus* - **country a.**: *Gloriosa superba* - **European a.**: term includes following spp.: *Aconitum napellus* (**garden a.**), *A. cammarum*, *A. lycoctonum* & c. - **Indian a.**: *A. laciniatum* & several other *A.* spp. - **Japanese a.**: *A. fischeri*; *A. uncinatum* - **Manchurian a.**: *A. kusnetzoffii* - **Nepal a.**: *A. laciniatum* - **Russian a.**: prob. *A. orientale* (rich in aconitine) - **Soviet a.**: *A. karakolicum* (etc.) - **A. tuber: Aconiti Tuber** (L, Phar. Int.): Aconitum: the tuberous rt. of *A. napellus* - **wild a**: 1) *A. vulparia*; *A. uncinatum* 2) country a.

aconitic acid: pl. acid occ. in *Aconitum* spp. +; prepd. by dehydration of citric acid (less 1 mol. H_2O); u. in mfr. itaconic acid (u. mfr. plastics).

aconitine: typical alk. of *Aconitum napellus* & some other *A.* spp.; classed as ester alk.; u. antipyr., card. depressant (dangerous); in exptl. pharmacol.; formerly off. in USP - **amorphous a.**: mixt. of NLT 8 alks. fr. *Aconitum napellus*; incl. aconitine, neopelline, 1-ephedrine, hypaconitine, & sparteine; dangerous poison.

Aconitknollen Ge): **aconito** (Sp, Por, It): *Aconitum* (tuberous rts.).

Aconitum: 1) genus (Ranuncul.) (nTempZone) of ca. 100 spp.; fls. showy; rts. often poisonous & source of several important drugs 2) (specif.) (L.): dr. tuberous rts. of *A. napellus*, off. in several phar. - **A. anthora** L. (sEu, cAs): wholesome aconite; rt. u. as bitter ton. in over-fatigue (Chin) - **A. balfourii** Stapf (Ind, Nep): c. pseudoaconitine; u. as adulter. of *Aconitum napellus*; rather toxic drug - **A. X bicolor** Schult. (eEu): possibly a form of *A. X cammarum*; common ornam. hb. - **A. X cammarum** L. (*A. napellus* X *A. variegatum*) (Euras, Hung): purple wolf's bane; rt. sometimes u. in med., sometimes as adulter. - **A. carmichaeli** Debx. (cChin): senbushi (rt.); c. aconitine, talatizamine, senbusine, & other alks.; rts. u. int. in cholera, kidney dis. with back pain, edema, dys.; pd. rt. mixed with egg white u. for boils; exter. appl. to athlete's foot - **A. chasmanthum** Stapf (A. laciniatum Stapf) (nInd, Pak): Nepal (or Indian) aconite; banbalnag (Kashmiri); rt. c. indaconitine; similar u. to regular aconite (*A. napellus*) but weaker (936); ex. neuralgia; one of oldest Ind remedies; strongly toxic; there are several other Indian aconites (incl. *A. ferox*, *A. palmatum* D. Don, *A. heterophyllum*) - **A. chinense** Paxt. (A. fortunei Hemsl.) (eChin, Indoch): rt. u. to treat rheum., numbness of limbs, abdom. pain, antipyr.; insecticidal - **A. chinense** Sieb. & Zucc. = *A. japonicum* - **A. columbianum** Nutt. (wUS & Cana, esp. in mts.): Columbian monkshood; all parts of pl. toxic (depressant) (106); in witchcraft (BC [Cana] Indians) - **A. coreanum** Rapaics (Kor, Chin): u. to treat the common cold - **A. deinorrhizum** Stapf (Ind, Pak.): rt. c. pseudoaconitine; u. adulterant of true aconite drug - **A. ferox** Wall. ex. Ser. (Ind, Pak, Nepal): Indian aconite; bish(ma); wild wolf's bane; Nepal aconite; rt. c. pseudoaconitine, &c.; most toxic of all aconites; u. to prepare bikh poison (for hunting) of Nepal; also u. in rheum. & neuralgia - **A. fischeri** Reichb. (eAs): Jap. a.; u. as adulter. and substit. for true (off.) aconite - **A. heterophyllum** Wall. (Ind, Pak -Himal): Himalayan aconite;

small rt. (called "Atis root"; atees), c. atisine, heterasine; u. antiperiodic. bitter ton. antipyr. aphrod.; in coughs, diarrh. &c. (suppl. fr. Nepal) - **A. japonicum** Thunb. (A. chinense Sieb. & Zucc.) (nChin, Jap, Siberia): Japanese aconite; decoc. u. in neuralgia, convulsions, beri beri, rheum., bronchitis (Chin); bruises & fractures (Vietnam); substit. & adulter. for off. aconite - **A. karakolicum** Rapaics = *A. tianschanicum* - **A. kusnezoffii** Reichb. (Kor, Manch, nChin): rt. specially treated by boiling with soy beans & licorice & u. int. in rheum., &c. - **A. laciniatum** Stapf (nInd, Tibet, Nep): Nepal aconite; source of bikh bish (Bengalese); toxic pl.; found admixed with *A. spicatum* & *A. ferox*; may be a var. of *A. ferox* - **A. ludlowii** Exell (Tibet): rt. c. 2.5% alks. calcd. as aconitine (108) - **A. luridum** Hook. f. & Th. (Ind, Nepal, Himal): bish (Bengalese); rt. anodyne; toxic pl.; may be a var. of *A. ferox* - **A. lycoctonum** L. = *A. septentrionale* - **A. manschuricum** Nakai (Chin): u. in Chin. med. - **A. moschatum** Stapf (eKashmir): toxic pl. - **A. napellus** L. (A. rotundifolia Kar. & Kir.) (Euras): monkshood; wolf's bane; mouse bane; the most important med. aconite, off. in numerous pharmacopeias (27); rts. & lvs. (hb.) c. aconitine, picroaconitine, benzaconine, aconine, &c.; u. anodyne, in trigeminal neuralgia, cardiac sed.; toxic (depressant); imp. fr. Ge, Hung - **A. orientale** Mill. (Caucasus, Crimea; Chin): Russian aconite; rt. c. up to 2.5% alks. (107); u. like off. a. - **A. palmatum** D. Don (Himal): c. palmatisine (non-toxic), aconitine, anthorine; u. ton. in diarrh., rheum.- **A. septentrionale** Koelle (A. vulparia Rchb.) (nEuras, esp. Nor, USSR, cAs, Himạlaya, Siber, Pak, Chin): (yellow) wolf's bane; y. aconite (SC yellow-flowered); source of Radix Ac. lutei (rt.), c. lycaconitine, lycotonine, myoctonine; *no* aconitine; rt. u. as adulter. & replacement for off. aconite; narcotic (Chin); fresh hb. u. in Homeop. med.; cult. orn. - **A. soongaricum** Stapf (Ind, Pak, Russia): u. similarly to *A. napellus,* of which it is at times considered a variety - **A. spicatum** Stapf (or Donn) (Ind, Nep): bikh; bish; rt. c. bikhaconitine; toxic - **A. X stoerkianum** Rchb. (*A. napellus* L. X *A. variegatum* L.) (cEu - Alps): close to & sometimes considered synonymous with *A. cammarum*; u. like & sometimes substit. for true aconite; old garden ornam. - **A. talassicum** M. Prop. (cAs, mts. of Talass-Alatau): rts. c. talatisine, talatisamine, &c. - **A. tianschanicum** Rupr. (A. karakolicum Rapaics) (USSR, cAs): Soviet aconite; c. 1.5-2% alks. (incl. aconitine), highly potent; formerly off. USSR Phar. - **A. uncinatum** L. (eUS [Pa to Ala]): (American) wild monkshood; a large type of aconite; u. anodyne, sed. - **A. variegatum** L. (Med reg., Alps): c. aconitine; activity comparable to that of *A. napellus* - **A. wilsonii** Stapf = *A. carmichaeli.*

acore (Fr): *Acorus calamus* - **a. faux**: *Iris pseudacorus* (false calamus) - **a. odorant** (Fr): **a. vrai** (Fr): calamus (root).

acorin: glycoside fr. *Acorus calamus*; bitter yellow subst.

acorn: frt. of the oak, esp. of *Quercus robur* & *Q. velutina* - **coffee a.**: *Q. robur* - **a. cups: a. hulls**: Valonia fr. *Q. aegilops* - **sweet a**: fr. *Q. ilex.*

acoro (Sp, It): **a. aromatico** (Sp): **a. nero** (It): *Acorus calamus.*

Acorus calamus L. (A. spurius Schott.) (Ar) (neAs incl. Jap & NA orig.; now nHemisph): sweet flag; s. sedge; source of sweet calamus (root) & VO (brown liquid) with isarone; acorin (bitter princ.); soft resin; u. arom. ton. stom. carm. in colic (Indians); flavor, masticatory; considerably u. in folk & home med.; anthelm.; in bitters (cordials), baths (perfume), rhiz. juice in inflamed eyes, &c.; popular food of muskrat; whole rt. insect repellant; prodn. (cultn.) Ge, Neth, Eng, USSR, US, Cana - **A. gramineus** Soland. (eAs): formerly much u. like *A. calamus*; adulter. of same; stom. ton. (Chin); anodyne, sed., vermif.; spasmolytic.

Acorus (Palustris) (L): **A. Vulgaris** (L): *Iris pseudacorus* (SC similar appearance to calamus & growing in similar areas).

acoschimperosides: series of glycosides occ. in *Acokanthera schimperi* - **a. P**: glycoside of oleandrigenin + L-acofriose (others of gp.: N,S,T,U+).

Acotyledones (Jussieu): acotyledonous plants: Cryptogamae; also sometimes applied to *Cuscuta,* &c.

Acourtia D. Don (Comp.) (NA): generally now SA taxa are separated into *Perezia* Lagasca (2588) - **A. adnata** (Perezia a. A. Gray) (Mex): pipitzahuac (root) (Mex): rhiz. & rts. u. as purg., &c.; esp. in typhoid fever - **A. alamanii** (DC) Reveal & King (Chiapas, Mex): papalohuiteconi; lvs. & rts. u. as antisep., purg. - **A. moschata** DC (Perezia m. Llave & Lex.) (Mex): yerba del zopilote; said to have stom. & strengthening properties - **A. multiflora** (Perezia m. Less.) (SA): fl. inf. diaph. - **A. nana** (Gray) Reveal & King (Perezia n. Gray) (sTex [US], Mex): desert holly perezia; u. like *A. adnata* - **A. oxylepis** Reveal & King (Perezia o. Sch. Bip.) (Mex): furnishes "Perezia root"; u. as purg., in hemorrhoids; as indicator (for alkalies) (also mentioned as u. in Mex med. are *A. rigida* DC and *A. wrightii* Rev. & King) - **A. runcinata** (Lag. ex. D. Don) B. L. Turner (Perezia r. Gray) (sTex [US], Mex): stemless perezia; piania (Mex); whole pl. u. for chest dis. (boiled & mixed with other herbs) - **A. thurberi** Reveal & King (Perezia t. Gray) (Mex): calzadilla; cola de zorra; pl. decoc. (concd.) & fomentations appl. to cicatrize wounds in man & lower animals; pd. lf. appl. usefully to syphil. sores; rt. decoc. an energetic emet. & drastic - **A. wrightii** (Gray) Reveal & King (Perezia w. Gray) (sTex [US], Mex): pink perezia; u. like prec.; lf. hemostatic.

acqua (It-f.): water.

acre: unit of area u. for land measurement; = 0.40 hectare; 43,560 sq. ft.; ca. 208 ft. sqd.

Acridocarpus natalitius Juss. (Malpighi.) (tropAf): mabophe (native languages): pd. rt. u. purg. in constip. & colic; in magic as charm, &c. - **A. plagiopterus** Guill. & Perr. (Sen): various parts of woody shrub u. in magic; inf. of leafy sts. u. in baths & drinks to strengthen infants & as aphrod. for adults.

acrinyl isothiocyanate: a. sulfocyanide: pungent liquid formed by breakdown of sinalbin of white mustard; chief active component of prepared mustard; u. as condim.

Acrocephalus capitatus Benth. (Lab.) (tropAs, Chin): hb. c. VO with marked anthelm. properties; decoc. to wash wounds - **A. indicus** Ktze. (Ind): ustukudus (Sindi): pl. u. as expect.

Acroceras macrum Stapf (Gram.) (As, Af): Nile grass; u. as pasturage (forage) & fodder (hay) (esp. eTransvaal); cult.

Acrocomia aculeata Loddig. (A. sclerocarpa Mart.) (Palm.) (WI-SA): grugru (Dominica); macaúba; mascasuba; coco de catarro; frt. & sd. both prod. FO, former u. in soap mfr.; sd. FO edible (mocaya oil [or butter] very much like coconut oil or even butter); u. in foods (cooking); frts. fleshy edible; sds. c. 9% resin, u. as purg. anthelm.; young lvs. u. as veget. - **A. eriocantha** Barb. Rodriguez (Br): mucaja(-mirim); frts. edible; sd. FO u. in rheum. & in making soap, &c. - **A. glaucophylla** Drude (Br): (coco de) catala; caatala; fleshy part of frt. & the sds. u. as purg. vermif. vermicide - **A. lasiospatha** Mart. (Guianas): mocaja; great macaw palm; source of mucuja (resin) - **A. mexicana** Karwinsky ex C. Mart. (Mex, Guat, Hond, Belize): cocoyol; rts. u. in diabetes (probably inefficacious): anti-atherogenic; frts. eaten raw or cooked - **A. microcarpa** Barbara Rodriguez (Br): frt. edible - **A. quisqueyana** Bailey (Hispaniola): corozo (criollo); apical bud (palm cabbage) furnishes liquid giving pleasant beverage (when fermented), said nerve stim., incr. fertility of women; frt. edible, supplies FO u. in cooking & to make soaps; wd. useful for cabinet-work; honey pl. - **A. sclerocarpa** Mart. = *A. aculeata* - **A. totai** Mart. (Bo, Arg, Par): macaw (or macuja) palm; Paraguay p.; believed chief source of mocaya (or Paraguay [coco] palm) oil (butter) (palm kernel oil), expr. fr. nut; u. to make soap (locally); frts. agreeably edible - var. *wallaceana* (Drude) Beccari (Br): u. similarly to *A. eriocantha* - **A. vinifera** Oersted (CA, Br?): palma de vino; mauriti palm; sds. yield 50% FO (muriti fat); frts. source of sugar, fermented to produce a palm wine ("sajetta"); sap diur.

Acrodiclidium Nees & C. Mart. = *Licaria.*

acrogens: plants wh. grow at the apex of st.; appl. to plants of highest gps. of cryptogams, incl. ferns & their allies, mosses & liverworts.

acrolein: oxidation product of unsaturated fatty acids.

Acronychia pedunculata Miq. (A. laurifolia Bl.) (Rut.) (Far East, esp. Chin): kramol (Cambodia); rts. u. as fish poison (sVietnam); resin fr. rts. as embrocation in chronic rheum.; hb. bitter, astr. u. in scabies, diarrh., colic, diur. (Java); young lvs. condim. (Java), rts. shoots, frt. u. to prep. stim. baths; wd. u. as anodyne, styptic in severe wounds (Chin); cult. orn.

Acroptil(i)on picris = *Centaurea p.* Pall.

Acrostalagmus Corda (= Verticillium) (Hyphomycetes - Fungi): c. acrostalic acid, antifungal antibiotic.

Acrostichum aureum L. (Adiant. - Pteridophyta) (Mal, Fiji, New Guin, PI): borete; forked fern; rhiz. grated u. as vulner. for wounds & boils; fronds emoll. appl. to syphil. ulcers (PI); very young lvs. eaten (Borneo).

Acrotome inflata Benth. (Lab.) (sAf): u. in "black-leg" in cattle.

Actaea (L): rt. of *Actaea rubra* or *A. pachypoda.*

Actaea pachypoda Ell. (A. alba auctt.) (e&cNA): white baneberry; w. cohosh; doll's eyes; rt. said eaten by black bear, hence called racine d'ours, bear root (Fr, Cana) & masqua-mitruin (=bear's food [Cree]); decoc. of rt. & spruce fir tops u. in gastric dis. (nCana); rt. u. emet., purg. - **A. racemosa** L. = *Cimicifuga r.* - **A. rubra** Willd. (e&cNA): red baneberry; r. cohosh; snakeberry; u. like *A. pachypoda* - **A. spicata** L. (nEuras, incl. Jap): herb Christopher; (black) baneberry; b. cohosh; rt. u. in rheum., phthisis, edema, pertussis, angina pectoris; St. Vitus dance; emet. purg.; hb. & rt. for asthma, esp. by Homeopaths; toxic.

actaplanins: glycopeptide antibiotics (6) prod. by *Actinoplanes missouriensis* Couch; u. as growth stimulants; incr. milk production in ruminants.

actée à grappes (Fr): Cimicifuga - **a. en épis** (Fr): *Actaea spicata.*

Acthar™: ACTH: adrenocorticotrophic hormone (Armour).

Actidione™: cycloheximide antibiotic fr. *Streptomyces griseus* (by-product) (isol. 1946); active against yeasts & other fungi.; u. as sd. disinf.; in pl. diseases; to prevent orange fall (abscission) fr. tree (Fla [US] (citrus indy.); rat repellant (1981); fungicide; in cryptococcosis (fungus dis.); retards cancer growth but toxic to humans.

actin: contraction protein found in muscle, &c.; also active in many functions (cell movement, &c.).

Actinea Juss. (Actinella Pers.) (Comp.): v. under *Hymenoxys/Helenium.*

Actinia equina L. (Actiniidae - Ord. Actiniaria [Coelenterata]) (Med Sea): sea anemones; invertebrates of sea bottom; c. curare-like compds., tetramines; u. by ancients for diarrh.

actinic radiation: "actinic rays" (less proper): radiation esp. ultraviolet.

Actinidia arguta Planch. ex Miq. (Actinidi./Dilleni.) (eAs, Jap, Kor): tara vine; bower actinidia; frts. & lvs. antipyr. to quench thirst; insecticide; vine juice u. expect. (Ainu); frt. ("tara") edible (Chin); dr. for winter use (Siber.); u. in pastry; pl. cult. as ornam.; may be a synonym of *A. callosa* Lindl. - **A. callosa** Lindl. (A. kolomikta Maxim.) (Chin, Jap): vine juice u. as expect.; cats very fond of pl. as food (claw sts. to pieces); cult. ornam. - **A. chinensis** Planch. (Chin): kiwi (frt.); Chinese gooseberry; frt. large, edible (*cf.* true gooseberry); u. arthralgia, cancer, diur. in fever, stone; raw frts. edible, meat tenderizer; rich in vit. C, delicious; in preserves; salads; wines; cult. Calif (US), NZ; exported to GB - **A. polygama** Maxim. (Chin, Jap): silver vine; lvs. boiled & eaten; salted frts. eaten in Japan; cats said fond of pl.; tranquilizer with mild hallucinogenic properties - **A. rubricaulis** Dunn (wChin): frts. flavorsome; cult. in Chin.

Actiniopteris australis (L.f.) Link (Actiniopteridaceae/Adiant - Pteridophyta) (Ind): rhiz. u. as anthelm. (probably with no value), in mal.; tonic; antifertility agt. - **A. dichotoma** Bedd. (Ind): morpankhi (Hind); pl. u. as styptic, anthelm.

Actinobolin: antibiotic prod. furnished by *Streptomyces griseoviridis* var. *atrofaciens*; isolated fr. Ga [US] soil (1958-); broad spectrum antimicrobial; u. to inhibit dental decay (exptl.).

Actinodaphne cupularis Gamble (Laur.) (eAs): tea fr. bruised lvs. u. in trauma (Chin) - **A. hookeri** Meissn. (Ind): tali (Tamil); lf. inf. u. in diabetes, urin. disord., sd. oil appl. to sprains - **A. pilosa** Merr. (Indochin): lf. inf. u. for stomachache.

Actinomeris Nutt. (Comp.): most or all taxa now transferred to *Verbesina.*

Actinomyces (Actinomycetaceae): genus of bacteria frequently causing disease; ex. *A. eriksonii* Georg & *al.* causes actinmycosis (lumpy jaw); (SC form characteristic radiating club-shaped formations, fr. actinus [L.]: ray).

Actinomycetes: "ray fungi"; generally regarded as order of filamentous bacteria; formerly thought fungi; some (Streptomycetaceae) furnish important antibiotics; others are disease-producing.

Actinomycins: antibiotics produced by several *Streptomyces* spp.; at least 12 forms are now recognized; some with cytostatic (neoplastic) activity; *cf.* Dactinomycin (A.D.)

Actinorhytis calapparia Wendl. & Drude (Palm.) (Mal, EI, PI): calappa (or galappa) palm; ripe nut u. for scurf (dandruff) (Sumatra) and chewed like betel (Mal, Java).

Actinospectacin™ = Spectinomycin.

Actinostemma lobatum (Maxim.) Maxim. (monotypic) (Cucurbit.) (Jap, Chin): gokisura (Jap); large sds. source of FO.

"activated" v. under charcoal; dehydrocholesterol; ergosterol.

activins: gp. of polypeptide hormones found in ovarian follicular fluid; stim. FSH secretions (1986-).

acuña (Br): *Derris floribunda.*

Acyclovir: acyclic purine nucleoside with activity against viruses of herpes, AIDS, &c. (1974-).

acyl: alkyl acid radical, ($C_nH_{2n+1}CO$-) or (RCO-).

adaga (Br): *Attalea compta* (*qv.*).

Adam and Eve: rt. drug (SC pair of rts. occur one above [Adam], one below [Eve]); u. as aphrod. ("to restore love"), appl. to 1) *Aplectrum* spp. such as *A. hyemale* 2) *Orchis latifolia, O. mascula, O. maculata* 3) *Pulmonaria officinalis*, &c.

Adam's apple: 1) kind of citron (*qv.*) 2) laryngeal prominence; projection of thyroid cartilage of larynx; more prominent in the male, hence common name - **A. needle (and thread)** (Md, La [US]): *Yucca filamentosa* - **A. pitcher**: *Sarracenia purpurea* - **A. weed**: shame weed: *Mimosa pudica: Schrankia* spp.

Adansonia digitata L. (Bombac.) (tropAf): baobab (tree); trunk huge (-9 m thick) (largest bole known); tree bears frt. of which the pulp ("monkey bread") is u. as food by natives; bk. & lvs. u. as tonic, in dys.; diaph.; FO fr. sd. u. (Madag); lvs. u. condim. in soups (*cf.* filet) - **A. gregorii** v. Muell. (Austral): bottle tree; (gourd-) gourd tree; sour g.; gouty stem; baobab; sds. eaten raw or roasted (aborigines, Queensland); water stored in large cavities in trunk for use during dry winter months; bk. u. antipyr.; frt. acidulous food - **A. madagascariensis** Baill. (Mad): bk. yields fiber;

frt. (cream of tartar frt.) eaten - **A. situla** Spreng. (seAf): u. med. (Moz).

adder('s) tongue: 1) *Ophioglossum vulgatum* 2) *Erythronium americanum* (also as **yellow a. t.**) - **a. wort**: *Echium vulgare*.

adelfa (Sp): *Nerium oleander.*

Adelia barbinervis Schlecht. & Cham. (Euphorbi) (Mex): rt. edible, u. in bee sting.

adelphia: union or combn. of stamen filaments into groups; term generally u. with prefix; ex. monadelphia (single group); diadelphia (two groups); polydelphia (3 or more groups).

Adenandra fragrans (Sims) Roem. & Schult. (Rut.) (sAf; Cape of Good Hope): klip buchu; u. arom.; tea substit. (natives); adulter. of long buchu; cult. as ornam (with anise odor).

Adenanthera L. (Leg) (tropAs, Austral, Pacif): bead trees; 4 spp. of trees; do not confuse with *Anadenanthera*! (2780) - **A. intermedia** Merr. (PI): wd. (tanglin or ipil-t.) has fine colored appearance & is resistant to termites & beetles; u. flooring, &c.; bk. & sds. u. for snakebite - **A. pavonina** L. (tropAs, spontaneous in Ind &c.; cult. in tropAm): (false) red sandalwood tree; bead t.; lf. decoc. u. in chronic rheum., gout, for inflamm. boils; sds. (Circassian beans) u. in necklaces, for weighing gold (uniform wt.), inf. in dys. (Cambod); bk. u. in native med.; rt. purg. emetic; wd. dyestuff; oil (coral tree wood oil) in comm.; lvs. stewed & u. for colds (Grenada); a street tree.

Adenia cardiophylla Engl. (Passiflor.) (seAs): u. to capture birds by poisoning (but flesh edible) - **A. cissampeloides** (Planch.) Harms (trop wAf): remedy in lumbago (sNigeria); rubbed on breasts to stim. lactation (Ghana); st. crushed & u. as fish poison - **A. digitata** Engl. (s&eAf): lemonade tree; frt. rapidly fatal (946); rt. & frt. u. in homicide & suicide; rt. c. HCN glycoside; modeccin (toxalbumin); rt. decoc. u. in leprosy, ulcers - **A. fruticosa** Burtt Davy (Af): elephant creeper; u. in tuberculosis - **A. glauca** Schinz (Af): lvs. c. HCN; frt. edible, delicious - **A. globosa** Engl. (sAf): tuberous st. u. as cattle med. (Masai) - **A. gummifera** Harms (seAf): rt. decoc. in mal. & leprosy (Zulu); lf. & st. emet. in biliousness; magic med. - **A. hastata** Schinz (Af): frt. edible, similar to granadilla - **A. kirkii** Engl. (eAf): u. in bronchitis (Moz) - **A. lobata** Engl. (trop wAf): many indications reported incl. icterus, headache, hemorrhoids, otitis, fainting; st. & lvs. as fish poison - **A. palmata** Engl. (Ind): mutakku (Malayalam); lf. & rt. juice u. in skin dis. (SriL) - **A. repanda** Engl. (Af): lvs. eaten by stock; appar. non-toxic - **A. singeporeana** Engl. (Mal): akar saoot; sacotan; rt. decoc. u. for ringworm (native med.); bk. fibers u. to make nets (wSumatra) - **A. wightiana** Engl. (Ind): pl. c. HCN - **A. venenata** Forst. (Nigeria): cult. for med. use - **A. zucca** (Blco.) Merr. (PI): vine pounded & made into decoc. u. to bathe swellings of body; rt. decoc. u. for stomach dis.

adenine: 6-amino-purine; "vitamin B_4"; occurs in many plants, animal tissues, human urine; extd. fr. tea; component of nucleic acids, ADP, ATP & other coenzymes, nucleotides (RNA, DNA), adenosine, some antibiotics (ex. Cordycepin); u. med. in granulo-cytopenia; tech. as defogging agt. - **a. arabinoside**: vidarabine.

Adenium boehmianum Schinz (Apocyn.) (e&swAf): c. echujin, ecubioside (cardioactive glycos.); u. in cardiac dis.; latex u. for arrow poison "echuja" - **A. coetaneum** Stapf (tropAf, Kenya, Nigeria): latex c. cardiac glycoside, echujin; u. as arrow poison of Watindigas; fish poison - **A. multiflorum** Klotzsch (e&sAf): c. cardioactive glycosides, hongheloside derivs.; pl. u. for fish poison, latex for arrow poisons - **A. obesum** Roem. & Schult. (A. honghel A DC) (tropAf, Nigeria): liar tree; toxic pl., c. honghelin (digitoxigen thevetoside) (cardioactive glycoside); u. as ordeal & arrow poison; appl. to bad teeth (Nigeria) - **A. oleifolium** Stapf (seAf, Kalahari Desert): c. hongheloside A, &c. (card. glyc.); rt. inf. u. ton. esp. in gastric dis., antipyr.; liniment for snake bite & scorpion sting - **A. somalense** Balf. f. (eAf): rt. & latex c. somalin (cardiac glycoside closely related to digitoxin); u. arrow poison.

Adenocalymna punctifolium Blake (Bignoni) (Mex, Guat): bulbs u. for asthma by Mayas (6) - **A. seleri** Loes. (Mex): op(p)olche; u. to cure carbuncles.

Adenocarpus commutatus Guss. (Leg.) (Sp): c. alk. adenocarpine, santiaguine; produces bradycardia in frog - **A. complicatus** J. Gaya (Sp): similar.

Adenocaulon bicolor Hook. (Comp.) (Cana, nUS, eAs): lvs. simmered in water to make tea as refrigerant - **A. himalaicum** Edgew. (A. adhaeresens Maxim.) (Jap, Kor, Chin): nobuki (Jap); dr. lvs. u. to antidote poison ivy dermatitis (Ainu).

Adenodolichos punctatus (Micheli) Harms (Leg) (c tropAf): rts. u. as remedy for abdominal pains (Congo); frt. edible - **A. rhomboideus** (O. Hoffm.) Harms (var. *lanceolatus* Wilczek) (Congo): rts. u. as remedy for cough.

adenohypophysis: anterior lobe of pituitary gland.

Adenolisianthus arboreus Gilg (Gent) (Br): said u. in prepn. Curare.

Adenophora capillaris Hemsl. (Campanul.) (eAs): u. for lung dis.; in weakness - **A. liliifolia** (L) Bess. (eGe, Russia, Siber): u. as nutrient; rich in sugars - **A. polymorpha** Ledeb. (Siber, Chin): c. saponins; u. as ton. aphrod. (Chin); rt. c. antibiotic - **A. potaninii** Korsh. (wChin): lady bell; rts. u. med. (55) - **A. remotiflora** Miq. (Chin, Jap, Kor): rts. u. to "promote vision"; antidote for poisons, snakebite, &c. - **A. stricta** Miq. (Chin, Jap): rts. u. in pulm. dis., sialagog.; female dis. ton. for weakness - **A. verticillata** Fisch. (Chin, Jap): u. similarly to *A. stricta*; expect. antipyr. stom.

Adenopus breviflorus Benth. = *Lagenaria b.*.

adenosine: adenine riboside (adenine combn. with ribose); glycoside-like compd., considered a nucleoside; widely distrib. in nature; obt. fr. yeast nucleic acids; plays very important role in energy metabolism of cell; u. in paroxysmal supraventricular tachycardia (PSVT) (arrhythmia); has been u. with nikethamide (Coramine) in angina pectoris, &c.; in brain tumors (orphan drug); vasodilator; a. triphosphate (ATP) is coenzyme instrumental in deposition of glucose as glycogen in liver - **a. arabinoside** (ara-A) (Vidarabine): fr. cultures of *Streptomyces antibioticus*; u. as antiviral agt., esp. to treat herpes, cardiac arrhythmias; encephalitis; may be useful in shingles, chicken pox, keratoconjunctivitis - **a. diphosphate (ADP)**: compd. found in muscle, yeast, &c.; important in devt. of energy by muscle, &c. - **a. monophosphate (AMP)**: adenylic acid - **a. triphosphate (ATP)**: Adephos (&c.): coenzyme wh. acts in transfer of phosphate bond energy (ATP —> ADP); occ. in muscle tissues; u. in biochem. research; to inhibit brown discoloration of pl. materials (ex. apples).

Adenosma bracteosa Bonati (Scrophulari) (Indoch): very arom. pl. u. by inhalation of aroma to relieve colds; associated with other pls. to wash hair (Laos) (947) - **A. caerulea** R. Br. (sChin to Austral): rt. decoc. of this ann. hb. u. for intest. dis., rheum. (Mal) - **A. indiana** (Loureiro) Merr. (A. capitata Benth.; A. bilabiata Merr.) (seAs from Ind to Indones, PI): hb. appl. as poultice for abd. pain (Mal).

Adenostemma lanceolatum Miq. (Comp.) (Malaysia-PacIs): amuel (Satawan Atoll); u. in leprosy; after-birth medication - **A. lobatum** (Maxim.) Maxim. (Jap, Chin): gokizuro (Jap); source of a cooking oil - **A. paniculatum** Maxim. ex Cogn. (Chin): rhiz. u. exter. in swelling of breast; in scrofula; for poisonous snake bite; wounds; astr. styptic - **A. platyphyllum** Cass. (Pan to Arg): lf. inf. drunk by aborigines for laryngitis (2263); lf. ash rubbed on head & ears for pimples (777) - **A. viscosa** Forster & G. Forster (A. lavenia O. Ktze.) (As, Austral, pantrop.) "flor de seda" (Pe); pl. has many uses incl. remedy at childbirth; in eye dis. (Pe); antispasm. (Réunion) in diarrh. mal.; rt. decoc. in gastralgia; often in combns.; lvs. in poultice for vertigo, headache, fever (Mal); lf. juice stim. sternutatory (Réunion); widely dispersed weed; hb. source of a blue dye ("indigo").

Adenostoma fasciculatum H. & A. (Ros.) (sCalif [US], nMex): greasewood; (redshank) chamise; shrub forms part of chaparral (or chamisal); u. to medicate cows (Coahuilla; Indians); lf. decoc. u. as wash for infected, sore, or swollen limbs (Calif [US] Indians) - **A. sparsifolium** Torr. (Calif & wMex): bastard cedar; greasewood; twig decoc. u. as emet., cath. & for many dis. (Am Indians).

Adenostylis Bl. Cass. (Comp.) (Euras) = *Zeuxine* Lindl.

adenylic acid, 3¹-: adenosine-3¹monophosphate; a nucleotide prepd. fr. yeast nucleic acid to give **5-adenylic acid**: adenosine-5¹-monophosphate; muscle adenylic acid; widely distributed in nature, thus in muscle & other animal tissues (energy source); may be prepd. by phosphorylation of adenosine; u. like adenine in microbiol. assays, research in cancerology, virology, &c.; combines with pantothenic acid to form co-enzymes A, which act with animal phosphorylase to form glycogen fr. glucose-1-phosphate; nutr., u. in bursitis, pruritus, &c.

Adeps (L): lard- **A. Anguillae** (L): eel oil - **A. Anserinus**: goose grease; u. emoll. - **A. Benzo(in)atus** (L): benzoinated lard - **A. Bovinus** (L): suet - **A. Gadi**: codliver oil - **A. Hominis** (L): human fat, u. in Middle Ages in med. (2091) - **A. Lanae: A. L. Anhydricus**: anhydrous lanolin; purified wool fat - **A. L. Hydrosus**: hydrous lanolin; lanolin; hydrous wool fat - **A. Ovillus: A. Ovinum**: prepared (mutton) suet - **A. Petrolei** (L): Petrolatum - **A. Suillus** (Eu Phar.): lard.

adermin(a): vit. B_6, pyridoxine (OW older term).

Adhatoda engleriana C. B. Clark (Acanth.) (trop eAf): rt. u. as purg. (Tanz, Kenya): lf. decoc. u. as purg. & to relieve pain of childbirth; in TB - **A. vasica** Nees (Justicia vastca L.) (sAs, esp. common in India): Malabar nut; kayagyi (Burma); lvs. & rt. c. vasicine (u. in bronchitis); antispasm. u. as expect. in resp. dis., for cough (TB); abortif. (*cf.* prostaglandins); lf. ext. antiseptic; lvs. u. in rheum., insecticide (Nep); lvs. smoked as cigarettes for asthma (Nep) (—> *Justicia*).

adiante (Fr): *Adiantum* spp.

Adianti aurei Herba (L): hair cap moss, fr. *Polytrichum commune*.

Adiantum aethiopicum L. (Adiant/Polypodi-Pteridophyta) (tropAf&As): large maidenhair (fern); inf. u. for pulmon. dis. (cough); caudex decoc. to promote childbirth; abort. (Kaffirs) - **A. capillus-veneris** L. (Euras, Af, Polynesia, seUS, WI, trop & subtrop Am): true (or southern or European) maidenhair fern; Venus hair-fern; entire pl. or fronds of this & other *A*. spp. sometimes u. as expect. emmen. stim. dem. (492); in rheum; hair restorant (pd. whole plant in olive oil) - **A. caudatum** L. (OW trop): fronds u. for coughs & fevers (Bourbon Is); extern. in skin dis.; int. in diabetes - **A. chilense** Kaulf. (Ch): culantrillo; u. diur. refrig. emmen. - **A. concinnum** H. & B. ex. Willd. (tropAm): u. med. (Col) - **A. pedatum** L. (Cana, nwUS, neAs): maidenhair (fern); American (or Canadian) m. f.; u. as expect. & antirheum.; astr.; u. like & sometimes preferred in med. (more aromatic & bitter) to *A. capillus-veneris*; var. *aleuticum* (Rupr.) Calder & Taylor (A. a. [Rupr.] Paris) (larger fronds) - **A. philippinense** L. (A. lunulatum Burm.) (PI & other trop): walking maidenhair fern; culantrillo; fresh lf. decoc. u. as antipyr. stom. diur., in dys.; sometimes administ. with *Centella asiatica* - **A. raddeanum** C. Presl (A. cuneatum Langsd. & Fisch.) (Ec): culantrillo del pazo; lf. decoc. u. for menstrual cramps [2449]) - **A. tenerum** Sw. (sFla [US], sMex, WI, nSA): (black-stick) maidenhair; slender m. h. fern; pl. decoc. u. as emmen. (Mex-Yuc); in chest complaints & catarrh (Cu) - **A. venustum** G. Don (nInd): fronds u. as ton. expect. diur. emmen. astr. emet.; to prevent hair fall (hakims).

Adina cordifolia Benth. & Hook. f. (Rubi) (Ind, Pak): Haidu (Hindi); rts. astr. u. in diarrh. dys. (Cambodia); bk. u. antipyr. antisep.; juice u. to kill maggots in sores; wd. ("modelwood") valuable article - **A. sessilifolia** Hook. f. (seAs): wd. inf./decoc. given after childbirth as lax. ton.; bk. astr. ton. & hemostat. to treat gingivitis, metritis, colds, diarrh., hemoptysis of TB, liver dis.

Adipatum: a mixt. of lanolin, petrolatum & ceresin; substit. for lard.

adipic acid = hexanedioic acid; HOOC-$(CH_2)_4$-COOH; occ. in beet juice, &c.; u. acidulant in baking pds., beer mfr., gelatin desserts (to confer acidity); in mfr. artificial resins, plastics (nylon); also prepd. by synth.

adjinimoto (Jap): subst. ext. fr. wheat gluten; c. mostly sodium glutamate; u. in flavor, seasoning.

adlai: adlay: *Coix lacryma-jobi* v. *mayuen*, kind of Job's tears; u. as food (seAs, Jap, PI).

Adleria gallae-tinctoriae Oliv. (Cynapidae - Insecta): wasp which produces galls in *Quercus*.

Adlumia fungosa (Ait.) Greene ex BSP. (monotypic) (Papaver.) (eNA): climbing fumitory; mountain fringe; Alleghany vine; pl. c. adlumine, adlumidine (phthalide-isoquinoline alk.); protopine; bicucculine; cult. ornam.

adonidin: mixt. of cardioactive glycosides of *Adonis vernalis* - **adonidoside**: amorphous cardiac glycoside of *Adonis vernalis*.

Adonis aestivalis L. (Ranuncul.) (s&cEu): pheasant's eye; hb. u. as digitaloid card. stim., diur., &c.; similar to *A. vernalis* but weaker; suggested as replacement (933) - **A. amurensis** Reg. & Rodde (Chin, Manch, Jap): hb. c. adonin (weaker card. activity than adonidin); off. in several Euras Ph.; u. in heart dis. - **A. annua** L. (Euras): digitaloid drug of lesser activity - **A. autumnalis** L. (may be synonym of *A. annua* L.) (Euras): autumn adonis; ornam. ann. hb. u. in med.; c. digitalis-like (cardiotonic) active princ.; formerly more often used in heart dis. - **A. microcarpa** DC (A. cupaniana Guss.) (Eu): cupana; c. cardiac glycoside; u. like *A. vernalis* (Sicily) - **A. mongolica** Sim. (Mongolia): barbad (Tibet); c. adonitoxin, cymarin, k-strophanthin, k-strophanthoside, &c.; with 1.2 x content of glycosides of *A. vernalis*; u. cardiotonic (digitalis-like) stim. - **A. sibirica** Patrin. (Siber): u. in folk med. - **A. vernalis** L. (c&seEu, esp. Ge, Balkans; Siber): (spring) pheasant's eye; s. adonis; false hellebore; hb. c. adonitoxin, (cardiac glycoside); adonital (sugar); adonidoside; u. cardiac stim. diur., in hyperthyroidism; in eye dis. (Russ); esp. popular in cEu; fresh pl. collected in spring u. in homeop. med. (more active than dr.), adulter. of *A. aestivalis* (1128); off. in several pharmacopeias (ex. DAB) - **A. wolgensis** Stew. (USSR): u. cardioton.

adonis, autumn: *Adonis annua* - **spring a.**: *A. vernalis* - **summer a.**: *A. aestivalis.*

adonivernoside: amorphous cardiac glycoside fr. *Adonis vernalis*.

adormidera (blanca) (Sp): opium poppy (white).

Adovern™: prep. of active glycosides of *Adonis vernalis*.

Adoxa moschatellina L. (monotypic) (Adox.) (Euras, natd. NA): moschatel; c. pseudoindican type of heteroside; u. in fever, hysteria, dystonia.

adragant (Fr): **adragante** (Por): tragacanth.

adrenal cortex extract: suprarenal c. e.; ACE; c. natural adrenocortical hormones (corticosteroids): 1) glycogen-forming (ex. hydrocortisone) 2) mineral corticoids (ex. aldosterone); prep. u. to treat adrenal insufficiency (esp. Addison's dis.); in acute stress (ex. serious dis., burns) - **a. gland:** living beef glands have been implanted in humans to furnish relief from chronic pain; secretes endorphin enkephalin - **Adrenalin™**: epinephrine - **adrenal substance, dried**: suprarenal; the dr. gland of animals u. as food by man; in this form, prod. now considered worthless in therapy; a gland transplantation (or fetal tissue injected) tried also for Parkinsonism

adrenochrome: pigment formed as metabolite by oxidn. of epinephrine; u. as hemostat.; has hallucinogenic properties; in chronic use esp. dangerous & may prod. schizophrenia (subtle and insidious in action) (*cf.* LSD).

adrenocorticotrophic hormone: ACTH: one acting to stimulate the adrenal cortex.

adriamycin: former generic name for Doxorubicin (*qv.*).

Adromischus mamillaris Lem. (Crassul.) (sAf): pl. non-toxic but may cont. cotyledontoxin; u. in epilepsy; cult. orn.

adrue: *Cyperus articulatus.*

adsorbate: substance that is attracted and concentrated on a surface by physico-chemical means.

adulteration: the addition (intentional or unintentional) of impurities or inferior products to a drug, food, etc., or the removal of valuable portions fr. such commodities.

ae: see also under "e."

Aechmea magdalenae Standl. (Bromeli) (CA, Co, Ve, Ec, Br): silk grass; lvs. furnish strong fiber u. for ropes & twines - pita fibre ("pita de madlena"); cult. ornam.

Aedes (Diptera - Insecta): genus of mosquitoes related to *Culex,* many of special importance as annoyances (bites, songs) but more importantly as disease vectors - **A. aegypti** (Eg & other trop areas): tiger mosquito; transmits yellow fever, dengue fever, &c.; other *A.* spp. transmit equine encephalomyelitis, filariasis, &c.

Aegiceras corniculatum (L) Blanco (Myrsin.) (Indomal, sChin, nAustral): u. as gargle (Vietnam); frt. & bk. c. saponin; u. as fish poison.

Aegilops cylindrica Host (Gram.) (Eu, natd. US): jointed goat grass; sds. eaten (Canary Is); pl. formerly u. in eye dis.; a serious weed - **A. ovata** L. (sEu, natd. US): sds. eaten - **A. triticoides** Bert. (Eu-natd. US): hard grass; u. similarly to *A. cylindrica*; closely related to wheat & u. in breeding wheat strains; weed.

Aeginetia indica L. (Orobanch.) (Taiwan to PI): u. ton. hematic in impotency, sterility.

Aegiphila Jacq. (Verben.) (tropAm, WI): trees & shrubs; lianas - **A. elata** Sw. (tropAm, WI, Guianas, CA): guairo santo (Cu, Pe); lvs. & sts. of shrub u. as inf. (without sugar) for ulcers, dys. diarrh., edema, tetanus (!?); lf. decoc. in edema (baths); for sores, wounds, & ulcers; int. for spasms (Cu) - **A. filipes** Mart. & Schau (Co, Pe): "arco sacha"; shrub; u. exter. for skin dis.; lvs. rubbed into skin - **A. graveolens** Mart. & Schau (Br): fleurs de bouc; fruta de sabia; lf. decoc. u. in skin dis. (exter.). in baths for arthritis - **A. hassleri** Briq. (A. poliantha Rejus) (Br, Par): frt. u. sudorific; dye - **A. integrifolia** Jacq. (Co, Ve, Guian, Pe, Ec, Bo, WI): tabaquillo; macerated lf. u. like soap for eruptions; oily sds. u. for itch mite lesions (Br) - **A. laevis** (Aubl.) Gmel. (Co, Sur, Br): smooth aegiphila; lvs. u. as insecticide - **A. lhotskyana** Cham. (Br-campo region): u. to furnish cork (layers 1 m thick) - **A. martinicensis** Jacq. (A. glandulifera Mold.) (WI, CA s to Co, Ve, Br): spirit weed; lengua de vaca; inf./decoc. of lvs. & sts. u. as diur. decoc. in baths; syrup u. in asthma - **A. mediterranea** Vell. (Br): pl. u. as insecticide (Espiritu Santo State) - **A. mollis** HBK (Co, Ve, Br, Par, CA): a tree; bruised fls. u. on bites; bk. u. as antidote for cobra bite (Br, Par); fls. u. in epilepsy (Ve Indians); rt. bk. u. as purg. - **A. obducta** Vell. (Br): c. alk.; pd. sds. u. ton. in diarrh.; bk. u. ton. diur.; FO of sds. used - **A. peruviana** Turck. (Br, Bo, Pe): arco sacha; bk. u. in poultices in inflamm. to treat inf. ulcers (Pe); for sunburn - **A. racemosa** Vell. (Pan, Guyana, Br): "cawuira"; u. in warm baths for nervous dis.

Aegle (L.) **marmelos** Correa (Rut.) (Ind, seAs): (Indian) bael; okshit (Burma); Bengal quince; small to medium sized tree; sacred to some Hindus; bael frt. pleasant tasting, popular with jaggery (sugar) as breakfast food (SriL); c. tannins, mucil. u. astr. in diarrh. dys., in prepn. lemonades, marmalades; bk. u. in digest. dis., source of a gum (good adhesive) (948); rt. u. in digest. dis.; wd. u. in building, making wheels, combs; fls. source of perfume; lvs. in ophthalmia; rt. bk. in fever; grows wild & cult. all over Ind & trop - **A. religiosa** Forst. f. = *Crataeva religiosa* Forst. f. (Capparid.).

Aegopodium podagraria L. (Umb.) (Euras, natd. eNA): goutweed; hb. u. mostly in fresh state (esp. in homeop.) to alleviate pains of gout & rheum.; in hemorrhoids; frt. u. to adulter. caraway; diur. sed.; lvs. u. as greens like spinach & in salads; exter. in fomentations; cult. ornam.

Aellenia subaphylla (Cam.) Aellen (Chenopodi.) (swSiber): solyanka malolistnaya (Russ); pl. u. in med. (Russ).

Aeolanthus canescens Guerke (Lab) (Af): decoc. in colds of children - **A. gamwelliae** G. Tayl. (Zambia): fls. c. VO rich in geraniol & esters; with aroma like rose - **A. helotropioides** Oliv. (trop wAf): hb. antipyr.; u. by natives to flavor soups - **A. pubescens** Benth. (tropAf): hb. u. as antipyr. & to flavor soups.

aerenchyma: pl. tissues suffused with air spaces; found in floating lvs., &c.

Aerides falcatum Lindl. (Orchid.) (Indoch.): crushed frt. u. for infected wounds, skin dis.; the small sds. (powder-like) are the "sulfonamides" of the country.

aerobiology: science of pollens, spores, microorganisms, etc., as carried in the atmosphere.

aerosol: gaseous suspension (producing mist) of very small particles; u. as surfactant, wetting agt. lax., in cosmetics, insecticidal & disinfectant sprays, &c.

Aerospora: *Bacillus aerosporus* Greer (2715): *Bacillus polymyxa* (Prazmowski) Macé.

Aerosporin™: streptothricin A; Polymyxin A; antibiotic prod. by *Streptomyces* sp. (soil microorganism); u. to eradicate Gm. negative organisms; in influenzal meningitis; u. in surgery (1946/7).

Aerva lanata Juss. (A. leucura Moq.) (Amaranth.) (sAf, Botswana, tropAs, SriL): lvs. u. as diur. in urin. dis., bladder trouble, GC, lithiasis; febrifuge; anthelm.; in cough, sore throat (2445); lvs. u. as spinach, shoots in curries (Ind); famine food (Ind); fls. to stuff pillows - **A. sanguinolenta** Bl. (A. scandens Wall. ex. Moq.) u. as lax. (Indones): diur. in hematuria - **A. tomentosa** Forsk. (tropAf near Red Sea): chaya; u. like *A. lanata*; in veter. med. (horses, camels).

Aeschrion Vell./Conc. = *Picraena; Picrasma.*

Aeschynanthus Jack (Gesneri.) (Chin, Indomal): blush-wort; many spp. cult. in greenhouse as ornam.; common names (US): lipstick plant; royal red bugler; basket vine - **A. marmoratus** T. Moore (Indoch): lvs. placed as poultices on boils - **A. poilanei** Pellegr. (Indoch): decongest. after childbirth (fumigant) - **A. radicans** Jack (Far East): appl. as poultice for headache.

Aeschynomene L. (Leg) (worldwide - trop & subtr): some 150 spp., about equally divided between Old and New Worlds; fr. ann. herbs to trees; st. pith. fr. eAs spp. u. for corks, stropping knives, to make artif. fls., floats for fishing nets, &c.; forage; honey pls.; soil nitrification; ornam. (2142) - **A. aspera** L. (Indoch): pith plant; pith-like wd. u. to make solar helmets of tropAs; insulator to keep beverages &c. cool in hot weather; st. pith u. as hemostatic in metrorrhagia - **A. fascicularis** Schlecht. & Cham. (Mex, Guat to nVe, Co): pega-pega; pl. decoc. u. to treat tumors (Yuc) - **A. filosa** Mart. (WI, Hond, nVe, Co, Br.): cama de venado; st. decoc. appl. as liniment to relieve rheum. pains - **A. indica** L. (Ind, Chin, natd. seUS, Af+): rt. decoc. u. to relieve swollen stomach; in GC; gastric dis., urin. dis.; u. —> charcoal - **A. leptophylla** Harms (Burundi, tropAf): umushi; bk. & lvs. mixed in food as abortient.

aesculetin: esculetin: the aglycone of aesculin - **aesculin: esculin**: the characteristic glycoside of the sd. of horse chestnut, *Aesculus hippocastanum.*

Aesculus californica (Spach) Nutt. (Hippocastan.) (Calif [US]): California horse chestnut (or C. buckeye); fresh frt. u. as fish poison; nut bitter but formerly u. in soups & gruels (Calif Indians); bk. u. in toothache - **A. chinensis** Bge. (Chin, Indochin): frt. u. in gastralgia, emetic; sds. u. in severe rheum. - **A. flava** Ait. (A. octandra Marsh.) (eUS): sweet buckeye; (large) buckeye; wd. u. occasionally; thus to make artificial limbs - **A. glabra** Willd. (e&cUS): (Ohio) buckeye (tree); fetid b.; bk. u. as ton. antipyr.; wd. u. for artificial limbs; powd. sts. & branches u. as fish poison; cult. as avenue tree - **A. hippocastanum** L. (Balkans, Iran, nInd; now cult. worldwide): (common) horse chestnut; English h. c.; marronier (Fr); bk. u. as astr. antipyr.; in prepn. aesculin; as tan; fls. as astr. ton. in gout, rheum.; lvs. in cough preps.; in hemorrhoids (suppositories); sd. kernels (Hippocastanum, L.) rich in & sometimes u. as source of saponins & starch; sd. as coffee substit. sometimes; ext. prevents anaphylactic shock (95); debittered meal as food - **A. indica** Hook. f. (Pak, Ind, Nep): Indian (horse) chestnut; frts. given to horses for colic; sds. u. as starvation food; cattle fodder; sd. FO appl. exter. in rheum.; fls. with very strong unpleasant odor; cult. ornam. (Pak) - **A. parviflora** Walt. (seUS, esp. Ala): white (or bottle-brush) buckeye; frt. with 1 or 2 chestnut-colored sds., considered edible & tasty - **A. pavia** L. (seUS n to WV): red (or little) buckeye; "flame b."; shrub or small tree (3-4 m); bk. ton.; fresh sds. u. as fish poison (Am Indians); fls. (as tinct. or oil ext.) in rheum. & gout; sd. was carried in pocket to prevent hemorrhoids (deep South) - **A. turbinata** Bl. (eAs): rt. decoc. u. to treat wounds (Ainu); frt. inf. astr. (Ainu); sds. u. in stroke, paralysis, neuralgia, rheum. (Indoch) - **A. wilsonii** Rehd. (Chin): frts. u. to treat rheum. & stom. illnesses (Chin).

aestivation 1) the way in which parts of flower are arranged before opening (see Appendix E) 2) unfolding process of flower parts from previously dormant bud.

Aether (Ge): ether; ester - **A. Petroleus** (L.): petroleum ether - **A. Vegetabilis**: acetic ether; ethyl acetate.

Aetheroleum (L.): volatile (or essential) oil.

Aethiops Mineral (L.): black mercury sulfide - **A. Vegetabilis**: seaweed charcoal.

Aethusa cynapium L. (Circuta minor Garsault) (monotypic) (Umb.) (Euras; natd. e&cNA): fool's parsley; dog p.; d. poison; garden (or lesser or small) hemlock; pl. c. coniine-like alk. ("cynapine"); VO; hb. u. in colics, diarrh. pollakiuria by homeop.; lvs. sometimes u. as substit. for conium, &c.; rt. & frt. also u. as diur.; chiefly important as poisonous pl.; ornam. - **A. divaricata** Spreng. = *Apium divaricatum* B. & H. (*qv.*).

Affenbrod (Ge): frt. of *Adansonia digitata*; ape's bread.

affenkruid (ape herb) (Belg): *Dictamnus albus*.

affinin: lipid amide occ. in *Heliopsis longipes*; serves as synergist to insecticides.

aflatoxins: mycotoxins (fungal metabolites) prod. by *Aspergillus flavus* (hence name) & *A. parasiticus*; several gps. designated as B (with blue fluorescence), G (with green florescence), & M (milk toxin; occ. in milk of cattle fed toxic meal); found in spoiled peanuts, corn, cottonseed, &c.; toxic to liver (hepatic cirrhosis) & may cause liver cancer; acts on DNA.

Afraegle paniculata Engl. (Rut.) (trop wAf): frt. + lvs. (+ bark) u. int. & as wash for rheum.; sd. FO u. in cooking & for food; lf. inf. u. as lotion.

Aframomum cuspidatum K. Schum. (Zingiber.) (trop wAf): (dadi) gogo (Guinea); u. purg. teniacide, in edema; lower parts of rts. u. in rheum. pulmon. dis. - **A. daniellii** K. Schum. (A. angustifolium K. Schum.) (trop wAf): bastard melegueta; rts. u. as vermif.; sds. u. med.; frt. pulp refreshing, u. to relieve thirst in fever - **A. granum-paradisi** K. Schum. (trop wAf): black amomum; paradise seed; arom. sds. u. in med. much like *A. melegueta*; pepper substit. & adulter.; rts. u. in roundworm; lvs. cooked with food to flavor; sds. not the true grain of paradise (= *A. melegueta*); u. in alc. drinks; sd. FO u. as body pomade - **A. korarima** (Pereira) Engl. (Amomum k. Pereira) (seEthiopia, Somalia): nutmeg cardamom; u. as spice (guragi spice) - **A. mala** K. Schum. (eAf, incl. Tanz): lvs. sds. rts. c. VO with cineole, terpineol, caryophyllene, &c; arom. frts. eaten as sweetmeat & to mask odor of hashish; condim.; sialagog; to stim. salivary flow; admixed with coffee (Saudi Arab); prepn. neck chains (Uganda) - **A. melegueta** K. Schum. (Amomum m. Rosc.) (SriL; trop wAf, Grain Coast): grains of paradise; Guinea (or melegueta) pepper (or grains); sds. c. VO with paradol (sharp taste), resin; u. as spice, stim., flavor in pungent liqueurs; VO in perfumes; during wartime, substit. for black pepper & as pepper adulter.; sds. u. in dys.; throat & tonsil inflamm., toothache - **A. sanguineum** K. Schum. (trop eAf): u. much like *A. mala* - **A. stipulatum** K. Schum. (trop wAf): frt. pulp edible; rt. u. for tapeworm & as abort. - **A. sulcatum** K. Schum. (tropAf): go go; gogui; u. similarly to *A. cuspidatum*.

afrecho (Sp, Arg): bran.

African: v. also under bdellium; chile; locust; rose.

African root: *Schizoglossum shirense*.

Afri-Cola™: caffeine soft drink popular in Ge (Buna Co.) (*cf.* Coca Cola).

Afrolicania elaeosperma Mildbr. (Chrysobalan./Ros.) (monotypic) (trop wAf): nikko(n) (nut) tree; sds. source of FO, "po-yoak oil"; u. as a hair pomade, said to incr. hair growth (Af women) (—> *Licania*).

Afrormosia angolensis Harms (Leg.) (tropAf): rt. cold inf. u. as ton. & in pulmon. dis.; wd. decoc. aphrod. - **A. elata** Harms (Congo); rt. u. as antipyr. (in combns.) - **A. laxiflora** (Benth.) Harms (trop wAf): false dalbergia; black gum (Liberia); lvs. & bk. analg. antipyr.; rt. fl. intoxicant; u. to increase intox. effects of palm wine; in arrow poisons; as ton.; wd. very valuable (termite resistant), for hoe handles, &c. (—> *Pericopsis*).

afs: galls of: 1) *Pistacia atlantica* coming fr. Gabel; product of *Pemphigus utricularius* 2) *Quercus* produced by *Cynips tinctoria*; imported into Tripoli fr. the East.

after-ripening: period of dormancy of sd. preceding germination & growth.

Afzelia africana Smith (Leg.) (tropAf): African mahogany; m. bean; aril u. for Guinea worm; bk. aphrod.; burnt pods u. in prepn. native soap; wd. valuable, u. in cabinet work - **A. quanzensis** Welw. (tropAf): pod mahogany; m. bean; rt. inf. u. in blepharitis; rt. decoc. in schistosomiasis, for stomach pains (Tanz); bk. in toothache; bk. & rt. in native magic; wd. useful for dug-outs, drums, &c.

Agalinis tenuifolia (Vahl) Raf. (Gerardia t. Vahl) (Scrophulari) (e&sUS, w to Mich & La): slender leaved gerardia; kidney weed; pl. decoc. u. diur. & for beverage tea (SCar [US]); (2451); said toxic to cattle & sheep (1992).

agalla (Sp): nutgall.

Agallochum (L.): 1) fragrant wd. of a tree, *Aquilaria agallocha* (*qv.*) also called aloes-wood (*qv.*) 2) (genus) = *Aquilaria*.

Aganosma calycina A DC (Apocyn) (Burma): malati (Sancrit); hb. ton., in bile & blood dis. - **A. dichotoma** (Roth) K. Schum. (Ind): malate (Hindi &c); lvs. u. as emet. in "biliousness;" fl. in eye dis.; antisep. - **A. marginata** G. Don (Ind+): rt. decoc. in urin. dis., ton. emmen.; lvs. fls. & frts. u. to make a digestive beverage; lvs. substit. for tea/coffee (Indoch).

Agapanthus africanus Hoffmgg. (Amaryllid./Alli.) (sAf): lily of the Nile; African l.; decoc. u. by pregnant female to ensure easy birth & healthy child; cataplasms for relief of abdom. pain; hot inf. emet.; ornam. (pot plant) - **A. umbellatus** L'Hérit. = *A. africanus.*

Agapetes saligna B. & H. (Eric.) (Ind): lvs. u. for tea in some parts of Ind.

agar: 1) dr. mucilaginous ext. of several genera of marine algae, mostly fr. *Gelidium* (esp. *G. cartilagineum*) and *Gracilaria* spp.; u. lax. emulsifier (for mineral oil, etc.); dem.; food component, & in industry; gives stiffness to ointments; a. broth u. in bacteriology; a. gel u. in electrophoresis; chief prod. in Jap, Chin, Calif [US] 2) (Hindi, Urdu): incense; made up of polyglucosans esterified with H_2SO_4 & combined with alkaline earth metals 3) agallochum. (prec.) - **agar agar**: synonym for agar - **American a.**: that prod. in US, esp. Calif, NCar - **Atlantic (coast) a.**: a. prod. in eUS (esp NCar) fr. *Gracilaria confervoides* & *Hypnea musciformis* - **a. attar**: VO of agar wd. - **Australian a.**: yielded by *Gracilaria confervoides* & prod. on that continent (935) - **Bengal a.**: Ceylon a. - **British a.**: not really an agar at all but carrageenan produced in GB - **Ceylon a.**: the dr. unaltered thallus (not ext.) of *Gracilaria lichenoides* - **Danish a.**: extractive prod. fr. *Furcellaria fastigiata* - **Indian a.**: that prepd. fr. *Gracilaria confervoides* (934) - **Japanese a.**: that prod. in Jap fr. *Gelidium elegans*, *G. japonicum*, *G. amansii*, etc.; represents biggest segment of world supply - **Pacific Coast a.**: prod. in US (Calif): fr. *Gelidium cartilag.*, *G. amansii.*

agari-bai (Arg): lvs. of spp. of Leguminosae u. in catarrh (949).

agaric (Fr): 1) mushroom 2) **a. blanc** (Fr): officinal white agaric: *Fomitopsis officinalis* (Fomes o.) - **a. acid: agaricic a.**: **agaricin** (comml. name): tricarboxy acid fr. *Fomes* spp. - **female a.**: Agaricus (2) - **fly a.**: *Amanita muscaria* - **larch a.**: Agaricus (2) - **a. of the oak**: *Phellinus* (*Polyporus*) *igniarius* - **purging a.: surgeon's a.: white a.**: Agaricus (2).

Agaricaeae: gill fungi fam. (Basidiomyc. - Fungi); the mushrooms & toadstools; constitutes the largest fungus fam. (c 5,000 spp.); prominent part is frt. body, often umbrella-like in shape & consisting of stipe (stalk), pileus (cap), & below the latter the lamellae (gills) bearing basidiospores; some spp. edible, some extremely poisonous; ex. fly agaric; field mushroom; *Coprinus comatus.*

agaricin (comml. name): tricarboxy acid fr. *Fomitopsis officinalis*; u. as antiperspirant.

Agaricus: 1) genus of Agaricaceae incl. several spp. of edible mushrooms 2) (L.) (specif.): the fruiting body of *Fomitopsis* (*Polyporus*) *officinalis* (Polyporaceae) - **A. bisporus** Pilat (Agaric. - Basidiomyc. - Fungi): (temp zone, worldwide): culti-

AGARICUS GAMPESTRIS L. (COMMON MUSHROOM)

vated mushroom; industrially cult. common mushroom, fruit-body widely u. as food (raw, fresh, cooked, canned); u. in homeop. med. (also known as **Agaricus brunnescens** Peck) (close to *A. campestris*) - **A. bitorquis** Sacc. (A. rodmani Pk.), (Af, Sahara region; NA): frtg. bodies (basidiocarps) esteemed as food by Arabs; edible and choice; gills pink to blackish brown (old); good keeping qualities - **A. campestris** L. ex. Fr. (Psalliota c. [L] Kummer) (widespread): (common) field mushroom; meadow m.; white m.; much cult. as popular food, the young sporophores ("buttons") being u. generally (fresh, dr., canned); close to *A. arvensis* Schaeff. (horse mushroom) with sporophores of fine quality - **A. chirurgorum** (L.): *Fomes fomentarius* - **A. emeticus** Schaeff. = *Russula emetica* Fr. = **agaric émétique** (Fr) - **A. muscaria** L. = *Amanita m.* - **A. silvicola** (Vitt.) Pk. (Eu, NA): forest agaric; sylvan (or wood) mushroom; a choice edible mushroom; goes well with meat; best mushroom for catsup.

agar-wood: agallo chum.

Agastache Clayt. ex Gronov. (Lab.) (As, NA): giant hyssop; gen. of ca. 30 spp. of perenn. hbs. with VO - **A. foeniculum** (Pursh) Ktze. (A. anethiodora Britt.) (wUS): (blue) giant hyssop; anise h.; lf. decoc. u. as beverage by Am Indians - **A. pallidiflora** Rydb. (Utah-Ariz [US]) subsp. *neomexicana* Lint & Epling (A. neomexicana Standl.) (NM [US]): lvs. with minty odor, c. pulegone, no menthol (950); u. as food flavor (Am Indians) - **A. rugosa** O. Ktze. (Ind, Indochin): "Korean mint"; pl. c. VO with methyl chavicol; u. as carm. stom. antiemet. antidiarrh. in colds, flu, fever, headache (Chin, where cult.); antivinous; flavor for food; for yeast cakes - **A. urticifolia** Ktze. (BC [Cana]-Calif [US]): nettle-leaved horse mint; sds. eaten raw or cooked.

Agathis alba (A. B. Lambert) Foxw. (Dammara orientalis Lamb.) (Araucari) (Indones, Austral, PI, Moluccas, Mal): Amboina pitch tree; very large & old tree (-65 m. high; 2 m. diam.; 500 yrs.); prod. white (or East Indian) dammar (resin); Manila (manila) copal (gum) eAsiatic c. obtained by tapping tree; u. to prep. shellacs & varnishes; in plasters, dental cements; in linim. (Mal); attached to feet to prevent leech bite - **A. australis** Salisb. (Damara a. Lamb.) (NZ, Austral): kauri (pine); wd. of this huge tree furnishes abundant resin (wood kauri); also semi-fossil ("recent fossil") fr. the soil; called (New Zealand) kauri copal (gum); agatho c.; much u. in mfr. of varnishes & linoleum; u. as masticatory by aborigines; good solid wood - **A. dammara** (Lamb.) L. C. Rich. = *A. alba* - **A. labillardieri** Warb. (Indones, PapNG, Austral): source of copal (damar putih); trunk furnishes excellent wd. - **A. loranthifolia** Salisb. = *A. alba* - **A. moorei** (Lindl.) Warb. (Oceania): u. in med. (New Caled) - **A. orientalis** Lamb. = *A. alba* - **A. ovata** Moore (New Caled): source of New Caled (or East Indian) kauri copal, furnished by tree trunk - **A. robusta** (F. Muell.) F. M. Bailey (Austral, Oceania): South Queensland kauri pine; source of both fossil & recent copal (fossil considered superior); excellent wood.

Agathisanthemum bojeri Klotz. (Rubi.) (eAf): lvs. c. alizarin; u. for snakebite; arom. rt. chewed or inf. drunk for pain in left side (Tanz); sds. u. by smoking (natives) in respir. complaints (coughs &c.) - **A. globosum** Klotz. (Af): remedy for hemorrhoids (Rundi) & dys. (Liber).

Agathodes Reichb. = **Agathotes** D. Don = *Swertia* L.

Agathosma Willd. (Rut.) (sAf): sea buchu; genus of some 200 spp. of upright shrubs very close to *Barosma*; ornam. - **A. apiculata** G. F. W. Mey. (Barosma a. Eckl. & Zeyh.): (s&eAf): sea buchu; "e-buchu" (natives); source of VO c. benzyl pentenyl disulfide, pinenes & other terpenes; with odor of asafetida; u. antipyr. diaph. stom.; cult. - **A. betulina** (Berg) Pill. = *Barosma* b. - **A. crenulata** Pillans = *Barosma c.* - **A. gnidioides** Schlt. (sAf): lf. decoc. u. stom. antipyr. antispasm. sed. nauseant; for persistent cough - **A. variabilis** Sond. (sAf): aniseed buchu; lvs. with anise-like odor, u. to replace buchu lvs.; several other *A.* spp. (sAf) are similar: **A. cerefolium** Bartl. & Wendl. and **A. ciliata** Link.

Agati grandiflora Desv. = *Sesbania g.* - *Agati* Adans. = *Sesbania.*

Agauria pyrifolia DC (l) (Eric.) (Réunion): mapou; similarly u. to *A. salicifolia*; toxic - **A. salicifolia** Hook. f. (Af, Madag): arrow poison antidote; u. in arthritis (lf. ash), pain; toxic plant; lf. u. as insecticide (212).

Agave L. (Agav./Lili.) (tropAm): false aloe; genus of over 100 spp.; pl. habit resembles that of *Aloe*; pl. dies after flowering; rather important economically; members c. saponins & steroid hormone raw materials; fls. edible; some spp. u. for fibers (lasso ropes, sisal, henequen); intoxicating liquors (mescal, pulque) prod. fr. saccharine juice of pl. - **A. amaniensis** Trel. & Nowell (origin unknown); may have originated in Tanzania as hybrid; source of superior fiber (eAf); lf. inf. u. as insecticide (PI) - **A. americana** L. (Mex, spread to SA, natd. Med reg): American agave; A. aloe; century plant (SC thought to bloom at ca. 100 yrs. of age; actually occ. variably at 5-60 yrs.); maguey; fresh juice lax., diur.; exter. in rheum.; lf. insecticidal (for termites); lvs. u. as food, fodder; cut buds furnish sugary liquid which is fermented into pulque and this distd. to prod. mescal (kind of brandy); pita fibers fr. lvs.; thickened rt. juice u. like honey; rt. u. as substit. for Sarasparilla - **A. angustifolia** Haw. (A. endlichiana Trel.; A. ixtle Karw.) (Mex, CA): American aloe; st. & lf. bases rich in sugars, u. to prep. the alc. spirits mescal & pulque; this called "mescal bacanora" (Sonora) with an "outstanding" flavor; lf. fibers considerably u.; pl. u. to make fences; mature pl. c. hecogenin, &c.; widely cult. in warm countries & Eu - **A. atrovirens** Karw. (sMex): maguey javalin; pulque agave; rt. u. in diarrh. & as vulner.; tapped for agua miel & pulque - **A. brachystachys** Cav. = *Manfreda b.* - **A. bracteosa** S. Wats. (neMex): amole de castillo; u. similarly to other *A.* spp.; cult. ornam. - **A. cantala** Roxb. ex Salm-Dyck (A. vivipara auct. non L.) (Mex, WI, natd. Ind, cultd. Indones, PI+): lf. juice for scabies; Koeki Indians; rt. decoc. in syph.; dys.; as vulner.; furnishes fine quality of fiber (cantala fiber) u. in hard twines; source of diosgenin (2220) - **A. chrysantha** Peebles (A. palmeri Engelm. var. c. Little ex Benson; A. repanda Trel.) (s&cAriz [US]): mescal; caudex & emergent scape roasted & eaten by Am Indians; rt. u. vulner.; ornam. - **A. cocui** Trel. (Ve): cocuiza (dispopo) (Ve); crushed lvs. u. as poultice on tumors; juice u. to prep. alc. beverage ("cordial") called cocuy (or cocqui) - **A. decipiens** Bak. (sFla [US]): false sisal; ornam. in formal plantings - **A. fourcroydes** Lem. (sMex, Yuc): henequen; Mexican sisal; "cahum" (natives); lf. fiber u. as binder twine, ship cordage (formerly); by-prod. sapogenins, esp. hecogenin (u. in synthesis of sex hormones) - **A. lecheguilla** Torr. (wTex [US], n&cMex): tula; ixtle (istle) (appl. to pl. & fiber); very common sp.; fibers u. in cordage, brushes,

&c.; pl. rich in saponins (toxic to livestock); u. as precursors in synth. of cortisone, &c. (juice); brown juice u. as auto radiator cleaner - **A. lophantha** Schiede (A. heteracantha Zucc.) (seTex, Mex): lechuguilla; lvs. u. to prep. fiber (ixtle de Jamauve; rula i.) - **A. lurida** Ait. (A. veracruz. Mill.) (sMex; natd. Med reg): maguey or pulque plant; st. juice c. polyfructosans & simple sugars (951); rt. u. in dys. & as vulner. - **A. mexicana** Lam.: may be *A. lurida* or may refer to *A. fourcroydes* - **A. palmeri** Engelm. (Ariz, NM [US], Sonora (Mex): lechuguilla; after removal of lvs. young tender flowering shoots eaten by Indians; also u. to make mescal & fiber; lvs., crown, & rt. u. as food (pitbaked); and fiber (baskets, ropes, cordage, mats; sandals, &c.); & to make mescal (spirit) (fr. boiled crowns, fl. stalks); sapogenins in small amt. or lacking, hence food rel. sweet to taste; lvs. c. subst. wh. causes erythema & itching - **A. parryi** Engelm. (Tex, NM, Ariz [US], Mex): Parry agave; mescal; u. medl.; ornam. - **A. potatorum** Zucc. (sMex): similar uses; u. to make mescal - **A. salmiana** Otto (cMex): maguey de pulque; chief source of pulque; lvs. u. for forage; fiber also u.; mature pl. c. hecogenin, gitogenin, tigogenin (2260); pl. as a windbreak - **A. schottii** Engelm. (nMex, sAriz, NM [US]): amole; u. for washing clothes like soap; inflorescence c. antitumor agents; young & med. aged pls. c. tigogenin & hecogenin, u. to prep. steroids; juice u. to prep. pulque, also as purg. (Mex Indians) - **A. sisalana** Perrine (A. rigida Mill.) (Mex, CA, introd. into Fla [US], thence to WI, Br pantrop): sisal (or green) agave; sources of sisal hemp & s. wax; sisal fiber one of chief natural hard fibers of world; u. for binder twine; pl. ext. u. in mfr. of steroids fr. hecogenin; lf. comml. source of Na pectate (u. to replace gums, agar, &c.) (952) - **A. tequilana** Weber (sMex): st. u. in prepn. tequila (spirit) (u. in making margarita cocktails); source of a fiber (by-prod.) - **A. toumeyana** Trel. (cAriz [US]): furnishes raw material for mfr. cortisone (493) - **A. virginica** Rose = *Manfreda v.* (eUS).

Agelaea emetica Plan. (Connar.) (Mad): lvs. c. VO; u. emet. - **A. lamarckii** Plan. (Mad): Lamarck agelaea; cephan-mahi; lvs. u. in GC - **A. obliqua** Baill. (trop wAf): frts. u. to rub teeth; ornam shrub - **A. trifolia** Gilg (trop wAf): wd. u. as chewstick (sNigeria); bk. decoc. as witch ordeal - **A. villosa** Pierre (wAf): lvs. u. in dys.

agenized process: bleaching of flour with NC_{13} (now prohibited in US); gluten so treated causes canine hysteria (running fits), esp. with animals on vit. A-deficient diet.

Ageratum conyzoides L. (Comp.) (pantropics): (wild) ageratum; floss flower; billy goat weed (Austral); curia; lvs. c. VO with caryophyllene; u. purg. antipyr.; as dressing for wounds & ulcers (Br); boils; in skin lotions; rt. u. in colic (tropAf); lvs. or entire pl. u. in syph.; lvs. u. for beverage tea & to adulterate true tea; in chronic cystitis (GC) ("gravel weed") (Salv.) - **A. corymbosum** Zucc. (wTex [US], Mex, Hond): flattop ageratum; mota (morada); pl. decoc. to expel intest. worms - **A. echioides** (Less.) Hemsl. (Guat, sMex): "frijilito"; pl. decoc. u. in mal., also in baths for same - **A. gaumeri** Rob. (Yuc): flor de San Juan; lvs. bound to temples to stop nosebleed - **A. houstonianum** P. Mill. (A. mexicanum Sims) (cMex, CA, WI, nSA; natd. Fla [US]): garden ageratum; pl. inf. diaph. for resp. dis. (head colds & hay fever); antirheum.; rt. decoc. in baths for rheum.; ext. has antijuvenile (antiallatotropic) hormone activity; c. coumarin, phytomelane (1983); precocenes; cult. as ornam. - **A. littorale** Gray (var. *hondurense* Rob.) (Mex): hauayche; u. in ringworm.

agglutination: clumping of bacteria & other m.o., red blood cells, &c.; may be accomplished by a specific agglutinin, agent wh. is supposed to act as antibody to the bacteria, &c., the latter acting as antigen.

aggregate fruit: many frts. collected into & having the appearance of a single frt. & prod. fr. a single fl.; e.g., blackberry (composed of many drupelets [ovaries]).

aggressin: subst. prod. by pathogen wh. aids in invasion of host tissues & dissemination of bacteria in host organism.

agialida: *Balanites roxburghii.*

agila wood: aloes wd.

Agkistrodon (*Ancistrodon*) (Crotalidae [Viperidae] - Ophidia - Reptilia) (NA, SA, As): genus of venomous snakes; incl. **A. contortrix** (L.) (A. c. mokasen Dauder) (eUS): copperhead (of NA); bite sometimes fatal to children - **A. piscivorus** (Lac.) (seUS swamps): cottonmouth (water mocassin of NA); venom c. hemolysins & hemorrhagins; u. as hemostatic; in prepn. of antivenins (both spp.); var. *laticinctus* (scUS): broad-banded copperhead - **A. rhodostoma** (Boie) (Mal): Malaysian pit viper; venomous snake; venom c. anerod.

Aglaia baillonii Pellegr. (Meli.) (Cambod): decoc. antipyr.; u. in dys. hemoptysis; as dental powd. - **A. glomerata** Merr. (tropAs, PI+): bk. appl. for pain in xiphoid process (lower sternum) & for anasarca (generalized edema) (PI); frt. pulp edible - **A. odorata** Lour. (Mal, Ind. +): priyangu (Ind.);

fls. u. to perfume tea & clothing; deriv. appl. to body after childbirth (folk med.); rts. & lvs. cardiac stim., antipyr.; in leprosy; comm. article; in joss sticks; cult. ornam. - **A. roxburghiana** Miq. (Ind, SriL): frt. astr. in inflammns. leprosy - **A. saltatorum** A. C. Smith (Fiji): langakali (Tonga); fls. u. in leis (Tonga) (86) & to scent coconut oil; med; cult. ornam. (Niue.).

Aglaonema Schott (Ar.) (Indomal): with 21 spp., mostly cult., ex. *A. commutatum* Schott (speckled lvs.) - **A. costatum** N. E. Br. (Indochin): u. antipyr. in respir. dis. - **A. modestum** Schott (eAs, Mal): "Chinese evergreen"; hb. (-7 m high) grown in house in water without soil & very little light; entire pl. ground into paste & u. to poultice dog bites - **A. oblongifolium** Kunth. (A. marantifolium Bl.) (eAs): lvs. warmed & bound on swollen joints (Mal): juice of pl. rubbed on body affected by "skin scale" (PI); ext. for sores (Java) (953) - **A. siamense** Engl. (Indochin): decoc. u. as lax.

aglucon(e): **aglycon(e)**: the non-sugar portion of a glycoside which appears on its hydrolysis ("y" pron. as "i" in ice).

Agnus Castus (L) (lit. spotless lamb): 1) *Vitex agnus-castus* 2) frt. of same (Semen Agni-casti) 3) castor bean (**A. Christus**, lit. Christ the Lamb) - **A. Scythicus** (L): Tartarian lamb (*qv.*).

Agomensin(e)™: water-soluble hormone fr. whole ovary.

Agonandra brasiliensis Miers ex Benth. (Opili./Olac.) (Br): sandalo; sds. source of FO, ivory wood seed oil (Pão Manfin oil), semi-drying, with indust. usage; u. as illuminant (natives); VO in bronchitis - **A. excelsa** Griseb. (Arg); sombra de toro; frts. yellow drupes, popular food of SA Indians & poor people; frt. c. VO u. in bronchitis - **A. racemosa** Standl. (Mex): "selva"; "margarita"; u. as physic (triturated with sugar); fls. u. as anti-inflammatory; sap u. in urin. retention; ground lvs. in alc. to heal contusions in man & beast.

agoniada (Br): *Plumeria lancifolia.*

Agonis flexuosa (Willd.) Lindl. (Myrt.) (sAf, Austral): willow myrtle; c. antibiotic; inhibits *Staph. aureus* growth; cult. orn.

agouman (Fr): pokeroot.

agoursi: ridge cucumber (*Luffa*); u. in Russian soup.

agra (CR): a wild grape - **Agracejo** (Sp): *Leonurus cardiaca.*

agrião (Por): water cress - **a. do Pará** (Br): *Spiranthes oleracea.*

agrifoglia (It): *Ilex aquifolium.*

Agrimonia capensis Harv. (Ros.) (sAf): lf. decoc. u. antipyr. - **A. eupatoria** L. (A. officinale Lem.; A. pilosa Hook. f.) (Euras, esp. Balkans; Ind, Pak, natd. NA): (common) agrimony; harvest lice; hb. c. tannins, VO, glycoside, bitter princ.; old English herbal remedy; u. astr. in gastric, gallbladder, liver, & kidney dis.; esp. in gall & kidney stones (colics); in bed-wetting; astr.; yellow dye - **A. gryposepala** Wallr. (A. hirsuta Bicknell) (Cana & nUS): stickseed; stickweed; flg. hb. u. astr. antiscorb. "tenicide" - **A. parviflora** Ait. (sUS, Mex): many flower groovebur; rt. u. stom. (Mex Phar.) - **A. pilosa** Ledeb. (Euras): in Chin rts. u. as cath. anthelm.; pl. astr. tonic, hemostatic, heart tonic; in diarrh. dys.; young lvs. u. as pot herb; dr. sed. mixed with noodles - **A. repens** L. (A. odorata Mill.) (c&sEu, wAs, nAf): similar to *A. eupatoria* & approved for such use (Austr Phar.) - **A. striata** Mx. (Cana, n&wUS): u. astr. stom.

agrimony, common: *Agrimonia eupatoria* - **a. (hemp)**: *A. striata*; *Eupatorium cannabinum*

agrio de guinea (PR): roselle.

agripalma (Sp, Por.): **agripaume** (Fr): *Leonurus cardiaca.*

agrobiology: science of living things produced in farming.

agrol: agricultural alc. blended with gasoline.

Agropyron cristatum (L) Gaertn. (Gram.) (Euras-Russ; cult. & adventive in nwUS): crested wheat grass; cult. in nGreat Plains for forage & fodder - **A. intermedium** Beauv. (Eu, cult. & adventive in US): intermediate wheatgrass; good pasture grass; drought resistant; now planted in nwUS - **A. repens** (L) Beauv. (Elymus repens var. r.) (Euras, natd. NA, SA): dog-grass; quack-grass; quitch g.; couch g.; an important weed in cult. grounds; rhiz. (Radix Graminis) c. triticin, sugars; u. dem. diur. emoll. lithontrip. (supplies fr. Ge, Hung) (1901); u. as starvation food; rhiz. substit. for coffee; rhiz. u. as alternative to hay in fodder - various *A.* spp. called wheat grass; u. by some to treat cancer.

Agropyrum Roem. & Schult.: 1) = *Agropyron* J. Gaertn. 2) (L): BPC 1934+ title for Triticum, *Agropyron repens.*

Agrostemma githago L. (Lychnis g. Scop.) (Caryophyll.) (Eu, natd. Af, eNA): (common) (corn) cockle; pl. u. as famine food (young lvs. with vinegar & bacon); rt. astr.; sds. toxic, c. githagin (prosapogenin) & other glycosides; u. exter. in skin dis., int. as diur. emmen. purg. anthelm. (for roundworms) (folk med.); weed in grain fields.

Agrostis L. (Gram) (widely distr. in temp regions OW, NW): bent grass; ca. 220 spp., perenn. grasses u. as forage & fodder (hay); lawn grasses - **A. alba** L. (A. gigantea Roth) (Euras, natd. NA): redtop bent (grass); white top; red t.; u. for lawns, pastures, meadows, sports turf; hay grass; soil building - **A. capillaris** L. (A. tenuis Sibth.) (Eu, natd. & cultd. NA): colonial bent (grass); forage grass; for lawns, sports turfs - **A. scabra** Willd. (A. hiemalis BSP) (e&cUS): creeping bent (grass); silk g.; pasture grass; ornam. (dr. bouquets) - **A. stolonifera** L. (A. palustris Huds.) (NA, Euras): creeping bent (grass); fiorin (GB); u. as lawn grass, for golf greens; valuable forage.

agrostology: science of grasses (Gramineae).

agrumen: citrus fruits (fr. It **agrumi**, acid fruits).

Agt(e)stein (Ge): amber.

agua (Sp) (*cf.* aqua, L.): water - **a. de anis** (&c.): anise (&c.) water - **a. de colonia** (Por): cardamon water.

aguacate (SpAm, Br): avocado.

aguamiel (swUS, Sp): 1) unfermented juice of Agave or maguey; u. by Indians of swUS sometimes as emet. etc. 2) (Mex): mead 3) (Ve): guarapa de cana.

aguapa (NM [US], Sp): cattail.

aguaraquilo (Por): *Solanum nigrum.*

aguardiente (Sp): brandy, sometimes whiskey.

ague: old term for mal. - **a. grass**: *Aletris farinosa* - **a. root**: *Aletris* - **A. weed**: 1) *Gentianella quinquefolia* 2) *Eupatorium perfoliatum.*

aguedita (Cu): *Picramnia pentandra* Sw.

Agurin™: theobrominc sodium acetate.

Ahnfeltia plicata (Huds.) Fr. (Ahnfelti. - Rhodophyta - Algae) (all oceans): chief raw material for Russian agar ("Sakhalin agar"); some other *A.* spp. also u. to furnish US Pacific coast agar (wNA Arctic to Mex, wSA).

ahoehoete (ahoehoetl) (Mex): *Taxodium distichum.*

Ahorn (Ge): maple, *Acer.*

ahuacatl (Mex): avocado.

ahuahtli (Mex): "Mexican caviar"; consists of eggs of flies fr. fresh water pools or lakes (Texoco Lake); u. human & canary food (94).

ahuhuete (Mex): *Taxodium mucronatum.*

ahuyama (Co): squash.

aiara: tucum nut.

Aidia cochinchinensis Lour. (Rubi) (Indochin, Micronesia): u. as antipyr.; in syph, TB; lf. decoc. u. as beverage tea.

aigremoine (Fr): *Agrimonia.*

ail (pl. aulx) (Fr): garlic.

Ailanthus altissima (Mill.) Swingle (A. glandulosa Desf.) (Simaroub.) (Chin, Jap, natd. NA, Af): (Chinese) tree of Heaven; (Chinese) sumach; copal tree (US); heaven tree; paradise t. (Utah [US]); "tree of poverty" (Philadelphia, Penn [US]); common tree of big city streets (954); pistillate trees very showy; male flowers bad smelling; bk. of trunk & rt. u. as bitter, anthelm. stom. ton. in dys. diarrh.; in epilepsy; (fresh fls. & shoots in homeop.); lvs. astr. (Chin); insecticide; wd. for paper pulp; young lvs. eaten (Chin); lvs. as adulter. of belladonna & senna lvs. (2639) - **A. excelsa** Roxb. (Ind) maharukha (Hind, Marathi); bk. u. as bitter ton. antipyr. expect. antisep. in chronic bronchitis, asthma, astr. in diarrh. dys.; lvs. ton. - **A. triphysa** (Dennst.) Alston (A. malabarica DC; A. fauveliana Pierre) (Ind, Indochin): perumaram (Tamil); bk. u. for bitter ton., for dys.; incision of bk. yields gum resin ("mutty pla") u. as incense in Ind temples, also in dys., bronchitis.

ailanthus family: Simaroubaceae.

ailanto (Eng, It, Sp, Por): *Ailanthus altissima.*

Aiouea brasiliensis Meissn. (Laur.) (Br): lf. u. as vulnerary - **A. tenella** Nees (Br): bk. u. for diarrh. colic, &c. (generic name of interest because it contains all of the vowels [except y]).

aipo (Por): *a. bravo: a. dagua: a dos pantanos: a. inculto: a. odorante: a silvestre*: celery.

Aira L. (Gram.) esp. **A. praecox** L. (Eu, nNA): (early) hair grass; foraged by animals.

airelle (Fr): *Vaccinium myrtillus* & other *V.* spp.

Airosperma trichotomum A. C. Smith (Rubi) (PapNG, Oceania): u. as cathartic (Fiji).

airplane plant (US): 1) *Chlorophytum* spp. 2) *Crassula falcata.*

air-plants: *Tillandsia* & *Catopsis* spp.; *Kalenchoe pinnata*; *Chlorophytum* spp. (epiphytes).

Aitkin's glycerite: g. of iron, quinine & strychnine phosphates - **A. (tonic) pills**: mild p. of iron, quinine, strychnine & arsenic - **A. tonic**: elixir of iron, quinine & strychnine phosphates.

Aizoaceae (Dicots.) carpet weed family; mostly sAf; c. 130 genera; & ca. 2,400 spp.

Aizoon canariense L. (Aizo.) (nAf, CanaryIs): hb. u. as food (Tuareg); fish poison (Cape VerdeIs).

ajeel (Iran): collection of nuts roasted & salted before eating (ex. pistachios, hazelnuts, &c.).

ajenjo (Sp): wormwood, *Artemisia absinthium*; *A. mexicana* (latter also **a. del pais**) - **a. menor**: wormwood, may be *A. pontica* - **a. officinal: a. maritimo: a. mayor**: *Artemisia maritima* - **a romano**: *A. pontica, A. absinthium* - **a. verde** (Sp): wormwood.

aji: 1) (SpAm): winter's bark 2) (Co, Mex): red pepper (also **a. caribei**).

ajinomoto (Jap): sodium glutamate.

ajipa (Pe): yam bean.

ajmaline: rauwolfine: alk. fr. rts. of *Rauwolfia serpentina*; u. as antihypertensive, antiarrhythmic. (disc. 1931) (S. and R. H. Siddiqui, India).

ajo (Sp): garlic - **a. porro** (Cu): leek (**a. puerro**, Sp).

ajoene: occ. in *Allium sativum* (garlic); reduces human platelet aggregate in blood; antithrombic; antifungal.

ajonholi (SpAm): *Sesamum indicum.*

ajowan (fruit): **a. "seed"**: *Trachyspermum copticum.*

Ajuga L. (Lab.) (OW temp): bugle; ground pine; ca. 50 spp. of herbs - **A. bracteosa** Wall. ex Benth. (Pak+): c. heart stim. alks.; u. stim. diur. - **A. chamaepitys** Schreb. (Eu, Levant, natd. NA): ground pine; bugle; hb. c. VO u. stim. diur. aper. emmen. in rheum., gout, edema, arthritis, intermittent fever, bil. dis. (off. Ajuga (L.) in several Eu Phar.) (*A. chamipitys*) - **A. decumbens** Thunb. (eAs): analg. & to reduce blood clots (Chin); st. & lf. decoc. (hot) u. to bathe rheum. & neuralgic painful areas (Jap) - **A. genevensis** L. (Eu, wAs, Levant): erect bugle; u. vulner.; in liver & lung dis.; ornam. - **A. iva** Schreb. (Med reg. incl. Crete & Gr): "Herb Eve"; muskey bugle; perenn. hb. c. VO (with odor of musk), NBP; u. like *A. chamaepitys* in mal., diarrh., vulner. (Tunisia); ornam.; in painful menstruation - **A. ophrydis** Burch. (sAf): rt. decoc. for rashes; in painful menstruation; to incr. fertility - **A. pyramidalis** L. (Eu): mountain bugle; in mal. diarrh., hemoptysis - **A. reptans** L. (Eu, natd. seUS+): bugle (weed); b. herb; sicklewort; common bugle; carpet b. weed; pl. u. astr. bitter, arom. vulner. in stab wounds & lung bleeding; rheum., &c.; cult. ground cover.

ajugarins: series of diterpenes fr. lvs. of *Ajuga "remota"*; with insecticidal values.

ajwain: ajwan: ajowan.

akage tree (Ituri Forest, Af): wd. u. produce poison for arrow points (955).

akalaka (Ind): astr. prepn. of *Acacia* spp. fr. Red Sea region.

akanga: *Strychnos icaja.*

akar lanyere: a. langir: *Securidaca inappendiculata.*

Akebia lobata (Houtt.) Dcne. (Lardizabal.) (Jap): woody vine; frts. eaten in some places; dr. young lvs. u. as beverage tea & in med. - **A. quinata** Dcne. (neAs): akebi (Jap); deciduous woody climber; diur.; wd. decoc. diaph. in fevers, stom. lax. vulner. anodyne (Chin, Jap); frts. rich in mucil. eaten, u. in lumbago, rheum. - **A. trifoliata** Koidz. var. *australis* Rehd. (Chin): st. ext. remedy for mechanical injury; insecticide.

akee fruit (Jam): *Blighia sapida.*

akene: achene.

Akonit (Ge): aconite.

aksamitink (Pol): *Calendula.*

akuammine: alk. fr. sds. of *Picralinia klaineana.*

Alabama wool (US slang): cotton.

alabaster: hardened plaster of Paris, $CaSO_4$.

alachofa (Sp): artichoke.

ala de murcielago (Pan) (lit. bat wings): *Passiflora* spp.

Alafia lucida Stapf (Apocyn.) (Ghana, Congo): latex u. to poison arrows - **A. multiflora** Stapf (tropAf, Mad): sds. u. to adulter. strophanthus sd. - **A. pauciflora** Radl. (Mad): u. med. in Madagascar - **A. perrieri** Jum. (wMad): alafy; latex rich in saponins u. in place of soap; rt. inf. u. against jaundice; lvs. u. for measles & eruptive fevers - **A. thouarsii** Roemer & Schultes (Mad): often u. to make straps & withes substituting for osier (willow) wickerwork; u. med.; fls. decorative, pl. cult. in greenhouses in Eu.

alamo (blanco) (Sp): **a. branco** (Por): *Populus* (*alba*); cottonwood tree - **a. preto** (Por): *P. nigra* - **alamoen** (Surinam): citrus frt. - **alamogordo** (Sp): round (or fat) cottonwood tree (*Populus* spp.).

alangina: *Alangium salviifolium.*

Alangium chinense Harms (Alangi) (eAs): lf. decoc. ton., to cure pain & itch; rt. bk. for phthisis, fever (Chin) - **A. faberi** Oliv. (Chin): lf. inf. expect. - **A. platanifolium** Harms (Chin): rt. u. in rheum. - **A. salviifolium** Wangerin (A. lamarckii Thw.) (tropAs, Ind, Af, PI, PapNG): sage-leaved alangium; ankota; ankol(a); evergreen tree or shrub; c. alangine; bk. bitter, emet. u. as ipecac substit.; much u. in Ind. med.; off. in several As Phar.; rt. juice u. as purg. anthelm. diaph., febrifuge, hypotensive; in rheum. skin dis.; dipth.; diur.; for animal bites, colic; frt. u. to stop hemorrhage; tabs. of dr. fls. & frt. with water u. extern. for eye sores (124); wd. valuable (also of other *A.* spp.); cult. orn.

alanosine: antibiotic fr. *Streptomyces alanosinicus* Thiemann & al. (1966-): antimicrobial (antiviral), antineoplastic; inhibits insect reproduction.

Alant (Ge): *Inula helenium.*

alant-camphor: alantolactone: helenin - **Alantwurzel** (Ge): inula root.

alantin: alant starch: inulin.

Alaria esculenta Grev. (Alari. - Phaeophyc. - Algae) (Atl & Pac Oceans): murlins; "honey ware"; this kelp u. in Scotland, Iceland, US, in fried form (badderlocks); also eaten raw; high in vit. B_6 & K; cult.

Alayam™: products of the sweet potato ("yam") (prod. in Ala [US]).

Alazopeptin: antibiotic fr. *Streptomyces griseoplanus* Backus & al. (fr. Iowa [US] soil); exptl. study 1956-; possible anti-cancer (antineoplastic) agt. (1993).

alazor (Sp): false saffron, Carthamus.

albahaca (Mex): *Ocimum*; (CR): *O. basilicum.*

Albamycin™: novobiocin (generic name); antibiotic; u. for dis. with pulmonary abscesses (494); &c.
albarcoque (NM [US], Sp): apricot (hence, possibly name of city, Albuquerque).
albarello (It, Eng): 1) drug jar of fine porcelain, usually of majolica ware; generally cylindroid, often with concave sides 2) *Populus alba.*
albaricoque (Sp): apricot.
albarillo (Sp): *Exostemma caribaeum.*
albarrana (Sp): white squill.
albedo: white inner portion of citrus peels.
alberchigo (Sp): peach - **albercogue** (NM [US]): albarcoque.
albergo (SpAm): **alberja** (SA): pigeon pea.
albero (-i) (It): tree.
albina (Por): *Turnera ulmifolia.*
Albiz(z)ia adianthifolia Wight (A. fastigiata Oliv.) (Leg) (tropAf, Madag): flatcrown (albizzia); bk. lvs. wd. c. tannins, u. diarrh. ophthalmia; bk. & rt. inf. u. for snakebite & for infest. parasites (Liberia); trunk source of sassa (or red) gum (similar to gum acacia); gastralgia; boutonneuse fever - **A. amara** Boiv. (Af, Ind): lallei (Bombay); u. emet. in malaria; oil in leprosy (Ind) - **A. anthelmintica** Brongn. (Leg.) (Ethiopia): musenna (bark); bk. c. musennin (saponin glycoside); u. purg. anthelm. tenicide; some other *A.* spp. u. similarly (formerly, *Rottlera schimperi* Hochst., thought source of musenna bk.); bk. u. as sex stim. (Kenya); sd. pods antiprotozoal - **A. antunesiana** Harms (Af): kawizi; bk. inf. as lotion to cuts; sap fr. st. as eye drops in ophthalmia; rt. inf. in GC, aphrod., prophylactic against colds; timber much used - **A. brachycalyx** Oliv. (Af): mumeya; bk. macerate u. in rheum., astr. - **A. chinensis** Merr. (Ind, Chin): chaku (Bengal): tree u. to shade tea shrubs in plantations; c. gum, saponin; bk. inf. u. lotion for cuts, scabies, skin dis.; pl. fish poison - **A. coriaria** Welw. (Af): bk. u. tanning leather; excellent wd.; u. med. (Angola) - **A. falcataria** (L.) Fosberg (A. molluccana Miq.) (Malaysia): wd. for tea chests (batai wood) & bk. technical uses; similar med. uses to *A. chinensis*; extremely fast growing tree - **A. ferruginea** Benth. (Af): c. saponosides; u. in burns, GC, infant convulsions - **A. harveyi** Fourn. (trop eAf): trunk yields gum u. intest. troubles (Mozam) - **A. julibrissin** Durazz. (As from Iran to Jap, esp. Thailand): mimosa (tree); silk tree; cult. ornam. tree of sUS (also natd.); gum u. dem., med. properties similar to those of *A. lebbeck* - **A. kalkora** Prain (Chin): fl. decoc. u. to improve vision & for int. injuries; also gum - **A. lebbeck** (L) Benth. (*A. lebbek*) (tropAs [Ind], Af, nAustral, natd. Fla [US+]): lebbek; woman's tongue (tree) (SC dr. pods clatter in wind); shirisha (Sanscr) lvs. & sds. u. eye dis.; sd. FO in leprosy; timber tree ("[East] Indian walnut"); cult. in trop for fodder (lvs.) & shade; lvs. u. to treat atopic allergy (anaphylaxis); in syph. (Madag); u. in asthma, insect sting; value attributed to action on immune system; bk. for boils (Burma); good timber tree - **A. lebbekoides** Benth. (Indo-Chin): bk. astr. u. in colic - **A. lophantha** Benth. (Paraserianthes I. (L.) Nielsen) (strop Af): bk. & rt. rich in saponins; rt. decoc. after childbirth for pain & discomfort; rt. soap substit. - **A. maranguensis** Taub. (trop eAf): mruka; bk. cont. saponin (musennin) u. cough, nail cleaning; wd. useful - **A. myriophylla** Benth. (Ind, Mal+): lvs. u. to make lotion; rt. inf. u. in combn. for fevers; bk. (with licorice flavor) u. in cough, bronchitis; u. in prepn. of press cakes of yeast for manuf. of rice alc.; lvs. crushed or minced appl. to wounds to stop blood flow (Indochin) - **A. odoratissima** Benth. (Ind, Burma, seAs): thitmagy (Burmese): lvs. u. as expect. in cough; bk. rich in tannin, u. in ulcers, leprosy; in prepn. fermented drink ("basi") (PI) - **A. petersiana** (Boll.) Oliv. (Tanz): bk. chewed for cough; in tapeworm; rheum. (cold water inf. bk. drunk t.i.d.) - **A. procera** Benth. (trop As, Austral, Af): lvs. as poultice to ulcers; insecticide; inner bk. fish poison; tree source of a gum; wd. valuable, in veneers; cult. ornam. - **A. saponaria** Bl. (Chin, PI, PapNG): bk. u. for washing hands, healing venomous stings of wasps, etc.; wd. & bk. also u. as detergents - **A. versicolor** Welw. ex Oliv. (tropAf): rt. bk., u. as headache remedy; purg. enema, rt. anthelm. & soap substit.; bk. inf. for skin rashes, int. for "puberty troubles"; wd. u. to make furniture & drugs; arrow poison - **A. zygia** Macbride (trop e&wAf): pd. bk. u. as dressing in yaws; lf. in diarrh.; wd. u. to make planks; bk. potion u. as "truth serum" & ordeal poison (Nigeria).
Albolene™: purified petroleum jelly similar to white petrolatum - **Liquid A.**: purified petroleum (mineral) oil.
Albomycin: antibiotic product obtained fr. *Streptomyces subtropicus*; discovered in Russia, 1951-; bright orange color (due to Fe incorp. in mol.); 6 fractions recognized; antimicrobial active in pneumonia, dys., measles, meningitis; circulates in blood as protein complex of serum; do *not* confuse with Albamycin.
alboroto (SA): sorghum.
Albromol™: proprietary contg. bromelin (enzyme).
Albuca cooperi Bak. (Lili) (sAf): lotion u. for washing wounds in animals - **A. major** L. (sAf): "soldier in the box"; edible; st. mucilaginous & chewed to quench thirst - **A. setosa** Jacq. (sAf):

tuber & foliage c. 1% hemolytic sapogenins (tigogenin, gitogenin) but no alks. or tannin; antibiotic activity negative; pl. u. like other *A.* spp. as ornam. - **A. trichophylla** Bak. (sAf): u. to treat GC.

Album Graecum (L): Stercus Canis; dog feces (SC white colored); u. by Ancients with honey in sore throat, dys. colic; extern. deterg. for ulcer, warts, &c.; made up mostly of Ca phosphates - **A. Nigrum** (L): rodent feces; dark excrement of rats & mice.

albumen: endosperm of sd. with storage foods - **vegetable a.**: plant tissue protein, esp. as found in seeds & incl. aleurone (*qv.*).

Albumen Ovi Siccum (L.): egg albumin, dried; many uses, incl. antidote in Hg poisoning.

albumin, blood: serum a.: crystallizable a. or mixt. composing over 50% of blood serum proteins; hence the chief protein fraction; synthesized in liver; maintains osmotic press. of blood; normalizes blood vol.; u. as substit. for blood plasma (E) (2091B); long history of use (2091B).

albumin, dry: protein prepared fr. egg white (ovalbumins) or fr. human blood plasma (serum albumins), the latter prod. is u. IV in 25% soln. to treat shock, hypoproteinemia; v. egg albumin.

albuminoid = scleroprotein.

albuminous seeds: those which c. endosperm (albumen) or storage food outside of the cotyledons: ex. castor bean.

albumins: group of proteins found widely distd. in living organisms, incl. ovalbumins, lactalbumins.

albumoses: primary cleavage prods. of polypeptides formed by hydrolysis of simple proteins; noncoagulable but can be salted out.

alburnum: sapwood or living part of wd. of tree trunk; occurs in circle around heartwood.

Albutannin™: albumen tannate.

alcachofa (Sp): artichoke.

alcacuz (Po): licorice.

Alcaligenes faecalis Castellani & Chalmers (Bacterium f. Mez) (Bacteria) (intest., decomposing veget. matter): produces biodegradable plastic subst. in cell (1991-); DNA code transferred to *E. coli*; potential source of good type (environmentally) of plastic.

alcanfor (Sp): camphor.

alcanna (root): *Alkanna tinctoria.*

alcaparra (Sp): caper.

alcaravea (Sp): caraway (seed).

alcaucil (Sp): (wild) artichoke.

Alcea rosea L. (*Althaea r.* [L.] Cav.) (Malv.) (AsMin, now c&sEu, Chin+): (English) hollyhock; garden h.; stockrose; rose (or garden) mallow; fls. (Flos Malvae arboreae) c. mucil. tannin; u. as inf./decoc. as antispasm., expect. & dem. in coughs, hoarseness, gargle in mild sore throat, in GIT inflammn.; exter. for sores; rts. & sds. also u. med.; now mostly disappeared from pharm.; marketed with and without calyces; to give color in wines, frt. juices, liquors, &c.; much cult. in Ge, Fr, Eng, Turk, US as ornam. (esp. var. *nigra* hort.) (967).

Alces alces L. (Cervus alces) (Cervidae) (Eu): (European) elk; elk horns u. medic., considered valuable (American elk, *Cervus canadensis*) - **A. americana** (nUS, sCana): (American) moose (deer); large animals, wt. to 650 kg; male with very large antlers; females antlerless; hunted for meat & antlers (ornam.) - **A. gigas** (Alaska [US]): Alaskan moose; wt. to 800 kg; u. as preceding.

Alchemilla alpina L. (Ros.) (Eu mts., Greenland; natd. eCana): alpine lady's mantle; Silbermaenteli; hb. (Herba Alchemillae alpestris) c. tannins, lecithin, resins; u. in popular folk med. as diur.; during female climacteric (Switz); may incr. milk secretion in cattle (Swiss cowherders); antipyr. vulner. - **A. arvensis** Scop. (Aphanes a.) (Euras, nAf, natd eCana): parsley piert; parsley breakstone; hb. (as Herba Percepier; "Fire Grass") formerly u. as dem. diur., in strangury; stim. - **A. conjuncta** Babingt. (eAlps, Eu): ladies' mantle; u. similarly to *A. alpina* (Herba Alchemillae) - **A. hoppeana** Dalle Torre (Eu, Alps): Sibermaenteli (Swiss, Ge): u. similarly to prec. - **A. woodii** O. Ktze. (sAf): smoke fr. burning pl. inhaled for headaches - **A. xanthochlora** Rothm. (A. vulgaris L. s. l.) (Euras, natd NA): (our) lady's mantle; lion's paw; hb. Herba Alchemillae; u. astr. diur. stypt. vulner. in hemorrhages, diarrh., excessive menstr.; in abdom. debility.

Alchornea castanaeifolia Juss. (Euphorbi.) (Ve): sauzo; bk. u. in cataplasm for treating sting of s. ray - **A. cordifolia** (Benth.) Muell.-Arg. (tropAf): "Christmas bush"; leafy twig decoc. u. for chills, rheum.; lvs. u. as hot press for feet (Sierra Leone); lvs. u. for sores, purg.; VD; dr. lf. u. in ulcers, yaws, bk. u. stom.; pith in diarrh. - **A. floribunda** Muell.-Arg. (tropAf): rts. c. alks., yohimbine; has hallucinogenic effects like Cannabis, u. aphrod.; may c. hypotensive constits. - **A. hirtella** Benth. (trop wAf): u. as remedy in toothache (Sierra Leone) - **A. iricurana** Casar (tropAm as Br): u. med.; wd. comm. timber - **A. latifolia** Sw. (WI, tropAm): achiotillo (PR): bk. u. bitter ton., c. alchornin (bitter subst.); u. in resp. dis.; wd. commonly u. for boxes, crates, &c. - **A. laxiflora** Pax & K. Hoffm. (trop eAf): ash fr. pith of pl. appl. to cuts in neck to treat sprained or dislocated neck - **A. rugosa** Muell.-Arg. (Indochin): sds. u. as purg.; rts. & lvs. decoc. u. for fever & ague (Mal) - **A. villosa** Muell.-Arg. (Malaysia): rts. & lvs. u. for itch; sap fr. young lvs. in fever; bk. bast fibers a ramie substit.

alchornin: amaroid fr. *Bowdichia virgilioides*.

alcino (Sp): wild basil, *Clinopdium vulgare*.

alcohol: aliphatic hydroxyl compd.; unqualified, generally means ethanol (<Ar. al-kohl); known for many centuries due to availability fr. fermented & distd. alc. beverages; much now prepd. by synthesis; u. as anti-infective, antisept., stim.; depressant; to destroy nerve tissue; topically in hay fever (1994); forms: 95%; dilute (50%); anydrous ("absolute") u. as dehydrating agt.; intoxicant (≤ 50% solns.) - **alcoholism** (dypsomania): fetal alcohol syndrome (FAS); cause of many social & econ. problems - **power a.**: industrial a. - **rubbing a.**: ethanol wh. has been partially replaced by adulterants ("denaturants"), such as methanol, isopropyl alcohol (sometimes largely replacing ethanol), camphor, &c.; u. locally to reduce fever (evaporative effect), disinfectant, to reduce local irritation, &c.; toxic to greater or lesser degree; must not be used internally.

alcoholate: 1) metal compd. with alcohol 2) alc. distillate of drugs.

alcornoque (Sp): alcornoco.

alcoolat (Fr): (distilled) spirit - **alcoolature** (Fr): alcoholic tincture of fresh pl. ("green tincture"); generally employed with drug pls. which readily deteriorate on drying; ex. aconite lvs. - **alcoolé** (Fr): simple soln. in alc. of drug or chem. entirely soluble in it.

alcornoco (Sp): *Bowdichia virgilioides*.

alcornque (Sp): cork oak; *Byrsonima crassifolia*; *Bowdichia virgilioides*.

alcotán (CA): 1) various pl. spp. u. in snakebite incl. *Cissampelos pareira* 2) (Guat): *Chondrodendron tomentosum*.

alder: *Alnus glutinosa* & other *A*. spp. - **(American) black a.**: 1) *Ilex verticillata*; bk. & berries cath. 2) Frangula 3) *Alnus glutinosa* (& perhaps *A. incana*) - **a. buckthorn**: *Rhamnus frangula* - **common (American) a.**: red a. (lvs.) - **common (European) a. (bark & leaves)**: **(European) black a.**: *A. glutinosa* - **English a.**: green a. - **false a. (bark)**: *Ilex verticillata* - **green a.**: *A. glutinosa* - **hazel a.**: red a. - **hoary a.**: *Alnus incana* & subsp. *rugosa* - **mountain a.**: *A. viridis* subsp. *sinuata*; *A. incana* - **red a.**: *A. incana* subsp. *rugosa* - **russet a.**: mountain a. - **smooth a.: speckled a.**: red a. - **spotted a.: striped a.**: *Hamamelis virginiana* - **swamp a.: tag a.**: *A. serrulata*, *A. incana* subsp. *rugosa*, *A. tenuifolia* (wNA) - **white a.**: hoary a.

Aldina insignis End. (Leg) (Guiana): dakampalli; u. in starch prodn.

aldolase: zymohexase; an enzyme wh. catalyzes the aldol condensation reaction important in the reformation of glycogen fr. lactic acid - **aldose-reductase** (AR): enzyme in blood stream active in eye tissues; produces retinopathy, cataract.

aldoses: collective name for monosaccharides which have an aldehyde gp. (CHO): incl. trioses, tetroses, pentoses, hexoses; commonest one: glucose.

aldosterone (earlier electrocortin): hormone (a mineral corticoid) of adrenal cortex; 25-30 X as strong as desoxycorticosterone.

alecost: *Balsamita major* - **alecrim** (Por): rosemary (lvs.).

Alectoria jubata (L) Ach. (Usneaceae; Lichen.) (nTemp Zone): **rockhair alectoria**; found on bk. of trees; u. as food of men (Sikkim); brown dye.

Alectra parasitica A. Rich. (Scrophulari) (Ind): parasite on *Vitex negundo* rts.; u. rhiz. in leprosy; no toxic reactions.

Alectryon macrococcum Radlk. (Sapind.) (Hawaii): mahoe; red frts. (sds.) edible (both aril & kernel).

alegar (Sp): alc. vinegar; prepd. by fermenting ale, esp. with raisins.

alehoof: *Glechoma hederacea*.

aleli (Mex): *Melia azedarach* - **a. amarillo** (Sp): *Cheiranthus cheiri*.

aleluya colorada (Nic): roselle.

Alepidea amatymbica Eckl. & Zeyh. (Umb) (sAf): kalmoes; bitter rhiz. & resin u. in gastralgia, ton. in larger doses as purg. in colic; mashed hb. appl. to wounds as vulner., hemostat. - **A. ciliaris** de la Roche (sAf): berg-kalmoes; rt. decoc. u. in chest pains - **A. longifolia** E. Mey. (Af): rt. decoc. in coughs; boiled lvs. u. as veget. - **A. setifera** N. E. Br. (sAf): u. similarly to *A. ciliaris*.

Aleppy: place in Ind; not to be confused with Aleppo (Syria).

alerce (Sp): larch tree, *Larix*, esp. *L. europaea*; (eChile): *Fitzroya cuppressoides*.

Aletris aurea (Walt.) (Lili) (eUS, coastal plains): yellow star root; rt. u. as bitter ton. - **A. farinosa** L. (Lili) (e&cUS, sOnt [Cana]): (white-tube) star grass; unicorn (roots); crow corn; star root; colic root; mehlige Aletris (Ge); alétris farineux (Fr) (957); rhiz. & rts. c. diosgenin glucoside, NBF, starch; rhiz. & rts. (**Aletris**, L.) u.; as bitter ton. diur. emmen. in uter. dis., flatus, descensus vaginae (prolapse of vagina [uterus]) (Am Indians); lf. inf. in colic (Catawba Ind, SCar [US]); u. with *Trillium erectum* as poultice in scrofula, "glandular swellings," &c. (956); pl. sometimes confused with *Chamaelirium luteum* - **A. lutea** Small (seUS): u. much like prec.

Aleurites J. R. & G. Forst. (Euphorbi) (tropAs, Oceania): frt. & sd. of some spp. u. to furnish tung (or Chin wood) oil, u. for varnishes, &c. (at one time this was chief cash crop exported fr. Chin); FO u. to make soap; appl. exter. as vulner. analg. in lacerations, abrasions, physical trauma, sprains, contusions, swellings, low back pain, varicose ulcers, sinus abscesses, burns (1-3 degree), puncture wounds. lymphangitis; int. said to reduce hypertension; ingestion of frt. causes emesis, sometimes lethal; cult. seUS (958) - **A. fordii** Hemsl. (A. cordata R. Br.; Vernicia f. Airy-Shaw) (c&eAs, Chin, PacIs; introd. Fla+ [US]): Chin tung (oil) tree; Japanese wood oil tree; Chinese (or Japanese) varnish tree; (Chin) wood oil tree; candelnut tree; frt. source of Japanese wood/tung (oil) (FO with eleostearic acid); u. for itch, wounds, scabies (Ind); poisonous (495); said to c. hypotensive principle; sometimes *A. cordata* (prod. Japanese wood oil as a sort of tung oil with lower sp. grav.) is recognized as separate fr. *A. fordii* (prod. Chinese wood oil); cult. US, Jap, USSR, Br - **A. laccifera** (L) Willd. (Croton laccifer L.) (Ind, SriL, Mulucca, WI): resin said source of lac (shellac), u. in varnishes - **A. moluccana** Willd. (A. triloba Forst.) (Molucca Is, Oceania, Polynesia, Mal): candlenut tree, c. oil tree (since frt. u. as candle); ku ku; akrote; source of c. oil fr. sds. u. purg. (Ind, Mal) in ointm. soaps; for hair & skin (Surinam): sd. kernels said edible (care!); aphrod.; fls. nuts & bk. u. for exhaustion, asthma, sores, ulcers, swollen uterus, constip.; bk. u. in dys. (Indones) & as dye (Japan, Tonga); FO fuel in stone lamps, illuminant (496); as artist's oil; nuts u. as torches (Surinam) - **A. montana** (Lour.) Wils. (A. cordata Gagnep.) (seChin): Japanese wood oil tree; mu (oil) tree; sds. source of mu (or chinawood) oil (497); u. in ointm. for ulcers, furnucles; drying oil (u. as paint base) - **A. trisperma** Blanco (PI): soft lumbang; sds. source of (bagi) lumbang oil, similar to tung & linseed oils; drastic purg. (*cf.* castor oil); sap fr. bk. u. to treat scurf; FO insecticide - **A. vernicia** Hassk. = *A. fordii.*

aleuronat(e): **aleuron(e)**: a plant protein (albumen), by-produced in mfr. wheat starch; u. prep. diabetic flours, etc.

aleurone grains: microscopic particles of protein substances charac. found in sds.

Alexanders, golden: *Zizia* spp., esp. *Z. aurea* - **purple A.**: *Thaspium trifoliatum.*

alface (Por): lettuce, *Lactuca sativa.*

Alfacene™: prepn. of alfalfa seed; claimed useful in arthritis & rheum. (959).

alfalfa, (common): **Spanish a.**: **wild a.**: *Medicago sativa* - **alfalfon** (NM [US], Sp): sweet clover.

alfavaca campestre (Br): **a. do campo** (Br): *Ocimum micranthum*, &c.

alfazema (Por): lavender - **a de cabocla:** hyssop.

alfilaria (stork's bill): **alfileria**: *Erodium cicutarium.*

alfombrita (Mex): *Euphorbia rosea.*

alforfon (Sp): buckwheat.

Algae: Phycophyta: seaweeds; sea plants; one of the three chief subdivisions of Phylum Thallophyta; pls. c. chlorophyll & are self-sustaining; show great range in size fr. microscopic diatoms to giant kelps; of considerable economic importance; furnish large part of diet of Japan & other nations; source of some important pharmaceutical materials; ex. agar, alginic acid, carrageenan, dulse, kelp products, mannitol, algulose - **yellow-green a.**: Chrysophyta.

algalia (Sp): 1) okra 2) wild o.

Algarobia glandulosa T. & G. = *Prosopis g.*

algarobilla (Sp): **algarobilli**: 1) *Caesalpinia melanocarpa*, *C. brevifolia*, & other *C.* spp. 2) frt. of *Prosopis juliflora* - **algarobillo** (SA): *Caesalpinia brevifolia* - **algarroba**: algaroba.

algarobilo (SA): *Prosopis* spp.

algarobito (Sp): *Caesalpinia brevifolia.*

algar(r)oba: algarrobo: (variant spellings) (Sp): 1) *Ceratonia* (*qv.*) 2) *Pithecellobium* spp. 3) algarobito (*qv.*) 4) *Hymenaea courbaril* 5) *Samanea saman* 6) other pl. taxa esp. *Prosopis chilensis*, *P. julifora* & other *P.* spp. (with sweet pods) (**algarobo** = the tree) - **a. bean**: *Ceratonia.*

algarrobin(a): dyestuff fr. wd. of *Ceratonia siliqua.*

algarrobo negro (Arg): *Prosopis nigra.*

algerita (Tex, NM [US]): *Mahonia trifoliata.*

algin: 1) any phycocolloid extractive fr. kelps (sea weed colloid) 2) **soluble algin**: any water-soluble salt of alginic acid, esp. Na; chiefly fr. *Laminaria* spp.; forms gels, u. in mfr. of jellies, lotions, sizing, u. to take impressions, thickening agt., ointment base; hemostatic.

alginates: salts or esters of alginic acid; u. in gel formn., regulation of viscosity, thickness, stability, &c.; u. in textile printing, biotechnology, foods, pharmaceuticals, &c. (498) (960); prod. of Norway, US, Cana (coasts); ultra pure prod. very valuable; u. for encapsulating seeds; pancreatic cells (diabetes), &c.

Alginex: 1) designation for greaseless non-staining alginate-containing cream u. for external application 2) an algicide with quinonamide (Hoechst).

alginic acid: a colloidal weakly hydrophillic acid subst. (polymannuronic a.) extd. fr. brown seaweeds (Algae- Phaeophyceae) (such as *Laminaria* spp.); u. with gastric antacids; of special value in pyrosis (forms foam barrier on top of gastric pool); foaming agent (to form & stabilize emulsions) (ex. mineral oils); paper & textile size; sulfamated compds. u. as anticoagulants.

algodon (pl. **algodones**) (Sp): cotton; (NM [US], Sp): cottonwood tree; cotton field - **a. en rama** (Arg): raw cotton; cotton wool - **a. hidrofilo** (Arg): absorbent cotton.

algology: science or study of Algae.

algulose: insol. cellulose-like component (possibly cellulose) found in marine algae, thus as residue in prodn. of phycocolloids.

Alhagi maurorum Medik. (A. pseudalhagi (Bieb.) Desv.; A. camelorum Fisch.; Hedysarum alhagi L.) (Leg) (Eg & Syr, wAs, Ind, Pak deserts): camel thorn; twigs produce Alhagi manna; u. by Bedouin; chiefly composed of sucrose; u. antisep. lax. expect. diur.; rts. sometimes eaten by nomads; important cattle fodder Desv.; Persian manna tree; manna (sugary exudn.) may be "asphaltus" of Bible; u. sweetmeat; pl. decoc. astr. bactericid. in dys. - var. *turcorum* (wAs): fls. u. in hemorrhoids.

alhandal (Sp): colocynth.

alhena (Sp): **alhenna** (Arab): privet; henna.

alho (Por): garlic.

alhucema (Sp): lavender.

alhumajo (Sp): pine lvs.

alicyclic: ring compds. derived fr. aliphatic; ring saturated (ex. cyclohexane, menthol).

Alidase™: hyaluronidase.

alimento (Sp): food.

alimo (Sp, Por): alamo; poplar.

aliphatic: acyclic: open chain carbon compd.

Alisma canaliculatum A. Braun & Bouche (Alismat.) (Kor): tubers eaten to prevent beri beri - **A. orientale** Juzepczuk = **A. plantago** auct. = **A. plantago-aquatica** L. var. *o.* G. Sam. (Jap, Kor, Manch, eSib): water plantain; rhiz. (Alisma, L.) c. VO resin, choline, u. as diur. in edema, kidney dis., GC (Chin); akenes u. simil.; in Kor. med. (Kor Phar); lvs. (hb.) u. as counterirrit., vulner. lithontrip.; at one time u. in Russia for rabies (!); food (Chin, Mongolia); cult orn.

alismo (It): *Alisma p.-a.*

aliso (Sp): 1) *Alnus glutinosa* 2) *Platanus racemosa* (Calif [US]).

alizarin: dyestuff fr. madder rt. (*Rubia tinctorum*); aglycone of uberythric acid, a glycoside; later (1868) synthesized; u. as dye, indicator, &c.

alkali, vegetable: K_2CO_3.

alkaline elixir: Elixir Hydrast. Comp.

alkaloids: complex organic nitrogenous compds. basic in reaction, which have physiological action on the animal body & which occur usually in pls.; are especially common among the dicotyledons; usually occur as salts or as tannates; typically alks. are xall. in structure & composed of C, H, O & N; ex. cocaine, quinidine, coniine (60) - **ester a.**: those composed of fatty acid esters of a base; readily hydrolyze to release base & acid; ex. Solanaceous alk. (tropane alk.: atropine, &c.), aconitine (hydrolyzes to benzylaconine & acetic acid; benzylaconine in turn to aconine & benzoic acid - **peptide a.**: generally now classed as peptides or modified peptides; ex. pandamine; ergot "alkaloids" (lysergic acid derivs.) - **phenolic a.**: those with phenols as part of compn. (with free OH gp. on benzene ring) (ex. dopamine) (**non-phenolic a.** lacking such; ex. mescaline with 3 methoxy gps.) - **secale a.**: ergot a.: SC *Secale cornutum*, L., title of ergot sclerotium.

alkanet, (dyer's): *Alkanna tinctoria* Tausch, *Anchusa officinalis* (rt.) - **American a. root**: (Tex [US]): *Lithospermum* spp. - **mountain a.**: *Arnica montana* - **Syian a.**: *Macrotomia cephalotes*.

Alkanna tinctoria (L) Tausch (Boragin.) (sEu, esp. Hung, Turk; cult. Fr+): alkanet; Spanish bugloss; rt. c. alkannin, tannin; u. astr. in diarrh. abscesses, etc.; red coloring agt. for FO, cosmetics, liqueurs (port wine); prepn. of alkannin; u. diagnostic stain for fatty oils - **Japanese alkanna**: *Lithospermum erythrorhizon* - **Syrian a.**: *Macrotomia cephalotes* - **yellow a.**: *Alkanna lutea*.

alkannin: quinone pigment of Alkanna rt.; this coloring princ. also found in *Echium humile*, etc.; red dye; astr.

alkanols: general designation for aliphatic hydrocarbons with an alc. (OH) gp.; alkandiols (with 2), alantriols (with 3), &c. (*cf.* Alkanol™).

alkekengi (berries): *Physalis alkekengi*, common winter cherry.

alkyl: radical or org. compd. representing an aliphatic compd. less one H; ex. methyl, CH_3 (CH_4 less H).

Allamanda = *Allemanda*.

Allanblackia Oliv. ex Benth. (Guttif) (tropAf): FO of sds. u. as butter at times of scarcity - **A. floribunda** Oliv. (wAf, Congo, Cameroun): kisidwe; sds. c. 46-66% FO, called bouandja fat, rich in stearin & u. in prodn. of stearic acid - **A. oleifera** Oliv. (cAf): sds. —> kagné butter - **A. parviflora** A. Cheval. (Sierra Leone to Ghana): sds. c. FO & alk. similar to physostigmine - **A. sacleuxii** Hua (Af): source of kanye or kagne butter, u. as cooking fat & for lighting purposes - **A. stuhlmannii** (Engl.) Engl. (trop eAf): African tallow tree; fresh lvs. chewed & juice swallowed for cough; bk. —> red dye; sd. FO., mkani fat; u. for cooking, lighting, butter substit. (Ge army, WWI); swallowed for rheum.; as ointment for pain in limbs; in cosmetics & to make soaps; sd. decoc. rubbed on joints in arthritis (Tanz).

allantoin: glyoxyl diureide; occurs in maggot excretions, etc.; prod. in metabolism of purines; breakdown prod. of uric acid; occ. in urine of mammals; also widespread in plant kingdom (ex. *Symphytum*); u. to stim. growth of healthy tissues in psoriasis; u. vulner. for wounds, ulcers, etc.; to treat suppurating wounds; acts —> proliferation of tissues; in cosmetics (low concns.); ulcus ventriculi and duodenal ulcer; skin ulcers (rodent ulcers); osteomyelitis; Alphosyl™ (cream).

Allegheny vine: *Adlumia fungosa.*

allele: allelomorph: either of a pair of alternative contrasting Mendelian characters or of a pair of genes which give rise to other characters (ex. rough vs. smooth).

allelochemistry: science of compds. produced by one organism which stim. or inhibit other organisms (<allelo-: reciprocally).

allelopathy: harmful or toxic effects of one plant on others growing in the vacinity.

Allemanda blanchetii DC (Apocyn.) (Br): sap of shrub u. as cath. emet. (in small doses) - **A. cathartica** L. (Br, now found in sFla [US], WI, pantrop; much cult. in tropAm): common (or yellow) allamanda; canario (PR); wild senna; bk./lvs./sds./juice u. cath. emet.; bk. ext. hydragog. in constip. of Pb poisonings; rt. u. in mal. & spleen dis.; poisonous pl. (Fla); popular ornam. liane - **A. doniana** Muell.-Arg. (Br): cath. emet.

allergen: agent which produces hypersensitivity (allergy) in the individual; ex. ragweed pollen antigen - **allergenic preparations**: pharm. prod. u. to test for & desensitize allergic states, such as asthma.

allergology: science of allergy.

allergy: sensitivity developed to specific substances called allergens: such symptoms appear as urticaria, vasomotor rhinitis (stuffy nose), etc., in such symptom-complexes or diseases as pollinosis, asthma, etc.

Allerlei-Gewürz (Ge): allspice.

allerman's root: *Allium victorialis.*

allheal: 1) *Stachys palustris* 2) *Prunella vugaris* 3) (Jam): *Satureja viminea* 4) *Valeriana officinalis.*

Alliaria petiolata Cavara & Gran. (A. officinalis Andrz.) (Cruc.) (Euras, nAf, natd. NA): hedge garlic; garlic mustard; hb. c. mustard glycosides; u. for colds, asthma, anthelm., in diarrh. & as antiscorbutic (It); earlier flavor; extern. for wounds & sores.

alligator: reptile; *Alligator mississippiensis* (Fam. Alligatoridae)(seUS); **American a.**: (-5 m): hide & flesh utilized - **a. bonnet**: *Nymphaea* (*Castalia*) *odorata* - **a. buttons**: *Nelumbo pentapetala* - **a. grass**: *Alternanthera philoxeroides* - **a. pear (tree)** (WI): *Persea americana* - **a. pepper**: *Amomum granum-paradisi* - **a. tree**: sweet gum (SC rough scaly bk. of trunk) - **a. weed** (sLa [US]): a. grass.

Allisatin™: active principles of *Allium sativum* adsorbed on charcoal to render odorless (proprietary); u. sed. stom. hypotensive.

Allium L. (Lili.) (nHemisph): 1) genus char. by having strong characteristic odors; many members u. for food (bulbs, lvs.) 2) (L.) garlic (bulb), *A. sativum* - **A. ampeloprasum** L. (Euras, natd. seUS): wild leek; bulb with lemon juice & Epsom salts u. in hypertension (folk med., SCar [US]); crushed bulb with salt placed on sores (SCar) (2451) - **A. ascalonicum** L. (A. cepa L. f. asc. Baker) (AsMin, cult. Eu): "shallot(t)" (error); bulbs esculent, flavor (more delicate than of onion); u. as vulner., appl. with honey to burns (Chin); flavor in soups, stews, salads, meats; diur. u. in digestive disturb. - **A. bolanderi** S. Wats. (US: sOre, nCalif): wild onion; u. as food, diur. (Am Indians) - **A. canadense** L. (eNA): Canada garlic; meadow (or garden) leek; bulbs boiled or pickled & eaten (Am Indian) (also var. *mobilense* (Regel) Ownbey (seUS): eaten by various Am Indian tribes); when foraged by cattle, may impart unpleasant odor/taste to dairy prods. - **A. cepa** L. (wAs, Iran, Baluchistan; widely cult. worldwide): a cultigen; (garden) onion; bulb & tops u. as food, flavor; expect. in colds, bactericidal (961) in wounds; cardiotonic; raw bulb appl. locally for insect (esp. mosquito) sting; in TB, as diur. (Chin); fresh juice hypoglycemic; to stim. hair growth; claimed aphrod.; somnifac.; peels source of brown dye; numerous vars., incl American group (white, yellow, red); foreign group (Bermuda, Spanish, Italian, Egyptian); Aggregatum gp = shallot; Vidalia (sweet o.) (962); bulb & fleshy lf. bases c. allyl sulfides - **A. cernuum** Roth (sCana, US): wild onion; bulb u. as carm. expect. like *A. sativum*; in soups, as pickle - **A. chinense** G. Don f. (A. bakeri Regel) (eAs): bulbs u. to treat chest cold, coughs, "shortness of breath," gastroint. dis., incl. dys. - **A. falcifolium** H. & A. (coastal Ore, nCalif [US]): scythe-leaved onion; u. similarly to *A. bolanderi* - **A. fistulosum** L. (eAs): spring onion; Welsh o.; stone leek; lvs. u. to flavor foods (seasonings for soups, salads, pickles, sausages); thought by Ancients to give strength; in diphtheria; sd. u. as ton. (cult. It, Eg) - **A. grayi** Regel (A. nipponicum F. & S.) (Jap, Ryukus): whole pl. decoc. u. for blood dis., diarrh., nervine; to start menses; for "nerves"; food - **A. macrostemon** Bge. (Far East): u. like *A. chinense* - **A. odorum** L. (A. chinense Max.; A. tuberosum Roxb.) (Chin, Sib): rakkyo; Chinese chives; u. in diphtheria; as food flavor (salads); sd. u. as cardioton., in spermatorrhea (Ind); to "purify the blood" - **A. porrum** L. (Euras, Eg): (garden) leek; lvs. eaten; u. med. to aid digestion, formerly in diphth.; diur. anthelm. roasted pl. in edema & antibacterial (sAf): for sores; to flavor soups, stews, salads (esp. popular in Fr); natl. emblem of Wales - **A. sativum** L. (Euras, natd. ne&cUS): (white) garlic; true g.; many med. values claimed; bulb (Allium. [L.]) & juice u. as hypotensive (vasodilator) in arterial hypertension; to prevent atherosclerosis & stroke; ton., sed. vulner. expect. in asthma, hoarseness, colds & resp. dis. (syrup) (965); bactericidal antibiotic, antifungal ("Russian penicillin" [ext.]) in suppurating wounds; carm. in flatulence; "antitumor" (963); diabetes; allergy control (964); as memory aid (!); arthritis (local); important culinary herb; in vodka for gout, rheum., kidney stones; diur.; g. oil (VO) prod. Eg, Bulg, Chin, &c., by steam distn.; c. allicin, (antibacterial), alliin. diallyl sulfide; small amts. of elements germanium & selenium; u. to reduce cholesterol & triglycerides in blood; incr. HDL & decr. LDL cholesterol in blood; antithrombotic (1588); "anti-cancer" (allicin); u. to remove plantar warts; deworming of dogs; u. to treat wounds (Russ); protozocidal; phytoncide (963) (2562); imported fr. Mex, It, Chin, Ind; prod. US (963); var. *ophioscorodon* Doell. (sEu; cult.): giant garlic; cibo(u); rocambolle; u. as pickling spice - **A. schoenoprasum** L. (A. sibiricum L.) (Euras, cult. & natd. NA): chives; lvs. juice & rts. antibacterial; u. as anthelm. (Eu); lvs. eaten in salads, soups, omelets, &c.; flavor (esp. popular in Scotland) - **A. scorodoprasum** L. (Eu, Caucas reg, Syr, USSR): sand (or giant) garlic; bulbs u. to flavor foods; med. - **A. sphaerocephalon** L. (Euras, Cauc, AsMin, nAf): bullhead onion; Kopflauch (Ge); similar medl. u. as other *A.* spp.; bulbs eaten by natives near Lake Baikal (Siber); ornam. - **A. tricoccum** Ait. (e&cNA): ramp(s) (2533); 3-seeded (or wild) leek; u. much like *A. schoenoprasum*; bulbs much sought after by mt. people (Appalachians) (964) - **A. triquetrum** L. (Eu, nAf): bulbs c. alliin (cysteine deriv.); u. med. - **A. tuberosum** Rottler ex Spreng. (c&eAs): oriental garlic; Chinese chives; popular food; u. stim. ton. stom. carm.; antidiarrh.; in flatulence; to aid menstruation; bulb juice u. in hemoptysis, hemorrhage (c. antibacterial); crushed lvs. & bulb. u. extern. in wounds, cuts, insect bites/stings; epistaxis - **A. ursinum** L. (temp Euras [woodlands] to Kamchatka): wild (or bear's) garlic; ransoms; rams(ons); much u. (hb. esp. fresh) in hypertension, arterioscl., gastric dis., asthma, in gravel, &c. - **A. victorialis** L. (Euras-mts.): bulb (Radix Victorialis longa) u. formerly as diur. vermif. (folk med.); bulb integument u. for bleeding cut wound; hb. u. for colds (Ainu) - **A. vineale** L. (Eu, natd eUS): field garlic; folk med. uses (diur., stim.); serious weed.

allopatric: growing in mutually exclusive regions.

Allophylus (2786) **abyssinicus** Radlk. (Sapind.) (e tropAf): lusasari (Tiriki); grated rts. eaten with salt & lemon water for coughs, rheum.; crushed rt. appl. to ringworm - **A. africanus** P. Beauv. (trop wAf): banda; lvs. u. in persistent headache, conjunctivitis, rheum.; rt. edible; wd. in construction - **A. alnifolius** Radlk. (tropAf): banque; rt. decoc. u. in poultice for swellings (Tanz); rt. decoc. u. in GC (Tanz), "dulazi" (int. dis.) - **A. cobbe** (L) Raeusch. (A. racemosus Sw.; A. occidentalis Radlk.; Schmidelia o. Sw.) (Mex, WI, CA, nSA): cajobe palo; palo blanco (PR); u. med. like *A. cominia* - **A. cominia** (L.) Sw. (WI): palo de caja (Cu); branches, wd. & lvs. u. in tenesmus & other intest. upsets; diabetes; VD - **A. edulis** Radlk. (Arg): joco; c. alks.; u. hepatoton. antipyr. - **A. griseo-tomentosus** Gilg. (e tropAf): lvs. boiled u. as antipyr. - **A. lastoursvillensis** Pellegr. (Congo, Gabon): lf. juice u. as vulnerary & for chest dis. - **A. pervillei** Bl. (trop eAf): muhene (Digo): rt. decoc. u. for stomach pains; lf. u. as dressing for head pains - **A. rubifolius** Engl. (trop eAf): rt. u. to render birth easier; rt. decoc. given for diarrh. in small children or as mouthwash for toothache - **A. serratus** Radlk. (A. serrulatus Radlk.) (Indochin): rt. inf. u. as aperitive beverage; lvs. & wd. u. for *rykuna* (pain in leg with swelling) by pounding lvs. & applying with hot stone in cloth to affected part with rubbing (Caroline Is) - **A. subcoriaceus** Bak. f. (Congo, Uganda): u. for headache; wd. in construction, to make spears - **A. timorensis** (DC) Blume (Melanesia incl. Fiji): lvs. baked & applied to swellings (New Hebrides); bk. decoc. for general malaise (PI) - **A. warneckei** Gilg (Ghana): rt. bk. inf. u. in diarrh.

alloploid: allopolyploid: polyploid with chromosome sets fr. different spp. &c.

alloro (It): laurel.

Allosan™: allophanic acid ester of santalol.

allouya: *Calathea allouia.*

alloxan: single-ring N compd. occ. as breakdown prod. fr. purine bases in potato; in mucus secreted during dys. &c.; u. in animal expts. to produce hypoglycemia, then hyperglycemia (alloxan diabetes), the latter being the result of destruction of the isles of Langerhans.

alloxur bases: xanthine bases.

allspice: *Pimenta dioica* (SC taste resembles mixt. of clove, cinnamon, nutmeg) - **crown a.**: *Pimenta acris* - **Mexican a. = Tobasco a.**: *Pimenta dioica* var. with larger frt.

allyl: radical $-C_3H_5 = -CH_2CH{:}CH_2$; propenyl. - **a. isosulfocyanate: a. isothiocyanate**: C_3H_5-NCS, volatile mustard oil; prod. fr. black mustard by enzymatic hydrolysis of sinigrin; u. carm. counterirrit., etc.; must not be confused with mustard gas (war poison gas) - **a. mustard oil: a. sulfocyanate: Allylis Isothiocyanas** (L.): a. isothiocyanate.

allyl normorphine: deriv. of morphine; u. as specific antidote for poisoning by morphine, Methadone, Demerol, &c.; u. as resp. stim. for new-born; pharmacol. reagent.

almaciga (Sp): mastic - **almacigo**: 1) (Sp): mastic tree 2) (SpAm): *Bursera* (*Elaphrium*) *gummifera* 3) term appl. to a Philippine manila gum.

almadina gum: almeidina caoutchouc: products of *Euphorbia resinifera.*

almecegão (Br): **almecega gum: almecegueira gum** (or **resin**): a kind of elemi (both hard & soft var.) fr. *Bursera simaruba, Tetragastris* spp., etc.

almendra (Sp): almond fr. the a. tree, **almendro**.

almidon (Sp): starch.

almizcle (Sp): musk.

almond(s): *Prunus dulcis* - **bitter a.**: bitter-seeded var. of almond, u. for prep. both sweet & bitter oils - **bitter a. oil**: 1) VO c. benzaldehyde & HCN & obt. by fermentation & dist. fr. sd. (**natural**) or 2) benxaldehyde made by synthesis (**artificial b.a.o.**) - **earth a.**: *Cyperus esculentus* - **expressed a. oil**: sweet a. oil - **green a.**: pistachio nut - **Indian a.**: *Terminalia catappa* - **Jordan a**: *Prunus amygdalus* - **Malaga a.**: 1) (PacifIs): *Terminalia catappa* 2) finest var. of sweet almond - **a. meal**: ground cake after expr. FO fr. sd. (prod. It, Fr) - **a. paste**: marzipan: mixt. of pd. sweet almonds, sugar, egg whites, water, &c., mixed to give paste; u. to cover cakes; in macaroons, &c. - **snow a.**: blanched sweet almond sds. (done by removal of sd. coats after 4 days soaking in water) - **sweet a.**: a sweet-seeded var. of almond; cannot be used in prep. bitter oil; u. as food & in prep. sweet a. oil - **sweet a. oil**: FO fr. sds. of either sweet or bitter almond; u. food, in pharmaceuticals; comml. prod. of Calif, It, Fr, Sp) - **tropical a.**: Malaga a. (1) - **Valencia a.**: distinctive var. of sweet a. - **West Indian a.**: Indian a.

Alniphyllum fortunei Perk. (Styrac.) (China): cooked bk. or decoc. u. for cough (Hainan) - **A. hainanense** Hay. (sChina): lf. inf. appl. to wounds (Hainan).

Alnus Mill. (Betul.) (mostly of nHemisph): alders, trees & shrubs found growing along lake margins & water courses - **A. acuminata** HBK (Mex): abedul; u. med. (Mex Ph.) - **A. arguta** Spach (Mex): bk.u. for tanning leather; in med. - **A. glutinosa** (L.) Gaertn. (Eu, Af, As, natd. eNA): European (or black) alder; purple a.; trees (-25 m); bk. c. tannins (-20%), flavones; u. in folk med. as antipyr. in gout, rheum.; as gargle in pharyngitis; tan (e&sEu); brown & black dye; lvs. astr. in inflammns. - **A. hirsuta** Rupr. (eAs): bk. ext. u. as tea before & after childbirth (Ainu); var. *sibirica* Schneid. (A. s. Fisch.); dr. young lvs. u. as diur. expect. sed. - **A. incana** (L.) Moench (Euras, cult. & escaped NA): European (or white) alder; u. as hemostatic like *A. glutinosa* - subsp. *rugosa* Clausen (A. rugosa Sprengel; A. r. var. americana Fern.): speckled/gray/hazel/American/tag alder; bk. u. astr. ton. emet. for indigest.; for itching rash (ointment combn.) - **A. japonica** Sieb. & Zucc. (eAs): dr. young lvs. u. as diur. expect. sed. (Kor) - **A. jorullensis** HBK var. *ferruginea* (HBK) O. Ktze. (Ec): heated lvs. tied to legs (etc.) for rheum. (2449) - **A. rhombifolia** Nutt. (sBC [Cana]-nCalif [US], Baja [Mex]): white alder; decoc. of fresh or dr. bk. u. as diaph. in diarrh., stomachache, to facilitate childbirth; to check hemorrhage in TB; dyestuff (Indians, Mendocino Co., Calif) - **A. rubra** Bong. (A. oregona) (NA, Pac Coast): red (or tag) alder; Oregon a. Nutt.; trees (to 25 m); bk. u. ton. emet. astr. alter. in indig., dyspep., source of red dye (Am Indians) - **A. serrulata** (Ait.) Willd. (eNA): smooth/tag/common/hazel/black alder; bk. u. as alter. "for the blood" formerly (NCar) & like *A. incana* subsp. *rugosa* - **A. tenuifolia** Nutt. (A. incana subsp. t. [Nutt.] Breitung) (Alas [US], wCana, s to cCalif): mountain (or river) alder; bk. inf. u. in rheumatic fever; inner bk. orange dye (Am Indian) - **A. viridis** Lam. & DC (Eu): green alder; bk. u. as for *A. glutinosa* - subsp. *sinuata* Löve & Löve (A. s. [Regel] Rydb.; A. alnobetula K. Koch p.p.): Sitka (or wavy-leaf) alder; u. for sunburn, hemorrhoids, itch (Am Indian); wd. u. as fuel, to smoke fish.

Alocasia indica (Roxb.) Schott (Ar.) (Burma, Indochin to Indones): tuber with coconut oil rubbed on scabies (Indochin); rhiz. rubefac., with NaCl appl. to snakebite, malignant sores, herpes; tubers source of starch (apé s.) - **A. longiloba** Miq. (Indochin): hb. recommended for abdom. pains; sap appl. to suppurating sores of cattle - **A. macrorhiza** (L) G. Don f. (trop seAs, esp. Indones, Mal, Austral, Oceania): apé (pron. ah-pay)

(Haw [US]): giant alocasia, g. taro; kapé (Tonga); tubers u. as food, also sts. (Tonga); toxic, however, and needs careful prepn.; juice (latex) u. to relieve sting fr. giant nettle (*Laportea*); mostly u. exter. as poultice for swellings, cuts, bruises (PI), furuncles, glanders; antiparasitic; starch fr. tubers important food of natives; ornam. - **A. odora** (Roxb.) C. Koch (eAs, Ind, Taiwan, PI): pounded tuber appl. to reduce swellings; sts. & young shoots u. for stomachache, colic (Chin) - **A. wenzelii** Merr. (PI): bugiane; petiole juice u. for itch & on scorpion stings & centipede bites.

Aloë L. (Aloe/Lili) (OW, mostly Af): aloe(s); genus of some 200 spp., mostly trop. perenn. plants with stiff basal rosettes of lvs. fr. which arises the flowering stalk; products incl. Aloe (L, Eng+) (USP & many other Phar.) (Eng pron. al´-oh; Latin pron. al-oh-ee); inspissated juice fr. cut lvs. of *A. perryi*, *A. barbadensis* & *A. ferox* & other *A.* spp.; c. anthraquinone derivs. (incl. anthranols) both free & as glycosides (glycosides much more active than aglycones); aglycones incl. emodin, aloe-emodin; glycosides (aloinosides; pentosides) incl. aloin, nataloin; barbaloin, isobarbaloin, beta-barbalion; u. as cath. to drastic purg., bitter, anthelm., antibiot., exter. in treatment of periodontosis (degeneration of tissues surrounding tooth) (963A), stings of Hymenoptera (964), using 20% soln.; another prod. is fresh leaf gel, claimed to have "healing substance"; u. by Am Indians for blisters, burns (X-ray), &c. (965-A); now much u. for dermatoses, sores; in gum dis.; in cosmetics, &c.; int. for arthritis (value?); vars.: 1) sAf: Cape (strongest, but irritating); Uganda; Natal (weakest) 2) eAf: Socotrine (Sokotrine, Sokotra) (one of oldest drugs known); Zanzibar; Mocha 3) West Ind: Curaçao; Barbado(e)s; Aruba 4) Indian: Jaffarabad; Musabbar; Musumbra - **A. abyssinica** Lam. (trop eAf): Indian aloes; lf. juice u. to prep. Mecca or Moka (Mocha) Aloe; u. cath.; lf. substance emoll. - **A. africana** Mill. (sAf): one important source of Cape aloe - **A. arborescens** Mill. (sAf; cult. sRuss, WI, Br, Fla [US]): zabila macho (LatAm): tree (-5 m); inf. u. in pertussis (CR); source of aloin (Russ); ornam - **A. barbadensis** Mill. (A. vera Tourn. ex L.) (latter name often now given preference) (Med reg.; natd. sFla [US], WI+): Mediterranean (or Barbados) aloe; "medicine plant"; source of Barbados, Curaçao, Aruba Aloes; both fresh & dr. lf. juice (latex) u. med. (v. Aloe [gen.]); lf. c. 3% raw saponins; lf. pulp (jelly) u. for X-ray (orig. Fla [US]) & other burns, &c. to reduce topical irritation, itching (esp. Fla); in headache, neuralgia, &c. (Mayas) (6); for healing stomach ulcers, colitis (US); for frostbite, abrasions, wounds, skin dis., &c.; incorp. in many creams, lotions & other cosmetics (US); minimizes hyperkeratosis; this aloe (gum resin) chief one imported into US (fr. nWI, as Bonaire); pl. occ. wild in sFla, ex. Bamboo Key; cult. ornam. bibliography (632) (Other *A.* spp. with similar properties are *A. arborescens* Mill., *A. burgerfortensis* Reynolds, *A. davyana* Schonl., *A. fosteri* Pillans, *A. longibracteata* Pole Evans, *A. macracantha* Bak., *A. mutabilis* Pillans, and others) - **A. candelabrum** Berger (sAf): lf. ext. antibacterial; one source of Natal aloe; old dr. lf. ash mixed with pd. tobacco in making snuff (Zulus) - **A. chinensis** Bak. (A. barbadensis var. c. Haw.) (Ind, sChin, Taiwan): one source of Barbados & Curaçao type aloes - **A. ecklonis** Salm-Dyck (sAf): grass aloe; ext. purg.; cooked fl. eaten as veget. (Zulus) - **A. ferox** Mill. (sAf): bitter aloe; the chief source of Cape aloe; fresh lf. juice appl. to eye in ophthalmia; nectar said narcotic - **A. humilis** Mill (sAf): u. in the same way as *A. saponaria* - **A. kirkii** Bak. (Tanz, Zanzibar): rt. u. to make beer - **A. kraussii** Bak. (sAf): rt. decoc. u. to aid conception & during gestation; fl. u. as veget. (cooked) - **A. marlothii** Berger (sAf): chopped up lf. in water with sugar for tapeworm; lf. & rt. decoc. in roundworm; shoot decoc. for stomach dis.; pd. of ash in Af snuff - **A. perryi** Bak. (SocotraIs): the source of Sokotra (Socotra) aloe; also Zanzibar a.; prod. also in eAf & sArabia; formerly packed in monkey skins - **A. saponaria** Haw. (A. latifolia Haw.) (sAf): soap aloe; white spotted a.; lf. pulp & yellow juice u. for ringworm; u. for boils & sores if inflamed; dr. lf. juice to tan leather; lf. cut open & appl. to wounds; heated lf. appl. as dressing to cuts, blisters - **A. spicata** L.f. (sAf): a source of Cape aloe; c. aloin; u. purg. - **A. succotrina** Lam. (sAf): one source of Zanibar aloe - **A. tenuior** Haw. (sAf): rt. decoc. u. by natives for tapeworm; purg. - **A. tomentosa** Deflers (sArab, Yemen, Somalia): "korr" (Saudi Arab); c. antibiotic; u. for its healing properties - **A. vera** auct. mult. = *A. barbadensis* (however, some authors prefer *A. vera* (L) Burm. f. in place of *A. barbadensis*) - **A. vulgaris** Lam. = *A. barbadensis*.

aloe: common a.: = **ground a.**: inferior type prepd. by expressing juice fr. lvs., boiling in water & evapg. to solid - **aloe-emodin:** hydroxy-anthraquinone deriv. found in aloe, rhubarb, etc.

aloenin: chromanol found in Cape and Curaçao aloe; bitter substance (2021).

aloes (Eng, Sp, Por): **aloès** (Fr): aloe.

aloetic: refers to medicinals with aloe.

aloin(um): xall. princ. obt. fr. aloe; made up of pentosides (glycosides); varies considerably acc. to var. aloe u. for extn.; one of active components is the arabinose glucoside of aloe-emodin; widely u. in mfr. small laxative pills.

Alonsoa meridionalis Ktze. (Scrophulari) (Ec, Co): decoc. u. as bath for vermin-infested babies; appl. to cuts in paste with "bad" cheese (Ec).

Aloysia barbata (T.S. Brandeg.) Mold. (Lippia b. T. S. Brandeg.) (Verben.) (Mex): hb. inf. u. in colds, &c. - **A. gratissima** (Gill. & Hook.) Troncoso (Lippia ligustrina Kearney & Peebles; L. lycioides Steud.) (Verben.) (Tex, NM [US], Mex, Br, Arg, Par, Ur, Bo): cedrón; hb. u. as pectoral - **A. triphylla** (L'Hér.) Britt. (A. citriodora Ort.; Lippia c. HBK) (Mex, CA, Co, Ve, Ec, Br, Bo, Ch, Arg, Ur; widely cult seUS, Sp, Fr): cedrón (Ve, Ec); lemon (-scented) verbena; vervain; hierba (yerba) Luisa (Sp); herb Louisa; lvs. c. VO ("verbena oil") with c. 25% citral, geraniol, verbenone; very fragrant lvs. (Herba Lippiae mexicanae) (?); u. at first as panacea, inf. as nerve sed. in "heart" pain (heartburn), to reduce stomachache (at times with baking soda), digestive, carm. stom. to aid in expelling intest. gas (inf. pc.c. [93]); ton., diaph., antipyr. in diphth. & mal.; in sore throat & colds (Br); to calm hepatic colics, in liver dis. (in general); arom., source of "true" verbena oil (Fr); as substit. for tea (Ve, Ch cult); to give lemon flavor (liqueurs), ornam.

alpha-estradiol: the primary estrogen which occurs in the ovary, urine & bloodstream of female mammals; isolated fr. natural sources or synthesized fr. estrone & other steroids - **alpha-hypophamine**: pitocin or oxytocin.

Alphitonia excelsa (Fenzl) Benth. (Rhamn.) (Austral.): "red ash"; bk. & wd. decoc. drunk as ton.; gargled for toothache; young lf. tips chewed for "upset stomach"; lvs. crushed in warm water for bath in illness (headache, skin dis.) (2263) - **A. incana** Teijs. & Binn. (seAs): bk. chewed with pinanga "to make throat smooth and voice clear"; to treat swellings; rubbed on hands for sores or chafes (PapNG) - **A. philippinensis** Braid (PI): bk. chewed & saliva swallowed for cough & stomach dis. - **A. vieillardi** Lenorm. (Fiji): lf. with some antibact. activity; u. medl. - **A. zizyphoides** Gray (Oceania; Fiji, Samoa): bk. u. medl.; wd. u. for interior work, cabinet making (Fiji, Samoa): lvs. u. as soap substit. (Tonga) (86).

Alphonsea arborea Merr. (Annon.) (PI): sapiro; med. sized tree; frts. eaten; frt. decoc. u. in amenorrhea, boiled frt. appl. locally in fever; bk. decoc. with garlic u. in urticaria.

alpine plants: those found in mountainous regions (orig. the Alps of cEu), i.e., at elevations approaching the limit of perpetual snow; also appl. to plants of other places in the world where the natural place of growth is near snows that are never melted by sun beams; thus, the term does not indicate elevation but rather aver. temps. of the region.

Alpinia allughas Rosc. (Zingiber.) (Ind): taro (Bombay); inferior form of lesser galangal; similar values & uses to *A. galanga*; cult. in the Konkan (Ind) - **A. antillarum** Roem. & Schult. = *Renealmia a.* - **A. calcarata** Rosc. (Ind): rhiz. u. as spice like galangal - **A. chinensis** Rosc. (Chin, Indochin): ma lat; rhiz. u. as stom., remedy for abdom. pain, to stim. blood circulation; for sunstroke - **A. conchigera** Griff. (Indoch, Mal): rhiz. stim. diaph. bechic, in uter. hemorrhage, bronchitis, arthritis; frts. edible & medl. - **A. elegans** K. Schum. (Kolowratia e. Presl.) (PI): pounded lvs. mixed with salt rubbed on affected parts of paralytic; rhiz. decoc. for hemoptysis - **A. galanga** (L) Willd. (A. pyramidata Bl.) (Ind, Chin, Indones-PI): (greater) galangal; source of Siamese & Chinese gingers; rhiz. c. VO with eugenol, cadinene; resin; starch; u. in rheum., bronchitis, resp. diffic. (esp. of children); ton. hypotensive; carm. stim. aphrod.; in diabetes (native to Java, Sumatra; cult. sInd, eBengal) - **A. globosa** Horan (Indochin): rhiz. & stom. & for intest. pain, vomiting, diarrh. - **A. haenkei** Presl (PI): frt. or sds. u. as stom. & in abdom. dis. - **A. japonica** Miq. (Jap, Chin, Taiwan): frt. edible, u. as stom. & in abdom. troubles; rhiz. ext. in wine u. for abscess (Jap); spice - **A. katsumadai** Hay. (eAs): frt./sds. u. as for *A. haenkei* - **A. malaccensis** Presl (sAs, PI): frt. decoc. in gastralgia, puerperal fever, other fevers; rhiz. for sores (Java); spice - **A. officinarum** Hance (Chin, esp. Hainan; Thail, Ind): (lesser) galanga(l); East Indian root; rhiz. ("root") c. VO with cineole, pinene, eugenol; resin, galangin (flavone deriv.), tannin, kaempferide; u. stom. in anorexia, carm. in gastric dis., arom. flavor, sed. for motion sickness, stim.; condim. in mfr. liqueurs (ex. vodka, USSR), &c.; in syncope, hypochondria, vertigo - **A. oxyphylla** Miq. (Chin): ingredient in many Chinese medicines; frt. boiled to make soup u. to aid digestion (Hainan) - **A. siamensis** K. Schum. (Thailand): kah (Siamese); rhiz. arom. stom. carm. condim. - **A. zerumbet** Burtt & R. M. Smith (A. nutans Rosc.; A. speciosa K. Schum.) (As, esp. Mal & pantrop): shell ginger; lvs. u. as decoc. in baths for fever (PI); lf. & fl. decoc. u. for gastric gas & indig. (Dominica; Carib, Indians); petiole juice put in ear for earache (2448); cult. ornam. - **A. zingiberina** Hook. f. (Thail): rhiz. u. as spice.

alpiste (Sp): canary sd.

alquequenje (Sp): *Physalis alkekengi.*

Alraun (Ge): *Mandragora officinarum.*

Alseikraut (Ge): absinthium.

alseroxylin: fat-free alkaloidal fraction of *Rauwolfia serpentina* (+) rt. contg. at least 9 alk.; non-adrenolytic amorph. alk. (reserpine +); action of sed. -antihypertensive-bradycrotic alks.; dose 2-4 mg/da.; lmg ≠ 0.2 mg reserpine; u. tranquilizer.

Alsidium helminthochorton Ktzg. (Rhodomel. - Rhodophy. - Algae) (Atl Oc, Med reg.): Corsican (worm-) moss; the drug so known ([Muscus] Helminthochorton) may c. up to 80 other spp. of sea mosses collected with it; thallus c. mucil. alkaloid-like subst., I_2, Br_2, &c.; u. as anthelm. vermif.; formerly & in scrofula (goiter?) (I. content); most collections in Med Sea at Corsica, hence name.

Alsike: v. under clover.

Alsine media L.: *Stellaria media.*

Alsodeiopsis poggei Engl. (Icacin.) (Congo): rts. u. as aphrod.; wd. u. to make bows - **A. rowlandii** Engl. (Congo, sNigeria): rts. u. as aphrod. (natives).

Alsophila glauca J. Sm. (Cyathc. - Pteridophy.) (Indones): scales fr. frond stalks u. as styptic & as substit. for *Cibotium*; young lvs. eaten with rice & onions to arouse appetite - **A. lurida** (Br) Hook. (tropAs): a tree fern; one source of Paku Kidang (Paleae Stypticae): chaffy scales fr. frond bases u. as hemostat., esp. in dentistry - **A. myosuroides** Liebm. (sMex, CA): tree fern; spores (& scales?) u. at times for styptic (Mex natives) - similar uses reported for *A. contaminans* Wall., **A. tomentosa** Hook., &c. (—> *Cyathea*).

Alstonia R. Br. (Apocyn.) (Mal-PacIs): 1) genus of alkaloid-rich trees & shrubs 2) (L.): specif. Dita bark, fr. *Alstonia constricat, A. scholaris* - **A. angustifolia** Wall. (seAs): lvs. boiled & heated appl. to splenic area in remittent fever - **A. boonei** de Wild. (tropAf, Seneg, Nigeria): bk. u. as antipyr. antimal. in enteritis (Seneg); aq. ext. u. for fever by Yorubas (Nigeria); expressed lf. juice u. to disinfect wounds, bk. latex u. for various skin eruptions (children) (Seneg) - **A. congensis** Engl. (tropAf): pattern wood; bk. (dita bark) c. echitamine, echitamidine, amyrin, u. as antipyr. (eAf); rubber source; rt. & lvs. u. in rheum.; this sp. was formerly confused with *A. boonei* - **A. constricta** F. v. Muell. (Australasia & other trop): fever tree; quinine alstonia; bk. (Queensland [or Australian] fever bark) c. alstonine, alstonidine; reserpine, yohimbine, &c.; u. bitter ton. antipyr. in mal.; stim.; in diarrh.; in prepn. reserpine - **A. costulata** Miq. = *Dyera c.* - **A. grandifolia** Miq. (Sumatra): source of Gutta Malaboeai (similar to gelutong [484]) - **A. lenormandii** v. Heurck & Muell. Arg. (New Caledonia): strongly bitter bk. u. as ton. antipyr.; wd. useful - **A. macrophylla** Wall. (seAs, PI; natd. Fla [US] [965]); bk. febrifuge, ton. emmen. vulner. in dys. diabetes, anthelm., as hypotensive, &c. - **A. plumosa** Labill. (New Caledonia): bk. latex very bitter, powerful ton.; source of Fiji rubber - **A. reineckiana** Lauterb. (Oceania): latex u. in chewing gums & for eye dis. (Fiji) - **A. scholaris** (L.) R. Br. (Ind, SriL, seAs, Austral, PI): devil tree; pali mara alstonia; source of dita bark (Australian fever bark); c. diamine, echitamine (ditaine), echitenine, &c.; bk. u. as antipyr. in mal., ton. in gastric dis., diarrh., lax.; astr.; quinine substit.; latex appl. to open sores; bk. inf. for sexual dis. of male (2448); c. blood sugar reducing subst. (966); bk. u. in spear & arrow poisons (PI); wd. u. to make coffins (SriL) - **A. spathulata** Bl. (Mal, Indones): pulai paya; bk. ext. u. in febrifuge & vermifuge mixts.; latex appl. to sores & skin dis. - **A. spectabilis** R. Br. (Mal, Andaman & TimorIs): corkwood; very bitter bk. (known as poelé bark) c. alstonamine, echitamine; u. as antipyr., like *A. scholaris* - **A. theaeformis** L. f. = *Symplocos t.* - **A. venenata** R. Br. (wInd): addasra (Kanarese); ripe frt. u. in epilepsy, insanity, syph. & as ton. anthelm.; bk. c. alstovenine, reserpine, u. as remedy in stomach pain & intest. ulcers - **A. viellardii** v. Heurck & Muell. Arg. (New Caledonia): c. alk. ochropamine, vobasine, +; u. antipyr., antimal. - **A. vitiensis** Seem. (Fiji, New Hebrides): smoke fr. burning bk. (also latex) u. for eye dis. (SolomonIs); juice squeezed fr. bk. appl. to sore eye in Fiji; latex u. as chewing gum - **A. yunnanensis** Diels (Chin): chih ku; bk. & twigs u. in mal.

Alstroemeria aurea Graham (A. aurantiaca) (Amaryllid.) (Ch): Peruvian lily; contact allergen; cult. as ornam. - **A. ligtu** L. (Ch): liuto; tuberous rts. yield starchy meal (chuño de Concepción), u. in simple erysipelas & cutan. irritations; food.

altamisa (Mex): *Ambrosia elatior* - **a. de la sierra** (NM [US]): *Artemisia franserioides.*

altea (It): **alteía** (Por): **a. rosea: Althaea** (L): marshmallow rt. fr. *A. officinalis.*

Alternanthera philoxeroides (Mart.) Griseb. (Amaranth.) (La, Miss, Fla n to Va [US], CA, SA; natd. Chin, Taiwan): alligator weed; hb. u. for salad greens; fodder; med. (Chin); a bad weed in seUS, in La almost as bad a pest as water hyacinth (bayous); cult. orn. - **A. polygonoides** (L.) R. Br. (seUS, tropAm): smooth chaff-flower (US): "sanguinaria" (Cu); hb. decoc. u. in colitis, dys. urticaria - **A. porrigens** (Jacq.) O. Ktze. (Ec): moradilla rosada; fl. inf. u. for bad case of grippe ("la peste") (2449) - **A. pungens** Kunth (A. achyrantha R. Br.); A. repens L.) (pantropics): "sanguinaria" (Cu); yerba del pollo (Arg); pl. u. as diur. - **A. ramosissima** Chodat (sFla [US], WI, seMex to Br): ma(r)cela; pl. decoc. u. to relieve cough (Yuc); pl. inf. diaph. (Br) - **A. sessilis** R. Br. ex DC (A. nodiflora R. Br.) (pantrops): "bella Maria" (Cu); pl. u. as galact., cholag. antipyr. (Ind); hb. eaten as greens, in soup, with rice & fish (As, Af) - **A. tenella** Colla (widespread in warm areas incl. sUS): chaff flower; manjerico (Por); u. similarly to *A. sessilis*..

alternative medicine: modalities wh. differ fr. the regular medical pattern ("mainstream medicine"): these practices incl. acupuncture; massage; electrical current appln.; yoga; hypnosis; music therapy; naturopathy; chiropractice; faith healing; homeopathy; biofeedback; macrobiotics; dance movements; "guided imagery"; &c.

Althaea armeniaca Ten. (Malv.) (eEu, AsMin, Thrace): u. like *A. officinalis* (off. Russ Ph.); bast fibers jute substit. - **A. ficifolia** Cav. (Eu): figleaf hollyhock; Antwerp h.; cult. ornam. hb. (-2m) - **A. ludwigii** L. (Ind, Pak): hb. u. as aperient - **A. officinalis** L. (c&cEu, esp Fr, Ge, USSR; natd. NA; commonly grows in marshy places): marshmallow (SC habitat); wymot(e); rt. (decorticated) & lvs. u. as dem. sed. in resp. dis. as in cough, hoarseness, bronch., asthma, pulm. TB, also diur. in GU dis., as dysuria, cystitis; in diarrh. proctitis (clysters), in cataplasms for skin burns, rt. in making tabs. (homeop.), veterin. med., cosmetics, plastic masses; previously u. in mfr. of candy marshmallows (m. paste) (2093) (now replaced by mixt. of gelatin, powd. sugar, &c.); fls. (July-Sept.) c. mucil. asparagin; u. to aid expect. & modify cough; in gargles - **A. rosea** (L) Cav. = *Alcea rosea*.

Althaeae Folia (Folium) (L.): marshmallow lvs., fr. *Alcea rosea* - **A. Radix** (L.): marshmallow rt. - **A., Unguentum** (L.): (ointment) with finely powd. marshmallow rt. with pine resin, yellow wax, turpentine, olive oil (Neth & Sp Pharm.)

althea (root, leaves, flowers): fr. *Alcea rosea* - **"althea": shrubby a.**: *Hibiscus syriacus*.

Altingia excelsa Noronha (Liquidambar altingiana Bl.) (Hamamelid.) (Ind to Java): rasmala; (rasa)mals; "storax": r. wood source of odoriferous resins, r. resin, also of r. wood oil (VO); source of a white soft balsam with styrol odor (c. 50% cinnamic acid esters) and of a solid brown balsam with cinnamon odor (c. free cinnamic acid + 7% esters of same); u. as perfume; wd. a timber u. for house beams, flooring; resin u. in orchitis (Burma); ton. (Chin); stim. in combns. (Mal) esp. for chest complaints; lvs. u. med. (Indones) - **A. gracilipes** Hemsl. (Indochin, Chin): latex fr. tapping trunk much u. by Chin. pharmacists; spread on paper & appl. to chest for resp. dis.; rt. bk. astr. depur. antisyph.

alubakhara (Beng.): *Prunus domestica*.

alum, rock: lump alum as it occurs with the earth attached - **a. root**: 1) *Heuchera americana* 2) (NCar [US]): *Geranium maculatum* 3) *Sanguinaria*.

Alumen Romanum (L.): **A. Rubrum**: rock salt - **A. Rupeum**: rock alum - **A. Ustum**: exsiccated (or burnt) alum.

aluminium: European continental spelling of **aluminum**; metallic element (Al) - **a. chloride**: $AlCl_3$: u. in deodorants and antiperspirant prepns.; topical astr.; many industrial uses - **a. hydroxide**: $Al(OH)_3$: occ. in nature in form of several minerals, such as gibbsite; also may be derived fr. Al oxides, such as bauxite; u. as adsorbent, antacid, in gastric hyperacidity; ulcers; dentifrices; antiperspirants; detergents; emulsifier; filtering medium; vaccine adjuvant; gastrointestinal protection (dogs); enzyme carrier; many indust. uses (E) - **a. magnesium silicate**: occurs as such in several minerals (colerainite, &c.); u. as suspending agt., antacid - **a. phosphate**: u. in gels as antiacid (Aluphos, &c. - **a. plant**: *Pilea cadierei* - **a. salicylate, basic**: u. as antidiarrheal - **a. sulfate**: u. as antiperspirant; anti-infective.

alun (Fr): cranesbill.

Alvaradoa amorphoides Liebm. (Simaroub.) (sFla Keys [US], WI, Mex, CA s to CR): besini(c)k-che; palo de hormigas; bk. decoc. ton. to digest. tract; in skin dis. (locally); lf. decoc. in urin. dis., in baths for rheum. &c. (Mayas) (6).

Alveloz balsam: **A. milk**: prod. of latex of *Euphorbia heterodoxa* Muelll. Arg. obt. fr. the trunk sap; u. exter. for treating sores, cysts, cankers, fibromas and other tumors (as ointment); also u. intern. (liq. preps.); possibly co-carcinogen.

alverja: 1) (SA): pigeon pea 2) (Co): pea.

Alvesia rosmarinifolia Welwitsch (Lab.) (sAf): u. in popular med. in leprosy, resp. dis.; for salting food.

Alysicarpus glumaceus DC (Leg) (trop eAf to Ind): pd. lf. appl. as remedy to old wounds, burns, veld sores; for thrush (eAf); appl. exter. to swollen feet; rt. decoc. for cough; in threatened abortion, antidiarrh. - **A. longifolius** W. & A. (Ind): jangli gailia (Marathi); rts. u. as substit. for licorice - **A. rugosus** DC (Af): lf. & rt. u. for fever & cough;

subsp. *penerrirufus* Leon. (Burundi, Af); umushi; crushed lvs. & rts. vulner.; u. as inf., for infant coughs - **A. vaginalis** (L) DC (OW trop, esp. Ind & Indones): Alyce clover; rt. decoc. u. in cough (Java); to aid digestion (Taiwan); pd. sds. u. as inf. against dys. & colics (Vietnam); pasture, hay (fodder), cover crop - **A. zeyheri** Harv. & Sond. (s&eAf): uruzi (Burundi); hb. u. for impotence (natives, sAf); lf. juice, raw rt. & rt. decoc. u. for cough; fresh rt. appl. to snakebite.

Alyssum campestre L. (Cruc.) (Iran to Med reg.): Steinkraut (Ge); sd. rich in mucil. u. as expect., emulsifying agt.

alyssum, (sweet): *Lobularia maritima.*

Alyxia flavescens Pierre (Apocyn.) (sVietnam, Thailand): day sen; wd. burned to furnish fumigant for treating cephalgia; bitter sap emet. - **A. madagascariensis** DC (Mad): andriam bacifohy; lvs. & bk. u. in mfr. rum - **A. pisiformis** Pierre (Indochin): u. like *A. flavescens*; burned in pagodas to give odorous fumes (Ind) - **A. stellata** Roem. & Schult. (seAs): pulasari (Java); bk. pounded with onions, cooked on *Musa* lf. & juice appl. to thrush; bk. & lvs. in GC (Mal); in cosmetics; flavor in native med.; intest. dis. (Indones); in Pac area, bk. u. in chronic diarrh. & febrile states; bitter ton. (non-alk.); rt. inf. drunk for bloody stool (Fiji) - **A. tisserantii** Montrouzier (A. disphaerocarpa v. Heurck & Muell. Arg) (New Caledonia): arom. bk. u. in obstinate constipation.

amadou (Ind): *Fomes.*

Amalocalyx microlobus Pierre (Apocyn.) (Indochin): may yam (Laotian); frts. edible; latex believed toxic; u. med.

amaltás (Ind): pods of *Cassia fistula.*

amamelide (It): *Hamamelis viriniana.*

amande (Fr-f.): almond frt.; a sd.; any sd. contained in a putamen (stone) - **a. amère**: bitter a. - **a. de mahalem**: sd. of *Prunus mahaleb* (*qv.*) - **a. douce** (Fr): sweet almond - **amandin**: globulin occ. in almond & peach kernels.

Amanita caesarea (Fr): Schw. (Amanit. - Agaric. - Fungi) (temp zones): Caesar mushroom; Caesar's a.; royal agaric; frt. bodies edible, popular in sEu & in ancient Rome - **A. capensis** Pears & Stephens (sAf, Cape Peninsula): Cape death cup; very poisonous mushroom - **A. citrina** Schaeff. ex. S. F. Gray (A. mappa Quél.) (eNA, Eu): delicate amanita; napkin a.; med. uses obscure; toxic - **A. muscaria** (L. ex. Fr.) Hook. (worldwide temp zones): fly (poison) amanita; f. agaric (SC kills flies); bug a.; sporophore c. muscarine, muscazone; muscimol; ibotenic acid; u. in CNS disturb., climacteric diffic., chronic arthritis; insecticide (muscimol); decoc. with sugar u. to attract & kill flies; with juice of bilberry (*Vaccinium uliginosum*) for colic, &c. (Kamchatka); to enhance flavor of foods (Jap); hallucinogenic (euphoriant); frt. body somewhat poisonous (variable but best avoided!), edible if boiled; fumitory, masticatory, addictive, narcotic agt. (Siberia); may represent Soma (of India); in homeop. med. in night sweats - **A. ovoidea** Quél. (Med reg., c&sEu): white frtg. bodies considered fine quality food; sold in markets (It, Sp, Por, sFr) - **A. pantherina** Secr. (worldwide in temp zones): believed poisonous sp.; u. med. in Mex - **A. phalloides** (Vaill. ex. Fr.) Secr. (Eu, NA): (green) death cup; d. angel; deadly amanita; destroying angel; one of the most poisonous toadstools; u. in criminal poisoning; frtg. bodies c. phalloidine, amanitins (968); poisonous effects delayed hence may be difficult to trace source of homicidal crime; u. in homeop. - **A. rubescens** S. F. Gray (temp & subtr zones): blusher; frtg. bodies u. as food (esp. Eng); in catsup - **A. spissa** (Fr.) Kummer (A. excelsa Kummer) (temp zones worldwide): tall amanita; said edible, but best avoided - **A. verna** Vitt. (worldwide): destroying angel (SC pure white sporocarp); spring amanita; claimed "deadliest of all known mushrooms" with no known antidote; very close to *A. virosa* - **A. virosa** Secr. (Eu, NA): destroying angel; poisonous amanita; virulent a.; sporocarps are deadly poisonous.

amanita toxin: highly poisonous princ. of the toxic mushroom *Amanita phalloides; also A. muscaria, A. phalloides, A. verna, A. virosa, A. capensis, A. pantherina, A. brunescens.*

amanitas, toxic (lethal): chiefly represented by *Amanita phalloides, A. muscaria, A. verna, A. virosa, A. capensis, A. pantherina, A. brunnescens.*

amanitins: gp. of toxic compds. found in *Amanita phalloides* & related *A.* spp.; oral admn. does severe liver damage; often lethal; compds. not destroyed by boiling or drying nor by digestive tract proteases; u. as "tool" in molecular biology.

amapol(l)a (Sp): 1) *Plumeria rubra* (Ve) 2) *Hibiscus rosa-sinensis* 3) *Erythrina* spp. - **a. seeds** (Arg): *Papaver rhoeas* sds.

Amaracarpus solomonensis Merr. & Perry (Rubi) (Oceania): bk. decoc. u. for constipation (white tongue in children) (Solomon Is).

Amaracus (L): 1) (Lab): gen. close to *Origanum*; ca. 15 spp. of eMed reg. 2) sometimes appl. to *Majorana hortensis* (Origanum majorana) - **A. dictamnus** Benth. = *Origanum dictamnus* (—> *Origanum*).

amaranth: 1) *Amaranthus caudatus*, etc. 2) an anilin dye permitted for use in foods, drugs, cosmetics, etc.; extensively u. in pharm. as red color to replace cudbear (in short supply).

Amaranthaceae: Coxscomb Family; mostly herbs, a few shrubs; ca. 800 spp. in ca. 71 genera; most grow in warm or tropical areas of both hemispheres; fls. often showy; most taxa are u. for decoration & foods (grains) (cult.).

Amaranthus L. (Amaranth.) (worldwide, temp & trop zones): pigweeds (SC weedy, of low value); genus of coarse ann. weeds; some grown for greens, others for cereal grains (Af, Ind; oldest food grain in NA); flowering a frequent cause of pollinosis, esp. in wUS - **A. albus** L. (NA): pigweed; "tumble weed"; hb. u. vulner. detergent - **A. australis** (Gray) Sauer (Acnida cuspidata Bert.) (seUS): "careless"; edible - **A. blitoides** S. Wats. (wNA): prostrate amaranth; lvs. eaten as potherb; sds. u. for pinole (Am Indians) - **A. blitum** L. (A. silvestris Desf.) (Med reg.; now Euras & trop worldwide, eUS): (wild) blite; w. amaranth; hb. & fls u. emoll. dem., astr. as salad, potherb (Ind); sds. u. porridge; pig food; common weed but cult. - **A. caudatus** L. (cAs, Af, Am): Inca wheat; Spanish greens; "love lies bleeding"; quinoa (also applies mostly to *Chenopodium quinoa*); lvs. c. amaranthin, rutin, betaine; lvs. u. as snuff; abort.; astr., deterg.; frts. & sds. of nutritional value (cereal grains) (c. lysine) (Chin); u. in breads (SA Indians); lvs. in salads; sts. also eaten (c&wAs); sts. in water —> brilliant red (969); ornam. ("Jussel flower") - **A. cruentus** L. (A. paniculatus L.): believed close to *A. hybridus*; important grain crop & potherb (cult., Ind.) grains eaten - **A. frumentaceus** Buch.) (seAs, tropAf, Austral, Am): purple amaranth; quadihtla; staple food of Aztecs (high nutritive value sine proteins with abundant lysine; vit. C); sds. u. in cakes, lvs. also eaten as salad greens; considerable red color so liquid prepns sometimes u. to simulate blood; cult. Ind (where important grain crop & potherb); important food of prisoners of war (WW II, Singapore) - **A. dubius** Mart. ex. Thellung (Eu, natd. WI, tropAm): pl. eaten by humans (cooked) and cattle (raw) (*cf.* spinach) - **A. graecizans** L. (US): tumble weed; similar to *A. albus*; serious weed in many countries - **A. grandiflorus** J. M. Black (Austral): sds. u. as food by aborigines (cAustral) - **A. hybridus** L. (may be of tropAm origin [indig. to Co]; now worldwide): careless; "pigweed"; green amaranth; bledo (Co); lvs. astr.; lf. decoc. u. for liver dis.; indig. (Co); rt. decoc. diur.; young shoots & lvs. potherb; sds. eaten (esp. swUS, Ind) - **A. hypochondriacus** L. (A. hybridus L. var. *h. Thellung*) (tropAm; cult. Ind): prince's feather; huauhtli; lvs. astr. deterg. in hemorrhoids; an important grain crop among Am Indians before the Conquest (A. leucocarpus S. Wats.) - **A. lividus** L. = A. blitum L. - **A. mangostanus** L. = *A. tricolor* - **A. palmeri** S. Wats. (US, mostly sw; Mex): Palmer's amaranth; (red [root]) careless weed; important hay fever plant - **A. quitensis** HBK (Ec, Arg, Par): hb. c. saponins, u. in hepatic diffic. (Arg) - **A. retroflexus** L. (temp zone, esp. Eu, NA): pigweed; redroot amaranth; (wild) red root (amaranth); wild beet (book name); green amaranth; young pl. u. as potherb, frt./sd. food (Am Indians); antihemorrhagic (Ariz [US] Indians); frts. adulter. of fennel & marjoram herb; semicosmop. weed; (introd. into Eu) - **A. rudis** Sauer (US, esp. Great Plains): western water hemp; a major hay fever pl. (Okla [US], &c.) - **A. spinosus** L. (tropAm native; now pantrop/subtrop): thorny (or spiny) amaranth; s. pigweed; green careless (Miss [US]); rt. decoc. u. as diur. (Cu); cath. refrig., in GC, menorrhagia (with rusot), diaph. antipyr. galact. (Ind); emoll.; rt. inf. & poultice in eczema, bruises, inflammn.; cyanogenetic when green; weed; major hay fever plant; young pl. eaten as veget. like spinach; ash mixed with tobacco as snuff; lvs. fodder; seed as birdseed - **A. tamariscinus** (Nutt.) Wood (wUS, esp. Great Plains, SDak+): western water hemp; major hay fever plant (Okla &c.) - **A. thunbergii** Moq. (tropAf): bondue; u. as veget. (*cf.* spinach) c. much vit. C; u. oxytocic - **A. tricolor** L. (A. gangeticus L. - A. magostanus L.) (trop worldwide, esp. warm As, as Ind & Jap): Joseph's coat; pl./lvs. astr. in diarrh. dys. deterg. in colic, GC (Ind); lvs. appl. to ulcerated state of mouth/throat; wash for ulcers; boiled lvs. eaten as greens or after salting (Ind, Mauritius, BourbonIs); sds. eaten (brède de Malabar); powerful aphrodisiac; FO prepd. - **A. viridis** L. (trop Am [native]; now pantrop): green (or notched) amaranth; tupu'a (Tonga); pl. decoc./inf. u. as diur. galactag. (Br); lvs. emoll. (Antilles); rt. inf. stom. (nVe); rt. decoc. for bleeding dys. (Cu); u. as antipyr. (Ghana); young lvs. eaten as veget. (Tonga & elsewhere) (86).

amarasca (It): black wild cherry.

amarogentin: glucoside fr. gentian rt.; said bitterest subst. known (2001); gentiopicroside is present in larger quantity but much less bitter.

amaroid: bitter plant principle of definite chem. compositition but neither alk. nor glycoside; also known as (neutral) bitter principle; specific names end in "-in"; ex. jasminin.

Amaryllidaceae: J. St. Hill.: monocot fam. with ca. 85 genera & ca. 1,100 spp., mostly trop & subtrop; ex. *Zephranthes*; fam. is often lumped with Liliaceae.

Amaryllis bella-donna L. (monotypic) (Amaryllid.) (sAf, natd. WI): belladonna lily; c. bellamarin; u. emet., card. agt.; poison; cult. ornam. (rose-red fls. with yellow anthers) - **A. equestris** Ait. = *Hippeastrum puniceum.*

"amaryllis" (of greenhouse): pls. of *Hippeastrum*, *Lycoris* & other genera.

Amasonia campestris (Abul.) Moldenke (Verben.) (WI, Guanas, Ve): mendoca; "cola de gallo" (Ve); pl. u. for gastritis; as rinse (970); to treat "palikur sterile" (Fr Guiana) - **A. lasiocaulis** Mart. & Schau (Br): pl. u. by natives to treat stomach inflammn.

Amazonia: area surrounding the Amazon River (Br); largest river of the world; area very rich in flora, many of medicinal interest; forests now rapidly being destroyed.

ambaiba (Br): *Cecropia* spp.

ambar (Sp, Po): amber - **ambaree: ambari (hemp)**: *Hibiscus cannabinus.*

ambarella: *Spondias dulcis* - **s. seed**: ambrette sd.

ambauba (Br): **ambauva**: *Pourouma* spp.

ambay (Sp): *Cecropia adenopus.*

Ambelania markgrafiana Monachino (Apocyn.) (Col): latex u. to adulter. chicle; rubber; u. to paint legs & arms as protection against gnats ("jejenes") (Witotos); large arom. frts. edible (2303).

amber (L., Eng.): Succinum (L.): **a. gum; a. resin**; a fossil resin fr. *Pinites succinifer* & other (extinct) pls.; u. in violin varnish, etc.; jewelry (beads); med. in chronic catarrh., etc. (prod. Ge) - **a. acid**: succinic a. - **rectified a. oil**: VO dist. fr. a.; c. terpenes & resinous compds.

amberbell: *Erythronium americanum.*

Amberboa divaricata Ktze. (Comp.) (Ind): pl. ton. aperient, antipyr., mucilaginous & u. in coughs; c. alks.

ambergris: a pathological concreted secretion in the whale intestine which is greatly prized in the perfume indy. because of its fine value as a perfume fixative; c. ambrein &c.; found washed up on trop shores (found by beachcombers) or in butchering whales (**"body ambergris"**); now illegal prod.; v. *Physeter*; mode of formation still not fully understood (2214).

Amblyolepis setigera DC (Helenium s. Britt. & Rusby) (Comp.) (monotypic) (swNA): powerful sudorific; pl. with arom. odor (Tex [US]).

Ambomycin = Alazopeptin: antibiotic obt. fr. *Streptomyces candidus* subsp. *azaticus* Awaya & Hata; isolated ca. 1972 (Russ).

amboside: glycoside of sarmentogenin + D-diginose; occ. in *Strophanthus amboënsis.*

ambra (L.): amber.

Ambramycin™: tetracycline.

ambrein: subst. like cholesterol obt. fr. ambergris by digestion in hot alc.

ambretta seed: ambrette (seed): musk sd., *Hibiscus abelmoschus.*

Ambrina Spach = *Chenopodium* L. - **A. foetida** Moq. = *Chenopod. schraderanum.*

Ambroma L. v. *Abroma.*

ambrine: paraffin dressing.

Ambrosia L. (Comp.) (mostly NA, SA): the (false) ragweeds; genus of mostly ann. hbs., a few shrubs; staminate fl. heads in racemes or spikes, the pistillate at base of these in axils of upper lvs.; mostly important as hay fever plants (Ambrosia means "food of the gods") - **A. acanthicarpa** Hook. (Franseria a. Coville) (wUS, Cana): bur ragweed; a range forage pl. - **A. ambrosioides** (Cav.) Payne (Ariz [US]): lvs. & sts. with cancer inhibitory effect (cytotoxic) - **A. artemisiifolia** L. (NA; natd. Eu+): (common or short) ragweed; Roman wormwood; hog (or bitter) weed; frts. (achenes) yield FO of drying type, edible (ragweed oil); in homeop. & native med. (Mex) to cleanse wounds; hay fever cause - var. *elatior* (L.) Descourtils (A. el. L.) (eNA, natd. Eu): dwarf ragweed; pl. u. in inflammn., deterg.; pollen an important cause of pollinosis - var. *paniculata* (Mx) Blank. (A. p. Mx.; A. peruviana Willd.) (seUS, WI, nPe): artemisa (*sic*); wegwood; mugwort; lvs. & fls. in baths for rheum.; lf. decoc. astr. ton., in fevers; fl. decoc. emmen. - **A. bidentata** Mx. (c&sUS): lance-leaved (or southern) ragweed; u. as tea for pertussis; lf. rubbed on skin for rhus poisoning - **A. confertiflora** DC (Franseria c. [DC] Rydb.) (swUS, nMex): slender ragweed; hay fever pl. - **A. cumanensis** HBK (Mex, WI s to Pe): altamisa; hb. inf. for yellow fever, menorrhagia, epilepsy, emmen.; hb. juice for pleurisy (Trin) - **A. deltoidea** (Torr.) Payne (Franseria d. Torr.) (Ariz, NM [US], nMex, CA): rabbit bush; donkey's hay; pl. decoc. drunk for rheum. & leg sores (Papagos); also in sweat baths (Guat); hay fever pl. - **A. hispida** Pursh (Fla, sMex, WI, CA): baby thyme (b. time, Bah); bay t.; b. tansy; lf. decoc. for fever, pain, gastralgia; cath. anthelm.; inf. of dr. lvs. u. for resp. dis.; stim in anorexia (Bah); "baysurina" - **A. maritima** L. (Med reg): dams-sissa (Arab): flg. tops c. ambrosin & damsin; hb. decoc. u. stim. stom. cordial; for nasal bleeding, flatulence (Eg folk med.) - **A. psilostachya** DC (wNA): western ragweed; perennial r.; hambre (Calif, Sp): lf. decoc. to treat sore eyes; inf. to heal sores at childbirth (214); for bloody stools; tea u. in mal. (Am Indian); heated lvs. appl. to aching joints (Calif Indians); pl. allelopathric - **A. tenuifolia** Spreng. (Ur): u. in fertility control - **A. trifida** L. (NA, natd. Eu, Jap): giant (or great) ragweed; buffalo weed; pollen said chief cause (ca. 80%) of pollinosis (cUS) (971); pl. u. as astr. deterg. &c.; rt. chewed as sed. (Wiscon Indians [204]); achenes possibly u. as "starvation food" by Indians of midwest (204); very common ann. hb. (esp. in low moist ground); l-6 m high; fls. mostly Aug., Sept.

Ambrosiaceae: fam. sometimes segregated fr. Compositae.

Amburana cearensis A. C. Smith (A. cluadii Schw. & Taub.; Torresi c. Fr. Allem.); (Leg) (Br, Arg): umburana (or amburana) sds.; cumarú; false tonka; sds. very rich in coumarin (4%), FO (22%); u. perfume, in soaps; source of coumarin (972); substit. for tonka beans; wd. yellow, valuable (1984).

ameixa (Br): 1) plum or prune 2) *Ximenia coriacea.*

Amelanchier Medik. (Ros.) (nTemp Zone, esp. NA): serviceberry, June berry; shadberry; shrubs or trees with berry-like edible pomes; eaten when over ripe; some cult. as ornam.; ca. 10 spp. - **A. alnifolia** Nutt. (& var.) (w. Cana, nUS): Saskatoon (berry); western service berry; popularly u. for its edible frt.; u. in pies, &c.; to make pemmican (American Indians) - var. *semiintegrifolia* (W. J. Hook.) C. L. Hitchc. (wOre [US] to BC [Cana]): Pacific saskatoon; with edible frt. - **A. arborea** Fern. (e&cUS, e. Cana): (downy) service berry; u. as food (Am Indians) - **A. canadensis** Medik. (eNA): service berry; shadbush; sugar pear; tree (-20 m in US); fls. Mar-June; frt. edible; wd. hard & u. for tool handles, &c. - **A. pallida** Greene (nCalif, Ore): pale-leaved service berry; frt. edible; wd. (fore-shafts) u. to tip salmon harpoons (nCalif Indians) (973) - **A. sanguinea** (Pursh) DC (eNA): "Saskatoon berry" (?); round-leaf service berry; shore shadbush; "bear berry" (SC eaten by black bears); frt. sweet, juicy, deliciously edible; dr. frt. u. as currants (Alta) - **A. stolonifera** Wieg. (A. spicata [Lam.] K. Koch) (eCana, neUS w to Mich, Minn): low Juneberry; bear berry; frts. pleasantly edible - **A. utahensis** Koehne (nwUS): Utah serviceberry; shadbush; frts. u. both fresh & dried-preserved (winter food by Indians); given during labor & delivery; emetic (Navajo); makes good jelly.

amendoa (Por): almond - **amendoim** (Por): peanuts.

ament: catkin.

American: v. under adder's tongue; adonis; agave; alder; aloe; arbor vitae; arrowroot; arrowwood; ash; aspen; aspidium; barberry; bear's foot; blackberry; blistering beetle; burnweed; cannabis; centaury; china root; colombo; custard apple; dittany; fir (silver); frankincense; gentian; ginseng; hellebore; hemp; holly; ipecacuanha; isinglass; ivy; jute; kino; licorice; mastic; mint; mistletoe; nightshade; pawpaw; pennyroyal; pepper; polygala; saffron; sanicle; sarasaparilla; senna; spikenard; spindletree; storax; turpentine; valerian; veratrum; vervain; water hemlock; white ash; white hellebore; worm root; wormseed.

American gum lack: *v.* Larrea.

Amerikanische Faulbaumrinde (Ge): Cascara Sagrada.

Amerind(ian): American Indian (to distinguish fr. Asiatic Indians).

ameixa (Por): plum.

Amherstia nobilis Wall. (monotypic) (Leg) (Burma): "queen of flowering trees"; "pride of Burma"; cult. ornam. (974).

Amianthium muscaetoxicum Gray (Zigadenus m. Regel) (Lili) (eUS): "fly poison"; toxic (alks.); cattle poison. (—> *zigadenus*)

Amianthus (L.): asbestos.

Amicardine™: khellin.

Amicetins: antibiotics (glycosidal) obt. fr. *Streptomyces vinaceus-drappus* Pridham & al. & some other spp. isol. fr. soils near Kalamazoo, Mich (US); 1953-; antimicrobial active against some Gm. m.o., TB, &c.

amidon (Fr): **Amidum** (L): amilo (Por): Amylum (L): starch.

Amigen™: hydrolysate of casein; c. amino acids and polypeptides; u. in protein replacement or supplementation.

Amikacin = Novamin™: broad spectrum antibiotic of aminoglycoside gp.; semi-synth. deriv. of kanamycin A; u. for infections by *Proteus, Providencia, Serratia* (&c.) spp. (1972-).

amines: derivs. of NH_3 in wh. one or more H's have been replaced by org. gps.; ex. methylamine, CH_zNH_2; some play an important role in physiolog. process = the **biogenic a.**, incl. such important members as histamine, acetylcholine, dopamine, nor-epinephrine, serotonin, epinephrine; *cf.* catecholamines.

amino acid: an organic compd. of the general formula $NH_2R.COOH$ (R: alkyl radical); compds. represent end products of protein cleavage & serve as building blocks with which organism produces its own protein tissue - **essential a. a.**: once believed necessary to continued existence; arginine; histidine; isoleucine; leucine; lysine; methionine; phenylalanine; threonine; trytophane; valine; (some believe arginine & histidine are not essential if other a. acids are present in adequate amts.) - **aminophylline**: theophylline & ethylenediamine compd.; u. heart muscle stim. &

diur. - **amino-penicillanic acid, 6-**: 6-APA; obt. fr. special cultures of *Penicillium chrysogenum*; fr. *Pleurotus ostreatus*; 1959-; u. as intermediate inprepn. of semisynth. penicillins - **aminopterin**: a folic acid derivative found in butterfly wings, now synthesized; u. experimentally as antagonist to folic acid in leukemia, cancer, etc.

aminoacetic acid: glycine; the simplest & most common amino acid; occurs in gelatin, proteins of sugar cane, etc.; comml. source of synth. fr. chloracetic acid; the HCl salt u. in gastric acid deficiencies.

aminobenzoic acid: v. para-aminobenzoic acid.

aminobutyric acid, γ-: GABA: this natural amino acid serves as an important neurotransmitter; u. as anti-hypertensive.

aminoglycosides: antibacterial antibiotics (broad spectrum) incl. neomycin, streptomycin, kanamycin, Tobrumycin, gentamycin, Amikacin.

Amioca (starch): amylodextrin fr. waxy maize (corn) grain (*Zea mays*).

Ammannia auriculata Willd. (Lythr.) (Indo-Pak): pl. u. as blistering agt. - **A. baccifera** L. (paleotrop, esp. Ind, seAs): dadmari (Hind, Beng); weed in rice fields & marshes; fresh lvs. u. in skin dis. as rubefac. (Ind); lvs. u. as vesicant (PI); in rheum. pains (Ind); decoc. for biliousness (PI) - **A. senegalensis** Lam. (wtrop Af): hb. u. as irrit., blistering agt. (vesicant).

Ammi L. (Umb.) (Euras; Med reg; nSA): bishop's weed; some 120 spp. of ann. hbs. with finely divided lvs. (**Ammi**, L., frt. of *A. visnaga*), *A. majus*, &c.) - **A. copticum** L. = *Trachyspermum ammi* - **A. majus** L. (Med reg. n. to Belg; wAs, natd. nAf, seUS): bishop's weed; herb William; arom. frts. ("sds.") c. xanthotoxin; bergapten; u. ton. stom. carm. diur. in angina pectoris, bronchial asthma, vitiligo (975); cult. since Middle Ages & use widespread - **A. officinal(e)** = *Trachyspermum ammi* - **A. visnaga** (L.)Lam. (Med reg., esp. Eg, Fr; cult. Arg): kellah; khella; toothpick ammi; Spanish t.; herbe aux cure-dents (SC u. in toothache); rt. edible; frt. ("sds.") & juice c. several furonochromones (khellin [visammin]; visnagin; khellinone; khellinol); FO ("visnagol"), &c.; long u. in Orient, esp. Eg as diur. in urin. dis. (esp. for kidney & bladder stones); hypertension; vasodilator in angina pectoris; emmen. (976); gallbladder, colic, allergy.

Ammiaceae: Umbelliferae.

Ammobroma sonorae Torr. (Pholisma s. Yatskievych) (Lenno.) (swUS, Ariz, Calif [US]; Mex): sand food; a root parasite (without chlorophyll); large rhiz. u. as food raw, cooked, or as flour (Papago Indians, Ariz [US]).

Ammodaucus leucotrichus (Coss. & Dur.) (1) Coss. (1) (Umb.) (nAf): lf. inf. pectoral; condim.; cult. & sold in markets.

Ammodendron karelini Fisch. & Mey. (Leg) (Siber): u. in Russian med.

Ammoides verticillata (Desf.) Briq. (Ptychotis atlantica Coss. & Dur.; P. trachysperma Boiss.; P. verticillata [Desf.] Duby) (Med reg.): rianeddo; ripe frt. c. 0.8% VO with 43% thymol; related genera (*Carum* & *Trachyspermum*) are also rich in thymol.

ammonia: NH_3: colorless gas with pungent odor; aq. soln. = NH_4OH = ammonium hydroxide; chloride (NH_4Cl) known to ancient Egyptians; word derived fr. Ammon, oasis near temple of Ammon (Amen) in Libya; many NH_4 compds. known & used; said to enhance effects of nicotine, hence added to cigarette tobacco (also to enhance flavor); permits body to absorb more (triple) nicotinc - **spirits of a.**: c. 10% NH_3 in alc.; sometimes with anethole (anise odor); u. to stim. person in faint. by reflex stim. of resp. & circn.; also by local appln. to stings of Portuguese men-of-war; smelling salts u. similarly (inhalation).

ammoniac, gum (Ind, Iran, USSR): **Ammoniacum** (L.) fr. *Dorema ammoniacum* - **African a.: Moroccan a.**: fr. *Ferula tangitana*, *F. communis* vars. (c. & u. similar to **Persian a.**: Ammoniacum).

ammonium sulfo-ichthyolate: Ichthammol.

Ammophila arenaria Link (A. arundinacea Host) (Gram) (Eu-coastal zones; natd. Mass [US]): (European) beachgrass; sea sand reed; marram (grass); u. to bind sand; sts. u. in basketwork, chair seats, &c.; rt. stocks edible - similarly **A. breviligulata** Fern.

Am(m)ophos (Am): an almost complete fertilizer.

amniotin: mixt. of estrone & other estrogenic ketones obt. fr. mare pregnancy urine.

amole (plants): **amolé** (Mex): **amolli**: soap plants; name given to various pls. with bulbs or rhiz. u. like soap; at times frt. are u. also as fish poison; incl. *Agave lechuguilla*, *A. lurida*, *Yucca elata* & other *Y.* spp., *Nolina* spp., *Polianthes tuberosa*, *Chlorogalum pomeridianum*, *Sapindus saponaria* & other taxa; rts. u. as deterg. like soap by Am Indians (Calif [US], &c) (*cf.* igamole [Aztec]); amole (Sp): a kind of gruel - **a. amargo**: *Phytolacca americana* - **a. de bolita** (Mex): igamollin, *Sapindus saponaria* - **a. de raiz** (Sp): *Agave lurida* (Mex Phar.).

amolonin: steroidal saponin of *Chlorogalum pomeridianum*; hydrolyzes and produces tigogenin + glucose, galactose & rhamnose.

Amomum L. (Zingiber.) (OW trop): cardamoms; some 160 spp. of large rhizomatous perenn. hbs.

quite similar in habit to ginger; (also L, Gr anciently for an arom. shrub of undetermined origin) - **A. amarum** Lour. (Indochin; cult. Chin): frt. u. med. in Jap, Chin - **A. aromaticum** Roxb. (A. medium Lour.) (Ind): Bengal (or Nepal) cardamom; a very minor & unoff. prod., u. as substit. for "true" cardamom; c. VO with cineole - **A. cardamomum** L. = *Elettaria cardamomum* - **A. cevuga** Seem. (Oceania incl. Tahiti): u. to scent coconut oil (Fiji); in med. (Tahiti); for thatch & beds - **A. compactum** Sol. ex Maton (A. cardamomum Roxb. or Willd.; A. cardamom L.) (Mal): (Siam) cardamom; round c.; produces an inferior sort of cardamom, not off.; frts. called "camphor seeds," round (or Java) cardamoms; (frts) c. VO with borneol & D-camphor; not to be confused with true cardamom; frt. u. as condim. to flavor cakes; chewed to sweeten breath; ground rhiz. u. for coughs & colds - **A. costatum** B. & H. (seAs): u. like *A. aromaticum* - **A. cotifolium** (Af): source of Sierre Leone cardamom - **A. granum-paradisi** L. = *Aframomum gr.-p.* - **A. kepulaga** Sprag. & Berk. = *A. compactum* - **A. krervanh** Pierre ex Gagnep. (Thail, Kamp): rhiz. lax.; frts. condim., in curries; ton. stim. aperitive, emmen. u. in TB, colic, rheum., &c. (Chin med. & pharm.) - **A. maximum** Roxb. (Mal, Indones): Java cardamom; perennial hb., pl. u. as condim. (Mal natives), sds. carm.; cult. in Java - **A. subulatum** Roxb. (tropAs, esp. Ind): Nepal or Bengal cardamom; frt. u. as anodyne, in debility; native spice - **A. villosum** Lour. (Chin): frts. u. stom. carm. ton. in dys., anuria, constipation, abdom. pain - **A. walang** Veleton (Java): "daong walang"; lvs. rts. & sts. furnish VO, walang oil, with unpleasant odor reminiscent of bedbugs - **A. xanthioides** Wall. (Burma, Indochin): source of bastard (Siamese) cardamom; cult. Ind, u. similarly to *A. villosum* - **A. zerumbet** L. = *Zingiber z.* - **A. zingiber** L. = *Zingiber officinale.*

Amoora cucullata Roxb. (Meli.) (Ind): amur (Bombay); bruised lvs. appl. to inflammn. - **A. rohituka** W. & A. (Aphanamixis polystachya Parker) (seAs, PI): sds. c. FO (amoora oil) u. as illuminant; bk. in glandular swellings, liver & splenic dis.

amora (Por): 1) mulberry 2) blackberry.

amoreira negra (Por): black mulberry.

Amorpha canescens Pursh (Leg) (cNA): lead plant; shoestring plant; sts. attached to skin, ignited & allowed to burn down to skin (moxibustion) for gastric or rheum. pains (Am Indians); dys. typhoid; growth; believed to indicate presence of lead ores - **A. fruticosa** L. (e&cNA): bastard indigo; (false) indigo bush; former source of indigo; pl. (rt. bk. &sds.) c. rotenoid derivs., amorphin (glycoside), amorphigenin, rotenone (sds.); sts. as bedding material (Kiowa Indians); pl. with insecticidal props. (977); pls. u. to keep meat clean (Pawnee Indians).

Amorphophallus Bl. ex. Dcne. (Ar.) (trop Af, As): some 100 spp., with enormous lf. (-3 m) & inflor.; some ornam., some toxic (CaOx, coniine) - **A. campanulatus** Bl. (Ind to Indochin & NG): Telinga potato (Bengal); inflor. has fetid odor; cooked (prolonged boiling) tuber a food; consts. (2094); caustic raw tubers (corms) appl. as rubefacient poultices to rheum. (swellings) (PI); boils (sds. also u.); juice of inner petiole fermented, then drunk for diarrh. (NG) - **A. dracontioides** R. Br. (tropAf): rt. u. in prepn. arrow poison; med. (Guinea); starvation food (washed & boiled) - **A. gigantiflorus** Hay. (Taiwan): appl. to buboes, swellings; intern. for resp. tract dis. - **A. konjac** Koch = *A. rivieri* - **A. prainii** Hook. f. (Mal, Sunda Is): lekir; pl. juice u. for arrow poison, usually added to the more poisonous juice of *Antiaris toxicaria*; tubers boiled & roasted for use as food - **A. rivieri** Dur. (Hydrosme r. Engl.) (Indochin, NG; cult. Jap, PI, Taiwan): konnyaku (Jap); devil's tongue; konjac rt. source of konyaku (*qv.*) u. as food by Chin & Jap (in sukiyaki, food thickener [konjaku flour or powder]) (c. glucomannans); said useful for reducing wt. in obese persons; tuber very toxic (CaOx), hence needs careful prepn.; raw tubers u. for cancer, rodent ulcer (Chin); insecticide; inflor. (stalks) (black & smelly) u. antipyr. in ophthalmia; ornam. - **A. sylvaticus** Kunth (Ind): cajra-kanda (Sanscr); sds. u. for toothache & glandular enlargements - **A. titanum** Becc. (seAs): prod. the largest inflorescence of any herbaceous pl. known (2.4 m long); tubers (-23 kg) eaten by natives (care!).

Amoxicillin: a semi-synth. penicillin (broad spectrum) antibiotic, similar to ampicillin; (1971-); bactericidal; prod. higher blood levels than ampicillin u. to treat GC.

Ampelocissus acapulcensis Planch. (Vit.) (Mex, Salv): sold as "uva (silvestra)" ([wild] grapes) in markets around Yautepec (Morelos); u. to make a kind of confection (978) - **A. arachnoidea** Planch. (seAs): u. med. in Kamp & Vietnam - **A. araneosa** Planch. (Ind): kauraj (Hind); rt. cooling, astr. - **A. arnottiana** Planch. (Ind): rt. juice u. as aper. (with coconut kernel); rt. decoc. u. as diur. & blood purif. - **A. cinnamomea** Planch. (Mal): lf. decoc. & poultice u. in labor, on wounds; rt. poultice in orchitis - **A. grantii** Planch. (A. asarifolia Planch.) (Tanz): lvs. in form of hot poultices u. for sprains, cuts; cold poultice u. for boils; frt.

sweet & edible - **A. martini** Planch. (sVietnam, Kamp, Laos): day dac; large rts. u. as food (Kamp); frts. good for jelly - **A. ochracea** Merr. (Mal, PI): bk. & lvs. powd. appl. to swellings & boils; rhiz. med. - **A. polystachya** Planch. (Mal): boiled rts. u. as poult. to rheum. & lumbago.

Ampelodesma Beauv. (sphalm.): **Ampelodesmos** Link (Gram.) (monotypic) (Med reg.; natd. Calif [US]): represented by **A. mauritanicus** Dur. & Schinz (A. tenax Link) (nAf): dis(s) (grass); lvs. u. by natives to make ropes (Alg); paper; u. like esparto; young pl. fodder; source of ergot of diss.

ampelography: study of (a) the grape vine, or (b) vines in general.

Ampelopsis brevipedunculata Trautv. (A. heterophylla Sieb. & Zucc.) (Vit.) (Kor, Chin, Jap+): nobudo (Jap); rt. decoc. u. as wash for sore eyes; lvs. to treat wounds & sores caused by millipede bites (!) (Indochin) - **A. japonica** (Thunb.) Makino (Chin, Kor, Jap): rts. of climber u. as expect., bitter, cooling; to treat large boils, carbuncles, burns, as anodyne, anti-inflam.; cult. ornam. - **A. quinquefolia** Mx. = *Parthenocissus q.* (Virginia creeper).

Ampelozizyphus amazonicus Ducke (monotypic) (Rhamn.) (Br-Amazon+): saracuramira; palo de culebra (Ve); arom. bk. of this liane u. as soap substit. (Wai Wai Indians, Guyana); inf. u. for snakebite (Ve).

Ampfer (Ge): *Rumex* spp.

amphetamine: a synthetic amino-propane; known under many propr. names, such as Psychedrine; Benzedrine (sulfate); u. as CNS stim., anorexic, in narcotic poisoning; to aid individuals exposed to hazards, such as auto driving; airforce pilots (to keep awake & alert; for hyperactive children; closely allied with various natural products, ex. phenylethylamine ("*natural a.*"); non-addictive but sometimes subject to misuse when u. in large doses - **a. natural**: phenylethylamine.

Amphiachyris dracunculoides (DC) Nutt. (Gutierrezia d. [DC] Blake) (Comp.) (s&cUS): broom weed; August flowers; flg. tops as syr, u. in coughs & colds (Okla [US] as home remedy); c. VO with α- & ß-pinenes, limonene, cadinene, borneol.

Amphicarpa Ell. = **Amphicarpaea** Ell. (Leg.) (pantrop): hog peanut; some frts. develop underground like peanuts - **A. bracteata** (L) Fern. (A. monoica [L] Ell.) (e&cNA): ground peanut; hog pea (nut); frt. edible (Am Indians); formerly cult. (*cf. Arachis*) - **A. edgeworthii** Benth. var. *japonica* Oliv. (Jap): sds. of subterranean frts. u. as food (Ainu).

Amphicome emodi Royle (Bignoni.) (Ind): kaur (Kashmiri); c. bitter alk.; pl. u. as antipyr., substit. for Chirata. (—> *Incarvillea*)

amphigean: occurring in both hemispheres, as plants (ex. *Gossypium*).

Amphimas pterocarpoides Harms (Leg.) (trop wAf): bk. source of a red glutinous resin u. in Liberia in dys. (decoc.); wd. u. for walls of small houses.

amphion: opium.

amphiploid: organism with all its chromosomes from two distinct spp. added together; ex. *Crepis capillaris* (n = 3) X *C. setosa* (n = 4); hybrid with n = 7 chromosomes.

Amphipterygium adstringens (Schldl.) Standl. (Juliana a. Schlecht.) (Anacardi./Juliani.) (Mex): cuachalala; matixeran; shrub often confused with *Rajania subsarmanta* (*sic*); bk. astr. (c. tannins) decoc. u. to harden gums, to cure old wounds, gastric ulcers; bk. or wd. u. as antipyr. in mal., typhoid (inf. or decoc.) (47); in GI cancer; bk. u. in prepn. of red dye.

amphitropical: found in the tropics of both hemispheres, east & west.

Amphomycin: antibiotic fr. *Streptomyces canus* Heinemann & al. (NY [US] soils); u. topically in dermatoses (Gm. + microorgan.); antibacterial in pl. & animal dis. (1985); to improve food efficiency in ruminants.

amphoteric: subst. able to act either as a weak acid or as a weak base; ex. aluminum hydroxide.

Amphotericin B: Fungizone™: an antibiotic fr. *Streptomyces nodosu*; Trejo announced 1955-56; has been u. in cryptococcal meningitis & other systemic fungal infections (E).

Ampicillin: semi-synthetic penicillin; much u. & the drug most often prescribed under generic name (-1979-); antibacterial; reputed to cure GC in single doses (E).

amplosia: an unfermented grape juice.

Amrad gum: a form of Ind acacia.

Amritsar gum: an Ind acacia obt. fr. *Acacia modesta* (Pak).

Amsinckia Lehm. (Boragin.) (wNA, sSA): fiddlenecks; pls. often toxic to cattle because of high conc. of nitrates; at times, however, excellent forage for stock (205) - **A. douglasiana** DC (Calif [US]): fiddle-neck; Douglas' amsinckia; sapoquelite (Calif Indians); c. alks.; unspecified med. use (Costanoan Indians) (980) - **A. intermedia** Fisch. & Mey. (Wash, sCalif [US], nMex): (ranchers) fireweed; young growth edible; c. alks.; sds./nutlets cumulative poison.

Amsonia palmeri Gray (A. hirtella Standl.) (Apocyn) (NA): bluestar; cult. ornam.; c. 2-5% rubber -

A. tabernaemontana Walt. (US, nAs): c. indole alks.; acts as CNS stim.

Amura Schult. = *Amoora* Roxb.

Amygdala Amara & **A. Dulcis** (L): sds. of *Prunus dulcis* - **A. Amara** (L): bitter almonds fr. *Prunus dulcis* var. *amara* - **A. Dulcis** (L.): sweet almonds fr. *Prunus dulcis.*

Amygdalaceae: family sometimes segregated out of Rosaceae; now not generally recognized.

amygdalase: emulsin: enzyme mixt. which c. ß-glucosidase; splits amygdalin; found in many pl. prods. (ex. bitter almonds; lvs. of *Hydnocarpus kurzii*, &c.) - **amygdalin**: glucoside commonly found in Rose Fam. members; breaks down into HCN, benzaldehyde, & d-glucose; crude product called Laetrile (*qv.*).

amygdal(in)ic acid = mandelic acid.

Amygdalus L. = *Prunus* L. (generally) - **A. communis** = *Prunus amygdalus* P. Hill = *P. dulcis* (almond) - **A. leiocarpa** Boiss. = **A. persica** L. = *Prunus persica.*

amyl alcohol, fermentation: fusel oil.

amyl valerate: apple oil.

amylases: amylolytic (starch-splitting) enzymes occurring severally in saliva (**salivary a.**), pancreas, blood, muscle, bacteria, sweet potato, &c.; converts starch to glucose; u. in brewing, &c.

amylodextrin: the first recognizable hydrolytic prod. of starch; gives blue color with iodine; often called "soluble starch."

amyloid (substance): a.s. deposits in various tissues of (1) animal body, chief proteinaceous fibrils; occ. in tumors, arthritic joints, brain of aging person, &c. (amyloidosis); (2) plants, with converted starch accumulating - **beta a.:** occ. in brain in Alzheimer's disease.

Amylon (Ge): Amylum (L.) (*qv.*): starch.

amylopectin: soluble component of starch, composed of mostly branched chains of glucose; one fraction of starch (gel).

amylopsin: amylolytic enzyme; intestinal amylase; an "animal diastase" constituting part of "pancreatin," the mixt. of digestive enzymes of the pancreatic juice.

amyloses: gp. of carbohydrates made up of mainly unbranched chains of glucoses, including starch, (one component), dextrins, glycogen, celluose, &c.

Amylotheca insularum Danser (Loranth.) (Oceanica): kainikavea (Tonga); lvs. u. med. (Tonga) (86).

Amylum (L): starch; specif. USP former title for maize s.; fr. *Zea mays*; this is the most important starch u. in Am - **A. Ari** (L): starch fr. *Dracunculus vulgaris* - **A. Erythronii** (L) (Jap Ph.): starch fr. *Erythronium denscanis* - **A. Manihot**: starch fr. *Manihot esculenta* - **A. Oryzae**: rice starch - **A. Phaseoli** (L): starch fr. *Phaseolus vulgaris* (bean) - **A. Puerariae** (L): starch fr. *Pueraria lobata*, kudzu starch (Jap) - **A. Risi: A. Rizi**: rice starch, fr. *Oryza sativa* - **A. Sagittariae**: starch fr. *Maranta arundinacea*, St. Vicent arrowroot - **A. Secalis**: rye starch - **A. Solani**: (Irish) potato starch - **A. Tritici** (L): wheat starch - **A. Zeae**: maize starch.

amyrin, (α **and** ß): crystalline triterpenoid alcohols, $C_{30}H_{49}OH$; occurs in **Amyris elemifera** (hence the name) (American elemi) (979), also in Manila elemi, in the latices of many pl. spp. (ex. *Ficus variegatus*), &c.

Amyris balsamifera L. (Rut.) (WI, esp. Jam, Cu, SA (Ve, Co.), sFla [US]): balsam torchwood; t. amyris; bois des roses; Rhodeswood; bois chandelle (candlewood) (Haiti, Guadeloupe); wd. source of American elemi (arom. resin) & of West Indian sandalwood oil (rhodium oil) u. in perfumery, cosmetics, soaps; arom. rosewood or rhodium wd.; u. in incense; "balsam tree bark" - **A. elemifera** L. (A. plumieri DC) (sFla Keys, WI, CA): **sea amyris**; torchwood; palo de tea (Sp torch); candlewoods; source of American/West Indian/Yucatañ elemi (gum resin); u. in lacquers; u. med. (this elemi has been off. in 10 or more pharmacopeias) resinous wd. u. as fuel & for torches (night crawfishing); lf. decoc. u. as antipyr. in diarrh. - **A. madrensis** S. Wats. (Tex [US], nMex): Mexican amyris; barreta china; trunk u. med. - **A. simplicifolia** Karst. (nVe, NWI): candil (de playa); lf. decoc. drunk by women during menstr. periods - **A. sylvatica** Jacq. (Mex [Veracruz], CA, WI, nSA): "tea" (PR: means torch); wd. source of resin u. in tech.

Anabaena cylindrica Lemm. (Cyanophyta - Algae): aq. organism which synthesizes large amts. of vit. B_{12} (6.3 mcg/100 gm. dr. wt.).

anabasine: alk. fr. *Anabasis aphylla* & *Nicotiana glauca*; u. insecticide (hazardous!).

Anabasis aphylla L. (Chenopodi) (USSR-Russian steppes to Caspian): pl. somewhat like tumbleweed; c. anabasine (neonicotine), lupinine, aphylline, neonicotine; u. exter. in skin dis. (natives), & to obtain soda; insecticide - **A. articulata** Moq.-Tand. & DC (Af-Sahara reg.; Eg): pl. c. saponins & prosapogenins; sts. prod. a kind of manna; pl. food for camels.

anabolism: the synthesis or building up of tissues in the living organism; the opposite of catabolism, together with which it constitutes the metabolism; in the course of life, produces an increase in tissue mass.; anabolic hormones include androgens, somatotropin, estrogens (slight effect).

anacahuite wood: fr. *Cordia boisseri*.

Anacampseros albissima Malr. (Portulac.) (tropAf): rts. u. to make beer - **A. arachnoides** Sims. (sAf): u. in med.; said emet. - **A. rhodesica** Br. (sAf): u. to treat mal., blackwater fever, anthrax, dys.; in prepn. beer; pl. toxic - **A. ustulata** E. Mey. (sAf): u. to prep. a kind of yeast (farmers) for making pastry & beer (natives).

Anacamptis pyramidalis L. C. M. Rich. (Orchis p. L.) (monotypic) (Orchid.) (s&cEu, wAs, nAf): pyramid(al) orchid; tubers u. as source of salep, u. med.

Anacardiaceae: Sumac Fam.; trees, shrubs & vines, (850 spp.) mostly in trops of both hemispheres; infamous as one of chief pl. gps. causing dermatitis (ex. *Toxicodendron*, *Rhus*, cashew nuts).

Anacardium L. (Anacardi.) (tropAm): genus of 8 spp. of trees & shrubs, often cult. for the nuts - **A. occidentale** L. (neBr; now widely natd. SA, CA, Mex, WI; cult. As, Af): cashew (nut) tree; kidney-shaped frt. ("nut") (promotion nut, coffin nail) enclosing sd. with edible kernel, popular as cashew nuts, delicious after roasting but expensive; unroasted toxic (irritant principles destroyed with heat); frt. shell is source of caustic oil (FO) which c. anacardic acid (phenolic subst.), cardanol, cardol; this FO (crude oil = cacajou balsam) ("raw cardol") is u. as preserv. & waterproofer, to paint boats, nets, woodwork; epispastic, vesicant; distd. residue furnishes prod. u. in floor tiles, auto brake linings, insulation, varnishes &c.; the greatly enlarged red flesh frt. stalk or receptacle, a "false frt." ("cashew apple") (*qv.*), is edible, u. in beverages (cachou wine, *qv.*); trunk yields acajou gum (SA, WI), u. as acacia substit. in med. & bookbinding (repels insects); sds. vermif.; bk. astr. ton.; in diabetes, diarrh. dys., gargle in sore throat; lf. decoc. drunk for indig. & stomachache; syr. prepd. fr. fls. u. to allay cough; anacardic acid larvicidal (mosquitoes, snails) & may be useful against mal. & schistosmiasis; sd. kernel yields cashew nut oil (bland FO like sweet almond oil) (chief prodn. Ind, Tanz); kernel u. in coffee for asthma; rt. u. as purg. (sAf); ingredient of zombie poison (2494); wd. u. to make boxes, chests, cases, &c.; juice fr. trunk bk. u. as marking indelible ink (983) (1468) - **A. orientale** L. = *Semecarpus anacardium*, Oriental cashew nut (tree).

Anacharis Rich.: *Elodea*.

Anacolosa griffithii Mast. (Olac.) (Indoch): lvs. & bk. grated & mixed with tobacco, smoked for nasal dis. or nasal ulcers.

Anacyclus officinarum Hayne (Comp.) (Med reg): (German) pellitory; Bertram (Ge); rt. (Radix Pyrethri germanici) c. anacyclin & pellitorin (acid amides), acrid resin, VO (traces); home remedy for mouthwash & toothache (tinct.); stim. sialagog. u. in neuralgia, lumbago, diaph. in rheum.; in snuffs; cult. in cEu - **A. pyrethrum** Link (nAf, wAs): longwort; source of (Roman) pyrethrum (or pellitory) root; c. anacyclin & derivs.; u. stim. & sialag. in toothache, oral dis., mouth dryness (in gargles & mouthwashes); rubef.; in headache, rheum. neuralgia; stom.; digest. dist.; in snuffs; insecticide; hb. also u. med. - **A. radiatus** Loisel (Med reg): ox-eye; camomile de Valence; u. like prec.; off. Por Phar.; rt. vulner. aper.; source of yellow dye.

Anadenanthera Speg. (Leg. - Mimosoideae, close to *Piptadenia*) (Br): yopo; angico (981) (do not confuse with *Adenanthera*) - **A. colubrina** (Vell. Conc.) Brenan (A. macrocarpa Benth.; Piptadenia c. Benth.) (Br, cPe to nArg): bastard tamarind; a source of angico (branco) bk. & gum; bk. u. as expect. hemostat. vulner. in gout; in GC; gum u. like acacia, preservative; sds. strongly narcotic, hallucinogenic (bufotenin derivs.), ground up & u. as snuff, &c.; lvs. & bk. also u. sometimes (SA Indians) - var. *cebil* Altschul, u. as abort. - **A. peregrina** (L) Spegazzini (Piptadenia p. [L.] Benth.) (WI to sBr): c. methyl tryptamines, & derivs.; powd. roasted sds. furnish niopa (cohoba or yopo); u. like the other sp. of the gen. (112); bk. u. as sternutatory, in bronch. dis.; as tan; (angico [vermelho]; "tan bark").

Anagallis arvensis L. (A. coerulea Lam.) (Primul.) (Euras, cosmopol. thus natd. NA): scarlet (or common) pimpernel; waywort; poor man's weather glass (Eng) (SC-fls. close in bad weather); bird's eye; hb. c. toxic saponins (rt. c. cyclamine, primaverase); u. in home remedies as nervine, expect. stim. diaph. vulner.; inf. drunk for intest. worms (Ec) (2449); ornam. - **A. foemina** p. Miller (Eu): blue pimpernel; c. saponins, tannin; u. as diur.

anagyrine: alk. occurring in sds. of *Anagyris foetida*, *Ulex europaeus*, &c.

Anagyris foetida L. (Leg.) (Med reg.): Mediterranean stink bush; (stinking) bean trefoil; hb. c. cytisine, anagyrine, sparteine; u. emet. purg. (fresh flg. hb. u. in homeop.); sds. with similar const. u. as emet. lax. (care!) - **A. latifolia** Brouss. (Pak, Ind): source of strongly adhesive gum.

Anamirta cocculus (L) W. & A. (A. paniculata Colebr.) (monotypic) (Menisperm.) (Indo-Mal area): (Malay) fishberry; (Indian) cockles; Cocculus Indicus, Indian berries; poisonous frt. & sd. c. picrotoxin, FO; poisonous, u. as parasiticide, in scabies & other skin dis.; for mfr. picrotoxin; fish poison (but disallowed) (1469).

ananá (Sp): pineapple.

Ananas comosus (L) Merr. (A. sativus Schult. f.) (Bromeli.) (tropAm; widely cult. in trop): pineapple; piña; ananas (do mato) (Sp); frt. delicious food, antiscorb. refrigerant; juice c. vit. C, bromelin (bromelain; protease); citric acid, sucrose; u. in beverages prepn., vinegar; med. as vit. source, digestive in dyspepsia, gastric hyperactivity, anthelm. (fresh juice); disproven (2641); frt. canned (sliced, crushed, chunks, tidbits, spears) & frozen; juice u. to make wines & brandies; pressed peels & cores u. as fodder; lf. fiber u. in making nets, cords, hammocks (984) - **A. bracteatus** Schult. f. (A. sagenarius Schult. f.) (Br): bromélia; sds. & outer part of aggregate frt. (syncarp) u. as anthelm. (also expressed juice of lf. bases); frt. juice emmen. abort.; frt. ornam.

Anaphalis adnata Wall. (Comp.) (Chin): u. in ointment for skin dis. (Taiwan) - **A. margaritacea** Benth. & Hook. f. (neAs; natd. Eu, NA): (common) pearly everlasting; everlasting (flower); immortelle (blanche) (Fr, Cana) (Qu.); pl. u. in folk med. as antispasm. vulner. arom., to perfume alc.; in dys.; contusions, tumors; claimed anodyne (113); expect. - **A. neelgerryana** DC (Ind): Kat plaster (Nilgiris); lvs. appl. (bruised) to wounds & cuts.

anaphylaxis: condition of increased hyper-susceptibility to an antigen after a previous injection of same.

Anaptychia ciliaris Koerb. (Physci. - Lichenes) (Eu, NA): occ. on rocks & trees; source of Cyprus Powder (popular 17th cent.): scented with jasmine rose & other VO; u. as hair powder to scent & cleanse the hair; c. atranorin (antibiotic activity).

anasarca: generalized edema.

Anastatica hierochuntica L. (monotypic) (Cruc.) (Af, As deserts, Mor to Iran): rose of Jericho; u. in difficult labor (Ind); dr. pl. sold to tourists ("rose of Jericho").

anatabine: an alk. of tobacco, one of its most abundant minor alk.

Anatherum Beauv. = *Vetiveria* + - **A. muricatum** (Retz.) Pb. = *Vetiveria zizanioides*.

anatoxin: toxoid.

anatta: anatto: annatto. <*Bixa*

Anaxagorea luzonensis A. Gray (Annon.) (eAs, PI): bobonoyang (Cibu Bisaya, PI): fresh lvs. u. topically for arthritis.

anchico colorado (Sp): *Piptadenia rigida*; angico gum (1).

Anchietea salutaris A. St. Hil. (Viol.) (tropSA, esp Br): mercury anchietea (pirageia); cipo'suma; sipo carneiro; rt. bk. u. emet. purg.; in sore throat, eczema, scrofula; st. u. in hemorrhoids.

anchoas (Mex): ginger.

Anchomanes difformis Engl. (Ar.) (Sen, tropAf): rhiz. u. diur. oxytocic, galactagog, aphrod.; famine food; CaOx raphides may produce toxic reactions.

anchovy (plur. anchovies): *Engraulis encfasicholus* (Engraulidae - Pisces): this & some other small herring-like fishes u. to make anchovy sauce & other sauces; in pickling.

Anchusa azurea P. Mill. (A. italica Retz.) (Boragin.) (Med reg. to Iran; nat. nwUS): bugloss; sea b.; Italian alkanet; Italienische Ochsenzunge (Ge) (Ital ox-tongue); fls. u. as pectoral, diaph. diur. (Alg): u. in jaundice; ornam. - **A. officinalis** L. (Euras): anchusa; (common) bugloss; pl. (herb) c. cynoglossine; allantoin; u. in folk med. as expect. diaph. dem.; in diarrh.; in "melancholia"; young lvs. eaten as greens & salads (spinach-like pot herb); fls. (as Flores Buglossi) c. mucil. u. as expect. in colds; yellow dye (rouge); rt. & sds. u. like hb.; c. anchusin (alkannin) - **A. tinctoria** L. = *Alkanna t.* (L) Tausch.

Anchusae (off.) Radix (L.): rt. of *Anchusa officinalis.*

anchusin = alkannin.

Ancistrocarpus densispinosus Oliv. (Tili.) (Nigeria): juice u. to heal circumcision wounds.

Ancistrocladus ealaensis Leon. (Ancistroclad.) (Af-Congo): bk. c. alks. saponins, tannins; u. med. - **A. tectorius** Merr. (seAs): rt. boiled & u. in dys. & mal.

Ancistrodon = *Agkistrodon.*

Ancistrophyllum secundiflorum Wendl. (Palm.) (trop wAf): u. med. (Guinea); young shoots boiled & eaten; sts. u. to make fish traps, swagger canes, &c. —> *Laccosperma*)

ancoche (Arg): *Vallesia glabra.*

ancolie (Fr, It): *Aquilegia vulgaris.*

ancrod: defibrinating enzyme in venom of *Angkistrodon rhodostoma*; u. as anticoagulant in circulation.

Ancylanthos fulgidus Welw. (Rubi) (Angola): u. for chest (resp.) dis.; ornam. tree.

anda oil: fr. **a. seed**: fr. **anda-assú** (Br): *Anda brasiliensis = Joannesia princeps (qv.).*

Andira anthelmintica Benth. (Leg.) (Br): "cabbage tree"; bk. c. berberine & andirine; u. as anthelm. - **A. araroba** Aguiar (Vouacapoua a. Lyons) (Br): angeleen tree; Goa angelin t.; heart wd. cavities source of goa powder (crude chrysarobin); fr. which is obtained chrysarobin (araroba depurata) c. chrysarobin (per se); chrysophanol anthrone; emodin & its anthrone monomethyl ether; ararobinol; u. in skin dis., esp. psoriasis, pityriasis rosea; skin irrit. in arthritis, rheum.; u. for prodn.

of chrysarobin (=chrysophanol: chrysophanic acid); poisonous! (term "chrysarobin" applied to both mixt. and to pure compd.) - **A. galeottiana** Standl. (sMex): macayo; bk. effective vermif.; shade tree in coffee plantations - **A. inermis** (W. Wright) HBK ex. DC (A. excelsa HBK; A. jamaicensis [W. Wright] Urban; Vouacapoua americana Aubl.) (sFla (US), Mex, CA, WI, SA, wAf): cabbage tree; brown-heart; dog almond; huacapu (Pe); (cabbage) angelin; dr. bk. (Cortex Geoffroyae jamaicensis) & sds.; u. lax. vermif. antipyr. toxic (Mex Pharm.); c. andirine (toxic alk.) (best source); timber (partridge wood) useful for furniture, cabinet work; wd. ext. u. in conjunctivitis (Amazonia) - **A. retusa** HBK (Surinam - mt. woods): bk. = Surinam bark (Cortex Geoffroyae surinamensis) & sds. c. berberine & andirine (occ. as incrustation on inner bk.); u. anthelm. - **A. stipulacea** Benth. (Br): sds. vermif.; wd. u. for boat construction; c. berberine, andirine - **A. superba** (Arg): manduvirá; natives eat frt. after previous roasting - **A. vermifuga** Mart. (Br): Brazilian angelin tree; bk. u. as vermif. purg. emet.

andiroba: *Carapa guianensis.*

Andorn (Ge): horehound.

Andrachne aspera L. (Euphorbi) (Ind, Pak): Indian (or Pakistan) senega; u. as adulter. of official Senega (Polygala s.) rt. - **A. cordifolia** Muell. Arg. (Ind, Pak, Himalayas w to Murree): gurguli (Punjabi); pl. c. HCN, toxic to cattle; med. u. obscure.

Androctonis australis L. (Buthidae - Arthropoda) (nAf, Ind): thick-tailed scorpion; sting may be fatal (antiscorpion serum available in Eg, Br); u. med. like Veratrum (with great care).

Androcymbium gramineum McBride (A. punctatum Baker) (Lili.) (Eg-Mor): lofut (Arabic); corms (bulbs) sds. fls c. colchicine; atimitotic; u. as condim. by the Tuareg (Sahara) (!) - **A. longipes** Bak. (sAf): u. in ointment for sore eyes - **A. malanthioides** Bak. (sAf): u. as med. for many diseases (sSotho); toxic bulbs u. in ointment for sore eyes; bulbs, sds. & fls. c. colchicine.

androecium: male reproductive organ of flg. plants; gp. of stamens.

androgen: male sex hormone (generic term) having masculinizing (virilizing) & anabolic effects; sometimes useful in treating male homosexuality.

Andrographis echioides Nees (Acanth.) (tropInd): kalukariatum (Guj.); pl. juice u. in fever - **A. paniculata** Nees (Ind, SriL, Indones; cult. WI): kariyat; cr(y)eat; hb. u. domestic med., popular in Ind as bitter ton. stom. in dys. diarrh. diabetes; rts. fls. & frt. ton. alter. antipyr. anthelm.; for kidney stones; entire pl. except rts. boiled to produce bush tea (Jam) for colds.

Andromeda arborea L. (Eric.) = *Oxydendron arboreum* (sourwood) - **A. leschenaulti** Broughton (Ind-Nilgiri): lf. c. VO with mostly methyl salicylate; u. as antisep. - **A. polifolia** L. (nHemisp) (incl. *A. glaucophylla* DC) (nHemisp, esp. Euras, nNA): bog rosemary; wild r.; c. andromedotoxin (andromeda toxin); tannin; causes poisoning of grazing animals; honey fr. plant causes toxic poisoning in humans (ex. soldiers of ancient Greece in AsMin); lvs. to prep. arom. tea (toxin destroyed by boiling); lvs. & twigs in tanning & for black dye (USSR); lf. inf. u. in rheum., as substit. for *Ledum palustre*; as adulter. of true rosemary (985); cult. orn..

andromedotoxin: amaroid occurring in *Andromeda polifolia, Leucothoë axillaris*; *Pieris japonica*; *Chamaedaphne calyculata*; *Rhododendron barbatum* Wall. ex. G. Don, *R. cinnabarinum* Hook. f., *R. chrysanthum* Pall., *R. falconeri* Hook. f., *R. fulgens* Hook., *R. grande* Wight, *R. indicum* (L) Sweet, *R. maximum* L., *R. ponticum* L.

Andropogon L. (Gram.) (trop&subtrop worldwide): beard/broom grasses; sand grass (SC pl. grows in dry soils); broom straw; infrutescence appears cottony (downy) at a distance - **A. bicornis** L. (Mex, SA, WI): penache; barba de indio (Cu); u. as thatch for roofs (tropAm); raceme hairs u. to stuff pillows - **A. caricosus** L. (tropAs, Mascarene Is; natd. WI): jiribilla (Cu); good forage, makes excellent hay - **A. citratus** DC = *Cymbopogon c.* - **A. gerardi** Vitman (NA): big bluestem; "red hay"; one of the most widely spread forage grasses; u. for debility (Omaha Indians) - **A. ischaemum** (*Ischaemon*) L. = *Bothriochloa i.* (L) Keng - **A. laniger** Desf. = **A. martini** Wats. = *Cymbopogon schoenanthus* - **A. muricatus** Retz. = *Vetiveria zizanioides* - **A. nardus** L. = *Cymbopogon n.* - **A. scoparius** Mx. = *Schizachyrium scoparium* Nash - **A. squarrosus** L. = *Vetiveria zizanioides* - **A. virginicus** L. (seUS): broom sedge; b. grass; b. straw (SC u. to make yard brooms, seUS); u. similarly to *Schizachyrium scoparium.*

Androsace maxima L. (Primul.) (seEu, wAs, nAf): groesster Mannsschild (Ge); hb. formerly u. as diur. in GC leucorrhea; cult. orn.

androstenol: occ. in testes of boars, axillary sweat of human males; truffles (*Tuber melanosporum*); in boar pheromone (with musk-like aroma); u. to detect estrus in sow in artificial insemination procedures; may be enabling factor in pigs detecting truffles.

androsterone: male hormone found naturally occurring in male urine.

Aneilema beniniense Kunth (Commelin.) (wAF, Congo, Ghana+): pl. decoc. u. as lax. for children - **A. conspicuum** Kunth = *Dictyospermum conspicuum* - **A. lanceolatum** Benth. (wAf): eye dis. treated by crushing sd. & placing in eye (mucil.); pl. juice u. to cure water sores in feet (Ghana) - **A. lineolatum** Kunth = *Murdannia japonica* - **A. nudiflorum** Sweet = *Murdannia nudiflora* - **A. scapiflorum** Kostelosky = *Murdannia edulis*.

Anemarrhena asphodeloides Bunge (monotypic) (Lili) (nChin): u. antipyr. in influenza, pneumonia, morning sickness, measles; cult. orn.

Anemone altaica Fisch. (Ranuncul.) (Chin): appl. to rashes on skin; int. to revive a person; appetizer - **A. caffra** E. & Z. ex Harv. (sAf): rt. inf. u. as emet., as enema for biliousness (Zulu); pd. rt. as snuff for headache, for colds; for mental dis. - **A. canadensis** L. (NA): windflower; rt. u. as tea to wash sore eyes & as throat lozenges for clearing throat; appl. to wounds; highly esteemed by Omaha & Ponco Indians - **A. cernua** Thunb. = *Pulsatilla c.* (Thunb.) Spreng. - **A. chinensis** Bunge (Chin): hb. u. for amebic dys., persistent mal. - **A. coronaria** L. (Med reg): crown anemone; "lily of the fields" (982); with red corolla; const. & uses like those of *A. nemorosa* - **A. cylindrica** Gray (NA): candle (or Indian) anemone; dr. lvs. u. in "Indian snuff" for head cold, also smoked - **A. dichotoma** L. (Siberia, Chin): u. as sed.; to quench thirst; usu. in admixt.; u. med. in Indoch (imported from Chin) - **A. hepatica** L. = *Hepatica nobilis* Mill. var. *obtusa* - **A. hupehensis** Lem. ex. Boynton (Chin): u. as substit. for *Centaurea monanthos* Georg (Shensi) var. *japonica* Bowles & Stern (A. J. S. & Z. (Chin; natd. Jap): appl. to swellings, itch, &c.; int. ton. for hemorrhage & int. injuries - **A. ludoviciana** Nutt. = *Pulsatilla patens* subsp. *multifida* - **A. nemorosa** L. (Eu, Siber, NA): (European) wood anemone; w. anemony; (red) windflower; hb. c. protoanemonin, saponoside; fresh hb. u. as hyperemic, vesicant; in bronchitis (children); arthritis, pleuritis; ornam. perenn. hb. - **A. obtusiloba** D. Don (Pak, nInd, Nepal): rattanjog (Punjabi); rts. u. int. for contusions, exter. as vesicant; sds. emet. purg. (Lahul) - **A. patens** L. = *Pulsatilla p.* Mill. - **A. pratensis** L. (Euras) = *Pulsatilla p.* Mill. - **A. pulsatilla** L. = *Pulsatilla vulgaris* Mill. - **A. raddeana** Regel (Chin): appl. to swellings & skin rashes - **A. ranunculoides** L. (s&cEu): yellow wood anemone; hb. u. like *A. nemorosa* - **A. rivularis** Buch.-Ham. ex DC. (Ind, Chin): rhiz. purg.; stock poison - **A. sylvestris** L. (Euras): snowdrop (anemone); c. & uses as for *A. nemorosa* - **A. virginiana** L. (e&cUS): thimbleweed; Virginia anemone; wind bloom; rt. or hb. u. as poultice for boils (Menominees); smoke fr. heated sds. inhaled in catarrh. or unconsciousness (Meshwakis); rts. u. as tea for lung congestion; to relieve dental pain (Am Indians).

anemone, European wood: grove a.: *Anemone nemorosa* - **long-fruited a.**: *A. cylindrica* - **prairie a.**: *Pulsatilla patens* - **rue a.**: *Thalictrum thalictroides* - **Virginian a.**: *A. virginiana* - **wood a.**: *A. quinquefolia* L.

Anemonella thalictroides (L) Spach = *Thalictrum thalictroides* Eames & Boivin.

Anemonia sulcata Penn. (Anthozoa - Coelenterata - Animalia) (Med, Atl Oc): (Actinaria); (green) sea anemone; body c. neurotoxins (inhibit proteinases); in It & Fr eaten baked or raw (*cf. A. edulis* [found on markets of Venice & Trieste]).

anemonin: anemone camphor; pulsatilla c.; a lactone; occ. in *Pulsatilla vulgaris* & other *P.* spp., *Caltha, Clematis,* & other genera of Ranunculaceae (precursor: protoanemonin).

Anemopaegma mirandum Mart. (A. arvense [Vell.] Stellfeld var. *petiolata* Bur.) (Bignoni.) (Br): catuaba; catuiba; tupi; rhiz. of liana u. as CNS stim. ton. aphrod. nervine (also lvs. st. bk. & rt.); causes pupillary dilatation; u. in galenicals.

anemophilous: pollen transmitted by wind currents; the state is **anemophily**.

Anemopsis californica (Nutt.) Hook. & Arn. (monotypic) (Saurur.) (swUS & adj. Mex) - (wet, alkaline soils); yerba (del) mansa; Apache beads; low pl. with dock-like lvs. & conical spikes; entire pl. considered infallible remedy by SpAm & Mex folk for many dis.; arom., pungent, astr.; c. allyl veratrol, VO with thymol; rt. ("black sarsaparilla") u. for wounds & abscesses (knee); rt. decoc. (or chewed) u. in (nasal) catarrh, in diarrh. dys.; for colds, indigest., colic; lf. inf. to "purify the blood" (Mex); as tea & poultice in rheum. swellings, abrasions, burns, cuts; astr. in hemorrhoids (ointment), female dis. (NM [US]); in ointments with lard for sores; herb also used, gargle for inflamed throat; antiluetic; cult. orn.

anergy: reduced sensitivity to specific antigens; also increased resistance to disease or other harmful agents.

anet(h) (Fr): **aneto** (It): dill frt. - **anet doux** (Fr): fennel.

Anethi Fructus (L.): dill frt. ("seed").

anethol(e): anise camphor; methyl oxy-prophenylbenzene; u. in perfume ind.; represents active princ. of anise oil, from which obtained, also by synth.; flavor for elixirs, &c.; carminative.

Anethum graveolens L. (Peucedanum g. [L] Hiern.) (Umb.) (wAs, Med reg.; cult. in ancient Greece, Rome, Palestine; natd. eUs): dill; dilling (or soothing) herb; ripe frt. ("seed"; **Anethum** L.) & its VO u. as flavor, arom.; hb. ("dill weed") & frt. u. as flavor in dill pickles, fish soups, salads, sausages, pickling spices; carm. stim.; corrective; VO distd. fr. whole pl. when frt. ripe (dill oil); cult. USSR, Ge, Neth, Ind (Madhya Pradesh nr. Indore), US - **A. sowa** Roxb. ex Fleming (tropAs, esp. Ind): (East) Indian dill; frt. u. as arom. carm. (like prec.); frt. of Indian dill longer; "Indian womum"; u. similarly; as adulter. for caraway; lvs. coated with castor oil appl. to boils.

anetseed: dill (sometimes confused with anise).

aneuploidy: dysploidy: state of individuals which differ in having a single chromosome number not a multiple of the basic haploid number, a deviation from euploid numbers; unlike polyploids which double the number (ex. *Athyrium crenatum* with n = 41, while other *A.* spp. have the basic no. x = 40.

aneurin(e) (Eu): thiamine; vit. B1.

angel eyes (Tex [US]): *Hedyotis caerulea.*

angelic acid: an unsaturated fatty acid, C4H7COOH; isomer of tiglic acid; occ. in *Angelica* spp.

Angelica acutiloba Kitagawa (Umb.) (Chin, Jap): toki (Jap); rt. (often in wine) u. as ton. circul. stim., emmen. galact. (Jap); diur. (Indones to which imp.); flavor correctant; cult. for med. (Jap) - **A. anomala** Avé-Lall. (nChin, Kor, eSiber, Manchur, Jap): yoroi-gusa (Jap); triennial hb.; rt. u. for headache, toothache, boils, GC; skin dis. (Chin); diphtheria (Mongol); cult.; often compared with *A. dahurica*; tang kuei; rt. u. carm. in dysmenorrh. - **A. archangelica** L. (Archangelica officinalis Hoffm.) (nEuras to Siber; swampy places; natd. in GB): European (or garden) angelica; rt. hb. & frt. u. flavor (in wines, esp. Benedictine, Chartreuse liqueurs, cordials) also VO so u.; carm. diaph. diur. ton. stom.; candied angelica stalks u. in pastries, cakes; hb. u. as ton.; pl. (lvs.) u. as veget. (n areas) (cult. & prod. Belg; VO prod. Neth, Ger); candied peduncles u. as candies for cake decoration; peeled sts. covered with sugar u. to strengthen stomach; highly delicious, sweet tasting - **A. atropurpurea** L. (eCana, neUS): American (or great) angelica; Alexanders; masterwort; rhiz. & frt. u. as stim. arom.; u. similarly to *A. archangelica* & u. as substit. for same (but considered inferior & less arom.) (more compact & whiter) - **A. breweri** Gray (wNA: Calif, Nev): Brewer's angelica; rt. u. in tea for colds & chest dis.; frt. stim. carm. - **A. dahurica** (Fisch.) Benth. & Hook. f. (Umb.) (Kor, Manchuria, eSiber, Jap): pai-chi; rts. diaph. for headache, common colds, rheum., anodyne in neuralgia, carm., styptic for epistaxis; female dis., skin dis., esp. itching - **A. decursiva** Franch. & Sav. (neAs; imported into Indochin): nodake (Jap); rt. anodyne, antipyr. expect. in bronchitis, cough (leaf eaten after boiling); emmen. - **A. genuflexa** Nutt. (A. refracta F. Schmidt) (Jap, eSiber, wNA): kneeling angelica; dr. rhiz. & rt. u. stim., flavor in food (Ainu) - **A. gigas** Nak. (Chin, Kor): danggui; u. similarly to *A. acutiloba;* sed. analg., antispasm in arrhythoria - **A. glauca** Edgew. (Ind): chora; rt. u. stim. in constip. vomiting; as spice - **A. graveolens** var. *dulce* (Miller) Pers.: celery; petiole blanched popular raw or cooked; known in ancient Eg; frt. ("seed") u. as flavor; in celery salt - var. *rupaceum* (Miller) Gerundst?: celeriac; edible tuberous rts. - **A. grosseserrata** Maxim. (Manchur, Kor, Chin, Jap): u. empirically in swollen feet, apoplexy, leprosy, edema, headache - **A. hendersonii** Coult. & Rose (Pac US): u. in colics, colds, &c. (NA Indians) - **A. japonica** A. Gray (Jap, Ryukus): cult. for VO ("Japan angelica oil"); similar to that of *A. archangelica* but c. phellandrene - **A. lucida** L. (eNA): u. similarly to *A. archangelica* - **A. megaphylla** Diels (Szechwan, Chin): u. for pain in joints & back, headache, toothache - **A. miqueliana** Maxim. (A. koreana Maxim.) (Kor): rt. u. for common colds, rheum. lumbago (Kor, Chin)- **A. nemorosa** Ten. (It): rts. & frt. u. as folk med. - **A. polymorpha** Maxim (eAs): rts. u. in rheum. neuralgia (in alc. as rub), to contract uterus (Eumenol); diur.; as hypertensive - **A. pubescens** Maxim. (A. polyclada Franch.) (eAs): rt. decoc. u. as emmen. to treat hematuria; exter. in hemorrhoids & abscesses (Chin, Jap): rts. as ton. pectoral (Indoch) - **A. roseafolia** Hook. (NZ): rose-leaved anise; lvs. u. diur. in syph. - **A. sinensis** (Oliv.) Diels (A. anomala Avé-Lall. var. *chinensis*) (Chin, Jap): tang-ku(e)i; man-mu; Chinese eumenol angelica; rt. source of Eumenol (fl. ext.); u. emmen. in menstrual & puerperal dis. ("female dis."); skin dis.; in colds & fever, hemorrhage - **A. sylvestris** L. (Euras; natd. NA): wild (or woodland) angelica; gout weed; pd. frts. u. to kill head lice; rt. as carm. expect. in gastric cancer; young sts. & lvs. boiled & eaten as veget. - **A. tenuissima** Nak. (Chin, Kor): carm. & in colds - **A. triquinata** Mx. (neUS): similar to *A. archangelica* but with different aroma - **A. uchiyamai** Nak. (Chin): rt. u. for female dis. - **A. venenosa** Fern. (eUS): hairy angelica; perenn. hb.; lvs. u. to discourage tobacco habit - **A. villosa** BSP = *A. venenosa.*

"Angelica" (Ala [US]): "a. tree"; *Aralia spinosa* - **American a.**: *Angelica atropurpurea, A. lucida* - **European a. root**: *A. archangelica* - **filmy a.**: *A. triquinata* - **hairy a.**: *A. venenosa* - **Japanese a.**: *A. anomala* - **a. lactones**: arom. compds. occurring in Angelica, catnip VO, & also made by synth.; flavor, cardiotonic, & hypotensive properties - **northern a.**: comml. article, *A. atropurpurea* (?) - **purple-stemmed a.**: *Angelica atropurpurea* - **a. root**: *Ligusticum filicinum* - **a. seed**: frt. of *A. archangelica* - **Southern a. root**: 1) lovage 2) *Angelica* spp. - **a. tree**: *Aralia spinosa* (bk. & rt.) - **wild a.**: *A. sylvestris.*

Angelicae Fructus (L.): angelica frt. (seed) - **A. Radix**: a. root.

angelico: 1) *Angelica villosa* 2) *Ligusticum canadense* 3) *Levisticum* spp.

Angelika (Ge) (also in compounds as *A. samen, A. wurzel*): *Angelica archangelica.*

angelim (Br): **angelin:** trees of certain genra of fam. Leg. incl. *Andira, Copaifera* (at times), *Hymenolobium, Pithecolobium, Vatairea*, &c.

angeline: andirine.

angélique (Fr-f.): 1) *Angelica archangelica* 2) bonbons made using fresh green sts. of this sp., these being candied & u. to decorate cakes & candies (chiefly Fr) - **a. officinale**: *Ang. archangelica.*

angelon: *Angelonia angustifolia* Benth.

Angelonia angustifolia Benth. (Scrophulari.) (Mex, WI, Guiana): Fernandina (Cu); angelón; fls. u. expect. ton. & in nerve dis. (Yuc) - **A. salicariifolia** (*salicariaefolia*) H. & B. (WI, nSA): Fernandina (Cu); u. diaph. (Ve).

angel's trumpet: *Datura suaveolens* & other *D.* spp.; Brug - *mansia* spp.

angico (Por): 1) *Anadenanthera peregrina* (bk. gum) 2) Acacia & *Pithecolobium* spp. 3) *Stryphnodendron barbatimao* (bk.) - **a. gum: goma de angico** (Br): 1) brown-colored gum fr. angico (*Anadenanthera rigida*) u. as acacia replacement; c. tannin; hence also u. as tan. 2) gum fr. *Acacia angico* (Para gum) - **a. wood**: fr. *Acacia angico.*

Angiopteris ervecta Hoffm. (l) (Angiopterid. - Pteridophyt.) (Australasia): turnip fern; pith c. starch, food of aborigines (Queensland, Austral); pith sap u. to make intoxicating beverage; int. in mal. (PapNG); fronds with VO u. to perfume coconut oil (South Sea Is).

Angiospermae: Magnoliophyta; the flowering plants; one of the two subdivisions of Phylum Spermatophyta; distinguished fr. Gymnosperms in carrying their ovules (later seeds) in a closed ovary which at maturity is the frt.; this subdivision incl. the great majority of plant drugs (ca. 241,000 spp. in subdiv.).

angiotensin: angiotonin (formerly): hypertensin (formerly): peptide acting as hypertensive; released fr. blood plasma globulin (angiotensinogenin) by action of renin (of kidney); **a. II** stim. aldosterone secretion; amide u. as vasoconstrictor.

angle-pods: *Vincetoxicum* spp.

Angola copal: A. gum: a fossil resin (red & white vars.) fr. *Copaifera demeusei*, etc.

Angolam Adans. = *Alangium* Lamb.

Angophora lanceolata Cav. (Myrt.) (Austral): roughbark apple: gum myrtle; source of Australian Kino; gen. close to *Eucalyptus.*

Angostura trifoliata (Willd.) Elias (Cusparia t. [Willd.] Engl.; A. cuspare Roem. & Schult.) (Rut) (tropSA, esp. Ve, Co, WI): angustura tree; source of true angostura bark (cusparia b.) (Cortex Angosturae); bk. c. several alks (3%) (cusparine, galipine, galipoline, quinoline, cuspareine); VO (c. galipol, galipene, cadinene, &c. [989]); u. bitter ton., antipyr.; Cinchona substit., in dyspep. (low stomach acidity), diarrh. dys.; "paralyses"; prepn. VO (u. in homeopath. med.); bk. formerly in prepn. of Angustura Bitters u. in mixed drinks (pink gins), confectionary, &c.

Angostura Aromatic Bitters™: popular bitters & flavor for mixed drinks, foods, &c.; bitter principle said to be now gentian with other herbs; orig. in Venezuela but now made in Trinidad (2523).

Angraecum fragrans Thouars (Jumellea f. Schltr.) (Orchid.) (Réunion, Mauritius Is, Mad): bourbon tea (orchid) lvs. = (Folia) Faham; orchid tea; c. coumarin, coumaric acid; u. to prepare Bourbon tea (*qv.*) (Mascarene Is); u. as expect. in TB, stom. (indig. med.); substit. for Chinese tea; filler in cigars; flavor for ice cream &c..

Ångstrom unit (.Åu.): 0.0001 micron: 0.1 m μ

Anguilla vulgaris Turt. (Anguillidae - Pisces - Chordata) (Atlantic Oc): eel; serum u. in endocarditis, nephritis, oliguria; flesh delicacy (fresh, marinated, roasted).

Anguria spinulosa Poepp. & Endl. (Gurania s. Cogn.) (Cucurbit.) (Trin to Pe, Bo, Br): lf. decoc. u. for constipn. (Trin) - **A. umbrosa** HBK (Ve): rhiz. thought toxic (—> *Citrullus).*

Angustmycin: antibiotic: v. Psychofuranine.

angustura (bark): Cusparia; fr. 1) *Angostura febrifuga* or 2) *Galipea officinalis* Hancock (2523).

anhalamine: anhalonidine: anhalonine: alkaloids of mescal buttons, *Lophophora williamsii* (previously called *Anhalonium lewinii*, hence the names).

Anhalonium Lem. = *Lophophora*; *Ariocarpus*; *Echinocactus.*

anhydrohydroxy-progesterone: semi-synthetic agent with props. of progesterone; u. in corpus luteum-deficiency states.

anhydrous: with water absent (incl. w. of crystallization); ex. a. lanolin, a. d-mannitol.

Aniba canelilla Mez (Laur.) (trop Am, Ve-Br): "(casca) pretiosa"; bk. with cinnamon odor, u. as stim. tea (Peru), to perfume linen; VO c. eugenol, methyl e.; wd. ("sandalo brasileira") source of arom. VO, c. nitrophenyl ethane - **A. cedrata** Aublet (SA): achariari; source of karama resin - **A. coto** (Rusby) Kosterman = *Nectandra c.* - **A. duckei** Kosterm. (A. rosaeodora Ducke) (Amaz, Br. Pe): (Brazillian) rosewood; bois de rose; pau-rosa (Br); source of VO, oil of Brazillian rosewood (there are several other rosewood oils); u. perfume - **A. megaphylla** Mez (tropAm): rt. c. cytotoxic neolignans - **A. pseudocoto** Kosterm. (tropAm): apparently the source of false coto bark (Cortex Paracoto); u. like coto bk. in diarrh. dys. neuralg. anti-perspirant - **A. terminalis** Ducke (Amazon Basin, Guian): one source of rosewood oil.

anice (It): anise.

añi(e)l (NM [US], Sp): sunflower.

añil: 1) (Sp): indigo 2) (CR, PR, Cu): *Indigofera tinctoria, I. suffruticosa*; also **a. colorado** (Mex).

aniline: C_6H_7N: occ. in coal tar & (in traces) in many pl. products; u. in mfr. numerous synthetic medicinals, dyes, perfumes, &c.

animal drugs: an increasingly important category of the materia medica since endocrines & amino-acid therapy have become so important in the practice of medicine - **a. excreta**: feces & urine, once long ago u. in regular medicine (Dreck-apotheke), still occasionally u. in home or rural folk medicine - **a. oil**: bone distillate or bone oil (*qv.*) - **a. protein factor**: APF; combination of amino acids with several vitamins; prep. by fermentation; u. like fish meal for single-stomach animals, such as hogs, which cannnot synthesize certain of their needed amino acids; also in poultry feeds; acts like Vit. B_{12} - **a. starch**: glycogen.

animé: animi: hardened fossilized copal resin fr. *Hymenaea courbaril* (eAf coast).

animic gum: gum animi: Zanzibar g.; copal resin found both as fossil & recent product; fr. *Protium icicariba*, &c. (WI elemi).

anis: 1) (Fr): anise 2) (Qu): caraway 3) (Por): fennel - **a. verde** (Sp) = **a. vert** (Fr) = aniseed.

Anisacanthus wrightii (Torr.) Gray (Acanth.) (s&wTex [US], Mex): Wright's anisacanth; hb. inf. u. in colic (Mex Indians).

anisaldehyde: anise aldehyde: anisic aldehyde; active principle of anise VO; also occ. in many fungi; u. in perfumes.

anise: 1) *Pimpinella anisum*; yields **a. fruit** ("a. seed"); one source of anise oil 2) *Myrrhis odorata* (occas. applied) - **Alicante a.**: Spanish a. - **Chinese a.**: *Illicium verum* - **common a.**: *Pimpinella anisum* - **Florida a.**: *Illicium floridanum* - **a. flower oil**: VO distd. fr. immature frts. of Tonkin star anise - **a. oil**: VO obt. fr. either *Illicium verum* (Chin) or *Pimpinella anisum* - **purple a.:** Florida a. - **a. root**: 1) *Pimpinella anisum* 2) *Osmorhiza longistylis* - **Russian a.: Spanish a.** (resp.): aniseed fr. Russ & Sp; considered best quality - **star a. (Chinese)**: *Illicium verum* (**Japanese s.a.**): *Illiciam anisatum* - **sweet a.**: 1) fennel; 2) *Osmorhiza longistylis* - **a. tree**: *Illicium* spp. - **wild a.**: (Calif [US]): Carum Kelloggii.

aniseed: telescoped term for "anise seed" (term commonly u. in commerce).

Anisette: liqueur prepd. with anise as chief flavor.

Anisi Stellati Fructus (L.): star anise (frt.) - **Anissame** (Ge): aniseed.

anisillo (CR): *Piper auritum, P. sanctum, Peperomia rotundifolia.*

Anisochilus carnosus Wall. (Lab.) (Ind): karpuravalli (Telugu): hb. c. VO with thymol; u. stim. expect. in cough in infants (mixed with woman's milk and sugar); lf. juice cooling, with sugar for coughs & colds; as liniment (with sesame oil); in synanche (severe sore throat with choking, as in diphtheria).

anisodamine: alk. found in *Scopolia tangutica* Maxim. & extensively u. in Chin for disorders of the microcirculation, as in diabetes, pulmonary heart, toxic shock, glomerular nephritis; it inhibits synthesis of thromboxane B_2; has reduced mortality from fulminating meningitis (Chin); also useful in emphysema, hypertension, coronary heart dis., viral hepatitis, and enteritis necroticans (2297).

Anisodus Link & Otto = *Scopolia.*

anisole: methoxybenzene; u. in perfumes.

Anisomeles indica Ktze. (A. malabarica R. Br.) (Lab.) (Ind, Mal+): Malabar catmint; karpuravalli (Telugu); rich in VO; lvs. c. alk.; lf. inf. u. in colic (carm.). stom. in dyspeps. (PI), ton.; VO for uter. dis. rheum. (exter.), perfume, &c.

Anisomycin: antibiotic isolated fr. *Streptomyces griseolus* Waksman; u. in *Trichomonas* infections (*T. vaginalis*) (in vag. tabs.); amebiasis; isolated c. 1954; less toxic than fumagillin.

Anisophyllea disticha (Jack) Baill. (Anisophylle./Rhizophor.) (Mal, Indones): keribor; lf. decoc. u. in diarrh.; wd. u. for arrows, &c. - **A. laurina** R. Br. ex. Sabine (tropwAf): "monkey apple" (Sierra Leone); frt. u. as food (Senegal), in preserves; sold in markets; pounded kernel of sd. emet.; young lvs. in oil appl. to circumcision & other wounds; twigs u. as tooth-sticks; wd. useful for corner posts, &c. (said termite-proof).

Anisosperma passiflora Silva Manso (monotypic) (Cucurbit.) (Br): sds. with FO u. as drastic, emet.; VO u. as anthelm.

Anisostichus capreolatus Bureau = *Bignonia capreolata.*

Anisotes formosissimus (Klotzsch) Milne-Redhead (Acanth.) (Moz, Malawi): u. to treat leprosy; c. alks. rt. decoc. u. in porridge (purg. action); rts. are incinerated & ash is appl. to lesions (2445).

anistreplase: anisolated plasminogen streptokinase activator complex (APSAC); complex of streptokinase & plasminogen; u. as thrombolytic in cardiac infarction; coronary thrombosis.

Anisum (L.): aniseed (anise frt.) - **A. Stellatum**: star anise, *Illicium..*

ankat bark: Ankat Cortex (L.): Chinese shavings.

ankora (Hind) = **ankota** (Ind): *Alangium lamarckii* Thunb.

annatto: a coloring matter consisting of dr. pulpy aril surrounding sds. of *Bixa orellana*; u. to dye cloth, margarine, cheese, &c. - **a. family**: Bixaceae.

Annelida: phylum of segmented invertebrate animals typified by leech & earthworm.

Anneslea fragrans Wall. (The.) (trop eAs): bk. u. in combns. for dys. & as vermif.; fls. in complex prepns. for fever (Indoch); wd. valuable, u. for "fancy" articles.

Annona (Anona) arenaria Thonn. (Annon.) (Congo): u. for stomachache & diarrh.; hemostat. & cicatrizant (2457) - **A. cherimola** Mill. (tropAm, incl. Pe, Andes): custard apple; cherimoyer; cherimoya; frt. delicious, considered with pineapple & mangosteen the most attractive trop frt.; u. extensively to flavor ice cream, &c. (Pe); sd. FO (c. fat) u. for inunctions, liniments; pd. sds. insecticidal (Mayas) (6); lf. decoc. u. as lotion for burns (Yuc); sds. in ointment for body lice & other parasites; small tree widely cult. in trop. - **A. chrysophylla** Boj. (s&cAf): c. tannin, resin; u. in dys. diarrh.; rt. decoc. for swellings, boils; frt. eaten for snakebite (Tanz); widely cult. - **A. diversifolia** Safford (sMex, CA; introd. into sFla [US]): (h)ilama; frt. delicious, eaten raw - **A. glabra** L. (A. palustris L.) (sFla, WI, sMex, CA to Br & Ec, trop wAf): pond apple; water a.; cow a. (Jam); alligator a.; ripe frt. edible but of poor quality (better fully ripe), somewhat insipid, u. for prepg. jellies; decoc. of lvs. of small tree u. in dry cough (Nigeria), vermif.; bast fibers & FO utilized; cult. sFla, wAf; common in swamps; rt. bark u. as cork substit. (Jam) for vinegar, &c., but not for rum bottles - **A. glauca** Schum. & Thonn. (Senegal); dugur mer; rts. u. diur. in gonorrhea - **A. globiflora** Schld. (Mex): chirimoya; frt. u. in diarrh. - **A. jahnii** Saff. (Ve): manirito; bk. u. antidiarrh., in edema, snakebite - **A. muricata** L. (WI, sMex to Br): soursop; guanába(na); small evergreen tree; lf. decoc. popular "bush tea" (beverage) of Latin Am; to induce sleep; rt. alexiteric; sds. astr.; lvs. rt. sd. u. as anthelm. & insecticide; lvs. (inf.) sed. hypnotic to reduce blood pressure; crushed lvs. astr. poultice on wounds & sores; frt. edible, in flavors, beverages, jellies, &c.; antiscorb.; cult. Fla - **A. obtusifolia** DC (Br-Bahía): frts. ("fruta de Conde") popular food; sd. FO bland & odorless - **A. paludosa** Aubl. (Guyan): lvs. u. in folk med. - **A. purpurea** Moç. & Sessé (sMex, CA, Co, Ve, Ec): toreta; lf. decoc. drunk for skin eruptions; frt. juice for chills & fever (Mex); jaundice (CA); inner bk. decoc. u. in edema & dys. - **A. reticulata** L. (Neth. WI, sMex to Br & Pe): (common) custard apple (PR, NWI); bullock's heart; sugar (or bastard) apple; frt. edible but lacks flavor & inferior; ripe frt. pulp plastered on boils, abscesses, ulcers (home remedy); decoc. of lvs. green frt. & bk. u. in dys. vermif.; pd. sds. appl. locally on head to kill lice; lvs. & branches u. in tanning & dyeing; bk. with strong fiber; cult. Fla, OW trop - **A. scleroderma** Safford (sMex to Guat): posh-ta; frt. delicious, eaten raw or as preserves; cult. occasionally - **A. senegalensis** Pers. (trop Af, Seneg, Zambesia): (wild) custard apple; pomme cannelle du Sénégal; frt. edible (pineapple odor); bk. u. in diarrh. dys. Guinea worm; bk. chewed to incr. sexual potency (Zamb); lvs. & rts. u. antipyr. diur. antisep. in dermatoses, ulcers, rheum., &c.; lvs. boiled as culinary flavor; rt. decoc. for stomachache, "palpitations" (Tanz) - **A. squamosa** L. (WI [native], Fla Keys, sMex, Br.; widely cult. OW [sAf, Ind], NW) occurs throughout much of Ind; (scaly) custard (or alligator) apple; anon (SA); sweetsop; sugar/caneel apple; tree cult. in trop as frt. tree; frt. deliciously edible; u. to make "lemonades" (squashes); pd. sds. c. NBP, u. as insecticide, parasiticide (esp. lice) (Yuc) (6); vermicide; abort. sometimes with *Plumbago zeylanica* lvs. (124); sd. FO ("Sirikaya oil") u. to make soap; green frt. astr. in diarrh. dys.; pd. rt. drastic purg., insecticidal; lf. insecticide, fish poison, vermicide, cataplasms for tumors; bk. purg. furnishes inferior fibers; tan (WI).

An(n)onaceae: custard-apple family; shrubs, lianes, & trees; mostly of tropics (OW, NW); with many (ca. 2200) spp.; lvs. simple, alternate, with entire laminae; fls. perfect; many with edible, berry-like frt., clustered or aggregate; u. for perfumes; ornamentals; ex. *Asimina*.

annotta, (roll): **annotto**: annatto, *Bixa.*

Annulata: Annelida.

anny seed: *Pimpinella anisum.*

Anobium paniceum: *Sitodrepa panicea* (*qv.*).

Anoda cristata (L) Schlecht. (A. hastata Cav.; A. triangularis [Willd.] DC) (Malv.): (cUS, WI, Mex, CA s to Chile, Arg; Austral+): crested anoda; violeta del campo; violata; malvavisco; hala(n)che; lf. & fl. decoc. emoll. u. in resp. dis., incl. pertussis ("tosferina"); eaten for fever (Mex) (Mex Phar.); pl. u. as wash for VD (Guat, Hond); boiled fls. u. as antidiabetic; cult. orn. - **A. triangularis** DC (Mex): pl. eaten for fever.

Anodendron paniculatum DC (Apocyn.) (Ind): rt. with properties of ipecac - **A. tenuiflorum** Miq. (Indones): bk decoc. (old plants) u. to treat sores (Dyaks, Borneo).

Anodontocarya tripetala Diels (Menisperm.) (Co): bk. boiled with that of *Matisia cordata* H. et B. and frt. of *Capsicum annuum* & brew u. to expel intest. parasites (natives).

anodynin: hormone-like subst. in blood with long enduring pain relief.

Anoectochilus formosana Hay. (Orchid.) (Taiwan): whole pl. decoc. u. in chest. dis. & gastralgia; cult. orn. - **A. setaceus** Lindl. (SriL): u. med.

Anogeissus latifolia Wall. ex Guillemin & Perrottet (Combret.) (Ind): Bakli gum axlewood; prod. gum g(h)atti (Ind gum); u. pharmaceuticals, acacia substit., mostly in technology (ex. calico printing) - **A. leiocarpa** Guill. & Perr. (A. schimperi Hochst.) (tropAf, esp wAf): lvs. sometimes u. as vermif.; in rheum. stim. aphrod.

anomers: α and β stereoisomers configurated about anomeric carbon atoms (C_1) in cyclic sugars, such as glucose.

anomeric: variation of sugar in which the H and OH (or H and OMe) on the reducing C are reversed.

Anomospermum Miers (Menisperm.) (tropAm): u. in arrow poisons (curare); fl. decoc. for hair wash - **A. reticulatum** (Mart.) Eichl. (Co): bk. u. in arrow poisons (curare); fl. decoc. for hair wash - **A. schomburgkii** Miers (Pe, Bo, Guianas, Trin): huano; similar; also *A. chloranthum* Diels and *A. grandiflora.*

Anona Miller: 1) *Annona* 2) (specif.) *A. squamosa.*

Anonymos Walter: generic name applied to some 40 plant taxa for which generic placement was not possible; thus, some *Liatris* spp. were placed here (genus now abandoned).

Anopyxis ealaensis Sprague (Rhizophor.) (trop wAf): lvs. & astr. bk. appl. to sores & skin inf. (Liberia); rts. as enema (Ghana).

Anotis hirsuta Miq. (Rubi.) (Indones): kahitootan; bruised pl. with fetid odor is u. as medication for colic & admixed with other native medicines - **A. urophylla** Wall. (Chin): pl. crushed & appl. as poultice to sores.

anotta: annatto.

Anredera bogotensis (Basell.) (Co): babosa; lvs. stripped of epidermis & appl. to pustules, &c. to quickly "bring to head" - **A. cordifolia** (Ten.) Steenis (Boussingaultia baselloides HBK) (subtropSA, Mex to Chile): Madeira (or mignonette) vine; cult. as veget. (like spinach) (sEu); tubers eaten; pl. u. med. (Mex); ornam. liane - **A. leptostachys** (Moq.) Steenis (sMex to Co; WI): "sacasil"; "yedra (del bois pais)"; rt. in plasters u. for broken bones, sprains, dislocations, &c.; pl. edible; tuber decoc. taken for cardiac palpitations (Cu); in inflamm.; piece of tuber to soften corns and calluses; vine inf. u. for asthma, bronchitis, pulmon. lesions; other uses in WI; lf. juice for toothache; lf. decoc. as soothing, relaxing bath; ornam. - **A. vesicaria** (Lam.) Gaertn. f. (A. scandens Moq.; A. cumingii Hassk.) (Tex [US], Mex to Pe; WI; natd. PI): olibato (Tagalog); Madeira vine; "glycerine"; lf. inf. u. as "shampoo for horses" (Guat); tuberous rt. appl. topically to mature boils.

Ansellia gigantea Reichb. f. (Orchid.) (tropAf): rt. inf. in cough, st. aphrodisiac - **A. humilis** Bull. (tropAf): iphamba (Zulu); rt. inf. emet.; lf. & st. decoc. emet. in insanity; contraceptive; cult. orn.

Anser anser (L.) (A. cinereus Meyer) (fam. Anatidae; subfam. Anerinae - Ord. Anseres - Aves): (large) gray goose; marsh g., domestic g. (**Anser domesticus**); flesh eaten as popular food (Eu) such as at Christmas (*cf.* turkey in US); liver in paté de foie gras; body grease (FO) (Oleum Anseris) (goose grease) u. in med. as emoll.; inunction for chest colds (popularly); goose quill tooth picks (Chin); geese u. to eat weeds in cotton fields (sUS).

anserine: peptide contg. histidine; occ. in muscles of mammals & birds (geese+).

ansérine vermifuge (Fr): chenopodium.

ant: small pestiferous insect (fam. Formicidae - Hymenoptera); occ. in all continents in numerous genera & spp.; in size fr. 2-24 mm in length; social colonies of varying size; feeding habits variable; one gen., *Solenopsis* (fire ants) (SA, sUS) especially serious pest; venom c. antibiotic (1902) (2522); ants eaten daily as medicine by Chinese; in US, eaten embedded in chocolate (formerly) - **red a.**: *Formica rufa* L.; fr. which formic acid was first isolated; ant eggs are much u. as food for trop. fishes & rare birds in zoos - **a. acid**: formic acid (2558) - **a. oil:** off. in early Brit Phar.

Antelaea Gaertn. = *Melia*; *Azadirachta.*

antelope: various ruminant spp.of *Addax*, *Antilope* and other Bovidae genera with unbranched horns (Af, As); domesticated by ancient Egyptians; flesh eaten; flesh, fat, horns &c. u. in Unani med.; horns u. as aphrod. (Chin).

antennaire perlée (Fr): life everlasting herb, *Antennaria* spp.

Antennaria dioica (L) Gaertn. (Gnaphalium doiocum L) (Comp.) (Euras): mountain everlasting; life e.; fl. heads c. VO, resin, tannin; NBP hydrocarbon; phytosterol; u. arom. bitter, vulner. diur. expect. in bronchitis, in breast teas, in diarrh., bile duct. dis. (990) - **A. margaritacea** L. = *Anaphalis m.* - **A. microphylla** Rydb. (A. rosea Greene) (wNA): rosy everlasting; pusstoes; st. gum chewed; decoc. of st. u. in colds & coughs.

Anthemidis Flos (Flores) (L.): **Anthemis** (L.): Roman chamomile fls., *Anthemis nobilis*.

Anthemis arvensis L. (Comp.) (Eu; circum-Medit): corn chamomile; mayweed; hb. u. as adulter. for true chamomile, of no proven med. value; const. of secret remedies for abort.; fl. heads u. bitter ton. emet. - **A. cotula** L. (Maruta c. DC) (OW, natd. NA): dog's fennel; wild chamomile; strong smelling hb. (USP 1820-70) c. VO, resin, valeric acid, tannin, FO acrid princ.; u. bitter, sed. diaph. antisp. in colic (Calif [US]); in dysmen. (folk med.); u. to scent hair oil (Gilbert Is); fresh lvs. bruised & appl. exter to prod. blister; insecticide (2188) - **A. nobilis** L. (Chamaemelum nobile All.) (w&sEu; cult. NA; Ar): Roman c(h)amomile; garden c.; fl. heads c. VO, NBP, quercitrin, &c.; fl. heads u. considerably in Euas as bitter ton. carm. &c., much less so in NA; diaph. antispasm; pwd. added to hair shampoo powders (991) - **A. tinctoria** L. (Euras; natd. NA): golden chamomile; yellow c.; yellow fl. heads u. as vulner. in icterus; as anthelm.; as yellow dye (—> *Chamaemelum*).

anther: male reproductive organ of flowering plant; the generally apically enlarged portion of the stamen (usually capitate in form) which bears pollen sacs contg. pollen grains.

Antheraea Huebner (Lepidoptera): one group of silkworms producing a natural silk.

antheridiol: first sex hormone found in plants (occ. in male pl. of the fungus *Achlya bisexualis*): u. to control pl. fertility.

antheridium: male organ of reproduction in cryptogams, male gametangium; analog of anther in phanerogams.

Anthidium (Megachilidae - Insecta): mason bees; line mud nests with pl. material; source of a resin (Mex).

Anthistiria L. f. = *Themeda* (Gram.).

Anthocephalus chinensis (Lam.) Walp. (A. cadamba Miq.; A. indicus Rich.; A. morindifolius Korth) (Rubi) (Ind-Indochin-PapNG): kadamb(a) (Hind) tree; kedam; mau-let-tan-she (Burma); hb. u. ton. antipyr. bechic; in ophthalmia; rt. antisep.; lvs. u. as poultice for headache, gland swellings, & as gargle; fls. c. VO (distd.); fl. receptacle edible; frt. cooling, as paste, antipyr. (124); blood "purifier"; in colic; wd. u. in carvings, in paper mfr., in mfr. of matches & match boxes; tea chests.

anthochlor pigments: flavonoid gp. incl. chalcones, aurones, &c.; non-carotenoids.

Anthocleista nobilis G. Don (Potali./Logani) (tropAf): arbre chou; fresh bk. raspings appl. to abscesses & ulcerated wounds; rt. macerate u. as purg. diur. antigonorrheal; lf. ash u. to make soap - **A. procera** Lepr. (Senegal, Af): fafa (Wolof race); bk. of trunk & rts. u. purg.; in leprosy; as emmen. abort.

anthocyan pigments: anthocyanins: glycosides of pl. constituting the blue, red & violet coloring matters; responsible for the reds & purples of autumn leaves.

Anthodiscus obovatus Benth. ex. Wittmack (Caryocar.) (Br, Col): bitter bk. of small tree u. as decoc. for fever; lvs. u. as fish poison (also **A. peruanus** Baill. [Col]); u. with *Strychnos* sp. to prep. a type of curare (Tukano Indians) (2100).

Anthodium (L): flower head, capitulum - **A. Cinae**: Cina.

anthogen: pl. hormone (in lipid fraction) which speeds up flowering process even in low light intensity & gives fuller blossoming (1960-).

Anthonotha acuminata Léon (Leg) (Congo); **A. crassulifolia** (Baill.) Léonard (Senegal to Nigeria; Mali) - **A. gillettii** Léon (Congo); **A. macrophylla** Beauv. (Congo; Guinea; Angola): med. plants of Africa.

Anthophylli (L.): mother of cloves; clove frts.; fr. *Syzygium aronaticum*; inferior to clove, but u. as spice, in stomach dis., etc.

Anthos Flores (or **Herba**) (L.): rosemary fls. (or hb.).

Anthospermum pumilum Sond. (Rubi) (sAf): rt. u. in dysmenorrh. & during pregnancy - **A. rigidum** E. & Z. (tropAf): decoc. u. in toothache; pl. decoc. added to bath of convalescent; st. & rt. u. in "lung sickness" of horses.

Anthostema senegalense A. Juss. (Euphorbi) (wAf): mano (Bambara); u. drastic, in amenorrh., leprosy (latex mixed with bread or other foods); for intest. parasites; insecticide; latex u. as bird lime.

Anthostyrax: section of genus *Styrax*, members of which furnish Siam benzoin; formerly as gen. (= *Styrax*).

anthoxanthins: vascular yellow pigments of fls., flavonoid in nature (flavonols & flavones); occ. in cell saps.

Anthoxanthum odoratum L. (Gram.) (Af, natd. Euras; NA): sweet (scented) vernal grass; spring g.; la flouve (odorante) (Fr); herbage c. coumarin; addn. to baths, in herbal pillows; in mfr. special brandy (USSR, &c.); rt. carm., in snuffs; a forage grass, rich in coumarin; occ. early in spring; low nutritional value.

anthracene: $C_{14}H_{10}$: solid hydrocarbon obt. as distillate fr. coal tar; almost colorless solid crystals; water insol.; starting point for synth. of many dyestuffs; u. in scintillation counters; to make lampblack; to detect thieves (dusted on articles wh. dust cannot be seen; contaminates fingers of thief & lights up in UV light) (fluoresces. violet); non carcinogen (2015).

anthracite: hard coal; good fuel but with much less tar than bituminous (soft coal), hence preferred as fuel since less smoke; prod. Denn. NM (US), +.

anthraglycoses: anthranols, oxanthrones, dianthrones (all give Borntraeger reaction).

anthraglycosides: anthraquinone glycosides; made up of 4 classes: anthraquinones, anthranols, oxanthrones, dianthranols (all give Borntraeger reaction.)

Anthramycin: antineoplastic fr. *Streptomyces spadicogriseus* Komatsu.

anthranilic acid: o-aminobenzoic acid; u. as tracer or marker in squalene to determine lf. u. to adulterate olive oil.

anthranols: hydroxy-anthracenes; reduction products of anthraquinones or anthrones; occur in plants; actively purg., ex. chrysophanol, anthranol.

anthraquinones: ketone derivs. of anthracene: anthracene with a keto pg. on each of the two opposite positions on the middle ring; reduced in body to anthranols; ex. moridone.

Anthrarobin™: anthracene derivative (synthetic); u. much like chrysarobin.

Anthriscus cerefolium (L.) Hoffm. (Umb.) (s,c&eEu; wAs; natd. eNA): (salad) chervil; hb. u. kitchen spice (fresh); diur., flavor; "blood purifier" (spring cures); emmen.; food in salads, lvs. (odor of Amines) u. like parsley (but less pungent); occasional dermatitis reported; often cultd. in kitchen gardens - **A. sylvestris** (L.) Hoffm. (Euras; natd. eCana, NJ+ [US]): cow parsley; c. weed; wild chervil; Waldkerbel (Ge); hb. stim. uter. contractions, u. in childbirth, as abort. (Russ); fresh pressed hb. appl. to sores, &c.; lotion prepd. fr. pl. u. as refreshing bath; frt. as adulter. of caraway.

anthrones: partially reduced anthraquinones; occ. in plants (oxidized anthranols).

Anthurium Schott (Ar.) (tropAm, WI+): flamingo flowers; banner plants; over 700 spp.; ornam. (cut fls. last two weeks or more) (cult. Hawaii [US]); various *A.* spp. u. in heart dis. (hence "corazón") (Co) - **A. acutangulum** Engl. (CR, Pan): pl. decoc. in pertussis (Choco Indians, Pan) - **A. andraeanum** Linden & André (seAs): crushed lvs. appl. to sores caused by stinging caterpillars (Thail) (2448) - **A. infectorium** R. E. Schultes (Co): ripened frts. u. to paint teeth purplish or black for ceremonial appearance (Kubeo Indians) - **A. oblongocordatum** Engl. (Ve, Pe, Co): komebiaya; pl. decoc. antidote (for poison?) - **A. oxycarpon** Poepp. (tropAm): dr. lvs. with vanilla & musk odor mixed with tobacco for special aroma - **A. tikunorum** R. E. Schultes (Co): hb. u. for pyorrhea (Tikuna Indians) (reduces pus flow); frt. juice astr.

Anthyllis Flores (L.): woundwort fl. - fr. **A. vulneraria** L. (Leg.) (Euras): kidney vetch; woundwort; perenn. hb.; pl. u. styptic, astr.; renowned vulnerary; form. off. now little u. med.; pl. formerly u. as yellow dye; fl. gives blue color; sometimes u. as substit. for Chinese tea; prod. in It, Yug; forage; cult. ornam.

antiallatotropic: agent which acts on the corpora allata of young insects (producing juvenile hormone) in such a way that the female insect becomes sterile; useful as natural insecticide.

antiar: poisonous gum-resin of foll.

Antiaris africana Engl. (Mor.) (trop wAf): bark cloth tree; mayo (Sereng); bk. & rts. u. purg., in leprosy; latex very toxic; wd. u. as timber, for canoes; cloth bark; latex u. as adult. of Funtumia rubber - **A. toxicaria** (Pers.) Lesch. (Indones, Ind+): upas tree; upas antiar (arrow poison); prod. poisonous sap (gum-resin); c. cardenolide glycosides (antiarins, bogoroside); u. to prepare arrow & dart poisons; strong digitaloid heart activity; lvs. bk. sd. u. as antipyr. & in dys.; juice as mild circulatory & heart stim. (Ind); ordeal poison; valkala (Sanscrit); fables tell of death when a person approached tree; u. for heart dis. in same manner as Digitalis/Strophanthus; bk. fibers u.

Antiasthmatic Elixir: Euphorbia elixir comp.

antibacterial serum: a serum separated fr. the blood plasma of an immunized animal; c. antibodies specifically antagonistic to the bacteria which serve as the antigen or as source of antigen.

antibiosis: association of organisms of various spp. in which more harm is produced in one than the other sp.; a competition in which one(+) sp. gains advantages (unlike symbiosis where both or all gain); v. App. D (properties & uses of drugs) - **antibiotic**: product..

antibody: agent in blood serum consisting of immunoglobulins which react with specific or closely related antigens - **monoclonal a.**: MAB: homogenous antibody prod. by hybridoma (hybrid cell made by fusion of normal lymphocyte with cancer cell); u. in research & in cancer immune therapy; prod. by genetic engineering techniques; u. to detect cancers, heart dis., &c.

Anticharis glandulosa Aschers. (Scrophulari.) (sInd, Pak): pl. u. in diabetes).

anticoagulant heparin solution: sterile soln. of heparin sodium in NaCl injection soln. (v. heparin).

Antidesma alexiteria L. (Stilagin./Euphorbi) (Ind, SriL): lvs. u. in snakebite - **A. bunius** Spreng. (swChin, Indomal, Austral, PI): Chinese laurel; huni; lvs. c. friedelin, u. diaphor.; sour frts. eaten fresh & u. for jellies & liqueurs - **A. cochinchinense** Gagnep. (Indochin); lf. inf. u. as ton.; rts. grated on stone & drunk with water for quotidian fever (mal.) - **A. cuspidatum** Muell.-Arg. (Mal): lvs. chewed for flatulence - **A. ghaesembilla** Gaertn. (Chin, Ind, seAs): bk. ton. astr. in diarrh., emmen. (usually in combns.) - **A. henryi** Hemsl. (Chin): lvs. mixed with sour wine, heated & appl. to contusions to remove blood - **A. montanum** Bl. (Mal): young lvs. as poultice for headache; rt. raspings drunk in water for measles, chicken pox, mal., thrush - **A. phanerophlebium** Merr. & Perry (Oceania): rts. appl. to stomachache - **A. velutinosum** Bl. (Mal): has stupefying effect - **A. venosum** E. Mey. ex Tul. (Moz & dry parts of Af): filo; ripe frt. edible but not palatable; u. as fish bait; rt. decoc. u. for bilharziasis, VO for removing afterbirth; lf. decoc. u. for sharp stomach pains, toothache (rinsed through mouth) (Tanz).

antidiphtheric globulins: diptheria antitoxin.

antiemetic root: *Cyperus*.

antigen: any foreign substance which after inj. into animal body promotes the formation of a specific antagonizing substance called an antibody; ex. toxin.

Antigonon (*Antigonum*) **flavescens** S. Wats. (Polygon.) (Mex): coamecate; u. med. - **A. leptopus** H. & A. (Mex, CA, nSA, natd. Fla [US]): coral vine; coral bloom tree; corallita; mountain rose; Confederate vine (cult. sUS); pine vine; tubers edible, (nutty flavor) u. as decoc. for chronic diarrh.; honey plant.

antihemophilic factor (human): obtained fr. blood plasma - **a. f. A.**: = Factor VIII - **a. f. B** = Factor B.

Antilles: all of the West Indies except the Bahamas.

Antilope cervicapra (L) (Bovidae - Mammilia) (Ind): baraseenga; black buck; true antelope; considerably u. in Unani med.; thus, blood in cystic & renal calculi.

antimacassar: protective cloth against spotting by macassar oil (*qv.*); formerly placed on upholstered chair backs, &c.

antimeningococcic serum: a prod. derived fr. animal blood of organisms immunized with cultures of various types of meningococci.

antimony: Sb (Stibium) (L.), metal obtained fr. ores fr. Mex, Bo, Chin; source of compds. u. in med. (SbK Tartrate [parasiticide+]); SbNa Gluconate (antiprotozoal); SbNa Tartrate (anthelm.); $SbCl_3$, &c.); also several organ. compds.

antimycin A: antibiotic fr. *Streptomyces* sp.; also insecticidal & miticidal; fungicidal (A_1).

anti-opium plant: *Combretum sundaicum*.

antioxidants: factors believed to destroy free radicals (wh. oxidize & damage cells of body) supposed to reduce occurrence of cancers, cardiovasc. dis. (heart attacks); stroke; cataracts; &c.; act to neutralize certain enzymes (peroxidases; catalases); ex. pro-vitamin A (betacarotene); vit. C; vit. E.

antipasto (It): an appetizer (u. in sandwiches, &c.) contg. meats & fish combined with olives, tomatoes, hot peppers, cheese, &c., with oil & vinegar (<ante [before] + pasto [food]).

antirabies serum: sterile soln. of antiviral antibodies fr. blood of horse immunized with rabies vaccine.

Antirhea (Antirrhoea) **coriacea** (Vahl) Urban (Rubi) (Dominica+): bk. decoc. u. to wash sores (Domin).

Antirrhinum linaria L. (Scrophulari.) = *Linaria vulgaris* Mill. - **A. majus** L. (worldwide temp zone, esp. Med reg.; natd. NA): (common) snapdragon; dragon's mouth; dog's m.; c. rhinanthin (glycoside), anthocyanins (flavonoids); u. wounds, sores, &c.; popular cult. ornam. - **A. orontium** L. (Misopates o. [L.] Raf.) (Eu): lesser snapdragon; calf's snout; c. alk.; similar uses to prec.; u. in witchcraft.

antiscorbutic juice: zumo antiscorbutico (Sp Phar.): prepn. made up of juice fr. equal wts. of fresh horseradish rt., water cress lvs., scurvy grass (or bitter cress) & buckbean lvs.; u. to prevent scurvy.

antisera: blood sera u. in combatting dis.; incl. antirabies (hyperimmune) serum (E); antiscorpion serum (E); antivenom serum (antivenin) (several kinds, ex. coral snake, adder [viper], crotaline [rattlesnakes], &c.) (E).

anti-snakebite serum, North American: Crotalus antitoxin or Antivenin; u. to immunize against rattelsnake venom.

anti-stiffness factor: FO-sol. subst. wh. occurs in green vegetables, molasses, raw cane sugar juice,

&c.; prevents muscular degeneration & stiffness syndrome, &c. (1950).

antistreptolysin: antibody active against hemotoxin of hemolytic streptococci.

antitetanic globulin: synonym for tetanus antitoxin.

antitoxin: an agent which tends to neutralize the harmful effects of a toxin; the latter usually of bacterial origin; when injected in small doses, a. produces passive immunity; ex. tetanus antitoxin; now replaced by t. toxoid - **Antitoxinum Diphthericum** (L.): diphtheria antitoxin; now replaced (US) by d. toxoid (v. under gas gangrene & scarlet fever) - **botulism a.:** prod. in horses & isolated fr. blood serum; protects against 3 strains of toxin produced by *Clostridium botulinum* (van Ermengem) Bergey & al.

antitrypsin, a: major glycoprotein, a protease inhibitor formed in liver; shows variation as subject of genetic influences; u. to treat emphsysema when due to a deficiency of this subst.

anti-ulcer factor: substance found in juice of head cabbage (*Brassica oleracea*) wh. inhibits gastric ulceration.

antivenin: antivenom: anti-snakebite serum; a venom antitoxin prep. fr. blood of horses or other mammals which have previously received repeated inj. of diluted venoms of snakes (2571), spiders, etc. (E) - **Latrodectus a.**: prepd. fr. venom of *Latrodectus mactans* (black widow spider) (see *Latrodectus m.*) - **Micrurus fulvus a.**: sterile prepn. of specific venom-neutralizing globulins fr. the serum of healthy horses immunized against venom of eastern coral snake (*Micrurus fulvus*) (CDC, Atlanta, Ga [US]) (*qv.*) - **polyvalent (Crotalidae) a.**: prod. derived fr. venoms of *Crotalus* spp.: *C. atrox* (w. diamondback), *C. adamanteus* (e. diamondback), *C. durissus terrificus* (South American rattlesnake), and *Bothrops atrox* (South American fer de lance); acts to neutralize venom of these and related spp. (see also under *Crotalus* and *Bothrops*).

antivirin: AV; virus inhibitory subst. prod. by various cells of animal body (ex. human embyro lung); wide spectrum; distinctive fr. interferons.

Antizoma angustifolia Miers (Menisperm.) (s&swAf): rt. u. in gastrointest. dis. - **A. capensis** Diels (sAf): rt. decoc. u. in boils, syph.; weak tinct. of pl. u. in dys.; rt. emet. purg.; in mixt. for erysipelas.

Antonia ovata Pohl (monotypic) (Antoni./Logani.) (Guianas, Br): iñacu (Wapisiana); falsa quina (false cinchona); lvs. u. as fish poison (slow acting).

Antrocaryon klaineanum Pierre (Anacardi.) (trop wAf): frt. with edible acid mesocarp; wood in carpentry ("white mahogany") - **A. nannani** Willd. (trop wAf): bk. c. tannin; sds. source of non-drying FO ("Congo oil") said similar to olive oil; u. med. (Congo).

ant's wood: *Bumelia angustifolia*.

antuitrin™: stimulating hormones secreted by the anterior pituitary gland - **A. G.™ = a. growth**: somatotrop(h)in - **a. S.™**: chorionic gonadotropin.

anu (Pe): **anyu** (Pe): *Tropaeolum tuberosum*.

Anzucht (Ge): cultivation, breeding, raising, propagation, culture.

aouará (Fr Guiana): awar(r)a palm (Surinam): *Astrocaryum tucumoides*; source of sd. kernel oil and FO fr. frt. pulp of same sp.; oils with different properties **- a. oil: aourata (kernel) oil**: awar(r)sa oil; FO fr. frt. of *Astrocaryum vulgare* (tucum nuts).

Apache beads: *Anemopsis californica*.

apamate (Ve, Br): *Tabebuia rosea*.

apamin: polypeptide in venom of *Apis mellifera* (bee).

aparine: *Galium aparine*.

apasote (CR): **apazote** (Cu, PR): epazote (Mex); *Chenopodium ambrosioides*.

Apeiba tibourbou Aubl. (Tili) (sMex, CA, WI to Pe, Br): monkey comb; pachiote; fls. u. antispasm., lvs. & bk. mucilagin. pectoral; sd. FO (apeiba or burillo oil) inedible; rubbed into area for rheum., to stim. hair growth; wd. very light ("pao de jangada") u. to make rafts.

Apera spica-venti Beauv. (Gram) (GB, natd. NA): silky bent grass; veldtgras (Belg); u. med.; u. to dye wool green (Russ); weed; cult. ornam.

apes: large tail-less primates (OW), including gorilla, chimpanzee, orangutang, gibbon; some may be trained to carry out various routine tasks; also source of amusement in the zoo, or even in the home; articles of commerce in ancient times; term sometimes appl. to any monkey.

Apetalae: subdivision of dicotyledons, in which the petals (& sometimes also sepals) are lacking: ex. walnuts, birches.

apex: tip, often rounded - **root a.**: growing tip of rt.

Apfel (Ge): apple - **Apfelbaum** (Ge): apple-tree - **A.-kraut**: horehound - **A-saeure** (Ge): apple acid = malic a. - **A-sine** (Ge): sweet orange, *Citrus sinensis* - **A-sin(n)enschalen**: orange peels; pomegranate (frt.) peels.

Aphanamixis grandifolia Bl. (Meli) (Indones, Mal): pericarp c. aphanamixin (toxic); bk. decoc. ingested for chest pains in colds - **A. polystachya** Parker (A. rohituka Pierre) (Ind): tiktaraj (Bengal): bk. astr. u. in tumors; sd. semi-drying FO u. in rheum.; lvs. & frts. widely u. in folk med. -

A. tripetala (Blanco) Merr. (PI): bk. u. in fermentation of "tubu" (alc. bev.).

Aphananthe aspera (Thunb.) Planch. (Ulm) (eAs): lvs. rough, u. like sandpaper; source of timber - **A philippinensis** Planch. (PI): dr. branches burned & ash u. to smudge the body for mal.; lf. ashes mixed with coconut oil rubbed on wound from a mad dog (!).

Aphanes arvensis L. (Alchemilla a. (L.) Scop.) (Ros.) (nearly worldwide): lady's mantle; Ohmkraut (Ge); hb. c. tannins; u. as diur. & in urinary dis. (folk med.).

Aphania senegalensis Radlk. (Sapind.) (tropAf, esp. Seneg): cherry tree of Cayor; lf. decoc. vulner. to wounds (lotion), also int. for contusions, falls, various accidents; bk. decoc. as beverage in bronchitis & broncho-pneumonia; vermif.; lvs. & frt. said toxic (but frt. eaten); sds. very toxic; wd. u. to make paddles. (—> *Lepisanthes*)

aphanin: carotenoid of Algae, such as *Aphanizomenon flos-aquae*; has vit. A activity but less than that of beta-carotene.

Aphelandra squarrosa Nees c. var. *louisae* (Acanth.) (warm Am areas): zebra plant; showy ornam. house plants.

aphidicolin: antibiotic fr. *Cephalosporium aphidicola* Petch; antiviral, antimitotic; u. as biol. "tool."

Aphis chenopodii glauci (Aphidinae - Homoptera - Insecta): aphid; pl. louse; u. homeop. med.

Aphloia mauritiana Bak. (1) (Flacourti.) (eAf+) lvs. u. as tea; inf. in hematuria (Mascarene Is).

aphrodaescin: saponin-like compd. fr. unripe sds. of *Aesculus hippocastanum*.

Aphyllon Mitch. = *Orobanche*.

apic (Fr, argot): garlic.

apichu (Pe): sweet potato.

apiculture: science & art of beekeeping.

Apidae: Bee Fam.; insects of Order Hymenoptera; incl. honey bees, carpenter bees, etc.

apigenin: 5,7,4'-trihydroxy-flavone; occ. in fls. of *Matricaria* spp.

Apii Fructus (L.): celery frt. ("sd.") fr. *Apium graveolens* v. *dulce*.

apiin: glucoside which occ. in parsley & celery; *Anthemis nobilis*; a flavonoid made up of apigenin with (glucose + apiose).

apio: 1) (Sp): celery 2) (SA, Sp): *Apium montanum* or *A. ranunculifolium* - **a. criollo**: Creole celery, *Ciclospermum leptophyllum* - **a. de España** (Mex): *Apium graveolens*.

apiöenin: 5,7,4'-trihydroxy-flavone; aglycone of apiin; occ. in fls. of *Matricaria* spp.; u. as dye.

apiol(e), parsley: parsley camphor; apioline (993); phenolic stearoptene fr. parsley VO; u. emmen., somewhat toxic (993); distinguish fr. the isomeric **dill apiole** (dillapiol) fr. *Anethum graveolens* - **liquid a**: oleoresin fr. parsley frt.; u. antipyr., emm.

Apios americana Medik. (A. tuberosa Moench) (Leg) (eUS, Cana): Indian potato; potato beans; ground nut (do not confuse with *Arachis*); tubers c. starch; cooked, boiled or roasted and eaten (Am Indians); said to have sustained the Pilgrims for the first few yr. in New England (US); taste much like Irish potato & with similar nutr. quality; has been adulter. with *Liatris* rts.

Apis mellifera L. (*A. mellifica* L., usu. in Eu) (Apidae - Hymenoptera - Insecta) (orig. Euras, Af; now worldwide): the (common) honeybee; c. hive bee; abeille (Fr); long domesticated & now one of man's most useful servants; source of honey (*qv.*), beeswax, venom, propolis, royal jelly; chief value in pollination of crops (2495); complex social organization as in many other Hymenoptera into three castes: workers, queen, drones; bees wax (*qv.*) from the hive considerably u. in ointments, polishes, crayons, candles, &c.; royal jelly (*qv.*) nutrient for larva of queen bee, claimed of med. value; bee venom has been u. effectively in collagen dis., such as progressive polyarthritis (994); several other bee spp. have provided honey, such as the African(ized) b. ("killer b."), an aggressive insect spreading in the wHemisphere.

Apium australe Cav. (Umb.) (SA): apio (silvestre); c. isopimpinelline, bergaptene, myristicin; u. ton. antipyr., carm.; in gout & rheum., antiscorbutic (Ch) - **A. divaricatum** B. & H. = *Spermolepis divaricatus* - **A. graveolens** L. (Euras; cult. started in Med reg.; grown almost everywhere in Temp. reg., ex. Eng, Eg, Ind, NA): var. *dulce* (Miller) Pers.: celery; ache (Fr); lf. stalks & hearts (bleached or blanched) popular table veget. consumed raw, in mixts. (salads, &c.) or alone; cooked in soups; lvs. garnish & to "neutralize" odor of garlic/onion & "repeating" after eating same; frt. ("celery seed") (chief source Fr, Ge) c. VO, c. limonene, selinene, sedanoic acid anhydride (chief odor source); prep. celery salt; hb. rt. frt. VO u. diur. in dropsy, bladder dis., nephritis; antipyr. flavor, in arthritis (olden times); uter. stim.; VO flavor, with strong antibacterial props.; supposed CNS sed. antispasm.; aphrod. in spices; as fixative in perfumes (lvs. u. in ancient Eg) - var. *rapaceum* (Mill.) Gaudich.: turnip-rooted/German/French celery; bulb (or knip) c.; celeriac; thickened rt. eaten as veget. (céleri-rave) - **A. prostratum** Labill. (Austral, NZ-seashore): green (or Maori or sea) celery; prostrate parsley; u. as antiscorb., in vapor baths (998).

Aplasmomycin: antibiotic prod. by *Streptomyces griseus* strain fr. shallow sea mud; u. promotes growth in Ruminantia.

Aplectrum hyemal60e (Muhl.) Torr. (monotypic) (Orchid.) (e&cUS): Adam and Eve; (American) putty root; p. wort; corms u. as mucil. expect. (folk med.); cult. orn.

Aplona: prepn. fr. apples; u. in diarrh.

Aplopappus Cass. = *Haplopappus* Cass incl. *Acamptopappus*.

apoatropine: alk. found in rt. of belladonna; also prepd. by synthesis fr. atropine; u. spinal & medullary stim., esp. of respn.; antispasmodic.

apocin: ext. fr. *Periploca graeca*; was formerly u. in cardiac dis.

Apocynaceae: dogbane family; large fam. (2100 spp.) of hbs., lianes. shrubs & trees; mostly trop, but some taxa in temp. zones; lvs. mostly opposite, fls. regular 4- & 5-merous, frts. drupes or follicles; with milky juice; many members poisonous; ex. periwinkle, oleander, *Rauvolfia*, *Strophanthus*, *Aspidosperma*.

apocynamarin: apocynein: apocynin: apocannoside: compds. reported to occur in rhiz. & rt. of *Apocynum* spp.

apocynin: 1) amorphous cardioactive glycosidal subst. fr. *Apocynum cannabinum* & *A. androsaemifolium* (Schmiedeberg) 2) acetovanillone, arom. ketone; fr. *Apocynum* spp., *Boophone disticha* Herb., *Iris*.

Apocynum L.: 1) genus of Apocynaceae (nTemp Zone); dogbanes; hbs. with pink or white fls.; sds. hairy 2) (L.) (specif.): *A. cannabinum* rt. - **A. androsaemifolium** L. (Cana, US, nMex): (spreading) dogbane; dogsbane; American ipecac; honey bloom; rt. c. cymarin, androsin, k-strophanthin; resin; emetic. card. stim. diur. in heart dis.; in mfr. cymarin; emeto-cath. in gastrointest. irritn. - **A. cannabinum** L. (e&cUS, Cana): Canadian hemp; "East west" (995); dogbane; rt. c. cymarin, peruvoside, apocannoside (cytotoxic, u. for tumors); rt. u. like preceding; latex in rubber products (996) - var. *hypericifolium* Gray (A. sibiricum Jacq.; A. s. var. salignum Fern.) (BC to Man [Cana] to sCalif, e to Tex [US]): clasping-leaved dogbane; Indian hemp; reported to have similar hallucinatory effects as Cannabis (999); pl. formerly u. in basketry by Am Indians - **A. juventas** Lour. = *Streptocaulon j.* Merr. - **A. venetum** L. (Trachomitum v. Woodson) (sEu, As): Vendeisches Hundsgift (Ge); rt. very similar to that of *A. cannabinum* (can scarcely be distinguished); st. fibers u. to make strings, cloth; in paper making (Russ).

Apodanthera pedisecta Arn. (Cucurbit.) (Pe): hb. & frt. edible - **A. smilacifolia** Cogn. (Br-Minas, Ger, Bahía): azouíue dos pobres (mercury of the poor); "cipó azougue"; c. alk. (apodantherine) & glycoside; all vegetative parts (esp. rt.) of liane u. (folk med.) in skin dis., rheum. ("depurative"); lvs. & rts. in syph. - **A. undulata** Gray (Mex, sUS): u. med. (Mex); furnishes oil sd.

Apodytes dimidiata E. Mey. ex. Arn. (Icacin.) (hot Af): white pear; rt. bk. u. for intest. parasites (as enema) (Zulus); for pain in left abdom. (Tanz); lf. appl. to ear inflammn.; wd. useful.

apoenzyme: protein part of enzyme bearing free charge & attracted to substrate.

apomixis = apomyxis: reproducing without the sexual process; ex. parthenogenesis.

apomorphine: semi-synth. alk. of morphine, prepd. by treating m. with concd. HCl; actually a dehydration prod. resulting in marked change of properties; u. emet. expect.; in "curbed" emesis prior to obstetric analgesia; in symptomatic relief of schizophrenia; aversion treatment of alcoholism, homosexuality, &c. (2095); dopamine antagonist.

Aponogeton crispum Thunb. (Aponogeton. - Monocots) (tropics & sAf): tuber u. as famine food.

apopin oil: VO fr. pl. of Lauraceae (Taiwan) with camphoraceous odor ("schuyu").

Aporocactus flagelliformis (L) Lemaire (Cact.) (Mex, CA, SA): rattail cactus; floricuerno; dr. fls. ("flor de cuerno") household remedy for heart dis. (Mex); pl. juice u. as vermif. (danger!); exter. rubefacient; cultigen (ornam.).

Aporusa lindleyana Baill. (Euphorbi) (Ind): vittil (Tamil); rt. decoc. u. in icterus, fever, headache, insanity - **A. microcalyx** Hassk. (seAs): muat; u. med. (Laos); wd. u. to make rice pestles, &c.

Apostasia nuda R. Br. (Apostasi./Orchid.) (Mal): boiled rts. u. in diarrh.; frt. inf. u. as remedy for sore eyes.

apothecary: term u. in Eu & Am denoting a professional pharmacist, i.e., one conducting a drug shop of the highest category, presumably one in which nothing but drugs and sickroom supplies are handled; nowadays, the term "pharmacist" is generally preferred (US, Eng).

apothegmatic matter: apothem(a): apothem: more or less insol. brownish subst. deposited fr. pl. extractive prepns. (inf., decoc. tinct. &c.) when exposed to heat in the presence of air.

Apotheke (Ge): pharmacy; drug shop - **Apotheker** (Ge): pharmacist.

appio riso (It): *Ranunculus sceleratus*.

apple: *Malus sylvestris* & other frts. (& some other *M.* spp.) + - **a. acid**: malic a. - **Adam's a.**: 1) *Citrus medica* vars. 2) laryngeal prominence; projection of thyroid cartilage of larnyx; more prominent in the male, hence common name - **African star a.**: *Chrysophyllum africanum* - **alligator a.**: *Annona glabra* - **a. butter**: a. pulp boiled longer than for a. sauce, often with added cider or vinegar and spices, hence of a dark red color - **Carthaginian a.**: pomegranate - **Chinese a.**: *Malus spectabilis, M. prunifolia* - **a. cider**: liquid expressed fr. fresh ripe apples; not filtered or clarified, nor pasteurized; hence cloudy & with residue; distinguish fr. **a. juice**, which is filtered, clarified, & pasteurized & is considered less nutritious - **crab a.**: name appl. to several *M.* spp., such as *M. halliana, M. ioensis*, etc.; lvs. u. med. in Yugoslavia; term appl. to escaped forms in which the frt. is usually small and tart; many are essentially cult. ornamentals, generally hybrids - **custard a.**: *Annona cherimola* + - **egg a.**: mad a. - **elephant a.**: *Feronia elephantum*, frt. - **emu a.**: *Petalostigma quadriloculare* - **gold a**: tomato (2536) - **gopher a.**: *Chrysobalanus* spp. - **a. haw** (Miss [US]): *Crataegus* spp. - **hedge a.**: *Maclura pomifera* - **hog a.**: 1) *Podophyllum peltatum* 2) (seUS): *Crateagus* spp. - **honeysuckle a.** (US): *Podophyllum peltatum* - **horse a.**: 1) large green a. formerly popularly grown in La (US), &c.; now much less so (sUS mts.) 2) (Miss, Ala [US]): *Maclura pomifera* - **Indian a.**: 1) hog a. (1) 2) *Datura meteloides* - **Jamaica a.**: custard a. - **Jew's a.**: mad a. - **lady a.**: small esculent apples u. mostly for ornament - **a. lotus**: *Nymphaea lekophylla* - **love a.** (formerly): tomato - **mad a.**: 1) *Solanum melonagena* 2) (early NA pioneers): guilder rose, *Viburnum opulus* - **Malay a.**: *Syzygium malaccensis* - **mock a.**: hedge a. - **monkey a.**: *Clusia flava*, &c. - **oak a.**: nutgall. - **Peru a.**: stramomium (frt. & sds.) - **Phoenician a**: pomegranate - **pond a.** (Fla): alligator a. - **Punic a.**: pomegranate - **rose a.**: *Syzygium jambos* + - **a. de San Domingo**: *Mammea americana* - **satin a.**: velvet a. - **a. of Sodom**: *Solanum sodomeum, S. carolinense* - **sorb a.**: *Sorbus* spp. (sorb), esp. *S. domestica* (frt.) - **sour a.**: *Malus sylvestris* vars. - **star a.**: *Chrysophyllum cainito*; *Averrhoa carambola* - **sugar a.**: *Annona squamosa* + - **thorn a.**: *Datura stramonium* & other *D.* spp. (**t. a. [tree]**): *Crataegus oxyacantha*) (**hairy t. a.**: *D. metel*) - **velvet a.**: *Diospyros* spp. - **water a.**: alligator a. - **wood a.**: 1) elephant a. 2) *Aegle marmelos*.

apprentice: one receiving practical drug store experience under supervision of a registered pharmacist; term "pharmacy intern" now preferred (US).

Apramycin: broad spectrum antibiotic fr. *Streptomyces tenebrarius* Higgins & Kastner; antibacterial.

apricot: *Prunus armeniaca* (tree & frt.) - **African a.**: *Mammea africana* - **a. kernel meal**: fr. a. sd. kernel; u. in confections, cordials, etc. - **a. k. oil**: one of the two off. forms of persic oil; typical FO u. as emoll. in ointments, etc. - **San(to) Domingo a.**: *Mammea americana* - **a. sheets**: a. leather: puree of a. frt. saturated with sugar; dried & rolled into sheets - **a. vine**: *Passiflora incarnata*.

aprotic: subst. which can act neither as acid nor base (ex. benzene).

aprotinin: polypeptide fr. lung tissue & parotid glands of cattle; u. to treat pancreatitis, some hemorrhages (Trasylol) (pancreatic "trypsin inhibitor").

Aptandra spruceana Miers (Olac.) (Br): nuts yield 46% very viscous oil (Castanha de Cotia oil; Sando de Maranhäo); rt. bk. source of greenish sandal oil with sassafras odor; u. med.

apteka (Russ): apothecary; drug store.

Aptosimum calicynum: E. P. Phill. (Scrophulari) (seAf): lf. & st. inf. u. antipyr.; as enema in milk for gastric dis. in children (Zulus); finely pd. lf. facial cosmetic - **A. depressum** Burch (seAf): carpet plant; inf. u. as diur. in anuria & difficulty in micturition; gargle in diphtheria (Eu); decoc. as wash for lesions of impetigo & ringworm - **A. indivisum** Burch. (seAf): Karoo violet; inf. for gastric difficulties (Europeans in Af); ton.; exter. in eczema (Europeans & natives).

apuna rong (Hind): *Auricularia*.

apupu (Mex): chayote, *Sechium edule*.

apyrase: enzyme found in plants, esp. *Solanum tuberosum*; hydrolyzes acid anhydride links; coenzyme = Ca ion; substrate = adenosine di- or triphosphate.

Aqua (L.): water - **A. aerata (fixa)** (L): *A. aeris-fixa*: carbonated water; satd. aq. soln. of CO_2; basis of soda waters - **a. ardens**: alc. - **a. bulliens**: boiling w. - **A. Cerasorum**: dild. bitter almond water - **A. Coloniensis**: Cologne water - **A. communis**: tap (or common) water - **a. crotona: a. crotonis**: w. fr. Croton Lake, New York City (NY [US]) reservoir; hence hydrant or tap w. - **a. fervens**: hot w. - **a. flava**: yellow wash (or lotion) - **A. fluvialis: A. fluviatilis**: river water - **A. foetida antihysterica**: a distillate fr. asafetida. galbanum, myrrh., valerian, etc. - **A. fontana**: spring (or distilled) w. - **a. fortis**: HNO_3 - **A. gaseosa**: A. aerata - **A. gummosa**: Mucilago Acaciae (ratio 1 to ca. 64 water) - **a. kali puri**: liquor potassae: KOH soln. - **a. lulliana**: ether - **A. marina**: sea water - **A. m. artefacta**: artificial sea water - **A. Menthae**: peppermint w. - **a. metallorum**: mercury - **a. naphae**: orange fl. w. - **a. nigra**: black wash (or lotion) - **a. phagedaenica**: yellow wash (or lotion) - **a. p. ni-**

gra: a. nigra - **a. plagiari**: mixt. of alum., Tr. Benxoin Compd. with w., boiled together; u. in hemorrhage - **a. pluvialis**: rain (or distilled) w. - **a. pura**: distd. w.; purified w. - **a. regia**: nitrohydrochloric acid - **a. saturnia**: lead w.; dil. soln. lead subacetate - **a. sedativa**: camphorated ammoniacal lotion - **a. temperata**: ether - **a. vitae**: water of life; brandy; whiskey; or alc.

aquaculture: the operation of "farming" the sea or other natural bodies of water for shellfish, sea plants, bait fish, food fish, &c.; this includes simple propagation of oysters, Algae, &c.; mostly involves food materials (distinguish from aquiculture [hydroponics]); ex. catfish (Ala [US]); salmon, prawn, shrimp (Ala), eels, sushi fishes, pearls, nori (sea weeds), &c.

aquamarine: greenish precious stone (gemstone); form of beryl; c. beryllium; u. in med. AD 1546-1600; known as "Smaragd."

aquénio (Por): indehiscent one-seeded frts., esp. achene or mericarp.

aquiculture: hydroponics; soilless gardening.

Aquifoliaceae Bartl.: holly fam. (<aqua, water + folium [L] lf., since lvs. often shiny); trees or shrubs of trop. & temp. zones; chiefly made up of hollies (*Ilex* spp.).

Aquilaria malaccensis Lam. (A. agallocha Roxb.; Aloëxylum a. Lour.) (Thymelae.) (Chin, Ind+): eagle (or agar) wood; Agallochum; aloes (Bible); aloes-wood (*qv.*); alim (Mal); wd. attacked by bacteria; slivers of low grade wd. found alternating with sound wd. (latter often in thin slices); heartwood c. VO (with dihydroagarofuran & derivs.), resin; u. in prepn. of "agar attar" (Ind); u. as perfumes, as joss sticks (Chin temples); incense; powd. wd. u. antipyr. in smallpox (formerly), rheum., carm. in indig. & abdom. dis., heart palpitations, anasarca; u. in Thai med.; two sorts known: agar real & dhum agar (1000) - **A. sinensis** (Lour.) Gilg (sChin); u. in place of prec.; wd. or its resin u. in kidney or male dis., in nausea, gas, chills, weak knees, &c.; injured wd. forms aromatic subst. of high value; "incense wood of Hainan."

Aquilegia L. (Ranuncul.) (temp. zones of both hemispheres): the columbines; a genus of ornam. herbaceous perennials; distinguished by having corollas bearing spurs - **A. alpina** L. (It, Fr): ancolie des Apes (Fr); aquilegia maggiore (It): unlike *A. vulgaris*, this sp. is dangerously toxic - **A. canadensis** L. (e&cNA): (wild) columbine; "meetinghouses"; "honeysuckle"; hb. u. as diur. emmen. diaph. ton.; sd. u. as perfume, crushed for headache & fever, diur. diaph. (Omaha Indians); toxic pl. - **A. formosa** Fisch. (Alas-Calif, e to Utah [US]): northwest crimson columbine; rts. c. alks. incl. magnoflorine; sds. u. in biliousness (Piute Indians) - **A. shinnersi** (Mex): rt. decoc. u. for bruises - **A. vulgaris** L. (nAs, c&sEu; natd. eNA): garden (or European) columbine; ancolie commune (Fr); entire pl. (herb) u. in jaundice, scurvy, as diaph.; hb. u. in nervous states, insomnia, skin dis. extern. in lotions; diur; sd. u. as gargle for mouth sores, throat dis., fistulas, eye weakness; toxic pl.; orn.

aquilmacal (Mex): yam.

araban: polyose resembling hemicelluloses of wds. & gums in structure of pectin & other pl. gums (ex. plum); yields arabinose when hydrolyzed.

Arabasan™: a dried purified sanitary acacia gum with no active enzyme content.

arabic acid: arabitic a.: tetrahydroxy valeric acid; isolated fr. gum arabic solns.

Arabic, gum: acacia.

Arabidopsis thaliana Heynh. (Cruc.) (Euras, Af, esp. Med reg; natd. Austral, NA): mouse ear (or Thale) cress; u. to cure sores in mouth (Spain) (Sp Phar.).

arabinose: pectin sugar; simple sugar (monose) occurs widely in plant spp. (incl. Mycobacteria), as component of sugar complexes, also in gum arabic (after which named); u. in nutrient media to identify microorganisms.

Arabis L. (Cruc.) (nTemp. Zone, incl. mts. of tropAf): rock (or wall) cresses; cult. ornam. hbs. (rock gardens) - **A. alpina** L. (Eu, Af [mts.], As [Himal]): with white fls.; pungent hb. u. as antiscorbutic - **A. glabra** (L.) Bernh. (Euras, natd. NA): tower mustard; young lvs. serve as potherb - **A. lyrata** L. (neAs, Alas [US]): young lvs. eaten (raw or boiled).

arabita (Por): *Pithecellobium dulce.*

Araceae: Arum or Calamus Fam.; large gp. (2950 spp.) of mostly trop. monocot. hbs. in which the pl. may appear tree-like in form; many spp. furnish foods; most are ornam.; ex. taro, calla lily.

arachain: proteolytic enzyme fr. peanut, *Arachis h.*

arachic acid: arachidic a.: characteristic fatty acid in FO of peanut, *Arachis h.* (where discovered) & other FO.

arachide (Fr): **Arachis** (L.): **arachis seed** (Eng): peanut.

arachidonic acid: an essential fatty acid, formerly one of those called collectively, "vitamin F"; occ. in many animal FO & foods; u. because of unsaturated fatty acid nature to treat skin dis. (veter.); involved in synthesis of prostaglandins, thromboxanes & leukotrienes; nutr.; in dermatitis of dogs, hogs; first discovered in peanut oil (hence name).

Arachis hypogaea L. (Leg.) (Br; now widely cult. worldwide): peanut; ground (or earth) nut; g. almonds; "grubies"; cultigen (wild form no longer known); very valuable crop of trop. & subtrop., esp. seUS (1001); after fertilization of fls., flg. branch bends over & pushes developing frts. into ground; frts. mature underground & c. 1-3 sds. ("peanuts"), a useful & important raw food; dried often salted, boiled (in salt water), roasted ("parched") (one of cheapest food protein sources), in chocolates (substit. for almonds) & candy; c. 50% of an edible FO u. in foods, injectables; peanut butter (ground sds. & additives) as bread spread (popular in US, Neth) fr. grinding roasted & skinned sds.; "p. milk" as milk substit. (2273); oil cake in stock foods; alc.-sol. lipid fraction of sds. u. to reduce bleeding tendency (oozing); peanut flour & p. meal (fr. sds. after expressing oil) u. as hemostatic in hemophilia (maybe by inhibiting fibrinolysis) (2401); in breads (cEu), macaroons, confectionery; waste peanut hulls u. as cork substit. (gaskets, &c.) & for insulation; in textile fibers, plastics, &c.; herbage (hay) u. as fodder; chief prodn. US, Af, esp. Ind (1002) (v. under peanut oil).

Arachnoidiscus ehrenbergii (Bacillarieae - Algae): a diatom sp. typical of much agar; an indicator of genuine Japanese agar.

arad: frt. of *Acacia senegal*.

araignée (Fr-f.): spider.

Aralia chinensis L. (Arali.) (eAs): Chinese aralia or C. angelica tree; rt. u. antidiabetic; stom., in dysmenorrhea; rheum. syph. (Chin); c. saponins with aglycone taraligenin (=panaxsapogenin); very young lvs. eaten as veget. (Jap); cult. ornam.; bk. u. diur., mixed with sugar as dressing for abscesses; rt. decoc. for flatulence, in plasters as antisep. - **A. continentalis** Kitagawa (Kor): rts. u. as analg. for labor pains & headache - **A. cordata** Thunb. (A. edulis Sieb. & Zucc.) (Chin, Jap, Manch. Kor.): udo; tu tang kuei (Chin); rts. diaph., recommended as substit. for ginseng; rts. & blanched shoots (lf. stalks) eaten as veget.; cult. - **A. elata** Seem. (eAs, fr eSiber to Jap): frts. & rts. u. for cough, diabetes, gastric cancer (Kor, Jap); rt. stom. (Ainu) - **A. hispida** Vent. (e&cNA): bristly (or hispid) aralia; bristly sarsaparilla; dwarf elder; a low shrub 0.2-1 m high; horizontal creeping rhiz. u. as ton. emet. diur. "alt."; hydragog; in poultices for swellings & infections (Potawatomi Indians) - **A. japonica** Thunb. = *Fatsia j.* - **A. mandschurica** Rupr. & Maxim. (eSiber, Manchur, Chin): Manchurian angelica tree (or aralia); rt. c. aralosides (triterpenoid saponins); rt. (bk.) u. ton. diur. for stomatitis; CNS stim.; anti-hypnotic; for nocturnal enuresis - **A. nudicaulis** L. (Cana, nUS): small/American/false/wild sarsaparilla; white s. (NJ [US]); small spikenard; rhiz. u. lax., sometimes as substit. for *A. racemosa* & to substit. for true sarsaparilla in root beer; *A. cordata* also so used; coll. in NCar [US] - **A. papyrifera** Hook. = *Tetrapanax papyriferus* K. Koch. - **A. racemosa** L. (e&cNA): (American) spikenard; life of man; petty morrel; spignet; "Indian root"; Aralia (L) (rt. formerly off. in US [-NF XI); rhiz. & rt. u. as stim. ton. (Am Indians), purg. diaph. diur.; expect. alt. in gout, rheum. scrofula; fresh rt. u. by homeopaths; med. values questioned; com. article in slices - **A. schmidtii** Pojark (eSiber): Schmidt's aralia; sts. u. to heal wounds (Sakhalin aborigines); analeptic; officially approved in USSR for med. use (1003) - **A. spinosa** L. (e&cUS, esp. common in seUS): prickly elder; angelica tree; Hercules' club; devil's walking stick; shot bush; "prickly ash" (confusion) (1004); toothache tree; much branched panicles of shrub or small tree; bk. c. araliin (saponin); & araligenin (= oleanolic acid); berries, fleshy rts. & bk. (USP 1879) u. stim. ton. "alt." sial. sudorific in arthritis; adulter. of prickly ash bk; cult. ornam. (protects property) - **A. villosa** Walt. = *A. spinosa*.

Araliaceae: ginseng (or spikenard) Fam.; large gp. (ca 800 spp.) of both temp. & trop. zones of world, represented by hbs. shrubs, lianes, & trees; many spp. u. med. or as ornam.; stems often prickly; ex. ivy, devil's club, rice-paper plant.

arali(i)n: saponin glycoside (crystd.) fr. bk. + rt. of *Aralia spinosa*; aglycone = araligenin = oleanolic acid - **aralin**: glycoside fr. lvs. of *Fatsia japonica*.

arana (SpAm): name appl. to several pl. spp., esp. *Hibiscus schizopetalus*, *Nigella damascena*.

arancia (It): orange.

arancini (It): small frt. of *Citrus sinensis*; powd. u. to adulter. black pepper.

aranduno (Sp): *Vaccinium myrtillus*.

aranea (L.): spider; s. web.

Aranea avicularis (L): web of bird spider; u. in homeop. med.

Araneus diadematus Clerck (Argiopidae - Arachnida) (Euras, nAf, Am): garden/cross/hazel spider; venom u. analgesic in neuralgia (longer lasting than morphine); in hemorrhaging, neuritis, dentalgia, mal. colds, flu - **A. exobolus** Thor. (Eu): black night spider; u. in hypertension.

arar(a) (tree): *Tetraclinis articulata*.

araroba (Eng, Fr): prod. fr. *Andira araroba* (cavities in heartwood); v. Chrysarobin.

Ararut (Ge): **araruta** (Sp, Por): arrowroot (Maranta).

Araucaria angustifolia O. Ktze. (A. brasiliana A. Richard) (Araucari.) (sBr, Arg): Brazilian pine; Paraná p.; candelabra tree; source of a turpentine (resina de pinheiro); sds. edible, sold in Br. markets; tree furnishes good lumber (important economic prod.) u. in box-making - **A. araucana** K. Koch (A. imbricata Pavon) (sChin, Arg): monkey puzzle (tree) (1005); Chile pine (-45 m); sds. (achenes) (roasted) ("Chile nuts") important food of some SA Indians; with aphrod. props.; resin u. for contusions, ulcers, headache; wounds (Araucanian Indians); wd. valuable lumber, u. for interior finish of houses; ornam. tree (1006) - **A. brasiliensis** A. Rich. = *A. angustifolia* - **A. columnaris** Hook. (A. cookii R. Brown) (New Caled, Polynesia); wd. u. by carpenters (in native areas); med. - **A. heterophylla** (Salisb.) Franco (A. excelsa R. Br.) (Norfolk Is): Norfolk Island pine; star p.; large tree (-60 m); produces a gum-resin; sds. edible; wd. excellent but scarce; formerly in ship building; seedling valuable ornam. pl for house; cult. outdoors in Med reg.: also Co. Cornwall (Eng).

Araucariaceae: araucaria fam. (sHemisph): evergreen resinous cone-bearing trees, often of large size; close to Pinaceae & sometimes included therein; ex. *Agathis* (Dammara).

Araujia sericofera Brot. (Asclepiad) (SA - sBr): cruel plant (SC fl. captures moths): latex u. for warts; fiber u. in textiles; cult. ornam.; climber.

arbol (Sp): tree - **a. de orquidea** (Mex): *Bauhinia purpurea* - **a. del cielo** (Sp): 1) *Rhus vernicifera* 2) *Ailanthus* spp. - **a. del hule** (Sp): *Castilla elastica* - **a. del Peru** (Mex): *Schinus molle*.

Arbor (L): tree - **Arbor Vitae** (L, Eng) (lit. tree of life): *Thuja occidentalis* & other *T.* spp. - **American A. V.**: *T. occidentalis* - **Western A.V.:** *T. plicata.*

arbor viruses: arthropod-borne viruses (formed fr. first 2 words).

arborescin: subst. in *Artemisia arborescens* L. & *Matricaria globifera* u. as contraceptives (ancient Greece).

arbre (Fr): tree - **a. à chandelle** (Fr-WI): *Byrsonima* spp. - **a. à suif**: *Myrica cerifera*, etc. - **a. de lilas** (Fr, Qu): *Syringa vulgaris* - **a. de neige**: 1) *Chionanthus virginica* 2) *Viburnum opulus* - **a. impudique** (Fr): **a. indecent:** *Pandanus* spp. (SC rts. dropping down, appearance like penises).

arbusto (Sp): shrub.

arbute (Eng): **a. tree**: *Arbutus* spp.

arbutin: **arbutoside**; a hydroquinone glucoside (hydrolyzes with β-glucosidases —> hydroquinone & (beta-d)-glucose); fr. lvs. of blueberry, cranberry, pear, &c.; common among members of fam. Eric.; (*Arctostaphylos uva=ursi*; *Arbutus unedo*; *Vaccinium vitis-idaea*; *Bergenia crassifolia*); u. diur., urin. anti-infective; to stabilize color photographic image.

Arbutus L. (Eric.) (Euras, Am): strawberry trees; shrubs or small trees with dry berries - **A. andrachne** L. (native to Med reg., esp Greece): eastern strawberry tree; oriental s. (t.); u. med. in ancient Greece, called komaros (Dioscorides); lvs. browsed at times by goats & berries by partridges & by natives (AsMin); often cult. as ornam. - **A. menziesii** Pursh (Pacific US, BC [Cana] - nMex): (Pacific) madrone (tree); madroño (Mex); madrona (laurel); bk. (madrona bk.) exfoliates fr. smaller trunks of evergreen tree (-40 m); u. astr. in diabetes; as tan; lvs. rubbed on burns, rheum. pain; decoc. in colds, sore throat, stomachache; lvs. c. "madronin" (mixt. condensed & hydrolyzed tannins) which is antibacterial agent for Gram pos. & acid-fast bacteria; lvs. smoked like tobacco; bk. ext. to tan paddles, fishhooks; wd. u. to make charcoal for gun powder; useful for furniture; frt. edible (raw, boiled, stewed); most of these uses by Calif (US) Indians; ornam. tree - **A. unedo** L. (A. vulgaris Bub.) (Med reg., swIre): (European) strawberry tree; madrono (Sp); madrone (*cf.* madrona [Sp], mother who spoils her children); bk. & lvs. c. arbutin, tannin, u. astr. in diarrh. (formerly urin. inf.); mealy frts. (berries) edible (but somewhat intoxicant); source of wine (Corsica) & brandy (Ionia); in marmalades; water distd. fr. fls. u. diaph.; lvs. frt. bk. formerly u. as tan; bk. c. 45% tannin; (SC frt. similar to strawberry both in form & color) - **A. uva-ursi** L. = *Arctostaphylos u.-u.* - **A. xalapensis** Kunth (A. texana Buckl.) (CA, Mex, swUS): madrono; Texas madrone; lvs. u. to feed silkworms; wd. useful for several purposes.

arbutus, trailing: *Epigaea repens*.

Arbuz: prepn. with special activated form of papain; recommended for use in GIT dis. where proteolytic enzymes are deficient.

Arca granosa L. (Arcinae - Arcidae - Mollusca) (marine habitats, worldwide): ark shells; conkle s.; Conchae arca; walengzi; this bivalve shellfish with rock-boring habit is used in Chinese med. (shell).

arcachon (Fr): **arcanson**: colophony in the shape of bread loaves; prep. in Eu by melting galipot in boiler, letting cool in mold; u. only in mfr. plasters & ointments; term also appl. to ordinary rosin, coal tar pitch, and a mixture of pine pitch & rosin.

arcana: secret remedy; nostrum.

Arcangelisia flava (L) Merr. (Menisperm.) (trop seAs): rt. & st. of liana u. as inf. as emmen. abort.

antipyr. ton. expect.; in hepatitis; c. berberine, palmatine & other alks.; lvs. u. in head packs for headache; wd. pulp u. for stomach dis. (PI).

Arceuthobium Bieb. (Loranth./Visc.): dwarf mistletoes; aerial parasites on Pinaceae & Cupressaceae, widely distr. in NA, WI, As; produce serious damage in NA forests (1007) - **A. americanum** Nutt. (Cana, nwUS, s to cCalif): (lodgepole) pine (dwarf) mistletoe; berry decoc. contraceptive; pl. decoc. for hemorrhages in mouth & lung; TB, stomachache (571) - **A. occidentale** Engelm. (Calif [US]): "wild ginger"; pl. decoc. u. as tea for stomachache (Indians, Mendocino Co., Calif) - **A. vaginatum** J. Presl subsp. *cryptopodum* Hawksworth & Wiens (A. c. Engelm.) (swUS, nMex): southwestern d. m.; u. similarly.

Archaebacteria: methanogens; microorg. which utilize CO_2 + H_2 to produce CH_4 instead of using oxygen, thus with different chem. requirements fr. other plants.

archangel root: *Angelica archangelica.*

Archangelica Wolf v. *Angelica.*

archegoniates: members of Bryophyta & Pteridophyta characterized by presence of archegonium as female organ.

Archichlamydeae: a sub-class of Dicots in which the petals are separate & distinct (Polypetalae) (ex. Rose Fam.) or else entirely lacking (Apetalae) (ex. *Clematis*).

Archidendron clypearia (Jack) I. Nielsen (Leg.) (Ind to NG): lvs. u. to darken the skin; dr. & pd. lvs. u. to treat sores (Indochin).

archil: *Rocella tinctoria.*

Archimycetes = Myxochytridiales: primitive fungi roughly including the Plasmodiphorales & Chytridiales.

architectonic: relating to architectural structure in an organ or system or in tissue composition.

arcilla (Sp): argil; kaolin.

Arctium lappa L. (Comp.) (Euras, nAf; natd. eNA): (great) burdock; pl. classed as weed in most places, but cult. by Jap for edible rt. (gobo), a popular vegetable with them; young lvs. eaten as salad (Scand); rt. c. insol. gum/oleoresin/tannin (traces); VO (traces); sitosterol, stigmasterol, polyenes; fresh rt. u. antibact. in furunculosis, seborrhea, impetigo, psoriasis, eczema, acne; lvs. also u.; rubbed into scalp (prevent hair falling); dr. rt. u. in western world (as Radix Bardanae) as diur. diaph. stomach. in skin dis. depurat.; for burns, rhus poisoning ("cure") (per ora as fluid ext.), diabetic food (as cookies [1008]); frt. ("sd.") c. 15-20% FO (1009), u. lax. diur. & stim. appetite; for dry scalp; c. antibiotic subst.; lvs. & frt. diur. "alt." depur. - **A. majus** Bernh.: this sp considered conspecific with *A. lappa* - **A. minus** Bernh. (Euras; natd. eCana, US): common (or smaller) burdock; rt. u. med. much as *A. lappa*; for this sometimes cult.; fresh lf. placed on head for headache (Abenakis Indians); decoc. of young fl. heads & rts. u. in rheum., pl. in "shaking ague" (mal.) (Abenak); frt. ("seed") c. FO, u. "alt."; sometimes differentiated into flg. & non-flg. plants - **A. tomentosum** P. Mill. (Euras; natd. eCana, e&cUS): clotbur; filzige Klette (Ge); Filzklette (Ge); popularly u. in Eu as Radix Bardanae much like *A. lappa* - **A. vulgare** (Hill) Evans (A. nemorosum Lej. & Court.) (Eu; natd. eCana, neUS): burdock; bat weed; hare burr; u. similarly to *A. lappa* .

Arctopus echinatus L. (Umb.) (sAf): bear's paw; rt. decoc. u. with KNO_3 for epilepsy, syph. (Europeans in Af); natives use popularly as rem. for bladder ailments, GC, purg.; in hypochondria.

Arctostaphylos Adans. (Eric.) (NA): some 80 spp. of evergreen shrubs or small trees; mostly northern, 2 spp. circumpolar (1010) - **A. alpina** Spreng. (Arctous alpinus Niedenzu) (nEuras, nNA [New England]): alpine bearberry; black b.; lvs. & twigs u. as tan; berries black, edible, most valued frt. of Eskimos - **A. arguta** (DC) Zucc. (Mex): madrono borracho; frt. u. as mild hypnotic for alcoholics; nervous insomnia; slightly lax.; lf. decoc. narcotic astr., u. to induce sleep - **A. glauca** Lindl. (cCalif to nwMex): (big-berried) manzanita; great-berried m.; lvs. c. arbutin, tannin; u. astr. like *A. uva-ursi*; food (Am Indians); ornam. shrub - **A. glaucescens** Kunth (Mex): macuate; frt. of shrub u. as medicinal, also edible - **A. nevadensis** Gray (wUS): pinemat manzanita; frt. edible; dr. lvs. mixed with tobacco & smoked (Klamath Indians, Oregon) - **A. patula** Greene (Wash to Calif, e to Utah [US], nMex): greenleaf manzanita; berries edible, better cooked; lf. decoc. wash for poison ivy; lf. chewed as poultice for sores; lvs. smoked - **A. pungens** HBK (Calif, Utah [US], Mex): (Mexican) manzanita; pingüica; frts. of shrub small spherical checked drupes of brilliant coffee-brown color, very sweet (c. 41% sugars) called "manzanitas" (Hidalgo, Mex); frt. & lvs. reputed astr. & diur. u. in edema, bronchitis, VD, &c.; wd. u. to make pegs, spoons; ornam. shrub - **A. tomentosa** (Pursh) Lindl. (Calif): (shaggy-barked) manzanita; frt. important food (Calif Indians); u. to prep. cider (Indians and whites); lvs. smoked; u. in med. - **A. uva-ursi** Spreng. (nHemisph, cooler regions; nEuras, NA, s to Calif): (red) bearberry (SC frt. eaten by black bear); rockberry; lvs. c. arbutin, methyl a., hydroquinone, tannin; u. as diur. astr. disinfect. in urinary & kidney dis. (cystitis, urethritis, pyelitis, lithiasis), in diabetes; hemorrhages, diarrh.; sometimes as part of mixt. with tobacco called "kinnikinnic(k)" as fumitory; in leather & textile indy. as tan & blue dye; lvs. as tea ("Caucasian tea") (Russia); berries edible (prodn. Sp, It).

Arctotis L. (Comp.) (sAf, Austral): some 50 spp. of herbs; c. latex; hb. of some spp. u. as appln. to sore breasts, douche in uter. cancer; cult. orn. - **A. stoechadifolia** Berg. (sAf): with antibiotic properties.

Arctous alpinus Nied. = *Arctostaphylos a.*

Ardall™: sodium salicylate fr. pl. sources.

Ardisia colorata Roxb. (Myrsin.) = *A. stylosa* - **A. crenata** Sims (A. crenulata Lodd.) (seAs): mata ajam (Mal); tea with other herbs u. for bruises, broken bones, sprains (also extern.); frts. consumed by natives (Chin) & u. med. - **A. crispa** A DC (A. elegans Andr.) (seAs, Chin, Ind to Jap-Indones): rt. u. for fever & ptyalism (excessive salivation); to heal injuries (Chin); u. diur., antidote (Taiwan); rt. inf. for resp. dis. (Indochin); lf. sap u. for earache, rt. inf. in diarrh. & dys. (Mal) - **A. escallonoides** Schiede & Deppe (Icacorea paniculata Sudw.) (WI, Mex, sFla [US], Guat): marlberry; morita; wd. of small tree useful - **A. ixoraefolia** Pitard (Indoch): boiled fls. component of inhalations u. to relieve pain fr. decayed teeth - **A. japonica** Bl. (Chin): rt. decoc. u. for influenza, carm., poison antidote; leafy shoot decoc. u. to stop cough, uter. bleeding - **A. pardalina** Mez (PI-endem.): lvs. appl. to abdomen for amenorrhea (PI folk med.) - **A. pickeringia** T. & G. = *A. escallonoides* - **A. quinquegonia** Bl. (Chin, Indochin): lf. inf. u. to treat toothache, lumbago (Indochin); u. med. (Chin) - **A. solanacea** Roxb. (Ind): rt. u. antipyr. in diarrh. rheum - **A. stylosa** Miq. (Mal+): lampene gedi; lf. decoc. u. in colic; rt. in combn. as postpartum protective med.; lvs. rubbed over head & body for headache (PI) - **A. villosa** Roxb. (seAs): remedy for contusions, rheum. neuralgia (Chin); lf. decoc. in edema; rts. for jungle fevers (Mal) - **A.** sp. (Pan): iskarmas; u. med. by Pan Indians; berries u. to stain body.

Areca catechu L. (Palm.) (orig. Sunda Is; now sAs, esp. Ind, Indochin): betel (nut) palm; areca (nut) p.; sds. (areca nut, betel nut; pan [Ind]) c. arecoline, guavacine, & other alks., FO, VO (traces), tannin; u. as vermif. (tenicide) (mostly in veter. med.), also for mfr. arecoline & its salts; sd. mostly u. for chewing (1011); u. in hallucinogenic mixts.; natives believe prevents dys.; said u. by some 200 million people in Asia; chewed, mixed with pale catechu, tobacco, betel lvs., lime, &c.; chewing associated with high levels of oral carcinoma; sd. u. as dye & tan (as Bombay or palm cutch, also frt. & heartwood) (esp. in Chin) (shipped in 100 lb. burlap bags, as tends to heat) - **A. hutchinsoniana** Becc. (PI): pisa (Yakán); raw terminal bud given to children as vermif. - **A. triandra** Roxb. (seAs): sd. u. as chew by natives similarly to *A. catechu*; claimed stim., relieving fatigue & hunger, sialagog; tree often cult. in Cambodia.

areca nut: fr. *Areca catechu.*

arecaidine (arecaine): **arecolidine**: alks. of areca nut; nicotinic acid derivs.

Arecastrum romanzoffianum Becc. = *Syagrus romanzoffiana* Glassman.

Arecolinae Hydrobromidum (L.): arecoline HBr. - **arecoline**: characteristic alk. of areca nut; u. mostly in veter. med. as tenifuge (dogs, poultry); anthelmintic (cestodes in humans).

Areka (Ge): *Areca.*

Arenaria juncea Bieb. (Caryophyll.) (Chin): rt. u. in mal., fever, kala azar. - **A. lanuginosa** (Mx) Rohrb. (Ec): chinchi mani; u. for kidney dis. along with other drugs (2449) - **A. serpyllifolia** L. (Euras; natd. NA): (thyme-leaved) sandweed; hb. c. saponins; lf. decoc. u. in dys. & as eye wash; hb. u. in bladder dis. (Chin); weed; cult. orn.

arene: aryl compds.: hydrocarbon compds. contg. at least one arom. ring.; ex. anthracene.

Arenga pinnata Merr. (A. saccharifera Labill.) (Palm.) (Indochin, seAs): sugar palm; gomuti (sugar) p.; st. sap furnishes palm sugar (jaggery); u. to prep. palm wine (toddy); giving on distn. arrack; fresh sap purg., u. in sprue (Sunda Is); frt. juice toxic, c. Ca oxalate, u. criminally as poison in coffee (Mal); petiole fuzz u. hemostatic; rt. stom., in chest dis. (Cambodia); petioles diur.; st. pith source of kind of sago; lf. sheaths u. for fibers (indust.) ("vegetable horsehair"); wd. finest palm wood; sts. u. for water pipes; (cult. Ind, Fla [US]) - **A. tremula** Becc. (PI): abigi (Bikol); bud eaten to excess produces a kind of intoxication & deep sleep; frt. toxic (Ca oxalate raphides) - **A. westerhoutii** Griff. (seAs): rts. u. in lithiasis, chewed like betel (without bark); frt. toxic.

arenobufagin: cardioactive glycoside found in *Bufo arenarum* (toad of SA).

arenque (Sp, Por): herring.

Arethusa bulbosa L. (monotypic) (Orchid.) (seNA): dragon's mouth; wild pink; rose lip; bog rose; cult. ornam.

argalia (Co): wild okra.

Argania spinosa (L) Skeels (A. sideroxylon Roem. & Schult.) (monotypic) (Sapot.) (wMor): argan (tree); on expression, sd. yields FO (like olive oil)

u. in foods, to invigorate body, aphrod.; wd. ("ironwood") hard & durable; frt. eaten by cattle; source of brandy; valuable gum; sds. often secured fr. goat feces & dried out over open fire (2497).

argel leaves: fr. 1) *Solenostemma argel*; u. adult. of Alex. Senna 2) (Mex): *Tecoma stans.*

Argemone albiflora Hornemann (A. alba Lestib. f.; A. intermedia Sweet) (Papaver.) (c&sUS, Mex): (white) prickly poppy (or thistle); pl. sap u. for sore eyes (Am Indians); hb. & sd. u. as narcot.; sds. emet. cath. c. FO (purg.) - **A. glauca** (Prian) Deg. (Haw Is): prickly poppy; capsule juice u. in toothache, ulcers, neuralgia, &c.; st. juice u. for warts; sds. u. like those of poppy - **A. mexicana** L. (seUS, nMex, CA, SA; natd. in trop): Mexican (prickly) poppy; thorn p.; M. (or yellow) thistle; chicalote (Mex); cardo (Mex); a yellow-flowered thorny hb. (ann.); latex c. alks.; sanguinarine (sds. rt.); chelerythrine derivs.; protopine (all parts) &c.; latex u. exter. for corneal spots, ophthalmia, skin dis., warts, muscular pains; int. diur. in dropsy; hb. lax., narcotic, sed.; rt. u. in guinea worm infestation (Ghana); as substit. for ipecac; fls. pectoral & narcotic; sds. lax. c. 30% FO (argemone oil), c. linoleic & ricinoleic acids; u. in soap mfr., as fuel, illumin., drying oil (1013); u. cath.; in skin dis.; oil dangerous adulterant of cooking oils; press cake u. as fertilizer; bee plant; weed - **A. ochroleuca** Sweet = *A. mexicana* - **A. platyc-era** Link & Otto = *A. mexicana* - **A. subfusiformis** Ownb. (Arg): cardo amarillo; lf. u. as ocular antisep. (exter.); rt. ext. (alc.) u. cholagog, antipyr. (Arg).

argenteuil: a dish composed primarily of Asparagus.

argil: argile (Fr): **Argilla** (L.): clay (pharm.): alumina.

arginase: enzyme which hydrolyzes l-arginine into ornithine & urea; abundant in liver.

arginine: essential amino acid, important to formation of prolactin, insulin, glucagon, somatotropin; obtained by hydrolysis of gelatin, &c.; stim. growth hormone secretion; may aid in healing of wounds; building body tissues; aiding penile erection; u. in hepatitis (to detoxify ammonia); diagnostic aid (pituitary).

argol(s): Weinstein (Ge) (wine stone); crude potassium bitartrate (50-90%); wine lees; formed on sides & bottoms of casks in which wine is fermenting; c. 6-12% calcium tartrate; u. in mfr. tartaric acid & tartrates; as mordant; in fertilizers.

argon: Ar; element abundantly present in earth's crust & in atmosphere; one of the noble gases (unreactive); u. in fluorescent light tubes.

argousier (Fr): *Hippophaë rhamnoides.*

Argylia huidobriana Clos. (Bignoni) (Ch): triaca; u. as stom. stim. & in catarrhal states (folk med.).

Argyreia acuta Lour. (Convolvul.) (seAs): silverweed; whole pl. decoc. u. int. & exter. for edema, tympanites (Chin); diur. emmen. (Indochin) - **A. cuneata** Ker. (Ind): kallana gida (Kanarese): lvs. u. in diabetes - **A. nervosa** Boj. (A. speciosa Sweet) (orig. Ind, Malaysia; natd. WI): woolly asiaglory; elephant climber (creeper); "ipomea" (Cu); rt. u. in arthritis, nerv. dis.; lvs. cooling, emoll. in poultices for wounds & skin dis.; sds. u. aphrod., hallucinogenic (105); to moderate effects of aging process; ornam. (cult.) - **A. wallichii** Choisy (seAs): pl. decoc. drunk as galactagog; bk. wrapped fresh around painful legs (Thai) (2448).

Argyrolobium marginatum Bolus (Leg.) (sAf): silver leaf; rt. c. alk.; u. in hiccup, gastrit. - **A. transvaalense** Schinz (sAf): remedy in female sterility.

Argythamnia heterantha Müll. (Euphorbi) (Mex): azafrán de bolita; aperitive - **A. tricuspidata** (Lam.) Muell. Arg. (Chiropetalum tricuspidatum [Lam.] A. Juss.) (Ve, Chile): lvs. & sts. u. as carm. in flatus; stim. (Chile); c. blue coloring matter.

aribine = alkaloid, harman < *Arariba = Simira.*

aricine: alk.; 5-methoxy-ajmalicine; occ. in *Rauvolfia canescens, R. hirsuta, Cinchona ledgeriana, C. pubescens;* identical with heterophylline.

Aries (L): ram, male sheep.

Arikuryroba schizophylla Bailey = *Syagrus s.*

aril(lus): **arillode**: fleshy outer covering of some seeds (such as mace of nutmeg).

Arillastrum gummiferum Panch. (monotypic) (Myrt.) (New Caled): chêne gom; st. produces exudn. of resin (20%) with 80% tannin; bk. tannin rich (15%); wd. valuable for interior work.

arind (Pak): castor bean.

aringa (It): herring.

Ariocarpus retusus Schiedw. (Cact) (sTex [US], Mex): false peyote; "chaute"; c. N-methyl-tyramine & other alks.; u. as med. & narcotic tranquilizer (Indians); similarly, *A. fissuratus* (Engelm.) K. Schum. with hordenine (phenylethylamine deriv.).

Arion empiricorum Ferrusac. (Arionidae - Mollusca) (Eu): slug; night snail; u. for cough, to prod. bronchial mucus softening; u. in homeopath med. ("Limax ater").

Arisaema amurense Maxim. (Ar.) (nChin): tuber processed & u. in blood poisoning, stroke, childrens' convulsions - **A. atrorubens** Bl. = *A. triphyllum* Schott - **A. brevistipitatum** Merrill (Chin, Kwang Tung): med. value claimed - **A. consanguineum** Schott (swChin): tuber soaked & u. in apoplexy, stomachache, &c. - **A. dracontium** (L) Schott (e&cNA): green dragon; dragon root; lvs. u. in edema (Am Indians) for female dis. (Menominees); in homeopathy; tuberous rt. ("tuber") (as Radix Aronis dracontii), u. like *A. triphyllum*; u. in cosmetics ("Cupress Powder") - **A. heterophyllum** Bl. (eAs): corm u. as sialagog (Chin); in stroke, hemiplegia, epilepsy (w&cChin); mixed with aconite as dart poison (Ainu) - **A. japonicum** Bl. (eAs): tuber u. as ton. expect. sed. in convulsions, anodyne, for numbness after stroke - **A. lobatum** Engl. (Chin): pd. corm appl. as antisep. to malignant sores - **A. peninsulae** Nakai (eAs): raw-raw (Ainu); sd. u. in gastralgia - **A. propinquum** Schott (A. wallichianum Hook. f.) (Ind, Pak-Himal): c. alk.; u. in cancer therapy - **A. ringens** Schott (Kor): tuber u. as vermif., bechic. - **A. speciosum** Mart. (eAs-Himalayas): kiralu; rt. u. as antidote for snakebite; tuber u. for colic in sheep & worms in cattle - **A. tortuosum** Schott & Hook. f. (Ind, Nepal): cobra flower, c. plant (also appl. to *A. speciosum* Mart.) (fr. inflorescence appearance); toxic to cattle, u. by villagers to poison wild animals (cats, &c.); raw tubers contain sterols, Ca Oxalate u. for colds & cough (with black pepper); to deworm cattle; considered antifertility, insecticide, insect repellant; sds. u. for colic in sheep (with NaCl) - **A. triphyllum** Schott (eNA): (common) Jack-in-the-pulpit; Indian turnip; Canada (or wild) t.; dragon rt.; rt. u. in indigenous med. as carm. expect. & as food (Radix Aronis triphylli); pd. u. in cosmetics; rt. of this & other spp. c. acrid principle & Ca Oxalate, hence strongly irritant & poisonous if eaten raw (however, black bears said to eat); variable (subspp.: *triphyllum*; *pusillum* Huttleston; *stewardsonii* Huttleston).

Arisarum vulgare Targ. Tozz. (Ar.) (sEu, Eg): friar's cowl; corms u. as emet. (*cf. Arum maculatum*); as famine food.

Aristea africana Hoffmnsg. (Irid.) (sAf): hb. inf. u. in abdominal troubles.

Aristida adscensionis L. (Gram) (pantropics): six weeks three-awn; fl. ashes made into ointment with lard & appl. to ringworm, itch (Madag) - **A. purpurascens** Poir. (e&cUS, Hond): arrow feather; u. forage for stock (US) - **A. sieberiana** Trin. (A. longiflora Schum. & Thonn) (trop wAf): rts. u. as antipyr. antimal.; in conjunctivitis (as pomade in butter); for thatching & matting - **A. stipoides** Lam. (trop wAf): makir (Niominka race); gall growing on grass u. as vermif., in anuria; plaited to make mats - **A. stricta** Mx. (NCar-Fla-Miss [US]): "wire grass"; pine-land three-awn; u. as forage grass (SC arista [L]: awn).

Aristolochia L. (Aristolochi.) (trop & temp zones): some 300 spp. of hbs. & vines; several med. spp. (1988) - **A. albida** Duch. (trop wAf): bitter rt. u. as ton. & for Guinea worm - **A. anguicida** Jacq. (CA, WI s to Co, Bo): South American birthwort; source of guaco (Cartagena) (*cf. Mikania amara*); u. in stomachache; lf. & lf. juice u. for chronic rheum. pains (Mart); antidote in snakebite (1020); pl. u. to clean clothes in washing (Salvador) - **A. arborescens** L. (sMex to Co): Dutchman's pipe; mata; rt. in snakebite (Pan); lvs./rts. emmen. antipyr. diaph. in colds, dys. VD, abort - **A. argentina** Hesse (Arg): rhiz. c. aristolochine, aristoline; u. diur. emmen. diaph. - **A. bilobata** L. (Haiti, DominRep): bejuco calzon; rt. inf. u. as stim. ton.; decoc. as emmen.; pd. rt. appl. to ulcers - A. **bracteolata** Lam. (A. bracteata Retz.) (Ind, Pak, tropAs, Af): bracteated birthwort; a shrub; lf. inf. u. as strong anthelm. & purg. (local physicians); lvs. brewed with castor oil in psoriasis; lvs. & rts. for Guinea worm; rt. u. to cure skin dis. (EI), for scabies, anthelm. (Ind) - **A. brasiliensis** (Mart. & Zucc.) = *A. ringens* - **A. clematitis** L. (Eu, [Med. reg.] Caucas reg., AsMin): birthwort; heartwort; European aristolochia; Osterluzei (Ge); pl. associated with vineyards; rhiz. c. aristolochic acid, clematitin (NBP), 0.4% VO; hb. u. vulner. for sores; in surgical operations on neck, nose, ears; in vet. med. for treating sterility; long u. in med. as emmen. abort. diur., in arthritis - **A. cordiflora** Britt. (CA, Co): bastard contrayerba; with large, bad smelling fl.; remedy for female VD (Salv) & infant dys. - **A. cordifolia** Mut. (SA): u. med. in Colombia - **A. cymbifera** Mart. & Zucc. (SA): thousand men wort; guaco; rts. u. as emmen. in diarrh. asthma, as diaph. diur. aphrod.; stim. in snakebite (Br) - **A. durior** Hill = *A. macrophylla* - **A. fragrantissima** Ruiz (Pe, Ec. Amazonas): bejuco de la estrella; u. in stomachache, diarrh., colds; snakebite - **A. glaucescens** HBK (Guyana): yellow pareira; rt. u. as purg. diur.; long-held source of Pareira Brava, more recently u. as adulter. of same - **A. grandiflora** Swartz (Mex, CA; natd. Jam): guaco (huaco); pelican flower; p. plant; "cunt flower"; rts. u. sudorific, abort., emmen., reputed effective in snakebite, &c.; young shoots veget., older pl. toxic; st. juice vesicant; lvs. u. in bath for gout; rt. stock toxic, u. for snakebite (Jam); ornam. with enormous flowers (-2.5 m. long); cult. in greenhouses; has abominable odor - **A. indica** L. (sAs, esp. Ind.): Indian birthwort; sunanda (Sanscr); isharmul; a liane; rhiz. & rts. & st. u. bitter ton. stim. in dyspep. indig., fevers, scabies, worms; emmen., snakebite, "white leprosy" (BPC 1934; Ph. Indica); u. abortient; c. "aristolochine" [= curine]; VO. c. cischwarone - **A. inflata** HBK (CA): entire pl.

boiled & decoc. appl. to skin in snakebite (2444) - **A. kaempferi** Willd. (eAs): u. in hemorrhoids, ascites - **A. klugii** O. C. Schmidt (Co, Pe, Br): bejuco caner; curarina (Co); u. for stomachache, toothache; headache; in snakebite int. & exter. (as poultice) (Amazonian Indians) - **A. lagesiana** Ule (Pe, Br): sapo huasco; u. in acute indigestion, cholera; rhizome decoc. for fever, snakebite - **A. ligulata** Ule (Pe): u. for stomach- and headache - **A. littoralis** Parodi (A. elegans Mart.) (Ind, wild & cult. in Af): calico flower; rhiz. c. alks., incl. cryptopleurine (similar to colchicine) - **A. longa** L. (Med reg., AsMin): long birthwort; hb. & rt. u. as stim. ton. in generalized weakness, snakebite, in sores, in childbirth; depurat.; abscesses, as "blood purifier"; ingredient in the celebrated "Ointment of the 12 Apostles" (XVII cent.) - **A. macrophylla** Lam. (A. sipho L'Hér.; A. durior Hill) (eUS): Dutchman's pipe; pipe vine; c. aristolochic acid, VO; flavones; u. like *A. clematitis*; cult. in Eu (Ge): ornam. - **A. macrota** Duch. (Surinam): sts. decoc. in colds, stomachache, to facilitate delivery - **A. manshuriensis** Komarov (Kor, Chin): sts. & shoots u. as antipyr. diur., in diabetes - **A. maurorum** L. (Iraq, Iran): sts. c. aristolochic acid, u. vulner. for scab of sheep - **A. maxima** Jacq. (Mex to Ve, Co): guaco (del sur); sts. u. as antidote for snakebite; decoc. in severe diarrh. - **A. medicinalis** R.E. Schultes (Co; Amazonian reg.): astr. rt. of vine u. as very bitter & pungent inf. for periodic attacks (epilepsy?) among Kubeo Indians; u. with care since toxic - **A. mollis** Standl. & Stey. (Guat): hoja del aire; lf. decoc. u. in stomachache - **A. mollissima** Hance (Chin): hairy birthwort; rt. & branches (Radix et Ramus) u. as contraceptive in Chin med. - **A. odoratissima** L. (WI, sMex to Pe, Par): sweet-scented birthwort; junction vine; saragossa; decoc. of woody sts. ton. to stim. appetite, in colds; diabetes; vermif.; bk. inf. u. for aches & pains - **A. pandurata** L. (trop.): u. for snakebite, wound antisep. - **A. pardina** Duchartre (Mex): lvs. antipyr.; sts. for cordage - **A. pentandra** Jacq. (Fla Keys [US]l WI, Yuc): coastal a.; guaco (de Yucatán); rt. & sts. antipyr. in neuralgia, rheum. emmen. - **A. philippinensis** Warb. (PI): barubo (Negrito); rt. decoc. stom. emmen. - **A. pistolochia** L. (sEu): French (or Spanish) birthwort; rhiz. u. similarly to *A. longa* - **A. reticulata** Jacq. (sUS): Texas snakeroot; T. serpentaria; source of Red River snakeroot; rhiz. & rts. c. reticuline; u. as arom. bitter, diaph. expect. stom. ton. (in whiskey) - **A. ringens** Vahl (A. brasiliensis Mart. & Zucc.) (SA): jarrinha (Br): u. as spasmolyt. in allergic dis., stomach, intest. & gallbladder colic, angina pectoris, bronchial asthma; ton. diur. antisep. emmen. in amenorrh. postpartum uterine inertia; folk med. antipyr. in high fevers, exter. appl. orchitis, for ulcers - **A. rotunda** L. (Med reg.): round (rooted) birthwort; rt. u. bitter ton., arom., stim. diaph. emmen. - **A. rugosa** Lam. (Trin): rt. inf./decoc. for indiges. urinary burning, dysmen. jaundice - **A. sempervirens** L. (Med reg.): climbing birthwort; rhiz. u. ton. stim. - **A. sericea** Blanco (PI): entire fresh pl. u. as carm. emmen. antipyr. gastralgia, uter. ton., abort. - **A. serpentaria** L. (e&cUS): (Virginia) snakeroot; snakeweed; sangree root; rhiz. & rts. c. VO with borneol; aristolochic acid (aristolochin); aristolactone; u. ton. diur. diaph. in fever, typhus, expect. in bronchitis; stom., gastric stim. - **A. tagala** Cham. (PI): rts. of liana u. ton. carm. emmen. infantile tympanites (appl. to abdomen) - **A. triangularis** Cham. & Schlecht. (Arg): "mil hombres"; rt., rhiz. & flg. tops c. aristolochic acid; u. as diur. diaph. - **A. trilobata** L. (WI, Belize to Br, Ve): tobacco pipe; contrayerba; rt. decoc. u. in colic, fever; emmen. abort.; lf. decoc. u. in diabetes, hypertension, snakebite; difficult menstr. - **A. wrightii** Seem. (Tex [US], Mex): u. for ulcers.

aristolochia root: *Aristolochia serpentaria* - **European a. r.**: *A. clematitis*.

Aristolochiaceae: birthwort fam.; dicot. hbs. & shrubs, or vines, mostly of temp. zone; frequently c. VO; ex. wild ginger, serpentary.

aristolochic acid: aristolochine = curine: aristolochia yellow; gp. of phenanthrene alk. (with COOH gp.); acts to stim. phagocytic activity of leucocytes; drug of value in infections; u. against postoper. staph. infections; antineoplastic, occ. in many *Aristolochia* spp., incl. *A. clematitis*, *A. rotunda*, *A. longa*, *A. indica*.

Aristolochite Society: early title (1908-) for honorary prof. U.S. society in pharmacy, later (1924) known as Rho. Chi. (named after Aristolochia, meaning well-born) - **aristoloquia** (Sp): *Aristolochia clematitis*.

Aristotelia macqui L.'Hér. (A. chilensis [Mol.] Stuntz) (Elaeocarp.) (Ch): maqui; lf. juice u. for throat dis.; bruised lvs. u. as cataplasms for tumors & also appl. to back & kidney areas to alleviate fever; frt. u. diarrh. dys.; "dust" u. for wounds; c. several alkaloids; frt. edible, u. to color wine; named after **Aristotle** (384-322 BC), Greek naturalist; his ideas prevailed in Eu for many centuries - **A serrata** W. Oliver (A. racemosa Hook. f.) (NZ): wineberry; frt. eaten, u. to color wines; wd. u. for inlaying; to make charcoal (for gun powder).

Arizona gum lack: A. shellac: v. *Larrea*.

Armadillidium pictum Brandt (A. vulgare Latreille) (Armadillidiidae - Isopoda - Crustacea - Arthropoda) (Euras, now widely distr. in damp places): pill bug; "roly-poly"; wood louse; rolls into ball; has been u. as diur. in dropsy, asthma; bronchial catarrhs, blood spitting (alc. ext., powd., roasted & dropped in tea).

armadillo, (common) (Texas): *Dasypus novemcinctus* L. (Dasypodidae - Edentata - Mammalia) (Arg to Ala [US]): 9-banded a.; omnivorous burrowing animal -0.8 m. long; flesh eaten; armored "shell" u. sometimes as basket.

armamentarium: the equipment of a medical professional, incl. medicines, books, instruments, &c.

Armatocereus laetus (HBK) Backeberg (Cact.) (nPe): pishicol; with psychoactive props.; decoc. of "slabs" u. as hallucinogenic (1023) (—> **Lemaireo cereus**).

Armeniaca (L.) (of Armenia): apricot, *Prunus armeniaca.*

Armenian bole: an argillaceous earth.

armepavine: isoquinoline type of alkaloid; occ. in *Papaver armeniacum* (L.) DC, &c.

Armeria (DC) Willd. (Plumbagin.) (nTemp Zone; Andes): thrift; some spp. make beautiful masses of reddish flowers (spring, summer) - **A. maritima** Willd. (A. vulgaris Willd.) (Euras, coastal reg.; cult. & occ. natd. in US): sea pink; perennial hb. u to treat obesity, as diur. in epilepsy (secret remedy); ornam. in rock gardens - var. *elongata* Messart (Euras): thrift; ornam. - var. *californica* Lawrence = subsp. **c. Porsild** (Pacific US n to BC [Cana]): ornam.

Armillaria ventricosa Pk. = *Calathelasma v.*

Armillariella mellea (Vahl ex Fr.) Karsten (Armillaria m. Kummer) (Tricholomat. - Agaricales - Fungi) (cosmpol.): the honey fungus (or mushroom); shoestring m.; causes serious root rots of trees, &c.; has shoe-string rhizomorphs (thread-like structures of hyphae); c. antibiotics incl. erythritol; armillol (antibacterial); melledonal (antibacterial), melleolide (antifungal) (derivs. of protoilludene); edible after cooking, pickling, or salting (popular in eEu); other *A.* spp. edible.

armoise (commune) (Fr): mugwort, *Artemisia vulgaris.*

Armoracia rusticana Gaertn. Mey. & Scherb. (A. lapathifolia Gilib.) (Cruc.) (seEu; much cult. Fr, Ge & wild; natd. NA): horseradish; strongly rooted perenn. hb.; horseradish rt. (**Armoraciae Radix**, OL, **Amoracia** USP 1820-50) u. as relish for beef, oysters, &c.; condim., veget.; stom. diur. flavor, stim.; appl. to ear in deafness (cEu; folklore); rich source of peroxidase, other enzymes; c. sinigrin (glycoside), wh. releases allyl sulfocyanate (myrosin); vit. C; freshly grated rt. sold in jars in grocery, preserved with NaCl, sugar, vinegar; powder and sauce are also used; juice added to sugar to afford syrup u. in urinary "gravel"; in scurvy; for colds; lvs. also consumed; sd. u. pepper substit. (Low Countries, Germ occupation; WWII) (1037).

armuche (Sp): *Chenopodium bonus-henricus.*

army pills: compound cathartic pills.

arnatto: annatto, *Bixa.*

Arnebia benthami (Wall. ex G. Don f.) I. M. Johnston (Boragin.) (Himal, wInd, Pak): Kashmiri gaózoban (Punjabi) (nInd, nPak bazars); a pachycaul; rt. u. for boils, ulcers, in heart dis., (also lvs. in Kashmir); headache, fevers; rt. c. alkannin; expect. useful in dis. of tongue & throat (Chitral, Gilgit [Pak]) - **A. euchroma** (Royle) Johnst. (Chin, Pak-Himal); u. in toothache & earache (Afgh); rts. burned & ash appl. to eruptions; source of red dye - **A. guttata** Bge. (Lithospermum guttatum) (nKashmir, Ind, cAs, cRussia): pinyin; rt. c. alkannin (dyestuff); u. antipyr. diur. purg. in cough; u. for bug bites; wounds, burns, ulcers, chickenpox; rts. c. shikonin (red dye) u. as cough remedy (Ladakh) - **A. nobilis** Reichb. f. (Afgh): "ratanjot"; imported into India; c. arnebins with anticancer & antibacter. activity; proven to be antifungal; appl. as antibacterial to wounds, sores, burns; red coloring matter u. in foods & drugs.

Arnica acaulis (Walt.) BSP (A. nudicaulis Nutt.) (Comp.) (eUS): leopard's bane; "water root"; u. in medicine, drunk for "pain in back" (Catawba Indians); several cases of dermatitis venenata reported (1038) - **A. alpina** (L) Olin (Cana): arctic arnica; similar c. & u. as for *A. montana* - **A. angustifolia** Vahl (nHemisph): u. med. (this epithet has also been appl. to at least 4 *Arnica* spp. often to *A. alpina* (L) Olin (nEu, high mts, Arctic, Greenland; s tip of SA); u. like *A. montana* - **A. chamissonis** Lessing (wNA to Alaska): meadow arnica; u. like *A. montana*; off. DAB VII, DDR Phar., Russ Phar. IX - **A. cordifolia** Hook. (wNA fr Alaska to Mex, common in the Rockies): heart-leaf arnica; formerly an off. source of Am Arnica; important med. sp. of *A.* in US - **A. foliosa** Nutt. = *A. chamissonis* var. *andina* Ediger & Barkl. - **A. fulgens** Pursh (A. monocephala Olin) (wCana, nwUS): orange arnica; arctic a.; arctic leopard's bane; formerly an off. source of American a.; u. locally in liniments (Mont [US]); considered more effective remedy than *A. montana*; suspected of being a stock poison - **A. montana** L. (nEuras): **mountain arnica**; m. tobacco; fl. heads, lvs. (hb.) & rt. c. VO, arnidiol, faradiol (arnicin), FO with lauric & palmitic acids; choline, quercetin glycosides, sterols; rt. c. also isobutyric acid & its phloryl ester; tannins; fl. heads u. in US with other *A.* spp. mostly topically for wounds, sprains, bruises, counterirrit.; elsewhere (esp. Eu); rt. still u. int. as exp. antipyr. in rheum. furunculosis, angina pectoris, etc.; very important med. of earlier centuries, now much decr. in usage; sometimes produces toxic effects; u. as tincture & in bitters & liqueurs (small amts.) (prod It, Ge) - **A. sacchalensis** A. Gray (Russia): Russian arnica (Russkie arniki); u. in Russian med. like *A. montana* (one of several Russian med. *A.* spp.) - **A. sororia** Greene (wNA): American a.; one of the 3 formerly off. native *A.* spp.; u. much like *A. montana* (1039).

arnica: 1) *Arnica* spp. 2) *Hieracium aurantiacum* 3) (eMass [US]): *Leontodon autumnalis* 4) (Mex): *Heterotheca inuloides* (chiefly); *Tithonia diversifolia, T. rotundifolia*; *Trixis angustifolia*; *Mentzelia conzattii*; *Zexmenia pringlei* - **American a.**: *Arnica cordifolia, A. fulgens, A. sororia* - **a. da serra** (Br): *Heterothalamus brunioides* Less., inferior substit. for Arnica - **a. del pais** (Mex): *Helenium mexicanum* HBK; *Heterotheca inuloides* - **a. do campo** (Br): *Chionolaena latifolia*; u. substit. for *A.* fls. - **a. de la costa** (Mex): *Verbesina greenmanii* Urban - **European a.**: *A. montana* - **a. flowers**: fl. heads of *A* spp. u. as vulner., etc. - **orange a.**: *A. fulgens*.

Arnicae Flores (L.): arnica fl. - **A. Radix**: arnica rt. obt. fr. *A. montana*; u. astr. antispas.

arnicin: xall. bitter princ. found in Arnica.

arnidiol: arnisterol; pentacyclic triterpenoid alc. widely distr. in Comp. (*Arnica*; *Helianthus*; *Taraxacum*; &c.); rhiz. & rts.; bitter principle; isomeric with faradiol.

Arnika (Ge, Turk): Arnica.

arnique (des montagnes) (Fr): Arnica fl.

Arnoglossum atriplicifolium (L) H. E. Robins. (Cacalia atriplicifolia L.) (Comp.) (e&cUS): pale Indian plantain; lvs. u. as poultice for cuts, bruises, "cancers"; to draw out poisons.

arnotta: annatto, *Bixa*.

aro (Sp): *Arum maculatum*.

Arobon™: carob flour; u. infantile diarrhea.

aroeira (Br): *Schinus molle, S. terebinthifolia*; *Astronium* spp.; &c.

Aroideae: subfam. of Araceae: **aroids**: members include *Arum, Arisaema, Typhonium* &c.

aroma (CR): **aromo** (CR): *Acacia farnesiana*.

aromatherapy (self-defining): treatment of disease or disorder by the application of VO or other arom. materials; by local appln., inhalation, int., &c.

aromatic chalk powder: a combination of cinnamon, clove, nutmeg, & cardamom with prepared chalk & pd. sugar; u. carm. antacid - **a. elixir**: simple e. (*qv.*) - **a. powder**: a mixt. of cinnamon, nutmeg, cardamom & ginger.

Aronia Medik. (Ros.) (eUS): the chokeberries (SC astr. to throat when swallowed); regarded by some as subgen. of *Pyrus*; most spp. have now been transferred to *Amelanchier*, also *Prunus, Sorbus,* & *Pyrus* - **A. arbutifolia** (L.) Pers. (eNA, Fla to Newfoundland): red chokeberry; shrub (-10 m); choke-pear; dogberry; frt. u. in prepn. of pemmican (Am Indians); very astr.; cult. ornam. - **A. melanocarpa** (Mx.) Ell. (nAla-Minn [US], Newfoundland): black chokeberry; similar to prec. with black frt.; berry inf. for colds; as food (Potawatomi, Mich [US]); bk. u. astr. - **A. X prunifolia** (Marsh.) Rehder (eNA): considered hybrid of other 2 spp. or var. of *A. melanocarpa* Ell. (1040).

Aronstab (Ge): *Arum maculatum*.

arpillera (Arg): burlap, jute cloth.

arra diablo: *Cnidoscolus urens*.

Arrabidaea chica Verl. (Bignoni) (SA; Amazon area): carajuru; chica; cigana; lvs. yield red dye u. to paint skin (Am Indians); formerly sold as "chica red" (Vermillion Americanum); lf. u. as diur.; in erysipelas (Co); rt. syr. in alc. u. with meals for syph. (Br) - **A. xanthophylla** Bur. & Schum. (Co, Pe): kah-pe-ree (Tikana Indians); lf. inf. u. in (epidemic) conjunctivitis (as eye wash).

arracacha (Ve): *Arracacia xanthorrhiza* Bancr. (A. esculenta DC) (Umb.) (Co) - **arracachuelo** (Co): *Rumex crispus*

Arracacia atropurpurea B. & H. (Umb.) (Mex to Pe): rt. lvs. frts. u. for fevers - **A. vaginata** Coulter & Rose (Ve): micuy; rt. u. to aromatize "aji" (local type of spicy pickles, Ve) - **A. xanthorrhiza** Bancr. (A. esculenta DC) (SA-Andes): Peruvian carrot; apio; afio (Cu); Brazil earth pear; u. as potato substit. (Pe); cult. in Cu for edible tuberous rts. (cf. parsnip) (Peruvian parsnip); may be source of Bogotá arrowroot; young lvs. = condiment.

arracachuelo (Co): *Rumex crispus*.

arra(c)k: alcoholic liquor dist. fr. fermented rice, dates, palm juice (most), mahua fls., etc. - **Batavian a.**: rum (EI).

arraclan (Sp): Frangula.

arraroba: araroba.

arrayan (Mex): *Myrtus communis*.

arrowhead: *Sagittaria sagittifolia* & other *S.* spp.

arrowroot (starch): 1) various starchy products derived fr. several different pl. spp. & distinguished by geographic names (see Table I) 2) (specif): St.

Vincent Arrowroot; prep. starch fr. rhiz. of *Maranta arundinacea*; u. dem. nutr. (SC rt. appl. by NA Indians to wounds fr. poisoned arrows) (see Table II).

arrowroot: American a.: fr. *Maranta a.* - **Australian a.: Bahama a.**: fr. rhiz. of *Zamia pumila* - **Bogotá a.**: fr. *Arracacia* spp. - **canna a.**: fr. rhiz. of *Canna edulis, C. coccinea*, etc. - **Chinese a.**: fr. *Nelumbium nelumbo* - **Costa Rica a.**: fr. *Yucca gloriosa* - **English a.**: fr. potato, *Solanum tuberosum* - **a. family**: Marantaceae - **Florida a.**: *Zamia integrifolia* starch - **Guiana a**: fr. *Dioscorea alata, D. batatas* & *Musa paradisiaca* (plantain) - **Guinea a.**: fr. *Calathea allouia* - **Japanese a.**: fr. *Erythronium denscanis, Pueraria thunbergiana* (kudzu), etc. - **Portland a.**: fr. taro; tubers of *Arum maculatum* - **Queensland a.**: fr. *Canna edulis* & other *C.* spp. (2745); also appl. to *Macrozamia spiralis* starch - **sago a.**: fr. sago palm, *Cycas revoluta* - **West Indian a.**: fr. *Calathea allouia* (at times).

arrowwood (viburnum): *Viburnum dentatum* (**American a.**) & some other *V.* spp. - **(Indian) a.**: *Cornus florida*.

arroz (Sp): rice.

arruda (Por): rue.

arrugula (It): arugula: *Eruca sativa*.

arse smart: arsmart (US slang): *Polygonum amphibium, P. hydropiper, P. punctatum*.

arsenic: As; a greyish solid element with odor of garlic; occ. in the earth's crust & widely in universe; known since antiquity; found mostly as arsenide of true metals; u. med. also as poison; sometimes term "arsenic" refers to As_2O_3; many compds. have been u. in med. & technology: **a. pentoxide** (As_2O_5) u. as fungicide, herbicide; **a. triiodide** (AsI_3) (formerly u. in dermatitis); **a. trioxide** As_2O_3 u. as parasiticide; herbicide; rodenticide; formerly in asthma, arthritis; many other compds., some org., have been utilized (the arsphenamines are outstanding).

Artabotrys brachypetalus Benth. (Annon.) (e&sAf): tinta (Chopi); pl. u. to treat GC, fevers, cough, rheum., hematuria; sexual impotence; frt. edible, u. to make intoxicat. liquor - **A. hexapetalus** (L.f.) Bhand. (A. odoratissima R. Br.; A. uncinatus Merr.) (sChin, Ind; from there OW trop, ex. EI, Malang): climbing ylang-ylang; (fragrant)

TABLE 1
DESIGNATIONS OF MAJOR ARROWROOT TYPES FOUND IN COMMERCE

Canna edulis+	*Curcuma angustifolia* & other *C.* species	*Dioscorea alata*	*Manihot esculenta*	*Maranta arundinaria* (rhiz.)	*Tacca leontopetaloides*
African*	Bombay	Brazilian*	Bahía	African*	African*
Australian	East Indian**	Guiana*	Brazil(ian)*	American	East Indian**
East Indian**	Indian*	South Guinea	East Indian**	Antilles	Fiji
New South Wales	Malabar	South Seas*	Hawaii(an)*	Bermuda	Hawaiian*
Queensland	Tellichery		Pará	Brazil(ian)*	Mozambique
Sierra Leone	Travancore		Rio	East Indian**	Pia
Tous les mois	Wild			Eucano	Polynesian
				Indian*	South Seas*
				Jamaica	Tahiti
				Natal	Williams
				St. Vincent	
				"True"	
				West Indian*	
				"West Indies"*	

* In part from this source
** Name also applies to starches fr. *Curcuma angustifolia, C. leucorrhiza*, Canna, cycads, & palms of seAsia

TABLE II
ARROWROOTS: MINOR TYPES

Plant	Name	Plant	Name
Alstroemeria ligtu	Chile	*Nelumbo nucifera*	Chinese
Arum maculatum	Portland	*Pueraria lobata*	Kudzu (Japanese)
Calathea allouia	West Indian	*Solanum tuberosum*	English (potato)
Musa X paradisiaca	Guiana	*Yucca gloriosa*	Costa Rica
Myrosma cannifolia	"marble"	*Zamia pumila*	Florida

tail grape; arom. fls. u. to scent FO (Tonga) (86); & to prep. a stimulating tea-like beverage (Ind); frt. edible; c. yinghaosins A & B (sesquiterpenes with antimal. activity); u. in cholera; cult. ornam. - **A. intermedia** Hassk. (Indones, esp. Java): fls. prod. VO (minjak-kenangan oil); u. to adulter. or replace ylang ylang oil - **A. monteiroae** Oliv. (eAf): u. to treat purulent ophthalmia & GC - **A. modestus** Diels (trop eAf): lf. inf. u. in nausea & vomiting; rt. decoc. in stomachache, diarrh. - **A. suaveolens** Bl. (Ind, Mal to Mol, Af): bk. c. artabotrinine & other alks.; bk. or lf. decoc. or inf. u. in cholera; bk. & rt. decoc. emmen. & after parturition (PI) (1041).

Artanema sesamoides Benth. (Scrophulari.) (Ind): rt. decoc. in rheum., calculi, syph. ophthalmia; sds. in "biliousness"; to favor conception.

Artanthe Miq = *Piper.*

artar radix: rt. of *Zanthoxylum senegalense*.

artarine: alk. fr. bk. of *Fagara xanthoxyloides*.

Artemia salina (Anostraca - Crustacea) (worldwide in salt lakes): (common) brine shrimp; 4-10 mm. long; u. in lab bioassays (toxicity tests for pesticides, &c.); eggs u. as food for trop. fishes in fish ponds; said only animal found in water of Great Salt Lake, Utah (US).

artemisa (Sp): *Artemisia.*

Artemisia L. (Comp.) (nTemp Zone; SA, sAf, often in arid regions): wormwoods; some 200 spp. of herbs & shrubs; the lvs. or unexpanded florets of many taxa are u. as anthelm; some spp. are bitter arom. hbs. (wormwoods); some spp. important causes of pollinosis, esp. in wUS (1130); many are small shrubs (sagebrush) (SC **Artemisia**, wife of Mausolus) - **A. abrotanum** L. (sEu, As): southernwood; old man; abroa (Dan); a bright green much-branched ornamental sub-shrub; arom. parts (lvs. & flg. tops) c. alks. (abrotanine+); u. as tea, arom. bitter, astr. ton. stom. expect., in love philters (superstit.); vermif. condiment; for clothes moths (e.g., equal parts with wormwood & lavender) - **A. absinthium** L. (Euras, nAf, eNA): (common) wormwood; absinth(e); mustiarah (Pak); hb. (lvs. & flg. tops) c. VO with "absinthin" (NBP), thujone; reputedly with the most bitter taste (but pleasant aroma) of all the wormwoods; u. bitter ton. stom., stim., lvs. & rts. u. emmen., abort. rt. in epilepsy, ton. antispasm. anthelm.; hb. long known & u. in prepn. of liqueurs (vermouth) (2272), although absinthe (the chief one) is generally outlawed as dangerous (1056); in fumigants (for dis.) (Eg); in cosmetics (Near East); to protect garments against insects; pl. widely cult. - **A. afra** Jacq. (sAf): (wild) wormwood; hb. c. VO with camphor-like odor; one of most widely popular folk med. (both whites & blacks); inf./decoc. (often as syr.) u. in bronch. states, cough, colds; chills; dyspeps. colic, croup, gout; purg.; lf. poultice for neuralgia, swellings of mumps - **A. alba** Turra (A. camphorata Vill.) (sEu, Med reg.): camphor basil; u. med. in It ("canfora"); hb. c. VO with cineole, thujone, thujyl alc.; 0.2% santonin (225) - **A. annua** L. (Euras; natd. eNA): sweet (or annual) wormwood; qinghaosu (Chin); frts. ("sds.") u. to treat skin dis., night sweats (TB), eye dis. (Chin); diur. stom. (Indochin); hb. u. in mal. (Chin) for over 1,000 years; found curative (series of derivs.) (Chin, 1973-80) (schizontocidal); this valuable antimalarial c. annuin, artemisinine = qinghaosu (sesquiterpene = lactone) (claimed active principle), arteunate; lf. u. as antipyr., in chronic dys., cerebral thrombosis; exter. in abscesses; many other uses are recorded (225) - **A. anomala** Moore (Chin): u. to aid menstr. - **A. arborescens** L. (Med reg.): absinthe de Portugal; tree wormwood; hb. with blue VO, long known & used; fresh lvs. in tea in place of mint; inf. u. as appetizer, diur. anthelm. - **A. argyi** Léveillé & Van. (Chin): u. as insecticide; also u. like *A. vulgaris* - **A. aucheri** Boiss. (Pak): source of santonin - **A. austriaca** Jacq. (Euras, esp. Russ, Orient): wormwood; c. VO u. in folk med. - **A. barreliera** Bess. (Med reg., esp. Sp): VO with nearly 100% thujone; u. in colic in folk med. - **A. biennis** Willd. (US, Cana): biennial wormwood; lvs. & flg. tops u. antispasm. emmen.; an "aggressive" weed - **A. brevifolia** Wall. (Ind, Pak. (Gilgit), Tibet): Kashmir santonica; chaboo (Kaghan); hb. c. VO with santonin (rich source) (1%), brevifolin, 1-camphor; dr. lvs. & fl. heads described in Extra Pharm. 1958 (1057) - **A. californica** Less. (cCalif to nwMex): coast sagebrush; California wild sage; old man; lvs. & sts. u. in cholera (Mex); as insecticide - **A. campestris** L. (Euras, nAf): fl. heads (as Semen Artemisiae rubrae) (c. VO) u. as bitter, vermif. - subsp. **borealis** Hall & Clements (Cana, nUS): Cana (or sea) wormwood; bruised lvs. u. with *Angelica* sp. to make poultice for pains in body (Menom Indians) - **A. camphorata** Vill. = *A. alba* - **A. capillaris** Thunb. (A. parviflora Buch.-Ham.) (Chin, Ind, Pak. [Kurram]): hb. decoc. as antipyr., diur. in cystitis; in hepatitis; in cataplasms for headache (Chin) - **A. carruthii** Wood (A. wrightii Gray) (wUS): wormwood sagebrush; hb. c. santonin in small amts.; lf. inf. u. in fever, flu, cough, cold sores, &c. (Navaho Indians); frt. as food (Apache Indians) - **A. cina** O. C. Berg ex Polj (sRuss [Uzbek, Turkestan], Iran): (Levant) wormseed; Tartarian southernwood; capitula santonica; rich in santonin (1-3.5%, sometimes more); lvs. & inflorescence u. anthelm. (roundworm) & for mfr. santonin (important source); cult. (1059) - **A. douglasiana** Bess. (A. heterophylla Bess.) (wPac, Calif [US], nwMex): wormwood; Douglas' mugwort; "bronchitis plant"; lvs. u. in gastralgia, diarrh. colic; bronchitis, colds; juice in poison oak poisoning, vapor fr. steeping lvs. u. for baby's fever - **A. dracunculus** L. (A. dracunculoides Pursh) (sEuras; natd. wNA): (false) tar(r)agon; Indian wormwood; estragon; hb. (anise-scented) sold both fresh & dr.; hb. (as Herba Dracunculi) u. as tarragon vinegar (flavored with green hb., 1:800) & cooking oil; lvs. & florets (c. VO) u. as food (Am Indians); arom. bitter, condim. to stim. appetite (VO u. in Fr); diaph., emmen. sed. - subsp. **glauca** Hall & Clements (A. glauca Hall & Clements) with similar use - **A. dubia** Wall. ex Bess. (seAs): pl. boiled in water & steam (with VO) u. for rash (Thail) (2448) - **A. filifolia** Torr. - (US cPlains, Neb to nTex - Ariz, NM - sUt, s to Mex): sand sage(brush); sagebrush (Tex); pollen allergenic; med. pl. of the Hopi Indians & early settlers in area; u. vermif. & stom. & intest. dis.; lvs. chewed for indig. flatulence, "biliousness" (Tewa Indians, NM [US]) - **A. franserioides** Greene (w mts. US, nMex): ragweed sagewort; sage brush; u. ton. "life medicine" (Navaho Indians) - **A. frigida** Willd. (wNA, eSiber): (Colorado) mountain sage; prairie sagewort; hb. u. bitter ton. arom. diaph. stim. aper. antipyr. in mal., yellow fever, chills, scarlet fever; inf. said u. in spotted fever (Mont [US]); diur. in rheum.; for worms *per recto* - **A. fukudo** Mak. (Taiwan): entire pl. c. santonin; u. vermif.- **A. gallica** Willd. (SA, Sp, Fr, GB coasts): sanguerie (Fr); frt. rich in santonin (1.24%) (1060) - **A. genipi** Weber (cEu [Pyrenees, Alps, Carpathians, Appenines]): one source of Herba Absinthii alpini (black absinth) u. by mountain farmers like wormwood, also in folk med. as antipyr. for lung and chest dis. as stom. tonic, also in tea & liqueurs - **A. glacialis** L. (Eu-high mts., as Switz): silky wormwood; genippi; genepi (Fr); odor similar to that of absinth but finer; u. stim. bitter, antipyr.; sometimes in prepn. absinthe-like liqueur (genepi); flavor in gingerale - **A. herba-alba** Asso (Med reg, Sp, Eg to Mor, Pak.): (white) wormwood; shih (or chieh) (Arab); believed wormwood of Bible; u. antipyr. emmen. vermif. (inf.); pl. decoc. for colic (Eg); VO u. as insecticide, parasiticide (veter.), perfume; source of African (or Berber) levant wormseed with Santanin - **A. heterophylla** Bess. = *A. furcata* var. *h. Hultén* - **A. iwayomogi** Kitamura (eAs): hb. & flg. tops u. arom. bitter, for colds (Ainu) - **A. japonica** Thunb. (Jap, China, Kor, Manchuria): decoc. as refrig.; for intest. parasites (Taiwan); young lvs. edible; ton. (Indochin) - **A. judaica** Lour. = *Crossostephium chinense* (L.) Mak. - **A. judaica** L. (Med reg., Eg): semen contra; Judean wormwood; u. anthelm, antispasm. in colic (Eg); young lvs. in moxibustion; cultd. - **A. kurramensis** Qazilbash (Pak, Iran; Kazakstan, cult. in Jap): santonica of Kurram or art. of Kurram; "spirah tarkha"; important source of santonin, with content of 1.2-2.5%; cult. in Kurram Valley (K. district, NFWP, hence bot. name); santonin-free material fed to stock (veterin.) (1062) - **A. lactiflora** Walt. = *A. vulgaris* L.- **A. ludoviciana** Nutt. (A. gnaphalodes Nutt.) (wUS, Ala e to Indiana): western mugwort; sage brush; u. febrif.; food; - subsp. **mexicana** (Willd. ex Spreng.) Keck (A. m. Willd.) (swUS, Tex, Mex): Mexican sagewort (or sage brush); wormwood; estafiate; black sage; inflorescence c. 1.24% santonin; estafiatin (guaianolide); very commonly u. in Mex folk med. as bitter ton. stim. emmen. vermif.; in diarrh., gastric pain, colds; alc. ext. rubbed on for rheum. (swUS) (100) - **A. maritima** L. (Euras, incl. Pak [Chitral]): Levant (or sea) wormseed; old woman; wild cypress; important comm. source of santonin; both lvs. & inflor. u. as vermif., ton. condim.; unexpanded fl. heads (Santonica) for santonin extn. (1060) (1061) - **A. mexicana** Willd. = *A. ludoviciana* subsp. *m.* - **A. moxa** DC (eAs, Chin, Jap): Chinese moxa; C. wormwood; hairs fr. sts. & lvs. prepd. in small cones (moxa; moxe) which are burned on different parts of body for therapeutic purposes (moxibustions); cautery - **A. mutellina** Vill. (Eu-mts.): alpine wormwood; genippi; u. as sudorific, febrifuge (Alpine natives) like *A. glacialis* & *A. spicata* - **A. neomexicana** Greene (sNM [US], nChihuahua): said to c. santonin & estafiatin; sometimes treated as conspecific with *A. ludoviciana* - **A. nova** A. Nels. (wUS, Calif to NM, Mont): black sagebrush; hb. tea u. for coughs & colds; physic - **A. pallens** Wall. (sInd): davana; lvs. fragrant u. in floral decorations; c. VO (devana oil) popular in perfumery trade & expensive; cult. in Ind - **A. pauciflora** Weber = *A. stechmanniana* - **A. persica** Boiss. (cAs, nInd, Pak, Iran+): pl. decoc. u. for stomachache (Afgh) - **A. pontica** L. (Eu, It, Alps, Caucasus): (true) Roman (or pontic) wormwood; absinthe romaine (Fr); hb. c. VO sapphire blue in color; NBP; hb. u. bitter ton. & in liqueurs, incl. vermouth & absinthe; stom. arom. flavor - **A. princeps** Pampanini (eAs, Taiwan): lvs. u. for moxa (Chin) (2582); similarly *A. montana* Pamp.

- **A. ramosa** C. Sm. (nAf, Canary Is): Barbary santonica; hb. u. as adulter. of *A. cina*; c. santonin - **A. rupestris** L. (Ge, Switz): creeping wormwood; one of 5 *Art.* spp. (hbs.) composing Herba Absinthii alpini (Hb. Genipi albi); u. to "strengthen" the stomach; to prep. a liqueur - **A. sacrorum** Ledeb. (Russ, Tibet, Ind, Pak, Chin, Siber): given to horses for head affections; in moxibustion - **A. santonicum** L. = *A. cina* & possibly other *A.* spp. - **A. scoparia** Waldst. & Kit. (eEu, As): darwang; fl. heads & frts. c. scoparone (dimethyl oxycoumarin); cult. for VO prodn. (USSR); u. much like *A. sacrorum* - **A. sieberi** Bess. = *A. herba-alba* - **A. spicata** Wulf. (Alps of Eu): Alpenbeifuss (Ge); u. med. (in Fr & It Pharm.) - **A. stechmanniana** Besser (sRuss, Turkestan): Semen Contra; a rich source of santonin; also u. as anthelm. - **A. suksdorfii** Pip. (VI [BC] s to Calif): coastal mugwort; golden-red sage; lvs. steeped in boiling water & vapors u. for baby's fever (Paiute Indians) - **A. tilesii** Ledeb. (Alas, nBC [Cana], eAs): lvs. u. as poultice for arthritis, skin infec.; claimed to have sed. props. of codeine; pl. juice deodorant; as smoking material (Eskimos) - subsp. **gormanii** Hultén (Alaska): lvs. of this & other related spp. u. as poultice for eye dis. - **A. tridentata** Nutt. (wUS & Cana, nwMex): (common) sagebrush; big s.; black s.; blue s.; shrubby pl. covers thousands of square miles (esp. Wyo, Colo, nNM, Utah, Mont, Nev, Ida [US]): may be most abundant shrub in US; pl. c. VO with camphor 40%, pinenes 20%, cineole 7%, sesquiterpenes 12%, artemisal 5% (active characteristic arom. const.), u. in emulsions for gingivits, as cosmetic; pl. u. arom. bitter, diaph. antipyr. by Coahuila Indians for GIT dis.; sheep forage - subsp. **vaseyana** Beetle (var. v. *Boivin*) (wUS): purple sage (brush); similar properties (1042) - **A. umbelliformis** Lam.(A. laxa Fritsch; A. mutellina Vill.) (cEu, Alps, Pyrenees, Med reg.): Alpine wormwood; Herba Genipi (alba) u. as bitter ton. (tea) antipyr. vermif.; in lung & bronchial dis. ("Swiss absinth"); in prepn. Benedictine (liqueur); ornam. perenn. hb. - **A. vallesiaca** All. (A. vallesiana Lam.) (Eu Alps, sSwitz [Valais (Wallis canyon) Aosta valley], nIt [cult.]): Walliser-Beifuss (Ge); gentile (or Valois) wormwood; Assenzio gentile (It); source of black genip (Genipus niger); armoise du valois (Fr); u. similarly to *A. absinthium*; strong camphoraceous odor; similar to *A. genipi* Weber; u. in mfr. of vermouth & other liqueurs (off. Fr Pharm. I); cult. in other places - **A. vulgaris** L. (Euras, Af, natd. NA, SA; prod Ne, Swe, Ge, Hung): mugwort; common (or wild) wormwood; ghost plant; *motherwort*; hb. c. VO (with cineole), NBP; u. as emmen. nervine, antispasm. bitter ton. anthelm; cult. in Yuc; young pls. boiled & eaten (Jap); young shoots & lvs. condim. for goose meat (Eu) & to flavor rice cakes (eAs) - var. *indica* (Willd.) Max. (As): u. in amenorrh., abdom. pain, metrorrhagia, styptic, moxa cautery; source of Indian & Javanese wormwood oil - **A. wrightii** Gray = A. carruthii.

Artemisiae Herba (L.): mugwort hb., *A. vulgaris.*

artemisin(e): hydroxy-santonin; fr. *Artemsia cina*, *A. maritima* & other *A.* spp.; formerly u. in mal.

artemisinin: artemisine; QHS sesquiterpene lactone (isolated 1972) fr. *Artemisia annua*; lvs. & (qinghau-su) fls. tops useful in mal. (ancient use in Chin; u. there for two millenia); derivs. also incl. Artesunate, effective antimalarials.

arterenol: nor-epinephrine; fr. adrenal glands; u. pressor agt.

arthanite (Fr): *Cyclamen* spp.

Arthraxon ciliaris Beauv. (Gram.) (Austral, As, tropAf): turade (Bombay); c. alks.; pl. u. med. (Chin) - **A. hispidus** Mak. (Chin): st. & lvs. u. as bechic for chronic cough; in wash for abscesses.

Arthrocnemom indicum Moq. (Chenopodi) (Ind): umari (Tamil); a common mangrove; pl. ashes u. as alexipharmic for scorpion sting; hb. u. as salad.

Arthrophyta: *Equisetum* & related genera; in past ages, a large gp. of vascular cryptogams.

Arthrophytum wakhanicum Korovin (Chenopodi) (Russ, Georgia [As]): pl. c. alk. dipterine (N methyltryptamine) with hypertensive action; source of firewood.

Arthropoda: phylum of animals characterized by possessing in their bodies a series of segments, all or some of which bear jointed appendages; both body & appendages are covered by an exoskeleton consisting of chitin; includes many organisms of enormous economic importance, favorable or unfavorable to human society; ex. insects, spiders, shrimp.

Arthrosolen gymnostachys C. A. Mey. (Thymelae.) (s&tropAf): lvs. c. daphnetin glucoside; u. for headache (lvs. smoked) (folk med.) - **A. polycephalus** C. A. Mey. (s&etrop Af): January bush; rt. pieces chewed for asthma; decoc. u. for "lamsiekte"; important toxic pl. (stock); abort. (—> *Gnidia*).

Arthrostemma ciliatum Pavón ex. D. Don (Melastomat.) (sMex to Pe, Bo, WI): nitro (real); pl. decoc. u. as diur. & cooling drink (Guat); wayfarers chew sts. to produce salivary flow & relieve thirst; cult. orn. - **A. volubile** Triana (Co): u. to prep. refreshing drink for fever patients.

Arthrothamnus kurilensis Ruprecht (Laminari. - Phaeophyc. - Algae) (Pac Oc, Jap & Chin coasts):

raw material for kombu; may be source of iodine, K; food.

artichaut (Fr): artichoke.

artichoke: 1) *Cynara scolymus* 2) Jerusalem a.; *Helianthus tuberosus* - **Common a.: Chinese a.**: *Stachys floridana, S. affinis* - **French a.: garden a.: globe a.: green a.: ground a.**: *Cynara scolymus* - **Jerusalem a.**: *Helianthus tuberosus* (SC corruption of It name, girasole) - **wild a.**: 1) *Cynara cardunculus* 2) *Onopordum acanthium.*

artificial oil of sassafras: safrole - **a. o. of wintergreen** (or **sweet birch**): methyl salicylate.

artillery plant: *Pilea microphylla* Liebm.; *P. serpyllifolia*; or *P. hernariaefolia* (SC pollen grains discharged forcefully).

Artiodactyla: order of mammals made up of large herbivorous organisms, which are provided with more or less hair & an even number of toes on their feet; ex. cattle, pigs, musk deer.

Artischoke (Ge): artichoke - **Spanische A.** (Ge): *Cynara cardunculus.*

Artisone™: 21-acetoxy-pregnelone; u. for arthritis.

Artocarpaceae: breadfruit family; generally Moraceae now preferred classification, with subfam. **Artocarpoideae**.

Artocarpus altilis (Parkinson) Fosberg (A. blancoi [Elm.] Merr.; A. communis J. R. & G. Forst.) (Mor.) (Mal, Chin, Moluccas; now cult. in Polynesia & elsewhere in trops): breadfruit (SC frt. with consistency of bread); pana (cimarrona); tree -20 m; very fruitful; frt. usually seedless or nearly so; a few have sds. (u. in fever [Chin]); sometimes numerous sds. like chestnuts or peanuts in pulp (PR) & as nutritive as these (u. roasted); frt. flesh rich in starch; frt. staple of many nations when cooked like a veget. (baked or roasted; natives bury frt. & eat fermented) (sometimes raw); breadfruit starch (comm.) (Amylum Artocarpi) prepd. fr. frt. (u. like rice starch); bk. decoc. u. in strangury; tree trunk u. to make canoes (Micronesia); lf. decoc. u. for hypertension (Jam, Baham); in heart trouble (Virg Is); sap boiled & residue appl. to wounds (Co) - **A. blancoi** Merr. (PI) = *A. altilis* - **A. heterophyllus** Lam. (A. integrifolius auct.; A. integra Merr.; A. jaca Lam.) (cultigen; long cult. Ind, SriL; now cult. eAs, Oceania): Jack (or Jak) fruit tree; tree with milky juice; -25 m; frt. (large -11 kg) highly delicious but with horrible odor, rich in pectin; eaten fresh, boiled, fried; in preserves (young frts. eaten as soup); jak sds. (roasted) & fl. clusters (with syr. & agar [Java]) also edible; lvs. & sds. c. acetylcholine; st. latex u. as bird lime, source of "canoe gum" u. as cement for broken china; lvs. u. in diarrh., skin dis.; rt. (bk.) decoc. in diarrh. fever, asthma, skin dis. worms; wd. valuable ("Jack wood"), u. in cabinet making; to dye silk yellow (with alum mordant), u. for robes of Burmese Buddhist priests.

artosterone: claimed androgenic, isolated fr. latex of ripe frt. of *Artocarpus heterophyllus* ("Indian summer fruit"); derived fr. artostenone (sterolactone) (highly androgenic) (1043).

aruca: *Calea pinnatifida* (*qv.*).

arücine: alkaloid: 5-methoxy-ajmalicine; occ. in *Rauvolfia canescens, R. hirsuta, Cinchona pelletierana*; identical with heterophylline; not hypotensive.

arugula: *Eruca sativa* (*cf.* r(a)uke [Ge]).

Arum: 1) (L.): *Arum maculatum* 2) *Arisaema triphyllum* & other *A.* spp. (Indian turnip); - **A. dracontium** L. (Ar) = *Arisaema d.* - **A. dracunculus** L. = *Dracunuclus vulgaris* - **A. esculentum** L. = *Colocasia esculenta* (var. *antiquorum*) - **A. italicum** Mill. (Ar.) (Euras, Af): wake robin; italienischer Aronstab (Ge); u. like *A. maculatum*; rhiz. u. as source of Portland arrowroot; rhiz. edible after drying; ornam. perenn. hb. - **A. maculatum** L. (c&sEu, nAf): common/European/spotted arum; dragon root; Aaron's staff; cuckoo-pint; calf's foot; dr. rhiz. (corm: Radix Aronis) u. ton. stom. in chronic catarrhs (resp., stomach); chlorosis; paralysis; dr. "tubers" c. 33% starch (u. nutr.), saponins, Ca oxalate; tubers edible after prepn.; dr. tuber source of Portland arrowroot - **A. triphyllum** L. = *Arisaema t.*

arum: arrow a.: *Peltandra virginica* - **a. family**: Araceae - **water a.**: *Calla* spp.

Aruncus dioicus (Walt.) Fern. (1) (Ros.) (Euras; natd. eUS s to Fla): goat's beard; perenn. hb.; sds. c. saponin (traces), hb. c. HCN glycoside (traces), antibiotic; rt. lvs. fls. u. as ton. astr. antipyr. in gall bladder dis., dys. (with *Filipendula ulmaria*).

Arundinaria anabilis McClure (Gram) (subfam. Bambusoideae) (sChin): Tonkin bamboo (or cane); a cultigen; sts. very expensive, furnish best type of fly fishing rods; lvs. u. ton. stom. carm. antitussive; rt. ton. esp. in rheum. - **A. densiflora** Rendle (Chin): lvs. ton. stim. antivenomous; rt. antipyr.; sap u. for ulcerative mouth, toothache, ophthalmia - **A. gigantea** (Walt.) Muhl. (A. macrosperma Mx.) (seUS): giant/large/southern/switch cane; colonial grass; young shoots u. as pot herb; sds. edible (u. like wheat by Indians & early settlers); forage (young lvs. & sds. eaten by stock); cult. for culm (st.) u. to make fishing poles, pipe stems, &c.; "cane brake" fire (sUS) produces great popping & cracking. (2672)

Arundinella nepalensis Trin. (Gram) (pantropics): river grass; perenn. grass serves as forage; pl. u. to make lotion for bathing wounds & elsewise (Basutoland).

Arundo donax L. (A. sativa Lam.) (Gram.) (Med reg, wAs; cult. & wild in e&sUS, SA, sAf): giant reed; reed (of Bible); carrizo (Mex); Spanish cane; perenn. grass -7 m high; rhiz. u. as antigalact. uter. stim. diaph. diur.; c. donaxine (gramine); sts. u. as fishing rods, walking sticks, screens, mats; reeds u. for clarinets, organ pipes (Eu); in mf. rayon cellulose; to prevent wind erosion; ornam.

arus(h)a (Hind.): *Adhatoda vasica.*

arveja: 1) (Sp): vetch 2) (Arg, Co): pea - **a. velluda** (Arg): hairy vetch.

arvejilla comun (Arg): common vetch.

arviculture: crop science.

arvy (Urdu): *Colocasia esculentum.*

Asa Dulcis (L.): benzoin.

asafetida: gum a.: Anglicization of **Asafoetida**; fr. *Ferula foetida*, etc. - **milk of a.**: 4% emulsion of a. in water; u. sed. carm.

asafran (Sp): azafran.

Asagraea Lindl. = *Schoenocaulon*; (Sabadilla) - **A. officinalis** Lindl. = *S. o.*

Asahi-aji (Jap): mono-sodium glutamate.

Asarabacca (, swamp): *Asarum europaeum* & other *A.* spp.

asaresinol ferulate: phenolic ester constituting c. 17% of asafetida.

asaret du Canada (Fr): *Asarum canadense.*

asarinin: xanthoxylin S; lignin fr. *Asarum sieboldi* var. *seulensis*, &c.

asaro (It): *Asarum eurpaeum.*

asarones: asarin; asarum camphor; trimethoxy-propenyl-benzene; arom. component of *Asarum europaeum* & other *A.* spp., *Acorus calamus.*

Asarum L. (Aristolochi) (nTemp Zone): wild ginger; Canada snakeroot; heart leaf; asarabacca; spp. of seUS have been segregated as *Hexastylis* (1045) - **A. canadense** L. (eNA): Cana (wild) ginger; rhiz. (Rhizoma Asari canadensis; **Asarum**) c. acrid resin, VO (with methyl-eugenol, geraniol, &c.); rhiz. u. arom. bitter, carm. stom. diaph.; ton. stim.; ginger substit.; in headache (eTenn, folk med.); weakly antibiotic (1989); rts. placed in linen bag & immersed in wine casks to improve flavor of wine (Fr); suggested as replacement for Yerba Santa - **A. caudatum** Lindl. (wNA): western wild ginger; lf. decoc. ton., tea of Indians of wWash state (US); rhiz. (dr.) u. like ginger (114) - **A. europaeum** L. (Euras): European wild ginger; asarabacca; wild nard(us); rhiz. (Rhizoma Asari) u. as arom. emet. (after excessive wine, in Fr), diaph. emmen. cath. cure for habitual intoxication (Russ); sternutatory; hb (lf.) u. similarly in sl. larger doses; rat & mice poison - **A. sieboldii** Miq. (Jap, Chin): te tan; c. asarinin; u. in toothache (pd. or in alc.), sternutatory; represents Jap drug sai-sen (also *A. blumei* Duch. [Jap]) - **A. virginicum** L. = *Hexastylis virginica.*

asbestos: Amianthus; a mineral obt. largely fr. eCana mines; made up of native calcium & magnesium silicates; largest prodn. in US, Cana (Quebec), Russ; long u. industrial area, as insulation, for heat resistance in gloves, clothes, &c.; of fibrous nature & fireproof; hence often fashioned into threads, cloth, boards, etc. & u. for fireproofing in laboratories, etc.; in inert fillers, brake linings; potent carcinogen after long latent period; considered serious hazard to health when inhaled; dangerous since may cause asbestosis (pneumo coniosis), tumors, lung cancers, etc.; u. now limited in US.

ascaridole: an organic peroxide representing active princ. of chenopodium oil; nematocide.

asclepain: proteolytic enzyme fr. *Asclepias* spp.; nearly as effective as papain in tenderizing meats & for removing warts, &c.

Asclepiadaceae R. Br.: milkweed family; named after chief genus *Asclepias*; most plentiful in tropics; hbs., lianes, & shrubs with milky juice (latex); lvs. entire, opposite or whorled; usually fleshy; inflor. generally brilliant; a corona occurs between petals & stamens; frt. a follicle; closely related to Apocynaceae; ex. *Calotropis, Gonolobus.*

asclepiade (tubéreuse) (Fr): milkweed, *Asclepias tuberosa* - **asclepiadin**: extractive prod. fr. *Asclepias tuberosa* & some other *A.* spp.

Asclepias L. (Asclepiad.) (Am, esp. US): milkweeds; silkweeds; latex of spp. with proteolytic enzyme, asclepain; perenn. hbs. with many-flowered umbels (1046) - **A. acida** Roxb. = *Sarcostemma viminale* R. Br. - **A. asperula** (Dcne.) Woods. - subsp. **capricornu** Woods. (wUS mts.): immortal (Pueblo Indians); popular med. of Tewa Indians (NM [US]) - **A. cornutii** Dcne. = *A. syriaca* - **A. crispa** Berg (Af): rt. emet. purg. diur.; inf. or decoc. u. in edema; tinct. u. in colic; toxic to grazing stock - **A. curassavica** L. (sMex to Par, WI; natd. sFla-Tex, sCalif [US]): now wide spread in tropics); Indian root; cancerillo (CR, Mex); blood flower; bastard ipecac(uanha); swallow wort; matac; paina de sapo (Br); quebojo (Ve); lvs. of shrub c. calotropin (cardioactive glycoside) producing cytotoxicity; lvs. appl. to sores & tumors ("cancers"); latex ("juice") u. in folk. med. against cancers, warts; in honey for intest. worms (Ve); sternutarory; sed. in scorpion string; lf. decoc. u. to wash ulcers; cardiac stimulating like Digitalis; latex sprayed in eye may cause blindness (hence common name "mata-olho," or eye death); hb. u. alt. in pneumonitis; irrit. emet. (similarly rhiz. & rts.); rt. expect. in cough; substit. for & adulter. of ipecac; aperient (Trin); decoc. antipyr. (2444); fls. astr. anthelm. in GC; insecticide; weed; cult. ornam. (OW trop) - **A. decumbens** Dcne. = *A. asperula* subsp. *capricornu* - **A. eriocarpa** Benth. (Calif [US], nwMex): latex u. as chewing gum (Am Indians); st. fiber u. for cordage - **A. exaltata** L. (e&cUS): tall milkweed; rt. ingested for gastric dis. (Omaha Indians); emet. cath. - **A. fruticosa** L. (sAf, Austral; sporadically in Euras): c. cardiac glycosides, gomphoside, afroside; lvs. u. as purg. (also latex); sd. hairs u. as textile fibers - **A. galioides** HBK = *A. verticillata* - **A. geminata** Roxb. = *Gymnema silvestre* - **A. glaucescens** Kunth. (Salv): lvs. placed on boil to bring to a head - **A. incarnata** L. (NA except w. coast): flesh-colored asclepias; swamp milkweed; white Indian hemp; rhiz. u. cardioton. diaph. diur. expect. & in large doses emet. purg.; anthelm. (tapeworm remedy, Mishwakis); bees in search of honey sometimes become so coated with sticky "wax" as to cause their own death - **A. latifolia** (Torr.) Raf. (c&sUS); broadleaf milkweed; fresh sap of st. u. appl. to sores, warts, freckles ("complexion balm"); tender shoots said edible; pl. toxic to livestock - **A. linaria** Cav. (Mex): venenillo; teperromero; juice u. as drastic (purg.); against anginas; in Mex Pharm. - **A. lineolata** Schltr. (Gomphocarpus lineolatus Decne) (tropAf): uzara root; c. uzarin (cardiac glycoside); u. as stom. emet. expect. (Nigeria); narcotic to catch birds; anthelm. for roundworm; wd. sap appl. to skin for itch - **A. longicornu** Benth. (Mex to CR): latex u. for toothache (Yuc) - **A. mexicana** Cav. (Mex to CR): Mexican whorled milkweed; leafy st. decoc. u. for colic - **A. nivea** L. (WI): flor de calentura blanca (Cu); hb. u. emet. vermif. - **A. oenotheroides** Cham. & Schltr. (Mex-CR): angelito; latex u. for warts & toothache - **A. ovalifolia** Decne. (cNA): dwarf milkweed; perenn. hb. (-7 m); hb. u. diaph. - **A. ovata** Mart. & Gal. (sMex): dr. pd. rt. gives a violent sternutatory action - **A. procera** Ait. = *Calotropis p.* R. Br. - **A. rubra** L. (e&sUS): red milkweed; rhiz. u. in pleurisy - **A. setosa** Benth. (Mex): contrayerba (Mex); rt./rhiz. u. as diaph. antiperiodic - **A. speciosa** Torr. (w. half of NA - roadsides wild): showy (or Greek) milkweed; the most common m.w. in US Pac area; latex c. asclepain (proteolytic enzyme); buds boiled & eaten with meat (Am Indians); young shoots & lvs. eaten (Hopi); latex u. as chewing gum (Gosuite Indians); potential rubber source; rubbed on sores, warts; rts. mashed in water u. as poultice for rheum. (BC [Cana] Indians); silk of frt. u. as nesting material by small birds; fiber u. as pillow filler; string; sugar source; host plant for Monarch butterfly; inflorescence with strong sweet odor, thought narcotic to humans; stupefying to insects (often found beneath pl. in drowsy state) - **A. stellifera** Schltr. (sAf): rt. inf. u. in "stomach troubles"; as local appln. to protruding fontanelles; in making arrow poison; latex yields good natural rubber - **A. subulata** Decne. (Calif, Nev, Ariz [US], nwMex): desert (or rush) milkweed; skeleton m.; ajamete; latex u. emet. purg. (Mex); studied as possible comm. source of rubber - **A. syriaca** L. (c&eUS, Cana): (common) milkweed; (c.) silkweed (m.); wild cotton; rhiz. c. nicotine (only occurrence outside of Solanaceae); asclepion; lvs. very rich in vit. C; u. as greens; shoots formerly cooked; for effect of bees, *cf. A. incarnata* (1047); rhiz. u. diaph. diur. in edema, sed.; latex as vulner.; bee plant; latex milkweed rubber; sd. & floss ("vegetable silk"); to treat warts; formerly cult for same & as ornam. (fls) (budgerigar) (Switz., for st. fiber) - **A. tuberosa** L. (e, c&nUS, sOnt [Cana]; dry fields & roadsides): pleurisy root; butterfly weed; chigger flower; fls. July-Sept; rhiz. (Asclepias, LT) c. glycosides (asclepiadin, &c) resins, VO, caoutchouc, amyrin, a potent estrogen, heterosides; u. expect. in pulmon. dis., pneumonitis, pleurisy; with estrogenic & oxytocic properties; cath. diaph. diur. carm.; in large doses, emet.; shoots & pods have been eaten boiled; expect. action attributed to nausea fr. glycosidal resins - **A. verticillata** L. (s&cNA w to Ariz, incl. Mex): dwarf(ed) (or sweet) milkweed; (Eastern) whorled m.; horsetail m.; lechones; lechugas; lengua de vaca (NM); latex u. for facial pain (Am Indians), snakebite; hb. galactagog; rt. stim. sudorific, in snakebite (selnd); claimed to antidote insect sting; buds & pods edible; sd. fibers in textiles - **A. vincetoxicum** L. = *Cynanchum v.*

Ascomycetes: Ascomycotina: largest class of fungi; in which ascospores (usually 8) are contained in characteristic sacs called asci (singular-ascus); about 30,000 spp.; ex. *Aspergillus*, *Claviceps* (ergot), *Peziza*.

Ascophyllum nodosum Le Jolis (Fucus nodosus L.) (Fuc. - Phaeophyc. - Algae) (both sides of nAlt Oc such as off Norway & neUS seaboard, n of NY City): rockweed; bladderwrack; yellow wrack; yellow wrack; one of the spp. which supply the drug "Fucus" (formerly off. -NF 1926); thallus c. mucil. mannitol, iodine, bromine; thallus u. dem. alter. in obesity, simple goiter, for atherosclerosis, Basedow's dis.; ext. antibacterial (87); a chief source of alginic acid & alginates; has been u. in prep. iodine; green manure; fodder additive; adulter. of *Fucus* spp. (1065-A).

ascorbic acid: Acidum Ascorbicum (L.): cevitamic acid; vitamin C; obt. fr. *Citrus* frts. and green vegetables & by fermentation processes; besides med. u. much employed as antioxidant, reducing agt., preservative, etc.; thus to preserve fresh flavor or color of frt. (ex. pears); antiscorbutic, specific in scurvy (scorbutus, L.) (PO & parenterally); gives lower BP; incr. HDL cholesterol blood level; antimicrobial, antihistaminic values claimed; aids in treating leukemia, neoplasms; antiarthritic (2403); reduces severity of colds (0.5g/d); in schizophrenia (2426); reduces gum bleeding; reduces male infertility; claimed protective against atmos. pollution (N_2O, O_3) causing lung irritation; to incr. F storage in body; some consider natural product superior to synth. (E); sometimes u. in form of Ca ascorbate; vit C should not be used with aspirin while exercising in hot weather.

ascorbyl palmitate (NF): ascorbic acid palmitic acid ester; u. as antioxidant in pharmaceuticals & foods; to prevent rancidity; reduce browning of cut apples, &c.; preservation of flavor in canned or frozen foods.

ascosin: antibiotic fr. *Streptomyces canescens* Waksman; active on fungal dermatoses (superficial fungus infections); isolated 1952.

ascus (pl. asci): sac-like cell of perfect state of Ascomycotina; bears ascospores (mostly 8) in its cavity (anagram of "sacus").

Ascyrum hypericoides L. = *Hypericum h.* (L) Crantz.

-ase: suffix denoting enzyme; the syllable(s) preceding generally denotes the substrate; ex. pectase, enzyme acting on pectin.

aselline: alk. claimed found in decomposing codliver oil.

asfodillo (It): asphodel; daffodil.

ash: incombustible material which remains following incineration of drug material, &c.; detn. of content in drugs represents one of best tests of quality & uniformity of drug; material u. in folk med. (suspension) to hasten labor (2431) - **acid insoluble a.**: residue from boiling total ash with 10% HCl; indicates sand (soil) content.

ash: Fraxinus excelsior & other *F.* spp. - **bitter a.**: *Quassia* - **black a.**: *F. nigra* - **blue a.**: *F. quadrangularis* - **common a.: European a.**: *F. excelsior* - **green a.**: *F. pensylvanica* - **hoop a.**: *Celtis occidentalis* L. - **manna a.**: *F. ornus* - **mountain a.**: *Sorbus americana* & other *S.* spp. - **poison a.**: 1) (US): poison sumac, *Rhus vernix* 2) (US): *Chionanthus* spp. 3) (WI): *Amyris balsamifera* - **pop a.**: *F. americana* - **quaking a.**: *Populus tremuloides* - **sea a.**: *Zanthoxylum clava-herculis* - **shrew a.**: an ash tree with hole bored in trunk in which a shrew was placed & sealed up; twigs or branches of tree were later applied to limbs of cattle to relieve pains thought suffered by the running of shrew over the part (superstitious practice, Eng) - **thorny a.**: *Zanthoxylum americanum* & *Z. clava-herculis* - **wafer a.**: *Ptelea trifoliata* - **white (American) a.**: *F. americana* (bk.).

Ashbya gossypii (Ashby & Nowell) Guilliermond (Endomycet. - Endomycetales - Ascomyc - Fungi): causes dis. in cotton bolls (sAf), coffee, legumes, &c.; fungus u. in industrial prodn. of riboflavin.

ashok(a): bk. of *Saraca indica* L.

asiaticoside: glycoside fr. *Centella asiatica*; made up of asiatic acid with 3 sugar mols.; u. to promote wound healing (1470).

Asimina angustifolia Raf. (A. longifolia Kral) (Annon.) (coastal plain, Fla, Ga, Ala [US]): rt. decoc. one of "black drinks" of Indians (not *Ilex*); u. diaph., purg. - **A. incana** Exell (A. speciosa Nash; A. grandiflora Dunal) (nFla, seGa [US]): papaw; "custard apple"; polecat bush (farmers, nFla); frt. eaten by Fla Indians ("delicious") - **A. reticulata** Shuttlew. ex Chapman (Fla [US]): (common) pawpaw; dog apple; fls. of this & other Fla *A.* spp. u. as inf. (Seminole tea) in renal dis. (Seminoles) - **A. triloba** (L) Dunal (NA: NY-Fla-Tex-Neb [US] to sOnt [Cana], common in Ohio Valley): North American (or common) pawpaw; papaw; (American) "custard apple"; "poor man's banana" (Ozarks); false banana; shrub or tree (-13 m); frt. yellowish to brownish when ripe or blackish after first frost; (other spp. yellow-green); fls. maroon, showy; appear before lvs.; frt. 5-15 cm, edible but sickly sweet; largest native Am frt.; called "banana" (Indians); food of Am Indians, sold locally (Tenn, nNCar [US]); dr. frt. purg.; lvs. & bk. u. diur. & appl. to abscesses; sds. c. asiminine (alk.); sd. emet.; inner bk. c. anolobine (alk.); inner bk. fibers u. to string fish & make fishnets (Miss [US]+); easily cult. (Calif [US]) (Kew, Eng) (1048); ornam.; sometimes confused with *Carica papaya*.

Asiminaceae: Annonaceae.

asiminier (Fr): pawpaw tree; *Asimina* spp.

asiminine: alk. in sds. of *Asimina triloba*.

asp: 1) apsen 2) (sEu): venomous snake; *Cerastes cornutus*, horned viper (nAf); also *Vipera aspis*; u. in Unani med., thus blood to cure cataract 3) Egyptian cobra, *Naja haje*; bite supposedly killed Cleopatra.

Aspalathus cedarbergensis Bolus. (Leg.) (sAf): dr. lvs. u. for tea in parts of Cape Peninsula (may be cultd. form of the foll. (one of sAf bush teas) - **A. linearis** (Burm. f.) Dahlgren **A. contaminatus** Druce, (sAf): rooibos or red tea (rooibostee) bush; flg. twigs with low tannin, u. comm. for caffeineless beverage tea, also Hottentots (Kaffree Tea™); (3855); c. flavonoids; stom. - **A. mollis** Lam. (sAf): u. in whooping cough, pulmon. TB - **A. tenuifolia** DC (sAf): African broom; u. as beverage tea (bush tea, naaldtee).

aspalathus (Bible): probably *Alhagi camelorum* v. *turcorum*.

asparaginase, l-: enzyme wh. occ. in fungi, bacteria, (esp. *E. coli, Erwinia carotevora*) higher pls., animals; u. antineoplastic in acute lympholytic leukemia (ALL) (children); acts to reduce circulating asparagine (needed by cancer cells to survive); Elspar™ (US).

asparagine: Asn; one of the non-essential amino acids; aspartic acid amide; *l*-asparagine commoner in nature than *d*-asparagine; common in asparagus (first found here, hence name), vetch, lupine, potato tubers, pea, soy, etc., esp. in seedlings & shoots; u. diur.; in fluid retention, fatigue; reagent; ingred. in bacterial media (1048) (1049).

asparag(in)ic acid: aspartic acid: aminosuccinic acid; a non-essential amino acid; occ. commonly in pls. & animals, esp. young sugar cane & s. c. molasses; formed by hydrolysis of asparagine (L . form), u. as roborant in chronic fatigue (excitatory neurotransmitter); nutrient; antioxidant & stabilizer in soap; in org. syntheses; Mg & K aspartate (Spartase™, Aspartat™) u. in therapy to reduce fatigue (reputedly).

asparago (PR): **asparagos** (Br): **Asparagus** (L.): *Asperagus officinalis*.

Asparagopsis sanfordiana Harv. (Bonnemaisoni. - Rhodophyc. - Algae) (PacOc): limu kohu (Haw); u. foods with fish, meat, poi; made into "inomena" with other substances, u. as condim.; comm. article.

Asparagus acerosus Roxb. (Lili.) (Indochin): u. expect. diur. appetite stim. - **A. adscendens** Roxb. (Ind): rts. u. stim. diaph. demulc. galact., in diarrh., dys., general debility; rt. u. like salep & to substit. for same; pd. rt. u. like tragacanth. - **A. africanus** Lam. (s&trop wAf): mukushi (Moz); rhiz. u. diur.; remedy for leprosy, syph. rheum. chronic gout, in arrow poison; lvs. u. to treat wounds - **A. asparagoides** (L.) Druce (sAf): Cape smilax; rt. u. to make lotion for sore eyes; rt. decoc. mild purg. - **A. buchananii** Bak. (sAf): "fishbones"; rt. decoc. in GC - **A. burkei** Bak. (sAf): rt. inf. u. in pulm. TB; rt. decoc. in sore throat - **A. capensis** L. (sAf): u. in consumption; young shoots edible - **A. cochinchinensis** Merr. (A. falcatus Thunb.; A. lucidus Lindl.) (Chin, Indochin): hatawariya (SriL); tien tung (Chin); tubers or rhiz. u. as ton. for cough, hemoptysis, dry throat, constip., as diur. antipyr.; eaten as veget. - **A. filicinus** Buch-Ham. (Ind, Pak, Chin): rts. u. like prec.; astr. antirheum., vermif. (Assam) - **A. laricinus** Burch. (sAf): rt. u. as diur. in edema; gives urine peculiar odor; young shoots edible - **A. officinalis** L. (Euras, nAf; cult. & natd. NA): (garden) asparagus; "sparrow grass"; rhiz. & rts. cardiac stim. diur. in edema; "kidney remedy"; lax.; reputed aphrod.; fresh tender young shoots ("spears") deliciously edible; c. asparagine, tyrosine, succinic acid, &c.; odorous methyl mercaptan (or else asparagine deriv.) excreted in urine in some individuals; sprouts homeopathic rem.; st. pulp marketed as paste (in cans); juice u. in bacteriol. culture media; crushed sds. u. as coffee substit. (Eu); bee plant; pl. seeded by birds eating frts.; "white" a. blanched by burial in the earth!; cult. sometimes as ornam. (450) - **A. racemosus** Willd. (wAs, Ind, Pak, Af, Austral): satawar (Hind); pd. rt. in honey in dys.; rt. decoc. in impotency (124) (with long pepper); as sed. purg. emet. in bilharziasis (Ivory Coast); dem. diur.; frts. eaten (Sudan) - **A. sarmentosus** L. (Ind): rts. u. aphrod. nutrient - **A. scandens** Thunb. (sAf): burnt & crushed rt. u. in colic - **A. schoberioides** Kunth (eAs): sts. & lvs. u. in foment. for backache; rt. sap u. to cure fevers (Ainu) - **A. setaceus** Jessop (A. plumosus Bak.) (Af): a. fern; rt. inf. in pulmon. TB; in mal.; branches decoc. u. in resp. inf.; frt. & lf. in pneumonia (Tanzania); cult. as popular ornam., cut for decorations (US+); as diur. (wMex) - **A. stipulaceus** Lam. (sAf): rt. inf. u. to treat pulmon. TB; as "blood purifier" (Xhosa); bk. as emet. - **A. virgatus** Bak. (sAf): rt. inf. anthelm. (in combns.); lf./st. inf. to stop vomiting; in mal.

asparagus fern: *Asparagus plumosus* - **winter a.**: *Scorzonera hispanica*.

Aspartame™: sweetening agt. ca. 160-180 X as sweet as sucrose; synth. fr. aspartic acid.

aspartic acid: v. asparaginic a.; a non-essential amino acid found in plants & animals; esp. rich in sugar cane - L-form excitatiory neurotransmitter; u. roborant; K Mg salt (Spartase; Aspartat) reputed reduces fatigue.

Aspartocin: antibiotic fr. *Streptomyces griseus* subsp. *spiralis* Pridham & al., &c.; active against Gm. + bacteria.

aspen: *Populus* spp. (*P. tremuloides*; *P. grandidentata*); with thin lvs. wh. flutter in the wind - **American a.**: *P. tremuloides* - **European a.**: *P. tremula* - **quaking a.**: American a.

aspèrge (Fr): asparagus.

aspergillic acid: antibiotic comp. secreted by *Aspergillus flavus* Link; studied since 1943; said active against Gm.+ & TB organisms in vitro; suggested as hypotensive (1990).

Aspergillus Micheli (Euroti. - Hyphomycetes - Deuteromycet. - Fungi) (cosmopol): some 50 spp. of smuts & molds; of increasing importance in fermentation ind. in furnishing various org. acids, &c.; (SC aspergillum [L], brush or instrument u. for sprinkling holy water in some Christian churches; based on similarity in appearance of sporangium) - **A. flavus** Link: u. in fermentation to prod. sake (rice wine) & soy sauce; possible source of FO - **A. fumaricus** Wehmer: u. to prod. fumaric & malic acids - **A. itaconicus** Kinoshita (A. glaucus gp.) (Jap): mold producing itaconic acid (u. in plastics ind. as plasticizers), itatartaric acid, itaconitin, ergosterol - **A. niger** v. Tieghem: u. in mfr. citric acid, gluconic a., glutaric a., saccharic a., &c. - **A. oryzae** (Ahlb.) Cohn: fungus growing on rice; u to produce sake, beer, taka-diastase, kojic acid, &c. - **A. tamarii** Kita: u. (Jap) to prepare tamari sauce (cf. soy s.); prepn. ascorbic acid - **A. versicolor** (Vuill.) Tiraboschi: source of aveffin (quinone); can metabolize tartrates (wine ind.) - **A. wentii** Wehmer: mold u. in mfr. of a soy sauce (Chinese in Java), bean porridge.

Asperugo procumbens L. (monotypic) (Boragin) (Eu): (German) madwort; wild bugloss; pl. u. as diaph., vuln.; rts. substit. for madder (formerly).

Asperula odorata L. = *Galium odoratum* (L.) Scop. - **A. tinctoria** L. = *G. tinctorium* (L.) Scop.

asperuloside: glycoside with a cyclopentindenone aglycone formed in organs of Rubiaceae, Euphorbiaceae, &c.; playing important part in their metabolism (ex. *Rubia tinctorium*) (SC *Galium* [*Asperula*] *odorata*, where first found); with hypotensive & antiphlogistic action.

Asphalat(h)us Burm. f. = *Aspalathus* L.

asphalt: Asphaltum (L.): a bitumen in solid state formed by evapn. & oxidn. of natural components of many petroleums; evapn. of the solvent leaves behind the native deposits (ex. Trinidad [Pitch Lake] & Dead Sea [Asphaltite, hence name of product] ["lake asphalt"]) mined in such places; u. in highway construction ("black top") (since ca. 1000 B.C.); roofs; in cements, water-proofing (tanks, &c.), ringing & sealing micro-slides, as plastic, filling cracks in joints, &c.; formerly in embalming mummies; appl. in skin dis. of humans & of dogs (pruritus, eczema, inflamm., allergy, maybe mange) - **Egyptian a.**: natural prod. probably same as Syrian a. - **natural a.**: that found in geological deposits & made up almost entirely of bituminous matter - **Syriac a.: Syrian a.**: natural a., from the Dead Sea; that exported to Egypt was u. in preparing mummies.

asphaltite: natural asphalt of the Dead Sea.

asphodel: 1) *Narcissus pseudo-narcissus* or *N. tazetta* (of Fr. & Eng. poets) 2) *Asphodeline lutea* (**a. of the Ancients**) 3) *Asphodelus fistulosus*, *A. ramosus*, & other *A.* spp. (fr. Gr. asphodelios: pl. sacred to Proserpine) 4) *Narcissus poeticus* (Ancients) 5) daffodil (of poets) - **wild a.**: *Tofieldia palustris*.

Asphodeline lutea Reichb. (Lili.) (Med, sEu reg.; AsMin, nAf, Caucas reg.): asphodel of Ancients; king's spear; (yellow) asphodel; bulb u. like that of *Asphodelus* spp., as diur. exter. for sores, scabies; bulbs eaten like potatoes by natives of ancient Greece, Rome; cult. orn.

Asphodelus albus Mill. (Lili.) (Med reg., Balkans): white asphodel; bulbs u. in mfr. alc.; fls. symbol of grief since antiquity; cult. - **A. fistulosus** L. (Med reg., Canary Is., Saudi Arab, Ind): hollow-stemmed asphodel; onion a.; weedy perenn. hb.; bulbs c. anthraguinones, flavonoids, chrysophanol, boiled & eaten (Bedouin); sds. u. as diur. (Ind); appl. topically for fungal inf. (Eg) - **A. luteus** L. = *Asphodeline lutea* - **A. pendulinus** Coss. & Dur. (nAf, wAs): u. as diur. - **A. ramosus** L. (A. racemosus Link) (Med reg., sEu, nAf, wAs): asphodel; bulbs c. mucil. inulin, sugars, u. as diur., exter. for ulcers, boils; itch; source of gum u. as glue in bookbinding, &c. (Turk).

asphodil: *Asphodelus ramosus*.

aspic (Fr-m.): spike lavender.

aspidie (Fr): Aspidium.

aspidin: anthelmintic (for Cestodes) fr. *Dryopteris austriaca* - **aspidinol**: ester of butyro-phenone; occurs in Aspidium Oleoresin; u. anthelmintic (Cestodes).

Aspidistra elatior Bl. (Lili) (Jap, Chin): cast iron plant; haran; perenn. acaulescent pl.; popular cult. ornam. (house pl. of former days); pollinated by snails.

Aspidium: 1) old bot. genus name (Polypodiaceae) for ferns (Swartz), now generally split up among the following genera: *Cyrtomium*, *Dryopteris*, *Polystichum*, *Tectaria*. &c. 2) (L.) (specif.): male fern root, fr. *Dryopteris filix-mas* (**Am. Aspidium**) or *D. marginalis* (**European A.**).

Aspidoglossum E. Meyer (Asclepiad) (trop & sAf): c. digitaloid glycosides; candidates for cardiovascular therapy.

Aspidosperma Mart. & Zucc. (Apocyn.) (tropAm, SA, WI): peroba; ca. 80 spp. of trees, most with valuable wd.; bark (quebracha) u. in tanning (1050); many spp. c. yohimbine & related alks. - **A. cuspa** (HBK) Blake (tropAm): lvs. & bk. u. as antipyr. (Ve) - **A. discolor** DC. (Br): ca. 10 alks.; u. to treat mal. - **A. excelsum** Benth. (Br): paddle wood; yaruru; bk. c. methoxy-yohimbine, aspidexcine, aspidexelcine, u. as carm. stom. - **A. illustre** (Vell.) Kuhlm. (Minas Ger, Br): "quina (de camamu)" u. in mal. - **A. nitidum** Benth. ex Muell.-Arg. (Br): u. in mal. - **A. peroba** Saldanda da Gama (A. polyneuron Muell. Arg.; A. dugandii Standley) (CR, SA, esp. Br): peroba-rosa (Br), guayacan (Co); bk. c. harman ester; u. med. (Br, Co) - **A. quebracho-blanco** Schlecht. (Arg, Ch, Bo, sBr): (common white) quebracho; bk. c. many alks. incl. aspidospermine (comm. available), quebrachamine, yohimbine, eburnamenine; soothing remedy (bronch. sed., antispasm.) in bronchial asthma, resp. stim. in bronchitis, dyspnea; tinct. u. in bronchial dis. (Bracodin, McNeil Labs.); antipyr. (in mal.); bitter, analeptic; exter. in wounds; for inhalations; ext. for tanning leather (c. 20% tannin); prodn. alks. - **A. quebracho-colorado** Schlecht. = *Schinopsis q.-c.* - **A. tomentosum** Mart. & Zucc. (Br): peroba-do-campo; piquia peroba; wd. (lemonwood); u. in carpentry; bk. furnishes a cork similar to that of *Enterolobium ellipticum* Benth. (Br).

Aspilia latifolia Oliver & Hiern (Comp.) (tropAf) h(a)emorrhage plant; fresh lvs. & flg. tops u. as styptic ("extraordinarily effective"); rt. ext. drunk for lumbago, sciatica, neuralgia (Shambala); galactagog (lf./rt.) - **A. holstii** O. Hoffm. (tropAf): rt. ext. u. for lumbago, neuralgia, sciatica; in eclampsia; lf. galactagog.

aspirin (former TM, now common & off. name): acetylsalicylic acid; ASA; popular antipyr. & analg. said most often used drug; although a synthetic, it is closely related to salicylic & acetic acids (both of natural origins) and with similar properties to willow bark.

Asplenium L. (Aspleni./Polypodi.-Pterid.) (cosmop.): spleenworts; some 700 spp., many u. as ornam. - **A. adiantum-nigrum** L. (sAf, Ind): black spleenwort; pl. u. as diur. lax. in ophthalmia, spleen dis., icterus; frond decoc. as expect. emmen.; rhiz. anthelmint. (sAf) - **A. bulbiferum** Forst. f. (NZ): mauku; mouku; rhiz. inf. u. as wash for skin dis. & sore eyes; rhiz. also eaten & juice drunk - **A. ceterach** L. = *Ceterach officinarum* - **A. crinicaule** Hance (Indochin): u. as antipyr., by women after childbirth - **A. cryptopteron** Kunze (Ve): inf. u. as bath in fever & general "cure-all" (Ve Indians) - **A. cuneatum** Lam. (sAf): rhiz. u. anthelm., sometimes in enemata - **A. falcatum** Lam. (Ind): pl. u. for calculi, jaundice, mal., splenic enlargement - **A. hastatum** Klotzsch (Ve): u. like *A. cryptopteron* - **A. macrophyllum** Sw. (tropAs, to Polynesia): pako (Manobo+, PI); frond decoc. strong diur., esp. in beriberi - **A. monanthes** L. (sAf): smoked by natives for head & chest colds - **A. nidus** L. (OW trop, esp. seAs): bird's nest fern (SC lvs. form rosette in which humus collects, rts. ramifying into this to procure food & water); laumea (Samoa Is); young sts. cooked & eaten (1051); u. sed., depur.; vermifuge (Tahiti); rheum. (New Caled); ornam. - **A. obtusatum** Forst. (Austral, NZ): rt. u. in skin diseases (NZ) - **A. praemorsum** Sw. (A. furcatum Thunb.) (pantrop): rhiz. anthelm. (sAf); like *A. cryptopteron* in Ve - **A. pumilum** Sw. (Fla, WI, Mex to Br): culantrillo (Yuc); u. med. (Yuc) - **A. rhizophyllum** L. (eUS): walking leaf; w. fern; pl. u. as ton. astr. - **A. ruta-muraria** L. (ncEu; cosmopol): wall rue; white maidenhair; hb. (as Herba Rutae Murariae) u. as expect. & in folk med. - **A. scolopendrium** L. (Phyllitis s. Newman; Scolopendrium vulgare Sm.) (Euras; cult. & natd. in eNA): hart's tongue (fern) (Eu); fronds u. as expect. diur. in pulm., bladder & spleen dis.; lax.; as vulner., urin. lithotryptic; in snakebite (Greece); pl. ornam. (1052) - **A. serratum** L. (sFla, WI, Mex to Br): New World (or wild) bird's nest fern; lvs. & rhiz. in decoc. or sweet syr. for liver & spleen; stubborn diarrh., hysteria (Haiti); in nervous tics; in bronchitis (Cu) - **A. trichomanes** L. (n&cEu, wAs, natd. NA): maiden-hair (or common) spleenwort; fronds u. for colds, in jaundice, edema; crushed pl. soaked in wine u. in mechan. injury, lumbago (Chin).

ass: mammal related to horse & zebra; includes several spp. of *Equus*, esp. *E. asinus* L. (domesticated forms, donkey, burro) (u. for carriage; milk, meat); *E. hemionus* Pallas (As): *E. Kiang* (Mong, Tibet-As) (incl. kiang & onager + [Ind]), &c.; u. in Unani med. (milk, dr. lung & other parts) - **Wild a(ss):** feral animals of As & neAf (*E. hemionus*) (*v.* donkey).

assaçu (Br): *Hura crepitans*.

Assam: area of seAs now covering large part of Bangladesh; residual area a small state of India (extreme northeast) - **A. indigo:** fr. *Strobilanthes flaccidfolius*.

assa-peixe (Br): *Boehmeria caudata*.

assay: the determination of strength (often %) of active princ. or class of active ingredient present in a substance (mixture); carried out by physical, chemical, or physiological means.

assefétide (Fr): asafetida.

assenzio (It): absinthe - **a. pontico: a. romano**: *Artemisia pontica.*

assiminier (Fr): *Asimina triloba* (fr. assimin, Algonquin Indians, means stone frt.).

ass's foot leaves: *Tussilago farfara.*

astacin: carotenoid; tetra-keto-carotene; fr. Algae, also Crustacea, sponges & other animals; believed derived fr. astaxanthin (carotenoid which occ. in pls. & animals); prepd. fr. lobster shells.

Astacus fluviatilis Fabr. (Astacidae - Homaridae - Crustacea - Arthropoda): (common river) crayfish (Eng); crawfish (US); "crawdad" (slang, US); "fresh water lobster" (last 3 names appl. to Am spp.); typically lives in bank of freshwater streams; source of Lapides Cancrorum (Oculi Astaci) c. $CaCO_3$ (chiefly), Ca phosphate, Mg phosphate; astacin; u. in gastric dis. (heartburn), nettle rash; kidney stones; in inf. dis., allergies; homeop. using tinct. of whole animal; shells in tooth powders; ornam. on girdles (Jap); flesh sometimes u. as food, thus very popular among Creoles of La (US) - **A. nigrescens** (PacUS) similar; sometimes u. as food.

Aster L. (Comp.) (temp zone: Am, Euras, Af): aster; starwort; Michaelmas' daisy; ca. 300 spp. of leafy-stemmed herbs (most perenn.), some fleshy halophytes; some ornam. - **A. ageratoides** Turcz. (Jap, Chin to Ind): many vars.; rt. u. in mal., pulmon. dis. (Chin) - **A. amellus** L. (Euras, esp. in Himalay): (Italian) starwort; rts. u. in colds, cough, pulmon. dis.; in rhus poisoning; hemorrhoids (Chinese med.) - **A. bakerianus** Burtt Davy (sAf): u. in native med. in gastrointest. dis., snakebite - **A. concolor** L. (eUS): edgeweed; silverleaf aster; u. for snakebite in dogs (sUS) - **A. cordifolius** L. (eNA): blue wood aster; beeweed; u. as incense to attract deer (NA Indians); as nervine like valerian (pioneer times); arom. tonic, diaph. - **A. ericoides** L. (NA): scrub-brush; this & many other *A.* spp. u. in Am folk med. - **A. erigeroides** Harv. (sAf): lf. inf. u. in native med. as clyster for intest. parasites, abdom. pains, strong purg. - **A. fastigiatus** Fisch. (Chin): rt. u. to treat fevers, "plague," dys., epileptoid conditions - **A. filifolius** Vent. (sAf): u. with camphor by natives for tapeworms - **A. fordii** Hemsl. (Chin): decoc. of whole pl. u. as ton. for colds, in GC - **A. hispidus** Bak. (sAf): rts. & tubers c. very bitter resin; u. in native med. as purg. emet.; enema for colic (combn.) - **A. indicus** L. (Boltonia indica Benth.) (Indochin, Ind): ton. stom. antipyr. carm. diur.- **A. lateriflorus** (L.) Britt. var. *hirsuticaulis* (Lindl. ex DC) Porter (e. Cana, neUS): Michaelmas (or blue) daisy; pl. u. diaph. arom. ornam. - **A. ledophyllus** A. Gray (wPac NA): Cascade aster; lvs. boiled for greens - **A. longifolius** Lam. (A. junceus Ait.; "A. aestivus" A. Gray) (eUS): rheumatic weed; u. antirheum. antispasm - **A. muricatus** Less. (sAf): pl. crushed & inhaled for headache; ordeal "poison" (medication) - **A. nemoralis** Ait. (NA): bog aster; rt. ext. u. for sore ears (Chippewa Indians) - **A. novae-angliae** L. (nNA s to Ala [US]; Co): New England (or hardy) aster; smudge (smoke) of burning pl. u. as fumigant to revive unconscious person (Am Indians); decoc. u. int. & exter. for poison ivy/sumac (Am Indians) - **A. puniceus** L. (e&cNA): purple (stem) aster; swamp a.; rt. u. arom. stim. ton. diaph. antirheum. antispasm. - **A. serrulatus** Harv. (sAf): decoc. purg. u. in gastric dis.; antidote in poisoning (native med.) - **A. spinosus** Benth. (Chloracantha spinosa [Benth.] Nesom) (swUS, Mex s to CR): devil weed aster; Mexican d. w.; oscoba (Mex); u. med. by natives; weed but useful to control erosion - **A. tataricus** L. f. (Chin, Taiwan, Jap): rts. & rhiz. u. dem. in bronch. dis.; purg.; in dysuria - **A. tradescantii** L. (eCana, neUS) = *A. lateriflorus* var. *hirsuticaulis.*

aster, bluewood: *Aster cordifolius* - **China a.: Chinese a.: garden a.**: *Callistephus chinensis* Nees (C. hortensis Cass.) (orman. ann. hb.) - **golden a.**: *Chrysopsis* spp. - **Mexican a.**: *Cosmos* spp. - **New England a.**: *A. novae-angliae* - **prairie a.**: *A. turbinellus* - **red stalked a.: swamp a.**: *A. puniceus* - **white wood a.**: *A. divaricatus* L.

Asteracantha Nees = *Hygrophila* R. Br. - **Asteracantha** (L., Ind Pharm. Codex) = **A. longifolia** Nees = *Hygrophila spinosa.*

Asteraceae Dumont = Compositae (former sometimes excludes Cichorieae, &c.).

Asterias rubens L. (Asteriidae - Echinodermata) (eAtlantic Oc): (common) starfish; common crossfish; c. clionasterol, poriferasterol; u. for mammary tumors & inoperable carcinomata (as tinct., Homeop.) (name *Asterias* Borkh. formerly appl. to *Gentiana* taxa).

Asteriscus aquaticus (L.) Less. (Comp.) (Med reg., maritime rocks): rose of Sharon; dr. pl. when placed in water swells to original size & form; u. for mystic purposes; similarly **A. maritimus** (L.) Less.; cult. ornam.

Asterohyptis mociniana (Benth.) Epling (Lab) (sMex, CA): "verbena montes"; hb. inf. u. for bruises (Salv).

Asterostigma luschnathianum Schott (Ar) (Br, Bol, Pe): the tubers ("jararaca") u. for insect stings.

asthma weed: 1) *Lobelia inflata* 2) *Euphorbia pilulifera* (also **a. plant**).

Asthmador™: propr. remedy for asthma; c. belladonna & stramonium powd. intended to be burned & smoke inhaled (atropine, scopolamine); some misuse.

Astilbe longicarpa Hay. (Saxifrag.) (Taiwan): raw frt. eaten as cure for colds & influenza - **A. rubra** Hook. f. & Thoms. (Chin): rt. decoc. diaphor.

astragalosides: saponins found in *Astragalus membranaceus*+; with useful therapeutic properties.

Astragalus adscendens Boiss. & Haussk. (Leg) (wAs, incl. Iran, Afghan): one source of tragacanth; also source of Persian (or astragalus) manna; u. in making candy & confections - **A. adsurgens** Pallas (A. nitidus Dougl.) (wNA): u. as wash for rhus poisoning (Cheyenne Indians) - **A. amphioxys** A. Gray (Tex, Ariz, Utah [US], nMex): taper-pod milkvetch; u. in med. (Mex) - **A. asymmetricus** Sheldon (wCalif): horse loco; San Joaquin locoweed; causes nervous disorders, addiction & ultimately death to horses & cattle; absorbs selenium fr. soil; this is true for several other wAm *A.* spp. - **A. brachycalyx** Fisch. (AsMin): one of several sources of gum tragacanth - **A. canadensis** L. (Cana, e&cUS): woolly milk vetch; Canada m. v.; rts. eaten raw or boiled (Blackfoot Indians); u. as febrifuge for children (Dakota Indians) - **A. crotalariae** (Benth.) Gray (Calif [US] & adj. Mex): (Salton) locoweed; crazy weed (1054); pl. toxic to stock. producing spinal tetanic stimulation - **A. glycyphyllus** L. (Euras): milk vetch; Wolfsschote (Ge); lvs. u. as medicinal tea in rheum. & skin dis. (Russ); diur. diaph. hb. & rt. c. glycyrrhizin, sugars; cult. for fodder - **A. gummifer** Labill. (AsMin, Syr, Iran, &c.) (A. tragacantha L.): gum tragacanth plant; (great) goat's thorn (2692); gum exudation arising fr. breaks in st. is (gum) tragacanth, of wh. this is chief source; several forms on market (flake, vermiform, sorts); gum is complex mixt. of polysaccharides, incl. bassorin (60-70%), starch, cellulose; u. dem. emulsifier for FO (as codliver oil), VO resins; adhesive for tablet masses; dental plates (with chalk, boric acid &c.); suspending agt.; emulsifier in mayonnaises, cheese spreads, &c.; thickener in calico printing; cigar wrapping adhesive; in toilet creams; taste corrective; allergy developed in time (2404) - **A. henryi** Oliv. (Chin): u. in med. of Chin - **A. heratensis** Bunge (cAs): Indian tragacanth; source of a gum (katira gabina) similar to tragacanth; u. to glaze & stiffen fabrics - **A. leiocladus** Boiss. (AsMin): a source of gum tragacanth - **A. lentiginosus** Dougl. (Wash to Calif; e to Utah [US]): (mottled) rattleweed; Tehachapi loco; produces very common form of locoism in horses (205) - **A. membranaceus** (Fisch.) Bge. (Chin, Kor, eSib): rhiz. & rt. (huanggi) diur. (comparable in effect to hydrochlorthiazide), hypotensive; u. in chronic renal edema (Chin folk med.); ton. in pulm. dis., night sweats; hematuria; to treat heart failure; to offset aging process (mode of action: scavenging superoxide free radical O_2); c. astragalosides (Sabinin): Se indicaor plant - **A. microcephalus** Willd. (Iran): source of a high grade tragacanth (best in Shiraz region) - **A. mollissimus** Torr. (cUS, Wyo to Tex): woolly loco (weed); crazy weed; infamous for causing intoxication ("loco disease") among horses & other animals (1054) - **A. mongholicus** Bge. (nChin, Jap): rt. u. in Jap folk med. as ton. - **A. pycnocladus** Boiss. & Hausskn. (wRuss, Iran): source of a Persian tragacanth - **A. sarcocolla** Dymock (Iran): one source of gummy st. exudation called "sarcocolla"; rather like tragacanth, with sarcocollin, sweet princ.; gum resin u. in medieval med. (aper. emoll., anthelm., antirheum.); also prod. of other pl. spp. - **A. scaberrimus** Bge. (Chin): above-ground parts u. in syph. & gallbladder dis. - **A. sinicus** L. (A. lotoides Lam.) (Chin, Jap): genge; dr. lvs. with sugar u. to treat GC (Chin); cult. as soil improver (green manure) on rice soils - **A. strobiliferus** Royle (wHimal-Pak, Ind): source of Chitral (or hof) gum, a kind of Indian tragacanth; u. in pharyngitis, to produce emulsions & tabs. - **A. verus** Oliv. (AsMin): source of Asia Minor tragacanth.

Astral oil (Conn.): refined whale oil; SC used in the "Astral" lamp (Argand lamp) (with tubular wick); sometimes kerosene.

Astranthium condimentum De Jong (Comp.) (Mex): hb. u. as spice by natives (Michoacán) in soups & meats.

Astrantia major L. (Umb) (c&seEu, wAs, Caucas reg): (black) masterwort; great m.; grosse Sterndole (Ge); rt. of perenn. hb. u. (as Radix Imperatoriae nigrae) in med. (as purg.); astr.; as food; cult. (NA) as ornam. border pl.

Astripomoea malvacea A. Meeuse (Convolvul.) (tropAf): rt. inf. u. locally in ophthalmia; poultice of crushed rt. for inflammn. & swelling; orn.

Astrocaryum acaule Mart. (Palm.) (Br, Co): espina (Sp); ripe sds. u. to cap "silipu" or snuff tubes (Co Indians) - **A. aculeatum** G. F. W. Mey. (A. tucuma Mart.) (Palm.) (Guyana, Surinam, Trin, Br): tucum(a) palm; aguire (Belize); source of tucum kernels & t. oil (FO); frt. also with FO; u. in soap mfr. (1055), &c.; lvs. furnish excellent textile fiber - **A. ayri** Mart. (Br): airi; frt. u. in erysipelas; sds. furnish edible oil; lf. fibers u. to make ropes, hammocks, fish nets, &c.; st. of palm covered with prickles wh. cause wounds & inflamm. on contact - **A. jauari** Mart. (equatorial Br, Pe, Co, Guayana, Surin): (liba): awarra (Pe); frt. furnish FO, awarra palm fat, u. to make soap; refined, as cooking oil - **A. murumuru** Mart. (tropSA fr. Amazonas to São Francisco, Br): murumuru palm; frt. kernels a source of FO, murumuru butter or fat or oil; u. in soap mfr.; foods — in butter, margarine, cooking fats & as lard substit. (1056); frt. hulls edible - **A. tucuma** C. Mart. (Br, Pe +): tucuma; FO from kernel much like that of coconut o.; u. in cooking; pulp of frt. eaten - **A. tucumoides** Drude (neSA): award; lvs. u. to make mats - **A. vulgaris** Mart. (Br & Guianas): tucuma (bravo); t. palm; ext. of thorns u. in erysipelas (Br); sd. kernels furnish 40% FO (t. [kernel] or nut oil) u. to make soap; refined as cooking fat; frt. flesh source of a red FO (tucum or aouara oil) entirely different fr. t. kernel oil; source of very strong fiber; sd. also called cumare, cumari, curua, coqueiro, tucum, large Panama.

Astronidium victoriae A. C. Smith (Melastomat.) (Oceania): herbage of small tree made into ointment with oil; u. for headache (Fiji).

Astronium Jacq. (Anacardi.) (tropAm): trees source of hard wds. & mastic-like resins - **A. balansae** Jacq. (Par): wd. u. for many purposes; very durable in water - **A. fraxinifolium** Schott (Br, Ve): bois de chat (Fr); gateado (Br); guereroba preto (Br); bk. astr. u. in tanning (much tannin); balsam (mastic-like resin) obtained fr. cut bark; wd. fine timber - **A. graveolens** Jacq. (sMex to Br, Co, Ve, Bo, Par): star tree; glassy wood; bk. & lvs. astr., decoc. u. in bronchitis, TB; appl. to ulcers; wd called "gonzalo (goncalo) alves," u. for bridge building, &c. - **A. juglandifolium** Griseb. (Arg): source of valuable wd. u. for furniture, &c.; prepn. urunday extract. - **A. urundeuva** Engl. (SA, Arg): urunday; dark reddish wd. u. for heavy construction (piling; bridge timbers); firewood.

Astrophytum asterias (Zucc.) Lem. (Cact) (nMex): sea urchin cactus; peyote; heads called "peyote" & u. as hallucinogenic (similarly u are **A. capricorne** (A. Dietr.) Britt. & Rose; **A. myriostigma** Lem.; & **A. ornatum** (DC) Web. [star cactus]).

Asystasia gangetica And. (Acanth.) (Ind, Indochin, Af+): lvs. & fls. u. as intest. astr.; pl. juice anthelm. in rheum.; rts. given orally for swollen stom. ("bengo") in child (Tanz); pantrop. weed.

Atalantia monophylla DC (Rut) (seAs): marandier sauvage (wild Mandarin [orange]); lvs. u. in resp. dis.; lvs. & twigs are cut up into fragments and these roasted like coffee, then boiled, and the decoc. drunk (Annam); frt. FO u. in chronic rheum. - **A. roxburghiana** Hook, f. (seAs): lvs. u. in resp. dis.; lf. juice in combns. for gastralgia (Mal); art. oil u. as embrocation in rheum. - **A. simplicifolia** (Roxb.) Engl. (Indochin): lvs. u. to treat resp. tract dis.; frts. edible.

ate: *Protium heptaphyllum*; prod. resinous exudate u. like Burgundy pitçh.

atees: atis.

Ateleia glazioviana Baill. (Leg.) (Br, Par, Arg): timbo blanco; pl. insecticidal, toxic to cattle (Arg) (2143).

Ateramnus lucidus Rothm. (Gymnanthes l. Sw.) (Euphorbi) (sFla Keys, WI, Yuc, Guat, Belize): (man) crab bush (Bah) (SC land crab eats bk. & berries); tabaco(n); bitter bk. decoc. in toothache (Cu); lvs. as decoc. or chewed in stomachache ("upset stomach") & "griping"; antidiarrh. (—> *Sapium*)

aterrimins A & B: antibiotics isolated fr. *Bacillus subtilis* var. *aterrimus* (Lehmann & Neumann) Smith & al.; believed to have growth stimulation properties for animals.

Athamanta cretensis L. (Umb) (sEu): candy carrot; pl. u. in flavoring liqueurs; u. as diur., carm. for flatulence, nerve sed.

Athanasia amara L. = *Brickellia cavanillesii.*

Atherosperma moschatus Labill. (monotypic) (Atherospermat./Monimi) (seAustral, Tasman): Australian (or Victorian) sassafras; bk. c. isoquinoline alk. (**atherospermine**), aporphine alk. (isocorydine) +; VO; bk. u. in arom. teas (Tasmania), as diaph. ton., astr. diur. in bronchitis, rheum.; sds. called "plum nutmegs"; wd. useful, u. to make clothes pegs, &c.

Athrixia phylicoides DC (Comp.) (tropAf): Kaffir tea; rt. decoc. u. as purg. & cough rem.; lf. inf. u. as beverage tea; for boils, sores, &c.; lf. & rt. decoc. u. to bathe sore feet.

Athrosolen: error for *Arthrosolen.*

Athrotaxis selaginoides D. Don (Taxodi.) (Tasmania): King William pine; wd. of this & other *A.* spp. u. technically, as to make bent wd. work.

Athyrium esculentum (Retz.) Copel. (Aspleni.-Pteridoph.) = *Diplazium e.* (Retz.) Sw. - **A. filix-foemina** (L) Roth (Himal fr. Afgh to Sikkim, s. to sInd, wNA): lady fern; rhiz. u. tenifuge, anthelm.; adulter. of & substit. for male fern.

Athyrocarpus persicariifolius Hemsl. (Commelin.) (WI, CR to Guyana to Pe): cojite blanco; pl. decoc. u. as bath for joint pains (Ve, Ind). (—> *Phaeosphaerion*).

atis(root): *Aconitum heterophyllum*; c. **atisine** (alk.) (frt. rt.).

atmosphere: the gaseous mantle ("air") of a planet, usu. the earth; earth's a. c. O_2 (20%); N_2 (78%); smaller amts. of CO_2, Ar, Ne, He, CH_4, H_2, N_2O, Xe, O_3, H_2O; comp. distinctive for our planet (supports life).

atole (SpAm, esp. Mex): porridge or gruel made of maize (*Zea mays*) meal (word fr. Nahuatl).

atopen: allergen; antigen (antibody generator) in allergy.

Atractylis chinensis DC (Comp.) (Chin, Kor): hb. rich in vit. A, u. in night blindness; diur. - **A. gummifera** L. (Med reg., as Gr, Crete, It): masticogna (It); c. atractyloside (toxic); rt. u. in dropsy, urinary diff., skin dis.; prod. mastic-like gum; resin (as acanthus mastic) u. similarly to true mastich (called pseudo-mastich); ext. u. as adulter. of mastic & licorice juice (ext.) - **A. lyrata** Sieb. & Zucc. (Chin): u. antipyr. for heavy colds; in vertigo - **A. ovata** Thunb. (Chin, Jap): rt. & rhiz. antipyr. in colds, vertigo, ton; in diarrh; diaph.; diur.; in dyspep. (Chin); VO u. in perfumery - **A. tomentosa** Lam. (Med reg., Eg): c. tomentsides (cyclo-artane glycosides) of possible value in treating leukemia.

Atractylodes japonica Koidz. = *Atractylis lyrata* - **A. lancea** (Thunb.) DC (Atractylis l. Thunb.) (Comp.) (eAs): pinyin; important medl. pl. of Chin; rhiz. arom. u. in dyspepsia; stim. diur.; for bad breath; pai-shu(h); rhiz. u. for gastric dis., night blindness; diur. - **A. macrocephala** Koidz. (Chin): dr. rhiz. important Chin med. u. sed. during pregnancy; sl. lax.; anhidrotic; for eczema; cancer(!).

Atramentum (L): black ink - **A. indicum: A. Sinense**: India ink fr. *Sepia* (cuttle fish).

atriopeptin: atrial natriuretic factor; occ. in atrium of mammalian heart; controls blood volume & diuresis.

Atriplex L. (Chenopodi) (temp & subtrop zones): saltbush; orach; some 150 spp. of hbs.; some like sagebrushes; grow in alk & saline soils (or even some in sea water); lvs. of some spp. edible after boiling - **A. argentea** Nutt. (cNA: Man, Sask [Cana] to Tex, Calif [US]): (silver) saltbush; silver scale; silver tumbleweed; pl. inf. for diarrh.; lvs. u. for spider bite, as poultice for pain (Navaho); claimed useful in diabetes (1063) - **A. canescens** (Pursh) Nutt. (c&wUS, Mex): fourwing (or hoary) saltbush; chamiza; cenizo ("purple sage"; 1042); a kind of sagebrush; lvs. chewed with salt for gastralgia (NA Indian); "physic" (Shoshone Indians); u. med. (Mex); claimed of use in diabetes; sds. u. as food; pl. u. as cattle forage (Gosiute Indians, Utah) (1065) - **A. halimus** L. (Eg to Mor): sea orache; s. purslane; qataf (bahari) (Arab); rts. u. in dropsy; cut into long pieces as toothbrush; ash of pl. alkaline, u. in gastric acidity - **A. hortensis** L. (Euras, later natd. worldwide, incl. NA): (garden) orach(e); sea purslane; mountain spinach; butter leaves; salt tolerant; sds. u. for flour (rich in vit. A), for resp. & digest. dis., but emetic; lvs. u. as spinach; long use as such may result in toxic skin damage; to feed sheep (Israel); u. med. (Calif Indians); widely cult. as ornam. - **A. hymenelytra** S. Wats. (sCalif, Ariz, Nev, Utah [US]-nMex): desert holly; silvery lvs. & twigs of this evergreen shrub collected for winter bouquets, Xmas decor., &c., at times gilded or dyed (wUS) - **A. lentiformis** S. Wats. (sCalif to sUtah [US], nwMex): lens-fruited saltbush; quail brush; Sonora Indians u. pd. rt. to treat sores or ulcers - **A. rosea** L. (Eu, nAf, adventive in NA): frosted orache; red o.; a serious cause of some pollinosis (s&wUS); staple food of camels (Sahara) - **A. vestita** Aellen (sAf): u. as hot cataplasm for bronchitis (Europeans in Af) - **A. wrightii** S. Wats. (NM, Ariz, Tex [US]; Sonora [Mex]): annual saltbush; Wright s.; lvs. popular as potherb (Am Indians); "salt bush tea" in diabetes (value?).

Atropa L. (Solan.): genus of Old World hbs., mostly poisonous; bear purplish black berries - **A. acuminata** Royle ex Lindl. (Ind, Pak): Indian belladonna; a source of belladonna hb. & rt. in Euras; similar comp. to *A. belladonna* with somewhat higher tot. alk. content (0.8%); adulterated with *Phytolacca acinosa*; cult. in Kashmir - **A. belladonna** L. (Euras): belladonna; deadly nightshade; dwale; lvs. & rts. c. hyoscyamine (chiefly), atropine, scopolamine (small amt.); total alks. 0.40%; rt. similar but also has belladonine, cuskhygrine (belladadine); Greek rt. c. latter & helleradine (high content); Bulgarian rt. (Bulgarian belladonna or chipka root) as white wine decoc. useful in postencephalitic parkinsonism ("Bulgarian cure") (1066); frt. u. as coloring matter for leather; toxic (2553); sds. u. med. (in homeop. med.); lvs. & rts. u. extensively as sed. antispas. mydr. anodyne, to reduce secretions; frt. c. alks., hyoscyamine, atropine (prodn. US, Eng, Fr, USSR) - **A. caucasica** Kreyer (Caucas reg.): krasavka kavkazskaya (Russ); u. like belladonna in med. - **A. komarovii** Blin. & Schalyt (near Caspian Sea); u. as med. pl. in Russ - **A. lutescens** Jacq. (Ind, Pak): Indian belladonna; an off. source of belladonna in BP & elsewhere - **A. mandragora** L. = *Mandragora officinarum* L.

atropamine: apoatropine: alk. fr. belladonna rt.; also prepd. fr. atropine; u. antispasm.

atropine: characteristic alk. of Fam. Solan. & esp. of *Atropa* & *Hyoscyamus* spp.; formed during curing & extr. processes fr. hyoscyamine (orign. present in drug pl.) by process of racemization; obt. fr. *H. muticus*, *A. belladonna* & by semi-synthesis; parasympatholytic (cholinergic blocking agent); mydriatic; antispasmodic; nerve gas antidote (E) - **a. methylnitrate**: Eumydrin; u. as prec.

atroscine: inactive atropine: D, L-scopolamine.

atta (Urdu): wheat flour.

attacins: immune protein P5; obt. from pupae of silkworm.

Attalea compta Mart. (Palm.) (Br): pindoba (palm); palma almendron; inaja; coqueiro oacury; indaiá (do Rio); indaiacu; source of FO - **A. funifera** Mart. ex Sprengel (eBr): broom palm; piassava; piassaba da Bahía; Pará grass; frts. source of piassava fat, u. like babassu fat (oil); earlier handled with palm oil; ext. u. for erysipelas; lf. bases & petioles source of piassava fibers, u. for brushes, brooms, twine, ropes - **A. speciosa** Mart. (Orbignya s. Barb. Rodrigues; O. barbosiana Burret) (Br, Surin, Guyana): caruá (palm); babassu (palm); frt. of tall tree u. to produce FO, babassu oil; u. for soap, margarine, &c. - **A. spectabilis** Mart. (equator. Br, Surin): caruá (palm); c. piranga; sd. kernels rich in FO (62-65%), curua fat, also c. palm oil; u. similarly to other palm fats; lvs. u. as thatch.

Attapulgite: kind of fine greenish clay or Fuller's earth (*qv.*) (hydrated Mg Al silicate) superior to kaolin & halloysite; adsorbent of bacteria such as *Staphylococcus aureus*, &c.; in suspensions & tabs.; u. (activated) in diarrh.; u. in many industrial adsorption & absorption processes (high sorptive capacity); in pharm. prepns. (SC mined with large prodn. at Attapulgus [Indian name] in swGa & neFla [US]); deodorant u. for cat litter (bedding) (large usage), &c.

attar: 1) perfume obtd. by distn. of arom. herbs & esp. fls. over sandalwood oil; ca. 20 pl. attars are prepd.—chiefly rose, jasmine, kewda (*Pandanus tectorius*) (Indo-Pak) 2) (Urdu): perfumer or druggist (pharmacist) (Indo-Pak) (<atr [Arab] perfume, orig. appl. to rose oil) (**a. of roses**) (corruption as "otto").

attemperator: pipe carrying water at const. temp. u. to control temp.

attenuated: 1) (bot.) drawn out to a sharp point (ex. rose sepal) 2) weakened in virulence as appl. to bacterial & viral products.

attenuation: 1) weakening the toxicity of a microorganism or virus 2) lowering of sp. grav. of liquid through formation of alc. in fermentation.

attic salt: Attichsalze (Ge): Succus Sambuci, juice fr. (mostly) *Sambucus ebulus* frt. (SC Attich, Ge [from *S. ebulus]* < atticus [L], Athens).

Attivisha: tubers of *Aconitum heterophyllum* (*qv.*).

attritus: particles of vegetable matter in swamps, bogs, coals, &c.

atyeng: *Plectronia* sp.

Atylosia goensis Dalz. (Leg) (Ind): rts. u. in rheum. fever, TB, swellings - **A. scarabaeoides** Benth. (Ind, China): rt. decoc. u. to treat inflammn. of mouth, nose, throat; for diarrh. in cattle.

aube-épin(e) (noire): **aubépin(e)** (Fr): (black) haw(thorn).

aubergine (rouge) (Fr): *Solanum melongena.*

Aucklandia Falc. = *Saussaurea.*

Aucoumea klaineana Pierre (Burser) (trop wAf): source of Gaboon mahogany; this & other *A.* spp. are the source of elemi (*cf. Canarium*).

Aucuba chinensis Benth. (Corn.) (Chin): crushed frt. appl. to mech. injury (as poultice); pounded lvs. appl. to burns, scalds, swellings (Taiwan) - **A. japonica** Thunb. (Jap to Himalayas): Japanese laurel; shrub (-3 m); various parts source of aucubin (glycoside found in many other plants; antibiotic); pl. u. similarly to prec., also appl. to chilblains, clots; c. var. *variegata* D'Ombrain: (Himal to Japan) gold-dust tree (or plant); common popular ornam. (house or outdoors).

Audibertia Briq.: 1) *Salvia* 2) section of *Salvia.*

Aufguss (Ge-m.): infusion.

Augentrost (Ge): eyebright, *Euphrasia officinalis.*

au(l)ne (Fr): elm tree.

Aulomyrcia Berg (Myrt.) = *Myrcia;* spp. of this now defunct gen. source of antibiotic, termed **aulomycin.**

Aulospermum Coulter & Rose = *Cymopterus.*

aunée (Fr): inula.

aurantiamarin: bitter glycoside of *Citrus aurantium.*

Aurantii Amari Cortex (L): bitter orange peel - **A. Dulcis Cortex**: sweet orange peel.

aurantiin: naringin.

aurantiogliocladin: antibiotic fr. *Gliocladium roseum* (*rubrogliocladi*) (Link) Bainier, & other closely related *G.* spp. (Moniliaceae; Hyphomycetes - Fungi): antibact. & antifungal; *cf.* gliotoxin, viridin.

Aurantium (plur. - **ia**) (L): orange - **A. acidum** Rumph. = *Citrus a.* - **A. maximum** Burm. = *Citrus grandis.*

aureine: alk. fr. *Senecio aureus*, etc.

Aureolaria Rafin. (Scrophulari.) (eUS, Mex, WI, &c.): one of the segregates of *Gerardia*; Am parasitic hbs. often cult. for their showy fls. - **A. flava** Farw. (Gerardia flava L., G. quercifolia Pursh) (e&sUS): false foxglove; golden oak; hb. & st. u. as stim. diaph. sed. - **A. pectinata** Pennell (Gerardia p. Benth.) (seUS n to Mo): said to cause sun poisoning (1067) - **A. pedicularia** Raf. (Gerardia

p. L.) (e&cUS): louse-wort (foxglove); hb. u. as diaph. antipyr. sed.; emet. antiscorb. (seAm Indians) - **A. virginica** Pennell (Gerardia v. BSP): said useful antidote for venom of poisonous snakes.

aureolic acid = mithramycin (*qv.*).

Aureomycin™: antibiotic subst. prod. fr. *Streptomyces aureofaciens* Duggar; useful in amebiasis, pertussis, tularemia, psittacosis, GC, syph. lupus, meningitis, trachoma, actinomycosis, etc.; also potent growth stimulator.

Aureothricin: antibiotic fr. *Streptomyces* sp. active against Gm. pos. and neg. bacteria; isolated 1947; *cf.* Aureothin.

Auricula (L): *Primula auricula.*

Auricularia Bull. (Auriculari. - Auricularales - Hymenomycetes - Basidiomyc. - Fungi): several spp. eaten, esp. in Orient - **A. auricula** (Hook.) Underwood (A. auricula-judae Schroet.) (temp zone): Jew's (or Judas') ear fungus; Jews' ears; frt. bodies u. as food (Chin); appl. externally to eye inflammn.; in throat inflammn. (Chin) - **A. mesenterica** Pers. (cosmopol): jelly-like sporocarps formerly u. in edema & sore throat (Eng); collected in NZ & exported to Chin - **A. polytricha** (Mont.) Sacc. (As, Austral): cultured on oak poles in Chin, where much u. as popular food; ingred. in bah-mi (Chin food) & other oriental (seAs, PI) cooking (soups, &c.); dr. & exported as "black fungus" - **A. sambucina** Mart. (temp zone): u. as emet. purg. for inflammn. esp. soreness of throat; boiled in milk, steeped in beer, vinegar, &c.

Aurikel (Ge-f.): bear's ear; *Primula auricula.*

Aurodox: antibiotic fr. *Streptomyces goldiniensis* Berger et al.; u. to stimulate poultry growth.

austral: southern.

Australian adonis: *Adonis annua, A. microcarpa* - **A. bottle tree**: *Brachychiton rupestris* - **A. cowplant**: *Gymnema sylvestre*; sd. hairs u. - **A. fever tree (lvs.)**: *Eucalyptus* - **A. gum**: exudn. fr. *Acacia decurrens* (wattle gum) - **A. mint:** *Prostanthera* spp.

Austrocedrus chilensis Florin & Boutelje (Libocedrus c. Endlicher) (monotypic) (Cupress.) (Ch, Arg): incense cedar; Chilean c.; cedro; wd. u. in house building, furniture, &c.; ornam. tree. (—> *Libocedrus*)

Austroeupatorium inulaefolium R. M. King & H. Robin. (Comp.) (WI, Pan, Co, Pe, Br; natd. Indones, SriL): Christmas bush; pl. decoc. u. for coughs & colds; fresh juice in ophthalmia (Trin); lvs. poulticed for sores.

autochtonous: native, indigenous.

autoploid: autopolyploid: organism genome redoubling with simple doubling, tripling, &c. of haploid no. of chromosomes; without structural morphological changes.

autotroph: autotrophic plant: self-supporting (ex. algae) in contrast with heterotroph (ex. fungi).

autumn(al) crocus: *Colchicum autumnale* (pl. or corm).

auxanography (bacteriol.): method of finding best nutrient medium for a certain strain of microorganisms by detg. growth stimn. or inhibition.

auxin: growth substance; pl. hormone, chemical wh. produces straight st. length increase, rooting in pl.; both natural and synthetic auxins are known - **auxin a.**: first obtained fr. yeast & human urine - **auxin b.**: obt. fr. maize germ oil, malt ext., sunflower sd., &c. (*cf.* heteroauxin).

auxotrophy: in bacteria, differing growth factor requirements fr. those of the ancestral or prototype strain.

auyama (Ve): squash.

avara (Malayam): **avarai: avaram** or **avaran** (Tamil): *Cassia auriculata* (bk.).

avaremotemo, cortex (Br): bk. of *Stryphnodendron barbatimao.*

avel(l)ana (Sp): filbert; hazel nut.

Aveloz: *Euphorbia heterodoxa.*

Avena fatua L. (Gram) (Euras): wild oat(s); u. as fodder (hay) for cattle; natd. in US, esp. Calif [US]; grain (frt.) u. as food (Am Indian); u. as emoll. refrig. diur. (Eu) - **A. sativa** L. (Euras; cult. worldwide): (common) oat(s); (cultivated) white oats (**Avena**, L.); grains (frts.) rich in starch (43%), protein (avenin) (albuminoid), FO (6-9%), mucilage, trigonelline; meal (Avenae Farina, L.) u. as nutr., dem., in powders, tablets, pills, &c.; in technology; fodder; flour is antioxidant to FO; not suitable for bread (lacks leavening power), mostly u. in oat cakes, wafers, cookies, crackers; dietetic, roborant; stim.; bran effective in reducing high blood cholesterol, to stim. libido; u. in shampoos & mild soaps; mucil. in gastroenteritis, dyspeps.; alc. ext. of pl. once supposed useful to combat opium habit/addiction (Ind); to reduce cigarette craving (1068); oat straw, Avenae Stramen(tum) (L) u. in barley straw baths (gout, rheum. liver dis., skin dis.), source of furfural, furfuryl alc., hydrofuramide, &c.; oat prods. incl. groats (v. oat), cereals, provender, porridge, gruel, oat hulls ("oat bran" u. to provide fiber & roughage in diet), oatmeal (rolled oats) (cult. mostly in nEu, NA, also common wild).

avenacin: antibiotic fr. rt. tips of *Avena sativa*; protects oats against "take all" disease of wheat, barley & rice; *cf.* avenacein (fungal antibiotic).

Avenae stramen(tum) (L): oat straw.

avenca (Br): *Adiantum subcordatum* - **a. brasileira**: various ferns incl. *Adiantum* & *Asplenium* taxa of Brazil, the lvs. of which are u.

avenin: plant casein; globulin fr. *Avena sativa*; nutrient.

avens, common: water a.: 1) *Geum rivale* (rt.) 2) (occas.): water cress - **white a.**: *G. canadense*.

average dose: 1) amt. of drug giving desired effect; precisely detd. by pharmacologic studies for a specific animal sp. 2) usual dose; that amt. generally administered orally to an adult human being at one time; usually may be repeated several times a day.

avermectins: class of antiparasitic antibiotic fr. fermentation broth of *Streptomyces avermitilis* (Merck Co.); represent macrocyclic (16 membered) lactone disaccharides effective in very high dilution; involve GABA mechanism; u. as anthelmintics (nematocidals) (esp. for onchocerciasis [river blindness]), insecticides, acaricides (miticides); several of series known; Ivermectin is a semi-synth. deriv. u. in veter. med. as antiparasitic.

Averrhoa bilimbi L. (Oxalid.) (Moluccas, now pantrop, incl. Mal, EI, Chin, WI, CA, SA): bilimbi; bilimbing (tree); grosella (Chin); small tree (-15 m); frt. too acid to eat as such; u. in pickles, curries, jellies, preserves (Ind); frts. u. in scurvy, pertussis, bilious colic; frt. decoc. with rice to treat fever, diarrh.; juice for stain removal; fl. syrup for cough in children; lf. juice with honey for thrush & other throat inflammn.; lf. decoc. u. in postpartum recovery; boils; pounded lvs. appl. as poultice to wounds (Caroline Is); (antibiotic?); wd. u. for construction - **A. carambola** L. (As-Mal; cultd. pantrop): starfruit; caramb(ol)a; kamrakh (Ind) (114); small tree (-9 m); unripe sour frts. c. -1% oxalic acid with Na & K salts; these u. for conserves, sauces, pickles; only ripe sweet frt. eaten (cultd. Ind); frt. rich in vit. C, reducing sugars (50% dr. wt.); u. as antipyr., sed., diur. (Br); popular hand frt., ("star fruit") in jellies & preserves; fls. vermif.; cult. ornam.

Aves (L): birds (hence Eng avian, aviary, &c.).

Avicel™: microcrystalline cellulose; colloidal suspensions u. as dispersing agt., thickener.

Avicennia L. (Avicenni./Verben.) (pantrop & subtrop): mangroves; ca. 30 spp. of shrubs or small trees; lvs. fodder (ex. *A. marina*); bk. tan; resinous exudn. of bk. medl. (aphrod., in dentalgia, contraceptive); rts. aphrod.; frts. eaten (farmer's food); sds. boiled or roasted (food); unripe in poultices to bring boil to head, abscesses; other mangroves are spp. of *Ceriops*, *Laguncularia* and *Sonneratia* - **A. africana** P. Beauv. (tropAf; natd. SA): black mangrove; "sana"; bk. u. in itch, parasitic skin dis., lice &c.; bk. tan. - **A. alba** Blume (sAs): white mangrove; rts. u. for all sexual dis. (Solomon Is) (rt. wd. burned & ashes eaten [Kursche]); frt. & sds. eaten, sometimes sds. preserved - **A. germinans** L. (A. nitida Jacq.) (Mex, WI, tropAm): black buttonwood (Bah); black mangrove.; canoa; mangle prieto (Cu) (109 vernacular names recorded by Moldenke [115]); sprouts toxic raw, edible when cooked; bk. c. 12% tannin, astr. in GC, u. as tan.; rts. & unripe frts. u. med. (WI) as decoc. to bathe for arthritis; resinous gum fr. bk. (gum canoa) u. in resp. dis., throat dis., to treat ulcers, tumors, hemorrhoids (ointment); aphrod.; a honey plant; wd. hard & resistant, esp. to sea water, u. for dams & naval construction - **A. marina** (Forsk.) Vierh. (Af [Malagasy], sAs): white (willow) mangrove; rt. mucil. u. as aphrod. (Arabs); lf. decoc. for fever (Tanz); wd. u. as firewood, for stakes; var. *resinifera* (Forst. f.) Bakh. (NZ, Austral): (gray) mangrove; native m.; frts. cooked (roasted) & eaten; pl. supposed to prod. edible resin (prob. untrue); u. in leprosy ("best remedy"); bk. bitter, astr. u. antipyr.; emet., as tan for leather; resin in tinct. to prevent pregnancy - **A. officinalis** L. (sAs, Ind, Oceania): white mangrove (tree); ripe frts. u. for food; for cough, asthma, GC, diarrh.; green frt. u. poultice for boils, &c.; heart wd. u. emet. snake poison, rts. aphrod. (Ind); bk. u. in tanning; wd. cinders (ashes) u. like soap to wash clothing.

Avicularia avicularia L. (Aviculariidae - Arachnida) (trop CA, SA): bird spider; mashed spider u. in gastrointest. dis. (Homeopathics).

avidin: protein of raw egg white & genital tract of hen which inactivates biotin (vit.).

avocado (pear) (tree) (US, WI): *Persea americana*; source of **a. oil**.

avoparcin: antibiotic fr. *Streptomyces candidus* (Krasnilikov) Waksman; antibacterial; growth promoter (livestock).

axenic cultivation: sterile cultivation of animals or plants.

axerophthol: antixerophthalmic vitamin; vitamin A.

axil: angle formed by two organs, typically lf. on stem.

axonge (Fr): 1) lard 2) other greases.

Axonopus compressus Beauv. (Gram) (sUS, Mex, WI, nSA): (American) carpet grass; a good turf grass; cultd. for forage (US, Austral); u. med. (Am Samoa); fls. chewed u. as facial poultice - **A. scoparius** (Fluegge) Kuhlm. (Paspalum s. Fluegge) (Mex to Pe): often cult. as green feed for cattle; source of pasta micael (Co), a paste originated by the Cauca Indian tribe of Co, which the people use to "cure" kidney dis.

Axungia (L.): hog lard - **A. Hominis**: human fat (in olden days, u. in therapy).

ayahuasca (SA—Jivaro Indians): *Banisteriopsis caapi.*

ayapan(a): *Ayapana (Eupatorium) triplinerve.*

ayfivin: bacitracin; antibiotic fr. *Bacillus licheniformis* Chester.

ayote (CR, SA): pumpkin - **ayotli** (Mex): squash.

ayua (varia) (Cu): *Zanthoxylum elephantiasis.*

Ayurveda: very ancient Indian system of treating various diseases & disorders of human body, chiefly with herbs, but also with animal prods. and minerals; continues to be practiced in India (2402) (*cf.* tibbi; Unani).

aza: prefix or interfix indicating replacement of CH in a cycle by N; ex. **azabenzene** = pyridine.

azadirach, Indian: Azadirachta (L): *A. indica.*

Azadirachta excelsa (Jack) Jacobs (Meli) (Mal, Indones): young shoots edible; old lvs. in folk med. (Borneo); timber source - **A. indica** Juss. f. (Melia a. L.; Antelaea (L) Adelbert) (Meli.) (OWtrop, Ind, Pak, SriL): "neem" (Bengali); n. tree; nim; Indian lilac; nim(b) (Hind); nimba (Sanscr); margosa; lvs., bk., berries u. raw & as aq. ext. for liver dis., sores, eruptions (Bengal); bk. inf. bitter ton., anthelm. astr. in dys.; antiparasitic, in mal. (before quinine introd.); lvs. stim. antisep., locally in ulcers; smallpox (formerly); pulped lvs. in psoriasis; scabies, other skin dis., sores (c. quercetin); lvs. & twigs as insecticide, insect repellant (locusts, grasshoppers [1069]); & in As libraries (lvs. placed between pages of books); gum exudn. fr. tree trunk u. dem. ton., antipyr.; sap (toddy) u. as cooling beverage; lvs. fodder; fls edible; frts. aper.; fls. stim. ton. stom.; rt. inf. ton.; sd. kernel yields FO with evil-smelling VO (garlic odor) admixed; this margosa (or veepa) oil u. as loc. stim. in rheum., skin dis. (parasiticide), anthelm. suppurating glands, mange (dogs), FO c. mostly glycerides with 2% bitter subst. c. nimbidin, nimbiol & nimbidol (keto phenols); azadirachtin; FO (margo oil) u. in soap mfr. (Neem toilet soap, Margo soap) (1070), hair tonics; dental pastes & lotions; lamp oil; also hydrogenated to furnish cooking fats; almost every house in rural Ind has tree in compound (*cf.* sUS with *Melia azederach* tree in back yard almost regularly); wd. good timber (mahogany substit.); fuel; frequent confusion in literature of *Melia azedarach* & *Azadirachta indica.*

azafran: 1) (Sp): true saffron 2) (Mex): *Carthamus tinctorius* 3) (Mex): *Buddleja marrubifolia* 4) (Sp): *Escobedia scabrifolia* - **a. estigmas**: saffron stigmas.

azafrancilla (flores) (Mex): *Carthamus tinctorius* - **azafranilla** (Sp): *Escobedia scabrifolia & E. laevis.*

azafrin: carotenoid carboxylic acid fr. rts. of azafranilla (*qv.*).

Azalea Desv.: formerly a good gen., now a sect. of *Rhododendron* (thus **Azalea japonica** Gray, is now recognized as *Rhododendron japonicum* Suringar).

azamboa (Sp): citron.

Azansa lampas Alef. = *Thespesia l.* Dalz. & Gibs. (Malv.).

azaro (Sp): *Asarum canadense.*

azaserine: serine diazo-acetate; antibiotic prod. by *Streptomyces* spp. (esp. *S. fragilis* Anderson & al.) (1954-); u. antineoplastic (tumor-inhibiting); inhibits growth of some bacteria, fungi, & viruses; inhibits rt. growth; is carcinogenic for pancreas of Wistar rats; acts synergistically with asparaginase.

azasteroids: sterane compds. with N as integral element of the ring or in side chain; ex. *Solanum* gly co-alkaloids.

azedarach (Indian) bark: *Melia azedarach.*

azelaic acid: occ. in rancid oleic acid; u. against acne vulgaris (topical antimicrobial).

azereiro dos danados (Por): *Prunus padus.*

azerole (Sp): acerola.

Azima tetracantha Lam. (Salvador.) (Ind): rts. u. in rheum.; lvs. stim. in rheum.; bk. expect.

azinho, lenhos (Por): Lignum limousianae, wd. of *Quercus ilex.*

azolitmin: blue colored princ. occurring in litmus & other related lichen pigments (preferred fraction); replaces litmus.

Azolla (Azoll. - Pteridoph.) (tropAs, NA): water fern; these harbor a primitive alga, *Anabaena azollae* Strasburger (Cyanophyceae) furnishing high N fixation; pl. u. as fertilizer (green manure) for N source in rice plantations in Chin; also weed suppressant; to control mosquitoes, by blocking water surface; (hence called "mosquito plant" or "m. fern;"); animal feed - **A. caroliniana** Willd. (seUS): mosquito fern; m. plant; mosquito plant; believed to drive away mosquitoes doubles wt. in 7 days - **A. pinnata** R. Br - **A. rubra** R. Br. (NZ): pl. chewed for sore mouth & tongue.

azomycin: antibiotic fr. *Streptoverticillium eurocidicum* Locci & al.; isolated in Japan (1953-); antibacterial; antifungal.

Azorella gilliesii Hook. & Arn. (A. caespitosa Cav.; Hydrocotyle gummifera Lam.) (Umb.) (Pe, Ch, Falkland Is+): balsam bog (Falk Is); yareta; pl.

produces bolax resin & b. gum; juice of perenn. hb. u. in kidney & liver dis., hypochondriasis; *cf.* yareta - **A. yareta** Hauman. (sSA): timiche; perenn. hb. forming "cushions"; resin u. as astr. & absorb. in native med.; u. as highly efficient fuel.

azota caballo (Sp): *Luehea divaricata.*

azote (Fr): nitrogen; symbol Az (Fr).

Azotomycin: antibiotic subst. derived fr. *Actinomyces cinereoviridis* Preobrazhenskaya & al.; antitumor properties; related to Ambomycin.

azougue de México (Mex) (Mexican mercury): *Hippeastrum reginae* - **a. dos pobres** (Br): *Apodanthera smilacifolia.*

Aztekium ritteri (Boed.) Boed. (monotypic) (Cact) (Mex.): "peyotl"; u. as hallucinogenic by Indians; may have similar alkaloids to those of *Lophophora williamsii.*

azucar (Sp): sugar-cane; sugar.

azulejo (Sp): *Centaurea cyanus* - **a. de montana**: *C. montana.*

azulene: caerulein; blue colored compd. found in some VO; also synth.; u. to dissolve gallstones & u. in brain hemorrhage; deriv. antacid.

azulenes: blue coloring matters of VO of chamomiles, wormwoods, &c.; thus of *Anthemis nobilis, Matricaria chamomilla, Andira excelsa, Inula helenium,* &c.; sesquiterpenes; higher b. pt. constit. (b.p. 170°); appar. derived fr. azulogens; as anti-inflammatory u. in cosmetics (1071) & medicinals.

Bb

babaçú (Br): babassu
babado (Br): *Datura suaveolens* .
babandi (Mex): *Chiococca alba* Hitchc. subsp. *parvifolia* var. *micrantha.*
babassú (nuts): *Orbignya oleifera, O. martiana, O. speciosa,* & other *O.* spp.
babes-in-the-cradle: *Rhoeo discolor.*
bablach pods: fr. *Acacia bamboiah.*
babonij (Arab): chamomile.
babool acacia: **babul a.**: *Acacia arabica*; source of **b. bark**.
baboons: apes (Af, sAs): *Papio* (*Cynocephalus* spp.) (Cercopithecidae - Anthropoidea - Primates); living hearts were u. in human transplants.
babul bark: *Acacia nilotica.*
baby's breath: ornamental flg. plants: 1) *Gypsophila paniculata, G. elegans* 2) *Galium sylvaticum* L. 3) *Muscari botryoides.*
bacalão (Sp): codfish; *Gadus* spp. - **higado de b.** (Sp): codliver.
bacalhaa (Por): **baccala** (It): codfish.
bacca (L): berry - **Baccae Spinae-cervinae** (L): buckthorn berries.
Baccaurea brevipes Hook. f. (Euphorbi) (Indones): lf. decoc. u. as emmen. - **B. cauliflora** Lour. (Indochin): frt., stom. & sl. anthelmintic - **B. motleyana** Muell. Arg. (Mal): rambai; liquid fr. pounded bk. for sore eyes; bk. prepn. appl. to skin dis. (Indones); edible frt. - **B. wilkesiana** Muell. Arg. (Fiji): lf. inf. strained through coconut cloth & drunk as "blood purifier."
baccelli (It): pods, beans.
Baccharis aphylla DC (Comp.) (Br): hb. u. in gastritis - **B. articulata** Pers. (Arg, Br): carqueja; entire pl. without rts. u. as hepatoton., in female sterility, impotence, antisep.; decoc. in rheum.; dyspep. formerly in brews for beer - **B. calliprinos** Griseb. (Arg): lf. inf. u. in colic; source of yellow dye - **B. confertifolia** Bert. (Ch): stom., ton. anti-spasm.; in baths for rheum. pains & urin. dis. - **B. cordifolia** DC (Arg, Br): mio-mio (hb.) earlier entire pl. as drug; c. several alks., FO (b. oil); toxic (1072) - **B. crispa** Spreng. (Arg): carquehilla; u. cholagog, digestive, aphrod. - **B. dracunculifolia** DC (seBr): vassoura; vassourinha; lvs. c. VO with limonene, nerolidol; hb. u. in gastritis; in perfumery - **B. fevillei** DC (Pe) common pl., coll. Jan. - May; u. antirheum. expect. aperitive; antiasthmatic - **B. floribunda** HBK (Pe): (ullcco) chilca; lvs. bound on cuts, bruises, &c. - **B. genistelloides** (Lam.) Pers. (B. venosa [R. & P.] Pers.) (Co, Ec, Bo, Pe, Br, Arg, Par, Ur): "carqueja (amarga)"; flg. pl. & st. u. as bitter ton. in bitter "species," Tr. Amara, &c.; antipyr. (Pe); in dyspeps., hepat. dis., diarrh. (Braz Ph. var. *trimera* Bak.) (dose 3-4 gm.); in impotence & frigidity (Arg); fertility control (Ur); anthelm.; in mfr. liqueurs (Br) (1074) - **B. glutinosa** Pers. (B. salicifolia Pers.) (sCalif, NM, Tex [US], Mex, to Bo, Arg): waterwally; seepwillow; jarilla (comun); jara dulce; chilca; lf. decoc. u. as eyewash; lvs. poulticed on sores; u. in baths; st. branches u. to cover rafters of houses & to make market baskets - **B. halimifolia** L. (seUS w to Tex, Okla, n to Mass [US]; Mex, WI): consumption weed; sea myrtle; (tree) groundsel bush; u. arom. dem. esp. in bronchial dis. - **B. latifolia** (R. & P.) Pers. (B. polyantha HBK; B. floribunda HBK) (Ve, Co, Pe, Bo, nArg): (ullcco) chilco; chilca; chilca negra (Hon); grd. rt. mixed with water, u. as shampoo

for dandruff (Ec); lvs. bound on cuts, bruises, &c. (Pe) - **B. linearis** (R. & P.) Pers. (Ch): romerillo; ash u. in gout, rheum. (there are at least 12 addl. *B.* spp. recorded for Ch u. in domestic med.) - **B. notosergila** Griseb. (Arg): carqueja; above ground parts u. as cholagog, digestive, antirheum.; in liver cancer (Mex folk med.) - **B. pilularis** DC (Calif, Ore [US]): (dwarf) chapparral-broom; coyote brush; kidney root; "cotton plant"; herva santa; decoc. u. as diur. in kidney dis. colds, fever, dys. (Jap) - **B. prunifolia** HBK (Co, Ve): chilco, lvs. u. for headache (Ve) - **B. salicifolia** Pers. (sMex, CA): chilca; u. similar to prec.; pl. u. in bath - **B. silvestris** DC (Br): alecrim do matto; u. as. stom. diur. - **B. tricuneata** Pers. (Co, Ve): "sanalotodo"; u. in various kinds of cancer (tinct.; inf.) (1073); var. *ruiziana* Cuatr. (Pe, Bo): contused lvs. appl to aid healing of wounds; decoc. drunk for dysuria - **B. tridentata** Vahl (Br): pl. u. diur, antipyr. - **B. trimera** DC = *B. genistelloides* var. t. - **B. trinervis** Pers. (sMex to Par): barba fina; lf. decoc. antipyr. appl. to swellings; as bath after childbirth (Guat): fresh lvs. as poultice for sores & ulcers - **B. vulneraria** DC (Br): herva santa; u. as vulner. - **B. wrightii** Gray (cUS, Mex): groundsel; inf. u. for sexual infec., emet.; snuff u. for head dis., esp. nasal.

bachelor's buttons: 1) *Centaurea cyanus*, *C. nigra* 2) *Ranunculus acris* 3) Nux Vomica 4) double fld. spp. of *Centaurea*, *Lychnis*, *Ranunculus*, etc.

Bacher's Pills: pills of hellebore & myrrh.

Bacillarieae: Diatoms (Algae); many spp. of unicellular pls. found in both fresh & salt waters; an important dietary of many small animals; all diatoms c. chlorophyll while living, sometimes also brown pigments; the organisms are protected by a transparent siliceous case made up of 2 dish-like "valves" which fit together; this silica covering is permanent & in masses collects on ocean floors to form huge deposits; some of these now being mined are far inland (ex., Nev [US]), marking the location of seas of enormous antiquity.

Bacillariophyceae: class of Algae; diatoms.

bacillin: bacilysin: antibiotic prod. by *Bacillus subtilis,* &c.; water-soluble (soil bacilus) (one of several prods. from this sp.); *cf* bacitracin.

Bacillol™ (Ge): prod. similar to Soln. Cresol Saponated (many pharmacopeia, formerly USP).

Bacillum: (L): 1) little staff, wand, stick; ex. Potassii Hydroxidi Bacillum (sticks of caustic potash) 2) bougie.

Bacillus Cohn (L): genus of Bacteria; a rod or stick-shaped micro-organism (fr. bacillum) - **B. brevis** (Bacillaceae): bacterial sp. fr. which is derived Tyrothricin (first isolated by Dubos) - **B. Calmette-Guérin:** BCG vaccine: a strain of *Mycobacterium tuberculosis* Lehmann & Neumann, causative organism of TB; u. as active immunizing agt. against TB (cattle); also for other infections, as leprosy (1473); anti-tumor value in Hodgkin's dis., inoperable lung cancer, &c. (E) - **B. polymyxa** (Prazmowski) Macé: fr. garden soil, air (Chin name); source of Polymyxins A, B (sulfate = Aerosporin), C, D, E, antibiotic complexes; ("P. B" - P. A + P. B.) (u. antibact. in surgery [1946-7]) - **B. subtilis** Cohn: hay bacillus; common soil-water saprophyte; may be occasional cause of conjunctivitis & other infections; spore-forming; carried in air; source of subtilin (antibiotic; food preservative) & bacitracin - **B. thuringiensis** Berliner: source of "biological insecticide" (BT) wh. produces death of larvae (caterpillars) of various Lepidoptera, incl. tomato hornworm, cabbage looper, & c.; spores u. for this are nontoxic to humans & other warm-blooded animals ("Dipel Dust™", Abbott).

Bacitracin: international free name for a polypeptide complex antibiotic found in nature (tissues debrided fr. compd. fracture of tibia) & obt. fr. *Bacillus subtilis* (Tracy strain) & *B. licheniformis* Chester; isolated in 1945; mixt. of 9 or more compds., mostly B.-A; antibacterial antibiotic mostly u. for gm. neg. organisms; mostly u. locally for furuncles, carbuncles, ulcers, wounds, &c.; also in enteric infec. (amebiasis, gas gangrene, &c.); promotes growth (B).

backache brake: *Athyrium filix-femina.*

backcrossing: crossing of hybrid with one of its parents.

Bacopa decumbens E. N. Williams (Scrophulari) (Af: Senegal): pl. not consumed by livestock & hence considered dangerous; c. hersaponin, saponin-glycoside. with cardioton. properties - **B. monnieri** Pennell (B. monniera [L] Wettst.; Herpestis monnieri Rothm.) (Ind, Indochin, SriL): water hyssop; "unuwila"; "bama"; hb. c. brahmine, baco-side A; u. as nerve ton. in mental dis., psychosis, epilepsy, &c. (Hindus); aphrod. mild purg. diur. (PI); hoarseness; lf. juice locally in rheum; lvs. for thread-worm; u. in aquaria.

ba(c)quois (Fr): *Pandanus* spp.

Bacteria: Schizomycetes: Schizonta: microscopic acellular organisms, sometimes classified with plants, sometimes with fungi, sometimes with an independent group, the Monera; acc. to food sources, 4 subdivisions are recognized: parasites, saprophytes, phagotrophes (ingest protoplasm or its products), and chemosynthetic organisms (obtain their nutrition from inorganic compds.); sometimes included with the blue-green algae in

Monera (Schizophyta; Prokaryota); Gm. positive b. (with thin walled cells) and Gm. negative b. (with complex thick walls) are recognized.

bacterial vaccine: **bacterin**: a suspension of killed pathogenic bacteria in sterile physiological salt soln.; ex., staphylococcus bacterin - **b. vitamin H**: para-aminobenzoic acid.

bacteriologics: **biologicals**: medicinal products derived fr. bacteria & other microorganisms & their products (ex. sera, vaccines, antitioxins).

bacteriophage: bacterial virus: virus which lyses bacteria; discovered by d'Herelle.

Bacterium lactis Lister = *Streptococcus lactis* (Lister) Lohnis.

Bactris gasipaes HBK (Guilielma g. Bailey; G. speciosa Mart. [vars.]; G. utilis Oersted; &c.) (Palm) (CA, SA [Br, Co]): pejibay(e) (palm) (Guat); peach palm; pupunha (Ve); frt. borne in enormous amts., considerably u. as food (boiled or roasted); looks like apricot & tastes like chestnut; in prepn. fermented drink; sds. source of FO (macanilla oil); rt. decoc. in colic (2444); as vermifuge; wd. u. in building & for bows (SA, Indians); apical bud u. as substit. for palm hearts (eaten); long & widely cultd. - **B. guineensis** (L.) H. E. Moore (B. minor Jacq.) (CA, Co): higuero de lata (Co); peach palm; frts. sold in market & edible (Co) - **B. maraja** Mart. (Br, Co+): espina; frts. edible with subacid flavor; u. to quench thirst.

baculum: penis bone found in *Canis*, &c. (2743)

badam (Urdu): almond.

badan: lvs. of *Bergenia crassifolia*.

Badeschwamm (Ge): sponge.

badger's grease: fat fr. *Meles meles* (L) (M. europaeus) (European b.) - **American b.**; *Taxidea taxus* Schreber: u. med. emoll.

badian: (Iran, Ind): **Badiane** (L., Fr): *Illicium anisatum*; anise.

Baeckea frutescens L. (Myrt.) (Chin to PapuaNG, Austral): tiongine; lf. inf. (tea) u. for sunstroke, fever; dr. lvs. insect repellant; fls. u. diur. & in headache, catarrh., rheum., colic, as emmen.; source of VO (essence de bruère de Tonkin); u. to perfume soaps (Fr).

bael (, **Indian**) (Hind): *Aegle marmelos* (frt.).

bagasse: 1) crushed st. of sugar cane or sorghum after expression of the saccharine juice; u. to make newsprint (Cu); source of cellulose, furfurol, aconitic acid; heat source 2) sugar beet pulp after being extracted.

bagh (Urdu): garden.

bagworm: *Thyridopteryx ephemeraeformis* (Psychidae - Frenatae - Lepidoptera): female caterpillar of this moth makes a case or bag attached to a twig (mostly on *Juniperus* & *Thuja* shrubs/trees) in which she lives to produce eggs; decoc. of bag u to help lose wt. (Seri, nwMex).

Bahía grass (Pe): *Paspalum notatum* - **B. powder**: goa.

baicalein: a flavone fr. rts. of *Scutellaria baicalensis*.

baie (Fr): berry - **b. de génièvre**: Juniper berry.

Baileya multiradiata Harvey & Gray ex Gray (Comp.) (swUS, nwMex): calancapatle (Mex); wild marigold; rt. u. as purg.; poultice for pain &c.; cult. orn.

bai-mo: *Fritillaria verticillata*.

bain-marie (Fr): water bath (Mary's bath): named by alchemists after Moses' sister, not the Virgin Mary!; conveys mild temp.

Baissea multiflora DC (Apocyn) (Senegal): eduden (Bassari); bk. & braches (rich in latex) recommended for edema of famine; frt. powd. in water for conjunctivitis; blended with foods in internal treatment of appendicitis; strong sts. u. to tie house roofs (Mali).

baits (for rat poisons, etc.): anise oil; artif. lovage oil; artif. rhodium oil.

bajra (Indo-Pak): **bajree: bajri**: spiked millet.

bakankosin: glycoside fr. sd. of *Strychnos vacacoua*.

bakas. *Adhatoda vasica*.

baker's salt: ammonium carbonate - **b. sugar**: corn sugar - **b. yeast**: a moist viable yeast, often compressed.

baking ammonia: baker's salt - **b. soda**: sodium bicarbonate.

BAL: dimercapto-propanol ("British Anti-Lewisite"); used in heavy metal poisoning.

Balaena mysticetus L. (Balaenidae - Cetacea - Mammalia) (Arctic Oc only): Greenland (or Arctic) (right) whale; right w. (meaning: right kind of whale to pursue for baleen); mammal 15-20 m. long; baleen (or whale bone) horny fringed plates on upper jaw; very long & finely fringed; formerly u. in corsets (stays), hair brushes, &c. then in high demand; body oil used (whale oil, superior) (Pacific right whale [18-22 m. long]); whaling now generally illegal.

Balaka seemanii (Wendl.) Burret (Palm.) (Oceania): lf. & rt. u. med. (Fiji Is); antimicrobial; sts. u. for walking sticks; ornam.

Balanites aegyptiaca Del. (B. roxburghii Planch.; B. zizyphoides Mildbr. & Schlecht.) (Zygophyll.) (nAf, Eg, Israel, Saud Arab, Syr. Ind, tropAf): desert date; Egyptian myrobalan; Jericho balm; agialida; Mexican (or Roxburgh) balanites; thorn tree; bito; subalive (Burma); zackum; frt. rich in saponins u. purg.; for intest. worms (Arabs); sd. kernel c. steroid saponin balanitesin (diosgenin); expect., sd. FO (zachun oil) u. mild purg., in sleeping sickness (Af); pl. c. resin (pseudo-balsam) ("Balm of Gilead"); lvs. sour but edible; unripe frt. emet. purg., added to drinking water as prophylactic to bilharziasis; bk. frt. lvs. u. as anthelm. (Burma); fls. & ripe frt. edible (cattle) ("desert date"; "slave dates"); wd. u. to make walking sticks, clubs (2770) - **B. maughamii** Sprague (Moz): torchwood; frt. lethal to snails & u. to eradicate bilharziasis; for this purpose, trees planted along water courses; sd. source of FO u. as lubricant, industrial purposes; rt. decoc. u. as emet.; frt. edible - **B. wilsoniana** Dawe & Sprague (trop eAf): pl. furnishes a gum; tests have shown the pl. of no value as antimal.

Balanocarpus heimii King (Dipterocarp.) (Malay): common lumber tree; yields resin, Damar penak (also called chengal); u. mostly in varnishes; similar to Ind dammar; resin burned with *Datura* sp. is stupefiant (—> *Hopea*).

Balanophora J. R. & G. Forst. (Balanophor.) (As, Austral): root parasites; said edible; u. for rheum; prod. a wax - **B. elongata** Bl. (tropAs, esp. Java, Austral): source of balanophor wax, u. to make candles - **B. involucrata** Hook. f. (seAs): pl. made into cakes with flour & u. for int. & ext. piles (Chin): pl. aphrod. (Mal); tonic ("valued") (Chin).

balata (Pe, Ve): (gum) chicle, sapote; zapote gum; non-elastic rubber; fr. *Manilkara bidentata* - **brittle b.**: - **Guiana b.**: **balata Rans:** exudation of *M. bidentata* (Rans: municipality near Penafiél close to Porto, Port).

Balaust(i)a (L.) (Balustius): pomegranate (several spelling variants).

bald-money: *Gentiana campestris*.

Baldrian (Ge): valerian.

baleen: whalebone; Barte (Ge-f.): Fischbein (Ge-n); transitional subst. between hair & horn, light, elastic, flexible properties; u. in bristle industry for street brooms, &c.; stiffening agt. in caps, brassieres, corsets (formerly), surgical appl., whip attachments, shoe manufacture, &c. (v. under "whalebone").

baleine: (Fr-f.): **Balena** (It); whale.

Baliosperum montanum (Willd.) Muell.- Arg. (B. axillare Bl.) (Euphorbi) (seAs as Cambodia): rts. u. in stomachache and to "warm the mouth"; rt., lvs., sd. & sd. FO u. as drastic purg.; lvs. u. as poultices (Mal); replacement for croton oil; raw rts. u. for intest. dis.

ball: frt. (berry) of potato pl., &c.

ballon (Fr-m.): round flask; carboy.

balloon vine: *Cardiospermum halicacabum*.

Ballota africana Benth. (Lab) (sAf): cat-herb; inf. of hb. u. for colds & flu, in thrush; locally for sores on head (lotions) (Europeans in Af); in severe colic and snakebite (Af natives); in asthma, bronchitis, hysteria (tinct.), neurasthenia, hoarseness, insomnia &c.; popular in Af folk med. - **B. foetida** L. = *B. nigra* L. - **B. hirsuta** Schult. (Med reg.) lvs. & flg. tops of perenn. hb. u. to adulter. hoarhound hb. - **B. lanata** L.= *Leonurus lanatus* - **B. nigra L.** (seEu, Med reg. nAf [Lib, Mor], AsMin, natd & US): black (or foetid) hoarhound; flg. hb. c. VO, NBP (marrubiin), tannin; u. as tranquilizer in neurasthenia, hysteria, hypochondriasis; carm., choleretic; exter. in gout; spasmolytic, emmen. ton.; for tinea; adulter. of hoarhound & *Melissa officinalis*; similarly var. *foetida* Vis. (B. f. Lam.).

Ballote (Ge): *Ballota nigra*.

balm: 1) fragrant healing ointment or to anoint with same 2) balsam 3) trees yielding balsams; ex. *Commiphora* spp. 4) (specif.): *Melissa officinalis* 5) (of Bible): gum fr. frt. of *Balanites aegyptiaca* 6) Mecca balsam - **bee b.**: 1) *Melissa officinalis* 2) *Monarda* spp. - **common b.**: mint balm - **field b.**: 1) *Nepeta cataria* 2) *Glechoma hederacea* 3) *Calamintha nepeta*; *C. sylvatica* subsp. *ascendens* - **b. gentle: lemon b.: mint b.**: *Melissa officinalis* - **mountain b.**: Yerba Santa; v. under "mountain" - **b. of Gilead** 1) balsam poplar, *Populus candicans* 2) *Commiphora gileadensis* 3) *Liquidambar orientalis* - **b. of life**: Tr. Benzoin Comp. - **red b.**: *Monarda didyma* - **sweet b.**: 1) *Cedronella triphyllum* 2) lemon b.

balmony (eUS & Cana): *Chelone glabra* (lvs. hb.).

balneology: science of baths; includes **balneotherapy** = therapeutic use of baths.

balsa (wood) (tree): *Ochroma lagopus* (& var. *occigranatensis*); *O. limonensis* (river rafts, Pe). (u. for building Kon-Tiki). (SC balsa, Sp raft).

balsam: 1) (generic): oleoresin or resinous prod. of two types: those with benzoic and/or cinnamic acid and/or the esters and those without; at times, specif. only for those of former type 2) tolu b. 3) *Abies balsamea*, tree + - **acajou b.**: v. under acajou - **b. apple**: *Momordica balsamina Echinocystis lobata* (wild)- **Arabian b.:** Mecca b. - **calaba b.**: fr. *Calophyllum jacquinii* - **Canada (fir) b.**: Terebinthina Canadensis: oleoresin fr. *Abies balsamea*; u. as cement for glass, etc.; mountant in making permanent micro-slides; occurs on market in 3 forms: natural, desiccated, as xylem soln.; prod. Quebec; u. clearing agt. - **capivi b.**: **copaiba b**: oleoresin fr. *Copaifera multijuga* - **Demerara b.**: fr. *Copaifera guayanensis* - **b. family**: Balsaminaceae - **fir b.: b. (of) fir**: *Abies balsamea*: the tree or its oleoresin - **friar's b.** : 1) Tincture Benzoin Compound 2) (strictly) = Balsamum Traumaticum - **garden b.**: *Impatiens balsamina* -

b. of Gilead: *Amyris gileadensis* - **Gurjun b.**: fr. *Dipterocarpus turbinatus* & other *D.* spp.- **Hungarian b.**: exudate fr. cut branches of *Pinus mugo* - **Illurin b.**: fr. *Oxystigma mannii*; *Daniella oliveri* - **Indian b.**: 1) Peru b. 2) *Abies balsamea* - **jungle b.**: FO fr. *Canarium* spp. - **Mecca b.**: fr. *Commiphora opobalsamum* - **b. needles**: lvs. of *Abies fraseri*; u. in med. pillows - **b. oil** (Pe): VO dist. fr. Peru b.; c. benzyl benzoate, b. cinnamate, cinnamic acid, vanillin, peruviol - **old field b.**: *Gnaphalium obtusifolium* - **Oregon b.** (fir): *Pseudotsuga menziesii*, Douglas fir (tree & oleoresin) - **b. pear**: *Momordica charantia* - **Persian b.**: friar's b. - **b. of Peru**: **b. Peru: Peru(vian) b.**: fr. *Myroxylon balsamum* var. *pereirae* - **pig's b.**: resin of *Tetragastris balsamifera* - **b. poplar**: *Populus balsamifera* - **Riga b.**: prod. of dist. fresh lvs. of *Pinus cembra*; generally an artific. arom. spirit - **small sea-side b.**: *Croton eleuteria* - **b. sulfur**: sulfurated linseed oil - **sweet b.**: *Gnaphalium obtusifolium* - **Thomas b.**: Balsam Tolu - **b. tips**: b. needles - **tolu b.**: fr. *Myroxylon balsamum* - **traumatic b.**: friar's b. - **b. tree**: 1) mastic 2) fir b. - **Turlington's b.**: **Wade's b.**: friar's b. - **b. weed**: *Impatiens biflora* - **white b.**: sweet b. - **b. wool**: fr. *Yucca* (*qv.*) - **wound b.**: friar's b.

Balsamea Gled. = *Commiphora* Jacq.

Balsamier (Fr): *Amyris balsamifera.*

Balsamina (L.): - **balsamine**: 1) *Impatiens balsamina, I. nolitangere* &c. 2) *Momordica balsamina.*

Balsaminaceae: Balsam Fam.; with about 900 spp. distributed over the globe; members of fam. are soft & succulent hbs.; lvs. simple, exstipulate, petiolate; chief genus, *Impatiens.*

Balsamita major Desf. (Chrysanthemum balsamita (L.) Baill.; Pyrethrum b. Willd.; Balsamita suaveolens Pers.) (Comp.) (sEu, wAs; natd. & cultd. e&cNA): balsamite; costmary; mint-geranium; hb. has an odor similar to spearmint & melissa; c. VO with L-carvone, L-camphor; hb. (as Herba Balsamitae) carm. sed. emmen., choleretic (in gallstones), diur. stom. astr., popular spice (Eng), to flavor alc. form.; cult. orn.

Balsamita: 1) (L., Sp): *Balsamita major* 2) (CR): *Palicourea crocea* (also balsamite) 3) *Tanacetum balsamita.*

balsamo blanco (Sp): white balsam; prep. by expression fr. frt. of *Myroxylon pereirae*; u. to treat ulcers, freckles, int. diur. stim. - **b. de Concolito** (Co): tolu balsam - **b. Negro** (Sp): Peru balsam.

Balsamodendron DC: *Commiphora* (*qv.*).

Balsamorhiza sagittata (Pursh) Nutt. (Comp.) (c&wNA): arrow-leaved balsam root; hb. inf. u. in gastralgia, fever, as "physic" (Cheyenne Indians); rt. chewed for sore throat, toothache; rt., pl. & sds. edible (raw or cooked) (Indians); roasted root u. as coffee substit. - **B. terebinthacea** Nutt. (nwUS: Idaho): turpentine balsam (root): rt. claimed useful in curing the tobacco habit (1075); the fleshy rts. (sometimes sprouts, sds.) of several wUS *B.* spp. were u. as food by Am Indians (**B. deltoidea** Nutt., **B. hookeri** Nutt.).

Balsamum Arcaei (L.): elemi ointment - **B. Canadense**: Canada balsam - **B. Carpaticum**: Riga balsam fr. *Pinus cembra* - **B. Catholicum**: Tr. of "*Hypericum vulneraria*" - **B. Cebur**: cebu(s) balsam fr. *Parameria barbata* - **B. Gileadense**: balm of Gilead fr. *Commiphora opobalsamum* - **B. Indicum**: peru balsam - **B. Judaicum**: balm of Gilead - **B. Mariae (Tacamacae)**: 1) fr. *Calophyllum tacamahaca* (tacamahac resin) 2) Resina Anime, fr. *Protium heptaphyllum* - **B. Nucistae**: ointment with expressed nutmeg oil; also the oil (mixed FO, VO) - **B. Opopalum: B. Peruvianum**: Peru balsam - **B. Styracis**: storax - **B. Terebinthina(e)**: turpentine oleoresin - **B. Tolutanum**: balsam of Tolu - **B. Tranquillans:** Hyoscyamus oil comp. - **B. Traumaticum** (NF. I-III): Tr. Benzoin Comp. - **B. Verum:**B. Gileadense.

balso (Mex): balsa tree or wd.

balucanat: balukanag: *Aleurites trisperma.*

balustier (Fr.): pomegranate.

bamboo: 1) giant woody grasses (Gramineae - subfam. Bambusoideae), growing in trop. & subtrop. areas; ca. 90 genera known; members of such genera as *Bambusa, Arundinaria, & Dendrocalamus*; in As, important source of timber, food, ornam.; etc.; wood u. in bldgs., furniture, fishpoles, ornam. items, phonograph needles, boxes &c.; st. segments with nodal septum at one end u. as water containers (Hond, Nic); bamboo shoots fed with eggs to cow with delayed parturition (St. Lucia, WI) (1076) 2) (Ala, Ga [US]): *Smilax* spp. - **b. briar root**: *Smilax rotundifolia, S. bona-nox* - **b. curare**: a type of curare which is packaged by the natives in bamboo sections - **giant b.**: *Bambusa vulgaris*; *B. arundinacea* - **b. grass:** *Arundinaria macrosperma* - **b. green brier**: *Smilax bona-nox* - **moso b.:** *Phyllostachys mitis* - **b. shavings:** bands or ribbons of the st. of 1) *Phyllostachys nigra* var. *henonis* 2) *Bambusa breviflora* 3) *Sinocalamus beechlayanum* var. *pubescens*; u. in resp. difficulties; antiemetic; inhibits some bacteria (Chin med.); u. since ca. AD 500 - **b. shoots**: young shoots of *Phyllostachys bambusoides* (chiefly); *Bambusa vulgaris,* &c.; used extensively in oriental (Chinese) cooking;

boiled & fried (*cf.* artichoke); canned - **b. vine** (US); *smilax* spp.

bambú (Sp) - **bambusa** (Mex): *Bambusa vulgaris.*

Bambusa Schreb. (Gram.) (trop&subtrop of both hemisph.): typical genus of the bamboos; celebrated for rapid growth (to 41 cm/day) & great height (to 40 m); b. sprouts u. in subgum chow mein & other Chin dishes - **B. arundinacea** Willd. (B. bambos Druce) (tropAs): spiny (or male) bamboo; buds popular food; also "Chinese bamboo shoots"; lvs. u. emmen. & in veter. med.; tea for mal. (Hainan); a b. manna (exudn.) ton. in fever, cough; silica cryst. secrns. ("tabashir") u. as aphrod.; sds. medl. food; wd. important; pl. sacred to Hindus - **B. multiplex** Raeusch. (As): shoots with pepper u. as inf. for inciting abortion - **B. spinosa** Roxb. (sChin to Moluccas): rts. & young lvs. u. for skin dis., fever; rt. decoc. in anuria - **B. vulgaris** Schrad. ex Wendl. (native to tropAs, cult. Java, &c.; natd. tropAm): (feathery) bamboo; common b.; rt. juice purg. (Trin); rt. decoc. diur. & for urinary stones (Cu); abort. (Trin); lf. decoc. or inf. antipyr. in fevers, flu, stroke, pneumonia (Trin); for stomach upsets, hemorrhoids (Br); shoots edible.

Bambuseae: sub-fam. of Gram. to which belong the bamboos, canebrakes, etc.; fls. at 150 yrs.+.

Bamfoline (Jap): ext. of bamboo lvs. u. in cancer.

banana (Eng, Por): 1) *Musa paradisiaca* v. *sapientum*, frt. + 2) (NA Indian slang): frt. of *Asimina triloba* - **Chinese b.: dwarf b.**: *Musa nana* - **b. figs**: dark brown leathery edible product made by sun-drying ripe pulp of bananas as aid to preservation (trops) - **poor man's b.** (Ozarks): *Asimina triloba* - **b. oil**: isoamyl acetate (synth.).

bandage, gauze: *Ligamentum Carbasi* (Absorbentis) (L): bleached cotton cloth of plain weave made in various weights & thread counts (Types I-VIII; Type I with closest weave); may be non-sterilized or sterile; may be folded flat or in tight roll; sometimes satd. with petrolatum ("p.g.") u. for sterile occlusive dressing of burns (reduces fluid loss; prevents adhesion); provides support, protection against mech. injury, insulation, &c., for wounds & burns; formerly some gauzes with phenol or iodoform - **adhesive gauze b.**: compress of 4 layers of Type I absorbent g. affixed to film or fabric with adhesive substance; sterile & may c. antimicrobial agt.; u. to apply as individual dressing.

Bandflechte (Ge): *Evernia furfuracea.*

Bandolin™: thick mucil. of *Chondrus* (or tragacanth), &c. scented; u. to stiffen ("dress") the hair; to stiffen silk.

baneberry: *Actaea* - **black b.**: *A. spicata* - **red b.**: *A. rubra* - **white b.**: *A. pachypoda.*

bang: bhang.

Banisteria auctt. = *Banisteriopsis* C. B. Rob. & Small.

banisterine: harmine fr. foll.

Banisteriopsis argentea Spring. (Malpighi) (tropAm): claimed to have hallucinogenic effects - **B. caapi** (Spr. ex. Griseb.) Mort. (Amaz basin, fr. Br to Co to Bo): caapi; yagé; ayahuasca; segments of lower part of st. of liana c. banisterine (harmine, telepathine), harmaline; u. hallucinogenic, narcotic; decoc. (ayahuasca) drunk; u. in post-encephalitic Parkinsonism; paralysis agitans; hypokinesis, muscle rigor; acts as strong sexual stim. - **B. cabrerana** Cuatrecasas (Co): liana st. the source of an hallucinogenic drink - **B. inebrians** Mort. (SA: westernmost part of Amazon basin; seCo, Br-terra firme): yagé (del monte); sts. of liana c. 0.15% harmine, lvs.in smaller amt.; harmaline; u. like caapi in narcotic drink (ayahuasca, yagé or pinde) acting as psychomimetic, delirifacient (*cf. B. caapi*); sometimes mixed with latter (1077); of considerable interest in exptl. pharmacology - **B. muricata** (Cav.) Cuatrecasas (eEc): bk. u. as hallucinogen by Waorani Indians - **B. quitensis** Niedenzu (Ec, Pe): another source of yagé, u. as delirifacient; claimed antimal.; "yagé cultivado"(?) - **B. rusbyana** Mort. (wAmaz, Co, Ec, Bo): ocó-yajé; lvs. c. N,N-dimethyltryptamine & other tryptamines; lvs. added to drink made fr. *B. caapi* or *B. inebrians* to lengthen and intensify the hallucinations sought after; may be identical with "Prestonia amazonica."

banjowan (Bengali): **banjwain** (Ind): *Seseli indicum* (perhaps).

banker (Newfoundland): fishing boat u. to catch codfish on the "banks" (seafloor elevation).

banko maria (Br): BM: Mary's bath = water bath of lab.

banks' oil: codliver oil.

Banksia L. F. (Prote) (Austral): honeysuckles; nectar u. by aborgines as food.

bankul oil: FO fr. *Aleurites moluccana.*

bannal: Scoparius.

banner: standard; large petal of fl. of Leg.

banyan (Hind): *Ficus benghaliensis.*

baobab (tree): *Adansonia digitata*; name first appeared in "De plantiis Aegypti Liber" (Alpino); AD 1592.

Baphia nitida Afzel. ex Lodd. (Leg) (trop wAf): (shiny) camwood; heart wd. source of red dyestuff, formerly popular & exported, u. in native cosmetics; redwood paste made up with shea butter, &c., & appl. to sprains or swollen joints; lvs. or lf. sap appl. to parasitic skin dis.; vulner.

for wounds; bk. & lvs. in enema for constipation; twigs as chew sticks; wd. (barwood) u. to make rafters, artistic cabinet work, violin bows - **B. obovata** Schinz (sAf): lvs. in face lotion for dizziness; pounded with salt & water for sores; comminuted rt. as soap substit.; wd. hard, u. to make gun stocks - **B. pubescens** Hook. f. (trop wAf): Benin camwood; bk. yields VO with santalol u. as urinary disinf. & expect. in bronchitis; appl. topically in lumbago; useful wd.

Baptisia alba R. Br. (Leg) (seUS): false indigo; white wild i.; lvs. & sts. c. cytisine, anagyrine, thermopsine; u. as substit. for *B. tinctoria* - **B. australis** R. Br. (eUS): (blue) false indigo; c. alks., flavonoids; u. as alter. - **B. bracteata** Muhl. (cUS): cream (or large bracted) wild indigo; black rattlepod; sds. in ointment base massaged into abdomen for colic (Am Indians) - **B. lactea** Thieret (B. leucantha T. & G.) (e&cNA, esp. se): Atlantic wild indigo; (large) white w. in.; prairie (false) indigo; hb. u. emet. cath.; for wounds, sores, eczema (Am Indians, settlers) - **B. perfoliata** R. Br. (seUS; natd. sAf): gopher weed; cat bells; lf. u. as diur. (sAf) - **B. tinctoria** (L.) Vent. (chiefly seUS, n & w to Minnesota): (yellow) wild indigo; rattleweed; horsefly weed; shoofly; honesty weed, c. cytisine, anagyrine; luteolin, &c.; rt. c. baptisin, pseudobaptisin; hb. u. lax., in large doses emet. & prod. diarrh; exter. as antisep. vulner. for sores; rt. u. in leucorrhea; pl. u. in country as dye & fly brush (seUS) (1078).

baptisin: glycoside with purg. action; hydrolyzes to baptigenin & rhamnose; also purified ext. (resinoid) (Eclectics).

barabang: barabong: dr. frt. of *Embelia robusta, E. ribes.*

barajo (CA): *Cassia alata, C. reticulata.*

Barba Caprinae Flores/Folia/Radix: fr. *Aruncus sylvester.*

barba de barata (Por): *Caesalpinia pulcherrima* (lit. cockroach beard).

barbado (Sp): st. transplanted with rts.

Barbado(e)s aloe(s): aloe fr. *Aloe barbadensis*; previously marketed in gourds - **B. flower-fence**: B. pride - **B. nut**: *Jatropha curcas* - **B. pride**: *Caesalpinia pulcherrima* - **B. tar**: petroleum (1).

barbaloin: an anthraquinone pentoside found in several types of Aloe.

Barbarea verna Aschers. (B. praecox R. Br.) (Cruc) (Eu, natd. NA): winter (or land) cress; Belle Isle (or early) c.; winter salad; lvs. u. as pot herb & for salads; pl. sds. source of FO - **B. vulgaris** R. Br. (Eu, natd. NA): winter cress; yellow rocket; spring mustard; hb. u. as balm for wound healing; lvs. in salads & garnishes; pl. cult. for sds. (source of FO); ornam. (biennial or perennial); weed.

barbasco: fish poison: 1) (SA): *Lonchocarpus nicou* 2) (wEc, Pe): *Tephrosia toxicaria* 3) (SA): *Jacquinia* spp. 4) CR): *Sapindus saponaria* 5) (Ve): Jamaica dogwood 6) (Mex): *Cracca* spp. 7) *Croton texensis* - **b. Negro**: (Pe): *Dictyoloma peruvianum.*

barbasso (It): mullen.

barbatic acid: depside (ester of phenylcarboxlic acids) found in lichens; originally found (1880) in *Usnea barbata*, hence name.

barbatimäo (Br): *Stryphnodendron barba-timão*, bk.

barbe pagnole (Haiti, Fr): Spanish moss.

barberry: *Berberis* spp., esp. *B. vulgaris* - **Allegheny b.**: **American b.**: *B. canadensis* - **common** or **European b.**: *B. vulgaris* - **b. family**: Berberidaceae - **India(n) b.**: *B. aristata.*

barbotine (Fr): Cina.

barbul bark: barbura b.: *Acacia leucophloea.*

Bardana (L., Sp): Bardanae Radix (L): **bardane** (Fr-f.): burdock, *Arctium lappa.*

Bärendreck (Ge) (bear's feces): *Glycyrrhiza* spp. incl. *G. echinata.*

Bärensaft (Ge): **Bärenzucker**: ext. ("juice") of licorice.

Bärentraube(n) (blätter) (Ge): uva ursi (lvs.)

Barfoed's test: for reducing sugars (glucose); Cu acetate with acetic acid.

barilla (Sp): 1) kali, *Salsola kali* 2) (Mex): *Calophyllum rekoi* 3) (CR, PR): *Batis maritima.*

barium sulfate: $BaSO_4$: occ. in nature as barite &c. (minerals); practically insol. in water; hence u. as X-ray contrast medium (in diagnosis of internal dis.); in making photographic paper, as radiation shield in concrete, &c. (E).

bark: 1) all that portion of a woody exogenous stem outside of the cambium layer; made up of an outer corky portion & an inner rel. delicate conductive & storage tissue mass, the phloem; ex. cinnamon, dogwood 2) (specif): (a) cinchona bk. (med., pharm.); (b) oak or hemlock bk. (tanning indy.) - **b. adstringens** (fr. Br): *Acacia jurema* - **Peruvian b.**: *Cinchona officinalis* - **sacred b.**: Cascara Sagrada - **scale b**: borke (ex. birch) - **shonny haw b**: *Viburnum nudum* - **sweet b.**: Cascarilla - **tree b.**: bk. of trunk or woody pl. stem. - **yellow b.**: *Cinchona calisaya, C. lancifolia*, etc.

barks, nine: *Physocarpus opulifolius* & other *P.* spp. (SC old bk. loose and separating in many layers) (also **nine bark**) - **seven b.**: *Hydrangea* spp. (SC exfoliation of bk. in thin layers).

Bärlappsamen (Ge): Lycopodium.

Barleria cristata L. (Acanth.) (eAs, PI): hb. crushed & appl. to malignant sores (Chin) - **B. prionitis** L.

(trop eAs, natd. Af): lvs. appl. as poultice for backache, rheum., lf. juice mild lax., appl. to skin of feet to prevent cracking; lvs. chewed for toothache; rts. antipyr.; lvs. cooked & mixed with maize flour for cough; lf. mixt. in water blown into penile urethra for syph; lvs. crushed and sap drunk for swollen testes & appl. as drops for sore eyes (Tanz); rt. appl. to boils.

barley: the pl. & frt. (grain) of *Hordeum vulgare*+ - **b. caustic, Indian**: sabadilla seed - **b. corn**: b. grain u. to prepare alc. beverages, hence "John Barleycorn," personification of malt liquors - **b. farina**: coarsely ground pearl barley - **b. flour, patent**: pearl b. - **b. grits**: b. farina - **b. malt**: Maltum = **malted b.**: fermented b. grains ("Scotch style" b.); coarsely ground b. - **pearl b., polished**: Hordeum Decorticatum: b. so milled as to remove chaff & most of bran; u. in puddings, breakfast foods (esp. popular among Jews); in Scotch broth, soups - **prepared b.**: *Hordeum distichon* - **b. sugar**: hard sugar made by boiling down with b. ext. added (formerly) - **wall b.**: **wild b.**: *Hordeum murinum* - **b. water**: beverage prepd. fr. pearl barley (GB); u. to give energy & dispel fatigue (Wimbledon tennis players).

barm (foam): yeasty foam rising to top of fermenting malt liquor; suspension of yeast in liquid.

barnacle: spp. of *Cirripedia* (subclass of Crustacea) (earlier classed with mollusks) found attached to marine rocks very tightly; c. adhesive material ("barnacle glue") with very strong bonding to substrate (thus when removed fr. steel, small particles of metal removed with organism); u. by dentists for bonding dental fillings to tooth.

barniz (Sp): 1) varnish 2) a juniper gum - **b. del Japon** (Sp): *v.* maque.

Barosma Willd. (Rut.) (sAf): buchu; some 25 spp. of arom. strong-scented evergreen shrubs or herbs of which 3 spp. are importantly used; often planted for ornam. (genus now generally transferred to *Agathosma*) - **B. betulina** Bartl. & Wendl. (Agathosma b. Pillans): short (leaf) buchu; bucco; (mountain) b.; round-leaf b.; "rounds"; lvs. c. VO with peppermint odor (c. buchu camphor [diosphenol], menthone, NBP); u. diur. antisept. in kidney dis. (given as b. brandy or b. vinegar, sAf); local appln. to bruises, relief of rheum. pains, gout (usu. in combns.) (1079); u. in syn. blackberry flavor - **B. crenulata** Hook. (B. crenata Kunze; Agathosma c. Pillans): oval (leaf) buchu; long (leaf) b. "longs"; similar to prec. - **B. pulchella** Bartl. & Wendl.: lvs. with odor of citronella; c. VO with citronellal, methyl heptenone - **B. scoparia** E. & Z.: lvs. known as "buchu"; VO c. no diosphenol - **B. serratifolia** Willd. (Diosma s. Curt.) = *B. crenulata* - **B. succulenta** L. var. **bergiana** H. & S. = *Diosma oppositifolia* L. - **B. venusta** E. & Z.: VO c. chavicol, myrcenol, myrcene, anethole, &c.

barra gokhru: *Pedalium murex.*

barras: *Pinus sylvestris.*

barrel tree: *Cavanillesia arborea.*

barrenwort: *Epimedium alpinum.*

barrica (Sp): cask, keg - **barril** (Sp): barrel, hogshead.

Barringtonia acutangula Gaertn. (Barringtoni. /Lecythid.) (Ind): hijjal (Beng); c. saponins, triterpenes (barringtogenols), triterpene dicarboxylic acids (acid A; barringtogenic acid); sds. u. for diarrh. & as emet.; domestic remedy in colds, as expect.; bk. ext. in fever, jaundice (Ind); various parts as fish poison (chiefly bk.); lf. juice for the itch (1080) - **B. asiatica** Kurz (B. speciosa Forst.) (sAs, Ind to Austral, Oceania): Indian barringtonia; bk. u. fish poison (Tonga) (86); formerly as homicidal poison (Gilbert Is); sds. c. FO, barringtongenol, a saponin, u. fish poison, FO as illuminant; pl. believed cardioactive; lvs. u. in prepn. varnish (Java); wd. for furniture mfr. - **B. excelsa** Blume (Malaysian Is): sds. c. chydenanthine & saponin; u. as fish poison - **B. racemosa** Roxb. (Af, As, Ind, Mal, Indonesia, PI): Rosenkranzjambusen (Ge); c. triterpenoids barringtongenol & barringtogenic acid (sapogenins); rt. & bk. u. stom. in skin dis.; frt. juice for eczema; sds. in eye dis. & as vermif.; young lvs. after debittering (lime water) eaten as veget.; bk. as tan; bk. & sds. fish poison.

barwood: *Pterocarpus soyauxii* (stated origin), *P. esculentus.*

baryta: **barytes:** barium sulfate.

basavana pada (Hind): *Gloriosa superba.*

bas du fleuve: gomme type of Acacia; hard gum; "down-river" grade "A."

Basella alba L. (B. rubra L.) (Basell.) (WI, SA, natd. in s&eAs): Ceylon (or red) vine or Indian spinach; Malabar nightshade; rt. lax., rubefacient, sap appl. to acne; lvs. diur. (also alkaline ashes); important cult. veget. food, eaten as spinach in Chin, Mal, tropAf, WI, tropAm, Haw (US); frt. juice u. to color food & agar.

basic number (chromosome) = x = lowest haploid no. in a series of polyploids (thus diploids = 2x).

Basidiomycetes: **Basidiomycotina** (more modern term): basidia fungi, a large group of pls. exemplified in the mushrooms & pore fungi; has **basidium** structure bearing 4 **basidiospores**.

basil: 1) *Ocimum basilicum* 2) *Satureja vulgaris* 3) *Clinopodium vulgare* - **bush (green) b.**: *Ocimum minimum* - **fever b.**: *Ocimum viride* - **Sweet**

(green) b.: *O. basilicum* - **wild b.**: 1) *Cunila mariana* 2) *Satureja vulgaris* 3) *Clinopodium vulgare* **basilic** (Fr): basil - **basilicon ointment: yellow b.**: rosin cerate (r. ointment) - **Basilienkraut** (Ge): *Basilienmünze* (Ge): basil.

Basiliken (Ge): sweet basil, *Ocimum basilicum*.

basionym (biology): oldest valid name on which all new names and combinations are based; the basionym becomes the synonym.

basket flower: *Centaurea americana* +.

Bassia All. (Chenopodi.) (Euras, Austral): hbs. or shrubs with narrow lvs. - **B. hyssopifolia** (Pall.) Ktze. (Euras, natd. ne&swUS): smotherweed; 5-hook bassia; small coarse hb.; a cause of summer pollinosis - **B. longiflora** L. = *Madhuca* I. - **B. scoparia** (L) A. J. Scott = *Kochia s.*

Bassia Koen. ex L. = *Madhuca* Gmel. & *Diploknema* Pierre (Sapot).

bassia tallow: FO of Madhuca indica & M. longifolia.

bassine (Dutch): palmyra fiber.

Bassora gum: false tragacanth; a tragacanth-like gum of unclear & variable origin; thus, from *Acacia* spp.; plum & almond trees; *Sterculia urens*; *Ailanthus excelsa*, etc., u. as lax. (1081); (SC Bassora or Basra, comml. city of Iran, n of Persian Gulf, sometimes an export point).

bassorin (fr. Bassora gum): pectin-like subst. insol. but swelling in water to form an irreversible gel; makes for high viscosity; occ. in tragacanth & gums & gum resins of many pl. such as *Dorema ammoniacum*, *Opopanax chironium*, *Boswellia carteri*, *Opuntia ficus-indica*, sarcocolla, frankincense, coconut gum, chagual gum, resina lutea, salep, &c.

basswood (bark): *Tilia americana*.

bast: phloem - **b. cell**: v. fiber, bast.

bastard dittany: *Dictamnus albus* - **b. feverfew**: *Parthenium hysterophorus* - **b. ipecacuanha**: *Asclepias curassavica* - **b. saffron**: *Carthamus tinctorius* - **b. wild rubber:** *Funtumia africana*.

Bastardia viscosa HBK (Malv.) (WI, sMex to Ec, Br): viscid mallow; escoton; lvs. crushed in wine appl. to fistulas; pd. lf. appl. to tumors, syphil. ulcers, lacerations; claimed useful in cancer.

basuco (Co): crude cocaine sulfate; prepd. & u. in Co; sniffed; considered dangerous.

bat: true flying mammal of Ord. Chiroptera (worldwide): with nocturnal habit, sleeping in caves, barns, &c.; consumes enormous amts. of flying insects; fecal matter accumulates in caves, collected as guano & u. as valuable manure; parts of animal u. in Unani med., incl. flesh (for arthritis), blood, brain & spinal cord, bile, milk.

batata (Sp, Po): potato - **b. da purga** (Br): *Piptostegia pisonis* - **b. de porco** (Por): (hog potato): *Boerhaavia diffusa* - **b. doce** (Por): sweet p. - **b. inglesa**: Irish p. - **b. purgante**: Brazilian jalap: *Ipomoea operculata* - **batatas canadiennes** (Fr): **b. de Canada**: *Helianthus tuberosus*.

batatinha do campo (Br): *Ipomoea operculata*.

bataua oil: FO fr. *Oenocarpus bataua*.

Batiator Radix (L.): **batiator root**: fr. *Vernonia nigritiana*.

Batis maritima L. (Bat.) (Florida (US), WI, Mex, tropAm, Hawaii): American saltwort; turtle weed; pl. decoc. u. int. & exter. for syph., skin dis., incl. psoriasis; for menstrual pains, GC, asthma; lvs. u. in salads (but salty); ashes u. to make soap; source of a crude soda.

batiste: thin fine cloth of cotton with fine weave; originally linen cambric.

batrachotoxin: said most toxic venom; fr. skin of kokoi frog of Co (SA); ca. 10x as lethal as terodotoxin (formerly considered most toxic).

battlefield flower: pansy.

batu gum: an EI damar.

Bauchspeicheldruese (Ge): pancreas

Baudouin's test: test for sesame oil using sugar & HCI reagent.

Bauhinia L. (Leg.) (trop regions): shrubs, vines & trees, with simple lvs. & flat pods (SC, John & Caspar Bauhin, Swiss botanists of 16th cent.) - **B. acuminata** L. (seAs): crushed leaves u. as poultice for nasal ulceration (Mal); cold inf. of rt. drunk for cough (Indones) - **B. bassacensis** Pierre ex Gagnep. (seAs): sd. macerate in alc. u. in stomachache (Cambod); lvs. u. as cigarette paper (Vietnam) - **B. bracteata** (Benth.) Baker (seAs): bk. inf. as antidote for poisoning by flesh of fish "tre prahing" (Cambod); also in diarrh. - **B. candicans** Benth. (Br, Ch+): pezuna de vaca (Arg); lvs. with alks. (small amt.), ß-sitosterol, u. anitdiabetic (Arg); lf. ext. reduced blood sugar significantly, cholinergic activity (1082) - **B. curtisii** Prain (seAs): sts. u. in beverage for dys.; st. fibers u. to make strong ligatures - **B. divaricata** L. (WI, sMex to Hond; cult. sUS): pata de cabra; bull hoof; Mayas u. fl. in decoc. with sugar as expect. in dry cough; boiled lvs. eaten for liver & kidney dis. (6); lf. inf. placed on sores on feet - **B. emarginata** Mill. (Pan, Co, Ve): urape; decoc. of lvs. & branch tips u. as remedy in diabetes - **B. esculenta** Burch. (swAf): sds. as Gemsbock (chamois buck) beans u. as food, source of FO u. as food oil; st. & rt. also eaten at times; FO in cosmetics - **B. excisa** (Griseb.) Hemsl. (WI, CA): chain-vine; bejuco de mono; decoc. of rt. & st. of vine u. for VD (Trin); st. decoc. emmen. & in yaws -

B. fassoglensis Kotschy (trop eAf): bean edible; bast fibers u. to make cloth & cords; astr.; tuberous rt. decoc. given as galactagog to cows - **B. forticata** Link (Br): c. insuloid prods. alks., glycosides, &c. u. in diabetes; lvs. u. astr. in gargles, &c. (2546) - **B. hirsuta** Weinm. (seAs Java): lvs u. for itch; bk. u. to wash (blanch) raw silk (Cambod) - **B. lakhonensis** Gagnep. (seAs): rt. decoc. u. for females after childbirth; u. in cough medicines - **B. macrantha** Oliv. (sAf): rt. u. in cough medicines - **B. malabarica** Roxb. (Ind, PI): alibangbang (PI); bk. ("Bauhinia bark") rich in tannins, u. in liver dis.; as tan; fls. u. in dys.; lvs. u. for culinary purposes; bk. strips u. as ropes; var. *acida*: fls u. for wounds (Indones) - **B. manca** Standl. (CR, Pan): escalera de mono; st./lf. inf. u. as astr. & diur. & in diabetes - **B. merrilliana** Perk. (Indones): u. for splenomegaly - **B. purpurea** L. (Ind, Burma, sometimes cult. as ornam.): purple mountain ebony; sona; bk. decoc. astr. u. in diarrh. (124); rt. carm. in fever (Laos); fl. lax.; sds. c. non-drying FO; fl. buds eaten (Sikkim) - **B. radiata** Vell. (Br): cipo de escada; st. u. as expect. - **B. reticulata** DC (tropAf): u. in mal., blackwater fever, blood poisoning, asthma; bast fibers in dyeing; lvs. to coagulate rubber latex - **B. rubro-villosa** K. et S. S. Larsen (Laos, Vietnam): sts. of liane u. in decoc. for bellyache - **B. thoningii** Schuw. (Piliostigma t. Milne-Redh.) (tropAf): camel foot; redhead; inner bk. roasted & u. as wash for ringworm; bk. chewed for cough (Tanz); in diarrh. (rt. decoc.) (Tanz); decoc. in milk or soup for GC (Masai); rt. ash with Fe dust in hookworm (Shambala); numerous other uses (129) - **B. tomentosa** L. (Ind, SriL, now in tropAf): St. Thomas' tree; fls. u. in diarrh., dys. (Ethiopia, 202); frt. as diur.; bk. & rt. bk. astr. for abscesses (also lvs. in latter); sd. furnishes FO "ebony oil"; edible; buds & lvs. u. in dys.; rt. bk. u. PO for abd. pains, liver dis.; sd. aphr. ton.; bk. fiber used - **B. ungulata** L. (sMex, Belize, Pan to Bo, Ve): pie de venado; p. de vaca; pata de venado; pl. decoc. u. diaph. purg vermif. (Mex); lf. decoc. u. for diarrh. & diur. in urin. dis (very active) (Mayas; 6); branches u. as poles in making huts (Yuc) - **B. variegata** L. (Chin, seAs, Indones): mountain ebony; bk. ton., rt. in dyspep.; bk. & fl. buds astr. in dys. (124); young fl. buds cooked & eaten; this & some other *B.* spp. produce gum - **B. viridescens** Desvaux (Burma, seAs. to Timor): pl. u. after delivery, in fever; young lvs. edible.

bauinia (Por): *Bauhinia forficata.*

Baum (pl. **Bäúme**) (Ge-m.): tree.

baume (Fr): 1) balm 2) spearmint 3) (Qu): *Mentha canadensis* 4) *Mentha* spp., esp. *M. rotundifolia* - **b. collodion**: gum damar in alc.; u. vulner. - **b. du commandeur**: Tr. Benzoin Comp. - **b. copalme de Mississippi**: American styrax - **b. de Capiri**: gurjun - **b. de Fioraventi**: complex formula chiefly Venice turpentine; also elemi, storax, etc. **b. de Pérou en coques:** balsam fr. *Myrocarpus fastigiatus* (in gourds) - **b. de sonsonate**: balsam Peru - **b. de vie**: comp. decoc. of aloe - **b. des champs**: wild mint - **b. des jardins** (Fr): 1) costmary 2) spearmint - **b. Maire**: Calaba balsam - **b. sauvage**: wild mint - **b. tranquille**: infused oil of hyoscyamus comp.; many different & complex formulas have been published - **b. vert**: spearmint.

Baumschule (Ge): (plant) nursery.

Baumwolle (Ge): purified cotton.

baunilha (Por): vanilla.

Baunscheidt Lebensweckers (Ge): apparatus for puncturing the skin in acupuncture counterirrit. therapy - **B. Oil**: euphorbium & mezereum extr. with alc., then ext. mixed with olive oil; u. as counterirrit. in **Baunscheidtism**: therapy of multiple puncture of skin (acupuncture) followed by appln. of irritants, esp. croton oil.

bauxite (pron. bow-zite; o as in "how," i as in "ice"), ferruginous $Al(OH)_3$.

bawa (Ind): *Cassia fistula.*

bawdmoney: *Meum athamanticum.*

bay: 1) (geog.): an ecological habitat being low land lying among hills 2) (bot.): spp. of *Laurus*, *Myrica*, *Gordonia*, *Magnolia*, &c. - **b. berry**: *Myrica pensylvanica*; *M. cerifera*, *M. californica* (name orig. with early Am. pioneers; also appl. to *Pimenta* spp.) - **bull b.** (Fla, Ala [US]): *Magnolia grandiflora* - **b. bush** (buds): *Myrica gale* - **California b.** (tree): *Umbellularia californica* - **Greek b.**: *Laurus nobilis* - **Indian b.**: *Persea indica* - **b. laurel: b. leaves**: laurel lvs., *Laurus nobilis* - **loblolly b.**: *Gordonia lasianthus* - **Mediterranean b.**: Greek b. - **b. oil** (WI): VO fr. b. rum leaf (also **sweet b. o.**) - **pyramid b.**: *Magnolia pyramidalis* - **red b.**: 1) *Persea borbonia* 2) *Gordonia lasianthus* - **b. rum** (St. Thomas, PR): Spir. Myrciae Comp. dist. (2344) fr. **b. rum leaf**; *Pimenta racemosa* - **stinking b.**: *Illicium floridanum* - **swamp b.**: 1) *Persea palustris* 2) *Magnolia virginiana* - **sweet b.**: 1) *Magnolia virginiana* 2) *Persea borbonia*, *P. palustris* 3) Greek b. (also **s. -scented b.**) - **tan b.**: loblolly b. - **West Indian b.**: *Pimenta racemosa* - **white b.**: 1) *Magnolia virginiana* 2) *Persea borbonia.*

baycuru radix (L.): **baycuru root**: fr. *Limonium brasiliense.*

bazna tree (Pak): *Zanthoxylum badrunga.*

bazora pretoe: *Scoparia dulcis.*

BCG vaccine: *Bacillus Calmette-Guérin* (vaccine fr. TB organisms).

Bdellium: (African) oleo-gum resin; u. as adulter. of myrrh; perhaps fr. *Commiphora africana* & other C. spp.; u. like myrrh; anthelm. - **Indian b.**: fr. *Commiphora mukul* - **opaque b.**: hard opaque masses; rich in tannin; often with small pieces of bk.; u. as adulter. of Bdellium (36).

bead bush (NCar [US], etc.): Hamamelis - **b. tree, common**: *Melia azedarach.*

beads (Eng): *Margaritana margaritifera* (Mollusca) - **prayer b.**: often are sds. of *Abrus precatorius* (SC u. in rosaries, &c.)

beaker cells: sclereid cells of endocarp palisade layer (as seen in frt. of *Piper nigrum*, &c.) - **b. plants**: cup plants: members of Fam. Nepenthaceae, Sarraceniaceae, & Cephalotaceae.

beam (tree) (, white): *Sorbus aria.*

bean: 1) generic name of *Phaseolus* spp., specif. *P. vulgaris* 2) loosely appl. to any sd. or pod resembling same - **Adzuk(i) b.**: *Vigna angularis* - **Algaroba b.**: *Prosopis juliflora* - **Bengal b.**: *Mucuna pruriens* var. *utilis*; *M. aterrima* - **black b.:** 1) sd. of *Phaseolus* sp. black colored, popular in SA, now in US 2) *Dolichos lablab* - **black-eye(d) (pea) b.**: sd. of 1) *Dolichos sphaerospermus* 2) *Vigna sinensis* **Bonavist b.**: *Dolichos lablab* - **Boston b.**: *Phaseolus vulgaris* - **broad b.**: 1) *Vicia faba* 2) *Canavalia gladiata* - **Burma b.:** *Phaseolus lunatus* - **bush lima b.**: *Phaseolus limensis* var. *limenanus* - **butter b.**: *Phaseolus limensis*; sometimes of smaller size than usual Lima b. - **b. cake**: *v.* Soy bean, *Glycine hispida* - **Calabar b.: b. of Calabar**: *Physostigma venenosum* - **China b.**: soy b. - **cluster b.**: *Cyamopsis psoralioides* - **common b.**: Boston b. - **cow b.**: *Vigna sinensis* - **b. curd:** b. cake; tofu; dou fu (Chin); prod. usu. fr. soy (also at times fr. mung bean, &c.); soaked beans ground to paste, strained —> bean curd residue & soybean milk (filtrate); $CaSO_4$ soln. added to latter —> ppt. "curd"; u. in skin conditions, sores, ulcers (local appln.); int. for intest. bleeding; common cold (1926) - **Duffin lima b.:** *bush l. b.* - **dwarf b.**: common b. - **dwarf lima b.**: bush l. b. - **fava b.: fave b.**: *Vicia faba* - **field b.**: common b. u. in the dried state, i.e., removed fr. the husk when mature & dr. - **flat b.** (GB): lima b. - **b. flour**: pd. dr. ripe common beans - **French b.**: common b. - **fuzzy velvet b.**: *Mucuna pruriens* - **Garbee b.**: *Entada scandens* - **garden b.**: 1) common b., esp. those u. in the fresh state as distinct fr. field b. 2) *Vicia faba* - **Goa b.**: *Psophocarpus tetragonolobus* - **great brown sea b.**: Garbee b. - **Great Northern b.**: common b., cultivar - **green b.** (wMidland, US): common b.; snapbean - **haricot b.**: common b. - **horse b.**: 1) *Canavalia ensiformis*, *C. gladiata* 2) *Phaseolus lunatus* 3) fava b. 4) *Parkinsonia aculeata.* - **hyacinth b.**: *Lablab purpyreus* - **Indian b.**: *Catalpa bignonioides* - **Italian pole b.**: common b. - **Jack b.**: *Canavalia ensiformis, C. gladiata, C. plagiosperma* - **Jambie b.** (BWI): 1) *Leucaena glauca* 2) *Phaseolus lunatus* - **jumping b.**: v. under "jumping" - **June b.**: *Canavalia limata* - **kidney b.**: a field b. - **kintoki b.**: fr. Chin, Jap; u. in chile con carne - **lablab b.**: Bonavist b. - **Lima b.**: *Phaseolus lunatus* - **locust b.**: *Ceratonia siliqua* - **lubia b.**: *Dolichos lubia* - **Lyons velvet b.:** v. b. - **Madagascar b.**: *Phaseolus lunatus* - **marrow b.**: field b. with sds. 1-1.5 cm. long & more than half as thick - **Mauritius b.**: *Mucuna* spp. - **medium b.**: type of field b. - **Mexican jumping b.**: v. under "jumping" - **multiflora b.**: scarlet runner b.- **mung b.**: *Vigna radiata*; u. for making bean sprouts for Chinese foods; v. also under "mung" - **Nankin b.**: peanut - **navy b.**: common b., ex. "Michigan Robust" (1083) - **nicker b.**: *Caesalpinia bonduc* - **ordeal b.**: Calabar b. - **oxe-eye sea b.**: *Mucuna* spp. - **pea b.**: any of various kidney b. cultd chiefly for the small white sds. chiefly u. dry in baking (ex. in Boston baked beans) - **pinto b.**: common b. var. developed in New Mexico (cult. in Ariz [Cantillo], Tex [US]); a speckled tender bean - **b. pod** (or husk): of *Phaseolus* spp., esp. *P. vulgaris*; commonly u. in med. (Eu) - **pole b.**: common b. with strong vine growth - **potato b.**: yam b. - **prayer b.**: *Abrus precatorius* - **princess b.**: Goa b. - **Rangoon b.**: lima b. - **rattle b.: rattlesnake b.** (Ala [US]): *Sesbania drummondii* - **red b.**: 1) *Abrus precatorius* 2) *Vigna catjang v. unguiculata* 3) kidney or **red kidney b.**: common b. (also white, blue-black & speckled vars.) - **rice b.**: *Vigna umbellata* - **sacred b.**: *Nelumbo* spp. - **sallet b.**: string b. - **St. Ignatia b.: St. Ignatius' b.: b. of St. I.**: *Strychnos ignatii* - **St. John's b.**: locust b. - **scarlet b.**: *Erythrina* spp. - **scarlet runner b.**: *Phaseolus coccineus* (*P. multiflora*), multiflora b. - **screw b.**: frt. of *Prosopis glandulosa* (curly var.) - **Sieva b.**: *Phaseolus lunatus* - **snap b.** (Va, NCar [US]): string b.- **soja b.**: soy b. - **b. soup** (USN): emulsion of soy b. proteins highly effective for extinguishing fires - **soy(a) b.**: *Glycine max* - **b. sprouts**: fresh young shoots of mung b. (or soy b.) picked before lvs. develop; u. in chop suey & other Chinese-Am dishes - **state b.**: pea b. - **stick b.** (Alabama): pole b. - **string b.** (New Eng, nUS): common b., grown for the edible immature pod - **sugar b.**: lima b. - **sword b.**: Jack b., *Canavalia gladiata* - **tepary b.**: *Phaseolus acutifolius* v. *latifolius* - **tonca b.:**

tonka b.: *v.* tonka - **towe b.:** lima b. - **velvet b.**: *Mucuna* spp., esp. *M. pruriens* (*M. deeringiana*) - **b. vine**: *Phaseolus polystachyus* - **white b.**: lima b. - **wild b.**: *Phaseolus sinuatus* - **Windsor (broad) b.**: *Vicia faba* - **winged b.**: Goa b. - **yam b.**: *Pachyrrhizus erosus, P. tuberosus; Sphenostylis stenocarpa* - **yard-long b.**: *Vigna unguiculata v. sesquipedalis* - **yellow eye b.**: var. of marrow b., *Phaseolus vulgaris* - **Yokohama b.**: *Mucuna hassjoo* Mansf. (cult. for the sds.).

bear: mammals of Fam. Ursidae (*Ursus, Euarctos*, &c.) - **b. berry**: 1) uva ursi 2) (Michigan): *Amelanchier* spp. - **black b. b.:** *A. alpina* - **red b. b.**: uva ursi) - **b. wood:** Cascara Sagrada.

beard tongue (US): *Pentstemon* spp.

bear's ear(s): 1) *Arctotis* spp. 2) *Auricula* spp. 3) *Saxifraga sarmentosa* - **b. foot (root)**: *Helleborus foetidus* - **b. foot (American)**: *Polymnia uvedalia* - **b. grape:** *Arctostaphylos uva-ursi* - **b. grease**: FO fr. *Ursus fuscus*; u. emoll. - **b. weed**: Eriodictyon - **b. wort**: *Meum athamanticum*.

Beaucarnea Lam. = *Nolina* (Agav).

Beaumontia grandiflora Wall. (Apocyn.) (Ind, Him): herald's trumpet; Nepalese trumpet flower; sds. c. oleandrin, beaumontoside; u. as floss.

Beaumont's root: Bowman's root.

beauty berry (US nurseries): *Callicarpa americana* (SC attractive clusters of mauve berries).

beauties (Cana): *Claytonia* spp.

beaver: **Castor canadensis** Kuhl & other *C.* spp. (*qv.*) - **b. poison**: 1) *Cocculus carolinus* 2) *Cicuta maculata*. - **b. tree (bark)**: **b. wood:** 1) *Magnolia virginiana* 2) *Celtis occidentalis*.

bebeerine: pelosine: chondodendrine: alk. obtained fr. the bk. of bebeeru bark (Nectandrae Cortex, L.) fr. *Nectandra rodiaei* & fr. rt. of *Chondodendron microphyllum, C. tomentosum, C. candicans*; u. as antimal., pacifier & analg. for small animal surgery - **b. sulfate** is actually a mix. of 6 alk. (sulfates).

bebeeru (bark): *Nectandrae rodioei*.

beccabunga: *Veronica b.*

bec-de-grue (tacheté) (Fr): *Geranium maculatum*.

becerro: (Sp): calf; calfskin; kip.

bêche-de-mer (swPac Oc): a marine invertebrate (sea cucumbers; Echinodermata) u. as food (trepang); also name for the *lingua franca* of swPacific since it is so widely traded there.

Becium kenyanum G. Tayl. (Lab) (swAf): inf. (cold) of rt. u. for indig.; lf. to restore hair (Zulu) - **B. obovatum** (Benth.) neBr. (sAf): Zula hair restorer.

bedbug: *Cimex lectularius* L. (Cimicidae-Hemiptera-Insecta): (cosmop. temp. & subtr.): (common) bedbug, temporary ectoparasite on man & many other animals; with nocturnal habits; sucks blood of host; u. in homeopath. med. & in Unani med.; dr. animal ground & appl. as paste to scalp to stim. hair growth in alopecia.

bedeguar: *Rosa canina*.

bedroega verdadeira (Por): *Portulaca oleracea*.

bedstraw: *Galium aparine, G. boreale*, & other *G.* spp. - **catchweed b.**: *G. aparine* - **ladies' b.: maiden's b.**: *G. verum* - **stiff marsh b.**: *G.tinctorium* - **sweet scented b.**: ladies' b. - **yellow (ladies') b.**: *G. verum*.

bee: the insect, *Apis mellifera* - **b. balm**: 1) *Monarda fistulosa* 2) *Melissa* - **b. bread:** mixt. of pollen, honey, & bee secretions; in form of masses stored in hive for nutrition of larvae; c. antibiotic (against *Salmonella*), growth stimulant (1817) - **bee-glue**: Propolis (*qv.*): resinous prod. fr. *Apis mellifera* - **b. plant**: 1) *Cleome* spp. 2) *Scrophularia* spp. - **Rocky Mountain b. pl.**: *Cl. serrulata* - **b. sting: b. venom**: prep. of the b. stings extr. fr. *Apis mellifera*; u. treatment rheum. chilblains - **bee tree**: 1) *Tilia* spp. 2) any tree visited by bees in search of nectar.

beech: *Fagus* spp., esp. *F. grandifolia* (in US), *F. sylvatica* (in Eu) - **b. bark** (lvs.): fr. *F. sylvatica* - **blue b.**: water b. - **copper b.**: *F. sylvatica v. atropunicea* - **b. drops** (seNA): *Epifagus virginiana* - **European b.**: *F. sylvatica* - **b. family**: Fagaceae - **Indian b.**: *Pongamia pinnata*, source of a b. oil, another b. oil: fr. *Fagus sylvatica* - **purple b.**: copper b. - **southern b.**: *F. grandifolia* - **water b.**: *Carpinus caroliniana* - **white b.**: *Carpinus betulus*.

beedi: bidi - **beedwood**: *Hamamelis virginiana*.

beef: muscle flesh of fully grown bull, ox, steer or cow, *Bos taurus*; source of beef extract, formerly much u. as convalescent aid, such as in b. tea, value considerably overrated; popular form, Liebig's Extract (2619) - **b. jerky** (wUS): combn. of b., salt, malt, *Piper nigrum*, sugar, MSG, &c.; thoroughly dehydrated; u. as masticatory - **b. suet: b. tallow**: internal fat of cattle u. as ointment base to raise melting pt. in hot climates.

beefsteak plant (Florida): *Perilla frutescens* - **b. (saxifrage)**: *Saxifraga stolonifera*.

beefwood: *Casuarina* spp. +

beer: 1) alc. beverage prepd. fr. grain, malt & hops in water with low (2-5%) alc. content (Bier, Ge; bière, Fr); found per some sources to aid in preventing heart attack when u. in moderation (2 bottles/day) 2) liquid part of antibiotic fungal culture contg. most of the antibiotic - **birch b.**: prepd. fr. *Betula lenta* - **b. bush**: hops - **b. seed, California**: sort of wild yeast (possibly fr. Ky & Tenn [US]) formerly widely u. in US (ca 1900); may be ordinary beer yeast or dr. mother of vinegar in some cases; occ. in white particles approx. size of radish sd.; placed in water to furnish ferment for beer of good quality (flavor) and low alc. content (also called ginger beer plant, beer bees).

Beere:(Ge-m.) berry.

bee's nest: *Daucus carota*, wild carrot - **beeswax, crude** (US, Br, etc.): the unrefined prod. fr. the honeycomb of the bee, *Apis* spp., u. by aborigines as cement for arrows, spear-heads; for lighting; for writing tablets (Ancients); embalming; sculpting; when purified, this constitutes the off. Cera Flava, Yellow (bee's) wax; when latter is bleached, the off. Cera Alba, White Wax, White beeswax; u. in prepn. ointments, cerates (fr. Cera), plasters, &c.; in protective coatings for airplane wings, ammunition shells, &c.; as cement, in surgery (bone wax), to prevent hemorrhage (cranial surgery); lubricant (large prodn. Ethiopia) (marketed as "bricks" & "slabs").

beet (, common): **(red) garden b.**: **beetroot** (Eng.) *Beta vulgaris* - **sea b.**: *B. maritma* - **silver b.**: *Beta vulgaris* var. *cicla* - **b. sugar**: sucrose obt. fr. *B. vulgaris* v. *altissima* - **wild b.**: sea b.

Beet (Ge-f.): **Bett** (Ge-m.): bed (as for plants).

Beetroot (GB): beet.

beetle: insect of order Coleopters - **b. bung tree** (Cape Cod, Mass ([US]): *Nyssa sylvatica* - **Japanese b.**: *Popillia japonica*, a leaf chafer which devours foliage of *Rosa* spp. & other plants (also fls. & frts.); larvae feed on grass roots; a bad pest of neUS.

Befaria Mutis ex L. = Bejaria.

beggar lice: b. ticks: 1) *Bidens frondosa* & other *B.* spp. 2) *Desmodium canadense* 3) *Lappula virginiana* 4) *Agrimonia gryposepala* 5) *Galium aparine*.

beggar's blanket: *Verbascum thapsus* - **b. buttons**: burdock - **beggarweed:** *Desmodium tortuosum.*

Begonia L. (Begoni) (trop & subtrop, esp. of Am): begonias; elephant's ear; some 1,000 spp. mostly perenn. hbs. with thick rhizomes (1084); of chief interest as ornam. (many hybrids) - **B. augustae** Irmsch. (PI; PapuaNG): lf. sap. appl. to relieve itching - **B. bahiensis** DC (Br): hb. u. as antipyr. - **B. carolineifolia** Regel (Guat, Mex): acid sap u. for fever & to quench thirst - **B. convallariodora** C DC (Mex & Guat+): lf. decoc. u. for stomach ulcers - **B. cucullata** Willd. (B. semperflorens Link & Otto) (eBr): hb. u. as diur.; incl. var. *hookeri* L.B. Sm. & Schub.; u. med in Par (Par Ph.) - **B. dominicalis** A DC (Dominica, Guadeloupe) lvs. & fls. u. as tea for colds (Caribs, Creoles) - **B. fimbristipula** Hance (Chin): decoc. u. in TB - **B. gracilis** HBK (Mex): carne doncella; rt. u. in folk med. as emeto-cath.; cult. ornam. - **B. grandis** Dry. var. *evansiana* Irmsch (B. evansiana Andr.) (Chin): u. stom., may be somewhat toxic; rts. diur. galact. purg. - **B. humilis** Ait. (WI, Mex to Br, Pe): Trinidad begonia; hb. inf. ("tea") u. popularly in chest colds, coughs (Trin); TB, fever - **B. luxurians** Scheidweiler (Br): lvs. u. in decoc. for fevers - **B. muricata** Bl. (B. tuberosa Lam.) (Moluccas, Indones): lvs. eaten by natives as sorrel, also cooked with fish & in sauce (484) - **B. rex** Putzeys (Ind, Himalaya): pl. u. as rhubarb substit.; juice toxic to leeches; popular ornam. in gardens & homes - **B. rossmanniae** A DC (Ec): lf. inf. u. for itch; pl. decoc. for eye wash in conjunctivitis - **B. rotundifolia** Lam. (Haiti): oseille marron; decoc. of entire pl. sweetened & u. as refreshing emoll. drink in inflamm., bilious fever; lvs. as poultice to tumors - **B. sanguinea** Raddi (Br): shrub; lvs. u. as diur. - **B. semiovata** Liebm. (Nic, CR, Pan): succulent lvs. & sts. cooked & eaten for vomiting & diarrh. (2444) - **B. stigmosa** Lindl. (sMex to Co): juna; juice squeezed fr. sts. u. to alleviate eye ailments (Ve Indians)

behen nuts: nuces behen (L.): fr. *Moringa oleifera* & *M. arabica* - **b. oil: ben o.**; FO fr. b. nuts; c. esters of **behenic acid** (satd. fatty carboxy acid; $CH_3(CH_2)_{20}COOH$; occ. also in many other sd. FO & animal fats).

behenic acid: docasanoic acid; occ. in seed fats, milk fats, &c.; solid at room temps.; u. in lubricants; to retard evapn.; antifoam agt.; &c.

Beifuss (Ge): mugwort, *Artemisia vulgaris*, *A. absinthium*, &c.

Beilschmiedia giorgii Rob. & Wilcz. (Laur.) (Congo): djombi; lvs. u. to prepare pomade ("ngula") u. by natives to rub on their bodies - **B. mannii** Benth. (Tylostemon m. Stapf) (w tropAf): spicy cedar; tola seeds; often confused with bitter kola nuts; however, tola sds. have 2 readily separable cotyledons; frts. u. in dys. (Guinea); sds. eaten raw or in soups, with rice, &c.; fragrant fls. u. to flavor rice, &c. - **B. pahangensis** Gamble (Mal): hb. decoc. u. in stomachache, digest. dis., dys. & postpartum.

Beinschwarz (Ge): bone black; animal charcoal - **Beinwurz** (Ge): comfrey.

Bejaria congesta Fedtsch. & Basilev. (Eric.) (Co): bk. decoc.with sugar u. for relief of severe coughs - **B. resinosa** Mutis ex. L.f. (Co): fls. u. as tea for expect.; "antitussant."

bejuco (Sp): rattan; liane; cane - **b. blanco: b. bravo de estrella**: *Paederia diffusa* - **b. de goma** (PR): *Cryptostegia*.

bel: 1) **Bela** (L.): **Fructus Belae** (L.): **Belae Fructus**: 1) bael: *Aegle marmelos*, frt. 2) Arabian jasmine, *Jasminum sambac*.

bela dama (Por): belladonna (beladona).

Belamcanda chinensis DC (Irid.) (Chin, Jap): blackberry lily; rhiz. important in Chin materia medica; chief remedy in tonsilitis; u. in chest & liver dis., added to tonics; u. in GC (Mal); in purges; emmen.; u. with caution; widely cult. as ornam. - **B. flabellata** Gray (Chin): similar.

beldroega (pequena) (Por): *Portulaca oleracea*.

beleño (Sp): henbane.

belladona (Sp, Po): **belladone** (Fr): **belladonio** (Por): **Belladonna** (Eng, L, Ge, Finn, Nor, Swed, Dan, Urdu, etc.): *Atropa belladonna*, of which lf. & rt. are commonly u. in med. - **b. del pais**: 1) (Cu): stramonium (name reflects common error in SpAm of confusing belladonna & stramonium) 2) (Mex): *Nicandra physalodes* - **Indian b.**: *Atropa acuminata*; off. in BP along with *A. belladonna* - **Japanese b.**: *Scopolia japonica* - **Russian b.**: Scopolia.

Belladonnae Folium (L.): belladonna lf. - **B. Radix**: belladonna rt. - **belladonnine**: alk. prepd. fr. *Atropa belladonna* & related Solanaceae; but probably formed during extn. (i.e., not native but a conversion prod. fr. atropine); also made by a half synthesis; isomer of apoatropine; u. Parkinsonism.

bellary leaf oil L.: VO fr. lvs. of *Neolitsea zeylanica* Merrill.

belle dame (Fr): belladonna.

bellflower (or **bell flower**): *Campanula edulis*, *C. laciniata* (Levant) - **b. family**: Campanulaceae - **b. (olive) tree: b. wood**: *Halesia carolina*.

bell-fungus: *Fomitopsis officinalis*.

Belliolum haplopus A. C. Sm. (Winter.) (Solomon Is): lvs. pounded & rubbed on skin for skin dis. in pigs. (—> *Zygogynum*)

Bellis perennis L. (Comp) (wEu temp, As, natd. Am): (English) daisy; garden (or true) d.; fl. heads & hb. u. as expect. in pulmon. dis., hemoptysis, vulner.; in skin dis., furunculosis, leukorrh., hematuria, hypotension (Eu); cultd. ornam. perenn. hb.

bellwort: *Uvularia perfoliata* & other *U.* spp.

beluga: 1) *Acipenser huso* (sturgeon) 2) white whale, *Delphinapterus leucas*.

ben nuts: behen nuts - **b. oil**: FO fr. behen nuts.

bendee: *Hibiscus esculentus* - **bendy tree**: *Thespesia populnea*.

Benedictendistel (Ge): *Cnicus benedictus* - **Benedictum Vinum** (L.): antimony wine.

Benedictine: a liqueur prepd. by Benedictine monks - **B. root**: *Geum urbanum* (rhiz.).

benefing oil (tropAf - Niger; Madag): *Hyptis spicigera (qv.)*; sd. source of drying FO (can replace linseed oil); in Madag u. as edible FO.

Bengal: v. under agar; ginger; kino.

Bengal beans: *Mucuna* spp. - **B. quince: Bengalische Quitten** (Ge): bael.

beni (seed): benne - **b. shoga** (Jap): unpeeled ginger rt. preserved in vinegar, brine & coloring matter.

Benincasa hispida (Thunb.) Cogn. (B. cerifera Savi) (1) (Curcurbit.) (trop As, Af, pantrop): wax gourd; petha; frts. u. in curries, as (canned) sweet pickles, preserves (Ind); soup vegetable (Chin); hb. u. for fever & colds; frts. sds. u. in diabetes, kidney dis., anthelm. (also FO); frts. covered with wax (hence common name); sds. u. diur in dysuria; lax, bk. antipyr. for hemorrhoids; frt. juice to wash inflamed eyes (PI).

Benjamin bush: *Lindera aestivale* - **B. tree:** *Styrax benzoin; Ficus benjamina* - **B. salts**: benzoic acid.

benjui: 1) (Sp): benzoin 2) (SA): exudn. fr. *Styrax pearcei*, *S. camporum*, *S. ferrugineus*, *S. reticulatus*; prod. actually more like styrax than benzoin.

benne plant (seed, oil, leaves): benni: *Sesamum indicum*, sesame.

bennut: *Juglans regia*.

benthic: bottom-living freshwater organisms; ex. some diatoms.

bentonite: Bentonitum (L.): native colloidal hydrated aluminum silicate (clay); highly plastic; swells greatly in water; forms gels, magmas; u. mud baths, beauty clays, to clarify wines, etc. (mined in Calif, Mont, Utah, Wyo [US]) (1085) - **fused b.** sulfur as a topical medication (1086).

benzaldehyde: C_6H_5CHO; important component of bitter almond oil, etc.; u. flavor, perfumes; most made by synthesis - **b. cyanohydrin**: mandelonitrile; combn. of *b.* & HCN - **b. spirit & comp. spirit**: u. as flavors like Spir. bitter almonds, but nox-toxic as free fr. HCN.

benzaline seed: behen nut, fr. *Moringa oleifera*.

benzene: 1) hydrocarbon, C_6H_6, distd. fr. coal tar 2) (NZ+): gasoline.

benzin(e): 1) **petroleum b.**: low b. pt. petroleum fraction; however, term somewhat indefinite & confusing; may be similar to petrol. ether (ligroin) (*qv.*); mostly butane C_4H_{10} to octane C_8H_{18}, but sometimes appl. to far less volatile solvent fractions; variously called **purified p. b.**, p. naphtha, &c.; do not confuse with benzene, which is a single compd. not a mixt. of hydrocarbons as is benzin 2) gasoline (NZ, &c.).

benzoaric acid: ellagic acid.

benzoates: salts or esters of benzoic acid with antisep. & irrit. properties & which are u. as food preserv., intest. & urin. antisep., expect., antirheum.;

for healing of small wounds, skin roughness, etc., ex. sodium b., benzyl b.

Benzoë (Ge): benzoin.

benzoic acid: C_6H_5COOH; simplest aromatic carboxy acid; occ. in nature free & combined, esp. in frts.; gum benzoin (after which acid is named) has up to 20%; component of most balsams, free or esterified; u. preservative, antifungal (E).

Benzoin aestivale: Nees = *Lindera benzoin*

benzoin (gum): 1) a balsam obt. fr. *Styrax benzoin, S. tonkinensis*, etc.; u. exp. arom. stim; inhal. in laryngitis, bronch.; cosmetic exter. in skin dis., antis.; added to spirit varnishes 2) synthetic org. compd. $C_6H_5CHOH.CO.C_6H_5$ 3) (Br): *Pamphilia aurea*; *Styrax ferrugineum*; *S. pohlii* - **block b.**: c. only benzoic acid not cinnamic - **Bogota b.**: exudn. fr. trunk of *Pamphilia aurea* - **b. bush**: *Lindera benzoin* - **Calcutta b.**: block b. exported through Calcutta - **flowers of b.**: benzoic acid - **French b.**: *Peucedanum ostruthium* - **Palembang b.**: inferior sort produced in sSumatra; c. only benzoic acid & esters; very low density; u. in prepn. "natural benzoic acid" by sublimation; u. as source of this acid (named after Palembang highlands in Sumatra) - **Penang b.**: derived fr. *Styrax subdenticulatum*; c. only cinnamic acid & esters - **Siam(ese) b. (gum)**: prep. fr. *S. tonkinensis*; u. perfume agt. in incenses, etc. - **Sumatra b. (gum)**: prep. fr. *S. benzoin* & some other *S.* spp.; u. exp., etc. - **b. tincture compound**: Friar's Balsam; c. benzoin, aloe, styrax, tolu balsam, peru balsam, myrrh, angelica, alc., &c.; u. as vulner.

Benzoinum (L.): benzoin.

benzol: benzene, C_6H_6; prod. of coal tar distn.

benzoline: ligroin: naphtha: solvent n.; petroleum fraction of same nature as petrol benzin but with higher density, higher boiling range (b. pt. 35-80°) & higher flash point (b. pt. 130- 50°); with persistent odor; same as "mineral spirits" (US).

benzoresinol: important constituent of resin in Benzoin.

benzoylaconine: picro-aconitine.

benzyl alcohol: peruvin: $C_6H_5CH_2OH$; occ. as such in VO of many flowers, ex. jasmine, hyacinth; but first discovered combined in Balsam Peru; u. as loc. anesthetic to mucosa & skin surfaces; for latter use lotion with alc. & water (for itching, ex. pruritus); bacteriostatic agt.; perfume, flavor, solvent - **b. benzoate:** arom. ester constituent of Peru and Tolu & other balsams, but not styrax; solid or liquid; u. in lotions as acaricide (scabicide, pediculicide); also u. as antispasm. in pertussis (worthless); in vet med. for mange; solvent for camphor, &c.; perfume fixative; in chewing gums, confectionery, &c. (1973) (E) - **b. cinnamate**: cinnamein: a chief component of storax, Peru & Tolu balsams; u. in flavors, perfumes, as fixative; lung collapse (1971) - **b. penicillin:** penicillin G: u. as salts of calcium, potassium, &c. (E).

Berberidaceae: Barberry Fam. (temp zone): over 600 spp. in 15 genera; perennial hbs. or woody pls. (bushes, trees); lvs. simple or compd.; spp. c. alk. rather similar to those found in Ranuncul.; berberine is found in almost every sp. of *Berberis*; subdivisions: subfam. (1) Podophylloideae: lvs. never pinnate (ex. *Podophyllum*): subfam. (2) Berberidoideae: lvs. always pinnate (ex. Oregon grape).

berberine: alk. characteristic of *Berberis* spp., also found in several related genera; thus, *Mahonia, Hydrastis, Coptis, Xanthorrhiza, Thalictrum, Adonis*; occ. in fam. Ranuncul., Annon., Papaver. Berberid., Rut.; has fungicidal/bactericidal value to plant; u. in eye prepns., GI dis., dysmenorrhea; as fluorescence producer; in syntheses; cheapest & most widely available alk. in world.

Berberis L. (Berberid) (native to every continent exc. Austral; most spp. in nHemisph & Andes); barberry; large genus (ca. 500 spp.) (1087); at one time, name covered the two subgenera of *Euberberis* (or *Berberis per se*) & *Mahonia* (or *Odostemon*); most writers now inclu. in *Berberis* those spp. formerly placed under *Euberberis* & make *Mahonia* (or *Odostemon*) a separate genus: *Berberis* spp. are armed & have simple leaves while *Mahonia* spp. are unarmed & have pinnately compd. lvs; shrubs, either evergreen or deciduous; fls. wd. & inner bk. yellow; "Berberis" is u. as title for the rhiz. & rts. of *B. vulgaris* (eHemisph) and *Mahonia* (*Berberis*) *aquifolium* & *M.* (*B.*) *nervosa* - **B. amurensis** Rupr. (e coasts of USSR): Amur barberry; lf. tinct. u. in obstetrics - **B. aquifolium** Pursh = *Mahonia a.* - **B. aristata** DC (Ind, Nepal, SriL): chitra; sumlu; kasmal; tsema; (hill) barberry; closely related to *B. vulgaris* (1088); rt. bk. a practical source of berberine; u. bitter ton., antipyr. in mal.; in eye dis.; as fish poison - **B. asiatica** Roxb. (Ind, Himalaya): chitra; kilmora; maté-kissi; &c.; commonly sold in bazars; like *B. aristata*, rt. bk. source of rusot (ext.); st. & rt. both rich in berberine - **B. baluchistanica** Ahrendt (nPak): shrub; c. isoquinoline alks., pakistanimine, zpakistanine, baluchistana-mine; these u. as starting pt. for synthesis of several isoquinoline alks., also berberine - **B. boliviana** Lechl. (Bo, Pe): chechejche; u. med. (a source of comml. Berberis rt.); (exports thru Lima, Peru) - **B. buxifolia** Lam. (Arg, Ch): calafate; michay (Ch); c. berberine (1089); u. in fevers, inflammns.; gentle purg. (Ch) - **B. canadensis** Mill. (se&cUS): (not in Canada!): American (or Alleghany) barberry; frt. edible, u. in jellies; in acidulous drinks in fever; as stom. to stim appetite; alternate host for cereal rusts - **B. darwinii** Hook. (Arg): michay; rts. reported to c. over 11% berberine (30-yr.-old plant) (1090); no HCN (1091) - **B. empetrifolia** Lam. (Arg, Ch): zarcilla (Ch): sacha michay; u. indig. diarrh. (Ch) - **B. flexuosa** R & P (Arg, Pe): sacha uva; huaccampe (Pe); u. like *B. buxifolia* - **B. hakodate** Hort. (Romania, cult.): c. berberine, berbamine, &c.; choleretic, cholekinetic, antibiotic activity demonstrated - **B. heterophylla** Juss. (Arg): calafate; similar to *B. Darwinii* - **B. heteropoda** Schrenk (Russ in As): rt. bk. with 4% berberine, &c.; u. med (Russ) - **B. japonica** R. Br. = *Mahonia j.* DC - **B. julianae** Schneid. (Chin): wintergreen barberry; u. as substit. for Coptis; cultd. ornam. (US) - **B. laurina** (Billb.) Thunb. (sBr): St. John's root (raiz de Sao Joao): rt. of shrub rich in both berberine (2.5%) & hydrastine (-1.4%); u.in mfr. berberine, medicinals, dyeing; substit for *Hydrastis canadensis* (1092) - **B. linearifolia** Phil. (Ch): calafate; u. like *B. buxifolia*; showed antitumoral props. (human naso-pharyngeal carcinoma) - **B. lycium** Royle (Ind): kashmal; chotra; ambar-báris, &c.; u. to prep. rusot (ext.) fr. rt. - **B. nervosa** Pursh = *Mahonia n.* - **B. pinnata** Lag. = *Mahonia p.* - **B. repens** Lindl. = *Mahonia r.* - **B. rigidifolia** HBK (Co): u. med. - **B. ruscifolia** Lam. (Arg): calafate da Patagonia; (raiz de) quebrachillo; rt. bk. c. berberine - **B. sibirica** Pall. (Siber): rt. u. as ton., for low blood pressure; in folk med for gastointest. dis.; icterus - **B. teeta** Wall. (Ind, Himal): mamiran (also appl. to other pls.); u. like *B. lycium* - **B. thunbergii** DC (Jap): Japanese barberry; megi; bk. antisep. anthelm; st. rt. lvs. decoc. u. as stom. (Chin); eyewash (Indochin); gargle for dental caries (Indochin); frt. & lvs. rich in alks. (1093); rt. bk. c. under 1% berberine; cult. ornam. - **B. trifoliata** Moric. = *Mahonia t.* - **B. vulgaris** L. (Euras, natd. e&cNA): common barberry; kashmal (Ind); bedana (Ind); rt. esp. bk. rich in berberine (1.07-1.3%) (1094); rt. & rt. bk. u. as hepat. ton. astr. bitter, &c.; rt. ext. u. for ulcers, eyelid dis., pimples, &c.; decoc. u. int.; astr. in cholera; also for injuries of cattle & people; frt. eaten, mostly in form of syrups, preserves, &c.; bk. u. as bitter ton.; rt. favored as tan for skins, &c.; frt. red berries u. acidulant, refrigerant; pl. is host of wheat rust (*Puccinia graminis*) hence the object of attempts at extirpation in many areas of eUS where pl. is now a weed pest; yellow dye (exp. of b. bark fr. Hung, Yug before WW II) - **B. wallichiana** DC (Chin): st. inf. u. to treat dental troubles (Chin) - **B. wilsonae** Hemsl. (Chin): u. as coptis substit.

berberis berries: b. fruit: zarishk; fr. *B. lycium*, *B. vulgaris* & other Ind spp. - **b. root**: fr. *Mahonia aquifolium*, *M. nervosa* - **b. sts.**: darhalad (Ind) - **b. sticks**: pieces of rt. 30-45 cm X 3 cm. (Kangra) (1095).

Berberitze (Ge): **Berberitzenwurzel** (Ge): berberis rt. (bk.).

berberry: *Berberis* & *Mahonia* spp.

Berchemia lineata DC (Rhamn.) (Chin, Ind): angari (Jaunsar); pl. u. as cure for fever - **B. scandens** (Hill) K. Koch (B. volubilis DC) (seUS): supple jack; rattan vine; rt. of twining shrub u. as antisyphil.

berengena (Mex, Por): **berenjena** (Sp): egg plant. *Solanum melongena* var. *esculentum*.

bergamot: 1) *Citrus bergamia* (Eu) 2) (NC etc.): horsemint, *Monarda fistulosa* - **red b.**: *Monarda didyma* 3) *Citrus aurantium* subsp. *aurantium* cv. Bouquet (US) - **wild b.**: 1) *Mentha canadensis* 2) *Monarda fistulosa*.

bergamota: 1) sweet lime 2) (Br): *Citrus nobilis* var. *deliciosa*.

Bergenia cordifolia Sternb. (Saxifrag.) (Siberia; high mts. of wHimal): lvs. u. as tea substit. (Chaga tea) (cAs) - **B. crassifolia** (L.) Fritsch (Saxifraga c. L.) (Siber, nMongol, Atlas Mts): rock foil; badanian rts. ("badan") (c. 20% tannin, arbutin) u. as antisep. int. for diarrh.; as tan for sole leather & Russian leathers; lvs. u. as astr. & to prep. "Tschagoric" tea, tea substit.; cult. ornam. - **B. ligulata** Engl. (Saxifraga l. Wall.) (sEuras, Ind, Him): breakstone; rts. u. in fever, diarrh., lung dis., scurvy, applied to boils; furuncles; eye inflamm. (Ind) - **B. purpurascens** (Hook f. & Thoms) Engl. (Chin): rhiz. u. hemostyptic; ton.

Bergia decumbens Planch. (Elatin.) (sAf): fresh frt. inf. u. for kidney dis., stomachache; crushed pl. as dressing for wounds (2145) - **B. odorata** Edgew. (Ind, Pak): lvs. rubbed in water & appl. as poultice to sores; u. to clean teeth - **B. prostrata** Schinz (sAf): whole pl. u. as inf for diarrh. (Ovambos) (Namibia) (2145).

Bergpech (Ge): asphaltum - **Bergpolei: Bergpoley** (Ge): *Teucrium polium* - **Bergt(h)ee** (Ge): wintergreen.

Berkheya carlinoides Willd. (Comp.) (cAf): rt. decoc. u. to dissolve vesical calculi; tinct. diur. - **B. discolor** O. Hoffm. & Muschl. (sAf): decoc. for sed. during illness; young lf. eaten as veget. (cooked) - **B. montana** Wood & Evans (sAf): lotion fr. rts. u. for bruises - **B. onopordifolia** O.

Hoffm. & Burtt Davy (sAf): u. for fever & in leprosy (with *Euphorbia clavarioides*) - **B. radula** Willd. (sAf): u. for pain over kidneys; rubbed on swollen testes - **B. seminivea** Harv. & Sond. (sAf): u. in GC, emet. (Zulus) - **B. setifera** DC (sAf): rt. decoc. for colds & coughs & related - **B. speciosa** DC (sAf): inf. u. for abdom. pain after eating.

Berlandiera lyrata Benth. (Comp.) (Tex [US], Mex): lyreleaf greeneyes; coronilla (Mex); rt. dec. & inf. u. for stomach dis.; sold in drug stores (Mex).

Berlinia auriculata Benth. (Leg) (trop wAf): lvs. u. in intest. troubles; in constipation (enema); bk. sap dropped in ears to stop discharge (Liberia); wd. u. locally - **B. globiflora** Hutch. & Burtt Davy = *Julbernardia globiflora* Troupin.

Bermuda: v. under arrowroot; grass.

Bernstein (Ge): amber.

berro (Sp): watercress.

berry: frt. characterized by having thin epicarp, thick fleshy, pulpy mesocarp (sarcocarp), & with small hard sds. loosely embedded in the latter; ex. grape, tomato, capsicum - **blue b.** (Tex[US]): *Smilax* spp.; *cf.* blueberry - **cat b.** (seUS): *Smilax* spp. - **fish b.**: Cocculus - **French b.**: *Rhamnus infectoria* - **Indian b.**: fish b. - **Jesuit b.**: partridge b.- **miracle b.**: *Sideroxylon dulcificium* - **oso b.**: *Oemleria cerasiforme* - **partridge b.**: *Mitchella repens, Gaultheria procumbens* - **snow b.**: **wax b.**: *Symphoricarpos albus* (etc.).

Bersama abyssinica Fres. subsp. *a.* (Melianth.) (trop eAf): c. cardioactive subst., hellebrigenin, &c.; with cytotoxic action for human nasopharyngeal carcinoma - subsp. **paullinioides** Verdc. (tropAf): "bitter bark"; bk. decoc. u. purg. anthelm.; lf. & rt. c. insecticidal principle - **B. lucens** Szysz. (sAf): bk. u. for menstrual pain, in impotence & sterility (Zulus) - **B. tysoniana** Oliv. (sAf): bastard sneezewood; bk. u. as antipyr. & in hysteria; bk. decoc. given to cattle for "gall sickness."

berseem: (Arab) *Trifolium alexandrinum.*

bertalha (Por): *Basella alba.*

Berteroa incana DC (Alyssum incanum L.) (Cruc.) (Euras, natd. Cana, nUS): hoary alyssum; h. madwort; sd. FO, unsatd., of possible value as drying oil; u. med.

Bertholletia excelsa Humb. & Bonpl. (Lecythid.) (trop nSA, Br, Guianas): Pará nuts; Brazil nut tree; bears large frts. each with c. 18-24 nuts; well-known as the edible "Brazil nuts" (sds.); nuts roughly triangular in x.s.; flavor of nuts resembles a blend of walnut and hazel nut; with considerable Ba (toxic), hence proposed to stop importation of nuts into Germany; source of FO u. in foods, soap mfr., illuminant; frt. shells u. to smoke rubber latex (coagulation); bk. u. to caulk boats - **B. nobilis** Miers (tropSA): another source of pará nuts and P. nut oil.

Bertiera racemosa K. Schum. (Rubi.) (trop wAf): lf. prepn. u. for teething babies; bk. appl. locally in snakebite; sds. boiled & sprinkled on yams.

Bertramgarbe (Ge): *Achillea ptarmica* - **Bertramblüten** (Ge): pyrethrum fls. - **Bertramskraut** (Ge) 1) *Achillea ptarmica* 2) *Anacyclus pyrethrum.*

Berula erecta (Huds.) Coville (Umb.) (Sium erectum Huds. (Umb.) (Euras, natd. NA): water parsnip; hb. u. earlier as diur. - **B. malacophylla** (USSR): hb. c. coumarins (1096) - **B. thunbergii** Wolff (Sium t. DC) (sAf): water parsnip; rt. analg., held in mouth or chewed for toothache; pl. inf. u. to wash body for headache; suspected of being toxic to livestock (a factor in "vlei" poisoning of cattle).

berza (Sp): cabbage.

Besenginster (Ge): scoparius.

Besenheide (Ge): *Calluna vulgaris.*

Besleria ignea Fritsch (Gesneri) (Co): epiphytic pl.; lvs. ingested as strong purg. (natives of La Pedredra) (2299) - **B. leucostoma** Hanstein (Co): mata de congo; crushed lvs. appl. as plaster to painful bites of the conga ant (Tikunas) (2299).

Besseya rubra (Dougl.) Rydb. (Scrophulari.) (nwUS): "kitten tails"; strong tea prepd. fr. fresh or dried rts. u. for colds & as physic (Flathead Indians, Mont [US]). (258).

be-still nut: *Thevetia nereifolia.*

Beta vulgaris L. (Chenopodi) (w&sEu, orig. wMed reg.; now widely cult., worldwide in temp. zone): (sea) beet; biennial hb.; subsp. *maritima* Doell. (B. maritima L.) wild sea beet; ancestor of other subspp. & vars.; subsp. **vulgaris** var. *conditiva* Alef.; (var. *esculenta* Sal.); red beet; betterave (cultivée) (Fr, Qu); betterave de table (Fr); common (red) garden beet; beetroot (GB); table beet; tuberous rt. cooked as veget., as salad, canned; lvs. eaten as pot-herb; rt. juice u. in prepn. borsch (*qv.*); to make beet wine (vin de bette), vinegar; in anemia (Fe, Cu); in nerve & liver dis.; b. root juice in cancers (Balkans) - var. *altissima* Doell.; sugar b.; rt. c. 20% sucrose; source of about half of sucrose prodn. (beet sugar) (big producers: US, Ge, USSR); b. wine; roasted rt. as coffee substit.; subsp. **cicla** (L.) Koch (Eu): leaf b.; (Swiss) chard; with lvs. greatly developed, esp. thickened; withstands much summer heat; sometimes u. as pot herb (blade & petiole); sometimes raised as ornam.; u. in ancient med.; f. rapa Dumme ("B. alba") (var. macrorhiza) (Eu): mangel (wurzel);

mangold; fodder b.; popular for stock feeding (Eu) (1097).

betabel (Mex): beet.

beta-hypophamine: pitressin or vasopressin obt. fr. posterior pituitary.

betaine: lycine; glycine betaine; occ. in many pl. parts, thus beet sugar juice & molasses (where first discov.), wheat germ, cotton seed, animal tissues, &c.; alkaloid-like compd. is a quaternary ammonium basic deriv.; u. by organism for choline synth., reduces blood cholesterol level, normalizes fat metabolism (lipotropic agt.); u. in atherosclerosis, hypertension, liver dis., obesity, muscular weakness (cardiac failure); in multivitamin formulations - **b. HCI**: u. in acid deficiency in stomach (as carrier); also u. in soldering, &c.

betalains: plant (fl. & frt.) pigments contg. dihydropyridine; include betacyanins (red), betaxanthins (yellow); occ. in Centrospermae (dicot order incl. some 18 fams., such as Cact.); one of these, *betanin* (beet red), is permitted in foods & cosmetics; is glycoside of betanidin. (—> *Beta*)

betarraga (Sp): **betarava** (Bo): beet - **betasterol**: a phytosterol fr. sugar beets.

betel (leaf): fr. *Piper betle* - **b. nut (palm)**; **b. n. tree**: *Areca catechu.*

beth root: *Trillium erectum*; *T. purpureum*.

beterraba (Bo, Br): beetroot.

bétoine (Fr- f.): betony - **b. des montagnes**: arnica.

Betonica L.: 1) genus *Stachys* 2) (L.) (specific.): *S. officinalis* (rts. &c.).

betony: 1) *Lycopus virginicus* 2) wood b. (2) 3) *Teucrium chamaedrys* L., *T. canadense* L. - **Betsy bug**: Bessie b.: horned beetles, spp. of Passalidae; blood u. in otalgia (seUS folk med.) - **betterave** (Fr): beet; sugar b. - **b. fourragère** (Fr): mangel - **marsh b.**: *Stachys palustris* - **b. sucriere** (Fr): sugar b. - **wood b.**: 1) *Pedicularis canadensis* 2) *Stachys officinalis*.

Betula L. (Betul.) (nTemp & Arctic zones): birches; trees hardy & lovers of cold, thus inside Arctic circle & high on mts.; wood valuable, u. for cabinet work; distn. products (1098); special purified charcoal (1099); bk. marked by prominent lenticels & often occurring in papery plates; u. to make canoes, boxes, shoes - **B. alba** L. = *B. pubescens* - **B. alleghaniensis** Britt. (eCana, US): yellow/silver/gray (swamp) birch; has been u. like *B. lenta* for prodn. of methyl salicylate (said richer source than latter); wd. useful - **B. latifolia** Tausch (Chin): bk. u. in icterus, cancerous mammae, suppuration - **B. lenta** L. (sCana, neUS s to nAla): sweet/cherry/black birch; twigs & bk. c. betulin (monotropitoside), source of natural methyl salicylate (sweet b. oil) by ferment. & distn.; bk. u. arom. antirheum.; buds arom. diur.; trunk tapped in early spring & sap fermented to give birch beer - **B. lutea** Mx. f. = *B. alleghaniensis* - **B. nigra** L. (e&cUS): red/river/black birch; buds boiled to syrup & u. with sulfur as ointment for ringworm, sores (Catawba Indians); bk. & small twigs inf. u. as arom. diaphor.; another source of sweet birch oil - **B. papyrifera** Marsh .(Cana, nUS s to wNCar): paper/canoe/white birch; pliable but sturdy bk. u. by American Indians to make canoes, baskets, cups, bags, &c.; to tan "Russian leather," wd. u. for woodenware, pegs, &c. - **B. pendula** Roth (Euras, natd. eNA): European (white) birch; bk (Cortex Betulae) & twigs (bk) source of birch tar (Pix Betulae) & Rectified Birch Tar Oil (Olem Rusci) by destructive distn.; tar u. in skin dis. (8% ointment), for rheum., gout, bronchitis; parasites on animals; in mange; vulner., prepn. Russian leather; in rum essence; oil u. counterirrit., parasiticide in skin dis. (also bk. decoc.), esp. eczema; insect repellant on skin; (also lvs.) in veter. med. for colics, worms; lvs. u. diur. in gout, edema; for falling of hair; with alum, source of green dye, with chalk yellow; other pl. parts brown & red dyes; bk. decoc. in baths for foot sweating, skin eruptions; tan (Scand, Russ): prepn. b. bark oil; buds u. arom. diur; choleretic; hair washes (chief prodn. USSR, Scand, CSR, Pol) - **B. populifolia** Marsh. (sCana, neUS): gray/white/fire birch; u. med. in TB (Creek Indians) - **B. pubescens** Ehrh. (B. alba L.) (n&cEu, natd. neNA): downy birch; white b. (of Europe); u. much like *B. pendula*; lvs. as diur. in rheum., gout, edema, "blood purifier," hair loss (Pol, CSR); adult. at times with lvs. of other birches - **B. pumila** L. (Cana, nUS): low/dwarf/swamp birch; (also var. *glandulifera* Regel): strobiles laid on plate of hot coals as incense for catarrhs; as tea for women in their menses, "strengthening"; branches as matting for bedding (Cana Indians) - **B. verrucosa** Ehrh. = *B. pendla.*

Betulaceae: birch fam. (temp. zones of both hemispheres); group of monecious shrubs & trees; ca. 150 spp.; lvs. simple, alternate; stipules deciduous; ex. alder; hazelnut; birch, ironwood, hornbeam (also called Corylaceae).

betulin(ol): betulol; birch camphor; triterpene; dihydric alc. in colorless crystals obt. fr. *Betula* tars & barks; also component of *Rosa* spp., *Zizyphus* spp. &c.

beurre de cacao (Fr): cocoa butter.

bezards: bezoar (stone): concretion fr. stomach or intest. of ruminant animals; often fr. goat; of various types (some hairy = "hair balls"); superstitiously thought of value to antidote poisons & for other therapy (listed in early Brit Phar.); as panacea; at one time among crown jewels of England - **b. de Goa**: an artificial comp. of pipe clay, amber, musk, etc. - **occidental b.**: fr. *Lama vicugna* - **oriental b.**: fr. *Gazella dorcas, Capra aegarus* - **b. stones:** b. - **western b.**: Occidental b.

bezoar root: Bezoar Radix (L): *Dorstenia brasiliensis.*

bhaji (Indo-Pak): spinach; sometimes appl. to any vegetable.

bhang (a) (Hind): cannabis; hemp. &c. dr. lvs. & young stems in combn. with sugar, milk & water; u. as popular hallucinative drug in cAs.; adm. perorally; term also means broken, destroyed, defeated - **b. ghotna:** stick with which hemp lvs. are rubbed down in producing beverage, bhang.

bhilawan: bhilwa: *Semecarpus anacardium.*

bhindi (Urdu; Hindi): okra.

biada (It): wheat.

Biancaea sappan Todaro = *Caesalpinia s.*

bibernel, small (herb): *Sanguisorba minor.*

Biberu-rinde (Ge): bebeeru bk.

bible frankincense: exudn. fr. *Boswellia carteri* - **b. leaf**: costmary.

bicho: 1) (Co): *Cassia occidentalis* 2) (CA, SA): *C. bicapsularis*, etc.

bichy nuts: Kola.

Bicillin™ (Wyeth): dibenzylethylene diamine dibenzyl penicillin; benzathine pencillin G; antibacterial; cure for syph. in single injection (claimed).

bicuiba (Br): *Myristica bichuhyba* - **b. redonda:** *M. fragrans.*

biddy's eyes: pansy

Bidens L. (Comp.) (cosmop.): tick-seed; weedy ann. or perenn. hbs., some 250 spp.; rarely cult; (1100) - **B. andicola** HBK (Ec): nache; fls. inf. u. to sedate nervousness, mania; mild spasms (also var. *decompositus* O. Ktze.) (2449) - **B. aurea** (Ait.) Sherff (Mex): té de coral; hb. decoc. u. as lax - **B. bigelovii** Gray (sUS): (Bigelow) beggarticks; flgs. tops u. for palatable bev.; ton. carm. diaph. - **B. bipinnata** L. (tropAm, e&cUS, semicosmop): Spanish needles; warm juice of fresh pl. u. in ear drops, in conjunctivitis, hemostyptic (Arg); rt. & sd. u. as emmen. expect. in asthma; pl. decoc. in diarrh. (Taiwan) - **B. cernua** L. (Euras, natd. NA): nodding beggar ticks; another source of Herba Bidentis tripartitae besides *B. tripartita* (Russia) - **B. connata** Muhl. (eCana, US): stick tight; beggar ticks; hb. u. as expect. - **B. cynapiifolia** HBK (WI, tropAm, Hawaii, Ind): romero amarillo (Cu); hb. inf. u. for fever (Trin): for stomach (Cur) - **B. frondosa** L. (NA+): beggars' tick(s); sticktight; cow lice; hb. u. as expect. emmen. - **B. leucantha** Willd. = **B. pilosa** L. (tropAm, Mex, sFla [US], natd. trop eAs): acahual (blanco); pacunga (Pe): shepherd's needle; (hairy) beggarticks; picão (de campo) (Por); lvs. pounded to give poultice appl. to eyelids for eye dis. (Chin); rt. inf. u. to cleanse "dimmed" eyes (Indochin); lvs. chewed or rubbed on gums for toothache; young hb. in diarrh., colic, as clyster in intest. dis.; lf. juice u. for eye irrit. (Domin); as diur. in diabetes, pectoral (Mex); rheum., in tonsilitis; lvs. u. for invigorating tea, fls. to stim. appetite (Haw [US]); fls. with boiled rice to make wine (PI); u. med. (Br) for pain (Pe); pl. inf. for cough (Mal), inf. substit. for Chin tea (Co) - var. *minor* Scherff (sAf): remedy for heartburn, icterus, cough, "fits" (children); crushed lvs. as poult. for boils (PI); var. *radiata* Schultz-Bip. (sMex): mozote amarillo; lvs. boiled with marigold fls. & decoc. u. in jaundice - **B. tetragona** DC (Mex): hb. u. to prep. a tea ("té de milpa") - **B. tripartita** L. (Euras - marshes; natd. Am): water agrimony; bur marigold; "shepherd's needle" (Bah); hb. (under Herba Bidentis aquaticae) (Herba B. tripartitae ([Russ.]) as decoc. or inf. as diur. in kidney and gallstones, expect. emmen.; diaph.; for fever, gas; decoc. for chronic dys., heart dis., refrig.; wash for chronic eczema; for hemorr.; tumors (folk med.); sd. (frt.) u. similarly.

bidet (Fr, Eng): **b. de toilette** (Fr): oblong bowl mounted on legs & u. for washing pelvis, etc.; on it one can sit astride; a form of sitz bath, popular in Euras.

bidi leaf (Indo-Pak): lf. of *Grewia microcos*; a cheap rough cigarette without paper (**bidi**) is prepd. fr. this; mfd. in Ind., exported to Pak & SriL, constituting **b. tobacco**.

bidon (Sp) (gallicism): steel drum, large tin can, carboy.

Biebersteinia odora Steph. ex. Fisch. (B. emodi Jaub. & Spach) (Gerani/Bieberstein.) (nIndo-Pak, Kashmir): hb. appl. to wounds of horse due to heavy loads.

Biene (Ge-f.): bee.

Bienenkraut (Ge) (= bee herb): thyme.

Bier (Ge): bière (Fr): beer.

bifarious: distichous: arranged in 2 separate ranks (such as leaves).

Bifora radians Bieb. (Umb) (c&sEu): hb. with intensively smelling VO; fr. u. as spice.

"big H": Heroin.

big root: *Echinocystis fabacea* & *Marah* spp. (cCalif [US]) 2) - **b. trees** (US slang): sequoias & redwoods (Calif), specif. *sequoiadendron gigantea.*

bigarade (Fr -f.): bitter orange, *Citrus aurantium* (*C. bigaradia*) - **bigaradier**: tree of same.

Bigelowia veneta A. Gray (Haplopappus venetus Blake) (Comp) (Mex): "Damiana"; appl. to wounds.

Bignonia alliacea Lam. (Bignoni.) (monotypic) (tropAm) = *Adenocalymma alliaceum* Miers; (u. as vermif.) - **B. capreolata** L.(Anisostichus crucifera Bureau) (seUS n to Illinois): cross-vine (SC XS shows pores in distinct cross pattern); bk 4-ribbed; trumpet vine; smoke vine; rt. u. like sarsaparilla; alter. deterg.; porous sts. u. by boys as fumitory like cigars; cult. ornam.; liane - **B. chica** Humb. & Bonpl. = *Arrabidaea c.* Verl. - **B. crucigera** L. = *B. capreolata* L. - **B. indica** L. = *Oroxylum indicum* Vent. & *Pajanelia longifolia* (Willd.) K. Schum. - **B. leucoxylon** L. = *Tecoma l.* L. - **B. ophthalmica** Chisholm (Guiana) = **B. quinquefolia** Vell. = *Cybistax antisyphilitica* Mart. - **B. radicans** L. = *Campsis r.* - **B. sempervirens** Ait. f. = *Gelsemium s.* - **B. stans** L. = *Tecoma s.* - **B. unguis-cati** L. = *Macfadyena u..-c.*

bikhaconitine: typical alk. of *Acontium spicatum.*

biladur (Persia): marking nut.

Bilderdykia Dum. = *Fallopia* Adans. (Polygon).

bile: liquid secreted by liver & poured into small intestine as digestant, &c.; c. bile acids: taurocholic, glycocholic, cholic; u. as choleretic, cholagog, cath. (usually as ext. of bile) - **b. de boeuf** (Fr): **Bilis Bovinum** (L.): ox bile; ox gall (obt. fr. Arg) - **fish b.**: **f. gall:** u. to anoint blind eyes to give sight (*sic*) (1478).

bilimbo (tree): *Averrhoa bilimbi.*

bill (obsolete): prescription.

bil(l)berries: frt. of *Vaccinium myrtillus*; exp. fr. USSR.

bil(l)berry (leaves) (Eu): blueberry, *Vaccinium* spp., esp. *V. myrtillus*, *V. vitis-idaea*, *V. uliginosum*, name also refers to several other genera of pl, esp. *Ribes* & *Rubus* spp.; *Viburnum nudum* - **red b.**: *Vaccinium vitis-idaea.*

billet: a small, short, rounded thick stick of wood, usually several cm. in diam. (fr. Fr **billette**, fr. bille, long round stick, fr. ML billa): ex. sandalwood billets.

billion: 10^9 (US, Fr); 10^{12} (UK, Ge).

billy-goat weed: *Ageratum conyzoides.*

bilma (Sp): = bizma.

bilocular: bicellular: 2-chambered.

Bilsenkraut (Ge): henbane.

bindle (US slang): small packet of narcotics; represents one ration - **b. stiff** (US slang): narcotic addict.

bindweed (common): 1) *Convolvulus arvensis* & other *C.* spp. 2) *Ipomoea pandurata* 3) *Polygonum convolvulus* - **common b.: bracted b.**: *Calystegia sepium* - **European b.: field b.**: *C. arvensis* - **great b.: hedge b.**: *C. sepium* - **b. knotweed**: *Polygonun convolvulus* - **lesser b.: small b.**: field b. - **tropical b.**: *Strychnos malaccensis.*

binomial: two names (genus, species) applied to organism.

Binse (Ge): rush (*Juncus* spp. +).

biocatalyst: enzyme.

biochanin A: genistein 4'-methyl ether; olmelin; isolated fr. red clover; *Lupinus termis* (sds.); *Cicer arietinum* L. (sds.) (1101); estrogenic - **b. B**: formononetin: a flavone; chief estrogenic factor in Leguminosae (soy bean; clovers, as *Trifolium pratense*).

biocides: subst. u. to destroy living organisms of various types, incl. pesticides, some antibiotics, &c.

biocolloids: colloids originating in pl. or animal tissues.

biocytin: complex of biotin; occurs in yeast.

bioengineering: genetic engineering, using recombinant DNA technology.

bioflavonoids: vitamin P complex: a group of compds. fr. *Citrus* spp. & other sources; antiinflam.; said to maintain normal blood vessel conditions by reducing fragility & permeability; ex. Rutin. (1182).

biogas: Methane.

Biogastrone™: carbenoxolone (*qv.*).

biological products: biologicals: prepns. made fr. microorganisms incl. vaccines, sera, antitoxins, &c., classes: 1) u. for passive immunization (to treat acute disease; for temporary immunization): antitoxins; serums (sera) 2) for active immunization (prophylactic): toxins, toxoids, toxin-antitoxin mixtures, tuberculins, bacterial vaccines (bacterins), sensitized bacterial vaccines (serobacterins), viruses 3) proteins (for foreign protein therapy, usu. nonspecific) 4) diagnostic reagents (some toxins, tuberculins, proteins, &c.).

biological warfare: modern plan of conflict based on large-scale attack on humans, other animals, crops, &c., by the use of living microorganisms or their derivs. (toxins, &c.), pl. or animal substances, &c., as opposed to explosives, gas (chemical warfare), &c.

bio-organic chemistry: the biochemistry of living organisms.

Biophytum petersianum Klotsch. (Oxalid.) (trop cAf): decoc. of peeled rts. u. purg. for infants - **B. sensitivum** DC (B. apodiscias Edgew. & Hook f.) (trop w&cAf, pantrop): lal chana (Ind); crushed roots appl. to skin dis. (Burundi); in stomachache; lf. inf. expect., decoc. in diabetes; diur. in lithiasis; to induce sleep (children) (124); pl. chewed

—> sterility in male (Ind); rt. decoc. for GC, bladder stones (PI); powd. sds. appl. to wounds (Ind); pl. u. for asthma (Br); TB. (lvs. sensitive).

Bios: growth-promoting factors which stimulate yeast growth; roughly equiv. to vit. B complex; incl. Vit B_1, inositol (**Bios I**), biotin (**Bios II**), pantothenic acid, &c.

biosynthesis: formation of compds. as they occur in pl. or animal tissues (in vivo).

biota (**plural biotas**): flora & fauna; pl. and animal life of a region.

Biota Endl. = *Thuja*.

biotechnology: genetic engineering.

biotin: vitamin H (<Haut, Ge, skin, SC u. in skin dis.); anti-egg white injury factor (since combines with avidin (of raw egg white) to become inactivated); compd. occ. in all living cells & regarded as constituting part of vit. B complex; chief sources: liver, yeast, kidney; known essential to growth of certain plants; u. in scaling skin dis. (ichthyosis), baldness (value unproven), seborrheic conditions; to treat **biotindiase deficiency dis.**

bipindoside: glycoside of bipindogenin with L-2, 6-deoxy-talose; occ. in *Strophanthus thollonii*.

biquico (Sp): bechic: expectorant.

birch: *Betula* spp. - **ashen leaf b.**: gray b. - **b. bark**. (of commerce): 1) (eNA): fr. *Betula lenta* 2) (Eu): fr. *B. pendula* - **b. beer**: sweetened effervescent beverage like root beer, contg. birch oil (methyl salicylate), &c., prepd. by carbonating or fermenting; formerly made by fermenting birch tree sap with grain - **black b.**: *B. lenta* (bk.) - **b. camphor**: betulin - **canoe b.**: *B. papyrifera*: bk. u. by Indians of North America to make canoes - **cherry b.**: black b. - **b. family**: Betulacese - **gray b.**: *B. populifolia* - **mahogany b.**: black b. - **paper b.**: canoe b. - **red b.: river b.**: *B. nigra* - **silver b.**: *B. papyrifera* - **sweet b.**: black b. - **b. tar oil** (USSR, Scand): rect. VO from bk. & wd. of white b. - **b. water**: unfermented liquid fr. sap of *B.* spp.; u. in OW as cosmetic for promoting skin beauty - **white (American) b.**: silver b.: gray b.- **white b.** (European): *B. pendula*, etc.; source of rect. b. tar oil - **b. wine:** prepd. by fermenting white b. sap - **yellow b.**: *B. alleghaniensis*.

bird bone: cuttle bone - **b. chili:** b. pepper - **b. of Paradise**: *Strelitzia* spp.; *Caesalpinia* spp. - **b. of P. clover**: *Strelitzia reginae* - **b. of Paradise** shrub *Poinciana gilliesii* - **b. of Paradise shrub:** *Poinciana gilliesii* - **b. of Paradise tree**: *Strelitzia nicolai*.

bird-eye: *Callicarpa americana*.

bird-lime: birdlime: prep. of sticky viscid nature, appl. as paste to posts, stumps, etc., u. to trap birds; made in various ways, as by boiling linseed oil several hrs.; using decoction fr. *Viscum album*, *Loranthus europaeus*, *Ilex opaca*, *I. aquifolium*, etc. - **b. pepper: b. pods**: *Capsicum annuum* var. *glabriusculum*.

bird's knot-grass: *Polygonum aviculare* - **b. nest (weed)**: *Daucus carota* (wild).

biri: variant of bidi.

Birke (Ge): birch - **Birkenrindenoel** (Ge): b. bark oil.

Birkenöl (Ge): birch tar oil.

Birne (Ge-f.): pear.

birth root: *Trillium erectum*.

birthwort fam.: Aristolochiaceae.

bisabol: v. bissabol.

bisabolene: monocyclic series of sequiterpenes c. 9-C side chain; occ. in bergamot oil.

bisam (Ge): musk.

bisbee: *Castela tortuosa*, *Castela texana*.

Bischofia javanica Bl. (Euphorbi.) (seAs, Oceania, Austral+): frt. nutr.; lvs. appl. to ulcers & boils; rt. diur.; bk. u. to dye tapa (Tonga) (86); bk. boiled with salt water & appl. to cuts (New Hebrides); inner bk. rubbed on skin irrit. by nettles; lf. eaten & juice drunk for tonsilitis (Fiji); wd. useful.

biscuit root: 1) (US): *Lomatium ambiguum* & other L. spp. (wUS); 2) (wUS) *Camassia* spp. (SC rt. eaten by Indians) (**Indian b. r.**).

bishop's weed: *Aegopodium podagraria*.

bisnaga ("barrel") (Sp): visnaga; barrel cactus; members of genera *Ferocactus* & *Echinocactus*.

Bison bison bison L. (Bovidae-Ruminantia): American bison; "buffalo" (do not confuse with the buffalo of OW); denizen of NA plains states; immense herds of early 1800s reduced by wholesale slaughter; meat, horns, bones, & hides u. by Am Indians & later by white man; in pemmican; gall stones u. med.

bissabol (myrrh): perfumed bdellium; a var. of myrrh obt. fr. *Commiphora kataf*, *C. erythraea*, etc.; u. in perfumery, as adult. of true myrrh.

bissy bissy (nut) (tropAf): bussi: *Cola acuminata*.

bister (Eng): **bistre** (Fr, Eng): brown pigment for water colors prepd. fr. charred wood.

bistort (root): **Bistorta** (L.): *Polygonum bistorta*.

Bistorta Adans. = *Polygonum* L.

bitch: *Solanum caroliniaum* - **b. wood**: *Piscidia erythrina*; *Lonchocarpus latifolius*.

biting stonecrop: *Sedum acre*.

Bitis gabonica (Dumeril & al.) (Viperidae: Ophidia) (wAf, rain forests): Gaboon (or giant) viper; largest of all vipers; large fat bodies from abdomen u. in med. concoctions (2023); venemous snake.

bito: *Balanites aegyptica*.

bitter: v. also almond; aloes; bugle; candy tuft; cassava; chocolate; clover; cress; cucumber; kola; milkwort; orange (oil); polygala; quassia; wintergreen.

bitter apple: Colocynthis - **b. ash**: *Picrasma excelsa* - **b. ball** (NM [US]): *Tagetes micrantha* - **b. bark**: 1) *Pinckneya* 2) *Alstonia constricta* 3) Cascara Sagrada (bk.) - **b. bone**: cat bone taken fr. boiled cat; u. in voodoo practice; supposedly has bitter taste - **b. chips**: *Quassia* - **b. cups**: b. wood cups: quassia cups turned fr. quassia wd. - **b. gourd:** *Momordica charantia* - **b. herb**: *Centaurium minus* - **b. nut:** *Carya cordiformis* - **b. pine**: Jeffrey pine - **b. plant**: *Aletris farinosa* - **b. root**: 1) *Apocynum androsaemifolium* 2) *Lewisia* spp. esp. *L. rediviva* 3) Gentiana - **b. salts**: Epsom salt - **b. stick(s)**: chirata - **b. stomach drops**: Tr. Amara (NF V) - **b. top** (Ala [US]): - **b. weed** (seUS): sneezeweed, *Helenium tenufolium*; formerly *Ambrosia* spp. - **b. wood**: *Picramnia, Quassia*, &c. - **b. wort**: *Gentiana lutea* & other *G.* spp.; *Taraxacum officinale.*

Bittere Fiebernuss (Ge): *Strychnos ignatii* - **Bitterklee** (Ge): *Menyanthes* - **Bitterkraut** (Ge): Absinthium - **Bittermandeln** (Ge): bitter almonds.

bittersweet (twigs) (hb.) (sts.): fr. *Solanum dulcamara* - **American b.: false b.: shrubby b.**: *Celastrus scandens* (rt. bk. berries); sometimes in error called simply "bittersweet" - **true b.:** fr. *Solanum dulcamara* .

bitterworm: *Menyanthes trifoliata* - **bitterwort:** *Gentiana lutea;* dandelion - **Bitterwurzel**: (Ge): gentian.

bitume de Judée (Fr): chapapote.

bitumen: pl. **bitumina**; asphalt; Asphaltum; tarry subst. originating fr. evapn. of petroleum under natural conditions (ex. Dead Sea) (2742) or prod. of the petroleum industry prepd. by distn. & mixing solid & semisolid petr. hydrocarbons (v. asphalt) - **blown b.**: mixt. of hydrocarbons of natural or pyrogenous origins; u. as coatings & binders - **b. de Judea** (Mex): chapapote (*qv.*) - **b. judaicum** (L): asphalt, orig. fr. Dead Sea - **b. sulfonate**: sulfonated bitumen (*qv.*).

Bixa orellana L. (l) (Bix.) (SA, WI, Ind): anatto tree; bixa, arnatto; pulpy aril around sd. u. as red, yellowish-red, or yellow dyestuff & flavor for coloring butter, cheese, FO (esp. in tropAm); spice; also to dye cotton, silk, wood, etc.; sds. u. by Mayas in measles, asthma, gastralgia; lvs. u. cephalgia (6); u. body & face paint (Am Indians) (Ec+) (serves as sun screen); u. in sores, burns, tumors (Mex); prevents scarring; in dys.; antidote to *Jatropha curcas*, manihot (HCN), &c.; insect repellant; source of orlean (annatto; roucou); u. to clear skin blemishes (Ec); veg. lipstick; flavor in stews; to color dairy prods. (butter, cheeses (Edam), noodles, &c.); rt. tranquilizing; cult. SA (Br), CA, PR, Indones - **B. urucarana** Willd. (Br): u. like *B. orellana.*

Bixaceae: Bixa (or Annato) Fam.; shrubs & trees with simple, alternate lvs.; composed of 3 genera (tropics) & 16 spp.

bixin: carotenoid carboxy-acid, yellow coloring matter fr. sds. of *Bixa orellana*; u. in cheeses, etc. - bixine: anatto.

bizma (Sp): plaster u. for strengthening & comfort; often made of sheep skin or linen.

biznaga (Sp): bisnaga.

black: v. also under alder (bark); ash; baneberry; birch; bryony; catechu; cherry; cohosh; currant; dogwood; elder; gentian; ginger; haw; hellebore; henbane; ipecac; licorice; mint; mustard (seed); myrobalani; nightshade; oak; pepper; pitch; poplar (buds); poppy; raspberry; resin; sanicle (root); sassafras; snake-root; spleenwort; spruce; tang; teas; thorn (blossoms); walnut; willow - **b. caraway:** *Pimpinella saxifraga* - **b. mulberry:** *Morus nigra*

black balsam: Peruvian balsam - **b. beetles** (GB): roaches (esp. *Blatta orientalis*) & some other insects - **b. boy gum**: acaroid resin - **b. b(r)ush** (Tex [US]): *Coleogyne ramosissima* - **b. cantharis**: *Cantharis atrata* (wUS) - **b. caps: blackcaps**: b. raspberry - **b. caraway:** *Pimpinella saxifraga* - **b. cumin:** Niqella - **b. doggert**: birch tar oil - **b. drink: b. draft**: beverage fr. *Ilex vomitoria* u. by seUS Indians in ceremonials & med. - **b. draught**: Infus. Senna Comp.; do not confuse with b. drop - **b. drop**: Vinegar of Opium; has been in use for some 500 yr. - **b. dye of the Shans**: *Diospyros ehretioides* - **b. Indian hemp**: *Apocynum androsaemifolium* & *A. cannabinum* - **b. jack**: 1) *Quercus marilandica* (**forked-leaf b.**: *Q. laevis*) 2) *Bidens* - **b. juice** (comm.): dr. aqueous ext. of licorice made by boiling rts. in water & evapg. decoc. to dryness - **lamp b.**: carbon or soot deposit fr. combustion of oils or tars - **b. mulberry:** *Morus nigra* - **narrow-leaved b.**: *Q. cinerea* - **b. root**: 1) Leptandra 2) *Aletris farinosa* 3) *Pterocaulon undulatum* 4) *Vernonia acaulis* - **roundleaf b. jack**: *Q. marilandica* - **b. salve**: *Unguentum Nigrum* (4); not the same as Ichthyol Ointment - **b. Samson**: Echinacea - **b. sulfur**: impurities formed during the purification of sulfur - **b. thorn**: sloe, *Prunus spinosa* - **b. wash**: b. lotion.

blackberry: certain members of genus *Rubus*; subgenus *Eubatus* composes the true blackberries;

specif. may be appl. to *R. fruticosus* (frt., fresh & dried); **b. bk. of rt**. is obt. fr. spp. of *Eubatus*; incl. *R. villosus*, *R. allegheniensis*, *R. cuneifolius* - **American b.**: *R. villosus* - **bramble b.**: *R. fruticosus* - **creeping b.**: *Rubus trivialis* - **b. leaf**: v. *Rubus caesius* - **b. lily**: *Belamcanda chinensis* - **low b.**: *R. canadensis*, *R. trivialis*, etc. - **white b.** (Miss [US]): frt. white with bluish tinge (13), as *R. allegheniensis* forma *albinus*.

blackbirds: merles (European b.); include *Icterus* & several other genera (Fam. Icteridae) (mostly of NW): males are chiefly black; some with addl. colorations (ex. redwing b.); birds cooked without eviscerating & after removing feathers, & eaten; in a related fam. Dicruridae, (mostly OW): black drongo (jhanpul) (Ind): whole bird boiled in sesame oil. u. for alopecia (Unani med.).

blackeyed Susan: 1) *Rudbeckia* spp., esp. *R. hirta* 2) *Thunbergia alata* 3) *Calochortus nuttallii* & *C. gunnisonii* (Colorado) - **b. s. seed** = sd. of *Abrus precatorius*.

blackhaw (viburnum): *Viburnum prunifolium*.

Blackstonia perfoliata (L) Huds. (Chlora p. L.) (Gentian.) (c&sEu, wAs, nAf): (common) yellow wort; pl. c. gentiopicrin; u. bitter ton. like *Centuarium erythraea*; yellow dye.

blackwood: *Haematoxylon campechianum* - **blackwort**: *Symphytum officinale*.

bladder nut: 1) *Pistacia vera* 2) *Staphylea trifolia* - **b. senna**: *Colutea arborescens* (lvs.) - **b. wrack**: *Fucus vesiculosus* & other *F.* spp.

bladders: dr. urinary receptacles fr. cattle, *Bos taurus*.

bladderwort: *Utricularia vulgaris* +.

blaeberry (Scot & New England): whortleberry (fr. blae, Scotch, blue) (1102).

Blainvillea gayana Cass. (Comp.) (Af: Seneg): decoc. or macerate of leafy sts. u. antisep.; inf. of seeds after crushing & sieving u. in eye infec.

Blakea subconnata Berg ex Tr. (Melastomat.) (wEc, Col): mucilaginous liquid fr. buds, inflated calyx u. in kidney trouble (Ec); frt. edible.

blanc de troyes (Fr): prepared chalk - **blanc fixé** (Fr): permanent white; $BaSO_4$; u. as pigment & paper glaze - **blanc-mange** (Fr, Eng) (pron. blahmonzh) (lit. white food): a jelly-like dessert food or pudding; made in Ire fr. soaking Chondrus (or Irish moss) & elsewhere similar materials (gelatin; isinglass; arrowwroot; starch; ground rice, &c.) cooked with milk & flavor - **transparent b.m.**: clear flavored jelly.

blanco (Sp): white - **palo b.** (Española): *Ilex* spp. esp. *I. macfadyeni* Rehder.

blanket leaf: b. weed: mullen.

Blasenstrauch (Ge): *Colutea* spp. - **Blasentang** (Ge): *Fucus vesiculosus*.

blasse Rose (Ge): *Rosa centifolia* (literally pale r.).

Blastmycin: antimycin A_3; antifungal & antiviral antibiotic fr. *Streptomyces blastmyceticus* Watanabe (Jap).

Blastomycin: sterile filtrate fr. culture of *Blastomyces dermatitidis* Gilchrist & Stokes (fungus) (mycelium) u. as diagnostic aid (skin tests) to detect NA blastomycosis (prod. scrofuloderma).

Blastophaga grossorum (Hymenoptera): a small wasp which pollinates the domestic fig.

Blatt (Ge): leaf - **B. kruell** (Ge): crumpled lf. material (sweepings) - **B. majoran** (Ge): *Majorana hortensis*.

Blatta orientalis L. (Blattidae - Orthoptera - Blattaria - Insecta) (Eu, NA): Oriental (cock)roach; common cockroach; Kuechenschabe (Ge): cucaracha commun (Sp); c. blattic acid, periplanetin; earlier dr. insect u. as diaph. diur.; in edema (Russ); in homeop. med. (512) *cf. Periplaneta*.

Blau (Ge): blue - **B. mohnsaat** (Ge): blue poppy sd.

blawort (Scot.): fl. of corn blue bottle, *Centaurea cyanus*.

blazes, blue: *Liatris squarrosa*.

blazing star (root): 1) *Liatris* spp. 2) *Aletris farinosa* 3) *Heleonias* 4) *Chamaelirium luteum*.

blé (Fr): wheat - **b. de Turquie**: maize.

bleaching powder: chlorinated lime.

Blechnum chilense Mett. (Blechn./Polypodi.) (Ch): palmita; appl. to wounds in the umbilical cord - **B. fraxineum** Willd. (tropAm): inf. u. as. cure-all (baths) - **B. hastatum** Kaulf. (Ch): palmilla; u. as emmen. abort.; rts. u. as anthelm. - **B. occidentale** L. (sLa, nFla [US], Mex to Arg, WI): doradilla; helecho; lf. decoc. u. in nervous conditions, hysteria; in constipation; kidney & bladder stones - **B. orientale** L (Chin): young shoots appl. to boils, swellings - **B. punctulatum** Swartz (sAf): rhiz. ingredient in anthelm. enemata.

Blechum brownei Juss. (B. pyramidatum Urb.) (Acanth.) (Mex, WI, CA, SA; natd. in Far East): to treat blenorrhea; crushed lvs. as vulner. (PI); lvs. & fl. tops u. as diur.; in night sweats, cough, fever, &c. (Yuc) (6); in amebic dys. (CR, Pan).

bleeding heart, little: *Dicentra pusilla* - **squirrel corn b. h.**: *D. canadensis*.

blennostasine: claimed deriv. of a cinchona alk.; u. hay fever, colds, influenza.

Bleomycins: group of antibiotics (Jap) fr. *Streptomyces verticillus* Takita; glycopeptides; useful as anti-neoplastic in various cancers; for plantar warts (E) (1129).

Blepharis buchneri Lindau (Acanth.) (cAf): ashes of pl. sprinkled on sores or stings; a pinch is taken for migraine, asthma, bronchitis (Burundi) (1103); for leprosy, scabies, GC (2445) - **B. capensis** (L.f.) Pers. (sAf): u. in anthrax by both natives & Europeans (tinct. int.), seems of value; lf. appl. in toothache & wounds; in snakebite & toxic insect bite (sting) (several other sAf *B.* spp. u. similarly) - **B. edulis** Pers. (Ind): sds. c. bitter glucoside, blepharin; sds. u. mucil. diur. aphrod.; lvs. vegetable - **B. linariaefolia** Pers. (Pak): sds. u. as "cure" for headache (Las Bela) - **B. maderaspatensis** (L) Roth. (tropAf): pl. ash in peanut oil u. as poultice for sore legs; rt. & lvs. u. for eye pain; rts. & lvs. pounded, soaked in water & u. as eye drops for infection (Tanz) - **B. pungens** Klotzsch (Moz, Namibia, Zimbabwe): u. in arthritic pains, esp. of legs; rt. incinerated with that of *Solanum incanum* L. & the ash inoculated for local pain (2445).

Blepharispermum zanguebaricum Oliv. & Hiern (Comp.) (c. & sAf): rt. prepns. u. for edema, hydrocele.

Blepharocalyx gigantea Lillo (Myrt.) (Arg): horcomolle; u. for asthma - **B. tweediei** Berg (Arg): anacahuita; lf. u. as astr. ton. in pulm. dis.

Blepharodon mucronatum Dcne. (Asclepiad.) (Mex, CA, nSA): chununa de caballo (Pan); tietie; pl. juice u. as antisep. (Yuc) (6); crushed lvs. u. as poultice for swellings, snakebite.

Blephilia hirsuta Benth. (Lab.) (eUS): woodmint; Ohio horsemint; hb. (Blephilia) u. arom., ton., carm., stim.

blessed thistle (herb): 1) *Cnicus benedictus* 2) *Silybum marianum* 3) *Cirsium lanceolatum.*

Bletilla striata (Thunb.) Rchb. f. (Blettia s. Druce; B. hyacintha R. Br) (Orchid.) (sChin, Jap mts.): rhiz. u. as lung ton., in TB, hemoptysis, exter. in swellings, dry cracked skin, chilblains; crushed tuber appl. to burns; rhiz. decoc. u. expect. dem.; appl. to malignant ulcers; insecticide; mucil. of tubers is superior to gum arabic & starch paste; u. as emulsifier & as binding agt. in tablet mfr.; good substit. for acacia & tragacanth.; popular ornam.

bletting: process of ripening or rotting of frt. (ex. medlar frts.); frt. placed in dry cool room after freezing; may be change in consistency of frt. without rotting.

bleuet: bluet (Fr): corn flower, *Centaurea cyanus* (v. bluet).

blewits: blewitt: blue leg: *Lepista saeva* P. D. Orton (edible mushroom).

Blighia sapida Koenig (Sapind.) (trop wAf): akee apple; a. fruit; savory akee tree; native to Guinea, wAf; cult., PR, sFla, natd. Jam; unripe arils c. hypoglycins A & B (with hypoglycemic activity); believed cause of "vomiting sickness" (Jam) following consumption of unripe or improperly prepd. frt.; may be fatal; floral water (distd.) u. in cosmetics; ripe frts. fried with salted fish (popular food in Jam); sd. arils edible, source of FO; wd. excellent building timber; u. for paddles, oars (named after Capt. Bligh of *H.M.S. Bounty* who introduced the tree to Jam) - **B. unijugata** Bab. (wAf): rt. u. as antipyr. purge; sds. frt. in nausea & vomiting, fish poison - **B. welwitschii** Radlk. (tropAf): "bate"; crushed frt. & juice u. to stupefy fish; wd. excellent; lvs. u. as cholagog. (Nig).

blindweed: *Capsella bursa-pastoris.*

blister(ing) beetle: insects of genera *Cantharis, Meloë, Lytta, Mylabris, Cerocoma, Zonitis,* etc. (Fam. Meloideae); specif. some common sp. such as *Cantharis vesicatoria* - **American b. b.**: spp. of *Lytta, Epicauta, & Macrobasis* - **blistering cerate**: cantharis c. - **b. collodion:** cantharidal c. - **b. fly, Chinese:** *Mylabris cichorii* & other *M.* spp. - **b. plaster**: cantharis p.

blisterweed: *Ranunculus acris.*

blood: chief body fluid of animal with respiratory, nutrient, excretory, body temperature-regulating, & protective functions; c. water, proteins, FO, carbohydrates, salts, endocrines, antibodies, &c., these composing the serum; along with floating or suspended corpuscles (RBC, WBC, platelets), the entire complex being whole blood - **b. bank**: organization which procures & stores (refrigerated) blood and/or its components - **b. cells, red** = erythrocytes - **b. cells, white** = leukocytes - **b. group:** class of human erythrocytes distinctive genetically and immunologically; depending on specific antigens (agglutinogens) present; antigens recognized: O, A, B, AB; also RH+ and RH- - **b. grouping serums:** sterile prepns. containing specific blood group antibodies derived from blood plasma/serum of humans; u to det. blood types (mostly types A, B, & AB) - **blood group specific substances A, B, and AB:** sterile isotonic solns. of polysaccharide-amino acid complexes capable of neutralizing the anti-A and anti-B isoagglutinins of group O blood; subst. A is obt fr. hog gastric mucin and subst. B and AB fr. horse gastric mucosa; u. with microscopic techniques to det. blood type of individual before giving a blood transfusion - **b. picture**: b. cell formula - **b. platelets**: thrombocytes, which function in agglutination, formation of thromboplastin, & blood coagulation - **b. transfusion**: passage of blood fr. one individual into another; storage for emergency use customary; in Russia, cadaver b. has been u. with some success (1930-) - **b. unit**: convenient vol. for administration; in US 500 ml., varies somewhat with country; large volumes (15) may be u. in open heart surgery, &c. - **umbilical cord b.**: u. in leukemia.

blood-berry: 1) *Rivina humilis* 2) b. lily - **b. flower**: *Asclepias curassavica* - **b. lily:** *Haemanthus coccineus* - **b. root**: *Sanguinaria canadensis* - **b. stone**: a mineral - **b. weed**: 1) b. flower 2) *Ambrosia trifida* 3) b. lily - **b. wood tree**: *Haematoxylon* spp. - **b. wort** (US): yarrow **(striped b. w.**: *Hieracium venosum*) (plus 12 other pls.).

bloodynose (US): Indian paint brush, *Castilleja* spp.

bloom: 1) flower or blossom (*cf.* Blume, Ge.) 2) heavy seasonal growth of algae produced by mineral enrichment; considered sign of environmental pollution.

blooming flowering spurge: *Euphorbia corollata* - **b. Sally**: *Epilobium angustifolium*.

blowball: dandelion.

blowfly (domestic): members of *Sarcophaga* & other genera (Sarcophagidae, large fam. of Diptera (Insecta); cause of myiasis in man & lower animals; maggots (larvae) u. in folk med. in osteomyelitis, &c. (bactericidal to several Gm. pos. bacteria) (2524).

blubber: FO fr. whale & other sea mammals; represents subdermal insulating layer.

blue: v. also ash; centaury; cohosh; copperas; devil; flag; galls; gentian (root); grass; gum tree; mallow; oil; ridge tea; rocket; vervain; violet (flowers); vitriol; weed.

blue bell family: Campanulaceae - **b. bells**: 1) *Polemonium reptans* 2) *Campanula* spp. 3) *Mertensia virginica* 4) *Muscari botryoides* 5) various other genera & spp. of plants with blue corollas (**English b. b.**: *Endymion nonscripta*; **b. b. of Scotland**: *Campanula roundifolia;* **Virginia b. b.:** *Mertensia virginica*) - **b. blossoms**: *Ceanothus thyrsiflorus* - **b. bonnets** 1) (Tex [US]) *Lupinus subcarnosus*, *L. texensis* (also Texas b. b.) 2) *Centaurea* spp. esp. *C. cyanus* 3) *Jasione* spp. 4) *Scabiosa* sp. - **b. bottle flowers**: 1) *Centaurea cyanus* 2) spp. of *Centaurium*, *Campanula*, *Muscari*, *Scilla* 3) (Ala [US]) *Callicarpa americana* - **b. butter:** stronger mercurial ointment - **b. curls:** *Trichostema lanceolatum* + - **b. John** (Ala negroes): po' John; *Plantago aristata*.- **b. mass**: mass of mercury - **b. mountain tea**: *Solidago odora* (lvs.) - **B. Ridge Mountains** (North Carolina): important area of crude drug collection, n of Asheville - **B. Ridge tea**: herb tea formula (SBP) u. as lax., diur. - **b. rocket**: aconite - **b. sailors:** chicory - **b. scarlet pimpernel:** *Anagallis arvensis* - **b. stone:** copper sulfate, popularly u. in voodoo amulets - **b. top:** *Verbena officinalis* - **b. vine:** *Clitoria ternatea*.

blueberry: *Vaccinium crassifolium*, *V. corymbosum*, *V. angustifolium* & other *V.* spp.; *Daniella nigra*; + - **b. barrens**: large areas of seMaine [US] where extensive cultns. of low-growing spp. of *Vaccinium* are made - **garden b.**: frt. of *Solanum nigrum* - **highbush b.**: *Vaccinium corymbosum* - **b. leaves**: fr. *V. myrtillus*, *V. angustifolium* - **low (bush) b.**: *V. angustifolium* - **rabbiteye b.** (wFla [US]): *V. corymbosum* (cult.) - **b. root**: *Caulophyllum thalictroides* - **white b.**: 1) whitish to pinkish frts. of *Vaccinium* spp. 2) albino forms; may be due to fungus growth.

bluet 1) *Houstonia caerulea* 2) *Vaccinium angustifolium* var. *myrtilloides* (also frt.) + (*cf.* also "bleuet") 3) *Centaurea cyanus*.

Bluete (Ge): flower - **Bluetengrus** (Ge): ray & disc fls. as distinct fr. whole fl. heads (ex. chamomile).

Blumea arfakiana Martelli (Comp.) (PapNG): lvs. & rt. u. for stomachache (1104) - **B. aurita** DC (B. viscosa Badell.) (tropAf, Ind, natd. WI): lvs. & st. u. in dyspep., sprue (Indones); pl. inf. in rheum. - **B. balsamifera** (L) DC (Pluchea b. Less.) (Indones, Indochin, Mal, sChin, PI): sambong (PI); lvs. & sts. c. VO with (-)-borneol, (-)-camphor (ai or ngai c.); lf. ext. u. as stom. expect in catarrhs, cough, antispasm; bath after childbirth (PI); diur, diaph. in tympanites, mal. fever; sed.; in lumbago, arthritis, rheum., locally as styptic to wounds; Indian sprue; may have antimutagenic activity; one of components of "Singapore Universal Tonic"; source of ngai camphor - **B. chinensis** DC (B. pubigera Merr.) (seAs): katarai; rt. decoc. u. in colic (Mal): for cough; substit. for *B. balsamifera*; stom. antispasm. diaph. (Mal) - **B. lacera** DC (Indones, Ind, Mal, Af): Malay blumea; hb. rich in carotene, vit. C, VO, with blumea camphor (66% cineole), fenchone; pl. juice u. as vermif. for threadworms; pl. crushed & u. by spreading on water to stun fish; fl. decoc. for bronchitis; to purify water; in scurvy (Angola) - **B. lanceolaria** Druce (B. myriocephala DC) (seAs): pl. or lvs. u. in asthma, mal., bronchitis, sudorific; lvs. as cataplasm for arthritis (Mal); young lvs. condiment with fish.

Blumenbachia insignis Schroed. (Loas.) (SA): ortiga (Arg); with stinging hairs; u. as rubefac. (Arg); c. quercetin, caffeic & ferulic acids.

Blumenkohl (Ge): *Brassica oleracea*, cauliflower.

blushwort: *Centaurium* spp. as *C. erythraea*.

Blutegel (Ge): leech - **Blutholz** (Ge): logwood - **Blutstein** (Ge): 1) bloodstone, manganese oxide 2) hematite, Lapis Haematitis - **blutstillend** (Ge): hemostatic; styptic - **Blutwurzel** (Ge): sanguinaria.

bo tree (Ind): *Ficus religiosa* (Budda).

boai (Br): *Buxus sempervirens.*
bobas (Br): *Ticorea jasminiflora*, baobas (Lindl.).
Bob's root: *Psoralea glandulosa* (etc.).
Bocconia L. (Papaver.) (warm As, WI): celandine tree; tree c.; plume poppy; shrubs, sub-shrubs, or hbs. with yellow or orange latex - **B. arborea S.** Wats. (c&sMex to Pan): celandine tree; llora sangre; chicolate (de arból) (Mex); c. opiate alks. bocconine, sanguinarine, allo-cryptopine, chelerythrine, &c.; u. by Mex surgeons by inj. for local anesthesia; bk. analg. (said similar action to morphine); bitter sap u. as purg. vermif. for toothache, vulner. (Mex Ph.); cooked lvs. as poultice on ulcers & tumors; cult ornam. (Guiana) - **B. cordata** Willd = *Macleaya c.* R. Br. - **B. frutescens** L. (WI, CA, Mex to Pe): parrot weed; John Crow bush; sarno; similar to *B. arborea*; bk. & wd. c. fumarine, chelerythrine, &c.; sap (latex) purg. vermif.; in bronchitis; for stomach cramps & cancer pains; appl topically to ulcers, warts, chiblains, callouses, corns, for skin parasites (Co); lvs. & sds. u. med.; rt. decoc. antipyr. liver remedy; kidney dis. (2444) in poultices for fungoid ulcers; & other uses - **B. integrifolia** Humb. & Bonpl. (sMex): u. similarly to *B. arborea* - **B. latisepala** S. Wats. (Mex): mala mujer; u. similarly to *B. arborea* - **B. vulcanica** Donn, Smith (Guat): cerbatana; sds. or frts. inserted into tooth cavity for pain.
Bockshornsamen (Ge): fenugreek.
bodan: badan.
bodark (bo'dark) (corruption of bois d'arc) (Miss [US]): *Maclura pomifera* ("bodock").
boe(a) gum: fossil resin of the manila copal class; fr. *Agathis alba*; u. in varnishes.
boegroemakka (Sur): *Astrocaryum murumuru.*
Boehmeria blumei Wedd. (Urtic.) (PI): lvs. pounded as poison for crabs - **B. caudata** Sw. (tropAm): source of fiber of poor quality (Mex); tinct. of drug a substit. for arnica tinct. (Br) - **B. irrusta** L. (Ind): reputedly useful for stomach dis., intermittent fevers; dys. - **B. nivea** (L) Gaud. (s&cChin, sJap; cult. Ind, PI & other subtrop): ramie, rhea; Chin grass; cult. for fiber mostly in Bengal; said the toughest, longest, & most silky veget. fiber; rich in pectocelluloses; important comm. product; rt. u. GC, fever, thirst, appl. to swellings; lvs. & rt. u. in urogenital inflammn., rectal prolapse, erysipelas; pounded lvs. as poultice for maturing boils - var *chinensis*: Chinese (or white) ramie (fr. swChin); var. *indicum* (B. tenacissima) (Gaud.) Min.: Indian (or green) ramie; rhea; rt. u. as emoll.; ingested to treat hematuria, restless fetus; appl. to swellings & boils; & many other uses; fibers of this & other spp. u. in mfr. textiles, ropes, (1105) - **B. platyphylla** D. Don (Oceania): "pute"; bk. yields fine strong fiber u. for fish lines (Marquesas); pl. u. in med. (New Caled) - **B. virgata** (Forst. f.) Guillem. (Oceania): lvs. appl to boils (Fiji).
Boenninghausenia albiflora Reichb. (monotypic) (Rut) (cAs-Chin+): rt. decoc. u. int. for malaria; crushed lvs. placed in nostrils for same; twigs in brooms u. to drive away fleas (Pak); cult. ornam.
Boerha(a)via diffusa L. (B. caribaea Jacq.; B. coccinea Mill.; B. repens L.; B. mutabilis sensu auct.; B. hirsuta Willd.; B. paniculata L. C. Rich.) (Nyctagin) (sFla, sTex [US], WI, Mex, CA to Pe, Arg; now trop & subtrop of NW & OW incl. Ind): (smooth) hogweed; spreading h.; erect sand verbena (Tex); yerba blanca (Mex); erva (de) tostão (Br); toston; punarnava (Sansc+) panarnaba (Bengali); gada (Bihar, Ind); rt. st. lvs. of this common weed c. punarnavine (alk.); boerhavin, starch (16%); pl. (as FE) u. as diur. in edema. ascites (1106), GC; in chronic alcoholism (Ind); hepatic cirrhosis (Br) emet. expect. in asthma; juice popular home remedy (Br); hb. u. as cataplasm for liver complaints (Br); in Ind to check bleeding (with chhatank lvs.) (1124); pl. decoc. u. to wash sores, to bathe erysipelas; as antispasm; in toothache (Br); imbibed in epilepsy, chorea (Yuc); rt. u. as emet in icterus (Ang); as poultice to wounds (Br); mal. (Br): rt. & lvs. stim. ton. aphrod.; cath vermif. in facial neuralgia (Haiti); rts. u. as antipyr. anthelm., in mal. lvs. u. as potherb or in soups; favorite food of pigs; common prostrate weed - **B. erecta** L. (sFla, swUS, WI, Mex, CA, to Pe, now pantrop): (smooth) hogweed; pl decoc. u. to bathe sores, erysipelas; also u. int. asthma, in epilepsy, chorea; emet.; rts. u. for stomachache (Tanz) - **B. plumbaginea** Cav. (Med reg., tropAf): lf. decoc. u. as vulner & for jaundice - **B. scandens** L. = *Commicarpus s.* Standl.
Boesenbergia rotunda Mansf. (Gastrochilus panduratus Ridl.) (Zingiber.) (Indochin, seAs): rhiz. c. VO with methyl cinnamate & cineole; rhiz. u. as stom ton. carm emmen; cholagog; in coughs (Mal); for "ague"; in lotions for rheum. (Mal) or muscular pains; in ovaritis; rhiz. & rts. as spice for foods.
boeuf (Fr): bull; ox; cattle; beef.
bog: a poorly drained depressed area filled with sphagnum moss (bog-moss), &c. & feeling spongy when walked upon.
bog-bean: *Menyanthes trifoliata* - **b. bugle**: *Sarracenia* spp. - **b. moss**: *Sphagnum* spp. - **b. myrtle**: bog-bean. (*Myrica gale*)

bog-butter (Irish): butyrelite, a soft, "fatty" substance (mixt. of fatty hydrocarbons) found in marshes & peat bogs & of unknown origin (20).

Bohne (Ge): bean - **Bohnenkraut** (Ge): *Satureja hortensis* (20).

bois (Fr): wood - **b. à envirer**: Jamaica dogwood, *Piscidia, piscipula* - **b. chandelle**: *Amyris balsamifera* - **b. cochon** (Fr): *Tetragastris balsamifera* - **b. d'arc** (seUS) ("Bow wood"): *Maclura pomifera* - **b. de betterave** (Cayenne - Fr): rosewood - **b. de buis**: dogwood - **b. de campêche**: logwood - **b. de chien**: Jamaica dogwood - **b. d'enfer** (Qu): *Rhus radicans* - **b. de rose (oil)**: (Mex, Br): lignaloe - **b. de rose (de la Jamaïque)**: *Amyris balsamifera* - **b. de rose femelle** (Fr): *Ocotea caudata* - **b. des roses** (Fr): b. de rose - **b. de sang:** logwood - **b. doux**: licorice - **b. du Brésil** (Fr): wd. of *Caesalpinia sappan* - **b. dur**: ironwood - **b. envirant**: **b. envirer: b. ivrant**: Jamaica dogwood - **b. pourri** (Qu): rotten wood.

boistroside: glycoside of corotoxigenin with D-digitoxose; occ. in *Strophanthus boivinii.*

bok choy (Chin): pak choi (preferred form): *Brassica chinensis.*

bolas: *Prunus domestica* var. *insititia.*

bold: 1) clean, bright, distinct, high quality of crude drug 2) large & clean pieces of resin, etc.

boldine: an aporphine alk. characteristic of boldus lf.; resembling bulbocapnine in physiolog. effects (suppresses but does not invert hypertensive action of epinephrine) (1107).

boldo (leaves): Boldus - **boldo-glucin**: glycoside of boldus (1108).

Boldus Schult.: *Peumus (boldus).*

bole: - 1) tree trunk 2) var. of clay (argillaceous mineral) u. in med. (terra sigillata) & as pigment - **Armenian (red) b.**: red b.: a natural iron-contaminated aluminum silicate orig. obt. fr. Armenia (hence name); u. in cements; as pigment, adsorbent in diarrhea, etc. - **French b.:** various forms of argillaceous earth - **white b.**: Kaolin.

boleko: *Ongokea* sp.

boleo: pennyroyal.

Boletus Fr. (Bolet. - Boletales - Hymenomycetes - Basidiomyc. - Fungi) (cosmopol): boletes; terrestrial; some frt. bodies edible - **B. crocatus** Batsch (As): u. int. in diarrh. & dys.; ground to paste & appl. to gums in excessive salivation - **B. edulis** Bull. ex. Fr. (worldwide): "Polish mushroom"; attractive food item (fresh, dr., pickled); however toxic if injected; crude exts. of dehydrated sporocarp retard growth of implanted tumors in mice (1109) - **B. fomentarius** L. = *Fomes f.* Kickx. - **B. igniarius** L. = *Phellinus i.* Quél. (Fomes i. Kickx.) - **B. laricis** Jacq. = *Fomes officinalis* Faull. - **B. luteus** L. - *Suillus l.* S. F. Gray - **B. satanas** Linz (Eu, NA): c. muscarine; toxic & not to be eaten; u. in homeop. med. - **B. suaveolens** Bull. = *Trametes s.* Fr. - **B. subtomentosus** Fr. (Euras, NA): yellow-cracked boletus; frtg. bodies u. as food in some countries of Euras - **B. sudanicus** Har. & Pat. (cAf): frt. bodies popularly edible by natives (hegba mboddo) - **B. versipellis** Fr. (Leccinum testaceoscabrum Sing.) (Eu): Rotkappe; popular food in many parts of Eu; frt. bodies sold in markets, made into pickles, "krassny grib" (red mushrooms) (USSR); to replace truffles (Ge) in making truffle liverwurst.

boll (cotton): rounded capsule of cotton, flax, &c. (fr. ball, bowl).

bolsa (Sp): bag; sack - **b. de pastor** (Por): shepherd's purse.

bolted: sifted by use of a bolter; a means of both assorting & refining; ex. separating bran fr. flour.

bolter: cloth or other fabric for separating finer & coarser particles.

Boltonia indica Benth. (Aster indicus L.) (Comp.) (seAs): u. ton. antipyr. diur. carm.

bolus (L.): 1) large pill 2) bole - **B. Alba:** kaolin - **B. Armena: B. Armenia(e)**: Armenian bole, aluminum silicate - **B. Rubra**: Kaolin with reddish admixture.

Bomarea edulis Herb. (Amaryllid./Alstroemeri) (WI, CA, SA): "gloriosa del pais" (Cu): rts edible, eaten raw as galactagog (sMex) (79) - **B. salsilla** Herb. (Ch): zarcilla; rt. u. like sarsaparilla as diaph. & stom.

bomba: 1) (Mex): *Datura candida* 2) (PR): *Calotropis procera.*

Bombax aquaticum (Aubl.) K. Schum. (Bombac.) (Br Guian): sd. kernels c. 55-60% FO, "mamurana fat"; u. to make soaps - **B. ceiba** L. (B. malabaricum DC) (Indo-Mal, nAustral): (red silked) cotton tree; 5-leaved silk c. t.; leptan (Burma); source of a reddish floss (fiber) fr. frt. (capsules) u. as stuffing material in pillows, cushions, &c., in bandages; gummy exudn. of trunk called Mocharas(a) resin, u. astr. in diarrh., dys., menorrhagia; aphrod.; against tumors (Ind); rt. u. as ton. diur. emet.; juice fr. rts. u. as antipyr. (Indones); bk. (ceiba bk.: corteza de ceiba [Pe]): u. as emet. in dys., swollen lymphatic glands; hemostat; antipyr; sd. c. FO similar to cotton sd. oil (19% linoleic acid); wd. of tree (50 m. high; 4 m. girth) u. to make dugout canoes, coffins, &c.; wd. pulp for paper mf.; called semal or cottonwood - **B. costatum** Pellegr. & Vuill. (tropAf): silk cotton tree; trunk & rt. bk. diur., emoll.; u. in GC diarrh., emmen.; wd. used - **B. ellipticum** HBK = *Pseudobombax e.* Dug.

"bombers" (US slang): strongest type of marijuana cigarettes with 95% cannabis & 5% tobacco (56).

bombilla (Arg): metal tube u. to suck the beverage maté.

Bombyx: 1) cotton 2) silk worms genus (Bombycidae-Lepidoptera): *B. mori* L. is the comml. silk worm or mulberry s. w.; chrysalis ("silk pod") u. as styptic, ton. astr. in leukorrh., chronic diarrh.

Bonafousia sp. (Apocyn.) (SA): morcrerin; lvs. rubbed on wounds esp. when between toes; removes itching, irritation. (—> *Tabernaemontann*)

bonbon: candy with center of frt., nut, &c. coated with chocolate or fondant (melted sugary paste) (fr. Fr, "good").

bonduc (bonduk) seed (or **nut**): *Caesalpinia bonduc.*

bonducin: earlier recognized glycoside (alk?) in *Caesalpinia bonduc* and *C. crista*; u. as quinine substit. in mal. ("poor man's quinine").

bone: Os (L., USP 1830-1870): hard skeletal tissues u. as transplants & grafts in surgery (Surgibone); bones of vertebrates thoroughly dr. & u. in prepn. bone-meal (ground up) & b. black (heated) - **b. ash:** crude calcium phosphate - **b. black**: purified animal charcoal prep. fr. mammalian bones & purified with dil. HC1 - **b. char**: animal charcoal; prepd. by heating bones with limited access of air, boiling with HCl, washing, &c.; u. in decolorizing & for adsorption; activated charcoal now usually preferred - **b. distillate**: b. fat: b. oil: Dippel's oil: animal oil: (Oleum Animali Dippelii): black liquid mixt. obtained by dry distn. of bones & other animal subst.; c. pyridine bases, picolines, lutidines, ammonium salts; u. to denature alc. (imp. fr. Arg) **- b. dry**: expression sometimes u. in drug trade to describe appropriate desiccation of pl. materials - **fatty b. oil**: clear liquid prepd. by pressure fr. bones; u. as watchmaker's lubricant - **b. marrow**: soft tissue in medullary canals of long bones; said to give protection fr. radioactive emanations - **b. marrow, red**: **Medulla (Ossium)** Rubra (*qv.*): prepd. fr. bones of calves; u. to aid in blood formation in microcytic anemia, debility, osteoarthritis - **b. meal**: ground whole dr. vertebrate bones; u. med. as source of Ca & phosphate, administered in tabs. or as powd.; u. also as fertilizer (generally as steamed b. m.) - **b. oil**: b. fat.

boneset (herb): *Eupatorium* spp. esp. *E. perfoliatum* (SC decoc. thought to alleviate aching bones) - **common b.**: *E. perfoliatum* - **purple b.**: *Eupatorium purpureum* - **spotted b.**: *E. maculatum.*

bonetero (Sp): *Euonymus europaeus.*

bonga (trees) (PI): bunga; *Areca catechu*, betel palm.

Bongardia chrysogenum (L.) Griseb. (B. rauwolfii C.A. Mey.) (monotypic) (Leontic./Berberid.) (Pak to Greece): u. as cure for sore eyes in horses (Baluch.); lvs. & rhiz. edible.

boniato (Cu): sweet potato.

bonina: (Por): *Mirabilis jalapa.*

bonnets (sUS): *Nuphar lutea* & other *N.* spp.

Bontia daphnoides L. (Myopor) (monotypic) (CA, WI, nSA, Guianas; Ve): buttonwood; kidney (or olive) bush; lvs. emollient, insect repellant; inf. u. for fish poison in Virgin Is (10); lf. decoc. for colds, coughs, asthma (Cur.); sometimes boiled with coffee beans in making coffee; lf. decoc. in hypertension, kidney dis. (Trin); to soak swollen feet; as bath to remove unpleasant odors (u. by aloe cutters); resin fr. trunk for parasitic skin dis.; cult. ornam.

boog(en)hout (Dutch): *Maclura pomifera.*

boojum (tree) (US slang): *Idria columnaris.*

bookoo: buchu.

boom (Neth): tree.

boon docks (PI): bamboo thickets; hence, common use (US) as "backwoods."

Boophane disticha Herb. (Amaryllid.) (s&cAf): candelabra flower; Kaffir onion; bulb scales appl. locally for urticaria, burns, boils; dressings for sores, whitlows, septic cuts (Europeans in Af); in rheum. pains, abrasions, minor traumata; fresh lf. appl. as styptic; dr. lf. moistened with milk to treat skin dis., varicose ulcers; component in poison arrows; bulb in mal., headache; weakness, eye conditions; bulb decoc. as enema —> suicide; c. buphanamine; lycorine; haemanthine; lf. generally considered non toxic.

Boquila trifoliata (DC) Dcne. (monotypic) (Lardizabal.) (Ch): pil-pil voqui; appl. to eye infections.

boracic acid: boric acid.

borage (common): *Borago officinalis* - **b. family**: Boraginaceae.

Boraginaceae: Borage Fam. (mostly Euras.) hbs, trees, or shrubs; rough, hairy, lvs. sl. bitter; with simple, alternate, exstipulate, & pentamerous reg. or irreg. fls., usually of a brillant characteristic blue color; ex. comfrey; forget-me-not ca. 2500 spp.

Borago officinalis L. (Boragin.) (AsMin; cult. Eu, nAf, NA): borage; hairy ann. with blue fls.; fl. hb. & lvs. u. exp. aperitive, emoll. pectoral, dem. flavor; lvs. eaten sometimes as salad & to flavor beverages; sds. u. as bird sd.; diaph. diur.; hb. rich in vitamins, esp. C; mucil; u. astr. anti-inflamm.; kitchen spice; fls. & shoots u. in teas & syrups; in claret. (negus); hydrolate (Fr Phar.); fls. u. to color vinegar (cultd. Ge, &c.); useful bee plant; fls. stim. (*cf.* amphetamines) (Duke); claimed to c. prostaglandin-like subst.; inactivates serum gonadotropin.

borajo (Guat): borraja: hb. u. in fevers.

Borassus L. (Palm) (As, Af): the fan palms; with c. 10 recognized spp. - **B. flabellifer** L. (B. flabelliformis Murr.) (Ind to PapNG, tropAf): Borassus palm tree; lontar p.; palmyra p.; African fan p.; black rhun p.; sap a pleasant beverage, source of palm sugar & p. wine (toddy); palmyra fiber fr. lf. stalk u. for rope, twine, &c.; inflorescence petioles decoc. u. diur. anthelm.; juice of spadices stim. in toothache, dys.; young seedlings eaten, source of good flour; frt. juice as beverage; lf. u. as writing paper fr. ancients times; pl. produces palmyra gum; trunk u. as lumber; for well sweeps, &c.; all parts of tree u. for some purpose or other (1110) (var. *aethiopicum* Warb. [B. a. Mart]).

borax: sodium borate.

Borbonia cordata L. (Leg) (sAf): old remedy for respir. congest., bronchitis, (whooping) cough; in erysipelas (Cape of Good Hope area); in bush teas (stekeltee) - **B. parviflora** Lam. (sAf): in bush teas (skagaltee) for asthma, hydrothorax; decoc. as diur. - **B. undulata** Thunb. (sAf): astr.; u. in bush teas. (—> *Aspalathus*)

bordalesa (Arg): large cask or barrel.

Boretsch (Ge): borage.

boric acid: boracic a.; H_3BO_3; occ. in nature as sassolite (mineral); obt. fr. Italy (Tuscany) (lagoons); California, &c.; colorless crystals or white powder; u. as antisep. antibact. antifungal in powders, aq. solns. for athlete's foot, also fungus inf. of ear; wounds & sores; ointment for shingles; not for int. use (toxic by ingest.); insecticide for roaches, black carpet beetles, &c.; many indy. tech. uses.

bork: Borke (Ge, Eng): outer dead corky layers of bark which slough off when the phellogen forms successively deeper & deeper in the outer layers of the stem-root axis; also called rhytidome, scale bark, &c. (ex. seen in birch bark).

Borneo tallow (or oil): fr. *Shorea aptera*.. & *S. palembanica*; also prod. fr. *Isoptera borneensis* (Dipterocarp.) (Indones); *Madhuca sebifera*; &c.

borneol: bornyl alcohol: an aromatic alc. found in many pls., incl. *Dryobalanops aromatica* of Borneo (hence the name); also prep. by synthesis.

bornesite: mono-methyl inositol; compd. occurring in Borneo caoutchouc (fr. *Urechites suberecta*); & in Madagascar caoutchouc (fr. *Landolphia madagascariensis*).

Bornträger's test: ammonia test for aloes & for senna (anthraquinone deriv.).

bornyl acetate: ester of borneol & acetic acid; found in many VO, incl. rosemary oil - **b. alcohol**: borneol - **b. chloride**: pinene HCl; "turpentine camphor"; the earliest artificial camphor; u. antisep. (formerly) - **b. isovalerate**: **b. isovalerianate**: occurs in VO valerian; u. CNS depressant like valerian; or menthyl valerate.

Borojoa patinoi Cuatr. (Rubi) (Co): borojoa; frt. pulp u. to make iced frt. drinks; this drink esteemed to be aphrod. (Chocó) (2218).

Boronia Sm. (Rut) (Austral): spp. are source of essential oils u. in perfumes - **B. alata** Sm.: c. 5-6% rutin; u. as source - **B. megastigma** Nees ex. Bartling (wAustral): source of *boronia oil* (VO) u. in perfumes fr. fls. (concrete otto); resembling violet & cassie; rich in rutin; cult. Victoria (Cana) - **B. rhomboidea** Hook.: anthelm.; chopped up with fodder & given to horses for worms - **B. serrulata** Sm.: "native rose"; lvs. rich in rutin.

borra de vino (Sp): wine lees.

borracha (Br): 1) *Hevea brasiliensis*, &c. 2) borage - **b. bravo** (Por): **borragem b.**: *Heliotropicum indicum*.

borrachero (SpAm): *Datura arborea*.

Borrago auct. = *Borago*.

borraja: borrage: (Sp): borage, *Borago officinalis*.

borrajón (Mex): *Heliotropium* spp., such as *H. indicum*.

Borreria G. Mey. (Rubi) (worldwide, trop & subtr.): buttonweeds; many taxa are now distributed in *Spermacoce* - **B. articularis** F. N. Williams (B. hispida Schum.) (Rubi) (tropAs+): bk. u. as febrifuge (Taiwan): emeto-cath. (Indochin); appl. as poultice for headache, wounds (Mal); in eye infections, lf. inf. for gallstones & kidney stones (Indones); lf. & bk. decoc. u. as emmen. (PI); lf. decoc. astr. for hemorrhoids (PI); sold in Chinese herbal shops - **B. capitata** DC (nVe, Pe, Br): poaya (do campo); poaia da praia; poya; rt. c. emetine (traces); u. as substit. for ipecac (unofficial i. or poaja); u. for constip. in infants (Ve); emet. - **B. centranthoides** Cham. & Schlecht. = *Spermacoce c.* - **B. compacta** K. Schum. (sAf): lf. paste u. as rubefac. skin counter-irritant much like mustard plaster; in skin rashes - **B. emetica** Mart. (Br): u. like ipecac; poaya da hasta comprida - **B. laevis** Griseb. (sFla, sLa [US], Mex, WI, CA, SA to Pe): button bush; b. weed; pl. decoc. in colds (Jam, Domin); emmen. in delayed menstr; crushed pl. appl. to cuts, burns; pl. decoc. u. to bathe skin, as in sunstroke - **B. latifolia** (Aublet) Schum. (Mex): u. in kidney dis. (Mex); in snakebite (Belize) - **B. natalensis** K. Schum. (sAf): rt. inf. u. as antiemet. (Zulus); in stomachache (Swaziland); rt. u. in enema for "gangrenous rectitis" - **B. ocimoides** DC (Spermacoce o. Burm. f.) (sFla [US], WI, tropAm, pantrop): "water breaker"; ponwatia (Ghana); juana la blanca (Domin Rep); pl. decoc. soporific, in chills, flu, rheum. (Guat); to relieve bladder dis. (Br), diur., hence common name (Virgin Is); crushed lvs. appl. to wounds (Indones) - **B. setidens** Ridl. (Mal): hb. u. as poultice for headache - **B. verticillata** G.F.W. Mey. (Spermacoce v. L.) (sMex, WI, CA to Par): wild scabious; pl. decoc. appl. to ulcers & itching skin dis., int. as purg. & in kidney dis. (Mex); antipyr., antidiabetic, emmen. (Trin); emet., leprosy (Af).

Borrichia arborescens (L) DC (Comp.) (WI): "samphire"; lf. inf. u. in poisoning fr. certain kinds of fish; inf. for colds, pertussis (Bah) - **B. frutescens** DC (seUS, Berm, Bah, WI, Mex): sea-ox-eye (daisy); lf. decoc. u. for asthma, chest dis., mal. (SC) (2451); lvs. raw in salads, &c. (WI, SA).

borsch(t) (Russ): borsht: soup or ragout made of sour cream &c., & colored with red beet juice (fresh or fermented).

borsello del pastore (It): shepherd's purse.

borsos menta (Hung): peppermint.

borussic: 1) Prussian 2) hydrocyanic.

Bos taurus L. (Bovidae - Ungulata - Mammalia) (orig. Euras, now worldwide): domestic cattle; the source of many important prods. such as meat (beef, veal), beef ext., liver, bone marrow & other bone prods., spinal cord, ox gall, neatsfoot oil, cattle ear hair, gall stones (very valuable as aphrodisiac - As), peptone, tallow, stearic & oleic acids, dr. blood and blood serums, hides (leather), hide glue, gelatin, endocrine glands, & endocrines (ex. insulin), products u. in culture media viz., beef, veal, beef heart, b. liver, calf's brain, blood, (dr. b.), serum, peptone, ascitic fluid, creatinine (ref. standard); gall stones u. aphrod. (Jap med.); cow arteries u. in reconstructive surgery; in Unani med. they are u. in blood, brains & spinal cord; bile, &c.; the cow gives milk fr. which come lactose, casein, cheese, butter; the Asiatic Indians u. these, also cow feces & urine in their med (*cf. Bos indicus* L. [As.]): Asiatic domestic cattle, (such as zebu, Brahmin): dung u. as fuel (2515), fertilizer, in building; animal u. for transport & traction (bull, cow, calf, bullock, ox, heifer).

Boschniakia rossica (Cham. & Schlechtd.) Fedtsch. & Flerov. (Orobanch.) (Jap, eSiber, nwNA): aq. ext. of this rt. parasite u. locally for wounds & bruises (Ainu).

Boscia albitrunca Gilg & Benedict (Capparid.) (Af): caper bush; shepherd's tree; rt. u. as coffee substit.; youngs buds u. as substit. for capers; lf. inf. u. for inflamed eyes; frt. edible; rt. decoc. u. in hemorrhoids; unripe frt. in epilepsy - **B. angustifolia** A. Rich. (Senegal, Sudan, Af): nos (Wolof): lvs. u. cholagog, neuralgia, headache, kidney dis., stiffness of nape of neck - **B. foetida** Schinz (Af): old woman's bush; pl. decoc. emmen. & in back pain (Hottentots); toxic to sheep - **B. rehmanniana** Pest. (Af): wonderboom (Dutch); rt. u. as chicory substit.; wd. u. to make spoons - **B. salicifolia** Oliv. (Af): starchy rt. starvation food; lf. u. to treat "chiufa"; bk. u. as soap - **B. senegalensis** Lam. (Senegal, Af): rts. bk. & lvs. u. in colic (without purging); lf. decoc. in ocular pruritis, in bilharziasis, tranquilizer in mental dis. (Af): starchy rt. starvation food; lf. u. to treat "chiufa"; bk. u. as soap - **bosque** (Por): wood (land).

Bosqueia angolensis Fic. (Capparid./Mor.) (trop wAf): okure (Ghana) lf. bk. twigs, decoc. u. diarrh.; latex serves as indelible ink; wd. practically useful. (—> *Trilepisium*)

Boston fern: *Nephrolepis exaltata* Birdw. v. *bostoniensis.*

Boswellia frereana: Birdw. (Burles.) (neAf, Somalia, Eg, seArabia [Oman]): East African elemi is prod. fr. this sp.; possibly also other types of incense; ex. luban mayeti; u. masticatory, incense - **B. neglecta** S. Moore (nSomaliland): produces an inferior sort of Olibanum - **B. papyrifera** Hochst. (Eth): source of a kind of frankincense - **B. sacra** Flueckiger-Dupéron (B. bhai-dajana Birdw.; B, carteri Birdw.) (Eg. nSomalia, seArabia [Oman]); frankincense tree; source since antiquity of (true) Frankincense (Olibanum), oleogum-resinous exudate of tree (bark shaved to stim. flow); prod. somewhat like myrrh; greatly esteemed in ancient times (llll) and still u. as altar incense; smoking frankincense may —> hashish-like effects; fumigation; (in combns.); in embalming of corpses (ex. King Tutankhamen); resin ("luban") also u. as pectoral (in chronic bronchitis, &c); exter. vulner. antisep. stim.; partly replaced by cheaper synth. prods. (1946-), packed in cases covered with burlap & bound with fiber rope - **B. serrata** Roxb. ex. Colebr. (Ind): Salai (tree); trunk wounds produce an oleogum-resin "Salai gugul" or Indian frankincense, (c. 9-12% VO [*cf.* turpentine oil], 57% resin [*cf.* rosin], 23% gum) u. fumigant (incense), in rheum. pain (indigest.) in ointments for sores & tumors; oleoresin u. like Cana balsam (microslides); resin u. in mfr. varnishes, lacquers, perfumes, adhesives, &c.; lf. juice appl. to cracks in feet; given in skin dis. (124); claimed hallucinogenic; wd. u. for chests, charcoal - **B. socotrana** Balf. f. (Af): source of a frankincense-like oleogum-resin; of minor economic importance.

Botany Bay kino: Eucalyptus gum.

bote (Sp): druggist jar, can or pot for medicines; hence botica (Sp) (drug shop) & boticario (Sp) (pharmacist).

Bothriochloa ischaemum (ischaemon) (L) Keng (Andropogon i. L.) (Gram) (Euras, Af, natd. US):

woolly andropogon; zacatow; the peeled root is used; cult. as pasture grass (986) - **B. odorata** A. Camus (Andropogon o. Lisboa) (eInd): hb. u. as food ("*via-digavat*"); VO smells like pine needle oil.

Bothrops atrox (L.) (Crotalidae: Ophidia) (SA): fer-de-lance; highly venomous snake; u. to make antivenin; in hypertension - **B. jararaca** (Wied.) (SA moist places): jarara (Arg); jararaca dormideira (Braz.): SA pit viper; venom releases bradykinin fr. plasma globulins, converts Angiotensin I to Angiotensin II; inhibits Angiotensin converting enzyme (ACE), producing widening of gall bladder passages; proposed for use in thromboses - **B. insularis** (Amaral) (SA): said to have the most toxic venom of all known (smallest MLD).

botica (Sp): 1) apothecary shop 2) potion given sick person.

boticario (Sp): pharmacist.

botogenin: steroidal sapogenin fr. *Dioscorea composita*.

bo-tree: *Ficus religiosa* (Buddha).

Botrychium ternatum Sw (Ophiogloss.-Pterid.) (eAs): hb. u. in stomachache & dys. (Taiwan); fronds eaten as veget. (Jap) - **B. virginianum** Sw. (worldwide): rattlesnake fern; pubic weed; Virginia grape fern; fleshy root applied to cuts & bruises (Am Indian); u. emet., diaph., expect. (seUS Indians).

Botrytis cinerea Pers. (Moniliaceae-Hyphomycetes-Fungi): (worldwide) common grey mold; some Eu grapes infected with this organism furnish wine with special bouquet (Sauterne type).

botterdel: old (XVII C.) popular beverage in Eng made of sugar, cinnamon, butter & beer brewed without hops.

bottlebrush: *Melaleuca leucodendron*; *Calestemon* spp.

bottle tree: barrel tree, *Brachychiton* spp. - **Australian b. t: 1)** *Sterculia rupestris* (9) 2) *Brachychiton rupestris* 3) *Adansonia digitata* ("b. baobab").

botulinum (or botulinus) toxin: botulism toxin type A: Botox: exotoxin prod. by *Clostridium botulinum* Bergey & al. (Bacillus botulinus Ermengen): harvested fr. lab cultures; responsible for systemic effects of botulism in man; useful in essential blepharospasm (very small amts. injected into eyelid [1985-]); in strabismus. in facial hemispasm; in spasms of voluntary muscles of vocal cords, arms, &c.; to remove wrinkles (very small doses; lasts for several months).

botulism: poisoning by an exotoxin fr. *Clostridium botulinum* Møller & Scheibel. (soil bacterium) wh. acts to produce botulism (paralysis of voluntary muscles, dysphagia, &c., often fatal); treated with **b. antitoxin** (b. antiserum) (prepd. fr. blood of immunized horses) given parenterally in large doses; also **b. immune globulin** now approved; u. in blepharospasm.

Boucerosia W. & A.: *Caralluma*.

Bouchea fluminensis (Vell.) Mold. (Verben.) (Br): gervão de folha grande; g. falso; hb. u. as antiemet.; stim. to digest. tract.

bouchon (Fr-m.): cork, stopper.

Bougainvillea glabra Chois. (Nyctagin.) (Br): rambling liana; ornam. u. med. (Mex) - **B. spectabilis** Willd. (seAs): tuber decoc. u. to give strength (Thail) (2448).

bougie (Fr, candle): smooth slender flexible cylinder u. to dilate or open urethra, rectum, esophagus, in stricture, etc. (SC, Ar. Bijiyah, town of Algeria fr. which certain wax candles were exp. to Eu) - **b. à boule:** bulbous b.

bouiller (Fr): to boil.

bouillon (Fr): beef soup or broth; u. as bact. culture medium - **b. blanc** (Fr-m.): mullen.

Bouin's fluid: hardening or fixative reagent in microtechnique; sat. soln. picric acid with HCHO & glac. acetic acid.

bouleau (Fr-m.): birch.

bouncing Bet (leaves; root): *Saponaria officinalis*.

bouquet garni: mixt. of hbs. u. in flavoring; ex. parsley, celery sd., thyme, sage, bay leaf, cloves, black pepper.

Bourbon whiskey: corn w. usually made with other grains admixed (SC first mfrd. in B. County, Kentucky [US]) - **B. tea**: *Angraecum fragrans*.

bourdaine (Fr): frangula.

bourgeons des pins (Fr): pine buds.

bourrache (Fr-f.): borage.

Bourreria ovata Miers. (Boragin./Ehreti) (WI): "strong back" (Bah); lf. u. for colds, flu; popular in combns. (Bah) - **B. pulchra** Millsp. (Mex): scraped bk. appl. to wounds as antis. (Mayas) (6) - **B. spathulata** Hemsl. (neMex): zapotillo; pl. u. for cough.

bourse à pasteur (Fr): shepherd's purse.

Boussingaultia Kunth = *Anredera* Juss.

Bouteloua hirsuta Lag. (Gram) (cNA, Mex): hairy grama; forage grass (1112).

bouton d'argent (Fr): life everlasting.

bouture (Fr-f.): slip, cutting (appl. to young plants).

Bouvardia erecta Standl. (Rubi.) (Mex, wTex): "hierba de San Juan"; u. med. - **B. longiflora** HBK (Houstonia l. A. Gray) (Mex): rosa (or flor) de San Juan; fls. very fragrant u. to scent ointments, etc. - **B. ternifolia** (Cav.) Schlecht. (wTex [US], Mex): trompetila; mirto (Mex); scarlet bouvardia; trumpet flower; often cult. as orn.; rt. cardiac stim. in dys. heat exhaustion, etc.; styptic; stems & branches u. by natives for rabies(!); claimed of use in infantile paralysis; boiled with cinnamon and u. for heart (Mex).

Bovidae: Cattle Fam.; made up of ruminants which bear hollow unbranched horns & have complex stomachs; ex., sheep, bison, buffalo, goat.

bovina (Sp): cow.

bovine growth hormone: BGH = bovine somatotropin; prepd. by genetic engineering (1989-); being u. to incr. milk prodn. in cows.

Bovista plumbea Pers. (Lycoperdaceae-Gasteromycetes-Basidiomycotina) (worldwide): a puffball; spores u. styptic; in folk med. (Perm, USSR).

Bovril™: a solid beef ext. or concentrate rich in amino acids & vitamins; at one time very highly rated for invalid food (fr. bos, bovis [L., ox.]).

Bowdichia virgilioides HBK (Leg.) (SA): alcornoco; alcornoque; cork tree; rt. bk. a comml article (not cork) called Cortex alcornoco hispanicus, cork (tree) bark; tradesmen distinguished male and female cork; bk. u. antipyr.; wd. very heavy, u. wheel hubs.

bowels of Christ (US Negro): *Salvia* spp. (appar.).

Bowiea kilimandscharica Mildbr. (Lili.) (eAf): u. like the foll. - **B. volubilis** Harv. ex Hook. f. (s&seAf): umagagana; c. large amts. of cardiovascular glycosides, incl. bovoside A; corm u. purg. (natives), in edema, female sterility; at times fatal; decoc. u. for inflammed eyes; occ. cult. as ornam. hb.

Bowlesia flabilis Macbride (Umb./Hydrocotyl.) (Ec to nArg, Mex): opuisoro (Pe): u. for pulmonary dis., styptic.

bowman root: Bowman's root: 1) *Veronicastrum virginicum* 2) *Gillenia trifoliata* (danger of confusion with (1) 3) *Apocynum cannabinum.*

bowstring fiber: b. hemp: *Sansevieria* sp.

box (bark): *Buxux suffruticosa, B. sempervirens* - **black b.**: one of several *Eucalyptus* spp. (ex. *E. bicolor)* (SC dark foliage) - **common b.:** *Buxux sempervirens* - **b. holly:** v. holly - **mountain b.**: *Arctostaphylos uva-ursi* - **b. myrtle:** v. myrtle.

boxberry: *Gaultheria procumbens.*

boxing: old fashioned method of turpentining, in which cavity is cut into tree trunk; into this the oleoresin runs & is collected: has been replaced by the superior modern cup & gutter system.

boxtree: (bark): *Buxus sempervirens.*

boxwood (bark, flowers): 1) (eUS): *Cornus florida* 2) *Buxus sempervirens* 3) *Amelanchier canadensis* 4) *Vitex umbrosa* - **false b.** (NCar [US]): *Cornus florida* - **South American b.**: *Vitex umbrosa* - **West Indies b.:** *Casearia praecox.*

boysenberries: fr. *Rubus ursinus v. loganobaccus*; closely allied to loganberry.

brab tree: *Borassus flabellifer.* (sds.).

bracelet wood: *Jacquinia barbasco.*

Brachiaria mutica Stapf (Gram.) (widespread): u. med. in Fiji.

Brachychiton (Sterculia) acerifolium F. Muell. (Sterculi.) (NSW, Austral): flame tree; ornam. - **B. populneum** R. Br. (Austral): kurrajong; cult. for fodder; sds. toxic; ornam. tree (-20 m.) - **B. (Sterculia) rupestris** K. Schum. (Austral): bottle tree; trunk swollen with remarkable similarity in shape to soda water bottle; refreshing sweet mucilaginous drink prepd. from trunk (holes bored); fiber u. to make nets.

Brachyclados stuckerti Spegazzini (Comp.) (Arg): hb. & rt. (falsely called "punaria tea") u. as fumitory in asthma (like Stramonium); int. in rheum.

Brachyglottis repanda J.R. & G. Forst. (Comp.) (NZ): puka-puka; c. senecionine, senkirkine; fresh hb, & fls, u. in homeop. med.; cult. orn.

Brachylaena discolor DC (Comp.) (tropAf): Natal silver leaf; u. ton. in diabetes, renal dis.; lf. for roundworms - **B. elliptica** Less. (trop & sAf): firesticks, bitter blaar; inf. of decorticated rt. u. emet, side pains; lf. c. tannins; decoc. topically for sore throat; mouth ulcerations; diphtheria, in diabetes (considerable reputation among white settlers), pneumonia - **B. hutchinsii** Hutch. (Af): muhugu tree; muhuhu; mububu; wd. distd. to yield VO similar to cedarwood oil.

Brachystegia spiciformsis Benth. (Leg) (c tropAf): bk. dec. u. as eyewash in conjunctivitis (Tanz); lvs. u. as disinfect. for wounds (var. *latifoliata*); bk. u. for making cloth, sacking (Moz).

bracken (fern): brake(n).

Brackenridgea zanguebarica Oliv. (Ochn.) (Tanz): bk. boiled & eaten in porridge for jaundice.

bract: reduced lf. esp. one which subtends a flower - **b. scale**: *Atriplex bracteosa.*

bradykinin: a plasma kinin (endogenous peptides) (SC brady, slows action): vasodilator; hypotensive; mediates pain & inflamm.; consists of chain of 9 amino acids (nonapeptide) liberated by action of cryst. trypsin or some snake venoms (ex. *Bothrops jararaca* Wied., jarara) on plasma globulins (1114); indications: colds, burns, allergies, &c.

Bragantia wallichii R. Br. (Apama siliquosa L.) (Aristolochi.) (Indo-Mal): alpam root bragantia; rts. u. in diarrh. dys. cholera; pl. in ointment for carbuncles, ulcers. (—> *Thottea*)

brai (Fr): resin, rosin.

braid root: *Rhinacanthus communis.*

brake(n): common b.: 1) *Pteridium aquilinum* or other *P.* spp. 2) dense thicket of undergrowth, esp. shrubs (ex. cane b. [seUS]) - **backache b.**: *Athyrium filix-femina* - **b. fern**: 1) *Pteridium* & other related fern genera 2) *Aspidium* (misnomer) - **brakeroot**: *P. lypodium vulgare* - **stone b.**: *Polypodium vulgare.*

bramble: *Rubus fruticosus+* - **b. brier** (Ala [US]): any woody vine or shrub with strong spines.

bran: Furfur (L): the coarse outer hulls (pericarps) fr. various cereal grains obtained during milling; ex. wheat b. (rice, maize); u. dietary, lax. for fiber value in preventing GIT dis., such as diverticulosis, cholelithiasis, hemorrhoids, etc. (1115); u. as such in cereals or in tablets (GB) - **diabetic wheat b.**: b. washed free fr. starch - **oat b.**: this and rice b. u. for hypocholesteremic effect - **b. oil**: furfural - **wheat b.**: b. washed free fr. starch.

brandy: Spiritus Vini Vitis: the distilled liquor made fr. wine by steam distn.; origins: grape (chief), peaches, cherries, apricots, apples &c.

Brasenia schreberi J. F. Gmel. (1) (Nymphae./Canbomb.) (worldwide): watershield; lvs. astr. u. dys. phthisis (NA); young lvs. eaten (Jap)

brasil: brasilleto; brazil; brazil wood: red wd. of various *Caesalpinia* spp. esp. *C. brasiliensis* (**b. tree**); (u. for cabinet work); Brasil (Brazil) the nation was named after these plants.

Brasil(ian): Brazil(ian) - **brasilin:** brazilin. - **B. waxes** produced fr. taxa of *Copernicia, Euterpe, Syagrus.*

Brassica L. (Cruc.) (nTemp Zone): large genus of ann. bienn. & perenn. hbs.; lvs. of following spp. u. as food: *B. chinensis; juncea* (esp. var. *crispifolia* Bailey); *nigra*; *oleracea*; *pekinensis*; *rapa;* also *Sinapis alba* - **B. alba** Rab. = *Sinapis a.* - **B. arvensis** Rab. = *Sinapis a.* - **B. besseriana** Andrz. (USSR; tropAs): "Russian mustard"; ann. hb.; sds. u. to prep. Serepta mustard; a brownish product; cultd. US (named after region of USSR) - **B. campestris** L. (Euras): field mustard; weed in US; cult. abroad for oil-rich sd. - **B. cernua** Forbes & Hemsl. (seAs): much like *B. nigra*; sds. made into ointment & u. for neuralgia, arthritis, pneumonia; appetizing, lvs. u. diaphor (Indochin) - **B. c.** var. *oleifera* DC; sds. one source of colza oil; u. for cooking (Ind, Pak) - **B. chinensis** L. (eAs): Chinese cabbage; colza; pak choi (Chin); pakatkào, pakaktao (Siam); white lettuce, garden crop (Thail+); pot herb, also pickled; antiscorb., antiarthritic. (Mal) - **B. fimbriata** DC (Euras.): Scotch kale; u. food, ornam - **B. hirta** Moench = *Sinapis alba* - **B. juncea** (L.) Czern.& Coss. (As): leaf mustard; Chinese m.; India(n) m.; fls. bright yellow; source of "mustard greens"; u. like spinach (var. *foliosa* L. H. Bailey); also southern curled mustard (var. *crispifolia* Baill.) (B. j. var *japonica* Hort.); black mustard sd. u. flavor, condim.; yields FO (mustard oil) - **B. napobrassica** Mill. rutabaga; yellow turnip; cult. for turnip-like tuberous rts; u. food; c. anti-thyroid factor (goitrin); superior for human food since sweeter and keeps better (this species sometimes equated to or made var. of *B. napus* L.) (Swedish turnip = "Swedes") - **B. napus** L. (var. *napus*) (cult. worldwide): (winter) rape; oil seed rape; colza; u. forage (rt. as stock food); fls. u. to improve tea; sd. source of rape or colza oil (FO); u. to make green soup; in cooking by poor people of India (2518) - **B. n.** var. *glauca* (Roxb.) O. F. Schulz (B. g. Roxb.) (Ind): Indian rape; has been source of FO and VO from sd. - **B. nigra** (L) Koch (Euras): black (or brown) mustard; weed; also cult. for sds. which represent chief source of table mustard; u. condim. flavor; source of VO & FO; FO (prod. Fra [Alsace, Flanders)] Eng) sds. c. sinigrin (—> allyl isothiocyanate + $KHSO_4$ [=volatile mustard oil]), sinapine, myrosin (enzyme); u. arthritis (tsp. whole seed) to reduce pain (1892); mustard foot bath popular in folk med. for treating colds - **B. oleracea** L. (Euras; now worldwide); cabbages (collectively) - **B. o.** var. *acephala***:** collards; (common) kale: boreocole; lvs. produced on long stalk (-22 cm.+) with lvs. loosely & openly arranged; not apparently affected by frost; some cultivars with colored lvs. are cult. as ornam. (hardy); highly nutritious; mostly u. as pot herb (1117); ext. of red cabbage u. as dye ("caulene"); common near Negro farm houses (seUS); said useful for opposing age-related macular degeneration; B. oleoracea var *acephala* subvar. *laciniata* L.: curly kale; c. greens; u. very popular vegetable as greens; to prevent pellagra - **B. o.** var. *botrytis* L (wEu): cauliflower (with large fl. heads) & broccoli (with small fl. heads) (2519); characteristic structures representing monstrous growths of stalks & undeveloped fls. & bracts; table delicacy; very nutritious (vit. & minerals): c. anti-cancer substance - **B. o.** var. *capitata* L.: cabbage: var. with very large lvs. u. as popular tasty pot herb; also widely u. folk med. as antidiabetic; raw juice or concentrate u. in peptic ulcers (anti-ulcer factor = healing); also in liver & pancreas disturbances; sds. c. antibiotic substance; claimed prophylactic to cancer (2400); lf. bruised & appl. as poultice for "erysipelas" (Ala [US] folk med.); eaten raw and u. in prepn. sauerkraut (1116) (in hypoglycemia) (1118) & cole slaw; crushed lvs. u. to treat bruises, &c. (in antiquity) (Cato, 147 BC) (sprouts = buds) - **B. o.** var. *gemmifera* Zenker:

Brussels sprouts; with many small edible heads: stem bears small lateral branches or "sprouts" (buds) of great delicacy as pot herb (SC long grown near Brussels) - **B. o.** var. *gongylodes* L. (B. caulorapa Pasq.): kohl-rabi; tuber above ground esculent vegetable - **B. o.** var. *sabauda***:** savoy (cabbage); white s. cabbage; with finely crinkled lvs.; u. as potherb - **B. pekinensis** (Lour.) Rupr. (Chin): Chinese (or celery) cabbage; Peking cabbage; pe-tsai; forms elongated heads u. for greens, salads; depur. (Chin); cult. nUS (1119); - **B. rapa** L**:** turnip; enlarged tuber which lies above the slender tap-root is eaten raw & boiled as table vegetable; lvs. (turnip greens) popular in seUS; pl has antithyroid activity; sd. u. condim. (Ind); a source of colza oil - subsp**. sarson** (Prain) Denford: Indian colza: brown sarson: var. *silvestris* Briggs; colza; turnip rape; var. *trilocularis* (Roxb.) Kitam. (Pak): yellow sarson - subsp. **oleifera** DC (B. campestris L.): entire pl. eaten.

brassica sterol: tetracycle; occurs in *Brassica napus, B. rapa, Pongamia pinnata.*

Brassicaceae Barnett: Cruciferae.

brassinosteroids: BS: brassins: steroids first (1965) found in pollen of rape (*Brassica napus*); with plant growth-promoting activity; 29 known compds. now known to occur in many other plants (dicots, monocots, gymnosperms, pteridophytes, algae); also synth.; one of best known = brassinolide; most interest in insect control (antiecdysteroid) (Jap); to incr. crop yields; kind of pl. hormone.

Brauneria Neck. ex Britt.: *Echinacea* Moench.

Braunwurz (Ge): *Scrophularia* spp.

Brausepulver: (Ge): comb. of $NaHCO_3$, tartaric acid, & pd. sugar; Seidlitz powders.

Brayera (anthelmintica) (L.): *Hagenia abyssinica.*

Brazil (Brasil) nut: *Bertholletia excelsa* - **B. wax**: v. wax - **B. wood:** 1) (Br): *Caesalpinia echinata* & other *C.* spp. (pau Brasil): this product gave its name to the nation 2) *Biancaea sappan.*

brazilette: *Haematoxylon brasiletto* - **braziletto wood**: *Caesalpinia brasiliensis.*

Brazilian gum (copal tree): 1) Courbaril copal (hardened resin fr. *Hymenaea courbaril*) 2) angico gum: type u. in tanning - **B. g. anime:** Courbaril copal - **B. waxes:** produced fr. taxa of *Copernicia, Euterpe, Syagrus.*

Brazil(ian): Brasil(ian)**:** v. cacao.

brazilin: xall. compd. fr. Brazil wood or Sappan wood.

brea gum (PI): fr. *Canarium luzonicum, C. edule* (1427) - **b. trees** (Mex): turpentine pines; fr. brea, tar, possibly resin residue after distn. of turpentine gum; u. to make torches, soap, &c.

bread, Boston brown: c. rye, Graham flour, corn meal, molasses &c. - **b. root**: *Psoralea esculenta.*

breadfruit: *Artocarpus altilis* - **African b. f.**: *Treculia africana.*

breast tea: Species Pectorales; pectoral species; mixed hbs. u. as tea for coughs, colds, bronchitis & other chest dis.

Brechnuss (Ge): Nux Vomica.

Brechwurzel (Ge): ipecac.

Breitlippe (Ge): *Lagochilus.*

Brennkraut (Ge): *Acalypha indica.*

Brennessel: (Ge): stinging nettle (lit. burning thistle).

Brenzkatechin (Ge): pyrocatechol - **Brenzkatechincarbonsaeure** (Ge): proto-catechuic acid.

Brenztraubensaeure (Ge): pyruvic acid.

bretones (NM [US], Sp): Brussels sprouts.

bretonica morada (Mex): *Melochia tomentosa.*

Brevoortia Mitl. (Pisces): menhaden; pogy; alewife; basis of important fishing industry (in wt. caught in US); body FO (fish oil) (highly unsaturated FA) u. med. (for atherosclerosis); bait, fertilizer, fish meal (after expressing oil) (rich in vit. B_{12}); chief source Gulf of Mexico; billions of pounds harvested annually - **B. tyrranus** (Atlantic coastal waters): menhaden; moss-bunker; fish -35 cm.; found in large schools.

brewer's flakes: wheat, barley, or corn flakes u. by brewers to replace malt (to reduce costs) - **b. yeast**: yeast formed during the process of brewing in which nutriment is composed of cereal grain ext. with hops; u. to make b. dried yeast, y. ext., etc.

brewing: mfr. of beer by mixing malt with water to give mash; fermentation to give maltose (&c.) (wort); addn. of hops, boiling, cooling, adding yeast to ferment, producing alc. & CO_2.

brewster: *Magnolia glauca*

Breynia fruticosa (L) Hook. f. (Euphorbi) (Chin, seAS): decoc. of leafy branches u. for swellings, abscesses, suppurating sores (Chin); rt. with fermented rice u. as galact.; lf. decoc. antisep. for cuts, in toothache (Indochin) - **B. rhamnoides** Muell. Arg. (Mal+): pounded lf. as poultice for swellings, headache, mumps (Mal peninsula); lf. inf. for stomachache (PI).

brezo comun (Sp): *Calluna vulgaris.*

briar (*cf.* also brier): 1) *Rosa rubiginosa* 2) (Ala [US]): blackberry 3) (Ala [US]): any thorny bush - **bamboo b.**: *Smilax bona-nox* - **black b.: blackberry b. (root)**: 1) *Rubus* sect. *Eubatus* 2) *Randia aculeata* - **bull b. : China b.:** *Smilax rotundifolia* **-** **b. bush**: *Rosa arvensis*, etc. - **(common) green b.**: *Smilax rotundifolia, S. hispida*, & some other *S.* spp. -**b. hips**: *Rosa canina* frt. - **holly b.**:

Ruscus aculeatus - **b. pipes**: made fr. rt. wd. of *Erica arborea* (fr. bruyère, (Fr) heath[er]) - **b. root**: 1) (US); *Smilax china* & other spp. 2) (US): *Rhododendron* spp. 3) *Erica arborea* - **b. root** 1) (Jam) *Cissus sicyoides.* 2) (US) *Smilax china* - **b. root, African** - *Erica arborea* - **sand b.**: *Solanum carolinense* - **saw b.** (seUS): *Smilax* spp. - **sensitive b. (, pink)**: *Schrankia* spp. - **stone bruise b.** (seUS): *Smilax bona-nox* of large size - **sweet b.: wild b.**: *Rosa canina.*; *R. rubiginosa.*

brick (gum) (US slang): gum opium - **b. dust: pd. brick**: u. in the voodoo med. of sUS Negroes, also as detergent.

Brickellia californica A. Gray (Comp) (swUS, Mex): California brickellbush; pl. u. as "pachaba" by rubbing into scalp for headache (Am Indians); lf. inf. u. for cough, fever (Navaho Indians); sds. in flour meal (Gosuite Indians) **- B. cavanillesii** A. Gray (Mex): atanasia amarga; "immortelle"; prodigiosa; atanasia (extranjera); shrub; lvs. u. bitter ton. antiperiod, tenif. (with kousso), to aid gastric digestive secretions; antipyr. in diarrh., in cholera, gallbladder dis. (Mex Ph. IV-VI [1906-52]) **- B eupatorioides** (L) Shinners (Fla [US]): (prairie) false boneset; pl. u. as bitter ton., diaph. - **B. veronicaefolia** A. Gray (Mex): lvs. u. stim. nervine, antirheum., in stomachache, atonic dyspepsia (Mex Phar.).

bridal wreath: *Spiraea prunifolia*; *Franco ramosa.*

Bridelia atroviridis Muell.-Arg. (Euphorbi.) (Af): umurama; u. med. (Congo); in Burundi, bk. maccrated several days and ext. u. for cough - **B. cathartica** Bertol. f. (trop eAf): u. as purg. (Zambia); bk. decoc. for cough; lvs. & sts. boiled with beans as aphrod.; lvs. partly burned & eaten with porridge for upper stom. pains; rts. in stomachache with sesame oil (Tanz); when lvs. chewed, water tastes sweet - **B. micrantha** Baill. (tropAf): Natal mahogany; lf. sap appl. to sore eyes (conjunctivitis); rt. u. for severe epigastric pain, purg.; black sweet frt. edible - **B. retusa** Spreng. (Ind, Pak): Kessai; bk. with sesame oil, astr. (124); in rheum. lithiasis, infertility: sweet frt. edible - **B. scleroneura** Arg. (Tanz): lvs. u. to treat headache; rts. u. for sores - **B. scleroneuroides** Pax (Burundi, Af): umunemberi rt. inf. u. as rinse for mouth for gum sores (Tanz); for abdom. pain, indig. (Tanz); frt. edible; lvs. boiled in water u. as vermif. & for GC.

Bridgesia incisifolia Bert. (monotypic) (Sapind.) (Ch): rumpiata; inf. u. for flatulence; loc. astr. & vulnerary.

brier: *v.* also briar - **box b.** (Jam): *Randia aculeata* - **bull b.: China b.**: *Smilax bona-nox* - **(common) green b.**: *Smilax rotundi-folia* - **b. hips**: *Rosa canina* frt - **b. root** (Jam): 1) *Cissus sicyoides* 2) *Smilax china* - **sweet b.: wild b.**: *Rosa eglanteria.*

brierberry: *Rubus cuneifolius* - **brierwood**: rhiz. of *Smilax rotundifolia*; u. to make pipes.

Brigham Young weed: *Ephedra* spp. of wUS.

bright: term referred to a drug conspicuously clear, clean & attractive in appearance.

brighteyes (US): wild violet.

brimstone: crude sulfur as obt. fr. volcanic deposits; roll sulfur.

brine: a liquid prepn. with NaCl, vinegar, lactic acid, water, sugar &c.; u. to prepare and preserve pickles, sauerkraut, olives, fish &c.

brinjal(l) (Indo-Pak): egg plant.

brinton root: Leptandra.

brio de estudiante (Por): *Caesalpinia pulcherrima.*

brionia (Sp): *Bryonia dioica.*

brisling: small European fish rather like sardine; packed in FO for food.

British gum: dextrin - **B. industrial spirit** (BIS): alc. 19 parts mixed with wood naphtha 1 part - **B. oil**: formerly popular rubef. linim c. linseed oil, amber oil, juniper oil, petroleum, mineral tar, etc.; u. exter. in rheum - **b. tobacco**: farfara.

brittle bush: incienso: *Encelia farinosa* - **b. stem (sarsaparilla)**: *Aralia hispida.*

Briza L. (Gram.) (Eu): quaking grass; ca. 20 spp., 3 cultd. in US; forage and ornam.; includes **B. maxima** L. (Euras): big quaking grass.

broad leaf: Plantago major: v. bean, dock, laurel.

broccoli: *Brassica oleracea* v. *botrytis.*

Brodiaea (Lili.) (wNA): firecracker flower; cluster lily; small bulbs roasted & eaten (Am Indian); esp. **B. douglasii** S. Wats. (B grandiflora J. F. Macbr.).

Brombeer(en) (Ge): blackberry(ies).

bromelain: bromelin.

Bromelia antiacantha: Bertoloni (Bromeli) (sBr): banana-do-mato (bush banana); pl. u. as emoll. & pectoral (popular med.): cowboys swig the juice of frt. (refrigerant); sds. anthelm. in constip.; vulnerary; frt. juice u. in thrush (mycotic stomatitis); ornam. & grown as a living hedge (2732) - **B. chrysantha** Jacq. (Ve): basal bracts of lvs. u. to disinfect wounds; pulp of berry pleasantly edible - **B. fastuosa** (Lindl. (sBr): lf. fiber u. by Br Indians under names of caraguata, gravata, pita - **B. karatas** L. (WI, Br. Co): wild pineapple; kurucujarro (Ven): lvs. u. antipyr.; frts. diur.; sd. anthelm; Mayas cook frts. as food either alone or with other foods (6); brownish hairs at base of fruit placed on burns - **B. pinguin** L. (tropAm WI): maya (fruit); juice of edible frt. c. pinguinain (protein-splitting enzyme); u. anthelm.; for this, juice of mature frt. mixed with coconut milk; esp. effective for trichuriasis; juice said more effective than leche de higuerón as anthelm; also u. diur.; pingui source of fiber - **B. sylvestris** Willd. (Br; sMex): u. similarly to *B. karatas*; lf. fibers long used.

Bromeliaceae: pineapple fam.; large monocot fam. with ca 2200 spp.; tropAm; many spp u. for fibers (*Aechmea magdalenae, Ananas* spp. *Bromelia sylvestris, Neoglaziovia* spp., *Pseudoananas sagenarius*).

bromelin: proteolytic enzyme of pineapple frt.; u. as skin clarifier (in cosmetics u. to remove dead cells of outer skin layers); meat tenderizer; anti-gellant (keeps gelatin-contg. drinks free-flowing); anthelm.; in mfr. cheese; to purify beer (chill-proof); in blood compatibility test; frt. b. differs fr. stem b. in remaining in frt. on ripening.

bromocriptine: ergot alk. deriv., in Parkinsonism; prolactin inhibitor.

bromonaphthalene, 1-: microscope immersion oil (mixed with castor oil).

Brompton's Mixture: B. cocktail: **B. syrup**: comb. morphine, (orig. heroin) cocaine, alc., etc. (many variants); u. max. pain relief in bone cancer and other painful malignancies.

Bromus catharticus Vahl (Gram) (SA; natd. US): brome; rescue grass; pl. decoc. purg. emet u. for winter forage - **B. hordeaceus** L. (Euras): blubber grass; u. forage - **B. mango** E. Desv. (Ch): u. purg.; formerly grain u. as cereal (bread) - **B. mollis** L. (Eu, Af, nAs, NA): (soft) chess; weed in grain fields - **B. ramosus** Huds. (B. asper Murr.) (Euras, nAf): wild oats; hb. u. for forage - **B. secalinus** L. (Eu, introd. US): chess; cheat; forage; u. to make hay; weed, esp. in grain fields.

bronchitis plant: *Artemisia heterophylla*.

brook lime: *Veronica beccabunga* - **brookweed**: *Samolus* spp.

broom, (common): *Cytisus scoparius* - **b. corn**: *Sorghum vulgare* v. *technicum* (SC long stiff panicle rays are u. to make brooms) - **English b. (flowers, herb, tops)**: common b. - **b. grass** (Mex): *Epicampes macroura* - **jointed b.**: *Genista sagittalis* - **b. rape**: *Orobanche americana* & other *O* spp. - **b. root**: zacatón (Mex): *Epicampes macroura* & other *E.* spp. - **b. sage** (seUS): b. sedge - **(brown) b. sage** (wUS): *Chrysothamnus graveolens* & related *C.* spp. - **Scotch b.**: common b. - **b. sedge**: 1) sedge grass, *Andropogon scoparius, A. virginicus* & other *A* spp. 2) arrow grass - **Spanish b.:** weaver's b. - **sweet b. root**: *Ruscus aculeatus* - **b. tops**: young twigs of *Cytisus scoparius* - **weaver's b.** - *Spartium junceum* - **yellow b.**: *Baptisia tinctoria*.

brose: dish (prepd. mostly in Scotland) made by pouring boiling water of other liquid on meal (ex. oat, buckwheat) & stirring (ex. beef brose) (2016); also potatoes ("bruisy").

Brosimum alicastrum Sw. (Mor.) (Mex, WI, CA, Ve): (Ramón) breadnut (tree); ramón (de México); (oh)ox; sds (achenes) ("bread nut") cooked & eaten (WI), sometimes ground & mixed with corn meal; u. as galactagog by Mayas & Sp of Yucatan; roasted as coffee substit.; lvs. forage for cattle; latex u. as sed. in asthma; toothache; as galactagog (also lvs. & frts); latex sold as "vegetable milk"; furnishes kind of rubber; wd. u. in carpenter work (1365) - **B. aubletii** Poepp. & Endl. (Guyana): tree; wd. c. black markings like hieroglyphs; very expensive wd. (letterwood) u. for drum sticks, fishing rods, arrows, etc. (Am Indians) - **B. galactodendron** G. Don (B. utile Pittier) (CR, Pan, Guian to Ve, Ec): (American) cow tree (palo de vaca); cow bread nut tree; on puncturing st., latex exudes resembling cow's milk, esp. rich in sugars & phosphates; c. wax-like FO, "galactin" (resinous subst.), rubber; latex u. like cow's milk, in bev. (*cf. Lacmellea utilis*); not as nutritious as once supposed; cow-tree wax obt. by boiling latex, like beeswax; to make candles; (1366) bk. u. for clothing - **B. gaudichaudii** Trec. (Br Minas Gerais): mama cedela; rt. of tree u. to treat dermatitis.

Brosme brosme Muell. (Gadidae - Pisces) (nAtl Oc - coasts on bottoms): cusk; torsk; lumb; this fish is a bottom feeder like cod; flesh eaten; liver furnishes FO similar to codliver oil.

brosses de bruyère: *Polytrichum juniperinum*.

brota (Sp): shoot, bud.

brotas de pinheiro (Br): balsam gilead buds.

Broussaisia arguta Gaudich (Hydrange.) (Hawaii [US]): "Kanawao shrub"; reddish frt. u. to aid conception.

Broussonetia kazinoki Siebold (Mor.) (Jap): Japanses paper mulberry; kozo; cultd. for paper mfr.; u. frts. for sexual impotence - **B. papyrifera** Vent. (Chin, Jap, Mal, natd. eUS): paper mulberry; inner bk. of shoots u. in paper and cloth mfr.; tapa (Tonga) = Kapa (Hawaii [US]) u. for loincloths, shirts, etc.; shade tree (-18m); ornam.; cause of hay fever; u. in building Jap houses; making umbrellas; clothing; handkerchiefs; bags; lvs. diur. diaph. lax. for dys. GC; bk. hemostatic, in dys. edema, menorrhagia, rt. galactagog.

Browallia demissa L. (Solan.) (Co): u. against mange (Tuma); u. as ornam. flg. plants.

Brown Algae: Phaeophyceae - **b. Indian hemp:** *Hibiscus cannabinus* - **B. Mixture** (not Brown's M.): Licorice Mixt. Comp.

Brownea grandiceps Jacq. (Leg) (Mex, SA): rosa de montaña, r. de cruz (Mex.), fls. & wd. u. med. in sterility & impotence (Ven Ph.); in birth control (Ec) (2470) - **B. latifolia** Jacq. (WI): mountain rose (Trin); fls. inf. u. colds, amenorrhea, bk. appl. to bleeding wounds.

browneyed Susan: *Rudbeckia hirta*; *Gaillardia aristata* Pursh.

browntop millet (or grass): *Panicum ramosum* L., *P. fasciculatum* Swartz.

bruca (Mex): *Cassia occidentalis*.

Brucea antidysenterica Roxb. (or J. F. Mill.) (Simaroub.) (eAf): melita; frt. u. in dys. & diarrh., colic, cancer (Eth) (also sd. rt. lf) - **B. javanica** (L.) Merr. (B. amarissima Desv.: B. sumatrana Roxb.) (eAs, Australasia, natd. Cey, wAf): ya-tan-tzu; kosam; macassar kernels c. yatanine; sds. very bitter, u. in dys. & malaria (splenomegaly) (Chin) (85); effective in amebic dys.; claimed specific; (1120); FO u. warts, condylomata acuminata: bk. u. antipyr. (Af) (1367); sd. decoc. u. to kill lice, fleas; to kill int. & exter. parasites in cattle (Chin); rts. for bodily pain, fever.

bruceantin: quassinoid fr. *Brucea anti-dysenterica* (and other Simaroub. [?]); (Kupchan, 1975-); antileukemic, antimalarial, anti-tumor, antiprotozoal.

brucine: alk. fr. nux vomica sd.; much more bitter relatively, less toxic than strychnine; detected at 1 ppm.; CNS stim.

Brugmansia arborea (L) Steud. (Datura a. L.) (Solan.) (Ec, Pe, Ch): angel's trumpet (tree): tree stramonium; moon plant; floripón (Ve); lvs. & sds. c. O. 5% alks. (hyoscyamine, scopolamine); lvs. u. fumitory, masticatory; by the poor (Cu) in asthma, hemorrhoids, vermicide; as cataplasm to relieve pain & accelerate suppuration; regarded as too toxic for narcotic (hallucinogenic use [Indians]); cult. ornam. (handsome flowers) - **B. X candida** (**B. aurea** Lagerheim); Methysticodendron amesianum R. E. Schultes) (mutant) (wSA, esp. Co): tree datura; huanto; culebra borrachero; closely related to *Datura* and sometimes included in that gen.; intoxicant (of Kamsa & Inga Indians); chewed for hallucinogenic effects; in treatment of arthritic pains; formerly anesth. for religious sacrificial victims (1121); may be dangerous; lvs. c. scopolamine, atropine, &c.; lvs. & fls. heated in water & appl. as plaster to reduce tumors & swellings of joints; warm decoc. of lvs. & fls. u. as bath in fever; Safford (Datura c. [Pers.] Saff.) (Mex, CA): floripondio (blanco); heated lvs. appl. to swellings (79); lvs. & fls. smoked for asthma; fls. placed on pillow for insomnia (Salv.); u. as card. stim. (poor people) (2434); very showy ornam. pl. with white fls. which are highly perfumed - **B. sanguinea** (R. & P.) D. Don (Co-Ch): u. as hallucinogenic (SA Indians); to restrain unruly children; supposed to allow communications with the dead - **B. suaveolens** Bercht. & Presl (Datura s. Humb. & Bonpl.) (seBr, Mex; cultd. NW crop): angel's trumpet plant; Ku-a-va-i (Kofane Indians, Ec); tree-like shrub -5 m. high; c. scopolamine, hyoscyamine; similar to *B. arborea* & u. by Indians of Latin America in similar way.

Bruguiera gymnorrhiza Lam. (Rhizophor.) (tropAs, Af, Oceania): one of the mangroves; bk, u. astr. in diarrh., antipyr.; tanbark; fl. ext. u. as dye (red, brown, black) (Tonga) (86); pith & sd. edible.

brujería (Sp): witchcraft, practice of **brujo**: Indian medicine or witchman of wSA (Pe & Co) who endeavors to harm through use of lethal drugs, psychism, &c.

Brunella R. & P.: *Prunella*.

Brunelle (Ge): 1) *Prunella* spp. 2) *Nigritella nigra* (Orchid) (*cf.* Prunelle).

Brunsfelsia americana L. (Solan.) (WI, cult. PI): lady of the night (night fragrances); frt. astri. ton.; syr. u. in chronic diarrh. (Martin) - **B. chiricaspi** Plowman (wAmazon: Co, Ec, Pe): u. to treat fevers; hallucinogenic, sometimes added to *Banisteriopsis* prepns. - **B. grandiflora** D. Don (tropSA): chiricaspi; (chiric) sanango; rt. bk. inf. u. hallucinogen (strongest of genus); or added to hallucinogens such as caapi; u. antipyr. aphrod. abortif. in arthritis, rheum. (Co) - subsp. **schultesii** Plowman (eEc): u. in arthritis, rheum (1124); hallucinogenic, antipyr. (777) - **B. hopeana** Benth. = *B. uniflora* - **B. hydrangeaeformis** Benth. (seBr): Mire raintree, u. like foll. - **B. mire** Monachino (Bo): miré; u. like curare (voluntary muscle paralyzant) (1112); for cutaneous parasites; gives profuse perspirn. (203) - subsp. **schultesii** Plowman (eEc); u. to expel cutaneous parasites - **B. uniflora** (Pohl) D. Don (B. hopeana Benth.) (equatorial SA esp. eBr): manaca (root); vegetable mercury; u. as antirheum. in arthritis, diur. (Br Ph.); purg. emet. abort.; cold inf. u. for VD, esp. syph.; bk. c. manacine, hopeanine; scopoletin, manacein, aesculetin (glycosides); horses browse but pl. believed toxic (nArg); sold in markets; ext. in arrow poisons; possibly hallucinogenic.

brusco (Sp): **Bruscus** (L): *Ruscus aculeatus*.

brush, headache: a hairbrush provided with coarse whalebone bristles; supposed to banish cephalgia.

Brussels sprouts: *Brassica oleracea* v. *gemmifera*.

Brustbeere (Ge): jujube frt.

brut (Fr): **bruto** (Sp): gross (wt.): crude; raw.

bruyère (Fr): *Erica arborea* - **b. (commune)** (Fr): *Calluna vulgaris*.

Brya ebenus DC (Leg) (WI): ebony cocus wood; Jamaica (or American) ebony; wd. of tree u. as lumber; in many musical instruments.
Bryamycin: thiostrepton (Bristol); antibiotic prod. of *Streptomyces azureus* (soil fr. NM [US]).
Bryatae: Musci, true mosses & liverworts.
bryology: science of mosses.
Bryonia L. 1) (Cucurbit): genus of vines 2) (L.) (specif.): title of tuber of *Bryonia alba* & *B. dioica* - **B. alba** L. (Euras): white bryony; tuber c. **bryonin** (glycos.), alk; u. strong purg. emet. diur., in gout, rheum. skin dis., etc. (poisonous) (1123) - **B. dioica** Jacq. (Euras, nAf): redberry bryony; u. similar to prec.; purple dye.
bryonidin: glycoside in *Bryonia* rts.; acts on nerv. system.
Bryonopsis affinis (Endl.) Cogn, (Cucurbit.) (Oceania): lf. & rt. decoc. u. for diarrh. (Solomon Is). (—> *Kedrostis*)
bryony (root): *Bryonia alba, B. dioica* - **black b.**: 1) *Tamus communis* 2) (less commonly) *Bryonia alba* - **white b**: *B. alba, B. dioica.*
Bryophyllum calycinum Salisb. = *B. pinnatum* Kurz = *Kalanchoë p.* - **pinnatum** Kurz = *Kalanchoë p.*
Bryophyta: Bryophytes: the phylum of mosses & liverworts; low-lying pls. growing on the ground, tree trunks, rocks, etc.; having primitive organization of tissues with more prominent gametophyte than sporophyte; ex. Sphagnum mosses.
Bryoria fremontii (Tuck.) Brodo & Hawksw. (Lecanorales*cf.* Alectoria) (Lichenes) (NA, wAlta, BC [Cana] e to ND [US]; cool habitats): black tree lichen; b. t. fungus; "tree ears"; c. vulpenic acid; u. med. (wNA Indians) dr. pd. lichen mixed with grease & rubbed on navel of newborn baby (anti-infective); after weaning, baby given mixt. of "sirup" of the lichen with saskatoon berries; reduces blood clotting (2606); baked & eaten by wIndians; cooked in pits & prepd. for highly valued food (2607); u. in Chinese cooking, often with tiger lily stem.
B. T.: *Bacillus thuringiensis.*
Bubalus bubalis L. (Bovidae) (sAs): water buffalo (sAs): common b.; carabao (PI); u. as draft animal, for milk (rich in butter fat), meat.
bubble gum: v. gum.
bubby: *Calycanthus floridus* var. *laevigatus.*
bucare (Mex): *Erythrina poeppigiana* or *E. velutina.*
bucchu: bucco (Fr): **bucha**: Buchu.
bucha (dos paulistas) (Por): *Luffa aegyptica.*
Buchanania arborescens Bl. (Anacardi) (Indones. PI): u. for diarrh. (Sabah); in stomachache (PI); lvs. cyanophoric - **B. lanzan** Spreng. (B. latifolia Roxb.) (Indochin, Burma, Ind): chirauli nut; cheroonjee oil plant; lumbo (Burm): thitsibo (Burm); small tree source of pial (or peal) gum (like bassora gum; ca. half sol. in water); popular adhesive; rt. & lvs. u. in diarrh.; sds. eaten like almonds; sd. kernel yields FO u. in ointments; bk. & frt. prod. a natural varnish; wd. inferior building material.
Buche (Ge-f.): beech (cognate to book [Buch]), SC bk. frequently used as record for initials, &c.
Buchloe dactyloides (Nutt.) Engelm. (monotypic) (Gram) (wUS, nMex): buffalo grass; a forage grass; source of hay; for erosion control; sod u. for s. houses (early settlers); lawn grass.
Buchnera cruciata Buch.-Ham. (Scrophulari) (seAs): pl inf. u. in cough (Vietnam); in epilepsy (sChin) - **B. elongata** Sw. (sFla [US], WI): bluehearts; inf. u. as emet. for headache, dizziness - **B. lithospermifolia** HBK (Hond, Guat, Pan): tronera del monte; hb. heated and appl. to forehead for headache.
Buchnerodendron lasiocalyx (Oliv.) Gilg (Flacourti) (tropAf, Tanz): muponono; pd. lf. strewn on yaws ; lesions; rt. raspings appl. to old wounds as vulner.
bucho da China (Por): *Murraya paniculata.*
buchu (leaves): *Barosma (Agathosma) crenulata, B. betulina, B. serratifolia*; u. diur. in GU dis. - **b. gin**: vin de buchu (Fr Ph.) (buchu 3% in white wine); prepd. by distn. of gin over b.; u. diur.- **Karoo b.:** lvs. of *Diosma succulentum* Berg var. *bergianum* (adulterant of long buchu) - **long b.:** *Barosma serratifolia* - **oval b.:** *Barosma crenulata* - **round b.: short b.:** *Barosma betulina*
buck: 1) *Artemisia* or specif. *A. vulgaris* 2) buckwheat 3) *Fagus sylvatica* - **b. brush** (Ozarks, wUS): 1) any of several spp. of shrubs u. by sheep, etc. for browsing (SC eaten by goats); specif. *Ceanothus velutinus* & several other *C.* spp. 2) *Symphoricarpos albus* v. *laevigatus* & *S. occidentalis* - **b. horn**: *Lycopodium clavatum*; *Osmunda regalis*; etc.
buckbean: *Menyanthes trifoliata* - **buckberry:** *Vaccinium stamineum* +.
buckbush: 1) *Salsola kali* 2) *Symphoricarpos occidentalis*; *S. orbiculatus* (also buckbrush) 3) *Ilex verticillata.*
buckeye: *Aesculus* spp. esp. *Ae. pavia* - **fetid b.**: Ohio b. - **Mexican b.**: *Ungnadia speciosa* - **Ohio b.**: *Aesculus glabra* (named by early American settlers) - **b. state**: Ohio (US) (SC fr. common occurrence of trees there).
buckhorn: 1) b. plantain 2) *Coronopus squamatus* - **b. brake**: *Osmunda regalis* - **b. plantain**: *Plantago lanceolata* (also **buck's horn plantain**) (8).
buck's horn tree: *Rhus glabra.*

buckthorn alder: v. alder - **b. (bark)**: 1) *Rhamnus frangula* 2) *R. persica* 3) *Prunus spinosa* - **b. berries (, purging)**: *Rhamnus cathartica*; u. as syr. for purg. - **b. brake**: *Osmunda regalis* - **Carolina b.**: *Rhamnus caroliniana* - **glossy b.**: *Rhamnus frangula* - **sea b.**: *Hippophaë rhamnoides*.

buckwheat (, common): pl. & frt. of *Fagopyrum esculentum* - **California b.**: *Eriogonum fasciculatum* - **climbing b.**: *Fallopia* spp. - **b. family**: Polygonaceae.

bud: an embryonic or undeveloped stem or s. branch (shoot) surrounded by protective b. scales; ex. poplar bud; usually proliferate ("swell") in the spring - **flower b.**: b. which develops into fl.; ex. clove, (unexpanded fl. bud) - **leaf b.**: one which develops into a leafy branch.

Buddleia: Buddleja americana L. (Logani) (Mex, WI, CA, SA): tabaquillo (Pan); lvs. bk. rt. u. diur. esp. in edema (as decoc.); vulner. & in rheum. & uter. dis.; in erysipelas (Pe); in Co appl. to forehead for headache; hypnotic; analg. - **B. asiatica** Lour. (Chin, PI): poisonous plant; lvs. & rts. appl. to tumors on head; rt. decoc. in mal.; abort., in skin dis. (PI); fish poison - **B. brasiliensis** Jacq. f. (Br): "barbasso"; vassoura; verbasco do Brasil; lf. inf. u. to bathe swellings; u. chiefly in treatment of pulmon. dis.- **B. globosa** Hope (Ch): helps heal ulcers - **B. marrubifolia** Benth. (Mex, sTex [US]): butterfly plant; asafrán del campo; shrub; pl. u. diur. aper., condim. (Chihuahua); lf. decoc. as dye for butter, pasta, &c. - **B. perfoliata** HBK (Mex): salvia de bolita (Mex); hot tea fr. lvs. u. for neuralgia & stim., antiperspirant esp. in TB - **B. scordioides** HBK (Mex): consound; "comfrey"; inf. u. in indiges. - **B. sessiliflora** HBK (sAriz, sTex [US], Mex): Rio Grande butterfly bush; tepozán (verde); tea fr. lvs. rt. & bk. u. indig. intest. inf., &c.; lvs. boiled with salt u. in baths for inflamm. & appl. locally to sores, wounds; cuts (Mex) - **B. verbascifolia** HBK (Ve): salvio; Santa María; crushed lvs. u. in cataplasm for rheumatic pains; wd. as firewood.

buena (Mex): *Euphorbia pulcherrima*.

Buettneria Murr.: v. *Byttneria*.

bufadienolides: scilladienolides found in plants and in toads (*Bufo*): cardioactive.

bufalaga (Sp): *Scrophularia auriculata*.

bufalin: aglycone steroid with cardioactive effect, occ. in toad skin (*Bufo bufo*).

buffalo: *Bubalus bubalis* L. (Bovidae) & several other *B.* spp. (As Oceania) water/common/Indian buffalo (Ind, SriL, later distrib. in Euras & Far East): large animals (-4 m. long) u. chiefly as draft animals, also for milk (very high butter fat content), meat (like beef) (u. in Unani med.); hide (good heavy leather); horns; dry feces ("b. chips") u. as fuel on treeless plains; carabao (PI) (domesticated), also wild b. (American bison (*Bison bison* L.) & European bison (*Bison bonasus* L.) very similar to each other (dorsal hump) are sometimes called "buffalo."

buffalo, American: bison: *Bison bison* - **water b.**: *Bubalus*.

buffalo berry: *Shepherdia argentea & S. canadensis* - **b. berry, Canadian**: *Shepherdia canadensis* - **b. b., thorny**: *Shepherdia argentea* - **b. bunch grass**: *Festuca scabrella* - **b. bur(r)**: *Solanum rostratum* - **b. grass**: *Buchloë* spp.- **b. herb:** alfalfa tea - **b. nut**: *Pyrularia pubera*.

Bufo bufo L. (Bufonidae - Anura - Amphibia): (nEu, As, nwAf.): toad; c. bufotalin, bufotoxin, &c.; with complex action (raise BP stim. uterus, cardioactive agent) - **B. b. gargarizans** Muell. & Hellmich. (eAs.): ch'an su (Chin.): toad venom (skin secretion) long u. in cardiac dis.

bufotalin: steroidal genin found in venom of *Bufo bufo* (*B. vulgaris*) (common European toad); cardiotonic.

bufotenine: 5-OH-N, N-dimethyl tryptamine; N, N-dimethyl-serotonin; occ. in many *Bufo* spp. (toad skin), mushrooms; *Adenanthera* (*Piptadenia*) *peregrina, A. colubrina* (cohoba snuff); u. hallucinogenic.

bufotoxin: cardiotonic conjugated compd. found in skin glands of *Bufo vulgaris* (Eu); similar to digitoxin.

bug: term commonly u. in English-speaking areas as equivalent of insect or other invertebrate animals of small size; properly applied to a member of Order Hemiptera (ex. stinkbug) (fam. Pentatomidae).

bug, lightning: firefly (*qv*).

bugbane: Cimicifuga.

bugle: *Ajuga* - **bitter b.**: *Lycopus europaeus* - **common b.: European b.:** *A. reptans* - **sweet b.: water b.: b. weed**: *Lycopus virginicus*.

buglosa (Sp): *Anchusa azurea*.

bugloss (, common): 1) *Anchusa officinalis* 2) *Borago officinalis* 3) *Echium lycopsis*- **Spanish b.**: Alkanna - **viper's b.**: *Echium vulgare*.

bugula (Sp): *Ajuga reptans* - **bugwort**: *Veratrum viride* (cimicifuga).

buisson (Fr-m.): bush, shrub.

buka: buku: buchu.

bukaryo: sclerotia (resting mycelial masses) of fungus *Pachyma (Poria) hoelen* (Polypor.); u. med. (Orient).

bulb: a short, thick, vertical subterranean stem surrounded by numerous lf. scales, the whole actually representing a kind of bud serving for reproduction; ex. squill, onion.

Bulbine alooides Willd. (Lili.) (sAf): rooiwortel (sAf Dutch): dr. bulb u. diarrh. dys. abdominal ailments - **B. asphodeloides** R. & S. (Af) "copaiba"; lf. juice appl to wounds like tinct. iodine; crushed lf. dressing for burns; bulb decoc. u. diur. in uremia, diabetes, VD dem. urinary antisep. - **B. latifolia** R. &. S. (sAf): rt. decoc. u. in rheum. blood disorders, diarrh; rt. & stalk decoc. to prevent miscarriages - **B. narcissifolia** Salm Dyck. (sAf): snake flower; fresh plant sap appl to wounds to hasten healing; appl to ringworm, gravel, rash, wounds; int. for rheum.; fresh lf. inf. for purg. - **B. rostrata** Willd. (sAf): smoked for cold in the head - **B. tortifolia** Verdoorn (sAf): lf. juice appl. to wounds & sores.

bulbocapnine: alk. found in *Corydalis* spp., *Dicentra canadensis*; sometimes u. in treatment of Ménière's dis., etc. (hypokinetic for tremor); (1234).

Bulbostylis barbata (Rottb.) Kunth - **B. globulosa** = *Fimbristylis b.* (Cyper.) *F.g.*

bulgar: American rice: dr. & cracked wheat which has been soaked & precooked; eaten with soups, meat, &c.; excellent dietary values; from Far East, this it the first processed food known.

bulgosa hierba (Sp): *Anchusa officinalis.*

bulgur: burgoo: cracked wheat.

bullace (seUS): muscadine grape, *Vitis rotundifolia.*

bullet wood (Jam): wd. of *Couroupita guianensis*; *Manilkara zapota, M. balata, M. multinervis.*

bul(l)berries (Mont.): *Shepherdia argentea: Vaccinium myrttllus.*

bull-head: *Tribulus terrestris.*

bull-nettle (seUS, Ala): any large sp. of nettle esp. *Cnidoscolus stimulosus.*

bullock's heart: *Annona reticulata* - **bull's foot (leaves)**: Farfara.

bully tree: *Sapota achras.*

Bulnesia arborea (Jacq.) Engl. (Zygophyll.) (Co, Ven): vera; palo santo; wd. (Maracaibo big vitae; vera wd.) u. same as guaiac wd. & *B. sarmienti* ; wd. bk & lvs. inf. u. as sudorific; bk. macerate said to kill fleas - **B. retama** (Gill) Griseb. (B. retamo?) (Arg): jarilla; very resinous shrub source of retamo wax; u. in shoe polishes, &c.; u. antipyr.; wd. as "hard as iron" - **B. sarmientoi** Lor. ex Griseb. (WI, Arg, Par, Br): palo santo; p. balsamo; wd. (Paraguay lign. vit.) similar to guaiac wood; u. in wood turning; as source (distn.) of "guaiac wood oil"; u. adult. of rose oil, perfume fixative, in artif. tea rose scents (VO), etc. (1365); wd. stim. diaph.; u. for rheum. arthritis, gout, dermal dis.; syph.; known and u. in Eu for several centuries (2605).

bulrush: 1) *Typha angustifolia* (var. *brownii*), *T. latifolia* 2) *Scirpus* spp. (1125) - **great b.**: *Scirpus validus* 3) (Bible) papyrus.

bulto (Arg): package, parcel.

Bumelia americana (Mill.) Stearn (B. retusa Sw.) (Sapot.) (WI, Bah, CA): cocuyo (Cu); bo-hog (Bah) "milkwood"; lf. decoc. u. (in mixt.) postpartum; edible frt. chewed as chewing gum (Bah) - **B. lanuginosa** (Mx) Pers. (Kan-Fla, Tex-Ark [US]): shittemwood; woolly buckthorn; tree 20 m.; st. rarely almost 1 m. thick; black frts. eaten by Am Indians; bk. (c. mucilage) pounded u. as chewing gum (NM [Kiowa Indians] [US]); source of orange dye (1800s) - **B. lycioides** Pers. (Ill-Fla [US]): American iron wood; fr. u. in diarrh. - **B. spiniflora** DC (sMex): saffron plum; downward pl; edible frt. eaten as aphrod. (Tamaulipas, Mex).

Bunchosia glandulosa (Cav.) DC (Malpighi.) (WI, Mex, Br): weakness bush (Bah); zipche (CA); lvs. in combn. decoc. u. as female remedy; rt. decoc. (tea) u. to increase female fertility; lf. branch decoc. u. for urinary dis. & to bathe body in epilepsy & rheum.

bund (Indo-Pak): embankment bordering field, ditch, canal, &c.

bundu: *Strychnos icaja.*

bunduk: (Arab): bonduc.

Bungarus (Ophidia - Reptilia) (seAS): krait; venomous snakes (-3 m) - **B. candidus** (seThail, Mal, Indones): exceptionally toxic snake - **B. multicinctus** (Far East): Chinese krait; u. curare-like action in neurophysiology.

Bunium bulbocastanum L. (Carum b. Koch) (Umb.) (c&wEu, As, Pak - old fields): earth (chest)nut; pignut; tubers eaten by swine & children (some believed eating too many would cause head vermin); lvs., fls. & frts. u. as condim.; frts. as carm.; tubers u. med. as astri.; substit. for caraway; form. cult. - **B. incrasatum** Battandier (Eu): "earth chestnuts"; prod. increases sexual desire.

Buphane Herb. = *Boöphone.*

Bupleurum chinense DC (Chin): "chai-hu"; dr. rts. u. in traditional med. for fever, flu, infectious hepatitis (h. A) - **B. falcatum** L. (Umb.) (Euras, Him reg. & s; widely distr.): hare's ear; c. triterpenes saikogenins A, B, C, D; spinasterol; stigmasterol; rutin (lvs. & sts.); u. in ancient Chinese med. & still one of most important Chinese drugs; u. antipyr.; rt. u. in gout (Manchuria); in mal. (Kor, Taiwan); var. *scorzonerifolium* (Willd.) Ledeb. (B. s. Willd.) (As, esp. Chin): in gout, prolapsed uterus; antipyr. in fevers, flu, mal., rheum.; hepatitis, rhinitis, deafness, dizziness (vertigo), &c. - **B.**

fruticosum L. (Euras): (shrubby) hare's ear; source of VO (Sardinian & Russian oils); cult. ornam. shrub (Med reg. Iran; Caucasus) - **B. ginghausenii** Y. Li et J. Y. Guo (Chin): u. similarly to *B. chinense* - **B. rotundifolium** L. (Euras; Caucasus reg. natd. NA): thorow-wax (or thorough wax) (lit. growing through); hb. (c. VO) arom.; sd. u. in rupture; decoc. of hb. u. as vulner. for broken bones & for dissoln. of goiter; u. for 15 centuries - **B. scorzonerifolium** Willd. (Chin): chai-hu; dr. rt. u. in traditional med. for fevers, flu, infectious hepatitis (h. A) - **B. stewartianum** Nasir (Swat, Pak): hb. u. in stomach & liver dis.

bur: hops - **b. dock**: burdock - **b. marigold**: *Bidens cernua* & other *B.* spp - **Paraguay b.**: *Acanthospermum* spp.

burbot liver oil: FO fr. liver of *Lota maculosa.*

burdock: *Arctium lappa, A. minus.*

Burgunder Harz (Ge): **Burgundisches Pech**: foll.

Burgundy pitch: oleoresin fr. *Picea abies* (*qv.*) - **American B. p.: artificial B. p.:** rosin & mineral oil, &c. u. for friction tape & rubber compositions; not medicinal B. p. - **domestic B. p.**: fr. *Tsuga canadensis.*

buriti: burity: *Acrocomia vinifera.*

Burkea africana Hook. (1) (Leg) (Tanz): rts. in warm water eaten for stomachache; bk. sap & rt. decoc. drunk for phlegmon & bacterial abscesses; bk. decoc. in cough; GC (in combns.); tanbark.

burlap: cheap coarse fabric prepd. fr. jute or hemp; u. to make bags, often u. for collection & storage of crude drugs.

burn plant - (Fla, Ala [US]): aloe (SC leaf u. for burns).

burnese (US addicts): crystallized cocaine.

burnet, garden: *Sanguisorba officinalis* - **b. (salad)**: *Sanguisorba minor.*

burning bush: 1) *Euonymus atropurpureus* (scarlet aril & red sds.) 2) *Dictamnus albus* var. *ruber*: pl. emits vapor wh. can be lighted (red fl.) 3) *Kochia scoparia* (mature lvs. bright red) - **b. fluid**: mixt. of 3 measures of alc. & 1 of purif. turpentine oil.

burnt alum: exsiccated a. - **b. hartshorn**: incinerated deer horns; fr. Cornu(s) elaphus (L.), antler of *Cervus elaphus*, red deer - **b. sugar (coloring)**: caramel.

burnweed, American: *Erechtites hieracifolia.*

burr, bathurst: cockle b.: *Xanthium spinosum* - **b. weed.**: *Galium aparine.*

bur-reed: *Sparganium* spp.

burro (SpAm): small donkey; usually pack animal - **b. weed**: 1) *Haplopappus fruticosus* (57) 2) *Allenrolfea occidentalis* 3) (Calif): *Ambrosia dumosa* (Bassia p. Don).

Burroughsia fastigiata (T.S. Brandeg.) Moldenke (Verben.) (wMex): "damiana"; u. in teas as reputed aphrodisiac (to prep. "creme de damiana") (Guadalajara), formerly large export; also in stomach dis. and to aid conception.

Bursa pastoria (L): *Capsella b. p.*

Bursaria spinosa Cav. (Pittosporaceae) (Austral): Australian black horn; native box; lvs. c. esculin. (prodn.); cult. orn.

Bursera Jacq. ex L. (Elaphrium) (Burser.) (tropAm): Jamaica birch trees; torchwood; resin-contg. passages occ. in tissues of st., hence name (bursa, L., small leather sac or purse) - **B. aloexylon** Engl. (sMex): linaloé (Puebla); VO fr. dist. of wd. or frt. u. mfr. perfume - **B. delpechiana** Poiss. (Mex): Mexican linaloe; wd. & VO (Mex Ph.) (linaloes aloe) c. linalool (60-70 [-90]%), linalyl oxide, geraniol, terpineol; VO fr. frt. = (Mexican) linaloe oil, valuable in perfum. (lily of the valley odor) u. in soaps & cosmetics; Indian linaloe oil; fr. pls. cult. in Ind - **B. fagaroides** Engl. (Mex): gum u. in toothache - **B. glabrifolia** (HBK) Engl. (Mex): linaloe bursera; wd. (lignaloe wd. - fr. Lignum Aloe) odoriferous, astr. bitter; c. arom. resin & VO (linaloe oil), smelling like caraway - **B. grandifolia** Engl. (Mex): palo mulato; hb. decoc. in fevers; gum fr. trunk u. to caulk boats & glue furniture - **B. jorullensis** Engl. (Elaphrium jorullense HBK; B. glabrescens Rose) (Mex): copal (blanco); copalli; reddish resin obt. by incisions in bk. of tree; resin u. to treat uter. dis. & in ointments; smoke inhaled for headache; makes transparent glossy varnishes; substit. for elemi (Mex) - **B. microphylla** Gray (Ariz [US], Mex, SA): elephant tree; torote; copal; gum resin (cuajiote [colorado]): u. as remedy for scorpion stings; VD (inf.) int. drastic purg. (*cf.* gamboge); lvs. & twigs with potent antitumor subst.; bk. u. tan, dye - **B. penicillata** (DC) Engl. (Mex esp. s. Chihuahua): herbage u. for catarrh; resin fr. st. u. as remedy in toothache; arom. wd. u. as substit. for aloe wood; much cultivated in Ind for prodn. of VO, linaloe oil ("Indian lavender") (one source) - **B. simaruba** (L) Sarg. (B. gummifera L.; Elaphrium simaruba [L] Rose) (tropAm, incl. sFla [US], WI, Bah, Mex, trop SA & CA): (West Indian) birch; incense tree; birchwood; gumbo limbo (tree) (Florida, [US]); almácigo tree; chaca(h); gommali (Bah); trees with smooth red bk.; "gum" (balsamic resin) = gum chib(o)u (caraña gum [wrapped in palm lvs.]); a gum elemi, is attributed to this sp.; gum, lvs. & rt. are u. as purg., diaph., diur., expec.; in dys.; VD; "male debility"; edema, &c.; mastic substit.; yellowish arom. resin was much u. by Mayas of Yucatan (6); u. to cure ulcers & sores

(CR) in mucous membr. dis.; to extract spines; bk. u. in measles; med. (PR); incense; to make balsams, pills, elixirs; glass & china cement; frt. u. for stomachache (Salv); lvs. tea substit.; lf. decoc. u. for rash, blood poisoning., back pain, &c. (Bah); sds. in rheum (Salv): (Tamaulipas, Yuc, Veracruz; resin prodn. WI, CA, Ve) - **B. tomentosa** Tr. & Pl. (Curaçao, nVe, SA): balsamo incienso (Ve):"tacamahaca"; resin u. in plasters for pain of waist area (centura); ton. astr. antirheum. diur. in gout.

Burseraceae: Myrrh Fam.; trop trees & shrubs, lvs. alternate, small fls. in thynses or racemes (1126), many with arom. gum-resins; ex. Myrrh.

bursting heart: *Euonymus atropurpureus.*

burstwort: *Herniaria glabra.*

burtse (Baltistan): wild *Artemisia* spp.

burweed: *Xanthium strumarium*; *X. spinosum*; *Triumfetta semitriloba*; *Ambrosia acanthicarpa*; *A. confertiflora*; *Medicago* sp. (?).

burwort: *Ranunculus acris.*

Busch (Ge-m.): bush, shrub - **Buschtee** (Ge) (bush tea): *Cyclopia latifolia* - **Buschwindroeschen** (Ge): *Anemone nemorosa.*

bush: a shrub or cluster of shrubs or a thick dense shrub; low shrub branching from ground (-5 ft. high) (1746); however, often u. as equivalent to shrub - **b. rope**: 1) Brazil plants; representing large woody vines (lianas) of the trop. forest, often of very large size; inclu. *Chondrodendron spp.*, many *Strychnos* spp. 2) (specif.) *Cissus* spp. (wild vine) - **sweet b.:** *Comptonia asplenifolia* - **b. teas**: (WI+) inf. or decoc. of various pls. u. chiefly as remedies of dys. diarrh. colds, influenza, sore throat, hemorrhages, &c.; ex. *Quercus falcata* var. *pagodaefolia.*

busserolle (Fr): *Arctostaphylos uva-ursi.*

butane: C_4H_{10}: colorless flammable gas; occ. in petroleum, natural gas, &c.; asphyxiant; u. as aerosol propellant (NF, XV)

butcher's broom - *Ruscus aculeatus.*

Bute Inlet wax: large masses of wax found in water at Bute Inlet, BC [Cana], believed prod. of pollen of *Pinus contorta* (1369).

Butea monosperma (Lam.) Ktze. (B. frondosa Roxb.) (Leg.) (Indo-Pak to Burma): dha(w)k tree; d. palm; dr. sap (gum) constitutes Bengal Kino **(butea gum; b. kino** palas k.): tannin-rich exudn. fr. bk. following injury) u. as astr.; gum u. substit. for true kino (med., dye, tan): (palas gum); fls. crushed in milk for intest. worms; bk. & lvs. pd. appl. to warts (124); lvs. fls. u. diur. astr. in diarrh., dys, aphrod.; sd. pd. u. purg. rubef. vermif; sds. source of moodooga oil; u. anthelm. as replacement for santonin; tree yields large proportion of sticklac of commerce; wd. (bastard teak) very durable; source of charcoal (for gunpowder); fls. u. as yellow/red dye; bk. furnishes palm fiber u. for sails - **B. parviflora** Roxb. = *Spatholobus p.* - **B. superba** Roxb. (seAs): palasa tree; potash tree; prod. an Indian Kino (Pauknive gum), also yellow dye (tesu); rt. c. estrogenic subst., also toxic princ. (2642) - **Buteae Gummi** (L.): Kino fr. *Butea monosperma* & other *B.* spp.

Butia Becc. (Palm) (SA) ef. *Syagrus.*

Butomus umbellatus L. (monotypic) (Butom.) (nEuras, natd. NA): flowering rush; pd. rhiz. u. for making bread; cult. ornam.

butter (, fresh): - milk fat, FO separated fr. milk of *Bos taurus* & less commonly other mammals; popular spread on bread; u. in folk med. in many parts of world; mixed with soot for burns to "draw out the fire" (Miss [US]) - **b. and eggs**: *Linaria vulgaris* - **b. color**: annatto - **b. of cacao: cocoa butter**: theobroma oil - **b. tree**: *Combretum* spp.

buttercup: 1) *Ranunculus* spp. 2) (Tex [US]): *Oenothera* spp. (improper) 3) (Ala [US]): daffodil & jonquil (improper) - **acrid b.**: *R. acris* - **(bulbous) b.**: *R. bulbosus* - **corn b.**: *R. arvensis* - **creeping b.**: - *R. repens* - **b. family** - Ranunculaceae - **field b.**: corn b. - **tall b.: upright b.** - acrid b. - **wild b.**: *Sarracenia flava.*

butterflower: *Ranunculus bulbosus.*

butterfly banner(s): *Dicentra cucullaria* - **b. weed**: *Asclepias tuberosa*; **b. pea:** *Clitoria ternatea.*

buttermilk: sour liquid remaining after fermentation of milk with certain microorganisms, such as *Streptococcus lactis* & *Leuconostoc citrovorum*; *cf.* **Bulgarian b.** using *Lactobacillus bulgaricus* (special culture).

butternut: 1) *Juglans cinerea* 2) *Caryocar* & *Euphorbia* spp. (less commonly).

Buttersaeure (Ge): butyric acid.

butterweed: *Senecio glabellus* - **butterwort**: *Pinguicula alpina, P. caerulea, P. vulgaris.*

button: popular term for spherical fls. or inflorescences (fl. heads) of *Platanus occidentalis*, *Cephalanthus*, *Liquidambar*, &c - **b. ball (tree)**: **b. bush:** b. will: *Cephalanthus occidentalis* (SC frt. form) - **b. nuts**: *Hyphaene crinita* - **b. wood** (or **buttonwood**): 1) (New Eng, Penn, NJ [US]): *Platanus occidentalis* 2) *Conocarpus erectus.*

button-tree: *Conocarpus* spp. - **buttonweed**: *Diodia teres*, etc.

butua (Sp, Por): pareira (brava).

butyric acid: Butansaeure (Ge): $CH_3 (CH_2)_2 COOH$; vol. fatty acid found in butter, Limburger cheese, feces, sweat, VO of *Heracleum giganteum* (as esters), & many other plants.

Butyrospermum parkii Kotschy (Bassia p. Don) = *Vitellaria paradoxa.*

Butyrum (L.): butter - **B. Majoranae**: ointment of sweet marjoram - **B. Nucistae** (L.): expressed nutmeg oil (c. VO & FO).

buxine: alk. fr. bk. of *Buxus sempervirens* (not identical with bebeerine of nectandra bk. as previously held).

Buxus sempervirens L. (Buxaceae) (Euras, nAf): (common) box; boxwood; a long-lived (300 yr.) shrub or tree; lvs. toxic with alkaloids, buxine +; buxine u. for fever, malaria (Chin); bk. u. in skin dis., gout, mal.; fresh twigs & lvs. off. in Ge Homeopath. Phar. - **B. s.** var *suffruticosa* L. (Euras): (dwarf) box; hedge shrub; lf. sprigs u. ornam.; bk as dye - **B. wallichiana** Baill. (Nepal, nInd): papar; wd. very useful; bk. u. febrifuge.

byne (bynin) (Eng): malt & malted prepns.

Byrsonima Rich. ex Kunth (Malpighi.) (NW trop): maricao (PR); ca. 150 spp. some u. for tanning - **B. brasiliensis** Jacq. f. (Br): "barbasso"; lf. inf. u. to bathe swellings - **B. bucidaefolium** (SpAm): frt. sour but edible; sometimes eaten cooked or spiced (Yuc) (6) - **B. cericaefolia** DC (B. chrysophylla H. & B.) (Pe): "indiano"; frt. edible, u. in fevers, dys. & in gargles; wd. & bk. astr. u. to stop blood flow fr. wound - **B. ciliata** Cuatrecasas (Co): tea fr. dr. lvs. u. for diarrhea (Kubeo Indians of Rio Vaupés) (2299) - **B. coriacea** DC (B. spicata DC) (WI, Pan, to Pe, Br): locust berry tree; bk. decoc. astr. u. in stomach dis., sd. decoc. in diarrh.; frts. eaten for sl. purg. action (FrWI); wd. u. for fancy furniture - **B. crassifolia** HBK (Mex, CA, WI, nSA): nance tree (n. amarillo); alcornoque bark; muruxi bark; inf. u. in fevers, dys. (Hond, Nic); lung abscesses, snakebite, antiscorbutic (Mex); inner bk. pounded to pulp & applied to wounds (vulnerary) to cattle & as poultice to humans; liquid gives effective healing; pl. astr. u. in folk med. for fever, colds, bronchitis, snakebite; calyces u. stom.; frt. edible; frt. juice as beverage, fermented & unfermented - **B. lancifolia** Juss. (Br): rt. decoc. u. in rheum.; fat edible - **B. lucida** DC (B. cuneata P. Wilson) (WI, sFla [US]): locust berry; rt. bk. & frt. u. as strong astr. (Cu) in gum sores, &c. - **B. poeppigiana** Juss. (Br, Pe): u. as vaginal wash & to combat TB (Pe); frts. edible (?).

byssinosis: cotton dust asthma; lung dis. ("black lung") caused by inhalation of plant fibers (cotton, linen, &c), leading to chronic bronchitis, emphysema; many cotton mill workers affected (SC byssos (Gr): cotton).

Byttneria herbacea Roxb. (Sterculi) (Ind.): idul sindur; ground rootstock rubbed on swellings of legs; (with *Aegle marmelos* frt.) in female dis. (124).

Cc

NOTE: The letters "c" and "k" are often interchangeable; it might therefore be advisable to sometimes look under each letter. (German & Spanish words beginning with "Ch" are often rendered in English as "Qu.")

caapi (Jivaro Indians, Pe): drug fr. *Banisteriopsis (Banisteria) caapi* and several other *B.* spp., now incorporates tobacco - **caapia** (Br): **caapiam**: *Dorstenia* spp. - **caapim de Angola** (Br): capim; *Panicum spectabile*.

caaponga (Por): *Portulaca oleracea.*

caatinga (forest): catinga (*qv.*).

cabaca (Por): **c. amargosa: c. de colò: c. de cuio: c. marimba: c. para vasilhas: c. purunga: cabaceiro:** *Lagenaria siceraria.*

cabac(h)inha (Br): *Cayaponia diffusa.*

caballon (Pr): **cabalonga** (Cu, etc.): **habas de cabalongas** (Ec): *Thevetia peruviana*, Thevetia nuts - **cabalongo** (Sp): calabar beans.

cabbage: *Brassica oleracea* v. *capitata* - **andira c.**: *Andira inermis* - **celery c.**: *B. pekinensis* - **Chinese c.**: *B. pekinensis* or less commonly, *B. chinensis, B. rapa* - **palm c.:** apical bud of various palm taxa: u. in salads - **c. palmetto**: *Sabal palmetto* - **c. p. gum**: prod. somewhat like tragacanth. - **Peking c.**: *Brassica pekinensis* - **Savoy c.**: Savoy, *Brassica oleracea* var. *sabauda*; types of *c.* with very wrinkled or blistered lvs., very dark green; considered a superior quality - **Shantung c: Siam c.**: *B. oleracea* v. *gongylodes*; kohl-rabi grown in US under names Pe-Taai, Pak Choi, Wong Bok, &c., *B. rapa* - **skunk c.**: **swamp c.**: 1) *Lysichton* spp. 2) *Symplocarpus foetidus* 3) heart of palm - **c. tree:** 1) *Sabal palmetto* 2) *Andira inermis* 3) many other plants with apical lf. masses - **c. tree bark, brown**: *Andira retusa* - **c. tree bark, yellow:** *Andira inermis* - **wild c.:** *Brassica oleracea* - **winter c**: one of several kinds of *c.* surviving the winter, growing outside in the southern US, esp. *Savoy c.*; incl. collards.

cabeca de carneiro (Por): **c. de cavalo:** *Cicer arietinum.*

cabeza (cabesa) de negro ("negro head"): **c. negra**: 1) (Mex): rt. of *Annona muricata*; less commonly appl. to *Dioscorea* spp. & to 8 other pls. 2) (Co, Pan): *Dioscorea alata*; *Phytelephas macrocarpa* 3) (Pe, Co): frt. of *Phytelephas seemanni.*

cabezuella (Sp): **cabo:** fl. head, capitulum.

Cabralea canjerana Sald. da Gama (C. polytricha A. Juss.) (monotypic) (Meli) (Br): "cangerana"; "canchalagua"; rt. bk. u. in popular med. as panacea, antipyr.; insecticide; wd. useful timber (1508).

cabreuva (Br.): 1) Tolu balsam 2) copriuva (wood oil): fr. *Myrocarpus fastigiatus* (M. frondosus) oil.

Cabucala cryptophlebia Pichon (Apocyn.) (Mad): c. cabucine (alk.) a sympatholytic agt. (Fr & Ge patents) - **C. erythrocarpa** Markgraf (Mad) u. in stomachache; bk. c. bitter princ. u. in alc. beverages; frts, edible - **C. madagascariensis** Pichon (w coast of Mad): c. 10-methoxy-ajmalicine & other alk.; wd. & lvs. u. as vermif., stom. & antirheum. in folk med. - **C. oblongo-ovata** Markgraf (Mad); lvs. u. in liver dis. - **C. striolata** Pich. (Mad): c. alks. with antiviral activity esp. against influenza; frts. edible.

cabulla tabuya (Domin Rep): henequen.

cabureiba balsam: prod. similar to Peru & Tolu balsams; fr. same pl. as cabreuva oil.

cacahaté (Sp): **cacahueté** (Fr): cacauatl (Aztec): peanut.

Cacalia: (Comp.); gen. often abandoned; some spp. are referred to *Parasenecio* (OW), *Arnoglossum* (NW); also at times to *Senecio* (sect.), *Emilia, Adenostyles, Crassocephalum,* &c. - **Cacalia atriplicifolia** L. = *Arnoglossum a.* - **C. decomposita** Gray (Senecio grayanus Hemsl.) = *Psacallium d.* - **C. hastata** L. = *Senecio sagittatus* Sch.-Bip.

cacao (beans): cocoa (beans): sd. of *Theobroma cacao* - **Brazilian c.:** Guarana - **c. butter**: Oleum Theobromatis; principally made up of glyceryl esters of arachic, stearic, palmitic & oleic acids, also of theosterols; when properly irradiated, these sterols furnish vit. D: hence the antirachitic value of such cacao products' - **c. de la tierra** (SA): **c. mani** (CR, Nic): peanut - **c. praeparatum** (L.): prepared cocoa (cacao) - **c. shells:** sd. coat of cocoa beans, consisting of inner pericarp, spermoderm, perisperm; used to dilute cocoa - **c. volador** (Sp): (flying c.): *Licania arborea* - **cacau** (Br): cacao.

cachibou: 1) Chibou 2) *Calathea lutea.*

cachito (SA, Guianas): pigeon pea.

cachou (Fr, Eng): 1) catechu 2) a tablet or pill for sweetening the breath - **c. fruit**: acajou or "yellow fruit" u. by Br Indians - **c. jaune** (Fr): gambir - **c. wine**: prep. fr. cashew frt. by Br Indians; rich in vitamin C; prod. on large scale in São Paulo State (Br).

Cactaceae Juss.: Cactus Fam.; mostly indig. to Am; many grow in arid to semi-arid areas (deserts); characterized by spiny & leafless fleshy stems in leaf-like segments; with showy fls.

cacti: plur. of cactus: term often u. by layman for almost any succulent pl. of arid regions - **cactier à grandes fleurs** (Fr); *Selenicereus grandiflorus.*

Cactinomycin: antinomycin C.: antibiotic complex obt. fr. *Streptomyces chrysomallus* Lindenbein; 1949-; with antineoplastic properties.

cacto (flores) (Br): *Selenicereus grandiflorus*, fls.

Cactus: correctly appl. to a member of Fam. Cactaceae, particularly to *Mammilaria* spp.; laymen misuse the word in designating almost any arid region succulent, thus incl. pls. from such diverse families as lily, euphorbia, amaryllis, asclepiad, sunflower, & others; the old genus of *Cactus* L. has now been split up into several other genera; important genera: *Cereus* with *C. grandiflorus, Opuntia, Mammilaria, Melocactus,* &c. - **cactus, angel-wing's:** *Opuntia microdasys* Pfeiff. - **barrel c.**: term appl. to ca. 40 cylindraceous spp. of cactus incl. *Ferocactus & Echinocactus* spp. u. as source of water in the desert - **beavertail c.:** *Opuntia basilaris* Engelm. & Big. - **brittle c.:** *Opuntia fragilis* - **bunny ears c.**: *Op. microdasys* - **button c.**: mescal. - **c. candy**: candied cubes mfrd. in Tex, NM, Ariz [US] fr. visnaga (barrel cactus) - **candy c.:** compass c.: *Ferocactus* spp.- **cholla c.:** *Opuntia* spp. - **Christmas c.**: **crab c.**: *Schlumbergera bridgesii* - **compass c.: devil's head c.**: *Ferocactus* spp. - **fishook c.**: *Echinocactus & Echinocereus* spp.; *Ferocactus*: *wislizenii*; *F. recurvus*, & *F. alamosanus*; also *Mammilaria* spp. - **giant c.**: **C. grandiflorus**: L. title for, also bot. synonym of *Selenicereus g.* - **grizzly bear c.**: *Opuntia erinacea* Engelm. & Big. - **healing c.**: *Aloe* spp. - **hedgehog c.**: *Echinocactus & Echinocereus* spp. (SC spines are long, stout, & sharp like hedgehog or porcupine quills) - **Mexican dwarf tree c.**: *Op. villis* Rose - **night-blooming c.**: **n. flowering c.:** *Selenicereus grandiflorus* Britt. & Rose - **old man c.**: *Cephalocereus senilis* Pfeiff. (SC long silky hairs covering plant) - **organ c.**: **organpipe c.**: *Cereus thurberi* Engelm. - **pin cushion c.**: *Mammillaria grahami* Engelm.; *Coryphantha vivipara* Britt. & Rose var. *desertii* W. T. Marshall - **rainbow c.**: *Echinocereus pectinatus* var. *rigidissimus* Engelm. - **rattail c.**: *Op. whipplei* Engelm. & Big. & *Aporocactus* (*flagelliformis* Lem.) - **snowy c.**: *Mammilaria nivosa* - **strawberry c.**: *Mammilaria dioica* K. Brandeg. (& *Echinocereus engelmannii*; *E. triglochidiatus* Engelm.; *E. fendleri* Engelm.; etc.) - **sweet potato c.**: *Cereus greggii* - **sword c.**: *Selenicereus* spp. - **vanilla c.**: *Selenicereus grandiflorus.*

cacueiro (Por): *Theobroma cacao.*

cacur: *Cucumis myriocarpus.*

cade: frt. of *Juniperus oxycedrus* - **c. berry oil**: VO fr. cade; yields 1.3-1.5%; c. pinene & myrcene-like compds. - **c. oil**: Juniper tar (*qv.*).

cadillo (Ve): *Urena.*

Caepa (L): *Cepa* = *Allium.*

caerulein: peptide fr. the skin of an Australian frog, *Hyla caerulea;* similar to cholecystokinin-pancreozymin (CCK-PZ), acting to contract the gall bladder (*cf.* CCK), increase intestinal mobility (peristalsis), and increase pancreatic enzyme secretion (*cf.* PZ).

Caesalpinia bonduc (L) Roxb. (C. bonducella Flem.) (Leg.) (cosmop. trop, esp. tropAs): nicker nut, bean, or seed; fever (or bonduc) nut; sd. c. FO; u. in techn. & for fever & edema (as Semen Bonduc); u. asthma, colic (mostly sInd) general ton.; lvs. u. anthelm. emet.; lf. inf. u. in sore throat; pd. sds. u. vesicant, c. bitter subst., bonducin ("the quinine of the poor") - **C. brasiliensis** L (tropAm, incl. WI): brazilwood; (páu brasil); peachwood (páu rosado [Por]); heart wd. u. for red dyes & inks; other parts of tree u. for similar purposes - **C. brevifolia** Baill. (Chile): algarob(ill)a; frts. (pods) rich in tannin (40-60%); u. as tan, in med. (Chin) (Chin Phar.) - **C. cacalaco** Humb. & Bonpl. (sMex): cascalote; frt. tannin-rich, u. as dye; decoc. in diarrh.; sds. eaten; bk. u. dentalgia; cult.; shade tree - **C. chinensis** Roxb. = *C. crista* - **C. coriaria** Willd. (WI, Mex, CA, Co, Ve, nBr) (cult. in Ind, &c.); divi-divi; libidivi; cascalote; small trees with C- or S- shaped pods; c. 40-50% tannin; with ellagic & gallic acids; very cheap & popular comm. tan in Eu (gives light colored leather); also intest. astr. in severe diarrh.; ton. (Jam); cancer "cure" (Domin Rep); in prepn. dyes & inks; source of gallic acid; bk. also comm. tan; sd. u. astr. in popular med., ex. hemorrhoids; inf. u. to strengthen the teeth (Ve) - **C. crista** L. (trop, esp. Jam, Br): kalein (Burma); lvs. rts. sds. u. ton. antiperiodic in fever, anthelm. (Burma); sds. u. colic (Ind); FO fr. sds. (bonduc nuts) u. in cosmetics; one source of Brazil wood or Pernambouca wood (this sp. sometimes included under *C. bonduc*) - **C. decapetala** (Roth) Alston = *C. sepiaria* - **C. digyna** Rottl. (Ind): frt. called tari (*not* tara); c. caesalpinine, 30% tannin; u. tan; st. u. to treat "yellow eyes" (Laos); ground bk. u. (as paste) as fish poison (sVietnam) - **C. echinata** Lam. (Br, CR, etc.): (prickly) brazil wood; peach wood; páu brasil; heartwood is a heavy dye-wood, Lignum Fernambuci (Pernambuco Wood); u. for violin bows, to dye red & as tan; astr. in diarrh.; tree (also *C. sappan*) a natl. emblem of Brazil and said to have given that country its name - **C. eriostachys** Benth. (Mex-CR): palo alejo; bk. u. to stupefy fish (Mex) - **C. gilliesii** Wall. (Arg, Ur, cultd. widely in hot areas, ex. sAf): "bird of paradise"; lvs. u. in folk med. to dissolve kidney stones; sds. & pods toxic; bk. rts. & fls. u. in pulmon. & skin dis.; frts. u. as tanning agt.; stamen filaments u. as saffron substit.; protein compds. with antitumor activity; lvs. u. purg. (Libya); stamens u. as saffron substit.; pl. considered toxic, lvs. give off strong odor, not eaten by stock animals; cultd. as ornam. shrub or tree (because of beautiful fls. & lvs.) - **C. major** (Medik.) Dandy & Exell (OW, NW, trop): scorched & milled sds. u. like coffee for resp. dis. (Camb) - **C. melanocarpa** Griseb. (Br, Ch, Arg): guajacán; algarob(ill)a; lvs. & pods u. as tan (21% tannin) & black dye - **C. paraguariensis** Burk. (Chaco, Arg): guayacán; st. bk. c. phenols, flavonoids, saponins, tannins, steroids (traces); sd. u. antiphlogistic (ext.) - **C. praecox** Ruiz & Pavon (Arg, Ch): brea; source of gum brea; u. like acacia - **C. pulcherrima** (L) Sw. (Poinciana p. L.) (trop, esp. Ind, natd. sFla [US]): (dwarf) poinciana; flower fence; bird of paradise; "Barbados pride"; peacock flower; shrub or small tree; lvs. (inf.) & sds. u. antipyr. antiperiod. in mal. ton. stim.; bk. & wd. emm. (Indochin), abort. purg. (*cf.* senna). for stomachache; frt. tannin-rich; sds. c. 35% mucil., galacto-mannan polysaccharide with gelling proprs. like those of *Ceratonia siliqua* sd.; rt. bitter, poisonous; fls. inf. in bronchitis, cough (Mex); pulm. dis. pneumonia; lvs. u. for diarrh. (Af); cult. ornam. - **C. sappan** L. (Ind, Mal, Java): ("European") sappanwood; brasileiro; brasilwood; heartwd. u. to furnish red dye; much less popular than formerly since dye is fugitive; wd. (decoc.) u. styptic, astr. in diarrh.; for the blood (Laos); for backache (nVietnam); emm.; bk. u. ton.; cult. ornam. - **C. sepiaria** Roxb. (As, incl. Indonesia): Mysore thorn; espino do cerea (Por); a drug of the old Chin pharmacopeias; rt. u. as purg., to help remove bone fr. throat; sds. u. as astr. anthelm. antipyr. antimalar.; bruised lvs. appl. to burns; fls. hallucinogenic (occult med.) (Chin) - **C. spinosa** (Mol.) Ktze. (SA, esp. Pe, Cu): tara; frt. tannin-rich (-60%), u. tan, indicator dye, inks; tannin source (1131); (cultd. worldwide, in trop & subtrop) - **C. tinctoria** Benth. (CA, Br, Pe): source of tara (pods); source of gallic acid (35%); u. tan. black dye; in home med. as astr. - **C. vesicaria** L. (Br): palo campêche; prod. a brown brazil wood.

Caesalpiniaceae: group of Fam. Leguminosae; sometimes treated as a separate fam., (senna fam.) at other times indicated only as one of 3 subdivisions of Fam. Leg. (subfam. Caesalpinioideae).

Caesar's amanita: *Amanita muscaria.*

cafard (Fr): (cock)roach.

café 1) (Sp. Fr): coffee 2) coffee house (hence restaurant) - **c. au lait**: c. crême - **c. Bonpland** (Arg): fr. *Cassia occidentalis* - **c. complet**: coffee with milk, rolls, & butter - **c. crême**: coffee decoc. with cream or milk added - **c. diable** (devil coffee) (Fr restaurants): mixt. with cinnamon, bay lf., cloves, candied orange or lemon rind (or orange juice), rum, cointreau - **c. de minas** (Por): **c. do mato grosso:** *Mucuna pruriens* - **c. du Soudan**: kola - **c. rhum** (Fr): coffee with rum - **c. sauvage** (La [US]): *Rhamnus caroliniana* - **c. vert**: green, unroasted coffee.

cafeana: *Tachia guianensis.*

caféier (Fr): coffee shrub.

cafeto (Sp): coffee tree; coffee.

caffeine: alk. derived fr. several different pl. spp. (coffee, tea, mate, etc.) & also now made by synthesis; u. central nervous system (CNS) stim. in lethargy, fatigue, headache, migraine; card. & resp. stim.; diuretic; large doses occas. lethal.; some consider mutagenic - **citrated c.:** caffeine

mixed with equal amt. citric acid - **c. hydrobromate: c. hydrobromide: c. bromide**: $C_8H_{10}N_4O_2HBr.2H_2O$.
Caffur: Kafur.
cafier: caféier.
cafta: kat, *Catha edulis*.
cahinca (root) = cainca (Br): *Chiococca alba* Hitchc.
caiapia (Br): caapia - **caiapó** (Br): cayapo
caimito (Sp, Am): cainito: star apple; *Chrysophyllum cainito*.
cainan (Br): **cainca**: (Br): cahinca (also other variants).
Cajanus cajan (L) Millsp. (C. indicus Spreng.) (Leg.) (OWtrop now pantrop): pigeon (or Congo) pea; dhal; ca(t)jang; cadjan pea; perennial shrub bearing sds. eaten by humans (after boiling) & stock, esp. poultry; pods & lvs. u. fodder; crop as green manure; lvs. chewed for cough; rts. boiled with flour u. for syph. (Tanzania); lf. inf. u. to bathe swellings & sores (Br).
cajaput: cajeput: cajuput (tree); cajuputih (Mal): white wood.
cajenne pepper: *Capsicum frutescens*.
cajeputol: eucalyptol.
cajú (Cu, Br): cashew, esp. *Anacardium occidentale* - **c. gum**: fr. 1) *Spondias venulosa* 2) *Anacardium* spp. esp. *A. occidentale* - **c. manso** (Por): cajuput - **c. resin**: acajou r. fr. frt. of *Anacardium occidentale*.
cajuco (Sp): liane; reed.
cajueiro (Br): **cajuil** (Dom): cashew tree.
cajuput: *Melaleuca cajuputi*; *M. quinquinervis* - **cajuputol** = eucalyptol (SC first discovered in cajuput VO).
cake seed: *Abutilon theophrasti*.
Cakile Mill. (Cruc.): sea rockets; several spp. u. as antiscorbutic, diur.; purg., in foods (lvs. in salads) - **C. edentula** Hook. (eNA; natd. wNA, Austral): American sea rocket; u. as pot herb & as cattle food - **C. lanceolata** (Willd.) O. E. Schultz (Fla & Keys [US]; shores of Caribbean Sea): hb. u. med., lvs. in salads; starvation food; pot herb; also rts. - **C. maritima** Scop. (Oceania): sea rocket; u. med. in New Caled.
calabacera (Co): gourd.
Calabar: river & towns of wAfrica (sNigeria); see C. bean.
calabash, **(Persian) (gourd)**: 1) *Crescentia* spp. 2) *Lagenaria* spp. - **sweet c.**: *Passiflora* spp. (WI) - **C. tree:** *Crescentia cujete*.
calabaza (Sp): pumpkin; gourd; calabash, &c. - **c. chata**: 1) (Mex): gourd 2) (Ve): *Luffa acutangula*.
calabazilla (swUS, Sp): *Cucurbita foetidissima*
calabazo (Pan): towel-gourd; *Luffa* spp.
Caladium bicolor Vent. (Ar) (tropAm): this & other spp. with fancy colored lvs. are popular ornamentals ("caladium"); toxic (because of CaOx); lvs. u. with olive (or other) oil to treat wounds & injuries; rts. also used; cult. ornam. - **C. esculentum** Vent.: *Colocasia esculenta* - **C. pictum** Lodd. = **C. seguinum** Vent. = *Dieffenbachia seguine* Schott.
calaguala (Sp): callahuala (root): rts. of ferns (*Polypodium*, etc.).
calalu: *Amaranthus spinosus*; *A. viridis* (foods).
Calamagrostis canescens (Web.) Roth (C. lanceolata Roth) (Gram) (n&cEu, Siberia) (moist places): reed bent grass; rt. c. saponin; u. diur. in rheum.
calamansi (Pl): *Citrus mitis*.
calamine: impure zinc carbonate (comml.); u. emollient (in ointments) - **prepared c.:** zinc oxide with Fe_2O_3 (0.5%); a zinc ore but also may be prepd. by mixing ZnO or $ZnCO_3$ with Fe_2O_3; mild astr. & antisep. u. to heal wounds, &c.; u.; as powd., ointment, &c.; skin protectant (E).
Calamintha alpina Lam. (Lab.) (sEu, alpine reg., Med reg.): "Alpenquendel"; hb. source of Schweizertee (Swiss tea) - **C. clinopodium** Benth. = *Clinopodium vulgare* - **C. graveolens** Benth. (Med reg. Transcaucas.): sds. u. as stim. & aphrod. - **C. macrostema** Benth. (Clinopodium macrostemum Ktze.) (Mex): tabaquillo (one of several plants so called); decoc. for GI dis.; lvs. as Chinese tea substit.; VO u. as replacement for peppermint oil (Mex) - **C. nepeta** (L.) Savi (Satureja n. Scheele) (and subsp. *glandulosa* P. W. Ball [S. calamintha Scheele]); (Euras, esp. Balkans (mts.) nAf, natd. e&cUS): calamint; field balm; marjolaine; (Wald) Quendel (Ge); hb. c. VO. with menthone, pulegone, isomenthone, pinene; 6% tannin; u. arom. medicine; epispastic; in urin. dis.; stom.; condim. - **C. sylvatica** Bromf. subsp. *ascendens* P. W. Ball (C. officinalis Moench; Satureja calamintha (L) Scheele var. *sylvatica*; S. nepeta (L) Scheele) (Euras [Balkans] nAf, natd. eUS) (common or wood) calamint; calamint (balm); field b.; basil thyme; mountain c.; flg. arom. hb. c. VO. with pulegone & other ketones; 6% tannins; hb. u. diur. epispastic; stom.; spice; proposed as substit. for Melissa (Fr Phar.).
calamo (It): calano aromatico (Por): Calamus.
Calamondin (orange): *Citrofortunella microcarpa*.
Calamus: 1) (L.): reed; writing reed or pen 2) genus (Palm.) (As, Af, Austral): the cane palms & rattan p.; largest palm genus (ca. 380 spp.); st. of various spp. u. entire or split to give canes, basketwork, furniture, novelties, &c. (esp. EI) 3) (L.) (specif.): rhiz. of *Acorus calamus* - **C. draco** Willd.: *Daemonorops draco* - **C. rotang** L. (EI): cane; "chair

bottom cane"; rts. u. in chronic fevers, snakebite; lvs. u. in biliousness; wd. vermifuge (Ind) - **C. scipionum** Lour. (Sumat, Cochin-Chin): source of Malacca canes (*qv.*); may be one source of dragon's blood - **C. viminalis** Willd. (EI, Mal): u. much like *C. rotang*.

calamus (root): **sweet c.**: sweet flag, *Acorus calamus*; prod. = **c. oil**.

Calandrinia ciliata DC (Portulacc.) (Pac coast NA): red maids; hb. grown in flower garden; potherb; garnish. (also var. *menziesii* Macbr.) - **C. barneoudii** Phil. (Chile): pata de huanaco; u. for headache; blows & wounds - **C. discolor** Schrad. (Chile): renilla; c. quercetin & kaempferol glycosides; u. for headache, rheum. & wounds (baths & massages) - **C. grandiflora** Lindl. (Chile): doquila; c. flavonoids; u. similar to *C. barneoudii*.

Calappus: coconut pearls; concretions rarely found in kernel of *Cocos nucifera* sd.

Calathea allouia Lindl. (Marant.) (tropAm, ex. Trin): allouya; topinambour; edible calathea; sweet corn root; rhiz. source of West Indian (or Guinea) arrowroot (starch); u. for scrofula & as antidote to manzanillo (Cu) - **C. cassupito** Meyer (tropAm): lvs. u. diur. - **C. lutea** (Aubl.) G.F.W. Mey. (CA, WI, nSA): cachibou; lvs. u. as thatch for roofs (Carib Indians); baskets; causassú wax fr. lvs. of hb. u. like carnauba wax - **C. tuberosa** Koern. (Br): tubers u. in treating wounds & and as food - **C. zebrina** Lindl. (sBr): zebra plant; lf. decoc. u. diarrh.; rt. stock as food.

Calatola columbiana Sleumer (Icacin.) (Ec): lvs. chewed by Indians to blacken teeth & lips (cosmetic).

Calavo™: California avocadoes.

Calcatrippae Semen (L.): sd. of *Consolida* spp. (*Delphinium*).

Calceolaria L. (Scrophulari) (tropAm): slipperwort; shoemaker flower; hbs. & shrubs mostly cult. as ornam. - **C. cuneiformis** Ruiz & Pavon (Pe, Ec): puru-puru; u. diur. & by natives for uter. dis. - **C. thyrisiflora** Grah. (cChile): hierba (and palo) dulce; palqui; lvs. u. as vulner. for stomatitis in lip & throat disorders, aphtha, stim. (gargles); exter. for wounds; small shrub c. anthraquinones, saponins, flavonoids, coumarins - **C. trifida** R. & P. (Pe): u. as antisep. & antipyr.

calcicole: pl. thriving in chalky soils ($CaCO_3$, &c.).

Calciferol™ (Brit): viosterol; irradiated ergosterol; vitamin D_2; mfrd. by u. v. irradn. of ergosterol in proper solvent; u. antirachitic. stim. in hypoparathyroidism, psoriasis, allergy, pemphigus, arthritides, rickets, etc.; of little value for poultry, which require vitamin D_3.

Calcii Amygdalas (L.): calcium mandelate.

calcite: Iceland spar (2568).

calcitonin: hormone of mammalian (pork & human) & fish (salmon) thyroid wh. lowers Ca level of blood; polypeptides derived fr. pork or salmon (preferred [2570]: Salcatonin); u. hypercalcemia, osteoporosis (post-menopause), in severe HBP; congestive heart failure.

calcitriol: steroid hormone representing biol. active form of vit. D_3 in calcium metabolism (bone, &c.); dihydroxy vitamin D_3; calciferol deriv.; u. as calcium "regulator"; originally obt. fr. chicken intestine; furnishing intestinal Ca transport & bone Ca resorption; however, now considered unsatisfactory for osteoporosis.

calcium: Ca: alkaline earth metal commonly found in large amts. in earth's crust, sea water, &c.; always in form of compds.; an essential element and constit. of bones, teeth, shells, &c.; chief common natural source, limestone; best dietary sources: cheddar cheese (0.67-0.93%); canned fish with bone (ex. sardine), greens (collard, kale, mustard, turnip), milk, soybean & soy flour; &c.; present in large proportions in bone & connective tissues; requisite in blood coagulation, neural transmission, &c.; administered in osteomalacia, osteonephropathy, tetany, inadequate uter. contractions in childbirth (often with oxytocics), &c.; in bone building, works in collaboration with vit. D & parathyroid hormone; claimed to reduce blood pressure (2472); Ca deficiency said often associated with hypertension; Ca salts with many indust. uses (2525) - **c. carbonate:** chalk; $CaCO_3$; occ. in nature in various minerals, as calcite, chalk; limestone, marble; &c.; prepd. chalk form sometimes preferred in med. (antacid); obt. fr. oyster shells (1241); (natural) & precipitated chalk (synth.) u. as antacid; dietary supplement (1241); antidiarrheal; u. in osteoporosis (also other Ca compds.); claimed to reduce risk of colonic cancer; dose 1-2 Gm. 3x/d (E) - **c. caseinate**: combination of casein & calcium salt; u. with food in infant diarrh. - **c. citrate**: u. as antacid; food stabilizer; to reduce hypertension in selected cases (also c. carbonate) (1973) - **c. gluconate**: u. as calcemic; less irritant than $CaCl_2$, &c.; may be used parenterally (E) - **c. hydroxide:** slaked lime; $Ca(OH)_2$; u. as astr.; mostly in building materials (concrete, plaster, &c.); pesticides; to destroy infection in wells; prepd. by addn of water ("slaking") to calcium oxide or quicklime (CaO; "lime"; burnt lime; Calx), in the process of which much heat is evolved (exothermic) - **c. mandelate**: a salt u. as urin. antisep. - **c. oxalate (neutral)**: a compd. of common occurrence in pls.; highly insol. in water, solns. of alkalis & alkal.

earths, acetic acid; sol. in dil. HCI or HNO_3; because of its polymorphy (5 xal forms) of much diagnostic value in microanalysis of pd. drugs - **c. oxide**: quicklime - **c. phosphate:** occ. in nature in various minerals & representing 3 compds.: **tribasic c. phosphate:** $Ca_3(PO_4)_2$ (NF) u. as calcium replenisher; antacid; in dental powders, &c.; **c. sulfate:** $CaSO_4$; occ. naturally as gypsum, alabaster; selenite, &c.; u. as diluent in compn. tabs.; plaster casts; pharmaceuticals; **calcium ursolate**: deriv. of ursolic acid, prep. fr. cranberries (*Vaccinium oxycoccus*); u. emulsif. agt.

calden(el) (Por): *Prosopis caldenia.*

caldera (Sp): retort, boiler.

Caldesene™: Ca salt of undecylenic acid.

caldo (Sp): broth.

Calea pinnatifida Banks (Comp.) (Br, Ur): aruca; jasmin de mato; hb. c. arucase; u. int. as amebicide in popular med.; lf. inf. u. as stom.; exter. as anti-inflamm. (wounds & bruises) - **C. zacatechichi** Schlecht. (Aschenbornia heteropoda Schauer) (Mex, Guat): Mexican calea; xicin; (falso) simonillo; bitter hb. u. in treating affections of stom., mal., cholera; emet., cholagog, dyspepsia; to remove scorpions; hallucinogen.

Calendula arvensis L. (Comp.) (Euras, Med, nAf) (c&eEu): (pot) field marigold; hb. & fls. (**Calendulae Flores**, L.) c. VO, resin, bitter subst. u. as hypotensive, emmen.; antibacterial, diaph.; at times for dyeing butter, cheeses, &c.; u. flavor, arom., condiment; anti-inflamm., diaph. in gastric dis. (choleretic), urin. dis. emmen.; CNS. sed.; for chilblains. warts; now important ornam. in fl. garden; pl. repellant to nematodes (1574) (prod. Fr) - **C. officinalis** L. (cEu): marigold; common, pot, or Scotch marigold; ruddles, u. med. for chilblains, warts, &c.; u. to color butter & to thicken soups; important cult. ornan., with many cultivars.

cali: *Salsola kali* - **c. beans**: **c. nuts: Semen Cali** (L.): Nuces C.: pseudo-calabar beans: sds. of *Mucuna urens* (*qv.*) (*Physostigma* & *Dioclea* spp. also suggested by some).

Caliciella corynella (Ach.) Vain, var. *chlorina* (Ach.) Räs. (Lepraria c. Ach.) (Lichenes) (widely distr.): brimstone lepraria; u. dyestuff.

calico bush: *Kalmia latifolia.*

Calicut: city of Malabar coast, sInd; now known as Kozhikode; important exports of drugs, &c.

California: v. under laurel, pepper tree, shellac - **C. bees**: **C. ferment:** beer seed (*qv.*) - **California gum plant**: *Grindelia* - **C. oak balls**: galls fr. *Quercus lobata.*

Calioben™: calcium iodobehenate.

Calisaya (**bark**): yellow cinchona, *C. calisaya* - **flat C.: table c.**: bk. pressed flat & dried (Bo).

calla, common: c. lily: **florist's c.**: *Zantedeschia aethiopica* - **true c.**: *Calla palustris* L. (monotypic) (Ar.) (e&cNA): u. ornam.

Calliandra anomala Macbride (C. grandiflora Benth.) (Leg) (Mex, Guat): (herba de) canela; u. in fevers (esp. malaria); rt. u. to retard fermentation in "tepache" (drink made from pulque & coarse sugar); as tan - **C. balakeana** Pittier (SA): clavellina (Ven); u. to produce tannins - **C. californica** Benth. (Baja Calif [Mex]): false mesquite; shrub with bright scarlet fls.; u. med. - **C. houstoniana** Benth. (Mex, WI, CA): rt. bk. called pambotano bark; c. alk. glycos. saponin; u. antiperiodic; in malaria (as FE & Elix.); bk. chewed to harden gums - **C. portoricensis** Benth. (tropAm); cojobillo; source of copaltic gum; u. in schistosomiasis (WI) - **C. tweedei** Mandar. (Br, São Paulo): u. emetic.

Callicarpa (means "beautiful frt.") (Verben.) (Am, As, Af): beauty berries; genus of ca. 140 spp. with attractive fls. in cymes, juicy frts. - **C. acuminata** HBK (Mex, CA): zacpucin; lf. u. diaph. purg.; decoc. in dys. diarrh. (sometimes cultd.) - **C. americana** L. (seUS, WI): French mulberry; beauty fruit; u. edema; frt. astr. acidul. (frt. typically mauve but white in one form); lvs. u. edema, malaria (Ala [US]) (sweat bath); dys., colic (Choctaw Indians [US]); rt. decoc. in gastralgia; antiviral activity in measles, polio, &c.; fish poison; mauve frt. clusters u. decor. (Fla [US]) - **C. arborea** Roxb. (seAs, Ind, Assam): bk. u. in leprosy & other skin dis.; ton. carm. - **C. candicans** (Burm. f.) Hochr. (sAs, Ind to Austral, PI+): arusha; hoary callicarpa; katoenapang; shrub; lvs c. callicarpone (kills fish) (callicapone has equal potancy to rotenone); lf. decoc. u. abdom. troubles; poultice for wounds & boils; expressed frt. juice u. for eyes (Caroline Is); rt. bk. antipyr. in lung and liver dis.; arrow poison ingredient; fish stupefying (Sumatra, PI); fiber u.; lf., st., rt. decoc. u. as appetite stim. after delivery (Indochina); wd. & lf. u. for headache (Caroline Is) - **C. elegans** Hayek (PI): "dup"; lvs. u. for treating headache & stomachache (exter. as plaster); lvs. & coconut oil appl. to wounds; frts. edible; pl. as fish poison (PacIs) - **C. erioclona** Schau. (IndoMal, PI): ligatar (Truk); lvs. u. for boils (Truk, Caroline Is) - var. *paucinerva* (Merr.) Moldenke: wd. & lvs. u. in headache, gastralgia (PI); lvs. put in mouth to produce emesis; lvs. appl. to wounds w. coconut oil; frts. edible; pl. fish poison - **C. formosana** Rolfe (Chin, Taiwan, PacIs): anoyop; Formosa beauty berry; poison b. b.; lvs. & fl. buds

u. for poultice (pounded) for cuts and bruises (Chin med.); lvs. & twigs pounded & u. to stupefy fish (PI) - **C. hitchockii** Millsp. (WI): boarhog bush; u. tonic (Bah) - **C. lancifolia** Millsp. (Bah): boarhog bush; lvs. u. ton. (local med.) - **C. longifolia** Lam. (Mal, Chin, nInd, Pak, Austral, Java): longleaf beauty berry; rt. decoc. diarrh., syph.; lvs. as poultice for wounds, treating hard tumors (Indonesia), decoc. in colic (Mal), fevers; to check dysem.; rt. lvs. bk. u. in sprue; lvs. fish poison; cultd. Pak - **C. longissima** (Hemsl.) Merr. (Taiwan): long-leaved callicarpa; Folium C.l. (Chin Phar.) - **C. macrophylla** Vahl (sChin, nInd, NG, &c.): "urn fruit tree"; large leaved c.; wd. rubbed against stone with water to form paste u. for oral sores (Ind); bk. as ton., carm., poult. for headache (Assam); rt. arom. oil u. in gastric dis.; lvs. warmed & appl. to rheumatic joints (Bengal); vulnerary; analgesic in neuralgias; shrub cultd. as hedge; in Chin med. (Chin Phar.) - **C. madagascariensis** Moldenke (s&eAs, Austral): u. fish poison - **C. nudiflora** Hook. & Arn. (Chin, cultd. Ind): u. for injuries (entire pl.?) - **C. pedunculata** R.Br. (As): frt. chewed with betal nut & spat into mouth of sick baby - **C. rubella** Lindl. (Chin, Burma, Assam): beauty berry; bk. & rt. chewed like betel nut; drug plant - **C. tomentosa** (L.) Murr. (eInd, SriL; Thail): "tondi"; c. mucil u. native med. int. & exter.; bk. chewed like betel (Ind, SriL); bk. & rt. decoc. in fevers, skin dis., liver dis.; lvs. boiled in milk & u. for oral aphtha; wd. carved - **C. vestita** Wall. (Ind): bk. chewed like that of *Carya arborea* (by Duff); lvs. rt. in skin dis. (Ind, Cey); juice u. in gastralgia (seAs).

callicrein: callikrein: enzyme fr. pancreas, lungs, & urine of mammals; hypotensive; acts on globulin of blood serum to form bradykinin, a nonpeptide with vasodilator activity (capillaries) (*cf.* vit. P.) (1611).

Callirhoë alcaeoides (Mx) A. Gray (Malv.) (2595) (central plains, Ill-Tex [US]): light poppy mallow; said to have been in the drug trade in the past (2512); rt. edible - **C. involucrata** (T. & G.) A. Gray var. *involucrata* (T. & G.) A. Gray (c plains, Ill-NM, [US]): purple (poppy) mallow; dr. pd. rt. burned & smoke inhaled for head colds; rt. decoc. ingested to alleviate int. pain; rt. edible (Plains Indians) = **C. i.** var. *lineariloba* (T. & G.) A. Gray (Tex, Okla [US], nMex): cowboy rose; rt. u. to cure "chipil" (sibling rivalry); rts. hung around neck of baby, who chewed on same (pacifier) (Kickapoo Indians); rt. eaten.

Callistemon citrinus Stapf (C. lanceolatus Sweet) (Austral) (Myrt.): lemon bottlebrush; red b.; lanceolate-leaved b.; shrubs u. as ornam.; u. med. (Ind).

Callistephus chinensis (L) Nees (C. hortensis Cass.) (monotypic) (Comp.) (Chin, Jap, Siber): Chin (or Chinese) aster.; rt. cooked with pork —> broth u. as ton. for general weakness; highly variable cult. ornam.

Callitris drummondii F. von Muell. (& some other C. spp.) (Cupress.) (Austral): cypress pine; lvs. & frt c. VO, callitris oil - **C. endlicheri** (Parl.) Bailey (C. calcarata [Cunn. ex Mirbel] F. Muell.) (seAustral): black cypress pine; furnishes good lumber; resinous exudation ("Australian sandarac") u. in varnishes - **C. quadrivalvis** Vent. (Tetraclinis articulata [Vahl] Masters) (nAf): "cedar"; arar tree; juniper gum tree; trunk exudes sandarac(h) gum (or pounce) (pd.); this resin u. for temporary dental fillings; dental cement; for incense; local stim. in rheum.; "risings"; techn. in shellacs & varnishes; prepn. plasters, retouching shellacs (photography); u. by Romans & ever since (chief prodn. Mor. export port, Mogador [Essaouira]) - **C. rhomboidea** R. Br. ex Rich. & A. Rich.) (C. cupressiformis Don) (seAustral): dune cypress pine; Oyster Bay pine; Illawara p.; Port Jackson p.; source of Australian (or Tasmanian) sandarac (pine gum) (considered less valuable prod.); cult. ornam. - **C. verrucosa** R. Br. (NSW, Austral): turpentine pine; produces the less valuable Australian sandarac ("pine gum").

Calluna vulgaris (L.) Nutt. (1) (Eric.) (Eu): Scotch heather; fl. (as Flores Ericae) c. methyl arbutin, quercitrin, ursolic acid; myricitrin, carotene, citric & fumaric acids, arbutase, gallo- and catechintannins, VO (with "Ericinol"); arbutin ("ericolin"); u. in colds, diarrh., insomnia, nervousness, female dis.; hb. (as Herba Ericae) u. astr. in diarrh., urin. calculi.; added to numerous teas; for insomnia; as yellow dye; hb. u. in beer ("heather ale"); fls. source of comm. honey (heather h.) to flavor "Drambuie" (malt Scotch whisky); young shoots & sds. eaten by game birds; sts. formerly u. to build cottages (mixed with peat, mud, straw; still u. for shed walls); lvs. in roofs (116); orn.

callus: callose: carbohydrate; polymer of glucose residues occ. in sieve tubes & other cells.

calmie (Fr): *Kalmia latifolia.*

calmodulin: Ca-dependent regulating protein (CDR); ubiquitous in pl. & animal cells; prepd. fr. bovine brain; Ca functions as cell regulator when bound with calmodulin.

Calocarpum sapota Merr. = *C. mammosum* (L) Pierre = *Pouteria sapota* Moore & Stearn = *Manilkara zapota* (*qv.*) - **C. viride** Pitt. = *Pouteria viridis* Pitt.

Calocedrus decurrens Florin (Libocedrus d. Torr.) (Cupress) (Ore, Calif [US]): Calif. incense (or post) cedar; (bastard) cedar; lf. decoc. u. for stomach distress (2441); wd. very durable, u. for shingles, &c.; ornam. tree.

Calochortus Pursh (Lili) (wNA): sego (lily) (Utah Indians); mariposa lily (Calif [US]); globe tulip; "black-eyed Susan" (error); small corms u. as "delicious: food (NA Indians), famine food (Mormons); orn. - **C. gunnisonii** S. Wats. (nwUS, mt. states): "mariposa"; dried corms pounded into flour to make porridge or mush (Cheyenne Indians) - **C. macrocarpus** Dougl. (BC [Cana]-nCalif [US]): green-banded star tulip; desert lily; mashed bulbs appl. directly to skin as antidote in poison ivy dermatitis; eaten raw or boiled with sweetish starchy taste - **C. nuttallii** T & G (wNA): sego lily; "children's food" (Indians) (SC children eat corms while playing); corms roasted or boiled as food; emergeny food; state fl. of Utah (US).

Caloncoba echinata (Oliver) Gilg (Flacourti). (tropAf): gorli; bk. & lvs. u. to treat cutaneous lesions of leprosy; &c.; sds. prod. gorli oil (g. fat) with hydnocarpic, gorlic, & chaulmoogric acids; u. like chaulmoogra oil - **C. welwitschii** (Oliv.) Gilg (etropAf, Tanz): mutire: decoc. of rt. with roasted squash sd. drunk for epilepsy; lvs. u. in dental caries; sd. FO u. like that of *C. echinata.*

Calonyction Choisy = *Ipomoea* - **C. aculeatum** House = *I. alba.*

Calophyllum brasiliense Camb. (Guttiferae/Clusi) (Amazon, Br, WI, CA, sMex): (balsam) maria; "river marble"; mangue; trunk & branches yield a balsamic resin ("jacareuba") u. as epispastic (in plasters); decoc. u. as diaph. in pulmon. dis.; fls. or bk. u. as pectoral; sd. FO (guanandy fat) u. as vulner. in skin dis.; cult.; var. *rekoi* Standl. (C. r. Standl.) (Mex, CA): barillo; cimarrón; branches yield resinous sap or latex-like chicle; u. by Am Indians as "Leche Maria" to heal infant's navel; sd. FO u. as vulnerary for burns, &c.; excellent wd. like mahogany; orn. - **C. calaba** L. (C. antillanum Britt.; C. brasiliense var. a. (Britt.) Standl.) (WI): (Santa) Maria; resinous balsam u. as vulner. on bites, wounds, & burns (Cu); c. bioflavonoids, triterpenes; FO (sd); inf. of bk. or fls. u. as pectoral; & diaph. - **C. inophyllum** L. (tropAf, As [Ind, SriL], Oceania, SA): Alexandrian laurel (tree); kamani; undi; sticky fragrant resinous sap fr. incisions in branches called (East Indian) "tacamahaca resin" (c. brasiliensic acid); u. exter. in plasters; for ulcers; as fumigant; resin fr. frt. u. as emet. purg; sd. FO (called pinnay, poon or domba oil) u. exter. in rheum. (liniment) itch; leprosy, antibacterial props. (u. to treat boils, sores) (Solomon Is) demonstrated (40); CNS depr. (1132); in candle & soap mfg., as lamp oil; bk. u. as diur. emmen. tan (PI); lf. in eye dis.; sds. bk. & lf. as fish poison; frt. oil u. med. & to scent coconut oil (Fiji); frt. edible; valuable wd. ("E. Indian mahogany") - **C. tacamahaca** Willd. (C. inophyllum Lam. non L.) (Madag, Bourbon Is): source of Réunion tacamahac (African or Bourbon t.) (or Mary Balsam) u. in wounds & sores.

Calopogonium mucunoides Desv. (C. orthocarpum Urban) (1) (Leg) (SA, WI+): jicana; a climbing perenn.; u. green manure, cover crop; stock poisoning pl.; juice squeezed fr. young lvs. & shoots u. for general debility (with coconut water added) (Caroline Is).

calorie: **calory**: heat unit, being quantity of heat needed to raise temp. of 1 g. of water 1 degree C - **Calorie: Calory: Kilogram c.: large c.**: = 1000 calories; unit u. in nutritional studies; av. cal needed (human) = 3500-4000/day.

calorific wool: capsicum wool.

Calotropis gigantea (L) R. Br. (Asclepiad.) (sAs, natd. Af, deserts): madar; mudar; giant milkweed; shrub produces silk cotton fiber fr. sds. (akon fiber) (*cf.* kapok) (floss u. to stuff pillows, mattresses, &c.); also bast fiber = jetee or yercum (calotropis floss): all parts u. in Ayurvedic med.; latex, fl. bk. of rts., lvs. rt. bk. u. diaph. alter.; lf. & fls. u. in asthma & as digestive; mudar gum obt. fr. latex, rubber-like; c. uscharin & uscharidin; latex c. papain-like enzyme; latex u. in eye dis., abortive; cardioactive, to stim. blood flow to surface wounds (Ind); in cancer; lvs. heated & appl. to insect bites (Indochin); rt. bk. emet. (Burma); fls. decorative - **C. procera** (Ait.) R. Br. (tropAf, As, natd. Latin Am, bunds & roadsides): auricula tree; akund; Sodom apple; us(c)har plant; botón de seda; rt. bk. called mudar bark c. phytosterol-like alc., taraxasterol. lineolone; rt. bk. u. in leprosy, syph. ton. diaph. emetic, in chronic rheum. (Ind); lvs. tied on head for headache (Bah); appl. with olive or coconut oil to inflamm. (Ve); latex c. calotropin; calotoxin, uscharidin, &c.; (cardioactive steroid); u. as purg. sternutat., arrow poison (Af); akon wax c. melissyl alc., phytosterol; floss u. to stuff pillows, mattresses, &c.; rt. bk. or latex u. in liquor for suicide or homicide (1207).

Caltha palustris L. (Ranuncul.) (Euras, NA): (common or wild) marsh marigold; u. pot-herb (greens) (also other spp.) (cook thoroughly!); pl. u. as exp.; yellow dye; pl. somewhat toxic (all parts) c. jervine (1509); palustrolide; (fresh flg. pl. employed in homeopath. med.); orn.

caltrop(s): *Tribulus* (the caltrop was an offensive weapon of metal u. in ancient times; when thrown around castle, one point always stood up & could cripple men or horses; SC rigid thorns on plant) - **c. family**: Zygophyllaceae.

Calumba: *Jateorrhiza palmata*, columbo (*qv.*) - **false c. (root): Mariettac: Spurious c.:** *Frasera Carolinisnsis* - **c. wood**: *Coscinium fenestratum.*

calumet: peace pipe of Am Indian u. for tobacco ceremonials.

calvacin: mucoprotein antibiotic active against animal tumors; prod. by *Langermannia gigantea* Rostk. (segregate of *Calvatia g.* Lloyd [Lycoperd.] "giant puffball" [widespread]; edible with delicious flavor) (1510).

Calvatia bovista Kambly & Lee (Lycoperdon b. Pers.) (Lycoperd. - Gasteromycet. - Fungi) (cEu, NA): giant puffball; ripe sporophore formerly u. as stypt.; anesth.; edible when unripe; spores; in Negro voodoo - **calvatic acid:** antibiotic fr. *Calvatia lilacina.*

Calx (L.): lime - **C. Viva**: quicklime.

Calycanthus floridus L. (Calycanth.) (seUS): common sweet shrub (SC sweet-scented); rt. u. emet.; bk. arom. stim.; bk. heated with lard & elder bk. to form "fire grease" u. for burns (Ala [US]); fls. arom. frag.; var. *laevigatus* T. & G.; (seUS): sweet-scented shrub; sd. c. calycanthine u. against intermittent fevers (mal.); uterine ton. (Am Indian); cattle & sheep poison; cultd. as ornam.

Calycogonium squamulosum Cogn. (Melastomat.) (WI, PR +): jusillo; wd. considerably u. in furniture, boat parts, &c.; aerial parts active against carcinosarcoma but richness in tannins precludes use in therapy (2276).

Calycophyllum acreanum Ducke and **C. spruceanum** (Benth.) Hook f. (Rubi.) (Pe): bk. u. for fungus infections (upper Amazon reg.); also for "sarna negra" (arachnid parasite under skin); &c.

Calycosia petiolata A. Gray (Rubi) (Fiji): tarutaru; bk. u. for toothache.

Calylophus serrulatus (Nutt.) Raven (Onagr.) (Tex [US]): close to *Oenothera*; half-shrub sundrop; wire primrose; u. in eye dis. & sore throat; ornam.

Calypso bulbosa (L) Oakes (C. borealis Salisb.) (1) (Orchid.) (Euras, wNA): fairy slipper; calypso; monotypic; found in cool bogs & damp woods; corms ("bulbs") eaten by Creek Indians; ornam.

Calyptocarpus vialis Less. (Comp) (sLa - sTex [US]; Mex s to CR, WI): pl. inf. u. for diarrh.

calyptra: 1) calyx in some pl. which is carried off in toto (ex. *Eucalyptus*, *Eschsholtzia* 2) covering of moss sporogonium 3) rt. cap 4) corona separating entirely.

Calyptranthes multiflora Poeppig (Myrt.) (Co): fls. of small tree u. to prep. tea u. for abnormally swollen breasts (given orally occ. for 2 days) (Taiwano Indians) (2299) -**C. paniculata** Ruiz & Pav. (Co): tea fr. red berries of treelet given to increase flow of milk from swollen breast (Barasana Indians) (2299); buds & frts. u. as condim (Pe).

calyptrogen: meristematic tissue fr. which root cap is formed.

Calyptrogyne ghiesbreghtiana H. Wendl. (Palm.) (Chiapas, Mex to Pan): guanito-talis (Mex); lvs. u. to cover huts - **C. sarapiguensis** Wendl. (CR): coligallo; lvs. u. for roofs & walls of houses; orn.

Calystegia sepium (L) R. Br. (Convolvul.) (Euras): bracted bindweed; wild morning glory; correguela major (Sp); bearbind (Eng); hb. & rt. u. as strong cath. (gum-resin) ("German scammony"); antipyr. anthelm. (folk med.); young lvs. edible - **C. soldanella** R. Br. (Convolvulus s. L.) (Med reg., on sea coasts worldwide): seashore glorybind; rt. c. resin, flavonoids; lvs. & rts. u. drastic; diur.

calyx: outermost whorl of fl. made up of segments called sepals.

calzoncillo (Mex): (lit. underpants) *Passiflora mexicana.*

camacarlata (CA): *Passiflora pulchella* +

camanulas (PR): *Coix lachryma-jobi* L. (Poaceae).

camara (Br): 1) *Acrodiclidium camara* 2) cambará.

camas(s) (fr. Nootka Indians, camash, chamass, meaning frt., sweet): *Camassia* spp. - **death c.: white c.**: *Zigadenus* spp. of wUS esp. *Z. venenosus*, *Z. nuttallii.*

camaso (Mex, Sp): *Lagenaria siceraria.*

Camassia quamash E. Greene (C. esculenta Lindl.; Quamasia q. Colville) (Lili) (wNA): common cam(m)as; blue c.; quamash; wild hyacinth; bulbs popular food of Piute Indians & of Indians near Vancouver (BC [Cana]), eaten raw (or best cooked) (*cf.* onion) when plant is in flower (safe); made into meal & cakes (Pl. sometimes confused with death camus, *Zigadenus* spp.).

cambará (Br): *Lantana* spp.

cambium: meristematic tissues of perennial Gymnosperm & Dicotyledon pls., i.e., actively reproducing & growing soft st. & rts., tissues found as (1) fascicular (inter-fascicular) cambium in the fibro-vascular bundles, from which arise phloem & xylem tissues; and (2) bark cambium (phellogen) fr. which arise phellem & phelloderm (atypically, Monocotyledons have vestiges of cambium in st. & rt.) (L. = exchange).

Cambodia (L.): **Cambogia**: gamboge.

cambric tea (NA): beverage made with little or no tea, but with hot water, sugar, & esp. with milk (flavored with coffee Mississippi [US] 1500).

cambur (Ve.) banana.

camedrio (It): **camedrois** (Sp): *Teucrium chamaedrys*.

camel: *Camelus* spp. (Camelida-Tylopoda-Artiodactyla-Mammalia): herbivorous domesticated animals of hot dry regions; Arabian or dromedary c. is *C. dromedarius* while the Asiatic or Bactrian c. is *C. bactrianus*; the former 1-humped, swift and u. in nAf and Arabia, the latter 2-humped and u. in cAs (Chin+); characteristic humps c. adipose tissue; u. as beasts of burden in desert regions (can go several days without water); flesh (meat); hair (c. hair brushes); milk; med. (Unani; flesh u. in TB of hip joint, as sexual ton.; ash of hairs styptic, &c.; milk, urine, &c. also u. med.) (other members of c. fam. include llama, alpaca, vicuña, guanaco, &c.).

camel-thorn: 1) *Alhagi camelorum* 2) *Acacia giraffae*.

Camelina sativa (L) Crantz (Cruc.) (sEu, cultd. Belg, Holland, USSR): "cheat"; gold of pleasure; (big seed) false flax; a fiber pl.; sds. yield FO, German sesame oil; u. soap making (formerly); culinary oil, illuminant; hb. & sds. (cameline) u. as soothing agent in cataplasms; sd. adulter. linseed; lvs. vermifuge; hb. & sds. were food of Iron Age man (Eu).

Camel(l)ia drupifera Lour. (C. kissi Wall.) (The.) (Chin, Himal reg., Indoch, EI): **Himalayan camellia**; sds. c. 28-35% FO; tea-seed oil (oil of camellia); u. food oil, cosmetics (hair oil, Jap) - **C. japonica** L. (Chin, Jap): common camellia; japonica (shrub); sds. prod. FO, camellia oil; fls. are u. folk med. for gastrointest. dis. & epistaxis; lvs. cardioactive drug; cult. for showy fls. (1133) - **C. oleifera** Abel (Chin): sds. represent comm. source of tea (seed) oil - **C. sasanqua** Thunb. (Chin, Jap): sds. prod. FO tea-seed oil; culinary oil, in hair oils, as lubricant for precision machinery; fls. u. to aromatize tea; showy ornam. - **C. sinensis** (L) Ktze. (C. Thea Link; Thea sinensis L.) (Chin, Ind-cult.): st. tips u. since antiquity for stim. beverage (inf.) (black, green, oblong); young shoots with bud & lvs. clipped, dr., rolled, fermented, dr. again & sorted (Pekoe, Souchong, Orange Pekoe, etc.); popular all over the world (most used caffeine beverage) — hot, iced, with cream, sugar, lemon, etc. (1578); sd. FO is nondrying, important industrial prod; u. soaps, food, illuminant; blow oil u. in textile indy.; pd. lvs. (tea dust) u. in prodn. caffeine, also theophylline; prepn. fungus tea (Combucha); hedge plant; var. *assamica* (Masters) Kitam. (*T. assamica* Mast.) (Chin): Assam tea; sds. c. FO (tea-seed oil) (1577), saponin subst.; other vars.: *v. bohea*; (Thea b. L.); *v. cantoniensis*; *v. viridis* (Thea v. L.) (green tea).

Camelliaceae: Tea Fam., Theaceae (*qv.*).

camellias, silky: *Stewartia malachodendron*.

Camelus bactrianus L. (eAs) (Camelida - Tylopoda - Artiodactyla - Mammalia) (Chin, Pak, Iran, USSR): Asiatic (or Bactrian) camel; 2-humped beast of burden adapted to hot desert transport; most domesticated animals (wild sp. in Chin, Mongolia) - **C. dromedarius** L. (Arabia, nAf, Austral - deserts; extinct in wild): Arabian (or dromedary) camel; herbivorous; the hump with much fat, serves for water storage; hair (camel hair brushes), flesh, milk employed; products considerably u. in Unani med.; flesh (for TB), lungs, fat of hump, milk, curd, rennet, urine, &c.

camomila (Sp): **camomile** (Eng, Fr): chamomile (*qv.*) - **golden camomile**: yellow c. - **rock c.**: *Anthemis montana* - **yellow c.**: *Anthemis tinctoria*.

camomilla (It): **camomille** (Fr): chamomile.

camote (SpAm) = camotti (Mex) = *Ipomoea batatas*.

camotillo: 1) *Exogonium bracteata* 2) (SA): food prepd. fr. tuberous rts. of *Mahihot esculenta*.

campana (Cu): *Datura arborea* - **c. blanca** (white bell) (PR): *D. metel* - **c. de pascua** (PR): *Solanum wendlandii* - **c. de Paris**: *D. suaveolens*.

campanelo (Sp): 1) *Thevetia peruviana* 2) *Datura candida* (Guat).

campaneta (Sp): sweet acorns.

campanilla (Mex): 1) *Datura candida* 2) *Ipomoea caerulea*.

Campanula L. (Campanul.): bell flower; hare bell; ca. 300 spp. mostly of nTemp Zone, esp. Eu; many planted as ornam. - **C. latifolia** L. (Euras): hask work; bell flower; breiblaettrige Glockenblume (Ge): fls. u. at times, as emetic. exp.; ornam. - **C. medium** L. (sEu; nSA): Canterbury bells; Coventry bells; hb. u. in rabies (?!) - **C. rapunculoides** L. (Eu, NA): rt. c. inulin; lvs. & rts. eaten; bad weed - **C. rapunculus** L. (Euras, nAf): rampion; young tuberous rts. edible, u. with lvs. as salad; cultd. - **C. rotundifolia** L. (Euras, NA): (Scottish) bluebell; harebell; much variation; fls. u. in epilepsy; ornam. garden hb. (Song of Solomon 4:13).

Campanulaceae: bluebell or bellflower fam.; (worldwide) large group (mostly herbs) with some 2,000 spp., incl. genera *Campanula, Specularia, Adenophora, Jasione, Lobelia*; notable for value in ornam. garden.

Campeachy wood: campêche (Fr): **Campecheholz** (Ge): *Haematoxylon*.

Campelia zanonia HBK (Commelin.) (Salvador): caña de Cristo; u. GC (—> *Tradescantia*).
campesino (Sp): small farmer.
campesterol: sterol fr. FO of soy bean, rape, wheat germ, &c., & mollusks.
camphene: terpene hydrocarbon (3 isomers; α- and ß- are solids); widely distr. in pl., ex. camphor oil, fennel oil, & numerous other VO (1511).
camphine 1) oil turpentine, rect. 2) decinene.
camphire: camphor.
camphor: 1) (Jap, Ind): dextro-camphor: concrete ketone (stearoptene) obt. by dist. all parts of tree of *Cinnamomoum camphora*, camphor oil being obt. at same time (the eleoptene fraction); many u in med., pharm., indy.: anti-infective (local); antipruritic (local); stim. carm.; antitussive; expect.; counterirrit. (local); nasal decongestant; in mfr. plastics, esp. celluloid; varnishes; moth repellant; preservative (pharm., cosmetics), &c. 2) *Dryobalanops aromatica* (VO) 3) generic term for vol. arom. stearoptenes which are solid crystals at room temp. & closely associated with & related to terpenes - **c. basil**: *Ocymum kilimandscharicum* - **Blumea c.**: obt. by sublim. fr. *Blumea balsamifera* - **Borneo c: borneol c.: d-borneol**: obt. fr. *Dryobalanops aromatica* (VO) - **buchu c.**: diosphenol - **crude c.**: prod. of first sublimation, commonly shipped fr. Chin & Jap & which must be resublimed & cleaned before being u. as a medicinal - **c. gum: gum c.: c. - c. ice**: comp. c. cerate - **c. herb:** *Camphorosma monspeliaca* - **c. in oil**: sterile soln. of c. in a bland FO (such as olive oil) for parenteral u.; distinguish fr. camphorated oil - **Japan c.: c. - c. julep**: c. water; sweet mucilaginous mixt. with c. - **c. laurel**: c. tree - **levo-c.**: obt. fr. VO of *Matricaria* spp. *Ocymum canum, O. kilimandscharicum* - **C. liniment**: camphorated oil: soln. of 20% camphor in olive or peanut oil (or other suitable FO); u. exter. as rubefacient & mild analg., counterirrit. in fibrositis, neuralgia, &c. - **Manipuri c.** (Ind): fr. *Blumea* spp. - **mint c.:** menthol - **monobromated c.:** compd. of Br with c. (1:1) - **natural c.**: c. obt. by sublim. fr. the tree as distinct fr. that made by synthetic means; d-rotatory; remains an important commodity; offered as slabs, tablets (1 oz.) & pd. (prodn. Jap, Taiwan, Chin, Ind) - **Ngai c.**: Blumea c. - **c. oil (white)**: 1) the VO obt. in the process of c. prodn.; rich in safrole; very important prod. in Far East; u. in embrocations, as solvent, &c. (chief prodn. Jap, Taiwan, Ind) (also yellow & blue oils) 2) VO of *Dryobalanops aromatica* - **peppermint c.:** menthol - **Samatra c.**: Borneo c. - **synthetic c.**: a prod. made by chem. processes fr. (1) pinene or (2) cyclopentadiene; is racemic; said sl. more toxic than natural; with slightly unpleasant aroma (1226) - **tar c.**: naphthalene - **thyme c.:** thymol - **tonka bean c.:** coumarin - **c. water**: sat. soln. of c. in dist. water; u. in eye preps. etc. - **c. weed**: *Trichostema lanceolatum*; *Pluchea camphorata*.
Camphora (L.): camphor - **Camphora officinarum** Nees in Wall.: *Cinnamomum camphora*.
camphorated mother plaster: Emplastrum Fuscum Camphoratum - **c. oil**: c. liniment.
Camphorosma monspeliaca L. (Chenopodi) (Med reg.): camphor herb; u. diur. diaph arom.
camphre 1) (Fr): camphor 2) henna.
campion: *Silene* spp.; *Lychnis* spp.
Campomanesia aromatica Griseb. (Myrt.) (WI Guianas): frt. pleasantly edible, resembling strawberry in flavor - **C. guaviroba** B. & H. (sBr); Pará guava; guabiroba; frt. yellowish, edible - **C. obversa** Berg (Par): med. u. (Par Phar.).
campos (Br): open grasslands.
Campsis grandiflora K. Schum. (C. chinensis Voss) (Bignoni.) (Chin, Jap): Chin trumpet creeper (or flower); all parts of pl. u. emmen. leucorrh., fever, diur., diabetes (!), appetite loss (infants) (oriental counterpart ["geminate sp."] of & also hybridizes with *C. radicans*) - **C. radicans** (L) Bureau (Bignonia radicans L.) (Fla-NJ-Ill [US], natd. Conn, Mich [US]): (common) trumpet creeper; cow itch; trumpet vine; rt. u. diaph. vulner (folk med.); cult. ornam. but weed in seUS; sometimes considered sensitizing; non-toxic.
Camptosema pinnatum Benth. (Leg) (Br, esp Minas Gerais): bk. macerate u. to treat scabies.
Camptotheca acuminata Dcne. (l) (Nyss.) (Tibet, wChin): wd. bk. frts. **c. camptothecin(e)**, non-basic alk.; very high activity against leukemia & some other malignancies (throat cancers); toxic.
camwood: Lignum Baphiae: fr. *Baphia nitida*.
can (US slang): ounce of morphine.
caña azucar: sugar cane - **c. brava**: *Bambusa vulgaris* (Cu): *Chusquea* spp. (CR); *Gynerium sagittatum* (Mex, Pe) - **c. dulce** (SA): sugar cane - **c. fistula** (Mex, CR): 1) *Cassia fistula* 2) (Br) *Cassia grandis* & other *C.* spp. (also **caña fistula** 3) *Peltophorum dubium*.
Canada balsam: the tree *Abies balsamea* or its oleoresinous prod. - **C. hemlock:** *Tsuga canad.* - **C. liniment**: Linimentum Opii Comp.
Canada: v. also fir; fleabane; hemp; moonseed; pitch; snakeroot; thistle; turpentine.
Canadian: v. also hemp; tea; puccoon - **C. fir balsam**: *Abies balsamea*.
canadine: alk. found in *Corydalis cava, Hydrastis canadensis*, etc.

cañagria (SpAm): 1) *Rumex hymenosepalus* (& possibly other *R.* spp.) rt. (**Canaigre**) 2) *Costus villosissimus* ("canagree") (canaigre) 3) (Mex): Mexican scammony rt.

canaline: cleavage prod. from **canavanine:** amino acids fr. *Canavalia ensiformis,* soybeans, &c.; inhibits growth of many m.o. (antimetabolite); toxic to cattle; induces wilting in some plants.

cañamo (Sp): 1) hemp 2) Cannabis - **c. indico**: Indian hemp.

cañandonga (Cu): *Cassia grandis.*

Cananga Hook. f. & Thoms (NL fr. Malayan name) (Annon.) (Mal): 1) genus of large trees valued for their fine fragrance 2) (seAs) tainghe; rt. or wd. u. for fevers, esp. malar.; fls. very fragrant, u. as substit. for ylang ylang - **C. latifolia** Fin. & Gagnep. (seAs): tainghe; rt. or wd. u. for fevers, esp. malar.; fls. very fragrant, u. as substit. for ylang ylang - **C. odorata** Hook. f. & Thoms. (seAs, Mal, Burma, PI): bois de banane (Maurit. Réunion); ylang-ylang; commonest *C.* sp.; large fls. are source of distd. VO called **c.** (or ylang-ylang) **oil**; greatly prized for its delicate perfuming properties (also terpeneless); lvs. crushed & appl. to boils; sds. u. exter. for intermittent fever, diarrh. (Indones); lvs. & bk. u. med. (Tonga) (86); frt. edible; FO u. to prep. macassar oil; tree planted for shade & orn.

canapa (It): hemp, Cannabis.

Canariese Semen (L.): canary seed, from foll.

Canarium album Raeusch. (Burser.) (PI, Moluccas): Java almond canary tree; arbol à brea; trunk bk. source of Manila elemi (resina elemi), an oleoresin which c. VO 10%, amyrin, elemic acid, resenes, etc.; Resina Elemi (or gum e.) is u. as stim. irrit. in plaster & ointment compositions, tech. in varnishes & lacquers; source of edible pili nuts (seeds) (Elemi is prod. by pls. of several other genera) (*Protium, Bursera, Boswellia,* &c.) - **C. asperum** Benth. (Oceania): ton. (in combns.); drunk during the puerperium (PI) & by convalescents - **C. commune** L. (C. indicum L.) (Moluccas): Java almond; sds. edible (eaten in pastries), source of FO (u. food, soaps, cosmetics, illuminant); pl. source of an oleoresin (Manila elemi); u. incense, perfume fixative; in shellacs, varnishes; VO fr. this resin u. perfumes, soaps, varnishes - **C. edule** Hook. f. (Angola+): frt. edible; bk. source of bulungu resin (oleoresin) u. as perfume & fumitory (incenses); in skin dis., for chiggers - **C. euphyllum** Kurz (Andamans): dhup; Indian white mahogany: one source of Manila Elemi - **C. hirsutum** Willd. (Indones - Solomon Is +): balsam appl. to wounds (Amboyna); kernels of nuts eaten by natives (Solomon Is) (also var. *leeuwenii*); rt. decoc. good for stomach dis. (Pe) - **C. luzonicum** (Bl.) A. Gray (seAs): Java almond; sds. one source of Manila Elemi; sds. ("pili seeds") edible; gum elemi u. in varnishes, &c. - **C. oleosum** (Lam.) Engl. (Oceania): gum u. with coconut oil as skin lotion & hair oil (PapNG); oil fr. oleoresin u. on wounds & for itch (Solomon Is) - **C. ovatum** Engl. (PI, Oceania): nuts purg., source of FO; u. food (roasted), in confectionery - **C. paniculatum** (Lam.) Benth. (Maurit.): source of Mauritius elemi (similar to Manila e.); c. amyrin. - **C. pimela** (L.) Leenh. (Chin): Chinese olives; a source of Elemi - **C. polyphyllum** K. Schum. (Mal to PapNG): frts. (sds) eaten, c. large amt. FO (Java almond oil; canarium oil); sd. presscake fed to cattle (rich in protein) - **C. samoense** Engler (Samoa; Tonga): makai (Tonga); u. to scent FO (Tonga) (86) - **C. schweinfurthii** Engl. (tropAf): incense tree; source of African elemi (Uganda or Cameroun e.) derived from trunk incisions; u. for roundworm, colic, dys. GC; to fumigate dwellings; incense in Uganda churches; mixed with fat & rubbed on body as perfume; frt. condim. - **C. strictum** (seAs): source of Black Dammar, also perhaps Alribe Resin - **C. villosum** B. & H. (PI): trunk produces sahing resin; u. as fuel, illuminant, caulking material, &c. (Philipp. Nat. Form.) - **C. vulgare** Leenh, (Java, Melanesia) tree -45 m.; Java almond; st. bk. source of Java elemi; sds. edible; u. as substit. for almonds; u. as shade and often planted along avenues or boulevards; trunk furnishes an elemi; sds., source of FO - **C. zeylanicum** (Retz.) Bl. (Af): tree another source of African elemi (Elemi commune).

canari(um) oil: FO fr. sd. of *Canarium luzonicum*: "Java almond oil."

canary grass (or **c. seed**), (**wild**): **c.-seed grass**: *Phalaris canariensis.*

cañatillo (NM [US], Sp): *Ephedra trifurca.*

Canavalia ensiformis DC. (Leg.) (tropAm, sFla [US] +, WI, PR, cult. Haw, Guam), jack (or sword or sabre) bean; Jamaica horse b.; young pods eaten but immature sds. thought toxic; meal u. as reagent (canavalia urease); widely cult. for seeds & as cover crop; chief u. as food (unripe tender pods & sds.) for humans & livestock (herbage also) (2026); also as source of urease; sd. coffee substit. (Negroes); sd. c. canavanine, concanavaline (*qv.*); poisonous saponin present in sd. should be destroyed before eating; green manure; as ornam. (at times) - **C. gladiata** DC. (pantrop, Far East, esp. Ind): sword bean; horse b.; broad b.; with red sds.; pods & sds. u. ton., stom. bechic (Chin); in rheum., sore throat (Congo); sds. c. canavanine (antibiotic useful against some pneumococci); roasted & eaten as delicacies, also boiled lvs. rubbed on urticaria (Congo); rt. decoc. u. for stomachache (Tanz); pods yield a strong mucilage - **C. lineata** DC. (tropAs as Indochin): June bean: arom. fl. u. spice (Mal); sds u. porridge (starch, pectin, no alk. nor HCN); good nutr. - **C. maritima** (Aubl.) Thou. (pantrop-beaches of S Seas, WI, &c.): "horse bean"; pl. u. in folk med. (Mex +); fls. u. to make leis; some consider conspecific with *C. rosea* - **C. microcarpa** (DC) Piper (Oceania): u. for headache & pains (Marshall Is) - **C. obtusifolia** DC (cosmop. trop): (purple-flowered) seaside bean; lvs. u. exter. as vulner.; sds. adulter. of Physostigma; rt. inf. u. for colds (2263) - **C. rosea** DC (EI, WI-coastal sands): jack bean (Austral): wild pl. considered toxic-narcotic (Java); sd. u. aphrod.; for baby being weaned (Haiti); adulter. for Physostigma (WI); rt. inf. u. for colds (2263).

canavalin: antibiotic subst. found in *Canavalia* spp; u. lobar pneumonia.

canaveral (NA Indians): cane brake (*cf.* Cape C., Fla [US], SC sugarcane grew there long ago).

cance (Fr): *Aira* spp.

cancer (L): crab - **c. drops** (US): *Conopholis americana* (c. powder) - **c. root** (US): 1) *Conopholis americana* 2) *Epifagus virginiana* 3) *Orobanche unicolor* - **c. weed:** *Prenanthes alba.*

Cancer irroratus (Cancridae - Crustacea - Arthropoda) (Atl NA): rock crab: meat popular sea food; shells u. in fish meals; soft vs. hard shell crabs; crab u. for chronic cough, TB (Unani med.).

cancerillo (Mex Sp.): *Asclepias currasavica.*

cancerwort: *Linaria spuria; L. elatine.; Veronica* spp.

canchalagua (Pe): *Schkuria octoaristata* (Comp.), u. in Pe folk med., esp. in malaria as substit. for canchalagua originating fr. Ch.

Cancrorum chelae (L): crab's claws - **C. Lapilli** (L.): crabs' stones; concretions fr. stomachs of crawfish; c. Ca CO_3 & Ca Phosphate; u. antacid.

Candelae Fumales (L.): fumigating pastils.

Candelariella vitellina (Ehrh.) Muell. - Arg. (Lecanor. - Lichen.) (cosmop. temp zone; on stones & walls): thallus u. to dye woollens yellow (Swed).

Candelilla Cera (L): **candelilla wax** (Mex): wax fr. stems of *Pedilanthus pavonis, Euphorbia antisyphilitica*, etc. (1575).

candicidin: macrolide antibiotic, mixt. of antifungal subst. obt. fr. *Streptomyces griseus* (strain) & related spp.; similar in properties to Nystatin but with greater effects against *Candida albicans* (&c.); u. chiefly in C. infect. of vagina (ointm. suppos.); very toxic (S. B. Penick & Co.) (A & B).

Candida albicans Berkh. (Cryptococc. - Hyphomycetes - Blastomycetes - Fungi): this yeast is a cause of candidiasis, cosmop. disease of man (thrush, &c.) and lower animals, moniliasis, & other mycoses (fr. L. candidus, light, clear, white, fr. appearance) - **C. utilis** (Henneb.) Lodder & Kreger-van Rijn) (Torulopsis u. [Henneb.] Lodder): food yeast; source of "vitamin T" (Torula Dried Yeast).

candleberry: 1) bayberry, *Myrica* spp.; 2) *Amyris balsamifera* - **c. nut (oil)**: *Aleurites moluccana, A triloba* (FO) - **c. tree:** *Byrsonima* spp.

"candle of the Lord" (swUS): *Yucca* spp. - **c. tree** (Pan): palo de velas: *Parmentiera cereifera.*

candlewood: 1) bayberry 2) **black c.: white c.:** *Amyris balsamifera.*

candurango: condurango.

candy: sweetmeats (fr. Persian) Arab: gand (sucrose); types: hard, cream, &c. - **rock c.**: sucrose in the form of hard, large, transparent crystals; popular as syr. in mixed drinks, folk remedy to treat colds, &c.; made by slow crystallization on thread or string.

candy tuft: *Iberis* - **bitter c. t.:** *I. amara.*

cane: 1) slender woody stem of grasses, palms, &c., often u. for wickerwork 2) *Arundinaria gigantea* - **c. brake:** 1) cane 2) dense growth of same - **c. cactus** (NM, Ariz [US]): *Opuntia imbricata* - **giant c.**: *Arund. gigantea* - **Indian c.:** *Rumex acetosella* - **ribbon c.** (seUS): *Saccharum officinarum* - **sweet c.:** Calamus.

canela (Sp): 1) cinnamon 2) winter's bark, *Canella winterana* (also **canella [bark]**) 3) (wMex - Baja Calif): *Pluchea odorata* - **c. amarilla** (Sp) *Nectandra lanceolata* - **c. guaiaca** (Sp): *Ocotea puberula, O. sassafraz, A. gardnerii*, &c. - **c. preta** (Br): *Nectandra mollis*, &c. (bk) - **C. sassafraz** = C. guaiaca.

Canella dulcis (L.): bk. of *Canella w.* - **C. zeylanica** (L): *Cinnamomum zeylanicum* - **C. winterana** (L.) Gaertn. (C alba [L.] Murr.) (monotypic) (Canell.) (sFla [US], Bahamas, WI [Virgin Is]): wild (or white) cinnamon; source of canella bark (Canella Alba); u. ton. stim. to flavor tobacco; condim.; fish poison (PR); lvs. also u.; water boiled with lvs. u. int. & exter. as diaph. in grip (10); timber u. as "Bahama whitewood." (2779)

Canellaceae: Canella Fam. (tropAm, Af): all spp. are trees with entire punctate lvs.; frt. a berry; incl. 5 genera - **canelle blanche** (Fr): - **canelo** (Sp): canella.

canelonie à l'italienne (Fr): a tube of wheat pastry stuffed with meat (SC appears like quill of cinnamon) (*cf.* canneloni).

canescens: hoary gray surface of pl. because of trichomes; *cf.* villous, strigose.
canescine = recanscine = deserpidine.
canferol (Por): kaempferol.
canfora (It): camphor.
canihua: *Chenopodium pallidicaule.*
Canis familiaris L. (*C. lupus* L.) (Canidae - Carnivora - Mammalia): dog; domestic (or house) d.; domesticated from small wolf (*C. lupus* L.) (gray wolf) in remote times (Euras); life span to 20 yrs.; body fat (dog fat) was u. in TB (Adeps Canis); dog milk (Lac Caninum) u. formerly in female dis. (homeopathy), arthritis; in Unani med.: prods. used (meat, liver, bile [u. in eye dis.], rennet, fat, ashes of tongue, bones); draft animal (Belg+); household pet; "man's best friend" (companionship, defense); to aid the blind (seeing eye d.); to smell & detect cocaine, cannabis, explosives &c.; 350 vars. ("breeds") (Chihuahua to Great Dane) (related also to jackal, coyote [*C. latrans*], &c.); flesh (meat) popular food in seAs (1668); u. in religious ritual sacrifices; meat u. in Unani med. in mental dis. (wild d.); ashed bones & teeth u. for piles, gout; dog bone, Os Canis (L.) = penis bone, baculum; u. in med. (Chin); other *C.* spp. include *C. nubilus* (timber wolf) (NA); *C. latrans* (northern or prairie coyote) (NA).
canistel: *Pouteria campechiana* (FO).
canja (Br): chicken soup or broth with rice, colored with curcuma, *Bixa orellana*, or saffron.
canjerana (Br.): *Cabralea* spp.
canker: 1) generally appl. to fetid sloughing ulcer of mouth; gangrenous stomatitis 2) (formerly) cancer - **c. berry**: **c. root**: *Coptis trifolia* ssp. *groenlandica* - **c. weed**: *Prenanthes alba, P. altissima.*
Canna coccinea Mill. (Cann.) (WI, CA, nSA): wild canna; "arrowroot"; lf. inf. u. diur., rhiz. decoc. diur. emm.; starch extd. fr. rhiz. u as food for invalids - **C. edulis** Ker-Gawler (tropAm, esp. Br): tous-les-mois; achira; gruya; edible canna; rhiz. source of Australian (or Queensland) arrowroot (or **canna starch**) (2749); this sp. may have been derived fr. *C. indica*; starch u. by invalids and children; black seeds u. as beads; now cult. worldwide in tropics - **C. X generalis** Bailey (SA. natd. Fl. Gulf coast): "canna lily" (misnomer); exudation on flg. stalks called "canna gum"; non-toxic, efficient emulsifying agt.; potential substit. for imported gums (1872) - **C. gigantea** Desf. (Br): rhiz. diur. & diaphor. - **C. indica** L. (C. orientalis Rosc.) (tropAm, cultd. Euras, sUS): biru manso (Por); Indian shot; chancle; rhiz. c. much starch, source of an arrowroot; rhiz. decoc. u. GC, amenorrh. (Chin); starch appl. to swellings, skin dis., bee stings; rhiz. decoc. u. yaws (Indochin) & other skin dis.; rhiz. decoc. diur. (PI), diaph., rt. juice u. Hg poisoning; fiber a possible jute substit. (often mixed with jute), in paper pulp & viscose mfr.; black sds. c. red dye; u. as beads; cult. ornam.
cannabinoids: monoterpenoid compds. characteristic of cannabis; some 50 compds. known; of potential med. value.
cannabinol: an active princ. of cannabis.
Cannabis: 1) formerly considered a monotypic gen. with the single sp. *C. sativa* L.; now, however, often considered to comprise at least 3 spp. (Mor./Cannabid.) (temp. sAs; now widely distributed over world); hemp 2) (L) (specif.): pistillate tops of *C. indica* or *C. sativa* (drug/hallucinogenic); habituating agt. - **C. Americana** (L.): **American c.**: American-grown *C. sativa* (2303) - **C. indica** Lam. = C. sativa var. i. Wehmer): **Indian c.**: I. hemp; marijuana; marihuana; (pot (US); and (true) dagga (1208); bhang; very many other common names (2599); considered to have high intoxicant potential; plants of intermediate stature (1-2 m); pistillate (female) plant tops c. Δ^9-tetrahydrocannabinol (THC) and its isomers (chief component with hallucinogenic effect); cannabinol (physiol. inactive); cannabidiol (physiol. inactive); many minor constits. incl. flavonoids, sterols, choline, VO, piperidine (giving characteristic odor to the plant & drug) (1135); urease; u. as "recreational drug" (illegal) (fumitory; masticatory; hallucinogenic & euphoric; orally eaten or drunk [hashish]; stim. aphrod. [supposed assn. with sexual crimes]) (illegal possession, US+) (1209); u. asthma (bronchoconstriction), glaucoma, polio & MS (value?), antinauseant in chemotherapy of cancer; surgical "anesthetic"; c. antibacterical subst. (1874); in epilepsy, HBR; for this & other *C.* spp., sd. u. as bird sd.; sd. source of FO (hemp oil) (indust. formerly u. in soaps); (chief prod. of drug: Turkey, Leb, Pak (Chitral state (district]); Afgan, Nepal, & sCalif, Mex, Col, Af, Mor) (hemp is one of oldest known cultigens, fr. ca. 8500 BC), and now one of the most valuable (2514) - **C. ruderalis** Janisch. (cAs, USSR): small weedy pl., may have derived from *C. sativa*; little intoxicant content - **C. sativa** L. var. *vulgaris* (widely spread fr. cultns. for fiber): tall plant (-5.5 m); limited intoxicant activity; u., mostly for fiber & sd., Eu); akenes edible & important fiber, hemp (chief prodn. Br, US [formerly one of greatest crops] [1512], Cana, USSR); source of paper (2590) - **synthetic c.**: TMC.
canne (Qu): *Typha latifolia.*
cannella (It): **cannelle** (Fr): cinnamon.

cannelon (Fr): fluted mold u. for shaping cheeses, ice cream bombs, confectionery, &c. (SC shape of cinnamon quills).

canneloni (It): It. pastry in form of tubes larger than spaghetti.

cannibal fish (Lake Superior [US]): c. liver FO rich in vitamins A & D.

cannodexoside: glycoside of cannogenin & D-digitoxose; occ. in *Pachycarpus distinctus.*

cannonball tree: *Couroupita guianensis.*

canola oil: rape (seed) oil; FO with high value of unsaturated fatty acids; u. as food (SC Canadian Oil).

Canscora diffusa R. Br. (Gentian.) (tropAs, Af): u. ton.; adaptogen; in stomachache; tea substit. (Pl).

cantala fiber: fr. *Agave c.*

cantaloupe (as generally u. in US): kind of muskmelon; *Cucumis melo* convar. *cantalupa* (Pang.) Greb. sometimes appl. to other vars. - **c. noir des carmes** (Fr): *Cucumis melo* convar. *cantalupa* (or var. *cantaloupensis* Naud).

Canterbury bell(s): *Campanula medium.*

cantharides: plural of Cantharis - **brown c.: Chinese c.**: *Mylabris vesicatoria* - **green c.**: Cantharis.

cantharidin: active vesicant princ. of *Cantharis* & other genera of Meloideae (blistering beetles); repellant to predators; attractant for mating; u. sex. stim. (for female animal); lethally dangerous (1810); solid subst.; u. like Cantharis; also to remove warts (2601).

cantharidism: unhealthy state prod. by local use of cantharidin.

Cantharis (L.): a beetle, *Lytta vesicatoria* Fabr. (Meloideae-Coleoptera): Spanish (or Russian) fly; blistering beetle; cantharis; c. cantharidin; u. vesicant, counterirrit. veter. aphrodisiac; irrit. (coll. mostly in Spain, USSR & Hung); canth. also obt. fr. *Mylabris* & *Epicauta* spp., sometimes mixed with *Meloë* spp. - **C. atrata**: = *Lytta a.* - **C. vittata** (USP. 1820-50); American Blistering Bettle; potato fly; old fashioned p. beetle; *Epicauta vittata* (= *E. solani*) (Meloidae).

canthaxanthin: b-carotene-4.4' -dione; pigment occ. in plants, fungi, bacteria, algae, some animals; u. as coloring addn. in foods; as tanning agt.

Canthium aciculatum Merr. (Rubi.) (Mal): melor hutan (Mal); lf. decoc. u. in hiccup in small children; rt. decoc. in stomachache - **C. afzelianum** Hiern (Rubi.) (Af): shrub u. astr. in swelling of legs & knees (Senegal) - **C. arnoldianum** Hepper (OW trop): u. med. (Congo) - **C. ciliatum** O. Ktze. (sAf): lf. & bk. in enema as purg.; charm in witchcraft - **C. dicoccum** Merr. (C. didymum Gaertn.) (Ind, Mal.): rt. decoc. to lessen diarrh.; pd. rt. rind u. as plaster for broken bones - **C. glabrifolium** Hiern (Plectronia glabriflora Holland) (trop wAf): bk. c. calmatambetin & calmatambin (glycosides) u. in cough & for certain mental dis.; wd. brown u. for carpentry - **C. horridum** Bl. (Ind, seAs): lf. bk. rt. u. for dys. (Indoch), worm rem.; frt. edible - **C. inerme** O. Ktze. (trop & Af): Cape date; lf. u. diarrh. dys. (inf. in milk) (Zulus) - **C. lanciflorum** Hiern (Plectronia lanciflora Benth. & Hook.) (sAf): frt. edible & considered one of best frts. of the area - **C. venosum** Hiern (wAf): lf. & wd. ash mixed with oil & appl. for pain in side & as poultice to boil - **C. zanzibaricum** Klotzsch (Tanz): rt. inf. u. in stomachache (Tanz); rts. chewed with water for ankylostomiasis; arom. (*Canthium* is close to *Plectronia*).

Cantua quercifolia Juss. (Polemoni.) (Ec): u. as remedy for mal. by natives (2449).

canutillo (NM [US], Sp): *Ephedra trifurca* Torr.

caobo (Br): *Swietenia macrophylla.*

caoutchouc: (India) rubber.

capacho (Mex): *Canna indica.*

capansa (Por): *Ryania speciosa.*

Cape jasmine (or **C. jessamine**): *Gardenia augusta.*

capelvenere (It): maidenhair fern.

caper(s): Capparis: fl. bud of the caper tree (or shrub), *Capparis spinosa* - **c. euphorbia: c. spurge**: *Euphorbia lathyrus* - **c. family**: Capparidaceae.

capeticobo (Por): **capicobo:** *Polygonum hydropiper.*

capilaria (Sp): *Adiantum capillus-veneris.*

capile (Br): **capillaire**: beverage (non-alcoholic) made with juice of *Adiantum* & other fern genera and syrup; of med. value.

capillaire 1) (Fr): *Chiogenes hispidula* 2) **c. du (de) Canada: capillary berb**: *Adiantum pedatum, A. capillus-veneris* (sirop) - **c. (de Montpellier):** *A. cap.-veneris.*

capim (Br): caapim de Angola - **c. cheiroso** (c. limão): *Vetiveria zizanioides* (&c.) - **c. gordura** (Arg.): molasses grass.

Capita Papaveris (L.): poppy heads.

capitulum: fl. head; the char. of most striking importance in Fam. Comp. but seen also in some other fam., ex. Leg.; ex. *Arnica, Matricaria, Trifolium.*

capivi (Br): copaiba.

capote de lac (Qu): *Sarracenia purpurea.*

Capparaceae = Capparidaceae: large fam. (ca. 700 spp.) of mostly warm areas; hbs. & shrubs.

Capparis aphylla Roth (C. decidua Edg.) (Cappar.) (nAf, wAs): karil (Pak, Panjab); shoots & lvs. u. as plaster for boils & swellings; pd. to relieve toothache; bk. lax. diaph. anthelm.; in cough; frt. in cardiac dis.; food of nAf tribes; wd. makes wooden combs (Pak) - **C. citrifolia** Lam. (sAf): Cape caper; rt. decoc. u. as emetic; u. as charm in superstitious practices; claimed u. in Voodoun rites (Haiti) (2487); rt. decoc. for gall sickness (cattle) (Xhosa) - **C. coriacea** Burch (Pe, Bo): simulo caper; frt. (as Fructus Simulo) u. in hysteria, epilepsy, chorea, as sed. - **C. cynophallophora** L. (sFla [US], WI, CA, Ve): wild orange, black willow (Bah); Jamaica caper; dog phalllus; rts. & lvs. u. med. ("burro") (PR) in edema (Mex); wd. u. lumber; street tree - **C. elaeagnoides** Gilg (Tanz. Rwanda): rts. crushed & u. as poultice for testis pain, boils (Tanz) - **C. flexuosa** L. (tropAm): limber (or bayleaf) caper; pan y agua (Ve): rt. inf. u. edema, emm.; lf. decoc. u. skin dis.; frt. sed. antispas.; bk. diur. emm. - **C. micrantha** DC (seAs): rt. grated with water & drunk in mal. (Indochin); rts. diur. uter. ton. (decoc.) for cough (Indochin); frts. edible - **C. sepiaria** L. (Asia, Austral+): rts. u. for boils (Hainan, Chin); antipyr. ton. in skin dis. (Ind) - **C. spinosa** L. (sEu, nAf, Med reg.): (common) capers; c. tree; citree; a shrubby pl.; fl. buds u. as flavor, appetizer, spice, esp. for pork sausage; in form of pickles (in vinegar & NaCl soln.) & caper sauce; frt. u. int. as decoc. for sciatica (Alger) - **C. tomentosa** Lam. (tropAf): purg.; rt. inf. for painful eyes; rt. appl. as paste with heat for swellings (Tanz); in chest pain, for cough; for impotency & barreness; & many other uses (129) - **C. zeylanica** L. (C. horrida L.f.) (Ind to Indochin): lvs. u. counterirrit., cataplasam for boils, swellings (Ind); rt. bk. sed. anhydrotic; for gastritis.

cappucini (It): espresso coffee mixed with milk or cream (SC Capuchins, religious order, the members of which [monks] wore white hoods).

Capra hircus L. (Bovidae - Ruminantia): (domestic) goat; like sheep but lighter build; source of meat, milk, cheese, hair, hide; massage with goat's fat to relieve pain (Unani med.); milk as dietary supplement; other *C.* spp. include ibex, tur, markhor, &c.

Capraria biflora L. (Scrophulari.) (sFla [US], WI, tropAm): goatweed; "Venezuelan tea"; flowers all the year; lf. inf. as eyewash in ophthalmia (Trin) (683); hb. decoc. u. in diarrh. & other intest. dis. (Ve+); decoc. for colds (Cu) (1513); emmen. (Cu).

Capr(e)omycin: Capastat Lilly: cyclic peptide antibiotic prod. by *Streptomyces capreolus* Higgens; *cf.* viomycin; several forms known; u. in TB (along with PAS, streptomycin, isoniazid) (IM); not for pediatric use.

Capricornus (L.): with goat's horn; sign of Zodiac.

caprification: the process of artificially pollinating figs by having certain wasps (fig wasp) which grow in the wild fig (caprifig) transfer pollen to the cultivated plant; also by hand pollination.

Caprifoliaceae: Honeysuckle Fam. (temp zone): mostly shrubs with a few hbs.; lvs. entire. opp. exstipulate; ex. honeysuckle, elderberry, *Weigelia*; *Diervilla*; *Triosteum*; *Viburnum*.

Capriola Adans. = *Cynodon* - **C. dactylon** (L.) Ktze.: Bermuda grass, *Cynodon dactylon*.

caproic acid: hexanoic acid; occ. in palm oils, coconut oil, milk fat, &c.; characteristic goaty odor; u. intermediate in mfr. artificial flavors.

capr(y)aldehyde: caprinaldehyde: decanal; in VO of lemon, lime, tangerine, grapefruit, sweet orange (Guinea) (traces), *Citrus nobilis* var. *deliciosa*, lavender; also synth.; u. as aroma in perfumes.

caprylic acid: oct(an)oic acid: occ. in lemongrass oil, coconut oil & many other plants, both free & as ester; rancid odor; insecticide; antifungal against spp. of *Trichophyton, Candida, Monilia* (formerly); intermediate in prepn. perfumes (fruit aromas), dyes, &c. (*cf.* undecylenic acid).

capsaicin: capsacutin: capsaicitin: active pungent crystalline princ. of Capsicum; taste burning, pungent; appl. to skin produces erythema and inflammation but not vesication; local anesthetic (1136); potent topical analgesic; lax., believed to reduce blood cholesterol, increase cholesterol excretion, reduce blood clotting capability; for pain of shingles & arthritis; peripheral diabetic neuropathy (chronic pain).

Capsella bursa-pastoris Medic. (Cruc.) (n&cEu, now worldwide): shepherd's purse; a widely distrib. weed; hb. c. glycoside; resin, choline bases; prosapogenins, sapogenin, 2 acid saponins; diosmin; histamine; sds. c. cardenolides; u. as ergot substit. hemostat. (2413); vasoconstrict.; diur. emm.; in eye dis. (Chin); antipyr.; in dys.; diarrh. (1475); antitumoral; lf. rosettes eaten as spring vegetable by Chinese (hemostat. effect attributed by some to fungi found commonly growing in pl.) (esp. popular with homeop.).

capsicin: capsaicin.

Capsicum (L.) 1) genus (Sol.) (tropAm); peppers, bearing frts. with greater or less pungency 2) (L.): Cayenne pepper; *C. frutescens*; u. flavor, condim. (biggest use); stomach. (in atonic dyspepsia); vesic.; exter. irrit. rubefac. counterirrit. (chief prodn.: Ind, Jap, Af, SA, NA) - **C. annuum** L. (CA, SA): in temp zone an ann. hb. but in trop grows as a woody shrub of good size; (sometimes) placed as var. under *C. frutescens*; c. 0.013-0.1-0.22% capsaicin (the hot pungent principle) in oleoresin; var. *annuum* cultivar 'Ancho': (chile) poblano; (fresh) green capsicum; popular for centuries in México; cult. mostly in Mex, less so in US (NM, Tex, Calif); eaten green mostly; sometimes dr. (sometimes such eaten like peanuts with beer) (SC ancho = broad; lge) - **C. a.** var. *conoides* Mill. (var. *tabasco* auct.): Tabasco (red) pepper, with pods c. 40 x 9 mm - **C. a.** var. *longum* (DC) Sendtn. (C. longum DC): Louisiana [US] long pepper (or long red pepper); var. *glabriusculum* Heiser & Pickersgill (C. minimum Roxb.); bird p.; (BPC) (2781) - **C. chinense** Jacq.: one of the cultivated forms; close to *C. frutescens* & u. in same way - **C. frutescens** L. (C. baccatum L.) (warm Am): (bush) red pepper; chile; source of African Chillies or Cayenne Pepper; u. flavor; cond. (paprika, pimento); lf. decoc. in coughs, colds, asthma; lvs. appl. to ulcers & headache (Pl); frt. decoc. gargled for sore throat (Jam); frt. in salads; u. in prepn. catsup, tamales, chile con carne, pickles; u. diur. (Cu); frt. stim. to improve digestion & dispel throat congestion; in throat troches for coughs & hoarseness; in liniments, dental poultices (1210); analgesic locally (cream) for pain incl. rheum.; shingles (post herpetic neuralgia); u. for internal bleeding, fibroid tumors; to make Tabasco sauce (Louisiana [US]) (C. fastigiatum Bl.); frt. very hot (claimed hottest red p.); grinding red pepper in small bldg. was most severe punishment in Austral for convicts; Cayenne pepper u. by criminals to throw bloodhounds off trail; oleoresin u. as repellant ("pepper") (acts rapidly) (like Mace); var. *cerasiforme* Bailey (CA, SA): (hot) cherry peppers; frt. small but very pungent (red, yellow, purple forms) u. to treat hemorrhoids (paste, &c.); subsp. ***longum*** Sendt (var. *longum* Bailey): long peppers; Turkish paprika; u., for tumors (Aztec [NM] Indians); var. *grossum* C. Bailey: bell pepper; sweet p.; pimento; fleshy mild frt. u. in fresh salads or filled with chopped meat; green, yellow, red vars.; rhiz. said toxic (prodn. Sp. Por, Hung, Turk, etc.) (some bot. authors [ex L. H. Bailey] place all 30 or so spp. as vars. under *C. frutescens*).

Capsicum Baccatum (L. not BO): *Pimenta officinalis* - **c. cumarim** (L.): pimenta de cheiro - "**c. oil**": **c. oleoresin**: a galenical oleoresin of *c.* made by percolation with ether, then distilling off ether fr. ext. - **c. wool**: absorbent cotton treated with soln. of c. oleroresin; u. in rheum.

capsique (Fr): capsicum.

capsula (Sp): capsule; adrenal gland.

Capsulae Papaveris (L): opium poppy heads.

capsule: dry frt. of more than one carpel, typically opening at maturity along one or more lines of dehiscence; ex. poppy.

captopril: mercapto-proline compd. wh. acts as enzyme inhibitor inhibiting angiotensin I-converting enzyme (ACE) thus to prevent formation of angiotensin II (vasoconstrictor, raising BP); obt. fr. venom of pit viper (Br); u. for heart failure with high BP; Capoten Tabs. (oral use) (Squibb); disc.; ca. 1987.

capuchina (Sp): capucine (WI): *Tropaeolum majus* - **capulin** (Sp): choke-cherry or choke-berry - **capulinar** (NM [US], Sp): choke-cherry grove - **capulin** (Mex); *Prunus capuli* (**c. cherry**) - **capulincullo** (Sp): *Rhamnus californica.*

Caput nigri (L.): cabeca de negro (Br); tayuya.

caqui (Br): 1) *Mimusops kaki* (kaki) 2) persimmon.

cará (Br): *Dioscorea sativa* - **caracolillo** (Arg): type of coffee.

Caragana Fabr. (Leg.) (cAs): (Siberian) pea tree; shrub pea; a genus of ca. 80 spp. representing shrubs to small trees; commonly cult. for hedge, windbreak, or ornam. purposes; grown in dry places for shelter belts - **C. ambigua** Stocks (Baluch): lvs. eaten as veg. - **C. arborescens** Lam. (Siber): rts. & bk. formerly off. in some pharmacopeias - **C. chamlagu** Lam. (C. sinica Rehd.) (Kor, neChin): branch, shoot & rt. u. for beriberi, osteomyelitis, & various malignancies; edema; syph.; fls. edible - **C. pygmaea** DC (nChin, Siber): tartaric furze; rt. u. like licorice - **C.** sp. (Turkestan): sd. mixed with that of an *Ervum* sp. & eaten under name of Adas (Jasmuk).

caraguatá (Br): *Aloe vera, Bromelia serra* (fiber).

carainin (Irish): carageen.

Caraipa fasciculata Camb. (Clusi/Gutt.) (Guayana): tree yields caraipa balsam u. for wounds; FO u. in rheum. & skin dis.; sds. c. tenicidal subst. - **C. grandiflora** Mart. (Br): bk. u. as astr. & vulner.; latex as wound balsam & for rheum. - **C. piscidiformis** Ducke (SA, Br, Bo, Ve): source of caraiba balsam, very similar to copaiba; u. in skin dis.; sds. c. resin u. tenicide & in skin dis. - **C. psidifolia** Ducke (SA, Br, Ve, Co): tamaquare; a tree of lower Amazon; sd. u. as coloring & flavoring of foods, liquors, tobacco, med.; trunk source of caraipa balsam, (*cf.* copaiba); u. skin dis. (Br) (1134); ophthalmia; frt. 2-seeded; hard timber.

Caralluma edulis Benth. (Boucerosia stocksiana Boiss.) (Asclepiad.) (Ind, Pak): pl. u. anthelm., antirheum. in leprosy & blood dis.; food of natives - **C. russelliana** Cufod. (eAf): succulent; pl. flesh is repellant to goats who will not browse on it - **C. tuberculata** N.E. Brown (B. aucheriana Hook. f.) (Pak): u. as ton. stom. carm.; succulent cult. ornamental.

carambola: *Averrhoa c.* (star apple).

caramel: burnt sugar (coloring); "sugar color"; occurs as dark brown amorph. pd. or as paste; u. as coloring & flavoring of foods, liquors, tobacco, med.; prepn. by heating sugar into brown mass; involves many trade secrets.

caramuru (Br): catuaba.

caraña gum: caranna (gum): fr. *Protium guianense*.

caranda(y) (palm): *Copernicia alba* (source of **c. wax**).

caraota (Ve): common bean.

Carapa guianensis Aubl. (Meli.) (Lesser Antilles, CA, Ve, Br): carapa blanc; bois carapat (Guiana); crabwood; bk. decoc. antipyr., in rheum. vermif., to bathe syphilitic sores; in diarrh.; sd. yields thick brown FO called c. fat (*qv.*); andiroba oil; u. in soap mfr.; bk. as febrifuge; anthelm.; fresh lvs. u. as epispastic; lf. decoc. u. to bathe itching skin; tree furnishes valuable lumber (Brazilian or bastard mahogany); u. to make furniture & in general constr.; FO formerly u. to shrink heads (2782) - **C. moluccensis** Hiern (C. granatum Alston) (eAf, Ind, Cey, Molucc): Molucca crabwood; tree classed as a mangrove; sds. prod. **carap oil**: bk. astr. u. in dyes (Guam); tree yields resin; valuable wd. (in veneers, &c.) - **C. procera** DC (Guinea, Senegal [trop wAf] & As): cola amarga (bitter cola); sds. prod. carapa or toulocouna fat; sd. u. for roundworm, purg.; bk. c. tulucunine; u. antipyr. in malar., dys.

carap(a) fat: c. oil: andiroba oil; FO fr. sds. of *Carapa* spp., esp. *C. guianensis*; u. mfr. soap; natives (SA) u. lubricate skin in skin. dis.; insect repellant; illumin.

carapia´ (Br): *Dorstenia multiformis*.

caraway (seed): *Carum carvi* - **black c. (seed)**: *Nigella sativa* - **ergotized c.**: an adult. disting. by its dark color - **levant c.**: Carum fr. Turkey, Asia Minor, Greece, Malta, &c. - **Mogador c.**: c. fr. Morocco - **c. oil**: VO dist. fr. frt.; important producers: Netherlands, Pol.

carbamide: urea.

Carbasus (L.): gauze.

Carbenia benedicta Adans. = *Cnicus benedictus*.

Carbenicillin: α-carboxybenzyl-penicillin; semisynth. antibiotic; antibacterial for *Pseudomonas aeruginosa* & *Proteus* spp. inf.

Carbenoxolene™ (sodium): glycyrrhetic acid hydrogen succinate, a natural glucocorticoid derived fr. licorice; u. as antiulcerative for gastric (not duodenal) ulcer.

carberry: *Ribes uva-crispa*.

Carbo Activatus (L.): activated charcoal; commonly made fr. woods (esp. willow, poplar, linden); often now also made fr. other organic materials; u. in GIT dis., ex. dyspepsia, flatulence, diarrh.; antidote in poisoning (adsorption); appl. topically to wounds (E) - **C. Animalis Purifactus**: purified animal charcoal; made by carbonizing bones, &c.; c. 10% C. and 90% Ca_3 $(PO_4)_2$, &c.; u. decolorization - **C. Ligni: C. Vegeta(bi)lis**; wood charcoal, usually prep. by carbonization of wds. of *Salix*, *Populus*, & *Acer* spp.

carbohydrases: enzymes wh. aid in hydrolysis of carbohydrates; incl. diastase (malt, taka-d., &c.), ptyalin, amylopsin, amylases.

carbohydrates: complex, inert, naturally-occurring substances with the general formula $C_xH_{2n}O_n$, which c. aldehyde-alcohol or ketone-alcohol gps.; two general classes are recognized: the sugars (xalline, soluble, sweet); the polysaccharides (generally non-xalline, insol., tasteless; incl. starches, celluloses, gums, etc.).

Carbomycin: macrolide antibiotic fr. *Streptomyces halstedii* Waksman & Henrici, u. for infect. with gm. + bacilli, some rickettsiae, and large viruses; u. in penicullin-resistant cases; v. under Magnamycin.

carbon 1) the charact. element of organic compds.; occurs free in nature in 3 forms: diamond, graphite, charcoal 2) (Sp): charcoal 3) (Arg): coal - **c. black**: lamp black; u. to color licorice candy, &c. - **c. de leña** (Arg): charcoal - **c. de piedra** (Sp): coal (mineral) - **c. dioxide**: carbonic acid gas; CO_2; inhaled to stim. respn.; to extinguish fires, &c.; solid CO_2 = "dry ice" (u. to remove keratoses (warts), nevi, &c.; packaged in cylinders - **c. vegetal activado** (Sp): activated wood charcoal.

carbone (It): coal.

carbonitrile: CN.

carbonization: process of heating an organized substance in the absence of air until charred; the volatile prods. are driven off, leaving the carbon & ash remaining behind.

carbono (Sp): carbon.

carbooja (Urdu): melon.

carboxy-glutamic acid: gamma-carboxy-L-glutamic acid: GLA (Gla); amino acid found in blood coagulation proteins, plasma proteins, &c.; *Spirulina* &c.; claimed of therapeutic benefit.

carcadech = carcadet.

carcadet (Eg): *Hibiscus* fls.

carcinogen(ic): giving rise to cancerous growth.

carcinophylin = carzinophylin.

carciofolo (It): *Cynara scolymus*.

Cardamine amara L. (Cruc.) (Eu &wSiber): bitteres Schaumkraut (Ge); (large) bitter cress(es); u. as stom.; lvs. as salad - **C. bonariensis** Pers. (nwSA): u. to "cure" liver & kidney dis. (Ec) - **C. pratensis** L. (Eu, NA): cuckoo flower; young lvs.

eaten as watercress; hb. formerly thought card. agt.; antiscorb. deobstr. stim. diur. diaph.

cardamom (seed,) bastard: fr. *Amomum xanthioides* - **Bengal c.**: fr. *Amomum aromaticum* - **Cameroon c.**: fr. *Aframomum hanburyi*; *A. daniellii* - **Ceylon c.**: fr. *Elettaria cardamomum*; often green; sometimes wild, sometimes cultd. - **Chinese c.**: fr. *Costus speciosus* - **C. Hills**: Kerala State, S. Deccan, India: much c. cultn. here - **lesser c.**: frt. (capsule enclosing seed) of *Elettaria cardamomum* - **Madagascar c.**: fr. *Aframomum angustifolium* - **Oriental c**: *Amomum* spp. (mostly) fr. China, Indoch, Siam, &c. (1291) - **round c.**: fr. *Amomum cardamomum* - **wild c.**: 1) *Zanthoxylum capense* 2) *Aframomum xanthioides* 3) *Elettaria cardamomum* v. *major*.

cardamomes (Fr): **Cardamomi Semen** (L.): cardamom seed(s) - **cardamon**: less preferred English title of cardamom: **cardamono** (Por).

Cardaria draba (L) Desv. (1) (Crucif) (Med region, wAs, US): (pepperweed) white top; bijindak (Afgan); hoary cress; pl. u. antiscorbutic, vegetable (Afghan); sd. u. condiment (pepper substit.) for flatulency, fish poison; bad weed (US).

cardenilo (Por): verdigris.

cardenolide: aglycone (genin) of cardioactive glycosides, representing combns. with sterane alcohols of rare sugars; these compds. are also cardioactive, ring E = 5-membered lactone; incl. digitoxigenin, strophanthidin, strophanthidiol, periplogenin, cannogenin, gitoxigenin, gitaloxigenin, oleandrigenin, uzarigenin, corotoxigenin, tanghinigenin, ouabagenin, &c.

Cardiaca (L.): **cardiaire** (Fr): *Leonurus* spp.

cardinal flowers, blue: *Lobelia syphilitica* - **red c. f.**: *L. cardinalis* - **c. spear**: *Erythrina herbacea*.

Cardinophyllum = Carzinophyllin.

Cardiospermum grandiflorum Sw. (Sapind.) (tropAm & Af): heart seed; rt. & lf. inf. u. for cough & bronchitis; sds. u. as beads; pl. veget. - **C. halicacabum** L. (pantrop, esp. tropAm; native to seUS): balloon vine; heart pea; "heart seeds"; badhi (Bomb.) pl. decoc. u. in arthritis; kidney dis.; in eye dis. (also lf. juice) (Af); lvs. ("bich" grass) (Trin) u. as inf. for "bich" (lower intest. dis.) (Trin) and indigestion; rt. (& frt. ([Mex]) u. diur. emet. purg. diaph. rubef. in rheum.; bad weed (Ala [US]); cultd. (some areas).

cardo: *Cynara cardunculus* - **c. alcachofero** (PR): artichoke - **c. benedetto, erba** (It): blessed thistle hb. - **c. corredor** (Sp): *Eryngium campestre* - **c. santo** (Sp): 1) blessed thistle 2) (Pe): *Argemone platyceras* 3) (Mex, Ve, WI): *A. mexicana* 4) (SA): *Caraguara* spp. (ornam.) - **c. s. amarillo** (CR): *Argemone mexicana* - **c. s. blanco** (CR): *Cnicus benedictus*.

cardol(um): **cardoleum**: an irrit. prod. (**crude c.**: *cf.* cantharidin); representing resorcinol derivs., found in mesocarp of frt. of *Semecarpus anacardium* (10% in cashew nut shell oil); u. as vesicant like cantharidin; similar prod. found as phenolic oily liquid ext. fr. frt. shell of *Anacardium occidentale* (comm.) with similar uses; take care! - **raw c.**: acajou balsam.

cardon 1) (Br, Eng): cardoon 2) (Canary Is): *Euphorbia canariensis*.

cardoon: *Cynara cardunculus*.

Carduaceae: segregate of Comp. Fam.

Cardui, Wine of: popular proprietary c. *Cnicus benedictus* (Carduus b.)

cardus plant: blessed thistle.

Carduus L. (Comp.) (Euras): plumeless thistles (*C. acanthoides* L. &c.); *cf. Cnicus* monotypic with *C. benedictus* and *Cirsium* (plumed or common thistles, close to *Carduus*) - **C. benedictus** Brunsf. = *Cnicus b.* - **C. getulus** Pomel. (Eg, delta region): pl. ext. c. alks., triterpenoids; u. as hypotensive; has moulting activity against cotton leaf worm - **C. lanceolatus** L. (Cirsium vulgare Tenore) (Euras, Af, natd. NA): common or bull thistle; rt. u. for stomach cramps; also eaten raw; fls. chewed to mask unpleasant med. taste (Am Indians); hb. tea u. emetic (Navajo Indians) - **C. marianus** L. = *Silybum marianum* - **C. nutans** L. (Euras, natd. NA, SA): musk thistle; Scotch t.; a biennial hb.; dr. fls. u. to curdle milk; pith edible after boiling; fls. u. antipyr. & to "purify blood" (Ind) - **C. spinosissimus** Walt. = *Cirsium horridulum* - **spotted carduus:** *Cnicus benedictus*

careless (weed): 1) (seUS): *Amaranthus spinosus*, *A. hybridus*, etc. 2) (seUS): *Acnida cuspidata*, *A. cannabina* - **green c. (weed):** *Amaranthus spinosus* (Miss [US]), *A. retroflexus* - **red careless (weed)** (seUS): 1) *Polygonum muhlenbergii* 2) *Amaranthus hybridus*, red form: *A. spinosus* (also **spiny c.**) - **white c.**: *A. retroflexus*.

Carex (Cyper.) (cosmopol, esp. in temp marshes): sedges; c. 1,200 spp.; lvs. & sts. u. to weave baskets; sds. of some spp. food (Am Indians) - **C. arenaria** L. (n&cEu): German sarsaparilla; rhiz. u. diaph. dem. exp. diur.; in rheum. - **C. hirta** L. (Euras, Alg, NA): hairy sedge; sha t'san; rhiz. u. as sarsaparilla substit.; diur. - **C. macrocephala** Willd. (Chin, NA): u. ton.

Careya arborea Roxb. (Lecythidaceae) (Ind, SriL., Bangladesh): bk. c. 19% tannin; u. astr.; as tan & for slow matches or fuses; substit. for bidi lvs. (process of boiling & drying); sds. edible.

Carica mexicana (A DC) L.O. Williams (Jacartia m. A DC) (Caric.) (Mex, CA): papaya; custard tree; papaya orejona; soft-wooded tree; frt. edible, made into sweetmeats - **C. papaya** L. (tropAm; natd sFla; now pantrop.); papaya; pawpaw; melon tree; actually tree-like herb, trunk rising to ca. 9 m without lateral branching; frt. rich in pectin, ca. 3 dm long, green, yellow, orange, darkening to black on ripening; yellow meat very sweet (generally eaten with lemon juice); also canned; flavor in ice cream, etc.; green frt. eaten boiled as vegetable (Dominica, Ind); split green frt. appl. hot as poultice, inflamm.; much cult. for its delicious frt. eaten as melon & u. for gastric dis. (PR); also for the dr. latex, with papain (*qv.*); frt. u. to make cough syr.; dr. pd. sds. u. to kill intest. worms, also latex (1519); latex u. to kill niguas (trop chiggers); papaya soap (Cu); papaya workers said to have good skin (hands); lf. & rt. inf. u. for TB (Caroline Is); many other uses (117); extensive plantations in Tanz, Popone, SriL + - **C. peltata** H. & A. (CA): lvs. decoc. u. to bathe erysipelas (extensive cultn. in Tanz, Popone, SriL).

Caricaceae: Pawpaw Fam.; small trees with milky latex found in tropics of Am & Af; general habit palm-like; frt. a large fleshy berry.

Caricae (L): figs, *Ficus carica.*

caricine (Fr): papain.

carinañolas (Pan): fleshy rts. of *Yucca* spp.

Carinta W. F. Wight = *Geophila.*

cariofilata (Sp); *Geum urbanum* - **c. acustica** (Sp): *G. rivale.*

Carissa bispinosa (L.) Brennan (Apocyn.) (sAf): pl. c. sterols & tannin; u. native med. as aphrod.; witchcraft drug; frts. edible tart - **C. campenonii** Palacky (Mad): frt. edible; cult. since ancient times by the Malagasy royalty on their palace grounds - **C. carandas** L. (Ind to Mal): karunda; rts. u. stom. anthelm.; lvs. for fever; rts. c. hypotensive agent - **C. edulis** Vahl (sAf [Natal], Eg, wAs): rt. c. **carissin**; cardioactive glycoside; u. in indigenous med; frt. ("Natal plum") edible, u. ton. in cough; abortient - **C. grandiflora** A DC (sAf): Natal plum; shrub u. as hedge pl.; frt. eaten raw & u. for preserves, jellies, etc. - **C. ovata** R. Br. (Austral): frts. edible, rts. c. odoroside H., cardioactive compd.

Carlina acaulis L. (Comp.) (s&cEu, mts.): **carline thistle**; rt. c. VO, tannin; u. diur, diaph. anthelm., emm., antipyr., emet.

Carlsbad salt: a salt said prep. fr. C. Springs water (Karlsbad, Bohemia, CSR) - **Artificial C. Salts**: a similar factitious mixture (sod. sulfate chiefly).

Carludovica palmata R. & P. (Cyclanthaceae) (CA, Pe, Ec, Co, Bo): Panama hat plant (or palm); said chief source of fiber for Panama hats; brooms (Col), matting &c.; for sacks, baskets, brooms, umbrellas, sails, paper, &c.; med. (Col); orn.

Carmania: Gum Carmania: Carmanian gum: like Bassora Gum; exudation of almond & plum trees; u. confectionery; adult. of Ghatti Gum (SC Carmania, ancient territory n of Persian Gulf; *cf.* modern Kerman).

carmin (CR): *Phytolacca octandra.*

carmine: aluminum lake of carminic acid (etc.) obt. fr. cochineal; u. as red dye in inks, stains, pharmaceuticals, frt. juices, lip sticks, eye make-up, etc. - **carminic acid**: glucosidal coloring matter of cochineal; aluminum lake is carmine.

carnation: *Dianthus caryophyllus* - **Spanish c.**: *Poincinana pulcherrima.*

carnauba wax (pron. carn-a-ú-ba): Brazil w.; waxy exudate scraped fr. the lvs. of the wax palm, *Copernicia prunifera* (**carnaubeira** [Br]): said the most important wax in industry (types: No. Country; Parnaiba; &c.)

carne (Sp): meat - **c. porcina**: pork - **c. vacuna**: beef - **extracto de c.**: beef ext.

Carnegiea gigantea B. & R. = *Cereus giganteus* Engelm.

carnitin(e): "vitamin B_T" (levo form) (levo-c.): trimethyl-hydroxy-butyrobutane; this essential amino acid deriv. is normal constit. of mammalian muscle, fish tissue, liver, cuttlefish, whey, & produced by liver & renal synthesis; chief source is meat ext.; stim. gastric & pancreatic secretions; u. po as HCl salt to produce wt. increase in pediatrics; in rickets; in carnitine deficiency (genetic impaired) (carnitine essential for normal FO utilization & energy prodn.); u. to incr. efficiency of athletes; in muscular deficiency; in hyperlipoproteinemia; in cardiomyopathy (as in valproic acid toxicity); to inhibit thyroxin action (SC carnis [L], flesh, meat); an orphan drug (USP XXI).

Carnitor™: carnitine.

Carnivora: order of Mammals of high intelligence, which have furry integuments & subsist on a meat dietary; aggressive active animals; ex. otters, wild cat, lions.

caroa (Br): *Neoglaziovia variegata* (fiber).

carob (bean): locust bean, *Ceratonia siliqua* - **c. gum**: locust (bean) gum: water dispersible gum fr. carob.

caroba: carob; *Cystibax* spp. - **c. gum:** polysaccharide fr. *Jacaranda caroba* - **c. leaves**: *Jacaranda tomentosa* (sBr), *J. copaia*; *Cybistax antisyphilitica.*

caroban: carobin: carob flour; *Ceratonia* meal; grd. kernel (endosperm); also polysaccharide of *Jacaranda caroba.*

carobin(h)a (Br): *Jacaranda & Memora* (esp. *M. nodosa*) spp. (other than *J. caroba & J. semiserrata*) - **carobon**: resinous constit. of lvs. & bk. of *Jacaranda copaia, J. macrantha*, etc.

carôco (Por): sd. endocarp (kernel).

Carolina thistle: *Carlina acaulis* - **C. vanilla:** *Trilisa odoratissima*; v. also cocculus, moonseed, pink.

Caroline: v. also jasmine.

carony bark: fr. *Angostura.* spp.

Carota (L.): carrot, *Daucus c.* - **carotene: carotin**: pro-vitamin A; so-called as first disc. in carrot; occ. in many frt. & vegetables; ß-form ca. twice as active as ∂-form; u. nutrient; to color butter, cheese, etc., yellow; claimed anticancerogenic (ingestion said to reduce risk of breast & kidney cancer) - **carotte** (Fr-f.): carrot.

carotenoids: carotene-like pigments (yellow, orange, red, purple) found in plants & animals (FO); highly sol. in fat solvents.

caroube (Fr): carob (frt.) - **caroubier** (Fr): carob tree.

caroy (PI): cashew.

carpaine: alk. of *Carica papaya*; u. as cardiac stim.; in amebic dys. (1479).

Carpathian balsam: fr. *Pinus cembra.*

carpel: pistil (unit); gynoecium (See Fig. in App.).

carpenter's herb: *Prunella vulgaris* - **c. square herb** (NCar [US]): *Scrophularia nodosa* & *S. marilandica.*

Carpesium abrotanoides L. (Comp.) (Eu, tempAs): nodding starwort; frts. u. as tenifuge, vermifuge (Chin) - **C. cernuum** L. (eAs): pl. u. for lung dis., emet. expect. lax.; bruised lvs. placed in nostrils for nosebleed (Chin).

carpetweed: *Mollugo* spp.

Carphephorus corymbosus (Nutt.) T. & G. (Comp.) (Fla, Ga [US]): "paintbrush" (1137) - **C. odoratissimus** (J. F. Gmelin) Herb. = *Trilisa o.*

Carpinus betulus L. (Betul./Coryl.) (Euras): European hornbeam; lvs. u. astr. in diarrh., relaxed mucosa of throat ailments, etc. (decoc.); yellow dye; orn. - **C. caroliniana** Walt. (C. americana Mx.) (US e of Miss River, sCana, Mex, CA; in moist areas typically along streams): American hornbeam; blue (or water) beech; ironwood; trunk resembles a sinewy arm; lvs. with low serrations (teeth < 1 mm high); u. med. (Mex); wd. useful (ex. tool handles); ornam. (-34 m).

Carpobrotus edulis (L.) N. E. Br. (Ve, Br) (Mesembryanthemum edule L.) (Aizo.) (sAf; cult. & escape in Calif [US]): Hottentot's fig; many med. uses in Af; lvs. c. 20% catechol tannins (comm. source); pl. juice appl. to ringworm; burns; eczema; lf. decoc. gargle in diphth.; frt. edible, lax.; pl. u. as soil binder; ornam.

Carpopogon Roxb. = *Mucuna.*

Carpotroche amazonica Mart. (Flacourti.) (Co): nina caspi; bk. u. as caustic - **C. brasiliensis** Endl. (Br, cult. Matto Grosso): sapucainha; sd. FO (carpotroche fat) u. as substit. for chaulmoogra oil with similar compd. (hydnocarpic, chaulmoogric, gorlic acids); no vit. A, D (1206); u. in leprosy, skin dis., TB (incl. TB meningitis) (1205); in soap mfr. (1139).

carque(i)ja (Br): *Baccharis genistelloides* & other *B.* spp. - **c. amar(g)osa**: *B. triptera.*

car(r)agaen: carrageen: (Scotch, Gaelic) **c. moss** (comml.); Chondrus (several variant spellings, such as **carraigein**) - **carrageenin**: purified extractive of Chondrus (or Irish moss) u. for peptic ulcer.

carrageenan: carrageenin: purified extractive of Chondrus (Irish moss); complex polysaccharide made up of polymers of galactose & anhydrogalactose; u. as gel, stabilizer, emulsifying agt., viscosity builder in many foods & medicines; dem. (1313).

carrapato (Por): *Ricinus.*

carré latin: Latin square: a square array of n letters, each repeated n times in different order and arranged so each letter occurs only once in each row and column; u. for exptl. design (for planting, &c.).

carreta: wooden cart yoked to oxen; u. to haul sugar cane, &c. (LatAm) (1576).

carrion flower: *Nepeta hederacea.*

Carron oil: lime liniment (equal parts lime water, linseed oil); u. to treat burns (SC first u. at Carron Iron Works, England).

carrot, common (or **garden**): *Daucus carota* (cult. form) - **c family**: Umbelliferae (Apiaceae) - **c. fl. oil; c. juice**: *v.* under *Daucus c.* - **wild c.**: is of same sp.

carruba (It): frts. of carrubo = carob tree.

Cart(h)agena bark: a yellow cinchona bk. fr. *Cinchona lancifolia* - **Carthaginian apple**: pomegranate.

carthamin: plant red fr. Carthamus.

Carthamus: L. 1) genus of ca. 20 wild spp. (Comp.) (Euras, nAf) 2) (specif.) (L.): fl. head of *C. tinctorius* - **C. helenoides** Desf. (Comp.) (OW as Alg): sometimes occurs as adult. of stramonium - **C. lanatus** L. (Euras): pl. u. as sudorific, antipyr., anthelm. (Fr) resolvent - **C. oxyacantha** Bieb. (Ind, wPak near Lahore): a bad weed with yellow fl. - **C. tinctorius** L. (Euras, Af, Eg; cult. wUS, esp. Calif): (American) safflower; American saffron; a cultigen; fl. heads yellow, orange, red, white; c. carthamin, safflower yellow; dr. fl. heads long u. as drug (lax., diaph., esp. in measles & scarlet fever); safflower oil considered of special value for burns: appl. to the burned area (reduces water loss; incr. speed of skin regeneration); red, yellow dye agent; sds. purg., considerably u. in rheum. (Ind); in pet foods (US); source of FO, safflower oil, important article of commerce (USP XXI); light-colored, c. more polyunsatd. fatty acids (essential f.a., esp. linolic ac.) than any other comm. oil (also lowest calorific value); hence healthful oil in dietary, said good for hair; u. in paints, varnishes; seedcake or sd. meal u. food for livestock (pl. also u. cattle feed); widely cult. Ind, Med, US (Calif 84,000 acres [1956]).

cartilage: white non-vascular connective tissue; often precursor of bone; appl. directly to lesions or wounds, psoriasis, eczema; c. poly- or acetyl-D-glucosamines; in combn. with bone u. as implants in grafting (calf bone).

caruba (Sp): carob.

Carum (L.): frt. of **Carum carvi** L. (Umb.) (Euras): caraway frt. (seed); source of VO, caraway seed oil; u. flavor, cond. carm. eupeptic (stomach); frts. often u. in rye bread to mask strong flavor & as carm. (popular in Ge); frt. whole & pd. u. culinary (as in goulashes) & meat flavors; fls. in confect. cheese (Neth); VO with carvone (50%+) u. in. perfuming soaps, &c., liqueurs (Kummel, aquavit); prepd with coffee (dalleh) (Saudi Arabia); plants claimed repellant to nematodes; rts. eaten in nEu (better than parsnip); young lvs. in salads, larger lvs. u. as spinach; supposed aid to memory (1140) (prod. Mor, Ne, Pol, Turk, Russ, Ind) (1140) - **C. ajowan** Baill. = (L.) Benth. = *Trachyspermum ammi* (L.) Sprague = *T. copticum* (L.) Link. - **C. bulbocastanum** Koch (Pak: Baluchistan, Kashmir): drought-resistant pl.; VO c. aldehydes; u. like *C. carvi* (for wh. at times substit.); tuberous rts. u. in salads, also cooked - **C. petroselinum** B. & H.: *Petroselinum crispum*.

caruncle: sd. outgrowth from integument occurring near micropyle.

caruto (Hon): *Genipa americana* var. *caruto*.

carvacrol: a phenol obt. fr. thyme & origanum VO, u. to perfume soaps.

carvi (Fr): Carum.

carvol: carvone: a ketone found in a number of VO; ex. spearmint, caraway, etc.; u. in perfumes, soaps.

Carya Nutt. (Hicoria Raf.) (Jugland.) (eNA, Chin): hickory; valuable trees for lumber (tough strong wd.) & for edible nuts; differs fr. *Juglans* in having fewer leaflets to leaf; Am Indians prepd. FO fr. nuts by boiling in water & skimming off oil - **C. cordiformis** K. Koch (C. amara Nutt.) (eNA): bitter nut (hickory); bitter h.; bk. u. cath. diur, "for simple sicknesses" (Am Indian) - **C. glabra** Sweet (e&cUS): pignut (hickory); hognut; nuts edible, sometimes sweet and pleasant, sometimes astr. & acrid; wd. valuable for tool handles, fuel - **C. illinoensis** K. Koch (C. pecan Engl. & Graebn.) (se&cUS, Mex; natural distrn. sWisc-wAla [US]): pecan (pron. pee-can or pee-cahn; deriv. fr. Am Indian paccan); sds. a popular nut; u. mfr. candy (pralines); FO like olive oil, u. by Am Indians as seasoning; pollen fr. catkins a cause of pollinosis; also cult. as nut & shade tree (orchards); grows to 50 m & 100 years; frts. to 90 mm - **C. laciniosa** Loud. (e&cNA): (big) shellbark h.; nuts u. for food & marketed; bk. astr.; wd. useful - **C. olivaeformis** Nutt.: *C. illinoensis* - **C. ovalis** Sarg. (C. glabra var. odorata Little) (e&cNA): sweet (or red) hickory; s. pignut (hickory); bk. u. to make hickory bark syrup (Mo [US]) (1514); wd. (of this & other *C.* spp.) u. for smoking hams, cheese, bacon, &c. (US) - **C. ovata** K. Koch (C. alba Nutt. p.p.) (eNA: Quebec [Cana]-Minn-Okla-cGa [US], Mex): shagbark (or shellbark) hickory; bk. ([white] walnut bark) u. as mild astr.; sd. kernel sweet, esculent; wd. strong, u. to make agric. implements, baseball bats, valuable for smoking meats (hams, bacon, &c.) - **C. porcina** Nutt. = *C. glabra* - **C. tomentosa** Nutt. (C. alba K. Koch) (e&cUS): mockernut (hickory); white-hearted h.; shagbark h.; shellbark h.; wd. very useful; small shoots placed on hot stones & fumes inhaled for convulsions (Chippewa Indians); bk. u. mild astr.; sd. kernel sweet, esculent; eaten.

caryer (**blanc**) (Fr): hickory.

Caryocar L. (Caryocaraceae) (tropAm, esp. Br): pequia (trees); bk. u. diur. antipyr.; frt. & sd. source of valuable FO - **C. brasiliense** Cambess. (Br, Guianas) (common pl. of cerrado): pequi, piqui; small tree; frt. (drupe) incl. sd. popular food among Brazil Indians (Minas Gerais, &c.); frt. (sex symbol) c. 10% starch; frt. u. to prep. liquors & condiments; sd. furnishes edible FO u. to make skin supple (Br Indians); shell u. as fuel, in prepn. charcoal; wd. u. in bldg. - **C. glabrum** Pers. (Guianas): barbasco propio; inner bk. u. for washing clothes (saponins); sds., souari nuts, edible (comm.) ("butter nuts"); frt. pulp & rind u. as fish poison; useful (durable) wd. - **C. gracile** Wittmack (Col, Ve, Br): barbasco; parts of tree u. as fish poison (2300) - **C. microcarpum** Ducke (Br, Guianas, Ve, Col): barbasco (del rio); c. tannins, saponins (oleanane skeleton); u. as ichthyotoxic agt. (Col) (2300) - **C. nuciferum** L. (nBr,

Guiana): Suari nuts; bk. u. febrifuge, diur.; kernels of Saouàri nuts edible fat, culinary use.

caryology 1) science of nuts or kernels 2) branch of cytology specially devoted to cell nucleus.

Caryophyllaceae: Pink Fam. (mostly temp zone): a large gp. of mostly hbs., a few undershrubs with lvs. opp. often joined at base; fls. regular, sepals & petals 4-5; incl. many attractive garden flowers, some u. in med.; ex. carnation, sweet William, campion; some 2,070 spp.; incl. genera *Stellaria*; *Cerastium*; *Silene*; *Dianthus*; *Lychnis*.

Caryophyllatae Cortex (L.): bk. of *Dicypellium caryophyllatum* - **Caryophyllus** (L.): **C. aromaticus**: clove, *Syzygium caryophyllata*.

caryopsis: dry, indehiscent, 1-celled frt. of cereal grasses; ex. oat.

Caryopteris Bunge (Verben.) (As): shrubs or small trees; some spp. u. as aphrod.; some ornam. (hedges) - **C. chosenensis** Mold. (Jap): "gio"; lf. decoc. u. for beri beri; has reputation of clearing the urine & rendering eyesight clearer; c. 3 insect antifeeding diterpenoids (caryoptin+).

Caryota mitis Lour. (Palm.) (tropAs): sago palm; juice u. with toad ext. as poison (Kelantan, Mal); fiber burning u. to cauterize wounds (moxas) - **C. monostachya** Becc. (eAs): sds. chewed with betel - **C. plumosa** Hort. (Ceylon): caryote; "fish tail" palm; cola pescado; popular ornam. - **C. urens** (Ind, Cey): toddy (fishtail) palm; source of kittul (or kittool) fiber (fr. lf.); jaggery (palm sugar), toddy (palm wine), & a kind of sago fr. the trunk's pithy center; timber useful.

Carzinophilin: anti-tumor, antibact. antibiotic isolated fr. *Streptomyces sahachiroi* Hata & al. (1954); clinically useful in cancer.

casa banaya: cassabanana.

casabe (WI, etc.): bread or biscuit prep. fr. grd. or grated yuca (Fecula de c.).

casanthranol: mixt. of anthranol glycosides of Cascara Sagrada & almost free of anthraquinones; 2 fractions A & B; u. purg. ("Peristim Forte").

casca (Por): bark; shell; peel; husk - **c. d'anta** (Br): **c. de anta** (Br): *Drimys winteri* & var. *granatensis* - **c. da mocidade** (Br): **c. da virginidade** (Por): bk. of *Stryphnodendron barbatimao*.

casca red water tree: *Erythrophloeum guineense*.

cascabel: SA rattlesnake, *Crotalus durissus*; antivenin prepd. fr. venom.

cascalote (Mex): *Caesalpinia coriaria*; divi divi.

cascara (Sp): bark; in Anglo-Saxon drug trade, usual short form of Cascara Sagrada - **C. Amarga** (L.): (lit. bitter bk.) (1142); Honduras bark 1) *Sweetia panamensis* 2) *Picramnia pentandra* - **C. Sagrada** (lit. holy bk.) (L., nomen vulgare): bk. of *Rhamnus purshiana*.

Cascarilla magnifolia Wedd. (Cinchona m. R. & P.) (Rubi) (Bol, Pe, wVe): bk. u. as bitter agt., cinchona substit. (monographed in pharmacopeias of mid-1850s & later) less effective (ex. USP 1820-1890) - **c. (bark, quills)**: 1) *Croton eluteria* 2) (Pe): false cinchona; *Pogonopus tubulosus* ("true cascarilla") - **c. oil**: VO of *Croton eluteria* (1-3% yields) (prod. Bahama Is).

cascarin: name u. by Leprince (1143) & others for xanthorhamnin (glycoside) found in Cascara Sagrada; same as frangulin.

cascarones (PI): coconut sweets.

Casearia esculenta Roxb. (Flacourti. or Samyd.) (tropAs, esp. Ind): rt. u. as astr. cath.; decoc. in piles, diabetes (Ind); c. hypoglycemic principle; in hepatic coma - **C. fuliginosa** Blco. (PI): rts. appl. to ulcers - **C. graveolens** Dalz. (Ind): churchi; pd. bk. with sugar for ton. (children) (124); frt. in edema; as fish poison - **C. grewiaefolia** Vent. (seAs): bk. ton, u. in metritis, metralgia, after childbirth; generally in combns. in snakebite (PI) - **C. nitida** Jacq. (Mex, WI, CA nSA): cafetillo; frt. eaten by birds; Mayas bathe in lvs. decoc. to treat bile and spleen dis. (6) - **C. praecox** Griseb. (Gossypiospermum p. [Griseb.] P. Wils.) (WI, tropSA, esp. Ve): West Indian boxwood; zapatero; jia (Cu); lvs. u. tonic, antipyr.; wd. u. to make engravers' blocks, carvings, rulers, etc., where a specially hard wd. is required; now mostly replacing the older boxwood fr. *Buxus sempervirens* - **C. richii** A. Gray (Fiji): lvs. chewed for thrush- **C. spiralis** Johnston (sMex, CA, nSA): aril around sds. edible; frt. said lax. - **C. sylvestris** Sw. (Mex, CA, WI, SA): sarnilla; lvs. c. VO, alk, saponin; lvs. & rt. bk. & sd. FO popularly u. as antilepric (*cf.* chaulmoogra), antisep. in skin dis. (Br); antipyr., stomach, antiprur.

casein: characteristic and principal phosphorus-rich protein of cow's milk & cheese; obt. by action of rennin or acid on milk; insol. in water; u. nutr. in special diets, in mixts. & hydrolysates; reagent; in mfr. paints, plastics, adhesives (Hammersten's C.).

caseinogen: precursor of casein (*cf.* fibrinogen & fibrin).

cashaw (US): *Cucurbita moschata*, large-fruited running-vine type of squash, frt. smooth with neck curved or straight.

cashew apple: "frt." of *Anacardium occidentale*; actually the fleshy edible receptacle, yellow or red, ca 1-1.5 cm long; u. as juice or syrup in beverages, in candy, jam, halva, chutney, preserves, pickles, &c. - **common c.:** *Anacardium occidentale* - **c. family**: Anacardiaceae - **c. leaves**: claimed u. (SA) as dentifrice, to preserve teeth

(1144) - **c. nut (kernels)**: 1) *Anacardium occidentale* 2) *Semecarpus anacardium* (roasted kernels of both spp. eaten) - **c. nut (shell) liquid**: fr. *Anacardium occidentale*; u. in mfr. resinous varnishes (1145) - **oriental c. (nut)**: *Semecarpus anacardium* - **West Indian cashew (tree)**: common c.

cashoo: cutch.

Casimiroa edulis Llave & Lex. (Rut.) (Mex, Guat): zapote blanco (white sapota); bk. lvs. frt. & sd. u. hypn. sed. analg. in rheum.; frt. (Mex apple) of tree c. casimirine (glyco-alk.), casimirol (alc.), sd. c. casimiroin, FO, resin; frt. long u. as fine food (bitter sweet); lvs. u. diarrh., anthelm, vulner.; frt. said hypnotic; sds. toxic or even lethal (in large doses) (resp. paralysis), bk. fish poison; cult. - **C. pringlei** Engl. (Mex): u. like **C. pubescens** Ramírez (Mex): frt. edible, u. as vulnerary, hypotensive; in arthritis, rheum. - **C. sapota** Oerst. (Mex, CA): sapote; frt. edible - **C. tetrameria** Millsp. (Mex, Guat, CR): matasano; frt. edible; u. like *C. edulis*; sds. c. soporific subst.; lvs. u. in mild cases of diabetes; cult. for delicious frt; a bee tree - **C. watsonii** Engl. (Mex): u. like *C. pubescens.*

casings: dr. mammalian intestine; u. sausage mfr.; largely imp. fr. Iraq, Iran.

caspi (Co): 1) *Mauritia heterophylla* 2) *Rhus juglandifolia.*

Cassaba melon: *Cucumis melo* var. *inodorus.*

cassabanana: *Sicana odorifera.*

cassada (WI): mandioca.

Cassandra calyculata D. Don = *Chamaedaphne c.*

cassareep: syrupy prod. made by macerating rts. of *Manihot esculenta* in water for a day then boiling down to syrup; has antisep. props. & is u. as base for many sauces, to preserve meat, etc.

cassaú (Br): *Aristolochia cymbifera.*

cassava (meal or **flour)**: **bitter cassava: cassave** (WI): mandioca (manioc) fr. *Manihot esculenta* (1640) - **c. starch**: prod. u. by dextrin mfrs. u. in yeast cakes, tapioca; as sizing to glaze twine (prod. in NEI, Nig) (1641) - **sweet c.**: *Manihot esculenta.*

casse (Fr): *Cinnamomum cassia* - **c. en batons**: cassia pods - **cassena**: *Ilex cassine, I. vomitoria.*

Casselia integrifolia Nees & Mart. (Verben.) (Br): flor de natal; rt. decoc. u. in rheum., lax. - **C. mansoi** Schau (Br, Bol): cha de Minas; u. diur.

Cassia L. (incl. *Senna* Miller) (Leg.) (cosmopol., trop & subtrop, not Eu): senna; hbs. shrubs & trees; sometimes specific. appl. (L) to *C. marilandica* - **C. abbreviata** Oliv. (tropAf): rt. decoc. u. for "sharp pains" in stomach (Tanz) - **C. absus** L. (Ind, trop, cosmopol), chaksu: chaksa; semen chichmae; chichim seeds; c. chaksine (alk.) (releases histamine to tissues); u. aphrod., in ringworm (Ind); in conjunctivitis, trachoma (Pak) - **C. acutifolia** Del. (Eg): Alexandria(n) senna; leaflets & pods c. emodin glycosides incl. sennosides, emodins &c.; u. lax. cath.; antibiotic; resin (greenish prod. extd. with alc.; c. myricyl alc. as tartr. & palmit., phytosterol; griping princ.); action somewhat more griping than that of Tinnevelly s. *(C. angustifolia)* (1515) - **C. afrofistula** Bren. (cAf): rt. decoc. u. for stomach pains, exter. for sores (Tanz); rt. inf. u. GC (Tanz); bk. u. in mal. & dys. (Moz); frt. purg. (Moz) - **C. alata** L. (Br + all tropics): ringworm cassia; dartrial; leaflets u. as prec.; for sores (Dom) & in skin dis. (Tahiti, PI.); general restorative (Tanz); rt. u. (Guat, Hond) as drastic & in rheum. (inf.); fresh bruised lvs. (or inf.) u. as diur. & in skin dis. (itch, ringworm); lf. u. in steam baths for women after delivery; st. u. in fever (Laos); cult. - **C. angustifolia** Vahl (As, Af) (cult.): Tinnevelly s. (TV senna); lvs. & pods u. as in *C. acutifolia*; action of this senna milder, less griping than that of Alexandrian (*C. acutifolia*) (1516); mostly fr. Yemen - **C. auriculata** L (Indochin, Chin +): palthe; tanner's cassia; tarwar; lvs. u. med.; tan; rt. bk. (as avaram bark) u. tan(s); sd. comm. article u. in ophthalmia &c.; decoc. fever. diur. for kidney stones (SriL) - **C. australis** Sims (Austral): Australian. senna; lvs. good s. substit. - **C. beareana** Holmes (eAf): said to bear largest of the cassia pods (-80 cm.); rt. bk. c. tannins, u. malaria, (esp. blackwater fever); ulcers (topically) - **C. bicapsularis** L. (tropAm): b(r)icho (Mex); shrub to 3 m; lvs. purg.; wd. u. in paper mfr. (Br) - **C. brachiata** (Pollard) J. F. Macbr. = *Chamaecrista fasciculata* (var. *brachiata*) - **C. brasiliana** Lam. = *C. grandis* - **C. chapmanii** Isely (C. bahamensis Mill.): stinking pea (Bah): lf. heated & appl. to draw boils; in bath water for all complaints (Bah); frt. said toxic - **C. didymobotrya** Fres. (c. tropAf): wild senna; lvs. are boiled in a rolled banana lf.; lf. decoc. u. as antipyr. purg. (Tanz); emet., also against infant verminosis (Burundi); rt. decoc. strong purg., lf. decoc. milder (Tanz): lethal to stock - **C. emarginata** L. (tropAm): herba de jolote; lvs. u. purg., appl. locally to allay insect bite pain, shredded & inhaled by Mayas for epistaxis (6) - **C. falcinella** Oliv. var. *parviflora* Stey. (tropAf): rt. inf. u. in diarrh.) - **C. fasciculata** Mx. = *Chamaecrista f.* - **C. fistula** L. (*Cathartocarpus f.* Pers.) (Ind+; seAs): purging cassia; golden shower tree; pudding pipe tree (Ind); frt. variously called legume & loment; (50 cm); frt. pulp u. as pleasant lax.; sometimes eaten with fish (Indochin); pulp added to tobacco (Ind); sds. laxative; rt. bk. u. as purg. & rubbed on snake & scorpion stings; cult. - **C. floribunda** Cav. (tropAf, As): frt. u. as coffee; given to babies with convulsions, also crushed lvs. rubbed on face (PI) - **C. fruticosa** Mill. (Mex, WI, CA, nSA): Christmas bush (Trin); quelite (Tabasco, Mex); fls. tea u. in diabetes (Trin) - **C. garrettiana** Craib (Indochin & Thail): fl inf. u. in fever & as "depurative"; lvs. & wd. u. in tetter; rt. decoc. u. in diarrh. - **C. grandis** L.f. (tropAm): cañafistula; cañandonga (Cu); frequently reported as *C. fistula* but pod of *C. grandis* considerably larger (3.8 cm X 60 cm); u. mild lax., esp. in veter. med.; antipyr. in colds (Guat); lvs. cut up & appl. to sores & wounds (Br); in lard for mange (dogs); fls. & sds. also u. (Br) (1227); frt. pulp edible (Hond); packed in baskets (c. 1946); - **C. hirsuta** L. (tropAm): inf. of ground sds. u. as antipyr. & coffee substit. (Domin, Caribs) - **C. hispidula** Vahl (Mex, CA, SA): nahuapate (CR); sds. c. abrin; lf. inf. u. analg.; u. in treating VD (CR) - **C. holosericea** Fresen. (Ethiopia): furnishes Aden senna; u. lax. - **C. italica** (Miller) Spreng. (C. obovata Collad.) (OW trops): furnishes "dog senna"; Italian, Spanish, Syrian, Aleppo, or Sudan s.; u. lax.; weaker than the off. senna (may be anthraquinone-free); occasionally occ. in commerce as Folia Sennae Italicae; sometimes served as adult. of official Alexandrian s.- **C. jahnii** Rose (Adipera j. Britt. & Rose) (Ve - Andes): fresh fls. (uruamco): u. as lax. by natives - **C. javanica** L. (Indochin, Indones, PI): ornam.; subsp. **nodosa** K. & S. S. Larsen (Ind, Burma, seAs): frt. purg. ; wd. decoc. u. for women following labor (Vietnam); sd. sometimes added to betel chew - **C. laevigata** Willd. (Mex, CA, widely pantrop): frijolillo (Hond, Guat): herb or shrub; frt. & bk. purg. emm.; hb. steamed & appl. to body for female dis.; emm (Mex); sds. u. coffee substit. - **C. marilandica** L. (C. medsgeri Shafer; Senna m. (L.) Link.) (e&cUS): Am (or wild) senna; lvs. said dr. & made into kind of frt. cake with lax. props.; senna substitn.; sds. u. coffee substit. - **C. medicinalis** Bischoff = *C. angustifolia* - **C. mimusoides** L. (cAf +): rt. decoc. u. as purg for abdom. pain; "deep" cough (Tanz); milk decoc. u. in dys. & in pneum. & other chest complaints; lvs. as tea substit. (Chin); lvs. chewed & appl. as poultice for scorpion sting (Tanz) - **C. montana** Heyne (Ind): u. adulter. of true senna - **C. moschata** HBK (tropAm): cañafistola (Ve); smaller pods than in *C. fistula* (-1.3 cm. diam.); u. mild lax - **C. multijuga** Rich. (Java & pantrop): lf. inf. in asthma; an ink prepd. fr. pounded beans (Br) - **C. nictitans** L. = *Chamaecrista n.* - **C. nigricans** Vahl (trop cAf): pl. u. for fever; rt. ext. for diarrh. (Tanz); lvs. decoc. u. antipyr. diaph. to control cough; rts. u. as vermif. (Seneg) - **C. nodosa** Buch.-Ham. (wMal): soothing; u. much like *C. javanica*; lvs. purg. antipyr.; bk. rubbed on skin for itch; rts. u. detergent for laundry (soap substit.); wd. u. for handles, &c. - **C. nomame** (Sieb.) Honda (Jap, Chin, Kor): rt. & lf. diur.; boiled young lvs. eaten (Jap) - **C. obtusifolia** L. (Senna o. [L.] Irwin & Barneby) (Tex [US], Mex): sicklepod (senna); coffee weed; fantupa seed (EI, Molucc, Sunda); lvs. & fls. u. as aperient & in skin dis. (antibacter.) - **C. occidentalis** L. (Senna o. (L.) Link) (trop&subtrop of world, se&cUS orig. SA): coffee senna; styptic weed; coffee weed; roasted sds. u. beverage (coffee substit.); lf. inf. u. cath. in arthritis (SA), intest. dis., stomachache (sAf+); "cure all"; for coughs & colds (Dominica); in yellow fever (wAf); to stim. appetite & for cramps (VirginIs); in skin dis. (Sabah); in ringworm, eczema, etc. (locally); lf. poultice for stiff muscles (Ec); headache (PI); rts. u. to prevent vomiting (Ec); abortif. (off. Br Phar.); fl. inf. u. in pregnancy; oxytocic; crushed lvs. u. for skin dis.; rts. reputedly purg. & antipyr. (Indochin); anthelm. - **C. pringlei** Rose (sMex): "rompebota"; shrub or small tree (lvs. purg.); u. in dys.; diur. diaph.; in ringworm & for its tanning properties - **C. quinquangulata** Rich. (Br): fedegôso grande; u. as antipyr. - **C. reticulata** Willd. (Mex, CA [CR], SA): Central American s. (leaves); lvs. c. rhein (antibiotic); purg. (CA); u. in arthritis, irreg. menstr. (CA) - **C. sericea** Sw. (tropAm): lvs. u. poultice for wounds; rts. u. in edema; sds. coffee substit. (Br) - **C. siamea** Lam. (eAs, PI, tropAf): kee lek (Siamese, Laos); wd. u. in liver dis., urticaria, anorexia, rhinitis; itch (Camb); lvs. fls. frt. u. to make soup for ton. & gastralgia; lvs. decoc. malaria - **C. singueana** Del. (C. goratensis Fres.) (hot Af, incl. Zimbabwe): rt. u. in VD; frt. pulp a galact.; rt. decoc. for stomachache, as purg.; in syph.; lf. decoc. with pepper u. for cough; as hot compress for fever (malarial) (Tanz); spasmolytic - **C. sophera** L. (C. lanceolata Forsk.) (pantrop, esp. Ind, Mol, Sunda): c. emodin & chrysophanic acid; lf. decoc. exter. for ringworm; pl. decoc. in acute bronchitis; abdom. pains; for reducing wt.; pd. sd. in diabetes (124); young leaves as veget.; *cf. C. occidentalis* - **C. spectabilis** DC. (cAf, NG): lf. decoc. for jaundice (Tanz); fls. c. pyrones; antiallergic - **C. spiciflora** Pittier (Ve): urumaco; lf. & fl. inf. u. as wash for fevers; lax. - **C. surattensis** Burm.f. (tropAs): scrambled eggs; common wild shower; rt. decoc. & bk. & lvs. u. GC (Indochin); fls. purg. (Indoch); bk. u. diabetes

(Haw); lvs. u. in bruises, dys. (Indochin) - **C. timoriensis** A. P. DeCandolle (Ind across seAs to Austral): raw frts. vermif. against *Ascaris* (Camb); macerated bk. (in oil) u. to treat the itch (Laos) - **C. tora** L. (considered by some identical with *C. obtusifolia* L.) (e&cUS, cosmopol. in trop): sickle pod; coffee weed; fantupa seed (EI, Molucc, Sunda); Folium Ketsumi; lvs. c. kaempferol sophoroside; lvs. ground with lemon juice & appl. to ringworm & scabies (Ind); cataplasm for skin dis. (Camb); sds. u. for eye dis., skin dis. such as trichophytosis (antifungal); in insect sting, lax. (Jap); hypotensive; sds. produce panwar gum (like acacia); young lvs. eaten as vegetable (eAf); bad weed - **C. villosa** Mill. (Mex): lvs. u. to treat dermal irrit., fungus growths; ash of lvs. appl. locally (6).

cassia: 1) cinnamon bark 2) wild senna, *Cassia marilandica* & related spp. 3) any cinnamon bk. (trade) esp. *C. loureirii* - **c. bark (tree): bastard c.**: *Cinnamomum cassia* - **Batavia c.**: *Cinnamomum burmanii*; u. adult. of Saigon & Chin cassias - **c. buds**: dr. fls. of *Cinnamomum loureirii* & *C. cassia* (Chin, Indochin) - **Canton c.**: **Chinese c.: Chinese cinnamon**: *Cinnamomum cassia* - **C. Caryophyllata** = *Dicypellium caryopyllata* - **fagot c.**: Batavia c. - **"c. fistula"**: 1) *Cassia fistula* 2) (Mx.) *Cassia grandis* (large); *C. moschata* or *C. nodosa* (small) - **c. flower tree**: *Cinnamomum loureirii* - **horse c.**: *C. grandis* - **Indian c.**: *Cinnamomum tamala* - **C. Lignea** (L.): Batavia c.: wood cinnamon; hard bk. fr. *C. cassia* & *C. burmanii* - **manna c.**: *C. fistula* (1480) - **Maryland c.:** *Cassia marilandica, C. fasciculata, C. nictitans* (1229) - **c. mata**: article claimed discovered in Java by Jap 1945 & u. as quinine substit. (1480) - **c. oil** (Chin): Oleum Cinnamomi fr. *C. cassia* - **Padang c.**: Batavia c. - **c. pods: purging c.**: *C. fistula* - **ringworm c.**: *Cinnamomum alata* - **Saigon c.**: *Cinnamomum loureirii* - **c. stick tree**: *C. fistula* - **Tanner's c.:** *Cassia auriculata* - **C. Vera** (L.): Chinese c.

Cassiafrüchte (Ge): *Cassia fistula* - **cassie**: *Acacia farnesiana* (VO).

cassie: *Acacia farnesiana* (with **c. oil**).

cassine: *Ilex c.*

Cassine croceum Ktze. (Celastr.) (sAf): saffronwood; pl. has digitaloid action; u. by witches for trial by ordeal; toxic; wd. useful - **C. velutinum** Loes. (sAf): rt. bk. inf. u. diarrh. dys.

Cassiope mertensiana D. Don (Eric.) (wNA): western mountain heather; ornam. shrub - **C. tetragona** (L) D. Don (wNA, polar s to Ore [US]): lapland cassiope; u. as summer fuel by Eskimos (1146).

cassis (Fr-m.): 1) cassie 2) *Ribes nigra*, black currant & its liqueurs (cordials).

cassumun(i)ar: *Zingiber cassumunar.*

Cassytha filiformis L. (Laur.) (sInd+): love vine (Bah); akas-borel (Beng); pl. a parasite (*cf. Cuscuta*); u. for hemorrhoids, bil. dis.; postpartum depr.; mixed with sesame oil to stim. hair growth (*sic*); juice u. in ophthalmias; u. rheum., aphrod.; frt. eaten with betel for colds (Solomon Is.).

Castagna (It): chestnut.

Castalia Salisb. = *Nymphaea.*

Castamargina™: claimed active princ. of *Castela tortuosa*; u. perorally in amebiasis.

Castanea Mill.: 1) genus (Fag.) of deciduous trees (chestnuts) (few shrubs); not to be confused with horse chestnut (*Aesculus*) of a different fam. 2) (L.) (specif.): chestnut lvs. fr. *C. dentata* - **C. chrysophylla** Dougl.: *Castanopis chrysophylla* - **C. crenata** Sieb. & Zucc. (Jap, Chin): Jap chestnut; sds. food of Chin & Jap; fls. remedy for TB, scrofula; wd. as timber - **C. dentata** (Marsh.) Borkh. (eUS & Cana): Am chestnut; lvs. tannin-rich; u. astr. & as tan; lvs. u. as tea as antiemetic in pertussis; nuts edible (superior); wd. formerly much u. (furniture, crates); wd. chief US source of tanning material in recent past; bk. u. to retard rancidity (1147); trees formerly grew to large size and great age; tree is now largely exterminated due to c. blight dis. (anthracnose) - **C. mollissima** Bl. (Chin, Korea): Chin chestnut; freq. cult. now in US for food since immune to chestnut blight dis. - **C. philippensis** (PI): bk. u. in asthma, headache - **C. pumila** (L.) Mill. (seUS): dwarf chestnut tree; chinquapin; Allegheny chinkapin; bk. astr. ton. antipyr.; nuts pleasantly edible but small - **C. sativa** Mill. (C. vesca Gaertn.; C. vulgaris Lam.) (Euras, nAf, Med reg.): Spanish (or European) chestnut; bk. u. astr. ton.; tan; much imported into US (2nd most important tan = bk); lvs. (as Folia Castaneae vescae) u. for cough (mostly as F. E.); sds. the large sweet European chestnut, a popular food esp. It & Sp; c. flour u. in sEu (mts.) in place of cereals in breads, fls. ornam. but with strong odor like semen; bk. source of starch (1756); for soup thickening, polenta (*qv.*); wd. useful - **C. vesca** Gaertn.: **C. vulgaris** Lam.: *C. sativa.*

castanea (comml.): **castanha** (Br): Brazil nut.

castañeda (Sp): chestnut grove.

castaño: 1) (Sp): chestnut 2) (Guat): *Quercus castanea* - **c. de indias**: *Aesculus hippocastanum.*

Castanopsis acuminatissima A DC (Fag.) (NG): bk. eaten to aid conception in woman who has never borne children (2261) - **C. chrysophylla** A DC (swUS): Calif (or giant) chinquapin (chinkapin);

golden chestnut; ornam. tree char. by lvs. bright golden yellow on lower surface; sds. edible - **C. philippensis** (Blco.) Vidal (Pl): "gemulaun sub"; bk. u med. for asthma, headache - **C. sempervirens** Dudley (Ore, Calif [US]): bush chinquapin; sds. (nuts) eaten (much like chestnut).

Castanospermum australe A. Cunn. & Fras. ex Hook. (l) (Leg.) (NZ, Austral, natd. trop&subtropAs): bean tree; sds. edible when ripe; roasted: ("Australian chestnuts"); pods astr.; cultd. as shade tree (Ind); wd. very valuable (carving).

Castela erecta T & G (Simaroub.) (tropAm incl. Col): goat bush; amargoso (Mex); pl. c. castelamarine, casteline; u. antiamebic; fevers, eczema (1148) - **C. texana** Rose (C. tortuosa Liebm.) (swTex [US], nMex): (chaparro) amargoso; allthorn castela; entire pl. (mostly u. tendril bk.) u. amebiasis, antipyr. astr. ton.

Castela, Herba (L.): **Castelae Planta**: *Castela texana*.

Castilla costaricana Liebman (Mor) (CR): a chief source of CA or Panama rubber - **C. elastica** Cerv. (Mex, CA): Panama gum tree; hule (Mex); a tall tree; latex fr. stem source of Panama (or CA) rubber (elastic); frts. & sds. u. galactogog. - **C. fallax** O. F. Cook (CA): source of "gutta percha" of inferior type (tunu gum, gum falax perhaps): hardy stick gum mass - **C. ulei** Warb. (Amaz, Br): "cauchu"; latex source of rubber.

Castilleja Mutis ex. L.f. (Scrophulari.) (NA, SA): (Indian) paint brush; painted lady; ca. 250 spp.; semihemiparasites - **C. canescens** Benth. (Mex): castilleja (Mex); u. med. (Mex Phar.) in dig. dis. (to incr. saliva & bile); as diur. - **C. linariifolia** Benth. (Calif, Ore [US]): painted cup; desert paintbrush; pl. decoc. u. in dysmenorrh., contraceptive, stomachache (Am Indians); rt. decoc. cath. emetic - **C. patriotica** Fern. (sUS, Mex): yerba del sapo; u. in urin. dis.

Castilloa Endl. = *Castilla* Cerv.

cast-iron plant: *Aspidistra elatior*.

castor: 1) Castoreum 2) (Fr): beaver - **c. bean**: sd. of c. oil plant - **c. grease:** Castoreum - **c. meal:** c. pomace - **c. oil fish**: *Ruvettus pretiosa* (Pisces) (trop seas): oil fishes; mackerel-like fishes u. for food; body FO c. ricinoleic acid (as wax); with lax. effect - **c. oil plant**: *Ricinus communis* (lvs. & sd. u.) (SC sd. u. prodn. c. oil) (**false c. o.**: stramonium) - **c. pomace**: residue after expr. FO fr. c. sd. - **c. sugar**: pd. sugar (SC can be u. in castor [= shaker]) - **c. wood**: *Magnolia glauca*.

Castor canadensis Kuhl (Castoridae) (NA, SA): American beaver; Canadian b.; large rodents u. as source of valuable fur and of Castoreum (L.), (beaver) castors (dr. glandular secretion of preputial follicles of sac) (musk); formerly u. as sed. ("nervine") in hysteria & analeptic (1149); menstrual difficulties (dysmenorrh.) (1517); also to promote healing of wounds (BC [Cana], Indians) (2306); u. by trappers to bait traps; perfume fixative (mfr.) (this sp. sometimes treated as *C. fiber* L. var. *canadensis*); chief food bk. & cambium of *Salix* & *Populus* (aspen, cottonwood) spp. - **C. fiber** L. (Eu): (European) beaver u. as preceding.

Castoreum (L.): castor; the dr. preputial follicles of beaver, *Castor fiber*; u. fixative in perfume mfr.; sed. (1149).

Casuarina equisetifolia Forst. f. (Casuarin.) (Ind to Polynesia): Australian pine; A. she-oak; horsetail tree; beefwood; bk. c. tannin; ext. u. hospitals (Ind) as substit. for Krameria, etc.; bk. u. diarrh. dys.; lf. decoc. in colic.; cult. ornam.; natd. in sFla [US], WI +; excellent firewood.

cat: *Felis catus* L. (*F. domestica* L.) (Felidae) (cosmop.): small carnivorous mammal, long domesticated; serves as pest-destroyer (rats, mice); pet; meat as food (seAs); u. in Unani med. (flesh, ribs, brain) (1668).

cat: 1) cutch 2) (Mex): *Parmentiera edulis* (Yuc) 3) khat (tea) - **cat bells** (Ga [US]): *Crotalaria* spp. - **c. claw** (US): *Acacia greggii* or other prickly plants, such as *Pithecellobium unguiscati.*

catabolism: breakdown of living tissue substances in organisms.

catachysis (L.): douche - **cataclysma** (L.): clyster.

catalase: enzyme wh. decomposes H_2O_2 to give H_2O, sometimes utilizing the freed O to oxidize alcohols to aldehydes; occ. richly in body tissues, as liver, RBC, &c.; cont. Fe.

Catalpa bignonioides Walt. (Bignoni.) (e&cUS, orig. seUS): **common catapla**; (Indian) cigar tree; bean t.; Indian b.; fl. white with yellow stripes & purple spots; shaped rather like foxglove fl.; frt. antispas. sed. in asthma; emet. ton. antisep. antipyr; bk. rich in tannin, has been u. in making leather; bk. anthelm. alt.; trees often planted (seUS) for "c. worms" (caterpillars) popular among amateur fishermen (1150); capsules "smoked" by children (US) (Am Indian word); wd. for cross ties - **C. bungei** C. A. Mey. (Chin): pounded lvs. u. as dressing on sores of pigs; sds. appl. to boils & scabies; twigs u. in kidney dis. - **C. ovata** G. Don (Chin): pl. insecticidal; frt. diur., in edema, peritonitis - **C. speciosa** Engelm. (eUS w to Mo): hardy (or northern) c.; Western c.; Indian bean; Shawnee wood; similar u. to prec. - **c. sphinx** = **c. worm** (Ala [US]): **catalpas** (Ala [US]).

catalpa (tree) (English fr. NA Indians): **bignonia c.**: **southern c.**: *Catalpa bignonioides* - **c. sphinx**: c. worm (Ala [US]): ("catalpas" ([Ala]); a striped caterpillar ca. 8 cm. long, which thrives on c. tree; much u. for fishing bait (seUS) - **catamenia** (L.): menstruation.

cataphyllary (leaves): early form of lf. (ex. cotyledon, rhizome scale).

Cataputiae Minoris Semen (L.): sds. of *Euphorbia lathyrus.*

Cataria (L.): catnip, *Nepeta c.*

Catathelasma ventericosa Sing. (Tricholomat. - Fungi): (NA, Jap): large-stemmed armillaria; sporocarps u. as food; sold in Jap markets.

catawba (seUS): 1) *Catalpa bignonioides* (c. tree) + 2) catalpa "worm" (caterpillar).

cat-bells (Ga [US]): *Crotalaria* spp.

cat-brier: *Smilax rotundifolia +.*

catch-fly: *Silene virginica* & other *S.* spp.

catchup = catsup.

catchweed: *Galium aparine.*

catechin: catechuic acid isolated fr. pale catechu (gambir), cutch, &c.; u. to fortify antihistaminic preps.

catechol: pyrocatechol.

catecholamines: amine derivs. of pyrocatechol; ex. epinephrine.

catecholase: tyrosinase.

catechu (gum): 1) **pale c.**; **true catechu**; gambier; obt. fr. *Uncaria gambier*; ext. fr. lvs. & young shoots; ext. in form of quite regular brown cubes; c. catechin & catechu-tannic acid (Indones, Borneo) 2) **black catechu**: cutch (Ind, Burma): ext. fr. heart wd. of *Acacia catechu, A. suma*; prepd. with boiling water; similar constits. to true c.; u. as natural dyestuff for textiles, tan; pharm.; somewhat similar are exts. of wattle bk. ("wattle gum") fr. *Acacia decurrens*, &c. - **gambier c.**: Gambir cutch: Gambier. - **Pegu c.**: black c.

catface (swGa [US]): side of pine tree which has been turpentined; (SC appearance of old style cutting [blaze] of trunk); blemish (esp. knot) in lumber.

catfoot: *Gnaphalium obtusifolium* (*G. polycephalum*).

catgut: 1) sheep intestine u. in sutures, strings of musical instruments, etc. 2) *Tephrosia* spp., esp. *T. virginiana.*

Catha edulis (Vahl) Endl. (monotypic) (Celastr.) (Saudi Arab, Ethiopia, Kenya, Somal, Yemen) (cult.): Abyssinian tea; k(h)at t.; lvs. c. cathine, cathinine, cathidine, no caffeine; shoots with leaves (preferably fresh) (esp. in Yemen+): fresh lf. chewed daily (air freighted fr. eAf [Kenya chief source] to Muslim nations); & as beverage (Khat) (Yemeni & Arabs); CNS stim. (euphoric *cf.* amphetamine); thought habituating; u. for reducing wt. (Eu); insomniac, anorexic, aphrodisiac; bronchodilat., hypertensive, astr.; u during Ramadan (Muslim month of fasting) (2600); use illegal in some eAf countries (1151) - **cathine**: d-nor-isoephedrine; (d-norpseudoephedrine) alk. with stim. properties somewhat like those of caffeine & cocaine.

Catharanthus lanceus Pich. (Lochnera lancea Boj.) (Apocyn.) (Mad): rts. c. ajmalicine, yohimbine (quebrachine); u. in dys. (sAf); as galact., emet. bitter, astr.; in folk med. (Mad); antineoplastic - **C. ovalis** Markgraf (Mad): u. as vermifuge for calves; ordeal poison (formerly) - **C. roseus** (L.) G. Don f. (Vinca rosea L.) (orig. Madagascar; now pantropic as Br [coast]): periwinkle; Madagascan p.; chatilla (Guat); fls. pinkish, violet, or white; c. very many (70+) alks., incl. catharanthine, vincaleucoblastine (vinblastine), vincristine, ajamalicine, akuammine, lochnerine, reserpine, serpentine, vindoline, yohimbine, etc. (90); some of alk. u. in acute leukemia, Hodgkin's dis., lymphogranulomatosis, lymphoreticulosarcoma, chorionepithlioma; in folk med. reputedly of use in diabetes (Domin+); dyspeps. (as tea) in mal. (1518); u. in med. in Tonga (86); to overcome primary inertia in childbirth; cult. ornam.; much exported fr. Ind; u. hypotensive (noted in 17th cent.); mashed between teeth to calm sensations of hunger & fatigue - **C. trichophyllus** (Baker) Pichon (Mad): rt. has great reputation as stim. (bitter ton.); in blenorrhagia & its sequelae.

Cathartocarpus = *Cassia* (*fistula*).

cathedral plant (US - popular): agrimony.

cathepsin: mixt. of proteolytic enzymes (peptidases & proteinases) found in crude pepsin & in many animal tissues; esp. rich in spleen, kidney, liver, gastric juice; acts to hydrolyze (autolyze) dead animal tissues.

Cathomycin™: novobiocin, antibiotic.

cathine: amphetamine-like compd. (D-norpseudoephedrine) (*qv.*) found in *Catha edulis;* action like that of caffeine or cocaine.

catinga (Br): region of stunted thorny vegetational growth, esp. low open forest in drought-stricken areas on sandy soils (seBr, Rio Negro region).

cativa: **cativo (balsam)**: **c. gum:** balsamic resinous exudn. fr. *Prioria copaifera.*

catjang: *Vigna cylindrica; Cajanus cajan.*

catkin: special type of inflorescence with several fls. on single stalk, usually all pistillate or all staminate fls.; ex. willow (ament).

catmint: **catnep: catnip**: *Nepeta cataria* +- **false catnip**: *Mentha rotundifolia.*

catrup: catnip.

cat's eggs: *Xanthium spinosum* - **c. faces**: pansy - **c. foot**: 1) c. paw (*qv.*) 2) *Antennaria* spp. 3) *Asarum canadense* & other *A.* spp. - **c. hair** (sUS): *Euphorbia pilulifera* - **c. heads**: *Tribulus terrestris* - **c. paw**: 1) ground ivy, *Nepeta glechoma* (lvs.) 2) *Trichinium spathulatum* (Tasmania) 3) *Anigozanthus humilis* (Haemodor.) (Austral): ornam. cult. for its woolly fls. - **c. paws**: fl. heads of *Antennaria* spp., *Trifolium arvense* - **c. tail**: *Equisetum arvense*; *v.* below..

catsup: food dressing or sauce made up of tomatoes, vinegar, sugar, spices; sometimes with oyster puree, onions, green walnuts, &c. (orig. of spiced mushrooms).

cat-tail: 1) *Typha angustifolia* & *T. latifolia* 2) *Equisteum arvense* - **common c**: 1) *T. latifolia* 2) (Eng): timothy - **cat-thyme**: *Teucrium marum* - **cat-tree**: spindle-tree.

cattle: animals of genus *Bos*, incl. *B. taurus* (European c.), *B. indicus* (humped zebu, etc., mostly tropical); and *B. bubalus* (usually now as *Bubalus bubalis* (water buffalo) - **c. foods**: commonly include besides grains & their husks, grasses, etc., various articles of pharmacognostic interest, such as fenugreek.

Cattleya Lindl. (Orchid.) (CA, SA): most are epiphytes; the most popular orchid for bouquets, etc.; chiefly *C. bowringiana* O'Brien & hybrids (over 100); mostly purple.

catuába (Br): 1) *Erythroxylon nitidum* 2) *Anemopaegma mirandum*.

cauaçu (wax) (Br): **cauassú wax** (Br): prod. readily obt. fr. underside of lf. of cauassú, *Calathea lutea* (Chenopodi) (*cf.* carnauba wax).

caucho (Sp): rubber - **c. del Pará** (Cu): Pará rubber tree.

caudex: 1) axis of woody pl. (tree trunk, shrub st.) or woody base of perennial hb. 2) st. of tree fern.

caudle: warm gruel with spice, sugar & wine; formerly adm. to invalids.

caudoside: glycoside of caudogenin & L-oleandrose, occ. in *Strophanthus caudatus*.

caujaro (Mex): *Cordia toqueve*.

caule (Por): = stem (<**caulis** [L]).

caulene: dye prep. fr. red cabbage (u. Germany in World War II) - **cauliflower**: *Brassica oleracea* v. *botrytis*.

Caulophyllum thalictroides Mx. (Berberid) (eUS): blue cohosh; a per. hb.; c. lupine alk. (anagyrine+), aporphine alk. [magnoflorine]); rhiz. cult. orn. u. antispas. emm. dem. diur.

caupi (SA): cow-pea.

caustic barley: Cevadilla.

cavalo (Por - slang): **cavalo** (It): Heroin; cabbage.

Cavanillesia platanifolia (Bonpl.) Kunth (Bombac.) (tropAm, esp. CA): wd. (quipo or bongo wood) soft & light u. for canoes & rafts; as substit. for balsa wd.; sds. edible - **C. umbellata** R. & P (Pe): sds. edible, sd. FO u. in ointments, &c.

cavete: cayate.

Cavia L. (Caviidae - Rodentia - Mammalia) (SA, Andes): cobaya (Br); cavy; guinea pig (improper name due to confusion of Guiana with Guinea [Af]); domesticated animal of Incas before Spanish conquest; of diverse colors; u. in lab. expts. extensively; u. as food (SA) - **C. aperea**; **C. porcellus**.

caviar(e): the salted & seasoned roe (fish eggs) of sturgeons (or other large fish) eaten as a delicacy; nutr. of high protein value; 3 chief vars. of Russian, including black (superior) and Beluga; exported fr. USSR & Iran; the Beluga (*Acipenser huso*) is found in the Caspian & Black seas, grows to very large size (-650 kg); (flesh) put up in cans; expensive ($60/oz.); considered a great delicacy, eaten often with champagne - **Am c**. (Ala [US], &c.).

cavolo (It): cabbage; (slang) Heroin - **c. flore** (It): cauliflower.

cayapia (Br): *Dorstenia* spp.

cayapó (, **purga de**) (Br): *Cayaponia diffusa*, *C. pilosa*.

Cayaponia cabocla Mart. (C. globosa Manso) (Cucurbit.) (Br): purga do gentio; silver naso; frt. u. purg. in edema, snakebite; sd. c. FO (14%) - **C. citrullifolia** Cogn. (Par): rt. u. as "blood purifier" - **C. diffusa** Manso (Br): frts. & rts. ("cayapó") drastic in edema; emmen. - **C. espelina** Cogn. (Br): tomha; rt. u. as bitter ton. for "blood purification," diur., purg., emet.; for asthma, epilepsy, syphilis - **C. glandulosa** Cogniaux (Co, Pe): frt. decoc. u. for liver complaints; lvs. & young sts. of vine dr. & u. as insect repellant dust for hammocks & clothes (2302) - **C. ophthalmica** R. E. Schult. (Col): soft green bk. u. to treat conjunctivitis (wash) - **C. pedata** Cogn. (C. pendulina) (Br): tayuya; strong purg. - **C. pilosa** (Br): **cayapó**; frt. & rt. u. drastic purg. emm. in snakebite - **C. racemosa** Cogn. (Mex, Salv, Pan): "camara"; u. med. (Mex Phar.); said toxic, esp. to cattle - **C. tayuya** Cogn. (sBr): taiuiá; taioia; u. in popular medicine as antidiab., antiarthritic, in rheum., "depurative" in syph. - **C. triangularis** Cogn. (Br, Amazon): purga de gentio; frts. & rts. u. as strong purg. (2302).

cayate: *Omphalea diandra*.

cay-cay fat: fr. **cay-cay nuts**: fr. *Irvingia oliveri*.

Cayenne (**pepper**): *Capsicum frutescens* (SC Cayenne, city of French Guiana exporting).

Cayratia trifolia (Domin) Quis. (C. carnosa Gagnep.; Vitis t. L.). (Vit.) (eAs): amarlata; decoc. for fever; lvs. u. as poultice for ulcerated nose (Mal); swelling of legs & mumps (124); lvs. appl. for dandruff (Indones); lf. decoc. or sap u. as antiscorbutic (PI); rt. ground with black pepper to apply to boils (Ind), carbuncles, swellings (Pak).

CCK (179): Hydergine (*qv.*).

Ceanothus L. 1) genus (Rhamn.) (mostly Pacific coast of NA) of shrubs or small trees; lvs. alternate or opp., fls. white or blue; sepals, petals & stamens in 5's; style 3-lobed; frt. separates at maturity into 3 nutlets; fls. c. saponin; u. by Am Indians & pioneers as soap substit. 2) (specific.) (L.): **C. americana** L. (e&cUS): (New) Jersey tea; dr. bk. of rhiz. & rt. u. hemostat. astr. & blood coag. (SC lvs. u. as tea replacement in Revolutionary War period) (1152) - **C. cuneatus** (Hook.) Nutt. (Calif, Ore, [US]): (common) buck brush; lvs. & fls. boiled to prep. tea (Am Indian) - **C. integerrimus** H. & A. (Calif [US]: deer brush; wild lilac; rt. c. integerressin, integerrin, integerrenin; rt. bk. decoc. u. in mal. jaundice, catarrh; ton.; bk. inf. emet. ton.; berries, sds. lvs. edible; shoots in basketry; fls. rubbed in water as soap - **C. reclinatus** L'Hér.: *Colubrina reclinata* - **C. sanguineus** Pursh (c&wNA): wild lilac; buck brush; shrub 1-2 m. high; sapwood dr. pd. & rubbed on wounds & sores (in grease); dr. bk. pd. & appl. as poultice for burns (Okanagan [BC] Indians) - **C. thyrsiflorus** Esch. (Calif, Ore, [US]) "blue blossom"; shrub or small tree; u. med. (1153); ornam. - **C. velutinus** Dougl. (NA: Pac. coast & Rocky Mt. area): sticky laurel; mountain l.; m. balm; "greasewood" (Wash [US]); snowbrush; deer bush; tobacco b(r)ush; soapbloom; evergreen shrub (1-2 m) branches green; decoc. u. int. & exter. for rheum.; VD (GC) (decoc. of branches); lvs. u. as tobacco substit. by Am Indians.

çeara wax: fr. *Copernicia cerifera.*

cebada (Sp): barley - **cebadilla** (Ve): *Schoenocaulon officinale.*

Cebar: chemical, biological & radiological weapons of offense/defense.

Cebatha Forsk. = *Epibaterium* = *Cocculus* (*C. carolinus,* &c.).

cebil: *Piptadenia cebil.*

Cebione™: vitamin C.

cebola (Por): onion - **c. albarrã:** squill.

cebolla (Sp): onion - **c. albarrana:** squill.

cebolleta (Sp): 1) tender onion 2) chives - **cebollin** (PR): leek.

ceci (beans): ceseron (Fr): cézé (Fr): *Cicer arietinum* (chick peas).

cecidiology: the science of galls, animal & pl. - **cecidium** (L.): a pl. or animal gall.

Cecropia adenopus Mart. (Cecropi) (Br, Arg): ambay lvs.; lvs. u. commonly as exp. in asthma; to reinforce digitalis action; frt. prod. wax similar to carnauba (1477) - **C. candida** Sneth. (Br): imbaúba; pith of long petioles u. in histological labs like elder pith - **C. hololeuca** Miq. (Br+): imbaúba; lf. buds u. in diarrh., diur.; lf. bud juice exter. for sores, int. in hemoptysis; rt. bk. in TB (pulm.); trunk bk. c. tannins, u. as astr. ton. - **C. latiloba** Miq. (nSA): lvs. ashed & added to coca lf. pd. before placing in mouth & swallowed, u. 3-4 times/day - **C. obtusa** Trécula (SA, Cu): medicinal pumpwood; juice u. caustic for warts; treat dys. & VD (SA); lf. decoc. u. edema, asthma, liver dis., cardioactive agt.; lvs. & fls. u. by Mayas in colds, Pinto dis. (6) - **C. obtusifolia** Bertol. (C. mexicana Hemsl.) (non C. obtusa Trécula) (sMex to Pan): trumpet; guarumo (morado); lf. decoc. u. in asthma, edema, liver dis., diabetes; sap of trunk u. on warts; forage, fodder - **C. peltata** L. (WI, nSA): West Indian snake (wood) bark; trumpet tree; imbaúba (Br): yagruma; lvs. c. steroid glycosides; u. exp. astr. antidys.; bk. u. in asthma; latex scraped fr. boiled bk. caustic u. to destroy warts, galls; in poultice for infected eye (2430); cardiotonic, diur., tops & latex u. exter. for pain (Cu); diaph. emm. ton. antipyr.; may be adult. of Cascara Amarga; water boiled with lvs. u. as bath water for rheum.; wd. u. to make rafts (*cf.* balsa); paper pulp, cellulose, lignin (1154); orn. - **C. surinamensis** Miq. (C. obtusa Trécul.) (Guyana, Ec, Pe): trumpetwood; medicinal pumpwood; lf. juice u. as caustic for warts, juice of new shoots in GC, leucorrh.; to treat dys. (SA) ; lf. decoc. & rt. sap strong diur.; cardioactive agt. in edema; in asthma, liver dis.; decoc. of fallen lf. u. in albuminuria, kidney dis.; lvs. & fls. u. in colds, Pinto dis. (6) - **C. tolimensis** Cuatrec. (Col): inf. rts. u. in fever.

cedar: term appl. to many different genera (16 or more) of shrubs & trees; such as *Juniperus, Cedrus* (the true cedar) (esp. *C. libani*), *Cedrela, Chamaecyparis, Thuja* (2592), *Hymenodictyon excelsum* (Rubi) - **c. apple:** 1) frt. of *Cedrela odorata* 2) - **Atlantic c.:** *Cedrus atlantica* - **c. ball** (sUS): **c. berry:** fungal-induced excrescences growing on red cedar, *Juniperus virginiana* - **deodar c.:** *Cedrus deodara* - **Eastern red c.:** *J. virginiana* - **c. family:** Cupressaceae - **c. gum:** fr. *Cedrela odorata,* etc. - **Himalayan c.:** deodar c. - **Japanese c.:** *Cryptomeria* spp. - **c. leaf (oil):** *Thuja occidentalis* - **Lebanon c.:** *Cedrus libani* - **Mexican c.:** *Cedrela* spp. - **mountain c.:** *Juniperus mexicana* - **c. oil:** VO fr. *Juniperus virginiana* - **old world c.:** *Cedrus* spp. -

Oregon c.: Port Orford c. - **pencil c.:** *Juniperus virginiana* & other *J.* spp. - **Port Orford c.:** *Chamaecyparis lawsoniana* - **prickly c.:** *Juniperus oxycedrus* - **red c.:** *Juniperus virginiana* (VO) - **rock c.**: *Juniperus mexicana* - **salt c.:** 1) *Tamarix gallica* 2) *Monanthochloe littoralis* - **Siberian c.:** *Pinus cembra* - **Spanish c.:** 1) *Cedrela odorata* (neither a cedar nor Spanish!) 2) *Juniperus oxycedrus* - **stinking c.** (Fla, [US]): *Torreya taxifolia* - **summer c.** (sUS): *Eupatorium capillifoium* - **true c.:** *Cedrus* spp. - **Washington c.:** *Sequoiadendron gigantea* - **West Indian c.:** 1) *Chamaecyparis thyoides* 2) *Cedrela odorata*, etc. - **white c.:** *Chamaecyparis thyoides* (**American w. c.**), *C. lawsoniana*; *Libocedrus decurrens, L. chilensis, Chukrasia* spp. - **C. wood:** 1) *Cedrus libani* 2) *Hymenodictyon excelsum* 3) *Juniperus virg.* 4) *Toona* spp. **- c. wood oil:** VO fr. *Juniperus virg.* (wd.) - **yellow c.:** *Chamaecyparis nootkatensis.*

cedrat(e): citron (*qv.*).

cèdre (Fr): cedar - **c. blanc:** *Thuja occidentalis.*

Cedrela odorata L. (C. mexicana M. J. Roem.) (Meli) (Mex, WI, CA, nSA to Br): Spanish (or West Indian or Mexican) cedar; cedro (blanco); kulche (Maya); grown as avenue trees; wd. & bk. u. in fever & epilepsy (arom. rt. bk.); sds. as vermif.; fragrant wd. much valued for furniture, cigar boxes, &c. (wd. floats in water); wd. inf. (in white wine) u. in mal. (Haiti); gummy adhesive resinous exudation ("sap") ("goma de cedro") fr. trunk u. as mucilage (acacia-like), in eyewashes to clarify vision; in chest dis.; frt. u. med. (1521) - **C. toona** Roxb. (Toona ciliata Roem.) (Ind, Himal to Austral): hill-toon; Indian mahogany; cult. as avenue tree; bk. u. in med. & as tan; prod. cedar gum (like acacia); fls. u. emmen., to dye yellow; valuable wood takes high polish - **C. tubiflora** Bertoni (Arg, Par, Br): cedro; wd. light (s.g.0.65).

cedro (Mex): *Cedrela* spp.; *C. odorata* (Hon) (**c. hembra**) - **cedrol**: cedar (or cypress) camphor; tricyclic sesquiterpene alc.; occ. in VO of *Juniperus virginiana, J chinensis, Cupressus sempervirens,* & other pls.; u. in perfumes; for soaps, &c.; occ. as crystals; cedryl acetate also occurs on market - **cedron**: 1) (Sp): *Quassia cedron* (sd.) 2) (Pe): *Lippia ligustrina* 3) *Aloysia triphylla* - **cedroncillo** (Pe).

Cedronella canariensis (L.) Webb. & Berth. (C. triphylla Moench) (monotype) (Lab.) (Canary Is, Madeira): "thé des Canaries"; perenn. hb.; lvs. u. to make tea; u. med. - **C. mexicana** Benth. (Mex): toronjil; lvs. & flg. tops u. in the form of inf. ("tea") as antispasm. stom. (fresh pl. preferred); med. props. attributed to VO (Mex Phar.); orn.

Cedrus Trew. (Pin.): the true cedar; 4 spp. - **C. atlantica** (Endl.) Carr (nAf, Alger; Atlas Mts): Atlantic or Atlas cedar; wd. u. as lumber; tree ornam.; VO u. in bronchitis, TB, gonorrh., skin dis.; cult. orn. - **C. deodara** (Roxb.) G. Don f. (Pinus d. Roxb.) AsMinor, Himal, Ind, Assam, Pak, Alg): deodar (tree); Himalaya cedar; Indian c.; source of resin, VO; VO u. as aid to digest.; antisep. in vet. med. & in skin dis. (like *C. libani*); wd. important u. furniture, bldg. (SC native Indian word meaning the gift of God [*cf.* Theodora]) (1154A); orn. - **C. libani** A. Rich. (Pinus cedrus L.) (AsMinor, Leban, Cyprus): cedar of Lebanon; wd. c. oleoresin; VO c. borneol; u. as percutaneous expect., arom.; digestive aid; useful wd. since days of ancient Egyptians (2275); source of resin u. in embalming (mummies); ornam. tree.

cee (US addict slang): cocaine.

Cefamandole: cephalosporin antibiotic; semi-synth.; wider spectrum of activity than others of series; low toxicity; u. GC+.

Cefoxitin Sodium: semi-synth. antibiotic, deriv. of cephamycin with strong activity in some bact. inf.

ceiba (bark): fr. *Bombax ceiba.*

Ceiba aesculifolia Britt. & Baker (Bombacaceae) (sMex, Guat, CR): pochote; young frts. eaten boiled (believed feminizing); sds. edible roasted; sds. c. FO, u. med. (Am Indians); for industrial purposes (like cottonseed oil); cotton fr. sds. u. in pillows (6); introd. into nSA: pochote (de pelota); inf./decoc. of bk. u. with sweetened water as diur.; young pods cooked as veget. - **C. blanca** (CA): *Hura crepitans* - **C. leucanthemum** DC (Eriodendron 1. DC) (Br - Mex): juice fr. bk. & thorns u. in eye dis. (Mex) - **C. pentandra** (L.) Gaertn. (tropAm, also trop As, Af): kapok tree; "ceiba" (Mex); silk cotton tree; corkwood; sd. furnishes silky fiber (sd. hairs) (Japanese kapok tree cotton) widely u to stuff mattresses, life preservers (1522), pillows, cushions, &c.; insulation (sleeping bags); lvs. & fls. & young frt. food; bk. appl. to wounds, decoc. u. int. as emet. diur. (rt., WI), antisp.; for "heart pain" (Tanz); edible oil (kapok oil) expr. fr. sds. (98), u. for soaps & foods; dark gum (like trag.) is "Malabar gum" u. as styptic, in diarrh. (wAf); shade tree (-70m); also to support Derris, pepper, and Vanilla plants; wd. to make matches, &c. (corkwood).

celandine, common: garden c.: greater c. (herb): *Chelidonium majus* - **lesser c.:** *Ranunculus ficaria* - **c. poppy:** *Stylophorum diphyllum* **- c. tree: tree c.:** *Bocconia cordata* (frutescens) - **wild c.:** *Impatiens pallida.*

Celastraceae: Bittersweet Fam.; shrubs, vines or trees; fls. small perfect; perianth 4- or 5-merous;

frts. often fleshy drupes; widely distr. except in frigid zones.

célastre grimpant (Fr): *Celastrus scandens.*

Celastrus paniculatus Willd. (Celastr.): panicled bittersweet; staff tree; melkangu(ni); lvs. c. non-toxic alk., u. antidote in opium poisoning; sds. prod. celaster oil (FO) u. nerve tonic, rheum., skin dis. (antisep.) for beri beri; sds. emet. stim. aphrod. lax.; bk. abortifacient; orn. - **C. richii** A. Gray (Oceania): lvs. chewed for toothache; bk. or lf. crushed in water, drunk for stomachache (Fiji) - **C. scandens** L. (SCar, Ala, Ga, La, Tex, [US]): false bittersweet; climbing b.; bk. u. emet. discut. alt. diur.; frtg. branches popular decoration; suspected poisonous pl. (frt. orange not red as in *Solanum dulcamara*) (other differences: Am b. has diff. kinds of lvs.; smaller greenish or whitish fls.).

céleri (NM, Sp): celery - **celeriac:** *Apium graveolens* v. *rapaceum* - **celery** (SC rapidity of therapeutic benefits fr. celer (L) speed; *cf.* celerity): *Apium graveolens* var. *dulce*; source of c. frt. (sd.); **c. (seed) oil** (VO) - **celery(-cabbage), Chinese:** *Brassica pekinensis; Oenanthe javanica* - **German c.** = turnip-rooted c. = celeriac - **perennial c.:** *Levisticum officinale* - **C. salt:** NaCl (80), pd. celery (19); celery VO (1); u. flavor, condim.

céleri-rave (Fr): **celeriac** (French celery: improved form cultd. (Fra.).

celidonia (Sp): garden celandine.

Celite: infusorial earth (TM, Johns-Manville).

cell: the characteristic unit of living tissues in both pl. & animals; consists of protoplasm, made up of cytoplasm surrounding the nucleus & usually surrounded by a cell wall - **c. inclusions**: ergastic substances (*qv.*) - **interstitial c.:** connective tissue c. - **stone c.:** sclereid.

Cellocidin: acetylenedicarboxamide; antibiotic fr. *Streptomyces chibaensis* Suzuki & al. (soil organism) (Japan); antibacterial.

Celloidin: soln. of pyroxylin u. as embedding agt. in histology.

Celloneze™: cellulose acetate - **Cellophane™:** viscose (cellulose soln.) solidified into transparent sheets (water permeable or impermeable); u. to wrap drugs, foods, candies, dressings, etc.

cellulase: enzyme wh. catalyzes hydrolysis of cellulose to cellobiose & sometimes even to glucose (may be due to presence of cellobiase); occ. in fungi, sds., malt, protozoa, many invertebrates, ruminant mammals.

cellulite (US popular term): waffle-like fat of legs (thighs) & waists of women, also buttocks, upper arm, &c.; do not confuse with cellulitis.

celluloid: synth. plastic prepd. fr. nitrocellulose & camphor mixt.

cellulose: $(C_6H_{10}O_5)_n$: a water-insoluble carbohydrate forming the basis of pl. fibers - **c. carboxymethyl** (sodium) **c.**: Carmethose™; cellulose deriv. u. as suspending agt., in adhesives, prepar. emulsions; drilling "muds"; cath. antacid; ophthalmic lubricant for dry eye (1% soln.) - **methyl c.**: c. ether forming a gel in water; u. as dispersive agt., bulking lax. - **microcrystalline c.:** white pd. made up of partly depolymerized c. prepd. by hydrolysis with mineral acids of purified wd. c.; u. suspending agt., filler, binder, moisture control, lubricant (tablet mfr.), disintegrator (in compressed tabs.), & reducing diets (appetite depression) (Avicel) - **oxidized c.: cellulosic acid:** a compd. made by treating c. with N_2O_4, rendering it absorbable by the body tissues - **powdered c.**: prepd. fr. alpha (wood) cellulose made by mechanical disintegration; u. as absorbent, tablet disintegrant, & suspending agt.

Celosia L. (Amaranth.) (trop & temp): coxcomb; Pampas plume; ca. 60 spp.; colorful orn. - **C. argentea** L. (tropAs, now widely distr. [weedy]): cockscomb; popular in Far Eastern med.; pl. with red fls. preferred; fls. u. dys. hemorrhoids, menorrh, leucorrh. (Chin); sds. u. diur; to strengthen eyesight & hearing; pl. u. as emmen. (Laos); lvs. eaten for beri beri & pellagra (Siam); as food (eAf); in blood dis.; beautiful ornam.; var. *cristata* (L.) Ktze. (C. cristata L.) decorative (tropics esp. eAs): cockscomb; keito (Jap); fl. comb dried & u. in dys., bleeding fr. lungs & intest.; diur.; tops appl. as poultice to painful bones (2448); many cvs; cult. ornam. - **C. trigyna** L. (trop eAf, Saudi Ar, Madag): igiri (Bukoba); c. kosotoxin; young shoots u. in tape- and roundworm (Tanz, Ethiopia); fls. & lvs. in diarrh. menorrh. & as partur. (Af); in skin eruptions; eaten as veget. (cultd.); rts. boiled in porridge for syph. (pregnant women) (Tanz); much cult.

celosia dos jardins (Por): *Celosia argentea* var. *cristata.*

Celsia cinnamomea Lindl. (Scrophulari) (Ind): bk. c. skatole; u. as "blood purifier" in skin eruptions - **C. coromandeliana** Vahl (Ind): lf. juice u. as sed. astr. in diarrh. & dys.; in fevers & skin eruptions; celsioside (saponin) with anti-cancer activity. (—> *Verbascum*)

Celtis L. (Ulm.) (nHemisph, sAf): hackberry; some spp. trees with much the habit of elms; frt. a drupe, often sweet (hence "sugar berry") - **C. brasiliensis** Planch. (Br, Arg): tala (Arg); bk. u. as antipyr.; wd. u. for agr. tools, paper pulp - **C. iguanaea** Sarg. (Tex [US] to tropAm): granjeno; juice u. for sore eyes (Mayas); lvs. u. for bathing in fever (decoc.) (6); frt. edible - **C. laevigata**

Willd. (C. mississippiensis Bosc.) (se&cUS, Mex, WI): sugarberry; Mississippi hackberry; bk. of lge. trunks (1-1.2 m diam) covered with corky warts (2.4-5 cm. high); frt. sweet hence name; ornam. tree (-35 m) - **C. occidentalis** L. (eNA): hackberry; bk. anodyne, refrig.; frt. pulp c. sugar & flavoring principles, sweet & edible; pollen important cause of hay fever - **C. pallida** Torr. (Fla, Mex): spiny hackberry; lvs. u. for head & stomach pains; frt. yellow; edible - **C. tala** Gill. (SA): tala (Arg); tree lvs. inf. u. for indig. (Arg, Ch).

Cenchrus L. (Gram) (trop & subtrop): sand spur; sand bur; cock spur; bur grass; some 25 spp., often pests <cenchra (Gr) = *Setiria italica* - **C. biflorus** Roxb. (C. barbatus Schum.) (C. catharticus Delile) (Af [Sahara region]; Ind, n&eAf, Austral, natd. NA): kramkrum; chevral; karindja; ann. grass; grains u. to make a flour fr. which a porridge is made; u. as a medico-magic subst. by secret societies (Senegal); rts. u. as vulner. after fights; buffel grass; sandbur; young pl. good forage (high protein); frt. (grains) decoc. u. diur. pectoral (Ind); eaten as porridge (Af); bad weed (US) esp. in sheep herds - **C. ciliaris** L. (Af, As, Austral): anjan grass; u. as fodder; frts. (grains) eaten - **C. echinatus** L. (tropAm: sUS): bur(r) grass; West Indian b. g.; frts. (caryopses) eaten & u. to make alcoholic beverage (poor people); vicious annual weed throughout the Americas & elsewhere - **C. incertus** M. A. Curtis (C. pauciflorus Benth.) (Fla [US], Mex): coast sandspur; rt. tea drunk by puerperal women (Mex) - **C. setigerus** Vahl (Af, wAs+): birdwood grass; grains u. as famine food.

Cenia hispida B. & H. (Comp.) (sAf): rt. decoc. drunk for relief of nausea.

cenizo (swUS): 1) *Atriplex canescens* 2) (Mex): appl. to 3 genera, esp. *Leucophyllum*; mostly *L. texanum*, *L. ambiguum*.

cenoura (Br): carrot.

Centaurea americana Nutt. (Comp.) (sUS, Mex): basket flower; med. (Mex Phar.); ornam. hb. - **C. behen** L. (Iran, Syr, Ind): suffed bahman (Indian bazars); rt. ("white rhapontic") u. blood "purifier"; aphrod., nerve sed. - **C. benedicta** L. = *Cnicus benedictus* L. - **C. calcitrapa** L. (Euras, natd. NA): corn flower; star thistle; fl. heads u. ton. stim.; young leafy sts. eaten (Eg); hb. rt. & frt. u. as diur. in intermittent fevers; pl. juice u. in eye dis. - **C. cyanus** L. (Euras, esp. Balkans, natd NA): cornflower; corn bluebottle; "blawort" (1481); ann. grain field weed; fls. c. tannins; u. (as Flores Cyani) diur. antipyr. added to fumigant powders; to improve teas - **C. diffusa** Lam. (Euras, natd w&cNA): tumble knapweed; allergenic (also other *C.* spp.); spines of this & some other *C.* spp. may cause mechanical damage to animal; this & other knapweeds allelopathic pernicious weeds resulting in great economic losses (Wash, Ore, Mont, Iowa [US], Cana) - **C. jacea** L. (Euras): brown knapweed; rt. u. antipyr., hb. in baths for rickets; gargle; fls. u. diur. hemostatic - **C. maculosa** Lam. (Eu, natd. NA): spotted knapweed; hb. c. cystall. antibiotic subst. - **C. montana** L. (Eu): Bergflockenblume; pl. weakly cyanogenic; fls. u. diur. in eye waters, for cough, as bitter ton.; for diarrh. (Romania) - **C. nigra** L. (Eu, natd. NA): (black) knapweed; star thistle; per. hb. u. bitter ton. diaph. diur.; supposed equal to gentian as ton. - **C. picris** Pall. (C. repens L.) (Azerbaidjan+): Russian knapweed; pd. hb. put in water for worms & wounds of sheep (Ind); horses sicken after use of pl. in encephalomyelitis - **C. scabiosa** L. (Euras: natd. US [Great Plains]) greater knapweed; hard head; greater centaurea; fls. c. cyanidin diglucoside; lvs. c. alk.; u. in folk med., esp. in Russia as diur.; also in scabies - **C. seridis** L. var. *maritima* (Sp): reputed antidiabetic; value thought due to presence of ß-sitosterol (2488) - **C. solstitialis** L. (sEu, weed in seUS): yellow centaurea; y. star thistle; spina giallo (It); fls. antipyr.; rt. stom. - **C. transcaucasica** D. Sosn. (USSR): has hypotensive properties - **C. urvillei** Stepposa Wagenitz (Turk): c. flavonoids (apigenin, &c.) with anticancer properties.

Centaurium beyrichii B. L. Robins. (Gentian.) (Tex [US]): quinine weed; mountain pink; early settlers in Tex used hb. for reducing fever - **C. chilensis** (Willd.) Druce (Gentiana peruviana Lam.) (Mex to Chile, high mts.): canchalagua; hb. u. as ton. stom. anthelm. antipyr. emmen. for snakebite (folk med.), kidney dis. (tea) (C. canchanlahuen Robins.) - **C. erythraea** Rafin. (C. minus Moench; C. umbellatum Gilib.; Erythraea centaurium Pers.) (Euras, Af, natd. NA): common centaury; centaurea menor (Sp); c. minore (It); (lesser) centaury herb; bitter-herb; c. VO, bitter princ., glycoside; u. (simple) bitter ton. stom. antipyr. diaph. in dyspep. gastric weakness, in liver & bile dis.; in bitters (liqueur industry); imported fr. Ge, Yugoslavia, Mor (1155) - **C. pulchellum** Druce (Euras, nAf): centaury; hb. found as adult. of *Swertia chirayita* (prod. Fr, Ge) - **C. venustum** (Gray) B. L. Robins. (Calif [US], nwMex): wild quinine; (beautiful) centaury; canchalagua; pl. with bright pink fls.; common on hills; hb very bitter, u. folk med. ton. in fever, ague, pneumonia.

centaury, American: *Sabatia angularis* - **blue c. flowers:** *Centaurea cyanus* - **common c.: European c.: foreign c.: lesser c. (tops): minor c.:**

Centaurium erythraea - **red (American) c.:** American c. - **yellow c.:** *Blackstonia (Chlora) perfoliata.*

centella (Sp): 1) *Caltha palustris* 2) (Co): *Cleome hassleriana, C. serrata* & several other pl. spp.; anti-fertility agt. (As).

Centella asiatica (L) Urban (Hydrocotyle a. L.) (Umbell.) (pantrop. OW, NW, esp. sAs, Ind; wet places): Indian pennywort; water navelwort; gotu kola; lvs. & sts. c. asiaticoside (glycoside) hydrosotyline (alkaloid); VO chief component of Fo-ti-tiang (Chin); u. sed. in insomnia; diur., Hansen's dis. (leprosy), psoriasis (VO); supposed to confer long life; hb. or lvs. (as Hydrocotylis Folia, Herba Hydrocotyles asiaticae) u. as astr. antispasm. stim. "nerve tonic," aphrod. alter. narcotic, arom. emet. alexipharmic; in "cancer"; TB; syph. skin dis.; pruritus; chickenpox; anorexia; dyspep. stomatitis, card. dis., worms in children; as gargle; lvs. eaten in curries & salads; to aid digest. (Hainan); for fever; exter. as anti-inflamm. (Tahiti); antifertility agt. (As); cicitrizant (Af): lvs. as poultice (Tonga); triterpenoid fraction favorable to wound healing; med. values claimed are disputed (1131-A); triterpene derivs. on market - **C. erecta** (L.f.) Fern. (C. repanda [Fern.] Small) (Fla to Md to Tex [US]) (grows in moist soils, mostly sandy, also saline [tolerant]); u. like *C. asiatica* (several *C.* spp. are toxic but also u. in med., ex. **C. glabrata** L., **C. montana** Donin.)

centeño (Sp): rye.

Centipeda cunninghamii (DC) A. Braun & Aschers. (Comp.) (tropAs & Austral) sneezeweed; u. like Arnica & as sneeze powder & in purulent ophthalmia - **C. minima** A. Braun & Aschers. (A. orbicularis Lour.) (eAs, Oceania, Austral): spreading sneezeweed; inhaled & rubbed on nose for colds; sternutatory (PI); in headache & colds in the head; in fevers (Ind); med (Chin) - **C. thespidioides** Muell. (Austral): desert sneezeweed; decoc. drunk for colds, sore throat, sore eyes, sternutatory; poultice of pl. appl. to sprains (2263).

centipede: *Scolopendra* spp.; large centipedes of many areas; member of class Chilopoda (Arthropoda) (worldwide; terrestrial); many fams. & genera; dr. centipede u. for pain, colds (Chin); as poultice for wounds, TB, tetanus (-1986-); oil in which boiled removes hair (2269).

Centotheca lappacea Desv. (C. latifolia Trin.) (Gram) (Oceania - New Hebrides Is+): now-now; pl. chewed & appl. to burns; fodder grass; lf. juice u. for finger "sickness" ("thava") (Fiji); lf. decoc. u. for vaginal bleeding (Samoa); weed.

Centranthus (Kentranthus) **ruber** DC (Valeriana rubra L.) (Valerian.) (UK, Med reg., wAs): red valerian; c. valepotriates; u. as sed.; quite popular in Ge (1523); cult. ornam.

Centratherum anthelminticum (Willd.) Ktze. (Vernonica a. [L] Willd.) (Comp) (tropAm): purple fleabane; sds. stom. ton. anthelm. diur.; exter. in skin dis. incl. pediculosis; frt. & lf. u. in colics, dropsy; exter. in rheum, gout - **C. punctatum** Cass. (CA, WI, SA): perpetua roxa do mato (Br); and subsp. **australianum** Kirkman (Austral, PI).

Centrolobium robustum Mart. ex Benth. and **C. tomentosum** Benth. (Leg.) (Br): arariba; bk. rich in tannins, u. as astr.; lvs. u. as plasters to cover wounds or contusions & for skin parasites (zebra wood).

centromere: point where 2 chromatids are fused; appears as transparent constriction.

Centropogon ferruginea (L. fil.) Gleason (Campanul.) (Co): bitter tea fr. lvs. u. in dys.

Centrosema plumieri Benth. (Leg) (tropAm): "patitos" (Mex); u. green manure & cover crop.

Centrostachys aquatica Wallich (monotypic) (Amaranth.) (Ind, Java, nAf +): u. like *Celosia trigyna* (*qv*.).

century = centaury = Centaurium - **American c.:** Sabatia angularis - **blue c.:** Centaurium erythraea - **c. plant:** *Agave americana, A. parryi.*

Cepa 1) (L) (Sp): onion, *Allium cepa*, bulb 2) strain, stump, stock.

cephaëline: one of chief alks. of Ipecac (*Cephaëlis* spp.); may be converted into emetine (the most important ipecac alk. therapeutically) by a process of methylation; more emetic, more toxic, & less expectorant than emetine (2541).

Cephaëlis Sw. (Rubi.) (trop): shrubs or small trees with c. 150 spp.; bot. synonyms incl. *Evea, Uragoga, Psychotria* - **C. acuminata** Karst. (Uragoga granatensis Baill.) (Co, Nic): Cartagena Ipecac; Nicaragua I.; Johore (or East Indian) I.; this sp. has rel. much less emetine than Rio I. & rel. much more cephaëline; hence previously regarded as much less valuable than the Rio, but now, with better methods of methylating, much less differentiation on price basis & most mfrs. prefer Cartagena; Johore with lowest alk. cont. (1.4-1.7%); Cartagena with the highest (commonly 2.1-2.8% up to 3.2%) (two kinds of Cartagena i. reported: (1) red brown rt. (2) gray brown rt. [richest in alk]) u. emet. exp. amebicide; in toothpastes; for mfr. alk.; rt. u. as expect. in bronchitis, bronchial asthma; emet. (in larger doses) in poisoning (presently syrup of i. considered the most appropriate emet. for use in the household) (disadv. nauseous; do not confuse syr. with F. E.); to empty stomach in pyloric stenosis; in amebic dys. (prolonged use may lead to chronic poisoning);

exter. in treating pustules; in folk med. for coughs & colds; tonic (tea); for teething (Ve); source of emetine & cephaëline - **C. emetica** Persoon (Co, Pe): greater striated ipecac; false (or black) i.; u. emet.; adulter. of Cartagena ipecac - **C. ipecacuanha** (Brot.) A. Rich. (Br, Bo): Rio ipecac; the preferred sort is Matto Grosso fr. that state, in 14-16° Lat. S. (1894); practically all fr. wild-growing pls.; another popular sort is Minas Gerais (after state of origin); Rio & Para types refer to port of export; also Bahía, which is inferior in quality; this sp. has rel. high emetine cont. but lower total alk. content compared to *C. acuminata*; u. similarly to *C. acuminata*; also u. as activator of tanning agt. - **C. muscosa** Sw. (WI): West Indian ipecac; u. like true ipecac but less potent - **C. tomentosa** (Aubl.) Vahl (Psychotria t. Hemsl.; Eva t. [Aubl.]) Standl.) (Trin, Hond, Nic, Guat, Belize, Mex): "false ipecac" (Rusby); "cresta de gallo" (Mex Tabasco): boca de sapo (Co): rt. c. little emetine (0.106% alk.) u. like true ipecac (1895).

Cephalanthus occidentalis L. (Rubi.) (NA, CA, eAs): (common) button-bush; honey-balls; popular in folk med.; bk. u. ton. alt. in home medicines; in cough med. (with syrup & paregoric) (nAlas [US]); st. & rt. bk. u. ton. antipyr. in dys. (decoc) (Choctaw Indians); in toothache, sore eyes; preps. incl. "globe flower syr.," "honey ball mixture," etc. (1156); cult. orn.

cephalins: kephalins: group of glycerophosphoric acid esters with fatty acids & ethanolamine; occ. in brain tissues; u. hemostatic; reagent.

Cephalocereus senilis (Haw.) Schumann = *Cereus s.*

cephaloglycin: semi-synth. antibiotic, deriv. of cephalosporin.

cephalosporins: antibiotics usually derived fr. *Cephalosporium* spp.; several (at least 9) known, incl. semi-synth. cephalothin, cephaloridine, cefoxitin, cephapirin.

Cephalotaxus fortunei Hook. f. (Cephalotax.) (sChin): plum yew; c. harringtonine & other alks.; antileukemic properties; u. to treat tumors - **C. harringtonia** (Forbes) K. Koch (C. drupacea Sieb. & Zucc.) (Jap, Korea, nChin): Jap. plum yew; sds. of evergreen shrub or small tree c. harringtonine, with activity against leukemia (L-1210); known only as cult. pl.; cult. Eu & US (not known in the wild) - **C. mannii** Hook. f. (Chin, nInd, Indoch): bk. has alkaloids charringtonine, &c.) u. in leukemia; cult. orn.; bk. lf. decoc. in stomachache & resp. dis.

cephamycins: group of antibiotics produced by several *Streptomyces* spp.; antibacterial.

cepharant(h)ine (Jap): pl. alk. discov. c. 1945; prepd. fr. tubers of *Stephania cepharantha* Hay. & *S. sasakii* Hay; appar. useful in TB & leprosy.

Cera (L) (It.): wax - **C. Alba** (L.): white bee's wax obt. fr. yellow by bleaching processes; preferred to yellow in most cosmetics, etc. - **C. Chinensis**: Chinese (insect) wax; nearly pure cerotic acid esters; obt. through action of insects, *Coccus ceriferus* Fabr. (Ericerus pe-la Kuenck), & *C. pe-la* Westwood) (on *Fraxinus chinensis*); u. in candles & med. - **C. Flava:** yellow bee's wax - **C. Japonica**: Japan wax - **C. Mineralis (Flava):** ozokerite; ceresin - **C. Sigillata**: sealing wax.

cêra de ouricury (Br): wax fr. *Syagrus coronata.*

cerasin: 1) metarabic acid fr cherry, beet, & other gums 2) incorrect spelling of ceresin(e) (*qv.*).

Cerastium arvensis L. (Caryophyll.) (NA, Euras): hb. u. in hemorrhoids, ophthalmia, &c. - **C. vulgatum** L. (Euras, natd. NA) (common) mouse ear (chickweed): hb. u. nutr. (in salads); refrig.

Cerasum Rubrum Acidum (L.): sour black cherry, frt. of *Prunus cerasus* (Ge Phar.)

Cerasus acidus Gaertn.: appl. to group in *Prunus* (sometimes as sect.) = *Prunus cerasus.*

Ceratanthera Hornem. = Globba - **C. beaumetzii** Heckel (Zingiber.) (wAf, trop): type specimen said mixt. of *Globba heterobracteata* Schum. & *Aframomum melegueta* Schum.; fresh rhiz. u. as purg.; tenifuge; c. tannin, gum.

cerate: pharm. prepn. based on wax (cera) and fat; sufficient wax is u. to render prod. solid, not to melt at body temp.; for exter. use - **compound camphor c.:** camphor ice.

ceratin: keratin.

Ceratiola ericoides Mx. (monotypic) (Empetr.) (seUS, esp. Fla): "rosemary"; a heath-like shrub; has needle-like evergreen lvs. with "peculiarly pleasant aroma"; u. folk med. (1158); cult. orn.

Ceratoides lanata J. T. Howell (Eurotia l. Mq.) (Chenopodi.) (c&wNA, Man [Cana] - NMex [US]): winter fat; white sage; hot decoc. of pl. u. to rid hair of lice; hair & scalp ton. (wIndians); to cure baldness (!), prevent graying of hair; decoc. of st. & lvs. for eye soreness (Nev Indians [US]). (—> *Axyris*)

Ceratonia (L.): frt. of **C. siliqua** L. (Leg.) (Med reg, Sp to Isr): carob (bean); locust seed; St. John's bread; figueira de egito (Por); bean u in mfr. drugs, syrups, as lax., confection of poor children (Eu); flavor in chewing tobacco, &c.; crushed bean in stock foods (calf & swine meals); frt. (bean) c. free butyric & isobutyric acids, pentosans, proteins; very rich in sugars (8% reducing sugars, 44% sucrose); concanavalin A.; sd. c. mucilage; pulverized & sold as carob gum (many TM names, such as Tragasol), sds. (karati) u. as weights; in textile industry as dressing agt. (to replace Trag.); frts. sold as dried, roasted, kibbled; (pods said toxic when young); boiled pods u. to treat "catarrhal" dis., cough; gum u. as stabilizer in freezing mixts.; to incr. viscosity of syrups, binder in baking industry (*cf.* chondrus, cydonia, tragacanth); in detection & detn. of borates (1211); to treat habitual vomiting (added to milk, drunk slowly), digest. disturb. (1173); juice of fresh frt. u. as mild lax. (Syria, Lebanon); astr. for diarrh. (1212); substit. for cocoa, coffee, & chocolate (2526); pd. sold in groceries (high K, low Na) (2737); useful shade & ornam. trees.

Ceratoniae Fructus (L.): frt. of prec. - **C. Gummi**: gum of prec.

Ceratophyllum demersum L. (Ceratophyll.) (NA, natd. As, now cosmopol.): coontail; hornwort; submerged water pl. of marshes (Rawlings, Cross Creek [72]); pl. u. in "biliousness"; scorpion sting (Ind); aquarium pl.

Ceratum (L.): cerate (ointment, the base of wh. includes wax; the prep. is stiffer & harder to melt) - **C. Citrini:** rosin cerate - **C. Epuloticum** (L.): calamine ointment - **C. Galeni:** cold cream.

Cerbera madagascariensis = *C. venenifera* - **C. manghas** L. (Apocyn.) (seAs along the sea coasts, Oceania): Schellenbaum; sds. constitute typical digitaloid drug (cardenolides & cardioactive osides); lf. & bk. purg.; lf. sap u. for temporary tatoo; frt. in cataplasms; sd. FO u. as vermifuge, illuminating oil (odolla fat) - **C. odollam** Gaertn. (islands of Indian & PacOc, [Polynesia & seAs], Mal): small tree or shrub; very toxic; frts. & sds. u. as fish poison, in suicides, as ordeal poison; c. cardioactive glycosides incl. cerberin; fls. u. for leis (welcome wreaths); lf. & bk. purg.; frt. in cataplasms; sd. most potent, cardioton.; sd. FO (odolla fat) rubefac. in arthritis; anthelm.; caustic latex emet.; locally for sores, in stingray poisoning; appl. to tattoo to aid in penetration of pigments; this sp. often confused with *C. manghas* but is distinctive - **C. venenifera** Steudel (Tanghinia v. Poir.) (Madag): Madagascar poison nut; tanguin (of Madagascar); sds. very poisonous, formerly u. as ordeal poison ("judgment of God"); c. glycosides with cardiotropic activity like digitalis (tanghinin); 75% FO; u. cardioton.

cerberoside: Thevetin B. = acetylneriifolin; do not confuse with cerebroside.

Cercidium floridum Benth. ex. Gray (C. torreyanum Sarg.; Parkinsonia f. S. Wats.) (Leg.) (sCalif, Ariz, Tex [US], nwMex); (blue) palo verde; tree usually found along water courses; sds. ground & made into cakes (Ariz Indians [US]) eaten with honey; also made into beverage; green pods covered with meat - **C. microphyllum** (Torr.) Rose et Johnst. (Parkinsonia microphylla Torr.) (seCalif, sAriz [US]; nwMex): yellow (or foothill) palo verde; sds eaten; fls. eaten fresh or cooked; juice expressed & drunk with honey; sds. in necklaces; wd. as fuel - **C. praecox** (R. & P.) Harms (C. spinosum Tul.) (Pe, Co, Br, Ve, Ec, Arg, Mex,): yabo; palo brea; cuica; source of cuica (or quika) resin (formerly collected) ("gum quick") (1157) u. as strong glue.

Cercis canadensis L. (Leg.) (eNA): (eastern) redbud; Judas' tree; bk. astr. said u. in old days by colonists as cinchona substit. in malar.; fls. & buds u. in salads, pickles, as antiscorb.; state tree of Okla [US] (pron. sersis) - **C. chinensis** Bge. (eAs): bk. & wd. u. to treat abscesses & bladder dis.; bk. antisep - **C. occidentalis** Torr. (wUS): western redbud; Judas tree; buds, fls. & young pods edible (fried).

Cercocarpus Kunth (Ros.) (Mex to Ore [US]): mountain mahogany; evergreen shrubs or low trees - **C. ledifolius** Nutt. (Calif to Wash [US]): curl leaf m. m.; toobe (Piute): important medicine of wAm Indian, esp. for pulmonary dis. (TB) (bk. decoc.); lf. decoc. for heart dis.; bk (or inner bk.) for blood tonic - **C. montanus** Raf. (C. betuloides Nutt. ex T. & G.) (c&wUS): hardtack m. m. (NM [US]); bk. u. dye; lf. decoc. lax. (Tewa Indians).

cerda (Sp): horsehair bristle - **c. de cola**: horse tail hair.

cerdo (Sp): hog.

cerea, erba da (It): *Monotropa uniflora* (1159).

cereal beverage (US): near beer; beer with c. 0.5% alc.

cereals: the fruits of grasses; also the grasses which produce them.

cerebral hypophysis = pituitary gland.

cerebrins = cerebrosides: complex glycosides found in nerve tissue (esp. brain); cont. N base, hexoses, & fatty acids; ex. phrenosin (fr. beef spinal cord).

Cerefolium (L.): **Chaerefolium cerefolium** (L.) Sch. & Thell.: chervil, *Anthriscus cerefolium* - **Spanish c.:** *Osmorhiza longistylis*.

Cerefolius (Ge Homeop. Phar.): Cerefolium.

cerejas (Br): cherries.

Cerelose™: dextrose.

cereo (It): *Cereus* - **C. a gran fiore odoroso:** *Cereus grandiflorus*.

ceresin(e) wax: earth wax; an impure yellow or white paraffin wax made by processing ozokerite; u. bee's wax substit. (prod. USSR, Galicia; Utah, Tex [US] fr. mineral deposits) (*cf.* cerasin).

cereus: common name appl. to many different cacti, such as *Echinocereus reichenbachii*; *Cereus greggii* (**desert c.**) (**c. root**) - **night-blooming c.:** 1) *Cereus grandiflorus* 2) *Cereus undatus* 3) *Cereus greggii* 4) *Epiphyllum hookeri* (Lk. & al.) Haw + (Cact.).

Cereus bonplandii Parm. (Harrisia b. [Parm.] Britt. & Rose) (Cact.) (Arg, Par, Pe): c. alk. (caffeine ?); u. in homeop. med. like *C. grandiflorus* - **C. deficiens** Otto & Dietr. (Lemaireocereus d. B. & R.) (Ve coast): breva; frts. edible, spineless in some cases at other times spiny pl. body c. K & has been exported to US recently to be u. as fertilizer; hedge pl. - **C. donkelaarii** Salm-Dyck (Selenicereus d. Britt. & Rose) (Yuc, Mex): pitaya; fls. u. as cardiac stim.; frts. u as food - **C. dumortieri** Scheidweiler (Lamaireocereus d. Britt. & Rose) (cMex): u. med. (Mex) - **C. giganteus** Engelm. (Carnegiea gigantea [Engelm.] Britt. & Rose) (sAriz, seCalif [US], nwMex): saguaro; sahuaro; giant cactus; the largest of the cacti; up to 20 m high, tons in wt., & centuries old; red sweet frt. ("apples") eaten fresh, as preserves, candy, molasses; sometimes dr. & stored; made into syr. & intoxicant beverage (ceremonial wine); ground into paste & u. as type of butter on tortillas, &c.; spines u. as pins (Am Indians); ribs of st. u. as lumber, ceilings of rooms (Am Indians) (1138) - **C. gracilis** P. Mill. var. *simpsonii* (Small) L. Benson (sFla & Keys): Simpson's prickly apple; frt. sometimes eaten - **C. grandiflorus** P. Mill. (Selenicereus g. [L] Britt. & Rose) (Mex, WI - Jam, Cu, natd. Latin Am+): night blooming cereus; Cactus Grandiflorus (L): "Cactus" (Mex); flor da noite (Br); fls. with vanilla-like aroma; fls. & succulent sts. u. sed. hypnotic, diur. rheum. (Mex); in neurasthenia, nicotinism; was brought in fr. Mex preserved in alc. in glass jars; u. in several functional heart dis. (valvular defects, muscular weakness, cardiac neuroses, myocarditis); nearly obsolete (1160) - **C. greggii** Engelm. (Peniocereus g. Britt. & Rose) (swUS, nMex): night blooming cereus; la reina de la noche (queen of the night); deerhorn (or desert or sweet potato) cactus; chaparral c.; large tuberous rts. (2.5-70 kg) have been u. as food & med. (Am Indian) (*cf.* turnip); slices bound to chest for congestion (364); lge. fls. odorous at half mile (said); frts. edible (delicious) - **C. gummosus** Engelm. (Lemaireocereus g. Britt. & Rose) (nwMex [Baja Calif]): very widely distr. (2061): pitahaya (agría); p. dulce (agriá); crushed rts. c. 24% saponins; u. to stupefy fish; source of resin u. as varnish & to calk boats; frt. sweet & tart, popular, delicious; said most valuable frt. of Lower Calif - **C. pentagonus** L. (CA, Mex, Tex [US]): pitahaya; frts. edible - **C. schottii** Engelm. (Lophocereus s. Britt. & Rose) (sAriz [US], nwMex [Sonora]): senita (cactus); sinita; rich in alks., lophocerine, &c.; lophenol; decoc. of pl. u. against cancerous tumors; frts. edible; sts. u. for fences - **C. senilis** DC (Cephalocereus s. (Haw.) Schumann) (cMex): old man cactus (since pl. covered with a beard-like growth of long silky white hairs); cabeza de viejo ("old man head"); vegetable wood; pl. grows to 15 m; pl. apex has odd appearance like face of old man; u. med. (Mex); fibers u. at times; pop. cult. as unusual ornam. - **C. spinulosus** DC (Selenicereus s. Britt. & Rose) (seTex [US], eMex): exotic ornam. - **C. striatus** Brandeg. (Wilcoxia s. Britt. & Rose; W. diguetii Peebles); (Mex [lower Calif, Sonora]): pita(ha)yita; saramatraca; sacamatraca; tuberous rts. crushed & juice appl. with cloth to chest for pulmon. dis., swellings; rts. eaten (Seri Indians) (2483); this cactus with nocturnal fls. - **C. thurberi** Engelm. (Lemaireocereus t. Britt. & Rose) (nwMex, sAriz [US]): organ pipe (cactus); pitahaya (dulce); frt. edible, important food of Papago Indians, u. to make candy, preserves, dr. sts. as fuel & for torches - **C. undatus** Haw. (Hylocereus u. [Haw.] Britt. & Rose) (Mex pantrop cultn.): pitahaya (orejona); chacoub; lge. frt. excellent food; st. juice u. exter. & int. as vermicide (care!); cult. as hedge plant; the finest of the night-blooming cereus.

Cerevisia (L.): beer - **Cerevisiae Fermentum** (L.): formerly USP tablets of yeast (Saccharomyces c.)

cereza (Sp): sweet cherry; sour c. - **cerezo** (Sp); cherry tree or wd., *Prunus cerasus* - **c. silvestre:** wild cherry tree or wd.

cerfoglio (It): chervil.

cerides: waxes.

ceriman: *Monstera deliciosa.*

cerin: ceresin.

Cerinthe glabra Mill. (Boragin.) (seEu, AsMin): honeywort; rhiz. c. mucil. reducing sugars, sucrose; u. in ointments for wounds; vegetable - **C. major** L.: (Med reg., sEu): honeywort; wax plant; u. similarly to prec.; cult. ornam - **C. retorta** Sibth. & Sm. (Med. reg.): cult. as ornam. ann. hb.

Ceriops Arn. (Rhizophor.) (tropics): mangroves; shrubs or small trees bearing aerial roots; growing on margins of sea; bk. u. tan. - **C. roxburghiana** Arn. (Pak): goran; bk. of bole & twigs with 50% ext. c. 66% tannin (of catechol type); lvs. also c. tannin - **C. tagal** C. B. Robins. (C. candolleana Arn.) (Ind to Austral): tangal; tengah; goran; bk. decoc. u. diabetes (*sic*); hemostatic (Ind); in childbirth (Mal); shoot decoc. u. substit. for quinine

(Af); bk. decoc. astr. to stop hemorrhage, appl. to malignant ulcers (PI); bk. c. 45% tannin: source of mangrove cutch, tengah extract, Bakau cutch (c. 58% tannin) u. tan., batik dyeing; wd. good for bldg. - **C. timoriensis** Daen. (New Caled): u. med.

cerisier (Fr): cherry tree - **c. de Virginie** (Fr): *Prunus serotina.*

Cerocoma (Meloideae-Coleoptera-Insecta): one of the blister beetle genera.

cerón (Fr): suron: comm. term for kind of bag or bale covered with a fresh cow hide, with the hair left on it; u. for some drugs fr. SA.

Ceropegia bulbosa Roxb. (C. acuminata Roxb.) (Asclepiad.) (Ind): tuberous rts. u. ton. digestant; food - **C. candelabrum** L. (Ind): lf. u. exter. for pains in limbs - **C. tuberosa** Roxb. (Ind): tubers u. in dys., diarr.; eaten.

cerotic acid: CH_3 $(CH_2)_{24}COOH$; occ. as esters in bee's wax (fr. cera, L. wax), carnauba wax, &c.

Ceroxylon alpinum Bonpland ex DC. (C. andicola Humb. & Bonpl.) (Palm.) (SA, sCo, Andes): common wax palm; palma blanca; "palm wax" deposits on trunk; u. much like carnauba wax (*qv.*).

cerrado (Br): dry prairies dotted with woods composed of stunted twisted trees and gnarled bushes, cattle grazing land; dry season 3-5 mo./yr. (Matto Grosso).

ceruletide: c(a)eruletin: decapeptide found in skin of *Hyla caerulea* (amphibian) (Austral): stim. GIT esp. gastric secretions; diagnostic aid.

cerveza (Sp): beer - **flor de c.** (Sp): hops.

Cervus (Cervidae-Ruminantia) (worldwide temp.): deer - incl. red d. (Eu), American elk, maral (As) - **C. canadensis** (US, Cana): (American) elk; wapiti; animal slaughtered for meat, fur. - **C. elaphus** L. (Eu): red deer; hart; horn (harts-horn) u. med. as raspings ($CaCO_3$, Ca phosphates); cartilage; u. addn. to pectoral teas; glue; fertilizers for flowers; horns in velvet (Typha Cornu Cervi); hart's suet (FO) in ointments.

ceryl alcohol: cerotin; found as cerotate in Chinese wax.

cesium: Cs: alkali metal occurring in earth's crust. - 137_{Cs} (isotope) u. to treat cancer.

Cestrum L. (Solan) (tropAm): poison berry; some 125 spp. of shrubs & small trees; several u. med., lvs. & sts. u. (seUS) emoll, analg. in urine dis.; several cult. as ornam. (fls. attractive & usu. fragrant) - **C. aberrans** Macbr. (Pe): laplacata; pl. poisonous to domestic animals - **C. aurantiacum** Lindl. (Guat): c. steroidal saponins; sterols, flavonoids; beta-amyrin, beta-sitosterol; u. primitive med., ornam. shrub - **C. conglomeratum** C. & P. (Pe, Col): lf. inf. added to chicha to make more soporific - **C. diurnum** L. (Mex, WI, SA): Juan de noches (Yuc); toxic plant; u. much like *C. nocturnum* - **C. dumetorum** Schlecht. (c&sMex to CR): galan; palo hediondo; decoc. of lvs. & new shoots appl. to skin dis., esp. pimples (granos) (Guat, Mex); rt. ext. u. for "fits"; lf. placed in ear for pain - **C. laevigatum** Schlecht. (coastal Br, Minas Gerais; natd. sAf): dama de noite; Coerana (Braz Phar I.); green berries c. gitogenin; frt. bk. & lvs. c. cestrumide (NBP); galenic prepns. u. as sed, antisep. emoll. in liver dis. diur., as cataplasm to reduce pain; cannabis substit.; much feared stock poison - **C. lanatum** Mart. & Gal. (sMex, CA): zorillo (CR); chacuaco; wd. decoc. cath. antipyr.; lvs. insectifuge (placed in hen nests); black frt. u. black dye - **C. latifolium** Lam. (Nic to Br, WI): jasmin (bois); wild jasmine; lady of the night (Trin); lf. decoc. emoll. u. to bathe skin rashes, mange, boils, marasmus (Trin) - **C. nocturnum** L. (sMex, WI, CA): night (blooming) jessamine (or jasmine); jasmin de nuit; huela de noche; frt. lvs. & st. ext. u. sed. antispas. in epilepsy; hysteria; analg. in urin. dis.; lvs. c. nicotine, nor-nicotine, yuccagenin, tigogenin; lf. decoc. u. as lotion in skin eruptions; pl. toxic; ornam. shrub (cult sFla [US]) - **C. parqui** L'Hér. (Ch, Pe): hediondilla (Arg); willow-leaved jessamine; durazuillo negro (Ur); toxic pl.; c. tannins, alks; solasonine; gitogenin & digitogenin glycosides; lf. u. in hemorrhoids (exter.); fr. u. antitumor (exter.) rt. bk. in edema, purg. antispas. (Arg); stock poison esp. frt. - **C. strigillatum** R. & P. (Ur to Col); ucha panga; lf. inf. u.; ton.; fls. with jasmine odor.

Cetacea: whales & related; mammals which spend their entire life cycle in water, usually marine; the anterior limbs are flipper-like; posterior have usually disappeared; formerly source of important fat & wax prods.; ex. whale, dolphin; fishing for whales is now largely prohibited by civilized nations; the products have disappeared from the markets.

Cetaceum (L.): spermaceti; a waxy brilliantly white solid removed fr. the cranial cavity of the sperm whale, *Physeter catodon*; c. mostly cetyl palmitate (m.p. 40-50°C); u. dem. emoll. in toilet creams, soaps, important ointment base; for candles; int. in diarrh. (Eu); formerly (ca. 1600) appl. to bruises & painful places (p. o.); product not now seen on market since killing of whales prohibited by most nations (1875) (1162).

Ceterach officinarum DC (Asplenium o. L.) (Polypodi) (Euras): rock fern; scale f.; fronds u. dem. diur. astr. in splenic dis. (Ind). (—> *Asplenium*)

cetico (Pe): *Cecropia peltata.*

cetin: cetyl palmitate - **cétine** (Fr): spermaceti.

cetology: science of whales (& related mammals).

Cetraria islandica (L.) Ach. (Parmeli. - Lichenes) (nEuras, nNA - arctic, subarctic, nTemp Zone): Iceland moss; I. lichen; Lichen Islandicus; Cetraria (L.); dr. foliaceous lichen; c. lichenin, isolichenin (galactosans), "cetrarin" (old name) (cetraric acid + lichen stearic acid + thallochlor [green pigment]); thallus u. to make bread (after removal of bitter princ. by water maceration); dem., bitter ton. lax.; substit. for ointment bases, in culture media; in fabrics for weaving; source of alc.; as brown dye, &c.; vehicle (disguises taste of nauseous med.); comm. source Scand, Siber, nAustr, Ge, Sp, Switz, Fr, NA: none fr. Iceland!) (1161) - **C. juniperina** (L.) Ach. (nTemp Zone): cedar lichen; juniper moss; u. to dye woollen goods brown; u. to poison wolves (*sic*) - **C. tenuifolia** (Retz.) Howe (oceans): often occurs mixed with *C. islandica* & u. similarly.

Cetratherum anthelminticum (Willd.) Ktze. (Vernonia a. Willd.) Comp) (Ind, seAs); purple fleabane; kali (Hindi); sds. u. as stom. ton. anthelm diur.; exter. in skin dis., to destroy pediculi (lice); in asthma, antipyr. lf. u for nasal discharges; similarly u. is *C. punctatum* Cass. (Ind).

cetyl alcohol: n-hexadecyl alc.; chief component of spermaceti; u. emoll. emulsifying agt., stiffening agt., lubricant, softener (cosmetics); medl. prod. (NF) with <10% stearyl alc. - **c. palmitate**: a characteristic ester. constit. of spermaceti.

cevadilla (seed): sabadilla - **cavadine:** veratrine; alk. fr. sd. of *Schoenocaulon officinale*; u. insecticide; poisonous.

cevitamic acid: ascorbic acid.

Ceylon: Sri Lanka (island): v. also cardamom; moss.

Ceylon cow plant: Gymnema lactiferum - **C. leadwort:** *Plumbago zeylanica.*

chá (Br): tea - **c. de maté** (Por)**: c. de paraguai:** maté, *Ilex paraguariensis* - **c. (de) pedestre:** *Lantana pseudothea.*

chabucano (Mex): apricot.

chaca(h) (Mex)*: Bursera gummifera.*

chacara (SA): a clearing with cultivated plants.

chadlock: Sinapis Nigra.

Chaenactis DC (Comp.) (wNA): false yarrow; some spp. with antibiotic props.; some spp. cult. ornam. herbs.

Chaenolobus Small = *Pterocaulon.*

Chaenomeles japonica (Thunb.) Lindl. (Ros.) (Jap native): "japonica"; (dwarf) Japanese flowering quince; shrub (1 m) cultd. for red fls. of spring - **C. sinensis** Koehne (Cydonia s. Thouin) (eAs, widely cultd.): big quince; Chin q.; produces lge, woody frt. (-2.7 kg); cut up & boiled (delicious flavor); Chin u. frt. to perfume room; u. in sweetmeats; frt. u. in jelly - **C. speciosa** Nakai (C. lagenaria Koidzumi); large-flowered Jap. quince; cultd. shrub (3 m); frts. u. in diarrh., emesis; arthritis; muscle cramps; frts. u. for jelly (2767)

Chaerofolium cerefolium (L.) Schinz & Thell. = *Anthriscus c.*

Chaerophyllum bulbosum L. (Umb.) (Euras, natd. US): turnip-rooted chervil; cultd. for thick rts., boiled & eaten; may c. vol. alkaloid - **C. procumbens** (L) Crantz (e&cNA): spreading chervil; rt. toxic, u. as emet.

Chaetachme aristata Planch. (Ulm.) (sAf): bk. u. to treat hemorrhoids; pd. rt. dental anodyne - **C. microcarpa** Rendle (cAf): buhonga; lf. inf. purg.; wd. to make musical instr.

Chaetocalyx latisiliqua Benth. (Leg) (CR-Ec): chupa-chupa (Ec); bruised lf. u. for skin eruptions (Ec) (2144).

Chaetoptelea mexicana Liebm. (l) (Ulm.) (Mex-Pan): bk. astr. u. for cough; wd. a lumber (cross ties).

Chaetospermum glutinosum Swingle = *Swinglea glutinosa.*

chaff: the glumes, paleas, lodicules, rachis, rachilla, & bristles of a grass spikelet (or fl. gp.) collectively.

chagas: chagueira (Por): *Poinciana pulcherrima.*

chagruna (leaves) (Pe): *Psychotria* spp., subgenus *Mepouria.*

chaguel gum: v. under gum.

chai (Russ): tea - **c. khan:** tea house.

chala (Arg): corn husk - **chalchay** (Yuc): **chalche** (Guat): *Pluchea odorata.*

chalchapi (Guat): rt. of *Rauwolfia heterophylla.*

chalchuan (Mex): *Erigeron affinis.*

chalchupa (Mex, Guat): *Rauwolfia tetraphylla.*

chalcones: chalkones: gp. of flavonoids (phenyl styryl ketones) with an allyl bridge (-CH:CH.CO) separating the 2 arom. rings; yellow/orange colored; xall. occ. in Rutaceae; pigments related to the flavones.

chalgooja (Baluch.): sds. of *Pinus gerardiana.*

chalk: one form of calcium carbonate - **aromatic c. powder** (NF): a mixture of cinnamon, nutmeg, clove, cardamon, prep. chalk & sucrose - **blackboard c.**: Ca phosphate - **compund c. powder** (NF): a mixture of prep. chalk with acacia & sucrose; u. gastric & intest. antacid - **French c.:** a soft native talcum (high grade) u. to remove oil spots (cleaner) & mark clothes (tailor) - **precipitated c.** (Eng): a synthetic form of $CaCO_3$, formed by ppt. of $CaCl_2$ with sod. carbonate; u. in dentrifrices - **prepared c.** (NF): a natural or native form of $CaCO_3$ which is mined & processed; actually represents the fossilized shells of spp. of

Foraminifera (protozoan animals); freed fr. most impurities by elutriation; u. antacid in gastric hyperacidity & diarrh.; exter. as dusting prd. & in skin lotions.

challote (Tex [US], Mex): chayote: peyote, *Lophophora williamsii.*

chalones: glycoproteins found naturally in mammalian body (liver, lungs, &c.) wh. act to hinder cell growth, hence to slow down process of aging of cells, possibly inhibit cancer growth; are tissue-specific but not specific for gen. or sp.

chalva (Gr): sweetmeat made fr. sesame sds.

Chamaebatia foliolosa Benth. (Ros) (Calif [US]): mountain misery; lvs. u. ton. antimal (inf.), orn.

Chamaecrista Moench (Leg.) (seUS): partridge pea; formerly made a separate genus, usually a subgenus of *Cassia*; u. bee plant; analogous to *Cassia marilandica* v. *Cassia fasciculata* - **C. fasciculata** (Mx.) Greene (Cassia chamaecrista L.) (se&cUS, WI): partridge pea; prairie senna; locust weed; sensitive senna (WI); lvs. u. cath. much like *Cassia marilandica*; honey & forage pl.; sds. serve as quail food; cult ornam.; var. *brachiata* Pullen ex Isely (swFla, sAla [US]): cold water inf. of pl. u. for nausea (Seminole Indians) - **C. nictitans** (L.) Moench (e&cUS): wild sensitive plant; lvs. reputed cath. vermif.; weed (—> *Cassia*).

Chamaecyparis lawsoniana Parl. (Cupressus l. A. Murray) (Cupress.) (PacNA): Lawson (or Port Orford) cypress; wd. of tree u. for many purposes; ornam.; resin a strong diur. - **C. nootkatensis** (D. Don) Spach (Cupressus n. D. Don) (nwPac coast NA): Nootka cypress; Sitka c.; Alaska cedar; yellow cedar (or cypress); excellent wood. (c. carvacrol & Me ester; nootkatin); bk. fibers u. clothing, fish nets, ropes, daily utensils, bailer for canoes (nwIndians); ornam. tree (-40 m) - **C. obtusa** Endl. (Jap, Formosa): Japanese cypress; VO & resin fr. wd. u. to treat GC (Chin, Jap); wd. useful; var. *formosa* Rehder (C. taiwanensis Masamune & Suzuki) (Taiwan): hinoki; a large tree, wd. source of VO (with chamenol, cadinene, ß-cresol); lvs. c. flavonoids (1204); cult. orn. - **C. thyoides** BSP (eUS): (southern) white cedar; "juniper" (seUS); lvs. u. stim. arom. diaph. stomach.

Chamaedaphne calyculata Moench (monotypic) (Eric.) (nEuras, natd. NA): leather leaf; u. diaph. in folk med. (toxic); fresh & dr. lvs. u. to make tea (Obijway Indians); rock garden ornam. shrub; c. andromedotoxin.

Chamaedorea Willd. (Palm.) (tropAm): young unopened spathes or young fl. clusters eaten as vegetable (*cf.* asparagus) (Mex & CA) - **C. elatior** Mart. (Mex): guayata; in Mex Phar. - **C. erumpens** H. E. Moore (Belize, Guat): cluster palm - **C. tepejilote** Liebm. (Mex, Col): "chimp."

Chamaedrys (, German): *Teucrium chamaedrys.*

chamaelirin: saponin fr. foll.

Chamaelirium luteum (L) Gray (monotypic) (1591) (Lili.) (eUS): helonias; false unicorn (root); rhiz. & rt. c. chamaelirin, starch, etc.; u. uter. ton. emm. diur. anthelm. alt. (1163).

Chamaemelum nobile All. = *Anthemis nobilis* L. (Comp.).

Chamaepitys Hill: *Ajuga c.*

Chamaerops humilis L. (Palm.) (Med reg.; cult. Ind, &c.): dwarf palm; Mediterranean palm; the only Eu palm; the young leaf buds of this small palm are eaten; lvs. source of fiber much u. for cordage ("vegetable hair") (crin végétale), ropes, sandals, sombreros (Sp), &c.; u. med. (Mex Phar.) - **C. ritchieana** Griff. (Ind, Afgh+): lvs u. in dys. diarrh.; fl. buds eaten.

Chamaesyce albomarginata (T & G) Small (Euphorbia a. T & G) (Euphorbi.) (swUS, nwMex): rattlesnake weed; golondrina (Mex); a prostrate hb. u. to treat snakebite; as emet. - **C. buxifolia** Small (Euphorbia b. Lam.; C. mesembryanthemifolia Dugand) (sFla, Mex, CA, WI): decoc. in diarrh.; latex u. for warts & bleeding wounds by appln. to same (Cayman Is); dr. lvs. u. for tea (Bah Is) ("bush tea") - **C. cinerascens** (Engelm.) Small (sTex [US], Mex): ashy euphorbia; latex obtained by boiling plant & appl. to boils & pimples - **C. densiflora** (Klotzsch & Garcke) Millsp. (sMex): golondrina; u. to wash wounds - **C. fendleri** (T. & G) Small (Tex, Mex): Fendler euphorbia; yerba de la golondrina; bruised fresh hb. u. for snakebite; or dr. hb. steeped in urine for same - **C. hanceana** Hemsl. var. *laureola* Gagnep. (Indochin): rhiz. u. to prep. tonic decoc. (Cambodia) - **C. hirta** (L) Millsp. (Euphorbia pilulifera L., E. hirta L.) (Ind, Austral, natd. seUS, Mex, WI, Fiji & pantrop.): pill-bearing spurge; "milkweed" (Bah); asthma plant; hb. c. antispasmodic princ. (1482); flavone glycosides; hb. u. against many dis. incl. asthma (expect.) (chief u.) (Hazelton); bronchitis, intest, dis.; dys. (New Caledonia) latex u. for worms (Bah); locally in ringworm (Ind); fungus inf. betw. toes (777); u. to cure boqueras (escoriations at the angle of the lips); in mouth ulcers of animals; hence called "hierba de boca" (Ve); depur. expect. (esp. in asthma), antihistaminic, anthelm., diur.; c. no alks.; tea u. to eliminate vaginal discharge (Bah); for dysm. (Domin); expressed juice u. for sore eyes (styes) (Domin, PI) - **C. hypericifolia** (L.) Millsp. (Euphorbia glomulifera Millsp.) (WI, seUS [Tex], Mex, tropAm): milkweed (Bah); u. as purg. in intest. fever; tea u. to eliminate vaginal discharge (Bah), for dysmen. (Domin), expressed juice u. for sore eyes (Domin) - **C. hyssopifolia** Small (sFla [US], WI, tropAm): pl. inf. for colds (Jam); diur. emmen. (Cu); inflamm. eyelids & poultice in erysipelas (Br); latex appl. to warts, calluses, ringworm - **C. maculata** (L) Small (Euphorbia m. L.) (NA, Fla to Calif [US] to Ont [Cana], Mex): large spurge; hierba de la golondrina (Mex); formerly u. to reduce spots & films on cornea (leukoma) after smallpox; latex u. for scarring of cornea; decoc. to bathe wounds (Mex) - **C. ocellata** Millsp. (Euphorbia o. Dur. & Hilg.) (Calif): yerba golondrina; latex appl. to skin lesions, infections, & black widow spider bites (Calif, Sp) (205) - **C. prostrata** (Ait.) Small (Euphorbia p. Ait.) (Fla to Tex [US], WI, Cu, trop. cont. America, natd. As, Af [Eg]): yerba de la niña; in rattlesnake bite (1350); c. flavonoids. (—> *Euphorbia*)

chambira: tucum nuts.

chamico (Cu): *Datura tatula.*

chamois: soft prep. skin of a wild Alpine goat (*Rubicapra.* sp.); u. as special cleaning cloth for glass, etc.; also name of wild goat.

chamomel: common chamomile.

chamomile (flowers): camomile (see this also): terms appl. to members of various genera of Comp.; If not otherwise qualified, applies specif. to German c.; or less commonly to Roman c.; sometimes appl. also to *Anthemis cotula & Achillea ageratum* - **anise c.:** *Matricaria matricarioides* - **common c.: corn c.: English c.:** *Anthemis nobilis, A. arvensis* - **f(o)etid c.:** *Anthemis cotula* - **German c.: Hungarian c.:** *Matricaria chamomilla* - **low c.:** common c. - **c. oil:** VO fr. German c. - **red c.:** *Adonis autumnalis* - **Roman c.:** common c. - **Spanish c.:** *Anacyclus pyrethrum* - **wild c.:** fetid c., also *Chrysanthemum parthenium* - **yellow c.:** *Anthemis tinctoria.*

Chamomilla recutita (L.) Rausch. = *Matricaria r.*

chamomille commune (Fr): Matricaria - **C. des champs** (Fr): *Anthemis arversis.*

champa: champaca (Por): *champak: Michelia champaca* - **c. wood oil** (misnomer): VO fr. *Bulnesia sarmienti*; in commerce called "Guaiac Wood Oil."

champagne: famous sparkling white wine prod. in Champagne region of neFr (ca. AD 1700-) (2520) - **C. fruit:** *Strychnos spinosa.*

Champereia manillana Merr. (C. griffithii [Hook f.]) (1) (Opili.) (Burma, Mal): u. to prevent colds & fevers (Taiwan); pounded lvs. appl. to headache & stomachache (PI); young lvs. u. in splenomegaly (PI); lvs. & frt. said edible, for ulcers.

champignon (Fr): field mushroom, *Agaricus campestris* (fungus).

champurrado (Mex): a beverage of chocolate mixed with corn, boiled & pounded or mixed with dough.

chana (Urdu): **chanaka** (Sanscr): *Cicer arietinum* - **chanal: chañar:** *Gourliea spinosa.*

chancho (Arg): pig.

chandam: (Hind, Sanscr): Red Saunders.

chandana (Sanscr.): 1) saffron 2) *Santalum album.*

chandani (Urdu, Hindi): *Ervatamia coronaria.*

chandelle (Fr): 1) *Arum maculatum* 2) *Amyris balsamifera.*

Chanel No. 5: most famous perfume in the world, in international usage (1922-); developed by Gabrielle B. Chanel (1883-1970); c. oils of rose, jasmine & orris, methyl ionone, vetiverol, &c.

chaney (berry) tree (rural people, seUS): chinaberry (tree).

chan s(h)an: chang shan.

chang shan (Chin): 1) *Dichroa febrifuga* 2) *Hydrangea macrophylla* - **chan-hsiang** (Chin): *Aquilaria agallocha.*

chanterelle (Fr): *Cantharellus* spp. (mushrooms).

chanvre (Fr): cannabis; hemp - **c. indien:** Indian hemp.

chapapote (Mex): kind of asphalt u. by NA Indians as masticatory, fuel & to prepare paint for ironwork.

chaparral (Sp, Eng) (chaparejos, Sp): 1) mixed thicket of low shrubby pls., often with brambles, esp. dwarf evergreen oaks, such as *Quercus garryana* var. *breweri*; also *Anenostoma fasciculata, Ceanothus cuneatus, & Arctostaphylos* spp., *Fremontia californica, Rhamnus californica* subsp. *tomentella,* mesquite, &c., as found in Calif, Mex, etc.; important fire hazard; riding horseback thru these thickets requires protection of legs with leathery strips called "chaps" 2) specif. the shrub *Condalia hookeri* 3) (specific.) *Larrea tridentata* (recently).

chaparro: 1) (Sp) *Quercus ilex* 2) (US, Mex) *Curatella americana* 3) any dwarf tree - **c. amargosa** (Tex, Spanish): *Castela tortuosa* - **c. prieto** (Sp): 1) *Condalia obtusifolia* 2) *Acacia amentacea, A. constricta* 3) *Mimosa* spp.

chapat(t)i (Indo-Pak): unleavened wheaten bread in thin, flat griddle cakes; prepared on short order & u. almost everywhere in nIndo-Pak.

chapeo (chapeu) de couro (Br): *Echinodorus grandiflorus.*

chapeu de Napoleão (Br): *Thevetia peruviana* (etc.) - **c. de sol:** *Terminalia catapea.*

chaplakha (eBengal): *Atrocarpus* sp. eaten by elephants (small frt. size of orange).

Chapman's mixture: Mistura Copaibae & Opii Comp.

chapopote (Mex): 1) kind of wax 2) *Casimiroa edulis* 3) chapapote (*qv.*) 4) (Tex): *Diospyros texana*.

chapparal: chaparral.

Chaptalia nutans (L.) Hemsl. (or Polak) (Comp.) (Arg, Co, Pe+): "valeriana"; juice u. in malaria; u. by Am Indians for infections (lvs. appl. with oil) & in resp. dis.; rt. decoc. u. in pulm. dis. bronchitis, GC (Hond, Guat).

char, coconut shell: this charcoal considered best absorbent filler for gas mask canisters.

Chara (Characeae-Algae): stoneworts; feather beds; common genus in brackish waters; useful in preventing devt. of mosquito larvae; hence of use in malaria control (1164); food of carp; no known med. use; (one of the 4 genera of Algae in Linnaeus).

charas (Pak): cannabis.

charbon animal ordinaire (Fr): animal charcoal.

charcoal: Carbo Ligni; a form of carbon prepared by incomplete combustion of wood & other org. materials - **activated c.:** Carbo Activatus (*qv.*): u. as antidote in poisoning (adsorbent) adm. orally or thru nasogastric tube - **animal c. (purified):** Carbo Animalis Purificatus (*qv.*) - **decolorizing c.:** activated c.; much more effective in this form in removing color than the ordinary, nonactivated c. - **medicinal c.:** activated c.: now the official form - **powdered c.:** Carbo Ligni Pulveratus; as distinct fr. c. in sticks; u. to color licorice extr., &c. - **wood c.:** Carbo Ligni, made fr. various inexpensive soft woods, as spp. of *Salix, Populus, Acer, Tilia*.

chard: 1) bleached stalks of artichoke, cardoon, &c. 2) **Swiss c.:** *Beta vulgaris* var. *cicla* (with large broad ribbed lvs.).

chardon (Fr-m): 1) thistle, incl. spp. of *Carduus, Cnicus, Dipsacus,* &c. 2) (specif.) *Carduus* spp. - **c. bénit:** *Cnicus benedictus*.

charlock: *Brassica kaber, Sinapis arronsis* - **jointed c.:** *Raphanus raphanistrum*.

charm (seUS Negro): hand (*qv.*).

charme (Fr): *Carpinus betulus* - **c. de Virginie:** ironwood.

charol (Mex): maguey.

charpie: lint.

Charta Sinapis (L.): mustard paper.

chartreusin: antibiotic prod. by *Streptomyces chartreusis* Leach & al.; has antitumor activity; (leukemia, &c. 1953-).

Chasalia curviflora Thw. = *C. chartacea* Craib. (Rubi) = *Psychotria curviflora* Wall.

chat (Somalia+): Khat.

chataignier (Fr-m.): chestnut tree.

chaucha (beans) (SA): pod of green immature beans found in Quichua region (Peru s to Arg).

chaudron (Qu): *Cirsium arvense*.

Chaulmestrol™: ethyl chaulmoograte.

Chaulmoogra: *Taraktogenos kurzii, Hydnocarpus wightiana, H. anthelmintica* - **C. oil:** FO obt. by expression fr. sds. of prec., also *Carpotroche brasiliensis*; has been u. in Hansen's disease (leprosy); now also in TB (combined with streptomycin & PAS) - **false c.:** *Hydnocarpus venenata, H. Wightiana*; *Gynocardia odorata*.

chaulmoogric acid: chief component of chaulmoogra oil; u. exter, in Hansen's dis. (leprosy) (ointment).

chaux (Fr): 1) chalk 2) limestone.

chavalongo, yerba del (Ch): natri (*qv.*).

Chavica Miq. = Piper. - **C. officinarum** Miq. (Piper o. DC) = *Piper retrofractum* Vahl.

chaw-sticks: chewing sticks (*qv.*).

chay: *Oldenlandia umbellata.* (root)

chaya: (Hindi) tea.

chayote (Sp): 1) squash, *Sechium edule* 2) challote (*qv.*).

cheat: *Bromus* spp. esp. *B. secalinus*.

checkerberry (leaves): 1) *Mitchella repens* 2) *Gaultheria procumbens* - **oil of c.:** methyl salicylate fr *Gaultheria procumbens* (**wintergreen c.**).

cheese: caseus (L.): tyros (Gr): solid or semisolid food mass made from casein of milk (mostly cows') by clotting (to form a curd) which is pressed, salted, & cut to various forms; u. in ancient Gr med.; considered diur., formerly thought to cause stone in bladder & recommended not to be eaten by religious in monasteries (2587); u. in Unani med. as stom., kidney stim., in biliousness, eye dis., to improve skin texture; types include Cheddar, American, Cream, Cottage, Edam, Gorgonzola (sometimes fr. goat), Limburger, Swiss, Roquefort (fr. sheep) & many others - **c. cloth:** thin muslin with coarse weave - **head c.:** prepd. fr. head & feet meat (mostly pork) finely cut, seasoned, boiled & pressed - **c. plant**: **Swiss c. p.:** *Monstera deliciosa* - **sage c.:** Cheddar cheese treated with sage; formerly popular in Eng & Am (ex. u. by Geo. Washington).

cheese coloring: annatto. - **c. log:** *Malva rotundifolia* - **c. rennet:** *Galium verum*.

cheeses: *Malva sylvestris, M. rotundifolia* & frts., also *Lavatera* spp. (SC frt. shape; eaten by children).

cheetah: *Acinonyx jubatus* Schreb. (Felidae) (Eg to Ind): hunting cheetah; resembles a spotted panther; domesticated (ancient Eg up to the present) & u. for hunting gazelles, &c.; very fast for short runs; fat u. in Unani med. as a sex stim.; urine u. as rat repellant.

Cheilanthes (Cheilanthus) farinosa (Forssk.) Kaulf. (Adiant.) (As; Pak to Burma): silver fern; c. cheilanthnatriol (rester-terpenoid) u. med. - **C. fragrans** Webb. & Bert. (sEu & nAs): u. emm. antiscorb.; tea substit.

Cheiranthus annuus L. = *Matthiola incana* (c. cheirotoxin) - **C. cheiri** L. (Cruc.) (cEu): common wallflower; hb. u. med.; fls. (as Flores Cheiri); c. cheirotoxin (digitaloid glycoside); VO (sulfurated); u. lax. in jaundice (folk med.); sed.; ornam. hb.

cheiro (Por): *Petroselinum crispum.*

cheirotoxin: glycoside of strophanthidin & D-6-deoxygulose & glucose; occ. in *Cheiranthus cheiri.*

chekan: checken (eugenia): *Eugenia chequen.*(lvs.)

Chelae Cancrorum (L): crab's claws; occasional use as antacid ($CaCO_3$); <kele (Gr): claw.

chelating compounds: c. complexes: inner complexes in which a metal atom is inserted like the joint of a scissor or is embraced like a claw; ex. chlorophyll, hemoglobin.

chelidonine: alk. of *Chelidonium majus*; also found in *Eschscholtzia californica, Dicentra formosa, Stylophorum diphyllum,* & other spp.; u. antispas. (somewhat like papaverine in its effects) (1213).

Chelidonium majus L. (monotypic) (Papaver.) (cEu, natd. NA): great(er) celandine; swallow-worts; hb. & pl. c. chelidonine, berberine (1214); u. diur. purg. irrit., claimed of use in infectious jaundice & in hepatic, gallbladder, & stomach dis.; fresh juice appl. to eye dis., warts. corns, bruises, eczema, etc.; dye-stuff (yellow); occas. cultd.

chellah (Arab): Kella, *Ammi visnaga.*

chellin: khellin.

Chelonanthus acutangulus (R. & P.) Gilg. (Gentian.) (Pe): u. for cuts - **C. chelonoides** Gilg (Co): pd. lvs. good insect repellant (Indians of upper Rio Vaupés) (2295) - **C. elatus** (Aubl.) Pulle (Mex, Co, Pe): rt. decoc. u. as antipyr. in cramps, indigest., to treat sores in mouth; for intestinal obstructions; pd. lvs. & fls. insectifuge; remedy for worm-infested wounds (cattle) - **C. uliginosum** Gilg (Co): rt. of herb u. as tea for curing stomach discomfort (2295).

Chelone glabra L. (Scrophulari) (eNA): balmony; rts. u. cath. ton. - **C. obliqua** L. (seUS): red turtlehead; hb. u. ton. cath. in liver dis. (homeopathy); vermifuge.

chemical radiation: c. rays: actinic radiation; u. v. radiation - **c. wood** (NC)**:** wd. of oak & several other hardwood trees; u. to produce acetic acid, methanol, &c. by destructive distn.

chemism: chemical properties or relationships.

chemist: 1) (Eng): professional man similar to but with somewhat inferior rank to pharmacist 2) (US, Eng): person who practices some branch of chemistry as in analytical work, synthesis, etc.

chemotherapy: treatment of disease with chemical substances.

chemurgy: appln. of farm (agricultural) products to chem. manufg. processes; ex. soybeans in plastics indy., licorice in fire extinguishers (fr, Gr. *chymeia,* mingling; *ergon,* work).

ch'en hsiang (Chin): chan hsiang.

chenar: chinar.

chêne (Fr.): **oak - c. d'ilex:** *Quercus ilex.* - **C. Liège:** cork oak - **C. velani**: *Q. aegilops.*

chenodeoxycholic acid (CDCA) (Chendol [Weddell], GB): chenodiol; bile acid normally present in bile (low titer) u. to reduce cholesterol secretion in bile & dissolve cholesterol biliary calculi (gallstones) (Eng, 1975-), thus avoiding surg. operation; in gall bladder dysfunction; *cf.* urso-deoxycholic acid.

Chenopodiaceae: Goosefoot Fam.; ann. & perenn. hbs., some xerophytes, some halophytes; many occur as weeds in many lands (ex. Russian thistle); most members are unattractive pls.; incl. beet, spinach, chard.

Chenopodium album L. (Chenopodi.) (Euras, natd. NA); lamb's quarters; common pigweed; fat hen; young pls. boiled & u. as greens; (esp. common wUS): old lvs. u. to relieve stomachache - **C. ambrosioides** L. (incl. var. *anthelminticum* Gray) (US, tropAm, trop worldwide, pantrop): (American) wormseed; Jerusalem tea; Mexican tea; sweet pigweed; Jerusalem oak; pico(ba) santa (Ve); decoc. for worms, stomach discomfort (Ve); ton. diur.; flavor for beans; tropical anemia; inf. of rts. & lvs. u. for colic (Mex, Ec); cough remedy (Br); frt. u. vermifug. (poultry) c. VO, with ascaridol (70%), cymol; frt. hb. u. to prepare chenopodium oil (American wormseed oil); u. for roundworms, chiefly hookworm, ascarides (1217) in teas (Mex-Dominica); sds. u. to prepare flour (Am Indians); hb. said to supress estrus cycle; contraceptive; pl. c. HCN; prepn. stock feeds; common weed but cult. at times (esp. Carroll Co., Md [US]) - **C. bonus-henricus** L. (Eu): English mercury plant; hb. c. kaempferol, caffeic acid, ferulic acid, histamine saponins; eaten as spinach; hb. & rt. formerly u. as vulnerary & in inflamm. & swellings - **C. botrys** L. (Euras, Af, adv. in NA): Jerusalem oak; wormseed; feather geranium; vermifuge (Mayas); ornam. ann. hb. - **C. californicum** S. Wats. (Calif): soap plant; rt. u. like soap (Am Indians) - **C. foetidum** Lam. = *C. vulvaria* - **C. fremontii** S. Wats. (wNA): pls. raw or cooked food of Am Indian; sds. food (mixed with corn meal & salt) - **C. gigantospermum** Aellen (Cana, nUS): maple-leaved goosefoot; sds. u. for ages by Am Indians - **C. graveolens** Willd. var. *bangii* (Murr.) Aellen (Arg.): larca yuyo; yerba del arca; u. as antisep. for skin maculae, sed. in digestive allergy - **C. hircinum** Schrader (Par, Br): quino quino; u. for inf. of intest. regions; vermifuge - **C. hybridum** L. (Euras, nAf, Haw.): sd. u. food (Am Indians) - **C. multifidum** L. (Arg, Bo): paico (Arg); paigo (Bo); rt. u. as vermif. emmen.; carm. in stomach dis., indig. dyspep.; diaph.; diur. in calculi; heart palpitations; chorea; with muna verde in appendicitis - **C. murale** L. (Med, cEu, swAs, Russia, NA): nettle-leaved goosefoot; mauern Gaensefuss (wall goosefoot); sds. u. for "Russian hunger bread" (during famines) (1524); lvs. u. as spinach (sAf); u. as wash for some skin inf. (Ec); for "eruptions on the head" (Mex) - **C. opulifolium** Schrad. (c&sEu, wAs, nAf, natd. NA): rt. decoc. u. in difficult childbirth (Tanz); u. med. (Rwanda); sd. & lf. edible - **C. pallidicaule** Heller (SA - high mts.): cañagua; sds. (cañagua) food of Indians (Bol, Pe); perenn. hb.; grain crop in highlands; u. to produce flour, middlings, bran - **C. purpurascens** Moq. (Pak): rai'han; considered medicinal - **C. quinoa** Willd. (Pe, Ch, Bo, etc.): quinoa; an ann. hb.; sds. u. food (rich in amino acids); (ground to a flour) by millions of SA natives; also u. med.; ash mixed with coca lf. to improve flavor; lvs. u. like spinach - **C. schraderanum** Schultes (C. foetidum Schrad.) (sMex, natd. eAf): hb. u. to eradicate ants, &c.; in folk med. for cough, digestive, abdom. gas.; VO u. as vermif. for round worm (Mex Phar.) - **C. vulgaris** Gueldenst. = **C. vulvaria** L. (Euras, incl. Pak) (natd. & weed in NA, Austral, NZ): stinking goosefoot; stinking orache; hb. (Herba Vulvariae) u. as sed. in hysteria; antispas. emmen. for abdom. cramps; hb. unpleasant odor ("stink") appar. fr. NH_3, propylamine, mono-, di-, & tri-methylamines (odor carries for long distances); saponin; sd. u. to prep. porridge (earlier by Incas).

ch'eno hsiangpi (Chin): dr. orange peel fr. *Citrus nobilis* vars. such as tangerine.

cheres (NM [US], Sp): cherries.

cherimoia (Sp)**: cherimolia: cherimoya:** *Annona cherimola.*

cherry: term denotes trees of 3 sections of *Prunus* (Ros.), viz., *Cerasus* (umbellate c.), *Padus* (racemose c.), & *Laurocerasus* (laurel c.); also appl. to frt. - **Barbados c.:** *Malpighia glabra & Bunchosia* spp.- **c. birch**: *Betula lenta* - **bird c.:** 1) *Prunus avium* 2) *P. pennsylvanica* 3) *P. padus* - **black c.:** 1) *Prunus serotina* (bk.) (also **wild b. c.**) 2) belladonna - **cabinet c.: choke c.:** *P. virginiana* - **cocktail c.:** Marasca c. - **Dalmatian c.:** Marasca c. - **English bird c.:** *P. padus* - **Florida c.:** Surinam c. - **flowering c., Japanese:** many *Prunus* taxa so designated, esp. *Prunus serrulata* (short-lived), *Pr. yedoensis* (long-lived), *Pr. subhirtella* v. *ascendens* & v. *pendula* ("spring c.") - **glace c. (cherries)** coated with icing; candied; glazed (marrons glacés: frts. &c. also frozen) - **ground c.:** *Physalis* spp., esp. *P. virginiana* - (**sweet g. c.**: *P. alkekengi*) - **c. gum**: exudn. fr. *P. cerasus*, etc.; u. thickening agt., sizing (Eu) - **Indian c.:** 1) *Rhamnus caroliniana* 2) *Amelanchier canadensis* - **Japanese c. (tree):** ornam. *Pr.* spp.; v. under flowering c. - **Jerusalem c.:** *Solanum pseudo-capsicum* - **c. juice, (sour)**: fr. *P. avium*; u. flavor, to make c. syrup, flavor ext., in proprietaries, galenicals, etc.; marketed as concentrate - **c. kernel** (or **seed**) **oil**: FO fr. kernel of *Pr. cerasus* sd. - **laurel c.: c. laurel (leaves), European**: *P. laurocerasus*. (**c. l., American:** *P. caroliniana*) - **c. leaves**: fr. *P. avium* - **Leyden c.:** with caraway, cloves - **Marasca c.**(It.): - **Maraschino c.** (US): *P. cerasus* v. *marasca*, red c. flavored with wild or m. c. flavor (M. cordial) - **Morello c.**: *P. cerasus* var. *austera* - **oriental c.:** Japanese c. (also simply "cherry" u. during WWII, Wash, DC) - **Puerto Rican c.:** West Indian c. - **rum c.**: *P. serotina* - **shrub c.**: winter c. - **sour c.**: *P. cerasus* - **c. stalks:** "**c. stems**" (US comml.): frt. peduncles of *P. cerasus* & other *P.* spp.; u. flavor - **Surinam c.:** *Eugenia uniflora* - **sweet c.**: *P. avium* - **c. syrup:** c. juice, sugar, alc.; popular soda water flavor - **C. water** (Gr Phar.): prepd. fr. bitter almonds - **West Indian c.:** *Malpighia glabra* - **whiskey c.**: *P. serotina* - **wild c. (bark)**: *P. virginiana, P. serotina* - **winter c., (common)**: 1) Jerusalem c. 2) *Physalis alkekengi.*

chervil: *Anthriscus cerefolium* lvs. & sds. - **Spanish c.:** *Myrrhis odorata* - **turnip-rooted c.:** *Chaerophyllum bulbosum.*

chervin (GB): *Sium sisarum.*

chervis (Fr): caraway, skirret.

Cheshire gum: locust bean gum (imported via Eng).

chesnut: chestnut: *Castanea* spp.- **American c. (leaves):** *C. dentata* - **c. blight:** *Endothia parasitica* (*Cystospora*) destroyed most of populations of *Castanea dentata* (US) - **European c.:** *C. sativa* (chestnuts largely imported fr. It) - **horn c.**: Chinese water c. - **horse c.:** *Aesculus hippocastanum* (**Indian [horse] c.:** *Aesculus indica*) - **c. leaves**: fr. *C. dentata* (US) - **Spanish c.:** *C. sativa* - **sweet c.:** any edible sds. of *Castanea* spp. - **water c.: Chinese water c.:** *Trapa* spp., esp. *T. natans & T.*

bicornis; *Eleocharis tuberosa* tuber; this latter sp. mostly u. in prepn. c. flour.

chess: *Bromus* spp. - **soft c.:** *Bromus mollis, B. secalinus.*

chewing sticks: twigs or sts. u. in many Af countries to clean teeth, remove dental plaque, &c.

chia (seeds) (Mex, CA, to Calif [US]): *Salvia columbariae & S. hispanica* - **chian** (Mex): mucilage, also bev. prepd. fr. chia - **chiantzoli** (Fr): chia.

chiba (chibu) (PR): product like gum thus (gum turpentine).

chibou(e): chibu gum: a gummy resin somewhat like elemi, for which sometimes substit. ("American elemi."); said to come fr. *Bursera Simaruba*; u. in mucous membrane dis.

chibouk (Turk): huqua; hookah; water-pipe.

chica (SA): *Arrabidaea chica.*

chicalote (Mex) *Argemone mexicana.*

chicha (SA): native wine or beer of Ch, Pe, &c.; made under primitive conditions (originally) fr. sugar cane, quinoa, raw sugar, &c. by fermentation.

chicharo (Sp): *Dolichos lablab.*

chichigua bark: *Aspidosperma febrifuga.*

chicken v. *Gallus* - **C. berry:** *Gaultheria procumbens, Mitchella repens, Callicarpa americana*

chickenpox: varicella; contagious dis. caused by a virus transmitted through resp. system; symptoms incl. dermal eruptions (with pruritus); years later may produce herpes zoster (shingles); with pain & aggravated itching; vaccine developed (1994) (Vacilex).

chickenweed (seUS): *Asclepias tuberosa.*

chickweed: 1) *Stellaria media* & other *S.* spp., *Alsine, Arenaria,* etc. spp. (Caryophyll.) 2) *Cerastium vulgatum* 3) *Montia* spp. - **small c.: common c. w.:** *s. media* - **c. w. family:** *Caryophyllaceae* - **gravel c. w.:** *Scleranthus annuus* (tea of different spp. said useful med. [Ala (US)]; u. frequently as potherb).

chicle (gum): dr. latex fr.: 1) *Manilkara zapota* (sapodilla) 2) *Mimusops balata.*

chico (zapote) (Sp, PI) *Manilkara zapota.*

chicory, common: *Cichorium intybus* (rt.).

chiendent officinal (Fr): Triticum, dog-grass.

chigger (mites) = red bugs; larvae of mites causing aggravating itching.

chiggerweed (seUS): *Asclepias tuberosa.*

chigoes: small trop. fleas ("sand fleas") also called "chigger" but not the same as the chigger or red bug of US (a mite).

chijol (Mex) Jamaica dogwood.

chikoo (Chin): drug of fam. Araceae believed to aid conception, producing male children.

chiku (Hind): *Manilkara Zapota.*

chilam (Ind): elongated funnel-like earthenware pot u. for smoking cannabis.

chilbinj: *Strychnos potatorum* - **chilcuan** (Mex): *Erigeron affinis.*

chile: *Capsicum frutescens* v. *longum* (1525) - **c. con carne** (Sp): mixt. meat cooked with small capsicum frt. (grd.), beans, &c. ("chili"), actually a US not a Mexican dish - **c. head** (US) (slang): people who consume large amts. of capsicum - **c. mill:** wooden mortar & pestle (noted in Harlem, New York City [NY]) - **c. powder:** compn. with capsicum, garlic, cumin, oregano, salt, &c. - **c. releno** (Sp): green pepper stuffed with minced meat, coated with eggs, & fried - **c. vinegar**: prepn. of capsicum in ordinary vinegar.

Chile saltpeter: crude $NaNO_3$ extd. from guano (excreta) of seabirds (cormorants) & bats living on the dry w. coast of Chile; deposits are of great antiquity & great size (Guano Islands); u. as fertilizer, source of nitrates.

Chilean pine: *Araucaria araucana.*

chilingo (Guat+): *Rauwolfia tetraphylla.*

chili: chile: (SC origin believed in Chile).

chilipucas (SA): lima bean.

chillie(s), African: red pepper, *Capsicum frutescens* - **Mombassa c.**: chiltipiquín (Mex): *Aulospermum purpureum.*

Chilopsis linearis (Cav.) Sweet (C. saligna D. Don) (monotypic) (Bignoni.) (swUS, nMex): desert willow; fls. dec. u. in coughs & bronchitis; antipyr; cardiac stim. (Mex); forage; cult. ornam. (pretty white & pink fls.); frt. look like string beans; u. for soil erosion; branches u. for baskets; shrub or low tree.

chilpanxochitl (Mex): *Lobelia laxiflora.*

chilte (Sp): *Cnidoscolus elasticus.*

chimaja (Sp): *Aulospermum purpureum.*

Chimaphila maculata (L.) Pursh (Pyrol.) (e&cNA, Mex): spotted wintergreen; ratsbane; hb. u. diur., urin. antiseptic, ton. alt.; boiled lvs. u. as kidney remedy (hierba del higado) (Mex) (1168); in rheum. gout; weak in kidney dis. (Gaspe Peninsula, Queb [Cana); in cough syrups - **C. umbellata** (L.) Bart. (Euras, NA): (common) pipsissewa: hb. c. ericolin, arbutin, **chimaphilin**, tannin: u. diur. ton. astr. (2777)

chimbolo verde (CR): *Dolichos lablab.*

chimbombo (Mex): okra.

chimó (Ve, Co): kind of thick jelly or paste made with chewing tobacco (plug) previously boiled with "natron" (crude hydrated sodium carbonate) (sal de urao) or $NaHCO_3$; coffee sweepings, ash, tonka; &c.; chewed & licked by rural people (Latin Am).

Chimonanthus praecox Link (Calycanth.) (Chin): fl. decoc. drunk for sore throat; fls. in oil u. for burns & as sialagog, to perfume linen.

china: v. also wax.

China: 1) (Ge): cinchona 2) (Sp): china root 3) (PR): orange - **C. bark:** *Quillaja* - **C. bell berries,** frt. of **C. berry (tree):** *Melia azedarach* - **C. brasiiliensis: C. californica:** *Symplocos racemosa* - **C. cinnamon:** *Cinnamomum cassia* - **C. clay:** kaolin - **C. morada**: *Pogonopus tubulosus* - **C. nova:** 1) *Symplocos racemosa* 2) *Croton eluteria* (bk) - **C. n. brasiliensis:** *Ladenbergia hexandra* - **C. oil:** 1) Balsam Peru 2) (less correctly): C. wood o. - **C. paraquatan:** *Symplocos racemosa* - **C. paya** (Pe): 1) *Pectis sessiliflora* 2) *Zinnia* sp. - **C. root** (Jam & other WI): 1 *Smilax china* & other *Smilax* spp. 2) *Dioscorea* 3) *Cissus sicyoides* 4) *Poria cocos* - **C. root, American:** *Smilax hispida, S. pseudo-china* - **C. root:** *Poria cocos* (Fungi) - **C. tree:** C. berry - **C. tubers, Occidental:** *Smilax pseudo-china, S. tamnoides* - **umbrella c. (berry tree)** (seUS): *Melia azedarach* var. *umbraculifera* - **West Indian C. root plant:** *Cissus sicyoides* - **C. wood oil:** fr. *Aleurites* spp., as *A. montana* - **Chinae Cortex** (L.) cinchona bk.

chinar (Urdu): oriental plane tree, *Platanus orientalis.*

Chinasaeure (Ge): quinic (kinic) acid.

chinchi (Pe): red pepper.

Chinchona: 1) bot. synonym for *Cinchona* 2) **corteza de c.** (Ec): cinchona bk.

chinckeri(n)chee (sAf): *Ornithogalum thyrsoides* & other *O.* spp. (vernac. name) (many variant spellings).

Chinese: v. also cabbage, isinglass, galls (nutgalls), ginger, potato, rhubarb, sugarcane, sumac, tallow tree (& vegetable t.), varnish tree.

Chinese bowl: opium - **C. gooseberry:** *Averrhoa carambola* - **C. lanterns:** *Physalis alkekengi* - **C. lantern tree** (Ala [US]): *Sapium sebiferum* - **C. nut:** litchi. - **C. oil bean:** soy - **C. root** (Jam): *Smilax china* (rt.) - **C. rose:** *Hibiscus rosa-sinensis* - **C. scholar tree:** *Sophora japonica* - **C. wood oil:** tung oil - **Chinesischer Zimmt** (Ge): Chinese cinnamon.

Chinese ink: India Ink; special grade of lampblack with binder in cakes or in suspension; u. for drawing or lettering - **c. shavings:** bamboo s.; ankat bark; u. to glaze furs, in hair waving (1526) - **c. yellow:** As_2S_3.

chini badam (Urdu) (lit. Chinese almond): peanut.

Chinium (L.): **Chininum**: quinine.

chin-jill (Chin): common name for "chin chiao": dr. ripe frt. of *Zanthoxylum* spp., prob. *Z. alatum* Roxb. or *Z. piperitum* DC or *Z. simulans* Hance.

chink(s): *Gaultheria procumbens.*

chinkapin: chinquapin - **water c.:** *Nelumbium nelumbo, N. pentapetalum.*

chino-: combining form of Chinium (quinine) (*cf.* quino-) - **chinoline:** quinoline: artif. alk.; u. antipyr.

chinquapin (Eng, fr, NA, Ind): *Castanea pumila.*

Chio (Chian) turpentine: *Pistachia terebinthus*; u. as abort. (Chios = Scio).

Chiococca alba (L.) A. S. Hitchc. (C. racemosa L.; C. brachiata R. & P.: C. anguifuga Mart.) (Rubi) (Tex, Fla [US], WI, Mex): David's root or milkberry; snowberry; cainca (Br); snakeroot (Fla); frt. juice & decoc. u. as purg.; c. cahincic acid, caincin (glycoside), bitter princ.; rt. (bk.) (Radix Caincae) u. purg. emmen, anti-asthm.; sometimes u. for snakebite; diur. in edema; tonic, astr. emet. exp. in VD (syph.); rheum. decoc. of fls. with fls. of *Petrea kohautiana* u. as abort. (Dominica); lvs. u. as poultices for sores, &c.; rt. is Radix Caincae, David's root.

Chiogenes Salisb. ex Torr. = *Gaultheria.*

chionanthin: phyllirin: forsythin; saponin-like glucoside fr. trunk & rt. bk. of foll.; hydrolyzes to resin & dextrose; not hemolytic (sometimes claimed hemolytic).

Chionanthus virginicus L (Oleac.) (eUS): (white) fringe tree (bark); old man's beard; rt. & st. bk. c. chionanthin, tannin; u. alt. bitter ton. aper. hepat. stim. diur. antipyr.; popular with Homeopaths & Eclectics.

Chione venosa (Swartz) Urban (C. glabra DC) (Rubi) (WI): martin avila (PR); "fat pork" (Montserrat); bois anda (Domin); violette; smooth chione; palo blanco (PR); bk. c. VO (chione oil) with hydroxy-acetophenone (wd. & bk. with fecal odor); rts. aphrod. (Grenada) small evergreen tree; furnishes useful timber (Grenada).

Chionia (Peacock Chem. Co. [d] Peacock Sultan) St. Louis, Mo [US]); later Natcon Chem. Co., Bethpage, Long Island, NY [US]): popular proprietary made fr. *Chionanthus virg.* (-1917-1965); u. as hepatic ton., stim.: "bile persuader"; in constip. jaundice, dyspepsia (alc. 18%); proposed for use in foods (Jap).

Chionolaena latifolia Baker (Comp.) (Br): arnica do campo; shrub; u. as anti-inflamm., substit. for arnica.

chipper: one who cuts V-incisions in trunk of pine (turpentining); the operation is called **chipping** (seUS).

Chiranthodendron pentadactylon Larréat (C. platanoides Humb. & Bonpl.) (Sterculi) (monotypic) (Mex, Guat): mapasuchel; mano de león; hand flower; tree; often cult.; fls. inf. u. for ophthalmia & piles (Am Indians).

Chirata (L.): Chirayita: Chirayta: Chiretta: chirette (Fr): *Swertia chirayita.*

chiratin: reputed bitter princ. of *Swertia chirayita, S. japonica,* & other Gentianaceae = impure amarogentin.

chirere (Mex): *Capsicum frutescens.*

chirimoya (Pan, Pe+): *Annona cherimola* - **chirivia** (Sp): parsnip.

chironja (PR): hybrid fr. cross of grapefruit X sweet orange.

chironji(i): *Buchanania lanzan* (FO).

chirua (Beng): *Poa annua.*

chiso (Jap): feast; dainty dishes; delicacy; vinegar.

chitin: horny polysaccharide subst. related to cellulose; occ. in exoskeletons (shells) of arthropods (esp. richly in crabs & other Crustacea), fungal cell-walls, &c.; has been called the second most abundant natural polymer, biodegradable; non-allergenic; u. to incr. shelf life by coating; antioxidant; u. for wound healing; u. hemostatic agt., anti-hypercholesterolemic; to kill nematodes (interferes with egg production in female).

chitosamine = glucosamine (*qv.*).

chitosan: deacylated chitin, prepd. fr. shells of certain shellfish (chiefly crab & shrimp) (Norway) fr. offal. of seafood factories; u. in wound-healing prepns.; in water purifying systems; drug delivery systems, &c.; in photo emulsions; to treat fabrics, &c.

Chitral gum (Pak): fr. *Astragalus gummifer.*

chittam bark: chittem b.: Cascara Sagrada - **chitam-wood: chittim-wood**: *Cotinus americanus.*

chive(s) (onion): *Allium schoenoprasum.*

Chlora Adans. = *Blackstonia.*

chloramphenicol: generic name for Chloromycetin, most popular brand.

Chloranthus elatior R. Br. (C. officinalis Bl.) (seAs, Oceania): lvs. bruised & appl. to ulcers, burns; lvs. heated & appl. for arthralgia; rts. or lvs. u. as inf. for fever as sudorific (Mal); mixed with cinnamon u. as antisp. during parturition - **C. spicatum** Mak. (seAs): diaph. excitant; for malaria; fls. & lvs. tea for cough; bruised rt. appl. to carbuncles & boils.

Chlorella pyrenoidea Chick. (Oocystaceae - Chlorophyta - Algae): rich source of protein with lysine, arginine, leucine (amino acids); cultured in water on large scale in some places as food source (Jap comm. product); c. **chlorellin** (antibacterial [1944-]).

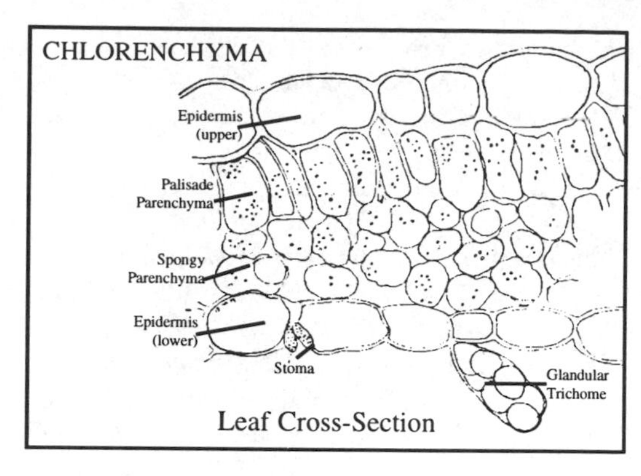

chlorenchyma: chlorophyll parenchyma, tissue of lvs. & other green parts of pl. serving mostly for assimilation (in lf. composed of palisade and spongy parenchyma) (see fig.).

Chloresium™: water-sol. derivs. of chlorophyll-α; u. in therapy.

Chloris gayana Kunth (Gram) (sAf, natd. US & tropAm): Rhodes grass (<Cecil R.); cult. for forage, meadow or pasture grass; u. med. (New Caled).

Chlorocodon whytei Hook. f. = *Mondia whitei* Skeels.

Chlorogalum pomeridianum Kunth (Lili) (Calif [US]): amole; common soap plant; soap root; Indian potato; bulb c. steroidal saponin (amolonin) (7%); u. detergent, hair washing, soap substit., fish poison; as poultice & antisep. for sores, bruises, boils; diur. lax. for flatulence; in poison oak rash; bulbs as food (roasted): industr. uses.

chlorogenic acid: condensation prod. of caffeic & quinic acids; widely distr. in pls; ex. apples, coffee, beans, tobacco, *Ricinus,* new wine, carrots, oranges; allergen at times; blackening of potatoes due to Fe compd. formed with c. a.

Chloromycetin™: Chloramphenicol: antibiotic subst. obt. fr. *Streptomyces venezuelae*: also by synthesis; useful against a wide variety of disease organisms, incl. many spp. of bacteria, viruses, rickettsiae, & protozoa; cases of agranulocytosis reported; preferential use in resistant staphyloccus inf.; in typhoid fever (E) (1891).

Chlorophora excelsa (Welw.) Benth. (Mor.) (tropAf): iroko; odum; African teak; termite-proof wd. u. for furniture; substit. for oak & teak - **C. tinctoria** Gaud. (WI, CA, Br): fustic tree; wd. u. brown, yellow dye (old fustic) & for construction; Mayas appl. sap with cotton to dental caries to relieve toothache (6); frts. u. for dental pain & gum for hemorrhage (Co); bk. u. tan, calking boats; wd. c. morin (flavone) (Calico Yellow R., Geigy) (khaki yellow); palo moro; moral de clavo; vermifuge (Br); frt. edible, sweet; somewhat milky juice u. for toothache (Br).

chlorophyll: green pl. pigment essential for photosynthesis; mfrd. fr. wild nettle (Eng) (1219): alfalfa, oats, clovers (US); u. vulnerary in wounds, ulcers, treatment of burns, skin dis., osteomyelitis, Vincent's angina; deodorant; natural green coloring matter in foods, drugs, etc.; antibact. (1166); in geriatrics.

chlorophyllins: Na & K salts of chlorophyll, water-soluble; u. to incr. rate of healing of wounds (acute, chronic); inhibits fibrin formation; deodorant.

Chlorophytum Ker Gawl. (Lili) (Af, Ind, Austral/SA): spider plant; popular house plants, often variegated ("airplane plant") - **C. arundinaceum** Baker (Ind+) safed-musli (Hindi): rt. u. ton. - **C. capense** Ktze. (C. elatum R. Br.) (sAf): spider plant; popular house plant; said useful in removing CO and NO_2 from air of room (var. *vittatum*) (with striped lvs.) - **C. comosum** (Thunb.) Jacques (tropAf): spider plant; tuber u. as purg., esp. for children; u. after childbirth; cult. ornam. - **C. drepanophyllum** Bak. (tropAf): Cape spinach; lvs. eaten like spinach.

chloroplast: chloroplastid (less preferable): particulate bodies of cytoplasm which c. chlorophyll-α, chlorophyll-β, carotene, & xanthophyll; play important role in photosynthesis; site for same; sometimes regarded as descendants of Cyanophyta cells which were first endosymbionts and later became subcellular organelles.

chlorosis: chlorophyll deficiency in pl. parts normally green.

chlorothymol: mono-c.: whitish xals. with characteristic odor; u. germicide (higher phenol coefficient than thymol), in antisep. solns.

chlortetracycline: generic name for the important antibiotic, aureomycin™ (*qv.*).

Chlumsky's solution (RB. III): camphor 600; phenol 300; alc. 125. (tot. 1000).

chnouf (Pe-slang): Heroin.

choc(h)o: *Sechium edule.*

choclo (Arg.): green corn ear.

chocolade = chocolate (*cf.* orangeade) (2278).

chocolate: (fr. Aztec chocolatl) - plain c. - **bitter c.: c. liquor:** natural unsweetened c. - **c. lily:** *Fritillaria lanceolata* - **milk c.:** prod. made by admixture of pd. (or condensed) milk with sweet c. - **c. nuts:** *Theobroma sylvestris* - **plain c.** solid mass made by grinding sds. of *Theobroma cacao* without removal of any appreciable quantity of FO; see serials List (App. C-1) - **c. root:** *Geum rivale* - **sweet c. (coatings):** c. mixed with sucrose & sometimes with spices & flavors - **white c.:** made up of cocoa butter (30%) c. some theobromine, lecithin, vanillin, & other flavors & colors, with sugar (50%) and non-fat milk (20%); u. to coat chocolate bars.

choke berry: *Aronia arbutifolia* & other *A.* spp. (SC astr. to throat when swallowed) - **c. cherry (bark):** *Prunus virginiana* (& var. *demissa*).

chola: *Cicer.*

cholecalciferol: vitamin D_3.

cholecystokinin: CCK: hormone of duodenum wh. stimulates gallbladder contraction & stim. secretion of pancreatic enzymes: recently marketed (US) as anorectic but no evidence of value when used p. o.; not allowed by US FDA.

cholera: infectious dis. of man caused by *Vibrio cholerae* Pacini (Spirillaceae - Bacteria); prod. profuse watery diarrhea; resultant extreme dehydration chief cause of death - **c. vaccine:** sterile suspension of killed c. organisms in isotonic NaCl soln.; u. to give active immunity.

cholesterin: cholesterol: the chief sterol of the higher animals; esp. rich in brain, spinal cord, animal fats; comml. source: spinal cord of cattle; fish meal, lanolin; (wool grease) also widely distr. in plant kingdom - algae, fungi, seed plants (ex. pine tree needles) (formerly thought restricted to animal tissues); proposed max. blood level in health: (c. index) 180 mg. % (humans) (1562); formula (2573); u. lanolin substit. (NF XVI).

cholette: *Arum* or *Coriandrum.*

cholic acid: cholalic acid: occurs combined with glycine or taurine in the bile of most vertebrates; comml. source beef bile by extn.; combn. with glycine —> glycocholic acid, an unconjugated bile acid related to betaine & muscarine.

choline: gossypine: basic component of lecithin; found in many pls. & animals; ex. belladonna, brain; regarded as member of Vit. B complex; u. in vit. deficiencies, esp. in lipid metabolism abnormalities, often with inositol; found of value in treating tardive hypokinesia (an effect of antipsychotic therapy) (2474).

choline esterase: enzyme wh. hydrolyzes the ester acetylcholine to c. chloride & acetic acid; occ. in blood cells, brain & other nerve tissues of animals (acetylcholinesterase is inhibited by nerve gases & P-contg. insecticides).

cholla (Sp skull) **(cactus)**: any of various spiny cactus taxa, ex. *Opuntia parryi* - **buckhorn c.:** *Opuntia acanthocarpa* - **c. gum:** fr. *Opuntia fulgida.*

Chomelia L. = *Tarenna.*

chondocurarine, d-: alk. fr. *Chondrodendron* spp., isomer of tubocurarine, with which it is found associated; c. 3-4 X as active.

Chondodendron Ruiz & Pavon = *Chondrodendron* Ruiz & Pavon.

Chondria armata (Kuetz.) Okamura (Rhodomel. - Ceramiales - Rhodophyta) (Pacific Oc): domoe (Jap); c. domoic acid (related to kainic acid); u. as vermif. esp. for ascarids, hypotensive - **C. crassicaulis** (Jap coasts): c. chondrine, amino acid - **C. littoralis** (WI coasts): shows antibacterial & antifungal activity (chonalgine) - **C. oppositiclada** Dawson (Pacific Oc): c. cycloeudesmol, a sesquiterpene with strong antibiotic activity; chondriol, with antiviral properties (anti-herpes).

Chondrodendron microphyllum (Eichl.) Moldenke (Menisperm.) (Br): pl. c. D-bebeerine; source of curare & of pareira brava - **C. platyphyllum** Miers (Br): (a)bútua (Br): bitter yellow rt. of liana one source of pareira brava; c. chondrofoline, curine; u. prepn. curare (chief source) (1278); diur. antipyr.; in edema (Indians) - **C. polyanthum** Diels (Br): one of the chief sources of curare - **C. tomentosum** Ruiz & Pav. (Br, Pe+): (white) pareira brava: rt. c. bebeerine, d-tubocurarine; ext. of rt. called curare (arrow poison); dr. rt. u. diur. bitter tonic (formerly off.: Pareira Brava); natives u. in menstrual dis., antipyr.; in mfr. of tubocurarine (*qv.*) (1170) - **C. toxiferum** (Wedd.) Kruk. & Moldenke (Br, Col): quina; c. curine, tubocurine; st. wd. crushed to make curare (arrow poison) (1268).

chondroitin sulfate (A, B, C): mucopolysaccharide mixt.; a major component of cartilage & skeleton (connective tissues); c. glucuronic acid (chondrosamine) (galactosamine) + H_2SO_4; related to hyaluronic acid & dermatans; split by hyalauronidase; obt. fr. nasal septa, aorta, &c.; u. antihyperlipoproteinemia (< chondros (Gr): cartilage) (2227).

Chondrus (L.): carageen moss; fr. *Chondrus crispus & Gigartina mamillosa* (Gigartinaceae) (shores of nAtlOc); mucilaginous, u. mostly for colloidal props. as aid to blood coagulation: dem. emoll.; moisture retention (ex. tooth pastes), tablet binder; emulsif. agt. for mineral oil, etc.; suspending agt. in foods & drugs; c. carrageenin (mucil.), antithrombic agt., anti-ulcer (inhibits proteolysis) (1222); for dental impression; agar substit. (1224); ice cream stabilizer; marbling paper; filler in canned meats (1221), giving body to confections; etc.; many TM preps. on market; mucil. not pptd. by alc.; alkali halides; protein, etc. - **C. crispus** Stackh. (e&w margins of nAtlOc) (prod. Algae, PEI [Cana], Fr, Ire): Irish moss: this sp. chief Am source of *C.*; popular as food in Ire (*blanc mange*); entire dr. clarified thallus also u. med. (1225); mostly imported fr. Eu; u. as described under Chondrus - **C. ocellatus** Holmes (Jap, Korea): tsunomata; false Irish moss; u. to prep. carrageenan, thallus u. to size cloth, for calcimines, plaster for house walls, &c.

chop (seUS): to hoe, as to "chop cotton" - **c. nut:** *Physostigma venenosum.*

Chorda Serica Chirurgicalis (L.): silk suture; u. for sewing up surgical incisions.

Chordata: the highest phylum of the Animal Kingdom, whose members possess a dorsal nerve-tract column (or notochord), typically protected by a series of bones called vertebrae, collectively the spinal column; in some lower forms among chordates, the notochord is present only at an early embryonic period; the chief classes are the sharks & allied fishes, reptiles, birds, & mammals.

chorionic gonadotrop(h)in, human: hcG; glycoprotein hormone isolated fr. placental chorionic tissues & pregnancy urine (chief source); u. in cryptorchidism; to stim sex glands; male hypogonadism; to induce ovulation & pregnancy; strong aphrod.; v. also under gonadotropin.

Choripetalae: subdiv. of Dicots in which the fls. possess separate & distinct petals: Polypetalae.

Chorisia insignis HBK (Bombac.) (wSA): palo borracho; produces useful fiber (*cf.* kapok) - **C. speciosa** A. St. -Hil. (Latin Am: Arg, Br) paina (de seda) (Br): floss-silk tree; silky fiber (floss) (paneira) of sd. capsules u. to stuff pillows, cushions (*cf.* kapok); sds. c. 14% FO, u. food purposes, also in industry (1171); wd. c. mucilage; ornam. tree (CR) with large pink fls. (1751).

Chorizanthe palmeri Wats. (Polygon.) (Calif [US]): cultd. as ornam. pl.

chorology: science of geographic distribution of organisms.

chou (Fr-m.): cabbage. - **c. -fleur** (Fr): cauliflower - **c. pomme = pomme de c.:** head cabbage.

chow-chow: mixt. of pickled vegetables, mustard, &c.

chows (, dog): receptacles of *Rosa canina* & other *R.* spp.

Christmas (Eng): holly (esp. *Ilex aquifolium*) & other evergreens u. for decoration at Christmas time (ex. Cornwall) - **C. bells:** *Malvaviscus arboreus* var. *penduliflorus* - **C. berry**. *Photinia arbutifolia* - **C. rose:** *Helleborus niger* - **C. star:** *Euphorbia pulcherrima* - **C. tree** (Eng-speaking countries): small tree usually *Pinus, Picea* (*P. excelsa*) (fir), etc.; u. at *C.* period (orig. in Ger).

Christopher root: *Actaea spicata* - **American Herb C.:** *A. alba.*

christophine: *Sechium edule.*

Christophskraut, traubenförmiges (Ge): Cimicifuga.

Christ's tears: *Coix lachryma-jobi.*

christyoside: glycoside of corotoxigenin & D-digitalose in *Strophanthus speciiosus,* &c.
chromatid: one of the pair of identical chromosome strands before a nuclear division.
chromatogram: record made by sepn. of compds. in chromatography.
chromatographie (Fr): **chromatography:** analytical procedure in which a mixture of substances is resolved into its components by selective adsorption in a column of adsorptive material; sepn. of gases (vapors), liquids, and solids by diffusion in appropriate media; incl. paper, column, thin layer, gas, etc. - **gas c.:** GC: separations in which a gas is the mobile phase passed through a column having a fixed adsorbent phase & u. for volatile constituents, ex. VO - **gel c.:** technique of sepn. of subst. on packings of granulated gels (ex. Sephadex) based on difference in sizes of molecules; u. both for analysis & prepn. - **paper c.:** distr. on special filter paper; the paper can be cut into parts for convenient elution to sep. compds.; useful in sepn. of amino acids (1-dimensional or 2-d.) - **partition c.:** chromatographie de partage (Fr); method in which equilibrium develops between 2 liquid phases, one held in form of a gel - **thin layer c.: TLC**; technique involving thin layers of asdorbent (such as silica gel, cellulose) instead of a column.
chromatographie à phase gazeuse: CPG (Fr): gas c. - **c. sur plaque** (Fr): thin layer c.
chromatomap: the distrn. of materials on a sheet of filter paper after 2-dimensional paper chromatography.
chromium: Cr; occ. in earth's crust as ores (chromite); considered a trace element in human nutrition; found in heart & other tissues of human body; in sea water; u. in heart dis. (2486).
chromosome: particular structure of cell nucleus wh. bears genes for inheritance of characters of cell & of organism; splits in mitosis.
chronotropic (+): to incr. heart rate.
Chrozophora senegalensis A. Juss. (Euphorbi) (Senegal, wAf): m'bolo; rt. decoc. drunk for green diarrh. (infants).
Chrysactinia mexicana A. Gray (Comp.) (Tex, NM [US], Mex): mariola; much u. in domestic med., as antisp. aphrod. (rts. & lvs): diaph. diur; for fevers & rheum.
chrysalis oil: FO fr. silk worm chrysalides; u. mfr. soap, paints, &c. (Shikim).
Chrysanthemum (Comp) (worldwide): formerly ca. 200 spp., now mostly transferred to other genera (2584); cult. often as ornam.; resin toxic - **C. balsamita** (L.) Baill. (Tanacetum b. L.; Pyrethrum b. Willd.) = *Balsamita major* Desf. - **C. cinerariaefolium** Vis.: (Yugosl): Dalmatian Pyrethrum fls.; as chief active princ. c. Pyrethrins I & II; VO. (0.5%); alk.; pyrethrosin (inert); u. insecticide, scabicide, in pediculosis - **C. coccineum** Willd. (Tanacetum c. [Willd.] Grierson): (Iran, Caucasus): Persian (or florist's) pyrethrum (fls.); painted daisy; one source of Pyrethrum Fls.; sometimes allergenic - **C. coronarium** L. (sEu. (Med reg.); natd. As, NA): garland chrysanthemum; (old) garden c.; crown daisy; fls. eaten by Chin, u. as digestive (Ind) & as emoll.; in icterus, GC; u. by ancient Eg for wreaths on the dead; repellant to fleas & bedbugs; in Dalmatian Powders (Crete); u. like chamomile (Ind), cult. Eu; var. *spatiosum* Bailey: cult. for edible young lvs.; stom. expect. - **C. frutescens** L. (sAf; Teneriffe): marguerite; Paris daisy; fls. c. frutescin & derivs.; rts. c. isobutylamides; fls. u. as insecticide; rt. u. rheum, toothache - **C. indicum** L. (C. japonicum Thunb.; Dendranthema indicum [L.] Des Moul.) (Jap, Chin): fls. & lvs. c. VO (kiku oil) with chrysanthenon, chrysanthemin (cyanidin-3-glycoside) (1172); fls. decoc. u. as bitter, lax.; in digestive dis., GC, boils, mumps (Jap) (lvs. in migraine; Parkinsonism; circul. ton., in dizziness, stom., carm.; to make tea [Chin]): in Chinese wine; keeps hair from turning gray (!?); to treat prostatitis (Chin); insecticide; pl. as a stock for many winter aster hybrids - **C. leucanthemum** L. = *Leucanthemum vulgare* Bernth. - **C. marschallii** (Iran, Caucasus): Caucasian (or Persian) insect fls.; u. like prec. as insecticide, parasiticide - **C. X morifolium** Ram. (C. sinense Sabine; C. japonense Nakai; Dendranthema X grandiflorum [Ram] Kit.) (Ind, Chin, Jap): (common) (cultivated) Chin chrysanthemum; pl. u. in circulat. digestive dis. as stom. purg. ton, in Parkinsonism, nervous ailments (headache, tinnitus, night blindness), rheum. lumbago; to keep hair fr. falling out & graying (?); exter. for skin dis., boils, inflamm.; juice as vulner.; "chrysanthemum wine" (fls. in wine) u. for digestive, circul. & nervous dis.; lf. & fl. decoc. u. for stomachache (PI); "antivinous"; pillow filled with lvs. or fls. u. for colds, headaches, inflamed eyes; substit. for chamomile; widely cult. ornam. - **C. parthenium** (L.) Bernh = *Tanacetum p.* - **C. roseum** Weber & Mohr = *C. coccineum* - **C. segetum** L. (Euras, natd. SA, orient, Af; cult.): corn marigold; c. chrysanthemum; "tung-has" (Chin); lvs. u. exter. for headache (Taiwan); young shoots u. as veget. (Chin); weed - **C. vulgare** Bernh. = *Tanacetum v.* (—> *Balsamita*; *Dendranthema*; *Leucanthemum, Tanacetum*+).
chrysanthemum, gilt (Chin): *Chrysanthemum balsamita* (possibly *Anthemis* spp.).

chrysarobin: powd. obt. by extn. of goa powder (Araroba) (*Andira araroba*) with benzene & evapn. to dryness; c. anthranols, anthranones; u. in psoriasis; disadv.: stains, causes irritation (formerly chrysophanic acid).

chryseoside: glycoside of chryseogenin (sarverogenin) & D-diginose; occ. in *Strophanthus intermedius.*

chrysin: dihydroxy-flavone; occ. in *Salix* buds, *Pinus,* Fag., Ros., Bignoni.

Chrysobalanus icaco L. (Ros) (trop&subtropAm, WI, Mex, SA, wAf, natd. Ind, Fla [US]): (h)icaco; coco(a) plum (tree); sd. FO u. in candles; frt. astr.; u. in diarrh.; edible, u. in preserves; bk. rts. lvs. astr.; sds. poisonous.

Chrysocoma coma-aurea L. (Comp) (sAf): Goldenhaar; goldylocks; u. as bitter (cEu): alexipharmic.

chrysophanic acid, medicinal: Chrysarobin.- **chrysophanol:** hydroxyanthraquinone compd. occurs free & as glycoside in cascara sagrada, senna, rhubarb, & therap.-related drugs.

Chrysophyllum africanum A DC (Sapot.) (tropAf): tree; frt. popular ("odara pears"); sds. rich in fat; u. med. (Guinea); wd. much valued - **C. cainito** L. (WI, CA, nSA): star apple; caimito; bk. decoc. u. in dys. (Mex); bk. inf. ton. refriger. (CR); latex u. drastic cath. (Br); sd. u. in diarrh.; round frts. very sweet, delicious, eaten raw or as preserves (cultd. sFla, PI, Indones); wd. useful (groves in PEI [Cana]); cult. ornam. - **C. glycyphloeum** Casar. = *Pradosia lactescens* - **C. lucumifolium** Griseb. (SA): star apple tree; wd. used; frts. edible; u. med. (Par Phar.) - **C. sanguinolentum** Baehni (Co): white latex appl. to open wounds as vulner. (2295).

Chrysophyta: golden (brown) algae; fresh & salt water habitats; incl. diatoms.

Chrysopogon aciculatus Trin. (Gram) (tropAs, Austral): bhatui grass; a perennial grass; rts. or hb. u. in decoc. in diarrh.; ashes int. for rheum.; pl. cult. Vietnam for rts. exported as Chiendent grenille à brosse; cattle fodder (PI).

Chrysopsis Nutt. (Comp.) (NA): golden aster; 10 spp. - **C. mariana** (L.) Ell. = *Heterotheca m.*

Chrysothamnus nauseosus Britton (Comp.) (wNA): (rubber) rabbit brush; grease bush; brown broom "sage"; yellow bush; yellow fl. heads; latex c. 2.-2.9% pure caoutchouc; source of Chrysil(l) caoutchouc; Am Indians chewed latex as "gum"; var. *graveolens* Piper (C. graveolens Greene) (US Rocky Mt. & Great Plains): heavy-scented rabbit brush; hb. u. for yellow dye (Nav. Indians); administered to parturient women to aid contraction of uterus; u. as emet. antipyr. dental remedy; in chicken pox, measles (Navahos) (1174); pl. toxic to humans & livestock - **C. vaseyi** (A. Gray) Greene (wUS) (Ut, Wyo [US] &c.): Vasey rabbit brush; u. to improve teeth - **C. viscidiflorus** (Hook.) Nutt.: mountain (or sticky-flowered) rabbitb(r)ush; fls. boiled with lichen u. as yellow dye for leather, woollens; inf. of tops rubbed on body to dry up lesions of chickenpox & measles.

Chrysothemis friedrichsthaliana Moore (Tussacia f. Hanst.) (monotypic) (Gesneri) (Pan): wilkwa (Cuna Indians, Pan); low hb. u. as antidiarrh., also in snakebite.

chua (Ind.): *Shorea robusta* (VO).

chuchau (Pe): *Furcraea* - **chuchupate** (NM [US], Mex): (Mexican) southern masterwort: *Ligusticum porteri.*

chuchu (Br): *Sechium edule.*

chuchuara (Pe): **chuchuhuasha**: *Maytenus* spp.

chuchupaste (NM [US], Mex): (Mexican) southern masterwort: *Ligusticum porteri.*

chufa (seed): *Cyperus esculentus.*

chukar: chakor (Hindi); *Alectoris graeca* Meisner (Phasianidae) (OW, esp. nInd, cAs, now natd. nwUS): chukar (or rock) partridge; important in game hunting; flesh eaten, delicious; u. in Unani med. (blood instilled into eye for e. dis.); many subspp.

Chukrasia tabularis A. Juss. (Meli.) (eAs): bk. u. as astr. & for diarrh. (Indochin); wd. valuable timber ("Chittagong wood"); yields gum.

chunine tree: v. chang shan (or ch'ang shan).

Chuquiraga insignis H. & B. (Comp.) (SA): u. as febrifuge (Ec).

church steeples (Eng): agrimony.

churras (cAs): the resin scraped fr. pistillate tops of *Cannabis sativa* subsp. *indica* & formed into balls; u. as fumitory; continued use produces a characteristic obstinate cough.

chutnee: chutney: chatni (Ind): pickle preserve: essentially comb. of sweet & acid frt. (ex. currants) with seasoning (ex. ginger, garlic, mustard, chillies, vinegar); frt. principally u. are green mangoes; also raisins, limes, lemons, tamarinds, mint, banana, &c.).

chymopapain: important enzyme component of crude papain (latex of *Carica papaya*); similar to papain enzyme (*per se*); made up of 4 components; proteolytic agent u. to tenderize meat; injection for shrinking slipped disc (herniated lower back) & relieving low back pain; some hazards in use.

chymosin: rennin (rennase).

chymotrypsins: major proteolytic enzymes of pancreatic juice; crystallized fr. ext. of pancreas of *Bos taurus* L.; Chymar™; u. topically for enzymic débridement of tissues, such as secretions of upper respiratory tract (asthma, sinusitis, rhinitis, &c.); in ophthalmology to facilitate cataract removal by causing lysis of the zonules (small bands holding lens); kerotitis, &c.

Cibotium barometz J. Sm. (Thyrsopterid. - Filicopsida - Pteridophyta) (seAs): tree fern; golden moss; Tartarian lamb; fibers at base of stipes of this fern formerly u. hemostat. (penghawar djambi); stem scales u. for filling pillows; wd. u. for sculptures - **C. glaucum** H. & A. (Hawls [US]): pengawar djambi; underblue cibotium; scales u. to stuff pillows, &c.; pith eaten; "pulu fiber".

cibo(u)l(e): cibo(u)ll(e): (Fr): various *Allium* spp., esp. onion, shallot, leek, chives.

Cibus Deorum (L.) (food of the gods): Asafoetida.

Cicada (Hemiptera): insects with loud characteristic song; shells u. in Chin med.

cicely, sweet: 1) *Myrrhis odorata* 2) *Urospermum claytoni* 3) *Osmorhiza longistylis* & other *O.* spp.

Cicer arietinum L. (Leg.) (prob. native wAs): very popular in Indo-Pak, also Af, NA (Calif, [US] Mex), SA, etc.; (Mexican) chick pea; garbanzo bean; Egyptian pea; garvanze; horse gram; young sds. eaten raw, older boiled (like green peas); roasted & u. as nutr. & in mfr. ersatz coffee (esp. popular among Spanish peoples); good protein source (1563); c. pangamic acid (antistress, stamina-building, antihyperlipidemic properties).

Cicerbita alpina (L) Wallr. (Lactuca a. R. & H.; Mulgedium a. Less.) (Comp.) (Eu, mostly in mts.): Alpen-Milchlattich (Alpine milk lettuce); European (or alpine) sow thistle; sts. are eaten raw; orn. - **C. rapunculoides** Beauv. (Lactuca r. C. B. Clarke; Mulgedium r. DC) (Pak, nInd - Kashmir): lf. & bk. appl. to sores.

Cichoriaceae: chicory family; fam. split fr. Comp. by some botanical taxonomists.

Cichorium L. (Comp.): chicory; hbs. with showy fls. - **C. endivia** L. (Med. reg.): endive; juice c. bitter princ. lactucopicrin; (also foll.) (1176); bleached tops eaten like celery, in salads; rt. lax. diur. ton.; var. *latifolium* Lam. - **C. intybus** L. (Eu; NA, wild & cult. US, [Mich, etc.]: (common) chicory; young rts. eaten like parsnips; older rts. (Cichorium, L.) u. as beverage flavor in coffee ("French coffee"), coffee adult. (roasted) (1175); u. bitter, deobstruent; soaked in water overnight & inf. drunk in morning for fever (Baluchist); for skin dis.; bitter lvs. u. as greens & in salads; attractive blue fls. - var. *foliosum* Hegi (Eu): (Brussels) witloof; lvs. in heads u. in fancy salads; "French (or Belgian) endive"; first u. near Brussels; cult. for centuries in Eu - var. *sativum*: Brunswick var. grown for rts. u. as coffee substit. (roasted & ground) (2223).

Ciclosporin: Cyclosporin A.

Cicuta (L, Sp, Por): 1) *Conium maculatum* (lvs., hb., & sd.) 2) (It): *Cicuta virosa* - **C. maculata** L. (Umb.) (e&cNA in damp meadows): (American) water hemlock; spotted cowbane; beaver poison; musquash; cicuta maceolata (It); lvs. &c. u. as alt. loc. analg.; c. citutoxin (very toxic); most poisonous pl. of NA; commonly causes death (of stock & children) (rt. often mistaken for parsnip); frt. c. cicutine (*cf.* coniine) & VO (*cf.* oil cumin); rt. chewed as contraceptive; u. as abort. (Am Indian) - **C. virosa** L. (nEuras): European water hemlock; hb. c. cicutoxin; cicutol (toxic phenol); u. alt. local analg. in cramps, rheum; u. exter. for cirrhous carcinoma (Eu); frts. u. in epilepsy; poison!

cicutilla (Mex): *Parthenium hysterophorum.*

cicutine: vol. toxic alk. fr. *Cicuta virosa* - **cicutoxin:** picrotoxin type compd. occurring in *Cicuta virosa*; resp. stim.; very toxic.

cider (, apple): cidre (Fr): unstrained unfermented expressed juice of apples (= **sweet c.**); delicious beverage (do not confuse with apple juice); fermented a. juice (= **hard cider**); "cider" (GB) (alc. beverage); products of cider incl. apple jack, vinegar, apple butter.

cidra (Sp): citron: (Eng) *Cucurbita ficifolia.*

cidrão (Por): *Aloysia triphylla.*

cidrón (PR): *Phyla nodiflora* var. *reptans.*

cigar tree (, ladies): Indian c. t.: *Catalpa bignonioides, C. speciosa.*

cigarette plant: *Kalanochoë.*

cigarillo (Sp): *cigarro* (Sp); cigar.

cigarrito (Mex): cigarette, esp. one prep. with cornhusk paper.

cigue officinale (Fr): *Conium maculatum* - **c. vireuse:** *Cicuta virosa.*

cila (Por): scilla.

cilantro (SpAm): 1) coriander 2) *Eruca sativa.*

Cimex: bedbug, &c.

Cimicifuga foetida L. (C. europaeum Schipcz.) (Ranuncul.) (Euras, NA): stinking bugwort; hb. u. for edema; rhiz. u. climacteric difficulties (Homeop.) - var. *simplex* (Wormsk.) Huth (C. simplex Wormsk.) (Jap, Kamschatka): perenn. hb.; rhiz. u. as heart ton. (Ainus), bitter, carm.; antipyr. in measles, &c. - **C. racemosa** Nutt. (eNA) black cohosh; b. snakeroot; rhiz. c. cimicifugine; 15-20% resins & bitter subst. ("cimicifugin"); also hormone-like steroids (*cf.* estrogen), acetylacetеol, cimigenol; FO; racemoside (anti-ulcer properties); u. stomach, ton. in rheum. & neuralg.; astr., sed., in chorea, chron. rheum., menstrual diff. (1177).

cimicifugin: 1) resin of cimicifuga 2) racemosin; xall. bitter princ.

Cina (L.): Levant wormseed, *Artemisia maritima* & other *A*. spp.

cina-cina (Arg): *Parkinsonia aculeata*.

cinamômo (Br): *Melia azedarach* (118).

ciñar (Hind): chinar.

Cinchona (L.) (fr. Countess Chinchon, Spain, first European said cured of malaria with this bk.): the dr. bk. of several *Cinchona* spp. which c. quinine & many other alk.; u. to treat malaria (1277); sometimes specif. appl. to *C. calisaya* & its hybrids; *C.* spp. native to SA; cult. Java, Sumatra, Guat, Mex, Mal, Congo, eAf; C. alk. are classed into crystallizable (quinine, quinidine, cinchonine, cinchonidine; hydrocinchonine, hydrocinchonidine, hydroquinine, hydroquinidine; quinamine, conquinamine; paricine; javanine) and amorphous (quinoidine; dicinchonine, diconquinine) (1236; 1275) (for serials re. *C.* see Appendix C-1) - **C. boliviana** Wedd. (Rubi.) = **C. calisaya** Wedd. (Bol): yellow (bark) cinchona ("Calisaya Morada" [m.=purple]) shows great variation in alk. content (0.64-12%); but u. rel. low in alk. - **C. Flava** (L.): yellow cinchona, obt. fr. *C. calisaya* - **Florida c.**: *Pinckneya* - **gray c.**: (quin)quina gris (Sp): *C. peruviana* & some other *C.* spp. (SC exter. color of bk. silvery grayish) - **C. hybrida** Hort. ex Sasaki (PI): u. med. (PI) (1483) - **Indian c.**: c. prod. in Ind; now mostly u. in that country (with world's greatest malaria incidence) - **Java c.**: c. cult. in J., largely for mfr. of quinine & other alks.; mostly exported - **C. lancifolia** Mutis (Co, Ve): Colombian (or Bogota) C. bk.; alk. content variable, in general low; tree cult. Ind - **Ledger bark c.**: **C. ledgeriana** Moens (Pe, Bo): yellow cinchona; tree most frequently cult, in Indones; alk. content varies considerably (1-14%); generally rich in quinine & richest of all cinchonas in tot. alk. (1276) - **C. macrocalyx** Howard (C. heterophylla Pav.; C. stupea Pav.) (Latin Am): bk. has been included in several natl. pharmacopeias; u. as bitter, &c. like other c. barks - **C. micrantha** R. & P. (Pe, Bo): gray bark; silver b.; tree source of Huanuco cinchona (yellow cinchona type); Lima bark; c. quinine, cinchonine, cinchonidine; bk. sometimes classed as a var. of Calisaya Bark - **C. nitida** R. &. P. (Pe): source of some Gray C. & some Loxa (pseudo-Loxa) *C.* - **C. oblongifolia** Mutis = *Cascarilla magnifolia* - **C. officinalis** L. (Pe, Ec): Loxa c. (bark); Crown Bk.; pale c.: C. Pallida; quinine cont. varies fr. 1.4-9.1% (271) - **C. ovata** R. & P. (Bol, Pe): source of Naranjada bark with 5% alks. & traces of quinine - **C. peruviana** How. (Pe): Gray C. (chief sp. of this); bk. exter. silvery gray - **C. pitayensis** Wedd. (Pe): source of a low grade (low alk. content) bk. - **C. pubescens** Vahl. (C. pelletierana Wedd.) (nAndes reg., Ven, Co, Pc, Bo; also CA): Cusco bark; source of Cortex Chinae flavus (yellow cinchona bk.) or Cusco bark; quin(a) (morada); c. cuscamine, cuscamidine, cusconine, cusconidine, aricine (closely related to *C. cordifolia* Mut.) ("wild cinchona"); bk. decoc. u. for fever (Ve); quinine source - **red c. (bark)**: *C. succirubra* - **South American c.**: that prod. in SA; for many years sJava prod. c. 85% of c. used, while SA prodn. dwindled; SA prod. revived during WWII due to Jap occupation of NEI - **C. succirubra** Wedd. (Ec): red (bark) c.; one of the off. *C.* bks.; earliest cult. in eAs; considerable variation in alk. cont. usually 7-10% (1276); the various bks. u. ton. antimal. astr. antipyr. stomach. & in mfr. alks - **C. uritusinga** Pav. (C. macrocalyx DC; C. heterophylla Pav.; C. condaminea H. & P. ?) (Peru): tree furnishes Loxa bark (China Loxa) off. in several pharmacopeias; trees almost exterminated by fungus following bad practice of leaving trees standing after bk. removal - **yellow c.**: *C. ledgeriana*, *C. calisaya* & hybrids.

cinchonidine: one of the less valuable alks. of Cinchona u. to treat malaria.

cinchonine: alk. stereoisomeric with cinchonidine; esp. rich in *Cinchona micrantha*.

cinchotannic acid: a phloba-tannin found in Cinchona; combines with alkaloids to produce a tannate salt.

cineole: eucalyptol.

Cineraria (L.): 1) *Senecio cineraria* (**C. maritima** L.) 2) *S. cruentus*.

cinerins: a group name for insecticidal compds. of pyrethrum fls.; rapidly destroyed in air by oxidn.; represent cyclopropane carboxy derivs.

cinnamaldehyde: cinnamic (acid) aldehyde charactertistic component of cinnamon oils; also made by synthesis (1564) - **cinnamein:** benzyl cinnamate; found in Balsams Peru & Tolu - **cinamic acid:** free & esterified org. acid in Styrax & other balsams; in cinnamon oil, etc.; mostly u. in perfumery & cosmetic trades.

Cinnamodendron corticosum Miers (Canell.) (WI): mountain cinnamon; bk. u. as spice ("false Winter's bark") - **C. axillare** Endl. (Br): source of paratodo bark; u. in scurvy, fevers, as stom.

Cinnamomum 1) C. Schaeffer generic name for true cinnamons (seAs, Austral) 2) (L.) (specific) bk. of *C. loureirii*, Saigon c. - **C. burmanni** Bl. (C. kiamis Nees, C. chinense Bl.) (Laur.) (Sumatra, Hainan, Timor): fagot (or Batavian) cinnamon; bk. u. flavor, adult. of better quality cinnamons; spice (much u. in Neth); twigs & frt. u. in incense pd.; older bk. called Massoy bark; u. in diarrh. stomach cramps; imported fr. Indones - **C. camphora** T. Nees & Eberm. (Chin, Jap, Taiwan): camphor tree; source of gum c. & c. oil (see also under "camphor"); cult. & natd. sTex, Fla (US) - **C. cassia** Bl. (L. & BO) (C. aromaticum Nees) (sChin, Vietnam, Indones): cassia cinnamon; Chinese cassia; source of VO, cinnamon oil; chief constit. cinnamaldehyde; u. flavor, condim.; perfume; in kidney dis. (1696) (2591); lvs. source of cinnamon leaf oil (Ceylon) (1284) - **C. culilawan** Bl. (Moluccas; Chin): culilawan (or lawang) bark; pleasantly odorous bk. with 4% VO, c. eugenol; u. puerperal eclampsia - **C. iners** Reinw. (Malay Archipelago): Siamese cinnamon (or lawang) bk. u. flavor, much valued condim.; c. much mucil.; cinnam. substit. arom. stim., carm., diur., galactag.; lvs. u. as condim.; VO distd. fr. lvs. u. stim.; sds. in honey for dys., cough - **C. javanicum** Bl. (Mal): sintoc (or sintok) bark; cortex sintoc; bk. somewhat like culilawan bk. - **C. loureirii** Nees (Indochin, Chin): Saigon cinnamon; found only under cult.; u. flavor; condim. stim. carmin., the chiefly u. medicinal cinnamon (1285) - **C. massoia** Schewe (PapNG): tree bk. ("Massoi[a] Bark") u. as spice; adulter. of/or substit. for Ceylon cinnamon; this epithet may represent a synonym of *C. burmanii* - **C. mercadoi** Vidal (PI): VO c. considerable amt. of safrole; bk. u. for flatus; expect., in headaches & rheum. (bk. & lvs. appl.); chewed for gastric dis.; in TB; proposed for use in root beer - **C. oliveri** F. M. Bailey (Austral): Oliver's bark; (black) sassafras; white s.; bk. has strong sassafras aroma; rich in camphor, methyl eugenol, safrole; u. med. (local): cinnamon substit. - **C. pallidum** Gillespie (Fiji Is): bk. u. to aromatize coconut oil - **C. pedatinervinum** Meissn. (Fiji Is): wild cinnamon; macou; mou; rt. VO c. safrole 50%, linalool, eugenol, &c.; u. to aromatize coconut oil - **C. pedunculatum** J. S. Presl (C. japonicum Sieb.) (sJap): sds. source of wax u. in mfr. candles; sd. oil u. in med.; bk. roborant - **C. sintoc** Bl.: *C. javanicum* - **C. solomonense** Allen (Solomon Is): bk. appl. for headache - **C. tamala** T. Nees & Eberm. (sAf): Indian cassia; (esp. rt.) bk. c. VO rich in eugenol, phellandrene; u. flavor in cooking, med. (GC); combined in an aromatic prod. called trijataka (with cardamom & Ceylon cinn.) - **C. zeylanicum** Bl. (C. verum J. S. Presl) (SriL, sInd, cult. WI, SA, Indones [esp. Sumatra & Java], Jap): Ceylon cinnamon; inner bk. with small percent of VO but of excellent quality; u. carm. flavor, condim. stim.

cinnamon (bark): bk. of *Cinnamomum* spp. usually *C. loureirii* (US), *C. cassia* or *C. zeylanicum* (GB) - **bastard c.:** *Cinnamomum cassia* - **Batavia c.:** *C. burmanii* - bitter c.; *C. culilawan* - **black c.**: *Pimenta racemosa* - **c. buds**: fr. *C. loureirii* - **c. canella:** *Canella winterana* - **Cayenne c**.: bk. of *C. zeylanicum* cult. in Guyana, Br, WI: in larger pieces & therefore inferior to true Ceylon c. - **Ceylon c.:** *C. zeylanicum* - **Chinese c.:** *C. cassia* - **culiban c.: culilawan c.:** *C. culilawan* - **fagot c.:** Java c. - **God's c.:** Saigon c. - **Himalaya c.:** *C. tamala* - **Java c.:** *C. javanicum*; *C. burmanii* - **c. leaf oil:** VO fr. lf. of *C. zeylanicum* & its v. *seychelleanum* (Seychelle Is) - **Malay c.:** *C. burmanii* - **massoia c.:** *C. massoia* - **c. oil:** c. bark VO fr. 1) *C. cassia* (US) 2) *C. zeylanicum* (GB) - **c. root**: sassafras - **Saigon c.:** *C. loureirii* - **Siamese c.:** *C. iners* - **sweet c.:** Calamus - **true c.:** Ceylon c. - **c. vine:** *Dioscorea batatas* - **white c.: wild c.:** 1) bk. of *C. pedatinervinum* 2) *Canella winterana* - **c. wood (oil):** sassafras.

cinnamyl cinnamate: styracin: ester found in balsams (Peruvian; Styrax) - **C. cocaine:** alk. found in coca leaf; hydrolysis yields cinnamic acid.

cinobufagin: digitaloid princ. fr. parotid secretions of a Chinese toad (really a toad venom) - **cinobufotoxin:** digitaloid princ. of same source & similar nature to prec.

cinquefoil (herb; root): *Potentilla* spp., esp. *P. reptans* - **Norway c.:** P. *monspeliensis* - **shrubby c.:** *P. fruticosa*.

cintul (Mex): *Veratrum frigidum*.

cioccolatta (It-f.): chocolate.

cipó (Br): liana - **c. azouque:** *Apodanthera smilacifolia* - **c. cabeludo**: *Mikania hirsutissima* - **c. caboclo**: 1) *Davilla rugosa*, *Echites peltata* - **c. cravo:** *Tynnanthus elegans* - **c. da praia** (Br): *Ipomoea pes-caprae* - **c. suma:** *Anchietea salutaris*.

cipolla (It.): onion - **C. marina** (Mex.): squill.

cipollina (It.): chives; shallot, &c.

cipress (Mex): *Cupressus sempervirens*.

circadian: diurnal but covering ca. 24 hrs. continuously (ex. 7 am - 7 am) (<circa, approx. + dies = day).

circulin: polypeptide antibiotic, fr. *Bacillus circulans* Jordan; Gm. neg. m.o. (tap water, soils) (1948-) (*cf.* polymyxins) (A & B).

cire blanche (Fr): white bee's wax - **c. jaune:** yellow wax.

ciriegia (It): cherry.

Cirramycin: series of macrolide antibiotics prod. by *Streptomyces cirratus*; forms A & B recognized (plus fractions); 1963- (Jap); active against Gm.+ & Gm.- bacteria, mycoplasma.

Cirsium Mill. (Comp.) (nTemp Zone, Med reg., NA, SA): closely affined with *Carduus*; many taxa have been transferred from latter (ex. *C. horridulum* Mx. from *Carduus spinosissimus* Walt.) - **C. arvense** (L) Scop. (Comp.) (Eu, natd. NA): Canada thistle; creeping t.; bad weed in NA; c. 1 m high; fl. diecious, purple, or rarely white; fl. heads ca. 2 cm high; rt. u. ton. diur. hepat. in bil. dis.; rt. decoc. for worms in babies (Ashenakis); starvation food in NA during depression of the 1930s - **C. eriophorum** Scop. (Cnicus eriophorus Roth.) (Eu): friar's crown; spear's thistle; hb. u. diaph. diur. in skin dis.; expressed juice (herbal j.) u. for cancer (1830s); VO c. polyacetylenes - **C. horridulum** Michx. (Carduus spinosissimus Walt.) (seUS): yellow thistle; chardon (Fr-La [US]); pl. inf. u. in colds, dyspeps., diarrh. - **C. japonicum** DC (Jap): rts. & lvs. u. as hemostat. (Chin); antipyr., ton., diur. (Jap); cult. for cut fls. - **C. muticum** Mx. (e&cNA): swamp thistle; biennial hb.; this & other *C.* spp. u. contraceptive (tea) (Zuñi, Chippewa Indians) - **C. neomexicanum** Gray (wUS): cold inf. u. as "life" med. (Navajo Indians); eyewash from rt. - **C. pumilum** (Nutt.) Spreng. (Carduus pumilus Nutt.) (seUS, Mex): bull thistle; biennial, growing to 8 dm; bud eaten by Am Indians - **C. segetum** Bge. (nChin, Kor): c. alk.; u. as ton. astr. "antidote" (Kor); for wounds, poisonous bites (Chin); otherwise like *C. japonicum* - **C. tuberosum** (L.) All. (w&cEu): Knollendistel (Ge); rts. u. as winter vegetable & stored through the winters; med. (the Lukakantha of Ibn el Baithar) - **C. vulgare** Ten. (C. lanceolatum Hill) (Euras, Af, natd. NA): common (bur) thistle; bull. t.; "true" thistle (of Scotland); spear t.; source of industr. alc. (1484): bad weed.

ciruela (Sp): plum - **ciruelo** (Sp): plum tree.

Cissampelos fasciculatus Benth. (Menisperm.) (Nic to Ec): st. juice u. orally for snakebite (Co) - **C. pareira** L. (warm As [Ind, SriL+], eAf, WI, SA): velvet leaf; false pareira (brava): picamano; rt. bitter, c. bebeerine (pelosine) & other alk.; u. diur. in lithiasis, emm. exp. antipyr. ton. astr. in diarrh.; substit. for Pareira Brava (of which formerly this sp. was thought the origin); rt. st. inf. u. to dissolve calculi of bladder, in snakebites (1566); for fertility control (Ur); frts. & sprouts also u.

Cissus alata Jacq. (Vitis sulcicaulis Baker) (Vit.) (cBr): mãeböa; lvs. u. med. in galenicals (Br Phar.) - **C. buchananii** Planch. (tropAf): rt. decoc. u. for pains in region of navel (Tanz) - **C. discolor** Bl. (tropAm): st. u. for baskets; in water produces a lather, u. for washing - **C. glauca** Roxb. (Vitis g. Wight & Arn.) (Ind, EI): lf. cooling; bk. resolvent; rt. u. for toothache - **C. incisa** Des Moulins (se&cUS): possum grape; marine ivy; expressed juice u. as mild expect. (1178) - **C. pallida** Planch. (Vitis pallida W. & A.) (tropAs, Ind, Af): kondage (Kanarese); bruised rts. appl. to leg ulcers, rheum. swellings, mucilag. prep. of rt. u. as anthelm. - **C. quadrangularis** L. (Vitis q. Wall.) (cAf, India, hotter parts): edible-stemmed vine; c. steroids; rt. & lf. c. tannins; u. as anthelm.; exter. as antiphlog. (Af); hb. exts. stim. healing of broken bones; st. fiber u. to stop bleeding (Digo, Kenya); rt. u. for myalgia; rt juice for earache; lvs. & young sts u. as stom. & for dig. disturb. (Ind) - **C. repanda** Vahl (Ind): bodlar lati; rt. pd. appl. to wounds, fractures (124) - **C. repens** Lam. (Ind+): pounded lvs. appl. as poultice to boils (to ripen); u. for stomachache (Mal) - **C. rhombifolia** Vahl (NA, SA): in Yuc, the Mayas wash wounds & sores with crushed bk. in water, then after cleansing wound, apply pd. bk. - **C. striata** R. & P. (Ch): "parrilla"; aerial parts u. as astr. in dys. hemorrhages; lvs. u. as cataplasm & as pomade for skin eruptions; c. triterpenes, sterols, alk. (traces), flavonoids, coumarins, phenolic compds.; quercetin, sitosterol - **C. tiliacea** HBK (Vitis t. Hemsl.) (Mex): tripa de Judas; lvs. u. in eye dis. - **C. trifoliata** (L) L. (C. acida L.) (sFla, tropAm, pantrop): sorrel vine; lf. decoc. u. to provoke vomiting in snakebite (PapNG) - **C. triloba** Merr. (eAs): rt. u. for headache, paralysis(!) (Indochin) - **C. verticillata** (L.) D. H. Nicols & Jarvis. (C. sicyoides L.; Vitis s. Miq.) (sFla [US], WI, Mex, tropAm): possum grape; China root; lvs. appl. to sores, inflammn., swellings; st. decoc. u. for rheum., certain mental dis. (2444); rt. u. ton.; substit. for adhesive plaster (Jam); sts. for cordage, to make baskets.

Cistaceae: Rock-rose Fam.; shrubs or woody hbs., mostly of nHemisphere; lvs. simple, entire; ex. *Helianthemum, Cistus.*

Cistanche phelypaea (L) Coutinho (Orobanch.) (swEu, nAf): thick lower part of pl. axis u. in diarrh., as aphrod. ton. in impotence; dr. pl. mixed with camel's milk u. as cataplasm against contusions (nAf) (2508); st. eaten like asparagus (wSahara reg.).

ciste (Fr.): *Cistus ladaniferus.*

Cistus L. (Cist.) (Med reg.): rock rose; holly rose; genus of low shrubs, mostly of importance as ornam. - **C. albidus** L. (Med reg., sEu, nAf): gum cistus; as adulter. of marjoram, inf. of lvs. & young shoots mixed with beverage tea u. as digestant ("Arab tea"); shrub 1-2 m high - **C. canadensis** L. = *Helianthemum canadense* - **C. cyprius** L. (AsMin, Crete, Cyp): rose of Crete; gland. hairs of lvs. prod. ladanum oil & resin - **C. ladanifer** L. (swEu, Med reg.): common cistus; (rose of) Crete; chief source of la(b)danum resin (*qv.*); "gum" ladanum obt. by boiling lvs. & tops in water; yield ca. 7% VO - **C. monspeliensis L.** (seEu, nAf, Med reg.): narrow-leaved cistus; tuzelt (Alg); lf. inf. as beverage tea; decoc. of flg. branches in asthma (Alg); folk med. in cancer; arom. in perfumes & incenses - **C. salviifolius** L. (se&cEu, nAf): sage-leaved cistus; holly rose; lvs. c. VO, much tannin; u. astr., adult. of Marjoram; tea substit. (Gr) - **C. villosus** L. (var. *undulatus* Gross) (C. incanus L.). (Med reg.+): rose of Crete; source of la(b)danum (resin); u. as expect. & in diarrh.; in perfume indy., to scent soaps; in leprosy, &c. (Br); stim. emmen.; pl. is source of Flores Cisti maris.

Citharexylum affine D. Don (Verben.) (sMex): alacate; "coral" lvs. appl. to swellings; emmen.; in chest dis.; frts. edible; cult. - **C. berlandieri** B. L. Robinson (Mex, Tex [US]): negrito; lf. decoc. u. to treat colds (Mex); sometimes cult. - **C. caudatum** L. (WI, CA): pigeon feed; upright fiddle wood; inner bk. u. emmen. abort.; frt. (drupes) eaten (Jam), u. in beverages; fls. pectoral & in diarrh.; lvs. inf. for hoarseness, sudorific (216) - **C. dentatum** D. Don (Peru): frts. u. to make marking & writing inks (black) - **C. ellipticum** Sessé & Moç. (Mex, Cu): anacahuita; pl. u. medicinally - **C. fruticosum** L. (Fla [US], WI, nSA): (Florida) (black) fiddlewood; u. med. (216); useful timber - **C. myrianthum** Cham. (Br): tarumá; u. as substit. plywood - **C. poeppigii** var. *margaritaceum* Poepp. & Moldenke (Peru): facedera; lvs. u. antipyr. (216) - **C. schottii** Greenm. (Yucatan, Mex; CA): palo de violín; "tatakache"; shrub/tree u. to treat and heal old "sores" (2490); "swellings"; crushed lvs. for cramps; young lf. inf. int. for asthma, coughs; crushed rt. syph. sores, erysipelas (217, 218) - **C. spinosum** L. (Cey, Haw Is [US], &c.): tree; drupes sometimes eaten by Negroes (Jam); useful wd. - **C. subflavescens** Blake (Col, Ve): "salvio"; tree -30 m; pl. toxic to skin (gives dermal swellings, itching) (*cf.* poison ivy); fls. with jasmine odor.

citral: citraldehyde: neral; present in VO of lemon, lemon-grass, orange, etc.; u. to flavor; strengthen lemon oil; mfr. raw material in perfume ind.

citric acid: characteristic org. acid of *Citrus* frts. (hence the name); also found in many other pl. & animal tissues; comml. prod. now (US), mostly obt. by fermentation of crude sugars with *Aspergillus niger* (1811); in sEu largely mfrd. fr. lemon culls; u. acidulant in beverages, pharm. preps.; mfr. citrates; as reagents; sequestering agt. in oil well operations; to disguise saline tastes; to beautify the complexion, etc.

citrin = vit. P; mixt. of citroflavonoids fr. lemon peel; hesperidin important component.

Citrine (Zitrine) (Ge): lemon; citron.

Citrinin: antibiotic subst. prod. by *Penicillium citrinum* Thoms, *Aspergillus niveus* &c. (1946-); active againt Gm. pos. m.o.; too toxic to be of practical importance (1812) (SC. lemon yellow color).

citrino (Por): citrine: citron tree (or frt.).

citroen (Du): lemon (1896).

citron: 1) *Citrus medica* 2) *Citrullus vulgaris* v. *citroides* 3) color of lemon.

citronella (grass) (oil): 1) *Cymbopogon nardus* 2) *Collinsonia canadensis* (also **c. horse balm**) - **c. leaves:** *Pimenta racemosa* - **c. plant** (NJ [US]): *Collinsonia canadensis.*

citronellol: acyclic terpene alc.; occ. rose & geranium oils (with rose-like odor); u. perfumes, cosmetics; in place of citronella oil.

citronnelle (Fr): 1) *Melissa officinalis* 2) vervain.

citronnier (Fr): lemon tree.

citropten: limettin; a coumarin fr. *Citrus aurantifolia.*

citrouille (Fr): squash.

citr(o)us products: v. under **Citrus**.

citrovorum factor: folinic acid.

Citrullus 1) (L.): watermelon sd., *Citrullus vulgaris* 2) genus *Citrullus* Schrader, with foll. spp. - **C. colocynthis** (L) Schrad. (Colocynthis citrullus O. Ktze.) (Cucurbit.) (As, Af): bitter apple; pulp of frt. c. colocynthin; elaterin, citrullol (alc.), resin (20-40%), FO; u. hydragog cath., drastic, considered of value in diabetes (US Negroes) (1372); to destroy bed bugs; imp. fr. Sudan via Egypt; pl. cult. to bind tops of sand dunes (Pak, Ind) - **C. lanatus** Mats. & Nak. (C. vulgaris Schrad.; Cucurbita citrullus L.) (trop&sAf, cult. all over globe): watermelon; cult. for delicious frt. & for dr. sd. (black & white) (wFla [US]); sd. ext. u. anthelm. diur. in hypertens., leukemia (exper.); also u. in mfr. FO; avg. melon bears some 1,500 sds., sds. roasted & cracked —> kernel (eaten); frt. eaten for hypertension (Sand Mt. area, Ala [US]; juice of frt. u. in water as refrigerant beverage (Pe); watermelon seed tea u. in prostatitis (Miss [US] Negroes); frt. eaten for typhoid (Indones); rind candied & u. as glazed frt.; sds. sucked to assuage thirst; one midget var. popular as "ice-box" watermelon; lvs. & frt. c. cucurbitacin E.; var. *citroides* Bailey (with hard white flesh); pickle melon; pine m.; preserving m.; "citron"; stock m.; rind u. as pickle, in conserves; inedible raw but slices fried, pickled, preserved; u. as cattle feed.

Citrus L. (Rut.) (orig. eAs; spp. now cult. in all trop & subtrop regions): important genus; many spp. bear esculent frts. of considerable economic worth (1181); **citr(o)us products** incl.: fresh frt.

peel; juice; pectin; pectic acids; galacturonic acid; bioflavonoid complexes; hesperidin; naringin; flowers & their VO - **C. acida** Roxb: *C. medica* var. *acida* - **C. aurantifolia** Swingle (C. lima Lunan) (cult. Fla, Calif [US], Mex, WI): (Persian) lime; acid l.; bigarade; shrub or small tree native to Ind & seAs (1362); frt. pulp very acidulous (too sour for eating); juice c. ca. 10% citric acid, u. antiscorbutic, flavor, refrigerant; as or in beverages; frt. sucked for seasickness (transAtlantic voyages); VO prod. WI, Mex, Ind (hand-pressed); fl. buds u. as soporific (Trin) - **C. aurantium** L. (C. vulgaris Risso) (seAs): sour orange; bitter o.; frt. peel c. aurantiamarin, hesperidin, acids; u. flavor, ton., bitter stom.; dr. peel u. in mfr. orange marmalade (GB, Calif [US]+); candied orange peel; fls. distd. to produce orange oil & o. water; u. perfume, in tisanes; beverage (SA); fresh peel source of VO, Neroli oil (bitter orange oil); u. arom. flavor, fls. decor. for weddings (symbol of purity); this is the first orange brought to Eu (now cult. It, Sp, WI); juice u. as vinegar (Yuc) (6); lvs. u. ton., bitter, arom., stom. sed.; lvs. (Folia Aurantii, Ge Phar.; Folium A., Swiss Phar.) (v. under "orange") (1814); subsp. *aurantium*: cultivar 'Bouquet'; bergamot orange of US - **C. bergamia** (Risso & Poit.) Wight & Arn. (C. aurantium subsp. b. [Risso & Poit.] Wight & Arn. ex Engler) (Eu, esp sIt): bergamot (orange); juice very bitter; peel u. prep. VO (b. oil); peels sold in markets of sInd for medl. use (1364); VO u. in hair tonics (Ala [US] Negroes); also for various perfumes (1813); u. in prep. Cologne Water (chief prodn. It) - **C. bigaradia Risso** = *C. aurantium* Thunb. - **C. decumana** L. = *C. maxima* - **C. depressa** Hayata (Taiwan, Ryukus): flattened lemon; frt. u. to neutralize jellyfish or sea nettle stings; juice in beverages; sds. in necklaces; tree ornamental - **C. grandis** Osbeck = *C. maxima* - **C. hystrix** DC (PI, Mal, Burma): cubuyao; "Mauritius papeda"; small thorny tree; frt. juice u. in "ades"; frt. peel to flavor meat (fish) (Java) & to perfume *Entada phaseoloides* (hair prepns) (PI); candied; in washing hair & cleaning clothes (1183); peel ton., for worms; headache; VO u. in perfumes; juice stim. - **C. jambhiri** Lush. (nInd, carried to Af, then Eu, then NA): rough lemon; mazoe; u. for frt. & juice in Haw+; juice very sour; frt. u. in marmalades, &c.; pl. serves as rootstock for other *C.* spp., esp. lemon tree (—> immunity to pl. dis.) - **C. japonica** Thunb. = *Fortunella j.* - **C. limetta** Risso (C. medica var. 1. Brandis) (cult sEu): sweet lime; (French) lime (tree); sweet l. (of India); true l.; source of Italian lime oil; frt. edible; sometimes confused with the very sour *C. aurantifolia*; frt. (flavor variously sweet, insipid sl. lemon flavor), juice & VO u. as other *C.* spp. - **C. X limon** (L.) Burm. f. (sEu, It, Sp; WI, Calif [US], Br): lemon; source of l. peel; fresh peel source of l. oil (VO), etc.; frt (c. pectin) u. flavor for Chartreuse & Curaçao liqueurs; juice u. for lemonade; popularly u. in arthritis & gout (in 1950s) (1179), juice & pulp (pulp with highest content in rag) strong in citric acid (ca. 7%) (may etch teeth); u. flavor, acidifier, antiscor.; peel candied & u. flavor, stim. stomach; as pickle; fresh juice canned, bottled, frozen - **C. X limonia** Osbeck (US): hybrid of *C. limon* X *C. reticulata;* Mandarin lime (Rangpur [lime]), Otaheite orange; (Ind, now widespread) - **C. maxima** Merr. (C. grandis Osbeck) (Mal, Polynesia, sChin): shaddock; pummelo (not pomelo); valuable frt. large, delicious, divisible into many segments or sections; popular food in sAs; sds. u. for cough, lumbago (Mal); peel u. as pickle; sometimes but wrongly called "grapefruit," which may, however, have derived fr. it (1389) - **C. medica** L. (sAs): citron; large oblong frt., resembling an oversized lemon (1361); with thick peel & scant acid pulp; mostly cult. for thick peel, which is candied & glacéd & u. in confect. & cakes as flavor; in Jewish ceremonials (ethrog. Heb.) (var. *ethrog* Engl.) also prod. VO u. in perfumes; this was the first citrus frt. introduced to the west (1180); mostly cult. for peel also prod. VO; inf. of fresh shoots u. as appetizer, in stomachache, as vermifuge (Mal); lge. prodn. in PR; var. *acida* Brandis - **C. microcarpa** Bge. = *Fortunella japonica* Swingle (hybrid) - **C. mitis** Blco. (C. fortunella Blco.; C. fortunella M. Ingram & Moore) (eAs, PI, cult. US, Haw): calamondin (orange); lf. o.; orangequat; calamansi; kalamondin; Panama (or Chin) orange; a cultigen (hybrid); food of natives, refreshing but bitter & sour (1183); in kidney dis.; u. by alcoholics (juice & pulp. u.); bleach; in marmalades; juice u. to disinfect shallow waters (typhoid, dys., cholera); & in prepn. of beverage (tea); rt. ingred. in childbirth med. (Mal); very beautiful tree with bright orange frt. - **C. natsudaidai** Hayata (Jap, Taiwan, Ryukus): Watson pomelo; natsu-mikam (Jap); frt. eaten as "summer oranges"; for marmalade, sweet pickle; peel med. u. bitter, flavor - **C. X nobilis** Lour. (C. reticulata X C. sinensis) (Cochin Chin.): king orange; tangori; frt. esculent; peel c. VO. similar to It orange oil) (prod. PI) - **C. X paradisi** Macfad.: (C. maxima X C. sinensis); grapefruit; pomelo (not pummelo); "forbidden fruit" (Paris); much cult. US for ripe fruit (rich in vit. C) which is popular breakfast food; VO important market commodity (Fla, Calif [US]; Israel) (1357); in perfumes; peel glacéd; slows down loss of drug activity (due to naringenin, bitter princ.) (recent finding) - **C. reticulata** Blanco (several vars.) (C. nobilis var. deliciosa Swingle) (seAs, PI, Ryukus): mandarin (orange); tangerine (orange); satsuma orange; pon-kan (Jap); sweet citrus; small frts. with loose, smooth skin, sweet pulp, popular & deliciously flavorsome, keeps well; peel c. VO similar to It orange oil (prod. Fla [US], PI, Br); u. as arom. to prep. floral formulas, eau de Cologne; as flavor, etc. (origin. cult. Indochin; now cult. Eu, US, SA, Af) - **C. sinensis** (L) Osbeck (C. aurantium Risso var. sinensis L.) (prob. origin. Chin; widely natd. in trop & subtrop (incl. sFla [US]): sweet orange; most popular sp. for edible frt. peel glacéd; lvs. u. carm. (Guat); appl. locally (Mex); VO expressed fr. fresh peel (yield ca. 0.1%) (1897) (Neroli oil); u med. as flavor, arom.; in perfumes, soaps; stomach.; frt. juice popular as beverage (US); distillate of fls. & of lvs. u. as beverage in Yuc (6); frt. often artific. colored with (1) dyes, (2) ethylene exposure; tisane made fr. o. leaves u. as kidney stim. in yellow fever (New Orleans, LA [US]); frt. source of bioflavonoids (1182); (large prodn. US, Sp, Port, It, Turk, WI, Af); many vars. incl Valencia, Jaffa, Maltese or blood (var. Maltense?), Washington Navel (or Bahía), Pineapple, Lue Gim Gong (1345); (var. *brasiliensis*); (v. under "orange") - **C. taitensis** Risso (C. X limonia forma) (Tahiti): Otaheite (orange); resembles lemon & may be a hybrid; hardy; frt. lemon-shaped; lvs. crenulate, with narrowly winged petiole; greenhouse and house (pot) plant as ornam.; frt. has insipid rather nauseating taste, non-acid - **C. taiwanica** Tanaka & Shimmada; (Taiwan): spiny shrub with edible frt. like lemon; may be a hybrid (1567) - **C. X tangelo** J. Ingram & H. Moore (C. paradisi X C. reticulata) (hybrid first made in Fla [US] 1897); cult. sFla: tangelo; 8 cultivars (or more) are recognized for this sp. incl. Ugli (Jamaica) - **C. trifoliata** L. = *Poncirus t.* - **C. unshiu** Marcovitch (C. nobilis Andr. var. unshiu) (Jap, cult. Taiwan): Satsuma (orange); cultigen, juice concentration, dehydrated pulp, frt. esculent; chief Jap orange; hardiest of sweet oranges (formerly grown in sAla, sGa [US]; Cairo [Eg]) (for serials re Citrus v. Appendix C-1).

citrus (bio)flavonoids: vitamin P complex; CVP; a series of compds. (incl. rutin) found in citrus & other frts., which are claimed to maintain normal blood vessel health by decreasing capillary permeability and fragility (for instance, reducing bruising tendency); comm. extn. fr. rinds of oranges, lemons, limes, grapefruit, &c.; these act synergistically with vit. C & are claimed to have other values (2446) - **c. canker:** *Pseudomonas citri* (Bacteria); caused very serious dis. in Fla [US] orchards, ca. 1910-20.

Cium = Sium.

Civet (fr. Arabic): preputial glands fr. civet cat, *Viverra civetta* - **civetone:** active constit. of civet; has arom. & fixative props. - **Civetta** (L.): civet.

clabberspoon (seUS): *Clitoria mariana.*

cladistics: grouping of organisms and ranking them on the basis of the latest branching point in phylogeny.

Cladium mariscus (L) Pohl (C. germanicum Schrad.) (Cyper.) (Euras, Af, Austral, Am): (marsh) sawgrass; u. diarrh. & metrorrhagia; sts. u. for thatch; brooms (rather too brittle); mfr. cheap paper - **C. jamaicensis** Crantz (seUS<WI, tropAm): sawgrass; "sticky grass" (Bah); this sedge is the characteristic plant of the Fla Everglades [US]; u. fr. cheap paper; lvs. for baskets; lf. decoc. u. as lotion in chickenpox eruptions (Bah); lf. base emergency food.

Cladonia Hill (Cladoni. - Lichenes): some spp. c. antibiotics (Jap) - **C. coccifera** Willd. (Arctic & subarctic reg., also SA): herbe du feu; thallus (Herba Musci pyxidati) c. barbatic acid; u. in pertussis - **C. didyma** (Fée) Vainio (US, growing on old wd.); said to c. antibact. subst. active against the TB organism - **C. fimbriata** (L.) E. Fries (temp. zone): trumpet lichen; u. to dye woolens red; c. protocetraric acid &c.; (Herba Musci pyxidati) - **C. mitis** Sandst. (circumpolar, arctic to temp): lichen des rennes (Fr); thallus c. rangiformic acid; polysaccharides show pronounced antitumor activity (Sarcoma 180) (2304); antibiotic - **C. pyxidata** (L.) E. Fries (temp. zone): "cup moss"; inf. with honey u. for pertussis & convulsive coughs; u. as dye to give woolens ash-green color; source of "Lichen pyxidatus" - **C. rangiferina** Web. (Arctic & nTemp Zone; over extensive areas): reindeer moss; pl. serves as caribou fodder & starvation food for man; source of usnic acid (antibiotic); prodn. alc. (Scand, nRussia); in resp. dis.; for bandages - **C. tinetoria** Raf. (eAs): yellow wood; u. as yellow dye.

Cladostemon kirkii Pax & Gilg (l) (Capparid.) (eAf): u. med. by natives (Moz).

Cladrastis kentukea Rudd (C. lutea [Mx. f.] Raf.) (Leg) (s&seUS): yellow ash; wd. heavy & u. for gunstocks; source of yellow dye - **C. tinctoria** Raf. (eAs): yellow wood; wd. u. for yellow dye.

clam: bivalve shellfish (Pelecypoda-Mollusca) of salt water wh. burrows in the sand; flesh popular food, often as **c. chowder**, stew with c. as chief component.

clammy sage: clarey, common: (common) clary: clarry (sage): *Salvia sclarea* - **wild clary:** *S. verbenaca.*

Claopodium crispifolium (Hook.) Ren. & Card. (Thuidi. - Bryophyta): (nTemp Zone): has very promising antitumor activity.

Clarase™: comm. prepn. of polygalacturonase; u. to clarify fruit juices (hence "the filtration enzyme"); gives "sparkle" to jellies.

Clathrotropis brachypetala (Tul.) Kleinh. (Leg) (SA, WI): pl. decoc. u. to destroy vermin.

Clausena anisata Hook. f. (Rut.) (tropAf): bois d'anis; lvs. with anise-like odor u. to repel mosquitoes; rt. decoc. u. for convulsions in children; in impotence, pain in testis (Tanz) - **C. anisum-olens** (Blco.) Merr. (C. excavata F. -Vill.) (seAs, PI): kayumanis (Tagalog); u. for indigestion (Hainan); rt. & frt. decoc. u. in coughs with fever; lf. decoc. for nausea during pregnancy; lvs. stuffed in pillows soporific; in baths for rheum. (PI) - **C. excavata** Burm. f. (Ind, seAs): good remedy for stomach dis. (Burma); rt. decoc. sudorific (Taiwan); bitter ton. astr. emmen. (Indochin); lvs. appl. to wounds (Sabah, Borneo) - **C. lansium** (Lour.) Skeels (C. punctata Rehd. & Wils.; C. wampi Bianco): uampi (sChin, PI): cult. in Chin & Haw (US) for edible frt.; dr. frt. given in bronchitis; stom., anthelm - **C. pentaphylla** (Roxb.) DC (Ind+): bk. u. in veter. med. for wounds (pd. appl. in FO) & sprains; spasmolytic (due to clausmarin [a coumarin]).

Clavacin: Clavatin: v. Patulin.

clavalier (Fr): prickly ash.

clavatoxine: alk. of *Lycopodium clavatum.*

clavel (Sp): carnation, pink - **c. de España** (Mex): Cryptostegia.

Claviceps purpurea Tul. (Clavicept.) (Euras, NA): ergot of rye; horn seed; the dr. sclerotium found growing on rye pls. & replacing the grain; c. ergonovine, ergotoxine, ergotamine, FO (50%), ergosterol; u. emm. hemostat. abort. ecbolic (3rd stage of parturition); (prod. Por, USSR, Pol, Ge, Hung, US).

Claviceptaceae (Claviceptales - Pyrenomycetes - Ascomycotina - Fungi): parasitic on pl. & animals; have well-developed stroma or sclerotium; incl. ergots; members incl. *Claviceps, Cordyceps.*

Clavija R & P. (Theophrast. - Dicot.): ca. 60 spp. of palm-like trees (tropAm); kuandu; frts. eaten; pl. u. med. (Cuna Indians, Panama).

clavine alkaloids: one of 2 series of alk. found in ergots; simple derivs. of dimethylergoline; also found in *Ipomoea & Rivea*; thought to be precursors of lysergic acid derivs.; a series of these alks. appear in ergot cultures, ex. fumigaclavine.

clavo de especia (Sp): cloves - **C. de pozo** (Mex): *Jussiaea suffruticosa.*

clavulanic acid: antibiotic prod. by *Streptomyces clavuligenus+*: antibacterial activity, (Gm. pos. & neg.), opposing bacterial resistance; lactamase inhibitor; u. as K clavulanate & as esters.

clay: China c.: white c.: kaolin (*qv.*) - **c. eating:** consumption of earths with high clay content, popular in Ala, Miss [US], &c. among rural Negroes; also elsewhere, ex. among Hond & Nic Indian women & children; with med. merits, reducing GI irritn., supplying minerals, to keep teeth clean, &c. (1536).

clays: plastic soft-colored native earths with high Al content; represent native Al silicates, u. in pottery; as carriers for insecticides & fungicides (ex. continental clay; Veegum [Vanderbilt]; and others), u. as emulsifying, suspending & thickening agts. (*cf.* Attaclay [attapulgite]); colloidal clay (bentonite) u. in facial masks.

Claytonia L. (Portulac.) (wNA): spring beauty; genus of small, mostly succulent herbs, some with large fls. & tuberous rts.; starchy corms of some spp. eaten by Am Indians - **C. lanceolata** Pursh (wNA): pigeon rt.; Indian (wild) potato; lvs. u. antiscorb. in salads; corms eaten raw or cooked (Am Indians) - **C. perfoliata** Donn (Montia p. Howell) (wNA, natd. GB): miner's (or Indian's) lettuce; winter purslane; lvs. edible, u. like spinach & for salads by Am Indians; ornam. ann. hb. - **C. virginica** L. (e&cNA): spring beauty; musquash; starch-rich corms u. food (Am Indians); cult. ornam. hb.

clayweed: coltsfoot.

clear eyes 1) clary 2) *Salvia verbenaca.*

cleavers: cleaverwort: *Galium aparine.*

Cleghornia dongnaiensis Pierre (Apocyn.) (Indoch): bk. u. remedy in dys.

Cleidion spiciflorum Merr. (Euphorbi.) (Ind to NG): lf. decoc. abort. (PI); bk. decoc. u. stomachache; sds. u. for constipation.

Cleistanthus collinus (Roxb.) Benth. (Euphorbi) (Ind): u. astr. as immunostim., fish poison; toxic pl.; c. cleistanthin (lignan lactone) (may protect bone marrow fr. cytotoxic subst.); dr. frt. criminal poison.

Cleistocalyx: mostly referred to *Eugenia & Syzygium* - **C. operculatus** Merr. & Perry (Eugenia operculata Roxb.) (Myrt.) (Chin, seAs): yethabye (Burma); an evergreen forest tree; lvs. c. arom. VO rich in terpene esters; u. in fomentations; rts. u. in embrocations; lvs. in jaundice (Chin); frt. antipyr. ton. in rheum.; sometimes eaten; bk. & lf. decoc. u. as wash for leprosy, scabies, vermifuge; wd. u. for canes, tools, &c.; bk. u. as tan.

Cleistochlamys kirkii (Benth.) Oliv. (l) (Annon.) (eAf, incl. Moz): rt. inf. u. in hemorrhoids, rheum.; in mixt. for TB; frt. of shrub/small tree edible. (2445).

Cleistopholis patens (Benth.) Engl. (Annon.) (tropAf): ofu; st bk. u. to treat dis., such as infectious hepatitis, skin dis., &c., & as antipyr.; lvs. & rts. also med.; c. sitosterol, campesterol, stigmasterol, VO with terpinene. (2427); fiber, timber.

Clematis L. (pron. Clém-atis) (Ranuncul.) (cosmopol, most temp): "climber"; some 300 spp., most climbing shrubs - **C. angustifolia** Jacquin (Chin): wei-ling-hsien; rt. u. as antipyr., antirheum. antineuralg.; c. VO, alk. sitosterol (1184) - **C. baldwinii** T & G. (Fla [US]): pine hyacinth; hb as cold inf. u. for sunstroke (Seminole Indians) - **C. chinensis** Franch. (Chin, Vietnam): Chinese clematis; pinyin; bago; rt. u as diur. carm, in arthritis, dysm.; fever, lumbago, headache; entire pl. said remedy for cancer (anti-cancer activity demonstrated) - **C. dioica** L. (cMex-Pe; WI): virgin's bower; crespillo; lvs. u. as rubef. & to remove freckles & skin blemishes; fresh lvs. vesicant, u. by beggars to disfigure skin & evoke sympathy; rt. u. purg.; for urinary obstructions; pl. u. rheum.; similarly u. is *C. grossa* Benth. - **C. gouriana** Roxb. (cAs): lf. & stalk u. diaph. diur. (Chin); pounded lvs. appl. to open wound (PI, Taiwan); rt. in goiter & frt. ton. (Indoch) - **C. hirsuta** Guill. & Perr. (cAf): rt. decoc. in milk u. for fever (Kenya); appl. exter. for sores (incl. leprotic); rt. pulp u. as rubef., esp. as cataplasms for resp. tract (Senegal) - **C. javana** DC (PI): pounded lf. or juice appl. to open wounds - **C. ligusticifolia** Nutt. (wNA): (western) white clematis; lvs. & branches decoc. u. shampoo; lvs. rubbed in water, lather u. as soap & for sores & boils; st. & rts. u. contraceptive; children smoke the stems (BC) - **C. loureiriana** DC (C. smilacifolia Wall.) (Indoch): rt. decoc. u. for lumbago; decoc. of pl. for itch; juice vesicant, u. for horse distemper - **C. meyeniana** Walp. (C. virginiana Lour.): (Chin, seAs, PI): "virgin's bower"; lvs. u. to make decoc. for women after childbirth - **C. minor** DC = *C. chinensis* - **C. montividensis** Spr. (Arg): herba de viejo; u. extensively in rheum. (fls. & lf.) & skin dis. (Arg) - **C. papuasica** Merr. & Perry (PapNG, Solomon Is): lvs. crushed & appl. to head for fever - **C. recta** L. (C. erecta All.) (sEu): virgin's bower; hb. u. diur. in kidney dis. & blood dis. - **C. sericea** HBK (NL, Mex): barbas de chivo; c. clementine (alk.); fresh. pl. u. rubefac. caustic. (Mex Phar.) - **C. viorna** L. (eUS): leather flower; viorna; ornam. vine - **C. virginiana** L. (eNA): devil's hair; ornam. woody vine; u. in domestic med. as diur. diaph. in chronic rheum., indolent ulcer, "palsy" - **C. vitalba** L. (Euras, nAf): traveller's joy; old man's beard; climber with sts. up to 30 m long; hb. u. diur.; young shoots food; pl. c. antibiotic.

Clematopsis scabiosifolia Hutch. (Ranuncul) (Af): pl. u. decoc. for chest dis., esp. in hemoptysis.

clémentines (Fr): 1) *Clematis* spp. 2) (Sp) "Algerian tangerine"; small tangerines; somewhat like tangelos; may be hybrid of tangerine & orange.

Cleome DC (Capparid.) (mostly trop e&w Hemisph.): herbs or small shrubs; fls. of varying colors, typically with strongly projecting stamens & pistil; bee plants - **C arabica** L. (Sahara area): spider flower; agasei; sd. u. in sauces - **C. brachycarpa** Vahl ex. DC (Arab, wAf, Sudan, Ind): hb. u. by natives for abdom. pain; pl. u. scabies, rheum., inflamm.; lvs. u. in leucoderma (Ind) - **C. droserifolia** Delile (Oman): zaaf (Jibbali): sprigs u. as deodor. & perfume; pl. juice painted around sore or painful eyes; u. in migraine, &c. - **C. gigantea** L. (SA, Br, WI): mussambé; rt. & lvs. u. bronchitis & asthma; rts. & lvs. ("mussanti"); rubef. (folk med.); c. kaempferol, ferulic acid; alkaloid-like subst.; sds. c. FO with oleic & linoleic acids - **C. gynandra** L. (Af, As, warm areas, natd. in US, WI, Mex, nSA) (Gynandropsis g. Briq.): cat's whiskers; akaja (Af): sd. decoc. anthelm. (*Ascaris*); exter. rubef.; lf. sap dropped in ear for earache (also pd. sds); various pl. parts as poultice; diaph.; for boils; inf. for colds; sd. FO u. as hair oil to kill head lice, &c.; lf. u. for rheum. pyorrhea, stiff neck; neuralgia; sds. u. like mustard; pl. fish poison.; lf. put in nostril for cold; lf. rubbed on temples for headache (Tanz); entire pl. u. for scorpion sting; lf. in otitis; skin dis.; vermicide; sd. for cough, antipyr. for fatigue in old men - **C. hassleriana** Chod. (C. pungens Willd.) (Arg. to Br.): mismia (Co); centella; u. as insecticide; odoriferous agt.; for stomach weakness - **C. monophylla** L. (tropAs, Af): hurhura (Bengal); lvs. u. as vegetable (Af); fls. & shoots also eaten; pounded rt. placed on lips to restore consciousness after a faint (Santal); chewed for cough (Tanz); u. as anthelm. (EI); sds. u. as mustard substit. - **C. serrata** Jacq. (SA [Co]) u. arom.; insecticide - **C. serrulata** Pursh (c&wNA): Rocky Mountain bee plant; spider plant; guaco (Navajo Indian); food pl. (young plants); boiled pl. as dumplings with meat or tallow & kept for winter food; fl. rich in nectar, source of honey; u. in gastric dis. (NA Indians); for insect stings; rts. u. as dye (pottery decoration) - **C. spinosa** Jacq. (tropAm now pantrop.): (giant) spider flower; u. as prec.; stom. vulner. (Br); bruised lvs. appl. to head for headache (Mal); ornam. - **C. viscosa** L. (C. chelidonii L.f.) (tropOW, adventive NW, natd. Mex, WI): "wild mustard"; hurh-hurhe (Beng); rt. u. vermifuge (Mal, Pl); stim. to give appetite; pods made into pickles; inf. of whole pl. u. skin dis. (Indoch); pl. u. in rheum.; rts. sds. & lvs. u. as mustard substit. condim. carm.; lvs. u. as ton., ear dis., headache (poultice); fish poison.

Clerodendrum (Clerodendron) aculeatum (L.) Schlecht. (Verben.) (WI, Mex to Ve; Tonga, Mal, PI): glory tree; madampōlam (Surin.); wild coffee (St. Vincent); lf. in poultice (house remedy, St. Thomas [VI]); to treat GC (Brit Virgin Is "privet wood"); lf. decoc. cough remedy (also for var. *gracile* Griseb. & Mold.); u. to bind soil; cult. as hedge plant - **C. adenophysum** H. Hallier (Borneo): med. pl. of the natives (2002) - **C. aucubifolium** Hemsl. (Madag): extd. juice u. med. - **C. barba-felis** H. Hallier (Indones): badi; lvs. prepd. in coconut oil for skin dis. - **C. bethuneanum** Low (seAs, PI, Indones): "oyang"; rt. decoc. u. to treat delayed menses - **C. bracteosum** Kostel. (Ind): bitter lvs. u. to make ointment for headache, eye dis., &c.; rt. aids digestion; pleasant smelling bitter frt. u. as lax. & diur. - **C. buchanani** Walp. (seAs) (cult. in some areas): "ti(n)tinga"; crushed lvs. u. in dys.; rt. u. in snakebite; lf. paste u. for burns, wounds, tumors - **C. bucholzii** Guerke (c&wtrop Af): "fafe"; lvs. mixed with food to stim. appetite; warm crushed lvs. rubbed on body for rheum.; snakebite; lf. inf. u. for colds in chest - **C. bungei** Steud. (Chin to nInd; widely cult.): glory bower; g. tree; lf. ext. u. in ankylostomiasis, arthritis, hives, filariasis, gastritis, exter. as antiphlogistic; pl. decoc. for cough, general weakness (Chin); rt. decoc. u. int. to treat weak muscles of ankles & legs - **C. buruanum** Miq. f. **lindawianum** Bakh. (Nig): lvs. rotted & rubbed on body of children with skin dis.; lf. decoc. appl. to sore legs - **C. calamitosum** L. (Br, seAs, PI, Indones, widely cult.): hurtful clerodendrum; u. in med. for kidney stones (Java); lvs. chewed before glass chewing by performers (Java) - **C. capitatum** Schum. & Thonn. (trop c&wAf): kaka (Tagoouana); rt. purg. in stomach pain; leafy sts. u. as antipyr. (decoc.); as gargle for toothache; in orchitis, elephantiasis of scrotum; in edemas - **C. cochinchinensis** Dop. (Indochin): cay leo trang; u. in childrens' med. - **C. colebro(o)kianum** Walp. (Pak to Ind, Mal, Indochin): "pelka"; young lvs. of shrub eaten by Lepchas (Ind); rt. lf. bk. decoc. u. for mal. fevers (2228); rt. decoc. drunk for stomachache (2448) - **C. confusum** H. Hallier (Oceania): sappy part of fresh inner bk. appl. to sores; lf. decoc. appl. to sore legs; lf. mash (lvs. rotted in water) appl. to skin dis. (Bougainville) - **C. cordifolium** A. Rich. (tropAf): yakuzum (Zaire); lvs. u. for snakebite; rt. inf. for liver & spleen dis. (children) - **C. cyrtophyllum** Turcz. (Chin): Chinese glory-bower; lf. decoc. u. to treat "night sickness" (*sic*), to treat pox & itch; antipyr. diur.; to treat tonsillitis, pharyngitis, mumps, insect bites; lvs. eaten as veget. - **C. deflexum** Wall. (seAs, Thailand, Indoch, Mal): nodding witch's tongue; rt. ext. u. as antipyr. & in intest. dis. as diarrh. (Mal) - **C. dekindtii** Guerke (tropAf): lvs. u. to treat headaches (Angola) - **C. discolor** (Klotzsch) Vatke (tropAf, incl. Sudan): rt. ext. u. in blenorrhea (Zaire), snakebite antidote (Tanz) - var. *oppositifolium* Thomas (tropAf): rt. decoc. as enema for constip.; cold water inf. u. for stomachache; lvs. u. in campfire to ward off mosquitoes - **C. disparifolium** Bl. (C. acuminata Wall.) (Mal, Indones, Indochin, Thail, Greater & Lesser Sunda Is): pounded rts. appl. locally for pain; packed into caries for toothache; wd. tar u. to blacken teeth (Mal); lvs. in combns. as ton. in constip.; for snakebite (Indochin) - **C. emirnense** Bojer ex. Hook. (Madag, Maurit, cult. Eu, US): "small-flowered (Madagascar) clerodendrum"; sasandroy (Malagasy); u. in cooling herb teas (tisanes) & fumigants; in blenorrhagia; may be very toxic - **C. erectum** DeWild. (tropAf, Zaire+): u. med. (Congo) (1839) - **C. excavatum** DeWild. (tropAf, Zaire+): hollow sts. u. to make pipestems & whistles - **C. floribundum** R. Br. (Austral.): lolly bush; wood decoc. drunk by Bushmen (or Aborigines) for aches & pains (2263) - **C. fortunatum** L. (Ind, seChin, Indochin, Indones): devil's lantern; hb. u. to treat bruises (local herbalists) ("Ramie Clerodendri fortunati"); antipyr. (Chin) - **C. fragrans** (Vent.) Willd. (Chin, Mal): rts. medicine taken to strengthen leg muscles; rt. decoc. as wash in skin dis.; lvs. in wash for furnuncles; lvs. as foment. for rheum. & in childbirth - var. *pleniflorum* Schauer (WI): fragrant clerodendron; inhalation of aroma cure for headache (Dominica) - **C. glabrum** E. Mey. (e&sAf): weeping tree; lf. decoc. with milk u. as purg. & vermif. for calves (sAf); for fever, patients bend over steaming inf. of lvs. & twigs (Tanz); lvs. as cough & fever remedy & in tapeworm & roundworm (Zululand); lf. inf. in colic & appl. to wounds to inhibit maggot growth (Sotho); rt. inf. u. in snakebite & as vermif. (donkeys); insect repellant (bees, beetles); wd. excellent timber, var. *vagum* (Hiern) Mold. (cAf, esp. Angola): rts. pounded into a mash & fed as porridge in diarrh., purg. for calves - **C. heterophyllum** (Vent.) R. Br. (Maurit. Réunion; cult. Ind+): various leaved c.; c.

VO; u. as antisyph. - **C. hildebrandtii** Vatke (Eth, Somal, Tanz): rt. decoc. u. stomachache, heartburn; fish poison - **C. indicum** Ktze. (C. siphonanthus Clarke) (Ind, Indochin, Mal, natd. in warm Am, &c.; cult. Fla [US]+): bharangi; ngayanpadu (Burma); ganja (ganja); tube-flower; pounded rt. given with ginger for asthma & other pulmon. dis. (bronchial cough, fever); lf. juice u. as vermif., bitter ton.; lvs. in ghee u. exter. in skin dis. (124), asthma & other pulmon. dis., cough; also exter. in herpes, &c. (2270); lvs smoked like Cannabis & as Opium substit.; sds. & wd. pieces u. in necklaces as charm (Ind) - **C. inerme** (L) Gaertn. (pantrop): seashore tube-flower; lf. juice u. as febrif. in mal. ("garden quinine") (Ind); fumigation with lvs. in steaming hot water u. for eye dis., incl. blindness (!) (Solomon Is); rt. boiled in oil for rheum.; frts. in dys. (Java); fls. source of perfume (exquisite); lvs. u. to make poultice (PI) for swellings; in skin dis. (Hong Kong); many other uses (2294) - **C. infortunatum** L. (C. viscosum Vent.) (SriL, Thail+): pinna; bhant: lf. paste appl. in headache (124); rts. & lvs. antipyr. (Burma); lf. inf. in angina; lf. decoc. in leukorrh.; lvs. anthelm. (SriL); considerable confusion in this sp.; widely cult. - **C. intermedium** Cham. (Indones, Taiwan, PI): lvs. with coconut oil appl. as plaster for headache (PI); dr. young shoots & fls. adminst. to counteract fetal death in womb; lvs. crushed & u. in water to bathe babies who are very weak; withered lvs. u. in poultice to relieve pains following childbirth, neuralgia, colic; rts. purg. (PI) - **C. japonicum** Sweet (seAs, Chin; widely naturd.): Japanese gloryberry; lvs. appl. to boils; lf. pulp given in edema; inflorescence decoc. u. in GC, hemochezia, epistaxis; frt. edible; cult. ornam. - **C. johnstoni** Oliv. (trop eAf): shimbo; lvs. or bk. u. as expect., in dyspepsia; lf. juice anthelm. - **C. kaempferi** (Jacq.) Sieb. (Ind to sChin, Indones, PI): scarlet (or scaled) clerodendrum; rt. decoc. u. in chest dis., TB; inf. of pl. in vinegar u. in GC (Indones); lf. juice u. as lotion (India) - **C. lanuginosum** Bl. (PI, Indones, Molluccas, NG): paiton; lvs. taken for splenomegaly; scraped bk. appl. to forehead for headache (PI) - **C. ligustrinum** (Jacq.) R. B. (Mex to Pan): snake tree; boiled lvs. u. as wash for snakebite; diur. in kidney stone (Mex); u. to season fish, camote stew, &c. (Mex) - **C. lindleyi** Decaisne ex. Planch. (sChin, Ind, Burma): Lindley's clerodendrum; well dr. rt. u. internally for strengthening leg muscles (Chin); to brew a tea; frt edible (also f. *albiflorum* Mold.) - **C. macrostegium** Schau. (PI): "volcameria de flores grandes"; u. as cancer "cure"; decoc. of pl. appl. as wash or lvs. as plaster; ornam. - **C. makanjanum** H. Winkler (eAf): u. in epilepsy by natives - **C. mandarinorum** Diels (Chin, Indochin): u. to treat rheum. (native herbalists) - **C. minahassae** Teijsm. & Binn. (Indones, PI): Hugo's clerodendrum; young lvs. of lge. shrub u. as greens & to treat stomachache (Indones); widely cult. - **C. myricoides** (Hochst.) Vatke (trop cAf): butterfly bush; rt. bk. u. for fever in cattle; pd. bk. u. for snakebite; pl. u. dysuria, sterility (female), impotence, cough, rheum. (broth fr. rt.), furunculosis, splenic enlargement; var. *camporum* Guerke (tropAf): u. medicinally for sick cows (Zaire); as tinderwood (fire by drilling) (Zululand) - **C. nereifolium** Wall. = *C. inerme* - **C. paniculatum** L. (Ind, seAs): (red) pagoda flower; inf. u. as purg., for leucorrhea (Indoch); as ceremonial; elephant med.; oil as "elixir"; cult. ornam. - **C. philippinum** Schau. f. *multiplex* Mold. (PI, WI, Indones.): fragrant clerodendrum; fragrance of fls. inhaled to cure headache (*sic*) (Domin) - **C. phlomoidis** L. f. (Indo-Pak, seAs): lvs. u. in syph.; in veterin. med.; rt bitter ton. - **C. rotundifolium** Oliv. (or Baker) (trop cAf): rt. inf. u. in diarrh. (Tanz) - **C. rubellum** J. G. Baker (Malagasy): lf. decoc. given as drink to child with marasmus - **C. serratum** Spr. (seAs, esp. Ind): barangi (Hindi); u. diur., pl. decoc. for colic; rt. chewed to "clear the voice"; pl. poultice for yaws, skin dis., headache & persistent fever; rt. & lvs. analg. & ton.; lvs. & fls. eaten - **Y. yunnanense** Hu (Yunnan, Chin): tien ch'ang shan; twigs of shrub u. in malaria.

Clethra acuminata Mx. (ceUS): sweet pepperbush; white alder; inner bk. inf. u. as emet. to "eliminate bile," in fever (with *Prunus virginiana* bk.) (neUS Indians) - **C. alnifolia** L. (C. tomentosa Lam.) (seUS): sweet pepperbush; summer sweet; rt. decoc. u. as foot bath for "ground itch" (SCar [US]) (2451); lvs. & fls. diaph. stim. antipyr. for pulmon. dis.; lf. u. for inflamn. & swellings (2425); ornam. shrub.

Clibadium asperum (Aubl.) DC (Comp.) (Co): guaco; cult. & u. as fish-poison - **C. sylvestre** Baillon (WI, Ve to Br): kunámi; bundles of lvs. pounded & tossed into river pools to kill fish (fish actually leap fr. water & are captured) (Domin, Caribs); other spp. u. similarly; u. ext.—> temp. heart arrest.

Clidemia setosa Gleason (Melastomat.) (sMex to Nic): "hembra"; decoc. ("potage") u. for sterile females (Guat).

clifers: *Galium aparine* - **cliff weed:** *Heuchera villosa.*

Cliffortia ilicifolia L. (Ros.) (sAf): pl. u. in preps. as emoll. & expect. in cough - **C. odorata** L. f.

(sAf): strong tea u. for colds, hemorrhoids, sore throat, croup, excessive menses - **C. ruscifolia** L. (Af): climber's friend; grazed by livestock.

Cliftonia monophylla (Lam.) Britt. & Sarg. (1) (Cyrill.) (seUS but not Fla penin.): titi; buckwheat brush; shrub or small tree; fls. important honey source.

climacterium: climacteric: menopause.

climax (ecology): the nearly stable population of pls. representing the final outcome fr. the evolution of the vegetative life of an area (climactic: adj.).

climbing entada: *Entada phaseoloides* - **c. staff tree:** *Celastrus scandens.*

clin, yerba (Sp, CanaryIs): *Ajuga iva*; u. in teas.

Clindamycin: semi-synth. deriv. of Lincomycin; antibiotic (Cleocin); u. in acne, bromidrosis.

cline (ecology): a local group or colony of organisms within a population of a single sp. or other taxon wh. shows gradations or intermediate stages of a character because of geographic or environmental factors (often with only gradual changes).

Clinopodium L. (Clinopodion) (Lab.) (nTemp Zone): savory; closely related to *Calamintha* & *Satureia*; herbaceous plants - **C. calamintha** O. Ktze. (Med reg., Atl, Eu, natd. NA): calamint; mountain mint; calament(a); nepetella; hb c. VO with pulegone, menthone, isomenthone; hb. c. caffeic acid & its depside; u. as stim.; cult. in gardens - **C. hortense** O. Ktze. = *Satureia h.* - **C. laevigatum** Standl. (Mex): lvs. u. with sugar as popular tea (wcoast of Mex) - **C. macrostemum** Ktze. (Mex): hierba del borracho; tabaquillo grande; té del monte; té nurite; lvs. u. to combat GI dis., as tea, for gastric pain; bile dis.; stomachic, aid digestion - **C. montanum** Ktze. = Satureia m. - **C. oaxacanum** (Fernald) Standl. (Mex): té del indio; u. as stom., similar to C. macrostemum - **C. repens** Benth. = *Satureja umbrosa* - **C. vulgare** L. (Satureia vulgaris Fritsch): (sEu; also sAf, As, NA): wild basil (herb); Wirbeldose (Ge); hb. u. arom., stim., diaph., carm.; substit. for Chinese tea.

Clintonia borealis (Ait.) Raf. (Lili) (neUS, eCana): corn lily; dragoness plant; "bilberry"; very young lvs. u. as potherb (Maine [US]), salad; similarly - **C. umbellulata** Morong (eUS); cult. ornam.

Clionasterol: α-sitosterol.

Clitocybe (Agaric. - Hymenomycetes - Basidiomycotina): most of the ca. 270 spp. edible, but a few are toxic, maybe hallucinogenic - **C. fragrans** (Fr.) Kummer (Eu, NA): edible fruiting body with strong flavor of anise; antibiotic activity (due to HCN) - **C. gigantea** Fr. = *Leucopaxillus giganteus* - **C. nebularis** Kummer (Eu, wNA): graycap; edible; c. nebularin (antibiotic); said to be tuberculostatic.

Clitoria L. (Leg) (tropics, esp. Am): butterfly pea; some 70 spp.; keel of fl. small, incurved. (SC fr. Gr. kleitoris, fr. similar appearance of keel to human female clitoris or fr. resemblance of form of fl.) - **C. arborescens** Ait. (pantrop): c. traces of curare-like alk. - **C. macrophylla** Wall. ex Benth. var. *laureola* Gagnep. (Indochin, Thail): rhiz. decoc. u. as ton. (Cambodia) - **C. hanceana** Hemsl. var. *laureola* Gagnep. (Indochin, Thail): rhiz. u. to prep. a tonic decoc. (Camb) - **C. macrophylla** Wall. ex. Benth. (Indochin, Thail, Burma): rhiz. decoc. u. as ton. (Camb) - **C. ternatea** L. (sAs, tropAf, native to NA): blue clitoria; b. pea; butterfly p.; Kordofan p.; rts. & lvs. u. lax. diur. emetic, antiperiod. (Ind); for infant diarrh. (Laos); bluish fls. used to color foods & drinks; hb. u. in skin dis., gout; sds. purg. emmen. but edible; cult ornam. climber.

clivers (Eng): cleavers.

Clivia miniata (Lindl.) Bosse (Amaryllid.) (sAf): St. John's (or Kafir) lily; c. lycorine, clivinine & other alks. with antiviral activity; rt. u. in snakebite (Zulus), in fever & to facilitate childbirth where onset retarded; ornam. cult. for brilliant fls.

Clomocycline sodium (also **c. calcium):** antibiotic; deriv. of tetracycline; claimed less deposited in bones & teeth.

clon(e): a group of pls. or animals propagated only by non-sexual means fr. an original stock, eventually fr. a single inividual; and not grown fr. sd.

Clostridium bifermentans Bergey & al. (Bacillacease-Schizomycetes): one bacterial sp. widely found in nature; causing gas gangrene; other causative spp. represented in Pentavalent Gas Gangrene Antitoxin are: **C. botulinum** Møller & Scheibel: anerobic organisms, the toxin of which causes botulism, an intoxication often fatal; mostly derived fr. home canned vegetables (peppers, string beans, beets), smoked fish, sausage; "ext." u. in essential blepharospasm (type A); **C. histolyticum** Bergey & al. (feces, soil); **C. oedematiensis** Bergey & al. (C. novyi Bergey & al.) (manured soils); **C. perfringens** Hauduroy & al. (feces, sewage & soil); **C. septicum** Ford (manured soils; animal intestines); **C. tetani** Holland: soils, feces, &c.: chief causative organism of tetanus; does injury through an exotoxin.

clot bur(r): *Xanthium*, Lappa - **little c. b.:** *Xanthium strumarium* - **spiny c. b.:** *X. spinosum.*

clou aromatique (Fr): **clou de girofle**: 1) clove (Eng): fl. bud of *Syzygium aromaticum* 2) garden **c.:** *Dianthus caryophyllus.*

cloudberry (bush): "yellow berries" (Cana): *Rubus chamaemorus, R. flagellaris* (rt. & rt. bk.); *cf* raspberries; grown in peat bogs (NA).

clove bark: bk. of 1) *Pimenta racemosa* 2) *Syzygium aromaticum* 3) *Cinnamomum culilawan* 4) *Dicypellium caryophyllatum* - **blown c.:** prod. of c. fl. which has opened & dropped its petals; much inferior to bud - **Brazilian c. bark: c. cassia (bark): c. cassia of Brazil: c. cinnamon:** fr. *Dicypellium caryophyllatum* - **c. bud:** c. - **c. cigarettes:** tobacco & clove mixt. made in Java (fr. Zanzibar c.); very popular in seAs - **c. fruit:** nearly ripe frt. of *Syzygium aromaticum* (detected by absence of starch in ripe); considered over-mature & inferior cloves - **c. garlic:** *Allium sativum* - **c. heads:** *Pimenta dioica* - **c. herb:** *Geum urbanum* - **mother of c.:** c. fruit - **c. nuts:** *Ravensara aromatica* - **c. oil:** VO fr. true c. - **c. pepper:** c. heads - **c. phenol:** eugenol - **c. pink(s):** *Dianthus caryophyllus* - **red c., flowers of:** c. pink - **c. root:** *Geum urbanum* - **c. stems:** peduncles of c. flower buds; u. as adulterant of c. - **synthetic c. oil:** eugenol - **c. tea:** concd. c. infusion (20%) (BPC) (cloves macerated in 25% alc.) - **c. tree (, aromatic**): *Syzygium aromaticum* - **water c. root:** *Geum rivale* - **wild c.:** *Eugenia acris, Pimenta acris* - **wild c. (tree):** *Pimenta racemosa* - **c. wood:** *C. cassia.*

clover (or **clover tops**): *Trifolium* spp. - **Alsatian c.: Alsike c.:** *T. hybridum* - **bear c.:** *Chaebatia foliolosa* - **c. belt:** Black Belt of Ala (US) - **bitter c.:** *Sabbatia angularis* - **Bokhara c.:** *Melilotus alba* - **buffalo c.:** *T. reflexum* - **bur c.:** *Medicago arabica. M. minima* - **bush c.:** *Lespedeza* spp. - **button c.**: *Medicago scutellata* - **Chilean c.:** 1) *T. pratense* 2) alfalfa - **cow c.:** *T. pratense* - **crimson c.:** *T. incarnatum* - **Dutch c.:** *T. repens* - **Egyptian c.:** *T alexandrinum* - **heart c.:** *Medicago arabica* - **hop c.:** 1) *T. agrarium* 2) *Medicago lupulina* - **Hubam c.:** *Melilotus alba* var. - **Japanese c.** *Lespedeza striata* - **King's c.:** *Melilotus leucantha* - **lappacea c.:** *T. lappaceum* - **low c.:** *T. procumbens* - **meadow c.:** cow c. - **Mexican c.:** *Richardia scabra* - **Negro c.** (Miss [US]): *Cassia fasciculata* - **old field c.:** *T. arvense* - **prairie c.:** *Petalostemum purpureum* - **purple c.:** cow c. - **rabbit foot c.:** *T. arvense* - **red c.:** cow c. - **smaller hop c.:** low c. - **spoiled sweet c.:** fermented *Melilotus* sp. herbage; formerly comm. article as anticoagulant (US-Wisc drug stores) - **stone c.:** red c. - **strawberry c.:** *Trifolium fragiferum* - **subterranean c.:** *T. subterraneum* - **sweet c.:** *Melilotus* spp., esp. *M. officinalis* - **c. tops:** cow c. - **c. tree** (NSW): *Goodia lotifolia* - **white (Dutch) c.:** Dutch c. - **white sweet c.:** *Melilotus* alba - **winter c.:** *Mitchella repens* - **Wood's c.:** *Dalea alopecuroides* - **yellow c.:** *Trifolium aureum* - **yellow sweet c.:** *Melilotus officinalis.*

clown heal herb: *Stachys aspera; S. palustris.*

club moss: clubfoot m.: *Lycopodium clavatum* Jacq.

Clusia columnaris Engl. (Gutt.) (Br, Ve, Co): "copey"; yields milky latex & resin; resin of bk. appl. to sprains & cuts (Co) - **C. flava** Jacq. (Mex, Jam): chunup (Mex); monkey apple; lvs. u. syph., for protruding navel (Mayas); antirheum; st. source of hog gum (latex) u. to heal wounds; yellow latex on cotton appl. to dental cavities to ease pain; also u. to adulterate chicle - **C. odorata** Seem. (Oceania, tropAm): copey; fls. u. expect.; latex u. in ointments for ulcers & skin inflamm.; purg. - **C. rosea** Jacq. (WI, Mex, SA, HawIs): copey; cupey (vernac.); lvs. u. in bath, for chest dis.; fls u. in pectoral tea (CR); frt. edible, exuding a "gum" u. as resolvent in treating fractures, dislocations, incipient hernias; bk. astr. in diarrh.; exudes latex (gum-resin) u. as cataplasm in pain (Trin), drastic, vulnerary; wd. useful & widely u.; thick, smooth lvs. u. as writing paper (sPan) & for playing cards - **C. utilis** S. F. Blakc (Guat): quiebramuela; yellow frt., juice u. for toothache (appl. on cotton); frt. u. for permanent marking of clothing (1537) - **Clusiaceae** Lindl. = Guttiferae.

Clutia (Cluytia) abyssinica Spach (Euphorbi) (eAf): rt. u. as ascarifuge, to prevent abortion; splenomegaly, influenza remedy - **C. hirsuta** Muell. Arg. (Af): lf. u. in combns. for anthrax, antipyr., ton. (alk.?) - **C. similis** Muell.-Arg. (sAf): pl. c. cluytianol (sterol), chrysophanol; u. anthrax antidote; rt. u. in snakebite; lvs. in indigestion, diarrh.

Clypea Bl. = *Stephania.*

Cneorum tricoccum L. (Cneor.) (Med reg.): spurge olive; all parts of shrub strongly purg. rubefac.; dr. branches u. fuel (in stoves, Narbonne, France).

Cnestis ferruginea Vahl ex. DC (Connar.) (wAf): bk. u. in general treatment of leprosy (2549); pd. bk. mixed with gum & appl. to swelling of gums in pyorrhea; rts. purg., used in dys. syph. GC, migraine, to prevent abortion; lvs. u. as abortif.; frt. decoc. in bronchitis, TB; toxic.

Cnicus L. (monotypic) (Comp.) (Eu): blessed thistle; much confusion with *Carduus* & *Cirsium* in past - **C. arvensis** Hoffm. = *Cirsium arvense* - **C. benedictus** L. (Eu, now worldwide): blessed thistle (SC supposed healing value); hb. c. cnicin (u. in gout), ceryl alc.; tannin, VO; u. bitter ton. stom. diaph. in menstrual dis.; antipyr. gastric ulcers, indig., liver & lung dis.; claimed useful in diabetes; cult. for use in Wine of Cardui (Ga) (US) (1538); frt. FO source (chief prodn. Eu, US, SA [cArg], sAf) - **C. eriophorus** L. = *Cirsium eriophorum* - **C. japonicus** Maxim. = *Cirsium japonicum* DC - **C. mexicanus** Hemsl. (sMex): u. in colics.

Cnidium dubium Thell. (Selinum venosum Prantl) (Umb.) (Eu, wAs): frts. c. cnidin (with spasmolytic activity), cnidicin, & other coumarins; u. as spasmolytic - **C. monnieri** Cusson (Manchuria to Indoch; cult. Jap): frt. u. ton. diur. emmen. resolvent; antirheum. sed.; aphrod; sd. decoc. u. prolapse of rectum, scabies, pruritus, pustules; insecticide (Chin); VO fr. sd. c. pinene, borneol (& esters) camphene; u. antiviral (flu) - **C. officinale** Mak. (Jap, Kor. [cult]); rhiz, c. ligustlide u. female dis., headache; rhiz. oil u. nerve & circ. stim.; rhiz. decoc. roborant, sed., anodyne, resolvent of blood clots; anemia (—> *Selinum*).

"cnidium root": *Peucedanum officinale.*

Cnidoscolus chayamansa McVaugh (Euphorbi.) (sFla, sMex, WI): chaya(col); lf. decoc. u. as lax. diur.; lvs. cooked & eaten as spinach (Fla+) - **C. elasticus** Lundell (wcMex, near coast): chilte; rubber frt.(?); latex obt. by tapping trees u. as elastomer (rubberoid) ("highland chilte rubber") - **C. multilobus** (Pax) I. M. Johnston (Mex, Guat): fl. u. in toothache; rt. in urin. dis. VD sores, decoc. in contact dermatitis; burns; sap as contraceptive (Mayans, sMex); in excessive menstr. (2594); fls. fried with eggs as food (Mex) - **C. palmeri** (S. Wats.) Rose (Mex): mala mujer; bush foliage u. as fish poison; latex prod. rash; roasted tubers eaten. - **C. stimulosus** (Mx.) A. Gray (Jatropha s. Mx.); J. urens L. var s. Muell. Arg.; Bivonia stimulosa Raf.) (seUS, n to Va; Mex): tread softly, pinpine; spurge (or bull) nettle; rts. u. in VD, psoriasis (1539); stomach ulcers, &c. - **C. texanus** Small (Okla, Ark, Tex [US]): bull nettle; rt. c 22% starch, 12% sugars; little or no alk. or glycosides; u. folk med. - **C. urens** Arthur = *Jatropha curcas.*

coacervate: stable liquid prod. formed by mixing 2 hydrophilic colloids of opposite reaction; may form a separate phase; represents intermediate stage between sol. & gel.

coacervation: coagulation.

coakum: *Phytolacca americana.*

coal: Lithanthrax (L.): the fossilized carbonized remains of pls. of a remote geological past; chiefly represent tree ferns, giant horsetails, tree lycopods, etc.; prod. of carbon chiefly with high energy value & source of coal tar; 3 chief types: anthracite (*qv.*); bituminous (*qv.*); lignite - **bituminous c.:** soft c.: burns with a sooty flame & much smoke, hence hazardous to environment; use sd. for heating, prohibited in some areas (Pittsburgh, Pa [US]); product forms over geological time from lignite; u. mostly for prodn. electricity, coke (residue), coal gas, coal tar (prime source of large no. of org. compds.) - **mineral c.:** pit c. - **c. oil:** 1) petroleum or a refined oil made from it 2) (c&sUS, wCana): kerosene; higher boiling point fraction in distn. of petroleum; u. solvent, vehicle; ex. in pyrethrum sprays - **pit c.:** mined c. as distinct fr. charcoal - **c. tar:** Pix Carbonis (L.): prod. obt. in the dry distn. of bituminous coal; u. source of benzene, toluene, naphthalene, etc., thus ultimate source of thousands of synthetic dyes, medicines, explosives & plastics; u. exter. for chronic dermatoses (mostly for psoriasis; often with X-rays) (E) soln. & ointment form. off. - **C. t. creosote:** phenol - **C. t. naphtha:** benzene - **C. t. pitch:** residue in still after distn. coal tar; u. in roofing, waterproofing, wd. preservation, &c.

cobalamin: generic term for vit. B_{12} group, ex. cyanocobalamin (SC c. Co) (c. E - as hydroxy c.).

cobalt: Co.; widely spread in nature, incl. ores such as cobaltite; element essential to life; aids erythropoiesis (mfr. of blood constits.); some labeled compds. u. in med. in place of Radium.

cobbler's pegs: *Erigeron canadensis.*

Cobione™: xall. vit. B_{12}.

cobnut: *Corylus avellana* (v. *grandis*).

cobra: *Naja naja* L. (& other *N.* spp.) (Elepidae - Ophidia - Reptilia): (eAs, Af): spectacled serpent; hooded snake; highly venomous snake (-1.8 m); venom with c. venom factor (CVF) u. in rheumatoid arthritis (acts on immune system); in heart attacks to reduce size of necrotic area in heart; cobratoxin (chief toxic protein; polypeptide with 62 amino acid residues in chain) (neurotoxin); stabilized inj. c. 1,500 mouse units/ml. u. analgesic for severe pains, neuritides, neuralgias (esp. trigeminal neuralgia), tabetic crisis, pain fr. malignant inoperable tumors, migraine, myocarditis, &c., (Nyloxin); gallbladder of snake u. in folk med.

cob web: v. spider web.

Coca: *Erythroxylon* spp., esp. *E. coca* & *E. truxillense*; source of **c. leaves,** which c. cocaine; small bushy pls. (c. shrub) - **Bolivian c.:** *E. coca*; said to have highest content of cocaine - **C. Cola™:** most popular carbonated non-alc. beverage; formula said developed at Columbus, Ga [US] by Pemberton ca. 1885; c. decocainized coca lf. ext., cola nut (ext.), caffeine (16 mg %), caramel (color), VO of lime, coriander, neroli, sweet or bitter orange, nutmeg, cinnamon; ginger essence; lime juice, glycerin, phosphoric acid (2.8%), sugar (12%), CO_2, water (2277); stimulating beverage; orig. sold as medicinal - **c. family:** Erythroxylaceae - **Huanuco c.:** a type of Bolivian c. - **Java c.:** *E. novo-*

granatense - **Peruvian c.:** *E. coca* - **Truxillo c.:** *E. truxillense.*

cocabutong: addictive beverage composed of absinthe fortified with coca.

Cocaina (L): **cocaine:** the characteristic alk. obt. fr. the fermentation & extn. of coca lvs.; u. as stim. of mental faculties (1815), "recreational" drug u by sniffing (snuff), by inj. (SC, IM, IV); (IV esp. dangerous) - **"crack" c.**: one of the most commonly abused drugs; C solns u. in topical anesthesia (oldest. in use); efficient for mucosa; rubbed on penis —> erection (*sic*); u. for nosebleed; most imports fr. Co, but chief prodn. of lf. & crude cocaine are Pe (60%), & Bo (30%), Co (10%) - **C. plant: c. tree:** *Erythroxylon coca*, etc.

cocales (Sp): coca plantations.

cocarboxylase: thiamine pyrophosphate ester chloride; represents the coenzyme (prosthetic gp.) of yeast enzyme, carboxylase (protein + apocarboxylase + c.); u. in acidosis (as in diabetic coma), toxemias, herpes zoster (2221), &c.

cocash root: *Aster puniceus.*

Coccidae: Cochineal Fam.; coccids; part of Order Homoptera of Insecta; scale insects; "tortoise scales" (hard scales); spp. have one pair of wings; female without wings or mouth, produces viviparous larvae; many spp. protected by a roofing of scales they produce; ex. shellac insects; some secrete wax, honey dew (some = economic prods.).

coccidiodomycosis: San Joaquin Valley fever; resp. dis. (chronic suppurative lesions) prod. by inhaln. of spores of *Coccidioides immitis* Rixf. & Gilcher. (fungus, Hyphomycetes) - **coccidioidin:** an antigen u. for skin testing.

Coccinella (Coccinellidae - Coleoptera): lady bird; l. bug; l. beetles; useful insect as it devours aphids (plant lice) (*C. septempunctata*).

Coccionella (L.) (Eu Phar.): **Coccionelle** (Fr): cochineal.

Coccinia cordifolia Cogn. (C. indica W & A; Cephalandra i. Naud.) (Cucurbit.) (eAs, esp. Ind): pakkep (Siamese); pl. u. for scabies (Indochin); fresh rt. u. for ointment for back pain fr. fever; hb. (Herba Cephalandrae) juice u. lax antidiabetic; young shoots & frts. eaten (1185) - **C. grandis** (L) J. Voigt (Tanz): lf. decoc. u. for delayed menses; lvs. eaten as vegetable; frt. eaten.

Coccognidii Semen (L): mezereon seed.

Coccoloba (Coccolobis) cerifera Schw. (Polygon.) (Minas Gerais, Br): waxy coating on lvs. may be u. in industry - **C. diversifolia** Jacq. (C. floridana Meissn.) (sFla [US], WI): pigeon (or dove) plum; (SC frts. eaten by birds); tie tongue; plum bush (Bah); mixt. rts. with other herbs u. for backache, strain fr. lifting, bedwetting, &c.; bk. chewed for gastralgia - **C. schiedeana** Lindau (Mex, Guat): palo de carnero: bk. u. as mouthwash in gingivitis (79); frt. edible; wd. useful - **C. urens** Gilg (wAf): sds. gently boiled in sugar soln. u. as med. for stomach; frt. burned & ashes u. in toothache (Zaire) - **C. uvifera** (L.) L. (sFla [US], WI, Mex, SA): (common) sea grape; seaside grape; g. tree; Jamaica Kino; pl. (esp. frt. & bk.) u. astr. ton. in diarrh. dys. VD; bk. source of West Indian (or Jamaica) kino; a tannin-rich drug; "gurn" u. as varnish on wooden objects; edible frts. u. to make jelly; beverage; ornam. cult. pl. (edible frt. also in other *C.* spp.).

Coccothrinax Sargent (Palm) (WI): silver palms; (yura) guano; some spp. cult. as ornam. - **S. argentata** (Jacq.) Bailey (sFla & Keys [US]; Bahamas, Cu): Florida silver-palm, Biscayne palm; silvertop (Bah); the only *C.* sp. occ. in the US; frts. eaten; terminal buds u. like cabbage; tough young lvs. u. to make baskets, hats, brooms, &c.; bee pl. (2544) - **C. barbadensis** Becc. (C. martinicaensis Becc.) (Lesser Antilles. Domin, Mart.): thatch palm; latanier (Domin); wd. u. to make bows (Caribs); lvs. u. to cover huts & shelters; for bags & sacks; similar uses to *C. argentata.*

cocculin: picrotoxin.

Cocculus carolinus (L) DC (Menisperm.) (se&cUS to Kansas): coral vine; Carolina moonseed; coral beads; weedy vine sometimes (rarely) becoming woody; u. ton. diur.; sd. u. fish poison; berries u. decorative; c. benzylisoquinoline deriv. alks.; cult. orn. - **C. diversifolius** DC (Mex, wTex, sAriz [US]: similar to prec. - **C. laurifolius** DC (Ind, Pak, Jap): c. coclaurine (& other tert. bases), laurifoline (quaternary base); lvs. c. diosmin; u. healing agt. in folk med.; ornament. shrub - **C. leaeba** DC = *C. pendulus* Diels - **C. ovalifolius** DC = *C. sarmentosus* - **C. pendulus** (Forst.) Diels (C. leaeba DC) (nAf, Arab, Ind, Pak): rt. of vine u. as diur., antipyr.; ton (Ind); frts. fermented —> "khumr vol majnoon" (beverage) (Arab countries); c. alks.: oxyacanthine, palmatine; cocculine (hypotensive; said active against carcinoma of nasopharynx), in intermittent fevers (mal.); cholagog; in serious dis. (Senegal); viral hepatitis; icterus; "biliousness"; a daily dose of decoc. in the morning on an empty stomach supposed to increase longevity (old fetishists, Senegal) (with *Tinospora bakis*); in leprosy, syph.; Newcastle dis. virus; invigorating, aphrod.; lvs. as vermif.- **C. sarmentosus** Diels (C. trilobus DC) (Malaysia, Chin, Taiwan, Jap): "fang-chi" (Chin); evergreen shrub; sts. (tea) diur., rts. c. alks. (menisine, &c.), decoc. u. diur., antisp. in epilepsy (Chinese); rts. u. as poison; in insecticides; alks. with neuromuscular blocking activity, hypotensive - **C. villosus** DC (C. hirsutus [L] Diels) (Ind, Af, [Transvaal - mts]): spasmolytic in chronic rheum., fevers, dig. dis; in making baskets (eAf).

Cocculus Indicus (L.): fish berry, *Anamirta cocculus.*

coccus: 1) (L) cochineal (insect): Dactylopius coccus 2) type of bacterium of spherical shape; formerly u as generic name, now invalid (SC kokkos, Gr, kernel, grain).

Coccus cacti L. = Dactylopius c. Costa - **C. ceriferus** Fabr. & **C. pe-lu** Westwood (Coccidae - Homptera - Insecta) (Ch, Jap): occ. on *Ligustrum lucidum & Fraxinus chinensis*; secretes Chinese wax (Cera chinensis), considerably u. in plasters, for candles (Chin); waxing of paper & cotton; c. ceryl cerotate - **C. lacca** Kerr = *laccifera lacca* Kerr - **C. ilicis** = *kermes vermilio* - **C. manniparus** Ehrenberg = *gossyparia mannifera* (Hardw.).

cochenille (Fr): **cochineal:** Coccus.

Cochin (China): Incochina, esp. South Vietnam.

cochineal: *Dactylopius coccus*, "Coccus" - **c. cactus: c. fig:** *Opuntia cochenillifera.*

cochitzapotl (Mex): *Casimiroa edulis.*

Cochlearia (L): Coclearia (Rom.): *Cochlearia officinalis.*

Cochlearia armoracia L. = *Armoracia rusticana* Gaertn, Mey. & Schreb. - **C. officinalis** L. (Cruc.) (nBoreal reg., nEuras): (common) scurvy grass; s. weed; spoonwort; flg. hb. (as Herba Cochleariae) lvs. & sds. c. VO, vit. C; u. antiscorb.; juice also u.; hb. sometimes consumed as salad (with watercress) (cult. Fr, Ge, Switz, Hung, Swed). (2719)

Cochlospermum gossypium DC (C. religiosum Alston) (Cochlosperm./Bix.) (Ind): yellow (or white) silk cotton (tree); frt. source of kapok-like fiber; one source of Sterculia (karaya) gum (1540); cult. orn. - **C. vitifolium** (Willd.) Spreng. (Mex to nSA): "wild cotton" (Belize); nunununa (Cuni Indians); "pumpoflor"; u. kidney & liver dis., to hasten childbirth (Guat); fls. u. med. (Cuna Indians); stamens u. to adulterate true saffron; bk. of tree u. for asthma & kidney dis. (Guat).

cocillana bark (Bo): guapi bk. fr. *Guarea rusbyi.*

cocinic acid: cocinic ethers: names formerly appl. to mixt. of lauric & palmitic acids fr. coconut oil.

cocinin: impure laurin (glyceride) of coconut oil; (chief component, ca. 40% of oil).

cockle: 1) term appl. to spp. of at least 6 genera of plants, esp. *Agrostemma & Lychnis* 2) edible bivalve mollusc with heart-shaped shell - **corn c.:** *Agrostemma githago.*

cocklebur: *Xanthium* spp. esp. *X. strumarium.*

cockles: *Cocculus indicus.*

cockroach, American: *Periplaneta americana (Orthoptera)* - **Oriental c.:** *Blatta orientalis* - **c. plant:** herba de cucaracha (Sp); *Halophyton cimicidum, H. crooksii.*

cockscomb: coxcomb - **cock-up-hat:** *Stillingia sylvatica.*

cocksfoot: *Dactylis glomerata.*

coclaurine: benzyl isoquinoline alk. found in *Cocculus laurifolius* (Menisperm.), *Persea japonica* (Laur.); & several other plants with curare activity.

coco (Sp, Por): 1) cocoanut palm; coconut, *Cocos n.* 2) (Por slang): cocaine - **c. de Bahía** (Por): coconut - **c. de mer:** *Lodoicea maldivica* - **c. fiber:** fiber of frt. of coconut - **c. grass** (Ark & other sUS states): *Cyperus rotundus* (SC tuber has 3 "eyes" like coconut endocarp) - **C. mama** (Hon): *Quararibea fieldi* - **c. yam:** *Xanthosoma* spp., &c.

cocoa (bean): 1) ripe sd. of *Theobroma cacao* after curing, roasting & powdering 2) **breakfast c.:** prepared c. - **c. butter:** FO of c. - **c. cake:** a refined form of c. expeller cake (residue after FO expressed fr. c. nibs) - **c. fiber: c. grass:** coco g. - **c. nibs:** embryo of sd. of c.; the part processed to make chocolate - **c. seeds:** after removal fr. shell, these roasted before separation of the nibs - **prepared c.:** cocoa, a pd. prod. representing a largely defatted milled c. nib, marketed as beverage & flavor - **c. shells:** represent dr. inner pericarp, spermoderm, perisperm.

cocoanut: coconut.

cocobolo: 1) (Pe): Guaiacum 2) (Nic): *Dalbergia* spp.

cocomero (It): watermelon.

cocona (SA-Andes): frts. of *Solanum hyporhodium & S. sessiliflorum.*

coconut: *Cocos nucifera* - **c. butter:** c. oil (SC because solid in cool weather) - **c. milk:** liquid remaining within (green) sd., gradually disappears; popular beverage (Ind+) (delicious, cooling); has been suggested for use in emergency when pure glucose IV solns. not available (pyrogen-free; c. 3-6% fructose & glucose; non-hemolytic); u. in tissue cultures & to preserve bull sperm (2779) - **c. oil:** FO of c. meat; u. cooking oil, mfr. soap, margarine; hair dressing (Ind) - **c. shells:** c. endocarp - **shredded c.:** desiccated c.: dr. meat of seed (mostly endosperm) cut into shreds; popular in baking (cakes), candy, frosting.

Cocos L. (Palm.): monotyic genus; previously thought made up of several spp. - **C. coronata** Mart. = *Syagrus coronata* - **C. nucifera** L. (prob. origin Af coast; now distr. in all warm ocean areas): coco(a)nut, the "tree of life" (SC many uses & essentiality); the common palm along trop sea coasts; said the most extensively utilized tree in the world; trunk of coconut palm furnishes material for house building; wd. ("porcupine wood") for cabinet work; young lvs. food; mature lvs. for thatching, mats, baskets; fl. spikes incised give juice, a source of palm sugar, toddy (palm wine), dist. to give arrack ("tuba"), vinegar (fr. wine); frt. (coconut) furnishes milk (latex); said highly nutritious, medl. (anthelm.); st. lax. husk of frt. fibrous, u. to prep. coir (*qv.*), u. in mfr. brushes, door mats, floor mattings, mattresses, pillows, cushions, wall board; source of charcoal for gas mask filters; oiled coco fiber u. to make ropes, cordage (As); coir dust c. 8-9% tannins (u. tan); shell u. to make ladles & other household articles (1266); sd. kernel (meat) u. fresh & desicc. (shredded coconut), fermented to give copra (*qv.*); expr. FO popular in shampoos & soaps; to administer medication orally; bitter subst. (WWI) but not popular; chief producers: PI, Oceania - **C. plumosa** Hook. = *Syagrus romanzoffiana* Glassman (queen palm) - **C. yatay** Mart. (Arg) = *Syagrus y.*

cocos preparados (Mx): beverage c. coconut milk, lime juice, sugar, ice.

cocostearic acid: old name for solid fatty acid mixt. (laurin, stearin) in coconut fat; actually not a specific acid - **cocostearin:** glycerides = "Nucoa butter."

cocotero (Sp): coconut tree.

coco-yam: *Colocasia esculenta*, *Xanthosoma spp.* +

cocuite (Mex): Jamaica dogwood.

cocyl: cocin.

cod: 1) c. fish 2) (O Eng, also sUS): testis; also (dial) scrotum - **c. fish:** *Gadus morrhua*, etc. - **c. liver oil:** FO expressed fr. liver of *Gadus morrhua, Pollachius virens* L., &c.; rich in vitamins A, D, E; u. as nutrient, restorative (esp. for children); in rickets; pulmonary TB, as vulnerary, in burns, sores, tissue injury; large doses in rhinopharyngitis chronica sicca (chronic inflamm. of nose & pharynx where tissues dry), installed into nose often with vit. A. palmitate (chief ester in FO; Arovit); reduces blood cholesterol levels (3574); prod. Mass (US), Scand. - **non-destearinated c. l. oil:** crude form of prod. as imported fr. which the destearinated or refined is made by chilling to ppt. the stearin; crude is u. for poultry feeding - **c. oil:** 1) body FO of cod (not liver FO) 2) codliver oil fr. deteriorated or low-grade livers.

cod, black: *Pollachius virens* (also *Anopoplonia fimbria*) - **gray c.:** Pacific c. - **ling c. (lingcod)**: *Lota lota* (also called burbot, fresh water ling, cusk) - **Pacific c.** *Gadus macrocephalus* (popular fresh, dried) - **red c.:** *Physiculus bachus* (NZ) - **rock c.:** *Eleginops maclovinus* (Atl, Pac, SA) (Scorpaenidae).

codamine: xall. alk. of opium; isomeric with laudanine & laudanidine.

codeine: an alk., methyl morphine, extr. fr. crude opium or (mostly) synthesized by methylating morphine; u. analg. exp. cough sed. in colds, bronchitis, asthma, narcolepsy (E) (1984).

Codiaeum luzonicum Merr. (Euphorbi) (Pl): rt. decoc. u. for stomachache - **C. variegatum** (L) Bl. (Moluccas) (incl. var. *reticulata*): pl. u. as abort., lvs. chewed & swallowed as contraceptive (Solomon Is); fresh rt. of green form chewed as purg. (Mal); juice of pounded lvs. & sts. u. for fever (Caroline Is); cult. orn.

codling: 1) *Gadus callarias*, kind of cod 2) small or inferior apple.

codoil: rosin oil.

Codonanthe calcarata Hanstein (Gesneri) (Ve): rts. u. to "cure" wounds & sores (Waika, Ind) - **C. crassifolia** (Focke) Morton (WI): hb. tea u. in coughs & colds (Trin) - **C. dissimulata** H. E. Moore Codonanthopsis d. (H. E. Moore) (Wiehler) (nSA): lvs. of this epiphytic hb. pounded & boiled: decoc. u. for toothache (held in mouth) & headache (liquid taken through nose using spoon) (777) - **C. uleana** Fritsch (Col): lvs. of this epiphyte appl. as plaster to slow healing wounds & infections (Tikanas) (2299).

Codonocarpus cotinifolius F. Muell. (Gyrostemon.) (Austral): bell fruit; medicine tree; bk. u. med.; rts. u. food (aborigines).

Codonopsis javanica (Bl.) Hook. (Campanul.) (seAs): tuberous rts. (juice) u. to "overcome vitamin deficiency;" as galactagog (2448) - **C. lanceolata** (S. & Z) Trautv. (Chin, Jap): chung yao chih; rts. u. galactag. (Ainu); general ton. & stim.; aphrod., sialagog, to treat palpitation of heart; substit. for *C. tangshen* Oliver (Chin) - **C. pilosula** Nannfeldt (nChin, Manchuria +): tan(g)-shen; rts. u. much like those of *C. tangshen*; tonic in weakness - **C. tangshen** Oliver (Chin): dangshen; tuberous rts. u. as ton. in pulmon. & gynecolog. dis. (Korea); stim. to give strength; sialagog, aphrod.; in breast cancer, substit. for expensive ginseng (Chin) - **C. ussuriensis** Hemsl. (Chin, Jap, Kor): u. med. (55); food - **C. viridiflora** Maxim. (Chin, Jap, Kor. +): one of most valued Chin med. (55).

coearana (Br): *Cestrum* spp.

Coelocarpum humberti Mold. (Verben.) (Madag.): rombaoola (Malagasy); u. in tisanes.

co-enzymes: substances found associated with & activating enzymes; separable by dialysis (ex. adenylic acid, corcarboxylase); several vitamins are components of c. - **co-enzyme Q10:** CoQ_{10}: ubiquinone (*qv.*).

coerana (Br): *Cestrum* sp.

coeurs-saignants des bois (Qu) (bleeding hearts of the woods): *Dicentra canadensis; D. cucullaria.*

Coffea arabica L. (Rubi.) (trop eAf, Ethiopia) (1270): (Arabian) coffee; sds. source of roasted coffee; u. beverage (1186); sweepings is important source of natural caffeine; 2 chief types; Br (incl. Santos, Rio) and mild (incl. Colom, Bogotá, Guatemala, Jam, Mex, Mocha, PR, Maracaibo, &c.) (Br chief prodn. area); roasted c. has caffeine, theobromine, niacin; coffee lf. juices for upset stomach (Guat) (1269); coffee enema u. in cancer (folkloric); overuse may be harmful to cardiovascular app. (2280); sds. orig. eaten grd. - **C. canephora** Pierre ex Fröhner (C. robusta Lindl.) (Congo orig.; now widely cult.): Congo (or robusta) coffee; often planted in areas where *C. arabica* is attacked by a rust disease; imported & used mostly blended with coffee mixes, instant c., for beverage use; considered inferior type - **C. liberica** W. Bull. ex Hiern (trop wAf; now cult. in Ind, Java, Br): Liberian coffee; beverage use; bitter taste.

Coffea Tosta (L.): roasted coffee.

coffee (fr. Arabic): dr. cotyledons of *Coffea arabica, C. liberica*, or *C. robusta*; also the beverage prep. fr. roasted ground sd. (note similar names in many languages; ex. café (Fr, Sp); cafea (Romanian); kia-fey (Chin); Kaffee (Ge) (for Serials, see Appendix C-1) - **c. adulterants & substitutes**: incl. chicory; coffee hulls; leg. sds. such as peas, beans, soy beans; cereals (ex. corn meal, sorghum, barley, rye, wheat) after parching; dandelion rt.; dr. carrots; dr mugwort lvs.; ephedra hb.; figs; okra sds, asparagus, chestnuts, dates, horse chestnuts, beechnuts, dandelion rts. - **American c. bean** (or **tree):** *Gymnocladus dioica* - **Arabian c.: Arabic c.:** *Coffea arabica* - **c. and chicory:** popular mix (seUS) - **c. bean:** 1) sd. of *Coffea* spp. 2) (US): *Gymnocladus dioicus* 3) *Glottidium vesicarium* - **c. bean oil**: FO fr. *Coffea*; non-saponif. matter c. sitosterol, kahweol (1187) - **c. berry:** *Rhamnus purshiana* - **black c.:** beverage prep. without cream or milk - **California c.:** *Rhamnus californica* - **c. char: c. charcoal:** formerly u. med. (1188) - **common c.:** *C. arabica* - **c. concentrate:** instant c.: popular ext. of c. (US) - **decaffeinated c.: decaffeinized c.:** c. with ca. 95% of the caffein removed; popular where insomnia is a problem; however, use said to raise cholesterol serum level - **French c.:** c. and chicory - **glazed c.:** c. beans coated with sugar before roasting (ex. Fr. roasted c.) - **green c.:** c. which has been freed fr. pulp & sun dried but which has not yet been roasted; best preserved in this form until just prior to use - **c. houses:** popular shops in Eu cities in past centuries where c. was consumed (later called café) - **c. hulls:** remains of frt. walls surrounding c. seed; u. as adulterant - **instant c.:** ext. of c. freeze-dried, readily dissolves at short notice; considered inferior in flavor - **Java c.:** good grade, quite similar to Mocha c. - **Kentucky c. tree:** Am c. tree - **Mocha c.:** a good grade of c. which originates fr. the Red Sea region - **Mogdad c.: Negro c.**:fr. *Cassia occidentalis* sds. (1281) - **oak c.:** toasted frts. (acorns) of *Quercus robur* - **c. plant**: 1) buckthorn 2) (eUS) *Oenothera biennis* - **raw c.:** dr. c. sds. before grading, garbling or roasting - **roasted c.:** raw c. submitted to sufficient heat to darken sd. color & develop aroma flavor, coffeol & part of the caffeine - **robust(a) c.:** *C. canephora* - **Sacca c.: Sakka c.**. (Ge.): **Sultan c.:** roasted frt. wall (esp. sarcocarp) of c. u. as c. surrogate (Ethiopia); sometimes u. in place of chicory (with coffee) - **c. senna:** *Cassia occidentalis, C. obtusifolia* - **Sudan c.:** kola - **c. syrup** (New Eng [US]): u. to make c. milk (popular locally) - **c. tree:** *cf.* c. bean - **c. weed:** c. senna - **white c.:** c. beverage with added milk or cream - **wild c. root:** *Triosteum perfoliatum.*

Coffeinum (L.): caffeine - **coffeol:** constit. of roasted coffee supposed to give its chief flavor & aroma.

cognac: 1) a high-grade Fr brandy made fr. wines produced in or near Cognac, Fr 2) any brandy of Fr origin - **c. essence: c. oil:** ethyl oenanthate.

cogon (PI): *Imperata* spp.

Cogswellia Roem. & Schult. (Umb.): invalid gen. replaced by *Lomatium; Peucedanum* - **C. cous** M. E. Jones = *Lomatium cous.*

cogumelo dos alpes (Por): white agaric.

cohoba (WI, Mex, SA): narcotic snuff made fr. sds. of *Anadenanthera peregrina*, u. as hallucinogenic, aphrod., c. bufotenin & other tryptamine derivs.

cohobate: in distn., residual water from condenser sepg. from the VO wh. has been distd. over.

cohobation: process of repeating distn. by returning separated water in distillate back into the retort, thus enriching the further yield.

cohoche bleu (Fr): **cohosh b.:** Caulophyllum.

cohombrillo amargo (Sp): *Ecballium elaterium* - **cohombro** (Mex, Ec): cucumber.

Cohosch, schwarze (Ge): Cimicifuga.

cohosh 1) NA Indians for dye; appl. to several NA pl. spp., esp. of Fam. Ranuncul. 2) (specif.) *Actaea spicata* (etc.); *Caulophyllum; Cimicifuga* - **black c.:** *Cimicifuga* - **blue c.:** *Caulophyllum* - **c. bugbane:** black c. - **false c.:** *Caulophyllum thalic-*

troides - **red c.:** *Actaea rubra* - **white c.:** *Actaea pachypoda.*
cohuaco (Pe): squash - **cohune (nuts)** (Guat): *Orbignya cohune* - **cohush:** *Actaea pachypoda.*
coing (Fr): quince.
Cointreau ™(Fr): type of sweet colorless orange-flavored liqueur mfrd. in Fr.
coir (Ind, Pak, PI): elastic but stiff fibers of outer husks of coconut; u. mfr. ropes, (door) mats (**c. yarn**), brushes, carpets; u. to string hops (wUS).
Coix lacryma-jobi L. (Gram.) (eAs, Mal, natd. NA): Job's tears; tall grass cult. as cereal; has large hard pearly white grains (frt.) (v. adlai); u. fodder; diur. lithontriptic; anthelm.; u. to make beads; orn. pl.
cojoba = cohoba.
coke 1) dry black solid residue fr. destr. dist. of coal; u. as fuel & in smelting Fe 2) (US, slang): cocaine or c. HCl 3) (US slang) Coca Cola™ beverage.
cokehead (US slang): **cokie:** cocaine addict.
cokomo (US slang): cocaine addict.
col (Sp): cabbage - **c. de Bruselas** (Arg): Brussels sprouts - **c. rabano:** kohl-rabi.
Cola Schott & Endl. (Sterculi.) (tropAf): cola; kola; gp. of over 125 spp.; fl. with campanulate calyx, no corolla; ovary 3-10-celled; sd. in various spp. with 2-6 cotyledons - **C. acuminata** Schott & Endl. (Sterculia a. Beauv.) (Sudan): (Abata) Cola nut; evergreen tree c. 14 m high; sd. ("nut") c. caffein, kolatin (glycoside), theobromine; VO, FO, tannin, sugar, starch, cola red; cotyledons u. nerve stim. ton., in neuralgia, migraine; heart ton. (galenic prepns. as Kolanin); sl. euphoric; in beverages, coffee substit. (fresh prod. often preferred, esp. by Af natives) (chewed, in drinks); widely cult. WI (hence commonly called Jamaica Cola) - **C. ballayi** Cornu (equatorial Af): Cola Gabon; u. much like *C. acuminata* - **C. nitida** Schott & Endl. (SA, Af): **Sudan cola nut**: u. stim. nervine, beverage, fumitory, masticatory; belongs to gp. of *C.* spp. generally having 2 cotyledons in sd. (some consider conspecific with *C. acuminata*).
cola: kola: carbonated beverage (soft drink) c. cola ext., pl. extractives, VO, &c.; Coca Cola™ (*qv.*) the first one (ca. 1890) (1816) - **c. de caballo** (Sp) (horse tail): *Equisetum arvense* - **c. de león** (Sp): *Leonotis nepetaefolia* - **c. de pescado** (Sp): isinglass (lit. glue of fish) - **c. de zorra** (Mex) (vixen's tail): 1 *Perezia adnata, P. moschata*, other *P.* spp. 2) *Lobelia fenestralis* - **fausse c.:** (Fr): **false c.:** *Garcinia cola* (v. also under kola).
coladera (Co): towel gourd, *Luffa* spp.
colamine: ethanolamine; occ. in bean (sds.), *Vicia faba*, soy, and many fungi & bacteria; was synthesized (Wurtz, 1862) before being discovered in natural products; u. as sclerosing agt. (oleate).
"colanin": cola catechol: cola tannin: constituent of cola nut; hydrolyzed with colazyme to caffein + glucose + cola red (hydroyzed to phloroglucin + cola dye).
colchiceine: alkaloid, deriv. of colchicine fr. which it may be formed during extn. by decompn. (1282); some think native alk., shows distinct. antibacterial activity.
Colchici Cormus (L.): corm of *Colchicum autumnale.*
colchicine: chief alk. of *Colchicum autumnale;* u. antirheum. alt., esp. in gout; in chronic back pain (1987-), in diskal & other spinal dis. (2429); for detached retina (1986); poisonous; in hepatic cirrhosis (2540); in exp. genetics, agr. to induce chromosome doubling (polyploidy) in pls. & animals (E) (682).
colchicoside: colchicum glucoside; u. in gout, gouty arthritis, allergy (only 10% of toxicity of colchicine); deriv. thiocolchicoside (u. tetanus+).
Colchico (Sp, It): **Colchicum** (L. Eng) (corm, seed): **Colchicum autumnale** L. (Lili.) (Med reg.): meadow saffron; (common) autumn crocus; c. alk., esp. colchicine, 0.5-1% & colchiceine, FO, phytosterol; corm & sd. as tinct. u. as diaph. diur. cath. antirheum., analg. esp. in gout; mfr. colchicine; sd. figure (1280) - **C. luteum** Bak. (Indo-Pak, Kashmir, Iran, Afgh): Indian colchicum; Indian corm; Colchici Indici Cormus; surinjan talakh (bitter var.); corms ovate, white, hard, entire, horny, semi-translucent; 3.5 - 5 cm long (1486); corm c. 0.8% alk., incl. colchicine; u. substit. for *C. autumnale* - **C. speciosum** Stev. (eEu, swAs): sts. & corms c. colchicine D (1280) colchamine; tubers known in Ind as Radix Hermodactyli; u. in gout; toxic - **C. variegatum** L. (C. tessulatum Mill.) (sEu, Med reg., AsMin): Hermodactyli; u. in gout; c. colchicine.
colchicum root: c. corm fr. *Colchicum autumnale.*
colchique (Fr): colchicum - **c. du printem(p)s:** *Colchicum autumnale* v. *vernum* - **c. jaune:** Sternbergia lutea.
cold cream: rose water ointment; this popular (& ancient) face salve is an emulsion of rose water with sweet almond oil, &c.; it is u. primarily as a cosmetic, but also med. as emoll. protective & cleanser - **c. seeds:** *Citrullus vulgaris* sds. - **greater c. s.:** sds. of pumpkin, gourd, muskmelon, cucumber; formerly so called as much u. for colds, also catarrh., fevers, GI & GU dis. - **c. turkey** (US & Cana slang): taking addict off Heroin (etc.) directly, producing acute withdrawal symptoms (*cf.* flu symptoms).

cole: 1) *Brassica* spp., esp. *B. napus*, *B. campestris* (also **C. cabbage**) 2) Crambe maritima ("cole" occurs as variant in many names for *Brassica* spp., ex. broccoli, cauliflower, collards, kohl-rabi, kale, colewort) - **red cole:** *Armoracia rusticana* - **C. seed:** fr. *B. campestris, B. oleracea* v. *acephala* - **c. slaw:** salad prepd. fr. cabbage (orign. kale) with vinegar, etc.

Colebrookea oppositifolia Sm. (monotypic) (Lab) (Ind): binda (Hind): bainsa; rt. prepn. u. in epilepsy; lvs. & fls appl. as paste to wounds & bruises (124).

Coleonema album B. & W. (Rut) (sAf): has been u. at times as "buchu"; ornam. shrub - **C. pulchrum** Hook. (sAf): similar.

Coleoptera: beetles; order of insects which bear armored wing covers (elytra) with head bearing well-developed jaws for biting; ex. cantharis, stag beetle.

Coleosanthus Cass. = *Brickellia* Ell.

Coleus amboinicus Lour. (C. aromaticus Benth.) (Lab.) (Ind, EI as Java): wild (or Spanish) thyme; oreganote (Ve); country borage; very arom. hb.; lvs. appl. as poultice in centipede & scorpion sting, headache; lvs. u. to scent laundry & FO (Tonga) (86); int. in asthma, bronchitis; kidney dis. (Ve); juice carm. in colic; culinary hb. u. like borage & thyme; cult. Ind, Bah, Fla [US]; natd. WI, Ve - **C. blumei** Benth. (Ind): lf. decoc. u. dyspep. (Mal); pounded lvs. cure for headache & healing of bruises (PI); ornam. hb. - **C. esculentus** (N. E. Br) Tayl. (Af): tubers source of coleus starch; u. food - **C. forskohlii** (Poir.) Briquet (Ind, eAf): c. forskolin, coleonol (diterpene, with spasmolytic & hypotensive activity); rts. u. as card. stim. (Ayurvedic med.); muscle relaxant (acts to stim. adenylate cyclase); u. in high BP, stomachache, insomnia, glaucoma; supposed to help nerve cells regenerate; rts. preserved & eaten (Ind) - **C. macranthus** Merr. (PI): lvs. pounded into "plaster" & appl. to wounds. - **C. scutellarioides** Benth. (eAs, Polynesia): u. in cough; lf. juice u. to make woman sterile (PapNG); sap appl. to sores & cuts (Solomon Is). (—> Solenostemon)

colewort: 1) v. of cabbage without compact head of lvs., esp. kale (1117) 2) any *Brassica* spp., esp. *B. napus*, *B. oleracea* 3) Crambe.

coliander: coriander.

colic root 1) *Aletris farinosa* 2) (NC): *Cimicifuga* 3) *Apocynum androsaemifolium*.

coliflor (Sp): cauliflower.

colliform: resembling the coli-aerogenes gp. of Bacteria.

colina (Sp): 1) rutabaga 2) cabbage sd 3) choline.

colinabo (Sp.): **colinabu**: rutabaga.

Colistin(um): cyclopolypeptide antibiotic subst. obt. fr. cultures of *Bacillus colistinus* (Aerobacillus c.) fr. Jap soils (A, B, C forms); 1952-; u. as intest. antibacterial; injectable form is the soluble Colistimethate Sodium; at least 11 forms known (A, B, C, plus subdivisions) (USP XIX).

coll, red: horseradish (red cole).

Colla Piscium (L.) (lit. fishes' glue): isinglass.

collagen: ossein; polypeptides fr. skin, connective tissues (white fibers, cartilage, bone, &c.) made up of 3 peptide chains and ca. 1,000 amino acids; prepd. fr. cow hide; u. fiber sutures, leather substit., coatings, dermatologic cosmetics (for removing facial lines & wrinkles, surgical scars, grafts for burned skin; breast filling in, acne), food casings, &c. (imp. fr. Ind) - **soluble c.:** u. as skin moisturizer - **c. sponges:** u. as patch for transdermal delivery of drugs.

collagenase: enzyme fr. filtrates of *Clostridium welchii* wh. hydrolyzes collagen, gelatin; believed to be K toxin of *C. welchii*.

collard greens: collards (seUS): v. of kale, *Brassica oleracea* v. *acephala*, that does not form head; may be the original precursor of cabbage; u. as spinach; collards are grown in the winter (sUS), very hardy, will not stand much heat (term also appl. to young lvs. of any type of cabbage) (possibly corruption of "colewort").

colle de poisson (Fr) (lit. fish glue): isinglass.

collection of drugs: important primary step in crude drug prodn.; requires knowledge of part(s) to be used, time of year & day; separation of unwanted parts, &c.; drying & storage follow, usually carried out by same operator.

collenchyma: earliest mechanical supporting tissue of pl., usually developed peripherally, typically in outer cortex, made up of thick-walled cells; often serve also for storage & assimilation.

collet (Fr): root neck = root crown; shoulder; neck. collum.

colleters: glandular trichomes secreting mucilage in buds of some plants.

Colletia paradoxa (Escal.) Exe. (Rhamn.) (SA): curro (Arg); bk. rts. u. febrifuge, purg. (Arg).

Collinomycin antibiotic prod. by *Streptomyces collinus* (1953-); identical with α-rubromycin; active against *Staph. aureus*.

Collinsonia canadensis L. (Lab.) (e&cNA): stone root; rhiz. & rts. (Collinsonia, L.) c. resins, mucilage, u. diur. diaph., astr., ton, exp.; in hemorrhoidal prepns. - **C. serotina** Walt. (seUS): horse balm; rich root; stone root; pl. lemon-scented; hence "citronelle" (Fr); lvs. & rts. c. VO, resin, mucil.; u. as tea ("French tea") with pleasant odor; med. - **C. verticillata** Baldw. ex Ell.

(Micheliella v. Briq.) (seUS): stone-root; pl. u. med.

collo (pl. colli) (It): bale.

Collutorium (L.): mouthwash.

Colocasia Schott (Ar) Indomal, Polynesia): elephant's ear; some 10 spp.; toxic plants (CaOx); ornam - **C. esculenta** Schott var. *antiquorum* Hubb. & Rehd. (C. a. Schott; Caladium esculentum Vent.) (PacificIs): elephant's ear; coco-yam; sato-imo (Jap); dasheen; eddoes; taro; tanier; tanyah; large ann. hb.; tubers cooked as edible veg., source of meal, starch, poi (HawIs); in Yuc, eaten with honey (6); young lvs. sometimes eaten; widely cult., also wild as in Fla [US], WI, Br, Cyprus, Canary Is, Caroline Is (Kolocasie of ancients) (1190) - **C. maclurei** (Chin): tuber u. for scrofula.

colocynth: c. apples: c. cucumber: Colocynthis (L.): *Citrullus colocynthis* - **c. pulp:** frt. deprived of epicarp & sds.; medicinal portion.

colocynthin: colocynth bitter; bitter princ. fr. *Citrullus colocynthis* (1373).

Colocynthis citrullus O. Ktze. = *Citrullus colocynthis.*

Cologania angustifolia HBK (& var. *stricta*) (swTex [US]), Mex): long-leaf cologania; "perrito"; rts. said purg.

Cologne spirits: deodorized alc. - **C. (water):** eau de Cologne; a toilet water u. to impart fragrance to skin, also as loc. stim., after shaving; after-bathing lotion; facial cleanser; deodorizer; relief of minor dermal irrit.; bath water addn., etc. (2222).

colombo (root): *Jateorrhiza palmata*; v. **Calumba - American c.:** *Frasera carolinensis.*

Colonial spirit: purified methanol.

colophane = colophony.

colophone (Fr): **Colophonium** (L.): **colophony**: resina, resin fr. *Pinus palustris*, etc.

coloquinte (Fr): **Coloquinthe** (Ge): **Coloquint(h)enapfel: coloquntida** (Sp): colocynth.

Colorado cigar: one which is medium in color & strength - **C. cold root: C. cough (& catarrh) root:** *Ligusticum filicinum* - **C. man root**: *Ipomoea leptophylla* - **C. mountain sage:** *Artemisia frigida* - **C. potato beetle:** *Leptinotarsa* (Doryphora) *decemlineata* (Chrysomelidae - Coleoptera - Insecta): yellow-black striped beetle; larvae (grubs) feed on lvs. of potato, &c.; u. crushed as toothache rem. - **C. root:** C. cough r.

colorless hydrastis: comp. hydrastine soln.

colostrum: first milk fr. mother's breast after childbirth; being marketed by natural food stores.

colpachi de Huasteca (Mex): *Croton draco.*

Coltricia cinnamomea Murr. (Polyporus c. Jacq.) (Hymenochaet. - Basidiomyc. - Fungi) (temp. NA): u. in a fl. ext. for bladder dis., dysmenorrh., & hemorrhoids.

coltsfoot: 1) *Tussilago farfara* (also **common c.**) 2) *Asarum canadense* 3) various spp., incl. *Piper peltatum, Petasites fragrans, P. frigidus* var. *palmatus, Ligularia grandis* - **coltstail:** *Equisetum arvense.*

Colubrina Rich. (Rhamn.) trop & subtrop of both hemispheres): snakewood; bk. of some spp. u. as purg. & of some as hops substit. - **C. arborescens** (Mill.) Sarg. (C. ferruginosa Brongn.; Ceanothus colubrinus Willd.) (sFla, Fla Keys [US], WI, Mex, CA): wild coffee; (WI) snakewood (tree) (or snakebark); soapwood (Bah); naked wood; bijáguara; bk. may c. alks.; u. as bitter, astr., antidys., antiscorb., expect., antispasm., u. bleaching & cleaning; in baths for germicide; to draw boils, reduce fever, &c. (Bah); lvs. u astr. in dys. diarrh., anthelm., emmen.; adulter. of Cascara Amarga bark; cult. ornam. - **C. asiatica** Brongn. (NA, Af, As): hoop withe; bk. & lvs. u. as soap substit. (lathers); lf. decoc. u. to alleviate skin irrit. & certain skin dis.; boiled lvs. u. as poultice for swellings, &c.; frt. fish poison, u. as abortif. (Mal); cicatrizant for wounds (Polynesia); lvs. boiled & eaten (PI, Sabah) - **C. elliptica** Briz. & Stern (C. reclinata Brongn.; Ceanothus reclinatus L' Hérit.) (sFla [US], WI, sMex, Guat, Salv, nVe): naked wood (Fla); soldier w.; smooth snake bark; bk. (as mabee or mabi bark) & lvs. u. as antipyr. in dys. diarrh. GC, syph., exp.; in milk as bitter ton.; in prepn. refreshing "mabi" beverage (WI, nVe); aids digestion (Domin Rep); wd. u. in construction - **C. glandulosa** Perkins (C. rufa (Vell.) Reissek) (tropAm): bk. (saguaragay) u. for fever (Br).

columba: columbe (Fr): calumba (*qv.*); columbo.

Columbian spirits: wood alc. - **columbine (flowers):** *Aquilegia vulgaris.*

columbine, wild: *Aquilegia canadensis.*

columbo: calumba (*qv.*) - **American c. (root):** Frasera corolinensis - **c. wood:** Coscinium fenestratum.

Columnea crassa Morton (Gesneri) (CR): lvs. & rts. of this epiphyte moistened & placed in nostrils to stop nosebleed or on forehead for headaches (2444) - **C. longifolia** L. (SA): wound wort; has been u. to treat tumors - **C. picta** Karsten (Ec): tea stim., u. for blood diseases - **C. rubriacuta** (Hond, Ec): u. for scalp eczema (Cayabi Indians) - **C. vilosissima** Mansfeld (Co): fleshy lvs. crushed & appl. as plaster for bite of deadly bushmaster snake (Mocoa) (2299).

Colusa natural oil: claimed a special unrefined petroleum oil u. in skin dis., esp. psoriasis (1191).

Colutea arborescens L. (Leg) (c&sEu, esp. Med reg.): bladder senna; lvs. & pods c. colutea-acid; sds. c. canavanine, FO; lvs. u. purg., chiefly in domestic treatment (inf.); substit. for senna; diur., blood "purifier"; sds. emetic.

colyone = chalone.

colza: (Euras) rapeseed, *Brassica napus* & v. *oleifera* (silvestre) also several related spp. of *Brassica* (incl. *B. campestris, B. rapa*) u. for mfr. FO = **c. oil:** sweet o.

coma: hairs at ends of some sds. (related to Komatis [Gr] [comet]) - **comate** (adj.).

Combretum Loefl. (Combretaceae) (As, Af): butter tree; frt. of spp. yield butter-like FO; lvs. often u. med. - **C. bracteosum** Brandis (Af); hiccup nut (frt.); u. to relieve hiccups (but causes hiccups!) - **C. butyrosum** Tul. (trop eAf): sd. prod. chignite butter (white fat); u. as anti-opium drug (1543) - **C. cacouciana** (Aubl.) Exell (CA s to Br): sds. toxic, said u. to kill bats (Surinam); fls. said toxic - **C. constrictum** Lam. (Angola): rt. u. remedy for worms in children - **C. erythrophyllum** Sond. (Af): bush willow; rt. u. purg. (toxic), produces persistent hiccup; trunk yields gum, u. tan; wd. useful for carving - **C. gossweileri** Exell (Af): heart rem. (Luvale) - **C. gueinzii** Sond. (tropAf): lf. as wound dressing; rt. to treat abortion, constip.; lf. decoc. wash for reducing fever temp.; inner bk. decoc. as enema; rt. u. for constip. GC.; locally to prevent scabies; on sores (Tanz) - **C. imberbe** G. Don (swAf, Moz): leadwood; smoke fr. burning lvs. u. for cough & other resp. affections (Moz); in fever - **C. latifolium** Bl. (C. extensum Roxb.) (Ind, seAs): green lf. placed over abscess to remedy (Ind); astr. frt. u. as ton (Indochin) - **C. micranthum** G. Don (C. altum Guill. & Porr.; C. raimbaultii Heck.) (wAf): kinkeliba(h); tree or shrub; dr. lvs. (esp. green) of climbing shrub u. by natives as prophylaxis for blackwater fever (mal.) (decoc.); tea also u. by Europeans in Af to help acclimate them to climate; twigs & sts. to control opium addiction ("Eumecon" anti-opium remedy) (1543); cholagog; in colics; diur.; rts. u. to incr. potency; c. catechol tannins; choline ("combretine"), vitexin; orientin (flavonoids), stachydrine; cult. in Vietnam - **C. phaneropetalum** Bak. (cAf): sds. u. in tapeworm; however, causes cramps - **C. pilosum** Roxb. (seAs): rt. u. in fever (Thail); lvs. anthelm. for Ascaris worms (Ind) - **C. quadrangulare** Kurz (Indochin): sds. u. as vermif. (ascarids, tenia) - **C. sundaicum** Miq. (Mal): jungle weed; lvs. & sts. formerly u. to combat opium habit; "specific" in black water fever; lvs. & frts. u. to combat headache; c. a green resin, no alk. - **C. ternifolium** Engl. & Diels (Af): rt. inf. u. in cough - **C. zeyheri** Wawra (seAf): Zeyher's bush-willow; pl. u. in diarrh.; pd. bk. to regulate menstrual flow; ash of bk. in ophthalmia (Moz); rts. cooked in porridge for ankylostomiasis; rts. chewed for bilharziasis (Tanz).

Combucha: Combuchu: Kombucha (*qv.*).

Comesperma Labill. (Polygal.) (Austral): ca. 30 spp.; said to have properties of Senega rt.

comestible: edible; for food use.

comfit (Eng): coriander encrusted with sugar; u. as candy.

comfort root (US voodoo): placed in perfume or oil, using drop as needed (word, corruption of comfrey).

comfrey (root) common: 1) *Symphytum officinale* 2) (Mex) v. consound.

comin (Eng, Du): **cominho** (Por): **comino** (Sp): cumin (at times appl. to caraway).

Commelina L. (Commelin.) (trop. & subtrop.): day flower; herbs cult. as ornam.; some with edible lvs. & rhiz. - **C. communis** L. (C. vulgaris Red.) (orgin. eAs, now trop cosmopol): common day flower; "Santa Lucia" (Bo); tsuyukusa (Jap); tubers rich in starch & u. as food; tubers u. as vulner hemost. in hemorrhoids (1194); lvs. styptic for surface wounds; young lvs. boiled & eaten with salt & miso (Jap); whole pl. decoc. u. for heart dis., diarrh., throat swelling, blood "impurity"; water collecting on lvs. u. for eye dis.; u. for general weakness (Caroline Is); u. for wounds & to remove particles fr. eyes (Fiji) - **C. diffusa** Burm. f. (C. nudiflora auct. non L.; C. caroliniana Walt.) (seUS, tropAm, now natd. As): spreading (or Virginia) day flower; spiderwort; expressed juice of pod (capsule) u. in ground itch, "sore head" (of poultry), &c. (1195) - **C. elegans** HBK (WI, sFla [US], CA, SA to Par): French weed; tropical day flower; water grass (Trin); pl. decoc. u. diur., VD, intest. dis. (Cu); urinary dis., hypertension, (Mex, WI); pd. lf. to stanch bleeding of surface wounds; hb. inf. u. boils, cystitis, fever (Trin), in stomach cancer (Mex) - **C. erecta** L. (e&cUS, nMex, SA): "spiderwort" (sphalm.); common dew flower; day f.; entire pl. u. ton. in malar. (inf.) (sFlorida [US]) var. *angustifolia* Fern. (C. ang. Michx.) (seUS): day flower; mucilag. sap (esp. fr. inflorescence) u. emoll. appl. to ground itch (Seminole Indians, Fla [US]); nectar u. for eye dis. (Arg) - **C. pallida** Willd. (Mx): hierba del pollo; hb. u. med (Mex Phar.) - **C. tuberosa** L. (Mex, Br): tuber day flower; yerba del pollo (Mex): tubers u. as food; hb. u. as hemostat. & vulner. - **C. virginica** L. (eUS, w to Kansas) = C. elegans (apparently).

Commersonia bartramia Merrill (Sterculi) (Australasia): rt. decoc. u. for diarrh. (Austral); pl. fodder in time of drought (Queensl).

Commicarpus plumbagineus Standl. = *Boerhavia p.* - **C. scandens** (L) Standl. (Nyctagin.) (WI): goma bush (Bah); pega pollo (PR); lvs. beaten in water & u. to wash cuts & sores; tea diur. (Bah); decoc. u. in diarrh. ("dejuco de purgación") (Cu); listed in earlier Br Phar. (genus now often assigned to *Boerhavia*).

comminution: process of reducing a drug (etc.) to finer particles by grinding, cutting, etc.

Commiphora abyssinica Engl. (Burser) (Af): (Abyssinian) myrrh tree; source of Arabian Gum Myrrh, a gum-resin wh. exudes fr. fissures in bk. of trunk; c. VO resins, gums, myrrhin (bitter princ.); u. perfume, flavor, astr. carm.; diaph. ton. diur. exp. emm. loc. stim. antisep.; in nephritis, dysuria; in incenses, anointing oils, embalming preps., dental prods., to perfume breath. (2465) - **C. africana** Engl. (Senegambia): African myrrh tree; may be source of Bdellium (Africanum), a gum-resin - **C. berryi** Engl. (c&eInd): prod. gum exudn. fr. st. called "mulukilavary;" c. resin, gum, VO; unlike myrrh - **C. boiviniana** Engl. (Tanz): rt. pounded & appl. fresh for sores. - **C. drakebrockmannii** Sprague (Somalia): habbak dunkel; oleo-resin u. to kill leopards - **C. erythraea** Engl. (Somalia): Bisabol myrrh tree; exudn. is African bdellium (Bisabol myrrh or female m.) (opopanax of commerce); also VO & resin - **C. gileadense** (L.) C. Chr. (Ar, Oman, Eth, Sudan): jibbali; tree or shrub; frt. edible, u. to quench thirst; for indig. flatul.; for cough; lvs. fodder, u. to make tisane; wd. u. as yellow green dye for cloth; under-bark as cosmetic by women to lighten, soften, & cleanse skin; also to perfume & color tea or as tea substit.; in med. to treat wounds, sores, skin dis. (eczema), shingles, disinfect. wounds; dead wd. u. to make charcoal, tobacco substit. as fuel; resin u. as depilatory; lf. tips deodorize; trunk exudation = opobalsamum (balm of Gilead) with scant prodn. & costly; u. in Muslim med. & as incense & perfume (2736); this sp. name sometimes considered synonymous with *C. opobalsamum* Engl. - **C. kataf** Engl. (Saudi Arabia): opopanax myrrh tree; may be source of arom. gum-resin called "opopanax," replacing "opopanax" of the Ancients; u. in incenses, perfume indy.; prepn. VO (u. as fixative); "perfumed bdellium"; much like myrrh fr. *C. molmol* & u. to "extend" same; c. oporesinotannol; ferulic acid; vanillin; bisabolenes; NBP; gum (bassorin type); santal - **C. molmol** Engl. (Eritrea, Ethiopia; Sudan, Somalia; swArabia): common myrrh tree; like *C. abyssinica*, source of Heerabol or true Gum Myrrh, a gum-resin; u. astr. vulner., purg. deodorant, etc. (prodn. Iraq, Somalia) - **C. mukul** (Hook.) Engl. (Balsamodendron roxburghii Stocks; B. m. Hook.) (Ind, Pak, arid regions): mukul myrrh tree; oleogumresin exudn. fr. trunk called gug(g)al gum (gug[g]ul g.); Indian bdellium (Bdellium indicum); mukul; gukkal; guggula opopanax; guggul (Sanscr) (false myrrh); c. resin 70%, gum 25-30%, VO; with considerable use in Ind, u. like myrrh, as dem. aperient, carm.; in virus dis., emmen. aphrod. rheum., leprosy, diur., in syph., skin dis, rheumatoid arthritis, obesity; vermif. (trop); gugglipid with hypocholesteremic & hypolipidemic props.; occ. in comm., sold in groceries (Ind); perfume; myrrh adulter. (1544); contained (1.8%) in iron compound mixt. (Griffith's mixture) (USP 1820-1900) - **C. myrrha** Engl. = C. molmol - **C. opobalsamum** Engl. (Balsamodendron o. Kunth): var. *gileadensis* Engl. (Amyris gileadense L.) (Saudi Arabia, Syr, Eg, Somalia): Mecca myrrh tree; Balm of Gilead; Gilead balsam or (true) Mecca b. produced as exudn. fr. wounds cut in bk. or by boiling twigs in water (2629); this oleogum-resin c. resin (burserin+), bitter princ., 10% VO; u. home med. as vulner., diaph. stom. antipyr., even for the plague; supposed useful for female sterility - **C. pilosa** Engl. (Tanz): rts. u. in snakebite - **C. pteleifolia** Engelm. (Tanz): rt. decoc. u. in GC, headaches - **C. schimperi** Engl. (eAf, incl. Somalia-Yemen): produces Gum Myrh (Arabian m.; Somalia m.) (Fahdli dist.).

common: v. also brake, bugle, cap moss, English sorrel, European ash, fennel (seed), larkspur, lilac, star grass.

Comocladia P. Br. (Anacardi) (WI, CA): guao; several spp. important because poisonous (*cf. Rhus* spp.) - **C. brasiliastrum** Poiret (SA): tree has toxic exhalations - **C. dentata** Jacq. (WI, SA): guao prieto (Cu); bastard Brazil wood; tooth-leaved maiden plum; guao (blanco); guau; latex very caustic, causing swelling and blistering of the skin (*cf. Metopium brownei*); u. by natives for ringworm, herpes & warts (dangerous); reported u. by American Indians of Hispaniola to whiten skin (women); bk. (Cortex comocladiae) u. as hypnotic, sternutat.; juice of edible frt. colors black; wd. dark red (called "bastard Brazil wood" since u. as dyestuff); pollen transmitted by specific fly to eyes of humans (blisters); several other *C.* spp. such as *C. glabra* with similar properties - **C. glabra** Spreng. (WI): latex (juice) toxic, caustic; u. to scarify skin (Haiti); employed in Vodoun practice; natives have superstitous dread of pl., considering even shade of tree & smoke toxic; contact with skin prod. inflammn., swelling, black coloration, itching, often requiring medical attention: antidote mudpack prepd. with ashes of *Aloe barbadensis*, &c. (2487); natives of Cuba combat irritation of this & other *C.* spp. by applying *Guettarda calpyptrata* A. Rich. (Rubi) & *Ouretia ilicifolia* (DC) Baillon (Ochn.) - **C. integrifolia** L. (C. pinnatifida L.) (WI, CA): day flower; bk. hypnotic; u. as sternutat.; juice dyes black (Martinique); frt. edible but unattractive - **C. platyphylla** Rich. (Cuba): guao blanco; gives poisoning like that of *Toxicodendron* spp. due to caustic latex - **C. repanda** Blake (sMex): tetlate (meaning "burning fire"); similar properties to prec.

Compactin (Jap): Mevastin.

companion cell: a cell closely related geneticallly & by position to the sieve tube; believed to furnish nutrition to the sieve tube with which associated.

Compari: liqueur prepd. fr. artichoke (mixed with orange juice).

compass plant: 1) *Lactuca scariola* 2) *Silphium laciniatum.*

Compositae: Daisy, Goldenrod, Sunflower, Dandelion or **Composite** Fam.; 2nd largest pl. fam.; (ca. 21,000 spp); outstanding character is combn. of fls into fl. heads (or capitula), surrounded by involucral bracts; mostly ann. & perenn. hbs., a few bushes & fewer trees; some spp. with milky juice; anthers fused at margins to form ring around pistil (syngenesy); most weeds causing pollinosis are fr. this family; Fam. often divided into 3 other fam.: Ambrosiaceae, Cichoriaceae, & Carduaceae; ornam. members incl. foll., genera.: Ageratum, Aster, Calendula, Chyrsanthemum, Coreopsis, Cosmos, Dahlia, Gaillardia, Gerbera, Helianthus, etc.; many medl. spp.

compost: mixt. of org. materials (such as lvs., banana peels, &c.) u. after aging as fertilizer.

compote (Fr, Eng): fruit boiled or stewed in syrup.

Compound Digestive Elixir: Elixir Pepsin Comp. - **C. E. (of Kendall):** Cortisone (*qv.*); steroid compd. u. in arthritis - **C. Elixir of Yerba Santa**: Elix. Eriodictyon Comp. - **(Kendall's) C. E.**: a sterooid compd., hydrocortisone - **C. Spirits of Lavender:** Lavender Tincture Comp.

comptie: *Zamia* spp.

Comptonia peregrina Coult. (incl. var. *asplenifolia* (L.) Fern.) (C. asplenifolia Ait.) (l) (Myric.) (eNA): sweet fern; arom. lvs. c. tannin, VO (with cineole, terpinene, caryophyllene), hemostatic; bk. u. diarrh; u. stim. astr. in colic, rheum., at childbirth (Am Indians); spice (Am Indians); ornam. shrub.

conalbumin: simple protein occ. (14%) with ovalbumin (65%) in egg white.

concanavalin A (conA): phytagglutinin (lectin) (carbohydrate-binding protein agglutinating certain animal cells) found as globoids in cells of jack bean (sd.), &c.; agglutinates mammalian red blood cells; said to restore growth pattern of tumor cells to that of normal ones; immunostimulating; has been tried in cancer; u. in exptl. studies.

conch (pron. conk)**:** large heavy marine mollusc (tropics & subtr.): "seasnails"; representing spp. of *Strombus* & related gen. (Gastropoda); or large bright shell of same u. as horn for blowing or as ornam. (cameos, &c.); calyx (calcined shell) u. for hyperacidity, colic, jaundice, eye dis. (Unani med.) (*Turbinella rapa* (vase shell); flesh often eaten (2529) - **Conchae** (Praeparatae): oyster shells (testa), u. as antacid; believed superior to chalk since more acceptable to stomach; in colic; pd. in eye dis.; dryaha (2585).

concombre (cultivé) (Fr): cucumber.

concrete: waxy essence of fls. etc. prep. by extn. & evapn. of solvent; pure fl. essence u. in pharm. & perfume mfr.

Condalia henriquezii Bolding (Rhamn.) (nWI, nVe): frt. eaten raw or cooked, in beverage; frt. juice as ink; stain (1545) - **C. hookeri** M. C. Johnston (C. obovata Hook.) (Tex [US], nMex [NL]): chaparral prieto; bluewood condalia; shrub/small tree; frt. edible, u. in jelly; suggested for use in cancer (Jan. 1969, Ohio [US]); wd. furnishes a blue dye - **C. lycioides** (Gray) Weberb. = *Ziziphus obtusifolia* Gray.

condit (Fr): conserve of fresh pl. decoc. in sugar - **condite** (adj.): pickled, seasoned.

condition(ing) powder: conditioner: u. to stimulate appetite of stock animals, esp. horses; kind of tonic (to fatten); kept in boxes where animal has free access; generally not considered advantageous (may be fr. Lat. *conditus*, seasoned, savory).

condom, skin: skins; prepd. fr. cecum of lamb; popularly u. as contraceptive sheath, also to prevent transmission of STD but not protective against AIDS; rather expensive; latex rubber c. generally preferred.

condor-vine bark: Condurango (bark): *Marsdenia cundurango*; 10-12 barks have been known under this name - **condurit(ol):** xall. cyclic sugar-alc. found in condurango bk. (1264).

cone: strobile - **c. flower:** 1) *Echinacea angustifolia*, etc. 2) *Rudbeckia* spp. - **purple c. f.:** *Echinacea purpurea* & other *E.* spp.

conessi (bark): *Holarrhena pubescens* (H. antidysenterica) - **conessine:** steroid alk. found in

Holarrhena febrifuga & other *H.* spp. (1196), oleander, &c.

Confectio Cardiaca (L.): arom. chalk powder - **C. Damacrotis:** comp. catechu confection - **C. Raleighiana:** aromatic confection.

confection, laxative: combn. of dates, prunes, raisins (all deseeded), fig. & senna.

Conferva L. = one of 4 genera of Algae of Linnaeus; this is a fresh-water gp. (Confervaceae) represented by genera *Tribonema, Alsidium,* &c. - **C. bullosa** L. ("bubbled cotton conferva"); duck mud - **C. helminthochorton** = *Alsidium h.; C. rivalis* was u. in "spasmodic asthma."

congealing point: highest temp. at which liquified subst. crystallizes when gradually cooled.

Congo bean: C. pea - **C. gum: C. copal:** a kind of copal obt. fr. Af, either recently prod. fr. *Copaifera demeusei* (a semi-fossil resin); or a fossil resin; mostly the latter; these are the hardest known resins; extensively u. in varnish indy - **C. pea:** *Cajanus cajan* - **C. root:** Samson's snakeroot, *Psoralea psoralioides.*

conhydrine: vol. alk. fr. poison hemlock, Conium; u. antispas.

coniceine, g-: major alk. of *Conium maculatum*; also found in some *Aloe* spp.; an oily liquid; 18 X more toxic than coniine

Coniferae (Coniferales): Pinopsida; order of Gymnosperms with 52 famillies, the members of which are conifers (or cone-bearers); incl. Pinaceae (chief one); Taxaceae, Cupressaceae, Taxodiaceae, Podocarpaceae, Araucariaceae, Cephalotaxaceae, &c.

coniferin: chief glucoside of Pinaceae taxa (ex *Pinus strobus*); on hydrolysis, yields coniferyl alc. (eugenol deriv.) & glucose; occ. in benzoin (esp. Siam), comfrey, &c.

coniferophytes: plants bearing cones (Coniferae; Humulus; etc.).

Conii Folium (L.): poison hemlock hb.

coniine: chief alk. of conium; liquid.

Coniogramme japonica (Thunb.) Diels (Adiant. - Pteridoph.) (China, Kor, Jap): feng ya chueh; rhiz. or whole pl. u. as decoc. for body aches & pains; bloodshot eyes; amenorrh., mastitis; TB; snakebite.

Conium (L.): frt. of **Conium maculatum** L. (Umb.) (Euras): poison hemlock; frt. c. coniine, coniceine, conhydrine; FO, VO, diosmin; u. prepn. coniine; rapid disappearance of coniine in drying hb. has diminished value & use of drug in med.; mentioned as hemlock or gall; in Bible u. analgesic (Matt. 27:31) (fr. koneion, Gr, fr. konos, cones; SC indented pinnatifid lvs. suggest pine cones); u. anod. sed. antispas. anticonvuls. poison (danger); lvs., hb. & juice also u. (1197).

coniz(z)a (It): fleabane, possibly *Blumea myriocephala* or *Conyza* spp.

conjure (man): (doctor) (seUS Negro) engaged in voodoo practice; voodoo (man) - **c. bag:** Jack - **c. bottle.**

conjugase: "vitamin B_c"; enzyme present in many animal & pl. materials which converts folic acid conjugates to folic acid; richly abundant in hog kidney, chicken pancreas, rat liver, potato, &c.

conk: hardened bracket-shaped fruiting body of some fungi on tree trunks.

conkurchine: alk. of *Holarrhena antidysenterica.*

Connarus africanus Lam. (Connar.) (tropAf, Guinea): sd. flour u. purg. (Sierra Leone); tenifuge (wAf) (st. bk. also u. thus) bk. c. tannin; astr. for bleeding wounds; decoc. of sds. called séribèle (1198) - **C. monocarpus** L. (Ind): sunder (Bombay): frt. pulp u. in eye dis.; rt. decoc. in syph.; bk. & wd. u. to treat ulcers, etc. - **C. schultesii** Standl. ex Schult. (Mex); very bitter red sap u. as tenif. by Chinantec Indians; large doses toxic - **C. sprucei** Baker (Co): bk. u. as fish poison; similarly u. are other *C.* spp.

connecting tissue (bot): special colorless tissues adjacent to veins in some lvs. (Solereder).

connective tissues (zool.): that derived fr. mesenchyma of embryo; incl. bone, cartilage, dermis, tendons, vessel walls, cornea, &c.; c. collagen, elastin, acid mucopolysaccharides (proteoglycans), & structural glycoproteins.

connina (It): goosefoot: *Atriplex* or *Chenopodium* spp.

Conocarpus erectus L. (Combret) (Fla [US], WI, tropAm): (black) button-wood; button tree (or mangrove); bk. u. astr. ton. antipyr.; bk. decoc. u. as lotion for prickly heat (Bah); wd. u. to make charcoal.

Conocybe siligineoides Heim (Agaric. - Basidiomycotina) (sMex): mushrooms u. as hallucinogens by Oaxaca Indians (ya'nte) (*C. cyanopus* [Atk.] Kuehner: c. psilocybin).

Conomorpha citrifolia Mez (Myrsin.) (Co): inf. u. as febrifuge; u. to flavor chicha - **C. lithophyta** R.E. Schult. (Col): da-pee-ka-hee (Kubeo Indians); lvs. crushed & u. as fish poison; also lvs. of **C. magnoliifolia** Mez (Surinam): ayaria (haiaria?); also powerful insecticide.

Conopharyngia elegans Stapf (Apocyn.) (sRhodesia): toad tree; latex fr. trunk u. styptic, adulter. of rubber latex; rt. u. in lung dis.; frt. edible - **C. pachysiphon** Stapf (e&c tropAf): pl. c. hypotensive alk. (R.A. Lucas); latex ("juice") u. to adulter. rubber latex; juice (latex) as bird lime; styptic, sore eyes; bk. fiber u. to make dodo cloth; lvs. source of black hair dye - **C. ventricosa** Stapf (tropAf): quinine tree; "fever t."; latex u. rubber, bird-lime; tests showed no antimalarial activity. —> *Tabernaemontana.*

Conopholis americana (L.) Wallr. (Orobanchaceae) (eUS s to cFla): squaw drops; s. root; spring flg.; parasite on rts. chiefly of oaks; pl. u. astr.

Conopodium majus (Gouan) Loret (C. denudatum DC) (Umb.) (s&w Eu): earth (or pig) nut; tuberous rt. edible after roasting.

conori (Pe): Pará rubber-tree.

Conringia orientalis (L) Dum. (Cruc.) (cEu, Jap, wAs, nAf): lvs. u. as veget. (cooked); sds. furnish FO u. cooking; rarely cult.

conserva (Sp): pickles, preserves, conserves.

consignment: goods with shipping tags made out to an agent or dealer.

Consolida ambigua P. Ball & Heywood (Delphinium ajacis L.) (Ranuncul.) (Med reg.): rocket larskspur; l. sds. c. alks. ajacine, ajacinine, ajaconine, FO (38%) (1198-AW); u. insecticide, parasiticide, esp. in pediculosis; poisonous; widely cult. - **C. regalis** S. F. Gray (Delphinium consolida L.) (Euras, occas. NA): forking larkspur; fls. u. as diur. anthelm., color for sweetmeats; prepn. eye waters; sds. (as Semen Calcatrippae) u. rarely as purg. diur. vermif. narcotic, analg. parasiticide; toxic. (2744)

consolidine: alk. fr. fresh pl. of comfrey.

consolva (Canary Is, Sp): *Sempervivum tectorum* - **consoude** (Fr-f.): comfrey - **consound** (Mex): *Buddleia scordioides.*

conspecific: synonymous.

consuelda real (Sp): *Consolida regalis.*

consumption bark: *Botrychium fumarioides* - **consumptive's weed**: Eriodictyon.

contamination of drugs: act or condition of being polluted by soil, chemicals, insects, &c.; care in both collection & garbling is requisite.

contee: conti: food fr. rts. of *Smilax pseudo-china* prep. by Seminole Indians.

contra-herva (Br): **contrajerva** (root): **contrayerva** (Sp): *Dorstenia contrajerva, D. brasiliensis*; *Aristolochia odorat.*- **contrahierba blanco** (Mex): *Psoralea pentaphylla L.* - **contrayerba** (NM [US]): *Kallstroemia californica* v. *brachystylis*; *Psoralea pentaphylla* (CR): *Dorstenia contrajerva* - **contra-semences** (Fr): *Artemisia maritima* (etc.).

convalaria (Sp): **Convallaria** (L.): *C. majalis* (fls. rhiz.) - **C(onvallaria) keiskei** Miq. (Lili) (Sakhalin, Sib): landitsch (Russ); u. med. (Russ Phar.); likewise **C. manschurica** Kom. (sSib, Manchur) - **C. majalis** L. (Eu, natd. NA): lily of the valley; the rhiz. & rt., fls. lvs. sts. & hb. c. convallamarin (chief glycoside); u. as card. stim. (like digitalis), diur. emet.; green dye; fls. u. perfume - **C. transcaucasus** Utk. = *C. majalis.*

convallarin: subst. isolated fr. Convallaria; term appl. to 3 components - **convallatoxin**: glycoside fr. *Convallaria* fls.

Convolvulaceae: Morning Glory Fam. (mostly trop): ca. 1,650 spp., many cult.; incl. hbs., vines, shrubs, & trees; corolla funnel-shaped, gamopetalous; ex. morning glory, sweet potato, bindweed, moonflower.

convolvulin: highly complex colloidal glycosidal subst. made up of some 16 compds. fr. *Exogonium purga*; with drastic activity - **convolvulinoleic acid**: dihydroxy palmitic acid.

Convolvulus althaeoides L. (Convolvul) (seEu, Syr+): mallow-leaved bindweed; rt. source of a resin u. as substit. for jalap resin - **C. arvensis** L. (Euras, natd. NA): (common, field) bindweed; rope weed; fl. & hb. c. acrid resin; u. as purg. diur. (obsolete drug); bad cosmop. weed - **C. erubescens** Sims (Austral): Australian bindweed; pl. decoc. drunk for diarrh., indig., stomachache (2263) - **C. floridus** L. f. (Canary Is): ann. herb; rhiz. & rts. one source of Rhodium Oil (gaudil) - **C. hamadae** (Uved.) Petrov (C. subsericeus subsp. h.) (Siber, Altai): c. hygrine (alk.); u. med. (cAs) - **C. hystrix** Vahl (Eg-Libya): shobroq (Arab); rts., &c. u. as purg. - **C. mechoacana** Vitm. = *Exogonium purga* - **C. nil** L. = *Ipomoea n.* - **C. operculatus** Gomez (Ipomoea operculata Mart) = *Merremia macrocarpa* - **C. pandurata** L. = *Ipomoea batatas* Lam. - **C. pluricaulis** Choisy (Ind): poprang; pl. c. sankhpuspine VO; hb. u. nerve ton. in insanity, sexual neurasthenia (Ind); sed. in epilepsy, insomnia, hallucinations - **C. pseudocantabrica** Schrenk (cAs): sds. c. convalamine (veratryl tropine); toxic - **C. purpureus** L. = *Ipomoea purpurea* - **C. scammonia** L. (seEu, AsMin, Iraq, sRuss): true (or Levant) scammony; Asiatic s.; rt. c. scammonium (scammony resin); tuberous rts. of vine c. scammonin (component of resin [3-13%]), complex glycosidal subst. with convolvulinoleic acid and jalapinoleic acid as aglycones; toxic (2343); u. drastic purg.; since WWI generally unobtainable & replaced by Ipomoea; much of s. gum resin (so-called Aleppo) is heavily adulter. with rosin, &c. - **C. scoparius** L. f. (Canary Is, WI): shrubs; wd. of rts. called rosewood (smells like roses & musk); source of rosewood or rhodium oil (generally adult. with santal or cedar wood oils); u. perfumery, etc. - **C. soldanella** L. = *Calystegia s.* - **C. turpethum** L. = *Ipomoea t.*

convolvulus, white: *C. arvensis.*

Conyza canadensis (L) Cronq. (Erigeron c. L.) (Comp.) (NA, natd. Eu+): horse weed; (Cana) fleabane; "fireweed"; vergerettes du Canada (Fr); lvs. & tops u. diur., ton., astr. which is u. as hemostat. in uter. hemorrhage; decoc. in diarrh. (Pe); hb. source of VO (with limonene, terpineol, cuminaldehyde, &c.), u. in bronchitis, tumors; hemorrhages; in soy diarrhea (1503); popular in eclectic medicine; common weed in mint fields - **C. dioscoridis** Rauw. (Near East, Eg): u. by natives to treat colds; lvs. & rt. excitant; arom.; palliative - **C. filaginoides** Hieron. (Mex, SA): simonillo; zacacjoccjoc (Mex) hb. u. med. (Mex Phar.); pl u. bitter ton. lax. cholag. - **C. lyrata** HBK. (Mex, WI, CA, nSA): lf. decoc. u. in mal. (Salv); u. med. (Mex) - **C. marilandica** Mx. = *Pluchea camphorata* - **C. parvifolia** Walt. (Mex): yerba del historico; pl. u. for gastralgia; bitter ton. - **C. pyrifolia** Lam. (tropAf): rt. eaten as porridge to aid conception; smoked for cough & chest dis. (Tanz); exter. in leprosy (Lib); in headache, worms; many other uses.

coolwort: *Tiarella cordifolia.*

coon (swVa [US]): puccoon - **c. root**: 1) *Sanguinaria* (US Negro) 2) *Hepatica triloba* (prob. from puccoon) (also red c. r.) - **c. tail** (Fla [US]): *Ceratophyllum.*

coondi oil: carapa oil *Carapa guianensis.*

coontee: coontie: *Zamia integrifolia.*

Cooperia drummondii Herbert (Amaryllid.) (Tex [US], Mex): rain lily; evening star (Tex); reputed toxic pl.; c. bulb (1548) (—> *Zephyranthes*).

copahiba (Br): **copahu, (baume de)** (Fr): **copahyba** (Br): copaiba (balsam).

Copaiba (L.): **balsam c.**: oleroresin (not balsam) fr. *Copaifera* spp. such as *C. bijuga, C. officinalis, C. reticulata* (Pe, Br); prod. collected in much the same way as pine turpentine; u. diur. in bladder calculi (1200); stim. disinf. vulner. (exter.) to treat GC (prod. Ve, Br, Bo, Co) - **African c.**: oleoresin prob. fr. *Hardwickia manii* - **Brazilian c.**: 1) Pará 2) Bahía - **East Indian c.**: Gurjun balsam - **Maracaibo c.**: (Bahía): thick liquid type - **Maranham c.** (Pará): thin liquid type - **c. oil**: VO dist. fr. c., the residue representing **c. resin** - **solidified c.** = c. mass - **Venezuela c.**: Maracaibo c.

Copaifera L. (Leg.) (tropics): some 30 spp. of large trees; some SA spp. produces copaiba (oleoresin), some Af spp. produce copals (resins); good wd. ("purpleheart") (1201) - **C. bijuga** Willd. (Br, Ve, WI): one source of Brazilian Balsam Copaiba - **C. conjugata** Milne-Redhead (C. gorskiana Benth.) (Guibourtia c. Bolle) (eAf, Moz, Nyassa, Jam, Singapore): Inhambane (copal) tree; one source of the semifossil Inhambane (or stake) copal - **C. copallifera** Milne-Redh. (Guibourtia c. Bennett) (tropwAf): West Africa copal tree; Congo copal tree; origin of both a semifossil & recent copal; Sierra Leone c. - **C. coriacea** Mart. (eBr): (Pará) copaiba; oleoresin fr. trunk = balsamo capivi (oleoresin of copaiba); off. Br Phar. - **C. demeusei** Harms (Guibourtia d. [Harms] Léonard) (wAf, Zaire): Congo copal tree; African rosewood; source of Congo (or Cameroon) Copal (fossilized, semi-fossilized, recent); Congo copal is the most important comml. copal - **C. guianensis** Benth. (or Desf.) (Guyana; Surinam; Col, Br, CA): prod. Surinam balsam (copaiba) - **C. jacquinii** Desf. = *C. officinalis* - **C. lan(g)sdorffii** Desf. (Br): balsam copal tree; another source of Maranham copaiba; bk. tannin-rich - **C. mopane** J. Kirk (stropAf): Rhodesian mahogany; mopane copal tree; prod. inhambane or red Angola copal - **C. multijuga** Hayne (Br): tall tree; trunk source of copaiba, variously called "balsam" (really oleoresin), "copahiba angelin," "c. marimary," copaiba angelin - **C. oblongifolia** Mart. (Br+): another source of copaiba; off. in Br Phar. - **C. officinalis** L. (Ve, Co, WI, Guyana, Br): copaiba copal tree; source of copaiba ("Maracaibo resin") - **C. reticulata** Ducke (Pe, Br): chief source of copaiba (Guayana balsam) (oleo vermelho); u. vulner. in skin dis., GC.

copaiva: copaiba.

copal (, gum): a class of hard transparent resins derived fr. (1) recent exudations of many different living trop tree spp., esp. of *Copaifera* (Legumin.); (2) semi-fossil deposits; (3) fossilized remains of great age; copals are u. in mfr. varnishes, lacquers, paints, linoleums, protective coatings, plastics, etc.; in med. plasters, dental cement; prod. in Af, SA, WI, Mex; classes of copal: **accra** (wAf): fossilized fr. *Daniela ogea* & other *D.* spp. - **accroides** (sAustral): v. under accroides - **American c.**: Mexican c. - **Angola c.**: v. Angola = **Angola c., red**: fr. *Copaifera mopane* - **A. c., white**: fr. *C. demeusei* - **animi**: v. under animi - **Benguela c.** (wAf): semifossil c. fr. *Tessmannia moesiekei* Harms. (wAf); type of Congo c. - **Brazil(ian) c.**: a type of Congo c.; sub-fossil, said obt. fr. *Hymenaea martiana, H. courbaril,* & other *H.* spp., also *Macrolobium, Trachylobium* & *Cynometra* spp.; c. 2-5% VO - **Cameroun c.**: fr. *Copaifera demeusei* - **Colombia c.**: fr. *Hymenaea* spp. - **Congo c.**: fr. *Copaifera demeusei, C. mopane,* etc. - **Demerara c.**: South American c. - **East African c.**: almost entirely fr. *Trachylobium verrucosum* (T. hornemannianum) some fr. *Copaifera mopane* - **East India(n) c.**: fr. *Agathis* *dammara* - **Inhambane c.**: fr. *Copaifera conjugata* - **Kauri**: under K. - **Kongo**: Congo - **Madagascar**: East African - **Manila**: East India(n) (hard and soft vars. recognized) - **Mexico, True**: fr. cMex; little known yet about this var. (types: Sonora; Yucatan) - **Mozambique c.:** Zanzibar c. - **red c.**: fr. *Vouapa phaseolocarpa* - **Sierra Leone**: fr. *Copaifera copallifera* - **South American**: chiefly fr. *Hymenaea courbaril* - **stake c.**: Imhambane c. - **West African**: fr. *Daniella thurifera* & other *D.* spp.; *Copaifera copallifera* - **W. Indian**: fr. *Hymenaea courbaril* - **Zanzibar c.**: fr. *Trachylobium verrucosum,* East African c.

copalchi (Mex): *Coutarea latiflora* (SpAm); *Croton glabellus, C. eluteria, C. guatemalensis, C. niveus, C. reflexifolius, C. schiedea, C. tonduzii,* &c. (bk.).

copalm balm (or **balsam**): fr. *Liquidambar styraciflua* - **copalquin** (Mex): *Coutarea pterosperma.*

Copelandia (Agaric. - Basidiomyc.) (warm areas): mushroom found growing on cow dung; c. psilocin & psilocybin; u. hallucinogen (introd. into Hawaii; illegal there to have in possession) (**C. cyanescens** Singer).

Copernicia alba Morong (C. australis Becc.) (Palm) (SA, Par [Gran Chaco]): caranda (branca); yields hard veg. wax (carandaly wax), imported as Paraguayan carnauba wax (1202) - **C. molineti** Léon, **C. fallaense** Léon: **C. humicola** Léon: **C.** x **textilis** Dahlgren & Glassman: **C. berteroana** Beccari (all of Cuba): yarey (de Cuba); all used to make sombreros (hats), baskets, &c. (Cu, PR) - **C. prunifera** H. E. Moore (C. cerifera Mart.) (Br): carnauba (pron. carna-óo-ba) (wax) palm; carnauba wax is scraped fr. lvs. of this tall, handsome tree; u. mfr. varnishes, protective coatings, phonograph records, waxes for polishing, candles; rt. u. depurat. ton. in rheum. skin dis.; sarsaparilla replacement (decoc.); "to clean the blood" (Br); wax c. FO with cerotic acid esters; white sap u. to make beverage like milk; frt. edible; sds. u. as coffee substit.; wd. u. in building houses; lf. fiber u. to make baskets & fish nets; lvs. u. to thatch roofs & make hats & brooms - **C. tectorum** (HBK) Mart. (Co, Ve): palma (de) cana; p. de sombrero; p. de cobija (roofing palm); p. llanera (lowland palm); frt. edible.

copey (CR): *Clusia rosea.*

copper: Cu: reddish metal, essential to nutrition; occ. in rocks as ores, such as chalcocite, malachite; constituent of many proteins, esp. enzymes (ex. cytochrome c. oxidase); found in blood pigments of invertebrates; deficiency rare, cause of some types of anemia (v. under cupric, cuprous).

copperas, blue: $CuSO_4$.

copperhead (snake): *Agkistrodon contortrix*: & vars., incl. *mokasen* (e&sUS): semiaquatic venomous snake; less toxic than rattlesnakes; venom preferred for dissolving blood clots.

coppice: dense growth of small trees grown from cut stumps of trees; process called coppicing; increases shade.

coppicing: process of cutting off tree trunk above ground, permitting several stems to develop fr. lateral buds, replacing the original single stem & thus increasing prodn. of bk., etc.; u. for cascara sagrada; Ceylon cinnamon, *Castanea, Corylus*

copra (Sp, Po): derived fr. Khipra (Hindu) = coconut; as used (US), copra is fermented coconut meat; u. food, expr. FO (coconut oil) ("copra oil"); press cake serves as cattle feed (exp. fr. SriL, PI, Af); fungi (substrate).

Coprinus (Agaric.; Basidiomyc. - Fungi): coprinus, inkcap; some 110 spp.; some are edible; lamellae deliquesce producing black liquid, bearing spores - **C. comatus** S. F. Gray (temp zone, NA+): shaggy mane; s. cap; s. beard; this unique mushrooom is edible when fresh; as the pileus (cap) matures, gills turn darker until liquid formed like India ink; u. as ink (for retouching photos) - **C. micaceus** Fr. (temp zone): common ink cap; frt. bodies u. as food, with delicate flavor - **C. quadrifidus** Pk. (temp zone): scaly ink; frt. bodies c. quadrifidins, antibiotics.

copriuva (wood) oil: VO fr. *Myrocarpus fastigiatus.*

Coprosma australis Robinson (Rubi.) (NZ): Tasmanian native currant; bk. inf. u. for itch, scabies, bruises, stomachache, cuts, fractures; lf. inf. u. for fever, cuts, festered sores, kidney dis.; lf. inf. as poultice for broken limbs; pl. juice u. skin dis. - **C. lucida** J. R. & G. Forst. (NZ): (Otago) "orange leaf"; u. dyestuff.

coptide (Fr): coptis.

coptine: characteristic alk. of *Coptis* spp. (toxic).

Coptis anemonaefolia S & Z (Ranuncul.) (eAs, esp. Jap): rhiz. c. berberine; off. Jap Ph. IV - **C. chinensis** Franch. (Chin): rhiz. u. to treat fevers, nausea, thirst, conjunctivitis; antibiotic, u. for bac. dys., gastroenteritis, diabetes(!), TB; cult. - **C. deltoidea** Cheng & Hsiao (Chin): comparable use to other *C.* spp.; monographed in Chin Phar. 1977 - **C. japonica** Mak. (Jap): rhiz. u. bitter stom., astr., in violent headache; ointments u. for eczema marginatum (Jap); ophthalmia (Mal) - **C. occidentalis** (Nutt.) T. & G. (nwUS): "snow flower"; u. much like *C. trifolia* - **C. tectoides** C. Y. Cheng (Vietnam): u. similarly to *C. teeta* - **C. teeta** Wall. (Chin, Ind, Assam, eHimal): mishmea teeta (Ind) source of mamira or mishmee bitter; rhiz. c. 8-9%

berberine; potent bacteriostatic hb. u. bac. dys.; bitter ton. antipyr. diur. lax. aphrod. stim. for depression (Burma); drug u. since ancient Gr (1203) - **C. trifolia** Salisb. subsp. *groenlandica* (Oeder) Hultén (C. groenlandica Hultén) (Alaska, neAs): Coptis (L); (common) gold thread; small hbs.; dr. pl. c. alks. coptine, berberine; u. astr. bitter ton.; rhiz. also marketed & used; decoc. with *Asarum candense* u. for colds (Abenakis).

CoQ_{10}: v. coenzyme Q_{10}; claimed essential to cardiac function & to immune system; u. in periodontal dis. (dose 100 mg/d).

coque du Levant (Fr): Cocculus Indicus - **coquelicot**: corn poppy - **coquelorde** (Fr): Pulsatilla - **coqueluchon** (Fr): aconite.

coquito (palm): *Jubaea spectabilis, J. chilensis.*

coral (SpAm): 1) invertebrate animals (polyps) occur. in marine areas (Phylum Coelenterata); the hard exoskeleton ($CaCO_3$) forms coral reefs & later islands; of increasing interest as to chem. pharmacol. & eventual therap. utilization 2) term appl. to many different spp. of pls., ex. in Liliaceae - **c. beads:** *Cocculus carolinus, Erythrina* spp. **c. bean**: *Sophora speciosa* - **c. berry**: *Symphoricarpos orbiculatus* - **c. moss**: *Corallina officinalis* (a red seaweed) - **c. root** (US): *Corallorhiza odontorrhiza* & other *C.* spp. - **c. snake**: *Micrurus* spp. (*qv.*) - **red c.** *Corallium rubrum* (*qv.*) - **c. tree**: *Erythrina herbacea* + - **c. vine**: 1) (Fla [US]): *Antigonon leptopus* (also c. bloom) 2) (seUS): *Erythrina herbacea* - **white c.:** *Amphihelia oculata* (Madripora o.) (ord. Madreporaria - Anthozoa - Coelenterata) (Med reg.); largely composed of calcium carbonate - **wild c. tree.**: *Symphoricarpos orbiculatus, Erythrina corallodendrum.*

coralillo (SpAm): *Satyria warscerviczii.*

Corallina officinalis L. (Corallin. - Rhodophyc. - Algae) (nAtl Oc, Med Sea): coral moss; thallus c. glyceritol (glycoside), antimicrobial subst., formerly u. as vermifuge.

corallito (Mex): *Hamèlia patens.*

Corallium rubrum L. (C. nobile) (Coralliidae - Gorgonaria -Anthozoa - Coelenterata) (trop seas worldwide, esp. Med Sea; Canary Is; Indian Oc; Indian & Jap coasts): (red) coral; moonga (Unani med.); pd. u. in resp. dis. coughs, retronasal catarrh., whooping cough (Homeopath); skin dis., esp. psoriasis; ash in tooth powders (improve gum & teeth health) (2447); in obesity (Eu, 16th century) (another Gorgonian found in Gulf of Mexico c. 1% prostaglandins); for millenia fragments u. as jewelry often in combn. with diamonds, sapphires, & turquoises; often u. as amulet against harm & danger; c. Ca carbonate, $CaSO_4$, $MgCO_3$, I-org. compds. (the related true or stony corals form large reefs in tropical seas).

Corallorhiza odontorhiza Nutt. (Orchid.) (eUS, s to Guat): (late) coral-root; crawley rt. (Corallorhizae Radix, L) u. diaph. antipyr.; sed. (Eclectics); cult. ornam.

corazón (SpAm): 1) (Trin): *Annona* spp. 2) (WI): *A. glabra* 3) (PR): *A. reticulata* 4) (Salv): *Echinodorus macrophyllus* 5) (Co): *Anthurium* spp. (flor de c.).

corchorin = strophanthidin.

corchoroside A.: glycoside of strophanthidin & D-boivinose; occ. in *Corchorus capsularis*, &c.

Corchorus L. (Tili) (warm areas of world): ca. 40 spp.; some spp. furnish the important textile fiber jute (Indian flax) or gunny fr. st. (75% world prodn. fr. Indo-Pak); u. mfr. gunny sacks, burlap, hessian, sand bags, sugar bags, etc.; nalte palta; dr. lvs. u. stomach; sds. - **C. aestuans** L. (C. acutangulus Lam.) (Ind, China, PI): nalte palta; young lvs. made into soup to drink as cooling beverage in hot season; bruised lvs. with sugar for abscesses; dr. lvs. u. stomach. - **C. capsularis** L. (Ind, cult. Indo-Pak; now pantrop): jute plant; source of important fiber; (jute, gunny) sds. c. corchoroside A (cardioactive glycos.); u. in dys.; poultice for boils & abscesses; sds. ton. antipyr. carm.; fls. for epistaxis; FO fr sd. u. food, lighting (chiefly shipped fr. Calcutta [Bengal]); some prodn. SA (1547) - **C. olitorius** L. (Ind, Bangla Desh): potherb jute; Jew's mallow; Naltu (jute); one of 2 important sources of jute; much u. for twine; lf. c. pectin, corchorgenin & other digitaloid cardioactive compds.; u. as potherb (Tun, eAf) ("meloukhia"); u. digestive & ton. in pulm. dis., constip., galactogog., antipyr. (prodn. best in Bangla Desh; wBengal [Burdwan dist.]), Assam, &c.) - **C. siliquosus** L. (Mex, WI): broom weed; malva (de cubierta); lvs. eaten by stock; Mayas u. to relieve bite of pic bug (6); antipyr.; lf. decoc. as enema or poultice; as infant enema (79); lvs. substit. for Chinese tea - **C. trilocularis** Burm. f. (C. tridens L.) (Af to Ind): "al moulinouquia"; sometimes cult. as veg. (near Timbuctu); sds. antipyr.; mucil. as demulcent (Pak).

cordão de frade (Br): 1) *Leonotis nepetaefolia* 2) *Borreria verticillata.*

Cordeauxia edulis Hemsl. (Leg) (Somalia, Eth.): gud(a); yehab; geeb; frt. & sds. edible; one of chief foods of poor people where growing wild; claimed to have med. properties also; lvs. tea substit.

cordero (Arg): lamb.

Cordia alba R. & S. (C. toqueve Aubl.) (Boragin.) (Mex): toqueve; cebito (Salv): lvs. & fls. emoll. dem. u. in resp. dis.; fl. decoc. u. as diaphor.; frt. edible, juice u. as adhesive for cigars; wd. u. in carpentry - **C. alliodora** (Ruiz & Pav.) Oken (LatinAm): arbre à l'ail (Fr) (garlic tree); picana negra (Br); "laurel"; cyp(re); lf. decoc. u. stom. ton., in inflammns. (swellings, bruises); pd. u. in ointment in skin dis. (WI); lvs. heated & placed on wounds (Salv); frt. eaten (Indian tribes, Mex); wd. u. for cabinet work, &c.; in plantations - **C. bahamensis** Urban (Bah): granny bush; decoc. u. for postpartum dis. - **C. boissieri** A DC (Tex [US], Mex): rasca vieja (Michoacán, Mex); anacahuita fls.; Mexican olive; bk. & lvs. u. antirheum. in pinta rabies(!); to treat fractures; frt. pleasantly edible but said intoxicating; source of anacahuite wood; frt. syr. u. in gastric, intestin. & bronch. dis., incl. TB (phthisis), pneumonia, mal. - **C. brittonii** (Millsp.) Macbride (Bah): u. like *C. bahamensis* - **C. curassavica** (Jacq.) R. & S. (WI): black sage (Trin); lf. decoc. u. in pneumonia, insomnia, cough; lf. juice in mal.- **C. dentata** Vahl (LatinAm): caujara; fl. inf. diaph.; wd. u. for barrel staves, fence posts, &c. (2435); frt. edible (Mex) - **C. dichotoma** Forst. f. = *C. obliqua* - **C. gerascanthus** L. (WI, SA): prince wood: Dominican rosewood; Spanish elm; timber u.; frts. u. cough; bk. antipyr.; st. gives "Narawali Fiber" - **C. globosa** (Jacq) HBK (Mex, CA, WI): papita; lvs. & shoots u. astr. hemostatic (Cu) in baths, to season meat (6) - **C. graveolens** A DC (Surinam): lvs. u. to treat GC (folk remedy) - **C. morelosana** Standl. (sMex): "macahuite"; fls. recommended for colds & sold in markets in Izucar de Matamoras (Puebla State, Mex); boiled with lemons u in pneumonia (Mex); fls. inf. for "stitch in the side" - **C. myxa** L. (Br, Ind, seAs, Austral): sebesten (plum); "plum"; bk. u. "mild ton."; astr.; green frts. likewise; ripe frts. edible; mucilage-rich; u. emoll. dem. lax.; in bird lime; lf. decoc. u. to bathe limbs in arthritis (NG) - **C. obliqua** Willd. (cAs, WI+): lasoori (Tibet); gout tea (Cu); sea coast t.; for bad colds, &c. (Tibet); bk. antipyr.; useful fiber - **C. ovalis** R. Br. (Tanz): bk. u. for swollen spleen in children; lf & bk. u. for leprosy (lf. decoc. in bath) & crushed bk. rubbed on area; frt. edible - **C. reticulata** Vahl (Lesser Antilles; Ve): coco poule (Dom); lvs. placed on forehead for headache (Dominica Caribs) - **C. sebestena** L. (Fla [US], Mex, WI): Geiger tree; frt. edible, emoll. dem. antipyr.; pl. u. gastroint. dis. & bronchial troubles; cult. ornam.

cordial: liqueur (from cor, cordis, L., heart): sweet alcoholic beverage drunk as a stim. (heart warming) in small amts. usually at end of meal; ex. arrak; crème de menthe - **c. flowers:** borage; rose; violet; *Anchusa azurea* P. Mill. ("American puccoon").

Cordillera (Sp): mountain range.

cord-oil: rosin oil.

cordoncillo (blanco): *Piper aduncum.*

Cordyceps (Fr) Link (Clavicept. - Claviceptales - Ascomycet. - Fungi): caterpillar fungi; "vegetable caterpillars"; found growing on fungi (*Elaphomyces*, &c.), insects, &c.; ca. 100 spp. - **C. capitata** Link (Mex): hombrecitos; u. as hallucinogen by native Aztecs - **C. cicadae** Sing. (As): on nymphs of cicadae; u. med. (Chin) - **C. gunnii** (B.) B. (Euras, Oceania): occ. on larvae of Lepidoptera; edible, medl. (Chin) - **C. martialis** Skeg. (Euras, SA): on pupae of Lepidoptera; med. (Chin) - **C. militaris** (L.) Link (Euras, NA): found growing on pupae of Lepidoptera; edible; med. (Chin) - **C. sinensis** (Berk) Sacc. (Chin, Austral): ch'ung tsao; u. ton. in anemia; stim. for convalescents; aphrod.; cure for opium "eating" & antidote in opium poisoning (Chin); powd. u. in culinary delicacies (119).

Cordyla pinnata· Milne-Redh. (Leg) (wAf): bush mango; pods fleshy, edible; bk. u. anthelm.; lf. macerate u. in anorexia; rts. cholagog, anthelm. (bilharziasis); oxytocic; aphrod. (Senegal).

Cordyline australis (Forst. f.) Endl. (Lili/Agav.) (NZ): palm lily; lvs. source of fiber u. to make paper, formerly for twine; c. high fructose sugar; lf. c. steroid saponins; u. as emulsifier; ornam. tree - **C. fruticosa** (L.) Goeppert (C. terminalis Kunth) (Ind, Oceania, Malaysia, Polynesia): palm lily; red ti; green ti; drac(a)ena; ti palm (Haw Is); "ric" (Caroline Is); tuberous rts. starch-rich; starvation food (Tonga); u. as sugar source (fructose), to prep. intoxicating beverage (Haw Is); lf. c. 2.3% raw saponin; bud lvs. u. stomachache, rhiz. in constip. (Caroline Is); in liver dis. (Java); rt. u. dys. diarrh.; heated bk. rubbed into gum where aching tooth (PNG); lvs. u. to wrap fish (Tonga); cult. ornam. (worldwide) - **C. indivisa** Steud. (NZ): lvs. antiscorb.; lf. fibers u. to make clothing (Maoris) - **C. roxburghiana** Merr. (tropAs): rhiz. u. in cough; as purg.; ton. antipyr. (Hindu med.); lvs. source of strong fiber; cult.; pl. u. to separate cult. land fr. wild jungle (Papua NG).

Coreopsis leavenworthii T & G (Comp.) (Fla [US]): tickweed; tickseed; pl. u. exter. by Seminole Indians for heat prostration (hot inf.) - **C. major** Walt. (seUS, n to Ohio): starry coreopsis; u. to dye eggs with lvs. (sUS).

Corhormon: heart muscle ext.; prepd. fr. hearts of calf embryos; u. as coronary vasodilator.

coriamyrtin: sesquiterpene toxic bitter princ. fr. *Coriaria myrtifolia & C. japonica*; has convulsant action like picrotoxin.

coriander (fruit) ("seed"): fr. *Coriandrum sativum* - **Bombay c.: Indian c.**: an inferior var. of *c.* with ellipsoid frts. (rather than subspheric) and low in VO.

coriandre (Fr): **coriandro** (It) coriander - **coriandrol:** d-linalool; first disc. in coriander frt. VO.

Coriandrum (L.): frt. of **C. sativum** L. (Umb.) (sEu, eMed reg.): coriander; dried ripe frt. & VO u. flavor (gin, bread, confect.), arom. carm. stim.; spice u. chiefly in sausage, curries (2530); lvs. u. as flavor; young lvs. eaten; (exp. fr. Mor, Ind, USSR, Romania); cult. in Calif & green hb. sold in markets there (cilantro, SpAm); u. for salads (raw) or as greens (boiled).

Coriaria myrtifolia L. (Coriari.) (sEu, nAf): tanner's sumac; lvs. c. gallic acid type tannin; st. frts. c. coriamyrtin (narcotic poison); hallucinogen; fresh lvs. u. homeop. med.; substit. for sweet marjoram; lvs. & bk. tan. - **C. nepalensis** Wall. (Nep, Ind, Pak): balel (Kashmiri); blue-black berries of shrub/tree purg.; lvs. c. 20% tannin; poisonous in large amts. - **C. thymifolia** H. & P. (NZ, Ve, Pe): ink plant; pl. c. tutin, tannin; u. mfr. inks (juice); poisonous.

Coridothymus Reichb. f. (Corydothymus) = *Thymus*.

corinths: corinthes (Fr): **Corinthian raisins**: corintos (Sp, Por): currants; a black, small-fruited seedless var. of *Vitis vinifera* grown in Gr since antiquity; u. like ordinary raisins; grades incl. Amalias, Patrás, Vostizza (prod. mostly in Ionian Is off Greece, thus Zante [Zakinthos]; name fr. Korinthos, near-by).

Coriolus biformis Pat. (Polyporus b. Fr.) (Polypor. - Basidiomyc. - Fungi) (cosmopol.): mushroom c. biformin (polyacetylene alc.) with antibacterial props.; frtg. body very toxic but u. in med.

cork: Suber (L.): outer protective layers of bark of older stems, roots, &c.; specif. of the trunk of the **cork oak**, *Quercus suber*; repr. the phellogen & its prods.; sometimes occurs adventitiously in patches on lvs. frt., &c.; typically in many layers; cells usually tabular, colored brownish & with suberin, wh. reduces water & gas penetration of tissues - **c. tree**: 1) c. oak 2) (As): *Phellodendron amurense* 3) (WI): bendy tree, *Thespesia populnea* 4) *Erythrina vespertilio* 5) *Ochroma lagopus* - **c. wood** (or **corkwood**): 1) *Annona glabra, A. palustris* 2) (US): *Leitneria floridana* 3) *Quercus suber* 4) *Duboisia myoporoides* 5) balsa, *Ochroma pyramidale* 6) *Alstonia spectabilis* 7) *Sesbania grandiflora* 8) *Pterocarpus draco* 9) taxa of at least 7 other genera.

corm: a solid bulb; an excessively thickened, firm, vertical, subterranean stem, with bud or buds at apex & fibrous rts. below; ex. colchicum corm, crocus.

cormier (Fr): mountain ash.

cormophyte: flowering plant or fern; incl. phanerogams & vascular cryptogams; plant with rts. sts. & lvs.

cormus: root & stem with leaves in higher plants; hence group Cormophyta as distinctive fr. Protophyta (Thallophyta).

corn: name appl. to most used (popular) cereal grain of a country, thus, in US, Cana = *Zea* (maize); in Eng = wheat; in Scotl. & Ireland = oats - **c. beard**: c. silk - **c. belt:** an area of the US where the chief crops are maize and maize-fed stock; mostly Iowa & Illinois, but extending fr. wOhio to Neb - **chicken c.**: *Sorghum bicolor* subsp. *drummondii* - **c. cob**: the core of an ear of c., technically the thickened rachis bearing the pistillate fls., later the frts. (or grains); reduced to *cob meal*, u. in stock feeds, chem. mfr. of lignin, xylose, furfural, dextrins, etc., as adult. etc.; granular cob u. abrasive for metal parts; whole cob u. mfr. pipes, fuel, etc. - **c. cockle**: *Agrostemma githago* - **c. collodion**: comp. salicylic c. (u. to treat corns) - **dent c.**: *Z. mays* v. *indentata*; the chief var. grown in US; a tall pl.; SC marked depression at apex of grain - **c. dodger**: cake of c. meal baked hard under embers; sometimes fried or boiled - **Egyptian c.:** Jerusalem **c.:** *Sorghum bicolor* subsp. bicolor; sometimes Jerusalem c. is distinguished as var. *durra* (Forsk.) Hubbard & Rehd. & Egyptian c. as var. *roxburghii* (Stapf) Haines - **c. ergot**: c. smut - **field c.**: any of several vars. u. principally as livestock feed - **flint c.:** *Z. m.* var. *indurata*; med. size pl.; SC grain hard & smooth - **c. flower (, blue)**: 1) *Centaurea cyanus* 2) (NY+) *Cichorium intybus* - **c. gromwell**: *Lithospermum arvense* - **Guinea c.**: *Sorghum vulgare* v. *drummondii* (or v. *caffrorum*) - **c. husks**: shucks; foliaceous bracts enclosing the pistillate inflorescence & infrutescence (cob with grains); u. (dr.) to cover tamales - **hybrid c.**: progeny of cross-breeding of inbred strains of *Zea mays*; pl. & grain usually of superior quality - **Indian c.**: the original Am name for *Zea mays*, now contracted in NA to "corn" - **Jerusalem c.**: *Sorghum bicolor* var. - **kaf(f)ir c.**: *Sorghum vulgare* v. *caffrorum* - **c. liquor**: c. whiskey - **c. meal**: the prod. of grinding (milling) corn grain less germ & most of bran; makes poor loaf bread (since low gliadin content); is u. to prep. c. pone, ash cake; batter bread (egg bread, spoon b.), griddle cakes (types: water ground [more nutritious because of germ]; new process [keeps better]; white [popular sUS]; yellow [pop. nUS]); u. by sUS. belles (early XIX cent.) by rubbing into body to help keep cool on hot days - **c. m. moth**: *Tinea zeae*, common enemy of miller & baker - **c. mint oil**: 1) dementholized Japanese mint oil 2) VO of *M. arvensis* - **c. oil**: the FO obt. by expression at high temp; fr. embryo of *Zea Mays* grain; u. culinary & salad oil; mfr. soaps, paints, grinding oil (with linseed oil); in vulcanizing rubber; solvent for viosterol; healthful in foods (lowers blood cholesterol) - **parched corn**: scorched c.; roasting corn; ear heated in ashes in husk or boiled in oven (sUS) - **pod c.**: *Z. m.* var. *tunicata*; each grain enclosed in its husk; believed close to primitive type - **c. pone** (sUS): c. meal cake baked with hot ashes; no milk or eggs u.; made in irregular oval masses shaped by hand - **pop c.**: *Z. m.* var. *everta*; grains hard & pointed; when heated, grain explodes & contents are everted; popular as food at circuses, cinemas, etc. - **c. silk:** the long styles with stigmas of the pistillate fls. of *Z. mays* (It, Turk); u. diur. (best fresh, most u. dried) - **c. smut**: *Ustilago maydis*, *qv.* - **c. snakeroot**: *Liatris spicata* - **soft c.**: *Z. m.* var. *amylacea-saccharata*; grown for silage or fodder - **c. stalk plant:** *Dracaena fragrans* (*Dr. massangeana*) - **c. starch**: Amylum (L.): derived fr. *Zea mays* grain endosperm; u. culinary agt. (to thicken gravy, soup, sauces; as pudding; blanc mange); to stiffen laundered cloth; in mfr. artificial gums, dextrins - **c. steep liquor**: a chief component of culture medium u. in growing *Penicillium* spp. for penicillin prodn.; u. with lactose - **c. sugar**: glucose - **sugar c.: sweet c.**: *Z. m.* var. *rugosa*; endosperm somewhat translucent; sweetish; a small pl. - **c. syrup**: liquid glucose - **c. tassels**: staminate inflorescence; u. as livestock & poultry feed; rich in vitamins - **Turkish c.**: Indian c. - **c. whiskey**: liquor made fr. c. meal mash; basis of Bourbon whiskey ("corn") - **Yankee c.**: flint c.

Cornaceae: Dogwood Fam.; shrubs & small trees of nHemisph.; small fls. often surrounded by petal-like bracts; incl. genera *Cornus*, *Corokia*.

cornegrasse (Fr): *Apera spica-venti*.

cornel (tree): *Cornus*.

cornel-cherry: frt. of *Cornus* spp. (**Cornelian cherry**), esp. *C. mas*.

cornezuelo de centeño (Sp): ergot of rye.

Cornicularia aculeata Ach. (Cetraria a. E. Fries) (var. *spadicea*) (Usne. - Lichenes) (temp. zone): cornicularia lichen; brown prickly l.; u. as brown, purple, orange dye (SC horn-like branches of thallus); in med. like *Cetraria islandica*; c. protolichesterinic acid (no usnic acid); grows in Ge & Swiss Alps; US mts.

corniculate: bearing a little horn or horns.

cornouille (Fr): berry of *Cornus*.

Cornu Cervi (L): hart's horn (USP 1820-30); ammonium carbonate. - **C. C. Ustum Nigrum** (L): bone black.

Cornus: 1) genus of some 45 spp. (Corn.) (Euras, NA): <L. cornus (horn), SC hard horny texture of wd. (*cf.* cornute [horned], kranos [Gr]), cornel 2) (L) specif. *C. florida* bk. - **C. alternifolia** L. f. (eCana & US): pagoda dogwood; bk. u. astr. antipyr. diaph. - **C. amomum** P. Mill. (C. sericea L'Hér.) (eNA): silky (or swamp) dogwood; bk. ("American red osier bark") "red willow"; kinnikinnik; u. diarrh. edema, expect. dyspep., pd. bk. in tooth powd. (to preserve teeth) (Homeopath.); inner bk. as tobacco; frt. edible (Am Indian); ornam. shrub - **C. canadensis** L. (nNA, neAs): bunchberry (cornel); dwarf cornel; herbaceous; bk. of rhiz. u. similar to that of *C. alternifolia*; frt. edible, tea fr. rt. u. for colic (babies) (Am Indians); pl. u. cardiac dis., boiled with wintergreen for colds; whole pl. in "fits"; orn. - **C. circinata** L'Hér: *C. rugosa* - **C. florida** L. (e&cUS): flowering (or American) dogwood; cornel; boxwood (1375); rt. bk. c. cornin (NBP), resin, tannin; u. antipyr. ton. astr.; formerly as substit. for cinchona (Old South); antiperiodic (in mal.) (Civil War); fls. formerly u. to make ague (malaria) pills (Ala); carm.; wd. to mfr. spindle & shuttle blocks (for cotton mills) (hence [American] boxwood) (1132); to make violins (Ala [US]); boiled wd. u. as bait for muskrat traps; bunched saplings u. to make "brush broom" to sweep yards of farm houses (seUS, Ga) - **C. foemina** P. Mill. (C. stricta Lam.) (seUS nw to sIll): stiff (cornel) dogwood; swamp d.; shrubby pl.; scrapings of rt. or bk. u. for decoc. for fever & mal. (Houma, Indiana, La [US]); cult. ornam. - **C. mas** L. (Euras): male dogwood; Cornelian cherry; fls. & lvs. c. rutin, quercetin; lvs. u. as tea sustit.; frt. edible (1132) (1375), u. to make frt. juices (rob), syr., wines to quench thirst in feverish patients; marmalades, preserves; bk. astr. in diarrh. dys. fever (folk remedy); cult. ornam. - **C. nuttallii** Audub. ex. T & G (Calif, Ida [US], BC [Cana]): Pacific (or mountain) dogwood; (white) d.; western flowering d.; shrub/tree; bk. u. as antipyr. (like quinine), in chronic diarrh., sores; wd. valuable; frt. eaten raw or cooked; cult. ornam. - **C. officinalis** Sieb. & Zucc. (Kor, Chin, cult., Jap): frt. edible (1375); nutr. roborant; incr. hormone secrns.; anthelm, antilithic, impotence, backache; bk. astr. ton. - **C. rugosa** Lam. (C. circinata L'Hér.)

(e&cNA): green osier; round-leaved dogwood; bk. u. bitter, astr. antipyr. - **C. sanguinea** L. (Eu, wAs): (bloodtwig) dogwood; dogberry; wd. useful for bobbins ("pegwood"); frt. juice u. in feverish dis.; in marmalades; sd. furnishes FO wh. is u. for soap mfr. & illumination; ornam. shrub - **C. sericea** L. (C. stolonifera Mx.) (NA-wet places): red (osier) dogwood; hart's poison; "red willow" (settlers, BC); western (or American) dogwood; bk. of shrub decoc. u. as antipyr. (120); rt. bk. in diarrh. dys.; inner bk. as dye & lvs. in smoking (Am Indians).

Cornutia latifolia (HBK) Mold. (Hosta l. HBK) (Verben.) (Mex, CA): pl. boiled with oil & honey & u. in asthma (121) - **C. microcalycina** Pavón & Moldenke (SA): Santa María; lf. u. in headache (Ve); var. *pulverulenta* Moldenke (SA, Co): med. pl. of Bayano Cuna Am Indians; u. to drive away lice fr. chicken nests - **C. odorata** (Poepp. & Endl.) Poepp. (Pe, Co, Ec): aquera (Peru); small tree; lvs. macerated in water & u. to wash head to treat earache, headache; rheum. pains (Ec); sapwood heated in water & water dropped in eye (for eye pain) (Pe); lvs. mixed with *Banisteriopsis caapi* to give stronger hallucinations - **C. pyramidata** L. (WI, CA): "salvia lugo"; blue frts. of tree are source of dye & ink (Domin, Yucatan); bk. u. as syr. in rheum.; lvs. ingredient of tea u. for retroversion of uterus (Domin, Caribs).

cornutine: alk. fr. ergot.

Corokia buddleoides A. Cunn. (Corn.) (NZ): lf. ext. u. in gastric pain.

corolla: floral whorl lying within the calyx and external to the essential floral organs (stamens & pistil); made up of segments, the petals, either fused or separate.

corona: outgrowth ("crown") fr. corolla seen in some fls.; ex. daffodil.

Coronilla scorpioides Koch (Leg.) (c&sEu, wAs): **scorpion coronilla**; hb. & sd. c. **coronillin** (digitaloid glycoside); sd. c. pseudo-cumarin, 4% FO; hb. u. cardiotonic, diur. for prep. coronillin; poisonous - **C. varia** L. (seUS; orig. Eu): crown vetch; devil's shoe string (sUS, Negro); c. digitaloid glycoside (coronillin); u. as purg. diur. digitalis substit.; cult.; u. in Negro folk med. & voodoo (in "hands").

Coronopus didymus (L) Smith (Cruc.): wart (or swine) cress; calachin (Arg) quimbe (Arg): hb. u. expect. antiscorb. antimal. (Arg) - **C. squamatus** Aschers. (Carara coronopus Medic.; Senebiera c. Poir.) (Eu, natd. eNA, Arg.): buck('s) horn; sowgrass; wart cress; hb. u. in scurvy; ashes u. against urinary stones.

corossol: *Annona squamosa.*

coroza (nuts): **corozo(n)**: 1) sd. of ivory palm, *Phytelephas macrocarpa* 2) *Elaeis melanococca* nuts 3) *Orbignya* spp. nuts.

Corpora Lutea (L.): pl. of Corpus Luteum - **corporin**: progesterone.

Corpus Luteum (L.): the product occupying the cavity left by the Graafian follicle after it breaks fr. the surface of the ovary; secretes progesterone; where pregnancy occurs, the C. L. remains active, otherwise it atrophies.

Corrigens: Eriodictyon.

Corrigiola telephiifolia Pourret (Caryophyll.) (Mor): towsergent; serghina (Arab); sarghine (maigre) (Fr); rt. u. in mixts. for cough, as diur. aphrod., in skin dis.; in cosmetic powders; fumes inhaled for flu, colds, as ton. fortifier (2336).

corrozo: coroza.

Corsican moss: *Alsidium helminthochorton.*

Cortaderia fulvida (Buchan.) Zotov (Gram) (NZ): toetoe; large coarse grasses with large plumes (compd. inflorescence) (u. to stop blood flow); pl. ashes appl. as poultice to burns; young lvs. eaten for diarrh.; juice of lower st. u. to clear infants' tongues; st. chewed & swallowed for kidney dis. - **C. selloana** (Schultes & Schultes f.) Asch. & Graebn. (Br, Arg, Ch): pampas grass; u. for paper mf. (SA); lawn grass; dr. inflorescence (ornam.); much cult. - **C. splendens** Connor (NZ): kakaho (Maori); u. much like *C. fulvida.*

Cortés, Sea of: Gulf of California.

Cortex (L.) 1) rind or outer portion of frt.; ex. lemon peel & other citrus rinds 2) bark (of trunk or root or branches) - **C. Alstoniae**: bk. of *Alstonia constricta* - **C. Angusturae verae**: bk. of *Cusparia trifoliata, C. febrifuga* - **C. Avaremotemo**: bk. of *Pithecolobium avaremotemo* (Por Phar.) - **C. Betulae**: birch bk. - **C. Caryophyllati**: clove bk. fr. *Dicypellium caryophyllatum* - **C. Chinae**: bk. of *Cinchoma succirubra* - **C. Chinae fusc.** (v. *provincianae*): C. Fuscus - **C. Colpachi**: fr. *Croton niveus* - **C. culiliban**: culilawan bk. - **C. Evonymi**: fr. *Euonymus atropurpureus* - **C. Fructus phaseoli:** bean pod - **C. Fuscus**: brown cinchona bk. (*C. nitida*; may also be *C. officinalis* or *C. micrantha*) - **C. Hippocastani**: fr. *Aesculus hippocastanum* - **C. Mezerei:** fr. *Daphne mezereum* - **C. Musannae: C. Musennae**: fr. *Albizzia anthelmintica* - **C. Nucum juglandis**: husk of frt. of *Juglans regia* - **C. Oliveri**: bk. of *Cinnamomum oliveri* - **C. Pareirae: C. Pereirae**: bk. of *Geissospermum laeve* - **C. primarius** (L): primary bk.; all basic tissues betwn. the outer layer & the peripheral vascular bundles - **C. Quercus**: fr. *Quercus robur, Q. petraea* - **C. Q. Ilex**: fr. *Q. ilex* - **C. Quillaiae**: Quillaja - **C. Radicis mali**: apple tree (root) bk. - **C. Rhois aromaticae**: fr. *Rhus aromatica* - **C. Thymiamatis**: bk. of *Liquidambar* sp.; used in prepn. Styrax Calamita - **C. vouacapouae**: bk. of *Andira inermis, A. retusa* - **C. Winterianus**: bk. of *Drimys winteri.*

corteza (Sp): 1) bark 2) peel, rind (of. frt.) - **C. de Carony: C. Orajure**: bk. of *Cusparia.*

cortiça (Por): cork.

corticcia (It): bk. - **c. Peruviana** (Chin): Cinchona.

corticoids: **corticosteroids**: adrenal cortex active compds. serving as endocrines (2281) (incl. some synth. derivs.) (types: mineral c.; gluco-c. (*qv.*) - **mineral c.: mineralocorticoid:** adrenal corticoid wh. chiefly regulates mineral metabolism, increasing Na retention; ex. deoxycorticosterone - **corticosterone**: a steroid hormone of the adrenal cortex, very similar to Cortisone (glucos-corticoid type); u. to improve muscular efficiency, carbohydrate metabolism, endurance to stress (poisons, cold, heat, etc.), electrolyte metabolism; essential to life.

corticotrop(h)in: adrenocorticotrop(h)ic hormone (ACTH); acts as adrenal gland stim. (glucocorticoid) incr. no. of RBC in body; effects pigmentation; u. in diagnostic tests; same indications as cortisone, &c. - **c. releasing hormones:** c. r. factor; CRF: secretion of hypothalamus (interbrain) wh. stim. secretion of ACTH & endorphin; prod. fr. pituitary of ovine origin.

cortin: adrenal cortex extract (natural); also u. as generic name for adrenal cortex hormones; formerly u. in literature (Fr, &c.) but not currently.

cortina (mushroom): shreds of inner veil left at margins of pileus like a thin curtain.

Cortisol: hydrocortisone.

Cortisone: Cortone™: endocrine compd. of adrenal cortex; as acetate, u. in state of adrenal cortex insufficiency: to treat inflammations, allergy (as asthma), psoriasis, gout, burns, snakebites, postoperative collapse, etc. (1132).

corvicide: crow-killing agt.

corydal, yellow: *Corydalis lutea* - **corydaline:** alk. chiefly fr. *Corydalis cava.*

Corydalis DC. (Papaver.) (nTemp Zone): wild fumeroots (SC rts. emit nitrous fumes); c. 340 spp., mostly perenn. hbs. (1233), specif., (L); dr. tubers of *Dicentra cucullaria, D. canadensis*; some 40 spp. u. in Chin med. - **C. ambigua** Cham. & Schlecht. (Manch.): u. for analg. in internal pain; antirheum.; dysmenorrh.; boils - **C. aurea** Willd. (NA): golden corydalis; colic weed; c. protoberberines, corpaverine (mixt.); rts. placed on hot coals, the smoke inhaled for "clearing the head" - **C. bungeana** Turcz. (eAs esp. Korea): u. in folk med. as antipyr. anti-infect.; in boils, carbuncles - **C. cava** Schweigg. & Koerte (C. tuberosa DC) (sEu): round birthwort; tubers c. bulbocapnine; u. vermifuge, depur. discut.; emmen.; ext. or bulbocapnine u. in hyperkinetic conditions, Ménière's dis., postencephalic Parkinsonism; chorea; muscle tremor; writer's cramp; sed. or hypnotic (with morphine & atropine in place of scopolamine) for pre- & post-narcotic agitation; source of bulbocapnine - **C. falconeri** H. & T. (nIndo-Pak): wash prepd. fr. this pl. u. to induce hair curliness (Nil, Kashmir) - **C. filistipes** Mak. (eAs): corms u. to treat dysmenorrh., lumbar pains, carm. - **C. formosa** Pursh: *Dicentra formosa* - **C. lutea** (L) DC (sEu): yellow corydalis; u. ton, yellow dye - **C. racemosa** (Thunb.) Pers. (Kor+): decoc. appl. to abscesses; int. in fever, bleeding piles; infections, jaundice - **C. ramosa** Wall. (Ind, Pak): mamiran (Kurram valley, Pak): pl. sap u. in eye dis. (Lahul, nInd) - **C. sempervirens** (L.) Pers. (eNA): pale corydalis; hb. ("Roman wormwood"); rt. u. bitter astr. anthelm. emmen.; ornam. hb. - **C. stewartii** Fedde (nPak, Ind): mamiri; pl. juice put in eye with surma (Changla) - **C. tuberosa** DC: various *C.* spp. so referred to; ex. *C. cava, C. pumila* (Host) Rchb., etc. - **C. yanhusuo** W. T. Wang (Chin): Chinese fumewort; rhiz. u. in indig. med. as hypnotic, sed.; in cardiovasc. ailments.

Corydothymus (*Coridothymus*) **capitatus** Rchb. = *Thymus c.*

Corylaceae Mirbel: Betulaceae.

Corylus americana Marshall (Betul.) (e&cNA): (American) hazelnut; nuts of this shrub commonly eaten; nut said diur. anod. anthelm. source of FO, hazelnut oil - **C. avellana** L. (Eu): (European) filbert; hazelnuts; European hazel; the frt. & sds. larger than those of prec. & foll.; the common filbert of commerce (cult.); nut edible & nutritive; nuts eaten salted, roasted, spiced, &c.; u. in chocolates; candy, cookies, cakes, nut breads, salads, &c.; source of FO (50%); widely cult. (USSR, Spain [Barcelona]) (considered best) - **C. cornuta** Marsh. (C. rostrata Ait.) (NA): beaked hazel(nut); California hazelnut; shrub; nuts edible & nutritive food; sometimes ground to give flour to make bread; source of "nut oil" u. in eye dis. (Abenakis) - **C. tubulosa** Willd. (C. maxima Mill.) (Euras): (Lambert's) filbert; sds. of shrub or tree (-10 m) edible ("Lambert's nuts").

corymb: a fl. cluster somewhat like a raceme but with longer pedicles on those fls. originating lower on main axis, resulting in a rather flat-topped broad inflorescence; ex. cherry, *Kalmia.* (v. appendix)

Corynanthe johimbé K. Schum. (Pausinystalia j. Diels) (Rubi.) (wAf): source of Yohimbe Bk., c.

yohimbine; u. antipyr. aphrod. - **C. pachyceras** K. Schum. (tropAf, esp. Congo): pseudocinchona; bk. c. corynanthine, corynanthidine, &c.; source of a false yohimbe bark; u. aphrod. local anesth. (less effective than cocaine), antipyr. in cough, leprosy; in arrow poisons; wd. used.

Corynebacterium diphtheriae (Kruse) Lehmann & Neumann (Mycobacteriaceae): the microorganism causative of diphtheria; straight to sl. curved bacilli.

Corynostylis arborea (L) Blake (Viol) (sMex, CA, nSA, WI): "monkey apple" (Belize); lvs. u. in fever (SA); rt. u. emetic (SA) - **C. volubilis** L. B. Smith & Fernández Pérez (Col, Pe): bk. decoc. (tea) u. for intestinal parasites; stim.

Corypha cerifera Arruda da Camara = *Copernicia prunifera* - **C. elata** Roxb. (Palm) (Ind, Burma, Mal, PI): u. source of jaggery, (sugar), toddy, vinegar, starch; fibers of petiole u. for mfr. hats, etc.; decoc. of young plants u. for colds ("febrile catarrh") - **C. palmetto** Walter = *Sabal p.* - **C. utan** Lam. (C. umbraculifera L.) (sInd, SriL, Mal): talipot palm; large tree (-24 m high); rt. decoc. u. for slight diarrh.; rts. chewed to relieve cough (also gum fr. tops); lf. juice emet.; pounded young frts. u. as fish poison (stuns); apical bud ("cabbage") eaten; lvs. u. as umbrellas, thatching, writing material (paper for Buddhist books, using metal stylus).

cos: c. lettuce (GB): **c. salad**: romaine lettuce; *Lactuca sativa* var. (<Cos or Kos (Khios) island) (Greece).

Coscinium fenestratum Colebr. (Menisperm.) (SriL, Ind): wd. c. 3.5 % berberine; wd. & rt. = False Calumba; rt. known as Pareira (not true Pareira); rt. u. bitter ton. in weakened states, fever, dyspeps.; furnishes dye like turmeric - **C. wallichianum** Miers (Ind): wd. & rt. c. berberine, palmatine; u. as arrow poison.

cosecha (Sp): harvest time (for frt. or grains).

Cosmegan™: Actinomycin D. (Merck [US]) = Dactinomycin.

cosmetology: the science & art of cosmetics, or materials appl. to improve ("beautify") the physical appearance of a person, esp. the face.

Cosmoline™: petrolatum usually white.

Cosmos caudatus Kunth (Comp.) (WI, Mex to Co; natd. OW trop): Spanish needles; mapola; u. to remedy muscular strains & spasms (Virgin Is) - **C. parviflorus** (Jacq.) HBK (sTex, NM [US], Mex): southwest cosmos; amores (Mex); dr. fls. inf. u. as purg.; in chest colds, pertussis (NM) - **C. peucedanifolius** (v. *tiraquensis*) (Kze.) Schott (Pe): panti (panti); fls. u. pulmonary dis. (decoc.), diaph.; rts. u. in pleurisy; cult. ornam. - **C. pringlei** Robins. & Fernald (nwMex): bavisa; rt inf. u. in headache; intestinal dis.; rt. as poultice for sores - **C. sulphureus** Cav. (Mex to Guat; natd. WI, CA, SA, sFla): "sunflower"; morocota; fl. decoc. u. in scorpion sting; card. sed.; to heal sores (decoc. to bathe skin) - **C. tiraquensis** (Kee.) Schott: flores de panti-(panti); u. pulmon. dis.

cosmosin: cosmetin: apigenin-7-0-glucoside; flavonoid found in *Bellis perennis, Alstroemeria pulchella*, &c.

costa de balsamo (lit. Balsam coast): **c. de Sonsonate** (Sp): area along coast of Salvador & Nic where Peru Balsam is chiefly collected.

Coster's paste: iodine diss. in light oil of wood tar; u. in ringworm.

costmary: *Balsamita major.*

Costus L. (Cost. or Zingiber. - Monocots) (tropics): costus; "wild ginger (of the West Indies)"; pinecone lily (Fla [US]); putchuk; some 100 spp., some cult. as ornam., esp. fr. SA; species u. similarly to *C. spicatus* include *C. glabratus* Sw., *C. mexicanus* Lieb., *C. ruber* Griseb., and *C. spiralis* Rosc.: the common name Costus has also been referred to *Saussurea* (incl. *S. costus*) and to *Cordiopatium corymbosum* - **C. afer** Ker (cAf, Senegal): rt. u. in constipation; rt. exter. for rodent ulcers (skin cancer) & abscesses; the sap appl. to wounds gives a sharp sensation of heat; diur.; for coughs, rheum; rts. in paste u. exter. for phagedenic ulcer (bacterial inf.); st. prepn. appl. to sores - **C. amarus** (L.): *Mussaenda landia* (M. stadmanni Mich.) - **C. corticosus** (L.): **C. doux: C. Dulcis**: *Canella winterana* - **C. root**: rt. of *Saussurea lappa*; source of the valuable c. oil - **C. lucanusianus** A. Chev. (Nigeria): juice u. for cough, bronchitis, rheum. - **C. mexicanus** Lieb. (Mex): caña dejabali; u. diur. (effective); rt. chewed (or decoc. drunk) for urinary calculi & in GC - **C. pictus** Don (Mex): u. as diur., in kidney cancer - **C. puberulenta** Presl (Mex): u. similarly to *C. mexicanus* - **C. speciosus** Sm. (C. arabicus L.) (eAs): cane reed; Arabian costus; c. plant; sap fr. crushed st. u. for diarrh.; eye trouble (Laos); lf. decoc. u. antipyr. (Mal), fresh rhiz. (Radix C. Dulcis; Kust root) juice purg.; antispas.; sds. c. diosgenin, FO; cult. ornam. - **C. spicatus** Swartz (C. cylindricus Jacq.): (sMex to Br, WI; cult. in sFla [US]): cañita agria (Tabasco, Mex); "pinecone lily" (Fla); st. & rts. u. in hepatic congestions; sap u. stim. emmen, diur.; appl. to skin to relieve burns (2444); rhiz. decoc. & inf. for VD (Trin) (1235) - **C. Spurius** (L.): *Opopanax hispidum* - **C. villosissimus** Jacq. (Mex, Guat to

Guyana, Pe): cañagria; st. decoc. u. kidney inf., VD.

cotarnine chloride: the HCl of a synthetic alk. prep. fr. narcotine (of opium).

cotinine: nicotine matabolite, pyrrolidine deriv., found in sputum of individual exposed to tobacco smoke (incl. non-smokers); u. antidepress.

Cotinus coggygria Scop. (Rhus cotinus L.) (Anacardi.) (Euras): (common) smoke tree; Hungarian fustic; wd. source of "young fustic"; lvs. c. tannin; u. brown & yellow dye; astr. - **C. obovatus** Raf. (Cotinus americanus Nutt.) (seUS): (American) smoke tree; yellow wood; to 10 m high; wd. yields Young Fustic; u. astr., yellow dye.

coto bark: c. fino: c. piquiante (Bo): **true c. (bark)**: fr. *Nectandra* (*Aniba*) *coto* (*cf.* paracoto); sometimes claimed as the bk. of *Drimys winteri* - **false coto: coto ordinario** (Sp): *Nectandra pseudocoto.*

coton (Fr): cotton - **cotonnier** (Fr): cotton pl.; c. products, etc.

Cotoneaster Medik. (Ros.) (nTemp Zone): rose-box; ca. 60 spp.; pls. resemble *Abelia* - **C. bacillaris** Wall. (Indo-Pak +): stolons u. astr. (Indoch) - **C. horizontalis** Decne. (Chin): lf. decoc. expect. - **C. microphylla** Wall. (Indo-Pak, Nep, Him): lvs. c. prulaurasin (producing HCN); stolons u. astr. (Indoch) - **C. racemiflora** Koch (C. nummularia Fisch. & Mey.) (nAf, cAs): pl. aper. stom. (Lahul); expect.; pl. source of Chir-Kist, manna-like subst. with sugars eaten in Iran, Ind.

cotton, absorbent: Gossypium Purificatum (OLT): the purified, bleached & sterilized trichomes of the sd. of *Gossypium hirsutum* & other *G.* spp.; consists of almost pure cellulose; u. absorbent, protective; sutures; color generally whitish, but sometimes other (ex. brown Peruvian c.) (cotton <Arab) - **American upland c.**: upland c. - **c. "buds"** (wGa [US]): c. seeds - **c. bush**: pl. of *Gossypium* spp. - **dead c.**: hairs of immature sds. of c. - **c. gin:** machine u. to separate seeds & lint of cotton bolls fr. the field; word is corruption of "engine" - **c. gum**: *Nyssa aquatica* - **hydrophilous** (or **hygrophilous**) **c.**: absorbent c. - **levant c.**: *G. herbaceum* - **c. oil**: cottonseed o. - **Piura c.**: a long-stapled form fr. P. Dept., Peru - **purified c.**: absorbent c. - **c. root bark**: freshly coll. air-dr. bk. of *G. barbadense*, *G. herbaceum* & other *G.* spp.; c. resins, FO; u. emm. oxytoccic, abort. - **Sea Island c.**: fr. *G. barbadense*; superior because long-fibered c. - **c. shell seed**: *Cochlospermum religiosum* - **silk c.**: sd. hairs of 1) *Asclepias* spp. 2) *Ceiba pentandra* (**5-leaved silk c. tree**) 3) *Eriodendron anfractuosum* - **c. thistle**: *Onopordum acanthium* - **c. tree**: *Platanus occidentalis* - **upland c.**: *G. hirsutum*; generally inferior to Sea Island c. mainly because of its short fiber - **wild c.**: *Apocynum cannabinum & A. androsaemifolium.*

cottonseed: dr. ripe sd. of *Gossypium hirsutum* & other *G.* spp.; flour has been u. as nutr. but hazardous as gossypol may be present; c. cakes u. in prodn. penicillin (Chin); FO as **cottonseed oil**: FO obt. by expression fr. sds. of *G. hirsutum* & other *G.* spp. (yield c. 15%); one of the cheapest & most readily available FO in US; nutr., rich in tocopherols; u. culinary substit. for & adulter. of olive oil; hydrogenated to make margarine & lard substit. (2575); soaps; med. as other bland FO; in packing sardines & other canned fish; to reduce cholesterol level in blood (Duke U. [NCar, US] research); allergy to sd. & FO occasionally seen.

cotton weed: *Gnaphalium* spp.

cottonwood (SC cottony tomentum of sds.): 1) *Populus deltoides* & some other *P.* spp. (2773) 2) (Ga [US]): *Paulownia tomentosa* - **valley c.** (NM [US]): *Populus wislizeni.*

Cotula (L.): May-weed; *Anthemis cotula* - **C. aurea** L. (Comp.) (Med. Sea region): fl. heads u. ton. diaphor. anthelm. antipyr. (Sp) - **C. coronopifolia** L. (sAf, originally; now widely distr.): buck's horn; Kuhdille; c. several polyenes; u. med. as substit. for chamomile fls. - **C. multifida** DC (sAf): u. in rheum. inflamm. & exanthemata.

cotyledon: 1) sd. leaf; the primary leaf of the embryo, serving as source of stored nutrition for the developing seedling, later as an early foliage lf. with photosynthetic functions - **Cotyledon** (L.): pennywort, *Umbilicus pendulinus, U. rupestris.*

Cotyledon orbiculata L. (Crassul.) (s&swAf): pig's ear(s); fresh lf. juice u. in epilepsy; fleshy part of lf. appl. to corns for easy removal; heated lf. u. as poultice for boils, earache, otitis media; widely u. in many dis; toxic.

couch (grass): **English c.**: *Elymus repens* - **Indian c.: plain c.**: *Cynodon dactylon* - **red c.**: *Carex arenaria*

Couepia floccosa Fritsch (Ros./Chrysobalan.) (CR): olosapo; tree bears delicious arom. frts. - **C. polyandra** Rose (Mex): zapotillo; zapote amarillo; frt. edible.

cough root: 1) *Trillium pupureum* 2) osha - **c. wort**: coltsfoot.

coulevrée (Fr): Bryonia.

Couma Aubl. (Apocyn.) (neSA): frts. palatable, u. in desserts, beverages; latex drunk directly or mixed with coffee or cassava porridge; u. as vermifuge (with sds.), as cement, varnish, &c. - **C. macrocarpa** Barb. Rodr. (Br, Co): latex source of Perillo (or Sowa) gum, u. as chewing gum base; berries eaten (natives); wd. in interior carpentry, ship masts; resin derived fr. tree u. as varnish in ceramics, to caulk canoes (Amazon Indians); pd. frts. u. for skin parasites - **C. utilis** M. Arg. (nBr, esp. Amazonas & Pará): **couman** (Br): cherry-sized savory; sorveira; latex mixed with castor oil or HOH makes good anthelm. for livestock; also u. to caulk canoes, boats, etc.; resins u. in varnishes, for ceramics (sorva gum); frt. edible.

coumarin (Cumarin, Ge): class of compds. lactones of o-hydroxycinnamic acid (coumarinic acid) found in nature; 238 shown in Devon Scott listing; sometimes occ. as glycosides; specifically, the lactone of coumarinic acid, sweet-smelling constituent of tonka, sweet clover, new mown hay, woodruff, etc.; reminiscent of vanilla; u. perfumes; flavors for medicines, tobacco; food flavor use discountenanced (1953) since chronic poison to laboratory animals - **c. bean**: *Dipteryx odorata*.

coumestrol: a coumarin isolated fr. alfalfa & strawberry clover.

coumou oil: fr. *Oenocarpus bataua*.

Coumarouna odorata Aubl. (&c): *Dipteryx o.* (&c.).

countercurrent distribution: process u. for solvent extn. involving multiple automatic repeated extn. operations with vol. solvent; developed by Lyman C. Craig ca. 1944 (122).

courage pills (US addict's slang): Heroin tabs.

courbaril resin: fr. *Hymenaea courbaril*.

Courbonia Brongn. = *Maerua*.

courge (Fr): gourd - (**c.**) **potiron: courgette** (Fr): squash, pumpkin, *Cucurbita* taxa.

couronne de moine (Fr): dandelion.

Couroupita guaianensis Aubl. (Lecythid.) (tropAm, WI): cannon-ball tree; frt. flesh yields a cooling beverage.

Coursetia glandulosa Gray (Leg) (Mex): broken branches secrete a gum, goma de Sonora; eaten with chile said good for stomach; lac fr. st. u. as adhesive for hafting harpoon heads &c. (Calif Baja Indians).

court plaster: isinglass plaster.

couscous (nAf): dish of crushed & steamed cereal grains (ex. rice) with meats & vegetables.

cousso: kousso.

Coutarea hexandra Schum. (Rubi) (sMex to SA): quin(it)a; bk. u. much like Cinchona in mal. but not as effective, known as "quinquina de Cumana"- **C. latiflora** Moç. & Sessé (Mex, Guat): copalchi; (falsa) quina; bk. u. as febrifuge (esp. mal. fevers); ext. (Sucontral) hypoglycemic in mature type diabetes (1265); bk. c. quinine, quinidine, flavonoid glycosides - **C. pterosperma** Standl. (Mex): copalchi; copalquin; bitter bk. u. for fevers, esp. malar. & pulmonary dis.; bk. ext. has been u. to adult. & sophisticate smoking opium.

Coutinia Vell. Conc. = *Aspidosperma*.

couve (Por): collards, rape, cole, &c. - **couve-flor** (Por): cabbage - **c. repolho** (Por): *Brassica oleracea* var. *capitata*.

cover crop: crop planted on land to prevent erosion & weed growth & to improve soil conditions.

Covillea A. M. Vail = *Larrea*.

cow bean: c. pea - **c. lice**: *Bidens frondosa* - **c. pea:** *Vigna unguiculata* - **c. plant (Australian): c. plant of Ceylon**; *Gymnema sylvestre* - **c. tree**: *Brosimum utile* - **c. wood** (NM [US]): dr. cattle manure chips u. as fuel.

cow: v. also parsnip; pea.

cowage: hairs fr. pods of *Mucuna pruriens* - (**spotted c.**: *Cicuta maculata*, less commonly & less correctly *Conium maculatum*) - **cowbane:** *Cicuta maculata, C. virosa* - **cow-berry**: *Vaccinium vitis-idaea* - **cow-cockle**: *Vaccaria pyramidata* - **cow-cumber tree** (Ala [US]): cucumber tree - **cowflop**: 1) *Digitalis*, esp. *D. purpurea* (also cowslop) 2) (Cornwall, Eng): *Cicuta virosa* (?) - **cowgrass**: *Trifolium pratense* - **cowhage**: cowage - **cowherb**: *Vaccaria pyramidata* (SC said favorite forage of cattle) - **cowitch (vine)** (prob. corruption of cowage): 1) *Mucuna pruriens, M. sloanei* 2) *Bignonia radicans* 3) *Decumaria barbara* 4) *Rosa canina* .

cow's lips: *Caltha palustris* - **c. milk**: the lacteal secretion of female animals of *Bos taurus*, secreted during the months after delivery of calves; u. nutr. bev., mfr. ice cream, cheese, casein, lactose, etc.

cowslip (flowers): 1) (GB): *Primula veris* 2) (US): *Caltha palustris* 3) *Mertensia virginica* (**Virginia c.**).

coxcomb: *Celosia* spp.; also several wild pl. spp.

Coxe's hive syrup: Syrup Squill Comp.

coyote: *Canis latrans* Say (Canidae): orig. wNA, now found NA to PR; fat cooked & swallowed for chest colds & tired muscles; furs marketed.

cozticpatli (Mex): *Thalictrum hernandezii*.

crab: one of numerous genera & spp. of Crustacea (ord. Decapoda); crabs' eyes & "claws" were off. in the earliest Brit Pharm. - **king crab:** 1) *Limulus polyphemus* 2) large crab of Pacific (popular food); v. also c. louse; c. thistle.

crab apple: crabapple - **c. (wood) oil**: carapa o. - **crabapple**: *M. sylvestris, M. angustifolia, M. baccata* & several other *M.* spp.; some cult. for frt., others ornam. only; some edible, some not - **crab-eye: crab's eye**: *Abrus precatorius* - **crabwood**: *Carapa guianensis*.

Crabbea velutina S. Moore (Acanth.) (eAf): lf. inf. u. as anthelm. for ascaris infections (Moz); fresh lvs. crushed & vapor inhaled for headache (Kenya); for pain in eyes, boil the rts. & allow vapor to make contact (Kenya); for skin eruptions, powder dr. pl. & apply to affected part over several days (Kenya).

crab's claw: *Polygonum persicaria*; *Stratiotes aloides*.

crab's eyes: c. stones: Lapis Cancri (*qv.*).

Cracca Benth.: an important synonym for some *Tephrosia* spp.

crack (cocaine) (US slang): pure form of cocaine; smoked; very dangerous (depressing) (1985-); prod. orgasmic-like sensations of pleasure; said "instantly" addictive; at times u. with Heroin (by needle or smoked).

cracker berries (Cana): frt. of *Empetrum nigrum*.

crackling (seUS): bit of pork meat with lard fried out.

craie lavée (Fr): prepared chalk.

Crambe abyssinica Hochst. ex. Fr. (Cruc.) (Ethiopia): crambe; colewort; sds. c. ca. 45% FO with ca. 60% erucic acid (*cf.* rape oil); u. as substit. for rape(seed) oil in indy. (plastics, synth. fibers); vol. isothiocynanates & non-vol. thiooxalzolidone constits. make toxic; may be detoxified to furnish nutritional feed (meal) for stock (1549) - **C. cordifolia** Stev. (Ind): pl. u. for itch ; cult. - **C. maritima** L. (Euras, esp. wEu seacoasts of Atl Oc & Baltic seas, nAf): sea kale; a shrub (1 m); sds. rich in FO (**Crambe oil**), with high content of erucic acid glycerides, u. lubricant in jet engines; polymers in mfr. plastics (Nylon); sd. meal (fr. press cake) may be u. (in blends) for cattle fodder; lvs. u. as potherb (Eng); rhiz. u. as potherb (Chin).

crameria (Mex): *Krameria*.

cramp bark (, true): *Viburnum opulus*.

cranberry: name appl. to several different pl. spp. incl. *Arctostaphylos uva-ursi* (mountain c.); *Chiogenes hispidula* (white c.); *Hibiscus sabdariffa* (Florida c.); chiefly *Vaccinium oxycoccos*, *V. macrocarpon*. & several other *V.* spp.; *Viburnum opulus* - **American c.:** large c.: *Vaccinium macrocarpon* - **European c. (bush):** small c. - **high bush c. (bark)**: bk. of *Viburnum opulus* v. *americanum* - **mountain c.: rock c.**: 1) *Vaccinium vitis-idaea* L.; 2) *Arctlostaphylos uva-ursi* - **small c.**: *Vaccinium oxycoccos* - **c. tree**: *Viburnum opulus*.

cranes: graceful water birds of genera *Grus* & *Balearica* (Gruidae) (nHemisphere & tropAf): superficially similar to herons; incl. *Grus grus* L. (common or eastern c.) (Euras, migrating to tropAf); notable for its dignified migration & mating behavior; domesticated by ancient Egyptians (for meat) & Romans (ornam.); u. in Unani med. (flesh in colic); fat, eggs, brains, testes, bile, &c.; also u.; *Grus virgo* L. (demoiselle c.) (Euras, nAf): meat u. in Unani med. as aphrod.; whooping & sandhill c. (NA) close to extinction.

crane's berry: *Vaccinium oxycoccus* - **cranesbill**: *Geranium maculatum* +.

Craniolaria annua L. (Pedali.) (nSA): rt. decoc. u. mild lax.; roots preserved in sugar, delicacy of WI natives.

crap: 1) buckwheat 2) **c. grass** (seUS): 1) crab grass 2) perennial rye g., *Lolium perenne* (also **crappe grass** (related to carpere [L.]: to pluck or rob, since hurtful to wheat field).

crapaud (Fr-f.): frog.

crape myrtle = crepe m.

Craspidospermum verticillatum Bojer (monotypic) (Apocyn.) (Madag): bk. u. as ton. antitussive, expect., in pulmon. dis. (folk med.); var. *petiolare* DC (Madag): workmen crushing the lvs. to ext. them have allergic symptoms (headache, &c.); attributed to condylocarpine.

crassine acetate: ester of base found in certain gorgoneans (corals) of Caribbean Sea; with antitumor activity.

Crassocephalum subscandens S. Moore (Senecio s. Hochst.) (Comp.) (Af, Ethiopia): lf. u. in abdom. pain, convuls., fever, toothache, and cancer(!) - **C. vitellinum** S. Moore (trop eAf, esp. Ethiopia): lvs. u. for GC, dermal suppurations, dentalgia; antipyr., antispasm. in colic.

Crassostrea virginiana (Ostreidae - Bivalva - Mollusca) (eNA coast): American oyster; shell 5-10-30 cm long; meat delicious, considered by some aphrod.

Crassula arborescens Willd. (Crassul.) (sAf): jade plant; j. tree; shrub (-3m); lf. juice appl. to lesions (US); popular house pl. (US); pl. u. as epilepsy remedy & corn "cure" (Cape); food of Hottentots - **C. argentea** Thunb. (sAf): "Japanese laurel" (misnomer); lf. juice u. astr. in diarrh; nectar u. as purg. - **C. rubicauda** Mey. (sAf): pl. juice in water u. as nasal douche in flu; reputed "cancer cure."

Crassulacean acids: malic, citric, & iso-citric acids; found originally in members of Crassulaceae; now known from 18+ families; in Crassulacean acid metabolism (CAM), these acids accumulate during the night & are converted to sugars during the day then metabolized.

Crataegus L. (Ros.) (nTemp Zone): hawthorn; (red) haw; hog apple; a large & complex genus of large shrubs & small trees; mostly with thorns (representing modified branchlets); cult. for ornam. hedges & edible frt. (pomes) (jellies), some of which are marketed; several spp. are u. medicinally - **C. aestivalis** (Walt.) T & G (wet woods, swamps, seUS): (apple) haw; may haw; frt. u. to prep. jelly, like guava in quality (2652) - **C. altaica** Lange (swSiberia): Russian med. plant - **C. azarolus** L. (eMed zone; cult. Ge): Welsh medlar; fr. rich in vit. A; u. as "stomach strengthener" and against dys., diarrh. - **C. coccinea** L.: now represented by the spp. *C. chrysocarpa* Ashe & *C. pedicellata* Sarg. - **C. crus-galli** L. (e&cNA): cockspur thorn (or haw); rt. u. aper. deobstr., diur. (Mex); frt. u. as exp., in preserves - **C. dahurica** Koehne (eSiber) med. plant of Russians - **C. flava** Ait. (seUS): summer haw; yellow berried thorn; frt. u. for prep. jellies - **C. intricata** Lange (C. coccinea auct. non L.) (eUS): haw; shrub to small tree (-3 m); u. ornam. - **C. marshallii** Egglest. (M. apiifolia Mx.) (seUS n to Mo): (parsley-)leaved (haw) thorn; parsley haw; said to be the closest Am sp. to the European *C. oxyacantha* - **C. mexicana** Moç. & Sess. (Mex): tejocote; frt. u. as preserves & jellies; frt. decoc. u. for cough; rt. decoc. u. as diur. - **C. mollis** Scheele (e&cUS, Cana): red (or downy) thorn; u. diur. heart ton., neuro-vasc. sed.; jelly prepn. - **C. monogyna** Jacq. (Eu, NA): English (or European) hawthorn; close to *C. oxyacantha* & sharing its props; lvs. fls. & frts. u. - **C. opaca** H. & A. (sUS): western (or riverflat) mayhaw: apple haw; frt. much u. for conserves - **C. oxyacantha** L. (Eu, nAf): European (or English) hawthorn; white (haw) thorn; red thorn; lvs. (hb.) & frt. c. glycosides, esp. oxyacanthin; saponins, flavonoids, crataegolic & ursolic acids (1238); frt. u. as astr.; fls. hypotensive u. in arterioscl.; (hypertonia); popular in Eu for cardiac & circulatory problems; diur.; redn. BP due to coronary vasodilation; in angina pectoris (1237); arthritis; nerve dis.; tincture fr. juice of berries sometimes with lvs.; for hedges (Ire+); common ornam. (Calif [US]) - **C. pedicellata** Sarg. (C. coccinea L. p.p.) (NA; cult. Eu): scarlet haw; u. as heart remedy; parts of this small tree u. as nerve sed., astr. in diarrh., emet.; hypotensive - **C. pinnatifidus** Bunge (Chin, Jap): frt. c. flavonoids; u. in cardiovasc. dis.; dyspepsia., as lax., gastric cancer; eaten; carbonized frt. u. in indig. diarrh - **C. pyracantha** Medik. = *Pyracantha coccinea* - **C. rivularis** Nutt. (C. douglasii Lindl. (wUS) (cult.): thorn tree; thorns said poisonous; frt. eaten by NA Indians - **C. sanguinea** Pall. (Russ & eEu): Russian drug plant - **C. succulenta** Schrad. ex Link (C. macracantha Lodd.) (e&cNA): long-spined thorn; large-thorned white thorn; succulent hawthorn; "thorn apple"; frt. rich in sugars (1% simple sugars); frt. u. for juice, jellies, jams, &c.; u. to reduce blood pressure; to combat coronary thrombosis; in various cardiac conditions - **C. uniflora** Muenchh. (e&cUS): dwarf thorn; unripe frt. of small tree u. astr. in bladder dis.; cardiotonic (Meskwaki Indians) (111).

Crat(a)eva nurvala Buch.-Ham. (C. magna DC) (Capparid.) (Ind to sChin, &c.): bk. juice lax. & to stim. appetite (Mal); lvs. appl. to skin as rubef. (as good as mustard plaster) to gums near aching teeth; bk. inf. appl. exter. for fever, cramps, &c.; lvs. u. to treat mental illness (Indones); fls. astr. cholag.; frt. lax. - **C. odora** Buch-Ham. f. *axillaris* (Presl) Jacobs (PI): garlic pear; lvs. u. as stom. emmen.; bk. sap for convulsions - **C. religiosa** Forst. f. (Ind, Mal, PI +): 3-leaved caper; bar(u)na (Hind); bk. of small tree c. saponins, u. stom. diur., lax., lithontriptic; frt. & bk. u. in embrocation for rheum.; fresh lvs. & rt. bk. u. rubef. vesicant; heated & appl. to ear for earache (Solomon Is); lvs. u. stom., menstr. dis., ton.; frt. mixed in mortar to produce strong cement - **C. roxburghii** R. Br. (Ind+): c. thiodiamine; bk. sed. ton. in cholera; lvs. & rt. bk. rubef.

Cratoxylum (Cratoxylon) blancoi Bl. (Guttif.) (Mal, Borneo, PI): bk. decoc. galactagog; bk. + lvs. decoc. abort.; lvs. crushed & placed on forehead & chest for colds; wd. useful - **C. celebicum** Bl. (Indones, PI): young lvs. chewed for cough & TB; decoc. of lvs. rt. & bk. u. as protective after childbirth - **C. formosum** (Jack) Dyer (C. cochinchinense (Lour.), Bl) (Indones, Mal): bk. & lvs. pounded with coconut oil, appl. to skin dis.; bk. decoc. in colic; bk. resin u. for itch; lvs. chewed for med. effect (Mal, Indones) - **C. prunifiorum** Kurz (C. prunifolium Dyer) (Indochin): decoc. of fresh or dr. lvs. u. as ton. stom. diur.; c. anthraquinones, flavonols (quercetin esters, incl. hyperoside).

craw-craw: obstinate dermatitis (Af) caused by a nematode (*Onchocerca*) - **c. c. plant:** *Cassia alata.*

crawfish: 1) *Panulirus* spp.: sea crayfish; spiny lobster; flesh u. as delicious food; offals u. in prodn. c. meal (highly nutr. protein & mineral cont.); u. stock food (prod. Belize; Fla [US]) (do not confuse with crayfish) 2) (US) crayfish (variant).

crawley (root): *Corallorrhiza odontorrhiza* (perhaps corruption of "coral root").

crayfish: "crawdad"; small freshwater crustacean, *Astacus fluviatilis* Fabr.; *A. nigrescens* (Pac US); *Cambarus diogenes* (cUS, &c.); etc.; u. as food with wild rice; med. in infectious dis., icterus, nettle rash (Homeop.); several *Astacus* & *Cambarus* spp. in NA u. similarly; thus *C. affinis;* popular as food in US (esp. La).

crazy (oats) weed: loco weed, *Astragalus crotalariae*; *A. mollissimus*, &c. - **c. weed:** *Oxytropis* spp.

CRC tablets ("pills"): formerly in popular demand as lax.; cath.; c. calomel; rhubarb; colocynth.

crea monda (**cerea monda**?) (It): identity of hb. not established; prob. member of Fam. Lab.; cut hb. u. in Vermouth herb mixtures (SBP).

cream nuts (Eng): Brazil nuts.

cream of tartar: potassium bicarbonate; $KHCO_3$ - **c. of t. tree**: *Adansonia digitata.*

creasote: creosote.

creat: kreat.

creatine: Kreatin (Ge): N-rich compd. found in muscle tissues; forms **creatinine** on dehydration; also found in blood, urine, also plant materials; admin. IV in diabetic coma.

creek gum: *Eucalyptus globosa.*

creeper: trailing pl. taking root at nodes along length of st. - **Virginia c.**: *Parthenocissus quinquefolia.*

creeping Charlie (NCar [US]): *Glechoma hederacea* - **c. Jenny**: 1) *Lycopodium clavatum*; *L. complanatum*; etc. 2) *Convolvulus arvensis* 3) *Lysimachia numimularia*- **c. Sarah**: *Cymbalaria muralis.*

creeping: v. also blackberry.

crème de menthe (Fr): mint cream; sweet green or "white" liqueur flavored with peppermint or other mints; u. in GI colic; flavor for ice cream.

cremocarp: a type of frt. found in & typical of Umb.; it consists of two achenes or mericarps borne on a forked stalk, the carpophore; ex. aniseed, caraway.

Cremor Tartari (L): cream of tartar: $KHCO_3$.

crenotherapy: **crounotherapy**: science of medical treatment by int. use of mineral waters.

Creolin™: cresols in soap soln. - **Creosotal**™: creosote carbonate.

creosol: methoxy methyl phenol; liquid found in beechwood tar & creosote; do not confuse with cresol.

creosote (créasote, Fr): a mixture of phenols obt. by distn. fr. wood tar, esp. that of beech; good expectorant ("brings up phlegm") - **beechwood c.**: c. phenols & cresols - **c. bush**: *Larrea tridentata* - **c. carbonate**: a mixture of the carbonates of constits. of wood creosote - **coal tar c.**: a mixture of phenols fr. coal tar; u. disinf. insecticide, germic., wd. preservative (impregnation under pressure) - **c. oil**: heavy o.; obt. in dist. of coal tar; c. phenols & cresols - **c. plant: c. shrub**: c. bush - **wood c.**: c.

crèpe myrtle: *Lagerstroemia indica* & other *L.* spp.

Crepis L. (Comp.) (worldwide temp & Af): hawk's beard; c. 240 spp. of ann., bienn., or perenn. hbs. (1239); mostly weeds, but one (*C. rubra* L.) (eEu) a showy ornam. fl. - **C. acaulis** Hook f. (Ind): rt. eaten raw for urin. dis.; baked lvs. or rt. mixed with goat's milk & given as galactagog - **C. japonica** (L.) Benth. = *Youngia j.* (L) DC.

Crescentia alata HBK (Bignoni.) (Mex, CA, cult. PI+): morro; frt. & rts. u. domestic med.; lvs. u. to promote hair growth (decoc.); sds. u. to make cooling beverage (Nic); decoc. for colds, pneumonia (Mex); gourds u. to make cups, etc. sold in Mex markets - **C. cucurbitina** L. = *Enallagma latifolia* - **C. cujete** L. (tropAm): calabash or gourd tree; cara (Mex); frt. shell (rind) u. for watergourds, cups, dippers (*jicara*), scoops, domestic utensils, hats, etc.; frt. pulp. u. as exp. dem. lax. astr. vulner.; frt. pulp boiled & decoc. u. in resp. dis. (VirginIs); ingred. of poultice for skin dis. (Domin, Caribs); med. uses similar to those of *C. alata*; frt. may be abortient to cattle; fl. juice u. for earache; gourd filled with mescal & drunk in small doses for pulm. dis. (Baja Calif [Mex]); sds. cooked & eaten; widely cult.

cresol: mixture of the 3 isomeric cresols (o-, m-, p-), usually with some phenol; obt. fr. coal tar; u. germicide, antisep. (GI, GU), disinf. parasiticide; u. as diluted saponated aq. soln. (1240); mfr. synth. resins, etc. (do not confuse with creosol).

cress: *Lepidium sativum* - **bastard c.**: *Thlaspi arvense* - **bitter c.**: *Cardamine* spp. - (**cuckoo b. c.**: *C. pratensis*) - **field penny c.**: bastard c. - **garden c.**: c. - **Pará c.**: *Spilanthes oleracea* - **penny c.**: bastard c. - **tongue c.: town c.**: c. - **water c.**: *Nasturtium* spp. (*v.* under "water"); term also appl. to *Apium* sp., *Arabis*, *Cardamine amara*, *Sisymbrium*, &c. - **winter c.**: *Barbarea* spp., esp. *B. verna.*

Cressa cretica L. (Convolvul) (nAf): rosin weed; cresse (Fr): hb. c. saponins, tannin, flavonoids; u. ton (Sudan); dr. lvs. crushed with sugar as emet. in icterus, followed by milk diet (Eg); anti-inflamm.

cresson (**de fontaine**) (Fr): water cress.

Creta (L, It): chalk - **C. Laevigata** (L.): **C. Praeparata**: prepared chalk - **C. Praecipitata**: precipitated chalk.

cretin (individual): **cretinism** (state): severe hypothyroidism; believed due to iodine deprivation (SC Crete (island) where some natives had dis.).

crex: *Prunus institia.*

crinotherapy: therapeutic use of glandular products of animal body or their secretions.

Crinum amabile J. Donn. (Amaryllid.) (widely cult. tropAm, Sumatra): giant lily; u. digitaloid cardiotonic; fl. decoc. chest dis. - **C. americanum** L. (seUS): southern swamp lily; bulb u. exp. stim. - **C. asiaticum** L. (C. latifolium L.) (Ind, Mal, PI): (Asiatic) poison bulb; giant crinum; bulb & rts. emet. diaphor.; ton. lax. expect.; bulb mixed with lime to treat wounds (PI); bulbs c. lycorine, crinidine; sds. purg. emmen.; lvs. expect. exter. for skin dis. & inflamm.; juice antidote for bites and arrow poisons; cancerostatic; substit. for ipecac (Ind) - **C. defixum** Ker-Gawl. (eAs): bulbs (tubers) u. as emet. diaphor.; topically for burns, nail abscesses; as poultice (Mal), ton. in impotence (Ind); juice in skin dis. (124) - **C. delagoense** Verdoorn (sAf, Zimbabwe; Moz): veld lily; (li)conwa; pieces of bulb appl. locally for furuncles; as galactagog (local appl.); for swelling of body (2445) - **C. erubescens** Ait. (Mex, CA, WI, nSA): lirio (silvestre); bulb juice u. in eye drops; cooked bulbs as poultice to "resolve tumors"; lf. juice & vinegar appl. to hemorrhoids; fl. decoc. u. as pectoral; & to bathe body (Ve) - **C. jagus** (Thomps.) Dandy (C petiolatum Herb.) (pantrop.): sap u. as disinfect. for trop. sores, cuts, lesions (Af); bulbs u. as charms (Br; Negroes); oncostatic - **C. latifolium** L. (eInd): c. latisoline, u. as immunostim.; bulbs & lvs. u. exter. in sores; crushed & washed bulbs u. as rubef.; lf. given in ear pains; ext. of tubers u. to inhibit tumor growth - **C. macrantherum** Engl. (eAs): bases of lower lvs. u. to cover arrow & stab wounds & burns.

crinum lilies: *Crinum* spp. - **southern swamp c.**: *C. americanum.*

Crisco™: the first hydrogenated cottonseed oil u. as lard substit. (shortening).

Crithmum maritimum L. (l) (Umb.) (Eu coasts, cEu+): (sea) samphire; lvs. with salt u. as condim. in capers & pickles; hb. u. digestive, diur. anthelm. (fleshy pls. collected fr. sea cliffs: notably hazardous occupation).

Crocanthemum Spach (Cist.): close to & usually reduced to *Helianthemum.*

crocodile: Crocodylus (Reptilia): (trop OW & NW) dangerous to man since aggressive; often large animals (-12 m) - **C. niloticus** Laur. (Af, esp. Eg): musk fr. m. glands u. in perfumery (ointments) esp. by natives; kidney u. as aphrod.; fat u. for promoting hair growth (with that of other animals) (ancient Eg).

Crocus aleppicus Bak. (C. gaillardotti Maw) (Irid.) (wAs): rose of Sharon (Song of Solom. 2:1) (Israel); autumn crocus; corms u. as food - **C. sativus** L. (C. longiflorus Rchb.) (seEu & As; cult. worldwide): true (or garden) saffron; dr. stigmas (Crocus, L.) c. crocin, picrocrocin, VO; u. flavor, coloring agt., stim. antispas. anodyne, diaph. emm. carm. stomach; perhaps the most outstanding drug in medieval Eu; sprinkled in Roman baths & Greek theatres (1242); important condim. (the most expensive of all spices), color & flavor in such foods as soups, pastries, cakes (saffron cake & bread popular in Cornwall, Eng) (2776); candies, chicken & yellow rice (pilau); yellow dye for butter, etc.; silk, cheese (Parmesan c. of It); chief prod. Sp, It, Fr (Crocus fr. krokos, Gr name of pl.) (trade vars: Mancha superior, Rio selecta, &c.); a cultigen - **C. vernus** (L) Hill. (sEu): pl. u. in chlorosis, epilepsy; corms u. as food.

crocus 1) *C. sativus* 2) *Anemone patens* (Ont [Cana] farmers) 3) *Iris pumila* 4) kind of coarse heavy cloth - **(common) autumn c.: fog c.: meadow c.: purple c.**: *Colchicum autumnale* (pl. or corm) - **c. sack** (seUS): burlap or coarse hemp or gunny sack - **saffron c.**: *C. sativus* - **wild c.**: *Anemone patens.*

croker sack: croaker s.; prob. corruption of "crocus s." (*qv.*).

crop, catch: cover c.: one planted to protect the soil in winter, to fix nitrogen in soil, etc. (ex. clover) - **smother c.**: one u. to suppress persistent weeds (ex. soybean).

crore (Indo-Pak): 10 million.

cross cells (Querzellen, Ge): found in endocarp (inner epidermis) of umbelliferous frts. (ex. Carum), forming a transverse layer; formed by 5-10 divisions of mother cells, the narrow cells produced lying side by side; good means of distinguishing Anise & Fennel - **c. vine**: 1) (seUS): *Bignonia capreolata* 2) (eUS): *Campsis radicans.*

Crossopteryx febrifuga Benth. (C. kotschyana Fenzl.) (l) (Rubi) (cAf; ex. Sierra Leone): African bark; "bembée"; bk. u. antipyr. (great reputation in Senegal); bk. decoc. for pruritus; u. fever, mal. (1257), diarrh. colic, intest. worms, ophthalmia, in cough (Sierra Leone+); emmen. abortif. (Tanz); rt. decoc. for syph. & TB (Tanz); wd. used.

Crossostephium chinense (L.) Makino (monotypic) (Comp.) (eAs): soft woolly mass of hairs (trichomes) u. as moxa; like cotton; lvs. & sds. u. as ton. anthelm. (Indochin).

crosswort (US): *Eupatorium perfoliatum.*

Crotalaria L. (Leg) (trop&subtr): some 700 spp. of herbs; many taxa rich in pyrrolizidine alks. (1243), often toxic (2282) - **C. acicularis** Hamilton (Ind, Indoch.): fl. inf. u.in local med. to combat lameness (lumbago?) (Laos) - **C. assamica** Benth. (Ind, seAs, Pl, Haw Islands): rt. decoc. u. to treat bladder calculi (Laos); pl. u. to enrich soil - **C. bracteata** Roxb. (sChin, seAs, Burma, Pl): rt. macerate u. as antipyr. & for herpes - **C. brevidens** Benth. var. *intermedia* (Kotschy) Polhill (C. i. Kotschy) (tropAf +): lf. applied to sore feet (wAf); st. u. to make string (eAf) - **C. brevifolia** (Chin): c. hypotensive alk. (also *C. agatiflora* Schweinf. ex Engl.) - **C. fulva** Roxb. (Ind, seAs, natd. in tropics as WI): consumption bush; pl. decoc. popular u. as medicinal "bush tea" (Jam); however, toxic, a cause of veno-occlusive dis.

(Jam), due to monocrotaline (alk.) - **C. incana** L. (pantropics, incl. sFla, WI, Mex to nSA): hoary rattle-wort; rattle bush; lf. inf. u. as bath for boils & rash; c. toxic alkaloids; stock poison; rt. tea u. for rash, yellow fever (Trin) - **C. juncea** L. (Ind; Austral, natd. WI [Jam, Hispan+] [pantrop]): shone (Beng): sds. u. depur. (folk med.); edible; (pods & lvs. emergency food); roasted to give coffee-like beverage; st. source of (Benares) Sunn fiber or Sunn-hemp; cult. for fiber, also as green manure; toxic to cattle - **C. longirostrata** Hook. & Arn. (sMex to CR): "chop"; lvs u. as purg. & emet.; cooked greens soporofic; stock poison - **C. maritima** Chapm. = *C. rotundifolia* - **C. nitens** HBK (CA, SA): appl. with salt to abscesses or swellings (Pe) - **C. pallida** Aiton (C. mucronata Desv.) (pantrop): fls. consumed as veget.; hb. u. to reduce fever (Laos); ripe sds. grilled & roasted u. as coffee substit. (non-stim.); rts. chewed with betel (Vietnam) - **C. quinquefolia** L. (Ind, seAs): frts. u. for serpent & centipede bites (Indoch); soil fertilizer - **C. retusa** L. (As, Af, natd. Fla [US]): wedge leaf rattle pod (Austral); pl. (hb.) c. pyrrolizidine alk. incl. monocrotaline (with tumor-inhibiting activity); lvs. u. in folk med. as febrifuge (decoc.); paste appl. to sores (Tanz); rt. for colics; pl. u. dye; lvs. eaten as veg.; u. to make tea; st. source of fiber u. for canvas, cordage; pl. chief cause of Kimberley Horse (or "walk about") dis. (Austral); pl. u. in local med. for certain pulmon. dis. (Vietnam); decorticated sds. edible - **C. rotundifolia** Poir. (C. maritima Chapm.) (seUS, WI); rabbit bells; pods inf. u. for sore throat (Seminole Indians) - **C. sagittalis** L. (eUS): rattlebox; rt. u. as narcotic; "venereal aid"; blood purif. by Am Indians - **C. spectabilis** Roth (OWtrop, now cult. & natd. seUS n to Va & Mo): showy crotalaria; c. alks. incl. monocrotaline (1244); hb. u. to treat scabies, impetigo (Ind); planted for erosion control, fodder, ornam. fiber - **C. stipularia** Desv. (US, Mex, SA): u. against snakebite - **C. tetragona** Roxb. (Ind, seAs): ground & appl. to dis. of tongue & gums (Indoch) - **C. verrucosa** L. (pantrop): rts. u. in local med. for fevers (Laos, Cambodia); abdominal pain (Vietnam).

Crotalidae (Ophidia - Reptilia): rattlesnake fam.; pit vipers (SC pit in head sensitive to infra-red waves so snake can detect presence of warm-bloooded animals); incl. *Crotalus* (rattlesnake proper); *Bothrops* (SA); incl. *B. atrox* (fer de lance, lit. iron point of spear); *B. jararacussu*, *B. jararaca*, *Agkistrodon* spp. (water moccasins, copperheads).

Crotalinae: sub-fam. of Vipers, representing the pit-vipers, to which belong the rattlesnakes, copperheads, & water moccasins.

crotaline: pertaining to the Crotalinae.

Crotalus L. (Crotalidae - Reptilia): rattlesnakes; 2 gps: NA & SA - **C. adamanteus** Beauvois (seUS): eastern diamondback (rattler); grows to 2.8 m long; found in wet woods, &c. - **C. atrox** Baud & Girard (wUS): western diamond rattlesnake; diamond back; "coontail rattler"; 0.9-1.2-1.8 m. long; cause of most severe envenomation (Tex [US]); FO (fat) u. to soothe nursing baby's mouth; cooked fat put on festering sores; meat & bones u. in TB; skin wrapped around arm for internal pain (folk remedies) - **C. durissus terrificus** Laurenti (nwBr): cascavel de quatro venta; venomous serpent; vemon u. in epilepsy, chronic bronchial catarrh.+ - **C. horridus** L. (e&cUS): timber (or canebrake) rattler, banded r.; timber rattlesnake; large snake (-2.1 m in Fla [US]); venomous snake, an antitoxin is available to antidote the venom (c. crotoxin); venom u. in small doses for blood disturbances, inf. dis.; cholescystopathy, myocarditis, resp. organ. dis., asthma, hay fever, epilepsy, polio; rattlesnake oil u. for rheum. (rubbed into skin) (1551); meat eaten (canned); leather fr. skin (2305) - **C. viridis** (Raf.) (wUS, swCana): prairie rattler; -1.5 m.

Crotalus antitoxin: an antivenin for rattlesnake venom; adm. intram. & subcut. following bite by any sp. of this venomous reptile.

Croton bonplandianus Baill. (Euphorbi) (Arg): nogal del zorro (Arg); hb. c. alkal., flavonoids; u. exter. as antiseptic - **C. capitatus** Mx. (NA): hb. u. in mal. - **C. ciliato-glandulosa** Orteg. (Mex, CA, WI): picosa (sMex); latex much u. in fever, purg.; for scorpion sting; said to incr. milk of goats foraging on pl. - **C. crassifolius** Geisel. (sAs): rts. in wine u. in scrofula, to heal wounds (Hainan) - **C. dichogamus** Pax (Tanz): decorticated rt. decoc. u. in syph.; rts. dr. & u. as snuff in head colds; dr. lvs. burned & smoke inhaled for fever & in chest dis.; lvs. chewed for chest & stomach dis.; chopped rts. in goat's meat soup as ton. - **C. dioicus** Cav. (Mex): "rubaldo"; rosval; yerba del zorillo; rts. sd. & hb. u. drastic purg. lax.; pl. u. in hysteria; in baths for rheum.; sds. claimed anti-syph. - **C. draco** Schlecht. (Mex, CA): dragon croton; sangregado; sangre de drago; u. in fevers, ulcers; astr. to harden gums; tree sap source of red dye (prod. dragon's blood); u. in hoof dis. of horses & burros; for vilmas - **C. elliottianus** Baill. (sAf, Mad): sd. c. nonvesicant FO; sd. u. as purg.; bk. mixed with curdled milk as purg. (Masai); diur.; sd. & FO anthelm. (1246) - **C. elliottii** Chapman (Flora s. U. St. 407) (seUS): sds. c. purg. oil (1245) - **C. eluteria** (L) Benn (Bah, CR): sweet(wood) bark; Chin Nova; source of cascarilla (bark); flavor in liquors, arom. in tobacco; fumigant; formerly USP & NF; c. VO with eugenol, limonene, vanillin, cascarillin (NBP); arom. bitter ton.; spice - **C. flavens** L. (WI): rock balsam; cough bush (Bah); white maron (Virgin Is); lf. u. as poultice in earache; inf. for sore throat (Trin), decoc. in cough; in bladder dis. (Virgin Is) - **C. flocculosus** Geisel = *C. flavens* - **C. glabellus** L. (Co): with hypertensive & cholagog properties; adulterant of *Croton tiglium* - **C. gossypifolius** Vahl (& var. *hibiscifolius* Muell. Arg.) (Ve, CR, WI): sangrito; candle tree; dark reddish resinous sap ("sangre de drago") u. as styptic on cuts, appl. to teeth & gums - **C. guatemalensis** Lotsy (sMex to Hond): copalchi(llo); bitter bk. decoc. u. in intermittent fever ("chills") (Guat), as ton. in rheum. pains - **C. gubouga** S. Moore (Transvaal, Af): lowveld croton; Transvaal c. (bark); sd. & bk. u. in malar. (both natives & Europeans); sd. drastic; wd. useful - **C. humilis** L. (sFla, Tex [US], WI, sMex): small seaside balsam; icaban; said poisonous, prod. blindness by mere contact with branches; fleas swept fr. house with broom of branches (Yuc) - **C. linearis** Jacq. & Benn (WI, Bah, sFla [US]): muckle (Bah); "marigold" (Bah); pl. decoc. in dysmenorrh., worms, after delivery; lvs. for beverage tea; for fever, colds, colics (Jam); u. like cascarilla bk. - **C. lucidus** L. (Bah Is): bk. u. as adulterant for cascarilla bk. - **C. macrostachyus** Hochst. ex. A. Rich. (Ethiopia, Tanz): frts. c. crotepoxide (carcinostatic activity (animal expts.); rt. decoc. anthelm for tapeworm, purg.; ash fr. burnt lvs. sucked for cough; juice fr. boiled lvs. u. for mal. & VD; lvs. boiled in pot & steam inhaled for cough (Tanz) - **C. malambo** Karsten (nSA, esp. Ve, Co): **malambo bark croton**: source of Melambo (Mateas, Matias) Bark; u. antipyr.; u. for colds (Co & Ve, Indians), as arom. (spice) & stom. (1247); appears as false cinchona bk. (obs.); odoriferous wd. (bk ?) u. in stom dis. - **C. morifolius** Willd. (Mex+): palillo; lf. inf. u. for stomachache; VO appl. exter. for neuralgia - **C. niveus** Jacq. (Mex, CA, WI, nSA): copalchi; Copalchi Bark u. ton. in malaria, etc.; arom. bitter; in diabetes, as vulner. for gastrointest. dis.; substit. for cascarilla bk.; as cinchona substit. (Mex) - **C. oblongifolius** Roxb. (EI, Burma, Bengal): sds. rt. & bk. u. as drastic (due to FO); hot decoc. of lvs. u. in scabies (Indochin); pd. sds. u. insecticide & fish stupefiant (Ind) - **C. ovalifolius** Jacq. (sMex, CA, nSA): quina (blanca); chul; bk. u. as ton. in mal., astr. in hemorrhoids, chronic diarrh. dys.; rt. inf. u. for "trista" (skin yellowing & fever) - **C. penduliflorus** Hutchin. (Sierra Leone): lf. inf. u. as lotion in fever; in biol. assay of garlic; wd. u. for rafters - **C. populifolium** Mill. (WI, Ve): bk. macerated in alc. u. in rheum. (Trin) (Ve) - **C. pottsii** Muell.-Arg. (C. corymbulosus Engelm.) (wTex to Ariz [US], Mex): flg. tops u. to make inf. or tea with diaph. stim. carm. & ton. properties (Mex, Am Indians) ("chaparral tea") (2230) - **C. reflexifolius** HBK (sMex to CR): "sasafras"; copalchi; bk. u. antipyr. ton. in colds (Guat), mal., &c. - **C. repens** Schlecht. (Mex, CA): chacotote; rt. u. gastric dis., dys. - **C. schiedeanus** Schlecht. (Mex to Pan, Co, Pe): "wild cinnamon"; cache; resinous sap to cauterize wounds; lf. decoc. for biliousness - **C. sparsiflorus** Morong (Ind): this weed c. alks. with hypotensive properties - **C. texensis** (Klotzsch) Muell-Arg. (c&sUS): skunk-weed; hb. tea u. for upset stomach (Zuñi Indians); eyewash, emetic, insecticide (NM [US] Indians) - **C. tiglium** L. (tropAs, often cult., Cey, Ind): purging croton; sds. source of FO, croton oil 30-50%, c. "crotin" (mixt. of 2 phytotoxins, very poisonous; has tumor inhibitory compd.); resin c. phorbol, tiglic acid; sd. c. crotonoside; FO u. drastic cath.; epispastic; u. by malingerers to simulate skin dis.: rt. u. cath.; wd. also u. med.; tree produces a kino; u. esp. in veterin. med. (danger!) for worms (sds. crushed & appl.); pustulant (leucotaxic effect: leucocytes collect in areas of injury & inflamm.) (*v.* croton oil) - **C. tonduzii** Pax (CR): quina amarga; bk. inf. as antipyr. - **C. wilsonii** Griseb. (Jam): pepper rod; decoc. to relieve colds - **C. xulapensis** HBK (sMex, CA): u. in toothache & sore gums; sap u. to treat wounds & in dressings (Guat).

croton oil: FO fr. **c. seed**: Grana Tiglii (L.): dr. ripe sd. of *Croton tiglium*; u. drastic purg., esp. formerly for acute hypertension & for the insane; exter. as epispastic, vesic. counterirrit. corrosive; dangerous; avoid contact with eye.

crotonic acid: $CH_3C_2H_2COOH$: occ. in clay soils (Tex [US]); in distn of wood; u. in prodn. synth. vit. A, threonine, &c.

crow: *Corvus brachyrhynchos* L. (Corvidae) (NA): common American crow; kau(w)a (Urdu, Unani): large black birds (not true blackbirds) (worldwide); this & other members of the fam. ("corvids") are among the most intelligent of birds (with largest cerebra); live in flocks; some believe they communicate by "crowing" (E. T. Seton); sometimes pestiferous; u. in Unani med.: flesh, blood (for piles); bile; ash of feathers (u. in eye dis.) (*cf.* other fam. members: *Corvus corax* L., common raven (Euras, NA); magpie).

crowberry: *Empetrum nigrum*.

crow-corn: *Aletris farinosa* - **crow-fig**: Nux Vomica - **crowfoot**: 1) *Ranunculus* spp., specif. *R. bulbosus* 2) *Geranium maculatum* - (**bulbous c.**: *R. bulbosus* - **celery-leaved c.**: **cussed c.**: *R. sceleratus* - **creeping c.**: *R. repens* - **marsh c.**: cussed c. - **tall** (or **upright**) **c.**: *R. acris*).

crowfoot family: Ranunculaceae.

crown: junction of stem & rt. in higher pls. - **c. bark**: fr. *Cinchona officinalis*.

crow-silk: agar.

crow-weed (NM [US]): *Croton texensis*.

Cruciferae: Mustard or Cross-bearing Fam.; mostly hbs. (many typical weeds), few shrubs; ca. 3,000 spp.; mostly of temp. regions; lvs. alternate, usually simple, exstipulate; fam. readily recognized by fls. which are mostly regular, 4-merous, with 4 sepals & 4 petals placed crosswise; 6 stamens, 4 larger, 2 smaller; 1 pistil; frt. a silique (capsule opening lengthwise along 2 sutures); many members are of economic importance; ex. cabbage, turnip, mustard, cress, stock, radish, wall-flower.

crude drugs: medicinal materials found in the raw form; term generally implies products fr. the pl. & animal kingdoms (ex. rhubarb rt., arnica fls., cochineal); however, sometimes (as in Eu) applied to incl. also pharm. products fr. the mineral kingdom in original or crude form & not necessarily of organic origin, such as bentonite, kaolin, &c.; crude drugs offer the chief resource fr. which active principles are obtained; such incl. alks., glycosides, VO, FO, waxes, &c. there is considerable variation in quality, so that various types of identity, purity, quality, & assay tests must be applied with care in their evaluation.

cruel man of the woods (US Negro): *Peltandra alba*.

cryoscopy: measurement of freezing or solidifcation point & study of physical or chemical changes occurring at low temps.

Cryptantha Lehm. ex G. Don f. (Boragin) (Pacif Am): ca. 120 spp. of low erect herbs (1248) - **C. crassisepala** Greene (swUS, Mex): thick-sepal cryptantha; u. to treat furuncles (Hopi Indians) - **C. fulvocanescens** (S. Wats.) Payson (Ut, NM, Colo, Ariz [US]): yellow-hair cryptanth; rt. inf. u. for many illnesses (Navaho Indians).

Crypteronia paniculata Bl. (Crypteroni.) (Oceania): bk. appl. to skin eruptions (PI); young shoots eaten with rice; timber useful.

Cryptocarya aromatica (Becc.) Kosterm. (C. massoy(Oken) Kosterm.) (Massoia aromatica Becc.) (Laur.) (seAs, native to NG): arom. bk. or sap eaten (2062); added to tonics & cigarettes (Mal) ton, u. by women after childbirth (Mal); for diarrh. & bowel spasms (Indones); for fever, TB (NG); headache; in many combns. for various complaints; spice, flavor(VO) - **C. bourdilloni** Gamb. (Ind): hypotensive; c. lactone with antifungal properties - **C. cordata** Allen (New Britain, Solomon Is): tembu (Solomon Is); frt. u. as relish with certain foods; lvs. heated & appl. to sore eyes - **C. glabella** Domin (Austral): poison walnut; touching bk. may cause severe illness; c. alk. - **C. glaucescens** R. Br. (eAustral): silver sycamore; brown beech; source of good timber - **C. hornei** Gillespie (Fiji, Tonga): motou (Tonga); bk. u. to scent FO (T.) (86) - **C. laevigata** Bl. var. *bowiei* Kosterm. (C. australis Benth.) (Austral): gray c.; bk. c. cryptocarpine, poisonous alk. (resp. depr.) - **C. latifolia** Sonder (sAf): source of ntonga nuts, u. for FO - **C. moschata** Nées & Mart. (Br); frt. called Brazilian nutmegs; c. cryptocarine, FO, VO; u. spice; very similar to true n. - **C. multipaniculata** Teschn. (Oceania); lvs. heated & appl. to sore eyes (Solomon Is) - **W. wightiana** Thw. (Ind): pd. bk. & lf. u. (with sugar) to remedy swellings & rheum.; grd. lvs. boiled in FO u. in elephantiasis.

Cryptococcus (Cryptococc. - Cryptococcales - Blastomycetes - Deuteromyc. - Fungi): **C. neoformans** (Sanfelice) Vuill. (Debaryomyces n.): causes cryptococcosis (or torulosis) in man (widespread); lungs, skin, &c. affected; may be fatal.

Cryptocoryne cordata Griff. (Ar) (Mal): crushed lvs. appl. to head for vertigo or headache - **C. retrospiralis** Wydler (Ind, Chin): crushed lvs. appl. to heads of children with high fever - **C. spiralis** (Retz.) Wydler (Ind, SirL): "Indian ipecac"; u. as calamus substit.; orn. in aquaria.

Cryptogamia: cryptogams: those plants which develop fr. spores rather than sds.; include all pl. groups except the Spermatophyta (seed plants): classical subdivision into Thallophyta, Bryophyta, Pteridophyta; formerly subdivided into acrogens (ferns & mosses) & thallogens (fungi, algae).

Cryptogramma R. Br. (Adiant. - Pteridoph.) (Eu, As, Am): rock-brakes; low ferns - **C. crispa** R. Br. ex. Hook. (Euras, NA): mountain parsley; p. fern; u. as anthelm. like Aspidium; orn.

cryptograndoside A: glycoside of oleandrigenin & D-sarmentose; occ. in *Cryptostegia grandiflora*.

Cryptomeria japonica (L. f.) D. Don (l) (Taxodi.) (Jap, Chin): Japan cedar; Japanese c.; peacock pine; tree -42 m; lvs. u. for incense sticks; wd. u. for many purposes; when buried, attains greenish color & popularly used; resin component of med. ointment; VO u. in GC; ornam.

cryptopine: alk. of opium, *Dicentra* spp. & *Corydalis sempervirens*; has narcotic properties.

cryptospecies = genetic sp.; a sp. of organism with polyploid no. of chromosomes, indistinguishable in morphology & other characters fr. original diploid (Löve proposed these should be considered distinct spp.).

Cryptostegia grandiflora R. Br. (Asclepiad./ Periploc.) tropAf & Ind, cult. Pl+): (India) rubber vine; this liana yields Palay rubber (used during WWII); pl. u. against vermin & in criminal poisoning (Madag); ornam. - **C. madagascariensis** Bojer (Madag): tombiro; rubbervine; latex of liana furnishes rubber; bk. u. in manuf. rum.

Cryptotaenia canadensis (L) DC (Deringa c. Ktze) (Umb) (eNA, natd. seAs, Austral, cult. Jap): mitsuba; young lvs. & rts. eaten, fried as veget. with agreeable flavor; lvs. prepd. like spinach; whole pl. decoc. u. for colds, rheum. tubercular glands, to stop diarrh. (Chin); hb. u. like *Anthriscus cerefolium* (diur. expec.).

Cryptotethya crypta (Spongidae - Porifera) (oceans): sponge; c. vidarabine; spongosine, &c. (thymine nucleosides); has been u. to treat leukemia.

cryptoxanthin: carotenoid pigment with effects similar to those of vitamin A; obt. fr. *Physalis* spp. petals & frt. but also found in many other pls. & animals.

Crystalli Tartari (L): $KHCO_3$.

crystal tea: *Ledum palustre* - **C. Violet**: synth. dyestuff u. for silk & wool; was u. as antisep. for malignant growths formerly, Gentian Violet.

CSF = colony stimulating factor.

Ctenium aromaticum Wood (Gram) (seUS): toothache grass (SC); cattle forage; u. as sialagog.

cuajiote (Mex): gum-resin of *Bursera microphylla.*

cuapiá (Br): *Dorstenia* spp.

cuas(s)ia (corteza) (Sp): quassia (bark); sometimes confused with Cassia.

cuastecomate: cuatacomate: *Crescentia alata.*

cuatomate (Mex): tomato.

cuauchichic (Mex): *Garrya laurifolia.*

cuba (Sp): barrel, cask, tub.

Cuba wood: fr. *Chlorophora tinctoria.*

cubé root: *Lonchocarpus floribundis, L. utilis.*

cubeb (berries): *Piper cubeba* (frt.) - **African c.: Congo c.:** *P. clusii* - **Guinea c.**: *Piper clusii, P guianense.*

Cubeba (L., Por.): **cubebe** (Fr): **cubebs** (Eng): cubeb, *Piper cubeba.*

cubeseed: *Iris prismatica.*

cuca (Pe): coca shrub.

cucaracha, yerba de (Mex): *v.* cockroach herb.

cucklebur (seUS dial): cocklebur - **cuckold**: *Bidens frondosa.*

cuckoo bread: *Oxalis acetosella* - **c. flower(s)**: 1) *Cardamine pratensis* 2) *Lychnis* spp. 3) *Orchis* spp. 4) *Stellaria* spp. 5) *Arum maculatum* 6) *Oxalis* spp. 7) wild hyacinth, &c. - **c. pint**: *Arum maculatum.*

cuckoo's cap: aconite.

cucumber: *Cucumis sativus* - **bitter c.**: colocynth - **Chinese c.:** 1) *Trichosanthes* spp., esp. *T. anguina* & *T. kirilowii* 2) *Luffa* spp. - **Indian c.**: *Medeola virginiana* - **one-seeded bur c.**: star c. - **squirting c.**: *Ecballium elaterium* - **star c.**: *Sicyos angulata* - **c. tree (bark)**: fr. *Magnolia* spp.: *acuminata, ashei, fraseri, macrophylla, pyramidata, tripetala* - **wild c.**: *Echinocystis lobata, Sicyos angulatus*; *Anemone quinquefolia, A. nemorosa*; *Ecballium elaterium.*

Cucumeris Fructus (L): cucumber frt. - **C. Semen (Semina):** c. sd.

Cucumis anguria L. (Cucurbit.) (sUS, Mex, WI): West India gherkin; produces a small cucumber with warty surfaces; frt. dem. refrig.; boiled pulp. u. cataplasms; frts. u. in pickles; rt. u. for edema; stomach dis.; lf. juice appl. for freckles; pl. cult. as ornam. - **C. citrullus** Ser. = *Citrullus lanatus* - **C. ficifolius** A. Rich. (C. pustulatus Hook. f.) (swAf): pl. c. cucumin, u. in homicidal poisoning; rt. u. as purg. (Eth) - **C. melo** L. (cAs, Af.): muskmelon; delicious frt. eaten; enzyme of frt. u. USSR in place of rennet in clabbering milk; sd. FO u. in cooking, soaps, sds. u. refrig. diur. (Ayurved. med); vars. include cantaloupe (var. *cantalupensis* Naud) (muskmelon, most popular name; melon [Eu]); cassaba melon, etc.; convar. *cantaloupa* (Pang.) Greb.: (cantaloupe melon); convar. *cassaba* (Pang.) Greb. (AsMin): winter melon; cassaba melon (in US, "cantaloupe" refers to any musk melon [SC Cantaluppi near Rome, It, where first grown fr. swAs stock]) - **C. myriocarpus** E. Mey ex Naud. (sAf): cacur; gooseberry gourd; frt. considered toxic, c. resins, myriocarpin; u. emet. purg. - **C. sativus** L. (cAs): cucumber; lvs. c. cucurbitacin C (steroid = aglycone); frt. culinary item., eaten raw or cooked; u. to prep. pickles; sds. c. FO; juice & sds. u. dem. emoll.; in cosmetics; tincture of frt. u. to depigmentize skin discolorations; also smoothes skin (2738) - **C. trigonus** Roxb. (Ind, SriL, seAs, Austral): wild cucumber; frts. u. as diur., antiphlogistic.

cucurbit: member or frts. of fam. Cucurbitaceae (esp. *Cucumis* & *Cucurbita* spp.).

Cucurbita foetidissima HBK (C. perennis Gray) (Cucurbit.) (sw&cUS, nwMex): wild squash; mock orange; buffalo gourd; unripe frt. pulp u. detergent as soap, food; rt. u. in hemorrhoids, purg.;

for skin irritations, etc.; is oxytocic (Ferguson, 1955) (c. cucurbitacin E); sds. edible (Am Indian) - **C. lagenaria** L.: *Lagenaria siceraria* - **C. maxima** Dcne. (tropAm, cultd. NA, Eu): (winter) squash; "hundred pound gourd"; fls. & frt. u. as food (esp. It; many vars.: ex. Hubbard, Turban, banana, mammoth, &c.); largest frt. of all plants (-85 kg) ("mammoth chile"); sds. anthelm. (Malta) - **C. moschata** Dcne. (tropAm, sAs; widely cult.): cushaw; cheese squash; large greenish frt. ripens in fall; u. as food; "pumpkin" seed u. to treat acute schistosomiasis - **C. pepo** L. (Mex, Tex [US], cult. Eu & natd.): (field) pumpkin; also called pie & cattle frt. orange, often of very large size (1250); in fall, u. food; sds. c. 36-57% FO, VO, active princ. unknown (1552); FO u. culinary; press cake u. fodder (esp USSR); fresh sds.as tea, u. tenifuge, dem., prostatic dis.; med. appl. to "stubborn" ulcers; lamp oil; varieties: *medullosa*; (vegetable) marrows; *melopepo*; scallop(ed) squashes; summer & autumn s., bush s., pattypans (or cymlings), round s.; *ovifera*: small inedible gourds (1299); yellow fl. gourds (1299); *torticollis*: long-necked squashes, warted s.; summer crooknecks; cv. zucchini; zucchini (squash) (zucca melon, wCana [sBC] of large size) (1251) - **C. radicans** Naud. (Mex): sanacoche (Mex): rt. u. to wash clothing in saline waters.

Cucurbitaceae: Gourd Fam.; ann. & perenn. hbs. climbing by tendrils; 760 spp.; mostly in trop of both hemisph.; lvs. alternate, broad; fls. unisexual; 5-merous; frt. typically a pepo (berry-like with thick rind); ex. luffa, balsam-pear; bryony; pumpkin, squash, melon, cucumber.

cucurbitacins: very bitter principles of many Cucurbitaceae & some other fams.; tetracyclic terpenes; variously designated A- Q (17) best known: B, E); antineoplastic (antitumor) (inhibit growth of solid tumors), & appl. to "stubborn" ulcers; anti-gibberellic activity; u. as purg., anthelm, & antimalarials; very toxic; cucurbitacin E said most bitter subst. known (detected at 1 ppb); formerly some cucurb. frts. considered dangerous to consumer.

cucurbocitrin: ext. of watermelon sds.; u. in hypertension (1376).

cudbear: Persio (L.); ext. fr. *Roccella* (esp. *R. tinctoria*) & *Lecanora* spp. (Lichens); u. coloring agt.; prod. in Swed, Switz, Nor, Neth, Madagas, Mozamb.

cuddy's lugs: *Verbascum thapsus*.

cudweed (seUS): *Gnaphalium* spp.; *Filago* spp. - **fragrant c.**: *Gnaphalium obusifolium*; *Anaphalis margaritacea* - **marsh c.**: *G. uliginosum*.

cuenca bark (Ec): pale cinchona.

cuica resin: fr. *Cercidium spinosum*.

cuichunchulli root: *Ionidium microphyllum*.

cuje: *cuji:* (Guat, Hond): *Inga spuria,* +.

cuke (US, slang): cucumber.

culantrillo (CR, Mex): *Adiantum capillis-veneris* - **culantro** (Sp): coriander.

culen: *Psoralea glandulosa*.

culilawan (bark): *Cinnamomum culilaban, C. iners*.

cull: to remove defective frts., nuts, &c., because small, unripe, deformed, discolored, or otherwise unsuitable for sale for direct consumption; generally u. in mfr. by-products (ex. cull lemons for citric acid; cull walnuts for shelled w.); also separate inferior but acceptable products.

cullen: culen.

cultigen: sp. of organism known only in cultivation; not known in the wild state, at least now; presumably originating as cult. pl. fr. less valuable wild growing sp., which subsequently became extinct in the wild; ex. *Erythroxylon coca*.

cultivar: cultivated variety (abbr. cv.).

Culver's physic: C. root: C. weed: *Veronicastrum virginicum*.

cum (Mex): pumpkin, squash - **cumahy** (Br): **cuman**: *Couma utilis* - **cumara** (Pe): sweet potato - **cumarin**: coumarin - **cumarú** (Br [Amazon]): *Dipteryx odorata* (oil).

cumbungi (Austral): *Typha latifolia*.

cumin (seed) (pron. come-in) (Eng. Fr): *Cuminum cyminum*; source of c. oil - **black c.; c. noir** (Fr): frt. of *Nigella sativa* - **Roman c.: sweet c.**: Anisum.

cuminal: cuminaldehyde: gives essential aroma to VO of cumin where first discovered.

cuminho (Por): cumin.

Cumini Fructus (L.): **cumino** (Sp): **Cuminum** (L.): frt. ("seed") of foll.

Cuminum cyminum L. (C. odorum Salisb.) (Umb.) (Turkestan [origin.] now seEu, Sic, Turk, Gr; cultd. NA, Chin, Iran, Ind): cummin (seed) (actually frt.); frt. c. VO (0.25%) with traces of perillaldehyde; u. flavor, stomach.; prod. VO; cond. in curry powders (Ind); mfr. pickles, in sausages; liqueurs; largely replaced by caraway; in Mex cooking (cakes); carm. stim. arom. (prod. Mor, Cypr, Malta, Sic, Turk, Chin, Iran, Ind). (2608)

cummin (seed): cumin.

cumquat: kumquat.

cundeamor (de Yucatan) (Mex): *Momordica charantia* L.

cundurango: condurango.

Cunila origanoides (L.) Britt. (C. mariana L. var. o.) (Lab.) (seNA): (American) dittany; Virginia dittany; Maryland d.; common d.; stone mint; hb. (Herba C. marianae) c.VO, u. as gentle stim.;

pleasant beverage (inf.) (smells like thyme) diaph. emmen., in fever, snakebite, u. for colds (Am Indians & settlers); to adulter. Spigelia; cult. ornam. (border pl); culinary hb. - **C. microcephala** Benth. (Br): poejo brasil.; lvs. & flg. tops u. for coughs, pharyngitis & pulmonitis.

Cunonia capensis L. (Cunoni) (sAf): red alder (of the Cape); useful wd. fr. tree, with rich red color & high polish.

cunshu-huayo: *Chondodendron toxiferum.*

cuoxam: cupric oxide (Cu_2O) dissolved in ammonia water; u. as microtest for sugars, &c.

cup: receptacle attached to pine tree trunk for reception of wound flow (turpentine); made of clay, tinned sheet iron, plastics, aluminum, &c. - **c. plate** (agar): method u. to assay antibiotics in which a cylindrical hole is cut or formed in agar after being melted & poured into Petri dish & left to harden (Lochplatte, Ge); the antibiotic soln. is then poured into this hole or cup & its effect observed on the m.o. grown on the rest of the plate.

Cupania americana L. (C. tomentosa Sw.) (Sapind.) (WI, Ve): guara comun; lf. & bk, u. as astr. in blenorrhea & cystitis; frt. & sds, edible, much esteemed; sds. u. for hemoptysis & diarrh.; wd. u. in naval construction; cult. - **C. racemosa** Radlk. (Br, São Paulo area+): sds. ("camboata(n)") appl. exter. to sores & rheum.

Cupaniopsis concolor van der Ham (Mischocarpus guillauminii Kaneb) (Sapind.) (Caroline Is): bk. inf. sprinkled over patient for debility - **C. leptobotrys** (A. Gray) Radlk. (Fiji): bk. inf. u. for stomach troubles.

cupey: copey.

cup-flower: *Nierembergia hippomanica* - **cup-fungi**: Discomycetes.

Cuphea carthaginensis Macbr. (Lythr.) (CA, SA): sds. rich in FO, u. as substit. for coconut oil; u. in soaps, lubricants, paints, medicines, &c.; c. mosquito repellant - **C. glutinosa** C. & S. (Br to Arg to seUS): lf. inf. u. as diur. purg. & to "clean" the blood - **C. lanceolata** Aiton or Dryand. (Mex): yucucacu; lf. u. as astr., in dys., in neuralgia(?); exter. as embrocation - **C. racemosa** Sprengel (Colo [US], Mex): pl. decoc. u. as diur. (Ingano Indians) (2303) - **C. strigulosa** HBK (Pe to Co): yerba de toro; inf. of pl. tops u. for upset stomach (Pe) - **C. viscosissima** Jacq. (CA, SA): pl. with digitalis-like activity; u. in homeop. med. (fresh pl.).

cupid's delight: *Viola tricolor*, pansy.

cupping: dipping of fresh semi-liquid turpentine.

cup-plant: *Silphium perfoliatum.*

cuprammonium sulfate: ammoniated cupric sulfate: prepd. by dissolving $CuSO_4$ in ammonia water & pptg. with alc.; u. as "solvent" for cellulose (actually only a suspension, pectic materials being dissolved); in textile printing; fungicide.

cuprea (bark): *Remijia (Ladenbergia) pedunculata, R. purdeana* - **cupreine**: an alk. (hydroxy-cinchonine) obt. fr. bk. & sds. of Cuprea.

Cupressus arizonica E. Greene (Cupress.) (swUS, nMex+): Arizona cypress; cedro (de la sierra); wd. u. for fuel & general construction; med. use in Hond ("pinabete") - **C. lawsoniana** Murr. = *Chamaecyparis l.* Parl. - **C. lusitanica** Mill. (Mex, CA, SA): ciprés; branch tips soaked in alc. & inf. u. for coughs & colds; cones boiled & decoc. appl. to hair & used for baths; wd. u.; var. *benthamii* Carrière (C. b. Endlich.); bk. u. astr. - **C. macrocarpa** Gord. (Monterey Peninsula, coastal Calif [US]): Monterey cypress; widely cult. ornam. tree; trees growing in Monterey area, often with picturesque distorted forms - **C. sempervirens** L. (Med reg, Euras): Italian (or European) cypress; lvs. c. procyanidin, u. as cardiotonic; wd. & frt. u. astr. bitter, insect repellant (esp. moths); prodn. VO, u. pertussis; generally treated by dropping VO on bed linen, thereby reducing the no. of paroxysms; scent soap; in bromidrosis; bk. u. antipyr. bitter astr. (Alg, Fr), wd. for bldg. (2545) - **C. thyoides** L.: *Chamaecyparis t.*

cupric oxide: CuO; black copper oxide; u. as complex with strong NH_4OH as "Cuoxam" reagent (dissolves cellulose) - **c. sulfate:** $CuSO_4(5\ H_2O)$: bluestone; blue vitriol; formerly u. as emetic (caution: if dose does not produce vomiting, second dose should be administered, since compd. is toxic); caustic (escharotic); in granulomas, &c.; hematinic; algicide; fungicide; anthelmintic (sheep); insecticide (prepns. Bordeaux mixt., Paris green); several other Cu compds. u. in therapy.

cuprous oxide: Cu_2O; "red copper oxide" (color varies from red to brown & yellow, depending on mode of formation); acts as fungicide, antiseptic, antifouling (in paint for ship bottoms); red pigment; &c.

cupulaire (Qu, Fr): *Adiantum pedatum.*

Cupuliferae: earlier name for Oak (or Beech) Fam.; now generally broken down into two fam.: Fagaceae & Betulaceae (*qv.*).

Curaçao (pron. curas-sow): 1) island of N WI after which are named several products, such as a type of aloe, a Dutch liqueur (but mostly mfrd. in Hamburg), bitter orange, etc. (*cf.* curassao, a gallinaceous bird) 2) elixir of bitter orange peel (similar to **curaçao liqueur**) - **c. peel**: rind of *Citrus aurantium.*

curara (L.): **curare: curar(e)i**: ext. fr. *Strychnos castelnaei, S. toxifera, Curarea, Chondodendron tomentosum* & other pl. spp.; u. arrow poison (SA

Indians): the last-named sp. is the one now u. in prep. of med. curare; 3 types (named for container): calabash; tube; pot (much fr. *Strychnos* sp.); formerly c. concentrate marketed (1267).

Curarea tecunarum Barn. & Kruk. (Menisperm.) (Amazonas, Br): st. crushed & u. in cold water inf., drunk by both sexes as contraceptive (Deni Indians); u. to prep. curare - **C. toxicofera** (Weddell) Barneby & Krukoff (Chondodendron t. [Wedd.] Krukoff & Mold.) (Br): bicava (Jamamadi Indians): st. bk. ingredient in arrow & dart poison, curare; antipyr.

curarine: alk. fr. curare; extremely poisonous parenterally.

Curatella americana L. (Dilleni.) (CA, Mex, SA, Cu): lengua de vaca (Salv); raspa-quacal; "rasca (de) vieja"; rough lvs. c. silica u. to polish wood or metal, sandpaper substit. (Hond); lvs. u. tanning; lf. decoc. u. to treat wounds (Br); grd. sds. mixed with chocolate to flavor; said "specific" in syph.(!!) (Mex).

curbana (Cu): *Canella winterana.*

Curcas purgans Endl.: = *Jatropha curcas.*

Curculigo brevipedunculata Elm. (Lili/Hypoxid.) (PI): st. juice rubbed on place where centipede has bitten (for pain) - **C. disticha** Gagnep. (Indochin): rts. & st. bases u. in decoc. against throat pains - **C. latifolia** Dryand. (Ind to Indoch, Indones): frts. edible, u. to stim. appetite; rts. & fls. decoc. stom. diur. in GU dis.; rhiz. decoc. in ophthalmia; fibers u. for fish nets - **C. orchioides** Gaertn. (seAs): kali musli; rhiz. important Ayuvedic & Unani drug; ton. in fever, pain, cuts; tuber u. for asthma, impotence (Ind) (124).

Curcuma (L.): *C. longa*, rhiz.

Curcuma amada Roxb. (Zingiber.) (Ind): mango ginger; pl. somewhat larger than *C. longa*; rhiz. white with very strong flavor, somewhat unpleasant; u. carm. stom. cooling; appl. to contusions, sprains; sometimes u. in pickles, curries, &c. - **C. angustifolia** Roxb. (seAs): tubers source of East Indian arrowroot (starch); dem. nutr. (Ind) - **C. aromatica** Salisb. (Ind): jedwar (Hind); rhiz. u. condim. flavor, stim, arom, yellow dye - **C. caesia** Roxb. (Ind): rhiz. c. VO with camphor; u. arom. stim. carm. & esp. in arthritis; exter. for sprains & bruises; in cosmetics (some cultn.) - **C. kwangsiensis** S. Lee & C. F. Liang (Chin): Radix Curcumae; R. Zedoariae (Chin Phar. XI) (Peking); c. germacrene, linderazulene; u. like other turmerics; choleretic, analg. sed.; in hepatitis, &c. - **C. leucorrhiza** Roxb. (Ind): tikan; rhizome one source of Bombay or East Indian arrowroot - **C. longa** L. (C. domestica Val.) (seAs, esp. EI, Java; cult. Ind, Chin, Br, WI, Af): temoe lawak (Java); (common) turmeric (root); prepared rhiz. (Rh. Curcumae javanicae; Rh. Zedoariae) of 2 kinds (round, long) c. curcumin (coloring matter), VO (with camphor, cineole), resins, starch; macerated rhiz. u. with mustard oil & salt in first-aid treatment of broken bones & open wounds (seAs); u. specif. in liver & gallbladder dis. (Indones), anthelm., antisep. depur. ton. in skin dis. condim.; lvs. u. med. (Tonga) (86); to dye faces of Hindu women; u. like *C. aromatica*; in mfr. pickles, curries, mustards, prepn. curcuma oil; tech. to dye wool & silk yellow (1283); yellow dye in carpets, &c.; vars. Aleppo ("Alleppy fingers") (superior); Madras (fingers); Rajapore - **C. xanthorrhiza** Roxb. (C. javanica) (trop sAs; widely cult. EI & Oceania; chief sources sChina, Java): temu lawak (1283); u. in disorders of gallbladder & ducts in cholelithiasis, cholangitis; stom. carm. dyspep. icterus - **C. zedoaria** Rosc. (C. pallida Lour.) (Ind, SriL; widely cult. Ind, Oceania): (round) zedoary; rhiz. c. 1% VO, resins, FO; u. flavor, perfumery, carm. stomach, spice in curries; u. in colics, cramps, as corrective, stim., bitter ton, in liqueurs; mostly rhiz. cut in discs.

curcuma, wild: Hydrastis. - **curcumin**: turmeric yellow, a coloring princ. fr. curcuma; anti-inflammatory.

curds, milk: casein.

cure all: *Melissa officinalis.*

curing (of drugs): process of proper storage so as to permit the enzymes present to act upon the constituents & transform them & thus improve the quality of the material; ex. coca, tobacco, vanilla beans.

curly cup: *Grindelia squarrosa.*

curo (Co): avocado.

currant(s): 1) *Ribes rubrum*, *R. nigrum*, etc. 2) small dr. grapes; v. corinthes - **American black c.: buffalo c.: clove c.**: (golden-flowered) c.: *Ribes americanum, R. aureum* - **European b. c.**: *Ribes nigrum* - **dried c.: grocer's c.**: c. (2) - **Indian c.** 1) *Symphoricarpos orbiculatus* 2) *Callicarpa americana* (sometimes) - **Missouri c.**: buffalo c. - **Zante c.**: corinth(e)s (*qv.*).

curry: 1) popular oriental (seAs, EI) food; essentially spiced & garnished rice 2) (Tamil): condim. u. in such foods; c. red pepper (also black, white), capsicum, turmeric, coriander, fenugreek, cumin, ginger, lime, white mustard, cardamom, fennel, cinnamon, mint, parsley, clove, allspice, mace (curry powder) (Ind) - **c. leaf tree**: *Murraya koenigii.*

Curtonus paniculatus Br. (Irid.) (sAf): remedy for diarrh. (humans, cattle) (—> *Crocosmia*).

curupay bark: *Piptadenia macrocarpa* v. *cebil.*

cusco bark: c. cinchona: *Cinchona pelletierana*; *C. scrobiculata* v. *delondriana*.
cus-cus (Ind): *Vetiveria zizanioides*.
Cuscuta L. (Convolvul./Cuscut.) (trop & temp zones): dodder; "scald"; common parasites & bad weeds in grain fields, clover patches, etc. some ca. 200 spp. of total parasites; often a serious pest (1221) - **C. americana** L. (sFla [US], tropAm): dodder; love vine; pl. u. ton. astr. alt.; in Virgin Is, pl. (as Yellow Love) u. with *Tecoma stans* in "fever tea" - **C. campestris** Yuncker (C. arvensis auct. non Beyr.) (tropAm, now worldwide incl. Fiji): "wabosutha" (Fiji); lf. squashed & appl. to wound as styptic (Fiji) - **C. chinensis** Lam. (Chin, Ind): sds. u. diaphor. in GC & urinary incontinence; for impotency, leucorrh.; vine exter. vulner., in eye dis. (Mal) - **C. corymbosa** Ruiz. & Pavon (Peru): u. cataplasms; for fire wounds; var. *stylosa* (Mex): u. in icterus (decoc. internally & as bath - **C. epithymum** L. (Euras, nAf): dodder of thyme; hb. u. cath. - **C. europaea** L. (Euras, nAf): greater dodder; devil's guts; hb. u. ton. astr. alt. antiper.; mostly in folk med.; juice purg. deobstr., for itch - **C. glomerata** Choisy (e&cUS): American dodder; hb. u. bitter ton. sl. astr. antiperiod - **C. hyalina** Wight = *C. hygrophilae* Pears. (OWtrop) - **C. planifera** Tenore (NCar [US]): u. stim, incisive, aperient - **C. reflexa** Roxb. (Ind, Nepal): pl. u. as purg. antipyr., exter. for itch; inf. as wash for sores; st. u. in bilious dis.; sds. u. as carm. anthelm - **C. umbellata** HBK (sFla [US], WI, nSA, esp Br): cipó chumbo; u. vulner. fresh juice u. in angina, hemoptysis.
cushaw: *Curcurbita moschata* (*C. argyrosperma*).
cushion plant: *Culcitium argenteum*.
cusk: *Brosme brosme* (fish).
cuskohygrine: alk.; identical with bellaradine; occ. in coca lvs.
cuso: kusso.
Cusparia macrocarpa Engl. (Rut.) = galipea (Br): angostura; lvs. & sts. c. evolitrine (alk.); u. in folk med. - **C. officinalis** Engl. (C. febrifuga Humb.; C. trifoliata Engl.; Galipea officinalis Hanc.) = *Angostura trifoliata* - **Cusparia bark:** *Angostura* or *Galipea*.
cusso: kusso.
Cussonia arborea Hochst. (Arali.) (sAf): cabbage tree; rt. u. in GC, ingredient in vapor baths; emmen. - **C. spicata** Thunb. (eAf): frt. stem basal thickening u. as decoc. for biliousness; rt. antipyr.; succulent rt. at times eaten; bk. in mal.
custard apple: *Annona muricata, A. glabra, A. squamosa* - **American c. a.**: papaw.
cut: cutch: Bengal c.: Bombay c.: Indian c.: dark prod. prepd. fr. heartwd. of *Acacia suma,* catechu gum: *A. catechu* & *A. sundra*; black catechu (when "cutch" is u. alone, refers to Indian c.) - **Borneo c.:** mangrove c.: fr. tengah bark (*Ceriops* spp.) - **Gambir c.**: Gambir - **Indian c.:** Bengal cutch. - **pale c.**: gambir.
cuticle: a structural waxy layer which originates fr. the epidermis (including epicarp of frts.) & forms a continuous cover over many of the aerial parts of plants (lvs., young sts., etc.) except at the stomata; protects fr. excessive water loss.
cutin: waxy component of cuticle of lvs. & frts.; mixt. of condensation prods. of free satd. and unsatd. fatty acids (esp. hydroxy-fatty acids) with some esters (leaf cutin mostly C_{16} fatty acids with 2-3 OH gps.); impregnates outer epidermal walls; broken down by fungi & pollen by **cutinase**.
cuttle (fish) bone: Sepia (L.): (qv.) calcareous body inside the muscular mantle (or covering organ) of the c. fish, *Sepia officinalis*; c. $CaCO_3$ & Ca phosphate, gluten; u. tooth powders, polishing pd., abrasive, cage birds (cleans beaks, supplies Ca) (chiefly fr. Eg, Fr). (off. in older Brit Phar.). (cf. *Sepia*)
cutweed: *Fucus vesiculosus*.
cuum (Mex): cushaw.
cuyabra (Co): gourd.
Cyamopsis tetragonoloba (L) Taub. (C. psoraloides DC) (Leg.) (Ind, Afgh, Pak, Iraq, cult. US): guar; cluster bean; cult. as livestock fodder (Pak), solid builder; green manure; pods & sds. eaten like string beans; sds. much u.; grd. endosperm —> guar gum or g. flour, c.8% protein (Indian Standards, 1958); gum largely composed of galactomannans (polysaccharides) with galactan (galactose) & mannan (mannose) units combined by glycosidal linkages (1487); u. paper size, thickener; suspending agt. in cheese, ice cream, salad dressings; in pharm. indy. as binder for tablets in toothpastes; emulgent; claimed to lower blood glucose in type II diabetes & to reduce cholesterol in serum; paper, dye indy. (pl. rel. drought-resistant; cult. for sds. & forage).
Cyani Flos (Flores) (L.): **cyani flowers** (comml.): *Centaurea cyanus*.
cyanin: 3, 5-diglucoside of cyanidin; fr. *Centaurea cyanus*.
cyanocobalamin: xall. vitamin B_{12}; hydroxycobalamin now preferred in therapy.
cyanogen = 1) CN radicle-$(CN)_2$ 2) NC-CN (a gas).
cyanogenic glycosides: those producing HCN, such release producing toxic effects; some 1,000 pl. spp. known to be cyanogenic; CN glycosides incl.: amygdalin; prunasin; sambunigrin; prulaurasin; vicianin; dhurrin; phyllanthin (taxi-

phyllin); zierin; linamarin; lotaustralin; acacipetalin; gynocardin (*Gynocardia odorata*; *Pangium edule*).

Cyanophyces = **Cyanophyceae**: blue-green algae; included with Bacteria in the Schizophyta; classification aided by fatty acid occurrence (1252).

Cyathea arborea (L) J. E. Smith (Cyathe.) (Mex, WI, Ve): tree-fern; "Tasmanian cup-fern"; woody caudex (st.) dr. & u. as tinder; also u. to carry fire fr. place to place (Carib Indians of Dominica); cult. ornam. - **C. australis** Domin. = *C. medullaris* - **C. brunoniana** C. B. Clarke & Baker (Nep, nInd): pith starch-rich & u. for food by aborigines - **C. divergens** Kunze (Ve): u. as remedy for backache during menstr. period - **C. medullaris** Swartz (Austral, NZ): sago fern; tall or black-stemmed tree fern; caudex medulla (pith) with abundant starch, eaten by aborigines; young lf. stalks roasted & eaten as ton. during convalescence - **C. mexicana** Schlecht. & Cham. (sMex, CA): ocopetate; scales fr. fronds u. to arrest hemorrhage (Veracruz) - **C. moluccana** R. Br. (Mal): poultice made fr. lvs. appl. to sores on legs - **C. usambarensis** Hieron. (trop eAf): u. as tenifuge (Tanzania). (care!)

Cyathula achrysanthoides (HBK) Moq. (Amaranth.) (CA): lvs. u. to stop bleeding (Pan, Cuna Indians) - **C. capitata** Moq. (Chin): rts. u. as emmen. hematic; in liver & kidney dis.; ton.; for pain in back - **C. officinalis** Kuan (Chin): in traditional Chin medicine, u. as ton.; c. insect molting hormones - **C. prostrata** (L) Bl. (trop. incl. PI): (pantropic): dayang (PI); pl. decoc. u. for cough (Mal); in cataplasms (for skin eruptions [Laos]); rt. decoc. in dys. (Mal); lvs. burnt & appl. to ringworm (PI): similarly for scabies, craw-craw (wAf); arthritis, dys. (Kamerun) - **C. spathulifolia** Lopr. (sAf, Moz): nama; pd. st. frts. & sds. u. in castor oil base locally for leprosy; scorched frts. as pd. appl. to lesions in heads of children (2445).

Cybistax antisyphilitica (Mart.) Mart. ex DC. (Bignoni) (tropAm): atunijanqua (Col); lvs. u. to furnish blue dye for cloth; u. med. in Col.

cycad: *Cycas* spp.

Cycadaceae: cycad family; (OW trop): "fern palms" (**Cycadales** with 3 fams.) only one gen., *Cycas* (ca. 20 spp.).

Cycas circinalis L. (Cycad.) (sInd, SriL, Sum, Java, Taiwan, PI): **crozier cycas** (or cycad); "sago palm" (but not a palm!); green sago; prod. tragacanth-like prod., **cycas gum**; kind of sago prep. fr. starch fr. trunk pith; sds. u. starvation food; lvs. inflor. sds. u. med. (nEI); fresh lf. juice u. in gastrointest. dis.; crushed frt. spadix u. as poultice for kidney pains (Af); cult. as ornam. (subsp. *madagascariensis* Schuster; Tanzania) - **C. inermis** Lour. (Indoch): sd. starch popularly u. (much finer than rice starch); must sep. fr. toxic substances (alks.): may be carcinogenic - **C. revoluta** Thunb. (eAs, esp. sChin, Jap; cult. Java, PI, Br): **sago cycas**; "sago palm" (but not a palm); Japanese sago (starch) prod. fr. trunk; frts. ton. nutr. emmen. expect. astr. in rheum. (Chin); sds. c. inhibitor of malignant growth; starch believed rejuvenating; lvs. popular for funeral wreaths; ornam. - **C. rumphii** Miq. (Malaysia, Austral): source of sago (starch); frt. poisonous.

cycasin: toxic glycoside fr. *Cycas revoluta, C. circinata*; believed carcinogenic.

Cyclamen persicum Miller (Primul.) (se&cEu, nAf [*not* Iran]): c. saponin ("cyclamin") (with activity against some cancers); rts. emet. emmen. purg. diur., as fish poison (Ind); cult. as "Alpine violet" (florists) - **C. purpurascens** Mill. (C. europaeum L. emend. Ait.) (c&sEu, As): (European) cyclamen; (European) sowbread; fresh tubers c. cyclamin (triterpene saponin) (strongly hemolytic); floral VO u. in perfume indy. (2706); rhiz. u. drastic cath. emet. emm., in colic, gout, rheum., toothache (considered too toxic for regular use in med. but much u. in earlier centuries [Arthanita Ointment]; u. now by homeopaths); cult. by florists as ornam. (SC Gr kyklos, circle, because of circular way in which fl. stalks are curled before & after flowering).

cyclamin: (o)enin: monoglucoside of malvidin HCl (an anthocyanidin); occ. in *Cyclamen* spp. incl. *C. persicum*; term also appl. to a saponin (do not confuse with **Cyclam(yc)in**, an antibiotic) (Troleandomycin; acetyl deriv. of Oleandomycin).

Cyclea barbata Miers (Menisperm.) (seAs): rt. decoc. antipyr. (Indones); in stomachache (Chin); c. cyceline (cardiac poison) - **C. laxiflora** Miers (Mal): rt. decoc. u. to treat fever, hemorrhoids, asthma; lf. decoc. u. for vermifuge for children - **C. peltata** Hook. f. & Thoms. (Indoch): rts. diur., in liver dis.; hemorrhoids; ton.

cyclitols: cyclic sugar alcohols; ex. viburnitol.

cycloheximide: Actidione: antimicrobial agt. prod. by strains of *Streptomyces griseus*; antifungal; u. in mycotic pl. diseases.

cyclol: cyclic peptide wh. occurs in some ergot alks.; diketopiperazine.

Cyclolepis genistoides Don (monotypic) (Comp.) (Arg): paloazul; lf. u. med.

Cyclopia buxifolia (Burm. f.) P. Kies (C. latifolia DC) (Leg) (sAf): bush tea; Boer tea; lvs. us as "honey tea" for pulm. dis., tea substit. (this sp. one of several *C.* spp. so u. is said to have the finest aroma) - **C. genistoides** R. Br. (sAf): Cape tea; honey tea; bush tea; c. VO tannin, no alk.; u. by early Af colonists as restorant, black tea substit.; expect. in bronchitis & pulmonary TB - **C. subternata** Vog. (sAf): Cape tea; (common) bush tea; hunger t.; lvs. u. as substit. for beverage tea ("Caspa tea" [comm.]); u. in chest. dis. by indigenous people, like *C. genistoides*.

Cycloserine: Oxamycin; antibiotic prod. by *Streptomyces garyphalus* v. *orchidaceus*; u. as antibacter. tuberculostatic.

cyclosis: movement of protoplasm inclusions within the protozoan cell.

cyclosporin(e)s: metabolites of fungus *Tolypocladium inflatum* Gams (Trichoderma polysporum Rifai) & other Deuteromycotina (Hyphomycetes; Fungi); isolated fr. soils of sNorw; series of cyclic peptides with antifungal & antiphlogistic properties; most importantly as immunosuppressive agts. to suppress immune response by suppressing helper T-cells (*cf.* immune system) [1970-]; in organ transplants to reduce or prevent rejection (liver, kidney, heart); in AIDS; juvenile-onset diabetes (type 1); maturity-onset diabetes (type II); prevent or reduce rejection by suppressing helper T-cells (of immune system); & other autoimmune dis. (uveitis, multiple sclerosis[?], schistosomiasis); myasthenia gravis (exptl.); in treatment of severe keratoconjunctivitis sicca; gives increased hair growth; 9 of gp. known (A-I), C most active, with A showing some activity.

Cydista diversifolia Miers (Pleonotoma diversifolium Bur. & Schum.) (Bignoni.) (sMex to Ve, Cu): rabo de mico; tsolak; lf. decoc. u. antispasm. to prevent premature childbirth (Yuc, Indians); in diarrh.

Cydonia oblonga P. Mill. (C. vulgaris Pers.; Pyrus cydonia L.) (Ros) (monotypic) (*cf. Chaenomeles*); (swAs, Iran to Turkestan; widely cult. Eu & worldwide): (common) quince; pineapple g.; frt. tart; rich in pectin; u. astr. mostly in jellies, preserves, marmalades, jams; eaten cooked (1553); as heart stim. (Unani med.); sds. c. mucilage, amygdalin, emulsin, FO; u. much as vehicle in skin lotions, hair setting preps., hand creams, emoll. for skin chapping, burns, etc.; emulsifier; dem. in cough, eye waters, etc.; decoc. of fresh or dr. frt. u. diarrhea, hemoptysis; in folk. med. for leucorrh., amenorrhea (1253); may represent "golden fruit of the Hesperides"; tech. u. in textile indy. (chief prodn. Sp, nAf, Iran).

Cydoniae Semina (L.): sds. of **Cydonium** (L.): quince, *Cydonia oblonga*.

Cylindropuntia Engelm. (Cact.) (NA): subgenus of *Opuntia* with cylindrical sts.; "chollas"; ex. *Opuntia imbricata* Haw.; *O. bigelovii* Engelm.

cymarin: glycoside occurring in Strophanthus, the aglycone being strophanthidin; u. digitaloid card. stim. diur.

Cymbalaria muralis Gaertn. B. Mey. & Scherb. (Linaria c. Tourn. ex. Mill.) (Scrophulari.) (Eu, natd. eNA): Kenilworth ivy; pl. u. in diabetes; ground & wall cover.

Cymbidium Sw. (Orchid.) (trop As, Austral): cybidium; ca. 50 spp. - **C. aloifolium** Sw. (C. finlaysonianum Lindl) (seAs): pl. decoc. u. to bathe sickly children; for irreg. menstr. (Malacca); pseudobulbs (salep) u. for cuts, sores, burns; nutr. demul.; as supernatural agt. (Malacca) - **C. ensifolium** Sw. (Ind, Chin, Indoch, Indones): rhiz. combns. u. stomachache, VD; fls. inf. u. for eyes; lvs. diur.

Cymbopetalum penduliflorum Baill. (Annon.) (Mex, Guat): orejuelo (Mex); a shrub or small tree; arom. petals long u. with vanilla to flavor chocolate (1254); fls. u. in asthma, sold in Guat markets; many other uses; cult.

Cymbopogon citratus Stapf (Andropgon c. DC) (Gram.) (origin unknown; cult. pantropics & subtrop., (As, sInd, Java, Vietnam, SriL, WI, seFla [US]+): lemon grass; "coal g." (cFla [US]); lvs. c. VO, with arom. lemon-like odor, c. geraniol, citronellal, citral; lvs. u. as tea in flatulence, intest. spasm, gastric irritation, as stim. ton. diaph. diur. (inf. or decoc.) (Mex Phar.); antipyr. (wAf); in mal. (1257); tea for "curing" colds, acid stomach, hay fever (dom. med. Fla); prodn. VO (lemon grass oil); u. in mfr. ionone, synth. vit. A; arom. in perfumes, sachet powders, soaps, &c.; in cuisine (Ind) forage for cattle (after pl. being exhausted by distn.) (sFla); ornam. (sFla, cultd. Ind [chief prodn.], Guat) (said plant was brought from WI to Fla in early days by Spanish colonists) - **C. flexuosus** W. Wats. (trop & subtrop worldwide): Cochin grass; lemon g.; Cochin lemon g. (broad-leaved type); source of VO, East Indian Lemon-grass or Malabar oil; u. perfume, vet. linim. (cult. Ind, Vietnam+) - **C. iwarancusa (jwarancusa)** Schult. (Ind-Himalay): Iwarancusa grass; similar to *C. schoenanthus*; c. 1% VO with piperitone & carene; u. flavor; pl. u. in cough, chr. rheum. cholera; arom. ton. in dyspep.; stim. & diaphor. in gout, fever; fls. u. styptic.; rt. is vetiver rt. (name also appl. to *Vetiverria zizanioides*), cuscus rt. - **C. martinii** W. Wats. (var. *motia* Burk. & var. *sofia* Burk.) (Ind, widely cultd. seAs): ginger grass; rosha g.; prod. VO of 2 vars.: 1) motia: palmarosa oil (prod. Turkey) 2) sofia: gingergrass oil; VO u. to give honey flavor - **C. nardus** (L.) Rendle (SriL, cultigen; widely cult., esp. SA, Java, Chin): Citronella grass; source of c. oil (VO) of 2 vars.:

1) Ceylon c. oil 2) Java c. oil; c. citronellol, geraniol; u. insect repellant; perfume (in cheap soaps, etc.) (1256); u. in mfr. synth. menthol; rt. u. in fevers, stomachache (tea) (Br) (2506); gastric dis., mal. (steam bath) (Hond) - **C. schoenanthus** Spreng. (deserts of nAf, cAs): camel hay; capim de São Jose (Br); a desert pl., the chief food of the camel; dr. hb. u. as diur.; rt. u. as diur. stim. carm. as tea in rheum. (PI); grass inf. as emmen. (Vietnam); for fevers, colds (Br), GC (wAf); source of camel grass oil; cosmetics; lf. bud centers u. in Ceylon curries; rt. med.; dr. hb. with c. 1% VO with phellandrene, piperitone - **C. winterianus** Jowitt (C. nardus var. mahapengir) (Java, SriL, Malacca): old citronella grass; maha pengiri ([or pangiri] Singhal.); sometimes united with *C. nardus*; prod. Java citronella oil, regarded as finer than the Ceylon c. o.; chief const. geraniol & citronellol (1255).

cyme: 1) an inflorescence in which the individual fls. develop. & open fr. the top down or fr. the center of the cluster out; ex. Linden (see App.) 2) cynee: senna (1752).

cymene: cymol: iso-pr. toluene; occurs in VO of thyme, *Monarda punctata*, etc. (SC sp. name of *Cuminum cyminum*, cumin, where first reported).

Cymopterus purpureus S. Wats. (Aulospermum purpureum C & A) (Umb.) (wUS): "wild celery"; lvs. & fls. u. in debility, gastric dis.; as seasoning in foods, liqueurs, &c.; rts. lvs. eaten by Am Indians.

cyna: Cina.

cynanche: severe sore throat with imminent suffocation.

Cynanchum acutum L. (Asclepiad.) (Euras, nAf): French scammony; latex c. cynanchocerin, cynanchin; u. sometimes like true scammony - **C. amplexicaule** (Sieb. & Zucc.) Hemsl. (eAs): decoc. of bitter sts. u. in fever & hematuria - **C. atratum** Bge (Vincetoxicum a. Morr. & Decne) (eAs): dog's bane; rt. u. during pregnancy & at parturition; for fever and treating wounds - **C. caudatum** Maxim. (eAs): rt. u. for all kinds of dis. & in magic med. (Ainu); rt. decoc. diur.; toxic pl. - **C. glaucescens** (Dcne.) Hand.-Mazz. (Chin): u. as expect. for cough; in asthma; in pneumonia - **C. japonicum** Morr. & Decne. (Chin): u. in bad colds for cough & chest discomfort; acute bronchitis - **C. stauntonii** (Dcne.) Hand.-Mazz. (Chin): u. similarly to *C. glaucescens* - **C. vincetoxicum** Pers. (V. officinale Moench) (Euras, Af): (white) swallowwort; rhiz. (as Radix Hirundinariae) c. vincetoxin & other glycosides; VO, asclepiac acid (saponin), resin; u. emet. cath. diur. diaph. chiefly in veter. med.; rt. u. in scrofula.

Cynara cardunculus L. (Comp.) (wMed reg.): cardon; cardy; vegetable artichoke; cardoon; pl. 6-18 dm. high; cult. as vegetable (rt. & thickened petioles & midribs eaten) & ornam.; rts. u. aper. diur. aphrod.; fls. u. to coagulate milk (Fr); sds. c. FO (cardy oil); u. in salads, soups, stews (cult. Fr, Sp, It, also US); lvs. blanched & eaten as celery - **C. scolymus** L. (orig. Ethiopia, now cultd. sEu+): garden a.; globe a.; alcachôfra (Br); (French) artichoke; a garden vegetable collected just before unfolding; fleshy receptacle of capitulum & thickened base of involucral bracts (artichoke hearts) are eaten; also blanched central lf. stalk; boiled in olive oil (It); after eating a., foods taste sweeter (123); laxative; aphrod. (reputed), reduces blood sugar (1505); bitter lvs. u. in edema, rheum.; decholesterolinization; in arteriosclerosis; in diabetes (1258); rt. u. cholagog. choleretic, depurat.; involucre, bracts, & stem c. inulin, inulase, etc.; fl. bracts. c. labenzyme **cynarase** (1488).

cynarine: active compd. (caffeic acid ester of kinic acid) of *Cynara scolymus* u. in hyperlipidemia & as choleretic.

Cynips tinctoria L. (Cynipidae; Ord. Hymenoptera): the gall-fly; sting causes formation of nutgalls on young twigs of *Quercus infectoria*.

cynocannoside: glycoside of cannogenin & L-oleandrose; occ. in *Apocynum cannabinum*.

Cynodon dactylon (L.) Pers. (Gram.) (probably fr. Bengal, now common in Euras, WI, sUS & many other regions) (many vars.): Bermuda grass; Bahama(s) g.; doob; Gramen (L.): Rhizoma Graminis (italica) (L); grama (Por); rhiz. c. saponins, starch, cynodin (neutral subst.); in med. (as Radix Graminis italici) somewhat like Triticum & Sarsaparilla; ext. of fresh pl. antimicrobial (bacter., fungi); u. in skin dis. (scabies, &c.); dys., fever, epistaxis, bleeding piles, fatigue, &c.; adult. for Triticum (1259); this grass (up to 40 cm high) considerably u. for pasture, hay, soil binding, lawns (popular in sUS); fls. u. prep. pollen ext.; u. in treating hay fever of which this sp. is a common cause in summer (seUS) - **C. incompletus** Nees var. *hirsutus* (Stent) de Wet & Harlan (C. hirsutus Stent) (e&sAf): red quick (grass); rt. decoc. u. diaph. diur.; for indiges.

cynoglossine: toxic alk. fr. fresh hb. or rt. of *Cynoglossum officinale*, *Echium vulgare* & some other spp. of Fam. Boragin.

Cynoglossum officinale L. (Boragin.) (Euras, NA): (common) hound's tongue; dr. hb. c. cynoglossine, consolidin, FO, choline; hb. u. dem. sed. in neuralgia, emoll., protective, antispas., for cough;

extr. astr. in diarrh. hemorrh., inflamm, ulcers, & tumors; to treat slowly healing sores; once thought of use in cancer; rt. u. astr. & analges. (exter. & int.) in neuralgia (exter. appln.); rat & mouse repellent; salad (young lvs.).

Cynometra cauliflora L. (Leg) (Ind): nam nam (SriL); sds. yield FO appl. exter. to cutaneous dis., leprosy; may yield a copal resin - **C. ramiflora** L. (sAs, Austral): oringen (PI); lvs. antiherpetic; rt. purg.; sd. FO u. lotion in skin dis. - **C. zamorana** R. E. Schultes (Col): "coca"; dry sweet pulpy aril of sd. eaten; reddish brown frt. edible; trunks of large tree (-33 m) u. to make rafts (2303).

Cynomorium coccineum L. (Balanphor./ Cynomori.) (nAf, Chin): pd. rts. u. as condiment (Tuareg); remedy for kidney dis.; in impotency, sterility, constip., sexual ton., purg., aphrod. - **C. songariorum** Rupr. (Chin): u. med.

Cynosbatae (L.): **Cynosbati Fructus**: hips; frt. of the dog rose, *Rosa canina*.

Cynoscion regalis (Sciaenidae - Pisces) (Atl Oc, trop & subtrop mostly; off eUS): (gray) weakfish; (gray) sea trout; edible marine fish; air bladder u. to produce fish sounds.

Cyperus L. (Cyper.) (trop & subtrop worldwide): galingale; 600 spp.; rt. stocks of some spp. u. as food - **C. alternifolius** L. = *C. involucratus* - **C. articulatus** L. (Nigeria, tropAm): funcia (Ve); adrue; rhiz. c. VO; u. with *Xylopia aethiopica* frt. pod as cough mixt.; sed. u. to reduce vomiting (as in pregnancy); u. in toothache (Salv); ingestion said to cause drunkenness (Ve) - **C. brevifolius** Endl. (Kyllinga b. Rottb.) (Indomal, PI, Austral): polell (New Hebrides); in splenic dis. (NH); whole pl. decoc. antipyr.; rhiz. to poultice sore legs (Mal); sds. u. in pneumonia (Nepal) - **C. bulbosus** Vahl (Maldive Is): gok kulandura; rts. & tubers u. as incense (very arom.) (Mald Is); nutrient - **C. cyperinus** (Retz.) Sueing. (Chin): decoc. u. for boils - **C. cyperoides** Ktze. (Indoch, PI): ungit ma; sd. inf. u. in toothache; vermifuge - **C. difformis** Bco. = *Scirpus grossus* - **C. difformis** L. (OW trop: now also in NA, [Mex], CA): small-flowered umbrella plant; dirty Dora (Austral): one of the world's worst weeds - **C. diffusus** Vahl (pantropical); rt. u. to treat singao (disease, PI); rts. eaten raw - **C. elegans** L. (US to CR, WI): doll grass (Bah); green parts boiled to give decoc. u. for colds - **C. esculentus** L. (Nile region, Med reg, now pantropics, esp. Mex, Br; cult. Sp, seUS): chufa (Sp): earth almond; tiger nut (Af); xosca; xosa in Dawa; var. *sativus* Boeck.; tubers u. human & hog food; as almond & coffee substit. (2225); rich in FO, which is sometimes expressed & u. as food & in soap mfr.; since antiquity a flour & beverage prepd. fr. tubers (1260); chewed for indigestion (esp. with bad breath); stim. ton. sed. stom. (Chin); bad weed - **C. involucratus** Rottb. (tropAf, natd. WI, SA): umbrella plant; common ornam. perenn. hb. (pot & porch) - **C. ligularis** L. (seUS, Af, CA): ceda; arom. rt. u. for debility & stomachache (Cuna Indians, Pan) - **C. longus** L. (Indochin): u. ton. arom. emmen. diur. - **C. malaccensis** Lam. (Indones): rhiz. & rts. u. as diur. - **C. monocephalus** (Rottb.) F. Muell. (seAs): decoc. u. as cooling beverage (Chin); rhiz. decoc. u. as astr. in diarrh.; for measles (Indones); rhiz. dococ. as diur.; exter. in oil in dermatoses (PI); juice mixed with lime for ringworm (PNG); u. med. (Guam) - **C. obtusatus** L. (Presl) Matt. & Kirkenthal (Arg): hb. u. as diur. digestive, antisp. (Arg) - **C. odoratus** L. (cosmopol): chintule (Mex); rhiz. u. stomach. in colic; ton.; tea u. as hallucinogenic (Ec, Pe, Amazonian Indians); similarly u. are *C. articulatus* L., *C. prolixus* HBK; u. to give luster & fragrance to hair (Mex, Indian women) - **C. papyrus** L. (neAf, wAs, sEu): (Nile) papyrus; bulrushes; st. pith the source of the first paper (papyrus) (ancient Eg); pl. -5 m; flourishes along the Nile (Sudan); u. in eye dis.; rt. edible - **C. polystachyos** Rottb. (Indochin): tubers u. as stom antidysenteric - **C. prolixus** HBK (Mex to Arg): rhizome in aq. prepn. u. for anemia (777) - **C. rotundus** L. (Med reg., Ind, Jap, sAf, natd. WI): nut grass, coco g.; false galanga (comm.) ajo cimarron (Cu); motha (Hind); tubers c. VO; u. diaph. diur. dem. stim. galactagog (Ind); eaten as foods, tuber chewed for toothache (Marianas Is); u. in perfumery (for hair oils [Ind]); in prep. of "lavender water" (Eng); u. in Cu (+) to make horchata, cold drink (1261); bad weed ("world's worst") - **C. scariosus** R. Br. (Ind, damp places): nagar musta; rhiz. u. arom. stom. diur. diaphor. cardiac remedy; decoc. in GC, syph.; prepn. hair oils, leather perfume; VO (cyperus oil) u. substit. for patchouli oil (u. perfume fixative & in hair tonics) - **C. sesquiflorus** (Torr.) Matt. & Kuekenthal (Arg): pl. u. diur. antisp. carm. (Arg) - **C. sexangularis** Nees (tropAf): water grass; rt. u. in enema for lumbago & back pain, stomach troubles - **C. tegetiformis** Roxb. (As): cult. for fiber u. for mats, &c. (Chin).

Cyphomandra crassicaulis (Cortega) Ktze. (**C.** betacea Sendt.) (Solan) (CA, SA, WI; cultd. worldwide): tree tomato; grenadilla; frt. esculent raw, rather like tomato; also in preserves &c.; tree-like shrub - **C. crassifolia** (Ort.) Macbride (Co): lf. tea u. to expel intest. parasites (Mitu region Indians); lvs. u. to color clay pots black (Kamsa Indian) (445) - **C. dolichorachis** Bitter (Co): lf. decoc. u.

as vermifuge (Kamsa Indian medicine men [445]).

cyprès jaune (Fr): *Chamaecyparis nootkatensis.*

cypress: name appl. to at least 8 genera of evergreens - **c. (bark)**: *Cupressus sempervirens*; prods. incl. c. nuts, c. wood, c. oil (VO fr. lvs. & young twigs); other prods. - **bald c.**: *Taxodium distichum* - **black c.**: *Taxodium dist.*; prob. also *T. ascendens* - **European c.**: *Cupressus sempervirens* - **c. family**: Cupressaceae - **flowering c.**: *Tamarix* spp. - **garden c.: c. herb**: *Santolina chamaecyparissus* - **pond c.**: *Taxodium ascendens* - **red c.: river c.**: Virginian bald c. - **c. vine:** *Ipomoea quamoclit.*

Cypridedium: 1) gen. (Orchid.) (nTemp Zone): lady's slipper (orchids); c. 40 spp. 2) (L) (specif.) rhiz. & rts. of *C. calceolus* - **C. acaule** Ait. (c&eNA): pink lady slipper; moccasin flower; fine fibrous rts. u. nerve sed. antisp. promote sleep, in epilepsy, tremors, nervous fevers, sometimes preferred to Opium; "love medicine" (Meskwaki Indians); in stom. & urin. dis. (Queb Indians); ornam. hb. - **C. calceolus** L. (e&cNA): (common) lady's slipper; yellow Indian shoe; rhiz. sed. u. in nervous hyperesthesia, esp. pathol. sensitivity of eye; to sedate hyperirritability fr. coffee or tea drinking; antisp. in nerv. intest. spasms - **C. candidum** Muhl. (e&cNA): white lady slipper; ornam. perenn. hb. - **C. hirsutum** Mill. = *C. calceolus* L. var. *pubescens* Correll. - **C. parviflorum** Salisb. (e&cUS): small yellow lady's slipper; American valerian; u. formerly in US med. (US Ph. V-VIII) - **C. pubescens** Willd. (eNA & cUS): (large) yellow lady's slipper; golden s.; rhiz. u. in nervous dis., joint pain (analg.); GI worms - **C. reginae** Walt. (e&cNA): showy lady's slipper; rhiz. & rts. u. similarly to prec. spp.

Cyprinus (Cyprinidae - Pisces) (As origin; now cult. in ponds, Eu, NA): (common) carp; flesh u. food (fresh; fermented [Jap]; smoked, canned [often steamed with soy bean paste]); considered a pest as fish roots in bottom of ponds producing muddy waters.

cypsela: achene in Compositae &c. (where ovary inferior).

Cyrilla racemiflora L. (monotypic) (Cyrill.) (seUS, WI, CA, nSA): (black) titi; ironwood; leatherwood; fls. of shrub or small tree furnish excellent honey (superior as it does not crystallize) (nFla, sAla [US]).

Cyrtandra auriculata C. B. Clarke (Gesneri) (PI): u. for fungus infection of feet; frt. pounded & this rubbed on affected parts - **C. cumingii** C. B. Clarke (PI): bk. scraped into water & given to woman after birth of child to control bleeding.

Cyrtanthus obliquus Ait. (Amaryllid.) (sAf): sore-eye flower; fire lily; bulb u. as ingredient in med. for scrofula & in chronic cough; bulb decoc. u. for tubercular cough; dark part of rt. u. as snuff for headache; in cystitis & leprosy.

Cyrtomium Presl (Polypodi. - Pteridophyta): segregate fr. *Polystichum* - **C. falcatum** Presl (eAs): cold water inf. of rhiz. drunk for summer fever & to purify the blood (Chin) - **C. fortunei** J. Sm. (eAs): rhiz. anthelmintic; styptic; insecticide (Chin); rhiz. decoc. for nervous stomach, vermifuge. (—> *Phanerophlebia*)

Cyrtosperma merkusii (C. chamissonis Merr.; C. edule Schott.) (Ar) (sPacific Is; Melanesia; cult. PI, PNG, EI): puna; swamp taro; frt. & fl. stalk pounded with coconut milk for children's med.; tubers edible after boiling (Caroline Is); with much starch, u. as meal ("maota"; "baba") palauan; fl. cluster (spadix) u. emmen. ecbolic; huge tubers (-60 Kg) often staple food, also starvation food.

cysteine: Cys (abb.); amino acid produced by hydrolysis of proteins; cont. sulfur; u. mostly as HCl salt, to stim. granulation of wounds (wet dressings); detoxicant; as dough conditioner; to reduce side-effects from use of cytostatics in cancer; in musc. dystrophy; in dermatoses (endocrine orig.); to protect the liver; inactivates some antibiotics of plants such as *Allium sativum* (1893).

cystine: amino acid importantly present in keratin & hair (-16%); u. in cosmetics (1950-); based on nails & skin rich in c. (219); said of use in nail fragility, seborrheic states, alopecia, onychodystrophy (natural form, l-cystine preferred).

cytisine: alk. fr. Scoparius (*Cytisus s.*) sd. of *Laburnum anagyroides* & other spp. of fam. Leg.

Cytisus canariensis Ktze. (Genista C DC) (Leg) (CanaryIs): "Genista" (florists); cult. as ornam. (1263); formerly thought source of Lignum Rhodii - **C. fragrans** (L) Lam. (Teneriffe, one of CanaryIs): ornam. shrub - **C. laburnum** L. = *Laburnum anagyroides* - **C. purgans** Spach (Genista p. L.) (sEu, nAf): lvs. & fls. u. in folk med. as lax.; replacement for senna lvs. - **C. scoparius** (L) Link (orig. c&sEu; now widely distr. to NA, Jap, sAf): common or Scotch broom; broom tops (young twigs) c. cytisine, sparteine, genisteine; u. diur. sed.; fls. u. card. sed., diur.; cath.; as yellow dye; moldy fls. have hallucinogenic effects; rt. (as Radix Genistae) u. similarly; sd. u.. as card. & nerv. remedy, diur.; sds. roasted & u. as coffee surrogate; (prod. Ger, GB, sEu, US); fiber; bee pl.; u. to bind sand.

cytochalasins: mold metabolites produced by *Helminthosporium dermatioideum* (A, B, F),

Metarhizium spp. wh. have an inhibitory effect on cell processes, movement &c.; six known (A - F, of which B is best known); inhibit thyroid secretion, growth hormone release.

Cytochrome A, B, C: iron porphyrin protein found in nearly all living matter: somewhat like hemochromogen; isolated fr. horse & cattle hearts, yeast, algae, etc.; a reducing agent; forms part of a respiratory enzyme system; suggested for use in angina pectoris, etc. - **C. C.**: compd. of hematin & protein; u. experimentally in peripheral vascular dis., degenerative dis. where sclerosis & anoxia are developing.

cytogenetics: study of the genetic, phylogenetic, & evolutionary relationships of organisms based on micro study of the cells, esp. chromosomes; of special interest in detg. origin of spp.

cytomegalovirus: salivary gland v.; host-specific herpesvirus infecting man & some other animals; an immune globulin is available.

Cytomel™: triiodo-thyronine sodium.

cytosine: a pyrimidine; fundamental unit of mucleic acids; DNA component.

cytotaxonomy: study of pl. & animal spp. & other taxa as related by chromosome no., morphology & behavior.

Cytospora (Sphaeropsidales - Endothia parasitica): the chestnut blight organism; ravaged the forests, NY (US) 1905.

czar nuszka (Pol): char nuski; sd. of *Nigella sativa.*

Dd

Dacrydium Lambert (Podocarp.) (Oceania, NZ, Indomal, Ch): rimu; some 25 spp. of evergreen trees; some spp. produce manool, u. as substit. for ambergris - **D. cupressinum** Soland. ex. Lam. (NZ): "red pine"; rimu (dacrydium): source of rimu resin (fr. heartwood) c. podocarpic acid with estrogenic properties; VO fr. lvs. & twigs; furnishes valuable lumber; bk. in tanning - **D. franklini** Hook. f. (Tasman): Huon pine; planted in English (Cornwall) gardens; excellent wd.; wd. VO mostly methyl eugenol; u. in soaps; suggested as replacement for pine oil - **D. laxifolium** Hook. f. (NZ): mountain rimu; one of smallest known conifers -7 cm; u. to control erosion - **D. pierrei** Hickel (Indochin, Thail): rimu (tree); u. med. (Indochin).

Dacryodes excelsa Vahl (D. hexandra [Hamilt.] Griseb.) (Burser.) (WI, esp. Domin Rep): tabonuco ("tabonico"); tree trunk exudes resin called "Dry Elemi"; u. in technology; u. to make torches & to light fires; with shark oil & cotton lint, &c.; u. to caulk canoes; as masticatory, to allay "shortness of breath"; appl. to aching tooth; trunk for dugout canoes, &c. (Domin, Caribs) (2328); earlier this sp. was regarded as source of Caraña Resin - **D. longifolia** H. J. Lam (seAs): decoc. u. as bath for fever (Mal).

Dactinomycin: actinomycin D: antibiotic subst. (5 cyclic) prod. by *Streptomyces parvulus*, Waksman & Gregory, *S. chrysomallus* Lindenbein; &c.; u. as antineoplastic in melanoma, &c. (E).

Dactyli (L.): dates; frt. of *Phoenix dactylifera* - **D. acetosi** (L): tamarind (lit. vinegary dates).

Dactyliophora salmonia (Loranth.) (Solomon Is): lvs. heated and rubbed on sore legs.

Dactylis glomerata L. (monotypic) (Gram) (Euras, natd. US, Cana): orchard grass; cock's foot; u. hay (fodder), forage, frts., nutr. med. (lax); pollen important cause of pollinosis.

Dactyloctenium aegyptium (L) Beauv. (Gram) (OW trop, Sahara & Sudan, introd. to US & tropAm): mahra (Hind); ann. hb. u. fresh or dr. in decoc. for dyspep. hemoptysis, to relieve pain in kidney region (Af); lvs. & rt. ext. in ulcers; grains u. food (natives).

Dactylopius coccus Costa (D. cacti; Coccus cacti L.) (Coccidae [Dactylopiidae] - Homoptera - Insecta) (sMex, CA [Hond], Pe, Ch, Canary Is, Teneriffe): cochineal (insect); lives on *Opuntia cochinellifera*, *O. tomentella* & other Cactaceae; the female insect c. carminic acid, coccerin (wax); red coloring matter (alkaline carminate); u. coloring agt. in foods, esp. candy, drugs, lipstick; also textiles (wool, silk); in mfr. carmine; in folk med. u. in pertussis, cough, chronic bronchitis, nephritis, nephrolithiasis; in homeop. for kidney & bladder dis.; revival of dye use (cosmetics, foods, textiles) in recent years.

Daemonorops (*Calamus*) **draco** (Willd.) Becc. (Palm.) (Indochin, Moluccas, Sunda Is): source of Sumatra dragon's blood (called "gum" but really resin); u. coloring & astr. agent in tooth powders, plasters, etc.; also in techn. as in wood stains, violin shellacs, etc.; also source of dragon's blood - **D. dracon(c)ellus** Becc. (Borneo): dragon's blood palm; slender palms another source of dragon's blood (*qv.*) - **D. propinquus** Becc. (Mal): another source of dragon's blood (*qv.*); exudn. of ripe frt.; u. med. as astr. sed. ton. to treat wounds

(Chin); to remove phlegm (Indochin); in indig. med. for hematuria, sprue, stomach pain (Mal).

daffodil: *Narcissus pseudo-narcissus* - **d. family**: Amaryllidaceae - **French d.**: *Narcissus tazetta* - **Peruvian d.: sea d.**: *Hymenocallis calathina*; *H. narcissiflora.*

daffydown-dilly (wNCar [US]): *Trillium erectum.*

dagga: 1) (sAf): cannabis 2) lvs. of *Leonotis nepetifolia* (Sudan), *L. leonitis* (**Cape dagga**) - **dagged: dagget: dagutt oil**: birch oil (empyreumatic) (b. tar oil).

Dahlia pinnata Cav. (D. variabilis Desf., Georgina purpurea Willd.) (Comp.) (sMex): **dahlia** (pron. dah'-lia; fr. Dahl, pupil of Linné) - **Aztec** (or **common**) **dahlia**; **garden d.**: tuber c. VO (dahlia oil), levulose; u. in med.; tuber food crop (Eu); source of inulin & phytin; as coffee alternate ("Dalopa") (Cal); inulin, source of fructose; cult. ornam. (2643).

daikon: *Raphanus sativus var. longipinnatus* (*qv.*).

daime: caapi.

daiquiri: cocktail prepd. fr. rum (or tequila) & frt. juices.

daisy (origin: fr. day's eye, SC closed at night): 1) *Bellis perennis* (also **English d.**) 2) (sUS): *Houstonia minor* 3) (seUS): *Chrysanthemum leucanthemum* - **African d.:** 1) *Gerbera* spp., esp. *G. jamesonii* 2) *Arctotis* spp. (Comp.) - **dog d.: field d.: Maudlin d.: moon d.: oxeye d.:** Shasta d. - **Michaelmas d.**: *Aster* spp., incl. *A. tradescantii* L., *A. novae-angliae* L., *A. novae-belgii* L.; *Chrythsanthemum parthenium* (NCar [US]) - **sea d.**: *Armeria maritima* - **Shasta d.:** (hybrids) *Leucanthemum* x *superbum* - **Transvaal d.**: *Gerbera jamesonii* - **true d.**: *Bellis perennis* - **white d.**: Shasta d.

dak: dhak.

dal (Ind): cajan; kind of pulse; chiefly appl. to *Cajanus indicus*; u. food (1554).

Dalbergia candenatensis Prain (Leg.) (As, Austral, Oceania): poisonous sds. u. as beads (Tonga) - **D. ferruginea** Roxb. (PI, Oceania): st. rt. decoc. u. emmen., in large doses, abort. - **D. glabra** (Mill.) Standl. (Mex): useful in treating dys. (Mayas) - **D. latifolia** Roxb. (eInd, Bangla Desh): Malabar rosewood; shisham (Urdu); blackwood (tree); wd. u. in cabinet making, veneers; in skin dis. (cattle); firewood - **D. melanoxylum** Guill. & Perr. (Af): granadilla; African granadilla wood; rt. c. tannin, VO; u. colic, toothache, said narcotic; wd. u. making walking sticks - **D. mimosella** (Blco.) Prain (Celebes; PI): bk. & lvs. pounded & juice placed on wound; lvs. crushed & appl. to abdomen for dys. (PI) - **D. nigra** Allem. (Br, esp. Bahía): jacaranda; tree yields Rio (or Bahía or Brasil) rosewood, u. for fine work, such as radio cabinets; bois de rose source of rosewood oil - **D. pinnata** Prain (Assam, eInd): daman; rts. u. as masticatory & anthelm. (Indoch); lvs. u. to poultice varicose veins & in nerve dis. (Mal); poultice for itch (Indones) - **D. retusa** Hemsl. (CA): excellent wd. appears in commerce; may be u. as dyewood (Nic); med. (Hond) - **D. sissoo** Roxb. (cAs, esp. Indo-Pak): sissoo; Indian rosewood; wd. raspings u. in native med.; otherwise for fine wood carvings (most popular furniture wd. in Pak), floor parquetry; fuel; empyreumatic VO fr. wd. u. skin dis., rheum.; rts. & lvs. u. cataplasm (Ind) - **D. sympathetica** Nimmo (wInd): pentagul; bk. u. by natives to remove pimples - **D. stipulacea** Roxb. (PI): wd. & rts. inf. u. emmen.; large doses abort.; bk. & rts. u. as fish poison - **D. vaccinifolia** Vatke (trop eAf): rt. decoc. purg., u. for asthma (Tanz) - **D. volubilis** Roxb. (Ind): bhatia (Hind), batvia; alei (Bombay); lf. juice u. as gargle in sore throat, appl. to aphthous stomatitis; rt. juice in GC; u. in bowel troubles.

Dalby's carminative: Carminative Mixture (NF VIII); c. caraway oil, etc.

Daldonia concentrica (Bolton) Ces. & de Not. (Xylari. - Sphaeriales - Ascomyc. - Fungi) (cosmop.): produces "calico wood" in *Fraxinus*; c. pl. growth inhibiting subst.

Dalea domingensis DC (Leg.) (swTex [US], Mex, WI, nSA): ruda cimarrona; pl. u. in arom. baths for rheum.; alc. inf. rubbed on painful rheumatic areas - **D. emoryi** Gray (Calif, Ariz [US], nwMex): pl. u. dye, med and st. for basket making (Am Indians) - **D.** (**Parosela**) **formosa** Torr. (wUS, Mex): yerba de Alonso Garcia; int. & exter. to treat rheum & rickets (NM [US]) (58) - **D. leporina** Bullock (D. alopecuroides Willd.) (NM, Ariz [US], Mex to Guat): Wood's clover; foxtail; u. in folk med. (crushed lvs. for poultice [Am Indians]); valuable soil builder - **D. purpurea** Vent. (cNA): (purple) prairie clover; cult. ornam. - **D. polyadenius** Torr. = *Psorothamnus p.* Rydb.

Dalechampia scandens L. (Euphorbi.) (WI cont., tropAm): ortiga (Nic); lvs. provided with stinging hairs; with many unrelated uses: in gout, rheum. dys. eczema, etc.; lvs. rubbed on cheeks for toothache (CR).

Dalmatian: relating to Dalmatia, a coastal area of Yugoslavia along Adriatic coast (approx. 200 mi.); previously a kingdom; produces D. pyrethrum fls.

damar (**resin**): resinous exudate of trees of Fam. Dipterocarpaceae; esp. of genera *Shorea* & *Hopea*; readily sol. in ether, chloroform, CS_2, etc.; c. VO, resins, bitter princ.; u. for enamels, spirit varnishes, lacquers, spraying on citrus frts. (gloss), specimen preservation, plasticizer, in water emulsions, for ointments, adhesives, polishing waxes, etc. - **Batavian d.**: fr. *Shorea Wiesneri* - **black d.**: fr. *Canarium strictum* - **East Asiatic d.: East Indian d.**: *Agathis alba* & other *A.* spp.; also *Shorea* & *Hopea* spp.; not a true d. - **green d.**: fr. *Shorea* spp. - **gum d.: Malaya d.**: Batavian d. - **mata puching d.**: fr. *Hopea micrantha* - **Padang d.: Pedang d.**: fr. Sumatra, believed fr. *Shorea* spp. - **Penak d.**: fr. *Balanocarpus heimi* - **Penang d.**: fr. Malaya; a type of Singapore - **sal d.**: fr. *Shorea robusta* - **Singapore d.**: Malaya d. - **d. temak**: fr. *Shorea crassifolia, S. hypochra* - **white d.**: *Vatera indica.*

damasco (Sp): apricot; damson.

damask: figured lustrous fabric u. for household purposes (Damascus) - **d. rose:** *Rosa damascena.*

dambonite: dimethyl-i-inositol; found in Gaboon rubber; u. in mfr. artificial resins.

dame's rocket: d. violet: *Hesperis matronalis.*

damiana (Mex): 1) *Turnera diffusa* 2) *Haplopappus venetus* (much sold one time as substit. for [1]) 3) *Burroughsia fastigiata* (Mex).

Dammara: dammar: green damar; hard resin u. in varnishes, &c.; fr. 1) *Agathis* 2) *Protium* 3) *Shorea* (chief source) - **d. batu:** Manila copal; "white dammar" (false name); fr. *Vatera indica* - **d. (resin)**: **gum d.**: damar.

Damnacanthus indicus Gaertn. f. (Rubi) (Chin): fls. recommended to treat rheum., bleeding piles, headache.

Damson: an oval purple plum native to Syria; *Prunus domestica* v. *insititia* - **mountain D.**: *Simaruba amara.*

Danagar™: Danish agar fr. *Furcellaria fastigiata.*

dandelion, blue: *Cichorium intybus* - **(common) d**: *Taraxacum officinale* or *T. laevigatum* - **dwarf d.**: *Krigia* spp. - **European d.**: d. - **fall d.**: *Leontodon autumnalis* - **potato d.**: *Krigia dandelion* Nutt. - **d. root**: the dr. tap rt. of d. (prod. Hung, Ge) - **true d.**: *Taraxacum offic.*

dander: hair, skin or feathers in particulate form; many produce asthma (allergy).

Danewort: *Sambucus ebulus* (rts. & berries).

Daniellia ogea Rolfe (Leg.) (wAf as Nigeria): tree trunk source of Benin (Ogea, Accra, West African) copal; u. in varnishes, med. - **D. oliveri** Hutch. & Dalz. (D. thurifera Bennett) (wAf, Congo, Togo): prob. chief source of resinous exudate of tapped trees called illurin balsam, West African copal, (Sierra Leone) frankincense (ogea "gum") or African copaiba balsam; u. as body perfume ("balsam of copaiba"), sometimes u. as substit. or adulter. of copaiba; u. in GC dys.; bk. u. in migraine & fever; lvs. also u. med.

Danosanum: proprietary (Ge) representing pd. hb. of *Galeopsis ochroleuca*; u. to "purify blood."

Danube: v. grass.

Daphnandra micrantha Benth. ("Sassafras daphnandra") (Monimi.) (Austral): socket wood; bk. c. daphnoline & other alks.; with strychnine-like activity - **D. repandula** F. Muell. (D. aromatica Baill.) (Austral): northern sassafras; bk. c. daphnandrine & other alks., VO toxic to animals.

Daphne cneorum L. (Thymelae) (Eu): Heideroeschen (little heath rose); lf. buds c. 22% daphnin; bk. u. as for *D. mezereum*; fish poison - **D. composita** (L.f.) Gilg (sChin, Burma, seAs): u. in traditional med. (Vietnam); lvs. & sts. —> VO - **D. genkwa** Sieb. & Zucc. (Chin): c. genkwanin (flavonoid) & glycosides; rt. bk. u. in ascites; schistosomiasis; vesicatory (1016); safe abortient - **D. gnidium** L. (sEu, Med reg.): spurge flax (daphne); bk. (as Mezereon or Cortex Gnidii); c. daphnin (glycoside); u. alt. acrid irrit. in plasters - **D. laurelola** L. (Euras): spurge laurel (daphne); u. as prec. - **D. mezereum** L. (Euras): February daphne; mezereon; wild pepper; bk. u. as preceding; frt. u. to adulterate black pepper; (prod. Fr); has been lethal - **D. miyabeana** Makino (Jap): karasu-shikimi; entire burned pl. appl. to bruises (Ainu) - **D. oleoides** Schreib. (Indo-Pak): kutti; popular in folk med.; rts. purg.; bk. & lvs. in skin dis.; lvs. (inf.) in GC & abscesses - **D. papyracea** Wall. (Nep): Nepa(u)l paper shrub; u. as bitter, purg., antipyr. - **D. striata** Trattin. (Chin): in Chin folk med., as abort. (lvs. & frt.).

Daphnia: "water flea"; small transparent fresh water crustacean; u. food for trop birds & fishes; pharmacolog. test animal (as in Veratrum Viride assay), esp. *D. magna.*

Daphniphyllum luzonense Elm. (Daphniphyll.) (PI): frt. & lf. decoc. u. for stomachache.

Daphnopsis Mart. & Zucc. (Thymelae.) (Mex, WI, SA): bk. of some spp. u. like Mezereum - **D. americana** J. R. Johnston subsp. **caribaea** (Griseb) Nevl.) (DC Griseb.) (PR, Lesser Antilles): maho; piment; lf. inf. u. for "spells"; tea brewed fr. bk. + *Inga laurina* bk. u. to induce lactation; fiber of bk. u. to make inferior rope & as caulking material (Domin, Caribs) - **D. brasiliensis** Mart. & Zucc. (Br): bk. of young pl. u. as drastic, & in erysipelas, psoriasis - **D. fasciculata** (Meiss.) Nevl. (sBr): "beta"; shrub; bast u. for cords for fastening sacks, &c.; pl. toxic (*cf. D. racemosa*) - **D. macrophylla** HBK (Ec): ripe frt. strong purg.; use care in dosage - **D. pseudocalyx** Domke (sBr): embirabranca; shrub; bast u. as cord for binding sacks & in making strings u. in country inns - **D. racemosa** Griseb. (sBr): imbero; shrub; bk. decoc. u. as abortient, in edema; bast u. for cords in home, &c. - **D. salicifolia** Meisn. (Mex): "ahuejote" (Veracruz); bk. & lvs. drastic, acrid; lvs. u. as blistering agt. like mustard plaster.

darnel(l) (grass) (US): *Lolium temulentum, L. perenne*, & other spp. of grasses (*Bromus hordiaceus, B. seclinus*).

dartrial (Br): *Cassia alata.*

darudi (Hind): *Argemone mexicana.*

Darwinia fascicularis Benth. (tropAs): source of VO u. in the perfume indy.; similarly with *D. taxifolia* Muell.-Arg. & *D. grandiflora* Benth.

dasheen: *Colocasia esculenta* (tubers).

Dasylirion longissimum Lem. (D. quadrangulatum S. Wats.) (Agav.) (Mex): junquillo; starvation food (leaf, pith, lf. bases); also u. to prep. sotol & in mfr. alcohol; med.; trunk interior as forage; useful fiber; lvs. u. to thatch houses (Queretaro) - **D. texanum** Scheele (Tex [US], nMex): (Texas) sotol; bear grass; central bud portion roasted & eaten (sugary) by Am Indians (antiscorb.); u. to prep. sotol (mescal) (beverage); perenn. hb. serves as emergency forage & fuel in arid areas - **D. wheeleri** S. Wats. (wTex-Ariz [US], Chihuahua): sotol; "desert spoon" (SC spoons made fr. leaf bases); head of caudex u. in mfr. alcohol; roasted heart u. as food, also to prep. "sotol" (beverage); forage.

Dasymaschalon clusiflorum Merr. (Annon.) (Indochin, PI): rt./dr. lvs. decoc. u. diur.; lvs. fresh or ashed u. as cicatrizant (PI).

Dasyphyllum diacanthoides (Less.) Cabr. (Comp.) (Ch): palo santo; p. blanco; lvs. & rts. c. terpenes, steroids; beta-amyrin, friedelin; u. in liver dis. (Ch), abscesses; diur. purg. astr. balsamic; bk. regulates blood sugar.

Dasystoma pedicularia Benth. = *Aureolaria p.* Raf.

date: *Phoenix dactylifera* (palm tree & frt.) - **camel date**: dry d. - **Chinese d.**: *Zizyphus jujube* & other Z. spp. - **desert d.**: frt. of *Balanites aegyptica* - **dry d.**: desiccated frt. is ground into flour & u. as general food article - **d. plum**: *Diospyros* spp. - **d. (tree) sugar**: palm sugar - **wild d.**: *Yucca mohavensis.*

dates: 1) frt. of *Phoenix dactylifera* (date palm, tree) 2) (NM [US]): frt. of *Elaeagnus angustifolia* (SC flavor & shape of drupe).

dathyl: datil: dattler: var. red pepper.

dátil (Mex): date, *Phoenix dactylifera.*

Datisca cannabina L. (Datisc. - Dicots) (seEu, cAs, Ind): lvs. & sts. bitter, purg. antipyr. in fevers (Iran); diur.; rt. sed. in rheum.; appl. to carious teeth; entire pl. u. as yellow dye, formerly u. as silk dye; has sometimes been mistaken for *Cannabis sativa*; cult. for ornam. foliage - **D. glomerata** (Presl) Baill. (wUS): Durango root; said to have anti-tumor activity (125).

datiscoside: cucurbitacin glycoside with new unique sugar fr. *Datisca glomerata.*

Datura L. (Sol.) (circumglobal): genus of poisonous & medl. pls.; usually strong-smelling - **D. alba** Nees = *D. metel* var. *alba* - **D. arborea** L. = *Brugmansia a.* L. - **D. candida** (Pers.) Saff. = *Brugmansia c.* - **D. discolor** Bernh. (Mex): similar to *D. inoxia* except that sds. black & fls. larger; u. as hallucinogenic, often in form of beverage - **D. fastuosa** L. = *D. metel* - **D. inoxia** Mill. (D. innoxia [in error]; D. meteloides DC) (Mex, swUS, now in all tropics): "scopolamine plant"; toloache (Mex); mad (or Indian) apple; sts. & lvs. hairy; lvs. c. scopolamine, hyoscyamine, atropine; pd. lvs. u. for hemorrhoids, boils, etc.; intoxicant (NA Indian): rts. u. narcotic, anod. anesth.; fls. & rt. ext. u. extern. for wounds & bruises; int. for asthma, inflammn. (Ve); scopolamine source; Indian ceremonial use (Calif [US] (1555) - **D. insignis** B. Rodr. (Br - Amazon reg.): toé; tohe; hb. c. high levels of scopolamine (0.3%); u. hypnotic; to produce trance (natives); fls. of great beauty - **D. metel** L. (sAs, now pantropic): recurved thornapple; metel nut (frt.) (jous-methel, Arab); pl. glabrous; fls. colored (v. *alba*: white fl.); lvs. c. 0.55% scopolamine (one of richest sources); sds. c. hyoscyamine; u. scopolamine prepn.; narcotic intoxicant; u. in Parkinsonism - **D. quercifolia** HBK (swUS, Mex): lvs. rts. & sts. c. scopolamine & hyoscyamine (eq. parts), u. folk med. - **D. sanguinea** R. & P. (Brugmansia s. Don) (Pe, Ec, Col): floripondio rojo (fr. *incarnanda*); dark red datura; huaca; sed. hallucinogenic, in witchcraft (assoc. with death); ointment of crushed lvs. appl. to genitals, legs, & feet; sds. u. to prepare ointment for med. u.; cult. as ornam.; poisonous; sds. u. narcotic, in intoxicating beverages called "Tonga"; may produce insanity (*sic*) - **D. stramonium** L. (OW trop, natd. SA, NA): Jimson weed (datura); found in richly manured soils; hb. with rank odor and showy whitish fls.; source of off. Stramonium; u. narcot. sed. fumitory, masticatory, lvs. also smoked in asthma (powders & cigarettes), often with KNO_3 (also relaxant), also in pertussis, cough; sometimes mixed with lobelia, cubebs, swamp cabbage; ointment for piles (formerly); u. hypnotic (Eg); hallucinogenic; (prod GB, US, Ind, Pak); rt. scraped & smoked with tobacco for asthma (Pe); var. *tatula* Torr. (f. tatula) (D. tatula L.) (trop): purple thorn-apple; only a color variant of *D. stramonium*; ext. u. in neuralgia, tic douloureux (sUS); aphrod.; lf. decoc. in leprosy (sUS); similar to *D. stramonium* - **D. suaveoles** Humb. & Bonpl. = *Brugmansia s.*

datura, black: d. d'Egypte (Fr): *D. metel* - **d. herb (d. leaf)**: dried lvs. & flg. tops of *D. metel* or its var. *fastuosa*; chief active const. scopolamine; u. in parkinsonism - **Indian d.: white d.**: *D. metel* var. *alba*.

daturine 1) mixed alkaloids of Stramonium; viz., hyoscyamine, atropine & scopolamine 2) (Fr Codex): atropine.

daturism: poisoning by *Datura* spp., esp. *D. stramonium*.

Daubentonia DC = *Sesbania*.

Daucus carota L. (Umb.) (Euras, nAf): wild carrot; Queen Anne's lace; ripe frt. (as Semen Dauci silvestris) (Carota USP 1820-70) c. VO (carrot oil); u. anthelm. (like santonin), flavor, stim. emmen., diur.; hb. u. folk med.; var. *sativa*: cultivated (or garden) carrot; source of c. fl. oil (u. special unique perfume note, adjuvant & medium for perfume agts.; arom. in condiments & prep. foodstuffs); sometimes u. to dye butter; c. juice expressed fr. rts.; popular nutrient (also inspissated juice) (1015); fleshy tap rts. well-known vegetables; eaten raw or cooked; rich in carotene (red color); rt. scrapings as local stim. for wounds (folk med.); said to c. substance inhibiting cancer; frt. contraceptive (Rajasthan, Ind), abort. - **D. pusillus** Mex. (NA, Pac, BC [Cana] to lower Calif [US]) also e to Fla): rattlesnake weed; "r. bite cure" (1017); biennial hb.; rts. u. raw or cooked by Indians as food; much u. as poultice for rattlesnake bites & infections; promotes wound healing.

Daunomycin: Daunorubicin: antibiotic of rhodomycin gp.; occ. in fermentation broth of *Streptomyces peucetius* Grein & al.; glycoside formed fr. daunomycinone (tetracyclic quinone) with daunosamine (amino sugar); antineoplastic u. in acute leukemia (children), soft-tissue sarcoma (children).

Dauria (L) = **Davuria** (L): region of seSiberia.

Davallia Sm. (Davilli/Polypodi.) (OW, warm): carrot fern; epiphytes; often cultd. - **D. fijeensis** (Fiji): u. med. - **D. solida** Swartz (Fiji): u. med.

davana oil (Tamil, Ind): VO fr. hb. of *Artemisia pallens* Wall.; u. arom. in perfumes.

David's root: Radix Caincae: rt. of *Chiococca alba*.

Daviesia latifolia R. Br. (Leg.) (Victoria, Austral): "native hop bush"; lf. inf. u. as ton. diur. & for hepatic dis., for hydatids (cysts), "low" fevers; c. VO, resin, dibenzoylglucoxylose (bitter subst.), phytosterol, quercetin glucoside (rutin?) (3564).

Davilla brasiliensis Mart. (Dilleni.) (Br.): u. exter. to soften crusts of leprosy ulcers - **D. nitida:** Kubitzaki (Co): lf. decoc. of shrub u. to cauterize bleeding wounds; others put lf. ashes into gashes made by machetes to stanch blood flow & hasten healing process (Tikuna Indians) (2299) - **D. rugosa** Poir. (CA, WI, SA): cipó vermelho (Br): lvs. & tops astr.; mucil. u. as cataplasm to treat ulcers, orchitis; scrotal tumor; rts. u. purg.; sds. u. emet. cath.; poisonous.

dawa: 1) (Urdu): remedy 2) (Fiji Is): edible frt. of *Dracontomelum dao*.

dawa dawa: frt. and seed of *Parkia filicoidea* (Af).

dawakhana (Urdu): dispensary for Unani or indigenous med. (mostly pl. drugs) (Pak).

daw-sun (Chin): *Zizania caduciflora* (grain).

day flower, common: 1) *Commelina communis*, *C. tuberosa* 2) *Tradescantia* spp.

daylily: 1) *Hemerocallis* spp. 2) *Lilium candidum* - **tawny d.**: *Hemerocallis fulva*.

DCA: deoxycortico-sterone acetate.

dead: d. Borneo: gutta gelutong - **d. camas**: *Zigadenus* spp. - **d. cup**: *Amanita phalloides* - **d. man's fingers**: 1) rt. of *Habenaria fimbriata* (great fringed orchid) 2) *Orchis*; *Alopecurus*; *Arum*; *Lotus* 3) (Ga [US]): a fungus - **d. m. hand**: *Laminaria digitata* - **d. men's bells**: Digitalis - **d. nettle**: v. under n. - **d. tongue**: *Oenanthe crocata*.

deadly: v. also under nightshade.

deadly amanita: death cup: *Amanita phalloides*.

deal (GB): pine or fir wood.

death camas(s): *Zigadenus* spp. - **d. cup**: *Amanita phalloides* - **d. head moss**: *Usnea hirta* - **tall d.**: *Amanita excelsa* - **d. of man**: *Oenanthe crocata*.

death's herb: belladonna hb.

debbil dour (Surin.): doberdua.

Decalepis hamiltonii W. & A. (Asclepiad.) (Ind, Deccan Peninsula, wGhats): rt. u. as appetizer, blood "purifier."

Decaschistia harmandii Pierre (Malv.) (Indochin): petals & tuberous rts. u. to treat plague - **D. parviflora** Kurz (Indochin): rts. given to antidote food poisoning, esp. that caused by toad meat.

Decaspermum coriandri (Bl.) Diels (Myrt.) (Oceania): pulped frt. in water kept in mouth for toothache (Solomon Is) - **D. fruticosum** (Bl.) Diels (Forst.) (eAs, Austral, PI): daniri (PI); terminal shoots u. as condim.; frt. sweet, black, eaten; frt. u. for stomachache (PI); chewed with betel for dys.; astring. lvs. chewed to make the teeth "firm."

Decavitamin Capsules: D. Tablets (USP): mixed vitamins with A (ergocalciferol or cholecalciferol), C (ascorbic acid or Na ascorbate); Ca pantothenate (or panthenol); B_{12} (cyancobalamin); folic acid; niacinamide; B_6 (pyridoxine HCl); B_2 (riboflavin); B_1 (thiamine HCl or mononitrate); & E (alpha-tocopherol); a multivitamin prepn. intended to supply the minimum daily requirements of vitamins per day for adults, using one unit per day.

deck (US slang): one ration of addictive narcotic, often in paper.

decoction (Decoctum, L.): liquid ext. of crude drug made by boiling with water for 15 min., then straining.

Decoctum Album: sort of chalk mixture - **D. Diaphoreticum: D. Lignorum**: Dec. Guaiac. Comp. - **D. Zitmanni**: Dec. Sarsaparill. Comp.

decorticated: with bk., husk or peel removed.

decoside: glycoside of caudogenin & L-oleandrose; occ. in *Strophanthus* spp.

dedifferentiation: loss of properties produced in the differentiation process; ex. tannin cells become meristematic cells in explants of *Juniperus communis* phloem.

dedit (L): he (or she) gave; refers to donors of plant specimens in herbarium.

deer: ruminant mammals of several genera of Fam. Cervidae (Euras, NW); horns of *Cervus* & *Elaphodus* spp. u. in Chin med.; deer are reared in fenced enclosures for horns, hide, venison, &c. (NZ; BRD) (domesticated in ancient times); for musk d., v. *Moschus moschiferus*; other taxa: *Axis axis* L. (speckled deer of Ind.); *Cervus* (red deer) (*C. canadensis* L.) (Am elk or wapiti); *Dama dama* L. fallow deer (Eu); *Odocoileus* (US) (black & white tailed deer); &c.; many parts of axis deer u. in Unani med. (flesh, blood, fat, horns, gall stones, &c.), often stag (hart) (male) or doe (female) specified - **d. apple** (Grove Hill, Ala [US]): *Aesculus pavia* - **d. berry** 1) *Gaultheria procumbens* 2) *Polycodium stamineum* - **d. bowl** (Fla [US]): *Liatris tenuifolia* - **false tongue d.:** *Liatris graminifolia* - **d. plum**: v. under plum - **d. tongue (leaves)**: 1) (seUS): *Trilisa odoratissima* 2) *Erythronium americanum*, &c. - **d. vine**: *Linnaea borealis* - **d. weed**: (Calif [US]): 1) *Lotus scoparius* 2) *Polygonum aviculare* - **d. wood**: *Ostrya virginiana* - **deerwort boneset**: *Eupatorium urticaefolium*.

Deeringia amaranthoides Merr. (Amaranth.) (eSolomon Is, Asia, Austral): sap of vine u. to wash skin of hogs with skin dis.; leafy shoots eaten with rice (Indochin); rt. juice with vinegar u. to clear congested nasal cavities; lf. decoc. in dys. (PI); lvs. appl. to sores (Indonesia) - **D. polysperma** Moq. (PI): young lvs. appl. during puerperium; rts. u. to prevent drunkeness.

deer's tongue: deer tongue.

deferoxamine: natural biol. subst. isolated fr. *Streptomyces pilosus* Ettlinger et al.; u. as chelating agt. to remove Fe & Al fr. body tissues by forming complexes; u. to treat Cooley's anemia; antidote in Fe poisoning.

Deflersia Schweinf. = Erythrococca.

defoliant: agent which removes lvs. usually by chemical means.

Degeneria vitiensis I. W. Bailey & A. C. Sm. (monotypic) (Degeneri) (monotypic) (Fiji Is): masiratu; large tree (-30 m) (largest in rain forest of Fiji); wd. useful; type specimen coll. by Otto Degener, 1941.

degras: purified wool grease; u. in prepn. of lanolin; c. lanoceric & lanopalmitic acids, lanosterol.

Deguelia: genus cognate with *Derris*.

deguelian: ketone representing an active constituent of derris & cube rts.

Dehaasia triandra Merr. (Laur.) (seAs PI): lvs. appl. to ulcers (Mal).

dehydrated: water removed fr. drug, frt., &c., to aid in preservation & to reduce transportation costs.

dehydrocholesterol, 7-: provitamin D_3; obt. fr. tissues of higher animals, etc. - **activated 7-d**: vitamin D_3 (*qv.*) - **dehydrocholic acid**: a steroid acid obt. fr. ox-bile & u. as oral tabs. & inj. in treating liver dis. & esp. gall stones - **dehydrocorticosterone**: a steroid comp. of adrenal cortex with endocrine function; u. much like corticosterone - **dehydro(epi)androsterone** (DHEA): Prasterone; endocrine secreted by adrenal glands; isolated fr. human male urine; secretion greatest early in life, later reduced; masculinizing effects; said to reduce obesity & oppose cancer & the aging process; may reduce heart attacks; antistress; improves muscular ability, less joint discomfort, sounder sleep - **dehydrogenase**: dehydrase enzyme which catalytically incr. the oxidation of a substrate by hydrogen removal; some act directly, some through a hydrogen acceptor - **dehydromorphine**: pseudomorphine obt. fr. Opium.

Deianira nervosa Cham. & Schlecht. (Gentian.) (Br): centaurea do Brasil; with high bitter value, u. successfully as substit. for *Centaurium erythraea*; flowering parts preferred.

Deinbollia borbonica Scheff. (Sapind.) (eAf): rt. decoc. u. in splenic pain (Tanz); stomach complaints; abdominal pain - **D. oblongifolia** Radlk. (sAf): rt. inf. u. in diarrh. dys.; pd. bk. rubbed into scarifications on forehead for headache; frt. may be edible.

Delmarva: peninsula between Chesapeake Bay & Atlantic Oc; comprises much of Del, Md, & Va (US).

Delonix elata Gamble (Leg.) (native to WI; natd. sAs, tropAf): flame tree; leb(i); sd. inf. u. as purg.; wd. useful; trunk yields gum like acacia; lf. u. in rheum. (WI, Indians) - **D. regia** Raf. (Madag, sAf): royal poinciana; flamboyant (tree); source of a mucilaginous gum u. in As; cult. for ornam. & shade; fls. & buds sometimes u. like spinach (Ind).

Delphinapterus leucas Pallas (monotypic) (Delphinidae - Monodontidae - Odontoceta) (circumpolar; Arctic Oc & subarctic): beluga; white whale; w. porpoise; this small (-4 m) toothed cetacean is utilized for its hide, FO & meat (Inuits); sometimes exhibited in oceanaria; belongs to same fam. as narwhal.

delphinin: glycoside made up of delphinidin & glucose; occ. in *Delphinium, Salvia, Verbena*, &c.

delphinine: toxic alk. found in *Delphinium* spp., esp. in stavesacre sd.

Delphinium (Ranuncul) (OW, NW): larkspurs; a genus of hbs. with irreg. spurred fl. (SC delphinos, Gr porpoise, because fl. sometimes so shaped) - **D. ajacis** L. = *Consolida ambigua* - **D. consolida L.:** *Consolida regalis* - **D. denudatum** Wall. (Ind): nirbisi (Hind): rt. bitter ton. stim. "alterative" (Lahul); in toothache; adulterant of aconite - **D. elatum** L. (Euras): bee larkspur; c. alkaloids delpheline, condelphine, deltamine; has curare-like action; u. emet. diur. anthelm. insecticide - **D. exaltatum** Ait. (D. urceolatum Jacq.) (w&cUS): tall larkspur; serious cattle poison (as others) - **D. peregrinum** L. (eMed reg., wAs [AsMin+]): u. for intermittent fever (mal.) (Nepal) - **D. staphisagria** L. (Euras, nAf): stavesacre; sd. c. many alk. incl. delphirine, staphisagrine; insecticide, esp. for pediculi; int. in neuralgia (danger!); poisonous - **D. tricorne** Mx. (e&cNA): ornam. perenn. hb.; c. delavaines - **D. zalil** Aitch. & Hemsl. (Iran): zalil; fls. u. ton., yellow dye (popular in Punjab).

Delphinus delphis L. (Delphinidae - Mammalia) (warm & temp seas; common in Med Sea): common dolphin; delphin; dolphin oil (Oleum Delphini) obt. fr. fat of animal; similar to whale oil; one of finest lubricating oils; killing of dolphins forbidden by the Soviet govt. because of the "humanoid" brain of this mammal (!).

delsoline: alk. fr. sds. of *Consolida regalis*; forms glycosidal dyestuff.

demanyl phosphate: metabolite of *Neurospora crassa* Shear & Dodge strain; u. as psychotonic.

Demeclocycline: Declomycin: 6-demethyl-7-chlorotetracycline: antibiotic fr. mutant strains of *Streptomyces aureofaciens;* u. like tetracycline but more stable.

demecolcine: alk. of *Colchicum autumnale*; antineoplastic; u. in leukemia; less toxic than colchicine.

Demerara (sugar): crude s.; named after river in Guyana.

demersal: bottom dwelling marine organism (ex. some fish).

demigod's food: *Chenopodium ambrosioides*.

denaturant: subst. added to another one to make latter obnoxious to taste or smell; esp. u. for alc. - **Alc. denaturants**: acetone, methyl alc., isopropyl alc., alks. (brucine), phenols, diethyl phthalate, &c.

denaturation: alteration in essential compn. of a protein or other org. substance by chem. action, boiling, &c.

denature: denaturize: change nature of subst. (specif. alcohol), rendering it unfit to eat or drink.

Dendrobium alpestre Lindl. (Orchid.) (nInd, Nepal): pseudo-bulbs u. as general ton. (Himal natives) - **D. herbaceum** Lindl. (As): st. u. in fever; appl. to forehead for headache; rheum. (124) - **D. nobile** Lindl. (Chin): u. stom. antipyr. sialag. analg.; hypotensive - **D. purpureum** Roxb. (eAs): u. as poultice (Mal); crushed st. pulp cooling & u. as maturative against whitlow (Indones).

Dendrocalamus strictus Nees (Gram) (Ind, Java): male bamboo; lvs. secrete b. manna (no mannite); wd. much u. (Ind, Pak); in paper pulp.

dendrochronology: science of dating events by careful study of annual ring sequences of tree woods, recent and older.

Dendrolobium lanceolatum (Dunn) Schindler (Leg) (sChin, Indochin): rt. decoc. u. in rheum. (Laos) - **D. triangularis** (Retz.) Schindler (sAs): lf. decoc. of roasted lvs. u. for cough (Camb) pl. u. in leucorrhea (Laos); rts. cut up & macerated in alc. —> bev. reputedly aphrod.

dendrology: science of botany of trees.

Dendropanax arboreus Dcne. & Plach. (Arali.) (tropAm, WI, Mex to Bol): angelica tree (Jam); lvs. & sts. decoc. u. in home med. (PR); honey pl.; wd. u. as lumber in general carpentry.

Dendropemon falcata (L. f.) Etting. (Loranth.) (nwInd): bandha; pl. decoc. u. as antifertility agt. in women - **D. purpureus** Krug & Urban (Loranth.) (WI+): "mistletoe"; u. tea for colds, worms, pain and "all sickness" (Bah) (similarly u. is **D. emarginatus** Steud., "proud tree" [Bah]).

Dendrophthoe falcata (L. f.) Etting. (Loranth.) (nwInd): bandha; pl. decoc. u. as antifertility agt. in women.

Dendropogon Rafin. = *Tillandsia*.

denizen: pl. or animal which has become naturalized in a region.

Dennettia tripetala Bak. (monotypic) (Annon.) (Nigeria): u. as antipyr. in folk med.

dental plaque: dental film; tartar; "coating slime" (Leber); "feltlike mass" (Williams, 1897); c. dextran formed by m. o. fr. sugars; may be mechanically removed from teeth or with dextranase.

Dentaria diphylla Mex. (Cruc.) (eNA): toothwort; crinklewort; crinkle root; rts. eaten raw or boiled by Indians; u. prepn. crinkle pickle (*cf.* mustard); cult.; at times ground with meat like horseradish; put in vinegar (Ashenakis); sometimes u. like mustard - **D. heptaphylla** Vill. (Cardamine pinnata R. Br.) (cEu): pepperroot; gefiederte Zahnwurz (Ge); tuberous rt. (off. in Pharm. Gall. I) u. in colics, &c. - **D. laciniata** Muhl. (eNA) pepperroot; u. similarly to prec. - **D. pinnata** Lam. = *D. heptaphylla* Vill.

dent de léon (Fr): dandelion.

dentelaria da india (Por): *Plantago indica.*

Dentella repens J. R. & G. Forst. (Rubi.) (Mal): boiled lvs. appl. as poultice to wounds (Mal).

Dentellaria root: *Plumbago europaea.*

Denver mud™: clay poultice, Cataplasma Kaolini.

deodar: *Cedrus deodara* (SC native Indian word meaning [like Theodore], the gift of God).

deoxy-: v. also under desoxy.

deoxycorticosterone: steroid occurring naturally in adrenal cortex; useful as mineralocorticoid (also d. acetate, compd. chiefly u.).

deoxyribonuclease: pancreatic desoxyribonuclease: enzyme capable of hydrolysis of highly polymerized DNA; as debridement agt. for removal of clots and exudates following injury.

deoxyribonucleic acid (DNA): coiled 2-stranded mol. (double helix) with genetic code detd. by sequences of 4 chem. bases (guanine, adenine, cytosine, thymine); attached to deoxyriboses; sets of these nucleotide polymers represent genes; DNA serves as template in RNA synthesis; occ. in cell nuclei of all living tissues; esp. abundant in animal body in thymus & spleen; u. for convalescence & weakness (Nuclifort [Fral]).

depsides: natural arom. compds. representing esterlike condensations or fusions of 2 anhydrous (poly)phenols, formed by combinations of COOH gp. of one with OH gp. of other; with important tannin-like properties; very common in lichens & valuable for taxonomy; some have been synthesized; ex. digallic acid.

Deravine: a mixt. of 5 alkaloids of *Veratrum viride*; u. in hypertension (1954-).

Deringa Adans. = *Cryptotaenia.*

dermatogen: outer meristem of pl. st. believed to give rise to epidermis.

Dermatomycetes: dermatophytes; fungi which attack skin, hair, nails, &c.

Dermochelys coriacea (L) **schlegelli** (Garman) (Dermochelidae; Reptilia) (trop Pac & Ind Oceans): leatherback (turtle); largest pelagic turtles (-700 kg; -2.7 m long); flesh fatty & oily, inedible; said poisonous; u. in Unani med. (flesh antiphlogistic; blood; eggs; bile, in eye dis.; bones; fat; ash); egg u. as aphrod. (Virgin Is), remedy for resp. dis.; FO u. to treat boat lumbers (Ind).

dermostatin: antibiotic subst. fr. *Streptomyces viridogriseus* Thirum. (Ind); with antifungal properties.

Derosne's salt: noscapine.

Derris spp. (Leg.) (PI, Mal, Ind, Jav, EI): woody vines; several spp. u. as insecticides (tuba root); "derris root" usually appl. to *D. elliptica* (other vernacular names incl. timbo, barbasco, haiari, nekol, cube) - **D. cebuensis** Merr. (endemic to PI): bk. appl. for internal pain; fish poison - **D. chinensis** Benth. (Far East): rt. u. fish poison, insecticide - **D. elliptica** Benth. (seAs, Oceanica): tuba root; jewel vine; rts. c. large amt. rotenone (commonly above 5%); (ya)nou (Mal); frt. & st. pulped up to produce fish poison (1024); lvs. u. for oxyuriasis; sts. appl. to impetigo (PI); cult. PR (1018); "rotenone resin" = deguelin & toxicarol - **D. floribunda** (Benth.) Ducke (wBr, Ind): acuna; u. fish poison (Paumari Indians of Br) - **D. latifolia** HBK (Br): u. fish poison - **D. malaccensis** Prain (Far East): Malacca jewel vine; tuba; rt. c. insecticidal principles; u. fish poison; insecticide (1014) - **D. polyantha** Perk. (Far East): u. similarly - **D. trifoliata** Lour. (D. uliginosa Benth.) (eAs, Oceania, Austral, eAf): rt. c. rotenone, anhydroderrid; u. fish poison (stupefiant), insecticide, antipyr. (NG); arrow poison; legumes u. as astr., in dentifrices, for leukorrhea; to prep. ink; fresh if inf. u. as purg.

desaroside: glycoside of sarmentogenin & D-digitalose; occ. in *Strophanthus vanderijstii.*

Descantaria elongata Brueckner v. *diuretica* C. B. Clarke (Commelin.) (SA, Br, Pe, Guat): trapoeraba (Br) (1019); entire pl. u. as diur.

descarozado (Sp): dr. frt. with rind & core of frt. removed (fr. carozo = frt. core).

Descurainia pinnata Walt.) Britt. subsp. **helictorum** Detling (Cruc.) (sTex [US], Mex): sds. in sugared water u. for liver dis.; boiled lvs. eaten as greens, purchased at pharmacy (Mex) - **D. sophia** (L) Webb (Sisymbrium s. L.) (Euras, nAf, natd. NA): tansy-mustard; flixweed; hb. u. astr. vulner. deterg.; frt. u. as purg. (Turkestan); sds. pungent, made into a kind of mustard; former Herba & Semen Chirurgorum (surgeon's hb. & sd.).

Desenex™: undecylenic acid & its Zn salt.

deseret (US-Utah): honeybee; original name of state of Utah; symbol of industry (Book of Mormon).

deserpidine = (re)canescine; fr. rts. of *Rauvolfia canescens*; u. sed. in hypertension; counteracts tachycardia fr. cigarette smoking; Harmonyl™.

desert black bush: *Larrea tridentata* - **d. finger**: *Cereus giganteus* - **d. holly:** v. holly - **d. post: d. sentinel**: *Cereus giganteus* - **d. spoon**: *Dasylirion wheeleri* - **d. tea** (Cal): *Ephedra* spp. - **d. tombstone:** d. post.

Desfontainia spinosa R. & P. (Potali.) (Pe, Ch): chapico (Ch): evergreen shrub; excels holly in foliage but not in frt.; c. alkaloid; lvs. u. for dyeing; showed some activity in lymphocytic leukemia; lvs. u. as stom (Ch); lf. tea u. as hallucinogenic agt.; u. in magico-medical practice of medicine men (Co) (borrachero de paramo).

desiccants (for crude drugs): driers such as silica gel (put up in perforated metal receptacles); Na_2SO_4; Drierite (anhyd. $CaSO_4$); activated Al_2O_3, etc.

desiccation: process of depriving solid of water at low temperatures.

deslanoside = de(s)acetyllanatoside C; purpurea glycoside C; fr. *Digitalis lanata* lvs.; a valuable cardiotonic.

Desmanthus illinoensis (Mx.) MacM. (Leg) (e&cUS): spider bean; prickle-weed; lvs. u. for itch (Pawnees).

Desmarestia Lamouroux (Desmaresti./Phaeophyc. - Algae): marine algae (Pac & Atl Oceans) - **D. aculeata** Lamour. (Atl & Arctic Oceans): Stacheltang; the thallus c. laminaran, mannitol, sulfuric acid (solid. subst. ca. 21%) (220); antimicrobial subst.

Desmodium Desv. (Meibomia) (Leg) (trop & subtrop): beggar's tick(s); beggar lice; beggarweed ("beggar" = segment); ticktrefoil; stick tights; tick-clover (SC appearance & adhesive nature of frts.); typical habitat; dry sandy open pine woods - **D. adscendens** (Sw) DC (WI): wild ground nut (Trin); "amor seco" (Cu): hb. decoc. u. in VD (Trin) - **D. auriculatum** DC (Indochin, Mal, Indones): lvs. u. vulner. - **D barbatum** (L.) Benth. (Tanz): rts. pounded & eaten with salt for ankylostomiasis - **D. canum** S. & T. (D. supinum DC) (WI, tropAm, pantrop): "amor seco" (Cu); mozote (CA); wild ground nut (Bah); rt. decoc. u. dys. & tenesmus (LatinAm); GC (Guat, Hond); as diur. stom. (Orient); in liver dis., as hemostatic; vulner. (Cuban war of independence ca. 1890); decoc. for dysmenorrh. (Bah); pl. u. as fodder - **D. caudatum** (Thunb.) DC (s&eAs): rts. medicinal u. in form of decoc.; pl. insecticidal, disinfect., u. in pickles (misos) to discourage growth of maggots (Jap) - **D. diffusum** SC (s&eAs): rt. decoc. u. in stomachache of infants (Laos) - **D. erythrinaefolium** DC (SA): rts. u. astr. in diarrh. dys. &c. (Br) - **D. gangeticum** DC (OW trop, esp. Ind): bhidi; salaparni (Sanscr) lvs. u. diur. in catarrhs., lithiasis, to expel stones fr. bladder; rt. u. in diarrh. (Mal); dys. (*cf.* ipecac); as one of "5 minor roots" (of SriL), decoc. for fever, catarrh. (1020) - **D. heterocarpum** (L.) DC (trop & subtrop of As, Af): lvs edible; u. in childbirth of young girls; good cattle forage; (trèfle de pays [Indochin]) - subsp. **angustifolium** Ohashi (seAs): crushed sts. appl. to broken bones & snakebite (Camb) - **D. incanun** DC (D. supinum DC) (CA, WI): Mozote, amor seco (dry love); collant; inf. u. in GC (Guat, Hond); considered an undesireable weed but u. as fodder - **D. latifolium** DC (tropAf): lf. decoc. u. in dys. lvs. appl. to skin (adhere) u. for diarrh. (Sen); in compns. for leprosy - **D. laxiflorum** DC (Ind to Taiwan, Mal+): rt. decoc. u. at time of puerperium (PI) - **D. microphyllum** (Thunb. ex. Murr.) DC (s&eAs): entire pl. u. in local medicine (for eyes, headache) (Indochin) - **D. molliculum** (HBK) DC (Ec): yerba del infante; rt. inf. u. to assist childbirth; u. for "hysterics" - **D. polycarpum** (Poiret) DC (Ind, Chin, Jap, seAs): "salaparni"; pl. decoc. u. as ton. & in cough; for convulsions & fainting - **D. rotundifolium** DC (Meibomia michauxii Vail) (eUS): dollar leaf; flux vine; much u. in folk med. (Negroes, Tallahassee, Fla, ca. 1940; Miss [US]) - **D salicifolium** DC (cAf): rts. pounded & u. in water orally to treat bilharziasis (Tanz) - **D. scorpiurus** Desv. (WI, CA): amor seco; pega pega; hb. inf. u. antipyr. (Hond, Guat) - **D. stipillosum** Schindler (Indochin, Burma): macerated rt. crushed & appl. as cataplasm to swelling of limbs - **D. styracifolium** (Osbeck) Merrill (s&eAs): all parts u. in local pharmacopeias; thus in colic (Malaya) - **D. tortuosum** (Swartz) DC (D. purpureum F. & R.) (seUS, WI, Col): "amor seco" (Cu); junquillo (PR); lvs. purg.; green manure - **D. triflorum** DC (sUS, WI, CA, world trop); trèfle noir (Fr) (black clover); plati; hb. inf. u. in gastric dis., dys. (Guat, Hond); decoc. for rheum. (Haiti); lvs. sts. rts. u. atonic dyspepsia, colics; fresh pl. appl. to abscesses & wounds that do not heal well (Co); lvs. eaten as veget. (Burma); a common weed - **D. triquetrum** DC (seAs, Austral): gerdji (Java); lvs. u. lax. diur. refrig. in hemorrhoids (1556) - **D. umbellatum** (L) DC (Andaman Is, Fiji): pl. u. for fever (And Is) (2226); young tender tips chewed with betel nuts & placed in mouth of sick baby (Solomon Is)

- **D. velutinum** (Willd.) DC (OW trop): squeezed lf. chewed with salt for diarrh. (PapNG).

desmolases: outmoded term for enzymes which catalyze formation or breaking of C-C bond, but not by hydrolysis, oxidn. or reduction; ex. zymase, alkolase, catalase, &c.

Desmos chinensis Lour. (Annon.) (Mal, Indochin): u. to treat colds; rt. decoc. u. in dys. & vertigo - **D. cochinchinensis** Lour. (Indochin): lf. inf. u. galactag. - **D. hancei** Merr. (Indoch): galactag.; branches & lvs. u. to prep. beverage; frt. edible.

Desmostachya bipinnata Stapf (monotypic) (Gram) (OW trop, EI, Ind, trop eAf): dab (Hindi); d. grass; culms. u. diur. stim. in dys. menorrh.; st. u. to mfr. carpets (Nile area of Af) (1495).

Desmotrichum fimbriatum Bl. (Orchid.) (Ind): jivanti (Hind, Sanscrit): pl. u. stim. ton. demul.

desoxy-: v. also dexoxy-.

desoxycholic acid: fr. ox-bile; found in bile of human, dog, &c.; u. as choleretic. in cholelithiasis & infections of bile passages.

desoxycorticosterone: steroid first isolated fr. adrenal cortex of mammals (also found in lower animals); but now synthesized fr. stigmasterol; potent hormone (mineralocorticoid) u. in form of the acetate ester in Addison's dis. (best treatment) & in other adrenal insufficiencies; adm. by injection or implantation.

de(s)oxyribonuclease: streptodornase - **pancreatic d.:** Dornevac™: enzyme fr. beef pancreas; u. as in débridement; in abscesses & hematomas; to reduce viscosity of mucous secretions in pulm. dis. (aerosol).

desoxyribunucleic acid: v. deoxyribunucleic acid.

destomycin A: antibiotic subst. fr. *Streptomyces rimofaciens* Niida (Jap); u. as broad-spectrum antimicrobial & anthelm.

destroying angel: *Amanita phalloides.*

detailist: detail-man: representative of a manufacturing pharm. firm who calls principally on the "profession" (practitioners) but also sometimes on the "trade" (pharm. retailers, other manufacturers, wholesalers).

Detarium microcarpum (G & P) G. & P (Leg) (wAf - Senegal): wonko; rt. decoc. u. as bev. in syph. - **D. senegalense** Gmel. (Leg.) (trop wAf): detar; dita (Wolof); tallow tree; bk. c. glycosides, u. in serious enteralgias (colic, intest. obstruction); frt. (pod, pith) edible, also u. to relieve cough; lvs. u. in conjunctivitis; bk. inf. used as ordeal poison in criminal trials; in arrow poisons; rt. bk. frt. u. in fever, anemia, coughs.

detoxin complex: subst. prod. by *Streptomyces caespitosus* var. *detoxicus* (Jap); u. to decrease phytotoxicty of blasticidin S to rice plants.

Deutscher Tee (Ge): German tea; many formulas have been given, these chiefly prepd. fr. mixts. of native lvs., such as strawberry, black raspberry, coltsfoot, currant, etc., generally rich in vit. C; some mfrs. are said to think of this as a profitable way of getting rid of their drug scraps.

developer (chromatog.) = **developing solvent**: u. to produce a spread of compds. in the chromatogram.

devil, blue: *Echium vulgare* - **d. in the bush**: *Erythrina herbacea* - **d. in the wood**: *Nigella damascena* - **d.-on-both-sides** (US): *Ranunculus arvensis*, corn crowfoot - **d. tree**: *Alstonia scholaris* - **devil(l)ed**: 1) grilled or dressed with hot condiments 2) chopped fine & highly seasoned with spices, &c.

devil's apple: 1) Podophyllum 2) (Miss [US]+): Stramonium - **d. apron**: Laminaria (seaweed) - **d. bit (herb)**: 1) *Succisa pratensis* (etc.) 2) Chamaelirium 3) *Liatris* spp., as *L. spicata* (**Ohio d. b.**) 4) Aletris (SC rt. appears as if bitten off) - **d. bite**: 1) *Veratrum viride* 2) *Liatris scariosa* - **d. bones**: Dioscorea - **d. chair**: *Fouquieria splendens* - **d. claw(s)**: 1) (US) *Martynia annua* 2) *Harpagophytum procumbens* - **d. club**: *Echinopanax horridum* - **d. cotton**. *Abroma augusta* - **d. darning needle**: esparto - **d. dung**: Asafetida - **d. fig**: *Argemone mexicana* - **d. finger**: *Echinocereus blankii* - **d. flight**: *Hypericum perforatum* v. *angustifolium* - **d. flower**: *Lychnis dioica* - **d. guts**: *Cuscuta americana* (dodder) - **d. herb**: *Succisa pratensis* - **d. knitting needle**: esparto - **d. medicine** (Ala [US] Negro): asafetida - **d. nettle:** milfoil - **d. paint brush**: *Hieracium aurantiacum, H. praealtum* - **d. pincushion**: *Leonotis nepetaefolia* - **d. plague**: *Daucus carota*, wild carrot - **d. plant**: *Abutilon theophrastii* - **d. plaything** (US): milfoil - **d. root**: 1) peyote (mescal) 2) *Succisa pratensis* - **d. shoestring(s)**: 1) *Tephrosia* spp., esp. *T. virginiana* 2) (seUS Negro): *Coronilla varia* - **d. snuff** (sUS Negro): spores of puff-ball - **d. snuff box** (sUS): puff-ball - **d. thorns:** *Tribulus terrestris* - **d. tongue**: *Opuntia vulgaris* - **d. trumpet**: *Datura meteloides* - **d. walking stick**: 1) *Ailanthus altissima* 2) d. club - **d. weed tea**: *Aster spinusus* - **d. wood**: *Osmanthus americanus*, *Sambucus nigra.*

dewberry: 1) name appl. to several *Rubus* spp. with frt. more delicious (sweet, juicy) than regular blackberrry; members of sect. *Flagellares*, in which the lvs. are not durable, evergreen, or glossy, the fls. & frt. on the trailing st. tend to hang towards the ground ("trailing blackberries") and the st. is not bristly; primocane leaflets are usually broad; rooting at nodes; fls. spring, frt.

early summer; ex. *R. flagellaris* (wild & cult.), *R. hisipdus*, *R. roribaccus*, *R. aboriginum*, *R. trivialis*, *R. canadensis*, *R. cuneifolius*, etc.; frt. & rts. u. 2) *Gaultheria procumbens* - **marvel d.**: *R. mirus* - **southern d.**: *R. trivialis* & other spp. of sect. *Verotriviales* (foliage evergreen, primocane leaflets longer than broad, often narrow, sts. bristly).

Dewee's carminative: mixture of magnesia, asafetida & opium - **D. tincture of guaiac**: guaiac tr. comp. (NF V).

dewflower (US): *Penstemon cobaea* - **dewgrass**: 1) *Drosera rotundifolia* 2) *Digitaria sanguinalis* - **dewplant**: *Drosera rotundifolia*.

Dextran: "fermentation gum"; high m. wt. glucose polymer; gummy carbohydrate found in unripe sugar beet, syrups, wines; etc.; prepd. fr. sucrose solns., molasses, beet juice, by action of *Leuconostoc mesenteroides* van Tieghem; formed through bacterial action; the cause of stringiness in wines; sterile soln. u. as blood plasma replacement; IV inj. for nephrotic edema; in soil conditioning (E) - **d. iron complex**: iron-d. complex: combn. of d. & $Fe(OH)_3$; u. as hematinic by parenteral route; for use only where peroral Fe is not possible; risk of anaphylactic shock, cancer (E) - **d. 40**: Gentran 40™: prepd. with *L. mes.* acting on sucrose; u. as blood flow adjuvant (m wt. 40,000) (E).

dextranase: enzyme wh. hydrolyzes the polysaccharide dextran yielding glucose; proposed for use (in sugar, dentifrices), for removing dental plaque, thereby combatting dental caries.

dextranne (Fr): dextran.

dextranomer: Debrisan™: 3-dimensional complex of dextran polymer; u. as vulnerary for suppurating wounds & ulcers.

Dextri-Maltose™: mixt. of dextrins & maltose obt. by treating corn flour with barley malt (with enzymes); sol. in water & milk; u. in infant feeding with milk ("carbohydrate modifier") & in gastrointestinal dis.

dextrin(e): carbohydrate intermediate prod. obt. by partial hydrolysis of (corn) starch - **white d.: Dextrinum Album** (L): pale colored form of dextrin u. as reagent, etc. - **yellow d.**: darker colored form of dextrin, u. as adhesive (for adhesive stamps, envelopes, &c.), emulsifying agt., size (textiles), thickener, excipient, etc.; prod. chiefly fr. maize (US), & potato (Ge), also cassava, &c.

dextrose: Dextrosum (L.): **d-glucose**: $C_6H_{12}0_6.H_2O$; a simple sugar which occurs as white or colorless small xals or powder; u. as nutr. esp. in shock fr. hemorrhage or insulin overdose, for edema (hypertonic solns.), when given by vein; also diluent, sweetener, energy source, in infant feeding, reagent; in bacteriological media, confectionery - **d. syrup** (Tariff Act): glucose.

dextrothyroxine (sodium): D-form of thyroxine with very little thyroid activity; chiefly u. as antihyperlipoproteinemic (orally) (Levothyroxine is active form of thyroxine).

dhak(i) (Hind): *Butea frondosa*+.

dhal (Urdu): *Cajanus cajan* (in split form); said sometimes appl. to other Leg. spp. sds.

dharmani: *Artemisia maritima*.

dhatura (Hind): *Datura*, esp. *D. stramonium*.

dhava: dhawa: Ind gum fr. *Anogeissus latifolia*.

dhupa: *Vatera indica*.

dhurra (Arab): Indian corn or millet.

dhutra (Ind): *Datura metel* (var. *alba*).

diacetyl morphine: Heroin.

diachenes = cremocarps (Umb.).

Diacodion: Diacodeum: Syrupus Diacodii: S. Papaveris: syr. of poppy capsules; u. in cough syrups; u. by Saml. Johnson "to improve his spirits" (per biography).

Diagridium = Diagryd(ium) (med. Lat): Scammony gum-resin.

Dialiopsis africana Radlk. (Sapind.) (sAf): rt. inf. u. dys.+; rt. & sd. toxic (—> Zanha).

Dialium cochinchinense Pierre (Leg) (seAs): (cay) xoai (Vietnam); bk. astr. u. in diarrh. & for some kinds of ringworm; wd. u. as tea for purg.; bk. masticatory & in urticaria (Indochin); frt. edible.

Dialyanthera otoba Warb. (Myristic.) (CA, nSA): otoba ("nutmeg tree"); sds. u. as antiparasitic & germicide for animals; sd. source of otoba fat (FO) = atoba (or Amer) nutmeg butter; u. in soap & candle mfr.; skin dis.

diamond: Diamant (Ge): adamas, adamantis (L) (hence adamanteu, adj., *cf.* adamant [Eng]): xall. form of carbon, occ. as a mineral (most fr. sAf); small d. made by synth.: hardest subst. known (No. 10 on Mohs scale); u. in jewelry of great value; for cutting glass; polishing, grinding; in semiconductor research; long ago u. in therapy of human dis.

diamorphine: Heroin.

Dianella ensifolia Redouté (Lili.) (seAs): "rat poison"; sisiak; toxic hb. u. to kill rats; st. inf. to overcome fatigue; rts. in dysuria; mostly now lvs. u. exter. as poultice.

Dianthera pectoralis J. F. Gmelin = *Justicia p.*

Dianthus L. (Caryophyll.) (mostly temp OW): clove pinks; a genus of pinks & carnations (lit. flower of the god, Zeus) - **D. barbatus** L. (Euras): Sweet William; u. like *D. caryophyllus* L. - **D. carthusianorum** L. (c&sEu): wild pink; fls. c. saponin; pl. fls. & sd. u. in folk med.; bee plant - **D.**

caryophyllus L. (Euras, Med reg.): carnation; (clove) pink; fls. (as Flores Tunicae hortensis) u. in tea mixt.; to aromatize herbal teas; fls.; antispas. sed.; syr. of fls. in pharmaceut. prepns. (17-18th cent. GB, Ge); fls. u. in perfumery; u. in jams, &c; good fixing value; much variation; cult. ornam. (red carnation; emblem of Kappa Psi fraternity) - **D. chinensis** L. (Ind, cultd. in Eu): rainbow pink; "Chinese clove"; numerous forms; c. saponins; u. in Chin as diur. anthelm. abortient; bienn. ornam. hb. - **D. deltoides** L. (D. crenatus Gilib.) (Euras, Af, natd. NA): maiden pink; rt. inf. u. emetic - **D. micropetalus** Ser. (sAf): u. to alleviate muscular & body pain - **D. superbus** L. (Eu, cAs, Chin, Jap): oeillet; sts. lvs. fls. u. as diur. vermif. abort. GC; dr. fls. & buds for indig., in difficult labor, diur.; lvs. in soups for intest. affections of infant (Chin); crushed old lvs. u. for eye dis. (Chin); cult. ornam.

diaspore: disseminule; any reproducing part of plant; ex. spore, seed.

diastase (of malt): a mixture of amylolytic (or starch-splitting) enzymes obt. fr. a malt inf.; digests 50x its own wt. of starch; u. digestant, predigestant; mfr. soluble starch - **taka d.**™: an enzyme prod. by fungal action; u. digestant; prep. sake; predigestant; reagent; in textile indy.

diatomaceous earth (, purified): diatomite; siliceous e. (*qv*.)

diatoms: uni- or multicellular algae with silica shells; constitutes large part of plankton (Bacillariophyceae - Chrysophyta - Algae).

dibble stick: u. for planting sd.

dibididibi: dividivi.

Dicentra Bernh. (Fumari.) (NA, As): genus of perenn. hbs. of ornam. & drug interest; c. alks. - **D. canadensis** (Goldie) Walp. (e&cUS, Cana): variously known as squirrel corn, bleeding heart, turkey corn; tubers (Corydalis) alt. bitter ton. diur. in chronic skin dis.; syph.; c. 6 alks. incl. bulbocapnine; u. to treat Ménière's disease & muscular tremors; as pre-anesth. by veterinarians; this & other spp. important stock poisons - **D. cucullaria** Bernh. (e&cNA): Dutchman's breeches; similar to prec. - **D. formosa** Walp. (wNA): Cal. bleeding heart; u. ton. - **D. peregrina** Mak. (D. pusilla Sieb. & Zucc.) (Jap, eSiber): little bleeding heart; koma-kusa (Jap); hb. c. dicentrine, protopine, monomethyl-quercetin; u. to relieve abdom. pains.

dicentra, showy: *Dicentra formosa*.

Dicerocaryum zanguebaricum Merr. (Pedali.) (Af): boot protector; lf. inf. u. to expel retained placenta; parturient (cows); for hydrocele; in GC; pl. formerly u. as soap substit.

Diceros bicornis (L) (Rhinoceros b. L.) (Rhinocerotidae - Mammalia) (tropAf, Kenya+): African (or black) rhino(ceros); animal killed often for horn, a very valuable article; u. as aphrod.; animal protected by law but poachers active (2232).

Dichaea muricata (Sw.) Lindl. (Orchid.) (Ec): wash prepd. fr. pl. u. to treat eye infections (Kofans).

Dichanthium annulatum Stapf (Andropogon a. Forsk.) (Gram) (worldwide trop): source of manna with about 75% mannitol - **D. gerardi** Vitman = *Andropogon g.*

Dichapetalum cymosum (Hook.) Engl. (Dichapetal.) (sAf): "poison leaf"; lvs. c. fluoracetic acid; one of most toxic plants (dangerous for stock) - **D. toxicarium** Baill. (Chailletia toxicaria Don) (trop wAf): ratsbane; "broken back"; c. fluroracetic acid; very toxic; u. by natives for arrow poisons; fish stupefiants; in homicides; for headache (1497); rat poison - **D. vitiense** Engl. (Tonga, Fiji): tuamea (Tonga); pl. said poisonous.

Dichondra J. R. & G. Forst.. (Convolvul.) (trop & subtrop): some spp. used at times for lawn plants (Jap tea garden, San Jose, Calif [US]) - **D. argentea** H. & B. (swUS, SpanAm): u. med. (Mex) - **D. micrantha** Urban (eAs, widely natd.): grown as substit. for lawn grass - **D. repens** Forst. & Forst. (tropics, subtrop, Austral, NZ): cult. as lawn or ground cover (creeping pls. forming mats) (San Francisco Bay area, Calif); u. med. Arg (antisep, anthelm, vuln, extern) - **D. sericea** Sw. (D. repens Choisy var. *sericea)* (Mex to Arg, WI) (1557); oreja de dato; above-grd. parts u. intern. as anthelm., exter. as antisep. vulner.

Dichorisandra hexandra (Aub.) Standl. (Commelin.) (CR): succulent sts. & lvs. of herb appl. to skin for burns & snakebite (2444).

Dichroa febrifuga Lour. (Hydrange) (Chin, eAs, PI): chang-shan; rt. & lvs.; c. febrifugine, (coccidiostat); u. antimalar., antipyr. (u. for 2000+ yrs.); purg.

Dichrocephala latifolia DC (Comp) (Ind, Indones+): tender shoots u. as poultice, in GC, insect stings (Camb); fl. bud decoc. u. diaphor. & diur. (Java).

Dichrostachys cinerea (L.) W. & A. (D. nutans Benth.) (Leg) (tropAf): rt. u. for intest. worms; bk. tan; wd. very heavy & strong - **D. glomerata** Chiov. (Leg) (Af, natd. WI): u. in tooth dis. (gingivitis), GC (lf. decoc.), arthritis (massage with mixt. of lvs. + salt + mineral oil); furuncles; leprosy.

Dicksonia blumei Moore (Dicksoni.) (Mal archipelago): a tree fern; scales u. as styptic.

Dick test: a skin test for determining if immune to scarlet fever, involving intracutaneous inj. of

Scarlet Fever Streptococcus Toxin (named after the Doctors Dick).

diclinous: diecious: unisexual, with female & male fls. on different plants.

Dicliptera unguiculata Nees (Acanth.) (SpanAm): sornia; u. in treating amebic dys. (CR).

Dicoma anomala Sond. (Comp.) (sAf): uzara (claimed); rt. c. digitaloid glycosides; u. colic, dys.; in anal & vag. suppos.; syph.; diarr., intest. worms; gall sickness of livestock - **D. speciosa** DC (sAf): rt. u. exp. - **D. tomentosa** Cass. (wAf): pl. u. topically on infected wounds.

Dicot: contraction of foll.

Dicotyledoneae: one of the 2 classes of Angiosperms, the other being the Monocotyledoneae; Dicotyledons ("Dicots") have 2 opposite cotyledons in the sd.; net-veined lvs.; annual rings developing fr. cambium layers in st. & rt.; fl. parts mostly in 4's or 5's - **dicotylendonous**: having the characters of Dicotyledoneae (**Dicotyledons**).

dicoumarin: dic(o)umarol: bis-hydroxy-coumarin; a compound found in fermented sweet clover; also prep. by synthesis; u. as anticoagulant to treat or prevent thrombosis, such as after operations.

Dicranopteris linearis Underw. (Gleicheni. - Pterid.) (As, Hawaii): uluke; false staghorn; bitter juice of st. & lvs. u. emet. & lax.

Dictamnus albus L. (monotypic) (Rut.) (Euras): (European) dittany; gas plant; itamo real (Mex); burning bush (SC ignition of gas given off by bush); fls. rose-red, rarely white (despite bot. name); rt. c. VO resin, bitter princ.; limonoids (limonin, fraxinellone); u. antipyr. diur. anthelm. in hysteria; uterine stim.; lvs. u. by Homeopaths; for beverage tea (Siber); occasional ornam. (US); photosensitizing.

Dictyophora indusiata Ad. Fischer (Phall. - Gasteromycetes; Basidiomyc.) (trop): sporocarps powd. mixed with *Hibiscus rosasinensis* fls. & placed on tumors - **D. phalloidea** Desvaux (trop): sporocarps ingested by curanderos to increase "divinatory" powers during use of *Psilocybe* spp. (sMex).

Dictyospermum conspicuum (Bl.) Hasskarl (Commelin.) (Ind, seAs): pl said u. as emmen. (Mal) (—> *Aenilema*).

Dicumarol = Dicoumarol.

Dicypellium caryophyllatum Nees & C. Mart. (monotypic) (Laur.) (Br): (Brazilian) clove bark; pão cravo; bk. in quills ("Cassia caryophyllata") with clove odor; u. as flavor, perfume, spice; bk. & lvs. u. as ton. & GI stim.; clove substit.; timber u. ("rosewood"); in mfr. "liqueurs" (2644).

Didymocarpus atrosanguineus Ridl. (Gesneri) (Mal): rt. decoc. drunk as protective during childbirth - **D. pedicellata** R. Br. (Ind): lvs. u. to "cure" stone of kidney & bladder - **D. reptans** Jack (Mal, Indones); lf./rt. decoc. u. in dys., colic; constipation.

Didymopanax morototoni Decne & Planchon. (Arali.) (WI, CA, nSA): matchwood; yagrumo macho; wd. u. in carpentry, interior finish, boxes, match sticks (Ve).

diecious: with staminate (male) and pistillate (female) fls. borne on separate individual pls.

Dieffenbachia costata Klotzsch (Ar) (Co, Pe): sap u. astr. to skin surfaces; lf. warmed & appl. for rheum. (Pe) - **D. seguine** (Jacq.) Schott (D. picta Schott) (WI, SA, esp. Br): dumb cane; arum des Antilles; rabano (cimarrón); yerba del cancer; ornam. pl.; toxic; juice of pl. u. antirheum. aphrod. (when chewed causes anesthesia & swelling of tongue [due to a protein & CaOx raphides], thus preventing speech); to produce impotence; as contraceptive (WI); u. antidote for tobacco smoking; combats loss of memory(?); u. in curare of Tecuna Indians; said to produce sterilization of humans(?); pl. u. in some sugar factories to facilitate crystn. of sucrose; dr. lvs. wrapped around dropsical or varicose legs to "draw out the water" & in local treatment for yaws (Domin, Caribs).

dienoic: fatty acid with 3 double bonds present (ex. linoleic acid).

diente de león (Sp): *Taraxacum officinale.*

Dierama pendulum Bak. (Irid.) (sAf): Zuurberg harebell; corm decoc. u. strong purg. enema (sSotho) - **D. pulcherrimum** Baker (sAf): similar to prec.

Diervilla Mill. (Caprifoli) (NA): genus of shrubs called "bush honeysuckle" - **D. florida** Sieb. & Zucc. = *Weigela f.* A DC - **D. lonicera** P. Mill. (D. trifida Moench; D. canadensis Willd.) (e&cNA): dwarf bush honeysuckle; rt. c. fraxin (glucoside); rts. lvs. & twigs u. diur. astr. alt.; alexipharmic. in mal., syph.; GC; ornam. shrub.

Dietes vegeta N. E. Br. (Irid.) (sAf): inner corm inf. u. per ora or per recto in dys. (Zulus).

diffusionists (taxonomy): "lumpers" (221); practicing aggregation of spp., etc., into fewer genera; opposite of "splitters."

"dig." (MD): Digitalis.

Digalen (Roche)TM: a formerly popular cardiotonic drug; mostly contg. gitalin and digitalein.

digallic acid: 1) tannic a. 2) (specif.): an ester of gallic acid occ. in Aleppo gallotannin.

Digenea Agardh. (monotypic) (Rhodomel). (Algae-Rhodophyceae): seaweed of trop & subtrop marine waters; u. as Fucus Digeneae - **D. simplex** (Wulf.) Agardh (all oceans); tanaka (Jap); thallus c. digenic acid (digenin; kainic acid); u. anthelm.

for *Ascaris* spp. (round worms) (100% decoc.) (esp. in Jap, but also US); thallus c. agaroid subst.; inferior substit. for agar.

Digestive Elixir: Pepsin Elixir Comp.

digger pine (Calif [US]): *Pinus sabiniana* - **digger's weed**: 1) *Lomatium nuttallii* 2) *Anethum graveolens*.

diginose: monosaccharide, $C_5H_{13}O_3CHO$, found as part of diginin (in *Digitalis* spp.) & of odoroside B (of *Nerium indicum)*.

digistroside: glycoside of digitoxigenin & D-sarmentose; occ. in *Strophanthus vanderijstii*.

digital (Sp): **digitale** (Fr, Sp): *Digitalis (purpurea)* (d. pourprée, Fr).

digitalin(e): a generic term for various extracts or principles of *Digitalis purpurea* - **French d.**: digitoxin - **German d.**: purified mixt. of glycosides of digitalis sd.; rel. inactive - **Homolle's d.**: an impure digitoxin - **Nativelle's d.**: digitoxin - **Schmiedeberg's d.: true d.**: mixt. of glycosides fr. digitalis sd.

Digitalinum Verum: glycoside of gitoxigenin & D-digitalose & glucose; occ. in *Digitalis* spp.

Digitalis L ("a" may be pronounced as in "hay") (New Eng [US]): 1) genus of splendid flowering biennial hbs. (Scrophulari.) (Euras): foxglove; several spp. are u. in med. 2) (L.) (specif.): dr. lvs. of *D. purpurea* - **D. amandiana** Samp. (seEu): lvs. show a higher cardioactivity than those of *D. purpurea* (1558) - **D. ferruginea** L. (sBalkans, AsMin, Caucas, nIran): rusty foxglove; lvs. c. lanatosides A, B, C; acetyldigitoxin B; u. cardiac dis., strongly potent - **D. grandiflora** Mill. (D. ambigua Murr.) (s&cEu, wAs): (pale) yellow foxglove; highest digitoxin content; considered equivalent of *D. purpurea*; u. to prep. digitalis compositions (USSR) - **D. laevigata** Waldst. & Kit. (sEu): Danube f.; u. cardiotonic, more potent than *D. purpurea*; adult. of same - **D. lanata** Ehrh. (sEu, Gr: Danube region): Grecian f.; st. & calyx woolly, with long hairs (but not the petals) (hence, the name); cardioactive; added to FrPr IX (1980); source of digoxins, lanatoside - **D. lutea** L. (Eu): yellow (or straw) f.; some say equivalent of *D. purpurea*, others that potency is variable - **D. X mertonensis** Butx. & Darl.: hybrid of *D. purpurea & D. grandiflora* - **D. purpurea** L. (wEu; much cult.): common foxglove; source of official crude drug & of digitoxin (coll. mostly Pacific coast US where natd.); primary glycosides A (hydrolyzed to digitoxin, this then to digitoxigenin) & B (hydrolyzed to gitoxin this then to gitoxigenin); gitalin; sometimes assayed on humans (1025); marketed as tabs., capsules, tinct.; fluid ext. (cult. Ge, Hung, Eng, Ind+) (2572) - **D. sibirica** Lindl. (USSR): Siberian foxglove; pharmacolog. studied (1558) - **D. thapsi** L. (Sp): Spanish digitalis; cardiotonic action c. 3x as strong as off. digitalis.

digitalis (foxglove) (unqualified): *Digitalis purpurea*; lf. u. heart stim.; also so u. its active principles, notably digitoxin - **d. pulverata** (L.): powdered d.; the pharmacopeial form intended for internal administration - **Spanish d.**: *D. thapsi*.

digitaloid: referring to glycosides with cardioactive properties such as those of digitalis; ex. in Scilla.

Digitaria eriantha Steud. (Gram) (Af): (Pongola) finger grass; minor cereal; hb. yields hydrocyanic acid - **D. exilis** Stapf (w tropAf): monco; a small millet cult. in dry areas of As; a minor cereal (Hausa natives); combn. with *Strophanthus sarmentosus* rts. u. in treating intest. meteorism with constipation; also diur. - **D. iburna** Stapf (Af): similar to *D. exilis* - **D. sanguinalis** (L) Scop. (Eu; now widely distr., incl. US): (hairy) crab grass; finger g.; sd. u. as rice substit. (CA); hb. u. as forage & fodder (hay) (as also other members of gen.); bad lawn weed.

digitately branched: 1) with divisions arranged like fingers of hand (*cf.* palmately) (ex. *Cannabis* lf.) 2) having projections like digits (fingers).

digitenolides: glycosides with sterane structure but physiologically inert; ex. digitonin (fr. *Dig. purpurea*).

digitonin: cryst. steroid saponin (glycoside with 6 sugar residues) occ. in *Digitalis* lvs. & sds.; with almost no typical digitalis action; u. as reagent in detn. of cholesterol.

digitoxin: Digitoxinum (L.): glycoside of Digitalis; supposed to represent most fully the action of d. leaf (E); made up of digitoxigenin (aglycone) combined with 3 mols. of digitoxose.

digoxin: Digoxinum (L.): cardioactive glycoside of *Digitalis lanata*; u. like digitoxin, but more rapid & transient in action (1559) - **d. immune fab (ovine)**: Digibind (BW): antibody (fragments) (binding part of immune globulin) produced in sheep by inj. of digoxin; u. specifically to antidote digitalis/digoxin poisoning (1986-).

dihydrocodeine: mfrd. by reduction of codeine; u. cough sed. - **dihydrocodeinone**: synth. fr. codeine; u. much like codeine but strongly addictive - **d. enol acetate**: u. analg. cough depressant, addictive - **dihydroergocornine**: synth. deriv. of ergocornine (alk. of ergot); adrenergic-blocking action.

dihydromorphinone HCl: hydromorphone; Dilaudid™: prepd. by reduction of morphine; strong analg., about 5x as active as morphine - **dihydroquercetin:** taxifolin (in wd. of *Pseudotsuga taxi-*

folia, now *P. menziesii*); distylin (in *Distylium racemosum*) - **dihydrostreptomycin**: occurs in *Streptomyces humidus*, *S. griseocarneus* Benedict & al. & *S. subrutilus* Arai + al.; usually prepd. by reduction of streptomycin; deriv. of streptomycin, with much the same action, but rather less toxic to the nervous system (usually as sulfate).

dihydrotachysterol: vitamin D_4; Hytakerol; AT-10 (anti-tetany); obt. by irradiating ergosterol (redn. of tachysterol); has vit. D activity; u. for blood Ca regulation.

diiodotyrosine: iodogorgoic acid; found in skeletal proteins of corals (ex. *Gorgonia cavollini*), sponges, & other marine animals; u. as thyroid inhibitor.

dika fat: FO of *irvingia gabonesis.*

dikmali gum: fr. *Gardenia gummifera* and *G. lucida.*

diktamo(n): **diktamnon** (Gr): 1) *Origanum dictamnus* 2) *Ballota acetabulosa.*

Dilaudid: dihydromorphinone.

dill: *Anethum graveolens* & its frt., d. fruit; Anethi Fructus (L.); "d. seed" - **d. pickles**: cucumbers pickled with addn. of dill, pimenta, black pepper, coriander & bay leaf - **d. water**: prep. fr. caraway; u. in beverages & med.

Dillenia indica L. (Dilleniaceae) (tropAs): karmal (Hind): thabyu (Burma); frt. bk. lvs. u. astr. antipyr. (Burma); petals attract insects; sepals edible; frt. pulp u. to cleanse hair; phloem fibers u. as cordage; lvs. under thatch of roofs; wd. u. in building boats.

dillesk: dillisk (Ireland): dulse.

dilling herb: dill.

dillweed: 1) dill 2) *Anthemis cotula.*

dilly: 1) *Anethum graveolens* (lvs. & frt.) 2) (US slang): daffodil.

diluent: dilutent (rare): a diluting agt.

dimer: a polymeric compd. (usually org.) with the same mol. compn. as another but doubled, and with twice the mol. wt. (ex. dimeric alkaloids).

dimethyl glycine: amino acid; compd. found in some samples of pangamic acid (vit. B_{15}).

dimethyl sulfoxide: DMSO: occ. as by-product in mfr. of paper pulp; developed at Crown Zellerbach plant, Camas, Wash (US); solvent, antifreeze, u. as topical anti-inflammatory; analgesic, to aid absorption of other drugs; proposed use in arthritis, bursitis, scleroderma, &c.

dimethyltryptamine (DMT): occ. in several hallucinogenic plants; ex. *Prestonia amazonica*; with psychotomimetic activity.

dimethyl-tubocurarine iodide: deriv. of d-tubocurarine; longer lasting action & less tendendcy to prod. respir. paralysis than latter.

dimethylxanthine: theobromine (3, 7); theophylline (1, 3).

Dimocarpus longan Lour. (Euphoria longana Lam.) (Sapind.) (Chin): longan; juice frt. delicious drink, ton.; frt. u. for nerv. dis., lf. decoc. cooling drink; fl. inf. u. in kidney dis.; rt.in GC; dr. frt. exported fr. Chin; cult. tree.

Dimocillin™: methocillin sodium; semi-synth. antibiotic derived fr. penicillin; u. in resistant staph. inf.

Dimorphandra mora Benth. & Hook. (Leg) (Guiana, Trin): mora; sd. "false cola nut" (no caffeine); u. to adulter. Cola (Domin Rep) (-1896-); wd. decay-resistant, u. for building.

Dinochlora andamanica Kurz (Gram) (eAs): young shoots u. as vermifuge; watery sap of st. u. for scurf (Mal) - **D. scandens** Ktze. (Indones): roasted young shoots u. vermifuge; st. watery sap u. for eye dis. & to improve vision.

Dinophora spennaroides Benth. (Melastomat.) (trop wAf): locally u. for sore mouth. (Lib.).

Dioclea Kunth (Leg) (tropics): cluster pea; Cali-nuts - **D. apurensis** HBK (Co, Ve): balu(y); sds. u. as food (*cf.* ordinary beans) - **D. paraguariensis** Hassl. (tropAf+): u. med. (Parag Ph.) - **D. reflexa** Hook. f. (Amtrop, incl. WI, Nig, Congo): marbles vine; sds. with alk. similar to those of Physostigma; u. ton. stim. (Lagos); rt. decoc. for coronary pain; to kill head lice (Sierra Leone).

Diodia teres Walt. (Rubi.) (US, Mex, CA): Poor Jo (Tex/US)+); Pó Jó; poor weed; (rough) buttonweed; u. folk med. - **D. orientalis** Koen. (Hydrophylax maritima L) (Malabar, Ind): bk. c. red dye - **D. sarmentosa** Sw. (tropAf, Am): u. in traditional med. of Rwanda.

dioecious: diecious.

Dionaea muscipula Ellis (monotypic) (Droser.) (seUS; pine barrens): fly-trap; carnivorous pl. with trap-like lvs. which fold over the prey; novelty pl.; u. in med. like *Drosera* spp.

Dionin™: ethyl morphine HCl.

Dioön (Dion) edule Lindl. (Cycad.) (Mex): sd. decoc. u. in neuralgia; sds. rich in starch, ground into meal & baked; often cult.

Dioscorea L.: 1) genus of trop. vines (Dioscoreaceae), representing the yams; most spp. of Af origin, where one of oldest cult. food pls. (tuberous rt. u.); often confused by layman with sweet potato 2) (L) (specif.): *D. villosa*, wild yam root; u. med. formerly NF - **D. aculeata** L. = *D. esculenta* - **D. alata** L. (D. atropurpurea Roxb.) (Austral, Polynes): white/water/winged/greater yam; inhame bravo; tubers excellent food; eaten during convalescence fr. phthisis, kidney & spleen dis.; appl. as poultice to mature boils; widely cult. Af,

SriL, Med, Mex (chief food yam) - **D. batatas** Dcne. = *D. oppositifolia* - **D. bulbifera** L. (D. sativa Thunb.; D. crispata Roxb.) (OW moist trop, esp. Ind, natd. sFla [US]): air potato; aiyam; lvs. appl. to pinworm (PI); u. galactag. (Upper Burma); tubers c. 7.5% crude saponins (hemolyzing) u. int. & ext. for sore throat, boils, swellings (Chin); against dys. diarrh. syph.; tubers toxic but u. as popular food (after repeated washing & cooking) (tubers aerial; -0.3 m long); toxic var. u. to poison baits for noxious animals; in criminal poisonings (cult. Br, Mex) - **D. cayennensis** Lam. (WI, Guyana); yellow (guinea) yam; name amarillo (Cu); lf. tea u. in fever (Domin) - **D. composita** Hemsl. (Fla [US], CA, sMex): Mexican (or tropical) yam; "barbasco (de camote)" (Mex); tubers esp. rich in diosgenin (13%), corellogenin, yamogenin (1044); u. mfr. cortisone fr. botogenin present; cult. Fla (US) (import. fr. Durham, sAf) - **D. convolvulacea** Schlecht. & Cham. (sMex to Pan): tuberous rhiz. boiled & eaten in some areas - **D. deltoidea** Wall. ex Kunth (Ind, Mal): rhiz. c. saponins; u. to wash hair (kills lice), silk, wool, etc.; in med. - **D. divaricata** Blanco = *D. opposita* - **D. esculenta** Burkill (seAs, esp. Ind, Indochin, Mal, Thail, PI, Polynesia, cult. in trop worldwide, ex. Mex): Asiatic yam/potato/Guinea/common; inhame (Negro, Pl); tugi (Tagalog, Pl); tuber c. diosgenin, decoc. u. in kidney dis. (diur.), rheum. (Indochin); beri beri (Chin); scraped rhiz. appl. raw as poultice to swellings of all kinds (throat, fingers, toes) (Indones); cooked tuber (whitish flesh) eaten like potato, one source of Guinea arrowroot - **D. floribunda** Mart. & Gal. (sMex, CA): "barbasco"; source of sapogenin - **D. hispida** Dennst. (D. daemona Roxb.) (eAs): intoxicating yam; tuber prep. u. as wash for skin dis., to matur. boils (Chin); cure for myiasis (126); tubers eaten esp. in famine yrs. (in ordinary yrs. *D. bulbifera* & *D. pentaphylla* are preferred); higher protein content than in potatoes; fresh tuber gratings appl. to corns, calluses, whitlow; arthr. & rheumatic pain; toxic, has been u. in criminal poisoning - **D. mexicana** Guillemin (D. macrostachya Benth.) (sMex to Pan): "cabeza be negro" (Mex); tuberous rhiz. rich in steroid saponins, a source of cortisone, &c.; cult. in Oaxaca, Veracruz - **D. nummularia** Lam. (PI, Fiji): frts. u. for intest. pain (PI, Calcutta) - **D. opposita** Thunb. = **D. oppositifolia** L. (D. batatas Dcne.) (Jap, Chin, PI, cult. Br): (Chinese) yam; cinnamon vine (ornam. pl. name); Chinese potato; tubers on st. c. starch (Guiana or Dioscorea arrowroot) (or starch); tuber eaten (*cf.* potato), also u. anthelm.; ext. u. on sores; lf. u. in snakebite, scorpion sting (Chin, Jap) - **D. paniculata** Mx. = *D. villosa* - **D. pentaphylla** Willd. (Ind, Polynes, jungles): "lena" (Tonga); tubers edible but make poor food; tubers u. to disperse swellings; ton; in cough, asthma - **D. pozucoensis** Knuth (Per): u. med. ("Zarzaparilla") - **D. quaternata** Walt. = *D. villosa* - **D. sativa** L. = *D. esculenta* - **D. tokoro** Mak. (eAs): diur. in arthr. & rheum.; to resolve blood clots; c. hemolytic saponin; fish poison; c. dioscin (saponin), diosgenin (hydrolytic prod.); yonogenin - **D. trifida** L. f. (Mex, Caribbean area+): yampi yam; cush-cush (yam); each pl. produces several tubers; tubers eaten ("vegetable relish"); cult. - **D. villosa** L. (e&cUS): wild yam; rheumatism root; tuber c. steroid saponins, resin; u. exp. diaph. in biliary colic, rheum.

Dioscoreaceae: Yam Fam. of Monocotyledons; widely distr. vines, often with tuberous rts. or tubers; fls. with 6 perianth segments; ex. *Tamus, Rajania.*

Dioscoreophyllum cumminsii Diels (Menisperm.) (tropAf): serendipity berry; "West African wild red berries"; frt. c. monellin (a protein); natural sweetener 3000x as sweet as sucrose; said sweetest natural compd. known (1972-).

dioscorine: isoquinolidine alk. obt. fr. rhiz. & rts. of *Dioscorea* spp. incl. *D. hispida, D. villosa, D. composita,* &c.; *Solanum* spp., *Agave* spp.; these c. tropane deriv.

diosgenin: steroid sapogenin found in *Dioscorea* spp., incl. *D. composita* & *D. tokoro*, *Trillium erectum*, *Solanum* spp., *Agave* spp.; starting material in synthesis of cortisone, &c.

Diosma 1) genus of shrubs (Rut.) (sAf) with aromatic lvs. & small pink or white fls. 2) (specif.): Buchu (BP Ne. Phar.) - **D. oppositifolia** L. (D. succulenta Berg; Barosma succulenta L. var. *bergiana* H. & S.) (sAf): Karoo buchu; false b.; lvs. with peppermint odor; lvs. u. for urinary dis., colds & tea substit. (folk med.); considered inferior to true buchu (var. *longifolia* Sond. called "wild b.").

diosmin: barosmin; flavone glycoside of diosmetin (flavonol) with rutinose (sugar) (7-rutinoside); occ. in dr. peel of ripe lemons, *Zanthoxylum avicennae*, *Sophora microphylla*, *Mentha pulegium*, *M. piperita*, *Hedeoma pulegioides*, *Dahlia pinnata*, &c.; marketed & u. as bioflavonoid (sometimes subst. called "hesperidin" actually represents diosmin).

diosphenol: barosma camphor; cyclic stearoptene found in VO of buchu.

Diospyros L. (Eben.) (mostly As): the persimmons; genus of trees, many of which yield ebony wood - **D. chloroxylon** Roxb. (Ind): "yllinda"; "kosai";

source of green ebony wood; wd. c. diospyrine, isodiospyrine, xylospyrine; u. for musical reed instruments (ex. clarionet); wd. loses its natural oil, replaced by soaking in "bore" oil (expressed fr. green grenadilla wood; linseed odor); frt. edible, peppery (*cf. Tecoma leucoxylon*) - **D. crassinnervis** (Krug & Urban) Standl. (WI): feather bed (Bah); rt. or lf. decoc. u. for bedwetting; lf. for female med.; male aphrod. (Bah) - **D. decandra** Lour. (Indochin): thi; frts. u. in insomnia, as emmen. (Cambod); lvs. anthelmint - **D. ebenum** Koen. (D. ebenaster Retz.) (EI, Mal, natd. tropAm, as Med): ebony; black apple; tree cult. in the Americas for frt.; pulp eaten with orange or lemon juice; source of brandy; formerly u. in leprosy, ringworm, itch; frts. & bk. fish "poison"; one source of ebony wood - **D. ferrea** Bakh. (Ind to HawIs): heartwd. u. int. & exter. for flatulence (children) (Indochin); bk. u. med. (Tonga) (86) - **D. haplostylis** Boiv. (Madag): one source of a Madagascar ebony or African grenadilla - **D. kaki** L. f. (Jap, Chin): kaki; Japanese persimmon; juice of unripe frt. rich in tannin, u. to impregnate & preserve fish nets, paper, etc.; arom. fls. u. in prepn. var. of Chinese teas; ripe frts. edible but insipid; juice squeezed fr. unripe frt. reduces high BP; reduces congestion of hemorrhoids; lax.; frt. stalk antiemetic, bechic; lf. decoc. bechic & antipyr.; cult. (sub)tropics - **D. lotus** L. (eAs): Japanese persimmon; date plum; frt. edible, antipyr., promotes secretions (Chin), cult. US+ - **D. loureiriana** G. Don (tropAf): rts. u. to cleanse teeth & impart a red color to them - **D. malabarica** (Desr.) Kostel. (D. embryopteris Pers.) (Chin, Indochin, Burma): gob (Hindi); gaub frt.; glutinous frt.; u. to caulk boats; astr. u. in diarrh., chronic dys., snakebite; mal.; dil. ext. u. vaginal discharge; frt. juice u. to treat sores & wounds; bk. u. in fever, stomachache, diarrh. (Burma); ulcerations; lvs. u. as wrappers for bidi cigarettes; young lvs. u. in aphtha, opthalmia; sd. FO u. purg. in diarrh.; wd. a fine "ebony" - **D. maritima** Bl. (seAs, Austral): frts. & bk. u. to stupefy fish (sVietnam) - **D. melanoxylon** Roxb. (Ind, SriL): frt. germicidal; sds. u. in dys.; lvs. u. as cigarette papers - **D. mespiliformis** Hochst. (tropAf): monkey guava; lf. inf. u. in fever, dys. in wound dressings (hemostatic); rt. bk. decoc. in leprosy, skin eruptions; anthelm.; pl. is antibacterial & insecticidal - **D. microrhombus** Hiern (Madagas): source of African grenadilla or Madagascar ebony - **D. mollis** Griff. (Thail, Burma): ma-gluca; maklua; small tree; frt. ext. or seed u. as anthelm. (c. diospyrol, active against hookworm, tapeworm); lvs. & fls. u. in dyeing black (also frt.) - **D. texana** Scheele (Tex [US], nMex): black (or Texas) persimmon; chapote (Mex); unripe frt. u. to dye sheepskins, etc.; wd. u. to make tool handles, engraving blocks &c.; frt. pleasantly edible - **D. tricolor** (Schum. & Thonn.) Hiern (trop Af): st. bk. antibiotic, u. in leprosy; anthelm.; active on Staphylococci, Streptococci, diphtheria bacillus - **D. virginiana** L. (e&sUS): common (or American) persimmon; bk. u. antipyr. tannin substit.; lvs. rich in vit. C, u. prep. of a good tea; popular in folk med.; ripe frt. eaten for hemorrhaging, diarrh. dys. (sUS); dr. frt. boiled in water u. for worms; frt. edible, very rich in sugar; it is popularly held that the astrigency of the frt. is lost with onset of frost in the fall; however, frequently frts. are sweet before the onset of frosts; u. in prepn. alc. & nonalc. beverages; blossom decoc. u. in sore throat ("plaguemma" - French Louisiana [US]) (ripe frt. tastes very much like dates) (1560); sds. c. FO (semi-drying, with oleic, linoleic, stearic, palmitic, myristic, arachidic acids; like sesame oil; potentially useful; wd. ("Oklahoma ebony") u. for golf-club heads, cotton spindles (shuttle blocks) - **D. wallichii** King & Gamble (Thail): lvs. appl. as poultice on yaws; frts. u. as fish poison.

Diotis maritima L. (D. candidissima Desf.) (*Otanthus*) (Comp.) (Ireland, Cornwall, Med reg., nAf, It, Fr): cotton weed; diotide (It); hb. u. as vermifuge, emmen. (Arabs).

Dipcadi erythraeum Webb. & Benth. (Lili) (Arab, Eg, Ind): bulb u. like Indian squill, with action of digitalis; in cough, expect.

Dipel v. Bacillus thuringiensis.

diphtheria: acute infectious dis. caused by *Corynebacterium diphtheriae*; may be lethal either fr. systemic effect of endotoxin (on heart, &c.) or by choking (suffocating) fr. leather-like pseudo-membrane formed by organisms in pharnyx; acute inf. treated with **d. antitoxin** (E), prepd. in horses which have been injected with d. toxin; since the antitoxin only neutralizes the toxic effects of exotoxin & does not affect the organism, it may be necessary to perform a tracheotomy or cut through pseudo-membrane to save life; before devt. of biologicals, disease frequently fatal, esp. for children - **d. toxin**: purified prod. fr. cultures of organism; u. in high dilution by intracutaneous inj. to det. if person immune or susceptible (redness = sign of no immunity); sterile aq. soln. treated with HCHO —> **d. toxoid**; u. to produce active immunity by SC inj. **- adsorbed d. t.**: when pptd. or adsorbed on alum, $Al(OH)_3$ or Al phosphate (slower absorption, gives fewer reactions & requires only 2 inj.); combinations: **d.** &

tetanus toxoids (also **adsorbed**); **d. & tetanus toxoids & pertussis vaccine (DPT)** (also **adsorbed**) for active immunization of infants (adv.: requires fewer inj.) (E); tetanus & d. toxoids (adsorbed) for adults c. 10x dosage of d. toxoid & must not be adm. to children under 6 yrs.

Diphysa carthagensis Jacq. (Leg) (CA, Mex, Co, Ve): u. in sores, wounds; sap u. in dys. & "chiclero" ulcer (Mayas); heart wd. source of yellow dye; sometimes u. to make live fences - **D. robinioides** Benth. (Mex, CA): palo amarillo (Guat): guachipelin (CR); macano; fine hard durable wd.; cult. ornam. (small tree); yellow dye; u. in chills. mal. (Salv) - **D. suberosa** S. Wats. (Mex): "palo santo"; pd. bk. remedy for catarrh (Jalisco).

Diphyscium foliosum (Hedw.) Mohr (Buxbaumi. - Bryophyta) (NA): moss ext. active against P388 lymphocytic leukemia.

Diplachne dubia Scribn. (Leptochloa d. (HBK) Nees) (Gram.): (sUS, Mex to Arg): green spangletop; important perenn. forage grass & for hay.

Diplacus Nutt.: *Mimulus.*

Diplazium esculentum (Retz.) Sw. (Aspleni. - Pteridoph.) (eAs, PI): pakoas; rhiz. & young lvs. u. in decoc. for cough, hemoptysis; young fronds u. as leaf vegetable (Sikkim).

Diploclisia glaucescens Diels (Menisperm) (Ind): lvs. admin. in milk u. in syph., GC, biliousness; lvs. c. saponins, mucil.

diploicin: antibiotic fr. *Buellia canescens* (Dicks.) De Not. (Diploicia c.) (lichen).

diploid (number): no. of chromosomes in the somatic or body cells of pl. or animal, representing a doubling of the haploid no. found in gametes; designated "2n."

Diploknema butyracea H.J. Lam. (Madhuca b. [Roxb.] MacBride; Bassia b. Roxb.; Illipe b. Engl.) (Sapot.) (tropAs, esp. Ind, Tibet, Nepal; tropAf): Indian butter tree; sds. source of tallow-like fat called fulwa butter, f. fat, ghee b.; u. similarly to FO of *Madhuca longifolia.*; incl. soap making; ointment for rheum., chapped hands - **D. sebifera** Pierre (Indochin, Indones [Borneo; Sunda], Moluccas): sds. prod. FO (Borneo tallow of comm.); better grades u. in foods, poorer u. for soaps, &c.

Diplopterys cabrerana (Cuatr.) Gates (Malphigi.) = *Banisteriopsis rusbyana* Morton.

Diplorhynchus angustifolia Stapf. (D. condylocarpon Pichon) (monotypic) (Apocyn.) (tropAf): Rhodesian rubber tree; rt. decoc. u. for blackwater fever (Europeans); remedy for screw-worm; diur.; pounded bk. wound dressing; latex u. as birdlime; rt. as gruel for cough.

Diploschistes scruposus (Schreb.) Norm. (Urceolaria s. [Schreb.] Ach.) (Lichens) (cosmopol): jagged (or rock) urceolaria (lichen) u. brown dye (woolens, Eng), red, orange, purple dyes; pl. collects Zn (-10% in thallus); highly resistant to gamma radiation.

Diplospora malaccense Hook. f. (Rubi) (Thail, Mal): lf. inf. u. as tea (Perak, Malacca).

Diplotaxis tenuifolia DC (seEu, Med reg.): wall rocket; sds. u. in folk med. (Fr Phar.); bee plant.

Diplotropis brachypetalum Tul. (Leg) (Guiana): bk. decoc. u. to kill vermin (Ind); wd. u. to make house frames.

dippel's oil: rect. animal oil, obt. by destr. distn. of bones, horns & other animal subst.

"dipper": operator who carries out **dipping**; removing crude gum fr. a "box" or cup attached to a turpentined pine tree.

Dipsacus (Dipsacaceae) (Euras): teasels; genus of coarse tall hbs. superficially resembling thistles - **D. asper** Wall. (Ind, Chin): rts. u. as ton. in breast cancer; sds. u. as emmen. (Chin); rts. exported fr. Indochin into Chin (formerly) - **D. fullonum** L. (Eu): fuller's teasel; rt. u. diur. diaph. stomach; frt. burrs u. in woollen mfr. to card (raise nap); pls. cult. NY & Ore (US) largely for export to GB - **D. japonicus** Miq. (Chin, Jap): rt. u. in Chin med. for dysmenorrh., hemorrhages, hemorrhoids, breast cancer - **D. sativus** (L) Honck. (sEu): fuller's teasel; fruiting heads u. in preparing cloth; rts. u. in syph.; fls. & sds. u. as antidote against rabies (!!!) (formerly); bee plant; cult. - **D. sylvestris** Mill. (Euras, natd. eNA): common (or Venus cup) teasel; rt. (Radix Dips. sativae s. cardui veneris) u. against chest dis. (phthisis); also as prec. sp.; fls. & sds. u. in rabies (folk med.).

Diptera: insect order which includes the flies & relatives (mosquitoes, gnats, etc.) which are characterized by typically having 2 wings, suctorial mouth parts & larvae in form of maggots.

Dipteracanthus prostratus Nees (Acanth.) (Ind): lvs. given with liquid copal in GC; remedy for ear dis. - **D. suffruticosa** Voigt (Ind): chaulia (Santal); rts. u. GC, syph. renal dis.; for abortion; prep. for sore eyes.

Dipterocarpaceae (Ind): Dipterocarpus Fam.; mostly of trees notably tall, evergreen, VO and resin-bearing; incl. *Dryobalanops, Shorea, Vateria.*

Dipterocarpus Gaertn. f. (Dipterocarp.) (sAs): several spp. yield gurjun balsam (an oleoresin) wh. exudes fr. trunk; u. in many varnishes; wd. valuable - **D. alatus** Roxb. (sInd, Burma, Thail): yang tree (Thai); source of Gurjun Balsam; u. GC (natives), in varnishes & to adulterate copaiba & Maracaibo balsams; bk. & rt. in hot effusions &

baths for rheum. - **D. glandulosus** Thwait (sAs, esp. Cey): resinous oil (Dorana oil) u. to treat Hansen's disease - **D. kerrii** King (Mal): yields oleoresin called minyak keruong, u. locally in med.; to caulk boats - **D. pilosus** Roxb. (Burma, Mal, Ind): bollong (Assam); balsam u. in GC, gleet, other urin. dis. - **D. tuberculatus** - Roxb. (Burma, Assam): oleoresin appl. to ulcers - **D. turbinatus** Gaertn. f. (eBangladesh +): common gurjun oil tree; trunk source of gurjun balsam (or EI, Copaiba Balsam) & g. oil; oleoresin appl. to ulcers, ringworms, skin dis., diur. in GC; u. like copaiba, exter. in skin dis. leprosy; substit. for copaiba; in mfr. shellacs (Gurjun: Bombay name).

Dipteryx Schreb. (*Coumarouna* Aubl.) (Leg) (SA, esp. Guianas): genus of trees, some with valued timber; some 10 spp., found along rivers & in woodlands - **D. odorata** (Aubl.) Willd. (Guianas, Br+): tonka (bean); (Dutch) t. bean; cumaru; sds. black, with resemblance to prune, coumarin-rich, c. FO, sitosterol; FO sometimes appl. to oral ulcers; u. TB; sd. u. as flavoring, perfume, source of coumarin (Pará tonka bean), in pertussis; TB (1561); sometimes coumarin crystals collect on surface of bean ("crystallized t. b.") ("frosted Pará"); bk. when cut gives rise to a kino-like secretion; bk. c. lupeol, betulin, methyl palmitate & methyl esters of 5 other fatty acids - **D. oleifera** Benth. (Hond - Mosquito Coast): eboer nuts; similar to but less useful than Tonka beans - **D. oppositifolia** Willd. (Guianas, Br, Ve): British (or English) tonka (bean); u. spasmolytic, odor corrective, to aromatize tobacco & spirits; in perfume indy.; formerly in prepn. coumarin - **D. panamensis** Record & Mell (CA): (small) tonka bean; almendro; eboe (CR); raw sds. u. as candlenuts; roasted sds. eaten (Pan); FO fr. sds. u. to prep. toilet soaps; wd. hard.

Dirca palustris L. (Thymelaeaceae) (n&cNA): (Atlantic) leatherwood; drago (Mex); a shrub of damp woods; bk. irrit. vesic rubef. esp. emet.; frt. poisonous; tough bk. u. to make thongs (NA Indians).

Dirty John: *Chenopodium vulvaria.*

disaccharides: class of sugars yielding 2 monosaccharides (simple sugars) when hydrolyzed; ex. sucrose.

Discaria longispina (N & A) Miers (Rhamn.) (Arg): coronilla; pl u. febrifuge; rt. u. in skin dis. (Arg).

Dischidia acuminata Cost. (Asclepiad.) (Indochin): pl. inf. u. in GC - **D. nummularia** R. Br. (seAs): pl. u. to allay pain of wounds fr. spines of certain fishes; latex & lvs. u. to correct sprue (children) - **D. platyphylla** Schltr. (Oceania): lvs. pressed to make juicy, then sap appl. to boils (PI) - **D. purpurea** Merr. (PI): kalipkip; crushed lvs. appl. as poultice or pomade (cooked in coconut oil) for eczema & herpes - **D. vidalii** Becc. (PI): lvs. crushed with salt & appl. to goiter.

disciplina de monja (lit. nun's discipline) (Arg): *Caesalpinia gilliesii.*

dishrag (sUS): sponge gourd, *Luffa* spp.

disinsection: removal of insects (*cf.* disinfection).

disparlure: sex pheromone of *Lymantria dispar* (gypsy moth).

disseminule: diaspore.

dissepiment: partition.

Dissochaeta celebica Bl. (Melastomat.) (seAs, Oceania): lvs. appl. for pinworms (PI); given after childbirth (Mal).

dissolution (Fr-f.): solution.

Dissotis canescens Triana (Melastomat) (sAf): lf. inf. u. as enema for diarrh., dys. - **D. grandiflora** Benth. (tropAf): juice squeezed fr. rt. u. as sweetener; source of fermented beverage ("biti") - **D. princeps** Triana (sAf): royal dissotis; rt. starvation food; aphrod.; used as soap - **D. rotundifolia** Triana (tropAf): lvs. u. anthelm. & in dys.; TB & yaws remedy; whole pl. u. in rheum., yaws; anthelm. (tropAf).

Distichlis spicata Greene (D. maritima Raf.) (Gram) (wUS): yerba del burro; seashore salt grass; u. for GC (combns.) (NM [US]); var. *stricta* Beetle (NA, Mex, WI, SA): u. as forage grass.

distillation: process of heating a liquid substance (in retort), then cooling the vapors in a condenser & collecting such distillate product - **destructive d.: direct d.: dry d.**: process of distilling without adding water to product in retort so as to destroy original substance in large part, the vapor of newly formed substances being condensed & collected - **water-steam d.**: process where steam is passed into retort contg. water suspension of subst. being distd.; the retort may or may not be heated.

Distylium racemosum Sieb. & Zucc. (Hamamelid.) (Chin, Jap): source of "bud" or Chinese galls c. about 38% tannin (Fr Phar.): wd. in commercial u. in Jap; ash fr. wd. u. in glazing.

dita (bark): **ditta**: *Alstonia constricta, A. scholaris.*

ditali (It-m.): Digitalis.

Ditaxis tinctoria (Millsp.) Pax & Hoffm. (Euphorbi) (tropAm): "azafran"; lf. decoc. u. to bathe babies affected by evil eye of a drunken person; yellow dye.

dithiol-thiones: gp. of compds. wh. occ. in cabbage, &c. (Cruc.); protective agts. against radiation injury & chemotherapy.

Ditremexa Rafin. = *Cassia.*

dittamo (It): **dittander: dittany**: *Dictamnus albus.*

dittany, American: common d.: *Cunila origanoides, C. mariana* - **Cretan d.: d. of Crete**: *Origanum dictamnus* - **domestic d.**: American d. - **European d.**: *Dictamnus albus* - **mountain d.**: common d. - **white d.**: *Dictamnus albus.*

ditto bark: dita.

Dittrichia Greuter - v. *Inula.*

Diuretin™: theobromine sodium salicylate.

divaricoside: glycoside of sarmentogenin & L-oleandrose; found in *Strophanthus divaricatus.*

divi-divi: *Caesalpinia coriaria, C. tinctoria.*

divinum remedium (L): *Peucedanum ostruthium.*

division: phylum (of plants).

divostroside: glycoside of sarmentogenin & L-diginose; occ. in *Strophanthus divaricatus.*

djenkolic acid: amino acid fr. *Pilhecellobium jiringa* (djenkol bean) (*qv.*).

DNA: deoxyribonucleic acid (*qv.*).

doberdua (Surin.): *Strychnos* sp. (bushy vine) of wh. bk. used.

dock: 1) *Rumex* spp. (dock means sharply cut off & may refer to truncate lf. base of some spp.) 2) also other plants, ex. *Tussilago, Malva, Verbascum* & *Petasites* spp. - **bitter d.: blunt-leaved d.: broad leaf d.**: *Rumex obtusifolius* - **bur d.**: burdock (*qv.*) - **d. cress**: *Lapsana communis* - **curled d.: curly d.**: *R. crispus* - **dove's d.**: *Tussilago farfara* - **English d.**: bitter d. - **fiddle d.**: *R. pulcher* - **mulle(i)n d.**: *Verbascum thapsus* - **narrow (leaf) d.**: *Rumex crispus* - **d. nettle**: *Urtica urens* - **patience d.**: *Rumex patientia* - **red d.**: *R. aquaticus, R. sanguineus* - **round d.:** *Malva sylvestris, M. rotundifolia* - **d. sorrel**: *R. acetosa* - **sour d.**: *R. acetosa, R. acetosella, R. crispus, R. obtusifolia* - **spatter d.**: *Nuphar advena* - **swamp d.**: *Rumex verticillatus* - **tanner's d.**: *R. hymenosepalus* - **velvet d.**: *Verbascum thapsus* - **water d.**: *R. aquaticus, R. hydrolapathum* - **white d.**: *R. pallidus* - **yellow d.**: *R. crispus* & some other *R.* spp. - **y. rooted (water) d.**: *R. britannicus.*

doctor's gum: 1) gum tragacanth. 2) gum resin fr. *Symphonia globulifera* 3) exudn. fr. *Rhus metopium.*

Doctrine of signatures: the belief that Nature has endowed pls. with symbols indicative of their value in med. (ex. *Iris* lvs., sword-shaped, hence u. as vulner.).

dodder: *Cuscuta* spp. - **lesser d.**: *C. epithymum.*

Dodonaea lanceolata: F. Muell. (Sapind.) (Austral): hop bush; decoc. drunk for various pains; also in warm baths for pain (2263) - **D. thunbergiana** E. & Z. (sAf): bosysterhout: inf. u. expect. in pneumonia, TB; pl./wd. (c. saponins) decoc. purg. in fevers; young twigs u. ton. (1031) - **D. viscosa** Jacq. (trop): candlewood; varnish leaf tree, gansies, kankerbos (Afrikaan); chopi (Af); a' ali's (Haw); lvs. c. saponin; crushed & inf. appl. to rashes & itches (Haw); lf. ext. int. for fever; mal., antirheum.; antipyr., sore throat (Madagas); hemorrhoids; tease out twig as toothbrush; lf. chewed as stim. like coca (Peru); bk. of shrub u. in astr. baths; frts. u. like hops (Austral); gastric dis. (decoc.)

dog: *Canis familiaris* (qv.)

dog chamomile: *Anthemis cotula* - **d. grass**: doggrass - **d. lichen**: ground liverwort - **d. parsley: d. poison**: *Aethusa cynapium* - **d. standard**: *Senecio jacobaea* - **d. tree**: *Cornus florida* - **dog**: v. also dog's; rose; grass.

dogbane (root): *Apocynum androsaemifolium* - **(Canadian) d.: hemp d.**: *A. cannibinum* - **spreading d.**: d.

dog-fennel: *Eupatorium capillaceum* (usually); *Anthemis Cotula*; *Peucedanum palustre.*

dogfish: Pacific shark; *Squalus acanthias, S. suckleyi.*

doggelt: birch tar oil - **doggrass**: *Agropyron repens.*

dog-peter (Ga) [US]) : *Leonitis* spp., probably also *Leonurus* spp.

dog's bane: dogbane - **d. buttons**: Nux Vomica - **d. grease**: FO of body of *Canis familiaris*; u. folk med. - **d. mercury**: *Mercurialis perennis* - **d. tongue**: 1) *Cynoglossum officinale* (lvs. & rt.) 2) *Trilisa odoratissima* - **d. tooth violet**: *Erythronium americanum* & other *E.* spp. (SC rts. look like d. t.).

dogwood: 1) *Cornus florida* 2) Jamaica d. 3) (Cu): *Randia mitis* 4) (Cu): *Lonchocarpus latifolius* (rel. abundant) - **American d.**: *Cornus florida* - **black d.**: Frangula - **bloodtwig d.**: *Cornus* - **common d.**: *Cornus sanguinea* - **English d.**: *Philadelphus coronarius* L.+ - **flowering d.**: American d. - **Jamaica d.**: *Piscidia Piscipula* (bk.) - **poison d.**: poison sumach, *Rhus vernix* - **round-leaved d.**: *Cornus rugosa* - **silky d.: swamp d.: sweet d.**: *Cornus amomum* - **Tallahassee d.**: American d. - **white d.**: *Viburnum opulus.*

dohang (Hung): tobacco.

Dolde (Ge): umbel.

Dolichandrone falcata Seem. (Bignoni.) (tropAs): lf. decoc. u. abortient; bk. u. as fish poison - **D. spathacea** K. Schum. (tropAs): lf. cold inf. u. as mouthwash for thrush (Indones); pd. sds. u. for nervous complaints (PI); fresh lvs. & bk. poultice appl. to relieve flatulence after childbirth; red capsules (dr. frt.) chewed with lime; lf. & frt. substit. for *Piper betle* in chewing betel; bk. u. in *Framoebia* infest. (Micronesia).

dolichol: very long chain unsatd. isoprenoid alc. with 80 C chain; occ. in plants (tobacco); kidney tissue, &c.

Dolichos biflorus L. (Leg) = *Vigna unguiculata* (L.) Walp var. - **D. katangensis** De Wild. (Congo, Zambia): u. as remedy for headaches (Congo) (2502) - **D. lablab** L. (OWtrop): hyacinth dolichos; pois contour; sem; shim; hyacinth bean; black-seeded kidney beans; sds. recommended as food (bean soup); raw sds. said toxic, must be cooked; veg. gum mfrd. fr. sds.; forage pl.; ornam - **D. lubia** Forsk. (Eg, Israel): lubia bean(s); u. as food in Israel - **D. pruriens** L.: *Mucuna p.* - **D. sphaerospermus** DC (WI, US): black-eyed peas; sds. popular food (seUS, WI, esp. Jam): esp. for New Year's Day meal (with hog jowl); really a bean, thus "rundsamige Bohne" (Ge) (round-seeded bean).

Doliocarpus dentatus Standl. (Dilleni) (Co): bejuco de agua; water fr. vine u. to allay after-effects of mal. (Kubeo Indians) (2299).

dollar leaf (US Negro): *Pyrola rotundifolia* - **d. weed**: 1) *Rhynchosia simplicifolia* 2) *Hydrocotyle* spp. (sUS).

dolloff: *Spiraea ulmaria* - **Dolly Varden**: *Tephrosia virginiana.*

dolomite: form of limestone; crude $CaCO_3$-$MgCO_3$; dolomitic limestone, fr. ancient marine deposits; with some Fe, Mn (multi- mineral chelated prod.) (Ca: Mg. = 2.1); u. as tablets in arthritis, food allergy, &c.; in stock foods; also u. as bricks lining furnaces, in paper making, &c. (SC mineralogist D.S.G.G. de Dolmieu).

dolphin: delphin: 1) *Delphinus delphis* L. (American or saddleback d.) & *Tursiops truncatus* (bottle-nosed d.) (Delphinidae - Cetacea - Mammalia) (worldwide) in warm temp. seas, sometimes in large rivers; carnivorous with numerous teeth; body 3-9 m long; with large brain & high intelligence; often trained to perform; source of FO 2) a true fish, *Coryphaena hippurus* (common d.) (warm seAs); emergency food for sailors (congregate around boats; up to 2 m long.

dolphin oil: 1) Oleum Delphini (L.): FO of d. liver; u. much like halibut liver oil 2) blubber oil similar to whale oil; c. spermaceti (wh. can be separated); u. as fine lubricating oil.

dom: doom (2) - **d. nuts** (Eritrea): sds. u. to make buttons; doom (2) (*qv.*).

Dombeya rotundifolia Planch. (Sterculi.) (tropAf): plum blossom tree; bk./wd. inf. u. for intest. ulceration; bk. decoc. for irreg. menstr.; abortient; rt. for colic; wd. useful - **D. wallichii** Benth. & Hook. f. (Af-Madag): dombeya tree; modeste (Maurit); tree -10 m; cult. ornam. (2340).

Dominican viper tail: *Urechites suberecta.*

Don Juanita (Mex): cannabis (*cf.* Maria Juana, marijuana).

Donax canniformis K. Schum. (Marant.) (tropAs, Oceania): lf. juice u. for eye dis. (Indones); rt. decoc. in snakebite, insect sting; inf. of young shoot (soaked for 1 hr. in water) u. as antipyr. (PI).

doncel (Co): *Berberis rigidifolia.*

dong quai (Chin): dongo yam: combn. of Spirulina, "bee" pollen, tienchi, fo-ti.

dongio yam.

donkey: domesticated form of *Equus asinus* L. (*cf.* burro); u. for transport, milk, meat; male bred with mare to produce mule; various parts u. in Unani med. (thus fat, liver [in mal.], lungs, penis, milk, urine, feces, &c.) (v. ass).

donkey's eyes: *Mucuna pruriens.*

doom: 1) doon 2) **d. (palm)** (Eritrea): *Hyphaena crinita*; sds. u. to make buttons (Eritrea).

doon (bark): fr. *Erythrophloeum guineense* - **doon-head clock** (NC): dandelion.

DOPA: dopa: di(hydroxy)-phenyl-l-alanine (formerly called dioxy-phenylalanine); L-form (levodopa) naturally occurring in *Vicia faba* (velvet bean), *Mucuna pruriens* (&c.); u. anticholinergic; to treat Parkinsonism; encephalitis (sleeping sickness); to replace depleted brain dopamine.

dopamine: natural endogenous catecholamine of body (adrenal glands), precursor of epinephrine &c; isolated fr. *Mirabilis alipes* (*Hermidium alipes*) (Nyctaginaceae); acts as neurotransmitter (brain); u. in shock syndrome due to myocardial infarction; trauma, open heart surgery, &c.; adm. IV; introd. 1974 (Intropin) (E).

dopastin: metabolic prod. of *Pseudomonas* spp.; acts as dopamine hydroxylase inhibitor.

dope (US slang): opiate or addictive narcotic.

doping: u. of medicinals by athletes to incr. strength, speed, endurance, &c.; the drugs are steroid anabolics (incl. testosterone), beta-blockers, peptide hormones, &c.; formerly strychnine, cocaine, &c.; term also appl. to treating race horses, &c. to improve performance.

Doppelzentner (Ge): 100 Kg (Dz.).

dora (Arg): dura.

doradilla (CR): *Ceterach officinarum.*

Dorema ammoniacum D. Don (Umb.) (cwAs, esp. Iran): giant fennel; ushak; source of gum-resin ammoniacum (Gum Ammoniac; Bombay Sumbul); feshook (Mor); Persian ammoniac; obt. by puncturing stem; c. VO (1%); resins (65-75%); soluble gums (20%); u. exp. stim. (do not confuse with African Ammoniac); distinguished fr. African a. by absence of umbelliferone; u. in plaster masses; also prod. by **D. aucheri** Boiss. & **D. aureum** Stokes (1496).

dormideira (Por): poppy, specif. opium poppy capsules - **d. de laguna** (Mex): *Neptunia oleracea.*

Dorn (Ge): thorn, spine.

Dornapfel (Ge): Stramonium.

Dornavac: deoxyribonuclease (pancreatic) (MSD).

Doronicum orientale Hoffm. (Comp.) (seEu): leopard's bane; Gemswurz; pl. & sds. u. in folk med.; cult. - **D. pardalianches** L. (Eu): leopard's bane; rt. u. as cardiac ton., in nervous depression, melancholia, scorpion sting; like Arnica; adulter. of & substit. for Arnica; ornam. perenn. herb.

Dorsch (Ge): cod (fish).

Dorstenia L. (Mor.) (Br): trop hbs.; several spp. yield medl. roots - **D. alta** Engl. (tropAf): rasped rt. bk. u. in tonsillitis, stomatitis; chewed for cough & catarrh.; powd. in meal porridge as aphrod.; decoc. in ankylostoma infec. - **D. asper** Wall. (Ind, Chin): rts. u. as ton. in breast cancer; sds. u. as emmen. (Chin); rts. exported fr. Indochin into Chin (formerly) - **D. brasiliensis** Lam. (SA): contrayerba (Arg); lf./rt. u. diur. diaph. (Arg.); rhiz./rt. u. stim. diaph. ton. purg., emetic; emmen. in diarrh., wounds, typhus (Br); fertility control (Ur) - **D. contrajerva** L. (Mex, WI, Pe): barbudilla; rts. (Radix Contrajervae); contrayerva torus herb; if decoc. u. ton. stim., in cough & gastric dis. (Hond, Guat), against wound infections, poisonous animal bites; diarrh., dys.; tumors; headache; in fertility control (Ur); antipyr (CR); rhiz. u. to flavor cigarettes - **D. convexa** De Wild. (tropAf): hb. u. as vulner. by natives (Congo) - **D. drakena** L. (Mex, Guat, Hond): another source of contrayerba root; rt. boiled & ingested for fever; in gastro-intest. dis. (inf.) (Mayas); lf. u. in alcohol overindulgence - **D. klainei** Heck. (tropAf, Gabon): rt. bk. c. coumarin-like smelling substit.; red rts. u. as perfume (natives), as potion, gargle - **D. multiformis** Miq. (Br): carapia; rhiz. u. as diaph. expect. diur. stim. antidote; in wounds, typhus, dys., diarrh., intermittent fever, mal., &c. - **D. psilurus** Welw. (s&eAf): u. in folk med.; contused rt. macerate u. for mal. (infants).

Doryphora sassafras Endl. (Monimi.) (NSW Austral; New Caled): "sassafras"; bk. u. as tonic; wd. fragrant, u. for insect-proof boxes, &c.; c. alks.

dosage form: a medicament in finished state ready for admn. to the patient; ex. capsules, tincture, ampuls, etc.; often mfrd. on large scale in factory, sometimes by the prescriptionist in his pharmacy - **dose, loading**: primary d.

doss: dyewood fr. *Ilex mertensii* - **Dost** (Ge): *Origanum* spp. (wild majoram), &c.

doub: v. grass.

double: v. tansy.

Douglastanne (Ge): Douglas fir.

doum palm: doom palm.

doundake (leaf): *Nauclea latifolia.*

doura(h) (corn): dura.

douradinha do campo (Br): *Psychotria rigida* var. *aurata.*

Dovyalis abyssinica Warburg (Flacourti.) (tropAf): frt. edible - **D. caffra** Warb. (tropAf): edible ripe acid frt. u. in pickles, preserves, compotes; unripe frt. u. to make jelly - **D. hebecarpa** Warb. (Ind, SriL): Ceylon gooseberry; berries with edible pulp u. in jellies & preserves - **D. xanthocarpa** Bullock (c tropAf): mgola; rt. decoc. & lf. juice drunk for purulent hardened abscesses.

down tree: balsa.

Doxantha Miers = *Macfadyena.*

Doxorubicin: Adriamycin: antibiotic fr. soil fungus, *Streptomyces peucetius* - subsp. **caesius** Arcamone & al. (It): important chemotherapeutic antineoplastic agt. in cancer (much used), leukemia, lymphomas, sarcomata, carcinomata of breast, bladder, thyroid, ovary, stomach; represents streptomycin deriv. (*cf.* carminomycin [USSR]); said most commonly u. anti-cancer drug (US); tetracycle, represents streptomycin deriv. (*cf.* carminomycin [USSR]); HCl salt called Adriamycin (E) (SC Adriatic Sea where organism discovered [1974]).

Doxycycline = Vibramycin (*qv.*).

Draba muralis L. (Cruc.) (Euras.): u. as antiscorbut. (Spain). - **D. nemorosa** L. (Euras, nAf, NA): whitlow; sds. c. sinapin; u. in Chin med. esp. for pleurisy & leukemia - var. *hebecarpa* Ledeb. (eAs): frtg. stalk & sds. u. as diur. & bechic (decoc.); also for edema, cough, nausea (Chin, Jap).

Dracaena Vand. ex L. (Agav.) (OW, warm areas): dragon trees; shrubs or trees often cult. for ornam. - **D. angustifolia** Roxb. (Pleomele a. N.E. Br) (Ind, sChin, seAs, EI): rt. decoc. u. in stomach dis. (PI); lf. decoc. in dys. leukorrh. blenorrh. (Vietnam); galactagog - **D. cinnabari** Balf. (Socotra): source of Socotra dragon's blood (gum), either spontaneous exudn. or through incisions - **D. draco** L. (Canary Is): dragon tree; prod. Canary Dragon's Blood, which c. red coloring matter; many tech. uses such as in topping powder in zinc engraving; trees of great age (hundreds, not thousands of years) - **D. fragrans** Ker. Gawl. (upper Guinea): "corn plant"; u. in stomach dis., flatulence, rheum. complaints; arborescent pl. -7 m; ornam. in malls & greenhouses - **D. ombet** Kotschy & Peyr. (Eg, Ethiopia): close to *D. draco*; source of dragon's blood.

dracocéphale (Fr): *Lallemantia.*

Dracocephalum canariensis L. (Lab) = Cedronella c. - **D. moldavica** L. (Euras, adv. in cUS): dragonhead; hb. c. VO, moldavoside (flavonoid); sds. ton. carm. demul. in fevers; pl. astr. ton. vulnerary; adulter. or substit. for Melissa officinalis; ornam. hb.

Draconis(,) Resina (L.): **D. Sanguis**: dragon's blood.

Dracontium asperum C. Koch Ar (Arg; nBr): jacaraca taia; tubers u. for snakebite (value?); against asthma - **D. polyphyllum** L. (Guiana; Surin; cult. Ind): jacaraca merim; large tubers u. as food (Am Indians); in asthma, hemorrhoids; as emmen.

Dracontomelon (Dracontomelum) dao Merr. & Rolfe (Anacardi) (eAs): frt. u. for sore throat, as depurative in dermatitis (Chin); to expel the afterbirth (Indones); juice of young lvs. u. as cooling drink in fever (Indones); frt. edible & sour relish; fls. to flavor food - **D. sylvestre** Bl. (D. mangiferum Blume) (Mal, Indones): sap u. to treat thrush (mouthwash) - **D. vitiense** Engl. (Fiji): med. use.

Dracunculus vulgaris Schott (Ar) (Med reg.): dragon plant; dragonwort; rootstock u. in folk med. for intest. parasites, rheum. extern. for sores, neuralgia, poisonous animal bites; as emmen.; boiled tubers as food; toxic (CaOx); ornam. cult. perenn. hb.

dragante (It): tragacanth.

dragée (Fr, Ge): 1) sugar-coated pill, bolus or tablet 2) sugar almond - **d. d'Inde** (Fr): croton sd.

Dragendorff, J. Geo N. (1836-98): outstanding Ge phytochemist; pharmacognosist - **D. reagent**: u. alk. test; $KBiI_4$.

dragon, green: *Arisaema dracontium* - **d. root**: fr. *Arisaema dracontium* (chiefly), *A. triphyllum*; *Arum maculata* (v. also dragon's root).

dragon's blood: 1) Draconis Sanguis (L.): dark red resinous subst. derived fr. the covering of ripe frts. of *Daemonorops draco* or other *D.* spp. (**True, Sumatra, Borneo**, or **East Indian d. b.**); marketed in thin cylinders wrapped in papyrus lvs.; u. coloring agt. (mahogany stain) in varnishes; astr. mouthwashes, &c.; voodoo med. (seUS); resin fr. *Dracaena ombet*, *D. cinnabari* (**False** or **Socotrine d. b.**); gummy secretion fr. *Croton draco* (**Mexican d. b.**, a false d. b.); (Mex): v. sangre de draco; *Geranium robertianum* & other plants (Mex) - **d. b. root** (Voodoo): Sanguinaria - **American d. b.**: *Pterocarpus draco* - **dragon's claw**: *Corallorrhiza odontorrhiza* - **d. root** (Mex): sangre de drago (*qv.*) - **d. tooth** (sUS): *Erythrina herbacea.*

drêche (Fr): malt.

Dreck (Ge): feces; filth - **D. Apotheke** (lit. fecal medicine): therap. use of feces & urine practiced in cEu in 1600s & earlier, up to ca. 1850 (Ge).

Dregea rubicunda K. Schum. (Marsdenia r. N.E. Br) (Asclepiad) (trop eAf): sd. c. strophanthin-like glycoside; toxic; rt. u. as expect. purg.; chewed as aphrod.; to improve flavor of honey beer; rt. decoc. for indigest. (Kenya); pd. rt. dusted in hair as perfume - **D. volubilis** Benth. (eAs): rts. emet. (Indochin); u. med. (Chin) (1032).

Drepanocarpus lunatus G. F. W. Mey. (Leg) (tropAf): leafy sts. u. purg.; urin. dis. (incl. GC); aphrod.; **bk.** u. int. for leprosy (—> *Machaerium*).

Drimia ciliaris Jacq. (Lili) (sAf): "itching bulb"; bulb u. as emet. expect. diur.; c. CaOx crystals, sap blisters skin - **D. neriniformis** Bak. (sAf): ashed rt. rubbed into exter. tumors after lancing (some place this gen. under *Urginea*).

Drimys aromatica F. Muell. (Winter./Magnoli) (Austral): pepper tree (of Tasmania); Tasmania pepper (plant); dr. frt. arom. (Tasmania pepper); bk. sometimes substit. for Winter's bark (SC drimus, L., acid, since bk. & lvs. pungent tasting) - **D. axillaris** Forst. (NZ): pepper tree; p. shrub; horopito; bk. arom. ton. astr.; wd. u. in inlays; analogous to *D. winteri:* cult. Calif (US) - **D. insipida** Druce (Austral): pepper bush; strongly arom. frt. & bk.; analogous to Winter's bark - **D. winteri** Forst. (Mex, CA, SA mts): Winter's bark (drimys): a fragrant evergreen tree; arom. pungent bk. c. drimenin (naphthol furanone); u. ton. antiscorb. in dys. gastric dis. (Br); chewed in dentalgia (CR); condiment (Mex); bk. & lvs. u. for itch, cattle ringworm, etc. (Ch); as blood coagulant (Co); var. *granatensis* Eichler (D. g. L. f.) (Br): u. in dys. yellow fever - var. *uniflora* (Ven): cupido; cubi (Lara); bk. with peppery taste; u. toothache, vermicide for cattle (with NaCl) (92); var. *chilensis* (DC) A. Gr. (Ch): canelo; for tumors & cancers (Indians).

Drogenkunde (Ge): pharmacognosy (lit. drug knowledge).

Drogenladen (Ge): drugstore - **Drogist** (Ge): druggist.

drogue (Fr): 1) drug 2) trash, worthless material (*cf.* "drug on the market").

Dronabinol: synth. delta-9-tetrahydrocannabinol (THC); identical with the natural compd. in *Cannabis sativa*; u. severe nausea and emesis fr. cancer chemotherapy, &c.

drop (Dutch): licorice - **d. chalk**: prepared chalk.

dropflower: *Prenanthes* spp.

drops, Jesuits': Friar's balsam.

dropsy plant: *Melissa officinalis.*

dropwort, hemlock: water d. (**crocus**): **d. w. hemlock**: *Oenanthe crocata.*

Drosera (Droser.) (worldwide): sundews; small insectivorous pls. growing in bogs; c. resin, peptidase, org. acids; some u. as purple dyes, some in med. as ton., stim., reputedly useful in pertussis - **D. anglica** Huds. (D. longifolia L.) (Euras, NA, HawIs): English (or great) sundew; u. like *Drosera rotundifolia* and to extend this hb. (often found admixed in the comm. article) - **D. intermedia** Hayne (c&nEu; high moorlands, natd. e&cNA): long-leaved sundew; narrowleaf s.; u. to extend hb. of *D. rotundifolia*; u. med. - **D. ramentacea** Burchell ex. Harv. & Sond. (D. madagascariensis A DC) (trop&seAf, Madag): hb. u. antispas. for chest dis., antitussive, to relieve cough; in pulmon. dis. (1845); c. plumbagin (antibiotic); u. in larger dose than *D. rotundifolia*; in dyspeps., to preserve teeth (Madag) - **D. rotundifolia** L. (c&eEu, natd. nNA peat bogs): dew plant; sundew; u. expect. in dry cough, whooping cough, asthma; bronchitis; in folk med. as diur. spasmolytic, aphrod. liver dis., arteriosclerosis; exter. for warts, corns, freckles; c. hydroplumbagin glucoside (cult. sGe, Neth).

Droseraceae: Sundew Fam; cosmopolitan hbs.; basal lvs. with elevated fls.; grow in moist places or swamps; insectivorous pls.; ex. *Dionaea.*

drosère (Fr): *Drosera.*

drosophilin = pleuromutilin.

dross: 1) refuse; worthless waste or foreign materials mechanically separated fr. superior portions; often left as residues or impurities after subst. has been processed or used; chaff; lees; dregs, etc. (ex. mustard d.) 2) (Ga): juice fr. the old turpentine still's skimmings; good for starting a fire (222).

drug: crude medicinal substance; usually appl. to dried **crude drugs** (plant & animal); prob. fr. Dutch droog (= to dry); to the layman, usually term means addictive narcotic - **d. fiend** (Americanism): narcotic addict - **d. store** (Americanism): apothecary shop or pharmacy; properly connotes establishment of lesser ethical status than these, however).

drugget: kind of coarse cloth (SC fr. drogue = trash).

druggist: pharmacist; however, often with understanding of second-class rating, one with limited qualifications.

drumstick: *Centaurea nigra* - **d. tree**: *Cassia fistula.*

drunkards: *Gaultheria procumbens* - **drunken plant** (US slang): darnel grass.

Drupaceae: plum fam.; now a part of Rosaceae (subfam. Prunoideae Focke).

drupe: type of 1-celled frt. in which mesocarp is more or less succulent but endocarp is stony or leathery, & encloses 1 sd.; ex. peach, cherry.

druse: globoid (spherical) cluster of crystals (commonly CaOx) occurring in pl. cells; rosette; excretory form.

dry flower: *Xeranthemum annuum* L. (sEu).

Dryas octopetala L. (Ros.) (Euras, NA; mts.): mountain avens; u. as tea substit., Schweizertee or Kaisertee (Alps); in folk med. as astr. & in diarrh.

drying: process of removal of water as appl. to drugs; necessary for their proper conservation; thoroughly dr., the activity of most drugs is retained for long periods of time; also reduces costs of shipping.

Drymaria (**Drymeria**) **cordata** Willd. (D. diandra Bl.) (Caryophyll.) (Ind, Nepal+): laijabori (Nep); u. in asthma (Nep); to stop vomiting (Vietnam); pl. juice as lax. antipyr. hb. eaten as cooling salad (Fr Guiana); lvs. u. for curing earache (Pe): in kidney & liver dis. (Ec) - **D. crassirhizoma** Nakai (nJap): rhiz. & stipes u. for stomachache & to relieve bruising (Ainu) - **D. gracilis** Cham. & Schlecht. (Mex): decoc. u. as wash for "yaza" - **D. villosa** Cham. & Schlecht. (Salv+): "poleo"; remedy for cough; pl. decoc. u. for dys. (Guat).

Drymoglossum carnosum Hook. (or J. Sm.) (Polypodiaceae - Pteridoph.) (Nepal, Sikkim, Bhutan): fronds u. as pectoral, poultice for whitlows; diur. astr.; in Chin u. for urinary calculi, rheum., hemorrhages - **D. heterophyllum** (L) Presl (tropAs, Oceania): ground fronds u. as styptic; in eczema (—> *Pyrrosia*).

Drymonia coriacea (Oerst. ex Hanst.) Wiehler (Gesneri) (Co, Ec): lf. decoc. gargled warm & kept in mouth 5 mins. for toothache and mouth ulcers (777).

Drynaria fortunei J. Sm. (Polypodi., Pteridoph.): rhiz. preps. u. as hemostatic, antirheum.; kidney ton. in hemoptysis, tinnitus (Chin) - **D. quercifolia** J. Sm. (Mal, PI): lvs. pounded & appl. as poultices to swellings; dil. juice sprinkled over head of a fever patient (Mal); rhiz. decoc. astr. anthelm. antibacter. u. in hemoptysis.

Dryobalanops aromatica Gaertn. (D. camphora Col.) (Dipterocarp.) (Mal, EI): hard camphor tree; oleoresin c. VO with borneol (Borneo or Sumatra camphor): known since the Middle Ages in Eu but supplanted since ca. 1600 by the ordinary camphor; tree also source of B. c. oil; u. to treat liver & splenic dis., tumors; as diaph. antisep. stim.; in dysmenorrh.

Dryopetalon runcinatum Gray (Cruc) (Tex, Ariz [US], Mex): crushed sds. in grease rubbed on back for back pain; lvs. u. as greens (Mex).

Dryopteris Adans. (Aspleni. - Pteridoph.) (worldwide): wood ferns; shield f.; found in both trop & temp zones; fronds compound; sori with or without indusia - **D. athamantica** O. Ktze. (Aspidium athamanticum O. Ktze.) (Eu, Af): panna-panna; rhiz. vermifuge esp. good for tapeworm (tenicide); some claim more reliable than male fern; in GC; frond bases & rhiz. c. pennaric acid (butanone-phloroglucide) flavopannin (phloroglucin glycosides); u. veter. med. anthelmintic; adulter. of Male Fern - **D crassirhizoma** Nakai (Chin, Kor, Jap): rhiz. u. as anthelm. (tenifuge; for schistosomiasis [liver flukes]); in dys. - **D. filix-mas** (L) Schott (cEu, esp. Fr, Switz, As, nAf, NA, SA): male fern (aspidium); oak fern; rhiz. & stipes off. & source of off. oleoresin; u. tenifuge; also for intest. & liver flukes (as clonorchiasis); c. filixic acid, filicinc acid (rhiz. best preserved in sealed dry dark containers; held for not over 1 year; toxic) (2612) - **D. marginalis** (L) A. Gray (e&cNA): leather wood fern; American aspidium; off. & u. in same way as prec.; toxic - **D. spinulosa** Watt (D. carthusiana H. P. Fuchs) (sEu mts., nAs; natd. NA, SA): spinulose wood fern; broad buckler f.; stronger & more toxic properties than male fern; rhiz. & stipes u. as adulter. for Male Fern (detected by numerous glandular trichomes on ramenta margins) - **D. villarii** Woynar (Eu, CA): rhiz. u. in CA as anthelm. like male fern.

Drypetes arguta Hutch. (Euphorbi) (tropAf): frt. edible & u. to prep. an intoxicating beverage - **D. gossweileri** Moore (tropAf): u. in traditional med. of Congo natives - **D. perreticulata** Gagnep. (eAs): frt. edible & medicinal (Chin) - **D. vitiensis** Croizat (Fiji): "meme"; lvs. squashed to furnish juice u. in nose & ears for headache.

Duazomycins (A, B, C): antibiotic fr. *Streptomyces ambofaciens* Pinnert-Sindico;1959-; is active against tumors (mice), bacteria, yeasts, &c.

Duboisia R. Br. (Solan.) (Australas): with 3 known spp. of alkaloid-rich shrubs & trees - **D. hopwoodii** v. Muell. (Austral): pituri pl.; c. nornicotine (1%), piturine; u. (as pituri) as narcotic poison, stim., fumitory, masticatory; emu poison; insecticide - **D. leichardtii** v. Muell. (Austral): pl. rel. rich in scopolamine; (1-3%); also c. hyoscyamine, atropine; u. mfr. alk. - **D. myoporoides** R. Br. (Austral, New Caled, NG; cult. US): corkwood (duboisia); lvs. c. duboisine (hyoscyamine [chiefly]), scopolamine, norhyoscyamine [pseudohyoscyamine], atropine, etc.), tigloidine (*qv.*) (tot. alk. 1.9-2.2%) (1035); u. mfr. scopolamine; fumitory u. like tobacco.

Duchesnea indica Focke (Fragaria i. Andr.) (Ros.) (sAs, natd. US): mock strawberry; snakeberry; st. & lvs. decoc. u. in rheum. for cough, ton, in TB, to reduce swellings; juice u. to treat fever, as antidote for arrow poison, antisep.; frt. edible but insipid; cult. (US) for ground cover & in hanging baskets.

duck: *Anas platyrhyncha* L. (Anatidae - Anseriformes - Aves) (Euras): water swimming bird, domesticated since ancient days; cited sp. is common mallard d. (chief d. shot in sport); many other wild spp.; in Unani med. (flesh; blood) (in cystic calculi); eggs; egg shell (pd. in dys.); brain; excreta; feathers (ash); preen gland grease (base of tail) (u. to grease feathers to prevent water logging & loss of plumage brightness); u. in ichthyosis; in cosmetics; ext. of heart & liver u. in homeopathic med., "A. narbaria" for flu symptoms (chills & fever, body pains).

duck's foot: Podophyllum.

duckweed: *Lemna minor* & other taxa of Lemnaceae.

duct: vessel.

Dugaldia hoopesii Rydb. (Helenium h. A. Gray) (Comp.) (Ore, Calif, e to Wyo [US]): orange sneezeweed; owlsclaws; South American rubber plant; c. dugaldin (toxic glycoside) causes "spewing sickness" (sheep); u. as dye (Navaho); u. to stop vomiting; rt. furnishes a chewing gum (—> Helenium).

Dugong dugong Erxl. (Ord. Sirenia - Mammalia) (OW, Af, As, Austral, coastal waters): dugong; sea cow; FO u. med. as substit. for codliver oil (*cf.* manatee).

Duguetia asteratricha Diels (Annon.) (SA): constit. of arrow poisons - **D. latifolia** R.E. Fries (Br): u. in arrow poisons - **D. pauciflora** Rusby (Ven): arairra-yek (local) (Ptari-Tepui); bk. decoc. u. as anthelmint.; also furnishes cords; branches u. for fishing poles (92) - **D. quitarensis** Benth. (Guiana, WI): (Guiana) lancewood; Jamaica l.; Cuba l.; bois de lance; Lanzenholz.; wd. useful, called yar(r)i yar(r)i.

duhan (Croat): tobacco.

Dulcamara (L.): bittersweet, *Solanum dulcamara.*

dulce: 1) (Sp): candy (sweet) confection 2) (Nic): false ipecac 3) (Atl seaboard, NA): *Rhodymenia palmata* (*cf.* dulse).

dulcite: 1) Madagascar manna fr. *Melampyrum* spp. 2) dulcitol.

dulcitol: **dulcose**; galactitol; an hexose sugar alcohol fr. dulcite (manna), spp. of *Euonymus*, *Melampyrum*, &c; u. as sweetner for diabetics.

dullings: dulse: *Rhodymenia palmata*; red seaweed often u. as food (also **Irish d.**) - **red d.**: *Iridaea edulis*.

Dumoria africana Dubard (Sapot.) (tropAf as Congo): yaka-yaka; sd. furnishes FO (yaka-yaka-butter) u. for soap mfr. - **D. heckeli** A. Chevalier (w tropAf): sds. source of Dumori butter (40% of kernel), u. as food (Af natives) (but press cake toxic); excellent wood (African mahogany, Makore wood) (—> *Tieghemella*).

dum palm: doom p.

Dunalia arborescens Sleumer (Acnistus a. Schlecht.) (Solan.) (nSA, CA, WI): wild tobacco; quiebra ollas (SC wd. often breaks when burned); lvs. u. with lard as emoll. & in rheum. (Pe); lf. u. in hemorrhoids.

dunbible weed: *Ambrosia elatior.*

dundaki: doundake.

dung: fecal matter; solid excrement of (higher) animals - **d. mushroom**: one growing exclusively on d.; ex. *Psilocybe cubensis.*

Duomycin: generic or non-proprietary name for Aureomycin (*qv.*).

Duotal™: guaiacol carbonate.

dura (Arg): kind of millet (*Pennisetum glaucum*); *Sorghum bicolor,*

duramen: the heart-wd. of a dicot. tree; characterized by its hardness, toughness & darker color fr. the younger "sapwood" (SC *durus*, L., hard).

Durandea monogyna (Linaceae) (NG): bk. decoc. given to dogs before a fight to make them more fierce (Zimakani tribe) (2261).

Duranta repens L. (Verben) (tropAm, cult. As, PI): golden dew drop; frt. toxic but has been u. as antipyr. & stim.; lf. inf. diur.; cult. as hedge pl. (Ind); ornam. shrub.

durazno (Sp): peach.

Durchwachs Oel (Ge): infused oil of hyoscyamus - **Durchwachsdost** (Ge): *Bupleurum rotundifolium.*

durian: Durio zibethinus Murr. (Bombac.) (seAs): tree produces large, delicious, edible but foul-smelling frt.; frt. latex said aphrod.; rt. decoc. antipyr.

Duroia hirsuta K. Schum. (Rubi.) (Co): bk. of small tree u. to bind on arms for cicatrization (Putamayo Indians); c. vesicant principle, leaving brown stain lasting several months (31); pl. toxic to other plants.

durra: dura.

durum wheat: *Triticum turgidum.*

Durvill(a)ea antarctica (Cham.) Hariot. (D. utilis Bory) (Durvillae. - Phaeophyc. - Algae) (antarctic regions; coasts of Austral, NZ, SA): bull kelp; cochayugo; "sea moss"; "South American Sea pods" (large kelp-like pl. -10 m long); "cochayugo" (Ch); large kelp-like pl (-10 m); raw material for prepn. alginic acid (c. 30%); u. fertilizer (promotes growth); u. for scabies (Maoris); u. for making jellies (Ch).

Dussia cuscatlanica Standl. (Leg) (sMex - CR): cashal (Salv); important lumber tree - **D. macrophyllata** Harms (sCR, wPan): citron; red sap of bk. u. as purg.; red frt. peel u. as antipyr. & sold for this in native drug shops (Salv).

dust devils (Ark [US] slang): puff balls.

dusting powder, absorbable (USP): cornstarch + MgO (2%); u. as lubricant for surgical gloves &c.

dusty miller: 1) *Senecio cineraria* 2) other genera & spp. incl. *Artemisia stelleriana*; also appl. to spp. of *Centaurea, Primula, Cerastium* 3) taxa of *Primula, Cerastium Centaurea* (SC white woolly yellowish fl. heads).

Dutch drops: D. oil: Haarlem oil - **D. rushes**: *Equisetum* taxa - **D. tea**: var. of St. Germain tea, Laxative Species

dutchie (Jam - Rastafarians): Cannabis.

Dutchman's breeches: Corydalis, *Dicentra cucullaria* - **D. pipe**: *Artistolochia reticulata, A. serpentaria, A. sipho.*

dwa (Urdu): dawa.

dwale: dwayberry.

dwarf: v. elder, larkspur, nettle, palmetto, pine.

dwayah (Urdu): drugs, medicines.

dwayberry: *Atropa belladonna.*

dye: dyestuff: coloring matter which binds to the substrate giving a degree of permanence; ex. cochineal; there are many natural dyes, such as the red dyes obt. fr. roselle, red cabbage; *Synsepalum dulcificum*, &c.

Dyera costulata Hook. f. (Apocyn.) (Mal, Borneo): **jelutong dyera**; latex fr. wounded bk. prod. part of gutta gelutong or g. pontianak ("dead Borneo"); low grade of gutta percha (1499); a mixt. of resins & caoutchouc formerly u. in mfr. rubber; now as chicle substit., base of bubble gums - **D. laxiflora** Hook. f. (Indochin) & **D. lowii** Hook. f. (Borneo): both u. as prec.

dyer's broom: *Genista tinctoria* - **d. bugloss** : *Alkannna tinctoria* - **d. greenweed**: d. broom - **d. weld**: 1) d. broom 2) *Reseda luteola* - **d. woodruff**: *Asperula tinctoria* - **d.** *v. alkanet*: madder; oak; safflower.

dynamite tree (nSA): prob. *Hura crepitans* (SC frt. explodes) - **wild d.**: Derris.

dynorphin: very potent neurotransmitter obt. fr. pork pituitary & duodenum; "slow reversing endorphin"; has regulatory role in analgesia.

Dyschoriste linearis Ktze. (Calophanes l. Gray) (Acanth.) (cTex [US]): narrow-leaf d.; "snake plant"; entire pl. chewed (or bruised in water) & appl. to snakebite, at the same time being eaten or inf. taken; Mex troops were ordered to carry small package of drug in pocket for this purpose (2230).

dysentery root: *Lappula virginiana* - **d. weed**: 1) d. root 2) *Gnaphalium uliginosum* 3) *Cynoglossum morrisoni*.

Dysophylla auricularia Bl. (Lab) (tropAs): sts. u. to treat fever (Indochin); pounded pl. u. as poultice for stomach dis., urin. dis. of children; in headache, diarrh., worms, sore throat (Mal); decoc. to wash wounds (Chin).

Dysosma pleiantha Woodson (Podophyllum versipelle Hance) (Berberid.) (Chin): rhiz. ground into paste u. as antisep. in syph.; decoc. antidote for poisonous snakebite; exter. as linim.

Dysoxylum (*Dysoxylon*) **acutangulum** Miq. (Meli) (Far East) (Sumatra, Java): sd. coats c. nerve poison; wd. excellent u. for furniture, coffins - **D. forsteri** C. DC (Tonga): mo'ota (Tonga): lvs. u. med.; wd. widely u. (Tonga) (86) - **D. fraserianum** Benth. (Tonga, eAustral): rosewood; Australian mahogany; useful wd. - **D. gaudichaudianum** Miq. (D. decandrum Merr.) (PI): bk. decoc u. as emetic (NG), emmen. antipyr. bechic - **D. hornei** Gillespie (Fiji): fresh lvs. squashed in water & drunk for diarrhea & bloody stools - **D. loureiri** Pierre (eAs): wd. VO u. diaphor. cardiotonic, diur., antipyr. (Indoc).

dysphoria (opposite of euphoria): unpleasant feeling as after termination of effects of cocaine.

Dyssodia acerosa DC (Comp.) (Tex [US]): prickleaf dogwood; hb. decoc. u. purg. - **D. anomala** (Canby et Rose) B. L. Robins. (Sonora, Mex): u. int. for stomach dis., catarrh; cataplasm to reduce "el baso" (Mex) - **D. gnaphaliopsis** (Hymenantherum g. A. Gray) (Mex): "lepiana" (Mex); hb. u. in "catarrhs" (AmIndians) - **D micropoides** (DC) Loes. (sTex [US], Mex): woolly dogweed; pl. with lemon odor; u. as poultice for wounds (Mex) - **D. papposa** (Vent.) Hitchc. (sUS, nMex): Mayweed dogweed, u. in headache (snuffing up inf. in nostrils); fever, pain, &c.; lf. inf. u. diarrh., stomachache, colic - **D. pentachaeta** B. L. Rob. var *puberula* Strother (D. thurberi (Gray) Woot. & Standl.) (Ariz, Tex [US], nMex): common dogweed; paraleña (Mex); inf. u. as physic, for constip.; decoc. u. for stomach pains, cough &c. (Mex) - **D. setifolia** B.L. Robin. (sUS, Mex): paraleña; pague; u. in diarrh. constip.; emesis; cough, stomachache; taken in water before meals.

Ee

eagle vine bark: Condurango - **e. wood**: agarwood, *Aquilaria agallocha*.

ear: frtg. spike of cereals (Gramineae), including grains as chief components.

earth, Japan: catechu - **e. nut**: peanut - **e. wax**: ozokerite.

earth: v. also almond; diatomaceous; infusorial; shellac.

earthworm: *Lumbricus terrestris* (*qv.*).

East India: collective name for India (more common previously), Indochin, & Malaya - v. under arrowroot; fish poison; kino; myrrh; nutmeg; sandalwood (oil); tamarind (incl. East Indian) - **E. I. gum**: fr. *Feronia elephantus* - **E. I. fish poison**: Derris root - **E. I. root**: Galanga - **E. Tennessee pink root**: *Ruellia ciliosa*; u. as adult. of Spigelia - **"east west"** (common name nr. Statesville, NCar [US]): *Apocynum cannabinum* (SC rts. supposed to grow in east-west direction).

East Indian: 1) native of EI 2) Asiatic Indian to distinguish fr. Am Indian.

East Indies: term formerly appl. to all of seAs, now generally refers to Malay archipelago (politically, Indonesia).

Easton's syrup: Syr phosphates of iron, quinine, & strychnine (NF V).

eau (Fr-f.): water - **e. de broccoherai**: creosote soln. - **e. de Cologne**: Cologne water; mixt. of aromatic VO, etc. (ex. lemon, orange flowers, rosemary, bergamot, petitgrain, orange); u. as toilet water - **e. des carmes**: melissa water comp. - **e. des créoles** (FrWI): distillate fr. fls. of *Mammea americana*; u. refreshing beverage - **e. sédative de Raspail**: ammoniac mixture camphorated.

eau-de-vie (Fr): brandy.

Ebers' papyrus: Papyrus Ebers: the oldest known (ca. 1500 BC) Egyptian papyrus dealing entirely with pharmacy & medicine; give many formulas utilizing crude drugs; named after the modern discoverer, author-Egyptologist-physician, Dr. Georg Ebers.

Eberthella typhi Schroeter = **E. typhosa** = *Salmonella typhi*.

ebonite: vulcanite; rubber combined with about 30% S, giving a hard black plastic insulation boarding.

ebony: *Diospyros ebenum*+ - **American e.**: wd. of *Brya ebenus* - **e. oil**: FO fr. sds. of *Bauhinia tomentosa* - **Oklahoma e.**: wd. of persimmon tree - **e. wood of Senegal**: *Dalbergia melanoxylon* (African e.).

Ebuli Radix (L.): dwarf elder, *Sambucus ebulus*, rt.

Ebur Ustum (L.): bone black.

ecbalatericin: fr. *Ecballium* sp. of Middle East; u. male contraceptive; lowers pH of sperm fluid medium.

Ecballium elaterium (L.) A. Rich. (*E. agreste* Reichb) (monotypic) (Cucurbit.) (Med reg): squirting cucumber; the only sp. of the genus; fresh frt. juice is obt. elaterium; cucurbitacin E; u. hydragog cath.

Ecbolium viride Merr. (Acanth.) (Ind): rts. u. in icterus, rheum.; amenorrh.; pl. u. in gout, dysuria (—> *Justicia*).

ecdysones: insect moulting hormones; metamorphosis hormones; control insect pupation; also found in some plants (ex. *Polypodium vulgare* [phytoecdysones]); ecdysterone (ß-ecdysone); have a steroid structure (**ecdysteroids**); stim. protein synthesis.

ecesis: establishment of a plant or animal (i.e., naturalizing in a new habitat or geog. range); there is need of proper growth, survival & reproduction.

ecgonine: base fr. which cocaine, truxilline & other alks. may be derived.

Echeveria secunda W.B. Booth (Crassul.) (Mex): hens & chickens; cult. as ornam.; u. by direct appl. of lf. juice to sores, bruises, corns (NM [US] folk practice).

Echinacea (L.): dr. rhiz. & rts. of **E. angustifolia** DC (Comp.) (cNA): (purple) coneflower; echinacea rt. u. alt. diaph. depur.; smoke treatment for headache (Plains Indians); rt. appl. for toothache; sore throat; to induce salivation; non-specific (general) value in TB; for reabsorption of cataract (1288); formerly u. in snakebite, rabies(!); syphilis; skin dis., esp. furnucles; said to be u. by Am Indian for more ailments than any other pl. - **E. pallida** Nutt. (cUS): pale e.; Sampson root; u. similarly to prec. - **E. purpurea** (L) Moench (cUS): purple coneflower; u. similarly to prec.; rt. & rhiz. u. exter. & int. in various "septic" states as furunculosis, carbuncles, paronychia, &c.; for cough, dyspep. (tincture); also in mastitis; cancer(!?); stim. immune system; hb. also u. (Eu Homeop.); said to prevent cataracts; cult. as ornam.

Echinocactus Link & Otto (Cact.) (swUS, Mex, SA): barrel cactus; spine c.; viznaga; ca. 10 spp. (originally some thousand spp. so classed are now transferred to different genera, ex. *Ferocactus*); pls. u. as source of water in desert; sts. and/or frts. of some spp. edible - **E. hamatacanthus** Muhlenpfordi = *Ferocactus h.* - **E. ingens** Zucc. (Mex): hedgehog cactus; Mexican giant barrel (cactus); body cut into sections like American cheeses & sent to candy factory, where boiled with sugar to produce "dulce" (like preserved citron) (similarly *E. horizonthalonius* Lem.); frt. eaten with sugar & cream - **E. lewinii** Schum. Hennings = E. williamsii Lem. = *Lophophora williamsii* - **E. oxygonus** Link & Otto (sBr, nArg, natd. sAf): "prickly-pear" (Af); slices (after removal of peel & spines) appl. to burns, scalds, ulcers; soothing effect; draws lymph fr. blisters (Af) - **E. texensis** Hopffer (Homalocephala t. Britt. & Rose) (Tex-NM [US], nMex): horse crippler cactus; devil's pincushion; devil's head cactus; manca caballo (N.L.) (Mex); farmers u. pl. to build fences; cult. as oram. ("curiosity").

Echinocereus blanckii Palmer (Cact.) (sTex [US], nMex): ornam. cult. for its large attractive fls. - **E. chloranthus** Engelm. (Mammillaria chlorantha Engelm.) (Utah to Calif [US]): fishhook cactus; raw frt. edible - **E. engelmannii** Parry (swUS, nwMex): golden hedgehog cactus; slab salted, roasted over fire until juices run out, then wrapped in cloth & appl. to painful areas (Seri Indians) - **E. enneacanthus** Engelm. (sTex, NM [US], nMex): strawberry cactus (var. *stramineus* L. Benson [E. stramineum Engelm.]); cob c.; frt. edible, said to be delicious as strawberries (223); eaten raw & to make jams (sTex) - **E. reichenbachii** (Terscheck) Haage (Cereus caespitosus Engl. & Gray) (sUS, nMex): lace echinocereus; lace cactus; cob cactus; u. med.; frt. edible as for many other *E.* spp.; cult. as ornam. - **E. salmdyckianus** Scheer (Mex): u. med. (Mex) - **E. triglochidatus** Engelm. (wTex, NM, Colo [US]): pitajaya; fls. u. for edema, swellings (decoc.) (NM [US]).

Echinochloa crus-galli (L) Beauv. (Gram) (eHemisph, warm regions, as Ind; NA): barnyard grass; cult. for green feed, hay; pl. u. to check hemorrhages & in splenic dis. (Ind); ointment u. for cancer (Wales) - **E. frumentacea** Link (widespread): Japanese (barnyard) millet; cult. for forage, fodder, silage, hay ("billion dollar grass"); cereal grain (As); pl. u. in constip., biliousness (Ind) - **E. crus-pavonia** Schult. (WI, Mex-Arg): capin (arroz); u. in seasickness (Ve): forage grass.

Echinocystis lobata (Mx) T & G(l) (Cucurbit.) (NA, incl. sCalif): wild cucumber; "creeper"; vine rts. u. as poultice in headache (Am Indian); decoc. as bitter ton.; in "love potions"; screen & verandah vine - **E. marah** Cogn. = *Marah gilensis* - **E. oregana** Cogn. = *Marah oreganus*.

Echinodorus grandiflorus Micheli var. *clausenii* (Alismat.) (Br): burhead; lvs. u. antiarthr. antirheum. diur. in skin dis. - **E. macrophyllus** (Kunth) Micheli (seUS, WI, Mex, CA, Ec, Guianas, Br): water plantain; chapeo de couro (Br); lvs. u. med. similar to preceding.

Echinomastus Britton & Rose = *Neolloydia*.

Echinopanax Decne. & Planch. = *Oplopanax*.

Echinops L. (Comp.) (Euras, Af): glove thistles; with coarse, spiny pinnatifid lvs. - **E. amplexicaulis** Oliver (eAf): rt. c. mucilage; appl. topically to chest for cough - **E. echinatus** Roxb. (Ind): pl. u. as diur. ton. in hoarse coughs, hysteria, dyspepsia, scrofula, ophthalmia; rt. u. against lice, maggots - **E. latifolius** Tausch (E. dahuricus Fisch.) (Chin): u. in gout, new growths, anthelm. against ascarids; rt. u. emmen. (Indochin) - **E. persicus** Stev. & Fisch.(Iran): source of trehala manna; prod. by action of parasitic beetles; c. sugars, starch - **E. ritro** L. (Med reg.; Russia): frts. c. echinopsine (*cf.* strychnine); sd. source of FO, echinops oil - **E. spinosissimus** Turra (E. viscosus

DC) (Med reg.): source of gum ("angadeo mastiche") u. for chewing (Greece).

Echites P. Br. (Apocyn.) (trop NW): some 30 spp. of woody climbers; milky juice acrid, narcotic, toxic; u. in med. - **E. longiflora** Desf. (Br, Ur): sd. hairs as vegetable silk - **E. peltata** Vell. (Br): cipo caboclo; paina de pen(n)as (Br); lvs. & tops u. in orchitis; source of a manila-like fiber - **E. religiosa** Teijsm. & Binn. (Thailand): a source of indigo - **E. tubiflora** Mart. & Gal. (Guerrero, Mex): hierba de la cucaracha; u. to kill lice.

Echium plantagineum L. (sAustral. introd. Calif): viper's bugloss; salvation Jane; c. alk. said toxic to sheep (Austral) - **E. vulgare** L. (Boragin.) (Euras, natd. in NA): blue weed; pl. u. emoll. protective; diur. exp.; to "purify the blood"; bad weed.

echujin: glycoside of digitoxigenin with D-cymarose & 2 glucose.

Ecklonia kurome Okam (Alari.-Phaeophyc.-Algae) (wPac Ocean near Chin, Jap): seaweeds served for human nutrition (several other *E.* spp. similarly u. in Orient).

Eclecticism: an Am medical system which supposedly selected the best from all other systems of medicine; it treated dis. by the use of single medicines ("simples") mostly representing indigenous Am plants, such as Stillingia; eclectic remedies were at times called "specifics."

Eclipta prostrata (L) L. (E. alba Hassk.; E. erecta L.) (Comp.) (trop worldwide): American false daisy; maco; inf. u. for whooping cough, cough (Trin); rts. u. as food (SA); lvs. as vegetable; boiled with congee (rice gruel) for indig. (Chin); lf. decoc. to dye hair; in diarrh., tonic in skin dis., toothache.

ecology, plant: relation of plants to conditions under which they grow (effects of soil, climate, other organisms on external form, internal structure & physiology of plant).

écorce (Fr): 1) bark 2) cortex 3) peel; rind (of frt.) - **é. elutherienne** (Fr): cascarilla - **é. sacrée** (Fr): cascara sagrade.

ecotype: ecological race: group of individuals of a sp. fitted to survive under a certain environment (climatic, edaphic).

ecytonin: antibiotic subst. isolated fr. *Microclona prolifera* ("red-beard sponge") (1807).

edaphic: influenced by soil conditions as opposed to climatic.

edaphology: soil science.

eddas: eddo (pl. eddoes) (WI): *Colocasia esculenta,* &c.

Edelsorte (Ge): domesticated, cult. vars.

edera terrestre (It): ground ivy.

edestin: globulin (protein) occ. in linseed, hempseed, cottonseed, &c.

edgeweed: *Oenanthe aquatica*; *Aster concolor.*

Edgeworthia papyrifera Siebold & Zucc. (E. chrysantha Lindl.) (Thymelae) (Chin; cult. Jap; natd. seUS): paper bush; bk. u. to produce high grade paper (Jap) (checks, &c.) (Japanese paper).

edo: eddo.

eels: fishes of several genera incl. *Anguilla* (fresh water), *Conger* (marine): *Muraena muraena* L. (Muraenidae) (Med Sea): Roman eel; moray; with voracious predatory habits; domesticated by ancient Romans & u. for pets & as food - **Anguilla anguilla** L. (A. vulgaris Turt.) (Anguillidae - Pices) (Euras): Europe eel; u. as food; in Unani med. as nutr. ton. & in TB (2763).

eelgrass: *Zostera marina.*

effluent = eluatc.

egg albumen (albumin): fresh e. white is u. to antidote mercuric chloride poisoning; dr. e. albumen (prepd. by desiccation) is u. for clarification, mfr. cements, plastics resembling ivory; reagents; constitutes ca. 68% of the protein portion of the egg white - **e. dye** (Ala [US]): *Coreopsis major* - **hen's egg**: ovum gallinaceum (L); the recently laid egg of this domestic fowl, *Gallus domesticus*; is u. as nutr., however, because of cholesterol content in yolk, feared by some as causative agt. of hypercholesterolemia; however, this may depend on mode of prepn. (1818); egg shell u. as antacid (formerly); egg white nutr. & in Canada Liniment; e. yolk u. as emulsifying agt., prepn. St. John Long's Liniment; fertile eggs (6- to 7-day-old embryos) u. in prep. various vaccines; may be preserved; culinary item, emulsif. - **e. oil**: FO fr. egg yolk; c. fat glycerides, cholesterol, lecithin; u. in ointment bases (med. & cosmetic formulations) (1796) - **e. plant**: 1) *Solanum melongena* (SC shape of frt.) 2) *Symphoricarpos* spp. 3) *viburnum opulus* (early Am pioneers) - **e. shell**: the calcareous covering of the hen's egg; c. $CaCO_3$; u. antacid, food supplement - **vegetable e.**: *Lucuma nervosa* - **e. white injury**: vitamin deficiency evoked by ingestion of excessive raw egg white.

eglantine (Eng fr. Fr): wild rose, *Rosa canina*; R. eglanteria L. (red e.).

ego oil (Jap): FO fr. *Styrax japonicus.*

Egyptian: v. under grass; gum; privet; thorn.

Ehretia laevis Roxb. (Ehreti/Boragin.) (nwInd): parki; c. tannin, saponins; rt. ext. u. as antipyr. - **E. microphylla** Lam. (Ehreti./Boragin.) (seAs, PI): Philippine tea; santing (Sulu); lf. decoc. u. beverage tea substit. (PI); for diarrh., dys., stomach dis.; for cough - **E. navesii** Vidal (PI): u. as antipyr. for intest. pain; bk. decoc. for cough; appl.

to athlete's foot; lvs. appl. to ringworm - **E. philippinensis** A DC (PI): alibungog (Cebu Bisaya): bk. & lvs. u. anti-inflamm. (locally); intern. for diarrh. dys. (tenesmus) - **E. saligna** R. Br. (wAustral): false cedar; coonta; wd. decoc. drunk for aches & pains - **E. tinifolia** L. (Mex, CA, WI): roble (prieto) (Cu); lf. decoc. u. in mouthwashes for pyorrhea; to bathe wounds & sores; frt. edible; shade tree.

Ehrlichin: antibiotic from *Streptomyces lavenduae* Waksman & Henrici; u. against viruses (Waksman, 1951) (named after Paul Ehrlich).

Eibe (Ge): *Taxus baccata.*

Eibisch (Ge): marshmallow.

Eiche (Ge): oak.

Eichelkaffee (Ge): oak (acorn) coffee (substit.).

Eichenmistel (Ge): *Loranthus europaeus.*

Eichhornia crassipes Solms-Laub (Pontederi.) (orign. SA, natd. trop & subtrop OW, NW incl. seUS n to Mo & Va): water hyacinth; green parts & inflorescence cooked & eaten; fls. u. for dis. (horses) (Java); the floating hb. a troublesome weed in Fla (US), Java, Bengal (Ind), Austral, Af, &c.; cattle fodder; to purify sewage water (SDak, Calif [US]); suggested as source of cellulose paper, paper board & biogas; hogfeed (Indones); forms islands in the water & is very bad pest to water navigation in some areas (seUS, tropAf); c. 3.7% org. matter, 95% water; ornam. floating plant.

eicosanoids: certain polyunsaturated essential fatty acids found in FO (as peanut o.), eggs, dairy prods., &c. (3/5 series in codliver oil, other fish oils, linseed oil) (<eicosanoic acid = arachidic a.).

eicosapentaenoic acid (EPA): analog of arachidonic acid; occ. in marine FO (fish, esp. mackerel & marine oils) (may originate fr. phytoplankton); u. to reduce plasma cholesterol & inhibit platelet aggregation to prevent coronary thrombosis (1290); derived fr. prostaglandins; Greenland eskimos consume large amts. of saltwater fish & show low incidence of cardiovascular dis.

Eierfrucht (Ge): egg fruit.

Einkorn: grain of *Triticum monococcum.*

Eisenhut (Ge): **Eisenhuettel** (Ge): aconite - **Eisenkraut** (Ge): *Verbena hastata.*

Eiweisz (Ge n): albumen; egg white protein.

ejoo fiber: ejow; fr. *Arenga pinnata.*

ejote 1) (Sal): common bean 2) (Mex): string bean.

Ekebergia capensis Sparrm. (Meli) (sAf): dog plum; bk. u. emet. tan; expect.; rt. u. in dys.; as expect. - **E. meyeri** Presl (tropAf): Cape ash; bk. u. in dys. heartburn; rt. decoc. for chronic cough, emet.; headache; lf. u. annthelmint. in scabies - **E. senegalensis** A. Juss. (trop wAf): bk. sold in markets for GC, stomach.; rts. for sterility; antisyphil. antilepric; aphrod.

elachi (Ind): cardamom.

Elaeagnus L. (Elaeagn.) (Euras, NA): silver berries; frt. of some spp. (ex. *E. umbellata*) edible; ornamental shrubs - **E. angustifolia** L. (E. hortensis Bieb.) (Euras): oleaster; Russian olive; Persian dates; small tree; fls. antipyr. frt. yellowish brown, -1 cm long ("dates," NM [US]) (flavor); pulp of some vars. edible ("Trebizond dates"); flesh sweet, mealy, sl. acid; odor prune-like; flesh dr. & made into flour & bread; FO (Oleum sanctum) u. med. (probably the "oil tree" of Old Test.); frt. c. tannin; u. diarrh. antiphlogistic (vet. med.) (homeopath. med.); ornam. shrub or small tree, cult. for fruit (Kashmir) - **E. commutata** Bernh. (E. argentea Pursh) (NA): silverberry; wolf willow; lemon-yellow fls. with heavy scent; silver-colored dry mealy berry with large stony sd.; frt. edible; ornam. shrub - **E. latifolia** L. (eAs): frts. & fls. u. as astr. (Burma) - **E. umbellata** Thunb. (Jap+): frts. scalded & eaten (Ainu); fls. stim. cardiac, astr.; sds. u. as cough stim.; FO of sds. u. in pulmonary dis. (Ind); ornam. shrub.

Elaeis guineensis (L) Jacq. (Palm.) (w tropAf, natd. Br): (African) oil palm; wine p.; occurs in many vars.; tender parts of st. yield juice from which are produced palm sugar, toddy (palm wine [malovu]), & arrack; frts. are source of 2 different kinds of FO; one fr. frt. pericarp called palm (or pericarp) oil ("p. butter"), the other fr. sd. called palm seed or kernel (or white) oil; latter u. for mfr. margarines ("palm" or "plant butter"), as liniment, & salve base in suppositories, for culinary uses, nutr. (children); prepn. stearin; hydrogenated u. in making candies, esp. caramels; former u. for lubrication (crude cheap) (1806), mfr. curd soap, in tin plate indy., candle making, also as fuel in int. combustion engines (diesel); in treatment of keratomalacia; this red orange mass or liquid has also been u. as source of carotene & vit. A & vit. D (codliver oil substit.), in mfr. chocolate, folk medical practice (Zaire); rts. u. diur. (wAf) & to relieve intestinal pain (Senegal); fresh st. juice u. as lax.; kernels of some vars. edible & u. in mfr. of flour (1568); lvs. & lf. stalks u. in building & thatching native houses; lf. ash u. to furnish salt (Ubangi); u. med. & as condim.; trees are cult. in Congo, Mal, Br+ - **E. melanococca** Gaertn. (Af): dende di Oara; palmeira caiaue; similar properties of oils; u. locally. - **E. oleifera** (HBK) Cortes (sometimes wrongly called *E. melanococca* auct.) (SA, CA): (American) oil palm; noli palm; batana (Mesquite Indians); corrozo (Colorado) (CA, Ve) (name also appl. to some other palms, such as

Phytelephas); u. similar to prec.; lf. fibers much u.; sd. kernels furnish FO u. for hair oil; frt. u. to make chicha; frt. flesh yields ca. 29% FO (palm oil) & c. 45% palm kernel fat (noli palm fat or cayaut oil); refined oil u. in margarine mfr.; fatty material fr. frt. u. for dandruff, also as scalp ton. & to prevent hair fr. falling out or turning grey (Co).

Elaeocarpus angustifolius Blume (Elaeocarp.) (Indones, Oceania): hando (New Caled) sds. u. to make necklaces & Hindu rosaries - **E. copaliferus** Retz. (OW): sds. c. 50% FO, similar to Malabar tallow - **E. dentatus** Vahl (NZ): bk. (20-22% tannin) u. as tan.; source of blue-black dye - **E. graeffei** Seem. (Fiji): lf. ext. u. for stomach dis. - **E. grandiflorus** Sm. (Indochin, Burma, Indones [Java], PI): bk. of large trees u. as poulticing for ulcers, women's dis.; lvs. & sts. u. intcr. as tonics; sds. said to cont. a heart poison - **E. lancaefolius** Roxb. (Ind): u. med. - **E. oblongus** Gaertn. (Ind): frts. u. as emet., in rheum. pulmonitis, swellings, leprosy, edema; some consider conspecific with *E. serratus* - **E. serratus** L. (Ind): frt. (Ceylon olive) edible; u. in curries; u. in dys. diarrh.; lvs. rich in vit. C; u. in rheum. & antidote - **E. sphaericus** (Gaertn.) K. Schum. (E. ganitrus Roxb.) (As-eNepal to Pl): East Indian lime tree; acid frt. flesh u. in epileptic seizures & cranial dis.; frt. stones u. in necklaces (1529); c. several alks. (2731) - **E. tuberculatus** Roxb. (Ind): rudraksha (Sanscr); u. in cholera, prolonged fevers (enteric fever) (Ind); sds. for rosary beads.

Elaeodendron glaucum Pers. (Celastr.) (Ind): crushed rts. as cold water inf. u. emet.; rt. bk. in paste appl. to swellings; pd. rts. as snuff in headache; ornam. tree (—> *Cassine*).

Elaeophorbia drupifera Stapf (Euphorbi) (wAf): pl. decoc. u. as ordeal poison in criminal trials (Af); fish poison.

Elaeosaccharum (L.) oil sugar; sucrose flavored with a VO.

Elaeoselinum asclepium Bertol. (Thapsia ascl. L.) (Umb.) (sEu, Med reg.): marsh parsley; "Aesculapias' panacea"; lf. juice u. as anthelm.; dye plant(?).

el(a)eostearic acid: fatty acid found in sd. FO of *Aleurites fordii* (70-80%) & other *A.* spp.

elaion (Gr): oil.

Elaphoglossum petiolatum Urb. (Polypod-Pteridoph.) (sAf): rhiz., decoc. ingested for colds & sore throats.

Elaphomyces cervinus Schroeter (Elaphomycet./Tuber. - Ascomycot.) (Eu mostly, esp. sAustr, Pol, CSR; NA-woods): hart's truffles; rut of harts; false truffle; lycoperdon nut; parasite on tree rts. (pine, beech, oak); mycorrhizal; resembles truffles in appearance; sporophores u. as aphrod. (cattle & swine); galactagog; styptic; narcotic (formerly), ecbolic; spores u. in form of electuary or mixed in beverages - **E. granulatus** Fr. (Lycoperdon solidum L) (Eu, chiefly s; NA): deer balls; u. similar to prec.

Elaphrium Jacq. = *Bursera* Jacq. - **E. simaruba** Rose = *Bursera s.* - **Elasmobranchii: Elasmobranchs**: cartilaginous fishes; chondrichthyes; class of Chordates similar to the true fishes but having a cartilaginous rather than bony skeleton & with different types of scales; ex. sharks, rays.

elastase: proteolytic enzyme of pancreas wh. digests **elastin** of arterial wall; has wide appln. since acts on fibrin, hemoglobin, albumin, casein, etc.; elastin is an insoln. protein found in animal connective tissues (fibers, &c.) u. in skin & hair conditioners (cosmetics).

Elastica (L.): rubber - **elastomer**; rubber-like substance, incl. natural & synth. rubbers.

elaterin: Elaterinum (L.): neutral principle obt. by crystallization fr. soln. of **elaterium**; dr. sediment fr. juice of green frt. of *Ecballium elaterium*; greenish to gray masses; active principle elaterin (20-30%); u. drastic cath. (danger!) - **English e.: white e.: Elaterium album**: dr. sediment fr. juice - **elatine** (Fr): elaterin - **B-e**: cucurbitacin E.

Elatostema bahamaense C. B. Robinson (Urtic.) (WI): mashed lvs. sts. & frt. u. for body itch (Bah) - **E. platyphyllum** Wedd. (Ind): u. as antiemetic - **E. repens** Hall. f. (eAs): u. as poultice for swellings, boils, painful abdomen (Mal) - **E. umbellatum** Bl. (& its var. *majus* Max.) (E. involucratum F & S) (Jap): midzu-na (Jap); lvs. u. as culinary veget. (Akita, Jap); rt. prescribed for colds, flatulence; worms (Chin).

elayotechnia (Sp): technology of oils.

elbow bush: *Cephalanthus occidentalis*.

elder, American: *Sambucus canadensis* - **American red e.**: *S. pubens* - **elder bark, e. berries**: *Sambucus* spp., usually *S. nigra*, also *S. canadensis* - **black e.**: Frangula - **black-berried e.**: German e. - **box e.**: *Acer negundo* - **common e.**: *S. nigra* (Eu); *S. canadensis* (NA) - **dwarf e.**: 1) *S. ebulus* 2) *Aralia hispida* - **European e.**: *Samb. nigra* - **e. flowers**: 1) American: fr. *S. canadensis* 2) (Ge): fr. *S. nigra* - **marsh e.**: *Viburnum opulus*; *Iva* spp. - **poison e.**: *Rhus vernix* - **prickly e.** (US): *Aralia spinosa* (bk.) - **red e.**: 1) (US): *Alnus rugosa* (misnomer for "alder") 2) *Cunonia capensis* 3) (extra-US): *S. nigra* - **rose e.**: *Viburnum opulus* - **scarlet e.**: *S. pubens* - **stinking: e.: swamp e.**: any *Sambucus* spp. (RMH) - **sweet e.**: *S. canadensis* -

water e.: marsh e. - **wild e.**: *Aralia hispida* - **yellow e.**: *Tecoma stans*.
eleboro (Sp): hellebore.
elecampane: *Inula helenium* (fr. elena, L. *Inula* + campana, L., field).
electre: Electrum (L.): (fossil) amber.
electricity: a stream of electrons in a conductor actuated by a difference of potential; serves as source of power - **electrical stimulation:** u. for pain in leg or low back pain (Univ Neb [US], Coll. Med.).
electrology: technique of using electric current appl. with electrodes to remove unwanted hair, warts, moles, birthmarks, &c.
electromicroscopy: the technique of using the electron microscope, in which images are formed by means of electrons rather than light beams & focused by electron lenses; very high magnifications may be attained.
electron (Gr): **electrum** (L.): amber (< electra, brightest).
Electuarium (L.): confection, a soft saccharine prep. made fr. dr. medl. agts. - **E. Dentifricium**: toothpaste.
electus, a, um (L): select, chosen (*cf.* Eclectic med.).
eledoisin: undecapeptide found in posterior salivary glands of some octopi; acts as hypotensive, vasodilator, smooth muscle stimulant.
elegueme (CR): *Erythrina corallodendrum*.
elemene: sesquiterpene hydrocarbon in *Calamus* and juniper VO.
elements, minor: trace e.: those chem. elements present in animal tissues in very small amts., either essential (ex. Cu, Mg, Mn, Co, Zn) or harmful (ex. Hg, Pb.).
elemi: collective or generic term for many different oleoresins of varying origins; originally pale liquids, they usually become hard on standing, although some may remain quite soft (elemi fr. spp. of *Icica*, *Bursera*, *Pachylobus*, *Ancomea*, *Amyris*, etc.); the most important one is Manila e. - **African e.**: fr. *Boswellia frereqaana*, *Canarium* spp. - **Almessega**: Brazilian - **American** (SA, WI): *Amyris elemifera*, *A. balsamifera* & related *A.* spp.; *Bursera gummifera* - **American Yucatan**: Yucatan - **Brazilian: e. of Brazil**: fr. *Protium heptaphyllum* - **Cameroon**: African - **carana** (Ve, nBr): **caricari** (Br): fr. *Protium carana* - **colophonia**: Mauritius - **dry**: fr. *Dacrodex hexandra* - **East Indian** (Moluccan): fr. *Canarium commune* (perhaps) - **gum**: fr. *Canarium luzonicum* (*C. commune*) - **Kamerun**: Cameroon - **Manil(l)a**: gum - **Mauritius**: fr. *Bursera paniculata* - **Mexican**: fr. *Amyris* spp., incl. *A. elemifera*; also some fr. *Bursera* spp. - **occidental**: Rio oil (PI): VO fr. e. gum - **Rio** (WI, Br): fr. *Protium icicariba* - **tacamahac(a)** (PI): fr. *Canarium villosum*: some types of unknown origin - **Uganda**: African - **West Indian**: Rio - **W.I. Yucatan: Yucatan** (WI): fr. *Amyris elemifera*, *A. plumieri* & *A. dacryodes*.
elemicin: phenolic compd. very similar to myristicin; occ. in elemi resin (fr. *Canarium commune*); VO of *Cinnamomum glanduliferum*, *Cymbopogon* spp., *Myristica fragrans*, &c.; with hallucinogenic props.
elenolic acid: carboxy acid found in aq. exts. of various parts of pl. of *Olea europaea*; Ca elenolate is virucidal; lactone ester, **elenolide**, is hypotensive.
Eleocharis acicularis Roem. & Schult. (Cyper.) (Chin): decoc. of pl. with brown sugar drunk for measles - **E. dulcis** Trin. ex. Henschel (E. tuberosa Schult.; Scirpus esculentus L.) (eAs): (Chin) water chestnut; water nut; grows in water; corm u. in eye dis., measles; perenn. hb. u. hemostatic, pith as diur.; more important as food; source of starch; solid corm much eaten by Chin (cult.); cut into slices (disks) in chow mein (600) and chop suey - **E. rostellata** (Torr.) Torr. (NA): beaked spikerush; u. as emet. (Navaho Indians) - **E. sphacelata** R. Br. (Austral): maya; spherical tubers eaten raw by aborig.
eleoptene: liquid part of VO (rm. temp.).
eleostearic acid: unsatd. fatty acid fr. FO of sds. of *Elaeococca vernicia*; forms occur in tung oil.
elephant: *Elephas* sp. (**Indian e.**, &c.) & *Loxodonta* sp.) - **African e.**: *Loxodonta* - **e. louse**: cashew nut - **e. root**: *Erigeron canadense*.
Elephantopus elatus Bertol. (Comp.) (seUS): elephant's foot; lactones elephantin & elephantopin carcinostatic to human nasopharynx carcinomas (224) - **E. mollis** HBK (Mex, natd. Oceania): "contrayerba"; lvs. appl. to wounds as vulner.; pl. diur. antipyr. (PI) - **E. scaber** L. (seAs, Oceania, seUS, tropAf): elephant's foot; lvs. u. vermif. aphrod., diur., antipyr. ton. in dys.; pl. a bad weed; tea u. in headache (Veadeiros, Br); pounded rts. appl. to boils (Tanz); lvs. u. in roundworm; c. elephantin, elephantopin (sesqueterpenoid tumor inhibitors) - **E. spicatus** Juss. (Pseudelephantopus s. [Juss.] Rohr) (sMex to Ve, nCh; WI): bulltongue bush (Virgin Is); devil broom (Trin); canasacanga (Cuna Indians, Pan); decoc. u. for fever, gastralgia, VD (Pan); for colds & heart trouble (Virgin Is); rt. inf./decoc. in diarrh. dys. (Salv, Trin); crushed leafy parts as poultice for swellings (Pan).
Elephantorrhiza burkei Benth. (Leg) (sAf): elandsboontijie (Afrikaans); rt. bk. u. as astring., rt. as tan. - **E. caffra** (sAf): remedy for GC, earache - **E. elephantina** Skeels. (tropAf): rt. u. dys. di-

arrh., fevers; cardiac remedy; inf. u. in hemorrhoids.

elephant's ear: *Begonia grandiflora* & other *B.* spp. (rt.); also appl. to *Colocasia (esculenta), Philodendron hastatum, P. pertusum; Enterolobium cyclocarpum*, &c. - **e. foot**: 1) *Elephantopus scaber* & other *E.* spp. 2) (Ind): *Amorphophallus campanulatus* 3) *Testudinaria elephantipes.*

Elephas maximus L. (E. indicus Cuvier) (Elephantidae; Proboscidea; Mammalia) (Ind e to Borneo): Indian elephant; domesticated animal u. in labor, esp. lumbering operations; for transport; in circuses; tusks (ivory) u. in ornam. work, very valuable; these & e. bones were distd. to produce oil u. as linim for spasms, paralyses, sprains, nervous weaknesses, &c.; calcined bones considered superior to calcined hart's horn; e. milk u. med.; skin u. as plaster (Chin); teeth u. against anuria & polyuria (Far East) (2349) (to reduce slaughter of elephants in Af, US & other nations are now prohibiting importation of tusks).

Elettaria cardamomum (L.) White & Maton (Zingiber.) (s&wIndia; cult. in Ind [1292], SriL, Jam, Guat): cardamom; cardamon (seeds); frt. a capsule enclosing sds.; sd. c. VO (4%), FO (1-2% sd., 10% frt. wall); sd. & VO u. flavor, carm. condim. stim. arom. stomach; vars. *minuscula* Burk (a-minor): smaller frts: b major Sm.: larger frts; wild Ceylon cardamom; following are trade vars.: Alleppy; Bengal; Ceylon-Malabar; Ceylon-Mysore; Guatemala; Java; Malabar; Mangalore; round; r. Chin; wild Ceylon; winged Java; bleached; unbleached (green); sds. are off. but must be recently removed fr. frt., hence generally entire capsules are encountered in retail trade but sometimes decorticated; VO obt. by steam distn. fr. sd. has penetrating arom. odor & taste; u. stim.; flavor (cakes), to flavor coffee (Bedouins, neAf); condim. (curries); perfume (in lily-of-the-valley type aromas), incenses, fumigating powders.

Elettariopsis sumatrana Valet. (Zingiber.) (Indones; Mal): u. for flavoring (Sumatra); juice u. for scorpion sting.

Eleusine coracana Gaertn. (Gram) (Ind, Af; cult. NA): African millet; ragi; cereal grain eaten (rich in protein); u. in prepn. of alcoholic beverage - **E. indica** (L) Gaertn. (pantrop as Ind, US): goose grass; fowl foot (Trin); lf. & st. decoc. u. diur. in cystitis, pneumon. (Trin, Nepal); poultice fr. crushed pl. with salt appl. to dislocations, sprains (Dominica); dys. (PI); entire pl. sudorific (Cambod); famine food; common weed.

Eleutherine bulbosa Urban (E. palmifolia Merr.); (Irid.) (tropAm, incl. WI, SA; natd. PI, cult. sAf): "lagrimas de la virgen"; bulbous hb.; bulb decoc. diur. (PI); vermif. (Ec); bloody diarrh., insect stings, boils, wounds, stings of poisonous fishes; hallucinatory(?); at menopause; macerated bulbs appl. to stomachs of children for gas pains (PI), "charley horse" (PI); inf. drunk with ext. of *Dichromena ciliata* (Scrophulari) to reduce fertility (Pe); bulbs u. for irreg. menstr., menopause; inf. - **E. palmifolia** Merr. (E. plicata Herb.) (tropAm, natd. PI): bulb decoc. diur. (PI), for menorrhagia; macerated bulbs appl. to stomachs of children for gas pains (PI), "charley horse" (PI), for rheum (Mart); insect stings, boils, wounds, sting of poison fishes.

Eleutherococcus senticosus Maxim. (Arali.) (As, Far East, Siber, near coast; nChin, Jap, Kor): Siberian ginseng; eleuthera (ginseng) (2324); devil's bush; rt. c. eleutherosides (A-G) (saponin glycosides); rt. lvs u. stim. ton. adaptogenic (to increase body resistance by activating immune systems); to treat carcinoma & diabetics (1293) (popular in USSR).

Elfenbein: 1) ivory 2) kind of vegetation in alpine regions - **E. palme** (Ge) (ivory palm): *Phytelephas macrocarpa.*

elfin-tree: dwarfed tree of **elfin-wood**; alpine forest where trees develop much wood & little foliage; incl. many different pl. spp., ex. *Rhododendron, Pinus*, etc.

elfwort: Inula.

Elionurus: *Elyonurus.*

elixir: sweetened aromatized liquid, usually with alc., generally u. as vehicle or base for medicines (fr. al-iksir, Arab, essence; philosopher's stone) - **e. adjuvans**: licorice e. - **e. ad longam vitam** (L.): Tr. Aloe & Myrrh. - **e. Aperitivum**: Tinct. aloe comp. - **e. calisaya alkaloidal**: = **e. cinchona**: c. alkaloids e. - **e. corrigens**: aromatic eriodictyon e. - **Daffy's e.:** Tinct. senna comp. - **e. gentianae ferratum**: gentian & iron phospate e. - **e. of long life**: modifn. of Swedish bitters (Tinct. aloe & myrrh.) - **e. of vitoril**: sulfuric acid e. arom. - **e. Paracelsus: e. proprietatis**: Tr. Aloe & Myrrh - **e. Purgans**: Tinct. jalap comp. - **red e.:** red aromatic e. - **e. salutis**: senna tinct. comp. - **e. Stomachic**: Tinct. gentian comp. - **e. Viscerale**: Tinct. orange peel + Tinct. gentian comp.

Elizabetha princeps Schomb. (Leg) (SA-Amazonia): bk. ashes u. with hallucinogens fr. *Virola* & *Justicia* spp.; may contribute to activity (1903); prepn. "yopo" (rapá de India).

elks: taxa of *Alces* L. ruminants (Cervidae) (n part of nTemp Zone): heavy animals with large flat antlers borne only by male (very different from horns); u. med. by Chin; best known is *A. alces* L. (nEuras): European elk; domesticated in earlier times (Eu) & u. for riding, pulling heavy loads & milking; very similar to *A. americana* L. (NA), American moose, large animal (-630 kg); *A. gigas,* Alaskan moose (nwNA) still bigger (-820 kg) (do not confuse American elk, a large deer, *Cervus canadensis* L.) (some authors include all 3 true elks under the name *A. alces*).

elk bark: fr. *Magnolia glauca* - **elk's claw**: fr. *Cervus canadensis*; u. folk med.

ellagic acid: a lactone found free & combined (**ellagitannins**) in nutgalls, divi divi, walnut lvs., uva ursi, jambul frts., tamarinds, etc.; less HOH-sol. than tannin; u. intest., astr. hemostatic.

ellebore vert (Fr): Veratrum Viride - **elleboro** (It): hellebore.

ellipticine: anti-tumor alkal. isolated fr. *Ochrosia elliptica.*

elm: *Ulmus* spp. - **American e.**: *U. americana* - **e. bark**: inner bk. fr. *U. rubra* - **bastard e.**: yellow e. - **Chinese e.**: *U. parvifolia* - **cork(y) e.**: *Ulmus alata, U. thomasii* - **Dutch e. disease**: lethal infection of American e. tree by *Graphium ulmi* (SC first discovered in Netherlands) - **English e: European e.**: *Ulmus procera* (bk.) - **false e.**: *Celtis laevigata* - **e. family**: Ulmaceae - **Indian e.: moose e.: piss e.** (seUS colloquial) (SC: sap spurted when wood is cut): **red e.**: *U. rubra* - **rock e.**: 1) *U. thomasii* (usually) 2) (Tenn [US]): *U. rubra* - **slippery e.: sweet e.**: red e. - **swamp e.**: *Ulmus thomasii* - **water e.**: *Planera aquatica* - **white e.**: American e., sometimes red e. - **winged e.**: cork e. - **witch e.: wych e.**: *U. glabra* - **yellow e.**: *Celtis occidentalis.*

Elodea (Helodea) (Anacharis) (Hydrocharit.): water weeds; water-thyme - **E. canadensis** Mx. (Anacharis c. Planch.) (Br, Arg, natd. NA, Eu): ditch moss; aquat. ornam. in aquaria & garden pools.

elsdizzato (It): hallucinated, as fr. LSD.

Elsholtzia Willd. (Lab.) (Euras, Af): hbs. or small shrubs; usually aromatic (spelling variants: *Els(ch)holzia, Elscholtzia, Elsholtzia,* etc.) - **E. blanda** Benth. (Nagaland, neInd): lf. juice u. for kidney & urinary bladder dis. (1508) - **E. ciliata** Hyland. (E. cristata Willd., E. patrinii Garcke) (Euras, esp. Chin, natd. NA): entire pl. formerly u. for abdom. pain, kidney dis. (Jap); leafy st. carm. stomach. diur. antipyr.; decoc. in epistaxis, VO (c. elsholtzia-ketone) u. in perfume indy.; edema, nausea; decoc. for alcohol hangovers (Ainu).

eluate: liquid washes obtained fr. ppt. by solvent.

eluant: eluent: eluting agent: washing agent.

elution: removal by means of appropriate solvent of substance separated out fr. a mixt. in chromatography.

elymoclavine: alk. fr. *Elymus mollis* ergot; first isolatn. 1955.

Elymus L. (Gram) (in temp zone, SA): lyme grass; wild rye; cult. for ornam. & sand binding - **E. arenarius** L. (Euras): sea lyme-grass; soil binder (sand dunes); forage grass; cult. in Jap to use for mats, ropes, paper - **E. canadensis** L. (E. robustus Scribn. & Smith) (NA, esp. Pac NW): Cana wild rye; good forage & cover crop; u. to rebuild sand dunes; make hay - **E. cinereus** Scribn. & Merr. (wNA): (gray) ryegrass; Cana r. g.; forage & fodder; frt. edible & u. as food by Indians; rts. mashed & steeped in water u. for GC; u. several days; claimed curative (Okanagan Indians); decoc. u. to stop internal bleeding; to wash hair to stim. growth; sts. u. to make coiled baskets (Salish Indians) - **E. virginicus** L. (NA): Virginia wild rye; forage grass.

Elyonurus (*Elionurus*) **argenteus** Nees (Gram) (sAf): lemon-(scented) grass; silk g.; wine g.; u. in colic - **E. latiflorus** Nees (Par, Br): source of espartillo oil; with odor like carrot seed oil; pl. u. by natives to "cure" leprosy (lf. inf.); "espartillo quazu."

Elytraria squamosa (Jacq.) Linden (Tubiflora s. Ktze) (Acanth.) (Tex [US], Mex, CA, nSA): coquillo; pl. decoc. administered to women with VD; bronchitis, pertussis.

"em" (US slang): morphine.

embalaje (Sp): packaging, baling.

Embelia (BPC 1934): E. Ribes Fructus (L.): barabang (frt.); dr. frt. of *E. ribes* - **E. kilimandscharica** Gilg (Myrsin.) (tropAf): bk. & berries u. as purg.; rts. & esp. dr. sds. important anthelmint. in Kilimanjaro area; lethal at times when overdosed - **E. kraussii** Harv. (sAf): frt. & lf. u. as tapeworm remedy (Zulus) - **E. laeta** Mez (Indochin): frts. u. as tenifuge - **E. ribes** Burm. f. (Ind, seAs, eAf): frt. c. embelic acid ("embelin"), FO of frt. very arom. u. alt., carm., diur. anthelm. adult. of black pepper, cubeb; to replace santonica; dyestuff - **E. robusta** Burm f. (Indies): u. as *E. ribes* - **E. schimperi** Vatke (Af): frts. or rts. u. as anthelm.; lvs. & frts. eaten; frt. (berries) best known anthelm. in nTanzania & is sold on the market - **E. tsjeriam-cottam** A DC (E. robusta Roxb.) (Ind - Malabar coasts): baberang; frts. anthelm. (round worms), spasmolytic, carm. adulterant of pepper; bk. of rt. u. for dental pain (124).

"embelic acid": embelin: phloroglucinol deriv., occ. in *Embelia ribes* frt.; *Maesa lanceolata, M. picta*; u. as anthelm. esp. for tapeworm; ammonium deriv. chiefly used.

Emblic myrobalan: Emblica officinalis Gaertn. = *Phyllanthus emblica.*

Embothrium coccineum Forst. (Prote.) (sChile): notru; bk. & lvs. u. in toothache & glandular dis.; decoc. for acceleration of healing of wounds.

embryo sac: large oval cell in nucellus of ovule where fertilization of ovum occurs (flg. plants).

emetamine: alkaloid of ipecac, found in small amts.

emetic herb: e. weed: *Lobelia inflata* - **e. root**: ipecac rt.

emetina (L.): **emetine**: an alk. occurring in ipecac rt. & in part extr. therefrom; a larger proportion is prepd. by methylation of cephaeline (1255), also occurring in ipecac; mostly u. in form of soluble salts, such as **e. HCl**; u. acute amebiasis (not chronic).

Emex australis Steinh. (Polygon.) (Af, Austral): Cape spinach (Austral); lf. u. as spinach; pl. suspected of being cause of poisoning of sheep (CaOx) - **E. spinosa** Campd. (Eu, hot Af): boiled lf. u. in dyspeps. & anorexia; purg. diur.; decoc. for threadworm in horses; pl. said u. as spinach.

Emilia coccinea G. Don (Comp.) (Zambesia+): mucilag. juice of pl. u. in ophthalmia; decoc. u. for colic & ulcers; syph.; lf. is appl. locally to contusions; swollen throat - **E. sagittata** DC (tropAs, Af): rt. u. in colic of babies; for fever, asthma, in abdom. dis.; juice for eye conditions; lf. appl. locally to contusions - **E. sonchifolia** DC (tropAs, Af): rt. u. in diarrh., lf. juice in night blindness, ophthalmia, sore ears; pl. decoc. antipyr. infantile tympanites (1295); appl. to skin eruptions, wounds (PI).

Eminium spiculatum (Bl.) Ktze. (Ar.) (Med reg., Iran): tubers consumed as food in parts of Eg.

emicymarin: glycoside of periplogenin with digitalose; has digitaloid action; occ. in several *Strophanthus* spp. incl. *S. eminii*.

emmer: chaffy wheat fr. *Triticum dicoccum*.

emmet (obsol.): ant.

emodin: an hydroxy-anthraquinone derivative found mostly as the glycoside (rhamnoside) in Rheum. Cascara Sagrada, Frangula, etc.

Empetrum nigrum L. (Empetr.) (temp Euras, NA, incl. Arctic reg.): (black) crowberry; curlewberry; edible; made into beverages with sour milk (Iceland); eaten with seal fat (Greenland); u. diur. (earlier); antiscorb.; above-ground parts astr. ton.; honey of flg. pl. toxic; lvs. c. tannin, resin, wax, fructose, rutin, benzoic acid - **E. rubrum** Vahl (Ch, FalklandIs): berries eaten in Antarctic regions; supposed to be tonic.

empirics: school of medical practitioners in which drugs are u. on the basis of practical experience; mostly crude drugs.

Emplastrum Lyttae (L.): cantharis plaster.

Empleurum serrulatum Ait. (*E. ensatum* Eckl. & Zeyh.) (Rut.) (sAf): lvs. u. as substit. for long buchu.

emu: *Dromiceius nova-hollandiae* (Casuariiformes - Aves) (Austral): lge. swift running flightless bird, -2 m; "thunder chicken"; FO u. by aborigines for dry skin; meat healthful & flavorsome; skin u. for leather; proposed agr. crop for US (like ostrich).

emulsin: mixt. of enzymes found in *Prunus amygdalus* (almonds), *P. laurocerasus* (lvs.). &c.; mixt. incl. esp. b-glucosidase (hydrolyzes b-glucosides, as salicin), amygdalase (acts on amygdalin), oxynitrilase (which acts on mandelonitrile); this term u. by Liebig & Woehler (1837) is being dropped as too indefinite.

Emulsio Communis (L.): almond emulsion (mixt.).

E-Mycin™ (Upjohn): erythromycin (enteric-coated tablets).

Enallagma latifolia Small (Crescentia cucurbitana L) (Bignoni) (sFla & Keys [US]; WI, Ve): small-fruited (or black) calabash; maguira; frt. u. emoll. exp. dem. lax. astr. vulner. in resp. dis. rheum.; pulp mixed inside shell with gin & consumed; rhiz. u. med.; properties much like those of *Crescentia cujete* (—> *Dendrosicus*).

Enantia chlorantha Oliv. (Annon.) (tropAf): African yellow wood; Af white wood; bk. exts. u. as antipyr., appl. to ulcers & hemostatic to wounds (Nig); arrow poison; components of yellow dye - **E. kummeriae** Engl. & Diels (tropAf, incl. Tanz): yellow wd. u. for cuts (Wasamba); wd. source of red dye - **E. polycarpa** Engl. (tropAf, incl. Nig): bk. c. berberine(?), palmatine, quinidine, hydroquinidine; u. like *E. chlorantha*; bk. ext. u. in ointments for sores (Sierra Leone); ulcers; leprosy (Ivory Coast).

Encelia farinosa Gray (Comp.) (swUS, nMex): (white) brittle bush; incienso; resinous exudn. (gum fr. plant) chewed by Ariz Indians (US); smeared (melted) on body for pain in side; pd. resin sprinkled on sores (Seri Indians); resin u. as incense in churches in Baja Calif (Mex) (1296) - **E. mexicana** Cass. (Mex): u. in folk med.

encens (Fr): incense; frankincense.

Encephalartos kaffer Miq. (Zami.) (sAf): Hottentot breadfruit; bread tree; the farinaceous pith of the palm-like pl. u. as food (sago) by the natives (Kaf[f]ir bread) - **E. poppgei** Aschers. (Congo): piondo; st. sap u. in prepn. glue; st. pith source of food (natives of Kasai).

enchilada (Am, Sp): tortilla (pancake) of maize stuffed with meat or cheese with spices incl. chiefly chile (Capsicum).

enchytraeids: Enchytraeus (Annelida): oligochete worms grown in the lab for feeding to birds, fishes, &c. (*E. albidus*).

encina, corteza de (Sp): bk. of evergreen oak, prob. *Quercus ilex*.

encino (Sp): oak tree - **e. blanco: e. colorado**: *Quercus velutina*.

endemic: growing in limited area; ex. on a single island.

Endiandra R. Br. (Laur.) (As, Austral): furnish valuable colored wds. (u. in panelling, &c.); ca. 80 spp., such as *E. palmerstonii* C. White (neAustral); Queensland walnut, with variegated wd.; *E. sieberi* Nees.

endive (garden) (Fr, Eng): *Cichorium endivia*.

endocrine glands: structures in the vertebrate body which manufacture hormones & release them directly into the circulatory system & not through ducts which open directly or indirectly to the outside of the body; ex., pituitary, thyroid.

endoderm(is): starch sheath; the innermost layer of the cortex of roots & many stems, lying against the stele; function uncertain; some believe it to represent a seal to protect fr. water loss out of the stele into the cortex, since the tangential outer & side walls are impregnated with lignosuberin.

endoenzyme: endocellular or intracellular e.; one remaining within the cell; ex. zymases.

Endomyces (Endomycet. - Ascomyc. - Fungi): u. in prodn. of fat (*E. vernalis* Ludw.) during World War I (Ge); furnished 25% of dr. wt. in FO.

Endomycin: polyene antibiotic fr. *Streptomyces endus* Anderson & Gottlieb cultures; antimicrob.

Endopleura uchi Cuatracasas (monotypic) (Humiri.) (Amazon reg.): sd. FO similar to olive oil (1492).

endorphins, esp. **beta-endorphin**: normal components of mammalian body which reduce pain; isolated fr. sheep brain but actually secretion of anterior pituitary gland; act like opioid analgesics (said 20-40x as effective as morphine); polypeptide (31 amino acids); reduces depressive moods & schizophrenia; possibly addictive; relieves symptoms of morphine withdrawl (1977-); incl. the enkephalins.

endosperm ("albumen"): portion of sd. external to embryo wh. furnishes nutrition to embryo; often lacking.

Endospermum formicarum Becc. (Euphorbi.) (eAs): u. as oral contraceptive (NG) - **E. moluccanum** Becc. (eAs): wd. decoc. u. for ulcers; bk. decoc. u. as diur.; young lvs. cooked as veget., sl. purg.; older lvs. strongly purg.; sap causes hair falling out - **E. peltatum** Merr. (PI): bk. decoc. u. as "alterative"; fresh bk. poultice for diarrh. (appl. to abdomen).

endotherapy: treatment of dis. in plants by applns. either to above-ground parts (before or after infection) or to the roots (for absn. & systemic distribution).

endotoxin: a toxic prod. of a bacterial (or other) cell which remains within the cell wall & is not diffused out into the surrounding medium, as is an exotoxin.

Endymion non-scriptum Garcke (Scilla non-scripta Hoffmgg. & Link) (Lili) (Eu, Af, sAs, Austral): wild hyacinth; blue bell (of England); pl. u. expect. emet. (Homeopath) (—> *Hyacinthoides*).

eneldo (Sp): *Anethum graveolens*.

enfleurage (Eng, Fr): process of absorbing a VO or aromatic princ. into a bland FO, esp. a solid fat (**enfleurage greases**), subsequently extracting the arom. prod. with a solvent, like alc.; operations conducted without heat; ex. rose oil.

Engelhard(t)ia roxburghiana Lindl. ex Willd. (Jugland.) (sChin, Burma, seAs): lvs. & bk. toxic; u. as fish poison; repulsive for leeches; bk. u. to make straps - **E. spicata** Bl. (Mal, Ind): soeval tree; lvs. prod. resin u. med. & as incense; bk. c. 20% tannins, u. as tan., as fish poison and timber.

Engel-wurz(el) (Ge): *Angelica archangelica*.

Engleromyces goetzii Henn. (Hypocre. - Ascomycet. - Fungi) (tropAf): dr. sarcocarps cooked & u. for stomachache by natives.

Englische Gewuerz (Ge): allspice.

English: v. alder; chamomile.

Englishman's foot: *Plantago major*.

eng-me-doo (Chin): frt. of *Kadsura* spp.; u. med.

engrais vert (Fr): green manure.

En(h)ydra fluctuans Lour. (Comp.) (eAs): lvs. lax., condim., used in skin dis., demulc., hypertension.

Enicostema littorale Bl. (Gentian) (trop): bitter hb. u. as stom. ton. lax. (Sudan) - **E. orientale** A. Payal (Slevogtia orientalis Griseb.) (Ind): hb. c. ophelic acid; u. as bitter ton. (Chirata substit.) - **E. verticillatum** Engl. (Java; Madura; WI): quinine bush (Trin); pl. decoc. u. as antipyr.; lf. & rt. u. for dys. malar. (Trin).

enkephalins: two opioid peptides each with 5 amino acids found predominantly in brain & spinal cord, &c.; found with the endorphins and with similar analgesic properties; both opioids; occ. in both neuron and endocrine cells.

Enkleia siamensis (Kurz) Nevling (Thymelae) (Burma, Indochin, Thail): lf. decoc. appl. to eyes in compresses; frt. purg.; bk. fibers u. as cords.

Enneastemon fornicata Exell (Annon.) (Tanz): pd. lf. u. in snakebite (—> *Monanthotaxis*).

enniatine: multiple antibiotic (A, B, C) fr. *Fusarium orthoceras* v. *enniatinum* & other *F.* taxa.

enrich: addn. of nutritional elements to food stuff to produce a more normal healthful food; ex. vit B_1 added to flour.

Ensete edule (Gmel.) Horan (Musa ensete Gmel.) (Mus.) (e tropAf): banana plant with reddish bitter milky juice; rt. pounded, warmed slightly, juice strained, salted & u. by drinking as galactagog (Tanz); good lf. fiber u. esp. in emergencies - **E. ventricosum** Cheeseman (trop eAf): wild banana; inflorescence stalk & sds. cooked & eaten; stalk yields good fiber.

ensilage: process of preserving or storing fodder in air-tight silo (or pit) with partial fermentation (anerobic f.); product not previously dried; also product of same; ex. green corn ensilage.

Entada abyssinicca Steud. (Leg.) (eAf): sea bean; rt. c. entadasaponin; u. in rheum. - **E. africana** Guill. & Perr. (E. sudanica Schweinf.) (w&cAf): bk. c. saponins, u. abortif.; lvs. c. rotenone, tannin; u. fish poison; lvs. & bk. u. wound dressing; tree yields a gum resembling acacia and tragacanth; bk. decoc. u. in constipation (Congo) - **E. gigas** (L) Fawcett & Rendle (CA, WI, SA, w&cAf): bk. decoc. u. in sitz baths or in enemas for women recently delivered & in anemia; for GC; inner bk. after retting releases fibers u. in cords (Gabon) - **E. phaseoloides** (L.) Merr. (E. scandens Benth.) (sAs, wAf, tropAm): Garbee (or sea) bean; St. Thomas (or nicker) b.; matchbox (or snuffbox) b.; gogo (PI); frt. (Fructus Entadae) largest pods in fam. c. 1.7 x X c. 1.8 dm wide; this liana c. saponins (esp. rich in bk.) with anti-tumor activity; u. fish poison; rts. macerated in H_2O to wash hair (Vietnam); hair ton. (Marianne Is) very hard, chocolate-colored beans roasted & eaten; adulter. of calabar beans; sd. kernels u. as antipyr. ("to cool blood after childbirth," Vietnam); young lvs. eaten as veget.; bk. fibers u. for cordage; bk. c. tannins, u. astr. for wounds; wd. u. skin dis.; wd. & bk. poisonous; "drift sds." carried to wEu by Gulf Stream) (1297) - **E. polyphylla** Benth. (Co): sd. decoc. u. as gargle in nasal-pulm. congestion - **E. polystachya** DC (tropAm, incl. Co): ojo de aquila (Mex); rts. of this shrub u. (decoc. & inf.) in VD (syph.) (Trin); st. u. for cord making & soap substit.; grd. rts. u. as hair tonic (Mex) - **E. pursaetha** DC = *E. phaseoloides*.

Entenfuss, schildblaettiger (Ge): Podophyllum.

enteramine: serotonin.

enterobacteria: enteritis bacteria; *Salmonella* spp.

enterococci: bacteria of GI tract, representing *Streptococcus* spp.; blood infections with these are esp. dangerous, often lethal, since may be resistant to all antibiotics.

enterocrinin: hormone fr. intestinal mucosa acts to stim. glands of small intestine.

enterogastrone: hormone fr. upper intest. mucosa; inhibits gastric secretions & movements.

enterokinase: activator in intestinal mucosa (esp. of duodenum) wh. catalyzes reaction: trypsinogen —> trypsin.

Enterolobium cyclocarpum (Jacq.) Griseb. (Leg.) (Mex, CA, WI, nSA): "parota"; guanacaste (blanco); bk. u. in colds (syrup); tumors; tree furnishes gum, goma de Caro, u. in bronchitis; bk. & frts. c. saponin with spermicidal activity; u. as soap substit.; pods edible after cooking (eaten raw by ruminants) - **E. timbouva** Mart. (WI, Br): frts. (c. saponins) u. as anthelm.; bk. (Cortex Pacarae, c. tannins) u. for tanning leather.

entheogenic: feelings of the supernatural experienced under influence of some hallucinogens.

entire (bot.): with margin of lf. (etc.) continuous & unbroken, without indentation large or small; ex. iris lf.

entomophilous: pollen transmitted by insects (as opposed to wind) (**entomophily**, state).

entraña (NM [US], Sp): *Opuntia imbricata*.

enula campana (It): elecampane.

envasar (Sp): to can, bottle, package - **envase** (Sp): container (bottle, can, etc.).

envenomation: bitten by a poisonous animal (esp. snake).

Enydra Lour. = *Enhydra*.

Enzian (Ge.): gentian.

enzyme: ferment: a complex protein material which acts in the living cell (plant or animal) as a catalyst to the reactions which proceed there or outside of the cell; the specific names for enzymes end in "-ase" or "in"; ex. emulsin, lipase - **amylolytic e.**: carbohydrase - **angiotensin-converting e.:** ACE: tonin: u. for both renovascular and essential hypertension (usually with diur.) - **extracellular e.**: exoenzyme - **intracellular e.**: endoenzyme - **lytic e.**: participates in degradation of subst. as opposed to synthetic e. - **old yellow e.: yellow e. of Warburg**: yellow oxidation ferment; phosporic acid ester of riboflavin (co-enzyme) united with a specific protein (apoenzyme); occ. only in yeast - **yellow e.**: dehydrogenase wh. removes H from the substrate, thus facilitating oxidn.; represent flavoproteins, based on riboflavin; several are known.

epaulette tree: *Pterostyrax hispida.*

epazote (several spelling variants) (SA): *Chenopodium ambrosioides* or *C. vulvaria* (also *e. de zorrillo).*

epena (SA, Indians): intoxicant snuffs.

éperonelle (Fr): larkspur.

épervière (Fr): *Hieracium,* hawkweed - **é. de murailles:** *H. murorum.*

Ephedra (pron. éf-e-dra) (Ephedr.) (OW, NW): joint firs; alkaloid-rich pls. commonly occurring in semi-arid areas; many spp. c. ephedrine; some u. in med. & for alk. extrn. (1352) - **E. alata** Decne. (nAf, Alg to Eg; Sinai Desert): ephedra; c. pseudo-ephedrine (1493); pl. u. in hypotension, asthma, astr.; sympathomimetic; branches chewed for headache, cooked in butter & eaten by women for miscarriage; bronchodilator (1494); u. for tanning in Sinai region - **E. americana** H. & B. ex Willd. (Pe, Ch, Ec, Bo, Arg,): pingo-pingo; pinco-pinco; cola de caballo; hb. c. no alk.; u. in domestic med. as diur. mouthwash for treating pyorrhea & gingivitis; var. *andina* Poepp. (E. a. Poepp.) (Andes esp. Peru): pingo-pingo; diur. u. for pain fr. kidney stones; styptic - **E. antisyphilitica** Berl. ex C.A. Mey. (swUS, Mex): squaw tea; twigs u. astr. in GC (chiefly), diur. alt. antisyph.; inf. fr. hb. of this & other wNA spp. u. as tea (hot or iced with sugar); sold for this purpose by "health food stores"; substituted by the Mormans for Chin tea, which they cannot drink - **E. californica** S. Wats. (sCalif [US], nwMex): California ephedra; miner's tea; hb. u. to prep. refreshing beverage, medl. - **E. distachya** L. (E. vulgaris L. C. Rich.) (s&cEu, Ind, nChin): ma huang; hb. c. ephedrine & pseudoephedrine; rel. low alk. yields; u. mydriatic, vasoconstrictor; in asthma, hypertension; gout (Russia); rheum.; frt. edible; u. in cough; earlier as febrifuge; u. med. by ancient Greeks; subsp. **helvetica** Aschers. & Graebn. (E. h. C. A. Meyer) - **E. equisetina** Bunge (cAs): Mongolian e.; rich in ephedrine (1.8%; u. as prec. & in mfr. alk. ephedrine (OW); u. for urinary dis. (Chin) - **E. foliata** Boiss. & Kutschy (E. ciliata Fisch. & Mey. ex C. A. Mey.) (wInd [Rajasthan], sPak desert plains): kutcher; tot. alk. content low (0.03%); u. med. - **E. fragilis** Desf. (eMed reg.): u. for hemorrhoids; var. *campylopoda* (C. A. Mey.) Aschers. & Graeb. (E. c. C. A. Mey.) (eMed reg., seEu, AsMin): low alk. content, higher in Jap (-1.25%) (cult.) - **E. gerardiana** Wall. (Ind-Pak) (E. vulgaris Hook. f.): one of most used spp.; shrub c. -3% alk., usually 0.5-2%; ephedrine, pseudo-ephedrine, methyl ephedrine; norephedrine, &c.; tannins, VO; u. in bronchial asthma, nasal congestion, spasmodic cough, diur. analeptic, anti-allergic; formerly in acute & chronic rheum. (1351); mydriatic; u. to clean teeth; ash as snuff (Himal) - **E. intermedia** Schrenk & Mey. = *E. distachya* - **E. monosperma** Gmelin (Chin, Siberia): u. in resp. dis., asthma paroxysms - **E. nebrodensis** (Tineo) Stapf (Sp, wAs): c. over 1% ephedrine; var. *procera* Stapf (E. procera Fisch. & Mey.) (Pak, esp. Baluchistan w to Gr): "golden herb" of Baluchistan; large amts. collected for prodn. of ephedrine; at one time, this indy. was main support of the Pak forestry dept.; high yields (2.8% tot. alks. with 1.9% ephedrine) - **E. nevadensis** S. Wats. (swUS, nMex): Nevada ephedra; Mormon tea; hb. u. (twigs & branches) in treatment of VD, esp. GC; to stim. urination (2293); as poultice for sores; u. for wt. reduction; ton.; hb. c. much tannin, VO; rich in vit. C; sometimes combined with *Gilia congesta, Larrea divaricata*; u. much like *E. antisyphilitica* (folk med.) - **E. pachyclada** Boiss. (Afghan, Tibet, nPak, Iran): uman; sam; "houma" (sacred Avesta); twigs u. to tan mashk leather; fuel; ashes added to tobacco & chewed (CNS stim.); c. pseudoephedrine - **E. pedunculata** Engelm. (wTex, Mex): canatilla; stems u. astr. diur. in VD, renal dis., in pleurisy, pneumonia - **E. procera** Fisch. & Mey. (Pak, esp. Baluchistan, Turk): "golden herb" of Baluchistan; large amts. collected for mfr. ephedrine; this indy. at one time a main support of the forestry dept. - **E. rupestris** Benth. = *E. americana* - **E. sinica** Stapf (c&sAs): Chin e.; rich in ephedrine (1.1-1.6-3.3%; varying with season, &c.); pseudoephedrine; L-norephedrine; one of chief sources of drug; u. in bronchitis, asthma, cough, as diur. - **E. torreyana** S. Wats. (swUS, nMex): popotillo; cañatillo; u. in folk med. like *E. antisyphilitica* - **E. trifurca** Torr. (swUS, nMex): three-forked ephedra; joint fir; tea fr. sts. popular as "teamster's tea" or "desert tea"; Mexican tea; mountain rush; u. folk med. like *E. torreyana* as ton. - **E. tweediana** C. A. Mey. (Arg, Ur): pico de loro; c. resins, peroxide, no alk. - **E. viridis** Coville (Calif, Nev [US]): green ephedra; Mexican tea; canyon tea; Mormon tea; u. astr. in renal & liver dis. (Indian folk med.); Apache & Navaho Indians u. in cough, GC; dyestuff (twigs & lvs.); st. inf. for arthritic pain - **E. vulgaris** L. C. Rich = *E. distachya.*

ephedra, Mongolian: *Ephedra equisetina.*

Ephedraceae: monogeneric fam. with *Ephedra*; closely allied with Gnetaceae; lvs. scale-like.

Ephedrina (L.): **ephedrine**: an alk. obt. fr. herbage of various *Ephedra* spp. (As or Eu) (*E. gerardiana* chief source of natural e.) or now mostly in US, prod. synthetically by fermentation process; the bulk of that prod. abroad is by extn. of ephedra; occurs in anhydrous & hydrated forms; sol. H_2O, alc., ether, etc.; tends to deteriorate in soln., developing garlicky odor; sympathomimetic agt. u.

mydr. vasoconstrictor (shrinks nasal passage mucosa); cerebral & medull. stim. (E) (1353).

Ephemeroptera: order of Insecta made up of mayflies & related gps.; chief gen. *Ephemera*; SC as mature adult form lives for only a few hrs. or days.

ephenamine penicillin G: generic title for antibiotic marketed as Compenamine; now in disuse.

Ephestia kuhniella (Phycitidae - Lepidoptera; Insecta): Mediterranean flour moth; larva attacks cacao, peanuts, cottonseed, almonds, capsicum, various crude drugs, &c.

Ephetonin™: racemic syn. ephedrine HCl.

Ephydatia fluviatilis L. (Spongidae - Porifera) (Euras, NA): freshwater sponge; u. in glandular swellings, eye inflammn., hay fever, cough, rheum. pains, arthritis, paresis (sic).

epiblema: periderm(is) of rt.

Epicampes macroura Benth. (Gram) (Mex): Zakaton grass; zacaton; "broom root"; rts. called zakaton or rice root; u. in mfr. brushes & brooms; exported to US; u. med. (987).

epicarp: exocarp: outermost (epidermal) layer of fruit; usually rich in waxes.

Epicauta tomentosa Maki (Meloideae - Coleoptera - Insecta): blister beetle; attacks hibiscus & veget.; c. cantharidin in free & bound forms; other spp. cont. cantharidin, such as *E. pestifera* Werner (margined blister beetle) (NA), with 1.09%; *E. vittata* (Fabricius) (E. solani) (old-fashioned potato beetle) (do not confuse with Colorado p. b.).

épicéa (Fr): spruce, esp. *Picea abies*.

Epich (Ge): celery frt.

Epidendrum bifidum Lindl. (Orchid.) (SA): this epiphyte is used against tapeworm & other intest. parasites - **E. brachyphyllum** Lindl. (Ec): "flor de Cristo"; fl. inf. u. as diur. - **E. cochleatum** L. (Mex, CA, WI, nSA): canuela (Cu); pseudo-bulbs u. as source of mucilage (Indians) - **E. pastoris** LaLlave (Mex): pseudobulbs u. like salep (*Orchis* spp.).

epiderm(is): outer protective cell layer of lvs., rts., soft sts. (before appearance of cork); cells rather flattened (tabular), often elongated, outer surface often convex, often covered by cuticle; the pattern by which the epidermal cells surround stomata is often of diagnostic value.

Epifagus (*Epiphegus*) **virginiana** (L.) Bart. (1) (Orobanchaceae) (eNA): (Virginia) beech drops; parasitic on beech tree rts.; rt. u. astr. vulner., esp. by Homeop. for ulcers, cankers.

Epigaea repens L. (Eric.) (eNA): **epigée rampante** (Fr): trailing arbutus; gravel plant; lvs. c. tannin, arbutin; u. like uva ursi, as diur. astr.; much prized ornam. shrub (1370).

epigean: epigeous: above-ground parts of pl.

Epigynum maingayi Hook. f. (Apocyn.) (Mal): u. as galactagog.

Epilobium angustifolium L. (Chamaenerion a. [L.] Scop.) (Onagr.) (Euras, NA): (great) willow herb; fireweed (SC commonly appears on burnt-over areas; ex. London Fires of 1666 & 1940); hb. u. ton. dem. tea adult. or substit., often with apple lvs. (Den, USSR = "Siberian tea"); sts. bruised & eaten with butter; rts. produce tisane (Abenakis); lvs. & sts. rich in vits. A, C; mucilage, pectin; u. potherb (Alas [US]) (201); fireweed honey highly rated (BC [Cana]) - **E. hirsutum** L. (Af, natd. NA): willow herb; lf. c. antibiotic principle; u. to cleanse foul ulcers; for warts; charms; toxic - **E. palustre** L. (Euras, NA): marsh willow herb; rt. u. homeop. med.

Epimedium alpinum L. (Berberid.) (Switz, nIt): (Alpine) barrenwort; Sockenblume; hb. u. astr. esp. in Middle Ages; in breast tumors; diaph.

épinard (Fr): spinach - **e. fraise**: strawberry s.

epinephrine: hormone secreted by the medulla of the adrenal glands (epinephros); occurs as microxalline pd.; 1:100 to 1:1000 solns. u. as vasoconstrictor, hemostat. styp.; to prolong loc. anesth.; circ. stim.; in anaphylaxis (inhalation), allergies, asthma (short acting but more effective in severe asthma); in heart stoppage; oil solution u. in bronchial asthma (constrictive action); inactive orally, hence, u. parenterally (E); **e. bitartrate** (2613).

epinet (WI, St Lucia, Trinidad): *Zanthoxylum* spp. (red & white) - **épinette** (Fr): spruce - **épéneux** (Fr): *Zanthoxylum* - **épinne-vinette** (Fr): *Berberis vulgaris*.

Epipactis gigantea Dougl. ex Hook. (Orchid.) (wNA incl. Mex): giant orchid; chatterbox; hb. inf. u. to treat insanity (Calif Indians).

Epiphegus: *Epifagus*.

epipodophyllotoxin: podophyllinic acid lactone; deriv. fr. *Podophyllum peltatum*; u. in cancer.

Epipremnum giganteum Schott (Ar) (Mal, Moluccas): pl. juice u. in dart poison (with that of *Amorphophallus prainii & Dioscorea hispida*) - **E. pinnatum** Engl. (E. mirabile Schott) (seAs, Oceania, Austral): (centipede) tonga vine; "tonga plant"; fue lau fao (Am Samoa); rt. u. in combination with bk. of *Premna taitensis* to form the drug Tonga; u. anod. in neuralgia & rheum.; lf. ext. u. for itching; rheum. (Chin); st. pith u. as compress for sprain (Indones); lvs. vermifuge (horses); ornam. climber (774).

epirenamine (Jap Phar.): epinephrine.

Episcia melittifolia (L.) Mart. (Gesneri) (WI), Domin, Guadeloupe): l'oseille (Domin); lvs. & pink fls. u. to make tea for colds (Domin, Caribs).

episperm: perisperm.

epitamo (It): **épithyme(e)** (Fr): **epittamo** (It): *Cuscuta europaea.*

Epithelantha micromeris Weber (Cact.) (sUS, Mex): button cactus; u. as stim., to "prolong life"; as hallucinogenic (Tarahumara Indians).

epithelium: 1) single layer of flattened cells bordering an oleoresin secretory canal (in Umbelliferae, &c.) 2) in animals, the peridermis or outer skin layer.

epithem: internal hydathodes (water excretion structures) in leaf mesophyll, representing non-chlorenchyma.

epithet (biol.): a scientific name for a genus, sp., or lower taxon (subsp., var., subvar.); ex. *Rosa chinensis longifolia,* in which *"chinensis"* is the specific e. while *"longifolia"* is the varietal e.

Epsom salts: orig. obt. fr. **Epsom springs; E. waters**: natural mineral springs in Surrey County, Eng; disc. as medl. waters 1618; most popular spa in Eng; waters rich in $MgSO_4$, hence this salt called **Epsom salts** (2246).

épurge (Fr): spurge.

Equisetaceae: horsetail family, one of three in **Equisetales**, order of Pteridophyta (ferns).

equiseto (It): **Equisetum** (L.) (Equiset.) (worldwide except Austral): horsetails; perenn. ferns, found mostly in swampy places; c. equisetine (palustrine) nicotine (alks.), aconitic acid; several u. in med.; some with sts. abrasive (c. silica) u. to polish wood, etc. - **E. arvense** L. (nEuras, NA): (field) horsetail; hb. c. equisetonin (saponin, 1%) as glycoside, equisetogenin; hb. u. diur. astr. ton.; pl. poisonous to horses (1298) - **E. bogotense** HBK (SA): limpiaplata (Ch): u. diur. (Ch) - **E. giganteum** L. (Arg): cola de caballo; pl. u. as diur. astr. (veter. med.); toxic; rich in silicic acid - **E. hyemale** L. (Euras, NA): scouring rush; shave grass; polishing rush; hb. (as Herba Equiseti majoris) u. diur. in liver & kidney dis., as cystitis, urethritis, urinary incontinence, noctornal enuresis; in support of digitalis therapy; eye dis. (Chin); st. decoc. for head lice; with willow lvs. for irreg. menstr.; heads of reproductive shoot ingested in diarrh. (nwAm Indians); epidermis c. silica, hence u. in Colonial days to clean pots & pans; Salish Indians (BC [Cana]) u. st. as file for wood & to shape artifacts of steatite (soapstone) - **E. laevigatum** A. Br. (NA): joint weed; dr. pl. ground & made into mush (NM [US] Indians) - **E. telmateia** Ehrh. (E. majus Gars.) (Euras, nwAf, wNA): (giant) horsetail; freshly pressed juice u. in metrorrh. & tubercular hemoptysis; hemostyptic; in Ukranian folk med.; earlier u. for polishing tin - **E. variegatum** Schleich. (Calif [US] to BC [Cana]): variegated scouring rush; inf. u. for sore eyes (Calif Indians); siliceous sts. u. like sandpaper.

Equus asinus L. (Equidae - Perissodactyli) (Af): (common) African wild ass; many domesticated forms (donkey; burro); u. as beast of burden; in breeding mules (hybrid with female horse) and hinnies (progeny of male horse with female donkey); flesh sometimes eaten (tough); milk u. as feed, med. in meningitis, sexual ton. (Unani); in baths (Nero's wife); dung (in classical medicine & Unani med.); many other parts of animal u. in Unani med. (ex. liver); hide u. to make parchment - **E. caballus** L. (orig. Euras): horse; many vars. (ex. Shetland pony [small]; Percheron [very large]); u. for transportation; source of meat & milk (many countries); hide; glue; serum; the horse was introduced into NA by European (Span) immigrants.

érable (Fr): maple; *Acer* spp. - **sirop d'é.**: m. syrup.

Eragrostis Wolf (Gram) (cosmopol., chiefly subtrop): lovegrass; some are u. as fodder grasses, some in soil & water conservation; recommended for lawns (NM [US]) - **E. chloromelas** Steud. (sAf, introd. swUS): Boer lovegrass; drought resistant, u. for erosion control; grain famine food, to make bread & beer (sAf) - **E. curvula** (Schrad.) Nees (sAf; natd. swUS): weeping lovegrass; forage (drought-resistant); for "stilling" sand; ornam. perenn. grass - **E. japonica** Trin. (eAs, PI): infused lvs. appl. as poultice in headache (PI) - **E. lehmanniana** Nees (Af introd. swUS): Lehmann lovegrass; Eastern Province vlei grass; remedy in diarrh., colic, typhoid fever (Europeans in Transvaal) - **E. pilosa** Beauv. (Eu introd. NA, SA): India lovegrass; effectively u. for contusions - **E. plana** Nees (hot Af): jongosgras (Afrikaans); rt. decoc. u. for profuse menstr., impotence, female sterility; ton.; to make hats, baskets - **E. tenella** R. & S. (Oceania): u. to treat scabies (Solomon Is) - **E. tef** Trotter (E. abyssinica Link) (Af): teff; frt. millet-like cereal, u. for prepn. bread (neAf); forage crop (Af+); cult. ornam.

Eranthemum roseum R. Br. (Acanth.) (Ind): rt. u. in leukorrh.; to promote growth of fetus (cattle) - **E. viscidum** Bl. (Mal, Indones): juice fr. leafy tops u. as eye drops in ophthalmia - **E. whartonianum** Hemsl. (seAs): rt. u. in indigest. & to remove object stuck in throat (Solomon Is).

Eranthis hyemalis (L.) Salisb. (Ranuncul.) (cEu): winter aconite; c. furanochromones; eranthin (bufadienolide?); appar. no alks.; u. antispas. for coronary vessels, bronchitis, GI tract, gallbladder (*cf.* Helleborus); cult. ornam.

erba (It): herb; grass - **e. amara** (della madonna) (It): *Balsamita suaveolens* - **e. anagallide**: chickweed - **e. colombina**: vervain - **e. di pilatro:** *Hypericum perforatum* - **e. di San Giovanni**: 1) *Hypericum perforatum* 2) *Salvia sclarea* 3) *Sedum telephium* 4) *Glechoma hederacea* - **e. pignola**: *Sedum album* & other *S.* spp. - **e. turca**: 1) *Arenaria serpyllifolia* 2) *Cnicus benedictus* 3) *Verbena officinalis* 4) *Spergularia rubra.*

Erbse (Ge): common pea & spp. of *Pisum*, also spp. of *Cajanus*, *Cicer*, *Dolichos*, *Lathyrus*, *Lotus*, *Vicia*, *Vigna*.

Erdbeere (Ge-f.): (wild) strawberry.

Erdbirne (Ge): Jerusalem artichoke - **Erdeichel** (Ge): peanut - **Erdkresse** (Ge): *Barbaraea* - **Erdmandel (gras)**: *Cyperus esculentus* - **Erdnuss** (Ge): peanut.

Erechtites hieracifolia Raf. (Comp.) (eNA): pilewort; fireweed; borraja (Mex); hb. u. emoll. astr. ton. emet. alt. in hemorrhoids; cough (Salv); pl. infests peppermint fields in Mich (US) - **E. praealta** Raf.: *E. hieracifolia.*

Eremocarpus setigerus (Hook.) Benth. (L) (Euphorbi) (Pacific US; nMex): turkey mullein; u. with *Aesculus pavia* and *Chlorogalum pomeridianum* pounded together & mixed to give a stiff paste to kill (or catch) fish (wNA Indians).

Eremochloa ophiuroides (Munro) Hack. (Gram) (seAs, esp. Chin; cult. seUS): centipede grass; lazy man's g.; popular lawn grass (esp. in Fla [US]) giving a good turf; prevents soil erosion; will grow on poor soils.

Eremophila longifolia (R. Br.) F. Muell. (Myopor.) (Austral): berrigan; decoc. u. for colds; (aborigines) (2263).

Eremothecium ashbyi Schopf (Saccharomycet. - Ascomycet.): yeast u. in fermentation process to produce riboflavin.

erepsin: mixt. of enzymes fr. intest. & pancreatic juice, mostly peptidases.

ergastic substances: passive secondary non-living components of protoplasm with inclusions; ex. starch grains; CaOx, $CaCO_3$, FO, proteins, VO, glycosides, alks. purines, SiO_2, resins; gums & mucil.; tannins, &c.

ergine: lysergic acid amide; product of hydrolysis of ergot alks. with lysergic acid nucleus; ex. ergotoxin, ergotamine.

Ergoapiol™: proprietary contg. apiol & "ergotin"; formerly popular in amenorrh. & dysmenorrh.; supposedly aid to abortion.

Ergobasine (Ergometrine; Ergotocin; Ergonovine; Ergotrate): simplest of ergot alk.; can be hydrolyzed to lysergic acid & 1, 2-amino-propanol; oral admin. produces rapid uterine contractions; u. as hemostatic after childbirth (not before or during) & in removal of afterbirth.

ergocalciferol: calciferol (vit. D_2) made fr. ergosterol - **ergocornine**: alk. fr. ergot.

Ergocristine: natural ergot alk.; lysergic acid deriv. of ergotoxin group.

ergocryptine: ergokryptine: alk. of ergotoxine gp.; occ. in ergots; 2 isomers, (a, b); u. like ergot; an important deriv. is bromocriptine (2-bromoergokryptine); u. as enzyme inhibitor (prolactin), in Parkinsonism, "genuine" aphrod.; b. methanesulfonate = Parlodel.

Ergoklinine™: **Ergometrine**: ergonovine.

ergoline: 4-ring basic (nuclear) system of D-lysergic acid alkaloids of ergot; ex. ergometrine.

ergonovine (US): generic name for most active alk. of ergot; other names are *Ergotrate*, *Ergotocin*; *Ergobasine*; etc.; u. oxytocic (3rd stage of delivery), hemostat. (E) - **Ergonovine maleate**: the ester most in medl. use.

ergosterol: sterol orig. disc. in ergot, but now obt. mostly fr. yeast, fungi, etc.; irradiated with UV light, it becomes vit. D_2 (1299) - **Ergostetrine**™: ergonovine.

ergot (of rye): **e. claviceps: e. de seigle** (Fr): dr. sclerotium of *Claviceps purpurea* fr. rye; c. ergonovine, ergotoxine, ergotamine; u. emm. partur. hemostat. ecbol. abortif.; pd. defatted e. claimed superior in therapy to e. alkaloids (2244) (chiefly prod. Port, Sp, USSR, Ge, US) - **domestic e.**: that prod. in Wisc, Minn (US), etc.; sclerotia somewhat thicker than in European; yields larger amts. alk.; most prodn. during war years - **Durum wheat e.**: domestic type growing on *Triticum durum*; rel. poor in alk. content; seen as adult. in regular ergot sometimes - **European e.**: the prod. generally seen during times of peace; rel. inferior to but much cheaper than domestic e. - **e. of diss.**: e. growing on *Ampelodesmos mauritanicus* Dur. & Schinz (A. tenax Link) (Gram) (Alger): rel. thin & long; highly active - **e. of oats**: u. med. in Algeria - **e. of wheat**: is occ. seen; has been u. med. in Fr - **prepared e.**: defatted, partly dehydrated, standardized, & pd. e. - **Russian e.**: that prod. in the USSR; usually considered inferior to Spanish type - **rye e.**: e. of rye - **Spanish e.**: represents good grade of European e.; sometimes exp. thru Portugal.

Ergota (L.): ergot.

ergotamine: an amide compounded of lysergic acid, proline, phenylalanine & pyruvic acid; found abundantly in NZ tall fescue ergot; less richly in ergot of cEu - **e. tartrate**: the tartaric acid salt. of ergot; occurs as microxalline pd. or colorless xals.; u. uter. stim. & uter. hemostat.; to control nerve & GIT effects of "shell shock" (1490); to relieve migraine (E).

ergothioneine: thiolhircinine: first noted in ergot but has since been found in animal tissues, incl. blood, semen, liver, &c.; u. antithyroid.

Ergotin Bonjean: aqueous ext. of ergot purified with alc. - **ergotinine**: alk. isomeric with ergotoxine - **Ergotocin(e)**™: ergonovine - **ergotoxin**: alk. of ergot made up of lysergic acid, dimethylpyruvic acid, proline & phenylalanine joined with amide linkages; u. oxytocic; before ergonovine was discovered, this was the most popular of the ergot derivatives - **Ergotrate**™: ergonovine.

Eria pannea Lindl. (Orchid.) (eAs): decoc. u. as med. bath for ague (chills & fever, usually mal.).

Erianthus ravennae Beauv. (Gram) (sEu, cult. sUS): plume grass; Ravenna g.; -4 m; ornam. (2645) (—> *Saccharum*).

Erica arborea L. (Eric.) (Med reg.): (tree) heath; bruyère (sFr); white Mediterranean heather root; rts. (stocks) c. ericolin; u. arom. astr. in bladder stone, snakebite; in mfr. of "briar" pipes (fr. bruyère [briarwood], Fr) (important ornam. gen.) - **E. multiflora** L. (Med reg., seEu, nAf): bruyère multiflore; urin. antisep.; u. in bladder stone; astr. (Tun) - **E. scoparia** L. (sFr): heath; bruyère; rts. u. for briarwood pipes - **E. scoparius** Koch = *E. s.* Thunberg = *E. arborea* - **E. tetralix** L. (Eu, esp. nGe): broom heath; cross-leaved h.; fls. u. in folk med. as antipyr. in fevers, expect.; source of yellow dye u. in Scottish Highlands - **E. vagans** L. (wEu): Cornish heather; similarly.

Ericaceae: Heath Fam.; found mainly in the temp. regions throughout the world; subshrubs, shrubs, & rarely small trees or some herbs; rather variable usually; includes many attractive ornam. & perfume plants; pls. characterized by their hairy, leathery lvs.; frt. berry, capsule or false drupe, often attractive (red, orange, etc.); mostly of value as ornam. pls.; ex. heather, bilberry, cranberry, strawberry tree (*Arbutus*), trailing arbutus, wintergreen, rhododendron, mt. laurel (Vacciniaceae).

Ericerus pela Chavannes (Coccidae - Hemiptera - Insecta): Chin wax scale; secretes white wax (Chin) = Chin wax (Cera Chinensis, *qv.*); formerly u. in special candles (Chin, Jap) as illuminant (but kerosene cheaper), in plasters, for waxing paper & cotton.

ericolin: glycoside or mixt. (may be impure arbutin, *qv.*); occurs in various Ericaceae spp.; ex. *Arctostaphylos uva-ursi, Ledum palustre, L. groenlandicum, Pyrola rotundifolia, Chimaphila umbellata*; *Calluna vulgaris, Rhododendron arboreum* & other *R.* spp., *Vaccinium myrtillus, V. oxycoccus*, &c.

Erigeron: 1) (Comp.) (worldwide): fleabanes; ann. & perenn. hbs. 2) (specif.): E. *canadensis* - **E. acris** L. (Eu, natd. NA): u. like *Conyza canadensis*; earlier u. in urinary tract dis., colds, pyrosis - **E. affinis** DC = *E. longipes* - **E. annuus** Pers. (E. heterophyllus Kunth & Bouché) (e&cNA): annual fleabane; daisy f.; "scabious"; source of Herba Asteri (L); u. diur. diaphor., stom. in edema - **E. canadensis** L. = *Conyza c.* (*qv.*) - **E. longipes** DC (E. affinis DC) (sMex to Nic): chilcuan rt. (raiz de piretre) of perennial hb. u. for dental pain (toothache); dentifrice; sternutat., gum stim. (1300), neuralgia, rheum.; insecticide (in place of pyrethrum fl.) - **E. philadelphicus** L. (NA): Philadelphia fleabane; c. tannins, VO; u. diur. ton. diaph. - **E. uniflorus** L. (E. patentisquama J. F. Jeffrey) (nInd, Pak, Himal): fls. in mixt. u. as poultice to bring boils to head; in stomachache.

Eriobotrya japonica Lindl. (Ros) (wChin, now pantrop): loquat; nispero (del Japon) (PR); "Japan plum"; tree -7 m; large lvs., yellow frt. edible, delicious, rich in pectin; frt. eaten fresh or cooked, made into jelly; frt. u. to allay thirst & vomiting; fls. expect.; lf. c. saponins; inf. in diarrh., dys. (Chin); diabetes, rheum. (Af); lf. decoc. u. in cough & nausea (Hong Kong); sd. c. HCN; ornam. (cult. Jap, Indo-Pak, USSR, Eu, US).

Eriocaulon L. (Eriocaul.) (worldwide): pipeworts; hat pins; bachelor buttons (wFla [US]); these monocots occur mostly in wet places and have crowded basal lvs. with long flowering scape; some 500 spp. - **E. bifistulosum** van Heurck & Muell. Arg. (Zimbabwe): water pipewort; aquatic sp. floating on water; lvs. u. as food (natives) - **E. buergerianum** Koern. (seAs): fl. heads & hb. sold separately in markets (225); fl. heads prescribed for headaches (hemicrania), cataract, toothache, sore throat; fl. head in eye dis. (doctrine of signatures); astr. in nosebleed (Chin) - **E. setaceum** L. (seAs): pl. boiled in oil & u. for scabies (itch) (Chin, EI) - **E. sexangulare** L. (E. sinicum Miq.) (sAs): rumput butang (Mal); pl. u. sed. anodyne, antipyr. cooling, diur.; in nosebleed; drug u. in fresh state (Chin).

Eriocephalus africanus L. (Comp.) (sAf): estragon du Cap (Fr); Cape of Good Hope shrub; source of an attractive VO; hb. u. as diaph. diur. - **E. ericoides** Druce (sAf): u. like prec. - **E. punctulatus** DC (sAf): hb. u. to fumigate hut of child with cold or diarrh. - **E. racemosus** L. (sAf): u. diaph. diur. - **E. umbellulatus** DC (sAf): wild rosemary; u. as diur. in edema (dropsy); in colic; tinct. u. in heart troubles.

Eriocereus platygonus (Otto) Ricc. (Cereus p. Otto) (Cact.) (SA): frt. eaten (—> *Harrisia*).

Eriocoelum microspermum Radlk. (Sapind.) (trop cAf): kaboli; bk. u. for coughs, VD, flatulence.

Eriodendron = *Ceiba* (*E. anfractuosum* DC = *Ceiba pentandra*).

Eriodictyon californicum Torr. (E. glutinosum Benth.) (Hydrophyll.) (Calif, Ore [US], nMex): (Calif) yerba santa; lvs. c. VO, eriodictyol (1301), resins; u. bitter ton. to mask bitter tastes like strychnine, expect. in resp. dis. as asthma & allergies, bronchitis; appl. to warts, wounds; has antibacterial properties (226) - **E. crassifolium** Benth. (*E. californicum* Decne.): pl. has antibiotic properties (Salle, 1951) - **E. tomentosum** Benth. (Calif): said to sometimes furnish Yerba Santa, but generally regarded as adult. of this article.

Erioglossum rubiginosum Bl. (Sapind.) (seAs+): rt. decoc. u. antipyr., for cough (nBorneo); lvs. for poultices for skin dis. or in embrocations; young shoots eaten as veget. to induce sleep (Mal); sd. decoc. in pertussis (Ind); frt. edible (—> *Lepisanthes*).

Eriogonum corymbosum Benth. (wNA): boiled lvs. eaten with maize (corn) meal by Ariz (US) Indians - **E. fasciculatum** Benth. (Polygon.) (Calif, Nev, NM [US]): Calif (or wild) buckwheat; boiled pl. u. in witchcraft (Navaho Indians) (127); frt. edible - **E. inflatum** Torr. & Frém. (Calif & lower Calif; e to Utah): desert trumpet; rhiz. u. for pickles; tender sts. eaten raw - **E. latifolium** Sm. (wNA as Calif): "sour grass"; naked stem eriogonum; decoc. of lvs. st. & rt. u. for stomachache, headache, female complaints; rt. decoc. u. for sore eyes; young sts. eaten as delicacy (Indians, Mendocino Co., Calif) (2441); edible - **E. longifolium** Nutt. (US - Great Plains s to Tex, Fla): Indian turnip; long-leaved eriogonum; rts. eaten by Kiowa Indians - **E. microthecium** Nutt. (wNA): wild buckwheat; b. brush; rts. edible; tea fr. rts. & tops u. in TB cough (Nevada Indians) - **E. nudum** Dougl. (Calif): tibinaguq; young sts. chewed for several days as spring ton. (Calif, Hispanics) (205) - **E. umbellatum** Torr. (wUS: Wash-nCalif, e to Colo): sulfur flower(ed umbrella plant); rts. eaten cooked or raw before flowering - **E. umbellatum** var. *stellatum* M. E. Jones (E. s. Benth.) (sOre, sNev, adj. Calif [US]): cottony lvs. placed on burns to ease the pain (Klamath Indians, Ore) (3583).

Eriolaena quinquelocularis Wight (Sterculi) (Ind): kattale (Kanarese): rt. decoc. in fevers; sd. decoc. in pertussis; lvs. & rts. u. for poulticing (Mal).

Eriophorum L. (Cyper) (nTemp Zone): in some spp., hairs of inflorescence u. to stuff pillows, &c.

Eriophyllum confertiflorum Gray (Comp.) (sCalif [US]; nwMex): yellow yarrow; hb. c. eriofertopin, a germacradienolide, with active tumor inhibition (128) - **E. lanatum** Forbes (E. caespitosum Dougl.) (wNA): woolly sunflower; lvs. rubbed on face to prevent chapping; dr. fl. mixed with grease & appl. as "love charm" (nWash [US] Indians); lvs. appl. as poultice to painful aching parts.

Eriopsis sceptrum Reich. f. & Wars. (Orchid.) (Br, Col): basal st. of this epiphyte boiled & mucilage extd.; appl. to sores of gum & oral mucous membr.

Eriosema chinense Vogel (Leg) (eAs): edible tuberous rt. u. as expect. in coughs; kidney dis. - **E. chrysadenium** Taub. (w&c tropAf): ikoma (Ruanda); var. *intermedium* Hauman (nCongo): macerated rt. u. for bellyache - **E. diffusum** (HBK) Don (Mex, CA, Col): "guapillo" (Guat, Hond): pl. inf. u. in female dis. (Guat) - **E. glomeratum** Hook. f. (tropAf): pd. sd. u. in ophthalmia; pd. lf. u. to remove warts, vulner. as dressing for sores, boils, styes; pounded lf. fish poison; native vegetable - **E. griseum** Bak. (w tropAf): kabala; hb. u. to reduce swellings (Congo natives) - **E. montanum** Bak. f. (tropAf): hb. u. as vulner. vermif. purg. - **E. psoraleoides** (Lam.) Don. (s&cAf): lf. & rt. u. to treat syph.; rt. decoc. aphrod. for impotence, stomachache (Tanz); lf. rubbed on dog for control of lice; fls. & frts. as calming inhalations & for headaches; rt. u. as vulner.; fish poison - **E. salignum** E. Mey. (sAf): rt. u. as expect. diur.; stim. to bull (129) - **E. tephrosioides** Harms (var. *angustifoliolatum*) Hauman (Congo): lf. inf. as remedy for relapsing fever (2502).

Erisimum = *Erysimum*.

Erithalis fruticosa L. (Rubi.) (sFla, WI, Mex to Hond): vibona (Cu); "black torch" (Bah); lf. decoc. u. for sores, measles (Bah).

Eritrichium gnaphaloides A DC (Boragin) (Pe, Chin): u. stomach digest. diur.

Erlangea moramballae S. Moore (Comp) (trop eAf): lf. u. in meteorism, habitual abortion; abdom. troubles esp. after childbirth (Tanz) - **E. tomentosa** S. Moore (Tanz): "sulla"; rt. decoc. u. for stomachache; rt. & lf. ext. drunk as anti-emetic; lvs. crushed in little water & paste put on sores caused by skin maggots (kills); raw roots chewed for sore throat.

Ernodea littoralis Sw. (Rubi) (Fla, WI, Yucatan): "cough bush" (Bah); inf. u. for coughs.

Erodium L'Hér. (Gerani.) (worldwide): heronsbills; storksbills; crowfoots (Austral); low hbs. rather like *Geranium* spp. - **E. cicutarium** (L) L'Hér. (Eu, esp. Med reg., As, now worldwide in temp reg., incl. NA): stork's bill; crane b.; red stem filaree; wild musk; alfilaria; alfilerilla (Calif); weedy pl.; hb. u. as diur. astr. hemostat. in diarrh. dys. (Af); menorrhag.; sds. u. as pasty poultice in tophi (gout) (Bo); decoc. u. for gargle (Mex); uter. stim. (debatable); young pl. u. in salads; pl. rel. drought-resistant; u. hay, forage crop (1302); style remains in frt. very hygroscopic, curls in accordance with humidity, u. in some hygrometers - **E. moschatum** L'Hér. (Med reg., Euras, Austral, now widely distr. thus in US, chiefly Pac): musk clover (pl. smells like musk); musk heron's bill; musky crowfoot; pl. formerly u. (as Herba Geranii moschati) as diaph., antipyr. vulner. in intest. dis., esp. dys. (1303); inf. u. for gas on stom. (Ec); ornam.; forage crop - **E. stephanianum** Willd. (eAs): whole pl. u. for pains, numbness of limbs, rheum., nerv. dis. (Chin).

ers: *Lathyrus niger*.

Ersatz (Ge): substit., artificial.

Ertron™: vitamin D.

Eruca vesicaria (L.) Cav. subsp. *sativa* (Mill.) Thellung (Cruc.) (Med reg., sEu [wild & cult.], As, now introd. in Am, Austral): garden rocket; rocket salad; hb. u. arom. aphrod. (1307), stim. diur. antiscorb.; lvs. u. salad; sd. c. FO; u. in prepn. mustard; important oil sd. of Indo-Pak; jamba oil (semi-drying) furnished.

erucic acid: important component of *Eruca* & Colza oils; C_{22} with one =; cardiotoxic (injures heart muscle); now mostly fr. rapeseed oil (Cana) (40-50% of total fatty acids); u. in mf. plastics (nylon); lubricants.

erva (Por): herb (v. also herva) - **e. bicha**: *Aristolochia longa* - **e. cidreira: e. de bugre: e. de lagarto** (Br): *Melissa, Casearia sylvestris* - **e. de bicho** (Por): *Polygonum hydropiper* - **e. de lombrigueira**: *Chenopodium ambrosioides* - **e. de mac(h)e** (Br): *Leonurus sibiricus* - **e. de passarinho**: *Phoradendron* & *Struthanthus* spp. - **e. de rato** (Br): *Psychotria marcgravii* lvs. - **e. de Santa Maria** (Br): *Chenopodium ambrosioides* var. *dentata* - **e. de São João**: *Glechoma hederacea* - **e. doce**: anise; fennel - **e. malamiente** (It): (bad hb.): poison oak (or p. ivy) - **e. tostão** (Br): *Boerhaavia*.

ervagem (Por): herbage; pot herb.

ervanario (Por): herbalist.

Ervatamia bufalina (Lour.) Pichon (Apocyn.) (eAs): rt. chewed to cure sore throat (Hainan, Chin); boiled crushed lvs. u. for boils, swellings (Hainan); bitter rts. u. for stomach dis., tonsillitis (Indochin); latex emoll. - **E. coronaria** (Jacq.) Stapf (Tabernaemontana c. Jacq.) (Apocyn) (Ind, Oceania, Austral): Ceylon jasmine; grape jasmine; East Indian roseberry; crape-jasmine; rt. anodyne, chewed for toothache; rt. bk. c. 1.5-2% alk., tabernaemontanine, coronarine, dregamine, resin alc., &c.; rt. bk. u. as vermifuge, in diarrh. dys. cataract; fls. c. steroid-like compds.; u. cardiac dis.; frt. pulp as dye; wd. u. as refrigerant, incense; in perfumery; latex in eye dis.; pl. toxic & lethal - **E. cylindrocarpa** King & Gamble (eAs): lvs. pounded with turmeric & rice u. as poultice in itch, eczema, beri beri (Mal) - **E. dichotoma** (Roxb.) Bl. (Ind): lvs. bk. latex & sds. purg. cath.; rt. & bk. u. as analg. in scorpion sting - **E. hainanensis** Tsiang (Chin): tan-ken-mu; rt. c. alks., u. to produce marked decrease in serum cholesterol & triglycerides, incr. in high density lipoproteins (HDL) - **E. hirta** Hing & Gamble (eAs): steam fr. boiled pl. parts inhaled for nasal ulceration; in dart poison - **E. liguana** Boiteau & Allorge (New Caled): u. to purge the new-born baby; to aid in cicatrization of umbilical cord (—> *Tabernaemontana*).

ervilha (Por): *Pisum sativum*.

Ervum L. = *Vicia* &c. - **E. lens** L. = *Lens esculenta*.

Erycibe malaccensis C. B. Clarke (Convolvul) (Mal): lvs. appl. as poultice to sores & to head for headache - **E. paniculata** Roxb. (Ind): kari; bk. & frt. pd. u. in cholera (124) - **E. rheedii** Bl. (Indones): rts. boiled in oil rubbed into abdomen to hasten delivery in labor.

Eryngium aquaticum L. (Umb) (e&sUS): button (or corn) snakeroot; eryngo; rhiz. & rt. (Eryngium, USP 1820-60) have minor uses as diaph. diur. exp. emm. sial., to dispel flatulence, emet. stim. GC (seAm Indians) (popular in Civil War) (1305) - **E. campestre** L. (Euras): field eryngo; basal lvs. eaten (Greece); rt. decoc. u. in hysteria (WI); in gravel, diur. in dropsy, jaundice (1804); ton. - **E. carlinae** Delwar (Mex, CA): escorzionera (Guat); in domestic med. in diarrh. & female dis. (Guat) - **E. foetidum** L. (WI [Jam]; Mex, Bo, Br; natd. Af, Java+): fitweed; "walung doeri"; culantro (Guat); lvs. u. condim. snakebite, hydrophobia(!), emmen.; pl. u. diarrh., diaph., antipyr. in severe fevers; aphrod. exp.; in edema; stim. in flatulent colic, hysteria (decoc.) (Cu); tea for "any" sickness (Dom Rep); frts. edible (Chin); cult. in Fla - **E. giganteum** Bieb. (Caucasus, As-Minor): Elfenbeindistel; c. sapogenins (1306); cult. ornam. - **E. heterophyllum** Engelm. (sTex [US], Mex): Wright eryngo; yerba de sapo; tea u. for kidney dis. (Mex) - **E. horridum** Malme (SA): turututu (Arg); rhiz. u. purg. (Arg) - **E. maritimum** L. (Eu coasts): sea holly; (sea) eryngo; rt. u. tonic, arom. diur., exp., formerly candied ("candied comfits"); reputed aphrod. (Eng folk med.) - **E. planum** L. (Euras): rt. u. for coughs, diur. - **E. yuccifolium** Mx. (NA, SpAm): button snakeroot; rattlesnake master; r. weed; water eryngo; w. snakeroot; u. as diur. expect. larger doses almost emetic (natives of Arg; Nuttall, ca. 1820); diaph. in snakebite (spurious claim); var. *synchaetum* Gray (E. s. Rose) (Fla-Tex, Ga [US]): rt. u. in making "black drink" (parde) (decoc.) of seAm Indians (Yuchi; Seminole) (also u. cassine).

eryngo: *Eryngium* spp., sometimes specif. *E. campestre* - **water e.**: *E. aquaticum*.

Erypar™: erythromycin stearate; u. in resp. inf., VD (where allergy to penicillin); u. parenteral (hence name).

Erysimum L. (Cruc.) (Euras, Temp reg.): blister cress; treacle mustards - **E. barbarea** L.: *Barbarea vulgaris* - **E. cheiranthoides** L. (Euras, nAf): treacle mustard; "wormseed"; hb. u. stom. vermifuge; in cardiac insufficiency - **E. crepidifolium** Rchb. (c&eEu): Sterbekraut; hb. c. helveticoside u. as cardiac with digitalis action - **E. diffusum** Ehrh. (E. canescens Roth) (cAs, also steppes of eEu): c. erysimoside, helveticoside (cardenolides) u. for cardiac insufficiency - **E. helveticum** DC (Eu): hb. c. helveticoside, cardioactive glycoside; u. heart dis. - **E. officinale** L.: *Sisymbrium o.* - **E. perofskianum** Fisch. & Mey. (Afghan, Baluchist): hb. c. several cardioactive glycosides: perofskoside, kabuloside, corchoroside A, &c.; u. for cardiac insufficiency.

Erytaurin: a glycoside, active princ. of centaury, *Centaurium erythraea*.

Erythea aculeata Brandegee (Palm) (Baja Calif [Mex]; cult. sCalif [US]): (Sinaloa) hesper palm; roasted sds. eaten; cult. as ornam. - **E. armata** Watson (Baja Calif, Mex): palm branch blanca; (Mexican) blue hesper palm; cult. ornam. (sCalif, Sonora) (227) - **E. edulis** Watson (Baja Calif [Mex]): frt. pulp sweet & edible; buds also eaten; u. med. (Mex) (227) (—> *Brahea*).

erythorbic acid: isoascorbic acid: isovitamin C: prod. by *Penicillium* spp.; u. as antixoidant; antimicrobial (in foods, canned meats, &c.): u. as Na salt, Na erythorbate, with only ca. 5% of vit. C activity.

Erythraea Borckh. = *Centaurium*.

Erythrina L. (Leg) (mostly in warm temp & trop areas of world): coral tree; about 110 spp. of herbs, shrubs & trees; sds. c. alk. often with a curare-like action (2504) but have tertiary bases not quaternary-like curare alk. (ex. erythroidine, erysodine), suggested of use in med., hence intensive study of alks. (much by Folkers at Merck) (1940s); many spp. of Mex u. as follows; stem & bk.: fish poison; bk. hypnotic: rts. diaph.; lvs. emm.; hypnotic; fls. in pulmon. dis. (decoc.); st. juice in scorpion stings; inf. u. in canine dis. (Seminole Indians) - **E. americana** Mill. (E. cornea Ait.) (sMex): colorin; pito; bk. u. diur. sed.; sds. u. intoxicant ("colorines"); fls. edible - **E. arborea** Small: *E. herbacea* var. *a.* (but sometimes *E. arborea* is regarded as a good sp.) - **E. berteroana** Urban (tropAm, sMex to Co, WI): coral bean; machette; parsu (Cuna Indians, Pan): u. for female dis. (Cuna Indians, Pan); pinon de pito (Cu); u. for female dis. (Cuna Indians); sed. hypn. antipyr. diur. purg. (WI); twigs, buds & young lvs. eaten as veg.; branches as fish poison; wd. u. as fuel & cork substit. (occasional misspellings: *E. berteroi*, *E. broteroi*) - **E. caffra** Thunb. (sAf): Kaffirboom; lvs. c. FO, VO, tannin, u. chancroid (cult. as ornam. for brilliant fls., Am); wd. very light - **E. corallodendrum** L. (As, WI, nSA, esp. Br): coral tree; immortal wood; coral bean tree; mulungu; bk. u. anod. esp. dyestuff; lvs. u. diur. lax.; in tumors; wd. is cork-like "coral wood"; cult. ornam. tree for showy fls. (sUS), and as living fence posts (Domin) - **E. coralloides** Moç. & Sessé (Mex): coral tree; "colorin" (Puebla, Mex): st. u. for scorpion stings, in epilepsy - **E. costa-ricensis** M. Micheli (CR): poró; u. popular med.; fls. believed soporific (inf.); lf. decoc. appl. to wounds or leishmaniasis (2444); female dis. (Pan); furnish living fences - **E. cristagalli** L. (Br; cult. New Orleans, La [US]): cockspur coral-tree; "cry-baby tree" (La); bk. u. astr. in diarrh., anthelm.; sed.; poisonous; toxicity stand in way of med. use; ornam. tree; planted for shade in coffee plantations - **E. edulis** Triana (SA): frts. edible; u. as shade tree - **E. flabelliformis** Kearney (sNM [US], wMex): chilicote; wd. u. in carvings; sds. u. for toothache, intest. disorders; in poultices for eye ailments - **E. folkersii** Krukoff & Moldenke (Mex to Co): tiger tree; sd. u. as diur. to cure styes (eyelids) (per ora); fls. in resp. dis.; rt. diaph.; fls. sed. - **E. glauca** Willd. (Br, Ve, WI): piñon; suinan (Br); fresh lvs. u. diaph. anthelm.; alks. patented - **E. guatemalensis** Krukoff (Guat): pito; bk. fls. lvs. hypnotic, u. in asthma, hysteria; sds. in divination - **E. herbacea** L. (seUS): cardinal spear; red cardinal; Cherokee bean; imortelle; cox's comb; coral plant; c. tree; (eastern) c. bean (Fla); mamou (La-Fr); cry baby tree; rt. u. diaph. anticoag.; var. *arborea* Small (Fla, s tip & Keys): herb, shrub, woody vine or tree; sometimes cult. as also the sp.; sd. an article of commerce among Indians formerly - **E. indica** Lam. = *E. variegata* - **E. lanceolata** Standl. (Hond): "pito"; tea u. for kidneys - **E. pallida** Britt. & Rose (WI, Ve): lvs. u. as poultice for sores (Trin); lf. decoc. u. to bathe same - **E. poeppigiana** (Walp.) O. F. Cook (tropAm, Pan-Bol; sometimes cult.): mountain immortelle; bucare (Mex); bk. twigs & sds. u. med. & as fish poison; paste fr. bk. appl. hot to strained ligaments (Ec); fls. eaten in soups & salads (CR); coffee plantation shade tree (Col) - **E. polyanthes** Hassk. (Java): bk. c. poisonous alk., less in lvs. - **E. rubrinervia** HBK (CA): gallito (Pan); u. hyp. poison; fls. & buds eaten cooked (Pan); lvs. in soups (Salv); trees planted to shade coffee trees - **E. sacleuxii** Hua (Tanz): hb. decoc. u. in GC by natives - **E. senegalensis** DC (w tropAf): arbre corail; rt. decoc. u. colitis, colic, renal pain, tympanites, dys.; trunk bk. u. in amenorrh., mal.; ext. in collyria for eye dis. - **E. standleyana** Krukoff (Hond): bk. eaten to treat intest. parasites (2395) - **E. suberosa** Roxb. (Indo-Pak, Burma): madar (Ind); lvs. u. as anthelm., cath., emmen., galact., exter. for ulcers, rheum.; bk. lax. emmen. in snakebite; bk. & lvs. u. sed. hypnotic, anticonvuls.; sd. c. erysodine, hypaphorine (latter also in lf.); wd. u. as furniture; bk. u. as cork - **E. subumbrans** (Hassk.) Merr. (seAs to PI): bk. decoc. u. for splenomegaly (Mal); steamed lvs. eaten in salads (PI) (2475) - **E. variegata** L. (Ind, Mal, Oceanica; cult. US): Indian coral-tree; dadap tree; bk. astr. anthelm. as collyrium in ophthalmia; lvs. lax. diur. anthelm. alt. galact. emmen.; lvs. & rts. antipyr.; lf. juice cath. vermif.; lvs. as cataplasm for chest inflammn. (Vietnam); tree sap u. in cough medicine (Solomon Is); shade ornam. tree - **E. verna** Vell. (E. mulungu Mart.) (Br, Pe): mulungu; trunk bk. u. as hypnotic, mild sed. (Br Phar.); anthelm. in croup; sds. u. as beads; cult. (Rio de Janeiro); trunk bk. important potential source of cork for use in stoppers, insulation, &c. - **E. vespertilio** Benth. (Austral): bat's wing coral tree; corkwood; lf. decoc. sed.; wd. u. to make shields (aborigines of cAustral).

Erythrite: Erythritol: **Erythrol:** a polyatomic alc. found in some algae, lichens, fungi, grasses; nitrate deriv. u. as vasodilator.

erythrocentaurin: aglycone of erytaurin (fr. *Centaurium erythraea* & *Swertia japonica* (swertiamarin) (u. bitter ton.).

Erythrochiton brasiliensis Nees & C. Mart. (Rut) (Ve, Br): u. in sore throat; as anthelm.

Erythrocin™: erythromycin.

Erythrococca anomala Prain (Euphorbi.) (w tropAf): lvs. u. analg. antisep. for ear- and headache, sore eyes; believed specific in tapeworm (Nig) - **E. bongensis** Pax (e tropAf): ngesa (Padhola); lf. juice u. in cough; lvs. pounded & cooked for stomach trouble (with peanuts) (130) - **E. fischeri** Pax (trop eAf): mtotwe (Shambaa): pd. rts. for chest complaints; boiled rts. u. anthelm. infertility, GC; lf. u. snakebite - **E. menyharthii** (Pax) Prain (Tanz): pounded rts. eaten with honey for cough; lf. sap u. for snake poison in eye; lvs. u. as vegetable.

erythrocytes: red blood cells; RBC; "packed (= concentrated) red blood cells (human)"; these corpuscles are sepd. by centrifugation fr. blood plasma for use as blood quality replenisher in certain anemias; with white blood cells present; may be u. like whole blood (latter preferred in acute hemorrhages).

erythrogenic acid = isanic a.; found in trop FO, ex. *Ongokea gore*.

erythroidine: alk. fr. *Erythrina herbacea, E. alloides, E. berteroana,* & 5 or more other *E.* spp. with curare-like action (neuromuscular & ganglion blocking agt.); short acting; active p. o. (curare not); not yet u. in therapy (hazards); suggested for use in dystonia & spastic states.

Erythromycin™: antibiotic subst. for soil fungus sp., *Streptomyces erythraeus* Waksman & Hernrici; active against Gm.+ organisms & a wide range of other pathogens incl. viruses & rickettsiae, esp. effective in intracellular inf. (E); most effective therapy known for Legionnaire's disease.

Erythronium albidum Nutt. (Lili) (eUS & Cana): white dog tooth's violet; fawn lily; corms eaten raw in spring by Am Indian children (18) - **E. americanum** Ker-Gawl. (E. flavum Sm.) (eCana & US): yellow adder's tongue; dog tooth violet; trout lily; tarmia (Mo [US] Indians); bulb pulp appl. as poultice to boils; u. alt. emet., colchicum substit.; breast complaints (wash) (Mo Indians) (16); lvs. eaten as greens; cooked, aperient: antibiotic; rts. nutr. (NA Indians) - **E. dens-canis** L. (Euras, Manch, Chin, Jap to Mal): rt. source of starch (prepd. in Jap) u. in pasta, cakes; lvs. boiled & eaten with milk (2517) (Mal, Jap Phar.); fresh tubers u. as aphrod. anthelm. antiepileptic, for abscesses, freckles, "ulcers"; snake & insect bites; tubers often eaten with milk (Mal, Jap Phar.) - **E. giganteum** Lindl. (sOre, nCalif [US]): giant fawn lily; crushed corm appl. as poultice to boils; eaten at times (2441) - **E. grandiflorum** Pursh (wNA): yellow fawn lily; corms edible, boiled or dried, thus by Thompson Indians (17) - **E. japonicum** Decne (E. dens-canis L. var. j. Bak.) (Jap): corm eaten as ton. for enteritis, diarrh. (Kor) remedy for emesis (child); for snakebites; lax. (Jap, Korea); corm c. 40-50% starch, source of same (Jap); Amylum Erythronii (Jap Phar. I-IV); in pasta, cakes.

erythrophleine: highly toxic alk. obtained fr. *Erythrophleum* spp.

Erthrophleum coumingo Baill. (Leg.) (Madag, Seychelles): bk. c. erythrophlein (cardiac glycoside), couminine (0.6%), coumingaine (1307); u. ordeal poison (if suspect vomits & recovers, considered innocent) - **E. fordii** Oliv. (sChin, Indochin, wAf?): bk. has been u. by criminals in robbing people; u. suggested for dental surgery & eye treatment; wd. of tree of good quality - **E. Suaveolens** (Guill & Perrot) Brenan (E. guineense S. Don.) (trop wAf, Sierra Leone): red water tree; source of Sassy Bark; c. erythrophlein (card. glycoside), cassaidine (1308); u. emet. narc. astr. purg. anthelm. digitaloid card. agt., loc. anesth., arrow poison, appl. as ordeal poison (*E. fordii*, *E. ivorense* A. Chev. [tropAf] & *E. laboucherii* F. Muell. [Austral] said also to prod. sassy bk.; an unknown sp. believed source of "Muava" bark).

erythropoietin (Ep, Epo): factor stimulating erythropoiesis (blood mfr.); hormone isolated fr. human urine, produced chiefly by kidneys; high titer in anemic animals (when caused by chronic kidney failure); c. sialic acid, hexoses, hexosamine, glycoprotein; u. to treat other anemias (such as HIV-related); made by recombinant technics (Epoetin Alfa—USP XXII).

Erythroxylaceae: Coca Fam.; tropics; c. 260 spp. of woody pls. mostly shrubs; 4 genera (*Erythroxylum*, *Aneulophus*) characteristic alks. shown so far only in *Erythroxylon* spp. which also prod. the sole economic commodity of importance, Coca Leaves - **erythroxyline**: cocaine.

Erythroxylum P. Br. (**Erythroxylon** L.) (Erythroxyl.) (trop & substrop SA chiefly): some 250 spp. of shrubs; many u. med. (*E. cuneatum*, *E. ecarinatum*, *E. kunthianum*, *E. rotundifolium* & others [1311]) - **E. acuminatum** Walp. (sInd, SriL): bk. & lf. juice u. anthelmint - **E. coca** Lam. (wSA, esp. Pe, Ec, Bo): coca (shrub); Huanuco cocaine tree; known only as cultigen; source of coca lvs., cocaine; u. fumitory, masticatory (used fresh or dried) by SA Indians; beverage (Andean areas); flavor in colas (u. decocainized lf. ext.); loc. anesth. stim. nervine; as stimulating tea; to allay thirst (SA Indians) (1311); cocaine exported worldwide fr. Co to furnish addicts (2493); many vars. recognized; ex. narrow-lvd., broad-lvd., etc.; var. *ipadu* Plowman (wAmazon, Co, Ec, Pe): Amazonian coca; (i)padu; cultigen - **E. dekindtii** O. E. Schulz (Angola): lf. & rt. decoc. u. antipyr. - **E. ferrugineum** Cav. (Madag): lf. decoc. u. in diarrh.; wd. very tough - **E. havanense** Jacq. (WI, Mex): jiba; no alk. present; u. hemostat., popular hb. in Cu (yerberos, herb shops) - **E. hypericifolium** Lam. (Mascarene Is): frt. juice in diur. & depurative syr.; wd. u. for cabinet work - **E. laurifolium** Lam. (Mascarene Is): lvs. astr. diur.; frt. lax.; u. for yellow fever (Reunion): wd. in cabinet making - **E. mannii** Oliv. (Sierra Leone to Cameroon): lf. & st. decoc. u. antipyr.; bk. in liniment for pleurisy & intercostal pains (with lemons & sds. of *Aframomum melegueta*) - **E. monogynum** Roxb. (Ind, SriL): wd. & bk. inf. u. stom. diaph.. diur.; bk. u. ton.; considered useful in cholera & prolonged fevers (Chopra); lvs. cattle fodder & starvation food - **E. myrtoides** Boj. (Madag): lf. decoc. & inf. u. carm. diur. in cystic & renal calcuil; lf. decoc. in indigest. & GC - **E. neo-caledonicum** O.E. Schulz (Melanesia): u. ton. (New Caled) - **E. nitidum** Spreng. (Br): catuaba; u. med. - **E. novogranatense** (Morris) Hieron (nSA): Java cocaine tree; cocaine plant; sometimes interpreted as a var. of *E. coca*; u. similarly; var. *truxillense* Plowman (E. t. Rusby) (nwSA): Peruvian (or Truxillo or Truillo) coca; similar uses to *E. coca* - **E. pervillei** Baill. (Madag): pl. decoc. u. in lenitive (demulcent) baths - **E. retusum** Baill. (Madag): lf. decoc. u. as vermifuge - **E. rotundifolium** Lunan (sMex, Guat, WI): bo-hog (Bah); rat-wood; lf. decoc. u. with salt as ton. for run-down system (Bah) - **E. sechellarum** O. E. Schulz (Seychelles Is): pl. u. against colics - **E. steyermarkii** Plowman (neVe): hayo; lf. c. 0.1% cocaine; lf. chewed by natives as mild stim. & med. (2486).

erythrozyme: glucosidase (enzyme) fr. *Rubia tinctorum* (madder); decomposes rubian (glycoside).

Erzengel (Ge): *Angelica*, sometimes specif. *A. archangelica*.

escabiosa (Sp): *Scabiosa officinalis* - **escamonea** (**, rais de**) (Sp): levant scammony - **escaramujo** (Sp): *Rosa canina* - **escarola** (It, Por, Catalan): **escarole** (Fr): endive.

Escallonia illinita Presl (Escalloni./Grossalari) (Chile): barroco; u. in folk med. for liver dis. - **E. paniculata** (R. & P.) R. & S. var. *floribunda* (HBK) Macbr. (Ve): jarilla; lf. ext. u. in chimó (plug of tobacco made into a paste with "natron" and chewed by Andean natives); wd. hard but easy to work; u. in plows, lintels, &c. - **E. pulverulenta** Pers. (Ch): mardono; u. in popular med. for liver dis.; inf. u. against cough, bronchitis, asthma, as expect.; diur. urinary disinf. - **E. revoluta** Pers. (Ch): l(i)un; u. much like prec.; also **E. rubra** Pers. (nipa) and **E. virgata** Pers. (meki) both of Ch.

escareol: **escarolle:** endive.

Esche (Ge): ash, *Fraxinus* spp.

Eschschol(t)zia californica Cham. (Papaver.) (Calif, Ore [US]): Calif. poppy; chamiso; c. alks. & glycosides; rutin; fresh rt. pulp inserted in dental cavity to ease pain; dr. rt. u. by gamblers to make opponents less keen (228); rt. ext. u. as wash for headache, supporating sores, to reduce milk secretion in women; int. emet. & to treat TB; lf. ext. in stomachache; lvs. sometimes eaten (with care) (2441); hb. u. hypn. analg.; Calif state fl.

Eschweilera C. Mart. ex DC (Lecythid.) (tropAm): boiled bk. of some sp. rubbed into skin of patient with fever (nwBr Indians)

escila (Sp, Po): squill - **e. roja** (Sp): red s.

escin: aescin; mixt. of saponins fr. sd. of *Aesculus hippocastanum*, horse chestnut; 2 major glycosides deriv. of aglycone, protoescigenin (triterpene); u. to treat peripheral vascular dis., hemorrhoids, varices (varicose veins), thrombophlebitis, &c.

escoba (SpAm): *Sida carpinifolia* - **e. amargo(sa)**: *Scoparia dulcis*; *Parthenium hysterophorus* - **e. amarilla**: *Sida rhombifolia* - **e. blanca** (CA as CR): **e. de castilla**: *Scoparia dulcis* (CA), *Sida carpinifolia* & other Malv.

Escobedia linearis Schl. (Sciophulari) (Mex): probably u. in same way as *E. scabrifolia*; rts. of this & other spp. u. for dyeing - **E. scabrifolia** R. & P. (CA, Mex, Par, Ch, Pe): Peru turmeric; azafran; palillo (Pe); azafranillo; rts. u. for dyeing conf. & foods yellow; dye princ. called "azafrin" (-8%); considerable exports fr. Pe to Ch (*E. laevis* Cham. & Schlechthd. (Br): u. similarly.

escobo (SpAm): 1) escoba 2) *Corchorus siliquosus*.

Escontria chiotilla (Schum) Rose (1) (Cact) (sMex): edible frts. sold in markets as chiotolla, geotilla, tuna, &c. (Tehuacán) (229).

esculent: edible; fit for food use.

esculin: aesculin.

esculoside: glycoside of esculin; occ. in bk. lvs., sd. of horse chestnut, *Aesculus hippocastanum*; u. as sunscreen agt. (lotion); in hemorrhoids (suppos.); in increased capillary fragility (*cf.* vit. P).

esdragol(e): estragole.

Esenbeckia febrifuga A. Juss. (Rut.) (Br, Par, Arg): "quina"; bk. (Cortex Angusturae brasiliensis) u. as substit. for cinchona in mal.; lvs. as diaph.; VO of frts. u. as vermifuge; wd. made into dishes, &c.; bk. c. esenbeckine, evodine, bitter princ. - **E. flava** HB & K (Hon): palo amarillo; u. med. in CR - **E. grandiflora** Mart. (Br): bk. powerful anthelm.; lf. decoc. u. med. - **E. leiocarpa** Engl. (Br, Ch): diaph.; oil of frt. u. as vermif.

esencia (Sp): VO.

esere nut: sd. of *Physostigma venenosum* - **eserine**: physostigmine.

Eskel (SKF): khellin.

espalier trees: frt. trees trained on a railing, trellis, etc.

Española (, La): Hispaniola (island of WI); occupied by Haiti & Dominican Rep.

esparceta (Sp): *Onobrychis viciaefolia* - **esparrago** (Sp): asparagus - **espartillo** (SpAm): 1) Chasmanthium laxum (*Uniola laxa*) 2) Schizachycium gracile (*Andropogon gracilis* var. *firmior*) 3) *Sporobolus* spp. 4) various other grasses & sedges - **e. oil**: VO fr. *Elionurus latiflorus* - **esparto** (grass) (nAf): 1) *Stipa tenacissima* 2) *Lygeum spartum* 3) *Ampelodesma*.

esparto wax: Spanish grass w.; wax extd. fr. Med reg. grasses, *Stipa tenacissima* L. & *Lygeum spartum* L. (Gram.); u. as substit. for carnauba wax; esp. for cosmetics, carbon papers, &c.

especias (Sp): spices.

especia (Sp): **especie** (Sp): species.

espelina (Br): *Cayaponia espelina.*

espinaca (Sp): **espinafre** (Br): **espiniche** (NW, Sp): spinach.

espinar (Sp): thorn forest; area of scanty rainfall bearing scattered small thorny trees; mostly in tropics (ex. Venezuela).

espinheira santa (Br): *Maytenus ilicifolia.*

espinheiro alvar (Por): *Crataegus oxyacantha.*

espino (Sp): 1) thorn 2) *Crataegus* spp. - **c. blanco**: *C. oxyacantha* - **espino cerval** (Sp): *Rhamnus cathartica* - **e. negro**: buckthorn (spp. of *Rhamnus*, *Prunus spinosa*).

espliego (Sp): *Lavandula latifolia*, *L. angustifolia* (L. spica).

espondilia (Sp): *Heracleum sphondylium.*

esrar (Turk): cannabis prepn.

essence: 1) (Fr): VO, except for a few foreign preps 2) (US, Eng): flavoring extract, usually made by dissolving a small amt. of VO in alc. - **e. de térébenthine** (Swiss Phar.): turpentine oil - **e. of lemon**: Tr. Lemon Peel - **e. of mirbane**: nitrobenzene; artif. bitter almonds oil (poison!) - **e. of neroli**: VO of orange fls. - **e. of peppermint**: Spt. of Peppermint - **e. of pepsin**: Elix. pepsin & rennin comp. - **e. of roses**: *Rosa moschata* VO - **e. of rum**: ethyl formate; u. to flavor rum - **e. of spruce**: *Picea mariana*; VO - **e. of turpentine**: t. oil.

essencia (Sp, Por): **essentia** (L): VO - **e. bina** (Sp): caramel.

essential oil: essenza (It): VO.

Essig (Ge): vinegar.

Essigbaum (Sp): *Rhus typhina.*

estafiata: estafiate (Mex): *Artemisia mexicana* & sometimes other *A.* spp. - **estafisagria** (Sp): stavesacre.

estearina (Sp): stearin.
estepa blanca (Sp): *Cistus albidus.*
ester: organic salt formed from an alc. and acid - **cyanacetic e.**: ethyl cyanacetate - **e. gum**: product made by combining resin of acid type with an alc. or phenol; resembles the hard natural resins; u. in varnishes & paints - **malonic e.**: ethyl malonate.
esterase: enzyme which splits (hydrolyzes) esters into alc. and acid; or synthesizes esters from alc. and acid; many specific esterases are known; richly present in liver; ex. lipases, chlorophyllase, tannase.
esterois (Por): sterols.
estilos de maiz (Sp): maize styles.
estivation: 1) arrangement of fl. parts in bud 2) survival by organism of heat & drought.
Estivin™: liq. prepn. of *Rosa gallica*; u. as eye drops for conjunctivitis in hay fever.
estoraque: 1) (Sp): storax 2) (Br): Tolu balsam, Peru b. (*cf.* benjui).
estradiol: the primary & most potent estrogenic hormone, which breaks down to give estrone; u. as for estrogen - **e. benzoate: Estradiolis Benzoas** (L.): ester which is FO-sol & may be adm. in such form parenterally - **e. dipropionate**: u. medicinally similar to prec. but with advantage of a longer continuance of action.
estragole: methyl chavicol; l-ally-4-methoxy-benzene; isoanethole; active constituent of VO of tarragon, fennel, basil, star anise; avocado bk. - **Estragon** (Ge): tarragon.
estramonio (Sp): Stramonium.
estrellamar (Sp): *Convallaria majalis.*
estriol: estrogen formed in body fr. estradiol; obt. fr. pregnancy urine, human placentae & by synthesis; occ. in many plants; function there unknown; some consider less likely to cause breast cancer.
estrogen: estrogenic hormone: steroid compd. which acts thru the blood stream to develop the typical female sex characters (1899), stim. proliferation of endometrium & cornification of vaginal mucous membrane; the "female hormone" secreted by ovary & placenta but some synthetic e. are non-steroidal (ex. diethylstilbesterol, dienestrol); also u. in prostatic carcinoma, hypermasculinity; very small doses reduce cholesterol in blood of male; for postmenopausal symptoms (locally in transdermal patch); for postmenopausal osteoporosis (per ora) (FDA approv. 1986); best known representatives are: estradiol, estriol, estrone - **conjugated e.**: mixt. of Na salts of sulfate esters of estrogens of type excreted in pregnant mare urine (PMU) (estrone, equilin); u. orally; being u. (often with progestin) in coronary heart dis. & in hormone replacement therapy (HRT) & to prevent osteoporosis - **esterified e**: mixt. of Na salts of sulfate esters similar to conjugated e. but with more estrone and less equilin; u. in same way.
estrogenic substances, water insoluble: Amniotin; conc. soln. of estrone with smaller amts. of other estrogenic ketones; obt. fr. mare pregnancy urine (PMU) - **water soluble e. s.**: Premarin; naturally occurring HOH-sol. conjugated forms of mixed estrogens fr. mare pregnancy urine (PMU).
estrone: Estronum (L): theelin; hydroxy-estratrienone; keto-hydroxy-estrin; metabolite of estradiol with much less activity; obt. fr. pregnancy urine (mares', womens'), human placenta; stallion urine; occ. in palm kernel oil, & other pl. substances; synthesized fr. ergosterol; u. as estrogen & in synth. of 19-nor steroids.
estropajo (Cu, Co): towel gourd, *Luffa* spp.
Ethal: Ethol; cetyl alcohol.
ethanol: ethyl alc.; CH_3CH_2OH said first org. compd. synthesized (Henry Hennel, 1826); occ. in frt. of *Heracleum giganteum, Pastinaca sativa*, orange juice, etc.; prep. by distn. of fermented saccharine liquids; u. fuel, preservative, solvent; sedative, hypnotic, aphrodisiac(?); delirifacient; quantities present in beer 2-5%; wine 7-13%, fortified wines (like sherry) 17-20%, whisky, brandy, rum, gin, etc. ("spirits") 40-45%; often abused (2283); mixed with gasoline (10%) as motor fuel (Br, US, Eu) - **dehydrated e.** = anhydrous e.
ethanol & dextrose (glucose) **injection**: soln. of 5-10% ethanol with 5% dextrose in water; u. in emergency situations to incr. caloric intake; in acute methanol poisoning; as pre-surgical analgesic.
ether-alcohol: mixt. of alc. & ether often u. in analyses & chem. prepn. as solvent, ppt. reagent; sometimes official as Spiritus Aethereus or S. Aetheris; 1 pt. ether & 3 pts. 95% alc.
ether soluble extractive: determined as follows: small amt. of pd. drug is dried over conc. sulfuric acid, then submitted to solvent action of anhydrous ether in a continuous extractor (Soxhlet apparatus) until completely extracted: the ether soln. is then allowed to stand in an open dish to evaporate spontaneously; weigh, then heat to 105° to constant wt.; the difference in wts. in the **volatile e.s. e.**; the residue left finally is the **non-volatile e. s. e.**
etheral oil: ethereal oil: etherische Oel (Ge): 1) mixt. of ether, light & heavy oils of wine (ethyl pelargonate, et. caprate, et. caprylate), & alc.; u. as base in mfr. Spt. Ether Comp. 2) VO.
éthérat: éthérolat (Fr): counterpart of alcoolat; class now abandoned - **etherolature** (Fr): counterpart

of alcoholature; ethereal trs. of green drugs - **éthérole** (Fr): simple soln. of drug materials in ether.

ethical pharmacy (etc.): one which adheres strictly to the pharmacists' code of ethics; term sometimes not in good form since reflecting unfairly on competitive units in the profession - **e. proprietaries**: those which are promoted at the professional level & not to the layman; ordinarily sold only on prescription or physician's order.

ethinyl estradiol: one of the most potent natural estrogens known; u. perorally; in birth control agts. (combn.) (E).

ethnopharmacognosy: science of crude drugs utilized by ethnic groups (folklore), a division of general pharmacognosy.

Ethulia conyzoides L.f. (1) (Comp.) (tropAf, As): hb. u. for intest. parasites, colic, abdom. dis. (Zulus); counterirrit.; for roundworm & ophthalmia (Madag); lvs. to prevent abortion (Liberia); pressed on fresh cuts; lf. juice u. for sore bloodshot eyes (Tanz); weed.

ethyl alcohol: ethanol - **e. chaulmoograte**: ethyl ester of chaulmoogric acid; u. parenterally in treating Hansen's dis. (leprosy) - **e. oenanthate**: wine oil - **ethylhydrocupreine hydrochloride**: Optoquine (Optochin) HCl; Numoquin HCl; semisynth. alk. made fr. quinine; also occurs naturally; locally u. for pneumococcus infections of eye; appl. locally to eye as loc. anesthetic; antimalarial; fraught with some danger (blindness).

ethylene: $CH_2=CH_2$: pl. hormone; gas used to ripen & color citrus frts., &c.; occ. in many pls. (esp. ripening frt.), petroleum (chief comm. source), natural gas; u. as inhalation anesthetic; u. to make polyethylene plastics, antifreeze, &c.

ethylmorphine HCl: Dionin: ethyl derivative of morphine; analg. antitussive; mydriatic.

etiology: cause of a disease/disorder.

etoposide: semi-synth. deriv. of podophyllotoxin; antineoplastic (*cf.* teniposide).

etrog (Hebrew): small unripe frt. of *Citrus medica* var. *ethrog* Engl.; u. in Jewish feasts; the first citrus introduced into Eu.

Euadenia eminens Hook. f. (Capparid.) (s tropAf): rts. u. as sex stim. aphrod. (Ghana).

Euarctos (Ursidae) (nTemp Zone): bears - **E. americanus** Kellar (most of NA, woody places): Am black bear (also brown & cinnamon color vars.); wt.-500 lb.; u. for fur, meat; killed by poachers for gallbladder u. folk med. (considered very valuable) (eTenn, wNCar [US]).

Eubalaena japonica (Lacepede) (Balaenidae - Cetacea - Mammalia) (Vanc Is-Aleut, USSR coasts): Pacific right whale; -21 m long; formerly used in whaling industry.

Eubatus: section of genus *Rubus*; preferred as source of tannin-rich medl. rt. bks.

eucalypt: eucalypto (Sp): **eucalipto** (It): eucalyptus, esp. *E. globulus*.

Eucalypti Gummi (L.): 1) eucalyptus gum 2) gummy exudation of *Eucalyptus* spp. (esp. *E. camaldulensis*).

eucalyptol: cineole: an arom. lactone repr. the chief compound of official eucalyptus oil & also present in rel. large amts. (70-75%) in several unoffic. VOs ex. *E. bakeri* Maiden, *E. bebriana* F. Muell., *E. bridgesiana* R. T. Baker, *E. cambagei* Maiden, etc.; u. as exp. to treat bronchial & pulm. dis. (Am).

Eucalyptus L'Hérit.: 1) genus of trees (Myrt.) (Austral): the principal & characteristic trees of Austral; some spp. very tall 2) (L.) (specif.): lvs. of *Eucalyptus globulus* - **E. amygdalina** Labill. (Tasmania; cult. Zaire, Georgia [USSR], Sp, Ch): willowleaf eucalyptus; black peppermint; peppermint gum; brown p. tree; one specimen claimed highest tree in world (150 m); lvs. c. tannin, VO (largest yield of VO in genus: 4.2%) with phellandrene, cineole (often cult.); var. *australiana* (E. a. Bak. & Sm): narrow-leaved peppermint; lvs. c. VO with cineole, terpineol, etc.; often cult. (2614) - **E. calophylla** R. Br.: red gum; marri; bk. c. tannin prod. kino; u. astr. tan. - **E. camaldulensis** Dehn. (E. rostrata Schlecht) (Austral): red (river) gum (tree); river (or Murray) red gum; longbeak eucalyptus; trunk source of "red gum" (Australian Kino) (creek g.); with 46% tannin; exudation u. astr. in cough, colds, dys. relaxed throat; diarrhea (children); wd. useful for ship building & construction; galls with 43% tannin (cult. sFra, Por, swAf): lvs. & VO c. valeraldehyde, cineole (77%) - **E. citriodora** Hook. (NSW, Queensland [Austral]): lemon (spotted) gum; l. scented g.; much cult. as source of valuable VO (distd. fr. lvs.), rich in citronellal, u. in perfumery, org. synthesis; cultd. in many warm countries - **E. cladocalyx** F. Mueller (Austral, Af): sugar gum; lvs. c. prionarin, mandelic acid nitrile glucoside, sumbunigrin, free HCN; pl. toxic; ornam. tree (-40 m) - **E. cornuta** Labill. (s&wAustral): yate tree; has useful wd.; ornam. - **E. deglupta** Sm. (eMal): kamerere; wd. u. in constr.; fallen pieces of bk. u. in med. (chewed with sirihpinnang); planted in Mal - **E. dichromophloia** F. Muell. (Austral): gum-topped bloodwood; source of a kino u. in native med. for pulmon. dis. (decoc. with sugar drunk), also for colds, flu, as general ton.; supposedly good in cardiac dis.; honey fr. flower u. for coughs & colds (2263) - **E. diversicolor** F. Muell. (wAustral): karri (gum); wd. termite-resistant, u. for construction; ornam. tree (Calif+) - **E. dives** Schauer (Austral): broad-leaf peppermint (eucalyptus); lvs. rich in VO; 3 vars. of tree recognized: **A, B,** & **C**, with varying VO, but all rich in piperitione; no menthol; in fever, lvs. burned & person held in smoke (2263) - **E. ficifolia** F. J. Muell. (Austral): scarlet (flowering) gum; flaming eucalyptus; a small tree; source of eucalyptus kino - **E. globulus** Labill. (Austral, cultd. Calif, Sp, Balkans): blue gum-tree; Tasmanian blue gum; the official eucalypt; one of the tallest *E.* trees with stems over 90 m high; the *E.* sp. mostly cutl. in NA; lvs. & VO u. antiperiod, antipyr. exp.; toxic by oral ingestion; said useful as aid for blocking smoking addiction; source of a kino; flea repellant (1809); wd. u. in shipbuilding & for wd. pulp (also several other spp. so u.); e. oil prodn. in Austral, Br, Chin, Af - **E. gunnii** Hook. f.: cider gum; bk. & lvs. tannin-rich; prod. eucalyptus manna (c. mostly melitose) - **E. leucoxylon** F. v. Muell.: (white) iron bark tree; bk. very rich in tannin; lvs. c. 9.5% tannin; wd. excellent lumber; prod. kino with 41% tannin, very much like Malabar kino - **E. loxopheba** Benth. (wAustral): york gum (tree); useful wd.; ornam. tree (133 m high!) - **E. macarthurii** Deane & Maiden (Austral): Camden woolybutt; source of a VO contg. geraniol & geranyl acetate; u. to denature alc. intended for use in perfumery (Austral) - **E. macrorhyncha** F. Mueller (Victoria, NSW, Austral): (red) stringybark; stringbark (tree); trunk source of ruby-colored kino (with 79% tannin), easily friable; u. med. (Vietnam); lvs. c. VO (with cineole, phellandrene; 12-24% rutin (comm. source), tannin (19%); ornam. tree (cult. Calif); honey tree - **E. maculata** Hook. (Austral): spotted gum; source of a brownish kino; wd. u. in construction of bridges, shingles, &c. - **E. mannifera** Mudie (Austral): manna e.; brittle gum; Australian (or e.) manna exudes fr. insect punctures; known in Eu since 1829; c. melitose; sometimes u. as food - **E. marginata** Sm.: jarrah; lvs. c. VO with much cymol, cineole, aromadendral, small amt. pinene; timber said resistant to Teredo (ship-worm) - **E. microcorys** F. Muell. (seQueensland, NSW): tallow wood; wd. much u. for telephone poles, railroad ties, &c.; source of VO - **E. microtheca** F. Muell. (Austral): coolibah; bk. placed in bag & u. as fish poison (2263); pollen/honey tree - **E. naudiniana** F. v. Muell. (E. deglupta Bl. or Labill.) (PI, Indones, PNG, New Caled); tree of rapid growth (trunk 6 ? in height & 8 cm diam. in 12 mo.); bk. u.. extern. for open carbuncles & wounds (PapNG); - **E. obliqua** L.'Hérit. (E. falcifolia Miq.) (eAustral): (messmate) stringbark; very large tree; a source of Australian kino; wd. & bk. much used, latter for roof thatch, &c. - **E. occidentalis** Endl.: flat-topped yate; Western Australian e.; source of mal(l)et(o) bark; inner bk. c. 35-52% tannins; u. tan; tannins similar to those of quebracho wd.; lvs. c. VO with much d-pinene, cineole, sequiterpenes - **E. odorata** Schltdl. (Austral): peppermint box; medl. use (Mex Ph.); source of VO - **E. phellandra** R. T. Baker & H. G. Smith: narrow-leaved peppermint (tree); lvs. with 3-4.5% VO with cineole, pinene, phellandrene, terpineol, geraniol - **E. piperita** Sm.: peppermint stringy bark; p. tree; frt. u. spice; kino prod. fr. stem wounds; source of oldest known (1877) e. oil; VO prod. fr. lvs. & twigs c. phellandrene, piperitone, cineole - **E. polybractea** R. T. Baker: blue mallee; shrub; VO c. 80-85-94% cineole, some pinene, similar in odor to VO of *E. odorata*; prod. on large scale - **E. polycarpa** F. Muell. (Austral): longfruit bloodwood; kino u. to relieve pain fr. dental caries by plugging hole in tooth (2263) - **E. pruinosa** Schauer: silver-leaved box; inner bk. is wound dampened around chest for rheum. - **E. regnans** F. Muell.. (Victoria, Tasman, Austral): swamp gum; "mountain ash"; largest & most valuable member of gen.; trees up to 96 m high (1320); wd. much u. - **E. resinifera** Sm. (eAustral) red mahogany; r. stringybark; source of Australian kino; strong wd. much employed, thus for posts & rafters - **E. robusta** Sm. (NSW): swamp messmate; s. mahogany; white mahogany; source of "ribbon kino" with tannin; galls c. 43% tannin, u. in sea sickness; lvs. c. VO; lf. juice u. as ton. antisep. antiperiodic; pl. u. insecticide (Chin); wd. u. in general building; ornam. tree - **E. rostrata** Schl. = *E. camaldulensis* - **E. sideroxylon** Cunn. ex. Woolls (Austral; cult. Mor, &c.): mugga; red ironbark; bk. rich in tannin; kino used in popular med.; very hard wood, u. in house framing, &c.; honey tree - **E. smithii** R. T. Baker (Austral): gully gum; VO rich in cineole; u. med. (BP 1973) - **E. tetradonta** F. Muell. (Austral): Darwin stringybark; inf. of crushed inner bk. u. in diarrh.; tree a bad pest; lf. decoc. drunk in flu (2263).

eucalyptus gum: e. kino: the dr. gummy exudate fr. *E. camaldulensis, E. leucoxylon, E. gunnii, E. piperita, E. macrorhyncha, E. amygdalina*, etc.; c. 46-47% kino-tannic acid, kino-red, pyrocatechin; u. astr. in laryngitis, pharyngitis, tonsillitis (as troches) - **e. oil**: VO of *Eucalyptus globulus* or *E. dives* or of some other *E.* spp. rich in cineole (eucalyptol); u. stim. antisep. in nasal preps., etc.; prod. Chin.

Eucarya acuminata (R. Br.) Sprague (Fusanus spicatus R. Br.) (Santal.) (Austral.): South Australian

Sandalwood Tree; source of sAustral Sandalwood oil (rose-scented); red frt. pericarp. u. to make jellies; preserves, &c.; (guan)dong nuts edible; wd. valuable cabinet wood - **E. spicata** (DC) Sprague & Summerh. (*Santalum spicatum.* DC; Fusnus spicatus R. Br.) (s&wAustral, New Caled; Af): wd. (West) Australian (or Swan River) sandalwood) distd. fr. trunk wd. & rts. for VO (Oleum Santali spicatae) u. med. & in incense; c. santalos (fusanols) bisabolene, farnesol &c; comml. oil u. like true sandalwood oil; very valuable wd. u. in cabinet work.

Eucerin: comml. name for ointment of wool fat alcohols u. for pharmaceuticals & cosmetics; 1912.

Eucheuma J. G. Agardh (Solieri. - Rhodophyt. - Algae) (warm seas, wPacOc, &c.): genus of red algae fr. which agar may be produced; source of Malayan agar & source in part of Japanese or Chin gelatin or isinglass; carrageenan - **E. muricatum** Web. van B. (E. spinosum J. Ag.) (wPacific, Indian Ocean): "Singapore weed"; u. as source of a very high grade agar (Austral).

Euchlaena mexicana Schrad. (Zea m. [Schrad.]) Ktze. = Z. m. L. subsp. m. (Schrad.) Iltis (Gram.) (Mex): teosinte; a tall grass, the nearest relative of corn, *Zea mays*, with which it can be crossed; grain rich in starch (37%), pentosans, etc.; fodder pl. (—> *Zea*).

Euchresta horsfieldii Bennett (Leg) (Ind, eAs, esp. Java, Thail, PI): palakija; very bitter rts. chewed in snakebite (value?) (Taiwan, PI): sds. bitter, u. as expect. in chest dis., ton., aphrod. (cytisine); lf. decoc. u. to facilitate childbirth (Thail).

Euclea daphnoides Hiern (Eben.) (sAf): bk. inf. u. as enema in painful menstr. - **E. fructuosa** Hiern (tropAf): rt. decoc. u. ankylostomiasis; pd. rt. in yaws; bk. decoc. u. for splenic enlargment; Af women chew rt. to impart red color to mouth (*cf.* betel) - **E. lanceolata** Mey. (tropAf): bush gwárri; rt. u. as purg. in biliousness; frt. so strongly purg. as to be considered toxic; pl. u. in diabetes (Europeans) - **E. multiflora** Hiern (trop eAs): u. for yaws; rt. vermifuge; branches u. as toothbrush - **E. natalensis** DC (Transvaal, sAf): inf. u. as purg. & in abdom. dis.; rt. u. in leprosy, ankylostomiasis, scrofula; toothache; rt. ext. u. for ulcers (Malawi); rt. bk. red dye - **E. pseudoebenus** E. Mey. (sAf): Orange (red) River ebony; Cape e.; frt. edible but astr.; wd. excellent, u. inlaying, &c.

Eucomis bicolor Bak. (Lili) (sAf): u. as colic remedy - **E. punctata** L'Hérit. (tropAf): krulkop; bulb decoc. u. in rheum.; enema given to children while teething - **E. regia** Ait. (sAf): rem. in VD, rt. astr. in diarrh. & to prevent premature childbirth - **E. undulata** Ait. (sAf): pineapple lily; wilde pynappel (Afrik, wild pineapple): core of bulb in decoc. u. for abdom. pain; in treating urinary. dis.; bulb decoc. in syph. (sSotho); sometimes cause of poisoning.

Eucommia ulmoides Olive. (1) (Eucommi.) (Chin): Chin elm; C. rubber tree; bk., lvs. &c. rich in caoutchouc, u. as source of rubber; gutta percha (1315); bk. acts on liver & kidneys, u. as ton. (Chin) (85); "antihypertonic."

Eucryphia cordifolia Cav. (Eucryphi. - Dicot.) (Ch): ulmo; bk. c. tannins, u. astr., &c.; c. azaleatin (quercetin methyl ester).

euforbio (Sp, Por, It): **euforbiu** (Rouman.): Euphorbium.

eufrasia (Sp): Euphrasia.

Eugenia acris Wight & Arn. (Pimenta a. Wight) (Myrt.) (EI, WI, Ve, eAf): wild clove tree; lvs. distd. to produce true bay oil (Oleum Myrciae) (not VO fr. *Pimenta racemosa*) u. in prepg. bay rum (u. in hair care, as mosquito repellant, for skin rashes); frt. u. as substit. for pimenta - **E. apiculata** DC (Ch): arrayan; lvs. u. for diarrh. & lung dis. - **E. aromatica** Baill. = *Syzygium aromaticum* - **E. axillaris** (Sw.) Willd. (sFla [US], WI): (white) stopper (Bah); decoc. u. for diarrh. & blood improvement, lf. & berries for grippe (Bah) - **E. capuli** (Schlecht. & Cham.) Berg (sMex, Guat): "capulin"; frt. edible, u. in gastralgia & to furnish black dye; wd. u. as fuel - **E. caryophyllata** Thunb. = **E. caryophyllus** Bullock & Harrison = *Syzygium aromaticum* - **E. cauliflora** DC = *Myrciaria c.* - **E. cheken** Hook. & Arn. = **E. chequen** Mol. (Ch): cheken; chekan; lvs. u. arom. astr. ton. exp.; source of c. lf. oil, very similar to myrtle oil (c. d-pinene, 75%; cineole 15%) - **E. florida** DC (Co): lf. inf. drunk for chest pains (Makuna Indians) (2299) - **E. jambolana** Lam. = *Syzygium cumuni* - **E. malaccensis** L. = *Syzygium m.* - **E. myrcianthes** Ndz. (Arg): igb jai; yellow juicy frt. eaten by SA Indians & poor people; as lax. & refrig. - **E. operculata** Roxb. (Cleistocalyx operculatus [Roxb.] Merr. & Perry - Syzygium operculatum Gamble) (trop As, as Burma, Ind, Chin): yethabye (Burma); lvs. c. VO rich in esters, u. in jaundice (Chin), to aid digestion (Indochin); bk. & lvs. decoc. u. as wash (fomentation) leprosy, scabies; rt. decoc. boiled down to consistency of syr. u. as embrocation in arthr. (Chin), vermifuge; bk. of old tree in formula for indigestion & cholera; frt. u. in rheum. antipyr. ton., edible; wd. u. in building, for canoes, agr. implements, &c. (medium-sized tree of savannah forests) - **E. patrisii** Vahl (Co): tea of lvs. twigs, & frts. reputed valuable remedy for persistent cough & other resp. dis. (Barasana Ind.) (2299) - **E. pitanga** (O.

Berg) Kiaersk. (Br, Arg): nangapire; lvs. u. digest. carm. source of VO; c. citronellol. geraniol, cineole - **E. pungens** O. Berg (Br, Arg): frts. delicious food - **E. rariflora** Benth. (Syzygium rariflorum) (Fiji): lvs. u. for thrush in babies; frt. edible; small tree - **E. smithii** Poir. = *Acmena s.* - **E. strigosa** Arech. (E. dasyblasta) (Ch, Ur): pitanga; pitanga; u. folk rem. (Ur); fresh or dr. lvs. as stomach. carm.; ripe frts. macerated in dil. alc. & sugar to prod. arom. liqueur u. digest. stim. - **E. temu** Hook. (Ch): temu; bk. & lvs. astr. u. in chron. diarrh., wounds, esp. in veter. med. - **E. uniflora** L. (Br. cultd. sAf, Fla [US]): Surinam cherry; pitagne; pitanga; evergreen bush or small tree; lvs. u. for rheum., gastric dis.; frts. esculent, in marmalades, preserves, to flavor sugar cane.

eugenol: allyl guaiacol; active princ. of clove oil; also present in many other VOs, etc.; u. perfumery; raw material in mfr. synth. vanillin; replacement for clove oil, etc.; exter. as local anesth., in red bug (chigger) bite; antisep., esp. in dentistry; int. to aid digestive process (1965), in diarrh.; in baiting insects.

eukaryote: eucaryote: organism with cells provided with well-defined nucleus; representative of most higher organisms.

eulachon: candlefish; rich in FO, burns like candle when ignited.

Eulicin: peptide antibiotic fr. *Streptomyces* sp. related to S. *parvus* Waksman & Henrici; active against fungi, incl. yeasts; very toxic; 1955-.

Eulophia campestris Wall. (Orchid.) (Ind): East Indian salep plant; rhiz. u. as aphrod. (Cashmere salep); in poultice for running sores (Indones); cardiac troubles (Ind) - **E. epidendrae** Fischer (Ind): tubers u. as vermifuge - **E. herbacea** Lindl. (Ind): tubers u. for tumors, bronchitis, vermifuge - **E. nuda** Lindl. (nwInd); landigadda; tuber edible; u. as antitumor.

Eumenol: prep. fr. Chin drug, tang-kuei (*qv.*).

Euodia = Evodia.

euonymin; comml. product representing alc. ext. of Euonymus.

Euonymus L. (*Evonymus*) (Celastr.) (NA, Euras, Austral): spindle tree; shrubs to small trees with opposite lvs., with cymes of tiny flowers; 177 spp., some spp. toxic; much u. as ornamentals; 5 spp. cult. US Pacific NW - **E. alatus** Sieb. (eAs): razor blade bush (NJ [US] vernac.): small sts. with broad corky wings ("winged spindle tree"); twigs hemostatic, to reduce blood pressure, in abdominal pain after childbirth (Chin, Korea); in chronic malaria (Chin); commonly cult. ornam. shrub - **E. americanus** L. (eUS): strawberry (shrub); burning bush; pl. has pink frt.; u. like foll. - **E. atropurpureus** Jacq. (eUS & Cana): (Eastern) wahoo; bk. of rt. c. VO resins, evonynol (alc.) FO, evonysterol; eu(v)atromonoside (cardioactive glycoside; digitoxigenin + arabinose); evonymin (glycoside); rt. bk. u. bitter ton. lax. cholag. hepat. stim.; inner bk. decoc. drunk for uterine trouble (Winnebago squaws) - **E. cochinchinensis** Pierre (Indochin): lo maly; tree bk. u. in alc. as aperient (Cambodia) - **E. europaeus** L. (Euras): **European euonymus**: spindle tree; frts. c. orange-red FO, phytosterol; juice (called "manna") c. mannite; sds. c. NBP resin, FO, emulsin, yellow coloring matter; several cardiac heterosides, evonside (digitoxigenin + rhamose + glucose), evomonoside, evobioside, etc.; furan-3-carbonic acid; evonic acid; sds. u. purg. (1471), emet.; cardioactive drug (1354); insecticide; yellow dyestuff; adult. for *E. atropurpureus*; a source of gutta percha - **E. japonicus** Thunb. (eAs): Chin box; spindle tree; ornam. shrub; bk. said ton. antirheum., anhidrotic; lvs. u. in difficult delivery (Jap) - **E. tingens** Wall. (Himal reg.): rt. bk. cardioactive, u. as diur. in cardiac edema; mild lax., to stimulate bile secretion, in digest. dis., in eye dis.; pl. u. as yellow dye (tika dye) - **E. verrucosa** Scop. (Eu): warty spindle tree; lvs. c. kaempferol glycosides, incl. astragalin; rt. bk. c. gutta, culcitol; u. as a source of gutta percha (Russia) (1355) - **euparin**: bitter princ. fr. *Eupatorium purpureum* (1325).

Eupatorieae; a tribe of fam. Comp. which is characterized by possessing discoid heads, stigmatic surfaces of pistil only on lower inside parts of upper working branches of style; representative genera: *Ageratum*, *Carphephorus*, *Conoclinium*, *Eupatorium*, *Garberia*, *Kuhnia*, *Liatris*, *Mikania* - **eupatorin**: 1) glycoside fr. *Eupatorium perfoliatum* or *E. cannabinum* 2) very sweet princ. of *Stevia rebaudiana*.

Eupatorium L.: 1) (Comp) (mostly trop & subtrop Am): bonesets; genus of herbaceous shrubby pls. with arom. & bitter princ.; inflorecences corymbs of numerous heads made up of rel. small fls. 2) (L.) (specif.): lvs. & fl. tops of *E. perfoliatum* - E. **adenophorum** Spreng. (Austral-NSW; Ind): Crofton weed; hb. u. as antisep. coagulant with stim. properties - **E. africanum** Oliver & Hiern (tropAf): hb. c. coumarin; u. med. (Congo) (1326) - **E. ageratoides** L. f.: *E. rugosum* - **E. aromaticum** L. (eUS): white snakeroot; pool-root; rhiz. arom. diur. antispas. in rheum.; hb. c. coumarin (1328); mildly hypoglycemic (1343) - **E. azureum** DC (sTex [US], Mex, CA): blue eupatorium; u. as astr. poultice (Mex) - **E. bahamensis** Northrop (Bah Is): cat tongue; lf.

decoc. u. with salt for stomachache; lf. juice with salt u. for intest. pain, &c. - **E. berlandieri** DC = *E. herbaceum* - **E. cacabioides** HBK (Ec): "guichilica"; lvs. tied around head for headache - **E. cannabinum** L. (Euras, nAf): hemp agrimony; water hemp; hb. c. euparin, eupatoriopicrin (NBP), resin, saponins (also called "eupatorin"); hb. chiefly in folk med., ague weed; u. as diur. in edema, scurvy, ulcers, swellings; antipyr. exter. for tumors, ulcers, skin dis., etc.; adult for hemp sd. (20); rhiz. & rts. u. for liver & gallbladder dis. - **E. capillifolium** (L.) Small (seUS, WI): dog fennel (confusion perhaps with *Anthemis cotula*); c. VO with much phellandrene; fresh juice u. insect bites; hb. strewed on floor as insectifuge (Calif [US]); lvs. collected in late summer u. for arthritis (folkloric) - **E. collinum** DC (Mex, CA): yerba del angel; hb. off. as Hierba del Angel (Mex Phar. 1904); lvs. u. diarrh., rheum. (Hond); bitter arom. in liver dis. (inf.); antipyr. as subsitit. for hops in beer mfr. - **E. compositifolium** Walt. (seUS): dog fennel; very close to *E. capillifolium*; by some regarded as identical - **E. cuneifolium** L. (Fla to NCar [US]): hb. c. eupacunin (germacranolide): has significant antileukemic and tumor-inhibiting properties - **E. dalea** L. (WI): vanilla (Cu); hb. with coumarin very fragrant; u. as vanilla substit. - **E. fistulosum** Barratt (Eupatoriadelphus fistulosus King & H. Robins.) (Fla-Tex, Ia-Maine [US]): queen-of-the-meadow; Joe Pye weed; tops & rts. u. for kidney dis. (Miss [US]); tea u. for polyps - **E. foeniculaceum** Willd.: *E. capillifolium* - **E. formosanum** Hay. (Chin): above grd. parts c. eupatolide, cytotoxic; antitumor, u. antipyr. antileukemic, antiinflammatory - **E. fortunei** Turcz. (E. stoechadosmum Hance) (Kor, Jap): rt. decoc. u. as antidote to poison; to regulate menstr.; cult. (Jap) - **E. glutinosum** Lam. (Mex): sometimes adulter. of Matico lvs. (appar. due to confusion of names) (1331) - **E. gorgonis** Berger (sAf): latex u. to remove warts, appl. to sores, cancerous skin conditions, toothache; styptic - **E. herbaceum** Greene (E. berlandieri[i] DC) (w mt., US, Colo to Calif): desert eupatorium; hb. arom. (c. coumarin); lf. u. by Apache Indians as tobacco substit. (smoke rather acrid); narcotic (prod. nervous tremor) (1329) - **E. hyssopifolium** L. (eUS): hyssopleaf thoroughwort; hempweed; lvs. u. alexiteric in snakebite (Tex); pl. decoc. for fever, colds (SC) - **E. incarnatum** Walt. (eUS, Mex): hb. c. coumarin; was (formerly) added to pipe smoking tobacco by Mexicans (NM [US]) ("mata") (1334) - **E. indigofera** Parodi (SA): source of indigo (other spp. of SA wh. prod. indigo are *E. laeve* DC [Par] and *E. lamiifolium* HBK) - **E. ingens** E. Mey. (sAf): candelabra tree; latex very irrit., toxic; u. in cancer; fish poison; drastic (danger!) - **E. inulaefolium** HBK (Trin): Christmas bush (Trin); lf. inf. u. for colds & coughs - **E. javanicum** Blume = *Vernonia arborea* var. *javanica* (DC) C. B. Clarke - **E. lancifolia** Schlecht. (Mex, Guat to Hond): ixbut; besmut; lvs. & sts. u. as galactagog (res. 1949) - **E. leptophyllum** DC (seUS): "fennel"; occurs in damp soil; u. folk med. - **E. leucolepis** T & G (e&seUS): justice weed; hb. u. as alexiteric in snakebite - **E. macrophyllum** L. (WI, Mex to Par): "aranda"; "boneset taffy"; formerly often u. in place of beverage tea (inf.); rt. u. in tisane for liguria, fever (Trin); in influenza, colds, icterus, general debility, esp. at onset of colds (snuffles, headache, cough) - **E. maculatum** L. (Cana, nUS): smoke weed; Joe Pye weed; marsh milk weed; sometimes confused with *E. purpureum*; lvs., fl. tops & rts. u. astr. in kidney dis., "polyps," etc.; lvs. u. as aper. diur. in jaundice, antipyr. (1805), rheum, &c. - **E. odoratum** L. (Chromolaena odorata [L.] King & Robins) (sFla, sTex [US], WI, tropAm, pantrop): Christmas bush; rt. u. emm.; lf. decoc. in gastralgia, nephritis; rts. boiled in salt water & u. as purge; GC, malar.; spasmolytic; substit. for mint lvs.; queen-of-the-meadow - **E. pallescens** DC (tropSA): u. as source of indigo (cult. Jam) - **E. perfoliatum** L. (e&cNA): boneset; thoroughwort; the chief med., usually as decoc. or inf. of lvs. & fl. tops; u. stim., ton., diaph., lax., emet., bitter, antipyr., quinine substit. in malaria, typhoid, yellow fever, dyspep., pneumon., rheum., sometimes a stock poison - **E. pichinchense** HBK (Ec): "pedorrera"; u. in diarrhea with flatus - **E. pilsoum** Walt. (e&seUS, incl. Fla): wild horehound; lvs. & flg. tops u. as ton. diaph., in mal.; diur. aperi. in colds & fevers (Am Indians) - **E. purpureum** L. (Eupatoriadelphus p. King & H. Robins.) (sc&sUS): Joe Pye weed; trumpet weed rt.; abela; gravel (or kidney) rt.; hb. c. euparin, coumarin; hb. & rts. u. diur. in calculus (kidney dis.); aphrod. (1338) - **E. pycnocephalum** Less. (Guat, Hond): te; lvs. u. diaph. (tea); in erysipelas, eye dis. (Salv) - **E. rebaudianum** Bert. = *Stevia rebaudiana* - **E. rotundifolium** L. (sUS): false hoarhound; hb. c. glycoside; u. TB (folk med.); c. tumor-inhibiting sesquiterpene lactones, incl. eupachlorin; eupatorin acetate (1339); var. *saundersii* Cronq. (E. teucrifolium Willd.; E. verbenifolium Reich.) (eUS, New Eng to La): rough thoroughwort; hb. (lvs. & flg. tops) u. ton. diaph.; diur. aper. in colds & fevers, TB; USP 1820-30 - **E. rugosum** Houtt. (E. urticifolium Reich.) (eCana & e&sUS): white snakeroot; Indian (or white) sanicle; hb. c. tremetol (toxic princ.), cause of "milk sick" or m. sickness (in humans) or trembles (lower mammals), SC poison carried in cow's milk & causes an ague (caused death of Lincoln's mother); rt. u. diur. diaph. antispas. arom. vulner - **E. sanctum** Moq. & Sessé (Mex): off. as Hierba del Angel (Mex Phar. 1904); u. as vulner. & in diarrh. - **E. semiserratum** DC (seUS): hb. c. antileukemic sesquiterpene lactones, incl. eupaserrin, in significant amts.; also inhibitory to human carcinoma - **E. serotinum** Mx. (c&eUS): silver rod; hb. c. VO - **E. sternbergianum** DC (Pe): u. as stom.; flavor - **E. subhastatum** H. & Arn. (Arg): pilarcito; u. antiphlog, vulner (exter. & int.) (Arg) - **E. teucrifolium** Wild. = *E. verbenaefolium* - **E. triplinerve** Vahl. (E. ayapana Vent.) (Br): aya-pana; lvs. (off. Fr Codex) with coumarin-like odor, bitter arom. taste; u. ton. stim. (in large doses, lax.), u. antipyr. & in gastric dis. (Domin Rep), hot inf. diaph., stomachache, emetic, in dyspepsia, cholera, ulcers, chronic diarrh., snakebite (as inf.); lf. juice styptic for tumors; u. as substit. for coffee & tea (1333) - **E. urticifolium** Reich. = *E. rugosum* - **E. verbenaefolium** Michx. = *E. pilosum* - **E. veronicaefolium** HBK: *Brickellia veronicaefolia* - **E. villosum** Sw. (sFla, [US], WI): tribulillo (Cu); bitter b(r)ush; b. sage; hb. u. in cholera, diarrh., diur. (Jam); also other local *E.* spp. u. similarly (Jam); lf. or twigs ext. u. in fever (1342); lvs. chewed to relieve flatulence (Bah) - **E. wrightii** A. Gray (NM, Ariz, Tex [US], nMex): snake weed; oregano (Mex); u. folk med.

Euphausia (Crustacea): (oceans) krill; chief food of some whales; small size (-3 cm) (*cf.* shrimp); red k. colors water red.

euphorbain: proteolytic enzyme fr. latex of *Euphorbia lathyris*, etc.

Euphorbia L. (Euphorbi.) (worldwide): spurges; hbs., shrubs and small trees, sometimes cactus-like; lvs. simple; latex-bearing; genus split up by some authors (ex. J. K. Small) into several genera (*Chamaesyce*; *Galarhoeus*; *Lepadena*; *Poinsettia*; *Tithymalopsis*, etc.); (specif.) hb. of *Euphorbia pilulifera* (= *Chamaesyce hirta*); now often transferred to *Nephelicum*- **E. amygdaloides** L. (c&sEu): wood spurge; hb. (& rt.) u. as antipyr. carcinostatic - **E. antisyphilitica** Zucc. (Tex [US], Mex): wax e.; this desert bush represents one source of candelilla wax; u. much like carnauba w. - **E. atoto** Furst. f. (Ind, seAs, PI): latex u. emmen. (Indochin); appl. to sores - **E. bupleurifolia** Jacq. (sAf): pl. u. as emet. purg. (danger); latex u. appl. to cancerous sores, cracked skin of feet, various skin dis. - **E. californica** Benth. (nwMex): "candelilla"; shrub; u. as spurge - **E. campestris** Cham. & Schl. (Mex): yerba del coyote; flgs. tops u. as purge - **E. canariensis** L. (Teneriffe): cardon latex c. euphol; dr. latex off. in several pharmacopeias - **E. clavarioides** Boiss. (sAf): latex u. for cancerous sores, warts, leprosy; as glue - **E. coagulum** = **E. gum** = Resina Euphorbii = Euphorbium - **E. collectoides** Benth. (nwMex): milky latex u. for eyes - **E. corollata** L. (eNA): (large) flowering spurge; snake milk; rt. emet. cath. irrit. diaph. (USP 1820-70) (1344); cult. ornam. - **E. cyparissias** L. (eEu, natd. NA): cypress spurge; hb. c. euphorbon; often cult. around cemeteries; latex u. earlier as emetic, in folk med. as purg. (also lf.) diur., rhubarb substit., skin irritant (exter.) - **E. dracunculoiodes** Lam. (tropAf, As, as Bengal): "sudab"; pl. c. euphorbol (triterpene), glyco-alkaloids; ext. u. for cholingergic activity; entire pl. u. in Unani med.; FO u. as lamp oil, paint base - **E. drummondii** Boiss. (Austral): caustic weed; pl. u. GU dis. dys. & fevers (by natives); claimed value of "drummine" as loc. anesth. not supported; stock poison (may be due to CaOx); latex appl. to sore eyes (2263) - **E. esula** L. (Eu): leafy spurge; livestock poison; milk juice c. caoutchouc; resins; flavonoids; sterols; cycloartanol deriv.; serious weed in NA; rt. u. int. for edema; exter. in scabies, herpetic eruptions - **E. geniculata** Ortega (Med reg., SA): source of milky latex similar to Euphorbium; c. quercitrin, kaempferol - **E. grantii** Oliv. (Tanz): rts. u. as strong lax.; emet. in snakebite; toxic plant - **E. helioscopia** L. (worldwide) (Eu, natd. Cana, nUS, Af; Austral): cat's milk; wartwort; wartweed; sun spurge (Eng); bk. hb. rt. u. med.; latex ("juice") appl. as caustic to warts, corns; int. strong cath. & in syph. (folk med.); VO = 0.1% with elemol & ß-eudesmol - **E. heterodoxa** Muell. Arg. (Br): alveloz (aveloz, arveloz) plant; produces alveloz balsam (a. milk) (seiva de alveloz), u. by natives to treat sores, skin cancer of lips, nose, eyelids - **E. heterophylla** L. (Br, now widely found in tropAm, eUS, Af, PI+): "avelot plant"; avelo (Br); hb. or shrubby pl.; Mexican fire plant; u. pulm. dis. (Guat, Hond): rt. & bk. decoc. u. malarial chills (Mal); st. acrid latex ("juice") against erysipelas (Yuc); substit. for castor oil (Amboina), lvs. c. red (violet) dye, poncetin; pl. u. for cancer (reduces tumor size [sic]); in chest, dis. (3584); pl. juice lactogenic (Guat); latex appl. to wounds (Salv) & to bad eyes (Mex) - **E. hirta** L. = *Chamaesyce h.* - **E. intisy** Drake del Cast. (Madag): intisy; herotin; large shrub or tree; formerly source of intisy rubber, c. caoutchouc; u. mfr. rubber products - **E. ipecacuanha** L. (eUS):

ipecacuanha spurge; American or Carolina ipecac; bk. of rt. u. like *E. corollata*, also as diur. in edema exp. emet. - **E. lancifolia** Schlecht. (Guat): ixbut; lvs. & st. u. galactagog (2142) - **E. lathyrus** L. (Euras, natd. nwUS [Calif, Ore], SA): caper spurge; mole plant; gopher plant (Calif); sds. (as Semen Cataputiae minoris) u. diur. drastic cath. yield FO like croton oil; strongly drying, u. in soap mfr.; ipecac substit.; latex proposed as source of crude oil (1316) - **E. marginata** Pursh (cUS): snow-on-the-mountain; sd. FO source of stand oil, superior to that fr. linseed oil (1347); ornam. - **E. mauritanica** L. (sAf): u. as arrow & fish poison; latex c. rubber, u. as adhesive; pl. foraged - **E. mauritiana** Webb = *E. piscatoria* - **E. milii** Desmoulins (E. splendens Boj. & Hook.) (Madag): svongo soongo; pl. c. sitosterol, cycloartenol; rt. c. alks.; u. as purge, vesicant, in homeop.; ornam. pl. (Eu) ("crown of thorns") - **E. misera** Benth. (sCalif [US], nwMex): shrub; rt. decoc. u. for stomachache, dys. VD) - **E. mitchelliana** Boiss. (E. mitchelli) (Austral): fls. chewed to stop diarrh. (2263) - **E. neriifolia** L. (tropAs, now in NW): common milk hedge; latex u. as drastic cath.; antipyr. exter. for sores; in otitis; lvs. diaphor.; expr. juice of lvs. for asthma; rt. u. in snakebite, fish stupefiant - E. **ocellata** Dur. & Hild. (Chamaesyce o. Millsp.) (cCalif): valley spurge; u. snakebite remedy (1348) - **E. oerstedianum** Boiss. (WI+): u. to cure styes (CA): lf. decoc. in oliguria (Trin) - **E. peplus** L. (sEu, AsMin; natd. e&cNA, CA): round-leaved (or petty) spurges; hb. u. as drastic purg. in Hippocratean medicine; in asthma (folk med.); exter. to remove warts & corns - **E. pilulifera** L. = *Chamaesyce hirta* - **E. piscatoria** Ait. (Madeira): u. fish poison - **E. plumeroides** Teysm. ex Hassk. (NG): bk. u. as purge, vermif., fish poison (chopped lvs. also) - **E. poissoni** Pax (wAf): latex u. purg. caustic in skin dis., leprosy, arrow poison - **E. prostrata** Ait. = Chamaesyce p. - **E. pulcherrima** Willd. ex Klotzsch (Poinsettia p. Grah.) (sMex, CA): Poinsettia (preferred pronunciation: second "i" silent); Weihnachtsstern (Ge [Christmas stars]); the well-known shrub, widely grown as ornam. house or greenhouse pl. in nUS, outside in sUS; bracts u. galactagog (Mex); lvs. in erysipelas, etc. (cataplasm): latex u. depilatory; brilliant colored upper (floral) lvs. (red, pink, white) account for its popularity as Xmas decor. (attractive contrast of red & green); floral parts small, yellow; lvs. st., & latex toxic (may prod. dermatitis); (named after J. R. Poinsett, US Minister to Mex.) - **E. resinifera** Berg (Morocco): source of dr. latex, euphorbium (*qv.*); u. purge, abort. in sciatica - **E. rigida** Bieb. (E. biglandulosa Desf.) (eMed reg, incl. Gr, Alb, Turk, &c.): narrow-leaved glaucous spurge; c. phorbol esters (2386); latex (rizziteddu [Sic.]); u. in cholera; as fish poison - **E. rosea** Retz. (eAs, Am): u. much like *E. thymifolia* L. (Indochin) - **E. royleana** Boiss. (Ind, Himalay): latex u. as anthelm. cath.; pl. u. as fish poison - **E. serpens** HBK (Arg, Par, Ch): yerba meona (Arg); latex u. as drastic; antidiarrh.; exter. in skin dis., antitumor; inf. as beverage, also to cure leukorrh. (Par); as general remedy (Arg) - **E. steyermarkii** Standl. (sMex, CA): latex u. for pain fr. broken bones (Guat) - **E. systyloides** Pax (Tanz): sds. rubbed into skin for ringworm - **E. terracini** L. (Med reg.+): c. quercetin, α-amyrin, ß-sitosterol; u. exter. to remove warts & corns; int. drastic - **E. thomsoniana** Boiss. (Pak, Tibet): rts. purg., &c.; detergent for washing hair; latex gives dermatitis - **E. thymifolia** L. (eAs, EI, Br): hb. astr. diur. lax. (Taiwan); anthelm. (Indochin); poultice for skin dis. (Mal); latex for eye infect. (Br); in coconut milk for leg & back pain (int. & exter.); pulped plant appl. to wounds; decoc. for diarrh. dys., prolapsed rectum (Indones); lf. decoc. in diarrh. (Trin) - **E. tirucalli** L. (eAf, Zanzibar, cult. Ind, eAs): milk bush; latex c. euphol, taraxasterin, tirucallol; irrit. toxic; u. fish poison, for homicide; emetic, antisyph.; lax; st. & bk. poultice to nose ulceration, hemorrhoids, swellings (Mal); skin cancer (sun) (2263); rt. in coconut oil for stomach locally (PI); antibiotic - **E. trigona** Haw. (eAs): juice of heated lvs. u. for earache; pounded lvs. appl. as poultice to boils (Indones); latex u. by natives for rheum. (EI) - **E. truncata** N. E. Br. (Af): latex u. in GC, acute appendicitis; juice fr. boiled plant u. as vet. anthelm., cough remedy (tropAf) - **E. verrucosa** L. (Eu): hb. c. flavonoids; euphorbon (rubber-like); sds. c. 25% FO - **E. xanti** Engelm./Boiss. (Mex): liga; latex very irritant to mucosa, sometimes acting as depilatory & even causing blindness.

Euphorbiaceae: Spurge Fam.; large gp. of hbs., shrubs, trees, ca. 7,950 spp. of all zones throughout the world; often pls. have latex (milky juice) in laticiferous tubes or sacs; with acrid properties; sds. often with FO; fls. apetalous & with essential organs surrounded by calyx- or corolla-like involucre; sometimes neither calyx nor corolla of various types; usually a 3-lobed capsule; some spp. have alks., but glycosides & VO are almost never found; poisonous protein substances are characteristic of several genera; therapeutic props; usually narcot. cath., emet., alt.; many important economic spp., ex. castor oil; Pará rubber, tung oil tree.

euphorbio (Por, Sp): Euphorbium.

Euphorbium (L.): **e. gum: Euphorbium Gummirestina** (L.): gum-resin obt. by desiccation of latex of *Euphorbia resinfera* which flows fr. incisions in sts.; a med. resin u. since the time of ancient Rome; dull yellow-brown tears; u. irrit., vesicant (veter.), drastic & in paint for ship bottoms to prevent fouling (barnacle growth); poisonous.

euphorbon: constit. of prec.; not a pure compd. but a mixture of vitorbol & novorbol & perhaps other substances as well.

euphoria: exaggerated sense of well-being, usually induced by drugs.

Euphoria longana Steud. (*E. longana* Lam.) (Sapind.) (Ind, Chin, Burma): longan; white frt. edible delicacy; frt. u. anthelmint. for nerve dis.; pd. kernel u. as styptic (Chin); sd. u. as alexiteric; fls. inf. u. for kidney dis., leukorrhea; rt. inf. to treat GC, glycosuria; fresh or dr. aril licked to stop hiccups; now often transferred to *Nephelium.*

Euphrasia L. (Scrophulari.): paleoarctic gen. of ca. 450 spp. - **E. nemorosa** Wallr. (E. canadensis auctt.) (Cana, neUS): eyebright; hairy e.; u. much like *E. officinalis* - **E. officinalis** L. (E. maculata Gilib.) (Eu): eyebright; pl. u. ton. astr. lax.; in ophthalmia, migraines, colds, asthma, skin dis. - **E. rostkoviana** Hayne (Euras, nNA): Augentrost (Ge): fig. hb. u. ton. in catarrhs., eye dis.; one source of Herba Euphrasiae (1317).

euploid: e. series: with chromosomes in multiples of n. (ex. 2n, 3n, 8n).

Euquinine™: quinine ethylcarbonate.

European: v. also alder, ash, arnica, aspen, aspidium, euonymus, goat's rue, holly, mistletoe, raspberry, rhubarb, white water-lily.

Eurotia Adans. = *Krascheninnikovia.*

Eurya acuminata DC (The.) (Chin): lvs. & frts. u. to promote digestion - **E. japonica** Thunb. (Jap): lf. poultice u. to treat skin dis.

Euryale ferox Salisb. (1) (Nymphae.) (eAs): sds. u. as stom., arthritis, ton.; in polyuria, spermatorrh., GC (Korea); and many other uses.

Eurycles amboinensis Loud. (Amaryllid.) (eAs): bulbs u. emeto-cath.; lvs. appl. in rheum. (—> *Proiphys*).

Eurycoma longifolia Jack (Simaroub.) (eAs): thonan; much u. in Vietnamese med.; bk. in indig., vermifuge, lumbago; rts. u. in jaundice, edema, cachexia (bitter); pain in bones; antipyr.; frts. in dys.; rts. in malar, ton., appl. to wounds & ulcers (Mal); lf. decoc. u. in skin dis. as itch (Sabah).

Euryops evansii Schltr. (Comp.) (sAf): st. smoked to relieve headache.

Euscorpius italicus Herbst (Chactidae - Scorpiones - Arthropoda) (Euras): scorpion; "Scorpio europaeus"; venom (c. neurotoxin) causes envenomation; u. in homeopath. med. (10% tinct.) in migraine, &c.; cancer.

Euspongia agaricina (Phylum Porifera): yellow sponge; sp. of toilet s. - **E. officinalis** L.: sp. producing finest bath sponges; source of burnt sponge, u. alt. in goiter, obesity (Homeop.) - **E. tubulifera**: glove s. - **E. zimocca** Schmidt: the hard or zimocca s.

Eustoma exaltatum (L) Salisb. (Gentian.) (WI, sUS, Bah, Mex): genciana del pais (Cu); mountain bob (Bah); lf. inf. u. for eye bath.

eutachin: v. *Lomatium dissectum* var. *pinnatifidum.*

Euterpe edulise Mart. (Palm.) (Br, Pará): assai palm; jussara; frt. edible & source of beverage; FO u. as cooking fat; wax; fresh pulp of trunk for dyeing; apical bud = palm cabbage (similar uses for *E. catinga* Wallace, *E. oleracea* C. Mart., etc.).

Euthamia Nutt. (Comp.) (Am): perennial hbs.; most now placed in sect. of *Solidago*; v. also under *Solidago* - **E. graminifolia** Nutt. ex. Cass. (Solidago g. Salisb.; S. lanceolata L.) (Cana, nUS): flat-top goldenrod; fragrant g.; hb. claimed useful in some dis.; rt. decoc. u. for chest pains (Chippewa Indians) - **E. tenuifolia** Greene (E. minor Greene; E. caroliniana Greene; Solidago m. Fern.; S. mexicana L.; S. remota Friesner) (seUS n to NS [Cana]; sandy soils): narrow-leaved bushy goldenrod; guobsque weed; rt. inf. u. in colds; ton. diur.; hb. c. VO with pinene, dipentene (1745).

Eutrema wasabi (Sieb.) Maxim. (Cruc.) = *Wasabia japonica.*

eutrophication: great increase in nutrients (NO_3, PO_4) in standing fresh water (ponds, &c.) resulting in overgrowth of plant life.

evatromonoside: glycoside of digitoxigenin & D-digitoxose; occ. in *Euonymus atropurpureus*, &c.

Eve and Adam root (sUS): prob. *Liatris* sp. (not *L. spicata*).

evening primrose: *Oenothera biennis* - **e. star** (Tex [US]): *Cooperia drummondii* (c. bulb) - **e. trumpet flower**: Gelsemium.

evergreen: 1) plant which keeps its green foliage while in a dormant state, ex. *Pinus, Magnolia* 2) (WI) *Ficus indica* - **e. box**: *Buxus sempervirens.*

everlasting, (life): *Gnaphalium* spp.+ - **pearly e.**: *Anaphalis margaritacea* - **sweet e.**: *G. obtusifolium* - **e. wood**: 1) cypress wood ("wood eternal") 2) sassafras wood (u. voodoo med.).

Evernia furfuracea (L) Mann. (Parmelia f. [L] Ach.) (Usneaceae; sub-phylum Lichenes) (Temp Zone, incl. seUS): oak moss; occ. on wood (tree trunks); oleoresin (resinoid) u. in perfumery (mousse d'arbre) (fixative); various chem. races recognized; pl. in diarrh. fever (bitter princ.); occurs in ancient Egyptian grave ornaments - **E. prunastri** (L) Ach. (eEu, NA): oak moss - **staghorn e.**: thallus c. VO; arom. substances; u. perfume fixative (p. base) in creams, soaps, etc. (mostly in form of Tr. or Ext.); in red, purple & orange dyes (chief prodn. Yug) (3582) - **E. vulpina** (L.) Ach. = *Letharia v.*

Eve's cups: *Sarracenia* spp.

evobioside: glycoside of digitoxigenin + L-rhamnose & glucose.

Evodia (Euodia) confusa Merr. (Rut.) (eAs+): bk. appl. to splenomegaly (PI); decoc. of lf. or leafy shoot mixed with liquor & ingested for hives (Taiwan) - **E. cucullata** Gillespie var. *robustior* A. C. Smith (Fiji): fresh lf. chewed for thrush; inf. u. for colds & headache (Degener) - **E. elleryana** F. Muell. var. *tetragona* (K. Sch.) W. D. Francis (Solomon Is): tree sap placed on boils by natives - **E. hortensis** Forst. (eAs): crushed lvs. u. as poultice for boils (Solomon Is) & in baths (Fiji) - var. *simplicifolia* K. Schum. (Pac trop): uhi (Fiji, Tonga); lvs. & bk. u. med. & to perfume coconut oil (Tonga) (186) - **E. lepta** (Spr.) Merr. (Chin): bitter evodia; rt. & lvs. u. in arthritis, rheum. fever, hemorrhoids, itch, flu, sore throat, trauma - **E. poilanei** Guill. (Indochin): frt. decoc. u. to treat resp. dis. - **E. radlkoferiana** Lauterb. (SolomonIs): heated lvs. u. as poultice for boils; lf. & bk. decoc. u. as lotion on skin dis. - **E. rutaecarpa** Hook. f. & Thoms (Chin, Jap): wu-chu-ya (Chin); frt. an old Chin. drug; c. evodiamine, rutaecarpine, evocarpine, limonin (bitter princ. of lemon); frt. u. antidiarrh. (1322), anthelm., in rheumatism; rt. in tapeworm; flg. shoots in indigest. and as diur. (—> *Tetradium & Melicope*).

evolution (biological): doctrine that all forms of life develop from earlier simpler forms, eventually starting with single-celled organisms.

Evolvulus squamosus Britt. (Convolvul.) (Virgin Is, Bah): candlegrass (Bah); pl. decoc. with salt drunk for fever or jaundice (Bah).

evomonoside: glycoside of digitoxigenin + L-rhamnose; occ. in *Euonymus europaea*.

evonoside: glycoside of digitoxigenin + L-rhamonse + 2 glucose.

Evonymus: name u. preferably to *Euonymus* by rule of priority (acc. to some) although less euphonious.

Exacum bicolor Roxb. (Gentian.) (Ind): pl. u. ton., stomach. substit. for gentian and chirata - **E. lawii** C. B. Clarke (Ind): pl. boiled with oil & appl. in eye dis.; pd. u. sed. in kidney dis. - **E. quinquinervium** (tropAf): juice u. for gnawing pains in stomach (Tanz); purge (Guinea).

exalbuminous: descriptive of sds. which do not c. endosperm (or extra-embryonic food materials, i.e., albumen); ex. bean.

Excoecaria acerifolia F. Didr. (Euphorbi) (Ind): lvs. u. in rheumatism - **E. agallocha** L. (Ind: Travancore, Cochin): gewa; geva; furnishes tejbala (resinous subst.) u. aphrod. ton.; lvs. u. epilepsy; gewa ws. u. as firewood; fumes of burning wd. u. for leprosy (Fiji).

excrescence: a pathological outgrowth on pl. or animal tissue such as a gall.

exile oil (Eng): FO of sd. of *Thevetia peruviana*.

Exocarpos aphyllus R. Br. (Santal) (Austral): decoc. u. for sores & colds; appl. as poultice on chest for "wasting" dis. (Aborigines) (2263).

exocrine gland: one which secretes externally directly or indirectly fr. the body through a duct. (ex. sweat g.).

exoenzyme: extracellular enzyme which migrates from cell where originated.

Exogonium Choisy: now usually placed in *Ipomoea* - **E. bracteatum** (Cav.) Choisy (Convolvul.) (Mex): "empanada"; bejuco blanco; tuberous rts. eaten raw or cooked (natives, wMex coast) - **E. purga** (L) Benth. (Ipomoea p. Hayne; I. jalapa Nutt.; Convolvulus mechoacana Vitm.) (Mex): jalap; (raiz de) jalapa; raiz de machoacan (michoacan); tuberous rt. c. 5-20% resins, with convolvulin (ether-insol.), jalapin (ether-sol.), both glycosidal; u. drastic, hydragog, lax. (in smaller doses), remedy in gastroint. dis., emmen., source of j. resin; component of many tablet compds. (1323) (—> *Ipomea*).

Exomis microphyllum Aellen (1) var. *axyrioides* Aellen (Chenopodi) (Af): milky decoc. of. lf. u. for epilepsy; aq. decoc. produces stupor.

Exostema (Exostemma) caribaeum (Jacques) Schult. (Rubi.) (Fla [US], Mex, Bah, tropAm): princewood; seaside beech; falsa quina (Mex); bk. (Jamaica b.) decoc. u. antipyr., substit. for cinchona; antiperiod. in malaris, emet.; wood u. for torches, in cabinet work - **E. cuspidatum** St. Hil. (Br): "quina do mata"; bk. u. in fevers; substit. for Cinchona & quinine - **E. sanctae-luciae** J. Britten (Lesser Antilles): quina; bk. decoc. u. as emet. purg.; in fevers (St. Lucia); in combns. to expel placenta (Domin, Caribs).

exotoxin: v. toxin.

expeller (press): equipment by which continuous prodn. of FO can be obtained; raw material is fed into press by a screw conveyor with gradually increasing pressure; feeding through a hopper above and release of cake at bottom proceeds throughout the operation (used for castor oil, etc.).

exsiccation: the process of depriving solids of their water at high temps.; ex. exsiccated alum.

extinction (European publns.): absorbance (US).

extract: the mixt. of substances (usually active) drawn out of a material by distn., soln., heat or other physical or chem. processess; ex. Gambir (71) - **bacterial e.**: aqueous solns. of protein materials fr. bacterial cultures u. for allergenic tests & desensitization - **crude drug e.**: evaporated decoc. of dr. drugs or fresh pl. materials; there are 2 main types: solid (or pilular) e. & powdered e. - **dry e.**: occur as powdered (more popular) & lump - **ferrated e. of apple**: crude iron malate - **fungus e.**: aqueous prepn. of allergenic subst. in various fungi such as *Monilia*, *Saccharomyces*, etc., u. for allergenic diagnostic tests & desensitization - **liver e.**: dry pd. prod. representing the soluble thermostable fraction of mammal liver; designed for use in treatment of permicious & some other types of anemia; acts to incr. the number of erythrocytes in the blood of the patient; some prods. are adm. orally, others parenterally - **ox-bile e.**: prep. made fr. ox-bile, the fresh bile of cattle; u. choleretic. - **pilular e.**: crude drug e. with rel. high water content u. in mfr. pills, ointments, etc. - **poison ivy e.**: allergenic preps. u. to desensitize individuals allergic or sensitive to poison ivy, *Rhus radicans* (etc.) - **pollen e.**: allergenic preps. intended to desensitize persons allergic to specific pollens & hence sufferers fr. pollinosis (hay fever) - **powdered e.**: dry form of ext. of crude drug u. in mfr. capsules, powders, etc. - **soft e.: solid e.**: pilular e. - **thin e.**: prepns. now rarely seen; usually syrupy in consistency - **vanilla e.**: Tr. Vanilla.

extraction: the process of separating the desirable parts of a crude drug fr. the inactive (or non-desired) parts by dissolving the former in an appropriate liquid solvent, as in percolation.

Extractum Carnis (L): beef extract; residue from beef broth obtained by extg. fresh sound lean beef by cooking with water; very popular formerly in med. practice - **E. Catholicum**: Ext. Rhei Comp. - **E. Faecis**: yeast ext.; u. as pill excipient - **E. Fellis Bovis**: ox bile ext. - **E. Pini**: prep. fr. lvs. (needles) of Euras. spp. of *Picea, Abies* & *Pinus*; u. in baths, for skin dis., rheum., etc. - **E. Sanguis**: dried blood.

Extralin™: liver-stomach concentrate; u. pernicious anemia.

eye: young bud, esp. in tubers, ex. potato - **e. balm: e. root**: Hydrastis.

eyebright (herb): 1) *Euphrasia* 2) *Lobelia* - **spotted e.**: *E. maculata*, *E. nutans*.

eyepiece, ocular: combination of lenses as removable unit set in upper part of microscope tube; sometimes provided with micrometer scale (oc. micrometer), allowing for measurement of objects in the microfield; sometimes with pointer for indicating items of interest.

eye's cup: *Sarracenia flava*.

eyestones: crab's eyes.

Eysenhardtia polystachya (Ortega) Sarg. (E. amorphoides HBK) (Legum.) (swUS, Mex): kidney wood; Lignum nephriticum; palo dulce; wd. inf. u. as diur. in kidney & bladder dis.; in chronic cough; aq. infusion of wd. shows unusual color changes & fluorescence; fls. furnish honey.

Ff

NOTE: The letters "f" and "ph" are often interchangeable; it might therefore be advisable to sometimes look under each letter.

fa-am tea: fr. *Angraecum fragrans.*

fab (abbr): antigen-binding fragments.

Faba: 1) (L.) bean, pod, sd. 2) formerly genus (1 sp.) now = *Vicia* - **F. Calabarica:** seeds of *Physostigma venenosum* - **F. Pichurin** (Port Ph.): sds. of *Licaria puchury-major* - **F. Porcina**: frt. of *Hyoscyamus niger* - **F. Purgatrix:** castor bean - **F. Suilla:** F. Porcina - **F. vulgaris** Moench = *Vicia f.* - **Fabaceae:** a fam. sometimes segregated fr. Leg. Fam.; the pea fam.; named after *Faba* Mill.; monotypic gen. with *F. vulgaria* Moench., now usually placed in Vicia (V. vulgaris L.) - **Fabae Ignatiae** (L.): Ignatia beans.

Fabiana imbricata R. & P. (Solan.) (SA, esp Pe, nCh): pichi-pichi; pichi lvs. & tops; Peru false heath; lvs. & twigs u. ton. diur. in kidney dis., bladder stones, enuresis; inflamm. of urinary tract, GC (rarely now); in vet. med. for liver fluke dis. (SA); wd. u. for kidney & bladder complaints.

fabrica (Sp): factory, mill, plant.

face: catface; blazed trunk u. formerly in turpentining pine.

factitious: artificial; ex. factitious styrax.

factor, animal protein: see under "animal" - **anti-alopecia f.**: inosite - **citrovorum f.**: part of Vit. B Complex; protective against inhibition of conversion of folic acid by aminopterin; u. in anemias; SC needed in growth of *Leuconostoc citrovorum* (bacillus) - **clotting f.:** factors which are essential to blood coagulation and normal hemostasis; often anisoylated; administered in AIDS - **colony stimulating f.**: CSF: natural endocrine-like subst.; occ. in bone marrow, &c.; increases prodn. of neutrophils (granulocytes); u. in transplants, cancer (exptl.), aplastic anemia, &c.; u. in combined inj. with inter-leukin-I; now prod. by genetic engineering - **cow manure f.**: vitamin B_{12} - **liver Lactobacillus casei f.**: folic acid - **Wills' f.**: unknown subst. in yeast & liver missing fr. or inactivated in highly refined liver extr. & required in cases of nutritional macrocytic anemia - **f. VIII**: blood coagulation f. VIII; protein found in human plasma/serum important to blood coagn.; u. for hemophilia; prod. by genetic engineering (E) - **f. IX**: blood coagulation f. IX; plasma thromboplastin antecedent; PTA; found in normal blood plasma; important to early phase of blood coagulation - **f. IX complex**: mixt. of factors II, VII, IX, X; hemostatic (E) - **f. XI**: blood coagulation factor XI; plasma thromboplastin antecedent; PTA; found in normal blood plasma; important in early phase of blood coagulation.

Faden (Ge): thread; fiber - **F. Hausenblase**: shredded isinglass - **F. klee**: *Trifolium filiforme* L.

fading purple test: test for strychnine using concd. H_2SO_4 & oxidizing agt. (dichromate, &c.) with characteristic color changes.

Fadogia agrestis Schweinf. (Rubi) (Nig): wood rootstock u. in Rx to increase virility - **F. erythrophloea** Hutch. & Dalz. (tropAf): rt. u. for constip. & intest. pain (powd. rt. added to food) (Senegal); lf. & berry u. as antidote to arrow poisons; decoc. emet. (Nig).

Faecula (L.) = Fecula: starch .

Faex Medicinalis (L.): dry yeast, *Saccharomyces* Siccum.

faftan: *Calotropis procera.*

Fagaceae: Beech (or Oak) Fam.; consists of ca. 400 spp. of monecious trees or shrubs, principally found in the nTemp Zone; chief subst. of chem. interest in fam. is tannin, found in nearly all parts of tree & in rich quantities (esp. bk. wd. frt.); sds. are starch- or FO-rich; sugar uncommon, VO little studied; alk., resins, & saponins are not found; best known members are the oaks (ca. 200 spp.), beeches, & chestnuts (SC Gr phago [to eat] since sds. of many spp. are edible).

Fagara L. (Rut.): ca. 260 spp. in tropics worldwide; most alkaloid-rich; c. DBA (dibenzanthracene) derivs.; u. in sickle cell anemia; rt. u. as toothbrush; taxa often now transferred to *Zanthoxylum* (*qv.*) - **F. chalybaea** Engl. (Zanthoxylum c. Engl.) (tropAf): rt. bk. (in ointm.) appl. to swellings; decoc. as gargle for weakness, and as emet. in fever; lf. decoc. drunk to treat edema in kwashiorkor; boiled lvs. taken for stomach pain, TB, fever; as poultice for toothache; rt. decoc. for the disease "litarangu" (Hehe); bk. decoc. for mal., colds, cough, dizziness; bk. chewed for toothache (Tanz); thin twigs as toothbrushes - **F. coco** (Gill.) Engl. (Bol, Arg): coco; cochucho (tree); lvs. & twig ends c. berberine, palmatine; bk. c. furoquinoline alks. (a-fagarine); skimmianine (b-fagarine), pro-topine alks. (fagarine-II, a-fagarine [allo-cryptopine]); quaternary aporphine alks. (N-methyl-corydine, N-methyl isocorydine) (1377); meranzin(?); u. in febrile ear dis.; cardioactive drug (*cf.* quinidine) - **F. leprieuri** Engl. = *Fagaropsis angolensis*- **F. macrophylla** Engl. (trop wAf): (st. &) rt. bk. c. resins, lupeol; alks.: angoline, angolinine, chelerythrine, skimmianine; u. expect. antiluet., local analg.; fish & arrow poison; insecticide; wd. c. fagaramide; furnishes African satinwood - **F. martin-icensis** Lam. = *Zanthoxylum martini-cense* - **F. octandra** L. = *Bursera tomentosa* - **F. olitoria** Engl. (Zanthoxylum olitorium) (trop eAf): lf. inf. u. as hair wash (for perfuming value); wd. in FO u. as body perfume; frt. edible; lf. u. to facilitate (difficult) labor; one of ingredients u. in bubonic plague - **F. pilosiuscula** Engl. (Congo, Angola): elondo; rt. decoc. u. as aphrod. - **F. rubescens** Engl. (tropAf): bk. u. in toothache - **F. xan-thoxyloides** Lam. (Xanthoxylum senega-lense DC) (trop wAf, esp. Nig): Senegal prickly ash; yields arta(r) rt.; rt. bk. (Fagara Bark) c. VO, fagaramide, zanthotoxin (fish poison); u. in native med. as arom. ton., in impotence; astr., diaph., antirheum., antipyr., in dental pain & caries, vermif.; rt. & frt. as fish poisons; lf. u. antipyr., flavor, arom.; sds. soaked in water u. in rheum. (inf.).

fagarine: alk. ext. fr. lvs. of *F. coco*; acts to reduce myocardial sensitivity; u. in auricular fibrillation, a. flutter, etc.; in some respects superior to quinidine.

fagaronine: alk. fr. *Fagara* (Zanthoxylum) *zanthoxyloides*: inhibits tumor growth.

Fagaropsis angolensis Mildr. (Rut.) (monotypic) (cAf-Ang): bk. u. useful in folk med.; "ikondu"; bk. decoc. u. to kill lice; rts. u. diur. purg. in GC, abdom. dis. (Senegal), intest. parasites; in sterility; exter. u. for kidney dis., arthritis (massage with pd. rt.) (Senegal).

fagiolo (It): **fagiuolo**: bean, *Phaseolus vulgaris*.

Fagonia arabica L. (F. cretica L.) (Zygophyll.) (Ind, wAs, Eg, swEu): pl. u. antipyr. ton. in edema, eye dis., hemiplegia; delirium; in mouth washes.

fagopyrism: reddish coloration of skin & mucous memb. of white rats, mice, etc., resultant fr. eating fl. or sd. husks of buckwheat.

Fagopyrum esculentum Moench (F. sagittatum Gilib.; F. vulgare Hill) (Polygon.) (c&eAs, now grown in many lands, OW, NW): (common) buckwheat; cult. nUS, Cana for grain (less important than formerly), u. in prepn. b. flour (u. in baking popular "buckwheat cakes") (batter c., flat c., pan c.); kasha (*cf.* groats) (u. in soups, gravies, porridge), b. starch; flg. hb. u. to mfr. rutin; fls. —> b. honey (delectable); fls. inf. u. in hypertension, headache; pl. as fodder; hb. c. fagopyrin (sun poisoning) - **F. tataricum** Gaertn. (Ind) India wheat; u. for edible grain.

fagot: v. cassia.

Fagraea fragrans Roxb. (Logani.) (Ind to seAs): bk. decoc. to wash scabies (Indochin), to assure person of long life (Campechy); against malaria (Mal); antipyr. (PI), lf. & twig decoc. where blood is passed in stool; lvs. & frts. against fever (nBorneo); wd. very useful - **F. gracilipes** A. Gray (Fiji): bk. or lf. inf. drunk to "purify the blood"; wd. useful - **F. racemosa** Jack (Ind to seAs, to Solomon Is): lvs. of tree appl. to boils on legs (condition commonly seen here); lf. decoc. u. as ton. after fever (Mal); lvs. smoked with Chinese tobacco for cold in head - **F. schlechteri** Gilg & Ben. (Melanesia): lf. in boiling water appl. to rash & skin irritations (New Caled).

Fagus L. (Fag.) (nTemp Zone): beeches; large heavy-trunked trees, often in stands, lvs. persistent thru winter; trees furnish heavy shade; bk. smooth & light colored, often carved by children & lovers (hence Buch [Ge] book); bk. & lvs. u. to make plaster appl. to relieve pain (Am Indians) - **F. grandifolia** Ehrh. (F. americana Sweet; F. ferruginea Ait.) (eNA, Cana to Fla & Tex [US]) ("southern b."): American beech; fresh nuts (beech nuts) of this tree u. as food; expressed FO u. as culinary oil; b. nut meal nutr.; bk. u. antisep. astr. ton.; wd. (beechwood) u. in hardwood dist. indy. to obt. wood alc., acetic acid, charcoal; wd. tar u. med. as exp. antisept. antipyr. in bronchitis, TB; tar c. creosote & guaiacol; mfr. paper; lumber water-resistant, hence u. for dams, mills, &c.; for veneers, flooring, spools, furniture, toys, crates, cooperage, for fuel; ornam. planted tree - **F. sylvatica** L. (cs&seEu): European beech; sd. edible delicacy; yields c. 20% FO (u. linim., etc.); (FO) u. as salad oil, bread spread (Ge); sd. roasted as coffee substit. (Fr); wd. source of beech wood tar (Pix Fagi) by dry distn.; c. phenols & derivs.; u. exter. in skin dis., rheum., gout; lvs. fermented & u. as tobacco substit.; as tea; bk. u. as tan (1472); ornam. tree.

faham: fahon: fahum (leaves): fa-am tea.

fain (Fr.): hay.

Fairbairn test: appl. for aloe types: char. xals form when a. is mounted in glycerin-alc. (2-1) mixt. (named after J. W. Fairbairn, Univ. London).

fairies' telephone (Eng): *Convolvulus* sp. with pink fls.

fairy bell(s): **f. cap: f. fingers:** *Digitalis*; *Schrankia* - **f.'s glove**: foxglove, Digitalis - **f. wand**: *Chamaelirium luteum*.

faitour's grease (means cheater's g.): spurge (SC acrid juice u. for malingering, ca. 1400).

falax, gum: prob. gutta percha of inferior form derived fr. *Castilla fallax*.

Falco: falcon: various spp. of *Falco* (Falcon-idae), esp. *F. peregrinus* Turnstall (peregrine; duck hawk) (Euras, introd. to NA): u. for hunting birds & mammals (for which domesticated since ancient times); falconry, the art of using falcons in hunting (Eu, wAs) (using hood, glove); gizzard, bile, excreta, &c., u. in Unani med.

Fallkraut (Ge): Arnica.

Fallugia paradoxa Endl. (Ros.) (swUS, Mex): lf. inf. u. to promote hair growth (Hopi Indians); small branches u. for brooms, larger arrow shafts.

"falsa": *Grewia subinaequalis* (G. falsa) frt.

false : **f. angustura**: 1) *Strychnos nux-vomica* 2) *Brucea antidysenterica* - **f. bay**: *Pimenta acris* v. *citrifolia* - **f. calamus:** *Iris pseudacorus* - **f. China bark: f. cinchona bark**: *Habenaria australis* - **f. cola nuts**: kola nuts, false - **f. dragon-head**: *Physostegia* - **false** also v. acacia, alder, bittersweet, calumba rt., (sweet) flag, hellebore, ipecac, kola nut, lily of the valley (*Maianthemum canadense*), mace, mitrewort, pareira, Peruvian bark, sarsaparilla, sunflower, unicorn, wheat, willow (look under each primary entry).

fame flower: *Talinum* spp.

fan palm: saw palmetto.

fancy: pansy.

fangotherapy: treatment of disease with an imported clay or mud (u. in arthritis, gout, etc.).

fanweed: *Thlaspi arvense*.

faperigua (Arg): *Cassia occidentalis*.

Faradaya amicorum Seem. (Verben.) (Tonga, Samoa): "afa"; frt. juice u. for children's fever (Samoa) - **F. splendida** F. Muell. (Austral, NZ): "buku;" middle layer of bk. u. as fish poison (saponins); frt. edible but acrid; rts. c. alk.; ornam. vine.

Faramea schultesii Standl. (Rubi) (sMex): frts. astr. said u. for sores of the mouth & tongue (Chinantecs, San Juan Lalana) (1780).

Farfara (L): coltsfoot, *Tussilago farfara*.

Farina (L): 1) flour (or meal) fr. cereal grains (usually wheat) or legume sds., specif. the gluten portion of kernels of spring wheat; u. in prepn. of macaroni, etc. 2) starchy or waxy deposit on lower surface of fronds of some ferns (mostly c. flavonoids) - **F. Amygdalae: F. Amygdalarum**: the press cake fr. extn. of FO fr. sds. of *Prunus amygdalus*; u. in cosmetic, perfume & liqueur indy. (both bitter & sweet vars.) - **F. hordei Praeparata**: barley flour heated in steam bath ca. 30 hrs. - **F. Lini**: linseed meal.

farine (Fr): farina.

farinha (Br): **farina** (Co): 1) Cassava (flour) 2) *Dimorphandra* sp.

farkleberry: *Vaccinium arboreum* (name may be corruption of whortleberry).

farmacia (Sp): 1) pharmacy (the subject) 2) pharmacy or drug shop.

Farnesian acacia: *Acacia farnesiana*.

farnesol: linear sesquiterpene alc.; occ. in VO of rose, orange, ylang-ylang, clover, etc.; u. in perfumes.

Faroa acaulis Fries and **F. graveolens** Bak. (Gentian.) (c tropAf): fresh or dr. flg. pl. u. as inf. with digestive & lax. effects; u. for abdominal pain.

Farsetia aegyptiaca Turra (Eg, As, incl. Punjab [Indo-Pak]): pl. said useful in rheumatism - **F. clypeata** R. Br. (sEu, Near East): upright madwort; used against rabies(!), skin eruptions (may represent "Alyssum" of Dioscorides) - **F. hamiltonii** Royle & **F. jacquemontii** Hook. f. & Th.: u. similarly to *F. aegyptiaca*.

faser (Ge): fiber.

fat: fatty or fixed oil, FO; usually appl. (by layman) to product solid at room temp; ex. lard; official fats are usually liquid (olive oil, &c.) (2492); FO needed by animal body to (1) furnish energy; (2) synthesize body tissues; (3) furnish fatty acids essential for many syntheses; (4) carry essential fat-soluble vitamins (A, D, K) - **monounsaturated f.**: one with fatty acids contg. a single double bond per mol. & 2 fewer H atoms per fatty acid molecule; ex. olein - **f. emulsions**: admin. in malnutr. - **fathead tree**: *Nauclea esculenta* - **fat hen**: *Chenopodium album* - **hard f.**: mixed glycerides of saturated fatty acids; u. as coating adjuvant: in tablet making; the coating is intended to allow slow release of drug particles (USP XXI) - **neutral f.** (fraction): 1) one made up of triglycerides 2) part of FO lacking in fatty acids: phospholipids (water-insol.) - **polyunsaturated f.**: one with 2 or more double bonds per fatty acid unit; ex. linolein, linolenin, arachidonin; some are now of the opinion that excess of these are not conducive to health - **saturated f.**: one with no double bonds in mol., ex. stearin; palmitin - **f. wood**: lighter wood (*qv.*).

Fatoua pilosa Gaudich. (Mor) (eAs, PI): lf. chewed raw in stomachache (Taiwan); rt. decoc. u. in fevers, irregular menstr. diur. (PI).

Fatsia horrida B & A.= *Oplopanax horridus* - **F. japonica** (Planch.) Decn. & Planch. (monotypic) (Aralia j. Thunb.) (Arali.) (Jap, sKor, Taiwan); yatsuda (Jap); lf. u. as expect. & in baths; bechic; toxic plant; lvs. c. fatsia-saponin & fatsin, aralin (saponin), oleanolic acid, hederagenin, idaein, quercetin; cult. as ornam. (1719); do not confuse with *Oplopanax japonicus*.

fattuchiere (It): woman practicing a simple kind of medicine.

fatty acids: carboxy acids characteristically present as esters of glycerin (glycerides) in fats; small amts. present free - **essential f. a.: unsaturated f. a.** (UFA): those with 1+ double bonds in chain; thought useful in reducing high blood cholestrol (2734) - **f. p-hydroxy**: occ. in fats, milk, &c.; u. in skin care prepns. (ex. Revlon); cannot be synthesized by human body, hence need of peroral intake - **f. oils**: FO: fatty acid glycerides; with one, two, or three fatty acid moieties; sometimes referred to as fats; classed as saturated, unsaturated, highly unsaturated (2617) (**Brazilian f. o.**: fr. genera *Acrocomia*, *Astro-caryum*, *Orbignya*, *Syagrus*.)

fatwood (seUS): resinous wd. fr. longleaf yellow pine (best).

Faulbaum (Ge): *Rhamnus frangula* - **Amerikanische F.**: Cascara Sagrada.

Faurea saligna Haw. (Prote.) (c&sAf): African beech; natives u. rts. for GC, dysentery, menstr. pains; wd. excellent for many uses, firewood; bk. tan. - **F. speciosa** Welw. (tropAf): broad-leaved boekenhout; rt. & lf. u. as appl. to ear; fine wood; bk. tan.

fausse orange (Fr): *Amanita muscaria*.

fauxréglisse (Fr): jequirity.

fava (Por): **f. beans**: 1) *Vicia faba* 2) common bean, *Phaseolus vulgaris* - **Greek f. b.**: *Vicia faba*.

favism: toxicity developed occasionally after ingestion of *Vicia faba* (broad bean) or by inhalation of pollen of same sp.; symptoms include acute hemolytic anemia with hematuria & jaundice; believed result of hemagglutinating antigen-antibody reaction.

fawn-lily (**, common**): *Erythronium (americanum)* (SC lf. spotted like fawn).

fazenda (Br): hacienda (Sp): plantation, esp. coffee p.

feather shank (swUS): Sabadilla.

featherfew: *Matricaria parthenium*.

Febrifuge (eAf): ext. formerly prep. fr. Cinchona; advantageously reduces transportation costs - **European F.**: former cinchona substit.: c. oak bark (24 pts.) nutgall (6), gentian (5), chamomile (4), Iceland moss (1).

febrifugine: antimalarial alk. fr. *Dichroa febrifuga* & *Hydrangea* sp.; too toxic for regular medicinal use.

feces: faeces (L.): plur. of faex, dregs or refuse, such as tartar pptd. fr. wine; now appl. to solid excreta of animals; *cf.* fecal medicine, involving use of feces in med. (Sterocora Medicata), practiced in cEu in past centuries; recognized in Pharmacopoea Württemberg 1786 (*cf.* Album Graecum [whitened dog feces] & Pavon Stercus [pea fowl feces]) (Stercus, stercoris, L.: dung, manure).

Fecula (L.): starch - **fécule de maïs** (Fr): corn starch - **f. de Barbade** = f. de Martinique = arrowroot - **f. de Tolomane:** Queensland arrowroot.

fedegoso (Br): *Cassia occidentalis*.

Federaloe (Ge): Aletris - **Federharz** (Ge): rubber.

fegato (It): liver.

Fehling's solution: copper test soln. for detecting, reducing sugars.

Feige(n) (Ge): fig(s).

feijão (Br, Por): bean; feijoca: thick bean - **feijãoado** (Br): beans with meat; the national dish of Br.

Feijoa (Myrt.) (trop&subtrop SA): showy shrubs & trees - **F. sellowiana** O. Berg (Arg, Br, Par, Ur): pineapple guava (SC frt. tastes like pineapple); "feijoa"; guayaba (del pais); shrub or small tree (-6 m); frt. green (sometimes red tinged), edible raw, also cooked; bk. c. saponins, tannin; u. astr. (20% decoc.); ornam. (cult. NZ+).

felandrio acuatico (Sp): *Oenanthe phellan-drium*.

Fel Bovis (L.): oxbile (or ox gall) fr. *Bos taurus*.

felce maschia (It): male fern, Aspidium.

Feldrittersporn(samen) (Ge): larkspur (seed).

fel(d)spar: mineral u. as antiskid; in mfr. glassware; light bulbs; white ware, &c.

Felis tigris L. (Panthera t.) (Felidae - Mammalia) (sAs, Siberia to Mal): tiger; source of t. fat, reputed as vulner. in ulcers & skin dis. (80); bones u. in arthritis (*cf. F. catus* L.) (F. domesticus) (domestic cat) (chiefly important as pet) (2501); *F. pardalis* L. (ocelot); *F. concolor* L. (puma, cougar, mountain lion); *F. leo* L. (lion); *F. bengalensis* Kerr (Bengal tiger); *cf. Panthera.*

feltwort: Verbascum - **felwort**: *Gentiana lutea.*

female agaric: v. under agaric - **f. nervine**: *Cypripedium hirsutum.*

feminine hygiene: euphemism for toilet articles designed esp. to serve the cleanliness of the female genital tract, incl. the contraceptive; sometimes specif. for the last-named.

feminize: produce characteristics of female, as by adm. estrogens.

Fenchel(samen) (Ge): *Foeniculum vulgare* - **fenchone**: arom. ketone of VO of fennel.

fennel, bitter: *Foeniculum vulgare* var. *piperitum* - **common f.**: *Foeniculum vulgare* - **dog('s) fennel**: 1) (sUS) *Anthemis cotula* 2) *Peucedanum palustre* 3) *Eupatorium capillifolium, E. compositifolium* - **f. flower**: *Nigella sativa* - **f. fruit** (**f. seed**): f. oil: chief products of f. - **German f.**: fennel with large green frt. - **giant Persian f.**: *Ferula persica* - **hog f.**: *Peucedanum officinale* - **Roman f.**: 1) *Foeniculum vulgare* var. *dulce* 2) *Pimpinella anisum* - **small f. flower seed**: wild f. - **sow f.**: hog f. - **sweet f.**: fennel & its var. *dulce* - **water f.**: *Oenanthe aquatica* - **wild f.**: 1) *Nigella arvensis* 2) *Foeniculum officinale.*

fenouil (doux) (Fr): (sweet) fennel - **f. bâtard** (Fr): dill - **f. de porc**: hog f. - **f. officinal**: fennel - **f. puant**: dill.

fenugreek: *Trigonella foenum-graecum.*

feral: wild, uncultivated.

Feretia apondanthera Delile (F. canthioides Hiern) (Rubi.) (tropAf): rt. decoc. u. for ureteritis, adenitis, GC sed.; frts. u. as cosmetic (marking face); sds. coffee substit.

ferment: an enzyme; agent which produces fermentation or enzymatic reaction; also as a verb, meaning to undergo enzymatic changes - **Fermentum Kefir** (L.): kefir (*qv.*).

fermentary: place where cocoa beans (etc.) are placed to ferment (ripen).

fern: plant of Phylum Pteridophyta, sub-phylum Filicineae, & particularly of Order Filicales (called the True Ferns); the typical shade-loving organism with prominent frond; ex. male f., Boston f., tree ferns - **asparagus f.**: *Asparagus plumosus* (not a fern at all: **"fern asparagus"** would be better name) - **basket f: beech f.**: *Phegopteris* spp. - **Boston f.**: *Nephrolepis exaltata* v. *bostoniensis* (1st coll. Fla Keys [US]) - **brake f.**: v. brake - **broad beech f.** (Ga [US]): *Dryopteris* sp. - **Clayton f.**: *Osmunda claytoniana* - **eagle f.**: *Pteridium aquilinum* - **female f.**: *Athyrium filix-femina* - **holly f.**: *Polystichum* spp. - **lady f.**: *Athyrium* spp. - **maidenhair f.**: *Adiantum capillis-veneris, A. pedatum, A. philippense* - **male f.**: *Dryopteris filix-mas* - **marginal f.**: *Dryopteris marginalis* - **meadow f.**: *Myrica gale* - **oak f.**: *Gymnocarpium dryopteris* Newman - **f. palms**: cycads - **shield f.**: oak f. - **sweet f.**: *Comptonia peregrina* - **sword f.**: brake f. - **tree f.: fern tree**: 1) many members of fam. Cyatheaceae & Marattiaceae (*Cyathea* & other genera); occ. in Austral; actually any fern of arborescent habit 2) *Jacaranda acutifolia.*

Fernambouc wood: Fernambuco w.: red dye wd. fr. *Caesalpinia echinata.*

Fernandoa adenophylla (Wall. ex G. Don) Steenis (Bignoni.) (seAs): bk. decoc. u. in local med. for childbirth (Laos).

Fernbach flask: reinforced Erlenmeyer flask u. for culturing microorganisms in antibiotic mfr.

Ferocactus (Cact.) (swUS, Mex): barrel cactus; u. emergency source of water in desert; mfr. cactus candy, fr. sts., etc. - **F. acanthodes** Britt. & Rose (Calif, Nev [US], Mex): u. in mfr. cactus candy - **F. covillei** Britt. & Rose (sAriz [US], nwMex): barrel cactus; bisnaga; buds & fls. eaten; emergency water supply; central spine u. to make face paint; as analg. (v. under *Pachycereus pringlei*) - **F. diguetii** (Weber) Brit. & Rose (nwMex): viznaga; st. source of water in emergency - **F. hamatacanthus** (Muhlenpf.) Britt. & Rose (Echinocactus h. Muhlenpf.) (Mex, Tex, sNM [US]): visnaga; bisnaga espinosa; Turk's head; spines (-8 cm long; 51,000 found on single pl.) u. as toothpicks (Mex); frt. u. in cooking in place of lemon (Mex) - **F. peninsulae** (Weber) Britt. & Rose (nwMex): similar u. to prec. & foll. - **F. wislizenii** (Engelm.) Britt. & Rose (swUS [Tex, Ariz], nwMex): st. pulp u. to make candy (sold commerc.); spines (-10 cm long) u. to make fish-hooks (Am Indians); trunk u. to make containers; source of emergency water; u. to make face paint (Am Indians); pulp & juice fermented.

Feronia limonia Swingle (F. elephantum Correa) (Rut.) (eAs, Ind to Java): elephant (or wood) apple; kath bel; frt. edible, c. much pectin; stom. ton. u. in dys., diarrh., in prepn. of jellies, sometimes u. in place of bael frt.; valuable; lf. decoc. carm. stom.; thorn inf. u. as hemostatic for metrorh.; bk. u. in biliousness; gummy exudn. fr. stem is like acacia, u. dem. (Tamil physicians), soothing rem.

in diarrh. & dys. (pd. mixed with honey); to prep. water colors; as adhesive; rt. u. med.

Feroniella lucida Swingle (Rut.) (Indochin, Java): frt. u. as vegtable; one var. said toxic - **F. pubescens** Tanaka (Indochin): frt. inedible by humans; rt. u. med.

Ferraria purgans Mart. (Irid.) (tropAm): rts. u. purg. ton. emmen.

ferric chloride: $FeCl_3$; occ. in nature as mineral (molysite); u. in med. as astr., styptic; in diagnosis of phenylketonuria; much u. in indy.; other salts u. in anemia (hematinic), food enrichment, etc. - **f. oxide, red**: jeweler's rouge; hematite is natural mineral source; u. as red pigment in Neocalamine, cosmetics, foods (GB) & many indust. products; to treat As poisoning; source of metallic iron - **f. o., yellow**: similar sources & uses (NF).

ferritin: crystal. org. iron compd. of animal body composed of apoferritin (protein) & colloidal $Fe(OH)_3$ (Fe up to 23%); source of blood iron in anemia.

ferrous salts: divalent iron compds. usu. green colored; u. as hematinics in iron-deficiency anemias; ex. ferrous carbonate; chloride, citrate, fumarate, gluconate, lactate, succinate, sulfate (E) (occ. in nature as such minerals as siderotil [hydrate]); sometimes u. in combn. with folic acid (E).

fertilizer: product u. to enrich soil (manure, lime, chemicals) often used along with insecticide, fungicide, etc.

Ferula alliacea Boiss. (Umb.) (eIran): source of hing asa (type of asafetida); this gum resin u. as carm., intest. antisep.; in hysteria, epilepsy; often transshipped via Ind (Bombay); fr. Khorosan (province) - **F. assa-foetida** L. (Iran, Afgh): asafetida (giant fennel); st. is cut down & living stem-bases & rhiz. & rt. crowns incised; exuding latex (oleo-gum-resin) dr. represents gum asafetida; c. VO with hexenyl sulfide & disulfide, cadinene, vanillin, resin with ferulaic acid (2459); u. sed., esp. in uterine irritation with threatened abortion; stim.; for colds; vermif. (Afgh); carm. lax., condim.; much u. in folk med. (sUS), thus in small bag around neck to keep away colds & other infections; in voodoo charms (1381) - **F. communis** L. (Med reg., n-e-s): giant-fennel; rt. u. in hemorrhages, hysteria; st. in hemoptysis, snakebite; sds. in colic; st. u. as tinder (Sicily); young inflorescence eaten; cult. ornam.; var. *brevifolia* Mariz. and var. *gummifera* Battandier are source of African (or Moroccan) Gum Ammoniac (*qv.*) (Bounafa); subsp. **glauca** Rouy & Camus (F. glauca L.) regarded as Narthex of Hippocrates - **F. foetida** Regel = *F. assa-foetida* - **F. galba-niflua** Boiss. & Bushe (sIran, Afgh, Turkestan): galbanum giant fennel; source of gum-resin galbanum; c. 60% resin, 25-30% gums, 9% VO; u. counterirrit. for rheum. inflamm., abscesses, catarrhs.; u. in plasters, cements; arom. expect antispasm - **F. jaeschkeana** Vatke (Iran, Afgh, Turkestan, Ind): hing; another source of gum asafetida; natives use fresh pl. in cooking (Aster); rhiz. inf. u. as diaph. & in rheum. - **F. marmarica** Aschers. & Taub (Cyreniaca, nAf): source of Cyrenian (or African) ammoniac gum (*cf.* Dorema) - **F. narthex** Boiss. (Baltistan, Pak, eAfgh): u. as spice by ancient Greeks (ca. 500 BC); reputed by ancient Romans in med.; another source of hingra asafetida; may be chief Afghan type; rt. u. in fever - **F. oopoda** Boiss. (Pak, Iran, USSR): eaten like staghåer (var. of asafetida) - **F. ovina** Boiss. (Baluchistan, Jordan): pl. ext. u. antispas. (uter. spasm); antioxytoccic - **F. persica** Willd. (swAs, esp. Iran, nIraq, eAfgh): Persian giant fennel; source of asafetida & of saga-penum; u. locally for rheumatism & lumbago - **F. reddeana** Regal (eAs, Turkestan [Turkkmen]): u. for lung dis. - **F. rubricaulis** Boiss. (Iran, Afgh+): a source of Galbanum - **F. schair** Borszczow (USSR: Turkestan): a source of Galbanum - **F. scorodosma** Bentl. & Trim. = *F. assa-foetida* - **F. suaveolens** Aitch. (Baluchistan [Pak], Iran, Afgh, USSR): chir; u. as vegetable & cattle fodder; may produce a kind of sumbul; probably chief source (1381); has sweet taste (faintly of angelica), then bitter - **F. sumbul** Hook. f. (cAs, Turkestan, eSib): source of true sumbul root; c. VO, resins with umbelliferone, valeric & angelic acids (1382); rt. u. sed. antispas. in hysteria, etc. (psychogenic value) - **F. szowitziana** (Iran): source of gum-resin Sagapenum, obt. fr. latex; c. VO, gum, resin; u. stim. antispas. discutient; ext. in plasters - **F. tingitana** L. (nAf): source of African ammoniac (fashook); some consider this sp. the source of an ancient drug, Silphium (*qv.*); obt. fr. latex; c. gum-resin (prod. exported); u. incense, int. as expect. diur. antispasm. (other species considered source of Gum Asafetida include: *F. caspica* M. B., *F. moschata* (Reinsch) K. Pol. (Kazakstan), *F. soongorica*, *F. pseudoreoselinum* L., & others) (all of cAs).

ferulic acid: caffeic acid methyl ether; occ. in many pls.; isol. fr. *Ferula foetida* (1866); u. as food preservative.

fescue (grass): *Festuca* spp.

feshook (nAf): gum ammoniac.

Festuca L. (Gram) (cosmop): fescue grass; ca. 90 spp.; incl. many good pasture grasses; fungal infection of some spp. causes poisoning of foraging animals; (<L., stalk, straw) - **F. elatior** L. (Eu,

natd. Cana, US): tall(er) fesue; meadow f.; in pastures - **F. rubra** L. (Euras, NA: Temp & Arctic zones): red (creeping) fescue; valuable grazing grass; u. for lawns - **F. scabrella** Torr. (Cana, nwUS): rough fescue; (buffalo) bunchgrass; useful forage grass; occ. in former buffalo range.

fetal tissues, human: now being u. exptl. in therapy (v. under fetus).

feterita (Arab): var. of grain sorgum (now cult. Calif, Ariz [US]).

fetid (or foetid): having offensive odor - **F. hellebore:** *Symphoricarpos foetidus* - **f. nightshade:** *Hyoscyamus niger.*

fetoprotein, α-: serum protein similar to serum albumin; present in large amts. in serum of fetus (prod. by liver) and in certain diseases of adults as hepatitis; in normal adult only traces; may have protective function; seems related in adult to carcinogenesis.

fetus: foetus (L): unborn offspring of viviparous mammals; the stage following the embryo; in humans, the stage 7 to 8 wks. after fertilization till birth; fetal tissues injected into the brain in Parkinson's disease (provides dopamine), Alzheimer's, diabetes, &c.; exptl. use shows relief.

Feuerschwamm (Ge): *Fomes fomentarius.*

feuille (Fr): leaf - **f. de chataignier**: chestnut (leaf).

Feulgen's test (reaction): nuclear stain in tissue sections testing as detector of DNA (aldehyde gp.).

fève (Fr): bean - **f. de Calabar**: Physostigma - **f. de St. Ignace**: Ignatia - **f. du Méxique**: cocoa bean.

fever bark: 1) *Alsotonia constricta* 2) *Pinckneya pubens* - **f. bush**: 1) *Benzoin aestivale* (lvs & bk.) 2) *Pinckneya pubens* - **California f. b.**: *Garrya fremontii, Garrya eliptica* - **f. grass** (Ga [US]): *Sisyrinchium* spp. - **f. nut**: bonduk sd. - **f. plant**: *Oenothera biennis* - **f. root**: *Tri-osteum perfoliatum* - **f. tree**: 1) *Eucalyptus,* esp. *E. globulus* 2) *Pinckneya pubens* (also Australian f. t.) - **f. weed**: 1) *Eryngium campestre* 2) Erigeron 3) *Verbena stricta.*

feverfew (chrysanthemum): *Tanacetum parthenium* (hb. & fl. heads) - **feverwort**: *Eupatorium perfoliatum*; variously appl. to *Centaurium minus, Eryngium* spp., and *Tri-osteum* spp. (especially *T. perfoliatum* L.).

Fevillea amazonica C. Jeffrey (Cucurbit.) (Co): sd. FO u. to hasten healing of serious burns (appl. tid for 10 days) (2302) - **F. cordifolia** L. (tropSA, WI): yabilla; antidote bean; a. cocoon; sds. (andhiroba seeds) furnish sekua fat (mandiro[ba] f.) u. as lax. (SA), also u. to polish sd. ornaments; in soap and candle mfr.; sds. (endosperm) u. as lax. emetic, vermifuge (Hond, Nic) (care!); formerly illuminant oil - **F. passiflora** Bell. (Br): jabitocaba; castanha de bugre; frt. & sd. u. purg. & stom. in gastric & liver dis.; sds. c. FO u. veter. med.; in colic - **F. trilobata** L. (SA, esp Br): sds. (nhandiroba seeds) c. 40-43% FO, u. in rheum. & skin dis.

Fiber zibethicus (Rodentia) (NA): the muskrat; source of so-called American Musk; u. as substit. for true musk; furs u. for coats; meat eaten.

fiber: a hardened sclerenchymatous elongated cell, generally with tapered ends; in pl. histology, 3 types are generally recognized: wood or xylem, phloem or bast (ex. linen), & pericyclic; *en masse* constitute heavily lignified supportive tissues; the term fiber is also loosely appl. to: trichomes (ex.

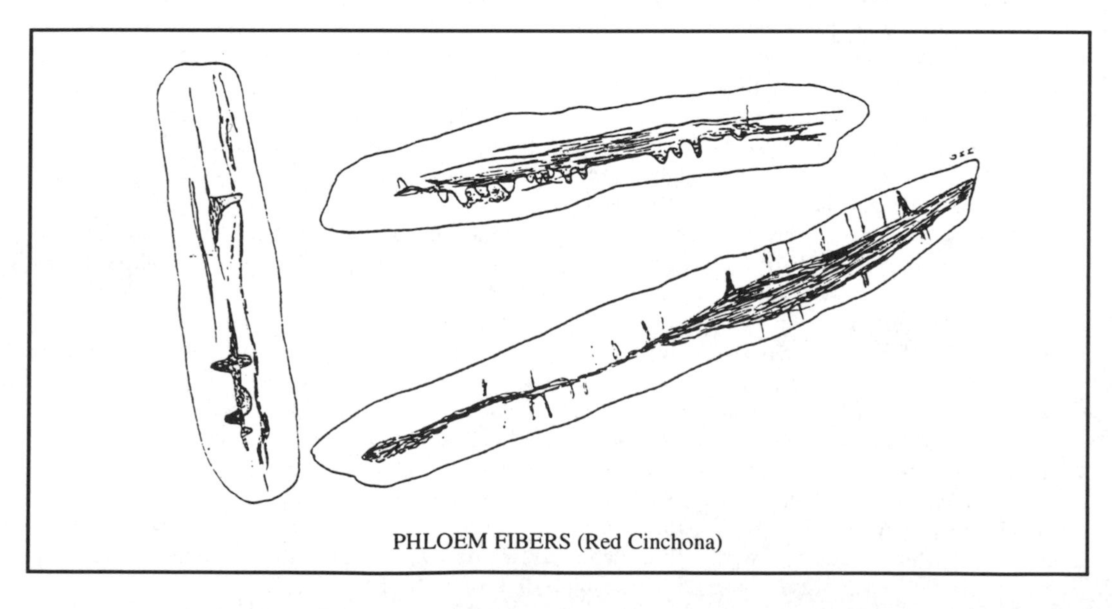

PHLOEM FIBERS (Red Cinchona)

cotton, kapok), lf. fibro-vascular bundles (ex. agave), collenchyma strands (ex. celery), shredded lf. & wd. tissues, sd. coat fragments, wd. cells besides wd. fibers, viz., tracheids (pine) (ex. paper pulp) - **crude f.**: in phytochemical analysis, represents cellulose with some lignin, being residue fr. treatment with 1.25-10% mineral acid, 1.25-2.50% KOH solution, & 1% H_2SO_4 & H_2O in succession; incl. cellulose, hemicellulose, pectin, lignin, cutin, suberin, etc. - **dietary f.**: remains of pl. cells resistant to digestive enzymes; plant f.; food roughage (ex. in frt. peels) believed with prophylactic values in gastroint. dis. (diverticulosis, colonic cancer, appendicitis, hemorrhoids, etc); found in frts. veget., cereals (2285) - **golden f.** (As): jute - **hard f.**: extending lengthwise in lvs. or stems of moncot pls. & very hard & wiry in nature; 3 major ones are: abacá (strongest vegetable fiber), henequen, sisal - **palmyra** (**bass f.**): fr. *Borassus flabellifer* - **phloem f.**: spindle-shaped often greatly elongated cells with cellulose walls, usually with higher lignin content & simple pits; furnish support & elasticity to st., esp. of herbs - **silver f.** (As): cotton - **soft f.**: rel. flexible & elastic fibers found in stems & inner bks. of dicots; ex. hemp, jute, ramie - **substitute f.**: f. of Sanio: cell intermediate between a wood fiber & wd. parenchyma cell - **wood f.: xylem f.**: similar to phloem f. but usually less elongated, more heavily lignified, with fewer pits.

Fibraurea chloroleuca Miers (Menisperm.) (Indonesia, Mal): rt. decoc. u. postpartum as protective; st. inf. u. as eye wash; in diabetes - **F. tinctoria** Lour. (F. recisa Pierre) (Indoch): rt. & lower st. u. as ton. diur. antipyr.; rts. very bitter & proposed as gentian substit.

fibrin: fibrous material of the blood formed during its clotting by the interaction of thrombin with fibrinogen - **f. film**: f. in thin sheets u. mainly in surgery to separate organs & prevent adhesions, also to replace dura mater of the brain - **f. foam**: prep. u. as hemostatic in surgery - **fibrinogen**: a protein constituent of the blood which by reaction with thrombin in the presence of Ca ions produces fibrin, which causes coagulation of the blood - **fibrinolase: fibrinolysin**: plasma trypsin: chief proteolytic (thrombolytic) enzyme of blood plasma which acts to digest clots in blood stream by dissolving fibrin, fibrinogen, prothrombin; exists as profibrinolysin which is activated to produce fibrinolysin (ex. by streptokinase [ca. 1960-]) - **fibroin**: characteristic protein of silk fibrin (insects, arachnids); hydrolyzed to glycine, alanine, etc. - **fibrous cell**: substitute fiber - **fibrovascular bundle**: structure made up of fibers (phloem) and vascular tissues (xylem) (together with accompanying tissues); several types are distinguished, of which the commonest is the collateral (xylem & phloem laterally separated by cambium); the term "vascular bundle" is now mostly preferred - **f. glue:** u. in reconstructive & blood vessel surgery; to stop bleeding; u. double-barrelled syringe with fibrinogen & plasma.

Ficago™: coffee substit. prepd. fr. figs.

Fichte (Ge): spruce - **Fichtenharz** (Ge): 1) colophony (resin) 2) resin fr. *Tsuga canadensis* - **gemeines F.**: *Pinus sylvestris* (resin).

ficin, crystalline: proteolytic sulfhydryl enzyme (the first to be crystallized and the first protease obt. fr. a plant source); obt. fr. latex of *Ficus glabrata* & other *F.* spp. (leche de higuerón) (*qv.*); close to papain & stem bromelain in props.; hydrolyzes peptide & amide bonds; acts on ascarids & other round worms like the latex itself; identical with papainase (1385).

fico (It): fig.

Ficus: 1) (Mor.) (trop&subtr of world): figs; vines; shrubs, & trees; some prod. edible frts. (ex. *F. carica, palmeri, pringlei*; latex of some spp. u. to remove warts; ca. 850 spp. (1883) 2) (specif.): fig, frt. of *F. carica* - **F. anthelmintica** Mart. = *F. glabrata* - **F. aurea** Nutt. (sFla [US]): strangler fig (SC destroys supporting trees by gradual envelopment); latex (in water) drunk to "cool the heart"; lvs. bruised & made into tea for cough & hoarseness (Bah) - **F. benghalensis** L. (F. indica L.) (Ind, Burma, Mal, PI): banyan; false b. tree; Indian fig; I. b. tree; "evergreen" (WI cult.); bk. c. tannin, sometimes u. as tan; rts. ton.; vulner. u. in GC; lvs. to treat dropsy; frt. starvation food; one of several *F.* spp. serving as hosts to the lac insect; tree sacred to the Hindus; tree yields India rubber; latex c. proteases u. exter. in rheum. & lumbago: bk. inf. astr. in diarrh. dys. diabetes; lvs. as poultice for sores; fodder; wd. of air rts. u. as posts & in paper indy. - **F. caballina** Standl. (Co): thick white latex of tree u. to set bones, forming hard mass (Tikunas) (2299) - **F. callosa** Willd (eAs): latex c. anticoagulase, no caoutchouc; appl. to cuts, sores, etc. - **F. carica** L. (eMed reg., Balkans): (common) fig; cult for its delicious sweet frt., eaten fresh, dr., or in confections, pastry, etc.; syr. medicinal; c. 50 invert sugars; wild (or Capri) fig develops in its fruit the fig wasp necessary for fertilization of the cult. fig.; fig collective frt.; cult. since 2400 BC v. a.; u. aper. corrective, diaph., aphrod.; as coffee substit. (roasted), in cataplasms, breast tea; fig poultice (plaster) u. for boils (Holy Land) (2508); latex of lvs. & twigs. u. to coagulate milk; lvs. u. to treat hemorrhoids;

also ointm. with latex (Chin); lf. inf. insecticide, latex u. for ringworm (spread on area); vars: Smyrna (best); Adriatic (on strings), Greek, Italian, Calimyrna (Calif grown fr. Smyrna stock), Istrian (small sweet), Mission Black (seUS) (1383) - **F. citrifolia** P. Mill. (Fla [US] to Par): suu (Pan, Cuna Indians): latex appl. to wounds & infections; seals wounds - **F. cotinifolia** HBK (Mex, CR): copó; latex (sometimes mixed with pd. bk.) u. vulner. appl. to wounds & bruises; latex u. for bowel ailments; frt. edible - **F. cystopoda** Miq. (Br) mercurio vegetal; bk. u. purg.; source of oil u. rheum. - **F. elastica** Roxb. ex. Hornem. (Ind, Mal): rubber plant; India rubber fig; Indian rubber tree; first comml. source of rubber; commonly cult., popular small tree in hotel lobbies - **F. erythrosperma** Miq. (SolomonIs): bk. inf. u. in GC - **F. eximia** Schott (Br, Par): latex with peptonizing enzyme; u. drastic, anthelm. esp. for hookworm; bk. also u. - **F. gemina** Ruiz (Co): white latex of tree u. as vermifuge, also spread on skin to relieve itching (fungal infections?) (Tikunas) (2299) - **F. glabrata** HBK (F. anthelmintica Mart.) (Br): higuerón; source of latex (leche de higuerón); u. vermicide (1504); purg. antipyr.; bk. u. in tertian mal. - **F. hispida** L.f. (seAs to Austral): lvs u. for reducing fever (Mal); bk. given for stomachache (Mal); latex u. for diarrh., appl. to cracks in soles of feet (Indones); frt. u. emet.; decoc. or crushed in salt water for diarrh. (Hainan); supposed to cause insanity when ingested (Ind) - **F. indigofera** Rechinger (Solomon Is): latex u. for sores - **F. laurifolia** (CA, SA) latex (as leche de higuerón) c. proteolytic enzyme (ficin), u. intest. anthelm. - **F. lyrata** Warb. (F. pandurata Sander) (trop wAf): fiddle-leaf (fig) tree (SC leaf shape) (tree -14 m): u. as popular indoor ornam. (very large lvs. shaped like violin body) - **F. maxima** P. Mill. (Mex to Amazonas): higuerón prieta; latex purg.; vermifuge (Br) - **F. nymphaefolia** Mill. (SA, sMex): Caracas fig; lf. inf. u. as bath for erysipelas; latex u. as cataplasm for pain - **F. obliqua** Forst. f. (Fiji Samoa to New Hebrides): ovava (Tonga); u. med for sore joints (Fiji); latex u. for bird lime; lf. ext. u. for carbuncles (Samoa); lvs. u. med. (Tonga [86]) - **F. padifolia** HBK (Mex, CA, Col, WI): American strangler fig; frt. (receptacle) edible; sold in markets; med. uses - **F. prolixa** Forster f. (Mariana Is): frt. mixed with vinegar for use in int. pain or locally on cuts; var. *carolinensis* (Caroline Is): bk. u. as pack for headaches - **F. pumila** L. (Jap, Chin, natd. NW): okque; st. & lvs. u. to bandage wounds, in rheum.; vine & latex antipyr., for swellings & suppuration fr. hemorrhoids; frt. u. in jellies (Jap); decoc. for hernia; rt. decoc. u. as diur. & in cystitis; frt. & lvs. in cancer; latex as anthelm. (1386) - **F. racemosa** L. (F. glomerata Roxb.) (Ind, Burma): cluster fig; pakri (Hind): frts. esculent, u. diabetes; rt. bk. u. astr. in gargles & vagin. inj. for leucorrhea (Ayurvedic med.); latex u. ton. locally in mumps, GC (Ind); latex in plasters appl. to temples for headache (PI) - **F. religiosa** L. (Ind): peepul; bo tree; banyan (tree); tree yields lac; bk astr. u. GC diarrh.; inf. for scabies; fls. u. as decoc. in cooling drink to lower temp of sick person; frt. lax. digest. nutr. dr. & pd. as antiasthma.; incr. female fertility (Ind); bk. fiber u.; a tree sacred to the Hindus; shade tree, planted close to every Buddhist temple & attracts almost as much veneration as the statue of Buddha himself; tree under which Buddha sat; hence bodhi tree (Telugu); sends many aerial rts. fr. its branches to ground (diff. fr. *F. benghalensis*); tree grows to huge size; shade tree - **F. retusa** L. (cosmop, trop): lvs. & rt. u. to treat wounds (Burma): bk. & aerial rts. u. as antipyr. in TB (Taiwan); bk. decoc. in the puerperium (PI) - **F. rumphii** Blume (eAs, esp Mal): bk. & lvs. decoc. u. to loosen phlegm, grd. bk. or frt. appl. to itch; resin with sugar for intest. worms; frt. edible (Bali) - **F. scabra** G. Forst. (Austral): latex u. as vulner. (aborigines) - **F. septica** Burm. f. (eAs+): bk. in formulas for arthritis; anti-emet. (Indones): latex on old painful sores (Celebes); latex for herpes (local); anti-rheum. (local) (PI); lvs. heated & appl. to bruises (Solomon Is) - **F. sycomorus** L. (nAf, Syria): "sycamore"; sycomore (fig) (tree); bk. c. tannin; u. diarrh., cough, scrofula; popular tree of ancient Egypt; ("tree of life"): "fig tree" of Bible (NT); bk. inf. u. in neck pains; frt. edible ("sycomore figs"); wd. u. in ancient Eg for burying mummies; frt. & lvs. galactagog (cattle); latex with peptonizing enzyme, pptg. casein; u. to coagulate milk - **F. tinctoria** Forst. f. (seAs, Oceania): rt. strong aperient (Burma) (also subsp. **gibbosa** Corn.); inner bk. u. as fish lure (Caroline Is) - **F. vogelii** Miq. (tropAf): latex u. as ton., styptic (locally); for dental caries; as wound dressing; bk. inf. u. for leprous ulcers; as purg. (Liberia); frt. edible - **F. yoponensis** Desv. (sMex to Pe): fresh latex of this huge tree ingested for diarrh. & intest. worms (777).

fiddle gum: tragacanth - **f. leaf (plant)**: *Ficus lyrata*; ("Fiddle Dee Fig").

fiddleheads (fern): *Osmunda cinnamomea* & other *O.* spp.

fiddlewood: *Citharexylum* spp. (SC Fr bois fidele, as faithful or trustworthy tree for furnishing wood in home construction): not used in mfr. fiddles or other musical instruments.

fide (L): on the authority of....
Fieberwurzel (Ge): Aletris.
fiel (Fr m): gall; bile.
fig: 1) pl. of *Ficus* spp. 2) fresh or dr. frt. of *Ficus carica* (var. Greek, Smyrna, etc.) - **Adam f.**: banana - **Barbary f.**: 1) *Opuntia vulgaris* 2) avocado - **caprifig** (frt.): 1) wild fig 2) *Opuntia* spp. (frt.) - **cluster f.**: *Ficus racemosa* - **f. coffee**: coffee substit. prepd. by roasting & grinding dr. figs, *Ficus carica* - **edible f.**: product of *Ficus carica, F. benghalensis, F. racemosa, F. religiosa, F. rumphii, F. sycomorus*, etc. - **golden f.**: *Ficus aurea* - **Hottentot f.**: *Mesembryanthemum edule* - **India rubber f.**: *Ficus elastica* - **Indian f.**: *Opuntia* spp. (edible frt.) (SC eaten by Am Indians) - **f. juice: f. latex: f. milk**: leche de higuerón (*qv.*) - **sea f.**: *Carpobrotus aequilaterus* (Aizoaceae) - **Smyrna f.**: finest var. of *Ficus carica* - **strangling f.: wild f.**: *F. aurea.*
figa (Pol): fig (tree).
figado de bacalhau (Por): cod liver.
Figue (Fr): fig - **figuera** (Sp): fig - **figueiro de inferno** (Br) (fig tree of hell): Stramonium.
figwort (herb): *Scrophularia nodosa, S. marilandica* (rt. also u.) - **f. family**: Scrophulariaceae - **water f.**: *S. aquatica* (rt.).
filament: stalk-like portion of stamen which supports the anther.
filament de maïs (Fr): corn silk.
filaree: filaria: *Erodium* spp. - **red-stem f.**: *E. cicutarium* - **white-stem f.**: *E. moschatum.*
filaria, red stem: *Erodium cicutarium.*
filbert: frts. of *Corylus* spp., esp. *C. maxima* (European hazel nut) (named after the celebrated & enterprising King Philibert of France) - **baked f.**: *C. cornuta.*
filé (La [US]): powd. young lvs. of *Sassafras albidum* u. to thicken soups or stews, such as gumbo (g. filé); thread (Fr).
filere: filaree.
filet (Fr): filament of stamen.
filicic acid: filicin: filicinic acid: mixt. of 6 homologs; u. mostly in vet. med. as vermifuge (1498); constituent of male fern, Aspidium.
Filicineae: sub-phylum of Pteridophyta, constituting the Ferns - **Filicis Liquidum Extrac-tum** (L.): oleoresin of male fern.
Filipendula rubra (Hill) B. L. Robinson (Ros.) (e&cNA): Queen of the prairie; u. as "heart medicine"; as "love potion" (Fox Indians) - **F. ulmaria** Maxim. (Ros.) (cEu, As, natd. NA): European meadow-sweet; queen-of-the-meadow; fls. (as Flores [Spiraeae] ulmariae) u. astr. in cough, grip, rheum., bladder dis.; as tan & black dye (Iceland); fls. c. free salicylic acid, methyl salicylate, salicin; gaultherin —> spiraein (glucoside) —> salicaldehyde; citric acid; u. as source of VO (c. heliotropin, vanillin); hb. u. antipyr., antispasm, diur. astr. styptic, in rheum, &c.; rt. also u. med. in gout, rheum. (c. methyl sal. +); ornam. hb. - **F. vulgaris** Moench (Euras, natd. eNA): dropwort; rt. & hb. c. gaultherin, vanillin, salicylaldehyde; u. in epilepsy, GC, bladder & kidney stone, edema, tapeworm, goiter; young lvs. eaten as veget.
Filipin: antibiotic complex fr. *Streptomyces filipinensis* Whitfield et al. (fr. PI soil): 1955-; made up of 3 pentaene components (II, III [most useful], IV); very active antifungal (in animals & plants).
Filix (Mas) (L.) (vulgar name): male fern; *Dryopteris filix-mas.*
filixic acid: filicic acid.
filler: liquid or solid u. to coat a surface or add to bulk or solidity of a substance.
filmaron(e): the chief active princ., an amorphous acid, of male fern; sometimes called "crude filicic acid."
filtrate factor (II): pantothenic acid.
fimble (hemp): staminate pls. of *Cannabis sativa* or its fiber.
Fimbristylis aestivalis Vahl (Cyper.) (eAs): u. in poultices for skin dis. - **F. asperrima** Boeck. (Mal): u. as postpartum med. - **F. barbata** Benth. (widely distr. eAs, Indones, PI): pl. decoc. u. for diarrh. (Sabah), dys.; in stomachache (PI); lvs. cyanophoric - **F. capillaris** (L) Gray (Bulbostylis c. [L.] C. B. Clarke) Gray (pantrop): patillo chico; u. in mouthwash for toothache (Par) - **F. diphylla** Vahl (Oceania; seUS): juice u. (with lime) for ringworm (NG) - **F. miliacea** Vahl (eAs): lvs. in poultices for fever (Mal) - **F. spadicea** Vahl (Mex to Pe: WI): esparrago; u. to heal umbilical cord in infants (Ve Indians) - **F. umbellaris** (Lam.) Vahl. (F. globulosa [Retz.] Kunth) (Chin, Mal, PI, Par): tingel sub; u. for splenomegaly; u. in female dis. & as "blood purifier"; cult. for weaving; sold in Asunción (Par) markets; u. to make baskets, &c. ("esparto mulato").
finfish: regular finny fish (ex. salmon) as distinct fr. shellfish.
finger flower: Digitalis - **f. of Hermes** (transl. of Lat.): Hermodactyla; *Colchicum autumnale*, &c. - **fingered kelp**: *Laminaria digitata* - **finger-grass fodder**: *Onobrychis viciaefolia* - **Fingerhut** (Ge): Digitalis - **f. root**: Digitalis - **fingers**: small pieces of rhubarb rt. originally made by trimming with knife, now often by compressing powd. rt.; formerly often carried in vest pocket for occasional chew.
finocchio (It): fennel - **finocchietto** (It): young fennel.

fiole (Fr-f.): bottle, flask (*cf.* Eng phial).
fiore (It-m.) (-i): flower.
fiorrancio (It): marigold.
fir: *Abies* (usually) - **American f., silver**: *A. alba* - **balsam f.: f. balsam**: **Canadian f.**: *Abies balsamen* (needle oil) - **Douglas f.**: *Pseudo-tsuga menziesii* - **joint f.**: cedar - **f. oil**: VO obt. fr. oleoresin vesicles in bk. of *A. religiosa* (Mex); u. as balsam in med. - **Prussian f.**: *Picea abies* - **Scotch f.**: *Pinus sylvestris* - **Siberian f.** = *Abies sibirica* (needle oil) - **silver f.**: *Abies alba* - **southern balsam f.**: *A. fraseri* - **f. wool**: pine wool: wood w. (1) (*qv.*).
fire bush: 1) *Pyracantha*, esp. *P. coccinea* 2) *Croton lucidus* (WI) 3) *Euonymus atropurpureus* 4) (Fla [US]) *Hamelia patens* 5) *Kochia scoparia* 6) *Streptosolen* spp. - **f. damp**: methane - **f. pink**: *Silene virginica* - **f. thorn**: *Pyracantha*, esp. *P. coccinea* - **f. weed**: 1) *Senecio hieracifolia* 2) *Erechtites praealta* 3) *Gutierrezia* spp. 4) *Epilobium angustifolium* 5) *Datura metel* 6) term appl. to any of various weeds growing in a burned-over area, or to various pls. which suggest the appearance of fire 7) *Datura stramonium* - **great f. w.**: *Epilobium angustifolium* - **f. wheel** (US): *Gaillardia* spp.
"firecracker" (localism-eUS): *Polygonum sagittatum* (SC stem joints explode when heated).
firefly: "lightning bug": nocturnal flying insects, occasionally emitting light, usually beetles of Lampyridae fam. (genera *Lampyrus*, *Photinus*). (v. under *Photinus*).
Firmiana simplex (L) Wight (Sterculi) (Chin; cult. US, as Ga, Ala): varnish tree; Chinese parasol tree; phoenix t.; crushed sds. u. in skin dis.; frts. cooked with meat —> ton. broth; rt. decoc. u. to reduce swellings (Chin); inner white bk. astr. to treat hemorrhoids; lvs. insecticidal (Chin); bk. c. useful fiber; wd. u. in furniture making (1387).
firm offer: definite, steady price put forward for a certain amount of a commodity.
Fischkörner (Ge): fish berries; cocculus.
fish berry: *Anamirta cocculus* - **f. bush**: *Hamelia erecta* - **f. bladders**: f. sounds - **f. maws** (comm.): f. bladders - **f. meal**: ground dr. fish & fish waste (rich in vit. B_{12}); u. animal food, fertilizer - **f. meat**: ingestion claimed of value in reducing coronary heart dis.; reduces cholesterolemia and hyperlipemia - **f. mouth**: *Chelone glabra* - **f. oils**: important FO of commerce; of hygienic value (unsatd. fatty acids reduce LDL cholesterol levels in blood); incl. fish liver oils but not whale oil; in prod. in Nor, US, Peru - **f. poison (tree)** (PR, Fla [US]): Jamaica dogwood - **f. protein concentrate** (FPC): "fish flour"; u. food supplement with 75%-90% protein c. essential amino acids; u. in soups, gravies, bakery goods, &c.; prepd. fr. any marine animals - **f. sounds**: air-bladders or swim-bladders of some fishes, serving as flotation organ, & u. as source of isinglass.
fish-catching coral tree: Jamaica dogwood.
fishing pole (wUS): *Equisetum* spp.
fishskin: 1) skins of various fish spp., chiefly cod, also haddock, pollock, hake, & cusk; u. in mfr. fish glue, also smaller amts. in prepn. leather (imports chiefly fr. Canada) 2) (slang) inner cecum lining of sheep u. as mechanical contraceptive sheath or condom; not protective against AIDS virus.
fission products: atomic fragments of all sorts resultant fr. nuclear fission; most are radioactive.
Fissistigma fulgens Merr. (Annon.) (Mal): lvs. as poultice to sore legs; lf. decoc. given postpartum - **F. kingkii** Burk. (Mal): fl. decoc. for gastric dis. - **F. lanuginosum** Merr. (Mal): similar to prec. - **F. manubriatum** Merr. (Mal): rt. decoc. in stomachache & as antipyr.
Fistulina hepatica Fr. (Fistulin. - Aphyllo-phorales - Basidiomycot.) (temp zone): beef-steak fungus; beef tongue; found on tree trunks; edible; produces valuable "brown oak" wd. (stained brownish).
fitch: *Vicia hirsuta*; *V. sativa* - **fitches** (Eng) (Bible): sd. of *Nigella sativa* (Isa. 28:35).
fitoatria (It): phytiatry.
fit root: *Monotropa uniflora.*
Fittonia albivenis Brummitt (Co to Pe): pl. decoc. u. exter. & int. for headache & muscular pain (777).
Fitzroya cupressioides (Molina) Johnst. (F. patagonica Hook. f.) (monotypic) (Cupress.) (sChile): alerce; Patagonia cypress; wd. of tree of good quality; u. in general carpentry; for musical instruments, shingles, pencils, &c.; resin & bk. u. med.; trees to 1000 yrs.
five-finger: 1) *Potentilla* spp. esp. *P. canadensis* (lvs. & rts.) (also f.-f. grass) 2) *Vitis* 3) *Primula elatior* 4) Virginia creeper - **dwarf f.**: *Potentilla pumila.*
Flachs (Ge): flax.
flacon (Fr, Ge): bottle.
Flacourtia indica Merr. (F. ramontchi L'Hér.) (Flacourti.) (seAs, tropAf, cult. WI+): grilled lvs. u. to make inf. given to female after childbirth; bee plant (Indochin); rts. eaten for leprosy (Tanz); red frts. edible; frt. u. in jaundice, splenomegaly (Ind); inf. of roasted lvs. given women postpart. (Indochin); frt. u. in jellies (vitamin-rich) - **F. jangomas** Raeusch. (F. cataphracta Roxb.) (seAs, esp. Ind, Mal): paniala (Beng): lvs. astr. u. in diarrh.; juice of crushed rts. appl. to herpes, residual

paste to wounds (Mal); frts. deliciously edible, u. in preserves; lvs. diaphor. (Burma); bk. ton. & stom. (Indoch); has been u. in leprosy - **F. sepiaria** Roxb. (tropAs, cult. tropAm): bonch (Beng); lf. & rt. inf. u. alexiteric in snakebite; bk. trit. in sesame oil u. in rheum.; frt. edible.

Flacourtiaceae: Flacourtia Fam.; mostly trop. trees; ca. 800 spp.; sds. characteristically c. FO; no alks. resins, saponins; very few have glycosides; ex. Chaulmoogra trees, Samaun tree.

flag, blue (**root**): *Iris versicolor, I. virginica*, & some other *I.* spp. - **cat-tail f.**: *Typha* spp. - **larger blue f.**: **f. lily**: blue flag (root) - **rattlesnake f.**: *Eryngium aquaticum* - **f. root**: 1) blue f. 2) calamus - **sweet f.**: **false f.**: calamus - **yellow f.**: *Iris pseudacorus.*

Flagecidin™: anisomycin Pfizer.

Flagellaria guineensis Schum. (Flagellari. - Monocots.) (tropAf): Kanot grass; a climbing plant; berry u. ext. in VD; hb. in skin dis. & persistent leg ulcers (Tanz); fl. st. lf. rhiz. u. as diur. (PI); lf. astr. & for hair washes (sAf); twigs u. in VD (Wasanya tribe, Kenya); (Flagellaria SC whip-shaped lvs. <L. flegra = whips with metal studs) - **F. indica** L. (seAs): lvs. & fls. decoc. u. diur., in urinary dis. & gravel (Mal); lvs. u. in hair wash (shampoo) to promote growth (Indones); contraceptive, to produce sterility in female (NG).

flags: *Iris* spp., *Acorus calamus*, etc.

flamboyan (Mex): **flamboyant-tree**: *Delonix regia.*

flame flower: *Kniphofia uvaria* - **f. of the forest** (Ind): *Butea frondosa* - **f. vine**: *Bignonia* spp.

Flamme (Ge-f.): *Iris germanica* - **Flammula Jovis** (L.): *Clematis recta.*

flamy: pansy.

flange: rim or projecting collar found at base of lf. blade next to lf. sheath & near the ligule; on outside of lf. not inside like ligule; frequently confused with ligule.

flannel leaf: *Verbascum thapsus.*

Flaschenkurbis (Ge); *Lagenaria siceraria.*

flat: shallow (wooden) box in which seedlings are started.

flat-top yate (**eucalyptus**): *Eucalyptus occidentalis.*

flavanone: a flavonoid cmpd. with pyranone group; ex. hesperetin; many combined as glycosides.

flavedo: the thin yellowish oily outer portion of the rind of a citrus frt.; ex. sweet orange peel (fresh).

Flaveria bidentis (L) O.K. (Comp.) (Arg, Ch): contrayerba; hb. inf. u. in indigestion, as arom. stim. vermifuge; inf. u. exter. in cataplasms for bites of poisonous animals, washing wounds; in cough - **F. trinervia** C. Mohr (seUS, WI, esp. Cu, Mex): "contrayerba"; hb. u. in gastrointest. dis. (Cu).

flavin: 1) class of yellow N-contg. pigments found in plants & animals 2) (specif.) riboflavin - **flavipucine**: antibiotic prod. by strain of *Aspergillus flavipes* Thom & Church (= glutamycin) - **flavone**: nuclear structure of flavonoids; very abundant in nature; phenyl-benzopyrone.

flavonoids: naturally-occurring compds. with the skeleton $Ph_2C_3Ph_2$; include the flavanones, flavanonols, flavones, flavonols, chalcones, aurones, isoflavones, &c.; possess useful therapeutic properties (antioxidant; antiviral; diur., spasmolytic; cytotoxic, &c.).

flax: linseed, *Linium*, esp. *L. usitatissimum* - **cathartic f.**: *L. catharticum* - **false f.**: *Camelina* spp. - **f. family**: Linaceae. - **mountain f.**: *Polygala senega* - **purging f.**: cathartic f. - **f. sd.**: the dr. ripe sd. of *Linum* - **toad f.**: *Linaria* spp. - **wild f.**: 1) *Linum virginianum* 2) *Camelina sativa* - **flaxweed**: *Cuscuta europaea* - **flax(wort) family:** Linaceae.

fleabane: members of *Erigeron, Conyza, Artemisia*, & some other genera of Comp. - **Canada f.**: **Canadian f.**: *Conyza canadensis* - **daisy f.**: *Erigeron annuus*+ - **marsh f.** (nFla [US]): *Pluchea* sp. probably *P. foetida* - **Philadelphia f.**: *E. philadelphicus* - **salt marsh f.**: **spicy f.**: *Pluchea camphorata*; *Plantago ovata* - **various-leaved f.**: *E. heterophyllus* - **white f.**: *Inula candida* - **flea husks** (Ind): fr. *P. ovata* sds. - **f. seed(s)**: 1) *Plantago arenaria* 2) *Nigella* sds. (white f. s.: *Plantago ovata*) - **fleaweed**: *Conyza canadensis* - **fleawort (seed)**: *Plantago arenaria.*

Fleckstorchschnabel (Ge): *Geranium maculatum.*

Flemingia chappar Hamilton ex Benth. (Leg) (Ind, seAs): inf. u. as remedy for white hair (Camb); lf. u. as beverage tea (Laos) - **F. macrophylla** O. Ktze & Prain (F. congesta Roxb.) (Ind, trop eAf): long-leaf wurrus; hairs fr. frt. (pods) constitute waras (false saffron); Ind red dyestuff (c. resins, flemingin) obtained by careful rubbing & shaking, u. for dyeing & in cosmetics; u. as anthelm. (tapeworm) (Arabia), in skin dis.; similar to kamala, for which sometimes u. as adulterant; rts. u. exter. as appln. to ulcers & swellings; husk eaten - **F. grahamiana** W. & A. (wAf, tropAs): similar to *F. macrophylla*; pl. hairs u. tenifuge, cath. - **F. stricta** Roxb. ex Aiton f. (Ind, seAs, Java, PI): remedy for scabies (Camb); lvs. u. as cigarette paper (Vietnam) - **F. strobilifera** (L) Aiton f. (Ind, SriL; seAs, PI+): lf. inf. u. as ton. after childbirth (Vietnam).

flesh-colored asclepias: *Asclepias incarnata.*

fleur (Fr): flower - **fleurs de muscade**: true mace.

fleur-de-lis (Fr): corrupted in Eng to **fleur-de-luce**: *Iris germanica, I. versicolor*, etc.

fleurs pectorales: breast tea.

Fleurya aestuans (L) Gaudich (Urtic.) (Cu, Bah, continental tropAm): red stinging nettle (Trin); rt. & hb. decoc. for oliguria, VD (Trin) - **F. interrupta** Gaud. (Laportea i. Chew.) (tropAf, As, Austral): lvs appl. locally to carbuncles; rt. decoc. u. diur., in asthma & coughs (PI).

Flieder (Ge): 1) *Sambucus* 2) *Syringa vulgaris* - **Fliedertee** (Ge): inf. of elder fls. - **Fliederbeere** (Ge): elder berries.

Flindersia maculosa Muell. (Flindersi./Rut./Meli.) (NSW, Queensland [Austral]): trunk & branches source of gum with ca. 80% "arabin"; u. in diarrh.

flixweed: *Sisymbrium sophia.*

"floating" (US slang): term describing intoxication by cannabis, etc. - **f. heart**: *Nymphoides peltatum.*

Flohsamen (Ge) (lit. flea seed): plantago sd. esp. fr. *Plantago arenaria.*

flor (Sp, Por): flower - **f. de la cruz** (Salv): Cryptostegia - **f. de noche (buena)**: **f. de pascua** (CA): poinsettia (Xmas fl.) - **f. do para** (Por): *Caesalpinia pulcherrima* - **f. de São João** (Por): *Artemisia vulgaris* - **f. de sol** (Sp): *Helianthemum nummularia.*

floral leaves: sepals & petals.

florefundia: florepondia: *Datura* spp., esp. *D. candida, D. arborea* (1389).

Florence oil: olive oil - **Florentine orris**: *Iris germanica.*

florepondia (Sp, Mex): *Datura candida.*

Flores (L.): plur. of Flos (*qv.*), flowers - **F. Africani** (L) (Eu Ph.): *Tagetes patula*; fl. heads - **F. Balaustiorum** (L): pomegranate fls. - **F. benzoes**: benzoic acid - **F. bismuthi**: bismuth oxide - **F. pectoralis**: pectoral species - **F. zinci**: ZnO.

floressence: the pure essence of odorous principles of fls.

florets: florettes (Fr, Eng): little flowers; ex. Roselle florettes, *Hibiscus sabdariffa*; specif. one of many small fls. which together compose the head in composite pls. (ex. daisy).

floricane: flowering & fruiting stems (*Rubus*, blackberries).

floriculture: the science & art of growing & processing flg. pls.; fls. are a commodity of increasing importance; new pl. vars. are constantly being sought out; of particular importance because of their unique beauty are the Orchid Fam. members.

Florida: v. under arrowroot - **F. bells** (Fla [US]): Turk's caps.

Floridin: floridine (Fr): Florida fuller's earth; a grayish white pd., mostly AlMg Silicate; u. in decolorization & cleaning; in insecticidal & fungicidal dusts and sprays.

floripondio (Mex): *Datura* spp. (various spellings, ex. florepondia, florefundia, floripondia) - **f. blanco** (Mex): *D. arborea.*

Flos (Flores) (L): flower(s) - **F. Acaciae**: 1) sloe fls., *Prunus spinosa* 2) fls. of *Acacia farnesiana* (true acacia f.) - **F. Amygdalae: F. Amygdalarum**: almond blossoms - **F. Anthyllidis**: fls. of *Anthyllis vulneraria* - **F. Cacti**: fls. of *Opuntia* spp. - **F. Calcatrippae**: fls. of *Consolida regalis* - **F. Caryophylli**: cloves - **F. Cassiae**: fr. *Cinnamomum cassia* - **F. Chamomilae Romanae**: *Anthemis nobilis* - **F. Cinae**: *Artemisia cina* - **F. Crataegi**: of *Crataegus oxyacantha.* & *C. monogyna* - **F. Cyani**: *Centaurea cyanus*- **F. Gnaphalii**: fls. of *Helichrysum* (*Gnaphalium*) *arenarium* = **F. Helichrysi** - **F. Lamii**: fls. of *Lamium album* - **F. Malvae Arboreae**: hollyhock fls. - **F. Naphae**: fls. of *Citrus aurantium* - **F. Oxyacanthae**: fls. of *Crataegus oxyacantha* - **F. Pedis Cati**: fls. of *Antennaria dioica* - **F. Persicae** (Por Phar.): peach blossoms - **F. Pruni**: fls. of *Prunus spinosa* - **F. Rhoeados**: fls. of *Papaver rhoeas* - **F. Robiniae**: fls. of *Robinia pseudoacacia* - **F. Spiraeae ulmariae**: fls. of *Filipendula ulmaria* - **F. Stoechados citrinae**: fls. of *Helichrysum arenarium* - **F. Tiliae**: linden fls. - **F. Verbasci**: mullen fls.

flos-adonis: *Adonis autumnalis.*

Floscopa scandens Lour. (Commelin.) (seAs): st. juice put in sore eyes (Ind); part of protective med. combination postpartum (Mal).

floss: fluffy substance chiefly kapok (silk-flower tree); fibers too short for textile use; also *Cochlospermum gossypium*, *Calotropis* spp., *Salix daphnoides*, *Beaumontia grandiflora*, *Populus ciliata*, &c. (*cf.* silk cotton; vegetable fiber).

flour: finely ground & bolted cereal grain with cellular tissues fr. embryo, endosperm, pericarp, &c.; when not otherwise qualified, usually refers to wheat (US, Eng); other common types are corn, barley, rye, buckwheat, etc.; distinction of flour fr. starch: 1) mix with water (flour gives sticky mass, starch a suspension) 2) micro study: trichomes in flour; c. chiefly starch, FO, protein, etc.; u. mfr. breads & other pastries; u. as placebo (flour pills u. in slavery days, sUS) - **gluten f.**: f. made fr. the protein portions of the cereal grain; should c. minimum amts. of carbohydrates; u. for diabetic foods - **Graham f.**: f. made by grinding the entire grain of wheat (etc.) after the chaff has been removed; named after its populizer, a Dr. Graham - **patent f.: white f.**: wheat flour less bran; the best wheat f. mfrd. fr. purified middlings (inferior grades are straight, clear, and red dog) - **whole wheat f.**: Graham f.

Flourensia cernua DC (Leg.) (sUS, nMex): "Mexican senna leaves"; hoja sé; h. sen; lvs. & buds u. purg. in indig. med. (decoc.).

flouve odorante (Fr): *Anthoxanthum odoratum.*

flower: a shoot (or branch) modified to serve as a means of propagation or reproduction, normally by prodn. of sd.; ex. lily, oak - **f. de luce**: *Iris sambucina* - **f. fence**: *Euphorbia pulcherrima* - **f. head**: capitulum (*qv.*).

flowering ash: 1) *Chionanthus virginicus* 2) *Fraxinus ornus* - **f. reed**: *Phragmites communis.*

flowers (chem.): sublimates; ex. **f. of sulfur**: sublimed s. (purified by sublimation) - **f. of benjamin: f. of benzoin**: benzoic acid.

Flueggea: v. *Securinega.*

flugsvamp (Vikings): *Amanita muscaria.*

fluoracetic acid: v. sodium fluoracetate.

fluorescence: phenomenon of absorption of light of one wave length & emission of same in another, hence with color change if in visible spectrum (ex. quinine solns., Chinese rhubarb suspension).

fluorides: salts (mostly Na) of **fluorine** (F): halogen element; composes considerable part of earth's crust; occ. as combined forms, esp. fluorite, etc.; sodium fluoride (NaF) u. as source of **fluorine**; to reduce tooth decay (prophylactic; fluoridation of water supplies practiced in many areas (one ppm. harmless); also u. in pediatric drops as preventative of dental caries in infants; u. widely as insecticide (roaches, ants), pediculicide, acaricide, anthelmintic; many indy. uses (commonly misspelled "flourine," "flouride").

flux: excessive or unhealthy discharge of body; ex. diarrhea - **bloody f.**: dys. - **f. root**: *Gentiana catesbaei* - **f. vine**: 1) (Miss [US]): *Desmodium rotundifolium* 2) (Indiana [US]): *Isanthus brachiatus.*

fly: v. under agaric, amanita, mushroom - **f. blister**: cantharides cerate - **f. catcher**: *Sarracenia* spp. - **Mexican f.**: v. under Mexican - **f. poison**: *Zygadenus* spp. - **f. poison plant** (Ala+ [US]): *Nicandra physalodes* - **Spanish f.**: Cantharis - **f. trap**: *Sarracenia flava.*

foalfoot: coltsfoot.

foam flower: *Tiarella cordifolia.*

fodder: herbaceous livestock food after being cured & dr.; ex. hay, silage.

Foehre (Ge): "fir"; *Pinus sylvestris.*

Foeniculum: 1) (Umb.) (OW): fennel; hbs. with small yellow fls., arom frts. 2) (specif.) (L.): frt. of *F. vulgare* - **F. vulgare** Mill. (F. capillaceum Gilib.) (Med reg.): (common) fennel; bitter f.; hinojo (Sp); perennial hb.; prods. frt. & VO (c. fenchone); inf. u. to bathe eyes, in skin lotions; for stomachache; lvs. u. in sausage; st. in salads; frts. u. flavoring, in confections, liqueurs (brandy); long u. popularly as estrogen; a perenn. hb.; condim; to mfr. VO (bitter f. oil) (prod. Ind, Eg, It, Fr, Ge+) - var. *azoricum* (Mill.) Thell. (Eu): cult. as vegetables - **F. v.** var. *dulce* (Mill.) Thell. (F. d. Mill.): sweet (or Roman or Florence) fennel; "sweet anise"; ann. hb. with 2% VO (light colored, rather sweet & not sharp tasting), c. much anethole, no fenchone; hb. u. for colds (cult. sFr); mostly as cremocarp (large & heavy) - subsp. **piperitum** (Ucr.) Presl Cout. (F. piperitum [Ucr.] Presl.) (It): (Sicilian) ass's fennel; bitter f.; perennial; frt. bitter, VO has little anethole; u. adult. of true f. (**F. v.** var. *v.*) (3587).

foenugreek (seed): **Foenum Graecum** (L): fenugreek, *Trigonella foenum-graecum.*

foglia (pl. **foglie**) (It): leaf - **f. farfarella**: *Tussilago farfara.*

foie (Fr): liver - **f. de morue** (Fr): cod liver.

Fokienia hodginsii Henry & Thomas (Pin.) (eChina, Indochin): nut has astr. & vasoconstrictor properties; VO c. fokienol, monocyclic sesquiterpene.

Folacin: Folaemin: Folettes™: folic acid.

folha (Por): leaf.

Folia (L): plural of Folium (*qv.*) - **F. Trifolii fibrini**: lvs. of *Menyanthes trifoliata.*

Foliandrin™ (Israel): the cardioactive xall. glycoside of oleander lf., *Nerium oleander.*

folic acid: pteroyl glutamic acid; vitamin B_c; member of the vitamin B complex; orig. isolated fr. green lvs. of spinach (hence name fr. folium, lf.), also brewer's yeast, liver, giblets, etc., & now synthesized; concerned with hematopoiesis; hence u. in pernicious anemia, nutritional macrocytic anemias, & sprue; said preventative of burning ("sore") mouth (E); to prevent spina bifida; often lacking in old age; deficiency may be associated with depression, paresthesias, walking difficulties, psychic dis.

folicoli (It): pods, husks.

folinic acid: active form of folic acid by conversion in body (ascorbic acid necessary factor); can be synth. fr. folic acid; factor for *Leuconostoc citrovorum* (bacillus) (hence called "citrovorum factor"); reverses action of aminopterin (& closely related methotrexate) in mice; u. in megaloblastic anemia (contraindicated in pernicious a.) & to antidote folic acid antagonists, such as methotrexate (blocks conversion of folic acid to folinic acid); u. in therapy in form of calcium folinate (c. leucovorin) (E).

Folium (L.): leaf; often u. equivalent to Herba - **F. Anthos**: Fol. Rosmarini - **F. Betulae**: lvs. of *Betula pendula* or *B. pubescens* - **F. Castaneae vescae**: of *Castanea sativa* - **F. Cathae edulis**:

khat tea - **F. Fraxini**: lvs. of *Fraxinus excelsior* - **F. Ilicis**: of *Ilex aquifolium* - **F. Juglandis**: of *Juglans regia* - **F. Liatris odoratae**: of *Trilisa odor-atissima* - **F. Mate**: l. of *Ilex paraguariensis* - **F. Matico**: l. of *Piper angustifolia* - **F. Murrayae**: lvs. of *Murraya koenigi* - **F. Myrtilli**: l. of *Vaccinium myrtillus* - **F. Oleandri**: l. of *Nerium oleander* - **F. Plantaginis**: plantain lvs. fr. *Plantago major, P. lanceolata* - **F. (Rhois) toxicodendri**: l. of *Toxicodendron toxicarium* - **F. Rosmarini**: l. of *Rosmarinus officinalis* - **F. Trifolii albil**: l. of *Trifolium repens* - **F. T. fibrini**: l. of *Menyanthes trifoliata* - **F. Uvae ursi**: l. of *Arctostaphylos uva-ursi.*

folk's glove: foxglove, Digitalis.

folle avoine (Fr): *Zizania aquatica.*

follicle: 1-chambered dry frt. that dehisces along only 1 suture; ex. prickly ash berries - **follicular hormone**: alpha-estradiol - **follicules de séné** (Fr): **Folliculi Sennae** (L.): senna pods - **folliculotropic hormone**: follicle-stimulating h.; FSH; extd. fr. urine of postmenopausal female or fr. pituitary; h. of the anterior pituitary which stimulates the sexual glands (testis, ovary) to produce androgen & estrogen (respectively); by inducing in the female Graafian follicle devt. & activatn. secretion of estrogen; u. for female infertility; and in the male, increased size of testis & stimln. of prodn. of sperm cells; formerly called Prolan A - **follutein**: chorionic gonadotropin.

Fomes auberianus Mont. (Polypor) (tropics): u. as contraceptive, abortient, intoxicant (NG) - **F. fomentarius** Fr. (Polyporus f. L.) (Eu, NA): surgeon's agaric; Fungus Chirurgorum; tinder fungus; bracket fomes; common parasite on beeches, elms, & oaks; has been/is off. in pharmacop. of 18 countries; central part of frtg. body c. fungus cellulose; FO, tannin, resins (with fomitin in fluid extract); fomentaric acid; u. styptic (formerly by surgeons); for slight hemorrhage; emet. (Am Indians); lax.; bitter ton., in night sweats of TB, fomitin (in bladder dis); tinder (after treating with KNO_3 [hence "touchwood" a synonym]).

Fomitopsis officinalis Bond. & Sing. (Fomes o. Faull; F. laricis Jacq.; Polyporus o. Vill.) (Siber, Eu-alpine areas, nAf, NA): (larch) agaric; white a.; growing on living or dead trunks of *Larix, Picea, Pinus*, & other conifers (397); frtg. body (as Agaricus [Albus][L]) c. fungus cellulose, resins with agaricin (agaricinic acid), FO, mannitol, gum, wax; u. bitter ton., quinine substit., irrit. anhydrotic, lax (Am Indians), for TB night sweats, rheum. edema, lung dis., gout; anthelmint.; prepn. agaricin; the natural & peeled (upper leathery & lower hymenial layers removed) forms occur on market (1392); dr. spongy part of sporocarp u. as moxa, styptic, tinder, formerly in dentistry as absorbent & styptic (also other spp.); substit. for hops in beer (NA); surface of sporocarp u. for art works (sketches, paintings); bitter in liqueurs; off. in 19 countries - **F. pinicola** Karst. (Fomes pinicola Cooke; Polyporus p. Sw.) (NA): red belt fungus; c. ergosterol, pinecolic acid; u. in homeop. med. (520).

fondre (Fr): to melt, dissolve, convert.

Fontanellkuegelchen (Ge): artificial issue peas (made of Iris rt., etc.).

food: nutrient taken into body of plant or animal organism (usually but not necessarily through oral feeding) for growth, repair, & metabolism - **cattle f.**: cereal feeds consisting of waste & inferior prods., such as bran, shorts (or middlings), screenings, etc. - **cereal breakfast f.**: popular food cereals generally served uncooked (or precooked) with cream or milk & sugar at breakfast; prep. fr. wheat, corn, oats, rye, etc., by such processes as rolling, crushing, steaming, puffing (exploding grains), caramelizing, etc. - **f., diabetic**: nutrients low in or free fr. sugars, starches, & other carbohydrates; u. to reduce titer of sugar in blood & excretion in urine - **f. of the gods**: Asafetida - **f. plant crops, primary**: rice (chief), wheat (2nd), corn (maize), potatoes, sweet potatoes, sugar cane, cassava, coconut, beans, bananas, oats.

fool (Urdu): phool (*qv.*) - **f. hay** (US slang): rough bent; witch grass.

fool's parsley (NA, Eu): 1) *Aethusa cynapium* 2) *Cicuta* sp.

foots: inferior grade of olive oil (etc.) prepd. by special treatment.

forage: herbage food for livestock as grazed in the field; pasture plants; green feed.

Foraminifera: marine Protozoa in which the protoplasm is covered with calcareous or chitinous shells; deposited over long periods of time, these animals' shells have built up massive formations in many parts of the world, constituting the large chalk cliffs of Dover (England) & chalk deposits in many other places (ex. Iowa, Ark, Tex [US]); important as source of prepared chalk; also as an indicator of petroleum deposits.

forb (ecol) (wUS): any hb. other than grass; ex *Liatris.*

forbidden fruit: sweet orange.

foreign organic matter: FOM: (USP & NF) requirement concerning content of impurities in the form of either materials from other plants or parts other than those represented; thus, belladonna root in a package of belladonna leaf is FOM.

foreskin: prepuce (human): preputium penis: skin covering the glans penis; u. as dressing for wounds; furnished after circumcision.

forest blanks (Ind): patches of open land in the forests where cultivation is carried out.

Forestiera neomexicana Gray (Ole) (swUS, esp NM): ironwood; wild privet; wd. u. for digging sticks (Hopi); frt. edible; u. in med. & ceremonial (Navaho Indians); lvs. emetic ("Evilway"); pl. u. for making "prayersticks of the south" (Nav); dye (Nav).

forget-me-not: *Myosotis,* esp. *M. sylvatica.*

Formalin: 40% aqueous soln. of formaldehyde (HCHO); u. antisept. preserv.

formic acid: HCOOH; obt. fr. *Formica rufa* (common red ant of Europe; hence the L. name); found in frt. of *Sapindus saponaria*; stinging nettles, etc.

Formica rufa L. (Formicidae - Hymenoptera) (Eu): (common) red ant; pismire; most spp. make nests of soil above ground; pupae called "ant eggs" u. as fodder for tropical fishes, exotic birds, etc. (imports fr. Finland, USSR); insect body c. formic acid; u. rheum., gout, nephritis (formica [L., It]: ant).

formula, constitutional: structural f.: chem formula showing all elements and bonds in graphic manner.

forskolin: diterpene fr. *Coleus forskohlii*; hypotensive, vasodilator.

Forsythia suspensa Vahl (Ole) (eAs, cult Eu, NA): forsythia; golden bell(s); Japanese gold-ball tree; frt. or frt. coat decoc. u. for furunculosis, skin dis., diur. emmen. vermifuge (in pinworms), lax; pl. decoc. u. for cancer of the breast (Chin folk med.); very popular med. in Orient; lvs. source of rutin; ornam. - **F. viridissima** Lindl. (Chin, cult. Jap+): Chinese golden ball tree; "jessamine" (sKy [US]); frt. u. as lax. ton.; early spring ornam. shrub.

Fortunella japonica Swingle (Citrus j. Thunb.) (Rut) (seChin; natd. seUS): (round) kumquat; "tropical orange"; small tree (often grown dwarfed); small frt. (approx. 25 mm diam.) sweet-sour, eaten fresh in entirety (except sds.); u. salads, preserves, marmalades, candy; considered stim. carmin. anti-phlogistic, antivinous, deodorizing (Chin); dr. peel u. as remedy in pulmonary dis. (Indochin); ornam. - **F. margarita** Swingle (eAs, esp. Jap): oval kumquat; frt. edible & esculent; pulp sour; rind sweet & pungent; VO used; frt. u. in jams, preserves, jellies; cult. Fla (US).

fossil(ized): refers to resins mined fr. the ground & fr. tree spp. which may or may not be extinct - **fossil alkali**: sodium carbonate - **f. dust: f. earth: f. flour**: infusorial earth (pd.).

Fossiline: Petrolatum.

Fossilite™: siliceous earth.

Fothergilla major Lodd. (Hamamelid) (Ga, Ala [US]): witch alder; ornam. shrub to 3.5 m.

fo-ti (Chin): *Polygonum multiflorum* Thunb.

fo-ti-tiang: f.-t.-tieng (Chin-TM): mixt. of *Centella asiatica* (chiefly), *Gillenia trifoliata*, & Kola; very popular herbal medicine.

fougère (Fr): fern - **f. mâle** (Fr): male fern - **f. musquée**: *Myrrhis odorata* - **f. odorante**: *Lygodium microphyllum.*

Fouquieria fasciculata Nash (Fouquieri) (Mex): alabarda; pulv. sds. u. in toothache (Durango); sts. for fences, huts - **F. formosa** HBK (Mex): palo santo (Puebla); st. u. depurative - **F. macdougalii** Nash (Mex): bk. u. soap substit., esp. good for washing woolens - **F. splendens** Engelm. (nMex, Tex to sCalif, NM [US]): coachwhip, slimwood; ocotillo; Jacob's staff; wild thorny desert or arid land shrub; rt. decoc. u. to relieve fatigue (Apache); fls. expect. remedy for cough (Coahuila); fls. & sd. pods eaten; prep. sweet beverage fr. fls. (Coahuila Indians); rt. u. for swellings (dust. pd. Am Indians); wax fr. bk. u. as belt dressing; plant u. for living hedge (Sonora state); shelters (1393).

four o'clock: *Mirabilis jalapa.*

fourmi (Fr): ant, pissmire.

fox: spp. of *Vulpes* (esp. *V. vulpes* L.) & *Urocyon* (*Canis vulpes*); common f.; red f. grown for pelt; believed domesticated in ancient times - **f. grease**: FO u. in folk med. (gout & earache [Unani]) - **f. lungs**: u. med. - **f. meal**: u. as food, aphrod. (Unani med.) - **f. tail**: 1) *Lycopodium clavatum* 2) (Vt): *Equisetum arvense.*

foxberry: 1) bearberry 2) mountain cranberry 3) (Ala [US]): common bramble 4) (Tenn Valley, Ala [US]): *Callicarpa americana* 5) *Mitchella repens.*

foxfire: forest fungus which glows in the dark (seUS); this luminescence noted especially in *Armillariella mellea.*

foxglove: *Digitalis* (usually); *Tecoma radicans*; *Verbascum thapsus*; also appl. to spp. of *Campanula, Impatiens, Penstemon* (Tex [US]), *Phytolacca, Sarracenia* - **common f.**: *Digitalis purpurea* - **Grecian f.**: *D. lanata* - **purple f.**: common f. - **straw f.**: *D. lutea.*

foxgrape: *Paris quadrifolia* (berries, lvs., rt.).

foxtail: *Alopecurus* spp.

fractional distillation: process in which the lower & higher boiling point portions of a substance being (re)distilled are vaporized at different temps. & are collected separately; commonly u. to separate components of VO, etc.

fracture, drug: the manner in which a drug breaks & the type of surfaces exposed in breaking when pressed between the hands; in many materials, this is characteristic: obviously it cannot be applied when the drug material is composed of pieces too small, too hard, or too soft & glutinous (plastic) to be broken; ex. of types of fracture: fibrous (blackberry bk.); short (prickly ash bk.); sharp (Cape aloe).

Fradiomycin: Neomycin, antibiotic.

Fragaria L. (Ros) (temp. reg.): strawberry; fresh ripe frt. u. nutr. flavor, delicacy; in preserves, jams, jellies, etc.; var. *ananassa* Bailey (wNA, wSA): **F. X ananassa** Duch. (F. chiloensis Dcne. var. anan. Bailey) (wNA, wSp): cultivated s.; this sp. constitutes largest part of cultivations; bulk of frt. ("berry") composed of fleshy receptacle; frt. deprived of epicalyx, calyx, & peduncle; u. as delicious dessert food, flavor in pharm. preps. as syrups; ripe frt. c. kaempferol, quercetin; much vit. C; u. in indig. diarrh., dys. & hemorrhages (Ch); c. 3-9% sugars, free acid 1%, pectins 2% - **F. indica** Focke = *Duchesnea i.* - **F. moschata** Dcne. (Eu): Moschus-Erdbeere; a small-fruited strawberry; edible but of decreasing popularity in Am - **F. vesca** Dcne. (F. elatior Ehrh.) (Euras, esp. It, Balkans): Hautbois strawberry; (common wild) strawberry, many forms are known, incl. the Alpine s.; frt. u. in gout; for syrups, liqueurs, species; hb. u. as emm. depur.; tea substit. & adult., hb. frt. rt. u. as astr. ton. diur; in diarrh., icterus; as component of "blood purification tea" (Kneipp); younger lvs. u. as tea substit.; subsp. **bracteata** (A.A. Heller) Staudt.: berries deliciously edible - **F. virginiana** Duch. (eNA): common wild strawberry (best known sp.); believed factor in genesis of cult. s., incl. subsp. **glauca** Staudt. (BC [Cana]): dr. lvs. appl. to navel of newborn baby to heal; also pd. lvs. appl. to sores as disinf. incl. baby's mouth when sore (Indians) - **F. viridis** Dcne. (F. collina Ehrh. (It, Balk): alpine strawberry; lvs. u. as with *F. vesca.*

fragarin: glycoside first found in lvs. of *Fragaria vesca*, raspberry lvs., etc.; relaxes uterus, etc.; u. in dysmenorrhea; presumed active princ.

fragrant: v. under cudweed, thistle - **f. tailgrape**: *Artabotrys odoratissima* - **f. weed** (US slang): tobacco.

frail (US slang): hickory or dogwood stick u. to knock down corn stalks; probably fr. flail.

frailillos (Sp): *Aconitum napellus*, etc.

fraise (Fr): strawberry - **f. des bois** (Fr): wild strawberries.

framboesa (Por): **framboise** (Fr-f.): **frambuesa** (Sp): raspberry, *Rubus idaeus*, etc.

Franciscea: *Brunfelsia hopeana.*

frangipani: *Plumeria.*

frangola (It): **Frangula** (L): *Rhamnus frangula.*

Frangula alnus P. Mill: = *Rhamnus f.*

frangulin: cascarin: glucosides (A, B) in *Rhamnus purshiana, R. alnifolia, R. cathartica*; u. cath.

Frankenia grandifolia Cham. & Schlecht. var. *campestris* Gray (Frankeni.) (Calif, Nev [US], nMex): salt grass (SC pl. grows on salty or alkaline soils); alkali heath; yerba reuma. ("rheumatism herb" [Ala]); hb. u. in catarrh, leucorrhea, GC, fever (before quinine), etc., in syph. (int. & exter.); mild astr. in diarrh, dys.; lvs. separate out hygroscopic salt mixture (NaCl, $MgCl_2$, etc.) on epiderm. surfaces.

Frankfurt black: said to be charcoal fr. grape & wine lees, peach kernels, etc., u. ink mfr.

frankincense: Olibanum, fr. *Boswellia carteri* - **American f.: common f.**: gum thus, a var. of crude turpentine (US).

Franseria Cav. = *Ambrosia.*

Franzosenkraut (Ge): *Galinsoga parviflora.*

frasco (Sp): small bottle, vial.

Frasera caroliniensis Walt. (F. walteri Mx.; Swertia c. Ktze.) (Gentian.) (eNA, Ont [Cana], s to Ga, La [US]): (American) (wild) columbo; Indian lettuce; tall bienn. or triennial hb.; rt. c. gentiopicrin, loganic acid, yellow coloring matter, no starch; u. bitter ton. stim; fresh rt. emet. & purg.; substit. for Calumba; lvs. cooked & eaten by Am Indians; in some areas treated as rare plant (Off. USP 1820-70).

Frauenhaar (Ge): *Adiantum capillis-veneris.*

Frauenminze (Ge): *Chrysanthemum balsamita.*

Fraxinella Mill. = *Dictamnus* L.

Fraxinus americana L. (*F. acuminata* Lam.) (Ole.) (NA): (American) white ash; bk. c. fraxin & fraxetin (glycosides), VO; u. astr. bitter ton., stim. emm. diaph. diur. cath., in mastitis, splen. enlargement, eczema, gout; sds. in obesity, lvs. thought repellant to snakes, relieve insect bites & stings; wd. hard & tough; u. to make implements; oars, wooden bowls, &c. - **F. berlandieriana** A DC (wTex [US], Mex): plumero; fresno; offered as drug item fr. Mex; planted as shade tree (neMex) - **F. chinensis** Roxb. (s&eAs): Chin ash; rt. c. sinine; bk. u. like cinchona (quinine) in malaria; in eye dis. (Chin); frt. carbonized & simmered in water & liquid u. as stom. (Chin); source of Chin (white or insect) wax (fr. cochineal attack); u. med. in plasters; for candles (Chin); to wax paper & cotton, patent leather; also a cochineal tree (2071) - **F. excelsior** L. (F. excelsa Thunb.) (Euras, esp. Balkans): European ash; bk. c. fraxinol (coumarin deriv.); fraxin & fraxidin (as glycosides) (bitter princ.); bk. lvs. st. u. much as for *F. americana*; also antipyr. in rheum., malar. diarrh. dys.; as yellow dye; believed wd. u. for staff of Aesculapius; tree believed involved in magic healing (2389) - **F. floribunda** Wall. (Ind, Himal): exudn. u. as manna substit.; u. for sweetening & mild lax. properties - **F. malacophylla** Hemsl. (Yunnan, Chin): pei chiang kan; c. glycosides (sinine) & VO but no alk.; u. in mal. said to be as effective as quinine in control; antipyr. - **F. nigra** Marshall (e&cNA): black (or water) ash; bk. u. astr. ton. in skin dis.; sts. u. to make baskets (Wisc [US] Indians) (2388) - **F. ornus** L. (sEu, wAs): flowering ash (tree); manna ash tree; source of manna (*qv.*); prod. It, Sic; bk. u. for wound healing (Bulg); in diarrh. dys.; value attributed to OH-coumarins, seco-iridoids - **F. pennsylvanica** Marsh (F. pubescens Lam.) (eNA): red ash; valuable timber (hard, heavy, strong), u. to make tool handles, etc. (2014) - **F. pubinervis** Bl. (F. bungeana sensu Hooper) (eAs): bk. astr. antipyr., to treat intest. inflamn., as diarrh. dys.; for eye abscesses (Chin) - **F. quadrangulata** Mx. (c&eUS, Cana): blue ash; wd. useful; inner bk. yields a blue dye - **F. rhynchophylla** Hance (Chin, Kor): bk. u. antipyr., for tremors; twigs to treat eye disease (folk med.) - **F. xanthoxyloides** DC (nwInd, Pak, Afgh): tree harbors an insect which prod. a kind of Chinese wax.

freebase (cocaine): crude cocaine treated with alkali ($NaHCO_3$) soln., boiling until whitish lump left; intended for smoking to give a stronger & more rapid effect; said strongly addictive.

freixo (Por): *Fraxinus excelsior.*

frejol (Sp): frijol.

Fremontodendron californicum (Torr.) Coville (Fremontia californica Torr.) (Sterculi) (Calif [US], nMex): flannel bush; "slippery elm"; mucilaginous inner bk. u. dem. for poultices; much u. like Ulmus.

French berries: *Rhamnus infectoria* - **F. chalk**: purified talc, a very soft mineral - **F. mulberry**: *Callicarpa americana* (neither Fr nor a mulberry) - **F. polish**: shellac in alc. soln. - **F. psyllium**: *Plantago indica* - **F. red**: rouge - **F. rhubarb**: *Rheum rhaponticum* - **F. rose**: *Rosa gallica* - **F. weed**: *Thlaspi arvense* - **F. white**: powd. talc. - **French** v. under artichoke.

frenching: disease of tobacco, etc., believed due to Th poisoning.

frêne (Fr-m.): ash.

fresa (Sp): strawberry (frt.) - **fresal**: (Sp): strawberry pl.

fresnillo (Mex): *Tecoma stans.*

fresno (Sp): Fraxinus (*cf.* Fresno, Calif [US]).

Freycinetia Guadichaud (Pandan.) (seAs, Indones, PI, Austral, NZ): u. as inf. in med. in many areas; fl. bracts & rts. eaten; aerial rts. u. to make cordage - **F. funicularis** Merr. (Moluccas): young inflorescences boiled & eaten as vegetable - **F. parksii** Martelli (Fiji): u. by healers; bk. leaf, rt. have some antibacterial properties.

friar's: v. balsam - **f. cap: f. cowl**: *Aconitum napellus.*

Friesodielsia obovata (Benth.) Verdc. (Popowia o. Engl. & Diels) (Annon.) (tropAf): for sores in mouth & throat, rt. powd. & incorporated in porridge; frt. edible.

frijol (plur. **frijoles**) (Sp): common bean - **f. caballero** (Cu): Lima b. - **f. de maïz** (Sp): cow pea - **f. de palo** (Salv): pigeon pea - **f. de vaca** (CR): catjang - **f. iztagapa** (Salv): Lima b. - **f. mungo** (Cu): mung b. - **f. verde** (NM [US], Sp): string bean.

frijoli (Mex): common bean.

frijolillo (Mex, Salv): Lima bean.

fringe tree (, white): *Chionanthus virginicus.*

frisol (Co): Lima bean - **f. de palo** (Co): pigeon pea.

Fritillaria cirrhosa D. Don (Lili) (wChin, mts.): dr. bulb or decoc. u. to "dissolve mucus" in coughs, asthma, bronchitis, hemoptysis, lung dis.; in breast cancer - **F. imperialis** L. (Iran): crown imperial; snake's head; bulbs u. in med. (Radix Coronae imperialis); poisonous but may be eaten after cooking; c. imperialine (alk) (2615) - **F. kamtschaticensis** (L.) Fisch. (Far East: Kamschatka, Sikhalin, nJap; NA [Cana]); bulbs c. starch & sugar (rice-like); u. as food often in place of bread; boiled & dr. for winter use - **F. lanceolata** Pursh (wNA): mission bells; bulbs edible - **F. meleagris** L. (Euras): snake's head; checkered lily; bulbs c. imperialine & other alk., starch; bulb juice u. locally in sores; fls. antipyr.; toxic pl.; ornam hb. - **F. pallidiflora** Schrenk (Chin): u. similar to *P. cirrhosa* - **F. pudica** (Pursh) Spreng. (wNA): gold bell; bulbs eaten raw or boiled (Indians); at times dried for winter use - **F. raddeana** Rgl. (cAs, Sib, Caucasus mt. of Turksmenistan): c. alk. raddeanine, verticilline, tritillarine; u med. toxic - **F. roylei** Hook. (Ind): bulbs boiled with orange peel u. in TB, asthma; galactagog; antirheum.; in cancer - **F. thunbergii** Miq. (F. verticillata var. t. Bak.) (Chin, Jap, Sib.): bai-mo: bulbs u. like those of *F. roylei*; c. verticine, fritillarine, verticilline, &c.; bulbs & sds. in u. lung dis.; rheum. & muscular pain, breast cancer (Chin med.); cult. for med. use in Jap gardens - **F. ussuriensis** Maxim. (Manchuria; nKorea): u. like *F. cirrhosa* - **F. verticillata** Willd. var. *thunbergii* Bak. = *F. t.*

fritillary (Crown), Imperial: *Fritillaria imperialis.*

Froelichia floridana (Nutt.) Moq. (Amaranth) (seUS): "cottonweed"; "wild cotton" (Ala [US]); u. folk med. (Redbay, Fla [US], 1940).

frog: tailless amphibian with smooth skin; the common genus *Rana* includes *R. pipiens* (NA), common leopard frog, commonly u. in lab. expts. & tests; formerly u. in med. (Ex. Brit Pharm. I); in Unani med. (*Rana tigrina* Daud) flesh & brains for burns; in coryza (surma); frog spawn (egg masses) u. in baths for rheum. (Eng gipsies) (2769); the slimy dermal secretions of a frog (CA) are rubbed into the skin to prod. hallucinations; dangerous, even lethal at times; skin secretions of an African frog are antiseptic, fungicidal - **arrow poison f.**: *Phyllobates (qv.)* - **f. bonnet**: *Sarracenia* spp.

froment (Fr): Frumentum (L); wheat.

Frosch (Ge-m.): frog.

frost flowers: *Verbesina occidentalis* - **f. plant**: f. weed - **f. proof**: to treat FO, etc., by cooling to low temp. (as outside winter weather) to ppt. stearin, etc; for soft drinks, beer, etc., papain also is u. to digest substances which ppt. when preps. are cooled - **f. weed: f. wort, (Canada)**: *Helianthemum canadense*.

Frucht (Ge): 1) fruit 2) wheat (commonly).

Fruchtstand (Ge): fructification; syncarp, such as cone.

fructans: fructosans: polyfructosans: fructose anhydrides; with 15-50 fructose units in chain; plentiful in grasses; important to human nutrition; hydrolyzed by stomach acids to fructose.

fructose: d-fructose: levulose: fruit sugar: "sweeter s." (c. 1.5 times sweeter than sucrose); occurs with glucose in frts., honey, etc.; prepd. by hydrolysis of insulin, esp. fr. *Helianthus tuberosus*; u. sweetener in diabetes mellitus (better tolerated than glucose) (2560); u. to prevent "sandiness" in ice cream; in bovine ketosis; u. in myopathy (coronary insufficiency); hepatic encephalopathy; gastric ulcer; ketosis of delayed labor; acute alcohol intoxication; DT; as donor in malnutrition, acute & chronic infections; in liver function testing.

Fructus (L) (sing. & pl.): fruit; sometimes u. in phar. as equivalent of Semen (seed) - **F. Anisi stellati**: star anise (frt.) - **F. A. vulgaris**: *Pimpinella anisum* - **F. Aurantii immaturi**: orange peas, o. apples, or o. berries - **F. Caesalpiniaei** divi divi - **F. Cardui Mariae**: frt. of *Silybum marianum* - **F. Cassiae (fistulae)**: pod(s) of C. f. - **F. cumini**: frt. of *Cuminum cyminum* - **F. Cynosbati**: rose haws - **F. Juniperi**: juniper berries - **F. Mariae**: f. of *Silybum marianum* - **F. Myrtilli**: f. of *Vaccinium myrtillus* - **F. Phellandri**: f. of *Oenanthe aquatica* - **F. Vanillae**: vanilla beans.

Fruehlingsteufelsauge (Ge): *Adonis vernalis*.

frugoside: glycoside of coroglaucigenin + D-6-deoxyallose; occ. in *Gomphocarpus fruticosus,* etc.

fruit: Fructus (L): 1) the product developed fr. the pistil or ovary of the flower of an angiosperm sp. after fertilization (normally), with or without additional structures from the other parts of the flower 2) to the layman, such of these as are edible; often called by other names, such as "vegetables," "nuts," etc. - **f. bodies**: spore-bearing structures of non-flowering plants, ex. sporophores - **collective f.**: simple frts. aggregated into a single mass; ex. mulberry; breadfruit; fig; hop; osage orange; banyan - **f. drugs**: quite a number are u. in med.; ex tamarinds, cassia fistula - **f. juice**: v. under Suessmost - **f. sugar**: levulose.

frustule: SiO_2 (silica) shell of diatom; covering made up of 2 valves (plates), one overlapping the other & enclosing the living organism; comparable to two parts of a petri dish (fr. frustum [L] piece; frustulum, small piece).

fruta (Sp): frt - **frutilla** (Sp): strawberry - **fruto** (Sp): 1) produce, staples 2) frt. 3) (pl.): seeds, grain - **frutta** (It): **frutto** (It): frt., frt. tree.

Fucaceae: Rockweed Fam.; of the Phaeo-phyceae (or Brown Algae); have perennial sporophytes, disc-shaped holdfasts, & flattened fronds; ex. Fucus.

Fuchsia L. (Onagr.) (CA, SA, NZ): c. 110 spp., popular ornam. shrubs or small trees (pron. fyoosh-ya) - **F. magellanica** Lam. (Ch): chilco; chilca; u. as diur. antipyr.

fucose: rhodeose: hexose sugar occ. in brown algae (esp. *Fucus* spp.), *Ascophyllum nodosum,* in intrinsic factor, &c.

Fucus crispus L.: *Chondrus c.* (*Fucus* one of 4 genera of Algae of Linnaeus) - **F. digenea**: *Digenea* spp. (*qv.*) - **F. helminthochorton**: *Alsidium h.* - **F. nodosus** L. = *Ascophyllum nodosum* - **F. serratus** L (Fucaceae - Fucales - Phaeophyta) (mostly nAtl off Eu coast): cartwrack rock weed; one of chief sources of Eu bladderwrack; u. as *Ascophyllum nodosum*; thallus c. laminarin (19%), mannose, fructose, vit. C, iodine, K, etc.; has been used in obesity, simple goiter, scrofula, hypertension, arteriosclerosis; has been used (ash) as source of I_2; livestock fodder; seaweed meal; fertilizer for crops (1399); in perfumery - **F. siliquosus** L. = *Halidrys siliquosa* - **F. vesiculosus** L. (coasts of nAtl & Pac Oc as Fr): rockweed; paddy tang; kelp ware; sea kelp; the commonest sp. & the chief source of Fucus (dr. seaweed), Aethiops Vegetabilis (*qv.*), & kelp (bladderwrack ash); thallus c. mucilage with alginic acid, fucose, laminarin, I, Br; u. much like *F. serratus.*

fuellin: *Peucedanum oreoselinum* (lvs. & rt.).

Fuerstia africana T. C. E. Fries (Lab) (trop eAf): u. as vermicide (in ankylostoma), purg. and malaria remedy; galactagog; to color butter; fls. & lvs. c. fuerstione (fuers-tiaquinone).

fugu: Fugu rubripes (Tetraodontidae - Pisces) puffer (fish); found in seas near Jap; swell fishes; inflate by gulping water or air to attain some 3x normal size; with sharp spines; popular food in Jap despite lethal poison carried in liver & other organs; very expensive; fish cultured in Jap; fugu poison represents tetrodotoxin (*qv.*).

Fuligo Ligni (L): soot - **F. Kali**: soot in alkaline soln. evapd. to dryness.

Fuller's earth (Chin clay): Floridin; non-plastic form of clay; highly adsorbent; similar silicate to kaolin, c. traces of Fe & Mg; much prod. in England; u. to purify fatty oils; u. dusting pd., in **fulling** cloth (process of thickening or felting woollen fibers by moistening, cleansing, scouring, pressing, heating, etc.); u. as dark protective covering over cheeses (mixed with oil).

fulmar: a sea bird, *Fulmarus glacialis*, u. by natives of St. Kilda formerly as food (meat with strong odor), lamp oil, feathers.

Fulvicin™: griseofulvin (*qv.*); PG (ultra-microsize; distr. fr. polyethylene glycol); UF (ultrafine; microsize) (Schering) sometimes preferred.

Fumagillin: Fumadil: antibiotic subst. prod. by *Aspergillus fumigatus* (this source of 4 other antibiotics); antiprotozoal; active against amebiasis, trypanosomiasis, & related; 1951-; u. to control *Nosema apis* inf. in honeybees; do not confuse with fumigatin.

Fumaria (L.): **F. capreolata** L. (Fumari.) (Eu, nAf): Rankenerdrauch (Ge): u. med. (Por Ph.) - **F. judaica** Boiss. (Eg, Libya): pl. u. in icterus, splenic dis., exter. in dermatitis - **F. media** Lois (Eu, Med reg.): bastard's ramping fumitory; u. much like *F. officinalis* - **F. muralis** Sond. (wEu): u. med. (Por Ph.) - **F. officinalis** L. (Eu, etc.): fume-root; fresh juice & hb. (lvs.) u. alt. discut. bitter ton. diur.; cholagog; choleretic; depur. diaph.; in asthma, cancers, etc. - **F. parviflora** Lam. (sEu, nAf, Near East, sAs): hb. & frts. u. lax, blood "purifiers," diur. (Iran); in carcinoma; sed. & tranquilizer (Ind); u. much like *F. officinalis* - **F. schleicheri** Soy. Will. (Euras): c. cheli-donine, protopine, sanguinarine; of value in hypertension (2390) - **F. vaillantii** Lois. (Euras, nAf, Sahara): alkafoun; hb. u. in liver & uter. dis.; green pl. u. to prep. sauces (Bilma, Niger) (it is believed that various *F.* spp. have similar properties; sometimes 2 or 3 spp. are mixed in collections & approved for such use).

Fumariaceae: Fumitory Fam.; hbs. found mostly native in nTemp. Zone; very closely related to Papaver. (Papaveraceae sometimes included in this fam.) but with watery instead of milky juice; many spp. have alks. (*Fumaria* fr. fumus [L], smoke, because of the plants' odor).

fumaric acid: COOH.CH:CH.COOH; isomer of maleic acid; obt. fr. *Fumaria officinalis* (hence name); fungi, &c.; now prepd. by fermentation of carbohydrates (1400); u. as substit. for citric acid in frt. juice beverages & for tartaric acid in same & in baking powders; antioxidant.

fumarine: (in *Fumaria*) protopine.

fumeweed: fumeterre (Fr-m.) **fumitory (herb,) (common)**: *Fumaria* - **fumitory fam.**: Fumariaceae.

Fumigatin: antibiotic fr. metabolic prods. of *Aspergillus fumigatus* Fres. (actually fungal toxin); 1938-; synthesized; do not confuse with Fumagillin.

Fumo (Por) = tobacco.

Funastrum clausum Schlechter (Asclepiad.) (Fla [US], WI, Mex, CA, nSA): petaquilla (Mex); "torsalo" (larva of fly) killed by acrid juice from crushed leaves appl. to skin (Mex) - **F. heterophyllum** Standl. (swUS, nMex): hortensia de guia (Jalisco, Mex); u. as remedy for snakebite (Sinaloa).

funcho (Por): fennel.

Fundstaette (Ge): **Fundstelle**: place of discovery, as of a pl. sp.

Fungi: a sub-group of Thallophyta, whose members are devoid of chlorophyll & hence must subsist on other organisms, living or dead; ex. fly agaric, *Penicillium*, diphtheria bacilli; term also u. to refer to organisms distinct fr. bacteria, yeasts, etc. - **Basidia F.**: Basid-iomycotina - **Gill F.**: Agaricaceae - **higher f.**: members of Ascomycotina, Basidiomycotina, Deuteromycotina - **F. Imperfecti**: Deuteromycotina: in this gp., the life history is imperfectly known or understood - **lower f.**: Myxomycota (formerly Myxomycetes); Mastigomycotina and Zygomycotina (formerly these 2 = Phycomycetes) - **pileate pore f.: Pore F.**: Polyporaceae - **Sac F.**: Ascomycotina.

fungicidin = Nystatin: antibiotic fr. *Streptomyces* sp. active against fungi.

fungimycin = Perimycin.

Fungus (L): sing. of Fungi - **F. Chirurgorum**: *Polyporus fomentarius* - **F. Cynosbati** = F. Rosaceus - **F. Laricis** (L): *Fomitopsis officinalis* - **F. Rosaceus**: gall formed on *Rosa canina* by sting of

Cynips rosae; c. tannin; u. urinary dis. (Eu) - **pineapple f.** (Cana): *Fomitopsis officinalis.*

funicle: a small cord-like structure, specif. the stalk of an ovule.

Funifera utilis Leandr. (Thymelaeaceae) (Br): embira sebo; fibers u. in mfr. of cords; juice of pl. u. strong tenifuge; emet.; intoxicant.

Funkia ovata Spreng. (Lili) (Jap): perenn. hb.; parts of petioles boiled & eaten (Ainu).

funori: funorin; funoran; seaweed glue; obt. fr. thalli of *Gloiopeltis tenax*, *G. coliformis*, etc. (Rhodophyta: Algae-Jap); composed of intertwined & mucilage-coated fragments of thallus; prod. in sheets & rolls; u. in weighting of silk & paper.

Funtumia africana (Benth.) Stapf (Apocyn.) (c&wtropAf): bobo(i) (Sierra Leone); false rubber tree; dr. lvs. u. for burns (dressings), lax.; rt. for weak bladder; st. to assist conception; source of Kicksia rubber; sds. adulter. of Strophanthus - **F. elastica** Stapf (Kitabalia e. Merr.) (trop wAf): w. African or silk rubber tree; most important rubber tree of Af; Cameroon r. tree; frt. & bk. u. in jaundice (bk. int.); piles (bk. beaten with spirits); latex fr. incisions in bk. source of (Lagos) silk rubber; sds. adulter. of Strophanthus; source of FO (cult.).

furanose: 5-membered ring sugar occurring in such forms as glucose, in contrast with the pyranose sugars (fr. furan, heterocyclic compd., C_4H_4O).

Furcellaria fastigiata (Huds.) Lamour (Furcell. - Rhodophyta - Algae) (nAtl Oc, Baltic Sea): "black carrageen"; Gabeltang (Ge); thallus u. to produce furcellaran = Danish agar, vegetable gelatin, jelly-like mucilage, sulfated galactoglycan; u. in food industry; chiefly for pudding powders; also in condim., ice cream, confectionary, baby & diabetic foods; ca. 1000 tons/ann. prod.; planned to cultivate in waste (polluted) water (BRD); u. under name Danagar (Denm); prepd. by drying aq. ext. of seaweed.

Furcraea agavephylla Brot. ex Schult. (Agav.) (Trin): rhiz. decoc. u. in rheum. - **F. foetida** Haw. (F. gigantea Vent.) (Agave foetida L.) (sBr, Bo, Ve): cocuiza; succulent pl. somewhat like Agave; lvs. source of fiber called Mauritius hemp; pl. cult. in Mauritius, WI, &c., for prodn. of this comm. fiber (u. in twine & cordage); introduced in many other trop. areas; rt. decoc. u. as diur. & in VD; lf. heated & appl. to tumor or swelling; boiled lf. juice appl. to ulcers, fistulas, etc. - **F. hexapetala** Urban (F. cubensis Vent.) (WI): "pita maguey" (Cu); rt. diur. & u. in hypochondria; cruched lvs. u. as cataplasm; fibers u. for twine, sacking, &c. - **F. humboldtiana** Trel. (Ve): pita; lf. u. for fibers & in alc. beverage - **F. tuberosa** Ait. f. (WI): langue boeuf; rt. "emulsified" to make tea u. in GC; lf. u. for fibers (Domin, Caribs) (2328).

Furcroya = *Furcraea.*

Furfur (L): bran.

furfural = **furfuraldehyde: furfurol: furol**: oily aldehyde occ. in many natural products, mostly as secondary deriv., thus, in lavender oil, marine algae; prepd. by distilling wheat bran, corn cobs, wd., &c.; u. to protect against fungal attack of wd.; starting material in mfr. nylon and other plastics (1402) (prod. in Fla, La [US]).

furze: *Ulex* spp. esp. *U. europaeus.*

fusain (Fr): *Euonymus atropurpureus.*

Fusafungine: antibiotic prod. by *Fusarium* spp., esp. *F. lateritium* Nees; u. in form of aerosol as antibacterial (1958-).

Fusanus R. Br = *Eucarya.*

fusaric acid: fusaric toxins: antibiotic isolated fr. *Fusarium heterosporium* Nees (1934-); also found in other *F.* spp. & *Gibberella fujikuroi* (Saw.) Wollenw.; wilting agt.; acts as hypotensive agt. (1876).

Fusarium Link (Hyphomycetes - Deuteromycotina): large gp. of parasites & saprobes; incl. **F. oxysporum** Schl.: produces enniatins, antibacterial antibiotics - **F. roseum** Link (universal): greatly facilitates healing of burns (humans) (1403).

fusel oil: a natural blend of amyl alcohols, succinic acid, glycerin, etc., obt. in process of fermentation prodn. of alc. fr. cereals.

fusidic acid: antibiotic substance (ramycidin) produced by *Fusidium coccineum* Fuckel (Cylindrocarpon) (Hyphomycetes - Deuteromycotina); with steroid structure; antibacterial antibiotic u. in TB, &c.

fusiform: spindle-shaped.

Fussblatt(wurzel) (Ge): mandrake, *Podophyllum peltatum* (rt.).

fustic (wood), old: *Chlorophora tinctoria* - **young f.**: *Cotinus coggygria.*

Futter (Ge): fodder - **Futterbrei: Futtersaft**: royal jelly; component of bee larvae food.

fybrene wax: an amorphous paraffin hydrocarbon; u. in polishes, ointments.

Gg

Gadidae: Cod Fam.; Class Pisces (Fishes); incl. many important food fishes: ex. cod, haddock, hake.

gaduol: an ext. fr. codliver oil; c. morrhuol.

Gadus: genus of the cods (codfishes) (Gadidae - Pisces) - **G. calarias** L. (*G. morrhua* L.) (nAtl Oc): (nAtlantic) codfish; cod; Dorsch (Ge); torsk (Nor, Dan, Swed); morue (Fr); codling (young cod); u. food (fresh, deep frozen, smoked, dr.) & as chief source of codliver oil (CLO) (*qv.*) (vit A & D); non-destearinated u. for prodn. destearinated CLO (by chilling); as vulnerary, incorporated in hemorrhoidal suppos., in veterinary med., codliver oil; ingestion reduces cholesterol level in blood serum (1569) (2374); caught mostly on the banks or shallows of the NA & Eu coasts (chief FO prodn. Newf, Nor, Icel, Ge, Eng) (2296) - **G. pollachius** L. (nAtl Oc; It to Nor): pollack; greenfish; fresh & salted fish u. as food, in fillets.

Gaensebluemchen (Ge): *Bellis perennis* fl. head.

Gaensefett (Ge): goose fat - **Gaensedistel**: sow thistle, *Sonchus*.

Gaensefinger (Ge): *Potentilla anserina* - **Gaensefingerkraut** (Ge): hb. of same.

Gaertnera cooperi Hutch. & M.B. Moss (Rubi) (Liberia): pounded bk. rubbed on swellings - **G. salicifolia**: Hutch. & Gillett (Liberia): frt. yields FO u. to treat craw-craw; wd. u. to make canoes - **G. vaginata** (OW trop): tree of Réunion; "wild orange"; sds. u. to make Mussaenda coffee (without caffeine).

Gaertneria Medik. = *Franseria*; *Ambrosia*.

gag root: *Lobelia inflata*.

Gageloel (Ge): VO of *Myrica gale*.

gaglee: *Arum maculatum*; source of Portland arrowroot.

gaiaba (Br): **gaiabera**: *Psidium guajava*.

Gaiadendron punctatum G. Don (1) (Loranth): (Ec) "violeta"; inf. of fls. u. in cough; constipation.

Gaillardia Foug. (Comp) (wNA): "brown-eyed Susans" (some spp., BC [Cana]); cult. as ornam. in fl. gardens - **G. aristata** Pursh (wUS & Cana): u. folk med. as poultice for pain (Okanagan) (in divination) of Thompson Indians (BC [Cana]) - **G. pinnatifida** Torr. (Colo to Tex, Ariz [US], nMex): blanket flower; Indian b.; slender gaillardia; coronilla; hb. u. as diur. (Hopi Indians); to aid conception, in anemia, &c.; lvs. added to warm water & u. int. & exter. for gout (Navajos) (1406).

galactan: hexosan which yields galactose on hydrolysis.

Galactia dubia DC (Leg) (WI): iron weed; lvs. u. as tea for relief of backache (Virgin Is); pl. c. latex.

galactic: lactic: concerning flow of milk.

galactose: "lactoglucose"; monosaccharide sugar obt. with glucose by hydrolysis of lactose; occurs as a part of molecule of many gums, mucilages, pectins, &c.

gala-gala: reddish brown to black brittle resin found in branches of various trees in wJava; natives believe ant secretion ("tai semut"); really a gum lac; in sticks; u. cement; in med. (powd. with coffee, nutmeg & sugar; for convalescents) (131).

galam butter: FO of *Madhuca longifolia*, also FO of *Butyrospermum parkii* - **g. gum**: 1) white acacia (gum) 2) Senegal gum.

Galanga (L): **galangal(e)**: **galangall**: dr. rhiz. of *Alpinia officinarum*.

Galanthus nivalis L. (Amaryllid.) (Euras): (common) snow drop; bulbs c. lycorine, galanthamine, & other alk.; u. emet., in paralysis & neuropathies; cult. as ornam. - **G. woronowii** Lozinsk. (USSR): Caucasian snow drops; c. galanthamine; u. med.

galanthamine: galantamine: alk. fr. rts. of *Galanthus woronowii* & other *G.* spp. & *Narcissus* spp.; u. as cholinesterase inhibitor (parasympathomimetic) in treatment of polio & other paralyses; u. together with exercises; also found in *Ungernia victoris* (USSR); "galantonin."

Galarhoeus Haw. = *Euphorbia.*

Galatella villosa Reichb. (Comp.) (Eu): inf. of young stalks drunk warm as expect.; in cough & throat ailments (—> *Aster*).

Galax Urceolata Brummitt (*G. aphylla* L.) L. (Diapensi.) (Va-Ala [US]): beetleweed; wandflower; galax(y); lvs. u. as ornam. by florists (Xmas evergreen); gathered for the trade (nCalif, Ky, around Evergreen, Ala [US]) (350).

galbano goma (Sp): gum-resin Galbanum.

galbanum (gum): gum-resin fr. *Ferula galbaniflua* & other *F.* spp.; source of VO, g. oil (Iran).

Galbulimima belgraveana Sprague (Himan-tandraceae) (PapNG): bk. u. int. for fever, skin dis. (shredded bk. taken with wild ginger) (2261) (2262); ingestion with lvs. of *Homalomena* spp. —> violent intoxication, hallucinations, & sleep.

galbulus: berry-like (cone) frt. of some spp. of Pinaceae; ex. Juniper "berry."

gale: sweet g.: *Myrica gale.*

Galega (L): hb. of **G. officinalis** L. (Leg) (sEu, wAs): (common) goat's rue; dr. flowering hb. c. **galegine** (bitter princ.), tannin; dr. flg. tops u. as mild astr. ton.; galactagog claimed (1404) incr. milk secrn. in cows for whom cult. as fodder pl. (sometimes); snakebite treatment; in diabetes; preps. for breast devt. (comml. supply thru Marseille, Liggorne).

galegine: alk. (guanidine deriv.) fr. hb. & sds. of *Galega officinalis*; acts to reduce blood & urine sugar & hence u. in diabetes (1405).

galenic(al): (**Galenik**, Ge): medicinal prepd. fr. plants & animals (in contrast with chemicals) (SC Galen [ca. 129-200] Gr physician).

galéope douteuse (Fr): *Galeopsis ochroleuca.*

Galeopsis ochroleuca Lam. (G. segetum Necker) (Lab) (c&sEu): hemp nettle; (common) dead-nettle; hollow tooth herb; source of Herba Galeopsidis (Herba Sideritis), u. folk med.; hb. u. as exp. in pulm. & bronch. dis., asthma, bronchitis, TB, diur.; exter. as vulner. (prod. Ge [Mosel], Austria) - **G. tetrahit** L. (Euras): hemp nettle; bee n.; stems swollen at joints; lvs. ovate to o.-lanceolate; corolla purplish or white; u. ton antiperiod; u. as adulter. of *G. ochroleuca.*

Galgant (Ge, Sw): galanga.

galingale: 1) arom. rhiz. of various pl. spp. related to ginger; ex. *Kaempferia galanga*; *Alpinia* spp. (u. med. purfumes) 2) *Cyperus longus* or any *C.* sp. (**English g.**) 3) *Dulichium arundinaceum.*

Galinsoga parviflora Cav. (Comp.) (tropAm as Mex, natd. NA, Euras, now cosmop.): small-flowered potato-weed; young pl. cooked as veg. (seAs); vulner. apply. locally to poison ivy, dermatitis; antiscorbutic; pl. occurs as common weed in both hemispheres; starvation food.

Galipea dichotama Fr. All. (Rut.) (Br): bk. very bitter u. as stom. - **G. febrifuga** St. Hil. (G. cusparia St. Hil. G. trifoliata Engl.; Cusparia f. Humb.; C. trifoliata Engl.) (tropAm, esp. Ve, Co, WI): may be an angostura bark (*cf. Angostura*) - **G. jasminiflora** Engl. (Rut) (SA): bk. u. as quinine substit. (Br); rt. decoc. u. for warts - **G. odoratissima** Lindl. (Br): u. similarly to prec. - **G. officinalis** Hancock (Cusparia o. [Willd.] Engl. [or Hanc.]): C. angustura Rich.; C. febrifuga Engl. [Humb.]; C. macrocarpa Engl.; C. trifoliata Engl.; Angustura cusparia, A. trifoliata) (Rut.) (WI, trop SA, esp. Guianas, Br, Co, Ve): u. similarly to foll.; angostura tree; bk. of twigs c. alks. (3%) (cusparine, cusparidine, galipoline; cuspareine, quinoline, galipine); VO (2%) (galipol, galipene, galipolene, cadinene); lvs. & st. c. evolitrine; source of True Angostura Bark (or cusparia b.); bk. u. in folk med. as bitter ton. in dyspeps., gastric subacidity & anacidity, antipyr. (febrifuge); cinchona substit.; in diarrh. dys., "paralyses"; u. in prepn. VO, bitter schnapps, pink gins, &c.; in homeop. med.; currently, the proprietary "Angostura Bitters" owes its properties to Gentian not Angostura Bark (as formerly).

galipine: galipoidine: alkaloids fr. *Cusparia trifoliata*; u. as arom. bitter; closely similar to cusparine.

galipot (resin) (Fr, Eng): 1) French turpentine; Bordeaux t.; the scrape resin obt. fr. *Pinus pinaster* (*qv.*) in sFr or less frequently fr. *P. sylvestris*, *P. halepensis*, etc. (quite similar to our "gum thus"); source of pimaric acid 2) a turpentine gum fr. *Picea abies*, *Abies alba* (prod. of latter also called Burgundy pitch) 3) resinous prod. fr. *Tsuga canadensis* 4) a small vessel as formerly u. by apothecaries for ointments, confections, etc.; hence, also an apothecary - **g. d'Amérique** (Fr): **American g.**: oleoresin fr. *Pinus strobus* & other *P.* spp.; very close to if not the same as our "gum thus" (*qv.*).

Galium L. (Rubi) (cosmopol): some 400 spp. of hbs. (SC gala, Gr, milk, since some spp. u. to aid curdling of milk [mentioned by Dios-corides]); "bedstraw" (SC formerly u. to make beds arom.) - **G. aparine** L. (Euras, natd. NA+): cleavers; clivers; goosegrass; catch weed bedstraw; hb. c. "galitannic acid," asperuloside (earlier called rubichloric acid), hb. decoc. u. since ancient times as diur. in fevers, edema, skin dis., gallbladder dis.; also expressed juice u. diur.; bad weed worldwide - **G. asprellum** Mx. (eNA): rough bedstraw; whole pl. diaph. diur. in measles (Choctaw Indians) - **G. boreale** L. (nTemp Zone): northern bedstraw; "savoyan" (Comanche Indians); u. diaph. diur. (Cree Indians); pl. decoc. as contraceptive; rts. u. as red dye (Am Indians) - **G. concinnum** T. & G. (n&cUS): shining (or pretty) bedstraw; hb. inf. u. in kidney dis. & edema - **G. cruciatum** Scop. (Ge, eEu): crosswort; pl. c. rutin; hyperoside; u. in epilepsy, gastric dis. - **G. lanceolatum** Torr. (eNA): (yellow) wild licorice; lvs. u. as greens; as flavor (like licorice) - **G. mexicanum** HBK (Mex): yerba de la pulga; sts. & lvs. u. against insects - **G. mollugo** L. (Euras, natd. eNA): white bedstraw; hedge b.; fl. tops antispas. in epilepsy; u. as substit. for *G. aparine* - **G. odoratum** (L) Scop. (Asperula odorata L.) (cEu, wAs, nAf, cult. orn. natd. US): (sweet) woodruff, woodroot; hb. c. coumarin (in glycosidal combn); asperuloside, NBP; u. diur. aper. ton. for liver dis. in breast teas; to prep. May wine (this mixed with champagne & other wines in alc. health tonics); to flavor tobacco; formerly in perfume indy. (now synth. coumarin gen. used); vulner.; in insomnia - **G. orizabense** Hemsl. (sMex): pd. lvs. & sts. u. for intest. parasites (Mazateca); antipyr. - **G. tinctorium** L.(Eu, NA): dyer's woodruff; stiff-marsh bedstraw; cojibwe (Am Indians); perenn. hb.; u. in resp. dis. (Am Indians), diaph.; rts. furnish red dye (Am Indians & early settlers) - **G. trifidum** L. (Euras, NA, Alas-Va [US], Mex): (small) bedstraw; pl. u. in inf. for skin dis. (Ojibwa Indians) - **G. triflorum** Mx. (NA [US, Mex], Euras): sweet-scented (or fragrant) bedstraw; hb. c. coumarin; u. diur. for bladder & kidney dis.; ton. decoc. for dys.; exter. for wounds & rheum; lvs. & sts. edible (Calif Indians) (152) (1570) - **G. umbrosum** Soland. (NZ): whole hb. u. in GC, esp. where urin. retention - **G. uniflorum** Mx. (eUS): whole pl. astr. u. diur. diaph. dye - **G. verum** L. (Euras, natd. Cana, US): yellow bedstraw; ladies' b.; hb. c. tannin, coloring matter; u. (as Herba Galii lutei) as diur., coagul. for milk; pectoral; red dye; whole pl. u. in skin wash for easier childbirth (folklore).

g-alkaloids: glyco-alkaloids.

gall: 1) nutgall 2) (ox) bile 3) gall bladder - **g. berry** (sUS): *Ilex glabra* - **Chinese g.** (or **g. nuts**): woo-pei-tzu; fr. *Rhus chinensis* - **marble g.**: nutgall - **g. oak**: *Quercus infectoria* & allied spp.

Galla (L): **Gallapfel** (plur. **Gallaepfel**) (Ge): **Galle** (plur. **Gallen**) (Ge): gall, nutgall, fr. *Quercus infectoria* - **G. Gallica**: fr. *Q. ilex*.

galla d'Alep (Fr): (Aleppo) galls - **G. Halepensis**: Aleppo gall - **G. Sinensis**: Chinses gall.

gallbladder: cholekystis (Gr): pear-shaped organ adjacent to liver; stores bile; exits through the cystic duct which leads to common bile duct; dr. beef g. u. in hepatitis (folk med.).

Gallerte (Ge): gelatin.

Gallesia gorazema Moq. (1) (Phytolacc.) (Br, Pe): wd. decoc. u. for intestinal worms; VO u. in GC.

gallic acid: trihydroxy benzoic acid; astr. compd. obtained by hydrolysis of nutgall tannin; also fr. *Caesalpinia tinctoria* pods (35% g. a.); u. formerly in hemorrhoids, diarrhea, &c.; now techn.

Gallidae: Phasianidae (Aves).

gallipoli: crude impure olive oil; u. in textile industry.

gallipot: galipot (*qv*.).

gallito (Hispaniola): *Aristolochia ringens*, *A. trilobata*; *Sesbania grandiflora*.

gallnut: Galla.

gall-of-the-earth: 1) *Prenanthes trifoliata*; *P. serpentaria* 2) *Pterospora andromedea*.

Gallogen™: anhydrous ellagic acid; astr. princ. of divi divi; u. intest. astr.

gallotannic acid: tannic acid.

galls: general definition: excrescences on leaves, stems, twigs, &c. of plant; they are produced following insect sting & usually harbor egg(s) or a developing larva (1407); specif. (oak) galls = galls on a Levantine oak, *Quercus infectoria*, the commonly off. nutgalls; classes in commerce: Aleppo, Mecca, Smyrna, Syrian, Turkey (galls are produced on other oak spp., such as *Q. alba*, *Q. coccinea*) (animal galls are simple sores due to friction, etc.) - **Aleppo g.**: one of chief comml. vars. exported thru Aleppo (Syria), Iraq, etc.; bluish; considered best grade - **American g.**: found on *Quercus coccinea*, *Q. imbricaria*, *Q. alba*, &c. - **artichoke g.**: fr. *Quercus robur* - **Bassorah g.**: tannin-rich g. formed on *Q. tauricola* Kotsch. by *Cynips insana* - **black g.**: blue g. - **blue and green g.**: best grades of med. gall - **California oak g.**: fr. *Quercus lobata* - **Chinese g.**: bud galls fr. *Distylium racemosum* (Hamamelid.), *Rhus semialata* (prod. in the Far East); generally unoffic.; c. 38% tannin - **crown g.**: 1) (Aleppo) from *Quercus* sp. by sting of *Cynips cerris* 2) bacterial pl. dis. caused by *Phytomonas tumefaciens*, producing tumors in crown & other parts of pl. (*Vitis*, *Rosa*, *Prunus*, &c.) - **French g. apples**: prod. by several *Q.* spp., esp. *Q. ilex* (galles de France) - **Hungarian g.**: fr. *Quercus robur* - **Istrian g.**: fr. *Quercus ilex* - **Japanese g.**: similar to Chinese g. - **Mecca g.**: Bassorah g. - **oak g.**: medicinal g. - **Smyrna g.**: comm. form of med. g., yellowish or reddish hues - **sumac g.**: fr. *Rhus chinensis, R. semialata* - **Syrian g.**: Aleppo g. - **Texas g.**: fr. *Quercus virginiana* - **Turkey g.**: regular med. g. fr. Turkey - **white g.**: lighter weight and color g. of inferior quality; product thus because of admission of air into interior due to egress of the contained gall-wasp (usu. wasp dies within gall); oxidation of the valuable tannins has taken place; the gall becomes light in color, wt., & value; not u. in med. or pharm. (large Hungarian g.).

galluba (Sp): gayuba (*qv.*).

Gallus gallus domesticus L. (Gallus gallus gallus L.) (Phasianidae - Aves): the hen, a domesticated fowl, of which the eggs, meat, feathers, etc., are u.; commonly but incorrectly called "chicken," a term which should be appl. to the young bird; meat u. to prep. chicken soup (c. broth) u. for colds ("Jewish penicillin") (u. by Maimonides c. 1200 A.D.) (2513) (v. also under egg).

gall-weed: 1) *Zygophyllum apiculatum* 2) *Gentiana quinquefolia* 3) *Linaria vulgaris.*

gallwort: *Linaria vulgaris*, toadflax.

Galphimia glandulosa Rose (Malpighi) (Mex): margarita; u. astr.; in rabies (!?) - **G. glauca** Cav. (Thyrallis glauca Ktze.) (Mex, WI, CA): rt. decoc. u. for urinary obstruction (Mex).

galuteolin: glycoside fr. *Galega officinalis.*

gama: v. grass.

gamabufotoxin: cardioactive glycoside occ. in *Bufo japonicus* & *B. formosus* (toads).

Gamander (Ge): *Teucrium chamaedrys* & *T. scorodonia*; also *Veronica* spp.

gambeer: gambier: gambir (gum): cube g.: **gambir cubique** (Fr): pale catechu, fr. aq. ext., *Uncaria gambir, U. acida, U. dasyoneura, U. lanosa,* & other *U.* spp.; c. catechin - **tanners g.**: cutch.

gamboge (gum): gum resin fr. *Garcinia hanburyi* & other *G.* spp.; u. drastic, c. gum. resin, gambogic acid; yellow dye for water colors; Buddhist priest robes, &c. - **g. butter**: FO fr. sds. of trees of *G. hanburyi* & *G. morella*; u. as food & soap oil, illuminant, in ointments - **g. family**: Guttiferae - **(false) g. tree** (Burm): *Thespesia populnea.*

Gambusia ("gamboozie"): top minnows; smaller fishes found in Everglades; feed on mosquito larvae, hence value in mal. control.

gambuzzo (It): lvs. & stems (twigs) of *Rhus coriaria*; tannin-rich.

gamene: madder.

gam-gun-far (Chin): *Lonicera japonica* or other *L.* spp. (fl. buds).

gamma-amino-butyric acid: GABA: non-protein amino acid; occ. in higher plants, microorganisms, animals; acts as neurotransmitter (inhibiting nerve messages); antihypertensive.

gamma globulins: one of the 4 globulins separable fr. blood plasma by electrophoresis; u. in measles, infectious hepatitis, anterior poliomyelitis - **gamma-linolenic acid**: GLA: gamolenic acid; essential fatty acid deemed of particular value in reducing cholesterol; occ. in colza, soy, borage, ex. evening primrose oils (7-10%), borage oil (18-26%), &c.; precursor of prostaglandins; u. nutr. in atopic eczema; hops; mother's milk, &c.

Gammexane™: hexa-chloro-cyclohexane; effective insecticide (c. 10x as lethal to house fly as DDT); parasiticide (for ectoparasites in vet. med.).

gamodeme: population of organisms so placed in space & time & so similar in morphology & physiology that they are capable of interbreeding within limits of sex differences, incompatibilities, etc.

Gamolepis pectinata Less. (Comp) (tropAf): said u. medicine (—> *Steirodiscus*).

gamopetaly: a state in which the corolla segments (petals) are sufficiently fused together (at least at the base) so that the corolla may be removed as a single structure; sympetaly; ex. *Mentha* - **gamopetalous**: corresp. adj. - **gamosepaly**: a state in which the calyx segments (sepals) are more or less fused together; ex. mallow - **gamosepalous**: corresp. adj.

ganado (Sp): livestock, cattle - **g. lanar** (Arg): **g. ovejuno**: sheep - **g. vacuno** (Arg): cattle.

gander vine: *Clematis virginiana.*

gandul (PR): pigeon pea.

gang flower: *Polygala vulgaris.*

ganja(h): ganz(h) (Indo-Pak): habituating & hallucinary drug Cannabis.

Ganoderma applanatum (Pers. ex Wallroth) Pat. (Ganodermat. - Polyporales) (worldwide): artist's conk; sporocarps with anit-tumor activity for sarcoma 180; low toxicity - **G. lucidum** Karst. (Fomes japonicus [Fr] Sacc.) (Polyporus tsugae [Murr.] Overh.) (worldwide, esp. Chin): lucid bracket; ling-chih; causes wood decay; u. in Chin (Chin Ph. 1977) med.; the mushroom of immortality of Chin & Jap legends; often grows to large size - **G.** sp. (seAs): parasitic on trunk of *Erythrophloeum fordii* Oliv.; supposed anesthetic, u. by criminals to aid in robbing persons; pd. rubbed on skin for skin dis. (Indochin); effect sometimes attributed to bk. of host tree.

garagaen = carageen.

garambullo (NM [US], Sp): gooseberry bush.

garance (Fr): madder.

garantogen: ginseng.

garapa (oil): sd. FO of *Tecoma stans*.

garb: 1) *Salix babylonica* (weeping willow) 2) (Ge) *Achillea millefolium*.

garbanzo beans (Sp): *Cicer arietinum*.

Garberia heterophylla Merr. & Harp. (G. fruticosa A. Gray; Liatris f. Nutt.) (monotypic) (Comp.) (peninsular Fla [US]): low shrub growing in sandy soils, with fragrant pale or purple or pink fl. heads; conspicuous in winter for brown pappus; fls. spring to fall; u. ornam.

garbling: picking over & separating by hand; classifying the prod. into desirable & undesired components; part of the processing of every crude drug.

Garcia nutans Rohr (Euphorbi) (Mex, CA, WI, Colombia): sd. kernels c. 37-54% FO (pale yellow); this oil purg., sometimes dangerous; sd. c. poisonous subst. somewhat like toxalbumoses of croton, tung, castor bean, & abrus.

Garcinia cambogia Desr. (Guttif.) (seAs): goraka; black dried segments of frt. very sour; u. to confer pleasant acidulous taste to Singhalese curries; source of gum-resin Gamboge - **G. cowa** Roxb. ex DC (eInd): cowa (Hindi); source of inferior type of gamboge, u. med.; arils & pericarps edible - **G. hanburyi** Hook. f. (Ind, Thailand, EI): Siam gamboge tree; pl. source of gum-resin gamboge; u. drastic cath., diur. anthelm.; yellow dye, for violin, candles, water colors, etc.; also prod. by other *G.* spp. (poisonous) - **G. humilis** Adams (Rheedia lateriflora L.; R. edulis Tr. & Pl.) (WI, CR, s to Bo, Br): hat-stand tree; jorco (CR); frt. edible & delicious; bk. decoc. u. to treat parasitic skin dis.; to make hat stands (FrWI) - **G. indica** DC (Ind+): kokam (Hind; Bombay); FO of sds. (kokum butter; k. oil) in soap mfr.; adulter. of butter; in skin dis.; frt. cooling, antiscorbutic, demulc.; frt. peel u. in curries - **G. kola** Heckel (G. cola) (wAf, esp. Sierra Leone): bitter kola; male kola; false kola (f. cola); bitter sds. c. flavonoids, xanthones, masticatory (chewed by natives like true kola nuts) as aphrod. (arom. astr.), vermifuge, in several dis., esp. cough, oral infect., liver dis., diarrh., dys; adjunct to kola as beverage (Nig); rt. ton. - **G. livingstonei** T. Anders. (eAf): deliciously edible frts.; lf. & fl. exts. antibacterial - **G. mangostana** L. (Mal): mangosteen (frt.); cult for delicious edible frt., & as substit. for bael frt.; bk. astr. in sore throat, diarrh, dys.; frt. antipyr; frt. peel u. for tanning & dyeing - **G. morella** Desr. (Indomal, Thail): kanagoraka (Singhalese); small hardly edible frt.; sticky gamboge-like resinous exudation of tree u. as favorite vulner. (for wounds); source of gamboge; pl. & exudation c. guttiferin, an active antibiotic; sd. c. morallin (*qv.*) - **G. multiflora** Champ. (Indochin): cay gioc (Viet); juice of frt. u. like lemon juice; sds. source of oil u. for illumination; yellow latex exuding fr. incisions u. as dressing for wounds - **G. myrtifolia** A. C. Smith (Fiji, Tonga): bk. u. to scent FO (Tonga) (86) - **G. oligophlebia** (PI): lvs. appl. for internal pain - **G. pictoria** Roxb. (G. roxburghii Engl.) (w coast of Ind): a source of gum gamboge - **G. pseudoguttifera** Seem. (Oceania): lf. squashed in water, strained, & drunk for body pains; frt. u. to scent coconut oil (Fiji); ripe frt. eaten (New Hebrides) - **G. sessilis** (Tonga, Samoa, Fiji): lvs. u. in domestic med. (86) - **G. sopsopia** (Buch-Ham.) Webb. (G. pedunculadta Roxb.) (eHimal-Tibet): cult. for deliciously acid flavor (malic acid) of frt. - **G. tinctoria** W. F. Wight (Ind & Indochin): one source of gamboge - **G. xanthochymus** Hook. f. ex T. Anderson (nInd); tapped—>gamboge.

garden angelica: *Angelica archangelica* - **g. burnet**: *Sanguisorba canadensis* - **g. dill**: *Anethum* - **g. gate**: pansy - **g. mint**: spearmint - **gardener's gate**: ribbon grass - **gardener's garters**: ribbon grass - **garden** v. under artichoke; carrot; cypress; endive; celandine; heliotrope; lavender; lettuce; marigold; marjoram; nightshade; parsley; purlsane; pursley; rosemary; rue; sage; sunflower; thyme.

Gardenia angkorensis Pitard (Rubi) (Indoch): lf. & st. decoc. u. diur. - **G. grandiflora** (Jap, Indochin): frt. called wongsky, Chinese yellow berries, or C. y. pods; u. as dye (comml. Chin) - **G. gummifera** L. f. (Ind, Chin): deka mali; tree produces gum-resin called dikamali (gardenia bud resin); u. carm. antisept., stim. anthelm. analges. tranquilizer; in dys. diarrh.; exter. in cuts & wounds - **G. jasminoides** Ellis (G. florida L.) (Chin, Indochin): widely cult. ornam; garden gardenia; "Malabar" (Mex) (var. *fortuniana* Lindl.); Cape jasmine; frt. c. yellow coloring matter, u. dye; cholagog for icterus (Jap); antipyr. in hemoptysis (Chin) (85); fls. emoll. in ophthalmia; bk. antipyr. ton. in dys. abdominal pain - **G. lucida** Roxb. (Ind): prod. decamalee (or dikkamale) gum (with odor of cat's urine); u. antispasm; c. VO, **gardenin** (a yellow amaroid); u. much like *G. gummifera* - **G. malleifera** Hook. (Af): "blippo"; bk. rich in tannin, u. (Senegal) to darken skin (natives) - **G. storckii** Oliv. (Fiji): cold water inf.

of rts. u. for lung dis. & constipn. (Degener) - **G. taitensis** DC (Polynesia, Fiji, Samoa): fls. u. to scent FO (Tonga) (86), in inf. for headache; lvs. u. in diabetes (int.), to "cleanse the blood" (prenatal care) - **G. thunbergia** L. f. (sAf): rt. bk. inf. u. in gallbladder dis.; as emet.; exter. & int. in syph.; rt. in leprotic skin dis.; lvs. as medl. in syph.; resin u. as purg.; frt. source of black cosmetic; wd. ashes u. in mfr. soap & in dyeing; fls. source of VO u. perfumery (native women of Sudan); wd. very hard, white, u. to make utensils - **G. turgida** Roxb. (Ind, Burma): rt. prepn. u. for indig. in children; source of gardenia gum (Punjab).

Gardoquia Ruiz & Pav. = *Satureja*.

garget: 1) a symptom (throat swelling) & a disease (udder inflammation) in cattle 2) *Phytolacca americana* (pokeroot & p. berries), **g. weed**.

garjan = gurjun.

garlic(k): pl. & bulb of *Allium sativum* - **bear's g.**: *Allium ursinum* - **crow g.**: *A. vineale* - **English g.**: g. - **field g.**: crow g.; *A. oleraceum* - **purple-flowered g.**: *A. rotundum* - **round-headed g.**: *A. sphaerocephalum* - **sweet-scented g.**: *Nothoscordum fragrans* - **three-cornered g.**: *A. triquetrum* - **true g.**: g. - **wild g.**: 1) *A. canadense*, *A. vineale* , *A. ursinum* (**wood g.**) 2) *Thlaspi arvense*.

garnet apple: pomegranate - **g. berries**: *Ribes rubrum* frt.

garofano (It): 1) clove 2) *Dianthus caryrophyllus*.

garou (Fr): *Daphne* spp., usually *D. gnidium*.

Garrya elliptica Dougl. ex Lindl. (Corn/Garry.) (Calif-Ore [US]): silk tassel tree or bush; fever b.; bk. or lvs. decoc. very bitter; u. ton. in malaria - **G. fremontii** Torr. (Calif, Ore [US]): (Calif) fever bush; lvs. c. garryine; u. ton. antiper. (Calif) - **G. laurifolia** Bentham (*G. racemosa* Ramsz.) (Mex, Guat): ovitano; bk. c. garryine; u. chronic diarrh. (Mex) - **G. longifolia** Hartw. (Mex): ovitano; bk. c. garryine, u. in diarrh. - **G. ovata** Benth. (*G. goldmanii* Dahling): Goldman silk tassel; lvs. & rt. may have been u. in malar.

garryfoline: diterpene alk. found in *Garrya* spp.; previously called laurifoline.

garryine: alk. fr. rt. of *Garrya fremontii* & fr. bk. of *G. laurifolia*.

Gartenbohne (Ge): *Phaseolus vulgaris* - **Gartenkervel** (Ge): chervil - **Gartenlauch** (Ge): garlic - **Gartenpappel** (Ge): *Althaea rosea* (garden mallow) - **Gartenthymian** (Ge): *Thymus vulgaris*.

garter pink: *Silene antirrhina*.

Garuga floribunda Dcne. (Burser.) (Indones): administered to lying-in women (bk. inf.) (Indones); rt. decoc. for pulmonary TB (PI) - **G. pinnata** Roxb. (tropAs, esp. Ind): koorak; juice u. to treat asthma (Burma): bk. with honey for asthma (Indochin); st. juice u. for opacities of conjunctivae.

gas: one of the 3 fundamental states of matter; a fluid which has no constant volume but expands to fill any confining vessel; loosely used to mean combustible gas (mixtures of hydrocarbons) - **g. black**: lamp-black - **natural g.**: mixture of hydrocarbons (methane [chiefly], ethane, propane, butane), N, He (the last 2 in Texas n. g. & u. commercially as source); produced from natural deposits, in Okla, Kans, NM, La, Tex [US], Mex & in many other parts of the world; chiefly for heating & cooking fuel, but also in mfr. H, CH_4, NH_3, &c.; potentially of value as automobile fuel; hazards: 1) gas in high concns. narcotic 2) with incomplete combustion, CO formed (frequent cause of death in rooms with defective gas heaters) 3) explosion (gas leak in room) - **g. liquor:** aq. prod. of coal distn. & in purification of coal gas; source of NH_3 & NH_4 compds. - **g. plant**: *Dictamnus albus*.

gas gangrene: an infectious dis. following massive injury of tissues & contamination with soil; characterized by presence of gas bubbles in tissues - **g. g. antitoxin**: a sterile soln. of antitoxic substances obt. fr. blood of healthy animals immunized against the toxins of *Clostridium perfringens* Hauduroy & al. & *C. septicum* Ford (bivalent), also sometimes that of *C. novyi* Bergey (*c. oedematiens* Bergey & al.) (trivalent), & at times additionally against those of *C. bifermentans* Bergey & al. & *C. histolyticum* Bergey & al. (pentavalent); u. parenterally proph. & therap. against wound infections.

Gasbrand (Ge): gas gangrene.

gasolene: gasoline: mixt. of saturated hydrocarbons of rel. low M.W. obt. by fractional distn. of crude oil; BP ca. 39-90-110°; incl. pentane, hexane, heptane, octane, etc.; u. solvent, cleanser, fuel.

gas(s)oul (blanc) (Arab): ghassoul: *Mesembryanthemum nodiflorum*; sometimes *Althaea officinalis*.

Gasteria Duval (Lili/Aloe) (sAf): xerophytes; succulents cult. for ornam.; ca. 15 spp. - **G. croucheri** Bak. (sAf): lf. pulped in cold water & washed on body of patient (esp. paralytics) as ton. - **G. disticha** (L) Haw. (Aloe lingua Salm.) (Cape, sAf): juice & lf. u. like aloe; was off. in Hung Phar. I.

Gastridium ventricosum Schinz & Thell. (Milium lendigerum L.) (Gram.) (Eu, natd. US): nitgrass ("millet"); st. u. as bougie for urethral stricture.

gastrins: hormones of gastric (antrum) mucosa; represent peptides, several gastrins are known; stimulate gastric secretion in low concn.; u. diagnostic agent.

Gastrochilus panduratus Ridl. = *Boesenbergia rotunda.*

Gastrococos crispa H.E. Moore (1) (Acrocomia lasiopatha Griseb.) (Palm.) (Cu, Br): great macaw palm; palma barrigona; source of mucuja resin.

Gastrodia cunninghamii R. Br. (Orchid.) (s&eAs): tubers roasted & eaten like potatoes - **G. elata** Bl. (eAs): above-ground parts u. for nervous dis.; tubers ton. & for common cold (Kor); tubers u. for headache, vertigo, lumbago, arthritis (Chin).

gateado (Mex): *Astronium graveolens.*

gates (slang): Cannabis.

Gatsch (Ge): solid white petrolatum; mfrd. from hydrocarbons (residues); can be u. in prepn. of solid fatty acids.

gattilier (Fr): *Vitex agnus-castus.*

gatu (Pl): coconut milk or cream.

gatuma (Sp): *Ononis spinosa.*

gauchamaca: *Malouetia nitida.*

gaultherase: glucosidase which catalyzes hydrolysis of gaultherin yielding methyl salicylate & sugars; occ. in *Gaultheria, Betula & Spiraea* spp. & *Monotropa hypopithys*, &c.

Gaultheria fragrantissima Wall. G, hispidula (Ind, Indones): c. methyl salicylate glycoside; lvs. u. to make arom. beverage (Indochin); source of Indian wintergreen oil - **G. hispidula** Muhl. (Chiogenes h. T & G) (Eric.) (nNA s to Catskills; swamps): creeping snowberry; Moxie plum; cancer wintergreen; pl. has wintergreen flavor & aroma; possible source of methyl salicylate; hb. u. for cankers, scrofula; hypnotic (Qu); pl. ornam. - **G. humifusa** (Graham) Rydb. (wNA): alpine wintergreen; red berries edible with spicy flavor - **G. oil**: VO fr. *G. procumbens*; natural methyl salicylate (also dist. fr. sweet birch); u. flavor. coloring syrups, & in perfumery; antispas.; exter. antisep. & rub in rheum. - **G. ovatifolia** A. Gray (wNA): Oregon wintergreen; frt. delicious - **G. procumbens** L. (eNA): wintergreen; checkerberry; fr. fermented hb. is dist. the arom. VO, chiefly methyl salicylate; lvs. u. flavor, to prepare pleasant beverage tea; diur.; in asthma, menstrual dis.; to prod. VO; berries u. to color distd. spts. - **G. punctata** Bl. (Indones): lf. inf. u. as medicinal tea by natives or lvs. chewed; VO (c. methyl salicylate) u. as embrocation in rheum. & on hair & as antisept. - **G. shallon** Pursh (wNA, Alas-Calif): salal; berries eaten raw or cooked; lvs. chewed & spit on burns & sores; lf. inf. for cough or TB; lvs. chewed for heartburn & colic; decoc. of lvs. u. in diarrh. (nwAm Indians) (230); important source of MeSalicylate.

gaultheric acid: methyl salicylate(!) - **gaultherilene**: terpene fr. wintergreen oil - **gaultherin**: glycoside fr. *Gaultheria procumbens*, black birch bark, & other natural prods.; on hydrolysis yields methyl salicylate as aglycone.

gauze: light loosely woven cotton fabric (name fr. Gaza, area sw of Israel, from which first imported); u. mostly in g. bandages (plain, often sterile); roller bandages (sterile, absorbent) (adhesive absorbent).

gay-do (Chin): *Lycium* spp.

gay-feather: *Liatris* spp. - **spike g.-f.**: *Liatris spicata* - **gay wings**: *Polygala paucifolia.*

Gaylussacia HBK (Vaccini. or Eric) (NA): true huckleberries; grow best on acid soils (2058) - **G. baccata** K. Koch (G. resinosa T. & G.) (eCana, US): black huckleberry (of stores); berries edible; u. in pies, etc. - **G. brachycera** (Mx) Gray (eUS): box (or Jerusalem) huckleberry; berries put up as preserves; also eaten raw - **G. dumosa** T. & G. (G. hirtella Klotzsch) (La-Fla [US]): gopherberry; dwarf huckleberry; woolly berry (local, Fla [US]); frt. edible - **G. frondosa** T. & G. (eUS): dangleberry; tangleberry; frt. u. nutr. diur. antiscorb. sweetening agt., in wine (NA Indians); pl. u. dyestuff (NA Indians) - **G. ursina** T. & G. (NCar, SCar [US]): bear huckleberry; buckberry; frts. edible.

gayuba (*galluba*) (Sp): uva ursi.

Gazania longiscapa DC (Comp.) (sAf): u. purg. (sSotho) - **G. pectinata** (Thunb.) Hartwig (G. pinnata Less.) (sAf): inf. u. to prevent miscarriages - **G. serrulata** DC (sAf): root decoc. (hot) held in mouth to relieve toothache; crushed pl. moistened with water placed in ear to relieve earache.

Gazella gazella Pallas (G. dorcas L.) (Antelope dorcas) (Bovidae - Ruminantia) (nAf, Algeria, Eg, Sudan, to sArabia & Ind): mountain (or dorcas) gazelle; "ravine deer"; rel. small graceful mammals; herded & may have been domisticated in ancient times; source of Oriental bezoar; skin furnished good leather; horns utilized; draft animal; astragalus (ankle bone) u. anciently as dice; excellent meat u. food; as cardioton. & antispasm. in Unani med.

gazon (Fr-m.): grass.

gean tree: *Prunus avium.*

Gebuesche (Ge): thickets; undergrowth; shrubbery.

gedda gum: var. of *Acacia.*

gederam (Dan): *Epilobium angustifolium.*

gee stick (US slang): opium pipe - **gee-yen** (US slang fr. Chin): residue of opium left in pipe bowl after smoking.

Gegengifte (Ge): antidotes.

Geigenharz (Ge) (violin resin): rosin.

Geiger tree: *Cordia sebestena.*

gein: spheroxals found in rt. of *Geum urbanum.*

Geisklee (Ge): *Galega officinalis*; *Cytisus* spp. - **Geisraute** (Ge): goat's rue.

Geissfuss (Ge): *Aegopodium podagraria.*

Geissorhiza bojeri Bak. (Irid.) (Madag): bulb u. as stom. & digestive.

geissospermine: alk fr. foll.

Geissospermum laeve (Vellozo) Miers (G. vellosii Allem.) (Apocyn.) (tropAm): bk. (Pau pereira) (pereiro bk.) (Br Phar.); u. in galenicals (F.E., Tr. Wine) against fever, GC, as purg.; bitter, antipyr. antiperiod., quinine substit. (c. alk.).

Geitonoplesium cymosum (R. Br.) A. Cunn. (Phlisi./Lili) (monotypic) (Fiji, Mal): lvs. u. in dis. similar to smallpox.

gel: colloid in more or less solid form; often jelly-like.

gelatin: an amorphous protein subst. obt. by partial hydrolysis of collagen present in skin (hog), bones (osseous g.), & other connective tissues of chordates; occurs in sheets (film), flakes, pd. (types: hard gel, soft gel); u. prep. nutr. culture media (bacteriolog; mycological); skin protective; emulsif. agt.; pill coating, plasma substit. (1821), blood replacement (4% aq. soln. of low m. p. product); ingestion (in frt. juice) claimed to prevent breaking and splitting of nails; hemostatic agt.; mfr. capsules (hard & soft); for photo films, etc.; (imp. fr. Neth, GB) (off USP 1900- NF 1985) - **Chinese g.**: agar - **food g.**: prepd. fr. beef, veal, or pork (Knox g. fr. beef; "Jello" fr. beef & pork); u. to dispel fatigue & furnish energy - **Japanese g.**: agar - **g. sponge**: sheet of g. applied to wound surface to stop hemorrhage (absorbable) (absorbed in ca. 6 weeks) - **vegetable g.**: agar.

Gelatina (**Alba**) (L): **gélatine officinale** (Fr): **Gelatinum** (L): gelatin for medicinal use - **gelatol**: ointment base, cosisting of FO, glycerin, gelatin, & H_2O.

gelbe Neisswurz (Ge): gold thread, *Coptis* - **gelbes Wachs** (Ge): yellow bee's wax - **Gelbwurz** (Ge): Curcuma - **Gelbwurzel, Canadische** (Ge): goldenseal.

Gelidiaceae: a fam. of Rhodophyta (red algae) fr. which agar comes.

Gelidium amansii Lamour. (Gelidi. - Rhodophyta) (Jap, Chin coasts): makusa (Jap); chief source of agar, much u. in Eu & As - **G. cartilagineum** Gaill. (e & w shores of Pac Oc, incl. Chin, Jap, Calif [US]): one source of common agar - **G. corneum** (Hudson) Lamour. (coasts of Eu, As, Af, NA, esp. swEu, Morocco): important source of agar; much u. for foods - **G. elegans** Benth. (Jap coast): "tengusa"; thallus u. in mfr. of agar, important source (Kinnhusa a.) - **G. japonicum** (Harv.) Okam. (coasts of Jap, Kor): onikusa (Jap); important source of agar of "hard" type; often in admixt. with other spp.; u. to make domestic homemade agar = "isinglass" ("Japanese i.").

géliform (Fr): jelly state; gelled.

gélin (Fr): agar - **Gellert** (Ge n): gelatin; glue; pectin; jelly; colloid - **gelling** (Dut): fimblehemp.

Gelonium multiflorum A. Juss. (Euphorbi) (Ind, Burma, Mal): bk. u. as ton. for gums in gingivitis; bk. u. with other drugs as gargle (Indochin); purg. (1739) - **G. zanzibarense** Muell. Arg. (tropAf): lf. & frt. u. to relieve abdom. pain; frt. said edible; rt. decoc. for bilharzia, GC (Tanz).

gelose: 1) gelatinoid carbohydrate (araban & galactan); chief component of agar 2) (Fr): agar.

gelsemicine: alk. of Gelsemium; motor depressant - **gelsemine**: xall. alk. of Gelsemium; u. in neuralgia & rheum. - **gelseminine**: amorphous alk. (or prob. mixt of alks.) fr. Gelsemium.

Gelsemium (L): dr. rhiz. & rts. of *G. sempervirens* - **G. elegans** Benth. (Logani) (Chin): allspice jasmine; u. exter. for boils, ulcers, ringworm, leprosy; int. in neuralgia (care!), pre-paralytic polio; lvs. u. (esp. among women) to commit suicide (3 lvs. & cup of water); in homicide; rich in alk. (1410) - **G. sempervirens** (L.) St. Hil. (seUS, eMex, Guat): (Carolina) yellow jasmine; **g. root** c. several alk. incl. gelsemine; u. sed. antipyr. antispasm., nervine, in neuralgia, infantile paral., tetanus (small doses), migraine, mydriatic, cramps, hysteria, asthma, etc.; ornam. vine; toxic!

gelso (It): mulberry.

gélule (Fr): gelatin capsule; now considerably u. to administer pd. crude drugs, &c.

gemein (Ge): common - **gemeiner Sauerdorn** (Ge): Oregon grape root.

gem fruit: *Tiarella cordifolia.*

Gemma (pl. **Gemmae**) (L): bud - **G. Betulae** (eEu, esp Pol, USSR): buds of various *Betula* spp. (Eu); u. arom. diur. - **G. Populi**: buds of *Populus nigra* & *P. tacamahacca* - **G. Ribis nigri**: buds of *Ribes nigrum*, u. arom.

gemme (Fr): turpentine (oleoresin).

Gemuese (Ge): vegetables.

genciana azul (Sp): *Gentiana acaulis.*

gene: unit of inheritance in the chromosome of cell nucleus - **artificial g.**: nucleic acids with some functions of true g. - **g. splicing**: recombination of genetic materials producing recombinant DNA.

genecology: ecology as related to genetics.

génépi (**blanc**): **génipe** (Fr): **geneppi** (It): wormwood, *Artemisia mutellina, A. spicata* - **g. des. Alpes**: Alpine wormwood; *Artemisia glacialis, A. laxa;* also liqueurs therefrom; *Achillea moschata.*

genest (Eng): **genêt** (Fr-m.): *Cytisus scoparius*, broom - **genêt à balais** (Fr): broom tops.

genetic engineering: technique of manipulating chromosomal genes as means of improvement in organism.

genetics: the science of heredity & variation in living organisms.

Geneva root (Qu): wild ginger, *Asarum canadense*; prob. SC gingembre sauvage.

genève (Qu): juniper.

genever (Dutch): juniper; may be source of word "gin."

Geneverwurz (Ge): pellitory root, *Anacyclus pyrethrum*.

genévrier (Fr): juniper.

gengibre (Hon): ginger.

genin: aglycone of cardiotonic heteroside or saponin.

Geniosporum prostratum Benth. (Lab) (Ind): pl. u. as antipyr.

Geniostoma vitiense Gilg & Bened. (G. rupestre Forst.) (Logani) (Tonga, Fiji): te'epilo'amaui (Tonga); pl. has unpleasant odor; bk. & lvs. u. med. (Tonga) (86); bk. ext. drunk for female dis.

Genipa americana L. (G. caruto HBK) (Rubi) (Mex, WI, CA, SA): caruto (Ve, Hond); genipa; marmalade box; jagua; frt. edible, very popular; u. along with pineapple & cashew pseudocarps to make a wine u. in dysen; bk. c. mannite; u. in syph. ulcers, granular pharyngitis; emet., etc.; tan.; frt. appl. to tumors, corns, u. in jaundice (Salv); juicy pulp with ashes of *Gynerium sagittatum* rubbed on body in yaws; rt. decoc. in GC; frt. juice dyes cloth & skin black; wd. u. for plows - **G. brasiliensis** Baillon (Gardenia b. Spreng.) (Br, Guiana, Par): frt. edible (genipa peira do mato), lax., u. for diarrh.; unripe frt. as astr. for washes & cataplasms, for syph. sores; lvs. & bk. c. glycoside genipin & mannitol; fls. c. VO ("gardenia oil"); wd. u. to make walking sticks, spears, etc. (Pe).

genipapo fruto (Br): frt. of *Genipa americana*.

génipis (Fr): *Artemisia glacialis*.

Genista germanica L. (Leg) (Eu): deutscher Ginster; u. like *G. tinctoria* - **G. sagittalis** L. (Eu): (dyer's) broom; flg. hb. u. as diur. lax. in homeopathy; u. like *G. tinctoria*; ornam. - **G. tinctoria** L. (Euras): dyer's greenweed; dyer's broom; hb. with fls. u. diur. cath. in gout, bladder dis., hemorrhoidal & depurat. teas; in yellow dye (Kendal green); c. genistin (glycoside) which hydrolyzes to give genistein (aglycone) which acts to reduce fertility (133) - **G. tridentata** L. (Br): u. for prepg. the VO (carqueja oil) with camphor-like odor; c. cineole.

genome (cytogenetics): basic haploid or single (n) set of chromosomes in cell nucleus (as found in gamete or spore); this has been termed the next higher genetic unit to the chromosome itself.

Genoscopolamine: scoplamine aminoxide hydrobromide (Lobica).

genotype: the expressed or latent genetic constitution of an organism in contrast to the phenotype (actual appearance).

Gentamicin: antibiotic mixt. fr. *Micromono-spora purpurea* Luedemann & Brodsky or *M. echinospora* Luedmann & Brodsky esp. u. for genitour. inf.; also in neonatal sepsis, bone inf., inf. by *Citrobacter* spp. = cat-scratch fever (or dis.), &c. (SC color of microorganism) (E) (usu. as complex sulfate).

gentian (root): *Gentiana lutea* - **American g.**: *Frasera carolinensis* - **black g.**: *Peucedanum cervaria* - **blind g.**: *G. andrewsii* - **blue g. (root)**: *Gentiana saponaria* - **bottle g.: closed g.**: blind g. - **English g.**: *G. campestris* - **five-flowered g.**: *G. quinqueflora* - **g. family**: Gentianaceae - **horse g.**: *Triosteum perfoliatum* - **fringed g.**: *G. crinita* - **Indian g.**: *Picrorhiza kurroa* - **red g.**: fermented dr. gentian roots u. in med. - **rose g.**: *Sabbatia* spp. - **southern g.**: blue g. - **stemless g.**: *G. clusii* - **g. violet**: synth. dyestuff u. in microstains, bactericide - **white g.**: 1) *Laserpitium latifolium* (rt.) 2) German rural domestic med. consisting of dr. & whitened dog feces 3) unfermented gentian rt. not much u. in med. - **yellow g.**: *G. lutea*.

Gentiana: 1) (Genti) (worldwide): gentians; small ann. or perenn. hbs. 2) (specif.) (L): a drug, rhiz, & rt. of *G. lutea* - **G. andrewsii** Griseb. (e&cNA): closed (or bottle) gentian; u. as simple bitter (to stim. appetite, aid digestion); for snakebite (Meskwaki Indians); u. similarly to *G. lutea*; formerly off. USP - **G. asclepiadea** L. (s&cEu, Caucas): Wuerger Enzian; rt. c. gentiopicrin, gentianin, etc.; u. to adulter. or as substit. for *G. lutea* (1411); earlier u. in ophthalmia, hematuria - **G. campestris** L. (nEu): field (or English) gentian; bitterwort; hb. u. ton. stom. - **G. catesbaei** Walt. (seUS): Sampson's snakeroot; u. much like *G. lutea* - **G. centaurium** L. = *Centaurium erythraea* - **G. chirayita** Roxb. = *Swertia chirata* - **G. clusii** Perr. & Song. (G. acaulis L. p. p.; G. elliottii Chapm.) (Eu, Alps): stemless g.; dwarf g.; hb. c. acaulin; u. as bitter ton., somewhat like generally off. Gentian; also antipyr. & in convalescence; ornam. perennial hb. - **G. cruciata** L. (cEu-mts.; As): cross g.; bitter-wort; hb. c. bitter princ., u. bitter ton. stom.; rt. u. as for *G. lutea* - **G. dahurica** Fisch. (G. olivieri Griseb.) (As, esp. cChin): rt. bitter, valued as medicine (55); u. esp. in arthritis, convulsions, numbness of limbs, edema, pulm. TB, dys. - **G. herrediana** (Pe): rt. looks like that of *G. lutea*; very bitter - **G. kurroo Royle** = *Swertia chirata* - **G. lutea** L. (Euras): yellow g.; bitterwort; u. as bitter by mfrs. as in Enzianschnapps; chief source of off. Gentian; rt. c. gentisin, gentianin, gentiamarin, aromatics; gentiopicrin, u. bitter ton., (2656); chewed to aid in breaking alcohol habit (chief prodn. Sp, Fr, Ge) - **G. nevadensis** Gilg (Ve): dictamo (riñon); potion u. for stomach dis. - **G. pannonica** Scop. (G. punctata Jacq.) (Eu-mts.): Hungarian g.; has been off. in 13 Phar., incl. Br Phar.; adult. & substit. for off. Gentian - **G. pneumonanthe** L. (Euras): lungflower; Calathian violet; hb. & rt. c. gentiopicrin; u. in pulmon. dis., spasmolytic; fls. give a blue dye - **G. puberulenta** J. Pringle (e&cUS): formerly off. USP; u. like G. - **G. punctata** L. (Eu-alps): dotted-flower g.; u. as g. root; adult of off. gentian - **G. purpurea** L. (Eu alps to sNor): purple (flowered) g.; u. like g.; adult. of & substit. for off. Gentian - **G. quinqueflora** Hill (NA): u. homeopathy (fresh pl.); bitter ton. - **G. saponaria** L. (eNA): soapwort g.; formerly USP (1820-1880); u. like off. G. - **G. scabra** Bunge (Chin, Mongol, Kor) (and var. *buergerll* [Miq.] Maxim.): rt. bitter, u. for liver dis., sore throat, pudendal itching; stom. ton. anthelm. antipyr. tumors; locally for skin dis., abscesses; adulter. of & substit. for off. Gent.; off. in Jap & other Phar. - **G. septemfida** Pallas var. *lagodechiana* Kusn. (G. l. [Kusn.] Grossh.) (Russ): pl. u. med. - **G. tibetica** King (Tibet): rt. u. for arthritis, convulsions, numbness of limbs, jaundice, edema; pulmon., TB, dys. - **G. verna** L. (Euras): spring g.; fls. c. gentianol (violet coloring matter); u. like off. prod. - **G. villosa** L. (G. ochroleuca Froel.) (US: Fla, La, sIndiana, NJ): Sampson's snakeroot; striped (or marsh) gentian; u. as stom. ton. digestive, antipyr. in gout, rheum.; rt. as anthelm. astr. in biliousness; substit. for other gentians.

Gentianaceae: Gentian Fam.; perenn. hbs. or sometimes shrubs, mostly in temp. regions; ca. 1,200 spp.; fls. regular, bisexual, pls. often c. bitter principles; ex. centaury, buckbean, yellow wort.

gentiane jaune (Fr): *Gentiana lutea.*

Gentianella amarella Borner (Gentiana a. L. p. p.) (Gentian.) (n&cEu, natd. Cana, nUS): bitterwort; felwort; rt. (& hb.) u. mostly in homeop. - **G. quinquefolia** Small (Gentiana q. L.) (eUS): gall of the earth, ague weed; u. antipyr.

gentianic acid: gentisic a. - **gentianin** = gentiin; aglycone of glycoside gentiopicrin fr. *Gentiana lutea*, etc. - **gentiopicrin**: bitter glycoside fr. Gentiana - **gentisic acid**: OH-salicylic acid; occ. in urine of dogs after salicylates; in gentian (hence name); u. as analg.

Gentianopsis crinita (Froel.) Ma (Gentiana c. Froel.) (Gentian.) (e&cNA): fringed gentian; ornam. perenn. hb.

gentianose: complex sugar (2 glucose + fructose) moiety of gentian glycosides.

gentisin: yellow colored xanthone occ. in fresh rt. of *Gentiana lutea*, etc.

Genussmittel (Ge): stimulant, esp. beverages, as coffee.

Geodorum dilatatum R. Br. (Orchid.) (nwInd): pd. tuber with ghee ingested in dys. - **G. nutans** Ames (eAs, PI, +): kula (PI); bulb u. as emoll. poultice (boils, abscesses); strong mucilage u. as glue for musical instruments (PI).

Geoffroea Jacq. (Leg.) (tropAm): bastard cabbage trees; close to *Andira* - **G. decorticans** Burk. (Ch, Arg): chanar; fls. u. expec. emoll. frt. u. expec. in asthma; st. bk. u. as antihemorrhag. (exter); lf. emoll.; fls. & bk. antitussive; pods edible; delicate food, pod u. in prepn. of bev. by fermentation; pl. u. as fodder - **G. inermis** Wright = *Andira inermis* - **G. spinosa** Jacq. (Ven): tague; frt. edible after cooking - **G. surinamensis** Pille (tropSA): bk. (Cortex Geoffroyae) u. as vermifuge.

Geoffroya bark: fr. *Andira inermis* - **geoffroyine**: andirine; alk.: N-methyl tyrosine; occ. in *Geoffroya* bk.

Geogenanthus ciliatus Bruckn. (Commelin.) (Co-Pe): pl. decoc. rubbed on swollen knee (777); cold water inf. given orally for intest. worms in infants (777).

geography, botanical: study of the range of pls. over the earth's surface; geographical distribution.

Geonoma linearis Burret (Palm) (Co, Ec): "calo" (Choco); pl. u. by "brojos" for pains in stom. (Cayapa Indians, Ec) (2482).

geophagy: the eating of various unusual substances (ex. clay, starch, etc.) engaged in by various peoples in various parts of the world (incl. Ala & Miss [US]); a form of pica.

Geophila herbacea L. (Carinta h.) (Rubi) (pantrop, ex. Cuba): lvs. u. med. (Tonga) (86) - **G. obvallata** F. Didr. (tropAf): fresh crushed lvs. analg. for pain in back, head, stomach, teeth (wAf); appl. to inflamed eyes (sNiger); cooked with food & u. for diarrh. (Liberia).

geophyte: perenn. hb. in which the overwintering bud is underground; ex. *Helianthus tuberosus.*

Georgia bark: fr. *Pinckneya pubens*, **G. fever tree**.

geosmin: volatile complex CHO compd. found in algae, fish, beef, ext., water, &c.; accounts for the earthy off-odor/taste of tap water.

Geraniaceae: Geranium Fam.; mostly small hbs., ann. or perenn.; bisexual, mostly regular fls., with 5 sepals, 5 petals, usually 5-10 stamens; pistil with 5 styles; ex. geranium, heronsbills.

geraniol: colorless fragrant liquid, chief constit. of rose geranium & palmarosa oils.

Geranium L. (Gerani.) (cosmopol., mostly temp.): ca. 320 spp.; some spp. with antibiotic properties (2396) - **G. carolinianum** L. (US, Mex, sBC [Cana]): crane's bill; Carolina geranium; pata de león (Mex); hb. u. as antipyr., astr. diur.; juice as purg. for children (not adults); decoc. to bathe rash of children; lf. inf. for dys.; rt. crushed & boiled to make syr. u. in same way; weed (US, Jap) - **G. dissectum** L. (Euras, Austral, NA): c. polyphenols, tannins, VO; u. astr.; weed (Austral) - **G. macrorrhizum** L. (Balkans): source of VO (Bulgarian g. oil; Zdravetz oil) with 50% germacrone; u. for unique exotic aroma in perfumery; aphrod. (Bulg) - **G. maculatum** L. (eNA): (wild) cranesbill; rhiz ("root") c. 10-28% tannin; commonly u. as astr. in intestinal dis. (esp. acute diarrh., dys.), styptic (esp. intest. & vesical hemorrhage); tan. - **G. mexicanum** HBK (Mex, CA): geranio (Mex); lf. decoc. (with *Gaultheria*) u. as ton., esp. in older patients - **G. molle** L. (Eu, wAs [Lebanon], NA [introd.], Chile, NZ): soft cranesbill; u. anod. astr. vulner. (Spain) - **G. pratense** L. (Euras): Wiesen-Storchschnabel (Ge); c. catechol tannins, VO, quercetin; u. as astr. chiefly - **G. pusillum** Burm. f. (non L) (Euras): small (flowered) cranesbill; hb. u. as astr. anodyne, vulnerary, styptic - **G. robertianum** L. (neNA, Euras, nAf): Herb Robert; Herb Robin; red robin; rhiz. c. 19-44% tannin; u. astr. in sore throat (gargles); formerly in veter. med. for hematuria, dys. diarrh; ornam. hb. - **G. sanguineum** L. (Euras): blood geranium; hb. c. tannin, u. astr. hemostat. as seasoning; rt. astr. ton for canker sores - **G. wallichianum** D. Don ex Sweet (Pak, Ind): hb. u. for tooth & eye dis. (Lahul); astr. rt. u. for tanning (32% tannin) & dyeing.

geranium: cranesbills, usually *Geranium maculatum* or *G. molle* - **cultivated g.**: *Pelargonium* spp. - **dove's foot g.**: *G. molle* - **East Indian g. oil**: fr. *Cymbopogon martinii* - **g. family**: Geraniaceae - **florist's g.**: *Pelargonium* spp. - **(greenhouse) g.**: *Pelar.* spp. - **hairy g.**: *Geranium dissectum* - **horseshoe g.**: *Pelargonium zonale* - **house g.**: *Pelargonium hortorum* (usually) (pot-plants) - **Indian g.**: East Indian g. - **Lady Washington g: Martha W. g.**: *Pelargonium domesticum* - **native g.**: hairy g. - **oak-leaf g.**: *Pelargonium quercifolium* - **g. oil**: VO fr. *Pelargonium graveolens, P. odoratissimum*+; important perfume oils; *Cymbopogon martinii* - **rose g.**: *Pelargonium graveolens, P. capitatum*, etc. - **spotted g.**: *G. maculatum* - **strawberry g.**: *Duchesnea indica* - **wild g.**: 1) *Erodium cygnorum* 2) *G. maculatum*.

géranium rosat (Fr): rose g.

Gerardia Benth. = *Aureolaria* (generally); *Agalinis*; etc. (—> *Stenandrium*).

gerbil = *Gerbillus iateronia* (Rodentia): small burrowing animal of Af & swAs; popular pet; u. in biol. exptl. studies.

Gerbera jamesonii Bolus. ex Hook f. (Comp.) (Af, Transvaal): Barberton (or Transvaal) daisy; African (or gerbera) d.; ornam. hb. - **G. kraussii** Sch. Bip. (sAf): lf. u. as anthelm. & to relieve stomachache (Zulu) - **G. piloselloides** Cass. (Af, Ind): rt. inf. in human urine appl. to ear for earache; rt. decoc. ton. & as milk decoc. for chest dis. (sSotho) - **G. viridifolia** Sch. Bip. (cAf): fresh rts. u. as cough remedy (esp. for childen).

gergelim (Port, Hind.): sesame.

gergelin (Ge): *Sium sisarum.*

geriatrics: science of treating diseases of old age.

germ: pathogenic bacterium & other microorganisms - **g. warfare**: form of biological warfare involving use of microorganisms.

germacrane: germacranolides; trivial names for class of sesquiterpenes, viz., cyclodecanes, 10-membered rings; common in Compositae; ex. liatrin; occ. in *Hedera helix, Heterotheca latifolia*, &c.

germacranolides: best known gp. of germacranes (lactone derivs.); named fr. *Geranium macrorrhizum*; strongly bitter, non-volatile; with outstanding physiol. properties (antibiotic; anthelm.; toxic).

German: v. fennel; pellitory; psyllium - **G. powder**: Licorice Powder Comp. - **G. tinder**: dr. thallus of *Fomes fomentarius*

germander: *Teucrium chamaedrys* but at times appl. to spp. of *Ajuga, Veronica*, &c. - **grey g.**: *T. racemosum* - **native g.**: *T. argutum* - **water g.**: *T. scordium* .

Germantown black: lampblack - **Germany**: *Anacyclus officinarum.*

germerine: alk. occ. in *Veratrum album, V. viride, V. nigrum.*

germidine: germine: germitrine: related alkamines of *Veratrum* & *Zygadenus* spp.; u. as antihypertensives.

Germisan™ (Ge): phenyl mercuric pyrocatechol; u. as fungicide.

gerontogenous: occurring in the OW (eHemisph).

Gerste (Ge): barley.

Gesneria allagorphylla Mart. (Gesneri) (Br): perenn. hb.; tuberous rts. u. as emoll. & ton.

getah rasamala (Indonesia): *Liquidambar altingianum.*

Get Away Powder (sUS voodoo): potion contg. Cayenne pepper, intended to get persons "off your neck."

Gethyllis afra L. (Amaryllid.) (hot Af): frt. edible, u. for colic & indigest.; to perfume rooms - **G. ciliaris** L. f. (tropAf): remedy in colic, flatulence; to perfume rooms of house - **G. linearis** L. (sAf): frt. inf. (in alc.) u. for digest. dis. (Cape colonists) - **G. spiralis** L. f. (sAf): alc. inf. of frt. u. for colic, flatulence; ripe frt. to perfume rooms & linen; fl. inf. u. for teething troubles (baby).

Getreide (Ge): cereal grains.

Geum L. (Ros.) (n&sTemp; Arctic): avens; perenn. hbs. with pinnate or lyrate lvs.; some 65 spp. (1413); fls. white, yellow, pink, purple - **G. canadense** Jacq. (eNA): white avens, rhiz. u. astr. stomach. tonic - **G. chilense** Balb. (Ch): rt. decoc. u. aper. diur. ton. "depurative"; u. for pain in molars (teeth) - **G. japonicum** Thunb. (Jap, Chin): decoc. diur.; frt. decoc. u. intern. for skin dis. (Indochin); rt. appl. to boils (Chin) - **G. montanum** L. (Eu-alps & lower alps): mountain avens; rt. u. in hemorrhages, leukorrhea - **G. rivale** L. (Euras, NA): water avens; rhiz. & hb. c. gein (glycoside); u. astr. ton. stomach (prod. in Balkans) - **G. urbanum** L. (Eu, wAs, Cauc): European avens; wood a.; clove root; rhiz. c. tannin, **geum-bitter**, gein (splits to give eugenol, etc.); hb. u. like prec., as stomach. for diarrh. in liqueurs; added to beer to give more pleasant taste & to prevent scouring (prod. in Balkans, incl. Yugoslavia) - **G. vernum** Torr. & Gray (e&cNA, s to Tenn): (early water) avens; u. similarly to *G. rivale* - **G. virginianum** L. (eUS, s to SCar): (white) avens; hb. u. in leukorrh., hemorrhages.

Geunsia farinosa Bl. (Verben.) (NG, PI, Mal): "tampang besi" (Mal); grd. bk. u. in vertigo, swellings - **G. paloënsis** (Elm) H. J. Lam. (C. pentandra var. paloënsis) (Solomon Is): bk. inf. u. in colds (—> *Callicarpa*).

geven (Turk): tragacanth.

Gevuina avellana Mol. (1) (Prote.) (Ch): avellano; chile nuts; u. in chronic diarrh., leukorrh., metrorrhagia, infections; nut edible.

gewa wood (Ind): a firewood fr. *Excoecaria agallocha.*

Gewaechs (Ge)**: gewas** (Neth): plant, herb - **G. Haus** (Ge): greenhouse.

Gewuerz (Ge): spice, seasoning, aromatic - **Gewuerznelken** (Ge): cloves - **Gewuerzepflanze** (Ge): spice plant - **Grosses Englisches G.** (Ge): large frt. allspice.

gezira gum: gum talha (Acacia).

ghati gum: ghatti g.: ghatty g. (Ind, SriL): mostly fr. st. of *Anogeissus latifolia*; at times fr. *Acacia arabica*; in acid hydrolysis yields arabinose, galactose, mannose, xylose, and glucuronic acid; u. as thickening agt., emulsif. for OW emulsions; in pharmaceut. & food indy.

ghee (Indo-Pak) = ghi: a clarified butter prepd. by melting & straining off; quite stable against rancidification; keeps much longer than butter, in place of which it is generally used in India.

gherkin: 1) *Cucumis sativa* 2) other immature small cucumbers u. for pickles.

ghezireh: gum talha (Acacia).

ghiandola (It): gland.

ghost plant: *Artemisia lactiflora.*

ghur (Urdu): crude sugar in lumps (*cf.* cane jaggery powder), palm ghur.

giant fennel: *Foeniculum; Ferula communis* - **g. reed**: *Arundo donax.*

Gibberella fujikuroi (Saw.) Wollenweber (Nectri. - Hypocreales - Ascomycot - Fungi): rice disease fungus; occ. on cotton, rice, maize, etc. (Jap+); source of gibberellins.

gibberellic acid (GA): gibberellin A_3 = most active of the known gibberellins; metabolic product of *Gibberella fujikuroi*, etc; acts as pl. growth substance; stimulates growth (esp. elongation) & early flowering/fruiting.

gibberellin: complex pl. hormone causing overgrowth & early maturation in higher pl.

gibbous: swollen on one side (as corolla).

gichero (It): *Arum maculatum*; *Arisaema atrorubens.*

Giesekia = *Gisekia.*

Gift (Ge): poison - **Giftjasmin** (Ge): yellow jasmine - **Giftlattichsaft** (Ge): wild lettuce juice, Lactucarium - **Giftsumach**: poison oak, ivy, sumach.

gigantin: active principle of *Calotropis gigantea, C. procera* (juice); with digitaloid action.

Gigantochloa verticillata Munro (Gram.) (trop eAs): whorled bamboo; a giant bamboo u. (Java) to build houses; young buds cooked & eaten; sometimes preserved in vinegar.

Gigartina acicularis (Wulf.) Lamour. (Gigartin. - Rhodophyta) (Eu, Af, NA): "Portuguese moss"; thallus source of a carrageenan; comm. prod. in Por - **G. mamillosa** Agardh (coasts of nAtl Oc [mostly] but also Jap): Irish moss; salt rock moss; chief Eu source of Chondrus; thallus distinguished by having sporangia projecting & borne on short stalks on surface; v. *Chondrus crispus*, the other sp. officially named as source of prod.; some consider conspecific under *G. stellata* (1415) - **G. stellata** (Stackh.) Batt. (nAtl Oc coasts): thallus u. as source in GB of carageenin -

G. teedii Lamour. (Jap, Eu): this seaweed is u. as a food in Jap.

Gigartinaceae: a fam. of Rhodophyceae (Red Algae) which originates Chondrus (*Gigartina mamillosa, Chondrus crispus*); thalli are more or less fleshy, with flattened, furcate forms; found in shallow marine waters.

gigas: gigantis: (L): giant (stature, morphology, etc.[botany]).

giggleweed (sAf): Cannabis.

giglio (It): lily - **g. de vallata** (It): Convallaria.

Gilia aggregata Spreng. (*Polemoni*) = *Ipomopsis a.* - **G. congesta** Hook. = *Ipomopsis c.* - **G. rubra** Heller = *Ipomopsis r.*

Gillenia stipulata Baill. = *Porteranthus stipulatus* Britt. - **G. trifoliata** (L.) Moench = *P. trifoliatus* Britt.

gill-go-over-the-ground: gill-(run)-over-the-ground: *Glechoma hederacea* (*Nepeta glechoma*).

gilliflower, wall: *Cheiranthus cheiri*.

gilly flower: primarily appl. to pls. or fls. with clove-like aroma; ex. clove pink, *Dianthus caryophyllus*; also to *Mathiola incana, Hesperis matronalis, Hottonia palustris, Armeria maritima* v. *purpurea*, etc.; also *Chionanthus virginicus*.

gilo (Br): jilo = eggplant frt.

gin: an alcoholic spirituous beverage, similar to Spt. Juniper Comp., made by distilling a fermented mash of malt, corn, & rye over juniper berries; also wheat, barley; u. hot, as an emm. stim.; a great var. of exts. may be added; ex. almond, angelica, anise, calamus, carum, cardamom, cinnamon, cloves, coriander, fennel, grains of paradise, lemon, licorice, nutmeg, orange, neroli, iris, sloe, turpentine oil - **ginepra** (It): juniper berry - **ginepro** (It): juniper tree.

ginestra (It): broom, genet - **ginestrone**: furze, gorse, whin.

gingelly (Ind): **gingili**: sesame, *Sesamum indicum* (FO).

gingembre (blanc) (Fr): ginger (Jamaica) - **g. sauvage**: wild g., Asarum.

ginger: 1) *Zingiber officinale*: pl. or rhiz. 2) contraction for Tr. ginger - **African g.**: that obt. fr. wAf: peeled on flattened sides & then treated with boiling water; largely u. in mfr. g. ext. and g. ale - **g. ale**: carbonated aq. beverage contg. g., capsicum, lemon juice or oil, caramel, sometimes sarsaparilla, etc. - **g. beer**: similar to g. ale but flavored with fermented g.; non-alcoholic or with very small amt. alc. (popular in UK) - **Bengal g.**: g. prod. in B. (eInd), partly or fully peeled product; considered the best sort - **black g.**: g. of unpeeled type which has been treated with boiling water & then rapidly dr. - **bleached g.**: peeled or unpeeled g. bleached with chloride of lime or SO_2, then dr. - **Borneo g.**: a trade name, not fr. Borneo - **g. bread**: sweet cake or cookie flavored with g., molasses, etc. - **Calcutta g.**: var. much like Af (broad lobed); exp. thru Calcutta - **Calicut g.**: produced in this center; much like prec. - **candied g.**: young fresh rhiz. boiled in heavy syrup after macerating fresh rhiz. in sea water; sugar xallizes on surface; u. confection (popular in Eng) - **Canton g.**: conserve prepd. fr. young green rhiz. boiled & cured in syrup & packed in jars or cups (Chin & WI) - **Chinese g.**: galanga - **coated g.**: unpeeled g. (such as Cochin g.), i.e., with cortex unremoved, altho may be bleached or whitened - **Cochin g.**: originating from this production center, both bleached (white) and unbleached vars. - **common g.**: g. - **crystallized g.**: candied g. - **dry g.**: the usual product fr. Jamaica or India, as distinct fr. preserved (or candied) - **East India(n) g.**: Calcutta or Calicut g. - **g. family**: Zingiberaceae - **g. grass**: v. under grass - **green g.**: prod. marketed before being dr.- **hand g.**: scraped g.; also term appl. to large portion of entire rhiz. as distinct fr. fingers (individual pieces of rhiz.) - **Indian g.**: that exp. fr. Ind - **Jamaica g.**: 1) that peeled, treated with lime juice, dr., & coated with $CaCO_3$ (SC island where prod.) 2) Tr. of ginger - **Japan(ese) g.**: *Zingiber mioga*; scraped or peeled; rather like Cochin g.; rarely seen in US - **leaf g.**: product made by shaving rhiz. into thinnish flakes - **lemon g.**: Calicut g. - **g. lily**: 1) *Hedychium* spp. 2) *Cymbopogon nardus* - **limed g.**: that treated with lime to preserve against insect attack & to give characteristic clean-looking prod. - **natural g.**: with cortex unremoved - **g. nuts**: g. bread nuts (Americanism): small rounded cakes c. molasses & flavored with ginger - **g. oil**: VO in commerce, prepd. by distn. of g. root (Chin, Ind) - **g. oleoresin**: a galenical prod. made by extn. of g. with acetone, alc., or ether, then evap. the solvent - **peeled g.**: g. with cortical layers removed; ex. Jamaica - **preserved g.**: candied g. - **Puerto Rican g.**: native var.; rather inferior grade; "many-fingered" - **race g.**: Calcutta g. - **ratoon g.**: an inferior kind of Jamaica g. made up of small incompletely peeled segments; made up chiefly of fibrous shoots of rhiz.; of a grayish color - **g. rhizome**: g. - **scraped g.**: uncoated g. - **shell g.**: *Alpinia tomentosa* (choice) - **Siamese g.**: Galanga - **g. tea**: inf. of g. rt. u. for colds - **uncoated g.**: with cortex unremoved; scraped - **unpeeled g.: unscraped g.**: natural g. - **wet g.**: candied g. as mfrd. in Hong Kong (casks with 180# g. & 44# sugar syrup) - **white g.**: 1) with cortex removed, as in Jamaica, African, &

Cochin g. 2) *Hedychium coronarium* - **wild g.**: *Asarum canadense.*

gingerbread nuts: ginger nuts - **g. tree**: *Hyphaene crinita.*

gingko: ginkgo: Ginkgo biloba L. (Ginkgo.) (monotypic) (eChin): maidenhair tree; sole modern representative of large geologically ancient gp.; distinguished as sole sp., genus, family, & order (Ginkgoales); lvs. notched wedge-shaped (hence called maidenhair tree); frt. u. to allay fatigue, as at long, drawn out dinners (Jap, Chin); lvs. off. in Pharm. Gall. (Fr Phar.) IX (1980); for coughs, colds, asthma, to dissipate phlegm, anthelm., in painful micturition; ginkgolides; bitter subst.; ext. u. in cerebral & peripheral vascular dis.; ginkgosides u. as pesticide; cult as ornam. (Chinese word) (1416).

ginseng: *Panax* spp. - **American g.**: 1) *P. quinquefolius* 2) *Rumex hymenosepalus* (wild red A.g.; wild red desert g.) (wrongly appl.) - **Asiatic g.**: Chinese g. - **Chinese g.**: *P. ginseng* - **dwarf g.**: *Panax trifolius* - **g. family**: Araliaceae - **horse g.**: *Triosteum perfoliatum* - **Japanese g.**: *P. repens* - **Korean g.**: *Panax ginseng* - **Sanchi g.**: *Panax notoginseng* - **Siberian g.**: *Eleutherococcus senticosus* - **spiny g.**: Siberian g. - **Tienchi g.**: Sanchi g. - **wild American g.**: *Panax quinquefolia* - **wild g.**: *Aralia nudicaulis* - **Wujia g.**: Siberian g. - **yellow g.**: *Caulophyllum thalic-troides.*

giotchoo: *Boerhaavia diffusa.*

gipsywort: *Lycopus europaeus.*

Girardinia leschenaultiana Decne. (Urtic.) (Ind): pl. u. for fiber; very close to foll. - **G. palmata** Gaudich. (G. heterophylla Decne.) (cAs, esp. Ind, Mal): Nilgiri nettle; rt. & basal st. decoc. u. in wine to treat malignant boils; broth (when lvs. cooked with pork) remedy for stomachache; diur. stinging hairs (c. enzyme); st. yields "vegetable wool," useful fiber, which makes a very strong cloth - **G. zeylanica** Decne. (sInd, SriL): lvs. u. for headache & swollen joints; decoc. antipyr.

girasol (Sp) (means turning with sun for plant compass property) (*cf.* Jerusalem artichoke): **girasole** (It): 1) sunflower 2) *Encelia mexicana.*

giraumon (Fr): *Cucurbita pepo.*

girofle (Fr): clove - **giroflée** (Fr): 1) wallflower, *Cheiranthus cheiri* 2) gilliflower, *Mathiola incana.*

Gisekia pharnacioides L. (Aizo) (Ind+): pl. u. arom. aper. anthelm., but edible.

gitalin: gitaligin: mixt. of cardioactive glycosides of *Digitalis purpurea* contg. gitoxin, gitaloxin, & digitoxin; considered superior drug for congestive heart failure.

gith: *Nigella sativa* (or possibly other *N.* spp.).

gitin: glycoside fr. *Digitalis purpurea.*

gitogenin: sapogenin occ. in green berries of *Cestrum laevigatum, Digitalis* spp. ("digin"), etc.; aglycone of gitonin.

gitomate (seMex): jitomate; tomato.

gitorin: glucoside occ. in *Digitalis purpurea, D. lanata*; with gitoxigenin as aglycone.

gitoxin: secondary glycoside obt. fr. *Digitalis purpurea & D. lanata*; prep. fr. purpureaglycoside B on hydrolysis, itself splits into gitoxigenin & digitoxose (sugar); has cardiac props.

glacé: candied, glazed (as cherries).

Glacies (L): ice (*cf.* glacier) - **G. Mariae**: isinglass-stone; Muscovy glass; a var. of native $CaSO_4$.

Gladiolus L. (Irid) (Euras, Af): some 180 spp.; lvs. rich in vit. C & recommended as good source of natural vit. (134) - **G. communis** L. (sFr, Med reg., Iran; often cult.): common sword lily; corn flag; bulb appl. exter. to wounds, int. in scrofula; in List D (Switzerland) - **G. dieterlenii** Phillips (tropAf): u. as remedy in headache & lumbago - **G. hortulanus** L.: garden (or florists') gladiolius; non-systematic name appl. to commonest cult. forms which cannot now be properly identified (hundreds of vars.); may be derived fr. *G. psittacinus*, &c. - **G. palustris** Gaud. (c&sEu): corn flag; corm (as Radix Victoralis rotunda) u. vulner., for scrofula; in veter. med. - **G. psittacinus** Hook. (tropAf): corm u. for colds & dys.

gladiolus (pron. glad-eye-olus) (pl. gladioli): *Gladiolus* spp.; mostly cult. as ornam. (SC dimin. of gladius. L., sword, because of lf. shape); layman contraction **glad**(s) - **large-flowered g.**: *G. grandis* - **wall g.: wild g.**: *G. cuspidatus.*

gland: secreting organ of pl. or animal; in the pl. there are numerous types, incl. the external glandular scales or trichomes (digestive, nectary, hydathode) and internal (cavities, resin ducts, oil ducts, laticiferous ducts, l. glands) - **ductless g.: endocrine g.**: secretes into the blood stream - **exocrine g.**: secretions reach the outside of body through a duct (ex. salivary g.).

Glandes Quercus(**tostae**)**:** (L): (roasted) acorn u. coffee substit.

Glandula Pituitaria (L): pituitary body - **G. Rottlerae**: Kamala - **Glandulae Lupuli**: Lupulin, glandular trichomes of *Humulus lupulus.*

glandular products: medicinal items representing or consisting in part or in whole of animal endocrine or ductless glands.

glassine: transparent cellulose prod. mfd. by heating & super-calendaring sulfate paper pulp; u. envelope windows, etc.

glasswort: *Salsola kali* - **jointed g.**: *Salicornia europaea.*

glaucarubin: antiamebic subst. obt. fr. meal of *Simaruba glauca* sd.

glaucine: alk from *Glaucium flavum*, *Corydalis* spp., etc.

Glaucium corniculatum (L) Curt. (Papaver) (Med reg., Canary Is): (red) horned poppy; in diabetes & neurasthenia (in form of FE); otherwise u. like *G. flavum*; sd. FO u. in food & soap mfr.; juice to adulterate opium - **G. flavum** Crantz (G. luteum Scop.) (Eu, Med reg., natd. NA): (yellow) horned poppy; sea p.; fresh hb. or juice c. sanguinarine, glaucine, protopine, chelerythrine; u. purg. sed. hydragog; sd. furnished FO u. food & for soap; diur. in urinary calculi; exter. for sores; juice has been u. to adulterate opium; border ornam. hb. - **G. f. picrin**: bitter princ. of *Glaucium flavum*.

Glechoma brevituba Kuprian. (Lab.) (Chin): hb. u. antipyr. diur. ton. for heart; in virus colds; gravel; insecticide - **G. hederacea** L. (Nepeta glechoma Benth.) (Euras, esp. Ge, natd. NA): ground ivy; field balm; hb. c. VO, with pulegone, menthone, menthol; tannin (3-7%), glechomin (bitter princ.), resin, saponin; expect. in lung dis., enteritis, kidney dis., antidiarrh., hysteria, neurasthenia; exter. for wounds, has been off. in 11 Phar.; ornam perenn. hb. (2543).

Gledits(ch)ia australis Hemsl. (Leg) (eAs): pods purg. emet. for chronic dys.; sd. u. to treat chronic TB, rectal cancer, difficult labor; powd. thorns u. for abscesses in mouth of child; decoc. to wash ulcers, skin dis.; deseeded pods macerated in water with citronella or vetiver u. as hair lotion (parasiticidal) (Indochin) - **G. japonica** Miq. (Far East): u. much like *G. australis* - **G. officinalis** Hemsl. (Far East): pods u. to treat cough & pulm. dis. - **G. rolfei** Vidal & Soler. (sChin, Indochin, PI): lf. decoc. purg. (Vietnam); pulp. fr. around sds. u. in local pharmacy (sVietnam) - **G. sinensis** Lam. (Far East): Chinese honey locust; u. much like *G. australis*. - **G. triacanthos** L. (e&cUS, cult. Eu): (common) honey locust (tree); sweet l.; "thorn tree"; pods c. sugar u. sweetener, food for catarrh.; st. produces "honey locust gum" as path. prod. fr. wounds in wood; c. Ca & Sr salts of glucuronic acid & sugars; bk. stim. ton.; pods u. formerly to make "locust beer" (US slaves); mixt. with sweet acorns —> dr. cake (eaten); bk. c. tannin, bitter princ.; sds. c. protein, FO; pods c. mucilag. (manno-galactan); wd. u. fenceposts, crossties; tree planted as ornam. wind-break; hedge (1417).

Glehnia littoralis Fr. Schmidt & Miq. (Umb) (Jap, Chin, Kor): rts. & rhiz. u. in Jap folk med. for common cold & lung dis. (TB, cough).

Glia gummifera Sond. (1) (Umb) (sAf): u. as diur. in edema, lithiasis.

gliadin: a protein occ. in gluten of wheat & rye; sometimes u. as film coating (medicinals); causes celiac dis.

Glimmentintling (Ge): *Coprinus micaceus*.

Glinus oppositifolius (L) A DC (Mollugo oppositifolia L.) (Mollagin.) (OW trop): grd. pl. mixed with water, boiled & appl. as paste to swollen fingers & toes (Tanz); cooked pl. (pot herb) u. as stom., appetizer; poultice for dyspepsia (children) (PI); common weed.

Gliotoxin: antibiotic (peptide) prod. by several fungal organisms, incl. *Gliocladium virens* Miller, Giddings, & Foster; *Aspergillus fumigatus* Fres.; *Trichoderma viride* Pers. (T. lignorum, Gliocladium fimbricatum Gilman & Abbott); & *Penicillium cinerascens* Biourge; 1932-; antibact. & antifungal but somewhat too toxic for human use; u. as seed dressing.

Gliricidia sepium (Jacq.) Walp. (Leg) (Mex-Col; Guianas): mother of cocoa; madre (de) cacao (Mex); St. Vincent's plum; lf. & rt. juice u. to cure itches & wounds (PI); lvs. crushed & u. as poultice (PR); to bring boils to head (Salv); honey pl.; u. to shade cocoa trees; bk. & lvs. u. to poison mice (not rats); fls. cooked & eaten (Salv)—> firewood.

Globba cambodgensis Gagnep. (Zingiber.) (seAs): u. to treat fever & rheum (Indochin) - **G. heteranthera** K. Schum. (w tropAf; PI): go-go bark; gogue; bk. c. saponin; u. hair ton. (claimed to remove dandruff); frt. u. as food; was formerly confused with *Ceratanthera beaumetzii* E. Heckel (in part; in part *Aframomum melegueta*) (231) - **G. pendula** Roxb. (G. panicoides Miq.) (seAs): rieng rung; pl. u. in fever, & arthritis (sVietnam, Mal); rhiz. u. poultice, vermifuge, fever, rheum. - **G. schomburgkii** Hook. (seAs): rhiz. macerate u. as antipyr., hemostatic.

globe amaranth: *Gomphrena globosa* - **g. artichoke**: v. artichoke - **g. flower**: *Eryngium aquaticum*.

globin: protein with 4 polypeptide chains obt. fr. conjugated proteins, such as hemoglobin - **g. insulin**: combn. of insulin with globin, has prolonged action - **g. zinc insulin injection** (GZI): i. with addn. of $ZnCl_2$ & globin (fr. beef blood); u. SC inj. for diabetes insipidus.

Globularia alypum L. (Globulari.) (Med reg, Arabia): (European) wild senna; lvs. (as Folia Alypi) c. globularin (alypin) (NBP), protocatechuic acid, mannitol, globular-iacitrin (rutin), cinnamic acid, no true phenols; lvs. u. as lax., substit. for Senna; poisonous properties - **G. aphyllanthes** Crantz

(Eu): similar c. & uses as preceding - **G. cordifolia** L. (Eu-alps): globe daisy; u. as purg. in edema, fever, &c. (254) - **globular-i(a)citrin**: Rutin.

Globuli Martiales (L): a crude iron tartrate in the form of balls or marbles.

globulins: class of simple proteins sol. in dil. NaCl solns. & heat-labile; incl. alpha-g. (most mobile in electrophoresis), beta-g. (intermediate), & gamma-g. (least mobile in electrophoresis); latter important in furnishing resistance or immunity to infections, antibodies (immunoglobulins, *qv.*) - **antihemophilic g.: coagulating g.**: blood fraction I; subst. in normal blood plasma which reduces the coagulating time in hemophilia - **human immune g.: Globulinum Immune Humanum** (L): sterile soln. of antibodies obt. fr. placentae & placental blood fr. healthy women; u. to prevent & modify course of measles (E).

glochid: glochis: 1) barbed hair (Cact., etc.) 2) bristle (animals).

Glochidion album (Bco.) Boerl. (Euphorbi) (PI): rts. boiled & rubbed on abdomen for indig. - **G. cauliflorum** Merr. (PI): rt. decoc. with rts. of cacao & coffee u. for quick delivery in childbirth - **G. concolor** Muell. Arg. (Tonga, Fiji): malolo (Tonga); lf. ext. u. med. (Tonga) (86); raw lf. eaten for stomachache & thrush & to stop diarrh. & bloody stool (Fiji) - **G. cordatum** (Muell. Arg.) Seem. (Fiji Is+): lvs. chewed for stomachache - **G. hohenackeri** Bed. (tropAs, Ind): kalchia; bk. given when stomach revolts against food - **G. molle** Bl. (eAs): lvs. appl. as poultice to abdomen in dys.; rolled lvs. as anodyne in toothache (Indones); bast u. for coarse cordage - **G. ramiflorum** Forst. f. (Oceania): inner bk. u. as tea, for cellulitis (Samoa); lvs. & bk. u. med. (Tonga) (86).

Gloiopeltis J. Agardh (Endocladi. - Rhodoph. - Algae) (coasts of eAs): funori (Jap); glue plant; source of funori (mucilage) u. as adhesive & sizing for textiles, esp. silk, paper, cosmetics (hair curling & dyeing); in water colors; food; u. anticoagulant, to reduce blood cholesterol; u. for centuries (Chin, Jap); funoran (mucil.) prepd. fr. thallus; cult. Jap (*G. furcata* Post & Rupr.) J. Ag.; *G. tenax* (Turn.) (J. Ag.; *G. complanata*) (1418).

Gloriosa simplex L. (G. virescens Lindl.) (Lili.) (tropAf, As): superb lily; glory l.; ihlamvu (Zulu); bulb c. colchicine, gloriosine, superbine; u. like *G. superba*; pl. decoc. drunk by childless couples to treat barrenness (Zulus), VD (1419); in impotency; juice to treat wounds (antisep.); homicidal & stock poison - **G. superba** L. (tropAf, As): climbing lily; maskaria (Ind); vine with yellow & red fls. c. 0.3% colchicine (gloriosine), resins; rhiz. c. colchicine & derivs. (1420); "superbin" (alk. NBP?); livestock poison, esp. to pigs, etc.; bulbs u. ecbolic, abortient; suicidal poison (Ind); girls crossed in love ingest rts. for suicidal purposes (Burma); rt. u. in colic, leprosy, piles; antiparasit. exter. in parasitic skin dis. (paste) (Ind); in ascites (some consider spp. conspecific).

Glossocardia bosvallia DC (Comp.) (Ind): seri (Hind); pl. u. as emmen. in female complaints.

Glossogyne pinnatifida DC (Comp) (Ind): rt. prepn. u. for snakebite & scorpion sting.

Glossonema varians Benth. (Asclepiad.) (Ind): frt. cooling & digestible.

Glossostemon bruguieri Desf. (monotypic) (Sterculi) (Iran, Iraq): moghat; c. mucil.; tubers source of a readily digestible meal, the true "Revelenta arabica."

Glottidium vesicarium (Jacq.) Harper = *Sesbania vesicaria.*

glow-worm: Gluehwurm (Ge): *Lampyris noctiluca* (Lampyridae - Coleoptera): wingless larva (female) glows; potential source of luciferin; some adults ("fireflies") also luminescent.

Gloxinia perennis (L) Fritsch (Gesneri) (trop, SA, widely cult. ornam.): pl. decoc. u. to bathe skin in area of boils (777).

Gloxinians: Gloxinoids: members of fam. Gesneriaceae, featuring members *Gloxinia* & related (*Sinningia*, synonymous); splendid ornam. floral pl. of tropAm.

glucagon: hyperglycemic-glycogenolytic factor; HGF; a straight-chain polypeptide hormone prod. by the alpha cells of the islets of Langerhans of pancreas; u. to treat hypoglycemia (by activating hepatic phosphorylase); also inhibits fatty acid synth. & stimulates ketone body production; antibodies may be formed; "gut g." that extd. fr. the intestine; useful in insulin-dependent diabetes (intranasal admin.).

glucides: carbohydrates & glycosides.

glucocorticoids: corticoids (adrenocortical hormones) affecting glucose & other carbohydrate metabolism; principally secreted by the adrenal cortex; u. as anti-inflammatory agts. as in allergy, arthritis; stress; promote glycogen storage; ex. cortisone, hydrocortisone.

glucokinin: 1) insulin 2) hormone in pl. tissue (1425).

glucomannans: polysaccharides in which mannose, glucose, etc. units participate; occ. in woods of many trees, ex. *Abies*, *Pinus*, *Picea* spp.; on hydrolysis yield such holosides as cellobiose, glucopyranose, mannopyranose, &c.; these are gum-like products; recently being "pushed" for body wt. reduction.

glucomycin: an antibacterial antibiotic; obt. fr. *Streptomyces flavogriseus* Waksman. (soil).

gluconic acid: $CH_2OH(CHOH)_4COOH$: compd. formed by oxidn. of glucose; actually prepd. by fermentation with *Penicillium chrysogenum*, *Aspergillus niger*, etc.; u. as acidulant in foods; source of Ca gluconate (Ca donation) (E) & glycogen deposit (1426).

glucosamine (aminoglucose): cyclic amino-sugar; occ in many foods; isolated fr. chitin; occ. in mucoproteins; u. pharm. aid; increases activity of antibiotics (ex. tetracyclines); in schistosomiasis; to improve intest. flora of infants; in arthritis; occ. usually as acetyl g.

glucose: 1) *d*-glucose; a simple monosaccharide sugar, grape sugar, dextrose (term of the pharmacists); common in both pl. & animals; $C_6H_{12}O_6$; made by complete hydrolysis of corn starch; u. for rehydration therapy in diarrh. fr. malnutrition (or with small amt. NaCl in water p. o.); u. in mfr. candies, syrup, bread spreads, jellies, preserves, etc. (advantage: does not crystallize) 2) **liquid g.: syrupy or plain glucose**: corn syrup: consists of a mixt. of ca. 1/3 dextrins, ca. 1/3 dextrose, ca. 10% maltose + 23% water (**Glucosum Liquidum** [L]) - **dried g. syrup**: dextrose.

glucosidase: class of enzymes which hydrolyze glucosidal linkages of certain glucosides & sugars, etc.; ex. alpha-glucosidase = maltase - **beta-glucosidase**: enzyme found in almond (best source) & other sds. of Rosaceae, some fungi (*Penicillium*, *Aspergillus*, *Polyporus*, etc.), intestine, liver & kidney; hydrolyzes amygdalin, arbutin, salicin, helicin, & other beta-glucosides.

glucosides: compds. which release simple sugars on hydrolysis; ex. sucrose; formerly appl. exclusively to what are now properly termed heterosides, in which besides sugar(s), non-sugar compds. are released; ex. salicin; also formerly restricted by some to compds. releasing glucose.

glucosiduronic acid: term preferred to glucuronide acid.

glucosinolates: group of volatile mustard oil or isothiocyanate glycosides, incl. sinalbin, sinigrin, erucin, iberin, glucotropaeolin & other ß-glycosides found mostly in Cruciferae.

glucuronic acid: carboxy acid formed by oxidn. of glucose; found in combined form in blood & urine.

glucuronidase, beta: enzyme prep. fr. calf spleen; u. in enzymatic hydrolysis of conjugated steroid hormones in the urine.

glucuronide: glycoside in which the glycone portion is represented by glucuronic acid instead of a simple sugar.

glucuronolactone: D-Gluconic acid delta-lactone; occ. in many plant gums & in animal tissues; prepd. by oxidn. of glucose; sweet tasting [soly. (H_2O) 59 g%]; u. tab. compression, diluent (1819); in cleaning compds. (1933-); detoxicant.

glue: an impure gelatin u. for strong adhesion of wood, etc.; now appl. to other plastic adhesive hardening materials.

glutamates: salts of glutamic acid; act as excitatory neurotransmitters in brain; it is thought excessive amts. may injure nerve tissue —> senility, dementia, &c., manifest in many of the aged.

glutamic acid: a nonessential amino acid, 2-amino pentanedioic acid; mfrd. mostly fr. waste vegetable materials by acid hydrolysis; u. in mental dullness (said to improve IQ in secondary mental deficiency), petit mal epilepsy, etc. (of questionable value); said to improve visual acuity (1421); as HCl salt to administer HCl in low or lacking gastric acidity; prep. of Na salt of L.-g. a. u. to give meat flavor (monosodium glutamate); to improve flavor of beer.

glutamine, l: modified amino acid; occ. both free GSH & combined in pl. & animal tissues; u. to control craving for sugar.

glutathione: a tripeptide chain made up of glutamic acid, cysteine, & glycine; found widely distr. in pl. & animal cells; important in metabolism; protective agt. against toxicity fr. arene oxides, etc. (2542); isolated fr. yeast; acts as detoxicant in heavy metal poisoning, liver disease, some skin dis., alcoholism; destroys free radicals, hence important to health; said to reduce cataract formation; dose: 300 mg p. o.; Glutathiol (propr.).

Gluten (L): 1) glue 2) mixt. of proteins; obt. fr. wheat & other cereal grains; u. adhesive, in place of flour, as nutr., etc. - **g. bread**: made of g. flour, with high g. & low starch content; u. mostly by diabetics - **g. flour**: v. flour - **whole wheat g.**: u. for suppositories.

glycérat (Fr): preps. based upon starch glycerite - **glycéré** (Fr): preps. of glycerin with starch, or with starch partly replaced by some solid or liquid (ex. tannin); also solns. in or mixtures with glycerin.

Glyceria canadensis (Mx) Trin. (Gram) (eNA): rattlesnake manna grass; palatable grass for livestock - **G. fluitans** (L.) R. Br. (Euras, natd. eCana, New Eng [US]): sugar grass; floating (manna) grass; m. g.; "Russia seed"; sd. u. as nutrient by Klamath Indians; in birdseed; palatable pasture grass.

glycerin(e): glycerol: a tri-OH alcohol universally found in combination with various fatty acids to form the FO; prep. fr. fats, also by synthesis fr. petroleum; an hygroscopic syrupy liquid; u. solvent, emoll., dem. sweetener, mild lax (in rectal suppositories); irritates & dehydrates mucosa; 10% aq. soln. u. to decrease intracranial hypertension in cerebral edema (osmotherapy); & intraocular pressure (glaucoma) in many industrial applns.; ex. nitroglycerin, glycerophosphates, &c. (1423) - **glycerinated gelatin**: a prep. made with glycerin, gelatin, & HOH heated together; u. as suppository base, semi-permanent mounting medium (microslides) - **glycerine tonic**: Elix. Gentian Glycerinated (NF, V) - **Glycerinum Papaini** (BPC, 1934): papain dispersed in glycerin, simple elixir, & dil. HCl - **glycerol**: glycerin - **glycérolé** (Fr): glycérat.

glycides: glucides.

glycine: 1) glycocoll 2) (TM): photographic developer; poisonous; must not be confused with (1).

Glycine max (L) Merr. (G. hispida Maxim.) (Leg) (Chin, Jap): soy(a) bean; this cultigen (China ca. 1000 BC) is one of the most nutritious pls. known; numerous vars. recognized; sds. c. 35% protein, 17% carbohydrates (sucrose, raffinose stachyose, pentosans; but no starch); 15-20% FO; lecithin, bitter princ.; u. nutr., in diabetic foods (canned); in prepn. FO, lecithin, leucine, seasonings (soy sauces, China); young beans eaten like green peas; soy bean flour (or bean meal) u. in many health foods, as soup noodles, in soy bread & component of other breads, cereals, muffins, semolina macaronis; almond paste substitutes; FO ("bean oil") u. as salad oil, mfr. margarine, soaps, to supplement linseed oil, medl. prepns.; proposed use as diesel fuel; bean curds in cheese (tofu); soy b. milk (miso); jellies, confections; in cooking; a source of vit. E; press cake good stock food; soy hay; sds. one of richest & most economic protein sources known (677); sds. c. canavaline (antibiotic) u. to treat pneumonia (2598); sds. u. to prep. artificial milk, u. for infants intolerant to cow's milk (1445); to inhibit cancer (claim, 1993); produces reduction in cholesterol in hypertension (claimed); also reduction in colon cancer (claimed); (pl.) widely cult., now a major crop eAs & US for forage, fodder, soil improvement (N fixation) - **G. soja** Sieb. & Zucc. (eAs): no-mane (Jap); wild growing pl.; sometimes considered conspecific with or the precursor of the cultigen *G. max*.

Glyciphagus spinipes & other *G.* spp. (Order Acarida, Class Arachnida): mite often encountered as infestants on crude drugs; such as cantharis.

glycocoll (= glycine): aminoacetic acid; the simplest & one of first amino acids isolated; NH_2CH_2COOH; obt. fr. gelatin (25%), silk fibrin; sweetish taste; only amino acid opt. inactive; u. myasthenia gravis; bladder irrigant; in some antacids; dietary supplement; serves as inhibitory neurotransmitter.

glycolic acid: $CH_2OHCOOH$; occ. in *Vitis* spp. (lvs., unripe frt.); in sugar cane juice; u. to improve skin texture & coloring after sun damage; to reduce skin wrinkling; appl. to "liver spots" (aging skin); represents cheap org. acid; many industrial uses (textiles, etc.).

glycolipids: complex lipids with carbohydrate residues.

glyconin: glycerite of egg yolk.

glycoside: a compd. which on hydrolysis (splitting with HOH), actuated by enzymes present or by reagents, breaks down to produce one or more simple sugars & a non-sugar component, the aglycone; found widely distributed in pl. kingdom; when formed sugar is glucose, compd. called a glucoside; chief classes differentiated by nature of aglycone present: thiocyanates, anthraquinones, HCN & benzaldehyde (cyanogenetic); ex. amygdalin (59), coumarin & oxycoumarin, phenols (ex. hydroquinone) - **coumarin g.**: combn. of a coumarin with glucose; ex. skimmin, fraxin, aesculin, daphnin - **pigment g.**: includes anthocyanins, anthoxanthins, anthraquinone glycosides, many flavones.

Glycosmis pentaphylla (Retz) Correa Ser. (Rut) (tropAs, esp. Ind, PI): Jamaica orange; inf. roasted lvs. u. as appetizer, aperient; juice of lvs. u. antipyr., liver dis., intest. worms; rt decoc. for bilious attacks (Indones); rts. appl. to chest pains & skin eruptions; lvs. u. in eczema; c. alkaloids, glycosine, glycosminine, glycosmine, skimmianine (penta-phylline); twig fibers u. as toothbrush (nInd) - **G. puberela** Lindl. (seAs): rt. & lvs. inf. u. as postpartum med.; poultice of bk., etc., for nausea.

glycuronic acid = glucuronic a.

glycyrrhetin: glycyrrhetinic acid: amorph. bitter subst. in licorice rt.; u. in Addison's dis.

Glycyrrhiza (L.) (lit. sweet root): 1) genus of 20 spp. of hbs. or sub-shrubs (Leg.) (worldwide): licorice, pinnate lvs, papilionaceous fls. 2) (L) (specif.): licorice rt., prod. of *G. glabra* vars.; common licorice; anti-inflamm., immunostimulating; estrogenic; antimicrobial; antiviral; antiulcer; hypolipidemic; in indig. med. u. to allay thirst; for fever, cough, pain & breathing distress - **G. echinata** L. (seEu, wAs): prickly-headed liquorice; u. similarly to *G. glabra* & vars. - **G. glabra** L. var. *glandulifera* Reg. & Herder (USSR, Syria, Iraq, Iran, Turk): Russian licorice; Persian l. (1448);

rel. large peeled segments of rhiz. & rts. often as small chips (cut l.); c. glycyrrhizin, di-gluco-liquiritoside (liquiriti-genin-rhamnoside-glucoside), glycyramarin (bitter princ., mostly in bk., hence more in Sp 1. which is unpeeled), mannite, desoxycorticosterone-like subst.; VO, etc.; aroma due to added anise oil; black color of ext. due to added charcoal; u. sweetmeat ("l. twist"); in liqueurs, porter, plug tobacco (572); in drops for voice (1820); licorice water sold on streets (Cairo); u. flavor, dem. in gastric ulcers (questioned by some), exp. mild lax., in Addison's dis.; pill & tablet excipient; to disguise unpleasant tastes; in fire extinguishers (foam prodn.); in mfr. l. juice (fresh); for uses of marc, *v.* rifatto; exhasted marc (fr. extn. of drug) u. as substrate for fungal cultures; as foaming agt.; in folk med. drug as diur. lax. emmen. aphrod.; roasted sds. as coffee substit. (also deglycyrrhinized licorice [DGL] [1978]) - **G. g.** var. *glabra* (Sp, It, Por, etc.): Spanish licorice; unpeeled, smaller rhiz. & rts.; usually whole; u. similarly to prec. var. - **G. g.** var. *violacea* Boiss. (AsMin): source of Persian licorice, mostly unpeeled - **G. lepidota** Pursh (w&cUS): American (or wild) licorice; u. like off. licorice; considerable cult. in Calif [US] - **G. malensis** (Chin): licorice of excellent quality; local use - **G. uralensis** DC (wSiberia, Chin): Manchurian l.; rt. somewhat like Russian l.; exp. fr. Inner Mongolia (55).

glycyrrhiz(in)ic acid: glycyrrhizin: a sweet water-sol. subst. fr. licorice rt.; a saponin of the sapotalin gp., representing a glycoside of glycyrrhetinic acid (sapogenin) & glucuronic acid (diglucuronide); occ. as K and Ca salt; occ. sometimes as ammonium glycyrrhizin (a. glycyrrhizinate).

glyoxyldiureide = allantoin.

Gmelina arborea Roxb. (Verben) (Ind to seAs): gamhar; gamari (Hindu); rt. & lf. u. in flatulence, gout; bk. antipyr.; lf. juice u. in cough, GC (124), for ulcers; rt. stom. (Burma); rts. diur.; bk. & lvs. as paste in scorpion sting (124); analgesic; tree fast growing; was cult. in Amazonia (Br) by Daniel K. Ludwig (3 million acres) for pulp & paper mfr.; trunk diam. to 0.4 m in 5 years, twice as fast growing as southern pine; frt. edible - **G. asiatica** L. (eAs): pl. inf. u. for yaws; rts. c. sweet mucil. u. for GC - **G. elliptica** Sm. (Mal): (common) bulang; rt. juice rubbed on head to prevent loss of hair; lvs. cath.; lf. & frt. sap dropped in ear for earache; appl. to wounds, as poultices for headaches, swellings; rt. u. for arthritic pains, nerve dis.; lvs. boiled & rubbed on gums for toothache (also frts.); frt. juice u. as syr. for TB, coughs - **G. philippensis** Cham. (G. hystrix Schult.) (Burma to PI): snapdragon tree; rt. u. for rheum. & purg. (int.); as poultice for throat to relieve cough (lvs. & rts. pounded with lime); juice of frt. for athlete's foot; lvs. appl. exter.

gnaphale des marais (Fr): mouse ear herb, *Gnaphalium uliginosum.*

Gnaphalium L. (Comp.) (worldwide): (life) everlasting; cud weeds; woolly hbs. with alternate lvs. - **G. antennarioides** DC (Ve): "lechuguilla"; entire pl. decoc. with milk & edible FO u. as refreshing drink for delicate children - **G. arenarium** L. = *Helichrysum a.* - **G. attenuatum** DC (sMex, Salv, Pan): gordolobo (Mex): boiled pl. u. as poultice for swellings (Salv) - **G. berlandieri** DC (Zacatecas, Mex): "sanguinaria"; whole pl. u. as emmen. - **G. bourgovii** Gray (sMex, natd. sAf): hb. decoc. u. for stomach dis. (Mex); general remedy for sickness (sAf) - **G. dioicum** L.: *Antennaria dioica* - **G. dombeyanum** DC (SA): u. in folk med. - **G. elegans** HBK (Ve, Ec, Andes): vira-vira; pl. decoc. in VD; branchlet decoc. u. in prostatism - **G. javanicum** Reinw. = *Ana-phalis javanica* - **G. luteo-album** L. (G. javaniucm DC) (worldwide, even to Antarctica): Jersey cudweed; pl. is antipyr., antimal., bechic; as remedy in lung dis., to reduce BP.; lvs. food - **G. obtusifolium** L. (G. conoideum HBK, G. polycephalum Mx) (eNA): sweet balsam; fragrant cudweed; (common) everlasting; (sweet-scented) life everlasting; rabbit tobacco; gordolobo (lit. fat wolf, Mex); hb. u. arom. anod. antispas.; antipyr. in colds (decoc. with syrup or lemon ext.) (SC); hayfever; u. diur. in balsam against painful tumors; in cataplasms for tympanitis (otitis media) (1424); c. gnaphaliin, obtusifolin; dr. lvs. u. masticatory & fumitory (young boys sUS: rabbit tobacco); decoc. u. to wash wounds (Mex); fomentation for bruises - **G. oxyphyllum** DC (Mex): manzanilla del rio; pl. u. as vulner. - **G. purpureum** L. (G. spicatum Lam.) (warmer regions of entire world): "lechiguilla" (Co); hb. u. as expect. for cough; u. in indigestion (Co) - **G. roseum** HBK (Mex): lf. u. to heal wounds - **G. semiamplexicaule** DC (Mex): "calampacate"; lf. appl. to wounds as emollient - **G. uliginosum** L. (Euras, natd. NA): March cudweed; life forever; sweet balsam; hb. u. in folk med. for diarrh.; in homeopathy for lumbago, neuritis - **G. vira** Mol. (nwArg, Ch, Bo): vira-vira; wira; eira; errcio-pelina; hb. inf. u. (with cocoa butter & oregano) for pleurisy; cough; pneumonia; & other chest dis.; TB (with molasses, cedar, cinchona ext.); diaph. antipyr.; to "purify" the blood; improve circulation - **G. viscosum** HBK (G. macounii Greene) (G. decurrens Ives) (wTex [US] to Hond; Haiti): flor de la seda; clammy cudweed (Tex); lf. decoc. u. in stomach cancer (with *Rhus terebinthifolia*); gastric ulcers, heartburn; for burns (Mex); lf. inf. u. for "coughing spells."

Gnetaceae: gnetum fam.; shrubs, lianas, or small trees; evergreen; fls. diecious; lvs. broad; opposite or ternate branches; frt. nutlets or false drupes; only one genus, *Gnetum*.

Gnetales (Gnetopsida) (Gymnospermae): order made up of Gnetaceae, Ephedraceae, Welwitschiaceae.

Gnetum africanum Welw. (Gnet.) (tropAf): bawale; u. med. (Congo, Mozamb); twigs u. for binding; sds. roasted & eaten - **G. campestre** Gamble (eAs): pounded frts. made into poultices for headache - **G. gnemon** L. (tropAs): bulso; u. as poison antidote (Indochin); in malaria (Indochin); u. med. (PapNG, Fiji); young lvs. eaten as veg. & for this cult.; frts. eaten cooked (PI); bk. source of fiber - **G. indicum** (Lour.) Merr. (eAs): sds. roasted & eaten (Hainan); st. with very strong bast fibers - **G. montanum** Mgf. (eAs): u. as remedy for poison; source of fiber.

Gnidia chrysantha Gilg (Thymelae.) (sAf): radjo; rt. & fl. strongly purg. & abort. - **G. eriocephala** (Wall.) Meisn. (Thymelae.) (Ind, SriL): bark u. to prep. paper; vesicant; u. as fish poison; lvs. in bruises; fiber & textile pl. - **G. glauca** (Fresen.) Gilg (Ind): pl. u. as vesicant; lvs. placed on contusions & swellings - **G. vatkeana** Engl. & Gilg. (tropAf, As): lf. u. as purg. (Tanz); rt. u. as vesicant (Ind).

gnoscopine: *d, l*-narcotine; alk. formed by racemization of narcotine.

go go bark: fr. *Ceratanthera beaumetzi*; c. saponin; u. hair tonic (*cf. gogo*).

Goa: former enclave of Port (1510-1961); on wcoast of India (Deccan Peninsula); prod. cocoanuts, timber &c.; now reincorporated into India - **G. butter**: FO of *Garcinia indica* sds.; u. to flavor curries - **G. powder**: Araroba - **G. tree:** deodar.

Goanese: v. ipecac.

goat: *Capra* spp., esp. *C. hircus* L. (Bovidae; Ruminantia); said the earliest ruminant to be domesticated (Euras, nAf); rather similar to sheep, but with sparse feeding habits (shrubs & trees); source of milk, meat, skin, hair, suet (fat) (u. in colds rubbed on back & chest) (folk med.) melted in hot milk p. o. in pneum, TB, coughs (135) - **g. bush** (Antigua): *Castela erecta* - **g. weed**: *Hypericum perforatum*.

goat's beard: 1) *Scorzonera* 2) *Aruncus sylvester* 3) goat weed - **g. cayenne**: Capsicum - **g. foot (vine)**: *Ipomoea pescaprae* - **g. rue**: *Tephrosia virginiana* & other *T.* spp. (do not confuse with foll.) - **European g. r.**: *Galega officinalis* - **g. thorn**: *Astragalus gummifer*, tragacanth.

gobernador (**, el**) (Mex): *Larrea tridentata* - **gobernadora de puebla** (Mex): *Brickellia veronicaefolia*, lvs.

gocci (It) (pl. **goccie**): drops.

gockroo: *Tribulus lanuginosus*

Godfrey's cordial: mixture of Opium & Sassafras.

Godmania aesculifolia (HBK) Standl. (Bignoni) (Mex, CA, Ve): corteza blanco (Salv); cacho de novillo (toro) (Hond); bk. u. med. (Hond).

godown: (As): warehouse.

God's grace: Haarlem oil.

goémon (Fr): various seaweeds incl. *Chondrus crispus*, *Laminaria* spp. etc.

gofruside: glycoside of corotoxigenin & D-6-deoxyallose; occ. in *Gomphocarpus fruticosum*.

gogo tree (Tobasco): *Salacia elliptica*.

gogru = gokhru (Pak): *Tribulus terrestris*.

gogue = go go.

goiaba (Br): guava - **goiabe(i)ra** (Br): guava tree.

goitrin: an antithyroid factor; C_5H_7NOS; found in sds. of various *Brassica* taxa.

goivo amarelo (Por): wall flower.

gold: Aurum (L): Au; noble metal of great intrinsic value; u. as monetary standard; foil placed on decubitus ulcer to promote healing; salts u. in rheumatoid arthritis (2701).

gold cups: *Ranunculus acris* - **g. of pleasure**: *Camelina sativa* - **g. root**: Goldwurzel (*qv.*) - **g. thread (, common)**: 1) *Coptis groenlandica*, *C. trifolia* 2) (seUS): *Cuscuta* spp.

Goldblume (Ge): calendula.

golden chain: *Laburnum anagyroides* - **g. herb** (Pak): *Ephedra* - **g. life root**: Senecio - **g. marguerite**: *Anthemis tinctoria* - **g. moss**: Pengawar djambi - **g. oil**: 1) sassafras oil (etc.) in alc. (Fenner's Formulary) 2) turpentine oil & few drops conc. sulfuric acid - **g. seal (root)**: *Hydrastis canadensis* - **(extra) large g. s.**: *Stylophorum diphyllum* - **g. senecio**: *Senecio aureus* - **g. shower**: 1) *Cassia fistula* 2) *Laburnum anagyroides* - **g. thread**: 1) *Cuscuta* spp. 2) (US): *Nigella* spp. 3) *Coptis trifolia* - **g. tree** (transl. of Fr, arbre d'or): mulberry tree (v. also g. aster).

Goldenhaar: Goldhaar.

goldenrod: 1) *Solidago* spp. (usually) (*S. leavenworthi* & *S. odora* are often supplied comm. on calls for g.) 2) *Neurolaena* & other plants - **anise-scented g.**: fragrant g. - **common g.**: *S. virgaurea* - **compass g.** (Manitoba [Cana]): *S. nemoralis* - **fragrant g.**: *S. odora* - **rayless g.**: *Bigelowia*, *Chondrophora*, *Chrysothamnus graveolens* - **rich g.**: *S. canadensis* - **sweet (scented) g.**: *S. odora*.

goldenshower (tree): *Cassia fistula* - **Goldpulver** (Ge): Tormentilla rt.

Goldhaar (Swiss Ge): *Chrysocoma* spp. incl. *C. coma-aurea* L. (also Goldblaetter); *Aster linosyris* (L) Bernhardi, *Polytrichum commune*, *P. vulgare*, etc.

Goldmelisse (Ge): *Melissa officinalis*; at times confused with *Monarda didyma* (1648).

Goldpulver (Ge): tormentilla rt., rhubarb (pd.), &c.

goldtree: *Tabebuia donnell-smithii*.

Goldwurzel: *Narcissus pseudo-narcissus*, bulb.

Golgi apparatus: G. body: G. complex: membranes, sacs, vacuoles, etc. (organelles) of living cell believed to aid formation of secretions in cell.

golondrina (Mex, CA): *Euphorbia pilulifera*.

goma (Sp, Por): 1) gum 2) (Arg) rubber 3) (PR) *Castilla elastica* - **g. adraganta** (Sp, Por): tragacanth - **g. arabiga** (Sp): acacia - **g. brea** (Arg): *Canarium luzonicum* (*cf.* brea gum) - **g. de mascar** (Arg): chewing gum - **g. de mezquite** (Sp): gum of *Prosopis* spp. - **g. de Pará**: Para rubbertree, *Hevea brasiliensis* - **g. elastica**: *Castilla elastica* - **g. galipodio (polvo)** (Urug): pine resin (fr. *Pinus sylvestris*, *P. pinaster*) - **g. grasilla**: sandarac - **g. guta** (Por): gambage - **g. liquida** (Arg): mucilage - **g. del pais** (Sp): g. de mezquite - **g. Rom**: gamboge.

gomart (resin): chibou.

gombo (leaves) (seeds): okra - **gomboes musc** (Trin): *Hibiscus abelmoschus*.

gomenol (Fr TM): **gomenoleo**: oleogomenol; 5, 10, 20, 33, & 50% solns. of niaouli oil; u. as substit. for eucalyptol; expect., anthelm.; fr. *Melaleuca viridiflora* or fr. cajeput oil.

gomma de batata: *Convolvulus operculatus*.

gomme (Fr-f.): gum - **g. adragante**: tragacanth - **g. arabique (vraie)**: Acacia - **g. bas du fleuve**: hard gum, Down River Grade "A"; type of Acacia - **g. blanc**: ghezireh gum - **g. de bassora**: bassorin gum - **g. du moyen fleuve**: Middle River Grade "B" (Acacia) - **g. du haut du fleuve**: Up River Grade "C" (Acacia) - **g. frialle**: Grade "D" (Acacia).

gommegutte (Ge, Fr): gamboge - **gommelin(e)**: a crude dextrin prep. made by heating starch in water.

Gomphia (Schreber) = *Ouratea*.

Gomphocarpus cornutus L. (Asclepiad) (Madag): rts. & lvs. u. emet. & in asthma - **G. fruticosus** R. Br. (Asclepias f. L.) (Af): hb. c. gofruside (cardioactive glycoside); lf. inf. u. for intest. dis. (children); in coryza (sSotho); snuff for pulm, TB, sternutatory; lf. & latex u. as purg. enema; glucoside with marked cardiac action, noncumulative; sd. floss u. to stuff pillows; rt. (uzara rt.) antispasm. u. in diarrh. dys. - **G. lineolatus** Decne. = *Asclepias lineolata* Schltr. - **G. rigidus** Decne. = *Pachycarpus r.* - **G. schinzianus** N.E. Br. = *Pachycarpus s.* N.E. Br. - **G. validus** Schltr. = *Pachycarpus v.* N.E. Br. (many *G.* spp. now transferred to *Asclepias*).

Gomphostemma crinitum Wall. (Lab) (seUS): one of the few rain forest taxa of fam.; rt. decoc. u. after confinement; lvs. pounded with camphor & appl. as poultice to groin swellings (Mal) - **G. philippinarum** Benth. (PI, endemic): lf. juice mixed with frt. juice of citrus called "Kubuyao" (Citrus hystrix DC) for warts - **G. phlomoides** Benth. (Mal): lvs. appl. to wounds.

Gomphrena decumbens Jacq. (Amaranth.) = *G. serrata* - **G. globosa** L. (eAs, sUS, WI, SA): "San Diego" (Cu); globe amaranth; bachelor's buttons (Trin); decoc. of flg. branch u. for leukorrhea, acute hemorrhage of women (Chin); fls. u. in catarrh (Cu); lf. inf. u. for hypertension, diabetes (Trin); ornam. - **G. martiana** Moq. (Arg): c. choline, betaine; u. med. - **G. nitida** Rothrock (Mex): rosetilla; good for baths - **G. officinalis** Mart. (Br): aconito do mato; rts. u. in fevers, snakebite; as panacea (Paratudo); ornam. ann. hb. - **G. serrata** L. (G. decumbens Jacq.) (sFla Keys [US], WI, SA): globe amaranth; "San Diego Cimarrón" (Cu); "retama" (Mex); u. diur. liver dis. (Cu); "internal diseases" (Mex) - **G. pulchella** Mart. (Arg): u. diur. purg. (Arg); ornam. ann. hb.

gomuti fiber: fr. lf. sheaths of *Arenga pinnata* - **g. sugar palm**: ejoo or ejou; *Arenga pinnata*.

gonad: sex gland producing male or female gametes (ovary, testis).

gonadotrop(h)in: substance secreted by anterior lobe pituitary (APL) or placenta which acts on sex glands; there are 2 types: follicle-stimulating hormone (*qv.*) and luteinising hormone (*qv.*); latter also called interstitial-cell-stimulating hormone (ICSH) - **chorionic g., human** (HCG): prepd. fr. human pregnancy urine; produced by placenta; u. in male to stimulate androgen secretion (for cryptor-chidism; aphrod.); in female in sterility, to induce ovulation, habitual abortion, secondary amenorrhea; to reduce wt. in obesity (with low-calorie diet) - **pituitary g.**: derived fr. anterior pituitary lobe (APL) of cattle; mostly FSH; u. only in vet. med., to initiate normal estrus cycle, in nymphomania (female); cryptorchidism (male); to stim. testosterone prodn., in cystic ovarian dis.; ex. Vetrophin™ - **g. releasing hormone**: Gn-RH: a decapeptide hormone prod. in the hypothalamus wh. stim. release of FSH & LH fr. APL; u. in gonadal dis. & in some forms of male & female infertility; in hypogonadism of both sexes - **serum g.**: prepd. fr. pregnant mare's serum (PMS) originates fr. endometrium (not APL); c. FSH (chiefly) + LH; combined g. has both chorionic g. + serum g. (Synapoidin, PD 1965-68-).

gonger (US slang): 1) opium addict 2) opium pipe.

Gonioma kamassi (camassi) E. Mey. (Apocyn.) (sAf): (African) boxwood; bk. of tree c. alk., kamassine (identical with quebrachamine) (curare-like); bk. u. to make a bitter like angostura (Cape); Kamassi wd. valuable for engraving; cabinet making, etc.; cause of frequent allergy (esp. sawdust) - **G. malagasy** Markgraf & Boiteau (Mad): bk. reputed toxic (c. alks.) & not u. apparently in prepn. bitter beverage (like *G. kamassi*); wd. excellent for fancy turnery.

Goniothalamus amuyon (Blco.) Merr. (Annon.) (eAs): liquid distd. fr. sds. u. to treat scabies (Taiwan); frts. stom. & for stomachache (PI); sd. cooked with oil u. as linim. in rheum.; decoc. in tympanites (PI) - **G. repevensis** Pierre (Cambodia, Laos): frts. popular, with flavor of grapes; u. med. - **G. tapis** Miq. (seAs): u. as abort. (early months of pregnancy); mosquito repellant (Borneo).

gonnyinnive (Burm): *Entada phaseoloides*.

Gonocaryum calleryanum (Baill.) Becc. (Icacin.) (PI): angkak (Bikol); u. for stomach dis. - **G. gracile** Miq. (Mal): frt. u. in poultices for headache - **G. subrostratum** Pierre (Indochin): do kom; boiled lvs. u. to treat beriberi; lf. prep. u. as drops for sore eyes; lvs. consumed as food, u. to prepare alc.

Gonocytisus angulatus (L) Spach (Leg) (Med region incl. Turkey); c. alks. (lupinine; anagyrine; cytisine, &c.); u. diur. antipyr.; toxic to livestock.

Gonolobus condurango (Rchbf.) Triana = *Marsdenia c.* - **G. edulis** Hemsl. = *Vincetoxicum edule*.

Gonystylus bancanus Kurz. (Gonystyl./Thymelae) (PI): bk. chewed & saliva swallowed for poisonous parasites fr. fish; wd. may be an adulter. or sophisticant of *Aquilaria* wood - **G. confusus** Airy Shaw (Mal): rt. decoc. u. after childbirth as protective - **G. macrophyllus** (Miq.) Airy Shaw (G. miquelianus Teijsm. & Binn.) (Indones): wd. VO u. as incense; smoke fr. burning oil u. to relieve asthma; useful wood.

goober (nut): g. pea (sUS): peanut (orig. Af; *cf.* nguba—Stanley's *Africa*).

Good King Henry: 1) *Chenopodium bonushenricus* 2) *Levisticum officinale*.

Goodenia ovata Sm. (Goodeni.) (Austral): hop goodenia; inf. of twigs & lvs. claimed antidiabetic (1573); believed stock poison.

Goodia lotifolia Salisb. (G. medicaginea F. Muell.) (Leg.) (Austral: NSW): goldentip; lvs. c. cyanogenetic glycoside; believed stock (cattle) poison (1428); cult. ornam.

Goodyera oblongifolia Raf. (Peramium decipiens Piper) (Orchid) (NA): (giant) rattlesnake plantain; Menzies r. p.; made into tea (Cowlitz Indians, wUS); fls. rubbed into body by women to make themselves attractive to husband; in snakebite - **G. pubescens** (Willd.) R. Br. (eNA): scrofula weed; ratsbane; (downy) rattlesnake plantain; cold lf. inf. u. in colds & for kidneys (Cherokee Indians); lvs. u. in scrofula; ornam. hb. - **G. repens** (L.) R. Br. (Eu-widely spread in NA): white plantain; kitten toes (Ala [US]): u. dem. to facilitate childbirth (Ala).

goofer (seUS) (Negro): jack.

goolwail: gulwail.

gooroo nuts: *Cola nitida*.

goose: water bird of fam. Anserinae, incl. taxa of *Anser*, *Chen*, & *Branta*; good fliers, with long migration routes, walking gait superior to that of ducks; *Anser anser* L. (Eu) & *A. indicus* L. (As) domesticated for several millenia; kept in herds; valued for food (incl. liver & eggs), fattened to improve liver since Roman days (*cf.* modern fattening to make paté de foie gras); pet; for sacrifice (ancient Eg); warning bird (cackle saved ancient Rome); down u. to stuff pillows; feathers for writing quills (since ca. 500 AD); flesh (gander, male) & eggs (cardiotonic) u. in Unani med.; lard u. in med. exter. for skin dis., int. in colic - **g. fat (g. grease, g. lard, g. oil)**: FO fr. *Anser anser*; u. emoll. popular in folk med. - v. under grass

gooseberry: 1) (nUS, Eu): *Ribes uva-crispa* 2) (seUS): *Vaccinium* spp. (esp. *V. stamineum*) - **Cape g.**: *Physalis* spp. - **Ceylon g.**: *Dovyalis hebecarpa* - **(common) garden g.**: *Ribes grossularia* - **Otaheite g.**: *Phyllanthus acidus* - **wild g.** (NA): *Ribes oxyacanthoides*, *R. hirtellum*, *R. setosum*.

goosefoot: 1) *Chenopodium* spp., esp. *C. ambrosioides* (v. *anthelminticum*), *C. album* 2) genera of Chenopod. Fam. 3) *Aspalathus chenopoda* (sAF) - **g. family**: Chenopodi-aceae - **nettle-leaved g.**: *C. murale* - **stinking g.**: *C. vulvaria*; perhaps also *C. ambrosioides*.

gopherberry: *Gaylussacia dumosa*.

gophertree: *Torreya taxifolia*.

gopherwood (Israel): *Cladrastis* sp.

goran (Pak): *Ceriops tagal*.

gordolobo (lit. fat wolf): 1) (Mex) *Verbascum thapsus* 2) *Gnaphalium viscosum*, *G. obtusifolium*, *G. luteo-album* 3) *Senecio barba-johannis* (Mex).

Gordonia lasianthus (L) Ellis (The.) (seUS): loblolly bay; red b.; tan b.; wd. of tree u. to make cabinets; bk. u. for tanning, etc. - **G. obtusa** Wall. (Ind): c. caffeine-like alk.; lvs inf. u. as stim. stom. appetizer - **G. penangensis** Ridl. (Mal): lf. inf. drunk for diarrh. & dys.

gordura (Ve): 1) *Melinis minutiflora, Gordonia luteo-album*; *Senecio barba-johannis* (Mex) 2) (Por, Sp): FO; grease.

gorlic acid: fatty acid, $C_{17}H_{29}COOH$, found in FO of sd. of *Oncoba echinata* (**gorli oil**).

gorse: *Ulex europaeus*.

Gossyparia manniparus (Coccus m. Ehrenberg; G. mannifera [Hardw.]) (Pseudococcidae - Homoptera - Insecta) (wAs): this coccoid prod. much exudate on *Tamarix* spp., the "honeydew" secreted ("man," Arabic, who use as food); believed manna of the Bible.

Gossypii (Radicis) Cortex (L.): cotton root bark.

gossypion (L not Gr): cotton.

Gossypiospermum praecox P. Wils. (Casearia p. Griseb.) (monotypic) (Flacourti.) (Cu, PR, Ve): (Jamaica) boxweed; lvs. bitter, antipyr.; wd. u. as engraving blocks, rulers, veneer, etc. (—> *Casearia*).

Gossypium L.: 1) genus of woody hbs. or shrubs (sometimes tree-like) (Malv.) (worldwide): cotton; many spp. u. for their sd. hairs (textile fibers); cotton rt. bk. = fresh (recently collected) air dried rt. bk. of *G. hirsutum* & other *G.* spp.; u. emmen. abort. (sUS Negroes) (1450); lf. & fl. inf. u. as gargle for sore throat (Mex); sd. meal fed to cattle; roasted sd. inf. u. in dys. (Mex); galactagog (1451); v. under "cottonseed" 2) (specif.) (L.): purified trichomes fr. sd. of *Gossypium hirsutum* & other *G.* spp. (G. Purificatum) (v. Cotton, purified) (for serials relating to cotton, v. App. C-1); for nomenclature, v. Fryxell, Nomenclator of Gossypium (136); books (1449) - **G. arboreum** L. (wAf, cult. Ind, Arabia, SriL): tree cotton; fiber u. - **G. barbadense** L. (tropAm): Sea Island cotton; prod. a superior long-fiber cotton; more vulnerable to boll-weevil; this & other spp. are source of cotton rt. bk., cotton, sd., &c. (widely cult. sUS, Eg, Pak, USSR) - **G. herbaceum** L. (wAs): Levant c.; prob. forms a component of some of the US short staple cotton var. - **G. hirsutum** L. (tropAm): upland c.; chief sp. cult. in seUS; mostly short staple; long staple in some hybrids with *G. barbadanse* - **G. nanking** Meyen (tropAs, Af): source of Nanking, Chinese, Siam, khaki and (tree?) cottons; with reddish brown fiber; formerly cult. by Negroes (US) ("Nankeen c.").

gossypol: toxic yellow pigment found in cottonseed (0.5%) representing a binaphthalene-dicarboxaldehyde (1452); found in cold pressed oil; exptl. studies for use as male contraceptive (Chin) (-1981-); said to reduce sperm count without affecting blood levels of androgen (2462) or potency of man; proposed for use as insecticide, rubber antioxidant, stabilizer, to destroy tapeworm larvae (cysticercosis).

go-to-sleep vine (Tex [US] children): *Schrankia* spp.

Gotu kola: *Centella asiatica* lvs. & sts.

Gouania leptostachya DC (G. domingensis Blanco) (seAs) (Rhamn.): lvs. u. in poultices for skin dis. (Burma); bk. in water u. as hair wash & to kill parasites in hair (Indones) - **G. lupuloides** (L) Urban (sFla [US], Mex, WI, CA): chew stick (SC pieces of stem chewed to clean teeth & heal gums); pd. st. u. to make dentifrice; st. u. as is hops in beer mfr.; st. decoc. to give strength after illness (2444); lvs. u. in edema & gastric dis.; pieces of stem frayed u. as toothbrush (1429) - **G. polygama** (L) Urban (tropAm): jaboncillo; stem. c. saponins; u. to launder clothes; vine inf. u. in VD; twigs u. to clean teeth (lather) (Hond); rt. for hypertension (Trin) - **G. richii** A. Gray (Oceania): lvs. of tree pounded in water drunk for stomachache (Fiji) - **G. tiliaefolia** Lam. (PI): rt. u. as soap substit. (PI); lvs. u. to wash ulcers (Ind).

goudron de houille (Fr): coal tar - **g. végétal** (Fr): pine tar - **goudronneux** (Fr): tarry.

goulac, gum™: a lignin pitch gum fr. NA *Pinus* spp. u. in book binding.

goulash (, Hungarian): gulyas: combn. of Capsicum with meat & vegetables.

goule: *Myrica gale*.

gourd: 1) *Lagenaria siceraria* (*L. leucantha*); also known as **bottle g.**; prods. gourds, g. seed 2) frt. of *Crescentia cujete* 3) *Cucurbita pepo*, generally of inedible type - **dishrag g.**: *Luffa* spp. - **g. family**: Cucurbitaceae - **leprosy g.**: *Momordica charantia* - **luffa g.**: *Luffa* spp. - **pipe g.: siphon g.**: *Lagenaria siceraria* - **snake g.**: *Trichosanthes cucumerina*; (less commonly) *Lagenaria siceraria* - **wax g.**: *Benincasa hispida* - **wild g.** (swUS): *Cucurbita foetidissima*.

Gourliea spinosa Sheels = *G. decorticans* Gill. ex H. & A. = *Geoffroea d.*

gousse (Fr-f.): legume; pod; husk; shell.

goutte (Fr-f.): drop.

gow (US slang): opium & other narcotics.

gowan (Scott.): daisy; *cf.* Old Eng golde (fr. gold) = marigold.

grabbling (potatoes, etc.): groping for in the soil.

Gracilaria Grév. (Gracilari. - Gigartinales - Rhodoph. - Algae) (temp & trop seas of world):

valuable agarophytes; 100+ spp.; thallus u. as food, fertilizer, source of agar or agaroids; harvesting a considerable industry - **G. confervoides** Grév. (Med Sea, Atl Oc, eUS coast): one of chief sources of agar (Atl Coast agar); food (934) (type sp. of gen.) - **G. lichenoides** (L) Harv. (trop Pac Oc, Indian Oc): Ceylon moss; Jaffna m.; u. in mfr. agar; u. in resp. dis., dys. (China); food; much exported; the dr. thallus not the extracted mucilage is given the name "Ceylon moss."

graciloside (odoroside F): glycoside of *Nerium odorum* consisting of digitoxigenin + D-digitalose-4-glucose.

Grad (Ge-m.): degree.

Graffenrieda rupestris Ducke (Melastom.) (Co+): lvs. rubbed on hands to relieve blisters (fr. paddling) (2299).

Graham, (Rev.) Sylvester (1794-1851): a miller who invented "**Graham bread**" (whole wheat flour with finely ground bran) (1837) & "**G. crackers**"; interested in vegetarianism, temperance; **Grahamism** (health fad) the beginning of the advocacy of fiber in diet.

Grahe's test: a color test for Cinchona; when bk. heated in test-tube, red sublimate on walls is positive indication.

grain: frt. of a cereal grass; caryopsis; ex. rye grain - **chaffy g.**: one in which glumes, paleas, lodicules, etc., closely surround frt. & remain with it when threshed; ex. oats, barley - **G. Coast** (wAf): along part of Gulf of Guinea (SC grains of paradise exported here) - **naked g.**: that in which chaffy structures noted are readily separated by the winnowing process; ex. wheat, corn.

graines d'Avignon (Fr): *Rhamnus infectoria* - **g. de canarie**: canary sd. - **g. de chapelet**: prayer beads.

grainetterie (Fr): sd. collection.

grains, chermes: *Coccus ilicis* - **coarse g.**: oats, barley, maize, sorghum, millets, etc. - **Guinea g.: g. of Paradise**: 1) frts. of *Aframomum melegueta* 2) *Amomum granumparadisi* (u. as substit. for [1]).

graisse (Fr): lard.

gram (Hind): pulses obt. fr. several Leg. genera, as *Cicer* (esp. *C. arietinum*), *Dolichos*, *Phaseolus* - **Bengal g.**: *Cicer arietinum* - **black g.**: *Vigna mungo* - **dew g.**: *P. aconitifolius* - **golden g.: green g.**: *Vigna aureus*; *V. mungo* var. *radiatus*+ - **horse g.**: *Vigna unguiculata* - **red g.**: *Cajanus cajan*.

gram(m)a (NM [US]): a wild grass, *Bouteloua* spp. - **g. alemana** (Mex): *Stenotaphrum secundatum* - **g. francesa** (Por): *Agropyron repens*.

gramicidin: antibiotic; originally obt. as fraction fr. tyrothricin fr. cultures of *Bacillus brevis*; 4 components (A, B, C, D); polypeptides with 15 amino acids; u. antibacterial.

gramigna (It): **gramilla** (Sp): doggrass, *Agropyron repens*.

Graminales: order of Monocots incl. Gramineae (grasses) & Cyperaceae (sedges).

gramine: alk. of barley, mustard, *Arundo donax*, etc.

Gramineae: Poaceae; Grass Fam.; consists of ann. or perenn. hbs growing worldwide; 9,000 spp.; one of most useful of all pl. gps. to man because of the widely u. cereal grains; lvs. typically with narrow blade, sheath at base, & ligule; stem hollow usually except at nodes; inflorescence a spike or panicle, bearing spikelets (small gps. of florets); frt. a caryopsis ("grain"); ex. wheat, rye, maize, *Perotis*.

graminin: an insulin-like carbohydrate found in *Phalaris arundinacea*, *Phleum pratense*, & many other grasses; later called phlein.

Grammatophyllum scriptum Bl. (Orchid.) (PI, PapNG+): letter-plant; epiphytic orchid; paste of pseudobulb appl. to sores; int. for worms & malignant tumors (!); diet in dys.

Grana (L.): pl. of Granum, sd.; cereal grain - **G. Paradisi**: grains of paradise, *Aframomum malegueta* - **G. Tiglii**: *Croton tiglium*, sd.

granada (Sp): 1) pomegranate (frt.) - **Cortex Granati** (L) (peel) 2) *Vitis* spp. - **granadilla** (Sp): grenadilla - **granadilla wood**: fr. *Brya ebenus* - **granado** (Sp): pomegranate tree - **granasang** (wNY [US]): ginseng - **Granatrinde** (Ge): **Granatum** (L.): pomegranate bk.

grand ciguë (Fr): poison hemlock.

grandaddy graybeard (sUS): **"gransy-graybeard"** (sUS): *Chionanthus virginicus*.

"grande flora" (Mex): *Tithonia* or *Encelia* spp.

grande gentiane (Fr): *Gentiana lutea*.

granejo (NM [US], Sp): mesquite bush.

Grangea maderaspatana Poir. (Comp) (eAs): lvs. u. as stom., sed. emmen.; lvs. u. as bechic & in antisep. fomentations (225).

Granola™: popular snack food, mostly cereals.

granulated: small sabulous (sandlike) pellets of sucrose, salts, or other crystall. material (granules or coarse powd. [smaller than 16 mesh]) prepd. by a modified crystalln. process, with added drying, screening, etc.

granulopoietin (GPO): subst. in blood useful in many infections.

Granum (L.): cereal grain; sd. (*cf.* Semen).

grape: Uva (L.): *Vitis* spp., vine or frt. - **European g.**: *V. vinifera* - **fox g.**: *Vitis labrusca* - **frost g.**: *Vulpina* (*V. cordifolia*) - **g. fruit (tree)**: *Citrus paradisi* - **green g.**: *V. vinifera* - **g. hyacinth**: *Muscari* - **Muscadine g.**: *V. rotundifolia* - **Muscat g.**:

various sweet spicy white grapes u. to make wine (**muscatel**) or raisins - **river g.**: *V. vulpina* - **scuppernong g.**: var. of *V. rotundifolia* - **sea(side) g.: shore g.**: *Coccoloba uvifera* - **g. sugar**: glucose - **table g.** (US): Am spp. of *Vitis* - **g. weed**: *Sargassum cuneifolium* - **wild g.**: any wild-growing *Vitis* sp.; *V. coignetiae* (Jap); *Coccoloba uvifera* (seaside g.); *Rhoicissus cuneifolia* (Am trop); *Chondodendron tomentosum* - **wine g.**: European g.

grapevine: *Vitis* spp.

graphite: "black lead"; form of native crystalline carbon, with iron traces; u. in "lead" pencils (pd. g. mixed with clay & hardened), as lubricant, stove polish, etc. (also amorphous g.).

grappa (It): 1) peduncle, specif. that of cherry (<Celtic, grap or crap) 2) branch (branca), claw, clutch 3) tendrils fermented & distd. —> brandy.

grappe rouge (Switz, Fr): *Lythrum salicaria*.

grappino (It): gin.

grappolo (It): bunch, cluster.

Graptophyllum pictum Griffith (G. hortense Nees) (Acanth.) (NG, cult. seAs+): lvs. emoll. u. as poultice for cuts, wounds, swellings, furuncles, etc.; lf. juice dropped into ear for earache (Mal); lf. inf. mild lax. for gallstones; weak diur.

Gras (Ge): grass.

grasa (Sp): fat, grease - **g. de cacao**: cocoa butter - **g. porcina comestible** (Arg): lard - **g. p. non-comestible** (Arg): hog grease.

grass: member of Fam. Gramineae (Poaceae) (common usage); also sometimes appl. (esp. by layman) to members of other families, esp. when the pl. is grass-like or of herbaceous habit (Ex. "grass," US slang for mint, also for *Cannabis*) (**non-Gramineae) - **alang g.**: *Imperata cylindrica* var. *koenigii* - **Algerian g.: ***Stipa tenacissima* - **alkali g.**: *Distichlis* spp. - **alligator g.**: *Aristida purpurascens* - **Arabian sea g.**: concocted name u. in proprietary formula (US) - **arrow g.**: *Aristida purpurascens* - **Bahiá g.**: *Paspalum notatum* - **bamboo g.**: b. briar: ***Smilax* spp. - **barnyard g.**: *Echinochloa crusgalli*; *Cynodon* spp. - **bear g.: bear's g.**: 1)***Yucca filamentosa* 2)***Dasylirion texa-num* 3) ***Xerophyllum tenax* - **beard g.**: 1)***Polygonum* spp. 2) (US) *Andropogon virginicus* - **bent g.**: 1) *Agrostis* spp. 2) sedge g. - **Bermuda g.**: *Cynodon dactylon* - **black oat g.**: *Stipa avenacea* - **blubber g.**: soft chess: *Bromus hordaceus* - **blue g.** (Ky [US]): *Poa pratensis* - **blue-eyed g.**: ***Sisyrinchium* spp. - **border g.**: ***Liriope* sp. - **bottle g.**: ***Trifolium arvense* - **bottle-brush g.**: *Hystrix patula* - **brome g.**: *Bromus* - **broom (sedge) (g.)**: sedge g. - **Buchloë g.: buffalo g.**: *Buchloë dactyloides* - **b. bunch g.**: *Festuca scabrella* - **buffel g.**: *Cenchrus biflorus* - **burr g.**: *Cenchrus* spp. - **canary g.**: *Phalaris* spp. (**toowomba c. g.**: *P. tuberosa*) - **carpet g.**: *Anoxopus compressus* - **cat ('s) g.**: ***Euphorbia helioscopia* - **cat-tail g.** (Ga [US]): timothy g. - **centipede g.**: *Eremochloa ophiuroides* - **China g.**: ***Boehmeria nivea* (fibers) - **citronella g.**: *Cymbopogon nardus* - **clayver g.**: ***Galium aparine* - **coal g.** (cFla [US], local term): lemon g. - **Cochin g.**: *Cymbopogon martinii* - **cogon g.**: *Imperata cylindrica* - **Colorado g.**: *Panicum texanum* - **couch g.**: *Agropyron repens* - **crab g.**: *Digitaria* spp. incl. *D. sanguinalis* - **crap(pe) g.**: 1) *Lolium perenne* 2) crab g. - **crawl g.**: 1) (US) couch g. 2) (US) ***Polygonum aviculare* - **crawly g.** (Tex [US]): *Sporobolus buckleyi* - **crop g.**: crap g. - **crowfoot g.**: *Dactyloctenium aegyptium* - **Cuba g.** (US): *Sorghum halepense* - **Dallis g.** (seUS): *Paspalum dilatatum* - **Danube g.**: *Phragmites maxima* - **deer g.**: ***Rhexia* spp. - **dew g.**: orchard g. - **dog. g.**: couch g. - **dog-hair grasses**: *Aristida* spp. - **doub g.**: Bermuda g. - **drop seed g.** (Tex [US]): *Sporobolus buckleyi* - **Egyptian g.** 1) crowfoot g. 2) Cuba g. - **elephant g.**: 1) *Pennisetum purpureum* (very valuable trop g.) 2) (at Pegu) spp. of *Coix*, *Saccharum*, *Andropogon*, *Phragmites*, etc., covering the ground 3) ***Typha elephantina* (*qv.*) - **fairway wheat g.**: *Agropyron cristata* - **g. family**: Gramineae - **feather g.**: 1) *Stipa* spp. 2) *Lep-tochloa* spp. 3) rot g. - **fever g.**: *Cymbopogon nardus*; *Pityopsis graminifolia* - **finger g.**: *Chloris* spp. - **foxtail g.**: *Setaria* or *Echinochloa* spp. - **gallow g. (hemp)**: ***Cannabis* - **gama g.**: *Tripsacum dactyloides* & other *T.* spp. - **gander g.**: ***Persicaria* spp. - **garden viper g.**:***Scorzonera hispanica* - **geranium g.**: rusa g. - **ginger g.**: *Cymbopogon martinii*, *C. nardus*, *C. schoenanthus*; *Setaria* spp. (VO) - **glow(-worm) g.**: ***Luzula campestris* - **goose g.**: ***Galium aparine* - **grama g.**: *Bouteloua hirsuta* - **green g.**: blue g. - **Guatemala g.**: *Tripsacum laxum* - **Guinea g.**: 1) *Panicum jumentorum*, *P. maximum*, & other *P.* spp. 2) *Sorghum halepense*, *Alfalfa* - **hard g.** (US): orchard g. - **herd g.**: timothy g. - **honey g.**: *Melica* spp. - **horsetail g.**:***Equisetum hyemale* - **Hungarian g.**: *Setaria italica* v. *nigrofructa* - **Indian g.** (Ala [US], Negro): ***Rumex acetosa* & related - **Jap g.** (swAla) = cogon g. - **Java g.**: 1) *Polytrias praemorsa* 2) ***Cyperus rotundus* - **Johnson('s) g.**: *Sorghum halepense* (W. Johnson, Ala [US], who introduced this g. ca. 1840) - **June g.**: *Koeleria cristata* - **Kentucky blue g.**: blue g. - **knot g.**: ***Polygonum* spp. - **Korean lawn g.**: *Zoisia japonica* - **lalang g.** (Mal): *Imperata* spp. - **lemon g.**: *Cymbopogon citratus*, *C. nardus* - **love g.**: *Eragrostis* spp., etc. (v. under l. g.) - **maha pangiri g.**: *Cymbopogon winterianus* - **Manila g.**: *Zoisia matrella* - **marsh g.**: *Spartina* spp. - **Mascarene g.**: *Zoisia tenuifolia* - **mat g.**: appl. to several spp. of *Ammophila, Cyperus, Elymus, Nardus* (chiefly in US) with sole sp. *N. stricta* - **meadow g.**: *Holcus lanatus* - **common** (or **English**) **m. g.**: blue g. - **Means g.**: Cuba g. - **molasses g.**: *Melinis minutiflora* - **mondo g.**: *Ophiopogon japonica* - **monkey g.**: 1) ***Liriope* spp. 2) ***Attalea funifera* - **mulga g.**: *Thyridolepis mitchelliana* (Nees) S. T. Blake (Austral): found in arid places; good fodder - **musk g.**: ***Chara* spp. - **muskus g.**: ***Erodium moschatum* - **Napier g.**: *Pennisetum purpureum* - **Natal g.**: *Rhynchelytrum roseum, R. repens* - **natural g.**: blue g.- **needle g.**: *Stipa* spp., less commonly *Aristida* spp. - **nut g.**: ***Cyperus rotundus, C. esculentus, C. strigosus* - **orchard g.**: *Dactylis glomerata* - **oyster g.**:** *Ulva* spp. - **panic g.**: *Panicum* spp. - **paper g.** (US): *Poa nemoralis* - **pepper g.**: ***Lepidium* spp. - **pigeon g.**: *Setaria glauca* - **pin g.**: ***Erodium cicutarium* - **plume g.**: *Erianthus* spp. - **porcupine g.**: *Stipa spartea* - **prim g.**: sweet vernal g. - **purple g.**: ***Lythrum salicaria* - **quack g.**: dog g. - **quaking g.**: *Briza* spp. - **quitch g.**: dog g. - **rat-tail g.** (Ala [US]): ***Plantago aristata* - **rattlesnake g.**: *Glyceria cana-densis* - **reed g.**: *Phragmites maxima* - **reed canary grass**: *Phalaris arundinacea* - **rib g.**: ***Plantago lanceolata* - **ribbon g.** (US): *Phalaris arundinacea* v. *picta* - **rice g.**: *Oryzopsis* - **rice-cut g.**: *Leersia oryzoides* - **rice-root g.**: *Epicampes macroura* - **rot g.**: *Holcus lanatus* & many other plants - **rusa g.**: *Cymbopogon martinii* - **rye g.**: *Lolium perenne, Hordeium murinum, H. secalinum* (**drunken r. g.** [US slang]: darnel) - **sage g.**: 1) *Lolium* spp. 2) sedge g. - **Saint Augustine g.**: *Stenotaphrum secundatum* - **Saint Mary's g.**: 1) Cuba g. 2) Guinea g. (1) - **salt g.**: 1) *Leptochloa* spp. 2)***Frankenia* - **sand love g.**: *Eragrostis trichodes* - **sandbar love g.**: *Eragrostis frankii* - **saw g.**: v. sawgrass - **scurvy g.** (, **common**): 1)***Cochlearia* spp., esp. *C. officinalis* 2)***Barbaraea verna* - **sea g.**: ***Zostera marina* - **sedge g.**: *Andropogon virginicus* - **shave g.**: ***Equisetum hyemale* - **shore g.**: **salt cedar - **silk g.**: ***Yucca filamentosa* - **sirkhi g.** (Ind): 1) *Latipes senegalensis* 2)***Bupleurum falcatum* - **sleepy g.**: *Stipa robusta* (*S. vaseyi*) - **smut g.**: *Sporobolus poiretii* - **snake g.**: spp. of 4 genera of Gram. Fam. esp. *Eragrostis cilianensis* (SC odor), also ***Myosotis scorpioides* (Boragin), ***Stellaria* spp. - **sour g.**: 1)***Rumex hastatulus* (Ala [US]), *R. acetosella* 2)***Xanthoxalis* spp. 3) *Andro-pogon pertussis* (Miss [US]) 4) *Trichachne insularis* - **sparrow g.**: ***Asparagus* - **garden s. g.**: ***Asp. officinalis* - **spear g.**: *Stipa spartea*, ***Ranunculus lingua*, ***R. flammula*, etc., & many other pl. spp. - **spring g.**: sweet vernal g. - **squirrel tail g.**: *Hordeum jubatum* - **star g.**: 1)***Chamaelirium luteum* 2)***Aletris* [also **common** (or **false**) **s. g.**] 3)***Hypoxis* 4)***Asperula odorata* - **stink g.** 1) (US): molasses g. 2) *Eragrostis cilianensis* - **Sudan g.**: *Sorghum vulgare* v. *sudanense* - **swamp g.**: smut g. - **sweet g.**: 1) *Hierochloë odorata* 2) *Acorus calamus* - **sweet vernal g.**: *Anthoxanthum odoratum* - **Syrian g.**: Cuba g. - **tear g.**: *Coix lacrymajobi* - **three awn g.: tickle grasses**: *Aristida* spp.- **timothy g.**: *Phleum pratense* - **toothache g.**: *Ctenium aromaticum* - **tree g.**: bamboos - **twitch g.**: dog g. - **vanilla g.**: *Hierochloë odorata* - **velvet g.**: rot g. - **water g.** (sGa): ***Stenophyllus barbatus* - **wheat g.**: *Agropyron* spp. - **winter g.**: *Poa annua* - **wire g.**: grasses with wiry culms or lvs. (ex. *Cynodon, Poa compressa*, etc,; esp. u. for *Cynodon dactylon*, Bermuda g. [wild growing], and *Aristida* spp. [*A. stricta*], *Eleusine indica*, & other grasses) - **witch g.** (US): 1) couch g. 2) *Panicum capillare* - **worm g.**: ***Spigelia* - **yard g.**: *Eleusine indica* - **zoysia g.**: *Zoisia* spp., esp. *Z. japonica, Z. matrella.*

grass-myrtle: *Acorus calamus* - **grass-spur** (Ala [US]): sand spur - **grass-tree**: *Xanthorrhoea* spp. - **grass-wrack**: *Posidonia oceani-ca* (Algae).

grasso (It): FO, fat - **g. di cacao** (Swiss Phar.): cocoa butter.

graticule: a network of lines or a scale engraved on a glass disc & placed at the focal place of a microscope (etc.) for making measurements.

Gratiola aurea Pursh (Scrophulari) (eNA, esp. seUS): hedge hyssop; rhiz. u. as drastic, anthelm.; in gout - **G. neglecta** Torr. (NA): clammy hedge hyssop; hb. cardiotoxic - **G. officinalis** L. (sEuras): hedge hyssop; c. gratioloside (gratiolin), cucurbitacin derivs. (toxic); hb. & rt. u. drastic, anthelm., diur. emet., in gout; cardiotonic - **G. peruviana** L. (Pe, Ec, Austral [NSW]): u. in "intestinal fevers."

Graupe (Ge): pearl barley.

gravel: sandlike granular material as found in renal or vesical calculi- **g. plant**: *Epigaea repens* - **g. root**: *Eupatorium purpureum* - **g. weed**: *Diervilla canadensis*.

gray: v. also grey - **g. moss** (Ala [US]): Spanish moss - **g. powder**: mercury with chalk.

grayfish: dogfish.

Grayia spinosa (Hook.) Moq. (Chenopodi) (wUS): spiny hop-sage; good browse pl. for livestock.

gray-land (sGa [US]): areas of gray sandy soils.

grayling grease: FO fr. *Salmo phymallus*.

grease: viscous animal fats; dark soft fatty waste product contg. lard, tallow, bone fat, horse f., fish f., stearin, &c.; u. as lubricant; occ. on market - **wool g.: Yorkshire g.**: crude fat-like product obt. by solvent extrn. fr. sheep's wool; u. mfr. degras & lanolin.

greasebush: greasewood, (Mexican): 1) *Larrea tridentata* 2) *Sarcobatus vermiculatus* 3) other swUS pls. such as *Grayia* spp., *Adenostemmo fasciculata, Audibertiella polystachya,* etc. (SC abundant FO or resin in wd., makes for good fuel).

Grecian: v. laurel.

grecua (Por): saffron.

Greek hay seed: fenugreek - **G. nuts**: sweet almonds - **Greek**: v. under sage, valerian.

green, bladder: sap g. - **g. endive**: Lactucarium - **g. gages**: *Prunus domestica* v. *italica* - **iris g.**: sap g. - **g. manure**: crop of Leg. pls. bearing rt. nodules, which is plowed under to supply nitrogenous & other materials to soil - **g. oil**: Baume Tranquille (Fr): Oil Henbane Infused Comp. - **sap g.**: juice of *Rhamnus cathartica* dr. with $Ca(OH)_2$ or other alkali; u. green pigment - **g. weed**: *Genista tinctoria* - **green**: v. under alder, ash, grapes, hellebore, osier, pepper, soap, vitriol.

greenbark: *Parkinsonia florida*.

greenbrier (, common): *Smilax rotundifolia*.

Greenea corymbosa K. Sch. (G. jackii W. & A.) (Rubi) (Indones): lf. decoc. u. as emet. & antipyr.

greeters (slang): *Cannabis*.

Gregory's Powder: Rhubarb Powder Comp.

grenade (Fr): pomegranate, frt.; also referred to *Diospyros haplostylis* & *D. microrhombus* - **grenadier** (Fr-m.): pomegranate - **grenadilla** (Sp): 1) (SA) *Passiflora quadrangularis* (**common g.**) & other *P.* spp. (SC similar to pomegranate frts.) 2) (Br) *P. edulis* (**purple g.**) 3) *Dalbergia melanoxylon* (also **African g. wood**) - **grenadille** (Fr): 1) *Passiflora* spp. 2) coc(c)us wood, *Brya ebenus*.

grenadine: 1) pomegranate 2) *Dianthus caryophyllus* (carnation) - **g. extract**: fr. pomegranate frt. - **syrup of g.: sirop de grenades** (Fr): pomegranate or red currant frt. juice, sugar, & water; u. as flavor or sweetener for carbonated & alc. beverages; also u. itself as beverage (with ice water); in US, an artificial syrup.

grenetina (Sp, It): **grenétine** (Fr): purest sort of gelatin; named after refined gelatin made by Grenet de Rouen to substit. for isinglass in phar. uses; still (-1951-) found in commerce.

grenouille (Fr-f.): frog.

Grensing (Ge): milfoil.

gresskar (Dan): pumpkin

Grevillea robusta A. Cunn. (Prote.) (NSW & Queensl [Austral]): silk(y) oak; disease causes gummosis with a gummy resinous exudate rich in Ca & Mg polyuronides (88); useful wd., ornam. shade tree, in coffee plantations.

Grewia acuminata Juss. (Tili.) (Oceania): rt. appl. to ulcers (PI); u. med. (Sumatra) - **G. asiatica** L. (G. subinaequalis DC) (Ind to seAs, PI): phalsa; falsa; falsé; frt. edible, sweetish, somewhat acid, astr.; u. to prepare beverage, spirit; bk. demulc.; lvs. appl. to eruptions (Burma); lf. fibers u. to make ropes & to clarify sugar; rt. bk. u. in rheum. (Santals) (cult. Ind, SriL) - **G. bicolor** Juss. (Tanz): rts. u. for chest dis., stomach pains, diarrh.; fibers u. to make ropes - **G. carpinifolia** Juss. (Ind, tropAf): hb. decoc. u. to wash hair to remove or prevent lice; oxytocic - **G. crenata** (J. R. & G. Forst.) Schinz & Guill. (Polynes): lf. decocc. drunk for cough, stomachache, & diarrh. - **G. elyseoi** Cavaco & Simões (Angola, wAf): u. oxytocic - **G. eriocarpa** Juss. (Chin): sd. & lf. decoc. u. for stomachache; frt. edible (Hainan) - **G. flava** DC (tropAf): brandy bush; frt. edible, u. to made beer & a spirit; lf. & rt. u. med. (Bechuanaland); lvs. fodder - **G. forbesii** Mast. (Tanz): u. similarly to *G. bicolor*; also remedy in lumbago, stiff neck - **G. latifolia** Benth. (Austral): dysentery plant; u. diarrh. (rt. decoc.) - **G. microcarpa** K. Schum. (Tanz): rt. decoc. u. for severe menstrual flow; bast fibers u. to make string - **G. microcos** L. (Ind+): myata (Burm); lf. u. to cover Burmese cigars, mfr. bidis (cheap Indo-Pak cigarette); branchlets u. to make "tooth-sticks" (a crude toothbrush); pl. u. in indig., dys. smallpox, eczema, itch - **G. mollis** Juss. (tropAf): rts. & bk. u. for persistent cough (Tanz); bk. & lf. u. for cuts & sores - **G. occidentalis** L. (tropAf): bruised bk. soaked in hot water & appl. to wounds; to facilitate childbirth; in impotency - **G. paniculata** Roxb. (seAs): rt. decoc. u. in cough (sVietnam); in fever (Mal); rt. inf. for abdomin. ailments - **G. platiclada** K. Schum. (Tanz): bk. u. as poultice for sores - **G. polygama** Mast. (nQueensland [Austral]): lvs. u. in diarrh. dys. - **G. salutaris** Span. (Lesser Sunda Is): rasped bk. u. to prep. paste (kajoo timor); u. for internal bruises, also as tea; in some areas u. for external bruises - **G. sclerophylla** Roxb. (G. scabrophylla Roxb.) (Ind+): petshat (Burm); rt. prepns. u. in cough, intest, & bladder irritat.; decoc. emoll. as enema; lvs. & rts. u. in leprosy (Hansen's disease) (Burma) - **G. venusta** Fres. (w tropAf): kee (Senegal); lf. u. for cuts, wounds, ulcers, fever; st. & twigs u. for colic (bk); lax (232) - **G. villosa** Willd. (tropAf): rt. u. as "strengthening" med. (Masai) - **G. vitiensis** Turrill (Fiji): lvs. crushed in cold water & drunk for hematemesis.

grey bark: quin(quin)a gris, gray cinchona.

Grias cauliflora L. (Lecythid.) (tropAm): anchovy pear; cult. in WI - **G. neuberthii** Macbride (Co, Pe): pear-sized frts. grated & mixed with water, u. as purg.; frts. also roasted & eaten - **G. peruviana** Miers (tropAm, incl. Col & Ven): cocota; scrapings fr. germinating sds. u. to induce vomiting (Indians).

griefo (slang): *Cannabis*.

gries: partially milled wheat, bran removed, & sifted.

Grieswurzel (Ge): velvet leaf, Pareira.

Griffonia simplicifolia Baill. (Leg) (trop wAf): boto (Ghana); lf. u. as purg. in diarrh., aphrod.; to kill lice in hen houses; delousing; sts. & twigs u. in soft chancre; purg.; sts. u. in cordage (2298).

Grifola frondosa S. F. Gray (Polyporus frondosus Dickson ex Fr.) (Polypor. - Basidiom. - Fungi): hen of the woods; "dancing mushroom"; maitake (Jap); edible but prod. hallucinations - **G. sulfurea** Schaeff. (Polyporus sulfureus Fr.) (Eu): Schwefelporling (Ge); sometimes sporocarp sold as "true larch agaric" (*Fomitopsis officinalis*); adulter. of same - **G. umbellata** Pilát (P. umbellatus Pers) (eAs, NA): many-capped polypore; u. diur. (Chin, Manch.).

Grifulvin™: griseofulvin (McNeil).

grigri: 1) (seUS Negro): gris-gris 2) *Astrocaryum* spp.

Grindelia camporum Greene (Comp.) (swUS, esp. Calif): field gumweed; gum plant; lvs. & fl. tops c. resins, grindeline, VO; u. stim. exp. in asthma & other pulm. dis.; has been u. in stomach cancer; for rhus poisoning - **G. glutinosa** Dun. (Mex): u. med (Mex Phar.) - **G. hirsutula** H. & A. (coastal Calif [US]): hirsute grindelia; perenn. hb. at times to 1.5 m (commonly 3-8 dm); u. med. - **G. humilis** H. & A. (Calif): marsh gumweed; similar c. & u. to *G. camporum*; off. grindelia hb. of several Ph. (1433) - **G. robusta** Nutt. (wUS, esp. Calif): gumweed; tarweed; gum plant; hb. u. in rhus poisoning (tinct); u. by NA Indians for colds, colic, etc.; in consumption, sores on horses, etc.; rubbed on legs for poison ivy (Am Indians); in home remedies, lvs. & flg. tops u. as mild stom.; to relieve throat & lung dis. (1432) (**G. Robusta** [L]: lvs. & flg. tops of *G. camporum* & other *G.* spp.) - **G. squarrosa** Dun. (w&cNA): curly cup gum-weed; yellow tarweed; snake-headed grindelia; rosin weed; yerba del buey (Mex); chewed as gum in early days; pl. u. in skin irritat. & stomach & kidney dis.; lvs. & flg. tops expect. antispasm. sed. u. in pertussis (1431); u. as is *G. camporum*; esp. in catarrhs.

griseofulvin: curling factor: Grisactin™; antibiotic compd. prod. by mycelium of *Penicillium griseofulvum* Dierckx & *P. janckzewskii* Zal.; isol. 1939; antifungal activity; u. to treat dermatomycoses (fungal skin dis.), such as athlete's foot, onychomycosis (tinea unguium), etc.; in angina pectoris (E).

Griseomycin: antibiotic fr. *Streptomyces griseolus* Waksman & Henrici, esp. active against *Neisseria* spp.

gris-gris (Fr): amulet against evil "spells"; bundle of odds & ends, such as snake skin, bird's claws, various drugs, etc.; "hand" or "jack."

grits (seUS): cracked maize grains; coarse grind meal popularly u. as food (occas. other cereal grains u.) (wheat, rice, &c.).

groats: 1) cracked wheat 2) edible parts of oats grain (v. oat-groats) 3) hulled & usually crushed grain (Eng) (oats etc.).

grocer (fr. Old Fr grossier <middle L., grossarius< L. grossus, large, since orig. a dealer on large scale or wholesaler): orig. these merchants sold spices & crude drugs in addition to foods, etc.

gromwell (leaves): *Lithospermum* spp., esp. *L. officinale*, *L. arvense* - **corn g.**: *L. arvense* - **false g.**: *Onosmodium* spp. - **seed g.**: *L. officinale*.

Gronovia scandens L. (Loas.) (sMex to Ec): pan caliente; trichomes sting, producing severe pain.

gros chardon (Qu): *Cirsium lanceolatum*.

groseille (Fr-f.): currant - **grossellero (negro)** (Sp): (black) currant, *Ribes (nigrum)*.

Grossularia (L): frt. of gooseberry bush, *Ribes uva-crispa* & some other *R.* spp.; delicious frt. eaten & u. in preserves, flavors, etc.

Grossulariaceae: gooseberry fam.; temp. regions worldwide; several genera, incl. *Ribes*; members: gooseberry, currant.

ground: v. under cherry; holly; ivy (herb); laurel; lemon; lily; maple; pea; pine; raspberry - **g. clematis**: *Clematis recta* - **g. hogweed** (BC [Cana]): *Pedicularis bracteosa*.

ground-nut: 1) peanut 2) *Apios tuberosa* - **Bambarra g.**: *Voandzeia subterranea* - **hausa g.**: *Kerstingiella geocarpa*.

groundsel: 1) *Senecio*—any herbaceous sp.; esp. *S. vulgaris* 2) g. tree - **g. bush**: *Baccharis halimifolia* - **common g.**: *S. vulgaris* - **silver g.**: *S. cineraria* - **g. tree**: *Baccharis* spp.

group (bacteriol.): rather indefinite meaning; may indicate gen. subgen., sp., subsp., var. or serological subdivision.

growth factor: g. hormone: Somatotropin (*qv.*).

grub oil: FO of body of earthworm, *Lumbricus terrestris* - **g. root**: *Chamaelirium luteum.*
gruener Germer (Ge): *Veratrum viride.*
Gruetze (Ge): grits, groats, porridge.
Grumilea capensis Sond. (Rubi) (sAf): rt. u. as emet. (Zulu); found not to be antibiotic.
Grus (Ge): garble; rubbish (ex. Bluetengrus).
Grusinian tea (Grusinischer Tee): Georgian tea; beverage tea; fr. Gruziya (Russ) (Georgia [USSR]).
guabirá (Co): **guabiroba** (Br): *Campomanesia* spp.
guacamole (Mex): avocado, mixed with tomato, spices, &c., u. as salad (sauce or dip).
guacamote (Mex): Cassava.
guacimo (Mex): *Guazuma ulmifolia.*
guaco (Sp, Mex): appl. in Am to various plants; u. to cure sores, for snakebite, rheum, etc., incl. 1) spp. of *Mikania*, esp. *M. guaco, M. cordifolia, M. cymbifera*, & related *M.* spp. *Aristolochia* (*Aristolochia grandiflora, A. odoratissima*, & related), *Eupatorium, Spilanthes* 2) (Mex) *Melicoccus bijugatus* 3) (NM [US]) *Cleome serrulata.*
guaguaci: guaguasi: guagusi (bark) (Cu): *Zuelania guidonia* & other spp. of *Z., Laetia, & Casearia.*
guaiac (gum): g. resin: **Guaiaci Resina** (L): **Guaiacum**: resin (over 10%) fr. wd. of *Guajacum officinale & G. sanctum* (1434); c. guaiaconic, guaiaretic, & other acids; VO; vanillin; u. diaph. exp., formerly in syph. rheum. as loc. stimulant; reagent in tests for occult blood, etc.; antioxidant - **g. wood: Guaiaci Lignum** (L): dr. heart wd. (usually as raspings) of *G. officinale & G. sanctum* - **g. wood oil**: VO in commerce.
guaiaco (Por, Br): Guaiacum - **g. wood oil**: 1) VO fr. *Bulnesia sarmientoi* 2) Guaiacum VO.
guaiacol: o-methoxy-phenol; xals. or liq. obt. fr. wood tar, esp. of beechwood; or by synth.; u. exp. in chronic catarrh, cough, etc.; now more u. in veter. med. - **g. carbonate**: this ester less irritant as exp. than g. (Duotal); odorless, colorless crystals or powd.
Guaiacum (Guajacum) (L): guaiac - **G. angustifolium** Engelm. (Porlieria a. Gray) (Zygophyll.) (sTex [US], Mex): soap bush; guayacán (Mex); decoc. u. as vascular stim., diaph. in rheum., amenorrh., VD (Mex) (2230); rt. bk. sold as "amole" (in balls) for washing woolen goods; trunks as fenceposts (long-lived) - **G. coulteri** A. Gray (Mex): guayacán; remedy for colds, &c.; fls. for pneumonia, cough, &c.; similar props. to *G. officinale* - **G. officinale** L. (WI, esp. Haiti, Hispaniola; nSA): (common) lignum vitae (tree); wd. (guaiac[um] wood) very heavy, almost like stone in consistency; c. resins, VO, saponin; rts. u. diaph. diur. in syph. rheum.; gout; skin dis., mild lax (Jam); in wd. carvings of ornaments, etc.; in making skittle balls, axle bearings; water lubricated (underwater) bearings (wd. almost impervious to water); source of resin (guaiac [*qv.*]) & VO; bk. also u. med. (drunk in gin as pain killer [Bah]); in gout & arthritis (Haiti); rubbed on teeth & gums to strengthen (Haiti); boiled lvs. abort. (Barbados) - **G. sanctum** L. (Antilles, Bah, sFla [US]): (hollywood) lignum vitae; holy wood; bk. boiled & whisky added to incr. appetite (137); c. & u. as prec.
guaiaretic acid: phenolic product fr. guaiac resin; antioxidant; a derivative, nor-dihydro-guaiaretic acid (*qv.*) formerly important antioxidant.
guiabeira = goiabeira, guava; **g. salad** also c. tomatoes, onion, etc.
guaicurú (Br): *Limonium brasiliense.*
guaifenesin: glyceryl guaiacyl ether; occ. in beechwood tar (distillate fr. creosote); claimed as only compd. in cough syrups which loosens phlegm & thins bronchial secretions (Robitussin™).
guairo (Pe): *Erythrina* sp. (búcaré) - **g. santo (de costa)** (g. comun) (Cu): *Aegiphila elata*, less commonly u. for *Citharexylum caudatum* (Cu) and *Pseudelephantopus spicatus.*
Guajacum = *Guaiacum.*
guajaca (Cu): Spanish moss.
Guajakharz (Ge): guaiac resin.
guaje (Sp): *Crescentia cujete* - **g. cirial** (Mex): **g. cirian**: *Crescentia alata.*
guama hediondo (Cu): Jamaica dogwood.
guana: guanito: guano: *Coccothrinax* spp. (as *C. argentea* Sarg.) (yuraguana, etc.).
guanabana: *Annona muricata.*
guanacasti (Sp, CA): *Enterolobium cyclo-carpum*; national tree of CR; name also appl. to mountains, parks, &c.
guandu (Co): pigeon pea, *Cajanus indicus.*
guanidine: $NH=C(NH_2)_2$; iminourea; occurs in rice hulls, mussels, etc.; mfrd. fr. guanine, which commonly occurs in animal & pl. tissues & in animal excreta & esp. in guano (hence the name).
guanine: one of four purine bases found in DNA/RNA; widespread in pls. & animals; first found in guano.
guano: bird manure (orig.) specif. dr. excreta of sea bird (guano) collected mostly in dry wCh; source of much sodium nitrate, iodides, etc; some is semi-fossil; also appl. to droppings of other gregarious animals, as bats (NM, Ariz [US] caves; Borneo), fishes, etc.; now commonly appl. to any fertilizer (even artificial manure) comparable to the bird guano of SA - **Peruvian g.** = popular fertilizer.

guanosine: guanine riboside; occ. in guano (bird excreta) (hence name); yeasts (chief source), pancreas, *Vicia faba*, etc.; component of nucleic acids fr. which prepd.; one of four repeating units of DNA.

guanylic acid: guanosine phosphate; widely distr. in nature (3'- & 5'-) forms isolated fr. yeast ext.; 5"-form = precursor of RNA/DNA; di Na guanylate occurs in meat ext., sardines, instant beef boullion; u. as flavor intensifier (cf. Na glutamate).

guapi bark: Cocillana.

guapilla (Mex): *Hechtia glomerata.*

guar(a): *Cyamopsis tetragonoloba* - **guaracaro** (Ve): Lima bean.

guaraná (**g. bread; g. paste; pasta de guaraná**): *Paullinia cupana* (SC Guarani = Guarany = Tupi; collective name for tribes of Indians living over a large part of Br) (1435)- **guaranine**: caffeine (SC first found in guaraná).

guarapo de caña (Ve): acidulous alc. beverage made in sugar mills fr. fermented cane liquors.

Guardiola mexicana Bonpl. (Comp) (Mex): u. med. (Mex Phar. 1974) - **G. platyphylla** Gray (sAriz [US], nMex): antimicrobial with demonstrated activity (1993 publ.).

guaré: Cocillana, *Guarea rusbyi.*

Guarea bangii Rusby (Meli) (SA): bk. occurs as adult. of Cocillana - **G. gomma** Pulle (Co): lf. tea very astr. u. to check diarrh.; rts. said toxic (Tikunas) (2299) - **G. guidonia** (L) Sleumer (G. guara P. Wilson; G. trichilioides L.) (WI, Pan, SA): guaré (Br); yamao (Cu); yamagua (Cu); lvs. & gum-resin exudn. fr. trunk bk. u. med. emetic (Guyana); gum-resin u. emet. drastic cath.; lvs. u. (decoc.) hemostat. in wounds, abdom. dis., intest. hemorrhage, hemophilia, eczema; for "lunacy" (lvs. rubbed on body) (Ec); emm. abort. antiperiod. in menorrhagia; chief source of sandalwood oil prepn. u. in GC; in worm infestations; antineoplastic in cancer (folk med.) (1436); poison (arrows); fleshy aril of frts. edible - **G. macrophylla** Vahl (Co): bk. u. as purg. (2299) - **G. rusbyi** (Britt.) Rusby (Syncarpus r. Britt.) (Boe slopes of Andes): cocillana tree; guapi (Indian natives); cocillana (or guapi) bark; dr. bk. c. rusbyine, resins, tannin; u. exp. resp. stim. esp. for asthma, in cough syr. drops, lozenges, pastilles; emet. emm.; replacement for ipecac rt.; packed with ox skin cover (also in strips & in burlap bags) (sp. named after H. H. Rusby, pharmacognosist, discov. of bk.) - **G. thompsonii** Sprague & Hutch. (trop wAf): wd. is Mbossé mahogany; u. for furniture, exported; c. dihydrogedunin, methyl ango-lensate (138).

guarumo de mejico (Mex): *Cecropia peltata* & other spp.

guassatonga (Br): *Casearia sylvestris.*

Guatteria calva R. E. Fries (Annon.) (Col): st. bk. of tree u. as ingred. in curare - **G. duckeana** R. E. Fries (Co): warm decoc. of lvs. rubbed on for rheumatic pains; c. alk. (2299) - **G. slateri** Standley (Hond Pan): u. med.

guava: *Psidium* spp. - **pineapple g.: purple g.**: *Feijoa sellowiana.*

guavacine: simple pyridine alk. $C_6H_4NH\text{-}COOH$; fr. *Areca catechu*; claimed to serve as growth factor for *Staph. aureus* in place of nicotinic acid.

guavacoline: methyl ester of guavacine; occ. in *Areca catechu.*

guaxima (Br): *Urena lobata* (**g. roxa**) & other pl. spp.

guayaba: *Psidium* spp.- **g. morati** (Par etc.): *Patagonula* sp. - **guayabí** (Arg): *Patagonula americana* - **guayac**: *Dipteryx odorata* - **guayacán** (Sp): 1) guaiac 2) Cascara Amarga 3) (Co, Pan) *Tabebuia rosea* 4) (Arg) *Caesalpinia paraguariensis, C. melanocarpa* 5) (Hond) *Pithecellobium tortum* 6) (Salv) *Pithecellobium microsta-chyum* (**g. negro**) 7) other pl. such as *Aspidosperma dugandi, Tecoma chrysantha* (CR, Hond) 8) various other Am pls., such as *Bulnesia arborea*, pareira, *Myrospermum frutescens, Porlieria* spp., etc. - **guayacine**: simple pyridine alk. $C_6H_4N\text{-}HCOOH$; fr. *Areca catechu*; claimed to serve as growth factor for *Staph. aureus* in place of nicotinic acid - **guayaco** (Sp): guaiacum - **Guayana** = **Guiana** = region in nSA; includes parts of eVe, Guyana, Surinam, French Guiana, and nBr; bounded by Amazon and Orinoco Rivers & Atl Oc - **guayule** (Sp, Am): *Parthenium argentatum*; *Guaiacum* spp., & others; some u. as rubber source, esp. the first named.

guayule: *Parthenium argentatum.*

guayusa (tea): beverage produced fr. lvs. of several spp. of *Ilex* in Ec; drunk like maté.

guaza: dr. fl. tops of cannabis, u. as hallucinating drug.

Guazuma polybotrya Cav. (Sterculi) (Puebla, Mex): guacima; bk. u. as vulnerary - **G. tomentosa** HBK (CA, SA, cult. Ind, SriL, Java): guasima (de caballo); guacimo (Mex); bk. u. in Hansen's dis., herpes; as diaphor. ton. demul.; frt. edible, in bronchitis; sds. (cf. Cacao) u. in stomach dis.; tree produces gum; bast fibers u. in rope-making; wd. u. in constructional ways, charcoal; honey pl.; juice of trunk u. in enemas for dys. & in burns - **G. ulmifolia** Lam. (Mex, CA, SA): guaci-m(ill)a; bastard cedar; pigeon wood; tannin-rich foliage u. astr. (tannin drug); frt. eaten, esp. when food scarce; as tea in kidney dis. (Mex); bk. u. in malar. syph. elephantiasis, chest dis., asthma, leprosy, to promote hair growth, etc. (some authors consider this sp. name a synonym of *G. tomentosa*).

gueso, hoja de (Mex): *Buddleja americana.*

Guettarda angelica Mart. (Rubi) (Br): rt. u. in veter. med. as vulner. astr.; diarrh. of cattle & horses (Br) - **G. elliptica** Sw. (sFla [US], Mex, WI): arbol sabroso; formerly u. alexiteric in snake & spider bite & in dys. (Yuc) - **G. speciosa** L. (orig. Indo-Mal; Af): powd. bk. u. for ulcers, abscesses, & wounds; astr. (Indoch); bk. decoc. (with *Artocarpus* sp. rt. bk.) u. in chronic dys. (Indones); lvs. in attar (sInd) as scent; fls. (Tonga); frt. epicarp u. with *Scaevola* rt. in VD (Gilbert Is) (103).

gugal (Ind): **guggal**: mukal (gum).

gugglipid: fr. *Commiphora mukul.*

Guhr (Ge): Kieselguhr.

gui de chêne (Fr): American mistletoe.

guiabo (Por, Br): okra.

Guibourtia coleosperma J. Leonard (Leg) (tropAf): bastard teak; rt. decoc. u. in VD (Luchazi); sd. edible, source of edible oil; excellent wd. (Rhodesian mahogany) - **G. copallifera** Benn. = *Copaifera c.*

Guiera senegalensis Gmel. (l) (Combret.) (Seneg, wAf): "cayor"; "nger"; lf. inf. u. for diur. & purg. in diarrh. dys. colic; antipyr. in cough; smoked with tobacco for therapeutic purposes (139).

Guilielma C. Mart.: *Bactris.*

guimauve (Fr-f.): marshmallow, *Althaea officinalis.*

guimbombo (CR): *Hibiscus esculentus.*

guindas de las Indias (Sp): *Solanum pseudocapsicum.*

guindo: *Prunus cerasus.*

Guinea grains: *Aframomum melegueta*, sds. - **G. (water) melon**: small nearly worthless m. growing up spontaneously in a cultivated field.

guineo: banana - **guingambo**: okra.

Guioa cambodiana Pierre (Sapind.) (Indoch): rt. decoc. u. in GC - **G. koelreuteria** Merr. (PI): salab (Tagalog); oil fr. sds. u. to treat various skin dis. (140)

guiro cimarrón (Cu): gourd, *Lagenaria leucantha.*

guisante (Sp): 1) pea, *Pisum sativum* 2) *Doli-chos lablab* - **guisquil** (Guat): *Sechium edule.*

Guizotia abyssinica (L.f.) Cass. (Comp.) (tropAf, As; cult. Ind): (Ethiopian) niger seed; sd. u. bird sd., source of FO (30% yield); the oil is u. as adulter. of mustard & rape oils; oil press cake u. as fodder (niger cakes); pl. grown as green manure; frt. furnishes FO u. as food oil & in soap mfr. (1437) - **G. scabra** Chiov. (e tropAf): ikizimyamuriro (Burundi); rhiz. furnishes a febrifugal decoc.

gulancha (Hind, Beng): guluncha: gulwail (gul-wel) (Bombay State); *Tinospora cordifolia.*

gulf weed: *Sargassum* spp.

gum: 1) amorphous transparent or translucent subst. derived fr. pls. thru disintegration of internal pl. tissues & ordinarily exuding through the bk.; these subst. swell in water to form a gel (mucilage) or dissolve to form a colloidal dispersal (sol); consists of glycosidal acids combined with calcium, potassium, or magnesium; on hydrolysis are partly converted to sugars; simple test for gums: add 95% alc. dropwise to 1% aqueous soln. of subst., gum ppts. 3 classes: A) arabin type: fully water-dispersable to a sol; ex. acacia B) bassorin type: partly water-dispersible to a sol; mostly gel; ex. tragacanth C) cerasin type: swell to form a gel; ex. cherry gum 2) pl. exudation of almost any sort; may be alc.-soluble, water-soluble, etc.; ex. gum myrrh; g. turpentine; g. acacia 3) gum-tree (*qv.*) 4) (US slang) gum opium — NOTE: the following are definitions (usually as name of originating pl. sp.) for a few gums; many more may be found under the specific name of the gum; thus, Indian gum under "I," zapote gum under "Z," etc. (for most names hereafter, the word "gum" may either precede or follow specific term) - **acajou g.**: fr. *Anacardium occidentale* - **African g.**: *Acacia* spp. - **agar g.**: a. - **anglco g.** = *Piptadenia rigida* - **anime g.**: *Hymenaea curbaril* - **apricot g.**: fr. *Prunus armeniaca* - **arabic g.**: Acacia - **artificial g.**: 1) *Acacia decurrens* 2) dextrin - **asafetida g.**: *v.* asafetida - **Australian g.**: *Acacia pycnantha* - **babul g.**: mostly *Acacia arabica* - **Bangalore g.**: fr. *Prunus armeniaca* - **Barbary g.**: *Acacia gummifera* - **Bassora g.**: *Acacia leucophloea*, etc., etc. - **batu g.**: an E. I. damar. - **benjamin g.**: benzoin - **black g.**: *Nyssa sylvatica* & v. *biflora* (tree) - **blue g. tree**: *Eucalyptus globulus* (lvs.) - **Brazilian g.**: anime g. - **Brea g.**: *Caesalpinia praecox* - **British g.**: dextrin - **bubble g. (base)**: v. under pontianak - **bush g.**: *Eriodictyon californicum* (pl.) - **canoe g.**: fr. *Artocarpus heterophyllus* - **Cape g.**: *Acacia karroo* - **carana g.**: **caranna g.**: *Bursera simaruba* - **Carmania g.**: Bassora g. - **carob (seed) g.**: fr. *Ceratonia siliqua* - **cashawa g.: cashew g.**: *Anacardium occidentale* - **cebil g.**: *Piptadenia cebil* - **cedar g.**: fr. *Toona cedrela* - **cellulose g.**: sodium carboxy methyl cellulose - **chaqual g.**: fr. *Puya chilensis* - **cherry g.**: *Prunus* spp. esp. *P. cerasus* - **chewing g.**: a masticatory mixt. based formerly on chicle g. with sugars, syrupy glucose, flavoring oils (typically mint & wintergreen); other bases u.: spruce g. (1438), paraffin - **chicle g.**: fr. *Manilkara zapota* - **cistus g.**: *Cistus* spp. - **coconut g.**: prod. separating fr. st.

of *Cocos nucifera* - **corn sugar g.**: polysaccharide g. obt. by fermn. of dextrose with *Xanthomonas campestris* (Pammel) Dowson (Bacter.) - **creek g.**: *Eucalyptus camaldulensis* - **crude g.** (US): turpentine oleoresin - **cycas g.**: *Cycas circinalis* - **damson g.**: that exuding fr. trunk, branches, & frt. of damson (type of plum) tree - **domestic g.**: sodium alginate - **dragon g.**: tragacanth - **dream g.** (US underworld): Opium - **Egyptian g.**: acacia; type u. in mfr. wallpapers; talha g. - **elastic g.**: rubber; first appl. to that fr. *Ficus elastica* (by Goodyear); later appl. to that fr. *Hevea guianensis*, etc.; *Bumelia lanuginosa* - **elemi g.**: *Protium heptaphyllum*, *Dacryodes hexandra* - **eucalyptus g.**: red g. (1) - **feronia g.**: *Feronia limonia* - **galam g.**: Acacia - **gedda g.**: var. of acacia - **Ghatty g.**: *Anogeissus latifolia* - **G. Goulac** (Puritan Co, Me [US]): trade name for concd. sulfate pulp process waste u. as binder - **guacama g.**: fr. *Protium* spp. (Co) - **guajacum g. : g. guaiac(um)**: guaiac - **guar g.**: fr. *Cyamopsis tetragonoloba* - **g. guggal** (**g. gugul**): gum-resin fr. *Commiphora mukul* - **guta g.** (Sp): **goma guti: gutti g.**: gamboge - **Haiawa g.**: *Protium heptaphyllum* - **hashab g.** (Ar & Sudanese): acacia fr. *A. senegal* (Sudan) - **hemlock g.**: resinous exudn. fr. *Tsuga canadensis* - **Hercules g.**: impure starch (mixed with dextrin ?); u. as cavity paste in embalming - **hog g.**: 1) *Clusia flava* 2) *Sym-phonia globulifera* 3) *Rhus metopium* 4) tragacanth 5) *Prunus* spp. 6) other pl. products - **Hyawa g.**: Haiawa g. - **India(n) g.**: 1) *Anogeissus latifolia* (true) 2) karaya (not true) - **Iowa g.**: Haiawa g. - **Jaguar g.** (TM): guar - **jeol g.: Jidda g.**: acacia var. - **kadaya g.: karaya g.**: 1) *Cochlospermum gossypium* 2) *Sterculia urens* - **katera g.: kathira g.: katila g.**: karaya - **Kelzan g.** (TM Kelco): corn sugar g. - **Kordofan g.: Kordophan g.**: Acacia - **kuteera g.: kutera g.: kutira g.**: karaya (g.) - **leaf g.**: flake tragacanth - **g. leaves**: *Eriodictyon* - **Lecca g.**: olive g. - **loba g.**: Manila copal; prod. of *Agathis alba* - **locust g.**: *Hymenaea courbaril*, *H. davisii* - **locust bean g.**: carob g. - **mango g.**: *Mangifera indica* - **mesquit(e) g.**: exudn. fr. trunk of *Prosopis glandulosa*+ - **Mogadore g.: Morocco g.**: fr. *Acacia* sp., perhaps *A. gummifera* - **Mukal g.**: g. guggal - **Niger g.**: Nigerian Acacia (3 grades) - **olive g.**: fr. *Olea europaea* - **Orenburgh g.**: *Larix europaea* - **pectoral g.**: jujuba, marshmallow, & Iceland moss pastes - **pendare g.** (Ve): chicle g. prod. in Ve (*Manilkara zapota*) - **Persian g.**: Tragacanth marketed through such Iranian cities as Shiraz - **pine g.**: crude g. - **g. plant**: *Eriodictyon* - **plum g.**: fr. *Prunus domestica*, *P. cerasus* - **quick g.**: acacia - **red g.** (**tree**): 1) *Eucalyptus camaldulensis*, *E. calophyllum* 2) *Liquidambar styraciflua* 3) exudn. fr. *Copaifera copallifera* - **red cashoo g.**: red var. of cashew g. - **sap g.**: sweet g. - **sapodilla g.**: *Manilkara zapota* - **sassa g.**: a red gum fr. *Albizzia adianthifolia* - **Seneca g.: Senegal g.**: Acacia - **Shiraz g.**: of unknown origin imported fr. Iran; may be acacia - **Sonora g.**: mesquite g. - **spruce g.**: *Picea mariana* (& *P. rubens* to lesser extent) - **starch g.**: dextrin - **starleafed g.**: sweet g. - **sterculia g.**: karaya g. - **Sudanese g. arabic**: fr. *Acacia senegal* - **sweet g.**: *Liquidambar styraciflua* (**Oriental s. g.**: *L. orientalis*) - **Syrian g.**: flake tragacanth - **talh(i)**: **talha: talka g.**: *Acacia seyal* - **thus g.** (Am & Br Phar): crude oleoresin fr. *Pinus taeda*, *P. palustris*, etc.; a crude nearly solid Am turpentine, which spontaneously exudes & hardens on the branches of pine trees; coll. mostly NCar; may also be appl. to frankincense - **tupelo g.**: *Nyssa aquatica*, tree - **turic g.: Turkey g.**: var. of Acacia - **vegetable g.**: dextrin - **West African g.**: Acacia - **xanthan g.**: corn sugar g. - **York g.**: fr. *Eucalyptus loxophleba* (tree) - **zapotl g.**: chicle.

gum: v. under acacia; accroides; almeceiga, almecega; amrad; angico; benzoin; boe(a); camphor; Congo; copal; gamboge; ghatti; Jesire; kino; myrrh.; opium; Pontianak; Senaar; South African; sterculia; Suakim; tragacanth; turpentine.

gumati fiber: gomuti f.

gumbo (seUS): okra; appl. to pl., frt. & soup made therefrom - **g. filet** (**g. filé**) (La [US]): soup flavored with sassafras lf. - **g. limbo**: *Pistacia simaruba* - **g. musque**: *Hibiscus abelmoschus* - **g. soil**: silty soil of s. & wUS which when wetted becomes extremely adhesive (also simply "gumbo").

gumbrin: Fuller's Earth.

Gummi (L): gum; most of the gums are under specific name or in prec. section under "gum"; thus for Gummi Breae, see "brea gum" **G. Acaroides: G. Acaroidis**: acaroid resin fr. *Xanthorrhoea* spp. - **G. Africanum**: Acacia - **G. Albizziae**: sassa g. - **G. Ano-geissi**: Indian g. - **G. Arabicum**: Acacia - **G. Bassorae**: Karaya - **G. Breae**: Manila elemi - **G. Cedrelae**: cedar g. - **G. Elasticum**: Indian rubber - **G. Eucalypti**: eucalyptus kino - **G. Guttae** (**G. Gutti**): gamboge - **G. Indicum**: ghatti g. - **G. Kino**: Kino (resin) - **G. Kuterra**: sterculia g. - **G. Mangiferae**: mango g. - **G. Mimosae**: Acacia - **G. Piptadenia**: Cebil g. - **G. Plasticum**: gutta percha - **G. Prosopis**: mesquite g. - **G. Rubrum Gambiense**: Kino - **G. Scorpionis**: Acacia - **G. Sonorae**: mesquite gum - **G. Sterculiae**: karaya gum - **G. Thebaicum**: Acacia - **G. Thus**: olibanum gum-resin, frankincense.

Gummigutt(i) (Ge): gamboge - **gummi-resina** (L.): gum-resin.

gum-plant: 1) *Grindelia camporum, G. humilis, G. squarrosa* 2) *Syringa vulgaris* - **broad-leaved g.**: *G. squarrosa.*

gum-resin: class of chem. constituents of pl. consisting of a mixt. of a gum & a resin forming a natural emulsion when triturated with water; ex. Euphorbium, frankincense.

gums, water soluble: as distinct fr. alcohol-soluble g., those which form hydro-sols or hydro-gels with water; ex. acacia, tragacanth, karaya, locust, ghatti, talka (comm. name).

gum-tree: generally trees of such genera as *Eucalyptus, Nyssa* (seUS), *Liquidambar* (seUS), *Sapium*, etc., which yield a gummy exudation; specif., *Eucalyptus* - **g. of Jamaica**: *Sapium laurifolium* - **Panama g.**: *Castilla elastica.*

gumweed: Grindelia.

gum-wood: Eucalyptus (wd.) (first used 1676) (141) - **g. w. tree**: *Liquidambar styraciflua.*

gun cotton: pyroxylin.

Gundermann, echte (Ge): *Glechoma hederacea.*

Gundlachia corymbosa (Urban) Britt. (Comp.) (WI): horse bush (Bah); inf. in rum or vinegar for backache, colds, etc. (Bah).

gunjah (Hind): Cannabis Indica.

Gunnera chilensis Lam. (Gunner./Haloragid.) (Chile, Ec): nalca; perenn. hb. with huge lvs. (2 m across); rhiz. c. tannins; u. in uter. hemorrhages, dys. diarrh. in tanning (Palo Pangue); lvs. u. in fever (cataplasms); in angina, mouth, & throat dis.; young lf. stalks eaten as veget. - **G. macrophylla** Bl. (Indones): frtg. panicle u. as stim. ton. refrigerant in native med. - **G. manicata** Lindl. (sBr): very large hb. with gigantic pedate orbicular lvs.; cult. as lawn foliage pl. in many places (ex. Vancouver & Lulu islands, BC [Cana]) (2476) - **G. perpensa** L. (sAf): river pumpkin; rt. decoc. for dyspepsia (white farmers, sAf); in brandy for kidney dis. esp., gravel; rt. u. for impotency (Zulu); to aid labor; rt. & lf. in colds (212) - **G. petaloidea** Gaud. (Haw Is): apéapé; large lf. u. as umbrella.

gunny: 1) strong coarse jute sacking 2) bag of same (also **g. sack**) (*Corchorus*).

guppy: small freshwater fish (warm Americas), *Lebistes reticulatus* (Poecillidae - Pisces); feed chiefly on mosquito larvae & pupae, hence useful in controlling malaria; aquarium fish.

gur: ghur.

Gurania acuminata Cogn. (Cucurbit.) (Col): lvs. of vine u. to make tea u. as vermifuge (La Pedrera region) (2302) - **G. bignoniacea** C. Jeffrey (Co): lvs. & fls. crushed & appl. to infected cuts & slow healing sores (Makunas); lvs. & rts u. in tea u. as vermifuge (Tikunas); lvs. rubbed on fungal skin infections (Colonos) (2302) - **G. guentheri** Harms (Col): lf. inf. u. as strong vermif. (Kofans) (2302) - **G. insolita** Cogn. (Col): crushed fls. appl. as poultice (cold) to boils & infected sores (Tikunas) (2302) - **G. pachypoda** Harms (Col, Pe): crushed lvs. u. as poultice to relieve headache (Tikunas) (2302) - **G. rhizantha** C. Jeffrey (Col, Ec): rts. & woody sts. of vine u. in tea for irreg. menstr. (Tikunas); lf. ashes spread on skin sores (Kofans) (2302) - **G. spinulosa** (Poepp. & Engl.) Cogn. (Anguria s. Poepp. & Engl.) (Trin to Pe, Bo, Br, Col): lf. decoc. u. for constip. (Trin); rt. inf. u. in faulty menstr. (Col).

gurjun: tung oil - **g. balsam & g. oil** (FO): fr. *Dipterocarpus* spp.

Gurke (Ge): cucumber.

guru: kola.

Gutierrezia sarothrae Britton & Rusby (Xanthocephalum s. Shinners) (Comp) (wNA, Sask [Cana], to n&cMex): (broom) snake-weed; matchweed; hb. u. as emmen. & in gastric dis. (NM [US]); pulp of chewed pl. placed on stings fr. ants, bees, wasps; on wounds; in snakebite (Navajo Indians); tops u. as yellow dye.

gutta: 1) (L.) drop of any liquid 2) trans-1,4-polyisoprene (a transisomer of natural rubber) (term u. in SA).

gutta gamba: gamboge - **g. gelutong**: g. Pontianak - **g. percha**: dr. exudn. fr. *Palaquium gutta* (*qv.*), *Payena leerii, Isonandra gutta* - **g. p. family**: Sapotaceae - **pseudo g. p.**: exudn. fr. *Mimusops elata* - **Malabuwai g.p.**: fr. *Alstonia grandifolia* - **g. Pontianak**: dr. latex of *Dyera costulata* - **g. taban**: coagulated latex fr. *Palaquium obovatum.*

Gutti (L.): gamboge.

Guttiferae: Gamboge Fam.; mostly trop. shrubs & trees (few hbs.), with simple opposite, exstipulate lvs., always with oil glands or passages; often lvs. show translucent spots; many prod. valuable resins, or more commonly gum-resins; sds. often c. useful FO; a few spp. with VO, bitter subst. & tannins; ex. gamboge tree; *Mammea.*

Guy's pills: Pil. Digit., Squill, Mercury (NF V).

Gymnadenia conopsea (Willd.)) R. Br. (Habenaria c. [L] Benth.) (Orchid.) (eAs): scented orchid; u. in Chin med (Chin Phar. 1977).

Gymnema acuminatum Wall. (Asclepiad.) (Mal): lf. poultice appl. to small sores - **G. lactiferum** R. Br. (Mal, SriL, EI): milk juice u. like cow's milk; lf. eaten as veget. - **G. latifolium** Wall. (Ind): lvs. c. HCN glycoside; toxic - **G. sylvestre** R. Br. ex Schult. (Af, Austral, Ind, SriL): "Australian cow plant"; bk. u. med; rt. emet., u. int. & exter. for snakebite (Ind); lvs. c. gymnemic acid, masks taste, esp. of sweet & bitter things (paralyzes taste buds); chewed for greater effectiveness in diabetes (Ind); lvs. rubbed into small skin cuts, for "stitch in side" (Tanz); pd. root made into ointment for boils (Bechuanaland); rts. cooked in food for epilepsy (Tanz); tabs. of powd. lvs. u. for paraguesia (perverted taste sensations) (lit. refs [634]) - **G. syringifolium** Boerl. (Mal): rts. finely ground in water u. for tightness of chest, shortness of breath; lvs. eaten as veget. - **G. tingens** Spreng. (PI): rts. u. as emet. (decoc., or powd.).

Gymnocladus Lam. (Leg.) (As, Am): trees with stout twigless branches, hence name (gymnos [Gr] naked + klados [Gr] branch) - **G. chinensis**: Baill. (cChin): soap tree; pods of tree u. in laundry esp. for fine fabrics (saponin); sds. u. in perfumed soap; in hair wash products - **G. dioicus** (L) K. Koch (G. canadensis Lam.) (eCana, neUS): Kentucky coffee tree; fresh sweet pulp of frt. (pod) u. by Eu Homeop.; lvs. cathart., u. as fly poison; sds. c. saponin, sed., formerly u. as coffee substit.; wd. is good timber; ornam.

Gymnogongrus flabelliformis Harv. (Phyllophor. - Gigartinales - Rhodoph. - Algae) (Jap): thallus u. as food - **G. griffithsiae** Mart. (nAtl & Pac Oc): a perenn. seaweed; u. in mfr. phycocolloids (Far East); med. (Hung & Mex Ph.) - **G. pinnulata** Harv. (Jap & China, coasts): alga eaten; made into a paste u. as hair dressing, shampoo; source of funoran.

Gymnopetalum cochinchinense Kurz (Cucurbit.) (China, Mal): u. in prepns. for women in confinement; lf. u. to treat tetanus(!); juice fr. pounded lvs. u. in eye for ophthalmia - **G. leucostictum** Miq. (Indones): lf. decoc. u. as appetizer, esp. after severe illness; crushed lvs. u. in prep. for rheum. - **G. pedatum** Bl. (China): decoc. u. as gargle as germicide for teeth.

Gymnosperma corymbosum DC (Comp.) (Mex): zazal; u. in uter. hemorrhage - **G. glutinosum** Less. (Selloa glutinosa Spreng.) (swUS to CA): mota (Mex); naked-seed weed; broomweed; tatalencho (NM [US]); jarilla; hb. decoc. u. in diarrh.; metritis; soln. of gum exudn. u. in rheum., ulcers, &c.

Gymnospermae: Gymnosperms: sub-phylum (or sub-division) of Phylum (or Div.) Spermatophyta; sds. borne naked on scales instead of in seed-vessels (pericarps).

Gymnosporia acuminata Szysz. (Celastr.) (tropAf): silk bark; lf. & bk. furnish gutta percha-like subst.; useful wd. - **G. buxifolia** Szysz. (tropAf): staff tree; bk. inf. remedy for amebic dys. & diarrh.; in chest colds & cough; good wd. - **G. marcanii** Craib (Indochin): water fr. macerated rts. appl. to chaps or cracks in skin - **G. putterlickioides** Loes. ex Engl. (tropAf): rt. u. as antiemet. (Shambala) - **G. senegalensis** Loes. = *Maytenus s.* - **G. undata** (Tanz): rts. & st. bk. boiled & u. for fever (—> *Maytenus*).

Gymnostachyum febrifugum Benth. (Acanth.) (Ind): rt. u. as antipyr.

gynaecium: gynecium: gynoecium: female organs (carpel) of pl. (collectively); portion of fl. wh. develops into frt. enclosing sd. (<gyne, female); pistil.

Gynandropsis gynandra Briq. (G. pentaphylla DC; Cleome gynandra L.) (Capparid.) (s&seAs, tropAf): bastard mustard; lf. inf. u. to ease childbirth (also rt.) (Tanz); bruised lf. & lf. juice u. as rubef., counterirrit.; lf. cooked & eaten as veget. (PI) & u. as flavor; sds. have hot taste like mustard & u. as carm. & in kidney dis. (Indochin); sd. pod inserted into ear to extract wax (Af); pl. u. as fish poison; VO with properties of mustard or garlic oil (142); sd. also cont. FO, edible & suitable for soap mfr.; lvs. u. for bilious disorders (PI) (—> *Cleome*).

Gynerium sagittatum (Aubl.) Beauv. (1) (Gram) (tropAm): wild cane; uva grass; st. u. as laths in construction; arrow shafts; to make hats, baskets; young shoots to prep. shampoo for hair (stim. growth); mashed st. piths appl. to fungal skin dis. (2444); ornam. (US).

Gynergen™: ergotamine tartrate; u. in migraine & as oxytocic.

Gynocardia odorata R. Br. (monotypic) (Flacourti) (nInd): chaulmoogra (Hind, Beng, Bomb); sd. FO u. in leprosy & other skin dis. ("false chaulmoogra oil"); frts. u. as fish poison.

gynocardic acid = oleic acid; formerly considered active principle of gynocardia oil.

Gynotroches lanceolata Bl. (G. axillaris Bl.) (monotypic) (Rhizophor.) (seAs, Oceania): lvs. appl. to ulcers (PI); lvs. appl. to head for headache & fever (Mal).

Gynura divaricata DC (G. ovalis DC) (Comp.) (Chin): u. to combat craving for opium ("o. habit"); u. as styp. & vulner. - **G. japonica** (Thunb.) Juel (G. pinnatifida DC) (eAs both wild & cult): fresh rt. u. as astr. for wounds & hemorrhages; juice u. for insect stings; also for scorpion & centipede (Chin); rt. cooked with chicken as ton. (Indochin) - **G. procumbens** (Lour.) Merr. (G. sarmentosa DC) (seAs): st. & lvs. antipyr. in eruptive fevers (Indochin); dr. lvs. rubbed with oil u. as salve for rash; in kidney dis. (Indones); lvs. as condim. (Mal) - **G. pseudochina** (L) DC (Ind, Chin, Thail): pl. u. as poultice for breast tumors, in erysipelas; sap of lvs. u. as gargle in throat inflamm.; rt. in dys. & inflamed wounds; tuber decoc. for fevers (Indochin).

Gypsophila L. (Caryophyll.) (Euras, nAf): chalk plants; ann./perenn. hbs. (SC fond of gypsum based on habitat) - **G. arrostii** Guss. (sIt, Sic): white soap root; rt. (as Radix Saponariae albae) c. much saponin & sapotoxin; u. diaph. diur. exp. in skin dis., icterus; prep. saponins - **G. oldhamiana** Miq. (eAs): rt. stocks u. to treat typhoid fever, rheum. lung dis. jaundice (Chin) - **G. paniculata** L. (Euras, natd. sUS: Ore, Wash): baby's breath; soap wort; rt. (as Radix Saponariae levanticae, etc.) similar to *G. arrostii*; u. detergent, spermicide - **G. perfoliata** L. (eAs): u. much like *G. oldhamiana* - **G. struthium** Loefl. (seAs [cult. Sp, Med reg.]): Levant soap root; soap wort; rt. (as Radix Saponariae hispanicae, etc.) u. much as *G. paniculata* with which sometimes confused; u. for certain tumors (folk med.); not now commonly seen; detergent.

gypsum: native calcium sulfate; plaster of Paris.

gyptol: sex attractant (pheromone) borne by female gypsy moth, *Porthatria dispar*; long-chain fatty acid esters with 2 or more unsatd. sec. alcs.

Gyrinops cumingiana Decne (Thymelae.) (Aquilaria c. Ridl.) (PI): bk. & rts. u. as hemostat.; bk. wd. frt. u. as quinine substit. - **G. vidalii** P. H. Hô (Laos): fibrous bk. u. to make bindings; in local med. mixts.

Gyrocarpus americanus Jacq. (Hernandi) (pantrop.): lvs. & bk. u. med (Tonga) (86); bk. ext. in cold water drunk for body aches (Hond).

Gyromitra esculenta (Pers. ex Fr.) Fr. (Helvell. - Pezizales - Ascomy. - Fungi): (Temp. zone worldwide; often under *Pinus*): (false) morel; "brain mushroom"; edible "mushroom"; however c. helvellic acid (non-toxic), gyromitrin (N-Me-N-formyl hydrazine) (prodn. benign & malignant tumors in mice); fatalities have been reported fr. consumption (61) hence bot. name is a misnomer and this fungus is often banned fr. markets; popular in parts of Eu as food when dried for winter use (innocuous when dried for 2 wks.); said safe after boiling & discarding the cooking water.

Gyrophora cylindrica (L) Ach. (Gyrophor. - Ascolichenes - Lichenes) (all continents): cylinder gyrophora or rock tripe; found growing on rocks; u. in kidney dis.; green-brown dye u. for woolens - **G. deusta** (L) Ach. (Temp. & Subarctic zones; on rocks): scorched rock tripe; u. like prec.; in nephralgia (1440); also u. in paints (Sweden) - **G. muehlenbergia** Ach. [Actinogyra m. (Ach.) Schol.] (NA, SA, nEu): source of a gel somewhat like that of *Chondrus crispus*; c. bitter purg. subst. removable by treating lichen with alkaline water, after which esculent; emergency food during wartime - **G. pustulata** (L) Ach. = *Lasallia p.*

Hh

"H" (Por slang): Heroin.

haabi: haabin (Mex): Jamaica dogwood, *Piscidia erythrina*

Haar (Ge, n): hair; trichome.

Haarlem.oil: old Dutch proprietary prep. c. linseed oil, bals. sulfur, turpentine oil, etc.; u. diur., in kidney dis. & skin dis.; vial, capsules (1441).

Haarstrang (Ge): **Haarwurzel** (Ge): *Peucedanum* spp.

haba (Sp): 1) broad bean, *Vicia faba* 2) (PR) Lima bean - **h. tonka**: tonka bean - **habalongo**: *Physostigma*.

habanera (Mex): good grade of rum prod. in Mex.

habb al han (Arg): cardamom.

Habenaria bifolia R. Br. (Orchid.) (Euras): butterfly orchis; hand-shaped under portions of tuber u. (doctrine of signatures) in nerve dis. fever, etc.; believed to be "Radix Satyrii" of Ancients - **H. commelinifolia** Wall. (Ind-Himal): rt. tuber u. as source of salep. - **H. dentata** (Sw.) Schltr. (Chin): rt. in broth u. as remedy in colic (Chin); rt. pounded & appl. to swellings & wounds (Taiwan) - **H. foliosa** Reichb. f. (tropAf): rt. inf. u. as emetic (Zulus).

habi (Mex): haabi.

habichuela (Sp): common bean, *Phaseolus vulgaris* - **h. mungo** (PR): mung b. - **h. soya**: soya b.

habilla (PR): yam bean.

habillo (Mex): *Hura crepitans*.

habitat: 1) (ecology): kind of environment in which a certain organism, pl. or animal habitually lives; ex., swamp, pine barrens, seaside 2) (as often understood in pharmacognosy): geographical area in which pl. or animal is found living originally or to which indigenous; ex. Calif (US) (for yerba santa); wUS (for Cascara).

habit(us): general appearance of pl. & manner of growth in the field, seen as a whole; often difficult to describe in words, better demonstrated with specimens, pictures, models, etc.; determined by such factors as color, size, branching, leaf arrangement, etc.

Hachettene: mineral tallow (1442).

hackberry: *Celtis* spp.

Hackelia virginiana (L) I. M. Johnston (Lappula v. Greene) (Boragin.) (eCana & US): stick weed; beggar's lice; burseed; rt. astr. demul.; rt. decoc. u. for the itch; to "improve memory"; in "cancer" (Cherokee Indians); pl. decoc. in kidney dis.

Hackelochloa granularis Ktze. (Rytilix g. Skeels) (Gram.) (tropics, Ind+; natd. sUS): pl. u. for enlarged spleen & liver (in olive oil; India); forage in swUS.

Hackfrucht (Ge): truck crop (lit. fruit of the hoe).

hackmatack (US slang): tacmahac.

hadacidin: subst. fr. fermn. fluid of *Penicillium frequentans* Westling; u. for adenocarcinoma (human) (inhibits adenylosuccinate synthetase activity).

hadrome: conducting portions of the xylem of stem or rt. (vessels, tracheids, xylem parenchyma).

hae-: v. also **he-**.

Haemanthus coccineus L. (Amaryllid.) (trop sAf): April fool; blood l. bulb (in vinegar) u. as diur. in edema; expect. asthma remedy; squill substit.; lf. appl. to ulcers & anthrax pustules; ornam. perenn. hb. - **H. multiflorus** Martyn (tropAf): African tulip; juice of bulb appl. to groin of cow to hasten parturition (Chagas); c. lycorine, haemanthidine,

& other alk. - **H. natalensis** Pappe ex Hook. (tropAf): blood flower; bulb decoc. u. as expect. in cough & emet.; dangerously toxic; c. natalensine (haemanthamine), coccinine, montanine, manthine, u. sometimes as homicidal poison.

Haematomma coccineum Koerb. (Lecanora h. Ach.) (Lecanor. - Lichen.) (Eu, NA): blood(y) spot lichen; u. to furnish red, purple, orange dyes - **H. ventosum** Mass. (Eu, NA): bloody spotted (or black) lecanora; u. to dye woolens red-brown (Swed).

Haematostaphis barteri Hook. f. (Anacardi.) (Nig): blood plum of Nupe; frt. edible; bk. c. gum, u. as purge & emet. in trypanosomiasis.

Haematoxylon (Haematoxylum) brasiletto Karst. (Leg) (Mex, CA, Co, Ve): brazilette; peachwood; palo de brazil; Brazil blood wood tree; wd. c. brazilein (dyestuff), u. red dye; wd. has antibacterial properties (89); in many parts of Mex wd. splinters added to drinking water & the water then u. as ton. & against infections - **H. campechianum** L. (Mex, CA, WI): logwood; campeachy (wood); wd. c. hematoxylin (12%); tannins; u. (as Haematoxylon) astr. antisep; purple dye & ink, histolog. stain, reagent (fermented wd. must not be u. in med.) (2646).

Haementeria (Hirudinea - Annelida): leeches, often of large size (giant l., Amazonia; Borneo) to 0.5 m long; with very large salivary glands secreting hementin (anti-thrombin activity; dissolving blood clots), orgelase (degrades hyaluronic acid) - **H. officinalis** (Mex): Glossi Phoniidae - **H. costata** (sRussia); similarly u.

Haemodorum coccineum Hook. (Haemodoraceae) (Austral): blood root; rts. u. by natives as aper. - **H. corymbosum** Vahl (Austral): blood root; u. like prec.; rts. eaten when roasted; stock poison - **H. spicatum** R. Br. (Austral): u. as dys. cure by aborigines.

h(a)emogallol: hemoglobin which has been deoxidized by pyrogallol; reduction prod. of hematin; thought to be more readily assimilable in regeneration of blood.

Haemophilus (*Hemophilus*) **influenzae** (Lehmann & Neumann) Winslow & al. (Pasteurell. - Bacteria): normal to resp. tract; formerly thought cause of pandemic influenza (now known to be a viral infection); is associated with bacterial pneumonias; bacterial meningitis (certain types) (type B) (Hib) in infants; Hemophilus b conjugate vaccine (meningococcal protein conjugate) u. for protection against meningitis.

Hafer (Ge): oats.

Hagebutte (Ge): hip, haw (wild rose).

Hagenia abyssinica J. F. Gmel. (Ros) (neAf, esp. Ethiopia): kousso tree; the dr. panicles of pistillate fls. called kousso or brayera; active toxic principle; protokosin (1443); kosotoxin (phloroglucin deriv.); kosin; 24% tannins; u. tenif. & anthelm. (as inf. or extr.) (2647).

Hahnenfuss (Ge): buttercup, *Ranunculus repens* & other *R.* spp. - **scharf H.**: *R. sceleratus* - **knolliger H.**: *R. bulbosus*.

hai thao (Chin): thao.

haiari(a) (root) (Guyana): *Lonchocarpus* spp. or possibly *Conomorpha magnifolia* - **white h.**: *L. densiflorus*.

hail (Arab): herb added to coffee (Bedouin) (*Scirpus*).

hair: 1) long, flexible, very thin appendage of skin of many animal spp. 2) outgrowth of pl. epidermal cell; better referred to as trichome; ex. cotton - **h. ball**: bezoar; mass in intest. made up mostly of animal hairs or pl. trichomes - **human h.**: much u. for wigs, toupees, etc.; exports through Marseille, France - **starred h.**: bundle trichome as in Verbascum lf.

hairbell: harebell - **haircap moss**: *Polytrichum juniperinum* - **hairstrong**: *Peucedanum officinale* - **sweet h.**: *P. palustre*.

hake: *Merluccius merluccius* - **h. sounds**: fish s.; u. in prepn. American isinglass.

hakim (Indo-Pak): medical man who sells drugs & operates a pharmacy, following the Unani or Ayurvedic system of native medicine.

hak-sen (Jap): *Dictamnus albus*.

hala (Haw): *Pandanus* spp.

haldu (Urdu, Hindi): *Adina cordifolia*.

Halenia corniculata (L) Cornaz (H. sibirica Borkh.) (Gentian.) (Sib, Mong, Manch): hb. u. as bitter. - **H. weddelliana** Gilg (H. antigonorrhoic Gilg) (sCo, Ec, Pe): cacho de venado; u. to treat VD (GC).

Halesia tetraptera Ellis (Styrac) (e&cUS): silverbell (tree); chittam wood; shittimwood (WVa [US]); sd. edible; ornam. tree (14-20 m high) (pron. hay-lee'-si[as in ill]-ah).

halfah (grass): *Stipa tenacissima*.

half-life: time needed for half of the amt. of drug introduced into an organism to undergo metabolization, excretion, etc.

halibut (liver oil): *Hippoglossus hippoglossus*.

Halidrys siliquosa (L) Lyngb. (Fuc. - Fucales - Phaeophyc. - Algae) (nAtl Oc mostly off Eu coasts): sea-oak; Schotentang; pl. thallus c. algin (16-17%); FO (ca. 2.2%) (as extd. with ether); u. as source of Fucus; drug; has antibiotic activity (143) (occurs in low littoral areas).

Halimium glomeratum Grosser (Cist.) (Mex to CR): Juanita; decoc. of subshrub u. for indig. & diarrh.; dr. plants sold in market.

Halimodendron halodendron Voss (H. argenteum Fisch.) (1) (Leg) (salt steppes of n&cAs, USSR): tschingil; salt tree; lvs. c. saponins; u. in erosion control; ornam. shrub; lvs. u. as purg. (*cf. Colutea nepalensis* Sims) (1444).

halite: rock salt, NaCl.

Haliver™: halibut liver oil, rich in vit. A & D.

hall (Ala [US]): frt. of haw, *Crataegus* spp. (**red halls, yellow h.**).

Halleria lucida L. (Scrophulari) (sAf): white olive; African honeysuckle; lf. juice u. in earache; regarded by Af natives as magic plant; frt. edible; cult. ornam.

hallucinogens: psycho(to)mimetics: use mostly learned from lore of native peoples, chiefly of SA; include both fungi and dicots.

Halogeton glomeratus C. A. Mey. (Chenopodi) (USSR; now a bad weed in NA): hb. c. oxalic acid; causes heavy fatalities among sheep (& some cattle) of some Rocky Mt. states (US, esp. Ida, Nev), where recently introduced - **H. sativus** (L) Mog. (wMed reg.): barilla; eaten as veg. & formerly burned for soda (Fr & Sp pharm.) (also called barrilla).

Halophila ovalis Hook. (Hydrocharit.) (Fiji): u. med. (Fiji) (144).

halophyte: pl. which grows in salt soil, saline marshes, etc.

Haloragis erecta Eichl. (Haloragid.) (SA, NZ): juice or inf. of pl. u. to treat scrofula (NZ).

Haloxylon multiflorum Bunge (Chenopodi) (Indo-Pak): lana (Punjabi); pl. source of a drug (145).

halva(h) (Turk Yiddish): confection representing paste of sesame sds., nuts, & honey (binder) (Arab halwa; halevah).

Halymenia (Cryptonemi - Cryptonemiales - Rhodoph. - Algae) (warm seas esp. of sHemisph): (common) dulse; some 30 spp.; u. in mfr. Russian agar; good source of carrageenan - **H. formosana** Harv. (Pacif Oc): thallus u. for food (PI) - **H. palmata** Ag.: dulse; a deep red seaweed; c. mannite, bromine, iodine; u. food, med.

Hamamelidaceae: Witch hazel Fam.; shrubs or trees (subtrop & temp zones); lvs. simple alternate; frt. a woody capsule; chem. constit. poorly known (tannins, etc.); incl. *Liquidambar*, *Fothergilla* - **Hamamelidis Folium** (L): witch hazel leaf.

Hamamelis virginiana L. (Hamamelid.) (e&cNA): (common) witch hazel (leaf, bk.); bk. & lvs. c. tannin, VO (trace), bitter princ.; u. astr. ton. hemostat. in hemorrhoids, hemorrhages, bruises, varicose veins, diarrh. vulner. in skin dis.; "extract" of twigs (steam distillate) (witch-hazel water) u. cosmetic & eye lotion; branch u. in water witching (US) (hence "witch" hazel).

hamamelis sugar: hamamelose: simple sugar (hydroxy-methyl-ribose) which is split out of hamamelis-tannin; a branched hexose.

Hamatocactus hamatocanthus Knuth (Cact.) (swUS, nMex): frts. u. in cooking as substit. for lemons (Nuevo León) (—> *Clacistrocactus* & *Ferocactus*).

Hamburg drops: kind of Swedish bitters - **H. tea**: like St. Germain tea but differently flavored.

Hamelia axillaris Sw. (Rubi.) (Mex to Br, Bol): decoc. fr. grated rt. taken for diarrh. & stomachache (777) - **H. nodosa** Mart. & Gal. (sMex, CA): u. for tanning hides (lvs. & bk.); dr. lvs. u. as tobacco (Hond); boiled lvs. u. for mange (dogs) (Pan); u. to wash infected wounds, scabies (Mart) - **H. patens** Jacq. (H. erecta Jacq.) (sFla [US], WI, CA, SA): firebush; chichi-pin(ce) (Hond, Guat); zorrillo (real) (CR); coral(illo) (Salv); crushed lvs. appl. to cuts, blisters, bruises, sores, scabies, eczema; lf. inf. u. to wash infected wounds (Pe); frt. edible, u. to make beverage, syrup u. in dys.; lvs. & sts. tan. (shrub to small tree).

Hamiltonia suaveolens Roxb. (Rubi) (Ind): rt. u. in courbature (muscle ache) (—> *Spermadictyon*).

Hamycin: polyene antibiotic complex formed by *Streptomyces pimprina* and *Actinoplanes caeruleus*; antifungal activity; 1961-; antitrichomonal, antiinflamm.; u. topically for fungal infect.; ex. tinea (Ind).

Hancornia speciosa B. A. Gomes (monotypic) (Apocyn) (Br): mangabeira rubber is prepd. fr. trunk latex; frts. edible u. to make marmalade.

hand: voodoo charm or amulet; often composed of small hand-like cluster of rts. or tubers, esp. one wrapped in red flannel (Am Negro, seUS) (1520).

Hanf (Ge): Cannabis.

han-ge (Chin): tubers of *Pinellia tuberifera*.

hanks: coils (or skeins) of yarn (*qv.*) esp. of given length (cotton: 7 hanks = 840 yds.).

Hannoa klaineana Pierre & Engl. (Simaroub.) (trop wAf): bk. decoc. u. in colic; wd. useful - **H. undulata** Planch. (wAf, esp. Mali): lf. appl. to bruises; bk. antipyr.; frt. for pediculosis capitis (—> *Quasia*).

Hansen's disease: preferred term for leprosy (named after G. H. A. Hansen).

Haplanthus tentaculatus Nees (Acanth.) (Ind): pl. u. as antipyr. - **H. verticillatus** Nees (Ind): u. similarly to prec.

Haplocarpha scaposa Harv. (Comp.) (Af): u. in later stages of mal.

haploid number: no. of chromosomes in gametes of pl. or animal, designated *n*.

Haplopappus angustifolius Reiche (Comp) (Ch): resin u. as stim. in indigest. - **H. baylahuen** Rémy (Chile, foothills of Andes): bailahuén; hb. u. in digest. dis., in liver & bile dis., lax. emmen. astr. - **H. heterophyllus** (Gray) Blake (swUS): rayless goldenrod; Jimmyweed; toxic to livestock; c. tremeton & toxol; early u. as remedy for animal dis. known as "trembles" or "milk sickness" - **H. multifolius** (Ok.) Reiche (Ch): hb. c. haplopine, haplopinol; u. to stim. stomach in indigest. & dig. debility; for liver; emmen., antisep. - **H. spinulosus** DC = Machaeranthera pinnatifida - **H. venetus** (HBK) Blake (Mex, sUS): (falsa) damiana; decoc. of shrub u. as bath in rheum.; confused with & often sold as substit. for Damiana.

Haplophyllum tuberculatum (Forssk.) A. Juss. (Rut.) (Pak, Saudi Arab, Eg): pl. u. as emmen. abort.; in rheum; fresh juice to increase hair growth.

Haplophyton cimicidium A DC (Apocyn.) (sAriz, Tex [US], nwMex, CA, Cu [Sagra]): roach herb (or plant); yerba de (la) cucaracha (cockroach plant); c. alk. (haplophytine, cimicidine, etc.) which act as both contact & stomach poisons for cockroaches; u. insecticide (146) (dr. lvs. mixed with molasses) - **H. crooksii** L. (Benson): cockroach plant; similar to preceding.

hapten: specific non-protein (often a synth. compd.) so formed that it can interact with specific combining gps. of an antibody; useful in detn. of which structural properties of antibody mols. enable them to combine with & defuse infectious subst. (antigen); incomplete antigen incapable of producing antibodies; may sometimes produce antibodies if linked to a protein ("half antigen").

hard pan: dense compacted soil stratum, often with clays mixed with sand or pebbles; may be formed by always plowing to same depth.

Hardenbergia monophylla (Vent.) Benth. (v. *violacea* Stearn) (Leg.) (Austral): false sarsaparilla; rt. u. as sarsap. substit., although ineffective; cult. ornam.

hardhack (leaves): *Spiraea tomentosa* - **hardheads**: *Centaurea nigra*.

Hardwickia mannii Oliv. (Leg) (wAf): source of (West) African copaiba (Balsamum Copaivae africanum); c. 44-45% VO, illurinic acid, caryophyllene; u. med. & techn. like copaiba but less commonly - **H. binata** Roxb. (tropAs, as Ind): source of Hardwickia balsam, similar to copaiba; c. VO, resins, hardwickia acid; u. as such, esp. in GC; u. to replace copaiba in shellac & porcelain industries (aids preservation); wd. heavy & useful; lvs. fodder; bk. tan; fiber.

harebell (, true): 1) *Campanula rotundifolia* 2) *Muscari* spp. (probably) - **hareburr**: Lappa - **hare's ear (herb)**: 1) *Bupleurum rotundifolium* & *B. fruticosum* 2) *Corringia austriaca, C. orientalis* - **h. grease**: FO fr. hare, *Lepus timidus* L.; u. folk med. as emoll.

haricot (bean) (Fr): *Phaseolus vulgaris*, common garden bean; also at times appl. to unripe pod (string b.) or sd.

harina (Sp): flour, meal - **h. de huesos**: bone meal - **h. de pescados**: fish meal - **h. de trigo**: wheat flour.

Harlem: v. Haarlem.

harmal: sds. of *Peganum harmala* - **harman**: indole alk. fr. *Simira* (Sickingia) *rubra* & *Symplocos racemosa* - **harmine gp. of indole alkaloids**: fr. *Peganum harmala*; incl. **harmine**, harmaline, harman, etc.; u. antithelm., hallucinogenic, quinidine-like action on heart (same as banisterine, yageine).

Harnkraut (Ge): *Reseda luteola*.

Haronga = *Harungana*.

Harpagophytum procumbens DC ex Meissner (1) (Pedali) (sAf): devil's claw, grapple plant, wool spider; perenn. hb.; secondary storage rts. (tubers) c. harpagoside, harpagide, procumbide (iridoid glycosides); u. (inf.) as antipyr.; decoc. in rheum., antiinflamm. oxytocic (Eu, Af); rt. u. to relieve pain of childbirth; ointment in ulcers, sores; FO purg.; value has been questioned; widely sold (US, Eu) as herbal tea in arthritis, diabetes, allergies (147).

Harpephyllum caffrum Bernh. ex K. Krause (1) (Anacardi) (s&eAs): Kaffir plum; Cape ash; bk. decoc. u. as emet. & to "purify the blood"; frts. arom. & acidulous, eaten; pulp made into jellies; pale reddish wd. u. industr.

Harpullia arborea Radlk. (Sapind.) (PI+): bk. & frt. u. to prevent leech bites; fish poison; sd. FO considered antirheum; bk. inf. drunk to allay pain (Solomon Is).

har(rar) (Hindi): *Terminalia chebula*.

Harrisia Britton (Cact.) (Fla [US]); WI, SA): "prickly apples"; frts. eaten; cult.; some spp. c. caffeine (ex. **H. adscendens** Britt. & Rose [Bahía, Br.]: slender-stemmed cactus) - **H. martinii** (Labouret) Britt. (Arg): declared noxious weed (sAf) (genus sometimes referred to as *Cereus* (2057).

Harrisonia abyssinica Oliv. (Simaroub.) (trop eAf): rts. u. as ingred. of remedy for bubonic plague; oxyuricide, ascaricide; lf. & twigs in hemorrhoids; lvs. in ghee poultice for boils (Tanz); carbuncles, abscesses; rt. decoc. with flour for stomachache (Tanz) - **H. perforata** Merr. (seAs): shoots at base (Thail), rt. bk. decoc. (PI), u. in diarrh. amebic dys.; in cholera; mal. (Chinese folk med.); lf. decoc. for skin dis. (Hainan).

Hartogiella capensis L. f. (l) (Celastr.) (sAf): ladlewood; lf. chewed to quench thirst; prevents fatigue; reduces appetite for food; wd. useful.

Hart Rotalge (Ge): agar.

Hartriegel (Ge): *Ligustrum vulgare.*

hartshorn: hart's horn: orig. prepd. fr. stag (*Cervus elaphus* L.) horn by incineration after boiling to remove gelatin; repr. nearly pure Ca phosphate, but modern meanings are 1) ammonia water 2) (in some localities) ammonium carbonate - **h. liniment**: Linim. ammonia - **salts of h.**: ammon. carbonate - **h. plant**: *Anemone patens* - **prepared h.**: made directly fr. stag's horn - **h. shavings**: parings fr. hooves of a deer, *Cervus elaphus* L.

hart's root: *Peucedanum cervaria* - **h. suet**: suet fr. deer; u. folk med. - **h. thorn**: *Rhamnus cathartica* - **h. tongue (fern)**: *Asplenium scolopendrium* - **h. truffle**: *Elaphomyces granulatus.*

hartwort: appl. to numerous pl. spp. incl. *Mentha cardiaca, Melilotus officinalis*, spp. of *Tordylium, Laserpitium, Levisticum.*

Harungana (*Haronga; Arungana*) **madagascariensis** Lam. ex Poir. (1) (A. paniculata Lodd.) (Guttifer.) (tropAf): "ogina"; uturu; bk. & lf. exts. u. in digest. disturb., pancreopathy, hepatic & gb. dis., dyspeps., meteorism, tapeworms; ringworm (with orange juice) (Lib); bk. u. for urinary fistulas, appl. as fumigant (Sen); dye; useful wd; antifertility agt.; abortif.; lf. antibact., antihepatotoxic.

Harveya speciosa Bernh. (Scrophulari) (sAf): u. as remedy for insanity (sSotho).

Harz (Ge): resin.

Haselnuss (Ge-f.): hazelnut, *Corylus* spp.

Haselwurz (Ge): *Asarum europaeum.*

hash (US slang): hasheesh.

hashab: Acacia (Kordofan).

hashberry: Haight-Ashbury area of San Francisco, Calif (US) (fr. hash [hashish] + Ashbury); area where the hippies met during the 1960s.

hasheesh: hashish: *Cannabis sativa* subsp. *indica* prepn. (2289) - **h. oil**: brown liquid concentrate of Cannabis (rich in THC) appl. to cigarettes to give hallucinogenic effect; u. like knockout drops in Near East (placed in sweet meats); much fr. Nepal, Pak ("hash oil").

hastacine: pyrrolizidine type alk. fr. *Senecio sagittatum* (Cacalia hastata).

hastula: terminal projecting part of petiole in palms (= ligule; flagellum) (see illustration).

hasty pudding: porridge of flour or oatmeal in water or milk; or corn mush (boiled corn meal).

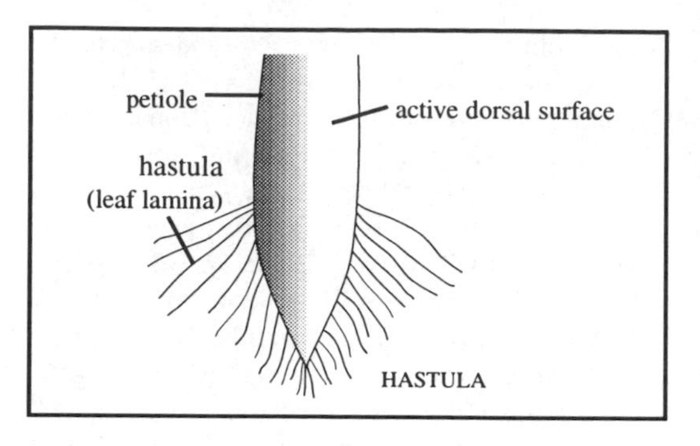

hasubanans: alk. with novel modified morphinan skeleton, having 5-membered heterocyclic ring skeleton; found in *Stephania japonica* (Jap: hasuno-ha kazura, hence name); ex. hasubanonine.

Hauhechel (Ge): *Ononis spinosa, O. repens*

haulm (halm): st. as of cereal grasses, potato plants, beans, etc.

Hausenblase (Ge): isinglass (lit. sturgeon bladder).

Hauya speciosa Donn. Sm. (Onagr.) (Hond, CR): supara (CR); esquisas (cabo de hacha) (Hond); u. med. (no details).

haw (abbreviation of mayhaw): frt. of 1) *Crataegus oxyacantha* & other *C.* spp. frt. 2) *Rosa canina* & other *R.* spp. - **apple h.**: may h. - **black h.**: *Viburnum prunifolium* & *V. rufidulum* (**Southern b. h.**) - **may h.**: *Crataegus aestivalis*+ - **mush h.**: *C. mollis* - **parsley h.**: *Viburnum apiifolia* - **pear h.**: various *C.* spp. with pyriform frts.; ex. *C. tomentosa* - **pond h.**: *Crataegus* spp. of seUS - **possum h.**: 1) *Viburnum nudum* 2) (less commonly) *Ilex decidua* - **red h.**: *Crataegus*; specif. *C. calpodendron* & *C. phaenopyrum* - **shonny h.**: *Viburnum nudum, V. cassinoides* - **sugar h.**: *C. spathulata* Mx (NA).

hawk bit: *Leontodon* spp.

hawk's beard: *Crepis* spp. - **h. weed**: *Hieracium* spp. (3 genera with name "hawk" are close together in fam. Comp.).

"hawk's vomit" (seUS Negro, slang): Nux Vomica.

hawthorn(e): *Crataegus intricata* & other *C.* spp. - **h. berries: h. blossoms**: fr. *C. oxyacantha* (chiefly) - **English h.: white h.**: *C. oxyacantha.*

hay: 1) grasses (etc.) mowed & dried for u. as fodder; pl. may sometimes be immature (ex. wheaten h.) 2) (US slang) Cannabis - **camel's h.**: *Cymbopogon schoenanthus* - **c. h. straw**: *Andropogon laniger* - **h. flowers**: v. Heublumen; mostly *Trifolium pratense* fls. - **h. grass**: *Andropogon caricosus* - **h. head** (US slang): Cannabis smoker - **h. process**: straw process (*qv.*).

haya (Sp): *Fagus silvatica.*

hayfever plants: mostly spp. which are anemophilous (wind-pollinated); 3 hayfever seasons of nHemisph: tree s. (spring) (incl. *Juglans*, *Acer*, Fag., Ole., Mor); grass s. (summer) (ex. *Poa* spp., *Gram.*); weed s. (fall) (ex. ragweeds [**h. weed**] & other Comp., Chenopodi, Polygon.); other important fam. in the etiology of pollinosis incl. Pin., Plantagin, Betul, Jugland, v. also pollinosis.

hayo: coca lvs.

hazel alder: *Alnus rugosa* - **h. wort**: *Asarum europaeum* - **h. (nut), snapping**: witch hazel - **hazelnut, American**: *Corylus americana* - **beaked h.**: *Corylus cornuta* - **European h.**: *c. avellana.*

Hazunta modesta Pichon (Apocyn.) (Mad): subvar. *velutina* Markgraf: lf. inf. u. to make body slender; fresh bk. very bitter, munched to overcome hunger pains (—> *Tabernaemontanum*).

HCG: human chorionic gonadotropin.

he-: v. also hae-.

headache crystals: menthol - **h. powders**: Acetanilid powder comp. - **h. weed** (Va [US]): *Caulophyllum.*

heal-all: 1) *Prunella vulgaris* 2) valerian - **h. herb**: *Collinsonia canadensis.*

healing herb: lilac - **h. shade**: *Plantago major.*

heart leaf: 1) *Medicago arabica* 2) *Nymphoides peltatum* 3) *Asarum* spp. - **h. nut**: cashew - **h. of the earth**: *Prunella vulgaris* - **h. trefoil**: *Medicago arabica.*

heart-leaved bixa: annatto - **heart-liver**: *Medicago arabica* - **heart-pea**: *Cardiospermum halicacabum.*

heart's ear: *Polygonum persicaria* - **h. ease: heartsease**: 1) *Viola tricolor, V. ocellata* (**h. pansy**) 2) *Polygonum persicaria* 3) *Prunella vulgaris* 4) *Gratiola* spp. (Austral) 5) other pl. spp. - **h. pansy**: pansy.

heart-seed: heart-pea - **heart-trefoil**: heart-liver - **heart-weed**: heart's ear - **heartwood**: duramen, the inner dead wood as distinctive fr. the sap wd.; heartwood is often rich in coloring matter, tannin & other pl. principles - **heartwort**: 1) *Melilotus alba, M. officinalis* 2) *Aristolochia clematitis*, & other *A.* spp. 3) mint spp., esp. *Mentha cardiaca* (obsolete).

heath: 1) *Erica*; *Calluna* 2) land on which heather grows - **common h.**: *Calluna vulgaris* - **h. family**: Ericaceae - **long h.: Scotch h.**: common h. - **sea h.**: *Frankenia campestris* - **white h.**: *Cassiope mertensiana.*

heather (common): *Calluna vulgaris.*

heaven tree: t. of h.: *Ailanthus altissima.*

heavy oil: higher boiling point fraction fr. distn. of coal tar.

Hebe parviflora (Vahl) Cockayne & Allan (Veronica p. Vahl) (Scrophulari.) (SA): u. in diarrh. dys. (2439) - **H. salicifolia** (Forester f.) Pennell (Austral, NZ & As): koromiko; often hybridizes; lf. inf. astr. in diarrh. dys. ton.; decoc. u. as gargle in ulcers, sore throat, VD; bruised lvs. u. as poultice in ulcers, headache; sores; lvs. chewed —> keen sense of hunger; lf. packs for sore skin in infants, other *H.* spp. u. similarly (2761).

Hebenstretia basutica Phillips (Globulari) (sAf): rt. inf. u. for indiges. & stomachache; pl. mixed with fat —> perfumed ointment - **H. dentata** L. (trop eAf): flg. pl. crushed & expressed juiced deposited in ear canal for earaches (Burundi); rt. decoc. u. for syph. in female (Tanz).

Hechtia glomerata Zucc. (Bromeli) (Mex): balsamo de hortelano; hb. u. as arom. antisep. (Mex Ph.).

Heckeria Kunth = *Pothomorphe*; *Piper.*

hectare (ha.): 1 sq. Hectometer; 10,000 sq. meters; ca. 2.47 acres; the common metric unit of area u. in much of world.

Hectorite: white clay (Hector, Calif [US]; Sweden; etc.): Mg Si hydrate; member of montmorillonite gp. of minerals; somewhat like bentonite; u. as suspending & dispersing agt., in ointments; emulsifying agt., plasticizer, esp. in cosmetics, since nearly pure white (purified).

Hedeoma nanum (Torr.) Briq. (Lab.) (Tex [US], Mex): hb. has arom. thyme-like aroma; herbal tea u. for stomachache (Mex) - **H. piperita** Benth. (Mex): tabaquillo (oloroso); hierbabuena; VO c. menthol; hb. u. much as peppermint (oil); also as source of menthol; tea u. in stomachache - **H. pulegioides** (L) Pers. (eNA): American (false) pennyroyal; Herba Hedeomae c. VO; dr. hb. u. carm. in colic, corrective, stim. emm. arom. abortif.; u. to repel fleas & other insects; hepatotoxic (in large doses); VO c. pulegone, menthone, isomenthone, piperitenone, etc. (1455) (do not confuse with European pennyroyal, *Mentha pulegium*).

hedeomal: methyl cyclohexanone: ketone in VO of *Hedeoma pulegioides*: liquid; may be effective in driving away mosquitoes & fleas.

Hedera helix L. (Arali) (Euras, nAf): (English) ivy; source of Herba Hederae helicis (hb. or lvs.); c. saponins; (hederin +); u. stim. for chron. catarrh, pertussis (2308), sores; "atrophy of children"; exter. parasiticide, insecticide; detergent; for corns; to counteract alc. effects; frt. cath. antipyr.; bk. rt. gum, wd. have also been u. med.; varnish resin; tan; dye; lvs. poisonous to some people by contact; berries toxic; hard wd. sometimes u. for engraving - **H. rhombea** Bean (Jap): Japanese ivy; st. & lf. juice u. to stop nose bleed.

Hederich (Ge): charlock.

hederin, alpha = helixin: saponin fr. *Hedera helix*; with strong antibiotic properties.

"hedge": 1) privet (h. row); 2) *Fouquiera splendens* (lower swUS) 3) **h. apple**: *Maclura pomifera* (wUS) - **h. balls** (Kans [US]): *Maclura pomifera* - **h. garlic**: *Alliaria officinalis* - **h. hyssop**: *Gratiola officinalis* - **h. maids**: *Glechoma hederacea* - **h. mustard**: 1) *Sisymbrium officinale* 2) *Erysimum officinale* - **h. orange tree** (cUS): **h. plant**: *Maclura pomifera* - **h. stachys**: *Stachys palustris.*

hediondilla (NM [US], Sp): **hediondo** (NM, Sp): 1) *Larrea tridentata* 2) (SA): *Paederia diffusa.*

Hedwigia Sw. = *Tetragastris.*

Hedycarya arborea Forst. (Monimi) (NZ): lvs. u. in vapor baths - **H. solomonensis** Hemsl. (Solomon Is): lvs. macerated & appl. to sores; lvs. rubbed on sore places.

Hedychium coronarium Koenig ex Retz. (Zingiber.) (EI, now pantropic, Ind, WI, Br): ginger lily; tubers u. in rheum., as arom. to scent oils (Tonga); in concoction for relief of int. pain (Domin, Caribs); lower part of st. u. for swollen tonsils & neck dis. (Moluccas); pl. decoc. drunk as "cure" for GC (Domin, Caribs) (2328); ornam. perenn. herb (1456) - **H. spicatum** Buch.-Ham. (eAs): "ginger"; ginger lily (this name appl. to ca. 40 other trop. spp. of *H.*); kur (Urdu); rhiz. important in Ind commerce; chief component of abir (scented pd. u. in Hindu worship); perfume; carm.; rhiz. u. as remedy for stomachache & toothache (Chin); fragrant stom. & spice (Taiwan); bitter ton. stim.; yields ginger lily fiber; in veter. med.

Hedyosmum brasiliense Mart. (Chloranth.) (Br): lf. decoc. u. as antipyr. sudorific, stom. diur. ton, & aphrod. - **H. toxicum** Cuatr. (Col): granisillo pequeño; lf. tea u. as stom. ton., purg., strong emet., intoxicant; never u. as beverage (2301) - **H. translucidum** Cuatr. (Col): granisillo del grande; lvs. arom., u. to prep. very hot decoc. useful as ton.-stim.; for stomach "upsets" (2301).

Hedyotis acutangula Champ. (Rubi) (Hong Kong): lvs. c. arborinine; u. by natives for sprains & contusions - **H auricularia** L. (Oldenlandia a. K. Schum.; H. umbellata L.) (Ind, Chin, seAs to Moluccas, PI): hb. ext. or decoc. u. in cholera, diarrh. dys. colitis, &c.; expect; prophylactic for enteric fever; lf. paste u. as emoll. appl. to wounds, abscesses; in deafness (Ind); lvs. eaten with rice (SriL) var. *melanesica* Fosberg (Fiji): lvs. u. in eye dis. - **H. biflora** (L.) Lam. (Oldenlandia b. L.) (Oceania): decoc. for diarrh. & dys. (PI); poultice for rash (Caroline Is), burns (Solomon Is), wounds (PI) - **H. caerulea** (L.) Hook. (Houstonia c. L.) (e&cNA): bluets; Quaker ladies; bk. u. as diur.; inf. u. for enuresis (Cherokee Indians); cult. ornam. - **H. corymbosa** (L) Lam. (Oldenlandia c. L.) (seAs): u. antipyr. (Indochin); stom. (PI); poultice for wounds & sores (Mal) - **H. costata** (Roxb.) Kurz (H. vestita R. Br.) (Ind to Mal; Moluccas, PI): rts. boiled to make lotion for rheum. (Mal); appl. to int. pain; powd. lvs. appl. exter. for headache (PI) - **H. diffusa** Willd. (Oldenlandia d. Roxb.) (eAs, Chin): u. for gastrointest. dis. (Chin); for stomach ulcers & dis. (Sing); in poultices for lumbago (Mal); pd. bk for general weakness; excessive thirst, fever GC (Ind); pl. decoc. as mouthwash in toothache (PI) - **H. fruticosa** L. (Ind): lf. decoc. u. for ophthalmia (SriL) - **H. glabra** R. Br. (eInd, Assam): hb. decoc. as blood "purifier"; poultice in headache & stomachache - **H. herbacea** L. (Ind): hb. u. as bitter ton; antipyr. in mal.; in bath for rheum., pd. lvs. in honey for rheum. fever & swellings; lvs. exp. in asthma, TB; rt. u. as dye - **H. hispida** Retz. (neInd): hb. decoc. u. in dys.; as poultice for headache - **H. parvifolia** Wall. ex. G. Don (seInd): hb. u. as poultice for headache(Mal) - **H. pinifolia** Wall. ex. G. Don (Ind): hb. u. as poultice for aching body parts - **H. scandens** Roxb. (Nepal to Assam): hb. u. in eye dis. & postpartum difficulties; rt. for sprains; capsules u. to blacken teeth (Bengal).

Hedysarum alpinum L. (Leg) (NA to Arctic Oc, cCana to Alas [US]): "licorice root"; rts. eaten raw or cooked in spring, an important & esteemed food of Indians of BC (Cana); hallucinogenic(?) - **H. boreale** Nutt. (nUS, sCana): northern hedysarum; rts. eaten; subsp. *mackenziei* (Richards.) Welsh (H. m. Richards.) (Cana Arctic, Alas): purple loment (SC sd. pod color); food of barren ground grizzly bear (149) - **H. coronarium** L. (wEu, Med reg.): French honeysuckle; sulla (Malta); grown for forage, fodder, to enrich (build) soil; cult. ornam. - **H. gangeticum** L. = *Desmodium g.*

heerabol: myrrh.

Heeria insignis O. Ktze. (Anacardi.) (tropAf): twig, bk. or lf. inf. u. in diarrh.; bk. decoc. purg.; bk. decoc. in diarrh. (Tanz); rt. galactagog (Tanz); wd. hard - **H. mucronata** Bernh. (e&tropAf): rt. bk. ext. u. in dys.; lf. sap for snakebite; lf. & rt. as tenicide (148) - **H. paniculosa** O. Ktze. (tropAf): resin tree; powd. bk. u. for acute inflamm. in chest, dys. - **H. pulcherrima** O. Ktze. (tropAf): powd. bk. inf. u. in abdom. pain, diarrh. - **H. reticulata** Engl. (tropAf): rt. ext. u. as galactagog & aphrod.; bk. decoc. for dys., remedy in "pink eye."

Hefe (Ge): dry yeast.

hegari: *Sorghum bicolor* var. (S. vulgare v. caffrorum); prob. type of Kaffir corn; mispron. heye gear (Ala [US]).

Heide (, **gemeine**): **Heidekraut** (Ge): *Calluna vulgaris*.

Heidelbeere (Ge): *Vaccinium myrtillus*.

Heilerde (medicinal earth) (Ge): clay, kaolin, Bolus Alba (150).

Heimia salicifolia Link (Lythr.) (wTex [US], Mex, CA, SA, WI): sinicuichi; willow-leaf heimia; lvs. c. NBP, much u. native med. as emet. antipyr. antisyph. (reputed); diur. lax. vulner. ton. strong diaph. astr.; fermented lvs. said to prod. pleasant intoxication with amnesia & yellow vision; ext. for Rhus poisoning (1459); fl. decoc.—> mild intoxication; mild hallucinog.; u. in fertility control; cult. ornam. shrub.

Hela cells: derived fr. cancer (carcinoma) of uterine cervix of woman; continuously grown in tissue culture (name derived from donor's name) (Henrietta Lacks) (Baltimore, Md [US], 1951); u. as culture medium for polio virus.

helado (Sp): ice cream.

heleboro blanco (Sp): *Veratrum album* - **h. fetido**: *Helleborus foetidus*.

helecho (Sp): fern - **h. macho** (Sp): male fern.

helenalin: bitter subst. found in *Helenium* taxa & other Helenieae; also called helenic acid.

helenien: xanthophyll palmitate.

helenin: white h.: elecampane camphor; alantocamphor; inula camphor; stearoptene fr. VO of rt. of *Inula helenium*; VO of *Artemisia maritima*, etc.; term was also appl. to inulin; alantolactone; isoalantolactone; mixture of these with alantic acid; benzoic acid; anthelmintic for nematodes (roundworms).

Helenium L. (Comp) (NA, SA): sneezeweeds; coarse ann. or perenn. hbs. with yellow or brown fl. heads (Helenium [L] also appl. to Radix Helenii, rt. of *Inula helenium*) (*qv.*) - **H. amarum** (Raf.) H. Rock (H. tenuifolium Nutt.) (seUS): bitter (sneeze) weed; hb. c. bitter glycosides (tenulin, isotenulin); u. errhine (Ind) (fine pwd.); deleterious to milk quality when foraged by cattle to whom occasionally poisonous (even lethal) (sesquiterpene lactones) (151); said useful in diabetes (1979); exter. in rheum. (2309); leaf decoc. in fever (2431); weed - **H. autumnale** L. (NA): common sneezeweed; sneezewort; hb. c. helenalin ("hele-nin") ("helenic acid") a sesquiterpene; u. errhine (Mo [US]); antipyr.; narcotic; cult. ornam. - **H. bigelovii** A. Gray (Calif, Ore [US]): Bigelow's sneezeweed; powd. fl. heads sniffed to produce sneezing to relieve head cold congestion (Indians in Butte Co., ncCalif [US]) (152) - **H. flexuosum** Raf. (Tex to Fla to WVa [US]): purple-head sneezeweed; u. as purple dye - **H. mexicanum** HBK (H. quadridentatum Labill.) (Mex City to Yuc, CR, WI): (yerba) manzanilla; m. montes; arnica del paiz; chapuz; powd. fl. heads (or lvs.) u. as sternutat., to kill lice, to cleanse infected ulcers; fl. head decoc., for flatulence, colic, emmen. in menstrual cramps, pain of tabes dorsalis; hb. decoc. u. as diur.; antipyr. in colic - **H. puberulum** DC (Calif): sneezeweed; rosilla; much valued folk medl. pl.; u. ton antiscorb., snuff in catarrh.; fl. heads & lvs. boiled for prep. u. in GC; ton. antiscorb.

Helianthemum canadense (L) Mx. (Halimium c. Grasser) (Cist.) (Cana, ne&cUS): frostwort; rock rose; sun r.; hb. (Herba Cisti canadensis) astr. alter. ton.; many others of the ca. 130 other spp. have similar properties - **H. corymbosum** Mx. (Fla-SCar [US]): rock rose; frostweed; ice plant; u. much like prec. (wounds, diarrh., dys.) - **H. glomeratum** Lag. (Mex): u. med. (Mex) - **H. nummularium** (L) Mill. (H. vulgare Gaertn.) (Med reg.): solflower; (little) sunflower; "garden s."; hb. astr. alt.; lvs. sd. also u. in Homeopathy (GB).

Helianthus annuus L. (Comp.) (wNA, Mex, Pe): common sunflower; 1-3-6 m high, lvs. mostly alternate, 10-30 cm long, long petioled, fl. heads -0.6 m wide; fresh fl. head & expressed juice of same u. med.; frt. akene ("seed") salted a popular nutrient nut in US, USSR+; fed to poultry, parrots, & stock; rich in FO (c. 50%) which is comml. article, u. foods, soap mfr.; pl. or lvs. & frt. u. diur. exp. dem. in malaria (Siberia) (substit. for Cinchona) (2491) (believed cult. in Am by Indians; the large-headed var. was developed in Eu, then reintroduced into US) (cult. NDak [US]) - **H. argophyllus** (Tex [US]): Texas (or silverleaf) s.; smaller than prec.; said rich source of furfural & cellulose - **H. giganteus** L. (e&cUS & Cana): tall (or wild or giant) sunflower; tuberous rts. edible; achenes powd. & mixed with corn meal to make bread (Indians) - **H. strumosus** L. (Atl NA & Mex): dahlia sunflower; tuberous rhiz. edible; u. in diabetic foods - **H. tuberosus** L. (thought native to c[&e]US & Cana; now found as various sorts in Eu [esp. Fr] & sAf): Jerusalem artichoke (corruption of girasole [di Canada] [It]); Topinamb(o)ur (Ge); patata del Canada; "Indian potatoes"(?); tall (2-5 m) weedy pl. (perenn. hb.); lower & middle lvs. opp., upper often alternate, winged petiole; rhiz. (tuber-like) c. inulin, levulose, sucrose; levulin (isomer of inulin); lectins; formerly eaten by Am Indians but rather insipid, slightly sweetish; u. raw in salads or as relish (unripe hazelnut flavor) (2535); boiled (1457); pickled, in soups, as nutr. u. as dry chips; meal, malt; steamed in their own juice (artichoke flavor) (*cf.* potato: 2 vars. red, white); diur.; in diabetic diets & other special diets; believed to have specific value in liver, glandular, & skin dis.; good source of industrial alc.; in mfr. fructose; u. to hinder crystn. in sweetmeats; tops (herbage) & tubers u. as forage, silage, fodder for livestock (esp. pigs); st. u. as fuel (Fr) (2649); occurs as weed (Iowa [US], etc.); also cult. for use or as ornam. (1589).

Helichrysum arenarium Moench (Comp.) (Euras): live ever; sandy everlasting; (yellow) immortelle; fl. heads (Flores Stoechados citrinae) c. bitter subst. (arenarine, antibact.) tannin, VO; formerly u. as diur. cholagog, choleretic, in gout, skin dis., as astr. & anthelm.; in gastric dis.; against moths (adulterant of Pyrethrum Fls.); filling in tea mixts.; an old-time drug; hb. (as Herba Gnaphalii arenarii) or whole pl. u. med. as emmen. (Mex) - **H. armenium** DC (Anatolia, Turkey): pl. rich in flavonoids (13 known); inf. u. as diur. in folk med., also to regulate bile secretion - **H. bracteatum** Andrews (Xeranthemum b. Vent.) (Austral): (golden) everlasting; c. several flavonoids (quercetin, luteolin, etc.); u. similar to *H. arenarium*, pl. used for cough (chewed) (NG); cult. orn. - **H. caespititium** Sond. (Af): decoc. drunk in GC (Kwena); rt. decoc. for nausea (sSotho) - **H. cochinchinensis** Spreng. (seAs): young twigs u. as condim., esp. with rice flour in dish called "bath khuc" - **H. crispum** Less. (sAf): Hottentot's bedding; u. for heart trouble ("heart weakness"), in backache, kidney dis.; inf. reputed useful in hyperpiesia (essential hypertension); coronary thrombosis; bladder dis. - **H. dregeanum** Sond. & Harv. (sAf): lf. smoked for cold in head (sSotho); inf. drunk for hiccup (Europeans in Af) - **H. foetidum** Cass. (sAf): lvs. make good dressing for festering sore; arom. astr. - **H. gerberaefolium** Sch. Bip. (eAf): rt. u. for coughs & colds; lf. as wound dressing - **H. imbricatum** Less. (sAf): "tea from the downs"; inf. ("tea") u. as dem. in coughs, lung dis., pertussis, bronchitis, etc. - **H. italicum** (Roth) G. Don f. (Med reg., esp. in e): u. as syrup or aerosol in chron. bronchitis, asthma, whooping cough, rhinitis, psoriasis; in burns; chilblains; headache; migraine, etc.; VO said antiviral - **H. krausii** Sch. Bip. (tropAf): lf. decoc. u. as wash for keloid scars; smoked for cough & pulmon. TB - **H. nudifolium** Less. (tropAf): lf. remedy for colds (Hottentots); in catarrhs, phthisis; wound dressing; as a tea - **H. odoratissimum** Less. (trop eAf): lf. c. flavonoids (antimicrob.), decoc. u. to relieve stitch & other abdom. pains; fresh lf. juice for heartburn; in coughs & colds - **H. pedunculare** DC (Af): said excellent dressing for festering sores, esp. following circumcision (sAf); rt. in cough & colds - **H. platypterum** DC (Af): rt. decoc. "to renew virility" of male - **H. sanguineum** L. (OW): (passion) everlasting (Israel); red cudweed; everlasting life; fls. sold in Jerusalem markets during Easter (symbolic) - **H. saxatile** Mor. (It): u. like *H. arenarium* - **H. serpyllifolium** Pers. (sAf): inf. of hb. u. as dem. in resp. dis.; inf. called "Hottentot tea," long popular in reg. - **H. stoechas** (L) DC (sEu): fl. heads u. like *H. arenarium* & to adulter. same - **H. vestitum** Less. (sAf): common everlasting; u. rem. for jaundice, biliousness; liver dis. croup.

Helicia artocarpoides Elmer (Prote.) (Sabah, former Brit, nBorneo, PI): rt. u. for sore eyes, ulcers, jaw.

helicin: hydroxy-benzaldehyde: glucoside obt. by oxidn. of salicin; very similar to or identical with spirein.

Heliconia bihai L. (Heliconi/Mus) (tropAm): falsa plátano; plátano (silvestre); p. de flor (Jalisco, Mex); cachipo; bastard (or wild or false) plantain; lobster claw, balisier; rhiz. decoc. u. diur.; lvs. wrapped around body to relieve stomachache (Choco Indians); lvs. & st. sap u. as healing agt. for burns, inflamn., etc. (Domin, Caribs); poultice of boiled shoots placed on foul ulcers resultant of snakebite; young shoots eaten as veget. - **H. brasiliensis** Hook. (Br; Col): rts. u. as antigonorrheal; sds. u. in diarrh. - **H. sp.** (Pan): susucan (Cuna Indians, Pan); young spears u. med. - **H. stricta** Huber (Pe, Ec): lvs. added to hallucinogenic mixts. (Amazonian Indians) - **H. subulata** Ruiz & Pav. (Pan): platavillo morado; rhiz. u. in skin cancer (plaster with oil).

Heliciopsis terminalis (Kunz) Sleumer (Prote) (Burma, sChin+): u. as antidote in local med.

Helicteres angustifolia L. (Sterculi) (sChina): rt. decoc. u. in mal., stomachache; pwd. lvs. u. as dressing for abscesses & boils - **H. isora** L. (Ind to Austral): (East Indian) screw tree; moraphal; rts. & bk. stom.; frt. (pods) u. intest. dis., diarrh. dys. colic, flatulence (c. tannin, mucil, sugars); syph.; appl. as liniment to ear sores; pd. lvs. appl. to wounds (Ind); branches furnish a coarse textile fiber (kaiva[n]); wd. u. as fuel & for charcoal.

Helinus ovata E. Mey. (Rhamn.) (sAf): soap plant; lf. c. scyllitol, aconitic acid, saponin (trace), wax (with ceryl alc.); rt. u. to treat hysteria (Zulu) (emetic); lf. juice soothes irritation of sand worm.

Heliobacter pylori (Marshall & Goodwin) Goodwin & al. (Bacteria): causes gastric ulcers; treated with antibotics & Bi subcitrate, etc.

Heliocarpus americanus L. (Tili.) (Mex, CA): jonote; powd. bk. or decoc. appl. to sores; bk. & lvs. in med. (decoc.) to overcome constip. - **H. donnell-smithii** Rose (Mex, CA): jolo-cin; bk. powd. appl. to sores; wd. soft & light & u. for bottle stoppers, etc; paper made fr. bk. also fr. wd. - **H. popayanensis** HBK (Mex, nSA): balso blanco (Co); bk. of st. u. to clarify sugars; to make cords - **H. terebinthinaceus** Hochr. (Mex): jonote (Oaxaca); pl. u. to make cordage, rope, paper; floats; resin fr. lvs. appl. to head to relieve sunstroke.

Heliocereus speciosus Britt. & Rose (Cact.) (Mex, CA): "Santa Maita"; flor espinosa; fl. decoc. u. against common cold; cult.

Heliopsis helianthoides Sweet subsp. **scabra** Fern. (H. scabra Dunal) (Comp.) (NA, incl. Mex): oxeye; u. as insecticide, esp. for common house fly; rt. c. scabrin (butyl amide of an 18 C unsatd. fatty acid); heliopsin - **H. longipes** (A. Gray) S. F. Blake (Mex): peritre del pais; c. N-isobutyl-2,6,8-decatrienamide; u. against flies & domestic household pests (scabrin); cult. ornam.

heliotrope, common: *Heliotropium europaeum* - **garden h.**: *Valeriana officinalis* - **h. oil**: VO fr. *H. arborescens*.

heliotropin: piperonal (qv.).

Heliotropium angiospermum Murray (Boragin.) (tropAm, incl. WI): "eyebright"; cat tongue (Bah); lvs. in ointment appl. to boils; hb. decoc. appl. to cuts, wounds, sores, u. to bathe irritated eyes (Virgin Is) - **H. arborescens** L. (Pe, Ec): common heliotrope (heliotropo, Mex); "cherry pie"; fls. arom.; ornam. cult. as border pl.; source of VO u. in perfumery; u. for cephalgia; in cancer, gangrenous wounds; for fever, vomiting, & urinary dis. - **H. bacciferum** Forssk. (H. ramosissimum Sieb.) (Saudi Arabia): ramram (Arabic); hb. u. in snakebite & scorpion sting - **H. bursiferum** Wr. ex Griseb. (Cu, PR): "alacrancillo blanco"; c. pyrrolizidine alks.; with antisep. props. (2485) - **H. curassa-vicum** L. (US Pac states; Mex): seaside heliotrope; wild h.; pond weed; pl. u. as source of a purple dye (Indians) (205); pl. decoc. u. in leukorrh. (Curaçao); u. as dry powd. appl. to sores & wounds (swIndians); widely u. med. (thus in Hawaii) (TB & bronchial infect.) - **H. europaeum** L. (Med Sea reg.): tournesol; herbe du cancer; hb. c. alk. (heliotrine, lasiocarpine, heleurine, cynoglossine, etc.); hb. u. against urinary calculi, worms, warts, cancer, etc; sds. antipyr.; cult. as ornam. - **H. indicum** L. (tropics of As, Austral, Af, Am): c. incidine (alk.); lvs. u. in folk med for swellings, burns, eye dis., insect stings; pl. decoc.; u. as diur. lax; refreshing drink for children (Br); in GC (Mex); lf. juice u. in ophthalmia (Trin) - **H. ovalifolium** Forsk. (tropics): dr. pl. mixed with ghee & u. as stick poultice for painful areas; also in fever (Tanz) - **H. polyphyllum** Leh. (Fla [US], SA): (pineland or yellow) heliotrope; fl. decoc. formerly u. for yellow fever; st. u. as depurative (Br) - **H. ramosissimum** Sieb. (Af, As): appl. topically for headache (nNigeria); inf. for GC & to incr. lactation; remedy in snakebite inter. & exter. (Ind, Arab) - **H. strigosum** Willd. (Ind, Pak, Austral): c. strigosine; u. as diur. lax. & against insect stings - **H. subulatum** (DC) Martelli (Tanz): lvs. u. as poultice for boils, yaws; in postpartum dis. in female - **H. supinum** L. (Euras e to Ind): hb. u. like *H. europaeum* - **H. undulatum** Vahl (wAf, nInd): c. toxic alk.; u. in veterin. med.

Helipterum DC (*Helipterus*) (Comp.) (sAf, Austral): c. 60 sp.; dr. fl. heads sometimes sold as "everlastings" - **H. eximium** DC (sAf): strawberry everlastings; in jaundice, biliousness, liver dis., in croup & diphtheria - **H. speciosissimum** DC and **H. variegatum** DC have similar properties.

helium: He: gaseous element found in natural gas fr. which obtained as comm. product; non-inflammable, hence u. to fill balloons, dirigibles, etc.; in lasers, etc.; in therapy u. to dilute gases (in anesthesia, etc.) mixed with oxygen; u. by patients with status asthmaticus, severe respir. obstructions, etc. to render respiration easier.

Helix pomatia L. (Helicidae - Stylommatophora - Mollusca) (Euras, Af): European garden snail; (edible) snail; vineyard s.; escargot (Fr); this land snail is u. in lung dis. (silicosis, pertussis), digestive dis.; warts; snail blood u. in blood gp. detns.; liver of snail with many enzymes (broad spectrum); u. in biochem. syntheses; to remove sugars fr. complex glycosides; popular food (cEu) (<fr. helix = 3-dimensional coiled figure [spiral: plane figure]).

hellardine: alk. in Greek belladonna rt. (= cuskhygrine).

hellebore, American (white): *Veratrum viride* - **bear's foot h.**: *Helleborus foetidus* - **black h.**: *Helleborus niger* - **European h.**: *V. album* (rt.) - **false h.**: 1) *Adonis vernalis* 2) *Veratrum* spp. - **f(o)etid h.**: *Symplocarpus foetidus*; *Helleborus foetidus* - **green h.**: *Veratrum viride*; *Helleborus viridis* - **stinking h.**: *Helleborus foetidus* - **swamp h.**: *V. viride* - **white h.**: European h.

hellebor(e)in: hellebrin: diglucoside of hellebrigenin (with glucose & rhamnose) (153); cardioactive.

Helleborus foetidus L. (Ranuncul.) (s&cEu): bear's foot (root); setter(s)wort; rt. c. quercetin glycosides; u. cardiac stim. (USP 1820); cath. (danger!); in veter. med; cult. as ornam. undershrub - **H. niger** L. (c&sEu mts.; often cult): Christmas rose; black hellebore; rt. c. glycosidal saponins, FO, glycosides; u. card. stim. diur. in menstr. dis., edema, skin dis. nervous dis. anthelm.; lax. (Hippocrat. med.); by Eu Homeopaths for kidney & urin. dis.; poisonous (take care!) (prod. sGe, Fr) - **H. orientalis** Lam. (Turk Caucasus reg.): eastern hellebore; hellebore of the ancients; rt. u. in edema;, epilepsy, melancholia, jaundice, intermittent fever; in veter. med. feeding aversion of calves - **H. purpurascens** Waldts. & Kit. (eEu-AsMin): u. similarly to *H. niger* & *H. viridis* but weaker - **H. viridis** L. (c&sEu; natd. US): bear('s) foot; Christmas rose; rhiz. c. alk. & glycosides; u. cardioactive drug often in combn. with *Rauwolfia serpentina*; earlier as emet. lax. anthelm.; component of sneezing powders; poisonous!

hellebrin: glycoside fr. rhiz. of *Helleborus niger*, representing combn. of hellebrigenin with rhamnose & glucose.

helmet flower: *Scutellaria lateriflora.*

helminth: intestinal worm.

Helminthostachys zeylanica (L.) Hook. (1) (Ophiogloss.) (Pteridoph.) (tropAs, Austral): a widespread fern of trop eAs & important village medicine; pl. decoc. appl. to boils, ulcers, "heat" hives; int. antipyr. (in mal.), stom. appetizer; rhiz. in dys. (Indones); aper anodyne in sciatica (Ind); ton. (Mal); cult. (lvs. eaten as salad).

Helmkraut (Ge): scullcap. *Scutellaria* spp.

Helodea: *Elodea.*

Heloderma horridum Wiegmann (H. suspec-tum) (Helodermatidae - Squamata - Reptilia) (swUS, esp. Ariz; Mex): Gila monster; "scorpion"; this poisonous lizard lives in hot deserts; venom u. in arthritis, cardiac weakness & collapse; neuralgias; paralysis agitans.

Helonias (L): rhiz. & rts. of *Chamaelirium luteum* (NF V VI).

Helonias bullata L. (monotypic) (Lili) (neUS s to Ga): swamp pink; often confused with *Chamaelirium luteum* & u. similarly; cult. ornam. - **H. dioica** Pursh = *Chamaelirium luteum* Gray. (rt.) (1590).

Helosis guyannensis L. C. Richard (1) (Balanophor.) (Co+): flor de terra; decoc. of entire pl. u. in diarrh., dys.

hematein: oxidized or fermented hematoxylin; u. indicator, lab. stain, textile dye.

hematin: ferriporphyrin hydroxide; complex Fe compd. found in body in pathological states (ex. pernicious anemia); an orphan drug; u. to treat hepatic porphyria (serious blood dis.).

hematite: blood stone; iron ore.

Hematoxylon: Haematoxylon.

Hemerocallis L. (Lili) (Euras): (common orange) day lily; some 15 spp.; mostly cult. as ornam. perenn. hb. (SC fls. last only one day generally) (1592) - **H. aurantiaca** Baker (Chin, cult. Jap): golden summer day lily; kanzo (Jap); rts. u. as diur. ton., to prevent jaundice (Chin); cult. for very large fls. (-15 cm across) - **H. flava** L. (**H. lilioasphodelusy** L.) (Eu, tempAs, Caucas area): lemon lily; fls. u. in folk med. as cardiac stim., in rheum. as spice for meat dishes (Caucasus) (2778) - **H. fulva** L. (As, natd. Eu; very common in US, thus wild along railroad, etc.): smaller pl. than *H. flava*; fls. u. in Chinese foods; cult. for this purpose (ex. farms in NJ [US]); possible source of carotene & colchicine; rhiz. & rts. diur. in edema, dysuria, lithiasis; fls. stom. diur. antispasm. sed.; antipyr. anodyne (Indochin).

Hemidesmus indicus Schultes (1) (Asclepiad) (Ind, SriL, seAs): "Indian sarsaparilla"; "vetiver(e)"; rt. popular drug in Ind; substit. for sarsaparilla (Ind); u. diur. ton. depurat. dem. cholagog. astr. antiperiod. in mal., galactagog; in ophthalmia, syph. (Indochin); rt. paste taken for impotency (124); cult. on small scale.

hemiparasite: pl. which contains chlorophyll but also is attached by rts. as a parasite, obtaining nutrition from both sources.

Hemiptera: the true bugs; an order of insects; these possess sucking mouth parts, wings (usually), & few antennae segments; ex. bed bugs, stink bugs, water bugs, lace bugs.

hemlock: 1) *Tsuga canadensis*; hemlock spruce; source of h. bark, gum, lvs. oil (VO) 2) *Picea* spp. (spruce) 3) poison h. 4) other Umb. genera, ex. *Aethusa* - **American h.**: A. water h.: water h. - **h. bark:** < *T. candensis* (ext. as tan) - **deadly h.**: *Conium maculatum* - **Eastern h.**: *Tsuga canad.* - **lesser h.**: small h. - **h. needle oil**: fr. *Tsuga candensis; Picea glauca, P. mariana,* &c. - **h. pitch**: Pix Canadensis - **poison h.**: *Conium maculatum* - **small h.**: *Aethusa cynapium* (lvs.) - **spotted h.**: 1) *Cicuta maculata* 2) *Conium maculatum* - **spotted water h.**: *Cicuta maculata* - **spruce h.**: Eastern h. - **water h.**: *Cicuta* spp. (**American w. h.**: *Cicuta maculata*) - **water fine-leaved h. seed: f.-l. water h.**: *Oenanthe phellandrium* - **western h.**: *Tsuga heterophylla.*

hemochromogen: compd. contg. heme & protein or N-contg. non-protein, ex. several respiratory pigments, as hemoglobin.

hemoglobin: characteristic pigment of the blood; prepd. fr. cattle blood; c. iron; u. in anemia, both as oxygenated & reduced compd., u. to make artificial blood for transfusion (oxygen carrying capacity) - **hemol**: compd. prod. by action of zinc dust upon hemoglobin; a reduction prod. of hematin; u. blood producer, ton.

Hemophilus = *Haemophilus.*

hemp: *Cannabis sativa* - **African bowstring h.**: *Sansevieria thyrsiflora* - **Ambaree h.: Ambari h.**: *Hibiscus canadensis* - **American h.: birdseed h.**: h. - **black Indian h.**: *Apocynum cannabinum* (or sometimes *A. androsaemifolium*) - **Bombay h.:** Sunn h. - **brown Indian h.**: Ambari h. - **bowstring h.**: African b. h. - **Canada h.: Canadian h.**: black Indian h. - **Chinese h.:** *Boehmeria niger* - **Deccan(i) h.**: *Hibiscus cannabinus* - **h. fruit**: frt. of *Cannabis sativa* - **Indian h.**: 1) *Cannabis sativa* (Ind-grown) 2) *Apocynum cannabinum* (frequent comml. name: misnomer) - **Madras h.** sunn h. - **Manila h.**: abaca, *Musa textilis* - **h. nettle** *Galeopsis* spp. - **Sann h.: h. seed**: h. fruit - **h. seed oil**: FO fr. h. frt. - **Siam h.**: Manila h. - **sisal h.**: *Agave sisalana* - **Sunn h.** (Indo-Pak): *Crotalaria juncea* & the fiber prod. fr. its stem - **white Indian h.**: *Asclepias incarnata.*

hena (Sp): henna.

hen and chickens: *Sempervivum tectorum.*

henbane: 1) *Hyoscyamus* spp. 2) (**leaves**): (specif.) usually *H. niger* - **black h.**: *H. niger* - **Cyprus h.: Egyptian h.**: *H. muticus* - **Japanese h.**: *Scopolia japonica* - **white h.**: *H. albus.*

henbit: *Lamium amplexicaule.*

henequen (Sp): *Agave fourcroydes.*

henna (leaves): *Lawsonia inermis* - **white h.**: $MgCO_3$ & solid oxidizing agt. (ex. ammonium persulfate) (or $MgCO_3$ appl. with H_2O_2 as pack) u. to bleach hair.

heno, flores de (Sp): hay flowers.

hepar (L.): liver (of sulfur, antimony, lime, etc.); also body organ. - **h. cum stomacho**: liver with stomach; u. in pernicious anemia (formerly).

heparin: natural derivative (sulfated mucopolysaccharide) of animal tissues (chiefly liver & lung tissues); prod. fr. intest. mucosa of cattle & hogs; u. anticoagulant; rapidly soluble & stable: hence form chiefly u. in therapy; u. in angina, etc. (E) - **h. sodium**: Liquaemin.

heparin lock (flush solution): h. sodium inj. in isotonic NaCl soln.; u. to maintain patency of intravenous inj. devices in hospital.

Hepatica nobilis Garsault. (H. triloba Gilib.; Anemone hepatica L.) (Ranuncul.) (Euras): (heart) liverwort; round-lobe(d) hepatica; liverleaf (hepatica); lvs. c. tannin, mucil. hepatrilobin (glycoside); u. ton. dem. "deobstruent"; astr., in liver dis., bronchitis, phthisis, influenza (temporary great popularity in NY in 1830 during flu epidemic) (1594) - var. *acuta* (Pursh) Steyermark (H. acutiloba DC); var. *obtusa* (Pursh) Steyermark (H. americana Ker-Gawl) (e&cNA): (heart) liverleaf; pl. inf. u. ton. in chronic bronchitis, liver dis. (folk med.); lvs. formerly coll. for drug market (Fol. Hepat.); ornam. hb.

hépatique (Fr): *Hepatica nobilis.*

hepatitis: liver inflammn.; of 2 types: **infectious h.** (caused by virus A; acute illness found worldwide; transmitted by oral ingestion of infected material) and **serum h.** (caused by virus B; acute dis; usually transmitted parenterally) - **h. B immune globulin** (USP) is u. as prophylactic in serum h. - **h. B vaccine**: u. as prophylactic.

hen plant: *Plantago major.*

Heptachlor™: hepta-chloro-tetrahydro-4,7-methanoindene; insecticide u. for many field insects (ex. boll weevil; webworm).

heptrec: *Rosa canina* (frt. & sd.).

Heracleum Meximum Bartr. (**H. lanatum** Mx.) (Umb.) (NA): (common) cow parsnip; berce laineuse (Queb [Cana]); rt. (Heracleum [L.] USP 1820-50) & frt. aromatic, bitter ton.; u. in dyspepsia; frt. for poultice in migraine, etc. (BC [Cana] Indians) (76); c. heraclin (=bergaptea) (NBP); entire pl. strongly musky-arom.; c. VO, acrid princ. rt. often presented for angelica or lovage rt.; vesicant, counterirritant, GI irrit.; young shoots eaten (Klamath Indians) (3583); rts. fls & sts. eaten; frt. spice (Sikkim) eaten by cow renders milk bitter; reputed —> dermatitis; pl. extraordinarily attractive to insects - **H. giganteum** Fischer (H. villosum Fischer) (Caucasus; cult.): frt. c. VO with ethyl & hexyl butyrate and free ethanol & methanol (distillate), the former mostly in ripe, latter in unripe frts.; pl. may produce light dermatitis fr. percutaneous photosensitization (129) - **H. rapulum:** c. coumarins with analg. & antifertility activity - **H. sphondylium** L. (Euras, nwAf, NA): cow parsley; hogweed; common (or European) cow parsnip; hb. (Herba Brancae ursinae [germanicae]) u. in muscle cramps, digest. disturb., diarrh. dys. gastritis; anthelm. emmen.; lvs. & frts. boiled in water u. to prep. alc. beverage (Slavic countries); VO c. octyl alc. u. in perfumery; rt. u. epilepsy; subsp. **sibiricum** simonkai L. (s&cEu to Siberia): Siberian bear paws.

herb: 1) the above-ground parts of a non-woody pl. which do not persist through the winter 2) dr. crude drug; general term appl. by laity to collections of herb doctor, etc.; strictly speaking, it should represent the above-ground parts and thus be distinct fr. a "plant" which incl. the underground organs 3) plants u. as flavor in cooking ("garden herbs"), perfume, &c - **beer h.** (Cornwall, Eng): *Achillea millefolium* (since u. in prep. beer) - **h. Bennet**: *Geum urbanum* - **h. doctor** (seUS): layman who prescribes & dispenses herbs, etc., for many kinds of indications; also "root doctor" - **h. of grace (& rue)**: rue, *Ruta graveolens* - **h. of the cross** (seUS Negro voodoo): *Verbena officinalis* - **h. Paris**: *Paris quadrifolia* - **herb-patience**: *Rumex, R. paticentia alpinus* - **H. Robert: h. robin**: *Geranium robertianum* - **h. strewer**: in medieval Eu post in courts, etc.; after scattering the herbs, people walked over them, bruised them & spread the aroma in the air - **h. trinity**: pansy - **h. vinegar**: prepd. by inf. of various hbs. (as dill, fennel, basil, mint, tarragon, garlic, burnet, chives, etc.); u. in cookery - **willow h., great**: *Epilobium angustifolium.*

Herba (L.): herb; *v.* also Folium (leaf) - **H. Abrotani**: fr. *Artemisia abrotanum* - **H. Absinthii vulgaris**: fr. *Artemisia absinthium* - **H. A. pontici**: fr. *Artemisia pontica* - **H. Anserinae**: *Potentilla anserina* - **H. Anthyllidis**: *Anthyllis vulneraria* - **H. Basilici**: fr. *Ocimum basilicum* - **H. Betonicae**: fr. *Stachys officinalis* - **H. Bursae Pastoris**: fr. *Capsella b. p.* - **H. Camphorata**(e): fr. *Camphorosma monspeliaca* - **H. Capilli Veneris**: fr. *Adiantum C. v.* - **H. Cardui benedicti**: blessed thistle - **H. Centaurii minoris**: centaury - **H. Cerefolii: H. Chaerophylli**: *Anthriscus cerifolium,* chervil - **H. C. hispanici**: *Myrrhis odorata* - **H. Chirettae indicae**: fr. *Swertia chirata* - **H. Dictamni oretici**: fr. *Origanum dictamnus* - **H. Dracunuculi**: fr. *Artemisia dracunculus* - **H. Farfarae**: coltsfoot - **H. Foeniculi**: fr. *Foeniculum vulgare* var. *dulce* - **H. Genipi**: fr. *Artemisia laxa, A. genipe, A. nivalis, A. nitida, A. glacialis* - **H. Grindeliae (campori)**: fr. *Grindelia camporum, G. humilis, G. squarrosa* - **H. Hederae Terrestris**: fr. *Glechoma hederacea* - **H. Herniariae**: fr. *Herniaria glabra* - **H. Hyperici**: fr. *Hypericum perforatum* - **H. Ivae moschatae**: fr. *Achillea moschata, A. atrata, A. nana* - **H. Judaicae**: fr. *Stachys recta*; (earlier) *Sideritis hirsuta* - **H. Kali (vulgaris)**: fr. *Salsola kali* - **H. Leontodontis**: fr. *Taraxacum officinale* - **H. Mari (veri)**: *Teucrium marum* - **H. Menthae foetidae**: fr. *Dysophylla auricularia* - **H. Musci acaciae**: oak moss - **H. Nasturtii**: fr. *Nasturtium officinale* - **H. Origani**: fr. *Origanum vulgare* var. *macrostachys* - **H. Pichi**: fr. *Fabiana imbricata* - **H. Plantaginis**: fr. *Plantago lanceolata, P. media* - **H. Pulmonariae (maculatae)**: fr. *Pulmonaria officinalis* - **H. Pulsatilla (nigricantis)**: fr. *Pulsatilla vulgaris & P. pratensis* - **H. Rutae hortensis**: fr. *Ruta graveolens* - **H. Sabinae** fr. *Juniperus sabina* - **H. Salicariae**: *Lythrum-salicaria* - **H. Santa**: fr. *Eriodictyon californicum* - **H. Scolopendrii**: fr. *Scolopendrium vulgare* - **H. Serpylli**: fr. *Thymus serpyllus* - **H. Sideritidis** (misnomer): *Galeopsis ochroleuca,* Sideritis hirsuta+ - **H. Solidaginis**: fr. *Solidago virga-aurea* (usually) - **H. Spiraeae ulmariae**: hb. of *Filipendula ulmaria* - **H. Tanaceti**: *Tanacetum,* tansy - **H. Tormentillae**: hb. of *Potentilla erecta* - **H. Vincae pervincae**: fr. *Vinca minor* - **H. Virgaureae**: fr. *Solidago virgaaurea* - **H. Vulnerariae (rusticae)**: *Anthyllis vulneraria.*

herbe à chats (Qu): catnip - **h. à coton** (Qu): *Asclepias syriaca* - **h. à. fièvre** (Fr): boneset - **h. à l'hirondelle**: *Chelidonium majus* - **h. à. pisser**: *Sarracenia purpurea* - **h. au mouton** (Fr, WI): *Leonitis nepetaefolia* - **h. aux chantres** (Fr): hedge mustard - **h. aux chats**: catnip - **h. aux mamelles**: nipplewort - **h. coeur**: lungwort (lvs.) - **h. d'eupatoire**: boneset hb. - **h. de lobélie enflée**: Indian tobacco - **h. parfaite**: boneset hb. - **h. de Provence**: combn. of dr. rosemary, thyme, orégano, chervil, violet fls. - **h. de pyrole ombellée**: Chimaphila - **h. de Ste. Christophe**: *Actaea alba* - **H. de Ste. Croix**: *Nicotiana tabacum* - **h. Sainte Jean** (Qu): *Artemisia vulgaris*

herbicide: agt. u. to kill weeds; weed killer.

herbology (inelegant layman's term): study & use of herbs, prepd. medicinal pl.

herboriste (Fr): dealer in crude drugs licensed in the same way as the pharmacist (state diploma).

Herbstzeitlose (Ge-f.): *Colchicum autumnale.*

Hercules: *Heracleum sphondylium* - **H. club**: 1) *Zanthoxylum clava-herculis* 2) *Aralia spinosa* - **H. tree**: *Aralia spinosa* (*v.* under gum).

Heritiera littorale Ait. (Sterculi) (OW trop): false cola nuts; looking glass tree; sd. c. FO, tannin, arom. subst. no caffeine; appar. source of balan oil (VO); sd. ext. u. for diarrh. dys. (Indones); u. anthelm., substit. & adult. for kola; bk. c. tannin - **H. minor** Lam. (Ind): sundri tree; source of charcoal; furnishes best fire wood in Calcutta; sds. eaten.

hermaphrodite (bot): perfect (fl.) in which both "male" (staminate) & "female" (pistillate) organs are present in same flower.

Hermannia betonicaefolia E. & Z. (Sterculi) (sAf): rt. & lf. inf. u. in treatment of resp. dis., esp. asthma; some 10 other *H.* spp. are u. med. in Af.

Hermas gigantea L. f. (Umb.) (sAf): tinder-brush; lf. u. as dressing for wounds; thick wooly indumentum of lf. u. as tinder.

Hermbstaedtia odorata (Burch.) T. Cooke (Amaranth.) (sAf, Moz, Zimbabwe): rt. inf. u. as diur. for children; as cleansing stomach "wash" (Zulu) (2445).

Hermidium S. Wats.= *Mirabilis.*

Hermodactyla (L): **Hermodactyli: hermodactyls**: corms of *Colchicum luteum* Baker, *C. variegatum* L., *C. speciosum* Stev., *C. illyricum* Stokes, etc., fr. e. extreme of Med region; at one time of unknown origin; chiefly with starch, no colchicine or veratrine; almost without activity; u. dem. lax.; formerly thought derived fr. rhiz. of *Iris* sp.; not the Hermodactyli of ancients (= *C. autumnale* u. in gout & rheum.).

Hermodactylus tuberosa (L.) Miller (Irid) (sFr-Gr): snakes head iris; tubers formerly u. med.; cult. ornam.

Hernandia moerenhoutiana Guill. (Hernandi.) (Tahiti+): u. med. & to scent coconut (Cook Is) - **H. ovigera** L. (H. peltata Meissn.) (Madag, Oceania, Antilles+): jack-in-the-box; st. wd. is "false camphor wood"; c. VO 1-2% with perillaldehyde (75%); sd. c. 1.4% VO; frt. & rt. wd. 0.5% VO, all with same comp.; pl. c. hernandione (mixt. of podophyllotoxins), hernovine, & other alk.; bk. lvs. & sd. u. purg.; u. during childbirth; juice of pl. strong depilatory (Ind); FO u. to restore hair & for dandruff (PI); fls. u. for stiffness of joints; grated frt. u. with grated coconut & turmeric for scalp treatment in women (Caroline Is); wd. u. lumber (termite-free); street tree (SC Hernandez, Am botanist) - **H. sonora** L. (tropAm, WI, sFla [US], Mex to Ec; also Ind, Java): jack-in-the-box; bk. c. bebeerine, thallicarpine, & other alk.; frt. & sd. c. FO, VO, bk. & sd. c. podophyllotoxin derivs., alk.; sap u. as depilat. for facial hair; bk. decoc. u. during childbirth (Mariane Is); shade tree.

Herniaria glabra L. (Caryophyll./Illecebr.) (Euras): (glabrous) rupturewort; common burstwort; Herba Herniariae c. glycosides, incl. quercetin arabinoside & q. galactoside; saponins; VO; u. diur. in chronic cystitis, bladder stone, urethritis, bladder tenesmus; exter. vulnerary (prod. Ge); formerly in "Hydroaogin" (Bischoff) - **H. hirsuta** L. (Euras fr. Canary Is to Pak, Ethiopia): hairy burstwort; props. & u. similar to prec.

herniary: *Herniaria.*

Heroin: diacetylmorphine; prep. by acetylation of morphine; u. very effective cough sed.; excellent analg., esp. in terminal cancer (1978); one of most widely u. & most pernicious addictive narcotics; addiction being treated with methadone (2456); off. in Brit Phar. & Por Phar.; synth. 1883 (2730) - **brown H.**: impure form fr. Mexico; much u. in wUS border areas - **China white h.:** imported fr. seAs; pure form, so pure that addicts accustomed to denatured or adulterated h. rec'd. overdoses, sometimes fatal.

Herpestis Gaertn. f. = *Bacopa.*

Herrania albiflora Goudot (Sterculi) (Col): cacao simarón; aril sweet u. like cocoa - **H. pulcherrima** Goudot var. *pacifica* R. E. Schultes (Col, Ec): cacao de monte; frt. edible, like cacao, sweet husks (2301).

Herreria montevidensis Smith v. *bonplandii* (Lecomte) Smith (Lili) (Arg): zarzaparilla (Arg); rt. u. as antivenereal (Arg) - **H. salsaparilha** Mart. (H. sarsaparilla Mart.) (Br): u. antisyph.

herring: *Clupea harengus* & other *C.* spp. (Clupeidae) (nAtl Oc) - **h. family**: smelt f.: includes sardines, smelt, menhaden, alose (alewives), shads, as well as herring - **Norway h.**: u. for canning - **h. oil**: FO (body) u. as food & in soap mfr.; most fr. Alaska h. (*C. pallasii*), & river h. (*Pornolobus aestivalis*, eUS); Scottish (1595) - **red h.**: fish cured by smoking - **salted h.**: mostly fr. Scotland (Scotch method of curing); popular among Jews (gefilte fish).

hertwort: *Fraxinus excelsior*; *Aristolochia* spp.

herva (Por) (v. also erva): 1) herb 2) (slang) Cannabis - **h. cidriera** (Br): *Melissa officinalis* - **h. doce**: anise - **h. de Santa Maria** (Por): *Chenopodium ambrosioides* - **h. mat(t)e**: maté - **hervaes** (Br): plantations of maté trees - **hervea** (Br): maté; *Ilex paraguariensis.*

Herzgespann (Ge): *Leonurus cardiaca*, motherwort.

Hesperaloë funifera Trel. (Agav.) (ne&cMex, sTex [US]): "samandogue"; cult. (Nuevo León) for fiber fr. lvs. (exported as "ixtle" or "Tampico fiber"); young pl. c. manogenin as steroidal sapogenin; with maturity, hecogenin, gitogenin, and tigogenin are formed (*cf.* **H. parviflora** Coult. [swTex (US), nMex]).

hesperidin: major flavonoid of lemon & other *Citrus* frts., esp. unripe frt; glycoside (hesperetin rhamnoglucoside); a bioflavonoid (vitamin P complex) - **phosphorylated h.**: antifertility factor; claimed useful as oral contraceptive; may reduce cholesterol in blood stream.

Hesperis matronalis L. (Cruc.) (Euras, natd. nUS, Cana): dame's violet; perenn. hb. 3-10 dm high; pl. u. diur.; sds. crushed —> perfume.

hessian: coarse sacking made of hemp or hemp & jute.

Hetaeria obliqua Bl. (Orchid) (Indones): lf. inf. u. to heal sores.

heta-starch: starch 2-hydroxy ethyl ether; a waxy starch, u. like dextran as blood replacement; plasma volume extender; cryoprotective agent for red blood cells (Voler); leukocyte harvesting.

heteranthy: occ. of different kinds of fls. on a plant.

heteratisine: alk. of *Aconitum heterophyllum* (rt.) (may be artifact fr. benzyl-heteratisine).

heteroauxin: growth hormone of lower plants; indoleacetic acid.

Heteromeles arbutifolia M. Roemer (H. sali-cifolia Abrams; Photinia a. Lindl.) (1) (Ros) (Calif [US] & nwMex): Christmas berry; toyon (tollon); lvs. colored; lvs. & frt. c. HCN-glycoside; frt. edible, raw or roasted; lvs. & bk. sometimes boiled in water & u. analg. incl. stomachache (Am Indians); cult. ornam. evergreen shrub.; *cf.* holly.

Heteromorpha trifoliata Ecklon & Zeyher (H. arborescens Cham. & Schlecht) (Umb) (sAf): live long; lf. decoc. u. in incipient & early nervous & mental dis.; intest. worms; antifungal; decorticated rt. decoc. u. cough, dys., threadworms in horses; inner bk. as tinct. in colic; anthelm.

Heteropappus altaicus (Willd.) Novopokr. (Comp) (Mongolia): aerial parts. c flavonoid glycosides, heteropappus saponins; u. in Tibetan folk med. for bronchitis, colds - **H. biennis** (Ld.) Tamamsch. (Mongolia): similar c. & u. to *H. altaicus.*

heterophyllin(e): alk. named fr. *Rauwolfia heterophylla* (= *R. hirsuta*); preferably called aricine with wh. identical.

Heteropogon contortus Roem. & Schult. (Gram.) (tropAf, Ind, Chin): black (or bunch) speargrass; fls. u. as dem. & antipyr. (Chin); decoc. u. for cough (Austral).

Heteropsis Kunth (Ar) (SA: Br, Ec+): **H. jenmanni** Oliv. (tropAm, Arg): ririmosirimi; cipo titica (Br); long aerial rts. u. to hold huts together; bk. of the liane is grated, mixed with water, & poured over head of patient for headache (Indians, 2506); other spp. in the literature are **H. helianthioides** and **H. longipes**; u. in folk med.

Heteroptera: subord. of Hemiptera (at times equated); incl. kissing bugs, bedbugs, assassin b.

Heteropteris (Heteropterys) argentea HBK (Banisteriopsis a. [HBK] C. B. Robinson) (Malpighi) (Guat-Pan, Col): climbing shrub; u. med. - **H. beecheyana** Juss. (Banisteria b. C. B. Rob.) (Mex, CA, Co): "margarita"; colors water rose-red; u. in folk med. to relieve person stung by scorpion - **H. macrostachya** A. Juss. (tropAm): sds. u. as tea for diarrh. (Taiwano Indians, Co): pl. considered toxic (Pan) - **H. pauciflora** A. Juss. (Br): u. aphrod. (Matto Grosso); rt. u. as ipecac adult.; c. heteroptin, no alk. - **H. riparia** Cuatrecasas (Co): tea fr. rasped bk. u. int. in GC (Indians of Rio Apaporis).

Heteropyxis natalensis Harvey (Myrt./Heteropyxid.) (Rhodesia, Af): lavender tree; "lemon verbena"; rt. decoc. u. to steam face in nose bleeding; lvs. u. to prepare medl. tea & as perfume (Zulus).

heteroside: glycoside which yields sugar(s) & aglycone (non-sugar compd[s]).

heterosis: hybrid vigor.

Heterosmilax gaudichaudiana DC (Lili/Smilax) (eAs): u. in med. (Indochin) - **H. japonica** Kunth (eAs): decoc. of tuberous rt. u. expect. diaphor.; to treat influenza, diabetes (Taiwan).

Heterothalamus brunoides Less. (Comp) (Arg-Br): u. stim. arom. antipyr. source of yellow dye and of rubber.

Heterotheca grandiflora Nutt. (Comp.) (Calif [US]): telegraph weed; st. u. to make arrow shafts (Luiseno Indians) - **H. inuloides** Cass. (sUS, Mex): arnica (del pais); falsa arnica; fl. heads of this & other spp. of wUS & Mex frequently occur as adult. of Arnica fl. (American); fl. heads c. resin, VO, tannin, bitter princ., FO; u. as alc. tinct. appl. to contusions, wounds, &c. with much the same value as Arnica (Mex Ph.) - **H. mariana** (L) Shinners (Chrysopsis m. [L.] Ell.) (e&seUS): golden star; tea u. as sed. for children (Del [US] Indians) - **H. subaxillaris** Britt. & Rusby (sUS, n&cMex): camphor weed; c. VO with alcohols, prob. borneol (37) - **H. villosa** Shinners (sUS): rosin wood; tea u. for chest pains; hb. heated & appl. to toothache (Am Indians); var. *foliosa* Harms (Chrysopsis f. Shinners) (wUS): u. as sed. hypnotic; supposed to kill ants (Am Indians) (*cf. Chrysopsis*).

heterotrophs (heterotrophic plants): not self-supporting (ex. mistletoe), sometimes dependent on other plants (saprophytes; parasites).

hetisine: alk. of *Aconitum heterophyllum* rts.

hêtre (Fr): beech.

Heublume(n) (Ge): a variable mixt. of hay flowers: 1) fls. of *Trifolium pratense*, *Andropogon caricosus*, *A. laniger*, *Cymbopogon schoenanthus*, *Genista sagittalis*, *Melilotus officinalis*, *Thymus serpyllum* & other spp.; popularly u. (as Flores Graminis) in Eu for herbal baths 2) the collected sd. shedding with smaller particles of hay; often u. to seed meadows; this material always however introduces weeds.

Heuchera americana L. (Saxifrag.) (c&eEu): (American) alum root; rt. u. astr. (similarly for other *H.* spp.); in scrofula, sec. syph.; var. *hispida* E. Wells (cUS): frt. inf. u. to stop diarrh. (Menomini Indians) - **H. bracteata** Ser. (Colo, NM, sWyo [US]): Navaho tea; lvs. u. for toothache, sore gums, indigest.; st. furnishes pinkish dyestuff - **H. villosa** Mx. (seUS): hairy heuchera; Am sanicle; u. astr. styptic; kidney remedy (sUS).

Hevea benthamiana Muell. Arg. (Euphorbi) (nBr): tree a source of Pará rubber - **H. brasiliensis** (A. Juss.) Muell. Arg. (Amazon reg., nBr, Pe, Bo-cult. Mal, Borneo): caoutchouc or Pará rubber tree; much cult. in both hemispheres as chief natural rubber source; latex is source of pará or fine rubber of commerce - **H. discolor** Muell. Arg. (Rio Negro, Br): less important source of natural rubber - **H. guianensis** (*guayanensis*) Aubl. (nSA, Guianas to Rio Negro, Br): India rubber tree; milk t.; latex source of crude rubber (caoutchouc; [American] rubber; "Resina elastica") fr. acrid latex; also of caoutchouc tree oil = siringa oil (of Brazil); sds. edible - **H. pauciflora** Muell. Arg. (Guyana): source of some natural rubber.

hexanedioic acid: adipic a.

Hexastylis Rafin. (Aristolochi) (NA): heart leaf; genus a segregate from *Asarum* (sUS spp.) (formerly sect. *Cerasarum*) - **H. arifolia** (Mx) Small (Asarum arifolium Mx) (eUS): lf. inf. u. for stomachache (Catawba Indians); rt. decoc. u. for coughs & colds, general remedy; emet. cath. (2431).

hexavitamin capsules & h. tablets (USP XIV): dispensing forms of a combination of 6 vitamins intended for daily usage: vits. A (natural), B_1, C, D, G, & nicotinamide; u. where several vitamins are thought needful, as in some avitaminoses; keep away from light.

hexham-scent: *Melilotus indica.*

HGH: Human Growth Hormone (v. Somatotropin).

Hibiscus L. 1) (Malv): genus of rose-mallows found throughout trop. & temp. zones of the world; lvs. of some spp. u. as potherb; mucilage hypoglycemic; bk. of some as fiber 2) (specif.) sd. of **H. abelmoschus** L. = *Abelmoschus moschatus* - **H. articulatus** Hochst. (Tanz): raw rts. chewed for cough - **H. bifurcata** Blanco = *H. surattensis* - **H. brasiliensis** L. (H. phoeniceus Jacq.) (WI, CA, nSA): (palo) peregrino (Cu); borrachita (Cu); mañanitas (Salv); altea (Cu); lvs. sour-tasting, u. for cooling drinks in "bilious fevers" (254); diaph. - **H. cannabinus** L. (eHemisph, tropWI, escape fr. cult.): kenaf; Ambari (or Deccan) hemp; furnishes important fiber of Old World; Indian h.; bimdipatam jute; replacement for jute; u. to make sacks; sd. FO u. food, illumination, & paint oil; st. u. for newsprint; young lvs. & sprouts as veget. (pot herb & in curries); fl. juice u. in gallbladder dis., lvs. as purg.; rt. u. med. (Br); cult. Ind, seEu - **H. esculentus** L. = *Abelmoschus e.* - **H. ficulneus** L. = *Abelmo-schus f.* W. & A. - **H. manihot** L. = *Abelmo-schus m.* - **H. micranthus** L. f. (Af, As, EI): rts. u. as purg.; for stomachache, cough (Tanz); poultice for sores - **H. moscheutos** L. (H.. palustris L.) (e&cUS): common (or swamp) rose mallow; sds. cordial, stom. dem. emoll., nerve sedative; rt. & lf. sedative - **H. physalioides** Guill. & Perr. (tropAf): lvs. in form of paste u. for boils (Tanz); fiber plant - **H. pilosus** Fawc. & Rendle (H. bancroftianus Macfad.) (WI, Mex to Pe): cupidito (Cu); oil u. in med. like *Althaea officinalis* - **H. rosa-sinensis** L. (tropAs, now widely cult.): Chin rose; rose of C.; Chinese (or red) hibiscus; shoe flower; fls. u. contraceptive, abort. oxytocic, emmen. (Indones); astr.; to treat cellulitis (Tahiti); juice for black dye; expect. emoll.; lvs. formerly in small pox (Cu); rts. u. like althea; cult. large shrub ornam. - **H. sabdariffa** L. (orig. tropAm. cult., WI, trop-As, Af): roselle, Guinea/Jamaica/(East) Indian/Florida sorrel; "Jamaica dogwood" (error) (160); Florida cranberry; rama; tall (3-5m) ann. hb.; fleshy red sepals & fl. bracts ("Jamaica flowers," Karkade flowers; "florets") u. as flavor, acidulous refrigerant, in agreeable astr. beverages (hibiscus tea; sorrel drink; Karkade tea; as tea substit.) (EI, WI), in jellies (*cf.* cranberry), sauces, tarts, tea mixts., cooked (as "cranberries"), diur. mild lax.; in vin d'oseille (ante 1793) (160); lvs. u. as salad (eAf) & in decoc. as sed. or emoll. (local); st. fiber makes good cordage ("kenaf") but less valuable than the kenaf fr. *H. cannabinus*; u. to make ropes, bags (esp. coffee b., CA), bagging, textiles, substit. for jute (adv: very durable, impervious to moisture); called roselle fiber (EI), r. hemp, vinagreira, (prod. Mex, Nic, Austral, USSR [nCaucas, cAs]); sometimes mistaken by the layman for marijuana (1596) - **H. schizopetalus** (Mast.) Hook. f. (eAs, eAf, now pantrop.): rose mallow; watutu (Cuna Indians, Pan); u. much like *H. rosa-sinensis* in Mal; fls. u. in eye med. (Pan); u. for hedges & ornam. (large shrub) - **H. surattensis** L. (OW tropics): lf. inf. u. in colds (Surinam); lf. decoc. u. in skin dis. (Mal); lvs. emoll. in cough; lf. or st. lotion for penile irritation (Zulus); inf. inj. into urethra & vagina for GC, &c.; lvs. edible - **H. syriacus** L. (native of AsMin, now common in eAs+): rose of Sharon; shrubby althea; Syrian mallow; fls. of shrub/small tree u. for dys. poor digestion, nausea, "regurgitation," leucorrh. (Chin, etc.); bk. (& rts.) decoc. u. for ascarids, dys. intest. dis., sds. emoll, to treat colds & headache (Chin); pl. insecticid.; rt. in poultices (Miss [US], Negro); stom. diur. expect.; the common garden hibiscus; splendid ornam. - **H. tiliaceus** L. (OW trop., now worldwide trop.): (coast) cotton tree; majagua (Cu); mahoe; rt. bk. u. antipyr. (Ind); lax. (Br); young lvs. & bk. u. in skin dis. (Tonga) (81); fls. rts. & bk. u. aper. emoll. diaph.; trunk bk. furnishes strong cord; wd. light, u. for "cork rose wood"; lf. decoc. in piles (Cu); buds u. for boils (Marianne Is); bk. starvation food (Cu); fls. for tea as diur. (Mal) - **H. trionum** L. (native to Med reg.; now Af, cEu, natd. NA): "flower of an hour"; bladder ketmia; hb./lvs.

u. expect. stom, diur. dem.; fls. in itching skin dis., diur.

hibiscus, wild: *Hibiscus lasiocarpus* Cav.

hiccup nuts: fr. *Combretum bracteatum* (SC prod. hiccups after ingestion).

hicker nuts (Ala [US] country): hickory nuts.

hickery-pickery: aloe & canella powder (hicra picra).

Hic(k)oria Raf. = *Carya.*

hickory (fr. Am Indians): *Carya* spp. - **h. bark borer**: beetle wh. penetrates bk. or wd. of hickory tree - **big shellbark h.**: *C. laciniosa* - **bitternut h.**: *C. cordiformis* - **mockernut h.**: *C. tomentosa* - **nutmeg h.**: *C. myristicaeformis* - **h. nuts**: fr. *C. tomentosa, C. ovata,* etc. - **pignut h.**: *C. glabra; C. ovalis* - **red. h.**: *C. ovalis* - **shagbark h.**: *C. ovata* - **shellbark h.**: *C. ovata* - **swamp h.**: *C. aquatica, C. cordiformis* - **white h.**: *C. tomentosa.*

hiedra (Mex): *Cryptostegia* - **h. terrestre** (Sp): *Glechoma hederacea.*

hiera (Gr. pl. of hieron, sacred thing): sacrificial offerings - **h. picra**: Powder of Aloe & Canella; the oldest known galenical preparation (commonly but incorrectly rendered "hicra picra").

Hieracium L. (Comp) (tempZone except Austral): hawkweeds; rts. of many spp. contain inulin - **H. aurantiacum** L. (Eu, Siber, esp. mts.; natd. eNA): orange hawkweed; devil's paint-brush; c. a yellowish pigment; hb. u. for cough; cult. ornam. - **H. gronovii** L. (e&cUS): vein-leaf; (hairy) hawkweed; hb. u. med. - **H. murorum** L. (Euras, natd. US): French lungwort; hb. (Herba Pulmonariae gallicae) u. in lung dis. (Fr Phar.), ton. - **H. pilosella** L. (Pilosella officinarum C. H. & F. W. Schultz) (Eu; natd. US): mouse-ear hawkweed; hb. c. antibiotic princ.; umbelliferone; luteolin; u. astr., diur., in grippe, bronchial colds, brucellosis, whooping cough, diarrh., worms, lvs. u. as tea substit., against dropsy; juice tumor-inhibiting agt. in folk med. - **H. polyodon** Fr. (sAf): in native med. u. as constituent in agent against sterility - **H. scabrum** Mx. (e&cNA): rough hawkweed; inf. or chewed for diarrh. (Am Indians) - **H. umbellatum** L. (Euras, natd. NA): narrow-leaved h.; pl. (hb.) u. in asthma, cough - **H. venosum** L. (e&cUS): poor robin's hawkweed; veiny h.; rattlesnake weed; pl. (lvs.) said alexiteric for snake bite; ton. astr.; for bowel dis. (Cherokee Indians) - **H. virosum** Pall. (Euras): u. as aper. vulnerary (Sp).

hierba (Sp): herb; also yerba (*qv.*) (latter considered less correct spelling) - **h. amargosa**: ragweed - **h. amarilla**: oxeye daisy - **h. bicha**: *Aristolochia longa* - **h. buena** 1) (Ve, CR, etc.) *Mentha citrata* 2) (Mex) *Hedeoma piperita* - **h. b. piperita** (Mex): *M. piperita* - **h. ceniza** (Mex): guayule - **h. centella**: *Anemone,* etc. - **h. de caballo** (NM [US]): *Senecio filifolius* - **h. de pasmo** (Mex): *Kallstroemia maxima* - **h. de San Antonio**: *Epilobium hirsutum* - **h. de San Juan**: *Artemisia vulgaris* - **h. de San Nicolas** (Mex): *Piqueria trinervia* - **h. de San Roberto**: *Geranium robertianum* - **h. de Santa Maria**: *Balsamita major* - **h. de la Cruz** (Mex): *Croton ciliato-glandulosus* - **h. de la golondrina** (Mex): *Chamaesyce maculata* - **h. de la orina**: *Herniaria glabra* - **h. de las reumas** (Mex): *Frankenia campestris* - **h. del Alacrán**: *Plumbago pulchella* - **h. del burro**: *Spigelia longiflora*- **h. del cancer**: *Plumbago* - **h. del gato** (Mex): *Valeriana procera* - **h. del hule** (Mex): guayule - **h. del negro**: *Malva* spp. - **h. del pollo** (Mex): *Commelina pallida* - **H. dulce de México** (Mex): *Lippia graveolens* - **h. gatera**: catnip - **h. Luisa**: lemon balm - **h. mala**: weed (*cf.* yierba mala) - **h. mora**: (Hond): *Solanum nigrum* - **h. velluda**: *Ranunculus bulbosus.*

hierbabuena (Mex): *Hedeoma piperita.*

hierbajes (Mex): herbs.

Hierochloë odorata Beauv. (H. borealis R & S) (Gram) (nEuras, NA [Alas-Ariz-Pa]): holy grass; sweet g.; Seneca grass; hb. c. coumarin; u. arom. stim.; in Am Indian medicine bundles; to make Indian baskets (Am Indians); perfume; incense (NM [US]); as spice (eAs); in liqueurs & tobacco - **H. redolens** R. Brown (Co): itámo real; valuable med. pl. u. for many dis., esp. in old age - **H. vulgare** L. (eAs): tea considered a remedy for many internal dis., esp. of old age (Indones).

higado (Sp): liver.

"high" (US slang): term descriptive of cannabis & other intoxications; a phase of euphoria.

"highbelia" (**"highbelier"**) (as opposed to "lobelia" [*L. inflata*]) (Cana): *Lobelia spicata* or *L. siphilitica.*

high bush blackberry: *Rubus allegheniensis, R. villosus* - **h. b. cranberry**: *Viburnum trilobum* - **h. gear** (Ala, Tex [US]): corruption of hegari.

High John the Conqueror (US Negro voodoo): piece of John the Conqueror (*qv.*) rt. carried in pocket against the evil eye.

higo (Sp): fig - **higuera**: f. tree - **higuerilla** (Sp): **h. blanca** (CR): **h. coloradá** (CR, Pe): castor bean.

higuereta (Cu): higuerilla **(Sp):** h. blanca **(CR): h. coloradá** (CR, Pe): **higurillo: higuero** (Salv): castor bean, *Ricinus communis* - **higuerón** (SpAm): fig - **h. latex**: latex of *Ficus* spp., esp. *F. laurifolia.*

hijjal (Bengali): *Barringtonia acutangula.*

hijlibadam: 1) *Semecarpus anacardium* 2) *Anacardium orientale.*

hikuli (Tarahumac Indians) (nwMex): hikuri: peyote & related cacti u. as hallucinogens.

hilo (Sp): thread, string, linen fiber - **h. sisal** (Arg): binder twine.

hilum: scar on sd. marking point of attachment to ovary (frt.) wall.

Himalayas (pron. Him-a-lias): large mt. chain of cAs ca. 2400 km long, with several of the highest peaks in the world (incl. Everest); hill stations in foothills (Simla, Murree); home of several important med. pl. spp. - **H. berry**: *Rubus procerus*, a large, very succulent blackberry.

Himatanthus Willd. = *Plumeria*; *Coutinia*.

Himbeere (Ge): red raspberry.

Himmelslilie (Ge-f.) (heavenly lily): *Iris germanica*.

hina (Ind): **hinna** (Arab): henna.

hing (Indo-Pak): asafetida, Afghan type fr. *Ferula narthex.;* Persian type fr. *F. asafoetida*.

hinojo (semillas) (Sp): fennel (frt.) (sds.); also **h. domestico**.

Hinterindien (Ge): Further (or Farther) India: = Indochin, Indones, Mal, Annam, Burma, Siam, etc. (*cf.* Vorderindien).

Hintonia latiflora (M. & S. ex DC) Bullock (Rubi) (Mex): "copalachi"; copalquín; bk. decoc. (in NaCl soln.) for fever; for gastric acidity; bk. soaked in mescal for mal., pinto; purg.; var. *leiantha* (Mex): lvs. u. int. in mal.; lf. decoc. as bath for pinto; bk. much u. in fever.

hip (pl. **hips**): frt. (enclosing the sds.) of *Rosa canina* & other wild *R.* spp.

hipericon (Sp): *Hypericum perforatum*.

Hippeastrum puniceum (Lam.) Voss. (H. equestre Herb.) (Amaryllid) (tropAm, now pantrop, seAs, Indones): Barbados or fire lily; "red spider lily"; c. alk. (lycorine; tazzetine, trisphaeridine); u. in poultices for quinsy & mumps (Java); bulb crushed in water, boiled or roasted & eaten for purg. in stomachache (777); cult. ornam.; this sp. may represent the true Amaryllis Belladonna - **H. reginae** Herb. (tropAm; wAf): azucena de México (Mex); very toxic bulbs u. by natives for arrow poisons - **H. vittatum** Herb. (trop&subtropAm, esp. Br): "Amaryllis"; c. alk. (lycorine, tazettine, hippeastrine); u. as antiasthm. (folk med.).

hippo (seUS, Negro, corruption of word): ipecac.

Hippobroma longiflora G. Don f (1) (Campanul.) (tropAm—>pantrop): juice skin irrit.; frt. toxic; lvs. drastic; cold rem. (Jam); to treat VD, bronchitis, asthma, rheum. (*Isotoma l.*, *Laurentia*).

Hippobromus pauciflorus Radlk. (1) (Sapind) (sAf): horsewood; lf. juice u. for eye inflamm. & corneal opacities; rt. u. in diarrh. dys. (Zulu); bk. u. as emet. & analg. for headache.

Hippocastani Semen (L): **Hippocastanum** (L): sd. of *Aesculus hippocastanum*.

Hippocrates (ca. 460 - ca. 377 BC): "father of medicine"; rationalized practice of healing; advocated good diet, fresh air, clean water, exercise, etc.

Hippoglossus hippoglossus L. (Pleuronectidae - Pisces) (nAtl & Pac Oc): halibut; liver is expressed for h. l. oil, with about 60x vitamin (A & D) of codliver oil; fish also furnishes important comml. meat.

Hippomane mancinella L. (Euphorbi) (along beaches, Fla Keys [US], WI, CA, Ec, Co, Br, seAs): manchineel (tree); manzanillo; manzanita (Mex); c. toxic princ. hippomanins A & B; fresh frts. c. physostigmine; latex is strong irrit. poison (—> dermatitis), violent emet. cath.; formerly believed dangerous (even lethal) to lie under tree, esp. when in flower; frts. eaten at times but toxic, sometimes leading to paralysis & death; frts. diur. purg.; latex u. in syph. & as arrow & dart poison (Carib Indians); equal parts lvs. bk. & frt. u. by homeop.; lvs. u. in paralysis, exanthemata, scabies; wd. valuable but hard to work because of blisters on hands (antidote, salt water); lvs. c. xanthoxylin (brevifolin); smoke irritant to eyes (2787).

Hippophaë rhamnoides L. (Elaeagn.) (Euras): sea buckthorn; purging thorn; Sanddorn (Ge); Seedorn (Ge); scheitbeziën (Neth); berries (scheitbessen, Neth) rich in vit C, edible, u. for jelly, sauce for fish, etc.; concentrate u. in med.; lvs. & fls. u. in rheum. gout, exanthemata; u. in stomachache (Tibet); wd. excellent; oil fr. frts. is u. exter. to protect the skin fr. sun ray injury (Russia) (154).

Hiptage benghalensis (L) Kurz (H. madablota Gaertn.) (Malpighi) (seAs, Ind, Burma to Timor): beinnwè; lvs. u. in skin dis.; bk. (with hiptagin) bitter (Burma); lf. juice appl. to scabies; insecticide.

Hiraea apaporiensis Cuatrecasas (Malpighi) (Co): tea fr. lvs. u. to treat conjunctivitis (Maku Indians); similarly u. is **H. schultesii** Cuatrecas (Makuna Indians of middle Apaporis [Co]).

Hirschfieldia incana (L.) Lowe (Cruc) (seEu): hoary mustard; pot herb eaten by ancients (Greece & Rome) (Lapsana); sds. u. as mustard substit.; mild lax. (*cf.* broccoli); diur.; bee pl.

Hirschhornsalz (Ge): ammonium carbonate; salts of hartshorn; u. in baking.

Hirschkolbensumache (Ge): *Rhus typhina*.

Hirse (Ge): *Panicum miliaceum*.

Hirtentaschlein (Ge): *Capsella bursa-pastoris*.

hirudin: dr. extn. prod. derived fr. buccal glands of leech; u. as anticoagulant; better yield fr. European l.

Hirudinea (Annelida): leeches (class); parasitic worms with anterior & posterior suckers; found throughout the world.

Hirudo (L): **H. medicinalis** L. (Sanguisuga officinalis Sav.) (Hirudinea-Annelida) (Eu): European medicinal leech; (blood-sucking) leech; green l.; blood sucker; the living animal u. to remove blood ("leeching") fr. bruises, inflammations, blood stagnation, hemorrhoids, etc.; recently to remove blood clots from tissue or organ transplants (exp. fr. Sp; Por; Hungary) (2228); many other spp. u. similarly; ex. *H. quinquestriata* Schmarda (Austral); *H. decora* (= *Macro-bdella decora*) (NA).

histaminase: enzyme extd. fr. hog kidney wh. inactivates histamine by oxidative deamination; formerly u. in prophylaxis & treatment of allergy (per ora) (Torantil).

histamine: primary amine found in normal tissues; esp. rich after protein putrefaction; made by fermentation of histidine, also by synth.; cause of food poisoning fr. spoiled fish, etc. (1460); loc. irrit. counterirrit.; u. for shock therapy in treatment of mental patients; u. in diagnostic tests for vascular sufficiency, gastric analysis - **h. phosphate**: the commonly u. salt.

histidine, L-: an essential (basic) amino acid derived fr. fish protamines, blood corpuscles, etc.; higher in potatoes (6%) & beans (3%) than many other foods; gelatin good source; formerly u. in gastrointest. ulcers; pain of endarteritis obliterans; rheumatoid arthritis (large doses) (1597).

histoplasmin: filtrate fr. culture of *Histoplasma capsulatum* (yeast-like fungus); u. in diagnosis of **histoplasmosis** (fatal dis. of reticuloendothelial system).

Hitchenia caulina Baker (Zingiber.) (Ind): source of Indian arrowroot.

hivedross: bee glue (*qv.*).

hive syrup: Squill syrup comp. - **h. vine**: *Mitchella repens*.

"hives" (US): urticaria, allergic reaction involving the skin & underlying tissues.

ho oil: a linalool-free fraction of hosho oil (v. *shiu oil*); u. perfumes, etc.

hoang-nan (Chin) = hwang-nao: chem. prepn. fr. bk. of *Strychnos gaulteriana*; c. strychnine; u. in syph. leprosy, rabies(!); & as arrow poison.

hoarhound: horehound.

hobble-bush (viburnum) (bark): *Viburnum alnifolium*.

Hobbock (Ge): large tin can for shipping materials, such as drugs.

hobo: 1) (Sp) *Spondias oxyacantha* (pl. & frt.) 2) (US slang): itinerant farm laborer.

hockelberries: fr. *Vaccinium myrtillus*.

Hodgsonia macrocarpa (Bl.) Cogn. (H. capniocarpa Ridl.; H. kadam Miq.) (Cucurbit.) (seAs): lf. decoc. for nasal dis., fevers (Sumatra); sd. FO (kadam oil or fat) mixed with *Kaempferia* lvs. & rubbed on swollen breasts (Mal); peeled sds. eaten; opium roasted in sd. FO; close to *Trichosanthes*.

Hoffman's balsam: Oleo-balsamic mixture (NF V).

Hoffmannseggia glauca (Ortega) Eifert (H. densiflora Benth. ex Gray) (Leg) (Calif adj. Mex): pignut; tuberous rts. roasted & eaten - **H. intricata** Brandegee (nwMex): rts. boiled —> red brown dye.

hog fish (US): appl. to many different kinds of fish; that u. in "fish & chips" (Fr fried potatoes) is cod - **h. grass**: *Richardia scabra* - **h. gum**: 1) tragacanth 2) gum fr. *Clusia flaba* 3) gum-resin fr. *Symphonia globulifera* 4) gum fr. *Metopium toxiferum* (also **false h. g.**) 5) gum-resin fr. *Moronobea coccinea* 6) carmania gum - **India h. g.**: karaya - **h. haw**: *Crataegus brachyacantha* Sarg. & Engelm.; also sometimes frt. of *C. oxyacantha* - **hog('s) hips** (swGa [US]): *Crataegus brachyacantha* (frt. dry; shrub grows in pasture, high lands) - **h. plum**: 1) *Spondias mombin* 2) *Ximenia americana* - **hog's bean**: henbane, *Hyoscyamus niger*.

hogla (Bengal, Ind): lvs. of *Typha angustifolia* - **hogweed**: 1) *Leptilon canadense* 2) *Ambrosia elatior* 3) *Eupatorium capillifolium* 4) Scoparius 5) *Lavatera plebeia* 6) *Heracleum sphondylium* 7) *Datura* 8) many other genera & spp. of plants - **hogwort** (sUS): *Croton capitatus*.

Hohlzahn (Ge): *Galeopsis* esp. *G. ochroleuca*, *G. tetrahit*.

hoisin sauce (Chin): red sauce with soy, garlic, etc.

hoja blanca (Hond): *Calathea lutea* - **hoja de poró** (CR): *Erythrina corallodendron* or *E. costaricensis*.

Holadin™: dr. ext. of entire pancreas; u. as starch & protein digestant.

Holarrhena africana A DC (Apocyn.) (w trop-Af): bk. (conessi bark) c. conessine, holarrhetine, holafrine, holarrheline; rts. u. antipyr. vulner. in diarrh. amebic dys.; floss fr. sds. u. to stuff pillows - **H. antidy-senterica** Wall. (Ind, seAs, w tropAf): kurchee bark; conessi bk., bk. & sds. (conessi seeds) c. conessine; rts. u. diarrhea & amebic dys.; sds. as antipyr.; important drug with many marketed prepns. (Ind); in trichomoniasis (trichomonal vaginitis) (Fr); as substit. for emetine (Ind); starting compds. in synthesis of steroid hormones using C_{21} alk. (holarrhimine, holarrhidine, etc.); lvs. u. to roll (or wrap) bidi cigarettes; said to reduce smokers' cough (Bibl. [635]) - **H. congolensis** Stapf (Congo, Af): c. conessine, holarrhenine; u. in dys. - **H. curtisii** King & Gamble (eAs): c. holacurtenine & holacurtin, with antibiotic & cytostatic properties; u. ameb. dys. (Camb) - **H. febrifugum** Klotzsch (trop s&eAf): kumbanzo bk. u. as antipyr. ton. (Mozambique); rt. decoc. u. in VD.

Holcus L. (Gram) (OW temp): broom corn; soft grass; c. 10 spp. - **H. halepensis** L.: *H. bicolor* = *Sorghum halepense* - **H. lanatus** L. (Eu, natd. NA): velvet grass; hay & pasture crop; recommended horse feed; esp. abundant on Pacific coast - **H. sorghum** L.: *Sorghum bicolor.*

holdfast: portion of alga stipe which appears to grip a stone or other substrate.

Holigarna arnottiana Hook f. (Anacardi) (Ind): bibu; trunk juice vesic., u. sometimes as varnish; bk. & frt. u. med. (156); wd. very useful - **H. ferruginea** March (Ind): juice acts like Cardol (*qv.*) - **H. longifolia** Roxb. (Indo-Mal): black varnish tree; juice u in toothache, sores; frt. against enteritis, furnishes a black varnish.

hollow tooth herb: *Galeopsis ochroleuca.*

holly (pl. hollies): *Ilex* spp., esp. *I. aquifolium* <: holm - **American h.**: *I. opaca* - **black h.** : *Ilex* spp. - **box h.**: *Ruscus aculeatus* - **desert h.**: *A triplex hymenelytra* (SC sp. has white holly-like lf.) - **ground h.**: *Chimaphila umbellata* - **knee h.**: box h. - **h. thistle**: blessed t. - **yaupon h.**: *Ilex vomitoria.*

hollyhock (flowers & root): **English h.**: *Alcea rosea*; *A. ficifolia.*

hollywood: guaiac (also holy wood).

Hollywood cup (TM?): coffee substit. made up of figs, barley, & bran.

holm(e) (Eng dialect): *Ilex.*

Holmskioldia microphylla Mold. (Verben) (Mad): lvs. of shrub u. as medicinal for eyes; wd. u. to make canoes, timberwork - **H. sanguinea** Retz. (Ind): rt. juice u. to relieve fever; u. with *Coelogyne stricta* for headache (rural Khasi tribes) (1980).

Holodiscus discolor (Pursh) Maxim. (Ros) (BC [Cana]-Calif [US]): ocean spray; "spirea"; cream brush; ironwood; frt. eaten raw or cooked; bk. decoc. ton. for convalescents & athletes; blossoms cure for diarrh.; inner bk. inf. for eye wash; sd. decoc. for chicken pox, black measles, etc.; wd. fire resistant, u. for many tools (w coast Am Indians) - **H. dumosus** Heller var. *glabrescens* Hitchc. (wNA-mt. states): desert ocean spray; lf./st. decoc. u. VD; stomach dis., diarrh. (Nev [US] Indians) (158).

holoenzyme: result of combn. of apoenzyme with coenzyme.

holoproteins: simple proteins & scleroproteins.

Holoptelea integrifolia (Roxb.) Planch. (Ulm.) (tropAf, Ind, Burma); Indian elm; kanju; bk. u. to treat rheum. (Burma); juice of boiled bk. in rheum. swellings (Ind); sd. edible, c. 50.5% FO, butter-like, mostly oleic acid ester; lvs. u. in ringworm, scabies, & other skin dis.; wd. of large tree important lumber, dry distd. —> charcoal, methanol, &c.

holoside: glycoside yielding only sugars on hydrolysis in contrast to heterosides; ex. sucrose.

holothurins: complex glycoside found in sea cucumbers (holothurians); very toxic to other sea animals; have digitaloid and antitumor activity; act as local anesthetic; detergent; said to control cancer growth.

holotype: single plant specimens designated by original author of a taxon to serve as the nomenclatural type of the taxon.

Holunder (Ge): elder, *Sambucus* spp.

Holy Cross (wUS): *Opuntia arborescens* - **H. Ghost root**: Angelica - **h. herb**: *Santolina chamaecyparissus.*

holy wood: hollywood.

Holzkohle (Ge): wood charcoal.

Homalanthus: *Omalanthus.*

Homalium fagifolium Benth. (Flacourti) (Chin, cult. Vietnam): astr. rts. u. to treat GC - **H. guianense** Warb. (H. racoubea Sw) (neSA): mavave; shrub; inf. of rts. u. in GC - **H. molissimum** Merr. (Chin): resin fr. bk. soaked in liquor (wine) & u. to heal cuts & wounds (Hainan) - **H. nitens** Turrill (Fiji): bk. of tree scraped & mixed with cold water for use as ton. (strengthening to children & old or weak people); fls. u. in necklaces - **H. racemosum** Jacq. (Mex to cSA; WI): white cogwood; rt. decoc. astr. u. in GC; wd. strong & heavy & u. in carpentry - **H. subsuperium** Sprague (sAf): pd. bk. u. in colic (Zulu) - **H. tomentosum** Benth. (Chin, seAs): rts. astr.; wd. u. to make charcoal for gunpower.

Homalomena Schott. (Ar) (tropAs&Am): various spp. u. in colic, fever, lumbago, wounds, sores, yaws, skin dis.; narcotic poisons (Indones); in cough, pain; rheum (PI) - **H. aromatica** Schott - u. in diarrh. (Indochin) - **H. caerulescens** Jungh.; u. to poultice sores (Mal); - **H. graffithii** (Schott) Hook. f.; decoc. u. to expedite childbirth (Mal).

homatropine: synth. alk. prep. fr. mandelic acid & tropine; u. in form of hydrobromide as mydr. &

cycloplegic; as **h. methylbromide** u. as antispas. in intest. spasm (colic).

hombre grande (CR) (big man): *Quassia amara* - **hombrecillo** (Mex): hops.

"home counties" (GB): those nearest London; formerly important for prodn. of crude drugs, VO, etc.

Homeria Vent. (Irid) (sAf): some 40 spp. of hbs.; causes "tulp" poisoning (digitaloid poisons) (129).

Hominidae: Fam. of Man; Order Primates (Mammalia): distinguished fr. apes by greater devt. of brain, lesser hair on body & limbs, & customary locomotion on rear legs; only one living sp., *Homo sapiens.*

hominy (Am Indian word): coarsely ground kernel of white maize, made fr. steamed softened prod. - **h. lye: big h.**: hulled corn made by soaking in lye.

homocysteine, S-adenosyl: SAH: amino acid which occ. in animal organs, esp. in the intersynaptic space; regulates membrane activity, stim. capture of serotonin & noradrenalin; u. in sleep disturbances ("paradoxical sleep"), depression; improvement of circadian rhythm; represents a new class of psychotropic agts. (1987).

homogenate: pl. or animal tissues reduced to a cream-like consistency in which individual cells are broken up.

homonym: exactly the same name for an organism previously appl. to a different organism.

Homo sapiens L. (Hominidae) (worldwide): human being; man; a primate now generally believed to have been evolved through ape-like forms representing common ancestors of both humans & anthropoid apes; several races distinguished primarily on basis of skin color; the source of several products (mostly biologicals) u. in med.; hair u. to make "hairpieces" (wigs); fetal tissues u. as grafts in exper. treatment of Parkinsonism, diabetes, &c.

Homonoia riparia Lour. (Euphorbi) (seAs): pl. a rheophyte (grows in running water); pd. lf. u. in skin eruptions (Thail); lvs. & frt. as poultice for skin dis. (Mal).

Homoptera (formerly classed as sub-order of Ord. Hemiptera - Insecta): order which includes scale insects, pl. lice, cicadas; possess 4 membranous wings; ex. cochineal.

Honckenya peploides (L) Ehrh. (Caryophyll.) (Arenaria p. L.) (nTemp & Arctic reg. of world): seabeach sandwort; lvs. eaten fresh, pickled, or in oil (Alaska [US] Eskimos).

Honduras bark: Cascara Amarga - **H. sarsaparilla**: fr. *Smilax regelii.*

honewort: name appl. to several Umb. spp., incl. *Cryptotaenia canadensis* (etc.), *Sison, Trinia,* & *Carum* spp. - **American h.**: *Cryptotaenia canadensis.*

honey: Mel (L); saccharine secretion of *Apis mellifera*, the raw material being obt. fr. fl. nectaries; c. equal parts of dextrose & levulose; VO, pollen grains; u. undiluted as antifungal & bactericidal agts. (acts by dehydrating effect); fresh floral h. generally liquid but on standing usually solidifies; coniferous honeys solidify less easily because of dextrin content; u. dem. in cough syrups, expect. nutr. mild lax. vehicle, sweetener, flavor; to "cure" ulcers; said better tolerated by diabetics (levulose); vulnerary appl. to wounds to prevent inf.; leading US producers: Minn, Calif, Ala, Wisc; clover & alfalfa products best known in US; other vars.: pine h. (NCar [US]); sourwood h. (Tenn, nGa [US]); gallberry h. (sGa [US]); mesquite h. (Ariz [US]); milkweed h. (Mich [US]); tupelo h. (non-solidifying) (Ala, Fla [US]) (2503); creamy blue curls h. (fr. *Trichostema lanatum)* (Calif [US]); various forms (2310); besides its use as a delicious food (later replaced by sucrose), h. is & was u. to make alc. beverage; mead (GB, nEu); also h. was/is added to alc. beverages; some h. are toxic (ex. Rhododendron); many folklore medical claims (157); given in small amts. of brandy or whisky for colds (children) (1461); sometimes with vinegar - **h. bloom (root)**: *Apocynum androsaemifolium* - **h. bread**: *Ceratonia*- **h. flavor oil**: tincture of styrax - **Grecian h.**: obt. fr. wild thyme - **h. locust**: *Gleditsia triacanthos* - **h. pod (tree)**: *Prosopis juliflora* - **Scotch h.**: obt. mostly fr. heather.

honeydew: 1) a sweet sugary liquid secretion formed by some lower plants (fungi) & lower animals (aphids, scale insects, etc.) occurring on fls. of seed pl., on epidermis of pine lvs.; also exudate fr. surfaces of some galls; incl. liquid secreted by sphacelial (conidial) stage of ergot fungus 2) saccharine secretion fr. lvs. & frt. because of excess water (when due to deficiency = gumming).

honeysuckle: *Lonicera* spp. esp. *L. caprifolium* (Appalachian, native); *Rhododendron* (*Azalea*) spp. - **bush h.**: *Diervilla lonicera* & some other *D.* spp. - **h. family**: Capri-foliaceae - **garden h.**: *L. caprifolium* - **Japanese h.**: *Lonicera japonica* - **Montana h.**: *Lonicera dioica* - **h. stick** (US slang): sugar cane - **trumpet h.**: *Lonicera sempervirens* - **white swamp h.**: *Rhododendron viscosum.*

honeywort: 1) *Cerinthe* spp. esp. *C. major* (SC yields much honey) 2) *Galium cruciatum* (SC fls. fragrant like honey).

honghelin: glycoside of digitoxigenin & D-thevetose, occ. in *Adenium honghel.*

hongheloside A: glycoside of oleandrigenin & D-cymarose & glucose; occ. in *Adenium honghel* - **h. G**: somalin.

Honig (Ge): honey.

hoodoo (doctor): **h. man** (seUS, Negroes): voodoo (*qv.*) practitioner.

hoodooism: voodooism but with less tendency to supernaturalism/superstition.

hoodwort: *Scutellaria lateriflora.*

hoof meal: horn meal.

hoopwood: *Ilex laevigata.*

hootch (US slang): intoxicating liquor (SC Hoochinoo, Alaska [US] Indian tribe, notorious abusers of alc.).

hop: 1) *Humulus lupulus* (hop flowers) 2) (US slang) smoking opium or other addictive narcotic 3) *Rosa canina* (frt. & sd.) - **h. head: h. hog** (US slang): addictive drug addict, esp. of opium - **h. hornbeam**: *Ostrya virginiana* (SC frt. resemblance to hops) - **h. powder**: lupulin - **h. tree**: *Ptelea* spp., esp. *P. trifoliata* - **h. vine**: pl. of *Humulus lupulus* - **wild h. vine**: *Bryonia alba*; *Humulus lupulus*; spp. of *Canavalia*, *Polygonum*, *Stachys*, etc.

Hope's camphor mixture: acid c. mixture.

Hopea Roxb. (Dipterocarp.) (seAs, esp. EI): furnishes damar; a good deal of confusion occurs in distinguishing copal & damar, since the natives call them by the same name & they look a good deal alike - **H. micrantha** Hook. (Java): source of damar mata kuching, best grade of gum damar; u. in med. plasters, shellacs, varnishes - **H. odorata** Roxb. (Burma, Indochin, Mal): thing wal; bk. u. in dental dis.; lotions for stiffness & pain; urinary incont.; hardens gums & u. in gingivitis, retards loss of teeth; damar sold in bazars, u. as vulner.; street tree.

Hopfen (Ge): hops.

Hoppea dichotoma Willd. (Gentian) (Ind): pl. u. in hemorrhoids, snakebite.

hoppie (US slang): **hoppy**: opium smoker or other drug addict.

hops: hop - **Spanish h.**: *Origanum vulgare.*

hop-toy (US slang): container to hold smoking opium.

hordas: figs of poor quality.

Hordei Fructus (L.): barley (grain).

hordein: characteristic protein of barley.

hordenine: alk. found in sprouting (germinating) barley (*Hordeum vulgare*); member of sympathomimetic group; u. diarrh. dys; cardioton (*cf.* epinephrine); usually as **h. sulfate**.

Hordeum distichon L. (Gram.) (Euras): 2-row(ed) barley; cult. for several millenia by man; in form of many vars. the most commonly cult. barley in US; important economic cereal grain, most u. in brewing beer (as malt) (rich in diastase) & liquor indy.; produces poor bread since inferior in rising power; other prods.: pearl barley (kernel) (decorticated grain, popular in home med. in mucosa inflamm. [1462], as b. water in gastroenteritis, diarrh.; for convalescents; as nutr. in soups & stews; roasted as coffee substit.); b. farina, b. flour (—> poor bread, since inferior in rising power), malt sprouts, brewer's grains, diastase, beer, whisky, etc.; cultd. in Eu, US (NDak & adjacent states) - **H. hexastichon** L. (Euras): 6-row(ed) barley; ancient type of barley found in antique tombs of Eg - **H. jubatum** L. (Euras, NA): wild barley; "foxtail barley"; squirrel-tail grass; stock poison (through mechanical injury); cult. as ornam.; troublesome weed in Alta (Cana), esp. in *Silene* glades - **H. murinum** L. (Eu; natd. wNA): mouse barley; barley grass; "Schlief-hannsel" (sleepy Hans); very common pl. in waste land; pl. u. in diarrh. dys. (sAf); antibiotic prop.; as "native coffee" (Calif [US]) - **H. spontaneum** Koch (wAs): wild barley; probably the ancestor of cult. spp. - **H. vulgare** L. (H. sativum Pers.) (Euras; a cultigen): (common) barley; 6- or 4-rowed b.; widely cult. over the world; believed oldest cult. cereal of man; u. food (breakfast cereals [nutr. flavor], soups, infant feeding), prep. malt (hence in beer, ale, stout, porter, & in whiskies & gins); starch (in powds., pills, technology, etc.); v. also under "barley," *Hordeum distichon.*

horehound (herb): *Marrubium vulgare* - **black h.**: *M. peregrinum, Ballota nigra* - **b. (stinking) h.**: archangel; *Lamium* spp., *Galeopsis; Stachys* - **foetid h.**: *Ballota nigra* - **water h.**: *Lycopus europaeus, L. americanus* - **white h.**: h. - **wild h.**: *Eupatorium rotundifolium* var. *saundersii.*

hormiga (Sp): ant.

hormiguillo (Mex): beverage of pd. biscuit, sugar, & spice boiled together to make a sweet syrup.

hormone: endocrine; a substance which is secreted by an animal tissue or organ into the blood stream, from which medium it exerts its effects on the body; pl. hormones of comparable nature are prod. by pl. organisms - **fertility h.**: gonadotropin releasing hormone (Gn RH) - **growth h.**: somatotropic h. - **h. (, ovine) corticotropin releasing h.** ([O]CRH): hypothalamus substance wh. stimulates secretion of corticotropin (ACTH) & endorphins fr. pituitary gland; isolated fr. sheep brain - **sanity h.**: serotonin - **thyrotropic h.**: one wh. stimulates the thyroid gland.

hormones, anabolic: u. to increase muscle mass, incl. androgens, somatotrophin, estrogens (slight effect) - **anterior pituitary h.**: a complex of endocrines, which includes Prolan A (folliculotropic hormone), Prolan B (luteinizing h.), prolactin, thyrotropic h., growth h., diabetogenic h. - (ovine) **corticotropin releasing h.** ([O]CRH): hypothalamus substance wh. stimulates secretion of corticotropin (ACTH) & endorphins fr. pituitary gland; isolated fr. sheep brain - **diabetogenic h.**: raises blood sugar by antagonizing insulin - **follicle stimulating h.**: appears to stim. follicles of ovary to secrete estradiol & in males to stimulate formation of androgen by the interstitial tissues of the testis - **folli-culotropic h.**: Prolan A (*qv.*) - **lactogenic h.: lactosecretory h.**: prolactin (*qv.*) - **local h.**: kinins - **luteinizing h.**: Prolan B (*qv.*) - **somatotropic h.: Somatotropin.**

hormonologie (Fr): **hormonology**: the science of endocrines.

horn chestnut: v. chestnut - **h. meal**: nitrogenous manure mfrd. by subjecting hoofs, horns, & claws to steam under high pressures, then drying & grinding the product; c. 14% N - **h. seed**: ergot, *Claviceps purpurea.*

hornbeam: *Carpinus betulus* (**European h.**); *caroliniana* (**American h.**) (SC u. to make yokes between horns of draft oxen) - **hop h.**: *Ostrya virginiana.*

hornet: *Vespa* (*qv.*).

horno (SpAm): round flat baked clay plates on which cassava, etc., is baked.

hornworts: Anthocerotae (closely related to the liverworts [Bryophyta]), sometimes specif. *Ceratophyllum (demersum).*

horopito: *Drimys axillaris.*

horse: 1) *Equus caballus* 2) (US slang): "the horse": Heroin - **h. balm (citronella): Canadian h. b.**: *Collinsonia canadensis* - **h. brimstone**: the impurities formed by the purification of sulfur - **h. foot (leaves)**: coltsfoot - **h. gentian: h. ginseng**: *Triosteum perfoliatum* - **h. gowan**: Matricaria - **h. physic** (early XIX cent.): aloe - **red. h.** (US slang): brownish heroin sometimes seen on the street - **h. salts(s)**: sodium sulfate - **h. sugar**: *Symplocos tinctoria* (SC lvs. sweet & very attractive to foraging horses) - **h. sulfur**: h. brimstone - **h. weed**: 1) (Indiana) h. balm 2) *Conyza canadensis* (also **h. w. fleabane**) 3) *Lactuca canadensis* - **horse**: v. under nettle, savin.

horse-chestnut (frt. & bk.): *Aesculus hippocastanum* - **horse-flesh mahogany**: *Lysiloma latisiliqua* - **horse-head:** *Inula helenium.*

horsemint: *Monarda* spp., esp. *M. fistulosa* & *M. punctata*, *Mentha longifolia* - **American h.**: *M. punctata* - **balm h.**: *M. didyma* - **common h.**: *M. fistulosa* - **dotted h.**: *M. punctata* - **European h.**: *Mentha longifolia* - **green h.:** *Monarda clinopodia* - **purple h.**: *Monarda fistulosa*+ - **round-leaved h.**: *Mentha rotundifolia* - **white h.**: *Monarda didyma* (white hort, var.); *M. fistulosa* var. *alba*; *Pycnanthemum montanum* (NCar [US]).

horseradish (root): *Armoracia rusticana* - **h. tree**: 1) *Moringa oleifera* 2) (Austral): *Codonocarpus cotinifolia* - **wild h.**: h.

horseshoe crab: members of Xiphosura - Merostomata (Arachnoidea - Arthropoda) (hence related to spiders, scorpions, mites, &c.); found on marine shores of tropics & temp zones; body shell distinctive with terminal telson at back; -40 cm (not incl. telson); *Tachyphleus* (e coast As); u. as aphrod. (2616); flesh & eggs eaten (Chinese); *Limulus polyphemus* (NA: NS [Cana]-Yucatán [Mex]).

horsetail (rush) (Eu): *Equisetum majus* - **h. tree**: *Casuarina equisetifolia* - **woody h.**: *Ephedra* spp. (v. under grass).

horseweed:. v. horse weed - **horse-wicky** (Fla [US]): *Pieris nitide.*

Horsfieldia glabra Warb. (Myristic.) (Indones): ki toongila; bk. & lvs. u. for intest. dis. - **H. iryaghedi** Warb. (Ind, SriL, Indones): fragrant fls. u. in perfumery (VO); wd. of good quality; sd. FO u. for candle making - **H. kingii** Warb. (eInd, seAs): sd. u. as substit. for areca; dr. juice of bk. (incisions) u. as substit. for Malabar kino; u. in mouth sores; frt. edible and intoxicating.

hortelã 1) (Por): mint 2) (Br) *Mentha piperita* - **h. comum** (Por): *M. spicata* - **h. pimenta**: *M. piperita* - **h. do Brasil**: *M. arvensis* var. *piperascens* - **hortelana** (Sp): *Mentha* X *verticillata* - **hortelão apimentada** (Por): peppermint - **hortensia** (Sp): *Hydrangea macrophylla.*

hôshô (Jap): shiu oil - **h. tree**: source of h. oil & of ho oil; subsp. of camphor tree.

Hoslundia opposita Vahl (Lab) (trop sAf): mvua (Swahili); rt. decoc. in cough; lf. inf. u. as wash in fever; rts. & lvs. inf. in fever & colds; rt. decoc. in TB, epilepsy (Tanz); u. in blenorrhea, cystitis; hookworm; abdom. pain (Shambala); VO with vanilla odor; frt. edible; var. *verticillata* Bak.: u. as antipyr. (lf. & rt.) (Tanz).

Hosta Trutt. (Lili) (Chin, Jap): plantain lily; several spp. eaten; popular garden ornam. - **H. lancifolia** Engl. (H. japonica Voss) (Jap): narrow-leaved plantain lily; var. *albomarginata* (Hook.) Onwi (Jap, popular cult. in US); rt. as tea for cough, hemoptysis (Clerhee Indians) - **H. plantaginea** Aschers. (Indoch): juice or fl. appl. to dress abscesses & skin dis.; pl. cooked with pork—> prepn. for cough, leucorrh.; decoc. for cough - **H. sieboldiana** Engl. (seAs, Jap.): seer-sucker plantain lily; crushed rt. appl. as plaster to cure abscesses; lf. cooked with pork eaten to relieve headache - **H. ventricosa** Stearn (Chin, Jap, Sib): blue plantain lily; Kau Tsit Lin; lvs. of this & some other spp. similar to those of *Convallaria*; u. as ornam. for both lvs. & fls.; u. med. (Chin).

Hottonia palustris L. (Primul.) (Euras): water violet; hb. u. as refrigerant in homeopath. med. (159); cult. aquarium pl.

hot drops: Tr. Capsicum & Myrrh (NF V).

houblon (Fr): hops.

Houmiriaceae Juss. (*Humeriaceae*): fam. (8 genera) of tropAm rain forests (one sp. in Af); closely related to Linaceae; *Houmiri* (Humiria) spp. prod. a balsam; some frts. of members edible; some frts. & sds. source of FO u. in domestic economy of SA; wd. u. in construction (97); genera with med. values: *Humiria, Humiriastrum, Sacoglottis, Schistostemon, Vantanea* (1463).

hound's tongue (common): *Cynoglossum officinale.*

houseleek, common: *Sempervivum tectorum* - **small h.**: *S. acre.*

Houstonia = *Hedyotis.*

houx (Fr-m.): holly.

Hoya australis R. Br. (Asclepiad.) (Austral+): waxflower; w. plant; hb. c. powerful cardiotonic glycoside; pl. u. for cough in children (Fiji); stock poison; white fls. worn as adornment; honey pl. (1598) - **H. carnosa** R. Br. (sChin, Indochin, Austral): wax pl.; hb. c. sterol glucoside; lvs. u. to hasten maturation of furuncles & anthrax (Indoch); honey pl.; cult. ornam. climbing shrub in greenhouses, etc.; several other *H.* spp. are u. in med., incl.: 1) **H. coriacea** Bl. (Mal): for coughs & asthma 2) **H. diversifolia** Bl. (Mal): lf. decoc. u. in hot baths for rheum. (Mal) 3) **H. imbricata** Dcne. (PI): u. like *H. carnosa;* charred lvs. powed. & mixed with coconut oil; appl. to old wounds & varicose ulcers to speed healing (161) 4) **H. rumphii** Bl. (Mal): latex u. to treat poisonous fish stings; lf. decoc. in GC.

huacacachu: *Datura sanguinea.*

huaco (Mex): v. guaco.

huage (Mex): *Pithecellobium brevifolium+.*

huajilio (Mex): *Leucaena esculenta.*

huang-yao (Chin): *Rhamnus crenata.*

huanto (SA): toxic bever. prep. fr. *Datura arborea.*

Huanuco: v. *Erythroxylon coca.*

huapi bark: h. corteza (Bo): *Guarea rusbyi.*

hubbly-bubbly (Ind): cannabis (may be SC hubab or habab, bubble, since often smoked in hookah or water pipe).

huckleberry (word formed possibly by corruption of "whortleberry"): term appl. to several bush frts., most correctly to edible black or dark blue frts. (more seeds than blueberry) of *Gaylussacia* spp., esp. *G. baccata, G. resinosa*; often also synonymous for blueberry (*Vaccinium* spp.) (incorrect usage) - **black h.**: *Gaylussacia baccata* - **blue h.**: 1) *Vaccinium angustifolium* 2) *Gayiussacia frondosa* 3) *Solanum melanocerasum* - **h. family**: Vacciniaceae (Ericaceae) - **garden h.**: frt. of *Solanum nigrum* - **hairy h.**: V. *hirsutum* - **he-h.**: *Lyonia ligustrina* - **high bush h.**: V. *corymbosum* - **h. paper**: Uffel-man's test paper made with ext. of h. frt.; indicates reaction; grayish blue (alkali) —> red (acid) - **squaw h.**: *Polycodium stamineum* - **tree h.**: *Batodendron* spp., esp. **winter h.**, *Batodendron arboreum.*

huckster: peddler.

Huechys sanguinea (Cicadidae - Homoptera - Insecta) (Chin, Jap, sAs): Chinese blistering cicada; chu-ki (red medicinal cicada); often found on *Ailanthus* trees (1599); u. vesicant.

huesillo (Mex): Cascara Amaraga - **hueso** (Sp): bone.

Huflattich (Ge): **Huflattig** (Ge): coltsfoot.

Hugonia mystax L. (Lin) (Ind, SriL): climbing flax; bruised rts. u. to reduce inflamm. swellings (exter.); antipyr. anthelm. elixiteric (int.) - **H. planchonii** Hook. f. (trop wAf): u. med. in dermatitis; compn. in treatment of abscesses, ulcers, & tumors of all kinds; lvs. u. as inf. or by chewing for chronic cough; pertussis, etc.; cult. ornam. (Liber) - **H. platysepala** Welw. (Congo+): u. in native med.; frts. considered lax. - **H. serrata** (Réunion Is): u. med.

Huhn (Ge): hen.

huile (Fr): fatty oil (usually), as in h. de cade, h. d'arachnide (peanut), h. de cottonier (cottonseed), h. essentielle (VO), h. de foie de morue (codliver), h. de maïs (corn), h. d'olive - **h. volatile de goudron**: pine tar oil - **h. v. de térébinthine**: turpentine oil.

huillca (SA): hallucinogenic snuff; *cf.* cohoba.

huisache: huizache (Mex): *Acacia farnesiana* & some other *A.* spp.

huitlacoche (Mex): *Ustilago maydis.*

hule: *Castilla elastica.*

hull = shell = shuck = peel (ex. walnut h.).

human (being): *Homo sapiens* L; many parts of the body have been u. in med. in earlier times or currently; blood & its fractions; menstrual blood; urine; feces; semen; hair; human fat; brain; milk; oil extracted fr. hair; ash of hair, &c. - **h. immune globulin**: v. Globulinum Humanum Immune - **h. measles immune serum**: sterile serum of healthy h. blood fr. persons who have survived cases of measles - **h. scarlet fever i. s.**: similar to prec. fr. individuals surviving attack of s. f. - **h. skull**:

compacted cranial bones of human; u. in med. several centuries ago (Eu).

Human Immunodeficiency Virus (HIV): discovered at the Pasteur Institute (Paris); develops into Aquired Immunodeficiency Syndrome (AIDS) dis.; an effective vaccine has not yet been discovered.

humic acids: complex org. compds., incl. carboxy acids, of unknown struct. formed in soil by breakdown by soil microorg. of org. subst. contg. CHON; occ. 1-10% in surface soils.

Humiria balsamifera St. Hilaire var. *subsessilis* Cuatrecasas (Humiri) (Co): dr. bk. appl. to cuts & wounds to hasten healing - **H. crassifolia** Mart. (Co): using lvs. & fls. similar to prec.; resin ("umiri") exudation u. as stim. diur. expect. tenif.

Humiriastrum piraparanense Cuatrecasas (Humiri) (Co): bk. chewed for toothache; inf. emet. purg. - **H. villosum** Cuatrecasas (Co); u. similarly to prec.

humoes(e): (Ge): abounding in mold or humus.

humulene: a terpene, chief constit. (60%) of VO of hops.

Humulin™: human insulin of recombinant origin (Lilly).

humulon(e): lupulone; alpha-bitter acid; antibiotic (antimicrobial) ketone of hops.

Humulus japonicus Sieb. & Zucc. (H. scandens Merr.) (Mor/Urtic/Cannabid) (Jap, Taiwan, eChin): kana-mugura; Jap hops; leafy shoots decoc. u. for coughs & colds; diur. & g. u. ton.; frt. bitter ton. stom. diur. in dys. (Chin); lf. decoc. u. as antipyr. & antidys. (Indochin); cult. ornam. - **H. lupulus** L. (eEu, later Euras, now on all cont. in Temp Zone): common hop(s); strobiles (Humulus [L], Strobili Humuli [L]) c. VO, resin (c. humulon), tannin, VO complex (with 96+ constituents) richest in glandular scales (called collectively Lupulin [*qv.*]); u. stom. nerve sed. anaphrod. diur. mild hypnotic, bitter ton. antispasm. in gout; in mfr. beers & ales & cereal beverages; gives sl. bitter flavor and distinct aroma & flavor to beer wort; is bacteriostatic & preservative; acts like yeast in bread-making; has estrogenic activity (1600); young tops u. in salads (in Roman days & still in Belg); u. in poultices (for swellings & bruises) (in Poloris dental poultices) & hot teas ("hop tea") (sed.) (Saaz [or Bohemian]); seedless hops = popular Eu var. fr. CSR; brown, yellow dye (1601).

humus: black or brownish amorphous subst. in soil; colloidal complex org. matter formed by decay of plant, (chiefly) animal, and microbial organisms resistant to decomposition; useful in soils, adds to fertility; absorbent, filler, reagent; source of humic acid.

Hundskohl (Ge): *Apocynum cannabinum.*

Hungarian: v. under balsam, turpentine.

Hunteria eburnea Pichon (Apocyn) (trop wAf): bark of small tree rich in many alks., incl. eburnamine; bk. in aging & wasting dis. such as myocardial weakness, hypertension, sinus tachycardia, arteriosclerosis, cerebral sclerosis, sleep disturbances, interrupted convalescence - **H. elliottii** (Stapf) Pichon (Picralima e. Stapf) (tropAf): locally u. for serious dis. characterized by hepatic dysfunction; action analg., lax.; frts. consumed during wintering (Seneg) - **H. umbellata** Hall. f. (wAf, esp. Nig, Ghana): u. med. under name "erin" - **H. zeylanica** Gardn. & Thw. (H. corymbosa Roxb.) (Mal): latex of tree smeared on sores of yaws (frambesia).

huntsman's cup: **h. horn:** *Sarracenia purpurea.*

huon pine: *Dacrydium franklinii.*

Hura crepitans L. (Eurphorbi) (tropAm, incl. WI; cult. Ind): sand-box tree; ceiba blanca (Mex); c. purgante (Ve); jabillo (tree) (Ven); sablier; habillo; huru; assacu (Br); sds. u. emet. drastic; latex u. anthelm. exter. for leprosy, vesicant (veter. med.); to protect against venom-poisoned arrows & spears (Amazonia); for toothache, syph. (Co); lvs. u. in baths & catapl. for rheum.; staminate fls. u. for furuncles; bk. (casca di assacu, Br) u. med.; fish & arrow poison (juice); "sand-box tree oil" (FO) u. as purg.; part of frt. u. to dress chronic ulcers (Indones); false reputation of lvs. & latex in leprosy; trunk u. for planks & dugouts (Amazonia) (name SC use of dr. rts. to hold sand u. to dry ink writing); frt. explodes with a loud noise - **H. polyandra** Bail. (Mex-CA): habal; habilla (de San Ignacio); "ava" (corruption); ovillo; arbol del diablo; frt. when ripe explodes scattering sds.; sds. drastic purg. u. to poison coyotes, etc.; latex u. fish poison; remedy in elephantiasis.

hurain: proteolytic enzume fr. *Hura* spp.

hurd: fiber residue left after hemp is processed; u. as poultry litter, etc.

hurr (Ind): *Terminalia chebula* - **h. burr**: *Arctium lappa, A. minus.*

hurrah bush (Ga): *Pieris nitida.*

hurricane plant: *Monstera deliciosa.*

hursing(h)ar: *Nyctanthes arbortristis.*

husk root: h. wort: *Aletris farinosa.*

huskinaugh: *Datura* sp. material u. in Indian ceremonies; noted by early Am settlers.

Huso = *Acipenser.*

Huxham's tincture (of bark): Cinchona Tr. Comp.

HVC: Hayden's Viburnum Compound: popular propriety with good reputation of relaxing uter.

cramps; u. in menstrual pain, threatened miscarriage, &c.; syrup c. bk. of Viburnum opulus, V. prunifolium, Dioscorea villosa (rt.) & Zanthoxylum americanum.

hyacinth: Hyacinthus orientalis L. (Lili) (Gr, Syr, AsMin): common h.; fls. c. VO, u. perfume previously, but h. oil rarely u. now because replaced by cheaper synth. compds. - **California h.**: *Brodiaea* - **grape h.**: *Muscari* - **h. paper**: indicator made by immersion of blue fls. in strong alc.; u. test for NH_3 (green) - **water h.**: *Eichhornia crassipes* - **wild h.** (Eng): *Scilla nonscripta* (misnomer).

Hyacinthaloë (Ge): *Aletris*.

Hyaenanche capensis Thunb. (H. globosa Lamb. & Vahl) = *Toxicodendrum globosum*.

hyaline: refers to a clear, homogeneous vitreous subst. often found in both normal & degenerated animal tissues (ex. mucin, glycogen, colloid of thyroid gland).

hyaloplasm: paraplasm; fluid part of protoplasm.

hyaluronic acid: HA: a mucopolysaccharide (N-cont.) found in many pl. & animal tissues (esp. connective t.); with the function of cementing cells & reducing spread of liquids; u. in cosmetics as emollient & greaseless lubricant; in matrix formulation; the Na salt mostly u. (NaHA).

hyaluronidase: enzyme in pathogenic bacteria, sperm cells, snake venom, poisonous insects, etc.; breaks down hyaluronic acid (found in protective N-contg. mucopolysaccaride cementing barriers, as in umbilical cord, around egg cells), decreases viscosity of mucins, allowing free invasion of underlying tissues (or cells) by the invading agt.; hence a "spreading factor"; prepd. fr. bull testes; u. to prevent formation of urinary caculi; to aid diffusion and absorption of injected materials (esp. hypodermoclysis solns.); to increase velocity of spread of local anesthetics; in glaucoma; assayed by redn. of viscosity or turbidity of tissue prepns.

Hybanthus calceolaria (L) G. K. Schulze (Viol) (Mex, CA, SA): poaya (branca); purga do campo (Br); u. as substit. for ipecac (c. emetine); exported to Eu for same purpose - **H. enneaspermus** (L) F. Muell. (tropAf to Ind to Taiwan to Austral): rts. u. diur. for urin. dis. & lvs. appl. exter. (Ind); sts. & lvs. u. to prep. cooling liniment for head (Ind); added to food of pregnant & parturient women to strengthen child and prevent after effects (1530) - **H. ipecacuanha** (L) Baill. (Ionidium i. L.; Calceolaria i.) (SA): false ipecac; rts. ("white ipecac[uanha]"; Ipecacuanha branca) u. as emet. expect. like true ipecac & to adulter. same - **H. lanatus** Taub (Br): pacacomha; lf. & rt. inf. extensively u. for cough & grippe (162) - **H. linearifolius** Urban (WI, Mex): naranjito; rt. inf. u. as purg. - **H. parviflorus** Baill. (Br, Co, Arg): "maytecillo"; maintecillo (Arg); u. emet. & as replacement for ipecac - **H. yacatensis** Milisp. (Mex, Guat): ta; rts. ground and mixed with pwd. rt. of *Pisonia aculeata* to apply to snakebite.

hybrid: offspring of animals or plants of differing races, forms, strains, lines, rarely spp. (interspecific) or genera (intergeneric); frequently intermediate in character - **h. swarms**: variable local populations where 2 interfertile spp. or subspp. overlap, with interbreeding and hybridization.

hybrimycins: new antibiotics produced by changing the genetic structure of the microorganism producing them (producing mutatants) or combinations.

Hydase™: brand of hyaluronidase.

hydathode: secretory organs (stomatal modifications) found on foliage lvs. (lf. tips), etc., which serve to separate water in drops; have been compared to the sweat glands of animals - **hydatophytes**: hydrophytes.

Hydergin™: CCK 179; hydrogenated natural ergot alkaloids (as methanesulfonates) vis., ergocornine, ergocristine, ergocryptine; u. as adrenergic blockers; in acute cardiac lung edema; to reduce appetitie (1980).

hydnocarpic acid: fatty acid, $C_{15}H_{27}COOH$, fr. chaulmoogra oil; cyclic f. a. with a single double bond; u. free & as esters in Hansen's dis.

hydnocarpylacetic acid = chaulmoogric a.

Hydnocarpus anthelmintica Pierre (Flacourti) (Thail, Camb, sVietnam): common chaulmoogra tree (or c. t. of Siam); source of **hydnocarpus** or krabao (or lucabro) **oil** expr. fr. sds.; c. 51% hydnocarpic acid; u. in treatment of leprosy, scabies, eczema, rheum., gout; bk. u. for fibers - **H. castanea** Hook. f. & Thoms (Burma): kalaw; FO u. in leprosy - **H. hainanensis** Merr. (eAs): sd. FO u. to treat leprosy & other skin dis. - **H. kurzii** (King) Warb. (Taraktogenos k. King, Hydnocarpus heterophyllus Kurz.) (Burma, Mal): kalaw seed; kalawso; sds. source of chaulmoogra oil (FO); c. chaulmoogric & hydnocarpic acids; u. in Hansen's dis., TB, skin dis. as anthelm.; bk. as antipyr., frt. pulp as fish poison (but fish flesh inedible) (Rangoon); sds. c. cyanogenetic glycoside (742); cult. in many areas, ex. PI, Sunda Is (1464) - **H. pentandra** (Buch-Ham.) Oken (H. wightiana Blume) (sInd, Chin): Wight chaulmoogra tree; sd. & FO ("Maratti oil") c. chaulmoogric & hydnocarpic acids; u. for leprosy, skin dis., rheum., to produce chaulmoogra oil & ethyl chaulmoograte - **H. venenata** Gaertn. (Ind, SriL): kaute; luhrabans sd. FO u. like chaulmoogra oil, in ointments, for Hansen's dis. & various skin dis.

Hydnophytum andamanense Becc. (H. formicarum Jack) (Rubi) (seAS, Indones, PI): swollen woody bases of pl. u. as decoc. for liver & intest. dis. (PI); pounded tuber u. as poultice for headache (Indones) - **H. formicarum** Jack (PI): lf. decoc. u. for treating boils; decoc. of swollen lf. bases u. for liver & intest. dis. - **H. stewartii** Fosb. (PI+): lvs. appl. to sore legs (Solomon Is).

Hydnora africana Thunb. (Hydnor.) (tropAf): kannip; parasitic pl.; u. as astr. in pharyngitis, tonsilitis - **H. longicollis** Welw. (Angola): u. as styptic; much tannin; u. as dye & to preserve fishing nets.

Hydnum L. (Hydnaceae - Hymenomycetes - Basidiom. - Fungi) (worldwide): a large bracket fungus causing heartwood rot in living tree trunks; "mushroom spawn"; some spp. edible, ex. *H. arinaaceus* Bull., *H. fragile* Petch., *H. repandum*, L. ex. Fr., etc.

Hydrangea arborescens L. (Hydrange/Saxifrag) (e&c-US): **mountain** (or **smooth** or **wild**): **hydrangea**; seven bark(s); rhiz. & rts. (Hydrangea, L.) c. **hydrangin** (glycoside), VO resin; u. diur. in kidney stones; cath.; said narcotic; old family remedy; popular ornam. shrub; subsp. *radiata* McClintock (H. r. Walt.) - **H. aspera** D. Don subsp. *strigosa* McClintock (Chin); crushed bk. of st. & rt. mixed with oil into plaster appl. to knife wounds - **H. macrophylla** Ser. (H. opuloides Steud.; H. hortensia DC) (Jap, Chin): greenhouse (or "changeable") hydrangea; common h.; garden Guelder rose; chang shan; stout bush 1-4 m high; fls. pink or blue, rarely white; often grown in greenhouses; fl. decoc. u. in mal. & some heart dis.; too toxic for clinical use (c. alk.) - **H. paniculata** Sieb. (var. *grandiflora* Sieb.) (Ind): lvs. smoked —> intoxication; u. as sweetener in beverages; cont. HCN - **H. serrata** Ser. var. *thunbergii* Sugimoto (H. t. Sieb.) (Jap, Chin, Kor): tea of heaven; u. med.; prepn. kind of tea (ex. Jintan: mouth freshener) (amasha, Jap.).

hydrangea, wild (seUS): wild-growing *H.* spp., esp. *H. quercifolia* Bartr.

hydrastine: h. white alkaloid (comml.): alk. typical of *Hydrastis canadensis*; mostly u. in form of **h. HCl**; as uter. hemostat., in GU, etc. - **hydrastinine**: deriv. of prec. by oxidation; u. oxytocic, uter. hemostat.

Hydrastis canadensis L. (Ranuncul.) (e&cNA): (2311): golden seal; rhiz. & rts. (*Hydrastits*, L.) c. hydrastine, canadine, & berberine; u. alt. antiper. hemostat. astr., ton., esp. for uter. & GIT mucosa; lvs. & sts. also u.; yellow dye; wild pls. largely exterminated, hence now mostly cult. (wUS & Cana).

Hydrilla verticillata Royle (monotypic) (Hydrocharit.) (Euras, Af, Austral, NA): aquarium pl. of the 1960s now a major weed problem worldwide; spreading through seUS (esp. Fla); submerged water hb. menace to navigation in rivers, lakes, etc. (fresh water).

Hydrochrite: barley water.

hydrocolloids: hydrophilic gums of pl. origin (often alginates) u. as stabilizers, emulsifiers, thickening agts., coatings, etc.; for taking impressions of dental inlays (1465) - **hydrocortisone: hydrocortone**: 17-hydroxy-corticosterone; occ. in adrenal glands; this rather than cortisone is believed by some workers to be the chief corticoid hormone which acts to reduce inflammation, itching, etc.

hydrocotarnine: alk. fr. opium; colored yellow by HNO_3; u. as cough suppressant, like noscapine (E).

Hydrocotyle asiatica L. (Umbell) = *Centella a.* - **H. bonariensis** Comm. ex. Lam. (Br, Co, Par, Arg, Pe, Ch, Urug, seUS; tropAf): sombrerito de abad (Pe); erva capitão (Br); juice u. as purg. emet. diur. (SA); as gargle in sore throat & thrush (Af); pl. aromatic; for cosmetic purposes (Br, Par); hb. u. in skin dis.; rts. u. in toothache (Pe) - **H. bonplandii** Rich. (Co, Ec): mixed with salt water to make a paste taken for animal bites (Ec) - **H. javanica** Thunb. (seAs, Austral): u. replacement for *Centella asiatica* (SriL); lvs. ton. diur.; local treatment of skin dis., digestive weakness, nervousness, dys.; inf. hypnotic (Vietnam); crushed pl. as fish poison (Indones+); grd. lvs. rolled up in banana or other lf. & appl. as hot fomentation for dog bite, &c. (Solomon Is); lf. juice & honey u. as wash for swellings (Taiwan) - **H. sibthorpioides** Lam. (H. umbellata Lour. non L) (tropAs to Austral, natd. NA+): pennywort; pl. u. to reduce phlegm, open sinuses, ease inflamm. (ears, nose, throat); appl. as poultice to bruises, styptic to cuts; decoc. as wash for itch; in dys. (Chin, Jap); in skin dis. (Mal); similar to *Centella asiatica,* cult. ground cover - **H. umbellata** L. (tropAm, WI, eUS, sAf, Madag): water navelwort; acarizoba (Br); juice of hb. u. aper. in liver & spleen dis., emet. (Mex); rhiz. diur. (Br); rheum., leprosy (*sic*) - **H. vulgaris** L. (Euras): shilling (or penny) grass; sheep's bane; earlier u. as diur. & vulner.; has hemolytic activity like that of senega rt.

hydrocyanic acid: HCN; prussic acid; natural compd., occurring free in *Pangium edule* sd., etc., also as glycoside (cyanogenetic); thus in lvs. of *Papaver nudicaule* (163); occurs in thousands of pl. spp. as cyanogenetic glycoside.

hydrogen: H: elemental state: H_2; this is the most abundant element in the universe; lightest gas known; formerly u. in balloons, dirigibles, &c., but now generally replaced by helium (non-flammable) which is nearly as light; potential auto fuel; in hydrogenation syntheses, hydrogenation (solidification) of FO, &c.; dangerous explosive (with O_2); H isotopes (deuterium, tritium) u. in medical tests - **h. acceptor**: subst. which takes H atoms fr. another subst. called an **h. donator** - **h. cyanide:** hydrocyanic acid, HCN - **h. peroxide**: H_2O_2; "hydrogen dioxide"; occ. in nature in small amts. in cells; however, prepd. by synthesis; strong oxidizing agt.; u. in therapy in diluted solns. (commonly 3%); u. antisep. antiinfective, antifungal, cleansing agt., thus frequently in mouthwashes.

hydrolase: enzyme wh. catalyzes hydrolysis of compd. by addn. & reaction with water.

hydrolat (Fr): a distilled water - **hydrolate**: water accumulating in condenser after distn. of VO = conobate - **hydrolé** (Fr): medicinal beverage with aqueous base & c. usually pl. medicinal agts.; ex. licorice, anise (*cf.* tisane).

Hydrolea zeylanica Vahl (Hydrophyll) (seAs): pulped lvs. u. as dressings for foul ulcers (Burma); antisep. cleansing; lvs. eaten as lablab with rice (Java).

hydromel (L): mixt. of honey & water; u. in Hippocratic med.

hydromorphone = dihydromorphinone.

Hydrophyllaceae: Water-leaf Fam.; widely distr. but found mostly in NA; many spp. cult. as ornam.; usually hbs., some spp. shrubby; very little known chemically, few hydrocarbons, FO, org. acids, most important economic sp.; *Eriodictyon californicum.*

Hydrophyllum L. (Hydrophyll.) (NA): (American) water leaf; rts. of some spp. u. in back or chest pains (Obijway Indians) - **H. canadense** L. (Atl NA): lvs. (John's cabbage) edible; u. against skin eruptions & snakebite - **H. occidentale** Gray (wUS): western squaw lettuce; lvs. u. as salad greens by some Am Indian tribes - **H. virginianum** L. (e&cNA): John's cabbage; Virginia waterleaf; lvs. earlier u. as cataplasm for festerings; in homeopath. med.; young tender shoots eaten (Am Indians).

hydrophyte: water plant; ex. water lily, cat-tail.

Hydropiper (L): *Polygonum h.*

hydroponics: soilless culture; s. gardening; water culture; pls. grown with rts. immersed in aq. soln. of nutrients, ex. Knop's soln.

hydroquinine: alk. of cinchona; more effective than quinine in malaria, but also more toxic (do not confuse with **hydroquinone**, synthetic photogr. developer).

Hydrosme Schott = *Amorphophallus.*

hydroxy-acids, α–: said to improve quality of skin (complexion); incorporated in some cosmetic ointments.

hydroxycobalamin: vit. B_{12a}; analog of cyanocobalamin in which CN gp. is replaced by OH; u. in therapy; has largely displaced cyanocobalamin.

hydroxy-corticosteroids: hormones similar to **hydroxy-corticosterone** having adrenal cortex activity.

hydroxystreptomycin: reticulin: antibiotic subst. fr. *Streptomyces reticuli* Waksman & Henrici; *Streptoverticillium griseocarneus* Baldacci & al. (Jap. soil), etc.; (1949-); u. much like streptomycin; active against Gm+ & acidfast organisms.

hydroxy-tryptophan, 5-: 5-HTP: occ. in many organisms (ex. sds. of *Griffonia simplicifolia*) & also synth.; precursor of serotonin; useful in postanoxic myoclonus (uncoordinated muscular movements); mongolism; its peptide (isolated fr. tomato rts.) acts as an auxin.

hyeble: *Aralia hispida.*

hyena: hyaena: *Hyaena crocata* Erxleben (Hyaenidae - Mammalia) (Af): spotted h.; dog-like carnivore; scavenges food, also kills prey; cave dwellers - **H. hyaena** L. (Ind, wAs, neAf): striped h.; formerly domesticated (Eg); u. for food & hunting; many parts (ex. flesh, blood, bone marrow, &c.) u. in Unani med.

hyena bane: h. poison: *Toxicodendron capense.*

hygrine: alk. of coca lvs., *Cinchona pubescens,* etc.; liquid volatile subst. decomposed by light.

Hygromycin: homomycin; antibiotic prod. by *Streptomyces hygroscopicus* (Jensen) Waksman & Henrici & other m. o.

Hygrophila guianensis Nees (Acanth.) (WI Mex to Arg): pl. decoc. u. to bathe skin (Pan) - **H. ringens** R. Br. (seAs): u. to treat wounds & edema; juice lightly salted u. as depurative & astr. (Ind) - **H. senegalensis** And. (tropAf): crushed sds. u. in treatment of various ocular dis. (Senegal) - **H. spinosa** And. (Asteracantha longifolia Nees) (Ind, SriL, trop sAf): asteracantha; star-thorn; rt. c. VO (small amt.), stigmasterol, wax, gum, lupeol (hygrosterol), sd. c. phytosterols (asterols); entire pl. u. diur. in dropsy; esp. in liver dis.; jaundice, rheum.; aphrod.; ashes u. diur.

hygrophytes: land pl. which have high water requirements; usually characterized by having weak rts., large thin leaf laminae, poorly developed woody tissues; leaves furnished with hydathodes.

Hygroryza aristata Nees (Gram.) (monotypic) (Ind, Mal): sds. u. cooling, astr. to urin. tract; in "biliousness."

Hylocereus napoleonis (Grah.) Br. & R. (Cereus n. Grah.) (Cact.) (WI, sMex): u. as cath. (St. Vincent); frt. called "mountain peas" - **H. undatus** (Haw.) Britt. & Rose (Cereus u. Haw.) (Mex, WI, CA, SA, orig. home unknown, cult., esp. in Chin, now worldwide): queen-of-the-night; pitahaya; night-blooming cereus (this is best known of various spp. so-called); u. med. (Mex & Col); cult. as hedge; popular ornam. (trop worldwide); frt. esculent, juice u. in cooling drinks & to color candy & pastry; vine (st.) crushed in water —> shampoo.

Hymenaea L. (Leg) (trop SA, WI): locust trees; source of copals, & balsams (animes); some copals are dug fr. ground (fossil or semi-fossil); u. in medicinals, jewelry, special varnishes; as incense before "gods" - **H. courbaril** L. (WI, sMex, CA, SA): courbaril; guapinole (tree); (SA) locust tree; source of hardened gum-resin called (South) American (or Colombia or Brazil) copal; jatoba resin (imp. into US); u. for plasters & dental cements, varnishes; in mask for making dental impressions; in incense in churches; also yields a balsam (Anime) and bk. (Cortex Lokri) (small piece of bk. soaked in Cochaca rum u. med. [Br]); frt. pulp around sds. edible & u. to prep. atole (gruel) & an alc. beverage; bk. inf. purg. & carm. (small doses), sap u. to control pain of ulcers (Sto. Domingo); lvs. u. to make medl. tea, also frt. peels; wd. useful - **H. martiana** Hayne (SA): source of Brazilian (or South American) copal; u. as shellac & in dye indy. - **H. oblongifolia** Huber (Amazonian reg, nSA): source of much Brazilian copal - **H. parvifolia** Ducke (Amazon area SA): resin u. to make microslide mounting medium - **H. stilbocarpa** Hayne (Br): source of copal similar to Courbaril copal.

Hymenantherum Cass. = *Dyssodia* Cav.

Hymenocallis arenicola Northrop (Amaryllid.) (WI): "lirio sanjuanero" (Cu); bulb u. as emet. & to prep. a cough syrup; ornam. - **H. caribaea** (L) Herb. (Pancratium speciosum L. f.) (WI): spider (or white) lily; bulb juice expressed & u. as emet. (Domin Caribs) (2328) - **H. harrisiana** Herb. (SA, CA, cUS): spider lily; bulbs c. lycorine, tazettine, galanthamine, galanthine, & other alk.; toxic pl. (164) - **H. lacera** Salisb. (H. mexicana Herb. App.): (Mex): u. as substit. for squill - **H. littoralis** Salisb. (seAs): bulbs c. lycorine, etc., toxic like squill; u. as vulnerary (Pl) - **H. tubiflora** Salisb. (nSA, WI): bulbs u. as astr. expect. diur.; bulb inf. in emesis, asthma (Trinidad).

Hymenocardia acida Tul. (Euphorbi) (tropAf): ashed rt. appl. to mouth infections; vapors from boiling lf. inhaled for headache; pd. bk. appl. to wounds as antisep. (Makua, eAf); bk. source of red dye.

Hymenocrater sessilifolius Benth. (Lab) (swAs): u. med. (Pak); lf. inf. u. as morning health drink for children; cooling for child with fever (Baluch); u. as spice (maybe replacing bay lf.); c. VO.

Hymenodictyon floribundum B. L. Robinson (trop eAf): tuber decoc. u. in threatened eclampsia (Tanzania) - **H. orixense** (Roxb.) Mabb. (H. excelsum Wall.) (Rubi) (trop seAs Pl): ulok (Siam); bhurkund; inner bk. u. antipyr. antiperiodic, esp. for tertian mal. ("ague") (with honey or sugar); valuable wd. ("cedar wood" = misnomer) - **H. parvifolium** Oliv. (tropAf): rt. u. as remedy in renal dis., for convulsions; in threatened eclampsia; lf. u. for inflamed eyes.

Hymenopappus filifolius Hook. (Comp) (wNA): rts. of bienn. hb. u. for chewing (Am Indians); u. med. - var. *lugens* (Nutt.) B. L. Turner (swUS nwMex): Columbia cutleaf; rts. u. by Hopi Indians as emet. & to treat toothache.

Hymenophyllum plumosum Kaulf. (Hymenophyll. - Filic. - Pteridoph.) (SA): inf. of fern u. as diaphr. & diur.

Hymenoptera: order of Insects, in which spp. have 4 wings, all membranaceous; social gps.; ovipositor modified for special purposes, such as boring; complete metamorphosis; ex. bees, wasps, ants.

Hymenopyramis cana Craib (Verben) (Indochin, Thail): kha pia; ashes mixed in dough u. in prepn. of cakes (biscuits), etc.

Hymenoxys argentea Parker (Comp.) (w mts. US): pl. u. for heartburn (Navaho); panacea (ingredient in "life medicine" [Navaho]) - **H. cooperi** Cockerell (Actinea biennis Ktze.) (swUS): rt. bk. made into chewing gum (NM [US] Indians) - **H. odorata** DC (Actinea o. Gray) (wTex nMex): bitter rubberweed; fragrant bitterweed; kills sheep; fl. head decoc. u. as beverage (Tex [US] Indians) - **H. richardsonii** (Hook.) Cockerell (swUS nMex): pingue; bk. & sts. ext. u. as substit. for chewing gum.

hyoscine: scopolamine, latter designation now preferred - **Hyoscyami Folium** (L): hyoscyamus; henbane leaf - **hyoscamine:** alk. fr. *Hyoscyamus niger* & other spp. of Fam. Solanaceae; readily converted into its optical isomer, atropine; mostly u. as HCl salt as hypn. sed.; in veter. med. (dogs).

Hyoscyamus L.: 1) OW toxic hbs.; frt. capsule 2) (L) dr. lf. sometimes with tops of *Hyoscyamus niger* - **H. albus** L. (H. varians Visiani) (Solan) (sEu, nAf): white henbane; hb. c. hyoscyamine, scopolamine, etc.; u. as drug similarly to *H. niger* since ancient times; u. cataplasm for breast tumors (Malta); adulter of *H. niger* - **H. aureus** L. (Med reg., incl. Crete, Rhodes): golden henbane; hb. has low alk. content; u. like prec.; cult. Hung. - **H. muticus** L. (Eg, swAs esp. Ind, Iran); Egyptian henbane; hb. c. 1.2-1.4%(!!) hyoscyamine; scopolamine; u. fumitory, masticatory, in Near East (some Bedouins smoke the hb. as a hallucinogen); adulterant of *H. niger* (1579); principally comml. source of atropine, etc.; cult. s.Yugo - **H. niger** L. (orig. home unknown; Euras, nAf; cult. and natd. all over the world) (1581): (black) henbane; hb. & tops c. atropine, hysocyamine, scopolamine (esp. in sds.), atroscine; u. sed. anodyne in neuralgia; hypn. mydr. carm. antispas. in convulsive coughs, asthma, stomach cramps, colics; source of alkaloids before *H. muticus* was used; poisonous (1580); masticatory, fumitory; rt & sd. u. similarly (Eu) (prodn. mostly fr. Hung, USSR, Mont [US]) - **H. pusillus** L. (AsMin): u. like other *H.* spp.; cult. in Hung - **H. reticulatus** L. (Baluchistan, Pak): kohibhang (Quetta); sd. appl. in toothache.

hyoscyamus, compound oil of: infused oil of h. to which have been added peppermint, lavender, rosemary, & thyme oils - **infused oil of h.** (NF V): h. macerated in ammonia water, alc., & sesame oil, warmed, strained, & expressed - **compound i. o. of h.:** inf. of narcotic materials in olive oil; appl. with friction.

hypanthium: receptacle or expanded torus below calyx of individual fl.

Hypericaceae: St. John's Wort Fam.; hbs. & shrubs mostly in temp. & subtrop.; lvs. smooth, simple, oval, opposite, exstipulate; fls. mostly yellow, with 4-5 petals & sepals; numerous stamens, these commonly arranged into 3 or more clusters; ex. St. John's wort; marsh wort.

hypericin: hypericum red: red coloring matter, a dianthrone, found in *Hypericum* spp; with antidepressant activity; active princ. of *Hypericum perforatum*; general ton.; u. in depressant states such as at menopause; dystonia, anemia, convalescence, also in atherosclerosis (aids capillary circulation); isolated by Brockmann (1942) (1602).

Hypericum androsaemum L. (Hyperic/Clusi./Guttiferae) (Euras nAf): all saint's wort; tutsan; hb. u. (park leaves) as diur. stom. (Por) - **H. calycinum** L. (Euras): St. John's wort; Bethlehem star; antimicrobial; c. no hypericins; ornam. subshrub - **H. connatum** Lam. (Arg, Par, Ur): inflor. apex u. as ton. vulner. - **H. crispum** L. (Euras): St. John's wort; u. like *H. perforatum*; in dys.; phthisis, hemoptysis, menstrual diff.; pl. has proven antimicrobial properties - **H. densiflorum** Pursh (seUS): bushy St. John's wort; American yellow heath; c. hypericin; may produce photosensitization in feeding animals - **H. denticulatum** HBK. (Mex): flg. tops u. med. (Mex Ph. 1874-1952-) - **H. ericoides** L. (wEu): st. lvs. fls. u. in folk med. (Valencia, Spain); c. several xanthones - **H. formosum** HBK (Mex): hipericon; flg. tops u. med. (Mex Ph. 1874-1952-) - **H. hypericoides** Crantz (Ascyrum h. L.) (US, CA, Mex, WI): St. Peter's wort; Poor (or Low) John; "olefield pine" (Oklahoma [US] Indians); hb. u. astr. (2431), diur. resolvent for glandular enlargement (promoting absorption); for kidney stones (1466) - **H. japonicum** Thunb. (trop&subtrAs): u. vulner. for wounds & leech bites (astr.); antimicrobial; antihemmorhagic for internal & uterine bleeding; to relieve intox. (Chin); appl. exter. for abscesses, fungoid skin dis. - **H. lanceolatum** Lam. (tropAf): curry bush (SC odor when crushed); is cause of photosensitization in sheep due to hypericin; source of balsam; u. med (Madag) (166) - **H. laxiusculum** St. Hil. (Rio Grande do Sul, Br): alecim cravo; u. in folk med. (Br) as vulner. in snakebite; hb. arom. stim. antispasm - **H. mysorense** Heyne (Ind): St. John's wort; lf. & st. juice u. in insanity, baldness, wounds; antifungal; much u. in folk med. - **H. patulum** Thunb. (Chin): u. to reduce ulcers (Chin); crushed lvs. liquid u. to wash blind eyes of poultry (Vietnam) - **H. peplidifolium** A. Rich. (tropAf): u. as indigest. remedy (Chagga) - **H. perforatum** L. (& subsp. *angustifolium* Gaud.) (Euras, nAf; now eAs, NA, SA, Austral): (common) (St.) John's wort; formerly very common in Klamath R. region (Calif [US]) (Klamath weed); fl. tops u. vulner. bitter; in intest. catarrhs, diarrh, liver & kidney dis, female conditions, a constit. of stomach, gall, reducing, & nerve teas (in hysteria) (2648); for prodn. of Ol Hyperici (VO with terpenes, cineole, myrcene, etc.); c. antibiotic (1604); livestock poison (Calif [US]); serious pest, since toxic to horses, etc. (goatweed); bad weed in Austral, etc. (destroys pasturage; spreads rapidly) (1603); yellow dye (prod. Ge) - **H. pratense** Schlecht. & Cham. (Mex): "sanguinaria"; tenchalita; u. in washing; claimed useful against uterine cancer - **H. punctatum** Lam. (e&cNA): spotted St. John's wort; u. in combn. to promote menstruation; in diarrh.; lvs. anthelm. (Am Indians) (167) - **H. pyramidatum** Ait. (neNA): great St. John's wort; rt. u. in pulm. dis (TB), kidney dis. (Menomini Indians); lax. alter. vulner. emmen. diur. expect. (white men); in TB, snakebite (Meskwaki Indians) - **H. teretiusculum** St. Hil. (Br): hb. u. as excitant, arom. emmen. - **H. uliginosum** HBK (Mex, CA): rabbit plant; tzotzil; ruda de monte (Guat); u. in diarrh. in domestic med. (Guat) (102); c. uliginosins A&B (antibiotics).

hyperin: hyperoside: quercetin galactoside hemipentahydrate; occ. in *Acacia melanoxylon* R. Br.; *Evodia hupehensis;* etc.; u. to decrease capillary fragility.

hyperisotonic: hypertonic.

Hyperoodon ampullatus Forster (H. rostratus Mueller) (Ziphiidae - Odontoceti - Cetacea - Mammalia) (oceans): (nAtl Oc); northern bottlenosed whale; one former source of Cetaceum (*qv.*)

hyphae: thread-like filaments, often branched, which constitute *en masse* the mycelium of a mold; also referred to hair-like growths in medulla of some algae.

Hyphaene crinita Gaertn. (Palm.) (SAf, Sudan, wAs): (South African) doum palm; doom p.; elalala p.; gingerbread tree; intoxicating alc. beverage made fr. sap; kind of rum (Madag); milk fr. sd. u. like coconut milk; kernel fr. sd. edible; sds. c. FO; fiber fr. husks & lvs. used (ropes, matting etc.); sds. like ivory u. to mfr. buttons; many other uses **- H. thebaica** (L.) Mart. (H. guineensis Schuman & Thonn) (n&w tropAf): (Egyptian) do(u)m palm, dum p. (*cf.* houm); gingerbread p.; orange frt. & sd. edible; rt. u in hematuria (at times due to bilharziasis); many uses reported for lvs. trunk, etc. (baskets, ropes, &c.); endocarp u. for buttons..

hypisotonic: hypotonic.

Hypnea musciformis (Vulf.) Lamour. (Hypne. - Rhodoph. - Algae) (Pac Oc, wcoast Atl Oc): sea weed u. as food in Chin (cooked with meat & vegetables, forming kind of jelly); source of phycocolloid, variously called hypnean, gelan, carrageenan, agaroid; thallus anthelm. u. as vermif. (Indones); one source of American agar - **H. nidifica** J. Ag. (Pac Oc): u. as food (Hawaii); fodder; in treatment of stomach dis. - **H. spicifera** J. Ag. (sAf coasts): raw material for African agar; u. as food.

hypo: sodium thiosulfate.

Hypochoeris glabra L. (Comp.) (Eur, nAf, As, incl. Ind): (smooth) cat's ear; rt. u. diur. ton. aper.; lvs. astr. hb. vulner.; weed - **H. tenuifolia** Griseb (Ch): escorzonera; rts. u. as diaphor. stim., diur.

Hypocreaceae: Fam. of Ascomycetes (Fungi) to which belong many pl. parasites, such as the ergots (now generally placed in Claviceptaceae); each perithecium encloses a cluster of 8-spored asci.

hypodermis: layers of parenchyma cells which lie directly beneath the epidermis & ordinarily separate the latter fr. the cortex.

Hypoëstes verticillaris R. Br. (Acanth.) (tropAf): cold water inf. u. to bathe children with fever (Tanz); pl. u. to drive away flies; u. as vegetable.

hypoglycin(e): amino acid (propionic acid deriv.) with hypoglycemic activity; forms A and B recognized; obt. fr. *Blighia sapida* sds.; reduces blood sugar.

Hypolytrum compactum Nees & Mey. (Cyper.) (Pl): salag salag sut; young sts. chewed with betel & appl. to ulcers.

Hypophaë: *Hippophae.*

Hypophysis Cerebri: Hypophysis Sicca (L): hypophyse (Fr): the whole desiccated pituitary gland; formerly quite popular, its potency in this form is very low & it is largely discarded in therapy, altho the fresh gland continues of much importance in the pharm. mfg. indy.

Hypoprion brevirostris Poey (Elasmobranchii - Selachoidei - Pisces) (WI, n to s coasts): source of a shark liver oil (FO with high vit. A content).

hypothalamus: part of interbrain wh. controls many autonomic functions.

Hypoxis aurea Lour. (Amaryllid.) (tropAs-Ind, Jap, PI): star grass; considered ton., rejuvenating, reconstructive, aphrod. (*cf.* ginseng) - **H. hirsuta** Cov. (e&cNA): stargrass; ornam. hb. - **H. nyasica** Bak. (tropAf): decoc. u. as cough rem., diaphor. - **H. rooperii** Moore (tropAf): purg. ton.; rt. stock juice appl. to burns; lf. u. to make strong cord - **H. villosa** L. f. (sAf): pl. u. as med. & charm (sSotho).

Hypserpa monilifera Diels (Menisperm.) (Solomon Is): bk. of vine pounded & appl. to head for headache.

Hyptis albida HBK (Lab) (Sonora, Sinaloa, Mex): Mexican bushmint; hoja de salvia mexicana; s. real; lvs. c. 1.6% VO, u. to flavor food; sold in Mex markets; u. in Sinaloa for ear dis. (by placing fls. in ears); in Guerrero decoc. of leaves u. as poultice for rheum. pains - **H. atrorubens** Poit. (WI, tropAm): "wild mint" (Trin); hierba buena (Ve); lf. juice u. in diarrh. dys. vomiting; lf. inf. u. in colds, indig. (Trin); hb. decoc. as hot bath for muscle soreness & congestion (Ve) - **H. brevipes** Poit. (tropAm, natd. eAs): lf. decoc. (Ve); given as postpartum protective med. (Mal); lvs. appl. to wounds & navel of newborn child to protect fr. infection (Java); in diarrh. (Par); lvs. as poultice for headache (Pe) - **H. capitata** Jacq. (tropAm, natd. seAs): botonera (Cerro Turumiquire, Ven); st./lf. decoc. u. in mal. flu, TB of lungs (Taiwan); lf. decoc. to cleanse wounds; rt. decoc. emmen.; pl.

cyanogenetic; lvs. boiled in water, then placed in sun & u. for fever (Ven) (92); lf. inf. in constip. (Trin); in diarrh. (777) - **H. emoryi** Torr. (Mex): fls. put in ear for earache; sd. u. for food - **H. fasciculata** Benth. (SA): hb. u. like melissa & origanum in med.; carm. diaph. anticatarrh. (Br) - **H. laniflora** Benth. (Baja Calif, Mex): "salvia"; pl. decoc. u. for fever - **H. multiflora** Pohl ex Benth. (Br): botaosinho; arom. hb. u. for gastralgia & indigestion - **H. mutabilis** Briquet (H. spicata Poit.) (CA, WI, Fla tropics): bittermint; lvs. c. VO (with menthone or pulegone) (233); lvs. as tea for stomachache - **H. pectinata** Poit. (pantrop as wAf, Fla [US]): lf. u. in roundworm, cough, antipyr.; poultice for boils (Tanz); source of resin, sometimes u. as incense (Af); lf. tea u. for stomach pains (WI); for diarrh. (Tanz); in prepn. of rum (Sakalava, Madag) - **H. rhytidea** Benth. (Mex): "salvia prieta," u. as remedy for fever (Sinaloa) - **H. spicigera** Lam. (CA, nSA, WI, tropAf, As): orégano (Cu); black sesame; beni; frts. & sds. source of FO (drying oil) (v. benefing oil); lvs. crushed appl. for headache & in baths; hb. insect repellant; fls. put into nostrils to combat headache & head cold; u. in embalming the dead and rubbing cadavers during night vigils (Diola, Senegal); sds. pulped & placed in stews & gravies (Sudan); sts. —> fibers - **H. suaveolens** Poit. (Mex, CA, Br-pantrop): bittermint; wild spikenard; "sandoval" (Cu); mastranto (de perro) (Br); chur (negro) (Mex); hb. c. VO (with menthol); u. to adulterate patchouli oil; carm.; pl. decoc. stom. antispasm. in resp. dis. (Cu); rt. decoc. u. as stom., emmen. in dys. (Mex); appetizer; in stomachache (Tanz); fungistat.; lvs. & tops u. in antirheum. baths (Pl); hb. u. diaphor. stim. (Chin); st. & lvs. appl. to skin for headache & skin dis. (Taiwan); for boils (Pl); wounds (Tanz); pl. as cataplasm for tumors (WI); galacatagog (Indones); lvs. as condim. (Burma); insect repellant (Pl); in mint-flavored beverage tea (wAf); c. suaveolic acid & suaveolol - **H. verticillata** Jacq. (WI, Fla, sMex to Co): "verbena" (Hond, Guat); "John Charles" (Jam); decoc. of shrub u. for indiges. colic; gastric dis.; u. as linim. for rheum. itch, insect bites; lax.; aphrod. (168); lvs. crushed & placed in hen coop to drive away fleas (Belize); antimitotic activity.

Hyraceum (capense) (sAf): prod. found on rugged mt. sides, supposed to represent desiccated feces & urine of *Hyrax capensis* (Hydracidae - Mammalia), klipdas (rodent-like ungulate); in hard dark masses with odor & taste resembling those of castor; u. like same in perfumes, etc.; *cf.* selajit (Indo-Pak).

hyssop (pron. iss'-up): 1) *Hyssopus officinalis* 2) *Capparis spinosa* - **Biblical h.:** 1) *Origanum maru* 2) *Capparis spinosa* - **giant h.:** *Agastache* spp. - **prarie h.;** *Pycnanthemum virginianum* - **wild h.:** *Verbena hastata.*

Hyssopus officinalis L. (Lab) (Med reg., c&sEu, nAf, cAs): (garden) hyssop; hb. u. since earliest times; lvs. & flg. tops c. VO, FO, hesperidin, diosmin (diosminoside) (flavonoid), marrubiin (bitter princ), sitosterol, choline; tannin (5-8%); VO c. pinocamphone (toxic), pinenes; hb. u. arom. stim. condim.; in night sweats, gall & kidney stones; in poultices, eye waters; hb. u. in medieval cuisine; in medicated wines & liqueurs; suggested for use in mfr. vermouth; expect. in coughs, hoarseness, tonsilitis, bronchial dis.; widely cultd. (prod. Ge, Fr, Balkans).

hysteranthous: lvs. produced after fls.

hysteresis: 1) (colloids) the reconstitution of a dr. gel (usually protein) by addn. of the requisite amnt. of water 2) (med.) delayed reaction or lag in gel formation as in retraction of blood clot.

Hystrix patula Moench (Gram) (eNA): bottlebrush (grass); some forage value; ornam. grass (bouquets).

NOTE: The letters "i," "y" and "j" are often interchangeable; it might therefore be advisable to sometimes look under each letter.

iamatology: materia medica.

iarovisation: vernalization.

Iberis amara L. (Cruc) (c&sEu): rocket (or bitter) candytuft; hb. & sds. c. sulfonated VO, a source of mustard; hb. u. antiscorb., antirheum.; sd. popular drug of Eu homeopathy; cult. ornam.; weed.

ibex: *Capra aegagrus* and other *C.* spp. (Euras, Afhigh mts.): formerly u. medicinally (Israel); a kind of wild goat.

ibira pita (Arg): *Peltophorum dubium.*

iboga (root): *Tabernanthe iboga.*

ibogaine: characteristic alk. of iboga (rt. & bk.); acts as hallucinatory (1582).

ibotenic acid: fr. *Amanita muscaria*; acts as psychomimetic agt., giving disorientation & deep sleep; related compd. muscimol.

Iboza riparia N. E. Br. (Lab) (tropAf): ginger bush; lf. inf. u. expect. in coughs & resp. troubles, emet. (Zulus); stom.; in edema; in mal; dengue fever. (—>*Tetradenia*)

Icacina senegalensis Juss. (Icacin. - Dicot.) (w tropAf): considered panacea (Casamancans, Senegal); lf. u. in the most diverse dis.; rt. very large & heavy, limiting use; considered invigorating, in rickets, anorexia, cachexia, senility, adynamic states; bucco-dental infections; purg. in meteorism, edema; rts. very rich in starch, but u. as food only in famines because of difficulty of extrn. of rt. fr. ground; sds. —> flour.

icaco (Mex): *Chrysobalanus icaco.*

Icacorea Aubl. = *Ardisia.*

ice: u. to reduce pain in rheumatoid arthritis (in plastic bags appl. locally), also to reduce inflammation, itching; u. crushed as packing in burns; reduces shock, pain, aids improvement; u. int. to cool beverages, &c.; (i. water); formerly harvested fr. rivers, lakes, &c. in winter time & stored in unheated bldg., pits, &c. until summer use; now mostly produced by freezing methods - **iceberg i.**: harvested fr. below glaciers in Alaska ("Alaska ice"); very pure, without contaminants; very old (thousands of yrs. at times); better tasting than ordinary ice (fr. tap water) (2398) - **i. acid**: glacial acetic acid.

Iceland lichen: I. moss: *Cetraria islandica* (misnomer as not prod. in Iceland).

ice-plant: *Mesembryanthemum crystallinum* - **ice-ribbons**: *Verbesina occidentalis.*

Ichnocarpus frutescens R. Br. (Apocyn) (trop s&eAs): rt. alter. ton. substit. for sarsaparilla; lf. & st. decoc. antipyr. - **I. oxypetalus** Pitard (seAs): lf. inf. u. for diarrh. (Indochin).

Ichthammol (L): generic name for sulfonated bitumen (*qv.*).

Ichthyocolla (L.): isinglass (*qv.*) - **I. in annulis**: ring i. - **I. in filis**: thread i. - **I. in folis**: leaf. i. - **I. in lyris**: ring. i.

Ichthyol™: sulfonated Bitumen (Merck).

Ichthyomethia = *Piscidia.*

Icica: *Protium.*

Ictodes foetidus Bigel. = *Symplocarpus f.*

id(a)ein: cyanidin chloride = galactoside found in *Vaccinium vitis-idaea.*

idioblast: a cell occurring singly or in gps. which differs in size, shape, cell wall markings, etc., fr. the cells around it; ex. the stone cell found in masses in the flesh of pears.

Idria columnaris Kellogg (Fouquieria c. [Kell.] Curran) (monotypic) (Fouquieri.) (Baja Calif & Sonora, Mex): boojum (tree) (slang); cirio (Mex); very odd-looking trees ("like nothing else on earth"); sts. u. to make corrals; as wattle for house walls & roofs; candle supports; fuel (poor); ashes u. —> soap; plant regarded superstitiously as harmful by Indians (*Fouguieria*).

if (Fr): yew, *Taxus*.

igamol(l)e (Mex): a sp. of amole.

ignace (Fr): St. Ignatius' bean.

igname(n) (Mex): *Dioscorea alata*; *D. batatas* (fr. Por. inhame fr. wAf native name of pl.); appl. to starch.

Ignatia (amara) (L): **I. Faba**: (St.) **Ignatius beans**: **Ignazbohne** (Ge): ripe sds. of *Strychnos ignatii*.

iguana (trop&subtropAm): large, harmless lizards usually greenish; female with eggs sold in markets (Belize); eggs eaten to increase fertility.

ikaja: *Strychnos icaja*.

ilama (Mex, CA): *Annona diversifolia*

ilang-ilang: *Cananga odorata*.

Iletin™: insulin.

Ilex L. (Aquifoliaceae) (worldwide): hollies; shrubs & trees, at times evergreen, many are ornam.; some med.; some beverage pls.; producing matte, quayusa. etc.; some spp. toxic - **I aquifolium** L. (Euras): English holly; lvs. c. bitter subst., mucil., coloring matter; u. ton. antipyr. in gout & cramps; fresh lvs. u. by Eu homeop.; frt. u. as lax. in colic; bk. u. to make birdlime - **I. aspera** Champ. (eAs): lvs. u. int. for GC, exter. for snakebite - **I. cassine** L. (seUS): yaupon; dahoon; lvs. c. caffeine 1%, tannin, VO; u. emet. beverage ("black drink") (NA Indians), med. as stim. - **I. chinensis** Sims (Chin): sds. & bk. soaked in wine & u. as carm. & ton.; in burns; ashes of lvs. u. in skin dis. - **I. cornuta** Lindl. (Chin): horned (or Chinese) holly; bk. & lvs. believed ton; tea of lvs. carm. contracept. abort.; evergreen lvs. given for TB cough, dizziness, ringing in the ears; headache - var. *Burfordii* DeFrance: Burford holly; popular ornam. - **I. decidua** Walt. (seUS): possum haw (holly); ornam. shrub or small tree - **I. glabra** (L) Gray (seNA): gallberry; inkberry; twigs for brushing grass-free (shaded) yards of (farm) houses ("yard brooms") (sUS); berries stain the fingers but do not furnish a satisfactory ink; lvs. c. caffeine(?), said u. to make a stim. tea; bk. u. astr. ton. antipyr. in colds (2431); honey plant - **I. guayusa** Loes. (eEc, Bo, Pe): lvs. c. caffeine, u. as beverage (guayusa tea) by the Indians as CNS stim., emet., also in cult practices; ritual psychotic (large does) in enemas & possibly snuffs; some other spp. also so used (170) - **I. integra** Thunb. (Jap, Kor): wd. u. for making furniture; also bk. —> Japanese birdlime; med. (169) - **I. laevigata** Gray (eUS): hoop-wood; bk. astr. ton antipyr. - **I. medica** Reiss. (Br): lf. inf. u. as stom. & diur. (171) - **I. mertensii** Maxim. (Japan - Bonin & Loochoo Is): tree yields dyewood "doss" (c. dossetin) u. as source of yellow dye - **I. opaca** Ait. (e&cUS): American holly; lvs. & bk. c. NBP u. bitter, antipyr. in malar. & rheum. (Miss [US] Negroes); lvs. u. as dem. exp. in coughs, colds (also rt. bk.); bitter frt. u. purg. emet. diur.; frtg. branches popular Xmas decor. - **I. paraguariensis** St. Hil. (Br, Par, Arg): maté (name first appl. to gourd fr. which beverage drunk, later appl. to both herb & drink) (927); Paraguay (or Brazilian) tea; yerba tea; lvs. c. 1.5% caffeine, 20% caffee-tannic acid, vanillin, VO; u. beverage (like tea); beverage is prepd. in Arg & Urug in gourd ("bomba") and sipped through a silver tube with spoon-like base with perforations (to keep out m. lvs.) ("cuia"); cerebral stim., ton. bitter, astr. diur. diaph.; in some medl. teas (*v.* mate) - **I. verticillata** (L.) Gray (c&cNA s to Ga [US]): American "black elder"; winterberry; Prinos; bk. lax. astr. bitter ton. alt.; wd. decoc. antipyr anti-inflamm. ton., in diarrh.; bk. & frt. antimalarial in place of Cinchona; emet. vermifuge; lvs. u. as tea substit. ("buckbush") - **I. vomitoria** Ait. (seUS): cassena; yaupon; lvs. c. caffeine; u. as ceremonial beverage ("black drink") by NA Indians; lf. decoc. u. in colds & fevers; emetic.; as beverage tea (2431); heart wd. valued; Christmas decoration - **I. yunnanensis** Franch. var. *ciliata* Hu (Chin): lf. tea u. as antipyr.; stim. (172).

Ilge (Ge-f.): *Iris pseudacorus*.

"Iliana": I. vine: "vegetable blood" transfusions u. by SA Indians (173).

ilima (Haw): *Sida cordifolia* & *S. fallax*.

Illicium (L): frt. of star anise, *I. verum* - **I. anisatum** L. (Illici./ Magnoli./Schisandr.) (Jap, Kor): shikimi frt.; sacred anise; Japanese (or bastard) star anise; sonki; cult. near Buddhist temples & u. in relig. ceremonies, hence synonym *I. religiosum* S. & Z.; arom. but poisonous (lvs. & frt.) (1583); c. shikimic & kinic acids; avoid confusion with *I. verum*; shikimitoxin; anisatin (toxic); u. stim. carm. (frt. decoc); wh. it is sometimes u. to adulterate - **I. cambodiana** Hance (I. cambogianum Pierre) (Tonkin, nVietnam): a harmless adult. of off. star anise - **I. floridanum** Ellis (Ga, Ala, Fla, Miss, La [US]): stink bush (SC); purple (or tree) anise; fls. smell like dead fish; bk. arom. ton.; lvs. & frt. toxic - **I. griffithii** Hook. f. & Thoms (Ind -

Himalaya reg.): frt. (false anise of Bombay); u. as arom. but toxic; harmless adult. or sophisticant of off. prod. - **I. japonicum** Sieb. = *I. anisatum* - **I. parviflorum** Mx. (Ga, Fla): "star anise"; yellow anise tree; bk. source of anise oil; rts. said like sassafras - **I. religiosum** Sieb. & Zucc.: *I. anisatum* - **I. verum** Hook. f. (seAs, sChin, Hainan Tao, Vietnam): true star anise tree; Chinese anise; source of s. a. frt.; c. -8% VO (with anethole); u. flavor., arom. stomach, carm. corrective, exp; for prodn. anise oil; mfr. liqueurs; condiment.

illipa: illipe (or **illipi**) **butter** (**tree**): *Madhuca indica*, etc., & *Shorea stenoptera*, etc. - **i. oil: illupi** (**oil**): FO obt. fr. sd. of illipe; solid at room temp.; u. food & illumin. (EI).

Ilotycin: erythromycin (Lilly).

imbauba: *Cecropia* spp.

imbu (Br): *Spondias tuberosa*.

imburzeiras (DominRep): *Spondias mombin*.

Immergruen (**evergreen**) (Ge): 1) periwinkle, *Vinca minor* 2) myrtle - **I. gewaechse**: Apocynaceae.

immiscible: liquid substances not readily held together when mixed; after shaking together, such components separate more or less completely; ex. chloroform & water.

immortal herb: Immortelle: (**Life**) **everlasting**: 1) *Xeranthemum annuum* 2) *Helipterum* spp. 3) *Erythrina* spp. 4) *Anaphalis* & several other genera of Fam. Comp. - **yellow i.**: *Helichrysum arenarium*.

immunity: state of being resistant to injury by (variously) viruses, toxins, parasites, etc.

immunogen: any substane wh. can stimulate an immune response.

immunoglobulin (**Ig**): any globulin of animal source with known antibody activity; glycoprotein; normally found in exter. secretions, such as tears; 5 classes (A, D, E, G, M); most result fr. reaction with antigen in tissues; ex. anti-D (Rh_o) (E); chickenpox; antilymphocytic; hepatitis B; Human Gamma (or Normal) (prevention of measles, polio, viral hepatitis A, etc.) (E); human measles; human tetanus; human antirabies; human antivaccinia (smallpox); human mumps; human pertussis; u. in primary immunodeficiencies; 5 classes recognized: IgM, IgG (gamma G globulin) (blocks allergenic reaction), IgA, IgD, and IgE (gamma E globulin) (excessive amt. gives allergic symptoms).

immunology: science of specific resistance to infectious disease, reactivity to allergen, etc.

impantine: callicrein.

Impatiens L. (Balsaminaceae) (worldwide): snapweeds; ann. or perenn. hbs. of considerable importance; c. 900 spp. - **I. balsamina** L. (Ind, Mal, Chin): garden balsam; spotted touch-me-not; pl. u. diur. emet. alt. cath.; lf. decoc. u. to wash hair to improve its growth; fls. u. to dye fingernails; popular ornam. (codded arsmart) - **I. capensis** Meerb. (I. biflora Walt.; I. fulva Nutt.) (e&cNA): (spotted) snap-weed; balsam weed; jewel weed; wild (or brook) celandine; quick-in-the-hand; orange (or wild) balsam; (spotted) touch-me-not; hb. aper. diur. emet., whole pl. for yellow orange dye (Am Indians); whole pl. inf. drunk for chest colds, stomach cramps; fresh juice reported useful as antidote in rhus (poison ivy) poisoning, a property not definitely proven to possess; in burns - **I. chinensis** L. (Ind): pylee (Madras State); pl. u. exter. for burns; int. in GC - **I. mooreana** Schltr. (NG): large petals & young lvs. rubbed into burns - **I. nolitangere** L. (c&wEu, n&eAs): hb. (Herba Balsamitae luteae) u. as diur. lax.; exter. as vulner.; juice to remove warts - **I. pallida** Nutt. (e&cNA): pale snap-weed; pale touch-me-not; jewel weed; hb. u. aper. diur. in edema, alt. in jaundice; juice & hb. decoc. u. locally in ringworm, tetter (psoriasis, eczema, etc.) - **I. platypetala** Lindl. (seAs): lvs. u. for skin eruptions & diur. (children) (Indones) - **I. scabrida** DC (nInd): pl. eaten as veget. (UP, Ind) - **I. walleriana** Hook. f. (I. holstii Engl. & Warb.) (trop eAf): rts. u. as abort.; st. for abdom. & liver pains; cult. ornam. (Eu).

Imperata cylindrica (L) Raeusch. (Gram) (sAf, seAs, trop & subtrop, both hemispheres incl. Fla [US]): alang-alang grass; cogon g.; cotton g.; lalung (Nig); rhiz. & stolons u. as diur. antipyr. hemostatic; for tumors (eAs); stom. (sAf); as ton. digestive aid, astr. in nosebleed (Chin); rt. u. for chest colds in children, for hiccups (sAf); frtg. spikes u. (decoc.) as vulner. sed.; grown for land reclamation (sAf); fairly good forage grass for water buffalo (cattle will not eat); u. for thatching; bad weed (Mal; Fla, Ala, Miss [US]) - **I. exaltata** Brongn. (I. brasiliensis Trin.) (As [home?] incl. Ind, Mal, tropAm [Br], PI, Sunda Is, New Hebrides): sapé (macho) (Br); startgras (Suriname); rhiz. u. as active diur. emoll. cholagog. lax. diaph. (Br folk med.) (Br Phar.); to make paper pulp; pl. c. arom. VO.

Imperatoria ostruthium L.: *Peucedanum o.*

imphee (**millet**): various millets of Af, forms of *Sorghum bicolor*.

imshay (Pers): "go away" = Pyrethrum flowers (*cf.* Flit).

Incaparina: combn. of high protein flour u. in CA (fr. I N C A P, Institución de Centro America y Panama).

incense (*broadly*): aromatic subst. such as wd. or resin wh. burns arom. smoke or fumes; (specif.) frankincense (gum i.) (2762) - **i. tree**: *Protium heptaphyllum*; *Bursera simaruba*; taxa of *Boswellia, Commiphora, Dacryodes.*

inchis (Pe): peanut.

incienso (Sp): *Encelia farinosa*; *Myrocarpus frondosus.*

incineration: process of burning organic substances to ash in the air.

incunabula: books printed fr. movable type fr. before 1501 (Am, later date).

indaconitine: alk. fr. rt. of *Aconitum chasmanthum.*

indaia do Rio (Br): indaia çu (Br): *Attalea compta.*

inDawa (Zulu): dawo; lateral tubers of *Cyperus esculentus.*

indene: indonaphthene, C_9H_8; found in coal tar, crude petroleum, some VO, etc.; u. in synth.

India, Farther: earlier term for area now comprising part of India e of Calcutta, Bangladesh (former E Pakistan), & Burma - **India**: v. under aconite, gum, rubber, tobacco.

Indian arrowwood: *Euonymus atropurpureus* - **I. barley-caustic**: sabadilla - **I. blanket** (seUS): *Gaillardia pulchella* & other *G.* spp. (SC color of fl.) - **I. boys & girls**: *Dicentra cucullaria* - **I. black drink**: *Ilex cassine* - **I. bread**: 1) tuckaho 2) corn bread - **I. b. root**: *Psoralea esculenta*; *Zamia integrifolia*, etc. - **I. cane**: sour dock - **I. clip**: *Sarracenia* spp. - **I. corn:** *Zea Mays* - **I. dye**: Hydrastis - **I. fennel**: *Foeniculum vulgare*, cult. in Ind; frt. is yellow - **I. geranium oil**: VO fr. *Cymbopogon martinii* - **I. gum tree**: Acacia - **I. hairtonic**: *Dracunculus vulgaris* - **I. hay** (US slang): cannabis - **I. laburnum**: *Cassia fistula* - **I. lettuce**: *Frasera caroliniensis* - **I. loaf**: *Poria cocos* - **I. millet**: pearl m.; durra - **I. nut**: *Pinus edulis* sd. - **I. paint**: 1) I. dye 2) *Sanguinaria* - **I. p. brush**: 1) (wUS) *Castilleja* spp. 2) (eAla [US]): *Spigelia marilandica* - **I. pear**: *Amelanchier canadensis* - **I. physic**: *Gillenia trifoliata* - **I. pink**: 1) *Spigelia marilandica* 2) *Lychnis floscuculi* - **I. plum**: *Flacourtia cataphracta*; *F. ramontchi* - **I. plume**: *Monarda didyma* - **I. red paint** (US): **red I. paint**: Sanguinaria - **I. red root**: *Lachnanthes tinctoria* - **I. rice**: wild r. - **I. root** (US): 1) *Aralia racemosa* 2) *Rumex hymenosepalus* 3) unknown Utah [US] pl. sp. possibly *Amaranthus retroflexus* 4) galanga - **I. shoe, yellow**: some *Cypripedium* spp. - **"I. soap"**: 1) fungus (perhaps *Polyporus* spp.) disintegrated by parasitic fungi 2) *Sapindus saponaria* - **I. type tea**: stated to be mixt. of ca. 10 drugs incl. gentian - **I. waterleaf**: *Hydrophyllum* - **Indian**: v. also under apple. azedarach; balsam; bean; berry; bdellium; cannabis; cherry; corn; cucumber; elm; fig; gum; hemp; ipecac; jalap; licorice; nard; olibanum; opium; pipe; plantago; podophyllum; poke; prickly ash; rubber; safflower; sage; sarsaparilla; senna; shot; squill; tobacco; tragacanth; turmeric; turnip; verbena; valerian.

indican (**, plant**): glucoside occurring in *Indigofera* spp., etc., apparent source of indigotin (indigo blue).

indicators: 1) reagents to determine acidity/alkalinity; many times inf. of plants, esp. petals, have been u.; ex. red cabbage, hollyhock, dahlia fls.; litmus 2) indicator plants: those wh. by their presence or state show nature of soil, etc. (ex. dwarfing of plant in dry soil).

Indicum (L): indigo.

Indien (Ge): 1) seAs (*cf.* Vorderindien, Hinterindien) 2) Indies (EI, WI).

indigenous: native to an area, country, continent, etc.; endemic.

indigo: 1) *Indigofera* spp. contg. indigotin, incl. *I. suffruticosa, I. tinctoria* 2) coloring matter, indigotin (considered oldest known coloring matter), indigo blue - **i. bush**: *Amorpha fruticosa* - **i. carmine: i. extract**: sodium indigotin disulfonate; u. dyestuff; now usually synth. - **false i.**: *Amorpha* spp. - **red i.**: cudbear - **synthetic i.**: indigotin. - **wild i.**: *Baptisia tinctoria.*

Indigofera arrecta Hochst. ex A. Rich (Leg) (Natal, sAf; Java): Natal (or Java) indigo; source of Java indigo (dyestuff); c. indigotin, indorubin, kaempferol, kaempferitrin (dirhamnoside); cult. as cover crop - **I. caudata** Dunn (Chin; seAs): sd. decoc. reputed tonic (Laos) - **I. dusua** Buch-Ham (Pak - Himalay reg.): useful fodder pl. fls. u. as potherb (Kangra) - **I. endecaphylla** Jacq. (trop Af, As): u. in prepn. indigo; fodder pl.; cover crop, to improve soils; med. (198); said to cause abortion in cattle - **I. gerardiana** Wall. (Ind, Himal reg.): kat(h)i (Punjabi); twigs u. to make baskets & rope bridges; fodder; shade tree - **I. glabra** L. (I. pentaphylla L.) (Ind): lvs. u. as bitter ton. antipyr, exter. appln. as emoll. sold in markets as sour component of curry - **I. heterantha** Wall. (Indo-Pak, Chin, Afgh+): "sakina"; fl. buds eaten as veget. for stomach dis.; lvs. furnish fodder for cattle & goats (175) - **I. hirsuta** (L f.) Harvey (eAf, As, PI, natd Fla [US]): hairy indigo; c. indican (glycoside); fresh lvs. decoc. u. as stom. in diarrh. as lotion in yaws; poultice for boils (Tanz); lvs. u. in chest dis.; cover crop; toxic (mech. injury) - **I. incarnata** (Willd.) Nak (Chin): indigo plant; leafy shoots u. in broth with pork; for cough, indigest, hemorrhage, &c. - **I. lespedezioides** HBK (Br, Ve, Guat, Hond): raiz de la virgen; cachicahua; hb. u.

for gastric dis.; rt. decoc. as antipyr. in mal. (given in syr. as very bitter); antidiarrh. stom.; rts. & lvs. as fish poison (Br) (176) - **I. linifolia** Retz. (eAs): pl. u. in fever eruptions, amenorrh. (Ind); vermifuge (Mal) - **I. spicata** Forsk. (Ind, Chin, trop Af): u. to improve soils; toxic plant - **I. suffruticosa** Mill. (I. añil L.) (tropAm, incl seUS Ve Pe): añil(ito); (WI) indigo; abugi (Cuna Indians, Pe); añil indigo plant; blue-black coloring matter "añil" (= indigo) extd. fr. lvs. for domestic use (Cu); pl. u. as febrifuge (baths with rts. & lvs.), vulner. purg. (with castor oil) (Sp Am in Calif); diur. antispasm.; rts. u. as insecticide for parasitic insects (Cu); ton. emmen. antiepilept. (formerly) (Mex); fish poison; cover or green manure to improve soils; lvs. u. as tea in snakebites; to "purify blood"; in treating "sarda"; cult. for indigo prodn. till ca. 1900 - **I. tinctoria** L. (sAs, tropAf, cult. in many trop countries, as Br & Salv): true (or East Indian) indigo; añil; lvs. c. indican (glycoside) & indigomulsin (enzyme) which produce indoxyl, this converted by O of air to indigo; u. for dyeing, printing, analy. reagent; u. against tumors (Chin, Ind); epilepsy, nerv. dis., bronchitis (ext. of pl.); exter. for wounds & hemorrhoids; rt. in hepatitis; Eu homeop. med.; formerly cult. seUS to mfr. indigo; prodn. in India (Madras); cover or green manure (to improve soil) - **I. trifoliata** L. (Ind): sds. u. as astr. aphrod. ton in rheum. & leucorrh. - **I. trita** L. f. (Ind, sAf): sd. u. as nutr. ton.; pl. indigo source.

indigotin: indigo; coloring matter of *Indigofera* spp., *Isatis tinctoria*, etc., also prod. synth.; insol. water, alc.; sol. chloroform (hot), aniline (hot); u. dye (usually as Na disulfonate), esp. for wool, rarely in med.

indio desnudo (Mex): *Bursera gummifera*.

Indochina: in recent decades separated. into Cambodia, Laos, Vietnam.

indoleacetic acid: heteroauxin: a plant hormone widely distr. in plant kingdom, recognized as chief auxin of higher plants; u. as pl. growth regulator; in exptl. study decreased blood sugar concn. in adult diabetics (Mirsky, A.; 1957).

indole-carbinol: *indoleethanol*; tryptophol; occ. in broccoli, &c.; said to protect against breast cancer.

inducer (biochem.): subst. wh. enables the synthesis of a specific enzyme (or enzymes) or enzyme sequence by combn. of a mol. (inducer) which deactivates the repressor & allows formation of the enzyme (or other mechanisms).

Indulin™: lignin derived fr. pine wd. in process of mfr. of sulfate paper pulp.

indumentum: pl. epidermal covering, esp. as describing the trichomes.

inertoside: glycoside of inertogenin (with D-diginose); occ. in *Strophanthus schuchardtii*, *S. intermedius*, etc.

infection: 1) disease communication 2) disease or condition caused by microorganisms - **infestation**: state where organisms or articles are beset with animals, esp. in case of drugs & foods by rodents & insects; the distinguishing hairs of various mammals (such as rats, cats, etc.) furnish the chief evidence of contamination; also fecal masses with embedded hairs; in insects & related forms, body parts & droppings are chiefly u. for detection.

Inflorescentia Helichrysi = Flos H. inflorescence of *Helichrysum arenarium*, &c.

inflorescense: fl. grouping or arrangement on st. & branches (determined by peduncle & pedicel attachment).

influenza: flu: acute respiratory infection caused by a virus or mixt. of viruses; manifested by rapid onset, fever, prostration; immunization sometimes possible using **i. (virus) vaccine** (viruses A, B, C) (chick embryo) (E).

infrutescence: cluster of frts. (parallel to inflorescence).

infusion (Infusum, L): 1) tea; prepn. with active principles made by extn. of drug with liquid, usually water; boiling water is usually added and left stand 15 min. but boiling is avoided then strained (2618) 2) slow injection into vein of medicament - **digitalis i.: Withering's i.**: digitalis leaf 1.5% in water with 10% alc. - **i. mug**: special utensil for making i.

infusorial earth, (**purified**): siliceous earth, purified (*qv.*).

Ingá affinis DC (Leg) (Arg, Br): ingá; frt. eaten by SA Indians; bk u. as tan. - **I. cinerea** = *I. saman* Willd. = *Pithecellobium* s. Benth. = *Samanea saman* - **I. edulis** Mart. (Mex to Pan & nSA): ice cream bean; mariá (Cu, Pan); guava (Pan); guayaniquil (CR); pulp of frt. eaten; u. med. as nerve sed. & in headache (Cuna Indians, Pan); this and other *I.* spp. grown as shade trees for coffee bushes & as windbreaks to protect cocoa plantations - **I. fagifolia** (L) Willd. (Amazonia, Br, Pan): "sweet pea"; guava (Pan); frt. pulp sweet, quenches thirst; believed to cause colic (Co) - **I. laurina** Willd. (WI): Spanish oak, guama de Puerto Rico; grown for shade in coffee & cacao plantations; good honey source; bk. u. in combns. (Pan) for erysipelas (2328) - **I. spectabilis** Willd. (CR-Pan): guava (real); guava de Castillo (Pan); monkey tambrin (WI); enormous pods sold in markets, Panama City (Pan) for edible pulp; the following *Inga* species have edible frt. pulp (sds. enveloped by a juicy pulp of sugary flavor, u. as food): **I. feuillei** DC (Pe vern. name, paccai; cult); **I. macrophylla** H. & B. (nSA); **I. paterno** Harms (Mex, CA) (paterno); **I. radians** Pitt. (sMex, CA) (guavo real); **I. spuria** H. & B. (Mex - SA) (guamo [Guat]) (nacaspilo [Salv]).

ingenio (de azucar) (Sp): sugar mill.

ingluvin: digestive enzyme fr. chicken gizzards.

Ingwer (Ge): ginger.

inhibin: polypeptide hormone secreted by special glands in gonads of both sexes wh. inhibit secretion of FSH, hence reducing available germ cells; suggested for use as potent contraceptive; diametrically opposite in action to activin.

inhibitol: constit. of non-saponifiable matter of pl. FO wh. delays oxidation, thus rancidity.

inhibitor: subst. that retards or reduces rate of chem. reaction; opposite of catalyst.

Injectio (L.): injection ampul; ampul-vial; dosage form for parenteral injection; ex. I. Insulini; I. Parathyroidei.

"Injun" turnip (eCana): *Arisaema triphyllum.*

inkberry (US): 1) *Ilex glabra* 2) *Phytolacca americana.*

inkroot: *Limonium carolinianum.*

inmortal (Sp, Pueblo Indians): *Asclepias asperula* subsp. *capricornu.*

Inocarpus fagifer (I. edulis) Forst. (Leg) (eAs, Malaya, Pac Islands, Polynesia, Tahiti, Samoa): Polynesian (or Tahitian) chestnut; ifi (Tonga); sd. kernels eaten boiled or toasted; med. (New Caled) (198); u. in treatment of injured eye; wd. useful.

Inocybe aeruginascens Babos (Cortinari. - Agaricales - Fungi): sporocarp c. psilocybin but not psilocin; u. as hallucinogenic - **I. patouillardii** Bres. (US): toxic to nervous system.

Inonotus obliquus Pilát (Poria obliqua Fr.) (Hymenochaet - Polypor. - Fungi) (US Eu+): clinker fungus; u. to inhibit cancer, decrease size of tumor; "rejuvenate" patient (Pol) (1585).

inosine: glycosidal base, hypoxanthine riboside; found in yeasts, sugar beets, animal tissues, etc,; activates cellular functions; u. in blood preservation; to reduce anxiety.

inosinic acid: nucleotide found in yeasts, *Penicillium chrysogenum*, meat exts., etc.; disodium inosinate & other salts u. as flavor intensifiers; *cf.* sodium glutamate.

inosite: inositol: muscle sugar; antialopecia factor; part of vitamin B complex; a saturated cyclic sugar alc. obt. fr. some green pls. (ex. Leg. spp. as *Vicia faba* unripe frt.), liver, muscle, lung, heart tissues (comml. source: corn steep liquor); essential to life (sometimes classed among the vitamins) ("non-essential" B complex); said formed by cyclization of glucose (cyclase, enzyme); when hydrolzyed, yields glucose; u. lipotropic in hypercholesteremia; u. treatment cirrhosis, fatty infiltration of liver; insomnia; often given with choline; i. hexaphosphoric acid $C_6H_6(OPO_3H_2)_6$ = phytic acid.

inotropic: affecting force or energy of heart or other muscular contraction (+ = stimulating; - = depressing).

in'sam (Jap): ginseng made into medl. beverage (best pls. for this grown in Korea, nChin, Manchuria).

insane root: 1) henbane 2) European mandrake.

insect flowers: pyrethrum flowers, fr. *Chrysanthemum* spp. - **Caucasian i. f.**: Persian i. f. - **Dalmatian i. f.**: *C. cinerariaefolium* - **Persian i. f.**: *C. coccineum, C. marschallii* - **i. powder**: 1) (US legally defined): I. flowers (1528) 2) insecticidal pd. material of various kinds & sources - **white i. f.**: Dalmatian i. f.

Insecta: Hexapoda: arthropod animals which possess 6 legs, 2-3 body sections (head, thorax, abdomen); undergo metamorphosis after egg is hatched; important mostly as destructive & disease-bearing pests.

insetto (It): insect.

inspissated: air-dried (referring to juices, etc.).

instant: prepn. very rapidly made; ex. **i. coffee**.

insulation: now u. at times to mean treatment with insulin (234), not conventional meaning.

insulin: Insulinum (L.): hormone secreted by isles of Langerhans of pancreas & which serves to aid metabolization of sugars & fats in the animal organism; m. wt. ca. 6000; polypeptide made up of 2 parts (chains A & B); i. crystals c. 0.4% Zn (+ other metals) altho Zn is not a component of the peptide mol.; lack of this hormone results in diabetes mellitus; u. in diabetic mellitus; diabetic coma, ketoacidosis, shock therapy, etc. (2249); extd. fr. pancreas of hogs, cattle, & humans (recently) (prepd. by genetic engineering or enzymic modifation of pork insulin) (1982-) (sometimes allergenic); major source of i. = pancreases of slaughtered animals, since synthetic is too expensive - **i. injection**: the basic dipensing form of i.; consists of acidified soln. of i. with preservative & put up in various strengths or potencies; generally adm. subcut. (sometimes continuous with pump) but occasionally IV - **i. isophane: isophane i. injection**: sterilized aq. suspension of Zn-insulin crystals modified by addn. of protamine sulfate in such proportions that solid phase of suspension is made up of crystals composed of i., protamine, and Zn. = **NPH i.** [N(eutral) P(rotamine) H(agedorn)] (in whose lab. 1st made); intermediate acting antidiabetic, with onset in 2 hrs. after SC inj., max in 10-20 hrs., duration 20-32 hrs. (E) - **protamine i.**: combn. giving longer action than i. alone; less sustained than with p. zinc i. - **i. zinc suspension**: Lente Insulin™: aq. suspension of i. with added $ZnCl_2$; permits longer maintenance of i. in blood.

intelligence plant: *Centella asiatica.*

interferons (IFN): group of small-mol. proteins (polypeptides), produced in infected cell cultures which inhibit multiplication of some animal viruses (by blocking virus replication) & thus provide resistance against viral infections to the host cell (1957-); production of i. in the animal tissues follows exposure to viruses, some bacteria, protozoa, etc., toxins, nucleic acids, agglutinins, antibiotics, etc.; they increase the body's immunity, also show antitumor activity (leukemia, solid tumors as ovarian cancer); multiple sclerosis; immunomodulator; u. in measles, chickenpox, meningitis, shingles (which is 50% fatal); inner ear infections, warts (multiple), to treat chronic granulomatous disease (CGD); nasal spray prophylactic to common cold (1985-) (to prevent spread in family gp.); various interferons are u. in hairy cell leukemia; chronic hepatitis B, Kaposi's sarcoma; condylomata aciminata; and various granulomatous dis. (SC inteferes with viral replication) - **gamma i.**: type with high antitumor activity.

Interleukin: group of protein factors (glycoproteins) produced by leukocytes (macrophages; T cells) in response to antigenic stimulation. - **i. 1** (IL-l) u. in therapy to increase immune capability; combined with colony-stimulating factor - **i. 2** (IL-2): deriv. of white blood cells (leukocytes): seems of value in treatment of cancer (1985-) - **i. 3** (IL-3): lymphokine-activated killer (cells) (LAK): a lymphokine prod. by antigen-activated T lymphocytes, wh. stim. blood-producing cells (granulocytes, &c.).

intermediate: compd. which is both synthesized & then further converted by organism into another product.

intermedin: melanocyte-stimulating hormone (MSH); polypeptide similar to ACTH; secreted by intermediate lobe of pituitary gland (APL, pars intermedia); stimulates pigmentation (melanin), retinal function, etc.

intermedioside: glycoside of sarverogenin & D-diginose; occ. in *Strophanthus intermedius.*

interoside: glycoside of sarverogenin & D-diginose & glucose; occ. in *Strophanthus intermedius,* &c.

interstitial tissue: that portion of testis which secretes androgen but not seminiferous cells, hence causative of devt. of secondary sexual characters & potency of individual but not related to fertility of male.

intisy: *Euphorbia i.*

intrability: ability to enter a cell or permeate through a cell wall.

intrinsic factor: obt. fr. hog stomach & duodenum; potentiates utilization of oral vit. B_{12}.

introgressing: back crossing of interspecific hybrid with one of the parental types producing new type; in true breeding results in a new sp.; intersp. hybrids become stabilized by this process, at the cellular level, the process in wh. genes fr. one taxon infiltrate genes of another to give modified characters.

introgressive hybridization: repeated back crossing of hybrids with one or other parent or both parents.

Intsia bijuga (Colebrooke) O. Ktze. (Leg) (seAs, trop islands): ipil (PI); frt. lax.; bk. u. in diarrh.; hard durable wd., u. in dyeing; sds. toxic but edible if soaked & heated; u. as starvation food.

Inula (L): rhiz. c. inulin (44%), VO, mucil; alantolactone (chief constit.) & isoalantolactone (mixt. called helenin or alant camphor [1587]), azulene, phytomelane (carbon encasing cells; useful for identification); rhiz. u. diur. emm. exp. stomach., anodyne., bactericide (*sic*); cholagog, anthelm. ton. against tumors; lvs. u. in epizootic dis. (horses) in mal., enteritis, dys., bronchitis (Chin, Jap); skin dis.; constit. (fresh) of diabetic foods (Eu); candied rt. formerly in cough; liqueur indy. - **I. britannica** DC (Comp.) (Euras): yellow starwort; rt. c. inulin; capitula sometimes found as adult. of arnica fls.; fls. u. as "depurative" (176); remedy for cough, vomiting, chest pain; in sore throat (Chin, var. *chinensis* Regel) - **I. conyza** DC (Eu, Med area, nwAf): ploughman's spikenard; hb. formerly popular drug (Herba Conyzae); u. diur. emm. rashes, stom. vulner.; lvs. to adulterate Digitalis; insecticide - **I. germanica** L. (c&sEu): Deutscher Alant; hb.u. against cachexia, scrofulosis, mucous membrane inflamm. - **I. graveolens** Desf. (Dittrichia g. (L.) W. Greuter) (sEu): herbe aux punaises (Fr); hb. rich in VO (with bornyl acetate) u. in colics, amenorrhea, asthma, "rheum." (watery discharge fr. eyes & nose) - **I. helenium** L. (Euras, natd. eNA, eAs; cult. Jap): elecampane; scabwort - **I. racemosa** Hook. f. (eAs-Himal): u. much like *I. helenium*; in cholera, diarrh. induration of organs - **I. royleana** DC (Indo-Pak, temp. wHimal): insecticidal, poisonous pl. (HCN glycoside); rhiz. u. to adult. kut rt. - **I. viscosa** Ait. (Dittrichia v. (L.) W. Greuter) (Prov., Med reg.): aromatic i.; herbo dey masquo (Fr); rt. u. in urin. dis., against tumors; in baths; flea repellants; weed.

inulase: enzyme which catalytically hydrolyzes inulin, irisin, & other D-fructose polysaccharides to D-fructose; occurs in fungi, snails, etc.

inulin: a starch-like carbohydrate found in underground parts of *Inula helenium*, dahlia, etc. $(C_6H_{10}O_5)_3$; hydrolyzed with acids into levulose; u. in diabetic flours; practically tasteless; as diagnostic reagent in kidney tests (1586).

Inventol™: mixt. of FO u. during World War II as olive oil replacement.

invernadero (Sp): covered place for protecting pls. fr. cold (hot house, green h.).

invertase: saccharase = sucrase; produces **invert sugar**.

involucre: whorl of bracts arranged closely below a cluster of fls. & behaving much as a calyx.

iodine: the least soluble of the halogen elements, obt. largely fr. Chilean nitrate (or guano) deposits (2316); u. antisep. alt. in hypothyroidism & palliative in hyperthyroidism - **"iodine"**: contracted form of Tr. iodinc - **i. disclosing solution**: I, KI, glycerin, HOH; appl. to tooth surfaces to show up bacterial & mucin films so that dentist can detect stains & decay areas - **i. swabs**: i. ampules, for first aid use in local appln.

iodoglobulin: thyroglandin: characteristic protein of thyroid contg. iodine; prepd. fr. thyroid gland.

Iodostarine™ (-1936-): I. addn. prod. of tariric acid (*qv.*) (hence name); formerly found on market.

Iodothyrin: 1) cleavage product of iodothyroglobulin, active constit. of thyroid gland 2) (TM) propr. prep. representing a thyroid ext. 3) triturate of (2) with lactose; formerly popular as dispensing form; supposed to represent equal wt. of fresh gland.

Ionidium Vent. = *Hybanthus* Vent.; *Calceolaria* Loefl.

ionone: Jonon (Ge): mixt. of ∂- & ß- ionones; occ. in VO of *Boronia megastigma*, henna, &c.; made by synth.; in dil. soln. is synthetic violet perfume; u. as arom. in perfumes, cosmetics.

Iowa gum: Haiawa g. fr. *Protium heptaphyllum* (exp. Georgetown, Guyana) (formerly Brit Guiana).

ipado: Coca lvs.

ipecac: abbreviated form of ipecacuanha; in common usage & considered proper; true i. has been variously placed in genera *Cephaëlis*, *Evea*, *Psychotria*, etc.; the first is most u. & recognized - **American i.**: 1) *Gillenia trifoliata*; less frequently, *G. stipulata* 2) *Euphorbia ipecacuanha* - **Bahía i.**: inferior sort of Rio i. - **bastard i.**: *Asclepias curassavica* - **black i.**: 1) (Hond): *Manettia cordifolia* 2) *Psychotria emetica* - **blue-flowered i.**: raicillo: *Polygala costaricensis* - **Brasil i.: Brazilian i.**: *Cephaëlis ipecacuanha*; 3 most important vars. are Rio (de Janeiro), Minas (Gerais), & Matto Grosso., rel. rich in emetine, hence previously rated superior - **Cart(h)agena i.: Carthaginian i.**: fr. *Cephaëlis acuminata*; altho with rel. less emetine & more cephaeline, often of higher tot. alk. content than Brasil, hence now preferred generally (cephaeline readily converted into emetine) - **Central American i.**: *Manettia cordifolia* - **Ceylon i.**: Indian i. (1) - **Colombia i.: Costa Rican i.**: Cartagena i. - **East Indian i.**: Indian i. - **false i.**: 1) *Manettia cordifolia* 2) *Richardia scabra*, *R. brasiliensis*, *R. rosea* 3) *Asclepias curassavica* 4) *Borreria poaia* 5) *Hybanthus ipecacuanha*, *H. poaya* 6) *Polygala angulata* - **farinaceous i.**: undulated i. - **Goa(nese) i.**: *Naregamia alata* - **greater striated i.**: *Psychotria emetica* - **Indian i.**: 1) *Cephaëlis ipecacuanha* grown in the Straits Settlements (Malay) (**Johore i.**) 2) *Tylophora asthmatica* - **lesser striated i.**: false i. - **Matto Grosso i.**: an excellent var. of Brasil i. (fr. State of Min Ger) - **Minas Gerais i.**: var. of Brasil i. (fr. State of Min Ger) - **Nicaragua(n) i.**: Cartagena i. - **North American i.**: *Gillenia* spp. - **Panama i.**: Cartagena i. - **Pará i.**: Brasil i. - **Peruvian i.**: black i. (2) - **Portuguese i.**: Goanese i. - **Rio i.**: Brasil i. - **Sabanilla i.: Savanilla i.**: Cartagena i. - **spurious i.**: term appl. to various i. adulterants or sophisticants; ex. *Richardia scabra*, *Calceolaria villosissima* (Br) - **striated i.**: greater s. i. - **substitute i.**: false i.; *Richardia scabra* - **Trinidad i.**: 1) *Asclepias curassavica* 2) *Psychotria tomentosa*, *P. verticillata* - **undulated i.**: *Richardia scabra* - **white i.**: 1) *Ionidium ipecacuanha* 2) *Richardia scabra* - **wild i.**: *Euphorbia corollata* - **wood i.**: *Triosteum perfoliatum* - **Yohore i.**: India i. (1).

ipecacuanha: 1) official L. title for ipecac - **Ipecacuanhae Radix** (Ph. Int.): rt. of *Cephaëlis i.* & *C. acuminata* 2) (PR) *Pedilanthus tithymaloides* & other *P.* spp.

"ipomea" (US): *Ipomoea alba.*

ipomée (Fr): *Ipomoea.*

Ipomoea (L): Mexican scammony, *Ipomoea orizabensis* - **I. alba** L. (Calonyction aculeatum [L] House; C. speciosum Choisy; Ipomoea bona-nox L) (Convolvul.) (trop Fla, swLa [US], WI, Mex, tropics incl. Br): moon flower; m. vine; nacta v.; "evening star"; bejuco de vaca; pl. u. for snakebite (Ind); lvs. as cataplasm for swellings; rt. bk. purg.; sds. eaten (Ind); latex u. (alk. decoc.) to coagulate fresh latex of *Castilla elastica* (Hond);

sds. roasted & u. as coffee substit.; ornam. night flowering pl. - **I. batatas** (L.) Lam. (CA; now natd. & cult. in all trop & subtrop) (677); sweet potato; patata (dulce); batata (Co); kumara (NZ Maoris); "yams" (La [US]) (incorrect); fleshy rts. c. 16% starch, sugars, vit. A, B complex, C; u. edible & nutritious root-vegetable throughout trop (2531); prep. many different ways; sometimes sun-dried & stored; comml. source of starch (Amylum Batatae) u. as food (low in calories) (flour, glucose, ex. pies, candy; prepn. "potato beer"); in pds., pills, etc.; techn.; sds. "parched" & u. as coffee substit.; vines u. as forage; lvs. u. in snake bite (Yuc); prep. hot potato poultices; "potato vine" tea for sore throat (Miss [US]); lvs. u. as food (Af); antidiabetic (PI); against laryngeal & other tumors (SA); liquor fr. whole pl. u. int. for "low" fevers, exter. in dermatitis (Maori, NZ); in prodn. alc. (2660) - **I. bracteata** Caven. = *Exogonium bracteatum* - **I. cairica** Sweet (tropAf & As+): mile-a-minute (Austral); mixt. of crushed lf. in liquid drunk for rashes, esp. when fever (Zulus); sd. purg. but sd. FO not; tubers edible; lvs. eaten throughout the year (natives) - **I. carnea** Jacq. (Mex, WI, CA, Co, Ve): campanilla (Mex); u. med. (Co Indians) (177) - **I. coccinea** L. (I. phoenicea Roxb.; Quamoclit c. [L] Moench) (sTex [US]; tropics): scarlet morning glory; star glory; rt. sternutatory; u. for intermittent fevers; coryza, some types of headache (Cu); lvs. u. as detergent; ornam. vine - **I. congesta** Br. (seAs; PacIs): u. as powerful cath. (Taiwan); med. in Fiji (199) - **I. digitata** L. (Operculina macrocarpa Urban) (cosmopol tropics, seAs, WI, cult. Cu): tuberous rts. u. as ton. aphrod. (Ind) to prevent obesity (boiled with sugar & water) (Taiwan) - **I. involucrata** P. Beauv. (trop wAf): pl. inf. drunk as stim. & to prevent fever; decoc. or freah juice u. for GC; inhibits growth of bacteria (Nigeria); lvs. u. in asthma - **I. jalapa** Schiede & Deppe = *Exogonium purga* - **I. leptophylla** Torr. (wNA): bush morning glory; Colorado man root; pd. rt. dusted on body with feather brush to alleviate pain or revive person; rt. remedy for nervous person (Pawnee Indians); rts. roasted & eaten (famine food) - **I. macrorhiza** Mx. (Mex): liane à minguet (Fr); u. med. (200) - **I. murucoides** Roem. & Schult. (Mex, Guat): palo del muerto; tree large or small; wd. decoc. u. in baths for paralysis; ashes u. in place of soap in washing clothes; juice is a latex - **I. nil** (L) Roth (I. hederacea Jacq.) (tropAm, WI, sTex, Fla [US], Ind, Chin): (blue) (white-edge) morning glory; Japanese m. g.; "aguinaldo azul claro" (Cu); ipomée du Nil (Fr); gaybine; fls. rose red, blue, etc.; sds. (Semen Kaladanae) drastic cath. (*cf.* Jalap); anthelm. (resin-active princ.); cult. ornam.; u. against cancer (Chin folk med.); sd. FO = Asaga oil; lvs. boiled & appl. to rashes & sores (Bismarck Archipelago); rt. cath. emmen. abort. - **I. operculata** Mart. = *Merremia macrocarpa* - **I. orizabensis** (J. Pellett.) Steud. (Mex): (Mexican) scammony; "Orizaba morning glory"; tuberous rt. c. glycosidal resins (-18%); u. drastic cath.; prep. resin (ipomoea or scammony resin) - **I. pandurata** (L) G.F.W. Meyer (eUS): man root; wild potato; huge tuberous rt. c. 5% resins (ipomein); u. mild purg.; rt. decoc. to expel gravel; vine inf. for mal. (2431) - **I. paniculata** R. Br. = *I. digitata* - **I. pescaprae** (L) R. Br. (trop & subtr of both hemispheres) (vine grows on coastal sand dunes): railroad vine; boniato de playa (Cu); beach morning-glory; entire pl. u. (Ind) exter. in rheum. (decoc.), in colic (Ind); lvs. c. VO, resin, FO, bitter princ., u. for bed sores (ungt.); drastic, detersive; u. against high BP; lvs. u. in combns. in baths for sickness (Domin Caribs) (2328) - **I. pisonis** Pittier = *Piptostegia p.* - **I. purga** Hayne = *Exogonium p.* - **I. purpurea** Roth. (tropAm, incl. Cu, tropAf): (common) morning glory; rt. & st. u. as purg. & antisyphilitic (Zulu); resin active component; ornam. vine - **I. quamoclit** L. (Quamoclit pennata Boj.) (seUS, WI, trop regions): cypress (morning) vine; cupid flower; cambustera fina (Cu); lf. rt. & sd. purg. u. in snakebite; cephal. stim. snuff; juice & pd. lf. sternutatory; u. for colds; lvs. detergent; weed; ornamental - **I. sagittata** Poir. (seUS, Mex, WI, Eu, Med): "snake grass"; rt. decoc. u. as "blood purifier"; in snakebite (Houma Indians, La [US]) - **I. sidaefolia** Choisy = *Turbina corymbosa* - **I. simulans** Hanbury (Mex): Tampico jalap; T. root; u. as drastic (similar to Ipomoea) & to prep. Tampico jalap resin (erroneously rendered at times as *I. stimulosa*) - **I. stans** Cav. (Mex): tuberous rt. u. in chorea & some forms of epilepsy; in homeopathy - **I. tiliacea** (Willd.) Choisy (Fla [US], WI, trop, continent Am): "marrullero" (Cu); sts. of vine are boiled & decoc. drunk in small amts. for parasites & in larger doses as purge (2444) - **I. tricolor** Cav. = *I. violacea* auct. div. - **I. triflora** Mar. & Velasco (Mex): purga de las animas (Mex); rt. u. drastic - **I. tuba** (Schlecht.) G. Don (Fla, WI, tropics): "flora de la Y" (Cu); te ruku (Gilbert Is); u. "as med. to attract the girls" (Gilbert Is) (103) - **I. tuberosa** L. = *Merremia t.* - **I. turpethum** R. Br. (Convolvulus t. L.; Operculina t. Silva Manso; Merremia t. (L) Rendle) (Ind, SriL, Austral, S Sea Is, worldwide): turpeth (root); t. pipes (2681); (East) Indian jalap; turbeth; rt. & resin u. drastic cath. vermifuge, in skin & liver dis.; adulterant of & substit. for jalap (also *I. violacea*); in abdom. tumors (folk med.); in prodn. turpethum resin (glycosidal resin); cult. ornam. - **I. violacea** L. (I. rubrocoerulea Hook.) (sMex, WI, tropics): morning glory; granni vine (St. Croix); ololioqui; piule; sds. c. 0.05% alk. (lysergic acid & clavine types), incl. ergine (lysergic acid amide), iso-ergine, ergonovine; sds. u. as hallucinogen (popularly u. by teenagers for "kicks" in '60s); ancient Aztec hallucinogen (Mex) (in religious practices of Indians [Zapoteca +]); 50-500 sds. produced hallucination lasting 5-8 hrs; not addictive but dangerous, esp. for neurotics & psychotics; chewing hard sds. bad for teeth; sds. extr. —> uter. stimln. (178) (Hortic. vars. favored [ornam] include "Heavenly Blue," "Pearly Gates," etc.).

Ipomopsis aggregata (Pursh) V. Grant [Gilia a. (Pursh.) Spreng.] (Polemoni.) (wNA): scarlet gilia; s. trumpet; timpiute; u. as emet. (Nev Indians) (1605) - **I. congesta** V. Grant (Gilia c. Hook.) (wUS): many-flowered gilia; hb. decoc. u. as blood tonic (Shoshone Indians, Nev [US]) - **I. rubra** (L) Wherry (seUS): standing cypress; ornam. perenn. hb.

ipyaku (Ec-Jivaro Indians): achiole, dye fr. *Bixa orellana*.

iraca (Co): *Carludovica palmata*, Panama hat plant.

irene: hydrocarbon, $C_{13}H_{18}$, derived fr. irone.

Iresine calea Standl. (Amaranth) (Mex-Calif): "amargosillo" (Mex); pl. decoc. u. in fevers (Puebla); diur. diaph. - **I. diffusa** HBK (I. paniculata Ktze.; I. celosia L.; I. celosoides L.) (seUS, WI, Mex to Pan; SA: Pe to Arg): blood leaf; bitter weed; juba (bush); bitter decoc. in coughs, colds, edema; pl. juice appl. to erysipelas; ashes with added lemon or lime u. as purg. - **I. herbstii** Hook. f. (tropAm, WI; natd. tropAs): "molleja" (Cu); lf. inf. u. for oliguria (Trin); med. (Vietnam); cult. - **I. spiculigera** Seut. (I. acicularis Standl.) (Guat to Peru & Arg): pie de paloma; inf. u. as purg. (Ve Indians).

iretol: part of aglycone formed in splitting iridin.

Iriartea *ventricosa* C. M. Mart. (Palm.) (Pan): ila (Cuna Indians); stilt palm; u. med.; outer part of trunk u. to make floors of houses; prop rts. u. to grate coconut (179).

Iridaceae: Iris (or Orris) Fam.; perenn. hbs. of temp. to trop. zones; usually low-growing, with sword-shaped or linear lvs.; fls. showy, bisexual, with a perianth of 6 segments (tepals) (3 sepals & 3 petals); frt. a 3-partitioned capsule c. few to many sds.; underground parts rhiz., bulbs, or corms, starch-rich, with resins, FO, VO, special carbohydrates; largely unknown chemically; ex. gladiolus; true saffron, Crocus.

Iridaea (Iridea) boryana (Gigartin. - Gigartinales - Rhodoph. - Algae) (coasts of Chile, Kerguelen, Crozet Is): thallus u. to make agar - **I. cordata** (Turn.) Bory (w coast NA): u. to produce iridophycan, a phycocolloid u. as mucilag. stabilizer, anticoagulant - **I. edulis** Harvey (Dilsea e. Stackhouse) (coasts of Scotland & Ireland): u. as food ("dulse") (but not true dulse, which has superior flavor) - **I. flaccida** (S & G) Silva (cCalif-Ore [US] coasts; also Chile, Jap): c. iridophycan (phycocolloid; *cf.* carrageenan); u. to stabilize chocolate milks, etc.; marketed as "kelp tablets"; u. as diaph., antipyr.

iridin: 1) xall. glucoside occurring in Iris rhiz. & splitting finally into d-glucose, iretol, formic & iridinic acids 2) resinoid fr. *Iris versicolor*; brownish red pd., sol. in alc.; formerly u. as purg. & cholagog (in pills); highly regarded by Eclectics - **irigenin**: glycoside, the first stage in enzymatic degradation of iridin.

iridoids: group of related structural-chemical monoterpenoid compds. similar to iridodial (named after *Iridomyrmex* [meat ant] where first found); often as glycosides; ex. oleuropein.

Iridophycus = *Iridaea*.

Iris L. (Irid.) (natives of nHemisph. only in temp. reg.): iris; flag; fleur-de-lis; flower of Louis, flower de luce (armorial symbol of kings, Prince of Wales); brilliant decorative pl. growing fr. rhiz. or bulbs (<iris, Gr., rainbow, SC diverse colors); Iris = L. title for rhiz. of *I.* X *germanica* (incl. var. *florentina*) & *I. pallida* - **I. decora** Wall. (I. nepalensis D. Don) (nInd, Pak, Nepal): rt. u. aper. diur., in bilious obstructions; exter. appl. to small sores & pimples - **I. douglasiana** Herbert (Ore, Calif [US]): coastal flag; Douglas' iris; fresh lvs. u. as cooling wrap for babies in hot weather; lf. fibers u. for deer snares - **I. ensata** Thunb. (cAs to Jap): irisa; sds. astr. hemostat. diur.; in hemorrhage, fevers, rheum., colic, epistaxis, wounds, ulcers, to relieve intoxication; rhiz. antipyr.; fls. anthelm. (Chin); fiber is u. - **I. florentina** L.: previously regarded as good sp., but now generally considered to be *I. germanica* L. v. *florentina* Dykes - **I. foetidissima** L. (wMed reg & wEunAf): scarlet seeded iris; stinking gladwine; lvs. rts. sds. c. glycosides, mannans, levulosans, VO; u. errhine hydrag. antispas. sial. astr. diur.; the dr. capsules in bouquets; popular ornam. - **I. X germanica** L. (nInd to Med reg., Mor): German iris; flag i.; common garden i.; source of Verona Orris Rt.; rhiz. c. iridin, FO, starch, resins, some tannin, VO; u. diur. cath. emet. stim. exp. corrective, er-

rhine; in perfumery, baby sachet, face, talcum, & tooth pds. (occasional allergy) & tooth pastes, shave creams, cleansing creams, cold creams, bath salts, dry shampoos; sometimes u. by bakers; pieces ("fingers") u. for teething children (masticatory); filler; formerly in adhesive plasters; var. *florentina* Dykes (seEu, AsMinor, nAf): Florentine iris; orris (plant); previously official as *I. florentina* L. (SC cult. widely in Florence, province of It); anciently u. for insomnia, cancers, freckles, etc. - **I. japonica** Thunb. (eAs): Chinese iris; rhiz. crushed into plaster to apply to mechanical injury; whole pl. decoc. u. to reduce swellings, rheum. (Chin) - **I. kaempferi** Sieb. ex Lemaire (eAs): Japanese iris; remedy for abscesses (Chin) - **I. kumaonensis** Wall. (Ind): piaz (Punjabi); rt. & lvs. given in fever; cult. ornam. - **I. lactea** Pall. (eAs): sds. u. in colics & diarrh. (176) - **I. lacustris** Nutt. (NA around Great Lakes): said u. similarly to *I. versicolor* - **I. macrosiphon** Torr. (Calif [US], Mex): rhiz. u. aphrod. emmen. (Mex) - **I. missouriensis** Nutt. (wNA): western blue flag; rt. u. in smallpox (NM [US]); adult. of blue flag rhiz.; pd. rhiz. mixed with tobacco & white camas & smoked at times —> nausea (Am Indians); stock poison - **I. nepalensis** D. Don (Ind): shoti; rts. u. as aper. diur. & in bile dis.; exter. for small sores, pimples, &c. - **I. pallida** Lam. (seEu, Balkans, Syria): sweet iris; a source of Verona Orris Rt.; may be chief source of official Orris Rt.; similar c. & u. to *I. germanica* - **I. prismatica** Pursh (Atl coast NA): slender blue flag; poison flag (root); has very slender rhiz.; said u. like *I. versicolor* - **I. pseudacorus** L. (Eu, nAf, natd. NA): yellow flag (iris); false calamus; rhiz. c. irisin, tannin; sharp (acrid) juice u. in med. anciently by Galen, Dioscorides; strong cath.; inf. in diarrh. dysmen. leucorrhea, dentalgia; in cosmetics; adult. of calamus (danger!); fresh juice as vulner., oil against cancer, sds. against flatulence; u. in veter. practice; rhiz. u. in tanning & dyeing (black & blue); sds. stomach, carmin.; roasted sd. as coffee substit. - **I. sibirica** L. (Euras esp. USSR): rhiz. u. as expect. & to destroy insect vermin; this & *I. pseudacorus* are the characteristic irises of Eu - **I. songarica** Schrenk (Baluchistan, Pak): rt. u. as antidiarrheal (pd. in curds) - **I. tectorum** Maxim. (eAs): rhiz. u. to treat marasmus & wasting dis.; as emet. & lax. (Chin) (SC grown on roofs long after edict forbade waste of ground space with ornam. pls.) - **I. tuberosa** L. (sEu): European orris root; rhiz. u. purg., for teething babies - **I. verna** L. (seUS): dwarf iris; rhiz. pungent & spicy; u. somewhat like foll. - **I. versicolor** L. (neNA s to Tenn [US] on margins of swamps, ponds, streams): (larger) blue flag (iris); beardless iris; said similar to European *I. pseudacorus* except for blue-violet color of corolla; rhiz. c. acrid resin, VO, tannin, starch; u. emet. cath. purg. cholag. alt. diur. in kidney dis., sial. vermif. much u. in cattle powders; poisonous; rhiz. esteemed by most NA Indian tribes, u. for gastric dis.; in earache; appl. in paste to sores & bruises; valued in chronic mal. (Creek Indians+); formerly highly adulterated & sophisticated (versicolor = changing color, due to varying colors of parts of fl.) - **I. virginica** L. (seUS s fr Tenn to Fla): (southern) blue flag; Virginia iris; c. & u. similarly to prec. with wh. has been confused.

iris, blue: *Iris versicolor* (etc.) - **i. de Florence** (Fr): orris - **i. family**: Iridaceae - **German i.**: *I. germanica* - **larger i.**: *Iris versicolor* - **Morocco i.**: rel. low quality of orris (Mogador; Saffi) - **i. oil**: VO of orris rt. - **i. peas**: artificial issue peas made fr. i. rhizome - **i. reagent paper**: filter p. impregnated with extractive of i. flower u. as indicator - **i. sauvage** (Qu): *I. versicolor* - **slender i.**: *I. virginica* - **i. varié** (Fr): *Iris versicolor* - **Verona i.: Veronese i.**: rel. low quality prod. - **wild i.**: *I.* spp. in US usually *I. versicolor*.

Irish bog-butter: unusual fatty subst. found in bogs (peat) of unknown origin; (180); butyrelite - **I. broom**: *Cytisus scoparius* - **I. daisy**: dandelion - **Irish**: v. also under dulse; moss.

irisin: a polysaccharide (carbohydrate) of same general type as inulin (fructosan); found in *Iris pseudacorus*, *I. sibirica*, & other *I.* spp. (rhiz.); may be identical with graminin - **iris-violet**: a synth. dyestuff.

iron: Fe; second most abundant element in earth's crust (Al = No. 1); red color of many soils, such as sUS, due to Fe minerals; valence 2, 3; several minerals incl. hematite, magnetite, siderite; many alloys (steels) with Ni, C, Cr, Mn, &c.; essential to animal metabolism; sometimes present in excess in animal body; many salts, ferrous (Fe^{+2}, ferric, Fe^{+3}) (2709) (v. under ferric, ferras).

iron bark tree: *Eucalyptus globulus*+ - **food i.**: iron available fr. more or less popular foods; best sources in decr. order of value: beef liver, fresh peas, raisins, bananas, apricots, cereal grains, lettuce, parsley, spinach (2597) - **i. pyrites**: native ferrous sulfide - **i. humate:** salts of iron with humic acid of soil; said good source of Fe for plants - **i. sorbitex: i. sorbitol**: complex of Fe, sorbitol, & citric acid with dextrin as stabilizer; u. by inj. as hematinic in anemias where Fe deficient - **i. soup** (Jam): water in which green bananas (plantains) have been boiled; thought rich in Fe & good blood tonic (21) - **i. tree**: 1) *Eucalyptus* (bk.) 2) *Metrosideros* spp.

iron-dextran complex: colloidal soln. of $Fe(OH)_3$ complexed with partly hydrolyzed dextrin in water; inj. u. as hematinic in iron deficiency anemias, esp. in pigs; carcinogenic.

irone: a cyclic methyl ketone, $C_{14}H_{22}O$; found as fragrant princ. in iris rt., violets, & other VO; u. in perfumery trade as a violet perfume; several isomers known; mostly present in *Iris* spp. as beta-irone, with smaller amts. of alpha- & gamma-irones present (latter with very little odor value).

ironweed: 1) *Lithospermum arvense* 2) *Polygonum aviculare* 3) *Vernonia* spp. 4) (Virgin Is) *Galactia dubia* 5) (US) *Verbena hastata* 6) (Ala [US]) *Sida rhombifolia* - **ironwood (bark)**: 1) *Ostrya virginiana* 2) *Carpinus caroliniana* 3) *Forestiera neomexicana* 4) 7 other pl. genera - **ironwort**: *Galeopsis ochroleuca*.

Irvingia gabonensis Baill. (Simaroub./Irvingi) (w tropAf): wild mango; African mango (tree); frt. edible but with turpentine flavor; sd. source of dika butter (or d. fat) (Gabon f.); food article, also u. in mfr. soap, candles; rather similar to cocoa butter but with some advantages (62); this solid fat u. as tablet lubricant; to give prolonged release in tablets & capsules; as sealant to microcapsules; roasted sds. u. in making dika bread; wd. very hard & heavy, u. to pave streets; wd. & rt. exts. not effective against exptl. mal. - **I. malayana** Oliver ex Bennett (Malacca; sVietnam): sds. yield FO, cay-cay butter, irvingia b., u. in prepn. candles - **I. oliveri** Pierre (Cambodia, Vietnam): sds. source of cay-cay butter (or fat), Cochin Chin wax; irvingia butter; techn. u. to make candles - **I. smithii** Hook. f. (tropAf): produces dika fat; hard lumber.

Iryanthera macrophylla Warb. (Myristic) (Pe): bk. u. to make narcotic pellets u. as hallucinogenic by Pe Indians - **I. ulei** Warb. (nwBr): u. to make hallucinogenic paste; c. dimethyltryptamine deriv.

isano: *Ongokea klaineana* (oil).

Isatis alpina Vill. (Cruc.) (wAlps, Eu): earlier u. to prep. indigo - **I. tinctoria** L. (c&sEu; As): (dyer's) woad; wade; formerly much u. as indigo source (cult. Fr, Ge); c. isatan (indoxyl gluconate), quercetin, kaempferol, etc.; u. as astr. in cicatrization of wounds, sores; at times against tumors incl. malignant type; blue dye, u. by early Britons (Celts) to paint their bodies blue.

Isertia hypoleuca Benth. (Rubi) (Co): lvs. u. as tea for menstr. irregularity (with sds. of *Carica papaya*); lf. decoc. rubbed on chest for pain (2299).

isinglass: Ichthyocolla (L): gelatin-like subst. fr. dr. swimming bladders ("fish sounds") of various fresh water fishes, esp. sturgeon (*Huso huso*, *Acipenser stellatus*, *A. sturio*, etc.); 3 types (curl, leaf, thread); u. to clarify (purify) beer, etc. - **American i.**: prod. of the hake, *Otolithus regalis* - **Bengal i.**: agar.- **Brazilian (shred) i.**: comml. prod., sort of fish glue, u. in techn. not pharm. - **Ceylon i: Chinese i.: Japanese i.**: agar; the last 2 largely fr. *Gelidium cartilagineum* - **leaf i.**: simplest form of prod. made by simply nailing the damp material to boards & letting dry out - **Russian leaf i.**: prod. of sturgeon fish sounds - **shredded i.**: prep. by rolling out the leaf i. & cutting into threads in a machine - **i. stone**: a mineral.

isoascorbic acid = erythorbic acid.

isobutyric acid: $Me_2CH.COOH$; occ. free in *Ceratonia siliqua* frt., *Arnica* rt. & esterified in *Pastinaca sativa*.

Isocarpha oppositifolia (L) Cass. (Comp.) (WI, Mex, CA): "white catnip" (Bah); hb. u. as ton. (manzanilla de la tierra) (Cu); rt. decoc. for chest colds (Bah).

isocitric acid: chief acid of *Kalanchoe pinnata* & other Crassulaceae; *Rubus fruticosus* (blackberry frt.); widely distr. in pl. kingdom (in traces in tissues); $COOH.CH_2.CH\text{-}COOH.CHOH.COOH$.

isodulcit(ol): rhamnose.

isoenzyme = isozyme.

isoephedrine: pseudoephedrine.

isoglucose (GB): high fructose syrup.

isohem-agglutinin: a fraction of the human blood serum; principally u. for typing bloods.

isolate: substance obt. in a pure form fr. a VO or other natural material; ex. eugenol.

isoleucine, 1-: essential amino acid; amino-methyl-valeric acid; essential in diet because not synthesized in human body; isolated from casein, ovalbumin, & other proteins; u. as nutrient; in amyotrophic lateral sclerosis (Lou Gehrig's disease).

Isoloma Dcne.: genus usually referred to *Kohleria* (*qv.*)

Isolona campanulata Engl. & Diels (Annon.) (Nig): bk. u. in coughs, fever, bilharziasis.

Isonandra gutta Hook. = *Palaquium gutta*.

Isopblad (Swed): **isopo foglia** (It): hyssop leaf.

Isoplexis isabelliana Webh. & Benth. (Scrophulari) (Canary Is): "digitalis of the Canary Islands"; lvs. with powerful cardiotonic activity (2061).

isoprene: unsatd. hydrocarbon; $CH_2{=}C(CH_3)$ CH $=CH_2$; distn. prod. of natural rubber (caoutchouc); occ. in plants, ex. *Hamamelis jelena* as VO product (181).

Isopyrum thalictroides L. (Ranuncul) (Euras): meadow rue; c. isopyroine, pseudoisopyroine; u. med. (Ind) (194); toxic pl.

isoquercitrin: isoquercitroside: trifoliin; quercetin-3-glucoside; flavone glucoside occ. in fls. of cotton, horse chestnut, garden nasturtium, *Arnica montana*, etc.

Isotoma (*Laurentia*) **longiflora** (L) Presl. (Campanul.) (Oceania-tropics): tibey (PR); c. isotomine (*cf.* lobeline); causes heart block; stops respirn.; very dangerous stock poison; produces excessive green saliva; latex u. med. (Cuna Indians, Pan); u. med. (Fiji) (199); finely pulped lvs. appl. to aching teeth (Indones); in Yucatán, vulner., in VD (syph.), rheum, epilepsy; bronchitis, antiasthm. (Br).

isotonic salt solution: physiological salt (NaCl) aqueous soln. (0.9%) u. in physiol. & pharmacol. experiments & in therapy, because with approx. the same tonicity as the blood serum.

isotype: typical specimen of organism taken from type series & type locality.

isozyme: one of the multiple (2, 3, or more) molecular forms of an enzyme found in a single sp. of an organism, which differ chemically, physically, & immunologically but still catalyze the same chem. reaction.

ispaghula seed: sd. of *Plantago ovata.*

ising(h)o: 1) (SA) *Gnaphalium* sp. (or related) 2) (Pe) various oily sds., incl. *Quercus, pucheri* (*Licaria puchury-major*), Peruvian oak.

issue peas: 1) small orange berries or round pills turned fr. orris rt. placed in a suppurating sore to prevent healing; formerly much u. in med. 2) *Cicer arietinum* - **artificial i. p.**: *Iris germanica* var. *florentina*; *Hedera helix* - **bitter orange i. p.**: small immature b. o. frt. u. in this way.

istle: fiber fr. *Agave heteracantha.*

itaconic acid: a dicarboxy aliphatic acid prepd. by fermentation using *Aspergillus terreus* Thom and *A. itaconicus* Kinoshita; u. in mfr. of plastics.

Italian juice root: licorice - **I. poplar:** *populus nigra.*

itamo real (Mex): *Dictamnus albus.*

itch(ing) powder: Juckpulver (Ge); novelty item based on pd. glochids of *Opuntia macrodasys* or mechanical trichomes of *Mucuna urens* (marketed Ge) - **i. weed**: *Veratrum album, V. viride.*

Itea chinensis H. & A. (Grossulari) (Chin): frt. decoc. u. to aid digestion (182).

Ithyphallus Gray (nom. rej.) = *Mutinus* (Fungi).

itr (Urdu): perfume (variant of attar).

Iva angustifolia DC (Comp.) (Tex): narrow-leaf sumpweed; native annual winter pl.; produces abortion in cattle - **I. annua** L. (I. ciliata Willd.) (seUS n to Indiana, Neb): sump weed; marsh elder; "careless weed" (prob. confusion); pl. cult. by Am Indians for edible sd.; pl. may cause contact dermatitis - **I. a.** var. *macrocarpa* R. C. Jackson (cUS): sump weed; large akenes former important food source of aborigines; displaced by maize; u. during 1st millenium BC (Ark, Ky, Mo, wNCar [US]); pl. now extinct - **I. axillaris** Pursh (wUS & Cana): marsh elder; poverty weed; pl. decoc. favorite remedy of several Am Indian tribes for stomachache or cramps, esp. in young children; claimed contraceptive; said u. in Chartreuse liqueur - **I. frutescens** L. (sNJ s to Fla [US]-coastal marshes): marsh elder; "horse-brush" (wFla); erect shrub; bk. u. as antipyr. - **I. xanthifolia** Nutt. (cNA, natd Eu): yellow sumpweed; burweed; cause of hay fever & dermatitis.

ivain: bitter subst. occurring in milfoil.

ivermectin: macrocyclic lactone related to the avermectins; fr. *Streptomyces avermitilis* (soils, Jap); parasiticide u. against onchocerciasis (river blindness) (wAf) (microfilaricidal agent); said superior to Suramin.

ivi apple: *Spondias purpurea.*

ivory: dentine of commerce, obt. fr. elephant, walrus, killer whale, & hippopotamus tusks; much of the so-called ivory of the modern day is a synth. plastic, esp. fr. casein - **i. black: black i.**: 1) animal charcoal fr. which inorg. compds. have been dissolved out with HCl 2) ivory fr. African elephant 3) (slang) Negro slaves (slave trade) - **i. nut: vegetable i.**: sd. of *Phytelephas macrocarpa* u. to mfr. buttons.

ivy: 1) *Hedera helix* + 2) (Appalachia [US], as nGa mt. people): mountain laurel, *Kalmia latifolia* - **American i. (bark)**: *Parthenocissus quinquefolia* - **i. berry**: *Gaultheria procumbens* - **climbing i.**: *Toxicodendron* spp. - **creeping i.**: *Toxicodendron radicans* - **English i.**: i. (1) - **ground i. (herb)**: *Glechoma hederacea* - **Kenilworth i.**: *Cymbalaria muralis* - **poison i.**: *Toxicodendron radicans* (eUS); *T. diversilobum* (NA-Pac coast); *T. radicans* subsp. *divaricatum* (R. greenei), & other *T.* taxa.

iwarancusa (jwarancusa): 1) *Cymbopogon i.* (*j.*) 2) *Vetiverria zizanioides.*

Ixora casei Hance (Rubi.) (Macronesia): lvs. u. stomachache, nausea; rts. u. as hemostat. in menstr.; cult. ornam. (Austral+) - **I. chinensis** Lam. (Ind-sChin): lf. decoc. u. for head- and stomachache (Indochin); rt. decoc. u. after childbirth & for urin. dis. (Mal); fresh fls. inf. u. for hemorrhage & incipient TB. (PI) - **I. coccinea** L. (Ind, SriL, cult. in the tropics): jungle flame ixora; flame (tree) of the woods; fls. u. rem. in dys. (Ind); dysmenorrh.; rts. for nausea, hiccups, anorexia (PI); cult. as ornam. - **I. cuneifolia** Roxb. (seAs): rts. u. to treat furunculosis, bubonic plague; abscesses; fl. inf. u. to treat fever & colic (Indochin) - **I. flavescens** Pierre var. *cambodiana* Pitard (seAs): inf. of lvs. st. fls. u. for fever & colic; decoc. to women at parturition - **I. macrothyrsa** Teijsmann & Binnenduk (Indones): fls. & bk. u. in med. - **I. nigricans** R. Br. (seAs): lvs. u. for dys. & at childbirth (Indochin).

ixtle: **istli:** *Avave lecheguilla*; *A. heteracantha.*

iztachtzapotl (Mex): *Casimiroa edulis.*

Jj

NOTE: "i" and "j" are often interchangeable in several languages (such as German); it might therefore be advisable to look under each letter.

jabilla (SpAm): javilla.

jaborandi (leaves): *Pilocarpus* - **Maranham j**.: *P. microphyllus* - **Paraguay j.**: *P. pennatifolius* - **Pernambuco j.**: *P. jaborandi*.

jaboticaba (seBr): tree; 1) *Fevillea trilobata* 2) *Myrciaria* spp. esp. *M. cauliflora* (sweet-sour edible frt.).

jaca (Br): jack: frt. of jaqueira.

jacape (Br): *Cymbopogon nardus*.

Jacaranda Juss. (Bignoni) (WI, CA, SA): some 35 spp.; carobinha; several spp. u. in syph. scrofula, buboes (Br) - **J. acutifolia** Humb. & Bonpl. (tropSA): "fern tree"; yaravisco (Pe); lvs. u. med. ("caroba" [Br]); bk. u. in skin dis. (palo de buba) (Pan); frt. u. in syph. buboes, astr.; wd. u. to make cups for drinking water in rheum. (*cf.* quassia wd.); frt. c. tannin, u. as astr.; lvs. appl. to sores as vulnerary - **J. aurantiaca** Ait. (CA): curobubo (Co); palo de buba (Pan); bk. u. to treat skin dis., VD (buboes) (Pan); wd. u. in light construction - **J. caroba** (Vell.) DC (Br): caroba (do campo); lvs. c. carobine (xall. alk.), resins; u. similar to *J. copaia*, in scrofula, buboes (folk med.); in amebiasis (2381) - **J. copaia** D. Don (J. procera Spreng.) (Hond to Br): boxwood; palo de buba; lvs. (as caroba or carobinha) c. carobine (alk.), resins, bitter princ.; lvs. (& bk.) u. as diur., diaphor. in GC, syph. buboes (baths); gout, rheum. tumors (Br); bk. emet. & cath. - **J. decurrens** Cham. (Br [Matto Grosso, Paraná, etc.]): carobinha (do campo); tree or shrub; u. syph. ulcers, skin dis. (aborigines); shown useful to control intestinal protozoan zoonoses, incl. *Entamoeba histolytica* (—> amebiasis), & *Giardia lamblia* (—> giardiasis), etc. (183) - **J. mimosifolia** D. Don (Br; tropics): frt. u. as antisyph. & astr. (sMozam) (some think this sp. identical with *J. acutifolia* H & P); wd. useful, formerly dyewood; cult. st. tree - **J. subrhombea** DC (Br): caroba preta; c. assu; bk. of tree considered diaphor.

jacaranda: *Machaerium villosum* & other *M.* spp.; source of Brazilian rosewood.

Jacaratia spinosa (Aubl.) DC (Caric.) (sBr): barrigudo; berries (yellow or orange) edible; frt. juice purg. deobstruent, vermifuge (large dose); st. pulp u. in prepn. of beverages & sweetmeats.

jack (seUS Negro): voodoo charm; sometimes consisting of a small hand-like cluster of rts. or tubers, often wrapped in red flannel; also called a charm, conjure bag, goofer, grigri, hand, luck ball, mojo, tobie, trick, tricken bag, & wanga - **j. bean**: *Canavalia ensiformis*; sd. mostly as the meal - **j. tree**: *Artocarpus integrifolia* (& j. fruit); also j. wood.

Jack-in-the-pulpit: *Arisaema triphyllum* (*Arum maculatum*) - **Jack-my-lantern**: *Clitocybe illudens*.

Jacksonia dilatata Benth. (Leg) (Austral): wd. inf. u. for stomach "upset."

Jacobinia incana Hemsl. (Acanth.) (Mex): muicle (NL); shrub; u. in colic (Tamaulipas) - **J. mohintli** Benth. (Mex): decoc. of lf. & fl. u. as ton.; lf. source of a purple dye u. as substit. for indigo - **J. sericea** Nees (Pe): u. for pneumonitis; to regulate menstrual flow; in dis. derived fr. VD - **J. spicigera** (Schlecht.) L. H. Bailey (J. mohintli Hemsl.;

Justicia s. Schlecht.) (Mex, CA): mohintle (Querétaro, Mex); tinta; decoc. of lvs. & fls. u. as ton. in dys. fevers, GC epilepsy, to prevent abortion (Guat), etc.; lvs. furnish dark blue dye ("Mexican blue") u. in laundering clothes; substit. for indigo; acts as indicator; cult. (Tex [US]) as "muicle" - **J. tinctoria** Hemsl. (Justicia t. D. Gibson) (CA, Co): azul de mata; decoc. of leafy branches u. in rheum. skin dis. (extern); contraceptive (CR, Ind); source of blue dye, u. to whiten clothes (—> *Justica*).

Jacob's chariot: *Aconitum napellus* - **J. ladder**: 1) *Polemonium occidentale* or *P. reptans* 2) *Smilax peduncularis* (sic).

Jacquemontia curtisii Peter ex Hallier f. (Convolvul.) (sFla [US]): "white jacquemontia"; whole pl. decoc. rubbed on ailing leg.

Jacquinia L. (Theophrast.) (tropAm): some 35 spp.; "barbasco" (SA) - **J. arborea** Vahl (J. barbasco Mez) (WI): cubé root; rich in rotenone; u. as insecticide & fish poison - **J. aurantiaca** Ait. (WI, Mex, CA): rosadilla; "knock me back" (Belize); bk. formerly u. in VD; sd. decoc. in headache, toothache; pd. bk. mixed with salt & appl. to sores of animals; crushed frt. or rt. u. as fish poison ("barbasco"); pd. sds. u. as vermifuge (Guat) - **J. pungens** A. Gray (Mex, Guat): "San Juan(ito)"; green frt. edible, fls. u. as dye (yellow); frt. u. as fish poison (stupefies).

jade: gemstone representing the minerals jadeite (chiefly) and nephrite; chiefly fr. Bermuda, also fr. Chin, NA, CA (CR); earlier u. in weapons, tools, jewelry, & med. (for kidney stone; colic; *cf.* name); for 7,000 yrs. worked into ornaments by Chinese; often very valuable (2521).

jade plant: *Crassula arborescens*; *C. argentea*.

jag: small load, portion, or quantity.

jagra: jaggary (Hind): **jaggery** (sugar) (Ind): ghur; raw brown sugar in lumps, aromatic (*cf.* maple sugar); finest flavored fr. *Caryota urens*; also prepd. fr. sugar cane, *Cocos*, *Arenga*, *Borassus,* etc.; rather crude only partially purified sugar; source of toddy, arrack, vinegar.

jak fruit: jack fruit.

jake(y): jakey gin (US slang): Fluid extract Jamaica ginger; **Jake paralysis** or **j. poisoning** was caused during prohibition days by tricresyl phosphate, a toxic solvent added to the Fluid extract by some unscrupulous manufacturers (Jake[y]: contrn. of Jamaica).

jalap (root): *Exogonium purga* (pron. ja´l-ap) (2317) - **Brazil(ian) j.**: rt. of *Merremia tuberosa* - **Brazil male j.**: Mexican scammony, *Ipomoea orizabensis* - **Indian j.**: *Ipomoea turpethum* - **j. léger** (Fr): *Ipomoea orizabensis* - **light j.: male j.: Orizaba j. (root)**: Brazil male j. - **j. resin**: a galenical gluco-resin ext. with alc. fr. jalap rt. c. jalapin & scammonin - **j. tubéreux** (Fr): Vera Cruz j.: j. - **wild j.**: 1) *Ipomoea pandurata* 2) *Podophyllum peltatum.*

Jalapa (L. Sp): jalapa root: 1) jalap 2) (occasionally) Mex scammony - **j. fusiforme** (Mex): Mex scammony.

Jalapenknollen (Ge): Jalapenwurzel: jalap - **jalapin**: alc.-soluble resin of jalap; glycosides of convolvulinic & ipurolic acids.

jalapeño (Mex, Sp): (pron. hal-apenyo): very hot Mex pepper, *Capsicum* sp. - **j. serrano (peppers)**: short, narrow, green pods, hotter than jalapeño *per se* - **Tampico j.**: *Ipomoea simulans.*

Jaltomata edulis Schlecht. (Saracha j. Schlecht.; S. edulis Tell.) (Solan.) (Mex): berries edible; Mex Phar.

jam: a sweet preserve made of various frts. with sugars & water; cheaper ingredients include apple juice, syrupy glucose, etc.

Jamaica: roselle, *Hibiscus sabdariffa* - **J. fire flower**: *Adenorima punicea* (with leaf-like bracts) - **J. fish fuddle tree**: *Piscidia erythrina* - **J. flowers**: **J. sorrel**: J.

jaman: 1) (Hind) jumbul, *Syzygium cumini* 2) (Punjabi) *Punica granatum.*

jambee beads (WI): sds. of *Adenanthera pavonina.*

jambo, wax: *Eugenia javanica.*

Jambolan: Jambos: Jambosa: jambosa (root): **jambi**: *Syzygium cuminii* - **j. assu** (Br): *Piper reticulatum* - **jambul (seed)**: jambosa - **jambulol**: ellagic acid - **jambun**: jambul.

Jamesonia Hook. & Grev. (Adiant/Gymnogramm.) (SA): xerophytic ferns; some spp. (doradillo) u. as decoc. of rhiz. for diarrhea; said very effective (Ve).

Jamestown weed: *Datura stramonium*; commonly contracted to "Jimson weed."

Janeway's Pills: Aloe & Podophyllum Pills Comp.

Janipha Kunth = *Manihot* Mill. - **J. manihot** Kth. = *Manihot esculenta.*

Jansbrood (Neth): John's bread.

jap drier: Japan.

japaconitine: alk. characteristic of Japanese aconite, *Aconitum fischeri.*

Japan (drier): varnish with a large percentage of resins - **J. camphor**: natural c. - Japan: v. under earth; isinglass; tallow.

Japanese lantern: *Hibiscus shizopetalus*; *Physalis* spp. - **J. salve** (US slang): menthol comp. ointment & similar prods. - **J. wax**: wax fr. the berries of *Rhus succedanea* - **Japanese**: v. under adonis; gelatin; pagoda tree.

japecanga (Br): *Smilax japicanga.*

Jap-menthol: menthol fr. Japanese (pepper)mint oil.

japmint: Japanese (pepper)mint (contraction).

japonica: 1) *Chaenomeles japonica*+ 2) *Camellia japonica, C. lagenaria* 3) catechu.

jaqueira (Br): *Artocarpus integra.*

jarabe (Sp): syrup.

jardiniere (Fr, Eng): stand or pot for ornam. flg. plant.

Jarilla caudata Standl. (Caric.) (Mex): jarilla; frt. u. for preserves & candies - **J. chocola** Standl. (wMex): frts. edible (lemon aroma); tuberous starchy rts. eaten baked; latex c. papain-like enzyme - **J. heterophylla** Rusby (Mex): "granadillo"; rts. & frts. u. as food or med. (Mex markets).

jarilla: 1) (Sp, Domin) *Hura crepitans* 2) (Mex) *Dodonaea viscosa*, etc. - **j. hierba** (Mex): *Larrea* - **j. pispa** (Arg): *Zuccagnia punctata.*

jarovisation: vernalization.

jarrah: *Eucalyptus marginata, E. rostrata.*

jasim de caiena (Por): *Plumeria alba, P. rubra* (also **j. de rabo**) - **j. do ceu** (Por): *Plantago ovata.*

Jasione L. (Campanul.) (Eu): sheep's bit; perenn. or bienn. hbs. - **J. humilis** Loisel (Sp, Pyren): perenn. hb.; cult. ornam. - **J. montana** L. (sEu, nAf, natd. US): sheep's scabious; u. astr. vulnerary - **J. perennis** Lam. (wGer, FrPyr): perenn. hb.; cult. ornam.

jasmine: 1) *Jasminum* spp. esp. *J. officinale* 2) *Gelsemium sempervirens* - **cape j.**: *Gardenia jasminoides* - **Caroline j.**: *Bignonia sempervirens* - **Catalonian j.**: *J. officinale* var. *grandiflorum* source of j. oil - **night j.**: *Nyctanthes arbortristis* - **night blooming j.**: *Cestrum aurantiacum*; *C. nocturnum* - **j. sauvage** (Fr) (wild j.): *Gelsemium sempervirens* - **Spanish j.**: Catalonian j. - **j. tree**: *Plumeria* spp. - **true j.: white j. (flowers)**: Catalonian j. - **yellow j.**: *Bignonia sempervirens.*

Jasminum anastomosans Wall. (Ole.) (eAs): sap of entire pl. u. as deobstruent both exter. & inter. - **J. angustifolium** Vahl (Ind): rt. u. in ringworm (exter. appln.) (with calamus, lime juice) - **J. arborescens** Roxb. (Ind): lf. juice u. as emet. & where bronchial tubes are obstructed with viscid phlegm (Ind) - **J. auriculatum** Vahl (tropAs): fls. given in TB (Ind); source of arom. VO u. perfumery (otto juhi) - **J. bifarium** Wall. (seAs): rt. decoc. u. antipyr.; cold water inf. in yaws; pounded lvs. u. as poultice in headache, vertigo (Mal); rt. decoc. u. as gargle for inflamed gums (PI) - **J. didymum** Forst. f. (Java to PI, Polynesia): frt. juice u. as ink; smoke fr. burning bk. inhaled to relieve headache (Solomon Is.) - **J. fluminense** Vell. (eAf): lf. & rt. bk. u. as snakebite antidote (Tanzania; Madag) - **J. fruticans** L. (Med reg., Albania): (wild) jasmine, shrubby j.; yellow j.; rhiz. & rts. u. adult. for Gelsemium - **J. grandiflorum** L.: *J. officinalis* L. var. *g.* Bailey - **J. humile** L. (As-Himal): Nepal yellow jasmine; fls. u. ton. for GI tract & heart; rts. in ringworm; latex to destroy accumulations in chronic sinuses & fistulas - **J. lanceolatum** Roxb. var. *puberulum* Hemsl. (Chin): rts. & st. u. as analg. - **J. laurifolium** Roxb. (Cambodia): fls. antipyr., antipruriginous; fl. juice u. in ulcers - **J. multiflorum** Andrews (J. glabriusculum Bl.) (seAs, Java): formerly u. in mal.; lvs. u. to treat ulcers (Burma); antipyr.; in dys. stomach ulcer - **J. nudiflorum** Lindl. (seAs): lvs. u. as diaphor. - **J. odoratissimum** L. (eAs, Taiwan, CanaryIs, Madeira): sweetest-scented jasmine; fls. u. to perfume tea (eAs); furnish VO u. perfumery; cult. - **J. officinale** L. (eInd Iran, Med reg.): common (white) jasmine; j.; fls. u. aphrod. antirheum; antispasm.; source of VO u. arom. in perfum.; var. *grandiflorum* Bailey (forma g. Kob.): Spanish jasmine; Catalonian j.; source of VO u. in perfumes (chief prodn. Fr, Mor) (184) - **J. pubescens** Willd. = *J. multiflorum* - **J. sambac** (L) Ait. (tropAs; widely cult. tropics): Arabian jasmine; motiá (Urdu); fls. white, double; u. to scent tea; rts. toxic, tincture u. as sed. anesth. vulner. (Chin); fls. in poultice for skin dis. & wounds (Mal); antipyr. - **J. simplicifolium** G. Forst. (Austral to Polynesia): lf. tea u. for sore throat & fever (Fiji); frt. juice u. as ink - **J. subtriplinerve** Bl. (Ind-Himalay reg.): u. to improve blood circulation; for eye dis. (Taiwan); lf. decoc. u. after partur. in quotidian fever (mal.) (Indochin) (the plant attributed to this name is considered by some to represent *J. pentaneurum* Hand.-Mazz. [185]).

jatahy (Br): jatii: *Hymenaea courbaril* & other *H.* spp.

jatamansi (Beng Sanscrit, etc.): *Nardostachys j.*

Jateorhiza Miers (Menisperm.) (Af): calumbo - **J. macrantha** Exell et Mend. (cAf): c. alk. of palmatine gp.; lvs. & rts. u. in snakebite; ulcers; bitter ton. hypotensive; rts. u. as food during times of want - **J. palmata** Miers (J. columba Miers) (seAf, esp. Mozambique; Ind, SriL; cult. Br): calumba (or colombo) root; ("Fraser root"); c. columbamine, palmatine, jateorhizine, columbin (NBP), no tannin; u. bitter ton., astr. in gastric & intest. dis., much u. in veter. practice.

jatoba resin: 1) fr. *Hymanaea courbaril* 2) resin resembling copal; of unknown origin.

Jatropha aconitifolia Mill. (Euphorbi.) (Mex, CA): chaya; bears stinging hairs; eaten boiled as a spinach; sap said u. in urin. dis.; shade tree - **J. calyculata** Pax & Hoffm. (Mex): "ortiga"; thick fleshy rts. u. in VD - **J. capensis** Sond. (sAf): strongly bitter juice u. in pulm. TB (sAf); & other pulm. dis.; against ringworm - **J. cardiophylla** Muell.-Arg. (sAriz [US], nMex): "torote" (Sonora); rts. c. 5% tannic acid; u. tan; in med. (Mex) - **J. cathartica** Teran & Berl. (J. berlandieri Torr.) (sTex [US], Mex): (Berlandier) nettle spurge; jicamilla; u. in med. by natives (Mex) - **J. chevalieri** Beille (Senegal, Af): popularly u. as vulner. (latex); rts. in leprosy, & diangara (Cayor); much u. by healers of leprosy (dists. of N'dande & Gueóul, Sen) - **J. cinerea** Muell.-Arg. (Mex): ashy jatropha; lomboi; sangregrado (Sinaloa, Sonora); astr. juice u. for sores, warts & sore throat; & to harden gums; decoc. u. as mordant in dyeing; sd. FO highly unsatd. (oleic ac. 65%, linoleic ac. 6.3%) (2003) - **J. cordata** Muell.-Arg. (nwMex): mata muchachos (Chihuahua); bruised lvs. appl. to sores; added to bath water of children to "strengthen" - **J. curcas** L. (sUS incl. sFla; WI, Mex-tropAm; tropics of world): Barbados nut; physic nut; sangre de drago (Mex, Tex); piñon (de purga); shrub to 5 m.; sd. c. FO, curcin (toxin somewhat like ricin); lf. decoc. u. as vulnerary wash (PI) & for dermatitis; sd. purg. for intest. parasites, etc.; source of FO (Internal Oil), u. drastic cath. (danger!); emet. (Hond, Nic); illuminant, lubricant, etc.; lf. decoc. u. in cough; "general decline" (Ve); lvs. appl. as poultice for eczema, scabies, local swelling, &c. (CR); snakebite (Tanz); lvs as fish poison (PI); latex appl. to sores & ulcers (Indochin) (large prodn. Madag); for toothache (esp. in molars); rt. to treat wounds (Bo); for sore eyes (Tanz); rts. appl. to sore gums (Tanz) - **J. dioica** Sessé (sTex, Mex): leatherstem; (sangre de) drago; "rubber plant"; u. diarrh.; juice of st. & rts. astr. u. to clean teeth & harden gums; also var. *sessiliflora* Hook. (sTex [US], Mex): leatherstem; "rubberplant"; tea (& juice) u. to harden teeth & cleanse & invigorate hair (Mex) - **J. gaumeri** Greenm. (nwMex): pomolche; rts. u. to treat snakebite; branches u. to make whistles - **J. glandulifera** Roxb. (Ind): pl. juice u. to "remove film from eyes"; rt. brayed with water & given to children with abdom. enlargements; purg.; sd. —> FO u. purg. in chron. ulcerations, wounds, ringworm, rheum. paralysis (!) - **J. gossypiifolia** L. (sFla Keys [US], Mex tropAm, tropAf, As): frailecillo; tua tua (Ve); belly ache weed (or bush) (SC—> gastralgia if eaten; poison); sds. c. FO with drastic & emet. properties; lf. decoc. u. in VD, emet. for stomach pain (Mex); lvs. as antipyr. in mal. (WI); rts. u. in Hansen's dis. (Ve); var. *elegans* Muell.-Arg. (tropAm): carretillo; with glabrous lvs.; c. jatrophine (inhibits tumors); resins; sds. u. as cath.; abort. & as emet. (WI); bk. u. in gastric dis., colics; lvs. (decoc.) lax. & for sores & ulcers (VirgiNIs); cult. widely - **J. hirsuta** Hochst. (sAf): pd. rt. u. as dressing for fresh wounds (Zulu) - **J. macrorhiza** Benth. (wTex [US], nMex): bigroot nettle-spurge; "jalapa" (Mex); rhiz. purg. emet.; sds. purg. (2230); var. *septemfida* Engelm. (J. arizonica I. M. Johnston) (sAriz, NM [US], nMex): u. as purg. (Mex) - **J. multifida** L. (sFla, tropAm, natd. pantropics): coral plant; physic nut; cabalongo (Veracruz); yellow sap u. as vulner.; roasted sds. for fevers & VD; sds. (bean magnum) purg. (*cf. J. curcas*); source of FO; rhiz. purg. emet. (nMex) - **J. platyphylla** Muell.-Arg. (Mex): u. med.; pl. cult. as hedge pl. - **J. podagarica** Hook. (Mex, WI, CA; pantrop): "ruibabro" (Salv); u. in schistosomiasis (wAf); in mal. (Nic); cult. (Salv) - **J. spathulata** Muell.-Arg. (sUS, Mex): telondilla (Mex); sangre de grado; sangregrado; pl. c. alks.; juice astr. u. to harden gums, for skin eruptions, sores, in dys. hemorrhoids, VD, as gargle for sore throat; wash to restore & give luster to hair; to remove tooth stains; bk. tan. & dye (red) (186); flexible sts. of shrub u. as whips & to make baskets - **J. stimulosus** Mx. = *Cnidoscolus s.* - **J. urens** L. (Mex, tropAm; pantrop): spurge nettle; mala mujer (Mex); purging nut tree; bed bug plant (wAf); quaritoto (Ve); fleshy rts. u. for VD (locally) (Mex) in dropsy (Br), &c.; considered very poisonous pl. (Af); produces fever, swelling, & discomfort in some people; toxic trichomes; sometimes equated to *J. curcas* - **J. zeyheri** Sond. (tropAf): tuber decoc. u. as "blood purifier," young shoots rubbed on sores (Zulu); tuber decoc. purg., remedy in headache (Lobedu); sap appl. to sores & burns; tuber with 20% tannin, u. tan.

Jatror(r)hiza Planch. = *Jateorhiza.*

jaundice root: Hydrastis.

Java almond oil: FO fr. *Canarium* spp. - **Java**: v. under grass; pepper; plum; tea.

javilla (SpAm, as CR): *Hura crepitans.*

javin (Mex): Jamaica dogwood.

jawar: jowar.

jecquirity: jequirity.

jecur (gen. **jecoris**) **aselli** (L): cod liver.

Jeera (Ind): cummin.

Jeffersonia diphylla (L) Pers. (Berberid.) (eNA): twin leaf; "yellow root"; hb. & rt. claimed by some to c. berberine; NBP, saponin, u. diur. alt. in arth. rheum.; adulter. of Hydrastis (hence "False Hydrastis"), Serpentaria (2027) - **J. dubia** Benth. & Hook. (eAs): rt. u. antipyr. ; decoc. of rts. & rhiz. u. as eyewash; bitter stom. (Chin, Kor): substit. for *Coptis*; c. berberine (1606).

Jello™: brand of flavored sweetened gelatin made usually fr. beef (occas. pork).
jelly: congealed sweetened frt. juices; gelling depends on presence of adequate pectins & acid; u. bread spread.
jelutong: *Dyera costulata.*
jeneverbes (Du): juniper berry (pl.-bessen)
jengibre (Sp): ginger.
Jennerian vaccine: smallpox v. (SC Edward Jenner, who introduced smallpox vaccination in 1796).
jequiriti: jequirity (seed): *Abrus precatorius.*
jequitiba (Br): *Couratari legalis.*
Jersey (US. slang): sweet potato; *Cariniana* spp.
Jerusalem corn: sorghum fr. the Nile region - **Jerusalem**: v. under artichoke; cherry; oak; sage; tea; thorn.
jervine: steroidal alk. found in *Veratrum viride, V. album,* & other *V.* spp.
jesaconitine: alk. fr. rt. of *Aconitum sachalinense, A. fischeri* (Jap. aconite).
jessamine: 1) *Jasminum* spp. 2) *Gelsemium* spp. - **Carolina j.: yellow j. (root)**: *Gelsemium sempervirens* - **night (blooming) j.**: *Cestrum nocturnum.*
Jessenia bataua (Mart.) Burret (Oenocarpus b. C. Mart.) (Palm.) (Br [Para], Pe, Co): patauá; patovouá; cuperi; FO prod. fr. frt. flesh; u. as substit. & adult. of olive oil (salad & cooking), for soap, in lamps, candles; sds. c. FO (comu oil), u. to prep. stearin; spines of lf. sheath u. as arrows (Am Indians); fresh frt. edible, source of delicious milk-like beverage ("bataua") (1607) - **J. polycarpa** Karsten (Ve, Co, Br): seje grande; milpesos; sds. yield FO with pleasant taste, seje oil, u. in consumption & other pulm. dis., chest pains; vermif.; frts. eaten by indigenous peoples for chest pains & TB (SA Indians) (187).
jesterin: emodin anthranol glucoside fr. *Rhamnus cathartica.*
Jesuit's balsam: Copaiba - **J. bark**: Cinchona - **J. drops**: Tr. Benzoin Comp., c. guaiac, sarsaparilla, & Peru balsam - **J. tea**: v. under tea.
Jesus's tree (Ala [US] Negro): *Albizzia julibrissin.*
jet: kind of very hard carbon with high luster; closely related to coal.
jetahy: jetai: Jetaica (L): anime gum.
jeti fiber: rájmahál hemp.
Jew plum: *Spondias dulcis.*
Jew-bird oil: goose grease, Adeps Anserinus.
jewel weed: *Impatiens fulva, I. biflora* - **j. vine**: tuba root, *Derris elliptica.*
jeweler's rouge: ferric oxide ground into an impalpable powder, u. as polish.
Jew's ear: *Exidia auricula* - **J. mallow**: v. under m.- **j. pitch**: asphalt; bitumen (SC obt. fr. Dead Sea in Israel).
jiba (Cu): *Erythroxylon havanense* - **j. de costa**: *Erythroxylon* spp.
jícama (SpAm): 1) *Pachyrhizus erosus* (j. *dulce*) 2) *Exogonium bracteatum* - **j. (de agua)**: (Mex): yam bean, *Pachyrhizus erosus* 2) several other pl. spp. - **jícara** (Mex): *Crescentia cujete, C. alata.*
jiconga nuts: sds. of *Telfairia pedata.*
jilli flower: *Entada scandens.*
jiló (Br): eggplant frt.
jimpson weed: jimson weed: *Datura stramonium,* also *D. meteloides.*
jiñicuite (CR): jiñocuabe (CR): *Bursera gummifera.*
jinko (Jap): aloes wood.
jinshang (US): ginseng (localism).
jintan (Jap): chewing pellet with gambir catechu & aromatics (ex. menthol).
jintawan: Borneo crude rubber.
jipijapa (Pan): Panama hat plant, *Carludovica palmata.*
jive (slang): Cannabis
Joannesia heveoides Ducke (Euphorbi.) (Br): frt. called castanha de arara; rich in FO - **J. princeps** Vell. (Br): anda-assú; frt. peel u. as fish poison; frt. u. diur. in liver dis., icterus; sd. (and nuts) purg., in edema; source of FO, rather similar to castor oil; u. in soap mfr. and varnish formulation; lax. in veter. med.; useful wd.; cultd. (2028).
jobo (Mex): *Spondias mombin.*
Job's tears: 1) *Coix lachrymajobi*; pl. or frts. or beads made fr. frts. (Ind) 2) (sUS): *Tradescantia* spp.
Jod (Ge): iodine.
Joe-pye (weed): *Eupatorium purpureum, E. perfoliatum, E. maculatum* (also spotted j. w.), *E. fistulosum, E. dubium.*
Johannesbrood (Dutch): John's bread.
Johanneskruid (Dutch): Herba Hyperici (L): *Hypericum perforatum.*
Johannisbeere (Ge): *Ribes* spp. - **J. tomate** (Ge): *Solanum racemiflorum* - **schwarze J.**: *Ribes americanum* (NA); *R. nigrum* (Eu).
Johannisbrot (Ge): John's bread.
johimbe: yohimbe.
John the Conqueror (Root): one of several different roots u. by Negroes (seUS, WI) as a talisman, in voodooism, etc.; term appl. sometimes specif. to high J. t. C. (2574) - **chewing J. t. C.**: galangal - **European J. t. C.**: natural beth root (Trillium) - **high J. t. C.** (in store): Jalap rt. - **low J. t. C.**: 1) Solomon seal 2) (as sold): galanga - **southern J. t. C.**: stripped beth root (Trillium).
johnin: paratuberculin: vaccine prep. fr. culture of *Mycobacterium paratuberculosis* Bergey et al., cause of Johne's dis. (kind of cattle enteritis).

Johnny Hump-up: j. jumper: pansy - **J. jump up**: *Viola pedata, V. cucullata.*

John's bread: Ceratonia - **Johnswort**: *Hypericum perforatum.*

joint (US slang): Cannabis cigarette - **j. fir (ephedra)**: *Ephedra* spp. esp. *E. distachya.*

jojoba (pron. ho-ho-ba) (Mex): *Simmondsia chinensis.*

Joneswort: *Hypericum* (St. John's wort).

jonquil: *Narcissus jonquilla.*

Joosia dichotoma (R. & P.) Karsten (Cinchona d. R. & P.) (Rubi) (Pe): "cascarilla ahorquillado"; bk. drunk in brandy as antipaludian (against mail.); bk. u. as Cinchoa substit.

jorojoro (Surinam): *Thevetia peruviana.*

jorra (Br: Minas Ger): *Datura arborea.*

Joshua tree: *Yucca brevifolia.*

joss stick: j. light: reed covered with paste made of dust of odoriferous woods, or a cylinder made up entirely of such paste, u. as incense before images, etc. (As) (joss = idol, god).

jowar(i) (Indo-Pak - nc & w): great millet, *Sorghum vulgare.*

Juanulloa ochnacea Cuatrecasas (Solan.) (Co): the trunk & lvs. of this liane are u. to treat wounds and as a narcotic (by addn. to ayahuasca drink).

Jubaea chilensis (Molina) Baill. (J. spectabilis HBK) (monotypic gen.) (Palm.) (wCh): Chilean palm; coquito (p.); syrup p.; wine p. (of Chile); honey p.; colihue stans (Ch); trunk yields sap, source of palm honey & wine ("guarapa"); sd. kernels rich in FO, edible, u. in confectionery (esp. fine flavor & odor); & marketed as coquitos (coquito nuts); lvs. u. in basketwork; cult.

jubilee (, fish) (Mobile Bay, Ala [US]): great concentrations of fishes, etc. along the shores; thought to be caused by mixing of fresh and salt water in certain proportions; possible wind & tide are factors; oxygen availability of water low.

juca (CA, etc): yuca, mandioca.

Juckpulver (Ge): itch powder (*qv.*); sold as novelty (joke).

Judas'(s) ear: *Exidia*, a fungus of elder trees - **J. tree**: *Cercis canadensis.*

Judean Pitch: *v.* under Jew's pitch - **Judenkirschen** (Ge) (Jew's cherries): *Physalis alkekengi* - **Judenpech** (Ge): mineral tar, *cf.* Jew's pitch - **judia** (Sp): common bean, *Phaseolus vulgaris.*

Juglandaceae A. Rich. ex Kunth: Walnut (or Butternut) Fam.; deciduous trees or shrubs of temp. zone; lvs. pinnately compd., exstipulate; frt. usually a nut; incl. walnuts, hickories; some spp. valuable for nuts (food, FO, confectionery), wood (furniture), and tannins (tanning & dyeing).

Juglans L. (Jugland) (mostly nTemp Zone, Euras, NA, nSA): walnut; 21 spp. of diecious trees; wd. u. for furniture, tool handles, etc.; often very expensive since no warping or cracking (as in oak); u. in antiques; green husks of frt. u. to kill fish (seUS); sd. kernels —> FO, much u. by Am Indians - **J. ailantifolia** Carr. (J. sieboldiana Maxim.) (Jap): Japanese walnut; Siebold's w. - var. *cordiformis* Rehd. (cult. Jap, NA): furnishes edible nuts ("heart nuts"); ornam. tree - **J. cinerea** L. (SC light gray bk.) (fr. cinereus, L., ash) (eNA, cult. Eu for long time): butternut (walnut); white w.; lemon w.; rt. bk. u. med. as cath. alt. anthelm. astr. in GI dis.; stem & twig bk. also u. but less popular; lf. u. astr. alt.; nuts u. by Indians as food item; sds. rich in FO vit. C (esp. green) (2029); sometimes pickled; frt. husks u. brown dye; wd. u. for furniture, gun stocks, tennis racquets, etc.; sap of tree claimed medicinal (New England [US]) (1608) - **J. fraxinifolia** Lam. (Caucas, wIran): lvs. & frt. husks u. as for *J. regia*; wd. useful - **J. major** Heller (J. rupestris var. *m. Torr.*) (swUS, nMex): Arizona walnut; nogal silvestre; nut u. as food; beverage made fr. lvs. (tea with sugar) (Chihuahua), decoc. u. as ton.; dye - **J. mandschurica** Maxim. var. *sieboldiana* Makino (Jap-Manch): Manchurian w.; aq. ext. of shell u. locally for athlete's foot (Ainu) - **J. microcarpa** Berl. (J. rupestris Engelm.) (NM, Ariz, Tex [US], nMex): nogal; little walnut; small nuts eaten; excellent shade tree; durable wd. - **J. mollis** Engelm. (J. mexicana S. Wats.) (Mex): "nogal"; lvs. heated & appl. for rheum.; frt. husks u. as dye; wd. valued - **J. nigra** L. (eNA): (Eastern) black walnut; bk. lvs. & frt. juice u. astr.; edible sds. rich in FO; frt. husks (or hulls) u. hair dye, carrier for insecticide, plastics; juice of unripe hulls u. as antifungal in ringworm (2031) (2431); rt. bk. u. as purg. in GI dis., vermif. (Am Indians); wild green frts. u. as fish poison (swGa [US]); yields valuable lumber - **J. regia** L. (sw&cAs, Balkans; cult. Fr, Eng, US): (common) English walnut (misnomer as not fr. Eng); European (or Persian) w.; welsh nut; furnish the walnuts of comm.; lvs. c. as chief constit. juglone (OH-naphthoquinone) (best yield in fresh lvs.) (hydrolyzes to ellagic and gallic acids); 9-11% tannins; u. astr. in diarrh., anthelm. in scrofula, alter., added to foot baths for frostbite; exter. in skin dis. vermin (scalding); ringworm; bk. c. 8% tannin, juglone, vit. C.; u. astr. to reduce perspiration, in baths & cataplasms, in liqueurs (stomach bitters), in rickets (folk med.); frt. rind (shell) u. fresh as hair dye (juglone) (in eq. pts. of lvs.), astr.; sd. FO bland, nutritious; young frts. rich in vit. C; pickled; shell flour u. as carrier for insecticides. fungicides, etc.; wd. valuable lumber, in veneers, etc.; often grafted on stocks of *J. nigra* (2620).

juglone: hydroxynaphthoquinone; coloring matter of various *Juglans* spp. (walnuts) esp. in fresh lvs. (2030); allelopathic; inhibits fungal growth; sed.; promotes tumor growth.

jugo (Sp): juice - **j. de piña**: pineapple j.

juhi (Hindi): *Jasminum auriculatum*

juice: common name for 1) pl. sap which flows in tree trunk in spring & fall 2) liquids of green tissues of pl. 3) milky secretion or latex which occurs in laticiferous ducts throughout pl. - **block j.: licorice j.: Spanish j.**: licorice ext.

jujuba (jujubae, pl.) (L.): **jujubae fructus: jujube** (pron. joo-joob): 1) *Zizyphus jujuba* 2) jujube (of candy store); lozenge of acacia or gelatin with special flavors (gum drops); previously in shops contd. jujube frt. juice or made to imitate j. frts. - **Indian j.**: *Zizyphus jujuba, Z. mauritiana.*

Julbernardia globiflora Troupin (Berlinia g. Hutch. & B. Davy) (Leg) (sAf): bk. decoc. u. as eye lotion in conjunctivitis (Mozamb); bk., rt., lf. u. in trial by ordeal (Lamba) (erratic behavior); in med. (Transkei); sts. comminuted & smoked in pipe for leprosy (Tanz); yields a kino; in tanning & dyeing - **J. paniculata** Troupin (sAf): bk. cold inf. u. as enema in dys. (Luvale); wd. u. for roofs.

Julocroton argenteus (L) Didr. (Euphorbi) (Arg): yerba mineral (Arg); rt. u. diur. (Arg).

Jumellea fragrans Schltr. (Orchid.) (Maurit. Réunion): lvs. (Faham tea; orchid tea); c. coumarin; u. as substit. for Chinese tea; u. in native med. against TB; u. in cigars as filler (*cf.* Trilisa).

jumping beans (, Mexican): frt. of *Sebastiania pavoniana, S. bilocularis*; *Sapium* spp.; c. moth larvae in 1 of 3 sd. chambers; this segment splits off & is actuated by spasmodic movements of the larva (2032).

Juncellus inundatus C. B. Clarke (Cyper) (Ind): pati (Hind, Bombay): tubers u. as ton. stim.

Juncus L. (Junc.) (cosmopol): rushes; fibrous parts u. to make baskets (Am Indians); fodder when grasses scarce; pls. grow in wet & cold places; rhiz. base of one sp. (sAf) eaten raw - **J. balticus** Willd. (Euras, NA): Baltic rush; u. as emet. (Navaho) (189) - **J. bufonius** L. (cosmop): (common) toad rush; hb. diur. & cath. (190) - **J. conglomeratus** L. (J. effusus L. var. *c. Engelm.*) (Euras, nAf, Newf, Br): (bog) rush; rashes; pith rush; rt. u. demulc. emoll. diur.; proposed for u. as textile fiber - **J. effusus** L. (cosmop-esp Euras, nAf, NA): bulrush; (common) rush; rice r.; rt. c. tannin, tripeptide; stalk medulla u. as mild diur. esp. in lithiasis, strangury; cath. (188); haulms u. fodder; scapes woven together to make floor mats & matting (Jap) (Jap mat rush); pith for candle wicks - **J. prismatocarpus** R. Br. (Ind): kundai; rt. & lvs. juice u. in earache (124).

Juneberry: pome of any of several shadberries (*Amelanchier* spp.); also any of the trees bearing this berry.

Junellia bryoides Moldenke (Verben.) (Ch): shrub; rts. u. for wood - **J. ligustrina** (Lag.) Mold. [Lippia l. (Lag.) Britt.] (swUS, Mex): u. in bladder dis., emmen., antispasm. (Mex); perfumery (sEu).

junga (Bo): semi-tropical valley.

Jungia floribunda Less. (Comp.) (Br): capitula u. as substit. for arnica fl.

jungleweed: *Combretum sundaicum.*

juniper: a common Eng word which in med. & phar. generally refers to *Juniperus communis* (ex. j. berries); the term has also been referred to the following tree & shrub genera: *Cedrela, Cedrus, Chamaecyparis, Cryptomeria, Libocedrus, Melia, Ostrya, Pinus, Retana, Taiwania cryptomerioides, Thuja*, & perhaps others - **black** (or **blueblack**) **j. berries**: the ripe frt. of *J. communis*, the source of **j. berries oil** - **j. cedar** (comml.): *J. scopulorum* - **coffin j.**: *J. recurva* - **j. gum: gum j.**: 1) sandarach gum 2) German sandarach, exudn. fr. stems & rt. of *J. communis* - **"j. juice"** (US slang): illicit liquor, perhaps formerly gin - **inspissated j. juice**: dr. expressed juice fr. j. berries (Ge & Swiss Phar.) - **j. leaves oil**: VO fr. tips of branches of *J. phoenicea* - **sacred j.**: *J. recurva* - **j. tar**: cade oil, fr. *J. oxycedrus* - **j. wood oil**: VO fr. j. wd. or twigs; u. only in veter. practice & exter. in home.

juniperic acid: fatty acid fr. *Juniperus communis*+; occ. in many other spp. of gymnosperms.

Juniperus L. (Cupress. - Gymnosperm.) (N Hemisph.): juniper (tree); evergreen needlebearing shrubs & trees; specif. (L) frt. of *J. communis* - **J. barbadensis** L. (WI, seUS): Jamaica cedar; West Indian red cedar; bk. u. med. arom.; cult. ornam. - **J chinensis** L. (temp. eAs; worldwide cult. as ornam. in numerous cultivars (Chinese juniper); lf. decoc. u. as ton. (Chin) & in bronchial hemorrhage; wd. u. in ointments, &c. (Indones); VO. of frt. & lvs. diur. aphrod.; wd. tar appl. to ringworm & eczema (Jap); arom. wd. u. for incense burning - **J. communis** L. (Euras, NA) (said widest distn. of all trees & shrubs): (common) juniper; ground j.; genèvre (Fr); source of j. "berries" (really galbuli), mostly u. as flavor (prod. It, Ge, USSR, US); in mfr. gin; for distn. VO (u. arom. stim. diur.) (prod. It, Ge, USSR, Fr, Hung); also med. as diur. (care in kidney dis.), in gout, bronchial dis., catarrhs, skin dis., flatulence, emmen., hemorrhoids, stomachache, in liqueurs, as condim., in foods (ex. fish conserves, sauerkraut, roast meat sauces); to combat "bad breath" (1823); fumitory; also source of inspissated juice; other prods. u.: leaves; tops (frt. & branches) (u. diur. depurat.); wood (c. VO resin); u. med. much like frt.; VO prepd. fr. dr. rt. & bruised wd. (2724); "gum" (resin) (German sandarac); smoke (with beechwood) u. to smoke hams; exudn. fr. old stems & esp. rt.; bark; tar; var. *saxatilis* Pallas (subsp. nana Syme) (Eu mts): mountain juniper; "mula pine"; pino macho; dwarf tree of semiarid (alpine) areas; lvs. u. antipyr.; hot tea of twigs u. gastralgia & colic (NM [US]) - **J. deppeana** Steud. (J. mexicana Schlecht.) (wTex [US], nMex, Guat mts.): Mexican sandarac (or alligator) juniper; sabiño; sabina; tascate; cidro chino; "Ozark white cedar"; lf. source of VO & wd. VO (both produced); latter u. cosmet. perfum. soaps; lf. VO u. rheum. neuralgia; bk. ashes u. to prep. corn for tortillas; wood a very useful lumber; fuel; frt. u. as food; ornam tree (Xmas tree); var. *pachyphloea* Martin. (swUS, Mex): alligator j.; lvs. u. rheum. neural. (2230) frt. food (Indians) - **J. excelsa** Biebers. (seEu, AsMinor. cAs): tall Crimean (or Grecian) juniper; lvs. & twigs source of Russian j. oil; frt. med.; wd. useful for building; this may represent the "cedars of Lebanon" of Bible rather than *Cedrus libani* (2725) - **J. foetidissima** Willd. (seEu, Turkestan): u. like *J. communis*; frts. u. to adulter. the genuine article - **J. horizontalis** Moench (wUS, common, esp. near Helena, Mont): creeping juniper; American savin; VO c. sabinene, limonene, &c.; cult. ornam. (many cultivars) - **J. macropoda** Boiss. (cAs, Himalay): dhup; shupa; shur; Himalayan pencil cedar; lvs. & VO u. med. (As); very large tree (-20 m.); brownish purple frt. (mature), very resinous; wd. u. in construction, for incense (2678) - **J. monosperma** Sarg. (wUS, nMex): cherrystone j.; frt. eaten by Indians & u. as diur.; staminate cones u. lax. emet. in diarrh., gastric ton., "cure-all"; bk. u. lax., rubbed on skin in spider bite; lvs. (decoc.) u. for pains after childbirth; gum for dental caries; (dr. grd. to meal —> cakes or mush) - **J. occidentalis** Hook. f. (wCana, US): Sierra juniper; western j.; Canadian j.; frts. u. food (Calif [US] Indians); wd. u. for fences, fuel - **J. osteosperma** (Torr.) Little (J. itahensis [Engler] Lemon) (wUS); Utah juniper; twigs decoc. u. for worms; heated twigs for sore throat, swollen tonsils, sore gums (locally); antisep. wash in measles (wAm Indians). (2700) - **J. oxycedrus** L. (Med reg., nAf, wAs, seEu): cade; prickly j.; prickly cedar; source of cade oil (or juniper tar), empyreumatic VO made by destr. distn. of woody parts (prodn. Sp, nAf); c. cadinene (sesquiterpene), phenols, cresol, acetic acid; u. strong antisep. in chronic skin dis.; parasiticide, ex. chronic eczema, psoriasis; in salves, hair waters; for gout, rheum.; in veter. med. to treat gale of sheep, etc; VO fr. "berries" also known; frt. & twig tips u. like *J. communis* - **J. pachyphloea** Torr. = *J. deppeana* - **J. phoenicea** L. (J. lycia L.) (seEu, Med reg.): Phoenician j.; French sade tree; twigs & lvs. source of VO; frt. also prod. VO rather similar to VO of *J. communis*; tops adult. of savin tops; bk. yields gum-resin. burned as incense ("Arabian olibanum," "A. frankincense") - **J. procera** Hochst. ex Endl. (Ethiopia, Kenya, Tanz [Usambara Mts.] Saudia Arabia): African juniper; eAf. cedar; heart wd. u. med. (Masai); source of eAf. cedar oil (VO); wd. u. for cabinet making, pencil casings, &c.; allergenic; lf. ext. active against TB (in vitro) - **J. pseudo-sabina** Fisch. & Mey. (cAs, Chin, Turkestan): Siberian savin; wd. source of incense; bk. u. to make pads for shoulders (191); furniture mfr. (Tibet) - **J. recurva** Buch-Ham. (J. squamata Ham.) (c&eAs, Himal): weeping blue juniper; drooping j.; scaly-leaved j.; decumbent or prostrate shrub or small tree; frt 1-seeded; smoke of green wd. u. as powerful emet. (Ind); wd. & twigs for incense; coffin wood; fuel; frt. red brown to purple (India[n] red [juniper] berry) (192)(?); said u. for yeast; prostrate ornam. shrub - **J. rigida** Sieb. & Zucc. (Jap, Kor): muro; frts. u. to treat neuralgia (Kor); VO fr. frt. & lvs. u. diur. aphrod. (Jap); wd. tar u. in ointments for ringworm & scaly eczema - **J. sabina** L. (c&sEu, cAs): savin(e) tree; savin (j.); u. by ancient Romans & in early Middle Ages extensively; fresh or young leafy twigs ("tops") & dist. VO u. uter. stim. emm. abortif. ecbol. diur. skin dis. (exter. in boils); insecticidal; in veter. med. for detaching afterbirth; frt. adult. of *J. communis* frt.; dangerously toxic (much local irritation); hence tree cannot be planted in public places (Ge); chief prod. Fr, It - **J. scopulorum** Sarg. (Sabina s. Rydb.) (wNA Rockies): Rocky Mountain juniper; (western) red cedar (term also appl. to *Thuja plicata*); "juniper cedar" (Mont [US]); bk. fr. near base of tree u. to make decoc. u. for colds & flu; branch tips inf. u. for int. bleeding; branches as poultice for skin sores & arthritis (Okanagan-Colville Indians, BC [Cana]); aq. inf. of branches with frt. u. as arrow & bullet poison (BC [Cana] Indians); berries thought toxic; u. in TB (BC [Cana] Indians) (1610) - **J. sibirica** Burgsd. = *J. communis* - **J. silicicola** Bailey (Fla & coastal areas of seUS): southern red cedar; similar to *J. virginiana* (differs in having smaller female cones & more slender twigs); lvs. c. silicicolin (lignan) wh. acts as antitumor agent (Sarcoma 37) (193); branchlet decoc. (in combn.) for colds; wd. formerly u. mfr. pencils - **J. thurifera** L. (Sp, Fr, Alg): Spanish juniper; incense j.; c. VO; tops u. as adult. of savin; var. *gallica* DeCoincy (sFr): source of sade tree oil (distn. twigs); twigs important component of comml. savin (which is a mixture) - **J. virginiana** L. (wide area of eNA esp. towards s): (eastern) red cedar; (American) pencil cedar; heart wd. (Lignum Cedri) bright red, u. mfr. "lead pencils" (casings), cabinets, closets (moth repellant); chests; shingles, etc.; dist. for VO (cedarwood oil) u. med. as disinf. techn. mfr. perfumes, soaps, polishes; inspissated VO u. as immersion oil (microscope); wd. fumitory, cigar box mfr.; frts. eaten by birds; cedar bk u. in voodoo (seUS) (finely grd. wd. u. for sachet fillers, etc.); young twigs as abort. (*cf.* sabin); also VO; galls on pl. (cedar apples, Fungus colombinus) u. as anthelm.; cult. ornam. tree.

"junk" (US slang): Heroin, etc.

junker (US slang): one who is an addict to narcotic drugs.

junket: dessert composed of milk curds (sepd. by rennet) with sugar & vanilla (or other flavor).

junky = junker.

Juno's tears: *Verbena officinalis*, hb. Herba Verbenae.

Jupiter's beard: J. eye: *Sempervivum tectorum*.

Jurinea himalaica R. R. Stewart (J. macrocephala Hk. f.) (Comp.) (nInd, Pak, Kashmir+): qugal; peeled rts. u. in fever, eruptions; as ton.; in incense (Hindus; Tibet).

jurubeba (Br): **jurumbeba**: *Solanum paniculatum*.

jus (Fr): juice (fr. L., broth).

Jusculum (L): broth (of beef, veal, turtle, etc.); convalescent food.

jusquiame (Fr): henbane - **j. d'Egypte** (Fr): *Hyoscyamus muticus* - **Jusquiamus** (L): *Hyoscyamus niger*.

Jussiaea L. (*Jussieua* Murr.) = *Ludwigia*.

Justicia adhatoda L. (Acanth.) = *Adhatoda vasica* - **J. bracteata** Ridl. (Mal): hb. u. to treat colic & diarrh. - **J. carthaginensis** Jacq. (CA, SA): hierba del susto (Salv.): u. for fits & spasms in children (Co) - **J. chlorostachya** Leonard (Co): pd. arom. lvs. u. as insect repellant; also appl. to fungal rashes of crotch region (Taiwanos) (2299) - **J. comata** (L) Lam. (Co): dr. pd. lvs. u. as insecticide or insect repellant (Tikunas) (2299) - **J. gendarussa** Burm. f. (e&seAs): natchouli; bitter rts. as decoc. u. for rheum. (Chin); pounded lvs. u. in chronic rheum (Ind); emet. (Reunion); antipyr. diaphor.; purg., etc. - **J. insularis** T. Anders. (lower Congo): young shoots & lvs. u. as salad by natives; rts. boiled & eaten - **J. neesiana** T. Anders. (eAs): lvs. pounded with camphor & rubbed over body for fever - **J. pectoralis** Jacq. (Stethoma p. Raf.) (Mex, WI, SA): garden balsam; water willow; tila (Cu); carpenter's bush (Trin); hb u. as pectoral, sed. expect. in cough syrups; in colds, flu, fever (Trin); diaphor. aphrod.; often in mixts. (syr.) (235); astr. vulner.; var. *stenophylla* Leonard (Co, Br): u. in prepn. hallucinogenic snuffs (virola, &c.); u. for washing (Pe) - **J. procumbens** L. (eAs): decoc. of vegetative parts u. in lumbago, backache, plethora, fever, dys. (Chin); lvs. u. exter. for skin dis. (PI); lf. juice u. in ophthalmia (Ind) - **J. ptychostoma** Nees (eAs): lvs. u. to cure colds; lf. & rt. decoc. given at childbirth (Indochin) - **J. secunda** Vahl (WI, Pan, nVe): (mata de) sangre; St. John's bush (Trin); hb. decoc u. as emmen. in amenorrh. (Trin); fl. inf. u. ton. & in hepat. dis. (Pan); pl. decoc. swished around in mouth to benefit gums & teeth; exter. to bathe skin (Ve) - **J. spicigera** Schlecht. (Jacobinia s. L. H. Bailey) (sMex, to Co): mozote; lvs. (of shrub) decoc. u. in dys. GC kidney inf.; as laundry "bluing" (235) & dye; stim. antipyr. in scabies; lvs. c. kaempferitrin; cult. as "muicle" (Tex [US]) - **J. tinctoria** D. Gibson (CA, Co): punciga; lf. decoc. for colds; rheum; exter. for skin irritations; contracept (CR); lvs. for bluing (hot water extn.) - **J. uber** C. B. Clarke (Indochin): lvs. in poultice for headache - **J. ventricosa** Wall. (eAs): st. & lvs. u. remedy for contusions (Chin) (*cf. Adhatoda*).

jute: yute: *Corchorus capsularis* - **American j.**: **Chinese J.**: *Abutilon theophrasti* - **Congo j.**: **Florida j.**: *Urena lobata* - **Indian j.**: jute produced in eIndia (Calcutta factories) - **j. stick** (E Pak): "j. butts"; lower st. said suitable as cellulose raw material.

juvenile hormones (Insecta): JH; hormones wh. prevent development of larva into adult; obt. fr. the male wild silk moth; u. in biol. control over insect populations by keeping insects in larval or pupal states; synthesized (Ayers; 1967).

jy-chee oil: FO fr. *Euphorbia dracunculoides*.

Kk

NOTE: The letters "k" and "c" are often interchangeable; it might therefore be advisable to sometimes look under each letter.

k: kilo: 1,000.

ka (Thai): galanga.

kaa-he-e: *Stevia rebaudiana.*

kabeljau (Ge): **kabeljauw** (Dut): codfish.

kabuloside: glycoside of strophanthidin & D-2-deoxygulose; occ. in *Erysimum perofskianum.*

kachoor (root) (Ind): **kachura**: Zedoary.

kadaya (gum): gum karaya.

Kadeoel (Ge): cade oil.

Kadsura Juss. (Magnoli/Schisandr.) (tropAs): drugs fr. this genus classed in 1st grp. (nonpoisonous drugs) of the famous Chinese work on materia medica, *Shan Nung Pen Ts'ao King* - **K. coccinea** A. C. Sm. (seAs): st. soaked in wine u. in lumbago; roasted frts. sl. hypnotic; sd. ton. aphrod. in bronchial dis. (Indochin) - **K. japonica** (L.) Dun. (eAs): frt. decoc. u. as stom. ton. in cough; in kidney dis. - **K. scandens** Bl. (seAs): rt. decoc. u. for general malaise; lvs. appl. to ulcers; med. for puerperium (PI).

Kaefer (Ge): beetle, chafer.

Kaempferia aethiopica Benth. (Zingiber.) (Senegal, wAf): rt. u. in beverages as purg. in intest. dis.; furnish ginger-like spice - **K. angustifolia** Rosc. (tropAs): rts. u. as masticatory; in cough (Ind); for veter. practice - **K. ethele** Wood (tropAf incl. Transvaal): shirungulu tubers; c. VO with cineole, pinene, linalool; u. in mal. menstrual pain (Swati); to treat "horse sickness" (Zulu) - **K. galanga** L. (tropAs, as Ind, Chin, Java; cult.): galangale; rhiz. ("Kentjoer") c. **kaempferia oil,** with ethyl cinnamate, cinnamaldehyde, borneol, etc.; VO & rhiz. u. as arom.; rhiz. has many med. uses as stim. stom. carm. in cholera, contusions, headache, constipation, etc. (Chin); chewed for cough; juice expect. (Mal); spice; lvs. u. as perfume in washing hair (PI) - **K. rotunda** L. (seAs, much cult.): round-rooted galangale; rhiz. appl. to reduce swellings (poultice), stom. (Ind); rhiz. is not the true round zedoary (Rhizoma Zedoariae rotundae); u. widely in treating GI dis. (Indones, PI); ext. mixed with oil as cicatrizant; arom. to flavor foods (Java).

kaempferol: tricyclic alc., flavonol, found in many plants, incl. *Rhamnus cathartica*, *Solanum* spp., etc.

kafenia (Gr): coffee house.

Kaffee (Ge): coffee - **Kaffeehaus** (Ge): coffee house, cafe - **Kaffeezeit**: period for coffee drinking.

Kaffir: Kafir: K. corn - **Kafir bread**: *Encephalartos* spp. (Cycad.), as *E. caffer*, & the starchy pith u. as food by the natives - **K. corn**: *Sorghum bicolor* - **K. tea: kafta**: *Catha edulis.*

kafur (Arab): camphor.

kaguno seed (Ind): *Setaria italica.*

kahlúa: lícor de café: Mexican liqueur with delicious sweet flavor and coffee bouquet; u. in mixed drinks often with milk or cream; c. 21% alc.

kahu (Hind, Bengal): *Lactuca serriola.*

kahuna: native medicine man of Hawaii (priest or sorcerer).

kail = kale.

kainic acid: digenic acid: compd. fr. *Digenea simplex* (*qv.*) (dr. thallus); u. to kill intestinal worms.

kairine: artificial alk. prep. fr. quinoline.

kaithi (wPak: Kurram valley): *Indigofera heterantha.*

kaivun fiber: fr. *Helicteres isora.*

kajal (Urdu): black dye for eye lids made up in homes in Indo-Pak; mostly by women & children.

Kakaobohnen (Ge): cocoa beans.

Kakaobutter (Ge): theobroma oil.

kaki (Br): persimmon.

Kalabarbohne (Ge): Calabar bean.

kaladana: *Ipomoea hederacea.*

Kalaharia uncinata (Schinz) Moldenke (l) (Verben.) (sAf, Namibia, Angola+): "kikosa"; pl. u. as remedy in headache by placing head in steam fr. decoc. of rt.; rt. decoc. u. for snakebite (violent emet.), to treat bilharziasis; u. as gargle in sore throat; dr. lvs. pounded with salt to treat chest colds; rt. inf. drunk twice a month as contraceptive (female); toxic to cattle, etc.

kalakinmum (Burma): *Sapindus.*

kalamondin: *Fortunella japonica.*

Kalanchoë crenata (Andr.) Haw. (Crassul) (w tropAf): lf. juice u. to discourage vomiting, sed. for intercostal & intestinal pain; lf. appl. to stop hemorrhage; decoc. to pregnant women to fortify (Ivory Coast); warmed lf. appl. to earache (236) - **K. daigremontiana** Hamet & Perrier (sMadag): pl. u. med. toxic to livestock & furbearers - **K. densiflora** Rolfe (K. glaberrima Volkens) (tropAf): warmed lf. appl. to contusions (Chagga); rts. & lvs. u. for abortion; lf. juice appl. to septic wounds - **K. integra** DC (As, Af): decoc. u. as eye-wash & styptic (contusions); emollient (Taiwan) - **K. laciniata** DC (trop seAs): tapak itek (Siam); crushed lvs. u. as poultice for coughs & colds (Mal); popular in combns. for smallpox; lvs. appl. to temples for headache (PI); poultice for ulcers (Ind, Indochin) - **K. lanceolata** (Forsk.) Persoon (tropAf): kuserwet; heated lvs. rubbed over body for rheum. & stiff joints - **K. pinnata** (Lam.) Pers. (Bryophyllum calycinum Salisb.; B. pinnatum Kurz) (pantrop, original home As, EI?): life plant (Bah+); air p.; everlasting; floppers; wonder of the world (Trin); prodigiosa; lf. emoll. u. int. in dys. diarrh. lithiasis, jaundice, dropsy, fever; cataplasm. in skin inflamm. (Bah); for earache, sprains, head colds (Trin); fls. in resp. dis. (WI); lf. juice with pronounced antibacterial props. (696); u. as diur. (wAf); ext. for snakebite, astr., styptic, in wounds, ulcers, etc.; popular ornam., cult. in greenhouses; natd. sFla (US); u. to demonstrate propagation of pl. fr. lf. since lf. (even when detached fr. plant) develops young plants at notches in leaf margin - **K. rotundifolia** (Haw.) Haw. (swAf): nenta; herbage boiled, cooled, & decoc. placed in ear to relieve children's fevers (2679) - **K. spathulata** DC (tropAs): "headache herb" (transl.); lvs. u. emoll.; in migraine; burnt lvs. appl. to wounds; ext. in med. (Ind); in cholera; juice u. purg.; lvs. insecticide (2019) - **K. thyrsiflora** Haw. (sAf): decoc. u. as anthelmintic enema; rt. decoc. drunk by pregnant women feeling unwell.

kaladana: sds. of *Ipomoea nil.*

kalaw tree (Mal): *Hydnocarpus*; source of chaulmoogra oil.

kale: *Brassica oleracea* v. *acephala*; *B. fimbriata*, etc. **- sea k.**: *Crambe maritima* (& other *C.* spp.).

kali (Fr, Eng): *Salsola k.*

Kalimeris incisa (Fisch.) DC (Aster i. Fisch.) (Comp.) (Jap, Kor, nChin, Siber): rt. & lf. diur. & u. for insect bite; young lvs. eaten cooked (Jap).

Kalium (Kalii) (L): potassium, potassa (**Kalium** [Ge]: K).

Kalk (Ge-m.): lime.

kallar (Urdu): a natural salt of the soil in Indo-Pak; c. Na_2SO_4, $MgSO_4$, lesser amts. NaCl, $MgCl_2$.

kallidin: kind of kinin liberated fr. blood plasma globulins by action of kallikrein - **k. I** = bradykinin; **k. II** = decapeptide made up of bradykinin with N-terminal lysine addn.

kallikreins = callicreins (*qv.*).

Kallstroemia californica Vail (Zygophyll.) (Calif, Ariz [US]; nMex deserts): California caltrop; var. *brachystylis* (Vail) Kearney & Peebles (K. b. Vail) (similar range): contrayerba (NM [US]); chewed lvs. placed on sore or swelling; rts. remedy for diarrh. (237) - **K. maxima** (L) H. & A. (sUS, tropAm, WI): abrojo (terrestre) (Cu); decoc. u. locally against various skin dis. (Cu); lv. mild purg.

Kalmes (Ge): **Kalmeswortel** (Dutch): Calamus.

Kalmia angustifolia L. (Eric.) (eNA): lamb-kill: sheep laurel; lvs. alt. (in exanthemata, syph. itch, etc.) (NA, Ind); sed. errh. astr. in diarrh.; poisonous; u. by natives as fish poison; u. by homeop. in rheum. neuritis; frt. esp. poisonous; var. *caroliniana* (Small) Fern. (K. carolina Small) (seUS): sheep kill; wicky; decoc. u. in colds, backache, poultice of lvs. for headache; lvs. in tea in gastric dis. (ulcers); said most toxic of gen. - **K. latifolia** L. (eNA): laurel; mountain (or American) laurel; calico bush; "ivy"; similar props. to prec.; c. andromedotoxin; lvs. u. exter. in skin dis.; u. in inflamm. heart dis. (homeopath.); astr. in diarrh. syph. itch; grd. sds. formerly put into mescal to prod. dizziness, stimn.; finally deep narcotic sleep (Tex [US]); cult. in Eu as ornam.; rt. ("burrs") u. to make tobacco pipes (*cf. Rhododendron maximum*) - **K. poliifolia** Wang. (NA): pale laurel; bog kalmia; b. l. swamp laurel; lvs. narc. errh., poisonous.

Kalmie (Ge, Fr): glands & hairs fr. frts. of *Kalmia latifolia.*

Kalmus (Ge): Calamus.

Kalopanax (Calopanax) septemlobus Koidz. (Arali.) (eAs) (monotypic): harigari; saponin protects wood against termites; u. med. (Chin) (95) (—> *Eleutherococcus*).

Kamala (L): glands & hairs fr. frts. of *Mallotus philippinensis.*

kamass (Chinook jargon): *Camassia* spp.

kamassine: alk. fr. *Gonioma kamassi.*

kameela: kamala.

kamilica (Yugoslav): **Kamille(n)** (Ge): chamomile.

Kampher (Ge): camphor.

Kampõ: Jap herbal medicine in the Chinese tradition; the practice goes back through the centuries (2621).

kampong (Java, Sumatra): native village.

Kampuchea (Campuchea): former Cambodia (seAs) 1975-.

Kanadisches Pech (Ge): Canada pitch; fr. *Tsuga canadensis.*

Kanamycin: Kantrex™: antibiotic subst. furnished by *Streptomyces kanamyceticus* Okami & Umezawa (fr. Jap soils); composed of 3 fractions, kanamycins A, B, & C; disc. 1957; u. antibacterial orally for intest. inf. (preoperatively), parenterally for lung (TB), urinary, blood, & other dis. (ex. pleuritis); toxic side effects; paste u. for gingivitis, plaque accumulations.

Kandelia rheedii W. & A. (Rhizophor.) (seAs): Kandal (Tamil); bk. in diabetes (mixed with ginger, etc.); tan bark.

Kaneel (Dutch): cinnamon.

kangaroo foot (Austral): **k. paw**: *Anigozanthos manglesii.*

kani (Senegal): *Xylopia aethiopica.*

Kaninchen (Ge-n.): rabbit; coney.

kaoliang (Chin): *Sorghum vulgare.*

kaolin (Chin, Eng): China clay; bolus alba; a hydrated aluminum silicate occurring naturally in deposits (mined in Eng [Cornwall], US [Ga]); u. adsorbent in gastric dis., emoll. (poultices) (2622); also u. in ceramics (porcelain) industry (paper making, cosmetics, paints, rubber) (1612); clarifying agts.

kaolinite: fine china clay ($Al_2O_3.2SiO_2.2H_2O$).

kapahosa (Br): *Ryania speciosa.*

kapheneion (Gr): coffee house.

kapok: "Java cotton" fr. *Ceiba pentandra.*

kaputu gedi (Java): sds. of *Brucea sumatrana, B. javanica.*

Kapuzinerbalsam (Ge): galenical prep. incl. both Peru & Tolu balsams.

Kapuzinerkresse (Ge): *Tropaeolum majus.*

Karachi gum: form of Indian gum arabic, claimed obt. fr. both *Acacia arabica* & *A. senegal*, chiefly the former (fr. Karachi, chief city of Pakistan).

karam (Hindi): 1) *Mitragyna parvifolia* 2) a vegetable - **karamkala** (Indo-Pak): cabbage.

karanja (monda): *Pongamia pinnata.*

Karatas plumieri Morr. (Bromelia plumieri [E. Morr.] L. B. Sm.) (Bromeli) (WI, CA, nSA): silk grass; pinuela (Cu); pita (Hond); frts. pleasantly edible; med. (Mex Phar.) (—> *Bromelia*).

karaya gum: Indian (or Sterculia) gum. fr. spp. of *Sterculia*, etc.; v. Sterculia.

Kardamomen (Ge): cardamom.

kardi seed (Ind): *Carthamus tinctorius* sd. (Ge): *Cynara cardunculus.*

Kardonen (Artischoke) (Ge): *Cynara cardunculus.*

Karfiol (Ge): cauliflower.

kari: 1) *Agelaia obliqua* 2) *Sterculia* gum (powd.).

karkade flowers: *Hibiscus sabdariffa*, u. to prepare **k. tea**.

Karo (Syrup)™: corn syrup.

karri (tree) (Austral): *Eucalyptus diversicolor* (**k. gum**).

karroo bushes (sAf): shrubs taking the place of grass on wide sandy plains - **K. thorn**: *Acacia horrida.*

Karwinskia calderonii Standl. (Rhamn.) (CA): guiliguiste (Hond); excellent wd. of many uses; u. med. (Hond) (196) - **K. humboldtiana** Zucc. (Rhamnus h. Roem. & Schult.) (wTex [US], Mex): tlalcapollin; cacachila (Chin); capulincillo (cimarrón) (Mex); coyotillo (Tex [US]); lvs. & rts. decoc. u. locally for fevers (Mex); hot tea held in mouth for toothache & neuralgia; anticonvulsive in tetanus & rabies(!); frt. edible but putamina (pits) are toxic; causes paralysis of voluntary muscles of limbs (FO non-toxic); toxic to livestock; decoc. exter. for infected wounds - **K. latifolia** Standl. (Mex): "margarita"; claimed useful in leprosy; decoc. antipyr.; purg.

karyosome: 1) chromosome 2) chromatin mass in nucleus distinct fr. nucleolus.

karyotype: 1) sum of specific characters of nucleus, as chromosome no., size, etc. 2) diagrammatic representation of nuclear makeup.

kasha: Jewish soup prepd. fr. buckwheat.

Kaskarillrinde (Ge): cascarilla bk.

Kassie (Ge): cassia.

Kastanienblaetter (Ge): chestnut lvs.

kat (tea) (Arab): *Catha edulis* (lvs).

katakar oil (Ind): FO of *Argemone mexicana.*

Katha: mixt. of several catechins prod. in Ind fr. heart-wd. of *Acacia catechu*; c. aca-catechin; u. to prevent FO rancidity.

Katzenkraut (Ge): *Calamintha nepeta.*

Katzenminze (Ge): catnip.

Katzenpfoetchen (Ge): *Antennaria dioica* (also **rote K., weisse K.**) - **Gelbe K.**: *Helichrysum arenarium.*

kauri copal: k. gum: resin fr. **kauri pine**, *Agathis australis.*

Kautschuk (Ge): rubber.

kaute (Ind): *Hydnocarpus venenata.*

kava: kava kava (root): *Piper methysticum*; also native drink brewed fr. the bruised rt.

keawe (SpAm): *Prosopis* spp.

kedlock: white mustard.

Kedrostis africana Cogn. (Cucurbit.) (Transvaal, sAf): "bryony"; c. cucurbitins; emet.; inf. in wine & brandy purg.; similar uses for **K. nana** Cogn. var. *latiloba* Cogn. (sAf) - **K. rostrata** Cogn. (Ind): rt. u. as electuary in piles; in pwd. as demulc., in asthma.

keef = kef = *Cannabis* flg. tops (mixed with tobacco); smoked or ingested.

kefir: k. grains: k. seeds: granules of yeast (Caucasian kefir fungi) & lactic acid bacillus organisms, which when cultured in cow's or goat's milk produce the fermented beverage also known as **k.**; latter an excellent ton. in bronchial, lung, & gastric dis.; convalescence, anemia, weakness, etc. (claimed) (c. 1% alc.).

Kelgin™: sodium alginate (<kel[p + al]gin).

Kellerhalsrinde (Ge): mezereon.

kelp(s): 1) brown seaweeds of genera *Laminaria, Nereocystis, Porphyra*; u. food, source of I_2, fertilizer, ind. applns. 2) ashes of seaweeds, former I_2 source - **broadleaf k.**: sugar k. - **giant k.**: *Macrocystis pyrifera* - **horsetail k.**: *Laminaria digitata* - **sugar k.**: *Laminaria saccharina*, source of alginates, iodine (cont. very variable), K. & Na also present in large amts.; livestock feed; u. in obesity, atherosclerosis; in K tablets (c. trace minerals) - **kelpwrack**: *Fucus.*

kemmel: kimmel.

kemp: 1) coarse animal fiber u. in mfr. of carpets 2) *Plantago major.*

kemps: *Plantago media.*

kenaf: kenaph: *Hibiscus cannabinus, H. sabdariffa.*

kendir: kendyr: textile fiber fr. *Apocynum venetum*, cult. & u. in USSR.

Kentish liniment: K. ointment: turpentine liniment.

Kentucky coffee tree: v. coffee.

keora (kaora): *Pandanus tectorius.*

kephir: kephyr: kefir (*qv.*).

Kepler™: appl. to malt & other prepns.

Keratan(e) sulfates: collective name for mucopolysaccharides; occ. in cartilage, bone, cornea (horny layer of eye), &c.

keratin: protein (of scleroprotein gp.) found in hair, wool, horn (ex. rhinoceros h.), fingernails, scales (alligator s.), beaks, etc.; insoluble in water & dil. acids (ex. gastric juice); u. to enteric coat tablets (dissolve in alkaline medium of intest); for prodn. of protein hydrolyzates; in skin & hair care (cosmetics).

Kerbel (Ge): Koerbel - **Kerbtier** (Ge): insect.

Kerfilz (Ge): notched & fitted together, ex. tomentum on some lf. surfaces, fungus mycelium, etc.

kerkadeb (Eg): *Hibiscus sabdariffa.*

kerlock: *Brassica nigra.*

Kermes vermilio Planchon (Coccus ilicis L.) (Kermesidae - Homoptera - Insecta): dr. insects furnishing a purplish red comm. dye.

kermes berry: Kermesbeeren (Ge): poke berries; frt. of *Phytolacca* spp., esp. *P. decandra.*

kernel: part of sd. within integuments (sd. coats) (<Kern [Ge]: little corn).

kerosene (kerosine): petroleum fraction; widely u. as lamp oil illuminant (early 1900s) - **deodorized**: highly refined petroleum fraction (b. pt. 175-325°F); prepd. by treatment with activated C or diatomite; comm. prod.; u. as solvent (vehicle) for insecticide sprays (ex. pyrethrum prepns.), treating skin dis., cleaning agt., etc.

Kerstingiella geocarpa Harms (Leg) (wAf cult.) (monotypic): geocarpic pl. (frts. produced underground, *cf. Arachis*); sds. eaten by natives (Nigeria, etc.) (—> *Macrotyloma).*

keruing oil (Mal): obt. by tapping *Dipterocarpus kerrii, D. cornutus, D. chartaceus, D. baudii*, etc.

kesso (root) oil: fr. *Patrinia scabiosaefolia.*

ketchup: catsup.

ketosteroids, (17-): natural steroids, chiefly products of metabolism of adrenal-cortical & gonadal steroids; have ketone substitution.

ketra gum = kuteera gum.

kewda attar (Ind): obt. by distg. fl. of *Pandanus tectorius* over santal oil.

key: samara or frt. of *Acer* (maple).

khagoor (Pak): *Phoenix dactylifera.*

khai (eAs, Laos+): mixt. of powd. residue fr. smoked opium, morphine, & aspirin; smoked by addicts.

khair (Ind): *Acacia catechu.*

khaki: yellow brown (dust colored) cotton or wool cloth; popular for use in army uniforms of British army, etc. (<khaki, Hindustani Urdu [mud.] <khak, Pers. (dust) (also name for color).

khandsari (Ind): open pan (raw) brown sugar prepd. fr. sugar cane (<khanda [(Sanscr]); partly refined molasses (hence word candy).

kharola (wPak: Kurram valley): *Salix tetrasperma.*

khas(h) (Urdu): **khas-khas**: *Vetiveria zizanioides* or its rts.

khat(tea): Khattee (Ge): kat (tea).

Khaya anthotheca C DC (Meli) (w tropAf): (smooth-barked) African mahogany; white m.; bitter bk. u. for fever (Angola); excellent wd. (exp. to Eu) - **K. grandifoliola** C DC (w&c tropAf, esp. sNigeria): African mahogany; broad-leaved m.; Benin m.; bk. (lower trunk) decoc. u. for postpartum pain; rt. bk. decoc. in blenorrhea; local appl. in dermatitis; calcined bk. for relief of hyperacidity; wd. approaching true mahogany in quality - **K. ivorensis** A. Chev. (trop wAf): red mahogany; African m.; wd. useful and was exp. to Fr; u. med. (197) - **K. nyasica** Stapf (e tropAf): brown mahogany; mululu; bk. in decoc. compn. for colds; FO fr. crushed sd. rubbed into hair to kill vermin; useful wd. - **K. senegalensis** A. Juss. (w tropAf): Gambia mahogany; dry zone m.; bk. c. tannin (10%), nimbosterol; calicedrine; astr. in dys. bitter ton., antipyr. in mal. ("Senegal cinchona"); anthelm. emmen.; bk. inf. in liver fluke of cattle; fls. given in gastric dis., syph.; sd. FO u to annoint the body (inunction); wd. valued.

khellactone: component of *Ammi visnaga* (Khella) with vasodilator effect.

khellin: glycoside fr. *Ammi visnaga*; u. antispas., coronary vasodilator; thought of possible benefit in effort angina pectoris; chronic pulmonary dis.; coronary insufficiency; asthma; ureteral lithiasis - **Khellinin: khellol-glucoside**: the oxyglucoside of visnagin; physiolog. inactive.

khing (Siam): *Zingiber officinale.*

khira (Ind): kind of slender cucumber, eaten fresh.

khobani (Urdu): apricot tree or its frt.

khus (Ind): *Vetiveria zizanioides* (rt. in screen or mat).

Kibatalia blancoi (Rolfe) Merr. (Apocyn.) (PI): laniting-gubat; bk. & lf. decoc. u. abort.; lvs. for covering of head for headache; bk. & lvs. fish poison.

kibble: break, bruise, or grind coarsely, as malt, beans, etc. (**kibbled**, p.p.) (may be cognate with "chip") - **kibbler**: machine which breaks up oil cake - **kibblerman**: tender of same.

kick stick: (US, slang): Cannabis cig.

Kickxia africana Benth. (Scrophulari) = *Funtumia a.* - **K. elastica** Preuss. = *Funtumia e.* - **K. elatine** Dum. (K. sieberi Dorfl.; Linaria e. Mill.) (Euras, natd. e&cUS, Austral): canker-root; "woolly toadflax" (Austral); u. in scurvy, wounds, bleeding, itch (1613).

kidneywort: *Hepatica americana.*

kief = keef.

Kiefer (Ge): *Pinus* (pine).

Kielmeyera coriacea C. Mart. (Guttifer.) (SA incl. Br): pau camp; p. santo; bk. u. as source of ground cork (may be substit. for true cork); may have insecticidal value (K. corymbosa Mart. & K. speciosa St. Hil. [Br] also u. as insecticides [Br]).

Kienoel (Ge): pine tar oil, VO fr. dest. dist. of rt. wd. of *Pinus sylvestris, P. palustris*, etc.; c. pinene, sylvestrene (prod. nEu).

Kieselguhr (Ge): **purified k.**: purified siliceous earth.

Kieselsaeuregel (Ge): silica gel.

kif: keef.

Kigelia aethiopica Decne. (Bignoni) (tropAf): (African) sausage tree; cucumber tree; frt. u. as intoxicant, aphrod. vulner. purg. in dys.; to restore the sense of taste (Ghana); added to beer to increase strength (*sic*); dr. bk. appl. to sores to heal; bk. in rheum. dys. (Ghana) - **K. africana** Benth. (K. pinnata DC) (w tropAf): (bologna) sausage tree; frt. u. purg. in syph. GC (nNigeria); in dys.; sold in native markets; ripe frt. in form of paste u. for appln. to malignant tumors; cult. Fla [US] (tourist attraction).

Kiggelaria africana L. (l) (Flacourti) (sAf): Natal mahogany; porkwood; wild peach; wd. useful in cabinet making, etc.; parts of tree as "medicine to protect the kraal" (238).

kikar (Pak): *Acacia arabica.*

killikinick: var. of kinnikinnik (mixt. of tobacco & sumac u. by NA Indians) (22).

kimchi: the national Korean dish, consisting of richly spiced foods, incl. ginger, garlic, capsicum, &c.

kimmel: caraway (fr. Ge. Kuemmel).

kindling (wood): v. lighter wood.

kinetin: furfuryl amino-purine; a pl. hormone said to stimulate cell division, probably thru effect on N metabolism; acts with auxin; occ. in plants, esp. yeasts.

king root (US): *Podophyllum peltatum* - **k. of the woods** (Am Negro): *Aralia racemosa* - **king's clover**: v. clover - **king's cups**: *Ranunculus bulbosus* - **k. fern**: *Osmunda regalis* - **k. spear**: *Asphodelus luteus, A. ramosus.*

Kingdom: a high division of living things, thus, the Plant Kingdom; considerable difference of opinion as to number of kingdoms; one popular treatment recognizes the following kingdoms: Plantae; Animalia; Fungi; Bacteria (or Monera); and Protista (Algae; Protozoa; Myxomycetes).

kinic acid: quinic a.: satd. cyclic carboxy acid, found in *Cinchona* bk. (quina) (hence name), tobacco, cranberry, & many other pls.

kinins: specific polypeptides wh. occur in animals as tissue hormones; act as hypotensives, contract some smooth muscle, and increase capillary permeability; ex. bradykinin.

kinkeliba: dr. lvs. of *Combretum micranthum.*

kinnickinic(k) (wNA Indians): kinnikinnik 1) *Cornus amomum, C. stolonifera* 2) willow bk. u. for smoking, etc., generally mixed with tobacco (several variant spellings) 3) *Arctostaphylos uva-ursi* (alone or in smoking mixt.).

kino (gum): 1) generic term for tannin rich resin-like prods. fr. spp. of *Butea, Eucalyptus, Pterocarpus,* etc. 2) (specif) the dr. juice of *Pterocarpus marsupium* - **African k.**: dr. juice of *Pt. erinaceus* - **American k.**: 1) *Geranium maculatum* (**A. k. root**) 2) *Dipteryx odorata* 3) *Coccolobis uvifera* - **Australian k.**: extr. fr. *Eucalyptus camaldulensis+* - **Bengal k.: Butea k.**: fr. *Butea frondosa* - **Cochin k.**: k. fr. *Pt. marsupium* - **k. de l'Inde** (Fr): **East Indian k.**: Bengal k. - **Eucalyptus k.**: Australian k. - **Gambia k.**: African k. - **Jamaica k.**: *Coccolobis uvifera* - **Madras k.: Malabar k.**: Cochin k. - **occidental k.** Jamaica k. - **red k.**: fr. *Eucalyptus calophylla* - **West African k.**: African k. - **West Indian k.**: Jamaica k.

kinoa: *Chenopodium quinoa.*

kip: undressed hide (cow, horse, etc.).

kiri (tree): *Paulownia tomentosa.*

Kirsch: 1) cherry 2) **Kirschwasser** (Ge): cherry water, c. brandy; liqueur; colorless liqueur made in Spain by fermentation of ripe cherries & distn.

kiskytom: *Carya ovata.*

kiss me: pansy.

Kisschen (Ge): cachet.

Kitamycin: Kitasamycin: Leucomycin: macrolide antibiotic prod. by *Streptomyces kitasatoensis*; similar in action to erythromycin.

kitol: chem. component of whale-liver oil & liver oils fr. salt water fish; u. mfr. vit. A.

kitool: kittool: kittul: 1) a palm, *Caryota urens* 2) a fiber yielded by petioles of this sp.; somewhat like horsehair; u. for brushes to polish linen, etc. (SriL, EI) 3) also prod. fr. gomuti palm; subst. for horsehair & oakum.

kitten breeches: *Dicentra cucullaria.*

kiwi berri (k. fruit): *Actinidia chinensis.*

Klamath weed: *Hypericum perforatum.*

Klatschmohn (Ge): **Klatschrose**: *Papaver rhoeas.*

Klauenoel (Ge): neat foot oil.

Klee (Ge): clover.

Kleie (Ge): bran.

Kleinhovia hospita L. (l) (Sterculi) (eAs, Oceania): arbre de Juda (Fr); lf. decoc. u. in skin dis. (scabies, etc.) (PI); lf. juice u. as eyewash; bk. source of cordage; pl. c. HCN.

Klettenwurzel (Ge): burdock; the FO was macassar oil, formerly made fr. burdock rt. (1) digested in sesame oil (10) (parts).

Klickum (New Orl, La [US]): cannabis impregnated with HCHO.

klip (p) fish (k) (Nor): cod (etc.) split, salted, & dried on rocks.

Kmeria duperreana (Pierre) Dandy (Magnoli) (Camb): bk. u. as antipyr.

Knabenkraut (Ge): *Orchis maculata.*

knackelbrod (Scand): rye hardtack (*cf.* Ry-Krisp).

Knackmandel (Ge): shelled sweet almonds.

Knaeuel: Knaul (Ge): 1) ball; knob; fascicle; tangle 2) *Scleranthus annuus* (Ruebeknaeuel).

knapp weed: *Liatris spicata* - **knapweed**: *Centaurea nigra, C. calcitrapa, C. americana* (NM [US]), & other *C.* spp.

Knautia arvensis (L) Coult. (Scabiosa a. L.) (Dipsac.) (c&sEu to the Caucasus; nAf, natd. eCana & neUS w to NDak): corn (or field) scabious; bachelor (or blue) buttons; hb. (Herba Scabiosae) c. NBP tannins (most in rt.), dipsacan (pseudoindican); u. in chronic skin dis., esp. eczemas, urticaria, sores; in cough, pulmon. & throat dis.; cystitis; fls. & rt. also u.; rts. to adulterate valerian; cult. orn. - **K. silvatica** Duby (Scabiosa s. L.) (s&cEu): Wildknautie (Ge); Waldgrindkraut; lv. fl. & rt. u. for scabies & other skin dis.

Kneipp: K. Heilmittel-Werk (Wuerzburg, Ge): mfr. med. chiefly of natural origin (ex. Species Laxantes, with 10 pl. constituents); popular on continental Eu.

Knema glomerata (Blco.) Merr. (Myristic.) (Burma, PI): rt. decoc. given during puerperium (PI) - **K. glaucescens** Jack (K. palembanica Warb.) (Indones): bk. scrapings decoc. u. for abdomin. illnesses; wd. easily worked, u. to make knife handles, etc. - **K. globularia** Warb. (K. corticosa Lour.) (seAs): muscadier à suif (Fr); arom. sds. rich in FO u. in ointments for skin dis., esp. scabies; & in medic. soap - **K. heterophylla** Warb. (seAs, PI): bk. inf. u. as gargle for sore mouth & throat (PI).

knight('s) spur: *Consolida regalis.*

Kniphofia rufa Baker (sAf): red hot poker; cult. as ornam. hb. (Tritoma Ker) - **K. uvaria** Hook. (Lili/Aloe) (sAf): red hot poker (plant); torch lily; poker plant; bulb & rt. crushed & u. as enema in dysmenorrh.; sds. c. antibacterial subst. (*cf.* rhein); cult. ornam. hb.

knob root: *Collinsonia canadensis.*

Knoblauch (Ge): garlic.

Knochenmark (Ge): bone marrow (*qv.*).

knockout drops: chloral alcoholate (chloral hydrate in alc.); sometimes other narcotics; slipped surreptitiously into beverages, esp. highballs, by criminals, to stupefy the victim.

Knoeterich (Ge): knot-grass.

knol-kohl (Indo-Pak): kohl-rabi.

Knolle (Ge): tuber; bulb.

Knopper(n) (Ge-f.): valonia, cup of *Quercus aegilops*.

Knop's solution: u. in hydroponics.

Knorpeltang (Ge): Irish moss.

knot grass (, **English**): *Polygonum* spp., incl. *P. aviculare* - **k. root**: *Collinsonia candensis*, stone root.

knotty break: Aspidium - **k. rooted figwort**: *Scrophularia nodosa.*

knotweed (, **common**): *Polygonum aviculare*+ - **Virginia k.**: *P. virginianum* - **water pepper k.**: *P. hydropiper*.

Knowltonia capensis Huth. (Ranuncul) (sAf): u. as remedy in sciatica & rheum. by local appln.; cantharis substit. (early colonists) - **K. gracisis** DC (sAf): lf. inf. u. for syph. by mouth or in enema (Zulus); smoke fr. burning lf. inhaled for headache - **K. rigida** Salisb. (sAf): u. to treat lumbago, rheum.; vesicant - **K. transvaalensis** Szysz. (sAf): crushed lf. inhaled for headache; lf. decoc. u. as lotion for festering wounds & for poisonous bites - **K. vesicatoria** Sims. (sAf): South African buttercup; blistering leaves; pl. u. as vesicant in rheum. & lumbago; hot inf. of rt. u. for colds & flu; diaphor.

Knoxia corymbosa Willd. (Rubi) (eAs): u. for dis. of excretory organs, edema (Chin); to improve fermentation of rice by rice alc. distillers (Indochin) - **K. valerianoides** Thorel (seAs): u. in rice ferment. like *K. corymbosa*, med. (Chin Phar.).

ko: kudzu.

koa = koi = *Acacia koa.*

koal (Dutch): cabbage.

kobe: var. of agar found on Jap seashores.

kobu = kombu.

Kochia scoparia Schrad. (Chenopodi) (Euras, natd. NA+): summer (or mock) cypress; "Belvedere"; lvs. & frts. cardiotonic (Ind); cult. ornam. for foliage (densely branched); incl. var. *trichophylla* Bailey: burning bush; (Mexican) fire bush; serious weed; hay fever pl. (2569) (—> *Bassia*).

Koellia Moench: *Pycnanthemum.*

Koelreuteria paniculata Laxm. (Sapind.) (Chin): golden rain; pride of India; fls. u. in conjunctivitis & epiphora (excessive lachrymation) (Chin); as yellow dye; lvs. with *Coptis* rts. u. in liquid for eye dis.; frt. u. in skin dis.; pl. produces a gum; sds. u. in necklaces; cult. ornam. tree with brilliant yellow fls.; var. *aculeata* Rehd. & Wils. (K. apiculata Rehd. & Wils.) (Chin, Kor, Jap): fls. of tree u. as yellow dye (Chin).

koemis koetjing (Java): *Orthosiphon stamineus.*

Koenigskerze (Ge): *Verbascum.*

Koenigspaprika: v. paprika.

Koerbel (Ge): 1) chervil 2) wheat, rye.

Kohautia amatymbica E. & Z. (Rubi) (sAf): remedy for sterility; inf. u. as emet. in lightning stroke; decoc. for washing children; general emet. (Transvaal) - **K. grandiflora** DC (Senegal): entire pl. u. as mastic. & in gargles in gingivitis, stomatitis.

Kohl (Ge): *Brassica oleracea.*

kohl: prep. of soot, Sb (pd.), burnt almond, shellac, olibanum, etc., u. by women of Eg & other Near East countries from the time of ancient Egypt as cosmetic to darken edges of eyelids (*cf.* kajal); esp. popular in Marrakesh (Mor) (the full name al-kohl was later related to alcohol, etc. (2284).

Kohle (Ge): coal, carbon, charcoal.

Kohleria (*Isoloma*) **deppeana** Fritsch (Gesneri) (Mex, Guat): tlalchi(n)-chinole; lvs. & fls. u. in GIT dis., esp. gastric ulcers, chronic diarrh.; vag. wash in leukorrh.; in asthma (decoc.) - **K. spicata** Hanst. (K. schiedeana Hanst.) (sMex to Ve): said important med. uses (chiefly in GI dis.) (Co); "glowing accounts of its medicinal properties" (230); also ornam.

kohl-rabi: *Brassica oleracea* v. *gongylodes* L. (*B. caulorapa* Pasq.).

koji ferment (Jap): diastase enzymes prod. by *Aspergillus oryzae.*

kojic acid: cyclic compd. prepd. by fermentation of wide range of C sources using *Aspergillis flavus* & other *A.* spp.; u. as intermediate in chem. syntheses.

Kokakauer (Ge): coca (lf.) chewer.

koke (US slang): 1) cocaine (HCl) 2) beverage, Coca Cola™.

Kokia drynarioides (Seem.) Lewt. (Malv) (Haw Is [US]): natives strip bk. to obtain brown sap u. to dye fish nets; lvs. foraged by cattle; rare.

Kokkelskoerner (Ge): *Cocculus indicus.*

Kokosnuss (Ge): coconut.

Kokoona zeylanica Thwaites (Celastr.) (sInd, SriL): bk. u. as cephalic snuff in headache; pd. bk made into cakes for use as soap substit.; sd. FO u. as protection against leeches.

Kok-saghys: *Taraxacum bicorne.*

kokum: **kokam**: *Garcinia indica.*

kola(nut): *Cola nitida* & other *C.* spp. - **bitter k.**: 1) *Garcinia kola* (& sd.) 2) *Carapa procera* - **false k.**: 1) *Heritiera littoralis* 2) *Dimorphandra mora* - **male k.**: 1) bitter k. 2) wild spp. of *Cola.*

kolanin: glucoside of kola sd. (fresh) (*C. acuminata*) wh. hydrolyzes —> caffeine, kola red, & glucose.

kolatin: d-catechin as found in Kola (240).

Kolbenhirse (Ge): *Setaria italica* (cereal).

Kolocassi = *Colocasia* (Arg).

Kolombowurzel (Ge): calumba rt.

kombológia (Greek): "conversation beads"; made fr. seaweed of Med Sea origin (Turkey) (2623); u. to reduce nervousness, as in place of cigarettes (SC kombologion [= beads-talk]) (2698)

kombu (Jap): solid seaweed prod. fr. *Laminaria* spp.; enormous useage in & export fr. Chin; various kinds; some prepd. fr. spp. of *Alaria, Arthrothamnus*, etc.; u. in Jap & Chin cooking, esp. with fish.

kombucha (combucha): *Kombocha*; Fungus japonicus: tea fungus, orig. fr. Japan, now worldwide; made up principally of yeasts: *Saccharomyces* spp.; *Pichia fermentans*; *Schizosaccharomyces pombe*; &c.; also fungi, such as *Torula* spp.; also cont. bacteria, incl. *Acetobacter aceti* (Pasteur) Deley & Frateur subsp. *xylinum* (Brown) Deley & Frateur, &c.; cultured in sweetened black tea inf. at room temp.; variable compn. incl. several enzymes; fruit acids, vit. C; vit B complex; u. folk med.; panacea for arteriosclerosis, to combat senility; in cancer; refreshing beverage; *cf.* kwass (241).

Kommisbrot (Ge): rye bread.

Kondurangorinde (Ge): condurango bk.

Koniga: *Lobularia*.

konjac(u): *Amorphophallus rivieri* (var. *k.*).

konna (Mal): **konnam** (or **konnan**) **bark** (Madras, sInd): fr. *Cassia fistula*; u. tan.

kon(n)yaku(ku) (Haw Is [US], Jap): devil's tongue, fr. *Amorphophallus rivieri* (Ar.); Konyaku; u. gelatin-like prod. fr. tubers; u. nutr. med. (1614).

koochla tree: *Strychnos nux-vomica*.

koolabah tree (Austral): native name for any one of several *Eucalyptus* spp.; ex. *E. microtheca*.

kool(e) (Dutch): cabbage.

kooso: kousso.

kootchie (Jam, Rastafari): *Cannabis*.

Kopal (Ge): copal.

Kopfduenger (Ge): top dressing with manure, fertilizer.

Kopfsalat (Ge): cabbage; lettuce.

Kopsia fruticosa DC (Apocyn.) (Burma; widely cult.): pounded rt. u. to poultice nose in tert. syph.; c. kopsine (lvs.); arrow poison (similarly u. are **K. larutensis** King & Gamble; and **K. singapurensis** Ridl.).

Korbbluetler (Ge): Compositae.

korinthin: raisin (grape) sugar; u. in Greece as sucrose subsit.

Korn (Ge): wheat; rye (in some places).

Kornrade (Ge): *Agrostemma (githago)*.

koromiko (NZ): 1) *Veronica salicifolia* 2) chondrus.

Korotrin™: anterior pituitary-like chorionic gonadotropin.

kosam seed: 1) *Brucea javanica* (ko-sam) 2) *Carthamus tinctorius*.

Koso(blueten) (Ge): **kosso: kousso (flowers)**: fls. of *Hagenia abyssinica*.

kosher: foods, &c., prepd. according to Jewish religious (Mosaic) law; thus, for instance, meat is obtained fr. animal slaughtered with minimum pain, drained (most of blood removed), soaked, & salted.

Kosteletzyka pentacarpa Ledeb. (Malv) (Am, sEu, Russia): rts. u. much like Althea - **K. depressa** (L) Blanch, Fryxell, & Bates (K. pentasperma Griseb.) (sFla [US], WI, Mex, to Ve): malva mulata; mashed lvs. appl. to cuts (Jam).

koumiss: koumyss: kumyss.

kouso: kusso.

Kowli seeds (Ind): fr. *Croton oblongifolius*.

krabao oil: FO fr. *Hydnocarpus abyssinica*.

Kraeuterkaese (Ge): sap sago cheese: a skim milk cheese with a powd. hb. (Trifolium?).

Kraeutertee (Ge): herb(al) tea or infusion; many formulas are prepared; more popular in Eu than in NA; ex. Breast Tea.

Kraftmehl (Ge): maranta starch.

krait: snake, a member of genus *Bungarus* (sAs); venom u. in multiple sclerosis (-1979); incl. some of the most venomous & dangerous serpents in the world; ex. **banded k., common k., Indian k.**

Krameria L. (Leg./Krameri) (Mex to Ch): rhatany; source of rt. called Krameria (L); fr. some 20 spp. (*K. triandra, K. argentea*, & some other *K.* spp.); c. tannin; u. astr. in tanning & dyeing (SC S. G. Kramer, Austria, -1720) - **K. argentea** Mart. (Br): Brazilian (or Para) rhatany; rt. c. 10-13% tannin; u. astr. in diarrh. hemorrhoids, etc., hemostat., ton.; to color wines; in mouthwashes, gargles, etc. - **K. cistoides (cistoidea)** H. & A. (Ch): (raiz de) pacul; strong astr. in diarrh. chronic dys.; in perspiration of TB, hemorrhages; in mouthwash to improve gums - **K. cytisoides** Cav. (Mex): donapé; raiz de rathanhia; rt. u. for dyeing wool - **K. ixina** L. (Co, Br, Guyana): Savanilla k.; u. as tan; similar u. to *K. argentea* - **K. lanceolata** Torr. (K. paucifolia Rose; K. secundiflora auct. non DC) (Fla, c&sUS, Mex): raiz de cuculillo (Mex); Texas (or American) rhatany; produces drug lacking in reddish color; rt. u. astr. - **K. parvifolia** Benth. (swUS, nwMex): pd. rt. u. to treat sores (Ariz [US] Indians) - **K. secundiflora** DC (K. pauciflora DC) (Fla, Ga, Tex [US]; Mex): Texas rhatany; "sandbur"; raiz de cuculillo; u. astr.; in chronic diarrh., hemorrhages; reddish dye for wool, inks, etc.; formerly in commerce; tannin different from that of *K. triandra* - **K. tomentosa** St. Hil. (tropAm): Savanilla rhatany; u. tan. - **K. triandra** Ruiz & Pav. (Pe, Bo, Br, Ch, Ec): Peruvian (or knotty) k. (or rhatany); red r.; ratanhia;

payta r.; as astr. &c.; to treat chilblains (frostbite); this is prob. the most popular sp.

krameria, Brazilian: *Krameria argentea* - **knotty k.**: *K. triandra* - **New Granada k.**: Savanilla k. - **Para k.**: *Krameria argentea* - **payta k.**: **Peru(vian) k.**: *K. triandra* - **Savanilla k.**: *K. ixina* - **Texas k.**: *K. secundiflora*.

Krameriaceae: Rhatany Fam.; one of the 4 families into which Leguminosae is sometimes split; distinctive in having non-legume type of frt.; now placed in Polygales (2624); only one genus (*Krameria*).

Krapp(wurzel) (Ge): madder rt. or m. dye.

Kraunhia Steud.: *Wisteria*.

Krauseminze (Ge): curl(ed) mint: crisp(ed) m.; *Mentha spicata* (usually); *Mentha aquatica* var. *crispa*.

Kraut (Ge-m): herb; cabbage; weed - **krautartig** (Ge): herbaceous.

Kraystay™: chondrus prepn. consists mostly of carrageenan.

kreat: *Andrographis paniculata*.

Krebiozen™: prepn. claimed of value in cancer (US) but worthless; claimed extd. fr. horse blood; actually mineral oil & creatine (242) (SC Krebs [Ge] = cancer).

Krebs' cycle: stepwise catabolic mechanisms of cell carried on in the mitochondria & representing the chief respiratory action of the cell.

Krebswurz (Ge) ("cancer wort"): *Scrophularia nodosa*.

Kreosot: (Ge): creosote.

Kreuzblumen: Kreuzblütler (Ge): Cruciferae - **Kreuzdorn** (Ge): buckthorn - **Kreuzkraut** (Ge): *Senecio*.

Kriebelkrankheit (Ge): ergotism.

Krigia virginica Willd. (Comp.) (e&cNA): dwarf dandelion; u. food, nutr., much like true dandelion.

krill: *Euphausia* (Arthropoda): u. food.

krokos (Krokus) (Gr): **Krokus**: (Ge): crocus, saffron.

Kronsbeere (Ge): *Vaccinium vitis-idaea*.

Krugiodendron ferreum (Vahl) Urban (l) (Rhamn.) (WI, Mex: Yuc): bk. & rts. u. locally for toothache, gum dis.; rts. (decoc.) purg.

kruizemunt (Dutch): spearmint.

Krumenduenger (Ge): vegetable mold fertilizer (black hearth); *cf.* top soil.

Krummholz (Ge): 1) elfin-tree 2) *Pinus mugo* (**Krummholzkiefer**) - **K. oel**: pine needle oil.

Kubeben (Ge): cubebs.

Kuchenschelle (Ge): Pasque flower.

Kuckucksblume (Ge): (Cookoo flower): *Orchis maculata*.

kudzu (vine) (Pron kood-zoo or cud-zoo) (seUS): *Pueraria lobata*.

Kueken (Ge): chick; stop-cock - **K. kuemmel** (Ge): *Thymus serpyllum* (hb.).

Kuemmel (samen) (Ge): caraway; u. in prepn. of kimmel, Kuemmel, liqueur.

Kuerbis (Ge): pumpkin, gourd.

Kuhnia eupatorioides L. = *Brickellia e*.

kuko (Jap): *Lycium chinense*.

kumara: 1) *Arbutus unedo*, frt. 2) sweet potato.

kumb(h)i (leaves) (Ind, Bangladesh): lvs. of *Careya arborea*.

Kuminon (Greek): cumin.

kumis kuching (Java): *Orthosiphon stamineus*.

kumquat: *Fortunella crassifolia*, *F. japonica*, *F. margarita*.

kumys(s): fermented milk, usually made fr. mare's milk, but often now fr. cow's or ass' milk; c 1.5-3% alc.; term also appl. to dist. intoxicating liquor made fr. it.

kundar: kindar: (Arab, Pers): frankincense; mastic.

kurchee: kurchi (Ind): rt. bk. of *Holarrhena antidysenterica*.

Kurkuma (Ge): Curcuma; turmeric.

kuro-moji (Jap): *Lindera umbellata*.

kurreh: a poor grade of tragacanth.

kurung oil: FO fr. *Pongamia pinnata*.

kusa (Sanscr.): *Desmostachys bipinnata* - **kusabura** (Jap): shiu oil - **kusam**: *Carthamus tinctorius* - **kusambi**: *Schleichera oleosa* - **kusan: kusin (root)**: *Sophora wightii*.

kushta (Ind, Pak): *Arisaema* spp. u. as remedy by hakims.

kusmus: *Shorea ciliata*.

kusso: *Hagenia abyssinica* - **kussum**: *v.* lac (resin).

kust (Kashmir): costus root, *Schleichera trujuga*.

kusu oil (Jap): FO of frt. of camphor tree.

kusum (Beng & Hindi): *Carthamus tinctorius* (sd.) - **k. tree** (Ind): *Schleichera trujuga*, source of resin.

kute(e)ra gum: 1) Karaya 2) Tragacanth.

kuth root: *Saussurea lappa* (*Costus*).

Kuzenelle (Ge): cochineal.

kuzu (Jap): kudzu.

kwan-nan (Chin): bk. of *Strychnos malaccensis*.

kwangoside: glycoside of sarmentogenin & K-diginose; occ. in *Strophanthus amboensis*, &c.

kwass: non-alcoholic refreshing beverage popular in Russia (dispensed fr. tank cars); prepn. by fermentation of meal, malt, or bread with yeast; looks like beer; rich in vit. B complex; often confused with Combucha.

Kyagutt™ (Ge proprietary): prepn. c. *Crataegus* frt. ext. u. in cardiac dis. edema.

Kydia calycina Roxb. (l) (Malv.) (Ind, Burma): pola (Hind); lvs. pounded & appl. as paste in rheum. &

lumbago; bk. fibers u. to make coarse ropes; mucilage fr. sts. u. to clarify sugar.

Kyllinga Rottb. (Cyper.) (trop & subtr, mostly Af): some spp. with aromatic rhiz. - **K. brevifolia** Rottb. = *Cyperus brevifolius* - **K. monocephala** Rottb. (K. triceps L. f; Cyperus m. F. Muell.) (pantrop, incl. sAs & Austral): decoc. u. as cooling drink (Chin); rhiz. astr. u. in diarrh. (Indones); rhiz. decoc. u. diur.; exter. in dermatoses (PI); folk med. for diabetes, &c. (—> *Cyperus*).

Kyor: *Sagittaria sagitifolia.*

Ll

labdano (Por): Labdanum.

Labdanum (L): gum l.; l. resin; oleogumresin, a prod. exuding fr. *Cistus ladanifer, C. creticus, C. cyprius, C. incanus, C. monspeliensis*+; obt. mainly by boiling lvs. & flg. tops in water; c. 1-2% VO, resin (acids, phenols, isohydroabietic acid methyl ester, labdaniol, labdantriol), paraffins; u. stim. expect. arom., in plasters (current use), perfumery, incense; cancer (folk med.); VO in soap mfr. chief prodn. Sp, Cyprus, Crete; compn. variable.

lab (enzyme): Labferment (Ge): rennin; chymosin.

labeled: contg. a radioactive atom; ex. labeled C-carboxyl acetate; CH_3COOH.

Labiatae: Mint Fam.; c. 5,600 spp. (in 221 genera) distr. widely over the world, predominantly in the temp. zone; mostly hbs., few shrubs, rarely trees; most spp. are arom., constituting a majority of garden hbs. u. in culinary arts; sts. quadrangular; lvs. opp.; fls. arranged in verticillasters; fls. bilabiate; frts. as 4 nutlets per fl.; arom. props. due mostly to epidermal glandular scales (occasionally on all parts of pl.); ex. *Salvia, Monarda, Satureja.*

lablab (bean): *Dolichos lablab.*

Lablab niger Medik. (Leg) = *L. purpurea.*

Labrador tea: *Ledum groenlandicum.*

Laburnum anagyroides Medik. (Cytisus laburnum L.) (Leg.) (c&sEu, cult. s&seUS): **laburnum tree**; bean tree; golden shower; g. chain; g. rain; common laburnum; sds. c. cytisine, laburnine; u. sed. hypn. aper.; all parts of tree are poisonous; fls. (as Flores Cytisi laburni) earlier u. as drast. emet. diur.; psychoactive action; anti-nicotine value; by Homeopath. as nerve ton.; sds. (as Semen C. l.) similarily u.; u. to prod. cytisine (u. in anxiety, asthma); ornam. shrub or small tree u. in Easter decorations; important poison in novel of DuMaurier.

Lac (L): 1) milk 2) shellac (also **gum l.**) 3) jujube; v. under Lacca - **button l.**: formed in thick pieces made by melting lac & then dropping on smooth surface - **l. sulfur(is)**: precipitated sulfur - **l. tree, Japanese**: *Rhus succedanea*; *R. vernicifera* - **l. tree, Malay**: *Schleichera oleosa* (or *S. trijuga*) - **L. Vaccinum** (L.): cow's milk - **L. Virginis** (L): Virgin's milk (*qv.*) - v. under **Lacca.**

lacamass: lakamass (fr. Fr, la camas, fr. Chinook): *Camassia* spp.

lacatan (US, Eng, CA): *Musa nana.*

lacayote (Pe): squash.

Lacca (L.): shellac; a resinous excretion which crusts over the bodies of the lac insect; *Laccifera lacca* & other *L.* spp.; various types of lac are marketed (ex. button, flake [orange flake shellac], kiri, seed, block, shellac, stripe, stick, shell); garnet lac, orange-colored prod. made by pouring lac into a mold - **L. Alba: Laccae Albae Gummi**: white shellack fr. *Aleurites laccifera*, bleached with alkalies - **L. Coerulea**: litmus - **l. de Arizona** (Sp): *v. Larrea* - **L. Japonica**: Japanese shellac fr. *Rhus verniciflua* - **L. Magnesiae**: milk of magnesia - **L. Musci: L. Musica**: litmus - **L. Virginis**: virgin's milk; Tr. Benzoin with rose water.

laccase: enzyme wh. occurs in latex of lac tree & in many fungi & other pls.; c. Cu; in presence of O_2 oxidizes o- & p-dihydroxy phenols (ex. hydroquinone; pyrogallol) to their dark oxidn. prod.; produces bluish Jap lac fr. yellow sap.

Laccifer lacca Kerr (Tachardia lacca Kerr) (Coccidae - Homoptera [Hemiptera] - Insecta) (seAs): lac insects; great numbers of insects (plant lice) found immersed in resinous secretion (lac) on young twigs of various tree spp., incl. *Croton lacciferus* L., fr. which they produce gum lac (Resina Laccae).

Lacerta agilis L. (Lacertidae - Reptilia) (Euras): common (or sand) lizard; u. in homeop. med.; flesh & fat u. as sex stim. in Unani med.

Lachesis muta L. (Bothrops sururucu Wagl) (Crotalidae - Ophidia - Reptillia) (CA, SA): bushmaster; very venomous large (-4 m) snake; venom u. in both human & animal med.; in mastitis, drusen, erysipelas, herpetic fever, epilepsy, also prophylactically in thrombophlebitis, badly healing wounds, endo- and myocarditis, hyperthyreosis; in Homeopath. med.; venom c. blood coagulating, proteloytic, & hemolytic enzymes.

Lachnanthes caroliniana (Lam.) (l) Dandy (L. tinctoria Ell.) (Haemodor.) (eNA, WI, low, damp places): red root; spirit weed; rt. & hb. u. ton. stim. hypnotic, expect. in cough, pnèum.; source of red dye.

Lacinaria Hill: **Laciniaria** Hill: bot. synonyms of *Liatris* (latter conserved).

Lack (Ge): Lacca.

Lackmus (L.): **Lacmus**: litmus. fr. *Lecanora tartarea* &c.

Lacmellea panamensis Monachino (Apocyn.) (CR): cerillo; latex u. to clean mouth & throat of infants with thrush - **L. utilis** (Arn.) Mgf. (Tabernaemontana utilis Arn.) (Guyana): latex ("hya-hya") u. as cow milk substit.; frt. edible (*cf. Brosimum utile*).

lacquer, Burmese: natural lacquer obt. fr. *Melanorrhoea usitata* - **Chinese l. Japanese l.**: fr. *Rhus verniciflua* - **natural l.**: a varnish which is collected as a liquid & appl. directly as a finish without adding a solvent or drier.

lactalbumin α–: the special form of albumen occurring in mammalian milk; considerably u. in pharm. mfr; supposed to have antineoplastic activity.

lactams, beta: cyclic amide gp. found in penicillins & cephalosporins; $C_{|}O\text{-}N_{|}$- made by removal of -CH–CH- H_2O fr. amino acid.

Lactarius (*Lactaria*) Pers. (Agaric. - Basidiomyc.) (mostly temp. zone): milk mushrooms (secrete milky exudate or latex); some are edible (ex. **L. deliciosus** Gray); some have sharp taste (due to sesquiterpenes) - **L. piperatus** (Scop.) Gray (Eu, NA): peppery milky cap; c. tyrosinase; earlier u. as diur.; edible after cooking - **L. rufus** (Scop.) Fr. (Eu, NA): red lactarius; slayer; frtg. bodies c. tarirufin (sesquiterpene lactone) with antibacterial properties) (243).

lactase: ß-galactosidase: enzyme wh. catalyzes the hydrolysis of ß-galactosides (ex. ß-lactose); occ. in alfalfa sd. (rich source), soy, almond; fungi; *Saccharomyces* (*Kluyveromyces*) *lactis* (yeast); propr. Lact-Aid (Sugar Lo).

lactene™: purified coconut oil.

lactic acid: hydroxy-carboxy fatty acid of syrupy consistence; occ. in numerous natural prods. (ex. tomato juice); prod. of many bacterial and yeast fermentations; u. as caustic, acidulant, GI, antisep.; dil. in cosmetics for chapped hands, scaled skin (v. sarcolactic acid).

lactin(e): lactose.

Lactobacillus acidophilus Hansen & Morquot (Lactobacill.): non-pathogenic rod; often formed into short chains; u. in making acidophilus milk (lactic acid); sometimes popular dietary; occ. in milk, saliva, feces (children), &c.; may enter into kefir culture; u. to restore normal intestinal flora (esp. after using antibiotics); functional diarrh. moniliasis, pruritus ani, proctitis, irritable colon, simple colitis, diverticulitis, functional constipation, &c. (2290) - **L. bulgaricus** Rogosa & Hansen: u. to ferment milk, produces Bulgarian buttermilk; now regarded as of less value in nutr. than acidophilus milk; u. to produce yogurt, in wh. it is the dominant organism - **L. casei** Hansen & Lessel: ("*L. casei* vitamin factor"); participates in mfr. of kumys along with a yeast sp.; a cheese bacillus wh. requires folic acid for survival, hence u. in assays for folic acid - **L. delbrueckii** Beijerinck: important in mfr. of fermentation lactic acid - **L. lactis** Bergey & al.; prod. l-lactic acid fr. lactose; u. as starter in prep. Swiss cheese; in Kefir; vit. B_{12} essential to growth (= *L. lactis* factor).

Lactococcus lactis Loehnis et al. (Streptococcus lactis [Lister] Loehnis) (Streptococc.) (Bacteria): said part of process of ripening of Cheddar cheese (occurs in Cheddar Caves, Somersetshire, Eng); also with involvement of *Lactobacillus casei* Hansen & Lessel.

lactoferrin: iron-binding protein of mammalian milk; deprives some bacteria of Fe needed for survival; hence acts as anti-infective.

lactoflavin (Eu): riboflavin.

lactone: inner anhydride: a cyclic compd. formed by removal of water from a straight chain compd.; ex. santonin.

lactose: Lactosum (L.): milk sugar, $C_{12}H_{22}O_{11}$; made up of D-glucose (1 mol.) + D-galactose (1 mol.); at body temp., lactose consists of 2 parts of α-lactose in equilibrium with 3 parts of ß-lactose; mfrd. fr. whey (by-prod. of cheese indy); u. nutritive, esp. for infants & convalescents; excipient, base, diluent; in culture media; aid to lactic acid fermentation; in chromatography; ß-lactose claimed superior nutritionally; harmful effects in older mammals (rats, men) have been noted due to inadequate secretion of lactase; the accumulation of the sugar in the body is thought to be responsible for eye dis., diarrhea, hair loss, etc.

lactoserum: milk serum; liquid sepd. fr milk in cheese making; microfiltration permits sepn. of major proteins—caseins (esp. ß-c.), a-lactalbumin, ß-globulin, ß-casomorphine (with morphine-like properties).

Lactuca canadensis L. (L. sagittifolia Ell.) (Comp) (NA-WI): (American) wild lettuce; wild opium; horse weed; trumpet weed; "fire weed"; latex ("juice") c. bitter subst. lactucin, lactu(co)picrin; with mild narcotic, sl. hypnotic props.; latex fr. rt. rubbed on warts to remove & for poison ivy (Am Indian); latex u. in prepn. Lactucarium Canadense; lvs. eaten in salads (Cherokee Indians) - **L. chinensis** Mak. (eAs): lax. refreshing food - **L. debilis** Benth. (Chin): whole pl. decoc. u. diur. antipyr. stom. for tinnitus; ext. appl. to boils, snakebite - **L. floridana** Gaertn. (e&cUS): "gall of the earth" (misnomer); bitter hb. u. in snakebite - **L. hirsuta** Muhl. (eNA): (hairy) wood-lettuce; similar props. to *L. canadensis* - **L. indica** L. (warm As, PI): lvs. ton. digestive; crushed & made into plaster or cataplasm to apply to skin to provoke discharge of abscess - **L. quercina** L. subsp. **chaixii** Hayek (Euras; cult. esp. near Clermont-Ferrand [Fr]; Eng): u. in prepn. of Lactucarium Gallicum (French L.) and L. Rossicum (Russian L.); latex c. hyoscyamine - **L. sativa** L. (Euras, now cult. worldwide): (garden) lettuce; common l.; salad; sallet; insalata (It); a cultigen: lvs. (head) u. salads (chief salad plant), nutr. in Homeop. med., juice u. as sed., mild hypnot.; in impotency (Homeopath.); var. *capitata* L. (head lettuce) u. to prep. Lactucarium Gallicum (French l.); prep. Lactucarium Gallicum (u. asthma, bronchitis); to prep. chlorophyll (US); latex u. occasionally for tumors; lvs. u. to make a nicotine-free cigarette (Bravo) (seUS) (926); var. *longifolia* Lam.: cos (or Roman) (or romaine) lettuce; spoon-shaped lvs. & columnar head & large midrib, popular form in salads (chichon, Fr) - **L. serriola** L. (L. scariola L.) (Euras, natd. NA+): prickly lettuce; compass plant; hb. u. as diur. against leukemia (eAs); latex source of Lactucarium Gallicum (French l.) (sometimes cult.) (prepd. in Fr); var. *laciniata* Hort. (chichons, Fr) - **L. thunbergii** Maxim. (temp Chin, Jap): u. as blood ton. & for contusions (Taiwan) - **L. virosa** L. (s&wEu, nAs, nAf) (cult. Eng, Fr, USSR): bitter (or wild or acrid) lettuce (of Europe); opium l.; flg. hb. u. sed., expect in laryngitis & tracheitis with severe cough; latex of this & some other *L.* spp. u. in mfr. Lactucarium Germanicum and L. Anglicum.

lactucario (Sp, Por): **Lactucarium** (L); lactucariu (Romanian): lettuce opium; dr. latex of *Lactuca virosa* & other *L.* spp.; u. sed. anodyne, hypnot. exp. in asthma, irrit. cough, etc. - **French Lactucarium: L. Gallicum**: that prep. fr. *Lactuca sativa* - **German L.: L. Germanicum**: prep. fr. latex of *Lactuca virosa*.

lactucin: lactucopicrin: bitter princ. found in Lactucarium fr. *Lactuca* spp., *Cichorium* spp., &c.

Ladanum: Labdanum.

Ladenbergia Klotz. (Rubi.) (SA): false cinchona; genus closely related to *Cinchona*, prod. barks alk.- & tannin-rich; bk. u. bitter, ton., antipyr. u. like Cinchona antiper. alt. - **L. dichotoma** Wedd. = *Joosia d.* (R. & P.) Karsten - **L. hexandra** (Pohl) Klotzsch (tropSA): bk. c. yohimbine & a related alk.; sold as "Quina do Rio"; has been sold to SA, etc. as "Cinchona" by American drug companies; u. like cinchona - **L. macrocarpa** Klotz. (Cascarilla m. Wedd.) (Ec, Co, Ve): white cinchona bark; payta alba; bk. c. paytine & isomer paytamine (data contradictory) - **L. oblongifolia** (Mutis) L. Andersson; Cinchona o. Mutis; Carcarilla magnifolia (R. & P.) (Weddell) (nCo, Pe, nBo); "cascarilla macho"; similar properties & uses.

ladies': *v.* also "lady's" - **l. glove**: *Digitalis purpurea* - **l. hat pin(s)** (Fla [US]): *Eriocaulon* spp. - **l. streamer** (Ga [US]): *Lagerstroemia indica* (fr. corruption of generic name) - **l. tresses**: *Spiranthes autumnalis*.

lads' love: *Artemisia abrotanum*.

lady's: *v.* also "ladies" - **l. blush**: carmine - **l. delight**: pansy - **l. fingers**: 1) okra (Euras) 2) *Anthyllis vulneraria* 3) grape var. - **l. laces**: ribbon grass - **l. mantle**: *Alchemilla vulgaris* - **l. thumb (knotweed)**: *Polygonum persicaria*.

ladyslipper (root): *Cypripedium* spp., esp. *C. calceolus* v. *pubescens* - **pink l.**: *C. acaule* - **showy l.**: *C. reginae* - **(small) white l.**: *C. candidum*.

Lady Webster's Dinner Pills: P. of Aloe & Mastic - **lady wrack**: *Fucus vesiculosus*.

Laelia Lindl. (Orchid.) (tropAm, WI): popular ornam. pl. with about 50 spp. (*cf. Cattleya*) - **L. autumnalis** Lindl. (Mex): flor de los muertos u. med. (Mex) - **L. speciosa** Schltr. (Mex): flor de todos santos; pseudobulbs furnish a greenish mucilaginous subst. molded into the form of animals, fruits, &c., used on feast days (an ancient art); *L. autumnalis* also so used.

Laetrile™: (fr. **lae**vo-mandelo ni**trile**): claimed derived fr. amygdalin fr. kernels of *Prunus armeniaca* (var. cult. in Calif [US]); represents mandelo nitrile-glycuronic acid; theory of action: amygdalin hydrolyzed by enzymes present in excess in cancer cells gives slow release of HCN, which does not injure normal cells because of special degrading enzyme not found in cancer cells; value in cancer not proven (242).

Laeusekoerner (Ge): stavesacre - **Laeusekraut (mexicanische)** (Ge): **Laeusesamen**: Sabadilla.

Lafayette Mixture: Copaiba Mixture (NF VIII).

lagaña de perro (Arg): *Caesalpinia gilliesii* (l. = slimy mucous secretion of eye).

Lagascea decipiens Hemsl. (Comp.) (WI, Mex): hb. decoc. u. as wash for insect bites, rattler bite (Mex) - **L. spinosissima** Cav. (Mex, natd. Ind): hb. c. alk. resin; u. med.

Lagenaria breviflora G. Roberty (Curcurbit.) (tropAf, eAs): rt. decoc. u. as purg.; trop wAf; sd. decoc. given to pregnant females; sds. chewed as intoxicant; u. to stupefy fish; frt. u. to dehair hides - **L. mascarena** Naud. (sAf): bottle-gourd; rt. or lf. inf. u. in stomachache (Zulus) - **L. siceraria** (Mol.) Standl. (L. vulgaris Ser.; L. leucantha Rusby; Cucurbita l. L.) (OW trop): bitter bottle gourd; "Calabash" gourd (although not a true calabash); white-flowered gourd; courage calebasse (Fr); some vars. have edible even sweetish frts., but these are inferior in other respects; bitter vars. c. cucurbitacins; frt. pulp near sd. purg.; often toxic or even lethal; young frts. sometimes boiled & eaten; frt. pulp. emoll. appl. to soles for burning feet (Ind); lf. decoc. in jaundice (with sugar); sds. in bladder dis. (nephritis); one of "four big cold seeds" (formerly u. med.); juice u. in asthma, ophthalmia & as drastic; cult. as ornam. vine (1430); lvs. ("patu leaves", "p. herb") marketed (Tanz, Sen) (diabetes [124]); popular u. in India for cancer; lvs. appl. to stomach of babies with diarrh. (Yuc).

Lagerstroemia floribunda Jack (Lythr) (seAs): bk. decoc. for diarrh. & poisoning; rt. decoc. for pain with fever (Indoch) - **L. indica** L. (eAs, neAustral): (common) crape (or crepe) myrtle; crape flower (white, pink, purple) ja(m)ponica; Indian lilac; rt. u. as astr. & gargle; rt., lf. & fls. purg.; bk. as stim. antipyr.; sds. said to c. narcotic princ.; pd. bk. u. as styptic & decoc. wash for abscesses (Chin); shrub or small tree cult. as ornam. for brilliant fls. - **L. loudonii** Teysm. & Binn. (Indones): lvs. ground with benzoin u. for skin eruptions, local appl. - **L. speciosa** Pers. (L. flos-reginae Retz.) (Ind, Mal, sChin, Indones, Austral): Queen('s) crape myrtle; pyinma (Burma); pride of India; with white or purple flowers; bk. rt. & lvs. c. tannins; rt. u. astr. stim. antipyr.; bk. & lvs. purg.; sds. narcotic; frt. c. 15-17% tannin, appl. extern. in thrush (aphtha; aphthous; stomatitis) (Andaman Is); bk. decoc. in hematuria; rt. decoc. u. in jaundice; during puerperium (PI); wd. u. for railway ties.

Laggera alata Sch. Bip. (Comp.) (tropAf, As): lf. paste put on sore eyes (Tanz); hb. u. as disinf. (Mad) - **L. brevipes** Oliv. & Hiern (s&wAf): rt. decoc. given to woman after childbirth; remedy for sexual disorders (—> *Blumea*).

Lagochilus inebrians Bunge (Lab) (USSR: Turkestan+ in cAs): "intoxicating mint" (transl. fr. Russian); hb. inf. u. as folk remedy for allergies, nervousness, hemostatic, antispasm., in glaucoma, skin dis.; for centuries, the Tartars have brewed the plant with honey & sugar to make an intoxicating beverage.

Laguncularia racemosa L. Gaertn. f. (Combret.) (tropAm incl. WI; Angola, Af): white mangrove; w. buttonwood; bk. astr. ton. for fever, aphthae, scurvy; lvs. boiled for greens (Scand, GB); appl. to wounds (vulnerary) (GB); all parts of pl. u. as tan. (esp. bk.); wd. of tree much u. (244).

Lagynias discolor E. Mey. (Rubi) (sAf): pd. lf. swallowed for diarrh. dys. (Zulus); frt. edible.

laiche (Fr-f.): sedge, esp. *Carex arenaria* - **l. du sable, racine de** (Fr): Rhizoma Caricis, rhiz. of *Carex arenaria.*

lait (Fr-m.): milk; latex - **laitue** (Fr-f.): lettuce - **l. vireuse**: *Lactuca virosa.*

lake weed: *Polygonum hydropiper.*

lakrids (Dan.): **lakritz** (Sw, Finn): **Lakritzen** (Ge): licorice or thickened juice fr. licorice rt.

Lakshadia indica: *Tachardia lacca.*

Lallemantia iberica Fisch. & Mey. (Lab) (cult. sRussia, eMed reg., Iran): sds. c. strong drying oil (lallemantia oil), u. as food oil, illuminant, in varnishes - **L. royleana** Benth. (Indo-Pak, Iran): tukkm-malangu (Punjabi): **lallemantia** (or tokmari) sd. u. as cooling & sed. remedy & to prepare mucilagin. beverage; sds. ("Adex psyllium") in flatulence, constip. (Ind); for cough, as aphrod. & cardiac ton. (Iran); as adult. or substit. for black psyllium sd. (2702).

Lama vicugna Mol. (Camelidae) (Ruminantia) (wSA, nAndes): vicuña; intestines source of occidental bezoar stones (*qv.*) (related to llama [lama]) (*L. glama* L. domesticated), alpaca (*L. pacos* L.; u. for wool; domesticated), guanico ("camels of the New World"); beasts of burden (*L. guanicoe* Muller: larger animal; beast of burden); "spitting" habit of this beast of burden actually emesis; wool u. for textiles.

Lamanonia speciosa (Camb.) L. B. Smith (Cunoni.) (sBr): guapêre; heart wd. dark yellow, much u. in furniture making; bk. approved for use on cuts - **L. tomentosa** Vell. (Belangera t. Camb.) (Br): cedro do campo; bitter bk. u. as ton. styptic in some parts of Br; also of *L. speciosa*; wd. u. in mfr. pencils.

Lamarckia aurea Moench (monotypic) (Gram) (Euras, Med reg., Ind, natd. swUS, nMex): goldentop; grass c. cyanogenetic glycosides & HCN; cult. for ornam.

lamb: v. under mint - **l. fries** (US slang): **l. testes**: eaten as delicacy; wrongly regarded as masculinizing.

lambkill: *Kalmia angustifolia; K. latifolia.*

lamb's foot: *Alchemilla vulgaris*; *Plantago major* - **l. lettuce**: *Plantago media*+ - **l. quarter(s)**: 1) *Trillium erectum* 2) *Chenopodium album* - **l. tail**: Lycopodium - **l. toes**: *Anthyllis vulneraria* - **l. tongue**: appl. variously to *Plantago lanceolata*, various other *P.* spp., *Erythronium americanum*, *Chenopodium* spp., *Mentha arvensis* & other pl. spp. - **l. wool**: soft form used as dressing, esp. for placing between toes for soft corns (with daily change).

Lamiaceae: Labiatae - **Lamiales**: mint order.

lamina: 1) leaf blade 2) flattened thallus of algae.

Laminaria bracteata Ag. (L. japonica Aresch) (Laminari. - Phaeoph. - Algae) (Jap & Siber coasts): makombu; u. as viscous soln. ("haitai") in menstrual dis. (sChin); food (As); source of kombu; source of iodine (USSR); to make sweets & flavor foods (Jap) - **L. bulbosa** (Atl Oc): a seaweed; thallus c. 0.2% iodine (dry wt.); u. formerly as alt. in scrofula & other glandular enlargements; food - **L. cloustonii** Le Joly (North Sea, nAtl Oc, esp. wScot, Ire, Nor to Sp; Mass [US]): laminaria; tangle (wrack); thallus c. algin. laminarin(e), mannite, Br, I, Na, Mg, etc.; u. to prepare Stipes (or Stipites) Laminariae, which serve as "tents" & dilating bougies for distending body cavities (os uteri, etc.); dr. stalks (laminaria stalks) (l. pins, Norway) (l. sticks) (plugs of dr. thallus wh. swell up to ca 5x original size when soaked in water) (inserted in cervix uteri to bring on abortion); also for l. powder u. in mfr. easily disintegrating tablets; as hemostat. (as pd. soln. or gauze); for mfr. of sodium alginate (algin) which has many techn. applns. (ex. artificial silk indy.); formerly alt. in TB - **L. digitata** (L) Lamour. (L. cloustoni Edm.) (nAtl Oc coasts, esp. wHemisph, Pac coast, Cana): fingered kelp; horsetail k.; sea tangles; s. girdles; a perenn. kelp; u. as source of sodium alginate & as prec. stalks; good material for dilating bougies & tents - **L. hyperborea** Foslie (coasts of nAtl Oc): tangle; chief source of material for prepn. alginates (coll on coasts of Nor & Scot); u. in making ice cream, jams, preserves, confections, custards, mustard, ketchup; in paper mfr., printed textiles, dental casts; in prepn. seaweed meal, in health foods, animal feeds, & fertilizer - **L. potatorum** Lab. (temp. seas): pls. important source of food among aborigines of Austral; elsewhere u. as food in form of French-fried potato-like product - **L. saccharina** Lamour. (nAtl, Pac, Arctic Oc; ex. off Vancouver Is, BC [Cana]): sugar wrack; "broadleaf (kelp)"; important source of alginic acid, alginates, kombu; u. against goiter (Ind), syphilis (Himalayas); skin dis.; stalks eaten (Orkney Is).

Laminariaceae: Kelp Fam.; sporophyte made up of lamina (or blade which floats), a stipe (or stalk) & holdfast (fixed to object on sea bottom); chief genus: *Laminaria* - **Laminariales**: Kelp Order; incl. Laminariaceae & Lessoniaceae - **laminarin(e)**: reserve carbohydrate of glucosan type, found commonly in brown seaweeds, recovered along with algin; mixt. of two substances, one forming aqueous soln., other hydrosol (pptd. by strong alc.); popular in GB as hemostat. (1% dispersion).

Lamium album L. (Lab) (Euras, native to NA): white (dead) nettle (SC no sting); dead n. (SC unpleasant odor); white archangel; fls. c. flavonoids (incl. glycosides, as rutin), mucil. catechol tannin; u. astr. in mild digestive disorders; as constip., diarrh.; bleeding piles; expector in respir. dis.; hemostat. in leukorrh., edema, kidney calculi; hypnotic (Eu); hb. lvs. rt. also u. (esp. folk med.); pl. decoc. u. for bleeding fr. uterus & nose (Ind); boiled lvs. u. as spinach - **L. amplexicaule** L (Euras, natd. US): dead nettle; henbit; winter annual; flowers Jan-June (Ala [US]); hb. u. stim. lax. diaph.; stom. ton. carm. - **L. galeobdolon** L. (Euras): (yellow) dead nettle; source of Herba Lamii lutei; u. in catarrhs. - **L. purpureum** L. (Euras, Alger, natd. eUS s to Ala, eCana): (red) (or purple) dead nettle; red archangel; juice of fresh pl. formerly u. to scatter "canker sores" (1825); hb. u. as styptic, for inflamm. swellings; as purg. diur.; believed more active than *L. album* (folk med.); in splenic dis., catarrhs, phthisis, as diur.

lammint: peppermint; spearmint.

Lampaya (Lampayo) medicinalis Phil. (Verben) (Chile - High Cordilleras): lampaya; arom. hb. u. as inf. to reduce fevers, esp. in rheum. & colds, esp. useful for liver dis.

lampayo: lamsiekte (Af): cattle dis. resulting fr. eating putrefying bone contg. botulism toxin (sAf).

lamp black: form of charcoal (soot) obt. by burning FO or by incomplete combustion of natural gas; this type of amorphous carbon is chiefly u. as pigment in printing, mfr. phonog. records, etc.

lampkill: *Kalmia latifolia* (corruption of lambkill).

Lana (L, Sp): wool - **L. Philosophica**: ZnO - **l. de cordero** (Sp): lamb's wool - **l. batu grass**: *Cymbopogon nardus.*

lanain(e): lanalin: lanolin.

lanatosides: primary glycosides of *Digitalis lanata* - **lanatoside A**: digitoxigenin(*) + digitoxose + glucose + acetic acid - **l. B**: similar but with gitoxigenin for (*) - **l. C**: similar but with digoxigenin for (*) - **l. D**: similar but with diginatigenin for (*).

lancewood: tough elastic wd. fr. various trees incl. *Duguetia quitarensis* u. for police truncheons (GB), &c.

land plaster: gypsum; u. as nutr. for peanuts & other crops.

Landolphia Beauv. (Apocyn.) (Af, esp. Mad, SA): African rubber trees; caoutchouc lianes; source of liane rubbers: *L. hispidula* Pierre; *L. mandrianambo* Pierre (source of one of best grades of rubber); *L. myrtifolia* Markgraf subvar. *myrtifolia* & subvar. *crassipes* Markgraf & vars.; *L. sphaerocarpa* Jumelle (white rubber); *L. trichostigma* Jumelle & Pierrier; frts. of some spp. edible - **L. amoena** Hua (tropAf): frt. edible, sap u. as drops in sore eyes (Tanz); source of rubber(?) - **L. capensis** Oliv. (tropAf): wild peach; frts. edible, u. in jelly, brandy, vinegar; aphrod. (male); frt. said toxic - **L. florida** Benth. (trop&sAf): latex the source of African & Madagascar rubbers; frt. edible - **L. gummifera** K. Schuman (Mad): st. furnishes a liana rubber called "black Madagascar rubber" - **L. huedelotii** A DC (Senegal): tol (Wolof); lf. decoc. u. for abdominal pain (not purg.); dental pain; frt. pulp edible & good for digestion; latex furnishes rubber, formerly exploited - **L. kirkii** Dyer (tropAf): source of African rubber - **L. madagascariensis** B. & H. (Malagasy): vine latex turns into rubber ("Madagascar rouge") - **L. nitens** Lassia (Mad): frt. astr. u. in chronic diarrh. - **L. owariensis** P. Beauv. (tropAf): white rubber (vine); u. med. (Congo, Guinea); source of rubber along with several other *L.* spp. (Accra rubber) - **L. petersiana** Dyer (tropAf): frt. & twig u. as snakebite antidote; rt. for colic; source of African rubber.

lanesin: lanolin.

Lanette wax™: emulsified mixture of cetyl & stearyl alcohols.

Langerhans, islets (islands) of: scattered gps. of cells in pancreas wh. secrete insulin (beta cells) and glycagon (alpha cells), also somatostatin; degeneration produces diabetes mellitus.

langsat: *Lansium domesticum.*

lang-de-beef: *Borago officinalis.*

langue de boeuf (Fr) (cow tongue): refers to several spp. as *Scolopendrium officinale*; *Arum maculatum*; etc.

lanichol: laniol: lanolin.

Lannea amaniensis Engl. (Anacardi) (eAf): bk. ext. u. for snakebite; rt. decoc. u. in stomachache (Tanz) - **L. ambigua** Engl. (cAf): crushed rt. u. as poultice for tumors (Tanz) - **L. discolor** Engl. (c&sAf): live long; grd. inner bk. u. in diarrh., lvs. remedy for boils & abscesses; frt. edible - **L. edulis** Engl. (eAf): wild grape; rt. inf. u. diarrh. rt. decoc. in blackwater fever - **L. fulva** Engl. (Tanz): u. med. (245); bk. chewed for cough; frts. edible - **L. grandis** Engl. (L. coromandelica Merr.) (Mal, Ind): gingelly; jingilli; goompany tree; wodier wood; bk. c. 9% tannins, astr. u. as lotion in impetigo, leprous ulcers; boiled lvs. appl. to swellings & pain; bk. decoc. u. in toothache; pd. bk. u. as toothpowder; useful wood; source of jingan gum. (fr. wounds in bk.); u. for contusions & sprains (with coconut milk) - **L. kirkii** Burtt Davy (cAf): baster-marvela; rt. u. for snakebite (Swahili); frt. edible - **L. nigritana** Keay (wAf): u. in abdom. dis., diarrh. (bk. decoc.), chest dis., & lumbago (Senegal) - **L. schimperi** (A. Rich.) Engl. (tropAf): bk. inf. u. as lax., for abdom. pains; diarrh.; inf. of entire pl. u. for cough; rt. decoc. for dental pain (2445); syph.; bk. u. for cordage - **L. stuhlmannii** (Engl.) Engl. (s&cAf): decoc. of lf. or rt. u. in colic, diarrh.; lf. as paste appl. as dressing to sores, boils, carbuncles & abscesses (cAf); decoc. of rt. bk. u. in GC, TB, sore gums, asthma, ulcer in nose (yaws); lf. inf. u. for paralyses (2445) - **L. velutina** A. Rich. (tropAf): bk. & rt. prepns. u. for diarrh., baths for children & adults with undefinable illness; in massage for muscular breakdown.

lanoceric acid: dibasic fatty acid found in wool grease; $C_{30}H_{60}O_4$.

lanolin: hydrous wool fat; Adeps Lanae (L); Lanum (L); purified wool "fat" (actually a wax) emulsified with water; represents sebaceous gland secretion of sheep epidermis deposited on wool fibers; complex mixt. of esters of many fatty acids & high mol. wt. alcs. (aliphatic, steroid, triterpenoid) (lanolin alcohols); u. as ointment base, healthful for scalp & hair, hence u. in many hair dressings; dressing for horse hooves (1826); source of **anhydrous l.** (wool fat) u. in ointments, with advantage it can take up large vol. of water - **l. alcohols**: wool wax alcohols: the unsaponifiable fraction of wool fat (lanolin) obtained after saponification of same; c. cholesterol, lanosterol, aliphatic C_{16} to C_{32}, univalent alcohols &c.; u. in treatment of skin keratoses; in cosmetics, as ointment base, &c.

lanosterol: kryptosterol: obt. fr. degras (wool fat of sheep).

lanopalm(it)ic acid: fatty acid occ. in wool grease.

Lansium domesticum Corr. Serr. (Meli) (eAs, Mal+): lanzon(e) (PI); langsat; lanseh tree; duku (Mal); bk. & lf. decoc. astr. in dys.; bk. u. in poultice for scorpion sting; raw frt. deliciously edible; peels c. lansic acid (triterpenoid), burned as incense (Java) to drive away mosquitoes; many other uses in orient; lansioside A (amino sugar triterpene glycoside; potent inhibitor of leukotriene-induced contraction of ileum [*in vitro*]).

Lantana (Verben.) (mostly tropAm, but also OW): "lantana"; ca. 85 spp.; toxic plants (all parts); several spp. commonly cult. ornam. - **L. armata** Schau (nSA): "sanguinaria" (Ve); decoc. of branches combined with that of *Commelina* sp. ingested for vaginal hemorrhage (Ve) (Falcon) - **L. bahamensis** Britton (Bah, Cu): (big) sage (Bah): "filgrana" (Cu); lf. decoc. u. in fever, measles - **L. brasiliensis** Link (SA, WI): hb. is "yerba sagrada" (Pe); c. lantanine, quinine-acting alk.; hb. u. antipyr.; lvs. u. as tea substit. (Br) & substit. for quinine - **L. camara** L. (tropAm, esp. Br, widely natd.): (Jamaica mountain) sage; s. tree; wild s.; orozuz; lvs. (as "Herba Camarae") u. arom. stomach. ton. stim. expect. in catarrhs. (as tea); in rheum.; decoc. u. as bath for babies, invalids, &c. (Mex), in rheum (Af); diaphor. for colds (Af); rt. cold water inf. u. for GC (Domin, Caribs; 2328); rt. decoc. u. to "purify the blood" (Salv); ornam.; bad weed (Austral: "lantana shrub"); var. *aculeata* (L.) Mold.: lvs. pounded into paste & appl. to stomachache; lvs. chewed for sore throat - **L. canescens** HBK (Ve): "orégano"; hb. u. as condim. for hogs - **L. demutata** Millsp. (Bah, wCaicos): (white) sage (Bah); pd. lf. sprinkled on bruises; decoc. u. int. & exter. for measles & chicken pox (to allay itching) - **L. fucata** Lindl. (Ve to Par): combara (rosa) (Br); fls. u. as syr. for cough, rheum. (Br); in mal de mer (Ve) - **L. involucrata** L. (Mex, sFla, Tex [US], WI, nSA): confite; fls. v. fragrant; pieces of lvs. or sts. put in ears for earache (246); decoc. u. as wash for insect stings (Mex) - **L. maxima** Hayek (Mex to Par): maiz de zorro (Ve); fl. decoc. u. in anuria (urinary stoppage) (Ve) (2435) - **L. mearnsii** Mold. var. *congolensis* Mold. (Cong): pl. boiled down for vapor baths; var. *latibracteolata* Mold. (Tanz): lvs. chewed for cough & sore throat - **L. ovatifolia** Britt. (seUS, WI): big sage (Bah); decoc. u. in measles - **L. pseudothea** A. St. Hil. = *Lippia p.* - **L. rugosa** Thunb. (s&eAf): bird's brandy; paste of lf. & st. appl. to festering sores (Xhosa); lf. inf. u. for bronchial dis.; cold inf. of lf. u. as nasal douche for coryza; in ear for earache; lf. as galactagog; var. *tomentosa* Mold. (cAf): boiled lvs. made into poultice for sores on lips; pounded lvs. & NaCl put on sore eyes of cattle (Tanz) - **L. trifolia** L. (SA introd. in As, Af): u. emmen. diaphor (Co); antirheum. (Br) - **L. viburnoides** Vahl (Af): lf. juice or inf. u. for cough; lvs. pounded & rubbed over body for persistent stomachache (Tanz).

lantanin: lantaden A: pentacyclic triterpene; u. as antipyr. in place of quinine.

lanthopine: one of the many alks. of Opium.

Lanugia N. E. Br. (Apocyn.) (eAf): rubber-producing trees (—> *Mascarenhasia*).

Lanum (L.): lanolin.

lanzone: langsat.

lapachol: hydroxynaphthoquinone deriv.; obt. fr. taxa of *Tabebuia* (lapacho), *Tecoma, Tectonia* (heartwoods) (1857-); starting point for new antimalarials; with antineoplastic activity (carcinomata).

Lapageria rosea Ruiz & Pav. (l) (Lili/Smilac) (sCh): copihue; rt. u. like sarsaparilla; diaphor.; frts. consumed for their sweet, refreshing taste; cult. ornam.

Lapides Cancrorum (L.): **Lapilli Cancrorum** (L.): Lapis Cancri (in Ge & Ne Phar. Suppl.): crabs' eyes; crab stones; eyestones; concretions found in the stomach of the European crawfish, *Astacus fluviatilis*; c. $CaCO_3$, Ca Phosphate; u. antacid; in toothpowders.

lapin (Fr-m.): rabbit.

Lapis (L.): stone.

Lapis Amianthus (L.): asbestos - **L. Baptista**: talc(um) (ancient term) - **L. Calaminaris**: prep. calamine, an impure zinc carbonate - **L. Cancri**: v. Lapides Cancrorum - **L. Haematitis**: hematite - **L. Lazuli**: ultra-marine - **L. Ophthalmicus**: Lapides Cancrorum - **L. pumicis**: pumice - **L. Smiridia**: emery.

Laportea Gaudich. (Urtic.) (As, Af, Austral, NA): lvs. armed with stinging spines, sometimes resulting in lingering pain of affected skin areas - **L. aestuans** (L) Chew (trop wAf): atro (Ghana); lvs. boiled & u. for constip.; dys. (lvs. as enema) (247) - **L. bulbifera** (S & Z) Wedd. (Jap, Chin): dr. ground herb u. as vulner. (Ainu) - **L. canadensis**

L. Wedd. (e&cNA): wood(land) nettle; Cana n.; rt. u. as diur. & in urin. incontinence (Ojibwe); sds. lax. & anthelm.; very strong fiber u. by Am Indians - **L. crenulata** Gaud. (L. stimulans Miq.) (Urtica s. L.) (tropAs): devil nettle; fever n.; devil's leaf; frt. juice u. in chronic fevers; sds. u. like coriander (Ind) - **L. decumana** Wedd. (seAs): legs of natives climbing high mts. stroked with bundles of lvs. to stim. circulation, also to relieve fatigue (Rumphius) - **L. gigas** Wedd. (Dendrocnide excelsa Chew) (Austral): Australian (or giant) nettle tree; n. t. (of Australia); stinging tree; source of Rangoon hemp; pl. bears very poisonous, painful spines (stinging may be dangerous); u. for rheum. by local appln. (aborigines) - **L. harveyi** Chew (seAs, Oceania [Tonga Is]): lvs. crushed in water to make paste u. to dress cleaned ulcers (Indochin); rt. decoc. diur. (PI); in asthma & coughs; fls. u. to treat sores on soles of feet (nwSolomon Is): - **L. meyeniana** (Walp.) Warb. (PI): shrub or small tree with similar properties to those of *L. gigas*; lvs. & rt. inf. u. as diur. in urinary retention - **L. photiniphylla** Wedd. (Austral, Pacif Is): shiny-leaved stinging tree; u. med. New Caled (248); fiber fr. inner bk. u. by aborigines for cords, fishing nets, &c.

Lappa (L.): burdock root, *Arctium l.*

Lapsana communis L. (Comp.) (Eu, NA): nipplewort; pl. u. bitter; appl. exter. to sore nipples.

laque (résine) (Fr): gum lac, shellac.

laranja (Por): *Citrus* spp., incl. orange; usually with qualifying term - **l. comum** (Por): *C. sinensis* - **l. de umbigo**: *C. sinensis* var. *brasiliensis*.

larch bark: fr. *Larix decidua* - **l. agaric: l. fomes**: *Fomitopsis officinalis* - **Larche** (Ge): larch, *Larix* spp. - **Larchenschwamm**: larch agaric.

lard: benzoinated l.: l. treated with Siam benzoin as preservative - **hardened l.**: partially hydrogenated; u. as margarine - **(hog's) l. : prepared l.**: purified internal abdominal solid fat of the hog, *Sus scrofa* v. *domesticus*; prep. by melting & straining ("rendering"); u. as base in ointments, etc.

Laretia acaulis (Cav.) Gill. & Hook. (Umb.) (Ch): llareta; u. as stim. stom. vulner.; in bronchial catarrh; for headache, also diur.; rt. juice u. sed. in mountain sickness, asthma; pl. u. fuel; dist. for med. oils, tars (for calking), source of laretia resin; u. substit. for galbanum, varnishes, illuminants (*cf.* yareta).

Larix Miller (Pin.) (colder parts of nHemisph): larch; evergreen trees with thick & furrowed bk.; lvs. deciduous, on short spurs; bk. u. med.; in prep. brown & green dyestuffs; furnish valuable lumber wood - **L. americana** Mx. = *L. laricina* - **L. decidua** Mill. (L. europaea DC) (n&cEu): European larch; common l.; bk. u. as bitter; source of Venice Turpentine (oleoresin), Terebinthina Laricina, through trunk incisions; u. antisep. diur. diaph. in gastric dis., tech. in shellacs & adhesives (coll. sTyrol chiefly); cult. ornam. tree (-50 m) - **L. kaempferi** Carrière (L. leptolepis Endl.) (Jap): Japanese larch; red l.; u. med.; ornam. tree; useful wd. - **L. laricina** K. Koch (nNA s to WVa): tamarack; American larch; tree to 27 m high; bk. u. ton. lax. diur.; in jaundice, hepatic obstruction, rheum. skin dis. (249); fresh small terminal branches with needles u. to make a tea; also u. without lvs. (since deciduous tree) (Qu [Cana]); tisane fr. bk. u. with *Epilobium*, &c., for persistent cough (Abenakis) - **L. lyallii** Parlatore (BC [Cana] to Mont [US]): Lyall's larch; alpine l.; tips of branches broken off & boiled to make strong tea u. as "blood purifier"; also as antisep. wash for wounds & sores; to soak arthritic limbs (250); pigment fr. pitch, &c., u. as cosmetic by Indian girls - **L. occidentalis** Nutt. (wNA): western larch; tamarack; larch sap exuding fr. tree hardens, can be eaten like candy; u. to make syrup; tops u. to make decoc. u. as bath for arthritis, severe skin sores; drunk for severe arthritis, cancer (250); wd. u. in building & interior finishes - **L. russica** (*rossica*) Sabine (L. sibirica Ledeb.) (USSR: neRuss, wSiberia): Siberian larch; notable as one source of Pix Liquida, pine tar; and as host for *Fomitopsis officinalis*; boiled inner bk. with flour u. to prepare a kind of leaven; inner bk. u. to make gloves; useful wd. (very hard).

larkheel: lark's claw: *Consolida ambigua*.

larkspur (seed): *Consolida ambigua & C. regalis* - **common l.**: *Consolida regalis* - **dwarf l.**: *D. tricorne* (poisonous) - **field l.: forking l.**: common l. - **l. lotion**: acetic tincture of delphinium formerly much u. for "crabs" (pubic lice) - **rocket l.**: *Consolida ambigua* - **Spanish l.**: *Gilia rubra* - **tall l.**: *Delphinium elatum, D. exaltatum*.

Larrea nitida Cav. (Zygophyll.) (Ch, Arg): jarilla; this shrub c. much resinous subst. wh. is recommended as a balsamic stim. & vulner.; emm. also u. as baths in rheum.; crushed lvs. cataplasm; digestive; toxic to animals (251) - **L. tridentata** (DC) Coville (L. mexicana Moric.; L. divaricata sensu auctt. subsp. t. [DC] Folger & Lowe; Covillea tridentata Vail.) (swUS; Mex): creosote bush; c. shrub; gobernadora; shrubby pl. growing in dry areas; pl. c. 17-20% resin, 12% mucil., resin acids (lignan type, esp. nordihydroguaiaretic acid [NDGA]), simple phenols (as guaiacol), VO; flavonol (O-dimethyl-morin); pl. u. in folk med. as antirheum. (baths) (NA Indian); in arthritic pain (recently); inf. u. for affections of pulm. tract (bronchitis), VD, diur. (Am Indian); lf. decoc. u. in dysuria (Mex); fumes fr. burning shrub u. to disinfect homes (Mex); antisep. lotion prepd. by steeping twigs & lvs. in boiling water, u. for sores, wounds; collar sores of draft animals; aerial parts u. as insecticide, fish poison (Tex [US]); fuel (lime kilns) (2230); dr. lvs. u. as dusting pd. for sores; rts. u. as "chapparral tea" as "blood purifier" in TB & locally to treat wounds & for pain; antiviral, anticoccidian; antiprotozoal; source of NDGA = antioxidant (u. fat preservative); source of kind of shellac prod. through action of insects, called American Gum Lack, Sonora Gum, &c. (1615); fl. buds pickled & eaten (*cf.* capers); strong decoc. u. to clean amalgam (miners); to dye leather red (Mex); resin u. to fasten arrowheads to shaft (Am Indian); large areas of sNM (US), &c., permeated by strong aroma of plant; much branched shrub, sometimes of great age (-17,000 yrs.; Mojave Desert).

Larvae Formicarum (L.): ant eggs, fr. *Formica rufa.*

lasagna (It): **lasagne**: wide flat kinds of macaroni or noodles or dish with same; u. in soups.

Lasallia pustulata (L) Mér. (Gyrophora p. [L] Ach.) (Umbelicari. - Ascolichenes): blister rock tripe; c. gyrophoric acid; u. in med.; as green & brown dye.

Laser trilobum Borkh. (Umb.) (sEu, As): dreilappiger Rosskuemmel (Ge); frt. (Semen Seseli) c. VO with 40-45% perillaldehyde; u. as carm. in colic, mostly in veter. med.; st. edible (639) (2632).

Laserpitium latifolium L. (Umb.) (Eu): white gentian; laserwort; woundwort; frts. u. stom. bitter ton.; frts. & rts. u. as diur. carm. in dysmenorrh.; decoc. in beer; spice (antiquity); veter. med.; rts. c. VO, resin, laserpitin (bitter princ.); gum resin ("laser") reputed purg., bitter (Hippocratic med.) - **L. siler** L. (s&cEu, rocky slopes): mountain laserwort; sermountain; sd. & rt. u. diur. for flatulence, toothache; in upper Austria u. as condiment & to prep. a liqueur.

Lasia aculeata Lour. (L. spinosa Thw.) (Ar) (Ind, seAs): kanta; rhiz. & frt. u. as paste for sore throat; rhiz. u. to induce abortion (124); for hemorrhoids; rt. & hb. decoc. u. in colic (Indones); med. for elephants (Indochin); in curries (SriL).

Lasiacis divaricata Hitch. (Gram.) (sFla, nMex to Arg, Pe): cane-grass; tibisi; rt. decoc. u. as diur. antisep.; boiled with rt. of *Chicocca alba* & decoc. u. for kidney pain - **L. procerrima** Hitchc. (Mex-Ve): wild rice; panatu; pl. decoc. u. as diur. in seasickness; in making fireworks (Co) - **L. ruscifolia** (HBK) Hitchc. (Mex-Arg): carrizillo; rt. & shoot decoc. u. in urin. dis., fever, stom. trouble.

Lasianthaea fruticosa (L.) K. Becker (Comp.) = *Zexmenia frutescens.*

Lasianthera africana P. Beauv. (Icacin.) (monotypic) (wAf): mundja; pl. u. by natives to drive away "spirits"; in med.; flexible branches u. to make arcades.

Lasianthus alatus Aubl. (Rubi) (Guyana): inf. of fleshy lvs. formerly u. for smallpox; also u. like Gentian, the drug - **L. fruticosa** (L.) K. Becker (Comp.) = *Zexmenia frutescens* - **L. hoaensis** Pierre (seAs): rt. inf. u. in lumbago, other pain & stiffness (Indochin) - **L. stipularis** Bl. (Mal): lf. decoc. u. with *Lindera* sp. for "noises in head."

Lasiocorys capensis Benth. (sAf): lf. decoc. u. for chest dis.; inf. of above-grd. parts for hemorrhoids; pl. eaten by stock (—> *Leucas*).

Lasioderma serricorne (Coleoptera): tobacco beetle; this beetle attacks drugs, spices & foodstuffs.

Lasiosiphon anthylloides Meisn. (Thymelae) (sAf): rt. u. snakebite remedy (Zulu; Ind) - **L. capitatus** Burtt Davy (sAf): rt. decoc. u. as emet., enema for backache; rt. decoc. or chewed for sore throat; for headache - **L. eriocephalus** Dcne. (Ind): powerful vesicant; bk. u. as fish poison; lvs. appl. to swellings & contusions - **L. kraussianus** Hutch. & Dalz. (sAf): yellow heads; chewed lf. appl. to burns (Lenga); decoc. u. to bathe wounds & bruises; for sore throat (Zulus); very toxic pl. - **L. meisnerianus** Endl. (sAf): u. for skin eruptions, sores; in "Karroo fever"; pd. rt. c. amorphous resin with phytosterol, oleic & palmitic ac.; sugars; tannins; inserted into tooth cavity for toothache; rt. inf. in mal.; for snakebite - **L. roridus** S. Moore (Af): remedy for mal (—> *Gnidia*).

Lasiospermum bipinnatum Druce (l) (Comp) (sAf): inf. of pl. u. for chest dis.; u. to fumigate sickroom.

latex: milky juice of pls. carried in special ducts called laticiferous tubules; ex. gutta percha, rubber, opium (inspissated latices) - **fig l.**: leche de higueron (*qv.*) - **liquid l.**: the fluid rubber latex fr. tree which has been left uncoagulated & is u. directly in mfr. soft rubber articles such as rubber gloves, condoms, etc.

Lathraea squamaria L. (Scrophulari) (Euras): (greater) toothwort; lungwort; total parasite on rts. of spp. of *Fagus*, *Corylus*, &c.; rhiz. c. aucubin; u. formerly in epilepsy, colic, & sterility; toxic.

lathi (Ind-Pak): bamboo canes u. by police as clubs.

lathyrism: food (& fodder) poisoning (intoxication) produced by ingestion of sds. of some *Lathyrus* spp., such as *L. cicera* L. (Singletary pea); *Vicia faba,* &c.; CNS disturbances; chief symptom: muscular paralysis (big muscles, esp. of leg).

lathyrogens: subst. wh. prod. the symptoms of lathyrism; ex. aminopropionitrile.

Lathyrus L. (Leg) (worldwide, nTemp Zone; mts. of sTemp Zone): vetchling; some 150 spp.; sds. of many spp. toxic, produce lathyrism (2682) - **L. aphaca** L. (Euras, Ind): rawari (Punjabi): ripe sds. narcotic; fls. u. to reduce morbid swellings - **L. cicera** L. (Med reg., As): dwarf chickling-vetch; cult. for fodder; sds. eaten but may —> lathyrism - **L. hirsutus** L. (Med reg. introd. into US): rough pea; Caley p.; poisonous to stock altho u. as forage - **L. japonicus** Willd. (L. maritimus Bigel.) (Euras, NA): beach pea; pl. & sds. u. in cataplasms; sts. & lvs. u. to promote gastric function - **L. latifolius** L. (cEu; natd. e&cUS): everlasting or perennial pea; u. as adulter. of herniary herb (Hung); ornam. hb. - **L. montanus** Bernb. (L. macrorrhizus Wimm.) (Eu): bitter vetch; tuberous rts. eaten like potates - **L. niger** Bernh. (Eu, nAf): black vetchling; ornam. perenn. hb. - **L. odoratus** L. (It, cult. worldwide): sweet pea; cult. for fl.; VO formerly u. in perfumery - **L. pratensis** L. (Euras): (bastard) vetchling; u. to dissipate inflamn. (Sp) - **L. sativus** L. (sEu, Med reg., As): chickling; commonvetch; grass pea (vine); vetchling; white (or wild) vetch; sd. u. Homeop. med.; in spinal cord dis., impotency; cult. for food (sd.) & forage & fodder (plant); *cf. Cicer* - **L. tuberosus** L. (Euras, introd. NA): earth chestnut; tubers boiled as veg.; fls. distd. for perfume formerly; fodder pl.; u. in Russian folk med. (241).

laticifers: cells or vessels in trees (&c.) bearing milky juice (latex), mostly dicots, most tropic; chief fams. Apocyn., Asclepiad., Caric., Comp., Euphorbi., Mor., Sapot.

Latin square: arrangement u. in expt. setups; an array of n different symbols arranged in n X n order, in which each symbol appears only once in each row & once in each column.

Latipes senegalensis Kunth (Gram) (tropAf): sirkhi grass; sd. of annual herb eaten by natives of desert; adulterant of zira (cumin) (Ind) (—> *Leptothrium*).

latour, écorce de (Fr): *Symplocos racemosa*, bk.

Latrodectus mactans Fabr. (Theriidae - Arachnida) (trop & temp reg. esp. Am): black widow (spider); bite (neurotoxin) may produce pain & other effects, occasionally death in young; anti-Latrodectus serum prod. in Bs. As. (Arg); venom neurotoxin u. in arthropathies, angina pectoris, cardialgia, &c.; spider web strands u. in eye pieces of microscopes, telescopes, &c., for cross lines; human hair is also so used.

Latschenkiefer (Ge): dwarf pine, *Pinus mugo*.

Lattich (Ge): **lattuga** (It): lettuce - **lattuga caprina** (It): **l. romana**: wild lettuce.

Latua pubiflora Baill. (l) (Solan.) (sCh): palo (or arbol) de los brujos (witches' tree); c. hyposcyamine, atropine, scopolamine, &c.; narcotic, toxic, may produce death; u. by Indian shamans as intoxicant, leading to hallucinations & delirium, insanity; aphrod.; juice fish poison (1616).

Laub (Ge-m.): leaf, foliage - **L. wald** (Ge): deciduous forest (angiosperms).

laudanidine: alk. of Opium; supposed l-laudanine - **laudanine**: xall. alk. of Opium; tetanic action - **laudanosine**: alk. of Opium; a derivative of papaverine.

Laudanum: Tr. Opium - **l. cydoniatum**: Vinegar of Opium (Van Helmont) - **laudenham, Sydenham's**: Opium 100, crocus 25, cinnamon 6, clove 6, dil. alc. qs. ad 1000.

Launaea intybacea (Jacq.) Beauv. (Comp.) (sMex, WI, CA, Ve): u. for throat dis. (Cur); in med. (Mex).

Lauraceae: Laurel Fam.; shrubs & trees of subtrop & trop generally; 2,200 spp.; simple lvs.; frts. a drupe or berry; aromatic pl. rich in VO, also often mucil.; incl. genera *Laurus*, *Persea*, *Sassafras*, *Cinnamomum*, *Cryptocarya*, *Benzoin*, *Litsea*, *Umbellularia*.

laurel (SpAm): term appl. to many different plants; ex. *Ficus retusa*, *F. religiosa*; *Ocotea* spp., *Nectandra* spp., *Persea* spp., &c. - **l. (leaves)**: 1) *Laurus nobilis* 2) (much less commonly) (Ga [US], etc.): *Kalmia latifolia* & other *K.* spp. 3) (Ala [US]): *Illicium floridanum* 4) (CA, SA): *Cordia alliodora* (Boragin.): *Laurelia aromatica* (Monimi.) 5) (Canary Is): *Laurus canariensis* 6) (US): *Prunus laurocerasus*, *P. lusitanica*, *P. caroliniana*, etc. 7) (nGa mt. people): *Rhododendron* (misuse of term) - **American l.**: *Kalmia latifolia* - **bay l.**: *Laurus nobilis* - **benzoin l:** *Styrax benzoin*, tree producing benzoin - **l. berries**: 1) *Pimenta racemosa* source of VO 2) *Laurus nobilis* frt. - **black l.**: 1) *Gordonia lasianthus* 2) *Kalmia* - **broad leaf l.: broad-leaved l.**: American l. - **California l.**: *Umbellularia californica* - **l. camphor**: *Cinnamomum camphora* - **l. canela**: *Nectandra* spp. - **l. cerezo** (Sp): **cherry l.: l. cherry**: *Prunus laurocerasus* - **l. de comer** (Mex): (l. for eating): bay l. - **l. family**: Lauraceae - **great l.**: *Rhododendron maximum* - **Grecian l.**: bay l. - **ground l.**: *Epigaea repens*, trailing arbutus - **l. hojas** (Canary Is): *Laurus canariensis* - **mountain l.**: 1) *Kalmia latifolia* 2) *Rhododendron maximum* - **narrow-leaved l.**: *Kalmia angustifolia* - **l. negro** (Sp): *Nectandra saligna* - **l. oil**: 1) VO fr. frt. of *Laurus* *nobilis* (prod. Fr, It) 2) FO fr. fresh frt. of same - **poet's l.**: bay l. - **sheep l.**: American l.; also *Kalmia angustifolia* - **stinking l.**: *Illicium floridanum* (SC pungent odor) - **swamp l.**: 1) *Kalmia polifolia* 2) *Magnolia glauca* - **sweet l.**: stinking l. - **l. tree** (Ga): *Persea borbonia* - **wild l.** (Ga): *Symplocos tinctoria.*

Laurelia aromatica Juss. (Monimi.) = *L. sempervirens* - **L. novae-zelandiae** Cunn. (NZ): pukatea tree; bk. (pukatea bk.) inf. u. in tubercular & chronic ulcers; fevers, gynecolog. dis., GC; c. pukateine, laurepukine - **L. sempervirens** (R. & P.) Tul. (cCh, Pe): laurel; lvs. c. saffrole liriodenine, oxylaur-eline; inf. u. to strengthen the nerves, improve paralytic dis.; in stomach dis.; as diaph. & expect.; lf. & frt. as spice & stom.

Laurentia longiflora (L) Peterm. = *Solenopsis l.*

lauric acid: satd. fatty acid ($C_{11}H_{23}COOH$), chief component of FO (laurel berry oil) of *Laurus nobilis*; also occurs in coconut & other fatty oils.

laurier (Fr): laurel - **l. de Saint Antoine**: fireweed - **l. des montagnes**: *Kalmia latifolia* - **l. rose** (Fr): *Nerium oleander.*

laurifoline: alk. of *Cocculus laurifolius*; previously appl. to alk. of *Garrya laurifolia* (term garryfoline now used for latter).

Laurocerasus (L.): cherry laurel lvs.

laurotetanine: alk. fr. various spp. of Lauraceae; powerful poison, like strychnine.

Laurus L. (Laur.): 1) laurel; sweet bay; formerly large genus, now with only 2 valid spp.; spp. were dispersed among following genera; *Benzoin*; *Cinnamomum*; *Ocotea*; *Persea*; *Sassafras*; *Machilus*; *Malapoenna*; &c.; trees of aver. size , with simple alternate lvs. & berry frts. 2) (specif.): *L. nobilis*, laurel leaves - **L. canariensis** Webb & Berth. (L. azorica Franco) (Laur.) (Canary Is, Madeira): Canary Is laurel; hojas de laurel; frt. ("lauro") produces pleasant smelling FO - **L. cassia** Nees = *Cinnamomum c.* - **L. cinnamoum** L. = *C. zeylanicum* - **L. indica** Lour. = *Persea indica* - **L. malabathrum** Burm. f. = *Cinnamomum javanicum* - **L. nobilis** L. (Med reg., orig. fr. Anatolia [Turk], Balkans): (Grecian) laurel; sweet bay, bay l.; "old world bay"; leaves u. flavor, arom., condim. (not eaten); in soups, broths, pork meat (Russia); of historical interest, as lvs. u. to wreathe the brow of victors & athletes in classical antiquity; frt. c. VO, FO, sugar; u. stomach, condim. to arouse appetite, in bladder & spleen dis., against noxious insects; FO with laurin, palmitin, olein, &c.; expr. fr. frts. c. 2.5% VO, chlorophyll, NBP; u. inunction for skin dis., swellings; mild skin irrit.; protective against insect stings; in veter. med.; ex. udder salves; in felt mfr. (prod. It, Fr, Gr, Turk, Moroc, Sp); actinodaphine (alk. in wd. & bk.) - **L. persea** L. = *Persea americana.*

lavanda (It): washing; lavender**: lavande** (Fr): lavender - **grande l.**: *Lavandula latifolia*, with coarse VO - **grosse l.**: lavandin (*qv.)* - **moyenne l.**: *Lav. officinalis* cult. at lower elevations - **petite l.** (small l.): *Lavandula officinalis* var. cult. at rather high elevations (best oil).

lavandin: hybrid of *Lavandula angustifolia* X *L. latifolia*; low-growing bush (Fr); prod. VO inferior to that of *L. angustifolia* with little cineole.

lavandola (It): lavender.

Lavandula L. (Lab) (sEu, nAf, swAs): 1) lavender; perenn. arom. hbs. or shrubs with fls. some shade of blue (fr. lavanda, It, washing, as u. to perfume baths & laundry) 2) (specif): *L. angustifolia* - **L. angustifolia** P. Mill. (L. vera DC; L. spica L. [pro parte]; L. officinalis Chaix) (wMed reg. & nAf): (true) lavender; English l.; source of l. fls. & l. oil; u. perf. arom. in sachets, cosmetics, better grade soaps, analg. in headaches, neuralgia, migraine; tea drunk at meals as carm.; antispasm., sed., cholagog; diur. cutan. irrit., corrective, to prevent cancer (claim); moth repellant; in prepn. of herb liqueurs; in hb. pillows (chief prodn. areas: Fr, Sp); formerly u. in perfuming baths, for body odor (<lavanda, It, washing) - **L. bipinnata** O. Ktze (L. burmani Benth.) (Ind): hb. u. med. antidote to snakebite - **L. hybrida** Rev. (L. intermedia Emeric) (Fr): lavandin (*qv.*); hybrid of *L. latifolia* X *L. angustifolia* - **L. latifolia** Medik (L. spica DC; L. spica L. pro parte) (Med reg.): spike (lavender); broad leaf L.; source of oil of spike, an inferior type of lavender VO u. arom. in cheap perfumes, soaps, varnishes; lvs. & fls. u. as ton. emmen. abortient - **L. spica** L.: now divided between *L. angustifolia* and *L. latifolia* - **L. stoechas** L. (Med reg.): French lavender; source of inferior VO, somewhat like that of *L. latifolia*; fls. u. (as Flores Stoechados purpurea); u. in lung & stomach dis., asthma; in diabetes (Spain); in medicated (herbal) pillows or cushions, sachets & fumigating powders, and by perfumers - **L. vera** DC = *L. angustifolia.*

Lavatera assurgentiflora Kell. (Malv.) (Ec): malva pectoral; decoc. of lvs., fls. & rts. u. as purg. (2449) - **L. kashmiriana** Bois (nPak, NWFP, Kashmir): resha khatmi; fls. lvs. & rt. u. emoll. cath.; rt. boiled in sugar soln. for cough & bladder & intest. irritability; rather like marshmallow in props - **L. plebeia** Sims (Austral): Australian hollyhock; "mallow"; rts. u. in med. by natives.

Lavendelblueten, echte (Ge): lavender fls.

lavender (SC It lavanda, washing, referring to use of pl. for perfuming the bath; word first u. [lauendre] ca. 1265 by Wr. Wuelcker; Vocab. Plants): l. flowers - **l. (flowers)**: 1) *Lavandula angustifolia* 2) (English l.) *Chrysanthemum balsamita* (name sometimes u.) 3) (Ala): *Vitex agnus-castus* - **French l.**: 1) *Lavandula stoechas* 2) *Santolina chamaecyparissus* - **garden l.**: *Lavandula angustifolia* - **Italian l.**: garden l. - **Roman l.**: wild l. - **sea l.**: *Limonium* spp. - **topped l.**: *Lavandula stoechas* - **true l.**: garden l. - **wild l.**: *Lavandula stoechas*.

laver (Engl.): *Porphyra laciniata*.

Lawsonia L. (Lythr.) (monotypic): **L. inermis** L. (Med reg.; As, n&cAf, as Eg; nAustral; cult. in Ind through Near East) (2 forms: f. *alba* Lam. & f. *spinosa* L.): henna (lvs.); "resada"; fls. pink, white, &c.; lvs. c. lawsone, a quinone; u. astr. in sore throat, intest. moniliasis (2312), mostly as coloring agt., orange dye for hair, nails, skin of hands & face & for other cosmetic uses; lends auburn (chestnut) tint to hair; VO of fls. u. as perfume, rubbed over body to keep the body cool; u. in attars ("otto hina"); diur., emmen., abort.; practically non-toxic; claimed u. by ca. 25% of world's population (mostly in Orient & nAf); sds. u. deodorant, u. in leukorrhea & vaginal discharges; sd. inf. u. in headache & wash for bruises; c. FO u. to anoint body (Uganda), has analgesic properties; pd. henna considered vulnerary, u. for humans & camels (foot pads); to dye hair when gray, also for men who have made the voyage to Mecca; cult. orn. shrub.

laxative elixir: Cascara Sagrada Elixir Comp.

layout (US addict slang): opium pipe, burner & pliers, &c.

lazy man's flower: day lily.

lead: Pb: plumbum (L.): heavy metal long known to man; occ. in earth's surface, combined as sulfide (galena, PbS), etc.; u. to shield radiopharmaceuticals; Pb salts u. as astr. insecticides, industrial (in solder, paints, lubricants); mostly important as environmental toxicant, carcinogen; Pb dishes u. in ancient Rome may have affected health of many.

lead & opium wash: Lotion of lead & opium (NF) - **black l.**: a more or less pure carbon, graphite - **l. number**: amt. metallic Pb in Gms. which is contained in 100 gm. (test of pepper) - **l. water**: lead subacetate dil. soln.

leader: 1) primary or terminal shoot of pl. 2) upper part (or apex) of primary axis of tree.

leadwort: *Plumbago europaea* - **Ceylon l.**: *P. zeylanica*.

leaf: Folium (l.): an expansion of the stem or branch of a pl., in the axil of which one or more branches arise; ex. of medl. lf.: Digitalis, Buchu - **l. & flowering** (or **fruiting**) **top**(s): the upper portion of an herbaceous pl. st. bearing lvs. (buds), fls. and/or frts. & also including st. tissues; ex. (med.): Cannabis, Peppermint - **l. plant, one**: young pl. fr. which all buds & all lvs. but one have been separated; preferable in pl. expts. for studying the chem. activity & influence on the rt.

leafcup: *Polymnia uvedalia*.

leaflet: division of a comp. lf.; ex. (pharm.): Pilocarpus, Senna.

leandro (It): oleander.

"Leaping Charley" (swEng) ref. (HSH): *Prunella vulgaris*.

leather bergenia: *Bergenia cressifolia* - **l. (cap) leaf**: *Chamaedaphne calyculata* - **l. flowers**: *Clematis viorna* - **l. meal**: nitrogenous manure obt. fr. scrapped l. by steaming, drying, then grinding: c. 6-11% N - **l. wood (, Atlantic)**: *Dirca palustris*.

leben (Euphrates River area): yoghurt, soured cow's milk (7).

Lebensbaum (Ge): arbor vitae, *Thuja* - **Lebensmittel** (Ge): foodstuff.

Leber (Ge): liver - **Leberbluemchen** (Ge): *Hepatica nobilis* & var. *obtusa* - **Lebertran** (Dorsch-l.) (Ge): codliver oil.

Leb-Kuchen (Ge): ginger bread.

Lecanora Ach. (Lecanor. - Gymnocarpeae - Lichen.): lichens found along sea coasts, u. in prep. cudbear, litmus, & other indicators & dyestuffs - **L. calcaria** (L) Smrft. (Urceolaria c. [L] Ach.) (temp zone): limestone urcelolaria; corkir lichen; source of cudbear (GB), a red-crimson dye u. to color woollens; also orange & purple (Sweden) - **L. cinerea** (L) Smrft. (Urceolaria c. [L] Ach.) (temp zone): greyish (or ashy) urceolaria; this lichen growing on rocks u. as red crimson dye for woollens (Eng); also orange & purple dyes - **L. esculenta** Eversm. (nAf, AsMin, Greece, wAs): manna lichen; thallus ground up & admixed with meal as food (Bedouins); was u. in med. diet (252); believed to represent manna of OT - **L. haematomma** Ach. = *Haematomma coccineum* - **L. parella** Ach. = *Ochrolechia p.* - **L. tartarea** (L.) Ach. = *Ochrolechia t.*

Leccinum boreale A. H. Sm. Thiers & Watling (Polyporus borealis Fr.) (Bolet. - Hymenomycetes - Basidiomyc. - Fungi) (Euras, esp. Siber, NA): "Agaricus femina"; thallus c. bacteriostatic agt.; u. as punk to stanch blood flow - **L. testaceoscabrum** Sing. (Boletus versipellis Fr. & Hoek.) (temp zone): orange cap boletus; food sold in markets; pickled in Russia; export to Ge where u. instead of truffles (truffle liverwurst).

leche (Sp): milk, milky juice.

leche caspi (Pe, Br): gum fr. *Couma* spp. - **l. de higuerón** (SpAm): h. latex; fig latex; latex obt. by incision of the frts. of certain *Ficus* spp. incl. *F. glabrata*, *F. doliaria*, *F. laurifolia*, etc.; c. ficin (enzyme); u. for several centuries past as vermifuge in intest. infestation by *Trichocephalus* (whip worm, nematode), *Ankylostoma* (hookworm), etc.; product must be refrigerated or stabilized - **l. Marfa**: latex of *Calophyllum rekoi* - **l. ojé** (SA): *Ficus* spp. u. vermifuge, esp. *F. glabrata* (latex) (1504).

Lechea villosa Ell. (Cist.) (e&cNA): pinweed; hb. u. ton. antipyr. antiper.

Lechee: lichi.

lechigilla: lechiguilla; also other variants; (Mex): term appl. to a score of pl. spp. incl. smaller *Agave* spp. (century plants), esp. *A. lecheguilla* Torr., *A. lophantha* Schiede, etc.; *Senecio vulneraria*, etc.

lechillo (Mex): *Carpinus caroliniana.*

lechosa (PR, Mex): **lechoza**: papaya - **lechuga** (Sp): lettuce - **lechuguilla**: lechigilla.

lecithin: (pron. less´-ith-in): **lecithol**: compd. of phosphoric acid with choline & glycerides of fatty acids (oleic, palmitic, linoleic, linolenic, stearic); occurs in nerve tissues, egg yolk, plants; comml. source soy beans & corn; u. dermatitis, eczema; lipotropic in hypercholesterolemia & atherosclerosis (value ?); in deficient memory; available in granules, wafers; appl. locally for diaper rash; gastric cancer; u. in various foods, as chocolate bars, margarines, &c.; emulsifier, surfactant; many industrial uses; liquid forms available.

lectins: phyto(hem)agglutinins (former name): sugar-binding proteins that agglutinate red blood cells; widely distr. in pl. & animal cells, esp. richly present in sds. (richest source, legume sds.) (ex. concanavalin A fr. jack bean); not an antibody but combines with antigen to give immune reaction; inhibits tumor cell growth; some toxic to mammals (ricin; abrin); role in nature unknown; 1945- (SC legere [L]: to pick out or choose since specific).

lectotype: cotype specimen of organism designated at date later than original description as a type of that taxon.

Lecythis ampla Miers (L. costaricenisis Pitt.) (Lecythid.) (CA-nSA): monkey pot (frt.); olla de mono; kernels much finer than Brazil nuts; supply small; when eaten in large amts., may prod. toxic effects (such as hair loss) attributed to Ba present; frt. & sds. u. by natives for pneumonia; biliary dis., diarrh. & to accelerate childbirth; bk. u. in tanning, as oakum, &c.; wd. of large trees u. in mfr. furniture for stakes, carts (23), etc. - **L. ollaria** L. (tropAm): coco de mono; sapucaia; inner bk. (kakarali) u. as cigarette wrappers; frt. & sd. u. as depilatory, hemostatic (253); sapucaia oil fr. sds. u. for illumination & soap mfr., also foods; sds. edible but toxic (Se); frt. sold as novelty (NA) after injecting meat with aromatics, such as rose oil - **L. paraensis** Huber (Br): sapucaia; nuts edible, sometimes sold as Para nuts; beautiful wood - **L. usitata** Miers (Br): sapucaia; nuts similar to Brazil nuts; useful wd. - **L. zabucajo** Aubl. (SA, esp. Br): paradise nut tree; similar nuts to Brazil nuts but with finer flavor; FO of sds. of this & other spp. u. lax., emoll. & to treat acne, psoriasis, etc. (prod. Guianas; Br).

Ledebouriella divaricata Hiroe (Umb) (neAs, s to Chin): dr. rt. antipyr. diaphor. expect. in colds, flu, headache; fls. in circul. disturbances (Chin, Jap) - **L. seseloides** Wolff (Manchuria, nChin): rts. sudorific prescribed in flu, headache, fever, joint pain, convulsions (Chin).

Ledum groenlandicum Oeder (L. latifolium Jacq.) (Eric.) (nNA-muskegs): hb. known as Labrador (or James) tea (thé sauvage [Anticosti, Qu]); (inf. formerly u. as tea substit.); c. VO 0.1% (dr. wt.) with ledol (ledum camphor), palustrol; hb. c. arbutin ("ericolin"), catechol tannin, quercetin; u. in chronic nephritis; bronchitis, whooping cough & like foll. sp.; some layman (Qu [Cana]) regard this sp. & *Kalmia angustifolia* as 2 phases of life cycle - **L. palustre** L. (nEuras, nNA): crystal tea (ledum); fls. & young shoots (sometimes known as Herba Rosmarini silvestris) formerly in popular u. as antipyr. emm. exp. diur. galact.; in diabetes; u. as tea substit. (Ainu); pl. c. arbutin (glycoside; earlier called ericolin), tannin, VO, wtih ledum camphor; wd. u. as summer fuel by Eskimo; distillate c. valeric. butyric. acetic, formic acids (prod. in Pol); subsp. *decumbens* (Ait.) Hultén (NA tundras): lvs. u. as tea by Eskimos (Cana; Alask [US]); tobacco substit.

Leea aculeata Bl. (Lee) (Indones, Oceania): malimali (PI); for intest. worms; vegetable; lvs. u. to "purify bad blood" (PI) - **L. crispa** L. (Ind): tubers u. as remedy for guinea worm; lvs. bruised & appl. to wounds - **L. macrophylla** Roxb. (or Hornem.) (tropAs, esp. Ind): dinda; astr. rts. u. for guinea worm (Burma); for ringworm (Hindu med.); cicatrizant for obstinate sores - **L. sambucina** (L) Willd. (L. indica Merr.) (trop seAs+): lvs. appl. as poultice for skin dis., caterpillar itch, headache; lf. rubbed on boils (Solomon Is); rubbed on chest for pulm. dis. (Fiji); eye med. (Sumatra) - **L. speciosa** Jacq. (Ind): juice u. for refreshing beverage; rt. bk. substit. for rhatany (Ind) (254).

leechee: leechi: lichi.

leeches: blood suckers; class Hirudinea (Annelida); some 800 spp.; found in fresh water, marine & terrestrial habitats; free-living or parasitic; some small (such as medicinal l.) (*Hirudo*, *qv.*), others of large size, such as some *Haementeria* spp. (*qv.*); oral secretons c. local anesthetic, anticoagulant; u. in microsurgery; ex. in reattaching fingers - **American l.**: *Macrobdella decora* (Hirudo decora) - **European l.**: *Hirudo officinalis* - **German l.: green l.: medicinal l.: speckled l.: Swedish l.**: *H. medicinalis* L. (Sanguisuga medicinalis Sav.: S. officinalis Sav.).

leek (, garden): *Allium porrum* - **Japanese l.: stone l.**: *Allium fistulosum* - **three-seeded l.: wild l.**: *Allium tricoccum.*

Leersia hexandra Swartz (Gram) (pantrop, seUS): herbe rasoir; u. in hemoptysis in depressed cachetic ill patients with frequent & severe fits of coughing (Senegal) (255) - **L. oryzoides** (L) Swartz (Eu, NA): rice cut-grass; forage, fodder (As).

lees: sediments, as fr. wine; v. under wine.

legen (pp. **gelegt**) (Ge): to sow; plant.

Legionella pneumophila Brenner et al: bacillus causing Legionnaire's disease, a kind of lobar pneumonia; first discovered in 1976, at American Legion convention; transmitted through air (air conditioning, showers, &c.).

legno (It): wood - **l. dolce**: licorice.

legumbre (Sp): leguminous pl., pulse.

legume: legumen, pl. **legumina** (L): 1) pod, leguminous frt. which opens along 2 sutures 2) (Fr): légume, any vegetable (*cf.* **legumi**, It).

Legumina Phaseoli: bean pods without the seeds.

Leguminosae: Bean, Pea or Legume Fam.; 3rd largest fam. of plants (16,400 spp.) now often broken down into Caesalpiniaceae, Krameriaceae (now often excluded), Mimosaceae, & Papilionaceae (Fabaceae) Fam.; commonest practice is to split Leguminosae into 3 subfams., Papilionoideae, Mimosoideae, and Caesalpinioideae; considerable variation within Leg. Fam.; fl. irregular (in Papilionaceae); usually monadelphous or diadelphous; usually perfect; frt. commonly a pod (legume) or loment; sds. usually exalbuminous; fam. of much economic importance (gums, tans, frts., alks., balsams, etc.); incl. many food pls. (ex. beans, peas, peanut, soy, locust, tamarind); forage pls. (soy, clover, alfalfa); med. pls. source of many important crude drugs; ex. acacia, copaiba, senna, Peruvian balsam, licrorice, Tolu, tragacanth, kino, &c.; source of pl. principles (*Sophora*, *Physostigma*, *Cracca*, *Lupinus*).

lei: flower wreath popular in Haw [US] & Oceania.

Lein (Ge-m.): linseed, *Linum*.

Leiocom: Leiogomme: dextrin.

Leitneria floridana Chapm. (1) (Leitneri) (seUS): (Florida) corkwood; wd. lighter than cork, u. for floats for fishing nets, stoppers.

Lelie (Dutch): *Lilium*.

lem-kee: gum opium.

Lemaireocereus deficiens Britt. & Rose (Cact.) = *Cereus d.* - **L. dumortieri** Britt. & Rose = *Cereus d.* - **L. gummosus** Britt. & Rose = *Machaerocereus g.* - **L. thurberi** Britt. & Rose = *Cereus t.*

Lemmaphyllum carnosum Presl (Polypodi) = *Drymoglossum carnosum* - **L. drymoglossoides** Ching (Chin): ext. prepd. by soaking pl. in wine for several weeks; u. to cure lumbago & other pains.

Lemna minor L. (Lemn.) (worldwide, incl. Eu, NA): (lesser) duckweed; pl. demul. exter. in cataplasms for carbuncles, eye dis.; int. for gout, rheum.; in Chin as diur. antisyph. antiscorb.; u. as fodder; to cleanse waste water (odor & sludge removed).

Lemnaceae Gray (Monocots): free floating water plants incl. the smallest angiosperms known; 7 genera; u. as food for many animals incl. dairy cattle; potential source of proteins & amino acids (L. L. Rusoff).

Lemnian bole: an argillaceous earth.

lemon: *Citrus limon;* frt. juice, peel, etc., u. - **l. acid** = citric a. - **l. bioflavonoids:** dr. l. peel infusion (Sunkist): complex contg. vit C, B complex, hesperidin, eriodictyol, &c. - **l. extract**: alc. prepn. of l. peel & VO (5% each); very popular culinary flavor - **l. grass**: v. under grass - **ground l.**: Podophyllum - **sweet l.**: pear-shaped l. of uncertain bot. origin (Gilgit, Pak) - **wild l.**: ground l.

lemonade bush (Cal): *Rhus integrifolia* - **l. berry**: frt. of same.

lengua de ciervo (Mex) ("deer tongue"): *Asplenium scolopendrium* - **l. de vaca** (Sp) (CR): *Elephantopus scaber* - **l. de v. grande** (Pe): *Miconia* spp.

lenhina (Por): lignin.

lenitive electuary: Confection of Senna.

leño (Sp): block of wood; log; firewood (*cf.* madera).

Lens culinaris Medic. (L. esculenta Moench; Ervum lens L.) (Leg) (s&cEu, As; widely cult.): (common) lentil(s); common subtrop. crop in OW, but not as popular in NW; one of most ancient cult. crops of man; sd. claimed exceptionally nutritious; source of lentil starch, u. sometimes as food, in soups, breads, &c.; for pds. pills, etc.; fodder.

lent (US): Japanese fibrous morphine; SC either as prod. lent (loaned) by Japanese or as less "rich substantial" narcotic than ordinary sort of m. (24).

lenteja (Sp): **lentil(s)**: *Lens culinaris*.

lentic: occurring in still waters, such as ponds.

lentil: lentille (Fr): lens, *Lens culinaris*.

lentilha da aqua (Por): *Pistia stratiotes*.

Lentinus edodes (B.) Pegl. (As [Chin], Af): shiitake mushroom; c. lentinan, a glucan with immunomodulatory activity; u. to treat CFS, low natural killer cell syndrome (LNKCS): edible - **L. tuber-regium** Fr. (Agaric. - Basidiomyc. - Fungi) (tropAf, As): sclerotia eaten (Af); u. in med. for fever, &c. (Mal); imported into Jap fr EI, Micronesia as substit. for drug "bukuryo" (256).

lentisk: lenticsco (It): lentisque (Fr): mastic.

Leo ferox: (L): lions now assigned to *Panthera leo*; some individuals regarded as maneaters; greatly feared; some parts u. Unani med.

Leocus africana Bak. (Lab) (monotypic) (trop wAf): shrub; decoc. of pl. u. in diarrh., dys., heart dis.

Leonotis dysophylla Benth. (Lab) (tropAf): minaret flower; lf. inf. u. before meals as ton. & for nerve "weakness"; lf. & st. decoc. u. for colds - **L. leonitis** R. Br. (L. ovata Spreng.) (sAf): lion's ear; klipdagga; like other spp. with golden flowers; u. much like *L. leonurus*; variously u. for itch, as purg. emmen., in pertussis; laryngitis; said to have some properties of Cannabis (257) - **L. leonurus** Ait. f. (L. oxymifolia warsson) (sAf): lion's ear; wild dagga; cape d.; minaret flower; lf. & fl. inf. u. as cath. tenifuge for tapeworm; in skin dis., esp. scabies; lf. decoc. purg. emmen.; tinct. u. in hemorrhoids; smoked like Cannabis or tobacco but gives off a nauseous odor (259); children suck the sweet juice (258) - **L. microphylla** Skan. (sAf): klipdagga; inf. appl. to hemorrhoids & painful places; inf. drunk for digestive dis., esp. when fever; in chest complaints; decoc. for cough & colds (Kwena) - **L. mollis** Benth. (sAf): "balm of Gilead"; lf. & rt. inf. u. as snakebite remedy, purg.; mixed with tobacco for smoking - **L. mollissima** Guerke (trop eAf): lf. u. for snakebite; rt. remedy for veld sores; pl. toxic to livestock - **L. nepetifolia** (*nepetaefolia*) (L) R. Br. (Af, natd. seUS, WI, SA, Oceania): lion's ear(s); cordão de frade (Br); widely u. in med. in eAf, Ind (where cult.); CA, Br; flg. pl. u. (Br Phar.); lvs. u. in many dis.; shown to have antitumor activity; smoking of dr. lvs. (as "dagga") common practice in parts of Af; fresh flg. pl. u. (in Syr) in kidney dis., rheum (Br); in Af u. in dysmenorrh., fever, worms; in Ind ash of inflorescence u. for burns, in worms, lvs. as ton.; lvs. in vagin. suppos. for uterine prolapse, as abort., inflor. in tea for mal., fever (Trin); poultice for fresh scratches (Tanz); in fish poisoning (Virgin Is); hb. decoc. in colds (Domin).

Leontice leontopetalum L. (Berberid.) (wAs, Med reg., seEu): rakaf; lion's turnip; tuber u. for epilepsy (Israel); u. to oppose effects of opium (opium smokers); u. detergent, to remove spots fr. cloth, as soap substit.; in snakebite - **L. thalictroides** L. = *Caulophyllum t.*

Leontodon L. (Comp.) (temp nHemisph): hawkbit; a genus close to *Taraxacum* - **L. hispidus** L. (Euras, natd. eUS): roasted rts. u. as substit. for coffee - **L. taraxacum** L. = *Taraxacum officinale*.

Leontopodium alpinum Cass. (Comp.) (cEu, Alps): (alpine) Edelweiss (Ge, noble white) (natl. fl. of Switz); fls. u. in diarrh. dys. colic (hence Bauchwehblumen, belly ache fls.); popular ornam. wild & in rock gardens - **L. ochroleucum** Beauv.: u. med. (USSR) (1289).

Leonurus (L.): **L. cardiaca** L. (Lab.) (Euras): (common) motherwort; hb. c. leonurin (bitter glycoside); alk. tannin; hb. (Leonurus, L.); u. arom. gastric complaints, flatus, stom. diaph. cardiac palpitation & other c. dis., as angina pectoris (esp. in women); climacteric dis. - **L. lanatus** (L) Spreng. (Panzeria lanata Bunge; Ballota l. L.) (Siberia, cAs, Balkans): woolly motherwort; hb. c. VO, NBP, tannin, resins; hb. (as Herba Ballotae lanatae) as folk rem.; vasodilator (USSR), in dropsy, rheum., gout; reputed ton. diur.; introd. to cEu as garden pl. (1829) - **L. sibiricus** L. (cAs, sSiberia, natd. SA): u. much like *L. cardiaca*; also bitter ton., antipyr. for mal. ("quinine of the poor") (Ind); emmen., frts. uter. ton.; juice of fresh pl. u. in hemoptysis (Br); in edema, rheum. gout.

leopard: *Panthera pardus* (Felis p.) (Felidae = black p.) (cAs): bones u. as aphrod. (Chin); fur very valuable; flesh, fat, brain; uter. secretions u. in Unani med. (blood u. in skin dis.).

leopard's bane: 1) *Arnica montana* & other *A.* spp. 2) *Doronicum pardalianches*.

Lepadena = *Euphorbia*.

Lepargyrea Rafin.: *Shepherdia* spp. esp. *S. argentea*, buffalo berry.

Lepechinia conferta (Benth.) Epl. (Lab) (Ec): malva pectoral; decoc. of lvs. fls. & rts. u. as purg. (2449): u. in rheum. (lf. inf. for int. use; hot compresses of lvs. locally [2449]) - **L. rufocampii** Epl. et Mathias (Ec): salvia real; inf. of hb. u. in rheum. (2449).

leper oil: chaulmoogra oil.

Lepidium africanum DC var. *capense* Thell. (Cruc.) (sAf): cape (or pepper) cress; tuber inf. u. for cough (Zulus); pl. u. as veget. - **L. bidentatum** Mont. (Polynes, Micrones+): hb. said u. as antiscorb. by Captain Cook in the South Seas (2307) - **L. draba** L. = *Cardaria d.* - **L. iberis** L. = *Iberis amara* - **L. densiflorum** Schrad. (L. apetalum Willd.) (cEu, eAs): pepperwort; sds. u. diur. purg. (eAs) - **L. latifolium** L. (Euras, nAf): dittander; rt.

& hb. u. for scurvy & abdomin. dis.; in skin dis.; pressed pl. juice u. for virulent tumors; to treat camel dis. (Sudan); salad pl. of ancient Greeks - **L. meyenii** Walp. (Pe, Bo): c. alks.; u. food (salad); said to incr. fertility of female - **L. micranthum** Ledeb. (eAs): seeds u. to incr. sexual power (Kor); as purg., diur., in pleurisy (Chin) - **L. ruderale** L. (eAs): pl. u. impetigo (Kashmir, Ind) - **L. sativum** L. (nwAf, swAs, now widely cult. c&sEu, NA, &c.): (common or garden) cress; bitter c.; hb. u. as kitchen hb. (salads) & folk remedy, antiscorb., in sciatica, antipyr.; hb. c. sulfonated VO; sds. u. ton. alt. diur., sds. c. FO u. in cooking; exter. for massage; "cress wine" aphrod.; rt. sometimes u. as condim.; decoc. u. against malignant tumors (breasts, uter.) (241) - **L. schinzii** Thell. (sAf): pepperwort; crushed lf. sniffed for headache; lf. juice dropped in eyes for ophthalmia; u. in GC; garden vegetable - **L. virginicum** L. (L. intermedium A. Gray; L. medium Greene) (NA, CA, SA, WI, natd. Eu): bird's pepper; (wild) p. grass; pungent hb. u. as antiscorb., diur. stim.; rt. decoc. u. with *Aloe barbadensis* rt. for coughs & colds (Mayas) (6); u. with apple vinegar & brown sugar as dope to paint ham & bacon slabs while smoking (Ala [US]) (260); hb. u. for acute & chronic diarrhea (Mex).

Lepidocaryum tenue Mart. (Palm.) (Co, Br): sts. u. in basketry.

Lepidoptera: order of Insects which incl. the butterflies & moths; many spp. are harmful to pls.; several moth larvae attack crude drugs; many spp. useful in pollination, etc.; highly ornam. group.

Lepidosperma gladiatum Labill. (Cyper) (Austral): sword sedge; u. in paper making & to bind sand dunes - **L. perplanum** Guillaumin (New Caled): u. med. (300).

Lepiota morgani Pk. (Chlorophyllum molybdites Mass.) (Agaric. - Basidiomyc. - Fungi) (Af & worldwide): green gill; the sporocarp (with green spores) is eaten raw or cooked; some tolerate & enjoy it, others suffer from emesis, diarrh., head & abdom. pains; shows antiviral activity.

Lepraria candelaris Fr. (L. flava Ach.) (Deuterolichenes) (Eu+): this lichen is said to c. antibiotic substances - **L. chlorina** (DC.) Ach. (temp. areas): brimstone-colored lepraria; u. as source of brown dye for woolens (Scandia.); likewise **L. iolithus** (L.) Ach.; viol massa lichen.

lepromin: subst. fr. leprosy bacillus analogous to tuberculin; u. in diagnostic tests (**lepromin test**).

leprosy: serious communicable disease now generally called "Hansen's disease" - **l. oil**: chaulmoogra oil.

Leptactin(i)a densiflora Hook. (Rubi.) (wAf): c. tannins, alk., VO (very pleasantly arom.); has antispasm. activity - **L. senegambica** Hook. f. (trop wAf): yombo (Swahili); drug pl.; source of VO u. in perfumery (gardenia odor).

Leptadenia hastata Decne. (L. lancifolia Decne) (Asclepiad.) (trop wAf): u. diur.; in veter. med. - **L. reticulata** W. & A. (Ind): dori (Hindi) pl. u. stom. ton. emoll. lvs. eaten (eAf); u. to make sauces; dr. frt. u. as tinder.

Leptandra (L.): **L. virginica**: *Veronicastrum virginicum* (*qv.*) - **leptandrin**: glycoside (or acc. to some a non-glycosidal amorphous bitter princ.) obt. fr. Leptandra.

Leptaspis cumingii Steud. (Gram) (PI): powd. rt. soaked in water & liquid appl. to wash hair & stop hair fr. falling out - **L. urceolata** R. Br. (Solomon Is): rt. decoc. u. as postpartum protective med. & ton. to induce fertility; lvs. heated & rubbed on legs to allow quick walking.

Leptilon Rafin. = *Erigeron; Conyza.*

leptin: hormone (protein) inducing wt. loss or reduction; fr. lentis (Gr [slender]).

Leptocarpha rivularis DC (monotyp.) (Comp.) (Ch): palo negro; u. stim. carm. stom.; in dyspepsias, indigestion, flatulence, &c.

Leptoderris fasciculata Dunn (Leg) (trop wAf): u. to treat asthenia, anorexia, adynamic states in course of fevers & infections (in baths & drinks); for prurigo - **L. nobilis** (Welw.) Dunn (Congo, Angola): kifuria; rt. decoc. u. to heal wounds, esp. fr. circumcision (natives).

leptome: sieve tissue of phloem, consisting of sieve tubes & companion cells.

leptoside: leptogenin glycoside with D-diginose; occ. in *Strophanthus intermedia*, &c.

Leptospermum Forst. & Forst. f. (Myrt.) (seAs, Australas): lvs. & young shoots u. as diur., urin., antisep.; *cf.* uva ursi - **L. attenuatum** Smith (neAustral): ti tree; shrub to 6 m; u. like other spp. of the genus - **L. ericoides** Rich. (NZ): heath tea tree; frt. inf. u. for diarrh. colic; as poultice for open wounds & running sores; gum u. for burns, scalds, cough; bk. inf. u. as sed. in dys. diarrh., skin dis.; sap for bad breath; lf. inf. for urinary complaints - **L. flavescens** Sm. (Burma to Austral): yellow tea tree; wild may (Austral); ti tree; lvs. u. for beverage tea, for fever & lassitude; to stim. appetite, in dysmenorrh. (Mal); lf. VO (tea tree oil) inhaled for bronchitis, in embrocation for rheum. (Indones) - **L. laevigata** F. Muell. (Austral): Australian tea (or ti) tree; coast t. t.; u. for reclamation of land ("sandstay"); VO u. urin. disinf. (phenol coeff. 11/13) - **L. lanigerum** Sm. (seAustral): pl. grows on wet semisaline places &

eliminates swamps, thus reducing mal. mosquito populations - **L. petersonii** F. M. Bailey (L. citratum Challinor, Cheel, & Penfold) (eAustral, esp. NSW): lemon-scented tea tree; VO. with lemon odor/taste; cult. often as ornam.; cult comm. in Kenya & Guat - **L. pubescens** Lam. (Austral): tee tree; u. similarly to other *L.* spp. - **L. scoparium** Forst. & Forst. f. (Austral, NZ): broom tea tree; pungent lvs. u. like Chin tea; wd. useful for cabinet making, &c.; bk. u. to cover hut roofs, natd. Scilly Is.

Leptotes bicolor Lindl. (Orchid.) (Br, Par): frts. u. to flavor ice cream & other foods.

Leptotaenia Nutt. (Umb.) = *Lomatium.*

Lepus europaeus Pallas (Leporidae - Rodentia) (Euras): (European) hare; body fat; hare's grease (Axungia leporis) u. for wounds & sores; flesh & fur also u.; ankle bones (hare's leap, Hasensprunge, Tali Leporum) u. in pleurisy - **L. timidus** L. (nEuras, Alas [US], tundras): mountain hare; hill h., alpine h., arctic h., varying h.; ears shorter than head; hair whitens in winter; source of hare's grease (FO); u. folk med. as emoll.

Lergotrile Mesylate™: ergot alk. deriv. close to bromocriptine; u. in Parkinson's disease.

Lespedeza Mx. (Leg.) (NA, As, Austral): bush clover; valuable forage, fodder & green manure; ca. 40 spp. - **L. bicolor** Turcz. (& var. *japonica* Nakai) (Chin, Kor, Jap): shrub lespedeza (-3 m high); u. for soil conservation & fodder for livestock; prod. strong uterine contractn.; c. alk. lespedamine & dimethyl-tryptamine - **L. capitata** Mx. (NA, cult. Eu): hb. u. in kidney insufficiency; anuria; nephrosclerosis; oliguria; uremia - **L. cuneata** Don (eAs, Jap): "sericea"; a perenn. sp.; u. as soil binder & for hay (forage) (sUS) - **L. X longifolia** DC (seUS): bush clover; u. (with other *L.* spp.) to furnish wildlife (mostly birds) with winter foods (sds.) - **L. sericea** Miq. (Kor, Jap): "sericea" (Ala [US]); whole pl. simmered in water & brew drunk as stom.; rts. in mal.; erosion control (sUS) - **L. striata** H. & A. (*Kummerowia s.* Schindl.) (Jap, Chin, natd. e&cUS): Japan(ese) clover; yankee c.; Confederate c.; German c.; little wild c.; "shamrock"; common lespedeza; king grass; annual sp.; decoc. for swellings & ton. for general weakness; in broth to improve appetite (Chin); pasturage for cattle (sUS); very common wild in US - **L. trigonoclada** Franch. (Chin): decoc. u. as ton.

Lesquerella S. Wats. (Cruc) (NA): 40 spp. recognized (261); bladder pod; in some spp., the sd. FO is similar to that of the castor bean in having fatty acids with the hydroxy gp. (lesquerolic and densipolic acids) - **L. douglasii** Wats. (PacNA): yellow "mouse-ear" mustard; rts. chewed one at a time for diarrh., the juice swallowed & the pulp spit out; also effective for heartburn (250).

lesser: v. under centaury.

Lessoniaceae (Phaeophyc. - Algae): a kelp fam.; spp. with branching stipes; incl. *Macrocystis.*

Lethane™: thiocyanate insecticide, u. esp. for bed bugs; low toxicity to humans.

Letharia vulpina (L) Hue (Evernia v. [L] Ach.) (Usne. - Lecanorales - Ascolichenes - Lichenes) (cosmop: arctic & alpine regions): wolfbane evernia.; lemon-yellow lichen; "yellow moss"; thallus u. as bedding material; to dry up running sores & relieve inflammn. (Indians, Mendocino Co., Calif [US]); perfume base (considered inferior to *Evernia prunastri*); yellow dye (to color woolens); wolf & fox poison (vulpinic acid).

let-pet: pickled tea u. as condiment in Burma & Thailand.

letterwood: *Brosimum aubletii.*

lettuce: 1) *Lactuca* spp. 2) (Eng): (specif.): *L. sativa, L. virosa* - **acrid l.**: *L. virosa* - **cabbage l.: Cos l.**: *Lactuca sativa* var. *longifolia* - **garden l.**: *L. sativa* - **l. juice**: Lactucarium - **lamb's l.**: *Plantago media* - **miner's l.**: *Claytonia perfoliata*; *Montia fontana* - **prickly l.**: *L. serriola* - **Romaine l.: Roman l.**: cos l. - **sea l.**: *Ulva* spp. - **strong-scented l.**: acrid l. - **wild l.**: 1) *L. virosa, L. canadensis, L. quercina, L. scariola*, etc.; lvs. have been comml. article 2) *Claytonia perfoliata* 3) *Prenanthes altissima.*

Leucadendron argenteum (L.) R. Br. (Proteaceae) (sAf): silver tree; bk. (silver tree bark) u. as tan.; silvery haired lf. u. for decorative purposes, for mats, book marks, paintings, curios, &c.; wd. u. for boxes - **L. concinnum** R. Br. (sAf): lvs. c. salicin, alk.; u. in malaria (antipyr.).

Leucaena collinsii Bieb. & Rose (Comp) (Chiapas, Mex to nGuat); sds. eaten; u. med. as antirheum. - **L. cuspidata** Standley (sMex): "huaxi"; "efe"; sds. eaten with tortilla; frt. sold in markets - **L. esculenta** (Moçiño & Sessé ex DC) Benth. (Leg) (sMex): quaje (rojo); quashi; guajul de campo; aachi; guaje barbaro; huassi; sds. (with salt) eaten by natives in tortillas as salsa or in cooked dishes (with pork); considered aphrod.; sold in markets; cult.; pods in commerce; u. med. to fortify the lungs; ground bk. u. for bad wounds; appl. directly; aphrod.; eupeptic (for stomachache); gall nuts of frt. ("tindes") eaten crude with frijoles (beans); to remove obstructions; **L. e.** subsp. *paniculata* (Britt. & Rose) S. Zarata (Mex): guaje barbera; guajal de campo; eaten; u. as depilatory, &c.; cult; **L. e.** subsp. *matudae* S. Zarata (sMex): u. as food, vulnerary; bk. mixed with honey u. to cure obstinate sores; locally u. as divinatory; bk. c. tryptamines - **L. glabrata** Rose (Mex): guaje; green pods & sds. u. as food; sold in markets; widely distr., considered a cultigen; "guaje blanco" g. verde; eaten fresh raw or cooked - **L. glauca** Benth. (Mex trop & subtrop Am & of world): wild tamarind; jumpy bean; ginkokai; lead tree; shack-shack (Tob); lvs. u. depilatory; pods & sds. edible (Java); sds. u. in prepn. starch; c. mimosine (toxic alk.); tree (or shrub) u. forage, fodder, but toxic to horses & cattle (causes loss of hair in tails & manes of horses; skin discoloration); wd. u. for timber, fuel, charcoal; planted for green manure; soil erosion; brown sds. u. in necklaces as ornam.; rt. & sd. decoc. u. as contraceptive & emmen., abort.; ornam.; planted by Japanese to camouflage gun positions (Bonin Is, Iwo Jima), now a pest (heavy growth) - **L. leucocephala** (Lam.) de Wit (sFla [US], WI, OW, NW, trop): aroma blanca (Cu); jum-bay (Bah); lf. decoc. u. for typhoid, intest. gas; rt. decoc. for fever (Bah); u. med. (Fiji) (199); lf. as paste appl. to poisonous bites & stings; sds. anthelm., in diabetes (Indones); roasted sds. u. emoll., rts. emmen.; sds. with 9% FO (2454); also toxic principle; tree rapid growing & recommended for use in backward countries for fuel, stock fodder, afforestation; "liliaque"; "guash"; u. med. (Yucatan); food (Puebla); as living ten (Verz Cruz); u. for removal of hair; sds. source of starch; cult. **L. l.** subsp. *glabrata* (Rose) Zarata = **L. g.** Rose - **L. pulverulenta** Benth. (swTex, Mex): "tepeguaje"; tree a source of lumber; u. med. (Mex) (262); "chichimeca"; "huache"; tender sts., fl. buds, & sds. eaten with enchiladas; lvs. u. med. with other pls.; bk. u. to harden gums.

Leucanthemum X superbum Bergmans & J. Ingram (L. maximum (Ramond) DC X L. lacustre Brot.) (Eu, esp. Por; cult. & natd. NA): Shasta daisy; popular cult. ornam; some races with double fls. - **L.vulgare** Lam. (Chrysanthemum l. L.) (Comp.) (Euras; natd. sAf, NA): oxeye daisy; marguerite; hb. u. as domestic remedy in "catarrhs." (syr., &c.); uterine hemorrhage (USSR); young lvs. u. exter, eaten as potherb, salad, emergency food foraged by sheep; adulter. of pyrethrum fls.

Leucas aspera Spreng. (or [Willd.] Link) (Lab) (sInd & seAs, Indones): pl. u. as antipyr., insecticide, lvs. in chronic rheum., fls. in colds, lf. juice in psoriasis, wounds, sores, & dermal eruptions (Ind) - **L. cephalotes** Spreng. (Ind): goma (Hindi); pl. u. diaph., antisep., insecticide; fresh juice u. in scabies; fls. in coughs & colds; scorpion sting (Ind); c. VO, alk. - **L. lavandulifolia** Rees (L. lavandulaefolia; L. linifolia Spreng.) (eAs, Ocean): u. similarly to *L. aspera;* as stom., for wounds, vertigo, anorexia, &c. - **L. martinicensis** R. Br. (tropAf, As+): manzana (DomRep): lvs. u. as poultice for cuts (Tanz); pl. inf. as bath for arthritic pains (Br); pl. burned to expel mosquitoes (Nig) - **L. zeylanica** R. Br. (seAs): lvs. u. to treat headache; juice of hb. u. for skin dis., esp. scabies; in colds; poultice for wounds; stom. (Pl), itch; vermifuge.

Leucenol = mimosine - alk. fr. *Leucaena glauca, Mimosa pudica;* u. depilat.

Leuchtstern (Ge): *Aletris farinosa.*

leucine, l-: Leu (abbrev.): essential amino acid, necessary for human nutrition & not synthesized in human body; isolated fr. casein, gluten, & other proteins; u. nutr.; in treatment of amyotrophic lateral sclerosis (also l-isoleucine l-valine).

leucoanthocyanins: condensed tannins; represent glycosides wh. on hydrolysis yield leucoanthocyanidins + sugars & in turn are converted into anthocyanidins; pigments.

leucocyte: v. leukocyte.

leucoenus (L): white wine.

Leucojum aestivum L. (Amaryllid.) (Eu & Levant): (mountain) snowdrop; snowflake; toxic; u. extern. as emoll. - **L. vernum** L. (cEu, mt. meadows): snowflake; bulb & lvs. c. cardioactive alk. lycorine, galanthamine, tazettine, &c.; fls. c. kaempferol glycosides, leucovernide, leucoside (241); bulbs u. exter. as emoll. int. as cardioactive drug (toxic!).

Leuconostoc (Lactobacteri.): saprophytic bacteria found in milk, sugar solns., &c.; u. to mfr. dextrans.

Leuconotis eugenifolius DC (Apocyn.) (Mal, Indones): u. in ringworm & other skin dis.; potential rubber source.

Leucopaxillus giganteus Sing. (Clitocybe g. Quél.) (Agaric. - Basidiomyc. - Fungi) (Eu, Am): (short-stem) giant clitocybe; sporocarp edible; c. antibiotic clitocybin B. (tuberculostatic).

Leucophyllum ambiguum H. & B. (Scrophulari) (Mex): shrub; u. in fever, liver dis., jaundice - **L. frutescens** (Berl.) I. M. Johnston (L. texanum Benth.) (sTex [US], nMex): cenizo; "purple sage"; Texas silver-leaf; pl. u. in fever & "ague" (lf. inf. u. as antipyr. by Am Indians) - **L. zygophyllum** Johnst. (Mex): tea u. for stomachache; decoc. in diabetes; as tea & hb. bath u. for bile dis.

Leucopremna Standl. = *Pileus* Ramirez - **L. mexicana** Standl. = *Pileus mexicanus* A DC.

Leucosyke capitellata Wedd. (Urtic.) (eChina, Oceania): alagasi (Pl); lvs. u. in poultice for swollen face; rt. decoc. "cure" for phthisis, coughs, headache, gastralgia (Pl) (301).

Leucothoë axillaris D. Don (Eric.) (seUS): fetter bush; drooping leucothoë; pl. u. as errhine (increasing mucus discharge fr. nose) (American Indian) - **L. racemosa** Gray (eUS): pepper bush (of North America); white ozier; sweet balls; ornam. shrub.

leucovorin = folinic acid.

leuenkephalin: one of the endorphins (neuropeptides); polypeptide with leucine at end of chain of amino acids; secreted by brain, &c.; reduces pain.

leukocyte: white blood cell; WBC; colorless ameboid cell of the blood stream, with nucleus & cytoplasm; various types distinguished by staining reactions; provide many of the natural defenses of the body against diseases; WBCs show two chief gps., granular (granulocytes) and non-granular (incl. lymphocytes & monocytes); lymphocytes of 2 types: B & T l. (or cells); B cells (source of body's immunity by prodn. of antibodies; mature in bone marrow); T cells (play a role in immunity); harbored in thymus gland; some are called Killer (T) cells (cytotoxic T lymphocytes); destroy tumors (incl. cancers); viruses, &c.; promote graft rejection - **l. typing serum**: derived fr. blood of persons or animals with antibodies, for identification of leukocyte antigens (in vitro).

leukoharmine: harmine.

leukotrienes: LT: biologically active compds. found in leucocytes of the blood, related to prostaglandins & thromboxanes; incl. l. A_4, B_4, C_4, D_4, E_4, (subscripts indicating no. of double bonds) formed in specific cells by stim.; act somewhat like histamine, esp. in bronchoconstriction; act —> tears, sniffles, inflammn., itching; found in anaphylaxis; represent long chain carboxylic acids formed as oxidized metabolites fr. certain polyunsatd. fatty acids, esp. arachidonic acid; LTB_4 stim. movement of leucocytes; important in mediating allergic reactions, esp. in bronchoconstriction.

Leuzea carthamoides DC (Comp.) (USSR): much u. as stim. & refreshing agt. (1618).

Levant: the Near East, specif. countries near the eMediterranean; the "Orient" (but not the Far East Orient as now usually understood) - **L. nut**: Cocculus indicus - **L. wormseed**: v. under wormseed.

levertraan (Dutch): **levertran** (Scand): codliver oil.

leverwood (Qu [Cana]): *Ostrya virginiana* (SC great tensile strength of wd., rendering it useful to make temporary levers).

Levisticum officinale W. Koch (l) (Umb) (swAs; ; cult. s&cEu, NA): (garden) lovage (root) tang kuei (Chin); rh. & rts. u. by ancient Romans & since as culinary hb.; rhiz. & rts. c. VO with terpineol, isovaleric acid, etc., resins; u. arom. flavor (together with fenugreek producing maple flavor) stim. carm. in flatus, diur. exp. "anti-melancholic," sed. in hysteria, reducing agt., constit. of many tea mixtures; rt. added to tobacco; frt. ("seed") & hb. u. stim. ton. diur., diaphor. (prod. Ge) - **levisticum** (US trade term): American Angelica.

Levodopa: L-dopa; hydroxy-tyrosine; naturally occurring form of dopa; precursor of the catecholamines; found in *Vicia faba*, *Mucuna pruriens*, &c.; u. to treat Parkinson's disease, herpes zoster (per ora) (E).

levose root: Levisticum.

levothyroxin(e): the levo isomer of thyroxin(e) with ca. twice the activity of the racemic.

levulose = fructose.

levul(in)ic acid: carboxy acid prod. by partial hydrolysis of starch or sucrose; u. acidulant.

levure (Fr): (dried) yeast - **l. de bière**: barm.

levuroids: yeasts or yeast-like organisms.

Lewisia rediviva Pursh (Portulac.) (Mt. & Pac US): bitter root (plant); spatlum (NA Indian); when boiled, rt. loses bitterness, is popular & nutritious food of NA Indians, preferably when first peeled; rts. & inner bark c. much starch; odor of boiled rt. like that of tobacco, hence called "tobacco root" (302); u. med.

ley: (<lea): meadow.

Leysera gnaphalioides L. (Comp.) (sAf): lf. inf. ton. to stim. appetite.

LHRH: luteinizing hormone-releasing hormone: hormone of hypothalmus; acts as male contraceptive.

Liabum verbascifolium (HBK) Less. (Comp) (Ec): inf. u. to bathe bruises, &c. (animals).

liana: liane: woody vine; some authors use term to incl. herbaceous vines as well as woody - **l. blance** (Fr): Pareira Brava.

Liatridis Radix (L.): **Liatris**: button snake root fr. *Liatris* spp., esp. *L. spicata*.

liatrin: 1) one of germacranes (sequiterpene lactones); fr. *Liatris pycnostachya*, *L. chapmanii*, &c.; shows tumor inhibitory properties (leukemia) (1828) 2) resinoid of *Liatris spicata* formerly mfd. by Lloyd Bros., Cincinnati (Oh [US]) .

Liatris Gaertn. ex. Schreb. (*Lacin[i]aria* Hill; *Serratula L.* [in part]) (Comp.) (NA e of Rocky Mts.): blazing star, etc.; genus of perenn. hbs. with about 40 spp. (incl. hybrids); simple st. with few branches, growing fr. a tap or tuberous rt., with simple, alternate lvs. & purplish to pink fl. heads; many publications on chem. have appeared in the last few years (2437); some spp. claimed arom. (*L. spicata*, *L. scariosa*, *L. elegans*; dr. specimens)

(SC liazo, I come forth, fr. early appearance of lvs. [303] [or fr. atros. Gr, invulnerable, since u. as vulner.]; generally name considered of unknown origin); many spp. are floral ornam. - **L. acidota** Engelm. & A. Gray (Tex, La [US]): sharp gayfeather; button snakeroot; c. heliangolides - **L. aspera** Mx. (sCana, s to SCar, w to Great Plains area [US]): tall gayfeather; c. germacranolides, salvigenin, eupatorin; lupeol; taraxasterol, &c. - **L. chapmanii** T & G (Fla, Ga-coastal plains [US]): Chapman gayfeather; pl. exts. showed significant *in vitro* inhibition of human carcinoma cells (263), attributed to chapliatrin (antileukemic) - **L. cylindracea** Mx. (sCana, neUS): cylindrical blazing star; rt. u. as diur. in GC, &c. - **L. densiflora** (Bush) Gaiser (L. mucronata DC) (Tex, Okla [US], Mex): c. lupeol, heliangolides, liscundin, liscunditrin, &c. (possibly *L. elegans* var., acc. to some) - **L. elegans** (Walt.) Mx. (seUS): gayfeather; hairy-cupped liatris; pl. said to be fragrant; popular among florists (cut fls.); rt. u. med. like others; diur.; c. ligantrol (diterpene); eleganin (cytotoxic sesquiterpene lactone); ornam. perenn. hb.; cult. as garden fl.; potted pl. - **L. gracilis** Pursh (Fla, Ga to Miss [US]—pinelands): blazing star; c. subst. active against lung cancer, leukemia, lymphocytic anemia - **L. graminifolia** (Walt.) Willd. (NJ s to Fla, Ala [US]): loose-flowered button snakeroot; grassleaf gayfeather; c. graminiliatrin (guaianolide) (cytotoxin); graminichlorin; u. for snakebite, hence called "Rattlesnake's Master"; diur. diaphor. ("sweating powd.") - **L. laevigata** (Nutt.) Small (penins. Fla [US]): c. sesquiterpene lactones (eudesmanolide) - **L. microcephala** K. Schum. (Ga, Ala, Tenn [US]): hb. c. euparin, cadinol (sesquiterpene alc.); taraxasteryl acetate; no sesquiterpene lactones - **L. odoratissima** (Walt.) Willd. = *Trilisa* (*Carphephorus*) *odoratissima*, deer tongue - **L. pauciflora** Pursh (NCar-Fla, Ala [US]): c. complex lactone mixts. - **L. platylepis** K. Schum. (eNA): c. heliangolides, toxol, elemanolide derivs., geranyl derivs. (304) - **L. provincialis** Godfrey (nFla [US]): c. provincialin (cytotoxic germacranolide) - **L. punctata** Hook. (cCana, cUS into nMex): dotted gayfeather; blazing star; (flor de) cachana (SpAm); okawa phah (Pueblo Indians, NM [US]) (264); c. liatripunctin (germacranolide); corm. u. for food by Tewa Indians (237); u. diur.; for "windstrike," headache (Pueblo Ind) (264); corm burned & inhaled for nosebleed, headache; sore throat; rattlesnake bite (Tex [US]); syph.; locally in tonsilitis; sold in Mex-Am stores (1619) with other *L.* spp. - **L. pycnostachya** Mx. (cUS, as SD): Kansas gayfeather; blazing star; c. pycnolide, spicatin HCl, & other germacranolides; u. in GC (Ind), pharyngitis - **L. scariosa** Willd. (eUS fr. Pa-Fla): devil's bit; colic root; rt. u. diur. diaph. stim. ton.; astr. in diarrh. of children, antisyphil., in snakebite; this & other *L.* spp. u. to adulterate or replace *L. spicata* (considered superior); lvs. fragrant, c. coumarin; a compass plant - **L. spicata** (L) Willd. (eUS, eCana): button snake-root; spike (or Kansas) gayfeather; commonest member of the genus; in fl. Aug, Sept.; purplish fl. heads; pl. 1-2 m high; c. spicatin (cytotoxic; sesquiterpene lactone); guianolide; euparin; flavonoid pigments (quercetin glycosides); tuberous rt. u. stim. diur. diaph. carm. in colic, "backache," bitter ton., in GC, etc.; resins; cult. ornam. (Eu); fresh rts. u. Eu Homeop.; lvs. c. VO (0.09%); coumarin (82) - **L. squarrosa** (L) Mx. (c&eUS): rattlesnake's master; backache root; blazing star; aerial parts c. sitosterol, ß-amyrin, desacetyl-spicatin, lupeol; u. similarly to prec., expec.; entire pl. u. as poultice for backache; ornam. pl. (cut fls.) - **L. squarrulosa** Mx. (L. earlei K. Schum [2477]; L. scabra K. Schum.) (eUS fr. Oh to nFla, La, Tex): c. eleganin (a heliangolide) (cytotoxic) & complex lactone mixts.; has been considered diaphor. diur. stim. ton. sialagog - **L. tenuifolia** Nutt. (SCar to sFla-Ala [US]): slender-leaved blazing star; backache root; "deer bowl" (Fla); c. chapliatrin (anti-leukemic germacranolide), spicatin (sesquiterpene cytotoxic lactones); ecuparin; u. diur., ton., stim. etc. (82); pl. blooms all year (sFla).

liber: phloem or bast; fr. **liber** (L), bark of trees (SC written upon, ex. *Fagus* bk., hence any bark or bark portion).

liberty caps (US): *Psilocybe pelliculosus*, sporocarps.

libi dibi: l. divi: l. libi: *Caesalpinia coriaria.*

libo (Arabic): pine tops (& buds); eaten with sugar (vit. C source) (Lebanon).

Libocedrus chilensis Endl. Cupress. = *Austrocedrus c.* - **L. decurrens** Torr. = *Calocedrus d.*.

libriform (cell): wood fiber.

Licania arborea Seem. (Ros. - Chrysobalan.) (Mex, CA, Br): cacahuananche (Mex); cacao volador; cacahoananche; sds. c. 30% FO; c. licanic acid; u. similarly to oiticica oil; u. to replace tung oil in varnishes, paints, &c., as polish for antique furniture, &c.; bk. inf. drunk for kidney dis. (305); u. as illuminant (candles); soap, lubricant - **L. cecidiophora** Prance (n&cPeru): insect galls on lvs. u. to make necklaces & mail-like capes - **L. hypoleuca** Benth. (Hond, Nic, Pan): tree furnishes useful wd.; u. med. (196) - **L. incana** Aubl. (Guianas): furnish edible sds. - **L. platypus**

Fritsch (Mex, CA): monkey apple (Belize); bears delicious sl. acidulous frt.; zapote (cabello) (Oaxaca; CR) - **L. rigida** Benth. (neBr, Pe, Guiana): oiticica; sd. kernel c. 60% FO; u. mfr. soaps, etc. (Oiticica oil, as previously labeled, originated also from at least 3 other plant genera) - **L. turiuva** Cham. & Schlecht. (Br, Guianas): sds. edible.

Licaria puchury-major (Mart.) Kosterm. (Acrodiclidium p.-m. [Mart.] Mez.; Nectandra p.-m. Nees & Mart.) (Laur.) (nBr, Amazon reg.): puchury; not pichurim (*Nectandra pichurim* Mez.); sds. (Faba Pichurim, Port Ph.) c. VO with safrole, eugenol, cineole, FO (with lauric acid, &c.); u. carm. stom. arom. in diarrh. dys.; stim. (1620); bk. u. in acute dyspepsia; vanilla substit. - **L. salicifolia** Kosterm. (WI): canela (de pais); dr. frts. given to check dys. (Fr, WI); wd. u. for posts - **L. triandra** Kosterm. (sFla, WI): (pepper-leaf) sweetwood; laurel blanco; bk. decoc. u. in stomach dis.; stim. astr.

lichenase: enzyme wh. hydrolyzes lichenin to cellobiose; occurs in *Aspergillus oryzae*, *A. niger*, sds., animal stomach, &c.

lichens: Lichenes: Lichenophyta: cryptogam symbionts, organisms made up of associated Algae & Fungi; most mycobionts are Ascomycetes, while most phycobionts are Cyanophyceae; found in many parts of the world & under varying conditions; variously colored; usually dry in texture; some are of food value, others are u. as med. (dem.); some have antibiotic properties; perfumes, dyestuffs, etc. (pron. leykens, litshins, [Eng]) - **L. Carrageen**: Chondrus - **dark crottle l.**: *Parmelia ceratophylla*, *P. physodes* - **l. d'islande** (Fr): Certaria - **evernia l., staghorn**: *Evernia prunastri* - **flowering usnea l.**: *U. florida* - **gyrophora l., cylindrical**: *Gyrophora cylindrica* - **Iceland moss l.: Lichen islandicus** (L.): *Cetraria islandica* - **light crottle l.**: *Lecanora parella* - **lungwort l.**: *Lobaria (Sticta) pulmonaria* - **map l.**: *Lecidea atroviridis* - (**mealy**) **ramalina l.**: *R. farinacea* - **oakrag l.**: *Sticta scorbiculata* - **orseille l.**: *Rocella tinctoria* - **parmelia l., sprinkled**: *Parmelia conspersa*, *P. perlata* - **pleated u. l.**: *U. plicata* - **L. Prunastri** (L.): oak moss, fr. *Evernia p.* - **L. Pulmonarius** (L.): lungwort l. - **L. Quercinus** (L.): **L. Quercus**: L. Prunastri - **ramalina l., ivory**: *Ramalina scopulorum* - **rock-hair l.**: *Alectoria jubata* - **sticta l., yellow**: *Sticta crocata* - **stone crottle l.**: *Parmelia caperata*; *P. saxatilis*; *Lecanora tartarea* - **umbilicaria l., blistered**: *Umbilicaria pustulata* - **usnea l., bearded**: *Usnea barbata* - **wall l.**: *Parmelia parietina.*

lichens, economic genera: *Alectoria*; *Cetraria* (Iceland moss), *Chondrus* (Irish moss), *Cladonia* (reindeer moss), *Evernia* (oak moss), *Lecanora* (litmus, &c.), *Lobaria*, *Ochrolechia*, *Parmelia* (dyes), *Peltigera*, *Ramalina*, *Rocella* (cudbear & orchil), *Umbilicaria*, *Usnea*.

lichenin: lichen-starch: moss starch; polysaccharide (reserve cellulose) found in *Cetraria* & a number of other lichen genera.

lichenology: the science of lichens.

lichi (**nut**): *Litchi chinensis*.

Lichtensteinia interrupta E. Mey. (L. pyrethrifolia Cham. & Schld.) (Umb) (sAf): rt. decoc. in milk for coughs & colds; antipyr. & in spleen swellings; to make a narcotic beverage & in snuffs (Cape); hb. & rt. u. for colds (natives) - **L. lacera** Cham. & Schlecht. (Transvaal): brushed hb. u. as styptic in wounds.

licorice (**root**) (corruption of Glycyrrhiza, Lat.) (*v.* also liquorice): 1) *Glycyrrhiza glabra* 2) (cCana to Alask [US]): *Hedysarum mackenzii*; eaten in spring by NA Indians - **American l.**: *Glycyrrhiza lepidota* - **Anatolian l.**: one of earliest types of Russian l. to be marketed; named after larger part of Turkey in AsMin - **deglycyrrhinized l.** (DFL): u. as antigastritis agt. in peptic ulcer - **l. fern**: *Polypodium occidentale* - **Indian l.**: *Abrus precatorius* - **l. juice**: made by decoc. & expression (preferably of fresh drug), then evap'n; dried aq. ext. of l. - **Oregon l. root**: l. fern - **Persian l.**: fr. *G. glabra* v. *violacea*; coll. in Iran & Iraq - **Russian l.**: *G. glabra* v. *glandulifera*; rhiz. & rts. left unpeeled - **Spanish l.**: *G. glabra* v. *glabra* - **wild l.**: 1) licorice (2) 2) *Glycyrhiza lepidota*.

licorne vraie (Fr): *Aletris farinosa*.

licuri: ouricuri.

"lidard knots": "lider" (lighter) **knots** (sUS slang): lightwood pine knots u. for kindling.

Liebermann-Burchard test (LB): reagent made up of acetic anhydride + H_2SO_4 (produces violet and green colors); u. as test for sterols & triterpenoids (glycosides).

Liebstockwuzel (Ge): Levisticum.

liège (Fr-m.): cork; c. oak.

lierre de terre: l. terrestre (Fr-m.): ground ivy.

Lieschkolben (Ge): cat-tails, *Typha* spp.

lievito (It): yeast.

life everlasting: *Gnaphalium polycephalum* - **l. of man**: *Aralia racemosa*; *Sorbus americana*; *Diervilla lonicera*; *Gnaphalium obtusifolium*; &c. - **l. plant**: *Kalanchoë pinnata* - **l. root** (**plant**): *Senecio aureus* - **sweet** (**-scented**) **l. everlasting**: *Gnaphalium*.

ligand: complexing gp. usually the source from which electrons are donated, often involving Fe or Si.

lighter wood (seUS): "fat" pine heart wood naturally impregnated with oleoresin & u. for starting fires, as torches, &c. (often pron. "leider" ["light'ard"]) best source = longleaf pine; "lighter knots."

lightwood: 1) *Amyris balsamifera* 2) lighter wood.

lignaloe(s): *Bursera glabrifolia, B. penicillata* (< lignum aloe; may also represent *Aquilaria agallocha* wd.) - **l. oil**: *v.* bois de rose.

lignans: natural phenolic resins closely related to lignin; occur often in pl. resins (in wd.); dimers of C_6H_3 units linked at p-position; widespread in pl. kingdom (resins of lvs., heart wd., sometimes as glycosides; ex. guaiaretic acid, cubebin, angelicin, podophyllotoxin).

lignin: amorphous subst. with glue-like consistency wh. by impregnating the cellulose of the cell wall of woody pls. confers the mechanical stiffness & rigidity needed to keep them erect; represents natural binding material of wd. products; polymerized coniferyl alc.; u. to make vanillin, vanillic acid, DMSO; imp. fr. Oslo, Nor; synthesized 1954.

lignite: brown coal: "wood coal"; intermediate prod. between peat & bituminous coal; source of **l. tar** - **l. wax**: high m. p. waxy prod. obt. by extn. fr. lignite coal (brown coal: an inferior type of coal); peats, etc.; of varying comp., usually esters of montanic acid; u. in place of carnauba, for mfr. candles, insulators, polishes, waterproof paints & varnishes, phonograph records, etc.; grades: crude (dark brown), bleached (white or nearly so); *cf.* Ceresin (prod. Ge fr. lignite; potential prod. areas: Okla, Tex [US]) - **ligno** (It): legno - **lignone**: lignite pitch; mfr. fr. sulfite waste liquors (of s. paper pulp indy.).

Lignum: (L.) wood - **L. Baphiae**: cam wood fr. *Baphia nitida* - **L. Braziliense:** L. Brisilium: Brazil wood fr. *Caesalpinia brasiliensis, C. sappan* - **L. Caeruleum**: logwood - **L. Casealpiniae**: peach wood fr. *Casealpinia echinata*- **L. Campechense**: **L. Campechianum**: logwood - **L. Fernambuci**: Brazil wd. - **L. Guaiaci**: L. Vitae (1) - **L. Juniperi**: wd. of *Juniperus communis* - **L. Nephriticum**: *Eysenhardtia amorphoides*; *Pterocarpus indicus* (306) **L. Pichi**: pichi - **L. Quassiae**: quassia wd. - **L. Quebracho colorado**: *Schinopsis* - **L. Rhois cotini**: young fustic fr. *Cotinus coggygria* - **L. Sanctum**: guaiac(um) wd. - **L. Santae Marthae**: L. Caesalpiniae - **L. Sappan**: European sappan fr. *Caesalpinia sappan* - **L. Tupelo**: wd. of *Nyssa sylvatica* - **L. Vitae** (L.): 1) guaiac wd., L. Sanctum 2) (Par): *Dasyphyllum dicanthoides* 3) (Par): *Bulnesia sarmienti.*

ligroin: solvent naphtha, b. pt. 30-145 degrees (close to petroleum benzine); sometimes equated to petroleum ether; solvent, extractive agt.; fuel.

ligulate floret: l. flower: small fl. with strap-shaped (ligulate) corolla, as found in Fam. Comp.; ex. *Arnica, Matricaria.*

Ligusticum acutilobum Sieb. & Zucc. (Umb) (Jap, Chin): *Angelica acutiloba* - **L. canadense** (L) Britt. (e&cUS): angelico; mondo; lovage; Southern masterwort rt.; u. med. as arom. - **L. canbyi** J. Coulter et Rose (wNA, Pac coast): still u. by the Flathead Indians (wWash [US]) for colds & sore throats - **L. filicinum** S. Wats. (wUS intermountain region): fernleaf ligusticum; fernleaf lovage; rt. u. to prep. cough remedy (Paiute Indians) (265) - **L. jeholense** Nak. & Kitag. (eAs): rhiz. & rt. administ. to relieve colds associated with headache; diarrh.; also appl. to skin dis. & boils (225) - **L. levisticum** L. = *Levisticum officinale* - **L. monnieri** Calest. = *Cnidium m.* - **L. mutellina** (L) Crantz (c&seEu - mts): alpine spicknel; lf. decoc. u. as stom.; lvs. in place of parsley; dr. hb. u. to make a tea - **L. pilgerianum** Wolff (wChin): very arom. pl.; valued med. (55) - **L. porteri** Coult. & Rose (Wyo to NM, Ariz, [US], nMex): (Colo) cough root (plant); oshá root (NA Indians); chuchupate; rts. (also lvs. st.) coll in wUS mts.; u. to treat coughs, colds, bronchitis, asthma, pulmonitis, pyrosis, indig., gastric dis., catarrhs. - **L. scoticum** L. (Euras, natd. eCana, neUS): (Scotch) lovage; persil (de mer) (Qu [Cana]); Alexanders; blanched pl. u. as potherb (coastal Scot, Hebrides); salad with olive oil (Qu); lf. stalks substit. for celery; young shoots candied like angelica; seasoning, esp. for fish - **L. sinense** Oliv. (eAs): rt. soaked in spirits 2 wks. & ext. drunk to relieve gout; also emmen. anodyne, sed. for headache; in flatulence - **L. striatum** DC (Chin): good for difficulties of menstr., circulation, headache, rheum.

Ligustrum japonicum Thunb. (Ole.) (eAs): (Japanese) privet; frts. u. ton. (esp. sexual) (Kor) or roborant (Chin, Jap); for tinnitus, irritability, insomnia, aid to bowel movements; roasted sds. u. as coffee substit.; boiled lvs. for swellings - **L. lucidum** Ait. f. (Chin): white wax tree; glossy privet; Chin p.; bk. & lvs. decoc. for diaphor.; lvs. in colds, congestions, dizziness, appl. to abscesses; frt. ton. in phthisis, edema; to "promote" longevity; wax prod. on branches by insect - **L. robustum** Bl. (seAs, esp. Bengal): wax tree; sts. u. to combat toothache & to treat sores (Vietnam); u. to aromatize tea (Java); c. alks, NBP & tannin; furnishes a useful hard wd. - **L. sinense** Lour. (Chin): Chin privet; decoc. of flg. branches remedy for oral inflammation - **L. vulgare** L. (Euras, nAf, commonly cult. seUS; readily natd.): European (or common) privet; primp; lvs. u. astr. stypt.; folk rem.; frts. dyestuff (illegal wine coloring); lvs. & fls. u. for treating oral & throat infections (Af); pl. long cult. as ornam. hedge.

lijnolie (Dutch): linseed oil.

likari: *Ocotea caudata.*

lilac (, common): *Syringa vulgaris.*

Liliaceae: Lily Fam.; ca. 4,500 spp., mostly of perenn. hbs.; st. usually a rhiz. or bulb or corm, sometimes aerial; fls. showy, with usually superior ovary, stamens usually 6 but occas. 3; of greatest importance as ornam.; a few genera furnish fibers (*Yucca, Sansevieria*); several med. (*Aloe, Colchicum, Veratrum*); Fam. is close to Amaryllidaceae & Juncaceae.

lilacine: terpineol.

Liliales: lily order; Liliiflorae, comprises Liliaceae & 8 other fams.

Lilie (Ge): lily, *Lilium.*

Lilium brownii F. E. Brown (Lili) (sChin, Indochin): lvs. kept in oil or alc. popular med. for treating cuts & bruises (Vietnam); bulbs eaten as veg. with meat; bulb scales sed. & ton., esp. for cough & lung dis.; in deafness, earache; carm. expect. - **L. bulbiferum** L. (s&cEu): orange lily; fire l.; fls. u. for lung dis. (folk med.) - **L. candidum** L. (wAs, esp. Syria, cult. Eu, Am): annunciation (or Madonna or Bourbon) lily; fresh & dr. fls. u. med.; fl. pollen u. epilepsy (Eu); fls. u. to prepare VO (concrete; absolute) (sFr) u. in perfumes - **L. columbianum** Hanson (sBC [Cana] to nCalif [US]): Columbia (or [wild] tiger) lily; cooked bulbs eaten, as seasoning (wNA Indians, as the Quinault tribe) - **L. convallium** Tourn. = *Convallaria majalis* - **L. lancifolium** Thunb. (L. tigrinum Ker-Gawl.) (Chin, Jap, cult. Eu, NA): tiger lily; spotted l.; bulb scales u. ton. for pulm. dis.; expect. for cough; in Eu homeopath. u. fresh fls. for female genital dis.; cult. for bulbs (Jap) - **L. longiflorum** Thunb. var. *eximium* Baker (Jap): Easter lily; popular ornam.; fls. u. for VO prepn.; *cf. L. candidum* - **L. martagon** L. (Euras): Martagon lily; turk's cap l.; bulbs eaten dr. or with milk (USSR, Mongol, Siber); in med. (266) - **L. michauxii** Poir. (seUS): Carolina lily; ornam. hb. - **L. philadelphicum** L. (eNA s to NCar): (wild) orange-red lily; wood l.; rts. u. for swellings, bruises, cough, fever, TB (1474) - **L. wallichianum** Schult. f. (Ind): dr. bulb scales u. demul. like salep in bronch. dis.

lily: *Lilium* spp.; less commonly appl. to other genera of Fam. Liliaceae & Amaryllidaceae - **alligator l.**: *Hymenocallis palmeri* - **annunciation l.**: 1) Bermuda l. 2) *Lilium candidum* - **avelanche l.**: glacier l. - **Bermuda l.**: *L. longiflorum* v. *eximium* - **blood l.**: *Haemanthus* spp., esp. *H. coccineus* (sAf) - **bluebeard l.**: *Clintonia borealis* - **Bourbon l.**: common l. - **butterfly l.**: *Hedychium coronarium* - **Calla l.**: *Zantedeschia aethiopica* - **checkered l.**: *Fritillaria meleagris* - **Chinese sacred l.**: *Narcissus*, esp. *N. tazetta* v. *orientalis* - **climbing l.**: *Gloriosa,* esp. *G. superba* - **common l.**: *L. candidum* - **day l.**: 1) *Hemerocallis flava, H. fulva* 2) common l. - **Easter l.**: 1) *Lilium longiflorum* var. *eximium* 2) *L. candidum* 3) *Zephyranthes atamasco* 4) (Cana, eUS): *Erythronium* spp. - **false l. of the valley**: *Maianthemum canadense* - **l. family**: Liliaceae - **fawn l.**: *v.* fawn-lily - **flag l.**: *Iris versicolor* - **ginger l.**: *Hedychium* spp. - **glacier l.**: *Erythronium* spp. - **gloriosa l.**: **Glory (Malabar) l.**: *Gloriosa superba* - **green l.**: *Schoenocaulon officinale* - **ground l.**: *Trillium* - **lemon l.** (BC [Cana]): orange l. - **liver l.**: *Iris versicolor* - **Madonna l.**: common l. - **Malabar glory l.**: glory M. l. - **Mariposa l.**: *Calochortus* spp. - **l. of the valley (root)**: *Convallaria majalis* - **orange l.** (BC [Cana]): *Hemerocallis* spp.; *L. bulbiferum* - **palm l.** (Australas): *Cordyline terminalis* - **park l.**: l. of the valley - **pineapple l.**: *Eucomis undulata* - **pinecone l.**: *Zingiber zerumbet* - **plantain l.**: *Hosta* spp. - **pond l., white**: *Nymphaea odorata* - **red l.**: *Lilium bulbiferum, L. philadelphicum* - **l. root, white**: of *Nymphaea odorata* - **royal l.** : *Lilian regale (Lilium reale)* - **sand l.**: *Mentzelia nuda* var. *stricta* - **sego l.**: *Calochortus nuttallii* - **shell l.**: *Alpinia tomentosa* - **southern r. l.**: *Lilium catesbaei* - **southern swamp l.**: *Crinum americanum* - **spider l.**: *Hymenocallis* spp. - **superb l.**: Glory Malabar l. - **toad l.**: white pond l. - **trout l.**: *Erythronium americanum* - **turf l.**: *Liriope* spp. - **valley l.**: l. of the valley - **water l.**: *v.* water-lily - **white l.**: 1) *Nymphaea alba* 2) *Lilium candidum* 3) 5 other genera, incl. *Trillium* - **white pond l. (root)**: *Nymphaea odorata, N. polysepalum* - **wood l.**: *L. philadelphicum*; *Clintonia* spp. - **yellow pond l.**: *Nuphar advena.*

lima: 1) v. under bean 2) (Sp, Por): lime 3) (Br): *Citrus limetta* - **l. bark**: *Cinchona micrantha.*

Limacia oblonga Hook. f. & Thoms. (Menisperm.) (Mal): rt. ext. appl. to sore eyes; c. berberine - **L. triandra** Miers (Indochin): ground rt. decoc. drunk for quotidian fever (Mal); bath fr. rt. in water u. for fever.

Limaciopsis loangensis Engl. (1) (Menisperm) (cAf): lf. & st. decoc. u. for stomach dis.

limão (Po): lemon - **limão (comum)**: **l. bravo** (Por, Br): *Siparuna* spp.

limãosinho (Por): Lignum Limosinae: 1) limousine wood; wd. of *Quercus ilex* 2) (Br): small lemon. (lit.); *Polygala brasiliensis.*

Limatura Ferri (L.): iron filings; u. folk med.

limber bush: *Jatropha spathulata* (SC long slender sts. can be tied in knots; various articles made fr. the very flexible sts.; "rubber plant" (Tex [US]).

lime: 1) *Citrus aurantifolia*, lime-tree 2) calcium oxide, quicklime 3) *Tilia* spp. (**lime tree**) - **acid l.**: *Citrus aurantifolia* - **finger l.**: *Microcitrus australasica* - **l. flour**: hydrated quick lime, slaked l. - **l. flowers** (USSR): *Tilia* spp. - **kãghzi l.**: *Citrus aurantifolia* - **key l.: Mexican l.**: acid l., the VO from the peel of this (*C. aurantifolia*) is **l. oil** (WI, Mex, Ind), usually hand-pressed fr. peel; formerly frt. u. in iced tea, &c. - **Persian l.**: similar to Tahiti l. - **slaked l.**: calcium hydroxide - **sour l.**: acid l. - **sweet l.**: *C. limetta* - **Tahiti l.**: type of acid l. looking more like small-fruited lemon, nearly seedless (cult. Fla, Calif [US]) - **West Indian l.**: acid l.

limeberry: *Triphasia trifolia* (VO) - **limero** (Sp): lime tree - **limestone**: calcium carbonate; u. building material; source of lime.

limequat: hybrid of kumquat X acid lime.

limetta: limette: 1) *Citrus limetta* 2) *C. aurantifolia* (C. medica var. acida).

limette, Southern European: *Citrus limetta*, Spanish orange - **West Indian l.**: *C. medica* v. *acida* ("lime").

limettin: citropten.

limey (US slang): British seaman; SC ration of lime or l. juice long served daily on board ship as prophylactic against rickets; facetious term also, generally for Englishman.

Limnanthemum S. G. Gmel. = *Nymphoides*.

Limnanthes alba Hartw. ex Benth. (Limnanth.) (Calif [US]): white meadow-foam; sds. c. 20-30% FO; potential oil sd. crop in Ore [US] - **L. douglasii** R. Br. (Pacific US; Calif): (common) meadow-foam; sds. yield highly unsatd. fatty acids; u. as raw material in lubricants, waxes, rubber additives, intermediates, plastics, &c.

Limnocitrus littoralis (Miq.) Swingle (Pleiospermium littorale Tan.) (Rut.) (Indochin): rts. u. in treating a skin dis. like ground itch or scabies found on the soles of the feet (natives); frt. edible (—> *Pleiospermum*).

limnology: sci. study of fresh waters (<limne [Gr]: marsh).

Limnophila aromatica Merr. (L. conferta Benth.) (Scrophulari) (seAs): "amba-wila" (SriL); remedy in menstrual pain & dis.; to treat intoxication (Taiwan); lvs. appl. as poultice to sore legs; rt. & lf. decoc. expect. antipyr. (Mal); chewed with betel (camphor flavor) (SriL) - **L. geoffrayi** Bonati (Thailand, Camb, Laos, Vietnam): strong coumarin odor; u. to perfume soups & other foods - **L. indica** Druce (eAs, Oceania): lf. inf. u. for dys. & dyspepsia - **L. rugosa** Merr. (seAs, Polynesia+): infused lvs. u. as diur. digest. ton. (PI); u. as arom. in cooking & to perfume the hair (PI); lf. decoc. u. in impotence - **L. villosa** Bl. (Mal): lvs. appl. to sores, ulcers, fever.

Limnophyton obtusifolium Miq. (monotypic) (Alismat.) (tropAf): ashes rubbed into cuts made in painful parts (Tanz); ash u. as source of NaCl (Liberia); fls. sl. arom.

limo (It): mud (sometimes as medicinal?).

limon (Fr, Sp): 1) lemon 2 (Ve): citron (also **l. criollo**) 3) limonin (*qv.*) - **l. dulce** (Sp): sweet lime - **l. real** (Salv): lemon.

limoncillo (Mex): *Triphasia trifolia.*

Limone: 1) (It): lemon 2) (Ge): citron, cedrat (saure L. = lemon) (suesse L. = lime).

limonene: terpene found in VO of lemon, orange, etc.

Limonenschale (Ge): lemon peel.

limonero (Sp): lemon tree.

Limonia acidissima L. (L. crenulata Roxb.) (monotypic) (Rut) (Ind, Jap+): elephant apple; wood a.; musk deer plant; tree source of a gum; wd. useful; frt. edible, u. as soap substit. (Jap); lvs. & fls. u. as stom.; bk. in perfumes (Thail).

limonic acid: citric a.

limonin: bitter princ. (earlier called limon) of sds. of orange; *Evodia* spp., &c.; heterocyclic lactone.

Limonis Cortex (L): lemon peel - **L. Succus** (L): lemon juice.

Limonium brasiliense (Boiss.) Ktze. (Statice brasiliensis Boiss.) (Plumbagin.) (Br, Ch, Ur, Arg): (Brazil) sea lavender; guaycuru; baycuru (Br); rt. of this halophyte c. baycurin; u. in bronchitis, pharyngalgia; glandular swellings; astr. for hemorrhoids; diarrh., dys., intest. hemorrhage; in edema; good vehicle for iodides, Hg, As; vaginal washes; may prod. uter. contractions; fls. ext. u. in Br - **L. carolinianum** (Walt.) Britt. (L. brasiliense Small; Statice brasiliensis A. Gray) (e&seUS, Mex, on sea coast): (Brazil) sea lavender; marsh rosemary; thrift; sea (side) thrift; rt. u. astr. for spongy gums (gargle), colds, sore throat, canker sores, in dys. diarrh. ulcers, &c.; hemostatic (SCar [US]); this sp. may have at times been u. as "Lobelia" in Thomsonian med. (267) - **L. gmelinii** O. Ktze. (Hung to Siber); rts. (Kermack rt.) c. tannin, u. in diarrh. & in gargles; tan. - **L. guaicuru** (Mol.) Gunckel (SA): natives u. as gargle, astr. - **L. latifolium** O. Ktze. (eEu, cAs, Caucasus): rts. (Kermack rt.) u. as astr., ton. styptic, tan. - **L. sinense** O. Ktze. (Chin): antipyr. hemostat. (Taiwan) - **L. suworowii** Ktze. (wTurkestan; cultd. NA): sea lavender; ornam. ann. hb.- **L. tetragonum** Bullock (Kor, Manch, Jap): rts. c. tannin; u. astr. - **L. vulgare** Mill. (Statice limonium L.) (wEu, NA - Atl coasts): European marsh rosemary; (common) sea lavender; hb. (Herba Statice) and rt. (Radix Behen rubri) u. as ton., astr., stypt. in diarrh., hemorrhages (TB) (1739), &c.; diur.

limonoids: bitter tasting compds. related to limonin; occ. in *Citrus* spp. (rinds, sds., juice of frt.); also in other Rut., Simaroub., Meli. spp.; some 100 known; act to stunt growth of cotton bollworm.

limousin(e) bark: Cortex Quercus ilex (**l. oak**) - **l. wood**: of *Q. ilex*.

limpiavista (NM [US], Sp): eye lotion.

Limulus polyphemus (Limulidae - Xiphosura - Arthropoda) (wAtl Ocean): horseshoe crab; king c.; blood blue, u. to prep. Limulus amebocyte lysate (LAL); shell u. med. as aphrodisiac (Mal), in mal.; as cattle food, manure (US); u. as diagnostic reagent: forms clot in presence of small amt. bacterial endotoxin, esp. gm. neg. m. o. (most sensitive test) & pyrogens.

lin: 1) (sUS): *Tilia* spp. (fr. linden) 2) (F-f.): linseed - **white l.**: *Tilia alba*.

Linaceae: Flax (or Linseed) Fam.; hbs. but sometimes sl. woody; fls. perfect, regular, carried in terminal racemes or corymbs; 5 or sometimes 4 sepals; hypogynous; petals of same numbers as & alternate with sepals; stamens likewise; ovary of 2-5 carpels; incl. genera *Linum*, *Reinwardtia*, *Hugonia*.

linaloe oil, Cayene: VO fr. wd. of *Ocotea caudata* - **Indian l. oil**: VO fr. *Bursera delpechiana*: rose-like aroma due to linalool; obt. fr. trees grown in Ind (Bangalore) (orign. fr. Mex) (**Mexican l. o.**: same fr. Mex).

linalol: linalool: linayl alcohol: aliphatic tertiary alc. (SC found in linaloe oil); also in sassafras; Ceylon cinnamon, orange fl., lavender, coriander, bergamot, jasmine, rose, thyme, & other VO; has rose-like aroma; u. perfumes in place of bergamot & Fr lavender oils - **linalyl acetate**: bergamol; this ester of linalool found in bergamot & lavender oils; much u. in perfume, cosmetics.

Linaria canadensis (L) Dumont (Scrophulari) (Mex, NA): (wild) toad flax; fls. lavender blue; hb. us. as diur. lax. in hemorrhoids (ointment fr. fresh hb.) - **L. cymbalaria** Tourn. ex. Mill. = *Cymbalaria muralis* - **L. vulgaris** Mill. (Antirrhinum linaria L.) (Euras, natd. NA): (yellow) toad flax; common t. f.; butter & eggs; hb. c. linarin, pectolinarin (flavonoid glycoside), aureusin (glycoside); peganine; formerly u. similarly to prec.; purg. in edema; insect poison; in homeopath. med.

linaza (Sp): linseed.

Lincomycin: antibiotic fr. *Streptomyces lincolnensis* Mason & al. var. *l.* (named after Lincoln, Neb [US], where first isolated) (*cf.* str. to *S. caelestis*); thioglycoside linked with propylhygric acid; u. (as HCl) antibacter. against gm.+ organisms; 1962-; u. in inf. resistant to penicillin & eyrthromycin.

lincture: electuary.

Lind, James (fl. -1747-): ship surgeon who first used oranges & lemons for preventing scurvy.

linda mañana (Cu) (lit. fine morning): *Hibiscus cannabinus*.

Lindackera Sieber = *Capparis*.

Lindackeria Presl (Flacourti) (SA, Af.): sds. yield FO similar to that of chaulmoogra; cf. *Oncoba* (do not confuse with *Lindackera* Sieber = *Capparis).*

Linde (Ge): Linden (768); *Tollia* spp. - **Lindenbluete** (Ge): **Lindenblumen**: lime fls. (*Tilia* spp.).

linden family: Tiliaceae - **l. (flowers)**: fr. *Tilia platyphyllos* - **l. blossm tea**: granulated l. fls

Lindenbergia indica O. Ktze. (Scrophulari) (Ind, Pak): juice u. in chronic bronchitis; with coriander oil appl. to skin eruptions.

Lindera benzoin Blume (Benzoin aestivale Nees) (Laur.) (seCana, e&cUS): spice bush; s. wood; Benjamin bush; shrub 2-5 m with alternate simple scented lvs.; frt. red drupes, obovoid; bk. & twigs (wd.) u. stim. diaph. antipyr. anthelm., as tea substit. (Ky [US]?); frt. condim. carm.; spicewood tea u. to "thicken blood" (like sassafras t.) - **L. caudata** Benth. (Chin): lf. decoc. drunk to relieve pain fr. mechan. injury - **L. communis** Hemsl. (Chin): u. for rheum. - **L. myrrha** Merr. (eAs): rt. inf. u. against various dis. (diur. emmen. anthelm., antisep.); VO fr. berries u. for scabies, pustules - **L. obtusiloba** Bl. (Korea): small branches (prepns.) u. to treat abdom. pain, cough, puerperal fever, as liver tonic - **L. sericea** Bl. (Chin, Jap): silky spice bush; lvs. & young twigs source of VO, kuromoji oil; bk. & rt. u. as stypt. as wash in skin dis. - **L. strychnifolia** Vill. (warm/temp, Chin, Jap): rts. c. linderoxide, linderane, linderene (azulogenic furanosesquiterpene), &c.; rt. u. for hernia, chest congestion & pain; in stroke; carm. for stomach spasms (Chin) - **L. umbellata** Thunb. (Jap, Chin): umbelled spice bush; a source of kuromoji oil; twigs. inf. u. in gastritis, flavor.

Lindernia ciliata Pennell (Scrophulari) (s&eAs): u. for menorrhagia (Taiwan); juice fr. crushed lvs. u. after childbirth - **L. crustacea** F. Muell. (trop

As+): u. for cleansing & healing power in herpes-like dis.; poultices for sores fr. "thicket-lice," infected sores, &c. (Indonesia).

linea fisural (Sp): fissural line seen in Leg. sds. (ex. *Calliandra* sp., in sds. of mature frts.), forms a fine bright line in the form of a horseshoe or completely closed; found at different stages of maturity (307).

Lineal (Ge): ruler.

ling (common): 1) **l. berry**: *Calluna vulgaris, Vaccinium vitis-idaea, Empetrum nigrum* 2) **ling-cod**:*Ophiodon* sp., large salt-water fish: a food delicacy.

lingen: lingon (Swed) berries: *Vaccinium vitis-idaea* (frt.).

lingomene (Por, Af): *Voandzeia subterranea.*

lingon(g): **lingos**: ling.

lingua franca: common language in Med Sea ports; combn. of It, Sp, Fr, Gr, Arab.

linguets: tablets intended for slow absorption fr. oral tissues (as under tongue); commonly hormones are so administered.

liniment, Canada: Liniment Opium Comp.

Linnaea borealis L. (l) (Caprifoli.) (Euras, Alask [US]): twin flower; u. in med. (Scand) (Swed & Dan Ph); ornam. subshrub.

lino (Sp, It): flax - **linseed**: sd. of *Linum usitatissimum* - **l. cake**: material left over after ground whole flax sd. (**l. meal**) has been expressed - **lint flowers**: linseed.

Linociera nilotica Oliv. (Ole.) (Tanz): rts. pounded & taken in water for enlarged spleen (in mal. in children); ornam. shrub - **L. oblongifolia** Koord. (Java): fresh frts. u. in combn. medicines (268).

linoleic (linolic) acid: essential fatty acid with 2 double bonds; found in linseed oil+ (2332); u. in atherosclerosis; hypercholesterolemia; skin dis. (linolenic acid u. similarly) - **linolenic acid**: essential fatty acid with 3 double bonds; in linseed oil, &c. u. in eczema.

linoleum: floor covering prepd. fr. linseed oil & pd. cork by treatment with SCl (sulfur chloride) and heat (<lin[um] + oleum).

Linostoma decandrum (Roxb.) Wall. ex Endlich. (Thymelae.) (Ind, seAs): dr. rts. u. as energetic purg. (Vietnam) - **L. pauciflora** Griff. (Thymelae.) (Burma): toxic pl. but u. medic. (308).

linoxyn: hardened (solidified) oxidized linseed oil; u. in mfr. linoleum & oil paints.

linseed: crushed l.: *Linumuti* u. as poultice (c. FO), shampoos, & wave-setting lotion.

linthead (Ala [US] slang): worker in cotton mill.

Linum: 1) genus (Lin.) (temp regions, mostly of Med area) of hbs. & subshrubs; flax 2) (specif.): dr. ripe sd. of *L. usitatissimum* - **L. catharticum** L. (Eurars): purging (or mountain or fairy) flax; hb. c. oleoresin, tannin, glycosides, linin (bitter princ.); u. as lax. (in form of tea), diur.; in folk med. as anthelm., in liver dis., &c. - **L. lewisii** Pursh (L. perenne subsp. l. Hultén) (w&cNA): western blue flax; sds. roasted, dried & ground for food use; st. fibers u. (Indians) - **L. perenne** L. (Euras): Siberian flax; perennial fl.; sds. emoll. (Lahul) (Chin): fresh hb. u. as decoc. in rheum. colds, coughs, edema (Eng) - **L. strictum** L. (Euras): (bab) basaut (Punjabi): sds. u. emoll. (Sp) - **L. usitatissimum** L. (wAs; cult. US, Cana, Arg, USSR+): linseed; (common) flax (seed); sd. c. FO (30%), protein, mucil., u. dem. expect. (flaxseed tea), lax. for inflamm., urin. passages; diur. in diabetic teas; antibiotic prop. (1827); sometimes u. as food (meal or grd. whole sd.) (ex. Roman meal, cereal meal), in chicken feeds, stockfeed; for prodn. FO (linseed oil), c. linolic & linolenic acids as glyceryl esters; the raw oil u. in cataplasms for pain, emoll.; fodders, veter. lax., in prepn. soaps, burn liniments (Carron oil), salves, hair prepns., depilatories; for poison ivy (neAla [US]); in hypercholesterolemia; in mfr. furniture polishes, printer's inks, as paint & varnish base (since a good drying oil), using the "boiled" oil (toxic) (2759); boiled oil u. also in mfr. oilcloth, imitation leathers, oilskin coats, rubber substit. &c.; fiber u. as important textile in prepn. linen (thread, cloth, twine, lace); found in ancient Eg mummy cloths; Belgian & Irish laces and linens are famous (boiled oil is toxic [c. Pb, Mn, &c., as drier] and must never be used internally!!) (Ind biggest exporter of l. oil).

lion: Leo (L): *Panthera leo* L. (Felidae) (tropAf, mostly; wIndia) (few) (P. leo persica); fat u. in Unani med. as sex stim. (by inunction); in rheum. (ointments); also u. bile; hide; previously, lion occ. in sEu (Gr); sometimes several hundreds u. in a single performance in ancient Roman Colosseum; still an important show (circus) animal.

lion's beard: *Anemone patens* (in frt.) - **l. ear**: *Leonotis leonurus* - **l. e. (of the Andes)**: *Culcitium argenteum* - **l. foot**: *Prenanthes alba, P. serpentaria, P. altissima* - **l. head**: *Dracocephalum* spp. - **l. mouth**: Digitalis - **l. tooth**: *Taraxacum.*

liothyronine: triliodothyronine; T-3; occ. in plasma and thyroid gland; considered the circulating thyroid hormone; u. as substit. therapy in thyroid deficiency (simple goiter, myxedema, &c.); reduces cholesterol level of bloodstream; mostly u. as Na salt; with 5X activity of L-thyroxine (Cytomel) (racemic form: Trionine).

Liotrix tablets (USP): mixt. of levothyroxine sodium and liothyronine sodium.

lipase (< lipos, Gr, fat): lipolytic enzyme; class of fat-splitting enzymes; occurs in pancreas (pancreatic lipase = steapsin), gastric juice, small intest., fatty tiss., castor bean (c. b. lipase), & other sds.

lipid(e): **lipin**: lipoid: fat or fat-like substance (sol. in various org. solvents not in water); found in cells of pls. & animals as metabolic prods.; included are fats —> fatty acids, waxes, lecithins (phosphatides), glycolipids, sterols, carotenoids, fat-sol. vitamins; occur in blood as lipoproteins (1621).

Lipiodol™: iodized oil; v. *Papaver somniferum*.

lipoclastic: lipolytic: descriptive of enzymes wh. hydrolyze FO to produce fatty acids & glycerin; ex. lipases.

lipio(n)ic acid, alpha: thioctic acid: growth factor essential to existence of many microorganioms; as carboxy acid with 2 S atoms in mol.; occ. as amide in pl. and animal tissues (ex. in vit. B complex); u. to treat liver dis., *Amanita* poisoning; heavy metal poisoning.

lipoprotein: conjugated proteins in wh. a simple protein is combined with a fatty acid - **high-density l.** (HDLP): **alpha-l.**: important in health since less risk of atherosclerosis for individual - **low-density l.** (LDLP): **beta-l.**: deposits cholesterol in arterial wall, more risk of atherosclerosis - **very low-density l.** (VLDL[P]): atherosclerosis promoted in presence of satd. fats.

lipotropin: *lipotropic hormone*: ant. pituitary hormone wh. stim. breakdown of fats —> f. acids; promotes formation of endorphins.

Lippia L. (Verben.) (subtrop & tropAm, a few in trop OW): 252+ taxa (spp. vars., & forms): hbs. u. as popular teas; as flavor, most orégano in US is fr. *Lippia* spp. (ex. *L. affinis* Schauer), some fr. *Origanum* (fr. Eu) (309); many arom. & flavor uses, ex. to deodorize corpses; u. as flea repellant - **L. abyssinica** (Otto & Dietr.) Cuf. (e&cAf): (Gambian) tea bush; shrub; dr. lf. inf. (decoc.) u. as beverage ("bush tea"), bathing for fever, as antipyr., diaph. in colds, lax. in colic, coryza, stim. in VD, dys. vermifuge; VO c. carvone, camphor - **L. affinis** Schauer (Br, Bol): orégano (de burro); camara; hb. u. as tea for dys. (Minas Giraes, Br) - **L. alba** (Mill.) N. E. Brown (Tex [US], WI, Mex, CA, SA, s to Arg; natd. As): white (or bush) lippia; wild sage; palisado; salvia (Arg); s. betonica (Mex); "poleo" (PR) (105 common names recorded by H. N. Moldenke); lf. decoc. or inf. (tea) expect. diaph. u. for flu, colds, cough (Guat, Trin) (310); sore throat; asthma (Ve); stom. in gastralgia, diarrh. dys. (Pan); antispasm. emmen., nervine; med. baths (PR); for hemorrhoids (exter.); u. as vegetable (Ind); like sage in cooking; as beverage (té del pais); VO c. piperitone, u. to make synth. menthol; cult. as medl. (Bayano Cuna, Pan; Tex [US]; Jam); forma *intermedia* Moldenke (LatAm): hierba luisa; u. for stomachache (CR), medl. tea (Haiti), for fever & grippe (Ec), forma *macrophylla* Mold.; cult. & u. to produce tea with orange (bon pom tea) (Haiti) (in hotels); var. *carterae* Moldenke (Mex): "salvia real"; u. for tea & in resp. dis.; var. *globiflora* (L'Hér.) Mold. (L. geminata HBK) (LatinAm): té del pais; pl. u. "for blood" (inf.), stim. beverage (Bol); tea & as antispasm. stom. carm. & to counteract effect of purgatives (Br); emmen. diaphor, cold rem. (Ur); u. med. (Par, Arg; cult. [Br]) - **L. burtonii** J. G. Baker (trop Af, Nigeria, &c.): u. in eye dis. - **L. callicarpaefolia** HBK (Mex): fls. u. for tea for inflamm. (Mex); inf. of lvs. & fls. stim.; estrogenic (*cf. Salvia officin.*) - **L. carviodora** Meikle var. *minor* Meikle (Kenya, Somalia): lvs. u. as substit. for tea (Somal) - **L. citriodora** Kunth = *Aloysia triphylla* - **L. curtisiana** Moldenke (Mex): "hierba de la mula"; u. as condiment at meals (Durango) - **L. dauensis** (Chiov.) Chiov. (eAf): "rehan"; VO used - **L. deserticola** R. A. Phil. = *Acantholippia d.* - **L. dulcis** Trev. (var. *mexicana* Hegi) = *Phyla scaberrima* - **L. geminata** HBK = *L. alba* - **L. graveolens** HBK (Tex [US], Mex, Nic) (cult. Mex): (Mexican) orégano; redbush lippia; hierba dulce (de México); lvs. u. for flavoring foods (CR); med. (CA); pl. decoc. stim. carm. in flatulence, meteorism, cholera morbus, muscular debility, as expect.; in gastralgia (Oaxaca); syrup fr. lvs. u. in coughs (Mex), colds, dys. diabetes; diarrh. (Guerrero); emmen. (Yucatan); condiment; furnishes part of the orégano of commerce; cult. ornam. - **L. grisebachiana** Moldenke (Arg): palo cenarillo; u. as blood "purifier" - **L. hastata** Jensen = *L. vernonoides* var. *attenuata* (Mart.) Mold. (Br, Bo) (market offering) - **L. hastatula** (*sic*) Hieron. = L. hastulata Griseb. = Acantholippia hastulata Griseb. - **L. integrifolia** (Griseb) Hieron. (Arg): yerba del Inca; u. as cardiotonic - **L. javanica** (Burm-f.) Spreng. (Ethiopia to sAf): linuñganuñga; lvs. u. to make tea for mal. ("fever tea"), coughs & colds; dys.; as anthelm.; colic (chewed); lvs. u. for sweeping, to repel fleas (269); smoked for resp. dis., in measles, urticaria, nasal hemorrhage, anthrax; gangrenous proctitis - **L. ligustrina** (Lag.) Britt. = *Junellia l.* - **L. mexicana** Cav. = *L. callicarpaefolia* - **L. micromera** Schauer (WI, Co, Ve): "false thyme"; Spanish t.; inf. u. for flu, baths for cough (Trin); asthma (Ve);

condiment in soups, &c. (*cf.* orégano) (Ve); to flavor soups & meatloaf (Haw [US]) (like sage); cult.; var. *helleri* (WI, CA): "mejorana" (PR) - **L. microphylla** Cham. (Guyana, Br): frt. u. as stim. & ton. - **L. multiflora** Moldenke (wAf): "tea bush"; ebolo; lvs. inf. u. as native tea, for fever; rt. decoc. u. as emet. in gastralgia, urticaria, colds; cult. for perfume oil (Congo) - **L. nodiflora** Rich. - *Phyla n.* - **L. oatesii** Rolfe (tropAf): "um susani"; branches u. to repel mosquitoes & snakes (Zambia) - **L. obovata** Sessé & Moç. (Mex): u. as condim. like Origanum. - **L. origanoides** HBK (Ve, Co, Br): oreganillo; "chara-ceur" (Roraima, Ve), lf. decoc. with sugar u. in colds; asthma or with *guarapo* (fermented cane liquor) in colds; lf. inf. for gastric dis.; lvs. as condim. & flavor for meat & fish dishes; cult. Co - **L. palmeri** S. Wats. (nwMex): "origano"; u. condim. ("orégano" of commerce); pot herb; forma *spicata* (Rose) Mold.: orégano; u. as spice - **L. pringlei** Briq. (Mex): "baca(r)ton"; a tree; sap u. in dentalgia (by chewing bark?) (Mex); lvs. decoc. u. in cough (mentholatum) for sores, bruises, headache, &c. - **L. pseudo-thea** (A. St.-Hil.) Schauer (Lantana p. L.) (Br, Ch): cha de pedestre; camará; lvs. of small shrub c. VO, make v. agreeable tea (tea substit., Br); flg. summits u. in arom. baths (Ch); in catarrhal dis., stim. diur. antipyr. diaph. (inf.); frt. edible - **L. rehmanni** H. H. W. Pearson (tropAf, As): u. diur. (Vietnam); lvs. cause poisoning of sheep; c. icterogenin; u. to produce artificial jaundice (u. therapeut. for tremor, improvement of psoriasis, mitigation of decubital ulcer) - **L. rondonensis** Mold. (Pe, Bo, Br): quicke orégano; hb. inf. u. for colic (Pe); in *Virola* snuff - **L. scaberrima** Sond. (sAf): "Benkie's bossie"; lvs. c. 0.25% VO (camphoraceous odor) with lippianol (alc.), glucoside (may be verbenalin); resin; dr. hb. hemostatic (native doctors), u. as ton. stom. (inf. with brandy) to treat sore eyes & in gastric dis.; in forage, taints cow's milk - **L. schomburgkiana** Schauer (Br): lf. inf. drunk for VO - **L. sclerophylla** Briq. (Par, Arg): ann. hb. u. in popular med. - **L. trifida** Remy = *Acantholippia t.* - **L. turbinata** Gris. (Arg): poleo; hb. u. as digestive; lvs. c. VO with limonene, cineol, lippione, lippianphenol.; forma *angustifolia* Osten (Arg, Ch): "Argentine pennyroyal"; shrub cult. in the Chaco for medl. use; hb./lvs. u. in inf. as diur. stom. (Bs, As) - **L. ukambensis** Vatke (Tanzania): lvs. c. VO with camphor, cineole, thujanol, &c.; u. as preservative to wrap meat; in abdom. dis. - **L. umbellata** Cav. (Mex): "hierba de mula"; decoc. of shrub u. in rheum, colic, night fevers, dys. - **L. vernonioides** var. *attenuata* (Mart.) Mold (L. hastata Jensen) (Br, Bo): tononjil; "salvia sija"(?); lvs. u. med. (as "Lippia leaves").

lippia leaves: lippie mexicain (Fr): *Phyla scaberrima.*

lippione: terpene ketone occ. in *Lippia turbinata.*

lipstick tree: *Bixa orellana.*

liquen (Sp): lichen - **l. de islandia**: Cetraria.

liqueur: sweet alc. liquor prepd. by admixt. of VO, syrups, flavors, &c. with alc. or alc. prepns.

liquid amber: sweet gum, *Liquidambar styraciflua* - **l. apiol**: oleoresin of parsley frt. - **l. aquilegius**: spirit of wine - **l. bismuth**: Sol. bismuth - **l. glucose**: syrupy g. - **l. smoke:** in dry distn. of wood, the first fraction is so named; c. mixt. phenols, cresols, tar, acetic acid (pyroligneous acid); u. to smoke meats (as preservative) - **l. storax**: v. under storax - **l. tar**: Pix Liquida.

Liquidambar altingiana Bl. (Hamamelid.) = *Altingia excelsa* Noronha - **L. formosana** Hance (sChin, Indochin): feng; "bastard chestnut"; balsam collected fr. incisions in trunk, dried, marketed in 3 forms; u. to aid circulation, antihemorrhagic, analgesic, in TB; u. in boils, carbuncles, toothache; appl. exter. & int. (pills); frts. u. in lumbago, arthritis, skin dis. (also resin); astr. bk. u. to wash skin dis. areas; lvs. & rt. in cancerous growths (Chin); balsam c. cinnamyl alc., cinnamic acid, & esters, borneol (1622); tree cult. in Jap, Taiwan; ornam. - **L. orientalis** Mill. (AsMinor; Dodecanese Is; cult. Med reg., esp. sFr): Oriental sweet gum (tree); source of Levant styrax; Styrax or Sweet Gum occurs both in the liquid form (Liquid Storax or Liquid Amber) & in solid form (Styrax Calamitus); Liquid c. cinnamic acid, cinnamein, styrol. vanillin, resins; u. int. as stim. exp. alt. masticatory, arom.; u. exter. as arom. vulner. parasitic in skin dis., esp. scabies; in perfume indy.; fat preservat. (natural & purified [waterfree] grades); solid c. 50% liquid storax, balance incl. resins (ex. olibanum), wood, bark, etc.; u. incense; usually marketed as granules or in pd.; synthetic prods. sometimes on market; bk. ("Cortex Thymiamatis") u. incense in perfume indy. - **L. styraciflua** L. (eUS, Mex, CA): (American) sweet gum tree; liquidámbar (Hond, CR); ocozote (Sp, Am); source of American styrax (liquid storax: Balsamum Styrax [liquidum]); prod. like prec. by incisions in tree trunk; similar props. (2552) & u. to prec.; u. in GC (int.); in dys.; in "fistula" (2313); antisept. for healing cuts; as incense (LatAm); in formulation of Whitehead's Varnish (comp. iodoform, paint); balsam chewed with Smilax frt. or blueberry to allow "penetration much like inhaled fumes" (Texas [US] Indians); wd. (satin walnut) useful as veneer, in cabinet making; "stobs" u. in "worm grunting" (Fla [US]); oleroresin u. as chewing gum & for diarrh.; to "desharpen" tobaccos of all kinds (chief US use); mixed with honey & consumed by women before & after childbirth; mixt. with garlic, onions, hot water u. for intest. worms (Honduras Indians) (2395); cult. Med reg., esp. sFr; ornam. tree (fall coloration; frt. often persistent through winter on leafless trees).

Liquiritia (L): licorice.

Liquor Carbonis Detergens (L.): coal tar soln. - **L. C. Hippocastanus**; prec. mixed with Tr. horse chestnuts - **L. Hepatis**: liver soln.; liquid liver ext.; prep. u. in pernicious anemia - **L. Potassa**: KOH Soln. - **L. Sedans**: popular and renowned propr. med. (c. black haw, *Piscidia*, *Hydrastis*, &c.); u. in dysmenorrh., menorrhagia - **L. Seriparus** (Russ Ph.): rennet prepn. - **white l.** (rural sUS—slang): alc.

liquorice (root): Brit spelling for licorice (*qv.*): *Glycyrrhiza glabra*; *G. echinata* - **block l.: stick l.**: soft ext. of *G. glabra* (prep. in It & Levant) - **wild l.**: *Aralia nudicaulis* (*cf.* licorice).

lirio (Sp, Por): 1) lily 2) *Iris germanica* - **l. acuatico** (Mex): *Nymphaea alba*; *Nuphar luteum* - **l. blanco** (Mex): orris - **l. convalla** (Por): **l. de los valles** (Sp): **l. dos valles**: Convallaria - **lirios** (Gr): lily.

Liriodendron chinenese Sarg. (Magnoli.) (eAs): bk. u. as antipyr (Indochin) - **L. tulipifera** L. (eUS): tulip tree; yellow poplar; tall tree with beautiful deciduous foliage & striking flowers (hence names); bk. bitter, ton. antipyr.; good lumber tree; wd. sometimes white ("white wood"), sometimes yellow; lvs. u. in stock feeds; bk. chewed as tea substit.; bk. ("poplar bark") u. in hysteria; wd. c. liriodenine (aporphine alk.; inhibits cancer cells); rt. bk. c. costunolide & tulipinolide (both with antitumor activity); inner bk. c. liriodendrin (diglucoside of lirioresinol); digitaloid subst. also present in bk., &c.; lvs. c. 0.6% tannin (SC lirion [Gr] lily or tulip + dendron [Gr]: tree).

Liriope Lour. (Lili) (eAs): lily turf; "monkey grass" (seUS); perenn. hb. with dark green lvs. in tufts, blue fls.; u. as border "grass"; hemiparasites; pot plant; *cf. Muscari* (grape hyacinth) - **L. graminifolia** Baker (L. spicata Lour.) (Jap, Chin): creeping lily turf; c. liriope; fls. purple or white; c. saponins, no tannin; frts. berry-like; tuberous rts. u. like those of *Ophiopogon japonicus*; ton. & aphrod. (candied); decoc. u. to fortify lungs in fevers, dys. coughs (Chin); border pl., good for shady locations - **L. muscari** Bailey (Jap, Chin): (big blue) lily turf; big blue lariope; bunching form (esp. var. *exiliflora* Bailey); cham (Jap); frt. black; ornam. border pl. (many vars., incl. var. *variegata* Bailey).

Liriosma ovata Miers (Olac.) (Br, Ve): source of drug Lignum muira-puama (wood); rt. bk. said most effective; u. aphr. stim. ton. in anter. poliomyeltis, dys., chronic rheum. neuralgias.

lis (Fr-m.): lily, *Lilium* - **l. des valées**: *Convallaria majalis* - **fleur de l.**: Iris - **l. martagon**: *Lilium martagon.*

Lisianth(i)us alatus Aubl. (Gentian) (Guyana): inf. of fleshy lvs. formerly u. for smallpox; also u. like Gentian, the drug - **L chelonoides** L.f. (nSA, Trin): ground itch bush (Trin); lf. decoc. or juice u. as bath for athlete's foot; lf. u. for worm-infested wounds of livestock (Pe) - **L. nigrescens** Cham. & Schlecht. (Mex): rt. decoc. bitter, u. in indig. heartburn; as febrifugal tonic; lf. u. as poultice for fungal inf. of feet & hands; black fls. u. as wash to remove ticks - **L. pendulus** Mart. (Guianas): u. as bitter ton, febrifuge.

Lissochilus arenarius Lindl. (Orchid.) (sAf): rt. decoc. u. for impotence & sterility (Zulu) - **L. krebsii** Reichb.f. (tropAf): rt. u. as sed. (tribal peoples) (—> *Eulophia*).

litchee: litchi: *Litchi chinensis* Sonn. (l) (Sapind.) (eAs, cult. trop Chin, Ind, Haw, &c.): litchi; lychee; claimed by some to bear the most delicious frt. known; eaten either fresh or dr. (Litchee "nuts") (*cf.* plum) or packed in syrup; frt. u. in Chin folk med. for sores & poisonous bites (snakebite); tumors; sds. anodyne, u. for orchitis; neuralgia; lumbago; ulcers (Chin); sds. macerated in alc. u. for intest. & biliary dis. (Indochin); aril u. antipyr.; inf. as cough remedy (Palau); not toxic; cult. (Chin, PI, US).

lite'ard (seUS slang): fatwood (*qv.*).

Lithanthrax (L.) (lit. black stone): coal.

Lithium: Li: one of the alkali metals with valence of 1; occ. in good quantities in earth's crust, as in mineral amblygonite ($AlPO_4.LiF$); various salts u. in med. - **l. bromide** (sed. hypnotic), **l. carbonate**, Li_2CO_3 (an antimanic [to treat manic phase of manic-depressive state] E); many ind. uses - **l. chloride**, **l. sulfate**, (antidepressant).

Lithocarpus cornea Rehd. (Pasania c.) (Fag.) (Chin): acorns edible, u. to make flour - **L. densiflora** (H & A) Rehd. (Calif, Ore [US]): tanbark oak; bk. u. for tanning; cult. ornam.; acorns edible after leaching (Am Indians); wd. hard, useful for implements (1625) - **L. elegans** Hatus. (Chin): frt. & lvs. u. in menorrhagia - **L. javensis** = *Quercus j.*

Lithophragma affinis A. Gray (Tellima affinis A. Gray) (Saxifrag.) (sOre to sCalif [US]): woodland star; rt. chewed to relieve stomachache, colds (2441).

Lithops N. E. Br. (Aizo.) (sAf): stone plants; windowed p. (some spp.); cult. ornam. - **L. hookeri** Schw. (sAf): hb. edible for humans & herbivora.

Lithospaston: hb. of *Plantago coronopus*.

Lithospermum L.: 1) genus of ann. & perenn. hbs. (Boragin.) (chiefly nHemisph.): gromwells 2) (specif.): *L. officinale* - **L. arvense** L. (Euras): corn gromwell; sds. formerly quite important drug as diur. lithont.; rt. u. source of red dye - **L. canescens** (Mx) Lehm. (e&cNA): (Indian) puccoon; hoary p.; yellow p. (corollas yellow); American alkanet (root); u. for face paint (Chippewa Indians) (2790); also u. as paint; vermifuge; hair dye (red) (270) - **L. carolinense** (J. F. Gmelin) MacM (e&cNA): (hoary) pucoon; yellow p. (corollas yellow); rt. u. red dye; for face paint (Chippewa Indians); also u. as vermifuge - **L. erythrorhizon** Sieb. & Zucc. (Jap, Chin): rt. is called Shikon; c. alkannin-like coloring matter, Tokio purple (shikonin); u. as folk med. agt. in tumors (eAs) - **L. guttatum** (Bge.) Johnst. (Chin): zicao; rts. u. exter. for cuts & burns & in chicken pox - **L. officinale** L. (Euras; introd. Chin, NA): (common) gromwell; stoneseed; frt. formerly u. as lithontriptic, diur. contraceptive, acts by inactivation of serum gonadotropin (pituitary hormone blocking effect); lvs. tea substit.; in emulsions; rt. dyestuff (also other *L.* spp.); hb. & fls. c. rutin, no alk. or glycosides; lvs. found in commerce under names Bohemian (or Croatian) tea; sds. u. in folk med. as laborpromoting agt., diur., kidney & bladder dis.; rt. u. in diarrh. (Am Indians) - **L. purpureo-caeruleum** L. (Euras): creeping (or purple) gromwell; lf. u. as emoll. & refrigerant - **L. ruderale** Douglas ex Lehm. (L. pilosum Nutt.) (wCana & US): western gromwell; Columbia puccoon; stoneseed; pl. u. as contraceptive (tea) by Nevada Indians (inactivates gonadotropin; inhibits ovulation); impairs devt. of gonads & secondary sex organs of both sexes, inhibits thyroid (271); rt. decoc. in diarrh. (Shoshone Indians).

Lithrea caustica Hook & Arn. (Anacardi) (Ch): litre; litri; lvs. as tincture u. as emet., skin irritant; frts. yield a kind of honey, u. in prepn. of maize brandy; wd. fuel but very irritant to skin; pl. c. oleoresin; cardol-like subst. is toxic princ. - **L. molleoides** (Vell.) Engl. (L. aroeirinha March.) (Br): "aroeirinha"; u. like *Schinus molle* but less strongly active; in folk med. a balsam exuding fr. the bk. is u. against swelling of legs (1631).

litmus: coloring matter obt. fr. *Rocella tinctoria*, *Lecanora* spp. & *Variolaria* spp. (Lichens); one of most popular indicators in cheml. laboratories, etc. (mnfd. in Neth); color change at ca. pH 7.

Litsea cubeba (Lour.) Pers. (L. citrata Bl.) (Laur.) (seAs, Java): berries carm. expect. in bronchitis, dyspep. (Chin); decoc. in vertigo, paralysis, melancholy, forgetfulness (Indochin); sd. u. as stomach tonic (Jap); bk. & lvs. u. in epidermophytosis & other skin dis.; bk. u. as perfume offered Budda (Indochin); frt. u. as substit. for Piper cubeba frt.; fls. for aromatizing tea; VO in perfumery; rt. u. after childbirth (Hainan); FO of sds. u. as illuminant; frt. edible - **L. glutinosa** C. B. Robins. (seAs, SA): bk. u. in dys. (Burma); garbijaur (Hindi); enteritis (PI); pounded bk. appl. as poultice to abscesses, & furuncles; rts. & lvs. u. on sprains, bruises & wounds (PI) - **L. hutchinsonii** Merr. (endemic to PI): bk. scraped & appl. to wounds - **L. luzonica** (Bl.) F-Vill. (PI): bk. appl. to wounds; lvs. appl. to pain in waist region; frt. edible - **L. odorifera** Val. (Indones): lvs. u. in biliousness, appl. to accelerate lactation - **L. papillosa** Allen (Solomon Is): com comu; frt. kernels eaten (272) - **L. perrottetii** F-Vill. (PI, Melanesia, Celebes): chewed lvs. appl. to boils (PI) - **L. quercoides** Elm. (PI, endemic): lvs. appl. for headache - **L. solomonensis** Allen (Solomon Is, New Guinea): arli-arli; bk. macerated in water & appl. to sore legs - **L. tersa** Merr. (L. chinensis Lam.) (s&seAs): bk. dem. u. in diarrh. dys. local antidote to toxic animal bites; styptic for wounds - **L. umbellata** Merr. (seAs, Indones): lvs. u. to poultice boils (Mal); lf. ext. with strong inhibitory action against sarcoma 180 (mice) - **L. vitiana** (Meissn.) Drake (Fiji): u. as stomach med. - **L. zeylanica** Nees = *Neolitsea z.*

"little round jug": (near Lookout Mt., Ala [US]): *Hexastylis* spp.

littorine: tropane-type alk. obtained fr. *Anthocercis littorea* Labill. (major component) & *Datura sanguinea* R. & P. (ca. 1968).

live for ever (long): *Sedum telephium* - **l. oak**: term appl. to several *Quercus* spp., esp. *Q. virginiana* (bk., lvs.).

liver, desiccated: now u. as ton. in macrocytic anemias; c. vit. B_{12}, folic acid, & other vit., Se, Cr, Fe, Cu; often u. in Fe combns. - **l. extract**: a dry, brownish pd. obt. by processing of mammalian livers; u. to treat pernicious anemia; formerly u. in treating pernicious anemia (oral, injectable forms); now largely displaced by vit. B_{12} but still marketed (**crude, refined**) (**l. solution**) - **kidney l. leaf:** *Hepatica americana* - **l. leaf**: 1) *Hepatica americana* 2) *Gaultheria procumbens* - **l. residue**: product of potential merit in prolonging survival - **l. soup**: ancient prepn. u. in pernicious anemia (China) - **sharp-lobe(d) k. l.**: *H. acutiloba* - **l. starch**: glycogen.

liverwort (leaves): **heart l.**: *Hepatica nobilis* & its var. *acutiloba* .

li-yuen: opium of high grade.

lizard's tail, common: *Saururus cernuus*.

llama: *Lama*.

llanten (CR): *Plantago major* (lvs.).

Lloyd reagent: a type of purified clay (aluminum silicate [hydr.]; fuller's earth) u. to adsorb alks.; u. analysis & mfr.

Loasa lateritia Gill (Loas.) (Ch): common Chile nettle; this & some other spp. sting - **L. tricolor** Ker-Gawl. (CA, SA): fresh pl. u. in homeopath. med. (273).

Lobaria pulmonaria (L) Hoffm. (Sticta p. Ach.) (Stict. - Ascolichenes - Lichen.) (c&nEu, nUS): oak lungs; "lungwort" (confusing name); l. moss; l. sticta; lobaria moss; thallus u. dem. astr. in pulm. dis. cough, bronchitis, rhinitis; hemorrhages, asthma; in brewing & tanning indy. (Ind); as brown, red, orange, purple dye (for woollens; Scand, GB); in domestic med. (Muscus pulmonaria) as bitter ton.; c. gryophoric and stictic acids; source of VO u. in perfumes; also f. *papillaris* (Del.) Heu - **L. scrobiculata** (Scop.) DC (worldwide in temp regions): pitted sticta; oakrag (lichen); "rag"; "raw"; warty leather lichen; u. as brown dye for woollens (Eng, Scot, Scand).

lobelanine: alk. next to lobeline the most abundant one in Lobelia.

Lobelia L.: 1) genus of ann. & perenn. hbs. (Lobeliaceae/Campanul.) (worldwide): often u. med. (named after Dr. Matthias de l'Obel [or Lobelius, L.], botanist to the court of King James I) 2) (specif.) (L): lvs. & tops of *L. inflata* - **L. cardinalis** L. (eNA): cardinal flower; red c. plant; r. lobelia; Indian pink; c. lobinaline+; hb. u. anthelm. (NA Indians); in syph.; considered similar to *L. inflata* but weaker - **L. decurrens** Cav. (Pe, Bo): soliman; u. to cure sores & for bites of poisonous animals; lvs. very hot appl. to swellings with brandy & salt; for warts; for pain in molars, liquid appl. directly; rt. inf. u. as purg.; in paralysis, appln. with massage or drinking small doses (274); resp. stim.; insecticide - **L. dortmanna** L. (n&wEu, nNA): German lobelia; water l.; u. as anthelm. toxic pl. - **L. excelsa** Lesch. (L. leschenaultiana Skottsb.) (sInd): lvs. insecticidal; as replacement for *L. nicotianaefolia*; acrid latex gives dermatitis; toxic pl. - **L. fenestralis** Gray (sTex [US], Mex): leafy lobelia; sold for heart dis. (Mex) - **L. inflata** L. (e&cNA): Indian tobacco; bladderpod lobelia; emetic weed; "low belia" (-0.6 m high); lvs. & tops u. as expect. antispas. sed. in asthma, whooping cough, croup, catarrhs.; hb. rich in alks; u. for mfr. lobeline (2291); smoked by Am Indian mixed with regular tobacco; sds. u. expect. antiasthmat. (coll. Va, NCar [US]; cult. US, Eu, Ind) - **L. laxiflora** HBK (Mex, Co): acrid rt. c. lobeline; u. like *L. inflata*; emet. expect. antiasthmatic, intest. ulcers (+ var. angustifolia DC) - **L. nicotianaefolia** Heyne (sInd, SriL): dhaval, dhayala (Bombay); lf. inf. u. as antispasm.; lvs. c. lobelanidines, &c.; u. like & as substit. for *L. inflata*; in asthma; insecticide; rt. u. for scorpion sting - **L. pinifolia** L. (sAf): stim. diaphor. diur.; hb. decoc. u. in skin dis., chronic rheum., gout - **L. radicans** Thunb. (L. chinensis Lour.) (Chin, seAs): hb. u. in fever & asthma (as snuff, Thail); rt. u. antisyph. antirheum. (Indochin); u. to treat ascites caused by liver fluke (Chin) - **L. salicifolia** Sweet (Ch): veneno del diablo; latex u. to subdue dental pains; fish poison; toxic pl. - **L. siphilitica** L. (e&cNA): (great) blue lobelia; "high belia" (1-1.5 m high); blue cardinal; hb. u. alt. diaphor. formerly in syph.; in cancer - **L. spicata** Lam. (e&cNA): pale-spike lobelia; "high belia" (-1 m high); u. as diur. (Am Indian) - **L. urens** L. (wEu): u. anthelm., similarly to *L. siphilitica*.

lobelia, great: (large) blue l.: *L. siphilitca*: - **red l.**: *Lobelia cardinalis*.

lobeline: α-lobeline: the most abundant & important alk. of Lobelia; u. resp. stim., analeptic in BA, poisoning, etc.; as tobacco cure; found effective in curbing tobacco smoking habit, esp. if subject has desire to stop (oral tabs) (403); u. for urticaria (402).

lobine: alk. fr. *Oxylobium parviflorum*.

loblolly: 1) marshy unfirm clayey ground (also in general a kind of mess), mud hole; swampy sterile soils of seUS 2) *Magnolia grandiflora* - **l. bay:** *Gordonia lasianthus* - **l. pine**: *Pinus taeda*.

lobo (swUS): gray wolf, much larger than coyote (Mex); *Canis mexicana* (loba).

lobster, spiny: crayfish.

Lobularia maritima (L.) Desv. (Alyssum maritimum Lam.) (Cruc.) (sEu, natd. eUS): sweet alyssum; panalillo (Mex); hb. u. as antiscorb., diur., astr. in GC (Sp); cult. ornam.

Lochnera Reichb. = *Catharanthus* - **L. rosea**: *Cath. roseus*.

loco (Sp): mad, appl. to horse crazed fr. eating **l. weed** (*Astragalus* & *Oxytropis* spp.) (wUS & Cana) (ex. *A. allochrous* & several other *A.* spp.) (term also appl. in error to *Datura stramonium*, *Crotalaria sagittalis*, *Cannabis sativa*, &c.) - **purple loco weed**: *A. mollissimus* - **white l. w.**: *Oxytropis lamberti* - **locoism** = loco disease.

locum pharmacist (Brit): man or woman who serves as relief pharmacist for 2-4 wks. - **l. tenens**: temporary substitiute (pharmacist, physician, et al.).

locus: position on chromosome occupied by a gene or its allelomorph in an assumed straight line arrangement.

locust (US): common l. - **African l.:** *Parkia biglandulosa* - **l. bean**: carob bean+ - **l. bean gum**: carob g., fr. *Ceratonia siliqua*; also *Prosopis glandulosa* - **l. beer**: prepd. fr. *Gleditsia triacanthos* pods (crushed, water added, left stand (sUS) - **black l. (tree)**: *Robinia pseudoacacia* (name fr. early pioneer days) - **l. bread**: l. bean - **common l.**: l. tree = *Robinia* spp. esp. *R. pseudoacacia* - **l. gum: l. bean g.**: carob gum, fr. *Ceratonia siliqua* - **honey l. (common)**: *Gleditschia triacanthos* - **pink l.**: 1) *Robinia pseudoacacia* v. *decaisneana* 2) *Sophora affinis* - **l. plant**: *Cassia marilandica* - **sweet l.**: honey l. - **l. tree**: black l. - **West Indian l. tree**: *Hymenaea courbaril* - **yellow l:** common l.

lodestone: magnetite; often u. in Voodoo med. (seUS Negro) as conjuring trick.

lodge: fall or lie down; ex. cereal grasses; *Mentha* sts.

lodicules: small structures at base of florets (in grasses) wh. absorb water & force floret open, disbursing the pollen grains.

Lodoicea maldivica (Gmel.) Pers. (L. sechellarum Labill.) (monotypic) (Palm.) (Seychelles Is mostly): coco de mer (sea coco) (SC floats on sea); double coconut; tree of life; double frt. very large (-28 kg); tree long-lived; has largest sd. of all pls.; frt./nuts u. ton. antipyr. aphrod. antidote to poisons; in nerv. dis.; endosperm astr. antiinflamm., aphrod. (Ind): nut shell u. for bowls; young lvs. u. for making hats.

Loeffelkraut (Ge): scurvy grass.

Loeselia caerulea (Lam.) Brand (Polemoni) (Mex): espinosilla (Mex); jarritos; lvs. & st. u. diaphor. antiinflamm., diur. for fevers, emet.; to preserve the skin; hair wash to discourage falling of hair; formerly u. as detergent like soap - **L. mexicana** (Lam) Brand (L. coccinea Don) (Mex): espinosilla (Mex); lvs. & st. u. diaphor. diur. for fevers, emet. (rt.); to preserve the skin; hair wash to discourage falling of hair; formerly u. as detergent like soap - **L. scariosa** Walp. (Mex): huachichile; pl. us. as remedy for "ague" (Coahuilla).

loess: wind-blown fine soil deposits (US: Mo River banks; Chin); believed esp. fertile.

loganberry: *Rubus ursinus* Cham. & Schlecht. var. *loganobaccus* Bailey; popular frt. (wUS).

Loganiaceae: Logania (or Nux Vomica) Fam.: incl. hbs. shrubs, vines & trees; found in warmer parts of globe; pls. often have bitter juice & c. alks.; lvs. simple, entire or dentate; fls. bisexual & regular; ovary superior; frt. capsule, berry, or drupe; Ex.: jasmine, pink root, nux vomica.

Loganobacci Fructus (L.): dr. loganberries, *Rubus ursina* var. *loganobaccus.*

log(arithmic) growth phase: stage where metabolism is most active; as with bacteria.

loggerhead turtle: sea t.: *Caretta* spp. (Chelonidae; Reptilia) (wAtlantic Oc): grows to 500 lbs.; meat & eggs as food; med.; exotic leather fr. flippers & neck skin; animal on course to extinction, efforts being made to conserve.

logwood (chips) (WI): fr. *Haematoxylon campechianum.*

Loiseleuria procumbens (L.) Desv. (monotypic) (Eric.) (circumpolar, s to Eu, Alps, US, sCana): creeping (or alpine) azalea; pl. u. astr.

lokao: *Rhamnus globosa, R. utilis*; source of Chinese green.

lokundioside: glycoside of bipindogenin + L-rhamnose; occ. in large quantity in *Strophanthus sarmentosus, S. thallonii*, &c.

Lolium multiflorum Lam. (Gram) (Euras, nAf, natd. NA): Italian rye grass; important forage & fodder grass; lawn g.; often a serious weed; rts. c. alks. anuloline, perloline (toxic) - **L. perenne** L. (Euras, Af, natd. NA): perennial (or English) ryegrass; the first meadow grass cult. in Eu & its most important forage grass; with detergent, antigangrene, antidiarrh. antiperiodic properties; c. perloline (alk.); sometimes toxic (HCN); lawn grass - **L. temulentum** L. (Euras, natd. US, Cana+): (bearded) darnel; (bearded) rye grass; poison darnel; cheat; "tares" (of Bible); u. in hemorrhages, urinary incontin., menostasis (Mor); sd. meal intern. in menopause, in poultry for skin dis., tumors (Eu); frts. narcotic, anod.; occas. poisonous (said due to temuline [base] produced by parasitic fungi); eaten in bread (mixed with wheat); u. by Eu homeopaths for rheum. pains, colic, dizziness, nose bleeding; weed.

Lomatia ferruginea R. B. (Prote.) (Ch): fuinque; u. in indigestion - **L. hirsuta** Diels (Ch): redal; lf. inf. u. in chronic bronchitis, bronch. asthma, expect. antispasm.

Lomatium Rafin. (Umb.) (wNA): Indian root; hog fennel; tuberous rts. eaten raw or cooked or as flour (Am Indian) - **L. ambiguum** (Nutt.) Coulter & Rose (Peucedanum a. Nutt.) (wNA, Wash east to Utah): "Wyeth's lomatium"; biscuit root; "Indian celery" (BC [Cana]); "cow-as" (Am Indian); tubers u. as food (aboriginies); inf. drunk for colds & sore throat (BC Indians); upper lvs. & fls. u. as flavor in foods (*cf.* parsley) - **L. cous** Coulter & Rose (Wash to Ore, Ida [US]): biscuit root;

cous (Am Indian); Am Indian u. starchy globose tubers (c. VO) for food - **L. dissectum** Mathias & Constance (Wash-Calif [US]): fern-leaved lomatium; var. *multifidum* Math. & Const. (Leptotaenia multifida Nutt.) (BC to sCalif; mt. states): cough root; Indian balsam; I. carrot; best known crude drug of Nevada, u. by Indians & whites; c. eutachin ("antibiotic"); boiled rt. u. for cough, colds, bronchitis; in pulmon. congest.; often u. in combns.; in GC, antisep.; trachoma; rheum.; distemper of horses (158) - **L. foeniculaceum** Coult. & Rose (cNA): tuberous rt. edible - **L. macrocarpum** (Nutt.) Coult. & Rose (c&wNA): prairie parsley; biscuit root; rts. eaten; mashed rts. u. as poultice for broken bones, &c.; soaked rts. chewed for colds, flu, & bronchitis (BC Indians) - **L. orientale** Coult. & Rose (cUS): whisk-broom parsley; eastern wild carrot; rts. u. for food (NM (US) Indians) - **L. suksdorfii** Coult. & Rose (Wash [US]): Suksdorf's lomatium; frt. c. VO having ascaricidal properties.

Lonchocarpus Kunth (Leg) (tropAm, Af, Austral): barbasco; cubé; rts. of some of the ca. 150 spp. u. as fish poisons; insecticides; pounded & pulp put in lagoons or ponds; c. rotenone (toxic to cold blooded animals); u. also by Indian brujos (medicine men); some common names same as appl. to *Derris* spp. - **L. bussei** Harms (tropAf): rt. u. as galactagog, in GC (Tanz) - **L. cyanescens** Benth. (Sudan, Sierra Leone, Lagos): liane à indigo; green lvs. c. indigotin; pl. u. as dyestuff "gara"; lvs. u. as dressings for tropical ulcer; lvs. u. as condim. for couscous (Senegal) - **L. densiflorus** Benth. (Guyana): (white) haiari(a) (root); vine u. to stupefy fish (f. poison); powerful insecticide - **L. eriocalyx** Harms (L. goossensii Hauman) (wCongo): wd. inf. u. as antipyr.; (Tanz): rt. u. as appln. to skin eruptions - **L. latifolius** HBK (WI trop, contin. Am, Mex): retama; lancewood; bitchwood; guama de costa (Cu); lvs. u. as purg.; rt. & frt. insecticidal; rt. bk. fish poison (Domin); wd. solid, compact, resistant, u. in construction & for equipment; med. (Col) - **L. laxiflorus** Guill. & Perr. (L. philenoptera Benth.) (trop Af): olwedo (Acholi); rt. inf. drunk for paralysis & backache; boiled bk. u. in indigest. (eAf); bk. & rt. u. general ton.; rt. u. for intest. pain, leprosy (Senegal); bk. with natron u. for flatus in horses, anthelm.; lvs. appl. to foot ulcers - **L. leucanthus** Burk. (Arg): rabo amarillo; prized as ornam. - **L. longistylus** Pitt (Mex, CA): balche; bai-ché (Yucatan); intoxicating beverage made fr. tree's bk.; bk. u. antipyr. ton. (Yuc) - **L. mexicanus** Pittier (Willardia mexicana Rose) (Leg) (sMex): nesco; bk. decoc. u. to destroy parasites on cows & horses; wd. u. for mining props, fuel, &c. - **L. nicou** DC (L. floribunda [Benth.] Killip) (Pe, Br, Guyana): Peru roots; nekoe; rts. c. rotenone (source), deguelin, tephrosin, rt. (barbasco; timbo; cubé) u. as insecticide, fish poison; cult. (SA) - **L. peckolti** Wawra (Br): timbo; c. tomboine (alk); timbo root u. for tumors & as fish poison (Br); bk. "timbo boticário" (Br) - **L. sericeus** (Poiret) DC (tropAf, rarely in Ve): Senegal lilac; bk. u. stom. in abdomin. complaints, purg. (children) (wAf); rt. decoc. lotion for wounds & as antiscorb. (eAf); for convulsions; locally for parasitic skin dis.; insecticide - **L. urucu** Killip & Smith (Br): crude drug u. in Jap; rts. suggested as rotenone source; u. insecticide - **L. utilis** A. C. Smith (L. nicou auctt. non DC) (Br [Amaz], Bol, Ec, Pe): rts. c. rotenone; u. insecticide; const. of curare; fish poison; also arrow poison - **L. violaceus** Kunth (WI Surinam): stinkwood; savonette; u. as emoll., fish poison.

longan(s): *Dimocarpus longana*.

long purples: *Lythrum salicaria*.

longwort: *Anacyclus pyrethrum*.

Lonicera caprifolium L. (Caprifoli) (s&cEu; occas. US): Italian woodbine; American w.; sweet honeysuckle; fragrant fls. & lvs. u. as diur. diaph. astr. against malignant tumors; in teas & perfumes; frt. diur.; sts. to "purify" the blood - **L. ciliosa** (Pursh) Poiret ex DC (Ore [US] - BC [Cana]): western trumpet honeysuckle; orange h.; edible frt.; branches u. to make med. for mother after childbirth (250) - **L. confusa** DC (eAs): lf. & fl. inf. in wine u. for itchy rash; to stop dys., appl. for boils (China); wash for wounds & rickets - **L. dasystyla** Rehd. (Indochin): kim ngan (Vietnam); u. similarly to *L. japonica* - **L. fuchsioides** Hemsl. (Ch): pinyin; u. as diaph. for cramps; &c. - **L. glauca** Hook. f. & Thoms (China); lvs. & fls. u. in VD - **L. involucrata** Banks (NA incl. Mex): (black) twinberry; twinflower honeysuckle; similar uses to *L. ciliosa* (BC Indians): frt. edible (Indian hunters, &c.) - **L. japonica** Thunb. (Jap, Chin, Kor; natd. e&cUS [Va-Fla-Miss.], sAf): (Japanese) honeysuckle; a twining & trailing shrub with fl. buds u. as diur. bactericide (Chin) (85); fls. white with purple, becoming yellow, very fragrant; c. luteolin & l. glycosides, saponins, tannin; fls. u. sed. refrig. in dermal dis. astr. diur. antipyr. stom. in tumors (esp. breast cancer); flg. tops u. depur. antisyph. as lotion for purulent abscesses, alexiter. for venoms, &c.; source of Jap drug, "nindo"; very common & destructive weed (esp. in seUS) since it overwhelms many hbs. & shrubs; a "romantic nuisance"; lf. u. as tea substit.; rt. inf. u. for gastric dis., diarrh., dys.; var. *chinensis* Baker (L. chinensis Wats.): corolla red outside; sts. & lvs. purplish - **L. periclymenum** L. (Euras): woodbine; common honeysuckle; u. as spasmolyt. in respir. difficulties; otherwise u. like *L. caprifolium*; st. u. as adulter. of Dulcamara; suspected poisonous pl. - **L. sempervirens** L. (e&cUS): trumpet (or coral) honeysuckle; corolla scarlet (rarely yellow) outside; infragrant; lvs. u. in asthma - **L. utahensis** S. Wats.: (wNA): Rocky Mountain honeysuckle; red twinberry; inf. of branches u. to wash sores, infections; lax.; ton.; frt. edible, source of water (emergency).

lontar palm: *Borassus flabelliformis*.

looch album: emulsion of almonds & almond FO.

loofah: vegetable sponge; the "skeleton" of the gourd (frt.) of *Luffa* spp.

loosener (US slang): compound cathartic pill.

loosestrife: *Lythrum salicaria* (also **purple l., spiked l.**) - **4-leaved l.**: *Lysimachia quadrifolia* - **yellow l.**: *Lysimachia vulgaris*.

Lopez root: 1) *Morus indica* 2) (Ind): *Toddalia asiatica*.

Lophanthus rugosus Fisch. & Mey. = *Agastache rugosa* - *L.* spp. Benth (NA) = *Agastache*.

Lophatherum gracile Br. (Gram) (eAs, Austral): tan chu; lvs. antipyr. healthful to lungs; diur. u. in hematuria (Chin); diur. in anuria.

Lophira lanceolata Van Tiegh. (L. alata van Tiegh.) (Ochn.) (tropAf): African oak; rt. decoc. drunk for menstrual pain, GIT dis., mal.; twigs inf. u. in fever, respir. dis., dys. (275); sds. furnish niam fat (n. oil); u. as food, soap mfr., hair dressing oil; wd. useful.

Lophocereus schottii Britt. & Rose = *Cereus s.*

Lophophora williamsii (Salm-Dyck) Coult. (Cact.) (swUS, c&eMex; nAf) (176): mescal button, peyote; cross cut slices of crown c. 30 alk., incl. mescaline, lophophorine, anhalonine, &c.; u. narcotic addictive (intoxicant), hallucinogen. —> colors, weightlessness; fumitory, masticatory; by NA Indians in ceremonials, feasts; sacramental fixture of native Am church (Indians); u. antipyr.; as heart regulation agt. in angina pectoris; fever, throat irritation, reduces hunger & thirst & keeps the mind awake (1626); added to "tizwin" (alc. beverage) to give added effect; when chewed called "dry whisky" (2230).

Lophosoria quadripinnata C. Chr. (1) (Cyath./ Lophosori.) (Ch): helecho arboreo; u. against hemorrhages & bruises.

loqhat = loquat: *Eriobotrya japonica*.

Loranthaceae: Mistletoe Fam.; hbs., shrubs, & trees; generally parasitic; with green lvs., small & inconspicuous fls.; lvs. opposite, sometimes appear as scales; frt. berry or drupe; ex. American mistletoe; *Phoradendron* spp.

Loranthus confusus Merrill: (Loranth.) (PI: Luzon): rt. decoc. u. in dys. - **L. dregei** E & Z. (sAf): pd. bk. in milk given as enema for stomach dis. in children (Zulus); sd. rubbed on stick to catch birds. - **L. europaeus** Jacq. (e&sEu, AsMinor): oak mistletoe; yellow-berried mistletoe; antique drug. u. somewhat like *Viscum album;* in tumors; berries the source of bird lime; the st. (Lignum Visci gerucini-Viscum quercinum) (cum foliis) u. like *Viscum album*; berries c. mucil., a resin, viscin (viscene) (as in *Viscum album*), choline; lvs. c. ceryl alc., phlobaphene, pectin-like mucil.; u. for tumors (homeop.) - **L. grandiflorus** King (tropAs): lvs. u. with turmeric & rice as poultice for ringworm - **L. marginatus** Desr. = *Struthanthus m.* - **L. uniflorus** Jacq. (Br): lvs. u. as cataplasms; frt. produce bird lime.

Lorbeer (Ge): laurel, *Laurus*.

lords and ladies plant (NCar [US]): *Arum maculatum*.

Loricaria ferruginea (R. & P.) Wedd. (Comp.) (Ec, Pe): chinchango (Pe); pl. decoc. drunk as strengthening tonic - **L. graveolens** Wedd. (Pe): dr. pl. sold at herb markets.

Loroglossum hircinum Rich. (Orchid.) (c&sEu): lizard orchis; tubers u. as aphrod. (—> *Aceras + Himantoglossum*).

losna (Por): *Artemisia absinthium*.

Losungen (Ge): feces, excreta.

lot (swGa [US]): 250 acres (50 X 50 chains).

Lota lota: **L. maculosa** (Gadidae - Pisces) (Euras, NA fresh water lakes): burbot; the only fresh water member of the fam.; common in Great Lakes (NA); livers are rich in vits. A & D; u. much as cod-liver & halibut liver oils.

Lotononis calycina Benth. (& other *L.* spp.) (Leg) (tropAf): u. as bronchitis remedy for children - **L. lanceolata** Benth. (sAf): rt. prep. u. orally or per recto for diarrh.

Lotur(r)inde (Ge): bk. of *Symplocos racemosa*.

Lotus arabicus L. (Leg) (nAf, Senegal; sArab): Arabian deer vetch; hb. c. lotusin, glycoside splits (with lotase) to prod. lotoflavin, HCN & maltose; poisonous; u. med. (Kuwait) (276) - **L. corniculatus** L. (Euras, n&eAf, Austral; natd. eNA): common bird's foot; bird's (foot) trefoil; formerly u. as emoll. for sores; now cult. chiefly for forage (worldwide crop), fodder (lea crop), ornam.; var. *hirsutus*: pile lotus; u. in hemorrhoids - **L. scoparius** (Nutt.) Ottley (Calif [US], nwMex): deer weed; d. lotus; foliage decoc. u. for cough; in house thatching (Costanoan Indians).

lotus, American: *Nelumbo lutea* - **l. bloom tea** (Vietnam): strong laxative - **East Indian l.: sacred l.**: *Nelumbo nucifera* - **l. tree** 1) *Zizyphus lotus* (Med reg.) 2) *Diospyros lotus* (cAs) 3) *D. virginiana* (US) - **water l.**: yellow l. - **wild l.**: American l.

louco (Por): *Plumbago zeylanica.*

loupe (Fr-f.): magnifying glass, lens, as u. by jewelers, pharmacognosists, et al. (fr. Ge. Lupe, microscope).

loureiro (Por): laurel.

louse (plur. lice; (*cf.* **lousy**; lousiness, infestation): broadly u. for insects wh. such blood (Anopoleura), incl. pl. juices; specif. *Pediculus humanus* (L.) (body or hair lice) & *Phthirus pubis* (L.) - **crab** (or **pubic**) **l.** (occ. in pubic regions) (both spp. of fam. Pediculidae): u. as antipyr. in mal. (Unani med.; taken live p. o. (Mallophagi, biting lice) - **l. powder**: pd. stavesacre - **l. seed**: *Delphinium staphisagria* - **l. wort**: *Pedicularis* spp.

lovage (**plant, leaves, root & oil**): *Levisticum officinale* - **l. root** (US): under this name, several Umbelliferae rts. have been sold (ex. *Heracleum lanatum*) - **Western l.**: *Ligus-ticum porteri*, osha.

Lovastatin™: Mevinolin: Mevacor; fungal metabolite isolated from *Aspergillus terreus* Thom (strain) & *Monascus ruber* v. Tieghem; u. to reduce hyperlipidemia (hypercholesterolemia) by inhibition of enzyme active in cholesterol biosynthesis; some by-effects & hazards in use (USP, 1988).

love apple: tomato - **love drug** (localism): pituri - **l. poison**: aconite (origin?) - **l. vine** (sUS): *Cuscuta americana* & other *C.* spp., dodder.

lovegrass: 1) *Eragrostis* spp. 2) (Cey): *Andropogon aciculatus* - **sand l.**: *E. trichodes* Wood.

love-in-a-cage: *Physalis alkekengi.*

love-in-a-mist: *Nigella damascena* - **love-lies-bleeding**: *Amaranthus caudatus*+.

Lovibond unit: u. in specifications of vit. A strength; 1 Lov. unit = 208 international (or USP) units.

low belia (US slang): *Lobelia inflata.*

Löwenzahn (Ge): dandelion.

Lox: smoked salmon (fr. Yiddish).

Loxa bark: *Cinchona officinalis, C. macrocalyx.*

Loxodonta africana Blumenbach (Elephantidae; Proboscidea; Mammalia): (tropAf): African elephant; generally not domesticated; larger body & much larger ears than in Indian e. (v. under *Elephas*); tusks u. in med., for carving, &c.; *cf. Elephas maximus* (sale of tusk ivory limited to that fr. dead (not killed) animals.

Loxosceles laeta (Scytotidae - Arachnida) (Pe, Ch, Arg, Ur, esp. near coast): spider wh. often produces death, esp. of children; anti-serum has been prepd. (1955).

LSD: lysergic acid diethylamide (Ge: Lysergsaeure Diaethylamid).

lubia (Iran+): *Phaseolus vulgaris*, kidney bean.

lucca oil: trade name for finest var. of olive oil; intended for table use.

lucern(e) (GB, Austral): Luzerne (Ge): alfalfa.

lucialima (Por): *Lippia citriodora* - **lucio marino** (Sp): *Glaucium luteum.*

luciferase: enzyme occurring in certain insect spp. wh. acts on luciferin (in presence of ATD + O_2) producing light (bioluminescence) & oxidized luciferin; u. to detect life.

Lucite™: polymer of methyl methacrylate (fr. *Anthemis nobilis*), a clear plastic u. in prepn. of permanent slides like Cana balsam (Fisher Sc.) (*cf.* plexiglass).

luck ball (seUS Negro): jack.

lucky nut: *Thevetia peruviana.*

Luculia pinceana Hook. (Rubi) (Indochin): lvs. u. to alleviate difficult childbirth (with lvs. of *Allophyllus glaber* Radlk.).

Lucuma bifera Molina (Sapot) (Ch, Pe): frt. eaten when very ripe - **L. caimito** Roem. & Schul. = *Pouteria c.* - **L. glycyphloea** Mart. & Eichl. = *Pradosia lactescens* **L. gutta** Ducke (Br): source of Abicurana Balata - **L. lasiocarpa** A DC (Br, Amazon.): abiorana; tree latex u. gutta percha substit.; bk u. as astr. (Amazonia) - **L. mammosa** Gaertn. = *Pouteria m.* - **L. nervosa** A DC = *Pouteria campechiana* - **L. obovata** HBK (Pe, Ch): lucumo; frt. yellow, edible; with high content of riboflavin - **L. salicifolia** HBK = *Pouteria campechiana.*

lucuma (SpAm): mamey apple, *Mammea americana* (25) - **L. gutta** Ducke (Br-Amazon): source of abicurana Balata - **lucumo**: tree of same.

Ludwigia adscendens (L) Hara (Jussiaea repens L.) (Onagr.) (e&seAs, trop): "false loosestrife"; pl. juice u. diur.; with rapeseed oil appl. to rheum., to wash wounds, fungous skin dis. (Chin); decoc. u. intern. for eye dis. (Taiwan); young shoots mixed with castor oil u. for ringworm & other scalp dis. (Indochin); poultice for skin dis. (Mal) - **L. alternifolia** L. (eNA, esp. seUS): (bushy) seed-box; rattle-box; hb. u. in chest dis., rt. decoc., emet. - **L. decurrens** Walts. (Jussiaea erecta L.) (As, Oceania): u. to dye pandanus lvs. (86) - **L. hyssopifolia** Exell (tropAs to NG): u. diarrh. dys. (Chin); antisyph. (rt.) (Mal) - **L. octovalvis** Raven (tropAs, Austral): lvs. as poultice for wounds; purg. vermifuge; mucil. lvs. appl. locally to headache, swollen glands (Mal) - **L. palustris** Ell. (NA, Mex): marsh purslane; phthisis weed; hb. u. to treat asthma, chronic cough, TB - **L. prostrata** Roxb. (eAs): for diarrh. dys., vermifuge; sds. for pertussis - **L. repens** Sw. (tropics): mururé; perenn. hb.; st. tops eaten as veget. (Indochin); hog food; lax(?) - **L. suffruticosa** Walt. (Jussiaea s. L.) (WI, tropAm): clavellina primrose willow; hb. u. for gastralgia, cephalic pain (Guat, Hond).

Luehea candida Mart. (Tili) (Mex, CA, Co): molenillo; dry frt. less sds. fastened to end of stick & u. to beat chocolate to make beverage light & frothy (CA) - **L. divaricata** Mart. (sBr, Arg): common whiptree; commonly wd. u. for diverse purposes, ex. shoe soles - **L. speciosa** Willd. (tropAm): fls. of this & other *L.* spp. u. (Br) for bronchitis; value attrib. to VO.

Luffa acutangula Roxb. (Cucurbit.) (OW trop): towel gourd; luffa gourd; vegetable sponge; dish rag gourd (sMiss [US] Negro); singkwa; young unripe frt. edible (like cucumber); ripe frt. emet.; dr. vascular skeleton of mature gourd u. as sponge; thus for washing dishes to filter oil; FO of l. seed u. as food oil (EI), substit. for olive oil; cath. in skin eruptions; sd. emet. diur. cath. tenifuge (u. with linseed tea & castor oil (*cf.* Pepo) (Ve) (26); rt. u. purg. diur.; lf. u. muscifuge, by rubbing on donkey's hide (Ve); sd. decoc. for hiccup (Indochin); finely pulped lvs. appl. to lesions of zona (herpes zoster) (Indochin); frt. u. to induce vomiting (Ind) - **L. aegyptiaca** Mill. (L. cylindrica Roem.) (OW trop, widely cult.): loofa towel gourd; sponge g.; wash-rags; (suakwa) vegetable s.; u. much like *C. acutangula*; frt. juice & pulp u. as bechic (for cough & phlegm), diur. (Chin, Jap) - **L. echinata** Roxb. (wInd, Pak): bristly luffa; pl. u. emet. anthelm. in icterus, phthisis, hiccup (Ind); frt. u. in edema, purg. as inf. in colic; frt. c. purg. bitter princ. luffein; cucurbitacins; sd. c. FO with much linoleic & oleic acids - **L. purgans** Mart. (L. operculata [L] Cogn.) (Mex, CA, nSA): koosia (Guyana); esponjilla (SA); sd. & frt. inf. u. in snakebite (Trin); frts. ("casadores") regarded as panacea, ex. purg. diur. antitumor (Br); homeopath. tinct. u. to treat colds, &c.

Luisia tenufolia Bl. (Orchid) (Ind): pl. emoll. u. as poultice to abscesses, tumors, boils.

lulu nuts (Sudan, Arab): sds. of *Butyrospermum parkii* c. 50-55% FO (or more) with large amt. free fatty acids (8-12%); roasted sd. kernels turn very dark chocolate color (1627).

Luma: v. *Myrceugenella.*

lumbang: *Aleurites triloba* (oil) - **soft l.**: *A. trisperma* (oil).

Lumbricus terrestris L. (Lumbricidae - Oligochaetae - Annelida) (worldwide): (common) earthworm; valuable in aeration of the soil, rendering it more permeable for water & minerals; FO u. in folk med. ("grub oil"); as aphrod. in ointment (Unani med.); u. in gourmet cooking (2633).

lump: v. under turpentine.

lumpers: those taxonomists who prefer to recognize larger groups (especially genera) with considerable diversity without naming such variants as species, &c.; i.e., they conserve genera, &c. as far as possible.

Lunaria annua L. (L. biennis Moench) (Cruc.) (Eu, natd. NA): dollar weed; d. plant; honesty; moonwort; satin-flower; "silver shilling" (Vancouver, BC [Cana]); silver dollar(s); penny flower (2662); ann. & bienn. forms; sds. earlier u. like mustard; rt. (of pre-flowering pl.) edible; cult. as ornam.; fls. often u. in dry bouquets - **L. rediviva** L. (Euras): satin flower; (perennial) honesty; rt. detersive; lvs. diur.; sds. acrid, formerly u. as mustard substit.

Lunasia amara Blco. (Rut) (seAs, PI): lúnas; bk. & lvs. c. lunasine & other alks.; u. for gastric dis.; reputed digitaloid action; shown to be hypotensive; c. some 10 alks., incl. lunasine, lunacrine, &c.

Lungenflechte (Ge): **lung moss**: lungs of the oak: *Lobaria pulmonaria.*

lungwort: 1) *Pulmonaria officinalis* & other *P.* spp. (lvs.) 2) *Mertensia virginica* - **American l.**: 1) *Mertensia virginica* 2) *Variolaria faginea* - **bullock's l.: cow's l.**: Verbascum - **French l.**: 1) *Hieracium murorum* 2) *Pertusaria faginea* - **maple l.**: 1) *P. Lobaria pulmonaria* 2) *Pertusaria faginea* - **l. moss**: *Lobaria pneumonia* - **Virginia l.**: American l. (1).

lupanine: alk. of some *Lupinus* spp.; strong antimalarial action.

lupe: *v.* loupe.

lupeol: tri-terpene (sterane deriv.); found in *Lupinus* sds., chicle, *Ficus* latex, &c.

Lupicaine: lupinine p-aminobenzoate HCl; loc. anesthetic.

lupin(e): *Lupinus polyphyllus* & other *L.* spp. - **blue l.**: *Lupinus subcarnosus* & *L. texensis* - **Egyptian l.: white l.**: *L. albus* - **yellow l.**: *L. luteus* - **lupinine**: alk. fr. various *Lupinus* spp.; has fruity odor.

lupinelli (It): wild lupine.

lupino (It): lupine.

Lupinus albus L. (Leg) (Eu, Levant): white lupine; often cult. to improve the soil (green manure); for fodder & human food (esp. It); sds. (roasted) as coffee substit.; debittered sd. added to wheat flour (10-20%) (Turkey); sd. u. anthelm. diur. emmen. abortient; in menstr. dis., tumors; in cataplasms - **L. angustifolius** L. (Med reg): blue-flowered lupine; in tumors; fodder; green manure; sds. as coffee substit. - **L. arboreus** Sims (NA: Pac coastal areas): yellow-flowered bush lupine; tree l.; cult. ornam. - **L. arbustus** Dougl. subsp. *silvicola* D. Dunn (L. laxiflorus Dougl. var. s. C. P. Sm.) (Calif, Ore [US]): hb. c. anagyrine; causes larkspur type of poisoning among cattle (28) - **L. littoralis** Dougl. (Pacifc NA, esp. BC [Cana]): chinook licorice; roasted rts. formerly u. as food by BC Indians - **L. luteus** L. (c&sEu; widely cult.): (European) yellow lupine; sds. c. alkaloids (sparteine, lupinine, lupanine, &c.); carbohydrates; FO; protein (valuable); good food if detoxified; coffee substit.; green manure; ornam. garden pl. - **L. mutabilis** Sweet (Ve, Pe): c. sparteine & lupanine; sds. eaten; u. as insecticide & fish poison; toxic pl.; however, cult. & sds. eaten - **L. perennis** L. (eUS): (wild) lupine; perennial l.; cold inf. drunk for hemorrhages & nausea (Cherokee) - **L. polyphyllus** Lindl. (wNA, cult. Eu): large-leaved lupine; lvs. c. estrogenic isoflavone, genistin; sds. c. alk.; u. as green manure, fodder, ornam. pl.; coffee substit..; many races; pl. decoc. u. as ton. (BC Indians) - **L. sericeus** Pursh (Pac Cana): sds. made into inf. for eye med. (BC Indians) - **L. subcarnosus** Hook. (Tex [US]): blue bonnets; ornam.; state flower of Texas (with *L. texensis* Hook.) - **L. termis** Forsk. = *L. albus*.

luppolina (It): **luppolo** (It): hops - **L. spagnolo**: Spanish h., *Origanum vulgare*.

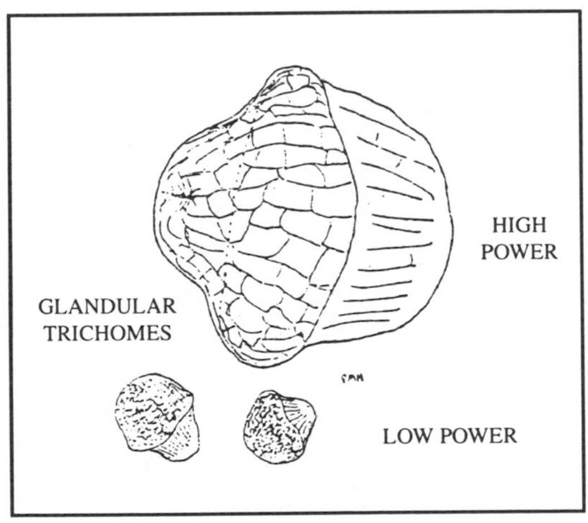

LUPULIN

lupulin: Lupulinum (L.): glandular trichomes (or scales) of hops, *Humulus lupulus*; u. sed. hypnot. anaphrod. bitter agent, antispas.; in some beers in place of hops (deteriorates with age).

lupuline: alk. fr. hops; liquid with coniine-like odor.

lupulo (Sp, Por): hops - **lupulon**: antibiotic fr. hops; u. against TB in lower animals (1830) - **Lupulus** (L.): Humulus, hops.

"lurk in the ditch" (Eng): **lurkey dish**: *Mentha pulegium*.

lust wort: *Drosera*.

lustrum (L): luster: 5-year period.

lutefisk: stockfish (codfish) treated with lye (alkali), boned & boiled; u. as food.

lutein: xanthophyll; yellow pigment found in egg yolks, corpus luteum, yellow flower petals (vegetable l.), &c.; antioxidant.

luteinizing hormone: LH (for female); interstitial cell stimulating hormone (ICSH) (for male): gonadotropic hormone of the anterior pituitary, which in the female stimulates the Graafian follicle to form the corpus luteum following extrusion of ovum (ovulation), with secretion of progesterone, also stim. (with FSH) secretion of estrogen fr. G. follicles; in the male, the interstitial cells of the testis are growth-stimulated with incr. secretion of androgens & devt. of secondary sex characters (formerly called Prolan B).

luteolin: a flavone dye found in glycosidal linkage in many plants; ex. in *Carthamus lanatus, Reseda luteola, Euphorbia cyparissias*; u. like other bioflavonoids; - **luteosterone**: steroid of corpus luteum: α, inactive; B & C are progesterone α & ß (resp.) - **luteotrop(h)ic hormone: luteotrop(h)in**: lactogenic hormone; anterior pituitary hormone which maintains corpus luteum in functional state & initiates & maintains secretion of mammary glands sensitized by estrogens; same as prolactin - **Lutocylin**™: progesterone.

lutoids: small bodies found in fresh rubber latex wh. act as stabilizers.

Luvunga scandens Buch-Ham. (Rut) (Ind, Mal): frt. u. in prepn. of arom. oil u. in Hindu med.; rt. inf. ingested as postpartum protective med. (Mal); u. in scorpion sting.

Luzula campestris (L.) DC (Junc.) (Ind - Himalay, Kashmir+, natd. NA): black-head grass; rhiz. diur. - **L. pilosa** (L) Willd. (L. acuminata Raf.) (Euras): (hairy) wood rush; stolons u. for kidney dis. & urninary calculi (925).

lycaconitine: alk. fr. rt. of *Aconitum lycoctonum*; ester deriv. of lycoctonine; aconitine deriv.

Lychnis: 1) *Silene dioica* (Lychnis d.); *Agrostemma githago* - **L. coronaria** Desr. (Caryophyll) (sEuras, Ind, Pak): rose campion; rt. decoc. u. for lung & liver dis. - **L. dioica** L. = *Silene dioica* - **L. flos-cuculi** L. (Euras, natd. NA): ragged Robin; rt. c. saponins, u. as "Saponaria Alba" - **L. githago** Scop. = *Agrostemma g.*

lycine: betaine (discovered in *Lycium chinense*).

Lycium L. (Solan) (temp. & subtr., many Am): some spp. furnish squawberries (Pima Indians, Ariz [US]), wolfberries, tomatillo - **L. andersonii**

Gray (swUS, nMex): water jacket; Anderson's desert thorn; frt. eaten fresh or in mush - **L. afrum** L. (Af, sEu, wAs): Kafir thorn; lvs. discutient, alt. in erysipelas, herpetic eruptions; young plant eaten; u. for hedge making (sAf) - **L. barbarum** L. (L. halimifolium P. Mill.) (Med reg., nAf, seEu, wAs): bastard jasmine; box thorn; tea tree; matrimony vine; c. traces of mydriatic alk. (hyoscyamine?); betaine, choline; hb. & rt. u. diur. purg. (Homeop.); lvs. u. in salads; cult. ornam. ("tea plant") - **L. chinense** Mill. (eAs): frt. & bk. u. to improve vision; frts. & sds. u. in diabetes (pl. cult. for this [55]); rts., sts. & frt. u. as ton., for chronic dis. & to improve blood & circ. (Chin); rt. bk. u. fever of pulm., TB (Chin) (85); frt. in nerv. dis. (Indochin); lf. inf. for toothache (Indones); lvs. food - **L. fremontii** A. Gray (swUS, nwMex): Fremont's desert thorn; one source of popular dessert frt. of Pima Indians (Ariz) (wolf berries) - **L. pallidum** Miers (swNA): rabbit thorn; frt. eaten raw or cooked by Am Indians - **L. ruthenicum** Murr. (cAs): khichar (Ladakh): pl. u. in ointments for blindness of camels - **L. vulgare** Dun = *L. barbarum*.

lycopene: major pigment of mycelium of *Podosordaria leporina* (Ascomyc. Fungi); also found in ripe frts., such as tomato; red carotenoid.

lycopenemia: variant of carotinemia, due to lycopene, often from ingestion of tomatoes (277).

Lycoperdon Pers. (Lycoperd. - Lycoperdales - Gasteromycetes - Basidiom. - Fungi.): common puff balls; immature sporophores of some spp. sometimes u. as food; at mature stages, enormous amts. of spores are released to the air - **L. bovista** L. = *Calvatia b.* - **L. candidum** Pers. (NA): edible mushrooms - **L. cervinum** L. = *Elaphomyces cervinus* - **L. gemmatum** Batsch = *L. perlatum* Pers. - **L. marginatum** Vitt. = L. candidum Pers. - **L. perlatum** Pers. (L. gemmatum Batsch) (NA, As): devil's snuff-box; gemmed puffball, spores u. as dusting powd. for burns, to stop bleeding (Ainus) - **L. pyriforme** Schaeff. (NA, SA): pear-shaped puffball; gem p.; spores u. as baby talcum to prevent or relieve skin chafing (Menominee Indians) - **L. solidum** Gron. = *Poria cocos* - **L. subincarnatum** Pk. (NA): pink puffball; powder (spores) u. as body powder to relieve skin irritation (Potawatomi & Rappahannock Indians) (278).

lycopersicin = tomatine.

Lycopersicon (*Lycopersicum*) **esculentum** P. Mill. (Solanum l. L) (sometimes treated as subgen. *Lycopersicon* of gen. *Solanum*) (Solan) (wSA, orig. Peru perhaps, now widely cult. worldwide, esp. US, Eu) (2383): (common) tomato; love apple (formerly); many vars. recognized; stem ext. c. tomatine (lycopersicin), glycoside of tomatidine, steroid with antiinflammatory props. equal to hydrocortisone (*sic*); with fungistatic/fungicidal potentiality; solanine; hb. bears popular esculent frts. (rich in vit. C); u. diur. nutr. antiscorb. aper.; u. to make tomato paste (It): prepns. of pulp freed fr. sds., core, & rind & condensed by evapn., u. to make sauces, soups, broths & with spaghetti, &c.; catsup (ketchup) (condiment contg. sauce combined with tomatoes) (1628); tops anodyne var. *Cerasiforme* Dunal & Gray (cherry t. [tom-a-toe] [small spherical]; Roma [u. for sauces for pastas, &c.]); t. seed oil = FO with semidrying nature, u. in foods, mfr. soap, &c.; tomato juice popular as beverage, "pick-me-up"; health drink; flavoring/masking agt. for administering KCl (&c.) solns. p. o.; possibility of overdosage of K, since t. juice is rich in same (2468) - **L. pimpinellifolium** Mill. (Peru; now widely cult.): currant tomato; small currant-like red frts.; lvs. c. tomatine & tomatidine, soladulcidine; frts. edible.

Lycopodiaceae: Club Moss Fam.; moss-like plants with small, almost scale-like lvs. & often prostrate sts.; genera; *Lycopodium*, *Phylloglossum*, &c.

lycopin = lycopene, carotenoid pigment.

Lycopodinae: former designation of an order in Pteridophyta; now supplanted by Lycopodiales, Lycopsida, &c; also in recent presentations made a division, **Lycopodiophyta**.

Lycopodium L. (Lycopodi - Lycopsida - Pteridoph.) (worldwide temp & trop): club mosses; some spp. c. alk., ex. *L. inundatum* L. - **L. annotinum** L. (Euras, NA): interrupted club moss; running ground pine; c. alk. incl. annotinine, annotine, annofoline; hb. & spores u. like those of *L. clavatum* (USSR Phar., Nor Phar.) - **L. cernuum** L. (OW, NW trop, particularly Br, Mal, Haw Is [US]): waiwaiole; hb. u. diur., in dys. gout; in skin dis., rheum., dys., colds, diarrh., tenesmus of rectum (sAf); spores u. like in *L. clav.* - **L. clavatum** L. (Euras, also worldwide in cooler areas): (common) club moss; running moss; creeping Jennie; lamb's (or fox) tails; lycopodium (moss); hb. c. alk. (lycopodine, &c.) (*cf. L. complanatum*); NBP; u. diur. emmen. antispasm., in folk med. in kidney & bladder dis.; spores = Lycopodium (L) (vegetable sulfur; v. brimstone), c. 40-50% FO, sporonin, resin, traces of alk., &c.; u. absorb. dusting pd., to prevent adhesion of suppositories, condoms, &c.; lenitive, to prevent piles; for athlete's foot; in tech. arts, to prevent adhesion of brass casting to mold ("dry parting compound"), in fireworks, as flash powder (theater, &c.); in microscopy (quantit.); to make finger prints visible (with carmine); former important allergen in theater (esp. opera), since u. on stage to indicate rain & fire (279); prodn. USSR, Ge, Switz; during WWII, good amts. prod. in US (New England) & Cana - **L. complanatum** L. (Eu, Siber, NA): creeping vine; trailing v.; hb. c. alk. (lycopodine, "complanatine" [mixt.]) (as in *Lycop. clavatum*, less toxic alk. than other spp.); hb. & spores (Lycopodium) u. as for *L. clavatum*; var. *flabelliforme* Blanchard (NA): c. clavatine, clavatoxine, obscurine - **L. saururus** Lam. (SA, Mauritius): hb. c. pilljanine, (very toxic alk.); spores u. as for *L. clav.* - **L. selago** L. (nearly cosmopol in mts.): fir(club) moss; tree m.; c. selagine & other alk.; alk. fraction u. as miotic for glaucoma in place of pilocarpine (Pol); hb. (Herba Selaginis) u. as strong emet. & drastic; earlier as enem. anthelm.; hb. decoc. for dyeing wool yellow; spores as for *L. clav.* - **L. serratum** Thunb. (eAs, Am): c. alk.; spores u. to prevent pills fr. adhering; hemostatic; for abraded skin, fungoid ulcers, &c.; pl diur. for nocturnal emissions; lax. (NG); antipyr.

lycopods: Lycopodiales, with single fam., Lycopodiaceae.

Lycopus europaeus L. (Lab) (Euras, natd. NA, Austral): gipsywort; water horehound; hb. astr. sed. diur. diaphor. u. in cough remedies, as bitter ton. in mal., metrorrhagia; in diabetes (unconfirmed value); antihormone effect: antigonadotropic, antithyrotropic; u. in milder cases of hyperthyroidism (incl. cardiac disturbances); antithyroid effect attributed in part to high F content, by some to lithospermic acid, "lycocyn" (311); furnishes black dye; said u. by gipsies to color complexions - **L. lucidus** Turcz. (eAs): u. as remedy for abscesses (Kor); circ. aid in edema; irreg. menses (Chin) - **L. virginicus** L. (eNA): (Virginia) bugle weed; betony; hb. u. ton. astr. sed.; in diabetes, hemorrhages; in simple & toxic goiters; in asthma, sometimes with Fumaria; antihormonal effect, as in hyperthyroidism (Grave's dis.; toxic goiter); cardiac dis., cardiac neuroses, &c.

lycorine: narcissine: galanthidine: pyrrolophenanthridine alk.; occ. in *Lycoris radiata*; *Narcissus tazetta*; *Buphane disticha*; *Crinum* spp.; *Galanthus nivalis*; *Amaryllis belladonna*, &c.; u. as substit. for emetine; with cytotoxic (1629), expect. emet. & antipyr. properties.

Lycoris aurea Herb. (Amaryllid.) (eAs): crushed bulbs u. for burns & scalds (Chin); cult. ornam. (fls.) - **L. radiata** Herbert (Chin, Jap): short-tube lycoris; poisonous pl.; c. lycorine, lycorenine, homolycorine, tazettine (sekisanine); bulb u. in decoc. to counteract poisons accompanying abscesses & ulcers; expect.; ext. appl. to ulcers & swellings; as plaster to burns (Chin); ornam. hb. - **L. sanguinea** Maxim. (Jap, Chin): bulbs. c. alks., u. in oint. appl. to swollen mammary glands - **L. squamigera** Maxim. (Jap, Chin): u. for tumors, abscesses, freckles; snakebite (Chin); ornam.; bulb. c. lycorine, tazettine, & several other alks.

Lycosa fasciiventris Duf.; **L. hispanica** Walck. (Lycosidae - Arachnida): "tarantula"; u. in nymphomania, hysteria, pruritus vulvae, multiple sclerosis (Homeopath.).

Lygeum spartum L. (l) (Gram) (nAf, Sp): esparto (grass); sts. source of fibers u. to make ropes, sails, &c.; for paper making; one source of esparto fiber (1312).

Lygodium circinatum Sw. (Schizae. - Pteridoph.) (tropAs, PI-Austral): stipe chewed & appl. to bites of poisonous animals; rts. & lvs. appl. to wounds (Indones); sap appl. to sprained limbs; rt. contraceptive (Solomon Is) - **L. japonicum** Sw. (Korea, Jap to Austral): decoc. u. diur. to treat GC, edema, vesical calculi; sed. in high fevers; circul. ton.; cath. - **L. microphyllum** R. Br. (L. scandens Sw.) (OW, NW trop): Japanese climbing fern; rhiz. astr. decoc. in dys. & spitting blood; lvs. in lotions & poultices to apply to skin dis. & swellings (Mal); sts. u. to make baskets - **L. smithianum** Presl (w tropAf): u. med. (Congo); sts. u. to make straps for fishing nets & traps - **L. volubilis** Sw. (nSA,WI): fiddle bush (Trin): rhiz. decoc. u. in VD.

lyme-grass: *Elymus* spp., esp. *E. arenarious* L. (SC planted in sand [arena (L.)] so that its rts. may keep this in place).

Lympha Pini (L): fresh sap fr. *Pinus pinaster*.

lymphocyte: type of leukocyte found in lymphoid tissues, blood, and lymph; subdivided into B and T lymphocytes, important to the immune process.

lymphogranuloma venereum: l. inguinale: an infectious systemic venereal dis. caused by *Miyagawanella lymphogranulomatis* Levaditi et al. (Chlamydiaceae - Bacteria); diagnosis effected with **l. v. antigen** (fr. organisms growing in chick embryo cells.)

lymphokines: low m. wt. protein factors prod. by activated T-lymphocytes (type of leukocytes of blood) (those differentiated largely in the thymus), producing stimulation of mitosis in other lymphocytes, immobilization of macrophages, lysis of other cells, &c.; incl. interferons, interleukin 2, tumor necrosis factor, &c.; important to the production of immunity.

lynn: *Tilia* spp. - **l. (wahoo)**: *Ulmus alata*.

Lyonia ligustrina DC (Eric.) (eUS): privet andromeda; maleberry; male blueberry; cult. ornam. shrub

(-4 m) - **L. mariana** D. Don (Andromeda m. L.) (e&cUS): stagger bush; toxic. fr. andromedotoxin; popular ornam. shrub; u. for ground itch, toe itch, &c. (country people, Ga, SCar [US]) - **L. ovalifolia** Drude (Ind to Jap): lvs. & frts. u. as ton.; however toxic (andromedotoxin).

lyophilization: removal of moisture by process of sublimation fr. the frozen state; appl. to sera, etc.; reduces danger of destruction by hydrolysis.

lypressin: vasopressin lysine; polypeptide isolated fr. hog pituitary glands, then purified; vasopressor, antidiuretic (*cf.* vasopressin).

lyre tree: *Liriodendron tul.*

lysalbinic acid: subst. prep. fr. casein or leather waste; u. protective colloid (metals), washing & filling agent for textile fibers.

lysergic acid: cleavage prod. from alkaline hydrolysis of characteristic alkaloids of ergot; high yields from *Claviceps paspali*; psychomimetic agent (controlled drug) - **l. a. diethylamide** (LSD 25): **lysergide**; deriv. of ergot with the basic str. of the various ergot; a monobasic carboxy acid having quinoline & indole nuclei; sythesized late 1954; psychomimetic, psychedelic; causes severe psychic disturbances; has been u. in study of mental disorders (1630); some simple l. a. derivatives have been u. therapeutically.

lysergol: alk. isolated fr. 1) saprophytic culture of ergot of *Elymus* type 2) sds. of *Rivea corymbosa*; u. in half synthesis of nicergoline (u. in myocardial infarction with diastolic hypertension) 3) other plants.

Lysichiton (*Lysichitum*) **americanus** Hultén & St. John (L. camtschatcense auctt. non L. Schott) (Ar) (Alask to Calif [US]): (yellow) (or western) skunk cabbage; rt. u. dermatits, esp. ringworm; u. as blood "purifier" (Makah Indians); abortient (by chewing raw rt.); small piece of rt. chewed for soothing stomach ("peppery"); rhiz. cooked & eaten (Indians); lvs. u. as poultice for headache & fevers (312); rhiz. eaten in spring by bears (ex. giant brown) - **L. camtschatcensis** (L) Schott (Jap, neSib, Sakhalin [Kuril Is]): chkau (Kamchatka): lvs. served as food to pigs (Sakhalin) (280); young "budlings" (shoots) edible (2480); pl. has arom. sweet scent; rts. food of bears.

Lysiloma acapulcensis Benth. (Leg) (Mex, Guat, Hond): tepeguaje; bk. astr. chewed to harden gums; gum exudate u. like acacia - **L. candida** T. S. Brandeg. (Mex): palo blanco (white wood) (SC trunks white); bk. u. to tan hides (27); export. fr. Mex - **L. latisiliqua** Benth. (L. bahamensis Benth.) (sFla, WI, sMex): wild tamarind; wd. of tree u. in boat making (WI).

Lysimachia chenii C. M. Hu (Primul.) (Vietnam): lf. decoc. u. in local pharmacopoeias in mixt. to facilitate cicatrization of certain sores on legs - **L. foenum-graecum** Hance (Primul) (Chin): pl. decoc. u. as mouthwash for toothache; u. to scent hair (odor of fenugreek); rts. chewed for fetid breath - **L. japonica** Thunb. (Korea, Chin, Taiwan, Mal, & Jap): decoc. u. diur. to treat GC, edema, vesical calculi; sed. in high fevers; circul. ton; cath.; decoc. of leafy shoot in honey u. as bechic (Chin); pl. appl. to swellings (Taiwan); st. heated & appl. to sore eyes (PI) - **L. nemorosa** L. (eEu): wood pimpernel; yellow p.; hb. (Herba Anagallidis luteae) astr. & healing agt. for wounds - **L. nemorum** L. (e&cEu, Caucasus reg.): wood pimpernel; formerly off. as Herba Anagallidis luteae; u. vulner. astr. - **L. nummularia** L. (Eu, Caucausus reg; natd. Jap, e&cNA): moneywort; creeping Charley; hb. c. tannin, saponins, primverase; ext. appl. to wounds, sores (vulner. astr.) & intern. in diarrh. & salivation; in Balkans, antipyr. for rheum. gout, & joint pain; sts. u. to make baskets; lvs. & fls. as tea; pl. has strong antibiotic activity - **L. paridiformis** Franch. (Chin: fl. top decoc. drunk as carm & diaph. - **L. quadrifolia** L. (4-leaved, but actually often up to 6-7 lvs. in whorl) (e&cNA): whorled (or 4-leaf) loosestrife; crosswort; perenn. hb. u. as expect.; as astr. & stom. by Am Indians - **L. sikokiana** Miq. (eAs): above-ground parts u. for stomachache (Chin); appl. to swellings & snakebite; spice; fumitory - **L. vulgaris** L. (Euras): yellow loosestrife (not related to purple l.: *Lythrum* spp.); (common) loosestrife; rhiz. c. saponin; rutin; u. astr. stomach. exp.; throat gargle; hb. & fl. previously u. int. & exter. as astr. in hemorrhages, diarrh., dys. fever, scurvy, etc.; smoke of burning hb. may kill or repel mosquitoes, flies, etc.

lysine, l-: diamino caproic acid; an essential amino acid; occ. free & as protein component, ex. serum albumin, fibrin, &c.; now prepd. by synthesis; growth factor in tissue synthesis; aids bone growth; produced by bacterial growth; proposed for use in herpes inf.; not synthesized by human; u. food enrichment (nutr.); may stim. appetite; often in mixts. with vitamins, minerals, &c.; to fortify wheat-based foods; u. proposed in shingles.

Lysionotus pauciflorus Maxim. (Gesneri) (Far East): tincture u. daily for lumbago (Chin).

Lysol™: prod. similar to Cresol Soln. Comp.

lysozyme: muramidase: enzyme wh. catalyzes lysis (or dissoln.) of bacteria by action on mucopolysaccarides of cell wall; occ. in egg white, nasal mucus, *Ficus* latex, &c. (1831).

Lythraceae J. St. Hil.: Henna (or Loosestrife) Fam.; hbs. shrubs & trees; mostly in tropAm; lvs. entire, stipules small or lacking; fls. bisexual, regular; ovary superior; frt. capsule; ex. crepe myrtle, henna; foll.

Lythrum alatum Pursh (Lythr.) (e&cUS, Cana, Mex): milk willow; yerba del cancer (de México); hb. u. astr. antiluet., in female dis.; against cancerous tumors (folk med., Mex) - **L. hyssopifolia** L. (Eu, neUS): hyssop-leaved lythrum; hb. u. astr. dem. in female dis. - **L. salicaria** L. (Euras, nAf; wetlands, natd. US): purple (or striped) loosestrife; long purples; handsome pl. with reddish-purple fls.; hb. (Herba Salicariae; H. Lysimachiae purpureae) c. tannin, salicarin (glycoside), lycorine; u. astr. in diarrh. dys., ton. dem. alter. styptic; fls. (Flores S.) & rt. (Radix S.) also u.; pernicious weed; bee plant; ornam.

Lytta atrata (Meloidae - Coleoptera - Insecta) (NA): insect u. like European cantharis; never official - **L. vesicatoria** L. (Cantharis v. Geoffr.) (c&sEu, fr. Spain to sRussia): chief source of Cantharis (Spanish fly); v. under Cantharis (1632).

lyxoflavine (, L-): analog of vitamin B_2 (riboflavin); occ. in human myocardiac tissues, also synthesized; growth stimulat. agt. in foodstuffs, feeds; mixed with fodders to give incr. growth; ineffective as antithyrotoxic factor; hypotensive.

Mm

ma huang (Chin): *Ephedra* spp. of Chin, yielding ephedrine; ex. *E. equisetina, E. sinica.*

Maackia amurensis Rupr. & Max. var. *buergeri* Schmid. (Leg) (eAs): bk. u. locally as fomentn. for bruises - **M. chinensis** Tak. (Chin): decoc. to bathe painful swellings.

maahlaa (Syr): mahaleb.

mabee bark: Colubrina.

maçá (Por): apple.

Macadamia integrifolia Maiden & Betche (M. ternifolia F. v. Muell.; M. tetraphylla L.A.S. Johnson) (Prote.) (Austral, esp. Qsld, NSW; cult. Austral, Haw, US, CA, SA, sAf): (rough shelled) macadamia nut; Queensland nut; sd. kernels delicious raw, boiled in oil, roasted, salted, canned; expensive delicacy (549).

Macaglia Rich. = *Aspidosperma.*

macal (Mex): yam.

maçaranduba = massaranduba.

Macaranga denticulata Muell.-Arg. (M. henricorum Hemsl.) (Euphorbi) (eAs): decoc. of wd. fr. tree stump u. to treat "paralysis" (Chin); juice decoc. u. after childbirth to prevent puerperal disorders (Indoch); bk. masticatory with betel - **M. gigantea** Muell.-Arg. (tropAs): rt. bk. decoc. u. in diarrh. dys.; tree sap u. as glue (Mal) - **M. grandifolia** Merr. (PI): with big lvs. 1-60 cm wide; resin fr. sts. u. as astr. gargle for oral ulcers; cult. ornam. - **M. harveyana** Muell.-Arg. (Oceania): lvs. u. med. for gastric dis. (infant) (Tonga) (86) - **M. heterophylla** (wAf): rt. decoc. u. as drink for amenorrh., emmen. abort.; various parts of pl. are purg. - **M. hypoleuca** Muell.-Arg. (seAs): u. as antipyr. expect. antispasm.; to kill insects & crawfish - **M. indica** Wight (Ind): gum exudn. ("Kino") appl. to sores - **M. involucrata** Baill. (Moluccas): bk. decoc. with spices u. as gargle for swollen tonsils; bk. decoc. u. GC; lf. sap u. for dys. - **M. kilimandscharica** Pax (eAf): rt. u. as remedy for biharziasis.; rt. decoc. in cough; lf. decoc. in stomach dis. (281) - **M. peltata** Muell.-Arg. (Ind): gum pwd. & appl. to venereal sores - **M. spinosa** Muell.-Arg. (Af, As): juice u. as vesicatory; fish poison - **M. tanarius** Muell.-Arg. (sChin to NG): bk. inf. for dys. diarrh. for prolapse of rectum; rt. inf. antipyr.; lf. ext. antibiotic to *Staphylococcus* spp.; bk. to make alc. beverage; source of binungia gum (Kino), u. as glue (PI); cult. as shade tree (Taiwan) - **M. triloba** Muell.-Arg. (eAs): large lvs. appl. as poultice to boils (Mal); lvs. & frt. u. for stomachache (Java) - **M. urophylla** Pax (SòlomonIs): lvs. macerated & appl. to sore legs - **M. vitiensis** Pax & K. Hoffm. (Fiji): lf. crushed & drunk with water for diarrh. - **M. winkleri** Pax & K. Hoffm. (EI): wd. decoc. u. for stomach, head & bone ache (Sabah [nBorneo]).

Macaronesia: region of Canary Islands (off Af); not to be confused with Macronesia.

macaroni: popular It food (pasta) prepd. fr. wheat flour, milk & water.

macaroons: biscuits prepd. fr. coconut (shredded). almond meal, egg white, sugar.

macassar kernels: sds. of *Brucea sumatrana* - **M. mace** & **M. nutmeg(s)**: *Myristica argentea*; substit. for true nutmeg - **m. oil**: old name for cananga oil *(Canagnga odorata)*; u. for hair dressing; name also referred to FO of *Schleichera aleosa*; *cf.* antimacassars (*v.* under Klettenwurzel) - **M. sandalwood**: *santalum album.*

macaw palm: *Acrocomia aculeata.*

mace: aril of frt. of nutmeg, *Myristica fragrans* (etc.); c. 30% FO, 15% VO, starch; u. spice (esp. for pickles), stim. carm. in GIT dis.; source of VO, FO (prod. Ind, SriL, Mal, Indones); hallucinogenic; (do not confuse with **MACE™**: chloroacetophenone; u. as riot control agent) (lacrimatory) - **false m.**: *M. malabarica* - **m. flowers**: mace - **Macassar m.**: **Papua m**: fr. *Myristica argentea* (u. as substit. for true nutmeg) - **true m.**: m. - **wild m.**: *Myristica sylvestris* Houtt. (Moluccas).

macédoine (Fr): hodge podge; jello - **m. de fruit**: frt. salad.

macela: 1) (Por): *Achillea ageratum* 2) (Br): *Matricaria americana*; uncata (Andrews) Sprague & Sandw. (Bignoni) (Pe) u. as "love charm" 3) (sBr): *Achryocline satureoides*

macene = pinene < nutmeg VO.

maceration: process of soaking drug in a menstruum to extr. or develop active principles by fermentation (enzyme action); often a process preliminary to percolation.

macerone (It): dandelion.

Macfadyena uncata (Andrews) Sprague & Sandw. (Bignoni) (Pe): u. as a "love charm," med. - **M. unguis-cati** (L) A.H. Gent. (Doxantha u. Miers) (Mex, WI, CA to Arg): cat's claw; catclaw (BWI); pago palo (PR, Domin Rep); bejuco (de) perdiz (Cu); young sts. & lvs. in bronchitis & catarrh (decoc. Mayas); enteritis; crushed fresh lvs. styptic for wounds (6); alexiteric; rts. & lvs. u. as antipyr. in mal. astr.; splenic dis.; possibly the widely touted aphrod. pago palo (1956).

Machaeranthera pinnatifida (Hook.) S. Shinners (Haplopappus spinulosus DC) (Comp.) (Mex, wUS): lvs. & rts. u. for toothache (Navaho).

Machaerium angustifolium Vog. (Leg) (Br): bk. u. as antidote for snakebite; tree produces a gum-resin ("dragon blood") - **M. macrocarpum** Ducke (Br): frts. u. as diur. - **M. robiniaefolium** Vogel (Ve): cascarón; bk. & resin astr. (282) - **M. villosum** Vog. (Br): Brazilian rosewood (jacarandá); useful wood.

Machaerocereus gummosus (Engelm.) Britt. & Rose = Cereus g. Engelm.

mache (Fr): *Valerianella locusta.*

machete (Mex) (pron. mash-ay´-tay): large heavy knife with wide blade (*cf.* corn knife of US) u. in Latin America in prepn. of many drugs, in agr. procedures, &c.

Machilus Nees = *Persea.*

machoacan, raiz de (Sp): *Convolvulus mechoacanna.*

macias (Sp): **macide** (It): **Flores Macidis** (L.): **Macis** (L.): **Macisbluete** (Ge): mace.

macin: proteolytic enzyme fr. *Maclura*; prob. same as pomiferin.

Macleania ecuadoriensis Horold (Eric.) (Ec): fresh frt. u. for lung trouble; frt. eaten.

Macleaya Betis (Blco.) Macbr. (Papaver.) (PI): bk. & lvs. u. to relieve stomach pain (children); latex appl. to abdom. to expel worms(!!); pd. bk. errhine; sd. FO u. as illuminant; wd. useful & durable - **M. cordata** (Willd.) R. Br. (Bocconia c. Willd.) (Chin, Jap): (pink) plume poppy; celandine tree; t. bocconia; all parts of lge (-2.5 m) perenn. hb. c. alks. esp. rts.; alks. incl. macleyine (= protopine) (homochelidonine), chelerythrine (*cf.* sanguinarine); juice u. caustic (as acrid); rt. u. bitter, sternut., red dye; bk. narcotic; ointment fr. decoc. of lvs. & sts. appl. to ringworms; lvs. & fls. insecticidal; u. to make a spirit (davu or darre, intoxic. bev.); cult. as medl. pl. in Russia; cult. ornam. - **M. longifolia** (Koenig) Macbr. (Bassia l. Roxb.; Illipe malabrorum [Koenig] Dubard) (tropAs, esp. Ind, SriL, Med, WI): mahua, mahwa (tree); mo(w)a; moha; mah(o)wa; mi; mowrah (butter tree); illipe b. t.; eloopa; u. to prod. FO, u. in margarines, &c.; fls. edible (raw, cooked) (c. sucrose 10-15%, in veget. sugar 45%); u. to prod. alc. bev. (davu or darre); sd. cake u. to kill worms in lawns(!!); bk. & lvs. astr. u. in chronic ulcers, gargle in tonsilitis & pharyngitis; fls. also u. in med. (expect. in chronic bronchitis, cardiac dis.); frt. u. in bronchitis; TB; sds. (illipe nuts) u. to prod. semi-solid FO, illipe oil or butter, mahwa b., eloopa (or bassia) oil (1634); u. as food (in margarines), in skin dis. chronic rheum. lax., in soap making, replacement for cocoa butter & tallow; sd. cake u. as emet. galactagog.; lvs. u. as tan (17% tannin), fish poison; in liniments for burns; wd. hard, useful, u. to make furniture; honey pl. - **M. neriifolia** (Moon) H. J. Lam. (M. malabarica [Bedd.] R. N. Parker; Bassia m. Bedd.) (Malabar coast, Ind, SriL): gammi (SriL); frt. u. in rheum., gall bladder dis., asthma, anthelm; sd. FO in rheum; fls. in kidney dis.

Maclura pomifera (Raf.) C. Schneid. (M. aurantiaca Nutt.) (monotypic) (Mor) (cUS esp. Ark, Okla, Tex, natd. in much of NA): Osage orange; bois d'arc; "hedge" (prairies); bow wood; a common diecious tree, sometimes considered a weed; dye fr. wd. (for uniforms; still in use) (2034); frts. ("horse" or "hedge apples") c. triterpene alcs., lupeol, lurenol (mixt. butyrospermol, lupandiol); wd. c. morin, maclurin; large green frt. (-8 cm diam) squirrel food; said to repel roaches; frt. ext. u. as card. stim. of great strength in card. collapse (2033); frt. c. latex rich in enzymes; wd. ext. u. as yellow or orange dye, tan; trunk u. for long-lived fence posts; planted for hedge; rt. bk. c. muritannic acid. u. leather tan; wd. hard, u. for tool handles (2034); branches u. for bows (hence bois d'arc); antioxidant (2033).

maconha (Br): Cannabis.

Macoubea guianensis Aubl. (Apocyn.) (Br): latex u. for pulmon. dis.; frts. eaten.

macroanalysis: chem. analysis with usual quantities, as in test tube amts.

Macrocystis pyrifera (Lessoniaceae Phaeophyta) (Pac coast NA): Calif [US] giant kelp; entire thallus u. as source of sodium alginate.

macrofungi: fungi with relatively large sporocarps.

macrolides: large ring lactone molecules found in some antibiotics (ex. erythromycin).

Macrolobium caeruleum Harms (Vouapa c. Taub) (Leg) (tropAf): wd u. for general carpentry; railroad ties - **M. phaseolocarpum** (Mart.) Steud. (nSA): resin u. as courbaril, American (Brazilian) copal.

Macromeria viridiflora DC var. *thurberi* I. M. Johnston (M. T. Gray) (Boragin.) (NM, Ariz, [US], nMex): smoked mixed with *Verbascum* for mental dis. (Hopi Indians).

macronutrients: chem. elements deemed very important for pl. growth, ex. Ca, Mg.

Macropiper excelsum (Forster. f.) Miq. (Piper) (NZ): pepper tree; kawa-kawa; u. in local med.; aphrod. (reputed); cult. ornam.

Macroscepis Kunth (Asclepiad.) (nSA): bk. of some spp. collected as Condurango Bark.

Macrosiphonia hypoleuca Muell. - Arg. (Apocyn.) (Mex): rosa de San Juan; hierba de la cucaracha; pl. decoc. u. for stomachache, toothache (appl. direct to tooth); ext. to inflamed eyes; powd. pl. with sugar u. to poison cockroaches - **M. longiflora** Desf. (Br): velame; flor de babado; decoc. of tree rt. taken with sugar for syph. ("depurative").

Macrosolen cochinchinensis Van Tieghem (Loranth.) (Indoch): lvs. u. to prep. tea-like drink; frts. u. for cough, appl. to head for h. ache; itch; st. juice drunk to help expel after-birth - **M. robinsonii** Danser (Indoch): lf. inf. u. as diur. & to combat enlarged abdomen - **M. tricolor** Danser (Chin, Indoch): pl. u. to prep. purg. drink.; rt. decoc. u. in gastralgia.

macrotin: cimicifugin: resinous prod. obt. by pouring strong alc. ext. of Cimicifuga into acidulated water & separating & purifying the ppte (< *Macrotys artacoides* Schmalz former B.O.).

Macrotomia benthami DC (Boragin.) (Pak, wInd): gaozaban (Punjabi): considered useful in tongue & throat dis.; said u. for cardiac dis.; expect. (Chitral, Gilgit) - **M. cephalotes** DC (Syr, Turkey, seEu): rt. (Syrian or Turkish alkanna) c. reddish pigment; u. as dye & as substit. or adult. of alkanna rt.

Macrotyloma geocarpum (Harms) Maréchal & Baudet (Leg) (wAf): ground bean; geocarpa ground nut; frt. buried in soil by st. like peanut; sds. eaten; other pulses of the gen. incl. **M. uniflorum** (Lam.) Verdc. (OW trop, esp. wAf): horse gram; u. as fodder, &c.; kooltee.

Macrotys: DC = *Cimicifuga.*

Macrozamia miquelii A DC (Cycad./Zami) (Austral): baked sds. (Queenland nuts) u. as food by aborigines - **M. spiralis** Miq. (Austral): kernels source of arrowroot (high quality); sds. eaten (aborigines).

mactins: anticoagulant subst. isolated fr. clams (*cf.* heparin).

macuja palm: macaw p.

mad apple(s): 1) nutgall(s) 2) stramonium - **m. (dog) weed**: *Scutellaria lateriflora.*

madagascoside: glycoside of uzarigenin + D-sarmentose; occ. in *Roupellina boivinii.*

Madagascar: Malagasy (Republic): large island opp. e coast of seAf, distinctive flora & fauna; former Port colony; prod. sugar, vanilla, coffee, &c. - **M. cardamom:** *Aframomum angustifolia* - **M. clove:** *Ravensara aromatica* - **M. fiber**: fr. *Aneilema fibrosa* (Dictyospermum f.) - **M. nutmeg:** *Ravensara aromatica* - **M. periwinkle:** *Catharanthus roses* - **M. plum:** *Flacourtia indica* - **M. rubber:** *Landolphia* spp.

madár (Hind) (**bark**): *Calotropis gigantea, C. procera.*

madder: **Dutch m.**: dark red var. of **common m.** (or **dyer's m.**) fr. *Rubia tinctorum* (rt.), made in the Netherlands - **m. family**: Rubiaceae - **Levant m.:** - **wild m.**: *Rubia peregrina.*

madeira: oleo (Po): tung oil.

madera (Sp): lumber, timber, wood.

maderwort: Absinthium.

Madhuca Gmelin (Sapot.) (seAs Malaysia): some 80 spp. of trees, several with economic values - **M. betis** (Blco.) Macbr. (PI): bk & lvs. u. to relieve stomach pain (children); latex appl. to abdom. to expel worms(!!); pd. bk. errhine; sd. FO u. as illuminant; wd. useful & durable - **M. butyracea** (Roxb.) MacBride = *Diploknema* b. H. J. Lam. - **M. indica** J. F. Gmelin (M. latifolia [Roxb.] F. Macbr.; Bassia l. Roxb.; Illipe l. Roxb.): moa tree; mahua (tree); butter tree (Indo-Pak); mahwa; bk. u. exter. as astr. in lotions for chron. skin ulcers; pruritus (itching); gargle (for bleeding, spongy gums, tonsilitis; pharyngitis): in rheum.; tan (17% tannin); lvs. u. exter. to treat burns & scalds, inflammn. (in ghee); fls. eaten; u. as expect. in chronic bronchitis; cooked in lard, u. in hemorrhoids; corollas distd. to give mohua liquor (intoxicant); u. as ton. astr. sds. yield FO (mahua butter); u. in skin dis.; fish poison; good wd. - **M. longifolia** (Koenig) Dubard (Bassia l. Roxb.; Illipe malabrorum Koenig) (trop. As, esp. Ind, SriL; Mad, WI); mahua (tree); mo(w)a: moha; mah(o)wa; mi; mowrah (butter tree): illipe b. t.; eloopa; sds. u. to prod. FO, u. in margarines, &c.; fls. edible (raw, cooked) & u. to prod. alc. bev. (davu or darre) (c. sucrose 10-15% invert sugar 45%); sd. cake u. to kill worms in lawns(!!); bk. & lvs. astr. u. in chronic ulcers, gargle in tonsillitis & pharyngitis; fls. expect. & in chronic bronchitis; FO u. as lax. & in chronic rheum. - (M. l. var. *Rom.* Chevalier) - **M. neriifolia** (Moon) H. J. Lam. = (M. malabarica (Bedd.) R. N. Parker; Bassia m. Bedd.) (Malabar coast, Ind, SriL) gammi (SriL); frt. u. in rheum. gall bladder dis., asthma, anthelm; sd. FO in rheum.; fls. in kidney dis.

Madia elegans Lindl. (Comp.) (nwNA, nOre [US] s to Baja Calif [Mex]): common madia; tarweed; magnonette vine (Eng); sds. u. for piole (also some other spp.) - **M. glomerata** Hook. (wNA, Calif-Alas [US]); tarweed; sd. u. as food (Am Indians) - **M. sativa** Molina (SA, esp. Ch, Arg; Calif to Alas [US]; cult. Eu, Ch, sAf): (Chile) tarweed; melosa (Ch); sd. source of madia oil, FO, 30-40%, semi-drying; u. as cooking oil (Ch), lamp oil, in soap mfr., paints, olive oil substit.; former (Ch) + potential future oil-seed crop; hb. as sheep fodder; pl. inf. u. in digestive dis., gout, ischialgia; frt. edible.

madnap: parsley.

madorius: *Calotropis gigantea.*

Madras gum: a very red form of Indian Gum Acacia.

madre dell'erbe (It): *Artemisia vulgaris.* - **m. selva** (Sp): garden honeysuckle, *Lonicera caprifolium.*

madrona: **madrone: madroño** (Sp): strawberry tree; *Arbutus menziesii* & *A. unedo*, the bk. of wh. c. **madronin** (antibiotic princ.).

madstone: concretion fr. gastro-int. tract of deer or other mammal; hair ball; supposedly of use to treat rabies ("madness"); *cf.* bezoar (2576).

madweed: 1) belladonna 2) *Scutellaria lateriflora* - **madwort**: *Camelina sativa*; *Asperugo procumbens*; *Alyssum* spp. (**German m.**); *Lobularia maritima.*

mäe boâ (Br): *Cissus alata.*

Maerua angolensis DC (Capparid.) (tropAf): frt. eaten but toxic; rt. u. in homicidal poisoning; lf. paste u. for lupus (Eth); purg.; placed on painful spot as analg. - **M. oblongifolia** A. Rich. (M. virgata Gilg; Niebuhria o. DC) (trop wAf, Ind): unripe frt. edible; rt. sweet, edible, u. as refreshing & fattening food; ton. stim.; u. with other drugs to treat syph. (Seneg; diangara Cayor) (255) - **M. siamensis** Pax (Niebuhria s. Kurz) (Indoch): rts. u. in vapor bath for swollen legs - **M. trichophylla** Gilg (sAf): roasted rt. u. as chicory substit. - **M. triphylla** A. Rich. (trop eAf): frt. edible; rt. softened in water u. as ton. in marasmus & in malnutrition in babies; in urinary dis.; sts. chewed for dry cough (Tanz) - **M. virgata** Brongn. (Courbonia virgata Brongn.) (c. tropAf, incl. Sudan): frts. c. betaine, betonicine, turicine, hydroxy-stachydrine; rts. c. tetramethylammonium iodide (tetramine), l-stachydrine; hence with strong fishy odor (1635); rt. appl. locally to chest for cough (Nigeria); bk. u. by Masai medicine men; toxic!

Maesa edulis C. T. White (Myrsin.) (PapNG, Oceania): u. to produce sterility in women (PapNG); for very sick person, lvs. heated & put in bed under person (Solomon Is) - **M. indica** Wall. (Ind): berries u. as anthelm.; rts. in syph.; lvs. as fish poison; condim. (Vietnam); c. saponins, maesaquinone, tannin; medl. & tech. potentials seem interesting (283) - **M. japonica** Mor. (eAs): u. as antipyr. (Taiwan), in mal. (Chin); lf. decoc. bechic (Indoch); rts. in blenorrhagia; c. maesaquinone - **M. lanceolata** Forssk. (eAf, Ethio): likoi (Hehe); frt. u. as purg. tenifuge, ascarifuge, in sore throat; inf. of decorticated rt. emet. (Zulus); rt. decoc. u. for lower abdom. pain during pregnancy, gastric. dis. (children); c. maesaquinone with antibiotic props. (cholera, diarrh., GU inf.); bk. & frt. u. to make stim. beverage ("tea") - **M. laxa** Mez (My) (PI): young lvs. u. in splenomegaly; pl. u. as fish poison - **M. parksii** Gillespie (Fiji): bk. crushed with water, strained thru coconut cloth & drunk for stomachache - **M. perlarius** Merr. (M. tonkinensis Mez) (Chin, Indoch): lvs. crushed & bound over broken bones; rt. diur. stom. (Indoch); c. alk. - **M. picta** Hochst. (Ethio); frt. u. for tapeworms & roundworms (*cf. Embelia*) - **M. ramentacea** A DC (s&seAs): pounded lvs. appl. to skin dis. (scabies+), eaten with rice for pain in region of heart (Mal); c. saponin - **M. subdentata** DC (Indoch): rt. inf. u. to expel leeches swallowed in water - **M. tetrandra** A DC (Indones): rt. u. as antipyr.

Maesopsis eminii Engler (monotypic) (Rhamn.) (tropAf-Nig to Angola): "musizi(i)"; bk. decoc. diur. purg. (Camer); lvs. drastic, emet. diur. (Nig); useful wd. in carpentry.

magdelate: Mg mandelate.

Magendie's solution: morphine sulfate 3.52 % soln. in water.

Magenkraut (Ge): Absinthium.

magey = maguey, *Agave* spp.

maggots: larvae of flies (ord. Diptera); in therapy sometimes were appl. directly to wounds of soldiers (WWI+) to accelerate their healing & to clean out debris; now in part replaced by less potent allantoin or urea (wh. are contained in the maggots); represents painless, inexpensive treatment replacing surgery; often represent larvae of horse bot fly, *Gastrophilus intestinalis*; maggots are an article of diet among many primitive peoples, being richly nutritious (1636).

Maghz (Urdu, Pers): seed; kernel; pith.

Magisterium Coccionellae (L.): carmine.

Magnamycin™: Carboymcin A; macrolide antibiotic prod. by *Streptomyces halstedii* (soil fungus); u. as antibacterial against resistant strains of Gm.+ organisms; spectrum rather like those of penicillin & erythromycin; withdrawn because of poor absn. fr. gastro-intest. tract.

magnesia: MgO: prepd. fr. magnesite, periclase; important in med.; u. antacid lax.; light and heavy forms recognized.

magnesium (Mg): an alkaline earth metal; several compds. u. med as cath., antacid, &c. (ex. m. hyroxide); low Mg levels of blood serum occur in cardiomyopathy; of value in mitral valve prolapse (2024) **- m. carbonate**; **m. chloride**; **m. citrate**; **m. phosphate**; **m. trisilicate** (& other m. silicates); combns. with aluminum hydroxide ("magaldrates") (do not confuse with magdelate = Mg mandelate) - **m. hydroxide:** m. hydrate; $Mg(OH)_2$; occ. in nature as mineral (brucite); u. in form of slurry, milk of magnesia; antacid, lax. (E) - **m lactate:** u. to supply Mg needs of body; sustained release tabs. reduces cathartic effects - **M. oxide**; magnesia (*qv.*); occ. in nature as mineral (periclase); prepd. fr. magnesite $((MgCO_3)Mg(OH)_2 5H_2 0)$; u. as lax. antacid; in hypomagnesemia - **m. stearate:** comm. prod. also c. some m. palmitate; u. in baby dusting powers; tablet lubricant - **m. sulfate:** Epsom salts; bitter s.; occ. in nature as mineral (epsomite) (fr. Epsom springs, Eng); cath. anticonvulsant; exter. for local inflammn., infected wounds; in general anesthesia (veter. med.); numerous indus. uses.

magnet: lodestone; body that attracts iron, &c. showing magnetic force; in antiquity pd. u. per ora for "gross humors" (Dioscorides); as rings for rheum. & gout; pd. u. in plasters to extract bullets fr. flesh; a physician, Trotula (ca. 1000), recommended tying a magnet around the leg of a pregnant woman as a means of drawing out the infant (**Magnet medicine**).

magnetite: $Fe_3 0_4$; lodestone; ore with magnetic properties.

magnoflorine: alk. (isoquinoline) orig. isolated fr. *Magnolia grandiflora* (1954); also occ. in *Cocculus* & *Thalictrum* spp., *Aristolochia clematitis,* &c. (284).

Magnolia acuminata L. (Magnoli) (eNA): cucumber tree; "cowcumber tree" (seUS); bk. u. bitter ton. antipyr. in rheum; vermifuge (eAm Indians); bk. u. for indigest. (chewed or as inf.) (Talladega Co., Ala [US]) (2035); wd. u. for flooring - **M. coco** DC (M. pumila Andr.; Liriodendron lililifera L.) (Indochin) buds, fls. & stalks u. as antipyr. stim. ton., in chronic rheum. - **M. denudata** Desr. (Chin; cult. Jap, Eu): bk. c. salicifoline, magnocurarine; buds & sds. u. as antipyr. in headache, as carm. diaphor. - **M. fraseri** Walt. (seUS): (large-leaved) umbrella tree; u. like *M. acuminata;* ornam. - **M. grandiflora** L. (seUS): bull bay; bk. u. in malar. rheum. as stim. arom. bitter ton.; fls. very arom.; popular shade & ornam. tree; frt. light pinkish, sds. bright coral red; st. bk. c. salicifoline, magnoflorine; diaphor. (Am Indians); hypotensive; sd. FO u. in soap mfr., wd. u. for furniture; bk. decoc. u. to bathe body to reduce itching of prickly heat (Choctaw Indians, [La]); cult. Ind, Jap); VO (2036) **- M. hypoleuca** Sieb. & Zucc. = *M. obovata* - **M. kobus** DC (Jap): bk. u. arom. ton; source of kobuschi oil (VO); fl. buds u. for headache & nasal congestion (Chin); wd. useful; cult. ornam. - **M. macrophylla** Mx. (e.US): (large-leaved) cucumber tree; large-leaved umbrella tree; u. med. (Mex); ornam. - **M. obovata** Thunb. (eAs, esp. Jap): honoki (Jap); large stately tree (much like *M. officinalis*); bk. c. VO with magnolol, machilol, u. stom. ton.; frt. rich in FO; wd. u. to prep. finest drawing charcoal; to make utensils; u. in engravings; young lf. & sd. pd. u. locally in fomentations & for colic (Ainu) - **M. officinalis** Rehd. & Wils. (wChina): bk. u. anorexia, nausea, dys.; anthelm. aphrod. diur. in pyrosis; fl. bud decoc. emmen.; said active on heart (313); expect. in cough; ton. stom. (85); considered toxic in Chin - **M. pyramidata** Bartr. (seUS): pyramid magnolia; "wood oread"; bk. ton. astr.; fls. arom. - **M. schiedeana** Schlecht. (Mex): "corpus"; fl. decoc. u. for scorpion strings (Tepic) - **M. tripetala** L. (eUS): umbrella tree; u. med. (formerly in USP); ornam. tree - **M. virginiana** L. (M. glauca L.) (eUS; in swampy places; cult. Eu): beaver tree; sweet bay; bk. (in brandy) bechic (colonial days); bk. decoc. in fever astr. for diarrh. dys.; colds, rheum., mal. ("intermittent fever"); other uses similar to those for *M. acuminata;* stim. (Ind); fls. popular homeopath. drug; wd. used; ornam. shrub or tree (-24 m); wd. u. for stakes in spitting meat (added sweetish flavor when tree in fl., otherwise very bitter) (Fla [US]) (2037).

magnolia: alley m: *Paulownia tomentosa** - **m. bark:** fr. *Magnolia acuminata, M. virginiana.* - **m. family:** Magnoliaceae - **Japanese m.** (swGa [US]): *Liriodendron tulipifera* - **laurel m.: small m.:** *M. virginiana* - **southern m.:** *M. grandiflora* - **swamp m.: sweet m.:** small m.

Magnoliaceae: Magnolia Fam.; shrubs & trees with alternate lvs., large fls. with numerous stamens; frt. collection of follicles (ex. yellow poplar tree, *Liriodendron*); punctate (pellucid spots due to oleoresin); some spp. arom.; some *Magnolia* spp. cult. as ornam. shrubs or trees; wds. of some spp. u. as timber - **magnolier glauque** (Fr): *Magnolia virginiana.*

maguey: (swUS, Mex): century plant with fleshy lvs.: *Agave americana* & other *A.* spp. - **m. manso** (Mex): *A. salmiana, A. potatorum* - **m. meco** (Mex): *A. americana* - **m. raiz** (Mex): rt. of *Agave* spp. formerly u against syph. (Mex).

maha pangiri grass: *Cymbopogon winterianus.*

mahaleb (cherry): mahalib (Ar.): *Prunus mahaleb* (sd.) - **mahalita** (king of bitters) (Ind): *Andrographis paniculata* - **mahlib: mahlab (seeds):** mahaleb.

mahmira: mamira.

mahoe, seaside: *Thespesia populnea.*

mahogany: **m. bush**: 1) *Ceanothus* spp. (US) 2) *Cercocarpus parvifolius* & other *C.* spp. - **m. family**: Meliaceae - **Indian m.:** *Toona surenii*; *Soymida febrifuga* - **mountain m.**: 1) *Betula lenta* 2) *Cercocarpus parvifolius* - **m. tree:** true m.: *Swielenai* spp. esp. - **West Indian m.** *Swietenia mahogani* - **white m.** (Calif [US]): *Anacardium occidentale*; *Eucalyptus acmenioides, E. robusta.*

Mahonia Nutt. (Berberid): sometimes treated as a genus itself, sometimes as sub-genus (or section) of *Berberis*; shrubs without spines; lvs. pinnately comp., evergreen - **M. aquifolium** Nutt. (Pacif NA): (tall) Or. grape; (Rocky) Mountain grape; tall mahonia; cooked frt. eaten by Am Indians; u. med.; cult. ornam. & widely natd. (Eu +) - **M. bealei** Carr. (Chin: lvs. & sds. u. as ton. in rheum., tinnitus; substit. for Coptis & *Ilex cornuta* - **M. fortunei** Fedde (Chin): whole pl./rt. decoc. u. as antipyr. & for "red eyes"; ornam. shrub (Eu) - **M. fremontii** Fedde (s&swUS): desert mahonia; Fremont m.; decoc. of branches u. as antipyr., in rheum. - **M. napaulensis** DC (Ind, Nepal): chatri (Nepal): berries diur. dem. in dys.; rts. of old plants richer in alk. (umbellatine) - **M. nervosa** Nutt. (Pac, Cana, & US): Cascade's mahonia; a source of Ore [US] grape rt.; frt. ("berries") edible; rt. u. occasionally as bitter; yellow dye for wool & silk - **M. philippinensis** Takeda (PI): bk. & rts. u. as ton. stim. antiperiod. stom.; in larger doses cath.; in dys. diarrh., fevers, splenic enlargement (fr. mal.) - **M. pinnata** Fedde (nMex): palo amarillo (Chihuahua, Mex) frt. edible, quite sweet; bk. u. med. (Mex); wd. u. purg. in coloring - **M. piperiana** Abrams (wNA): Piper's mahonia; berries edible - **M. repens** (Lindl.) G. Don. f. (Pac, Cana, & US.; Rocky Mt. states): Ore [US] grape; holly g.; rt. c. berberine; blue berries source of a fermented beverage; in jelly; shrub conserves soil; u. as decoc. for stomach dis. & to "purify the blood" (Indians, Mendocino Co., Calif [US]) - **M. swaseyi** Fedde (Tex): blue agarita; rt. c. berberine, berbamine (1637); Tex barberry; T. mahonia; berries effectively u. for jelly & wine; wd. & rts. u. as dye (Indians) - **M. trifoliata** Fedde. (Tex): agarita; algerita; pd. chaparral berry; "wild currant"; rt. u. as dressings for impetigo, ringworm, etc.; red berries edible, u. for jelly making; yellow wd. u. as dye (Tex pioneers).

mahonia, trailing: *Mahonia aquifolium* & some other spp.

mahua butter: m. (seed) oil: mahwa oil: FO tr. sds. of *Bassia latifolia, B. longifolia.*

Maianthemum bifolium F. W. Schmidt (M. dilatatum A. Nels. & J. F. Macbr.) (Lili) (nEuras, incl. Switz): May flower; M. lily; rts. & hb. c. digitaloid glycosides; u. as diur. in folk med. - **M. canadense** Desf. (eNA, Cana s to La & Ga): false (or wild) lily of the valley; Cana Mayflower; Massachusetts M.; pl. c. saponins, tannins, no alk.; incr. intest. tone, reduced BP (285); u. as hypotensive; hb. inf. drunk for headache (Indians) & sore throat u. as expectorant; smoke inhaled.

Maibowle (Ge): Maitrank.

maicillo (Sal): sorghum.

maiden cane (US): *Sorghum halepense.*

maidenhair (fern,) (European): *Adiantum capillus-veneris* - **American m.** (f.): *A. pedatum.*

maid of Kent (Eng): *Dianthus caryophyllus* L. (gillyflower; carnation).

Maieta guianensis Aubl. (Melastomat.) (Pe, Co, Ve, Br, Ec): lvs. chewed into ball wh. is appl. to wounds to stop bleeding (hemostatic); frts. eaten.

Maiglöckchen (Ge): lily of the valley.

maikoa (Ec): *Datura arborea, D. suaveolens* (narcotic prep.) (Jivaro Indian).

Mais (Fr, Ge): maize - **Maispistille** (Ge): corn stigmas, c. silk.

Maisena: maize starch.

Maispistille (Ge): corn stigmas, c. silk.

Maitrank (Ge): **Maiwein:** May wine; a sweet wine prep. with fresh hb. of *Asperula odorata* (just before fl.); long history of use, esp. around Vienna & in Rhine region.

maiz (Sp): Indian corn - **m. de guinea:** Guinea corn, millet - **maize (, Indian):** Indian corn, *Zea mays.*

majoo: maju.

makasar: v. macassar.

majoram: Majoran (Ge): **Majorana** (L): **Majorana hortensis** Moench = *Origanum majorana* L. - **M. onites** Benth. = *Origanum onites* - **M. maru** Briq. = *Origanum m.*

maju (Punjabi): frt. of *Juniperus macropoda & J. communis* - **majuelo** (Sp): *Crataegus oxyacantha.*

maki (Jap): *Podocarpus*

mala (Sp): pome; apple. - **m. mujer: m. suegra** (NM [US], Sp) tumbleweed.

Malabar: v. under kino.

Malacantha alnifolia Pierre (monotypic) (Sapot.) (wAf): hot crushed lvs. appl. to wounds caused by bullets; claimed to draw out the lead or iron projectile & aid in healing of wound (Senegal).

Malacca canes: sts. of *Calamus scipionum* Lour.: usually imp. into Ind after having been smoked to prod. rich brown color.

Malachra (Pan): kwala tumat (Cuna Indians): u. for stomachache - **M. alceifolia** Jacq. (Malv.) (WI, tropAm contin.): malva mulata (Cu); lf. in pouitices & baths for sores, gargle for sore throat, cough (Trin) - **M. capitata** L. (WI contin., tropAm + pantrop.): malva mulata (Cu); cult. ann. hb.; fiber u. like jute; u. as emoll. & pectoral (Réunion).

malacology: study of mollusks.

Malagueta (Sp): bay - **Malaguetta pepper**: *Amomum granum-paradisi.*

Malaisia scandens Bl. (Mor) (monotypic) (eAs, Austral Oc): crow ash (Austral): lf. decoc. u. in childbirth (PI); vine u. as crude rope.

malambo 1) matias bark, *Croton malambo* 2) (WI): winter bark, *Canella winterana.*

malanga: 1) *Xanthosoma* spp. 2) (PR, Cu): dasheen.

Malarene: 1) a standardized cinchona febrifuge 2) propr. consisting of amorphous cinchona alks. with arsenic compd.; u. malar.

malaria: a serious tropical infectious dis. endemic to Africa, Asia, Oceania, Central & South America, &c.; caused by the protozoan *Plasmodium* spp. (incl. *P. vivax*); very debilitating & harmful; sometimes deadly; has been treated with Cinchona (quinine) & derivs., synthetic compds., &c.; a vaccine using genetic engineering procedures is being researched; some Af natives have a partial immunity.

Malaxis acuminata D. Don (Microstylis wallichii Lindl.) (Orchid.) (nInd, Himal): pseudo bulbs u. as general ton. & spermatopoetic (2258).

Malaya: Malay Peninsula - **Malaysia:** Malay Archipelago (West M; East M= nBornes.

male fern (root): Aspidium. *Dryopteris* spp. - **m. jalap:** *Convolvulus orisabensis* - **m. nervine**: *Cypripedium parviflorum.*

malet bark: inner bk. of *Eucalyptus occidentalis.*

malevecino (Pan): Cascara Amarga.

malic acid: apple acid; dicarboxy acid, CH_2 $CHOH(COOH)_2$; occ. in frts., rhubarb, &c.; malus, apple (where first discovered); u. in foods & beverages, competitive with citric acid.

malkanguni (Hind) = *Celastrus paniculatus.*

mallein: bouillon culture of *Pfeifferella mallei* (filtrate sterilized); u. as in diagnosis of glanders (farcy).

Mallotus apelti (Lour.) Muell. Arg. (Euphorbi) (Chin): lvs. mixed with glutinous rice & appl. as salve to itchy rash - **M. auriculatus** Merr. (PI): rt. chewed with betel nut & appl. to ulcers - **M. lianus** Croizat (Chin): lvs. u. to heal cuts & similar injuries - **M. paniculatus** Muell. Arg. (eAs): lf. decoc. u. to cleanse wounds; rt. decoc. as postpartum protective med. - **M. philippinensis** (Lam.) Muell. (Arg, Ind, seChin, Burma, EI, Mal, Indones, PapNG, PI, Austral): kamala tree; frts. source of Kamala, Glandulae Rottlerae (the hairs & glands) separated fr. the capsules; u. vermif. for tapeworm (tenifuge, tenicide), roundworm, pin (ceca) worms, threadworm, esp. in poultry (2040) (use now discountenanced by USDA since toxic); but not always dependable; in veter. med. as vermif.; cath. (given with milk); sds. source of camul oil (FO) (Ind) ; c. kamlolenic acid; u. in ointments for skin dis. (scabies, ringworm), tumors; red dye fr. frt. u. in silk dyes, varnishes, paints, &c.; frt. & lf. decoc. u. in colds (PI); lf. decoc. u. in diarrh. (PapNG); sap appl. extern. to wounds (PapNG) - **M. poilanei** Gagnep. (Indochin): root ton. & aphrod.; in headache - **M. repandus** Muell.-Arg. (seAs): u. to stop itching, as insecticide (Taiwan); rt. decoc. for colds - **M. ricinoides** Muell.-Arg. (seAs): sap mixed with coconut juice for dys. (NG)

mallow: common name appl. to various taxa of Malvaceae (*Malva, Abutilon, Althaea, Hibiscus, Lavatera*, &c.) - **blue m.**: **common m.: dwarf m.:** *Malva neglecta, M. rotundifolia* (fls. lvs. hb.) - **m. family**: Malvaceae - **forest m.:** *M. sylvestris* - **Indian m.:** *Abutilon theophrasti* - **Jew's m.:** jute - **light m.:** forest m. - **low m.:** blue m. - **marsh m.:** *Althaea officinalis* - **Mexican m.: musk m.:** 1) (Eu) *M. moschata* 2) *Hibiscus abelmoschus* - **northern m.:** 1) blue m. 2) *M. borealis* - **poppy m.:** *Callirhoë digitata* - **rose m.:** *Hibiscus* spp. - **round-leaved m.:** blue m. - **running m.:** M.

neglecta - **tall m.:** light m. - **tree m.:** *Lavatera arborea* - **white m.:** marsh m.

Malmea depressa (Baill.) R. E. Fries (Annon.) (tropAm): box elemuy; rt. boiled with Zea silk decoc. u. GC & in kidney & bladder dis.

mal-me-quer do campo (Br): *Grindelia* spp.

malört weed: *cf.* malurt (Dan.): absinthe; u. in Malort Bitters.

Malouetia duckei Markgraf (Apocyn) (Col+): crushed lvs. & sts. u. as fish poison - **M. nitida** Spruce (CA, Br): guachamaca; bk. c. guachamacine, with curare-like action; pl. sap u. as arrow poison (Indians) - **M. tamaquarina** (Aubl.) A DC (nSA, esp. Co): cuchara-caspi (Pe, Co): "spoon tree" (SC utensils made fr. soft white wd.); ripe frt. chief food of pajuíl (bird); the bones of which are held toxic to dogs at time of treelet's fruiting (286); hallucinogen (danger!); latex u. to hasten wound healing; said adulterant of Hevea rubber (formerly).

Malpighia coccigera L. (Malpighi) (WI): frts. esculent, very rich in vit. C; palo bronco de pinar (Cu); frts. esculent, very rich in Vit. C - **M. glabra** L. (sTex [US]; WI, tropAm; cult. Ind): Barbados (or Jamaica) cherry; frt. delicious; somewhat like true cherry in flavor; u. in preserves, jams, etc.; beverage made fr. juice u. for sore throat; frt. u. astr. in dys. diarrh. liver dis.; pd. sd. mixed with resin fr. tree u. in pulmonary dis.; tree source of nance bark (astr.) very rich in vit. C; as tinct. u. as antipyr. - **M. polytricha** A. Juss. (Cu, Bah): palo bronco (Cu); touch-me-not (Bah); rt. decoc. u. diur. for bedwetting in children (Bah) - **M. punicifolia** L. (WI, CA, nSA): WI cherry; Puerto Rico c.; Barbados c.; acerola; cerezo (Cu); ripe frt. deliciously edible, said richest known natural source of vit. C (juice some 85-180 times as rich as fresh orange juice) (2042); u. for jellies, syr., tarts; bk. u. as red dye (manquitta bk), tan; juice u. in place of orange juice where individual allergic; shrub cult.; some consider conspecific with & a synonym of *M. glabra* - **M. urens** L. (tropAm): juicy frt. edible; called "Grosse cérise" (Martinique).

malt: (Eng, Fr) partly & artificially germinated & fermented grain generally of *Hordeum vulgare* (but also sometimes of rye or wheat), then dried; rich in m. diastase (*qv.*), also c. maltose, glucose, dextrins, protein bodies, salts, protease; u. food digestive aid, vehicle (CLO, &c.) - **m. blocks:** confection popular in Eu - **m. extract:** prep. made by extn. of m. with warm water, 50°C - **m. syrup:** prep. made by evapn. of m. extr. with hops; u. to make beer in home ("home brew").

Malt Soup Extract™: m. ext. with diastase removed & $KHCO_3$ added; u. lax. for infants: said useful in intractable cases of pruritus ani (itching & burning symptoms); acts by encouraging growth of aciduric bacteria.

maltase: α-glucosidase; enzyme wh. hydrolyzes α-glycosides (ex. maltose α-glucose); occ. in yeast, fungi, malt, small intest.; often found with amylase.

malted milk: combn. of malt (barley), desicc. milk, wheat flour, $NaHCO_3$, &c.; u. as nutrient, esp. in milk shakes.

"Maltese": v. under sponge.

Maltin: diastase of malt.

Maltine™: ext. of malted barley, wheat, & oats; u. nutr. vehicle.

maltobiose: maltose: malt sugar; a disaccharide sugar, $C_{12}H_{22}O_{11}$; in hard crystals; less sweet than sucrose; occ. in malt ext.; u. in baking ind.

maltol: hydroxy-methyl-pyrone; const. of arom. products found in roasting of bread, chocolate, chicory, &c., occ. in larch bark, *Abies* lvs., roasted malt, wood tars; prod. by dry distn. of starch & cellulose, u. flavor, to give "fresh baked" odor to bakery goods.

Maltum (Hordei) (L.): barley malt.

Malucidin™: antibiotic fr. yeast fraction wh. ends pregnancy in dogs, sheep, &c.; antimicrobial; antifungal; supposed to be anti-allergic (1957-) (1832).

malum (L): means both apple and evil (calamity, adversity, &c.), apparently with the idea that in the Garden of Eden, Eve tempted Adam with an apple.

malurt (Dan, Nor): absinthe - **strand-m**. (Dan, Nor): Santonica.

Malus Mill. (Ros) (nTemp reg.): apple; c. 35 spp., many vars.; chief frt. tree of temp. zone; often included under *Pyrus* (v. also under "apple") - **M. angustifolia** Mx. (Pyrus a. Aiton) (eUS): southern crab apple; frt. u. in jellies, preserves, pickles, cider; wd. useful; ornam. tree - **M. baccata** Borkh. (Pyrus b. L.) (nChin, Sib): Siberian crab (apple); Chin c. (a.); frts. eaten fresh, dr., or preserved; ornam. tree - **M. coronaria** (L) P. Mill. (eNA): (wild sweet) crab (apple); fragrant c.; Alabama c.; frts. inedible raw (as hard & sour); cooked to make jellies, preserves, marmalade; also for cider, vinegar; early US colonists buried apples over winter to reduce acidity; rt. u. as ton. in mal., TB (287); wd. hard, useful - **M. domestica** Borkh. = *M. pumila* - **M. fusca** Schneid. (Pyrus f. Raf.; P. diversifolia Bong.) (Aleutian Is, Alas to nCalif [US]): (Ore) (Pacific) crab apple; inner bk. decoc. u. ton. in rheum., cough; lax. diur.; juice fr.

bk. scrapings for eye dis. (Gitksan Indians, BC [Cana]); lvs. chewed for "lung trouble" (Makah Indians, BC); frt. eaten (sour) (BC Indians) (1768) - **M. halliana** Koehne (Chin, Mong, Jap): Hall crab apple; wild ornamental shrub; at times cult. - **M. hupehensis** Rehd. (Pyrus h. Pampan.) (Chin, Assam): Chinese crab apple; lvs. u. to make beverage tea (Chin); frt. edible; small ornam. tree - **M. X ioensis** (Wood) Britt. (cUS such as Minn, Kans): prairie crab apple; Iowa c.; small tree with double fls.; ornam.; several vars. & cultigens - **M. prunifolia** Borkh. var. *rinkii* Rehd. (neAs): Ringo crab apple; frt. bittersweet, edible; u. as carm. (Chin) - **M. pumila** Mill. (M. domestica Borkh.) (Euras; commonly cult. worldwide): (common) apple; fresh (cider) (apple juice, Succus Pomorum NF) or fermented —> (hard) cider; apple jack (higher alc. content); apple sauce; chewing apple claimed to cleanse teeth; frts. tonic for anemia; remedy in colic & bilious dis. (Chin); many vars.: McIntosh; Delicious (1900); Stark; Jonathan; Northern Spy; Winesap; Baldwin; Golden Delicious - **M. sieboldii** Rehd. (Pyrus s. Regel) (Jap, Korea): zumi; Toringo crab apple; bk. u. for yellow dye - **M. spectabilis** Borkh. (Chin): cultigen; frt. consumed - **M. sylvestris** (L) Mill. (P. malus L) (Euras-widely cult.): (common) apple (tree); very close to *M. pumila*; some consider *M. pumila* a var. of *M. sylvestris*; frt., dried pomace, & peel u. in European med.; thus the pomace in pediatry for dyspepsia, diarrh., skin dis.; peel constit. of many health tea mixts. (2479); sd. FO u. as food oil, formerly lamp oil, &c.; excellent wd. u. for rulers, knobs, saw handles, &c.; good fuel (1638); fresh frt. delicious, nutritive, flavorsome food (2318); stem. bk. u. tonic., antipyr.; ripe frt. source of apple juice; (fresh cider; apple juice, Succus Pomorum) (NF); (clear); cider (US) is turbid; (in US formerly & in GB, "cider" is fermented = hard c.) (c. 12% alcohol); distd. —> calvados (apple brancy) (or water may be frozen out); apple syrup (u. to humectate tobacco, cigarettes); vinegar, jellies; pulp for apple butter, pectin, apple powder (u. in diarrh. & other intest. dis.); peel u. as tea substit.; apple essence (VO); ursolic acid (fr. frt. peelings); yellow dye (2041).

Malva L. (Malv.): mallows (nTemp Zone): ca. 50 spp.; pls. generally rich in mucilage (2043); u. as dem. emoll. &c.; formerly u. in cancer (Eu, NA) - **M. alcea** L. (Eu, natd. NA): European (or vervain) mallow; u. as emoll. in folk med.; rt. adulterant of marshmallow rt. - **M. arborea** St. Hil. = *Althaea rosea* - **M. carolina** L. = *Modiola caroliniana* - **M. crispa** (L.) L. (Med reg.. natd. NA): **malva de castilla**: Castillian malva; curled mallow; hb. u. in folk medicine (NM [US], Mex) for sore throat, measles, wounds, boils, bruises, post parturition convalescence; cultd. for salads; pl. now grows wild over much of NA - **M. moschata** L. (Eu, nAf, cult. & natd. in NA): musk mallow; musk plant; u. arom., substit. for *M. rotundifolia & M. sylvestris*; hb. u. in folk med. as mucilaginous drug - **M. neglecta** Wallr. (M. rotundifolia sensu auct. non L.) (Euras, nAf, natd. Am, Austral): dwarf mallow; common m.; lvs. u. as dem. emoll. mild astr. in inflamn, gastroenteritis; in folk med. as vulner. in emoll., poultices; tea mixts.; weed - **M. nicaeensis** All. (Med reg., wAs, natd. NA): mallow; fls. u. in upper resp. dis., mild astr. pharyngitis, gastroenteritis; in folk med. in bladder dis. exter. as vulner.; as coloring matter in food indy. - **M. parviflora** L. (Med reg., AsMin, natd. Am, Austral): little (or Egyptian) mallow; small-flowered m. (Cana); cheeses; sds. rich in FO with biol. active fatty acids (314); lvs. mucilaginous u. as ton., vulner. in eye drops; lvs. & fls. as dem. in gargles, as emoll. in cataplasms (folk med.); rts. purg. (Arg); u. as foods for millenia (Near East) - **M. rotundifolia** L. (M. borealis Wallr.) (e&cEu; now widely distr. in temp zone): (common) mallow; dwarf (or low) m.; "cheeses"; hb. & lvs. (Herba Malvae minoris s. vulgaris); whitish fls. c. mucil. tannin; u. dem. in gargles, emoll. in cataplasms (folk med.), enemas (fls.); fls. as coloring agt. (prodn. Fr, Hung): various medl. uses (Lahul) (pl. lvs. sd.) thus diur. in kidney & bladder dis. (315) - **M. sylvestris** L. (Euras, nAf, natd. NA; cult. Ind): (common) mallow; fls. similar c. to & u. like *M. rotundifolia*; expect. in upper resp. tract, catarrhs., &c.; frts. edible - **M. verticillata** L. (Eu, natd. NA, SA, As): whorled (or curled) mallow; malva amargo (Pe); rt. decoc. u. as diur. antilithic; in diarrh.; decoc. for cough, general weakness, lactagog (Chin); in pertussis (Indochin); lvs. as poultice for suppurating wounds (Pe); sd. FO c. nutritive fatty acids like *M. parviflora*.

malva (SpAm): *Althaea rosea, Malva parviflora* (Guat); *M. rotundifolia*; *M. sylvestris* (CR); *Malvastrum vitifolium*; *Sida* spp., etc.; in Mex, 41 or more pl. spp. are referred to as **"malva"** - **m. blanca** (Cu): *Urena* - **m. flowers**: Malvae Flores (L.): 1) *M. rotundifolia* 2) *Althaea rosea*; *A. officinalis* - **m. grande** (Por): *Malva sylvestris* - **m. leaves**: Malvae Folia (L.): lvs. of *M.* spp.

Malvaceae: Mallow family; hbs. or shrubs of temp & trop climates of both hemispheres; lvs. simple; fls. usually 5-merous; calyx gamosepalous; stamens numerous, united in a column surrounding pistil; ovary 2- to many-chambered; frt. capsule or berry-like or separating into carpels (achenes); rich in mucilage; ex. *Gossypium* (cotton), *Hibiscus, Althaea, Abutilon, Malvaviscus, Pavonia.*

malvaisco (Por): *Althaea officinalis* (also **malvarisco**); *Alpinia officinarum*; *Urena lobata*; *Piper marginatum.*

malvarreal (Sp, Am): *Althaea rosea* - **malvavisco** (Sp): 1) *Althaea officinalis* 2) (Mex): *Sida triloba.*

Malvastrum coromandelianum Garcke (Malv.) (Mex, widely distr. trop, subtrop): lvs. appl. to treat carbuncles (PI); pl. emoll. bechic (WI); rt. decoc. drunk for menstrual dis. (Easter Is.); fls. diaph. pectoral - **M. lacteum** Standl. (Mex, Guat): pl. boiled & given for indig. in children; Mex med. (288) **M. peruvianum** (L.) Gray (Malva peruviana L.) (Ec, Pe): cuchi malva (blanca); rt. decoc. u. as purg. (2449).

malvavisco (Por): *Piper marginatum.*

Malvaviscus arboreus Cav. (Malv.) (sUS, Mex to Pe & Br; cult. US): sleeping hibiscus; Turk's cap; monacillo (Mex); wax mallow; fl. decoc. u. dys. diarrh. cystitis, for sore lips (Salv); lvs. fls., & rts. emoll. in pulm. dis. (cough, sore throat) (fl. decoc.); lf. decoc. to make hair smooth & lustrous (Salv); ornam. shrub; var. *drummondii* Schery (M. d. T & G) (seUS, Mex): frt. edible; ornam. shrub - var. *penduliflorus* Schery (Mex, SA): Turks's cap; ornam. shrub; u. med. (Mex); var. *mexicana*; u. for cough, pulmon. dis.; lf. in fever - **M. grandiflorus** HBK (Mex): manzanilla; mosilado; Turk's cap; fl. decoc. u. as gargle for sore throat; lf. & bk. decoc. for purpura; frt. sweet, edible; common cult. ornam. (seUS) - **M. pentacarpus** Moç. & Sess. (Mex): (malva visco); rt. u. like Althaea.

Malvella leprosa Krapov (Sida hederacea T. & G.) (Malv.) (seUS): rt. u. in diarrh. dys. burns (seIndians) (852).

malvita (Salv): *Urena* sp.

malvón (Mex): *Pelargonium inquinans.*

Malz (Ge): malt.

mama (Ec-Jivaro Indians): manioc.

mamáo (Po-Br): 1) papaya (tree or frt.) 2) pl. sucker.

mamee: mamei: **mamey** (**rojo**) (Pan): 1) *Manilkara zapota* (Mex, Cu): (also **m. colorado**) 2) *Mammea americana* (also **m. apple**) 3) mamou (*qv.*), & other plants.

mamira (Hind, Sind, Bombay): *Coptis teeta.*

Mammalia: order of chordate animals which suckle their young, possess hair, are warm-blooded & are generally viviparous; ex. elephant, man.

mamma teiga (Malta): *Hyoscyamus albus.*

Mammea africana Sabine (Guttif./Clusi.) (tropAf): African mammy apple; A. apricot; exudate, rts., bk. u. (resin) for parasitic skin dis., syph. as fish poison - **M. americana** L. (WI, CA, nSA): toddy tree (fr. fermenting sap); **mammee** (or mamey) **apple**; bk. decoc. u. against skin parasites (lower animals), in eczema; frt. edible; frt. juice as tincture u. for ulcers; lf. in mal.; liquor dist. fr. fls. u. refreshing beverage (eau de créole) as digest.; roasted sds. u. as insecticide (Ve); sap of trunk extern. u. to prevent insect bite; fermented to furnish toddy; resin exudn. fr. bk. u. for chigoes (niguas), red bugs, sand flies, &c.; frt. u. as preserves; marmalades; lumber wood (2044) - **M. harmandii** Kosterm. (Indochin): frts. eaten by women in childbirth - **M. longifolia** Planch. & Triana (Ind): fl. buds mild carm. astr. in dys. & hemorrhoids; frt. edible; fls. u. in perfumes; lf. buds dye silk red.

Mammillaria Haw. (Cact.) (swUS to Co): small plants as a rule; most popular gen. of ornam. cacti - **M. craigii** Lindsay (Chihuahua, nwMex): peyote de San Pedro; central part of pl. st. (after roasting) squeezed out into ear for earache, headache, deafness; hypnotic & hallucinogenic props. claimed (1639) - **M. grahamii** Engelm. var. *oliviae* L. Benson (nwMex): híkuri; frt. & st. eaten as sed. & hallucinogenic (1639) - **M. heyderi** Muehlenpf. (swUS, nwMex): biznaga; u. as hallucinogenic drug by the Tarahumara Indians (289); purplish frt. mashed & added to atole (maize) gruel (290); roasted center of "slabs" squeezed into ear for earache (Mex) - **M. magnimamma** Haworth (cMex): pincushion cactus; frts. eaten by nursing mothers to increase milk supply (Mex, Kickapoo Indians) (this an example of the doctrine of signatures, fr. resemblance of tubercles on st. to breasts; the sp. name means "large breasts") - **M. melanocentra** Poselger (M. runyoni Boed) (cMex): c. mamillarol (triterpenoid), steroid, acetovanillon (316); several spp. are reported u. med. in Mex: *M. geminispina* Haw. (M. bicolor Lehm.); *M. goodridgii* Scheer (strawberry cactus; near San Diego, Calif [US], Baja Calif [Mex]); and *M. obconella* Scheidw. (M. tetracantha Pfeiff.) - **M. microcarpa** Engelm. (swUS, nwMex): spines burned off, pl. boiled, peeled, and "fiber" (vasc. cylinder) crushed; resultant liquid u. as ear drops for earache (Seri Indians, Sonora, Mex) (*cf. M. heyderi*) - **M. pectinifera** F.A.C. Weber (Solisia pectinata Britt. & Rose); (sMex, esp. Puebla): peyotillo; cochiniti; u. in folk med., also u. like peyote.

Mammilopsis senilis Britt. & Rose (Cact.) (nMex): u. as hallucinogenic like Peyote (Tarahumara Indians).

mam(m)oneiro (Br): *Carica papaya* tree (SC fr. L. mamma, breast, fr. frt. shape).

mamona (Br): *Ricinus communis.*

mamou (La-Fr): mamu (rt.); *Erythrina herbacea.**

mamrah gum: Ind Acacia gum obt. fr. *Acacia arabica & A. modesta* (mixed).

man root: *Ipomoca pandurata*

Manaca: 1) **manaca:** raintree, *Brunfelsia hopeana* (rt.) 2) various palms: *Calyptrogyne dulcis*; *Euterpe*; *Attalea.*

manatee: *Trichechus latirostris* Harlan (*Trichechidae - Sirenia - Mammalia*): (Fla [US], CA - WI - coastal waters): sea cow; common American manatee; an herbivore -5 m long; meat eaten formerly by Indians; protected by law (US); 2 related spp. in SA.

manchineel: Mancinella (L.): *Hippomane mancinella*

mancona bark: fr. *Erythrophleum.*

mandarin (orange): *Citrus reticulata* (*C. mandurensis*); source of VO, mandarin oil (prod. It, Br)

mandarina (Sp): Mandarin orange.

mandelic acid: arom. carboxy acid; Ph-CHOH. COOH; occ. naturally in form of cyanide: mandelonitrile in bitter almond, &c.

Mandeln (Ge): almonds - **mandelonitrile glycoside**: active component of wild cherry bark, etc., which on hydrolysis yields HCN & benzaldehyde & dextrose.

Mandelzypergras (Ge): *Cyperus esculentus.*

Mandevilla steyermarkii Woodson (Apocyn) (nBr, Amazon) panacea of Vaupés Indians; latex u. for healing of sores & skin infect.; roasted lvs. added to porridge for diarrh.; fls. in chicha u. as aphrod. - **M. subsagittata** Woodson (Apocyn.) (sMex to Pe): flor de(l) mico; pl. u. for VD (Yuc); for female sterility (Ve) - **M. velutina** (Mart.) Woodson, Jr. (Br): rt. ext. u. as antidote for jararaca bites; blocks bradykinin receptors (reducing pain).

Mandioca (L, Por): maniock; tapioca (flour); Amylum Brasiliense (Brazilian Arrowroot) fr. *Manihot esculenta.*

mandola (Hung): **mandorla** (It): almond.

Mandragora autumnalis Bertol. (Solan) (Med reg.): autumn mandrake; source of m. root or Alraun rt.; long known & u. as sed. hypn. mydr. anesth. poison; mentioned in Ebers Papyrus - **M. officinarum** L. (Atropa m. L) (Med reg.; sometimes cult. cEu): insane root; rt. c. scopolamine, atropine; u. much as prec.; folklore; hb. (rt.) also important in ancient and medieval periods as a magic drug (fr. appearance of rt. to human form) & panacea; rt. u. as hypnotic, anesthetic (Dioscorides), aphrod. sed.; more recently in gastr. ulcer, obesity, colic, hemorrhoids, dysmenorrh. asthma, hay fever, pertussis; hb. u. in homeopath. for dystonia, gastritis, rheum., &c. (2045) - **M. turcomanica** Mizg. (swTurkmenistan): white mandragora; similar props. & uses to prec.; may represent the "soma" (khaoma) of central As.

mandrake (root,) (American): *Podophyllum peltatum* - **European m.**: *Mandragora offcinarum.*

manduba (Por): tapioca.

mandubi (Por): peanut.

manduvi (Pan): peanut - **mandwa** (Urdu): *Eleusine coracana.*

Manettia cordifolia Mart. (Rubi.) (tropSA): rt. of tree furnishes Radix Ipec. striatae minor, u. as adulter. of Ipecac rt.; cult. ornam. - **M. glandulosa** P. & E. (Ec): lvs. chewed to prevent decay of teeth; causes blackened teeth; mostly chewed by women who often chew "yuca" in making of chicha (2449) - **M. ignita** Schumann (tropAm): rt. u. to adulter. Ipecac rt.; c. 0.14% alk. incl. emetine - **M. reclinata** L. (M. coccinea Willd.) (WI cont., tropAm): "cambustera cimarrona" (Cu); nut u. to coat teeth black (Ec-Jivaro Indians).

Manfreda Salisb. (Agav./Amaryllid): genus of fleshy hbs. close to *Agave*; rt. of Mex spp. called amole; u. like soap & sold as detergent in markets (amole); ex. *M. jaliscana* Rose ("mescalito") - **M. brachystachys** Lindley (nMex, Guat): amole; soap plant (rt. u. like soap) (317) - **M. jaliscana** Rose (Agave j. (Rose) Berger (Mex: Guadalajara): rt. u. as detergent like soap; sold in local markets for this - **M. variegata** Rose (seTex [US], nMex): "huaco"; mottled manfreda; hb. u. as remedy in snakebite - **M. virginica** Rose (Agave v. L.) (eUS): rattlesnake master; false aloe; rt. bitter, u. as carm. in colic; antispasmod., stom.; rts. boiled in milk for rattler bite (eAm Indian) (852).

manga (Br): frt. of mangueira: mango, *Mangifera indica.*

manganese: manganum (L); Mn; essential element; with many compds. u. in med./pharmacy.

mangel: *Beta vulgaris* var. *macrorhiza* - **m. wurzel** (Dutch, Ge): *B. vulg.* v. *macrorhiza*; u. as cattle food (Eu).

mangeriçao (Por): sweet basil, Ocimum; *Alternanthera sessilis.*

mangerioba (Br): *Cassia occidentalis*; *C. sericea.*

mangerona (Br): *Majorana hortensis.*

Mangifera indica L. (Anacardi.) (Ind, cult. widely in trop&subtrop): (common) mango; a cultigen; ripe frts. edible & delicious (treasured); most popular

Indian frt.; large unripe frts. of some var. u. for conserves & pickles (eAs); young lvs. u. cholagog; bk. u. astr. resinous exudn. of tree (mango gum) mixed with egg albumen & opium as antidys.; diur. antirheum.; bk. c. tannin u. in combn. with others in diarrh. (Ind); c. tannin; nut kernels u. as starch source (c. 65-70%) (Ind) (v. mango) (2046); lvs. inf. with cooling effect (Chin); astr. u. in dys. (Nigeria); skin dis.; indirectly source of yellow dye (Indian yellow) (obt. fr. urine of cows fed on m. lvs.); oil fr. st. u. in syph.; gum-resin sold in Ind as "gum Arabic"; sd. kernels anthelm., roasted in diarrh.; frt. pulp juicy, u. to promote blood circ.; in heart failure, sunstroke; rind u. in tea, astr. in diarrh., dys. tan; sds. covered with fibers; u. in coughs & colds, diarrh.; dr. fl. spike (manjari) smoked (Ind); fingers rubbed over fls. then on skin, said to prevent pain & inflammn. fr. bee sting (2319); frts. & fls. regarded as having special beauty & u. as sacred pl. in religious ceremonies (Ind); shade tree; wd. u. considerably - **M. macrocarpa** Bl. (eAs): frt. hypnotic - **M. minor** Bl. (Oceania): bk. chewed & saliva swallowed for snakebite (NG); bk. with lime appl. to abdomen in infant with sick stomach (Bougainville); skin spots (Solomon Is) - **M. odorata** Griff. (Indones): bk. in mixt. appl. to treat epilepsy; frt. delicious; tree planted.

mangiferin: xanthone-C-glycoside (flavonoid) fr. *Mangifera indica* (rts. lvs. wd. bk. frt.).

mangle (Mex, CR): **mangericao**: *Rhizophora mangle* (also **m. rojo**) - **m. boton** (Mex): *Conocarpus erectus.*

mango: edible delicious sl. acid drupe frt. of *Mangifera indica*; c. vitamins, sugars, mineral salts, etc.; said of med. value as digestive aid, in syncope, heat apoplexy, heart failure(!?), etc.; many prepns. such as jams, salads, squashes, confections, preserves, chutneys; frt. 2nd most largely used for food in the world; sd. kernel nutritious (2636) - **mangold root** (GB): **mangold Wurzel** (Ge): mangel - **mangosta** (Por): **mangostao** - **mangosteen** (, **wild**): *Garcinia mangostana.*

mangosta (Por): **mangostao**: mangosteen.

mangostin: complex deriv. of xanthone; occ. in various parts of mangosteen tree; yellow coloring matter.

mangrove: trop maritime tree or shrub of gen. *Rhizophora*, or any plant resembling same in habit (some spp. of *Avicennia, Laguncularia, Ceriops, Sonneratia*); grows in mud at sea margin - **black m.**: *Avicennia germinans* - **m. (common) (red)**: *Rhizophora mangle* - **m. extract**: fr. *Rhizophora mangle, R. mucronata*; u. in tanning - **Indian white:** *Canarium euphyllum* - **Tengah m.**: *Ceriops tagal* - **white m.**: *Laguncularia racemosa.*

mangueira (Br): *Mangifera indica.*

mangue vermelho (Por, Br): **mangueiro** (Br): *Rhizophora mangle.*

mani (SpAm): peanut.

Manicaria saccifera Gaertner (Palm.) (Ve): bussu palm; frt. juice u. for fever; cough; potion fr. green juices of fresh eophylls of *Mauritia flexuosa* mixed with this frt juice for appl. in fever & cough; sd. FO (turluru fat) u. in soap-making; liquid endosperm u. for asthma & catarrh.; grated palmito in frt. water u. to facilitate breathing; sago prepd. fr. starchy pith of trunk; lvs. u. to make sails, hats (Warao Indians) (2507); sds. —> FO.

manicure (US dope pedlars): to strip lvs. & fls. fr. Cannnabis tops; also high-grade Cannabis.

Manihot dichotoma Ule (Euphorbi) (Br): manicoba brava; source of manicoba rubber - **M. esculenta** Crantz (M. aipi Pohl: M. dulcis [G.F. Gmelin] Pax; M. utilissima Pohl) (tropAm, esp. Br, but widely cult. in trop): (common) cassava; yuca (SA); tapioca plant; bitter cassava; Aipi cassava; sweet c. (Fla [US]; manioc(a); manihot (Sp); herbaceous shrub; -4 m; cultigen; many intraspecific variations; rts. edible after processing (one of staple crops of n&cSA [Amazon basin]) (1531); cassava (tapioca) eating thought protective against diabetes (2467); u. to mfr. starch = tapioca (manioc) (cassava starch or flour) (fécule de yuca) (Brazilian arrowroot) & dextrin; farinha (coarse meal) (Br) (one of cheapest forms of starch); u. in puddings; adhesive on postage stamps; u. to prep. alc. beverages; also —> acetone, sugar, &c.; u. in tablets, pds., &c.; fresh tuberous rts. c. cyanogenetic glycoside (linamarin) (with enzyme, emulsin); ext. u. commonly for poison darts & arrows (Indians) (1642); cassareep (*qv.*); lvs. eaten (eAf); young lvs. u. to "cleanse the blood" (Indones); tuber u. to dress ulcerous sores (Indochin); in erysipelas, eczema, whitlow (Ve); in diarrh. (SA) - **M. glaziovii** Muell. Arg. (Br): mandihoba; tapped st. yields Ceará rubber; sds. (manicoba nuts) are source of manihot (or Ceará) rubber oil (FO); genes source of resistance to drought & mosaic dis.; tree cult. in Java, SriL, Af (Nigeria+) - **M. palmata** Muell. Arg. (Br, widely cult. in pantrop): sweet cassava (plant); tuber source of food (Af, Bantu), however may be toxic; source of Rio arrowroot - **M. tristis** Muell. Arg. and subsp. **saxicola** (Lanj.) Rogers & Appan (tropAm): starch-rich tubers eaten (other *M.* spp. u. as food, incl. *M. caerulescens* Pohl [manicoba] and *M. catingae* Ule [manicoba brava]).

manila: hemp u. in ropes, cordage, cables, etc., fr. *Musa paradisiaca* v. *sapientum* - **boea m.**: boea gum - **m. copal**: prod. of *Agathis alba* (Mal) - **m. fiber: m. hemp**: fibers fr. lf. bases of abaca (*Musa textilis*) - **m. loba gum: l. m. gum**: half-hardened alc.-soluble resin fr. tapped trees of *Agathis alba* - **manilla**: 1) manila 2) (Guat): peanut.

Manilkara bahamensis (Baker) Lam. & Meuse (Sapot.) (WI): wild dilly (Bah); w. mango; lf. tea for grippe & fever (Bah) - **M. balata** (Aubl.) Dubard = *M. bidentata* - **M. bidentata** (DC) Chev. (Mimusops balata Gaertn. f.; M. globosa Gaertn.f.): (tropAm, CA, WI, nSA, esp. Ve, FrGuiana): (common) balata tree; bullet tree; bully; latex fr. trunk yields balata (gum) (rubber) of commerce, a prod. when dr. intermediate between rubber & gutta percha ("inelastic rubber") u. as substit. for gutta percha & chicle; for machine (driving) belts, shoe soles, &c. (easily vulcanized); earlier as dental cement (temporary); in plasters; wd. valuable as lumber (dense, elastic) ("bullet wood"), u. for shingles, bridge piles, railroad ties, ship propellors, &c. - **M. chicle** (Pittier) Gilly (tropAm): crown gum; u. as substit. for *M. zapota*; milky juice (latex) fr. incisions in the bk. boiled & allowed to harden in brick-shaped molds. - **M. elengi** Chev. (Mimusops e. L.) (Ind, Mal, SriL, cult.): (West) Indian medlar; bukul tree; rt. decoc. u. antipyr., as mouthwash; lvs. & frt. u. astr. in wounds & sores; unripe frt. chewed to fix loose teeth (due to HgCl); ripe frt. edible as such or in confections; u. in chronic dys.; pd. lvs. strongly irritant to nasal mucosa, u. for "head pains," snakebite; lvs. smoked for asthma (Indones); sds. lax. & u. to furnish FO u. in culinary processes; bk. source of Pagoda Gum (Madras); fls. u. as perfume (distd. water), yield attractive VO; bk. c. saponins (also in fls.), tannins (ca. 7%), u. astr., esp. as gargle with decoc. mostly in gum & tooth pains (Ayurved. med.); ton. antipyr., tan. - **M. hexandra** Dubard (Mimusops h. Roxb.) (Ind): palu; rather similar to *M. elengi*; bk. c. 40% tannin; u. astr. ton. dem. emoll. antipyr.; for tanning; frt. edible, sed. & tranquilizer, u. in alcoholism; sds. c. FO (rayan oil), u. to adulter. ghee (clarified butter); wd. (palla wood) u.; tree also source of a gum - **M. kauki** Dubard (Mimusops k. L) (seAs, Mal, Ind, Burma, cult. OW, NW trop, Af, Austral): sawo (Java); ka(u)ki rt. & bk. u. astr. in diarrh.; beri beri (also lvs.); fls. & sds. (c. saponin) u. ton. antipyr. anthelm (Mal); frt. edible; also source of a gum & durable lumber; pl. u. as stock for grafting *Manilkara zapota* (but dwarfed) - **M. obovata** (Sabine & G. Don f.) J. H. Hemsley (trop wAf:) African pear; mungambo (Swahili); rts. u. in med.; frts. eaten - **M. zapota** (L) P. van Roy. (M. achras [Mill.] Fosberg; M. zapotilla Gilly; Achras zapota L.; A. sapota L.; A. chicle Pittier; Sapota achras Mill.) (tropAm, chiefly sMex [Yuc], CA [Guat], WI, natd. sFla (US), Co, Ve, Ch, widely cultd. & natd. OW trop, esp. Java): sapodilla (tree); chicle; níspero; chico; sapoti; chicazapote; marmalade (or sapodilla) plum; naseberry; marmalade fruit; (mamey) sapote; zapote (Mex); beef apple; an evergreen tree (-22m) with edible frt. (ca. 10 cm.), delicious when ripe, one of best in tropAm (sapodilla plum); u. to make preserves; sds. c. HCN(!), but edible, u. diur. in bladder dis., lax. in catarrh. (CR); sd. u. as confect. (sapuyl); prepn. beverage, kola adulterant; to stop vomiting (Yuc); source of FO (zapoyol), u. in pop. med. as "cure" for colds (CR, CA); trunk bk. decoc. u. in diarrh. (Mayas) (6); ton. antipyr. to treat mal. (Chin), TB (Haiti); as cinchona substit.; c. tuberculostatic subst. sapotine (318); condensed latex fr. incisions in trunk or frt. chief source of gum sapota (sapotille gum); (Mexican) chicle (gum), best grade, u. as chewing gum base (as in Chiclets™) (chewing gum of the Aztecs): surgical materials, formerly in mfr. rubber products (*cf.* gutta percha); temporary filling of carious teeth (formerly); source of mispel wax; lf. (ca. 12 cm long) u. as inf. for rheum pains (Br); wd. good lumber; terminology confusing; chief prodn. area, sMex.

man-in-the-ground: *Ipomoea pandurata*.

manioc: Manioca (L.): **maniock**: *Manihot dulcis* or *M. esculenta*; tapioca.

maniqueira (Br): liquid squeezed fr. mandioca-water mixt.; c. HCN & is toxic; boiled to remove HCN, then u. as beverage.

manita de león (Mex): **flor de manita**(s): *Chiranthodendron pentadactylon*.

manjari: fl. spike of mango (Ind).

man-mu: *Angelica sinensis*.

manna: 1) saccharine exudation fr. *Fraxinus ornus*; c. 70-90% mannitol; u. mild lax. dem. in bronchitis, nutr. (children, convalescents); in bacteriolog. culture media; formerly for prepn. of mannitol 2) exudns. produced by 52 or more pl. genera (mostly Angiosperms, Gymnosperms) (ex. spp. of *Alhagi*, *Calotropis*, *Echinops*, *Salix*, *Lecanora*, &c. 3) (Bible): prod. by M. lichen (or otherwise) - **Arab m.**: prod. fr. 1) *Tamarix mannifera* 2) *Alhagi maurorum* - **m. ash**: *Fraxinus ornus* - **Astragalus m.**: m. secreted by *Alhagi maurorum* (SC gen. of Leg.) - **Australian m.**: mellitose fr. *Eucalyptus mannifera* (swAustral) - **bamboo m.: b. sugar**: Tabaschir (not true manna, since no mannitol) - **Bible m.**: 1) nutritive lichen (*cf.* Iceland and reindeer mosses); grows on arid steppes, food in times of drought & famine; c. starch & sugars (Algeria; cAs) (2047) 2) honeydew excretions of pl. lice & scale insects (*Gossybaria maniparus*), evapd. in dry air of desert; mostly sugars 3) exudation on *Hammada salicornica* (Moq.) Iljin. - **Boer m.**: *Setaria italica* - **M. Brianzona: M. Brigiantica**: m. fr. larch trees (Eu, Alps) - **California m.**: fr. *Phragmites maximum* - **common m.**: m. in fragments ("in sorts"), fr. lower part of ash trunk - **Douglas fir m.**: fr. *Pseudotsuga menziesii* (Wash [US], BC); c. melezitose; u. by Indians (stored) - **M. Eucalypti** (L.): Australian m. - **false m.**: Arab m. - **false m., American**: fr. *Pinus lambertiana*; sweet & cathartic - **fat m.**: m. in a viscid mass - **flake m.**: the best quality of m. fr. upper part of ash trunk - **m. gum**: fr. *Eucalyptus viminalis* - **Harlálu m.**: fr. Bihar, Patna, etc. (Ind) - **Hebrew m.**: Arab m. - **Indian m.**: fr. *Alhagi maurorum* (Afghan) - **Jew's m.**: Arab m. - **m. ladanifera**: fr. *Cistus ladaniferus* - **larch m.**: m. Brianzona; fr. *Larix decidua*; c. melezitose; toxic to bees - **m. lichen**: *Lecanora* spp. (still u. by Bedouins) - **Madagascar m.**: of unknown origin; represents crude form of dulcite - **M. Metallorum** (L): calomel - **m. of Moses**: Bible m. (1) - **m. of Sinai: Persian m.**: Arab m. - **m. sugar**: mannitol - **M. Syriaca** (L.): fr. cedars of Lebanon - **tamarisk m.**: Arab m. (1) - **Turkish m.**: prob. oak m. fr. *Quercus vallonea* & *Q. persica*.

mannan: polysaccharide made up entirely of mannose units.

manne (Fr): manna - **m. noire** (black m.): *Digitaria iburna* - **m. rouge** (red m.): *D. sanguinalis*.

Männertreu (Ge): **Mannstreu**: eryngo.

Manniophyton africanum Muell. Arg. (monotypic) (Euphorbi.) (tropAf): dr. lvs. put on sores (pd.) (Lib).

mannite: mannitol: polyatomic (hexahydric) sugar alc. $CH_2OH\ (CHOH)_4\ CH_2OH$ ($C_6H_{14}O_6$); earlier obt. fr. manna; also occurs in celery, sugarcane, lilac, rye flour, &c, now prod. largely fr. sucrose by electrolytic or enzymic hydrogenation; u. excipient, diluent (as for heroin), mild lax.; base for chewable tablets; appl. in glaucoma; diur. in kidney failure (intrav.) test for kidney function; in convalescent foods; in prepn. - **m. hexanitrate** (hypotensive); in technology (E).

mannose: simple sugar, $C_6H_{12}O_6$; widely distr. in nature, esp. in form of glycosides & complex carbohydrates; ex. ivory nuts (fr. wh. prepd. comm.).

manoba (Por): peanut.

man-of-the-earth: *Ipomoea pandurata* - **man-root**: 1) prec. 2) *Megarrhiza* (*Micrampelis*) spp.

mansanilla: 1) (Mex): *Matricaria* 2) (CR): Anthemis (*cf.* manzanilla).

Mansoa standleyi (Steyermark) Gentry (Bignoni) (Ec): lvs. & sts. of liana pungent, u. to treat fever, sore muscles, arthritis; lvs. crushed in hot water & appl. locally & decoc. drunk (emetic) (1643).

Mansonia altissima A. Chev. (Sterculi) (tropAf, as Nig): bk. c. mansonins A, B, D, E; u. in leprosy; as aphrod.; c. cardioactive glycoside & potentially of value in heart dis.; wd. termite-resistant & u. in carpentery (prod. dermatitis) (2361) - **M. gagei** J. R. Drumm. ex Prain (Burma): Kalamet; wd. fragrant, u. as cosmetics (pd).

manteca (Sp): butter; lard; fat - **m. de cerdo**: lard.

manuko: *Leptospermum scoparium* & possibly other *L.* spp.

Manulea crassifolia Benth. (Scrophulari.) (sAf): appl. to swollen umbilicus of infant; to relieve headache (bathing head with lotion fr. lf. & branches) - **M. paniculata** Benth. (tropAf): strongly emetic.

manure: ordure of animals, mixed with straw, etc.; valuable fertilizer, esp. for drug crops (types: farm [or barn] yard, stable, horse, cow, etc.) - **m. salt**: KCl & other potash salts u. as fertilizer.

manzana (Sp-f.): apple.

manzania (Calif [US]): *Mentha piperita* (110).

manzanilla (flores) (Sp): a term appl. to many spp. of pls., esp. many Comp. Fam. hbs. (ex. *Bidens* spp.), several Mex trees & shrubs (ex. *Quercus* & *Ximenia* spp.); a small, green olive, etc.; most commonly : 1) *Matricaria* & *Malvaviscus* 2) a pale dry sherry wine (Sp) 3) (CR): Anthemis 4) (Mex): *Matricaria grandiflorus*; *Helenium grandidentatum*; *Chrysanthemum parthenium* 5) *Crataegus stipulosa*- **m. amarilla** (Sp): *Dyssodia berlandieri* - **m. dulce**: Matricaria - **m. romana**: *Anthemis nobilis*.

manzanillo (Sp): 1) *Hippomane mancinella* (also **m. comun**) 2) manzanilla 3) (Mex): *Trixis* spp.; *Arctostaphylos pungens*; *Malvaviscus drummondii, Chrysanthemum parthenium* - **manzanita**; *Arctostaphylos* spp. (*A. pungens*) - (**great-berried m.** = *Chrysanthemum parthenium*, *Arctostaphylos glauca* - **point-leaf m.**: *Arct. pungens* - **manzanito** (Mex) (lit. little apple): common name for various pl. spp. which prod. small frts somewhat apple-shaped; ex. *Malpighia mexicana* Juss. (2048) (lvs. & berries) - **manzano** (Sp): apple tree.

manzoul: mixt. of Cannabis & Myristica; hallucinogen.

mapén (PR): **mapin** (PR): frts. of tree *Artocarpus heterohyllus.*

maple: *Acer* spp. - **ash-leaved m.**: *Acer negundo* - **big leaf m.: broad-leaf m.**: *Acer macrophyllum* - **black sugar m.**: *Acer nigrum* - **field m.** *A. campestre* - **m. family**: Aceraceae - **m. flavor, imitation**: formerly exts. of hickory bk. & vanillin; now ext. of fenugreek sd. - **great m.**: *Acer pseudoplatanus* - **ground m.**: *Heuchera acerifolia* - **hard m.** (wNew Eng [US]): *A. saccharum* - **Japanese m.:** *A. palmatum* - **Manitoba m.**: *A. negundo* - **mountain m.**: *Acer glabrum*; *A. spicatum* - **Norway m.**: *Acer platanoides* - **red m.**: *A. rubrum* (1834) - **rock m.**: (eNew Eng): *A. saccharum* - **scarlet m.**: red m. - **silver (leaf) m.**: *A. saccharinum* - **soft m.**: red m. - **striped m.**: *Acer pensylvanicum*; *A. striatum*, *A. saccharum* - **sugar m.**: rock m. - **swamp m.**: red m. - **sycamore m.**: *A. pseudo-platanus* - **m. vine**: *Menispermum canadense* - **vine m.**: *Acer glabrum*, *A. circinnatum* - **white m.**: silver m.

Mapleine™: artificial maple flavor; prepd. fr. fenugreek, &c.

Maprounea africana Muell.-Arg. (Euphorbi) (cAf): rt. decoc. u. as purg.; face lotion (Luvale); pl. sap u. to dress penis after circumcision (Luvale).

maque (Mex): 1) hard impremeable varnish prep. fr. Asiatic resins, pl. juices, etc. 2) Japanese sumac 3) charol, very lustrous & permanent varnish 4) tree (Ve) furnishing valuable wd.

maqui (Mex): 1) kind of ginger 2) *Aristotelia chilensis.*

Maquira calophylla (C. & E.) C. C. Berg (Mor.) (tropSA): bk. c. furocoumarins; latex caustic & toxic to skin (Indians, Leticia, seCo) - **M. sclerophylla** (Ducke) C. C. Berg (Br, Amazonas, Pará): rapé dos indios; hallucinogenic snuff prepd. fr. sds.; other *M.* spp. u. similarly.

mâquis: typical shrubby growth found on Corsica.

maraca (Pe, Br): precious balsam; term also appl. to container for same.

maracuja (Br): *Passiflora elata* & other *P.* spp.

Marah fabaceus (Naud.) Greene (Echinocystis fabacea Naud.) (Curcurbit) (Calif [US]): (California [or valley]) man-root; C. big root; yerba marra; u. as drastic (291); mydriatic; rt. c. fabacein, cucurbitacin B, saponin (with echinocystic acid) - **M. gilensis** Greene (seCalif) (1532) (Ariz - swNM [US]): bitter root; big (or man) root; wild cucumber; chili cajote; sds. & large rts. (-48 kg) u. in popular treatment of rheum. & VD; in mfr. of a Calif "bitters" - **M. macrocarpus** Greene (Echinocystis macrocarpa Nutt.) (sCalif [US], nwMex [Lower Calif]): taw man-root; chilicothe; rts. u. as purg.; large sds. roasted & taken for kidney dis. (toxic when raw), sds. source of red dye (Luisino Indians); cult. ornam. - **M. oreganus** T. J. Howell (E. o. Cogn.) (sBC [Cana] to Calif): (coast) man-root; (western) wild cucumber; frt. a kind of gourd, with weak spines; may c. saponins (1288) - **M. watsonii** Greene (sCalif, nwMex): chilicothe; rts. u. as purg.; sds. source of a red dye.

marañon (CR): *Anacardium occidentale.*

Maranta arundinacea L. (M. indica Tussac) (Marant.) (WI, CA, SA, now pantrop cult.): (Bermuda) arrowroot (plant); rhiz. source of starch known variously as Maranta (L), true, Bermuda, St. Vincent (main prodn. place), Jamaica, & West India(n) arrowroot; perenn. hb. (-2 m) (similar to Canna but with smaller lvs.); arrowroot u. as nutrient, esp. for invalids & children; in culinary delicacies (puddings [2049], biscuits, cakes, jellies); thickener for soups, broths, ice cream, &c.; in gastritis, enteritis; pd. rhiz. in beeswax appl. to sores, infected wounds (incl. arrow wounds) (hence name); size for bk. cloth (Tonga); in making "self" carbon paper, in computer paper.

Marantaceae: Arrowroot Fam.; perenn. hbs. with both radical & st. lvs.; inflor. a raceme or panicle with irreg. 3-merous fls.; ovary 1-celled; ex. genera *Maranta*, *Thalia*, *Calathea.*

marapuama (Br): muirapuama.

Marasca (cherry): a sour, wild cherry, *Prunus cerasus* var. *marasca*, processed into **Maraschino** (liqueur) in Zara, Dalmatia (Yugoslavia) (v. under cherry) - **Maraschino (cherry)**: Marasca cherries leached, bleached, boiled, dried, candied, & artific. flavored in red colored syrup; popular in mixed drinks - **M. cordial**: liqueur prepd. fr. Marasca cheries by ferment. & distn.

Marattia fraxinea Sm. (Maratti. - Pteridophyta) (sAf, Tanz, Mal): rhiz. u. as tenifuge in ankylostomiasis; st. eaten formerly by Maoris (NZ) & aborgines (Austral); cult. ornam - **M. novoguineensis** Ros. (NG): petiole juice is appl. to treat a kind of leprosy (319).

maravilla: 1) (Sp): common marigold 2) heliotrope 3) *Carthamus* (safflower) (usually) 4) (NM [US], Sp): *Mirabilis multiflora* (maravilha, Por) 5) other pls. as calendula, bindweed - **m. de Indias: m. de noche**: *Mirabilis jalapa.*

marble: Marmor (*qv.*) - **m. flower**: *Papaver somniferum.*

Marchantia polymorpha L. (Marchanti. - Hepat. - Bryophyta) (worldwide in moist places): (brook) liverwort; moss u. in liver dis. (in Homeopathy).

mare-blebs: mare-blobs: *Caltha palustris.*

Maredith root (secret comml. name?): Indian rhubarb.
mare's milk: lacteal fluid of lactating females of *Equus caballus*; u. nutr. in prepn. of kumyss, etc. - **mare-tail**: *Equisetum arvense*.
Mareya micrantha Muell. Arg. (Euphorbi) (tropAf): bitter lvs. u. as purg. (drastic), antipyr.; abort.; topically for pain in fractures, sprains, rheum. (Nig) - **M. spicata** Baill. (wAf): purg.; bk. rt. & st. u. exter. & int. in leprosy.
margaret: *Bellis perennis*.
Margaretta rosea Oliv. (monotypic) (Asclepiad.) (tropAf): lf. decoc. u. as male aphrod. (Tanz); c. digitaloid glycosides.
margaric acid: daturic acid; $C_{16}H_{33}C$ OOH; satd. fatty acid; rarely occ. in nature; found in mutton fat; repts. in early literature of plant analysis actually repr. mixts. of palmitic & stearic acids.
margarin(e): oleomargarine: synthetic butter or b. substit., mfrd. fr. hydrogenated vegetable oils (esp. cottonseed), beef fat, pork fat with added color, flavor, vitamin A, salt, &c.; at times lard(!) - **autumn m.** (Calif): m. colored with natural subst. such as carotene.
margarita: m. mayor (Sp): *Chrysanthemum leucanthemum* - **m. punzó** (Ur): *Verbena peruviana*.
margiapantes: *Peltodon radicans*.
marginal: v. under fern.
margosa (**bark**): fr. *Melia azadirachta*.
marguerite: 1) *Chrysanthemum leucanthemum*; *C. frutescens* 2) *Bellis perennis* - **golden m.**: *Anthemis tinctoria*.
Margyricarpus pinnatus (Lam.) Ktze. (monotypic) (Ros) (sSA - Andes); pl. u. in fertility control (Ur); cult. ornam.
maria mole (Por): **m. preta**: *Solanum nigrum*.
Mariani's tonic (Coca wine): Vin Mariani; French wine of coca leaf; u. as stim.; popular ca. 1890-1910.
marianum (**antigen**) **vaccine**: dil. culture of *Mycobacterium marianum* Suzanne & Penso (M. scrofulaceum) u. to treat leprosy (Hansen's dis.).
mariculture: growing plants, &c., on/in the sea.
Mariendistel (Ge): *Silybum marianum*.
marigold: *Tagetes erecta* - **African m.: Aztec m.: big m.**: *Tagetes erecta* - **common m.:** pot. m. - **corn m.**: *Chrysanthemum segetum* - **French m,**: *Tagetes patula* - **garden m.: marsh m.**: *Caltha palustris* - **pot m.**: **Scotch m.:** *Calendula officinalis* - **sweet-scented m.**: *Tagetes lucida*.
marihuana (Mex): **marijuana**: *Cannibis sativa* subsp. *indica* - **synthetic m.**: Δ^8-tetrahydrocannabinol (THC) (1970-).
marine acid: HCl - **m. oils**: FO derived fr. m. animals, as seals, whales, sardines, &c., when ingested, reduce serum cholesterol levels; hence considered healthful in diet - **m. salt**: NaCl.
marinobufagin: genin found in *Bufo marinus* (SA) & *B. vulgaris* (Eu) (toads); cardioactive.
mariola: 1) garden majoram, marjolaine (Fr) 2) (swUS): *Parthenium incanum* 3) (Mex): *Chrysactinia mexicana*.
mariposa: sea water drinking; ingestion —> death; however, when added 20% to fresh water is potable in small amts. (2658).
marjoram, (common): 1) *Origanum vulgare* 2) *O. majorana* - **garden m.: leaf m.:** *Origanum majorana* - **pot m.**: *Origanum onites* - **sweet (German) m.**: garden m. - **wild m.**: pot m.
Marjorana hortensis auct. = *Majorana h.*
Markhamia hildebrandtii Sprague (Bignoni) (cAf): bk. chewed for toothache; wd. u. to make bows - **M. obtusifolia** Sprague (tropAf): rt. u. for convulsions & sterility; rt. bark galactagog; fermented drink prepd. fr. rt. u. for sores of secondary syph. (Tanz).
marking nuts: frt. pericarp of *Anacardium semecarpus*; u. all over Ind in place of marking ink.
marlberry (US): *Ardisia escallonoides*.
Marlierea insignis McVaugh (Myrt.) (Co): frt. edible, u. to prep. tea snuffed hot into nostrils for sinus congestion (2299) - **M. spruceana** Berg (Co): hot decoc. (frt.?) u. to clear congestion in throat (Taiwanos) (2299).
marlo (Arg): corn cob.
marmalade: preserve made with peels of *Citrus* spp., esp. orange & lemon - **m. fruit: m. tree**: *Manilkara zapota*.
marmelo (Por): quince; hence marmelada (jam); (Eng marmalade).
Marmor (L): marble; native $CaCO_3$; sometimes u. as antacid.
Marniera macroptera Backbg. (Epiphyllum m. Br. & Rose) (Cact) (CR+): dama de la noche (Guat): the wide sts. are bound around fractured wounds like bandages to immobilize.
marram: *Ammophila arenaria*.
marri: *Eucalyptus calophylla*.
marron (Fr-m., It): sweet European chestnut, u. as such freshly roasted or preserved in syr. with vanilla flavor; candied chestnuts esp. popular in Fr - **m. d'Inde** (Fr): horse chestnut - **m. glacé**: m. coated with sugar icing - **marronier** (Fr-m.): chestnut tree.
marrow (**, vegetable**)**:** edible greenish squash. *Cucurbita maxima, C. pepo*.
marrube blanc (Fr): horehound - **marrubiin**: crystalline bitter princ. fr. horehound - **marrubio blanco** (Sp): **Marrubium** (L.): *M. vulgare*.

Marrubium incanum Desr. (Lab) (sEu): popular med. pl. (folk med.) in Yugoslavia - **M. incisum** Benth. (eAs): decoc. with fermented rice u. in amenorrh. (Chin); aids against suppuration (Tibet) - **M. paniculatum** L. (eEu): source of Herba Marrubii peregrini; u. bitter ton. in phthisis. - **M. peregrinum** L. (sEu, wAs): one source of Herba Marrubii peregrini (Hba. M. cretici); substit. for Herba Marrubii albi - **M. vulgare** L. (M. album Gulib.) (Euras, nAf, natd. NA): (common or white) hoarhound (or horehound); hb. (Herba Marrubii albi) c. marrubiin (0.34-1%) & other NBP; VO; u. as drug since antiquity; lf. inf. u. exp. in bronchitis, coughs, colds (horehound drops); hepat. stim. (choleretic); flavor (u. as tea, in liqueurs); in chronic diarrh., hemorrhoids; emmen.; ext. in skin dis. ton (2050); hypotensive (3590) (prod. Sp, Hung, Alg); common weed.

mars (gen. **martis**) (L.): iron (ex. Tinctura Martis; iron tr.).

Marsdenia condurango (cundurango) Reichb. f. (M. reichenbachii Triana) (Asclepiad.) (Pe, Ec, Co; cult. eAf): condor vine source of condurango bark (1546); c. condurangin & other glycosides, condurit(ol) (0.4-0.8%) (chief active princ.); oleoresin; u. astr. alt. stim. bitter ton. in certain chronic blood dis.; bk. said u. in cancer treatment (formerly, ca. 1870) (1644); sd. hairs u. as styptic - **M. erecta** R. Br. (seEu, wAs): source of Montpellier scammony; natural exudn. c. resin 16-21%, starch 50-60%, ash 11-18%: u. med. (Greek Ph.), latex strongly narcotic; epispastic; toxic to grazing cattle - **M. latifolia** Schum. (wtrop Af): hasoub (Sen); leafy sts. with lime juice in colic & intest. dis.; frts. as lax. (ex. Muslims during Ramadan); sts. chewsticks - **M. rubicunda** N. E. Br. (Dregea r. K. Schum.) (cAf as Kenya): rt. u. as purg. expect.; rt. decoc. to improve the flavor of beer; rt. chewed as aphrodisiac; sds. c. digitalis or strophanthus type of glycosides - **M. rubrofusca** Fournier (SA - Amazonia): c. alk; very toxic, u. in poison dogs - **M. tenacissima** Wight & Arn. (Ind): rajmahal hemp; bk. c. fibers u. to make strings; latex to make rubber; u. med. (Nepal) - **M. tinctoria** R. Br. (M. tomentosa Morr. & Decne.) (Indones, PI): condor vine; lvs. u. in stomachache; rubbed on head of small children to promote hair growth; c. indigo-like dye.

Marsden's paste: arsenic 6, acacia 3, cocaine HCl 1, HOH to make paste.

marsh: low treeless flat grassy land area inundated part of the time with fresh water or sometimes saline or brackish water; has grasses and sedges but no trees; *cf.* swamp - **m. marigold**: *Caltha palustris* - **marsh:** v. under parsley, rosemary, skullcap, St. John's wort, tea, trefoil, watercress, wort.

marshmallow: 1) *Althaea officinalis* (**common m.**) 2) (Austral) *Malva rotundifolia* 3) **m. candy**: confection orig. prepd. fr. *Alth. offic.*; now made fr. pd. sucrose, glucose, gelatin, &c. - **marshmelon** (Indo-Pak): muskmelon.

marshwort: *Vaccinium oxycoccos.*

Marsilea minuta L. Marsile. (&c.) (As): u. as sed. hypnotic - **M. quadrifolia** L. (Marsile. - Filicopsida - Pteridophyta) (Euras, natd. US): pepperwort; aquatic pl., edible, diur. antipyr.; juice appl. to abscesses & ulcers; decoc. u. to reduce fever & swellings; tonic in eye troubles - **M. vestila** Hook. & Grev. (Ind): u. as sed. & anticonvuls.

Marsypianthes hyptoides Mart. (Lab) (Br): "paracari"; lf. juice u. int. for snakebite or any poisonous bite; baths for arthritis.

Martinella obovata (Kunth) Bureau & K. Schumann (Bignoni) (Co): bk. inf. antipyr. (danger!); component of curare; rt. u. in eye inflammn.; rt. sap u. for eye infections, arrow poison (all Indian uses); cult. by Indians.

Martynia annua L. (M. diandra Glox.) (monotypic) (Martyni.) (Mex, WI, CA; natd. Ind [common in wild], to Mal): tiger claw; devil's claw; arana gato (spider cat) (Cu); ann. hb.; lvs. u. as anticonvuls. in epilepsy, appl. to tuberculous glands in neck; juice u. as gargle for sore throat; frt. in inflamm., for scorpion sting (Ind); sds. u. dem. for urinary dis.; frt. at times pickled - **M. louisiana** P. Mill. = **M. proboscidea** Glox. = *Proboscidea louisianica.*

marula: *Sclerocarya caffra* (**m. berries**).

Marum verum (Homeop.): *Teucrium marum.*

maruna (Pol): *Chrysanthemum parthenium.*

marupá (Br) *Quassia amara.*

Maruta Cass. = *Anthemis.*

marvel: horehound - **m. of Peru**: *Mirabilis jalapa.*

Mary bud: Marygold (flowers): Calendula, garden marigold - **marygold** (NC [US]): marigold.

Maryland: v. under pink.

marzipan: (sweet) almond paste; a. bread; prepd. fr. sweet almonds & sugar, etc.; represents an ancient confection & medicinal (Iran into Eu, medieval period).

masa harina (Pima Indians, Ariz [US]): ground maize paste; u. to make tamales.

Mascagnia glandulifera Cuatrecasas (Malpighi) (Co): poultice of crushed & boiled lvs. appl. commonly to boils & similar infections (Indians of middle Apaporis) - **M. macroptera** (Sessé & Moç.) Niedenzu (Mex, esp. nw): milky latex purg.; rt. said "specific" in syph.; rt. decoc. u. for colds, diarrh.; to strengthen after childbirth (Seri Indians, Sonora, Mex).

mascalage: removal of outer layer of bk. fr. cork oak; repeated on tree every few years.

Mascarene Islands: near Mauritius Is & Madagascar (off eAf).

Mascarenhasia arborescens A DC (Apocyn.) (Madag): latex formerly collected for rubber making (majunga noir); wd. white, useful - **M. elastica** K. Schum. (e tropAf): latex u. to produce Mgoa rubber - **M. lanceolata** Dec. (Madag): latex of trees formerly exported for prodn. of rubber - **M. lisianthiflora** A DC (Madag): latex formerly source of rubber; that from this sp. and *M. arborescens* furnished largest exports; similarly **M. rubra** Jumelle & Perrier (Madag).

mash (Urdu): 1) *Phaseolus mungo* v. *roxburghii* 2) muskmelon.

masopin (Mex): sweetgum tree (2510).

Massanalyse (Ge): volumetric analysis.

Massaranduba gum: a false gutta percha obt. fr. sap of Brasillian cow tree, *Mimusops elata*.

massasauga: pygymy rattlesnake, *Sistrurus* sp. (NA-SA fr BC [Cana] to Chile) (SC Missisanga R. Ont, Cana).

Massoia Becc.: = *Cryptocarya.*

massoia (bark): **massoy bark**: fr. *Cinnamomum aromatica.*

Massonia bowkerii Bak. (Lili) (sAf): Abraham's book; u. as ophthalmic appln. (with other pls.); as magic med.

massu (Bo-Jivaro, Ind): rotenone-bearing pl. u. as fish poison - **massur** (Indo-Pak): **masur** (Bengal): **masura** (Sanscr): kind of pulse, *Lens esculenta.*

mast: hard m.: collective name for frt. of beech, oak, chestnut, walnut, hickory & other forest trees; esp. when serving as swine food & wild life food - **m. tree**: 1) a tree prod. m. 2) cork tree 3) *Picea abies.*

masterwort (term often improperly u.): 1) (Eu): *Peucedanum ostruthium* (also **imperial m.**) 2) (Am): *Heracleum lanatum* 3) *Astrantia major, A. maxima.*

Mastfuetterung (Ge): fattening pasture (acorns & other nuts).

mast-holm: holm oak, *Quercus ilex.*

mastic(he): 1) **gum m.**: arom. oleoresin fr. **m. tree**, *Pistacia lentiscus* 2) any similar resin 3) cement of pasty consistence with pd. rock, asphalt, &c., u. in road making - **American m.**: resin fr. *Schinus molle* - **Bombay m.**: resin fr. *Pistacia atlantica* - **Chios m.**: the leading g. mastic; fr. C. - **East Indian m.**: Bombay m. - **Mastiche (Resina)** (L): Mastix (L): mastic gum.

Mastichodendron angustifolium (Standl.) Cronquist (Sideroxylon a. Standl.) (Sapot.) (Mex): tempisque; bk. u. to curdle milk (Sinaloa) - **M. foetidissimum** H. J. Lam. (Sideroxylon f. Jacq.) (sFla, Fla Keys [US], WI): wild olive; "ironwood" (SC strong wood); latex source of false mastic (u. in fresh hernias): u. in cabinet & boat bldg. (Fla [US]) - **M. sloaneannum** Box & Philipson (WI): Barbados mastic.

Mastix (Ge): Mastic.

Mastixiodendron flavidum A. C. Smith (Rubi) (Fiji): lf. exts. showed activity against *Proteus mirabilis*; u. med. (Fiji) (293).

mastransia (Sp): *Stachys bullata.*

mastranzo (Sp): mastranto (Pan): *Hyptis spicigera.*

mastuerzo (Mex): *Tropaeolum majus* - **m. retortano** (Sp): *Cardamine pratensis.*

mastworts: Betulaceae.

máta (Sp, Por): 1) shrub 2) forest of trees of same sp. or similar spp. 3) *Eupatorium incarnatum.*

máta (Czech, Bohem): mint - **m. peprna**: peppermint.

matalahu(e)ga (Co): **matalahuva**: anise.

mata-olho: *Asclepias curassavica* - **mata-ratón** (Mex): *Gliricidia sepium.*

mata-pasto (Por): *Cassia* spp.

matasano (CR): *Casimiroa tetrameria, C. edulis,* & other *C.* spp.

Matayba macrolelpis Radik. (Sapind.) (Co, Br): crushed lvs. & FO rubbed on testes to reduce swelling (Co Indians); c. cleomiscosin with anticancer activity - **M. purgans** (Amazonia): matayba seed & FO u. as purg.

mate: matte: hb. (lvs.) of *Ilex paraguariensis*; u. as beverage in Par, Br, Arg, & other SA countries; name first appl. to gourd fr. which drunk, later also to both hb. & drink; grades: green, toasted, etc.; stim. ton.; advantage; no tannin; u. also as source of caffeine - **m. de costa** (Cu, Dom): *Canavalia rosea.*

mateco: matico.

Matelea viridiflora (G. F. W. Mey.) Woodson (Trin): lf. u. as poultice on boils, carbuncles, gum boils (Trin).

Materia alba (L): debris deposited by tooth m. o. (bacteria, molds, &c).

Materia Medica (L.): lit. medical materials; a designation now largely in disuse because quite variously interpreted; sometimes as synonymous with pharmacognosy, sometimes as equivalent to pharmacology & its branches; at other times representing both of these fields - **M. Prima** (Arg): raw materials.

"maternity plant": *Kalinchoë* sp.
Mater secalis (L): ergot.
mathi (Pak): methi, fenugreek.
matias bark: fr. Croton malambo.
matico (leaves): 1) *Piper angustifolium* (Pe, Br) 2) group name for many *Piper* spp. u. in tropAm as vulner. & styptic 3) (Pe): *Salvia bridgesii.*
matière première (Fr): raw (or crude) material(s).
Matisia cordata H. & B. (Bombac.) (Br+): chupa-chupa (sapoté); sometimes cult. in Amazonian Pe for edible frt. pulp (30); frt. soft, brown (some claimed finest flavored frt. in world) (Co) - **M. paraensis** Huber (Br): cupuassurana; st. & branches source of good textile fiber (30).
mato (Por): bush, shrub.
matorral(es) (Sp): uncultivated lands with shrubs & weedy pls.
Matricaria chamomilla L. (M. suaveolens L.) (Comp.) (Euras, natd. NA; much cult.): German (or sweet false) c(h)amomile; fl. heads c. glycosides, VO with chamazulene, bisabolol, immunostim. polysaccharides; NBP; fl. heads & hb. u. arom. bitter, diaph., sed. in tisanes; antiphlogistic; antiallergenic; anthelm.; exp. in bronchitis; antipyr.; to aid digestion; to reduce inflammations, esp. in hemorrhoids, arthritis (2050), exter. in wet dressings, &c. for eczema, wounds, burns, sores; analg. in migraine; in hair preps. (1645); very popular in herbal teas (prod. Ge, Hung, Fr, Bel, Pol, USSR, Rom, Eg, Arg; VO [blue c.o.] prod. Ge, Neth, Bel, Hung, Eg) (literature) (636) - **M. discoidea** DC (cEu, Russia, NA) rayless chamomile; hb. with antispasm., antiphlogistic & anthelm. properties - **M. glabrata** DC (s&eAs): wild chamomile; active constit. VO; u. as spasmolytic; arom. bitter taste hence u. as remedy for dyspep. & in gastric dis.; exter. for inflammn. - **M. maritima** L. (w&cEu, natd. eNA): scentless (or wild) chamomile; corn mayweed; c. matricaria-ester; u. as substit. for true chamomile (but worthless) & as adulter. of Roman chamomile - subsp. **inodora** (L) Soó (eEu, Caucas reg.): scentless chamomile - **M. matricarioides** Porter (M. suaveolens [Pursh] Buchenau non L.) (orign. neAs now wCana, Alas [US], Mex, Eu): fragrant chamomile; pineapple weed (because of aroma); similar to *M. chamomilla* but very pleasant smelling, with stronger, more compact growth; u. as substit. for *M. chamom.* but inadvisable since VO does not cont. azulene & not antiinflammatory; fl. heads c. VO; inf. or pd. u. orally as spasmolytic, anthelm. (ascarids, oxyurids, *Trichocephalus*); or as enema or supp. per recto; insecticide - **M. nigellaefolia** DC (e&sAf): bovine staggers plant; u. in anthrax, skin dis. (inhalation of steam of boiling lvs. or appln. in ointment); in cattle causes "pushing disease" in which bovine pushes head against obstructions & refuses food - **M. parthenium** L. = *Chrysanthemum p.* - **M. perforata** Merat (M. inodora L.) (eEu): scentless (false) chamomile; VO c. matricaria ester; u. in folk med. (Perm, USSR); adulter. of true chamomile.
matrimony vine: *Lycium barbarum (L. halimifolium)* or other *L.* spp.
matriz (Sp): 1) flask (glass bottle with long neck) 2) mattress.
matstsoth (Heb): matzoth (qv).
matsu (Jap): *Pinus massoniana* - **m. oil**: VO fr. *Pinus massoniana* or *P. densiflora* - **m.-take** (Jap): mushroom, *Armillaria edodes*, growing in red pine groves - **m. -taku** (Jap): mushroom fr. pine tree cooked with sauce (with water cress, ginger, &c.), sometimes c. chicken, fish.
matta: the cortical portion fr. a millet sd. (var.) u. to adulterate pepper.
matte: maté.
Matteuccia struthiopteris Todaro (Aspidi. - Filic. - Pteridophyta) (nNA): ostrich fern; rhiz. u. med. (USSR) (320); ornam. pl.
Matthiola (*Mathiola*) **incana** R. Br. (Cruc.) (Med reg, Fr, Eng, Turk, natd. wUS): common (or Brompton) stock; gilliflower; u. as purg. stom. emmen. in hepatic cirrhosis; bitter ton. expect. diur. aphrod.; inf. in cancer (Indo-Pak med.); ornam. hb.
matzo(th) (Yiddish): matzos; motza; unleavened bread, usually wheaten, eaten as large biscuits mostly at Passover (2538).
Maulbeer(baum) (Ge): mulberry (tree).
maund (Indo-Pak): unit of wt. = 82.3 lbs. (c. 38 kg) (Ind); varies fr. country to country.
Mauritia flexuosa L. f. (M. minor Burret; M. vinifera Mart.) (Palm) (nSA, esp Br, Co): moriche (Ve); aguaje, canangucho (Co); bache (Fr Guiana); ita palm; burity (Br); frt. collected on large scale (Co) for sugary juice (also fr. trunk) to prep. beverages; fermented producing wine (chicha de canangucho) (31) also alc.; trunk source of starchy food (sago); fiber fr. lvs. u. for weaving; frt. pulp oil u. as strong vermifuge; lvs. u. as thatch (Co).
Mauritius: island in Indian Oc e of Mad; one of the Mascarenes; many medl. pls. are/were produced on the island, ex., cloves, *Kola, Cinchona, Cocos, Myristica, Vanilla*, calumba, &c. - **M. elemi**: *Canarium paniculatum* - **M. hemp:** aloe h.: *Furcraea foetida* (Ec).
mauve (Fr): mallow.

mavenolin: compd. isloated fr. *Monascus ruber* v. *Tieghem* (fungi) & *Aspergillus terreus* (fungi); inhibits enzyme needed in cholesterol synthesis; hence, indicated in hypercholesterolemia, &c.

mawah oil: 1) (Tanz) FO fr. *Madhuca longifolia* 2) fr. African geranium, *Pelargonium radula.*

maw seed: black poppy sd. fr. *Papaver somniferum.*

Maximiliana maripa Drude (Attalea m. Aubl.) (Palm) (Guianas, Surin, Co, Br): maripa; kokerite; frts. (oil seeds) source of maripa fat, similar to coconut oil, u. in foods, soap mfr.; frt. mesocarp mixed with water produces bev.; sd. endosperm roasted important food (Co): lvs. u. for thatch; cult. ornam. - **M. vitifolia** (Willd.) Krug & Urban = *Cochlospermum vitifolium.*

may (contraction of **m. thorn** or **m. bush**): 1) *Crataegus* spp. 2) *Spiraea* spp. 3) *Ulmus campestris* 4) various other pl. spp.

May apple root (or **rhizome**): *Podophyllum peltatum* - **M. blob:** *Caltha palustris* - **M. blossom**: *Convallaria majalis* - **M. flower**: term appl. to at least 13 genera, incl. *Epigaea repens* (also **Canada M. f.**), *Maianthemum canadense*, *Anemone patens* (Pulsatilla) (NA) - **M. haw**: v. haw & *Crataegus* -**M. leaf**: *Achlys triphylla.* - **M. lily**: M. blossom - **(stinking) M. weed**: *Anthemis cotula* - **M. wine**: Maitrank - **M. worm**: *Meloë majalis* (Coleoptera).

maya (Cu): *Bromelia pinguin* (**m. fruit**).

Maya leaf: *Cassia reticulata.*

mayberry: *Rubus palmatus.*

Mayer's reagent: alk. reagent consisting of mercuric potassium iodide soln.

Mayna amazonica (Mart.) Macbride (Flacourti) (Co): lvs. u. in bath for cramps; decoc. intern. for aching of legs, &c.; bk. of small tree u. as ingredient of curare - **M. grandiflora** (Spruce) R. E. Schultes (Br, Co): treelet; bk. said strongly poisonous to rodents, man; methyl salicylate odor (31) - **M. linguifolia** R. E. Schultes (Co): sd. FO u. for sores - **M. longifolia** Poepp. & Endl. (Co, Ec): sd. u. as emet.; poison - **M. toxica** R. E. Schultes (Co): sds. & bk. u. as poison for dogs - **M.** sp. source of FO like chaulmoogra oil.

mayonnaise: emulsion of olive oil (or cottonseed or other veget. FO) with egg (or e. yolk) as emulsifier; vinegar, spices; u. as salad dressing, sandwich spread, &c. (SC bayonnaise, adj. of Bayonne, Fr).

may-pop: pl. or frt. of *Passiflora incarnata.*

maytansine: alk. fr. *Maytenus ovatus* & *M. buchananii*; 0.7 ppm in wd. & bk.; active in leukemia; new type of ansa macrolide related to rifamycins & streptovaricins (antibiotics) (maytansinoids).

Maytenus boaria Mol. (Celastr.) (Ch): maiten; lvs. antipyr., u. to cure skin eruptions prod. by "litre"; active against lymphocytic leukemia (321); lvs. c. lupenone, spinasterol, daucosterol, dulcitol; sds. purg.; wd. u. for timber - **M. buchananii** R. Wilczek (Kenya): c. antileukemic ansa macrolides (maytanprine, maytanbutine); u. in leukemia, cancer (294) - **M. buxifolia** (A. Rich.) Griseb. (WI): spoonbush (Bah); carne de vaca (Cu); lvs. chewed in gastralgia; decoc. in colds, fever, after childbirth, &c. - **M. ilicifolia** C. Mart. (Br, Bo, Arg; Ur): mayteno; congorosa; mayten; sombra de toro (Arg); lvs. of tree c. tannin, NBP; u. as intest. disinf. (popular in gastric ulcer); analg. ton. astr.; contraceptive; stom. aper. cicatrizant, in skin cancer (Br); adulter. of maté; u. to revive drunken person (Arg); sds. c. caffeine-like alk. (2051) - **M. ovata** (Wall.) Loes. (eAf, esp. Ethiopia): shrub c. maytansine, maytanvaline (0.055 ppm), maysine, normaysine, maysenine; u. leukemia, not antitumor; toxic - **M. phyllanthoides** Benth. (M. texana Lundell) (sUS; nMex, WI): leatherleaf; guttapercha (or Fla [US]) mayten; verdolaga (Cu); mangle dulce (Mex); lvs. of shrub furnish a gutta percha; lvs. u. as remedy in toothache, scurvy (Mex); wd. u. as fuel - **M. pseudocasearia** Reiss. (Minas Gerais, Br): lf inf. u. for dys. - **M. senegalensis** (Lam.) Exell (Celasrus s. Lam; Gymnosporia So. (Lam) Loes.) (wtropAf, sAs): confetti tree; kulumbila; hazimo; Kisambila; umiviesa; Senegal pendoring; buds u. to treat GC; st. decoc. in colic, dys., diarrh. (children) (Senegal); green lvs. placed on sores (or appl. as poultice of pounded lvs.) (Tanz); in ulcers (Ind); rts. (bk.) u. chronic dys., diarrh. rheum.; cytotoxic to some cancers; inf. for wounds; lf. in mouthwash for toothache (Togo); ashes as salt substit.; lf. vermif. ; in fever; rts. chipped & put in beer as aphrod.; fire prod. by friction fr. sticks - **M. vitisidaea** Gris. (Arg, Ur): palta; u. in eye & ear dis. (folk med.).

mayweed (**camomile**): 1) *Anthemis cotula* 2) *Matricaria chamomilla* - **maywort**: *Galium cruciata.*

Mazola™: corn oil; the first brand marketed.

mazrhi: mazri (Pushto): *Nannorrhops ritchieana*; palm u. in Pak (esp. Baluchistan), Ind u. for making fiber sandals, fabrics, baskets, carpets, etc.

mazun (Armenia): a milk beverage made fr. buffalo milk treated with proper enzymes; acid arom. product similar to yogurt.

Mazus pumilus (Burm. f) Steenis (M. japonicus O. Ktze.; M. rugosus Lour. (Scrophulari) (Taiwan, Indoch): pl. inf. antipyr. ton. aper. (Indoch); to treat snakebites (Java); emmen. (Taiwan); natd. in lawns (US+).

mazzard: *Prunus avium.*

McMunn's Elixir: deodorized opium tr.

meadow: m. anemone: *Anemone pratensis,* Pulsatilla - **m. beauty**: *Rhexia* spp - **m. bloom**: *Ranunculus acris* - **m. cabbage (root)**: *Symplocarpus foetidus* - **m. foam:** *Limnanthes douglassii* - **m. rue**: *Thalictrum* spp. - **m. sweet**: *Gillenia trifoliata*; *Filipendula ulmaria* (also **m. wort**); v. also under crocus, fern, grass (English), rue (& early m. r.), saffron, sage, &c.

meal: a coarsely ground grain such as maize.

mealberry: Aretostaphylis *uva-ursi.*

mealies (sAf): Indian corn.

mealworms: larvae of *Tenebrio molitor* (& other spp.) (beetles); infest flour, some drugs, &c.; raised for feeding birds, zoo animals, &c.; occasionally as human food.

mealy: v. under starwort.

measles: rubeola: acute infectious viral dis. characterized by pinkish macular eruption, fever, mucosal inflammn., &c., treated with **m. convalescent serum** (E) - **m. prophylactic**: human m. immune globulin; is used for prevention - **m. (virus) vaccine (live)**: grown on chicken embryo primary cell tissue cultures (less popular is the **inactivated m. virus vaccine**); combns. also u.: **m. & mumps virus vaccine**, **m. & rubella virus vaccine** & combns. of the three.

Mecardonia procumbens (Miller) Small (Scrophulari.) (nwInd): pinjung; ext. u. as cardiac ton.

Mecca balsam (balsam of M.; Balsamum meccaense; B. de M.; Oleoresina de M.; Resina de M.): oleoresin of *Commiphora opobalsamum* - **M. galls**: fr. *Quercus infectoria*; similar to Bassorah (Basra) g. - **M. oil** (early term): crude petroleum - **M. senna**: *Cassia angustifolia* - **M. wheat**: *Zea mays.*

meconha (Por): Cannabis.

mechoacán (root) (CR+): **mechoacanna**: **mecoacan** (Por): *Convolvulus mechoacana.*

mechudilla (Mex): *Caesalpinia* spp.

Meconella oregana Nutt. (Papaver.) (Ore, Wash [US]): u. anod. sed. narcotic.

meconha (Por): Cannabis.

meconic acid: a dicarboxylic acid found in Opium; forms meconates; although previously controlled because derived fr. Opium (tax stamp, &c.), it has no narcotic effects.

meconin: lactone of meconinic acid; occ. in Opium & Hydrastis.

meconidine: amorphous alk. of Opium - **Meconium** (L.): juice of the poppy, Opium.

Meconopsis aculeata Royle (Papaver.) (Ind): kanta (Simla); pl. esp. rt. considered narcotic & dangerous; c. alkaloids (*cf.* Opium) - **M. nepaulensis** DC (nInd, Napal): rt. u. as narcotic.

Mecopus nidulans Bennett (monotypic) (Leg) (Ind, seAs, Java): u. in local med. to relieve certain pains, esp. in resp. dis.; to fertilize soil.

Medemia argun (C. Mart.) Wuerttemb. (monotypic) (Palm.) (ntropAf); frt. u. as offerings in tombs of ancient Eg (flavor of edible frt. superior after burying): frond u. as matting; endangered sp.

Medeola virginiana L. (monotypic) (Lili.) (eNA): cucumber (Indian) root; tuberous rhiz. u. diur. hydragog; food (tastes like cucumber or potato); eaten raw or cooked.

median: representing the value at central position of a series, not the average (ex. 3,4,5,7, 11: aver. = 6; median = 5).

medic, black: *Medicago lupulina* - **purple m.**: *M. sativa* - **spotted m.:** *M. arabica.*

"Medica": Haarlem oil.

Medicago arabica All. (Leg) (Euras, nAf): heart clover; spotted bur c.; pl. u. forage & for hay - **M. arborea** L. (s Euras, nAf): tree medic(k); pl. eccoprotic, galactagog, diur., vulner.; u. as fodder for livestock - **M. falcata** L. (Euras): yellow lucerne; yellow-flowered alfalfa; a fodder plant - **M. hispida** Gaertn. (M. denticulata Willd.) (Med reg., As): pl. considered diur. stom. and useful for intest. & kidney dis. (Chin); pasture & hay crop - **M. lupulina** L. (Euras, Af; natd. NA): black medick; hop clover; yellow trefoil; u. as green fodder, hay, green manure; med. (Ind) (2245); claimed the original shamrock - **M. minima** (L) Bartilini (Eu, natd. US): bur clover; "medicks"; with small yellow fls.; fodder - **M. sativa** L. (swAs, nAf, natd. Eu & NA): alfalfa; lucerne; herbage u. fresh & as hay as animal fodder; green manure; most important fodder crop in world; must be stored at low temps. (dehydrated, sundried, leaf meal, st. m.); dehydrated a. prevents scale in sea water evaporators (1646); for chlorophyll extn.; mostly dehydrated meal in folk med. as nutr. arom., "alfalfa tea," as beverage tea substit., etc.; rich in vits A & C; sd. c. FO (Alfalfa Sd. Oil) comm. available fr. American "Fancy Utah" var.; sd. u. arthritis & rheum.; rt. adult. of Bellad. Rt.; reduces blood cholesterol by reducing its absn. into the blood stream (295) (extravagant claims)- **M. scutellata** Willd. (Med reg, Eu, natd. seUS): snail medick; button clover; pasture crop (perenn.).

medical drugs (US): term u. to distinguish therapeutic fr. "recreational" drugs (hallucinogens, &c.).

medicament (Fr-m., Eng): medicine - **Medicamentum**: Haarlem Oil; Dutch Drops - **"medicine"**: Castoreum; secretion of beaver u. to trap same; sometimes other odorous substances are added - **medicine tree**: 1) (Austral): *Codonocarpus cotinifolia* 2) *Moringa oleifera* 3) (Calif): *Pinus* spp. (SC pitch u. burns & bruises) 4) (BC [Cana]): *Rhamnus purshiana*.

medick: medic.

medico (SpAm): physician (**medica**, fem.).

Medinella crispata Bl. (Melastomat.) (seAs): mucilagin. bk. & rts. u. as plasters for flesh wounds (Indones) - **M. heterophylla** A. Gray (Oceania): stalk grated & u. int. as abortif. (Fiji) - **M. rubescens** Merr. & Perry (Solom Is): lf. inf. ingested as remedy for sick stomach.

Mediterranean diet: supposedly healthful food regime commonly u. by natives of It/Gr/Sp +; this consists of pasta, olive oil, frts. & vegetables; wine; avocadoes; nuts; &c.; such foods have been associated with low rates of heart attack.

medlar: *Mespilus germanica* - **Mediterranean m.**: *Crataegus azarolus*.

medronheiro (Por): *Arbutus unedo*.

medulla (L, Eng): 1) rel. soft pith of pl. st. of endogenous type (ex. sassafras) 2) central portion of long bones of vertebrates, glandular organs, etc. - **m. bovis**: ox (bone) marrow - **m. lactis**: butter - **M. ostium rubra:** red bone marrow; human b. marrow u. to treat leukemia after removal from compatible host & transfer to invalid - **m. sambuci**: elder pith - **m. sassafras**: s. pith.

Medusanthera laxiflora Howard (Icacin.) (PI, EI, Oceania): yemollew; branches u. to make mallets to beat pulp of sago palm (Solom Is) (296).

meenchi (Quetti; Sinhalese): *Mentha spicata*.

Meerriche (Ge): Fucus.

Meerrettig (Ge): horseradish root.

Meerschaum (Ge, Eng) (lit. sea foam): hydrated magnesium silicate, like talc; u. for carvings, pipes, &c.

Meerschweinchen (Ge-m.): guinea pig.

Meerzwiebel (Ge): *Urginea maritima*.

meeting (house) seed: anise, dill, fennel, cumin, or caraway; SC in early days in New Engl [US] munched to counteract drowsiness at long church services.

Megarrhiza T & G = *Marah* - **M. californica** Torr. = *Marah fabaceus* - **M. gilensis** Greene = *Marah g.*

mega-vitamins: vitamins u. in large doses (sometimes defined as 150%+ recommended dosages); ex. vit. C.

megg (US addict slang): Cannabis.

megilp: magilp: combn. of boiled linseed oil, turpentine, & mastic varnish; u. in painting as vehicle for colors, furniture polish, &c. (SC McGuilp inventor).

mehedi (Hind): *Lawsonia inermis*.

Mehlblume (Ge): **mehlige Aletris**: *Aletris farinosa*.

Meibomia Fabr.: *Desmodium*

meimendro (Por): henbane.

Meiocarpidium lepidotum Engl. & Diels (wEquat, Af): (Annon.): c. alk. (phenanthrene type), polycarpol (triterpene); wd. u. to make ladles, paddles, oars (Gabon); med. (322).

meiosis: process of cell division & chromosome configuration in wh. the net effect is reduction of the chromosomes in egg or sperm from a diploid (2n) to haploid (n) no.; in this process, there is no chromosome splitting.

Meiran (Ge): majorana.

Meisterwurz (Ge): masterwort.

mejorana: 1) (Sp): marjoram 2) (Mex): same; also *Lantana* spp., *Brickellia veronicaefolia* 3) (Guat): *Mentha citrata* - **m. del pais** (Mex): *Salvia ballotaeflora*, *S. patens* - **m. de perro** (Mex): *Origanum* sp.

Mel (L.): honey (*qv.*) - **M. Depuratum**: purified honey. clarified h.

mela (It): apple.

Melagueta: v. *Aframomum m.*

Melaleuca alternifolia Cheel (Myrt.) (eAustral): (narrow leaf) Ai tree; VO c. terpinene, limonene, cymol, terpineol (tea tree oil), bactericidal, u. in furunculosis - **M. cajuputi** Powell (or Roxb.) (native to Molucc, esp. Buru Is, Ind, Vietnam, nAustral, NG, New Caled, often cult.): a source of cajeput oil; u. as heart stim. disinf. embrocation for rheum. & arthritis; local analg.; inhaled for rhinitis, colds; lung dis.; on sugar in colic, stom. indigest. (Chin); diaphor, antispasm for leg cramps; pain (Indones); rubbed on body in mal. ("the malaria tree") (NG); lvs. u. in asthma (PI) (1835); useful timber; ornam. - **M. leucadendron** L. = *M. quinquenervia* - **M. linariifolia** Sm. (Austral): narrow leaved tea (or ti) tree; VO u. antisep. in dermatitis (esp. fr. cutting oils) - **M. nervosa** (Lindl.) Cheel (Austral): u. by aborigines for colds (fresh lvs. crushed & inhaled, rubbed on forehead [2263]) - **M. preissiana** Schau. (Austral): mountain tea tree; shrub cult., usually planted to control soil erosion - **M. quinquenervia** (Cav.) S. T. Blake (M. leucadendran auctt.) (eAustral, Indomal, NG, New Caled); a source of cajeput oil; much like "Niaouli oil" (New Caled); punk tree; soaked bk. appl. to sores (NG) (2261); much cult.; natd. US, cultd. in Fla as windbreak, now a serious pest & displacing native plants there; cajeput tree (2053); ti; Australian pine; fr. lvs. VO ("tea tree oil"), chiefly cineole, also cajeputol, terpineol., limonene, pinene; VO u. exp. carm. counterirr. anthelm., exter. parasiticide, antisep. (strong), in skin dis., in otorrhea, otalgia, diarrh. (*cf.* eucalyptus oil) (765); wd useful - **M. uncinata** R. Br. (Austral): tea tree; broom honey myrtle; lvs. chewed for pharyngitis/laryngitis (natives) - **M. viridiflora** Soland. (New Caled): broad-leaved tea tree; niaouli tree; lvs. source of niaouli oil (VO), antisep. u. as substit. for eucalyptus & cajeput oils; u. in cough, neuralgia, rheum.; lvs. considered potent medicinal; oil u. in prepn. (oleo)gomenol (*qv.*), the oil itself sometimes being so called; VO c. Me cinnamate, ocymene, eugenol, linalool, terpineol.

melambo: malambo.

Melampodium camphoratum B. & H. (Comp.) (SA): u. as diaphor. (SC melampodion, Gr black hellebore fr. Melampous, legendary Greek physician).

Melampyrum nemorosum L. (Scrophulari) (Euras, Madag): Hainweizen (Ge); blue cow-wheat; semi-parasitic ann. hb.; source of "Madagascar manna" (c. dulcitol) - **M. silvaticum** L. (Euras): horse flower; wood cow-wheat; source of "blue bread"; bread baked with the meal turns blue, attributed to aucubin (rhinanthin); such bread causes headache & vertigo.

melancia (Por, Br): watermelon (vine or frt.).

Melanesia: island gp. of swPacific Oc (SC people dark-skinned, fr. mela [Gr] [dark, black], nesos [Gr] [island]); *cf.* Micronesia, Polynesia.

melanin: black pigment found in higher animals (ex. human skin); dark hair, retina; fungi, bacteria, & some higher plants (pods of *Vicia faba*); some are formed by oxidation of tyrosine enzymatically (tyrosinase).

melanocyte-stimulating hormone (MSH): secretion of pars intermedia of pituitary gland.

malanoma: malignant tumor chiefly of skin formed fr. the melanocytic system of the skin - **m. vaccine** (Malaccine); u to treat stages III-IV; classed as orphan drug.

Melanorrhoea inappendiculata King (Anacardi) (Mal, Ind): yields resin, source of indelible black ink; pl. u. sometimes for criminal poisoning - **M. laccifera** Pierre (Indochin): source of a lacquer; milky juice sap has caustic effect (counteracted with *Alocasia macrorrhiza* juice) - **M. usitata** Wall. (Burma): (black) varnish tree (of Burma); u. to prep. fine Burmese lacquer, for preserving woodwork; also as cement in Burmese glass mosaics.

melanosmegma: soft (green) soap.

Melanthera biflora Willd. (Wedelia b. DC) (Comp.) (seAs): pounded lvs. u. as poultice or vulner. for wounds, sores, insect bites, swellings, itch, festers; pl. decoc. antipyr. rt. decoc. diur. emmen. - **M. chinensis** Merr. (seAs): lvs. ton. u. in coughs, headache, skin dis., baldness (Mal); appl. to swellings with rice (Taiwan) - **M. gambica** Hutch. & Dalz. (cAf): in combns. u. for icterus; purg. diur. (323) - **M. scandens** Brenan (Equat, Af): lf. & st. inf. u. emet.; lf. & fl. hemostatic, induce rapid healing; lf. juice with lime j. u. as eye drops; love philtre (Hausa) - **M. strigulosa** B. & H. (seAs): antipyr. (sts.).

Melanthium virginicum L. (Lili) (c&sUS): bunch-flower; black f. (SC darker color of perianth after expanding); decoc. of toxic rts. u. for scabies; cult. ornam.

melanzana (It): **melanzane** (It): egg plant frt., *Solanum melongena.*

melão (Br): *Cucumis melo.*

Melastoma ackermanni Mure. (Melastomat.) (Co, SA): (South) American gooseberry; hb. c. VO; u. as antineuralg. (SA); in homeopathy - **M. decemfidum** Roxb. (Burma, Mal, Indoch): lf. frt. rt. u. astr. in diarrh. dysmen.; rt. u. in liver complaints with jaundice; stim. ton.; in malaise & vertigo - **M. malabathricum** L. (Ind, seAs, Oceania): "Indian rhododendron"; Malabar gooseberry; blue tongue; lvs. accumulate Al; lvs. (Folia Senggani) & fls. u. astr. in diarrh., dys.; lvs. chewed for "thrush" (Fiji); leukorrhea; bk. & rts. for wounds, skin dis. & gargles; frts. edible, produce a red dye; wd. tar u. to blacken teeth; foliage u. as food for silkworms - **M. polyantha** (Bl.) Volkens (M. polyanthum Blume) (Indones, PapNG, Mol, Ind, sChin, Mal, Sum, PI, Austral): taval (PapNG); senggani (PI); sap drunk for GC (Solomon Is) lvs. & rt. u. int. for diarrh. and at child-birth (PapNG).

melatonin: hormone of pineal gland; soporific; higher blood levels at night; represses sex hormones (antigonadiotropic) (potential contraceptive); may be involved in thermoregulation & circadian rhythm; u. to alleviate untoward symptoms of "jet lag"; also fatigue in psychiatry; aids sleep at night; antioxidant; possibly opposing Alzheimer's dis. & cancer; sold only at health stores

melegueta (Sp): *Pimenta officinalis.*

Meleguette pepper: v. Malaguetta p.

Meles (L.): **Meles taxus** (Mustelidae - Mammalia): (OW) ([gray]: earth pig); badger; burrowing carnivore; body fat. u. in med. prepns.; the common American b. = *Taxidea taxus.*

mélèze (Fr): *Larix laricina*, larch.

melezitose: a trisaccharide, representing a foreign sugar (probably a honeydew, excretion of aphids, &c.) occurring unpredictably in honeycombs in widely scattered parts of country; most obtained fr. Md & Penn (US); found in almost pure form in periods of drought ("pine honeydew") (2054).

Melia azadirachta L. (Meli) (Ind): = *Azadirachta indica* - **M. azedarach** L. (M. australis Sweet) (As-Himalay reg. to Austral.; cult. & grown wild in Med reg. Af & fr. Va s in US): pride of China; Chinaberry (tree); pride of India; bead tree (frt. u. as beads); Persian (or Cape or Barbados) lilac; dongoyaro; frt. pulp in tallow formerly u. for "scald Head" (seUS); frt. pulp (c. melianone, triterpene); u. for colic, in feed to make horses & mules shed hair in spring (swGa, US) & as vermifuge; supposed to have therap. benefits of neem tree; to cure coughs & colds (Bah); mal. (Nigeria); frts & lvs. insect repellant; sds. yield 50-60% FO, called Melia Oil; frt. said eaten by children with no bad effects (SC) (297); lvs. said insecticidal; bk. c. 4-7% tannin, u. as vermifuge (Chin) (85); lf. juice u. as vermif. diur., emmen.; rt. bk. (c. vanillic acid = anti-ascariasis component) u. cath. anthclm. (2634), tenifuge; wd. yields a gum; (Ga); juice of green frt. (with S & curds) u. in skin dis. & for sores infested with maggots (Ind); frt. pulp bitter, nauseous, u. for colic; anthelm. (324); lvs. appl. for int. pain (PI) inf. of lvs. fls. bk u. as antipyr.; inf. lvs. or frts. u. exter. for eczema, dermatitis; timber u. in construction ("white cedar"); useful in boat building, etc.; cult. ornam.; var. *umbraculifera* Knox (forma umbraculiformis Berckm.) sometimes recognized as distinct fr. sp. (Texas): (Texas) umbrella tree; (T.) umbrella china(berry) tree; tree with shape of umbrella; popular shade tree - **M. composita** Willd. (M. dubia Hiern non Cav.) - **M. dubia** Cav. = *M. azedarach* **M. indica** Brand = **M. japonica** Don (M. azedarach L. var. j. Makino) = *M. azedarach*..

Meliaceae: Mahogany Fam.; shrubs & trees of trop & subtrop regions of world; lvs. usually pinnnate, exstipulate; frt. a capsule, berry or drupe; sds. rich in FO (ex. *Carapa, Trichilia*); trunks furnish some valuable timbers (ex. true mahogany, Gambia m., "cedar" wood).

Melianthus comosus Vahl (Melianth.) (Sapind.) (eAs+): tuffed honeyflower; lf. as poultice for old sores; decoc. of pl. appl. to slow healing wounds; lf. paste to dress sores & swellings; rt. bk. emet. general ton.; lf. & st. decoc. for rheum., relieve foot problems; in skin dis. (often in combns.) - **M. major** L. (sAf): honey flower; krikkiebos (Afrikaans); lf. decoc. u. as lotion for sores, wounds, in tinea capitis, necroses, foul ulcers, & as gargle for sore throat & in gum dis.; appl. with flour to boils (Transval); pl. toxic, also honey fr. nectar said toxic; cult. ornam.

meliatin: glycoside fr. *Menyanthes trifoliata* hb.

Melica ciliata L. (Gram) (Euras, nAf): ciliated melic grass; melick; honey g.; c. cyanogenetic glycoside; cult. as ornam (Other *M.* spp. found in temp. zone except Austral).

Melicoccus bijugatus Jacq. (Sapind.) (sFla [US], WI, nSA): Spanish lime; "guaco"; bk. decoc. u. in dys.; frts. edible, juicy, sold in markets; wd. as timber.

Melicope monophylla Merr. (Rut.) (PI): lvs. appl. to ulcers, pimples; pounded rts. drunk with water for puerperium.

Melicytus micranthus Hook. f. (Viol.) (NZ): manakura; u. to scent titoki oil; bk. u. as decoc. in several complaints - **M. ramiflorus** J. R. & G. Forst. (NZ): mahoe; lf. decoc. u. to bathe parts affected by rheum.; boiled lvs. bandaged on surfaces affected with scabies; inner bk. frayed & appl. as pack to burns (1647).

melilot: *Melilotus* spp. - **white m.:** *M. alba* - **yellow m.:** *M. officinalis.*

melilotic acid: O-OH-Ph-$(CH_2)_2$COOH; occ. free in lvs. of *Melilotus* spp.

Melilotus alba Medik. (M. leucantha Koch) (Leg) (Euras sometimes cult., natd. Am, Austral): sweet (scented) clover; white melilot (clover); white sweet c.; bienn. hb.; flg. hb. u. in plasters & cataplasms for rheum.; constip.; sd. u. in gargles; u. to flavor cheese & tobacco; source of superior honey; green manure; livestock forage; (< mel [honey] + lotus = "honey lotus"); some consider conspecific with *M. officinalis* - **M. altissima** Thuill. (Euras): yellow sweet clover; similar c. & u. to *M. officinalis* - **M. indica** All. (Med reg.; now natd. in Af, Austral, nInd, NA, SA): stinking clover; sweet c. (US); scented trefoil; hb. c. coumarin, sitosterol; u. discutient, emoll. as hot poultices & plasters; powd. hb. in snuff (formerly); fl. inf. to flavor milk; stim (Mex); stock food but may be toxic - **M. officinalis** (L) Pallas (c&eEu, As, natd. NA): yellow melilot; y. sweet clover; hb. c. VO coumarin, melilotin, resin; u. bronchial dis., colics, antispas. stim.; in plasters appl. to wounds, glandular swellings, etc.; in cigarettes for asthma, bronchitis; in halitosis; fermented ("spoiled") hay formerly commodity sold in drug stores as anticoagulant, the active principle Dicumarol now replacing it (chief prodn. nIt) (2055); formerly u. in tobaccos (Bull Durham?) - **M. sicula** Vitm. (Sicily): c. spermicidal saponin - **M. suaveolens** Ledeb. (seAs): pl. c. coumarin; pl. decoc. u. as lotion to relieve eye dis. (Indoch).

Melinis minutiflora Beauv. (Gram) (Af, natd. SA, esp. Br, Ve, CA, WI, cult. Fla [US]): molasses grass; (capim) gordura (Br, Ve); source of Efwatakala Grass oil; arom. odor; u. as repellant to snakes & insects (mosquitoes, tsetse fly, ticks) (this may be without foundation); u. as high-value fodder in tropics (Austral, Af).

Meliomen (Ge, TM propr.): combn. of Hydrastis & *Potentilla anserina*; u. in spastic dysmenorrhea.

melisa (Sp): **Melissa** (L): hb. of *Melissa officinalis* - **M. calamintha** L. = *Calamintha officinalis* Moench = *C. sylvatica* Brosf. subsp. *ascendens* P. W. Ball - **M. officinalis** L. (Lab) (Euras, nAf, natd. NA): (common) balm; lemon b.; bee b.; b. mint; "honey plant"; Bienenkraut (Ge); hb. u. flavor, diaph. carm. stim. antipyr. sed. esp. in nervous states; digest dist.; u. as cooling beverage; one of oldest medicinal plants (2056); source of VO (prod. Fr, It) (< mel, honey) u. in perfumes, liqueurs - **M. parviflora** Benth. (Ind, in Himal reg.): lvs. & sts. u. antipyr. stom. in liver & heart dis.; for insect stings; frt. as ton.; hb. in mouth washes; substit. for *Melissa off.*

mélisse (Fr): Melissa - **m. citronné**: lemon balm hb.

Melissenkraut (Ge): *Melissa* hb.

Melissengeist (melissa spirit): patent med. (Ge) with high alc. content; u. as defense for failing alcohol test (Ge).

melissic alcohol = **melissyl a.** = myricyl a. (*qv.*).

Mellitis melissophyllum L. (monotypic) (Lab) (Eu): (bastard) balm; b. gentle; honey leaf; honey balm; b. (leaf); u. in folk med. as aperient, diur. antisp. hypnotic, emmen. antidiarrh.; for cleansing healing wounds; in Maiweins.

méllite (Fr): medicinal prep. based on honey; Mel.

Mellorine (US): kind of ice cream contg. vegetable FO (instead of milk fat) + vit. A & D (-1962-).

Melocactus maxonii (Rose) Guerke (Cactus m. Rose) (Cact.) (Guat): melon cactus; "chili"; frt. pulp edible, sweet pulp u. for gastritis; cult. ornam.

Melocanna baccifera Kurz (M. bambusioides Trin.) (Gram) (Bangladesh, Burma): terai bamboo; fleshy frts. recalling apple in size & appearance baked & eaten; wd. useful for building & mat making; u. med. (325).

Melochia corchorifolia L. (Sterculi) (pantropics): lvs. (c. mucil.) u. in poultices for sores, swellings, body pains; lf. decoc. in urin. dis. (Mal); for dys.; rt. bk. chewed for sore lips (Kordofan); lvs. u. as veget. (Ind); fiber obt. fr. st. - **M. nodiflora** Sw. (WI): black marshmallow; u. in salt & vinegar for wounds & sprains (Virgin Is) - **M. tomentella** Hemsl. (Mex): malva de los cerros; pl. source of good quality of fiber - **M. tomentosa** L. (tropAm into pantropics): bretónica (morada) (Ve, PR); "red rope" (Bah); frt. eaten & lvs. placed on forehead for headache (NG); rt. decoc. for back pain (Bah) - **M. umbellata** Stapf (Ind, Mal, Java+): betenook; lvs. macerated in water & appl. to sore back (Solomon Is); u. to make "tempe" (fermentation food of Mal natives); bk. source of cordage; wd. not durable but u. to make tea boxes & as fuel.

melocoton: 1) (Mex): cushaw 2) (Sp): peach.

Melodinus Forster (Apocyn.) (seAs, Oceania): some 80 spp.; many taxa have complexes of many interesting alks - **M. monogynus** Roxb. (Ind): u. as fish poison - **M. orientalis** Bl. (Indones): latex very poisonous but u. in exter. med. - **M. scandens** Forst. (New Caled): u. in otalgia (lvs. reduced to poultices, introduced into ears); also u. to calm cough; lf. decoc. drunk after confinement; frt. sweet & edible; u. to prep. fermented beverage - **M. suaveolens** Champ. (Chin): frts. u. for cough & to treat "glandular swellings" in neck; for contusions & gas pains (small intest.).

Melodorum gracile (Engl. & Diels) Verdc. (Annon.) (Moz): u. for headache, joint pains, fever, generalized edema; frt. of shrub edible.

Meloë (Meloidae - Coleoptera - Insecta) (dry temp & trop regions): blister beetles; genus with 146+ spp., separate fr. *Lytta*, *Mylabris*, *Epicauta* - **M. vesicatoria** L. (Cantharis v., Lytta v.) chief source of Cantharis; several *M.* spp. c. cantharidin (ex. *M. proscarabaeus* L.) (v. under Cantharis).

Meloideae: Blister Beetle Fam; blister flies; beetles with rather soft integument which commonly causes blistering when appl. to skin; incl. many thousands of spp.: ex. *Cantharis*, *Epicauta* (genera) (1633).

melon: 1) muskmelon, *Cucumis melo* 2) watermelon - **butter m.** (Ark [US]): yellow-fruited watermelon - **camel m.**: wild watermelon: *Citrullus vulgaris* - **Cas(s)aba m.**: winter m. - **citron m.**: *Citrullus vulgaris* var. *citroides* (rind candied) - **m. de agua** (Cu): watermelon - **musk m.**: *Cucumis melo* - **Persian m.**: derived fr. winter melon; light colored rind & orange flesh; orign. fr. Levant - **sugar m.**: **m. tree**: papaya - **winter m.**: *Cucumis melo* var. *inodorus*, with sweet light-colored flesh (fr. wAs).

Melonis Oleum: FO of sd. (**Melonium Semen**) of fruit (**Melonium Fructus**) of *Cucumis melo.*

Melophagus ovinus L. (Hippoboscidae - Diptera) (universal): sheep tick; ked; u. in folk med. to treat icterus (of no value).

Melothria heterophylla Cogn. (Curcurbit.) (eAs): "gulkakri"; rts. purg.; frt. decoc. u. in fevers (UP, Ind) - **M. indica** Lour. (eAs): hb. u. for cooling drink, to relieve inflamm. (Chin); frt. u. purg. (Indoch) - **M. maderaspatana** Cogn. (tropAf, As): fl. & rt. decoc. u. antidote for "poison"; anodyne (Taiwan); aper. in flatulence, diaphor. (w coast of Af) - **M. perpusilla** Cogn. (eAs): frt. diur. (Indones).

meloukhia (Tunisia): **muluhia** (Turk): *Corchorus olitorius.*

melt (sugar, &c.) (popular usage): dissolve.

membranate: like, pertaining to, or forming a membrane.

membrillo (Sp): quince.

Memecylon edule Roxb. (Melastomat.) (seAs): lvs. astr., bk. u. as fomentation (Burma); lf. inf. antipyr. (Indochin); furnishes yellow dye; frt. edible, pulp astr.; wd. heavy, useful.

Memora axillaris Schum. (Bignoni.) (SA): pl. u. in folk med. hemostat. in uterine hemmorhage; contracts uterus; lax. - **M. nodosa** Miers (Bigonia n. Manso) (Br): sena do campo; carobinha; caroba lvs.; lvs. u. as lax.

Menabea venenata Baill. (monotypic) (Asclepiad) (Madag): fiofio; kiso(m)pa; thangin des Sahalaves; rt. c. digitoxigenin glycosides; u. as ordeal poison in criminal trials (rt. chewed or decoc. drunk); suicide poison.

Menadione: synthetic vitamine K-like compd. of quinone base; prothrombigenic vitamin; this & its various salts u. in therapy of hemorrahges where due to low prothrombin titer of blood (monoxime, sodium bisulfite).

menaphthone = Menadione.

menformon(e): estriol; estrone; estradiol.

menhaden: fish of gen. *Brevoortia* (*qv.*).

menhdi: (Sanscript): henna.

meningococcal meningitis: meningitis prod. by *Neisseria meningitidis* (Wollstein) Murray ("meningococcus"); inflammn. of membranes of brain & spinal cord; prod. drowsiness, pain in limbs, sometimes death - **m. vaccine: meningococcus v.**: suspension of polysaccharides of m. organism u. to produce immunity (mostly u. in epidemics) (E); groups A & C (sialic acid polymer)

menisarine & **menisine**: alks. of *Cocculus sarmentosus* (mu-fang-chi) (Chinese drug) (326).

Menispermaceae: Moonseed Fam.; mostly vines, few shrubs, trees; mostly in trop.; frt. a drupe; ex. Moonseed; Indian Berry.

Menispermum canadense L. (Menisperm.) (e&cNA, Japan): (common) moonseed; Canada m.; yellow parilla; Texas sarsaparilla (241); Canada wormwood (241); vine maple (298); rt. c. dauricine, acutuminidine, magniflorine; rt. u. bitter, ton., alter. diur. stom.; value in doubt; cult. ornam. liane - **M. carolinum** L. = *Cocculus carolinus* - **M. cocculus** L. = *Anamirta cocculus* - **M. dahuricum** DC (e&cAs): c. alk. dauricine & tetrandrine; rt. u. as diur., for skin allergies; hypertension; pl. insecticidal.

menomin (NA, Indians): wild rice, *Zizania aquatica.*

menotropins: ext. of human female post-menopausal urine c. gonadotropin (FSH and LH) activity; u. to treat female sterility (by stim. growth & maturation of ovarian follicles); also in males to stim. fertility and masculine qualities (by stim. testicular interstitial cells [Leydig tissue], concerned with testosterone prdn. & devt. & maturation of spermatozoa); also u. in veterin. med. as gonad stimulant; u. by inj. (ex. Pergonal, TM).

menta (Sp, It, Hung, Roum): 1) mint, *Mentha* (also spelling in Old Latin) 2) (Salv): *M. citrata* 3) (Sp): *M. piperita* - **m. acuatica** (Sp): *M. aquatica* - **m. argentina** (Sp) = *Bystropogon* (401) - **m. inglesa** (Sp): **m. peperina** (It): **m. piperita** (Catalan): peppermint.

mentastro (Sp): *Mentha rotundifolia.*

Mentha L. (Lab) (nTemp Zone, sAf; Austral): genus of true mints incl. peppermint, spearmint, &c.; rich in VO, the arom. pl. u. in med. (flavor preferred by old people); perfumes, as spices, & for ornam.; widely u. in tisanes for colds & fevers; some 30 spp.; about half native to or natd. in NA - **M. alopecaroides** Hull = *M.* X *villosa* - **M. aquatica** L.(c&wEu, Med, As —> worldwide): water mint; marshmint; yerba buena (Arg) lvs. c. VO, tannin, NBP; u. stim. for cramps, flatulence; for "horseworms" of horses & humans (Ec) (2449); u. by ancient Romans as flavorings; this sp. a putative parent of peppermint; marshmint; yerba buena (Arg); **M. a.** var. *crispa* Benth. (Eu, natd. eUS): Krauseminze (Ge); Menthe crépue (Fr); crisped mint; hb. u. arom.; for gallbladder & stomach dis.; in flatulence; source of VO (poco oil, Chin) - **M. arvensis** L. (M. dubia Schleich.) (Euras, NA): corn mint; lamb's tongue; fieldmint; marsh (whorled) mint; pudina (Ind); poleo (Mex); yerba buena (PI); hb. (lvs. & young shoots) u. as carm. diaphor. (Chin); in colds decoc. heated with vinegar by rubbing on forehead (Mex); arom., antispasm. stom. emmen. diur. (Ind); bitter ton.; antirheum. (claim); insect repellant (pd. lf.) on fresh meat (Flathead Indians); pd lvs. appl. to wounds inflicted by bears; flavor in chutneys, &c.; popular in folk med.; hb. c. 1-3% VO with pulegone, menthofurane; hb. c. hesperidin - **M. a.** subsp. **haplocalyx** Briq. var. *piperascens* Holmes (Jap, Chin, Korea; cult. Eu, Br): Japanese (field) mint; J. peppermint; hortelã; poho (Jap); hb. very rich in VO with menthol (-80%); hb. u. flavor, carm. diaph. analg. (Chin); antiinflamm.; antipruritic (Ainu); pl. chiefly cult. for prodn. Jap peppermint oil & natural menthol (Jap, Arg); residual eleoptene fr. distn. called "cornmint oil" & u. in cheap flavors, adult. other mint oils - **M. a.** var. *glabrata* (Benth.) Fern. (Chin): Chin peppermint; hb. u. Chin folk med.; source of natural menthol - **M. canadensis** L. (M. arvensis var. c. [L.] Ktze.; M. a. var. villosa [Benth.] S. R. Stewart; M. a. var lanata Piper) (NA): (American) wild mint; meadow m.; Cana m.; wild peppermint; said the only native American sp.; VO c. pulegone; inf. u. with cedar & sagebrush lvs. in cough, colds, headache (Utah [US] Indians) (327); beverage tea, cold rem. (wWash, Ore Indians); hb. decoc. u. as beverage, for stomach pain (Bella Coola Indians, BC [Cana]); for fevers, pains, swellings, colics (children), "summer complaint" (Okanagan Indians) (250) - **M. X cardiaca** (S. F. Gray) Berard ex Baker (M. x gentilis L.; M. gracilis Sole) (GB, most cult. US, natd. NA): Scotch (spear)mint; heart mint; red m.; little-leaf m.; constit. & u. much as for *M. spicata* but with higher yield & superior VO; cult. since ca. 1910 (US); inflorescence of whorled fls. subtended by small lvs. (BO, SC kardiakos, Gr, for the heart or stomach, for which symptoms often confused [404]; vernac. name Scotch, SC original stocks brought from Scotland [405]); more commonly cult. & distd. for spearmint oil than the traditional *M. spicata*; VO with carvone, &c.; supposed hybrid of *M. spicata* X *M. arvensis* (NF.BPC) - **M. citrata** Ehrh. (M. x piperita nm. citrata) (Eu, natd. NA, SA): lemon mint; orange m.; bergamot m.; VO c. 84-90% linalool & linalyl acetate; VO with aroma like lavender u. as raw material in perfume mfr.; sometimes considered a var. of *M. piperita* - **M. cordifolia** Opiz (PI): Philippine mint; herba buena; some consider conspecific with *M. spicata*; VO with high proportion of piperitenone oxide; lfy. tops u. as carm. (decoc.); crushed lvs. with coconut oil appl. as poultice to tympanites; lf. juice for roundworms; abortient; ear drops for otitis; often in combn. with other herbs; fresh young tops to flavor tea (400) - **M. crispa** L. = *M. spicata* (some consider it to represent a var. of *M. aquatica* L.) - **M. cunninghamii** Benth. (NZ): Maori mint; hb. u. as diaph. - **M. X gentilis** L. (*M. arvensis* X *M. spicata*) (Eu; cult. & natd. e&cNA): whorled mint; American apple mint; perenn. hb. rather close to *M. cardiaca*; odor reminiscent of peppermint; u. similarly (*M. rubra* Sm. [red mint], a "luxuriant form" of *M. g.*) - **M. glabrior** (Hook.) Rydb. = *M. arvensis* subsp. **haplocalyx** Briq. - **M. gracilis** R. Br.: generally regarded as a slender form of *M. gentilis* (328); sometimes equated with *M. cardiaca* - **M. incana** Willd. = *M. longifolia* - **M. lanata** (Piper) Rydb. = *M. canadensis* - **M. longifolia** (L) Huds. (M. sylvestris L) (Euras [Kashmir], nAf, natd. US): horse mint; English h. m.; brook m.; Rossminze (Ge); "white mint"; VO c. carvone, menthofuran, dihydrocarveol, limonene, caryophyllene; quercitrin, tannin (c. 9%) in hb.; hb. inf. u. carm. stim. antisept.; for headaches, general pain (folk med.): eaten sprinkled on "mast" (curds) (condim.) (Afghan); often confused with spearmint - **M. microphylla** C. Koch (Eg): habag; hb. c. sitosterol, ursolic acid, VO with thymol, &c.; u. carm. (folk med.) (241); eaten sprinkled on most (= curds) (Afgh) - **M. nemorosa** Willd. (sometimes thought a form of *M. niliaca* or conspecific with *M.* X *villosa*) (Euras. natd. PR: cult. SA): yerba buena: (hoje) bueno; popular in WI for GIT disorders, as flatulence, stomachache, diarrh., emesis; generally in illness; to stim. appetite - **M. X niliaca** Jacq. (M. longifolia var. vulgaris) (Med reg., sEu): Egytptian (or Persian) m.; appar. hybrid of *M. spicata* X *M. rotundifolia*; itself regarded by some as one parent of triple hybrid *M. piperita*; VO c. dihydrocarvone; hb. u. by Ancients much like present day peppermint - **M. penardi** (Briq.) Rydb. = *M. arvensis* subsp. **haplocalyx** - **M. X piperita** L. (or Hudson) (L. & BO) (Eu): peppermint; menthe poivrée (Fr); Pfeffermuenze (Ge); French mint; hybrid (*M. aquatica* X. *M. spicata*); widely cult. mostly in US & Euras chiefly for VO which c. menthol, m. esters (menthyl acetate+), menthone; oil prodn. (US [50%]: Ore, Wash [Willamette; Yakima], "Midwest" [Mich, Indiana]; Br; Russia, Eng, Chin, Ind [Madras]; &c.) (also a "synthetic p. o." in comm.); hb. u. flavor (p. water); in pharmaceuticals, toothpastes, cough drops, etc.; perfume, arom. carm. stim. antispas. in colic (often as p. tea) for flatulence, diarrhea, cholera, cramps, headache, vomiting, anorexia, bile dis., menstrual derangements, antisep. in nasal prepns. (but menthol preferred); in liqueurs (crème de menthe), peppermint lozenges (p. drops), flavor in chocolates, tea, ice cream; cosmetics (esp. tooth prepns.); for masking bitter tastes (1st choice) (1836); exhausted "hay" (after distn.) u. as fodder (but not for milch cattle); 2 vars. recognized: nm. *piperita* (black pm), nm. *officinalis* (white p. m.) - **M. pulegium** L. (Med reg., now widely distr. Euras, NA [also cult.], SA, Arg [1053], sAf): pennyroyal (mint); pudding grass; poley; European pennyroyal; hb. u. as stim. carm. flavor, emmen. diur. for abdom. pains, liver & gallbladder dis. in flatulence, hysteria, nerv. dis., gout; misused as abort. (since ancient Greece); insect repellant; dyestuff (Turkey); beverage tea ("organ tea") at meals (Cornwall, England)*; hb. u. exter. as rubefacient, c. VO with 85% pulegone; u. in soaps - **M. rotundifolia** (L.) Huds. = *M. suaveolens* - **M. royleana** Wall. ex Benth. = *M. longifolia* - **M. sachalinensis** (Briq.) Kudo = *M. arvensis* var. *piperascens* - **M. sativa** L. = *M.* X *verticillata*; per some American authors = *M. cardiaca* - **M. saturejoides** R. Br. (Qland, New S.W., Austral.): "(native) penny royal"; "creeping mint"; with strong peppermint odor; u. like *M. crispa* (J. H. Maiden); suspected of stock poisoning - **M. X spicata** L. (M. viridis L.) (Euras, cult. & natd. NA, SA): (true or common) spearmint; green mint; lambmint; menthe verte (Fr); m. romaine (Fr); "American mint" (formerly; misnomer); yerba buena (Mex, Domin Rep); hb. c. VO with carvone, dihydrocarveol, phellandrene; hb. (lvs. & flg. tops comml. article); u. flavor, arom. coloring, masking bitter drugs, stim. carm. in colic (strong inf.); emmen. "antigalactagog"; garnish in salads, &c.; in alc. beverages (mint julep; crême de menthe); in tisanes, candy; macerated sts. u. in poultices for bruises (Chin); very popular as chewing gum (chief use of VO) & mint jelly flavor; prod. US, Chin - subsp. **crispata** (Schrad.) Briq. (M. crispa L.) (Eu): curled leaf mint; crisp(ed) m.; lvs. (Herba Menthae crispae) u. in stom. & gallbladder dis.; carm.; source of VO (cult. Ge, &c.) much like typical spearmint (some consider this taxon to be a var. of *M. aquatica*, viz., *M. a.* var. *crispa* [L] Benth. [Krauzeminze]) - **M. suaveolens** Ehrh. (M. rotundifolia (L.) Huds) (s&wEu; natd. eUS+): apple mint; round-leaved m.; woolly m.; this & spearmint are the most common mints of Eu; hb. sometimes u. for flavoring; u. for stom. dis. (Ch); sometimes cult var. *variegata* auct. (Eu); pineapple m.; hb. u. in mint sauce & jelly - **M. sylvatica** Host = *M. arvensis* - **M. sylvestris** L. = *M. longifolia* - **M. X verticillata** L. (M. aquatica X M. arvensis) (M. sativa L.) (Euras): marsh whorled mint; hb. with VO like that of lavender - **M. X villosa** (Hull) Briquet (M. X spicata X M. suaveolens) (Euras, widely natd.); hybrid of M. X spicata X M. suaveolens: nm. alopecuroides (Hull) Brig, (M. a. Hull): Bowles m.; claimed to have superior arom. properties; incl. nm. (nothomorph) *M. alopecuroides* Hull (Bowles's mint claimed to be one of best mints) - **M. viridis** L.: = *M. spicata* .

Menthaceae: mint fam., Labiatae.

Menthastrum: subgen. of *Mentha* - **M. virdis** L. = *Mentha spic.*

menthe (Fr-f.): 1) mint, *Mentha* 2) (Qu [Cana]): *M. canadensis* 3) peppermint - **m. anglaise: m. d'Angleterre: m. du pays: m. poivrée**: peppermint - **petite menthe** (Guat): *M. rotundifolia* - **m. silvestre**: *M. longifolia* - **m. verte** (Fr): spearmint.

menthol: peppermint camphor; $C_{10}H_{20}O$; obt. fr. Japanese mint or other m. oils; also by synth. fr. thymol by hydrogenation (inferior prod.); occ. as white (colorless) crystals; u. stim. loc. anod. in pruritus; antispas. carm.; formerly in cigarettes, cough drops; perfumes, liqueurs, nasal drops (for rhinitis sicca) (2320); ointments, confectionary, dentifrices (toothpastes), etc. (1374); clears nasal passages giving sensation of patency; (natural m.; fr. Chin, Jap, Ind, Br) - **spirit of m.**: 5% in alc.

Mentholatum™ Ointment: formulation with menthol, camphor, boric acid various VO., &c. in petrolatum base; popularly u. for colds & other resp. dis.

menthone: a terpene ketone occurring in VO of *Mentha* spp. (such as *M. pulegium, M. requienii*), *Pycnothymus muticus, Geranium* spp., etc.

mentruz (Bahia, Br): *Chenopodium ambrosioides.*

Mentzelia albicaulis T & G (Loas) (wNA inland e of Cascades & Sierra Nevada): white-stemmed stick-leaf; sds. ground & u. as food by the Hopi - **M. conzattii** Greenm. (Oaxaca, Mex): "arnica"; yaga-duchi; lvs. & rts. u. in itch & other cutaneous dis. - **M. cordifolia** Dombey (SA - Andes, slopes): u. in folk med. as inf. for stomach dis. (angura tea; Alsitan Soc.); cult. ornam. - **M. decapetala** Urban & Gilg (cNA): ten-petal mentzelia; overground hb. with demonstrated hypotensive activity; c. choline - **M. hispida** Willd. (CA, SA): pegajosa; rt. c. drastic of unknown compn.; u. for syph. (Mex); antibilious; "cures malignant pustules"; Homeopath.

Menyanthes trifoliata L. (Menyanth) (monotypic) (Euras, NA swamps): (common) bog bean; buck b.; marsh (or bitter) trefoil; Bitterklee (Ge); Menyanthes (L); Folium Trifolii fibrini (L): hb. c. glycosides trifolioside, menyanthin (bitter subst.), menthafolin, loganin, &c. of primary importance and alks. gentianine, gentialutine, &c. of sec. importance; catechol tannins; u. as bitter ton. antipyr. in (intermittent) fevers, rheum. skin dis. icterus, for migraine, nerve dis., dry coughs; in liqueurs (bitter schnapps; boonekamp); lvs. substit. for hops (beer); rt. (as Radix Trifolii fibrini) as bitter, stom. antipyr (prod Pol, CSR, Belg); pd. rts. in bread (nEuras).

meona (Sp): *Acalypha alopecuroides*, copper leaf.

mepacrine hydrochloride: quinacrine HCl, Atabrine™: synth. antimalarial, now being superseded by other synth. compds.

mephitic odor: one poisonous or offensive to the smell, such as mercaptan bisulfide.

Mephitis mephitis L. (Mustelidae - Mammalia) (all of NA): (common) (striped) skunk; pole cat; "stink martin"; secretion of perianal glands sprayed at enemy assailant; secretion u. therapeutically in cough with laryngospasm; in neurasthenia, insomnia (Homeop.); also by hunters to mask human smell on traps (bottles) (other skunks = *Conepatus & Spilogale* spp., swUS, Mex).

Meractinomycin = Dactinomycin.

mercadela (Mex): *Calendula officinalis.*

mercaptan bisulfide: occurs in *Allium ursinum, Capsella bursapastoris*; stench around a Kraft paper mill; skunk spray.

Mercenaria mercenaria (Veneridae - Mollusca) (NA; Atlantic coast; Eng, Fr) quakog; hard clam shell up to 15 cm long; edible gastropod; entire animal u. in Homeopathy.

mercure végétal (Fr): Manaca rt., genuine.

Mercurialis annua L. (Euphorbi) (Eu, esp. Med reg.): French (or annual) mercury; a field weed; hb. c. saponin, VO, cyanogenic glucoside, bitter subst.; u. rarely as weak purg., in folk med. in edema, syph., emmen. in amenorrh. expect., diur. (fresh hb.); ancient folk med. to inhibit milk secretion (juice, in It); in suppurative inflammns., in mucous obstructions; at times toxic to horses (2057); dye pl. - **M. perennis** L. (cEu & Med reg): (dog's) mercury; u. similarly to *M. annua* & often mixed with it (emt. purg. poison); natural dye.

mercury: 1) a liquid metal, Hydrargyrum (L.); symbol Hg; mined in Sp, It 2) *Mercurialis annua* (lvs.) 3) *Chenopodium bonus-henricus* 4) *Toxicodendron* spp. - **three-seeded m.**: *Acalypha indica.*

merenchyma: spherical-celled pl. tissue.

merey (Mex): *Anacardium occidentale.*

merisier (Fr): wild cherry tree.

meristem: plant tissue specialized to carry on active cell division, including the types: **apical m.** (that at tips of sts. & rts.), **interfascicular m.**; phellogen (cork cambium); **intercalary m.** (that at base of internodes of grasses).

Merluccius bilinearis (Atl. Oc.) (Gadidae - Pisces) whiting; New Eng. hake; flesh eaten; **- M. merluccius** (Gadus m.) (neAtl Oc, North Sea): (European) hake; merluce; sea pike; chief source of American isinglass; food fish; found as deep as 800 m in ocean - **M. productus** (Ayres) (Pac Oc off Calif to Alas): Pac hake; potential source of fish meal, fish oil, solubles, isinglass; rarely u. for food; c. antibiotic (1655).

merluzzo (It): codfish.

Merremia aegyptiaca (L) Urban (Convolvul.) (tropAs?): sds. are eaten - **M. distillatoria** (Blco.) Merr. (Convolvul.) (PI): fls. placed on boil to heal (folk med.) **M. emarginata** (Burm. f.) Hall.f. (seAs): u. for cough (with sugar) (Indones): lvs. & tops decoc. u. as diur. (PI); pot herb - **M. gangetica** (L.) Cufo. (Ind): u. diur.; in cough & neuralgia - **M. macrocarpa** (L) Roberty (Antilles to cBr; cult. Br): "Jalapa (do Brasil)"; Brazilian jalap; tuberous rt. occurs in slices; c. 20% resin; u. much like true jalap, as drastic purg., lax. - **M. mammosa** (Choisy) Hall f. (Ind, seAs to Polynesia): u. in throat & resp. dis. & GIT complaints; fever; burns, skin dis.; formerly in diabetes (proven of no value); mild purg.; cult. for med. uses (Java) & for edible rt. - **M. peltata** Merr. (Oceania): pounded lvs. put in breadfruit poi to improve taste (Caroline Is); lf. placed on skin sore (PapNG), swellings & wounds (Indones) on wounds & sore breasts (Mal) - **M. tridentata** Hall. f. (Ind): rt. u. as bitter astr. ton. in rheum. hemiplegia, swellings, hemorrhoids, urin. dis.; lax. - **M. t.** subsp. **angustifolia** Ooststr. (Af): decoc. of entire pl. u. as eyewash in conjunctivitis; cerebral stim. - **M. tuberosa** (L.) Rendle (Piptadenia pisonis Mart.) (Br, widely natd.): Brazilian jalap; similar c. &. u. to *M. macrocarpa*; adult. of jalap.; (Operculina t. (L.) Meisn.) (orig. tropAm, esp. Br; now pantropical); sliced tubercles furnish Brazilian Jalap; also resin (B. j. resin); c. operculinic acid; cult ornam. (yellow fls.) - **M. umbellata** Hallier f. (pantrop): lvs. u. as poultice for sores, burns, scalds (PI); pl. decoc. u. as diur. antirheum. in neuralgia, epilepsy; rt. decoc. for hematuria (PI) - **M. vitifolia** Hallier f. (seAs, Mal, PI): juice cooling u. in fever; lvs. for poulticing; for urethritis (Ind).

merry-bells: *Uvularia* spp., esp. *U. sessilifolia* (**little m.**).

Mertensia maritima S. F. Gray var. *asiatica* Welsh (M. asiatica Macbr.) (Boragin.) (eAs, nEu coasts): bluebells; gromwell; oyster leaf; rt. ext. u. in gastralgia (Ainu); rhiz. eaten by Eskimos - **M. virginica** Pers. (e&cNA): (smooth) lungwort; (Virginia) bluebells; cowslip; u. lenitive, dem. expect. in pulmon. dis. & catarrhs.

Meryta spathipedunculata Philipson (Arali) (Solomon Is): bk macerated in warm water & appl. to sore legs.

Mesadenia Raf.: now distr. in *Cacalia, Frasera, Senecio.*

mesbe: *Sida rhombifolia.*

mescal (Mex): 1) peyote 2) *Agave atrovirens, A. parryi*; u. to prep. 3) strong alc. liquor (brandy) made fr. *Agave* spp. by distn. of fermented sweetened juice of penco (*qv.*) like mescal but with numbing effect (alks.) - **m. buttons**: peyote, *Lophophora williamsii*.

mescaline: alk. fr. peyote with uses as psychotomimetic agt.

Mesechites trifida Muell.Arg. (Apocyn.) (Co): pl. latex u. to cauterize wounds & hasten healing of recalcitrant sores & ulcers (Makunas) (2299).

Mesembryanthemum crystallinum L. (Gasoul c. Rothm.; Cryophytum c. N.E. Br.) (Aizo.) (Calif [US] [natd.], sAf, Austral, Canary Is, Greece eMed reg. [natd.]): ice plant (SC lvs. covered by small glistening hairs); cult. for greens (like spinach) & in salads; & as window pl. (ornam. succulent); u. med. (as Herba M. c.) as diur. & in kidney dis.; formerly for obt. soda (Alicante s.); dr. lf. u. to remove hair from skins (Hottentots); in soap making (Ethiopia, Mor): antipyr. (Ethiopia) - **M. edule** L. (Carpobrotus e. (L.) Bolus (sAf): Hottentot fig; frts. edible; hallucinogenic; juice fr. pounded lf. u. as gargle in sore throat, thrush; in digest. dis., to treat diarrh. dys.; juice antisep. astr.; lvs. c. 14-19% catechol tannin; pl. grown as soil binder - **M. nodiflorum** L. (Gasoul n. Rothm.) (nAf): gas(s)oul (blanc) (officinel) (Mor-Arab, Fr): Egyptian (fig) marigold; u. similarly to *M. crystallinum*; hb. & sd. edible (mostly in famines) - **M. tortuosum** L. (Sceletium t. N.E. Br) (sAf): channa; kanna; pl. u. as narcotic & intoxicant; c. mesmbrine with cocaine-like action; lf. chewed for toothache & other pains; decoc. or tinctr. u. as sed. (farmers, Cape of Good Hope); addiction among the Nama reported; u. to relieve hunger (329).

meso-: optically inactive, as in meso-inositol (also myo-i.).

Mesoamerica: middle America; (s) Mex + CA (+WI according to some); that part of Am occupied before Spanish Conquest.

Mesona grandiflora Doan (Lab) (eAs): pl. decoc. appl. to scabies (Indochin) - **M. palustris** Bl. (eAs): pl. decoc. u. for dys. & hoarseness (Indochin); furnishes refreshing beverage & flavoring (Java).

mesophyte: pls. needing a fair amount of moisture.

Mespilodaphne sassafras Meissn. = *Ocotea s.*

Mespilus germanica L. (Pyrus germanica Hook. f.) (monotypic) (Ros.) (Euras): (German) medlar; thorn tree; frt. eaten after it is frostbitten & is on point of decay ("bletted"); close to *Crataegus* (hawthorns).

mesquit (Sp): **mesquite (beans)**: *Prosopis* spp. incl. *P. juliflora* v. *glandulosa*; source of m. pods - **m. flour**: m. gum: Sonora g. - **m. grass**: *Muhlenbergia porteri* - **m. gum:** Sonora g.

mesquital: area (swUS) in which mesquite is dominant pl.

messenna bark: musena.

mesta (Ind, Pak): fiber of inferior quality fr. *Hibiscus cannabinus*; mixed with jute (some consider as of equal quality to good jute); prod. in Bihar, Mysore, &c. - **m. pat** (Ind, Pak): fiber fr. *Hibiscus cannabinus*.

mestom(e): vascular bundle (not incl. fibers) - **m sheath**: outer lf. sheath of vascular bundle.

Mesua ferrea L. (Guttif.) (Ind, seAs, PI, tropics): na(gas tree); palo de fierro (Baja Calif, Mex); ironwood; fls. u. astr. stom. expect. in cough; for bleeding hemorrhoids, burning of soles; in perfumes for cosmetics; to stuff pillow (arom); sd. source of FO; bk. u. as pectoral & for snakebite (Mex); resin fr. unripe frt. bk. & rts. u. as replacement for Canada balsam, in varnishes; sds. edible sl. lax.; lvs. & fls. u. for insect bites; wd. source of valuable timber (extremely hard); ornam. tree revered in Ind.

metabolism: the total of all chem. processes in the living organism; made up of anabolism (building up) and catabolism (tearing down) processes of pl. or animal.

metabolite: a substance participating in the metabolic processes of pl. & animal, usually the product of catabolism (tearing down of tissues by the living body) but may be of anabolism (building up of tissues); or may be a product taken into the organism (ex. CO_2 fr. plant sources) and essential to the life of the organism - **secondary m.**: the synthetic compounds in the organism not directly related to life process products (**primary m.**), which include many substances of medicinal importance (ex. terpenes, alkaloids, &c.).

metaderm(a): non-corky tissue substituting for cork.

metallic pills (bitter): strong pills of iron, quinine, strychnine & arsenic.

Metamucil™: "psyllium hydrophilic mucilloid" obt. fr. the outer epidermis of blond Psyllium sd.; recently extolled for fiber content.

Metaplexis japonica Mak. (Asclepiad.) (eAs): crushed sds. appl. to wounds, ulcers, insect bites; pounded lvs. appl. to abscesses & swellings; sd. hairs styptic (Chin, Jap).

Metastelma schlechtendahlii Dcne. (Cynanchum) (Asclepiad.) (sMex): rt. decoc. u. as rinse for canker sores (Yuc).

metate: crude mortar u. by wNA Indians to triturate maize, cacao, &c. to make flour, meal &c., using a pestle (mano).

metathesis: double decomposition: d. replacement: d. displacement: chem. reaction with exchange of elements or radicals with no valence changes; ex. $AgNO_3 + HCl = AgCl + HNO_3$.

Metazoa: animals composed of many cells, as opposed to the one-celled animals or Protozoa; the M. incl. all the higher animals up to & incl. man.

Metel (Folia) (Folium) (L): *Datura m.* (lvs.).

met-enkephalin: peptide of endorphin class found in brains, pituitary glands & peripheral tissues of all vertebrates; methionine present at end of peptide chain; stronger than leu-enkephalin; analg. property.

methacrylic acid: simple carboxy acid; occ. in VO of *Anthemis* (Roman chamomile); methyl ester important for plastic mfr. (Lucite, Plexiglas).

methane: marsh gas; CH_4; the simplest org. compd.; occ. commonly in nature (decompn.), in atm., chief component of natural gas; prod. in marshes, rice fields, mines (cause of explosion); coal beds (Ala [US]), &c. (natural pollutant); u. in org. synthesis (2466); in gas heating; potential fuel for int. combustion motors.

methanogen: primitive anerobic organisms which produce CH_4 fr. CO_2+H_2.

methanol: methyl alc.; wood alc.; CH_3OH; mfrd. by dcstr. distn. of wood; widely u. in chem. indy. as solvent, denaturant, anti-freeze, preservative, etc.; potential fuel for int. combustion motors; poisonous, esp. to optic nerve; occ. in juices of some frts. & in various prepns. of plant juices; mostly now synth.; in chem. mfr.

methi (Urdu): fenugreek.

methionine: amino acid essential for human life; prep. fr. wool, gelatin, casein, etc.; along with cysteine provides chief source of S of proteins; u. in liver dis.(cirrhosis); benzene intoxication; urinary infections; diaper rash; slow healing wounds; arterial calcification (inhibits cholesterol in artery); thyroid deficiency.

methodology: sci. of arrangement, systematic description & evaluation of arrangement of materials of instruction.

Methonica Gagneb. = *Gloriosa* - **Methonicae Radix** (L.): bulb. of *Gloriosa superba*.

methoxy-psoralen: xanthotoxin: methoxsalen™: coumarin deriv.; occ. in *Psoralea corylifolia*, *Ammi majus*, *A. visnaga*, &c.; u. to increase dermal pigmentation in sun, repigment vitiligo (vitiliginous areas), thus permitting sun-sensitive persons to tan quickly to protect against sunburn; to treat psoriasis; u. locally & orally.

methscopolamine salts: synth. deriv. of scopolamine; u. as anticholinergic, in peptic ulcer, &c.

methyl cellulose: prod. made by methylating cellulose; u. substit. for tragacanth & other similar gums; in cosmetics; as bulk laxative, etc. (1081) - **m. chavicol**: estragole - **m. hydrate** (Cana): legal term for methanol - **m. mercaptan**: methanethiol; CH_3SH; occ. in rts. of radish; in coal tar, petroleum; Kraft paper pulp; mill odor; in urine after ingestion of asparagus (pronounced aroma); in GIT, source of bad breath (halitosis); u. as intermediate in synthesis of insecticides, plastics, &c. - **m. morphine**: codeine - **m. reserpate** = seredine (fr. *Rauwolfia vomitoria*) - **m. salicylate**: Methylis Salicylas (L.): ester either natural source (by ferment. & dist. of wintergreen, *Gaultheria procumbens* or sweet birch, *Betula lenta*) or synth.; occurs in smaller amts. in many other pl. materials, ex. *Polygala senega* & other *P.* spp.; *Lindera benzoin*; *Spiraea ulmaria*, etc.; prod. u. flavor VO, &c., antisp. arom. deodorant, stim. carm. antirheum. diur. solvent, penetrant; exter. analg. antisep. antirheum. counterirrit.; after oral ingestion, 15 ml may be rapidly fatal (shock); keep fr. eyes!; often in ointments with menthol; in fly sprays (2635) - **m. testosterone**: methyl ester of t., an androgen; this ester is active by oral admin., absorption either fr. gastrointestinal or buccal mucosa - **m. theobromine**: caffeine.

Methymycin: macrolide antibiotic prod. by *Streptomyces eurocidicus* Okami & al. &c.; fr. soil near Oswego, NY (US); u. antimicrobial; for Gm- bacteria.

Methysticodendron R. E. Schultes = *Brugmansia* - **M. amesianum** R. E. Schultes = *Brugmansia aurea* Lagerheim.

Methysticum (L.): kava.

Metopium brownei Urb. (Anacardi.) (WI, sMex, CA): guao (de costa) (Cu); cochinilla (Domin Rep); (black) poison wood; toxic (allergenic) pl. (*cf.* poison oak, Toxicodendron); lvs. u. (folk med.) to treat erysipelas, measles, inflammn., &c. (Yucatan) - **M. toxiferum** Krug & Urb. (sFla [US], WI): (Fla) poison tree; p. wood; sap strong contact poison (allergen); bk. of shrub or tree u. in compn. for abortions; put on aching tooth to remove it (Bah), trunk exudn. "doctor's" (or hog) gum: resinous, u. emet., drastic purg. (Mex); wd. u. to adulterate quassia (distinction: c. tannins); latex gives painful blister (1649).

metopium: (L.) expressed almond oil.

metrenchyta: vaginal inj., sometimes v. syringe.

Metrodin: folliculotropic hormone.

Metrosideros albiflora Soland. (Myrt.) (NZ): inf. of bk. of tree u. to relieve pain, wounds, & bleeding - **M. excelsa** Soland. ex Gaertn. (NZ): Christmas tree; C. rose; bk. inf. u. to treat diarrh.; fl. nectar u. in sore throat; cult. ornam. - **M. macropus** H. & A. (Hawls): ohia lehua; large trees but very variable; wd. u. in flooring, &c.; bk. & young shoots with red lvs. u. in med. (Haw natives) - **M. scandens** Soland. (NZ): rata vine; inner bk. of shrub st. u. as hemostat. to heal sores - **M. tomentosa** Rich. (NZ): bk. decoc. u. in diarrh. (Bushmen); wd. useful - **M. vera** Lindl. (Amboina; Celebes): bk. chewed or in inf. u. for leucorrh. diarrh. (sometimes with ginger, cloves, &c.).

Metroxylon laeve Mart. (Palm.) (Thail, seAs, Oceania, cult.): (smooth) sago palm; sago (starch) is obt. fr. the pith of the stem after felling, just before palm fruits & dies; sago u. as delicious nutr., diluent - **M. rumphii** (Willd.) Mart. (Mal, Indones+): (Rumpf) sago palm; (common) s. p.; sometimes considered the cultigen while *M. sagu* is the wild type; pith of tree yields sago starch (powd.) and pearl sago (granular); represents staple food in much of Oceania; fibrous residues of pith u. as hog food - **M. sagu** Rottb. (Mal, Moluccas+): sago palm; s. tree; considered the wild form fr. which the cultigen (*M. rumphii*) developed; source of sago; lf. juice appl. to wounds, also fresh lvs. (NG).

Metubine Iodide™: di-methyl ester of d-tubocurarine iodide.

metva (Croat): mint.

Meum athamanticum Jacq. (monotypic) (Umb.) (w&cEu mts.): (common) spignel; bearwort; bald money; meu; rt. c. oleoresin, gum, starch, FO; u. arom. carm. (in indigest. flatulence) stom. ton. sed. emmen. in delayed menses; in fever; leukorrhea; in veter. med.; in prepn. of a stomachic spirit (sGe); frts. u. to adulterate fennel; rt. eaten in highlands; cult. ornam.

mevalonic acid: theoretical basis of "active" isoprene & terpenoids; precursor of cholesterol, ergosterol, digitoxigenin, carotenoids, caoutchouc, &c., using appropriate enzyme systems; isolated fr. distillers' solns. (Folkers, 1956-).

mevastin: Compactin (Jap): fungal metabolite (fr. *Penicillium citrinum)*; reduces cholesterol formation by interference with enzyme activity; of potential value in hypercholesterolemia.

mévente (Fr): sale under cost price.

mevinolin: a fungal metabolite compd. isolated fr. *Monascus ruber* v. Tieghem & *Aspergillus terreus* (fungi); inhibits enzyme needed in cholesterol synthesis; hence u. in hypercholesterolemia, &c.; Lovastatin™.

Mexcal (Mex): mescal.

mexerica (nBr): *Citrus nobilis* var. *deliciosa*.

mexicain: Mexican papain: proteolytic enzyme fr. *Pileus mexicana* A DC (*Leucopremna mexicana* A DC Standl.) (330) *cf. Jacaratia*.

Mexican chicle: chicle - **M. fiber**: istle - **M. fire plant** (Mex): Poinsettia - **M. flies**: mixt. of insects obt. fr. fresh water ponds near Mexico DF; u. fish food & to feed rare birds in zoos - **m. poppy:** *Argemone mexicana.*

Mexican v. also under sarsaparilla; scammony; tea; thistle; vanilla.

Mexico seed: castor bean.

mezcal (Mex): mescal.

mezerein: acrid resin fr. **mezereon (bark)**: *Daphne mezereum: D. laureola; D. gnidium*; effective against leukemia.

mezereon (bark): *Daphne mezereum*; *D. laureola*; *D. gnidium* - **American m.**: *Dirca palustris* - **Mezereum** (L.): mezereon.

Mezia includens (Benth.) Cuatrecasas (monotypic) (Malpighi) (Co): much u. vine of Makuna Indians; bk. inf. in urinary dis., to stim. urination; tea for edema (swollen legs); lvs. strong emet.; placed as cataplasm over abdomen in jaundice; rts. soaked in water with flour u. as strong lax.

Mezilaurus Kuntze (Laur.) (Br): silvia; lumber trees; bk. a tannin source - **M. navalium** (Fr. All.) Taub. (Br): tapinhoã; wd. useful & used; lvs. u. as cataplasm for buboes.

Mezonevron (*Mezoneurum*) **benthianum** Baill. (Leg) (wAf): lvs. c. mucil. saponin; u. in hookworm, GC (Nig), in female sterility, impotence, lvs. for abdom. pain, lax. (Senegal) - **M. latisiliquum** Merr. (PI, Timor): young lvs. of vine eaten as salad; lf. decoc., for asthma - **M. sumatranum** W. & A. (seAs, Indones): lf. decoc. u. postpartum; to treat diarrh; as vermifuge (Mal).

mezquite (Mex, NM [US]): mesquite (m. gum).

mezros: "mez": popular term (ca. 1940s) for Cannabis.

mezz (US slang): Cannabis.

mgoma wood (sAf): wd. fr. this tree compressed, then soaked in water several hrs., u. as substit. for cork & rubber stoppers for bottles.

miata (Russ): mint.

Mica Panis (L.): bread crumb.

Micania D. Dietr. = *Mikania.*

micelle: any colloidal particle, now specif. appl. to a generally submicroscopic aggregate of micromolecules (ex. in strong soap soln.) taking on different forms, mostly long chains of units chemically joined & laid side by side to form bundles; not very stable; often broken down by changes in concn., pH, heat, reactants (emulsifiers, Tweens, &c.).

Michelia baillonii Fin. & Gagnep. (Magnoli) (seAs): bitter bk. stim. antipyr. (Indochin) - **M. champaca** L. (eAs, incl. Ind, Mal, Nepal, Sunda; cult. Ind, Br): champaca tree; golden champa; champak (Siam); pand-oil plant; arom. bitter bk. astr. antiperiodic, as adulter. of cinnamon; rt. (rt., bk. Siam) u. as purg. emmen.; fls. stim. ton. carm., yield valuable VO u. as perfume in prepn. flower scents; to treat leprosy; lvs. vermifuge, in colic, yield a lf. VO; sds. c. FO "michelia fat," u. in embrocations; wd. useful, "guaiac wood" ("champaca wood" of commerce) (said an imaginary name not really applying to this sp.); lvs. to feed silkworms - **M. figo** Spreng. (Jap, sChin; cult. Ind, Java+): bk. & lvs. u. as antipyr.; fls. in perfumery; to scent hair oils; to aromatize tea; wd. useful - **M. fuscata** Bl. (cAs [Georgia], sChin); lvs. u. as antipyr. hypotensive - **M. longiflora** Bl. (M. longifolia Bl.) (Siam, PI): champie; fl. buds u. arom. (Thailand); fl. buds inf. u. in sapremia following miscarriage; in perfumes (Mal) - **M. montana** Bl. (Ind, Java): bk. u. as bitter ton. antipyr.; wd. light, resistant, u. to make houses, bridges, furniture - **M. nilagirica** Zenker (Ind): bk. & lvs. u. as antipyr.; wd. valuable.

Micheliella anisata (Sims) Briquet = *Collinsonia serotina* - **M. verticillata** Briquet = *Collinsonia* v.

michoacán (Mex): 1) Jalap (Mex Phar.) 2) *Operculina* (Convolvulus) *alatipes* (*cf.* mechoacán) - **m. negro**: Jalap.

Mick(e)y (Finn) (US slang): **"Mickola"**: 1) knockout drops (chloral hydrate in whiskey, &c.) to render victim unconscious 2) croton oil, jalap, &c. in alc. drinks (preferably whiskey) to give intestinal pain; u. in order to get rid of an unwelcome visitor; or appl. in lactose ("shoo fly powder," New Orleans, La [US]) 3) narcotic, opiate.

Miconia Ruiz & Pavon (Melastomat.) (tropAm, WI): "lengua de vaca grande," genus of ca. 1,000 spp.; lvs. of some u. as tea substit., some in med., usually astr.; small frts. generally edible (bush currants); bk. of some c. tannin, u. dye - **M. holusericea** (L) DC (Br?): frt. edible ("bush currants"), u similarly to *M. rubifinosa* - **M. impetiolaris** D. Don (WI, Mex to Br): cristo verde (Cu); lvs. u. as arom. in baths (Cu) - **M. prasina** DC (WI, tropAm): "camasey" (PR); frt. edible; plant yields black dye - **M. rubifinosa** (Burpl.) DC (Br): lf. decoc. u. in bathes as sed; tranquilizer; in infantile "nervousness" - **M. willdenowii** Klotzsch (Br): bk. of shrub u. in "swamp fevers"; lvs. u. as tea substit. (0.2% caffeine).

Micrampelis Rafin. = *Echinocystis*; *Megarrhiza* (incl. man root).

Micrandra Benth. (Euphorbi.) (Br, Ve): spp. c. latex with rubber.

microbes: microorganisms, usually deemed harmful.

microchemistry: the study of chemical reactions & prods. observed under the microscope; a chief advantage is the ability to determine very small amts. of a subst. under examination.

Microcitrus australasica Swingle (Rut) (Austral): finger lime; a small spiny tree with small smooth elongate frts.; pl. u. in hybridization & as stock - **M. inodora** Swingle (Queensland): Queensland wild lime; frts. ("Russell River limes") u. like ordinary limes.

microclimate: c. of a restricted area, such as in a room.

Micrococcaceae: large fam. of Bacteria incl. genera *Micrococcus*, *Staphylococcus*, *Sarcina*, *Gaffkya.*

Micrococcus Haupt: gen. of bacteria (Fam. Micrococcaceae): Gm.+, in irregular masses (*cf.* staphylococci); not pathogenic; type gen. of fam. (1877); type sp. *M. luteus* (Schroeter) Haupt; many spp. now transferred to other genera (*M. catarrhalis* Frosch & Kolle = *Neisseria c.*; *M. gonorrhoeae* Schroeter = *N. g.*; *M. intracellularis* Migula = *N. meningitidis*; *M. melitensis* Bruce = *Brucella m.*).

Microcos paniculata L. (Tili) (seAs): lf. inf. u. to aid digestion & as cooling drink; decoc. of roasted lvs. u. as vermifuge (children); astr. lvs. in mal. & diarrh. (Indochin) - **M. tomentosa** Sm. (Indochin to Java): rt. decoc. for cough & fever; lvs. as poultice in itch; treatment for thrush (Indones) - **M. vitiensis** A. C. Smith (Fiji): crushed lvs. in cold water drunk for "spit blood" (hematemesis).

microcosmic salt: sodium ammonium phosphate.

Microcystis aeruginosa Kuetzing (Chroococc. - Algae - Cyanochloronta - Chroococcales) (cosmop.): colonies of this blue-green alga occur in fresh water to produce "water blooms"; c. "fast death factor" (toxin); may produce acute poisoning.

Microdesmis caseariaefolia Planch. (Euphorbi.) (seAs): fresh sap u. for dental caries (Indochin); frts. toxic - **M. puberula** Hook. f. (wAf): bk. u. in burns, pediculosis; if decoc. abort. emmen. (Liberia); sds. lax.; useful timber wd; u. for tool handles, combs, &c.

microfungi: fungi with small frtg. bodies.

Microglossa DC (many spp.) = *Conyza* L. + *Psiadia* Jacq. - **M. mespilifolia** Robinson (Comp.) (sAf): u. for fever & as ton. to cattle - **M. oblongifolia** O. Hoffm. (Tanz): in toothache; fever; lf. for eye trouble, appl. to sores - **M. pyrifolia** Ktze. (M. volubilis DC) (tropAf & other tropics): rts. u. for cough (smoked), chest dis., to aid conception (eaten as porridge) (Tanz); abort. (Burundi); headache, for worms (Liberia); for heartburn; in neuralgia & toothache (given as nasal powder) (Senegal): arom u. for fever with headache; lotion to cause perspiration; decoc. in labor; in edema, yellow fever; pd. rt. as snuff for colds; in ringworm of scalp (juice of warmed lvs.).

micrology: science of prepn. of microscopic objects for study, incl. sectioning, slide prepn., staining, etc.

Micromelum compressum Merr. (M. pubescens Bl.) (Rut.) (s&seAs, Austral, Polynes): thé de Lifou (New Caled); rt. & lvs. antipyr.; bk. appl. to wounds (PI); lvs. chewed for sore tongue (Fiji); young shoots boiled in oil u. for infantile convulsions; rt. inf. carm. lvs. crushed in coconut oil as liniment; for diarrh. (children), toothache; lvs. for tea (Fiji); lvs. appl. to splenomegaly (PI) - **M. falcatum** Tan. (seAs): lvs. u. as emmen. & to clean infected sores & insect bites (Chin, Indochin).

Micromeria abyssinica Benth. (Lab.) (eAf, Ethiopia): lf. u. in bronchitis, fever, cough; arom. - **M. biflora** Benth. (UP Ind, Himal, Arab, tropAf): whole pl. made into paste & appl. to cuts & wounds as antisep.; often mistaken for thyme - **M. brownei** (Sw.) Benth. (Satureja b. Briq.) (NA, SA): "poleo"; pennyroyal; West Indian thyme; pl. decoc. for stomach dis. (Yucatan); pectoral, (pulm. ton.) emmen. (Ven); abort. (Jam); inf. in liver trouble; lf. & fl. decoc. for pleurisy (2066); fresh lvs. appl. to chest for pleurisy; also u. in hysteria, flatulence, tinnitus (lvs. inserted in ear); nasal congestion; culinary (Co) incl. var. *pilosiuseula* Gray - **M. chamissonis** Greene = *Satureja douglassii* - **M. douglasii** Benth. = *Satureja d.* - **M. eugenoides**: (nwArg): muña muña; shrub; tea u. for stomachache - **M. juliana** (L) Reichenb. (Med reg.): Bartsaturei (Ge) (bearded savory); savory; lvs. u. like sage - **M. thymifolia** (Scop.) Fritsch (Eu - Balkan Peninsula): VO with strong antibacterial (and some antifungal) activity.

micrometer: 1) device for measuring very small distances with the microscope; the **ocular m.** is u. to measure directly by means of a ruled scale which fits into the eye-piece of the microscope; the **slide m.** (or **stage m.**) is a scale fitted to a micro-slide with markings of 10 microns & multiples & is u. to calibrate the ocular m. 2) one millionth of a meter; 0.001 mm; symbol: μ or μm. - **micrometry**: the science of measuring very small objects or distances by means of the compound microscope & appurtenances; an important technic of the microanalyst - **micron** (pl. **microns, micra**): 1/1000 mm.; symbol μ(m).

Micronesia: island gp. of Oceania in nwPac Oc (SC small islands) (incl. the Caroline, Marshall, Mariana & Gilbert Is & Nauru Is); many medl. pl. occur in area.

micronize: to pulverize into very small particles only a few (400-1,000) mikra in diameter; ex. mica, clay, crude drugs.

micronutrient: substance in trace amts. essential for health of individual; ex. vitamins, minerals.

Micropholis obtusifolia Baehni (Sapot.) tropSA (esp. Br); catuaba; u. as aphrod.

Microphyllophyta: primitive vascular cryptogams, the oldest of the gp. of vascular plants; current genera *Lycopodium* & *Selaginella* (lycopods).

micropyle: small pore in sd. coat by which water is admitted during germin.

microscope: an instrument for magnification of view of a relatively small object - **compound m.**: consists essentially of 2 converging lenses or lens systems (objective & ocular or eye-piece), u. for high magnification - **electron m.**: instrument with very high powers of magnification (1000-10,000 X that of optical m.) attained through the use of directed streams of electrons with much smaller wave length than light waves - **light m.: optical m.**: orginal old-fashioned type using light waves as in contrast to electron m. - **phase (contrast) m.**: one possessing greater contrast than usual light m. by utilizing differences in phase of light to give differences in intensity of light transmitted by or reflected fr. object - **simple m.**: magnifying glass or m. lens; a convex lens or lens system providing a virtual image; magnifies 5-20x - **ultra m.**: instrument for study of submicroscopic particles using high intensity light & a compound microscope to study the points of light.

microscopy: science of pd. drugs, foods & other materials as studied under the microscope, simple & compound; the operator is a microscopist - **micro-slide**: thin glass slide for mounting material for study under a comp. microscope; common size 1" x 3".

Microsechium ruderale Naud. (M. helleri Cogn.) (Cucurbit.) (Mex): chichicamole; rt. u. purg. in obstipation (chronic constipation); diur. (32).

Microsorium scolopendria Copel. = *Polypodium s.*

microsublimation: the process of isolating a substance, usually a comp., by heating a small amt. of source material to vaporize this subst., then collecting it by condensation upon a cooled microslide; examination is made of sublimate by means of the microscope.

Microtea debilis Sw. (Chenopodi) (WI, tropAm): hb. decoc. u. in cough, colds (Trin).

Microtoena patchouli (C. B. Clarke) C. Y. Wu & Hsuan (Lab) (tropAs); Chinese patchouly; VO u. much like true patchouly oil.

microtome: apparatus u. to cut very thin sections of material for micro study; hand or machine operated - **microtomy**: technique of same.

Micrurus corallinus Wied. (Elapidae - Ophidia - Reptilia) (CA, SA, esp. Br, WI): coral snake; c. viper; snake with very toxic vemon (c. enzymes), may cause death; u. to prepare antivenin; venom u. in hemiplegia, migraine, ozena, hemoptysis, TB, &c. (Homeopath.) (2321) - **M. fulvius** (L.) (US - Gulf states; nMex): American (or northern) coral snake; harlequin s.; highly venomous.

Mid(dle) East: Near E.: Orient (Ge): portions of wAs & nAf surrounding the eMediterranean Sea, ex. Jordan.

middlings: sd. tissues (embryo, endosperm) with some of frt. tissues (pericarp, sd. coat) fr. cereal grains such as rice - **wheat m.**: is u. as drug adult., also diluent in vet. pds, etc.

"midnight horror" (Mal): *Oroxylum indicum* (SC foxy smell of fls. at night).

miel (Fr, Sp): honey - **m. de Narbonne**: honeys of ePyrenees, light colored & of agreeable flavor - **m. de palma** (SA): syrup prod. fr. sap of *Jubaea spectabilis* - **m. de palo** (CR): wild honey - **m. de pierre**: rock honey; collected fr. holes in rocks & fr. caves (Ga [US]) - **m. de purga**: molasses - **m. de tuna** (Mex): syrup prepd. fr. frt. of *Opuntia* spp.

miele (It): honey.

mielga (Ec): alfalfa.

miet(k)a (Pol): mint.

mignonette: *Reseda odorata* & other *M.* spp. (VO): **m. tree:** henna - **wild m.**: *R. luteola.*

mijo (Co): millet.

Mikania Willd. (Comp.) (mostly tropAm): climbing hempweeds; herbaceous or woody vines with opp. lvs. & small fl. heads (discoid) in panicles; ca. 90% belong to Br flora; nearly all of the many spp. of Br are said u. in therapy (SC J. G. Mikan, Bohemian bot.) - **M. amara** Willd. = *M. guaco* - **M. capensis** DC (sAf n to Sudan; Mad): lf. as paste for local appln. to venereal sores (Zulus) - **M. cordata** (Burm. f.) Robins. (wAf [esp. Nig] seAs, Oceania+): u. by natives for insect & scorpion stings & snakebite; wounds; eye inflammn.; cough; in mal. (Af); has adaptogenic value - **M. cordifolia** Willd. (Cacalia c. L.f.) (Fla, La [US], Mex, Guat; SA, esp Br): hempweed; guaco de Tampico; climbing vine, commoner near coast (Fla), blooms nearly all yr.; corolla pink or white, 6 mm. long; hb. & young sts. c. VO; u. ton. diur. in albuminuria, antipyr., diaph. in snakebite & scorpion sting; VO similar to that of *Eupatorium triplinerve* - **M. guaco** H. & B. (H. amara Willd.) (sNA [Mex, CA], nSA [Co, Ve, Bo, Br+]): (Herba) Guaco; guaco (morado) (Ve), hb. (lvs. & sts.) c. guacin (resinous bitter subst.), tannin; u. for snakebite (Baird), scorpion sting; for cholera (Eu); antipyr. in nerve dis., diarrh. cramps; rheum., arthritis; stom.; insecticide; vulnerary; in eczema, syph.; for carcinoma (Br); often combined with condurango bk. (Br); in tetanus(?!), rabies(?!), lf. juice & inf. u. as emmen., in gout; in bath after fevers - **M. hastata** Willd. (WI): guaco; popular med. in Cu; rts. u. in diarrh., vomiting, rheum., lf. decoc. in cholera - **M. hirsutissima** DC (Br): cipó-cabeludo; flg. pl. u. as diur. for albuminuria (Br); cult. Br (São Paulo & Matto Grosso states) (331); off. Braz Phar. - **M. micrantha** Kunth or HBK (WI trop, contin. Am now worldwide): "Chinese creeper"; waku (Trin); guaco (blanco) (Ve); lf. juice rubbed on athlete's foot (Trin); crushed lf. poultice for sores (Mex); crushed lvs. exter. for hornet stings & to stop bleeding (Samoa) - **M. officinalis** Mart. (Br): fresh lf. inf. u. ton. antipyr. stom. - **M. parviflora** (Aubl.) Karst. (Co, Ve, Cayenne, Ec, Pe, Br.): u. med. like *M. guaco* - **M. scandens** Willd. (seUS n. to Mass, nMexBah): climbing hempvine; c. hemp-weed; guaco (Mex); sts. & rts. u. for bites of poisonous animals; u. like *M. amara*; to reduce tumors (Antilles) - **M. triangularis** Baker. (Br): guaco a folha fino (sBrazil); u. antiseptic due to pimaradienic acids (diterpenes); acute inflammatory states; antitumor; expect.

mikron (pl. **mikrons, mikra**): micron.

mil = 0.001 inch - **m. homens** (Br.): *Aristolochia brasiliensis* & other A. spp. (also **milhomens**).

Milch (Ge): milk - **M. cow:** producing milk for human consumption - **Milchstern** (Ge) (milky star): *Ornithogalum* - **Mikchzucker** (Ge): lactose.

mildews: pl. diseases where pathogen appears at surface of host; types: powdery, downy (**mildiou** Fr) - **black m.**: members of Meliolaceae (Ascomycetes); pl. parasites of trop.

milenrama (Sp): **milfoil**: *Achillea millefolium* (fls., lvs.).

milfoil (, Eurasian) water: *Myriophyllum spicatum.*

milho (Por): maize.

Milium effusum L. (Gram) (Euras, natd. e&cNA): (wood) millet; m. grass; fodder; grain has been u. for flour for bread; straw to weave mats; sometimes cult. as ornam. - **M. esculentum** Moench = **M. panicum** Mill. = *Panicum miliaceum* L. (**milium** reflects ancient L. name for millet).

milium solis: (L) *Lithospermum officinale.*

milk (Lac, L): 1) normal lacteal secretion of the mammae (breasts, udders) of various mammals, starting after parturition; when term unqualified, generally means cow's m., fr. *Bos taurus*; supposed to oppose insomnia because of tryptophan content; nutritive (*cf.* culostrum) 2) colored thick liquid of various pl. parts, ex. coconut m. (fr. sd.), latex of rubber trees, *Asclepias* spp., etc. - **m.**

TABLE OF MILLETS

Broomtop/Japanese	*Brachiaria ramosa*
Adlay	*Coix lacrymajobi*
Shama	*Echinochloa colona*
Little	*E. crus galli*
Barnyard/Japanese/Sanwa	*E. frumentacea*
Channel	*E. turnerana*
African/Finger	*Eleusine coracana*
Millet Grass	*Milium effusum*
Bread/Broom/Common/ French/German/Hog/Little/ Proso/Russian/True	*Panicum miliaceum*
Little/Sumatra	*P. sumatrense*
Kodo	*Paspalum scrobiculatum*
African/Bulrush/Great/Indian/ Pearl/Spiked	*Pennisetum americanum*
Russian	*P. glaucum*
Foxtail/German/Hungarian/ Italian	*Setaria italica*
Egyptian/Great/Milo	*Sorghum bicolor*

acid: lactic acid - **m. ball** (Ky [US]): frt. of *Maclura pomifera* - **condensed m.**: m. partly evapd. with sugar added - **m. curds**: casein - **fermented m.**: incl. such products of fermentation as buttermilk, Yogurt, kefir, &c. - **m. ipecac**: *Trillium erectum* & other pl. spp. - **malted m.**: mixt. of dr. milk (powd.), malted barley & wheat flour; usually mixed with milk & ice cream for beverage use - **m. of almonds**: emulsion of a. - **m. of asafetida**: emulsion of a. (formerly USP) - **m. powder**: prep. fr. cow's milk (etc.) by a process of spray drying; popular for prepn. of babies' milk formulas; frequently modified by addn. of various sugars, dextrins, proteins, etc.; mostly defatted - **skim(med) m.**: whole m. less butter fat - **m. shake:** mixt. of milk with ice cream, flavoring, sometimes malt ext.; pleasant beverage esp. popular in summer & often consumed with meals - **m. sugar: sugar of m.**: lactose - **m. thistle**: *Sonchus oleraceus* - **virgin's m.**: rose water rendered milky by the addn. of Tr. Benzoin - **m. weed** (or **milkweed**): 1) *Asclepias* spp. such as *A. syriaca*, *A. tuberosa*, *A. subulata* 2) *Euphorbia corollata* (seUS) - (**common m. w.**: *Asclepias syriaca* - **m. w. family**: Asclepiadaceae - **horsetail m. w.**: *Asclepias verticillata* - **orange m. w.**: *Asclepias tuberosa* - **red m. w.**: *A. rubra* - **m. w. root**: fr. *A. syriaca* - **swamp m. w.**: *Asclepias incarnata*) - **m. wort** (or **milkwort**): *Polygala* spp. - (**bitter m. wort**: *P. amara* - **m. wort family**: Polygalaceae - **orange m. w.: yellow m. w.**: *Polygala lutea*).

millefeuille (Fr-m.) **Millefolium** (L.): milfoil.

mille-pertuis (Fr-m.): *Hypericum* spp., esp. *H. perforatum*.

millets: *Panicum miliaceum,* &c. (See Table) - **African m.**: Indian m. - **bread m.**: common m. - **broom corn m.**: common m.- **bulrush m.**: *Pennisetum americanum* - **common m.:** *Panicum miliaceum* - **Egyptian m.** (US): *Sorghum halepense* - **finger m.**: *Eleusine coracana* - **foxtail m.**: Italian m. - **French m.**: common m. - **German m.**: 1) common m. 2) Italian m. 3) some other grasses - **great m.**: 1) Indian m. 2) *Sorghum bicolor* - **hog m.:** true m. - **Indian m.**: *Pennisetum americanum* & other grasses - **Italian m.**: *Setaria italica*.- **Japanese (barnyard) m.**: *Echinochloa frumentacea* - **kodo m.:** *Paspalum scrobiculatum* - **little m.**: 1) common m, *Panicum sumatrense* 2) *Echinochloa crus-galli* 3) *Panicum sumatrense* - **pearl m.**: Indian m. - **proso m.**: **Russian m.:** common m. - **spiked m.**: bulrush m.- **true m.:** common m.

Millettia W & A (Leg.) (OW trop&subtrop): spp. c. rotenone; u. insecticide, mfr. rotenone - **M. barteri** Dunn W & A (Leg.) (OW trop. & subtrop.): spp. c. rotenone; u. insecticide, mfr. rotenone. (trop wAf): st. ext. strongly virucidal to Newcastle's disease - **M. eriocalyx** Dunn (eAf): rt. appl. to skin eruptions - **M. grandis** Skeels (sAf): Kaffir ironwood; sd. u. as vermifuge (roundworms) resin exudate similar to guaiac; pl. u. as fish & arrow poison; wd. u. to make spokes, implements, walking sticks, &c. - **M. oblata** Dunn (tropAf): rt. u. as bladder remedy (Sukuma) - **M. pachycarpa** Benth. (sAs): rt. u. as fish poison; insecticide; u. as ton. & to induce growth of red blood cells (Chin) - **M. sericea** Benth. (Ind to seAs): lf. decoc. antipyr. urinary dis.; lvs. u. as poultice for sore eyes; rt. sap u. to heal wounds of horses & cattle (Timor); pd. bk. vermifuge; fish poison - **M. zechiana** Harms (trop wAf): duahoma (Ghana): pulped lvs. rubbed on painful parts; bk. pulped in sea salt & Guinea grains in warm water as gargle for rhino-pharyngeal dis.

milliard: 1,000,000,000 (in US = billion).

Millingtonia hortensis L. f (monotypic) (Bignoni) (Burma, Thail, Indochin): bk. & wd. u. in the local pharmacopeia against scabies (Indochin); fls. u. to perfume tobacco; lvs. u. like opium in cigarettes.

millo (Cu, Ve): millet.

milloside: glycoside of corotoxigenin & D-cymarose; occ. in *Strophanthus boivinii*.

milo: any of certain grain sorghums derived fr. *Sorghum bicolor*; closely related to durra; **m. meal**.

Milorganite: activated sewage sludge prepd. by treatment of sewage with microorganisms; prod. u. as fertilizer and as source of vit. B_{12}; first obt. from sewerage of City of Milwaukee, Wisc (US) (hence name); dust can irritate eyes & resp. tract; some think may cause amyotrophic lateral sclerosis (Lou Gehrig's dis.)

milpa (Sp Am): field (patch, tract) of farm cut out of forested land, u. for growing maize or other crop (esp. beans).

milpita (Mex): small patch of maize, &c.

mimbre (Arg): willow cane, osier twig.

Mimosa L. (Leg) (most in tropAm): hbs., shrubs & trees; some spp. with sensitive foliage; some Co spp. called "bambuces"; u. as antipyr. - **M. bahamensis** Benth. (M. hemiendyta Rose & Robins.) (sMex, Guat, Bah): pepinillo blanco (former bot. name *Mimosa julibrissin* [Durazz.] Scop.); bk. boiled with salt water, decoc. u. for coughs & colds (Yuc) - **M. catechu** L. f.: *Acacia c.* - **M. invisa** Mart. (sMex; Latin Am): dormilona (CR): lf. inf. bitter ton.; sds. u. sometimes as emetic; bad weed in sugar cane fields - **M. pigra** L. (Par. Arg. & Af); recently introd. into As): sensitive du Sénégal; lf. u. as bitter ton. purg. in diarrh.; GC; crushed & placed on boils to draw them (Tanz); rt. raspings u. for inf. instilled into nostrils for bad colds (Gabon); aphrod.; in leposy; sd. emet. expect. in dental dis.; u. in mental dis. (Toucouleur race, in Sénégal) - **M. pudica** L. (sMex, tropAm; now pantrop.): sensitive plant; humble p.; dormilona (CA); lvs. u. for wounds, rarely for ischias (pelvic pain); lf. inf. u. in GC (Hond, Guat); lvs. toxic but rts. of same pl. said antidote (Br); rt. rich in tannin; sds. u. as emet. & expect. & in urinary dis.; cult. ornam. - **M. sensitiva** L. (SA, natd. EI): lf. u. for ishchias; bk. as astr. rt. in hemorrhoids, urinary gravel; small doses for diarrh. & vomiting - **M. tenuiflora** (Willd.) Poir. (M. hostilis Benth.) (Br, Ve+): rt. c. dimethyltryptamine; rt. u. in prepn. narcotic beverage "wine of Jurema"; Pancara (Indians of Br); hallucinogenic; formerly u. before battle (hence bot. name) (Paneara Indians, Br).

mimosa: 1) lay name for *Albizzia julibrissin* (also **m. tree**) (former bot. name *Mimosa julibrissin* [Durazz.] Scop.) 2) *Acacia* spp. producing wattle bark ("mimosa bark") (Austral) (also florist's m.) 3) *Leucaena pulverulenta* - **m. family**: follows.

Mimosaceae: one of the 4 families sometimes split out of Leg. Fam.; mostly trop or subtrop; with regular fls.

Mimosoideae: subfamily (tribe) of Leguminosae; component genera incl. *Mimosa, Prosopis, Acacia, Albizia, Entada,* &c.

mimotannin: catechutannic acid.

Mimulus glabratus HBK (Scrophulari); (Andean SA): "violeta"; "yerba amarila"; fl. inf. u. for bronchitis, chest colds (Ec); urinary tract dis. (nephritis) (2449); tender shoots eaten; var. fremontii (Mex): emollient poultice - **M. guttatus** Fisch. ex. DC (wNA, Alas-Calif, now throughout US, Mex, natd. cEu): yellow monkey flower; fls. eaten raw in salads; homeopath. med. (GB) (332) - **M. moschatus** Dougl. (w&neNA, mostly BC [Cana]-Calif, cult. & now natd. in c&wEu): yellow flowering musk; m. flower; m. plant; pl. occasionally with musk-like odor (strong odor claimed lost for many years after ca. 1910 [2058]; a botanical "mystery" [84]); u. stim. - **M. strictus** Benth. (M. gracilis Hook. f.) (As, Af): monkey flower; u. as remedy for irregular menstrn., lotion for bathing tired & feverish patients (sSotho).

Mimusops balata Gaertn. f. = *Manilkara kauki*. - **M. coriacea** Miq. (Sapot.) (Br): frt. c. gutta percha-like subst. - **M. darienensis** Pittier (Pan): source of Panama balata gum or gutta percha; wd. very valuable lumber - **M. djave** (Lan.) Engl. (tropAf): adjab tree; African pearwood; bk. u. in mouthwash to relieve toothache (Nig); moumgou sds. (djave nuts) expressed for FO (false shea butter: adjab b.; mahagoni nut fat) u. as ointment base (Nig), cooking fat, soap ind.; press cake toxic; frts. or sds. u. as ordeal poison; wd. known as Congo or Cameroun mahogany - **M. elata** Allem. (Br): Brazilian milk tree; masseranduba; when bk. cut, exudn. is yielded in abundance, u. like milk (consistency of cream) & as glue; frt. edible; bk. tannin-rich; latex gives inferior rubber; tree furnishes valuable wood - **M. elengi** L. (Mal, Ind): Spanish cherry; bakul(a); bk. astr. ton. antipyr.; bk. decoc. u. in fevers, diarrh.; boiled lvs. appl. as cold compresses for headache (Java); expressed lf. juice dropped in eye for eye dis.; rt. in tonic mixts.; frt. pulp astr. u. in chronic dys.; c. saponins; u. in Unani & Ayurvedic med.; cult. for arom. fls. (Mal) - **M. henriquesii** Engl. & Warb. (Moz): source of an inferior gutta percha - **M. kauki** L. = *Manilkara kauki* - **M. kummel** Bruce (tropAf): source of balata (mimusops gutta [Ethiopia]); lvs. rts. & bk. vasoconstrictive, u. as sed. & heart remedy (Senegal) - **M. obovata**

Sond. (sAf): ukbomvu; roode melkhout (Afrikaans); frt. edible - **M. parvifolia** R. Br. (M. elengi var. p. H. J. Lam.; M. erythroxylon Boj.) (seAs, PI, Austral): bk. astr. in gargles, lotions (for ulcers); inj. in GC; frt. edible; fine lumber - **M. zeyerhi** Sond. (sAf): red milkwood; frt. edible (c. vit. C).

mina (Sp): spring; mine (as in Minas Girais, [Br]).

mince meat: combn. of minced meats, frts., wines, liqueurs, &c., u. in prepns. of foll. (meat, suet, & brandy now generally deleted) - **m. pie: minced pie**: shred p.; formulas indicate a large var. of meats; frts. (such as apples, dates, lemon & orange peel, raisins); suet; spices (such as ginger, coriander, allspice, cloves, nutmeg); wines & liquors, esp. brandy; sugars, molasses, cider, etc.

mineral glycerin: petroleum - **m. gum**: sodium (or potassium) silicate - **m. oil (white)**: paraffin oil; liquid petrolatum (US, USSR) (light s. g. 0.818-0.880; heavy s. g. 0.845 - 0.905 [av. 0.875]); u. as lax., emollient, dem., lubricant, solvent (1837) - **m. pitch**: Bitumen - **m. spirits:** petroleum s. - **m. wax**: ozokerite; v. under wax - **m. wool**: asbestos.

mineral waters: natural spring waters contg. 0.1% + salts, sometimes carbonated; very popular, esp. in Eu; several types recognized: 1) simple acidulous (CO_2 + salts) (ex. Apollinaris) 2) alkaline (ex. Vichy) 3) Glauber salts type (Na_2SO_4) (ex. Karlsbad, Marienbad) 4) bitter ($MgSO_4$ (ex. Friedrichshall, Epsom) 5) table salt (NaCl) (ex. Wiesbaden) 6) chalybeate or iron (ex. Spa; St. Moritz) 7) sulfur (H_2S, &c.) (ex. Aachen) (recommended for arthritis, gout) 8) earthy or chalky (ex. Wildungen) (useful for kidney dis.); the devt. of carbonated beverages culminated from efforts beginning in AD 1265 to imitate naturally carbonated mineral waters (2063).

mineralocorticoid: adrenocortical endocrine wh. regulates the electrolytes in the body; ex. DOCA, aldosterone.

minerals, essential: compds. or substances contg. such elements as Ca, Mg, K, Na, P, S, C, H, O, N, that are known to be requisite for maintenance of life, incl. trace elements (or **t. minerals**) (Fe, Cu, Mn, Zn, Se, Mo, Co, I, F, B).

mingwort: Absinthium.

mini-pill: contraceptive tablet c. progesterone; for oral use.

mink: *Mustela* spp. (Mustelidae): mink (body) fat ("mink oil"); u. to condition & waterproof leather; in cosmetics (2062); in dermatitis, eczema, burns, &c.; furs much sought after.

Minocycline: semi-synth. deriv. of tetracycline; antibiotic u. in prophylaxis against meningococcal infections; Minocin™ (M. HCl); in GC (male) & non-GC urethritis (NGU) (*Chlamydia* caused); also in resp. & skin dis.

minor forest produce: MFP: incl. such items as lumber, fibers, rubber, oil seed, medicinal pls., &c.

Minquartia guianense Aubl. (1) (Olac.) (WI, CA, nSA): (black) manwood; manu; bk. u. as fish poison; frt. edible; wd. useful timber; u. —> black dye.

mint: 1) members of the genus *Mentha* 2) members of Fam. Lab. (2608): 3) spearmints (incl. *M. cardiaca*) 4) peppermint (2450) 5) *M. canadensis* 6) a liqueur made or flavored with peppermint - **American m.**: 1) peppermint of NY State[US], &c. 2) *Mentha canadensis* 3) *M. spicata* (formerly) - **apple m.**: 1) *M. suaveolens* 2) *M. spicata* (**American a. m.**) (**golden a. m.**: *M. gentilis*) (**silver a. m.**: *M. suaveolens* v. *variegata*) (**striped a. m.**: *M. suaveolens*) - **balm m.**: 1) *Melissa officinalis* 2) crisp mint - **baulme m.**: *M. aquatica* - **bergamot m.**: 1) *Mentha citrata* 2) *Monarda mollis*, &c. - **bitter m.**: *Hyptis* spp. - **black m.**: Mitcham peppermint: *M. piperita* v. *officinalis* f. *rubescens* - (**black p. m.** [*M. x pip.* nm. *piperita*]) - **brandy m.**: peppermint - **brook m.**: *M. longifolia*; sometimes *M. aquatica* - **brown m.** (Ge): *M. spicata* - **m. camphor**: menthol - **Canada m.: Canadian m.**: *M. canadensis* - **candied m. (leaves):** dr. m. lvs. are dipped into stiffly beaten egg whites then into granulated sugar, left to dry thoroughly on wax paper; such stay green & sweet for a year - **cap m.**: *Satureja calamintha* (prob.) - **cardiac m.**: *M. cardiaca* - **cat m.**: 1) *Nepeta cataria* + other *N.* spp., catnip 2) *Satureja calamintha* (prob.) - **Chinese pepper m.**: *M. arvensis* v. *glabrata* - **cock m.**: *Chrysanthemum balsamita* - **common (garden) m.**: spearmint - **corn m.**: *M. arvensis* & vars. - **Corsican m.**: *M. requienii*; u. in juleps - **coyote m.**: *Monardella villosa* - **creeping m.**: 1) *M. satureioides* 2) *M. gentilis* - **crisp(ed) m.: crossed m.: curled m.: curly(-leaved) m.**: *Mentha aquatica* var. *crispa*; (*M. crispa*) - **dead m.**: *Lamium* spp. (SC pl. has the fetid smell of death) (also "dead nettle") - **downy whorled m.**: *M. gentilis* - **eau-de-cologne m.**: *M. citrata* - **Egyptian m.**: *M. niliaca* - **m. family**: Labiatae (Lamiaceae - Menthaceae) - **field m.**: 1) *Mentha arvensis* 2) *Nepeta cataria* (Feldminze, Ge) - **fish m.**: brook m. - **flea m.**: *M. pulegium* - **French m.: garden m.**: peppermint - **m. geranium**: *Chrysanthemum balsamita* - **Gibraltar m.:** flea m. - **ginger m.:** Scotch m. - **green m.**: *M. spicata* - **hairy m.**: *M. aquatica* - **h(e)art m.**: *M. cardiaca* (old name) - **horse m.**: 1) *M. longifolia* (etc.) 2) *Monarda* spp. - **Hungarian m.**: Egyptian m. - **Italo-Mitcham m.**:

Mitcham or black m. cult. in It - **Japanese (field) m.**: **J. pepper m.**: *M. arvensis* v. *piperascens* (**"Japmint"**) - **m. julep**: alc. beverage garnished with a sprig of spearmint - **ladies' m.**: **lady's m.** spearmint - **lamb('s) m.**: 1) spearmint 2) peppermint - **lemon m.**: 1) *Monarda pectinata* 2) *Mentha citrata* - **littleleaf m.**: *M. cardiaca* - **mackeral m.**: **mackerel m.**: spearmint - **marsh m.**: 1) *M. aquatica* 2) *M. sativa* - **m. whorled m.**: *M. sativa* - **meadow m.**: 1) *M. canadensis* 2) *M. gentilis* - **mountain m.**: 1) *Pycnanthemum* spp. as *P. incanum*, *P. montanum* 2) *Monarda didyma* 3) *Calamintha officinalis* 4) *Origanum vulgare* - (**wild m. m.** [Calif]: *Salvia* spp.) - **orange m.**: 1) *M. citrata* 2) *M. aquatica* v. *hirsuta* - **Oregon m., southern**: coyote m. - **Patagonia m.**: *M. suaveolens* - **pepper m.**: (v. also **peppermint**): *M. piperita* (usually) - **pineapple m.**: *M. suaveolens* hort. form *variegata* - **pink m.**: *Stachys drummondii* - **rattlesnake m.**: *Brazoria truncata* - **red m.**: *M. gentilis* - **rock m.**: *Teucrium scorodonia* - **Roman m.** (It): peppermint (formerly) - **roundleaf m.**: **round-leaved m.**: *M. suaveolens* - **Russian m.**: peppermint cult. in USSR - **m. sauce**: spearmint oil, sucrose, vinegar, water, u. as condim. - **Scotch m.**: *M. cardiaca* - **slender m.**: *M. gentilis* v. *gracilis* - **spear m.**: v. spearmint - **spire m.**: *M. x spicata* (old name) - **squaw m.**: *Hedeoma* - **state m.**: peppermint cult. in central NY State (US) (earlier) - **stone m.** (Md [US]): *Cunila origanoides* - **thick-leaved m.**: *Mentha arvensis* var. *penardi* Briq. - **tufted m.**: water m. - **tule m.**: Canadian m. - **water m.**: 1) *M. aquatica* 2) *M. longifolia* - **m. weed** (Austral): *Salvia reflexa* - **white m.**: 1) *M. piperita* v. *officinalis* f. *albescens* (w. pepper) 2) *M. longifolia* - **white horse m.**: *Pycnanthemum montanum** - **white leaf m.**: *Pycnanthemum albescens* - **white peppermint:** *M. x pip.* nm. *officinalis* - **whorled m.**: *M. gentilis* - **wild m.**: 1) *M. suaveolens, M. canadensis* (**American w. m.**), etc. 2) *Ajuga reptans* 3) *Teucrium scorodonia* 4) *Salvia reflexa** 5) *Stachys arvensis* - **wood m.**: *Blephila hirsuta* - **woolly m.**: *M. suaveolens, M. lopecuroides.*

mint-bush: *Prostanthera* spp. - **minthis** (Gr): mint - **mints**: originally referred only to pls. of genus *Mentha*; term now sometimes extended to any spp. of Fam. Lab. - **mintweed**: *Salvia reflexa.*

Mint(h)ostachys mollis (Kunth) Gris. (Mentha m. Benth.) (Lab) (c&nwArg+): (menta) peperina; (guarmi) poleo (Ec); "orégano" (Co); above-grd. parts u. as antispasm. digest. antidiarrh.; pd. lvs. rubbed on rheum. areas (2449); in bronchial affections (Co); to rid house of fleas (Co); spice (Pe); VO c. pinene, limonene, menthone, pulegone, &c. - **M. verticillata** (Griseb.) Epling (nArg, sBo): piperina; rich in VO with menthol mostly as esters, called "Argentine peppermint oil" (406); inf. u. in GI dis, diarrh.; emesis; in prepn. brandy, liqueurs (2065).

Minze (Ge): mint - **gruene M.** (Ge): *Mentha spicata.*

Mio (abbr. Ge): million.

mio (Mex colloquialism): corn - **m. m.**: *Baccharis cordifolia.*

miquit: mesquit.

Mirabella (L): **mirabelle (plum): Mirabellen** (Ge): Damson plums, *Prunus domestica* var. *insititia, P. cerifera.*

Mirabilis alipes Pilz. (Hermidium a. Wats.) (Nyctagin.) (wNA: Utah, Nev, Calif [US]): four o'clock; rt. decoc. u. as headache remedy (Piute Indians): pl. fr. which Dopamine origin. isolated (1650) - **M. jalapa** L. (tropAm; widely cult. worldwide): marvel of Peru; four o-clock (plant); belles-de-nuit (Fr); rt. (false jalap) strong cath.; all parts of pl. so u. (Am Indian) (SC formerly thought source of jalap); achenes somtimes filled with cocaine as surreptitious means of administering a single dose (size of #3 capsule) (333); rts. u. in edema; bruised lvs. u. as poultice for abscesses, boils, scabies; fls. steeped in water to give dye u. to color cakes & seaweed jellies (Chin); pd. sds. u. as cosmetic pd. (Jap); cult. ornam. - **M. longiflora** L. (Mex): four o'clock flower; sweet f. o. c.; rt. u. as drastic - **M. multiflora** Gray (wUS): many-flowered (or Colorado or wild) four o'clock; beverage tea made fr. pl.; poultice fr. rt. u. in swellings (edema, goiter) (407); tumors; in rheum. (408) - **M. nyctaginea** MacM. (Allionia n. Mx.) (cNA): u. antipyr. (Teton Dakota Indians).

miracle fruit: miraculous f.: 1) *Synsepalum dulcificum* 2) *Thaumatococcus danielii* 3) *Gymnema sylvestre.*

miramashikima: *Skimmia japonica.*

mirasol (Sp): sunflower.

mirbane: nitrobenzene (dangerous substit. for or adult. of benzaldehyde).

mire: *Brunfelsia hydrangeaeformis.*

mirlitono (New Orleans, LA [US]): *Sechium edule.*

mirto (Sp): *Myrtus communis.*

Miscanthus japonicus Anderss.(Gram) (eAs): shoot decoc. u. as rem. for cough - **M. sinensis** Anderss. (Jap, Chin; natd. eUS): eulalia (grass); juice of young st. u. for inflamm. & to incr. absn. of extravasated blood (Chin); juice or decoc. u. as dressing for bite of wild animals (Indochin); pl. grows so rapidly that it can be heard (sic); cult. ornam.

Mischocarpus sundaicus Bl. (Sapind.) (seAs): lf. decoc. u. to quiet headache (Indochin); rt. decoc. u. for deep-seated cough (Mal).

mishme teeta (Ind): **misimee teeta**: *Coptis teeta.*

mish-mish (Iraq): apricot.

miso (Jap): soy bean milk (or paste or sauce) prepd. by mashing beans in water with rice or wheat & salt; u. in milk sensitivity, by bakers, chocolate mfrs.

misquit: mesquit.

missey: moosey (NC): mountain ash.

Mistbeet (Ge): hot-bed.

mistletoe: parasitic pls. of Fam. Loranthaceae - **American m.**: *Phoradendron flavescens* & other *P.* spp. - **Christmas m.**: *Viscum* spp., esp. *V. album* - **European m.**: *Viscum album* - **m. family**: Loranthaceae - **oak m.**: *Loranthus europaeus.*

Mistura Acaciae (L): acacia mucilage of different strengths - **M. Fusca** (L.): brown mixture, comp. licorice mixture - **M. Oleo-balsamica** (L): alc. soln. of Peru balsam, amber oil & various VO - **M. Rhei et Sodae**: rhubarb mixture comp.

miswak (Urdu): stick u. in Indo-Pak to brush teeth (c. 2 dm X 2 cm diam.); sold in bazars.

Mitchella (L.): **M. repens** L. (Rubi) (eNA): partridge berry; squaw vine; pl. & hb. u. ton. partur. astr. diur., emmen. (Am Indians).

mite: small 8-legged organism of Class Arachnida, commonly found infesting many crude drugs, foods, etc.; ex. cheese mite, *Glyciphagus* sp. (common in Cantharis).

Mitella diphylla L. (Saxifrag.) (e&cNA): bishop's cap; lvs. u. diur. astr. (<mitella [L], dimin. of mitra [L], cap: little cap) - **M. pentandra** Hook. (BC [Cana], Wash, Ore [US]): bishop's cap; 5-stamened mitrewort; c. tannin, bitter princ.; u. styptic, &c. (2064) - **M. prostrata** Mx. (neUS): miterwort; u. ton. - **M. trifida** Graham (BC - Alta [Cana]-Calif [US]): Pacific ozomelis; hb. u. as ton.

Mithramycin: antineoplastic antibiotic fr. *Streptomyces argillaceus* Sabin & al., *S. plicatus* Pridhem & al. & *S. tanashiensis* Hata & al. have also some antibacter. activity; mostly u. in inoperable testicular cancer (IV) (Mithracin Pfizer); isol. 1953.

miticide: agent which destroys mites.

mitochondrion (pl.-**ia**): minute particles of variable shape (usually sausage-shaped) found in the cell cytoplasm of nearly all organisms; source of ATP, hence concerned with oxidative metabolic (respiratory) processes in cell.

Mitomycins: antineoplastic antibiotics produced in growth of *Streptomyces caespitosus* Hata & al. (soil, Japan) (1956-); **M.-C.** u. for solid tumors (as sarcoma) by IV inj., esp. in GI cancers (adenocarcinomata of stomach, colon, pancreas, rectum) (Mutamycin Bristol).

Mitracarpum scabrum Zucc. (M. verticillatum Vatke) (Rubi) (tropAf): ann. hb.; dr. lvs. u. to heal old ulcers; for leprosy (exter. & int.) (wAf); antidote for some arrow poisons; pd. rt. (flour) u. in GC erysipelas (Tanz).

Mitragyna africana Korth (Rubi./Naucle) (Equat, Af): bk. u. as antipyr. (Seneg): lf. u. for fever, GC, Hansen's dis., wounds; rt. in colic (wAf); this & other *M.* spp. much u. in wAf med. for colds, dys. syph., &c. (409) - **M. ciliata** Aubrév. & Pellegr. (wAf): bk. rich in alks. (ciliaphylline, rhynchociline) u. as antipyr. in dys. - **M. diversifolia** (Hook. f.) Havil. (Indo-Mal area) = *M. rotundifolia* - **M. inermis** O. Ktze (tropAf: Seneg to Sudan): xosse; bk. c. mitrinermine (rhynchophylline); wd. c. mitraphylline; lf. u. exter. & int. for fevers & chills, GI troubles; diur.; bk. antipyr.; both lf. & bk. in GC (decoc. with native natron); rheum pains (some consider conspecific with *M. africana* Korth.) - **M. parvifolia** Korth (Ind, Pak-Himalay reg.): kaim; kaddam; c. alk.; u. to stim. appetite (bitter green lvs.); good wd. u. for carving cups, combs, &c.; bk. has fiber u. in cordage - **M. rotundifolia** O. Ktze. (M. diversifolia Korth) (seAs, esp. Burma, PI): mambog; c. rhyncophylline, mitraversine, alk.; psychoactive drug; alk. tried as replacment for Opium; u. in folk med. - **M. rubrostipulata** Havil. (tropAf): c. mitraphylline; kinovic acid; bk. u. for dys., fever, roundworms - **M. speciosa** Korth (Mal, Borneo, NG, PI): lugub (Mandaya); beinsa (Burm); gralom (Siam); lvs. c. mitragynine; lvs. chewed or smoked like Opium (narcotic), for which it is u. as substit. (Iran) (danger!); claimed addictive (334); stim.; pounded lvs. appl. to wounds; cultn. & sale prohibited in Thail (335) - **M. stipulosa** Ktze. (M. macrophylla Hiern) (trop wAf): African linden; bk. c. mitrinermine, mitraphylline; lvs. u. vulner. antipyr.; decoc. with pepper u. internal dis. & pregnancy dis.; bk. antipyr. diur. alexiteric to poisons, body pains; in leprosy; rts. in colic; lvs. & bk. for cough, mal.; wd. very hard, durable, u. to make pestles, &c.

Mitraria coccinea Cav. (Gesneri) (monotypic) (Ch): botellita; voquivoqui; lvs. & bk. purg. (u. as inf. or decoc.); pd. hb. u. as pomade appl. to some skin dis.

mitrewort: *Mitella nuda* - **false m.**: *Tiarella cordifolia.*

mitsuba (Jap): *Cryptotaenia canadensis.*

mitsumata (Jap): *Edgeworthia papyrifera.*

moa tree: *Bassia latifolia.*

mocan: mocassin flower (prob.): *Cypripedium* - **mocassin plant**: *Cypripedium acaule* - **mocassin, water**: cottonmouth: toxic venomous snake: *Agkistrodon piscivorus* (eUS): venom u. to prepare antivenin.

Mocha (US slang): coffee; also icing, &c. flavored with coffee; <seaport of Yemen former exporting point for coffee.

Moderil: reserpinine (Pfizer).

Modiola caroliniana G. Don (monotypic) (Malv.) (SA, natd. US): Carolina bristly mallow; pila pila (Ch); emoll. sed.; u. in edema; country people u. for maladies of throat (Ch); cold water ext. u. in bath.

modus faciendi (L): **m. operandi**: method of manufacture or prepn.

Moehre (Ge): **Mohrruebe** (Ge): carrot.

moelle (Fr): pih, medulla - **m. de sureau**: elder pith - **m. noire**: *Solanum nigrum.*

Moenchspfeffer (Ge): sds. of *Vitex agnus-castus.*

Mogadore: Essaouira, Morocco; exporting port; v. under orris; gum.

mohair: long stiff hairs of Angora goat u. in mens' coats to give firmness & shape.

Mohawk-weed: *Uvularia perfoliata.*

Mohnkapseln (Ge): poppy capsules, p. heads - **Mohnkuchen** (Ge): poppy sd. cakes - **Mohnsaft**: (fresh) poppy juice (ancient article of medicine).

Mohrah butter: mahua butter.

Mohrenpfeffer (Ge): *Xylopia aethiopica.*

Mohria caffrorum Desv. (Schizae) (Pteridoph.) (sAf): scented (or frankincense) fern; frond smoked for relief of head & chest colds; dr. frond incorporated in ointment for cooling appln. to burns & scalds.

moisissure (Fr-f.): mold(iness), mildew, mold growth.

mojo (US slang): 1) narcotics, esp. morphine, heroin or cocaine 2) (Am Negroes): voodoo amulet; hand (*qv.*).

mokko-mokko: moko-moko (Surinam): *Montrichardia arborescens.*

molascuit (WI): mixt. of sugarcane molasses & bagasse (pith fiber) in small particles; u. stock feed of high quality.

molasses (US): a thick dark sweet liquor or treacle obt. in prepn. of sucrose as non-crystallizing residue; u. in cooking, nutr., in silage for fattening stock, prepn. of exts., base in mfr. of rum, &c. (2025) - **blackstrap m.: New Orleans m.**: very thick, dark type of m. rich in minerals; u. to mfr. sucrose & industrial alc.; u. as "cureall," lax. - **refined m.**: treacle; golden syrup; u. as bread spread - **sorghum m.**: syrup prepd. by boiling down juice of sorghum.

mold: mould; term appl. to many fungus spp., usually saprophytes but occasionally parasites, which cause moldiness on various organic materials, ex. *Aspergillus, Penicillium, Mucor* (**bread m.**).

Moldavian balm: *Dracocephalum moldavica.*

Moldavica Fabr. = *Dracocephalum.*

mole plant: 1) (seUS Negroes): *Ricinus communis* 2) *Euphorbia lathyrus* (sd.).

molineria (Sp): grinding; milling.

molle: 1) (Sp, incl. Ec): *Schinus molle* 2) (Arg): *S. dependens.*

Mollin™: K soap of coconut oil.

Mollinedia guatemalense Perkins (Monimi.) (Guat): lf. inf. u. for stomachache.

Mollugo cerviana Ser. (Aizo./Mollugin.) (Ind): pl. u. as antipyr. in GC & lochial discharges - **M. lotoides** Ktze. (M. hirta Thunb.) (pantrop): st. & lf. decoc. u. in dyspepsia; purg. (Punjab); in diarrh. (Sind); in itch & other skin dis. (Ind); tender lvs. as vegetable (Sudan); weed - **M. nudicaulis** Lam. (tropAf, Ind): lvs. appl. to draw out boils; bitter, pector. in athrepy (form of malnutrition) - **M. oppositifolia** L. (Glinus oppositifolius A DC) (OW trop): ladies' bedstraw; juice u. in itch & skin dis. (Puddokota); veg. eaten as appetizer & stom. (PI) - **M. pentaphylla** L. (M. stricta L.) (eAs, PI): hb. u. as bitter (Ind), made into soup to promote appetite; made into poultice for sore legs (Mal); stom. aper. emmen. antisep. (Ind) - **M. verticillata** L. (eCana, eUS, Mex, tropAm): carpetweed; Indian chickweed; hb. u. dem. in cataplasms; a common weed.

Molucca bean: *Caesalpinia bonducella.*

Moluccas: Spice Islands; now part of Indonesia; since ca. 1500, the source of many of the spices u. in the West, incl. nutmegs, pepper.

Mollusca: animal phylum in which invertebrate organisms are characterized by having neither segments nor appendages; a dorsal fold of body wall often bears a protective shell, while a ventral protrusion of body u. for locomotion is the foot; ex. clams, octopus, snails.

molly-pop (Ala [US] slang): *Passiflora incarnata*, frt.

Mombasa: island & port of eAf; v. pepper.

momira (Urdu): *Thymus* sp. u. as flavor & spice in curries (local name in Kurram Valley, NWFP, Pak).*

Momordica balsamina L. (Cucurbit.) (OW trop, esp. Ind, Chin, Pak; natd. seUS, LatAm): balsam apple; b. cucumber; sometimes regarded as identical with *M. charantia*; vine u. as stom. ton.; in children's colds (Domin); frt. drastic cath., antispasm., emet. but edible; exter. in linim.; often preserved in alc. & pulp appl. as vulner. (also sds.); lf. & rt. ext. u. as antidiabetic (Pak); rt. for jaundice, liver dis. (indigenous med.); decoc. abortif. (Penang); camel fodder (lvs. & sts.); vine is cult. for ornam. & edible frt. in US & elsewhere - **M. charantia** L. (OW trop as Ind, Af; natd. sUS, LatinAm): balsam pear; bitter cucumber, wild balsam apple; "art pumpkin"; (planta) cundeamor (PR); la-kwa (Chin); leprosy gourd; lvs. (as Folia Papari) u. as emet. purg. in colic, stomachache, gallbladder dis., vulner skin dis.; frts. & lvs. as anthelm. in icterus; these & rts. in diabetes (Ind, PR); frt. as condim. in rheum. gout; stom. antipyr. in stews for sick (Chin); sd. anthelm., sd. FO against hair loss; rts. in hemorrhoids, as aphr(od). (2069) - **M. cochinchinensis** Spreng. (tropAs): frt. & lvs. u. in lumbago, abscesses; young lvs. appl. to int. pain (Pl); sd. in cough, liver & spleen dis.; wounds & ulcers; rts. to clean clothes (lather) - **M. dioica** Roxb. (Ind): rt. u. as antisept. in hemorrhoidal bleeding, intest. & urin. dis.; sed. in fever; sd. FO u. in dye & shellac indy. - **M. grosvenori** Swingle (Chin): arhata fruit; rakanka (Jap); frt. & lvs. u. for bronchial & pulm. dis. (PR, Chin, Ph. ed. 11 [1977]) - **M. tuberosa** Cogn. (Ind): tubers u. as abortif.

Monadenium invenustum N. E. Br. (Euphorbi) (tropAf): lf. decoc. u. in fevers, chest dis. - **M. lugardae** N. E. Br. (eAf): ext. u. in GC; latex thought anesthetic; ash of pl. rubbed into scarifications to relieve pain in rheum.; rt. hallucinogenic (prophetic visions) - **M. schubei** Pax (eAf): mild purg. (mixed with food).

Monanthochloë littoralis Engelm. (Gram) (saline marshes - sTex, sFla, NM [US], Cu): salt cedar; key grass. shore gr; uses not found.

Monarda L. (Lab.) (NA, incl. Mex): horsemints; ca. 20 spp. of arom. showy hbs., several u. in med. (2081) - **M. citriodora** Cerv. ex Lag. (Mex, seUS, Tex): prairie bergamot; lemon bee balm; lemon mint; hb. eaten with meat (Hopi Indians); hb. c. much VO with thymol; hb. u. as diur. & bactericidal agt.; subsp. austromontana (M. a. Epling); "orégano" u. as insect repellant (crushed on skin) (Tex); hb. decoc. u. for unsettled stomach; hb. u. to season meats - **M. didyma** L. (eNA, sandy places; cult. SA as tea plant; cult. Eu as ornam.): (Oswego) bee balm; mountain b.; Oswego tea; flg. hb. u. arom. carm. diur. diaph. stim. ton. antipyr., "nervine"; as tea substit.; earlier as quinine substit.; in homeopath.; lvs. eaten with meat; rel. low yields of VO, with linalool & esters, limonene, camphene, &c.; no thymol (2082); u. in pomades - **M. fistulosa** L. (with var. mollis Benth.) (e&cNA): (wild) bergamot; (American) "horse mint"; hb. c. VO with thymol quinone, isovaleraldehyde; hb. chiefly u. as antipyr.; stim.; rt. decoc. for GIT pain (Ojibwas); in combns. for colds (Meskwakis Indians); "baume des prairies" (Creoles, Mo [US]); lvs. u. in lotion for pimples, &c. (410); otherwise u. like *M. didyma* (2083); *f. albescens* Farw. (var. alba Hort.): cult. as ornam. - **M. pectinata** Nutt. (wNA): pony beebalm; c. VO. with thymol, carvacrol, &c. (2433); hb. u. flavoring (Am Indians) (NM [US]) - **M. punctata** L. (e&cUS): spotted beebalm; dotted monarda; horse mint; monarde (Fr); lvs. & tops c. VO with 60% thymol ("monardin") & esters; hb. u. carm. diaph. emmen., diur., antispasm. in flatulent colic, in colds, fevers; in "Am bilious fever"; exter. stim. (linim.); counter-irrit; rt. decoc. emmen.; otherwise like *M. didyma*; formerly indy. source of thymol (2084).

Monardella lanceolata A. Gray (Lab) (sCalif-wNev, [US], nMex): mustang mint; lvs. & flg. tops u. as beverage tea; also in colds, headache, diaph., in colic (decoc.) - **M. odoratissima** Benth. (wUS, Cana): coyote mint; wild pennyroyal; western balm (var. *parvifolia* Epling); lvs. & st. u. as decoc. or inf. as beverage (tea) or to treat severe cases of common cold (BC [Cana]) (336) (Nevada Indians) (171); decoc. in indigest. - **M. villosa** Benth. (Calif [US]): coyote mint; flg. tops u. as beverage tea (Calif Indians); lf. & st. inf. u. in colic, as ton., in stomachache.

Monascus purpureus Went (Monasc. - Pezizales - Asomyc. - Fungi): source of angkak & other Asiatic foods & beverages; source of monascoflavin (monascin) (pigment).

monda: *Pongamia pinnata*.

Mondia whitei (Hook. f.) Skeels (Chlorocodon w.hyteiHook. f.) (Asclepiad.) (s&tropAf-Uganda): mundi; murundo; liana with latex; chlorocodun rt. u. for indig., abdom. dis., ton., in enteritis, purg., heartburn, impotence; st. fiber for cordage; rt. c. coumarin & methoxysalicylaldehyde, hence, smells like vanilla ; VO; glycosides, resin, FO; sd. u. in arrow poisons (?) (1169).

Mondo Adans. = *Ophiopogon* (Lili).

Monechma hispidum Hochst. (Acanth.) (Ghana, Nigeria): pd. dr. lvs. burned & smoke inhaled for head colds.

monecious: pl. spp. in which both staminate & pistillate fls. occur on the same pl., as distinct fr. where they occur on different pls. (diecious).

monellin: "chemostimulatory" protein fr. *Dioscoreophyllum cuminsii* (Stapf) Diels (Menisperm.); with very sweet taste (800-1500); X as sweet as sucrose.

monensin: complex (polyether ionophore) isolated fr. *Streptomyces cinnamomensis* Okami; u. as anticoccidial, antifungal, &c., additive to beef cattle feed, 1967-.

Moneses uniflora Gray (monotypic) (Pyrola u. L.) (Pyrol.) (cEuras, Jap, nNA): one-flowered pyrola; hb. c. arbutin, ericolin; u. diur. emet.; in homeopathy.

monesia (Br): *Pradosia lactescens* Radlk. (Lucuma glycyphloea; Chrysophyllum glycyphloeum).

money plant: *Lunaria annua* - **Cornish m.**: *Sibthorpia europaea.*

moneywort: *Lysimachia nummularia.*

Mongolian: pertaining to nation of eAs; v. under Ephedra

mongoose (plur.) mongooses: *Herpestes ichneumon* L. (Viverridae - Mammalia): (nAf, AsMin, Sp; later Ind+): first domesticated in ancient Eg, probably because of its facility in killing rats & snakes; still domesticated in trops.; u. in Unani med are the following parts: flesh; blood; leg bones; not allowed into continental US because of danger of transmitting rabies; other spp. of this animal are known.

moniliform: series of swellings on root, st., hypha, &c. resembling a string of beads.

monilla (nuts): sds. of *Ungnadia speciosa.*

Monimiaceae: Monimia Fam.; arom. shrubs & trees, mostly of sHemisph & chiefly of Oceanica; leathery generally opp. lvs.; fls. solitary or in cymes; ex. *Laurelia*; *Peumus.*

monkey bread: *Adansonia digitata* - **m. face**: 1) pansy 2) *Mimulus* spp. (also **m. flower**) - **m. pod** (tree): *Samanea saman* - **m. puzzle** (tree): *Araucaria araucana* (common name originated in Co. Cornwall, Eng.) - **m. weed** (BC [Cana]): *Clematis* spp.

monk's hood: monkshood: 1) *Aconitum napellus* 2) *Consolida regalis.*

Monnina R. & A. (Polygal.) (Mex to Ch): some 160 spp., many with constit. & applns. similar to those of Senega rt. - **M. phytolaccaefolia** HBK (SA): u. med. (Co) - **M. polystachya** R. & P. (SA, esp. Pe): rt. & bk. c. saponins - **M. salicifolia** R. & P. (Pe): iguila; rts. ground up & u. as soap, esp. to remove dandruff fr. head (2449).

monochloracetic acid: $CH_2ClCOOH$; u. mfr. dyes; as beverage stabilizer; for latter condemned as poison & deleterious substance (FDA, 33).

Monochoria vaginalis Presl (Pontederi.) (seAs): "sesela"; corms pounded & appl. to ulcer; lvs. eaten raw or cooked (Mal); tubers eaten (PI); bk. u. in asthma; subterr. parts u. for gastric dis. & toothache (Java).

monoclonal: v. antibodies.

Monocotyledoneae: Monocotyledons: familiarly known as Monocots; a subdivision of the Angiosperms; char. by having 1 cotyledon in sd.; parallel-veined lvs.; fls. with divisions of 3 or a multiple; ex. *Iris..*

monocrotaline: alk. of *Crotalaria spectabilis* +.

Monodora myristica Dun. (Annon.) (trop wAf fr Sierra Leone to Congo; cult. St. Thomé, Jam): calabash (or American or African) nutmeg; sd. (owere seeds) c. VO with limonene; u. condiment (*cf.* nutmegs), stim. arom.; frt. arom. pungent - **M. stenopetala** Oliv. (Moz. Malawi): rt. inf. u. in TB, intestinal worms; frt. edible - **M. tenuifolia** Benth. (trop wAf): sds. u. as in prec.; bk. med. for dogs.

monoenoic: fatty acid with only one double bond; ex. oleic acid.

monoglyceride: ester of glycerin, with one of its OH gps. attached to the fatty acid (di- and triglycerides also occur in natural FO).

monograph: written detailed description of some single topic or object or of a single class of objects - **official m.**: definitive treatment of a drug or medicine as presented in an official compendium (In US: USP, NF, or Homeopathic Phar.).

Monolepis nuttalliana (Schult.) Greene (Chenopodi) (wNA, incl. nMex; Siber): patata; rts. cooked with fat & NaCl u. as food (Ariz [US] Indians); sds. u. as pinole.

monomer(ic): relating to alks., &c., single molecule, often large, which can polymerize with other similar molecules.

monomial: designation of organism (&c.) with a single term; *cf.* binomial.

Monopteryx angustifolia Spruce ex Benth. (Leg) (SA): bk. of tree u. as vermifuge; sds. stored for emergency food use (Amazonas) - **M. uauca** Spruce (Co): uauça; very large tree; sds. u. as food (roasted or boiled); very rich in FO, u. as food & lamp fuel; to make soap; bk. for intest. worms.

monosaccharide: simple sugar, either pentose (ex. arabinose) or hexose (ex. glucose).

monosodium glutamate (MSG): sodium g. (*qv.*).

Monotes glaber Sprague (Dipterocarp) (sAf): yellow wood; rt. bk. cardiac rem. (Zezuru).

Monotropa hypopithys L. (Eric./Monotrop./Pyrol.) (Euras, NA): pine sap; false beech drops; yellow bird's nest; chlorophyll-less pls.; rt. parasites; c. gaultherin (& gaultherase) wh. hydrolyzes to methyl salicylate; u. bitter, diur., emet., mostly in veterin. med. (cough in sheep & cows); VO can be obt. fr. sprouts - **M. uniflora** L. (NA, CA, eAs): corpse plant; fit root; Indian pipe; p. plant; wood snowdrop; thick st. clasped by broad white pointed scales & terminated by snow white fls.; coll. in summer; turns dark when dried; rt. u. ton. sed. nervine, antispasm; in eye lotions (folk remedy).

Monotropaceae: Indian pipe fam.; close to Eric. with which sometimes combined; incl. *Monotropa, Epifagus.*

monotypic: genus having only one sp.; also appl. to other levels (ex. Fam.).

Monsonia biflora DC (Gerani) (Af): dysentery herb; pd. rt. u. in dys. diarrh.; hb. ("Slangebos") decoc. for varicose veins & ulcers; pl. with strong perfume property (VO) - **M. burkeana** Planch. (sAf): lf. & rt. u. in same way as prec. - **M. ovata** Cav. (tropAf, Ind): dysentery herb; u. astr. in dys. diarrh., decoc. diaphor. in feverish colds; in anthrax; sed. stomach dis.; whole pl. inf. u. abortif. (Af, Ind) - **M. senegalensis** Guill. & Perr. (wAf): rts. u. as emmen.

Monstera deliciosa Liebm. (Ar.) (Mex, CA [Guat]): "delicious monster"; "Swiss cheese plant"; cutleaf philodendron; ceriman; plant characterized by large lvs. (1 m long) with large holes through lamina on opposite sides of the midrib; large edible frt. (spadix) ("monstera"), esculent pulp, u. in beverages & desserts, such as ice cream; rts. u. to make strong baskets (Mex); u. med. (Homeopath); for arthritis (folk med.); ornam. vine; frt. should be ripe & soaked in water before consumption (CaOx needles, irrit.); lf. or rt. inf. u. for arthritis (Mex); rt. in snakebite (Mart); ("kusbubbla") - **M. pertusa** de Vriese (Lesser Antilles): mibi; lvs. u. as poultice for pain (Domin, Caribs); rts. in basket making.

montan wax: montana w.: lignite w. (*qv.*).

montane: rel. cool moist slopes; upland terrain below timber line with large evergreen trees as dominant life form.

Montanoa hibisciflora (Benth.) Schultz.-Bip. (Comp.) (tropAm & widespread): tree daisy, lf. decoc. u. to wash hair to reduce effects of sunlight & aging (Easter Is) - **M. tomentosa** Cerv. (Mex): zoapatle; (herba) cichoapatli; cihuapatli; hb. u. stom. diur. emmen. galactagog, parturient, abortive; in gastric & uterine tumors (411) (412); contraceptive (CA).

Montia L. (Portulacc.) (worldwide, temp): (water) blinks; some 15 spp. - **M. fontana** L. (cEu, natd. eCana, Mex): blinking chickweed; (water) blinks; miner's lettuce; lvs. eaten as salad - **M. perfoliata** J. T. Howell = *Claytonia p.* var *p.*

montmorillonite: chief component of bentonite (clay); u. in toilet soap mfr. (Ge).

Montrichardia linifera Schott (monotypic) (Ar) (Br, Ve): lf. with sharp tasting juice u. for sores; decoc. in gout (Pará, Pernambuco); fr. edible; lvs. appl. to back or side pains (Ve) (*Pleurospa* Raf.).

moocah (slang): Cannabis.

moonflower: *Ipomoea quamoclit*; *I. alba*; *Menyanthes trifoliata*; *Leucanthemum vulgare* (daisy); *Anemone nemorosa; Brugmansia suaveolens.*

moong: mung.

moonseed (, Canada) (or **Canadian**): *Menispermum canadense* (rt.) - **Carolina m.**: *Cocculus carolinus* - **m. family**: Menispermaceae - **heart-leaved m.**: *Tinospora cordifolia* - **moon-vine** (Miss [US]): *Dioscorea* (2517).

Moor (Ge): peat.

moorberry: *Vaccinium oxycoccus* - **moor-grass**: *Drosera rotundifolia.*

Moos (Ge): moss.

moosebush: *Viburnum alnifolium* (bush or frt.) - **moosewood**: 1) *Dirca palustris* 2) *Acer pensylvanicum* (bk.) 3) *Viburnum lantanoides.*

moot(ers) (US slang): Cannabis.

mora (It, Sp): blackberry; mulberry, *Morus* - **m. di rovo** (It): blackberry; bramble - **m. negra** (Sp): black mulberry.

Moraceae: Mulberry Fam.; hbs., shrubs, trees, vines; mostly trop.; with milky juice; ex. Cannabis, fig. Osage orange.

mor a grifa (US slang): Cannabis.

moral (CR): *Chlorophora tinctoria.*

morango (Par, Br): strawberry.

Morchella esculenta (L) Pers. (Morchell. - Pezizales - Ascomyc. - Fungi) (worldwide, temp.): common (or edible) morel; frtg. bodies delicious food; sometimes cult.

morel: *Morchella* species - **bell m.:** *Verpa* spp. - **common m.:** *M. esculenta* Pers. - **conic m.:** *M. conica* Pers. - **delicious m.:** *M. deliciosa* Fr. - **false m.:** *Gyromitra* esp. *G. esculenta* Fr. - **hybrid m.;** *M. semilibera* DC. ex Fr. - **thick stemmed** (or **t. footed**) **m.:** *M. crassipes* Pers. ex. Fr.

Morelia senegalensis Rich. (monotypic) (Rubi.) (trop wAf): wd. u. as fish poison; pl. u. to "poison water courses" (Senegal).

morelle furieuse (Fr): belladonna - **m. noire** (Fr): common nightshade.

morellin: antibiotic (antibact.); a xanthone found in *Garcinia morella* (pericarp & sd. coats) (yellow pigment).

morello: v. under cherry.

morera (Sp): mulberry tree.

Moricandia tortuosa Hook. f. & Th. (Crucif.) (Ind): fl. decoc. rubbed on for eczema.

morin: flavone pigment wh. occurs in *Chlorophora tinctoria, Maclura pomifera,* &c.

Morinda bracteata Roxb. (Rubi) (PI endemic): lvs. u. for cuts & sores - **M. citrifolia** L. (Rubi) (tropAs & other trop.): Indian mulberry; bankoro; "pain killer" (Virgin Is); soranjee; brimstone tree;

rt. rich in anthraquinone glycosides; u. hypotensive in arterial hypertension, anticonvuls.; diur. lax. antipyr. in gout; rt. esp. bk. rich in dyestuff (morindin —> morindone), u. yellow dye for fabrics; hair dye; fls. —> red dye; frts. c. VO (morinda oil); u. in urin. & spleen dis., dys., asthma; frts. heated & appl. to sores & inflamm. (Virgin Is); lvs. in diarrh. menstr. diff., colic, ton. antipyr.; lvs. as poultice wrapped around rheumatic joints (Domin), appl. extern. as decoc. (also frt. decoc.) to relieve pain of wounds & ulcers, boils (86); source of canary wood; frt. may be toxic but emergency food (Polynesia); var. *bracteata* (Roxb.) Hook. f. (Chin, se As Oceania) (PI, endemic) tinct. whole plant u. for aching bones; rheum.; bk anthelm.; lvs. u. cuts & bruises; swellings; backache; stomach pains, lf. juice in eye dis.; rts. red dye (basketwork) - **M. elliptica** Ridl. (Mal): widely u. in native med.: lvs. eaten for fever & to stim. appetite; in headache, cholera, dys. diarrh.; poultice or decoc. for hemorrhoids - **M. geminata** DC (tropAf): rt. bk. u. as vermifuge - **M. longiflora** G. Don (wAf): u. as lax., an important folk remedy in fevers; called the most valuable drug u. in wAf - **M. lucida** Benth. (cAf): brimstone tree; rt. bk. u. by natives as diur. purg.; hypotensive; for leprosy, mal., yellow fever, schistosomiasis (wAf); lvs. u. astr. in ulcers; bk. in diarrh. dys. - **M. officinalis** How (Chin): fleshy rts. u. as ton., to strengthen the kidneys; for impotency, infertility; to incr. mental power - **M. parvifolia** Bart. (eAs): "hong'zhu-teng" (Chin); rhiz. & rts. u. in bronchitis, pertussis (Chin. folk med.); purg. (Indochin); c. morindaparvin-A & antileukemic anthraquinones - **M. persicaefolia** Ham. (seAs): anthelmintic, hypotensive, but toxic; not antibiotic - **M. royoc** L. (CA, WI, Mex, seFla [US], nSA): yaw weed; red gal; red root (bark) (2067); rt. u. lax drastic, in diarrh., "blood tonic" (astr. sed.); as orange dye source; bk. antipyr. - **M. tomentosa** Roth (M. tinctoria Roxb.; M. coreia Buch.) (eAs): ach(u); (dyer's) Indian mulberry; source of morinda bark, u. astr.; rts. crushed in alc. u. to stop vomiting in cholera; antipyr.; rt. & st. bk. u. in urin. dis. (124); lf. & rt. decoc. u. in "ague" (Mal) - **M. umbellata** L. (eAs): lvs. & rts. administered in diarrh. dys. (Ind, Indochin); earlier u. as natural dye for wool & silk (red, yellow); rt. bk. (mang-koudo) c. morindin, glycoside wh. splits to give morindone, dyestuff - **M yucatanensis** Greenm. (sMex [Yuc], CA): hoyoc; piñuela; pl. juice rubbed on granulated eye-lids (Maya Indians); pl. decoc. ton. diur. lax.; frt. rubbed on warts to remove; rts. u. as dye.

Moringa concanensis Nimmo (Moring.) (Ind). small tree; u. like *M. oleifera* - **M. hildebrandtii** Engl. (Madag): sds. rich in maroseran oil, a bitter fat; pl. cult. as orman. - **M. oleifera** Lam. (M. pterygosperma Gaertn.) (Ind, widely cult. in trop, incl. Eg, Saud Arab, tropAm, as Haiti): (Indian) horseradish (tree); drumstick t.; anambo (Madag); rt. like horseradish u. as flavor, condim.; colic (St. Kitts, WI); lvs. (eAf) & frt. ("drumsticks") eaten as veget.; wd. (as Lignum Nephriticum) pulped —> cellulose; formerly u. in kidney dis.; rt. u. on skin like mustard plaster; diur. in edema; ton. emmen.; to prevent scurvy (seamen); lvs. diur. antibiotic; sds. (ben nuts) furnish behen (or ben) oil (FO) u. as purg., in gout, rheum (PI); techn. for artists' paints; for soaps; to lubricate watches (does not easily become rancid) (known to ancient Greeks); st. source of moringa gum (c. bassorin), u. med. (Ind, Java) - **M. peregrina** Fiori (M. aptera Gaertn.; M. arabica Pers.) (deserts of wAs, nAf, also WI, SA): ben tree; behen nut t.; Arabian horse radish t.; sds. source of FO, be(he)n oil; u. as watchmaker's oil; in perfumes of floral type; sds. u. purg. emet.; lvs. u. as dressing for contusions; trunk of large tree yields water-swelling gum, moringa gum.

morkov (Russ): carrot.

morning glory: *Ipomoea purpurea* & other *I.* spp.; also some *Convolvulus* spp. - **blue m. g.**: *Ipomoea hederacea* - **m. g. family**; Convolvulaceae - **seaside m. g.**: *Ipomoea pes-caprae.*

Moronobea coccinea Aubl. (Guttif.) (WI, tropAm): manil; source of a resin ("gum") (trunk exudative) called "mani" or "anani" u. by Indians of Co, Ve, Guianas as adhesive for arrow points, knife blades, ax heads, &c.; "Piraman" ("Paraman") resin ("gum pitch") u. to caulk canoes (Ve); in making torches; spoken of as one var. of "hog gum" (fr. *Symphonia globulifera*) - **M. riparia** Planch. & Triana (nSA): yields yellow latex, source of resin u. to make masks & for torches (Indians, Co); also u. like prec.; also var. *pirapanensis* R. E. Schultes - **M. rupicola** R. E. Schultes (Co+): resin u. to make torches.

morphia (obsolete): **morphina** (L.): **morphine**: chief alk. of Opium, with strong narcotic effects (analgesia, hypnosis, sedation, antispasmodic action) (E); to produce analgesia & euphoria in terminal cancer (heroin with no advantage here); dangerous addictive agt. (1651) - **diacetyl m.**: Heroin (*qv.*)

morphology: the science of form & structure of living organisms - **plant m.**: may incl. gross structure, organography, histology, embryological devt., cytology, etc., or term may be restricted to first two.

morphometry: measurement of external form of any object (< Gr. morphe= form + Gr. metreo = to measure).

morphotype: morph. differentiated individual organism of unknown species or of sp. of no taxonomic significance.

morrah butter: FO of *Bassia longifolia.*

Morrenia brachystephana Griseb. (Ascle-piad.) (Arg, Br): rt. c. morrenine, asclepiadin; u. galactagog - **M. odorata** (H. & A.) Lindl. (Arg, Br): tasi (Br): doca (Arg); u. galactagog (413).

Morrhua (L): codfish, *Gadus m.* and other *G.* spp.

morrhuate sodium: sodium m. (*qv.*); prepd. fr. mixed free fatty acids (mostly unsaturated) **(morrhuic acid)** obt. by hydrolysis of codliver oil; u. as sclerosing & fibrosing agt. (2265); inj. u. in hemorrhoids; hernial rings closure; removal of condylomata acuminata, &c.

morrhuine: "alkaloid" formerly claimed found in decomposed codliver oil.

morrhuol (Chapoteaut): gaduol; alc. ext. of codliver oil; formerly thought representative of oil's activity; actually of little value; adm. in capsules.

morso del diavolo (It) (lit. devil's bite or sting): *Succisa pratensis.*

morte aux poules (Fr) (hen poison): *Hyoscyamus.*

mortification root: Althaea.

morue, foie de (Fr): codliver.

Morus L. (Mor.) (OW, NW temp): mulberries; collective frt. (Syncarps) often eaten; lvs. food of the silkworm; inner bk. of some spp. furnishes strong fiber - **M. alba** L. (Ind, Chin; later Eu, cult. Med reg., natd. NA): (white) mulberry (tree); morera (Sp); frt. sweet, edible, but inferior to that of *M. nigra* & *M. rubra*; c. invert sugar, pectin; vit. C; rutin; dr. frt. medl., u. as winter frt. (Pak); in prepn. m. syrup; u. refrig. lax.; in cardiac dis., in hypertension; dropsy; neck pains; lvs. c. rutin, u. as antidiabet. (value disproven), to feed silkworms & domesticated animals; inner bk. of annual shoots u. paper mfr. & to make cords; wd. u. in making furniture, mosaics, &c. - **M. bombycis** Koidz. (eAs): bk. u. as antipyr.; in diabetes, diur. in edema; expector.; lvs. u. as fodder for silkworms (Chin, Jap); frt. u. to promote sexual functions - **M. celtidifolia** HBK (Mex; CA to Pe): mora; morera (Mex); "brasil"; frt. edible; wd. useful - **M. indica** L. (eAs): (East) Indian mulberry; bk. (Cortex Mori) u. diur. & in chest dis.; young lvs. as galactagog - **M. microphylla** Buckl. (swUS, Mex): Texas mulberry; "fruitless"(?) mulberry; produces much pollen; said to cause some 40% of hay fever of Arizona; frt. edible, acidulous - **M. multicaulis** Perrot (Chin): frt. edible; popularly u. dr. (cAs); much like *M. alba*; lvs. u. as silkworm forage; so used many yrs. ago in eUS (exptl.) (414) - **M. nigra** L. (wAs, Iran+; cult. sFr, It, swUS): (black) mulberry (tree); common m. (t.); rt. bk. u. as purg. tenifuge; rt. bk. in paper mfr.; frt. mild lax. expect.; in coloring of wine; earlier u. to make a natural rouge & lipstick; flavor; as food not sufficiently tart & better admixed with acid frt. (ex. apples) or cooked in pies, puddings, &c.; frt. juice off. in Fre Codex & several other Ph; wd. u. to make furniture, inlays & c. - **M. rubra** L. (e&cUS): (American) red mulberry (tree); American m.; frt. u. as *M. nigra*, also in prepn. wines; frt. with superior flavor; lvs. u. as tea for sore throat (US middle west); bk. vermifuge, cath.; inner bk. u. for strong fiber for clothing, &c. (Am Indians); lumber source; cult. ornam. tree - **M. tinctoria** L. = *Chlorophora t.*

Moscharia pinnatifida R. & P. (Comp.) (Ch): almizcle; pl. has odor of musk, hence name; u. as stim. antispasm. carm.; cult. ornam. annual.

Moschidae: mammalian fam. of small deer which are antlerless & bear musk glands (musk deer).

Moschosma multiflorum Benth. (Lab) (trop): rt. & lf. u. as cough remedy; rt. emetic & for flatulence - **M. polystachyum** Benth. (tropAf, As): pl. juice squeezed into nostrils of children for headache (Gold Coast).

Moschus (L.): musk; obt. fr. **M. moschiferus** L. (Moschid. Cervidae) (Chin, Indo-Pak, esp. Tibet, Nepal; in Himalayas above 1600 m): musk deer; small hornless deer, yellow to reddish brown with 2 white stripes; prod. represents dr. secretion fr. musk sac, the preputial follicles of male deer between navel & prepuce; c. muskone; u. perfume fixative; for card. weakness, impotence (esp. as Tr. M.) qualities: **M. cabardinicus**; **M. tonquinensis** also allied *M.* spp. (2583).

Mosch(us)wurzel (Ge): *Ferula sumbul* rt.

Moses-in-the-bulrushes: **M.-in-the-cradle:** *Rhoeo spathacea.*

Mosla Buch.-Ham. (includes *Orthodon*) (Lab) (eAs): ann. hbs.; some 10 spp. - **M. chinensis** Maxim. (Chin): hb. sudorific, diur., u. in headache, fever, diarrh., edema - **M. dianthera** Maxim. (Chin): decoc. u. as wash for parasitic skin dis.; carm., for headache - **H. hadai** Nak. (Jap): decoc. u. as vermicide - **M. japonica** (Benth.) Maxim. (Orthodon japonicum Benth.) (Chin, Jap, Korea): decoc. u. as ascarifuge, prepn., VO & thymol (comm.); germicide; VO c. thymol, carvacrol, terpinene.

mosquito plant: 1) **m. bush:** *Ocimum basilicum*, common sweet basil (SC reputed in SA to drive away mosquitoes) 2) *Azolla pinnata, A. caroliniana* (**m. fern**).

moss: 1) a pl. of the Phylum Bryophyta 2) (layman): pls. of many different classifications 3) (drug trade): Irish moss or Chondrus 4) (seUS): Spanish m. 5) (Ireland): bog - **m. animals**: Bryozoa - **black m.**: Spanish m. - **bog m.**: *Sphagnum* spp. - **cap m., common**: *Polytrichum commune* - **Ceylon m.:** dr. thallus of *Gracilaria lichenoides*. (Ceylon agar) - **Chinese m.**: agar - **club m.**: member of Lycopodiaceae - **Corsica(n) (anthelmintic) m.**: *Alsidium helminthochorton* - **dyer's m.**: *Rocella tinctoria* - **m. fiber**: fr. Spanish m. - **Florida m.**: **hanging m.**: Spanish m. - **Iceland m.**: *Cetraria islandica* - **Irish m.:** *Chondrus crispus, Gigartina mamillosa* - **Lapland m.**: *Usnea plicata* - **long m.**: Spanish m. - **lungwort m.**: *Lobaria pulmonaria* - **New Orleans m.**: Spanish m. - **pearl m.**: Irish m. fr. *Chondrus* - **peat m.**: bog m. - **reindeer m.**: *Cladonia rangiferina*+ - **salt rock m.**: Irish m. - **Spanish m.**: *Tillandsia usneoides* - **staghorn m.**: *Lycopodium* spp., esp. *L. clavatum* - **m. starch**: lichenin - **Swedish m.**: bog m. - **Tartarian m.**: *Lecanora tartarea* - **wall m.**: *Parmelia parietina* (lichen) - **wire m.: wiry m.**: form of Irish m. with narrow segments (coll. in Ireland).

mossambi (Ind): sweet citrus, *Citrus reticulata.*

"mossing": 1) (seUS): collecting Spanish moss 2) (Ireland): collecting & winning the turf for the year's supply of fuel (summer).

mostacilla (Mex): rocket salad.

mostarda (Por): **mostaza** (Sp): mustard.

Mostuea angustifolia Wernham (Logani) (tropAf): rts. u. as excitant, aphrod. - **M. batesii** Baker (M. stimulans Chev.) (w tropAf; Gabon): no common name(s) recorded; a spiny shrub; stim. (natives u. to stay awake during nights devoted to dances); aphrod.; said similar to *Schumanniophyton magnificum* Harms in properties - **M. thomsonii** Benth. (tropAf): similar to *M. angust;* may be hypotensive.

motes: particulate waste sepd. fr. cotton in process of ginning; u. as mulch, fertilizer (cottonseed m.).

mother heart: *Capsella bursa-pastoris* - **m. of rye**: r. ergot - **m. of thousands**: *Saxifraga sarmentosa* - **m. of thyme**: *Thymus serpyllum* - **motherherb**: *Artemisia* spp. - **mother-in-law plant**: *Dieffenbachia picta* (**m. tongue**: *Sansevieria* spp.) - **mother's drops**: Mutterkolikessenz (Ge): aromatic tr. 5, orange tr. 5, saffron tr. 2.5, Tr. Opii crocata 2.5, dil. alc. 10 - **m. plaster: m. salve**: brown ointment (or b. salve) - **motherwort**: 1) *Leonurus cardiaca*+ 2) *Artemisia vulgaris* 3) *Lysimachia nummularia* 4) *Chrysanthemum parthenium* - (**woolly m.**: *Leonurus lanatus*).

motia oil: VO fr. *Cymbopogon martinii.*

mouche (Fr-f.): fly.

mould: mold.

mountain ash: *Sorbus americana, S. aucuparia* (bk.) - **m. balm**: 1) *Eriodictyon californicum* 2) *Monarda* spp., esp. *M. didyma & M. punctata* 3) *Calamintha napeta* 4) *Ceanothus velutinus* - **m. fat**: paraffin wax - **m. leather**: asbestos - **m. misery** (Calif): *Chamaebatia foliolosa* - **m. oysters** (US slang): lamb testes (as food); reputed aphrod. - **m. weed**: *Eriodictyon* - **mountain**: v. also under alder, box; damson; dittany; flax; laurel; mahogany; maple; mint (wild m. m.); parsley; rhubarb; rush; sage; sorrel; tea; thyme; tobacco.

mouse (plur. mice): spp. of *Mus, Perognathus, Micotus* (Rodentia), &c.; distinction from rat by smaller size; some forms albino; *Mus musculus* (house mouse); various parts u. in Unani med.; thus, flesh u. int. & ext. in resp. dis., &c.; ash of excreta u. in cystic calculi; eaten in As (served as delicacy in Chinese restaurants); chief importance in destroying & spoiling food (excreta).

moura fat: FO fr. *Madhuca longifolia.*

mouse bloodwort: *Hieracium pilosella* - **m. seed**: oily sd. (linseed?) u. by fur houses to make furs glossy - **mouse-ear**: *Hieiracium pilosella* - **mouse-tail**: *Myosurus* spp.

moussache: *Manihot esculenta.*

mousse (Fr): moss - **m. d'arbre** (Fr) (lit. tree moss): **m. de chêne: m. des chênes**: oak moss, as *Evernia prunastri & Pseudevernia* spp. (etc.) - **m. de Chine**: Agar - **m. d'Islande**: Cetraria - **m. perlée**: Chondrus.

moutarde blanche (Fr): white (or yellow) mustard - **m. des moines**: horseradish rt. - **m. noire**: black mustard.

mouth root: *Coptis trifolia.*

moutillable (Fr): wettable.

mowah: mowra(h) seed: fr. *Madhuca latifolia, M. longifolia* - **m. butter** (**m. fat**): mahua b. - **m. meal**: granulated sd. of *Madhuca* spp.

moxa (<mogusa, Jap): 1) soft, woolly tomentum fr. young lvs. of *Artemisia moxa* DC, *A. chinensis* (As), *A. vulgaris* (Fr), *A. judaica*, or *Senecio* spp. or *Crossostephium chinense*; or use as cautery by burning on skin 2) other subst. thus u. as counterirritants to cure muscle stiffness, headache, etc.; sometimes by igniting cotton, sunflower pith, cork, etc. over an area of the skin, specif. called foll.

moxibustion: process of burning substances placed on skin for therap. purposes.

Mozinna spathulata Ort. = *Jatropha dioica.*

mozote de caballo (CR): *Malachra capitata.*
MSG: monosodium glutanate.
muava bark (Af): **muawin b.**: believed fr. *Erythrophloeum* spp.
mucca-mucca: moko-moko.
mucic acid: galactaric acid; a dicarboxy sugar acid; occurs in ripe frts. as cherries; prepd. fr. wood (sawdust); u. to replace KH tartrate in baking powd.
mucilage: pl. substance which forms in water a thick, adhesive liquid of high viscosity; ex. agar.
mucins, animal: glycoproteins representing major components of saliva, gastric & intestinal juice, &c.; serve as lubricants & protectives (buffers) to body surfaces, int. & extern.; prepd. fr. snails, &c.; u. as gastric dem.; adsorbent; in stomach ulcers - **gastric m:** prod. of stomach mucous lining fr. hog; u. dem. to treat peptic ulcers - **vegetable m.**: gum of *Cyamopsis tetragonoloba* or of okra frt.
mucinase: hyaluronidase.
mucoids: term u. to incl. gums, mucilages, &c.; also specif. gp. of glycoproteins with higher S content than mucins; occ. in connective tissue, as bone, tendons, & in some degenerating tissues.
mucopolysaccharides: combs. of mucin (protein) with carbohydrate; occ. in thickened skin of hypothyroidism, in atherosclerotic blood vessel walls, etc.
Mucor Mich. (Mucor. - Zygomycetes - Zygomycotina - Fungi): (cosmopol.); molds, often found in stored foods; grown in culture media & u. to convert starch to fermentable sugars; c. diastase - **M. mucedo** Pres. (cosmopol): a common bread mold; possibly of value in transforming steroids.
mucuja: mucuje: m. resin: fr. *Acrocomia sclerocarpa, A. lasiopatha*, etc.
Mucuna alterrima (Piper & Tracy) Holland (M. pruriens var. utilis Wahl) (Leg) (tropAs): Mauritius (or Bengal) bean; nescafé (Mex): cult. as green manure & soil cover crop - **M. altissima** DC (Br): ojo de Zamuro; mucuna; stinging hairs u. like those of *M. pruriens*; juice u. by natives in cancer (337); sds. sometimes u. antipyr. in inf. or macerated in spirits (34) - **M. bracteata** DC (Ind, Chin, seAs): rt. u. in med. (Laos) - **M. deeringiana** (Bort.) Small = *M. pruriens* var. *utilis* - **M. gigantea** DC (seAs): cowitch (Austral); hairs mixed with food to destroy rats & in criminal poisoning (Indones); sds. eaten, are said aphrod. (Guam); purg.; bk. u. extern in gout; st. source of a gum - **M. mitis** DC (trop): sds. in plasters to relieve pain of insect stings - **M. monosperma** DC (Ind): pod bristles irritant, u. as "poison"; bk. u. in rheum. appl. extern.; in combns. for acute spasms (Bombay); sds. expect. - **M. nigricans** Steud. (seAs, PI): sap of vines ingested as antipyr. (PI); pounded st. tied around waist of sick person(!) - **M. poggei** Taub. (M. rubra-aurantiaca De Wild.) (Tanz, Uganda): rt. ext. u. for ankylostomiasis, diarrh.; rts. crushed at edge of water to stupefy fish; lf. juice darkens in air to produce an ink - **M. pruriens** DC (Dolichos p. L.) (tropAs; EI, WI widely natd.): cowhage; a twining shrub; trichomes (setae) (2322) on pod penetrate epidermis easily & cause intense itching & transient pain; u. anthelm. (vermif.) (as electuary) esp. to de-worm stock; irrit. to increase absn. through skin (not proven useful; however recommended for scarification as in vaccin. & diagn. tests) (1652); in itch powders (practical joking); pods, rts. sds. also u. as aphrod. (Ind), hallucinogen; coffee substit. & food (boiled); sds. c. Dopa, lecithin, mucunine & other alks., FO with ß-sitosterol (2070); hairs in fat (such as butter) u. as anthelm. revulsive, counterirritant (2323); lf. inf. given in colic (Congo); var. *utilis* (Wallich ex Wight) Baker ex Burck (sAs, but plant described from specimens growing in Fla [US]): (Florida, fuzzy, or cowage) velvet bean; F. b.; Bengal b.; u. for stock feeding, forage, cover crop; green manure; pods glabrous; sds. occ. eaten by humans (after boiling a long time); c. Dopa (415); lvs. & hulls boiled & eaten - **M. sloanei** Fawc. & Rendle (Nigeria - indig. now pantrop, incl. Fla, Mex): horse-eye bean; sea bean; s. heart; cow-itch (vine); sds. u. diur. (Am); decoc. in hemorrhoids; sds. said to c. physostigmine; u. arrow poison, charm (Nigeria); control of micturition in children; pods c. very irritant hairs; sds. u. as ornam. (esp. watch charms); sds. transported by Gulf stream, &c., & deposited on beaches - **M. urens** Medik. (pantropics; wAf, WI, SA): cali beans (or c. nuts): pseudo Calabar beans; sea beans (since often carried up on beaches); as in the similar Calabar beans, sds. characterized by having hilum at margin for 3/4 of circumference; c. eserine; frt. bears irritating hairs (setae); sds. u. like & as a substit. for Calabar beans (2324); anthelm - *M.* spp. grown as green manure & cover crop.
mucunain: active pruritogenic protease of *Mucuna pruriens* (338).
mud: 1) a viscous adhesive semi-sold mass of great complexity, essentially consisting of earth particles (incl. many of colloidal size) in a water base; formerly u. in cancer treatment (c. 1870); in soaps 2) (US slang): opium 3) (US slang): Denver mud; Antiphlogistine™, a clay cataplasm - **m. dauber:** wasp of *Sceliphron* & other genera of Sphecidae;

nest mixed with vinegar & appl. as poultice; also decoc. intern. (Miss.).*

mudar bark: *Calotropis gigantea.*

Muehlenbeckia hastatula I. M. Johnst. (Polygon.) (Ch): mollaca; c. emodin & anthraquinone glycosides; u. as diur. & in liver attacks.

muérdago (Mex): *Phoradendron velutinum*; *Loranthus* spp. (Sp); *Viscum album* (**m. europeo**).

muestra (Sp): sample; specimen - **m. sin valor**: s. of no value.

mu-fang-chi (Chin): *Cocculus sarmentosus* (or perhaps *Pericampylus glaucus*, u. in rheum., &c.).

muggles (US slang): *Cannabis* cigarettes.

mughetto (It): convallaria.

muguet (Fr): **muguete** (Sp, Por): **m. de los valles** (Sp): *Convallaria majalis.*

mugwort: **common m.:** *Artemisia vulgaris*, *A. absinthium*, etc. (lvs. or hb.) - **Alpine m.**: wormwood.

Mugwurz (Ge): *Artemisia vulgaris.*

Muhlenbergia capillaris (Lam.) Trin. var. *filipes* (M. A. Curtis) Chapman ex Beal (M. f. M. A. Curtis) (Gram) (seUS): hair grass; "muhly"; u. as fodder & forage pl. - **M. dumosa** Buckl. (eUS. nwMex): otatillo; rts. u. to make tea for stomach cramps (Mex) - **M. huegelii** Tren. (Java): perennial grass much foraged & preferred by cattle - **M. porteri** Beal (wUS, nMex): black Grama; important range grass; hair grass - **M. tenuifolia** (Kunth) Trin. var. (M. monticola Buckl.) (Tex, Ariz [US], cMex): mesa muhly; zacate liso (Mex); u. in folk med. (but details are lacking)

muhly: *Muhlenbergia* spp.

muira puama (Br): wd. & bk. of sts. of *Ptychopetalum olacoides* or *P. uncinatum*; has also been attributed to *Liriosma ovata* & less commonly to "*Acanthea virilis.*"

muitle (Mex): *Jacobinia mohintle.*

mukal: mukul (Ind): oleo-gum-resin of *Commiphora mukul.*

muktak: whale skin & underlying flesh; rich in vit. C; much relished by the Inuit (Eskimo of NA); esp. in wintertime; that of forehead said best; (ex. narwhal).

mulathi: Ind. licorice.

mulberry: *Morus* spp., esp. *M. alba* - **black m.**: *M. nigra* - **m. "bush"**: actually tree spp. of *Morus* - **common m.**: *M. rubra* - **French m.**: *Callicarpa americana* (misnomer as neither Fr nor a mulberry) - **Indian m.**: 1) *Chlorophora tinctoria* 2) *M. indica* 3) *Morinda atrifolia* - **paper m.**: *Broussonetia papyrifera* - **Persian black m.**: black m. - **red m.**: *M. rubra* - **"smoking" m.**: paper m. (SC great clouds of pollen released by stamen mechanism) - **white m.**: *M. alba.*

mulch: any subst. such as mixt. of straw, lvs., pine needles, soil, etc., spread on the ground to protect pl. rts. fr. cold, reduce moisture loss, reduce weed growth, etc.

mule 1) sterile F_1 hybrid offspring of female horse (*Equus caballus* L) and male ass (*E. asinus* L.) (or other *E. sp.*); early crosses were made in past millenia in wAs (Med reg.); popular riding animals, & for traction; used in Unani med.: flesh (u. in rheum.); fat; heart; urine (abort.); hoof; ear secretions (hinny = cross of male horse with female ass; uncommon) 2) (US slang): alc. - **m. tail tea** (Miss, NM+ [US]): *Conyza canadensis.*

mulga grass: *Thyridolepis mitchelliana.*

mullein: mullen: *Verbascum thapsus* (lvs. fls.); also appl to *V. phlomoides*, *V. thapsiforme*, &c. - **m. dock**: m. - **great m.: grey m.**: *V. thapsus* - **moth m.**: *V. blattaria* - **velvet m: woolen m.**: *V. thapsus.*

mulligatawny (means "pepper water") (EI): soup contg. mango, coconut, rice, capsicum, meats (incl. chicken), vegetables, &c., highly flavored with curry pd.

"Mullins" Food™: dry ext. fr. inf. of wheat, malt barley & K_2CO_3; u. nutrient.

mullite: an argillaceous mineral found in Isle of Mull (Scotland); u. as abrasive, furnace linings, etc.

mulmul: myrrh.

multibiotic: combn. of streptomycin, bacitracin and polymyxin; supposedly destroys all microorganisms of open wounds.

multoside: glycoside with several sugar moieties.

mulungú (Br): 1) *Erythrina crista-galli*, *E. corallodendron* 2) *Piscidia.*

mummy: dr. tissues of human body (generally skin & muscle over bones) or of other animals, often of great antiquity, as fr. ancient Eg; u. med. in Eu (ca. 1600 - date) in epilepsy, TB, &c., also as pigment in art works, fertilizer, etc. (2397) - **m. apples**: 1) (Oceania): *Mammea americana* 2) papaya (339).

mumps: epidemic parotitis; communicable viral dis.; causes painful enlargement of salivary glands, also may prod. swelling & injury to testes, pancreas, &c.; several biologicals are available: **m. immune serum (human)**: sterile serum of healthy donors recovered fr. m.; u. both prophylactically & therapeutically; **m. skin-test antigen**: dil. form of vaccine; u. as intracutaneous test for susceptibility to m.; **m. virus vaccine**: prepd. fr. infected embryonated chicken eggs; the live attenuated vaccine (prefer. to killed) u. by SC or im. inj. as immunizing agt. against mumps (children; adults).

mums (US slang): *Chrysanthemum* spp. (ornam.) (contraction).

muña (verde) (Bo, Pe): *Minthostachys setosa* Epling.

muña muña (Co, Arg): *Satureja eugenioides* (*qv.*).

Mundia spinosa DC (Polygal) (sAf): decoc. of branch apex u. for phthisis, sd. in hysteria & insomnia; in stomachache, acidity; rt. in mal. diaphor.; frt. edible; st. gives buazi fiber (*Nylandtia*).

Mundulea sericea A. Chev. (M. suberosa Benth.) (Leg) (tropAf, As): fish bean; bk. c. rotenone, deguelin; u. fish poison, insecticide; also sd. lf. rt. so used; bk. u. ton. antidiarrh., astr.; for impotence, stomachache (Tanzania).

mung bean (, green) (pron. moong) (Urdu): *Vigna radiata*; small beans u. for sprouting (Chin foods).

Mungo 1) *Ophiorrhiza mungos* L. (rt.) 2) mung (bean).

munj fiber: of *Saccharum ciliare*.

Munronia Wight (Meli) (Indochin, Malaysia+): shrublets; some spp. u. med.

munt (Dutch): Muenze (Ge): mint; specif. spearmint.

Muntingia calabura L. (monotypic) (Elaeocarp.) (tropAm; natd. eAs): capulin; calabura; nigrio; jamaican cherry; buyam; fl. inf. u. for headache, incipient colds, antispasm.; diaphor. (PI); lf. decoc. u. in measles & smallpox (sMex); the sweet red or yellowish frts. pleasantly edible; cult. for firewood; cult. ornam. (sFla [US]).

Mupirocin : Pseudomonic acid A (antibiotic).

muramidase (lysozyme): mucolytic enzyme widely distr. in nature (ex. egg white, nasal mucus, serum, also some plants); active against some viral & bacterial dis. (ex. herpes zoster).

mûre (Fr): mulberry.

Murdannia edulis (Stokes) Faden (Commelin.) (Ind): rts. astr. ton.; rt. bk. in colic, piles, asthma; aphrod. - **M. japonica** (Thunb.) Faden (sAs): pl. u. as abort. (Mal) (bruised sts. u. as suppository) - **M. medica** (Lour.) Hong (Indochin+): tuberous rts. u. for child's ills caused by constip.; maturative poultice; for coughs, asthma, dysuria, &c. - **M. nudiflora** (L.) Brenan (Ind): hb. cooked in FO. u. for leprosy; juice fr. warmed hb. u. to wash ulcers; heated lvs. rubbed on fungus-infected areas; in dys., fever, &c.

Murex (Muricidae - Gastropoda - Mollusca) (seas in many areas): purple shellfish; "purple glands" secrete dyestuff; u. since antiquity (eMed reg.) to color textiles, parchment, cosmetics, finger nails (Marshall Is), very expensive dye (hence "born to the purple"); u. in gynecology for menstr. disturbances (dysmenorrhea), circulatory dis., &c. (homeopath.) (M. Purpureus) (**M. brandaris** L.; **M. cornutus** L.; **M. trunculus** L.).

muri muri kernels: *Astrocaryum murumuru* (source of **m. m. oil**).

murillo bark: Quillaja.

murlins: *Alaria esculenta* (kelp).

muromonab - CD3: monoclonal antibody derived fr. mice serum (1986); u. as immunosuppressant to reverse acute allograft rejection in renal transplant.

murrapo (Co): Panama hat plant.

Murraya crenulata (Turcz.) Oliv. (Rut) (New Hebrides): frt. eaten - **M. koenigii** Spreng. (As): (Indian) curry leaf tree (Ind); bk. lvs. & rts. of large shrub u. as ton. (natives); cult. in SriL & lvs. (fresh) marketed; lvs. boiled in rice (curries) to aromatize (VO with caryophyllene, sabinene) - **M. paniculata** Jack (M. exotica L.) (seAs, Austral, Melenesia+): orange jessamine; Chinese box; cosmetic bark tree; lf. decoc. u. for stomachache, chronic dys.; for bruises; in fungoid skin dis. (wash); lvs. mashed & put in alc. (or boiled); u. for snake bite (Ec); lvs ton. (Mal); emmen. (PI); remedy for tapeworm, herpes; bk. pleasant smelling, u. as cosmetic; wd. ("satinwood") fine hard, u. for knife handles, walking sticks (40); cult. ornam.

murta (Por): *Myrtus communis*.

murungu (Br): Jamaica dogwood.

murúré (Br): 1) *Jussiaea repens* 2) *Ficus cystopoda*.

murú-murú butter (or **fat**) (Br): veg. FO fr. muri muri kernels.

Musa L. (Mus.) (OW trop, orig. As): bananas; some 40 spp. & many vars. (340); pls. tree-like but the "stem" is actually a mass of overlapping leaf bases, hence pls. actually large herbs; stumps u. in mfr. kraft paper; frts. popular food (rich in K) - **M. acuminata** Colla (Java-NG): pisang jacki; frt. edible, consumed by natives; close to *M. nana* - **M. basjoo** Sieb. & Zucc. ex Linuma (M. japonica Veitch) (Jap): lf. decoc. diur. in edema; rhiz. inf. cooling, stom. & to stimulate appetite; lf. fiber u. to make cloths, sails, &c. (2074); cult, ornam. - **M. corniculata** Rumph (Réunion Is): horse banana; very large frts. eaten by natives; popular comm. sp. - **M. discolor** Horan. (New Caled): u. med; frt. consumed by natives; lvs. furnish fiber with local use - **M. ensete** Gmel. (Ensete ventricosa Cheeseman) (trop c&eAf): wild banana; Abyssinian b.; lf. & rt. stock edible; frt. famine food (insipid); furnishes good fiber; sd. endosperm eaten; stalk cooked & eaten: lvs. source of starch - **M. errans** Teodoro (eAs): young unfolded lvs. u. to cover wounds & ulcers; var. *botoan* Teodoro; lvs. appl. for chest pain; sap u. vulner., juice of corm ingested for TB (PI) - **M. fehi** Bert. (New Caled, Tahiti): aiori; frt. med. &

foodstuff of natives - **M. holstii** K. Schum. (M. ulugurense Warb. ex Moritz) = *Ensete superbum* - **M. nana** Lour. (M. cavendishii Lamb; M. chinensis Sweet) (sChina, Oceania): dwarf banana; Canary b.; Chinese b.; frt. small, seedless; of good quality, eaten; u. med.; hardiest sp.; cult. in Canary Is, sUS - **M. oleracea** Vieill. (New Caled): banana poreté; rhiz. eaten as vegetable; frts. eaten by natives - **M. X paradisiaca** L. (OW trop; cult. trop of both OW, NW): plantain (banana); cooking b.; large frt. seedless, edible after cooking (boiled, roasted, baked, fried); eaten like bread; difficult to distinguish fr. ordinary banana (but larger & green colored); pulp made into flour, u. to feed invalids, children; rts. u. for tumors, measles, sunburn; trunk juice (obt. by inserting bamboo tube) u. to cure headaches with fever, as hair tonic, emet.; powd. flower "relaxing" to heart; lvs. u. as packing material - subsp. **sapientum** O. Ktze. (M. X sapientum L.) (trop, much cult.): (common or honey) banana; pisang; tall perenn. hb.; fresh frt. an important food worldwide (dessert fruit) (sometimes fried); u. in many edible prod. such as bonbons, "figs," meal & flour (fr. unripe bananas); ice cream, cake, with cereals, &c.; starch; powder (dried & pd. ripe bananas) ("dehydrated"); starch also prepd. fr. stalks; pulp rich in vit C, has serotonin, noradrenalin, dopamine; rich in K; exts. with hypoglycemic activity, not hallucinogenic; u. for banana preserves, coffee substit., alc. beverage (b. wine); alc., vinegar; green frt. peel u. in hypertension; yellow peel u. as poultice in migraine (Trin); frt. seeded or seedless; frt. u. in general debility, asthma (Haw); nutrient for infants & older persons, in obesity (2072); in stomach dis. (1653); dr. peel smoked in asthma (Ala [US]); lvs. u. in prepn. coarse fiber (banana fiber); frt. pulp u. as absorbent in gastric dis.; boiled unripe frt. with salt astr. u. in acute dys. (174); ripe frt. lax.; wax fr. lvs. (Cera Pisang); many vars.; fig or lady fingers; Gros Michel (most popular); Bluefields (for export.); Bungulan, Colorado (red, morado) & many others; chief brands; Chiquita; Dole, Del Monte; the banana is said the most abundantly produced & heavily consumed food of the world & a chief carbohydrate source - **M. superba** Roxb. (OW trop): wild plantain; frt. st., bases, rhiz. eaten; sds. were u. as smallpox "prophylactic" & as "specific" in rabies (Ind); lf. fiber u. for well ropes, etc. (2073) - **M. textilis** Née (M. silvestris Colla) (OW trop, chiefly cult. PI): abacá; manila hemp; prepd. fr. lf. bases, u. for ship ropes, twine, for "paper" walls in Jap houses; sails, &c.; rt. u. as worm remedy (Annam); this & other spp. source of pisang wax fr. lf. surfaces; cult. PI, Indochin, CA (SC chiefly exported fr. Manila) - **M. tikap** Warb. (Caroline Is): lf. fiber u. by natives.

musana: musena.

Musanga cecropioides R. Br. (Cecropi.) (wAf, as Cameroon): umbrella tree; lf. u. to induce labor & to lower BP; wd. light ("corkwood") u. for floats, rafts, isothermic ceilings; source of charcoal; very fast growing cult. ornam. - **M. smithii** R. Br. (Mor) (tropAf): kombo-kombo; parts of small tree u. to prepare beverage (natives).

musaroside: glycoside of sarmentogenin & D-digitoxose; occ. in *Strophanthus divaricatus*.

musc (Fr): musk - **muscade** (Fr): nutmeg - **fleurs de m.**: mace.

muscadine (**grape**): *Vitis rotundifolia*.

muscarine: lethal alk. (quaternary ammonium base) fr. *Amanita muscaria*, some *Inocybe* spp. & other fungi.

Muscari Mill. (Lili) (Euras, Med reg.): grape hyacinth; g. flower; some 30 spp.; popular spring flg. ornam. esp. **M. botryoides** (L.) Mill. (sEu): (blue) g. h.; baby's breath; u. as diur. exp.; ornam. pl. - **M. comosum** Mill. (Med reg., wEu, nAf): purple tassel(s); purple grape (or t.) hyacinth; bulbs eaten esp. in early spring (Greece); u. diur. stim.; ornam. - **B. muscarimi** Medik. (Eu): cult. for arom. VO.

muscat nut: *Myristica*.

muscio (It): musk.

muscle sugar: inositol (SC sweet).

musco islandese (It): Cetraria.

Muscus (L.): moss.

musen(n)a: *Albizzia anthelmintica*, bk.

musgo comun (Sp): moss; spp. of *Bryum*, *Polytrichum*, etc. - **musgo de Corceza**: Corsican moss, *Alsidium helminthochorton* - **m. de Irlanda** (Sp): *Chondrus crispus*.

mushmelon (Ala [US] slang): musk-melon* - **mushroom**: sporocarps of members of the Agaricaceae (Fungi); same as toadstool; edible spp. eaten raw or in soups, gravies, sauces, &c.; as flavor - **common m.**: *Agaricus campestris* - **fly m.**: f. agaric. - **magic m.**: those contg. psilocybin (NA).

musk: 1) the **m. plant**, *Mimulus moschatus* 2) muskus grass 3) *Muscari botryoides*, *M. racemosum* & other *M.* spp. 4) *Hibiscus abelmoschus* 5) (Austral): *Olearia* spp. 6) *Erodium moschatum* 7) the animal product (*v.* Moschus) - **American m.**: m. obtained fr. the Am muskrat - **artificial m.**: synth. compds. such as trinitro-butyl-toluene; alc.-soluble not HOH-soluble - **Assam m.: Bengal m.**: a comml. var. of musk fr. Assam but exported thru Calcutta - **Canton m.**: an artificial mixture - **Chinese m.**: Yun-nan m. - **deer m.**: genuine m. (Moschus) (*qv.*) fr. the musk deer - **m. flower**: m. plant - **m. leaves**: fr. *Hibiscus esculentus* - **m. mallow**: *v.* mallow, musk - **monkey m.:** m. plant - **Nepa(u)l m.**: small pods (1/3 size of Tonkin) with dark skin (comm. sort) - **m. oil**: musk seed o. (probably) - **m. plant**: *Mimulus moschata* - **m. root**: 1) *Ferula sumbul* (also **muskroot**) 2) *Adoxa moschatellina* - **m. seed**: fr. *Hibiscus abelmoschus* - **sweet-scented m.**: monkey m. - **T(h)ibet m.**: trade v. of m. (2059) - **Tonkin m.**: **Tonquin m.**: the best grade of m. fr. *Moschus moschatus* (Tibet) (2060) - **m. tree**: *Olearia argophylla* - **vegetable m.**: 1) m. root (1) 2) m. plant 3) m. seed - **wild m.**: 1) *Erodium cicutarium* 2) *Malva moschata* 3) Nos. 1, 3, 4, 6 under musk - **m. wood**: Cocillana (Rusby) - **Yunnan m.**: next to Tonquin the best quality of m.; fr. Chin.

Muskambrett (Ge): *Hibiscus abelmoschus* - **Muskatblueten** (Ge): mace (lit. nutmeg flowers) - **Muskatenbalsam** (Ge): nutmeg butter - **Muskatnuss**: nutmeg - **Muskatoel**: nutmeg oil .

muskeg: bog (subartic, as in Newfoundland): mostly made up of *Sphagnum* moss deposits.

muskmelon: *Cucumis melo.*

muskone: cyclic ketone; active constit. of musk fr. m. deer.

muskrat: mushrat: *Ondatra zebethica* (L.) Link (Cricetidae) (NA): large rodent with similar habits to beaver; source of hide (fur), meat, musk (fr. perineal glands).

muslin: cotton fabric with plain weave (< Muslim).

Musops elengi L. (seAs): bk. astr., decoc. u. in fevers, diarrh.; boiled lvs. appl. as cold compresses for headache (Java); expressed lf. juice dropped in eye for eye dis.; rt. in tonic mixts. cult. for fragrant lvs. (Mal).

musquash: *Cicuta maculata* (rt.).

Musops elengi L. (seAs): bk. astr., decoc. u. in fevers, diarrh.; boiled lvs. appl. as cold compresses for headache (Java); expressed lf. juice dropped in eye for eye dis.; rt. in tonic mixts. cult. for fragrant lvs. (Mal).

Mussaenda afzelii Don (Rubi) (trop wAf): damaram; bk. c. tannins; u. as astr.; pl. (bk.) c. tannins, u. astr.; lf. juice u. in ocular inflamm., GC; as mouthwash - **M. anisophylla** Vidal (PI): decoc. of fresh lvs. u. for asthma - **M. arcuata** Poir. (Sudan): boiled lvs. appl. to leprous ulcers - **M. cambodiana** Pierre (Indochin): fls. diur. & adm. for cough, asthma, intestinal tumors; skin dis. - **M. elegans** Schum. & Thonn. (wAf trop): lvs. u. in elephantiasis, edema (decoc.) - **M. erythrophylla** Schum. & Thonn. (trop wAf): "Ashanti blood"; flame of the forest; rt. chewed as appetizer; with Guinea grains as expect.; cult. ornam. - **M. glabra** Vahl (SAs: Mal): lvs. u. to treat headache; rt. decoc. in cough; inf. of lvs. for cough, bowel dis., & fever - **M. hainanensis** Merr. (Chin): u. as cure for cancer (Hainan) - **M. landia** Poir. (*M. stadmanni* Mich.) (Maurit, Madag): quinquina de l'île de France, q. indigène; source of "Costus amarus" = bk. of this sp. (Cortex Belahe) (bela-aye) (416); c. tannin, u. astr. antimal - **M. philippica** A. Rich. (sAs, Oceania [PI]): juice of bk. u. for headache; tobacco substit.; bk. u. for wounds & ulcers (PI): rt. & lvs. in chest dis.; lvs. u. as emet. frt. in GC; rts. & white sepals u. in jaundice - **M. pubescens** Ait. f. (Chin): st. & lf. u. med. & to prepare cooling drink in hot season; tea substit. - **M. raiateensis** J. W. Moore (sPac Oc islands): st. grated & mixed with water u. for chest complaints & kidney dis.; rt. in asthma; bk. u. med. (Fiji) - **M. rehderiana** Hutchins. (Indochin): fls. diur. & in asthma - **M. theifera** Pierre (seAs): lvs. for fever - **M. villosa** Wall. (Mal): lf. decoc. for rheum.

mussel: marine bivalves shellfish found attached to rocks & structures such as pilings; chief genera *Mytilus* and *Modiolaria* (Mytilidae; Mollusca); *M. edulis* (cosmopol.) is u. as food; may develop toxicity with mytilotoxin (neurotoxic) formed; small blue mussel (of Maine [US]) source of natural glue—permits attachment of bones & metal prostheses, & of tendons to bones; also of gums to teeth (1986).

mustard: 1) *Brassica* spp. 2) prepared m. (*qv.*) - **artificial m.**: allyl isothiocyanate - **ball m.:** *Neslia paniculata* - **m. bath**: made by stirring half cupful of ground m. in water & adding to tepid bath; u. to give relaxation & euphoria by improving circulation - **black m. (seed)**: **Bordeaux m.**: **brown m.:** *B. nigra* - **m. cake**: ground m. sd. fr. which some FO has been removed - **Chinese m.**: *B. juncea*, one of sources of black m. - **corn m.**: *B. kaber* - **crambling m.**: *Sisymbrium officinale* - **Dijon m.**: white m. - **m. dross**: v. dross - **"Durham m."**: prepared m. (SC first made in County Durham, Eng, ca. 1720; with sepn. of sd. coats) - **m. family**: Brassicaceae; Cruciferae - **m. flour**: pd. m. sd. with hulls mostly removed; with or without FO - **French m.**: table m. + cloves + sugar - **garlic m.**: *Alliaria petiolata* - **m. gas**: $(ClCH_2CH_2)_2S$; u. poison gas; not related in any way to vol. m. oil - **German m.**: m. flour + laurel lvs., cinnamon, cardamom, vinegar & sugar - **m. greens**: lvs. of *Brassica juncea, B. hirta* (as in mustard & cress salad) - **ground m.**: prepared m. - **hedge m.**: crambling m. - **hoary m:** *Hirschfel-*

dia incana - **India(n) m.**: Chinese m. - **Japanese m.**: *Brassica cernua* - **m. leaf** (Eng): m. plaster - **Mithridate m.:** *Thlaspi arvense* - **m. oil**: 1) VO: allyl isothiocyanate 2) FO: mustard sd. oil - **m. oils** (Ind): FO. obt. fr. *Brassica campestris, B. juncea, & B. napus* - **m. plaster**: admn. form in which black m. is incorporated into the plaster base & gives counterirritation & stim. effects over several hrs.; u. backache, lumbago, rheum., etc. - **prepared m.**: paste made of mixt. of grd. m. sd., sometimes with m. flour & m. cake, also NaCl, vinegar, sugar, turmeric & other spices (Jamaica pepper), water, etc., often with horseradish; u. table condim. - **red m.**: black m. - **m. salad**: m. greens - **Sarepta m.**: fr. *Brassica besseriana* - **m. seed**: sd. of *Brassica hirta, B. nigra, B. juncea* & vars. (USDA) - **southern curled m.**: *B. juncea* v. *crispifolia* - **table m.**: prepared m. - **tower m.:** *Arabis glabra* - **treacle m.:** *Erysimum cheiranthoides* - **true m. oil**: volatile m. oil, allyl isothiocyanate - **tumble m.**: *Sismybrium altissimum* - **white m.**: (sd.) *Brassica juncea, B. hirta* - **wild m.**: 1) corn m. 2) hedge m. - **yellow m.**: white m.

Mustela (Mustelidae - Carnivora - Mammalia): incl. minks, weasels, stoats, ermines, ferrets; semi-aquatic animals, very fond of fish - **M. lutreola** (Euras): European mink; really a large weasel - **M. vison** (US, Cana): American mink; both spp. source of fine furs (most popular on market); now raised in large colonies in Cana, US, Scand (ca. 200,000 harvested per annum in US); FO u. in cosmetics for hand care.

mutagen: something wh. produces a genetic change or mutation; ex. teratogen.

mutah (US slang): Cannabis.

mutation: change in organism caused by alteration in heredity material; may be chief mechanism of evolutionary process.

Mutisia retrorsa Cav. (Comp.) (Ch): virreina; flor de Granada; fls. u. lax. diur.; lvs. in cataplasms for tumors; bk. against mal. - **M. rosea** Poepp. (Ch): u. med. in Ch - **M. viciaefolia** Cav. (Ch): vetch-leaf mutisia; shrubby pl.; lvs. c. saponin; fls. u. in cardiac diffic., hysteria, epilepsy.

Mutterkorn (Ge): ergot - **Mutterkraut** (Ge): 1) *Melissa officinalis* 2) *Matricaria parthenium* - **Muttersaft** (Ge) (lit. mother juice): first molasses.

mutton suet: suet.

myagre (Fr): gold of pleasure.

myall (trees) (Austral): *Acacia pendula* A. Cunn. (with useful wd.), etc.

myata perechnaya (Russ): peppermint.

Mycelin: antibiotic fr. *Streptomyces roseoflavus* Arai: 1952-; antifungal.

Mycelis muralis (L.) Dum. (Comp.) (Euras, nwAf, natd. NA): wall lettuce; ivy-leaved l.: milk juice c. lactucopicrin and an alk. with atropine-like action; latex of rt. u. for snakebite; lvs. as salad.

Mycetia javanica (Bl.) Korth. (Rubi) (PI, Mal, PapNG): rt. decoc. u. for colds (PI).

mycetismus: poisoning by mushrooms following ingestion.

Mycobacterium Tsukamura (Mycobacteri. - Actinomycetales - Bacteria): atypical mycobacteria; rod-shaped aerobic saprophytes & parasites; best known as cause of 2 obstinate dis., tuberculosis & leprosy - **M. leprae** (Hansen) Lehmann: Hansen's bacillus; cause of Hansen's disease (leprosy); vaccine sought; m. o. confined to skin, testes, & peripheral nerves; nerve injury paramount; 9-banded armadillo may furnish vaccine; therapy; rifamycin; sulfones, esp. Dapsone (but resistance developing) - **M. tuberculosis** Lehmamn & Neumann: the micro-organism causative of tuberculosis; 2 vars. are important in medicine; v. *bovis* Bergey & al. (M. b. Bergey & al.), causing cattle TB & also pathogenic for man; v. *hominis* (Bergey & al.), causing human TB.

Mycoblastus sanguinarius (L) Norm. (Lecide.) (Lichen) (nUS+): blood red lichen; u. red, purple, orange dyestuff.

mycology: science of fungi.

Mycomycin: antibiotic compd. produced by *Proactinomyces acidophilus* (Arai) Krasnilikov (*Nocardia acidophilus* Arai) (1947-); tuberculostatic but said too toxic for medical usage; also highly unstable.

mycophagy: eating of mushrooms.

Mycophyta = Fungi.

Mycoplasma (Mycoplasmat. - Cl. Mollecutes [parallel to Schizomycetes]): minute microorganisms lacking true cell wall (-300 nm); incl. pathogens of man & lower animals (resp. & GU tracts) and non-pathogens; related to Bacteria & Rickettsiae; occ. in body fluids of pls. or animals (parasitic); pleomorphic (1998) - **M. mycoides** Freundt: causes pleuropneumonia in some stock animals - **M. pneumoniae** Somerson & al: causes primary atypical pneumonia of man.

mycostatin = Nystatin.

mycosterols: sterols of fungi, incl. yeasts; ex. zymosterol, episterol, &c.

Mylabris bifasciata De Geer (Meloideae - Coleoptera - Insecta): c. 1% cantharidin - **M. cichorii** L. (nInd, Chin, EI): Chin Cantharides; C. (or Indian) blistering beetles; C. flies; brown cantharides; the dried beetle is rich in cantharidin & often u. in place of the sometimes official Cantharis; u. epispastic; in chronic GC, emmen.;

source of cantharidin - **M. pustulata** Thunb. (Ind): Indian blistering beetle; c. -2% cantharidin u. like prec. - **M. sidae** Fabr.; blister fly; props. & uses like prec.

mynte (Dan, Nor): **mynth** (Swed): mint.

Myocardone™: beef heart ext. u. cardiac dis.

myoctonine: alk. fr. rt. of Aconitum lycoctonum.

myoinositol = inosit(ol).

Myoporum laetum Forst. f. (Myopor.) (NZ): ngaio; frt eaten; source of manna; good wd.; shade tree.

Myoschilos oblonga R. & P. (monotypic) (Santal.) (Ch): codocoipu; lvs. digestant & lax.; emmen. & uter. ton.

myosin: muscle fibrin; an albuminoid (with properties of a globulin) found in muscle portions of lean meat of higher animals; also in maize, wheat, oats, grains, cannabis sd., etc.; u. to perfume tobacco (417) - **myositic**: relating to **myositis** (voluntary muscle inflammation) (do not confuse with myotic).

myosmine: alk. of tobacco; also synth.

Myosotis afropalustris C. H. Wright (Boragin.) (sAf): "forget-me-not"; decoc. ton. in hysteria, to aid memory; emet. - **M. arvensis** Hill (Euras): field scorpion-grass; rt. u. for eye dis.; decoc. or powd. u. in oral & genital cancers (411); hb. u. in homeop. & folk med. (Eu) - **M. micrantha** Pall. ex Lehm. (Euras): sand forget-me-not; u. in pulm. TB, chronic bronchitis, TB of intest. - **M. scorpioides** L. (M. palustris Hill) (nHemisph, esp. nEuras, in moist places): (true or water) forget-me-not; c. curare-like alk. (scorpioidine, symphytine, myoscorpine); u. like *M. arvensis*; also in lung dis.; fls. u. expect. - **M. sylvatica** Hoffm. (Eu, Siberia): wood (or garden) forget-me-not; rt. u. in eye dis.; ornam. hb.

Myosoton aquaticum Moench (monotypic) (Stellaria aquatica Scop.) (Caryophyll.) (temp Euras; natd. eCana, neUS): giant chickweed; sts. & lvs. eaten in appendicitis, gastric cancer (Jap); decoc. as galactagog (Chin); in ulcers, hemorrhoids; pl. u. for fistula (Nepal; Murree [Pak]).

Myosurus minimus L. (Ranuncul.) (Euras, nAf, eNA, Calif [US]): mousetail; hb. (Herba Caudae musinae) irritant, astr., u. in homeopath.

myrbane: mirbane.

Myrceugenella apiculata Kaus. (Luma a.) (Myrt.) (Ch): temo; lvs. & bk. astr. balsamic; u. in chronic diarrh. & infections; vulner. - **M. chequen** (Mol.) Kaus. (Luma c.) (Ch): chequen; arrayan blanco; astr., inf. u. for cough, aid digest., expect.; diur.; wd. with water in ophthalmia.

Myrceugenia exsucca Berg. (Myrt.) (Ch): patagua; popular med. for dermal dis. - **M. obtusa** Berg. (Ch): arrayan; astr. vulner. in diarrh. dys.

Myrcia acris DC = *Pimenta racemosa* - **M. citrifolia** (Aubl.) Urban (Myrt.) (WI): couroupoume; frt. of shrubby tree edible & u. as special tea to induce lactation (Domin, Caribs); lf. in combns. for diarrh.; wd. insect-resistant, u. to build houses (2328) - **M. salicifolia** DC (Co): dr. lvs. mixed with *Manihot esculenta* flour; astr. in diarrh.; emetic in excess (Taiwano Indians) - **M. splendens** DC (Co): bk. of small tree u. to paint gourds (*cuyas*) black.

myrcia oil: Oleum Myrciae (L): bay oil fr. lvs. of *Pimenta racemosa*.

Myrcianthes fragrans (Sw) McVaugh (Myrt.) (sFla [US], tropAm): naked wood (Bah); u. in combn. for influenza, &c. (Bah).

Myrciaria O. Berg (Myrt.) (trop SA, WI): igba puru; some 70 spp.; frt. of wild traees edible (jaboticabá), popular; u. by poor people of Arg to make refreshments (cool drinks, ice cream) - **M. cauliflora** (DC) Berg (Eugenia c. [Berg] DC) (seBr, cult. CA): jaboticabá; jabuticabá; tree (-17 m) produces grape-like purple frt. (berries), well flavored, deliciously edible, popularly eaten; frt. borne on short spurs on tree trunk or branches; sds. c. myrciarin; inner bk. u. in asthma (Br, Par); much cult. in sBr (frt); (cult. Palm Beach, Fla [US]).

Myrianthus arboreus P. Beauv. (Mor.) (wAf): bk. & lvs. u. in dys. & tapeworm (Nigeria); frt. edible; wd. very light; lighter than cork.

Myrica asplenifolia L. (Myric.) = *Comptonia a.* - **M. caraccasanna** H.B. (Co, Ve): bk. u. as astr.; prod. myrica wax - **M. cerifera** L. (M. carolinensis Mill.) (eUS sNJ-Fla-Tex, WI): southern wax myrtle; candleberry; bayberry; tallow shrub; "merkle" (SCar [US]); dr. rt. bk. (Myrica [L.]) u. astr. ton. stim. to ulcers, chronic pharyngitis; lax., in many comp. prepns.; dr. frt. (bayberries) u. source of bayberry wax (for candles chiefly) (2076) & bayberry soap (actually wax on frt. surface is a FO rich in palmitic acid) (yield 20-25%); lvs. u. arom., stim.; as tea; to drive away fleas, was kept in dog kennels (seUS); may be carcinogenic - **M. conifera** Burm. f. (sAf): washes; frt. eaten; pl. u. in dysmenorrh.; chewing lf. prod. intense headache & pain at back of nose - **M. cordifolia** L. (sAf): waxberry; waxbush; branches placed in boiling water & "(bay)berry wax" skimmed off, formerly u. to make candles; "wax" is true fat not wax, formerly u. as food (Hottentots); in ointments for wounds; to make soap; makes best coating for metal patterns (Fe, brass) - **M. esculenta** Buch.-Ham. ex D. Don (Ind, Chin, Jap): box myrtle; bark ("kaiphal") u. as ton. carm. antisep. astr. in asthma, colds; tan, dye (yellow); fish poison;

sweet-sour frt. edible - **M. gale** L. (Euras, neNA) (ex. bogs & moors of Scottish Highlands & Hebrides): (sweet) gale; gaul (Scotch); bay bush; Dutch (or bog) myrtle; lvs. & buds u. ton. alter. vulner. in dys. skin dis. as abort.; exter. antiparasitic, for clothes moths; formerly added to beer for its foaming & intoxicating effect; in liver dis., jaundice (US); twigs & bk. in tanning & as yellow dye; as replacement for *Ledum palustre* - **M. heterophylla** Raf. (M. caroliniensis P. Mill.) (e&seUS n to sNJ): wax myrtle; swamp candleberry; (evergreen/northern) bayberry; bk. astr.; frt. c. & u. as other *M.* spp. - **M. inodora** Bartram (Cerothamnus inodorus Small), (nFla, Miss. [US]); odorless wax-myrtle; shrub or tree; u. (ca. 1800) by inhabitants of sAla [US] for producing "wax" candles (2478) - **M. kandtiana** Engl. (Af): powd. rt. as restorative in fainting - **M. kilimandscharica** Engl. (tropAf): bk. emet. in mal.; rt. in indigest. (Sukuma); rt. tonic (Masai) - **M. mexicana** Willd. (Mex, Guat): arbol de la cera; arrayan (Guat); a source of bayberry "wax" u. to make candles (with arom. smoke), intern. for icterus, diarrh.; to adulterate beeswax, formerly to make phonograph records - **M. nagi** Thunb. (Podocarpus n. [Thunb.] Zoll. & Moritzi); (Ind, Pak, Chin, Jap; also cult.): box myrtle; kaiphal; u. prodn. kino c. 60-80% tannin; bk. u. arom. stim. astr.; a fish poison; c. myricetin (color) - **M. pensylvanica** Loisel. (e&cNA): (northern) bayberry; u. as *M. cerifera* - **M. pringlei** Greenm. (Mex): u. for its wax (candles) - **M. pubescens** H & B. ex Willd. (Latin Am): "roble"; encenillo; "laurel"; berries u. to make wax; bk. u. for tanning leather for sandals (cotizas) (Ve); ingred. of inf. u. to bathe women for 5 days after childbirth (2449) - **M. quercifolia** L. (M. laciniata Willd.) (sAf): waxberry; u. like *M. cordifolia* - **M. rubra** Sieb. & Zucc. (Chin, Jap, Kor, PI): Chin strawberry; nagi; frt. eaten raw or boiled; u. to prep. beverage; boiled to give myrtle wax (appl. to ulcers); frt. u. for dig. dis.; bk. decoc. u. in asthma, diarrh., fevers, pulm. dis., to treat As poisoning; bk. chewed in toothache; sd. aril u. carm.; bk. in oil for earache (tree cult. in Jap, Chin).

Myricaceae Blume: Wax Myrtle (or Bayberry) Fam.; arom. shrubs or small trees fr. temp. into subtrop zone; soft-wooded; ex. sweet fern; s. gale.

Myricaria elegans Royle (Tamaric.) (Ind): lvs. appl. to bruises - **M. germanica** Desv. (Euras): tamarisk; bk. u. as astr. (Homeop.).

myricin: myricyl palmitate; occ. in wax fr. bk. of *Myrica* spp.; beeswax.

myristicin: a phenolic ether; component of VO. of nutmeg & other pl. prods.

myricyl alcohol: triacontanol; occ. in cuticular waxes such as carnauba wax & beeswax - **m. palmitate**: melissyl p.; ester occurring in beeswax, lf. cuticular waxes, etc.

Myriophyllum aquaticum Verdc. (M. brasiliense Cambess., M. proserpinacoides Gill.) (Haloragid. - Dicot.) (SA+): parrot's feather; water ornam. pl. (in aquaria) - **M. spicatum** L. (cosmop.): (European) water milfoil; submerged water pl. with much divided leaf; bad pest in many areas now (ex. Okanagan Lake, BC); pl. u. to treat fever, relieve thirst, "fluxes"; sap of crushed lvs. u. in chronic dys. (Chin).

Myristica (L.): sd. of *M. fragrans* - **M. argentea** Warb. (Myristic) (New Guinea, cult.): Papua (or Macassar) nutmeg; sd. is narrower & longer than of true n. & inferior as to taste & odor; u. nutmeg substit.; nuts u. in diarrh., aphrod.; has replaced *M. fatua* in Java - **M. bicuhyba** Schott = *Virola b.* - **M. fatua** Houtt. (Moluccas): wild nutmeg; mountain n.; mace flowers; u. condim., substit. for true nutmeg, but inferior in quality - **M. fragrans** Houtt. (M. officinalis L. f.) (Moluccas, NG, now widely cult., ex. Grenada) (1838): (common) nutmeg (tree); source of nutmeg (sd. freed from aril & sd. coats) (sd. kernel) and mace (aril) (*qv.*) both u. as condim.; flavor carm. arom. stim.; u. in colics, dyspeps., arthritis; kitchen spice; sd. source of VO & FO (mixt.= nutmeg butter u. in ointments, for candles); VO c. myristicin, eugenol, camphene, borneol, geraniol; nutmeg & other prods. (esp. mace) hallucinogenic ("narcotic"), u. as substit. for cannabis & LSD; toxic (418); VO phytocidal. (cult. EI [Indonesia] [2077], Ind [Nilgiris], WI ["pale blade mace"]) (literature [631]) - **M. malabarica** Lam. (Ind): Bombay (or Malabar) nutmeg; almost odorless; hence of little value as spice; u. substit. & adulter. of true n.; u. for headache, aphrod. (Ind); bk. yields a kind of kino; sd. c. 30% FO (poondy oil); aril adulter. of Banda mace - **M. sebifera** Sw. = *Virola s.* - **M. simiarum** DC (PI): tanghas; sds. yield FO u. in itch, skin dis. (or pulped bk.) - **M. speciosa** Warb. (Moluccas): sds. considered about equal to those of *M. fragrans*, also those of *M. succedanea* Bl. (Indonesia).

Myristicaceae: Nutmeg fam.; trees & shrubs of tropics (mostly As); lvs. evergreen, simple, entire; fls. small with superior 1-celled ovary; frt. fleshy capsule, sds. with aril. ex. *Myristica, Virola, Otoba, Compsoneura.*

myristic acid: $CH_3(CH_2)_{12}COOH$; char. satd. fatty acid of Myristica FO; but occurs in most pl. & animal FO in smaller proportion; u. in soaps, lubricants, coating.

myristicin: a phenolic ether; component of VO of nutmeg & other pl. prods. as *Anethum graveolens*; u. as hallucinogen.

myristin: trimyristin: glyceryl trimyristate (in FO of nutmeg); spermaceti, etc.

myrobalans: Myrobalanum (pl. Myrobalana) (L): Baccae Myrobalani; dr. frt. of *Terminalia chebula, T. bellirica, T. citrina* & other *T.* spp. - **black m.:** *T. chebula* - **Egyptian m.:** *Balanites aegyptiaca.*

Myrocarpus fastigiatus Allem. (Leg) (Br, Par, nArg): cabriuva; copriuva; wd. source of VO (brown oil; cabriuva [wood] oil); with odor reminiscent of rose and sandalwood; also balsam prod. fr. trunk injuries (cabureiba or myrocarpus b.); packed in gourd; earlier u. med. - **M. frondosus** Allem. (Br): like prec., noteworthy for hard wd.

myronic acid: occurs in rape & black mustard sds. & horseradish rt.; present free in very small amts., mostly as K salt, sinigrin.

myrosin: enzyme also known as myrosinase, sinigrinase, & thioglucosidase; occ. in Cruciferae, &c.

Myrosma cannifolia L. f. (Marant.) (Guiana): rhiz. source of marble arrowroot.

Myrospermum frutescens Jacq. (Leg) (sMex, WI, Guat, CR, Co. Ve. & elsewhere in SA) balsamita; guatamare (Ve); guayacan; in folk med. sd. in rum u. for rheum., body pains, "bad food," &c. (Ve); vermifuge; antispasmod. in convulsions; tetany; exter. in liniments; trunk resinous exud. put in cavity of painful tooth; frt. c. oleoresin; fls. u. also.

Myrothamnus flabelliformis Welw. (Myrothamn.) (trop eAf): resurrection plant; fl. inf. drunk for colds; strong decoc. u. for backache, dysmenorr.; st. & lf. decoc. analg.; lf. chewed for halitosis, Vincent's angina; pl. ton.

Myroxylon balsamum (L) Harms (M. toluifera HBK) (Leg.) (nSA, esp. Co, Ve to Pe; cult. Guat, WI): balsam of Tolu tree; Tolu b. t.; balm t.; Tolu balsam (balm) tree; a tall tree, trunk of which is incised somewhat as in producing turpentine oleoresin (V-shaped incisions in bk.); Tolu balsam (estoraque [Br]) c. VO with benzyl benzoate & b. cinnamate; resin with tolu-resinotannol esters, free benzoic & cinnamic acids; u. stim. exp. in cough syr.; antisep. in ointments; flavor, adhesive for wounds, in mfr. cosmetics, perfumes, incenses, etc.; balsam formerly chewed (like chewing gum); wd. u. to make furniture - **M. pereirae** Kl. (M. balsamum var. pereirae Harms) (CA, esp. Salv): Peru balsam (balm) tree (Peru is misnomer); balsam obt. after mechanical & heat injury of bk. by letting soak into rags & then boiling out in water; c. cinnamein (VO with benzyl benzoate, b. cinnamate), vanillin, resin; u. parasiticide in ringworm; stim. exp. & in perfumes; in moustache wax (with wax & fats); in veter. med. (1654) - **M. peruiferum** L. f. (Pe, Co): quina quina (Pe, etc.); source of nearly solid balsam, balsamo peruviano solido (off. Por Phar.); u. somewhat like prec.; esp. in perfumes; bk. yields VO (2 types depending on sp. gr.).

myrrh (gum) (< Arabic): oleo-gum-resin fr. *Commiphora molmol* & *C. kataf*; some believe m. of Bible was a mixt. of the two & perhaps with gum-resins of other *C.* spp. - **African m.**: the best var. of the usual type of m. - **Arabian m.**: prep. fr. pls. cult. in sArabia - **Bisabol m.: Bisbadol m.: Bissabol m.: Bissobal m.: East Indian m.**: former names for gum-resin obt. fr. *Commiphora erythraea*, source of opopanax oil - **false m.**: Bdellium - **m. family**: Burseraceae - **Herabol m.**:<*Commiphora m.* - **Indian m.**: prob. Bissabol m., sometimes fr. *C. mukul.* - **Mecca m.:** < *C. gileadense* - **m. oil**: VO fr. true m. (prod. Somaliland, Arabia) - **Somali m: true m.**: fr. *Commiphora molmol* - **Turkey m.**: African m. - **Yemen m.**: the least arom. var. of m.; shipped fr. Sau. Arabia.

Myrrha (L.): myrrh - **M. Indica**: Indian m.

Myrrhis odorata Scop. (monotypic) (Umb.) (sEu Caucas reg., As): sweet (or Spanish) chervil; sweet cicely; garden myrrh; sometimes cult. for sweet aroma; rt. u. arom., carm.; in salads; hb. (c. VO with anethole, anisaldehyde) u. as expect. (sometimes smoked like tobacco) in asthma, difficult breathing; frts. in kidney dis.; potherb; u. in prepn. of Chartreuse.

Myrsine africana L. (Myrsin.) (s&eAs; Af): African myrtle; frt. c. embelic acid; u. like embelia for tapeworms; abortif., antifertility agt., u. in dysmen. - **M. rhododendroides** Gilg (eAf): u. similarly.

Myrtaceae Juss.: Myrtle Fam.; shrubs & trees; mostly occur in Austral & tropAf; lvs. simple, usually evergreen, entire, opp.; fls. perfect; frt. a berry, drupe, nut or capsule; many spp. have edible frt., many are arom.; ex. *Eucalyptus*, cloves, *Darwinia.*

myrte (Fr): *Myrtus communis* - **Myrtenharz** (Ge): **Myrtenwachs**: myrtle resin - **myrticolorin**: rutin; yellow dyestuff; xall. in needles; fr. *Eucalyptus macrorhyncha* (2078) - **myrtille** (Fr-f.;): bilberry - **myrtillin (concentration)**: Extractum Myrtilli (L.): obt. fr. dr. frt. of *Vaccinium myrtillus*; u. skin dis. burns, etc.; also in supp. or enema for hemorrhoids, small intest. inflammn.; ext. of blueberry lvs. (*V. corymbosum*, etc.) u. in diabetes.

Myrtillus (L): huckleberry frt., fr. *Vaccinium myrtillus.*

myrtle (< murtos [Gr] < Myrrha [nymph who was converted into this plant in mythology]; or < muron [Gr] perfume): 1) *Myrtus communis* 2) *Myrica cerifera*; *M. pensylvanica, M. gale* (lvs. hb.) 3) *Vaccinium* spp. 4) *Vinca minor* 5) *Umbellularia californica* - **black m.**: box m. - **bog m.**: *Myrica gale* - **box m.**: *Myrica rubra* - **California m.**: *Umbellularia californica* - **candleberry m.**: *Myrica* spp. - **crepe m.**: *Lagerstroemia indica* - **Dutch m.**: bog m. - **m. family**: Myrtaceae - **m. flag**: Calamus - **m. oil**: VO fr. myrtle (1) (Sp, It) - **Oregon m.**: California m. - **m. resin**: fatty exudn. fr. *Myrica* spp.; u. mfr. candles (pleasant smelling smoke); in mfr. soap, phonograph records, ètc. (2079) - **running m.**: *Vinca minor* - **sand m.**: *Leiophyllum buxifolium* - **sea m.**: *Baccharis halimifolia* - **sweet m.**: *Acorus calamus* - **m. wax**: m. resin - **m. w. berry: wax m.**: *Myrica cerifera*, etc. (bk.).

myrtlewood: California myrtle.

myrtol: gelomyrtol; fractional distillate of VO of *Myrtus communis*; b.p. 166-80 degrees; chiefly eucalyptol & d-pinene; u. to replace eucalyptus oil.

Myrtus arayan HBK (Myrt.) (Mex): arrayán; lvs. c. VO; bk. astr. ton. (Mex Ph.) - **M. communis** L. (Med reg., wAs; cultd.): (true) myrtle; known fr. classical days; lvs. fls. frt. c. VO, resin, NBP, tannin, myrtol (myrtle oil, decoder c. pinene, camphene, dipentene, cineole, geraniol, myrtenol); u. astr. in cararrhs., bronchitis; lung dis.; as balsamic agt.; fresh flg. twigs off. Ge Homeop. Phar; reputed abortif. (419); arom. stom. spice; source of myrtle oil (eau d'Anges, u. perfumes) (hb. distd. sFr, Sp); wd. u. to make furniture, walking sticks, &c.; bk. & rt. u. to tan leathers; cult. ornam.

mystery bark: Mezereon bk.

mytilotoxin: muscle poison; paralytic shellfish poison; saxitoxin; clam poison; specific neurotoxic poison produced in digestive glands of "toxic" mussels (&c.) infected by a dinoflagellate alga, *Gonyaulax catenella/G. tamarensis*; attacks Alaskan butter clam, scallop, &c.; said one of the most potent poisons known; thought related to the development of "red tides" where enormous numbers of the alga give the ocean water a distinct coloration; formerly supposed a "ptomaine" formed in mussels living in stagnant water and exposed to dryness or sun; produces muscular paralysis & death; proposed for use in paints to prevent barnacle fouling of ship hulls (WWII); u. in neurochem. research (2080).

Mytilus edulis (Mytilidae - Mollusca) (salt water Eu, NA): mussel; frequently eaten, esp. in Fr, Neth but sometimes danger of poisoning by mytilotoxin; inadvisable to eat this shellfish during the summer months; several other spp. so used (ex. *M. californicus*).

myxedema: state of hypothyroidism in which low metabolism, dry skin, loss of hair, low metabolic rate, mental dullness, anemia, slow reflexes, &c. are found.

Myxobacteria: Myxobacteriales: slime bacteria; order in Schizomycetes; flexible slender rods; these aggregate to form frtg. bodies ("fruiting myxobacteria").

myxology: the science of mucous membranes.

Myxomycetes = Myxomycophyta = Myxomycota = slime molds; may be confusion with foll.

Myxophyceae = Cyanophyceae: blue-green algae.

Nn

Nabalus Cass. = *Prenanthes.*

nabo: 1) (Sp, Por): turnip 2) (Mex): some *Opuntia* spp. (neMex).

nacaguita flowers: fr. *Cordia boissieri.*

nacasolo (Mex, CR): **nacascolote**: *Caesalpinia coriaria.*

nacker: *Margaritana margaritifera.*

nacre: mother of pearl (lining of some sea shells).

nacuma: Panama hat plant.

Nadelholzteer (Ge): pine tar.

nafta (Sp): naphtha; gasoline.

Naftalan (Oil) (Azerbaijan): black liquid c. hydrocarbons + tar; claimed of value in rheum., eczema, polyarthritis (in baths); nervous dis. (antisep. & analg.); c. "biostimulants" - **purified N.:** special crude naphtha + 5% med. soap; u. for eczema, skin parasites, &c.

nagas: *Mesua ferrea.*

nagbo (Pers): spearmint.

Naegelein (Ge): clove.

nagi (Jap): *Podocarpus nagi*; *Myrica rubra.*

naguapate (CR): naguapacle: **nahuapate**: *Cassia hispidula.*

naharan fat: FO fr. sds. of *Horsfieldia irya.*

Naias Juss. = **Najas** L. (Najad. - Monocot.) (cosmop.): naiads; ann. fresh water pls.; at times u. as fertilizer; dr. pl. as packing material - **N. flexilis** Rostk. & Schmidt (NA, WI, OW): (Canadian) water nymph; submersed pl. - **N. marina** L. (N. major All.) (OW, NW temp & trop): naiad; pl. u. as food (with salt) (Haw [US]) (*cf.* watercress); considered appetizing with raw water shrimps, crabs, &c.; sold in Honolulu [Haw] markets; cult. as ornam.

Naja: cobra (*qv.*).

naked boys: n. ladies: 1) *Colchicum autumnale* (also **n. lady**) 2) *Amaryllis* sp.

Nama hispidum Gray (Hydrophyll.) (swUS [sTex], nMex): rough nama; hierba de la ventosidad ("herb of flatulence") (Mex); hb. decoc. drunk for gassy stomach.

ñame: yam (pron. nyamaye).

ñamole (Mex): *Phytolacca americana, P. octandra.*

namur oil plant: *Cymbopogon martinii.*

nana (Pishin, Arab): *Mentha* spp., esp. spearmint.

nance bark: fr. *Malpighia glabra.*

nancite (CR): *Malpighia punicifolia.*

Nandina domestica Thunb. (Berberid.) (Chin, Jap, Ind): nanten; heavenly bamboo; sacred b.; lf. c. 0.25% HCN in labile form; rt., bk., st., frt. c. some 10 alks., incl. domesticine (nantenine), nandinine(?) (420) (rt. bk.), berberine; frt. ton. antipyr. useful in asthma, cough, sore throat, impotency, sed. to drunkards; pl. insecticide; rt. bk. (Cortex Nandinae) in many disorders; ornam. shrub.

nangca: nangka: jack tree & its frt. (*Artocarpus altilis*) or (in Guam) the related breadfruit (*A. heterophyllus*).

nanju (CR): *Hibiscus esculentus.*

Nannorhops ritchieana (Griff.) Aitch. (Chamaerops r. Griff.) (Palm) (Indo-Pak): maz(a)ri; chatai; young lvs. fls. & frts. u. as potherb (natives); lvs. made into baskets, sandals, brooms, brushes, fans, mats, ropes, &c.; sds. u. as beads; lvs. u. in diarrh. dys. purg.; mostly u. in veter. med.

nannyberry: *Viburnum prunifolium* - **nanny-bush bark**: fr. *Viburnum lentago.*

nanometer: nanon: nm. = 10^9 m.; 0.001 mikron.

nape: *Brassica napus.*

naphtha: a transparent colorless or pale yellowish volatile limpid liquid found abundantly in most samples of petroleum oil & obtained as one of the lowest boiling point fractions fr. that liquid (also obt. fr. coal tar distillates); mixt. c. pentanes & hexanes chiefly; about the same as petrol. benzin or petrol. ether - **coal tar n.**: benzene, C_6H_6 - **petroleum n.**: p. benzin - **wood n.**: methyl alc. fr. direct distn. of wd.

ñapinda (Arg): *Acacia bonariensis.*

Napol™: pure amylose (of starch).

Napoleonea heudelotii A. Juss. (Lecythid.) (w tropAf): gohemi (Sierra Leone): macerated frts. u. for inguinal hernia - **N. imperialis** P. Beauv. (trop wAf): frt. rind & bk. u. for cough & in criminal poisoning; c. saponins - **N. leonensis** Hutch. & Dalz. (w tropAf): chopped bk. chewed with cola nuts; also mixed with rice as food; inner bk. or twigs chewed for cough & asthma - **N. vogelii** Hook. & Planch. (w tropAf): bori-bori (Nigeria); rts. u. in fever; frt. pulp edible.

Napralert: Natural **P**roducts **Alert**: a computerized database for medicinal plants (some 500 genera) (chem. & pharmacology) kept at Univ Ill [US] Coll. Pharm. (Chicago) since ca. 1950.

naranj (Pers): **naranja** (Sp): orange - **n. agrio** (CR): **n. amarga** (Sp): *Citrus aurantium* - **n. dulce** (Mex): sweet o.

naranjilla (Sp): *Solanum quitoense* - **naranjita** (Sp): frt. of *Citrus nobilis* - **naranjo** (Sp): orange tree - **naranjón** (NM, Sp): grapefruit.

Narbe (Ge): stigma of pistil.

Narbomycin: antibiotic fr. *Streptomyces narbonensis* Corbaz & al. (soil, Cannes, France); macrolide structure; active against Gm.+ bacteria; 1955-.

narceine: xall. alk. fr. opium; narcotic, antispas.; governed by Harrison Narcotic Act.

Narcissus L. (Amaryllid.) (Euras): daffodil; 27 spp. of herbs; fls. with well-developed coronas; showy ornam.; bulbs toxic - **N. jonquilla** L. (sEu; nAf): jonquil; source of VO u. as perfume; bulbs & fls. emet. & u. in diarrh., epilepsy, & exter. in inflammn. - **N. poeticus** L. (sEu): poet's narcissus; pheasant's eye; VO u. in perfume; common ornam.; bulbs u. med. as for *N. pseudonarcissus* - **N. pseudonarcissus** L. (wEu): (common) daffodil; wild d.; lent lily; bulb & fl. u. emet. antispas.; has been u. int. for tumors, dys., ulcers, indurations, &c.; toxic; widely cult. ornam.; hydrolate (Fr Phar.) (1840) - **N. tazetta** L. (Euras, nAf): (French) daffodil; polyanthus (daffodil); narcissus; nargis (Punjab); bulbs c. tazettine, lycorine & other alks.; white fls. source of VO; bulbs u. as poultice for swellings (boils, ulcers, mastitis), emetic, to treat itching skin dis.; headache; bulb juice for eye dis.; cult ornam. hb. - **N. t.** var. *chinensis* Roem. (v. *orientalis* Hort.) (much cult. Chin, Jap, US): Chin (sacred) lily; bulbs &c. u. in Far East same as *N. t.*; esp. for tumors; attractive yellow fls. freely developed when bulbs kept in water.

narcosine: narcotine: now called noscapine (*qv.*) (since non-narcotic) (1656).

nard (celticus): *Valeriana celtica* - **Indian n.**: *Nardostachys jatamansi* - **nardine**: relating to or smelling like nard - **nardo** (Mex): *Polyanthes tuberosa.*

Nardostachys grandiflora DC (N. chinensis Batal.; N. jatamansi DC) (1) (Valerian.) (Chin): u. as ton. & stom. in Chin folk med.; as source of VO (do not confuse with Valeriana j. Wall.) (Ind-Himalay): (true) Indian nard; spikenard (of Ancients); rhiz. arom., very ancient & expensive perfume of Indians; c. 1% or more VO; u. arom. adjunct. in perfumes esp. for hair prepns.; said to incr. blackness & growth of hair (fr. Gr. nardos, valerian); rt. (nard) u. in epilepsy, hysteria, cramps, as diur. emmen. carm. lax.; to make arom. ointment; substit. for valerian.

Nardus Indica (L.): 1) (true) Indian nard, *Nardostachys jatamansi* 2) (ancients): perhaps *Cymbopogon nardus* - **N. Italica** (L.): *Lavandula stoechas*; the spikenard (Spica Nardi) of the Middle Ages.

Nardus stricta L. (1) (Gram.) (Euras, NA s to Mich [US]): nard(grass); mattgrass; rt. (Wendewurzel) u. stim. vulner. arom.; for mal., &c.

nardy: belonging to or redolent of nard.

Naregamia alata W. & A. (Meli) (Ind, EI, WI): Goa(nese) ipecacuanha; Portuguese i.; rt. & st. c. **naregamine**, FO, etc.; u. alt. in rheum., bilious dis., diur., diaph., exp., emet.; mostly as Tr. or naregamine (u. in dys.); action compared with that of cocillana (421).

narghil (Arab): coconut; hence **narghile**, water pipe, oriental tobacco pipe in which the smoke is filtered through a bowl cont. water, since bowl orginally made fr. coconut shell.

naringin: the glucoside, naringenin-7-rhamnoglucoside; found in fl. & frt. peel of lemon, grapefruit, & shaddock; u. bitter ("Amerin") in bev. & confections to confer piquancy.

Naristillae (L): nose drops.

narra wood: fr. *Pterocarpus vidalianus* Rolfe (PI) (dye).

Narthecium ossifragum (L) Huds. (Lili.) (Euras, natd. NA): (bog) asphodel; u. med.; yellow hair dye; saffron substit. (Shetlands).

narwhal(e): *Monodon monoceros* (Cetacea) (Arctic Oc esp. nr Greenland): "sea unicorn"; one of two teeth greatly extended in male 2-2.8 m in length (unicorn "horn"); medical uses in Eu ca. 1600, esp. u. to protect against poison (as antidote) and as aphrodis.; may sometimes have been rhinoceros horn (traded by Por fr. Ind) (1657).

naseberry: sapodilla, *Manilkara zapota.*

Nashia inaguensis Millsp. (Verben.) (WI): natives use lvs. of shrub as tea; in decoc. as febrifuge ("Mountain Tea") (Bah) (1658).

nasitort (Fr): **n. officinal**: Nasturtium.

Nasturtium R. Br. (Cruc.) (Euras, Af, NA): (water) cress; aquatic or marsh plants; some 10 spp. (the plant gen. called "Nasturtium"): *Tropaeolum majus* and other *T.* spp. (Tropaeol.) (origin of name: nasus [nose] + tortus [twisted], nose twister, referring to pungent properties of pl.) - **N. fontanum** Aschers. (N. officinale R. Br.) (Rorippa nasturtium-aquaticum Hayek) (Euras, Af, natd. Am) = *N. officinale* - **N. indicum** DC (Cruc.) (eAs): pl. u. as diur. stim. (Lahul); antipyr.; var. *apetalum* Gagnep. (Java): pl. remedy in leukorrhea - **N. montanum** Wall. (eAs): weed eaten as aid to digest carm., for aching bones, cough; summer or green watercress; Herba Nasturtii (aquatici) (important prodn. at limestone springs near Huntsville [Madison County], Ala [US]); grows in clear running water (2085); c. VO (sulfurated) glycoside, vits. C, E; u. diur. spring ton., antiscorb. in metabolic dis., catarrhs., eczema; as nutr., esp. in salads; for cough, kidney dis., asthma, coryza, contraceptive; today u. in chronic skin dis., gingivitis, periodontosis (328); has been u. in carcinoma (411); esp. as garnish in salads; substit. for lettuce; sd. condim. (like mustard); c. Fe (7-8 mg %) - **N. palustre** DC = *Rorippa palustris.*

nasturtium (, garden): *Tropaeolum majus.*

nataloin: glycosidal bitter princ. fr. Natal Aloe.

Natamycin: antibiotic (polyene type) fr. cultures of *Streptomyces natalensist+*; u. as antimycotic in epidermophytosis & candidamycosis; in ocular dis. as conjunctivitis, keratitis, fungal blepharitis; using 5% aq. suspension.

natema (Ec-Jivaro Indians): caapi (narcotic drug).

National Formulary (US): NF: 1880 (1888) to the present; originally published by the Am Pharmaceutical Assn., from ca. 1970 by the U.S. Pharmacopeial Convention, Inc., jointly with the USP; on an equal footing in the eyes of the law with the U.S. Pharmacopeia, this publication has often been misunderstood by professionals abroad as simply a formulary (or recipe book); for many years the NF was the depository of crude drugs deleted fr. the USP; originally it had been made up entirely of pharmaceutical preparations, hence the name.

native: 1) occurring or growing naturally or indigenously in a certain place or area; ex. plants, minerals, etc. 2) natural rather than artificial or synth.; ex. n. proteins, those ready-formed in tissues.

natob: a bush lespedeza u. for soil conservation.

natre (Ch): **natri**: *Solanum crispum, S. tomatillo, S. gayanum.*

natrine: alk. fr. natre; supposedly active as antipyr.

Natrium (L. Ge): **natrum** (Arab): **natrun**: sodium (Na) - **Natro-kali Tartaricum** (L.): sodium potassium tartrate; native Na_2C0_3 or $NaHCO_3$ or other mineral alkali, as salt deposit.

natron: 1) **natrun;** soda; trona; crude (native or natural sodium sesquicarbonate) Na_2CO_3., 10 H_2O or $NaHCO_3$; originally obtained in the Egyptian Desert; a native subst. found as the dihydrate in Owens Lake (US), Lake Magadi (Kenya); orign. fr. Natron Lakes w. of delta near Nile River, Eg; u. in mummification of bodies (esp. to fill in cheeks, thus giving a more lifelike aspect to mummy) (ancient Eg) 2) (occasionally) NaOH.

natural: 1) drug article in state in which collected but dr.; with rts., fibers, outer bk., etc. still attached & not removed 2) occurring in nature (pl. or animal) as opposed to that synthesized by man (esp. appl. to drugs & foods); however, the active principles of these may be synthesized.

naturalized: pl. or animal introduced by accident or intention fr. an area where native to a place in which it becomes so well adapted that it eventually grows wild and would appear to be native, were its history not known; ex. jimson weed in US, pokeroot in Eu.

nature's mistake (NC): *Cornus florida.*

Naturstoffreagenz (Ge): natural substance reagent u. for tests in chromatography.

Nauclea inermis R. Br. (Naucl./Rubi) (tropAf): "khoss"; bk. & lvs. u. as antipyr., analg. in childbirth; abort. (prob) (Seneg.) - **N. junghuhnii** Merr. (Sarcocephalus j. Miq.) (PI): Guinea peach; bk. decoc. u. for menstr. dis. - **N. latifolia** Sm. (N. esculenta Merr.; Sarcocephalus e. Afzel.) (trop wAf): Guinea (or African) peach; (Negro) p.; fathead tree; d(o)undake; Khusia (Ghana); pl. much u. in native med.; bk. bitter ton. antipyr. in mal. ("African quinine"), stom. antiemet., purg., emmen.; lf. & rt. u. fever, cough, dys., diarrh., nausea, indigest.; wd. (as njimo or njuno) c. indole alk., resin, sitosterol, pyrocatechol tannin; u. stim. ton. digestant; relief of pain fr. swollen spleen in mal. (Nigerians claim synth. antimalarials do not relieve this pain); frt. edible, juicy, u. emet.; rts.

with yellow dye, u. to make Morocco leather; st. u. for indigest., wounds; frt. in piles, dys., emetic; fl. heads eaten as veget.; tree gum u. (Senegal) (1532) - **N. officinale** (L.) Pierre (S. officinalis Pierre) (seAs): huynh ba; yellowish wd. & bk. (Camb) reputedly antipyr.; sold in Chin herb stores - **N. orientalis** L. (Austral, Polynesia, PI): canary cheesewood; bk. bitter ton., antipyr., reputed antimal, vulner. for toothache, diarrh. (soaked appl. to sores); lvs. appl. to boils, tumors, headache (PI) - **N. pobeguinii** Petit (S. diderrichii DeWild.) (Merr.) (trop wAf): West African boxwood; lvs. c. alk.; antipyr.; wd. u. for shuttles, sometimes causes fatal cardiac poisoning in workers.

naughty man's cherry: Belladonna.

nauli gum: oleoresin fr. *Canarium commune*; source of **n. resin**.

Nautilocalyx Linden (Gesneri) (tropSA): most spp. are known as ornam.; however, one sp. of Amazonia (unnamed) is u. to relieve severe cramps (esp. in the aged) (warm baths); formerly u. in one of the forms of curare; c. alks. (2733) - **N. melliti-folius** (L.) Wiehler (Gesneri) (Lesser Antilles, Dominica): fls. u. in tea for colds (Caribs, Dominica).

Navaho tea: Navajo tea: *Thelesperma subnuda.*

naval stores: products now defined as turpentine oil (gum & wood); rosin (gum & wood), pitch (SC highly essential to naval operations in the days of wooden-hulled vessels & calking) (2625); *Pinus* spp.

navelwort (, nodding): *Umbilicus pendulinus.*

navet (Fr): turnip.

nay-kassar: fl. of *Mesua ferrea.*

ndawa: *v.* dawa.

NDGA: Nor-DihydroGuaiaretic Acid.

Nealchornea yapurensis Huber (Euphorbi) (Co): lvs. of tree u. as fish poison.

neamine: antibiotic derived by acid degradation fr. Neomycin.

Neapolitan: pertaining to Napoli (Naples); v. under nuts.

neat (cattle): cattle of genus *Bos* (as distinct fr. horses, sheep, goats, & other cattle in a general sense) - **neat's foot oil**: Oleum Bubulum: FO prep. fr. feet tissues of *Bos taurus*: u. leather lubricant; formerly in med.

Nebennieren (Ge): suprarenal (adrenal) capsules - **N-mark**: adrenal medulla - **N-rinde**: adrenal cortex.

Nebenzellen (Ge, Eng): neighboring cells.

nebo (Arg): rape.

necklace weed berries: *Actaea americana.*

Nectandra Roland ex Rottb. (Laur.) (tropAm): c. 100 spp. of shrubs & trees; fls. perfect, perianth lobes fleshy; anthers 4-lobed - **N. acutifolia** Mez (Co): bk. c. alks., u. as warm tea to relieve excessive fatigue (Kubeo Indians) (2299) - **N. cinnamomiodes** Nees (Ec-Andes Mts) (tree cult. Ec): canello de mato; calyx & bk. u. as spice (*cf.* cinnamon); unripe frt. u. to adulter. cassia buds - **N. coto** Rusby (Aniba c. [Rusby] Kosterm.) (CA, SA, esp. Bo, Br): coto bark; source (long in doubt) of coto bark; c. arom. ketones cotoin, hydrocotoin, methylhydrocotoin, protocotoin, &c., VO, alk. (parostemeine) tannin; u. in diarrh. dys. prepn. cotoin; toxic (1659) - **N. cymbarum** Nees (Br, Ve): (pão) sassafras; rt. strengthening ton.; balsam obt. fr. incisions ("oil of sassafras") u. as diur., antispasm., emm., diaph., exter. appl. to sores; wd. appl. locally in rheum. (Ve) - **N. elaeophora** Barb.-Rodr. (nBr): (i)nhamui; bk. c. VO (like turpentine oil) (prod. by boring hole in trunk) with pinene, turmerone; bk. & beans u. as arom.; bk. is sometimes found in comm. coto bk.; VO u. in skin dis. & as insecticide - **N. globosa** Mez (Mex, WI, CA): aguacatilla (CR); u. med. (Hond) - **N. oleophorum** (nBr): inhamui; c. VO with limonene, pinenes; u. insecticide - **N. pichurim** (HBK) Mez (Ocotea p. Ruiz. & Pavon) (N. pichury-minor Nees & Mart.) (tropSA, esp. Br): pichurim (or purchury) (bean) ; dr. cotyledons very arom.; u. (as Fabae P.) as substit. for nutmegs & vanilla (mostly by natives); stim. ton.; also source of med. ext. called "pichuri"; u. to treat diarrh., dys., asthma; to regulate menstr.; appl. exter. to wounds & ulcers; do not confuse with *Licaria pichury-major* (Nectandra p.-m.) (*qv.*) - **N. pseudocoto** Rusby (Aniba p. Kosterm.) (Br. Bo): paracoto; bk. (paracoto bk., "false coto bk.") similar compn. to *N. coto*; c. paracotoin; VO; u. similarly to coto bk.; less irritant to mucosa; toxic - **N. puberula** Nees (N. amara Meissn.) (Br): bk. source of VO (aceite de sassafras); u. med.; off. as Canela preta (Br Phar.) - **N. puchury-major** Nees & Mart. = *Licaria p.-m.* Kosterm. - **N. rodioei** (variant spelling) Hook. (BrGuianas): itanha branco; bebeeru (tree); greenheart; a large tree, the wd. valued as lumber; bebeeru bark (greenheart bk. fr. Demerara).

Nectandrae Cortex (L.) (BP 1885): c. bebeerine (also in sds.); u. bitter ton. antipyr.; in mal., extern. in eye inflammn.; bk. & wd. tannin-rich, with groenhartin; u. dye.

nectandrine: "bebeerine tot. alks."; several alks. in mixture found in pareira rt. (*Chondodendron* spp.) & *Nectandra rodioei* bk.

nectar: 1) drink of the (Greek) gods 2) sweet liquid saccharine secretion of pl. nectaries, chief source of honey 3) sweetened prep. of juice & pulp of peach, apricot, or pear.

nectarine: smooth-fruited peach, *Prunus persica* v. *nectarina.*

Nectria ditessima (Necrtri. - Hypocreales - Fungi): organism paarasitic on trees; subc. inj. u. (sporocarp) in cancer (ca. 1900).

Neea parviflora Poepp. & Endl. (Nyctagin.) (Co, Pe): yano muco; lvs. chewed by natives to preserve & blacken teeth & gums; gives strong teeth with little decay (Br) - **N. theifera** Oerst. (Br): lvs. u. as tea.

needle: lf. of conifers (fam. Pinaceae).

neem: *Azadirachta indica*; sds. c. **n. oil** (FO) (Ind).

neesberry: Sapodilla.

Negerhirse (Ge): *Pennisetum americanum*, cereal.

neguilla (Sp): *Nigella damascena.*

negus: Negus; a mild punch of port (or other) wine with a little lemon & sugar (SC Colonel N., fl. 1710).

Negundo(L.): bk. of *Acer negundo* (Homeop.).

neighboring cells: epidermal cells which surround a stoma.

Neisosperma oppositifolia (Lam.) Fosberg & Sachet (Ochrosia o. K. Schum.) (Apocyn.) (Oceania: Micronesia; Soc Is to PI, SriL): bk. c. reserpinine, reserpilene, & other indole alks.; sds. edible; wd. u. in construction (Tonga) (86); bk. u. antipyr. bitter ton. stom.; antipyr. in anemia.

Neisseria intracellularis Buchanan & Murray: **N. meningitidis** (Wollstein) Murray (Neisseriaceae: Cocci): bacteria spp. which causes epidemic cerebrospinal meningitis.

neja nuts: sds. of *Pinus gerardiana.*

Nelke (Ge): 1) clove 2) carnation - **Nelkenpfeffer**: pimenta.

Nelsonia canescens (Lam.) Spreng. (Acanth.) (pantrop, esp. Af): pl. juice expressed into eyes for fever (Ghana); rt. decoc. drunk for bilharziasis; lf. inf. drunk for diarrh. (Tanz); lvs. scrubbed & pressed in a cloth and the mass formed dipped into water, drops put in ear for otitis (2445).

nelumbine: alk. fr. cotyledons & pl. juice of foll.

Nelumbium Juss. = *Nelumbo.*

Nelumbo lutea (Willd.) Pers. (N. pentapetala [Walt.] Fern.) (Nelumbon.) (e&cNA, WI, SA): water chinquapin; yellow nelumbo; "American lotus"; pond nuts; water chinkapin; sacred bean; frt. appears similar to a shower head (ornam. uses); rhiz. (rts.) st. lvs. & sd. u. as food (Am Indian); refrig. lax. diur. - **N. nucifera** Gaertn. (Nelumbium speciosum Willd.) (sAs, Australas, Ind; natd. eUS): (East) Indian lotus; sacred l.; lotas; with pink petals; stalks & rts. commonly eaten (also boiled young lvs.); sds. consumed as delicacy in Chin theatres, Kashmir; sd. & rhiz. frt. source of Chin arrowroot; rt. & sd. u. in diarrh. dys. hemorrhoids; st. juice antiemet.; fls. astr.; stamens u. to flavor tea (Indochin); pl. u. to control hemorrhaging (Chin); frts. ("Pythagorean beans") & rhiz. as food; sacred in Ind & Chin (Tibet); introd. into Egypt ca. 600 BC (Egyptian lotus) (u. in decorations, 5th dynasty) (2086).

nema: nematode: round worm of Phylum Nematoda; eelworms; cylindrical tiny worms living in soil or parasitic on animals (incl. man) & pls. (3562).

Nematoloma caerulescens Pat. = *Psilocybe cubensis* (Thailand, Camb) (2351).

Nemesia strumosa Benth. (Scrophulari.) (sAf): ornam. ann. bh.

Nemexia Rafin. = *Smilax.*

Nemopanthus mucronatus (L) Trel. (Aquifoli.) (e&cNA): mountain holly; berries edible with high carbohydrate cont.; cult. ornam.

Nemuaron vieillardii (Baill.) Baill. (N. humboldtii Baill.) (Atherospermat./Monimi.) (NewCaled): arbre absinthe; bois pernod; bk. u. chewed as ton. stim. back pain, kidney dis.; decoc. for stomachache, bronch. dis., colics, paralyses; large dose as abortif. (aborigines) (341).

nenufar (blanco) (Sp): *Nymphaea alba.*

Neoarsphenamine: Neosalvarsan™: antisyphilitic now almost entirely displaced by penicillin, etc.

Neocallitropsis araucarioides (Compton) Florin (monotypic) (Cupress.) (New Caled): foriage dist. —> "araucaria oil" (VO): classed as fixative oil u. to scent soaps.

Neocinnamomum poilanei Liou (Laur.) (seAs): lvs. & bk. u. to prep. a drink for abdomin. dis.

Neo-Digalen (Russ Phar.): aq. ext. of *Digitalis ferruginea*; u. by sc. admn.

Neogale: fresh milk.

Neoglaziovia variegata (Arruda) Mez (Bromeli) (Br, Arg): caroa (verdadeira); lvs. source of fiber u. like jute (nets, packing material) (422).

neohesperidin: sweet tasting compd., n-dihydrochalcone; found in *Citrus aurantium.*

Neolitsea umbrosa (Nees) Gamble (Laur.) (nInd, Nepal): frt. furnishes FO (48%) u. for heating & illumn.; med. for itch, burns, skin dis.; in food prepn. (1661) - **N. zeylanica** Merr. (Litsea z. Nees; N. cassia Kost.) (seAs): bellary (Ind); lf. provides b. leaf oil (VO); u. arom. in perfumery.

Neolloydia johnsonii L. Benson (Echinomastus j. [Parry] Baxt.) (Cact.) (swUS): frt. peeled & eaten (Piute Indians).

Neomethymycin: macrolide anitbiotic isolated fr. metabolic prods. of *Streptomyces venezuelae* Ehrlich et al.; also fr. mother liquors fr. methomycin; research of Djerassi et al. (1957); active against Gm.+ bacteria.

Neomillspaughia emarginata Blake (Polygon.) (sMex, CA): zac-itza (Maya); lf. decoc. for coughs & colds (Yuc); bee tree.

Neomycin: antibiotic prod. by *Streptomyces fradiae* Waksman & Henrici, *S. lavendulae* Waksman & Henrici, obtained fr. soils, & *S. albogriseolus* Benedict & al.; complex of A,B,C; isolated by Waksman & al. 1949; **N.-B** mostly used (2233) (Framycetin, &c.); u. to combat Gm.+ & Gm.- bacteria; to sterilize intest. before surgery; in ear inf., &c.; *Staph.*, *Proteus vulgaris*, *Pseudomonas aeruginosa*; topical not systemic; lowers cholesterol levels in blood (E).

Neonauclea formicaria (Elm.) Merr. (Rubi) (Borneo, PI): bk. decoc. drunk for mal., rheum., puerperium (PI).

Neonelsonia acuminata Coulter & Rose (1) (Umb.) (Co, Ec): wild arracacha; lvs. & sts. u. for intest. inflamm. & post partum (85).

neopelline: alk. isolated fr. *Aconitum napellus*.

neopine: alk. of Opium, isomeric with codeine.

Neoraimondia arequepensis Backeberg var. *roseiflora* Rauh (Cact.) (Pe): u. to make cimora, an hallucinogenic beverage.

Neorautanenia mitis (A. Rich.) Verdc. (Leg.) (tropAf): claimed to destroy snails (carriers of bilharzia) - **N. pseudopachyrhiza** (Harms) Milne-Redhead (across tropAf): rt. u. against scabies & as fish poison (Congo).

neotype: type specimen or illustration of pl. selected to replace lost previous type.

Neo-uvaria merrillii (Annon.) (nBorneo): frt. u. in fever in children.

Nepenthaceae: Nepenthes Family (Dicot.) (Indomal, Australasia, Oceania): pitcher plants; genera *Nepenthes* L. (ca. 70 spp.).

Nepenthes ampullaria Jack (Nepenth) (Mal): (bottle-like) pitcher plant; insectivorous; rts. boiled to make poultice for stomachache & dys.; st. decoc. for remittent fever - **N. boschiana** Korth. (Indones): water fr. unopened pitchers u. as eye wash for inflamed eyes.

Nepeta cataria L. (Lab) (Euras, natd NA, sAf): catnip; catnep; catmint; arom. lvs. & tops c. VO, with nepetalactone & several other lactones, carvacrol; tannin; inf. u. flavor, stim. anod. antispasm. (in gastric dis., infantile colic [pain] [with baking soda]), in chronic bronchitis, hysteria; neurasthenia; hallucinogenic; as tea to induce sleep (adults) (Ala [US]); ant repellant (strewn on trail); in folk med. u. very extensively; thus tender young shoots inf. for anemia in infants; excitant to feline (lactones of VO responsible for activity in cats & hallucinogenic for humans but not the subst. of therapeutic value) (tea) (329) - **N. c.** var. *citriodora* Dumoulin (N. c. Beck.): u. like Melissa; source of geraniol - **N. ciliaris** Benth. (Indo-Pak, Himal): pl. u. in fever & cough; administered in sherbet (phool) - **N. elliptica** Royle (Ind): sd. inf. u. in dys. - **N. glechoma** Benth. = **N. hederacea** Trev. = *Glechoma hederacea* - **N. hindostana** Haines (Ind): pl. u. in fevers, as card. ton. in GC; decoc. gargle in sore throat - **N. mussinii** Spreng. ex. Henk. (Euras, natd. US): "catmint" (but not true c.); u. in folk med.; fresh lvs. chewed for headache; inf. u. as arom. tea; & infant colic (area of Buffalo, NY [US]).

Nepetaceae: Labiatae or segregate fr. same.

Nephelium lappaceum L. (Sapind.) (Mal, cult. eAs): rambutan; esculent frt. u. as compote; sds. eaten (roasted, &c.); sds. source of rambutan tallow; frts. u. as carm. in dys., antipyr.; rt. decoc. in fever - **N. litchi** Camb. = *Litchi chinensis* - **N. longana** Camb. = *Euphoria longan(a)*.

Nephrolepis cordifolia Presl. (Davalli./Polypodi) (tropAf, As; natd sFla [US]): decoc. of tuber with pork eaten to strengthen kidneys & to relieve hernia; decoc. said antipyr.; decoc. of fronds bechic (PI) - **N. exaltata** Schott (trop & subtr OW, NW, ex. sFla [US]): sword fern; popular living room & conservatory ornam. fern, esp. var. *bostonensis* Davenport (trop): Boston fern; ornam. house fern - **N. hirsutula** Forst. (eHemisph): laufale (Tonga): various parts of pl. u. med. (Tonga); u. med. Fiji.

Nephroma parile Ach. (Nephromat./Peltiger.) (Lichenes) (temp zone; in US, New Eng-NY, Minn, Mont): chocolate colored nephroma, powdered Swiss lichen; u. to produce red, orange, purple dyes, for woollens (Scotland).

neps: small tangled masses of fibers in cotton visible to naked eye.

Neptunia oleracea Lour. (N. prostrata Baill.) (Leg) (trop): sprouts of this water pl. are eaten as potherb.

néré (Fr): *Parkia biglobosa*.

nereistoxin: neurotoxin fr. *Nereis* spp. (polychete worm); u. as insecticide for rice insects (Jap); "Padan" comm. article.

Nereocystis luetkeana Post. & Rupr. (monotypic) (Lessoniaceae) (algae) (Pac coast of NA): bladder kelp; a giant kelp (-40 m long) found growing in immense numbers; u. prodn. sodium alginate; during World War I, source of potash, halides, etc.; succulent parts u. as food by Orientals & Am Indians; fronds dr. & pd. u. in pills as mineral supplement; stalks u. to make fishing lines (Am Indians); mucil. for burns; candied stipes, &c. sold as "Seatron" (sweetmeat) (2087).

nerigoside: glycoside of oleandiginin + D-diginose; occ. in *Nerium oleander*.

neriifolin: cardioactive glycoside with digitoxigenin & L-thevetose; occ. in *Thevetia nereifolia*.

neriine: conessine.

Nerine Herb. (Amaryllid.) (trop & sAf): alkaloidal pls. often the cause of poisoning (chiefly stock) - **N. krigei** Barker (sAf): c. neronine, krigenamine (1662) - **N. sarniensis** (L.) Herb. (sAf): (common) Guernsey lily; c. nerinine & tazettine.

nerinchil (sInd): *Tribulus terrestris*.

neringe: nerunji (Tamil): *Tribulus terrestris*.

neritaloside: glycoside of oleandrigenin + D-diginose; in *Nerium oleander*.

Nerium indicum Miller (N. odorum Sol.) (Apocyn.) (Iran to Jap, cult. Ind, sUS, Jap): sweet-scented oleander; sts. & twigs c. cardioactive glycosides, odoroside A, B, G, H; lvs. c. oleandrin; rutin; rts. &. lvs. u. as cardioton.; ton. abortif. toxic; suicidal agt. (Turk); locally to reduce swellings; rt. bk. u. extern. in skin dis.; VO fr. rts. u. to treat chronic skin dis. (also bk. & lvs. mixed with oil appl. to skin eruptions [herpes] [PI]) - **N. oleander** L. (Med reg., Levant, nAf, As, natd. Fla [US] seaside, Ind): (common) oleander; rose bay; laurier-rose (Fr); lvs. bk. & fls. c. glycosides (2102) oleandrin (with very rapid onset of action); neriin; u. cardioton. agt.; lvs. also u. in cutaneous eruptions; poison(!!) (351); cult. ornam. shrub; highly toxic (sap, fls.); old lvs. rich in rubber (Edison).

neroli (flowers) (petals): orange fls.: source of **n. oil** (Fr) - **nerolin** (comml. name): beta-naphthol methyl & ethyl esters; u. as substit. for neroli oil (SC Princess of Neroli or Nerola [near Tivoli] who u. the oil to perfume her gloves ca. 1675).

nerprun (Fr): buckthorn.

nerve growth factor: NGF; occ. in various tissues of mammals; acts to stim. growth & differentiation of nerve ganglia; said to extend life span **- n. root: nervine**: *Cypripedium pubescens* & other *C.* spp.

Nesaea polyantha Guill. (Lythr.): (wAf); pl. u. in ophthalmia - **N. sagittifolia** Koehne (tropAf): u. for blackwater fever & other hematurias of natives - **N. salicifolia** HBK = **N. syphilitica** Steud. = *Heimia s.*

net (weight): **netto** (Sp): the wt. of an article aside fr. packing materials.

netrospin: antibiotic fr. *Streptomyces albus* Waksman & Henrici; u. fungicide.

nettle: 1) any stinging or prickly plant 2) *Urtica* spp. 3) *Amaranthus spinosus* (Clark Co., Ala [US]) - **blind n.**: *Lamium album* (fls., lvs.) - **bull n.**: 1) *Solanum carolinense, S. aculeatissimum* 2) *Cnidoscolus stimulosus* 3) *Jatropha urens* - **Carolina horse n.**: horse n. - **common n.**: *Urtica dioica* - **dead n.**: *Lamium amplexicaule* (**white d. n.**: *L. album*; **red d. n.**: *L. purpureum*) - **devil n.**: *Laportea crenulata* - **dog n.: dwarf n.**: *Urtica urens* - **dumb n.**: dead n. - **n. family**: Urticaceae - **gympie n.**: *Laportea moroides* Wedd. (Austral) - **hedge n.**: *Stachys arvensis* - **horse n.**: *Solanum carolinense* (berries) - **lesser n.**: dog n. - **mulberry n.**: gympie n. - **Nilgiri n.** (Ind): *Girardinia palmata* - **poison n.**: *Stachys sieboldii* - **n. rash**: urticaria: skin eruption like that from nettles; common allergic reaction - **sting n.**: *Cnidoscolus stimulosus* - **stinging n.**: 1) *Urtica dioica* 2) (NC): horse n. - **tall n.**: common n. - **n. tree**: *Celtis occidentalis, C. laevigata, C. australis*.

Neuburgia corynocarpa Leenh. (Logani.) (Solomon Is): macerated bk. appl. to skin dis. (Guadalcanal).

Neuracanthus sphaerostachyus Dalz. (Acanth.) (OW trop): u. ringworm (423).

neurohormones: endocrines produced by nerve tissue; ex. endorphins.

neurohypophysis: pituitary posterior lobe.

Neurolaena lobata R. Br. (Comp.) (CA, WI)): halbert weed; tespuntas (Guat, Hond); hb. u. antipyr. & in gastric dis. (inf. every morning) (Guat, Hond); in colic, diabetes (US, Virgin Islands); fish poison (Pan Indians); pl. juice rubbed on skin to repel ticks; bitter tea fr. lvs. u. for gastritis; fever, mal. (Cuna Indians, Pan); decoc. for intestinal parasites (2444).

Neurospora Shear & Dodge (Neurospor. - Sphaeriales - Pyrenomycetes - Ascomycot. - Fungi.): red bread molds; bakery m.; taxa much u. in biochem., genetic, & physiol. research studies - **N. crassa** Shear & Dodge: orange bread mold; various strains u. in bio. assays of vit. B complex, choline, amino acids, &c. - **N. sitophila** Shear & Dodge: closely related & similarly used.

neutral fats: triglycerides (believed significant in coronary heart dis.).

neutral mixture: potassium citrate soln. - **n. oils**: 1) lightest lubricating oils (fr. Am petroleum) 2) refined coal tar oils - **neutralizing cordial**: alkaline rhubarb mixt.

neverdie (Fr): *Moringa pterygospermum*.

Neviusia alabamensis A. Gray (monotypic) (Ros.) (seUS-Ala): snow wreath; cult. ornam. shrub (as far n as Mass [US]) (1841).

Newbouldia laevis Seemann (1) (Bignoni) (w tropAf): vulner. (topical use); in *Dracunculus* (threadworm) inf., abscesses; bk. u. locally after massage for arthritic pains; rheum. in lung tumors (411).

New Granada: old name for Colombia & Panama (earlier incl. part of Venezuela); this geographic designation is ocassionally u. in modern books!

New Jersey: v. under "tea."

New Zealand cabbage tree: *Cordyline (Dracaena) australis* - **N. Z. flax plant**: *Phormium tenax.*

nezareno (Mex): *Lonchocarpus* spp.

nguba (Af): peanut; appar. origin of "goober" (sUS Negro).

nhandiroba (Br): *Fevillea trilobata* sds.

niabi tree: *Mimusops djave.*

niacin: nicotinic acid - **niacinamide**: nicotinamide.

niam fat: FO fr. sds. of *Lophira alata.*

niaouli: *Melaleuca viridiflora*; source of **n. oil** ("gomenol" in dilution).

Nicandra physalodes (L.) Gaertn. (monotypic) (Solan.) (Pe, now natd. US & throughout trop): apple of Peru; shoo-fly plant (SC lvs. & sts. crushed & pulp placed in milk in shallow pans to kill flies); said fly repellant; u. diur. in kidney & bladder dis.; garden ornam. ann. hb. (fls. blue with white center).

niche (ecological term): location of pl. or animal existence defined by particular environmental conditions.

nickar bean: nicker sd. - **nicker tree** (Tenn [US]): *Gymnocladus dioicus* - **n. seed**: *Caesalpinia bonducella* - **nickers**: nicker seed.

Nicotiana alata Link & Otto (N. persica Lindl.) (Solan) (SA): lvs. smoked & u. as masticatory (SA) - **N. attenuata** Torr. (swUS, Mex): dr. & fermented lvs. of hb. u. in smoking (cigarettes made fr. corn husk [Indians]) - **N. bigelovii** Wats. (Calif [US]): Indian (or wild) tobacco; smoked by Indians like regular tobacco (205); toxic to stock; var. *quadrivalvis* Fast (N. q. Pursh) (2315) - **N. glauca** Graham (SA, orig. Arg, Bo & Mex, natd. Calif, Ore, Tex [US]): tobacco tree; lethal cattle poison; tree t.; Mexican tob.; tronadora; tabaguillo; palán palán (Arg.); occ. as tree living for 10 yrs.+ (Calif); lvs. (tree tobacco) c. anabasine (u. insecticide); appl. as poultice for relief of pain, esp. headache, arthritis; has been mistakenly eaten for wild cabbage; narcotic, toxic; lvs. u. for vulner. & anodyne poultices; anti-hemorrhoidal - **N. latissima** Mill. (US, cult. Eu): Maryland tobacco; u. like *N. tabacum* - **N. rustica** L. (US, Mex, cult. Af): common (or East Indian or Aztec) tobacco; Turkish t.; comm. source of nicotine sulfate 40% (factories in Uganda, Kenya: at least formerly); lvs. u. as fumitory, masticatory - **N. tabacum** L. (SA; widely cult. throughout world): (common) tobacco; true t.; the chiefly u. tobacco sp.; cultigen; c. nicotine, nicotianine, nornicotine, anabasine, &c.; chem. thoroughly studied, 210+ compds. have been reported; lvs. u. as popular fumitory (cigarettes, cigars, [pipe] smoking tobacco, &c.) (1665); masticatory (chewing t.), snuffing material (snuffs), dipping (snuff appl. between lower lip & gums, retained considerable time [Af, sUS]); sd. FO proposed food oil (1665); source of nicotine, &c.; formerly used as lax., sed., antiasthmatic, &c.; anthelmintic (Chin); juice appl. to bee stings; sts. u. to nest poultry (insecticide); poisonous; cigarette smoking is now considered the greatest public health hazard in the US; danger fr. smoke, not nicotine (424) (for serials re Tobacco, v. App. C-1) (344).

nicotinamid(e): niacinamide: $C_5H_4N.CONH_2$; one of vit. B complex factors; occ. widely in nature in both pls. & animals, both free and bound; u. in vitamin deficiency states, esp. pellagra; also in alcoholism; has been u. in schizophrenia; after gastroint. surgery; after chemotherapy; in tinnitus; certain skin dis., &c. (E) - **nicotine**: colorless liquid alk. char. of tobacco (2-8% in t.); u. insecticide (usually as Black Leaf 40™); parasiticide; pharmacol. reagent; strongly addictive & toxic (1664); active constit. of tobacco; causative of addiction to tobacco incl. smoking; n. tablets & gum & transdermal patches now being administered to satisfy cravings or in treating addiction; however, chief harm from smoking is from tar, &c. in smoke; u. to treat ulcerative colitis - **nicotinic acid**: niacin; Vitamin B Complex factor (Vit. B_3) found in liver, milk, animal & pl. tissues; important growth factor; mostly made by synth.; u. to prevent & treat pellagra; lowers serum cholesterol (1842) & triglycerides; to produce vasodilation (425); nutritional additive (ex. in enriched flour); its vasodilator effect is utilized for treating vasomotor head pains, tinnitus, vasc. disturbances of eye & coronary vessels; causes flushing of face, &c. (while niacinamide does not) - **n. a. amide**: nicotinamide - **nicotyrine**: alk. of tobacco; u. as insecticide.

Nic-wood: Nicaragua w.: fr. 1) *Dalbergia hypoleuca* 2) *Caesalpinia crista* L.

Nidorella hirta DC (Comp.) (sAf): u. to fumigate native hut when child is feverish.

Niebuhria DC = *Maerua.*

Niederschlag (Ge): lodging of herbaceous plants; ex. wheat.

Niemeyer's Pills for Dropsy: Pil. Digitalis, Squill & Mercury (NF. V.) - **N. P. for Phthisis**: Pil. Opium, Digitalis & Quinine (NF V).

Nierembergia hippomanica Miers (Solan) (Arg): (dwarf) cup flower; cult. as ornam. border hb. &

pot pl. (*cf. Petunia*); other spp. in trop & subtrop Am.

Niespulver (Ge): sneezing powder; snuff.

Nies(s)wurz, gruene (Ge): *Veratrum viride* - **schwarze N.**: *Helleborus niger.*

Nifext: nitrogen-free extractive.

Nigella L. (Ranuncul.) (Euras): fennel flower; 14 spp. (SC nigra [L] black fr. intensely black sds.) - **N. arvensis** L. (Euras, nAf): (field) fennel flower; sds. ("black cumin") smaller than those of *N. sativa*, for which. u. as adulter. - **N. damascena** L. (sEu, wAs): love-in-a-mist; (damask) fennel flower; small black sds. c. damascenine, damascinine (alks.), VO (with blue fluorescence & odor of pineapple); saponin, FO; u. like those of *N. sativa* but less popular; sometimes u. like poppy sds. on rolls, bread, & other pastries - **N. sativa** L. (Med reg., Balkans, wAs): Nigella (seed); black caraway (seed); b. cumin; (garden) fennel flower; "nutmeg flower" (seed); Semen Nigellae (L) = **Nigella (seed)**: black caraway; b. cumin; (garden) fennel flower; sd. mostly u. in US as flavor, spice; in Eu as diur. for flatulence, worms; emm. abortif. (reput.) after delivery to involute uterus (Ind), galactag.; in nasal catarrh.; esp. popular in veter. practice as diur.; repellant for insects which attack clothing; FO u. as food oil (EI); in fruit essence mfr.; VO with thymoquinone, nigelline (2098).

Niger gum: kind of Acacia - **N. plant: N. seed**: *Guizotia abyssinica* (FO).

nigger head: 1) (US slang): *Rudbeckia hirta* (also **n. nabel**) (navel) 2) (Mass [US] slang): *Plantago lanceolata* - **n. teats** (US slang): *Rudbeckia bicolor* - **n. toe** (US slang): Brazil nut.

night blooming cactus or **cereus**: *Selenicereus grandiflorus.*

nightshade (sUS): *Solanum nigrum* - **American n.**: 1) *Phytolacca americana* 2) *Penthorum sedoides* - **bitter n.: bittersweet n.**: *Solanum dulcamara* - **black n.**: *S. nigrum* - **black-spined n.**: *S. sodomaeum* - **broad-leaved n.**: *Nicandra physaloides* - **common n.**: *S. nigrum*; frequently confused with belladonna - **deadly n.**: 1) Belladonna 2) *S. seaforthianum* - **enchanter's n.**: *Circaea latifolia* & other *C.* spp. - **n. family**: Solanaceae - **garden n.**: bitter n. - **stinking n.**: henbane - **woody n.**: bitter n. - **yellow n.**: *Urechites lutea.*

night soil: human excrement collected at night & u. for manure (Orient).

Nigritella nigra L. (1) (Orchid.) (Eu-alpine reg., esp. Bavaria): Schwarzlein (Ge); fls. purplish black (with red dyestuff); with sweet vanilla-like odor (c. coumarin).

nihamanchi (Ec-Jivaro Indians): alc. beverage made fr. manioc; the root is chewed and the starch is hydrolyzed to sugars by means of human saliva.

nikethamide: deriv. of nicotinamide; u. orally & parenterally as resp. & circ. stim.; barbituric acid antidote.

nim(a): neem.

nimar: namur.

nimble Kate: *Sicyos angulata.*

nine-bark: *Physocarpus (Opulaster)* spp.

ninfea (It.): water-lily, esp. *Nymphaea alba.*

niño rupa grande (Arg): *Aloysia virgata.*

ninsin root: ginseng rt.

niopa: Niopo snuff: *Anadenanthera peregrina.*

Nipa = *Nypa* (**nipa palm**) (v. *Nypa).*

nipplewort: *Lapsana communis.*

Niptus hololeucus Falderm. (Coleoptera): a beetle doing considerable damage to crude drugs.

nisin: antibiotic subst. prod. by growth of *Streptococcus lactis* Loehnis; active against gm.+ bacteria & actinomycetes; preservative for cheese, &c.; antimalarial.

nispera: nispero: 1) (Sp): medlar tree 2) (Mex, Arg): loquat 3) (Co, Ve, Ch, WI): *Manilkara zapota* - **n. cimmarón** (PR): *Symplocos lanata* - **n. de invierno** (Ch): medlar - **n. de montaña** (Sal): *Manilkara zapota* - **n. gum**: one of the high resin wild rubbers; originates in Nic & CR.

Nissolia fruticosa Jacq. (Leg) (sMex to Ve): hierba del tamagas, h. de tamagaz (Salv); hb. u. as antidote for bite of "tamaga" (venomous snake) (Salv); fish poison (2140).

niter: 1) nitre 2) sweet spts. of nitre 3) "sugar sand" (*qv.*) - **n. cake**: $NaHSO_4$ - **nitre**: KNO_3 - **rough n.**: $MgCl_2$ - **nitrobenzene: nitrobenzol**: myrbane: artificial bitter almond oil; u. indy. technical processes; mfr. soaps & shoe polishes; poisonous!

nitrification: conversion of certain N compds. (ex. peptone) to nitrates.

nitrogen: N (elemental state as gas N_2); abundant in nature (78% of earth's atmosphere); indl. prod. by distn. of liquid air in mfr. oxygen; u. as inert gas to fill air space above liquid in packaging (ampuls, vials, bottles, etc.) to exclude air (O_2) (ex. storage of 7-dehydrocholesterol, FO, vitamins); N compds. highly essential to life; stored in metal cylinders - **liquid n.** is used to remove common warts, calluses (cauterizing); as coolant (liquid produces gas); u. as crop fertilizer - **n. mustards**: series of N, Cl org. compds. u. IV to treat chronic leukemia, lymphosarcoma, etc.; potential war gases; ex. chlorethylamine; $N(CH_2CH_2Cl)_3$ - **n. oxides:** 1) nitric oxide (said related to impotency) 2) n. dioxide: n. peroxide: NO_2 (or N_2O_4): brown irritant toxic gas 3) n. monoxide: nitrous oxide:

dinitrogen oxide (pref.): N_2O = laughing gas; dental gas; u. as dental anesthetic (self-administered); delirifacient - **Nitrophoska** (Ge): mixt. of KNO_3 & ammon. phosphate; an almost complete fertilizer.

Noah's ark: *Cypripedium acaule* & other *C.* spp. - **yellow N. a.**: *C. reginae.*

noble: v. under pine; yarrow.

Nocardia asteroides Blanchard (N. farcinica Trevisan; Actinomyces farcinicus Gasperini) (Actinomycetes - Bacteria): organism causes nocardiosis (pulm. dis.); bovine (cattle) farcy (form of glanders); mycetoma - **N. formicae** Harris & Woodruff: forms noformicin, antibiotic wh. inhibits TMV. & swine influenza virus - **N. uniformis** subsp. *tsuyamanensis* Aoki & al.: source of nocardicins, antibiotics (A - G); various *N.* spp. have produced nocardamin (antibiotic fr. bee hives) & **nocardic acids** (included among the mycolic acids).

nocardicins: series of antiobotics (A - G); derived from *Nocardia uniformis* subsp. *tsuyamanensis*, &c. active against gm.- bacteria (*Proteus* spp., &c.); A = most active.

nocciolo (It): kernel.

noce (It): walnut - **n. moscada**: nutmeg.

nocella (It): hazel nut.

nochestli (Sp-LatAm): *Opuntia cochenillefera,* cochineal pl.

nochizo (Sp): hazel-nut tree.

Noformicin: antibiotic prod. by *Nocardia formicae* Harris & Woodruff culture; pyrrole deriv.; u. antiviral in flu (or grippe), mumps; TMV; 1957-.

nogal (Sp): (black) walnut, *Juglans regia* (Mex); *J. mollis.*

noisetier (Fr): hazel (nut) shrub.

noisette (Fr): hazelnut.

noix (Fr-f.): nut; walnut - **n. d'arec**: Areca - **n. de vomique: noix-vomique**: Nux Vomica.

Nolina microcarpa S. Wats. (N. greenei S. Wats.) (Lili/Agav.) (NM, Ariz, Tex [US], Mex): bear grass; squaw flowers; small-seed nolina; soyate; zacate cortador; lvs. u. to make baskets & mats; fiber of comm. value, u. for hanging meats, brooms; sds. ground to flour & made into bread; pl. c. yuccagenin (a sapogenin) - **N. recurvata** Hemsl. (Mex): lvs. u. in making mats, baskets, &c. - **N. texana** S. Wats. (N. greenei S. Wats.) (swUS n to eColo): bear grass; bunch g.; similar to *N. microcarpa*; pioneers (Tex) u. lvs. for thatching, to make baskets, hang bacon; ornam. (borders); sds. ground to flour & made into bread.

nol-kol (Ind): noll-kholl; kohl-rabi.

Nolletia ciliaris Steetz. (Comp.) (sAf): lf. smoked for relief of headache.

Noltea africana L. Endl. (1) (Rhamn) (Af): soap bush; lf. decoc. u. for "quarter-evil" of stock animals; macerated lvs. & twigs u. to wash clothes; toxic to sheep.

no-name tree (Ga [US]): *Halesia carolina.*

non-descript: 1) neither readily nor easily classified 2) of no specific class or kind 3) not previously described.

none-so-pretty: pansy; *Saxifraga exurbium.*

ñongue (morado) (Mex): *Datura stramonium.*

nonpareil: kind of apple.

nopal (Mex): member of *Opuntia* subgen. *Platyopuntia* (345); ex. *Opuntia tuna.*

Nopalea Salm-Dyck = (near) *Opuntia* (cact.) (Mex, CA) - **ñopo** (LatAm Indians): cohoba; 10 spp.

nor-di-hydroguaiaretic acid: NDGA: Masoprocol™; compd. obt. fr. *Larrea tridentata*; guaiac, &c.; u. antioxidant in fats, such as lard, butter, etc.; later banned for int. use in US (346) (supposed kidney damage).

norepinephrine: noradrenaline: symphathomimetic hormone found in animals (incl. man); also occasionally in plants (bananas); serves as neurotransmitter; l-form = levarterenol, u. as symphathomimetic, vasoconstrictor, to raise b.p. in shock.

Norit(e)™: purified charcoal fr. birch-wood; u. in adsorptive decolorization of sugars, etc.

norm: average or typical state.

normal human serum: the sterile s. obt. by pooling the liquid portions of coagulated whole blood fr. 8 or more humans; occurs as liquid or solid (dr.); u. after extensive hemorrhage & to combat infections - **n. saline solution**: physiological salt soln.

normalisation: making a prod. conform to a norm or standard; standardization.

norpseudoephedrine: katine: amphetamine-like compd. found in *Catha edulis*, *Maytenus krukovii* A. C. Smith, *Ephedra* spp., &c.; u. as stim., euphoriant, anorexic, in org. synthesis, &c.

Norway: v. under spruce - **N. codliver oil**: FO obt. fr.. livers of *Gadus pollachius.*

noscapine: alk. of Opium representing 3-10% of its compn.; formerly called narcotine; controls cough in cats but non-narcotic.

"nose-bleed": *Achillea millefolium.*

nose candy (US - slang): cocaine.

nosology: classification of diseases.

Nostoc commune Vaucher (Nostoc. - Cyanochlor. - Algae) (worldwide: on wet rocks, fresh water, etc.): scoom; fallen stars; star-slime; ke-sien-mi (Chin); u. as food (Chin) (cooked, fresh, as soup; with high protein content) (also forma *flagelliforme*) - **N. pruniformis** (L) Ag. (fresh water, cosmop): beurre d'eau (water butter) (Fr); with antisep. properties; specimens kept in water in bottle in good condition for 10 yrs.

nostrum (L) (lit. our own): secret remedy usually of little real value or else promoted unethically or greatly over-priced (ex. Hadacol).

notatin: glucose oxidase: "Penicillin B" (formerly); enzyme found in mycelia of *Penicillium* & *Aspergillus* taxa (first found in *P. notatum* Westling, hence name); catalyzes oxidation of glucose to gluconic acid with reduction of O to produce H_2O_2; useful in food protection, thus to remove O_2 fr. canned foods, beer, &c.

notchwood: *Chenopodium vulvaria.*

Nothaphoebe umbelliflora Bl. (Laur.) (seAs): bk. c. toxic alk., laurotetanine, with similar action to strychnine but less active.

Nothofagus menziesii Oerst. (Fag.) (NZ): silver beech; bk. rich in tannin; weed tough, useful; several of the *N.* spp. (sHemisph) noted for their excellent timber, incl. **N. cunninghamii** Oerst. (Tasmania) ("myrtle tree"; "myrtle beech") - **N. procera** Oerst. (Ch): rauli.

Notholaena eckloniana Kze. (*Cheilanthes*) (Polypodi./Sinopterid. - Filic. - Pteridophy.) (sAf): resurrection fern; frond smoked for relief of head & chest colds - **N. tomentosa** Desv. (Ch): doradilla; u. as diur. & "depurative" - **N. velutina** C. Chr. & Tardieu-Blot (Indochin): heated fronds u. for respir. tract dis. (Indochin).

nothomorph: nm.: variation possibly due to hybridization.

nothospecies: hybrid sp.

Notonia grandiflora DC (Comp.) (eAs): lvs. u. in pimples or pustules; hb. stocked by Chin pharmacies —> *Kleinia*).

novilla (Sp): heifer - **novillo** (Sp): steer; young bull or ox.

Novobiocin: antibiotic prod. by *Streptomyces spheroides* Wallick et al. & *S. niveus* Smith et al.; stable glycosidal compd. u. as antibact. with little cross resistance to other antibiotics; binds to serum albumin; u. mostly for resistant mo & in pl. bacterial dis. (ex. fire blight in pears & apples); u. as salts of Na and Ca (Cathomycin) (1844).

Novo Galiciano: States of Jalisco & adj. Michoacán (Mex).

nox vomik(y) (Ala [US] slang): **nox vomit**: nux vomica.

noyal (ingles) (Sp): (English) walnut.

noyer (Fr-m.): walnut; tree or wd. - **n. blanc: n. gris**: butternut, *Juglans cinerea.*

nozes (Por): walnut.

nubbin (sUS slang): small undeveloped ear of corn.

nubs: pieces of resin not exceeding 2-4 cm in largest dimension (ex. **manila n.**).

Nuces (L. pl. of Nux): nuts; sds. (*cf.* Semina) - **N. Vomicae**: Nux Vomica sds.

nuclear magnetic resonance: NMR: kind of absorption spectroscopy; very valuable in detn. structure of org. compds.; depends on spin of protons in the nuclei of atoms, when magnetic field is applied (spin resonance); also u. in med. diagnosis.

nucleases: generic term for enzymes (in both plants & animals) wh. attack the nucleic acids or their decompn. prods.; incl. polynucleotidases, nucleotridases, nucleosidases (2363).

nucleic acids: polynucleotides; occ. in all living cells & control growth, reproduction, & metabolism; of pivotal importance in molecular biology; isolated 1868/9 by Miescher ("nucleins"); 2 main types (deoxyribonucleic acids [DNA] and ribonucleic acids [RNA]); made up of purine & pyrimidine bases + sugar + phosphoric acid.

nucleosides: glycoside-like complexes made up of either a purine or pyrimidine base combined with a pentose carbohydrate (ribose or desoxyribose).

nucleotides: nucleosides combined with phosphoric acid (as ester); include several coenzymes.

nucleus sheath: endodermis.

nuez (pl. **nueces**) (Sp): (wal)nut - **n. moscada** (Sp). nutmeg.

Nujol™: purified mineral oil.

Nulomoline™: invert sugar.

Number 6: hot drops; Tr. Capsicum & Myrrh Comp. (No. 6 of Samuel Thomson).

nummulaire (Fr): *Lysimachia nummularia.*

nuns: 1) *Houstonia caerulea* (NA) 2) small marine molluscs (Cornwall, Eng)*.

Nuphar japonicum DC (Nymphae.) (Jap, Chin & colder areas): rhiz. u. as ton. beneficial to digest., styptic; sd. inf. stom., for splenic dis.; lvs. u. as tea surrogate (Jap) - **N. luteum** (L) Sm. (Euras, e&cNA): (small) yellow water lily; European (yellow) pond lily; beaver root; brandy-bottle; rhiz. c. alks. thiobinupharidine, desoxynupharidine, nuphleine, & others; u. in many infirmities, as astr., dem. in disturb. of sexual system ("anti-erotic"), also as poultice for sores, swellings, cuts; as food (bread); fls. u. as aphrod. in impotency; spermatorrhea; fls. formerly popular under name of "white sea rose fls." in toothache, &c. - **N. l.** subsp. *variegatum* E. O. Beal (N. advena R. Br.) (eUS): common spatterdock; ornam. in water garden or greenhouse; rhiz. eaten raw or cooked; sds. in soups (Am Indian) - **N. l.** subsp. *polysepalum* E. O. Beal (N. p. Engelm.) (Calif, Ore to Alas [US]): western (or American) yellow pondlily; green p.; sds. starch-rich, parched (fried or roasted) & u. as food often as meal (Klamath Indian); tubers boiled & eaten -

N. pumilum (Timm) DC (N. luteum subsp. p.) (Euras): white waterlily; fls. u. as for *N. luteum*.

nushumbi (Ec-Jivaro Indians): nut of *Manettia coccinea*; juice u. to color teeth black.

Nuss (Ge): nut (pl. **Nuesse**) - **Nussbaum** (Ge): (lit. nut tree); appl. to *Corylus avellana* (European filbert) and *Juglans regia* (English walnut).

nussi (Ec): peanuts.

nut: a frt. char. by being dry indehiscent with 1 sd. (the kernel portion eaten if nut is edible) & with a hard, stony pericarp (or shell) - **Bichy n.**: *Cola acuminata* - **Brazil n.**: *Bertholletia excelsa* - **buffalo nut** (US): *Pyrularia pubera* - **butter n.** (US): *Juglans cinerea* - **n. butter**: coconut oil, FO of *Cocos nucifera* - **crazy n.**: *Pyrularia pubera* - **cream n**: Brazil n. - **earth n.**: peanut - **gooroo n.:** Bichy n. - **Greek n.**: bitter & sweet almonds - **Indian n.**: pine nut (v. under pine) - **ivory n.**: *Phytelephas macrocarpa*, sd. - **Jesuit('s) (water) n.**: water chestnut - **jur n.**: peanut - **kola n.**: *Cola* spp. - **Manila n.**: peanut - **Maranhäo n.**: *Pachira aquatica* (sds.) - **monkey n.** (Eng): peanut - **Neapolitan n.**: *Corylus tubulosa* - **oil n.**: *Juglans*: - **n. oil**: FO fr. various nuts, esp. filbert, walnut or tung - **Pará n.**: Brazil n. - **paradise n.**: *Lecythis allaria* & *L. zabucaje* (frts.) - **physic n.:** *Jatropha curcas* - **poison n.**: Nux Vomica - **purging n.**: physic n. - **n. shells**: hard pl. tissues, such as walnut shells; formerly u. to adulter. spices, &c. - **squirrel n.**: acorns (Bronx, NY [US]) - **stone n., (Brazilian)**: ivory n. - **tiger n.**: *Cyperus esculentus* (tubers) - **n. tree**: orchard tree bearing economically important nuts, such as walnuts.

nutgall(s): excrescences on *Quercus infectoria* (*qv.*; also *v.* galls).

nutlet: small indehiscent 1-seeded frt. of Labiatae.

nutmeg: ripe sd. of *Myristica fragrans*; prod. n. oil - **Banda n.**: the best grade (named after the B. Islands, a group in the Moluccas [Indonesia]) - **Bombay n.**: sd. of *M. malabarica* - **n. butter**: expressed FO fr. *M. fragrans*, mixed with smaller amt. of VO; u. exter. as counterirrit. in ointments & linim. - **Calabash n.**: *Monodora myristica* - **clove n.**: *Ravensara aromatica* - **East Indian n.**: good grade of prod., often limed; shipped through Java, Malaya, Chin - **n. flower (seed)**: *Nigella sativa* - **Grenada n.**: inferior grade fr. WI; incl. much bruised, odd sizes, etc. - **Macassar n.**: Papua n.: sd. of *Myristica argentea*: u. substit. for true n. - **male n.**: *Myristica castaneifolia* A. Gray (Fiji) (male; Fiji n.): *M. fatua* Houtt. (Moluccas) - **Papua n.**: Macassar n. - **round n.**: wild n. - **N. State**: Connecticut [US] (SC wooden n. made & sold here reputedly or because natives ate n.) - **West Indian n.: wild n:** n. - **wooden n.**: supposed mfr. & marketing of n. carved fr. wood (eUS); probably a myth (681).

nutria: coypu: *Myocastor* sp. (Rodentia) (SA): large rodent found in marshlands, now distr. in US (s&w); fur u. in clothing, made to resemble beaver fur; meat considered a delicacy (La).

Nux: Nux, Nucis (L): nut (v. Semen) - **N. Aromatica:** nutmeg - **N. Metella:** Nux Vomica - **N. Moschata: N. Muscata: N. Myristica: N. Nucistae**: nutmeg - **N. Vomica**: sd. of *Strychnos nux-vomica*.

Nuxia congesta R. Br. (Logani.) (tropAf): bogwood; bk. or lvs. chewed or u. as decoc. for indigest.; wd. u. as fuel - **N. floribunda** Benth. (trop eAf): mgandu (Shambaa); twigs with lvs. put under beds to repel mosquitoes; rt. u. to treat influenza; lvs. u. for "nyago" (Taita), a malignant childhood dis.; wd. u. for rural work (347) - **N. verticillata** Comm. ex Lam. (trop eAf): lambinana; lvs. u. to flavor betsabetsa, natl. alc. beverage (fr. sugar cane juice) u. in feasts (Mad).

Nyctanthes arbor-tristis L. (monotypic) (Verben.) (As, Mal, Java): sad tree; tree of sadness; night jasmine; yellow fls. u. in eye dis., as arom., antisep.; dyestuff for silks (shrub a holy pl. cult. near temples, India); lvs. u. as antipyr. in sciatica (Ayurvedic); in intest. dis. incl. amebiasis; in Indian folk med. for sores; ulcers.

Nylandtia Dumort: v. *Mundia.*

Nymania capensis Lindb. (1) (Meli) (sAf): Chin lanterns; u. to treat convulsions (Hottentot; European settlers, sAf).

Nymphaea L. (Nymphae.) (world; trop & subtrop): water lily; grows in shallow fresh water; sds. & tubers u. as food in CA, Austral, sAs, wAf) - **N. alba** L. (c&sEu, nAf): (European) white water lily; w. pond l.; water bells; rt. & rhiz. c. alks., gallic acid tannins, astr. u. in diarrh.; fls. astr. u. in hemorrhage, GC (folk med.); sds. u. med. as antiaphrod.; exts. show antitumor activity; furnishes brown dye - **N. ampla** DC (WI, Mex): ova blanca (Cu); has been u. as ritual narcotic & hallucinogenic; u. to mediate ecstasis in priests (psychodysleptic); thought to have been sacred to ancient Mayas (Mex) - **N. caerulea** Sav. (nAf-Eg): blue lotus; b. water lily; rt. & st. inf. u. as emoll. diur.; fl. decoc. narcotic & anaphrodisiac; fl. u. in cough; sd. in diabetes; rhiz. & sd. u. as foods; substit. for opium (Antilles?) - **N. elegans** Hook. (sTex [US]-Mex; lowlands): señorita (or white) water lily; small black tubers u. as potato substit. (Mex wcoast) (1666) - **N. lotus** L. (Nile reg. Af, now tropAs): Egyptian (or sacred) lotus; important in the culture of ancient Egypt; sacred pl. of Egypt, Ind, Chin; night blooming; rhiz. as tea u. in cystitis, nephritis, metritis, enteritis, fevers, insomnia (hypnotic?); lf. decoc. as intrauterine inj. for hysteria; fl. decoc. narcotic, sed.; juice u. in GC; st. & rt. inf. u. as emoll. diur.; sd. rich in starch, u. to make bread (Chin); famine food (Ind) - **N. mexicana** Zucc. (Mex): yellow waterlily; u. med. (Mex); cult. ornam. - **N. odorata** Ait. (Castalia o. Woodv. & Wood.) (eNA): (sweet-scented or European) white pondlily; fragrant water l.; American w. l.; spatter dock; rhiz. u. in med. like that of *N. alba*; esp. popular with Eu homeopaths - **N. pubescens** Willd. (N. nouchali Burm. f.) (s&eAs): Indian red water lily; fl. cardiotonic; rt. demulc. for piles; in dysp. & dys.; astr. pl. juice u. to treat GC; rubbed on forehead & temples to induce sleep - **N. rubra** Roxb. (Ind): red India waterlily; rhiz. u. dem. in dyspeps. diarrh. hemorrhoids; fl. decoc. for cardiac palpitations; cholagog, in diarrh.; rhiz. edible - **N. stellata** Willd. (eAs, tropAf): blue (Indian) waterlily; blue lotus; rhiz. pd. u. in dyspeps., diarrh., hemorrhoids; fls. u. in cardiac palpit.; sds. rich in starch, u. for food; rhiz. edible, u. in colic; macerated lvs. u. in cooling lotions for fevers.

Nymphoides indica O. Ktze. (Menyanth.) (eAs): u. in fever, jaundice; substit. for Chirata - **N. peltata** O. Ktze. (Eu, seAs): floating heart; pl. diur. antipyr., ton.; appl. exter. to furnucles, bruises; fresh lvs. for headache; ornam. floating hb. (aquaria) (natd. eUS).

Nypa (Nipa) fruticans Wurmb. (1) (Palm.) (sAs, Ind to Austral): nipa palm; fresh lf. decoc. u. to treat indolent ulcers; remedy in centipede bite; sap of inflorescence c. 15% sucrose; u. sugar source, mfr. alc. beverage "tuba" (or Philippine toddy).

Nyssa aquatica L. (N. uniflora Wang.) (Nyss.) (seUS): tupelo (gum); cotton gum; water tupelo; rt. u. med. esp. by homeop. of Eu; wd. u. for frt. & veget. boxes, &c.; rt. wd. (tupelo wd.) extraordinarily soft u. for net floats, formerly like laminaria pegs in surgery & gynecology for opening of wound canals, &c.; frt. u. in preserves - **N. ogeche** Bartram ex Marshall (seUS, SCar-Fla): Ogeechee plum (or tree); O. lime; tupelo gum; one source of tupelo honey; frt. flavorsome, mildly acid hence u. to make acid beverages & preserves - **N. sylvatica** Marshall (N. multiflora Wang; N. caroliniana Poir.) (e&cUS): black gum; (b.) tupelo; pepperidge; hornbeam; sour gum (tree); fls. source of tupelo honey (best pharm. honey as it does not crystallize out); frt. acidulous, edible (in preserves); popular food of black bear; rt. wd. u. for tents; to swell body cavity (compressed into smaller volume & u. for swelling sticks in abort., &c.) (*cf.* Laminaria pegs); twigs u. as "chewsticks," toothbrushes, snuff dipsticks; wd. u. in many ways, ex. crates - **N. s.** var. *biflora* Sarg. (N. biflora Walt.) (seUS: NJ, Fla.-Tex.): black (or sour) gum; pepperidge; swamp tupelo: bear gum; blue drupes popular food for bears (348); wd. u. as for *N. aquatica*; bk. u. in diarrh. as anthelm. in TB (Creek Indians); cancer (Ala Indians) - **N. uniflora** Wang = *N. aquatica* - **N. ursina** Small (wFla [US]): bear gum; very bitter drupes popular food of bears.

Nystatin: Mycostatin (Squibb): polyene antibiotic prod. by *Streptomyces noursei* Brown, Hazen & Mason, *S. aureus* Waksman & Henrici (from soils in Va); u. as antifungal; athlete's foot; vaginal & ear Candida infections, &c. (isolated by Hazen & Brown, 1950) (name from N[ew] Y[ork] Stat[e] In]stitute] Dept. of Health) (where isolated, 1950) (Fungicidin); u. to supress superinfections (esp. secondary inf. of intest. following tetracycline use) (moniliasis); topical antimycotic (cutaneous candidiasis, ringworm) (the first safe and effective antifungal marketed); to promote growth (lower animals) (E).

Oo

O (CA Maya dial.): avocado.

oak: member of genus *Quercus* (Fag.) - **o. agaric**: v. under agaric - **Australian o.**: *Eucalyptus regnans* - **o. balls**: nutgalls - **Bartram o.**: *Q. heterophylla* - **basket o.: cow o.** - **bastard o.** (Tex [US]): *Quercus glaucoides* - **black o.**: *Q. velutina* (bk.) - **black jack o.**: *Q. marilandica*, *Q. laevis* - **blue jack o.**: *Q. incana* - **bur(r) o.**: *Q. macrocarpa* - **California o: ***Q. lobata* - **C. tanbark o.**: *Lithocarpus densiflora* - **Ceylon o.**: *Schleichera oleosa* - **chestnut o.**: *Q. prinus* (SC): 1) lvs. resemble those of *Castanea*; or 2) lvs. in clusters of 5 commonly, resembling palmate leaflets of horse chestnut - **chestnut white o.**: white c. o.: c. o - **chinquapin o.**: *Quercus prinus* - **cow o.**: *Q. michauxii* - **o. currants**: galls fr. *Q. robur* - **dollar leaf o.**: *Q. marilandica* - **dwarf live o.**: *Q. virginiana* - **dyer's o.:** *Q. velutina; Q. infectoria* - **English o.**: *Q. robur* - **evergreen o.**: *Q. engelmannii*, *Q. ilex* - **o. galls**: v. under galls - **g. oak**: *Q. infectoria* - **high ground willow o.**: blue jack o. - **holm o.**: evergreen o. - **Jack o.**: black j. o. - **Jerusalem o.**: **o. of Jerusalem:** *Chenopodium* sp. - **kermes o.:** *Q. coccifera*; *Eucalyptus coccifera* Hook. f.- **laurel o.**: *Q. laurifolia* - **o. leaves** (med.): fr. *Q. virginiana* - **leopard o.**: *Q. rubra* (SC bk. mottled) - **live o.**: *Q. virginiana* (**dwarf l. o.**) & all evergreen oaks of NA - **lungs of o.:** *Lobaria pulmonaria* - **o. mistletoe**: *Phoradendron coryae* - **o. moss**: lichens, usually *Evernia prunastri* & *E. furfuracea*, growing on oaks (& other trees); source of a resinous subst. u. in perfume mfr. as fixative (chief source: Sp, Fr, Yugoslav) - **mountain o.:** *Q. prinus* - **northern red o.**: leopard o. - **o. of Jersalem**: *Chenopodium ambrosioides; C. botrys* - **overcup o.**: *Q. lyrata*; *Q. macrocarpa* - **pin o.**: *Q. durandii*; *Q. palustris* - **poison o. (leaves)**: *Toxicodendron diversilobum*+ - **post o.**: *Q. stellata*; *Q. wislizenusii* - **quercitron o.**: black o. - **red o.**: *Q. rubra* (bk.) (**northern r. o.**) - **sandhill post o.**: *Q. stellata* - **sawtooth o.**: *Q. muhlenbergii* - **scarlet o.**: *Q. coccinea* - **scrub o.** (wFla [US]): *Q. laevis* (usually); *Q. incana*; *Q. myrtifolia*; *Q. chapmanii* (u. at times for paper pulp) - **scrub live o.**: *Q. virginiana* - **sea o.:** spp. of *Fucus* & *Halidrys* - **she-o.:** *Casuarina* spp. - **shin o.**: *Q. annulata* - **silk(y) o.**: *Grevillea robusta* - **southern red o.**: *Q. falcata* - **o. spangles**: galls on lvs. of *Q. robur* - **Spanish o.**: scarlet o. - (**swamp Spanish o.**: *Q. palustris*) - (**swamp post o.**: *Q. lyrata*) - **stone o.**: 1) *Quercus javensis* 2) *Q. alba* - **swamp chestnut o.**: basket o. - **swamp o.:** *Casuarina obesa* - **sweet acorn o.**: *Q. ilex* v. *rotundifolia* - **sweet o.**: *Myrica* - **tan (bark) o.**: *Lithocarpus densiflorus* - **tanners' o.**: *Q. pseudocerris* - **Turkey o.**: blue jack o. - **valley o.:** *Q. lobata* - **valonia o.**: v. under valonia - **water o.**: *Q. nigra* - **white o.**: *Q. alba* (bk.); *Q. durandii; Q. lobata* (Calif [US]) - **willow o.**: *Q. laurifolia*; *Q. phellos* - **yellow (chestnut) o.**: chinquapin o. 3) *Chasmanthum latifolium* (eUS): all are u. for forage.

oakum: hemp or jute or Sunn fiber impregnated with a tar; u. to caulk or pack joints in timbers, as between boards of a boat's hull; types: marine (better grade); plumbers'.

oast: (Eng): kiln for drying (& storing?) hops, tobacco, &c (o. house, Kent).

oat(s): *Avena sativa* - **black o.**: (Calif [US]): *A. s.* var. *nigrescens* (cult. It) - **common o.: cultivated o.**:

A. s. - **o. groats**: oat kernels hulled (deprived of husks), either whole or coarsely ground - **o. meal**: like rolled oats, but may be either rolled or ground - **quick o.**: rolled o. which has previously been partly cooked (pre-cooked cereal); each groat is cut into several pieces before rolling; product is for quick cooking - **rolled o.**: oat kernels processed by rolling after removal of chaff (flattening with heated rollers) and steaming - **sea o.**: *Uniola paniculata* & other *U.* spp. - **o. straw**: Stramen Avenae (L): dr. stems of threshed grain; u. in herbal mixts. - **o. thistle**: *Onopordum acanthium* (vernac. name no longer u.) - **water o.**: wild rice, *Zizania* spp. - **wild o.**: 1) *Bromus ramosus*; *B. secalinus* 2) *Avena sativa* subsp. *fatua*.

obakhta (Pak): *Juniperus macropoda.*

obeah (Jam Negro): magic involving employment of such articles as camphor, myrrh, asafetida, etc.

Oberonia anceps Lindl. (Orchid.) (Mal): whole crushed pl. u. as poultice for boils - **O. longibracteata** Lindl. (Indochin): lvs. appl. to scorpion stings.

Obetia pinnatifida Bak. (Urtic.) (Tanz): lf juice u. for toothache (mouth rinse); lf. rat repellant (foot irritation).

obier (Fr): Guelder rose, *Viburnum opulus* v. *roseum*

objective: the mounting (which bears one lens or more) lying nearest in the optical system of a microscope to the object under examination; in this light rays proceeding fr. the object are brought to a focus.

oblet (veter. med.): large tablet.

obligatory: in ref. to organisms, required to live in a definite pattern.

Obolaria virginica L. (Gentian) (e&cUS): pennywort; rt. decoc. mixed with decoc. of rt. of sweetgum; mixt. appl. to cuts & wounds (349) (Choctaws; La [US]).

Obst (Ge): large edible frt. as distinct. fr. Frucht, the term for any fructification, principally appl. to grains.

obturator: 1) caruncle (*Ricinus* sd.); protuberance near sd. hilum 2) small bodies found in pollen masses wh. close the cavity of the anthers.

oca (SA): *Oxalis tuberosa.*

occurrence: as used in pharmacognosy, the geographic range, areas where most common, and environment (habitat) where an organism is found.

Oceania: large region of the western Pac Oc n of Equator wh. is arbitrarily divided into Melanesia, Micronesia, and Polynesia; sometimes, Austral, NZ & Malay are included.

Ochna arborea Burch (Ochn.) (sAf): (African) redwood; Cape plane; timber strong & heavy; u. in making handles, good fencing poles, &c. - **O. atropurpurea** DC (sAf): red pear; rt. decoc. u. for gangrenous rectitis in children - **O. integerrima** Merr. (seAs): bk. very bitter (cf. quassia); u. bitter ton.; frts. edible - **O. mossambicensis** Klotzsch (seAf): rt. decoc. (in porridge) u. as masculinizing agt.; frts. sometimes eaten; frt. FO u. in cosmetics - **O. pulchra** Hook. f. (sAf): morelle; FO fr. frt. pericarp & sd. u. to grease head (Bushmen); FO edible but with unpleasant odor of valeric acid; stock poison - **O. pumila** Buch.-Ham. (Ind): bhuin champa; rt. u. in menstr. dis., TB, asthma (124).

ochra: ocher: ochre: an earthy or clayey ore consisting chiefly of iron oxide; u. as yellow, red, or brown pigments (mined in Washington Co., Ala+ [US]) - **red o.**: hematite u. in paints, linoleum, &c. (1712).

ochratoxins: toxic metabolites of various fungi: *Aspergillus ochraceus, A. mellus, A. sulphureus*; *Penicillium viridicatum*; prod. injurious to liver, kidney, & intest.; occ. in moldy corn, &c.; 1st found in fermented frt. prod. (Japan, 1911).

Ochrocarpus A. Juss. = *Mammea.*

Ochrolechia pallescens (L.) Koerb. (Lecanora p. Roehl.) (Lecanor. - Lichenes) (cosmop.): pale lecanora; u. for red, purple, orange dyes - **O. parella** (L) Mass. (Lecanora p. L.) (Euras): crab's eye lichen; Lichen parellus; light crottle; parelle d'Auvergne (Fr) (oireille d'Auvergne); a source of cudbear red, purple, & orange dyes for wool (GB, Fr) - **O. tartarea** (L) Mass. (Lecanora t. [L.] Ach.) (nEuras, US, on trees & rocks): crab's eye lichen; stone crottle l.; a source of litmus, cudbear, & red & purple dyes; other *O.* spp. are u. in mfr. litmus.

Ochroma lagopus Sw. (O. pyramidale (Cav.) Urban; O. limonensis Rowles) (monotypic) (Bombac.) (sMex to Bo, Br; WI): cotton (or down) tree; balsa; balso (blanco) (Co); balsa wd. with extremely low density, (s. g. 0.11; 7-8 lbs./cu. ft.), lightest comm. wd. known, lighter than cork (hence [SA] cork wood); rt. bk. inf. u. as diur. aperitive; lf. & fl. decoc. emoll.; st. bk. diaph., sl. emet., formerly in syph.; fl. decoc, with honey u. to relieve colds, grippe, coughs, &c.; balsa wd. u. in mfr. life rafts, boats (Lake Titicaca); bolsters, airplane parts (formerly), military bridges, architectural models, in movie furniture (broken over head, &c.) (*Cavanillesia* = substit.); for insulation; pod hair ("cotton") u. in filling mattresses, pillows (*cf.* kapok), &c.; wd. impregnated with perfume carried in ladies purses; bk. fibers u. to make ropes; cult. as ornam. tree in Fla [US], Cu.

Ochrosia Juss. (Apocyn.) (Oceania, sAs, Austral): 23 spp., with yellow wd.; all parts rich in indole alk., esp. ellipticine - **O. elliptica** Labill. (Austral, New Caled; cult. Indones, Haw [US], PR, &c.): bk. inf. u. as antipyr. ("quinquina du pays"), diur. purg.; latex appl. to contusions & cuts; c. ellipticine, u. in leukemia - **O. glomerata** F. Muell. (PI): bk. u. to reduce swellings - **O. maculata** Jacq. (O. borbonica Gmel.) (Madag, Java, Oceania): bk. (mongumo bk.) c. methoxyellipticine, reserpine, reserpiline, &c.; u. antipyr. in anemia, bitter ton. stom. - **O. oppositifolia** Schum. (Neisosperma o.) (Indones): rt. as bitter, stom. carm., antidote to eating poisonous fish & crabs; sds. edible & medl. - **O. sandwicensis** A DC (Haw Is [US]): holei tree; rt. c. alk. with tranquilizing effect.

Ocimum basilicum L. (Lab) (sAs [Ind], Pac Is, tropAf; cult. cEu [Balkans, Fr], Ind, cMex, Sudan, &c.): (sweet) basil; Herba Basilici; (grand) basilic; lvs. & fls. c. VO (with linalool, methyl chavicol) (2099); compn. vary variable, tannin; u. flavor (*cf.* cloves), arom. stom. carm., ton. for nausea, antipyr.; alimentary (culinary), condiment hb. in pasta, pizza, &c. (often with tomatoes); sed. in catarrh., diur. in kidney dis., galactagog, in bronchitis, rheum; sd. mucilaginous, dem. in dyspeps. cough, GU dis. (GC); source of VO (basil oil); in mfr. Chartreuse; perfumery; stalks in house as insect repellant (esp. mosquitoes); weed in Gezira (Sudan) (prod. Fr, Ge, Hung) - **O. canum** Sims (O. americanum L. pp.) (As, esp. Ind, Austral, tropAf, NA, SA): hoary basil; fever leaf; "(pepper)mint" (Trin); hb. u. somewhat like *O. sanctum* (Ind); inf. of hb. in cocoa butter u. skin dis.; in bilharziasis & snakebite (Tanz); insectifuge (wAf): diur. diaph. (Br); smoke of burning pl u. for sore eyes, mosquito repellant (Tanz); hb. inf. in diarrh. indigest. (Trin); lf. juice u. in hair oil (Tanz); source of VO with d-camphor; source of latter (2100) - **O. caryophyllatum** Roxb. (As+): small basil; bush green b.; hb. u. arom. stim. sed. corrective (may be a form of *O. basilicum*) - **O. fluminense** Vell. (Br): u. as med. & spice - **O. gratissimum** L. (tropAs, Af [Madag])+): lf. c. VO with thymol, eugenol, ocymene; u. ton. stom. expect. diaph. lax. antispasm. antipyr. anthelm.; in stomachache, yaws (Guat) - **O. incanescens** Sims (Br): alfavaca de cheiro; hb. u. stim., carm. diur. diaph.; baths in rheum.; decoc. in genitourin. dis.; "remedio de vaquero" (cowboy medicine) - **O. kilimandscharicum** Guerke (e&sAf; cult. US, Br, Ind): camphor basil; VO with 50-70% camphor (d-camphor) (352); u. as source of VO & camphor; rt. decoc. u. in stomachache; lvs. chewed for colds (Tanz) - **O. micranthum** Willd. (sFla [US], WI, Bah, Mex, tropAm, esp. Br): albahacca (cimarrona); a. blanca; duppy basil; great marjoram; lf. decoc. stom. cold remedy, for fevers, in baths & poultices; pl. decoc. for stomachache & dys. (Yuc); applied to ears in otalgia (Salv); condiment; lvs. rubbed on horses to prevent bite of horseflies - **O. minimum** L. (tropAm, &c.): bush (or least) basil; hb. c. VO with eugenol; u. flavor (may be small form of *O. basilicum*); weed - **O. sanctum** L. (Euras, esp. cooler parts of Ind): holy (or monk's) basil; tulasi; lvs. sds. & rts. u. antimal. expect. dem. antipyr. for cough, pain in the side, oxytoc. (in Ol. Sesam.); sds. u. to make rosaries (Ind); said antidiabetic; source of VO (SC sacred to Hindus) (2557) - **O. suave** Willd. (cAf): VO c. eugenol, methyl-e., methyl iso-e., ocymene, cubene, bergamoptene, &c.; because of the very different nature of this oil, it is held that this is a valid sp. and not synonymous with *O. gratissimum*, as usually held (2359); u. stomachache, & for catarrhs.; lvs. u. for sharp pains in stomach & to treat swollen gums; lf. paste rubbed on sores (Tanz) - **O. viride** Willd. (trop wAf; cult. Ind): fever plant; f. basil; lvs. rich in VO with thymol, terpinene; hb. u. antipyr. diaph. stom. lax. prob. not insectifuge as claimed; rt. u. snakebite.

ocote (Mex): *Pinus teorote.*

Ocotea Aubl. (Laur.) (trop&subtrop): large genus (ca. 200 spp.) of arom. shrubs & trees - **O. bullata** E. Mey. (sAf): (black) stinkwood; African oak; pd. bk. sniffed for headache; in urin. dis.; bk. u. tan; wd. useful, dark brown, u. for gunstocks, doors, &c. - **O. caparrapi** Nates (Br, Guianas): VO (caparrapi oil or balsam) u. in bronchitis, laryngitis, for insect stings, skin sores - **O. caudata** (Nees) Mez (Guianas; Co): (Cayenne) linaloe wood; keretiballi (Arowaks, Surinam); Brazilian bois de rose; source of VO (oil of bois de rose, Brazilian; or rosewood oil); u. for floral type perfumes (as lily of the valley); wd. u. to adulter. sassafras wood - **O. costulata** Mez (Guianas, Br): "camphor laurel"; pao rosa (Br); wd. VO u. perfumery & to adulterate sassafras wd. - **O. moschata** Mez (WI, tropAm): nuez moscada (PR): frts. u. med.; wd. u. as fine lumber, termite-resistant - **O. opifera** Mart. (Co): dr. crushed frt. powd. & mixed with coca (Kubeos) (2299) - **O. pretiosa** Benth. & Hook. (Br): false cinnamon; "coto bark"; canela-sassafras; source of sassafras nuts; bk. (as "Priprioca" or "Preciosa") u. aperitive, stom. gout, edema, antirheum.; similar to *Dicypellium caryophyllatum*; wd. bk. & lvs. by steam distn. yield Brazilian sassafras oil, VO c. safrole, methyl eugenol; u. in prepn. heliotropin (*qv.*) - **O. pseudo-coto** Rusby =

Aniba p. = *Nectandra p.* - **O. rodiaei** Mez = *Nectandra r.* - **O. sassafras** (Meissn.) Mez (Mespilodaphne s. Meissn.) (Br): sassafras do Brasil; bk. c. VO with safrole; source of Brazilian sassafras wood; VO u. in arthritis, rheum. syph. (Br); in homeopath. med.; however not recommended for med. or food use, only in technical fields (US) - **O. usambarensis** Engl. (Tanz): East African camphor wood; decoc. of decorticated bk. u. in stomachache & gastric dis.; c. VO with myristaldehyde, cineole, terpineol; wd. comm. timber; produces dermatitis in some workers - **O. venenosa** Kosterm. & Pinkley (Co, Ec): frt. u. to make arrow poisons; c. rodiasine, demethyl-r. (1715).

ocotilla: corruption of **ocotillo** (Mex): *Fouquieria splendens*.

ocoxote (Mex): **ocozote** (Mex): *Liquidambar styraciflua;* sweet gum (tree).

octacosanol: cluytyl alc.: octacosyl alc.; $C_{28}H_{57}OH$; higher aliphatic alc.; occ. in waxes of apple peel, cotton, &c. & in FO (ex. olive); claimed to improve memory (*sic*).

octadecanol: stearyl alc.; n-octadecyl alc.; occ. in lard oil; u. as anti-foaming agt. in antibiotic mfr.

October flower: *Polygonella polygama*.

octoic acid: caprylic a.

Octopus L. (Mollusca - class Cephalopoda) (worldwide on sea bottoms): octopus (lit. 8-footed); devilfish; head, arms, body recognizable; size variable (3 cm-10 m diam.); arms popular as food (esp. Med reg.); another gen. *Eledonella* (small, ocean depths) secretes eledoisin (*qv.*).

ocular micrometer: scale engraved upon a small circular piece of glass for insertion into a microscope ocular for use in making measurements of microscopic objects; ordinarily divided into markings 100 mikra apart; the microscope with which it is u. should be calibrated by means of slide micrometer.

Oculi Cancrorum: Lapides C. (*qv.*).

Oculus Christi (L.): clary sage.

ocumo: *Xanthosoma sagittifolia* (*qv.*) - **o. culin** (Mex): *Colocasia esculenta* var. *antiquorum*.

Ocymum Wern.: *Ocimum*.

Odermennig (, kleiner) (Ge): *Agrimonia eupatoria*.

Oderminze (Ge): peppermint.

Odina Roxb. = *Lannea*

Odontocarya tripetala Diels (Menisperm.) (Pan to Pe): bk. u. in combs. for intest. parasites (Co).

odontology: science of teeth.

odorobioside G: glycoside of digitoxigenin + D-digitalose + glucose; found in *Nerium odorum*.

odoroside A,D,G,H: cardiotonic glycosides of *Nerium odorum*, contg. resp. various sugars.

Odostemon Rafin.: *Mahonia*.

odum (Jam-Negro): silk cotton tree (Ashanti).

Odyendea (Pierre) Engl. = *Quassia*.

oe-: look also under "e."

Oel (**Öl**) (Ge): oil - **Oelsuess** (Ge): glycerin - **Oelkurbis**: *Cucurbita pepo*.

oeillet (Fr): pink; formerly poppy.

Oemleria cerasiformis Landon (Osmaronia c. (T&G) Greene) (monotypic) (Ros.) (Pac NA): Indian peach (or plum); oso berry; bird cherry: skunk bush: close to *Prunus*; frts. eaten (fresh or dr. for winter use) by BC [Cana] Indians (flavor not too pleasant).

Oenanthe aquatica Poir. (O. phellandrium Lam.) (Umb.) (c&sEu; n&cAs): water fennel; (5-leaved) w. hemlock; (fennel-leaf or water) dropwort; frt. (Fructus Phellandri) c. VO, resin; u. as diaph. carm. diur. expect. in bronchitis, asthma; in veterin. med. (flu in horses); toxic; prepn. VO (yield 1.5-2.5%); prodn. in Balkans - **O. crocata** L. (swEu, GB, Ire; Alger, Mor-swamps): water hemlock; (yellow) (or hemlock) water dropwort; w. lovage; dead tongue; toxic plant; fresh rhiz. & rts. (c. oenanthotoxin, oenanthetol, oenanthetone) u. by Homeopaths (Eu) in epilepsy (also hb.), delirium, apoplexy, &c. (426) - **O. fistulosa** L. (Euras-Casp reg.): hemlock dropwort; pl. juice mixed with As compd. u. perorally in leprosy (1832) - **O. javanica** DC (O. stolonifera Wall.) (eAs to Austral): bamboong; consumed raw with rice; cult.; bruised pl. u. as dressing for abscesses, "horse bites," cancerous swellings (Chin; int. for fevers, colds; frts. carm. diur. - **O. sarmentosa** Presl (Pacif NA): water parsley; sweet tubers (rich in starch) boiled and eaten by nwAm Indians.

oenanthic ether: artificial cognac oil; ethyl oenanthate; wine oil (*qv.*).

oenin: enin: cyclamin(chloride): dye subst. (glucoside of malvidin Cl) fr. the skins of red grapes.

Oenocarpus bacaba Mart. (Palm.) (Br, Pe, Guianas, Co): bacaba verdadeira; frts. source of FO like olive oil; also beverage fr. frt. - **O. bataua** (Mart.) Burret = *Jessenia b.* - **O. distichus** Mart. (Br): bacaba de azeite; b. palm.; frts. u. to prep. beverages called bacaba branca & b. vermelha; u. alone or with sugar or manioc; frts. source of greenish FO u. in cooking - **O. mapora** Karsten (CR, Co, Ve, Ec, Br): seje pequeño; small palm furnishing ripe frts. u. to make milklike beverage (high protein content) & FO; sts. u. to make gates.

oenolé (Fr): medicated wine.

Oenothera (Onagr.) (tempAm): evening primrose; many mutations among members of gen.; fls. open & exude scents in evening, hence name; sds. FO c. gamma-oleic acid; important as source of various fatty acids & prostaglandin - **O. acaulis** (O. laciniata var. *pubecens*) (Ec): oleoresin u. in plasters; colsilla; u. in cases of ulcers & contusions - **O. affinis** Cambess. (Chile): u. as vulner. in wounds, contusions, &c.; for enteritis, stomachache, & "female diseases" - **O. biennis** L. (NA, now cosmop. - natd. n&cEu, As): common (or large) evening primrose; scabish; hb. u. in diarrh. skin dis.; rts. boiled to furnish nutritious & wholesome food; rt. inf./decoc. u. as antispasm., anticonvuls.; young shoots eaten in salads; FO fr. sds. (15-20%) high in γ–linoleic (58%) & linolenic (8%) acids (as glycerides) (similar to poppy sd. oil) u. as food oil & in dermatologic prepns.; in multiple sclerosis (Eu, Cana); oil sold in health food stores (capsules) for PMS, atopic eczema, &c. (2325); ornam. hb. - **O. caespitosa** Nutt. (Wash, Ore, Nev, Colo, Utah [US] n to Sask [Cana]): morning lily; several vars.; potential value in cancer treatment; c: gallic acid (chief cytotoxic agt.) - **O. kunthiana** (Spach) Munz (sTex [US], Mex): Kunth snowdrops; u. as poultice for bruises (Mex) - **O. laciniata** Hill (e&cUS, Mex, SA): cut-leaf evening primrose; amapola (Mex); grows to 1 m; u. med. (Mex); lvs. boiled & salted & eaten as greens or added to atole - **O. odorata** Jacq. (Arg, Chile): cult. as ornam. hb. - **O. rosea** Ait. (sTex [US], Mex, SA): rose sundrops; yerba del golpe; decoc. of whole plant u. for upset stomach; to "refresh" the intestines (Ec) - **O. speciosa** Nutt. (Hartmannia s. Small) (c&seUS, Mex): wild primrose; wind flower; "roadside poppy" (sAla [US]); evening primrose; ornam. hb. in flower garden & common escape; often seen in spring along highways: sd. FO rich in vit. A. - **O. stricta** Ledeb. ex Link (Chile): flor de San Jose; u. similarly to *O. affinis* - **O. tetraptera** Cav. (wUS [Tex+], Mex, Ve, Co, Ec): linda terre; lvs. ground & placed on wounds.

Oesipus (L): suint; impure wool fat; crude lanolin.

Oestrone = Estrone.

Oestrus: gen. of bot flies, wh. attack humans & sheep.

oeuf (frais) (Fr-m.): (fresh) egg.

offal: all waste (inedible parts) of an animal body, generally fr. dressing of fish; incl. heads, fins, tails, visceral organs except liver; carrion; refuse.

official: a subst. or prod. described in monographs or important textual subdivisions of the official compendia; in the US, the US Pharmacopeia, National Formulary, and Homoeopathic Phar. (in common practice, often understood as referring in the U.S. only to the first two of these) - **o. de sala** (Br): *Asclepias curassavica*.

officinal: article or prod. commonly carried as merchandise in the pharmacy or drug shop (officina, L.) but not represented by a monograph in the official compendia (in US: USP, NF, HP; last compendium commonly excepted); ex. Manna.; Valerian.

ogea gum: fr. *Daniella ogea*.

oignon (Fr): onion.

oil: Oleum (L.) (see also under specific name): a water-immiscible, usually liquid material fr. various sources; two main classes are the volatile (or essential) oils (which can be evaporated completely in a relatively short time) & fixed (or fatty) oils (which do not evaporate spontaneously); when unqualified, "oil" generally means a fatty oil (FO) - **ajowan o.**: a. seed o.: fr. *Carum copticum* - **almond o.**: sweet a. o. - **American o.**: 1) castor oil (Scandinavia, etc.) 2) rel. light mineral oils (in distinction fr. the heavy Russian oils) - **animal o.**: empyreumatic VO obt. by dist. of animal tissues such as bones, horns, etc., very similar to bone o.; c. pyridine bases, amines, etc.; u. alc. denaturant - **anise o.**: most obtained fr. star a. - **apple o.**: isoamyl isovalerate - **apricot kernel o.**: FO expr. fr. kernels of *Prunus armeniaca* vars. - **arbor vitae o.**: cedar leaf o.; VO fr. *Thuja occidentalis* (**American c. l. o.**: distd. fr. lvs. of *Juniperus virginiana*) - **arachidic o.**: peanut o. - **astral o.**: kerosene - **Australian sandalwood o.**: fr. wd. of *Fusanus spicatus* - **banana o.**: mixt. of ethyl butyrate & amyl acetate - **bay o.**: 1) (generally understood) fr. *Pimenta racemosa* (o. of bay rum) 2) laurel berries o. 3) *Myrica* spp. VO - **bead o.**: patented liquid for "beading" liquors - **behen o.**: **ben o.**: fr. *Moringa* sp. & not the same as oil of **benne** (fr. *Sesamum indicum*) - **betula o.**: sweet birch o. - **birch tar o.**: empyreumatic VO made by dry dist. of wd. of *Betula pendula* - **bitter almond(s) o., artificial**: benzaldehyde - **blue o.**: azulene - **boiled o.**: linseed oil wh. has been treated (heating with litharge, &c.) to shorten the drying period; is darker & thicker than refined (or raw) l. o.; u. in paints, varnishes, printing inks; toxic! - **bone o.**: 1) empyreumatic VO obt. by destr. distn. of bones, still with some FO; u. alc. denaturant; with pyridine, amines, hydrocarbons, pyrrole, quinoline, &c.; very bad smelling 2) liquid portion of bone fat; u. lubricant, neat's foot oil substit. u. in leather mfr. - **British o.**: composition of crude petroleum 35, mineral tar 105, crude amber oil 140, juniper oil 140, linseed oil 280, turpentine oil to 1000; u. rubefac. in linim. - **brominated vegetable oils**: formerly u. (US) to produce cloudiness or turbidity in soft drinks - **burbot liver o.**: FO fr. liver of *Lota maculosa* - **buchu o.**:

VO fr. b. lvs., a comm. article - **cabbage o.**: elder oil - **cade o.**: juniper tar - **camphor o.**: v. under camphor - **cassia o.**: cinnamon o. - **c. o., artificial**: prob. camphor o. or a neutral VO perfumed with c. o. or cinnamaldehyde - **o. of castor**: Castoreum - **cedar leaf o.**: VO dist. fr. twigs of *Thuja occidentalis* (American c. l. o.: distd. fr. lvs. of *Juniperus virginiana* - **checkerberry o.**: wintergreen o. - **cherry kernel o.**: FO fr. sd. kernel of *Prunus cerasus*, &c. - **China (wood) o.**: tung o. - **cinnamon o., artificial**: cinnamaldehyde - **cloves o., artificial**: eugenol - **coal o.**: 1) crude petroleum 2) kerosene - **coconut o.**: FO obt. fr. sd. kernels of *Cocos nucifera* - **cod o.**: FO of body of codfish, *Gadus* spp. - **c. liver o.**: FO fr. *Gadus morrhua* liver - **cognac o.**: prep. fr. press cake fr. expression of yeast left in wine vats by steam distn. & sepn. of uppermost layer of distillate - **colza o.**: rapeseed o.; FO fr. sds. of *Brassica rapa*, *B. napus*, *B. campestris* v. *oleifera*, & other *B.* spp.; u. for cooking, lubricant - **cooking o.**: cottonseed o. - **copra o.**: coconut o. - **coriander o.**: VO fr. *Coriandrum sativum* - **crab o.**: **crabwood o.**: FO fr. *Carapa guianensis* - **creosote o.**: fractions of coal tar distilling at ca. 200-270° C - **crude o.**: petroleum - **dagget o.**: empyreumatic birch oil (o. of Russia leather) - **dead o.**: 1) that fraction of shale o. distilling at high temperatures & in which most of paraffin is xallized out 2) creosote o. - **deciduous o.**: VO fr. pls. with deciduous lvs., thus excluding oils fr. pls. with evergreen lvs. (conifers, citrus, &c.); ex. lavender - **Dippel's (animal) o.**: rectified animal (or bone) o. - **drying o.**: unsatd. FO which forms hard films when spread in thin layers - **o. of eggs**: FO obt. by expr. of hard boiled egg yolks - **elaeococca o.**: tung o. - **elder o.**: infused oil of *Sambucus* sp. flowers prepd. by boiling elder fls. in olive oil - **essential o.**: **ethereal o.**: VO - **evening primose o.**: *Oenothera* - **fatty o.**: **fixed o.**: glyceride of 1, 2 or 3 mols. of fatty acid - **fleabane o.**: Erigeron o. - **floor o.**: cheap FO prepn. u. particularly to keep down dust & facilitate cleaning & polishing operations; u. on dust mops, etc. (types: mop, wax, concrete & cement, etc.) - **fuel o.**: indefinite term for any petroleum product u. for heating & power; several grades - **fusel o.**: mixt. of alcs., mostly isoamyl & amyl; solvent; in mfr. of perfumes; poison - **galangal o.**: VO fr. *Alpinia officinarum* - **gaultheria o.**: methyl salicylate - **gingelly o.**: sesame o. - **o. of grain**: fusel o. - **green o.**: olive o. - **halibut liver o.**: FO fr. *Hippoglossus hippoglossus* liver - **hardened o.**: hydrogenated FO - **hartshorn o.**: animal o. prep. fr. cattle horns, etc.; hence the name - **heavy o.**: creosote o. - **hydnocarpus o.**: chaulmoogra o. - **infused o.**: solns. in FO of active princ. of pls., esp. alk. & VO; ex. lobelia, stillingia, hyoscyamus - **Japanese mint o.**: **J. peppermint o.**: VO fr. *Mentha arvensis* v. *piperascens* - **juniper tar o.**: cade o. - **kusum o.**: Paka o. - **lard o.**: FO obt. fr. hog's lard; u. to control foaming in penicillin fermentation tanks - **laurel o.**: VO fr. *Laurus nobilis* - **leper o.**: **leprosy o.**: chaulmoogra oil - **Lima o.**: crude o. - **linseed o., sulfurated, terebinthated**: Haarlem o. - **Macassa(r) o.**: **Makassar o.**: 1) FO fr. *Schleichera trijuga* 2) cajeput o. (VO) - **margarine o.**: any edible FO u. to mfr. margarine - **Mecca o.**: crude petroleum - **melissa o.**: fr. *Melissa officinalis*; u. in perfumery & cosmetics (lemon odor) - **mineral o.**: 1) liquid petrolatum 2) (GB): petroleum - **mirbane o.**: mononitrobenzene - **mocayo o.**: fr. *Acrocomia aculeata* - **mountain pine o.**: dwarf pine needle o. - **mu o.**: fr. *Meum athmanticum* & other *M.* spp. - **muscle o.**: perfumed FO mixt. u. to massage sagging portions of face muscles, etc. - **mustard o.**: VO fr. *Brassica nigra* or *B. juncea*; or synth. allyl isothiocyanate; less commonly FO fr. sd. of various *Brassica* spp. - **myrcia o.**: bay o. fr. *Pimenta acris* - **neat o.**: common name for unrefined shark liver o. - **neatsfoot o.**: FO obt. fr. feet tissues (without hooves) of cattle (neat) (USP 1830-60); u. lubrication, softening of leather; formerly, in med. like cod-liver o. - **neroli o.**: bitter orange o. obt. by distn. - **n. portugal o.**: sweet orange fl. o. obt. by extn. (inferior prod.) - **niger o.**: FO of sd. of *Guizotia abyssinica* - **o. nut (tree)**: *Juglans* spp. - **nut o.**: fr. *Corylus avellana* - **oiticaca o.**: formerly u. for sd. FO fr. 3 different genera of plants; chiefly fr. *Licania rigida* - **orange peas o.**: petitgrain o. - **origanum o.**: VO fr. *Origanum vulgare* or *O. creticum* (**Spanish o. o.**: VO fr. *Thymus capitatus*) - **Paka o.**: FO of *Schleichera trijuga* - **palm kernel o.**: FO fr. sd. kernels of *Elaeis guineensis* - **palmarosa o.**: VO fr. above-ground parts of *Cymbopogon martinii* - **paraffin o.**: mineral o. - **peach kernel o.**: FO fr. sd. kernels of *Prunus persica* vars.; official with apricot kernel o. under the title Persic O. - **peanut o.**: FO of *Arachis hypogaea*; c. oleic. arachidonic & hypogaeic acids - **pear o.**: isoamyl acetate - **pennyroyal o.**: VO fr. *Hedeoma pulegioides* - **percomorph liver o.**: FO fr. livers of percomorph fishes; esp. tuna, swordfish - **persic o.**: apricot or peach kernel oils - **petitgrain o.**: v. under p. - **pimenta o.**: allspice o. fr. *Pimenta officinalis* - **pine o.**: **pinewood o.**: VO distd. fr. wd. of *Pinus palustris*, &c. - **pineapple o.**: ethyl butyrate - **pine tar o.**: empyreum. VO fr. distn. of pine tar - **Portugal o.**: sweet orange peel o. - **rapeseed o.**: colza o. - **red o.**: 1) red thyme o. 2) crude oleic acid - **Ricinus o.**: castor o. - **rock o.**: petroleum - **rose geranium o.**: VO fr. lvs. & fls. of *Pelargonium odoratissimum* & other *P.* spp.; comm. vars. incl. "African" (=Algerian) (fr. *P. roseum* or *P. graveolens*); (Belgian) Congo (fr. *P. capitatum* or *P. radula*); French; Morocco; Réunion (special race of *P. graveolens*); Spanish - **rosin o.**: empyreum. VO obt. by destr. distn. of rosin; usually 2 or 3 fractions are collected; all are acidic because free abietic acid is present - **Russia leather o.**: birch tar o. - **Russian o.**: heavy mineral oil - **salad o.**: any edible FO other than olive o.; esp. cottonseed o. - **o. of salt**: 1) Oleum Salis (L.) = HCl. 2) a propr. medicine - **santal (wood) o.**: VO fr. *Santalum album* - **sassafras o., artificial**: safrole - **o. seed**: niger sd.; sesame sd. - **Seneca o.: Seneka o.**: crude petroleum - **shale o.**: dark prod. obt. by destructive distn. of bituminous shales - **shiel-kanta o.**: FO fr. *Argemone mexicana* - **Siberian fir o.**: VO dist. fr. fresh lvs. of *Abies sibirica* - **o. of smoke**: creosote - **snake o.:** org. liquid of indeterminate & variable nature; sold by itinerant vendors (usually) with exaggerated claims of value; term may be derived fr. "Seneca Oil" (mineral oil) - **o. soap.**: neutral pure virgin o. jelly soft soap - **o. of soap**; glycerin - **sod o.**: degras, crude wool fat - **solar o.**: kerosene - **Spanish hop o.**: origanum o. - **sperm o.**: FO fr. cranial cavity of sperm whale, *Physeter catodon* - **spike o.**: crude o. of lavender - **star anise o.**: VO fr. *Illicium verum*, marketed as "anise oil" - **o. of stone**: crude petroleum - **succinum o.**: amber o. - **sulfonated o.**: pl. or animal FO treated with sulfuric acid, neutralized, & washed - **sulfur o.**: kind of olive o. extd. with CS_2 fr. press cake; bleached & u. in mfr. white soaps - **o. of swallow**: elder o. **sweet almond o.**: FO expressed fr. blanched seeds of *Prunus amygdalus* - **s. birch o.**: 1) teaberry o. 2) methyl salicylate - **sweet nut o.**: cottonseed o. - **sweet o.**: 1) olive o. 2) colza o. 3) other bland edible FO - **Swiss mountain pine o.**: dwarf needle pine o. - **table o.**: cottonseed o. - **talisay o.** (PI): FO fr. *Terminalia catappa* - **tall o.**: FO fr. pine wd. (tall = Swed. for pine; pron. "a" as in bad) u. in mfr. soaps, linoleum, paints, emulsions, cutting oils, &c. - **tanner's oil**: neatsfoot o. u. in dressing leather after coming fr. tan vats - **tar o., rectified**: pine tar o. - **o. of tartar**: K_2CO_3 soln. - **teaberry o.**: methyl salicylate - **teel o.**: **til o.**: sesame o. - **templin o.**: v. under t. - **thistle o.**: FO of *Argemone mexicana* - **train o:** whale oil: FO extd. fr. whale blubber, pilchards (Cornwall), &c.; mostly u. in crude lamps earlier - **o. tree**: 1) *Ricinus communis* 2) *Bassia latifolia* 3) *Elaeis guineensis* (o. palm) 4) candlenut 5) (Bible): *Elaeagnus angustifolia*; prob. not the pine tree as held by some (427) - **tropical o.**: such oils as coconut, palm, & palm kernel; considered nutritionally unsatisfactory because of saturated fatty acids - **tucum o.**: FO fr. sd. kernels of *Astrocaryum vulgare* - **tung o.**: FO fr. *Aleurites* spp. - **Turkey red o.**: sulfated castor o. or its Na salt - **turpentine o.**: VO distd. from pine tree exudn. (see under turpentine) - **vegetable o.** (common usage): bland FO fr. plants (usu. sds. or frts.) other than olive (usually soy o.); occasionally appl. to VO fr. plants - **(hydrogenated) v. o.**: refined, bleached, hydrogenated, & deodorized stearins (mostly triglycerides of stearic & palmitic acids) - **verbena o., Indian**: lemongrass o. - **vetiver o.**: VO steam-dist. fr. rts. of *Vetiveria zizanioides* - **o. of vitriol**: sulfuric acid (conc.) - **volatile o.**: v. under "oil," & under "volatile" - **watch o.**: FO fr. cranial cavity of black fish (or pothead); said most valuable lubricant o. - **Wesson O.**™: cottonseed o. - **whale o.**: FO rendered fr. fat layer of blubber of whales; fractions are u. in mfr. soaps, leather dressings, lubricants, illuminants; when hydrogenated u. in margarine & as cooking fat component; now illegal in US - **wheat germ o.**: FO expressed fr. fresh wheat embryos; rich in vit. E. - **wine o., light**: cognac o. = product of esterification of FO - **wintergreen o.**: methyl salicylate - **wood oil**: 1) Chinese/Japanese w. o.: tung o. fr. *Aleurites* spp. 2) Gurjun o. 3) various other oils obt. fr. wds. 4) oils obt. along with methanol in wd. distn.

oilcloth: cloth coated with colored linseed oil & dried.

oil-nut bark: bk. of *Juglans cinerea*

oil-stone: a bituminous schist of the Tyrol; u. as source of ichthyol.

ointment: 1) semisolid prepn. for appln. by inunction to skin as emollient & protective or for admin. of medication 2) (specif.): mixt. of wax & petrolatum forming a "**simple o.**" u. as emollient & vehicle (white & yellow) - **black o.**: black salve; ichthyol ointm. - **holy o.** (Bible): composed of myrrh, cinnamon, cassia, sweet almond, olive oil.

oiticica (oil): *Licania rigida* & other spp. (FO).

ojo de pollo (Mex): *Thunbergia alata* - **o. de zamuro** (Mex): *Mucuna pruriens.*

oka: (oca): *Oxalis tuberosa* (2nd most important root crop of Andes).

okan (Urdu): *Tamarix articulata.*

okolehao ("oke") (Haw Is [US]): distilled spirits.

okra: *Hibiscus esculentus* (lvs., young frt.) - **musk o.**: *Abelmoschus moschatus* - **Texas o.**: *Luffa* spp. - **vining o.**: *Luffa* spp.

ol (Beng): *Amorphophallus campanulatus* (rt.).

olaj (Hung): oil.

Olax dissitiflora Oliv. (Olac.) (Tanz): sap fr. crushed lf strained & mixed with coconut oil for u. as emet. - **O. nana** Wall. (Ind): frt. u. med. (428) - **O. scandens** Roxb, (Ind, seAs): bk. u. in fever (Burma); in prepn. for anemia (Ind) (428).

old field pine: v. under pine.

old maid (or **ol' m.**) 1) (Ala [US] slang): *Zinnia* - 2) **o. maid's flower**: *Catharanthus roseus* - **o. m. nightcap** (NCar [US]): *Geranium maculatum* - **o. m. pine**: *Bidens* spp. - **o. man (wormwood)** (Calif [US]): **o. man leaves**: *Artemisia abrotanum*; *A. californica*; sometimes *Clematis vitalba*; *Rosmarinus officinalis* - **o. man's beard**: 1) *Chionanthus virginica* 2) (NCar) *Fraxinus americana*, &c.- **o. man's love**: o. man - **o. man's pepper**: v. under pepper. - **o. wife's shirt**: *Liriodendron tulipifera* - **o. woman**: *Artemisia maritima*.

Oldenlandia affinis DC (Rubi.) (sAf): rt. decoc. u. for shortness of breath & cardiac dis. (Zulu) - **O. corymbosa** L. (Af, As): u. as abortif. (Haya); pl. juice appl. to palms & soles for burning sensations due to fever (UP Ind) - **O. johnstoni** K. Schum. (Af): lf. juice u. as diarrhea remedy - **O. nudicaulis** Roth. (nwInd): ext. u. in cough, asthma - **O. umbellata** L. (Hedyotis umbellata (L) Lam.) (seAs): Indian madder; chirwal (Hindi); lvs. u. as expect. (Indochin) in bronchitis, asthma; rts. ("chay") u. to prod. red dye like madder; in snakebite.

Olea cuspidata Wall. (Ole.) (As-Himalay reg., Baluchistan): Indian (or wild) olive; kao (Pak); frt. yields FO, rubefac.; lvs. & bk. astr. antipyr. & in debility; lvs. in GC; gum with Sb appl. to eye; wd. u. in furniture, &c., for making wooden combs. - **O. europaea** L. (O. ferruginea Wall.) (orig. eMed reg.; now widely cult. as in Calif [US], Ind, Jap, Austral, s&swAf): common olive (tree); (European) olive (tree); unripe (green) & ripe (black) frt. fresh & pickled u. as relish (green) in salads ("table olives"); pickled with NaCl (& vinegar) (using oil-poor frt.) (sometimes with condiments, ex. fennel); frt. expressed for FO u. food (mostly); considered very healthful ("liquid gold") (2460); also as lax. in hemorrhoids, colic; in salves, enemas, plasters; c. olein, palmitin, squalene, &c. (adulter. common during World War II, using cottonseed & teaseed oils, chiefly); olive o. of several types: It; Fr; Calif; virgin (superior) (with greenish cast), extra virgin, pure, foots (inferior grades prepd. by treatment with hot water & high pressure) (1716), u. in mfr. soaps, lubricants, the pressed cake (incl. endocarps [stones] u. in adulter. spices & drugs); olive juice c. bitter agent with foaming properties; bk. u. in mal.; resin exudn. (oleo-gum-resin; "ol. gum"), fr. bk. u. in fever & as vulner. & arom.; lvs. u. in mal. & wounds, hypotensive agt.; wd. u. in furniture making; huge trees (6 m diam. at rts. Syracuse, It; also in Holy Land) believed to be 1500-3000 years old (regions of cult. & prodn. Med, Sp, It, Por) (354); Arg, Pe (ancient cultn. in Holy Land) (355) (for serials on olive, v. App. C-1) (2101) - **O. glandulifera** Wall. (nInd): bk. & lvs. astr. u. in periodic fevers; kind of manna produced (c. 95% mannitol); bk. c. quercetin - **O. Pinguia** (L): FO - **O. Volatilia** (L.): VO (plur. of Oleum Volatilium).

Oleaceae: Olive Fam.; shrubs & trees of temp. & trop. zones (900 spp.), frt. a drupe, berry, capsule, or samara; lvs. opp. exstipulate, simple or pinnate; fls. usually in panicles; chief representatives: olive; others: ash, lilac, jasmine (true), privet, forsythia.

oleaginosos (Sp): oily sds.

oleander: *Nerium oleander* - **yellow o.**: *Thevetia peruviana* - **oleandrin**: glycoside fr. lvs. of *Nerium oleander*; cardiac poison.

oleandomycin: Matromycin: antibiotic prod. by *Streptomyces antibioticus* Waksman & Henrici; antibacterial activity similar to that of erythromycin; active against gm.+ m., enterococci, gm.- cocci; various pyogenic infections; generally u. as phosphate; adm. po, iv, im: deriv. = triacetyl-o.

Oleandra colubrina Copel. (Oleandr./Davalli - Filic. - Pteridoph.): (PI; Mal) decoc. of stipes effective emmen.; rhiz. for venomous snakebite.

oleandrin: neriolin: oleandroside of 16-acetyl-gitoxigenin; complex cardioactive glycoside; in lvs. of *Nerium oleander*; u. cardiotonic, diur., prodn. in Israel (429).

oleandrose: monosaccharide sugar, $C_5H_{13}O_3$.CHO: fr. hydrolysis of some cardiac glycosides.

oleanolic acid: caryophyllin: carboxy acid form of triterpenoid (sapogenin); ($C_{30}H_{42}O_3$); occ. in olive & centaury lvs., cloves, &c; often as glycoside to form a saponin.

Olearia argophylla F. v. Muller (Comp.) (Tasman): (silver-leaved) musk tree; -7 m high; lvs. & wd. with musky odor (2103), u. antispasm.

oleaster: *Elaeagnus angustifolia*+.

oleates: esters or salts of alks. & metals (resp.) with oleic acid; combns. of the following have been u. in medicine/pharmacy: aconitine; atropine; cocaine; mercury; morphine; aluminum; arsenic; bismuth; calcium (lubricant; claimed useful against viruses of colds, &c.): copper, iron, manganese, nickel, silver, sodium, acid sodium, tin.

oleic acid: 9-octadecanoic acid; $CH_3(C_{16}H_{30})COOH$; unsatd. fatty acid; orig. disc. in FO of *Olea europaea*, hence name; occurs in commonest pl. & animal FO, mostly as ester (glyceride), such as **olein** = triolein = glyceryl trioleate; u. as solvent; prepn. oleates, soft soap, &c. (2393).

óleo (Sp): **ôleo** (Por): oil - **o. de dende** (Br): FO fr. frt. pulp of *Elaeis guineensis* (oil palm) - **o. de pardo** (Br): VO fr. wd. of *Myrocarpus fastigiatus*; weakly arom. *cf.* myrrh.).

oleo-cerolé (Fr): cerate.

oleogomenol: v. gomenol.

oleo-gum-resin: natural combination of gum, VO, & resin; ex. Asafetida; forms natural emulsions with water & aqueous solns.

oleomargarin(e): v. margarin(e).

oleoresin: natural mixt. of VO & resin; ex: copaiba ("balsam" of c.); elemi; galbanum; olibanum - **o. turpentine**: gum dip.

Oleoresin of Mecca: oleoresina di Mecca: secretion of *Commiphora opobalsamum.*

oleo-stearin(e): stearin, beef.

Oleovitamin A: natural vit. A in FO; u. as food factor esp. for infants & mothers - **O. A & D**: composed of fish liver oil, alone or diluted with edible FO or a soln. of vit. A & D concentrates in f. l. o. or an edible FO - **O. D, synthetic**: soln. of activated ergosterol (vit. D_2) or of activated 7-dehydro-cholesterol (vit. D_3) in edible veg. FO.

olericulture: science of production, &c., of vegetables.

oleu (Roum.): oil.

Oleum (L.): 1) oil either fixed oil (FO) or volatile oil (VO) 2) crude animal oil 3) fuming H_2SO_4 (industry) - **O. Aleurites cordatae**: tung oil - **O. A. trilobae**: candlenut o. - **O. Amygdalae expressum**: sweet almond o. - **O. Animal Foetidum**: crude oil of hartshorn - **O. Anisi**: anise o. - **O. A. corticis**: VO fr. bk. of *Persea gratissima* - **O. Anthos**: rosemary o. - **O. Apopinense**: shiu o. (SC prodn. at Aupin) - **O. Arachidis**: peanut o. - **O. Avellanae**: FO fr. Corylus avellana - **O. Aurantii (amari)**: (bitter) orange o. - **O. A. Florum**: orange fl. o.; neroli o. - **O. Badiani**: star anise o. - **O. Balsamum** (oil of balsam): birch tar - **O. Ben**: behen o. - **O. Betulae**: sweet birch o. - **O. B. albae**: **O. B. empyreumaticum rectificatum**: **O. Betulinum**: birch tar o. - **O. Bubulum**: neatsfoot o. - **O. Cadinum**: Juniper tar - **O. Cajuputi**: cajeput oil - **O. Camphoratum**: camphor liniment - **O. Cari**: caraway o. - **O. Caryophylli**: clove o. - **O. Cassiae**: cinnamon o. - **O. Castoris**: castor o. - **O. Cedrae**: lemon o. - **O. Cedrii Folii**: cedar lf. o., fr. *Thuja occidentalis* - **O. Chaberti**: rectified animal oil - **O. Chamomi(l)lae citratum**: lemon oil distd. over charcoal - **O. Cinnamomi**: cassia o. - **O. Cocois**: coconut o. - **O. Coctum**: FO (usually olive oil) infused with a single herb of one kind or another - **O. commune**: olive oil - **O. Cornu Cervi (Foetidum)**: crude animal oil - **O. Crotonis**: FO fr. *C. tiglium* (Ol. Tiglii) - **O. Cupressi**: cypress o. - **O. Dipterocarpi**: gurjun o. - **O. Dryandrae**: tung o. - **O. Dulce Paracelsi** (old name): ether - **O. Empyreumaticum**: an empyreumatic VO, one produced by destructive distn. of plant or animal materials - **O. Erigerontis**: VO of *Conyza canadensis* - **O. Euphorbiae**: caperspurge o. - **O. Fermentationis**: fusel oil - **O. Foeniculi**: fennel o. - **O. Gaultheriae**: methyl salicylate obt. as VO of *Gaultheria procumbens* - **O. Gossypii Seminis**: cottonseed o. - **O. Graminis citrati**: lemon grass o. - **O. Harlamensis**: Haarlem o. - **O. Hippoglossi**: halibut liver o. - **O. Hyperici**: VO of *Hypericum perforatum* - **O. Illicii**: star anise o. - **O. Jasmini**: Spanish jasmine o. **O. Jecoris Aselli**: codliver o. - **O. Juniperi Empyreumaticum**: **O. J. Oxycedri**: juniper tar - **O. J. virginianae**: cedarwood o. - **O. Laurinum**: expressed FO of laurel berries - **O. Limonis**: lemon o. - **O. Linderae**: kuromoji o., VO fr. *Lindera sericea* - **O. Lini**: raw linseed o. - **O. Lumbricorum**: infused o. of earthworms - **O. Maydis**: corn o. - **O. Menthae piperitae**: peppermint o. - **O. M. viridis**: spearmint o. - **O. Morrhuae**: codliver o. - **O. Myrciae**: bay o.: VO fr. *Pimenta racemosa* - **O. Myristicae**: nutmeg o. - **O. Napi**: rapeseed o. - **O. Nervinum**: neats' foot o. - **O. Nucis moschatae**: **O. Nucistae**: expressed nutmeg o.; both VO & FO mixed - **O. Olivae**: olive o. - **O. Omphacinum:** impure olive o. - **O. Ovorum**: oil of eggs - **O. Palmae Christi**: castor o. - **O. Percomorphum**: percomorph oil; FO fr. livers of tuna fish, &c. - **O. Persicae: O. Persicorum**: peach kernel o. (FO) - **O Petrae**: crude petroleum (lit. rock oil) - **O. Philosophorum**: olive o. distilled over hot bricks; u. anciently - **O. Picis Rectificatum**: rect. pine tar o. - **O. Pimentae**: allspice o. - **O. Pinguium**: any FO - **O. Pini**: pine o., VO obt. by distn. of p. wood **O. P. pumilionis**: dwarf pine needle o. - **O. Portugallicum**: bergamot o. - **O. Ricini**: castor o. - **O. Rosae**: otto of roses - **O. Rosmarini**: rosemary o. - **O. Rusci**: empyreumatic birch o.; birch tar o. - **O. Sabinae**: savin o., fr. *Juniperus sabina* - **O. Salis**: salt o.: muriatic acid, tech. HC1 - **O. Santali**: sandalwood oil, VO fr. *Santalum album* - **O. S. Australiensis**: VO fr. *Fusanus spicatus* - **O. Sesami**: FO fr. *Sesamum indicum* - **O. Sinapis Volatile**: vol. mustard o. - **O. Templinum**: VO fr. pine cones & small branches of *Pinus mugo*, but usually fr. green cones of

Abies alba - **O. Terebinthinae (Rectificatum)**: (rectified) turpentine o. - **O. Terrae**: crude petroleum - **O. Theae Seminis**: teaseed o., FO fr. sd. of *Thea sinensis* v. *assamica*; *Camellia sasanqua* - **O. Theobromatis**: cacao butter - **O. Tiglii**: croton o. - **O. Vini**: wine oil - **O. Vitaminatum** (BP): refined shark liver oil (vits. A, D) - **O. Vitrioli Dulci** (old name): ether - **O. Volatilium**: VO or essential oil.

oleyl alcohol: found in fish oils; u. as surfactant (emulsion stabilizer, carrier for medication; wetting agt.); emollient (Novol).

Olibanum (L.): frankincense; oleoresin fr. *Boswellia carteri, B. neglecta, B. bhawdajiana*, etc. (430) - **Arabic O.**: fr. *Juniperus oxycedrus* - **O. gum**: O. - **Indian O.**: fr. *Boswellia serrata* - **O. oil**: VO fr. bk. of *B. carteri* (prod. Somal, Arabia) - **O. Sylvestre**: oleoresin fr. *Pinus sylvestris*; *cf.* Thus.

olie (Dan, Nor, Neth): oil.

Oligomycins: macrolide antibiotics prod. by an organism like *Streptomyces diastatochromogenes* Waksman & Henrici (Actinomycetes); isolated by E. McCoy et al. (Univ. Wisc, 1953); active against higher fungi (as in plant dis.) (esp. O-D), but not bacteria (mkt. by Serva, Heidelberg, BRD); compds. A-D known.

oligosaccharide: polysaccharide carbohydrate with fewer than 10 monosaccharide gps.

olio (It): oil.

olive: *Olea europaea* - **black o.**: *Laguncularia racemosa* - **Californian o.**: *Forestiera neomexicana* - (**C. wild o.**: *Umbellularia californica*) - **o. family**: Oleaceae - **o. gum**: **o. resin**: o. oleo-gum-resin fr. *Olea europaea* - **o. pits**: o. endocarps u. as pd. to adult. cloves & other spices - **Russian o.**: *Elaeagnus angustifolia* - **spurge o.**: *Daphne mezereum*; *Cneorum tricoccum* - **sweet o.**: 1) *Calycanthus* spp. 2) tea o., *Osmanthus fragrans* - **wild o.**: *Olea cuspidata*.

Oliver('s) bark: fr. *Cinnamomum oliveri*.

olivil: complex phenolic deriv. of furan; fr. gum-resin of *Olea europaea*.

olivo (Sp): **oliwa** (Pol): olive - **O. criollo** (Mex): *Bontia daphnoides*.

Olivomycin (A, B, C, D): antibiotic fr. the aureolic acid gp. derived fr. *Streptomyces olivoreticuli*; Arai & al. antineoplastic; toxic.

olja (Swed, Finn): **olje** (Nor): **oljy** (Finn): oil.

Olneya tesota Gray (1) (Leg.) (Ore, Calif [US], nwMex): (desert) ironwood; tree to 10 m, with pea-like fls.; bk. is drought-resistant; wd. heavy, u. as fuel, for arrowheads (Indians); sds. food of Pima & Mohave Indians; stored for winter use.

ololiuqui (SA Indians): *Ipomoea tricolor*; *Turbina corymbosa*: both u. as hallucinogens.

olutkombul: *Abroma angusta*.

Olyra latifolia L. (Gram) (tropAm; natd. wAf): lvs. u. in fomentations for abscesses & various cutaneous eruptions (Bainouk, Senegal).

Omalanthus *(Homalanthus)* **nutans** (Forst.) Guill. (Euphorbi) (Fiji, Samoa, Solomons+): latex drunk for GC; lf. inf. in stomach dis. (Solomon Is.); in yellow fever; rts. in diarrh.; wd. c. prostratin, of potential value against AIDS - **O. populneus** Pax (Mal & Indones): oiled lvs. heated & appl. to abdomen to treat fever (Mal); pulp of soft wd. smeared on painful joints (Indones); latex fr. inner integuments of st. mixed with water & drunk by pregnant woman to induce labor & ease delivery (PapNG); u. as hemostatic by Chin in Austral (2262).

omega-3 (fatty acid): designated unsaturated fatty acid found in fish oils (ex. mackerel) & some pl. materials (ex. soybean, wheat germ), in which an unsaturated bond occurs at the 3d carbon from the methyl end of chain; supposed to reduce risk of heart attack by reducing sticking together of platelets in blood, to reduce arthritic pain, &c.

Omphalea cardiophylla Hemsl. (Euphorbl.) (CA): sds. u. purg. anthelm.; ca. 55-58% FO, u. cath. diur. lubricant; possible substit. for castor oil - **O. diandra** L. (tropSA, CA, WI): cavetė; cayaté; similar to prec.; sds. toxic raw, edible after roasting, weakly lax.; u. for FO (ouabe oil) prodn.; u. for chest, intest. & kidney dis. (native med.); fls. astr.; latex c. caoutchouc. - **O. megacarpa** Hemsl. (WI, tropSA): climbing shrub close to *O. diandra*; large sds. ("Hunter's nuts") u. as food by natives; stim.; source of cayete oil, u. as purg. (*cf.* castor oil) & to make varnish; sds. also purg. - **O. oleifera** Hemsl. (CA): source of sd. FO, tambor oil - **O. triandra** L. (WI & tropAm): frts. edible; latex may be source of rubber (caoutchouc); sds. c. purg. FO.

Omphalina (Omphalia) lapidescens Schroeb. (Tricholomat. - Agaricales - Basidiomyc. - Fungi) (eAs): toxic mushroom; u. in marasmus, hookworm, tapeworm, roundworm (Chin).

Omphalocarpum elatum Miers (Sapot.) (trop wAf): sds. yield oily ext. u. to treat yaws & other sores; wd. u. for household utensils; latex a kind of gutta percha; u. to adulter. rubber - **O. procerum** Oliv. (trop wAf): frt. edible, u. as remedy for yaws; bk. fibers u. to make cloth.

Omphalotus olearius Sing. (Clitocybe illudens [Schw.] Sacc.) (Tricholomat. - Agaric. - Basidiomyc. - Fungi): (NA+): jack-my-lantern; false chanterelle; c. sesquiterpene alks.; has bacteriostatic & cancerostatic activity (356); phosphorescent; sl. toxic (emetic).

onagr(air)e (Fr): evening primrose, *Oenothera* spp.
onagrine: γ : u. to make skin supple (<*Onagra* Mill. = *Oenothera* L.)
Oncidium Sw. (Orchid.) (Fla [US] to temp. SA, WI): bulbs of one small sp. (SA) peeled & decoc. prepd. for kidney troubles (natives of Cuenca, Ec)
Oncoba echinata Oliv. (Flacourti) (tropAf): sds. produce gorli fat (g. oil), a type of chaulmoogra oil, with hydnocarpic, gorlic, & chaulmoogric acids; u. to treat skin lesions of leprosy, &c. - **O. glauca** (Beauv.) Planch. (trop wAf): props. & uses similar to those of *O. echinata* - **O. spinosa** Forsk. (tropAf): snuff box tree; sds. c. FO but without fatty acids of type of chaulmoogra oil; rts. & lvs. u. in dys. & bladder dis.; urethritis; frt. pulp edible; large frts. u. as snuff boxes & rattles for children - **O. welwitschii** Oliv. (wAf): FO u. like that of *O. echinata*; lvs. u. in dental caries.
oncolipids: lipid moieties of lipoproteins in plasma of persons with cancer; detected by magnetic resonance spectroscopy.
Onc(h)orhynchus (Salmonidae) (NA, As; chiefly Pacif Coast of NA) (some spp. landlocked): genus of fishes fr. the sperm or testes of which is prepared protamine u. in mfr. Protamine Zinc Insulin (injection).
oncornavirus: virus gp. wh. cont. RNA & produces tumors; causative of carcinomata & sarcomata.
Oncosperma horridum (Griff.) Scheff. (Palm) (Mal): rt. decoc. drunk for fever; apical bud is "palmito" ("cabbage") eaten (with pleasant flavor).
Oncovin™: vincristine.
one berry leaves: *Mitchella repens*.
Ongokea gore Engl. (O. klaineana Pierre) (Olac.) (trop wAf, esp. Congo, Gabon): isano; nut (ongueco [ongeko] or isano nuts) furnishes much FO = isano (ongokea or boleko) oil, highly unsatd., esp. isanic acid; u. in tech. & industrial areas, for coatings (polymers) fire retardant paints; not food; purg. action; substit. for linseed oil (2581).
onguent (Fr): ointment, often with resins, salts, extracts, etc. - **O. Nerval** (Fr): a veter. remedy c. althea, laurel oil, sage, lavender oil, etc. - **O. Nutritum**: **O. triapharmacum**: c. litharge 3, vinegar 3, olive oil 9; stiff as a plaster; u. as resolvent.
oni (Chama Indians, Pe): caapi.
onion: *Allium cepa* - **Bermuda o.**: a large mild tasting o., orig. fr. Bermuda; now also cult. US - **climbing o.**: *Bowiea volubilis*; possibly *Stropholirion californicum* - **garden o.**: o. - **sea o.**: *Urginea maritima* & some other *U.* spp. - **spring o.**: **Welsh o.**: *Allium fistulosum*.
Oniscus asellus L (O. murarius Cuv.) (Oniscidae - Isopoda - Crustacea) (Eu, NA): woodlouse; tiller's louse; like millipedes; u. as diur. in asthma, gout, kidney stones; leprosy (Chin); edema (Homeopath.).
Onobrychis viciifolia Scop. (Leg) (c&sEu, As): sain(t)foin; esparsette; good fodder, forage; hb. & sd. u. as diur., for tumors; in Ind as aphrod. (cult. pl. apparently introduced fr. seEu).
Ononis spinosa L. (Leg) (Euras, nAf): rest-harrow; hb. c. saponin-like glycoside (ononin), VO; u. diur. aperitive, to aid slow healing wounds; rt. (as Radix Ononidis [spinosae]) u. diur. exp. in chronic skin dis. & slowly healing wounds; u. in many tea mixt. (Eu) for arteriosclerosis, etc. (prod. Ge); the spp. *O. arvensis* L., *O. antiquorum* L., & *O. hircina* Jacq. are now regarded as subspp. of *O. spinosa*.
Onopordum (**Onopordon**) **acanthium** L. (Comp.) (Eu, esp. Med reg., nAf, Levant; natd. NA): Scotch thistle; silver (or cotton or Argentine or oat) t.; spina alba; lvs. very spiny (2286); fl. heads 3-5 cm. broad, fls. pale purple, all hermaphroditic & fertile; hb. u. for severe fevers (Eu homeop.); substit. for *Cnicus benedictus*; lvs. & rts. formerly u. in inoperable cancer (rts. per Linné), as diur., stom., lax., in GC; juice of fresh lvs. u. exter. in skin growths, facial cancer (appl. to lint of dressings), cancerous ulcers (folk remedy); florets have appeared as adulter. of Arnica & Saffron; rts. of 1st yr. pl. eaten as veget. (mostly It, Sp); FO obt. fr. frt ("seed"); pappus u. to make a kind of cloth ("étoffe de chardon") (Picardy); ornam. - **O. viride** Desf. (Euras): cotton (or tomentous) thistle; juice u. in cancers, esp. scirrhous carcinoma; at times entity treated as *O. acanthium*.
Onosma bracteatum Wall. (Boragin.) (Pak, Kashmir, Iran - wHimal): gaozaban (Bengali); has hypotensive & spasmolyt. effect; lvs. fls. & sds. u. in bronchial asthma, fever; in syph. leprosy; pharyngitis, diur. tachycardia, card. arrhythmias, hypertension (folk med.); fls. sold in As bazaars (guli-i-gao-zaban) as reputed mild ton. & alter. - **O. echioides** L. (sEu, As): ratanjot (Beng, Hindu); fls. u. as cardiac agt. & stim. in cardiac palpitations, rheum.; tonic (Unani med.); rt. u. as coloring matter for med. & toilet oils, fats, & woollens; bruised & appl. to eruptions; adulter. of alkanna rt.; pl. plentiful in the wHimalayas - **O. paniculatum** Bur. & Franch. (Chin): u. locally for cuts, ulcers, blisters.
Onosmodium virginianum (L) A DC (Boragin.) (eUS): false gromwell; wild Job's tears; gravel weed; rts. & sds. u. as diur. lithontriptic.

ontjom (Indones): fermented peanut press cakes after inoculation with *Neurospora sitophila* or *Rhizopus oligosporus*; popular food = lontjom.

ontogenesis: **ontogeny**: life history or development of individual plant or animal organism (as in contrast to phylogeny, the evolution of a group).

Onychium siliculosum C. Chr. (Adiant. - Pterid.) (tropAs): decoc. of fronds u. to remedy dys.; juice fr. crushed fronds u. to prevent falling hair (PI).

oolong (Chin): type of tea (like black dragon: color & strength).

Operculina turpethum (L) Silva-Manso = *Ipomoea t.*; many *O.* spp. are now referred to *Ipomoea* or *Merremia*, thus *O. tuberosa* Meissn. = *Merremia t.*

Ophelia: *Swertia* (in part).

Ophidia: the snakes (Reptilia).

Ophiodon (Hexagrammidae - Pisces) (world's oceans): lingcods; important comm. food fishes (with greenish flesh); liver oils rich in vit. A, D; such as *O. elongatus*.

Ophioglossum vulgatum L. (Ophiogloss. - Pteridoph.) (Euras, seUS): (southern) adder's tongue; hb. u. as emet. in edema, hiccups; fresh lvs. in poultices, ulcers, & tumors.

ophiology: science of serpents.

Ophiopogon japonicus Ker-Gawl. (Lili) (Jap, Kor): dwarf lily-turf; dwarf ophiopogon; mondo grass; Japanese snake's beard; lvs. to 30 cm, fls. hidden in lvs.; mucilag. tuberous rts. u. as expect. ton. in chest/abdom. dis.; nutrient; fine lvs. produce good lawn - **O. planiscapus** Nagai (Jap): u. similarly - **O. wallichianus** (Kunth) Hook. f. (Chin): plant crushed in water appl. as plaster for injury & inflammn.

Ophiorrhiza mungos L. (Rubi) (eAs, PI): earth gall; mungo; snake-root; mongoose plant; rt. u. to treat wounds & snakebite; has been u. in treating cancer; bitter ton.

Ophioxylon L. = *Rauvolfia* - **O. serpentinum** L. = *R. serpentina*.

Ophrys insectiferus L. (Orchid) (Eu): fly orchid; fls. so similar to insects that male insects often attempt to copulate with the fl.; tuber of this & other spp. u. as salep in prepns. of Turkish delight (gelatinous candy), &c.; cult. ornam.

opiase: peroxidase enzyme believed responsible for destruction of morphine in opium under certain conditions of prepn. & storage.

opiate: deriv. of Opium; distinguish fr. **opioid:** opioid compds. are synthetics with similar effects & values as morphine; ext. Demerol.

Opium (L.): gum o.: air-dried gummy latex of full-grown but unripe capsule of *Papaver somniferum* & v. *album*; c. 25 alks., the chief one morphine (10-13%), with smaller amts. of codeine, narcotine, narceine, papaverine, thebaine, etc.; u. narcotic, exp. antispas. (esp. for intestinal dis. as cholera, colic), analg. aphrod. (low dose), an aphrodisiac (last resort); to control diarrh. & intest. cramps (2326); prepn. *O.* ext. & various alks. (chief cult. & prodn.: Ind; Pak [Swat]; Eg; Bulg; USSR; Indochin; Iran; Turkey; Mex, nGr, Jap) (1717) - **abkari o.** (Ind): that used for local consumption - **Bengal abkari o.**: o. fr. Ind in flat cakes wrapped in oiled paper - **Chinese o.**: flat cakes of o. prod. & u. in Chin; c. 4-11% morphine - **Constantinople o.**: v. of Turkey o. rel. low in water & mostly u. in mfr. granular & pd. o. - **denarcotized o.**: **deodorized o.**: O. freed fr. narcotine & odor by treatment with petroleum ether; claimed to be free of side-effects due to narcotine - **druggists' o.**: that var. of Turkish o. which has low water content & is largely u. in mfr. of granulated & powd. o - **excise o.**: occurs in cuboidal cakes of ca. 1 kg made & u. in Ind - **granulated o.**: one of the 3 official forms of o.; the others being gum & pd. - **Indian o.**: that prod. mostly in state of Bihar (dists. Benares & Patna); usually in blocks often wrapped in white paper & tied with string; usually c. 9-10.5% morphine - **Iranian o.**: Persian o. - **lettuce o.**: Lactucarium - **licit o.**: legally imported opium u. to make morphine, codeine, etc., also now thebaine u. in mfr. of oxycodone, nalbuphine, naloxone, naltrexone, hydrocodone, &c.; best obtained fr. gum o. since poppy straw has only trace quantities of thebaine - **Malatia o.**: found in cakes of ca. 125 g; low morphine content - **Malwa o.**: Indian o. made into rounded balls - **manipulated o.**: that in more or less finished form with a fairly uniform content of alks. after adjustment - **medical o.**: type of Indian o. made up in blocks or flat cakes of ca. 1 kg & wrapped in wax paper - **natural o.**: crude o. as it comes fr. the orig. producer & before adjustment - **Persian o.**: Iranian o.: that prod. near Ispahan & Shiraz; in brick-shaped cakes wrapped in red paper; morphine runs 10-12.5% as a rule - **o. plant:** *Combretum sundaicaum* (SC u. to control opium addiction) - **o. poppy**: *Papaver somniferum* & v. *album* - **powdered o.**: one of the 3 off. forms of o.; ground to a very fine pd. (80 mesh): this is the form u. int. by preference - **provision o.**: an Indian o. worked into cakes weighing ca. 1.6 kg; u. mainly in alk. mfr. - **O. Pulveratum**: powdered o. - **Smyrna o.**: Turkey o. - **soft shipping o.**: rel. soft o. with 30% or more water; mostly imported by morphine mfrs. - **O. straw, concentrated**: ext. of above-ground parts of poppy pl. u. as source of o. alkaloids - **total o. alkaloids** (prepns.): Opium

Concentratum: Papaveretum (BPC); Opialum: HCl salts of opium alks. standardized for morphine, codeine, noscapine, papaverine; fewer by-effects than Opium; easier to administer, may be injected (solns.); may be given to infants & children (SC inj. or p.o.); preferred by many physicians; trademarked prods.: Pantopon; Laudanon; Pavon; Narkophen; Laudopan; Glycopon; Nealpon; Totopon; Holopon; Domopon - **Turkey o.**: occurs in subcylindrical cakes covered with pd. poppy lvs.; c. 10-15% morphine; chief subvars.: Balukissar, Boghaditz, Ghivez, Karahissar, Malatia, Salonica, Tokat; prod. by a government monopoly.

Oplopanax elatus Nak. (Arali) (Kor): twigs u. as ton. in progressive emaciation - **O. horridus** (Sm.) Miq. (Echinopanax horridum Dcne.; Fatsia horrida B. & H.; Aralia japonica Thunb.) (nwNA, Alas-Calif): (American) devil's club; Hercules c.; sts. & petioles of shrub covered with prickles (may cause irritation & mech. injury); rt. & st. c. VO with echinopanacene; rt. claimed useful in diabetes mellitus, this later disputed (1718); sts. heated & appl. to rheumatoid arthritis (BC [Cana] Indians) (77); young sts. boiled & eaten by Eskimos; bk. u. med.; cult. ornam. - **O. japonicus** Nak. (Echinopanax j. Nakai) (Jap): haribuki (Jap); lvs. c. aralin (saponin), idaein, quercetin; st. & lf. decoc. u. as antipyr. & bechic; in pneumonia; cult. often as ornam. (do not confuse with *Fatsia japonica*).

Opobalsamum de Tolu: Tolu balsam - **O. Liquidum**: Peru balsam - **O. Verum**: balm of Gilead (Mecca balsam) fr. *Commiphora opobalsamum* and var. *gileadensis* Engl. (also called simply "**Opobalsamum**").

Opodeldoc (liquid): soap liniment - **o. (solid)**: soap lin. camphorated; solid linim. soap.

opopanax (gum): **o. resin**: fr.: 1) *Opopanax* spp. 2) *Commiphora* spp., esp. *C. kataf* (present chief source: *cf.* Bdellium) 3) *Acacia farnesiana* 4) spp. of Fam. Umb. - **Opopanax chironium** (L) Koch (Umb) (Med reg., sEu, wAs, Iran+): rough parsnip; Hercules all-heal; a source of gum opopanax (surgeon's o.; o. resin; Hercules gum) = dr. juice obt. by incision of the rts. as oleo-gum-resin (tears or irregular lumps) u. as perfum. antispas. anti-inflamm. stim. (like ammoniac gum); u. in prodn. VO (**o. oil**); in incense, Fr perfumes; formerly u. in med. (from ancient times); still in homeopath. (Sudan gum) - **O. hispidum** Griseb. (Greece; AsMin): costus spurius; supposed to be one source of opopanax gum.

opossum: possum: *Didelphis virginiana* Kerr (Didelphidae - Marsupalia - Mammalia): (US): American (or Virginian) o.: the opossums are the only marsupials outside of Austral; nocturnal omnivorous animals with prehensile tail; young born in marsupium (pouch): u. as food (but fatty) (possum & sweet potatoes).

opotherapy: organotherapy.

optical activity: influence of a soln. or other liquid on direction of beam of polarized light, detd. with a polarimeter.

Optochin™: **optoquine**: ethyl hydrocupreine; prep. fr. cupreine (of Cuprea bk.); u. antiseptic.

Opuntia Mill. (Cact.)(NA, SA): prickly pears; chollas; large gen. (nearly 300 spp.) often found in xeric habitats; some said to prefer rocky soils; vary fr. small procumbent hbs. to large tree-like forms with flat stem joints; frts. of many spp. edible (1533); rts. of some spp. u. heated to make poultice for wounds & bruises (Tex [US]); feeding on pl. by animals may produce phytobezoars in intest. (431); many spp. are cult. as ornam. such as *O. microdasys* Pfeiffer (NM [US]) (very variable) ("bunny ears"), *O. acanthocarpa* Engelm. & Big. var. *coloradensis* L. Benson (buckhorn [cactus]) - **O. bigelovii** Engelm. (sNev, Calif [US], nwMex): teddy bear cholla; young sts. baked & eaten; rt. core boiled & decoc. u. as diur. (Seri Indians) (Sonora, Mex) - **O. cochenillifera** (L.) P. Mill. (Nopalea c. Salm-Dyck) (Mex, Jam, tropSA, CA): tuna (real) (Mex); nopal plant; cochineal plant; fleshy joints appl. as poultice (or cataplasm) to inflammn. swellings (Antilles, Haiti); rheum., earache, skin dis., &c. (432); chief sp. u. for growing coccus (cochineal insect) (*Coccus cacti*) (cult. Canary Is); fleshy segments considered specific in liver dis. (CR); shoots eaten as asparagus-like veget., considered anti-inflamm. - **O. elatior** Mill. (SA, natd. Ind - waste land); fls. c. opuntiol, ß-sitosterol; u. med. like *O. vulgaris* - **O. engelmannii** Salm-Dyck: 1) *Opuntia phaeacantha* Engelm. var. *discata* Benson & Walkington 2) var. of **O. ficus-indica** (L) Mill. (O. megacantha Salm-Dyck) (prob. orig. Mex, now natd. trop&subtrop, Carib & Med reg., sAf; widely cult. WI, Eu): Indian fig; tuna; nopal; cult. as "spiny mission cactus"; fresh sts. (slabs) & fls. u. in Homeop. med. (also alc. preserved); fl. yellow, ext. u. as diur.; frt. reddish with yellow to reddish flesh, delicious, u. in salads (*cf.* watermelon); in Fla (US [nr. Tampa]) & Calif (cult. in rows); pls. -5 m high; joints -5 dm X 1.5 dm; frts. -5 cm. diam.; some vars. almost spineless; fls. & slabs boiled with peppers & orange peel, vapors inhaled for colds; tender slabs (joints) u. as poultice for inflammation; sds. eaten (constipating if too many ingested); frt. u. in prepn. of intoxicating beverage; slabs fed to livestock; pl. source of tuna gum; host pl. for cochineal insects (dye); common weed (PR, Austral) - **O. fragilis** Haw. (Tex to Wash [US], BC [Cana]): brittle prickly pear; pigmy tuna (2104); claimed useful in diabetes; pl. reproduces by detachment of br., rarely fls. - **O. fulgida** Engelm. (swUS, esp. Ariz; nwMex): (jumping) cholla (cactus); source of c. gum; u. as food (toasted, or as beverage); for diarrh. & shortness of breath; fresh inner part of st. u. as decoc. for heart pain or heart dis.; rt. tea held in mouth for toothache; rt inner bk. + *Argemone* sp. lf. prepd. into tea u. as diur. for urinary dis. (Seri Indians) - **O. humifusa** (Raf.) Raf. (O. compressa (Salisb.) J. F. Macbride) (e&cUS): (eastern) prickly pear; frt. juice u. as diur. in urinary gravel; split lvs. u. as emoll. in acute rheum. (appl. direct); split lvs. heated appl. as poultice in gout, wounds, &c.; decoc. c. mucil. u. in lung dis., for tumors, &c. - **O. imbricata** Haw. (O. arborescens Engelm.) (wUS): cane cactus; choya; fls. u. as diur. (inf.); rts. u. in hair tonics; frts. u. emergency food (*cf. O. fulgida*) - **O. marenae** Parsons (Sonora, Mex.) u. like *O . reflexispina* - **O. megacantha** Salm-Dyck = *O. ficus-indica* (var.) - **O. phaeacantha** Engelm. (Tex-Ariz [US]; nMex): brownspine prickly pear; lf. pulp u. as poultice for ulcers, sores, bruises (Mex); said to stim. healing; u. to clarify water; food (boiled with salty water); ext. u. in diabetes (folk med.); frt. eaten; joints ("slabs") relished by cattle & sheep & as source of water. (2230) - **O. reflexispina** Wigg. (Sonora, Mex): thick rts. cooked in hot ashes & eaten for diarrh. - **O. soehrensii** Britt. & Rose (nArg, Bo, Pe): "ayrampo;" sd. u. as pomade for body pains, headache, and as inf. in stomachache; frt. has red dye u. in jellies, wines, &c.; cult. as protective - **O. stricta** Haw. (O. inermis DC) (sTex, Fla [US]; Cu): "tuna"; u. in diabetes (fl. ext.); frts. treated as "pear" in Queensland [NZ]; popular food, fresh or cured (*cf.* figs.) (Am Indians); hedge plant.; var. *dillenii* L. Benson (O. d. Haw.) (Fla, WI s to nSA): seaside tuna; raquette; frt. eaten & source of culinary dye; fleshy joints u. in treating enlarged spleen (Domin) (2328) - **O. triacantha** Sweet (Lesser Antil, Cu, PR): "cactus"; cult. - **O. tuna** Mill. (Jam; cult. sUS, Med reg., sAf, sAs, Austral): prickly pear; tuna; st. pulp emoll. u. as clarifying agt.; cattle feed on st. segments, may be lax.; pl. u. for hedges; weed; frt. with red pulp edible; host pl. of cochineal insects; red dye; one source of tuna gum - **O. vulgaris** P. Mill. (Br to Arg; cult. WI, NA, sEu): (common) prickly pear; Barbary fig; st. ext. u. as discut. in emuls.; exter. for gout & rheum., for softening skin cornifications; juice u. as purg. & anthelm.; support pl. for cochineal insects; frt. edible; fls. u. in treating enteritis - **O. villis** Rose (Zacatecas, Mex): prostrate plant; u. in inflammn.

or (Fr-m.): gold.

ora pro nobis (Por): *Portulaca oleracea* (L., pray for us).

oraca: a combination of orange & carrot juices; u. nutr., dietary vitamin source.

orach(e) (**, garden**): *Atriplex* spp., esp. **red o.**: *Atriplex hortensis* - **wild o.**: *A. hastata*.

Oral Rehydration Therapy (or treatment) (ORT): simple mode of treatment of diarrhea in babies, a very often lethal dis.; process combn. of sugar (or starch), salt, and water u. perorally in good amts. for dehydration (chief cause of death) fr. infant diarrhea; claimed the most important medical advance in the century (BMA); great reduction in mortality rates; as commonly prescribed in "Third World" (or developing) countries, a "pinch of salt, a fistful of molasses," with a cupful (+) of water; replaces the former sometimes difficult IV administration of fluid in such emergencies; discov. in Bangladesh epidemic (1950s); rice u. often in lieu of sugar (superior).

orange: 1) certain *Citrus* spp.; if unspecified, sweet o. is understood 2) a color representing a combn. of red & yellow; named fr. external rind color of frt. of same name - **o. apples: o. berries**: dr. immature frt. of bitter o. (those aborted & falling fr. tree early); u. flavor, perfum. stomach. bitter (chiefly prod. Sic) - **Batavia o.** (Madras): sweet o. - **Bigarade o.**: *Citrus aurantium* - **bitter o.**: *C. aurantium* (dr. peel & VO) - **blood(y) o.**: sweet o. with red or red-streaked pulp (sFr & It) - **o. blossom**: 1) fl. of sweet o., popular decorative theme at weddings (symbol of purity) & distd. for VO 2) *Trillium erectum* - **Chinese o.**: sweet o. - **Curaçao o.**: bitter o. - **o. flowers**: fls. of *C. aurantium*; dist. fresh for VO fr. orange fls. (chief prodn. Fr) - **o. flower water**: a satd. soln. of o. fl. oil, representing the cohobate fr. steam distn.; u. to prepare toilet waters, &c. (1720) - **green o.** (Pan): Osage o. - **horned o.**: bitter o. - **horse o.**: *Maclura pomifera* - **Kaffir o.**: *Strychnos speciosa* - **o. leaf**: *Coprosma lucida* - **lumilareng o.** (Malta, sEu): frt. of this citrus is sweet, while rind is green - **Malta o.**: sweet o. - **mandarin o.**: *C. reticulata*; small flattened frt. the peel separating easily, the pulp sweet & delicious; *cf.* tangerine - **o. milkweed (root)**: *Asclepias tuberosa* - **mock o.**: 1) horse o. 2) *Philadelphus* spp. 3) *Prunus caroliniana* (seUS) 4) *Bumelia lycioides* 5) *Styrax grandifolia* 6) *Buckleya* spp. 7) *Echinocystis lobata* 8) *Citrus maxima* 9) many other plants - **Natal o.**: *Strych-*

nos spinosa - **native o.**: *Maclura pomifera* - **navel o.**: seedless sweet o. with protrusion at one end representing supernumerary carpels - **Osage o.** (swUS): native o. - **Panama o.**: *Citrus mitis* - **o. peas**: smallest forms of o. apples; formerly u. as issue peas - **o. peel**: rind or cortex of frt., made up of outer yellow flavedo (rich in VO) & inner white albedo (rich in bitter princ., pectin) - **pineapple o.**: type of Med orange said developed in Fla [US]; thin-skinned, spherical, juicy (best for o. juice) with ca. 90 ml juice/orange); coated with blue or gray pigments - **o. root**: 1) golden seal rt., *Hydrastis* 2) *Asclepias tuberosa* - **sage o.**: Osage o. - **Satsuma o.**: *Citrus reticulata* (C. nobilis v. unshiu) - **Seville o.: sour o.**: *C. aurantium* - **Spanish o.**: *C. limetta* - **o. spoon**: special toothed or pointed spoon, for eating oranges or grapefrt. - **o. stick**: small cylinder of orangewood, u. to manicure nails - **sugar o.**: sour o. (SC had to be eaten with sugar) - **Sumatra o.**: *Murraya paniculata* **sweet o.**: *C. sinensis* (source of frt. peel, VO of peel) - **Temple o.**: hybrid, *C. reticulata* X *C. sinensis*; prob. orig. Jam before 1894 (1358) - **wild o.**: 1) bitter o. 2) (WI): *Drypetes* spp. 3) name appl. to many plants, incl. *Citrus* spp. growing wild; *Poncirus trifoliata*, *Aralia spinosa*; cherry laurel (sUS); *Drypetes glauca*; *Capparis* spp. (Austral); *Canthium oleifolium* (wild lemon, Austral), &c. - **o. wine**: prepd. by fermentation of peel or juice; deteriorates rapidly - **o. wood**: much u. for carving & wood turning.

orangelo: frt. of hybrid orange & pomelo.

oranger doux (Fr): sweet orange.

orangery: greenhouse or other enclosed place in cooler climates where oranges are cultd., usually on a small scale; popular in Eng, Fr (ca. 1500).

oranje (Dutch): **oranje appel** (Dutch): orange frt.

Orbignya cohune (C. Mart.) Standl. (Attalea c. Mart.) (Palm.) (sMex, CA to Hond, Salv): cohune palm; coco de aceite (Mex); corozo (Guat); manaca(nuts) (Guat); young buds u. as veget.; frts. for sweetening fodder; sds. (cohune nuts) (like ivory) to produce cohune oil (non-drying) (65-71% of kernel) (esp. Belize); u. as substit. for coconut oil; food, soap, illuminant; shell u. mfr. charcoal (for gas masks); st. juice u. to prep. wine - **O. martiana** Barb. Rodr. (Br): uaussú; coco naia; another source of babassú nuts; FO (70% of kernel) u. as cooking fat & in prepn. margarines, & in technology, similar to palm kernel and coconut oils; press cake u. as fodder; smoke of burning sds. u. to promote coagulation of rubber latex - **O. oleifera** Burret (Br): babassú; babacú; sds. u. as source of FO, b. oil & b. flour; the FO u. in mfr. of soaps, margarine, lubricants, fuel.

orcanette: alkanna.

orcein: dye made by oxidn. of orcinol.

orchata (Mex): cooling beverage of sugar & almond paste in water; orgeat.

orchic substance: testicular tissue of food animals of man; formerly marketed dried, in tablets (worthless since destroyed in stomach).

orchid family: orchis f.; **Orchidaceae**: possibly largest pl. fam. (ca. 17,500 spp.); hbs. found in all parts of temp & trop zones; exc. where arid; both epiphytic & terrestrial; fls. irregular, often grotesque; frt. capsule with many very small sds.; the most popular o. in US is *Cattleya*; ex.: *Paphiopedilum insigne*; *Cypripedium* spp. - **fringed o.**: *Habenaria* spp. - **great fringed o.**: *Habenaria fimbriata* - **Hooker's f. o.**: *Habenaria hookeri* - **marsh o.**: *Orchis latifolia* - **o. tree**: *Bauhinia variegata*.

orchil: orc(h)illa (Sp): **orchilla weed**: *Rocella tinctoria*, cudbear.

Orchis coriophora L. (Orchid.) (seEu, Gr, Turk): bug orchid; c. coumarin; rt. a form of salep - **O. incarnata** L. (Napal): tubers u. as ton. aphrod. - **O. latifolia** L. (Euras): marsh orchis; salap; source of a var. of salep tubers called Radix Palmae-Christae; rt. tubers u. nutr. in gen. debility, aphrod. in impotence, spermatorrhea; diabetes; (Orient); in convalescence, demul. in colitis; expect. (fr. orchis, L., testis, fr. shape of tuber); u. in chocolates & sweetmeats (Ind) - **O. laxiflora** Lam. (c&swEu): Jersey orchid; tuber u. as expect. astr. nutrient - **O. maculata** L. (Euras): spotleaf orchid; u. like prec. - **O. mascula** L. (c&sEu, As, nAf): male (or early purple or salep) orchid; a chief source of salep "root" (tuber); c. 50% mucilage, starch, protein; u. dem. esp. in diarrh. of children; vehicle for acrid drugs; nutr.; substit. for *O. latifolia* (exp. Iran [Tehran], Fra) - **O. militaris** L. (Euras): soldier o.; another source of salep - **O. morio** L. (Euras): salep orchid; another source of salep - **O. odoratissima** L. (Eu): fls. (Flos Palmae Christi) c. coumarin; u. in dys. - **O. pyramidalis** L. (c&sGe): salep plant; u. aphrod. (ancient times) - **showy o.**: *Galearis spectabilis* Raf. - **O. simia** L. (seEu, Gr; Turk): monkey orchid; fl. c. coumarin; rt. a form of salep; several **Orchis** spp. of Eu have furnished salep: **O. anatolica** Boiss. (sEu); **O. italica** Poiret (O. longicruris Link) (Gr); **O. majalis** Rchb.; **O. pallens** L.; **O. papilionacea** L. (Med reg.) (pink butterfly); **O. purpurea** Huds. (Eu) (lady orchid); **O. sancta** L (seEu, Aegean area) (holly orchid); **O. tridentata** Scop (sEu) (tootherd o.); **O. ustulata** L. (formerly off. in Ge Phar.).

orcil: orcillana (Sp): orchil.

orcin(ol): dihydroxy-toluene; this phenolic compd. & derivs. are important constits. of several lichens & lichen dyes (litmus, orseille); u. as reagent for sugars, lignin, &c. (microchemistry).

ordeal bean (of Calabar): *Physostigma venenosum* - **o. bark**: *Erythrophloeum guineense* - **o. poison**: toxic prod. (ex. *Parkia bussei*) administered to a suspected criminal or wrong-doer in order to determine guilt or innocence; u. in primitive societies believing in the supernatural where it is believed that divine intervention will permit the innocent person to avoid harm; there were also ordeals by fire, water (ex. Salem [Mass, US] witches), &c. - **o. tree**: the upas tree.

orecchio d'orso (It): bear's ear, *Primula auricula* - **o. di topo**: mouse-ear chickweed, *Cerastium vulgatume*.

oregão (Por): **orégano** (Sp): 1) (wild) marjoram, *Origanum* 2) *Lippia graveolens* (Mex) 3) (NM [US]): *Monarda* spp. - **o. de la sierra** (NM): *Monarda fistulosa* L. var. *menthaefolia* (Grah.) Fern - **o. del campo** (NM): *M. pectinata* 4) (as sold in US): botanical identity in doubt (357).

Oregon: v. under balsam - **O. grape (root)**: *Mahonia aquifolium*.

oreilles (Qu): *Asclepias syriaca*.

orellana (Sp, L): annatto - **orellin**: yellow & red coloring matter fr. *Bixa orellana*; related to bixin; u. to dye cheeses, etc.

Orenburgh gum: fr. *Larix europaea*.

Oreoherzogia fallax (Boiss.) Bent. (Rhamn.) (seEu): bk. c. various anthraquinone glycosides; u. purg.; decoc. in skin dis. (folk med.); to sophisticate Frangula bk. (Balkans) (—> *Rhamnus*).

Oreopanax capitatus Dcne. + Planch. (Arali.) (tropAm): palo de vela; sts. u. as candles (Indians); branches claimed snake repellent; unconfirmed by actual test; cult. orn.

organ: 1) *Mentha pulegium* (Eng) 2) in pl. or animal, organization of tissues for the purpose of carrying out a certain function or functions; the organ itself is usually one of several constituting a system; ex. stomach as part of digestive system = **vegetative o.**: one not concerned with spore or sex reproduction.

organelle: intracellular particle (ex. plastid; mitochondrion).

organic gardening: that carried out usually on a small scale using natural fertilizers & mulches (pl. & animal); and without synthetic pesticides, &c.

organography: the science of description of pl. & animal organs, esp. concerned with the internal structures; incl. under morphology.

organoleptic: as perceived by the sensory organs (eye, nose, tongue, ear, & sense of feel); making an impression upon an organ or upon the entire organism (odor, form, &c.); appl. to recognition or identification.

organotherapy: treatment of disease or disorder by the use of endocrines.

orge (Fr): barley - **o. monde**: b. grain denuded or deprived of its chaff - **orgead: orgeat** (pron. orzháh) (**syrup**) (Fr fr. orge): a heavy syrup formerly prep. fr. a barley decoc. (eau d'orge); now made chiefly fr. sweet almond emulsion with orange; u. flavor; beverage based on barley flavored with sweet almond; u. as summer drink & to flavor cocktails (rum) & foods - **o. perlé**: pearl b.

oriana (It): orleana.

oridigin: glycoside of gitoxigenin + 2,6-deoxyhexose + glucose; found in *Digitalis* spp.

oriental: v. under bezoar; cashew nut.

origan (Fr): *Origanum vulgare*.

Origanum creticum L.: *O. vulgare* v. *macrostachyum* - **O. dictamnus** L. (Lab) (Gr, Crete): (Cretan) dittany; hb. c. VO with pulegone; u. by ancient Greeks; now rarely u. as emm. stim.; in herb baths (Eu) (1721); as tea for GI dis.; in mfr. liqueurs (vermouth) - **O. heracleoticum** L. (O. hirtum Link) (wMed reg.): Spanish hops; hb c.VO with carvacrol, called "Trieste origanum oil"; u. veterin. linim. - **O. majorana** L. (Majorana hortensis Moench) (s&cEu, wAs, nAF): (sweet) marjoram; hb. (Herba Majoranae) c. VO. with arom. odor of cardamom; c. terpineol, carvacrol, terpinene, linalool, caryophyllene; culinary hb. u. as flavor, seasoning (esp. in sausages); stim. carm. emmen. (inf.); as arom., diaph. astr.; in prepn. VO (prod. Ge) (sweet German m.), Fr, Austr, Sp, Hung, CSR, nAf, (Eg), tropAm (Ch) (2038) - **O. maru** L. (Majorana m. [L] Briq.) (Syr, Isr): Egyptian marjoram; source of thyme-like VO (Syrian origanum oil); rich in thymol; probably the "hyssop" used by sprinkling in the Jewish & Roman Catholic rites ("Asperges me, Domine, hyssopo . . .") - **O. onites** L. (Majorana o. Benth.) (Euras.: Crete, Cyprus, Gr. Yugo.): pot marjoram; Spanischer Hopfen; foliage of dwarf shrub u. as arom. (as "Herba Origani cretici"), as spice for anchovies, &c. (fish industry); source of VO (Smyrna origanum oil) (c. carvacrol, linalool) - **O. smyrnaeum** L. = *Majorana onites* - **O. syriacum** L. (near East): "hyssop" (Bible); u. as spice; lvs. dr. malted & eaten with bread; u. in resp. dis.; u. in Roman Catholic rituals - **O. vulgare** L. (Euras, Med reg., esp. It, nAf, wAs; natd. NA): wild (or common) marjoram (marjoran); orégano; hb. (Herba Origani) is chief source of origanum oil

(VO with 50% thymol, carvacrol, cymol); also c. tannins, NBP; u. arom. flavor, spice (popular with Italian cuisine as in pizzas), stom. carm. diaphor. diur., in inflamm., gingivitis; to incr. bile secrn. (1846); in gargles, herbal baths (Eu); extern. in wounds; fresh lf. placed in ear to remove pain & dry up pus (UP Ind); pl. said ant repellant; with high antioxidant value (prodn. Sp, It, Mor, Hung, Russia) (2105); var. *macrostachyum* Brot. (var. creticum) (L.) Hal.) (sEu): Spanish hops; source of Cretan origanum oil; c. carvacrol, linalool; u. in toothache, urin. dis., arom.; hb. adulter. of Thyme; several *Origanum* taxa also u. under names "Spanischer Hopfe" (Spanish hops) & Herba Origani cretici.

origanum: 1) *Origanum vulgare* (source of **origanum oil**) 2) *Monarda punctata* .

Oritrophium peruvianum (Lam.) Cuatr. (Comp.) (SA, Andes): frailejoń morado; syr. u. in certain types of asthma.

Orixa japonica Thunb. (1) (Rut) (Jap, Chin): ch'ang shan; dr. lvs. c. VO, berberine; u. as antipyr. but not antimal.

orizaba (root, resin, &c.): *Ipomoea orizabensis*

orlean (Fr): **Orleana** (L.): annatto.

orme (Fr): elm (tree).

Ormocarpum cochinchinensis (Lour.) Merr. (O. sennoides DC) (Leg.) (Ind, seAs): rt. u. ton. stim. in lumbago (wInd); pulped, lvs. u. for boils (Indones); lvs. chewed & saliva swallowed for sore throat or sores in mouth (Caroline Is); to treat frambesia (Indones), &c; u. in local med. for relief of stomachache - **O. kirkii** S. Moore (tropAf): cooked rts. u. as poultice on sores; rt. flour placed on scabies, sores (Tanz) - **O. trichocarpum** Burtt Davy (tropAf): rts. cooked with fat meat & juice drunk for impotence; remedy for bilharzia (Tanz).

ormono (It): hormone.

Ormosia calavensis Azaolo (Leg.) (Pl, Indones, Austral): decoc. of bruised lvs. taken for abdominal cramps - **O. coccinea** Jacks. (SA, esp. Br, Ve): bead tree; sds. brilliantly marked red & black; c. physiologically inactive alk., ormosinine, dasycarpine, &c.; u. by indigenous peoples in med., as ornamental in necklaces, &c.; useful wd. - **O. fordiana** Oliver (seAs, incl. sChin): lf. inf. u. for certain skin dis. (Vietnam) - **O. krugii** Urban (WI): mato(s); palo de matos; wd. of value, u. for many purposes; ornam. - **O. ligulivalvis** Rudd (Guianas; Ve, Co, Br): cocho; bk. inf. u. to wash sores with pus; sds. toxic - **O. macrocalyx** Ducke (sMex-Br [Amazonia]): casique; sds. c. panamine with hypotensive activity (2141) - **O. macrophylla** Benth. (Co): "peonia"; ingred. in curare prepd. fr. *Strychnos* spp.- **O. monosperma** Urb. (O. dasycarpa Jacks). (Ve, Br, WI): peonia; necklace tree; jumbie beads; (red) bead t.; caconnier; frt. or sds. u. as *O. coccinea* (esp. Swed); u. occasionally in native med. (sds. decoc. for "heart pain"); c. ormosine, narcotic alk. somewhat like morphine with analg. properties (2062); ormosinine (isomer of ormosine) - **O. panamensis** Benth. (CR, Pan): tree with alks. having analg. & narcotic action; panamine is hypotensive.

ornithine: diamino valeric acid; non-essential amino acid found in animal proteins, esp. those of fish flesh; u. as anticholesteremic (best administ. on empty stomach); said to protect liver (named from ornis [L], bird, since 1st isolated fr. excreta of poultry [Jaffe, 1877]).

Ornithogalum caudatum Ait. (Lili) (sAf): star of Bethlehem; the squashed lvs. are laid over wounds; bulb u. as adulter. of squill - **O. lacteum** Jacq. (Af): milk-white chinkerchu; pl. toxic to stock; bulbs. long u. in Eu as squill substit. - **O. thyrsoides** Jacq. (sAf): star of Bethelem; lf. inf. u. by Europeans in sAf for diabetes mel. with contemporaneous restriction of carbohydrates; lethal poison to stock - **O. umbellatum** L. (Euras, nAf; natd. NA): star of Bethlehem; snowdrop; nap-at-noon; hb. after roasting u. as emoll.; hb. potent digitaloid cardioton.; bulb u. as dispersing agt., sometimes eaten; fresh lvs. u. in homeop.; ornam. shrub.

Orobanche L. (Orobanch.) (world, temp, & subtr): broom-rapes; like other members of fam., root parasites; chlorophyll-less - **O. americana** L. = *Conopholis a.* - **O. californica** Cham. & Schlecht. (wNA): succulent underground sts. u. as food by Indians (Calif, Nev [US]); likewise **O. fasciculata** Nutt. (wNA), and **O. ludoviciana** Nutt. (wNA) - **O. major** L. (O. elatior Sutton) (Euras, Ind): common parasite on *Centaurea* spp.; u. med. - **O. minor** Sutton (Euras, natd. eUS): hell root; clover broom-rape (parasitic on clover rts.); u. folk med. - **O. uniflora** L. (Aphyllon u. Greene) (Calif) cancer root: u. in intest. dis., formerly in ulcers & cancers (997).

Orontium aquaticum L. (monotypic) (Ar.) (e&sUS): golden club; tuckahoe; "never wet"; rhiz. & sds. former food (Am Indians); babies bathed in decoc. every month (Cherokee Indians).

Orophea polycarpa A DC (Annon.) (seAs): rts. u. as antipyr. bechic (Mal), diaph.

orosul (Nic+): *Phyla scaberrima*.

orotic acid: whey factor; milk component, pyrimidine precursor; pyrimidine carboxy acid deriv. (uracil carboxylic acid); sometimes claimed as vit. B_{13}; growth stim. for calves (put in feed); uricosuric agt.; claimed to reduce cholesterol in blood serum; but unproven (Univ. Ill. res.).

Oroxylum (Oroxylon) indicum (L) Kurz (monotypic) (Bignoni) (Indo-Malaysia+): India trumpet flower; midnight horror (SC bad odor of night-opening fls.); bk. bitter u. in gastric dis. (Java); as tannin drug (Ind); rt. bk. u. in local med. (Indochin); sd. u. to allay inflammn. & pain; int. in treatment of ringworm in cattle; lvs. emoll. dem. for ulcers; soft cambium tissue u. as styptic (Sumatra); sold in markets (433); pd. sds. (damree seeds) u. as purg., st. as paste for scorpion bite (124); inf. for tinea capitis; green lvs. fls. & frts boiled & eaten (Indochin).

orozul: orosul.

orozús (SpAm): orozuz (cimarrón): 1) (Hisp) *Phyla scaberrima* & others 2) **orozuz** = licorice (*cf.* sous. Arg).

orphan drugs: those agts. with therapeutic value, such as are used for rare diseases, but with limited comml. potential; incl. some crude drugs, ex. peg-1-asparaginase &c.

orrada (Mex): sabadilla.

orrice (It): **orris (root)**: rhiz. of *Iris* X *germanica* (& v. *florentina*), *I. pallida*, (SC corruption of Iris) (chief prodn. It, Fr) (*v.* under *Iris germanica*) - **o. fingers**: pieces of rhiz. trimmed & perforated to hang around an infant's neck for teething - **Mogadore o.**: rhizome originated mostly fr. *I. germanica* cult. in Morocco; of inferior quality - **o. oil**: VO preferably fr. *Iris pallida* v. *clio* - **Portuguese o.**: appeared on drug markets in 1940s; peeled rhiz. split up longitudinally - **Verona o.**: **Veronese o.**: prod. much like, but inferior to, Florentine o., usually former more yellowish, often with holes for cords.

orseille: lichens which after maceration in urine are u. for their violet tinctorial properties (orchil, cudbear); 2 chief groups recognized: 1) **o. de mer**: found at the seashore (esp. *Rocella* spp.) 2) **o. de terre**: found growing on the land (esp. *Variolaria* spp.).

orsella (It): uva ursi.

Orsinus orco (Cetacea) (seas worldwide; off nw coasts of NA, ne coasts NA, Pac Oc); killer whale; orca; black & white whales up to 10 m long; u. for oil prodn., protein source; popular at exhibitions.

ORT: oral rehydration treatment.

Orthodon chinensis Kudo (Mosla c. Max.) (Lab) (Chin): whole pl. u. as diaph. diur., in fever of colds, diarrh. - **O. grossiserratus** Maxim. (eAs): st. & lvs. u. carm. - **O. japonicus** Benth. (Mosla japonica Maxim) (Jap, sKor): yama-jiso (Jap); VO c. thymol; decoc. of lvs. & sts. vermifuge for ascarids; germicide for intest. ferment.; source of thymol. - **O. lanceolatus** Kudo (eAs): decoc. u. in colds, scabies, boils, hemorrhoids; VO u. in perfumes (Taiwan) (—> *Mosla*).

Orthosiphon pallidus Royle (Lab) (Indo-Pak; sArabia): pl. c. orthosiphonol, choline, betaine; hypotensive (434) - **O. stamineus** Benth. (O. spicatus Bak.; O. aristatus Miq.; O. grandiflorus Bold.) (seAs, Austral, tropAm): Java tea; kumis kuching; lvs. u. as diur. in kidney dis. ("Indian kidney tea") (esp. in Bright's dis.); albuminuria, chronic nephritis, cystitis, gout, rheum. dem.; hb. (off in Neth Phar.) c. orthosiphonin (glycosid. bitter subst.), saponins (orthosiphonosides A,B,C,D,E), many K salts (1283).

Orthrosanthus chimboracensis Baker (Irid.) (Co, Ec, Pe): amargoso; rt. u. to make brushes; source of fibers.

ortica: (It): **Ortiga** (Sp): nettle. - **ortiga (blanca)** (Sp): *Lamium album* - **o. di canapa** (It): *Cannabis*.

orto botanico (It): botanic garden.

oruga (Sp): *Eruca sativa*.

orujo (Sp): brown presscake residue fr. olive oil prodn.; c. 6- 15% FO, which is extd. with CS_2 producing "sulfur oil"; this is bleached & u. in mfr. of white soaps.

Oryza rufipogon Griff. (Gram.) (seAs): red rice; bad weed in rice fields (sAs; Miss [US]); red sd. coat; caryopsis nutritive, similar to common r.; some regard this as hybrid of *O. sativa* X *O. perennis* Moench - **O. sativa** L. (sAs, nAustral; cult. in most trop & subtr areas, esp. It, Sp, Por, Eg, Ind-Chin area; NA, seUS (Ark, La, Tex, Calif), SA): (common) rice; cultivated r.; source of chief cereal grain of 1/2 of world's population, most important (largest produced) food of world (2561); made into r. starch, r. flour (unsuited to bread making, since low leavening quality; hence u. in pastries, puddings, rice-macaroni) (Jap), pancake flours) (v. under "rice" for prods.); often combined with spices (as saffron), vegetables, cheese, &c; FO, & hard wax prod. fr. rice husks; beer, sake (rice wine, beverage of Jap); rice broth consumed for colds, diarrh. (Indones); ground rice grain u. dem. refrig. (with lint = hemostatic); bran u. to reduce hypercholesterolemia; rice husks u. mixed with cement in building blocks to afford insulation (poor mech. props.); husk ashes proposed for this use (2106); rice straw —> furfural.

Oryzopsis hymenoides (Roem. & Schult.) Ricker (Gram) (w&cNA, incl. nMex): Indian rice grass; silk g.; very palatable to livestock; sds. eaten by Indians (wUS) (meal in breads, gruel).

orzata (It): barley water, beverage dispensed by NY City (NY [US]) street vendors in Little Italy (mid-Manhattan, NY).

Os (L): (plur. **Ossa**) bone - **O. Sepiae** (L.): cuttle(fish) bone fr. *Sepia officinalis*.

Osage: v. under orange

Osbeckia chinensis L. (Melastomat.) (seAs): st. & lf. decoc. u. in dys. (Taiwan); rts. chewed & saliva swallowed for cough & hemoptysis (PI) - **O. crinita** Benth. (seAs): rt. inf. u. as stom., remedy in GC; dr. lvs. held in mouth for toothache (Indochin) - **O. nepalensis** Hook. (Ind+): fls. pounded & appl. to sores in child's mouth.

oseille (Fr): *Rumex acetosa* - **o. de Guinée** (Fr): *Hibiscus sabdariffa*.

osha root (Colo, NM, Ariz [US]): *Ligusticum porteri*.

osier: 1) *Salix* spp. esp. *S. viminalis* 2) **green o.**: *Cornus rugosa* (bk.) 3) **red o.**: *C. sericea* (bk) 4) **white o.**: *Leucothoë racemosa*.

Osmanthus americanus Benth. & Hook f. (Ole.) (eNA, esp. seUS): devilwood; American (or wild) olive; "Olea americana"; bk. u. bitter astr., lax. emoll.; frt. oleaginous; wd. hard & resistant, useful - **O. fragrans** (Thunb.) Lour. (Chin & adj.- Himalaya): tea olive; fragrant o.; lvs. finely toothed margin; fls. strongly & deliciously fragrant (hence name); lvs. (& fls.) u. to give aroma to beverage tea; frt. edible; lf. decoc. for cough; lateral rt. decoc. u. in rheum.

Osmaronia cerasiformis (T. & G) Greene = *Oemleria cerasiformis*.

Osmites dentata Thunb. (Comp.) (sAf): inf. u. for cough; exter. appln. in inflammn. (—> *Relhania*).

Osmitopsis asteriscoides Cass. (Comp.) (sAf): diaph. u. for body aches, influenza, rheum.; in gastric dis.

Osmorhiza aristata Mak. & Yabe (O. japonica S. & Z.) (Umb) (seAs): lf. decoc. drunk to regulate digest. tract function; fresh lf. sap antiemetic; ash of rts. mixed in oil for scabies (Indochin) - **O. claytonii** C. B. Clarke (wNA): rts. u. to gain weight (Wisc [US] Indians) - **O. longistylis** (Torr.) DC (e&cNA): sweet chervil; (wild) sweet anise; American sweet cicely; rt. (as "anise root") u. arom. carm. stom. (2107) - **O. occidentalis** Torr (wNA): sweet root; rt. decoc. u. for colds & other resp. dis. (sometimes pd. rt. smoked for same).

Osmunda cinnamomea L. (Osmund. - Pteridoph.) (e&cNA, cAs): cinnamon (or fiddle-head) fern; young fronds boiled & eaten in soup (Am Indians); ornam. perenn. hb. - **O. claytoniana** L. (As, e&cNA): interrupted fern; rhiz. sometimes seen as adult. of Aspidium - **O. fiber**: masses of reddish brown ramenta (scales) fr. the rhiz. of *Osmunda cinnamomea* or *O. regalis*; coll. in eFla (US); u. for packing pls., etc., in greenhouses (*cf.* Sphagnum moss) - **O. japonica** Thunb. (seAs): rhiz. anthelm. in tapeworm, liver fluke (sheep, goats); preventive measure in epidemics - **O. regalis** L. (Euras, nAf, natd. NA, SA): flowering (or regal) fern; buckhorn brake; fronds diur. u. in edema; purg.; rhiz. u. dem. astr. ton.; has been noted as adulter. of Aspidium; trichomes of young lvs. u. with wool to make a cloth & raincoat material.

Ossa Sepiae (L.): Os Sepiae.

ossein: collagen.

Osteophloeum platyspermum (Poepp.) Warburg (1) (Myristic.) (Br); sap of tree drunk to cure cough & colds (Maku Indians); smoke fr. burning lvs. u. for asthma. (1029).

Ostindien (Ge): 1) (East) India (*qv.*) 2) East Indies.

ostindischer Enzian (Ge): Chirata.

Ostrea edulis L. (Ostreidae - Bivalvia - Mollusca) (nAtl Oc coasts): (common) oyster; European flat o.; body a delicious food, eaten raw or cooked; c. antibiotic & antiviral subst.; considered aphrod.; calcareous valves or shells (Conchae Praeparatae) c. $CaCO_3$, calcium phosphates, silica; u. detergent, abrasive in toothpowders; source of Ca for bone building, ton.; gastric antacid; shell u. in osteoporosis & disturbances of Ca metabolism in climacteric (female) (homeop.); in chicken feed (to strengthen egg shells) - **O. gigas** (Jap): Japanese oyster; one of largest oysters (30 cm long); source of meat, shells, pearls; cult. BC [Cana], Wash (US) - **O. virginica** (eUS maritime areas): meat & shells u.; mostly coll. in Chesapeake Bay.

Ostreogrycin: antibiotic subst. obtained fr. cultures of *Streptomyces* spp. incl. *S. ostreogriseus* Shirling & Gottlieb; several known, designated by letters: **A** = Virginiamycin M_1; **B** = Mikamycin B; $\mathbf{B_1}$ & $\mathbf{B_3}$ = Vernamycin B.

ostrich: *Struthio camelus* L. (Struthionidae) (Aves) (Af, wAs): (Strausz [Ge]): largest living bird (-2.5 m tall, 135 kg wt.); terrestrial & flightless; lives in flocks; tamed in ancient Rome; domesticated in modern times (farms in sAf; Calif [US]); flesh eaten (low in fat & cholesterol); better tasting than chicken; display feathers (white) popular for decoration of hats; dusters, &c.; eggs considered valuable for shells, u. as vessel, &c.; flesh, fat, blood, gizzard, excreta (appl. to skin pigmentation), u. in Unani med.; skin u. for boot leather.

ostruthin: crystall. coumarin. deriv. found in rhiz. of *Peucedanum ostruthium*, &c.

Ostrya carpinifolia Scop. (Betul.) (Eu, Med reg. into Alps): (European) hop hornbeam; bk. c. 7% tannins (pyrocatechin type); lvs. u. med. (Eu); good hard wd., u. to make charcoal - **O. virgini-**

ana K. Koch (eNA): (American) hop hornbeam; ironwood; leverwood; bk. u. bitter ton.; wd. u. by homeop. physicians (Eu); tough durable wd. u. for tool handles, &c. (lvs. doubly serrate; frt. a strobile resembling hops).

Ostryoderris stuhlmannii Dunn (Leg) (cAf): lf. decoc. u. as cough remedy; lf. & frt. foraged; wd. u. to make railway ties (sleepers); bk. inf. u. for diarrh. (Sukuma tribe, Tanz).

Ostryopsis davidiana Dcne. (Betul.) (Chin): lvs. u. as tobacco substit.

Oswego: v. under tea.

Osyris alba L. (Santal.) (Med reg.): ancient hb. known to Dioscorides and Galen; tops u. diur., as substit. for Scoparius (2108) - **O. arborea** Wall. (Himalay, Ind): popli; lf. inf. u. as powerful emet. (Ind) - **O. compressa** A DC (sAf): bark bush; rt. decoc. u. in milk for GC, in honey for rheum., with fat as ton. & galactagog; tan; wd. of little significance; source of African sandalwood oil, substit. for true Indian s.w. oil; lvs. c. osyrit(r)in (= rutin) - **O. tenuifolia** Engl. (eAf, Ind, seAs): wd. yields VO, "East African sandalwood oil" u. in perfumes; wd. similar to Santalum Album; lvs. u. to prep. osyris tea; u. in native med. much like *O. compressa*; useful wd.

Otaheite: Tahiti (old name).

Otanthera cyanoides Triana (Melastomat.) (eAs): rts. grated in water u. to prevent abortion; frts. given to children to prevent bed-wetting (Indones).

Otanthus maritimus Hoffmanns. & Link (Diotis m. Desf.) (1) (Comp.) (wEu coasts to Cauc, nAf): cotton weed; sea cud-weed; flg. branches u. as antipyr. emmen. ton. tenif.; cult. orn.

otate, palo de (Mex): a wild bamboo, *Arundinaria longifolia*.

otoba: Virola.

Otolithus regalis Can. = *Cynoscion r.* (fish-croakers).

Otostegia limbata Hook. f. (Lab) (Ind): bui (Punjabi); lf. juice u. for ophthalmia (man & beast).

ottar: **otto**: fragrant essential oil or perfume mostly of floral origin (variant of attar).

Ottelia alismoides Pers. (Hydrocharit.) (tropAs, nAustral): pl. appl. as poultice to ulcers, abscesses, burns, scalds, cancers; lvs. u. as styptic; in hemorrhoids; sometimes u. as pot herb.

otto of Rose(s): rose oil fr. *Rosa damascena*, *R. centifolia*, etc.

ouabain(um): bushman's poison: G-strophanthin; glycoside obt. fr. *Acokanthera ouabaio* & *Strophanthus gratus*; also occ. in mammalian tissues; u. cardiotonic, pharmacol. reagent. (1847); topically in glaucoma.

oudoudalam (Tamil): *Mentha piperita*.

Ougeinia dalbergioides Benth. (Leg) (Ind): sandan (Hindi); bk. u. antipyr.; kino-like exudn. fr. incised bk. u. in diarrh. dys.; bk. fish poison (—> *Desmodium*).

our Lord's candle (NM [US]): Yucca - **ouricuri** (Br): **ouricury (oil)**: FO fr. *Scheelea martiana*, *Syagrus coronata*, (etc.) - **o. wax**: fr. *Syagrus coronaia*.

Ouratea parviflora Engl. (Gomphia p. DC) (Ochn.) (Br): pl. yields VO; sd. FO, batiputa (or bati) oil, u. in skin dis., Hansen's dis.

Ourouparia gambir Baill.: *Uncaria gambier*.

Ova Formicarum (L.): ant "eggs" (larvae), *Formica rufa*.

oval: shape of hen's egg as opposed to elliptical (latter is popular idea).

ovalbumin: egg albumin.

ovarian residue: pd. dr. ovary less the corpora lutea, obt. fr. cattle, sheep, & swine; previously u. for ovarian hypofunctioning & in menopause; now believed worthless in the desiccated state (oral use).

Ovarium (L.): **ovary**: pd. dr. **whole ovary** of animals u. as food by man; of little or no therap. value in this form, hence now largely replaced by the active constituents of the gland.

ovate: shaped somewhat like the outline of a hen's egg; said of a leaf, etc.

oveja (Sp.): ewe, sheep - **ovejunos, animales** (Sp.): ovinos (Arg): sheep in general.

overground parts: when non-woody, this constitutes the herb; otherwise, it would incl. woody stems, bark, etc., as well as lvs., fls. & frts., also small sts.; ex. Pulsatilla, Verbena.

Ovi Albumen Recens (L.): fresh white of hen's egg; u. nutr., ton., in reagents, bacteriolog. media - **O. Vitellum Recens** (L.): fresh yolk of hen's egg; u. emulgent; nutr.

Ovibos moschatus (Bovidae): (Arctic NA): musk-ox; long-haired wild ox, feeding on lichens, mosses, grasses; source of hair, hide, meat, &c. for northern Indians & Eskimos (< *Ovis*, sheep + *Bos*, ox).

ovine: relating to sheep.

Ovis aries L. (Bovidae) (Ovidae) (cosmop.): (domesticated) sheep; some 120 breeds; a ruminant mammal long domesticated & u. for its wool, wool fat, meat (mutton, lamb), suet, endocrine glands, etc.; many prods. in chem. mfr. such as cholesterol; clotting factor (blood); intestine u. to prepare catgut u. for surgical ligatures, strings of musical instruments, in tennis rackets, &c.; sheepskins (less wool) u. to make parchment (for diplomas, &c.) & in treatment of bedsores (Ala

[US]); sheep manure inf. ("Saffron tea") u. for measles (US s. mts); fat-tailed s. (As); source of food fat; ram's horn was u. as the first wind musical instrument (shofar) u. by the ancient Jews; now serves in ceremonials at Yom Kippur, &c.

ovoid: hen's egg-shaped; term appl. to solid articles, such as frts: *cf.* ovate.

ovomucoid: egg white.

ovule: immature unfertilized young seed in ovary; megasporangium of a seed plant.

Ovum (L.): egg, specif. hen's egg.

owala grains: fr. *Pentaclethra macrophylla*, **owala oil tree**.

ox: 1) various spp. of genus *Bos* or closely allied genera 2) castrated male *Bos taurus* (u. as draft animal).

Oxacillin: a semi-synth. penicillin; active against *Staphylococcus* spp.; 1961-.

oxadoddy: Leptandra.

oxalic acid: a di-carboxy acid $(COOH)_2.2H_2O$; commonly found in the pl. kingdom, generally as the calcium, sodium, or potassium salt; comml. source by a simple synth.; represents strongest (most highly ionized) org. acid; u. hemostat., many techn. usages; ex. of natural occurrence: in *Cicer arietinum* separates out with other org. acids on lf. surface, also present in stem trichomes; frequently occ. in Oxalidaceae; in *Averrhoa carambola* (-1% in unripe sour frt. var.; danger of large crystals penetrating stomach wall); K acid oxalate found in *Oxalis* & *Rheum* spp., etc.

Oxalis acetosella L. (Oxalid.) (Euras, nAf, NA): (common or European) wood sorrel; (common) field s.; "shamrock" (popular but incorrect); pl. antiscorbut. refrig. diur.; acidulous; poisonous (since c. potassium oxalate) (552), but u. in salads as ton. - **O. corniculata** L. (OW origin, natd. NA, pantrop): (yellow) wood sorrel; hb. u. antipyr. antiscorb. & exter. appln. for bruises (Chin); corns (Ind); lf. juice appl. to wounds to cleanse, also for itch (PI); lvs. u. med. in Tonga (86) - **O. latifolia** HBK (cMex, to Pe, Ve): acederilla; lf. decoc. u. as mouthwash in oral inflammn. - **O. oregana** Nutt. (Pacif US): redwood sorrel; with white or pink fls.; lvs. & sts. edible - **O. stricta** L. (NA): upright yellow wood sorrel; lvs. eaten raw or cooked (some Indian tribes); lvs. chewed to assuage thirst on long walks - **O. tetraphylla** Cav. (Mex): hb. u. med. in México - **O. tuberosa** Mol. (O. crenata Jacq.) (Co, Pe, Bo): "oca"; "oka"; cult. for millenia in SA for tubers u. as food ("forgotten crop" of the Incas; second most important rt. crop of Andes; of modern potential value); cult. somewhat in Eu; juice u. as astr. acidulant - **O. violacea** L. (e&cUS, Mex). violet (wood) sorrel; u. similarly to *O. acetosella*.

Oxamycin = Cycloserine.

Oxandra laurifolia A. Rich. (Annon.) (WI): bois pian; wd. at one time exported to GB as "lancewood spars"; sap said to have been u. to treat yaws ("pians") by Caribs of Dominica (2328).

ox-bile: **ox-gall**: fresh bile or secretion of the gallbladder of *Bos taurus*; v. under bile; c. potassium & sodium taurocholates & glycocholates, biliverdin & bilirubin, cholesterol, choline; u. choleretic - **o. b. extract**: prep. of prec. c. 45% or more cholic acid as Na & K salts (equivalent); u. cholag. cath. - **o. gall, inspissated**: dr. o. bile - **o. marrow**: marrow of long bones u. med.

ox-eye: *Helenium autumnale*, etc.

oxidases: class of enzymes wh. accelerate oxidizing processes; closely related to the dehydrogenases, which are sometimes included with them; found in fresh digitalis lvs., horseradish rt., gum acacia, &c.; classified into iron oxidases (catalase, peroxidase, cytochrome oxidase, &c.), copper oxidases (incl. tyrosinases, laccase, &c.).

ox-lip: *Primula elatior* - **ox-tongue**: *Anchusa officinalis*.

oxyacanthine: alk. fr. *Berberis vulgaris*.

Oxycoccos Baccae (L.): cranberries.

Oxycoccus Hill = *Vaccinium*.

oxyconiine: conhydrine.

Oxydendrum (**Oxydendron**) **arboreum** (L) DC (monotypic) (Eric.) (eUS): sorrel tree; sour wood; elk tree; lily of the valley tree; angel fingers; "titi" (misnomer); -15 m high; lvs. u. as tea; diur. ton. refriger.; to assauge thirst; in cardiac dis.; for pyrosis & flatulence (folk use); fls. source of well-known sourwood honey (seUS); cult. for ornam. (435).

oxygen: **Oxygenium** (L): O_2: gas composing 21% of atmosphere, 47% of earth crust; most abundant element on earth; prepared by fractional distn. of liquefied air; u. to relieve hypoxia (h. chamber); CO poisoning, at hyperbaric pressure (3X atm. press.) in cardiac surgery; anerobic infections; stroke, smoke inhalation, electrical, & crushing injuries, stim. in senility, &c.; cryotherapy (liquid); with inhalation anesthetics; many ind. uses (E); delivered compressed in metal cylinders; off. in nearly all pharmacopeias..

oxygenase: thermolabile enzymes or enzyme systems which permit the taking up of atmospheric oxygen to form peroxides, which are u. by the organism.

Oxygonum atriplicifolium Mart. (Polygon.) (cAf): lf. juice u. as cough remedy (Chagga); freshly roasted lf. u. in bronchitis to "loosen the phlegm"; food.

oxymel: oxymellite (Fr): very ancient gp. of prepns. (Hippocrates); base of honey boiled with vinegar; u. as pharmaceutical base.

Oxymycin: Oxypan: Terramycin.

oxynarcotine: an alk. of Opium.

Oxystelma esculentum (L.f.) Sm. R. Br. (Asclepiad.) (seAs: Camb, Vietnam): decoc. u. in gargles, latex in ulcers; rt. in jaundice (Indochin); rt. & st. "depurative"; frt. u. for aphtha (aphthous stomatitis) (Ind); frt. edible; famine food.

Oxystigma mannii Harms (Hardwickia m. Oliv.) (Leg) (Af): tree trunk is "boxed" to produce "illurin balsam" (African copaiba), an oleoresin, u. as replacement for copaiba (less effective, but more pleasant to take); the VO distd. off u. to adulter. peppermint oil (US); u. in porcelain ind.; & to make a kind of varnish.

Oxytetracycline: generic name for Terramycin - **O. HCl:** Oxycycline.

oxytocin: alpha-hypophamine: oxytocic princ. of the posterior pituitary (E).

Oxytropis lambertii Pursh (Leg.) (ncUS): white loco; crazy weed; hb. causes loco dis. in horses; u. med. for rheum.

oyster: **Ostrea** spp. (*qv.*) - **oyster green**: *Ulva latissima* - **o. plant**: 1) *Tragopogon porrifolius* 2) *Rhoeo discolor* 3) *Scolymus* spp. esp. *S. hispanicus* (also **Spanish o. p.**) - **o. shells, prepared**: fr. *Ostrea edulis* (dietary source of Ca) - **vegetable o.**: *Tragopogon porrifolius*.

ozokerite: earth-wax; an impure paraffin wax, brown to grayish; made up of mixed solid hydrocarbons.

ozone: O_3; has sometimes been u. in resp. dis.; disinf. for air & water; bleach; formed in upper atm. by action of UV(L) on oxygen (ozone layer, considered protective to human skin, reducing incidence of skin cancer.

Pp

PABA: para-amino-benzoic acid.

Pablum™: infant cereal food; c. corn meal (< pabulum [L] food).

Pachira aquatica Aubl. (Bombac.) (sMex, CA, SA): Guiana chestnut; provision tree; new shoots, bk. & unripe frt. u. for liver dis. (Guat); bk. decoc. antidiabetic (Pan); lvs. eaten as greens; sds. roasted & eaten; c. edible arom. FO (tree much cult. orn. Fla [US], tropAf) - **P. insignis** Sav. (Antilles): u. similar to prec. - **P. macrocarpa** Sch. & Cham. (sMex, CA): sapoton (CA); apompo (Vera-Cruz): fls. & lvs. emoll. u. in conjunctivitis; frt. size of coconut; sds. u. as food, cocoa substit.; source of FO.

pachomonoside: glucoside of strophanthidin + D-glucose; occ. in *Pachycarpus schinzianus*; sd. & rt. of *Parquetina nigrescens*; cardioactive.

pachycarpine HI: (+)-sparteine iodide; alk. of *Sophora pachycarpa*; u. as ganglion blocking agt.; oxytoccic.

Pachycarpus distinctus Bullock (Asclepiad.) (sAf): rts. c. cannodoxoside (cardioactive glycoside); u. med. - **P. rigidus** E. Mey. (sAf): uzara; pl. u. sed. antispas. against colic & intest. dis. (dys., diarrh.); rt. u. in dys. - **P. schinzianus** N. E. Br. (sAf): rt. c. digitoxigenin, pachygenin; u. in syph., sterility; rt. u. in hemorrhoids, edema; similar to prec. in dys.; in prepn. of uzara - **P. validus** N. E. Br. (sAf): maketula; rt. latex u. as purg.; similarly to *P. rigidus*.

Pachycereus marginatus (DC) Britt. & Rose (Cact.) (s&cMex): organ pipe cactus; organo; c. pachycereine, cereine (alks.); popular use to color skin dark; u. for skin swellings (place slice of st. previously heated); cult. to furnish living defensive fence - **P. pecten-aboriginum** Britt. & Rose (nwMex): cardon (barbon); (h)echo; sd. ground & mixed with corn meal; bristly frt. covering u. as hairbrush (Goldman); st. juice (also decoc. of branches) hallucinogenic drug of Tarahumara Indians (J. S. Bruhn) (1667); c. carnegine - **P. pringlei** (S. Wats) Britt. & Rose (nwMex): cardón; wd. u. to make walking canes, in house building, as fire wd.; sds. & frt. pulp u. to make flour for cakes & in tamales; slabs heated in hot ashes, wrapped in cloth & appl. to body for pain; formerly, juice boiled down to give kind of balsam appl. to wounds, &c. (2271).

Pachycormus discolor Coville (monotypic) (Anacardi.) (Baja Calif [Mex]): elephant tree; latex hardens to resin; bk. u. as tan (exp. to Eu).

Pachylobus edulis G. Don (Burser.) (w tropAf): bush butter tree; frt. cooked & eaten; bk. yields resin (elemi) u. for jiggers, skin dis., etc.

Pachyma Fr. = *Poria*.

Pachypodium lamerei Drake var. *ramosum* Pichon (Apocyn.) (Mad): fleshy pith filled with water serves as food for livestock; liquid expressed is a refreshing drink; cut into pieces, this pulp is appl. to face to give a refreshing sensation; ornam. in cultivation - **P. rosulatum** Baker var. *gracilis* Perrier (Mad): juice expressed fr. sts. considered powerful vulner.; instilled into wounds, it seems to protect fr. microbial infection - **P. rutenbergianum** Vatke (Mad): bk. fiber of good quality; u. as textile fiber like ramie.

Pachyrhizus ahipa Parodi (Leg) (Bol, Arg): ahipa; long cult. for tubers u. as food - **P. erosus** (L) Urb. (P. angulatus L.C. Rich.; P. bulbosus Britt.) (Mex [Vera Cruz], CA, trop-SA; cult. & natd. tropAs):

(Wayaka) yam bean; potato b.; sds. toxic; c. rotenone, pachyrrhizone; u. as purg. (emulsion), pediculoside (Tr), antipsoriatic; vermifuge (nViet-nam), insecticide (often with pyrethrum) for aphids, codling moth larvae, cabbage-worms, &c.; tuberous rts. boiled & eaten (Mex, Yuc) (Bengal: raw); young husks & sds. also eaten; cult. (Belize) - **P. palmatilobus** B & H (Mex): sds. edible, u. for cooling drink, in fever, &c.; rt. bk. u. in rheum., hulls for skin exanthemata - **P. rigidus** E. Mey. (Gomphocarpus r. Decne.) (sAf): uzara; u. sed. in intest. dis., dys. diarrh. colic, antispasm.; veget. - **P. tuberosus** Spr. (WI, CA, SA): (West Indies) yam bean; young frts. & rts. are eaten; crushed fresh pod mixed in lard u. to cure itch (Chin).

Pachysandra procumbens Mx. (Bux.) (eUS): Allegheny pachysandra; (A.) mountain spurge; A. spurge; fls. white or purplish, on spikes -18 cm long; thick exserted stamens; u. as ornam. perenn. hb. (ground cover) - **P. terminalis** Sieb. & Zucc. (Chin, Jap): Japanese pachysandra; fukki-so; (Jap); lvs. cath. stom.; frt. pleasantly edible (Ainu); evergreen sub-shrub often cult. in gardens (Jap, US+) as ornam. (fls. white in spikes 3-5 cm long).

Pachystachys coccinea (Aubl.) Nees (Acanth.) (Trin, Guianas; natd. WI): "coral" (Cu); "chandelier" (Domin); lf. inf. u. in headache.

paciencia (Sp): *Rumex patientia.*

Pacific coast: v. under adonis.

pacifier (US): violet.

packaged medicine: a household remedy ordinarily sold over the counter & not on prescription; SC ordinarily put up in cartons bearing fully descriptive data & directions in a layman's language; may have a proprietary title or be a common article, such as Epsom salts.

pacova (Br): *Renealmia exaltata*

Pactamycin: antibiotic subst. fr. *Streptomyces pactum* subsp. *pactum* Bhuyan et al. (1961); sepd. by chromatography; antitumor, active against human epidermoid carcinoma cells (*in vitro*) & many tumors in hamsters & mice (*in vivo*) (1878); u. exptl. in cancer patients (Nat. Cancer Center).

pacul (Ch+): *Krameria cistoides*; *K. triandra.*

pad: fleshy lf. of a water pl. - **p. joints** (US slang): places where cannabis is smoked by groups (SC smokers recline on mattresses).

paddock: fenced field or other enclosure - **p. pipes**: *Equisetum* spp., esp. *E. palustre* - **p. stools**: *Polyporus igniarius.*

paddy (As): rice.

Padina pavoni(c)a (L) Thivy (Dictyot. - Phaeophyc. - Algae): (Med Sea): peacock's tail (seaweed); c. mucilage; u. to make edible jelly.

pado (Por): *Prunus padus.*

Padutin™: kallikrein: callicrein; hypotensive enzyme found in animal body tissues, esp. pancreas & urine (chief sources); c. no insulin; u. as vasodilator.

Paederia diffusa Standl. (Rubi) (Pe, etc.): hediondo; vejuco; bejuco; u. folk med. - **P. foetida** L. (Ind, seAs, Java): gandhali (Hind); lvs. have fecal odor (indole); entire pl. u. in rheum.; lvs. eaten to aid digest.; lvs. in soup for invalids & convalescents; to treat anuria & fever (Indochin); rt. u. sed. nervine, ton. analg.; for toothache (blackens teeth) - **P. scandens** (Lour.) Merr. (sAs, Mal): pl. has fecal odor but lvs. u. as vegetable; lvs. in rheum. (Burma); to incr. fertility (PI); digestive aid (Chin); juice fr. bruised frt. u. rubbed into body for cold injury (Jap); lvs. ton. antipyr. (Indochin); many other uses.

Paederus fuscipes Curt. (Meloideae: Coleoptera; Insecta) (It): blister beetle; c. paederine = pederin (*qv.*).

paen (Ind): pan = betel

Paeonia L. (Paeoni./Ranuncul.) (Euras, wNA): peonies; some 40 spp. of showy perenn. hbs., often of horticultural value (358); < paeonia (L.) < paionia (Gr) since discovered by Paion (physician to the gods) (u. anciently in insanity, convuls.); specif. L. title for peony rt. fr. *P. officinalis* - **P. brownii** Dougl. (wUS: Calif, &c.): western peony; Christmas rose (Calif); u. against colds. (359) - **P. dahurica** Andrews (Turk): rt. c. paeonoside, paeonol; u. as anti-inflamm., analg.- **P. emodi** Wall. (Pak): rt. u. as ton. in bladder dis., nervous dis.; sds. emet. & cath.; dr. lvs. in diarrh.; tender shoots eaten as veget. - **P. japonica** Miyabe (Far East): rts. u. as circul. stim., diur. (Kor, Chin) - **P. lactiflora** Pall. (P. albiflora Pall.) (Siber, Manchur, nChin): rts. u. circul. stim., diur., in TB, eye dis., lumbar pain; has antibiotic properties; ton. analges. antispasm. in GIT dis.; chief source of many garden hybrid cultivars. - **P. obovata** Maxim. (Jap, Kor, maritime Chin): rt. of perenn. hb. u. for stomachache; pd. sd. u. for eye dis. (Ainu); astr., antipyr. aid circulation - **P. officinalis** L. (Euras, esp. Med & Balkan reg.): common peony; rhiz. c. antibiotic; source of FO (similar to linseed oil); u. antispasm. in cramps, epilepsy; fls. c. paeonidin; u. in epilepsy; in breast teas, fumigants; sds. u. in epilepsy; emet. emm. c. solid VO with much paeonol (arom. ketone, hemostatic); rt. uter. & intest. stim. - **P. peregrina** Mill. (eAs): ram's horn peony; u. similarly to *P.*

officinalis; lf. ext. shows anti-tumor activity - **P. suffruticosa** Andr. (P. moutan Sims) (Far East, esp. Chin): moutan; (Chin) tree peony; mau tan (Chin); rhiz. u. sed. in nervousness, emmen., female dis., caecum inflamm. analg. styptic; antispasm. - **P. veitchii** Lynch (wChin): u. anticoagul. emmen. antipyr. circulat. stim.

paeonin: peonin

Pagaea recurva Benth. & Hooker (Gentian.) (SA, incl. Co.): "hierba amarga"; tea fr. whole pl. u. for "forgetfulness" of elderly (natives) (2299) (—> *Irlbachia*).

Pagiantha cerifera Markgraf (Apocyn.) (New Caled): bk. inf. u. as drastic purg. (—> *Tabernaemontana*)

pagoda tree: 1) *Plumeria rubra* var. *acutifolia*, *P. alba*; *Sophora japonica* (also **Japanese p. t.**).

pahadi pudina (Bombac.): **paharipodina** (Punjabi): **paharipudina** (Pak, NWFP): spearmint.

paico (SA): epazote.

paigle tea: *Primula veris*.

paina (Br): complex of silky fibers fr. sds. of spp. of fams. Bombacaceae, Typhaceae, & Asclepiadaceae - **pain de sucre** (Fr): pineapple.

paineira (Br): *Chorisia speciosa* (tree).

paint brush: *Carphephorus corymbosus* - **Indian p. b.: scarlet p. b.**: *Castilleja* spp. - **"painted leaf"** (Mex): poinsettia.

pai-pu (Chin): *Stemona tuberosa* (rt.).

paishte: towel gourd, *Luffa*.

Pajanelia longifolia K. Schum. (P. multijuga DC) (monotypic) (Bignoni.) (Indomal): rt. bk. u. as ton. in dys. (Mal); lf. ext. as antipyr. (Indones).

Palaquium Blanco (Sapot.) (seAs, Malaysia, Pacif Islands): large laticiferous trees; some 120 spp., several yielding gutta percha (1995) - **P. grande** (Thw.) Engler (SriL): meeria-gass; sds. yield FO u. in med., for cooking (when fresh) - **P. gutta** (Hook f.) Baill. (P. borneense Burck; P. oblongifolium Burck): (Mal, Indones: Borneo, Sumatra): gutta percha plant; Malay g. p. nato tree; trunk tapped for latex (also bk. & medulla), generally after first felling tree (lvs. also c. latex); one of chief sources of gutta percha (coagulated dr. purified milk juice), considered of best quality; u. in prepn. plasters, dental cements, dentures, fracture splints, surgical insts., chewing gums, temporary dental fillings; ligatures, "traumaticin" (soln. in $CHCl_3$); for insulation (to cover underseas cables, impervious to sea water), centers of golf balls; in mfr. of rubber, &c. (1453); sd. prod. 33% FO, Njatu fat; u. in foods, soaps, candles (cult. Borneo, Sumatra, Java, Thailand, Mal) - **P. leiocarpa** Boerlage (Borneo, Celebes): latex u. to prep. rubber, sometimes mixed with superior types - **P. obovatum** (Griff.) Engl. (Ind, Mal, PI): another source of gutta percha; wd. keeps well under water; not much attacked by termites - **P. rostratum** (Miq.) Burck (seAs, Indones): frts. edible; sds. yield FO, edible; wd. u. for building purposes - **P. stellatum** King & Gamble (Mal, Sumatra): frt. edible; "vegetable butter" extd. fr. sds.; wd. for canoe making.

palas: *Butea monosperma*.

pale: v. under catechu; rose (leaves), &c. - **p. bark**: *Cinchona pubescens* (etc.).

Paleae Haemostaticae (L.): **P. Stypticae**: chaffy hairs fr. *Cibotium* spp. (SC u. hemostat. stypt.).

Palembang: v. under benzoin.

Palicourea crocea Roem. & Schult. (Rubi) (WI, contin. tropAm): ponasi (Cu); rt. u. emet. - **P. densiflora** Mart. = *Rudgea viburnoides* - **P. marcgravii** St.-Hil. (Br forests): "coto-coto"; erva de rato (rat plant); hb. u. diur.; berries & lvs. u. to poison small rodents (interior Br) (said highly efficient) (482); hb. c. strongly toxic monofluoracetic acid - **P. rigida** HBK (tropAm, esp. Br): douradinha; lvs. (& inner bk. of young branches) c. douradine (CNS paralyzant); u. as diur. diaph. cardiac agt., in skin dis., prostatism; in diarrh. (Br, Ve) & liver dis. (Br).

palillo: 1) (Pe): turmeric, Aleppy 2) (Sp): little stick, toothpick, &c. (fr. Croton, &c.) 3) (Mex): *Croton morifolius*.

palisade ratio: a figure representing fr. 4 determinations the average number of palisade cells lying beneath one epidermal cell (as seen in mount looking down on surface of cleared leaf); this is a constant for each sp. & is u. in identification of lf. materials - **p. tissue**: layer(s) of parenchyma cells beneath epidermis of lf. with cells lying at right angles to surface of lf.

Paliurus aculeatus Lam. (Rhamn) (sEu, wAs): Christ's thorn; paliure (Fr); frt. astr. u. in urin. dis.; rts. & lvs. anti-inflamm., in diarrh.; sds. u. in lung dis. - **P. inermis** Pur. = *Colubrina elliptica* - **P. ramosissimus** Poir. (Chin): rt. u. for sore throat; pulped lvs. dressing for abcesses; fls. diur.

pallao (Pak): dish somewhat like pilau (*qv.*).

pallar (Pe): lima bean.

palm: member (shrub or tree) of Fam. Palmae. - **African oil p.**: p. oil tree - **babassú p.**: *Orbignya* spp. - **Buri p.**: *Corypha elata* - **cabbage p.: c. palmetto**: *Sabal palmetto* - **p. cabbage**: *Roystonea regia* - **Chilean (nut) p.**: **coquito n. p.**: *Jubaea spectabilis* - **coco(a)-nut p.**: *Cocos nucifera* - **coquito p.**: *Jubaea chilensis* - **date p.**: *Phoenix dactylifera* - **doum p.**: *Hyphaena thebaica* - **p. family**: Palmae. - **fan p.**: any palm having palmate leaflets arranged in the shape of a fan;

ex. canyon palms (*Washingtonia* spp.) (Calif fan p.) - **feather p.**: any palm having pinnate leaflets; ex. coconut palm - **fishtail p.**: *Caryota mitis* - **gingerbread p.**: doum p. - **gomuti p.**: *Arenga pinnata* - **Guinea p.**: oil p. tree - **p. hearts: heart of p.**: apical stem bud of many palm spp.; u. as food ("cabbage") - **honey p.**: *Jubaea chilensis* - **jaggery p.**: *Caryota urens*; *Borassus flabellifer* - **p. kernel oil**: fr. p. oil tree - **needle p.**: *Rhapidophyllum hystrix* - **p. nut**: sd. of **p oil tree**: **oil p. tree**: *Elaeis guineensis*, source of p. oil - **palmyra p.**: *Borassus flabellifer* - **peach p.**: *Bactris gasipaes* - **p. real** (Cu): **royal p.**: *Roystonea regia* (Cuban); *R. elata* (Floridian) - **sabre p.**: cabbage p. - **sugar p.**: gomuti p. - **tagua p.**: *Phytelephas macrocarpa* - **toddy p.**: *Caryota urens* - **traveler's p.**: *Ravenala madagascariensis* - **Washington p.**: *Washingtonia* spp., esp. *W. filifera* - **wax p.**: *Ceroxylon andicola* - **wild date p.**: *Phoenix sylvestris*.

palma (Sp): 1) palm 2) *Yucca australis* - **p. almendron** (Br): *Attalea compta* - **p. Christi** (L.): *Ricinus communis* (FO) (**p. c(h)risti**: [Mex]) - **p. de aceite** (Mex): *Elaeis guineensis* - **p. de azucar**: *Arenga saccharifera* - **p. de bussu** (Br): *Manicaria saccifera* - **p. de coco**: coconut tree - **p. de miel** (Sp): coquito p. - **p. real** (Cu): royal palm - **p. rosa** (or **palma-rosa**): *Cymbopogon martinii*, *C. schoenanthus* (VO); *cf.* ginger-grass (oil).

Palmae (or **Palmaceae**): Palm Fam.; trop. & subtrop. shrubs & trees of both hemisph.; 2,650 spp.; chief characteristics: 1) no st. branches, lvs. growing direct fr. st. 2) lf. crown at apex of st. 3) fls. developed in lf. axils 4) lf. laminas pinnate (feather palms) or palmate-compound (fan-palms) 5) fls. trimerous, usually diecious 6) frt. a nut, drupe, or berry; ex. *Mauritia*, *Borassus*, *Bactris* (1848) - **P. of Amazonia**: furnish food, medicine, drink, shelter, clothing, illumination, fuel, materials for making fish nets, ropes, blowpipes, wax - **Palma nux**: palm nut; nut of *Elaeis guineensis*.

palmatine: alk. found in *Jateorrhiza palmata*, many *Berberis* spp., etc.

palmetto (Eng fr. Sp): fan palm, several spp. of *Sabal* & *Serenoa*; specif. saw p. or scrub p. - **cabbage p.**: **common p.**: *Sabal palmetto* - **dwarf p.**: 1) *Sabal minor* 2) saw p. (less commonly) - **dwarfed cabbage p.**: cabbage p. - **saw p.**: *Serenoa repens* - **scrub p.**: *Sabal etonia* - **swamp p.**: *Sabal minor* - **white p.**: common p.

palmiche (Co): Panama hat plant - **palmilla** (NM [US]-Sp.): *Yucca glauca* & other *Y.* & *Agave* spp. (*cf.* soapweed) - **palmita** (Sp): *Yucca rigida*, *Y. rostrata*.

palmitic acid: $CH_3(C_{14}H_{28})COOH$; satd. fatty acid occ. in palm oil (hence name), & many other fats & FO; m. pt. 64°, hence a solidifying component.

palmitin: glyceryl tripalmitate; chief component of palm oil (frt).

palmito (Sp): terminal bud of various palms, esp. *Euterpe edulis* 2) (Br) *Euterpe oleracea* 3) (Pan, Cu): *Roystonea regia*.

palmitoleic acid: physetoleic a.: $CH_3(CH_2)_5$-CH=CH. $(CH_2)_7COOH$; occ. in many marine oils (ex. whale blubber, codliver [zoomarinic acid]), but also in pl. FO, as palm, peanut, olive, *Argemone mexicana*, &c.

palmyra: palmyra (palm): p. tree: *Borassus flabellifer* (a source of **p. fiber, p. sugar):p. jaggery**: sugar (fr. sap of p. palm).

palo (Sp): wood - **p. amarillo**: *Mahonia fremontii* - **p. blanco** (Sp): 1) *Chione glabra* 2) (NM [US]): *Populus wislizeni* - **p. (de)-borracho**: *Ceiba* spp., *Chorisia insignis*, *C. speciosa* (Arg) - **p. de campeche** (Mex): **p. de tinte**: *Haematoxylon campechianum* - **p. del muerto** (Mex): *Ipomoea murucoides* - **p. de vaca** (cow tree): *Brosimum utile* - **p. de vela** (Pe) (candlewood): *Oreopanax capitatus* - **p. de velas** (CA): *Parmentiera cereifera* - **p. de zope** (fool tree) (Salv): Jamaica dogwood - **p. Maria** (PI, &c.): *Calophyllum* spp., esp. *C. inophyllum* - **p. mulato** (Mex): *Zanthoxylum pentanome* - **p. natri** (Ch): sp. of Fam. Sol.; good source of solanin; u. folk med. - **p. santo**: 1) (SA) (tree): *Bulnesia sarmienti* 2) (SA): *Triplaris* spp. 3) (Pan) *Rourea erecta* 4) various others: *Guaiacum sanctum*; *Porlieria hygrometrica*; *Swartzia tomentosa*; *Eriodictyon glutinosum* - **p. verde** (lit. green stem) (swUS): *Parkinsonia* spp. such as *P. florida*, *P. microphylla*.

palta (Pe, Ch, Ec): avocado, *Persea americana*.

Paludrin™ (Brit): recent synth. antimalarial.

palustric acid: component of pine oleoresin & rosin; belongs to abietic acid type of resin acids.

palynology: sci. of pollen grains & spores.

palytoxin: lethal marine poison found in *Palythoa* (Coelenterata) (Maui, Hawaii [US]), but produced by *Vibrio*-like microorganisms living symbiotically in tissues (virulence comparable to those of ricin & botulinus toxin).

Pamaquine Naphthoate: generic name of the first effective synth. antimalarial brought out by the German dye trust under TM name "Plasmochin"; has been effectively u. in combination with quinine, but quite toxic; u. is being abandoned.

pambotano (bark): fr. *Calliandra houstoniana*, *C. anomala*.

pampa (sSA): vast grassy treeless plains, mostly areas of cattle prodn. (Arg, Br, Ur).

pampelmusa (Arg, Co, Par): *Citrus grandis* (v. *pampelmusa*); shaddock.

Pamphilia aurea Mart. (Styrac.) (Br): source of a benzoin-like balsam ("Bogata storax") u. in plasters, as incense in churches, &c. (*cf. Styrax reticulatum*).

pan: 1) (Indo-Pak): popular masticatory of As composed of betel nut, lime, spices (such as cardamom, cloves, etc.), gold leaf, etc., wrapped in a betel leaf (*Piper betle*) (*cf.* pan supari) 2) (Sp): bread; wheat; food - **p. copal**: *Vouapa phaseolocarpa* - **p. supari** (Indo-Pak): masticatory of pan (betel vine lf.) & supari (betel nut), etc. - **p. yen**: Opium.

pana (PR): frt. of tree *Artocarpus altilis*.

Panacea (L.): 1) universal cure-all 2) *Arnica montana* - **P. Holsatica**: potassium sulfate (old name) - **P. Lapsorum** (cure-all for contusions): Arnica - **P. Mercurialis**: calomel.

panada: panado (SpAm): toasted bread boiled in water or milk to form a pulp, then flavored (ex. ginger); *cf.* bread and milk; intended for sick persons; sometimes u. as substit. for soups; USP 1850-60.

Panaeolus campanulatus Quél. (Coprin. - Agaricales - Basidiomyc. - Fungi) (eNA, SA): stained gill fungus; teonanacatal ("food of the gods"); oral ingestion of sarcocarps produces hallucinogenic state; utilized by curanderos of sMex; mushroom conceived as divine - **P. papilionaceus** Quél. (cosmopol.): butterfly panaeolus; fool's mushroom; psychotropic, leading to mental confusion; u. in love philters (Hung).

Panalba™: combn. of Novobiocin & tetracycline.

Panama bark: P. cascara (Sp): soap bk., Quillaja - **P. hat plant**: *Carludovica palmata* - **P. wood.**: P. bark.

panapen (PR): pana.

panaquilon: glycoside fr. ginseng rt. (foll.).

Panax ginseng C. A. Mey. (P. schin-seng Nees) (Arali.) (Kor, Manchur, Chin, Nepal): (Chin or Asiatic) ginseng; djen-shen; "ninsi"; tuberous rt. preferred by Orientals, but supply so limited that now largely superseded by the very expensive, imported Am sp.; c. ginsenosides (with aglycones panaxatriol, panaxadiol, saponigenins), very complex mixt.; u. ton. in states of exhaustion; such as low blood pressure, climacteric depression, memory loss; sed. in neurasthenia, stim., adaptogen; antipyr. in diabetes (2334) & cancer; use similar to that of *P. quinquefolius* - **P. japonicus** C. A. Mey. (P. repens Maxim.) (Jap): Japanese ginseng; decoc. or inf. of dr. rhiz. (c. saponins 21%), u. as expect., stom. antipyr.; ton. giving strength to convalescent people & women after childbirth; represents a panacea not specif. remedy; lvs. marketed in Chin & u. as expect. & emet. (cult. Chin/Indochin) - **P. notoginseng** (Burkh.) F. H. Chan (Chin-cult.): tienchi; sanchi; rt. u. much like that of "true" ginseng - **P. pseudo-ginseng** Wall. (eAs, Kor, Manch): rhiz. u. as ton. styptic, in nosebleed, astr., vulner., as aid in causing edges of wounds to stick together ("mountain varnish") (sometimes confused with *Gynura pinnatifida* DC) - **P. quinquefolius** L. (e&cNA fr. Qu, Man [Cana] to Ala & nFla [US]): American ginseng; sang; red berry; found in shady woodlands, but rapidly disappearing; also cult. (US & Cana [1710]); rhiz. u. mostly by Orientals as ton. stim. "reconstructive," aphrod. often valued very highly; adaptogen; Chin prefer wild pl. rt. because more irregular in shape, more like human figure (doctrine of signatures); marketed in many forms, ex. chewing gums, sodas, hair lotions, facial creams, &c.; market value very high - **P. vietnamensis** Ha & Grushv. (Vietnam): rt. given general stim. in physical, mental & sexual asthenia, incr. muscular strength & immunity, reduced fatigue from physical labor, to remedy hypotension, anti-arterioslcerotic. (2484).

panbotano (bark): v. pambotano

pancake plant: *Malva sylvestris*.

Panchina: total alk. prep. of Cinchona of Jap mfr. (formerly Formosa, ca. 1928).

pancreas: a gland which secretes digestive enzymes into the intestinal canal (exocrine) & hormones incl. insulin into the blood stream (endocrine); deproteinized pancreas ext. u. as antispasm. & vasodilator (ex. Depropanex™).

pancreatic dornase: deoxyribonuclease (pancreatic).

pancreatin: Pancreatinum (L.): a mixt. of enzymes, chiefly trypsin, amylase, & lipase, obt. fr. the fresh pancreas of hog or ox; assayed for digestive activity on starch & casein; u. in digestive derangements (tabs. capsules); as pre-digestant; in prepn. hydrolysates; formerly emulsifier; removal of dead growths in diphtheria, TB, cancer.

pancrelipase: pancreatic lipase: concentrate with mostly steapsin, but also some trypsin, &c.; u. as digestive adjunct.

pancreozymins: hormones fr. duodenal mucosa of pig; stim. pancreatic digestive secretions; may represent cholecystokinin; u. as diagnostic agt.

panda (bear) plant: *Kalanchoë* spp., esp. *K. tomentosa* (K. pilosa).

Pandaca retusa Markgraf (Apocyn.) (Mad): latex u. to prepare limed twigs ("bird lime") with which to capture small birds; aril of frts. sugared & u. to feed flying-foxes (fruit bats) & other animals (roussettes) (sharks).

Pandanus L. f. (Pandan.) (OWtrop): screw-pines; frts. of some spp. edible, lvs. u. for weaving (plaiting); for bags (for karaya gum, &c.); some with sweet-scented lvs. or fls. u. as ornam. (<pandan [Mal name of pl. < pandang Mal.] = conspicuous) - **P. candelabrum** Beauv. (trop wAf): screw-pine; u. med. (Guinea); lvs. u. in matting, for baskets, &c.; young leafy shoots supply a good fiber; stilt rts. u. to make brushes for whitewashing - **P. capusii** Martelli (Thail, Indochin): shrub; lf. inf. u. for fever in infants - **P. furcatus** Roxb. (Indones): young top lvs. eaten as antidote for poisoning; lf. sap u. against diarrh. dys.; unripe frt. emmen. - **P. klossii** Ridl. (Mal): decoc. of crown u. as post partum "protective" medicine - **P. luzonensis** Merr. (PI): tips of prop rts. decoc. u. as diur.; crushed fresh tips with mint lvs. u. stom. - **P. odoratissimus** L.f. (P. tectorius Parkinson) (Ind, seAs, widespread in Pacific Is): fragrant-flowered screw-pine; lvs. u. arom. bitter, pungent, in leucoderma, &c.; VO fr. bracts stim. antisep. in headache & rheum.; staminate fls. u. to perfume areca nuts, also in prepn. of a toilet water; frt. segments u. as starvation food & masticatory; lvs. boiled with brown sugar u. as sweetmeats (Indones); lvs. boiled with rice, &c. as flavor; lf. fibers (vae(o)ua) u. in weaving; source of kewda attar, u. in perfumery, cosmetics (VO with coconut oil) - **P. polycephalus** Lam. (seAs): frt. said to be abortif. - **P. tectorius** Warb. (seAs, Pac Is.): widely distributed sp. on or near beaches; screw pine; pandan(g); rt. decoc. + pineapple rts. u. as diur.; young undeveloped lvs. u. for dizziness; floral bracts laid over stab wounds (Banda); apical bud (cabbage) applied with *Fortunella japonica* juice as hot poultice to boils; aerial (prop) rts. decoc. u. as diur. & in blenorrhea.; var. *samak* (Hassk.) Warb. (P. samak Hassk.) (seAs): pandan samak in tikar; rt. u. in diarrh. - **P. utilis** Bory (Madag): screw-pine; fls. pleasantly odorous, edible; considered aphrod.

Pandigal (Eu): prep. of the pure *Digitalis lanata* glycosides for injection or oral u.

pandotan(d)o: common erroneous spellings for pambotano.

panela (Sp): brown sugar

Panellus stypticus (stipticus) Karst. (Panus, sFr) (Tricholomat. - Basidiomyc. - Fungi) (cosmopol.): bitter panus; fall & winter mushrooms; u. styptic (local appln.).

pangamic acid: vitamin B_{15}.

Pangestin™: pancreatin.

Pangium edule Reinw. (1) (Flacourti) (seAs, Mal, EI, Sunda Is): samaun tree; pangi; cold water inf. of fresh lvs. or sds. u. extern. as antisep. disinfect. antiparasitic; bk. u. as fish poison; lvs. & esp. sds. rich in cyanogenetic glycoside, also free HCN; dangerous poison; however, when rid of HCN (heat), sd. kernels eaten by natives (Mal, Indones); sd. FO (pitjung oil) u. as illuminant; lvs. wrapped around meat to keep fresh.

panhotano cortex: pambotano (misspelling).

pania (It): *Viscum album*.

panicle: type of inflorescence representing a comp. raceme (as in oats:—panicle oats) (*v.* App.F).

Panicum L. (Gram.): the panic grasses & millets - **P. antidotale** Retz. (Ind, cult. US): smoke of burning pl. u. to disinfect wounds - **P. crusgalli** L. = *Echinochloa c.* - **P. glutinosum** Sw. (WI, Mex-Par): ginger grass; rt. decoc. or syr. u. for pain in side & kidney region; to expel gravel & stones fr. bladder (Ve); pl. decoc. for sunstroke (as bath) - **P. maximum** Jacq. (P. jumentorum Pers.) (Af, now pantrop & subtr): Guinea (grass); Moha g.; most important forage (pasture) grass of tropAm; said shade-tolerant; hay & green feed; ripe sds. milled to make flour u. to make bread-soup (panade); vermifuge for children; tied around head for headache(!) (Uganda); rhiz. decoc. u. diur. & in bladder dis.; lvs. u. in mal. (Jam) with *Bambusa vulgaris*; inflorescence u. to make brooms (natd. sUS) - **P. microspermum** Fourn. (tropAm): arom. rhiz. of perenn. grass u. as emoll. stim. diur. (Br) - **P. miliaceum** L. (tropAs, esp. Ind, Eu, eNA): (proso) millet; broom-corn m.; cult. for grain u. for food (flour, bread) (Euras); cult. for forage fodder (US); u. to prep. braga (or busa), alc. beverage (Euras); grain fed to livestock (hogs, birds) - **P. repens** L. (trop&subtr worldwide): rhiz. decoc. u. in ovaritis & abnormal menstr. - **P. sarmentosum** Roxb. (tropAs): rts. chewed with betel as aphrod. (Mal); in medicinal baths (Indones); livestock food - **P. sumatrense** Roth (P. miliare Lam.) (trop-As, Af): little millet; ton. nerve stim. (Ind); lf. u. as antidote to poison fr. *Synadenium ballyi* & *S. volkensii*; food - **P. turgidum** Forssk. (nAf [Sahara], swAs): millet grass of extremely dry regions; grd. sts. u. for healing of wounds; frts. emergency food for porridge, breads; cattle fodder; sts. u. to make mats - **P. virgatum** L. (NA, Mex, CA): switchgrass; a perennial grass of econ. importance in prodn. alc.; good fodder (hay) & forage; proposed for use in producing alcohol for auto fuel (gasification).

panis (L.): bread; the staff of life; nutr., u. for heartburn - **P. St. Joannis** (L.): *Ceratonia*.

panocha (Mex): **panoche**: raw (brown) sugar loaves made in Lower Calif (Mex) (557).

Panopsin™: pancreatin.

Panosialin: antibiotic from Actinomycetes taxa, incl. *Streptomyces rimosus* sabin et al. *f. panosialinus,* &c.; u. as antiinfluenzal antiviral agt. (inhibits enzymes).

panstroside: glycoside of sarverogenin + D-digitalose; occ. in *Strophanthus courmontii*, &c.

pansy (herb): flg. hb. of *Viola tricolor* v. *hortensis* - **field p.: wild p.**: *Viola pubescens*; V. *arvensis*.

pantanals: areas in Matto Grosso (Br) where clay & arid grasslands of winter are converted into marshy plain & pools by summer rains.

Panteric™: pancreatin of triple strength.

Panthenol™: dexpanthenol; pantothenol: pantothenyl alcohol; d(+) form with vitamin activity; dietary source of pantothenic acid; u. in human & veterin. med. (intestinal atony; cholinergic).

Panthera leo L. (Felidae - Mammalia) (Af, swAs): (African) lion; body fat u. in Ayurved. med. as inunction to male organ of generation to incr. sexual power; bile for high fevers; smoke of burning hide inhaled in mal.(!!); other *P.* spp. u. in Ayurved. & Unani med.: **P. pardus** (black leopard; flesh & fat cooked & appl. in rheum.); **P. tigris** (tiger; flesh, fat, testis, tongue, bile, milk, excreta, u. in various ways in Unani med.) (1668).

panti (-panti), flores de (Pe): *Cosmos peucedanifolius* v. *tiraquensis* (fls.).

Pantopon™: prep. c. all the water-soluble alk. of Opium in the approx. ratio in which they occur in the crude drug; may be u. orally or hypodermically; thus as premedication before surgery.

pantothenic acid: vit. B_3; member of the Vit. B Complex; common in liver & molasses; also now prep. by synth.; dextrorotatory; u. in vit. deficiencies; to enrich flours, etc. (2109); administ. as Ca salt; helps restore normal bowel action (tone); may be helpful in allergy; important role in antibody formation (1669).

pantropic(al): found in tropics of both hemispheres.

Panus rudis Fr. (cosmopol, incl. NA) (Tricholomat. - Basidiomyc. - Fungi): frt. bodies u. in nCaucasus for prepn. of cheese from ewe's milk, "airan"; young frt. bodies sometimes eaten.

pan-yen (Chin): Opium.

Panzeria lanata Bunge: *Leonurus lanatus*.

pao (Br): seed, bread, &c. - **p. marfim** (Br): *Agonandra brasiliensis* - **p. parahyba** (Br): *Simarouba versicolor*.

paolins: antimicrobial & antiviral subst. fr. clams & oysters, also found in other animals (ex. calf thymus) & plants (ex. snow peas); original work on abalone (*Haliotis* spp., Haliotidae - Mollusca) (Chin, 1960) (canned a. juice); later u. ext. of clams; paolin I antibact., p. II antiviral (ex. herpes simplex, cold sores); tumor-preventive action claimed; (paolin is Chin name; *cf.* kaolin) (1670).

papa (Arg): potato; white potato (Sp) - **p. dulce** (NM [US]): sweet p. - **p. lisa** (Pe): *Ullucus tuberosus*.

papain: 1) digestive enzyme mixt. fr. latex of *Carica papaya*; incl. proteolytic enzyme, lipase, pectase, rennin-like enzyme, amylopsin; carpaine (alk.); c. chymopapains, payaya peptidases (hydrolases); u. to purify silk; protect wool in processing; in prepn. chewing gum (US); in remedying skin dis. (eczema, psoriasis); in dyspepsias & GIT dis.; in allergy (soln. sprayed into nose); to remove freckles; aid fat digestion & pancreopathies (2335); u. predigestant, esp. of tough meats; to chillproof beer, soft drinks, &c. (remove protein, &c.); to dehair hides (tanning indy.); int. to aid digestion (like pepsin); to tenderize beef (IV inj. before slaughter); for intestinal worms; to prevent abdomin. adhesions; appl. to stings of wasps, coral, jellyfish, Portugese man-of-war, &c. (often u. in form of meat tenderizer); for spinal disc herniation (prod. SriL, Tanz ["best"], Jap) (1849) 2) purified mixt. of proteolytic enzymes (papayotin; papoid) in ± pure form; u. similarly to 1).

Papaver argemone L. (Papaver.) (Eu): sand poppy; petals inf./syr. u. as diaph. (Sp) (similarly to *P. rhoeas*; c. rhoeadine); ornam (Eu) - **P. bracteatum** Lindl. (eEu, wAs): "codeine poppy"; great scarlet poppy; chief alk. thebaine (analg.) wh. can readily be converted into codeine but not with ease into morphine; hence recommended for cultn. in US (DEA) to furnish codeine (since morphine not produced in US; can be grown legally) - **P. dubium** L. (Euras, Med reg.): smooth poppy; Saatmohn (Ge); u. in homeop. med. - **P. nudicaule** L. (Arctic reg., e&wHemispheres-Himalay): Iceland poppy; arctic p.; yellow (arctic) p.; fls. white, red, orange, yellow; sds. c. opium alk.; u. for pain, tranquilizer, in headache (Kalmucks); fls. & capsules mildly diaph. (Ind); lvs. c. HCN-glycoside, cause of fatal sheep poisoning (Austral) - **P. orientale** L. (Euras to Iran): oriental poppy; c. many alks. chiefly isothebaine but no morphine; petals u. as diaph.; pl. has bright scarlet fls.; cult. as ornam.; dr. latex is opium-like, c. thebaine - **P. rhoeas** L. (Euras, introd. into NA, nAf, Austral, NZ): corn (or red or common) poppy; wild maws; cup rose; coquelicot (Fr); all parts of pl. c. alk. (fls. 0.12%), chiefly rhoeadine; also thebaine, protopine, coptisine, chelidonine (1671), not morphine (altho sometimes reported); petals rich in pigments, papaverrubins (4); petals u. in coloring medicines, wines, &c.; fls. u. in folk

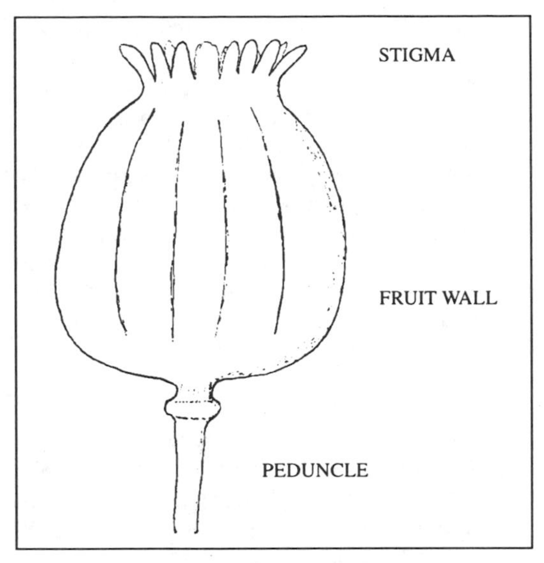

CAPSULE (FRUIT) OF OPIUM POPPY
(PAPAVER SOMNIFERUM)

med. as expect. & sed. (pectoral teas); sd. FO similar to that of opium poppy (2129); dem. anodyne, for cough; cult. ornam. - **P. setigerum** DC (wMed reg.): Riesenmohn (Ge) (giant poppy); besides *P. somniferum*, this is the only opium-producing poppy; latex bears similar alks. but in smaller amts.: however, this is not regarded as a comm. source of opium (low yields) - **P. somniferum** L. (Gr, Near East; eMed reg. where first cult.; cult. now in many subtrop areas): opium (or garden) poppy; (white) p.; chess-bowls; pavot du jardin (Fr); latex fr. full-grown but unripe capsules, removed & dr., constitutes Opium Gum; petal color varies: pink, red, purple (violet), white (var. *album*); gum opium rich in morphine (6-20%), codeine, papaverine, noscapine (narcotine), thebaine, narceine, in all some 30 genuine alks. + several arising during processing by oxidn., hydrolysis, &c.; alks. constitute 20-25% of wt. of opium; u. narcot. sed. hypn. analg. antispas. fumitory, masticatory (opium habit); opium in 3 forms (gum [crude], granulated, & powdered); Opium also produced by straw & green poppy methods (most for codeine) (1672); poppy capsules (Fructus Papaveris) (poppy heads) very small amt. alks. u. sed. in irrit. cough, insomnia, in eye waters, etc.; sd. (blue, white) (Dutch sds. best [blue]); Russian, Ind, Turkish (almost white); maw seed u. food (in pastries, etc. to ornament & flavor) (2130), source of FO; u. as any bland FO, for culinary & med. purposes, also in prepn. iodized p. sd. oil (1922-) u. as X-ray contrast medium (iodized oil inj.) (chief prodn. Neth, Turk, Chin, nAf, Co); only opium (not its alkaloids) can be imported into US, of this, 60% u. in mfr. of codeine; opium retained in national strategic stockpile; over the years, produced in small amts. in US; last grown legally in US in 1942; lvs. u. sed.; var. *album* DC: this & the typical var. are the off. sources of Opium; var. *nigrum* DC: black opium poppy.

Papaveraceae: Poppy Fam.; small gp. (210 spp.) of hbs. & few shrubs, mostly of nTemp Zone; possess milky juices (latices) which may be brightly colored (red, yellow, etc.); fls. regular with 2- 3 sepals, 4-6 petals, many stamens, 1 massive, rounded pistil (or 6 or more pistils); frt. capsule; pls. of fam. rich in alks.; ex. *Sanguinaria*; *Chelidonium*.

papaveramine: xall. alk. of opium - **papaverine**: alk. of Opium (0.8-1% of Op.); also synth.; HCl salt acts as antispasm. in angina pectoris, embolism, spasms, colics, vasodilator in stroke victims, org. brain syndrome (b. damage, senility); intracavernous (inj. into penis blood vessel rapidly active) to give erection (sometimes with phentolamine); impotence; non-addictive - **Papaveris Fructus** (L.): poppy capsules.

Papaveretum: "concentrated opium"; total alks. of opium; Pantopon (*qv.*) - **papavero** (It): opium poppy.

papaw: papaya (*cf.* pawpaw) - **p. family**: Caricaceae (or Papayaceae) - **northern p.**: *Asimina triloba*.

papaya: papayer (Fr): *Carica papaya* - **papayotin**: papain.

papel (Sp): paper - **p. higienico** (Arg): toilet p. - **p. secante** (Arg): blotting p. - **papier** (Fr, Ge): paper.

paper: a film or layer of cellulose fibers prepd. fr. cellulose pulp, wh. represents phloem or wd. fibers or tracheids fr. which lignin has been removed (or other fibers) - **p. shell**: term appl. to pecans, almonds, &c., denoting thin husks (shells) which can be broken between the fingers.

Papilionaceae: Legume, Pea, or Bean Fam.; by some authors segregated as a separate fam. fr. the Leg.; fls. irregular, often superficially similar to a butterfly, hence the name (papilion, Fr, butterfly); frt. a legume or loment; lf. blades compd. (Fabaceae).

papoid: papain.

papoose = pappoose.

papoula (Por): 1) poppy sd. 2) *Hibiscus rosasinensis* (also **p. dobrada**).

Pappea capensis E. & Z. (1) (Sapind.) (sAf): sand olive; sd. kernel c. FO non-drying, edible but cath.; formerly appl. to tinca capitis, alopecia, &c., & as hair restorative; u. as lubricant & in mfr. soap; tinct. of sd. u. in constip.; frt. ("wild plum") edible, u. in jelly making; rt. u. for ankylostomiasis.

Pappel (Ge): 1) poplar 2) mallow.

pappoose root: *Caulophyllum.*

paprika: *Capsicum frutescens* & other *C.* spp. bearing large frts. - **Hungarian p.**: large red conical pepper commonly cult. in Hungary & widely used as cond. - **King's p.: Koenigs-p.**: pd. prod. made fr. prec. incl. sds. & peduncles - **rose p.: Rosen-p.**: pd. prod. made by grinding select Hungarian p. after first removing sts., peduncles, & placentae - **Spanish p.**: pimiento: globoid frts. sweet & nonpungent; u. to stuff olives, etc. - **Turkish p.**: similar to Hungarian p.

paprikash (Hung): natl. Hungarian dish (cf. goulash) prepd. with paprika.

papyrus: *Cyperus papyrus*, u. to make Egyptian paper (origin of prod. & word "paper") - **p. Ebers**: *v.* Ebers papyrus.

para- (p-): prefix meaning opposite, beyond, irregular, wrong.

Pará: large Amazonian state of Br made up mostly of tropical rain forests. v. under arrowroot; cress; rhatany; rubber - **P. gum**: prods. of *Hevea brasiliensis*, *H. guianensis*; *Piptadenia rigida.*

para-aminobenzoic acid: PABA: Paba; part of vit. B complex; widely distr. in nature, in both pls. & animals; both free & in conjugated (combined) form; sources: meats (as liver), yeasts (a good source), vegetables, synth.; this & K salt u. in prophylaxis & treatment of rickettsial dis. (ex. typhus, Rocky Mountain fever, &c.); collagen dis. (ex. rheumatic fever); antidote in sulfanilamide poisoning; sunburn preventive (sunscreen agent); "anti-gray hair factor" (but only for rats); in eczema nasi (dogs); considered non-essential vitamin (part of vitamin B complex).

paracari (Br): **paracary**: *Peltodon radicans.*

paracasein (US): the casein of GB; mixt. of phosphoproteins found in milk, beans, nuts, &c.; strongly nutritious since rich in essential amino acids; u. in paints (water type), plastics, adhesives, coating papers, sizing of textile fibers, &c.

parachute flowers: 1) *Ceropegia sandersonii* 2) *Aristolochia tagala.*

Paracodin™: dihydro-codeine.

paracoto bark: fr. *Nectandra pseudo-coto* (Bo, Br).

paradise plant: *Daphne mezereum* - **p. tree**: 1) (Utah [US] Mormons): *Ailanthus altissima* 2) (Belize): chinaberry tree 3) *Simarouba glauca.*

Paradisi Granum (L.): sd. of *Amomum granumparadisi.*

Paradisia liliastrum Berth. (Lili.) (Eu-mts.): rt. c. cardiac poison; u. med.

paradol: yellow oily subst. conferring its pungency on Paradisi Granum.

paraffin: 1) **Paraffinum** (L) (US) : solid wax-like mixt. (**p. wax**) of hydrocarbons fr. petroleum; u. as stiffening agt. in ointments, &c. 2) (Eng): gasoline 3) (Ge): p. wax, p. oil, kerosene 4) (chem.): any saturated hydrocarbon of the methane series - **household p.**: kerosene fractions fr. crude petroleum - **p. dressing**: cotton or other gauze impregnated with mineral oil (Curatio Paraffini) - **medicinal p.**: liquid p.: **p. oil**: 1) liquid petroleum, liquid paraffin, mineral oil, Paraffinoel (Ge) 2) (GB): kerosene - **p. scale**: p. - **soft p: solid p.**: Paraffinum Molle.

Paraffinum Molle (L): petrolatum, petroleum jelly - **P. Perliquidum** (L,DAB): thin mineral oil, thin liquid medicinal oil, light mineral oil - **P. Subliquidum** (L,DAB): thick liquid p., heavy medicinal white oil, heavy mineral oil.

paragene: plasmid.

Paraguay roux: *Spilanthes oleracea.*

Parahancornea amapa (Hub.) Ducke (Apocyn.) (Surinam, Br): am(a)apa; latex u. as cicatrizant, in intest. dis., TB, &c.; painted on wounds; in general debility; frt. edible, savory, nutr. - **P. krukovii** Monachino (Amazonian Co): latex painted on wounds to hasten healing.

Paralstonia culsiacea Baill. (Apocyn.) (PI): bk. u. on swellings (—> *Alyxia*).

Parameria laevigata (A. L. Juss.) Moldenke (P. vulneraria Radlk.; P. barbata Schum.) (Apocyn.) (seAs): tagulaway (bk.); bk. inf. u. in mal. as antipyr. in rheum.; bk. inf. u. after birth to make uterus shrink (Mal); lf. & fl. inf. u. as emmen.; source of cebu(r) balsam, u. as vulner. & for scaly skin dis.

parameter: 1) independent variable thru functions of which other functions may be expressed (ex. event detd. by 4 parameters [3 = position determinants, 1 = time determinant]) 2) arbitrary constant each value of which determines the specific form of a system.

Paramignya scandens Craib (Rut.) (seAs): rt. decoc. taken post part. & for abdomin. complaints (Mal).

paramos (Sp, SA): high dry treeless grassy plateaux (plains) esp. of SA; ex. Ve, Br (< paramo [Sp]: wasteland, alpine scrub plains).

Parapiptadenia rigida (Benth.) Brenan: name now replacing *Piptadenia r.* (*qv.*).

paraplasm: hyaloplasm; the more fluid finely granular portion of cytoplasm.

Parartocarpus venenosus Becc. (Mor.) (Mal, New Britain): latex appl. to sores; ripe frts. pd. & u. for healing of wounds; frt. & sds. eaten.

parasympatholytic drugs: those wh. block the action of parasympathetic nerve fibers; ex. atropine - **parasympath(et)-omimetic drugs**: those wh. act as though the parasympathetic nerves are stimulated; ex. pilocarpine, physostigmine, arecoline.

Paratecoma peroba (Record) Kuhlm. (Tecoma p. Rec.) (monotypic) Bignoni) (c&sBr): peroba; bk. antipyr. u. in fever; wd. much u. for building; tree threatened with extinction.

parathyroid gland: Parathyroideum (L.): the dr. p. glands of animals u. as food by man to raise calcium level in blood; of no provable value when given by mouth; hence now u. in form of a sterile soln. inj. parenterally in tetany, chorea, epilepsy, allergy, etc.

paratuberculin: johnin.

paratudo bark (lit. good for all): fr. *Gomphrena officinalis*.

paratype: specimen or illustration cited in protologue wh. is neither holotype nor isotype nor syntype; when 2 or more specimens were simultaneneously designated as types.

parcha (Mex): *Passiflora caerulea, P. edulis* - **p. granadina** (Mex): *P. quadrangularis*.

parched (seUS): roasted, as peanuts.

parchita (Sp): passion flower or frt., *Passiflora foetida*.

parchment (, artificial): linen or other rag paper rapidly passed through 50% sulfuric acid soln., then washed & dried.

parde: black drink (sacred) fr. *Eryngium* or *Cassine* spp. (seUS Indians).

pardillo (Mex): *Cordia alliodora*.

paregoric: Opium Tr. Camphorated - **p. root**: *Osmorhiza longistylis*.

Pareira: pareira brava: *Chondrodendron tomentosum* (rt.) - **false p.**: *Cissampelos pareira* - **p. root**: p. brava - **white p.**: *Abuta rufescens, A. candollei* - **yellow p.**: *Abuta amara*; *Aristolochia glauca*.

parenchyma: relatively simple undifferentiated ("primitive") tissue in wh. the relatively large commonly isodiametric cells have a prominent living & active protoplast; this constitutes the bulk of living pl. tissues devoted to the functions of assimilation & storage - **conjunctive p.**: parenchyma lying between vascular strands of stele - **palisade p.**: v. under palisade - **spongy p.**: irregularly scattered parenchyma in leaf lying below the palisade layer(s).

parenteral: administration of drugs other than by g.i. tract, esp. by needle (IV, IM, SC).

parica (tropAm Indians): plant snuff derived fr. *Virola* spp.

pariétaire (Fr): *Parietaria officinalis*.

Parietaria debilis Forst. (Urtic.) (Par, Co): baico; u. as vulner. diur.; in "tropical anemia" - **P. officinalis** L. (P. erecta Mert. & Koch) (Med reg., esp. It, cEu): common (or wall) pellitory; hb. c. mucil. tannin.; u. as vulner. emoll. diur. in rheum. & dem. in coughs, hemorrhoids; in kidney & bladder dis.; inflammn. abscesses - **P. pensylvanica** Muhl. (NA, incl. Mex): (American) pellitory; u. med. (Mex) (Mex Ph.,1952) - **P. ramiflora** Moench (Euras): u. similarly to *P. officinalis*, of which it is possibly a var.

parilla: little p.: *Solanum dulcamara* (probably) - **yellow p.**: *Menispermum canadense* (rhiz. & rts.).

parillin: sarsapogenin with 3 glucose & 1 rhamnose units at C-3; fr. sarsaparilla rt.

Parinari (Parinarium) curatellifolia Planch. ex Benth. (Chrysobalan.) (tropAf): mbura; bk. with earth of termite nests (incl. fungal gardens) u. int. for hookworm; in mental dis.; pounded rts. eaten with porridge for edema; rt. juice for stomachache - **P. excelsa** Sabine (trop wAf): grey plum; u. like prec.; wd. like that of oak; frt. edible - **P. glaberrima** Hansk. (Java to Polynesia): u. med. by natives (Fiji); sd ("pipi") u. to scent FO (Tonga) (86) - **P. macrophylla** Sabine (wAf): tamba kouma (rotten plum); ginger bread p.; pomme de Cayor (Fr): frt. eaten (but of poor quality); tree will grow on saline soils; fresh frt. rind has pleasant odor, u. in ointments; bk. & lf. decoc. u. as mouthwash; lotion fr. bk. u. in ophthalmia; sd. kernel source of "neou" oil; wd. u. - **P. polyandra** Benth. (wAf): lf. u. like that of *P. curatellifolia*; wd. u. to build sheds, make charcoal (excellent); bk. tan.

pariparoba (Br): *Piper umbellatum* or a related sp.

Paris chinensis Franch. (Lili) (Chin): rhiz. decoc. u. in mental disorders; rhiz. ground in vinegar u. in skin dis., esp. ringworm - **P. polyphylla** Sm. (nAs): pd. rt. in water u. as ton.; for worms; epilepsy - **P. hainanensis** Merr. (eAs): u. med. to "remove toxins" (Chin) - **P. quadrifolia** L. (Eu, nAs): (herb) truelove; erba Paris (It) (herb Paris); one berry; entire pl. (Herba Paridis) c. the saponins paridin & parisstyphnin; rhiz. juice appl. to itchy sores (Chin); vulner.; for tumors; ophthalmia; in gout, rheum.; fresh pl. u. in Eu. homeop. practice for headaches, neuralgias (2131).

Paris black: animal charcoal; lampblack.

Parkia africana R. Br. (Leg) (tropAf, seAs): African locust; frts. of tree eaten; considered good nutrient for children; sds. food, aphrod. weakly purg., roasted ("parched") u. as Sudan coffee & in prepn. of dana-dana cheese; u. as condim. in colic; fish poison; source of "nete" meal; bk. tannin-rich; u. in fever, for toothache, tumors; tanning of animal hides (red-brown leather produced); wd. u.

in carpentry - **P. bicolor** Chev. (trop wAf): soma (Gold Coast); macerated bk. & lvs. u. to prep. eye lotion; powd. bk. u. as vulner. - **P. biglandulosa** Wight & Arn. (Mal): Malay nitta tree; sd. c. 18% FO = inga (or ingo) oil - **P. filicoidea** Welw. ex Oliv. (trop wAf): African locust bean; dawa dawa (frt. & sd.); "wo"; bk. u. ton. (inf.); frt. husks decoc. astr. in diarrh.; fl. buds u. in Hansen's dis.; frt. pulp edible; sds. after fermentn. u. as condim. - **P. roxburghii** G. Don (P. javanica Merr.; P. biglobosa Benth.) (tropAf, As): African locust bean; bk. hot decoc. u. as mouthwash & in dentalgia (wAf); frt. pulp mealy, u. diur. emoll. dem.; frt. & sds. edible; sds. u. as coffee substit.; sd. decoc. carm. in colic, general ton. & to remedy distended stomach (Mal); pd. sd. decoc. appl. to wounds & ulcers (PI) - **P. speciosa** Hassk. (Indones): sds. (petch beans) with odor of garlic u. as spice raw or cooked in rice; u. in diabetes & for worms (Java) - **P. sumatrana** Miq. (seAs; Sumatra): ta sek (Camb): bk. u. in baths; (macerated) young lvs. eaten (Camb, Laos).

Parkinsonia aculeata L. (Leg) (WI, tropAm, sUS - natd. eAf, As): Jerusalem thorn; palo verde (green stick, SC bk. green); cloth of gold (Tex); retama; espinillo (Ve); thorny shrub or small tree; lf. fls. sd. antipyr. in recurrent fevers & phthisis (tropAf); folk med. u. for fever & diaph. abortif. (Ve); epilepsy (Mex Indians); fls. eaten by children (Senegal); sds. edible; wd. u. in paper making; ornam. (natd. trop. & subtr. of world) - **P. florida** S. Wats. (Cercidium floridum Benth.; C. torreyanum Sarg.) (Calif, Ariz [US]; nwMex): (blue) palo verde; border p. v.; small tree; sds. ground to make cake u. as food by Am Indians; also in prepn. beverage - **P. texana** S. Wats. var. *macrum* Isely (Cercidium m. I. M. Johnst.) (sUS): Texas palo verde; pods & sds. edible (as meal); bee plant; forage plant (2132) (emergency food); eaten by jackrabbits.

parmaceti: corruption of spermaceti (= Cetaceum).

Parmelia caperata (L) Ach. (Parmeli. - Ascolich. - Lichenes) (cosmopol.): stone crottle (lichen); wrinkled parmelia; chalcacura; thallus u. for red, orange, purple dyes; c. polysaccharides wh. inhibit growth of tumors (Sarcoma 180) - **P. cirrhata** Fr. (widely distr.): dark crottle lichen; u. dyestuff; extensively studied - **P. conspersa** (Ach.) Ach. (temp zone, widely distr.): "sprinkled (parmelia) lichen"; c. usnic & salazinic acids; u. in Chin med.; source of yellow, orange, red, purple dyes - **P. esculenta** = *Lecanora e.* - **P. islandica** = *Cetraria i.* - **P. kamtschadalis** (*camtschadalis*) Eschw. (tropAs): gives pale rose dye u. to print & perfume calico cloth (Ind); smoke inhaled to relieve headache; pd. u. as cephalic snuff (Ind) - **P. omphalodes** (L) Ach. (cosmopol. in temp. zone): black crottle (parmelia); smoky shield lichen; u. for red, purple, orange, brown dyes (Scandia, Brit-Isles) - **P. parietina** (L) Ach. = *Xanthoria p.* - **P. perforata** (Wulf.) Ach. (cosmopol): u. diur. (Ind) - **P. perlata** (Huds.) Ach. (temp. areas of world): sprinkled (p.) lichen; c. atranorin, stictic acid; u. astr. carm. in dyspeps. tonic in amenorrh. in calculi; aphrod. diur. (Ind); purple dye - **P. physodes** (L) Ach. (P. ceratophylla Schaer) (worldwide): darks crottle (p.); d. c. lichen; horny p.; puffed shield lichen; said most commonly found of the ca. 800 *P.* spp.; c. lichen acids incl. physodic acid (fungistat.); u. as acacia substit.; cooked & consumed as soup (Wisc [US] Indians); source of yellow & brown dyes u. for calico printing (Indians); c. antibiotics (1673); u. int. in TB, acute enteritis; leucorrh.; exter. in inf. skin dis., cut wounds; hemostatic; vulnerar. in veter. med. - **P. saxatilis** (L) Ach. (Parmotrema s. (L) Choisy) (cosmopol., esp. Indo-Pak, nwHimal, wNA): stone crottle p.; u. as "cure" for epilepsy (Indians); u. for red, purple, orange, yellow dyes, u. for woolens (BritIsles) (ex. Harris tweed) & calico - **P. vagans** Nyl. (seRuss): c. usnic acid, salazinic acid; u. exter. as styptic, bactericide.

Parmeliaceae: Parmelia Fam.; a division of the Lichens which incl. a large number of spp. which can be u. for dyeing; genera incl. *Parmelia, Cetraria.*

Parmentiera aculeata (HBK) Seem. (Crescentia a. HBK) (Bignoni) (sMex): pepina; lvs. u. in retention of urine - **P. alata** Miers = *Crescentia a.* - **P. cereifera** (*cerifera*) Seem. (Pan): candle tree; palo de vegas; palo de velas; frt. resembles yellow candles, suspended 1-2 m long, with apple-like odor, good food for cattle; lvs. c. glycoside - **P. edulis** DC (Mex, Guat, Salv): guajilote; chote; sweet frt. edible, raw or cooked; frt. roasted & eaten for colds (Salv); colds, rheum. lumbago; rts. u. as diur. in edema; diabetes - **P. stenocarpa** Dug. & L. B. Smith (Pan): candle tree; arbol de la bujias; frt. pulp & sds. eaten & u. to make beverage (436).

Parmotrema perlata (Huds.) Choisy = *Parmelia p* - **P. saxatilis** (L) Choisy = *Parmelia s.*

Parnassia faberi Oliv. (Saxifrag.) (eAs): crushed pl. u. to poultice dog bites (Chin) - **P. palustris** L. (nEuras, nAf, Bering Sea, NA): grass of parnassus; white buttercup; hb. u. as astr. sed. in nervousness, epileptic seizures, decoc. of entire pl. u. in stomach pains & pyrosis (Swed); also u. in hemorrhages, eye dis., as diur. (cult. Ind med.).

paromomycin: antibiotic isolated fr. several *Streptomyces* spp. incl. *S. rimosus* subsp. *paromomycinus* Coffey et al. (1959); active against & u. for control of several bacterial spp. & *Entamoeba histolytica.*

Paronychia argentea Lam. (Caryophyll.) (Med reg., sEu, Mor): silver nailroot; hb. u. in fevers, grip, as diaphor. diur. aphrod. (Mor); tea substit. (Arabian tea) - **P. jamesii** T & G (wUS): James whitlowwort; hb. u. to prep. a tea (Kiowa Indians) - **P. serpyllifolia** DC (sEu): u. for sores at base of nails; hence common names, nailwort & whitlow-wort.

Paropsia brazzaeana Baill. (Passiflor./Flacourti) (tropAf): rt. inf. u. as gargle for toothache; in GC; frt. remedy for headache.

Parosela Cav. = *Dalea.*

parotin: endocrine of salivary glands; globulin type of protein; prod. by parotid gland (hence name); stim. dental calcification, but at same time reduces Ca level of blood. (1953).

parqueted cells: elongated cells in gps. with the cells in each gp. arranged parallel & at right angles fr. direction of cells in other adjacent gps.; ex. fennel endocarp.

Parquetina nigrescens (Afzel.) Bullock (monotypic) (Asclepiad.) (wEquatorial Africa): kunto; wd. c. strophanthidin & its glucoside; convallatoxin; u. in prepn. of arrow poison; natives use lf. latex against tumors, diarrh., as body paint; lvs. appl. locally for craw craw (skin dis.).

parra (SpAm): grapevine - **p. cimarrona**: *Vitis caribaea* (WI), etc. - **parra-vid**: grapevine.

parri (Br): *Arrabidaea chica.*

parrot's feather: *Echinocystis lobata*

parrot seed: parrot's corn: sds. of *Carthamus tinctorius.*

parsley: p. family: Umbelliferae (Apiaceae) - **common p.**: *Petroselinum crispum* (lvs. rt. frt. sd. oil) - **fool's p.**: *Aethusa cynapium* - **garden p.**: common p. - **Hamburg p.**: - *Petroselinum crispum* var. *tuberosum* Crovetto (form with edible rts.) - **marsh p.**: *Selinum palustre* - **milk p.**: *Peucedanum palustre* - **mountain p.**: *Peucedanum oreoselinum* - **p. piert**: 1) *Aphanes arvensis* (commonest) 2) *Erica aphanes* 3) *Scleranthus annuus* - **poison p.**: p. hemlock - **rock p.**: common p. - **p. seed**: frt. of common p. - **tulip-rooted p.: turnip-rooted p.**: Hamburg p. - **water p.**: *Oenanthe crocata* - **Welsh p.**: Cannabis - **wild p.**: 1) *Zizia aurea* 2) *Aethusa cynapium.*

parsnip: *Pastinaca sativa* - **early p.**: **(golden) meadow p.**: *Zizia aurea* - **cow p.**: *Heracleum lanatum* - **parsnip (root)**: *Pastinaca sativa* - **water p.**: *Sium* spp., esp. *S. latifolium* **(common w. p.)**: *S. suave,* &c. - **wild p.**: 1) *Conium maculatum* 2) *Pastinaca sativa* (wild growing) 3) *Zizia aurea.*

Parsonsia helicandra H. & A. (Apocyn.) (tropAs): pl juice u. int. in psychosis (Hindu med.); placed on leg swellings (Solomon Is).

Parthenium argentatum Gray (Comp) (swUS, Tex, adj. Mex): guayule (plant); Mexico rubber plant; low shrub - 1 m; silvery white lvs. (2133); lvs. rich in VO (c. pinenes, cadinene); bk. c. caoutchouc (polyterpene) (-30%); u. mfr. guayule rubber; formerly widely cult. (1674) - **P. hysterophorus** L. (eUS, Mex, WI, SA): bastard feverfew; "ragweed" (do not confuse with others); Santa Maria; altamisa del campo; yerba de la oveja (Arg.); pl. c. parthenine, parthenin, hysterin (sesquiterpene lactones), bitter glycoside; u. bitter ton. antipyr. exter. as emoll. vulner. (Arg); snakebite (Arg); analg. for headache, neuralgia; fever; arthritis, sores (Aztec & other Am Indians) (pd. lvs. on open sores-Bah); inf. of tops u. in dyspep. & as emmen.; in mal.; lf. decoc. spilled on floor to kill fleas (Trin); in place of hops in making beer - **P. incanum** HBK (Chihuahua, Mex): mariola; u. med. (Mex); entire pl. u. to ext. rubber - **P. integrifolium** L. (e&cNA): wild quinine; rt. decoc. u. in burns (Catawba Indians).

Parthenocissus quinquefolia (L) Planch. (Ampelopsis q. (L) Mx.) (Vit.) (eNA, Mex; natd. Eu): Virginia(n) creeper; woodbine; five leaves; American ivy; vina virgen; vigne vièrge (Fr); bk. young twigs, & lvs. (5 leaflets) u. ton. alter. expec. diur. astr. sl. vesicant when lvs. crushed on skin; formerly u. for rheum. & pain; often cult. to cover walls & arbors; toxic pl. (esp. berries [blue]) (some fatalities) (< parthenos [Gr]: Athene, virgin) - **P. serianaefolia** (China): u. as analgetic - **P. thomsonii** Planch. (Chin): poultice u. for sores & boils - **P. tricuspidata** Planch. (Chin, Jap, natd. Eu): st. lvs. frt. u. as ton. & to treat inflamm. swellings; frt. decoc. for obstinate nosebleed.

partridge: *Perdix* spp. (Phasianidae - Aves): small grayish birds, often sought as game, esp. *P. perdix*, European (or Hungarian) p. (Eu, now in w&nwUS, swCana); u. in Unani med.; v. also under berry; pea.

Pasionaria (Mex): *Passiflora incarnata* or *P. alba.*

pask flower: pasque f.

pasota: pasote (Sp): *Chenopodium ambrosioides* (corruption of apazote).

Paspalum conjugatum Berg. (Gram) (sFla [US], WI, tropAm-OWtrop): bed grass (Bah); fresh frts. u. in diarrh. (PI), sore throat (nwSolomon Is); in mixt. for clearing lungs in TB (Bah); in fever, flu, pleurisy (Trin); forage for livestock; common weed - **P. dilatatum** Poir. (tropAm, natd. sUS, sAf): Dallis grass; popular as forage, lawn, & pasture grass - **P. distichum** L. (tropAm, s to Arg, US, OW coasts): knot grass; pasture grass; u. to retain banks of streams & ditches - **P. laxum** Lam. (WI; Key West, Fla [US]): chien-dent; drink made by placing whole pl. in water (cold water inf.) given infants to hasten teething (Domin) - **P. notatum** Fluegge (eUS, Mex, WI, SA): Bahia grass; perenn. grass cult. as fodder (cattle) - **P. scrobiculatum** L. (As, now cosmopol., cult. in Ind): kod(r)a; Koda millet; millet-like cereal widely u. as food & fodder; rts. u. as remedy in childbirth (PI); believed to cause digestive disturbances (var. *commersonii*) (toxic principle?).

pasque flower: pulsatilla, *Anemone patens*, &c.

Passiflora L. (Passiflor.) (mostly Am, a few spp. in As, Austral, Af): passion flowers; granadilla; maracudja; climbing pls.; fls. often very beautiful; "passion" fl. SC symbolism of Christ's passion (signs of cross, scourges [tendrils], crown of thorns [corona], nails of crucifixion, wounds [5 stamens], disciples, &c.) (2337); u. in herbal teas - **P. alata** Ait. (Br, Pe): maracuja; grenadille; sds. u. hypnotic (in large dose); emmen. diaph. (claimed); roborant; lvs. u. mild hypnotic, sed. (Br); cult. for delicious frts. (u. in ice cream) & as ornam. (Br) - **P. alba** Link & Otto (Mex, Br): pasionaria; u. med. Mex - **P. caerulea** L. (sBr, Par, Ur, Arg, WI, Mex): blue passion flower (or vine); lvs. c. cyanogenetic glycosides; rt. u. as emet., lvs. vermifuge, entire pl. antispasm. sed.; in native med.; frts. ("les barbadines") esculent (grenadille) (1053); vine commonly cult. as ornam. - **P. coccinea** Aubl. (Pe, Guiana): rt. u. as abort. (Pe); frt. edible, antipyr. - **P. coriacea** Juss. (sMex, CA, SA to Bo): (ala de) murcielago (CR); media-luna; juice or latex of rt. appl. with cotton to eye (360); u. in domestic med. as diur. in urin. (kidney) dis.; sd. u. as insecticide for "chuchus"; lvs. with lard u. as poultice for wounds & swellings (Salv); lf. inf. in infections (Pe) - **P. cuprea** L. (WI): sabey de costa (Cu); lizard's tail (Bah); vine juice appl. to cuts (Bah); u. as tea (Bah) - **P. edulis** Sims. (sBr, CA, US, Burma, cult. Austral, sAf): (purple) granadilla; passion fruit; (yellow) lillikoli (Haw [US]); large pulpy frt. & juice of this & several other *P.* spp. u. esp. in Pacific area as stim. ton., food (raw) & flavor, in beverages (2136), ice creams, sherbets, candy &c.; sd. source of a drying FO - **P. ed.** var. *flavicarpa* Degener: comm. crop raised for frt. juices (cult. Calif [US], SA) ("Lillikoi" cult. var.) (Polynesia); only *P.* grown comm. in US.) - **P. foetida** L. (P. ciliata Dryand.) (Mex, WI, Br, Ve+): pasionaria hedionda (Cu); stinking passion flower; "lemon apple" (Haw); parchita; herbaceous vine; pl. u. as narcotic sed., esp. in insomnia & convulsions (children), hysteria; lf. inf. in cough, colds, intest. worms (Trin); earache (Pe); frt. edible in reasonable amts. (c. HCN) - **P. f.** var. *ciliata* (Mex, Guat, Jam): pochcac; u. sed., hypnotic - **P. incarnata** L. (seUS n to Ind, Bermuda, SA, natd. seAs): (white or wild) passion flower; maypop; passion vine; apricot v.; plant sometimes weedy; passion frt. size & shape of hen's egg, edible; herbage (coll. after frts. ripened) c. alks. (harman [passiflorine]), harmol, &c. (2134); flavone glycosides, sterols; catechol, gallic acid; pl. mildly antispas. sed. in neuralgia, etc.; hypn. in epilepsy (2135), neuroses, menopause; lvs. & fresh expressed juice of fls. also u. med. at times; mental anguish (1675); nervousness; insomnia; anxiety; analg.; rt. ext. u. to treat ulcers, hemorrhoids, etc.; frt. u. to make jellies; cult. ornam. - **P. laurifolia** L. (tropAm, seAs): water lemon; bell-apple; lvs. toxic (HCN), u. anthelm.; lf. decoc. u. to induce sleep (Kubeos, Co) (2299); frt. grenadille (Fr) edible; sds. u. cardioton. hypn. emmen. diaph. (Br) - **P. ligularis** Juss. (sMex, CA, SA): (sweet) granadilla; one of chief *P.* spp. of CA; esculent frt., sds. also eaten & u. in beverage - **P. lutea** L. (e&cUS): yellow passion flower; u. similar to *P. incarnata* - **P. manicata** (Juss.) Pers. (Ec): frt. edible; frt. juice u. to bathe nervous baby (by "gumming up" fleas & lice?) - **P. mexicana** Juss. (Mex): itamo; calzoncillo; decoc. u. in bronchial & pulm. dis. (361) - **P. mucronata** Lam. (Br): frt. edible; acid beverage made fr. frt. - **P. pulchella** HBK (CA): "calzoncillo" (Salv); u. as diur. - **P. quadrangularis** L. (tropAm, now pantropic): water lemon; (giant) granadilla (real); barbadine; maracuja (Br); tumbo; "apricots" (Berm, sCA); frt. deliciously edible, pulp acid u. to flavor beverages & ices; green frts. cooked as veget.; rt. vermifuge; hb. u. as narcotic & as given under *P. laurifolia* (several other *P.* spp. so u. in Br); fresh rt. narcotic & toxic; dr. rt. pd. & mixed with FO u. as emoll. in cataplasms (Co) (2137) - **P. racemosa** Brot. (Ind+): c. HCN glycoside; u. as sed. - **P. rubra** L. (WI contin., tropAm): "pasionario de cerca" (Cu); "trois quarts" (Domin); decoc. fr. sts. & lvs. drunk to induce lactation (Domin) (2328) - **P. salva-dorensis** J. D. Smith (sMex, CA): u. as purg. & in GC & urine retention (Salv).

Passifloraceae: Passion-flower Fam.; hbs. or woody vines of the warm temp. & trop. zones, esp. of SA; spp. often cult. for food or ornam.; fls. regular, showy; frt. a capsule or berry.

passion flower: *Passiflora incarnata* - **bay-leaved p. f.**: *P. laurifolia* - **wild p. f.**: *P. incarnata* - **p. fruit**

(**juice**): 1) any edible frt. of *Passiflora* spp., esp. *P. edulis* 2) mango (Hawaiian Is [US]) - **p. vine**: 1) p. flower 2) *Passiflora caerulea*.

passions: *Rumex patientia*.

Passiphen™: proprietary contg. exts. of Passiflora & Valerian (McNeil).

Passulae (majores) (L.): Uva Passa; raisins - **P. minores**: Corinthian raisins; grocers' currants.

Pasta (L.): 1) paste 2) (It): processed wheat flour paste representing a rich dough for cooking (ex. spaghetti, vermicelli) 3) (Sp) pulp (paper) - **P. Guarana**: guarana - **P. Picis Carbonis** (BPC): coal tar 34.3 with comp. ZnO paste qs. ad 1000.

paste: soft smooth mixt., such as **almond p.** - **guarana p.**: Guarana.

paste (pron. 2 syllables) (CR, Nic, Salv): towel gourd, *Luffa*.

pastel leaves: *Isatis tinctoria*.

pasteque (Fr): watermelon.

Pasteur Prophylactic: P. Treatment: rabies vaccine.

Pasteurella pestis (Lehmann & Neumann) Bergey et al. = *Yersinia p.*

pastille (dim. of panis, loaf): troche.

Pastinaca opaca Bernh. (Umb.) (Euras): source of furocoumarins, such as imperatorin, xanthotoxin - **P. Opopanax** L. = *Opopanax chironium* - **P. sativa** L. (Peucedanum sativum B. & H.) (Euras, cult. & natd. Am, Austral, NZ, sAf): (common, wild) parsnip; (garden, cultivated) p.; frt. FO & VO (apiol; butyric, heptanoic, & capronic acids & their esters, free & bound methanol & ethanol) (362); u. in stomach, bladder dis.; source of VO, xanthotoxin, apiol; rt. (fleshly tap r.) edible cooked (long used); c. FO, VO pastinacine (alk.), furocoumarins (bergapten; xanthotoxin; psoralen [?]); u. as diur. in fever, lithiasis, phthisis; fresh rt. u. in homeopath. med. in Eu; hb. u. as carm. arom. eaten as veget.

pasto ovillo (Ch): *Dactylis glomerata* (orchard grass).

pastrami (sandwich) (Romanian/Yiddish): made with specially cured beef (like ham) with much seasoning; sold in delicatessens.

Patagonula americana L. (Boragin./Ehreti) (sBr, nArg): guayabi; guayaba morada (morata); bk. u. med.; lvs. & buds in plasters for buboes; int. in traumatic fever; wd. valuable.

patata (Sp, It): **patate** (Fr-slang): potato.

pataua (Br): **patava**: *Oenocarpus bataua*, source of **p. palm frt. oil**.

patchouley (leaves): **patchouli: patchouly**: *Pogostemon cablin* - **p. oil**: VO fr. prec. & *P. heyneanus*.

patch test: appln. of subst. to intact skin, allowing contact for 48 hrs. to det. sensitivity.

pâté de foie gras (Fr) (lit. paste of fatty liver): prepd. fr. liver of goose specially fattened for that purpose; with truffles & spices; u. sandwich spread (now often c. pork); exported fr. Strasbourg, Fr, in little stone pots.

patent medicine: packaged drug prod. which has been patented in US Patent Office (or corresponding govt. agency abroad); relatively few in number as compared to TM prods. or proprietaries.

Paternosterbohnen (Ge): *Abrus precatorius*.

pathogens: microorganisms wh. produce disease.

pathosis: state of disease.

patience (Fr-f.): *Rumex patientia* - **wild p.** (Eng): *R. obtusifolius* (rt.).

patient: in pharmacy, term of reference for customer who has prescription or physician's order filled.

patilla (Co, Ve): watermelon.

pat(o)u (leaves): *Lagenaria vulgaris*.

Patrinia villosa Juss. (Valerian.) (Chin, Indochin): lvs. & frt. edible; white fragrant fls. eaten to remove poisons fr. body; decoc. emmen.

patriota (Pe, Mex): banana.

Patrisia Rich. = *Ryania*

patulin: antibiotic prod. by several microorganisms, such as *Aspergillus clavatus* Desmaz, &c., *Penicillium patulum* Bain. (= P. urticae Bain.) (hence name), *P. expansum* Link; antimicrobial.

patwa (Hind): Sunn hemp.

pau (Por): wood; stem; stalk (*cf.* bois [Fr]) - **p. brasil** (Br): *Caesalpinia echinata* - **p. bugre:** *Lithraea brasiliensis* March. (good resistant wd.) - **p. d'arco** (Br): 1) *Tecoma heptaphylla*, *T. impetiginosa* & others 2) *Ocotea pretiosa* - **p. de gafauhoto** (Br): *Simabo cedron* - **p. lepra:** *Pisonia tomentosa* Casar (source of cork) - **p. pereira**: pear wd., *Geissospermum vellosii* - **p. roxo** = red stick: pau d'arco (roxo) (red bow wood): *Tabebuia avellanedae (qv.)* - **p. santo:** *Kielmeyera coriacea* Mart. (source of cork) - **p. urubu:** p. lepra.

paulioside: glycoside of crotoxigenin + D-sarmentose; occ. in *Strophanthus boivinii*.

Paullinia cupana HBK (P. sorbilis Mart.) (Sapind.) (n&wBr, sVe, Amazon reg.): guarana shrub; bitter sd. of climbing shrub c. caffeine, small amts. of other xanthines, saponins, resin, tannin; u. to prepare guarana paste, u. to make stimulating beverage (*cf.* cocoa) (c. 3-5% caffeine & theobromine) & in many pharm. prepns. as stim. analg. in neuralgia, to suppress appetite, migraine, astr. in diarrh.; natives use solid mass (guarana) by rasping with tongue of piranucu (fish); marketed as stick & powder - **P. emetica** R. E. Schultes (Co): vine; lvs. u. as emet. - **P. maritima** Vell. (Br-Rio; Minas G): cururuape; sds. ext. with alc. u. locally for rheum. neuralgia, paralytic states, etc. (several other *P.* spp. u. similarly) - **P. pinnata** L. (WI, tropAm, esp. Br, now tropAf): cipotombo; timbo (cipo); rt. c. timboin (impure rotenone) u. fish poison (sometimes also frt. sd. whole pl. u.); insecticide (several other *P.* spp. u. similarly); u. as ton. exter. vulner.; lf. sap u. as vulner. anthelm. antimal. amaurosis (partial blindness) - **P. thalictrifolia** Juss. (Br): source of Lignum (wd.) Quamacai; u. for rheum., beri beri, inflammn. catarrhs.; in diaph. baths - **P. trigonia** Vell. (Br): timbo; sds. c. 27% FO; roasted sds. u. in diarrh. - **P. yoco** Schultes & Killip (Co, Pe): yoco; u. as stim. beverage (c. caffeine) (Putumayo Indians); expressed sap u. as stim. (Co); alk. chiefly in bk. (437).

Paulownia tomentosa Steud. (P. imperialis Sieb. & Zucc.) (Scrophulari.) (Chin, Jap, natd. eUS): princess (or empress) tree; karri t.; "alley magnolia"; royal paulownia; Paulownie (Ge): juvenile pl. very herbaceous (more so than most other woody pls.) with lvs. of enormous size (lamina 32 X 40 cm); fl. resembles that of catalpa tree, but with violet brown spots & yellow stripes; bk. c. syringin, u. astr. vermicide, to treat falling hair (also wd.); lf. decoc. u. as wash for sores; to restore growth & color of hair (Chin); sds. source of FO, toi (or abur) oil; close to *Catalpa*; cult. orn.

pa(u)pau: pawpaw: *Asimina triloba.*

Pauridiantha callicarpoides (Hiern) Brem. (Rubi) (c tropAf): u. med. (Congo) - **P. paucinervis** (Hiern) Brem. (tropAf): rt. decoc. drunk for headache & as purg.

Pausinystalia lane-poolei Hutch. (Rubi.) (w tropAf): bk of tree u. for yaws & ground itch (Liberia); wd. u. for poles & interior work - **P. johimbe** (Schumann) Beille (Corynanthe j. K. Schumann) (trop wAf) [Cameroon+]): source of Yohimbe bark (BPC); c. yohimbine, &c.; u. antipyr. aphrod. (yohimbine incr. sexual motivation in rats) - **P. macroceras** Pierre (tropAf): u. as adulter. of yohimbe bark (fr. *Corynanthe johimbe*), not distinguishable microscop. but only by lower content of yohimbine - **P. pachyceras** K. Schumann (Corynanthe p. K. Schum.) (tropAf): pseudocinchona; bk. c. corynanthine, corynanthidine, &c.; u. aphrod., local anesth. (less effective than cocaine); antipyr.; in cough, leprosy; to antidote arrow poisons; wd. u.; (tropAf, esp. Congo): source of a false yohimbe bark - **P. sankeyi** Hutch. & J. M. Dalz. (w tropAf): bk. of large tree u. antipyr.

Pavetta crassipes Schum. (Rubi.) (tropAf): lf. u. for GC (Tanz); acid inf. of lf. for cough; lf. foodstuff (wAf); sap coagulant of rubber latex - **P. indica** L. (Ind to seAs): rt. u. as aper. lvs. in fomentation (Burma); wd. chips decoc. in rheum. (Indochin); bk. decoc. u. for visceral obstructions; lf. decoc. for pain of hemorrhoids (PI) - **P. tomentosa** Roxb. (Mal): lf. decoc. for fever; rt. inf. to promote delayed parturition.

Pavonia hirsuta Guill. & Perr. (Malv.) (tropAf): rts. u. in mixts. for syph.; in diarrh. - **P. odorata** Willd. (Ind): perfumed rts. u. arom. refrig. stom. antipyr., often in combn. with other drugs (Ind); fls. & lvs. c. mucil. u. dem.; pl. "cure" for rheum. - **P. propinqua** Garcke (Pak, Af): lf. decoc. u. as wash to prevent abortion in syphilitic women (Somalia).

pavot (officinal) (Fr-m.): **p. somnifère, (blanc)**: (officinal or opium) poppy.

pawpaw (Eng fr. Am Indian): 1) papaya 2) *Asimina triloba* (also with following qualifying terms: **American**; **common**; **native**; **northern**) - **p. family**: Caricaceae.

paxte (Salv): towel gourd, *Luffa.*

Payena leerii Kurz (Sapot.) (Mal, Burma): source of a good grade of gutta percha called g. sundek - **P. lucida** A DC (Indomal, Penang, Singapore, Sunda Is): nyatok bunga (Mal); source of inferior gutta percha; sometimes u. to adulterate superior g. p.; sds. c. bassic acid; rt. decoc. u. in childbirth; wd. serves as lumber - **P. maingayi** Clarke (Malacca): source of useful gutta percha.

pazote: epazote.

pea: 1) name appl. to many spp. of Fam. Leg. 2) (specif.): *Pisum sativum*, common p. 3) (seUS specif.): cow p. - **asparagus p.**: *Psophocarpus tetragonolobus* - **beach p.** *Lathyrus maritimus* - **black-eyed p.**: 1) *Vigna sinensis* var. (really a bean) 2) *Dolichos sphaerospermus* (Jam); *D. lablab* - **butterfly p.**: 1) *Centrosema* spp., esp. *C. virginianum* 2) *Clitoria* spp. - **cadjan p.**: *Cajanus cajan* - **Caley p.**: *Lathyrus hirsutus* - **Canada white p.**: *Pisum sativum* var. *arvense*: much u. in Qeubec (Cana) & formerly in New Eng (US); sown with oats for forage or soil building; sds. light green to grey, lvs. with whitish bloom; now mostly replaced by sugar or French vars. - **chick(en) p.**: *Cicer arietinum* - **clay p.** (Ala [US]): red v. of cow p. - **common p.**: green p. - **Congo p.**: pigeon pea - **cow p.**: *Vigna sinensis*; *V. catjang* - **Crowder p.**: *Vigna unguiculata* - **English pea:** green p. - **field p.:** *Pisum sativum* var. *arvense* - **everlasting p.**: *Lathyrus latifolius* - **flat p.**: *Lathyrus sylvestris* - **garden p.:** green p. - **grass p.**: *Lathyrus sativus* - **green p.**: *Pisum sativum* - **ground p.** (seUS): peanut, esp. Spanish type (sGa [US]) - **hoary p.**: *Tephrosia* spp. - **Japan p.**: soy bean, *Glycine hispida* - **jequirity p.**: *Abrus precatorius* - **Jerusalem p.**: *Phaseolus trinervius* - **Pak(istan) p.**: *Pisum sativum* var. -

partridge p.: *Chamaecrista* spp. - **pigeon p.**: *Cajanus cajan* - **rabbit's p.**: wild p. - **rosary p.**: *Abrus precatorius* (sd.) - **rough p.**: Caley p. - **scented p.**: sweet p. - **Scotch p.**: prob. a field pea; recommended by Parkinson (AD 1629) - **self-perpetuating p.**: *Phaseolus angularis* - **sensitive p.** (US): some *Cassia* spp. sensitive to touch; ex. *C. fasciculata, C. nictitans* - **Singletary p.**: Caley p. - **snow p.**: Chinese p.: *Pisum sativum* var. *macrocarpum*; differs in having edible pods; this var. was developed in Chin centuries ago - **split p.**: green p.; pigeon p. - **sugar p.**: *Pisum sativum* (snow pea) - **sweet p.**: 1) *Lathyrus odoratus* 2) *Dolichos lablab* 3) green p. - **Tangier p.**: *Lathyrus tingitanus* - **p. tree, Siberian**: *Caragana* spp. - **turkey p.**: hoary p. - **vetchling p.**: *Lathyrus sativus* - **wild p.**: *Tephrosia virginiana* - **w. winter p.**: Caley p.

peach: *Prunus persica* (SC fr. Persia, once supposed place of origin; also origin of "peach") - **p. wood**: fr. *Caesalpinia bonducella* - **wild p.**: 1) *Prunus caroliniana* 2) *Eriodictyon* 3) *Kiggelaria, africana*.

peacock: common pea fowl: *Pavo cristatus* L. (Phasianidae) (Ind, Burma): domesticated for millenia; male bird (peacock) can expand a fan-like canopy of brightly colored feathers for sexual allure; female bird (peahen) lacks same; meat formerly popularly eaten but not very palatable; feathers u. as decorative article formerly in women's hats; bird valued in Unani med. (flesh, fat, blood, eggs, bile, excreta, ashed bones, feathers).

peanut: 1) *Arachis hypogaea* 2) (slang) something or someone small, insignificant, petty or worthless - **Russian salted p.**: salted sunflower sd.

pear: *Pyrus communis* (many vars.) & some other *P.* spp. - **prickly p.** (cactus): **tree p.**: *Opuntia* spp., as *O. monacantha, O. stricta* var. *dillenii* - **vegetable p.**: *Sechium edule.*

pearl: margarita (L); lu-lu (Arab); lustrous concretion occurring in various bivalve molluscs, that most prized fr. the oyster (*Ostrea & Crassostrea* spp.); mostly c. $CaCO_3$; produced by reaction to a foreign body, such as sand grain; u. for ornaments of great value (necklaces, &c.), the finest fr. pearl oysters of Persian Gulf; "cultured" pearls prod. in Japan; "artificial pearls" are in no way related; u. med. in ancient & medieval times, but still in Unani med., usually powd. (in hepatic & renal dis., palpitation; epilepsy; contraception [in form of pessary], &c.) (Unio[n] = large pearl) (2728) - **p. ash**: "pot ash"; impure potassium carbonate; fr. wood ashes - **p. barley**: grain of barley, *Hordeum vulgare*, husked, rounded, & polished; u. in infant foods - **cocoa p.: coconut p.**: found occasionally in sd. of *Cocos nucifera* - **p. essence**: aqueous suspension of fish scales fr. herrings, etc., u. to make artificial pearls - **p. flower**: fl. of *Exochorda racemosa* (p. bush); or of *Lithospermum arvense* (wild p. plant) - **p. millet**: *Pennisetum glaucum* - **p. moss**: Chondrus. - **mother of p.:** nacre; nacra perlarum (L.); Perlmutter (Ge); iridescent covering of inner surfaces of mollusc shells; with pearl-like brilliance; u. in making ornamental objects; in old-time medicine; off. in pharmacopoeias of 1700s - **p. shell**: p. oyster or other pearl-bearing shell - **p. tapioca**: a factitious tapioca.

pease meal: made by grinding peas.

peat: porous partly decayed pl. materials of considerable age & largely decomposed; made up of entire *Sphagnum* sp. plants; believed to represent early stage of lignite (hence of coal) formation; u. fuel; soil improving (humus); packing material; distd. for generation of electric power; u. as paste for gastritis ("peat drink cure": Moortrinkkur) (1676); peat baths for rheum. (USSR, Ireland, NA); mech. & athletic injuries; dressings for wounds (WWI) (antibiotic values); tolerated when heated to 45-48° C. (to reduce pain); female infertility; absorbent for oil spills; source of mineral wax; to treat eczema, ulcers, & burns; eye prepn. for near-sightedness; preservative. - **p. moss**: *Sphagnum* spp.

pebble plants: *Lithops* spp.

Pebermynte (Dan): peppermint.

peca: Podophyllum.

pecan (nut): *Carya illinoensis*; forms in wh. u.: plain (unshelled and shelled); p. halves; meal; salted & toasted; sugar-coated; sugared (plain & with orange peels or figs); pralines; p. oil (Am Indians; pron. pee-kahn´ or pee-can´).

Pech (Ge): pitch, tar.

peche (Fr-f.): peach (frt.) - **pecher** (Fr-m.): peach tree.

pechita (Mex, Sp): gum Sonora fr. *Prosopis* spp.

pechorim (Por Phar.): pichurim beans.

pecilocin: Variotin: antibiotic obtained fr. *Paecilomyces varioti* Bainier var. *antibioticus* Takeuchi et al. (1959); antifungal agt. very active locally against dermatophytes.

pecora (It): sheep.

pectase: pectinesterase = enzyme wh. catalyzes hydrolysis of pectin to give pectic acid & Me alcohol; occ. in higher pls.

pectin: Pectinum (L): complex carbohydrate mixt. consisting chiefly of partly methoxylated polygalacturonic acids; obt. fr. albedo of orange & lemon frt. (yield 2.5-5.5%) & fr. apple pomace &

cull apple peelings (yield 1.5-2.5%) chiefly, by extn. with dil. sulfuric acid; less commonly fr. cucurbitaceous frts., &c. (vegetable p.); u. to combat effects of hemorrhage, in (infant) diarrh.; for colitis (adm. in tabs & capsules); reduce serum cholesterol (1677); in prepn. jellies, jams, & other food prods. (added to prepns. of juices of strawberry, raspberry, pineapple, grapes, which "set" to a gel [jelly]) (2352); in mfr. mint jellies; candies (ex. gum drops); thickener & stabilizer in latherless shave creams, toothpastes, & other cosmetics; emulsif. agt. for VO, mineral oils, &c.; adhesive in pastes for skin dis.; stabilizer for ice cream, salad dressings, table sauces, &c.; antisticking or glazing agt. for dr. frts. Salts: Ca pectinate: film in surgical dressing; Ni pectinate u. in infected wounds, colitis, bacill. dysent. - **p. sugar**: arabinose.

pectinase: pectolase: polygalacturonase: enzyme prod. by most Thallophyta; hydrolyzes pectin & pectic acid; u. in wine making & to reduce viscosity of frt. juices.

Pectis augustifolia Torr. (Comp.) (US Rocky Mt states, Mex): limoncillo; pl. u. as food & for seasoning (Hopi Indians); prep. tea for gastralgia; prep. dye - **P. bonplandiana** HBK (Guat, Hond): culantrillo; inf. u. for female dis. - **P. ciliaris** L. (WI, nVe): donkey weed; a decoc. of pl. u. for colds & TB; to dispel flatulence; emmen. - **P. elongata** HBK (WI, CA, sMex, SA): cominillo; hb. decoc. u. as carm. for intest. dis.; for colds & fevers; pl. decoc. u. as bath water in fevers (Ve, Ind) - **P. febrifuga** van Hall (WI, CA, nVe): stink weed; tebink; pl. decoc. u. for colds & fevers; diaph.; in urin. irrit. & burning; decoc. as enema in intest. dis. - **P. leonis** Rydb. (WI as Cu): hb. popularly u. for intest. dis.; also u. in asthma, bronchitis - **P. linifolia** L. (sArg, Mex, CA, WI, Co): comino de monte; hb. decoc. u. for nausea & gastric upsets; in colics - **P. papposa** Harvey & Gray (swUS, nMex): fetid marigold; chinch-weed; capitula u. to season meat (NM [US] Indians); sugg. for lemon ext. substit. (also *P. angustifolia*) - **P. prostrata** Cav. (sUS, Mex, CA, WI, nVe): tomillo; u. like *P. ciliaris* - **P. sessiliflora** Sch. Bip. (Pe): china-paya (*cf. Zinnia*); u. for GC (folk med.).

pectolinarin: glycoside fr. fls. of *Linaria vulgaris*; hydrolyzes —> pectolinarigenin + glucose + rhamnose.

pectose: 1) raw material of frt. pulp fr. which pectin is obtained 2) sugar fr. which pectin is polymerized 3) protopectin (Tschirch).

Pedaliaceae: Pedalium Fam. (OW subtrop): mostly hbs., few shrubs & trees, closely related to Scrophulariaceae; ex. *Sesamum*.

Pedalium murex L. (monotypic) (Pedali.) (trop eAf, As): entire pl. u. diur. (in GC, Af); ext. of lvs. emoll.; sds. u. as diur. antispasm. dem. in GC urin. incontinence, impotence, cystitis; lvs. u. as veget.

Peddiea africana Harv. (Thymelae.) (tropAf): cord tree; u. med.; tough bk. furnishes cordage; u. in homicide.

pederin(e): toxic compound isolated from *Paederus fuscipes;* inhibits protein biosynthesis & mitosis.

pedicel: individual fl. stalk in inflorescene with 2 or more fls. (*cf.* peduncle).

Pedicularis L. (Scrophulari.) (nHemisph, esp. As): lousewort; some 350 spp.; semi-parasitic pls. - **P. canadensis** L. (eNA, incl. nMex): (common) lousewort; wood betony; ornam. hb. u. in rock garden; decoc. of entire pl. u. to reduce internal swellings; rt. poultice u. for exter. swellings (Meskwaki Indians); u. ton. sed. astr. vulner. (Potawatomi) (364) - **P. palustris** L. (Euras): hb. u. int. as diur. & in excessive menstr.; exter. in sores; insecticide - **P. pectinata** Wall. (Indo-Pak): pounded lvs. u. for hemoptysis; diur. - **P. Semen** (L.): stavesacre seed - **P. silvatica** L. (Euras, NA): hb. c. aucubin; u. similar to *P. palustris* - **P. siphonantha** D. Don (Ind): pl. u. as diur. (Punjab).

Pedilanthus bracteatus (Jacq.) Boiss. (P. pavonis Boiss.) (Euphorbi.) (Mex): candelilla (Sinaloa); wax slipper flower; one source of candelilla wax u. as substit. for carnauba wax; u. in furniture & shoe polishes, candle mfr.; pl. said purg. emet. emmen. antiluetic. (chief source of c. wax is **P. cymbiferus** Schlecht. [sMex]) - **P. tithymaloides** Poit. (P. grandifolius) (sMex to nSA-WI as PR): candelilla (Mex); Christmas candle; "ipecacuanha" (PR); latex caustic, irrit. emet. toxic, topically for warts, calluses, ringworm, cancers, umbilical hernias, dental pain (into cavities) (folk med.); pl. decoc. u. emmen., for menstrual pain; lf. decoc. for VD; rt. decoc. purg. & abortif.; subsp. *smallii* Dressler (sMex, WI, natd. sFla [US]): decoc. of sts. & lvs. u. emmen. & in syph.; subsp. *parasiticus* Dressler (P. itzaeus Millsp.) (sMex, WI, cult. Mex): mayorga; cult. as hedge, ornam. shrub.

Pediomelum canescens Rydb. = *Psoralea c.*

peduncle: stalk of inflorescence where 2 or more fls. are on stalk; or else where a single fl. occurs.

Pedunculi Cerasorum (L.): cherry stalks; "cherry stems"; u. flavor; commerl. product.

peel: the rind or epicarp (exocarp) fr. a fleshly frt.; ex. orange, lemon p.

peepal: peepul: pipal.

pega palo: drug advertised (1956) fr. the WI as the "most powerful" aphrod. in the world; identity uncertain as this name has been appl. to several spp. incl. *Macfadyena unguiscati, Dendropemon purpureus, Rhipsalis baccifera, Stellaria antillana, Rauwolfia pyramidalis, Aechmaea nudicaulis, Rhynchosia pyramidalis* (name usually cited) (2516), &c.

pega pollo (Cu): *Plumbago scandens.*

Peganum harmala L. (Zygophyll.) (sEuras, nAf, cTex [US]): Syrian (or wild) rue; har-mal; hurmal; Steppenraute (Ge); sds. c. many alks. (tot. 2-4-6%), incl. harmine, harmalol, harmaline, peganine, pegamine, peganol; rts. also c. alks. (1678); sds. u. anthelm. diur. diaph. emmen. eye dis.; exter. for wounds, head lice, fumigation; prepn. red coloring matter (harmala red or Turkey red, harmalol) (important in carpet mfr. before anilin dyes available); as spice (Near East); source of FO (Alg); prepn. of harmine & vasicine; poisonous pl.

Pegolettia senegalensis Cass. (Comp.) (wAs, Af): pl. given with ghee to children to make strong; juice appl. to wounds of camel (Baluchistan, Pak).

peh hsien (Kor): *Dictamnus albus.*

pejibay(e) palm (CR): *Bactris gasipaes.*

pekea: pequia.

pekoe: tea prepd. with first 3 lvs. of "spray" (twig) - **orange p.**: prepd. fr. 1st or 1st & 2nd lvs. of spray.

pe-la (Chin): Cera Chinensis (*qv.*).

pelagic: of the open oceans - **p. whaler**: factory ship.

Pelargonium L'Hérit. (Gerani) (trop&sAf; As, Austral): storksbill; "geranium" (of greenhouses); some 280 spp., many hybrids; many of comm. vars. are said to have been re-exported to sAf, fr. sFr where they had been developed by hybridization; "geranium" oils are among most important perfume oils, incl. Algerian (African), Moroccan, Egyptian, Chin, &c. - **P. capitatum** Ait. (sAf): rose geranium; "geranium rosat"; one source of geranium oil; rts. u. in diarrh. (Civil War) - **P. crispum** L'Hérit. (sAf): geranium; one of the sources of rose geranium oil (VO) dist. fr. the lvs. - **P. cucullatum** L'Hérit. (Af): decoc. orally or as clyster for nephritis, urin. suppression; c. VO mucil.; lvs. u. in colic & diarrh. (Cape Colony) - **P. domesticum** Bailey (sAf): showy geranium; ornam. - **P. graveolens** L'Hérit. (sAf): rose geranium (commonest form); apparently chief sp. concerned in most comm. r.g. oils (ex. Réunion; African; eAf; Algerian; Moroccan; French; Spanish) (438); cult. for its sweet aroma; VO u. in perfum. ind., as replacement for rose oil - **P. hortorum** Bailey (sAf): fish geranium; cultigen; the common geranium of windows; often overwintered in cellars; believed hybrid - **P. inquinans** L'Hérit. (Mex, CA, Af): malvón (Mex); u. med. (Mex); u. for colds & headache (tribal peoples, sAf) - **P. luridum** Sweet (sAf): rt. inf. adm. for diarrh.; rt. mixed with porridge for dys.; in colic & fevers - **P. odoratissimum** (L.) L'Hérit. (sAf; cult. e&nAf, Réunion; Fr, Corsica): sweet-scented geranium; nutmeg g.; a source of r. g. oil; cult. as arom. pl.; hb. u. in med. (homeop.) - **P. peltatum** L'Hérit. (sAf): geranium-lierre; juice u. as astr. antisep. in throat inflammn., aphthae - **P. quercifolium** Ait. (sAf): oak-leaved geranium; ornam. - **P. radula** L'Hérit. (P. radens H. E. Moore; P. roseum Willd.) (sAf): rose geranium; rough r. g.; g. r.; u. to produce r. g. oil; u. as soap perfume (cult. Kenya) (1679) - **P. roseum** auct.: sometimes u. as group name for various *P.* hybrids u. in prodn. r. g. oil - **P. triste** L'Hérit. (sAf): mourning geranium; the scarlet red edible rt. u. in diarrh. dys. anthelm. - **P. zonale** Ait. (sAf): horseshoe geranium; banded g.; petals c. pigment, pelargonin; source of VO; mostly cult. as ornam. hb.

pelargonium, rose: scented p.: *Pelargonium graveolens.*

Pelea anisata Mann. (Rut.) (Hawaii Is [US]): mokihana; capsules of tree strung into lei (garlands) with anise-like perfume (Haw); allergenic - **P. clusiaefolia** A. Gr. (Haw Is): sacred balm of the goddess Pele.

Pelecyphora aselliformis Ehrenberg (Cact.) (San Luis Potosí, Mex): pe(y)ote; peyotillo; hatchet cactus; this small cactus u. as stim. antipyr. inebriant, hallucinogenic (Am Indians); c. mescaline & derivs., pellotine, phenethylamines, &c. (2000); do not confuse with *Lophophora williamsii*, chiefly used.

pelican: taxa of *Pelecanus* (Pelecanidae) (marine areas of both hemispheres): large heavy gregarious birds with large pouched bill u. to scoop fish fr. water; wing spread to 3 m; *P. onocrotalus* L. (It, Med reg.): domesticated by ancient Egyptians for eggs; bird a religious symbol of self-sacrificing love to early Christians; whole bird (est.) (grey p.) & fat u. medicinallly in Unani med.

pelican flower: *Aristolochia grandiflora, A. serpentaria.*

pelitre (Sp): 1) pellitory, *Anacyclus pyrethrum* 2) pyrethrum fls. (*Chrysanthemum* spp.) (**p. de Africa**).

Pellaea andromedifolia Fée (Polypodi/Adianti.) (Ore-Calif [US], nwMex): coffee fern; "stick gut" (SC sts. may perforate intestine of foraging sheep); hb. officinal (Mex) - **P. atropurpurea** Link (sUS, Mex): cliff brake; u. as ornam. fern - **P. calomelanos** Link. (Ind, sAf): rhiz. u. as anthelm.; frond smoked for head & chest colds **P. involuta** Bak. (sAf): rhiz. u. for diarrh., bite of spiders; frond smoke inhaled for colds (sSotho) - **P. mucronata** DC Eaton (sCalif, nwMex): bird's foot cliff brake; b. f. fern (Calif); frond-tea; inf. u. in epistaxis, TB; spring ton. for feverish colds; pd. lvs. appl. to sores (Calif Indians).

pellagra preventive (P.P) **factor**: niacin, niacinamide.

Pelletierinae Tannas (L.): **pelletierine tannate**: a mixture of alks. ppt. with tannic acid fr. pomegranate bk. ext., consisting of pelletierine (per se), isopelletierine, pseudopelletierine, etc.; u. tenifuge (SC Pelletier, pharmacist, early alk. chemist); other p. salts are known, such as the sulfate.

Pellionia daveauana N. E. Br. (Elastostema repens Hall. f.) (Urtic.) (Ind, Indochin): cult. as ornam. (hot house); u. folk med. (lf.); as poultice for swellings, boils, & painful abdom. troubles (Mal) - **P. pulchra** N.E. Br. (Indochin): "satin pellionia"; lvs. u. in folk med.; pl. cult. as ornam. for colored lvs. (hot house).

pellitory: *Anacyclus pyrethrum*, etc. - **European p.**: *Achillea ptarmica* - **German p.**: *Anacyclus officinarum* - **p. of Spain: (Spanish) p. root**: 1) *Anacyclus pyrethrum* 2) *Achillea ptarmica* 3) *Peucedanum ostruthium* - **sweet p.**: *Tanacetum umbelliferum* - **true p.**: p. - **p. of the wall: wall p.**: *Parietaria officinalis*.

pellotine: alk. fr. mescal buttons; sed. props.

pelon (Arg, Ch): nectarine.

pelosine: berberine.

pelo de Indio (Mex): *Cynodon dactylon*.

Peltandra virginica (L) Schott (Ar) (eUS): (Virginia) tuckahoe; (green) arrow arum; corms with starch, u. as food (Ind); also spadix & frt. boiled & eaten; ornam. decorative hb.

peltate: state where lf. lamina is attached to petiole away from edge; ex. *Podophyllum, Nasturtium*.

Peltigera aphthosa (L) Willd. (temp. zone) (Peltiger. - Lecanorales - Lichen): "ground liverwort"; source of mannitol, mannose, d-galactose (Dan Ph.) - **P. canina** (L) Willd. (nHemisph): English liverwort; ash-colored ground l.; inf. of thallus as lax. ton.; in liver dis.; in rabies (*sic*) & icterus (Ind); source of iron-grey dye.

Peltiphyllum peltatum Engl. (Saxifraga p. Torr) (monotypic) (Saxifrag.) (Ore, Calif [US]): umbrella plant; sts. petioles eaten (Am Indians) (raw or boiled).

Peltodon radicans Pohl. (Clinopodium repens Vell.) (Lab) (Br, Par): paracary; margiapantes; juice fr. fresh lvs. & sts. u. to treat eczema & other dermatoses; flg. pl. u. asthma, bronchitis, obstinate cough; diur.; to treat flatulence, snakebite (2219).

Peltophorum africanum Sond. (Leg) (sAf): African (black) wattle; pd. decorticated rt. appl. to wounds; bk. inf. anthelm.; fresh bk. chewed for colic; bk. of st. & rt. u. in diarrh., dys. - **P. dubium** Taubert (Br, Par, Arg?): caña fistula; bk. source of tannin & dyes.

pelure (Fr): peel.

Pemba: island in Indian Ocean off coast of Tanz near Zanzibar; prod. cloves (chief supply); nutmegs; copra, &c.

pemmican: common food of the Am Indians, consisting of dr. ("jerked") meat (beef or bison) with fat & service berries or wild cherries incorporated; now often prepd. only of peanuts, raisins, dr. apples, soy oil, shortening, sugar, &c.

Penaea fucata L. (P. fucata Endl.; P. fucata Ait.; P. sarcocolla L.; Astragalus s. Dymock; Sarcocolla fucata Endl.) (Penae.) (sAf): source of a gummy subst. (Sarcocolla) like tragacanth with licorice taste; very similar is **P. mucronata** L. u. as aper., emoll. in chest dis., hemorrhages, antirheum., anthelm.; exter. in wounds; much u. in Middle Ages - **P. squamosa** L. (Sarcocolla s. Endl.; S. imbricata Benth. & Hook.) (sAf): source of the gum Sarcocolla, u. in chest dis., hemorrhages, exter. in wounds & in corneal maculae (spots); this prod. is distinctive fr. the Sarcocolla, long- known gum-like subst. formerly (in Middle Ages) considerably u.; this derived fr. *Astragalus sarcocolla* Dymock (Iran).

pencil flower: *Stylosanthes* spp.

penco (NM, Tex [US], Mex, CA): fleshy caudex of certain *Agave* spp. (*A. scabra, A. lurida, A. decipiens*).

pendare gum (Ve): a type of chicle fr. *Mimusops* spp.

Penfield's method: a means of distinguishing the anterior & posterior lobes of pituitary glands in pd. form.

pengawar djambi (Chin): "golden moss"; hairs fr. *Cibotium* spp.

penicillamine: dimethyl cysteine; most characteristic degradation prod. of penicillin-type antibiotics, from wh. obtained by hydrolysis; u. as chelating agt. (in Wilson's dis.), in rheumatoid arthritis, cystinuria; to stimulate hair growth; to remove excess Cu, Pb; disadv.: toxicity (E).

penicillin(um): the first practical & comml. antibiotic; prep. fr. *Penicillium notatum* & *P. chrysogenum*.; chief types: Pen. G & Pen. V.; penicillinase-resistant; the latter semi-synthetic (439) u. to treat many infectious bacter. dis. (ex. GC) (440); generally considered the greatest advance in treating infectious diseases - **buffered crystalline p.**: either potassium or sodium p. buffered with 4-5% sod. citrate aq. solns; is stable in refrig. for ca. 10 days - **p. calcium**: Ca salt of p.; sometimes preferred for tabs.. ointments, troches; sometimes u. for IM. inj. - **crude p.**: was sometimes u. locally; costs less (35) - **p. dental cones**: p. calcium with required binders, diluents, lubricants, etc.; u. with or without sulfanilamide &/or sulfathiazole for tooth socket infections after exodontia - **dibenzyl-ethylene-diamine p.**: stable p. salt (*cf.* procaine p.); u. orally, IM; rather more toxic than p. - **p. G.: p. II**: crystalline benzyl p.: the compd. now mostly prod. & the generally most potent prod.; is stable up to 3 yrs. without refrigeration; mostly furnished as peroral tablets or troches of potassium & sodium salts - **p. G procaine**: the procaine salt of the acid of p. G; u. orally but mostly in IM inj. or in repository form - **p. inhalation**: p. potassium, p. sodium, or p. procaine with or without harmless diluents; in very fine pd. for inhalation into nasal & upper resp. passages to combat infection there directly; or u. as atomized soln. - **p. injection in oil & wax**: sterile soln. of p. calcium in peanut or sesame oil with white wax (Romansky formula), an early form of depot therapy, now *passé* - **P. O.**: allyl mercapto methyl penicillin; less allergenic form - **p. ointment**: p. calcium in proper ointment base; common strength 1,000 units per Gm.; anhydrous bases preferred to lengthen useful life of prod.; however, diffusion rate fr. such bases rather low - **p. potassium**: K salt of acid of p.; commonly p. G; often called amorphous p.; usual IM dose 300,000 IU - **p. procaine**: a salt of p. with procaine which retains its potency in body for rel. long periods of time; in form of aqueous inj. a single dose of 300,000 units will maintain effective blood levels for 12-24 hrs., the same as the oil-wax mixt.; first u. for depot therapy; the oil inj. in the same dose acts effectively for 24-48 hrs. (with aluminum monostearate this is again doubled); hence this prod. has entirely displaced p. in oil & wax - **p. sodium**: generally u. in the xall. form, p. G sodium - **p. tablets**: p. calcium or p. sodium with a buffer, such as $CaCO_3$; this oral form generally less reliable than parenteral medication forms; best given remote fr. meal times or when fasting; sometimes estimate 20% of IM effectiveness - **p. troches**: similar to prec. except instead of being swallowed, these are chewed or left slowly dissolve in mouth to exert local effect on oral, pharyngeal, & other tissues; soluble type has base of agar, gelatin, sugars; chewing type has wax base; u. Vincent's angina, etc. - **p. V.**: methoxyphenyl penicillin; less inactivated by gastric juice & hence more absorbed by GIT (E).

penicillinase: enzyme wh. destroys penicillin; occ. in many bacterial spp. both Gm.+ and Gm.-; ex. *Escherichia coli.*

Penicillium chrysogenum Thom (Aspergill. - Hyphomycetes - Deuteromyc. - Fungi): bluish-grey mold (color variations); found growing widely; u. in mfr. of penicillins; sometimes allergenic - **P. notatum** Westling: original source of penicillin; c. notatin (glucose oxidase: u. as purifier of foods, stabilizer; antioxidant).

penicilloic acids: obt. by alkaline cleavage of various penicillins.

Peniocereus Britt. & Rose = *Cereus.*

penis: male organ of copulation & for transfer of urine to the exterior; homologous to clitoris of female; u. in Unani med. as aphrod. & to give potency (erection); obt. fr. bull, ass, zebra, tiger (Taiwan); urethra (tube traversing penis) u. similarily in Unani med. (fr. crocodile; bear).

pennawar djambi: pengawar d.

Pennisetum compressum (L) R. Br. (Gram) (eAs): pl. u. as ton. (Chin) - **P. glaucum** (L) R. Br. (P. americanum Schum.; P. typhoideum L.) (orig. prob. in Af, OW now cult.; mostly OW, tropAs, Af, limited cultn. in sUS): (Indian, African, pearl) millet; this cultigen u. as forage (stalks as cattle feed); cereal grain for humans (since time immemorial); grain u. as ton. appetizer, in heart dis. (*sic*) (Ind); to make flour; crop of special interest because able to thrive under severe drought conditions - **P. pedicellatum** Trin. (tropAf): u. as diur. & exter. hemostatic in popular medicine - **P. purpureum** Schum. (tropAf, cult. Fla [US], WI, SA): Napier grass; elephant g.; perenn. grass (-4 m) cult. as pasture (forage); excellent fodder (silage) for livestock; produces more biomass/unit area than any other hb. u. for feeding livestock; u. to improve soils; source of crude paper fiber; u. as esparto substit. - **P. spicatum** Koern. (Af): bulrush, cat-tail, or Kaffir millet; grain nutritive & in prepn. beer & brandy; in chest dis.; sometimes toxic to cattle (HCN) - **P. spicatum** R. & S. = *P. glaucum* - **P. subangustum** Stapf & Hubbard (wAf): this common weedy sp. is u. for impotence in medico-magic practice of the fetishists (Sen); furnishes good fodder (Sierra Leone; Gold Coast, Nigeria) - **P. typhoides** Stapf. & Thunb. (Af & As, cult.): poko grass; u. for beer making (Af) & brandy; ground up grain u. anthelm.; pl. u. as food; in chest dis.

Pennsylvania: v. under sumach.

pennyroyal (, American): *Hedeoma pulegioides* - **Argentine p.**: *Lippia turbinata* - **Brisbane p.**:

Mentha satureioides - **European p.**: *Mentha pulegium* - **false p.**: 1) *Isanthus brachiatus* 2) (wNY): *Prunella vulgaris* - **native p.**: Brisbane p. - **p. oil**: VO fr. *Hedeoma pulegioides* (Sp, Fr, US).

pennywort: 1) *Umbilicus pendulinus* 2) *Hydrocotyle* spp. - **stinking p.**: *H. laxiflora* - **water p.**: *H. umbellata* - **wild p.**: *Monardella* spp.

penole (Sp): parched Indian corn mixed with cold water.

pensamiento (Sp): pensée (Fr-f.): pansy; thought.

Penstemon (Pentstemon) Schmid. (Scrophulari) (mostly NA, esp. w): beard-tongue; some 250 spp. of perenn. hbs. with showy fls.; some u. in med., thus *P. campanulatus* Willd., of Mex - **P. barbatus** (Cav.) Roth (Mex, wTex [US]): beardlip penstemon; u. med. (Mex); subsp. **torreyi** (Benth.) Keck (P. t. Benth.) (Mex, wTex): Torrey p.; varas de San Jose; hb. u. as diur., diaph. in colds (Mex) - **P. bradburii** Pursh (P. grandiflorus Nutt.) (cUS): shell-leaf penstemon; wild foxglove; lf. decoc. u. as febrifuge in chills & fever (Pawnee Indians); rt. decoc. u. for toothache (Kiowa Indians) (NM [US]) (Pawnee Indians) - **P. cobaea** Nutt. (cUS s to Ark): "foxglove"; cobaea penstemon; u. for rattlesnake bite, dental analg. (Am Indians); ornam. hb. - **P. laxiflorus** Pennell (cUS): wild foxglove (Tex); "balmony" (early Tex settlers); u. as lax. (tea fr. lvs.) (Tex settlers).

Pentaclethra filamentosa Benth. (Leg) (nBr to Nic): source of paroacaxy (or pracaxy) nuts with ca. 50% semisolid fat and resin; FO u. in soap mfr. - **P. macrophylla** Benth. (Br, w&etrop Af): pauco nuts; oil bean tree; sds. (owala grains) (Gabon) c. 30-36% FO (in shell), u. for lubrication, soap & candle mfr.; sds. nutr. condim.; bk. c. 11% tannin, decoc. u. galactagog. astr. vulner. anthelm. lax.; crushed sds. u abortif. (with ants) (Gabon); fish poison; for Hansen's dis.; refined oil u. in cooking, in dika bread.

Pentacme A DC = *Shorea*.

Pentadesma butyracea Sab. (Guttif.) (trop wAf): butter (or tallow) tree; sds. source of edible fat, u. in cooking (Lamy, Kanga or Sierra Leone butter); also in mfr. soap, candles, margarine; frt. at times u. to adulter. cola nuts.

Pentagonia brachyotis (Standl.) Standl. (Rubi.) (CA): cuamtulo (Cuna Indians, Pan); u. medl. - **P. spathicalyx** K. Schum. (eEc): frt. edible, appl. to sting ray wounds (most painful of jungle wounds) (906).

Pentanisia ouranogyne S. Moore (Rubi) (tropAf): u. for abdom. pains & purg. (Shambala tribe) - **P. prunelloides** Walp. (sAf): wild verbena; enjoys considerable u. as household med. of natives; ex. for painful swellings, rheum., sprains, sores, fevers; rt. decoc. in influenza epidemic (1918).

Pentarrhinum insipidum E. Mey. (Ascepiad.) (s&cAf): lvs. u. as poultice for boils; rts. poultice for sores (Tanz).

Pentas bussei Krause (Rubi) (Tanz): lvs. bruised, mixed with water, & drunk for bloody diarrh., dys.; rt. decoc. u. in GC & syph. - **P. longituba** K. Schum. (eAf): rt. eaten raw as cure for fever; some of pounded rt. may be appl. to small incision in patient's chest - **P. purpurea** Oliv. (eAf): rt. decoc. in GC (Tanz); with sugar cane u. to bring on the menses; juice in headache, fever, rheum.

Pentatropis cynanchoides R. Br. (Asclepiad) (Indo-Pak): rt. decoc. astr. in GC.

Penthorum sedoides L. (Penthor./Saxifrag.) (e&cNA, eAs): ditch (or Virginia) stonecrop; hb. u. as astr. dem. lax.; in diarrh. hemorrhoids; in gastritis.

Pentopetia androsaemifolia Dcne. (Ascelpiad./Periploc.) (Madag): tandrokosy; c. periplocymarin (cardenolide); u. in heart dis. - **P. elastica** Jumelle & Perr. (Madag): source of Mavokeli rubber.

pentosan: complex carbohydrate yielding pentose sugar on hydrolysis; found in cell walls, etc.; ex. xylan - **pentose**: simple sugar with general formula $C_5H_{10}O_5$; widely distributed as glycosides in pl. kingdom; ex. xylose, arabinose, ribose - **pentoside**: glycoside which yields a pentose on hydrolysis; ex. barbaloin.

pen tsao (Chin): Chin pharmacopeia - **p. yen** (Chin): Opium.

Pentstemon = *Penstemon*.

Pentzia globosa Hutch. (Comp.) (sAf): hair karoo; lf. inf. in stomach diff.; vulnerary (exter.); woody parts chewed & saliva swallowed in erysipelas - **P. incana** O. Ktze. (Af): (common) karoo (bush); pl. bitter & ton.; similarly u. to *P. globosa* - **P. schistostephioides** M. R. F. Taylor (sAf): u. to treat duodenal ulcer - **P. suffruticosa** Hutch. (sAf): ton. antispasm. anthelm. in gout.

peonia del pais (Mex): *Cyperus esculentus*.

peonin: anthocyanin dyestuff of fuchsia, cyclamen, magnolia, &c.; petals; glucoside of peonidin.

peony: *Paeonia officinalis*+ (fls. rt. sd.).

pepe (It): pepper.

pepermuntblad (Dutch): peppermint lf.

Peperomia Ruiz & Pav. (Peperomi/Piper.) (mostly Am): pepper elders; over 1,000 spp.; popular ornam. f.i. in Haw.Is. with over 100 spp.; u. med.; as gray dye for tapa cloth (389) - **P. emarginella** (Sw.) DC (NA, SA, WI, CA): pl. inf. u. in colds, flu, fever (Trin); pounded hb. u. to poultice bites of cungamanda ant (Co) - **P. glabella** A. Dietr. var. *melanostigma* Dahlstedt (Co): hb. u. as remedy in conjunctivitis - **P. leptostachya** H. & A. (tropics): lf. juice u. in skin dis., eye infection & burns (trop. islands) - **P. obtusifolia** A. Dietr. (tropAm): "purslane"; crushed lvs. rubbed on rheumatic joints for pain (Co); cult. as ornam. hb. - **P. pellucida** (L.) HBK(orig. SA, now pantropics): shaving brush (Trin); hb. decoc. u. for flu, cough, diarrh., chest cold (Trin); crushed soaked lvs. poulticed on ulcers & wounds (Tikuna Indians, Co); sore throat (Domin); lf. juice u. for abdom. pain & colic (Indones) - **P. reflexa** Dietr. (China): pl. soaked in wine 10 days u. as vulner. in mechanical injury - **P. rotundifolia** (L) HBK (Piper rotundifolium L.) (WI; trop, contin. Am): anis de Sierra (PR); baume des chasseurs; lvs. with strong camphor odor; stim., u. in "catarrhs." (inf. or tr.), anaphrod., antispasm.; ornam. - **P. serpens** Loudon (Co): arom. lvs. & sts. u. to relieve irritant sting of conga ant - **P. victoriana** DC (Co): crushed lvs. rubbed on forehead for headache (Rio Vaupes natives) (2299).

pepino (Sp.): 1) cucumber 2) (SA) *Solanum muricatum* - **p. cohombro** (Co): cucumber - **p. de monte** (Mex): *Cucumis anguria*.

pepita (Mex-US groceries): pumpkin sd.

Peplium (L): *Euphorbia peplus*.

Pepo: 1) (L.): pumpkin sd. fr. *Cucurbita pepo*; also fr. *C. maxima*, *C. moschata* 2) type of frt. found in Fam. Cucurbit. (large berry with thick epicarp).

pepparmynt(h) (Swed): **pepparmynta** (Pol): peppermint.

pepper: 1) *Piper* spp., sometimes (specif.) *P. nigrum* 2) *Capsicum* spp. - **Acheen (Achin; Atschin) p.**: Penang p. (SC export town Acheen in nSumatra, now Koeta Radja) - **p. adulterants**: incl. p. hulls; Penang p.; sweet orange immature frts.; grains of paradise; coconut shell charcoal; grd. nutshells; cereals; mustard hulls; buckwheat hulls; sds. of Fam. Leg.; olive pits; sawdust; flaxseed; cocoa shells; p. stems; exhausted spices; etc. **African p.**: A. chillies: *Capsicum frutescens* grown in Af; formerly the only kind off. in US - **alligator p.**: *Aframomum granum-paradisi*, *A. melegueta* - **American p.: annual p.**: *Capsicum annuum* - **Anaheim p.**: *Capsicum annuum*: long rd & green chillies; said first peppers brought to NM (US) fr. Mex (ca. 1600); cult. later at Anaheim, Calif (US); pungency varies; u. to make chile powder - **Ashantee p.: Ashanti p.**: frt. of *Piper clusii* - **Australian p.**: *Piper novae-hollandiae* - **Australian p. tree**: 1) *Umbellularia californica* 2) *Echinus molle* - **bell p.**: *Capsicum* cultivar; with sweet thick flesh, eaten raw as veg.; u. in salads.; "green p." (but some vars. are red, yellow, orange, or brown) large bland broadly ovoid type of *C. frutescens*; u. in prep. "stuffed peppers" - **Bengal p.**: long p. - **Benin p.**: Ashantee p. - **p. berry**: *Capsicum* frt. (as distinct fr. Piper frt., which is a drupe) - **betel p.**: *Piper betle* - **bird p.**: *Capsicum frutescens* (with undeveloped fruits) - **bird's p.**: *Lepidium* - **black p.: b. African p.**: *P. nigrum* - **Bombay p.**: frt. of *C. annuum*, of large size, less pungent, & with thicker pericarp than in chillies, *C. frutescens* - **Brasil p.: Brazil(ian) p.**: *Xylopia frutescens*, *X. grandiflora*, etc. - **butcher's p.**: coarsely ground black p. - **California p. tree**: p. shrub - **candle p.**: *Capsicum annuum* var. *conoides* - **Cayenne p.**: 1) *C. annuum*, less commonly *C. baccatum* var. 2) at times but less correctly appl. to *Zanthoxylum danielii*, *Capparis sinaica*, *Vitex trifolia* (SC town in Fr Guiana) - **cherry p.**: *Capsicum cerasiforme*, *Capsicum annuum* (SC cherry size) - **chile p.**: v. chile - **clove p.**: Pimenta - **cockspur p.**: *C. frutescens* - **common p.: corn p.: p. corn**: black p. - **p. cress**: *Lepidium* spp., esp. *L. latifolium* - **cubeb p.**: *P. cubeba* - **Ethiopian p.**: *Xylopia aethiopica* - **p. family**: Piperaceae - **garden p. (, common)**: *C. annuum* - **p. grass**: *Lepidium intermedium* & other *L.* spp. - **green p.**: sweetbell p. collected before turning red; u. in salads, &c. - **Guinea p.**: 1) *Capsicum frutescens* 2) *Aframomum malegueta* 3) *Xylopia aromatica* - **p. herb**: 1) *Satureja hortensis* 2) *Lepidium latifolium* - **p. hull**: outer layers of black & white p. frts. - **Hungarian p.**: paprikas, fr. *C. frutescens*, *C. annuum*, & other *C.* spp.; large reddish frts. with only slight pungency - **Indian p.**: garden p. - **Jamaica p.**: *Pimenta officinalis* - **jalapeño p.** (pron. halapeyno) (Mex): very hot *Capsicum* spp.; u. spice, condim. - **Japanese p.**: *Zanthoxylum piperitum* - **Java p.**: cubeb p. - **king's p.**: Koenigs-paprika - **Kissi p.**: *Piper guineense* - **long p.**: 1) *P. longum*, *P. officinarum*, *P. reticulatum* 2) (US): *Saururus* sp. - **Louisiana long p.**: *C. annuum* v. *longum* - **L. short p.**: hybrid between Honka v. of Japanese capsicum & Old Louisiana Sport capsicum - **L. sport p.**: mutations arising in Louisiana Capsicum plants - **melegueta p.**: 1) *Aframomum melegueta* 2) alligator p. - **Mexican p.**: 1) *Schinus terebinthifolius* 2) *Capsicum* spp. cult. in Mex - **mignonette p.**: coarse p. - **Mombasa p.**: frt. of *C. frutescens* imported fr. Af; one of the best grades - **Moorish p.**: *Xylopia aethiopica* - **mountain p.**: *Daphne mezereum* - **negro p.**: Moorish p. - **Nyasaland p.**: similar to Mombasa p. - **old man's p.**: Achillea - **Penang p.**: superior v. of black p. exported through P. (off Mal) - **Peruvian p. tree**: *Schinus* *molle* - **pod p.**: *C. annuum, C. frutescens* - **p. pot**: dish origin. in WI, now popular in US; c. tripe, dumplings, pepper (Capsicum), okra, &c.; kind of stew - **red p.**: pod p. - **p. shell**: p. hull - **p. shrub**: *Schinus molle* - **shrubby p.**: *C. frutescens* - **Sierra Leone p.**: Af. chillies (*C. frutescens*) imported fr. that place (w tropAf) - **southern p.**: *C. annuum* - **Spanish p.**: red p. - **sport p.**: strain of *Capsicum* developed by mutation -**p. spray**: prepn. of *Capsicum* for use in personal defense against muggers, &c. - **spur p.**: *C. frutescens* - **p. substitutes**: 1) paradise sd. 2) sd. of *Vitex agnus-castus* 3) Moorish p. - **sweet(bell) p.**: *Capsicum frutescens* var. *grossum*; large mild-flavored fruits, green & red (ripe); both u. = bell p. - **Tabasco p.**: *C. annuum* v. *conoides* - **tailed p.**: cubeb p. - **Tasmanian p.**: *Drimys aromatica* - **Thevet p.**: *Eugenia pimentoides* - **p. tree**: 1) *Schinus* spp., esp. *S. molle* 2) Tasmanian p. - **p. t., Brazilian:** *Schinus terebinthifolius* - **p. tree, California: Peruvian p. t.**: *Schinus molle* (**p. t. of Tasmania: Tasmanian p. t.**: *Drimys aromatica, D. axillaris)* - **true p.**: *P. nigrum* - **Turkish p.**: *C. frutescens* - **turnip p.**: *Arisaema triphyllum* - **p. vine**: *Ampelopsis arborea* - **water p.**: *Polygonum hydropiper* (**mild w. p. [knotweed]**: *P. hydropiperoides*) - **white p.**: dr. ripe decorticated frt. of *P. nigrum* - **wild p.**: 1) *Solanum carolinense* 2) *Daphne mezereum* & other *D.* spp. - **wood p.**: *Zanthoxylum americanum* - **p. wort**: 1) p. cress 2) *Pimpinella saxifraga* (etc.) - **Zanzibar p.**: superior grade of Af chillies, *C. frutescens*, exp. fr. Zanz (trop eAf).

pepperette: powd. olive pits; u. to adult. pepper.

pepperidge: *Nyssa sylvatica.*

peppermint: 1) *Mentha* X *piperita* (lvs. hb. VO) (laymen often misapply this name to spearmint) 2) any sp. of *Mentha* which c. menthol in VO 3) *Mirabilis* spp. 4) several Eucalyptus spp. (Austral) 5) (Austral): *Agonis flexuosa* 6) (mostly in CA [Guat &c.]; also SA [Ec, Pe]): *Mentha citrata* - **American p.**: that cult. in such US states as Mich & Ore - **Australian p.**: *M. australis* R. Br. - **black p.**: 1) *M. pip.* v. *officinalis* f. *rubescens* 2) *Eucalyptus australiana* (cult. Austral) - **broad-leaved p.**: *Eucalyptus dives* - **p. camphor: p. crystals**: menthol. - **Chinese p.**: *M. arvensis* v. *glabrata* - **Eales p.**: *M. piperita* - **English p.**: Mitcham strains of *M. piperita* - **Essex p.**: *M. spicata* - **Japanese (field) p.**: *M. arvensis* v. *piperascens* - **native p.**: *M. australis* - **Sidney p.**: *Eucalyptus piperita* - **small p.**: *Thymus piperella* L. (Med reg.) - **p. tree**: *Eucalyptus* spp., esp. *E. microcorys* & *E. pauciflora* - **wild p**: *M. canadensis.*

peppermynte (Nor): peppermint.

pepperoni (peperoni) (It): beef & pork sausage highly seasoned with pepper, chillies, &c.; u. in pizzas.

pepperweed: *Lepidium* spp. - **pepperwood**: *Umbellularia* spp.; *Dicypellium caryophyllatum.*

pepperwort: pepperweed.

Pepsin(um): digestive enzyme wh. converts proteins into peptones; secreted by the glandular layer of the stomach obt. fr. the hog; u. digestive, predigestant of protein materials; for removal of necrotic cancer tissue; diphtheria membrane; in ulcers & wounds; to aid gastric digestion; acts in faintly acid medium (optimum pH = 1.5-2); types: scaly; spongy; saccharated (mixed with lactose).

pepsinogen: pepsin precursor converted into pepsin by HCl.

Peptenzyme™: Elixir of enzymes.

peptide: polymer with 2 or more linked amino acids, representing small monomers (building blocks); with ten or more am. acids, called "polypeptides" - **atrial natriuretic p.**: ANP: hormone prod. by heart wh. acts as antihypertensive, incr. excretion of Na & H_2O thru the urine; no by-effects - **peptide T**: octapeptide fraction of glycoprotein of HIV (human immunodeficiency virus) ("T" stands for the high threonine content); u. in rheumatoid arthritis, AIDS & other low immunity states; increases the level of immunity.

Peptone: Peptonum: mixt. of proteoses, polypeptides, & amino acids; obt. by action of pepsin on lean meats, blood, or casein; u. nutr. mostly for microorganisms growing on artificial media.

peptonizing powder: pancreatin powder comp.

pe(e)pul (tree): *Ficus religiosa.*

pequi: pequia (Br): *Caryocar* spp.

pera (Sp, Por): 1) pear 2) avocado 3) (Suriname): *Couma guianensis.*

peral (Sp, Por): pear tree.

Peramium Salisb. = *Goodyera.*

pérasse: *Tsuga canadensis.*

percomorph: Percomorphi: large order of fishes (Pisces) with ca. 80 fams.; found worldwide in fresh & salt waters; incl. Percidae (perch fam.) with perches, darters, saugers, &c.; the order incl. basses, mackerels, tunas, sword fishes, bonitos, sardines, groupers, weakfish, snappers, &c.; **percomorph (liver) oil** is very rich in vitamins A & D.

percoon (root) (US Negro): puccoon.

Perebea lecithogalacta R. E. Schultes (Mor.) (Col): latex source of rubber (caucho negro) (2303).

pereira (Por): pear tree; pomaceous tree.

pereiro (Por): 1) pear or apple or crab-apple tree 2) white quebracho - **p. bark**: fr. *Geissospermum* spp., esp. *G. laeve* (*qv.*); formerly thought to originate fr. *Picramnia ciliata*.

perejil (Sp): parsley.

perera (Catalan): pear tree.

Pereskia aculeata P. Mill. (Cact.) (Mex, WI & tropAm): West Indian gooseberry; Barbados g.; lemon vine; frt. eaten raw or preserved; lvs. u. as potherb - **P. grandifolia** Haw. (Br, WI): Barbadoes gooseberry; u. to make strong hedges; u. (Co) cardiovascular dis. (365) - **P. guamacho** Weber (Ve, Co): lvs. u. as med. (Co); pl. u. for hedges; wd. u. for posts - **P. nicoyana** Weber (CR): "mateares"; lvs. u. for cataplasms; small trees u. to form a "living fence"; u. in prepn. Ung. Althaeae.

Pereskiopsis porteri Britt. & Rose (P. brandegeei Schum.) (Cact.) (Mex): alcajer; frt. edible but very sour.

Perezia Lag. (Comp.) (sUS to Patagonia): herbs; some spp. said to prod. perezia root (radix pipitzahuac; r. pereziae) (*P. nana*, *P. oxylepis*, *P. rigida*); u. in cholera & other dis. as purg. (some have divided this group into *Acourtia* [for North American], Perezia [for South American] spp.) - **P. adnata** Gr. u. in Mex med. (perezia rt. decoc. as lax. & antiparasitic) - **P. atacamensis** Phil. (Ch, high peaks of Cordilleras): maranzel; c. balsamic resin; u. as teas in heights sickness, resp. difficulties at high elevations; much u. in chest dis. - **P. cuernavacana** Rob. & Greenm. (Acourtia c. Reveal & King) (Mex): rt. decoc. u. as lax., antiparasit.; c. perezone (2424) - **P. hebeclada** (DC) A. Gray (Mex): rt. u. as lax., effect due to hydroxyperezone monoangelate (2424).

perfect: of fls., when both stamens & pistil(s) are present in each fl.

perforator (Ge): continuously operating types of liquid extractors; ex. Soxhlet extractor.

perfume: VO or other volatile subst. found in nature or made by synthesis; mostly u. for body scenting or cosmetics; the perfumery industry is a large and complex one (word derived fr. per fumo [L], through smoke, with ref. to early efforts of man to contact the gods by means of fragrances emitted by heat; incense); voodoo prepns. incl. "enlightenment p." (lemon, jasmine), "success p. oil" (sandalwood, sage, petitgrain, magnolia), &c. - **p. plant**: *Matthiola bicornis* DC (Gr, AsMin).

Pergularia africana N. E. Br. (Asclepiad.) (tropAf): pl. ext. u. to adulter. dragon's blood of commerce - **P. extensa** N. E. Br. (P. daemia Chiov.) (Afghan, Ind, SriL): dr. entire pl. c. cardenolides ("calotropin" +), uzarigenin, &c.; u. expect. emet. (Ind); lf. juice in infantile diarrh., lax. in rheum., amenorrh. & dysmenorrh.; lvs. & fls. eaten (Ind) - **P. minor** Andr. (Telosma cordata Merr.) (Daemia cordata R. Br.) (Ind): c. pergularin (hydrolyzing to uzarin, digitoxose, glucose) with cardioton. properties; u. emet. expect. in asthma; anthelm.

Periandra dulcis Mart. (Leg) (Br): rts. (Brazil roots; raiz doce) u. as substit. for licorice (c. 4-8% glycyrrhizin) ("national licorice" [Br]).

periblem: part of meristem giving rise to cortex (bk.).

pericambium: pericycle of rt. - **pericarp: Pericarpium** (L.): ripened ovary wall or frt. wall; typically consists of 3 layers: epicarp (exocarp), mesocarp (sarcocarp), & endocarp.

Pericampylus glaucus Merr. (Menisperm.) (eAs): lf. inf. u. for coughs, asthma (Mal); lf. ext. appl. to head for falling hair (Indones).

pericycle: regarded as the outermost layer of the stele, ordinarily bounded on its outer side by the endodermis; function mostly of storage.

Perideridia Reichenb. (Umb.) (NA): yampah; 9 spp. of branching perenn. hbs.; c. flavonoids, anthocyans; rts. & sds. eaten; sds. as poultice for black eyes, sds. inf. u. in colds, indig.; rt. poultice for swellings, rt. chewed for sore throat (Calif [US] Indians) - **P. gairdneri** (H. & A.) Mathias (wNA, Calif [US] to BC [Cana]): squaw root; Indian/false/wild caraway; yampah (Klamath Indians); tuberous rts. (raw, preserved) important food of Am Indians ("Indian carrot"), u. as flavor or relish; sds. u. for seasoning; several other *P.* spp. u. likewise, incl. **P. kelloggii** Matthias (Carum k. A. Gray) (cCalif): Kellogg's yampah.

periderm: cork: phellogen (cork cambium) & its products, phellem & phelloderm; sometimes restricted to 2 latter.

perilla: 1) perillo 2) *Perilla* spp. - **Perilla frutescens** (L) Britt. (P. ocymoides L.; P. nankinensis Dcne.) (Lab) (nInd, Chin, Jap+): common perilla; "purple mint plant"; ann. hb. with strong balmy odor; when bruised, pl. gives off strong aroma; aboveground parts u. for cough, colds, indigest., &c. (Kor, Chin); sed. spasmolytic; diaph.; cold water inf. of pl. u. to wash feet with itch from wading in dirty water (Chin); may be toxic to cattle & humans (perillaldehyde antioxime) (sweetening agt., Jap); sd. source of perilla, yegoma, or zegoma oil (FO); a drying oil u. in oriental foods, in technology, as linseed oil substit.; in shellacs, prepn. of Japan wax; fish poison; var. *crispa* (Benth.) Dcne. (Far East): u. for coloring foods; ornam. ann. hb.; var. *nankinensis* Voss (P. n. Dcne) (eAs): source of FO; hb. u. for cough & lung dis.; u. to incr. appetite, remove phlegm; fls. eaten (Chin); ornam. (forma *viridis* Mak. u. as source of VO).

perilla-aldehyde: perillaldehyde: arom. component of many VO, incl. those of *Perilla frutescens*, esp. var. *nankinensis*, *Sium latifolium*, *Siler trilobum*, *Hernandia ovigera*, cumin, &c.; u. as sweetener (Jap) (2000x as sweet as sucrose) - **perilla-ketone**: furyl isoamyl ketone; pulmonary edematogenic agt.; pulm. toxin (produces emphysema in grazing cattle).

perillo (Co): tree of Fam. Euphorbi. with fig-like frts. u. for intest. dis.; latex u. as chicle (**p. gum**).

Perimycin: polyene antibiotic obt. fr. *Streptomyces coelicolor* (Müller) Waksm. & Henrici var. *aminophilus*; 3 components; isolated 1960; antifungal.

periodot: woodbine green; suggested as color of jackets for pharmacists.

Periplaneta americana L. (Blatta a. L.) (Blattidae - Orthoptera) (Am tropics & subtr., introd. into Eu last few centuries): (American) (cock)roach; light to dark reddish brown insect -5 cm; common in sUS; in Fla [US] lives outside in summer; in ancient med. folklore as diur. in edema (Russia); asthma; nephritis of scarlet fever; chronic nephritis; bronchitis; laryngeal croup; epilepsy (SCar [US]); now u. by homeop.

Periploca aphylla Dcne. (Asclepiad.) (Ind): hb. u. as antipyr.; pressed juice u. for swellings, tumors; bk. decoc. purg. - **P. forrestii** Schlecht. (eAs, Chin): c. cardioactive glycosides (241) - **P. glauca** (AsMinor): climbing dog's bane; wd. u. as cardiac med. - **P. graeca** L. (eMed reg. Gr, AsMin, Iran, tropAf): (Grecian) silk-vine; topi; hb. c. digitaloid glycosides; periplocin, periplocymarin; u. like digitalis, as dig. substit.; in prepn. periplocin; sometimes u. with Strophanthus; cult. ornam. - **P. laevigata** Ait. (nAf, Lib-Mor): hallaba (Arab): sd. decoc. rubbed on skin as local analg. (rheum., &c.) - **P. nigrescens** Afzel. = *Parquetina n.* - **P. sepium** Bge. (Chin): c. cardioactive glycosides (periplocin, &c.); rt. bk. (bei wujiapi of Chin med.) as card. stim.; in rheum., lumbago, & impotency.

periplocin: glycoside of periplogenin + D-cymarose—ß-O-glucoside.

perisperm: sd. storage tissue formed fr. the nucellus of ovule; lies directly beneath the sd. coat (spermoderm).

Peristaltin™: mixt. of anthranol glucosides (rhamnosides) of Cascara Sagrade (chrysophanic acid, frangula-emodin methyl ester, cascarol, &c.).

Peristrophe acuminata Nees (Acanth.) (Mal): lvs. u. in poultices for wounds, esp. snakebite - **P. roxburghiana** Bremek (P. bivalvis Merr.; P. tinctoria Nees) (eAs): lvs. u. for skin dis.

perithecium: hollow frtg. body of Ascomycotina.

peritre, raiz de (Mex): *Erigeron affinis* - **p. del pais** (Mex) (means native pyrethrum): *Heliopsis longipes*.

periwinkle: *Vinca* spp., esp. *V. minor*, *Catharanthus roseus* - **blue p.**: *V. major* - **cape p.**: *Catharanthus roseus* - **greater p.: large p.**: blue p. - **lesser p.**: *V. minor* - **pink p.**: cape p. - **small p.**: lesser p.

Perlgerste (Ge): **Perlgraupen**: pearl barley.

perlière rameuse (Fr): mouse ear herb, *Hieracium pilosella*.

Perlzwiebel (Ge): *Allium sativum* v. *ophioscorodon*.

Pernambuco wood: Brazil w., fr. *Caesalpinia echinata*.

Pernettya Gaudichaud (Eric) (LatinAm; NZ, Tanz): several spp. claimed to have hallucinogenic properties: **P. furens** (Hook. ex DC) Klotsch. (Ch); **P. parvifolia** Benth. (Ec); **P. prostrata** (Cav.) DC ("albricia negra") (Ve).

pero (It): pear tree.

peroba (wood) (SA): *Aspidosperma* spp. also appl. to *Tecoma peroba*, *Paratecoma peroba*, &c. - **perobo**: *Aspidosperma peroba* - **p. de campo** (Br): **perobinha (campestre)** (Br): *Sweetia elegans*.

perofskoside: glycoside of strophanthidin + D-2-deoxyglucose; found in *Erysimum perofskianum*.

Peronema canescens Jacq. (monotypic) (Verben.) (Mal archipel.): wd. useful & in demand; lf. decoc. u. as mouthwash for toothache.

Perotis latifolia Eckl. (Gram) (sAs, tropAf): hairy grass; rt. diur. & styptic; *cf. Imperata cylindrica*.

Perovskia abrotanoides Karel (Lab) (Indo-Pak): water-soaked fls. appl. to body of fever patient.

peroxidases: enzymes wh. catalyze oxidation of certain compds. by H_2O_2 (increases the oxidizing power of peroxides formed in living tissues through the activity of oxydases); form active oxygen; occ. in pl. & animal tissues but commonly obtained fr. pls. (horseradish rt., fig tree sap), higher Basidiomycetes; ex. *Poria*.

Perpolitiones Oryzae (L.): rice polishings.

perry: fermented beverage produced fr. pears (British); *cf.* cider.

Persea americana Mill. (P. gratissima Gaertn. f.) (Laur.) (CA, tropSA, now widely cult. Calif [US], Mex, Israel+): (American) avocado (tree); alligator pear (SC frts. green with rough peel); aguacate; cult. (since ca. 8000 B.C.) for edible nutritious frt. which is the driest of all fresh frts. with much protein (in compn. more like meat than frt.); highest caloric value (741/#) of all frts; pleasant nutty buttery flavor; c. vit. A, C, E; FO; u. in salads with sour juices such as lemon or lime; FO (prod. Haw [US]) rich in oleic acid; u. med. & in cosmetics (green soaps); refined for food use; said protective against heart attack; frt. pulp in folk med. on wounds, as emmen. aphrod.; sd. astr. u. (pd.) as rubefac. locally decoc. in caries; aphrod.; lvs. u. as diur. carm. stom. emmen. tea substit.; lf. decoc. abortif. (WI); expect. & in fever (Yuc); bk. anthelm. (Mex) (alt.); var. *drymifolia* (Schlecht. & Chamisso) Blake (P. d. Schlecht. & Cham.) (Mex): aguacate; u. similarly to type var.; sd. kernels u. in colics; for indelible inks; residue fr. expressing FO fr. frt. u. as cattle fodder (*Persea* < pera [pear] since frt. pear-shaped) (1680) - **P. borbonia** Spreng. (eUS): red bay; tree grows in swamps; lvs. c. VO with camphor, cineole; pl. ornam.; wd. excellent, u. in boat building; dr. lvs. u. to flavor soup (crab gumbo); condim. in stuffing fowl; var. *pubescens* Little (P. p. Sarg.) (seUS): swamp red bay; ornam. u. diaph. hydragog, in edema (852) - **P. indica** Spreng. (Canary Is): Canary wood; royal bay; small tree; ornam. pl. (sUS) - **P. lingue** Nees (Ch): lingue; bk. rich (17-25%) in tannin u. astr. in galenical prepns.; in prepn. of valdivia leather; decayed wd. strong astr. u. in leucorrh. - **P. meyeriana** Nees (Ch): u. like *P. lingue* - **P. palustris** Sarg. = *P. borbonia* var. *pubescens*. - **P. schiedeana** Nees (Mex): u. much like *P. americana* var. *drymifolia* - **P. thunbergii** Kosterm. (Kor, Jap): bk. u. for eczema, asthma; pd. bk. corrective for bad odors.

perseite: perseitol: sugar formed in flesh & lvs. of *Persea americana*; glycero-galacto-heptitol.

Persian berries (trade name): frt. of *Rhamnus infectoria*; chiefly *R. oleoides*; u. yellow dyestuff - **P. root**: type of Russian licorice.

persic oil: peach & apricot kernel oils (FO) - **Persicae Folia/Folium** (L.): peach lvs.

persicaire douce (Fr): **persicary, common**: *Polygonum persicaria* (heartsease).

Persicaria Mill. = Pollygonum L.: term often appl. by medical writers of the 16th cent. to *Polygonum hydropiper* (Wasserpfeffer) & *Pulicaria* spp. (Flohkraut).

persil (Fr-m.): parsley frt.

persimmon (Am Indian), **common**: *Diospyros virginiana* (frt., bk.) - **Japanese p.**: *D. kaki.*

Persio (L.): cudbear (*qv*).

Personae Formicarum (L.): ant larvae ("a. eggs").

pertuisane (Qu [Cana]): *Hypericum perforatum.*

Pertusaria concreta Nyl. (P. pseudocorallina (Sw.) Arn.) (Pertusari. - Lecanorales - Lichen.) (Eutemp. & subarct. regions): thallus u. for red & purple dyes for woolens (Nor, Swed) - **P. corallina** Th. Fr. (temp. zone): found on rocks; u. to give red-orange color to woolens (Scotland) - **P. pertusa** (L.) Tuck. (Porina p. (L.) Ach.) (NA): thallus u. as source of dyestuff.

pertussis: whooping cough; infectious dis. caused by *Bordetella pertussis* Moreno-Lopez; causes inflammn. of resp. passages, usually resulting in peculiar & characteristic cough (hence name); biological products: **p. immune globulin (human)**: sterile soln. of globulins fr. blood plasma of adult donors immunized with p. vaccine; u. prophylactically & therapeutically - **p. immune human serum**: fr. blood of donors who have recovered fr. attack of p.; u. both prophylactically & therapeutically - **p. vaccine**: sterile susp. in isotonic NaCl soln. of killed p. organism, *Bordetella pertussis* (cause of whooping cough); u. in active immunization of children; inj. to give active immunity; acellular p. v. now in use; **adsorbed p. v.** preferred; there are also combns. of p. v. with diphtheria & tetanus toxiods for multiple vaccination.

Peru Balsam: Peruvian B, v. under Balsam.

perussié (Fr): *Pyrus communis* var. *pyraster* L.

Peruvian bark, false: *Buena obtusifolia* - **red P. b.**: red cinchona - **yellow P. b.**: yellow cinchona.

Peruvian: v. under balsam; Krameria; rhatany.

Peruvianische Rinde (Ge): Cinchona.

peruvoside: glycoside of cennogenin + L. thevetose; occ. in *Apocynum cannabinum;* u. as cardiotonic.

pervenche (Fr-f.): **p. officinale** (Fr): **pervinca** (Sp): *Vinca minor*.

pesce (It): fish.

pessego (Br, Por): peach.

pesticide: agent u. in struggle against predators & unwanted pls. & animals; incl. insecticides, rodenticides, fungicides, herbicides, seed protectants, wood preservatives, &c.; important natural pesticides: pyrethroids; rotenoids; nicotine; quassia; sabadilla; hellebore; Ryania; limonoids; natural isobutylamides.

Pestilenzkraut (Ge): Galega.

Pestilenzwurz (Ge): *Petasites.*

petal: floral leaf or segment of corolla; ex. red rose p.

Petalonia fascia Kuntze (monotypic) (Scytosiphon - Phaeophyc Algae) (Pac Oc coasts): sun-dried thallus eaten (eAs) by rural populations.

Petalostemum (*Petalostemon*) **purpureum** Rydb. = *Dalea purpurea* (Leg.).

Petalostigma pubescens Domin. (P. quadriloculare F. Muell.) (Euphorbi.) (Austral): native quince; quinine berry; q. bush; crab tree; emu apple; frts. bitter ("quinine berries"); these & bk. reputed antimalar.; c. glycosides, no alk.; frts. u. vermif. fish poison, opium antidote.

Petasites albus Gaertn. (Comp.) (Euras): butterbur; c. albopetasin; lf. u. as expect - **P. frigidus** Fries var. *palmatus* (Ait.) Cronq. (P. speciosus [Nutt.] Piper) (circumpolar, nwUS-Alas; nEuras): (common) (or sweet) coltsfoot; butterbur; rt. u. as cough med., either eaten raw, boiled, or as tea; mashed rt. u. for swellings & sore eyes (Quinault Indians); in TB (hemoptysis); emet.; lvs. warmed & appl. in rheum.; lvs. u. as greens (Eskimos); sts. boiled & eaten (Muckleshoot Indians) (441); ash fr. pl. combustion u. as salt by some Indians - **P. hybridus** Gaertn., Meyer, & Scherb. (P. officinalis Moench) (Euras, natd. eNA): butter dock; b. bur; pestilence; hb. c. mucil. tannin, resin; u. ton. in anorexia; diur., anaticonvulsant, antipyr. diaph. anthelm.; fresh lvs. vulner.; substit. for or adulter. of Farfara; rhiz. c. eremophilene, furanopetasin; u. ton. diaph. exter. as vulner. for sores - **P. japonicus** Maxim. (Jap, Chin, Kor): lf. stalks & fl. buds eaten boiled as veg. (Jap); fls. as bitter & in colds; u. as expect. sts. inserted in uterus —> abortion; var. *giganteus* Nichols. (P. amplus Kitamura) (Jap & nearby): decoc. u. for colds & sed. in cough (Ainu) - **P. paradoxus** Baumg. (Eu-Alps & Pyrenees): u. like *P. albus* - **P. spurius** Rchb. (n&eEu): c. eremophilans; u. like *P. albus*.

Petersilie (Ge): parsley (frt.) - **Petersilienapiol: Petersiliencampher**: apiol.

petha (Hindi): *Benincasa cerifera*.

petit baum (Guad, WI): peppermint (as grown in gardens) - **p. ciguë** (Fr): *Aethusa cynapium* - **p. thé** (Anticosti, Qu): *Chiogenes hispidula* - **petite centaurée** (Fr): Centaurium - **petit-houx** (Fr): *Ruscus aculeatus* - **p. oseille** (Qu): *Rumex acetosella* - **petitgrain oil**: VO fr. *Citrus aurantium* (lvs. twigs, young frts.) (most prodn. Par, nArg) - **petitpois** (Salv): **petits pois** (Fr) (little peas): *Pisum sativum*.

Petiveria alliacea L. (monotypic?) (Phytolacc. or Petiveri.) (sFla [US], Mex, WI, tropAm, esp. Br, Co): polecat weed; (garlic) guinea hen weed; zorrillo; lvs. sts. & rts. c. vol. mustard oil (foul odor); after cow feeds on pl., much gas generated in milk (sufficient to blow the stopper out of container); gas formed is strongly malodorous (skunk aroma); source of "pipi" root (raiz de guini), u. as antispasm., anticonvuls., counterirrit. in dentalgia, &c.; in hysteria, rheum. cancer (2490): to promote menstruat.; aphrod.; in prostate hypertrophy; prepn. of kind of curare u. by Tecumas Indians; very popular in folk med. as parturient (Domin) abortif.; rts. placed among woollen goods to protect against insects (repellant), toxic to cattle (abort.).

Petradoria pumila (Nutt.) Greene (Solidago p. Blake) (Comp.) (Calif to Wyo to Tex [US]): rock goldenrod; pl. u. to alleviate chest pains (Hopi Indians) (1998); in folk med. u. as diur. in kidney & bladder stones; chronic nephritis; rheum.

Petrea kohuatlana Presl (Verben.) (Lesser Antilles,Ve): queen's wreath; fls. u. as tea with *Chiococca alba*; u. as abortive (DominRep); cult. in Eng - **P. racemosa** Nees (P. subserrata Cham.) (tropAm, natd. seAs): lvs. u. as diaphor. stim. (Vietnam, Laos) - **P. volubilis** L. (Am, tropWI, Java): purple wreath; "flor de papel" (Cu); "choreque" (CR); woody vine cult. as ornam.

petrochemicals: those prepd. fr. petroleum oil, natural gas, &c.; ex. ethylene, glycerol, styrene, &c.

petrographic microscope: instrument equipped with lenses & special equipment to permit the ready examn. of rock fragments in determining their optical properties, as a means of identification & determination.

petrol (Br): 1) gasoline, motor fuel, sometimes u. to ext. drugs 2) **p. ether**: petroleum ether.

Petrolatum (L., Eng.): (yellow) petroleum jelly; (yellow) Vaselin(e); mixt. of hydrocarbons, a semisolid prod. of distn. of petroleum; u. emoll., protective, lubricant, ointment base; by rubbing to one side of ball, u. by baseball players to "curve" the ball ("vaseline ball") - **P. Album**: white petroleum jelly; u. in light-colored ointments, etc., as in cosmetic preps. - **P. Liquidum** (L): mineral oil - **P. Spissum**: hard petrolatum; usually the yellow form.

petrolene: asphalt.

petroleum (, crude): 1) petroleum oil; "crude oil"; a blackish oily liquid, insol. in water & only partly sol. in alc.; formed in shale & sandstone deposits worldwide; possibly derived fr. plankton or fish remains fr. past geolog. ages; also prod. by distn. of some plants (ex. *Euphorbia lathyris*, milkweeds, &c.); consisting mostly of a natural mixt. of hydrocarbons & of these chiefly the paraffin & cycloparaffin compds., with some cyclic arom. compds. as well as sulfurated & oxygenated compds. in smaller amts.; u. in folk med. as expect. & vermif., exter. in skin dis.; in rheum; sores; also in war paint & to caulk canoes (Am Indians); distn. yields kerosene, gasoline, naphtha, petroleum ether, petrolatum, paraffin wax, lubricating oils & greases (b.p. c. 40°) &c.; produced US (Tex, Okla, Ohio, Ala), USSR, Ve, Mex, Rom, Iraq, Iran (Caucasian p. cont. members of ethylene series) (2755) 2) (Ge): mineral oil - **P. benzin: benzin(e)**: petroleum ether - **P. ether**: petrol. distillate, purified p. benzine, mostly pentanes & hexanes; with boil. range 35-80° (mostly 50-60°); much u. as solvent; in chem. & pharm. extns. & other processes; ethanol denaturant, in chromatography, &c. - **P. jelly**: Petrolatum - **P. naphtha** (US): gasoline, with b.pt. range of 30-220 C°.

petrosel(in)ic acid: $\Delta^{6,7}$-octadecenic acid; $C_{17}H_{33}$ COOH; occ. in frt. & sd. of *Petroselinum sativum*; FO of frt. & sd. of *Hedera helix*; frt. of *Picrasma quassioides*, &c.

Petroselini Herba (L.): **Petroselini Radix** (L.): hb. & rt. (resp.) of parsley (foll.).

Petroselinum crispum A. W. Hill (P. sativum Hoffm.) (& var. *latifolium* Airy-Shaw) (Umb.) (cEu, mts. - Calif [US], &c.): common (garden) parsley; cultivated p.; hb. (incl. lvs.) rich in vit. C (more than in water cress); hb. u. flavor, condim. arom. emmen. diur. carm.; garnish in soups, broths; suggested for use in salads & sandwiches (366); believed to neutralize odor of garlic; parsley frt. ("seed") (Petroselinum, L.) c. VO (-7%) (parsley oil, prod. Fr, Ge); also FO; frt. u. as diur. carm. prodn. VO, apiol (green & yellow); ext. & VO parasiticide; flavor; rt. u. as diur. esp. in edema, kidney, & cardiac dis., emmen.

pettitgrain oil: petitgrain o.

petty-mor(r)el: *Aralia racemosa.*

Petunga roxburghii DC (Rubi) (Indochin): decoc. of rts. u. for yaws - **P. venulosa** Hook. f. (Mal): pounded rt. u. formerly to poultice smallpox; boiled rts. in rheum. (—> *Hypobathrum*).

Petunia Juss. (Solan.) (mostly sNA, SA): popular ornam. flowering hbs. (Pe: one spp. hallucinogenic—> sense of flying).

Peucedanum ambiguum Nutt. (Lomatium a. (Nutt.) Coulter & Rose) (Umb.) (wNA): cow-as (Am Indians); tubers u. as food (aborigines) - **P. cervaria** Lapeyr. (s&cEu, wAs): hart's wort; much-good; broad-leaved spignel; hb. c. VO with coumarin; rt. & frt. u. for stomach dis., gout, in mal. & as diur. in edema, fevers; emmen.; formerly cult. - **P. decursivum** Maxim. (P. decumbens Maxim.; Angelica decursiva Maxim.) (eAs): "nodake" (Jap); young lvs. boiled & eaten; rt. febrifuge & expect. to stop cough; rheum. - **P. foeniculaceum** Nutt. = *Lomatiam f.* - **P. galbanum** Benth. & Hook. (Af): blistering bush; wild celery; u. as abortif., diur. in "gravel"; blisters skin on contact; to relieve suppression of menses - **P. graveolens** Benth. = *Anethum g.* - **P. japonicum** Thunb. (Jap, Chin, PI): gante; hb. u. as ton. diur. sed.; in mal.; u. in pregnancy prior to delivery (Chin); u. after delivery (Mal); antipyr. (mostly in mixt.) - **P. officinale** L. (w&cEu, wSiberia): (common) hog fennel; sulfur weed; s. wort; s. root; Ross Kümmel; rt. c. unpleasant smelling VO with peucedanin (NBP), resin, gum; u. in veter. practice (Eng); formerly as diur. stom. digestive; emmen. expect. antispasm., in mal. GC; scabies, scurvy - **P. oreoselinum** Moench (s&cEu): hb. source of VO (Ol. Oreoselini); frt. c. oreoselone, VO; rt. & half-ripe frt. c. athamantin (vasodilator); u. diur., in fever & jaundice; homeopath. - **P. ostruthium** Koch (Imperatoria o. L.) (c&nEu, natd. NA): masterwort (hog fennel); pellitory of Spain; rhiz. c. ostruthol, ostruthin, osthol (a coumarin), peucenin (a chromone); hesperidin (367); u. diur. diaph. arom. stim. stomach. digestive; in gout; fevers; in veterin. practice (Eng); acrid poison; hb. (Herba Imperatoriae) u. stom. condim. (prod. Austral, Ge) - **P. palustre** Moench (Euras): milk parsley; marsh smallage; marsh milkweed; rt. of bienn. hb. (Radix Olsnitii) u. as antiepileptic; ginger substit. (Slavic countries); hb. u. emmen. diur. antispas. antiepileptic - **P. sowa** Kurz = *Anethum graveolens* (*cf. A. sowa*).

Peucephyllum schottii Gray (monotypic) (Comp.) (Calif [US], nwMex): desert fir; pigmy cedar; claimed useful in kidney & bladder dis. (368).

Peumus boldus Molina (Boldu boldus [Mol.] Lyons) (1) (Monimi.) (Ch): boldu tree; boldo; lvs. c. boldine & other alks., "boldoglucin" (glycoside; not yet obt. in pure form); VO; u. arom. stim., diur. ("boldo tea," coffee/tea substit.), hepatic stim. (in liver dysfunction, cholelithiasis); in stomachache; sed. ton. anthelm., kitchen spice (Ch); frt. edible; bk. u. as dye, wd. hard, useful (prodn. Ch, Pe).

peuplier: 1) (Fr): poplar 2) (Qu [Cana]): *Populus tacamahacca.*

peyote (mexicano): **peyotl**: *Lophophora williamsii.*

Pfaffenroehrchen (Ge): dandelion.

Pfeffer (Ge): pepper - **schwarzer P.** (Ge): black p. - **Pfefferkraut** (Ge): *Lepidium latifolium*; *Satureja hortensis* - **Pfefferminze** (Ge): **Pfeffermuenze**: peppermint.

Pferd (Ge-n.): horse - **Pferdepulver**: condition powder.

pfirco (Pe): pfiuco: *Ephedra americana.*

Pfirsich (Ge-f. or m.): peach.

Pflanze (Ge): plant - **Pflanzenrot** (Ge): carthamin.

Pflaume (Ge): plum.

Pfund (Ge): pound.

ph: *v.* also under "*f*" (Spanish, Portugese, simplified spelling, etc.).

Phacelia neomexicana Thurb. ex Torr. (Hydrophyll.) (sNM, Ariz [US]): pd. pl. appl. locally for rash (Belize) (Maya Indians) - **P. palmeri** Torr. ex S. Wats. (Nev, Utah, Ark [US]): Palmer phacelia; u. as skin irritant (Navajo) - **P. ramosissima** Dougl. ex Lehm. (wPacif states; Nev, Ariz): pl. u. for greens (Calif [US] Indians) (442).

Phaenogams: Phanerogams.

Phaeocystis pouchetii Lagerh. (Phaeocyst. - Chrysophyta - Algae) (nAtl Oc, Arctic Sea): thallus c.

antibiotic; eaten by crustacean, later by antarctic penguin; intestinal contents rendered sterile.

Phaeophyceae: brown sea weeds, a class of marine algae wh. show great variation in form; usually fixed by holdfast to ocean shore bottom, the other end of the stipe being supported by a float provided with fronds; one of the most important groups are the Kelps, u. industrially on a large scale.

phage: bacteriophage.

phagocytes: leucocytes, &c., with the power of engulfing & digesting foreign particles, cells, &c.

Phalaris aquatica L. (P. tuberosa L.) (Gram.) (Eu, Med reg.): Harding grass; Peruvian g.; capim doce (Br): cult. in some countries as winter green grass; forage, fodder - **P. arundinacea** L. (NA, Euras): reed canary grass; lady's laces; grass with high water usage, thrives on 6 m. water/yr.; u. to purify waste waters fr. chicken processing; lvs. formerly u. in med.; pasture (cattle forage), hay; var. *picta* L. (NA, Eu): ribbon grass (blades white-striped), cult. ornam. - **P. canariensis** L. (wMed reg., nAf, Canary Is; natd. NA): (true) Canary grass; "bird seed grass"; dr. ripe sd. ("canary seed"; "phalaris") u. as bird sd. for cage birds, fodder, & as cereal (for bread baking in sEu); bears an ergot; diur. in bladder dis. (calculi); sd. c. FO; packing material for stick $AgNO_3$; the meal for sizing in cotton weaving; hb. as forage.

Phallus Pers. (Phall. - Gasteromycet. - [Basidiomyc. - Fungi]) (widespread): "stinkhorns"; SC resemblance to pen. erect. - **P. impudicus** (L.) Pers. (nHemisph): u. to treat gout; kidney dis.; aphrod. (Doctrine of Signatures); toxic.

phalsa = falsa.

Phanerogamia: Phanerogams: Spermatophyta: the phylum of Seed Plants as distinguished fr. the spore plants or Cryptogams (Thallophyta, Bryophyta, Pteridophyta).

Pharbitis Choisy: gen. name appl. to some spp. now conserved under *Ipomoea*; ex. *I. purpurea.*

pharmacal: concerning or dealing with drugs; pharmaceutical - **pharmaceutic(al)**: relating or pertaining to pharmacy - **pharmaceutical**: a pharmaceutic prepn., a medicinal - **pharmaceutically**: fr. the viewpoint of pharmacy - **pharmaceutics**: the science of pharmacy; that branch of the medical science which relates to the prepn., use & form of medicinal prods. - **pharmaceutist**: a practicitioner of pharmacy; a pharmacist; druggist (term popular in early 19th c., US, now rarely heard) - **pharmacian** (obsolete): **pharmacien** (Fr): **pharmacist**: a person skilled in the science of pharmacy; one who prepares &/or dispenses medicines; a druggist; a pharmaceutical chemist; in Brit, one who has passed the qualifying examns. of the Pharm. Soc. of Gt Brit (either the Chemist & Druggist or the Pharm. Chem. exams) - **pharmacize** (obsolete): to treat with medicines; to "physic" - **pharmack** (obsolete): a medicine.

pharmacobotany: the botany of drug pls. - **pharmacochemistry**: the chem. of drug pls. & pl. drugs - **pharmacodynamic**: relating to the effects of drugs - **pharmacodynamics**: the science of the effects of drugs on healthy animals, as determined by laboratory experiments (orign. AD 1829) - **pharmacoemporia**: the trade & commerce in crude drugs - **pharmacoergasy**: the science of drug pl. cultivation & harvesting - **pharmacogeography**: the geographical relationships of drug pl. growth & commerce.

pharmacognic: relating to pharmacognosy - **pharmacognocist**: frequently noted incorrect spelling of pharmacognosist - **pharmacognosia** (L.): pharmacognosy - **pharmacognosist**: a practitioner of prec. - **pharmacognistic(al): pharmacognostically**: pertaining or in relation to pharmacognosy - **pharmacognosia** (L): **pharmacognosis**: **pharmacognostics**: **pharmacognosy**: 1) the knowledge of drug materials or pharmacology (as a general term) (botany, chemistry, & pharmacy) 2) now generally restricted to the science of crude drugs, that branch of pharmacy relating to medicinal substances fr. the pl., animal, & mineral kingdoms in their natural, crude, or unprepared state, or in the form of such primary derivatives as oils, waxes, gums, & resins (2528) (**economic p.**: study or knowledge of pesticides) - **pharmacography**: a description of crude drugs - **pharmacohistory**: the history of crude drug usage, commerce, prodn., etc.

pharmacologist: a person trained in the techniques & theory of **pharmacology**, a term now restricted to the science of the action of drugs on the animal or pl. body (usually in abnormal state) (originally, **p.** was defined as the science of pharmacy, incl. prepns., actions, uses, etc.) (orig. AD 1641).

pharmacomania: a craze for trying or using drugs - **pharmacomathy** (obsolete): pharmacognosy - **pharmacometer**: vessel or device for measuring drugs - **pharmacomorphic**: relating to form or appearance of drugs - **pharmacomorphology**: the science of crude drug gross description - **pharmacopedia** (obs.): **pharmacopedics** (obs.): **pharmacopedy** (obs.): the total of scientific knowledge concerning drugs & medicinal preparations - **pharmacopeia: pharmacopoeia**: 1) a book or books c. monographs for drugs & medicinal prepns., with descriptions, tests, methods of preparation, &c., so that they may be manufactured & standardized in a uniform manner; published by official direction of a government or organization collaborating with governmental authorities; subject to more or less frequent revision 2) a collection of drugs (obs.) - **pharmacopoeist**: a compiler of a pharmacopeia - **pharmacooryctology**: the science of mineral or inorganic drugs (ex. kaolin, chalk) - **pharmacopoietic(al)** (obs.): pertaining to the compounding of medicinal prepns. - **pharmacopole** (fr. pharmacopola, L.) (obs.): **pharmacopolitan: pharmacopolist**: a seller or vendor of drugs & medicines; an apothecary - **pharmacopoly** (obs.): an apothecary shop - **pharmacosology**: study of undesirable reactions produced by drugs directly or indirectly (1965) - **pharmacosystematics**: the taxonomy of drug pls. - **pharmacotheon** (obs.): a divine medicine - **pharmacozoology**: the science of animal crude drugs & their derivatives - **pharmacus** (L.) (fr. farmakon, Gr: poison): poisoner, sorcerer.

pharmacy: 1) the science, art, & profession of medicinal agents; the collecting, preparing, compounding, examining, & dispensing of drugs & pharmaceutical prepns., involving selection, identification, combination, preservation, analysis, & standardization techniques 2) the occupation of an apothecary or druggist 3) an establishment where medicinal articles are stocked, compounded, & dispensed; a drug dispensary - **p. intern**: a prospective pharmacist who is working under supervision for a definite period of time (in most states, one year) in order to partially qualify for the requirements of full licensure.

Pharmagel™: a pure gelatin for pharmaceutic u.; **P. A.** is cationic & is u. with acid media, while **P. B.** is anionic & u. with rel. alkaline media.

pharmaka: pl. of **pharmakon** (Gr): medicinal agt.

Pharmakognoscht (Ge): **Pharmakognost** (Ge): pharmacognosist, student of Pharmakognosie (Ge) (pharmacognosy) - **Pharmazeut** (Ge-m.): pharmacist.

Pharnaceum lineare L. f. (Mollugin.) (sAf): inf. u. for pulm. dis., formerly for TB.

phaseolin: an isoflavanone found in sds. of *Phaseolus vulgaris*, *P. radiatus*, soy bean, etc.; u. as fungicide.

phaseolunatin: linamarin; HCN-producing glycoside.

Phaseolus L. (Leg) (trop. & subtrop., most Am): (kidney) beans; some 50 spp.; many taxa have been transferred to *Vigna* - **P. acutifolius** Gray (swUS [Ariz, NM], nMex+): sds. boiled, ground & added to soups (Uganda); beans soaked overnight to form gelatinous soup base; hay & cover crop; adapted to heat & drought; var. *latifolius* G. Freeman is cult. form; tepary bean; annual vine or bushy herb; sds. small, white or variously colored; popular in drier areas (cult. NM [US], Af) - **P. adenanthus** G. F. W. Meyer (pantrop.): decoc. u. in intest. dis., "strictures" (Ind); women express lvs. & drink juice for labor pains (Caroline Is) - **P. angularis** Wight = *Vigna a.* - **P. angulatus** Wight = *Vigna mungo* - **P. aureus** Roxb. = *Vigna radiata* - **P. calcaratus** Roxb. = *Vigna umbellata* - **P. coccineus** L. (Mex, CA, now widely cult.): scarlet (runner) bean; pods boiled & eaten; diur., reduces blood sugar (adjuvant in mild diabetes); in kidney & cardiac dis.; gout & rheum.; sds. food (ground to flour, Indones); much cult. for food; sds. u. in cataplasms for exudative eczema; for tablet binding; in prepn. bean starch & phytoagglutinin (u. in aplastic anemia); rts. claimed toxic - **P. formosus** HBK (Mex, Guat): both sds. & tubers edible - **P. limensis** Macf. (prob. origin tropAm): Lima (or butter) bean; sds. large, flat & somewhat kidney-shaped; larger than in similar *P. lunatus* (but maybe conspecific); popular food; frequently canned or frozen - **P. linearis** HBK (Ve, Br, CA): guaipanete (Ve); kind of pea native to LatinAm - **P. lunatus** L. (pantrop. esp. tropAm): sieva bean; civet b.; (small) Lima b.; Carolina b.; frijol (Sp): butter bean; sds. (beans) often large, popular as foodstuff; young pods & sprouts also eaten; u. folk med. (Yuc); sds. in diet foods in fevers; pods, sd. & st. decoc. u. in kidney dis., diabetes, dropsy, eclampsia (1682); var. *lunonanus* Bailey: dwarf lima bean - **P. max** L. = *Glycine max* - **P. multiflorus** Willd. = *P. coccineus* - **P. mungo** L. = *Vigna m.* - **P. polystachyus** (L) BSP (e&sNA): bean vine; wild b.; sds. popular food (dr. & cooked) (esp. Am Indian) - **P. radiatus** L. = *Vigna radiata* - **P. ritensis** M. E. Jones (P. metcalfei Woot. & Standl.) (swUS, nwMex): cocomelca (native name for sts.); rt. decoc. u. for stomach upsets (Tarahumar Indians), purg.; to prepare refreshing beverage to overcome pneumonia, aid to losing weight; diur.; to reduce intest. gas; as catalyst in fermentation of frt. juices in making tesguino (alc. beverage); rts. furnish a kind of glue; sds. roasted & eaten or u. in amole - **P. trilobus** Ait. = *Vigna trilobata* - **P. vulgaris** L. (prob. Am origin, now worldwide): kidney bean; common (garden) b.; red- or white-seeded climbing b., frijoles (Sp) popularly u. as a vegetable food (pods & sds.) (unripe pods eaten cooked as "string [or snap] beans"); bean husks u. as diur. depur. in pulmonary dis., arteriosclerosis; "kidney bean pod tea" u. in diabetes (reduces sugar) (Eu) (popular use) (2422), kidney stones (443); claimed healthful for heart; sds. u. in bean meal, bean starch, as nutr., for cataplasms, pill adhesive, &c.; sds. u. to produce phytohemagglutinins, etc.; bean hb. u. med. (Eu Homeop.).

Phasianidae: pheasant fam. of birds; ex. domesticated chicken, turkey.

Phaulopsis (Phaylopsis)falcisepala C. B. Cl. (Acanth) (trop wAf): fresh juice of pl. appl. to small sores; dr. hb. u. as dressing for wounds; c. lignins sesamin & eudesmin.

pheasant: *Phasianus* & other genera (Phasianidae - Galliformes - Aves): game birds; flesh a favored food; specially known is *P. colchicus* (As, now worldwide); ring-necked p.; flesh u. in Unani medicine.

pheasant's eye: *Adonis vernalis*.

Phelipaea (Phelypaea) annua C.A. Mey. (Orobanchaceae) (nwAs, Siber, Chin): a rt. parasite; hb. said one of most valued Chin med., also u. as food, fresh or salted (55).

phellandrene: cyclic terpene found in many VO, ex. fennel, lemon, star anise, *Eucalyptus* spp. (at times u. as adulter. of off. eucalyptus oil) (SC < *Oenanthe phellandrium*).

Phellandrium (L): frt. of **P. aquaticum** L. = *Oenanthe aquatica* Poir., water fennel.

phellem: layer of cork cells ("cork") prod. extern. by division of the **phellogen** (cork cambium layer); **phelloderm** is present on inner side of phellogen.

Phellinus igniarius Quél. (Fomes i. Kickx.; Polyporus i. L. ex. Fr.) (Hymenochaet./Polypor. - Hymenomycet. - Basidiomyc. - Fungi) (cosmop., esp. Eu): false clinker fungus; tinder f.; similar to *Fomes fomentarius* but sporocarps harder in texture & dark reddish (*cf.* cinnamon bk. color); c. polyporic acids (A, B, C); dr. sporocarp u. formerly as tinder or spunk (to light firecrackers, &c.); also source of brown dye; in med. as styptic, absorbent; for ornaments & pictures; incense (comminuted) - **P. pachyophloeus** Pat.: sporocarps ground up, cooked, & eaten (PI) - **P. rimosus** Pilát (Fomes r. Cke.) (widespread): cracked fomes; sporocarp ashed, mixed with salt, & taken for colds & coughs; ash alone sprinkled on wound to stimulate healing (sAf).

Phellodendron amurense Rupr. (Rut.) (neAs): (Amur) cork tree; Siberian c. t.; bk. & frts. in fever; frts. expector.; bk. u. as skin disinfect. (Ainu, Sakhalin); astr.; stom. antiphlogistic, antidote (Chin) (85); ext. insecticidal (for horsefly & mosquito control) (369); frt. edible; wd. valuable for furniture making; var. *lavallei* (Dode) Sprague (P. lavallei Dode) (Jap): insecticidal (369) - **P. chinense** Schneid. (cChin): c. alks., sitosterols; u. much like prec. - **P. wilsonii** Hay. & Kaneh. (Taiwan): cork tree; inner bk. & rt. decoc. u. stom. & in kidney dis.; wash for red eyes (N.B.: avoid confusing *Phellodendron* & *Philodendron*) (latter sometimes erroneously rendered *Phyllodendron* & *Philodendrum*).

Phelypaea = *Phelipaea*.

phenology: study of relation of climate (esp. of recurring seasons) to periodic phenomena in pls. (as flowering, fruiting, &c.) & animals (as sexual activity).

phenons: gps. of organisms which on the basis of numerical values approach the natural taxa more or less closely, but are not synonymous with taxa.

phenotype: the expression of plant or animal characteristics exhibited in its habit, structure, &c. as distinct fr. the genetic make-up (= genotype).

phenylalanine: essential amino acid found in pl. & animal tissues; nutr. sexual stim. (370) (1850); for depression.

phenylethylamine, beta-: Ph-$(CH_2)_2$-NH_2; phenethylamine; occ. in bitter almond oil, cheese, coffee, &c.; prod. by animal body; has amphetamine-like action; higher levels found in urine in mental dis., in cases of chronic paranoid schizophrenia, in stress & by persons in love(?!); skin irritant; may be carcinogen.

phenylhydrazine: reagent for detection & estimation of simple sugars through formation of crystall. osazones.

phenylpropane: C_6H_5-C_3; benzene with propane (C_3) side-chain (often isopropyl); the compds. as a group are called **phenylpropides** Ph-C-C-C & are related to phenylpropenes (with propenyl side-chain) & phenylpropynes (with acetylene side-chain); Ph. C = C-CH_3 or Ph. CH_2-C=CH; these compds. are basic to many important, mostly aromatic pl. constituents; ex. eugenol, thymol.

pheromones: ecto-hormones; vol. org. compds.; substances secreted by animals to affect the behavior of other animals of same spp. by communication, esp. relative to location; ex. sex attractants: represent volatile org. compds. u. by man to bait insect traps & to produce confusion in mating patterns of insects. (< pherein = carry + hormao = stimulate).

pheophytin: prod. representing chlorophyll with its Mg replaced by H; the Fe & Cu derivatives (**pheophytins** group) are u. in med. as anti-infectives.

Philadelphus coronarius L. (Hydrange) (Euras, as escape in NA): mock orange; English dogwood; in homeopath. med. for nerve dis.; bk. u. to treat mange in dogs (hence name); cult. ornam. shrub (white arom. fls.) - **P. coulteri** S. Wats. (Mex): cult. ornam. shrub - **P. lewisii** Pursh var. *gordonianum* Jepson (P. g. Lindl.) (BC [Cana] - Calif

[US]): Gordon's syringa; arrow wood (SC u. by Am Indians to make arrows); mock orange; wd. u. to make spears, digging sticks; rts. in basketry (Am Indians); attractive food for deer; cult. ornam. shrub - **P. mexicanus** Schlecht. (sMex): jazmin (del monte); Mexican mock orange; flowering branches distd. to make scented water (pre-Columb. Am Indians); lvs. in wine for colic & in plasters for strains (recorded in 1651).

Philibertia sp. (Asclepiad.) (Mex): lengua de vibora; sts. & juice u. for cholera morbus (Puebla state).

Phillyrea latifolia L. (Ole) (Med reg., sEu, AsMin): stone linden; mock privet; evergreen shrub or small tree; lvs. u. diur. emmen. astr. for mouth sores (gargle); in mal.; fls. in cataplasms for head pains (cephalic).

Philodendron Schott (Ar) (tropAm): large gen. with c. 500 spp. of climbers; u. as popular ornam. house pls. (usually) for large coriaceous lvs.; toxic (CaOx+) (SC "loving trees" since tree-climbing vines) (sometimes erroneously given as *Phyllodendron* or *Philodendrum)* - **P. bipinnatifidum** Schott (Br): juice of lvs. & twigs u. in rheum. & ulcers (folk med.); frts. edible; sds. u. as anthelm. for intestin. worms - **P. cordatum** Kunth (sBr): guimberana; lf. juice u. in eczema (mixed with soap) - **P. dyscarpium** R. E. Schultes (Co): u. as oral contraceptive (SA Indians) - **P. imbe** Schott (WI, CA): corde molle; lf. compresses u. for acute or chronic orchitis (Br); fresh lf. & st. decoc. u. in edema, rheum. & for boils (cataplasms) - **P. krebsii** Schott (P. hederaceum C. Wr.) (WI): "macusey hembra" (Cu); anthelm. - **P. lacerum** (Jacq.) Schott (WI): fomentation in oil u. for pain; exter. in dropsy (Cu) - **P. laciniatum** Engl. (Br): fresh lvs. covered with oil u. for rheum. pains (Br) - **P. latifolium** Koch (WI): lf. poultice u. for erysipelas, rheum. (Trin) - **P. pedatum** (Hook.) Kunth (nSA): u. in snakebite (Ve) - **P. radiatum** Schott (CA): lf. decoc. u. in baths for rheum. pain, rickets (Salv) - **P. scandens** K. Koch & H. Sello (WI, CA): macerate u. to produce sleep (CR) - **P. selloum** Koch (swBr): "lacy tree philodendron"; rts. drastic; frt. edible, u. as compote; important ornam. shrub - **P. speciosum** Schott (Br): aringa iba; fresh lvs. u. to treat boils; decoc. in rheum.; pd. sds. u. as anthelm. - **P. squamiferum** Poepp. (Guianas, Br): guiambe; crushed lvs. u. as cataplasm for edema.

philter: philtre: love potion or one intended for "magic" use.

phlein: inulin-like carbohydrate found in *Phleum pratense* & other grasses (earlier called graminin).

Phleum pratense L. (Gram) (temp. Euras, natd. Cana, n&cUS): timothy (grass); important forage & fodder grass; chief hay crop of US (esp. around Great Lakes & in mid-West [US]) (371); u. in mfr. of chlorophyll; pollen often causes hay fever.

phlobaphene: a red insol. compound derived by natural processes in the pl. fr. **a phloba-tannin**, one of the 2 classes into which tannins are sometimes divided (the other being pyrogallol tannins): occurs in Hamamelis lf., Kino, etc.

phloem: that portion of a stem or root which conducts the pl. sap & is generally also provided with a quantity of strong elastic protecting fibers; phloem sometimes also incl. secretory tissues.

Phlojodicarpus sibiricus K. Pol. (Umb) (Siberia): rhiz. & rts. u. in spasmolytic prepns.

Phlomis lychnitis L. (Lab) (sEu): lampwick (plant); lvs. sometimes found as adulter. of *Salvia officinalis*, true sage; other spp. also so occur (*P. fruticosa* L., *P. herbaventi* L., Jerusalem sage) - **P. maximowiczii** Regel (Chin): rhiz./rt. u. in paralysis & as ton. for women - **P. pungens** Willd. (Siber, Chin-mts.): hb. u. as folk remedy in bronchitis & colds; exter. in suppurating wounds - **P. tuberosa** L. (Euras): Flammenlippe; tubers c. alk., saponins - **P. umbrosa** Turcz. (Chin): rt. & st. in joint pains & to prevent chilling.

phlorhizin: phloridzin: phlorizin: glucoside fr. rt. bk. of pear, apple, cherry, plum, &c.; inf. u. to test kidney function; produces temporary glycosuria, hence u. in malingering (med. examn.) (1851); antimalarial; additive in lubric. oils; on hydrolysis gives phloretin & glucose.

phloroglucin(ol): tri-OH-benzene, isomeric with pyrogallol; an important aglycone quite abundant in the pl. kingdom; the active constits. of Aspidium are p. derivatives; **P.** is an important reagent in microscopy for detection of lignin, pentosans, pentoses, etc.

Phlox carolina L. (Polemoniaceae) (eUS): **thick-leaf phlox**; Carolina pink; rhiz. & rts. sometimes found as adult. of Spigelia; formerly stocked as "false pinkroot"; some have suggested that this sp. has some of medicinal props. of Spigelia (372) - **P. divaricata** L. (eNA): wild Sweet William; cult. ornam. perenn. hb. (many other *P.* spp. u. as ornam.) - **P. ovata** L. (eUS): mountain p.; u. similarly to *P. carolina.*

"phlox" (Ala [US]): *Dianthus barbatus.*

Phoebe henryi (Hemsl.) Merr. (Laur.) (eAs): lf. & frt. cooked for stomachache (Hainan); u. in incense powd.; c. alks.

Phoenix atlantica Chev. (Palm.) (nAf): false date; similar to *P. dactylifera* but frts. of poorer quality - **P. canariensis** Chabaud (Canary Is; natd. nAf, sEu; cult. Fr, Fla, Calif [US]): Canary Islands date (palm); "pineapple palm" (SC shape of trunk); frt.

edible (but scarcely so), sometimes canned; cult. as ornam. (2771) - **P. dactylifera** L. (nAf, Canary Is, wAs - Med reg.): date palm (tree); trees diecious; fresh or dr. ripe frt. edible & delicious; sweetmeat (safe to eat regardless of unsanitary conditions); unripe frt. u. astr.; ripe frt. source of date honey (exudn. of or expr. fr. frt.); d. syrup (Algiers), date wine (lakmi) prepd. fr. frt. juice (c. 55% sugars) or juice of end shoots; frt. u. in prepn. marmalade (jam), paste, d. pulp, d. pickles, &c.; in laxative prepn. (ancient Egypt); numerous med. uses; sds. roasted & ground to furnish date (seed) coffee; date (kernel) oil (FO) expr. fr. sd.; with pleasant flavor; d. butter; apical bud lvs. cooked & eaten (*cf.* palm cabbage); lf. fibers u. to make matting (poor quality) (Eg, Iraq); trunk & fronds u. to build huts (Arabs); tree cult. ornam. (sUS); useful for shade; cult. for frt. in Calif, Ariz (US) (generally inferior quality), swAs (largest prodn.: Iran, Iraq, Pak, Israel [cult. since 4000 B.C.]) (dates may represent the "lotus" of the little lotus eaters of Zerbi [island of Lotophagi]) (of antiquity); many cultural vars. are known, such as Halawi (chief var. of world), grown in Iraq, with firm, light-yellow flesh; Khadhrawi from Iran, with darker & softer flesh; Fardh, from Oman, with dark brown flesh; much exp. to Saudi Arabia (where eaten daily) & US (v. under "date") (1852) - **P. farinifera** Roxb. (P. pusilla Gaertn.) (Ind): frt. juice lax.; "gum" u. in diarrh. & GU dis. (Ind) (444); lvs. made into ornam. baskets, &c. - **P. reclinata** Jacq. (wAf): false or dwarf date palm; wild d.; may be a source of palm wine; frt. edible; u. med. by Zulu medicine men; terminal bud cooked as vegetable - **P. sylvestris** Roxb. (Ind, Pak, sIran): wild date (palm); common Indian date palm; cult. for juice (u. in prodn. of jaggery or date sugar & wine); frt. u. as nerve ton. & sed.; ornam. tree.

phoenix tree: *Firmiana simplex.*

Pholisma arenarium Nutt. ex Hook. (P. sonorae) (Lenno.) (Ariz, sCalif [US], nwMex): (scaly-stemmed) sand plant; sand food; native perenn. hb., parasitic on spp. of *Croton*, &c.; food pl. (shoots) of the swAm Indians; attempts to cultivate unsuccessful.

Phoma Sacc. (Coelomycetes - Deuteromyc. - Fungi) (widespread, some with marine distrn.): c. phomactins (sesquiterpenes), specif. antagonistic to PAF (platelet activating factor).

phool (Urdu): **phul**: fool (Urdu): 1) flower 2) kind of sweet dessert, like a sherbet; ex. mango fool (c. green raw mangoes, sugar & rich milk) (2138).

Phoradendron Nutt. (Visc.) (Am.): American mistletoes; some 200 sp. (mostly trop); parasitic; some spp. c. alk.; dr. fls. of CA spp. (flores de palo) sold as ornam. in markets; distinguish fr. *Viscum* (2729) - **P. californicum** Nutt. (Visc.) (swUS): mesquite mistletoe; Calif m.; dr. frts. u. as food by Papago Indians - **P. juniperinum** Engelm. (swUS, wMex): juniper mistletoe; sds. u. as coffee substit. (Hopi Indians) - **P. piperoides** (HBK) Trel. (tropAm): bird vine (Trin); lf. decoc. u. to bathe children with marasmus (wasting body tissues); boiled pl. placed on cuts; frt. u. for indigest. (Ve) - **P. serotinum** (Raf.) M. C. Johnston (P. flavescens [Pursh] Nutt.) (e&cUS): American (Christmas) mistletoe; frtg. shoots popular Christmas decoration (sometimes shot with gun fr. trees); pl. u. as antispasm. ecbolic; abortif. (inf., nwFla [Escambia County] [US]); uter. stim.; card. ton., hypertensive, nervine; all parts of pl. toxic, esp. frt. - **P. trinervium** Griseb. (WI, Key Largo, Fla): lf. inf. u. in colds (WI); decoc. u. for pre- and postnatal care (Bah) - **P. vermicosum** Greenm. (Yuc, Mex): caballero; pl. decoc. u. to expedite childbirth; in epilepsy, dementia, paralysis, various nerve disorders - **P. villosum** Nutt. (Ore, Wash, Calif [US], nwMex): hairy mistletoe; this & other spp. of *Ph.* cause abortion if eaten by pregnant cows; lf. inf. u. as abort., chewed for toothache (Calif).

phorbol: complex diterpene alcohol; derived fr. croton oil; the diester (not parent compd.) is a cocarcinogen & mutagen; u. in differentiation of human leukemia cells; termed TPA (**t**etra **d**ecanoyl **p**horbol **a**cetate).

Phormium tenax Forst. & Forst. f. (Lili/Agave) (NZ, cult. Chin): New Zealand flax (or hemp); hb. st. fibers u. in mfr. of clothing (Maori), matting, sacking, fishing nets, (binder) twine, & ropes (however weak); rt. & basal lvs. or pods boiled to furnish a dye; fresh rt. pounded & juice u. as lotion in burns, scalds, &c.; poultices made fr. roasted sts. u. in abscesses & swollen joints; gum fr. lf. in water u. for burns, scalds, old sores.

phosphatases: a large gp. of enzymes which serve to hydrolyze esters of phosphoric acid, esp. nucleotides —> nucleosides; it is subdivided into several classes (ex. phosphoamidases).

phosphate rock: natural deposit of calcium phosphate minerals u. as fertilizers.

phosphatides = phospholipids: esters of phosphoric acid contg. 1-2 mols. of fatty acids, an N base, & an alc.; ex. lecithin (base: choline; alc. = glycerin); sphingomyelin.

phosphorolysis: phosphorylation: process of esterification of glycogen with phosphate; glucose monophosphate.

phosphorylase: enzyme wh. effectuates transfer of P to glycogen; this breaks down to give glucose monophosphate.

Photinia Lindl. (Ros) (s&eAs, NA): popular ornam. shrubs; many spp. c. HCN glycosides (< Gr shining red lvs.) - **P. glabra** Maxim. (eAs): frts. u. in obstinate dys., piles, intest. worms, icterus - **P. serrulata** Lindl. (Chin, PI): u. similarly to prec., but more popular; often cult. ornam. evergreen tree (sUS).

Photinus spp., esp. **P. pyralis** (Lampyridae - Coleoptera - Insecta) (Am): (American) fireflies; glowworm (larvae): bioluminescence due to luciferin (firefly l.) (with luciferase); u. in assay of ATP; hospital tests for detection of heart attacks, muscle dis. (*Lampyris* spp., alternate source).

photosynthesis: process of formation of carbohydrates from H_2O & CO_2 in presence of light & chlorophyll; basic to independent pl. life.

Phragmites australis (Cav.) Staud. (P. communis Trin.; P. vulgaris Trin.) (Gram) (cosmpol., margins of rivers, ponds, seas): (common) reed (grass); American r.; giant r.; grass growing -8 m; sts. u. in building of houses (Iraq) (373); hb. & rt. stock c. saponin, u. as diur. diaphor., esp. in rheum. states; mucilag. juice of st. u. to soothe insect stings (Am, As); rhiz. c. 5% sugars, u. in diabetes, as food (seAs); entire pl. (esp. tender sprouts) u. in leucemia, breast cancer, as tea substit. (Chin); in salves for burns; somewhat bitter young sprouts u. as delicacy (Jap); pl. grown for land reclamation - **P. karkara** Steud. (Austral, Oceania): tropical reed; lf. decoc. u. int. to promote easy childbirth (NG) - **P. maxima** Blatter & McCann (Euras, Austral, Ind+): ditch weed (d. reed); rhiz. eaten; similar u. to *P. australis*; st. u. for plant stakes &c.; furnishes cheap paper; u. in place of lathing in walls (Swed) (1684).

Phryma leptostachya L. (monotypic) (Phrymat/Verben) (e&cNA, seAs): lopseed; lf. & rt. ext. insecticidal; c. phrymarol (sterol); pl. u. to kill house flies (Jap, Chin); entire pl. u. for preventing pruritus & bone-muscle dis. (Kor); smashed pl. in lard with black pepper u. as poultice for sores; crushed pl. may be placed over boils, carbuncles, cancers (Chin).

Phthirus pubis (Pediculus pubis) (Phthiridae - Anopleura - Insecta) (universal): crab/pubic/body louse (pl. lice); "crabs"; parasite on man, usually in pubic region, but also sometimes on armpits, eyebrows, &c.; conventional treatment of recent past "larkspur lotion" (acetic tincture of Delphinium); currently, parasiticides.

Phthirusa adunca (Mey.) Maguire (Loranth.) (tropAm): lf. decoc. u. as bath in marasmus - **P. caribaea** Britt. & P. Wils. (Dendropemon caribaeum Krug & Urb.) (WI): (West Indian) mistletoe; pl. decoc. u. as invigorating beverage & bath; pl. bad pest for woody crops.**phudina** (Sindi): **phudino** (Gujerati): spearmint.

phulwara (Ind): *Madhuca butyracea.*

phycobiont: algal component of a lichen.

phycocolloid: colloidal mucilaginous prod. fr. Algae; chiefly marine spp.; ex. agar, algin.

phycoerythrin: red pigment found in predominance in Rhodophyceae, or Red Algae - **phycology**: the science of algae - **phycophaein**: brown pigment found predominantly in Phaeophyceae or Brown Algae - **phycoxanthin**: yellow pigment occurring in some Algae, esp. Phaeophyceae.

Phycomycetes: rel. primitive gp. of lower Fungi; with non-septate mycelium; generally now included in *Mastigomycotina* (forming zoospores or mobile spores) (ex. *Phytophthora*) & *Zygomycotina* (mostly saprophytes or parasites on insects) (ex. *Rhizopus*); some taxa are now u. in biosynthetic mfg. processes.

Phycophyta: incl. all Algae except Cyanophyta.

Phyla nodiflora (L) E. Greene (P. chinensis Lour.; Lippia n. Mx.) (Verben.) (trop OW&NW; ex Tex [US]): cow slip (Bah); turkey tangle (Tex); cape weed; vervaine du pays; coronilla (Ve); u. diur. antipyr. in digest. disturb., pd. lvs. dusted on diaper rash, &c. (Bah); in erysipelas, sores; in Chinese med.; ext. antibacterial; frt. food (SriL); to replace lawn in dry countries; var. *reptans* (Spreng.) Moldenke (WI & trop) - **P. scaberrima** (Juss.) Moldenke (Lippia dulcis Trev. & var. mexicana Hegi) (Mex, WI, CA, Co, Ve): "menta" (Co+); fog fruit; orosul; dr. hb. or lvs. (Herba Lippiae mexicanae) ("L.M.") c. VO with ocymene, ocymenone, myrcenone (445); quercetin; u. dem. expect. in asthma, bronchitis, coughs; arom. (in perfumes) (2764); stom.; lvs. & fls. u. as sweetener (Aztecs) - *Phyla* is closely related to *Lippia* & some taxa have been transferred to that gen.

Phyllanthus acidus (L.) Skeels (P. distichus (L) Muell.) (Euphorbi.) (sAs, EI, natd. Fla [US], WI): (Otaheite) gooseberry (tree) (Ind); rts. & lvs. u. in med.; lvs. appl. in urticaria, poultice for lumbago & sciatica (Java); rt. boiled & steam inhaled for cough; u. for psoriasis of foot sole; rt. bk. juice u. in criminal poisoning, may be lethal; frts. edible, u. in prepn. wine, pickles, preserves, confectionary, pies, &c. (gooseberry substit.) - **P. acuminatus** Vahl (nwMex to nArg; WI): gallina; crushed lvs. appl. to breasts (&c.) to relieve inflammn. - **P. amarus** Schum. (orig. Am trop, now circumtrop.; sFla [US]): hurricane weed; hb. decoc. u. in VD, oliguria (Trin); lf. inf. to relieve pain in stomach (eAf) - **P. aspericaulis** Pax (Af): sap u. to treat wounds (Kenya); pl. eaten by livestock - **P. botryanthus** Muell. Arg. (DutchWI, nwVe): liki loki; juice of crushed lvs. appl. to skin inflamed by sap of *Hippomane mancinella*; formerly as remedy for abdom. dis.; witchcraft poison - **P. caroliniensis** Walt. (seUS: Pa, Mo to Fla; Mex, WI, CA, s to Arg): cababesinixte; whole pl. inf. u. as diur. (FrWI) - **P. corcovadensis** Muell. Arg. (Br): quebra-pedra (lit. break-stone); fls. c. VO, salicylic acid, lineol, methyl salicylate, cymol; u. diur. celebrated remedy to reduce kidney & gall stones; in bronchitis - **P. discoideus** Muell.-Arg. (cAf): Egossa red pear; bk. decoc. drunk for relief of postpartum pain; anthelm.; leaf & twig poultice u. for jigger sores (Swahili); frt. eaten with coconut; rt. decoc. for heartburn (Tanz); wd. dense & red - **P. elegans** Wall. (Indochin): rts. u. as antipyr.; lf. sap appl. to mouths of children with coated tongue - **P. emblica** L. (Emblica officinalis Guertn.) (Pak, Ind, sChin, seAs): emblic (myrobalans); (Indian) gooseberry (St. Croix); aonla (Ind); amla(ki) (Ind); zibya (Burma); shrubby phyllanthus; small tree; frt. edible, very acid, high in vit. C (c. 20x C of orange), usually in preserves & jellies; refrigerant, digestant, ton., diur. vehicle; frt. juice u. for eye inflammn. (Burma); lax.; young frt., bk. & lvs. u. in tanning & dyeing; astr.; lf. decoc. u. for fever (sChin); dr. frt. in diarrh. (constipating); typhoid; tree host for lac insects - **P. engleri** Pax (cAf): rt. bark smoked slowly as cough cure & for chest dis. (Tanz); rapid smoking to commit suicide (Zambia); lf. chewed for indigestion & constip. - **P. epiphyllanthus** L. (WI, Cu, Bah, Antilles): hardhead (Bah); panetela (Cu); lf. decoc. u. for colds, dentalgia, excessive menstr. (Bah) - **P. fraternus** G. L. Webster (Ind, Pak, natd. seUS): hb. u. for GC & urogen. inf.; young shoots in dys., icterus; lvs. in diabetes; c. flavonoids with hypoglycemic activity - **P. glaucescens** HBK (sMex, Guat): monkey rattle; pix-thon (Yuc); lf. decoc. u. folk med. as bath in pellagra (Yuc) - **P. lathyroides** HBK (Mex): lf. decoc. u. to wash infected eyes (conjunctivitis); moistened lf. u. as poultice for boils; lf. strong inf. u. as emet. - **P. leucanthus** Pax. (tropAf): rt. juice given in large doses to neonatal infants to hasten detachment of umbilical cord - **P. maderaspantensis** L. (tropAf, Ind): pl. ext. u. for scabies; lf. inf. in migraines; rt. decoc. for constip. (Tanz) - **P. muellerianus** Exell. (trop cAf): young lvs., twigs, & rts. u. as mild purg. in dys., urethral discharges; lf. juice in ophthalmia (Nigeria) (292); rt. in water for diarrh. (Tanz) - **P. niruri** L. (sTex [US]; nMex s to Arg & Br, WI; natd. seAs, Mal, Oceania+): Santa Maria; meniran; erva pombinha (Br); lvs. u. diur. in kidney dis. ton.; emmen., antipyr. (PI); lf. decoc. in colds, flu, fever, anorexia, gastritis; in dys. (Ind); in GC (Gold Coast); in jaundice; fls. ground to paste to soothe spider bites (Hond Indians) (2395) - **P. pentandrus** Schum. & Thon. (cAf): lf. juice u. for newborn baby (to aid suckling & act as mild lax.) - **P. piscatorum** HBK (Ve, Co, Guianas): shrub; macerated lvs. u. as fish poison (Guian); powd. lvs. insect repellant; sometimes cult. in small farms (Ve) - **P. reticulatus** Poir. (seAs, Af): lf. decoc. u. for sore throat; pin worm (PI).; diur. refrig. odontalgic; bk. inf. for dys.; rt. inf. in asthma (PI); headache (Tanz); male aphrod. to incr. virility (Tanz); powd. lvs. u. locally for sores, burns, suppurations, & skin chafing (sAf) - **P. rubriflorus** Beille (Indochin): u. to treat itch - **P. simplex** Retz. (P. anceps Wall.) (Chin): tea for cough & to remove phlegm; in rash; lf. juice dropped into sore eyes (PI) - **P. stuhlmannii** Pax (Af): appl. to suppurative swellings of leg, inflammn.; lf. for asthma - **P. tenellus** Roxb. (c&sFla [US], Br): quebrapedra; c. alks., flavonoids, VO, gallic acid; hb. u. as diur. litholytic (1685) - **P. urinaria** L. (seAs, EI, Oceania, eAf, Austral, WI, sFla): daoen menira; u. ton. diur. in gout & rheum.; as diaphor. emmen (Vietnam); antispasm. in hiccup, abortif.; in GC; has some antibacterial props.

phyllary: involucal bract (Compositae).

Phyllis Amara: bitter almond bran.

Phyllitis Hill = *Asplenium* - **P. scolopendrium** Newm. = *Asplenium s.* - **P. fascia** Ktz. = *Petalonia f.* (Algae).

Phyllobates (Dendrobatidae - Amphibia) (SA, WI): small tree frogs; some spp. called "arrow-poison" frogs (*P. bicolor* [Co]; *P. aurotaenia*; *P. chocoensis*) since Latin Am Indians ext. poison fr. skin & u. in blowgun darts; skin c. batrachotoxin & batrachotoxinins, steroidal alk., some of the most potent non-protein poisons known (446); potentially valuable pharmacolog. agents.

Phyllochlamys taxoides Koorders (Mor.) (seAs): boiled bk. u. as poultice for ulcers; smoke of burning bk. inhaled for common cold (Mal); rt. inf. u. to treat "anito" (PI) (—> *Streblus*).

Phyllocladus asplenifolia Hook. f. (Podocarp./ Phylloclid.) (Tasm): celery-leaved pine (tree); tree-like pl. bears phyllocladia; bk. (toa-toa) with 23-30% tannin; u. in tanning - **P. trichomanoides** D. Don (NZ, NG, Mal): celery (top) pine; tanekaha; bk. u. tanning material; excellent wd.; phyllocladia source of VO with myrcene, dipentene, pinenes; lvs. mixed with tobacco.

Phyllodium elegans (Loder.) Desvaux (Leg.) (seAs): leafy branches u. to provide hen nests; infused fls. u. to cure respir. dis. (Vietnam); rts. u. in treatment of edema in newborn infants (Camb) - **P. kurzianum** (L) Desvaux (tropAs, Austral): pl. u. in traditional med. for young mothers in childbed (Laos) - **P. vestitum** Benth. (Burma, Indochin): lvs. u. in local med. for wounds of cattle & other domestic animals (Camb).

Phyllonoma laticuspis Engl. (Grossulari) (Mex): herba de la viruela; reputedly of value in smallpox.

Phyllophora nervosa (DC) Grev. (Phyllophor. - Rhodoph. - Algae) (Black Sea): thallus c. phyllophoran (an agaroid); this & foll. sp. are u. as basis of Soviet agaroid industry; extractive has antilipemic but not anticoagulant activity (2004); source of iodine - **P. rubens** (L) Greville (Black Sea): thallus source of phyllophoran, Russian agaroid, u. in mfr. of agar & iodine.

Phyllostachys bambusoides Sieb. & Zucc. (Gram) (Chin, cult. Jap): (giant) timber bamboo; culm sheaths of new shoots u. as remedy in hematuria (Chin); shoots u. as food in Oriental dishes; sts. u. in building (2292) (2774) - **P. glauca** McClure (eAs): u. med. (Chin Phar.) - **P. nigra** Munro var. *henonis* Stapf (Kor, Chin): Henon (or black) bamboo; juice & lvs. u. in hemoptysis, pulm. atrophy, & other respir. dis.; bamboo cult. for young buds (Chin, Jap) u. as veget.; sts. u. as walking sticks (wanghee canes) - **P. nidularia** Munro (eAs): rts. u. to quiet uterus & in postpartum fever - **P. nuda** McClure (Chin): shoots edible; sts. u. as fishing poles, pl. stakes, &c. - **P. pubescens** Houz. de Lehaie (eAs): pan chu; lvs. u. in herbal med. for arthritis; culm sheaths u. in emesis & to quench thirst (Chin) - **P. sulphurea** Riv. var. *viridis* Young (P. mitis Riv.) (Chin, Jap): moso bamboo (Jap); edible shoots; shoots grow 6 dm in 24 hr. period (Calif [US]).

Phyllostylon rhamnoides (Poisson) Taubert (Ulm.) (Ve): baicoa; deciduous tree; sds. edible; good wd.

phylogeny: evolution of the race or gp. (ex. sp.) = genealogy & classification; treats of relationships of organisms.

phylum: division; there are 6 phyla (usually termed divisions) generally recognized in the plant kingdom, ca. 14-34 in the animal (classifns. vary).

Phymatodes Presl = *Polypodium*; *Microsorium*.

Phymatosorus diversifolius (Willd.) Pichi-Serm. (Polypodi) (OW trop.): sap squeezed fr. washed & scraped rts. & mixt. in water, drunk 3 times a day for rheum., asthma, headache, tonic in general weakness (Easter Is) - **P. scolopendria** (Burm. f.) Pichi-Serm. (OW trop) (cult.) oil u. to flavor coconut oil.

physal(a)emin: polypeptide found in skin of amphibian, *Physalaemus fuscumaculatus*; acts as potent vasodilator & hypotensive (kinin-like).

Physalis alkekengi L. (Solan.) (Euras, natd. NA): winter cherry; strawberry tomato; Chin lantern; bladder herb; frt. (Fructus Alkekengi) c. physalin (NBP), cryptoxanthin (hydroxy-carotene); diur. antipyr. in gout, rheum. icterus; local analg.; fresh berries edible; cult. ornam. - **P. angulata** L. (pantrop, CA, SA, tropAs): (tooth-leaved) winter cherry; ground c.; capuli cimarrón; tomatillo (Cu); hb. inf. u. antipyr. in mal. (Pe); frt. ton. lax. diur.; in sterility; fresh frt. soaked in vinegar & eaten; med. in Tonga (86) - **P. costomate** Moç. & Sess. (Mex): costomate; lvs. diur., rt. carm. & antidiarrh.; in colic (Mex Ph.) - **P. franchetii** Masters (eAs): rt. prod. uterine contractions; frt. may produce dizziness (may be due to physali[e]n [dihydroxy-carotene ester]) - **P. heterophyllus** Nees (e&cNA): (clammy) ground-cherry; husk tomato; yellow frts. of this perenn. hb. eaten raw or made into a sauce (Meskwaki & other Indians of Middle West); however, c. solanine & may poison or even kill (esp. children eating unripe berries) - **P. ixocarpa** Brot. (Mex): husk tomato; tomatillo (SpAm); Mexican tomato; frt. c. hygrine (alk.), edible; u. in stews; sold in markets, San Jose, Calif+ (US); cult. in Calif - **P. minima** L. (OW trop): sunberry; tepary; rt. u. as anthelm., antipyr., in diabetes; lf. decoc. u. with *Plantago major* against GC & as diur., frt. in diarrh. & in abscesses; frt. occ. eaten (Austral) but not pleasant-smelling or tasting but flavor in soft drinks (Ind); common weed (Ind); lvs. c. quercetin-3-0- galactosides with antiinflamm. activity; frts. u. diur. purg.; for splenic dis.; c. physalin X with abortif. properties - **P. peruviana** L. (earlier SA-now pantropics): Cape gooseberry; lf. & rt. diur.; white lf. u. as poultice for inflammn.; fl. inf. u. during parturition; sweet-acid frt. eaten raw, cooked, or in vinegar & to make preserves (said good; ex. "Inca Conserve"); lvs. u. med. (Tonga) (86); frt. lvs. & st. may be toxic to lower animals - **P. philadelphica** Lam. (sUS): green Spanish tomatoes; rt. u. as diur.; widely cult. for food (Calif to Mex+) (*cf. P. ixocarpa*) - **P. pruinosa** L. (e&cUS): strawberry tomato; husk t.; yelow frt. u. as food - **P. pubescens** L. (sUS, WI, Mex, CA to Peru): (downy) ground cherry; low hair g. c.; pop vine; whole pl. u. as diur. in edema, &c.; remedy in stomach dis. (PR); decoc. as antiemetic - **P. pumila** Nutt. (cUS): frt. eaten; subsp. *hispida* Hinton (P. lanceolata Mx) (US): ground cherries;

lantern plant; frt. edible - **P. virginiana** Mill. (e&cUS): ground cherry; wild c.; frt. eaten - **P. viscosa** L. var. *mollis* Waterfall (P. m. Nutt.) (Okla, Tex [US]): field (or large bladder) ground cherry; (smooth) g. c.; lvs. c. alk. & glucoside; u. as muscicide; chewed rt. u. to dress wounds (Omaha Indians) (1686).

Physaria newberryi Gray (Cruc.) (Ariz, NM, Utah [US], swNA): rt. decoc. u. as emet., esp. during the "Snake Dance" (Ariz).

physcion(e): 3-methyl ether of emodin; found in *Harungana* spp.; *Rhamnus* spp. &c.; cathartic.

Physeter catodon L. (P. macrocephalus L.) (Physeteridae - Cetacea - Mammalia) (all warm oceans): (great) sperm whale; cachalot (Fr); Spermwal (Ge.); cachalot whale; fr. head cavity is obtained sperm oil (mixt. of sperm whale oil & spermaceti (Cetaceum [*qv.*]); sperm whale oil is u. as lubricant, in mfr. soaps, cosm., in lamps, for hardening steel; in special cutting & lubricating oils (sulfurized); formerly only known subst. resistant to high temps & pressures found in much machinery (ex. automatic transmissions) (synthetic oil, Maysperm; 1972); comm. sp. oil may be 2/3 body oil & 1/3 head oil; inferior grades may be obt. fr. sperm whale blubber (1687); blubber eaten by Eskimos; flesh u. as nutr. (said indistinguishable fr. beef); teeth u. to make scrimshaw (drawings); liver furnishes vitamin A; fr. intestine is sometimes obt. ambergris (*qv.*); more often this is found as deposit on beaches; since catching of whales is no longer allowed by US & most other nations because of the danger of extermination of the animals, such products as spermaceti, sperm oil, &c., are now disallowed & no longer available legally (US) (*Hyperodon rostratus* Muller [bottle-nosed whale] is another source of spermaceti [BPC]; other whales have also been considered potential sources).

Physeteridae (Mammalia): Sperm Whale Fam.; monotypic with *Physeter*; fish-shaped mammals of the oceans; the largest of all known animals are included.

physic: laxative, also appl. to medicines or drugs in general (older term); a physician is one who applies medicines - **p. nut**: 1) *Jatropha curcas* 2) sd. of *Caesalpinia crista* 3) bonduk seed - **p. root (, common) Indian**: *Gillenia trifoliata* - **western Indian p. r.**: *G. stipulata* - **wild I. p. r.**: *G. trifoliata* & *G. stipulata*.

physiological modifiers: plant hormones.

physiology: the science of the function of living beings & their component parts (divisions: **animal physiology** & **plant p.**).

Physocalymma scaberrima Pohl (monotypic) (Lythr.) (Br, Pe): tulip wood; wd. yields VO (azeite de pau de o rosa, rose wood oil) u. perfum.; valuable wd.

Physocarpus Maxim. (*Opulaster*) (Ros.) (NA, neAs): nine bark; shrubs rather like *Spiraea*; bk. u. astr. ton.; cult. ornam. as door-yard shrubs with white to pink fls. in clusters - **P. opulifolius** Maxim. (e&cNA): (common) nine barks; bk. u. by exter. appln. to burns, swellings, wounds; int. as diur. astr. in female dis.; ornam. similar to spirea.

Physochlaina praealta (Dcne.) Miers (Solan.) (cAs, nInd, Pak, Turkestan, sChin): source of "Indian belladonna" (along with *Atropa belladonna* & *Scopolia lurida*); good source of atropine gp. of alkaloids.

physodic acid: a lichen acid extd. fr. *Parmelia caperata* & other *P.* spp.

Physostegia virginiana (L.) Benth. (Lab) (e&cNA): (false) dragon-head; "lupine" (sphalm.) (374); c. iridoids (rts. larger proportion than lvs.) (375); cult. as perenn. hb. in borders, &c.

Physostigma cylindrospermum (Welw.) Holmes (Leg.) (trop wAf): sds. c. physostigmine, &c.; u. similarly to *P. venenosum*, for wh. a substit. - **P. venenosum** Balfour (trop wAf, esp. in Calabar): Calabar bean; ordeal b.; chopnut; ripe seeds. (L.: Physostigma; Semen Physostigmatis) c. physostigmine, geneserine, physovenine; u. antispasm. sed., to counteract atropine alk., in postoperative paralysis of intestine; for intest. colics of horse (veter. med.); in mfr. of physostigmine; as ordeal poison (natives); very poisonous; sd. oil u. to kill mice (2502).

physostigmine: eserine: alk. fr. prec.; cholinergic (parasympathomimetic); myotic; in rheumatoid arthritis, bursitis (1689); in Alzheimer's dis. as memory restorative (1690); generally u. in therapy in form of its somewhat water-sol. (1:75) official salt, **p. salicylate**; u. myotic, in glaucoma; rheumatoid arthritis; antidote in atropine poisoning; in intest. atony.

Phytamin: plant hormone.

Phytelephas macrocarpa Ruiz & Pavon (Palm.) (CA, Co, Ve, Ec, Br): (common) ivory palm; vegetable ivory plant; jarina; frt. called "negro('s) head"; sd. tagua or corozo (corusco) or ivory nuts; sd. kernel (endosperm) constitutes vegetable ivory or palm i.; u. in diabetic flours; juice of unripe sds. & fr. flesh u. in prepn. beverages; sprouts as food; stone-hard sds. u. mfr. buttons & ornaments; import fr. Hond - **P. microcarpa** Ruiz & Pav. (Pe, Br): tagua palm; yarina (Br); sds. u. similarly to prec.; juice of unripe sds. for beverage; frt. pulp in prepn. of chicha de tagua (alc. beverage); sprouts as palm cabbage; ripe hard sd. kernels u. to make buttons, knobs for furniture, &c. - **P. seemannii** Cook (Par, Co): ivory palm; cabeza de negro; sds. vegetable ivory u. for making buttons, chessmen, inlays, &c.

Phyteuma hemisphaericum L. (Campanul.) (Eu Alps): Gras(blaettrige) Teufelskralle (Ge); rampion (Eng); perenn. hb. u. med.; this & other spp. popular in Ge med. recently - **P. orbiculare** L. (cEu, Alps): Kugel(ige) Teufelskralle; (horned) rampion; latex c. phytosterols; lvs. & rts. u. as veget. & in salads; fls. ornam.

phytiatrie (Fr): **phytiatry**: study of therapeutic management of pl. diseases; or the study of pl. enemies & the best means of combat.

phytin: inositol hexaphosphoric acid ester; found in many sds. (ex. hemp, sunflower, fenugreek, peas, beans); u. as Ca & Mg salts in rickets, anemia, TB.

phyti(ni)c acid: inositol hexaphosphoric acid; occ. in pl. sds.; cereal grains; u. as complexing agt. to remove heavy metals; hypocalcemic.

phytoalexins: psoralens & other defensive subst. of the plant; natural antibiotic (*cf.* phytoncides).

phytobezoar: a gastrointestinal concretion made up principally of pl. fibers (ex. persimmon) (1691) ("hair balls").

phytochemistry: the chemistry of plants, plant living processes, & pl. products.

Phytocrene blancoi (Azaola) Merr. (Icacin.) (PI, endemic): rt. decoc. u. in puerperium; for "black urine"; lvs. for pinworm; juice fr. cut sts. u. for pink eyes ("singam"); st. of Liane source of water.

phytoestrogens: pl. compds. with estrogenic activity; thus, found in peanuts, soy, carrots, cereals, *Digitalis*, *Rauwolfia*, *Cannabis*; said useful in osteoporosis.

phytohemagglutinins: complex mucoproteins extd. fr. sds. of *Phaseolus vulgaris*, &c.; u. in clinical tests, aplastic anemia.

phytol: long chain aliphatic alcohol; occ. as ester in chlorophyll; u. in prepn. of vit. E and K_1.

Phytolacca acinosa Roxb. (Phytolacc.) (Chin, Jap, Ind): Indian pokeberry; u. in ancient Chin med., diur. in edema, nephritis, pleuritis; fresh rt. appl. exter. as poultice for sores, swellings, &c.; toxic; frt. dye; lvs. & rt. as adulter. of *Atropa acuminata* - **P. americana** L. (P. decandra L.) (e&cNA, Mex; now cosmop.; cult. ornam. Eu, nAf): poke(weed); common pokeberry; pigeonberry; "American nightshade"; "cocum"; "coakum"; perenn. hb. up to 3 m; rt. c. mixt. of triterpenoid saponins, phytolaccatoxin (phytolaccagenin = aglycone); cystine-rich glucoprotein simulating mitosis, pokeweed mitogen (PWM) (formerly phytolaccine mentioned); phytohemagglutinin; rt. (**Phytolacca** [L.]: pokeroot) u. emet. purg. anthelm. in chronic rheum. (rt. steeped in whiskey) (folklore) (2519), arthritis; gout (using F.E.); cancers; obesity (nostrums for reducing formerly popular) (1693); syph.; exter. in skin dis. ("camp itch," Civil War); in sickness of cows & chickens (sUS); rt. on standing develops disagreeable odor; rt. & lvs. formerly sent to Eu to adulterate belladonna rt. & lvs. (376); lvs. u. like rt.; substit. for belladonna lvs.; dil. ext. u. against tobacco mosaic virus (Jap); vermifuge; to kill snails carrying bilharzia; as dye (1879); frt. becomes black & shriveled in late fall but juice still dark red; u. in prepn. of a juice (Succus P.); to color wines. alc. bever. (1694); u. for thrush ("thresh") (NC) (2547); frt. may be toxic if eaten raw, but harmless if boiled; young lvs. u. as potherb ("sallet greens," seUS) (spring tonic) but water in which boiled should be discarded; common weed (Eu) - **P. bogotensis** HBK (Co): lvs. & frt. toxic to cattle - **P. dioica** L. (Mex, Br+): bella sombre; ombu (Arg); frt. of evergreen tree u. as emet. purg. anthelm., food(!); lf. u. vulner. (exter.), purg. (Arg); popular shade tree (Arg) - **P. dodecandra** L'Herit. (P. abyssinica Hoffm.) (s&eAf, incl. Madag): endod; rt. u. for tapeworm; lf. juice as styptic; emet. purg. in epilepsy; berries c. saponins (endod); glycoside of oleanolic acid; u. as red dye in cosmetics (shampoos); in bilharziasis (molluscicide: kills snails also cercariae, small fish; biodegradable) (1692); rt. may be very poisonous, but young sts. & lvs. cooked & eaten as potherb (Eth); juice, &c. u. as homicidal poison (Af) - **P. esculenta** van Houtte (P. acinosa var. e. Max.) (Chin, Jap): rt. u. diur. (Jap); rt. c. cynanchotoxin (377) but this may have derived fr. replacement of drug by *Cynanchum caudatum*; u. in kidney dis., pleuritis, pneumonia; u. in med. since ancient times - **P. heptandra** (sAf): umbra tree; "wild sweet potato"; rt. u. as purg. emet.; in GC; toxic; for tapeworm; frt. edible - **P. insularis** Nak. (eAs): rts. appl. to eczema - **P. japonica** Mak. (Jap, Taiwan): hb. u. as diur. - **P. octandra** L. (P. icosandra L.) (pantrop, sMex, CA, to Co, Af, As, Austral): carmin (CR); calalu (CR); calaloo (Hond, Nic); young lvs. u. like spinach; dye plant; soap substit. (Mex); toxic; rt. emet., against tapeworm; frt. decoc. u. formerly in small pox; decoc. of lvs. in rheum. (WI); lvs. applied locally to pimples & sores - **P. rivinoides** Kunth & Bouché (WI, tropAm, as Co): altusa (Co); bledo carbonero (Cu); lvs. u. to stupefy fish & like soap in washing clothes; young lvs. boiled as edible greens & to reduce edema (Domin,

Caribs) (2328) - **P. thyrsiflora** Fenzl. (SA): u. drastic; frts. for dyeing; young shoots are eaten.

Phytolaccaceae: Pokeroot Fam.; hbs. shrubs & trees of trop & subtrop Am, As, Af; lvs. alternate, entire; fls. without petals; frt. a berry, samara, or capsule (ex. *Petiveria, Rivina*).

phytolaque (Fr): Phytolacca.

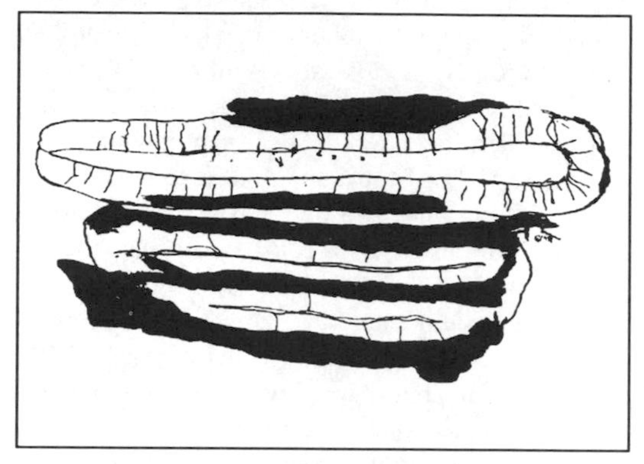

LIATRIS TENUIFOLIA, ROOT.
PHYTOMELANE ENCASING STONE-CELLS

phytomelane: unusually resistant carbon-rich blackish masses found especially often in the frts. of members of Fam. Comp.; appar. formed fr. the middle lamella of the cell partitions; often also found in rts.; ex. *Inula, Perezia, Liatris*, etc.

phytomenadione: phytonadione.

phytomucins: plant mucilages, such as found in *Cydonium*; *Dioscorea* tubers (1695).

phytonadione: vitamin K_1; phylloquinone; "blood-clotting vitamin"; widely spread in higher green plants, such as spinach, tomatoes, alfalfa (where first isolated); synthesized by bacteria in lower intest. (mammals); now mostly made by chem. synthesis; u. in hypoprothrombinemia; where clotting time is prolonged; in biliary fistulae & obstructive jaundice (lack of bile prevents absn. of vit. K); hemmorrhagic dis. of newborns; prophylaxis in labor & before surgery; antidote in dicoumarol poisoning; overdose harmless (E); inj. & tabs.

phytoncides: bacterial & protozocidal subst. found in VO of onion, garlic, &c. (96).

phytonyme (Fr): **phytonym**: plant name.

phytopharmacology: effect on pls. of various compds., esp. of natural compds.

phytopharmacie (Fr): **phytopharmacy**: study of products & their prepns. u. for protection & improvement of plants; also assimilating prods. (except fertilizers, &c.); esp. agts. for combatting pl. diseases; but also insecticides, herbicides, growth subst., fumigants, wetting agts., &c. (adj. phytopharmaceutical).

phytosanitaire (Fr). **phytosanitary**: related to fighting pl. dis. & animal enemies of plants.

phytosterol: generic term for 1) any plant sterol complex tricyclic alcohol wh. forms esters & occurs in plants 2) (specif.) isomer of cholesterol. (β-sitosterol-[α-phytosterol] is considered the commonest pl. sterol.).

phytosterolin: sterol glycoside of pl.; yields phytosterol on hydrolysis.

piantaggine (It): plantain.

piassava (Br): 1) the SA palms, *Attalea funifera* & *Leopoldinia piassaba* Wallace, *Raphia* spp., &c. 2) the strong coarse fiber fr. these trees (vascular bundles) u. to make cords, brushes, brooms, ropes, snares, &c. - **p. (fiber,) Bahía**: fiber fr. *Attalea funifera* - **Pará p.**: fiber fr. *Leopoldinia piassaba* - **p. tow**: waste matter accumulated in prepn. of p. fiber; u. in upholstery.

pica: eating of non-nutrients, ex. dirt, paint (Pb poisoning); clay; ice; starch (amylophagia) (SC pica, L., magpie).

picado (Sp): 1) minced, chopped, broken into small pieces 2) (Arg): damaged by weevils (grain, etc.).

picão (Br): *Bidens pilosa+*.

piccalilli (orig. EI) (piche): mixt. of pungent spices (incl. mustard) & veget., incl. green tomatoes; popular in New Eng [US] as pick-up (SC pickle & chile).

picciolato (It): petioled.

Picea A. Dietr. (Pin) (nTemp. Zone, esp. eAs): spruces; some 40 spp. of large to medium evergreen trees; with pendulous cones; trees with conical outline; lvs. keeled on both surfaces; lvs. of some spp. u. in making "pine needle extract" u. in baths for rheum., etc. (< L., pitch pine, < pix, picis, pitch) - **P. abies** (L) Karsten (P. excelsa Link; P. rubra A. Dietr.; P. vulgaris Link) (n&cEu; natd. neUS & seCana): Norway spruce; common (Norway) s.; mast tree; "white fir"; tree grows to 60 m (cEu); oleoresinous exudn. fr. trunk constitutes Burgundy pitch (or Jura turpentine), oleoresin u. in various medl. plasters & varnishes; young shoots & lvs. u. in mfr. spruce beer; u. as diur.; antiscorb. beverage; wd. is preferred source (Ge) of medicinal charcoal (v. Carbo Activatus); lumber; lvs. (needles) much u. in folk med. for coughs, air passage inflammn., gout, rheum., scurvy; source of VO u. as addn. to baths & for inhalation for resp. dis.; young shoots u. like pine tops; young tree u. as Christmas tree; spruce ext. fr. bk. u. as important tan. - **P. engelmannii** Parry (Cana, US-Rocky Mt. reg.): mountain spruce; "green s." (447); bk. inf. u. in TB & other resp. dis. (BC [Cana] Indians); decoc. of bk. spring ton.; rts. much u. by Pac Coast Indians, ex. for matting needles (ca. 1 m long); lf. decoc. u. as tea (BC Indians) - **P. glauca** (Moench) Voss (P. alba Link) (nNA): Canadian spruce; (western) white s.; trees -45 m high; lvs. u. in gastric dis., sores, in salves for cuts & wounds; tea u. as antiscorb. (for scurvy); gum (oleoresin) appl. to scabs, sores; u. to make spruce beer; u. as antiscorb.; tisane of green gummy cones u. for too frequent urination (Abenakis); rts. u. to sew canoes, baskets, &c.; useful wd. & —> paper pulp; cult. ornam.; u. similarly to P. engelmannii (BC Indians) - **P. mariana** (Mill.) BSP. (P. nigra (Ait.) Link) (nNA): black (or bog) spruce; (SC dark bark; dark green lvs.; dark purple cones); double s.; twigs tips & needles distd. —> spruce oil (similar to VO of *Tsuga canadensis*); young shoots to prep. American "spruce beer" (antiscorb); resinous exudn. similar to that of *P. rubens*; wd. u. to mfr. paper pulp; var. *semiprostrata* (Peck) Teeri (Qu): lvs. u. in prepn. spruce beer (c. molasses, raisins, yeast, spruce branches, water) (379) - **P. rubens** Sarg. (P. rubra Link) (eNA s to Tenn & NC): red (or yellow) spruce; chief source of spruce gum (oleoresin) formerly popular as masticatory & fumitory; wd. source of most NA paper pulp - **P. sitchensis** (Bong.) Carr. (NA Pac Coast, Alas to Calif): Sitka spruce; coast s.; oleoresin (exudn.) u. as chewing gum sometimes (378); rts. u. to make baskets (nw Indians); bk. inf. u. in TB & other resp. dis.; decoc. of branches u. as strengthening bath.

Piceae Resina (L.): spruce gum, obt. fr. *Picea rubens*, *P. mariana*.

picene: di-benzo-phenanthrene; occ. in petroleum distillate residues, lignite coal tars, etc. - **piceous**: resembling or relating to pitch.

pichão (Br): picão.

pichi leaves: p. tops: herbage fr. *Fabiana imbricata*.

pichurim (beans): Fabae P. (L.): fr. *Nectandra pichurim* (do not confuse with puchurim).

pickerel weed: *Pontederia* spp.

pickles: pickled cucumbers cured while green - **chow-chow p.: mixed p.**: mixt. of cucumbers (60-70%), cauliflower (20- 30%), onions (10%), green tomatoes (10%), few tabasco peppers - **dill p.**: cucumber p. flavored with *Anethum graveolens* - **p. spices**: combinations u. in making pickles; incl. various mixes of the following: bay leaves; cardamom; cinnamon; chillies; cloves; dill; fenugreek; coriander; black pepper; allspice; ginger; turmeric; mace.

pick-purse: *Capsella bursapastoris*.

picolines: methyl-pyridines; occ. in coal tar & bone oil; u. solvent; intermediate in mfg. (niacin, &c.); waterproofing, &c.

Picraena Lindl. = *Picrasma*.

Picralima kleiniana Pierre (P. nitida (Stapf) Th. & H. Durand) (monotypic) (Apocyn.) (wEquator, Af): panwe (Ghana); st. u. in fevers, gastric dis.; rt. in pulm. dis.; frt. in mal.; sd. in pneumonia; bk. c. alks., antipyr., vermifuge & in place of quinine (Ghana); green frt. u. to kill fish.

Picramnia andicola Tulasne (Simaroub.) (Veracruz, Mex): bitter bk. u. as quinine substit. (< pikros [Gr]: bitter) - **P. antidesma** Swartz (Mex, WI, CA): "bitter wood"; "quinina"; lvs. & bk. bitter (Cascara Amarga; corteza de Honduras) with licorice-like flavor; bk. formerly u. in VD & erysipelas (exp. to Eu); in intermittent fevers (mal.) (Belize, Hond) & gastro-int. dis. (WI); one of "major" bitters (Jam); subsp. *fessonia* (DC) W. Thomas (P. locuples Standl.) (Hond): "quina roja"; bk. u. as antipyr. - **P. carpinterae** Polak (Mex, CA, Guianas, Br, Pan): caregre (CR); tariri; sds. c. solid fat (semilla grassa) with tariric acid (*qv.*); source of tariri fat; decoc. of bitter bk. of trunk & rts. u. as ton. antipyr. - **P. ciliata** Benth. & Hook. (WI, Br): one source of Honduras bk.; formerly thought source of Pereiro Bark (do not confuse with Pareira Brava, rt.); bk. u. skin dis. - **P. lindeniana** Tul. (Mex, Guat): bk. u. like others of the gen. - **P. pentandra** Sw. (sFla [US], WI, tropAm): bitter bush (Fla); snake root (Bah); quinina del pais (Cu); shrub or small tree; bk. rt. & lvs. u. locally for fever (Fla); cinchona substit., as stom. & for dys. (Cu); bitter decoc. u. in colds, TB, menstrual pain (Bah); one source of Honduras bark (or Cascara Amarga) u. as bitter ton. stim. antipyr., in skin dis. - **P. pistaciaefolia** Blake & Standl. (Mex): cascara amarga; u. for syph. (Am Indians) - **P. polyantha** (Benth.) Planch. (Mex): bk. u. similarly to others of gen. - **P. tetramera** Turcz. (Mex): similar uses to other *P.* spp. - **P. tariri** DC = P. carpinterae Polak - **P. xalapensis** Planch. (sMex): bitter bk., u. similar to others of this gen.

Picrasma antillana (Eggers) Urban (Simaroub.) (Lesser Antilles): bitter ash; "simaruba"; cold water inf. of wd. chips u. for anorexia; sap expressed fr. bk. appl. exter. for yaws (framboesia) (Domin, Caribs) (2328) - **P. excelsa** Planch. (WI, esp. Jam, nVe): Jamaica quassia; "quina quassia"; one of the 2 off. bot. sources of quassia wood, the other being *Quassia amara*; wd. c. quassi(i)n (diterpenoid lactone) (= isoquassin), neoquassin; wd. u. bitter ton., vermifuge for thread worms, insecticide; in dyspep. intermittent fevers, anorexia; antipyr.; bk. u. in same way but less active; u. in brewing in place of hops (GB, earlier) (< pikros [Gr] bitter) - **P. javanica** Bl. (with var. *nepalensis* Benn. [P. n. Benn]) (nInd, Nepal+): bk. u. as antipyr.; lvs. on wounds; lvs. of var. u. as insecticide (Assam); bk. antidote after eating poisonous turtles (NG) - **P. quassioides** Benn. (P. ailanthoides Planch.) (Ind [Himal] Nepal, Kor, Chin, Jap): nigaki (Jap); Indian quassia; small tree; wd. c. quassi(i)n, neoquassiin, picrasins; u. antipyr. anthelm.; substit. for wd. of *P. excelsa*; bitter bk. u. stom., antipyr., parasiticide (lice & fleas) (Ainu); adult. of quassia wood (pd.) - **P. vellozii** Planch. (Picraena v. Engler; Picrasma crenata Engler) (sBr): "quassia from south Brazil"; bitter wd. u. to replace Jamaica & Surinam quassias; bk. u. in dyspeps. & mal.; antidiabetic.

Picria fel-terrae Lour. (Curanga f. Merr.) (Scrophulari.) (seAs, Oceania): "fel-terrae" (= earth gall): lvs. aper. emmen. intest. stim.; diaph. diur. (Indochin); decoc. anthelm. for children (Indones); lvs. crushed & appl. to snakebite & other wounds; alc. macerate u. as ton.; pl. often cult. & sold in markets.

Picris divaricata Vent. (Comp.) (eAs): u. bitter, diur. galactagog; in leucorrhea - **P. hieraciodes** L. (Euras, Austral): yellow succory; lvs. u. as antipyr. (mt. peoples, Indochin).

picroaconitine: benzaconine: alk. fr. tuber of *Aconitum napellus*.

picrocrocin: saffron bitter; glucoside fr. stigma of *Crocus sativus*; on hydrolysis yields safranal & glucose, all give characteristic odor/taste of saffron.

Picrolemma Hook. f. (Simaroub.) (Amaz, Br, & Pe): taxa u. in arrow poisons (Br) - **P. pseudocoffea** Ducke & **P. sprucei** Hook. f. (nBr): u. to adulter. simaruba barks.

Picromycin: Pikromycin: macrolide antibiotic (the first isolated) prod. by *Streptomyces felleus* Lindenbein (earlier referred to as *Actinomyces* sp.); active against Gm.+ bacteria, incl. mycobacteria.

picropodophyllin: constit. of the resin of various *Podophyllum* spp.; an isomer of podophyllotoxin.

Picror(r)hiza kurroa Benth. (P. lindleyana Wettst.) (Scrophulari.) (Pak-Himal): "kutki"; rhiz. u. bitter ton. cholagog, antiperiod. stom. mild cath. antipyr.; ground up & given in water for gastric dis. of children; dyspeps. (Ind); for fever, kala-azar, conjunctivitis (Chin) - **P. scrophulariflora** Pennell (eAs): rt. ext. u. as antipyr. & for stomach dis. (Sikkim); u. in Chin med.

picrotoxin: non-N bitter principle (non-alk.) fr. sd. of *Anamirta cocculus*; one molecule c. 1 mole of picrotin + 1 mole of picrotoxinin; u. CNS & esp. respir. stim.; antidote to barbiturate & other hypnotic poisoning; analeptic.

picry (US): poison ivy.

pie de cristo (Sp): *Potentilla reptans* - **p. de león**: *Alchemilla vulgaris* - **p. de perro** (Pe): *Rhynchosia* spp.; *Potentilla repens.*

pie berry (Eng): frt. of Solanum nigrum - **p. plant**: *Rheum rhaponticum.*

pied de chat (Fr): cudweed, *Gnaphalium* spp.

piel (Sp): skin, pelt, fur, hide.

Pieris japonica D. Don (Eric.) (eAs): Japanese andromeda; shika (Jap); lf. & st. decoc. u. as insecticide exter. to kill flies, maggots, skin parasites (incl. itch mite); vermifuge; very toxic; good shrub (1-2 m) for acid soils; ornam. - **P. formosa** D. Don (Chin): sap fr. leafy shoots smeared on ringworm 2-3 times to cure; insecticide.

pierre de vin (Fr) (wine stone): red argols.

Pierreodendron kerstingii Little (Simaroub.) (wAf): st. bk. c. antileukemic quassinoids, incl. ailanthinone, dehydroailanthinone, glaucarubinone, &c.; with significant activity against certain leukemias & carcinomata (2339).

pig: *Sus serofa* L.

pig family: Suidae.

pigeon: Gmelin (Columbidae - Columbiformes) (GB to Ind) (cult. worldwide): *Columba livia:* this, the wild rock pigeon, is ancestor to modern vars.; the bird has long been domesticated by man (since Roman times); symbolic of peace, love, faithfulness & fertility; housed in special bldgs. (dovecotes; columbaria); u. for meat (esp. young, squabs); in sacrifices (ex. Hebrew); as message carrier (c. pigeon) since Roman (ancient) times; u. in Unani med. (meat = tonic; hematinic); blood, eggs, gizzard, excreta, ash of bones ("Dove" is a term applied to smaller types of pigeon) (2769).

pigeon berry: 1) *Phytolacca americana* 2) *Rhamnus purshiana* 3) several other genera & spp. - **p. weed**: *Verbena officinalis.*

pigment: substance commonly insol. in water wh. colors a surface or a liquid, but is distinct fr. a dye, which penetrates & is fixed to the fiber or tissue; u. in exter. pharm. applns. (galenical class, Eng) (**Pigmentum**) - **Pigmentum Indicum** (L.): indigo - **P. Tinctorium** (BPC): paint of Brilliant Green & Xal Violet in water & alc.

pigna (It): pine cone.

pignolat: pine sd. put into dragees (*qv.*), also u. in ragouts.

pignole (nuts): **pignolia** (It): **pignolo** (It): pine nut (sd. of European pines, esp. *Pinus pinea*).

pignon (Fr): piñon - **p. de Barbarie** (Fr): castor bean.

pignut: 1) frt. & sd. of *Carya cordiformis* & *C. glabra* 2) *Conopodium majus* (Euras).

pig-turd (Virgin Is): *Andira inermis* - **pigweed:** 1) *Amaranthus* spp., incl. *A. graecizans* L. (prostrate p.); *A. hybridus* L. (green p.); *A. hypochondriacus* (Ind; natd. NA) (redroot p.); *A. retroflexus* L. (redroot rough p.; rough p.); *A. spinosus* L. (spiny p.) 2) *Chenopodium* spp., esp. *C. album* 3) *Portulaca oleracea* - **p. (redroot): rough p.**: 1) *Amaranthus retroflexus*; *Amaranthus hybridus* v. *hypochondriacus* 2) *Chenopodium album*, etc. 3) *Portulaca oleracea* - **pigwrack**: chondrus.

pikant (Ge): pungent, highly spiced, piquant.

Pikantheit (Ge): spiciness (*cf.* piquant, Eng).

pilaf(fe): pilau (Persian): dish of rice, chicken or fish, saffron, &c. (Eg, Ind, Pak, Turk; now in seUS).

pilchards: Cornish sardines: *Sardinia pilchardus*, marine fish (Clupeidae, herring fam.) (Atlantic Oc): popular food fish with highly unsatd. FO, supposed preventive of atherosclerosis; in the past was good-sized industry (Cornwall), other Eu countries (sAf, swAf, earlier US & Jap) - **South African p.**: *Sardinops ocellata* - **California herring (pilchard)**: *Sardinops caerulea*.

Pilea microphylla (L) Liebm. (Urtic.) (sFla, WI, tropAm): artillery plant (SC massive pollen grain discharge with little smoke-like puffs); "fire bush"; crushed pl. appl. to sores & bruises (Guat); hb. decoc. u. for fever (Guad); diur. & in liver & urinary dis.; antiasthmatic, in diarrh. (children); pl. placed under pillow to induce sleep (Curaçao); ornam. hb. - **P. pumila** A. Gray (eCana, neUS w to Ia): richweed; clear weed; hb. u. for urtication (prod. of wheals) - **P. rotundinucula** Hay. (eAs): appl. exter. for headache (Taiwan).

Pileus mexicanus A DC (Leucopremna mexicana Standl.; P. heptaphyllus Ramirez) (Caric.) (sMex, CA): kunche; bonete; semi-cultd. tree with conical trunk -12 m high; frt. a large berry, somewhat like a pepper, edible; eaten as salad, cooked, or made into a kind of custard; starch fr. trunk made into kind of tortilla; juice somewhat like that of papaya; c. mexicain (Mexican papain), a digestive enzyme; young shoots furnish drink u. for jaundice & "bilousness"; poultices on buboes; decoc. of trunk pith u. for pyuria; in skin dis. & abscesses (380) (—> *Jacaratia*).

pileus: cap-like upper part of basidiocarps & ascocarps; SC pileus: pileum (L): cap worn on heads of slaves & celebrants at feasts.

pileweed: *Ranunculus acris* - **pilewort**: 1) *Amaranthus hybridus* v. *hypochondriacus* 2) *Erechites hieracifolia* 3) (Brit) *Ranunculus ficaria* 4) *R. bulbosus*.

pili (nut) (PI): pleasantly edible sd. of *Canarium ovatum* & *C. luzonicum*.

Piliostigma reticulatum (DC) Hochst. (Leg) (trop wAf): rts. & bk. c. tannin, alk. (traces) u. astr. for wounds, diarrh.; frt. rich in & source of tartaric acid - **P. thonningii** Milne-Redh. (Equat, eAf): camelfoot; redhead; agrario di gondar (It, Ethio); bk. & lf. inf. u. in toothache; bk. & young lf. u. as expect. & in chest dis. (Ethio); bk. cold inf. u. in dys. diarrh.; bk. rt. & lf. u. as "cure" of leprosy & smallpox (formerly) (natives) (Ethio); bk. & smaller twigs yield 20% tannin, u. in tanning; rt. gives red-brown dye; lvs. purg. (Nig) (—> *Bauhinia*).

"pill" (layman): pill (rarely), tablet, or capsule - **p. evacuant**: laxative pill comp. - **pillenterry** (Va): *Zanthoxylum clava-herculis* - **pilli-pilli** (Pe): *Hypochoeris sonchoides* - **pillpod euphorbia**: *Euphorbia pilulifera* - **pills, C. C.**: compound cathartic p.; mild mercurous chloride pills compd. (formerly very popular).

pilo (Iran) (Eng pilau): staple Muslim dish, very fatty; *v.* pilau.

Piloblephis rigida Raf. (Pycnothymus rigida Small; Satureja r. Bartram) (Lab.) (peninsular Fla [US]): "pennyroyal"; small shrub flowering all year; u. to make a popular tea; med.(?) (Wewahitchka); hb. c. VO with alcohols (prob. thujyl alc.) & a ketone (prob. pulegone); u. as popular tea (—> *Satureja*).

pilocarpine: the characteristic alk. of Pilocarpus; occ. as oil or in xals; u. myotic, in glaucoma, diaph.; in impaction (horses); atropine antidote; to darken the hair (2147); cholinergic (parasympathomimetic) agt.; many uses now dropped; however, it remains important myotic (sometimes admixed with physostigmine); u. in xerostomia (dry mouth) (5 mg p.o., lasts ca. 3 hrs.); Mistura P. (Lederle): ophthalmic agt. (spray delivery system) (E).

Pilocarpus (L.): jaborandi leaf, fr. *P. jaborandi* & *P. microphyllus* - **P. jaborandi** Holmes (Rut.) (nBr): Pernambuco jaborandi; lvs. c. pilocarpine, isopilocarpine, pilocarpidine, VO; u. diaph. sial. myotic, sed.; hair stimulant (*sic*), in kidney & heart dis.; intestin. atony; mostly in the form of pilocarpine & its salts (1287) - **P. microphyllus** Stapf (neBr): Maranham jaborandi; little leaf p.; similar to prec. (2338) - **P. pennatifolius** Lem. (sBr, Par, nArg): Paraguay jaborandi; c. much less alk. than the preceding spp.; u. adult. of same - **P. racemosus** Vahl (WI, Cuba to Mart): Guadeloupe jaborandi; lvs. c. pilocarpine & isopilocarpine; similar uses to prec.; lvs. added to mash in making beer (Curaçao) (2325) - **P. spicatus** St. Hil. (SA, esp. Br): source of Arcati jaborandi; c. 0.16% alk. - **P. trachylophus** Holmes (nwBr): Ceara jaborandi; c. less than 0.5% alk.

pilón (PR): wooden or metal mortar & pestle; u. for comminuting onion, garlic, grains, etc., u. mostly in rural areas; also specific pestle.
piloncillo (Mex): raw brown sugar commonly u. in Mex.
Pilosella (L): piloselle (Fr): *Hieracium pilosella.*
Pilostaxis Rafin. = *Polygala.*
Pilulae ad prandium (L.): dinner pills; pills of aloe & mastic, NF VIII - **P. e. gummi**: pills of galbanum comp. - **P. metallorum amarae**: Metallic pills (*qv.*) - **P. rufi**: pills of aloe & myrrh, NF VII - **P. triplices** (sing. **Pilula triplex**): pills of aloe, mercury, & podophyllum, NF V.
pilule (Fr-f.): pill.
Pilz (Ge): 1) fungus 2) mushroom.
pimaric acid: cyclic diterpenoid; phenanthrene deriv.; isolated fr. Am rosin & oleoresin; Burgundy pitch; galipot resin; opt. active d- & l- isomers; stable to heat & acid.
pimbina (Fr-Cana): *Viburnum opulus, V. edule, V. trilobum.*
Pimelea trichostachya Lindl. (Thymelae) (Austral): spiked riceflower; "native flax"; "Australian flax"(?); toxic pl., producing coma & gastroenteritis.
piment couronnée (Fr): *Eugenia pimentoides*
Pimenta acris Kostel. (P. racemosa [Miller] J. W. Moore) (Myrt.) (WI, esp. VirginIs, Ve, Guiana, cult. Indones): bay(rum) tree; wild clove (tree); West Indian bay tree; lvs. & frt. u. as inf. for colic & stomachache & as shampoo to wash hair or body; heavy wd. u. to make pestles, housepiles, &c. (Domin) (2328); lvs. (Myrciae Folia) distd. to yield (true) bay oil; VO c. eugenol, chavicol; u. in perfumery, cosmetics (arom. soaps, hair prepns., toilet waters, &c.), to prep. cheaper types of bay rum; frt. bayberries (Kronpiment), u. as adulter. of true Pimenta (WI chief source); c. 37% VO with eugenol & phellandrene; var. *citrifolia* (P. c. Kostel.) (WI): false bay; limoncillo; lvs. source of VO (lemon-scented bay oil) with much citral, phenols; cannot be u. as bay oil (do not confuse with Bay Leaves fr. *Laurus nobilis*, u. for culinary purposes) (introd. EI, Indonesia, where cult. for VO) - **P. dioica** (L) Merr. (P. officinalis Lindl.) (sMex, CA, WI, nSA, introd. EI): allspice; pimenta tree; pimento; "Jamaica" (< Jamaica pepper tree); VO of rapidly dr. frt. c. 2-5% VO with much eugenol, similar to clove oil; dr. *unripe* frt. u. arom. flavor, ton. condim. digest. stim.; carm. in digestive diffic.; substit. for cloves; in mfr. sausages; wines source of VO (allspice oil) (WI); allspice ointment u. for earache (slavery days, sUS); decoc. of branch tips u. in diabetes (CR) (381); lvs. smoked like tobacco; lvs. at times u. as flavor (CA) (381); var. *tabasco* (P. t. Lundell): Mexican allspice; Tobasco a.; large-fruited form u. as adulter. of Pimenta.
pimenta, Mexican: Spanish p.; Tabasco p.; large frtd. form of *P. officinalis* - **p. dagua** (Por): **p. do brejo:** *Polygonum hydropiper* - **p. do reino** (Por): *Piper nigrum.*
pimento (pods) (oil): 1) Pimenta 2) (Sp): *Capsicum frutescens* with mild flavor - **p. dram** (Jam): ripe p. berry soaked in good rum; u. for abdominal pain, in liqueurs, etc. - **p. de Tabasco** (Por): Tabasco pimento - **pimentón** (Sp): Paprika - **pimienta** (Sp): pepper - **p. de agua** (Sp): *Polygonum hydropiper* - **p. malagueta** (Cu, PR): *Pimenta officinalis* - **p. negra**: black p.
pimpernel, blue: *Scutellaria lateriflora; Anagallis monelli* - **common p.**: *Anagallis arvensis* - **p. root**: fr. *Pimpinella saxifraga* & *P. magna* - **scarlet p.**: common p. (var. *phoenicia*).
pimpinela branca (Por): *Pimpinella saxifraga..*
Pimpinella anisum L. (Umb) (Levant; now widely cult.): anise; aniseed (plant); frt. ("seed") (aniseed) c. 1-6% VO FO; VO rich in anethole, anisketone; u. flavor esp. in pastry, also in chewing gum, dental lotions, &c., in liqueurs (anisette, Pernod, Goldwasser, distillate in ouzo (natl. Greek drink); condiment, expect. arom. stom. antispasm, carm. emmen. galactagog; analgesic (proven!); addn. to children's milk; aphrod. (folk med.); topically as parasiticide; anise oil u. on traps & fish hooks to attract animals; cult. Cu; component of many medl. teas (Species); for prepn. VO & anethole (residue u. as cattle fodder); rt. formerly off. USP; ground with sugar & water to furnish cooling draught (Ind) (2148); substit. for tea; for long time u. as estrogen; whole pl. sold in markets (PR); pl. known fr. very ancient times; prod. It, Sp, Tunisia (2149) - **P. diversifolia** DC (Ind, Chin): crushed pl. plastered on inner side of wrist in mal.(!) (Chin); hb. u. as carm. (Ind) - **P. major** (L) Huds. (P. magna L.) (Euras; natd. NA): greater pimpernel(l); great (burnet) saxifrage; a source of pimpernell root (together with *P. saxifraga*); u. as mild expect. (pastilles or powd.); in angina, pharyngitis, laryngitis, bronchitis, gout; urinary calculi; much u. in folk med. in hoarseness; asthma; in homeop.; in liqueur ind. (bitter schnapps & herbal exts.); exter. u. as masticatory; tinct. added to eye, mouth, & gargle mixts. - **P. pruatjan** Molkenb. (P. alpina Koord.) (Indones): arom. rt. u. as diur. aphrod. - **P. saxifraga** L. (Euras, natd. eNA; NZ): burnet-saxifrage; (saxifrage) pimpernel; bibernell; pimpinela branca (Br); rhiz. & rts. constitute Pimpinella (L.), Radix Pimpinellae (albae) (L.) (pim-

pernel or saxifrage root) c. VO resin, saponin, tannin; u. expect. diur. stom. carm. condim.; in inflammn., cough, menstrual & lactating problems, in gargles & mouthwashes; hb. (Herba P. albae et nigrae) u. in folk med. (Euras); young pl. u. in "spring herbal soups" & as spinach, in Hamburg eel soup; veterin. med.

pimpinella root: pimpernel rt. - **small p.**: *Pimpinella saxifraga.*

pimpinellin: a substituted coumarin; occ. in rt. of *Pimpinella saxifraga.*

pimprenelle (Fr): sanguisorba.

pin (Fr): pine - **p. blanc** (Fr): *Pinus strobus* - **p. d'Alep**: *P. halepensis* - **p. maritime**: *P. pinaster* - **sève de p.**: fresh sap of *Pinus pinaster.*

piña (Sp): 1) pineapple 2) pine cone - **p. colada** (Sp): combn. of coconut, pineapple juice, & rum, popular as beverage - **p. de ratón** (Cu): *Bromelia pinguin.*

Pinaceae: Coniferae: Pine (or Conifer) Fam.; the chief fam. of the Gymnosperms; found as shrubs or trees all over the world; somewhat similar anatomic. to Dicots; lvs. needle-shaped; fls. strobiles; frt. a cone or galbulus; important chem. compounds incl. oleoresins, balsams, tannins, FO, etc.; of greater importance in past geological ages than at present; ex. *Pinus, Abies, Pseudolarix, Cedrus, Juniperus, Picea, Taxus, Thuja* (2345).

pinal (Sp): **pinar** (Sp): pine tree grove.

Pinanga dicksonii Scheff. (Palm.) (Chin, Ind+): seeds c. tannin u. as stom. antidiarrhetic; frts. substit. for areca nuts (chewed) - **P. disticha** H. Wendl. (Mal): sds. u. as antidote to certain poisons - **P. duperreana** Pierre (Vietnam, Laos, Camb): sla condor; pith of trunk eaten by natives (Camb); frts. u. as masticatory like betel - **P. latisecta** Bl. (Indones): sds. & lvs. u. as astr. (int. & exter.).

pinard (Fr): popular name for wine.

pinaster: *Pinus pinaster.*

Pinckneya pubens Mx. (Cinchona caroliniana Poir.) (monotypic) (Rubi) (SCar, Ga, Fla [US]-very rare): quinine (or Georgia) tree; fever tree; bk. (Georgia [or bitter or quinine or fever] bark) c. glycoside; u. antipyr. ton. in malar. fevers; considerable use during Civil War (448); attractive shrub or small tree.

pincushion tree: *Hakea platysperma.*

pinder (Ala+ [US]): peanut.

pine: 1) member of genus *Pinus* 2) (Austral) *Araucaria, Callitris,* etc. - **air p.**: *Tillandsia* spp. - **Aleppo p.**: *Pinus halepensis* - **p. bark**: v. white p. - **bastard p.**: *P. virginiana* - **bishop('s) p.** (Cal [US]): *P. muricata* - **black p.**: 1) *P. rigida, P. serotina* 2) *P. taeda* (SC dark bk.) 3) other *P.* spp. (source of **b. p. needles**) - **black-bark p.**: *P. jeffreyi* - **blue p.**: *P. griffithii* - **bull p.**: *P. ponderosa* - **buttermilk p.**: one growing in the open with many branches down to ground (449); *P. taeda* (SC stump covered with milky exudation) - **Caribbean p.**: Cuban p. - **cliff p.**: bastard p. - **cluster p.**: *P. pinaster* - **Cuban p.: Dade County p.** (local name, Fla [US]): *P. caribaea* - **digger p.**: *P. sabiniana* - **p. drops**: *Pterospora andromedea* - **dwarf p. (needle oil)**: *P. mugo* - **Georgia p.**: longleaf p. - **gold p.** (Ga): *P. palustris* - **goose p.** (US Negro): spruce p. - **ground p.**: 1) *Lycopodium clavatum* 2) *Ajuga chamaepitys* - **p. gum**: turpentine (1) - **heart p.: longleaf p.**: *P. palustris* - **hill p.** (swGa [US]): *P. palustris* (SC clumps on hills survived fires, accounting for more longleaf pines on hills) - **Jack p.**: *Pinus banksiana* - **Japanese umbrella p.**: *P. densiflora* - **joint p.**: *Ephedra* spp. - **lodgepole p.**: *P. contorta* & vars. - **longleaf Indian p.**: *P. longifolia* - **longleaf (yellow) pine**: *P. palustris* - **p. needle oil** (USSR): fr. lvs. of Scotch p. - **Nepal p.**: blue p. - **nigger p.**: bastard p. - **noble p.** (NH [US]): *Chimaphila* - **Norfolk (Island) p.**: *Araucaria heterophylla* - **northern p.**: *P. strobus,* etc. - **p. nuts**: sds. of *P. pinea, P. edulis,* & 10 or more other *P.* spp. - **p. oil**: Oleum Pini: VO dist. fr. wd. of *P. palustris* & other se. US *P.* spp.; u. in disinf. soaps, insecticides, techn. prods. (distinguish fr. wood turpentine) - **old field p.**: *P. taeda* (SC grows up spontaneously as second growth in old fields) - **Oregon p.**: 1) *Pinus ponderosa* 2) *Pinus contorta* 3) *Pseudotsuga menziesii* - **Paraná p.** (Br): *Araucaria angustifolia, A. brasiliana* - **parasol p.**: umbrella p. - **pitch p.**: *Pinus palustris, P. caribaea, P. rigida, P. elliottii, P. ponderosa* - **p. pollen**: pollen grains fr. *P.* spp. sometimes found as adult. of Lycopodium - **pond p.**: *P. serotina* - **ponderosa p.**: *P. ponderosa* - **poorlan** (=land) **p.** (Miss [US]): *Ascyrum* sp. - **princess p.: Prince's p.**: *Chimaphila umbellata* - **pumpkin p.**: *P. strobus* (lumber indy.) - **resin p.**: Scotch p. - **riff p.** (sUS): p. lumber with straight grain - **rosemary p.**: *P. echinata* - **round timber p.** (sUS): lumber fr. tree never turpentined - **running p.**: *Lycopodium clavatum* - **p. sap**: fr. *P. pinaster,* etc.; prod. u. for sore throat (Anticosti, Qu [Cana]) - **Scotch p.**: *P. sylvestris* - **screw p.**: *Pandanus* spp. - **scrub p.**: *Pinus banksiana, P. virginiana, P. taeda* (382) - **sea p.**: 1) (SC) *P. elliottii* 2) (Eu) *P. pinaster* - **Siberian p.**: *P. sibirica* (source of needle oil) - **slash p.**: Cuban p. - **spruce p.**: *P. glabra* - **star p.**: *P. pinaster* - **stone p.**: *P. cembra*; *P. monophylla*; *P. pinea* - **p. straw**: fallen & dead p. needles u. as mulch for strawberry beds, on stable floors, etc. -

sugar p.: *P. lambertiana* (**huge-coned s. p.**) - **p. tar**: Pixi Pini: prod. of dest. distn. of wd. of *P. palustris* & other *P.* spp. (*v.* tar, pine) - **p. t. oil**: rectif. t. oil, VO fr. pine tar by steam distn. - **p. tops**: fr. *Tsuga (Pinus) canadensis* - **umbrella p.**: 1) *P. pinea*; *P. densiflora* var. 2) *Sciadopitys verticillata* (u. fir) - **Weymouth p.: white p. (bark)**: *P. strobus* - **wild p.**: Scotch p. - **p. wool**: *v.* wood wool (1) - **yellow p.** (wUS): older ponderosa pine (SC bark color) (—> *P. palustris*).

pineal body: p. gland: Corpus pineale; small cone-shaped endocrine body located near center of brain; at this site, melatonin is synthesized (this inhibits gonadal development & growth & effects estrus in lower animals); ext. has been u. to treat schizophrenia (ca. 1957).

pineapple: *Ananas comosus* - **p. fungus** (Cana): *Fomitopsis.*

Pinellia ternata Ten. (P. tuberifera Ten.) (Ar.) (Arg, eAs): tubers (hange) u. antiemetic, expect. diaph.; in mammary carcinoma (411); many other uses (450).

pinene: a terpene, $C_{10}H_{16}$; found in turpentine oil & many other VO (1853).

pinettes de prairie (Fr): *Liatris* spp. (Cana boatmen).

pinguedo (old Lat.): lard.

pinguica (Mex): *Arctostaphylos pungens.*

Pinguicula L. (Lentibulari.) (nTemp regions; Antarctic zone); butterworts (SC mucilag. secrn. by lvs.); 46 spp. of small insectivorous or insecticidal perenn. hbs. - **P. alpina** L. (n&cEu; Siber-mts.): mountain butterwort; white-flowered b. - **P. caerulea** Walt. (P. elatior Mx.) (NCar to Fla [US]): tat grass; fls. u. purg., deterg. vulner. - **P. vulgaris** L. (n&cEu, natd. nNA): (common) butterwort; bog violet; sheep ro(o)t; hb. or lvs. u. for pertussis, spasmodic cough, dry cough (esp. as F.E.); diur.; purg.; insecticide; fls. u. as in prec.; lvs. u. to curdle milk (Lapland).

pinguin (fiber): *Bromelia p.*; c. **pinguinain**: protein-splitting enzyme.

pinguoleum (L.): fatty (or fixed) oil.

pinha (Por): *Annona squamosa.*

pinh(a)o (Por): pine kernel (sd.) - **p. bravo: p. da india: p. das barbadas: p. de purga: p. do paragui:** *Jatropha curcas.*

pinheiro (Por, Br): 1) trees of Fam. Pin. & Araucariaceae 2) (specif.): *Araucaria angustifolia, A. brasiliana* - **p. broto**: pine bud - **p. do brejo** (Por): *Talauma* spp. (as *T. dubia*).

pinhões (Por): pine nuts ("fir kernels").

pini-pini (Sp): *Jatropha urens* (Br also pina pina); *Morinda royok*; &c.

pinite: pinitol: inositol methyl ether (a cyclitol or hydrated cyclic alc.); occ. in resin of "American manna" (exudn. fr. injured trunk of *Pinus lambertiana*); also in many other plants; very sweet.

Pinites succinifera Conw. (Pin.) (nEu, Baltic region): fossil sp. fr. the Oligocene; source of gum amber (resin), Succinum (*qv.*).

pink: *Dianthus* spp. - **Carolina p.**: 1) *Phlox ovata* 2) *Spigelia marilandica* (misnomer) - **Dutch p.: English p.**: yellow dye lake prepd. fr. Persian berries; u. by artists (yellow madder) - **East Tennessee p. root**: fr. *Ruellia ciliosa* - **p. family**: Caryophyllaceae - **grass p.**: *Dianthus plumarius* - **Indian p.**: 1) *Spigelia marilandica* 2) *Arethusa bulbosa* - **p. lady** (US slang): alc. treated with croton oil (denaturant) u. to fire torpedoes; has been u. as "beverage" by sailors, hence name - **Maryland p.**: p. root - **rainbow p.**: *D. chinensis* - **p. root**: 1) *Spigelia marilandica* (rts.) 2) fungous dis. of onions (esp. in Tex [US]) - **p. saucer**: color obt. fr. *Carthamus tinctorius* - **p. weed**: *Polygonum aviculare* - **wild p.**: *Silene virginica* & other spp.

Pinkneya = *Pinckneya.*

pino: 1) (Sp) pine 2) (NM [US], Spanish) any of the taller pines.

pinocchio (It): pine seed (p. nut).

pinole (swUS, Spanish): meal u. as food by Am Indians made fr. ground maize & wheat with mesquite beans & various dr. sds. as of *Chenopodium fremontii.*

pinolin: rosin oil.

piñon (pl. **piñones**) (swUS, Mex-Sp): pinyon, *Pinus edulis* & other low-growing nut pines (*P. quadrifolia, P. cembroides, P. monophylla, P. cembra*); tree & nut (sd.) (p. nut); term also appl. to *Jatropha curcas* (Mex) - **piñonero** (NM [US], Sp): one who picks piñon nuts.

Pinsel (Ge): 1) pencil, brush 2) penis of wild boar or whale - **P. schimmel** (Ge): *Penicillium.*

pinto (bean): bean mottled like some kidney beans (Colo [US], &c.).

pinuelita de paramo (Ve): *Paepalanthus schlimii.*

Pinus L. (Pin.) (nHemisph, both OW, NW): pines; 93 spp.; trees, often of large size; lvs. of 2 types: deciduous scale-like & permanent needle-like in clusters of 2-5; see many uses under "wood" - **P. abies** L. = *Picea abies* - **P. Alba** (L.): white pine bark fr. *P. strobus* - **P. aristata** Engelm. (wUS-Rocky Mts.): bristle-cone pine; wd. u. locally in mine construction & for fuel; trees of great age (ca. 2,000 yrs.) are found in eCalif (US) - **P. balsamea** L. = *Abies balsamea* - **P. banksiana** Lamb. (NA): jack (or black) pine; decoc. of cones u. as remedy in coryza; wd. u. as lumber, fuel, paper pulp mfr. - **P. canadensis** L. = *Tsuga candensis* - **P. caribaea** (seUS, WI, CA): slash pine; Cuban p.; an important source of turpentine oil & rosin - **P. cembra** L. (Euras): Swiss stone pine; Siberian cedar; S. fir; Zirbel (Ge); sds. (as Nuclei Cembrae) edible, c. resin 50% FO (cedar nut oil) rich in linolenic acid; formerly u. like codliver oil; as delicacy in baked goods (Russia); bird feed; distn. young twigs —> Hungarian (or Carpathian) turpentine oil; fr. trunk bk. Riga or Carpathian balsam, cedrobalsam (Balsamum [Carpathicum]) or galipot (*cf.* Canada Balsam, for which it is used); u. vulner. diur.); fr. lvs. a needle VO u. to make synth. camphor; bk. c. coloring matter u. to color beverages - **P. cembroides** Zucc. (swUS, Mex): Mexican pinyon (pine); piño prieto (Coahuila); sds. eaten & sold in markets; source of oleoresin; VO c. pinene, limonene, longifoline - **P. contorta** Dougl. (w&cNA, nwMex): lodgepole pine; shore p.; scrubby tree, usually up to 10 m (var. *latifolia* Engelm. ex. S. Wats. to 70 m); black pine; possible source of oleoresin; source of a manna; sap u. as food by Chilcotin Indians (BC [Cana]), either fresh or dr. (sapscraper [bone] u. to gather); var. *murrayana* Engelm. (Pac, NA): pitch u. for sore eyes (Calif [US] Indians) - **P. densiflora** Sieb. & Zucc. (Kor, Jap, eUSSR): Japanese red pine; resin & VO u. to treat chronic bronchitis, cystitis (Kor); pollen u. in folk med.; ornam. tree; cv *'Umbraculifera'*: umbrella pine (Tahyoshi pine) - **P. deodara** Roxb. = *Cedrus deodara* - **P. echinata** Mill. (e&cUS): shortleaf pine; yellow p.; occasionally source of Am turpentine (inferior); wd. u. as lumber - **P. edulis** Engelm. (P. cembroides var. e. Voss) (swUS, esp. NM; nMex): pinyon (pine); piñon (Sp); nut pine; tree bears edible pine nuts (sds.) ("Indian nuts"); u. nutr. delicacy; in confections; an article of commerce; oleoresin appl. to wounds & sores; by Am Indians in syph.; wd. fuel, house construction; state tree (NM [US]) - **P. elliottii** Engelm. (seUS, incl. lower Fla Keys): slash pine; an important source of Am turpentine oil & rosin; (P. caribaea Mor. [Cuban pine] is a distinctive sp. but was formerly confused with this sp.) - **P. gerardiana** Wall. (cAs, Himalay, Indo-Pak, Afgh.): chilgoza pine; neoza (or neosa) p.; large sds. edible (neja nuts); turpentine exudn. fr. trunk - **P. glabra** Walt. (seUS): spruce pine; cedar p.; black p.; source of various pine products - **P. grandis** D. Don = *Abies g.* - **P. griffithii** McClelland (P. excelsa Wall.) (cAs as Pak): blue pine; kail; turpentined; source of edible nuts - **P. halepensis** Mill. (P. maritima Lambert) (Med reg., wAs): Aleppo pine; sec. source of Eu (Greek) turpentine (t. oil, rosin) - **P. insularis** Endl. (seAs, PI): source of a turpentine oil u. in similar therapy as regular (PI); bk. u. in urinary dis.; resin in plasters for bone fractures (Ind) - **P. jeffreyi** Grev. & Balf. (wPac, US, nMex): Jeffrey pine; black p. (SC needles dark bluish-green); source of various pine products; its turp. oil is 95% n- heptane - **P. koraiensis** Sieb. & Zucc. (Kor, Jap, USSR): Korean pine; pollen u. as food & ton.; sds. are ton. - **P. lambertiana** Dougl. (swUS [Calif, Ore], nwMex): (giant) sugar p.; Lambert p. (Calif); largest of all pines; -80 m, cones 0.5 m long; juice & dr. exudation ("American manna") prod. fr. heartwd. injuries, c. oleoresin, pinite (*qv.*), very sweet; u. as sugar substit., cath., chewing gum; buds & dr. juice u. in homeop. med.; wd. valuable; nuts food (Calif Indians) - **P. maritima** Lambert = *P. halepensis* - **P. maritima** Poiret = *P. pinaster* - **P. massoniana** Lamb. (eAs, csp. seChin, nVietnam): Masson pine; a chief timber pine of eAs, valuable wd.; pl. much u. in med. in Chin (lvs. bk. of rt. [pd.]); resin; cones; sds. fls.; pollen u. intern. to incr. blood devt., to give strength, tinct. for violent headache; pollen appl. exter. to boils & carbuncles; pine knots u. in rheum.; decoc. of wd. shavings u. in colds; lvs. insecticidal; wd. for lumber, fuel, & charcoal source (Chin) - **P. merkusii** Jungh. & De Vriese (Chin, Burma, Indochin, Sum, PI): Merkus pine; u. for boils & skin dis.; source of turpentine; sds. edible; bk. u. as cork; wd. useful, does not rot easily in soil - **P. monophylla** Torr. & Frémont (P. cem-broides var. m. Voss) (Calif to Utah [US], nwMex): single-leaf pinyon; one-leaf pine; sds. edible; wd. u. to make charcoal for smelting, fuel; sds. furnish FO similar to olive oil, u. in cooking - **P. montezumae** Lamb. (Mex, CA): Montezuma pine; resin u. for healing open wounds; important source of turpentine (Mex); incl. var. *hartwegi* Engelm. - **P. monticola** Dougl. ex D. Don (wNA-mts.): (western) white pine; useful wd.; needles yield VO, with pinenes, phellandrene, borneol, azulene; sap edible, lax.; bk. decoc. u. for TB, cuts, sores, stomachache (Calif) - **P. mugo** Turra (P. montana Mill.) incl. var. *pumilio* (Haenki) Zenari (P. p. Haenki) (c&sEu): mountain pine; dwarf (knee) pine; Knieholz (Ge) (knee timber, elfin wood); source of VO, dwarf pine needle oil; u. antisep. exp., in nasal prepns., inhalants, disinf. for local appln. in rheum., in perfume indy. (prod. Fr, Hung); disinf. of sickrooms - **P. muricata** D. Don (Calif [US], nwMex): bishop pine; good lumber wd.; cult. ornam. tree - **P. nigra** J. F. Arnold (sEu, AsMinor): Austrian pine; (European) black p.; widely cult., good lumber; pine wand (scepter surmounted by a pine cone [thyrsus]) carried by Dionysius (shown on vases, &c. fr. ancient Greece); furnishes turpentine u. in ointments & plasters - **P. occidentalis** Sw. (P. cubensis Griseb.) (Cu): Cuba pine; bastard p.; pino (Cu); source of turpentine; roasted sds. eaten - **P. palustris** Mill. (P. australis Mx.f.): long-leaf (yellow) pine; southern (yellow) p.; pitch p.; slashes made in trunk yield crude turpentine oleoresin (semi-liquid: "dip"; solid [2167]: "scrape" if fr. incisions; "gum thus" if spontaneous exudn.); mostly distd. with steam or water to furnish (Am) turpentine oil, the residue, rosin (*qv.*); important folk med. usage (formerly); oleoresin appl. directly to cut & wrapped in handkerchief (383); VO ("turpentine") u. for sore throat, backache, kidney trouble; rheum. croup, pneumonia, earache, toothache, worms & colds (using sugar dampened with oil) (384); such widespread domestic use is hazardous (damage to kidneys) & is now legally controlled; rosin (*qv.*) (colophony) much u., mostly in industry; wd. source of (pine) tar (*qv.*), tar oil, tar pitch, tar water (u. in "nervous colic") pine oil (wood turpentine), &c.; many of these products are termed "naval stores" (chief source); wd. ("fat wood") ("kindling") u. for starting fires in fireplace, &c.; this wd. now sold in commerce; young shoots & needles at branch tips u. in decoc. as cold remedy, for menstrual cramps & as beverage (SC) (2431); young tree forms grass-like cluster (6-7 yrs.) & requires much sunlight for survival (2346) - **P. picea** L = *Abies alba* - **P. pinaster** Ait. (P. maritima Poiret) (Med reg., mostly Sp, sFr): French maritime pine (pin maritime); cluster p.; chief source of Eu turpentine (t. oil, rosin); trunk exudn. called Bordeaux t. (semiliquid), with resin acids (esp. d- and l-pimaric acid, also abietic acid), resenes, VO (15- 30%), u. as diur. & in varnishes; exudn. when dr. solidifies as "galipot" (*qv.*); trunk also source of Lympha Pini (Por Ph.), fresh sap or juice (sève de pin [Fr]) obt. by incisions in bk., u. for bronchitis & phthisis (syr. & F.E.); pine tops (Turiones Pini; sprouts of pine) u. as popular folk med. in bronchitis, gout, rheum., vit. C source; in baths to improve peripheral circulation - **P. pinea** L. (Med reg., incl. Sp): (Italian) stone pine; sds. edible (pignolia nuts) (Rome); recommended for high linolenic acid content in arteriosclerosis - **P. ponderosa** Dougl. ex Lawson & P. Lawson (wNA, nMex): ponderosa pine; (western) yellow p.; bull p.; important lumber tree; oleoresinous exudn. u. as masticatory & adhesive; source of turpentine oil (u. in bronchial dis.); fiber fr. lvs. u. to stuff mattresses, pillows (medicated); soft sapwood u. as famine food (Indians of wNA.); var. *arizonica* Shaw (P. a. Engelm.) (NM, Ariz [US], nMex): Arizona pine; turpentined in Mex - **P. pungens** Lamb. (Appalachian region, eUS): Table Mountain pine; tree occasionally turpentined; wd. u. for charcoal (parts of Pa [US]); fuel - **P. radiata** D. Don (cCalif): (small-coned) Monterey pine; a source of turpentine; wd. u. in cabinet work, matchboxes; cult. in GB, Austral, NZ, &c. (2442); var. *binata* Engelm (Baja Calif [Mex]) - **P. resinosa** Ait. (Cana, e&cUS): red pine; Norway p.; yields oleoresin with high m. pt.; useful wd.; source of a tanbark - **P. rigida** Mill. (eNA): pitch pine; "black p."; wd. lumber, fuel, charcoal; source of pine tar - **P. roxburghii** Sarg. (P. longifolia Roxb.) (Indo-Pak + mts.): chir (pine); cheer p.; tree is turpentined; wd. u. to make charcoal for Chin fireworks; VO (Indian turpentine oil) rich in carenes, longifoline, poor in pinenes; rt. & bk. u. diabetes (folk med.) - **P. sabiniana** Dougl. D. Don (Calif): digger pine; nuts u. for food; oleoresinous exudn. u. as protective for burns, sores, as poultice (mixed with corn flour) as counterirrit.; dr. prod. u. as chewing gum, & for rheum.; resin c. n-heptane (formerly "abietene") (hydrocarbon); VO fr. resin u. med.; bk. inf. in TB; bk. & lvs. u. in kind of sweat bath for rheum. & body bruises (Mendocino Co. Indians, Calif) (SC Digger Indians) - **P. strobus** L. (e&cNA, Mex, Guat): (eastern) white pine; northern p.; weymouth p.; source of white p. bk. (as Pinus Alba, dr. inner bk.) (formerly largest cash crop among crude drugs in US [-1952-]); u. expect. in coughs, grippe, colds, in cough syrups (esp. white p. syr. comp.); kidney dis. hemorrhage, scurvy (385); oleoresin ("gum") u. in cough prepns. (Abnakis Indians, eNA) (386); decoc. & plaster fr. inner bk. u. for colds (386); VO fr. lvs. & twig ends also u. - **P. succinifera** Conw. = *Pinites succinifera* - **P. sylvestris** L. (Euras, natd. eNA; cult.): Scotch pine; Scots' p.; source of some Eu common turpentine (mostly USSR, Ge) fr. which turpentine oil & rosin are prod.; one source of Galipot (US "scrape"), mostly rosin, little VO, u. skin dis. & in plasters, technol.; important lumber tree; popular Xmas tree; wd. source of wd. turpentine oil (distn. of rts.), known to ancients; tar (*qv.*) & pitch; paper pulp; wood wool (*qv.*); needles u. arom. in baths, for rheum. gout, source VO (pine needle oil); bk. u. tan.; buds (as Turiones Pini) u. diur. exp. antirheum., prodn. VO & ext.; sds. source of pine sd. oil (chief prodn. Swed, Fin, USSR, Ge); buds off. in Braz Ph. - **P. tabulaeformis** Carr. (Chin, Mongolia, Manchuria): Chin pine; much u. in Chin med.; pollen, resin, lvs. u. much like *P. massoniana* - **P. taeda** L. (e & esp. seUS): loblolly pine;

oldfield p.; (black [slash] p.); said the most abundant tree of NA; lvs. in 3's, 15-23 cm long); source of minor amts. of Am turpentine; exudn. low in terpenes, c. 90% resin acids (may be converted into steroids, incl. testosterone) (aryl 18-nor-steroids); tree grown for erosion control; lumber; pulpwood; pine tar (451) - **P. teocote** Schildl. & Cham (Mex, nGuat): pino real; source of turpentine, (VO) (trementina del pais) u. stim.; tar (brea) u. in soaps, for torches, &c. - **P. thunbergiana** Franco (P. thunbergii Parl.) (Jap): (Japanese) black pine; resin & VO u. to treat chronic bronchitis, cystitis - **P. virginiana** Mill. (eUS): Virginia pine; scrub p.; Jersey p.; black p.; hardened oleoresin made into pills for kidney dis. (Rappahannock Indians); lumber u.; ornam. tree, esp. for Christmas t. (cut in Ala [US], &c.) - **P. wallichiana** A. B. Jackson (P. excelsa Wallich) (temp. Himalayas; Afgh to Nepal): Himalayan pine; oleoresin obt. by tapping & fr. this turpentine oil & rosin are prepd.; rts. yield VO to anoint legs & arms of natives to keep off water insects while working in rice fields; a manna appears in dry winter seasons as lf. exudn., eaten by natives; tree resistant to blister rust; wd. u. in building, making planks, & for firewood.

pinweeds: *Lechea* spp.

pinyon (NM [US]): *Pinus edulis.*

pioche (Fr): kind of mattock u. to raise plants fr. ground (long-handled).

piolho de padre (Por): *Bidens pilosa.*

Pionocarpus madrensis Blake (Comp.) (Mex): rts. u. in rheum. &c.

pip: 1) sds., ex. in raspberries, citrus; apple 2) rootstocks of some pls., ex. *Convallaria* 3) bird dis. with thick mucus in throat (corruption of pituita, Gr, phlegm); facetiously, of some human dis.

pipal (Hind): *Ficus religiosa.*

pipara minta (Hindi): peppermint.

pipe gamboge: form of g. in cylindrical sticks made by pouring oleo-gum-resin into bamboo sections as molds - **p. gourd**: v. gourd - **Indian p.: p. plant**: *Monotropa uniflora.*

Piper L. (Piper.) (pantrop): peppers; c. 1,100 spp. mostly of climbing shrubs with frts. as berries; also (L) frt. of black pepper, *P. nigrum* - **P. abbreviatum** Opiz (PI+): lvs. appl. to treat splenomegaly; frts. arom. stim., carm.; u. in coughs & colds (Ind) - **P. acuminatissimum** C DC (CR): alkatan; dr. lvs. u. against diabetes (387) - **P. aduncum** L. (Guat, Hond+): cuturo; lvs. astr. stim. diur., tea for cough; inf. in mange of dogs; fl. spikes u. to flavor fish; frt. eaten - **P. aequale** Vahl (Ve, Co): ancillo (Ve); lf. decoc. u. in rheum. - **P. Album** (L.): dr. ripe drupe of *P. nigrum* fr. which outer portions have been removed; u. condim. stomach. ton.; in meat indy.; distinctive finer flavor than black p. but less pungent (2178); prod. Indonesia (Muntok) - **P. aleyreanum** C DC (Br-Amazon): rts. sts. lvs. decoc. in rheum. - **P. allenii** Trel. (Pan): rts. u. to deaden pain (Pan Indians); lvs. u. in snakebite - **P. amalgo** L. (WI): black wattle; fresh lvs. (inf.) u. for cough; dr. lvs. & inflorescence u. in prepn. alc. beverage; chewed lvs. placed on cuts to stop bleeding (PR); wd. lumber; rem. for infect. & inflammn. (Co) - **P. amazonicum** (Miq.) C DC (nAmazonia, SA): lf. decoc. antipyr. in high fevers; purg.; poultice for swellings (SA Indians) - **P. angustifolium** R. & P. (P. elongatum Vahl) (Bo, Pe, Ve, Br): matico (pepper); hb. & lf. u. arom. diur. vulner. aphr. astr. in GC, hemostat. stypt. stim. (2176) - **P. attenuatum** Miq. (seAs): crushed lvs. smeared on abdomen in fever & "bilious colic"; lf. sap in GC (Indones) - **P. aurantiacum** Wall. ex DC (China, Ind): pl. decoc. u. to bathe sores & swellings (China); frt. refrig. (Ind) - **P. auritum** HBK (sMex, CA, Co): "Santa Maria"; acoyo; lvs. u. to season tamales, &c. (Veracruz); fresh lvs. u. for headache (CR); hb. juice u. to remove leeches & ticks (Salv); in snakebite (Nic); var. *amplifolium* DC: lvs. have sassafras odor & u. to make beverage tea; fl. spikes to flavor fish (Guat, Hond) - **P. baccatum** Bl. (seAs): lf. juice drunk in cough, frt. native ton. (Indones); rt. decoc. for VD (Mal) - **P. bartlingianum** DC (Surinam): ingredient in arrow poisons (Waiwai Indians) - **P. betle** L. (seAs, Mal, Indones) (cult. in hot damp areas of Ind [Cochin, Bombay, Maharashtra]; WI; Madag): betel pepper; b. vine; pan (Hind); betel lvs. c. VO with chavicol, cadinene, u. masticatory in chewing betel nut by wrapping fresh leaf around mixt. with chalk, spices; acts as stim. euphoric (danger: carcinogenic); rts. u. in coughs, diphth. otitis media, in gastric dis. - **P. bogotense** C DC var. *ovalifolium* Trel. & Yunck. (Ve): u. in baths for animals with alunadura (inflammn. & sores [tumors] attributed to moonlight![?]) - **P. bredemeyeri** Jacq. (Ve): cordoncillo negro; lf. inf. u. as purg.; pl. decoc. in VD; lf. decoc. emmen.; rt. decoc. purg. in TB; decoc. to bathe swollen animals (enludados) (Ve) - **P. caninum** Bl. (seAs): lvs. chewed with betel quid for hoarseness (Java); lf. wash u. after childbirth - **P. chaba** Hunter (P. officinarum C DC) (Mal, PI): frt. u. as spice, sometimes called "long pepper"; rt. chewed & saliva swallowed for colic (PI); dyspeps. stomachache; *cf. P. retrofractum* - **P. chlorocarpum** C DC (PI, Solomon Is): frt. chewed with betel nut & lime to bring out stim. effects of betel

- **P. citrifolius** Lam. (Br, Fr. Guiana): poivrier à feuilles de citronnier; frts. u. like black pepper; lvs. u. as cataplasms in baths, rts. diur. diaphor. (rt. bk) - **P. clusii** DC (P. guineense Schum. & Thom.) (equatorial wAf): Ashanti pepper; fefe (Malinke); poivre du Kissiv; African black pepper; Congo cubeb; widely u. as condiment in wAf (natives); u. med. in Angola; frt. c. piperine, chavicine; u. as pepper & cubeb substit. - **P. crassipes** Korth. (seAs): sds. u. in mixt. for treating GC - **P. crocatum** Ruiz & Pav. (P. nubigenum Kunth) (Pe, Ec): inflor. & lvs. u. as arom. & to disinfect wounds; inflor. has yellow dye - **P. cubeba** L. f. (Java, Cey): cubeb (pepper); cubebs (frt.); unripe fully grown frt. u. diur. exp. sed. fumitory in asthma (cigarettes); u. in Ulmer Pfefferkuchen (elm gingerbread); in prepn. of a galenical oleoresin, VO; urinary antisep.; flavor (cigarettes) - **P. dariense** C DC (Pan): kana; u. in baths for colds & snakebite (Cuna Indians) - **P. erythroxyloides** R. E. Schultes (Co. Ve): chewed sts. produce anesthetic effect & sensation similar to that fr. coca chewing (hence name); thought to reduce dental caries; aphrod. - **P. geniculatum** Sw. (Artanthe geniculata Miq.) = *P. stamineum* - **P. guineense** Schum. & Thonn. (tropAf): Guinea/Ashanti/Benin/Kissi pepper; Congo cubeb; frts. u. as seasoning; as astr. in diarrh. carm. ton. (often with massage); in rheum.; as insecticide (house fly); u. as cubeb adulter.; VO rubefac. - **P. hancei** Maxim. (Chin): lf. decoc. u. as wash to reduce swellings & relieve inflammatory rheum. - **P. Hispanicum** (L.): Capsicum - **P. hispidum** Sw. (WI, cont. tropAm): platanillo (de Cuba); cordoncillo; arom. lvs. u. as astr. hemostat., in baths for rheum. (Cu); frts. u. to delouse dogs (Co); crushed in water to kill head lice (Ec); lvs. as tea in mal. (Ingas); fish poison (with *Phyllanthus*) (Co) - **P. jaborandi** Vell. (Br): (Brasil) jaborandi; falso j.; j. do mate; j. do Brasil; jambu assu; rt. (Radix Jaborandi; J [Raiz]) or rt. bk. u. as diur. diaph. sialagog; in catarrhs., toothache, angina, to incr. salivary secretion (*P. corcovadensis* [Miq.] DC similar) - **P. kadsura** Ohwi (eAs): frts. stom. expect. & stim.; similarly for **P. kawakamii** Hay - **P. kava** Royle = *P. methysticum* - **P. latifolium** L. f. (Austral Is, Caroline Is): lvs. cooked in coconut milk u. for cramps (Austral Is) - **P. lolot** C DC (eAs): dr. lvs. decoc. u. for maladies of the bones (Indochin) - **P. longum** L. (s&seAs): long pepper; consists of entire spikes of immature frts.; c. piperine; whole frt. u. in pickling; adulter. for black & white peppers; u. flavor, condim. carm. stim.; well-known drug in Ind since time immemorial; widely cult. in Ind (2177) - **P. lowong** Bl. (Java): false cubebs (frt.) c. piperine, pseudocubebine; u. cubeb sophisticant - **P. lucigaudens** C DC var. *alleni* (Pan): rts. u. to deaden pain; lvs. in snakebite - **P. marginatum** Jacq. & var. *catalpaefolium* HBK (WI, Br, CA): cake bush (Trin); lf. inf. u. to ease birth; lvs. as covering for sores; frt. in colic; rt. diaph. & to stim. saliva - **P. methysticum** G. Forst. (South Sea Is, Figi, Polynes, NG): kava (pepper); kava kava; rhiz. of complex comp. (c. kawain, methysticin [non-N lactones, probable active constits.]) u. to prep. stim. & intoxicating beverage "ava"; social intoxicant (South Pac) formerly chewed (by old women) to crush, then spat into common mixing bowl and left to ferment; now comminuted in mortars; also as fumitory & masticat.; diur. anod. in rheum. gout; fish poison - **P. molliconum** auctt. = *P. wrightii* - **P. nigrum** L. (tropAs, esp. Ind; often cult. seAs, tropAm): black pepper; "pepper"; unripe frt. (peppercorns) c. VO, piperine; piperidine; chavicine, methyl pyrroline, FO resin; u. condim. (most used of all spices), flavor, stim. stom. countrirrit. in dentalgia; prepn. of white pepper, Piper Album (*qv.*); in former times much sought after such as in war booty (at capture of Caesarea AD 1101, Genoese recd. as their portion over 160,000 lbs. of pepper [388]); comm. sorts; Malabar (best); Goa; Tellicherry; Singapore; Penang; Kompong; Sarawak; Brazil - **P. novae-hollandiae** Miq. (nAustral, esp. Queensland): Australian pepper; u. as med. in mucous membr. dis., esp. GC; tumor-inhibiting properties - **P. patulum** Bertol. (Salv, Guat): u. to repel wood ticks - **P. peepuloides** Roxb. (Ind-Himalay): frt. u. for dis. of resp. tract (Ayurvedic med.) (not *P. peepuloides* Wall. = mixt. of *P. retrofractum* & *P. betle*) - **P. peltatum** L. = *Pothomorphe p.* - **P. penninerve** C DC (PI): lvs. appl. to deep ulcers - **P. phytolaccaefolium** Opiz (SA: nAmazonia area): st. sap dropped into eyes for blindness (cataracts?); st. chewed for sore throat; teeth colored black (thought to preserve teeth); several other *P.* spp. of nBr u. similarly - **P. puberulum** Seem. (Fiji): bk. decoc. u. in fever; lf. decoc. in bloody stools; var. *glabrum* (C DC) A.C. Smith (Tonga, Fiji): lvs. u. med. (86) - **P. pyrifolium** Vahl (P. borbonense C DC) (Madag, Mascarene Is): cubèbe de pays; an adultern. of cubebs & black pepper - **P. quadrifolia** (L) HBK (Pothomorphe quadrifolium L.) (WI, Mex, nSA): "cuartillo" (Mex); u. as antilithic; in skin dis. - **P. reticulatum** L. (WI, Mex-Br?): wild jaborandi (pepper); false j.; similar to *P. jaborandi*; rt. has peppery aroma - **P. retrofractum** Vahl (P. officinarum C DC) (Indo-Mal, archipelago; cult. Java,

&c.): (Javanese) long pepper; rt. decoc. u. for stomach dis. (PI); dr. unripe frt. (in entire spikes) u. condim. flavor, as ton. stom. carm. stim. diur; in bronchitis, asthma, fever, TB; diur., diaph.; added to black pepper; in pickling, to season curries & pickles; fly poison; frt. c. VO piperine - **P. ribesoides** Wall. (Burma, Mal): frts. as "false cubebs" at times u. as adulter. cubebs - **P. rohrii** C DC (Mex, CA, esp. SA): cordon; u. to wash wounds (Hond, Guat) - **P. sanctum** Schlecht. (sMex): acuyo; tlanepaquelete (Mex); shrub lvs. u. as condim., to aromatize soups; flavor; lf. decoc. in indig. & abdom. cramps; folk use as stim. local anesth. toothache - **P. sarmentosum** Roxb. (Ind, eAs): rt. u. in wash for fungoid dermatitis of feet; lf. decoc. u. embrocation for "weakness"; with rice as antipyr. & to aid digestion (Hainan) - **P. stamineum** (Miq.) DC (P. geniculatum Sw.; Artanthe g. Miq.) (WI, Mex): rt. or rt. bk. c. piperine; u. drastic, emmen., may be admixed with curare (452) - **P. subpeltatum** Willd. (Af pantrop): lf. juice for intest. dis.; rt. decoc. for cough (Af), remedy in oxyuris (Haya); u. as antispasm. expect. stom. diur. (Br, Par) - **P. sylvaticum** Roxb. (Ind): frts. u. like "long pepper"; carm.; rts. u. in snakebite as "antidote" - **P. tricuspe** C DC (Pan): "chiti"; u. for headache & pain in ribs area; cult. - **P. umbellatum** L. = *Pothomorphe umbellata* - **P. volkensii** C DC (Af): rt. decoc. u. for sores, stomachache; raw or cooked rt. u. in impotency; frt. as cardamom substit. - **P. wallichii** Hand-Mazz. (P. aurantiacum Wall. ex DC.) (Ind, Chin): u. uter. stim. (with hypotensive properties); pl. decoc. u. to bathe sores & swellings (Chin); frt. refrig. (Ind) - **P. wrightii** C DC (Cu): another source of matico.

Piperaceae: Pepper Fam.; hbs., vines, shrubs, & trees of both hemisph., mostly subtrop. or trop.; lvs. alternate or rarely opp. or whorled; fls. very small; frt. small globose or ovoid drupe; members often aromatic, with VO or oleoresins; ex. *Piper, Peperomia.*

piperettine: alk. of frt. of *Piper wallichii, P. nigrum,* &c., chief source of pungency, less active than piperine.

piperidine: amine found in *Piper nigrum*; also formed by hydrolysis of piperine.

piperine: alk. fr. *Piper nigrum, P. clusii,* & other *P.* spp.; has been synth.; u. antipyr. carm.; to give pungency to brandies; as insecticide.

piperina (Arg): *Minthostachys verticillata.*

Piperis Nigri Fructus (L.): black pepper.

piperitone: terpene ketone with peppermint odor; occ. in many different pls., incl. peppermint (VO), *Eucalyptus dives* (VO); Japanese mint (VO) & other plants; u. to mask odors of dentifrices - **piperonal**: heliotropin; aldehyde found in VO of *Robinia pseudoacacia*, etc.; u. perfume (artificial heliotrope); as pediculicide in lousiness.

pipe-vine: *Aristolochia* spp.

pipi (raiz) (Por) (**root**): *Petiveria alliacea* - **pipian** (Salv): pumpkin.

pipit zahoac (Mex): *Acourtia media.*

pip-menthol: menthol fr. Am & Eu peppermint oils (as distinct fr. that of Japanese mint, which has different phys. props.).

pippins: 1) *Gaultheria procumbens* 2) kind of attractive yellowish apple 3) seeds.

pippul (tree): peepul; pipal (*qv.*).

pipsissewa, common: *Chimaphila umbellata* - **spotted p.**: *Chimaphila maculata.*

Piptadenia colubrina Benth. = *Anadenanthera c.* - **P. excelsa** (Griseb.) Lillo (Leg.) (SA): claimed to have hallucinogenic props.; excellent timber - **P. incanus** Wedd. (P. velutinus Wedd.) (Indones): bk. pounded & scraped & u. for boils, cascado, herpes, burns - **P. macrocarpa** Benth. (incl. var. *cebil*) (Bo, nwArg): cebil; curupay; red & white cebil bk. (vilca bk.) c. 15-18- 34% tannin, tryptamine derivs.; u. tan. & in med.; source of cebil gum (c. 81% arabin) ("Brazilian gum arabic") u. as replacement for acacia; excellent wd. - **P. peregrina** Benth. = *Anadenanthera p.* - **P. pisonis** Mart. = *Merremia tuberosa* - **P. rigida** Benth. (Parapiptadenia r. Brenan; Acacia angico Mart.) (sBr, Par, nArg): angico; bk. c. 22% tannin, u. as dyestuff & tan; in bronchial dis.; gum (Brazil[ian] g.; gomma angico [Angico gum]) c. bassorin; u. like gum arabic; lvs. & pods also u.; hard wd. comm. timber, u. in construction; mamaki (Haw); frts. of shrub. u. as lax. & in thrush; bk. u. to make tapa (or paper) cloth.

Piptadeniastrum africanum (Hooker f.) Brenan (1) (Leg) (tropAf: Sen-Sudan-Uganda-Ang): "dabéma" (comm.); bk. decoc. abortif. & in gargle for dental pains (Gabon) (2489).

Piptoporus betulinus Karst. (Polyporus b. Bull.) (Polypor. - Basidiomycet. - Fungi) (Euras, NA): birch polyporus; razor- strop fungus; c. polyporenic and tumulosinic acids; cancerostatic activity; corky white mass of frt. bodies u. to draw on with charcoal sticks; wd. attacked by fungus u. to burnish watches (Switz); frt. body u. to make razor strops formerly.

Piptostegia pisonis Mart. (Convolvul.) (2152) = *Merremia tuberosa* - **P. gomesii** Mart. = *Merremia macrocarpa* (—> *Operculina*).

Pipturus arborescens C. B. Rob. (Urtic.) (Borneo, PI): bk. scrapings u. in poultice for boils - **P. argenteus** Wedd. (NG): juice appl. locally to fresh wounds or aching tooth; var. *lanosus* Skottsb. (Tonga, Samoa, Fiji): alanga (Tonga); bk. & lvs. u. med.; bk. to make cordage fiber (Tonga) (86) - **P. gaudichaudianus** Wedd. (P. albidus Gray) (Haw Is [US]): mamaki (Haw); frts. of shrub u. as lax. & in thrush; bk. u. to make tapa (or paper) cloth - **P. incanus** Wedd. (P. velutinus Wedd.) (Indones): bk. pounded & scraped & u. for boils, cascado, herpes, burns.

piquant: pungent, spicy.

Piqueria trinervia Cav. (Comp.) (Mex, CA): (hierba del) tarbadillo; hierba de San Nicolas (Mex); lf. inf. u. in typhoid fever; at times pollen cause of hay fever; u. ornam. spray with cut fls.

piqui(a) (Br): *Caryocar villosum.*

pirageia: *Anchietea salutaris.*

Piraman gum pitch: fr. *Moronobea coccinea.*

pirarucu (Br): very large fish of major rivers of Br & Guianas; -5 m long; *Arapaima* spp.; guarani (Corrientes, Arg); tongue (ca. 5 cm X 20 cm) has very rough surface; u. to grate guaraná (SA Indians).

Pircunia Bert. = *Phytolacca.*

pirliteiro (Por): *Crataegus oxyacantha.*

piromen: pyromen: *Pseudomonas* polysaccharide complex: prepd. by proteolysis of organism & sepn. of complex by dialysis; u. as pyrogen.

Piru (Mex): *Schinus molle.*

Pirus auct.: *Pyrus.*

Pisangi wax: derived fr. various *Musa* spp. of Indonesia; c pisangceryl acid.

Pisces: class of True Fishes (Phylum Chordata); customarily, water-living & respiring through gills; ex. perch, salmon.

piscia (It): urine.

pisciacane (It): broom tops (SC diur.).

Piscidia piscipula L. Sarg. (P. erythrina L.; P. communis Blake-Harms) (Leg) (sUS, Mex, WI, esp. Jam, nSA): (Jamaica) dogwood; bitchwood; haabi; mulungu; Jamaica fish fuddle tree; (inner) bk. of rts. & sts. (latter chiefly now); c. jamaicin (isoflavone); saponin; (piscidiasaponin) tannin; resin; rotenone & rotenoids; glycosides; piscidic acid; piscidin (NBP) (c. 2 compds.); acts as uterine depress. with low toxicity; u. anod. in dentalgia; hypn. antispasm. diur. diaph. expect. in cough; for mange in dogs; fish poison; insecticide; fine wd.; bk juice & frt. u. for arrow poison (natives) (2153); (*P. carthagenensis* Jacq. similar) (other plants of Cuba have been called "dogwood," incl. *Lonchocarpus pentaphyllus* and *Randia aculeata.*)

piscio (It): piscia.

Piscipula Loefl. = *Piscidia.*

pisello (It): *Pisum sativum.*

Pisonia aculeata L. (Nyctagin) (trop everywhere, incl. Fla [US], tropAm): cockspur; lf. decoc. u. in rheum. & VD (Mex, Jam) (folk med. Yuc); lvs. & bk. counterirritant; fresh lf. decoc. as wash for scabies (PI) - **P. alba** Span. (P. morindaefolia R. Br.) (Ind, Indones): lettuce tree; milk tree; lvs. warmed with coconut milk appl. to swollen legs; dressing for corns, callosities; lvs. u. diur.; rt. purg.; lvs. eaten in salads - **P. capitata** (S. Wats.) Standley (Mex): u. for various fevers by native Mexicans - **P. longirostris** Teijsm. & Binn. (seAs): charred bk. mixed with FO appl. as salve to fresh wounds - **P. umbellifera** Seem. (PI): bk. decoc. given to newborn babies; u. in gastralgia; rts. u. for soup.

pissabed (Brit): usually dandelion, but also *Houstonia caerulea, Helenium autumnale* (SC diur.).

Pisselaeum Indicum (L): petroleum.

pisse-sang (Fr): *Fumaria officinalis* (bloody urine, SC colors urine blood red after ingestion) - **pissmire** (Eng): dandelion.

pista (Hind): **pistachio: p. nut**: fr. *Pistacia vera.*

Pistacia khinjuk Stock. (P. integerrima Stew.) (Anacardi) (Indo-Pak: Eg, Iran): kakra; kakkar; lf. galls (kaktasinghi) u. in tanning & dyeing (high [20-75%] tannin content); oleoresin (Bombay mastic) u. against cancer; valuable wd. - **P. lentiscus** L. (Med Sea reg., esp. Gr islands, incl. Chios): lentisk; mastiche tree; pistacia (Por); the brittle yellowish concrete oleoresinous exudn. is (Chios) mastic; c. 98% resin, VO, NBP; u. in temporary dental fillings, varnishes (surgical & in technology, as for paper, glass, porcelain), pill excipient, in plasters; stim. in bronch. dis., inflamm.; corrective in aloe laxatives, masticatory, fumitory; condim., adhesive; prepn. m. wine, in pastry; for embalming, incenses, reagents (2516) - **P. mutica** Fisch. & Mey. (P. cabulica Stocks) (eMed reg. Iran, Afgh, Baluch [Pak]): "turpentine tree"; source of mastic-like resin (North African mastic) u. as masticatory (Iran) & as adulter. of true mastic (Bombay m.); oleoresin u. against cancer (Ind) - **P. simaruba** L. = *Bursera s* - **P. terebinthus** L. (Med reg. fr. Canary Is to As-Minor-nAf): terebinth; incisions in bk. yield Chios turpentine (really kind of mastic) (mostly fr. C.) (known to ancients as "terebinthos") c. 10-12% VO (chiefly α-pinene, dipentene), c. 85-90% resin; oleoresin formerly considered specific for uterine cancer; u. in skin dis.; tree bears tan galls (pistacia galls, "Jew pods") with ca. 60% tannin; bk. with 14+% tannin; lvs. with 23% tannin u. as emmen in dys., albuminuria (1696), & as tan; frt. oil against gout (2154) - **P. vera** L. (w&sAs, esp. Syr, Iran; now cult. Med reg., Calif [US]): common pistache; pistachio; small tree with reddish wrinkled frt. bearing green or yellow sds. or kernels, pistachio nuts (nuts u. in ice cream, liver sausages, etc.); these are edible, rather like almonds (hence "green almonds"), source of FO, pistachio oil; u. as delicacy, flavor, in confectionary.

pistacia galls: found on *Pistacia terebinthus* (also *P. khinjuk).*

pistil: the individual carpel of a gynoecium; sometimes the entire g.

Pistia stratiotes L. (monotypic) (Ar) (trop & subtrop, as tropAm, Fla [US]): tropical (or great) duckweed; water lettuce; "dunsul"; free-floating lvs. appl. to ulcers, boils, in skin dis.; in headache; ashes appl. to scalp ringworm; rt. lax. emoll.; was u. in ancient Eg.

Pisum sativum L. (Leg) (orig. fr. Near East; now widely cultd. temp. areas worldwide): (garden) pea(s); petits pois (Fr); common p.; mattar; sds. (sometimes pods) u. nutr. in soups, as veget. (pulse); eaten cooked or uncooked; source of starch; earlier u. in med. (esp. folk med.), sd. meal in plasters for wounds, warts, & inflamed skin to prevent eruptions; int. in dropsy, jaundice; vars. early June (superior flavor) var. *macrocarpahser.*; sweet peas (more wrinkled, inferior); Pakistan pea (Pak, Nepal+); claimed contraceptive by peroral use; now claimed ineffective in reasonable amts. (453); var. *arvense* Poir. (Euras): field (or gray) pea; u. as (green) fodder (small pods & sds.).

pit (Eng fr. Dutch): 1) stone fr. such drupes as peach, plum, cherry; represents endocarp enclosing sd. 2) (histology) tiny pores or thin spots in cell wall of vessels, tracheids, &c. - **bordered p.**: one with marked appearance of a rim, due to overarching of secondary wall over pit cavity - **simple p.**: plain, without border - **vestured p.**: a bordered p. lined with projections from a secondary wall.

pita (Sp, tropAm): fiber or fibrous pl. (broadly u.) - **p. istle** (Mex): **p. ixtle**: fiber fr. various *Agave* spp., as *A. lecheguilla*, *A. endlichiana*, *A. gracilispina*, etc., u. for twine, brushes, &c.

pita bread (< pita: modern Gr): round, flat bread that can be slit open to form a pocket for meat, &c.

pitahaya: *Cereus pentagonus*; *C. undatus* (also **p. oregona**).

pitalla: pitaya: plant (cactus) or frt. (edible); name u. in tropAm for many spp. of various genera of Cact.; ex. *Cereus*, *Echinocereus (E. triglochidiatus)*, *Echinocactus*, *Ferocactus*, &c.

pitanga (Br): *Eugenia uniflora.*

Pitcairnia integrifolia Ker-Gawl. (Bromeli.) (WI): powd. fr. leaf under surface u. for rubbing on VD lesions in male (Trin).

pitch: Pix. (L): 1) dark viscous residuum remaining fr. the dry distn. of tar, wood, &c. 2) certain oleoresins, resins, asphalts, bitumens, &c., usually dark-colored (*cf.* expression, "dark as pitch") 3) (specif.): pine tar (misnomer) - **artificial p.**: mineral p. - **black p.**: residual prod. obt. fr. distn. wood tar of *Pinus sylvestris*, *P. palustris*, & other *P.* spp.; c. rosin, empyreumatic prods. of rosin; u. for adhesive tapes (in Eu; formerly US) - **Burgundy p.**: Pix Burgudica; actually oleoresin exudn. fr. *Picea abies* (chief prodn. Ge, It) - **Canada p.**: resinous exudn. fr. *Tsuga canadensis*; represents an oleoresin with only a very small amt. of VO; u. mild rubef.; in plasters somewhat like Burgundy p. - **coal tar p.**: residue in distn. of coal tar; u. roofing, paving, water-proofing, protective coating, etc. - **p. coke**: charcoal obt. by fractionating coal tar p.; u. fuel, electrodes, etc. - **domestic Burgundy p.**: rosin - **hard p.**: that var. of coal tar p. which is least plastic; u. paving, insulating, etc. - **hardwood p.**: residue in distn. of tars fr. destructive distn. of hardwoods - **hemlock p.**: Canada p. - **Jew's p.**: mineral p. (*v.* under "j") - **lignin p.**: prep. fr. wd.; u. as core binder, i.e., to strengthen sand molds, etc.; has glue-like props - **linseed p.**: residue fr. refining linseed oil - **liquid p.**: wood tar, pine t. - **medium p.**: that sort of coal tar p. which is intermediate between soft & hard; u. paints, roofing, metal protection - **mineral p.**: bitumen; asphaltum - **naval p.**: black p. - **palm p.**: black prod. obt. fr. distn. of p. oil - **petroleum p.**: tarry residue remaining after distn. of petroleum; u. road-making, roofing, as artificial asphalt, etc. - **pine (tar) p.: prime p.**: Pix Pini (L): residue remaining after steam distn. of pine tar - **soft p.**: that fraction of coal tar p. which is most plastic, u. protective, insulating, electrodes, etc. - **stearin(e) p.**: brownish to black residue fr. distn. veg. & animal FO, soap mfr., etc.; u. in floor coverings, insulators, paints, varnishes, etc. - **Trinidad p.**: asphalt fr. Trinidad (WI); mostly u. in road-making - **vegetable p.**: residue fr. distn. of vegetable materials; u. mfr. varnishes, etc. - **wood tar p.**: pine p. - **wool p.**: residue fr. distn. of wool grease.

pitcher plant: *Sarracenia* spp., as *S. purpurea* - **trumpet p. p.**: *S. flava* - **pitch-weed**: *Grindelia squarrosa.*

piteira (Por): *Agave americana.*

pith: medulla: central portion of st., consisting of soft (loose spongy) tissues, constituting a residual part of the ground meristem (storage parenchyma) - **ague p.**: fr. Sassafras st. - **p. helmet**: a light hat u. in sunny trop. areas (ex. Ind) to protect the head fr. sunburn; made of pith; mfd. in Calcutta, &c.; now often provided with waterproof cover.

Pithecellobium (*Pithecol[l]obium*) - **P. albicans** Benth. (Leg) (sMex, Belize, Guat): huisache; bk. with c. 18% tannin u. as astr. in diarrh. - **P. arboreum** (L) Urban (sMex, Belize, Guat, WI, CA): frijolillo (Veracruz; Hond); u. med. (Hond); excellent wd. - **P. auaremotemo** Mart. (Br): source of a gum (angico) resembling Acacia (1371) - **P. bigeminum** Mart. (eAs): kachlora; sd. u. in diabetes (Ind); lf. decoc. u. to stim. hair growth; pl. u. as fish poison - **P. brevifolium** Benth. (P. pallens Standl.) (swTex [US], Mex): guaje; guajolla; tenaza; huajillo (Tex); altho shrub is spiny, lvs. are foraged; wd. u. in various ways - **P. clypearia** Benth. (seAs): lvs. u. as poultice in cough, chicken pox, sore legs; rt. & bk. in itch; bk. scraped & appl. to snakebite wound (also lf.) (PI) - **P. dulce** (Roxb.) Benth. (Mex, CA, Co; natd. Fla [US], seAs): Manila tamarinds; guamuchil; white frt. pulp around sds. edible; also astr., u. in hemostat. in TB (Mex); bk. astr. in dys., as dye; gum (trunk exudn.) serves as good mucilage; lvs. for VD lesions (cataplasm); intest. indiges.; sd. juice insufflated into nose; pd. sd. u. for internal ulcers said esp. efficacious mixed with rue; bk. u. as fish poison (PI) - **P. flexicaule** (Benth.) Coult. (swTex [US], Mex): Texas ebony; ebano; maguacata; shrub or small tree; green pods cooked & eaten; sds. roasted & eaten as pistachios (Mex); sd. coats u. as coffee substit.; wd. useful - **P. jiringa** Prain (P. lobatum Benth.) (Indones): lvs. u. to poultice skin dis.; scorched lf. powd. sprinkled on wounds, esp. of circumcision; sds. eaten but somewhat toxic (c. djenkolic acid wh. can plug kidney tubules); pods u. to make purple dye - **P. keyense** Britt. (P. guadalupense [Desv.] Chapm.) (Fla Keys [US]; WI, Yuc): ram's horn (Bah); lvs. chewed or decoc. u. to stop menstruation in pregnancy (Bah) - **P. mindanaense** Merr. (PI, endem.): rt. decoc. u. for cough - **P. parvifolium** (Sw.) Benth. (WI, nSA): frt. c. tannin; frt. pressed —> algarovilla, orange-yellow dyestuff (copal) - **P. rufescens** Pittier (Pan, Co): jarino; coralillo; ina pisu; bk. decoc. in bath for "female trouble" (Cuna Indians, Pan) - **P. saman** Benth. (Samanea saman Merr.) (tropSA, WI, CA): rain tree (SC ejection of juice by cicadas on branches of tree); kala kokko; saman tree (SA); kindly tree (Virgin Is); monkey pod; cult. as large shade tree (-60 m) (Ind, Indones); bk. u. as astr. antisep.; sd. c. pithecolobine & other alks., chewed for sore throat (Virgin Is); fleshy pods serve for cattle fodder - **P. tortum** Mart. (Mex, WI, CA, SA): guayacan (Hond): u. med. (Hond) - **P. trapezifolium** Benth. (P. jupunba [Willd.] Urban) (Lesser Antilles; nSA): bois cicerou; bk. decoc. drunk for dys.; pounded bk. pieces u. as substit. fish poison; leaflets rubbed in water —> lather u. as soap; wd. u. for boards (Domin, Caribs) (2328) - **P. unguiscati** Mart. (Fla, sMex, WI, nSA): cat's claw; black bead; rolon (PR); bk. decoc. astr. u. in diarrh., as ton. diur.; in fevers & kidney dis. (black sds. resemble kidney; u. per doctrine of signatures); frts. edible.

Pithecoctenium echinatum Schum. (P. cruciferum Gent.) (Bignoni) (sMex to Pe): monkey comb; lf. decoc. u. in skin dis. locally; fl. inf. antipyr.; rough frts. u. as back scratchers & pin cushions.

Pithecolobium auct.: sphalm., *Pithecellobium* Mart. (*qv.*).

pitocin: alpha-hypophamine; an aq. soln. of the oxytocic hormone of the post. pituitary gland, which is u. in obstet. veter. practice for uterine inertia (in difficult labor, placental retention).

Pitraea cuneato-ovata (Cav.) Caro (monotypic) (Verben.) (SA): u. med.; tubercles edible.

pitressin: vasopressin: beta-hypophamine; an aq. soln. of the blood-pressure raising princ. (hormone) & of the diuretic-antidiuretic principles of the post. pituitary gland; u. in bladder atony, adynamic ileus (intest. obstruction due to paralysis), as antidiur., also in veter. med.

Pittosporum arborescens Rich. (Pittospor.) (Tonga, Fiji): masikona (Tonga); frt. u. as fish poison (Tonga) (86); lvs. in water for stomachache (Fiji) - **P. brackenridgei** Gray (Oceania): lf. crushed in water & drunk for cough (Fiji) - **P. coriaceum** Ait. (Br-much cult. as ornam.): sds. c. 3% FO, known in comm. as malhado oil - **P. ferrugineum** Ait. (seAs): lvs. & rts. (poultice) to treat mal. (Mal); lvs. & frt. u. as fish poison (Indones) - **P. floribundum** W. & A. (Ind): bk. u. as bitter, arom., in bronchial dis. as expect. antipyr., alexiteric; VO as stim., ton., locally in rheum.; int. in certain skin dis., syph., chron. rheum. leprosy - **P. nilghirense** W. & A. (Ind): pl. c. spermaticidal saponins; u. as spermaticide in fertility control - **P. pentandrum** Merr. (PI): powd. bk. antipyr. in bronchitis - **P. phillyraeoides** DC (Austral deserts): narrow-leaved pittosporum; poison berry; butter bush; sds. bitter but not toxic; pounded into flour (aborigines) - **P. resiniferum** Hemsl. (PI): frt. u. as panacea (by curanderos); frt. inf. to treat abdom. pain; oleoresin in skin dis. & as cicatrizant - **P. tobira** Ait. (Chin, Jap): Japanese pittosporum; bitter bk. u. to treat dys. & rheum.; decoc. to brush teeth (prevent decay); very popular ornam. shrub - **P. undulatum** Vent. (Austral, sAf): Victorian box; mock orange; pl. c. saponins; VO; u. as soap substit.; tree but usually pruned to a shrub (ornam.) - **P. viridiflorum** Sims. (sAf): Cape pittosporum; stinkbas (Afrikaans); bk. decoc. or inf., u. as emet., antipyr. sed.; roasted bk. for dys.; added to Kaffir beer in place of hops; cult. ornam.

pituicytes: characteristic cells of the post. pituitary gland.

Pituitarium Anterius (L.): anterior pituitary lobe; APL; an important gland of internal secretion; constitutes one of the 3 anatomical divisions of the pituitary gland, the anterior lobe (AP) (consists chiefly of the glandular part); the AP has a very complex & diverse function; secretes 6 or more hormones regulating body growth, & controlling other endocrine organs, viz., thyroid growth & function, gonads, suprarenal cortex, parathyroids, & lactation; the AP is regarded as the chief of all the internal secretory glands, but the dessicated or dr. prod. is considered of little value & is now u. very little - **P. Posterius** (L.): PPL: posterior pituitary; the posterior lobe is generally thought to incl. the pars neuralis & p. intermedia; endocrine secretions of the PPL affect smooth muscle contractility, renal function & blood pressure; u. to prod. contraction of uter. wall (oxytocic effect) & blood vessel walls (pressor); the desicc. prod. is now going into disuse, although there is perhaps some potency per ora - **P. Totum** (L.): (whole) pitutary gland; u. mostly in desicc. form; any effects are to be attributed to the poster. portions present.

pituitary body: p. substance: whole pituitary gland.

Pituitrin: obstetrical product, c. both oxytocin & vasopressin.

Pituitrin-S: surgical prod.

pituri (native Austral name): stim. drug fr. *Duboisia hopwoodii* - **piturine**: old name for alk. fr. prec.; has nicotine-like odor; may be nicotine or nornicotine (2458).

Pityopsis graminifolia (Mx.) Nutt. (Chrysopsis g. Ell.) (Comp.) (seUS, Bah): golden aster; fever/blue/silk/scurvy/silver grass; lf. decoc., u. in kidney & bladder dis., diur. in edema (NCar, SCar [US]) (2451), in poultices for sprains (Am Indians).

Pityothamnus Small = *Asimina.*

Pityrogramma calomelanos Link (Adiant. - Pteridoph.) (tropAm): white back fern (Trin); frond inf. u. in cough, flu, amenorrhea.

Pix (1.): 1) tar, pitch 2) pine pitch (P. Nigra) - **P. Alba**: P. Burgundica - **P. Arida**: P. Navalis - **P. Betulina** (Ge Phar.): **P. Betulinum**: rect. empyreumatic birch tar oil fr. *Betula pendula* - **P. Burgundica**: Burgundy pitch, oleoresinous exudn. fr. *Picea abies* - **P. Cadi**: P. Juniperi - **P. Canadensis**: hemlock pitch; exudn. fr. *Tsuga canadensis* - **P. Carbonis (Praeparata)**: (prepared) coal tar - **P. Fagi**: beech wood tar - **P. Juniperi (oxycedri)**: juniper tar; cade oil - **P. Liquida**: 1) P. Pini 2) P. Betulina (**P. L. Oxycedri**: P. Juniperi - **P. L. Pini**: P. Pini) - **P. Lithanthracis (Cruda)**: (crude) coal tar - **P. Navalis: P. Nigra**: black pitch; the residue left behind after pine tar has been dist. to prod. pine tar oil, etc. - **P. Pini**: pine tar; wood tar; obt. by destructive dist. of wd. fr. *Pinus* spp. of either hemisphere, esp. fr. *P. palustris, P. taeda, P. rigida, P. sylvestris*; sometimes fr. other conifers - **P. Solida**: P. Navalis.

pizza (It) (pron. peet-sa): pie-like pastry with spiced (esp. with orégano) mixt. of tomato paste, meat, & cheese (now very popular in US); also egg-shaped cheese.

Pizzle: penis, esp. of bull, formerly u. as whip - **p. grease** (Ala [US] slang): popular ointment made by boiling hog penis in lard.

placebo (fr. L., I shall please): usually inert med. given to satisfy (or deceive) patient (ex. bread pill).

placenta: afterbirth; organ developed in mammals during pregnancy wh. serves to provide substances incl. hormones to fetus (provides nutrition, respiration & excretion); extruded fr. uterus after fetus; u. in exptl. pharmacology.

Placenta (L.): cake - **P. Amygdalarum**: pressed cake of almond meal - **P. Seminis Lini**: linseed cake; residue fr. FO expression.

placental extract: human immune globulin; u. to prevent or modify the course of an attack of measles (SC obt. fr. expelled placenta or afterbirth).

placentation: mode of attachment of ovules to ovary wall in flower - **axile p.**: att. to central axis of ovary.

plague bacilli: *Pasteurella pestis* (Lehmann & Neumann) Holland: u. in prep. **p. vaccine**, highly effective in conferring active immunity against bubonic plague.

Planchonella firma (Miq.) Dubard (Sapot.) (Malaysia, Solomons): pinago (Sum); soft wd. chewed with betel for sprue; wd. u. to make furniture - **P. obovata** (R. Br.) Pierre (Ind, Pak, seAs, PI, Austral, Oceania): maha (Annam); lf. decoc. u. for stomachache & chest pains; bk. heated over fire chewed for sprue; crushed lvs. u. as poultice for lumbago; wd. u. in carving, cabinet making.

Planchonia papuana R. Knuth (Lecythid./Barringtoni) (Oceania): bark macerate & sap drunk for headache (Solomon Is).

plane tree (, American): *Platanus* spp., esp. *P. occidentalis* - **Oriental p. t.**: *P. orientalis* (chinar) - **Scotch p.**: *Acer pseudoplatanus.*

planer tree: Planera aquatica J. F. Gmel. (Ulm.) (seUS): water elm; a small tree.

planimetry: measurement of any plane surface, ex. leaf.
plank: thick, heavy wooden board.
plankton: floating life in the water, esp. of the oceans or large lakes.
plant: a member of the plant kingdom, Plantae; a living organism wh. generally mfrs. its own food fr. rel. simple compds. (see under individual [specific name: e.g., pie plant is under "pie") - **p. hairs**: trichomes - **p. red**: carthamin - **planta** (L.) (Sp, Por): plant - **p. de huevo** (NM [US], Sp): egg p. - **p. sensitiva** (Br): *Mimosa pudica*.
Plantae: the pl. kingdom, made up of four divisions (phyla): Thallophyta, Bryophyta, Pteridophyta, & Spermatophyta.
Planta Genista (L): yellow broom; hence Plantagenet, name assumed by royal family in Eng as symbol of humility (1300s- 1500s).
Plantaginaceae: Plantain Fam.; biennial & perenn. hbs. & under- shrubs; widely distr. in OW & NW; lvs. mostly basal, sometimes also cauline; small fls.; frt. capsule or nutlet.
Plantaginis Semen (L.): plantago seed, psyllium; sd. fr. *P. arenaria*, or *P. ovata* (NF).
Plantago L. (Plantagin.) (cosmop., mostly temp. zone): plantains; some 250 spp. of herbs; lvs. of some spp. (*P. lanceolata, P. major, P. maritima*) sometimes u. as salads (Fr+); mostly acaulescent herbs; at times short-stemmed; one pl. is a subshrub; some 42 spp. indigenous to or natd. in US (391); p. sd. c. psyllium hydrophilic colloids with bulk-forming lax. & dem. properties; claimed to lower elevated serum cholesterol and lipoprotein levels in blood serum (Metamucil; Citrucil) (Synonyms: *Coronopus* Ludw. non Gaertn.; *Coronopodium*; *Asterageum*; *Psillium* (in part)) (2150) - **P. altissima** L. (P. lanceolata var. a.) (scEu; Balkan Penin): rib grass; u. similarly to *P. lanceolata* - **P. amplexicaulis** Cav. (tropAs, esp. Ind): hb. astr. u. in pulmon. dis., fevers; ophthalmia - **P. arenaria** Wald. & Kit. (P. indica L., P. psyllium L., P. ramosa Aschers.; P. afra L.) (Euras, Med reg., Siberia): fleawort (plantain); sandwort; sds. (Spanish [or French]) psyllium (seed); whorled plantago; fleaseed; black psyllium (seed); sds. mucilage- rich (in sd. coats); u. lax. dem. emoll.; effects due to swelling of mucilage in water; dem. in enteritis, bronchitis; in cosmetics (setting lotions); as filler & glossy agt. in paper mfr.; in textile indy. (2169) (mostly cult. in Sp, sFr, US) - **P. aristata** Mx. (c&eUS): bracted plantain; Po' Jo'; Po' John; buckhorn; close to *P. patagonica* Jacq.; pl. supposed to be an indicator of poor soil; sds. rich in mucilage (454) - **P. cordata** Lam. (e&cNA): (water) plantain; heart-leaved p.; ribwort; lvs. u. astr. diur. anod. - **P. coronopus** L. (Euras, nAf, natd. NA, Austral): buckshorn plantain; crowfoot p.; buckthorn p.; xerophyte; ann. pl. u. diur. for kidney stones (folk med.) (claimed to dissolve renal calculi [Litho-spaston]); hb. u. as poultice; lvs. u. in salads - **P. cynops** L. = *P. arenaria* - **P. depressa** Willd. (Manchur, Kor): sds. u. dem. antipyr. diaph. diur. in diarrh. dys., rheum.; in GC - **P. indica** L. = *P. arenaria* - **P. lanceolata** L. (P. decumbens Bernh.) (Euras; natd. sCana, US, SA, Austral): ribwort (plantain); English p.; rib grass; leadwort; lvs./hb. (Herba Plantaginis (lanceolatae)) (L) & rts. u. hemostat. for coughs, pulmon. dis., asthma; vulner. for burns; to dress wounds; c. 7% tannin, astr.; sds. (German Psyllium Seed) mucilaginous, u. mild lax., psyllium sd. adult. (but with less mucil.); linseed substit.; folk remedy for *Rhus* poisoning - **P. loeflingii** L. (status?) (Med reg., esp. Sp): said u. in snakebite; sds. u. as lax.; mucilage in Mucilose™ (Winthrop) claimed fr. this sp. - **P. major** L. (P. erosa Wall.) (Euras, natd. NA, worldwide): (common) (greater) plantain; "lahuria" (Ind); broad leaf; Englishman's (or white man's) foot; cart track plant; way bread; lvs. (Herba Plantaginis [latifoliae]) c. glycoside (glucoraphenin), NBP alks. (plantagonine, indicaine +); u. vulner. (ungt.); diur. astr. intest. disinf.; calculi; resp. inflammn., facial neuralgia; exter. as poultice for bruises, dermal inflammn. ("Your plantain-leaf is excellent for . . . your broken skin" *[Romeo & Juliet])* (390); burns &c.; sds. mucil. diur. lax.; in dys.; substit. for Ispaghula; pl. ground up to paste, appl. to inflammn., impetigo; also to warts (1697); to promote growth & luster of hair (UP, Ind); pl. decoc. u. to wash sore or inflamed eyes (Domin, Caribs); food pl. (Chin); pot herb (US); fodder for cattle, pigs; lvs. chewed with betel & lime (pan) (Bhutan yak breeders); universal (lawn) weed - **P. maritima** L. (Eu, Cana, Alas [US]): goose tongue; sea(side) plantain; a halophyte; c. aucubin; hb. u. lithontriptic for urinary stones (folk med.); sds. mucilag.; supposed to effect disgust for tobacco (Homeopath.) - **P. media** L. (Euras, natd. eNA): hoary plantain; hb. u. like that of *P. major* - **P. ovata** Forssk. (P. isphagula Fleming; P. decumbens Forssk.) (Med to Ind, Pak, Iran): sds. (spogel seed); blond psyllium (seed); Indian plantago seed; ispag(h)ul(a) (sd.); ispagol; isabgol (Ind); psylla seed; the sp. now chiefly u. in US; husks are employed as lax. dem. in dys. diarrh. emoll.; mucilage in cosmetics; for obstipation, internal hemorrhage; inflammn. (GI, GU); ulcers; mucil. in fabric stiffening, dyeing, & finishing calicoes, &c.; in snakebite; lvs. sometimes u. as

salad (Fr); bruised lvs. u. for sprains, inflammn., &c. (folk med.); cult. Ind, Pak. (chiefly), sEu, nAf (455) - **P. paralias** Dcne. (Arg): llanten; u. as antiphlogistic; lf. hemostatic (exter.); sd. emoll. (exter.) - **P. patagonica** Jacq. (wNA): sagebrush plantain; pl. u. in bloody diarrh. (Zuñi Indians, NM [US]) - **P. psyllium** L. = *P. arenaria* - **P. rugelii** Dcne. (seUS): plantain; fresh lvs. u. for burn dressings & for other inflammn.; in poultices for swellings - **P. virginica** L. (US-Mex): pale-seed (or hoary) plantain; dwarf p.; u. symbolically by the Kiowa Indians; garlands tied around heads of old men as emblem of health; entire pl. u. as diur., vulner. alexiter(?).

plantago seed (NF): Plantaginis Semen (L) - **Indian p. (seed)**: *P. ovata* - **whorled p.**: *Plantago arenaria.*

plantain: 1) *Plantago* spp. (p. seed) 2) *Musa paradisiaca*, a kind of banana - **blond p.**: *Plantago ovata* - **broadleaved p.**: *P. major* - **broad leaf p.**: *P. ovata* - **branching p.**: *P. arenaria* - **buck's horn p.**: *P. coronopus*; English p. - **common p.**: **p. commun** (Fr): **p. des oiseaux** (Fr): **dooryard p.**: *P. major* - **dwarf p.**: *P. virginica* - **English p.**: *P. lanceolata* - **p. family**: Plantaginaceae - **flaxseed p.: flea-seed p.: fleawort p.**: *P. arenaria* - **grand p.** (Fr): **greater p.**: common p. - **heart-leaved p.**: *P. cordata* - **hoary p.**: *P. media*; *P. virginica* - **Indian p.**: 1) blond p. 2) *Cacalia* spp. - **large p.: p. leaves**: common p. - **narrow-leaved p.**: English p. - **rabbit p.**: *Antennaria* spp. - **rattlesnake p.**: *Goodyera pubescens* - **ribwort p.**: English p. - **rippleseed p.: roundleaved p.**: common p. - **sand p.**: *P. arenaria* - **p. seed(s)**: sds of various *P.* spp. c. protective colloid mucilages; for u. note *P. ovata* - **small p.**: various *P.* spp. - **snake p.**: English p. - **spotted p.**: rattlesnake p. - **water p.**: 1) *Alisma plantago-aquatica* 2) rattlesnake p. 3) *Plantago cordata* - **white p.**: *P. virginica.*

plante fraxinelle (Fr): dittany herb - **p. vivace** (Fr): perennial; hardy plant.

plantemal: combn. of pl. & animal cells made by fusion.

plant growth substance: org. substance which alters rate of growth; may be natural (ex. abscisic acid) or synthetic; phytohormone.

plant of peace (US Negro): *Arisaema triphylla.*

plantule (Fr-f.): plantlet; seedling; sprout; small plant.

plasma: 1) fluid part of the blood; after formation of the clot, the residue is called the serum; blood minus b. cells = **plasma**; plasma minus clot = serum 2) starch glycerite (Glyceritum Amyli) - **p. extender**: p. volume expander - **normal human p. (citrated)** (Plasma Humanum Normale (Citratum), L.): sterile p. obt. by separating blood cells by centrifugation or sedimentation; formerly obt. by pooling equal volumes of plasma from 8 or more healthy humans; now banned in US because of risk of transmitting hepatitis or AIDS; no risk in single donor use; X-ray irradiation to destroy microbes ineffective; fractions: portions obt. by fractional pptn. with variant pH; incl. gammaglobulins, prothrombin, fibrinogen, lipoproteins, blood coagulation factors V (to treat parahemophilia), VIII, IX; human antihemophilic globulin concentrates; &c.; forms: *liquid*: original form; may be stored at 15-30°; but limited life; *fresh frozen*: obt. by quickly freezing citrated normal human p. no longer than 72 hrs. after bleeding (life ca. 3 mos.); *dried p.*: made by freezing p. then desicc. at high vacuum at temp. below freezing (cryodesiccation) (5-8 yrs. life) - **p. protein fraction**: soln. of albumin & globulins (α, ß) u. like albumin to treat shock; to provide IV nutrn. - **p. substitutes**: diverse substances have been u., incl. dextrans; polyvinyl pyrrolidone; gelatin soln.; acacia soln.; Na arabinate soln.; pectin soln.; bovine soln.; &c. - **p. trypsin**: plasmin - **p. volume expanders**: blood volume replenishers; u. to restore blood volume in body, esp. when blood unavailable; may use physiological soln.; glucose soln.; various blood substitutes.

plasmapheresis: removal of blood, sepn. of plasma (or components), & reinjection of balance (RBC, &c.) into donor.

plasmids: 1) (generic term): intracellular inclusions; incl. bioblasts, plastids, viruses, &c. 2) paragenes; extranuclear extrachromosomal replicating units contg. nucleoprotein (c. DNA) involved in metabolic process in microorganisms (2139) (**resistance p.**: believed to cause resistance to antibodies of bacteria, &c.)

plasmin: fibrinolysin; proteolytic enzyme of plasma of blood; important in breaking down blood clots (fibrin skeleton).

Plasmochin™: Pamaquine Naphthoate.

plasminogen: profibrinolysin: precursor of fibrinolysin (*qv.*); occ. in blood plasma; activated by streptokinase, &c.; u. in wound débridement - **anisoyl plasminogen streptokinase activator complex**: APSAC: u. in single dose IV therapy of coronary thrombosis; thrombolytic & fibrinolytic in acute myocardial infarction - **tissue type p. activator** (TPA): forms plasmin; u. as thrombolytic in coronary thrombosis, &c.; adm. IV over several hrs. period.

plasmolysis: contraction of cytoplasm of living cell; induced by exosmosis, &c.

plaster, cantharides: vesicant p.: warming p.: Emplastrum Cantharidis.

plastids: any specialized organ of the cell besides nucleus & centrosome; ex. chloroplast(id), chromoplast(id).

platane (Fr): **platano** (Sp, Por): 1) plantain, *Musa paradisiaca* 2) *Platanus* spp. - **falsa p.**: 1) *Acer platanoides* 2) *Heliconia bihai.*

Platanthera bifolia (L) L. C. Rich. (Habenaria b. R. Br.) (Orchid.) (Euras, nAf): butterfly orchis; one of the many sources of salep root; u. mucilag. agt., aphrod. (doctr. sign.) (resemblance of rt. to testis); nutr. gel, for convalescents & children - **P. orbiculata** Lindl. (Habenaria o. Torr.) (nNA): round-leaved orchis; fringed o.; lvs. u. to treat blisters - **P. psycodes** Lindl. (Habenaria p. Spreng.) (neUS; eCana): small purple-fringed orchid; "wild hyacinth"; cult. as ornam. hb.

Platanus hybrida Brot. (P. acerifolia Willd.) (Platan.) (Euras): London (or bastard) plane (tree); platane (Fr); quite similar in const. & uses to *P. orientalis*; sap bitter; lf. c. HCN glycoside; allantoin; bk. u. in scurvy & rheum. as ton. nutr., lf. for conjunctivitis; rt. in hemorrhage; frt. for dys.; often cult. in parks, along streets, &c. (456) - **P. occidentalis** L. (e&cNA): (American) plane (tree); sycamore; buttonwood; button tree; frts. ("buttons") persistent through winter; bk. scales off in large pieces; lvs. formerly u. in eye lotions; antipyr. ton. nutr.; c. cyanogenic glycosides; bk. antirheum. antiscorbut.; wd. useful; ornam. tree (in parks, &c.) (*Plantanus* < platus [L] broad, wide leaf) - **P. orientalis** L. (seEu, wAs, Indo-Pak, nIran): oriental plane (tree); eastern p. (t.); sycamore; chinar (Punjabi, Kashm); bk. c. & u. similar to those of *P. hybrida*; bk. antipyr. (folk med.); in diarrh. dys.; frt. (capsules) u. int. for ophthalmia (Kashmir); tree grows to enormous size & great age.

platelet: blood p.: thrombocyte; spheroid to ovoid gray-colored body found in blood; measures ca. 1-2.5 mikra in diam.; essential part of blood clotting mechanism - **p. concentrate**: prod. by sepn. fr. blood plasma after centrifugal removal of red & white blood corpuscles; u. prophylactically prior to surgical operation to reduce cerebral hemorrhage in thrombocytopenia.

Platonia esculenta Rickett (P. insignis Mart.) (Gutt.) (Br, Co, Guian): pacouri; large fleshly frt.; popular food of SA Indians; yellow resin u. much like that of *Moronobea coccinea*; sds. yield Bacury kernel oil (non-drying FO); wd. u. in construction, flooring (parcouri wd.).

Platostoma africanum Beauv. (Lab) (wAf, Ind): u. for fevers, chills, rheum. (Nigeria); lvs. & sds. in children's fever & cough (Ghana); to prevent miscarriage (Ivory Coast).

Platycarya strobilacea Siebold & Zuccarin. (Jugland) (monotypic) (Chin, Jap, seAs): lvs. u. as pesticide; st. bk. u. for fibers; bk. & frt. u. for black dyes in textile indy.; cult. ornam.

Platycerium coronarium Desv. (Polypodi. - Pteridoph.) (seAs): elks horn fern; ashes of pl. rubbed over body for splenic enlargement (Mal).

Platycodon grandiflorus DC (1) (Campanul.) (eAs): Chinese (or Japanese) bellflower; balloon flower; rt. c. platycodin (saponin), polygalic acid, platycodigenin; u. as expect. in asthma, astr. carm.; in cholera, dys.; sed. vermifuge; replacement for ginseng; u. in ancient Chinese med. (Pen t'sao).

Platyhelminthes: division (phylum) of flat worms.

Platymiscium polystachyum Benth. (Leg.) (CA, nSA): quira; macawood; trunk furnishes good wd.; u. med. (Hond).

pleasure smoke (US slang): opium smoking.

Plectaneia elastica Jumelle & Pernes (Apocyn.) (Mad): this & other *P.* spp. u. as source of caoutchouc.

Plectocomia elongata Mart. & Bl. (Palm) (seAs): raw sap drunk for fever; appl. as lotion to heal wounds & burns.

Plectranthus elegans Britten (Lab) (Equat, eAf): lf. c. VO (Kaffir potatoes); u. for sore throat - **P. floribundus** N. E. Br (sAf): rhiz. u. as substit. for sweet potato; u. in native folk med. - **P. forsteri** Benth. (Oceania): laca (Fiji); sap mixed with sea water u. by natives for sores (New Hebrides) - **P. fruticosus** L'Hérit. (sAf): pl. u. in fever & cramps (Homeopath. med.); lvs. u. as insecticide; adulter. of patchouli lvs.; pl. cult. as room plant - **P. hirtus** Benth. (sAf): rt. decoc. or lf. inf. u. in cough & chest complaints; lf. rubbed into skin for scabies - **P. incanus** Link (sAf-Ind): dr. lvs. & flg. tops c. VO, resin, tannin; VO has antimicrobial activity; incr. depth of respiration, promotes bronchial dilation; spasmolytic - **P. japonicus** Koidz. (P. rugosus Mak.) (Far East): u. in stomachache; entire pl. decoc. u. as ton., to stop diarrh., "cure" convulsions; c. plectranthin, extremely bitter subst. - **P. laxiflorus** Benth. (sAf): pd. leaf in enema for flu, fever, abdomin. upset - **P. macranthus** Hook. f. (eAs): pl. crushed & appl. as plaster to sores; soaked in wine & u. for sprains (Chin) - **P. nummularius** Briq. (P. australis R. Br) (Austral, Oceania): Swedish ivy; ornam. vine u. in hanging baskets - **P. parviflorus** Willd. (P. graveolens Willd.) (Oceania): with *P. patchouli* Clarke, a source of kind of patchouli leaves.

Plectronia auctt. = *Canthium*; *Acanthopanax*; *Olinia*.

Pleioblastus amarus Keng f. (Gram) (Chin): bamboo lvs. u. for thirst, insomnia, painful eyes, laryngitis, alc. poisoning.

Pleiocarpa micrantha Stapf (P. tubicina Stapf) (Apocyn.) (Ghana, Nigeria): grd. rts. mixed with wine & u. as lax.; in anorexia, anthelm., for heart & stomach dis. - **P. mutica** Benth. (Congo): c. indole alk., rt. ext. u. for reducing blood pressure.

Pleioceras barteri Baill. (Apocyn.) (Ghana, Nigeria): frts. u. as emmen. & abort.

Pleiogynium solandri (Benth.) Engl. (Anacardi.) (Oceania): tree - 30 m.; bk. inf. drunk for thrush & stomachache (Fiji); wd. u. to make canoes.

Pleione pogonioides Rolfe (Orchid.) (Chin): pseudobulbs u. for asthma, TB.

Pleiospermum littorale Tanaka = *Limnocitrus littoralis*.

Pleomele Salisb. = *Dracaena*.

pleomorphism: structural changes in microorganism induced by environmental conditions such as light.

Pleonotoma diversifolium Bur. & Schum. = *Cydista diversifolia* - **P. melioides** (S. Moore) A. Gentry (Bignoni) (tropAm) u. in mal.

Plerandra solomonensis Philipson (Arali) (Solomon Is): inner sap drunk with water for constipation (—> *Schefflera*).

plerome: central zone of meristem of shoot apex wh. gives rise to the central cylinder (stele) of the pl. axis.

pleurisy root: *Asclepias tuberosa*.

Pleurococcus = *Protococcus*.

pleuromutilin: Drosophilin B: diterpene antibiotic fr. *Pleurotus mutilus* (Fr) Sacc., *P. passeckerianus* Pilát, *Psathyrella* spp., *Drosophila subatrata* (Batsch) (ex Fr) Quél. (Basidiomyc.-Fungi) (do not confuse with fruit fly, same generic name) (*cf. Psathyrella*) (Coprin.): active against gm.+ bacteria.

Pleuronectidae: Fam. of bottom-feeding flat-fishes to which belong the halibut, sole, flounder, & plaice.

pleurotin: antibiotic fr. *Pleurotus griseus (sic)*, &c.; with antitumor activity.

Pleurotus ostreatus (Jacq. Fr) Kummer (Agaric./Tricholomat. - Basidiomyc. - Fungi) (cosmopol): oyster cap; o. mushroom; damages wood; cult. on beech trunks & sold in Eu markets; popular food in Chin & exp. there fr. Siberia; has a hypocholesterolemic effect (2756).

Plexaura homomalla (Plexauridae - Gorgonacea - Anthozoa - Cnidaria - Coelenterata) (Fla - coral reefs of coast): colonial arborescent animals symbiotic with Algae (Zooxanthellales); this sea whip (gorgonian) is a source of prostaglandins (intermediate); collections fr. Grand Cayman, Bahamas.

Plocamium violaceum Farl. (Plocami. - Rhodoph. - Algae) (Pac coast of NA): active constit. of thallus is plocaneme B (cyclic monoterpene); shows marked insecticidal activity (for mosquitoes).

Pluchea camphorata (L) DC (Comp.) (se&cUS): camphor weed; stinkweed; camphor plant; hb. c. VO; u. similarly to other spp. - **P. carolinensis** G. Don (sMex, WI, CA, Ve): bitter tobacco; (big) saab (Bah); cough bush; considerably u. in rheum. & fevers; lf. & fl. inf. diaphor. & expect. in cough; mouthwash in toothache - **P. dioscoridis** DC (trop eAf, wAs): rt. decoc. u. for colds (Af); lf. & rt. stim., arom. "comforting medicine" (wAs) - **P. foetida** (L) DC (wFla to Tex, n to NJ [US], Mex, swamps & marshes): stinking fleabane; tea u. for fever (Choctaws) - **P. indica** Less. (sAs, Ind, Chin, Indones, Austral): "loentas"; lvs. eaten raw or cooked, diaphor. arom. vulner., in reducing teas; in lumbago; lvs. & rt. u. as antipyr. astr., in diarrh. (Ind) (457) - **P. lanceolata** Oliv. & Hiern (Indo-Pak, sInd, Bengal - Gangetic Plain): lvs. u. as lax., substit. for Senna; analg. sed. - **P. leubnitziae** N. E. Br. (sAf): pl. u. as remedy in bone dis., peptic ulcer, VD (folk med.); milk of cattle feeding on pl. has disagreeable flavor - **P. odorata** Cass. (US, Mex, WI, tropSA): salt-marsh fleabane; wild tobacco; "salvia (colorado)"; spur bush; lf. inf. in alc. u. in neuralgia, rheum.; hb. decoc. for bronchitis; lvs. analg. in lumbago; antipyr.; in menstr. & uter. dis. (Ve Phar.); u. for colds (St. Lucia) - **P. pinnatifida** Hook. f. (Pak): pl. rubbed on inflamed & wounded areas; sts. u. in anorexia, debility - **P. pteropoda** Hemsl. (Indochin): rts. u. as antipyr. - **P. rosea** Godfrey (seUS; Mex, WI, CA): may-grow; pl. decoc. u. as vermifuge (Bah); with other pls. after childbirth - **P. sagittalis** (Lam.) Cabr. (SA): lucera (Arg); VO c. pinene, d-camphor, borneol; inf. of lvs. & lower st. u. as carm. (Arg); & in prepn. of an apertitif - **P. sericea** Cav. = *Tessaria s.*

Plukenetia conophora Muell.-Arg. (Euphorbi) (trop wAf): sds. (owusa) source of N'gart or conophora oil, FO with 65-67% linolenic acid; u. in foods, as paint base; substit. for linseed oil.

plum: *Prunus domestica* & other *P.* spp. of Section *Prunophora* (Ros.), etc. - **American wild p.**: *Prunus americana* - **p. bark**: fr. *Cordia myxa* - **Bassett p.: beach p.**: *P. maritima* - **black p.** *Syzygium cumini.* - **Canada p.**: *P. americana, P. nigra* - **cherry p.**: *P. divaricata* - **Chicksaw p.**: wild p. - **Chinese p.**: 1) *P. japonica* 2) *Diospyros kaki* -

Damson p.: *P. domestica* or *P. insititia* var. - **date p.**: *Diospyros ebenaster* - **deer p.**: *Chrysobalanus* spp. - **greengage p.**: *P. insititia* var. *italica* (GB) - **hog p.**: 1) *P. umbellata* 2) *Spondias mombin* - **Japan**(ese) **p.**: 1) *P. japonica* 2) loquat - **Java p.**: jambul - **Jew p.** (Jam): *Spondias cytherea* - **marmalade p.**: 1) *Salicornia europaea* 2) *Manilkara zapota* - **Moxie p.**: *Chiogenes hispidula* - **Natal p.**: *Carissa grandiflora* - **prairie p.**: Canada p. - **sand p.**: beach p. - **sapodilla p.**: *Manilkara zapota* - **wild goose p.**: *P. americana*; *P. hortulana* - **wild p.**: *P. angustifolia* - **wild p. tree blossoms**: fr. *P. spinosa* - **yellow p.**: *P. americana*.

Plumbaginaceae: Plumbago (or Leadwort) Fam.; rel. small fam. widely distributed; hbs. & small shrubs often in salty or limey soils; pls. acaulescent; ex. *Limonium*.

Plumbago auriculata Lam. (P. capensis Thunb.) (Plumbagin.) (sAf): cape (or blue) leadwort; rt. u. int. as emet.; exter. for warts; decoc. in blackwater fever (int.) (folk med.); popular peroral u. (whites & natives) for GU infections (c. plumbagin = antibacterial) - **P. europaea** L. (sEu, wAs): common leadwort; toothwort; lvs. rt. & juice acrid; c. plumbagin, plumbagol (phytoncide); u. as epispastic, emet. in epilepsy (folk med.); rt. u. as Radix Dentariae - **P. indica** L. (P. rosea L.; P. coccinea Salisb.) (s&setrop As): rose-colored (or red) leadwort; this and *P. zeylanica* much u. in Ayurvedic med. (Ind); rt. (Radix Vesicatoria) u. as caustic, vesicant, in rheum.; narcotic (1698); hemostat.; generally with other drugs as digest. carm., incl. many popular prepns. for anorexia & dyspepsia; in hemorrhoids; abort. (locally & orally); substit. for cantharis - **P. pulchella** Boiss. (sMex): cola de iguana; u. to destroy ulcers; in toothache; popular rem. in rheum; Mex Phar. - **P. scandens** L. (sFla [US], Mex, tropAm, WI, CA, SA): pega pollo (Cu); devil's herb; turicua; zapatilla; crushed lvs. & rts. rubefacient & epispastic; lf. u. in itch, erysipelas, &c.; u. for worms (Mex); lf. inf. emet. purg. (danger!); rt. chewed sialagog, u. in toothache; lvs. u. to produce sores by beggars (creates pity) (458) - **P. zeylanica** L. (Ind, Af-Austral, Am. Oceania): Ceylon (or white) leadwort; similar to *P. indica*; rts. "chita(cra)" u. as for *P. auriculata*; to stim. appetite & regulate digestive process (dyspepsia; diarrh.) (normalization of intest. flora); diaph.; rubef. in skin dis. as itch, leprosy (eAf Indian natives); leukoderma (exter); in piles; int. & exter. as emmen. partur. abort. (fresh lvs. chewed) (hazard!) (683); lvs. vesicant vulner. (1699).

plumcot: a hybrid of plum X apricot (Burbank).

Plumeria L. (Apocyn.) (Carib reg. fr. WI to nSA; now cult. worldwide). frangipani; 57 spp., some cult. for ornam. (Muslim cemeteries), worship (Buddhist temples), perfume mfr. - **P. alba** L. (WI, Mex, Af): (West Indian) pagoda tree; jasmine t.; sap u. to reduce swellings (Yuc); latex u. to make rubber, to treat warts & skin eruptions (Af); arom. wd. u. in place of sandalwood; frt. edible - **P. bicolor** R. & P. (SA, Mex): u. in folk med. (Mex) - **P. elegans** (eAf): exudate u. to treat pulmon. hemorrhage - **P. lancifolia** Mart. (Coutinia l.; Himatanthus l. Woods) (SA, esp. Br, Rio de Janiero, Min Ge, esp. Sant): agonia(da) (tree); aconiada; sucuuba; arapue; bk. c. agoniadine (*cf.* quinine), agoniadin (glycos.); plumieride (indican glucoside); u. emmen. antipyr., in mal., purg.; lvs. u. in asthma (as F.E., cigarettes [459]); tree a source of rubber - **P. multiflora** Muell.-Arg. (tropAm): trunk bk. c. plumericin, broad-spectrum antibiotic (active against fungi, some bacteria) (392); u. against scabies & intest. worms - **P. pudica** Jacq. (Ve): amapola; bk. c. plumeride; decoc. u. to wash ulcers - **P. rubra** L. (tropAm, [Mex to Pan], eAs): frangipani plant; kalachuchi fr. ancient Nahuatl "Kalachuchl" (Mex); cacalojoche; rt. c. plumericin, fulvoplumierin; u. lax. in dropsy; in VD; warts; var. *acutifolia* Bailey (P. a. Poir.): (tropAm) frangipani; temple tree; wd. well chewed to control cough, vermif.; fls. antipyr.; u. to prep. perfumes; milky latex for itch (with santal oil or camphor); rheum.; rubbed into cracks in sole of foot; u. for aching, decayed teeth; frts. abort.; fls. called "the Malay flowers of the dead"; trees found in every Muslim cemetery fr. sInd to the Moluccas ("graveyard flower"); very fragrant fls. offered in Buddhist temples; much cult. in OW trop; in Hawaiian leis; bk. purg. in GC; tuberculostatic; lvs. u. in cataplasms for bruises - **P. sucuuba** Spruce (Br): sucuuba bk. c. plumieride (agoniadin) (dihydropyran glucoside); u. antipyr.

plumericin: antibiotic fr. *Plumeria multiflora* & *P. rubra*; u. fungicide, antibacter.

plumeride: agoniadin: cryst. bitter principle; a spiropyrane glucoside; occ. in *Plumeria lancifolia*, *P. rubra*, &c.

Plumiera = *Plumeria* L.

pneumonia serum: Antipneumococcic Serum, Type Specific; obt. fr. the blood of a mammal immunized to *Diplococcus pneumoniae* Weichselbaum, the organism of pneumonia; organism also causes other inf. dis. such as otitis media, meningitis, etc. - **p. weed** (**root**) (Ala [US], local, Negro?): *Ludwigia alternifolia*, &c.

po (Por): 1) powder 2) (slang) Heroin - **em p.**: in powder form - **p. de mico:** *Plumeria rubra.*

Poa annua L. (Gram) (orig. Eu, now cosmop): ann. blue grass; serves as forage & fodder; common weed - **P. caespitosa** F. Forst. (NZ): "native paper grass"; stems with strong fibers, u. in paper prodn. - **P. nemoralis** L. (Euras, natd. NA, esp. neUS): wood blue grass; meadow g.; u. as forage - **P. pratensis** L. (Eu, now widely cult. & natd.): Kentucky blue grass; important US forage, fodder (hay), & lawn grass; the chief pasture grass (US) since very nutritious.

Poaceae = Gramineae.

poaia (**de preto**) (Br): **poaja: poalla**: **poaya** (Br): Ipecac (rt.) (< Mata da Poaia, large forest where plant occurs) (Tupi Indians apply this term to other pl. spp.).

pocan: Phytolacca.

poción (Sp): mixture or inf. - **p. Jaccoud**: several formulas, of which the most popular is **p. alcoholica**: c. Tr. Canella, Ext. Cinchona, Syr. Orange-peel, red wine, brandy.

pock wood: *Guajacum officinale.*

poco oil, Chinese: VO of *Mentha aquatica.*

Podachaenium eminens Schultz (Comp.) (Mex to CR): tacote; fragrant lvs. u. as poultice to wounds (Sinaloa).

Podaxis pistillaris (L. ex Pers.) Fr. (P. carcinomalis Fr.) (Secoti. - Hymenomycetes - Basidiomy. - Fungi) (cosmop): desert coprinus; fruiting bodies u. in treatment of carcinomatous ulcers (Af); spores u. as face powder (Hottentot women, possibly also in ancient Rome); to treat skin burns; young frt. bodies eaten.

podina (Urdu, Punjabi, etc.): mint, *Mentha arvensis* or *M. spicata.*

podoberry: *Podocarpus* spp.

Podocarpus L'Hérit. (Podocarp.) (mostly OW): yellow woods; Ahern pines (Chin); some 100 spp., mostly trees with valuable timber; bk. & frt. of some u. med. (Chin) - **P. amara** Bl. (Java): black pine; wd. useful, u. for beams, &c. (142) - **P. ferrugineum** D. Don (NZ): miro; wd. u. in turnery, for cabinet work; gum u. in med. - **P. macrophyllus** (Thunb.) D. Don (Chin, Jap): Japanese yew; kusamaki; u. for hedges; ornam.; frt. decoc. drunk as ton. for kidneys & lungs (Chin); var. *maki* Endl.: bk. u. in ringworm; frt. u. to restore health of lungs, kidneys, heart; for stomachache - **P. nagi** (Thunb.) Mak. (Jap. Chin-Taiwan, Ryukus): juice u. as beverage; frt. cooked as food; ornam.; pot pl. for dwarfing; valuable timber - **P. neriifolius** D. Don (seAs n to Chin): thitman; lf. inf. u. to medicate bronchitis; frt. receptacles eaten (Nepal); wd. useful - **P. spicata** R. Br. (NZ): black pine; wd. of great value, thus in use for ballroom floors; bk. u. as tan.

podophyllin: Podophyllum Resin - **podophyllotoxin**: a lactone compd. found as active princ. in rhiz. of *Podophyllum* spp.; antineoplastic agt. u. in plantar warts (1700); condylomata acuminata (topical soln.—Podofilox).

Podophyllum hexandrum Royle (P. emodi Wall.; "P. indicum") (Berberid.) (As-Himal-nIndo-Pak, Chin, Afgh): Indian podophyllum; I. mandrake; bankakri (Punj); u. much like *P. peltatum*, but has stronger med. action; rhiz. & rts. powd. mixed with water & given for indig., as lax. drastic cath. (hepatic stim.); cholagog; in chronic obstipation; source of P. Resin; larger rhiz. has more resin (12%) with higher content (ca. double) of podophyllotoxin than *P. peltatum*; frt. edible (E) - **P. macrovaphalum** (Co): said active against Hansen's disease - **P. peltatum** L. (eNA): (American) mandrake; (American or common) mayapple; ground (or wild) lemons; umbrella plant; vegetable calomel; v. mercury; rhiz. c. resin (podophylloresin) with podophyllin (c. podophyllotoxin, α- and ß-peltatins); picropodophyllin (older lit.) (2175); u. cholag. cath. in chronic constipation; anthelm. diur. emet.; to treat carcinoma (Penobscot Indians of Mo [US]); decoc. in leukemia (Ohio); salve or paste for tumors in cattle (Indians); in veter. med. (for pole evil [horses] [eCana]); in mfr. P. Resin, (soln. with benzoin in alc.) u. in superficial epitheliomatoses, laryngeal papillomata, condylomata acuminata (venereal warts), lymphogranuloma inguinale, &c. (1710); in psoriasis.; poisonous! (E) - **P. pleianthum** Hance = *Dysosma pleiantha* - **P. sikkimensis** R. Chatterjee & S. K. Mukerjee (Himal-Sikkim): rhiz. & rts. c. resin with active tumor-necrotizing properties like other *P.* spp.; c. sikkimitoxin, quercetin, isorhamnetin, & hyperoside (quercetin galactoside) - **P. versipelle** Hance (Chin): u. like *P. peltatum.*

podophyllum: 1) *P. peltatum* 2) *P. emodi* 3) *Achlys triphylla* - **American p.**: *P. peltatum* - **Indian p.**: *P. emodi.*

podsolnechniki (Russ): sunflower.

Poecilia reticulata (Lebistes reticulatus) (Poeciliidae - Pisces): guppy; small freshwater fish (Ve, Barbados, Trinidad): feeds largely on mosquito larvae & pupae; hence useful in malarial areas; a related sp. of the same fam. is *Gambusia affinis* (se&sUS); very useful in mosquito eradication.

Poecilochroma punctata Miers (Solan) (Pe, Bo?): bitter lvs. u. as analg.

poejo (**do campo**) (Br): 1) *Hedeoma denudata* 2) *Cunila microcephala* 3) *Mentha pulegium.*

poele bark: fr. *Alstonia spectabilis.*

Poga oleosa Pierre (monotypic) (Anisophylle.) (wEquat, Af-esp. Gabon): ovoga; sds. u. as condim.; source of FO, oil of inoy kernel (pleasant flavor).

Pogogyne douglasii Benth. subsp. **parviflora** (Benth.) J. T. Howell (P. p. Benth.) (Lab) (Calif [US]): perenn. arom. hb. u. to flavor pinoles (Calif Indians) - **P. nudiuscula** Gray (Calif): "pennyroyal"; "pennyroyal tea" u. as folk medicinal agt. for various minor ills (393).

Pogonanthera reflexa Bl. (Melastomat.) (Moluccas, Indones, Mal, PI): lvs. appl. to wounds (PI).

Pogonatherum crinitum Kunth (Gram) (Chin): u. antipyr. - **P. paniceum** Hack. (seAs): decoc. for chest pains (Taiwan); paste of pl. or ashes u. in skin dis. (Mal).

Pogonopus febrifugus B. & H. (Rubi) (Pe, Arg, Bo): bk. (Cascarilla Morada) u. as diaphor., substit. for Cinchona; c. moradeine (alk.) & moradin (NBP) - **P. tubulosus** K. Schum. (nwArg, Br, Bo): quina (Chin) morada; st. bk. c. moradine, moradeine, steroids; u. antimal.

Pogostemon cablin Benth. (P. patchouli Pell.) (Lab.) (Indo-Mal region; much cult. over As, esp. Chin, PI, WI, Par): patchouly; fleshy lvs. after fermentation yield on distn. VO, the famous patchouli oil; u. perfum. in soaps (*cf.* sandalwood oil), to scent mattresses and carpets (Arabs); antisep.; to drive away insects; for clothes moths (prod. Sum, Java, Penang; Chin) (395) - **P. glaber** Benth. (Chin): lf. juice u. on mosquito bites to relieve itching & pain (Hainan) - **P. heyneanus** Benth. (Indones): India patchouli; u. much like *P. cablin*; rt. decoc. in edema, lotion in rheum. (Mal); a source of p. oil; lvs. appl. to wounds (PI).

poha: *Physalis peruviana*.

poi: 1) (HawIs [US]): native food paste made fr. taro corm (*Colocasia esculenta*); main dish at luaus (feasts) of natives 2) (Hindu): *Basella rubra*.

Poikilospermum suaveolens Merr. (Conocephalus s. Bl.) (Urtic.) (Ind to seAs+): lf. decoc., for kidney dis.; lf. juice to treat fever (Mal); sap u. to treat cuts (PI); a liquid exudate fr. cut st. u. for pink eyes, substit. for water (PI).

Poinciana auctt. = *Delonix* - **Poinciana** L. = *Caesalpinia* - **P. gilliesii** Hook. = *Caesalpinia g.* - **P. pulcherrima** L. = *C. p.* - **P. regia** Bojer = *Delonix r.*

poinciana, dwarf: *Caesalpinia pulcherrima* - **royal p.**: *Delonix regia*.

Poinsettia pulcherrima Grah. = Euphorbia p. Willd.

pois (Fr): pea(s) - **p. carré** (Maurit.): *Psophocarpus tetragonolobus* - **P. petits**: English peas.

poison: a substance in or acting on the blood stream which in relatively small dosage (ca. 60 grains = 4 Gm. or less) destroys life or impairs the body functions in a serious manner - **p. berry**: *Actaea rubra*, etc. - **p. creeper**: *Toxicodendron* spp. - **dog p.**: *Aethusa cynapium* - **p. dogwood**: p. sumac - **economic p.**: pesticide (preferred term) - **p. elder**: p. sumac - **p. nut:** 1) *Jatropha curcas* 2) nux vomica - **p. oak**: *Toxicodendron toxicarium* (**western p. o.**: *Toxicodendron diversilobum)* - **p. parsley**: *Conium* - **rat p.: p. root**: *Dianella ensifolia* - **p. sumac**: *Toxicodendron vernix* - **p. tobacco**: *Hyoscyamus* - **p. tree: p. wood**: *Metopium* spp.

poison: v. also ash, hemlock, ivy, nut, vine.

poisson (Fr-m.): fish.

poivre (Fr): pepper - **p. de guinée**: *Capsicum annuum* - **p. noir**: black p. - **p. sauvage**: mezereon bk. - **poivrette**: powd. olive pits; formerly u. as pepper adult.

poix (Fr): pitch - **p. blanche: p. de bourgogne: p. jaune**: Burgundy p.

po' Jo: poor Joe (*qv.*).

poke: *Phytolacca americana* (berries, lvs. rt.) - **Indian p. r.** (US): *Veratrum viride* (misnamed): - **p. salad: p. sallet**: young lvs. boiled for greens (early spring) (460) - **Virginia p.: p. weed: p. wood** (Ga [US]): p.

pokeberry: *Phytolacca americana* or its frt.

polarization: change in the nature of a light beam whereby motion of the transverse vibrations is restricted to one plane instead of occurring in all planes, as in ordinary light; phenomenon is appl. in microscopy.

polder (Neth): drained marsh.

polecat: 1) (NA) skunk 2) (Eu) black-footed ferret 3) other animals.

pole-cat weed: 1) skunk cabbage 2) (Ky [US]) *Pluchea camphorata*.

Poleiminze (Ge): poleo.

Polemonium boreale Adams subsp. **richardsoni** J. C. Anderson (P. r. Grah.; P. humile R. & S.) (Polemoni.) (Arctic: Alas [US], Greenl, nEuras): (blue) Jacob's ladder; skunkleaf; sds. c. saponin; u. expect. - **P. caeruleum** L. (temp. zone; Euras, esp. Russia, NA): (blue) Jacob's ladder; Greek valerian; moss pink; "charity"; rhiz. & rts. u. as expect. with analg. & sed. action; hypotensive; senega rt. substit.; c. spermicidal saponin; formerly in folk med. for rabies; hb. u. as diaph. astr.; fls. c. saponins (high cont.; 10%+, Russ Ph.) - **P. reptans** L. (e&cUS): abscess root; blue valerian; blue bells; Jacob's ladder; rhiz. of perenn. hb. u. as astr. diaph. expect. alter., &c., in coughs, colds, pulmon. dis., pleurisy, bronchitis, fevers - **P. viscosum** Nutt. (wNA, Siber): rt. u. as expect., may be hypotensive.

polenta (It): pudding or porridge (mush) of maize; formerly of chestnut flour.

poleo: 1) (Sp) pennyroyal 2) (Arg) *Lippia turbinata* - **p. de Castilla** (Arg): *Aloysia polystachya.*

Polianthes tuberosa L. (Agav.) (Mex; cult. in Far East): tuberose; tube rose; bulb u. for mal. (PI); b. scales antispasm.; bulb. decoc. in GC; cult. as ornam. also for VO (in perfumes); u. mfr. cosmetics, fine perfumes; fls. u. in veg. soups (China, Jap).

Polidase™: comm. amylase prepn.

poligala (Sp): *Polygala* spp., esp. *P. senega.*

poligono (It): knotgrass.

Poliomintha incana (Torr.) A Gray (Lab) (swUS): hoary rosemary mint; pl. u. as potherb; fls. u. as flavor; VO c. pulegone.

poliovirus vaccine: produced from poliomyelitis (infantile paralysis) virus (strains 1, 2, 3) grown in cultures of monkey kidney tissue; in 2 forms: 1) interactivated with HCHO (Salk vaccine) & assayed; u. to produce active immunity to polio (3 doses at intervals SC or IM); u. in combn. with DPT Vaccine (E) 2) live attentuated (oral) (Sabin) (polio) (vaccine); c. either one or all 3 of the polio strains; attenuated with light, heat, or chemicals; advantages as prophylactic for polio: a). easier to administer (p.o. not by inj.) b). longer-lasting, better-type immunity c). lower cost (E).

polipodio (It, Por): polypody rt.

polipor de guetite (CR): *Polypodium vulgare.*

polk, Indian: I. poke - **polkweed** (error): pokeweed.

Pollachius pollachius (Gadus pollachius) (Gadidae - Pisces) (nAtl Oc): pollack; green-fish; source of codliver oil (Nor); marketed as food: whole, fillets, salted in oil - **P. virens** (US): (green) pollack; coalfish; common American pollack or bluefish; important in comm. fisheries.

pollack: various fishes of genera *Pollachius* & *Theragra* (Gadidae) resembling cod; liver oil u. like cod's (Jap, US).

pollard: 1) to cut off apical part of tree trunk —> greater foliage growth (< poll [head]: since head of leafy branches produced) 2) mixt. of rice bran & polishings.

pollen: microspore of a Spermatophyte; germinates to prod. a male gametophyte which prod. sperm cells; "bee pollen" popularly u. in many "herbal" medicines, marketed in tabs., capsules; "health food" candy bars (rich in nutrition); **p. grains** are characteristic of a sp. & useful in microidentification; presence of p. g. in susceptible individuals produces allergy & irritation of the bronchial areas, called pollinosis or hay fever; specific pollens are collected from many different spp. for use in making antigens (ex. giant ragweed; alfalfa) to use in prophylaxis & treatment of pollinosis (*qv.*); some pollens are very expensive; sometimes exceeding the price of gold (ex. catalpa) (2151); pollen counts (atmospheric) are important in studying hay fever (2231) - **"bee p."**: p. collected fr. beehives; u. in therapy (colitis; to prevent premature aging [value ?]); claimed to c. antibiotic (antibacterial, antifungal), gonadotropin substances.

Pollia japonica Thunb. (Commelin.) (Chin, Jap): rt. carm. sed. stim. ton.; for colds, vertigo - **P. thyrsiflora** Endl. ex. Hassk. (PI, Mol): lvs. appl. for amenorrhea.

pollice (It): unit of length; = one inch.

pollinosis: allergic reaction to pollen in atmosphere, giving hay fever, rose cold, &c.

pollock: pollack, *Pollachius* spp.

polmonaria (It): Pulmonaria - **p. du montagna** (It): Arnica.

polo-polo (Ec): *Bombax roseorum.*

polpa (Por): pulp.

polvo (Sp): powder.

Polyalthia cauliflora Hook. f. & Thoms var. *beccarii* J. Sinclair (Annon) (eAs): pounded lvs. appl. as poultice to skin dis. (Mal) - **P. hypoleuca** Hook. f. & Thoms (seAs): rt. decoc. may be u. after childbirth - **P. longifolia** (Sonn.) Thw. var. *pendula* R. Br. (Bangladesh, SriL): debdara (Bengal); bk. of this small tree u. as antipyr.; aq. ext. lowers blood pressure & slows heart rate; antibact. antifungal; c. steroids, alk., diterpenes; cult. India - **P. suberosa** Thw. (eAs): karadia (Oriya, Orissa, Ind); frt. edible; fresh rt. decoc. u. as abort. (PI); wd. useful - **P. sumatrana** Kurz (Mal): frts. are intoxicating - **P. thorelii** Finet & Gagnep. (Chin, seAs): bitter bk. u. to treat stomach dis.

Polyanthes Hill = *Polianthes* - **Polyanthes** Jacq. = *Polyxena.*

polyanthus narcissus: *Narcissus tazetta.*

polyarch: referring to certain roots having many xylem strands or gps.

Polycarpaea arenaria Gagnep. (Caryophyll.) (Indochin): woody rts. u. to treat intermittent fever & night sweats - **P. corymbosa** Lam. (Ind-Indochin, Mal): fls. u. as emoll. astr.; pounded lvs. as poultice over boils & inflammns. (Ind); for venomous reptile bites - **P. gaudichaudii** Gagnep. (Chin, Indochin): inf. of rt. st., lvs. in wine u. for wounds in head.

Polycarpon indicum Merr. (Caryophyll.) (Ind, Indochin): roasted lvs. inf. u. for cough following fever, esp. measles.

Polycillin: ampicillin.

polyenes: long-chain hydrocarbons, as seen (e.g.) in carotenoids.

Polygala L. (Polygal.) (widely distr.): milkworts; some 600 spp. of hbs. shrubs, small trees; *cf.* therap. props. of *Comesperma* & *Monnina* of same fam. (611) - **P. alba** Nutt. (cUS, nMex): white senega; rt. c. saponins; u. as substit. for or adulter. of *Polygala senega* (2171) - **P. amaniensis** Guerke (Af): bruised lf. u. as wound dressing - **P. amara** L. (c&sEu): bitter milkwort; hb. c. saponins, esp. senegin; polygalin (NBP), VO with methyl salicylate, FO, resin; u. expect. bitter ton. stim. stom. galactag.; in mucous diarrh.; rt. u. bitter ton. expect. esp. in TB - **P. amatymbica** E. & Z. (sAf): stim.; u. as cattle med. - **P. angulata** DC (Br): source of poaya blanca (white ipecac); c. saponins; u. emet. like ipecac; & as adult. of same - **P. aphrodisiaca** Guerke (Tanz): rt. u. as aphrod. (cooked with young rooster) - **P. arenaria** Willd. (tropAf): crushed lf. appl. as poultice in smallpox formerly (Malawi); u. as purg. in dys., wound dressing (Nigeria) - **P. boykinii** Nutt. (seUS, w to Tex, Mex): false senega; rt. c. methyl salicylate; u. like true senega (Mex) (2170); cold water inf. u. as headwash to correct dizziness (Fla [US]) - **P. brasiliensis** L. (Br-São Paulo to Patagonia): "barba de São Pedro"; pl. c. as high content of saponins as rts. of *P. senega* (highest in fls.); folk. med. u. in GC & intest. dis. - **P. butyracea** Heckel (trop wAf): cultigen; sds. u. in soups; cult. for sd. oil, malukang butter - **P. chamaebuxus** L. (cEu-mts.): hb. c. saponins, u. in folk med. as expect. & galact. - **P. chinensis** L. (Ind, tropAs, Austral): Indian senega; gaiguru; rt. juice u. as antipyr. & for dizziness (124); exp. in bronchitis & asthma - **P. costaricensis** Chodat (sMex, Guat, Nic, CR): "ipecacuana (amarilla)"; "raicillo"; u. to treat amebic dys.; as emet. (2172) - **P. crotalarioides** Buch. Ham. (Ind): bijuri; pl. ext. u. for cough, goiter dis. (124) - **P. elongata** Klein (Ind): powd. lf. u. for constip. (124), biliousness; in snakebite - **P. glomerata** Lour. (seAs): rt. pith u. for fever, throat inflammn. spermatorrhea, hematuria (Indochin); lf. inf. u. for cough, asthma - **P. gomesiana** Welw. (Equat, Af): rt. u. as cough remedy (Tanz) - **P. hottentotta** Presl (sAf): rt. inf. in mumps (children) & anthrax - **P. japonica** Houtt. (Kor to Indochin): ton. for males & very old people (Kor); rt. decoc. expect., in TB; in amnesia, sexual impotency (Indochin) - **P. kilimandjarica** Chod. (Tanz): lf. u. for infertility; rt. in snakebite - **P. lutea** L. (eUS, esp. se-pine barrens & bogs): candyweed; orange (or yellow) milkwort; dr. blossoms mixed with hot water —> poultice for swellings (Choctaws-La [US]) - **P. oppositifolia** L. (sAf): rt. decoc. u. in edema; in TB; diaphor. (Zulus) - **P. paniculata** L. (Tex [US], to Br, Pe; WI): milkwort; lanillo; rt. decoc. diur. in urin. dis.; inf. in pulmon. catarrh., dry cough - **P. paucifolia** Willd. (e&cNA): flowering wintergreen; fringed polygala; cult. ornam. perenn. hb. - **P. persicariaefolia** DC (Af): aphrod., in snakebite; lf. edible - **P. polygama** Walt. (P. rubella Willd.) (eUS): pink milkwort; bitter milkwort (or polygala); u. as bitter ton., much like *P. amara*; cough med. (Montagnais Indians) - **P. pruinosa** Boiss. (Near East): rt. proposed to be u. like *P. senega* - **P. rarifolia** DC (P. rehmannii Chod.) (s&eAf): "senega"; rt. u. for indig. (Bantus); & in love philtres (sSotho) - **P. rugelii** Shuttlew. (sFla [US]): yellow bachelor's button; hot inf. of hb. u. int. for snakebite (Seminole Indians) - **P. senega** L. (e&cNA): seneca (or senega) (snake) root; rt. c. senegin (saponin), polygalic acid, FO, resin, methyl sal.; u. exp. diur. diaph. in catarrhs, colds, arthritic dis., snakebite (2174); Northern, Western & Southern vars. are recognized in commerce; imported fr. Cana prairie provinces (Man-Alta) - **P. serpentaria** Eckl. (sAf): pd. rt. purg. for children; as enema in milk (toxic: danger!); rt. as expect. - **P. sibirica** L. (P. tenuifolia Willd.) (tempAs, Siber): rt. u. diur. expect. in bronchitis, coughs & colds; substit. for Senega rt. - **P. telephioides** Willd. (Ind, Chin): cross-flower; pl. u. expect., in opium poisoning - **P. virginiana** = *P. senega* - **P. vulgaris** L. (Euras): milkwort; hb. u. in inflammn. dis. of respir. & urin. passages; hb. c. senegin, polygalic acid; adulter. of *P. amara.*

polygala, bitter: 1) *Polygala amara* 2) *P. polygama* (also **American b. p.**) - **p. de Virginie** (Fr): Senega.

Polygalaceae: Milkwort Fam.; hbs., shrubs, & trees; found in all parts of world exc. Arctica, Antarctica, NZ & Polynesia; simple entire lvs.; fls. perfect but irregular; frt. a capsule, nut or drupe; ex. *Xanthophyllum*; *Securidaca.*

polygalacturonase: enzyme wh. acts on de-esterified part of pectin molecule, reducing vicosity & increasing reducing gps.; occurs in fungi, malt, snails; thought responsible for flax retting & frt. autolysis (ripening).

polygalic acid: polygalin: senegic acid; acid saponin of *Polygala senega*, &c.; but name also appl. to one of prods. of hydrolysis of senegin & to polygamarin (NBP).

polygalitol: anhydride of mannitol, sweet tasting subst. found in Senega rt.

polygamous: sp. which bear on the same pl. staminate, pistillate, & hermaphroditic fls.

Polygonaceae: Buckwheat (or Dock) Fam.; mostly hbs. of nTemp Zone; pls. have prominent nodes with manifest stipules (ocreae) which usually clasp st. above lf. base; infl. racemose, involucrate; frt. usually a nut; ex. *Chorizanthe*, *Polygonum*, *Rumex* (SC *poly*: many, & *gonios*: angle; as st. has many angles & turns).

Polygonatum Mill. (Lili.) (nTemp Zone): Solomon's seal; perenn. hbs. (do not confuse with *Polygonum*, *Polygala*) - **P. biflorum** (Walt.) Ell. (e&cNA): small (or hairy) Solomon's seal; rhiz. coll. as "Solomon's seal"; "conquer John"; u. ton. expect.; emet. astr.; var. *commutatum* Schult. f. (P. c. (Schult. f.) A. Dietr.) (NA): great Solomon's seal since larger pl. (tetraploid); u. like typical sp.; for mouth sores (2155) - **P. cirrhifolium** Royle (Chin, Ind): "meda"; boiled rts. eaten to give strength; also in uter. dis. & postpartum; aphrod. (UP, Ind) - **P. falcatum** A. Gray (eAs): antipyr.; eating pl. supposed to prolong life - **P. kingianum** Collett & Hemsl. (Chin): rhiz. u. as ton. for heart & lungs; stim. saliva - **P. latifolium** Desf. (seEu, wAs, NA): another source of polygonatum rhizome - **P. multiflorum** All. (Euras; natd. NA): David's harp; u. like *P. biflorum*; rt. u. as diur., once highly reputed in gout, rheum. diabetes; dem. ton. in leukorrh. (fluor albus) (Ind) - **P. officinale** (L) All. (P. odoratum [Mill.] Druce) (Euras-Siber): (aromatic) white root; (true) Solomon's seal; rhiz. c. saponins, asparagin, allantoin, mucil.; mostly u. as diur. in urin. dis. (as Radix Sigilli Salomonis); hypotensive; antirheum.; ext. in contusions - **P. sibiricum** Red. (Siberia): u. to stim. digestive juices incl. saliva - **P. verticillatum** All. (Euras-Indo-Pak): rhiz. has onion-like aroma & mucilag. a bitter taste; c. & u. much like those of *P. officinale*.

Polygonum L. (Polygon.) (worldwide): knotweeds; fleece-weeds; mostly hbs.; much u. in folk med. (do not confuse with *Polygonatum*, *Polygala*) - **P. acre** HBK = *P. punctatum* - **P. amphibium** L. (P. muhlenbergii S. Wats.) (Polygon.) (Euras, NA): water (or mud) knotweed; rhiz. u. for "cleansing blood" (Homeop.) like sarsaprilla; sds. emet. & vermifuge (Indochin); hb. u. as diur. for kidney stones; red fls. u. as bait formerly (BC [Cana] Indians); young shoots eaten by Sioux Indians; fed to hogs (Miss [US]) - **P. amplexicaule** D. Don (Indo-Pak, esp. Kashmir): rts. u. in kidney dis. (Koelz) - **P. aviculare** L. (temp. zone worldwide): (male) knotgrass; prostrate knotweed; crawl grass; large variation (esp. in Eu); hb. c. tannin, org. acids; avicularin; quercitrin; u. as astr. in diarrh., in urin. dis. (flg. hb.), esp. nephritis (kidney teas, Eu); tea to prevent abortion (NA Indians); in lung therapy, as for TB (night sweats) (Ge popular) (2181); diur.; in diarrh. (2180); rhiz. u. in diabetes (probably worthless) (2179); as hemostyptic; sds. purg. emet. - **P. barbatum** L. (eAs+): sds. u. as arom. purg. emet.; in colic & enteritis (Ind); rt. u. for sores (Chin); pl. juice in baths for hemorrhoids & gout (Br) - **P. bistorta** L. (Euras, nNA): (European) bistort; English serpentary; snake weed; rhiz. (as Radix Bistortae) c. catechol tannins, gallic acid, tannin content 15-36% (depends on season); u. as astr. in diarrh., dys. ton. diur., hemostatic; ext. for sores, mucous inflammn.; hb. u. astr. hemostat. in kidney & pulm. dis. - **P. bistortoides** Pursh (Bistorta b. Small) (wNA): American (or western) bistort; rts. starchy, edible - **P. chinense** L. (trop&subtr As): pl. decoc., u. for dys. & dis. of large intest.; with sugar to relieve headache; rt. vermifuge (Chin) - **P. convolvulus** L. (ntemp. reg.-Eu, NA): (black) bindweed; c. emodin glycosides; u. lax. - **P. cuspidatum** S. & Z. (eAs, now natd. Eu, NA): cuspidate knotgrass; Japanese k.; hb. c. quercitrin, isoquercitrin, rutin; u. as lax.; yellow dye (eAs) - **P. divaricatum** L. (Russia): pl. with high vit. C content, also with carotene, rutin (hyperoside); rt. c. tannin (-20%); u. astr. in diarrh., antidote for some poisonings - **P. dumetorum** L. (nEuras): hb. c. emodin glycosides, rutin; u. as lax.; purg.; sds. as grits (porridge) - **P. fagopyrum** L. = *Fagopyrum esculentum* - **P. hydropiper** L. (Euras, NA, nAf): water pepper(wort); hb. c. VO with polygodial, flavonoids; u. diur., in rheum.; jaundice; pepper substit.; uter. stim. hemostat.; rt. u. med. for wounds (Chin) & sores on horses; lvs. eaten in salads (Chin); fresh flg. pl. off. in Ge Homeop. Phar. - **P. hydropiperoides** Mx. (NA, SA, Austral): tasteless knotweed; American water pepper; (wild) smart weed; hb. c. tannins, rutin (3% in lvs.), quercetin, kaempferol; u. like *P. hydropiper* (but with sharper taste); for intest. hemorrhages, metrorrhagia, in strangury, stim.; diur. (Br); emmen. (Mex); epispastic; cult. in Eu - **P. japonicum** Meissn. (Chin, Jap): flg. shoot decoc. u. to relieve atrophy, diarrh. (Chin) - **P. longisetum** De Br. (Jap, Chin): "inutade"; lvs. boiled & eaten when food scarce; whole pl. u. in gastric cancer - **P. minus** Huds. (seAs+): lf. decoc. taken after childbirth; also in indig. & in gastric dis. (New Hebrides) - **P. molle** D. Don (P. frondosum Meisn.) (Ind-Himalayas): pl. u. as astr. in diarrh.; young shoots dr. & eaten like rhubarb stalks (Sikkim) - **P. multiflorum** Thunb. (eAs): rhiz. c. emodin & chrysophanol; u. as ton., turns hair dark if u. for long time; u. for tumors, piles, colds, diarrh. (Chin); ext. influences alimentary hyperglycemia - **P. odoratum** Lour. (Indochin): st. & lvs. diur. antifebrile, antinauseant; anaphrodisiac

(u. by Buddhist priests [Manchuria]) - **P. orientale** L. (eAs): frts. u. for dyspeps., constip. (Manchuria); leafy branch decoc. u. for hernia (Chin), as carm.; in ophthalmia - **P. paleaceum** Wall. (Chin, Ind): rt. decoc. u. for wounds, hemorrhages - **P. pensylvanicum** Small (e&cNA): pinkweed; foxtail; Pennsylvania persicaria; lf. inf. drunk by women after childbirth (Menomini Indians); for hemorrhoids (Meskwaki Indians); has been u. for diabetes (inf.) (394) - **P. perfoliatum** L. (eAs, natd. neUS): lf. & st. decoc. u. in fever, diarrh. dys. ton. (Chin) - **P. persicaria** L. (temp&sEuras, natd. NW): lady's thumb; heart's ease; common persicary; hb. u. in folk med. similarly to *P. hydropiper*, as kidney stone remedy, in gout, rheum. hemorrhoids, icterus, exter. as vulner. on wounds & in eczema - **P. plebeium** R. Br. (cosmopol., trop & temp, esp. Austral): small knotweed; raiphur (Ind); folk med.; rt. u. in bowel dis.; pl. paste u. with *Ficus semi-cordata* frts. in diarrh. (124); pl. powd. u. inter. for pneumonia (Ind) - **P. polystachyum** Wall. (Ind): hb. c. quercitrin, isoquercitrin, quercetin, quercitol (461); u. med. (Ind) - **P. portoricense** Bert. (sUS, Mex, WI, CA, nSA): guinea hen bush; exhalation of crushed rt. inhaled for headache - **P. pulchrum** Bl. (Burma): rt. decoc. u. for gastric dis. of children; lvs. eaten as ton. (Mal) - **P. punctatum** Ell. (As, NA, SA): water (or dotted) smartweed; American s.; chilillo; chile de perro; hb. u. diur. diaph. emmen.; in rheum.; lf. juice u. for dog mange (inf.) (CR, Guat, Hond) - **P. sachalinense** Fr. Schmidt (Sakhalin & sKuril Is): Sakhalin knotweed; "o-itadori" (Jap); rather common pl.; lvs. u. as disinf. for boils (Ainu); young shoots eaten (Ainu) (rich in vit. A & C) - **P. salicifolium** Brouss. (sAf): knotweed; lf. paste appl. to sores - **P. senegalensis** Meisn. (P. glabrum Willd.) (Ind & Af): lf. inf. u. for colic pain; decoc. & washes u. as antipyr. in syph. (Senegal) - **P. tinctorium** Ait. (eAs): frts. u. as antipyr. - **P. virginianum** L. (Ind - Himal): pl. u. as dem. ton. astr. diur., in pulmon. dis. - **P. viviparum** L. (circumboreal-Alas [US], nNA; Ind-Himal): alpine bistort; rt. appl. to abscesses; decoc. inj. in GC, leukorrhea; lotion for ulcers; in diarrh.; lvs. rich in vit. A & C, boiled & eaten; rt. starchy, eaten raw or boiled (201) - **P. weyrichii** Fr. Schmidt (Russ): hb. c. quercitrin; u. fodder (vit. A, C).

polygony: *Polygonum* spp.

polygraph: "lie detector": instrument u. to record simultaneously physiological processes, as respiration, blood pressure, pulse, perspiration, &c.

Polymnia uvedalia L. (Comp) (e&cUS): bearsfoot; (yellow) leafcup; rt. c. resins, tannins, reducing sugars, &c., u. ton. stim. in rheum. arthr. gastritis, hepatic enlargement, cervix uteri hypertrophy (int. & exter.).

Polymyxin A: formerly but not now called Aerosporin - **P. B sulfate**: Aerosporin (*qv.*).

Polynesia: island gp. of e&sePacif Oc incl. Hawaii, Tonga, Samoa, Society Is, Cook Is, Easter Is (SC so many islands or so many different kinds of people inhabiting them).

Polyosma longipes Koord. & Val. (Grossulari.) (Indones): inner bk. scrapings u. for itch; lf. sap remedy for sore eyes - **P. mutabilis** Bl. (Taiwan): bk. u. in fever & edema.

polypeptides: protein derivs., consisting of 2 or more amino acids linked together (-CONH).

polypetalous (adj.): where corolla of fl. is made up of several separated petals, ex. rose (**polypetaly**).

polypharmacy: the use of many different drugs together, often producing disorder.

polyploid: having 3 or more sets of chromosomes in cell (do not confuse with polypoid = polyp-like, ex. polypoid rhinitis).

Polypodiaceae (Pteridoph.): Common-fern Fam.; most common low ferns of temp. areas belong here; widely distr.; ex. *Pteretis*, *Dryopteris*, *Pteris*.

Polypodium aureum L. (Polypodi. - Pteridoph.) (Mex, WI): haresfoot fern; calaguala (Mex); Rhizoma Calaguala (off. in Mex & Ve Ph.) c. mucil. NBP, tannin, FO, sugars, cyanogenic glycoside; u. decoc. as diaph. pectoral in chronic resp. dis., pertussis; vulner., styptic - **P. calaguala** Ruiz (Mex, Pe, Ch, occas. in Eu): echte Calaguala (Ge) (true calaguala); polypody; rhiz. u. as expect. diaphor. antisyph.; exter. for contusions - **P. californicum** Kaulfuss (Calif [US], lower Calif [Mex]): California polypody; juice fr. bruised rhiz. rubbed on sores to heal & on body for rheum. (Wailakis Indians, Calif); rt. ext. u. to wash sore eyes - **P. crassifolium** L. (WI, Mex to Ve, Br): rhiz. u. much like *P. calaguala* (Br); rhiz. rts. stipes, fronds & paleae u. as vulner.; in skin cancers (396) - **P. decumanum** Willd. (tropAm): u. med. (Mex, Co) - **P. filix-mas** L. = *Dryopteris f.-m.* - **P. hastatum** Thunb. (eAs): rhiz. & frond decoc. drunk for GC, as antipyr. (Chin, Jap, nVietnam); whole pl. for syph. (Manchuria) - **P. lanceolatum** L. (Mex, Af): lengua de ciervo; Hb (off. Mex Phar.) c. tannin; u. expect. for sore throats, colds (sAf); considered to have similar props. to *Phyllitis scolopendrium*; vulner. - **P. phymatodes** L. (sAs, Austral, Af): u. kidney dis. & to perfume FO - **P. plebeium (plebejum)** Schlecht. & Cham. (sMex, SA): u. as expect. & purg. (Mazatecs) - **P. polypodioides** (L) Watt (seUS, trop & subtropAm): resurrection fern

(SC unfurls from dormant state when placed in water); entire pl. u. for bronchitis, liver dis.; ornam. curiosity - **P. scolopendria** Burm. f. (seAs): u. as remedy in chronic diarrh. (Indochin) - **P. virginianum** L. (NA): rock polypody; r. cap fern; wall f.; u. in pleurisy (Maritime Indians); stomach dis.; ornam. fern - **P. vulgare** L. (Euras, Af): (common) polypody; wood fern; rhiz. (polypody root) c. catechol tannins, sugars, FO mucil. saponins, phytoecdysones, glycyrrhizin(?); u. mild purg. for liver dis., in colds, bronchitis; hb. u. in folk med. (Eu) for colds, hoarseness, TB; in liqueur indy. for bitter schnapps; var. *occidentale* Hook. (NA, nwUS, Br): "licorce fern"; c. polydin (glycoside); sugg. as licorice substit. since has similar flavor.

polypody: *Polypodium vulgare.*

Polyporaceae (Hymenomycetes - Basidiomyc. - Fungi): Pore Fungi Fam.; "tube-bearing fungi"; fruiting bodies commonly bracket-shaped, but at times resemble mushrooms, etc.; the pores are lined by the hymenium or basidial layer & here the basidiospores are prod. (1391).

polypore (Eng, Fr): a sp. or individual of genus *Polyporus+* - **p. du mélèze** (Fr): *Fomitopsis officinalis.*

Polyporus betulinus Bull. (Polypor.) (Euras, NA): = *Piptoporus b.* - **P. anthelminticus** Berk. (eAs): parasitic on bamboo; fruit bodies u. as anthelmin. for cattle (Burma) - **P. fomentarius** L. = *Fomes f* - **P. igniarius** Quél. = *Phellinus i.* - **P. officinalis** Vill. = *Fomitopsis o.* - **P. sulphureus** Bull. ex. Fr. (temp-worldwide): chicken fungus; sulphur polypore; lacks agaricin but occurs in comm. ("true larch fungus") as adulter. of *Fomitopsis officinalis* - **P. squamosus** Micheli (temp. zone): scaly polyporus; frt. bodies u. to make razor strops; when young, consumed as food - **P. tuberaster** Jacq. (P. tuckahoe Guessow) (temp. zone): tuckahoe (sporophore) eaten by Am Indians (nwNA).

Polypremum procumbens L. (monotypic) (Logani.) (Salv): u. as remedy in metritis.

polysaccharides: polysaccharoses: large group of natural complex carbohydrates with the general formula $(C_6H_{10}O_5)_n$ (n is a large number); insol., tasteless; easily hydrolyzed to simple sugars; ex. starch, cellulose, inulin.

Polyscias fruticosa Harms (Panax fruticosus L.) (Arali) (Indones, cult. trop): rts. & lvs. decoc. u. diur. in gravel, dysuria; powd. lvs. vulnerary (PI); lf. inf. with Acalypha u. for swollen testes & for stomach trouble (Fiji) - **P. nodosa** Seem. (Indones, Mal, Java): "deleg"; umbrella tree; lvs. (& resin) u. to stupefy fish (poison); wd. useful & u. to make vessels, knife handles, &c. - **P. odorata** Blanco (Schefflera o. Merr. & Rolfe) (PI): c. saponins; bk. u. as cough cure; lvs. antiscorb.; resin as vulner. - **P. scutellaria** (Burm. f.) Fosberg (seAs+): lvs. u. to treat inflammn., ulcers, paste smeared on scalp to prevent baldness; rts. diur. diaph. (Indones); lvs. heated appl. to abdomen to induce urination (PI).

polysepalous (adj.): showing **polysepaly**: the presence in the fl. of several & distinct sepals; ex. *Peony.*

Polysporin™: mixt. of polymyxin B sulfate and bacitracin Zn in lactose base; u. as powder, ointment, &c.

Polystichum Roth. (Aspleni. - Pteridoph.) (widely distr.): ca. 200 spp.; holly ferns; shield ferns - **P. achrostichoides** Mx. (e&cNA): Christmas (or dagger) fern; canker-brake; rhiz. u. as anthelm.; in stomach trouble (Am Indians); u. in hoarseness (Maritime Indians) - **P. aculeatum** (L) Roth (Ind.): hard fern; rhiz. u. to adulterate & sophisticate *Dryopteris filix-mas*; fronds u. to prep. curry - **P. adiantiforme** Sm. (trop & subtrop sHemisp): rt. (Radix Calagualae) u. in chronic pulmon. dis., asthma, & severe cough - **P. lonchitis** Roth (temp zone) (Euras, Greenl, NA): holly fern; fronds appl. exter. to wounds, also in splenic dis. (formerly) - **P. munitum** C. Presl (Alas-sCalif e to Mont [US]): (western) sword-fern; Christmas f.; large quantities of fronds collected each yr. to make wreaths, sprays, &c. (2341); rt. stocks boiled & eaten (banana flavor) by BC Indians (Vanc Is [Cana]) (1768) - **P. pungens** Kaulf. (sAf): rhiz. decoc. u. as enema for intest. worms; bots (horses).

Polystictus Fries = *Polyporus* &c.

Polytrias praemorsa Hack. (Gram.) (monotypic) (Java): Java grass; moderately good fodder for cattle; turf grass.

Polytrichum commune L. (Polytrich. - Musci - Bryoph.) (Euras, NA): rockbrake herb; (common) hair(cap) moss; bear moss; pl. (as Herba Adianti aurei) c. resin, tannin, FO, gum, wax; u. astr. in hemorrhages, menstr. diffic.; diaph. ton.; popular in folk med. - **P. formosum** Hedw. (Polytrichastrum f. G. L. Sm.) (worldwide): u. much like prec.; a source of Herba Adianti aurei - **P. juniperinum** Hedw. (NA & nHemisph): ground moss; u. diur.; exts. have weak tumor necrotizing effect - **P. strictum** Brid. (worldwide): u. similarly to *P. commune.*

polyunsaturated fatty acids: linoleic, linolenic, & arachidonic acids; u. in therapy to lower level of blood cholesterol (462).

Poma Aurantii (L): **P. Aurantiorum** (L): orange buds.

pomace: frt. or sd. pulp; ex. apple p., castor bean p., &c.

Pomaceae (Rosaceae - Pomoideae) (Pyroideae): pome frt. subdivision of fam.

pomade: Pomatum (L.): a perfumed ointment esp. for the hair.

Pomaderris elliptica Labill. (Rhamn.) (Tasmania): "kumarahon" (Maori); u. in asthma, cough, bronchitis, chest dis., rheum., kidney dis. (Maori).

pomander (ball): a kind of sachet (ball) made by sticking the entire surface of an apple (p. apple) or thin-skinned orange (Seville) full of whole cloves, then rolled in powd. iris & clove (eq. parts), then dried out; formerly u. to protect against disease & for odor; now hung in linen or clothes closet for long-lasting fragrance; term also appl. to box (case) to contain same (463).

pome: fleshy frt. with sds. but no stone; some Rosaceae; ex. apples.

pomegranate: *Punica granatum* - **p. family**: Punicaceae.

pomelo (Ind): *Citrus paradisi* (avoid confusion with pummelo: *C. grandis)* (*C. maxima*).

Pomeranzenschale (Ge): bitter orange peel.

Pometia pinnata Forst. & Forst. f. (Sapind.) (Malaysia; Oceania): bk. or lf. decoc. u. to bathe patients with fever; bk. u. in dressing for festering sores or wounds (Mal); frt. & sds. eaten; bk. in gastric dis. (Tonga) (86); wd. u. in house building - **P. tomentosa** Teysm. & Binn. (PapNG): timber (taun) extensively u.; med.

pommade (Fr): ointment (or salve) (often perfumed) which does not c. resin.

pomme (rousse) (Fr-f.): (russet) apple - **p. de Canada**: *Helianthus tuberosus*, tubers - **p. de Cythère** (Fr): *Annona squamosa* - **p. de pin**: pineapple - **p. de terre**: potato - **pomme-épineuse** (Fr): Stramonium.

pommier (Fr): apple tree.

pomo (Br): apple - **p. do diablo: p. espinhoso:** *Datura stramonium.*

pompelmous: *Citrus maxima.*

pompion (Fr): pumpkin sd.

Poncirus trifoliata (L) Rafin. (Citrus t. L.) (Rut) (n&cChin, natd. seUS): trifoliate orange; wild o.; frts. inedible, but at times u. in preserves; dr. frts. u. as digestive, for fever (Kor); stom. antidiarrh., lax., antispasm. diur.; in rheum. (Chin); bitter arom. stim. (Mal); thorns as toothache rem.; bk. for colds; VO c. limonene (1703); pl. u. as root stock to graft mandarin orange, &c.; as hybrid parent of Citrumquat (resists cold); ornam. in hedges; "specimen" tree (2411).

pond lily: *v.* lily, pond.

Pond's Extract™: a witch-hazel liquid ext., representing aq. distillate prod. of *Hamameltis virginiana* made near Utica, NY (US) (since ca. 1842); originator of "Double Distilled Witch Hazel" (popular skin treatment) by E. E. Dickinson; has been prepd. in Conn (US) since ca. 1835.

pone: corn p. (sUS): maize bread made without milk or eggs; *cf.* Johnny cake (< Am Indians).

Pongamia pinnata Merr. (P. glabra Vent.) (Leg.) (eAs, sChin to Polynesia; Ind, SriL): Indian beech; poonga oil tree; karum; karanj(a); thinwin (Burma); med.-sized tree in coastal forests; sds. c. thick yellowish or reddish bitter FO (pongam or beech oil) (2156); u. locally in scabies, eczema, prurigo, pityriasis, herpes, psoriasis; often mixed with mustard oil; in dermatol. cond.; mixed with kamala & camphor in scabies; int. stom. cholag.; techn. purposes (fuel); sd. u. antipyr. (2226); frt. pericarp u. pertussis, chronic bronchitis; rt. (juice) u. in ulcers, sores, wounds; to clean teeth (Ind); bk. in hemorrhoids; splenomegaly; lvs. (bath) for rheum.; wd. as timber for bldgs., oil mills, cart wheels; cult. roadside tree (—> *Millettia*).

Pontederia cordata L. (Pontederi. - Monocot.) (trop&subtr; tempAm): pickerel weed; wampee; frts. sometimes eaten (famine food); to prevent pregnancy (Maritime Indians); may be added to caapi; to control facial paralysis (Co) - **P. nymphaefolia** Kunth (Par) and **P. rotundifolia** L. (Par): fls. u. diur.

pontianak (gum): a gutta percha fr. latex of *Dyera costulata*; u. as base of bubble gums; named after Pontianak (island off w coast of Borneo) - **black p.**: *Shorea* spp.

pontico (Sp): *Artemisia pontica*, wormwood.

poon tree: *Calophyllum* spp.

poor Joe (seUS, Ga to Miss) (pron. pojo): **poor John** (Miss): *Plantago aristata*; *Diodia teres* (398) - **p. man's garlic**: garlic.

pop (seUS): frt. of *Passiflora incarnata.*

popcorn: *Zea mays* var. *everta*; grains of which explode on heating - **p. tree** (Ala [US]): *Sapium sebiferum.*

popinac, white: *Leucaena glauca.*

poplar: 1) *Populus* spp. 2) (sUS): *Liriodendron tulipifera* - **American p.**: 1) *Populus tremuloides* 2) (nCalif [US]): white p. - **balsam p.**: *P. balsamifera* - **p. bark**: fr. *P. tremuloides* - **black p. (buds)**: *P. nigra* - **p. buds**: Populi Gemmae - **Carolina p.**: *P. canadensis* - **p. leaves**: 1) fr. *P. tremula* 2) (comml.): *Liriodendron tulipifera* - **Lombardy p.**: *P. nigra* - **silver (leaved) p.**: *P. alba* (clone) - **tulip p.**: yellow p. - **white p.**: 1) *P. alba*; *P. tremuloides*; *P.* X *canescens* 2) yellow p. (bk.) - **yellow p.** (seUS): *Liriodendron tulipifera.*

popotillo (Mex): *Ephedra nevadensis.*

Popowia aberrans Pierre (Annon.) (sVietnam, Laos, Camb): arom. fls. u. to perfume lip paste (combn. with beeswax & *Sterculia lychnophora* FO) - **P. caffra** Baill. (trop eAf): u. med. (Moz); frt. eaten (Bantu); as charm.

popple (US slang, Great Lakes area): *Populus grandidentata & P. tremuloides* (*cf.* Ge Pappel).

poppy: spp. of *Papaver*, etc. - **American prickle p.**: *Argemone mexicana* - **black p. seed**: *Brassica nigra* (misnomer) - **California p.**: *Eschscholtzia californica* - **p. capsules**: the dr. ripe frt. of *Papaver somniferum* & v. *album* - **celandine p.**: *Stylophorum diphyllum* - **common p.: corn p.: field p: p. flowers**: *P. rhoeas* - **garden p.**: opium p. - **p. heads**: p. capsules - **horn(ed) p. (yellow)**: *Glaucium flavum* - **Iceland p.**: *P. nudicaule* (Arctic) not California p. - **Mexican p.: M. prickle p.: M. prickly p**: *Argemone mexicana* - **p. oil**: FO fr. sds. of Dutch-grown opium p. - **opium p.**: *P. somniferum* & its v. *album* - **oriental p.**: *P. orientale* - **plume p.**: *Bocconia cordata* - **prickly p.**: Mexican p. - **red p. (flowers)**: corn p. - **p. seed**: sd. of opium p.; var. blue (superior), white (**Dutch p. s.**: blue v., u. for FO formerly imp. by oil mills) - **p. straw, concentrated**: form of opium imported into US for manuf. of opium alks.; *cf.* "Straw" - **thorn p.**: Mexican p. - **white p.**: opium p. (sd.); sometimes u. to refer to fresh white latex (399) - **white prickly p.**: *Argemone alba* - **yellow p.**: 1) Mexican p. 2) celandine p. - **yellow-horn(ed) p.**: horn p.

Populi Gemma (L.): balm of Gilead buds fr. *Populus balsamifera.*

populin: glucoside fr. *Populus* spp. incl. *P. alba, P. nigra*; on enzymic hydrolysis yields saligenin & benzoyl glucose.

Populus L. (Salic.) (nTemp Zone): poplars; Espe (Ge); shrubs or trees with soft wood, resinous buds, diecious; some 40 spp. (2157); (New Lat) (2158); wd. u. to make boxes, excelsior, furniture, paper pulp; wood charcoal, plywood, &c.; trees popular as shade trees, windbreaks (esp. wUS); < arbor populi (L) = tree of the people; several spp. u. for prepn. yellow dyes; mfr. salicin (bk.); masticatory (464) - **P. alba** L. (Euras, cult. US, Cana): white poplar; silver p.; great aspen (tree -35 m) ("poplar tall," expression); abele; lvs. u. as decoc. for decayed teeth & bone necrosis (Chin); bk. c. VO, salicin (source), populin; tannin (10% in bk.); long u. as antipyr. arom. ton.; in bladder dis. sciatica; as dye; wd. u. as veneer material; makes good grade cellulose & charcoal; tree u. for erosion control; popular shade & ornam. tree, but rts. cause difficulty in clogging sewer lines, &c.; var. *nivea* Ait. (var. *argentea* Hort.): silver (leaf) p.; ornam. clone - **P. balsamifera** L. (P. tacamahaca Miller; P. candicans Ait.) (nNA, natd. Eu): balsam poplar; balm of Gilead; tacamahac; peuplier (baumier) (Qu [Cana]); source of poplar buds (lf. buds); c. VO with cinnamic acid, salicin, acetophenone, u. arom. stim. expect. in catarrhs., cough; GU dis.; in chronic polyarthritis; ointment u. for sores, burns, hemorrhoids; source of sort of shellac (by placing buds in alc.); u. as vulner. by covering wounds, sores (Anticosti, Qu) (64); good shade tree (most prodn. fr. US, Cana, Hung) - **P. canadensis** Moench (P. X eugenei Simon-Lewis) (NA): Carolina poplar; Eugene poplar; cult. ornam. tree - **P. ciliata** Willd. (Ind - Himal reg.): falsh (Kashm, Punj): black berries; bk. u. ton. stim.; source of a floss; wd. u. to make matches - **P. deltoides** Bartr. ex. Marsh (P. virginiana Desf.; P. sargentii Dode) (e&cNA): (northern) cottonwood; eastern cottonwood; u. for lf. buds; important lumber source; ornam. tree; subsp. *monilifera* Eckenwalder (P. m. Ait.) (Rocky Mt. area-NA): western (or Great Plains) cottonwood; bk. u. in mal. ("swamp fever"); pollen a frequent cause of hay fever; bk. u. as vermif. (Iroquois) (2160) - **P. euphratica** Oliv. (wAs, Pak): "weeping willow"; bahan; bk. u. as vermifuge - **P. fremontii** S. Wats. (swUS, nMex): Fremont (or Arizona or California) cottonwood; bk. decoc. u. as wash for cuts, sores, & bruises (Calif Indians); inner bk. u. in scurvy (Am Indians); twigs u. to make baskets; wd. u. for water troughs - **P. nigra** L. (Euras, natd. NA): black poplar; source of (balsam) poplar buds (European) u. in chronic polyarthritis; diur. emmen. expect.; in GU dis.; vulner., in ointment for burns, wounds, bruises (esp. Fr), hemorrhoids, inflamed breasts; gout; int. in colds, enteritis, cystitis (2161); cult. var. *italica* Moenchh. (P. it. Moenchh.) (cult. Eu, NA): Lombardy p.; tree with unique columnar habit; popular windbreak, ornam.; however tree banned in Atlanta (Ga [US]) since rt. system often blocks water pipes - **P. tremula** L. (Euras, nAf): (European) aspen; bk. & lvs. u. like foll.; bk. u. in strangury - **P. tremuloides** Mx. (Cana, US - Pac & mt. reg.; nMex): quaking aspen; (trembling) poplar; white p. (sometimes confusion of collectors with bk. of yellow "poplar" [*Liriodendron tulipifera* [1702]); bk. ("poplar bark") u. antiperiod. antipyr. (lvs. & bk. mixed in equal amts. u. by Eu Homeop.); bk. u. in colds & to stim. appetite (Maritime Indians); bk. tisane vermif. (Abnakis) (2759) - **P. trichocarpa** T. & G. (wNA-Pac & mts.): western balsam poplar; (black) cottonwood); lvs. c. trichocarpin (fungistatic); bk. u. as poultice for wounds; ton. (with p. buds) (BC [Cana] Indians) (75); catkins eaten (Am Indians); cult. shade & ornam. tree - **P. wislizeni** Sarg. (P. fremontii var. w. Wats.) (Colo, NM, wTex [US], nMex): Rio Grande (or valley) cottonwood; álamo; lvs. u. edema; bk. as splints for broken bones; ashes appl locally in boils.

popusa (Arg): *Zuccagnia punctata.*

porakary: paracary.

Porana volubilis Burm. f. (Convolvul.) (Indones): decoc. u. as postpartum stim.; lvs. eaten to remove nasty taste from mouth also in comb. as ton.

porcino (Sp): young pig - **porcinos** (Arg, etc.): hogs in general.

porcupine: *Erethizon* spp. (Rodentia): *E. dorsatum* (L.) Cuvier (Cana, nUS): Canadian (or gray) p.; slow-moving armored arboreal; serves as emergency food in woods; quills u. in Indian art work.

porcupine wood: wd. of *Cocos nucifera.*

pore (plant histology): 1) pit 2) tiny perforation of sieve plates - **p. fungi**: Polyporaceae.

Porfiromycin: antineoplastic antifungal antibiotic derived fr. *Streptoverticillium ardum* Locci et al. (*Streptomyces ardus* De Boer et al.) (*cf.* mitomycin).

Poria cocos (Schw.) Wolf. (Pachyma c. Fr.; P. hoelen Fr.) (Polypor. - Basidiomyc. - Fungi) (Jap, China): produces large masses of sponge-like sclerotia in rts. of pine trees, &c.; known as Hoelen (Chin med.), Chin root (term also appl. to *Smilax glabra* frt.), foe-lin; "Fuh-ling"; buyuryo; Indian bread; I. loaf; tuckahoe, &c.; c. pachyman (polysaccharide), ebicuricolic acid (triterpene); pachymic acid; u. as special ton. for children ("sushin"); food - **P. obliqua** Fr. = *Inonotus o.* - **P. weirii** Murr.: wood rotting fungus on *Pinus taeda* prod. high yields of peroxidase (different fr. horesradish p.); sugg. as comm. source.

Porifera: Sponge Phylum; invertebrate animals which are attached like pls. to a fixed position; possess no coelom (or body cavity); mostly marine; a few fresh water spp. known.

Porlieria chilensis Johnst. (Zygophyll.) (Ch): guayacán; inf. u. as diaph. stim. balsamic; in chron. rheum.; u. exter. for contusions.

poro: 1) (It) pig 2) (Pe) gourd.

porongo (Por): *Lagenaria siceraria..*

Porophyllum confertum Greene (Comp.) (nwMex): yerba del venado; lvs. & fls. astr.; tea u. in various intest. dis. - **P. gracile** Benth. (sCalif [US], nwMex): hierba del venado; odora; bitter lf. decoc. u. for intest. dis. (465); a popular deer & cattle forage - **P. porophyllum** Ktze. = *P. ruderale* - **P. pseudodynum** Robins. & Green. (Mex): maravilla; u. for headache - **P. punctatum** Blake (sMex, CA, to CR): pioja; pl. inf. u. in GC - **P. ruderale** (Jacq.) Cass. (P. ellipticum Cass.) (WI, CR, to Arg): guacamaya (Ve); lf. fl. decoc. u. antispasm. (Arg); in stomachache, severe eructations (Ve); VO has antimycotic activity (Br); rt. decoc. in erysipelas, joint pain (Br); pl. decoc. hemostat. (Co); much valued by brujas & herb doctors (Ve); subsp. **macrocephalum** R. R. Johnson (P. macroc. DC) (sUS, WI, Mex, to Bo): chipaca; one of the "bush tea" plants (Cur.) decoc. ton. beverage, stim. in flatulence, diarrh.; hemostat., in leprosy (Co); antipyr. (Aruba) - **P. scoparium** Gray (Tex, NM [US], nMex): jarilla; pl. u. in fevers, rheum.; gastroint. dis. - **P. seemannii** Sch. Bip. (nMex): maravilla; u. for mal.

poroto (Pe, Arg, Ch): common bean - **p. balin** (Arg): catjang.

Porphyra perforata J. Ag. (Bangi. - Rhodophyta - Algae) (Pac coast of NA, Alas to Calif): California (or purple) laver; plants are picked fr. rocks & dried in sun; eaten by Chin living in US, also exp. to Chin; ground up & made into a soy sauce (*cf.* sloke); algal harvesting indy. operated in Gulf Islands (BC [Cana]) before World War II by Jap (2410) - **P. umbilicalis** (L) J.G. Agardh. (L. vulgaris Ag.) (nAtl & Pac Ocs): common porphyra; laver (weed); limu (Haw Is [US]); this red seaweed is a very popular food (Brit Is); u. to make laver-bread (466); var. *laciniata* (P.l. [Lightfoot] Ag.) (Pac & Atl Oc): source of edible red laver, using gelatinous fronds (thallus) for popular Jap food; added to foods as "sea (or marine) sauce" (Eng) (since 5th cent.); ton. (Chin).

porphyrins: large-ring org. compds. with several pyrrole gps.; widely distr. in pls. & animals as well as microorganisms; strongly colored; incl. chlorophyll, heme (protein deriv. = hemoglobin); often assoc. with Fe, Cu, Mn, Zn, &c.

porphyroxine: xall. alk. of Opium.

porpoises (Delphinidae - Cetacea - Mammalia): small, gregarious, carnivorous, ocean-dwelling cetaceans found in most seas; like the dolphin but with blunt rounded head & no beak - **harbor p.**: *Phocaena phocaena* - **Dall's p.**: black & white p.; *Phocoenoides dallii*; 2-3 m long - **white p.**: u. for jaw oil & hide (467).

porro (Por): *Allium ampeloprasum.*

Porrum (L.): leek, *Allium p.*

Porterandia penduliflora Keay (Rubi) (Af): powd. lf. appl. to purulent wounds, eczema, & scratches.

Porteranthus stipulatus Britt. (Gillenia s. Baill.) (Ros.) (eUS): American ipecac (root); rt. c. gillenin (gillein) (glycoside); rt. (wild) Indian physic rt.; American ipecac rt. u. as emet.; rt. bk. u. formerly like ipecac - **P. trifoliatus** Britt.

(Gillenia t. Moench) (eNA): Indian physic; Bowman's root; rt. c. gillenin (bitter glycoside); rt. or rt. bk. u. as mild emet. (Am Indians), cath. diaph. expect. ton. (—> *Gillenia*).

Portia tree: *Thespesia populnea.*

Portugal oil: VO of sweet orange (SC frts. first imported fr. Chin by Portuguese).

Portulaca grandiflora Hook. (nSA; natd. NA): "portulaca" (fl. gardens); hb. u. diur. emoll.; ornam. hb. - **P. lutea** Forst. (Portulac.) (Polynesia): pokea; raw shoots eaten by natives; u. med. (New Caled) - **P. oleracea** L. (cosmop fr. As): garden (or common) purslane; common Indian p.; pu(r)sley; verdolaga (Pan); common universal weed (expression: "cheap as pusley") (2329); hb. u. vulner. antiscorb. diur. purg. pl. decoc. for worms; lf. poultice for backache or dysmenorrh. (Domin, Caribs); pot herb in CA (468) (Af, Austral); lvs. boiled as veg.; also lvs. eaten raw as salad (using for latter, var. *sativa* (Haw) DC = (kitchen) garden purslane; larger pl. not the wild-growing) - **P. perennis** (Pe): u. in urin. dis. - **P. pilosa** L. (WI, SA): "jump up and kiss me" (Trin); hb. inf. u. for intest. worms; lvs. u. as diur. &c., much like *P. oleracea* - **P. quadrifida** L. (s&eAs, Af, Ind): wild purslane; pusley; u. as remedy in lung, kidney, & skin dis. (Ind); anthelm.; GC (Af); emet. (Zulus); to prevent abortion; foodstuff (Af); lf. inf. drunk for abdom. trouble (Indones); lvs. u. refrig. antiscorb. (Ind) - **P. rubricaulis** HBK (P. phaeosperma Urban) (WI): "wild parsley" (Bah.); decoc. lax. & u. for baby's "tea" (Bah).

portulace (Por): purslane.

Posidonia australis Hook. f. (Potamogeton. - Posidioni) (Monocot.) (Austral): source of posidonia fiber, Cellonia, Lanamar; u. to make coarse fabrics, sacking, sometimes mixed with wool; st. u. for packing (glass) & stuffing - **P. oceanica** (L) Delile (P. caulini Koenig) (Med Sea): "alga" of Greeks; c. iodine; u. for goiter & skin dis.; produces Pilae marinae (Aegogropilae) (spherical masses of dead & living plants formed by the waves); a form of sea wrack.

posole (Mex): dish made of hominy & capisum spiced with oregano, mint, cloves, etc.

possau (Am Indians of seUS): a brew with emet. props. fr. *Erynigium* spp. u. in the physic dance of the Indians, esp. Cherokees (spring ceremonial).

possession vine: *Convolvulus arvensis.*

possum (sUS): common contraction for opossum - **p. berry** (nwFla [US]): prob. **p. haw**: 1) *Ilex decidua* 2) *Viburnum nudum* - **p. wood**: 1) *Halesia* spp. 2) (Fla): *Diospyros virginiana.*

posterior pituitary: *v.* under pituitary.

pot (US slang): *Cannabis.*

potable (Fr, Eng): drinkable, fit to drink (esp. re. water).

Potalia amara Aublet (Potali.) (monotypic) (tropSA): fresh lvs. chopped in water & taken int. for snakebite or attack of other venomous animal; said sed. & analg.; bk. inf. mild lax. (Ve) (906).

Potamogeton L. (Potamogeton. - Monocots.) (cosmop): pondweed; floating or submerged aqueous plants; pl. u. as fertilizer; most important food of ducks - **P. chamissoi** A. Benn. (Mauritius): u. in card. palpitations, sed. in nervousness - **P. javanicus** Hassk. (Taiwan): u. as antipyr. - **P. natans** L. (temp&subtr zones OW, NW): flutter-dock; tench-weed; ext. gives prolonged prothrombin time; lvs. u. in colic, exter. for sores & herpetic eruptions; in homeopathy; antipyr. (Taiwan) - **P. schweinfurthii** (Nile Valley, Eg): juice was inj. into uterus to cause uter. contractions (ancient Eg); lvs. & sds. often u. in therapy (ancient Eg) (2250).

potash: 1) potassium carbonate 2) KOH (**caustic p.**) 3) pot ash(es), wood ash, earlier prod. by combustion of plant materials (SC lye of wood ashes extd. & evapd. outdoors in iron **pot**) (baking p.: formerly prepd. in home) (2766) 4) any K salt, ex. chloride of potash; $KC10_3$; KI. (Egyptian kali, 19th cent.) - **potassa: potassium hydrate: potassium hydroxide**: KOH.

potassium: kalium (L., Ge.): K; an alkali metal richly present in earth's surface & salt lakes (Caspian Sea; Dead Sea); also in deposits (ex. Stassfurt, nGe); K is essential to life & often deficient in pathological states; best imbibed fr. frts. (ex. bananas) & vegetables (ex. potato); important med.-pharm. compds. include **p. acetate**; **p. bicarbonate**; **p. chloride** (see below); p. citrate; p. gluconate; p. iodide (E); p. permanganate; p. sodium tartrate; &c. - **p. chloride**: KCl; occ. in nature as mineral (sylvite); u. in med. as electrolyte replenisher, K donor (ex. Slow-K Tabs. propr.) (E); u. considerably in indy. - **p. citrate**: very sol. salt with slight alkaline reaction; u. as alkalinizer, diur.; reduces formation of kidney stones (Ca type), sometimes eliminates them (1984); approved (in wax base) for use in uric acid lithiasis (stones), Ca oxalate nephrolithiasis, and renal tubular acidosis with calcium stones.

potato: (pl. or tubers): 1) Irish p., *Solanum tuberosum* (or its tubers) 2) (seUS): sweet p. - **Bermuda p.**: sweet p. - **China p.: Chinese p.**: taro, 1) *Colocasia esculenta* 2) *Ipomoea batatas* - **common (Irish) p.**: Irish p. - **p. family**: Solanaceae - **Indian p.**: 1) *Apios americana* 2) *Dichelostemma pulchellum* Heller 3) *Dicentra canadensis* 4) (Fla

[US]): *Liatris* sp. - **Irish p.**: *Solanum tuberosum* - **Japanese p.**: *Stachys sieboldii* - **Kaffir p.**: *Coleus* spp. such as *C. esculentus*, *C. dysentericus* - **nettle p.**: *Stillingia sylvatica* - **red p.**: sweet p. - **round p.**: Irish p. - **Spanish p.**: sweet p. - **p. spirit**: fusel oil - **sweet p.**: 1) *Ipomoea batatas* (either entire pl. or tuberous rt.) 2) *Dioscorea* spp. (sphalm.) - **p. tree, Brazilian**: *Solanum macranthum* - **West Indian p.**: sweet p. - **white p.**: Irish p. (distinguish fr. white sweet p.) - **wild p.**: *Ipomoea pandurata*.

potency wood: v. Muira Puama.

potentiation: use of 2 drugs —> greater effectiveness than sum of action of the 2 drugs.

Potentilla L. (Ros.) (nearly cosmop.): cinquefoil; 5-finger; some 500 spp. of hbs.; (L): *P. erecta*; *P. palustris* (2162) - **P. anserina** L. (P. argentina Huds.): (orig. Euras, now cosmop): silverweed; (silver) cinquefoil; Herba Anserinae; c. tannins; u. for arthritic dis. & menstr. complaints (as spasmolytic), uter. sed.; colics; ton. (2163); rts. edible - **P. argentea** L. (Euras, natd. US): silver herb; u. in night sweats (Dragendorff) - **P. arguta** Pursh (US, Cana): u. as styptic, dry or moistened rt. (Ojibwa Indians) - **P. aurea** L. (Eu-mts.): Gaensblume(n) (Ge); golden yellow cinquefoil; hb. u. in Eu folk med.; fresh pl. u. in homeopath. med. - **P. canadensis** L. (eNA): fingerleaf; hb. c. tannin; u. astr. ton. in night sweats; constits. similar to those of potentilla rt. - **P. candicans** HBK (Mex): sinfito; hb. u. as astr. in diarrh.; rich in tannin; u. as substit. for European potentilla - **P. chinensis** Ser. (Kor, Chin): sts. & lvs. u. as expect., antipyr.; for ulcers & abscesses (Kor) - **P. discolor** Bge. (Chin): astr. rts. u. as aphrod. & hemostat.; in mal. - **P. erecta** (L) Raeusch. (Euras-natd. NA): tormentil(la); "Blutwurz"; rhiz. of perenn. hb.; c. 16% tannin; rhiz. & hb. u. as astr. ton. in gastroenteritis, colitis, obstinate diarrh. dys.; hemostat., in menstrual dis.; rhiz. u. in teas, in brandy for stomachache; exter. as vulner., in dusting pds. - **P. fruticosa** L. (nTemp. Zone): golden hardhack; shrubby cinquefoil; hb. u. like that of *P. anserina*; dr. lvs. u. as tea (Eskimos, Alas [US]) - **P. glandulosa** Lindl. (wNA): sticky cinquefoil; glandular 5-finger; edible rts. of perenn. hb. boiled or roasted & eaten; lvs. boiled —> beverage (BC Indians [Cana]) - **P. kleiniana** W. & A. (eAs): rts. & sts. appl. to snake & centipede bites (Indochin); boiled with egg u. for cough & int. injuries (Chin) - **P. nepalensis** Hook. (Ind): rts. u. as "depurative" (folk med.) & for coloring wd. red - **P. norvegica** L. (P. monspeliensis L.) (nEuras, NA): Norway cinquefoil; rts. u. astr. - **P. pacifica** T. J. Howell (P. anserina subsp. P. rousi) (wNA, eAs): hb. occurs sometimes as adulter. of *P. anserina* (36) - **P. palustris** (L) Scop. (Comarum palustre L) (Euras, natd. NA): marsh cinquefoil; hb. u. antipyr. astr. ton.; red, yellow dyes; hb. c. VO tannins; subterr. runners u. as decoc. for gout, rheum., styptic, vulner. to strengthen the gums - **P. pumila** Poir. = *P. canadensis* - **P. recta** (L.) Raeusch. (Euras, nwAf, natd. NA): tormentilla cinquefoil; rhiz. off. in various phar. as Rhiz. Tormentillae; c. 17% tannin; tormentilla red, quinovic acid; u. astr. substit. for Krameria; red dye - **P. reptans** L. (Euras, nAf, natd. eNA): (creeping) cinquefoil; 5- finger grass; hb. (as Herba Pentaphylli); c. tannin; u. astr. hemostat. antipyr.; in epistaxis, wound infections (esp. of horses), sterility - **P. rupestris** L. (Euras): lvs. u. in Siberia as tea ("Siberian tea") - **P. sylvestris** Neck. = *P. tormentilla* Neck. & Schrank = *P. erecta* - **P. thurberi** Gray (Mex+): clameria; yerba colorada; rt. decoc. u. for stomach ailments; rt. interior crushed & placed between cheek & gums for toothache.

potentille rampante (Fr): *Potentilla reptans*.

Poterium L. = *Sanguisorba*.

pot-herb: various plants u. for "greens"; lvs. & herbaceous sts. boiled & eaten like spinach or used as spice; the liquid may then be poured off & u. ("pot liquor").

Pothoidium lobbianum Schott (monotypic) (Ar) (PI, Moluccas): ayasun mung; caisimon; rts. bruised & appl. to part bitten by centipede; cyanophoric pl.; source of fiber u. for fish-corrals.

Pothomorphe peltata (L) Miq. (Piper p. L) (Piper.) (WI, CA): Santa Maria (Pan); patza (Cuna Indians, Pan); lvs. u. stim. as cold cure (Bayano, Cuna Indians); lvs. rubbed on body for swellings, ulcers, syph. sores; to exterminate lice, ticks (garrapatas); heated lf. as poultice in headache (Trin, Domin); rhiz. strong diur. - **P. sidaefolium** Miq. (Piper s. Link) (Br, Par, natd. trop): caapeba; lvs. u. in liver dis.; u. in asthma as expect. stom. diur. in dropsy; hb. eaten as veget. (Java) & appl. to wounds - **P. umbellata** (L) Miq. (Piper u. L.; Heckeria u. [L] Kunth) (CA, Br): periparoba; ca(a)peba; Yerba Santa Maria; all parts of pl. u. med. (frt. & rt. considered most effective); u. diur., in stomach & liver dis. (bil. stim.), in rheum.; rt. u. arom.; frt. rich in VO, u. as stom.; lvs. to destroy ticks; with chicken fat to reduce swellings (Salv).

Pothos repens Druce (Ar) (Indochin): st. juice rubbed on rheum. joints; lf. inf. emmen. - **P. scandens** L. (eAs): lvs. u. in fever in monsoon season (Ind-Indones); anthelm., anticonvulsive; ornam. vine (several other spp. also u. as ornam. vines).

potiguaya (Sp): crude marihuana.

potiron (Fr): pumpkin or squash - **p. commun** (Fr): *Cucurbita maxima* (with yellow frt.).

pot-liquor (sometimes as "pot-likker") (Ala [US], &c.): liquid poured off boiled greens or field peas & fat meat (hog's jowl); good for soups.

pot-pourri (Fr): 1) medley or mixt. 2) jar of mixed arom. fl. petals (esp. rose) or hbs. (as cloves), generally preserved with salt; u. to fumigate (or scent) a room; fumigating species.

potrero (SA): pasture land; cattle farm.

potstone: talc.

Pouchetia africana A. Rich. (Rubi) (wAf): leafy branches u. to prep. beverage u. for rheum., intercostal pain, lumbago; lf. decoc. in dys.

poudre (Fr slang): cocaine (lit. powder)

poule (Fr-f.): hen, fowl.

pouliot américain (Fr): American pennyroyal.

Poulssen units: Oslo units (in standardizing codliver oil).

Pourouma acuminata Mart. (Urtic.) (Br): frt. edible - **P. bicolor** Mart. (Br): similar - **P. cecropiaefolia** Mart. (Br): frt. edible with grape-like flavor; cult. in Amazon area - **P. mollis** Tréc. (Br): rt. bk. u. in dys. (Peckolt).

pourpier (Fr): purslane.

Pouteria caimito (Ruiz & Pav.) Radlk. (Lucuma c. A DC) (Sapot.) (Br, Pe, Co): abiu (egg frt.); abi(a); abieiro; frts. are eaten; latex of stem u. for abscesses; toasted macerated young lvs. appl. to wounds as disinf. (Wioto Indians, Co) (2299); frt. juice u. drastic, anthelm.; comm. timber; sds. c. saponins - **P. campechiana** Baehni var. *nervosa* Baehni (Lucuma n. A DC; L. salicifolia HBK) (sMex, nSA): caniste(e)l (lucuma); "egg fruit" (Fla [US], SC, yolk-like appearance of frt. flesh); zapote borracho (drunken sapota); frt. edible, as flavor (c. VO); said hypnotic; bk. reputed antiperiod. - **P. gutta** Ducke = *Lucuma g.* - **P. lasiocarpa** (Lucuma l.) (Br): abiorana; latex of tree u. as gutta percha substit. - **P. magalismontana** (Sond.) A. Meeuse (Af): frt. & rt. inf. u. for "fits"; pd. rt. rubbed into skin incisions for rheum.; frt. edible, very rich in vit. C - **P. melinonii** Baehni (Co): bk. inf. u. as strong purg. (Tikuna Indians, Co) (2299) - **P. sapota** H. E. Moore & Stearn (P. mammosa (L) Cronquist; Calocarpum sapota Merr.; Achras m. L.) (Guat; cult. CA & trop): mamee apple; zapote (colorado); mamey (sapote) (tree); marmalade plum; mamey colorado; large frt. with sweet red pulp edible; sds. with flavor of bitter almond, u. in beverages (false cola), candies; sd. kernel furnishes FO, u. mfr. soap, hair dressing (belief it prevents hair falling); for musc. pain, rheum., colds; colitis, cardiac dis.; bk. & lf. decoc. u. to relieve arteriosclerosis & to lower bl. press.; sap emet. anthelm.; wd. useful; ornam. tree - **P. ucuqui** Pires & Schultes (wBr): ucuquí; frt. u. as vermif. (popular med.); sds. u. to make toy whistles; frt. attractive to natives - **P. viridis** Pitt. (Guat, Hond, CR): green sapota; ingerto (verde); edible frts. sold in market, very popular; ripe frts. thought healthful but u. as vermicide; sds. u. in cocoa to remove phlegm & "bad humors" (2602).

Pouzolzia denudata De Wild. (Urtic.) (Congo): betaume; lvs. u. to heal wounds & for eye dis. in dogs; consumed as veget. - **P. hypoleuca** Wedd. (sAf): wood old fly; st. decoc. u. for biliousness (enema) (Zulu); bk. source of tingo fiber, u. fish nets - **P. viminea** Wedd. (eAs): fibers u. to make fish nets, &c.; paste of whole pl. appl. to itch (EI natives) - **P. zeylanica** Benn. (s&eAs): pl. or rt. pounded & appl. as vulner. to sores, abscesses, swellings (Chin); diur., cicatrizant for gangrenous ulcers (PI); galactagog (Indones).

poverty weed: 1) *Diodia teres* (398) 2) *Iva axillaris* & other *I.* spp. 3) *Monolepis* spp.

powder, digestive: pepsin pd. comp.

powderpuff flowers: of Mimosa tree; *Albizzia j.*

po-yoak oil: fr. *Afrolicania eleosperma.*

Prachtscharte (Gc): *Liatris scariosa, L. elegans*; *L. spicata* (lit. splendid saw-wort).

Pradosia lactescens (Vell.) Radlk. (Lucuma glycyphloea Mart. & Eichl.) (Sapot.) (Br, esp. Rio de Jan): monesia; m. bk. c. monesin (saponin) (32-62%), tannin, gallic acid, glycyrrhizin, NBP; u. astr. in enteritis, hemoptysis, GC; expect.; stom.; exter. for skin ulcers, ophthalmia, &c.; in tanning indy.

Praecocia: apricot, *Prunus armeniaca* (SC early fruiting) (Dioscorides).

Praequine (As): Plasmochin.

Prague powder: P. salt: curing pickle; c. NaCl, $NaNO_2$, dextrose.

prairie: extent of flat or rolling grassland enclosed by woodland; typical of cNA - **p. dock**: *Parthenium integrifolium* - **p. flower**: *Pulsatilla patens* - **"p. plant"**: *Achillea millefolium* - **p. wool**: upland grass.

prairie: *v.* also under anemone; quinine.

Pralin (Ge): dr. peeled almonds or hazel nuts roasted brown with sugar, chocolate, & finely pd.

praline (Eng, Fr): confection of dr. peeled nuts or nut kernels, often almonds (or hazel nuts), roasted in boiling sugar until crisp brown with added chocolate; more commonly & orginally (seUS, esp. La) a cake of brown sugar c. pecan kernels; chocolate cream (SC Marshal of Plessis-Praslin, fl. 1650, whose cook invented this confection); sometimes u. in puddings, ice cream, cremes.

Prangos pabularia Lindl. (Umb) (Indo-Pak, Iran): komal (Hind); frt. stim. carm. stom. diur. emmenag.; aid to expel fetus.

Praeparirte Kreide (Ge): prepared chalk.

prasterone: androsterone deriv., isolated from male urine; also synth.; androgen.

praticultura (Sp): art & science of cultivating fields, lawns, &c., hence of grasses, &c.

prawn: large shrimp-like crustacean (several genera incl. *Palaemonetes*); u. as popular sea food.

prayer beads: **p. beans: p. seeds:** *Abrus precatorius*, sds.

précis (Fr): resume; summary.

précurseur (Fr): precursor: any compd. (exogenous or endogenous) wh. can be converted into some other compd. or prod. by a living organism (*cf.* intermediate).

prednisolone (E) and **prednisone**: pregnadiene ketones prepd. from steroid precursors with microorganisms or by chem. synthesis; glucocorticoids u. as anti-inflammatory; anti-cancer agts.

pregnane: synth. prod. related to pregnanedione - **pregnanediol**: prod. of metabolism originat. fr. progesterone; occurs in pregnancy urine - **pregnanedione**: steroid compd. occurring in pregnant mare's urine.

pregnenolone: steroid compd. mfr. fr. stigmasterol; obt. fr. *Agave* & *Yucca* spp.; u. in rheumatoid arthritis; fatigue due to stress; nocardiosis.

Preis(s)elbeere (Ge): red whortle berries; lingon b.; *Vaccinium* spp. (Eu), esp. *V. vitis-idaea* L.

prêle (Fr-f.): horsetail - **p. des champs**: *Equisetum arvense*, field h.

Premarin™: water-soluble conjugated mixed estrogenic substances fr. pregnancy urine.

Premna arborea Farw. (Verben.) = *Premna taitensis* - **P. chrysoclada** Guerke (tropAf): lf. u. for ulcerations & inflammn. - **P. cordifolia** Roxb. (Mal): rt. & lf. decoc. antipyr (Mal); rts. for "shortness of breath" (Borneo) - **P. cumingiana** Schau. (PI): lf. inf. diur. in edema; for general malaise - **P. gaudichaudii** Schauer (Fiji, &c.): rts. u. with plant sap in fever with pain in limbs; lvs. warmed & placed on skin for muscle pains; fls. popular for decoration (Caroline Is) - **P. latifolia** Roxb. (sAs): paste of rt. u. as local appln. after childbirth (Burma) - **P. nauseosa** L. (P. integrifolia Blanco) (tropAs): headache tree; lvs. remedy in gastric diffic. (PI); lvs. chewed sometimes with betel nut in place of b. lvs. - **P. obtusifolia** R.Br. (P. corymbosa Miq.) (Af, As, Austral): arani (EI); decoc. of leafy shoot u. mal. (Chin); lf. decoc./inf. for cough (PI); u. med. in Gilbert Is (103); decoc. u. rheum. & neuralgia, lvs. bitter, carm; soup, pectoral (mixed with pepper) in fever; rt. powd. cordial, stim. in dig. weakness, in fevers (Malagasy) - **P. odorata** Blanco (P. tomentosa Willd.) (PI, now spread throughout eAs, Oceania): decoc. diaph. carm., in pulmonary dis., cough; fl. decoc. in dys. & abdom. pain; valuable wd. - **P. parasitica** Blume (Indones): vulner. after childbirth - **P. taitensis** Schau. (Tonga to Fiji): yaro; one of the constituents of the crude drug, Tonga; u. in neuralgia; lvs. u. med. (Tonga) (86).

Prenanthes L. (Comp.) (OW, NW): rattlesnake root; perenn. hbs.; some writers put spp. under *Nabalus* - **P. alba** L. (e&cNA): white lettuce; (w.) rattlesnake root; pl. u. bitter ton., famous snakebite remedy (Am Indians) - **P. altissima** L. (eNA-woodland thickets): wild (or white) lettuce; cankerweed; lion's foot; poultice of fresh pounded lvs. appl. to ulcers, sores, &c. (folk med.) (2165) - **P. serpentaria** Pursh (eNA): gall of the earth; lion's foot; hb. (as Herba Nabali serpentariae) c. tannin; lvs. & rt. u. as bitter ton. astr.; for snakebite; in homeopathy; in hypertension (US Negroes).

prenyl: 3-methyl-2-butenyl (isoprene) radical.

prepared: *v.* under chalk, hartshorn.

prescription specialties: TM medicinals intended primarily for sale on prescriptions; also known as ethical proprietaries.

preservation of crude drugs: best managed by thorough drying of drug (crisp dr.; so dry that part breaks when bent), glass or metal packaging, together with appropriate measures against insect & rodent attack, such as fumigation with chloroform, CCl_4, etc.

Prestonia amazonica (Benth.) Macbride (Apocyn.) (Br+): apparent source of hallucinogenic drug of Indians; "yage"; reputedly c. dimethyltryptamine; may have been confused with *Banisteriopsis rusbyana.*

prezzemolo (It): parsley.

Priapiscum (L): *Orchis pyramidalis.*

Priapus Caballi (L.): dr. horse penis; u. in med. (Jap) - **P. Cervi**: dr. deer penis; u. aphrod. (Jap) - **P. Ceti**: whale penis; u. med. - **P. Tauri**: bull penis; u. antiluetic (Jap.).

prick off: to transplant young pls. once or more, preparatory to final planting out in field.

prickly alder (bark): *Aralia spinosa* - **p. ash**: 1) *Zanthoxylum americanum* (northern); *Z. clavaherculis* (southern) 2) *Aralia spinosa* (confusion) - **Indian p. a.: Southern p. a.: yellow p. a.:** *Z. clavaherculis* - **p. leaf** (Miss [US]): *Solanum carolinense.*

prickly: *v.* also under poppy, lettuce, pear, saltwort.

pride of China: p. of India: *Melia azedarach.*

priest's crown: dandelion.

Primaquine(e): synth. antimalar.; quinoline derivative; prep. by Elderfield.

Primates: order of mammals, consisting of monkeys, apes, man, &c.; hands & feet are adapted for grasping; regarded as highest group in evolutionary devt.

primavera (Sp): **Primel** (Ge): **primèvre** (Fr-f.): primrose - **primola** (pl. primole) (It): primrose: *Primula veris.*

priming dose: loading dose.

primocane (blackberry, *Rubus*): 1st season's shoot.

primrose: 1) *Primula veris* & other *P.* spp. 2) *Rosa eglanteria* 3) *Oenothera* spp. (some) - **birdseye p.**: *Primula farinosa* - **evening p.: tree p.**: *Oenothera* spp., esp. *O. biennis* - **white e. p.**: *O. speciosa* - **p. willow**: *Jussiaea repens.*

Primula auricula L. (Primul.) (Eu-Alps): (garden) auricula; bear's ear; rt. c. saponins, paeonol, primveroside; hb. (Herba Auriculae muris) & rt. u. in folk med. for headache, vertigo, lung dis., TB, phthisis; expr. juice in chilblains, sores - **P. elatior** (L) Hill (Eu, e to Iran): oxlip; true cowslip; fls. (as Flores Primulae) c. VO resin, flavonoids (quercetin & kaempferol glycosides); u. diaph. expect. in coughs, colds, gout, rheum.; hb. u. similarly; rt. (as Radix Primulae) c. saponins, VO resin, volemitol; u. like senega rt. as expect. in pulm. dis., asthma, pneumonia, diur. migraine, card. weakness; like *P. veris* (off. several Eu Phar.) - **P. farinosa** L. (worldwide, nTemp zone - wSA - alpine regions): mealy (or birdseye) primrose; rt. c. saponins, flavones (incl. rutin); u. to facilitate resp. in mountain climbing - **P. obconica** Hance (Chin): top primrose; glandular trichomes (esp. of calyx & pedicels) secrete poisonous subst. (primin) to which some persons are sensitive (true allergy); mostly dermatitis produced; hb. u. in folk med. & homeop. for urticaria, eczema, &c.; cult. ornam. - **P. reticulata** Wall. (c&eHimal-Ind, Chin): hb. u. as anodyne (exter.) (Hindu med.); toxic to cattle - **P. sieboldii** E. Morr. (Chin): c. sakuraso acid (saponin); u. expect. (good) - **P. sinensis** Sabine ex Lindl. (Chin): Chin primrose; secretion c. flavones; u. med. - **P. veris** L. (P. officinalis Jacq.) (Euras): cowslip; primrose; fls. (with & without calyces) u. much like *P. elatior*; rhiz. & rts. (off. Austral, Hung, Swiss & 7 other Phar.) c. saponins, VO with methyl salicylate; tannin; u. expect. in pertussis, asthma, pulmon. dis., as snuff; analg. in rheum. gout, migraine, diur. in urin. inflammn. antipodagric.; hb. u. in card. weakness; as spring ton. (tea); substit. for Senega; potherb; salad plant (2164) - **P. vulgaris** Huds. (P. acaulis (L.) Hill) (Euras, nAf): common (or spring) cowslip; oxlip; primrose; lvs. & rts. u. like *P. elatior* & *P. veris* as emet. sternutat. anod.; pl. ornam. perenn. hb. (2017).

Primulaceae: Primrose Fam.; mostly arctic & temp. zone hbs.; ovary superior; unilocular with many ovules; ex. *Lysimachia*; *Cyclamen.*

Primulae Radix (L): primrose rt., of *Primula veris* &/or *P. elatior.*

Primycin: debrycin; macrolide antibiotic formed by *Streptomyces primycini* & *Micromonospora "galeriensis,"* occ. in intestinal tract of larvae of *Galeria melonella* (wax moth); active against *Staph.* spp. (boils) & mycobacteria; u. topically as antibact.; discovered by Valyi-Bagi et al., Debrecen, Hungary, 1954.

prince's (or **princess'**) **feather**: 1) *Amaranthus hypochondriacus* 2) *Polygonum orientale* - **prince's pine**: *Chimaphila umbellata* - **princess' pine**: *Lycopodium clavatum* - **p. tree**: *Paulownia tomentosa* - **princewood (bark)**: *Exostema caribaeum; Cordia* spp.

Pringlea antiscorbutica R. Br. (monotypic) (Cruc.) (Kerguelen & Crozet Is-sIndian Ocean near Antarctica): Kerguelen cabbage (shows habit of cabbage); u. as antiscorb. (seamen; 19th c.); coll. by J. H. Kidder (1874); nearly exterminated by rabbits brought to island ca. 1895; u. by islanders as salad (469).

Prinos verticillatus L.: *Ilex verticillata.*

Prinsepia utilis Royle (Ros.) (nInd, Pak): sd. FO rubbed or appl. exter. in rheum.

printanisation (Fr): vernalization.

Prioria copaifera Griseb. (1) (Leg) (Co, Ve, Pan, Jam): cativo; cativa; wounds in trunk prod. cativo balsam, similar to copaiba; "oil tree gum"; u. turp. & copiba substit.; u. as urin. antisep. in cystitis, for sores, tetanus(!?) (Resina Prioria); u. as adhesive, in prepn. resin soaps; binder for paper & stucco mfr.; wd. useful.

priprioca (Br): *Cyperus odoratus.*

prise d'essai (Fr): sample for analysis.

Prismatomeris tetrandra K. Sch. (Rubi) (seAs): wd. inf. ton. esp. at childbirth (Indochin); bruised lvs. as poultice for wounds (Mal).

Pristimera celastroides Miers (Celastr.) (CA, Mex): sds. crushed in lard & appl. for delousing (Salv) ("mata piojo").

Priva cordifolia Druce (Verben.) (sAf): cold inf. of lf. appl. to conjunctivitis; sd. paste appl. to sores & wounds - **P. curtisiae** Kobuski (Tanz, Kenya): "ikubya"; lf. juice squeezed into eye for ophthalmia; pd. lvs. & fls. as poultice for sores - **P. laevis** Juss. (trop): tubers edible - **P. lappulacea** (L) Pers. (P. echinata Juss.) (WI, tropAm cont., Java): styptic bur; "farolito" (Cu); lvs. u. as poultice in headaches (Sonora [Mex]), in pertussis, (Choco Indians); good vulner. for severe wounds (470); styptic (Jam) (also juice); in leukorrhea (Mex); lvs. in tea; medl. (Pan) - **P. meyeri** Jaub. & Spach (Sudan, Moz, sAf): sds. ground up & appl. to sores (Zulus); lf. inf. u. for inflamed eyes (Zulus).

privet (leaves): *Ligustrum vulgare* - **Egyptian p.**: *Lawsonia inermis* - **evergreen p.: Japanese p.**: *Ligustrum japonicum* - **wild p.**: *Forestiera neomexicana.*

pro anal.: pro analysi (L.): analytical quality of a chemical.

Probecillin (TM: Astra, Sodertolje, Sweden): combn. of penicillin & benamide (tabs.); supposed to increase blood level of p.

Proboscidea fragrans (Lindl.) Dcne. (Martynia f. Lindl.) (Martyni) (Mex): sweet devil claws; toloache; catachio; u. against "convulsions" (471); by Mexicans believed to produce permanent insanity (2230) - **P. louisianica** (Mill.) Thell. (Martynia p. Glox.) (c&sUS, nMex): (common) unicorn plant; devil's claws; "rat tags"; "proboscis flower"; frt. has 2 long, woody horns, made into craft objects (dolls, &c.); small immature pods made into pickles (*cf.* cucumbers); sds. u. dem. in urinary dis.; cult. Italy; both spp. & **P. parviflora** Woot. & Standl. widely cult. & u. by swUS Am Indians; for basket-making & food (frt. & seed) (1704); fls. pretty, orchid-like; cult. ornam.

procaine penicillin: *v.* penicillin.

Procardin™: proscillaridin.

"prodigiosa" (SA, Mex): *Brickellia cavanillesii.*

prodigiosin: antibiotic pigment fr. *Serratia marcescens* Bizio (Chromobacterium prodigiosum; Bacillus prodigiosus, &c.); antifungal; antimalarial; (but too toxic to use); antileukemic.

productos alimenticios (Arg): food stuffs - **p. de caucho diversas** (Sp): rubber sundries - **p. de lecheria** (Arg): dairy prods.

pro-enzyme: zymogen.

profibrinolysin: plasminogen.

progestin: progestogen: one of a class of endocrine agts. u. to promote the progestational part of menstrual cycle manifested by corpus luteum endometrial secretory activity; term sometimes appl. specif. to progesterone.

progesterone: corpus luteum hormone; has been synthesized fr. steroids; u. in threatened abortion to maintain pregnancy; menstr. dis. (IM inj. of FO soln.); helps overcome PMS; contraceptive - **p. root**: *Dioscorea* spp. (SC u. source of synth. p.).

pro-insulin: insulin precursor made up of single chains A and B plus a connecting polypeptide; the no. & sequence of amino acids is species-dependent; occ. in beta cells of pancreas, &c.

prokaryotic (procaryotic): anucleate organisms lacking a well-defined nucleus; ex. Bacteria.

prolactin: luteotropin: the lactosecretory hormone of the anterior pituitary, which initiates & maintains milk secrn. in female.

Prolan: confusing term; formerly appl. to follicle-stimulating hormone (A) and luteinizing hormone (B); later to human chorionic gonadotropin; in America also appl. to DDT deriv. (pesticide), & in Ge, to a concrete condensation agt.

proof spirits: 100-proof spirit is a liquid which c. 50% (V/V) of absolute ethanol.

propadrine: phenyl propanolamine; sympathomimetic amine; occ. in small amts. in *Ephedra* spp. (2676).

propagules: buds, stem cuttings, bulbs, gemmules, &c., capable of reproducing the pl. asexually.

propepsin: pepsinogen, precursor of pepsin.

properdin: serum protein factor believed useful in natural immunity against infect. diseases & possibly cancer (1954-).

propharmacien (Fr): **propharmacist**: a physician who sells or dispenses drugs to his own patients at times when or in places where a pharmacist is unavailable.

prophylactic ointment: calomel 30% ointment; u. to prevent infection by syphilis (& GC) by local apln. *post coito.*

Propionibacterium freudenreichii subsp. **shermanii** (van Niel) Krasilnikov (Propionibacteri.): microorganisms believed responsible for the spaces ("eyes") produced in Swiss cheese.

propionic acid: methyl acetic acid; CH_3CH_2COOH; occ. in many natural prods. (ex. dairy prods.; as esters in VO [ex. lavender oil], fly agaric; sargasso weed [alga]); synth. 1844 before being found in nature; u. generally as salt as antifungal; as esters in flavors; org. synth.; to preserve arterial grafts; various salts u. as antifungals, esp. sodium propionate (u. as fungicide, in dermatoses, conjunctivitis, vulvovaginitis, pruritis, otitis, eczema, burns, wounds, etc.).

Propolis (L): bee glue; red odorous resinous subst. collected fr. coniferous trees, sweet gum tree, &c. by honeybee & u. to fill openings (cracks, crevices) in hive walls; to seal wax cells; to embalm intruders; c. pinocembrine, galangin; claimed to have antibiotic (antibacterial) properties & sold in capsules (sometimes in eq. parts with pollen) in health stores (1705); anciently u. as cement; substit. for Galbanum.

proprietary: Trademark (TM); medicinal commonly sold OTC to the layman without supervision;

those advertised professionally are generally sold only on prescription or physician's order.

proscillaridin (**A**): glycoside obt. fr. *Urginea maritima, U. indica, U. burkei*; u. as digitaloid in congestive heart dis. AP.

prosenchyma: tissue consisting of elongated cells with tapering ends which overlap; ex. bast fibers.

proso (**millet**) (US): *Panicum miliaceum*.

Prosopis L. (Leg.): mezquite; screwpod; a desert & arid land shrub or small tree; source of mesquite gum; frts. food (2187), forage; prepn. alc. beverages; sds. for cakes; bk. for fuel, diapers (Ind); shade tree - **P. africana** Taub. (w&cAf): rt. u. in pop. med. as diur. & urin. disinf. (GC); vermifuge; lf. cataplasm for dermatoses, mycoses, filarioses; appln. of hot macerate of lvs. u. for head pain (earache, migraine, toothache); in dental caries; sialagog; wd. soporific - **P. alba** Griseb. (Arg): algarroba blanco; sweet pods u. as food (very important in nutrition of native poor of neArg); beans u. to prep. beverage like beer ("aloja de algarrobo"); dr. beans u. to prep. meal u. to make tortas (sweet) ("patay") (472); good forage for cattle; source of a kino; useful wd.; lvs. & frt. u. astr. (exter.) in ophth. dis., antilithic - **P. algarrobillo** Griseb. (Arg): algarrobillo; algarroba negro; source of a dye; wd. useful in construction - **P. algarrobo**: *P. alba* & *P. nigra* - **P. caldenia** Burkart (Arg): calden; frt. nutritive; wd. useful - **P. dulcis** HBK (SA): algaroba; frt. of tree edible; useful stock food; a source of mesquite gum - **P. farcta** (Banks & Sol.) Macbride (P. stephaniana [M. Bieb.] Kunth) (Pak, Syr, Iran, Saudi Arabia, Cypr): frt. (pods) have food value (free of HCN); long cult. (Mesopotamia) - **P. glandulosa** Torr. (Algaroba g. T. & G.) (swUS, nMex): (honey) mezquite; mesquite (Sp) (2110); largest tree of genus (-20 m; -1 m thick at ground); pods slender & straight, in clusters, 10-15 cm long; pods, lvs., small sts. u. as forage; trunks (wd) for fence posts, fuel, mfr. charcoal to prepare smoked meats; furniture, cross-ties, house foundations, paving streets; exudn. fr. trunk called gum mesquite (yellowish or amber); u. for fine mucilages; in confectionary; salad dressings (Calif [US]) (2111); u. as dye by Mexicans; rts. u. as fuel; fls. cause of pollinosis (swUS); shade tree; beans as flour in foods, fls. furnish honey; inner bk. pd. with salt & water as med. (Mex-children); frt. food, mfr. beer - **P. juliflora** DC. (P. chilensis [Mol.] Stuntz) (Mex-CA, WI, SA to Arg, Ch): mesquite; honey pod; cuji (yague); similar to *P. glandulosa*; sweet frts. (algaroba pods) (somewhat like carob beans) edible, pd. in fat appl. to boils to mature; sds. & bk. inf. gargled for bronchitis, pharyngitis, laryngitis; has antibiotic props.; pods (meal) u. as stock food; trunk secretes gum mesquite (Sonora or prosopis gum) u. as mucilage, in sore throat, colds; in foods, chewing gum, dye (2186); soln. of gum (or rt. inf.) u. astr. in diarrh. of children; tree cult. in trop. (as Ind); sometimes *P. chilensis* considered a good sp. - **P. microphylla** HBK= *Acacia tortuosa* - **P. nigra** (Griseb.) Hier. (Arg): algarrobo negro (Arg); frt. u. in ophth. dis. (exter.); eupeptic; in prepn. of alc. beverage; excellent wd. - **P. panta** Hieron. (Arg): algarobilla; cama tala; wd. source of valuable timber - **P. pubescens** Benth. (Tex, Calif [US]; Mex): screwbean; screwpod mesquite; tornillo; curly mesquite tree; pod tightly coiled, hard; nutr.; u. to prep. alc. beverage; wd. durable - **P. spicigera** L. (P. cineraria [L] Druce) (wInd, Pak, Iran): pod u. astr.; bk. in rheum., scorpion sting; ashes rubbed on skin to remove hair; heart wd. valuable - **P. strombulifera** Benth. (Arg, Andes): frt. c. tannin, u. astr.

prostacyclin: a prostaglandin wh. is produced by cells lining the arteries & wh. relaxes the blood vessels & inhibits blood platelet aggregation.

prostaglandins: PG; hormones prod. by many tissues of animal body incl. the prostate gland & semen; exert many physiol. actions such as stim. of smooth muscle; facilitates childbirth (relaxes uterine cervix); u. to prod. abortion, in amenorrhea ($PGF_{2\alpha}$); E_2 most potent of gp.; u. for induction of labor, menstr., abortion; to promote penile erection; group incl. prostacyclin, thromboxane (prod. by platelets; furnish nucleus of blood clot); PGF_2; (reduction prod. of E_2) gel u. to facilitate cervical "ripening" process (to aid initiation of childbirth); oxytocic suppository u. as abort.; so named since early observations indicated the prostate gland was the place of origin; several compds. recognized; also prod. by gorgonian soft corals (sea fans), as Plexauridae (perhaps by associated alga) (WI, Fla [US]).

prostanoids: collective term for prostaglandins, prostacyclins, & thromboxanes.

prostate gland: secretory gland of male sexual apparatus; related to potency in male; enlargement leads to benign prostatic hyperplasia (BPH), causing urinary dysfunction; one of chief sources of prostaglandins; ext. proposed for use in impotence.

prosthetic group: co-enzyme.

Protalba™: protoveratrine A.

protamine: simple protein yielding basic amino acids on hydrolysis & wh. occ. in fish testes & sperm (combined with nucleic acids); group of protamines, incl. salmine, iridine, &c., wh. inhibit inflammn. (cervical arthritis), tumor growth,

&c. - **p. sulfate**: anticoagulant but neutralizes heparin (also an anticoagulant); u. by intrav. inj. for overdoses of heparin, hemorrhagic states, &c. (E) - **p. zinc insulin**: combn. of $ZnCl_2$ with insulin & protamine (fr. sperm or testes of *Oncorhynchus*, *Salmo*, or *Trutta* spp.); adm. SC; advantages: more prolonged action; allows greater utilization of dextrose by body; requires fewer & smaller injections.

Protea gaguedi Gmel. (Prote.) (Af): sugar bush; rt. decoc. u. for pains in lower abdom.; for fainting - **P. grandiflora** Thunb. (tropAf): waggon tree; astr. bk. u. in diarrh., as tan, black inks; good wd. - **P. hirta** Klotz. (sAf): (white) sugar bush; decort. rt. inf. u. in dys.; bk. rich in tannins - **P. madiensis** Oliv. (cAf): crushed lf. u. as poultice; boiled lf. eaten as famine food - **P. mellifera** Thunb. (sAf): nectar condensed into a syr. (Syrupus Proteae) of value in pulm. dis. & cough; has been u. in diabetes (c. levulose, no dextrose); fls. source of fine honey.

protease: protein digesting enzyme.

protective foods: those fending off deficiency dis. ex. vitamins.

proteic acid: amino acid.

proteins: complex org. nitrogenous substances rel. richly present in meats, fish, eggs, cheese, legumes, &c.; of very high molecular wt.; characteristic of living matter & representing its bulk; made up of C,H,O,N (S,P); yield amino acids on hydrolysis; in pls. often found deposited in aleurone grains; ex. albumin, casein, &c. - **crude p.**: computed in analysis by multiplying N content by 6.25; incl. primary proteins, peptides, amino acids, &c. (simple p. soluble in water; globulin p. & sclero-p. insol. in water) - **p. hydrolysate**: sterile soln. of amino acids & short-chain peptides derived fr. hydrolysis of casein, lactalbumin, plasma, fibrin, &c.; must be rich in amino N; u. by inj. as nutrient & fluid replenisher - **p. sparers**: carbohydrates & fats eaten along with proteins; useful in furnishing energy sources & conserving the rel. expensive proteins for tissue building.

proteinase: proteolytic enzyme wh. hydrolyzes whole protein mols. esp. high m.wt. proteins, such as albumin —> amino acids; also involved in synthesis of proteins; found in all living tissues; ex. cathepsin.

proteoclastic: proteolytic: process of converting or ability to convert proteins into soluble peptones or amino acids usually by hydrolysis or decompn.; ex. papain, trypsin.

proteohormones: those made up of proteins or polypeptides, such as ACTH; injected into lab. animals said to promote faster learning.

prothrombin: subst. essential to blood clotting which occurs in plasma; converted to thrombin in presence of thromboplastin & Ca ions.

protirelin: thyrotropin-releasing hormone, TRH: occ. in hypothalamus.

Protium aracouchini March. (Burser.) (Guiana, Br): produces a balsam (Balsamum Acouchi or Aracouchini Balsam); u. as vulner. & antisep. (for teeth) (SA) - **P. carana** (Humb.) March. (nBr-Ve): tree the source of Hyowana resin (Carana, Hyowa, or Mararo Elemi) (Haiawa or Hyawa Gum) u. for wounds & contusions; also carana gum - **P. guaianensis** March. (Icica guyanense Aubl.) (tropAm-Guiana, Ve, Br): tree trunk source of a balsam (variously called Resin Tacamahaca from Cayenne; Elemi of Guiana, Cayenne incense, Balsamo real) (prod. wrapped in palm leaves); u. to treat GC (exter.); in lacquers, incenses (churches); in prepn. black paint - **P. heptaphyllum** (Aub.) March. (Icica heptaphylla Aubl.) (nBr, Co, Ve, Guiana, Par): incense tree; bastard cedar; Brazilian resin tree; balsam (oleoresin) forms fr. wounds in tree trunk = (Resina) Almecega or Elemi of Colombia (Venezuela), Tacama(ha)ca Resin, Brazilian Elemi; hiawe(resin); u. to perfume anointing oils (SA Indians); as vulner., against tumors; very arom. resin u. as cure for headache (Br); added to coca-ash mixt. (2510); mixed in clay in making pots; & many techn. uses - **P. icicariba** (DC) March (Icica i. DC) (WI, Br): furnish elemi resin (Rio or WI elemi; hard elemi [of Brazil]; Anime); u. exter. for wounds & sores, in coughs, & as incense - **P. insigne** Engl. (Ve): masramo tacamahaco; colorless oleoresin (exudation) u. in varnishes, to substit. for true elemi; u. in family med.; wd. u. in carpentery, for house beams, construction (473) - **P. javanicum** Burm. f. (Indones): lf. poultice appl. for abdomin. pain; rts. & lvs. in combins. for dys. - **P. octandrum** Swart. (nSA): oleoresin u. to treat ulcers, pimples, syph. (Co); for ringworm, tumors (Ve).

proto-alkaloids: simple type without indole nucleus; ex. mescaline.

protoanemonin: antibiotic princ. (lactone) of spp. of *Pulsatilla*, *Clematis*, *Ranunculus* (988).

protocatechuic acid: $Ph(OH)_2.COOH$; dihydroxybenzoic acid; occ. in many plants & pl. materials; ex. lvs. of *Knautia silvatica*, frts. of *Illicium religiosum*; also prep. synth. & u. in org. syntheses.

Protococcus viridis C.A. Agardh (Pleurococcus naegeli Chad) (Ctenoclad. - Chlorophyc. - Algae) (very widely distr., cosmop. on bk. or wd. & in soil); tree-scum alga; a unicellular alga; c. erythritol (a lichen sugar), a thermostable phytoncide active against *Strep.*, *Staph.*, & *E. coli* at dilutions of 1:1 million; u. in skin dis. (ointment), also in parasitic dis.; usually now recognized as close to *Chloroccum* sp.

protolog (biol.): original description of a taxon.

Protomycin: Streptimidone: antibiotic prod. by *Streptomyces reticuli* var. *protomycicus* Sugawara & al.; 1959-; active against yeasts; fungi; viruses; amebic dys. organism; etc.

protopapaverine: alk. of Opium related to papaverine.

protopine: basic alk. found in Opium, Sanguinaria, & other drugs of taxa of Fam. Papaver.; also in Fam. Fumariaceae.

protoplasm: the living matter of extraordinary complexity found in all living pl. & animal cells; this colloidal mixt. constitutes the seat of life.

Protorhus longifolia Engl. (Anacardi) (sAf): red Cape beech; gum exuding fr. bk. very adhesive; u. as glue, depilatory (smearing fingers & pulling hair out by roots) (Zulus); useful timber.

protoveratrines A & B: chief hypotensive alks. fr. rhiz. of *Veratrum album*; with local anesth. props.

Protozoa: the phylum of unicelluar animals; the simplest of all animal life; spp. occur all over the world, mostly in fresh & salt water, but also on solid objects, such as in wet soil; many spp. are int. or exter. parasites; some are loosely colonial (ex. *Volvox*), others free-living individuals (ex. *Paramoecium*; Foraminifera, *Stentor*).

Proustia cuneifolia DC (Comp) (Ch): lvs. c. flavonoids, terpenes, triterpenes; u. inf. or decoc. for rheum. gout (baths); bacteriostatic.

provenance: Provenienz (Ge): origin, source, place where found.

provender: dry fodder of horses & other livestock, made up of grd. oats & maize mixed, also hay; sometimes food generally.

pro-vitamin A: carotene, &c. - **p.-v. D**: precursor of vitamin D.

pruche (Fr): hemlock, *Tsuga* spp.

prugno (It): plum tree.

prugnolo (It): 1) sloe-bush, *Prunus spinosa* (also *p. spinoso*) 2) edible mushroom.

prulaurasin: glycoside occurring in lvs. of cherry laurel, *Prunus laurocerasus*; splits on hydrolysis into HCN, benzaldehyde, & dextrose.

Prumnopitys ferruginea (D. Don) Laubenf. (Podocarp.) (NZ): miro; gum medicinal; wd. u. in cabinet making.

prunase: enzyme in wild cherry bk. & cherry laurel lvs. which fosters hydrolysis of cyanogenetic glucosides - **prunasin**: cyanogenetic glucoside, 1-mandelonitrile, occurring in wild cherry bk.; is hydrolyzed (β-glucosidase) —> HCN + C_6H_5 CHO + β-glucose.

prune: 1) *Prunus domestica* (fresh or dr.) (blue plum) & other plums wh. may be cured & dr. without removing pit; **p. juice** popular 2) (US slang, Prohibition era): intoxicating fermented p. juice - **p. oil**: synth. combn. of ethyl acetate, acetaldehyde, benzaldehyde, glycerol, alc., u. as p. flavor, &c. - **pruneau noir** (Fr): prune - **p. pits:** p. stones or putamina enclosing sds. - **p. salt: Sal Prunella:** pellets of sodium nitrate - **p. seeds**: sds of sloe, *Prunus spinosa.*

Prunella vulgaris (*Brunella*) L. (P. asiatica Nak.) (Lab.) (Euras, natd. NA): (common) self-heal; wild sage; heal-all; hb. (**Prunella** (L.): Herba Prunellae) c. tannins, saponins, NBP, rutin, antibiotic; u. astr. in diarrh. GIT dis., hemorrhoids, gargle in quinsy, in coughs & colds; exter. as vulner. for neck & mouth sores; widespread weed (*P. grandiflora* Scholl. & Jacq. [Eu]: u. similarly ["istuchados," Turkestan] [474]).

Prunelle: 1) (Ge): prunes wh. have been peeled, destoned, & dr. 2) (Fr): sloe 3) *Prunella* 4) (Ge): *Nigritella nigra* (**Pruenelle** (Ge): olive green liqueur fr. plum juice, sugar, spirits, bitter almond oil, & prune water) - **Prunum** (L.): dr. prune; off. formerly NF; u. lax.; as breakfast food with cream.

Prunus L. (Ros) (temp. zone): incl. the stone-fruits plum, apricot, almond, peach, cherry, & cherry-laurel spp.; ca. 400 spp.; frts. in form of drupe; usually edible; often luscious; bk. exudes cherry (tree) gum (esp. fr. plum tree); u. in purple, yellow dyes; u. as thickening agt. in cloth printing; bk. u. to make baskets (nwAm Indians) - **P. africana** (Hook. f.) Kalkm. (Pygeum africanum Hook f.) (trop&sAf): red stinkwood; "bitter almond"; lf. bk. frt. c. amygdalin yielding HCN; lf. decoc. mixed with milk a good substit. for "almond milk"; wd. hard & useful for wagon making; sds. c. 46% FO, frt. wd. & fls. c. sterols incl. ß-sitosterol; liposoluble complex on market; u. in prostatic hypertrophy (adenoma) - **P. americana** Marsh. (ne&cUS; cult. Cana): (American) wild plum; yellow p.; rts. boiled & u. as vermifuge (Ojibway Indians); sds. source of plum kernel oil (FO); bk. comml. article; frts. eaten raw or cooked: much used to develop new vars. - **P. amygdalus** Batsch = *P. dulcis* - **P. angustifolia** Marsh. (seUS): Chickasaw plum; wild p. (Ala [US]); frt. edible & delicious, eaten raw or in jellies & preserves; sometimes cult. - **P. arborea** Kalkman (Pygeum persimile Kurz) (Mal): lf. decoc. u. to facilitate delivery in labor (Sakai people) - **P. armeniaca** L. (c&wAs): apricot; "moon of the faithful" (Pak); sds. u. to produce Persic Oil (Oleum Persicarum), edible FO, much u. in med.

& pharm., incl. cosmetics; & volatile almond oil (benzaldehyde & HCN), now chief source of bitter almond oil (prepd. fr. press cake after extn. FO); frt. eaten fresh & dr. (considerable part of diet in nwHimalay); important trade article fr. Afghan.; furnishes much of vit. C in winter (dr. frt.) (Pak); mostly u. for jams & dr. in cooking (Eu, NA); frt. juice marketed as beverage; kernels eaten, u. to make flour for cakes, bread, &c., very nutritious; pits (hulls, shells) u. for plastic insulation material in electric switches; widely cult. in warm temp. zones, as Mor, Syr, AsMin, Calif (US), China (cult. 4,000 yr.) (vars. Park; Blenheim); kernel paste u. as substit. for almond p.; var. *ansu* Max. (orig. fr. Armenia & regions around; hence L. name) - **P. avium** L. (Euras, natd. NA): sweet cherry; small c.; mazzard; frt. delicious, nutr.; c. malic & other acids u. in cherry juice & syr. for flavor improvement; in brandies & liqueurs (ex. Maraschino); stated source of "Laetrile"; lvs. u. as tea substit. in anemia; trunk & frt. source of cherry gum (u. like acacia); cult. in many vars. (546) - **P. capuli** Cav. (P. capollin Zucc.) (Mex, CA to Peru): capuli(n); cerezo; frt. delicious, often baked with corn meal; distd. water of lvs. u. antispasm.; bk. decoc. in dys., antipyr. - **P. caroliniana** (Mill.) Ait. (seUS): wild (or mock) orange; American cherry laurel; lvs. & bk. c. prulaurasin (—> HCN), poisonous to livestock; pl. u. as hedge; u. folk med. (2112) - **P. cerasoides** D. Don (P. puddum Roxb.) (Indo-Pak-Himal): panni (Burma); kernel & bk. c. amygdalin, u. in stone & gravel (lithontryptic) (2113); bk. u. to make cherry water; astr.; trunk source of cherry gum, u. as adulter. tragacanth (Kordofan); wd. hard & strong u. for furnit. walking sticks; handsome fls., tree often cult. as ornam. (Burma) - **P. cerasus** L. (AsMin & Transcauc; cult. Eu, NA): (wild) sour cherry; w. c.; Morello c.; Kirsch (Ge); frt. esculent; u. in prepn. cherry water (Kirschwasser; "Kirsch"; spirits made in Alsace fr. cherry pits; also sweet c.; 43-58% alc. 14-48 mg HCN/1.); c. juice (former. NF), c. syrup (form. NF); st. & frts. furnish cherry gum (glucuronides of mannose, arabinose, &c.; u. like acacia & as sizing); frt. flavor in pharm. (syr.), prepn. pies, &c.; cherry brandy (comm. source); cherry frt. stalks ("cherry stems") (peduncles); c. tannin, u. diur., in diarrh., obesity, in therapy of aging heart; condim. & flavor; FO expressed fr. sds. (25-35% yields) cherry seed oil of comm.; press cake c. amygdalin —> HCN (frt. pulp much less); VO prod. by fermn. & distn. of sd. cake (0.016% yield); bk. & lvs. u. in tea mixts. (Eu); var. *austera* L., Morello c.; with colored juice; var. *marasca* Schneid. f. (It): Dalmatian marasca; small bitter wild c.; u. to make maraschino, a liqueur - **P. ceylanica** (Wight) Miq. (Pygeum gardneri Hook. f.) (seInd to seAs): daka (Marathi); frt. kernel c. HCN, u. as fish poison; bk. u. for stomachache (Vietnam) - **P. communis** Huds. = *P. domestica* - **P. c.** Arcang. = *P. dulcis* - **P. divaricata** Ledeb. (wAs): mirabelle (plum), myrobalan cherry; m. plum (round cherry plum); u. much like prunes (Russ.) - **P. domestica** L. (wAs. now widely cult. Eu & wUS): common (or garden) plum; prune; source of dr. prunes (Prunum), popular, mildly lax. dessert frt. (as stewed or boiled prunes), also fresh, dr., canned; aids appetite; dem. refrig. (2119); u. to make brandy (Slibowitz), as pulp or jam; quality related to s. g. (475); p. juice u. as beverage (sl. lax.) (often with cranberry j.) (popular with elderly); flavor; in arom. prepns.; lvs. insecticidal; chief prodn. Pac coast US, esp. Calif & Ore (esp. famous Ore Petites); Fr, It; many vars. incl. var. *italica* Schneid. (swAs, It): green gage (plum); edible frt. dr. for use (SC Rev. Gage introd. to Eng); var. *insititia* Bailey (P. i. L.): Damson plum (or D. bullace); b. plum: alubakhara (Beng.) (2116); small dark frt. a favorite food, plum much u. in preserves; liqueur ("Mirabelle"); the indigenous Eu plum u. antipyr. (Pak); sds. with edible FO u. as substit. for sweet almond oil - **P. dulcis** D. A. Webber (P. amygdalus Batsch) (As, widely cult. Eu, wUS, as Calif): almond (tree); both sweet a. ("var. *dulcis*") & bitter a. ("var. *amara*") u. in prodn. FO ("sweet almond oil") but only bitter a. u. for bitter a. oil using sds.; sweet a. sds. much u. as nutr. flavor; in cosmetics, food, candies; emulsion for cough; roasted ("burnt") sweet a. u. in prepn. marzipan, pralines; FO much u. in cosmetics, foods; VO (bitter almond oil) u. to flavor med., respir. stim., in cough mixts., perfum. (danger: toxic) (*v.* under "almond"); almond meal (fr. both vars.) u. as cosmetic (skin improvement), diabetic food; fls. u. formerly; frt. pulp dry, nearly leathery, thus inedible; almond hulls u. cattle feed; good vars.: Jordan (Sp); Alicante; sweet a. prod. in Sp, It, Fr, Mor, Turkey, Calif [US] - **P. emarginata** Walpers (wNA): bitter cherry; frt. edible but very bitter; rt. & bk. decoc. u. in "heart trouble"; formerly u. contraceptive; fls. c. quercimeritrin (1706) - **P. japonica** Thunb. (Kor, Chin, Jap): endocarp u. as mild lax., diur.; sd. kernels dem., for fevers, rheum. - **P. laurocerasus** L. (Laurocerasus officinalis Roem.) (seEu to Iran): (common) cherry laurel; laurel (cherry); fresh lvs. c. prunasin & sambunigrin (cyanogenetic glycosides); u. bitter ton. respir. stim., sed. antispasm. arom.; in prepn.

VO & c. l. water; poisonous (2114) - **P. lusitanica** L. (Sp, Por): Portugal laurel; cult. as large ornam. shrub - **P. mahaleb** L. (eMed reg. to Pak): Mahaleb (or St. Lucie) cherry; perfumed c.; rock c.; lvs. & bk. c. coumarin; sds. c. amygdalin (—> HCN); u. in spirit indy. (mfr. cherry brandy) & perfum. mfr.; culinary material (Syr); sd. kernels u. for scorpion sting, ton. (Pak); sds. ground & mixed with milk —> jelly (eaten) - **P. maritima** Marsh. (neUS, eCana near coast): beach (or sand) plum; frt. sweetish, edible; u. in jellies & preserves; ornam. shrub - **P. mexicana** S. Wats. (sUS, Mex): Mexican plum; a source of cultiv. plums; perhaps cult. by Am Indians - **P. mume** Sieb. & Zucc. (Far East): creeping plum of the Japanese; smoked unripe frts. u. carm. antispasm. astr. for bacterial dis. of intest.; lvs. & fls. bechic - **P. myrtifolia** (L.) Urban (P. sphaerocarpa [Sw.] Roem.) (sFla, Fla Keys [US], WI): West Indian cherry; cuajani hembra (Cu); almendrillo (Cu); one of the cherry-laurels; u. similarly to *P. occidentalis* - **P. occidentalis** L. (WI, CA): almendro (macho) (Cu); cuajani (Cu); West Indian cherry laurel; sds. c. HCN glycosides; bk., lvs. & twigs u. fresh for asthma & cough (Cu) - **P. padus** L. (Euras to Jap): (European) bird cherry; cluster c.; bk., fls. & sds. c. amygdalin; lvs. & twigs c. prunasin; bk. u. much like *P. serotina*; antipyr. ton. in bronchitis & bronch. dis.; sed.; frt. u. in jams; wd. useful - **P. pennsylvanica** L. f. (Cana, nUS): bird/pin/fire/ (wild) red/sand/cherry; frt. edible; bk. u. for erysipelas; wd. u. for chafed skin, prickly heat (Maritime Indians); very hardy bush - **P. persica** (L.) Batsch (Amygdalus leiocarpa Boiss.) (eAs, esp. Chin, also wAs, cult. Eu, NA, SA): peach (tree); frt. esculent & succulent, diur. (vars.: freestone; cling [with pulp adhering to pit]; brandied peach); sds. (peach kernels) by-product of peach canning indy.; u. in prep. persic oil (FO - *cf. P. armeniaca*) & bitter almond oil; lvs. u. arom. in rheum., bladder weakness; purg.; to mitigate cough; in poultices for sores & wounds (476); fls. (Flos Persicae, Por Phar.) u. in prepn. of a syr. (mild purg.); bk. u. for roundworms & in fever (folk med.); gum exudn. (water-soluble) claimed used (2440); branches formerly popular for whipping children ("peach hickories"); Georgia [US], the "Peach State"; many vars., incl. var. *nucipersica* (Suckow) C. Schneider (var *nectarina* Maxim.) nectarine; frt. smooth like plum, usually smaller than ordinary p.; popular frt.; diur.; var. *compressa* Bean: flat peach - **P. pseudocerasus** Lindl. (Chin): lvs. stom. ton. & to retain virility - **P. salicina** Lindl. (Chin): u. for liver dis.; white bk. of rt. u. as antipyr. & to control nausea & sudden fever in children - **P. schlechteri** (Koehne) Kalkman (Pygeum s. Koehne) (Solomon Is): natives crush bk. & apply to aching teeth; hot water macerate of bk. appl. to sore legs - **P. serotina** Ehrh. (e&cNA, natd. nwEu): wild (black) cherry; Virginian prune; large tree (-33 m; 1.3-1.8 m diam.); bk. c. prunasin, emulsin; u. as expect. (syr.) in pulm. dis., coughs, sed., bitter ton.; dr. frt. u. astr. bitter ton. (in commerce); sprouts poisonous (HCN); bk. (inner) aromatic; represents Prunus Virginiana (USP) (2118) - **P. serrulata** Lindl. (Jap, China, Kor): Japanese flowering cherry trees; beautiful pink fls. (last 10-12 days); trees planted along Tidal Basin (Washington, DC [US]), ca. 1910 (1167); many vars. - **P. spinosa** L. (Acacia nostratas) (Euras, at times natd. NA): sloe (tree); black thorn; a small bitter wild plum; frt. c. sugars, tannin, acids, vit. C, pectin; frt. edible only after repeated freezing; u. in marmalades for stomach weakness, bladder & urin. dis.; flavor, astr.; black dye; sloe wine u. purg. diur.; in mfr. liqueur (sloe gin); frt. juice u. in bleeding of the nose, uterine prolapse; fls. c. quercetin, kaempferol; amygdalin (nitrile glycoside) u. mild lax. diur. diaph. expect. in cough; emmen.; fl. buds in homeop.; lvs. (as "Folia Acaciae" = misnomer) u. blood "purifying" teas; substit. for tobacco & black tea; thorns on sts. often cause flesh wounds & infections (due in Eu to shrikes hanging meat on them) (2117); prodn. frt. It, Bulg; bk. u. as antipyr. in diarrh., asthma - **P. triflora** Roxb. = *P. salicina* Lindl. - **P. umbellata** Ell. (seUS): black sloe; frts. u. in jellies & jams - **P. undulata** Buch.-Ham. (P. capricida Wall.) (Ind-Himal): frt., sds. & lvs. c. HCN; pl. toxic (cattle) - **P. virginiana** L. (Cana, nUS): choke cherry (SC frt. strongly astr.); black frt. & glabrous lvs., tree -8 m; var. *demissa* (Nutt.) Torr. (P. d. [Nutt.] Walp.) (Wash to sCalif [US]): wild (or choke) cherry; western c. c.; inner bk. not aromatic, u. as ton. in diarrh. & to control nervous excitability; frt. sour & astr. but edible if boiled; frt. eaten fresh or dr. (Indians, Mendocino Co, Calif [US]); lvs. c. HCN; var. *melanocarpa* (A. Nels.) Sarg.: western c. c.; inner bk. u. for diarrh. (Seneca Indians) (do not confuse with Prunus Virginiana [L]: wild cherry bk. fr. *P. serotina*) - **P. yedoensis** Matsum. (eAs): one of the flowering cherries popular in Jap; bk. decoc. u. as bechic (Chin) - **P. zippeliana** Miq. (P. macrophylla Sieb. & Zucc.) (eAs): lf. distillate u. as bechic & sed. to relieve asthmatic breathing.

Prussian powder: licorice pd. comp. - **prussic acid**: diluted (2%) hydrocyanic acid.

Psacalium decompositum (Gray) H. E. Robins & Brett (Senecio grayanus Hemsl.) (swUS, nMex):

matarique; "life root plant" (US); rhiz. & rts. u. for rheum. & neuralgia (extern. appln.); in skin irritation; aid to healing of wounds (cicatrization); purge; decoc. in urinary dis., diabetes, &c. (Mex); said one of 4 most useful med. plants of Mex mountains.

Psalliota (Fr) Kummer = *Agaricus.*

Pseudananas sagenarius (Arreda) Camargó (monotypic) (Bromeli) (Br+): gravatá de rede; fiber useful; cult. ornam.

Pseudarthria hookeri Wight & Arn. (Leg.) (tropAf): rt. decoc. u. for stomachache (Tanz); to cleanse teeth - **P. viscida** W. & A. (As): pl. u. for rheum. excessive heat, fever, diarrh. asthma. heart dis.

Pseuderanthemum bicolor Radlk. (Acanth.) (seAs): whole pl. decoc. u. to treat aphthae & to heal wounds & ulcers (PI) - **P. curtatum** Merr. (PI): pl. u. in asthma; lvs. appl. to boils - **P. graciliflorum** Ridl. (eAs): rt. decoc. given after childbirth - **P. hildebrandtii** Lindau (Tanz): lvs. ashed & ash rubbed into cuts for pains of arthritis; rt. rubbed on exter. for bronchitis, snakebite; rts. & lvs. u. for intest. pain; rt. decoc. for dys.

Pseudevernia spp. (Parmeli. - Ascolichen. - Lichenes): source of oak moss - **P. furfuracea** (L.) Zopf (Tatra Polan); ext. inhibits proteolytic enzymes (ex. trypsin; elastase); c. atranorin +.

pseudo-acacia: locust (tree) - **pseudo-aconitine**: alk. fr. tubers of *Aconitum ferox.*

Pseudobombax ellipticum E. Dug. (Bombax ellipticum HBK) (Bombac.) (Mex to CA [Salv]): cotton tree; cornsilk flower; bk. & rt. decoc. u. in toothache & to harden gums; fl. decoc. in fever, coughs; nectar as eye drops for inflamed eyes; cult. ornam. sFla [US], Mex - **P. septenatum** (Jacq.) Dugand (CA): bottle (or square) tree; urtuwala (Cuna Indians, Pan); prob. u. in med.; fiber around sds. u. to stuff pillows & mattresses; bk. for cordage (Pan).

Pseudocassine transvaalensis Bredell. (sAf): bk. inf. u. to "clean" the stom. & as enema in fever; cherry-like frt. edible; wd. useful.

Pseud(o)elephantopus spicatus (Juss.) Rohr = *Elephantopus spicatus* Juss. ex Abul (*qv.*).

pseudoephedrine: isoephedrine: stereo-isomer of ephedrine; occ. in *Ephedra* spp.; u. (mostly as HCl salt) for nasal congestion; somewhat less stim. to CNS & lower pressor effect than ephedrine; causes less insomnia.

pseudo-fruit: the so-called frt. of *Anacardium occidentale* (edible part) but actually the receptacle.

Pseudolachnostylis maprouneaefolia Pax. (Euphorbi) (Af): Kudu berry; yolia (Tanz); rt. u. int. & as inhaled smoke from burning rt. to treat pneumonia; rt. purg.; frt. edible; wd. a source of charcoal.

Pseudolarix amabilis (Nelson) Rehder (P. kaempferi Gord.) (monotypic) (Pin) (eChin): golden pine; g. larch; tree large, scarce, restricted in area of growth; valuable timber tree; very hard permanent lumber; bk. (Cortex Pseudolaricis) u. in Chin med.; ornam. tree (-45 m.).

Pseudolmedia laevigata Trécul (Mor) (Co): tea made fr. young twigs of small tree u. to stim. appetite (middle Apaporis, Indian tribe).

Pseudomonas radicicola (Beijerinck) Moore (Pseudomonadaceae): bacteria which inhabit the nodules of leguminous pls. & prod. nitrates fr. atmospheric nitrogen.

pseudomonic acid A: Mupirocin; prod. of fermentation of *Pseudomonas fluorescens*; antibacter. u. against gm.+ microorganisms; in impetigo; eczema; folliculitis, &c.

Pseudosmodingium perniciosum Engl. (Anacardi) (Mex): caujiote; gum-resin exudn. fr. trunk purg. diur., rubbed on skin to reduce pain of scorpion sting; poisonous pl. (*cf. Toxicodendron*).

Pseudotsuga menziesii (Mirb.) Franco (P. taxifolia Britt.; P. douglasii Carr.) (Pin) (Pac NA, incl. Mex): (common) (Ore) Douglas fir; (Ore) Douglas spruce; Oregon pine; large tree (-100 m+) (2259); trunk wd. source of a balsamic oleoresinous exudn., "Oregon balsam" (Balsam fir, Ore) obt. either by felling the tree & coll. prod. in special vessels fr. broken areas or by boring into heart wood & inserting a tube; u. techn. as substit. for Canada balsam; melezitose (rare sugar) exudes fr, branches sometimes in hot summers; formerly u. sweetener (BC Indians [Cana]) ("Douglas fir sugar"); decoc. of spring buds in VD (Calif [US]); decoc. of 1st year shoots u. for high fevers & anemia (emetic); bk. decoc. & lf. inf. (by steeping) u. med. (BC Indians); coffee substit.; lvs. in sweat bath cure for rheum. (Calif Indians); bk. potential source of cork (2121); very important lumber tree; bk. tan. (477).

pseudo-whorl: verticillaster.

Psiadia arabica Jaub. & Spach. (Comp.) (tropAf & As): u. for arthritis, neurological dis.; as poultice on joints & skeletal muscles for rheum., muscular atrophy (after polio); fuel; fumigant(?).

psicofuranine: psychofuranine: angustmycin A; nucleoside antibiotic fr. *Streptomyces hygroscopicus* Waksman & Henrici var. *decoyicus* (1958); adenine deriv.; with antibacterial & antineoplastic activity.

Psidium: former Lichen genus close to *Pertusaria*, now name invalid because u. for gen. of higher plants - **P. acutangulum** DC (Myrt.) (Co): lvs. very astr. & u. for hemorrhoid pains (Tikuna Indians) (2299) - **P. araca** Raddi (P. guineense Sw.)

(Br): Brazilian guava; lvs. u. as astr.; in bronchitis; rts. for menorrhagia; styptic for swellings - **P. cattleianum** Sabine (Br+): strawberry guava; small bushy tree (-8 m); lvs. opp., pinnately veined, veins not prominent; reddish frt. eaten raw & u. to make jellies, drinks; u. stom. antispasm.; widely cult.; var. *littorale* Fosberg (P. l. Raddi) (Br): strawberry (or Cattley) guava; araca era (Br); small trees with edible frt. - **P. densicomum** Mart. (Co): frts. often dr. & chewed as needed for mouth sores (astr.) (Rio Apaporis Indians) (2299) - **P. guajava** L. (P. pomiferum L.; P. pyriferum L.) (tropAm, esp. WI, SA, but now pantrop.): common (or white) guava; red g.; guayabo (Arg); tree (-10 m) bears edible yellow berries (-1 dm long) with delicate flavor & very high vit. C content; eaten raw & u. to prep. exquisite sweetmeats (neArg), jellies, preserves, & jams (Cu, PR, Br+) (1855); frt. rich in pectin (478); lvs. (as Folia Djamboe) c. VO (with eugenol), FO malic acid, ursolic acid & derivs.; inf. rt., bk. & lvs. u. as astr. in GIT dis.; as gastritis, enteritis, diarrh. (Domin, Caribs), &c.; inf. emoll. (Yuc); female dis. (Rota); vulner. antipyr. - **P. guianense** Pers. (Co): mature frt. eaten raw to control diarrh. (Tikunas) (2299) - **P. microcarpum** Camb. (tropAm): lvs. u. as astr. - **P. montanum** Sw. (Jam): mountain guava (tree); bk. decoc. u. for diarrh.; frt. eaten raw or u. in jams - **P. sartorianum** Niedenzu (sMex): pichiche; guayabillo; frts. eaten fresh or dr. & u. to make refreshing beverage; lvs. u. as astr., ton.; many folk med. uses lack rationality; bk. tan.

Psila spartioides (N. & A.) Cabr. (Comp.) (nArg): pichana; VO of shrub c. cryptone (5%), limonene, pinenes, citronellal, &c.; u. perfumery (—> *Baccharis*).

Psilocaulon absimile N. E. Br. (Aizo.) (sAf): asbos; ash rich in mineral salts (Na, K) & u. as lye in soap making; pl. c. oxalic acid, piperidine; cause of much stock poisoning - **P. parviflorum** Schw. (sAf): ash u. as lye in soap making & to soak grapes when making raisins.

psilocin: hallucinogenic tert. amine obt. fr. *Psilocybe mexicana* &c.

Psilocybe (Fr) Kummer (Strophari. - Basidiomyc. - Fungi) (widely distr.): sacred mushrooms (teonanacatl) (Mex); many spp. of these small mushrooms are u. as hallucinogenics; ex. **P. aztecorum** Heim; **P. bohemica** Sebek.; **P. caerulescens** Murr. + vars. + forms; **P. callosa** Quél.; **P. cyanescens** Wakefield; **P. cyanopus** (Atk.) Kuehner; **P. muliercula** Sing. & Smith; **P. semperviva** Heim & Cailleux; **P. zapotecorum** Heim; & some 73 other *P.* spp. - **P. baeocystis** Sing. & A. H. Smith (nwUS): sporocarp c. psilocybin, psilocin, &c.; shows psychotropic activity much like that of *P. mexicana* - **P. cubensis** (Earle) Sing. (Stropharia c. Earle) (WI, as Cu; Fla, Ala [US], Mex, CA, Thail, Kampuchea, Austral - grows on horse & cow dung): c. hallucinogenic compds. like *P. mexicana*; strong hallucinogen; u. in ceremonies (Mex Indians) - **P. hoogshagenii** Heim var. **convexa** Guzman (P. semperviva Heim & Cailleux) (Mex): hallucinogenic mushroom; this & *P. mexicana* believed most important mushrooms in magico-religious use in Mex - **P. mexicana** Heim (Mex, Guat): "flesh of the gods"; ingested orally for its hallucinogenic (psychomimetic) (dream-like) effects; u. in Mex Indian religious rites (hence "sacred mushrooms"); effects due to psilocybin (0.2-0.4%) & psilocin (in traces only) (*cf.* mescaline, LSD) - **P. pelliculosa** (A. H. Sm.) Singer & A. H. Sm. (wUS, nEu): "liberty caps"; similar c. & u. to prec. (479) - **P. semilanceata** (Fr.) Kummer (Eu, Cana, US, Br, Ch, Austral, Tasm): "liberty caps"; c. psilocybin (1%), baeocystin; u. hallucinogenic.

psilocybin: phosphorylated indole alk.; occ. in *Psilocybe* & some other genera of mushrooms; hydrolyzed to psilocin, similar to psilocybin but very unstable; a psychotomimetic agt.; u. in mental clinics; *cf.* LSD, mescaline; synthesis 1963 (2347).

psilocyn = psilocin.

Psilopeganum sinense Hemsl. (1) (Rut) (Chin): pl. u. as diur. in edema.

Psilotrichum ferrugineum (Roxb.) Moquin-Tandon (Amaranth.) (Indochin, Thail, Mal): u. for cough.

Psilotum complanatum Sw. (Psilot. - Pteridoph.) (Fiji, NA+): u. med. (Fiji) (480) - **P. nudum** Beauv. (eAs, Haw [US]): moa; decoc. of hb. u. as lax. & for thrush; ext. (pl. soaked in wine) u. for trauma (Chin); spores u. as kind of talcum powder to reduce irritation of loin cloth (Haw).

Psittacanthus calyculatus Don (Loranth.) (Mex, CA): ingerto; lf. & fl. decoc. u. as vulner.; distd. water of hb. u. as cosmetic - **P. collumcygni** Eichl. (Br): tonton; juice u. in hemorrhage - **P. cordatus** Blume (Par): u. in mal.

Psophocarpus tetragonolobus (L) DC. (Leg) (seAs, incl. PapNG): Goa (or winged) bean; princess b.; asparagus pea; "wonder bean"; pois carré (Fr) (square pea); ann. vine (- ca. 4 m); rt. u. as poultice for vertigo (Mal); lf. inf. wash in ophthalmia (Indones); lvs. & sds. eaten to cure skin sores (NG); all parts of pl. edible; pods (-0.6 m long) eaten raw or cooked; sds. (rich in protein, 42%), tuberous rts., fls. & shoots important food sources useful for providing protein-rich food to poor countries; lvs. rich in tryptophan & vit. A; widely cult. OW, NW,

Psoralea argophylla Pursh (Leg) (cNA): silverleaf scurfpea; tea of lvs. & sts. u. to reduce fever (Cheyenne Indians) - **P. bituminosa** (Med reg., wAs): bitumen trefoil; hb. u. as antisp., antiepileptic, in hysteria - **P. canescens** Mx. (Pediomelum c. Rydb.) (Ala, Fla to NCar [US]): buck (t)horn; u. in gastric dis. - **P. corylifolia** L. (Arab, Ind to Indochin): guar; babchi; lvs. u. in diarrh.; arom. fls. in local med.; frts. (Malaya tea; Psoralea frts. (or sds.) [bawchang seed]) c. psoralen, psoralidin, u. as stom. in gastralgia, ton. aphrod. lax. diur. exter. (cataplasms) in leprosy, psoriasis, leucoderma ("white leprosy") (FO); alc. macerate u. in arthritis, female dis. (Vietnam) - **P. eglandulosa** Ell. v. *P. psoralioides* - **P. esculenta** Pursh (cNA, Sask [Cana]-Mont [US]): common breadroot scurfpea; prairie turnip; Indian breadroot; u. for intest. infec.; tuberous rt. eaten (staple) by Am Indians (1534) - **P. glandulosa** L. (Ch, Pe): Jesuit's tea; culen (Ch); rt. u. as emet.; lf. inf. u. in indig. worms & to wash sores, antipyr., emoll. vulner.; sl. astr. & u. for stomach dis.; lvs. u. to prep. beverages - **P. hypogaea** Nutt. (cUS): smaller Indian breadroot; tuberous rt. u. as Am Indian food - **P. mexicana** Vail (tropAm): "culin"; "trinitaria"; inf. u. for diarrh. & stomachache (Ec); also var. *trianae* - **P. obliqua** E. Mey. (SA): leaflets serve to adulter. Buchu - **P. pentaphylla** L. (SpAm): contrayerba; contrahierba blanco; pl. inf. ("tea") u. for fever, flu, colds (nMex) (481) - **P. pinnata** L. (sAf): pin-wortel (sAf, Du); blue peas; cold inf. of rt. u. as emet. for treating hysteria (natives) - **P. polysticta** Benth. (sAf): rt. smoked for head colds - **P. psoralioides** (Walt.) Cory. (P. pedunculata Vail) (seUS): Sampson's snakeroot; Congo root; scurfy pea; rt. c. oleoresin, u. arom. bitter ton. stim.; commonly in Negro voodoo, thus in "hands"; similarly var. *eglandulosa* Freeman (P. e. Ell.) (—> *Pediomelum*; *Orbexilum*).

psoralens: furocoumarins; found in many members of fams. Rutaceae, Leguminosae, Umbelliferae, Moraceae; gp. incl. psoralen per se: latter occ. in *Psoralea corylifolia*; *Ficus carica*; &c.; with phototoxic, mutagenic, photocarcinogenic props.; not destroyed in cooking; members of series incl. xanthotoxin (methoxypsoralen), bergapten (acetyl deriv., in *Apium graveolens*); &c.; psoralen u. topically with UV light in psoriasis (PUVA = psoralen + UVL-A) (& in vitiligo ("white leprosy").

Psorospermum baumii Engl. (Gutt.) (Af): boiled rt. effective in scabies & lice; gives useful mucilage - **P. febrifugus** Spach (tropAf): claimed antipyr.; purg.; juice u. to treat constip.; bk. in leprosy & skin dis., also insect bites; "poison antidote" (Java) - **P. senegalense** Spach (P. guineense Hochr.) (wAf): shrub or shrubby tree; bk. or rts. u. in all kinds of dermatitis, as eczema, scabies, psoriasis; in syph. leprosy; pd. rt. for colics, rt. decoc. in GC; lf. decoc. in arthritis (Sen).

Psorothamnus emoryi Rydb. (Leg.) (sCalif, Ariz [US], nwMex): Emory indigo bush; found on sand dunes; st. & lvs. u. to dye yellow (baskets); twigs u. to apply face paint (Seri Indians); for roofing - **P. polyadenius** Rydb. (Dalea polyadenia Torr.) (wNA): smoke bush; tea made fr. perenn. hb. st. u. for coughs & colds (Nev [US] Indians); in pneumonia; flu; TB; VD; muscular pains - **P. spinosus** Barneby (Dalea spinosa Gray) (wUS deserts [Ariz, sCalif], nwMex): (desert) smoke tree; indigo bush; i. thorn; u. as flowering ornam.

psychosomatic therapy: beneficial effect of emotions & mental attitudes on course of a dis.; effective in some disorders, ex. warts (note folklore use of worthless materials in effecting disappearance); cancers.

Psychotria adenophylla Wall. (Rubi) (seAs): lvs. u. for chest dis. (Indochin) - **P. archeboldiana** Fosberg (Fiji): lf. inf. u. for TB - **P. cooperi** Standl. (Co): amarga (Choco); u. antirheum. - **P. cuernosensis** Elm. (PI, endemic): lvs. appl. to burns - **P. curviflora** Wall. (seAs): lf. inf. antipyr. (Indochin); lf. poultice appl. for headache, convulsions; as lotion to treat injury; lvs. u. in cough; lf. juice dropped into ear; lf. in coconut oil u. to dye hair black (Fiji); rt. decoc. u. in childbirth (Mal) - **P. emetica** L.f. = *Cephaëlis e.* - **P. insularum** Gray (Samoa): matalafi; lvs. u. as poultice in localized cellulitis - **P. ipecacuanha** Muell.-Arg. = *Cephaëlis i.* - **P. ligus-trifolia** (Northr.) Millsp. (Fla [US], WI): "wild coffee" (Bah); taburete (Cu); pl. decoc. u. as bath for dropsy - **P. luconiensis** F.-Vill. (PI): rt. inf. u. in dys.; bk. for internal pain; fresh lvs. laid on wounds; lf. decoc. to cleanse ulcers - **P. macrocalyx** A. Gray (Fiji): lf. juice put in eye for eye dis. - **P. manillensis** Bartl. (PI): rt. decoc. for excessive menstruation - **P. marcgravii** (St.-Hil.) Spreng. = *Palicourea m.* - **P. membranifolia** Bartl. (PI): lvs. appl. to treat splenomegaly - **P. mindorensis** Elm. (PI): lvs. appl. to ulcers; for eye dis.; skin eruptions - **P. microdon** Brem. (sMex, WI, Guiana, Ec): bacelac (Mex); tapa camino (Cu); lvs. u. in folk med. (Yuc); lf. decoc. to bathe infants with diarrh. - **P. montana** Bl. (seAs): boiled lvs. u. for sores, colds (Indochin); rt. as poultice for swellings & ulcers (Mal) - **P. nervosa** Sw. (P. undata Jacq.) (sFla [US], WI, sMex to Ve, Co): St. John's bush; wild coffee; crushed lvs. u. to stop bleeding; rt.

emet., lf. decoc. antipyr. in colds - **P. olivacea** Val. (Solomon Is): lf. decoc. appl. to sore legs; lvs. eaten with betel nuts to relieve stomachache; sap of vine u. for penis dis. (probably GC) - **P. psychotriaefolia** (Seem.) Standl. (Co, Ec): admixed with hallucinogenic yaje drink (to fortify & make it last longer) - **P. psychotrioides** Rob. (trop wAf): vapors of bk. being boiled u. to treat ocular maladies (Sen) - **P. rubra** Poir. (Chin & sAs): lf. & st. decoc. u. to relieve pain of bruises; rt. inf. in mal. (Indochin) - **P. sarmentosa** Bl. (Mal): lf./rt. decoc. u. to expedite childbirth; lvs. u. to poultice sores - **P. schmielei** Warb. (Solomon Is): decoc. of scraped bk. ingested for stomachache - **P. tephrosantha** A. Gray (Fiji): lf. decoc. u. for dis. of bowels & stomach ("nimbatha") - **P. tomentosa** Hemsl. (Co): rt. of hb. u. as emmen. & for asthma - **P. uliginosa** Millsp. (sMex, WI): tres cabezas; decoc. u. to treat asthma, dys. hemorrhoids, arterial hypertension - **P. uncinatum** Anselmino (SA): long u. for impotency (Br) - **P. undata** Miq. (SA): rts. emet. - **P. viridiflora** Reinw. (Mal): crushed lvs. u. for skin dis., bites of poisonous snakes & insects - **P. viridis** Ruiz & Pavon (Cu, CA, s to Amazon Basin): yaje(?); lvs. c. N, N-dimethyltryptamine; u. to fortify psychomimetic activity of *Banisteriopsis* spp. (Br) (483).

psychotrine: one of the 3 chief alks. of Ipecac; physiologically inactive.

psychrophilic: cold-loving; appl. to bacteria (usual opt. 15-20° C).

Psyllii Semen (L): psyllium sd. (orig. < psylla (Gr.), flea).

psyllium (seed): fr. *Plantago ovata* - **black p.**: *P. indica* - **blond p.**: *P. ovata* - **brown p.**: *P. indica* - **English p.**: *P. lanceolata* - **French p.**: black p. - **German p.**: English p. - **Spanish p.**: black p. - **white p.**: blond p.

Psylocybe = *Psilocybe.*

PTA: plasma thromboplastin antecedent: Factor XI (*qv.*).

Ptaeroxylon utile E. & Z. (P. obliquum Radlk.) (Ptaeroxyl.) (sAf): source of sneezewood or Cape mahogany; pd. wd. u. as snuff to produce sneezing to relieve headache; inf. of wd. u. for rheum. & heart dis., to rid cattle of ticks; in sinusitis; bk. as antipyr.; resinous juice appl. to lupus & warts; fish poison; wd. gives extremely durable timber.

Ptarmica: *v. Achillea p.*

Ptelea trifoliata L. (Rut) (e&cNA): (common) hop tree; wafer ash; skunk bush; potato chip tree; swamp dogwood; rt. bk. (ptelea or wafer ash bk.) c. alks. (pteleine, dictamnine, pteleatinium Cl., +) u. antipyr. stom. ton.; exts. anti-yeast, anti-TB; rt. decoc. u. for oral & throat dis. (inflamed uvula); lvs. & shoots anthelm.; arom. bitter winged frts. u. as hops substit.; eq. pts. of fresh lvs. & bk. u. by Ge homeop.; other *P.* spp. with similar props.

ptenophyte: intermediate pl. community.

Pteretis struthiopteris Nieuwl. = *Matteuccia s.*

Pteridium aquilinum (L) Kuhn (Pteris aquilina L.) (Polypodi. Pteridoph.) (nearly cosmpol.): brake (fern); bracken (fern); common b.; poor man's soap (Ala [US]); rhiz. u. in diarrh., anthelm., yellow dye; strongly carcinogenic; source of kelecho bread (Canary Is); psychosedelic; juice fr. rhiz. aphrod. (Af); source of starch (Jap); hb. (fronds) u. as tea for rheum.; vulner. (Af); diur. (horses; Calif [US]); stalk bases for tapeworms (Por); to adulter. med. fern hbs. (2122); has antihistaminic effect; hb. a stock poison (causes "staggers" in horses); young fronds c. HCN glycoside.

Pteridophyta: Phylum of ferns, club mosses, & scouring rushes; the sporophytic generation is more prominent, with a leaf-like frond & root-like rhizoids; incl. Filicales, Equisetales. (ex. *Adiantum*, *Polypodium*, *Aspidium*).

Pteris aquilina L. (Polypodi.) = *Pteridium aquilinum* - **P. buchanani** Bak. (sAf): rhiz. decoc. u. as vaginal inj. in cow to aid expulsion of placenta when retained - **P. dentata** Forsk. (Af-Tanz): rhiz. u. tenifuge - **P. ensiformis** Burm. f. (Ind-Polynesia): pl. decoc. u. as diur. & in GC, mal. & dys.; lvs. u. in bacill. dys. (effective in cases resistant to chloramphenicol & syntomycin); rt. juice u. for glandular swellings of neck (Mal); young fronds edible - **P. multifida** Poir. (Chin): u. in diarrh. dys. anthelm. - **P. tri-partita** Sw. (seAs, Taiwan, Polynesia, &c.): fronds u. in parturition (nw Solomon Is).

Pterocarpus angolensis DC. (Leg) (s&eAf): African teak; bloodwood; sd. ash appl. to bleeding gums & dermal inflammn.; sap fr. trunk controls ringworm (*sic*) & intest. parasites; rt. decoc. in GC vulner. dys. aphrod. analg. in headache (poultice) (Tanz); wood valuable ("muninga") for construction - **P. draco** L. (sMex, WI, CA-Ec): swamp bloodwood; kanirabe (Cuna Indians, Pan); red juice outflowing fr. trunk after incising hardens to form WI dragon's blood (Dragon's Blood of Cartagena; sangre de drago); with 34% tannin, 33% gum; bk. & lf. astr.; u. hemostat.; source of cork - **P. erinaceus** Poir. (Equat, wAf-Senegambia to Angola): source of West African (or Gambia) kino with 7.5% tannin, 24% gum; may be superior to Malabar kino; tincture does not gelatinize - **P. esculentus** Schum. (Sierra Leone, Af): wd. u. like that of *P. santalinus* - **P. flavus** Lour. (Chin, Annam, Moluccas): bk. u. in rheum.

jaundice, as ton. & diur. - **P. indicus** Willd. (P. dalbergioides Wall.; P. pallidus Blanco) (Indo-Mal, Chin, PI): wd. (pad-[o]uk or bar wood) u. like Santalum Rubrum; u. for urinary calculi; as red dye; also produces kino; red sap of bk. u. in toothache (Sabah); sap appl. to thrush (PapNG) - **P. lucens** Lepr. (trop wAf): bk. decoc. anti-diarrh. but not often u. in med. (Sen); wd. u. for furniture - **P. macrocarpus** Kurz. (Ind, Burma): source of (red or brown) pad(o)uk wood (bar wood) (of comm.); u. red dyes; for inlay work - **P. marsupium** Roxb. (Ind-Malabar Coast; SriL; Af, Br): Amboyna kino tree; bastard teak; vengai padauk; source of east Indian (or Malabar) kino; (black) kino (gum); the dr. gummy exudn. fr. bk. incisions; c. kino-tannic acid (23-80%) (kino red), gallic acid, gum; astr. (pd. or decoc.) in diarrh., intest. hemorrhage; hemostat. in gargles, to color mouthwashes; wd. u. for cups from which to drink water (cold water inf.) in diabetes (UP, Ind) (value?); kino u. earlier for coloring port & Burgundy wines; in dyeing, printing, tanning (expensive), & other trades (prodn. Ind, SriL) - **P. officinalis** Jacq. (sMex, to Br): latex u. as astr.; to firm up gums; bk. inf. hemostat.; formerly exp. to Eu - **P. podocarpus** Blake (Ve): sangre (de) drago; bk. with resinous exudn. as decoc. u. astr. (65) in diarrh., hemostat., to tighten loose teeth; shrink hemorrhoids - **P. santalinoides** L'Hérit. (w tropAf): source of barwood; sds. roasted & eaten but may be toxic (most poison in shells) - **P. santalinus** L.f. (seAs, SriL, PI): source of red sandalwood; r. saunders; caliatur wood; ruby wood; wood long known & u. in med.; ext. as analg.; astr. for cauterization; in plasters, tooth powders, mouthwashes, snuffs; coloring agt., u. for caste marks (Hindus); u. tech trades (ex. wool dyeing) (red) (2123) - **P. soyauxii** Taub. (trop wAf-as Gaboon): source of African red (or sandal) wood (barwood) u. as dye, in cosmetics (Af); in palm oil as body paint (ngula) (natives); similar to that of *P. santalinus*; grated bk. u. med. (Congo): wd. u. to make canoes - **P. subvirgatus** Malme (SA): lf. u. as digestive (Arg) - **P. tinctorius** Welw. (tropAf): rt. decoc. in fever (Tanz); wd. & rts. source of dye resembling that of *P. soyauxii*; u. to paint newly-born infants; to color feet of native women in imitation of shoes & slippers (484); bee plant - **P. ulei** Harms (Br): bk. ext. like kino; u. in ulcers, snakebite - **P. vernalis** Pittier (nwSA): sangre de drago(n) (Ve); tree exudn. u. locally for cuts, fever, infections fr. arrow wounds, &c. (Ec) (2430).

Pterocarya Kunth (Jugland.) (tempAs): deciduous trees with alternate pinnate lvs.; frt. small, winged nutlet, in pendulous racemes; pith lamellate - **P. rhoifolia** Sieb. & Zucc. (Jap): u. to cure nausea & vomiting in pregnant women; wd. u. for chopsticks, matches, &c. - **P. stenoptera** C DC (Chin): lvs. & bk. carm., anthelm.; lvs. as insecticide; var. *tonkinensis* Franch. as fish poison; tree u. in culture of stick lac. (Laos, Vietnam).

Pterocaulon pycnostachyum Ell. (P. undulatum C. Mohr) (Comp.) (seUS): (Indian) blackroot; rt. u. alt. narcotic, diaph. purg. stim. styptic. emm. oxytocic, abort. (Negroes, wFla [US]) - **P. redolens** F.-Vill. (seAs, Austral, PI): lf. decoc. u. in stim. baths - **P. serrulatum** (Montrouz.) Guillaumin (Comp.) (Austral): lf. decoc. drunk for colds, flu, & sore throat; lvs. crushed & stuffed up nose (2263) - **P. subvaginatum** Malme (Arg): yaguarete caa; pl. u. as insecticide (myiasis); emmen.

Pterocymbium tinctorium (Blanco) Merr. (Sterculi.) (PI): rts. appl. for ulcer; bk. & frts. toxic.

Pterolobium exosum Bak. f. (Leg) (cAf): redwing; lvs. eaten with butter for fever; roasted rt. squeezed between teeth for toothache; juice analg. - **P. stellatum** (Forsk.) Brennan (trop eAf): gara; lvs. rich in tannin (21%); inf. & decoc. u. for TB, &c. (485); to make inks (2559), in tanning; rt. decoc. for stomachache.

Pteronia camphorata L.(Comp.) (sAf): plaster u. for drawing boils - **P. onobromoides** DC (sAf): mixed with grease & u. med. by Hottentots; Hottentot name "buchu" appl. to this & other arom. pls. u. med. & in cosmetics - **P. stricta** Ait. (Af): lf. inf. u. for intest. dis.

pteropterin: pteroyl triglutamic acid; PTGA; isolated fr. cultures of *Corynebacterium* sp.; antimetabolite; u. antineoplastic (formerly).

Pteropyrum olivierii Jaub. & Spach (Polygon.) (Pak): pl. considered cure for sore throat; u. to disinfect milk, pots, &c.

Pterospermum acerifolium Willd. (Sterculi) (Pak to Java): fls. general ton., appl. locally for migraine; lf. pubescence u. as hemostat.; wd. much u. - **P. heyneanum** Wall. (Ind): lvs. u. in leucorrh.; smoked like tobacco - **P. niveum** Vid. (PI): bk. decoc. u. for puerperium & swelling - **P. suberifolium** Lam. (Ind): fls. in paste appl. for hemicrania.

Pterospora andromedea Nutt. (monotypic) (Monotrop./Pyrol.) (Cana, nUS): pinedrops; giant bird's nest; false crawley; a root parasite; u. diur. bitter, emmen.; st. & berries inf. u. to prevent bleeding fr. nose & lungs (Cheyenne Indians); pl. u. as dye (Navaho).

Pterostyrax corymbosa Sieb. & Zucc. (Styrac.) (Jap, Chin): asagara; prod. valuable heartwood.

pteroyl-(mono)-glutamic acid = folic acid.

Pterygium Endl. = *Dipterocarpus.*

Pterygota alata R. Br. (Sterculi) (Ind): sds. u. as substit. for Opium.

Ptinidae: beetle fam. to which belongs the common drug store beetle, *Sitodrepa panicea.*

Ptinus brunneus Dufts. (Ptinidae): spider beetle; causes some damage among drugs.

ptisan(e): tisane.

ptomaines: very poisonous org. compds. produced in animal protein putrefaction; most food poisoning, however, is not due to ptomaines.

ptyalin: amylase of saliva of man & some other animals; salivary amylase; has been crystallized; opt. pH = 6.8.

Ptychopetalum olacoides Benth. (Olac.) (Br-Amazonia, Suriname, Guiana): source of muirapuama (rt. & st. wd.; or rts.) (Braz Ph.), called potency wood, u. ton. aphrod. in impotency int. & exter. (genital baths); in nerve weakness (neurasthenia), rheum.; to stim. appetite & assist digestion; very popular - **P. uncinatum** E. Anselm. (nBr): u. in chronic rheum., nerv. dis., dyspepsia.

Ptychotis ajowan DC. (Umb.) = *Trachyspermum ammi* - **P. ammi** Halacsy = *P. atlantica* Coss. & Dur. = *P. trachysperma* Boiss. = *P. verticillata* (Desf.) Briq. = *Ammoides verticillata* (Desf.) Briq. - **P. saxifraga** (L) Lor. & Barr. (monotypic) (c&sEu): ann. hb.; ornam.

Pubes Dolichi (L.): Mucuna, trichomes.

puca (Br): *Cissus sicyoides.*

Puccinia malvacearum Mont. (Pucciniaceae-Basidiomycetes-Fungi): common rust infestation of mallow lvs.

puccoon: term appl. by NA Indians to pls. furnishing yellow or red dyes or to pigment; also medl.; specif. Sanguinaria and Hydrastis. - **Canadian p.**: *Sanguinaria canadensis*; perhaps also *Lithospermum* spp. - **false p.**: term may have been u. to differentiate *Lithospermum* fr. *Sanguinaria* - **hairy p.**: *Lithospermum gmelinii* - **hoary p.: Indian p.**: *L. canescens* or *L. carolinense* - **red p.**: *Sanguinaria canadensis* - **white p.: yellow p.**: *Hydrastis canadensis*; *Lithospermum canescan*, &c.

puchere = pucheri = puchuri (major) (Br): **puchurim: puchury** (Amazonia): *Licaria puchury-major* (do not confuse with pichury).

pudding pipe (tree): p. stick: *Cassia fistula.*

pudina (Hind, Beng, Bomb, Marathi, Punjabi, Telugu): **pudneh** (Pers): podina: spearmint; peppermint.

Pueraria lobata Ohwi (P. hirsuta Schneid.; P. thomsonii Benth.; P. thunbergiana Benth.; P. triloba Mak.) (Leg.) (Jap, Kor, Chin s to NG; introd. & natd. seUS; now very common): kudzu (vine); kuzu; tuberous rt. u. antipyr. diaphor. in mfr. superior starch (10%) (Yoshinokuzu) (Jap [arrowroot]) (2125); in wines; in diarrh., dys. chicken pox, flu; typhoid fever; alc. intoxication; in fomentation for bruises (Ainu); starvation food & hog food (Tonga) (86); tender lvs. eaten (Jap); fls. (buds) in dys. & intest. distention, esp. in alcoholics; vine ton. & cough remedy; vine (-20 m, L H Bailey) cult. as forage (partial feed), tender lvs. eaten (Jap); cover crop, to prevent soil erosion (soil conservation); oldest fiber pl. of Chin; ornam.; pl. now covers large areas of sGreat Plains (US) (2124) - **P. mirifica** Airy Shaw & Savat. (Thail): c. estrogenic subst., u. ton. rejuvenating; prod. abortion due to an estrogen, miroesterol - **P. phaseoloides** Benth. (trop seAs): lvs. u. as poultice for boils (Mal); decoc. drunk for foul-smelling ulcers; cult. as green manure & cover crop - **P. tuberosa** DC (Ind, Jap; cult. As & Am trop): ban kumra; rt. juice rubbed on body for rheum. & fevers (with hot mustard oil) (124); emet. ton. aphrod.; cardioton.; diur. galactag. contraceptive (anti-implantation) (2264); dem. (as cataplasms); rhiz. & rt. eaten.

puericulture: the science of the proper development of youth—physical, mental, & moral.

puerro (Sp): leek.

puffball: 1) *Lycoperdon* spp., such as *L. candidum* 2) *Taraxacum* spp.

puffer (fish): fugu (Jap) (*qv.*).

pukateine: alk. fr. bk. of *Laurelia novae-zealandiae*; aporphine type alk.

puke weed: *Lobelia inflata.*

pukikawa (NZ): *Drimys axillaris.*

Pulegium (L): *Mentha p.*

pulegone: menthenone: cyclohexanone deriv.; a terpene ketone found in VO of many Lab spp., notably the pennyroyals.

Pulicaria crispa Sch.-Bip. (Comp.) (tropAf, As): dr. hb. u. as vulner. for bruises - **P. dysenterica** Gaertn. (sEu, cAs, nAf): fleabane; (Grosses) Flohkraut (Ge); hb. & fl. heads u. in dys., hemoptysis; rt. u. to adulter. Arnica rt.; fl. heads u. like Arnica (for wh. substit. & sophisticant) - **P. glaucescens** Jaub. & Spach (Pak): pl. eaten by camels & believed to be strengthening (but purg.) - **P. scabra** Druce (sAf): inf. u. as lotion to bathe hemorrhoids; pd. leaf in paste appl. to vaginal tumors; pd. lf. u. for colds (Zulus) - **P. vulgaris** Gaertn. (Euras, nAf): hb. u. as insecticide (smoke); earlier in med. & witchcraft.

pull-and-haul-back: pull-and-hold-back: *Pisonia aculeata.*

Pulmonaria angustifolia L. (Eu, esp. mts.) (Boragin): narrow-leaved lungwort; hb. u. med. like that of *P. mollis* - **P. mollis** Wolff (Euras): hb. u. in lung dis., expect.; for adulter. & substit. of pulmonaria herb - **P. officinalis** L. (Euras): (common) lungwort; sage of Jerusalem; hb. (as Herba Pulmonariae [maculatae]) c. mucil., crude saponin, tannin, silicates, allantoin; u. dem. expect. ("pectoral"), astr. in diarrh. dys., hemorrhoids; in resp. dis.; exter. as vulner.; for tumors (Sp.); abort. (Caucas.).

pulp: Pulpa (L): fleshy tissues of frt. (sarcocarp); ex. colocynth; tamarind (**preserved p.**) - **p. chemicals**: by-products of the pulp & paper industry (esp. fr. sulfate & sulfite mills); incl. rosin, pine oil, α-pinene, turpentine oil, terebene, dipentene, terpin hydrate, terpinolene, camphene, methyl chavicole, &c. - **P. Tamarindorum** (L): tamarind pulp; prepd. by boiling frt. pulp until a sticky mass is formed, sometimes with addn. of sea water (as conserving agt.), then packed into sacks, bales, or casks for shipment - **P. T. depurata**: purified t. pulp; made by softening of crude p. with eq. amt. of hot water, rubbing thru a sieve, & heating in porcelain container until a thick ext. is obtained; 20% pd. sugar added & mixed.

pulque: alc. beverage (wine) obt. by fermentation fr. *Agave atrovirens* (**pulque agave**); sometimes distd. —> mescal.

Pulsatilla cernua (Thunb.) Spreng. (Anemone c. Thunb.) (Ranuncul.) (Chin, Kor, Jap): okinagusa (Jap); rts. u. for bone fractures; as carm. in gastralgia; dentalgia, sore throat; astr. styptic - **P. occidentalis** Freyn (Alaska-Alta-Calif [US]): (western) pasque flower; western anemone; Oregon (or desert) buttercup; mountain pasque; c. protoanemonin; with antibacter. properties (2126) - **P. patens** P. Mill. (P. ludoviciana Heller; Anemone ludoviciana Nutt.; A. patens L.) (Arctic Am, Cana,-nUS; nEuras): pasque flower; spreading anemone (or p. f.) "(American) pulsatilla"; wind flower; (prairie) crocus; blooms (purple) in early spring (601); hb. c. protoanemonin (anemone-camphor), VO u. as stom. counterirritant; antispasm. card. sed., expect.; crushed fresh lvs. appl. to rheum. (Am Indians); much like *P. pratensis* & *P. vulgaris* - **P. pratensis** (L) Mill. (Anemone p. L.) (nEu): (meadow) wind flower; m. pasque-flower; Easter flowers; hb. c. ranunculin (glycoside) which on hydrolysis releases protoanemonin; flavonoid glucosides, tannin; u. as sed. in painful spasms, neuralgias, rheum. migraine; diur. diaph. otitis, rhinitis, emmen. in menstr. disturbances - **P. vulgaris** Mill. (Anemone pulsatilla L.) (e,c&nEu): (European) pasque flower; c. & u. like those of *P. pratensis*; formerly off. (USP 1880-90) as Pulsatilla (L); rt. said eaten by black bear (486) (487).

pulse: edible sds. of pods of Leg. family members, mostly peas & beans, esp. of *Ervum lens* & *Vicia hirsuta* - **p. family**: Leguminosae.

Pulvis Aloes et Canellae (L) (USP 1820-70): Powder of Aloes & Canella; Hiera Picra; "hickory pickery"; claimed the oldest galenical (ancient Eg) - **P. Infantum: P. Puerorum**: dusting powder (exter.) - **P. Purgans** (L): Jalap Powder Comp. - **P. Sanguis**: dr. blood.

Pumex (L): **pumice (stone)**: a very light volcanic porous rock; c. 60-80% silica (kind of glass actually); full of small cavities; u. cosmet. to rub down warts & other skin growths; for polishing, cleaning, &c. (1707) (prod. Haw; Rheinland [Ge]; Lipari Is [It]) (2127).

pumilin: antibiotic fr. *Bacillus pumilus* Gottheil (soil; Nigeria); active against Gm.+ bacteria

pummelo: pompelmous.

Pumpernickel: coarse rye bread popular in neEu; sl. acid; "black bread" (US prod. c. also wheat flour, caramel, &c.).

pumpkin: *Cucurbita pepo* & its vars. (frt. pulp, sd.); *C. moschata*, &c.

punarnava: punarnavi (Sanscrit): 1) *Boerhaavia diffusa*; white-fld. form, sometimes considered a separate sp., *B. p.* 2) *Trianthema portulacastrum* - white fld.; both are u. in Ayurvedic med. (Ind).

punche: punchi (NM [US], Mex): local smoking tobacco; fr. *Nicotiana attenuata* (731).

puncture vine (Calif [US]): **p. weed**: *Tribulus terrestris*.

Punica granatum L. (Punic.) (sAs, Iran, Pak, Ind, Chin; now natd. in many trop&subtrop areas, esp. in Med reg. as Mor): (common) pomegranate; Carthaginian apple; rt. bk. (Cortex Rad. Gran.) (Balausta granati) stripped fr. trees no longer fruit-bearing; c. pseudo-pelletierine (chief alk.), isopelletierine; tannin; u. astr. anthelm. tenifuge; pomegranate frt. eaten, pulp around sds. refreshing (the Pomum punicum of Ovid); c. 1-5% citric acid; fermented —> grenadine cordial; frt. peel (rind) c. 19-28% tannin, u. astr. in diarrh. dys. colitis, tenifuge; frt. or sds. popularly eaten by women to prod. sterility (Ind); dye, tan (in prodn. fine leathers, as Morocco l.); fls. (blossoms) (Flores Balaustii); balaustion fls.; balaustias (Por) u. astr. in diarrh., in dyeing, decorations; frt. juice as syrup u. flavor, in beverages (488); shrub an ornam. (chief prodn. Fr) (2128).

Punicaceae: Pomegranate Fam.; shrubs & small trees of Euras; frt. a thick-skinned berry; only 1 genus, *Punica* (with 2 spp.).

punicine: pelletierine; alk. of pomegranate bk.

Punicum Malum (classical Lat.): pomegranate.

punk (fr. spunk, *qv.*): decayed wd. or other spongy pl. material u. as tinder, &c. - **p. tree**: *Melaleuca quinquenervia*.

punnag (seed & oil): *Calophyllum inophyllum*.

Pupalia lappacea Moq. (Amaranth.) (PI): lvs. or frt. u. for boils; rt. purg. in colic (499).

pupunha (Br): *Bactris speciosa* (Palmae).

purée (Eng, Fr): food boiled into a pulp or paste & then rubbed through sieve; ex. in thick soups (as tomato paste).

purga dos paulistas (Br): *Joannesia princeps*.

purgative lemonade: magnesium citrate soln.

Purgiercassie (Ge): **Purgirkassia**: *Cassia fistula*.

purging root: *Euphorbia corollata* (rt.).

purging: *v.* also under cassia; flax; nuts.

purified: *v.* under animal charcoal; cotton; infusorial earth; kieselguhr; siliceous earth.

purine: a double-ringed compd., $C_5H_4N_4$; closely related to uric acid; when oxygenated, forms the xanthine alks; its derivatives are quite common in nature; ex. uric acid, caffeine, guanine, etc.

Puritan (Salad) Oil™: cold-proofed cotton-seed oil; u. for cooking, salads, etc.

purity rubric: official rubric; the official (USP, or NF) standard of quality, strength, or purity, as specif. given in the off. monograph; stated in various ways, however, most commonly as a percentage representing the minimum allowable assay of the active constituent(s); ex. for Asafetida: shall yield not less than 50% alc. sol. extractive.

Puromycin: antibiotic prod. by *Streptomyces alboniger* Hesseltine & al. (soil actinomycete); disc. 1952; formerly called "Achromycin" (but no more); ribofurane-purine deriv.; antineoplastic (esp. mammary cancer); antimitotic; antiprotozoal (*Trypanosoma*) (amebiasis) (1708).

purple: *v.* under boneset; cone-flower; loose-strife; willowherb.

Purpurblume (Ge): *Aletris farinosa*.

purpurea-glycoside A: a primary glycoside of *Digitalis purpurea*; is hydrolyzable to digitoxin & dextrose - **p.-g. B**: a primary glycoside of *D. purpurea*, hydrolyzing to gitoxin & dextrose.

Purshia tridentata (Pursh) DC (Kunzia t. Spreng.) (Ros.) (BC [Cana]-Mont-NM-cCalif [US]): (northern) antelope bush; buck brush; rt. decoc. for cough & other resp. dis.; dried ripe frts. inf. u. as emetic (very bitter); stain fr. outer sd. coat u. to color arrows, bows, &c. (Klamath Indians) (3583).

purshiana bark: Pursh's bark: Cascara Sagrada.

purslain: purslane: *Portulaca oleracea* - **Florida p.**: *Richardia scabra* - **garden p.: green p.**: p. - **water p.**: 1) *Vernonia beccabunga* 2) *Ludwigia repens*, & other *L.* spp.

pursley: 1) *Portulaca oleracea* 2) (Ala, Miss [US] slang): *Richardia scabra* (corruption of purslane; confused at times with parsley).

purutu (Pe): common bean.

pusley, (Florida) (seUS): **pussley**: pursley, esp. *Richardia scabra* (*cf.* phrase: "as mean as pusley" [2329]).

pussur (Pak): *Carapa moluccensis*.

pussy paws: *Calyptridium umbellatum* - **p. toes**: *Antennaria* spp. - **p. willow**: *Salix discolor* (SC gray fur-like catkins (or aments) on leafless branches).

Putranjiva roxburghii Wall. (*Drypetes* sp.) (Euphorbi.) (Ind, Mal): lvs. frts. & putamens u. as decoc. for colds & fevers; rheum.; sd. kernels c. VO & FO; wd. shining, whitish; tree both wild & cult.

putrefaction: decompn. by bacteria of dead pl. or animal tissues, esp. appl. to protein decompn. with effusion of evil-smelling gases.

Putterlickia verrucosa Szyszyl (Celastr.) (sAf): c. antileucemic alk.

putty: mixt. of whiting & linseed oil u. to glaze windows, seal joints, &c. - **p. powder**: artific. prepn. of tin oxide, sometimes mixed with lead oxide; u. to polish glass, &c. - **p. root: p. wort**: *Aplectrum hyemale*.

putz pomade: a polishing paste made with ferric oxide & prob. pumice stone (pd.); u. for polishing.

Puya chilensis Mol. (P. coarctata Fisch.) (Bromeli) (Ch): **puya**; cardón; ft. spikes source of chagual (or maguey) gum (through insect injury to inflorescence) c. bassorin; u. med. & technol., in textile printing; decoc. of lvs. & sts. u. for tumors; chaqual gum u. exter. for bone fractures; young shoots eaten in salads; lf. fibers (cardón) u. to make rot-proof nets - **P. floccosa** (Ve): arua yek; boiled lvs. u. as purg. - **P. gummifera** Mez & Sodiro (Ec): achupalla; lf. bases eaten, said good for fodder - **P. lanuginosa** Schult. f. (SA): chagula gum produced by inflorescences.

Pycnanthemum Mx. (*Koellia* Moench) (Lab.) (NA): mountain mints; perenn. hbs. with small purplish or whitish fls. in heads or cymes (fr. pyknos [Gr], dense; anthos [Gr], flower, SC fls. in close heads) (500) - **P. albescens** T. & G. (seUS): (American) mountain mint; lf. decoc. of perenn. hb. u. as sudorific in colds (Choctaw Indians) - **P. flexuosum** BSP (seUS): American wild basil; hb. u. med. - **P. incanum** (L.) Mx. (eUS): Atlantic mountain mint; wild basil; pl. decoc. u. to stop nose-bleeding (Koasati Indians) - **P. lanceolatum** Pursh: *P. virginianum* - **P. muticum** (Mx.) Pers. (P. miticans Pers.) (e&cUS): horse mint; odor of pennyroyal; hb. c. VO with pulegone (predominating), menthol, menthone, limonene; u. arom. carm. stim.;

substit. for pennyroyal oil; source of synth. men thol - **P. virginianum** Durand & Jackson (P. lanceolatum Pursh) (e&cCana & US): mountain thyme; m. mint; comp. lf. decoc. u. as antipyr.; diaph.; rt. decoc. for stoppage of menst. period (Chippewa Indians); panacea for run-down patients (Fox Indians); lvs. as snuff as stim. (Fox Indians); antispasm.; VO similar to that of *P. muticum*; suggested for u. in tooth powders & pastes, &c. (501); honey plant.

Pycnanthus angolensis (Welw.) Warb. (Myristic.) (tropAf): false nutmeg; sap collected on cotton from incisions in bk. appl. to diseased teeth, buccal mucosa (for aphtha, thrush, &c.); instilled into ear for otitis; pl. decoc. u. in cough, diarrh., snakebite (Ivory Coast).

Pycnarrhena manillensis Vidal (Menisperm) (seAs): ambal; lf. inf. u. for high fever; pd. rt. u. as ton. in gastric dis. (PI); water & fl. squeezings form gelatinous mass appl. to head as remedy for falling hair (Indones); rt. vulner. esp. for snakebite.

pycnolide (pyknolide): $C_{20}H_{28}O_6$: long-chain compd. found in *Liatris pychnostachya* gum (Herz & Shamma); germacradienolide; supposed to have antitumor properties (2473).

Pycnoporus sanguineus Murr. (Polyporus s. L.) (Polypor.) (trop OW, NW; widely spread): u. as remedy in hemoptysis (Indones) (sympathetic medicine); c. polyporin, an antibiotic subst.; toxic; frt. bodies u. for intest. dis., VD, eczema (Ind) on market.

Pycnostelma paniculatum K. Schum. (Asclepiad.) (eAs): rt. u. to treat nerve dis. as vertigo (Chin); u. during pregnancy & during confinement (Kor).

Pycnothymus Small = *Satureja.*

Pycreus umbrosa Nees (Cyper.) (sAf): u. for chest colds; scented rhiz. put among clothes.

Pygeum Gaertn. = *Prunus*; thus, **P. africanum** Hook. f. = *Prunus africana*; **P. gardneri** Hook. = *Prunus ceylanica.*

Pygmaeothamnus zeyheri Robyns (Rubi.) (sAf): u. in ointment rubbed on scarifications of swollen legs to reduce swelling.

pylinyado (wood): *Xylia xylocarpa.*

pyocyanase: pyocyanin(e): one of earliest known antibiotics (1889); prod. by *Pseudomonas aeruginosa* (Schroeter) Migula (P. pyocyanea Migula); actually a crude material contg. this was introduced ca. 1900 in Germany for diphtheria & grippe; the variable compn. made its use unpredictable, it was rather toxic & not standardized, and soon faded away; active *in vitro* against Gm.+ organisms.

Pyracantha M. Roem. (Ros.) (seEu to Chin): fire thorn; evergreen thorny shrubs, with white fls. & red frts.; 6 spp.; cult. ornam. (genus close to *Crataegus* & *Cotoneaster*) - **P. angustifolia** Schneid. (orig. Chin; s&seEu): (orange) firethorn; thorny shrub -4 m high; frt. c. polycislycopene; toxic, even fatal, to children; cult. ornam. - **P. coccinea** M. Roem. (Crataegus pyracantha Medik.) (sEu, wAs): (red) firethorn; fiery (or everlasting or Christ's) thorn; shrub sometimes -7 m; lvs. c. rutin, chlorogenic acid, pyracanthoside; fls. c. amylamine; frt. u. as ton. astr. in diarrh. dys., vaginal discharges; kernels of sds. u. as coffee substit.; eaten by birds; frt. (non-toxic) (1709) often in marmalades; ornam. shrub (hedges) - **P. crenulata** Roem. (Ind, Chin): cult. ornam. shrub or small tree.

Pyrahexyl™: synth. compd.; euphoriant action as of Cannabis.

pyramid flower: *Frasera carolinensis.*

pyranose: cyclic form of glucose in contrast with furanose (acyclic).

Pyrenacantha scandens Planch. ex Harv. (Icacin) (sAf): chief source of edible spinaches (Natal) - **P. staudtii** Engl. (cAf): u. med. (Congo).

pyrenoids: small bodies found mostly in chloroplasts of Algae; protein center sheathed with starch; prob. serve in starch mfr., although some think they may represent reserve protein supply.

Pyrethri Flores (L.): pyrethrum (flowers): insect flowers; fl. heads of *Chrysanthemum cinerariaefolium, C. coccineum, C. marschallii* (chief prodn. areas: Kenya, Br, Ec, Jap, Dalmatia) - **pyrethrin I, p. II**: the most active insecticidal components of Pyrethrum Flowers; organic esters of rather complex structure; I is more strongly insecticidal than II.

pyrethrosin: one of active components of pyrethrum fls. (502).

Pyrethrum Zinn = *Tanacetum* (*chrysanthemum*).

Pyrethrum (L.): **Pyrethrum Flowers**: Pyrethri Flores: do not confuse with Pyrethrum root formerly off. under same L. title (Pyrethrum) - **p. flowers, exhausted**: the marc fr. extraction of p. f., c. residual pyrethrins & cinerins; u. along with other insecticides, mostly of synth. type, mostly in dusts (in fine pd.) - **p. root**: fr. *Anacyclus pyrethrum.*

pyridine: heterocyclic tertiary amine, C_5H_5N; occ. in belladonna, coffee, bone oil, & other natural sources; obt. fr. coal tar by fractionation; readily hydrogenated to form the alk. piperidine; very strong unpleasant odor; u. alc. denaturant, in org. synth.; formerly u. antisep., in asthma.

pyridoxine alpha - keto glutarate (PAK): said to help control diabetes.

pyridoxin(e) hydrochloride: vit. B_6 HCl; a component of the vit. B complex (one of its 6 known essential vitamins); in rats, represents the antidermatitis factor; occ. in wheat germ, liver, meat &c.; intake said to improve intelligence; u. for arthritis & other pains; sore throat; effective in carpal tunnel syndrome, trigger finger, &c.; given orally & parenterally.; u. food supplement; in vit. medication, parenterally for emesis gravidarum (pregnancy sickness) (E).

pyro: pyrogallic acid; pyrogallol.

pyrocatechin: pyrocatechol: dihydroxy-benzene (1:2); $C_6H_4(OH)_2$; a phenol wh. occ. in crude beet sugar & as degradate of many tannins (catechol type).

pyrogenus (L): alcohol.

pyrogen: fever-producing substances representing metabolic products of bacteria: mucopolysaccharides.

Pyrola asarifolia Mx. var. *purpurea* Fern. (P. uliginosa Torr.) (Pyrol.) (nNA s to SD, NM [US]): (pink) wintergreen; p. pyrola; u. in kidney dis., GC, hemoptysis (Maritime Indians); lvs. boiled in water panacea (Montagnai Indians) - **P. elliptica** Nutt. (nNA, similar range as preceding): shinleaf (SC u. like shin plasters [cheap brown paper for bruises] wrapped around sore legs); wild lily-of-the-valley; lvs. u. astr. (inf. u. as gargle for sores & cankers in mouth) (Mohegan Indians); diur. ton. (rt. decoc. u. for weakness & "back sickness" [Montagnais]); diaph. anodyne - **P. maculata** L. = *Chimaphila m.* - **P. rotundifolia** L. (Euras, natd. e&cCana): wild lily-of-the-valley; round-leaved (or false) wintergreen; round-leaf; lvs. c. ericolin, arbutin, tannin; u. as diaph. astr. anodyne; diur. antilithic; in ophthalmia; & as blisters; var. *chinensis* Andres (n&wChina): lvs. u. styptic; in rheum.; as tea by lumbermen - **P. secunda** L. = *Orthilia s.* - **P. uniflora** L. = *Moneses u.*

Pyrolaceae: wintergreen family; close to Ericaceae; recognized by some botanists to incl. saprophytes; otherwise usually placed under Monotropaceae, also sometimes genera placed under Ericaceae (*Pyrola, Chimaphila*, &c.; with polypetalous corolla; cold nTemp Zone & Arctic).

Pyroleum (L.): empyreumatic VO - **P. Juniperi**: juniper tar; cade oil - **P. Lithanthracis**: coal tar - **P. Pini**: rect. pine tar oil.

pyroligneous acid: crude acetic acid obt. by direct distn. of wd. & wd. tar (often pine); c. methanol, acetone, &c.

pyrolusite: native MnO_2.

pyrolysis: decompn. by heat.

Pyromen™ (Baxter): Piromen (Travenol): complex polysaccharide of bacterial orgin (*Pseudomonas*) prepd. by proteolysis & dialytic separation; a pyrogen; c. DNA, hexosamine; u. as antasthmatic, in allergies.

pyrones: derivatives represent many important pl. constituents, such as anthocyanidins, flavones, & other pigments; coumarin, umbelliferone, scopoletin, aesculetin, etc.

pyroxylin: gun cotton.

Pyrrhopappus carolinianus DC (Comp.) (s&cUS): (leafy- stemmed) false dandelion; "sunflower" (cLa) (since compass plant); rts. were eaten in fall (Kiowa Indians).

Pyrrosia adnascens Ching (Polypodi) (China): juice fr. pounded fronds u. for burns & as dyes - **P. lingua** Farwell (China): fronds astr. diur.; in GC & urin. dis.; rhiz. to treat wounds, carbuncles - **P. petiolosa** Ching (China): ext. of whole pl. in wine u. for lumbago (int.); fronds & rhiz. for pulmon. dis. - **P. sheareri** Ching (Chin): fronds & rhiz. u. diur.

Pyrularia edulis (Wall.) DC (Santal.) (tropAs, Ind-Himal): frts. edible; sd. source of FO - **P. moschifera** (Bl.) DC (seAs): u. as astr. in diarrh. dys. - **P. pubera** Mx. (P. puberula Mx. sometimes in error) (eNA): rabbit-wood; shrub bears edible frts.; FO of sds. ("oil nuts") said to be bad-smelling & poisonous (503).

Pyrus L. (Ros.) (orig. temp. Euras): pears &c.; frts. edible, sometimes delicious; pulp c. stone cell masses (diff. fr. *Malus* & *Sorbus*); some 20 spp.; many taxa originally assigned to this gen. have been transferred to others (*Malus*; *Sorbus*; *Aronia*; *Cydonia*; &c.) - **P. aria** Ehrh. = *Sorbus a.* - **P. aucuparia** Gaertn. = *Sorbus a.* - **P. calleryana** Dcne. (Chin): frt. u. to treat dys. - **P. communis** L. (P. domestica Med.) (temp. As; cult. Eu & worldwide in temp. regions): (common) pear (tree); pera (Sp); frt. esculent, u. nutr. antiscorb. astr. diur. lax. (syrup); sds. for adulter. qunice seeds; frt. eaten raw, cooked, dr.; as preserves, jam, pear butter; lvs. (heated at 100° C for short time to destroy enzymes) u. as urin. disinfectant, diur. & sed.; wd. u. in the technical. arts, in engineering, &c.; wd. excellent fuel; cult. in many vars. (Bartlett, Kieffer, Seckel, Vicar) - **P. cordata** Desv. (Eu): cult. in hedges & for its wd. - **P. cydonia** L. = *Cydonia oblonga* Mill - **P. lindleyi** Rehd. (Chin): fresh (or dr.) pulp & peel u. in dys.; frt. peel u. for colds, dry cough, vomiting, inflammation, pulmonary. dis.; rt. bk. astr. for wounds; lf. decoc. u. for urinary tract infections - **P. malus** L. = *Malus sylvestris* - **P. pyrifolia** Nakai (& var. *culta*) (n&cChin): sand pear; frt. excellent for preserves; pl. drought-resistant - **P. sitchensis** Piper = *Sorbus s.* - **P. sorbus** Gaertn. = *Sorbus domestica.*

pyxis: 1) capsule in wh. the upper part dehisces —> kind of lid 2) theca of mosses.

Qq

qahwa (Arab): coffee.
qat: *Catha edulis*.
quadang nuts: frt. of *Fusanus acuminatus*.
quadruplex pills: Pil. Iron & Quinine Comp.: Pil. Iron, Quinine & Nux Vomica.
quail: Several genera of small game birds, chiefly *Coturnix coturnix* L. (Phasianidae) (sEu, wAs, nAf): grey quail; bater; a kind of small partridge; excellent meat & hence much hunted; domesticated in ancient Rome; flesh & other parts u. in Unani med., thus blood instilled into ear for earache (*Perdicula asiatica*, jungle bush quail [As]: flesh u. for gastric & liver dis. & diur. in Unani med.) (2627).
quakenash (Penn): quaking ash: *Populus tremuloides*.
Quaker bonnet: Scutellaria - **Q. button**: Nux Vomica - **Q. laudanum**: Acetum Opii (38).
Quaker's black drop: Opium vinegar (Acetum Opii).
quamash: Quamasia Rafin. = *Camassia* Lindl.
Quamoclit Mill. = *Ipomoea*.
quantitate: measure; define in concrete terms.
quantitative microanalysis: determination with the microscope of the proportions of various drugs or other materials in a powder mixture; based upon a study under the microscope of powders with known quantities of a component with fixed & uniform comp. (such markers as Lycopodium, corn starch, etc.); can be u. with materials wh. possess uniform type of constituent, such as starch grains, pollen grains, spores, trichomes, etc.
Quararibea fieldii Millsp. (Bombac.) (Yucatan, Mex, Hond, Bel): maha; fls. u. to flavor chocolate - **Q. funebris** (Llave) Vischer. (Mex, Guat, Salv): rosa de cacao; dr. fls. u. to flavor chocolate; young shoots u. as utensils (frothing sticks) to beat chocolate to a froth; fl. decoc. in cough & to regulate menstr.; frts. antipyr. - **Q. putumayensis** Cuatrecasas (Co, Pe): rts. u. in prepn. of arrow-poison; frt. edible (Kofan Indians) - **Q. turbinata** Poir. (WI, contin. tropAm): swizzle-stick tree; whorled twigs fr. 3-5 forked branches u. as swizzle sticks for drinks, soups, &c.
quassation (phcog.): reducing tough rts. & bks. to smaller fragments.
Quassia africana Baill. (Simaroub.) (w & inner Af): nsimbikale (Angola); st. bk. & wd. c. quassin, these (& lvs.) u. as very bitter ton., appetizer, vermifuge, antipyr. (natives) - **Q. amara** L. (Guian, esp. Surinam, nBr to Pan, WI): (Surinam) quassia; heart wd. & bk. c. bitter princ. quassins, includ. quassi(i)n (per se), isoquassin (picrasmin), neoquassin, quassol; wd. ("bitter wood") u. as bitter ton., stom. antipyr. anthelm. (mostly for roundworms [incl. threadworms] [mostly as inf., tincture, ext. or fl. ext.]); insecticide (in pediculosis); as alc. denaturant; pesticide (for aphids, flies, caterpillars); formerly cups were made of the wood (on lathes) which when filled with warm water & left to stand rapidly furnished a bitter potion (see plate); in veter. med.; bk. also u. but less commonly, as anthelm., formerly as stom. ton. antipyr. adult. of q. wood (pd.); formerly in beer mfr.; rt. (Radix Quassiae amarae) also sometimes u. med.; in most pharmacopeias (incl. USP), two sources of q. wd. are indicated as alternatives: title sp. and *Picrasma excelsa* (USP, 1820-1930; NF 1930-60) (v. *Simaba*).

quassia bark: Cortex Quassiae (Jamaicensis): bk. fr. *Picrasma excelsa* & *Quassia amara*; u. bitter ton., like q. wood - **bitter q.**: *Q. amara* (wd. & bk.) - **q. chips**: wd. fragments fr. *P. excelsa* & *Q. amara* - **q. cup: q. goblet**: a medicinal container; cup turned fr. q. wd.; u. as bitter ton. draft by filling with water & letting stand few minutes before quaffing (see Frontispiece) - **q. do Brasil meridional** (Br): *Picrasma vellozii* - **q. family**: Simaroubaceae - **Jamaica q.: lofty q.**: *Picrasma excelsa* - **Surinam q.**: *Q. amara* - **q. wood: Lignum Quassiae**: fr. *P. excelsa* & *Q. amara*.

Quassie (Ge): **Quassienholz**: quassia (wood) - **quassiin: quassin**: bitter princ. found in wd. of *Quassia amara* (chief source) (68%), *Samadera indica*, *Simaba cedron* (all of Fam. Simaroub.); consists of mixt. of quassin (proper) & neoquassin; permitted as substit. for brucine in specially denatured alcohol (SDA); method of prepn. (504).

quebrachal (Arg): forest of quebracho trees - **quebracho**: *Schinopsis lorentzii*, *S. balsanae* - **q. bark**: fr. *Aspidosperma quebracho-blanco* (**q. blanco**) (quebracha) - **q. colorado: red q. (wood)**: *Schinopsis lorentzii*, &c. (**q. moreno; q. negro; q. prieto**) - **white q.**: q. bark - **q. wood**: fr. *Schinopsis* spp.

quebrachine = yohimbine.

quebra-pedra (Br)*Phyllanthus corcovadensis*, *P. niruri*.

Quecke (Ge): Queckenwurzel(stock) (Ge): doggrass (rt.).

Queen Anne's lace (seed): *Daucus carota* (wild) - **q. of the meadow(s)**: 1) *Stillingia sylvatica* 2) *Eupatorium purpureum* (hb. & rt.) 3) (Miss [US]): *Eupatorium maculatum* 4) *Filipendula* (*Spiraea*) *ulmaria* - **queen's delight: q. root**: *Stillingia sylvatica* - **Queen Elizabeth root** (US-voodoo): *Iris germanica* - **q. of the night (cactus)**: *Cereus grandiflorus*.

quelites (Sp): tender greens u. as food, usually annual lvs. but also fls., &c. of perennials; ex. *Phytolacca esculenta*, *P. octandra*.

Quendel (Ge): wild thyme, mother of t., *Thymus serpyllum*.

quequeste (Mex): dasheen.

quercetin: a flavonol pigment (a pentahydroxy flavone) representing the aglucone of quercitrin, rutin, & other glycosides; found in broccoli, onion, &c.; claimed anticarcinogenic; found in pls., esp. barks, rinds, etc.; ex. *Quercus* bks., Uva Ursi, Podophyllum, horse chestnut, *Allium cepa* bulb scales, *Forsythia*, *Lathyrus odoratus* - **quercetol** (Fr): quercitrin - **quercimeritrin**: quercetin glucoside; pigment occ. in fls. of *Gossypium herbaceum*, *G. hirsutum*, *Prunus emarginata*, *Helianthus annuus* - **quercitol, alpha**: a rare alc. rather like inositol or sorbitol; obt. fr. acorns (*Quercus* spp. frts.) - **quercitrin: quercitroside**: quercetin 3-rhamnoside; a glycoside found in quercitron bark (fr. *Quercus velutina*), *Vaccinium myrtillus* lvs., horse chestnut lvs., & many other pls. (2405); with same aglycone as rutin but sugar fraction of latter is rutinose instead of rhamnose; hydrolyzes to form quercetin & rhamnose; u. like rutin (med.); as yellow dye (textiles) - **quercitron: quercituron**: quercetin glucuronate: red dye; in *Phaseolus vulgaris*; term also appl. to *Quercus velutina* (quercitron oak).

Quercus L. (pron. kwérk-us) (Fagaceae) (cosmpol temp. to subtrop): genus of oaks; incl. the most majestic of forest trees; many spp. evergreen with thick glossy leathery lvs.; some 600 spp.; bks. commonly u. as tans, in folk med. for eczema, fungus inf., &c. (decoc.), bk. ext. u. in boiler compds.; lvs. u. as tannin drug; frt. acorns, important wildlife food (deer, squirrels, &c.), human food, in tanning; potential source of FO; valuable wd.; ornam. trees - **Q. aegilops** L. (Euras, esp. Gr): valonia oak; source of Valonia or Acorn Cups; very large acorns c. 30% tannin; u. tanning very fine leathers, dyeing; unripe acorns, Catmata, u. tan, dye - **Q. alba** L. (e&cNA): white oak (SC white bk.); tanner's o.; Quebec o.; bk. (Quercus [L.]) c. tannins, u. as astr. in tanning & med. in diarrh., dys., &c.; FO proposed for use (2406); inf. astr. wash in bleeding piles (Penobscot Indians); inf. of inner bk. u. as expect. (Meskwaki Indians); chief source of American galls (with only c. 10-20% tannin); Am Indians prepd. meal fr. acorns, washing with hot water to remove tannins, u. to bake bread, in children's food (Racahout); acorns (mast) relished by hogs, &c. (NCar [US]); rts. crushed in whiskey for liniment in rheum. (Houma Indians); wd. a fuel source - **Q. aquatica** Walt. = *Q. nigra* - **Q. ballota** Desf.: *Q. ilex* v. *rotundifolia* - **Q. bicolor** Willd. (Q. platanoides Lam.) (e&cNA s to Ga & Okla [US]): swamp (white) oak; econ. useful; strong hard wd. u. for making furniture, boats, cooperage, fuel, &c. - **Q. cerris** L. (Euras, Austral): turkey oak; manna (or mossy cupped) oak; source of Morea or crown galls; Gallic galls; with ca. 30% tannin; acorns & lvs. also utilized; acorns eaten like chestnuts (sEu); acorns u. to prep. flour; gum ("oak manna") u. to make sweetmeats; wd. useful - **Q. chrysolepis** Liebm. (Ore, Calif, Ariz, Nev [US], nwMex): (canyon) live oak; maul o.; economically important tree, with wd. very strong & much used, tan, &c.; acorns sometimes eaten (Am Indians) after reducing to meal & cooking - **Q. coc-**

cifera Lam. (Euras, nAf Med reg.): Kermes oak; shrubby; rt. bk. (gazouille) c. 15-25% tannin; u. in tanning; one source of Tripoli galls; host pl. for *Chermes ilicis*, source of kermes dye - **Q. coccinea** Muenchh. (e&cNA): scarlet oak; source of tanbark & "American nutgalls" (rich in tannin); wd. hard & durable - **Q. crenata** Lam. (Fr, Alg): false cork oak; source of tan bark; furnishes inferior kind of cork - **Q. durandii** Buckl. (s&seUS, neMex): Durand (or bluff) oak; Durand('s) white oak; medium-sized tree; acorns edible (505) - **Q. engelmannii** Greene (swCalif [US]): evergreen white oak; mesa o.; prob. the sp. referred to for which mold growing over mush (fr. acorns) was u. for boils, inflammn., &c. (crude antibiotic) (Luiseno Indians) - **Q. falcata** Mx. (Q. digitata Sudw.) (eUS): Spanish (red) oak; southern red o.; bk. astr. (6-8% tannin) (decoc. drunk for piles (Ala [US]); antisep. ton., put in poultry drinking water for "limber neck"; acorns fed to hogs; may be FO source; var. *pagodifolia* Ell. (eUS): cherry-bark oak; (swamp) red o.; wd. excellent; lumber equal to that of *Q. alba* - **Q. garryana** D. Dougl. (wBC [Cana]-Calif [US]): Oregon (white) oak; (Pacific) post o.; Garry('s) o.; an economic lumber sp.; u. for furniture, &c., fuel; acorns edible (Am Indians) (2330) - **Q. ilex** L. (Med reg, sEu, wAs): holly (or holm) oak; evergreen o. (tree); bk. (as Cortex Quercus viridis or Limousin Bark) (named after former province of Fr.) c. tannin (5-11%); u. decoc. in veter. med.; astr.; prepn. of cognac; to age whiskey; source of Tripoli (or Istrian) (or French) galls (Galla gallica); furnishes sweet acorns, with fine flavor, eaten in sEu; wd. as Limousin wood (Lignum Limosinae; limaosinho [Por.]) as grated wd. fr. staves of old wine or cognac barrels ("cognac wood"); u. arom. in spirits ind.; as tan; cork source; var. *rotundifolia* (West.) Rehd. (Q. ballota Desf.) (wMed reg.): frts. (acorns) sweet (*cf.* chestnuts), u. as strengthening nutr. - **Q. imbricaria** Mx. (e&cUS): shingle oak; source of American (nut) galls produced after sting of *Cynips aciculata*; wd. u. for clapboards, shingles - **Q. incana** Bartr. (Q. cinerea DC) (non Q. incana Roxb. of Ind.) (seUS): blue jack oak; upland (willow) o.; wd. useful (2708) - **Q. incana** Roxb. (w. Himal, Ind, Nepal, Upper Burma): grey (or ban) o.; trees sometimes in pure forests; acorns u. in indigenous med. as diur. & in GC; astr. in diarrh. indigest.; buried in soil, then extd. with lime water in prepn. of edible flour & starch (76%); sweet exudn. of tree u. as confection (Iraq & Iran) ("oak manna"); lvs. fodder; wd. to make agr. implements, construction; bk. tan. - **Q. infectoria** Oliv. (Q. lusitanica Lam. var. i. Oliv.) (eMed reg. to Iran): Aleppo o.; chief source of nutgalls, excrescences formed on twigs after sting by *Cynips tinctoria*, a gall-fly or wasp; galls c. tannin (-58%), gallic acid; u. astr. in diarrh. burns, gum dis.; as dusting pd.; tanning material, in dyeing; mfr. of tannin (Coll. Turk, Syr; sorts incl. Aleppo; Jerli; Sorian; Mossul; Smyrna; etc.) - **Q. kellogii** Newb. (Ore, Calif [US]): (California) black oak; acorns edible - **Q. laevis** Walt. (seUS): turkey (or scrub) oak; bk. u. as astr. (folk med.); wd. fuel - **Q. laurifolia** Mx. (seUS): (swamp) laurel oak; tannin-rich bk. and galls available (2189) - **Q. lobata** Née (Calif [US]): (valley) white oak; v. oak; roble o.; overcup o.; specimens in Calif largest of all known oaks; large galls ("California oak galls") are very rich in tannin; u. to make ink (permanent) (2436); inf. of fresh galls u. (diluted) as a wash for sore eyes (Indians, Mendocino Co., Calif); acorns previously the chief food of the Calif Indians - **Q. lyrata** Walt. (seUS): overcup oak; bk. decoc. astr. u. in dys. (Cherokee & Creek Indians), for stomachache (Choctaws); good hard wd., u. in cooperage - **Q. macrocarpa** Mx. (cNA): bur oak; inner bk. chewed for diarrh. (Am Indians); wd. strong, u. in boat building, &c. - **Q. macrolepis** Kotschy (eMed reg., Crete, Gr, AsMin): valonia oak; one source of valonia acorns; sometimes regarded as *Q. aegilops* var. - **Q. marilandica** Muenchh. (se&scUS): (black)jack (oak); barren o.; wd. u. to make charcoal; bk. in folk med. - **Q. nigra** L. (se&scUS): water oak; bk. u. astr.; tan. - **Q. numidica** (Alg): source of an inferior grade of cork - **Q. occidentalis** Gay (Med reg., swFr, Sp, Por): corkwood oak; biennial cork o.; important source of cork - **Q. oophora** Kotschy (Armenia): a source of valonia acorns - **Q. palustris** Muenchh. (e&cUS): Spanish (or pin) oak; bk. u. much like that of *Q. alba*; bears galls rich in tannin - **Q. pedunculata** Ehrh. = *Q. robur* - **Q. persica** Jaub. & Spach (Iran): acorns u. as food (Iran, Kurdistan) - **Q. petraea** Liebl. (Euras): (Ourts) durmast oak; similar in u. to *Q. robur* - **Q. phellos** L. (e&sc US): willow oak; source of tanbark; acorns eaten by birds & rodents; wd. u. in construction, as fuel; ornam. tree - **Q. platanoides** Lam. = *Q. bicolor* - **Q. prinus** L. (Q. montana Willd.; Q. michauxii Nutt.) (e&cNA): chestnut (or cow) oak (as eaten by cows); (rock) chestnut o.; basket o.; bears edible acorns; bk. important tan; u. astr. ton. antipyr. (2185) - **Q. pseudocerris** Boiss. (Q. valonea Kotschy) (AsMin): Gerbeiche (Ge.) (tan oak); tanner's o.; unripe cupules still contg. acorns called valonea (Wallonen; levant gallnuts), with tannin content of 30-42% u. in tanning (ext.);

Smyrna vallonia (exp. fr. Turkey) - **Q. pubescens** Willd. (s&cEu, AsMin): quercia (It); st. bk. decoc. u. as wash for skin inflamm.; produces galls - **Q. robur** L. (Q. racemosa Lam.) (Euras, nAf, natd. NA): English (or common) oak; bk. (as Cortex Quercus) c. 8-20% tannin; gallocatechol; quercitol; quercetin; resins; u. astr. in enteritis, hemostat.; ton. antipyr.; occas. in liver & gb. dis.; exter. in rashes, frostbite; in veter. med. for diarrh.; in tanning & dyeing (blue, brown); French wine is aged in oak casks (bois de chêne) with tendency for wine to have strong oak flavor instead of grape flavor (best wines); source of Serbian Nutgalls: lf. galls formed by sting of *Cynips lignicola* Hart., c. 30-40% tannin; frt. galls (Knoppern) formed on the cupules by the sting of *Cynips calicis* Burgsdorff, c. 24-35% tannin; u. techn. Acorns (as Glandulae Quercus) c. quercitol, FO, tannin, sugars; u. astr. in diarrh.; pd. acorn meal ("Farina") as nutr.; in chlorosis, menstr. dis., rickets; coffee substit. (roasted); timber formerly much u. to build ships for British navy - **Q. rubra** L. (Q. borealis Mx.) (e&cNA): (northern) red oak (< red petioles); inner bk. u. for sore eyes (Am Indians); bk. in dys. (Am Indians); important lumber tree (incl. var. *maxima* [Marsh.] Ashe) - **Q. rugosa** Née (Q. reticulata Humb. & Bonpl.) (swUS, cMex): ensina (de miel); netleaf oak; acorns u. as substit. for coffee (S. Luis Potosi) - **Q. sessiliflora** Salisb. = *Q. petraea* - **Q. stellata** Wang (Q. obtusiloba Mx.) (eUS w to Ia & Kans): post (or iron) oak; acorns u. in form of meal (Am Indians); coffee-like beverage; lvs. as cigarette wrappers (Okla [US] Indians); in peyote ceremony; heavy wd. u. for fencing, fuel, &c. - **Q. suber** L. (Med reg., wAs, nAf, cult. mostly in Port, sporadically in Ala [US], &c.): cork oak; c. tree; chief source of cork of commerce (outer bk. of trunk); 2 types: male (or virgin) cork, that originally formed; female, newly formed after removal of male bark; c. cellulose, triterpenoids (cerin, friedelin, betulin); lignin, suberin, tannin, sitosterol, cork wax, vanillin (paraffins); male cork u. for linoleum, life-belts; female cork for cork stoppers, soles, prepn. gaskets; suberic acid; much u. for insulation & in plastics, &c. (506-A) - **Q. valonea** Kotschy = *Q. pseudocerris* - **Q. velutina** Lam. (Q. tinctoria Bartr.) (e&cNA): black (or yellow) oak; dyer's o.; quercitron; bk. u. as astr. & tan.; inner bk. source of yellow dye in printing calico; acorns very bitter - **Q. virginiana** Mill. (se&sUS, neMex, WI (Cu): (Virginia) live oak; encina (Cu); wd. hard & tough; formerly much u. in shipbuilding; note live oak reservations in wFla [US] Panhandle held by US Navy; source of "Texas nut galls"; bk. & fls. antipyr., astr. antisep., also lvs. (Cu) (507).

Querzellen (Ge): cross cells; cells strongly extended across the long axis of lf., each bearing a papilla; occ. in lower surface of grass lf.

quetchalalatl (Mex): *Amphipterygium adstringens.*

quiabo (Por): okra.

quick, gum: may be quika g. (*qv.*) - **quickens**: *Agropyron repens.*

quihuicha (Pe): Amaranth.

Quiina amazonica A. C. Smith (Quiin.) (Co): lf. inf. u. to treat sores of mouth (Tikuna Indians) (2299) - **Q. leptoclade** Tul. (Co): lvs. u. as snuff for nosebleed.

quika gum (Co): **q. resin**: cuica r.; prod. of *Cercidium spinosum, C. praecox.*

quill: roll of dr. bk. (ex. Ceylon cinnamon).

Quillaiae Cortex (L.): quillaja bk. - **Q. Lignum**: q. wd.

Quillaja (L.): bk. of **Quillaja saponaria** Mol. (Ros.) (Pe, Ch, Bo): soap tree; s. bark t.; Panama bk. tree; dr. inner bk. c. saponins (incl. sapotoxin, poisonous); u. deterg. foaming agt. (not for int. prepns.), in shampoos (508); & other cosmetics; for polishing metals (such as Zn & Al) (2191); in some foreign countries, in soft drinks & beer (foam formation); exp. in chronic bronchitis, pharyngitis, etc.; decoc. formerly u. as gargle in emphysema (bronchial secretions made more fluid & abundant); in mfr. saponin; (cult. Calif [US], nInd; sEu) - **Q. smegmadermos** DC (Ch): bk. (écorce de quillaja savonneux) u. in same way as prec.; likewise **Q. poeppigii** Walp. (Ch): with bark c. saponin.

quillajasapotoxin: a neutral saponin; the chief active component of Quillaja; made up of a composition of quillaic acid, a sugar, & perhaps other substances.

quillay (bark): quillaya (bark): quilley bark: **quilleys**: Quillaja (L.): dr. inner bk. of *Q. saponaria.*

quilling (GB): 1) quill 2) *Cinnamomum verum.*

quimbombo (Co, Cu): okra - **q. chino** (Cu): roselle.

quina (Sp, Por): 1) Cinchona tree 2) *Exostema caribaeum* & other *E.* spp. 3) *Holarrhena febrifuga*, &c. - **Q. amarela** (Por): yellow cinchona bark - **q. amarilla** (Ve): **q. blanca** (Ve): *Galipea* spp. as *G. longifolia*: (Mex); *Exostema caribaeum* - **q. do campo** (Br): *Strychnos pseudoquina* - **q. gris** (Sp): gray cinchona, *C. peruviana, C. nitida*, etc. - **q. mineira** (Br): *Remijia ferruginea* - **q. morada** (Sp): 1) *Cinchona boliviana* 2) *Pogonopus tubulosus* - **q. de serra** (Br): q. mineira - **q. tabla**: tabla bk. (Cinchona) (*qv.*) - **quinacrine (hydrochloride)**: Atabrine (HCl); synth. antimalarial

quinamine: a secondary alk. of Cinchona bk. - **quina-quina**: Cinchona; *v.* Tolu Balsam.

quinbolite (Guat): Lima bean.

quince, Bengal: *Aegle marmelos* (frt.) - **common q.**: *Cydonia oblonga*; *C. rubra*; frt. sd.

Quinchamalium chilense Mol. (Santal.) (Ch): "quinchamali"; hb. aerial parts u. as panacea in liver dis., ulcers, antihemorrhagic, diur. emmen.; c. triterpenes, &/or sterols, alks. (traces), flavonoids, saponins, beta-sitosterol.

quinchoncho (Ve): pigeon pea.

Quinetum: Chinetum: mixt. of Cinchona alks. in varying proportions as they occur naturally in the bark of *C. succirubra* (Red Cinchona), incl. both xall. & amorphous alks.; u. mostly in form of sulfate, in malar. (1273).

quinic acid: kinic a.

quinicine: alk. of Cinchona; isomer of quinine & quinidine; u. as heat stabilizer & in plastic resins.

quinidine: Chinidin (Ge.): alk. of *Cinchona* & other Rubi. genera; stereoisomer of quinine; now made synth. by isomerization of quinine (chief US source); mostly u. med. as **q. gluconate** or **q. sulfate** in cardiac arrhythmias; antimal. (1274) (E).

Quinina (L.): **quinine**: bitter crystalline compd., $C_{20}H_{24}O_2N_2$; chief alk. of Cinchona; numerous salts are available; until ca. 1930 was the most effective antimalarial, now partly superseded in US & some other areas by synth. antimalar.; in treating nocturnal leg muscle cramps (often with vit. E); bitter; considered (1985) one of four essential antimalarials (1501); still u. in many medicaments as antipyr. (esp. in cold mixt.), analg. etc.; chiefly u. as q. sulfate (E) - **q. bark**: 1) Cinchona 2) *Pinckneya pubens* - **q. bark tree**: *C. calisaya*, &c. - **q. bush**: 1) *Sabatia angularis* 2) *Garrya fremontii* - **q. carbacrylic resin**: Azaresin (NF XIV): Diagnex (Blue); combn. of polyacrylic carboxylic acid resin with quinine; u. to det. acidity of stomach; assay of quinine in urine indicates presence or absence of free HCl in stomach - **q. conk: q. fungus**: *Fomitopsis officinalis* - **q. herb: q. plant**: *Sabatia elliottii* - **Indian q.**: aq. ext. of *Tinospora cordifolia* - **native q., Australian**: *Alstonia constricta* - **prairie q.** (Ia [US]): *Sabatia anguloris*; American centaury pl. - **q. tree**: 1) *Codonocarpus cotinifolia* 2) *Alstonia constricta* 3) *Pinckneya pubens*.

quino: 1) Cinchona 2) combining form of quinine - **q. do mato** (Br): *Exostema caribaeum* - **quinoa** (Bo): **quinua**: *Chenopodium q.*

quinoidine: Chinoidin (Ge): non-crystallizable alk. residue remaining after crystalln. of c. alkaloids; said to have 5- 10% therap. value of quinine (Rosin); salts used.

quinoline: chinoline; benzopyridine: occurs in coal tar; mostly prod. synth.; u. preservative, solvent, formerly in med. - **quinovin**: glycosidal compd. found in bks. of *Cinchona calisaya*, & spp. of *Ladenbergia*, & *Esenbeckia*.

Quinora: quinidine sulfate.

quinova bitter: quinovin: occ. in cinchona bk.; glycoside of quinovic acid; a pentacyclic triterpene; sugar moiety is **quinovose** (deoxyglucose).

quinquina (, corteza de) (Sp): 1) cinchona 2) *Quinquina* Boehm.; formerly u. by some botanists as generic name for what is generally called *Cinchona* - **q. Huanuco**: quina gris.

quintal (metric): 100 Kg.

quinua: quinoa.

qupita: quipito (Ve): *Croton* sp.(?); wd. u. in construction (at least formerly); distd. to give VO with turpentine aroma.

Quisqualis conferta (Jack) Exell (Combret.) (seAs): lf. decoc. or juice fr. pounded rts. u. as vermifuge (Mal) - **Q. indica** L. (Q. malabarica F.-Vill.) (seAs, Ind, PI, trop wAf): Rangoon creeper; frts. & rts. u. as vermifuge (As.); sds. c. FO sugar-like subst., gum, tannins, much K_2SO_4 (3.87%), alk. trigonelline; u. purg., decoc. to stop diarrh. in children (Mal); substit. for Santonica (1910); natd. in tropics (as ornam.); sds. & lvs. u. anthelm. (Ind); lvs. in oil appl. to inflamed ear, nose, & throat (Tonga); cult.

Quitten (Ge): quince.

Rr

rabanete (Por): radish.

rabanillo (Sp): **rabanito** (Arg): radish - **rabano** (Sp): 1) radish 2) horse-radish.

rabarbaro (It): rhubarb.

rabbit: grayish mammal found worldwide (Rodentia - Lagomorpha): most domestic vars. are based on *Oryctolagus cuniculus* (L.) (orig. Eu, nAf): European gray rabbit; rapidly progressing animals (hopping, running) with high reproductive rate; hence, often raised for meat & fur; popular vars. include Belgian Hare, Flemish Giant; albinos; u. in clinical test for pregnancy & pyrogens (bacterial endotoxins); many parts & products of rabbit u. in Unani medicine (ex. bile = sed.); related genera: *Lepus* (hares), *Sylvilagus*.

rabbit bells: *Crotalaria rotundifolia*; *C. spectabilis* - **r. berry**: *Shepherdia argentea* - **r. brush: r. bush** (wUS): *Chrysothamnus nauseosus* subsp. *graveolens* & related taxa (SC foraged by rabbits but toxic to livestock) - **r. faces** (Tex [US]): bluebonnets (*qv.*) - **rabbitwood**: *Pyrularia pubera*.

rabente (Por): cabbage.

rabies vaccine: anti-rabic v.; the uncontaminated suspension of the attenuated, diluted, dr. or dead fixed virus of rabies (or hydrophobia) u. to prevent the devt. of rabies in persons bitten by rabid animals (E); SC inj. in side of abdomen, now often given in combn. with **rabies anti-serum** or **r. immune globulin** (prepd. fr. blood of persons vaccinated with r. vaccine); the virus is obt. fr. the CNS of an animal infected with fixed virus rabies; the method of attenuation varies (E) - **adsorbed r. v.** (RVA): special vaccine adsorbed on $AlPO_4$; u. for pre- or post-exposure to infection.

rabo amarillo: *Lonchocarpus leucanthus* - **r. blanco**: *L. muehlbergianus*.

raceme: a fl. arrangement of simple type in which the fls. are arranged along a main axis on pedicels of about equal length; fl. opening proceeds upwards fr. the base of the inflorescence (v. Diagram in Appendix).

racemic acid: d,l-tartaric acid.

rachis: 1) axis of inflorescence 2) in pinnately compd. lvs., prolongation of petiole 3) (Zool) vertebral column.

racilla (Nic, &c.): raicilla.

racine amère (Fr) (bitter root): *Lewisia rediviva* - **r. blanche** (Qu [Cana]): rt. of *Peucedanum ambiguum* & other *P.* spp. - **r. brésilienne**: ipecac - **r. du St. Esprit**: angelica rt.

racoon berry: *Podophyllum peltatum*.

radica (It): root; briar(wood) - **r. di San Pietro**: *Succisa pratensis*.

radice (It): root; horseradish - **r. bulimaca**: *Ononis spinosa*.

radichio (It): radish; also rafano.

Radicula Hill = *Rorippa*; *Nasturtium* - **R. armoracia** Robinson = *Armoracia rusticana*.

radicula: radicle: young rt. of seedling - **r. byssoidea**: fungal mycelium.

radio (Sp): medullary ray.

radio-: as in radio-iodine; tagged or labeled element; isotope; generally u. as tracers.

radioimmunoassay (RIA): procedure of nuclear medicine in which *in vitro* tests are made with the help of specific antibodies and a radioactive indicator (marked antigen or second antibody) to det. amts. of hormone, vitamin, &c.; u. for precise &

accurate assays, such as for digoxigenin in crude exts. of digitalis leaf.

radiopharmaceuticals: radioactive materials u. to diagnose & treat many injuries, defects, diseases, &c.; chiefly of natural origin; incl. radioactive iodine, &c. (radioactive isotopes); strength of radiation measured in rads.; radionuclides emit alpha and beta particles and gamma rays; radionuclides u. in med., &c. (ex. cobalt-60; gold-198; carbon-14 [radiocarbon u. for dating in archaeology, geology, &c.]); these materials are extremely hazardous!

radis (It): **radish** (Eng): *Raphanus sativus* - **water r.**: *Nasturtium amphibium* - **wild r.**: *R. raphinistrum*.

Radium: Ra; radioactive alkaline earth metal; found in earth's crust; disintegrates spontaneously, releasing radon (dangerous gas found in many Am homes); formerly u. as antineoplastic, now superseded by radon, cobalt-60, &c.; chief prodn. Zaire, Cana.

Radix (L): root or rhizome (v. also under Rhizoma) - **R. Acori** (L): sweet flag; calamus rt. - **R. Angelicae**: rt. of *Angelica archangelica* - **R. Asphodeli**: rt. or bulb of *Narcissus pseudo-narcissus* - **R. Bardanae**: burdock rt. - **R. Belladonnae**: rt. of *Atropa belladonna* - **R. Caincae**: *Chiococca alba* - **R. Calamagrostidis**: rt. of *Calamagrostis lanceolata* - **R. Calami aromatici**: R. Acori - **R. Caryophyllatae**: rt. of *Geum urbanum* - **R. Chinae ponderosae**: rt. of *Smilax china* - **R. Colombo**: *Jateorrhiza palmata* rhiz. - **R. Consolidae**: rt. of *Symphytum officinale* - **R. Curcumae**: turmeric rt. - **R. C. americanae**: rt. of *Calathea allouia* - **R. dulcis**: licorice - **R. Ebuli**: rt. of *Sambucus ebulus* - **R. Fraseri**: Calumba - **R. Helenii**: rt. of *Inula helenium* - **R. Imperatoriae**: rhiz. of *Peucedanum ostruthium* - **R. Ipecacuanhae**: rt. of *Cephaëlis ipecacuanha* (&c.) - **R. I. Carthagena**: rt. of *Cephaëlis acuminata* - **R. Ivarancusae**: rt. of *Vetiveria zizanioides* - **R. Kava-kava**: rhiz. of *Piper methysticum* - **R. Liquiritiae**: licorice rt. - **R. Methonicae**: *Gloriosa superba* rt. - **R. Ononidis**: *Ononis spinosa*, rt. - **R. Pareirae**: rt. of *Chondodendron tomentosum* - **R. Pseudacori**: *Iris pseudacorus*, rhiz. - **R. Pyrethri**: rt. of *Anacyclus pyrethrum* - **R. Rhataniae**: krameria rt. - **R. Rhapontici: R. Rhei rhapontici**: rt. of *Rheum rhaponticum* - **R. Rhei**: *Rheum* spp. (medicinal) - **R. Saponariae albae** (or **magnalbae**): rt. of *Gypsophila arrostii*, *G. struthium*, or *G. paniculata*; also **R. S. hungaricae**; **R. S. aegyptiacae**; **R. S. turcicae**; **R. S. indicae**; &c. - **R. S. albae turkestanicae** (*v.* levanticae): *Acanthophyllum* spp.- **R. S. rubrae**: rt. of *Saponaria officinalis* - **R. Sarsae**: rt. of sarsaparilla - **R. Scammoniae**: rhiz. of *Ipomoea orizabensis* - **R. Serpentariae**: rt. of *Aristolochia serpentaria* & *A. reticulata* - **R. Spicae celticae**: rt. of *Valeriana celtica* - **R. Sumbul**: rhiz. of *Ferula s.* - **R. Tormentillae**: rt. of *Potentilla erecta* - **R. Uzara**: rt. of *Gomphocarpus* spp.

raevekage (Scand): Nux Vomica.

rafia: raffia: *Rafia*: *Raphia*.

Rafflesia hasseltii Suring. (Rafflesi.) (Mal): u. to expedite delivery in childbirth - **R. patma** Bl. (Indones): fl. bud inf. u. as post-partum cleansing agt.; astr.

Rafnia amplexicaule Thunb. (Leg) (sAf): veldtee (Du); rt. decoc. with licorice flavor u. as dem. in catarrhs & phthisis - **R. perfoliata** E. Mey. (sAf): lf. decoc. powerful diur. in edema; tea pl. (Hottentots & farmers).

rag: fibrous tissue of citrus frts.

ragi (Indo-Pak): small millet var., *Eleusine coracana*.

ragweed: 1) *Artemisia* spp. 2) *Ambrosia elatior*, &c. 3) *Parthenium hysterophorus* - **common r.**: **dwarf r.**: *Ambrosia elatior* - **giant r.**: *Ambrosia trifida* - **golden r.**: *Senecio aureus* - **great r.**: giant r. - **prarie r.**: *Iva* spp. - **short r.**: common r. - **yellow r.**: *Senecio jacobaea* - **ragwort**: *Senecio jacobaea*, *S. aureus*, & other *S.* spp.

raicilla (SpAm): **raicillo**: true ipecac; also appl. to false i. root, &c. (lit. little root).

raicilleros (Br): collectors of ipecac rt.

raifort (Fr): horseradish.

rain tree: 1) *Albizia samam* 2) *Laburnum*, gold chain.

raina (or **rayna**): pl. with insecticidal properties.

Rainfarn (Ge): tansy.

rainha (Por): queen - **r. do abisso** (Por) (lit. queen of the abyss): *Sinningia* spp. (Rechsteinera leucotricha).

raisin: 1) Uva Passa: dr. grape, frt. of *Vitis vinifera* 2) (Fr) grape (fresh) - **Corinthian r.**: *v.* corinthés - **r. d'ours** (Fr): bearberry - **r. (seed) oil**: grapeseed o.

raiz (Sp, Por): root.

raizilla (root): 1) (Co) ipecac (2) *Polygala costaricensis*.

Rajania cordata L. (R. pleioneura Griseb.) (Dioscore.) (WI): bihi; swollen rhiz. of vine u. as food (Carib Indians); cult. ornam. - **R. subsamarata** Moç. & Sessé (Mex): cuachalala; bk. c. tannins, u. as astr.; in homeop.

rajmahal hemp: 1) fibers of *Marsdenia tenacissima* Wight & Arn. 2) cortical fibers of *Calotropis gigantea*.

ramad: a fair quality of Karaya gum.

Ramalina calicaris (L) Roehling (Usne. - Lichen) (e&cUS): common twig lichen; c. lichenin, usnic acid; source of yellow dye for woolens (Eu); pd. pl. formerly u. to dye perukes & wigs; med. **- R. farinacea** (L) Ach. (widely distr. on trees): mealy ramalina; c. protocetraric acid (ramalinic a.); u. in cosmetics, to dye woolens brown (Eu); red, purple, orange (Ind) - **R. fraxinea** (L) Ach. (widely distr. ex. Himalaya, Austral on tree stems): lichen de frêne (Fr) (lichen of ash); u. as acacia substit. & in cosmetics & perfumes (nwHimal, Indo-Pak) - **R. pollinaria** Ach. (nHemisph. on trees, rocks, &c.): c. obtusatic (or ramalic) acid, supposedly with antibiotic properties - **R. reticulata** (Noedh.) Kremph. (Wash-Calif [US] on trees & old wood): "California Spanish moss"; c. antibacterial substit. (509), usnic acid - **R. scopulorum** (Retz.) Ach. (worldwide): ivory(-like) ramalina; c. lichenin, usnic acid; u. med.; for purple, red, orange dyes; yellow-brown dye for woolens (Scotland); on marine rocks (SriL).

rambutan (eAs): *Nephelium lappaceum.*

Ramenti Ferri (L): iron filings.

rami: ramie: ramio (SpAm): a fiber & the pl. producing it, *Boehmeria nivea.*

ramo (It): branch.

Ramona Greene = *Salvia.*

rampion: 1) *Campanula rapunculus* 2) horned r. - **German r.** (hort): *Oenothera biennis* - **horned r.**: *Phyteuma* spp.

ramps (wVa [US]): wild leek, *Allium tricoccum* - **r. of Europe**: *Allium ursinum* - **ramsen**: 1) *Allium ursinum* 2) *Sisymbrium officinale.*

ramste(a)d (corruption of ramshead?): *Linaria vulgaris.*

ram-teel: ramtil: Niger seed.

rancher's tree: *Prosopis glandulosa*, mesquite.

Randia aculeata L. (R. mitis L.) (Rubi) (Mex, WI, CA, Pe, Ve, Co): chill-busy (Bah); "café cimmarón"; "dogwood"; tintillo; lf. inf. in fever, as bath for skin infections; green frt. astr. u. for dys. (WI); for dyeing blue & black; produces galls; mature frt. edible - **R. annae** K. Schum. (Br): false cinchona bark; u. as bitter, quinine replacement - **R. cochinchinensis** Merr. (R. densiflora Benth.) (Indochin): bitter bk. u. to treat jungle fever; purg., against urticaria - **R. echinocarpa** Sessé & Moç. ex. DC (wMex): granjel; frt. u. in kidney, pulm., & circul. dis., incl. kidney stones, cough, diabetes (lvs. also u.); abortif. (2705) - **R. esculenta** Merr. (R. fasciculata DC) (seAs): rt. inf. in quotidian fever (mal.); wd. sed. to nervous system & anticonvuls. (Indochin) - **R. horrida** Schult. (R. longiflora Lam.) (Indochin): lvs. u. to make refreshing tea; bk. antipyr. & astr. - **R. longiloba** Hemsl. (sMex): canalkax; rt. u. in erysipelas (formerly) - **R. maculata** DC (trop wAf): frts. source of a dye u. to mark face blue-black, to color fibers; sts. u. for chewsticks (Ghana); likewise **R. malleifera** Benth. & Hook. - **R. nilotica** Stapf (tropAf): tsura (Hausa); rt. u. in VD formulations; rt. inf./decoc. drunk as arrow poison antidote; fresh frt. emet. u. as fish poison - **R. pubescens** R. Br. (Par, Pe): frt. emet. antidysent. for washing; fish poison - **R. spinosa** Poir. (R. dumetorum Lam.) (Ind, Chin, seAs, Af): Malabar ipecac; sethanbaya (Burma); frt. (gelaphal, chela frts., jela frts.) c. saponins, scopoletin; u. as emet. in dys.; fish poison; pericarp for washing (c. saponin); substit. for bela frt.; sd. c. FO, randia fat; bk. u. antipyr.; rt. expect. & spasmolytic - **R. taylori** S. Moore (trop eAf): pl. c. saponins; u. in schistosomiasis (Tanz) - **R. tomentosa** Bl. (seAs): frt. decoc. or macerate keeps hair glossy & supple - **R. uliginosa** DC (eAs): unripe frt. & rt. u. in dys. diarrh. - **R. vestita** S. Moore (tropAf): rt. bk. u. diur., in GC; purg.; frt. fish poison (saponins); u. med. (Wabena tribe); rt. for foam prodn. (in fire extinguishers) (comm.) (510) - **R. walkeri** Pellegr. (tropAf): u. for stupefying fish; tree cult. (*sic*).

Rangifer caribou (Cervidae - Ruminantia - Mammalia) (nNA): caribou; pd. horns u. as aphrod. (Korea) (valuable); antlers occur in both sexes; important characters maned neck, broad hooves - **R. tarandus** L. (Euras; introd. into NA): reindeer; important animal to nEu (Lapland); semi-domesticated; fur & meat source.

Rangoon creeper: *Quisqualis indica.*

ranoncule (Fr): buttercup; *Ranunculus* spp.

Ranunculaceae: Crowfoot (or Buttercup) Fam; hbs. or sometimes shrubs of temp. & arctic zones of nHemisph; lvs. mostly radical or cauline alternate; lamina simple or generally divided; fls. generally regular, cup-like, commonly with 5 sepals, 5 petals (sometimes lacking), many stamens; petals & sepals sometiems irreg., as spurred (ex. larkspur, columbine); pistils few to many; pistil typically drawn out to point; frt. achene, berry, or capsule; ex. *Clematis*, *Thalictrum.*

Ranunuculus L. (Ranuncul.) (cosmop. cold & temp. regions): buttercups; crowfoot; ann. or perenn. herbs; some 250 spp.; most bitter (alks.), saponins (2192), cattle will not eat even when pasturage is scarce (folk belief erroneous that cattle eat b. to improve their milk); many spp. are common weeds, others beautiful ornamentals; many are toxic - **R. abortivus** L. (e&cNA): kidney-leaf buttercup; chicken pepper; hb. u. as vesicant, counterirrit. - **R. acris** L. (R. acer L.) (Euras, nAf, natd. eNA): field (or acrid) buttercup; (tall) b.; a common weed; c. protoanemonin, rapidly dimerized to the less active anemonin; hb. u. epispastic, counterirrit., rubef. irrit. acrid for chronic skin ailments (as warts), rheum., gout, bronchitis, pleuritis; fl. & frt. buds (pd.) u. for headache (Algonquin); fresh fls. & lvs. crushed in hand & inhaled for headache (Abenakis); fresh hb. u. by Homeop. - **R. alpestris** L. (Eu): Alpine crowfoot; c. protoanemonin; pl. diur. purg. stim. - **R. aquatilis** L. (Eu, widely natd.) (white) water crowfoot; u. med (Sp Phar.) - **R. arvensis** L. (Euras, nAf, natd. NCar-Ga [US]): corn crowfoot; hard-iron(!); yellow-cups; pl. u. in intermittent fevers, gout, asthma (Eu); acrid poison (HCN) - **R. bulbosus** L. (Euras, natd. NA): bulbous buttercup (or crowfoot); frogwort; c. & u. much like those of *R. acris*; in neuralg.; grip, herpes zoster, &c.; poisonous - **R. capensis** Thunb. (sAf): weak decoc. u. for gastric dis., esp. with severe vomiting; milk decoc. of rt. u. for cough & sore throat; epispastic - **R. diffusus** DC (eAs): lvs. u. as medicine for eye (Chin) - **R. falcatus** L. (Ind, Pak): pounded pl. u. as vesicant - **R. ficaria** L. (Euras, natd. NA): lesser celandine; pilewort; figwort; fresh hb. (Herba Ficariae; H. Chelidonii minoris) u. for hemorrhoids; in folk med. in chronic skin dis. (warts) & to "purify the blood" (vit. C) (511); rt. u. diur.; juice of fresh rt. u. in hemorrhoids - **R. flammula** L. (Euras, natd. eCana, Ore [US]): (lesser) spearwort; u. in homeop. med. - **R. glacialis** L. (Euarctic & high mountains): glacier crowfoot; u. in folk med. as diaph.; homeop. - **R. hirtus** Banks & Soland. ex DC (Austral, NZ): pl. u. to treat toothache, ophthalmia, abrasions (NZ) - **R. hirtellus** Royle (Ind, Pak): "airu" (UP); whole pl. as paste u. by appl. to rheum. - **R. multifidus** Forsk. (s&eAf): (wild) buttercup; pd. pl. hb. u. for cough; weak decoc. for emesis, diarrh. dyspeps.; & many other uses - **R. multiflorus** Forsk. (Af): for coughs, colds, syph.; exter. in cancer, scabies (512) - **R. muricatus** L. (Pak, Kashmir Med reg.): spiny-fruited buttercup; u. in gout & asthma (Eu) (428) - **R. pensylvanicus** L. f. (Cana, Alas, nUS): bristly crowfoot; entire pl. u. as astr. (Potawatomis); blistering agt. (in Ind, where weed) - **R. peruvianum** Pers. (Col): rt. u. as stim. tea for weak & aged - **R. repens** L. (Euras, nAf, natd. n&wNA): creeping buttercup; c. (or common) crowfoot; sitfast; strong antibiotic activity reported; u. in Ukrainian folk med.; bad weed; ornam. - **R. rivularis** DC (Austral, NZ): small river buttercup; pl. sap u. for rheum. & arthritis; lf. inf. for quinsy (peritonsillar abscess) - **R. sceleratus** L. (Euras, natd. NA): cursed crowfoot; marsh c.; "water celery"; (cut-leaved) buttercup; hb. u. as irrit. for skin dis. (ex. scabies); leukoderma; fresh hb. or pl. popular with homeop. physicians; juice burns, blisters, & puckers the mouth; formerly, boys used hb. to make "tetters" around mouth so they would not have to go out; lesions u. by beggars to attract sympathy; toxic (2193); (sceleratus [L] = rascally; wicked [513]) - **R. ternatus** Thunb. (eAs): u. to erode abscesses (using paste of lvs. & sds.) but with care to protect skin with paper layer (Indochin) - **R. thora** L. (sEu-mts.): kidney-leaved crowfoot; erba tora (It); c. protoanemonin; very toxic when fresh; when dr. loses most of its activity (514).

Rapa: Rapa Semen (L); rape (seed).

Rapamycin: antibiotic of polyene type fr. *Streptomyces hygroscopicus* (Jensen) Waksman & Henrici; orig. found in Easter Is; antifungal; with immunosuppressive props.

Rapanea coriacea Mez (R. ferruginea Mez) (Myrsin.) (sMex, WI, CA, SA-Br): ratón (CR); azeitona do mat(t)o (Br); lvs. u. diur.; u. med. in Hond - **R. melanophloeos** Mez (sAf): Cape beech; decoc. of tannin-rich bk. u. expect. emet.; lvs. astr.; frt. c. embelin - **R. usumbarensis** Gilg & Schellenb. (eAf): rt. u. as purg.

rape(seed): *Brassica napus* var. *n.*, source of FO (1216).

Raphanus raphanistrum L. (Cruc.) (Euras, natd. eAs, NA): wild (or field) radish; a common weed; sds. similar to those of white mustard; source of sulfurated VO; pl. has antibiotic properties; sd. toxic (sinalbin), c. 15-40% FO u. as food oil (Balkans); sd. in mal. (earlier); ravison oil (Fr) is fr. a var. - **R. sativus** L. (Euras): (common garden) radish; garden (or cultivated) r.; small r. (var. *sativus*); rabano (Sp); a cultigen; juice u. for pertussis (It) (tops cut off r., center scooped out, & juice left collect in cavity) as relish, sauce; c. VO, vit. C; rts. u. diur. lax. for gall bladder dis., icterus; fresh in salads, curry; sds. c. FO u. lax. diur.; Japanese (or Chinese) radish ("Daikon") = *R. sativus* cult. var. *longipinnatus* Bailey (forma *gigantissimus*), giant whitish tap rt. with low pungency; (Sakurajuma) daikon; mula; with rts. to 20 kg.; cult. Japan & Cal (eaten cooked or raw in salads); also large rooted Chin var.; similar large rt. var. (noted in BRD): var *niger* S. Kerner ("black radish"), rt. black or gray; "Russian radish"; (Med Sea reg., As): garden vegetable; fresh radish juice u. in bronchitis, dyspep. liver dis.; cholagog in gall bladder dis.; in folk med. for coughs, obstipation, arthritis, preventing colds (claim) (uses also of type) (var. *sativus*).

raphe: ridge along ovule representing adhesion of part of funicle.

Raphia farinifera N. Hyl. (R. pedunculata Beauv.; R. ruffia Mart.) (Palm.) (Mad): raffia (palm); East African wine p.; raphia p.; lvs. source of raffia (fibers) (=bast fibers) popularly u. in basket making, by nurserymen; to make hammocks, curtains, mats, &c.; juice fermented to make palm wine with low alc. content (2-5%); believed galactagog; petiole, &c. u. to make piassava (fiber); raffia wax (fr. secretions on lf. surface); sd. decoc. u. to control hemorrhage (Mauritius); sago; nut source of FO, raphia butter; kernels food; terminal bud eaten as veget.; leaflets for roofing; tree cult. in eAf - **R. hookeri** G. Mann & H. Wendl. (tropAf): Ivory Coast raphia palm; wine p.; mimbo p.; sap of young inflorescence prepd. for palm wine (or mimbo); fiber u. to make cloth; lvs. u. for mats, thatching; midribs as poles, rafters, &c. - **R. sudanica** A. Chev. (tropAf): u. to make palm wine; lf. mid-ribs u. for bedsteads, &c.; frt. source of an oil wh. is emetic & u. exter. & int. for leprosy (Bassari, Senegal) - **R. vinifera** P. Beauv. (trop wAf): West African wine palm; wine raffia p. ; bamboo p.; cuticle of leaflets furnishes raphia bast (raffia piassave); frt. & st. juice (sap) u. to make wine (toddy); whole lf. for thatch; sds. (bambu nuts) kernels u. to make buttons; frt. shell with orange bitter FO u. med.; midrib of lf. u. to make arrow shafts, rafts, gates, &c.

raphides: aligned calcium oxalate needle-shaped (acicular) crystals wh. usually occur in **r. bundles** (ex. Scilla); these are discharged when the various pl. parts are moistened —> skin or mucous membrane irritation (< L. rhaphides).

Raphionacme hirsuta R. A. Dyer (Asclepiad.) (Perioplac.) (tropAf): large underground tubers u. to make highly intoxicating beer ("ghali"); also a med. for internal tumors; may c. depressant subst. - **R. utilis** Brown & Stapf (tropAf): Bitinga rubber plant; source of ecanda or marianga rubber (Angola).

rapper dandies: *Gaultheria procumbens.*

rap(p)s (Ge, Sp): rape seed.

Raputia sp. (Br): "orapoca de cheiro"; lvs. chewed as coco substit - **R. alba** Engl. (Rut.) (Br): bitter bk. of tree stim. antipyr.; as fish poison - **R. aromatica** Aubl. (Guian-Br): bk. of shrub u. stom., as antipyr. - **R. magnifica** Engl. (Br): bk. c. chrysophanic acid; u. as anthelm.

rasamala hars (Ind): *Liquidambar altingianum*; *L. orientale.*

Rasapen: glycosidal components of horseradish rt. u. as urin. antiseptic.

rascadera (Co): yautia.

Rasen (Ge-m): turf, lawn - **rasenartig** (Ge): grassy.

rasot: rasout: rusot.

raspberry: *Rubus idaeus*; *R. strigosus* - **black r.**: *R. occidentalis* in its several cultural vars. - **European r.**: *R. idaeus* - **ground r.**: Hydrastis - **r. jam tree** (Calif [US]): *Acacia* sp. - **r. juice**: the expressed juice of the fresh ripe frt. of *R. idaeus* - **red r. (juice)**: *R. idaeus.*

Rastafarian: cult of black religious nationalists in Jamaica who use Cannabis (home grown) as a part of their religious ritual; they consider Haile Selassi of Ethiopia as God (Ras Tafari).

rat: general term for medium-sized rodent of one of several genera (esp. *Rattus*; others: *Sigmodon*; *Dipodom*, & many other genera of several families (*Cricetidae*, *Muridae*, &c.); generally larger than mice but sometimes confused; white (or albino) rats (& mice) commonly u. in drug tests, demonstrations, assays, &c.; various parts u. in Unani med.; often u. as food, esp. in famines; chiefly important for their destructive habits.

ratafia de benzoin (Sp): Benzoin Tr. Comp.

ratan (SpAm as Ch): **ratan(h)ia**: (Br) *Krameria*, sometimes *K. triandra* - **r. de Nueva Granada: r. morena**: *Krameria ixina* - **ratanhine**: andirine.

ratbane: ratsbane.

Ratibida columnifera (Nutt.) Woot. & Standl. (R. columnaris D. Don; Lepachys c. Rydb.) (Comp.) (NA Plains: BC [Cana]-Tex [US]; natd. eNA): (upright) prairie-coneflower; inf. emet. (Oglala Indians); as beverage; in fever; chest pain; exter. snakebite, poison ivy (Am Indians) - **R. pinnata** (Vent.) Barnh. (prairies, NY-Fla [US]-Man [Cana]): cone-flower; u. for toothache (Meskwaki Indians) - **R. tagetes** (James) Barnh. (US plains, Tex-Ariz): (short-ray) prairie-coneflower; tea u. for cough, headache, fever, sex. infect., &c.

raticide: agt. u. to destroy rats (the most harmful & dangerous of the rodents).

ratifia de benzoin (Sp): Benzoin Tr. Comp.

Ratin (Ge): sd. of *Nigella arvensis.*

ratón (Sp): rat; also (Mex) *Gliricidia sepium.*

ratoon: stalk or shoot of a perennial pl. esp. fr. 2nd year's growth or later (ex. sugar cane; ginger).

ratsbane: 1) *Chimaphila maculata* (Ve) 2) *Dichapetalum toxicarium* 3) *Goodyera pubescens* 4) *Anthriscus cerefolium* - **rat-tail (plant)**: 1) *Plantago lanceolata* 2) *Aporocactus flagelliformis* (r. t. cactus) 3) *Peperomia obtusifolia* & other *P.* spp. 4) *Limonium suworowii.*

ratsvane: corruption of ratsbane.

rat-tags (US): *Martynia probiscidea.*

rattan (cane) (PI): solid cane u. to make furniture, such as chairs; baskets; obt. fr. **r. vine**, *Calamus rotang*, *C. javanensis*, *C. viminalis*, &c. (climbing palms); name also appl. to *Berchemia scandens, Daemonorops, Korthalsia* spp.

rattle root: Cimicifuga - **rattler** (US slang): rattlesnake, *Crotalus* spp. (Crotalidae).

rattlebox: *Sesbania drummondii* (Drummond's r.); *Crotalaria* spp.; *Rhinanthus alectorolophus*; *Silene vulgaris* - **rattlepods**: *Crotalaria* spp.

rattlesnake: *Crotalus* spp. (*qv.*) - **r. bite**: 1) r. fern 2) *Thalictrum* spp. - **r. b. cure** (Calif [US]): *Daucus pusillus* - **r. fern**: *Botrychium virginianum* - **r. master**: 1) *Eryngium aquaticum* 2) *Liatris scariosa* & *L. squarrosa* (also **rattlesnake's m.**) (SC thought to relieve rattlesnake bite) 3) (Fla [US]) *Trillium erectum* 4) *Agave virginica*, rt. - **r. oil**: FO of body; formerly common nostrum ("snake oil"), in proprietaries (claimed of value in rheum., eczema, &c.) - **r. root: rattlesnake's r.**: 1) *Goodyera pubescens* 2) Cimicifuga 3) Senega 4) *Trillium cernuum* 5) **r. root, white**: *Prenanthes alba* 6) (US voodoo) Serpentaria & Asarum - **r. weed**: 1) *Hieracium venosum* 2) *Daucus pusillus* 3) *Cimicifuga racemosa* (NCar [US]) (SC sd. pods rattle in wind) 4) *Eryngium aquaticum* 5) *Goodyera pubescens* 6) (Arg) *Chamaesyce albomarginata* - **rattlesnake**: v. also under flag; grass; plantain.

rattle-top: Cimicifuga - **rattle-weed**: 1) *Cimicifuga* spp., incl. *C. racemosa* 2) *Astragalus mollissimus* & other *A.* spp. with bladder-like pod (SC sds. give out sound when pods shaken) 3) *Crotalaria retusa* 4) (Mex): *Actaea racemosa.*

raubasine: ajmalicine.

Raudixin: Rauwolfia serpentaria (rts.) (Squibb).

Rau(l)ke (Ge): Ruke: *Eruca sativa.*

rauli (Pe): *Nothofagus procera.*

raunormine: deserpidine, alk. fr. *Rauwolfia canescens* (S. B. Penick & Co.).

Raute (Ge): *Ruta graveolens.*

Rauvolfia: *Rauwolfia.*

Rauwenhoffia siamensis Scheff. (Annon.) (Indochin, Thail): lf. decoc. u. as ton., also appl. to cuts (2448); supposed anti-venom.

Rauwiloid™: ext. or mixt. of alks. ("alseroxylon" fraction) fr. *Rauwolfia serpentina.*

Rauwolfia L. (preferred by taxonomists is *Rauvolfia* L.) (Apocyn.): genus with c. 135 spp. (2206) - **R. amsoniaefolia** DC (seAs, PI): lvs. lax.; new buds for gastric dis. of young babies (PI); bk. decoc. to treat yaws - **R. caffra** Sond. (sAf): quinine tree (Natal); c. rauvolfine (0.1%) with hypotensive activity; bk. u. as cinchona substit. (2204) - **R. canescens** L. = *R. tetraphylla* - **R. concolor** Pichon (Mad): sts. u. to poison rabid dogs; alk. with sympatholytic and hypotensive properties; var. *intermedia* Markgraf: lvs. u. as decoc. in stomachache; aphrod. (in alc. beverages) (var. *angustifolia* Markgraf with similar properties) - **R. confertiflora** Pichon (Mad): bk. of young branches & lvs. (crushed) appl. topically to eyes for viral conjunctivitis; bk. decoc. u. to destroy varmints; bk. shipped to Eu to make hypotensive & sympatholytic prepns. - **R. decurva** Hook. f. (Ind): rt. c. reserpinine, rescinnamine, reserpine, reserpiline, isoreserpiline, sarpagine (no ajmaline); close to *R. densiflora* - **R. degeneri** Sherff (Oahu, Lanai, HawIs): hao; devil hopper; c. reserpine, &c. - **R. densiflora** Benth. (Ind, SriL): c. ajmaline; adulter. of *R. serpentina*; can be distinguished fr. latter by chromatography - **R. inebrians** Schum. (tropAf): pl. & bk. intoxicant; bk. ascaricide; bk. astr. in colic; frt. & rt. added to pombe (narcotic beverage); wd. useful timber - **R. javanica** Koord. & Val. (Indones): lvs. vulner. (in ointment) - **R. ligustrina** Roem. & Schult. (sMex, WI, CA to Co, Br): frutillo; c. many alk. incl. reserpine, isoreserpine, raugustine; inf. or decoc. of rt. u. in snakebite & scorpion stings (Trin) - **R. media** Pichon (Mad): bk. u. to make hypotensive prepns.; shipped to Eu for this purpose - **R. micrantha** Hook. fil. (Ind-Malabar dist.): rt. u. as adulter. & sophisticant of *R. serpentina* - **R. mombasiana** Stapf (trop eAf): rt. c. reserpine, &c., u. as homicidal agt. & to kill dogs; added to pombe (beverage); to act as rem. in stomachache; rt. ext. hypotensive; powd. lvs. appl. as paste to breast abscess; rt. decoc. for asthma, TB (Tanz) - **R. natalensis** Sond. (sAf): African quinine; q. tree; c. 0.25% rauvolfine, reserpine, ajmaline, & other alk.; bk. antipyr. in mal., abdom. troubles (515) - **R. nitida** Jacq. (WI): milk bush; huevo de gallo (Cu); crushed rt. u. in erysipelas; lvs. for healing ulcers; latex emet. purg. - **R. obliquinervis** Stapf (eAf): rt. c. alks. added to alc. beverage prepd. fr. banana to make more potent; homicidal agt.; to kill dogs - **R. perakensis** King & Gamble (sAs-Indones): u. to treat inflamed eyes & suppurating eyelids; rt. ext. u. as antidote to snakebite; rt. u. to adulter. *R. serpentina* - **R. pyramidalis** (Domin Rep): pega palo; u. med (2208) - **R. rosea** K. Schum. (trop eAf): frt. strong purg.; u. as sex stim. & in VD (Tanz) - **R. sellowii** Muell. Arg. (Br): rt. bk. c. reserpine (small amt.), ajmalicine, aricine, tetraphylline, ajmaline, ajmalidine, tetrahydroalstonine; latex drastic (Arg) - **R. semperflorens** (Muell.-Arg.) Schlechter (New Caledonia): rt. c. semperflorine, &c.; u. med. - **R. serpentina** (L) Kurz (R. trifoliata Baill.) (s&seAs as Ind, Burma, Thail, Java): (East Indian) serpent wood; snakewood; Java devil pepper; insanity herb; bois de serpent (Fr); rt. (as Radix Mustelae)

c. reserpine, many other alks. incl. the following: ajmalicine (raubasine; δ-yohimbine), ajmaline (neo-ajmaline [Dun]) (= hypertensive), ajmalinine, "alkaloid C," isoajmaline (Dun), isorauhimbine, papaverine, deserpidine (516), pseudoyohimbine, corynanthine (rauhimbine), rauwolfine (= ajmaline), rauwolfinine, rescinnamine, *reserpine*, reserpiline, sarpagine (raupine), serpentine (Ind, card. stim.), serpentinine (= hypertensive), thebaine, yohimbine; also present: FO with oleic acid, phytosterol; oleoresin; unsatd. alcs., &c.; u. as hypotensive (chief use) sed. in insomnia (superior to barbiturates), insanity (mania, schizophrenia, paranoia, chronic alcoholism, hysteria), epilepsy; cerebral palsy; hypnotic in twilight sleep (2205); antihistaminic; in uterine inertia; in veter. med. (Java); in native med. (Ayurvedic) for snakebite, cholera; blindness; poisonous; several vars. are distinguished by geog. origin (Ind [Dehra, Dun], Bengal, Cochin, Bihar, Assam) (lit. ref. 637) - **R. sumatrana** Jack (R. samarensis Merr.) (Indones): bk. decoc. for mal., stomachache (PI); general antidote for poisons (Indones) - **R. tetraphylla** L. (R. heterophylla Roem. & Schult.; R. hirsuta Jacq., R. tomentosa Jacq.; R. canescens L.) (Mex, CA, WI, nSA): hoary-leaved milkwood; "palo boniato" (Cu); chalchupa (Guat); pinique-pinique (Co); chiolingo (Guat+); shrub or small tree; c. many alks. incl. ajmalicine (raubasine), aricine (heterophylline), corynanthine, ajmaline, deserpidine (recanescine), reserpine (0.03-0.05%), reserpiline, reserpinine, sarpagine (raupine), serpentine, serpentinine, isoreserpinine, isoreserpiline, isoyohimbine (rauwolscine, Penick), rauvanine, yohimbine, pseudoyohimbine, "chalchupine A & B"(?); (total alk. 6.5%); rt. decoc. antipyr. in mal. & snakebite (Guat); bk. decoc. in skin dis. (WI); antisep. for wounds (Guat, Mex); insecticide (ants, beetles, &c.); latex for sore eyes, granulated eyelids; cholera, mal. (popular med.); comm. source of alks.; frt. juice as ink & in dyes (2207) - **R. verticillata** Baill. (Chin+): rt. u. as hypotensive (Lo-fu-chia-su); believed to have less intense but more sustained action than *R. serpentina*; rt. decoc. insecticide; lf. & st. decoc. u. for bath for aches & pains & in scabies (Burm); cult. in Java - **R. viridis** Willd. (R. lamarckii DC) (SpAm): "bellyache" (VirginIs); lvs./decoc. u. in 3rd stage of parturition (VirginIs); med. (Co); latex powerful drastic - **R. vomitoria** Afz. (tropAf): swizzle-stick (tree) (SC twigs u. in mixing drinks); "asofeiyeje"; rt. c. most of the alks. found in *R. serpentina* (reserpine+); u. purg. anthelm. bitter (vermifuge) (schistosomiasis, wAf), emet. ton. in GC; bk. inf. antipyr, u. in indigest., scabies; lf. emet.; latex for colic, diarrh.; rt. bk. drastic purg. & emet.; lf. u. exter. with friction against vermin (head lice); frt. emet. expect.; lf. latex u. in parasitic skin dis., pediculosis; sed. anticonvuls., aphrod., in neuropsychiatric cases.

ravanello (It): radish.

ravasin (Eg): *Sesbania aegyptiaca.*

rave (Fr-f.): turnip, radish, rape, beet.

Ravenala madagascariensis Sonn. (monotypic) (Strelitzi./Mus.) (Mad): traveller's tree (SC source of water accumulated at lf. bases useful in drought periods); sds. rich in starch, nutr. (esp. for young); lvs. eaten like cabbage; petioles in making paper; sts. u. in construction of huts; widely cult.

Ravensara aromatica Gmel. (Agathophyllum aromaticum Willd.) (Laur.) (Madag, Réunion): Madagascar (or clove) nutmeg; frt. (ravensara) c. FO, eugenol; u. in kari pd.; sds u. as spice (Madagascar nutmegs); bk. u. in mfr. rum.

raw (as for linseed oil): not subjected to heat & chem. treatment.

rawa (Urdu): raw material for macaroni, vermicelli, etc.

ray: sheets of tissue which extend radially (hence name) through xylem & phloem tissues - **medullary r.**: older term for what is now generally called "vascular r."; SC supposed origin fr. medulla, acutally originates fr. phloem & xylem tissues - **phloem r.**: portion of a ray which lies in the midst of phloem tissue - **pith r.**: medullary r.; these terms are passé - **vascular r.**: newer term for tissues previously mostly designated medullary r.; consists of phloem & xylem r. - **wood r.: xylem r.**: portion of a ray which lies in the midst of xylem tissues.

reagent, universal: u. in microanalysis for detecting various elements, in bot. sections; incl. chloral, Fe, I, Sudan III, &c. (to detect cutin, suberin, lignin, &c.) (2195).

reagin: special antibody formed in response to an allergen acting as antigen.

Reaumuria hypericoides Willd. (R. cistoides Adam) (Tamaric.) (Pak, Iran, Afgh+): lvs. u. to treat prurigo & itch.

rebacillo (Arg): middlings (in flour milling).

receptacle: enlarged or elongated apex of st. or pedicel/peduncle in which fls. are borne (in Comp. often greatly expanded); torus.

receptor: site in cell wh. combines with drug or other substance to give change in cell function.

receta (Sp): prescription.

recino (Sp): ricino (Sp): *Ricinus communis.*

recombinant DNA technology: genetic engineering technology of new gene combinations u. to produce complex org. compds. or substances such as insulin, antihemophilic factor VIII, rabies vaccine, interferon, antibodies, &c.; involves linkage, insertion of genes, &c.

recreational drugs: those u. for non-medical purposes, often euphoriants (as cocaine), hallucinogens (as Cannabis), &c.

rectified: double distilled - **r. spirits**: ethyl alc. - **r. tar oil** rect. oil of pine tar.

red algae: Rhodophyceae - **r. argols**: argols; red sediment in wine barrels, c. chiefly crude potassium bitartrate (*cf.* wine lees) - **r. bark**: *Cinchona oblongifolia* - **r. berry**: 1) ginseng 2) uva ursi 3) wintergreen - **r. bud**: *Cercis canadensis* - **r. bug weed** (sUS): *Asclepias tuberosa* - **r. bush tea**: rooibos - **red cedar, eastern**: *Juniperus virginiana* - **r. c., western**: *Thuja plicata* - **r. chickweed**: *Anagallis arvensis* - **r. coral**: Corallinum Rubrum - **r. devils**: capsules of Calomel & Rhubarb - **r. gum**: eucalyptus g. - **r. head**: *Asclepias curassavica* - **r. ink berry: r. i. plant**: *Phytolacca americana* - **r. maids** (Calif [US]): *Calandrinia ciliata* - **r. oil**: 1) first distillate of thyme oil (fr. iron still u. in Sp. giving reddish tint) 2) crude oleic acid (comml.) 3) olive oil c. alkanet (*cf.* Turkey r. oils) 4) acid-treated petroleum distillate; u. as lubricant 5) crude cottonseed oil (fr. decorticated sd.) 6) Ol. Hyperici - **r. o., Turkey**: castor oil sulfonated with conc. H_2SO_4 - **r. pollen**: wintergreen - **r. robin**: *Geranium robertianum* - **r. rod**: *Cornus sericea* - **r. root**: 1) *Ceanothus americanus* 2) *Geum virginianum* 3) Sanguinaria 4) *Morinda roioc* (WI, CA) 5) *Lachnanthes tinctoria* 6) *Amaranthus retroflexus* - **r. Sally**: *Lythrum salicaria* - **r. scale insect**: cochineal - **r. shank**: *Geranium robertianum* - **r. shanks**: *Ceanothus americanus* - **r. stone** (Negro): red Fe_2O_3 (pd.) - **r. tartar**: argols - **r. top**: *Agrostis stolonifera*; *A. gigantea* - **r. tide: r. water**: discoloration of surface ocean water caused by a spreading of protozoa (dinoflagellates) through the salt water, poisoning enormous numbers of fish, turtles, molluscs, &c.; associated with rich mineral runoff fr. the land via rivers; common off wFla [US] coast; gas (NO_2) prod. —> cough, sneezing, nausea, eye-sting (case reported 1844) (2196) - **r. wood (tree)**: wd. yielding red dye or with reddish color; appl. to many plant spp. or their wd. (**African r. w.**: camwood); v. redwood - **Indian r. w.**: 1) Indian mahogany, *Toona calantas* 2) sappan wd. 3) *Soymida febrifuga* [*cf.* redwood]) - **red**: *v.* also under alder; balm; bearberry; bilberry; bole; cinchona; clover (blossom); cohosh; couch grass; elm; gum; indigo; mustard; oak bark; pepper; Peruvian bark; poppy (flower); puccoon; quebracho (wood); raspberry; rice; rose (leaves); sandalwood; santal; saunders; squill; tea; thorn.

redoul (Fr): *Coriaria myrtifolia.*

redox (reaction): reduction-oxidation reaction.

reductone: org. compd. with -CHO or -CO gp. with strong reducing properties; ex. vit. C - **reductonic acid**: glucic acid; formed fr. glucose by action of KOH; found in dehydrated foods, interferes with vit. C. detn. since reacts in same way.

redwood: *Sequoia sempervirens* (**coast r.**) - **Sierra r.**: *Sequoiadendron gigantea.*

reed, bur: *Sparganium* spp. - **common (water) r.: giant r.**: *Phragmites australis; Arundo donax.*

reefer (US slang): Cannabis (marijuana) cigarette.

refining: as applied to FO, generally represents solvent-extracted oil; (ex. coconut oil fr. copra) for food & drug use, free fatty acids removed with NaOH (—> soap), drawn off & put through filter press; may be later washed with water, decolorized (activated C), deodorized with superheated steam, dried, and cooled (ex. cottonseed oil).

refrigerant counterirritant: Dental Liniment of Aconite & Iodine Comp.

regaliz(a) (Sp): regoliz (Por) - **r. de Cuba** (Sp): *Lippia dulcis* var. *mexicana.*

régime (Fr-m.): diet.

réglisse (Fr-f.): licorice.

regressive therapy: returning to earlier types of therapy (ex. folk medicines).

Regulin™: Cascara Sagrada with agar.

Rehmannia glutinosa Lib. (Gesneri) (nChin, Inner Mongolia): rt. u. as antipyr. for cough, headache, vertigo; in aphrodisiac formulas; a very ancient drug - **R. lutea** Maxim. (Chin): hb. u. diur.; to induce labor.

Reichstein's "substance S" (or compd. S): 11-desoxy-cortisone.

reina de los prados (Sp): *Spiraea ulmaria* - **R. Luisa** (Cu): *Zinnia elegans* - **reine claude** (Fr): *Prunus domestica* - **r. des prés**: meadowsweet.

reindeer: *Rangifer tarandus* L. (with several vars.) (Cervidae - Ruminantia) (northern parts of Euras, NA): reindeer (Euras) is semi-domesticated, while caribou (var. *caribou* Gmelin) (R. c.) (North American woodland reindeer) (nNA) is wild; long a source of mest, milk, cheese, bone, sinews, hides, fat, antlers, fur, as well as of transportation & carriage; antlers exported fr. Siberia; u. as popular aphrodisiac (seAs).

reineckates: general insoluble ppts. produced by reacting soln. of Reinecke salt with dissolved amines, alks., proline, amino acids (some) (2198) - **r. salt**: NH_4 or K reineckate.

Reinwardtia indica Dum. (R. trigyna Planch.) (Lin.) (wInd, Pak, Chin): gudbatal (Punj.); yellow flax;

u. to treat "founder" (acute gastroenteritis) in cattle; pl. decoc. u. in dys. (124); ornam. subshrub.

Reineckea carnea Kunth (Lili) (Chin, Jap): rt. u. vulner. for contusions; in rheum. colds.

Reis (Ge): rice.

Reissantia indica N. Hallé (Celastr.) (Ind, Af): lf. & rt. decoc. u. antipyr. for enteritis, & for children's dis.

Reissekia smilacina Stendel. (monotypic) (Rhamn.) (Br): lf. tea u. as "blood purifier."

relaxin: releasin (former name): polypeptide (non-steroidal) hormone secreted by the corpora lutea of ovary & placenta of many mammals during pregnancy & presumed present in serum of pregnant female; it permits relaxation of pelvic ligaments (esp. of pubic symphysis) & reduces (relaxes) uterine contractions (softens cervix); widens vagina; first discovered in guinea pig & burrowing mammals (mole, pocket gopher); recommended to shorten childbirth (50%) & prevent prematurity in threatened miscarriage (2022).

Relbunium hypocarpium (L) Hemsl. (Rubi.) (Ch, Co): relvún (Ch); u. as astr. in diarrh., hemorrhages; to aid sight; lvs. & rts. good diur. & hypertensive.

relojes (Sp): *Erodium cicutarium*.

remède man (or **woman**) (La [US]): herb doctor (esp. voodooman).

Remijia ferruginea DC (Rubi.) (sBr): remijia bark (Cuprea bark; false cinchona bark) c. cupreine; u. as cinchona bk. replacement - **R. pedunculata** (Karsten) Flueck. (Pe to nCo): Cuprea bark (SC has coppery color); c. cupreine, quinine, quinidine; u. adult. & substit. for Cinchona; important comml. article in 1880s as chief source of quinine (1275) (1725) - **R. purdieana** Wedd. (Co): bk. earlier known as cinchonamine bark; adulter. cuprea bark; c. cinchonine, cinchonamine, hydroquinone, &c.; u. as substit. for cinchona & source of quinine.

Remirea maritima Aubl. (monotypic) (Cyper.) (tropics): rt. inf. u. as diaph. & diur. (Br, Guianas).

remolacha (Sp): **r. azucarera**: (sugar) beet (root).

Remusatia vivipara Schott (Ar) (Ind): rt. in ointment with Curcuma for itch; juice with cow's urine alixipharmic.

Renanthera moluccana Bl. (Orchid) (Moluccas): young lvs. pickled in salt water & vinegar & eaten as delicacy.

render: produce FO by melting out of an animal tissue; ex. lard fr. hog mesentery.

Renealmia asplundii Maas (Zingiber.) (Ec): st. inf. of this hb. drunk to reduce swelling fr. snakebite (fer-de-lance) (517) - **R. domingensis** Horan. (R. aromatica Griseb.) (Guianas, Br): chucho (Guat); sds. u. as emmen.; arom. hence surrogate for cardamom; pl. juice to treat hemorrhoids, emmen. (Maya) - **R. exaltata** L. f. (Mex to Br. Guian, WI): cardamomo do Brasil; pacová; cuité assú; pacoseroca; pako-kating; "made gwa" (Fr. Creole-Trin); rhiz. c. 0.4% VO; bk. decoc. u. for inflamm. contusions; rhiz. u. for urin. inflamm. (decoc); frt. & sds. (Semen Renealmiae) c. VO, u. arom. carm.; lf. as poult. for swellings, sprains, sores (Trin) (2199) - **R. nutans** Andr. = *Alpinia n.* - **R. occidentalis** Sweet (R. aromatica Griseb.) (sMex to SA; WI): co(j)ate; rhiz./sts. decoc. diur.; in rheum. dyspepsia; antispasm.; vermifuge - **R. pedicellaris** A. C. Smith (Guian): decoc. of entire pl. u. as bath for fever - **R. thyrsoidea** P. & E. (Ec): st. inf. u. for snakebite, to reduce swelling.

renin: proteolytic enzyme liberated fr. kidney (mostly) under anerobic conditions; acts on angiotensinogen globulin) of blood serum —> angiotensin I, this soon converted to angiotensin II, a vasoconstrictor peptide —> hypertension.

rennase: rennin: Renninum (L): the partly purified milk-curdling enzyme fr. the glandular layer & juice of the calf's fourth stomach (abomasum) (*Bos taurus*); coagulates casein of milk; as Junket tablets u. in prepn. custards made fr. milk; in mfr. cheese (curdles milk); int. as milk digestant (usually infants).

rennet: 1) prep. of the calf stomach lining u. as source of rennin 2) rennin itself - **vegetable r.**: *Withania coagulans*.

renostyptin: epinephrine.

renouée (Fr-f.): *Polygonum*.

renuevo (Sp): shoot; sprout.

replicate: duplicate or repeat or multiply precisely, as in expl. work.

repolho (Por): cabbage.

repollo (Sp): cabbage - **r. de Bruselas**: Brussels sprouts.

repository formula: composition allowing deposit of drug such as penicillin in the subcutaneous tissues to permit its slow absorption into the system; mostly u. for hormones & antibiotics.

répulsif (Fr-m.): repellant.

Reptilia: class of Chordates; cold-blooded, lung-breathing; scale- or bony-plate-covered animals, generally repellent to humans; ex. cobra snake; alligator; turtles; lizards.

resaca (Sp) (swUS, Mex): dr. channel of stream.

rescinnamine = reserpinine.

Reseda lutea L. (Resed.) (Euras, Med reg., natd. NA): base (or Italian) rocket; hb. u. diaph. diur. anthelm. - **R. luteola** L. (orig. Med reg., now s&cEu, wAs, sAf): (dyer's) weld; d. rocket; Gaude (Ge); gaude des teinturiers (Fr); pl. source of "wild extract" (liquid or solid); dyer's weed; (wild) woad; hb. c. gluco-nasturtiin, mustard oil & luteolin glucosides, barbarin; u. in dyeing yellow ("Dutch pink"), for silk; as diaph. diur. anthelm. (Af); sds. earlier for varnish mfr.; formerly much cult. - **R. odorata** L. (n&sAf, cult. Eu): (common or garden) mignonette; hb. c. VO, mustard oil glycosides; fls. with solid VO with farnesol; u. in perfum. indy.; aq. lf. exts. are bacteriostatic; formerly u. as tenifuge; rt. u. as diaph. lax. diur.; cult. ornam. - **R. phyteuma** L. (Med reg., AsMin): herbe au mort (Fr); pls. sometimes u. as veget.

resenes: complex neutral components of resins, which are chemically inert, do not form salts or esters, & are insol. in alkalies; ex. in Asafetida, in resin of *Plantago coronopus*.

reserpiline HCl: alk. fr. *Rauwolfia serpentina*; ester u. as antihypertensive.

reserpine: Serpasil™: the most important and chiefly u. alk. of *Rauvolfia* spp (2020); hydrolyzed (alkaline) —> reserpic acid; chiefly u. as sed. in psychosis, anxiety, &c., hypertension (E).

reserpinine: raubasinine; rescinnamine; alk. occ. in *Rauvolfia* spp., *Vinca major*, &c.; sympatholytic action like yohimbine; hypotensive; tranquilizing agt. in alcoholism (1722).

Residuum Ovarii (L.): ovarian residue.

resin: class of solid or semi-solid organic substances which are amorphous, often soften & melt on heating but not at a definite temperature; mostly of pl. origin, usually representing excretions & exudations; sometimes of fossil origin (fr. pls. of distant past); usually yellowish to brownish; sol. in alc., ether, &c., insol. in water; often transp. or transluc.; nonconductors; ex. Rosin - **balsamic r.**: balsam with rel. more resin content; with VO contg. cinnamic & benzoic acids - **black r.**: black rosin: dark solid resin obt. by distn. of oleoresin of *Pinus palustris* & other *P.* spp.; earlier, crude distn. methods resulted in dark yellow to black prod. (518); term may also be appl. to other dark-colored resins; ex. Benzoin - **r. copal**: *v.* Copal - **fossil(ized) r.**: r. dug fr. ground in areas where originating trees are no longer found (extinct) - **r. guajac**: guaiac fr. *Guajacum officinale & G. sanctum* - **Jatabo r.** (Bahía, Br): fossil copal fr. *Hymenaea* spp. - **r. mucuje**: fr. *Acrocomia lasiopatha* - **semi-fossil(ized) r.**: r. dug fr. ground where the originating trees still grow in area - **synthetic r.**: plastic made by chem. processes in laboratory or factory; with much the same physical properties as natural r. - **tanroutk rouchi r.**: *v.* Madagascar copal - **violin r.**: rosin fr. *Pinus* spp. - **r. wine**: resinata (*qv.*) - **yellow r.**: rosin fr. *Pinus palustris*, &c. - **resin**: v. also under "gum"; v. also under acajou; acaroid.

Resina (L.): 1) resin 2) (specif.) rosin; pine resin - **R. Alba**: resin fr. *Pinus sylvestris* - **R. Benzoë**: Benzoin - **R. Communis**: galipot - **R. C. Nativa**: R. Pini (Communis): R. Alba - **R. Drastica**: rt. of Mex. pl. sp. (Fam. Convolvul.) somewhat resembling *Ipomoea orizabensis* (Mexican scammony) or *Piptostegia pisonis* (66); product fr. *Ipomoea* sp., acts as drastic cathartic. (Resina Jalapae mexicanae) - **R. Elastica**: India rubber - **R. Empyreumatica Liquida**: tar - **R. E. Solida**: pitch - **R. Flava**: rosin (yellow) - **R. Jalapae**: fr. *Exogonium purga* - **R. Lacca**: u. in mfr. of shellac - **R. Lutea**: yellow acaroid resin - **R. Mastiche**: Mastic - **R. Nigra**: black resin (*qv.*) - **R. Oxycedri**: Thus Commune - **R. Pini**: 1) turpentine (crude gum) 2) rosin - **R. P. Communis: R. P. Nativa**: R. Alba - **R. Prioris**: cativa - **R. Scammoniae**: resin fr. *Ipomoea orizabensis* - **R. Thapsiae**: fr. *Thapsia garganica*.

resinata: a resin wine prepd. in Greece; v. retsina.

resinate: the salt of a resin acid mixt.; often combined with metals.

resinates, fused: combination of metal with resin acids of a resin; u. driers in varnishes; ex. manganese resinate; calcium r. - **resine d'euphorbe** (Fr): euphorbia resin - **resinoid**: a subst. with some resin-like properties; specif. of the Eclectic class of pharm. prepns., made by ppting an alc. extraction of a crude drug with dil. HCl, which pptes. resins & active princ. such as alks. or glycosides; ex. aletrin - **resinol**: resin alc. which does not give tannin reaction with ferric salts; ex. found in gutta percha - **resinotannol**: resin phenol; resin alc. which shows a positive tannin reaction with ferric salts; ex. found in Benzoin, Tolu Balsam.

resistin: subst. believed a ketosteroid (perhaps fr. adrenal cortex) found in liver & other organ tissues of mammals submitted to experimental procedures; has antihistaminic properties (67).

resorption: reabsorption = absorption into blood fr. intestinal tract.

restinga: 1) (Br) sand banks, coastal thickets 2) (Arg) *Cassia crassiramea*.

rest harrow: *Ononis spinosa*.

resurrection plant: *Selaginella lepidophylla* +.

retama (Sp): 1) *Cytisus fragrans* Lam. 2) *Parkinsonia aculeata* (**r. de cerda**, Mex) - **retamo** (Arg): *Diostea juncea* (also as retama) **r. wax**: fr. *Bulnesia retamo* - **retamón** (Sp): *Cytisus purgans*.

Retanilla ephedra (Vent.) Brongn. (Rhamn.) (Ch, Pe): retanilla; caman; rt. u. for indig. carm. astr.

reticulin: 1) hydroxystreptomycin 2) protein associated with connective tissues.

retine: subst. wh. inhibits the growth of cells, incl. those of malignant tissues (transplants in mice); occ. in urine, tissue exts. (esp. liver, thymus); disc. Szent-Györgyi; not present in Krebiozen (cancer nostrum) as claimed.

retinoic acid: vit. A acid; tretinoin; Retina-A; vit. A acid; occ. in nature; keratolytic; u. to treat acne, reduce pigmentation; removes wrinkles; aids hearing (claim); u. exter. & int.

retinol: 1) vit. A; u. as wrinkle remover 2) rosin oil - **rétinolé** (Fr): ointment c. resins, FO, &c.; (less commonly) plaster.

retrorsine: alk. with pyrrole nucleus; fr. *Senecio retrorsus*, *S. ilicifolius*, *S. isatideus*; toxic (39).

retroviruses: gp. of RNA viruses which replicate through a DNA intermediate called the "provirus" (SC carry reverse transcription).

retsina (Gr): **restinata**: the ordinary wine (delicious yellow wine of Gr) flavored with pine resin which is put into the wine when fresh as a preserv.; u. since very early times.

Rettig (Ge): radish.

retting: a process of rotting away of the soft tissues of sts., leaving the phloem fibers behind; the process is accomplished by laying the fresh material in water, where bacteria (such as *Clostridium felsineum*, *Plectridium pectinorum*, in case of hemp & flax) feed upon the tissues; involves action of polygalacturonase.

reuniol: mixt. of geraniol & citronellol found in rose geranium oil (fr. Réunion Is).

Réunion: island in Indian Oc close to Af; orig. Bourbon Is, renamed 1948.

rewind chini (Dukanais): Rheum.

Reynosia septentrionalis Urb. (Rhamn.) (sFla [US], WI): darling plum ("dollen p." Bah); lf. decoc. u. for "weakness"; frt. delicious food.

Rhabarbarum (L.): **Rhabarberwurzel** (Ge): rhubarb rt.

Rhamnaceae: Buckthorn Fam.; widely distr. in temp. & trop. zones; over 800 spp. of shrubs, trees, & a few woody climbers; lvs. mostly simple, stipulate; frt. a drupe or capsule; fls. small, regular (ex. buckthorn); many spp. have stout thorns; many spp. have methyl anthraquinone glycosides (red with alkali); ex. *Ceanothus*, *Rhamnus*, *Zizyphus*.

rhamnase: rhamninase: glycosidase; enzyme fr. *Rhamnus cathartica*, *R. frangula*; serves to hydrolyze rhamnosides.

rhamnetin: an ether of quercetin; found in *Rhamnus* spp. frts. - **rhamno-glucoside**: a glucoside which yields on hydrolysis both rhamnose and glucose besides aglucone; ex. glucofrangulin of Frangula - **rhamnose**: sugar, methylpentose; occurs in many glycosides, ex. glucofrangulin - **rhamnoside**: a glycoside which yields on hydrolysis rhamnose (methyl pentose) & an aglycone.

Rhamnoneuron balansae (Drake) Gilg (monotypic) (Thymelae.) (Chin, Vietnam): sd. u. in local med.; bk. u. in mfr. paper; cult in Vietnam.

Rhamnus alaternus L. (Rhamn) (Med reg., seEu): Mediterranean buckthorn; lvs. very rich in anthraglucosides (4.1%); bk. inf. u. in edema & icterus (Arabs); lvs. u. as gargle in sore throat - **R. alnifolia** L'Hér. (Cana & nUS-swamps): dwarf alder; buckthorn; low shrub (2-8 dm); bk.u . as lax.; adulter. of Cascara Sagrada - **R. alpina** L. (Oreoherzogia fallax Vent.) (s&cEu, esp. mts [It-nwAf]): Carniolian buckthorn; u. in folk med. as lax. like frangula; decoc. for skin dis.; to adulter. frangula bk. (Balkans) - **R. californica** Eschsch. (Calif [US], nMex): (Calif) buckthorn; coffee berry; wild c. (tree); bk. u. like Cascara Sagrada & at times substit. for it; irrit.; berries food, esp. for mammals, have been u. as coffee substit. (2200); with several subspp., incl. *tomentella* Wolf (519) - **R. carniolica** Kern = *R. alpina* - **R. caroliniana** Walt. (seUS): southern buckthorn; Indian cherry; yellow wood; bk. u. as lax; frts. at first red, finally purple black; frts. eaten by Am Indians (hence "Indian cherry") - **R. cathartica** L. (Euras, nAf) widely natd. (most prodn. Ge, USSR): (common, European, or purging) buckthorn; ripe frt. (b. berries) u. purg. in diarrh.; diur.; fresh frt. u. to obt. expressed juice; evap. to dryness with CaO or other alkali to prod. Succus Viridis (Sap Green) u. yellow, green, dye; also called bladder green (Blasengruen; vert de vessie) (SC evap. to thin paste & dr. in animal bladder by placing in chimney flue in old days); unripe frt. also u. purg., dye; bk. u. mild purg., diur.; popular as lax. for dogs - **R. crenata** Sieb. & Zucc. (Jap, Chin, Kor): huan-go-yo; rts. sts. lvs. c. rhamnin A & B; u. in parasitic skin dis.; for boils (sChin); for whitlow; insecticide; said poisonous pl., may cause death if ingested - **R. crocea** Nutt. (Calif [US]): buckthorn; red berry; (Calif) mountain holly tree; ext. of bk. dark brown with charact. red color & pleasant arom. odor; u. like Cascara Sag. (& u. as substit.) but milder in action; subsp. **ilicifolia** C. B. Wolf (R. i. Kellogg) (Calif [US]): holly-leaf berry; berries u. to make sour beverage, summer cooler (pioneers); sometimes u. as astr. addn. to gin drinks; berries u. in pemmican (Indians) - **R. dahurica** Pall. (*davurica*) (R. virgata Roxb.) (e&cAs, Kashmir): frt. bitter, emet. purg.; for spleen dis. (Ind); bk. u. med. (Chin) - **R. frangula** L. (Euras, nAf, natd. e&cNA): (glossy) buckthorn; bk. c. frangulin, frangula-emodin, chrysophanol, & other anthraquinone deriv.; well-known purg.; in hepatic dis.; hemorrhoids; emm.; bk. misused as abortient; should be aged one yr. before use; frt. ("French berries") u. purg. & as subst. for *R. cathartica* berries (as dye); wd. u. to make charcoal; chief prodn.: USSR, eGe, Pol, CSR - **R. globosa** Bunge (nChina): Lokao buckthorn; Chin green; u. green dye - **R. graecus** Boiss. & Reut. (Gr): source of Greek yellow berries; best qualities fr. Turk & Iran (520) - **R. humboldtiana** Zucc. = *Karwinskia h.* - **R. infectoria** (*infectorius*) L. (Med reg., esp. AsMin): Persian berry; source of the Ital, Fre ("graines d' Avignon") (Avign. berries), Levant, Span, Syr & Turk yellow berries of comm.; u. to make yellow, green dyes - **R. japonica** Maxim. (Jap, Chin): Japanese buckthorn; small tree; bk. & black frt. u. ton. lax. (Ainu); wd. u. to make small furniture - **R. lotus** L. = *Zizyphus l.* - **R. nipalensis** (*nepalensis*) (Wall.) Lawson (Nepal+): frts. pounded & macerated in vinegar for herpes (Indochin); var. *tonkinensis* Tard. - **R. oleoides** L. (Med reg., esp. AsMin): source of Persian yellow berries (or Avignon grains); lvs. u. in skin dis.; very similar to & confused with *R. infectoria*; may be the *Rhamnus* of Theophrastus; possibly the same as *R. graeca* - **R. prinoides** L'Hér. (SAf): "dogwood"; decorticated rt. decoc. u. as "blood purifier" & in pneumonia; wd. ext. u. in rheum.; added to wines & beers to give narcotic effect - **R. purpurea** Edgew. (Ind-Himal): frt. u. as purg. (UP) - **R. purshiana** DC (Pac coast NA, fr. BC [Cana] to Calif, Ida, Mont [US]; cult. US [1860], eAf]: cascara sagrada; c. buckthorn; the medl prod. is the bk. (Cascara Sagrada, sacred bark, chittem bark, bitter [or yellow] bark); c. anthraquinones (anthranols, anthrones mostly as glucosides), aloin, cascarosides A-D (aloe-emodin glycosides), chrysaloin, emodin, iso-emodin, tannins, VO; u. purg. in obstipation, pregnancy, obesity; digestive tract weakness (1723); many pharmacopeias require storage of bk. for one year or more (or heating at 100° for 1 hr. in some cases) in order to reduce griping effect; lvs. twigs & bk. u. as tea for emesis (Klamath Indians); bot. name sometimes u. as off. title of drug; many proprietary prepns. of drug; frt. edible (2197); u. to adulterate *R. cathartica* berries - **R. saxatilis** Jacq. (c, s &seEu, Med reg., Balkans): rock buckthorn; source of Persian berries, Hungarian yellow berries, u. as yellow dye; *cf. R. cathartica*; said *Rhamnos* of Galen - **R. sectipetalis** Mart. (Br): similar c. & u. to *R. purshiana*; during drying of bk., anthranols converted to anthraquinones by oxidn. (similarly to *R. pursh.*) - **R. serrata** Schultes (sTex [US], Mex): tlacapollin; u. med. (Mex) - **R. tinctorius** W. & K. (Balkans): may represent hortic. var. of *R. cathartica*; frt. u. as yellow dye - **R. tomentella** Benth. = *R. californica* subsp. *t.* C. B. Wolf - **R. triquetra** Brandis (Ind): antiinflamm., cardiac stim., antispasmodic - **R. utilis** Dcne. (Chin, Jap): Chin buckthorn; bk. c. anthraquinones & anthranols, cocaine (?) (*sic*) (520); small tree; c. rutin; source of dye (Chin green indigo) u. as direct dye for wool & silk - **R. wightii** Wt. & Arn. (Ind, SriL): bk. u. as ton. astr. lax.; much like Cascara Sagrada - **R. zeyheri** Sond. (sAf): red ivory (wood); red ebony; bk. inf. u. for pains & soreness in back; as enema for rectal ulceration in children; frt. eaten fresh & dr.; wd. red-colored, u. to make small items - **R. zizyphus** L. = *Zizyphus jujuba.*

Rhaphidophora (*Raphidophora*) **falcata** Ridl. (Ar) (Siam): pounded rts. u. for threadworm in children - **R. hookeri** Schott (Indochin): pl. inf. (alc.) u. as stim. aphrod. - **R. merrillii** Engl. (PI): reputed cure for snakebite; spadix u. as emmen; cult. ornam. - **R. minor** Hook. f. (Mal): lf. & rt. decoc. u. in "delayed confinement."

Rhaphiostylis beninensis Planch. (Icacin.) (trop wAf): macerated bk. u. as lax. for infants with chronic constip. (Senegal).

Rhapidophyllum hystrix H. Wendl. & Drude (1) (Palm) (seUS, Ga-Ala): needle palm; frt. edible; ornam. shrub; monotypic; spines at base of lvs. u. as needles for leather, &c. (Am Indians).

Rhapis excelsa (Thunb.) Rehd. (Palm) (eAs): lady palm; ground rattan cane; rt. decoc. u. as circ. stim. & in rheum.; ash of petiole base u. as hemostat (also ashed bk. or frt.) (Chin); cult. ornam. (dwarf form, Jap) (2407).

rhapontic root: fr. *Rheum rhaponticum* - **white r.**: *Centaurea behen* - **rhaponticin**: xall. glycoside found in prec.

Rhaponticum carthamoides (Willd.) Iljin (Leuzea c. DC) (Comp.) (USSR): rhiz. & rts. c. several flavonoids; CNS stim., hypotensive; u. as ext. ("saian") to "strengthen the nerves"; ton.; with invigorating properties for states of complete exhaustion (asthenia) of various origins (Siberia) (closely related to *Centaurea*) - **R. scariosum** Lam. (Eu Alps): Swiss centaury; rhiz. u. as replacement for rhubarb root.

rhatany (root): *Krameria triandra* & other *K.* spp. (v. *Krameria* also) - **Brazilian r.: Pará r.**: *Krameria argentea* - **Payta r.: Peru r.**: *K. triandra* - **Savanilla r.**: *K. ixina*, *K. tomentosa.*

Rhazya stricta Dcne. (Apocyn.) (wInd, Pak): vena (Punjabi); lvs. bitter ton., in low fevers, general debility, sore throat (inf.); frts. & lvs. in boils, eruptions, &c.; tree source of lacquer & wax (frts.) u. for candles (2201).

rheadin: v. rhoeadine.

Rheedia acuminata (Ruiz & Pavon) Planchon & Triana (R. madruno [Kunth] Planchon & Triana) (var. *bituberculata*) (Gutt./Clusi) (CA, Guianas, Co, Ve, Br.): madroño; latex (Brazilian tacamahac) u. for ulcers & sores; frt. believed useful in jaundice & cholera; frt. staple article of diet, with delicious taste; u. in preserves & beverages - **R. aristata** Griseb. (WI): manaju (Cu); yellow resin u. in field to cover injuries & prevent tetanus; reddish hard wd. very resistant & good for construction - **R. brasiliensis** Planch. & Triana (var. *salicifolia*) (Arg, Br): pacuri; bakupari; frt. of tree delicious, famous for flavor; u. in jams - **R. edulis** Triana & Planch. (sMex, CA): "ceró"; greenish-yellowish gum exudn. u. for wounds, cuts, &c. (Pan) - **R. lateriflora** L. (WI): hat stand tree (Trin); yellow latex of wd. u. to heal cracked or fissured feet (Domin, Caribs); u. in plasters; frt. edible (2328) - **R. magnifolia** Pittier (CR): with edible tasty frt.

rhein: aglyconelike aloe-emodin but with the CH_2OH gp. replaced by COOH gp.; occ. in senna, rhubarb rt.

Rhesus monkey: *Macaca mulatta* (common macaque) (c&nInd): the small monkey often seen in zoos & biol. labs - **r. factor: Rh factor**: one of several erythrocyte antigens originally found in blood of rhesus monkeys by Wintersteiner et al.; related to blood hemolysis.

Rheum L. (Polygon.) (temp & subtr As): rhubarb; c. 60 spp.; many spp. toxic (lf. lamina with Ca oxalate) (2408); perenn. hbs. v. under "rhubarb"; also (L) for med. rhubarb rt. fr. R. *officinale*, *R. palmatum*, &c. - **R. compactum** L. (Mongolia-Chin-Siberia): purg. ton. cholagog, &c.; adulter of true rhubarb rt.; forms hybrid with rhapontic r. - **R. cordatum** A. Los (As): rhiz. c. aloe-emodin, rhein, rheum-emodin, physcion, chrysophanol, rhaponticin, almost no tannin - **R. coreanum** Nakai (Kor): c. sennosides A & B; u. to promote digestion; popular in Jap med. (522) - **R. emodi** Wall. (R. australe Don) (Indo-Pak, Sikkim, Nep, Tibet, Himal reg.): (Himalayan or Indian) rhubarb; chief source of Indian Rhubarb; rhiz. & rts. u. tonic lax. (ca. 30% less effective than Chin r.) followed by astr. action; in chronic diarrh., acute bacillary dys. (esp. of children) (2210); dye; dusting pd.; etc. (70); lf. stalks eaten raw, boiled, dr.; lvs. & fls. also edible (2464) - **R. franzenbachii** Muenter = *R. rhabarbarum* - **R. hybridum** Murray (Mongolia, eAs, cult.): garden rhubarb; considered hybrid of *R. rhaponticum* X *R. palmatum*; stewed petioles u. as dessert & in pies; r. wine is prepd. fr. juice - **R. Indicum** (L): Indian (or Himalayan) rhubarb; rhiz. & rt. obt. fr. several *R.* spp. of Ind.; u. ton. lax. stom. - **R. leucorrhizum** Pallas (Chin): rts. u. as purg.; rts. u. as substit. for superior kinds of rhubarb - **R. moorcroftianum** Royle (nInd): c. & u. like *R. emodi*; a sp. of Ind rhubarb - **R. officinale** Baill. (wChin incl. Tibet): (medicinal) rhubarb; Chin r.; a less important source than rhiz. & rts. of *R. palmatum*; c. anthraglycosides; anthraquinone derivatives, incl. emodin, chrysophanic acid, rhein; u. purg. stomach, ton. lax., bitter, in acute bacillary dys. (83); less active than foll. - **R. palmatum** L. [incl. var. *tanguticum* (Maxim.) Tschirch] (Chin incl. Tibet): Chin rhubarb; (sorrel) r.; Shensi r. (best grade of med. r.); similar to prec. but richer in active princ. & therap. more effective (2212) - **R. rhabarbarum** L. (R. undulatum L.; R. franzenbachii Muenter) (cAs, as nwChin, seSiber, Turkestan, Mongol): Siberian rhubarb; an adulter. of Chin rhubarb; forms hybrids with *R. rhaponticum*; classed with & u. like latter - **R. rhaponticum** L. (R. sibericum Pallas) (Siber, nChin, Bulg, Thrace; cult. Eu, Fr, Eng, US+): (garden) rhubarb; rhapontic (r.); pie plant; Radix (Rhei) Rhapontici is off. in some phar.; u. mild lax. astr. substit. for & adult. of off. Chin Rhubarb (2211); considerably u. veter. med.; lf. stalks or petioles stewed as dessert (rich in vit. C) (pudding, pies, tarts); lf. lamina toxic (CaOx) - **R. ribes** L. (Iran, Afgh, Turkestan): current rhubarb; rts. u. as vermifuge for horses; sold in Tehran markets - **R. spiciforme** Royle (Himal-Tibet, Sikkim): rhiz. u. purg.; a source of Indian rhubarb - **R. tanguticum** Maxim. & Balf. = *R. palmatum* var. *t.* - **R. tartaricum** L. f. ("Tartary", USSR) (Iran, Afgh): fools' rhubarb; rewash-i-dewana; frt. decoc. u. as purg., considered stronger than the tuberous rt. & rhiz. which is also u.; rhiz. c. frangula-emodin, chrysophanol, physcion, &c.; u. as purg; large underground parts u. as fuel in Afghanistan, burning with beautiful blue flame (1737) - **R. tibeticum** Maxim. (Tibet, Kashmir): revand chini; one source of Indian R.; natives u. as vegetable - **R. webbianum** Royle (Nepal, Tibet, Kashmir-Himal): one of chief sources of Indian R. Root.

rheum-emodin: aglycone found in Polygonaceae members esp. *Rheum* spp.; identical with emodin, frangula-emodin, frangulic acid.

Rheum(a) (Ge, Eng): 1) rheumatism 2) rheum (watery discharge fr. nose) (catarrh) ("epidemic rheum" = influenza) (formerly).

rheumatism herb: *Frankenia campestris* - **r. root**: 1) *Jeffersonia diphylla* 2) *Dioscorea* spp. - **r. weed**:

Chimaphila umbellata - **rheumatology**: the science of rheumatism incl. arthritis (diagnosis & treatment).

Rhexia virginica L. (Melastomat.) (eNA): meadow beauty; vinegar wood; lvs. & sts. u. as "throat cleaner" (Maritime Indians); cult. orn.

Rhigiocarya racemifera Miers (Menisperm.) (Nig): lvs. u. for neuralgia; sds. u. as aphrod. (523).

Rhinacanthus calcaratus Nees (Acanth.) (Ind): lvs. u. as popular anthelmintic; juice of rt. aphrod.; prep with citrus juice & pepper u. in herpes (40) - **R. nasutus** Kurz (R. communis Nees) (eAs, Chin, Indochin, Java): nasuta shrub; rt. (ringworm root) u. as remedy for herpes, "cascado" (Moluccas) & other skin dis.; c. rhinacanthin (quinone-like); lvs. c. alk., coumarin; rt. & lvs. u. to treat ringworm; supposedly extraordinary aphrod. effect (sInd).

Rhinanthus alectorolophus Poll. (R. major Ehrh. (Scrophulari) (Eu): (yellow) rattle (box); cock-grass; pl. c. aucubin; insect poison; hb. u. in homeop. - **R. minor** Ehrh. (R. cristagalli L.) (Eu-Greenland; Cana, New Eng [US]): yellow rattle; rattle box; hb. c. aucubin (rhinanthin), dulcitol, mannitol; u. (as Herba Cristae galli) in homeop.; as insecticide.

Rhine berry: thorn berry: *Rhamnus cathartica* frt.

Rhinoceros unicornis L. (R. indicus Cuv.) (Rhinocerotidae - Mammalia) (Pak to nInd [Assam] s of Himalay): (Indian) rhinoceros (lit. horn-nose [Gr]); the horn is much u. (powd.) as antipyr., aphrod. in Chin & Kor med. & highly valued; greater value than gold or finest elephant ivory (said worth $15,000/lb [1988]); made up of keratin; formerly u. in Far East, cut into form of goblets to test drinks for poison (*cf. Diceros bicornis* L.); the white r., *Ceratotherium simum* Burchell (sAf) (gray, with 2 horns on snout) is similarly u.; horn also u. to form handles of daggers (jembia) u. by men of N Yemen (hence of great value).

Rhinopterys spectabilis Niedenzu (Malpighi) (trop wAf): rhiz decoc. u. as bath & in beverage as parasiticide, esp. for itch.

rhinoviruses: coryza v.; respiratory v.; one of several gps. of viruses responsible for the common cold; belong to the ECHO gp. (serum type); virus vaccines have been prepd.; belong to Picornaviruses.

Rhipsalis baccifera Stearn (R. cassutha Gaertn.) (Cact.) (trop&subtr Am incl. sFla [US]+): mistletoe cactus; "pega palo"; bejucco de quebradura (= rupture rattan) (SC u. to tie up fractures of man and horses); u. anthelm.

rhizobia: microorganisms wh. live symbiotically in nodules produced upon the roots of leguminous plants; *Rhizobium* spp. (Azotobacter. - Eubacteriales - Bacteria) (Nitrobacteriaceae); the very small rod-shaped organisms fix atmospheric nitrogen; ex. *R. radicicola* Hiltner & Stoermer, on soybeans & alfalfa.

rhizoderm(is): epidermal layer of rt.

rhizogenous layer (of root) = pericycle.

Rhizoma (L): rhizome; root stock; "root" (*v.* also Radix) - **R. Arnicae**: arnica rhiz. (rt.) - **R. Aronis**: fr. *Arum maculatum* - **R. Bistortae** (= Radix B): rhiz. of *Polygonum bistorta* - **R. Caricis**: rt. of *Carex arenaria* - **R. Curcumae domesticae**: *Curcuma longa* rhiz. - **R. Filicis**: rhiz. of *Dryopteris filix-mas & D. marginalis* - **R. Galangae minoris**: fr. *Alpinia officinarum* - **R. G. majoris**: fr. *A. galanga* - **R. Graminis**: fr. *Agropyron repens* - **R. Imperatoriae**: fr. *Peucedanum ostruthium* - **R. Podophylli indici**: Indian podophyllum, *P. emodi* - **R. Polygoni**: fr. *Polygonum bistorta* - **R. Rhei**: Rhubarb "root" - **R. Tormentilla**: fr. *Potentilla erecta* - **R. Zedoariae**: fr. *Curcuma zedoaria.*

rhizomatous: with the character of a rhizome.

rhizome: an underground st. of horizontal (i.e., creeping), vertical, or oblique growth; ex. ginger, blue flag, veratrum; in crude drugs, often associated with rts. as Rhiz. & Rts.; ex. gentian, licorice, rhubarb, ipecac.

rhizomorphs: thread- or cord-like structures of fungi made up of hyphae.

Rhizophora caseolaris L. = *Sonneratia c.* - **R. conjugata** L. (Bruguiera c. Merrill) (Rhizophor.) (tropAs, Af, Austral, coasts): mangrove tree; source of tanbark; brace rts. u. to make anchors, in fishing nets (Mal) - **R. mangle** Roxb. (sFla [US], w. Mex (coast), WI, SA, Af, trop Atlantic Ocean coasts): American or red mangrove; bk. (Cortex Mangles) c. 30-40% catechol tannins, u. astr. in fever, to stop hemorrhage; bk. syr. in asthma & "suffocation"; in diarrh., leucorrh., GC; Hansen's disease; nasal hemorrahges (Br); rts. u. like bk.; bk. ext. u. for tanning, dyeing; ext. said cheapest comm. natural tan; wd. u. for piling (not attacked by molluscs or Teredo); honey pl.; frt. edible - **R. mucronata** Lam. (sAs, Indones, eAf): (red) mangroves; mangle; bk. astr. in diarrh., dys.; c. pyrogallol tannins (4-48%); related to oak tannin; red dyestuff; u. as tan; in hypertension, diabetes, rheum. (Mauritius); ext. c. 70% tannin, u. as catechu substit.; frt. edible, sometimes fermented to give a mild wine; durable wd. u. techn.; lf. in fever, u. as poultice for injuries fr. "armored" fish; honey reported toxic - **R. racemosa** Mey. = *R. mangle.*

Rhizopogon rubescens Tul. (Rhizopogon./Hymenogastrales/Basidiomyc.) (widely distr.): subterr. frtg. hodies eaten (Jap).

rhizopterin: a folic acid factor with high activity in promoting growth of *Streptococcus lactis* u. in experim. med.

Rhizopus Ehrenb. (Mucor. - Mucorales - Zygomycetes - Fungi): widely distr. molds, some the cause of mucormycosis, incl. common bread mold (*R. stolonifer* Lind. [R. nigricans Ehrenb.]), common saprobe & facultative parasite on ripe frts. & veget.; u. comm. in mfr. of fumaric acid & in prepn. of cortisone; *R. oryzae* &c., u. in mfr. lactic acid; *R. arrhizus* u. in making many steroids; *R. japonicus* u. to make saccharified starches u. to prep. alcohol; for conversion of starches into sugars (Amylo process); source of amylase (diastase) (Eu); orig. isolated from Chin yeast; prepn. sufu & tempeh (Chin foods fr. soybeans using *R.* spp as agt.).

rhizosphere: soil surrounded & influenced by the rts. of a plant.

rho I: prolan A.

Rhodamnia cinerea Jack (R. trinervia sensu auct.) (Myrt.) (Mal to NG): lf. sap or inf. of lvs. or rt. decoc. u. as post-partum protective med. - **R. dumetorum** Merr. & Perry (Indochin): frt. decoc. u. to treat ulcers & gum dis.; bk. & lvs. vulner. & astr.; wd. useful, u. to make charcoal (Sumatra).

Rhodea Endl. = *Rohdea.*

rhodeose: fucose.

Rhodesia —> Zimbabwe

Rhodeswood: Guiana rosewood, *Amyris balsamifera.*

rhodexin B: glycoside of gitoxigenin & L-rhamnose; occ. in *Rohdea japonica.*

Rhodiola rosea L. Sedum r. (Crassul.) (nHemsph): rose root; lvs. edible.

rhodium oil, artificial (or **factitious**): mixt. of oils of copaiba (or cedarwood), sweet almond, rose (or rose geranium); or of sandalwood & rose (or rose geranium) oils - **r. wood**: 1) rt. wd. of shrubs of *Convolvulus floridus* & *C. scoparius*; source of rhodium oil, VO 2) wd. of *Amyris balsamifera*; oil u. as rat & fish lure (2666).

Rhododendron L. (Ericaceae) (nTemp Zone): rose bay; rhododendron; large genus (some 850 spp.) of deciduous & evergreen shrubs & small trees of cooler temp. regions; sometimes genus *Azalea* segregated from *R.*; phytochemistry of interest (524) (SC rhodon, Gr, rose; dendron, Gr, tree) - **R. arboreum** Sm (Ind, Nepal, Burma): medium-sized tree; pl. juice u. to treat diarrh. dys. (also fls); paste (dr. juice) u. in headache; fls. eaten by children (but intoxicating !); honey fr. fls. toxic; fish poison; toxic to cattle - **R. barbatum** Wall. ex G. Don (Nepal, nInd, Himal): evergreen tree; lf. ext. has ionotropic effect; fish poison; honey toxic, wd. u. as fuel - **R. brachycarpum** D. Don (Kor): lvs. c. ericolin; gathered in fall, u. as kidney ton. & for itching - **R. calendulaceum** (Mx.) Torr. (eUS, s to Ala): flame azalea; boiled twigs rubbed on rheumatic areas; inf. u. for female dis.; fungal growth ("apple") u. to alleviate thirst (Cherokee Indians) - **R. campanulatum** D. Don (Indo-Pak, Himal): simru(ng) (UP) twigs & pd. wd. u. in phthisis & chronic rheum; syph; sciatica; u. as snuff in colds - **R. canadense** (L) Torr. (neNA): rhodora; ornam. shrub - **R. catawbiense** Mx. (seUS): mountain rosebay; very showy shrub to small tree; lvs. c. arbutin, pyroside, rhododendrin; honey discarded each year since toxic; pl. toxic to cattle; many ornam. cult. vars. - **R. chrysanthum** Pall. (ne&cAs, Davuria [seSiber]; now cult. in eEu): Siberian (or yellow snow) rose; yellow (flowered) rhododendron; c. andromedotoxin (andromedol) (toxic!); lvs. u. diur., diaph., narcot. in neuralg., gout, rheum.; astr. in hemorrhages - **R. cinnabarinum** Hook. f. (Ind, Nepal, Himal): fls. u. to make a jam; honey fr. fls. toxic; wd. u. as fuel - **R. dauricum** L. (Chin): lvs. c. flavonoids, ericolin; VO c. menthol, juniper camphor; betulin - **R. falconeri** Hook. f. (nInd, Nepal): lvs. c. skimmiol, quercetin, &c.; u. as fish poison - **R. ferrugineum** L. (cEu-Alps): "Alpine rose"; dwarf rosebay; rusty-leaved alpenrose; lvs. & sts. c. ericolin (arbutin), rhododendrin; VO tannin; u. narcot. diur. in lithiasis, gout, rheum. sed. diaph. (danger!); in tanning - **R. fulgens** Hook. & **R. grande** Wight (both Himalayan India reg.): *cf. R. Cinnabarinum* - **R. hirsutum** L. (cEu, Alps): (hairy) alpenrose; lvs. & green sts. c. phlobatannins; rhododendrin, VO, FO, arbutin, &c. (2331); lvs. u. like those of *R. ferrugineum* - **R. indicum** (L.) Sweet (sJap, cult. Ind): parent of "Indian Azalea" of florists - **R. luteum** Sweet (R. flavum G. Don) (Caucas, cRuss): "Pontic azalea"; fls. & lvs. narcot.; honey fr. fls. said to cause intoxication, emesis, diarrh., insanity(!!) (Xenophon); pl. sometimes fatal to foraging goats - **R. maximum** L. (eNA): mountain (or great) laurel; "American rhododendron"; rose bay; shrub or shrubby tree (-10 m) with leathery lvs. wh. c. acetylandromedol (toxic, thus in honey); ericolin (arbutin); ursolic acid; tannin; lvs. u. diur. astr. narcot.; fish poison; sometimes in place of *R. chrysanthum*; wd. u. for tool handles; rt. wd. u. for tobacco pipe bowls (525) in place of African brier - **R. molle** G. Don (R. sinense Sw.) (Chin, Jap): lf. & fl. tincture u. in rheum. arthritis; fls. u. in marsh fever; insecticide;

toxic - **R. moulmainense** Hook. (eAs, Burma): honey fr. this pl. stupefying; white person becomes lethargic & quite ill, but not Tibetans - **R. periclymenoides** (Mx.) Shinners (Azalea nudiflora L.) (e&sUS): pinxter flower; purple (or bunch or wild pink) honeysuckle; election pink; galls called "May apples" (on lvs. & sts.) pickled with vinegar & spices & eaten (since early colonial days) - **R. ponticum** L. (Iberian pen., Thrace, AsMin, Leb, Cauc): common (or yellow) azalea; lvs. c. ericolin (arbutin), acetylandromedol (andromedotoxin); rhododendrin; fl. nectar c. sucrose, c. 0.1% andromedotoxin (hence honey poisonous) producing intoxication, vomiting, diarrh., &c. (Xenophon); fls. & lvs. narcotic; u. in gout, rheum. hypotensive (AsMin) - **R. seniavinii** Maxim (Chin): man-shan-bai; lvs. u. in bronchitis - **R. thomsonii** Hook. f. (cAs): with toxic honey (acetylandromedol); u. med. (Ind) - **R. vidalii** Rolfe (PI): "cure" for itching - **R. viscosum** (L) Torr. (neUS): clammy azalea; (white) swamp honeysuckle; cult. ornam. shrub.

rhododendron: spp. of *Rhododendron* & of *Azalea* (when latter sometimes recognized as separate genus) - **American r.**: NA spp. of *R.* cult. as ornam.; u. for making pipes in place of African brier root - **yellow flowered r.**: *R. chrysanthum.*

Rhodomela larix (Turn.) C. Ag. (Rhodomel. - Rhodoph. - Algae) (e&wcoasts of nPac Oc): annual seaweed; thallus c. antibiotic (63); alga rich in Br (3%) & a comm. source in Jap.

Rhodomyrtus Reichb. (Myrt.) (sAs, Austral, Oceania): c. 25 spp. of trees and shrubs - **R. macrocarpa** Benth. (Austral): finger cherry; f. berry; native loquat; toxic - **R. tomentosa** Wight (Ind, Chin, Indones, Jap, PI): hill guava; (Nilghiri) h. gooseberry; downy myrtle; lvs. styptic, antipyr. anodyne; rt. for cardialgia; frt. edible; sds. edible u. as digest. ton. (Chin); rt. decoc. u. for stomachache, diarrh. (Mal); bk. c. 19% tannin, poisonous.

Rhodophyceae: Red Algae; a large & variable class of the marine Algae (a few spp. grow in fresh water); color varies through purple, violet, reddish brown, & even green, incl. *Chondrus, Gracilaria, Rhodymenia.*

rhodopsin: "visual purple"; conjugated carotenoid important to vision; (acted on by light to give visual yellow); deficient in retinal rods.

Rhodora L. = *Rhododendron*; now treated as section in latter - **R. canadense** L. = *Rhododendron c.* (526).

Rhodothamnus chamaecistus Rchb. (monotypic) (Eric.) (seEu, wAs-hilly areas): Alproeslein (Ge): Alpenradrose (Ge, Alpine wheel rose); dwarf evergreen shrub; cult. ornam.

Rhodotypos scandens (Thunb.) Makino (R. tetrapetala Makino) (monotypic) (Ros.) (Chin, Jap): jet-bead; jet berry bush; toxic (HCN): cult. ornam. shrub.

Rhodymenia palmata (L) Grev. (Halymenia p. Ag.) (Rhodymeniaceae - Rhodophyceae - Algae) (1724) (nAtl Oc, esp. N Sea): (Irish) (or purple) dulse; varette (Qu [Cana]); sea kale; eaten raw for constipation (Qu); eaten by Canadian Indians (BC [Cana]); u. as vegetable, in ragouts (red in color, thicker consistency), relishes, condim., masticatory (Scotland); for intest. worms; prepn. iodine. (2667).

rhoea: *Boehmeria nivea.*

rhoeadine: non-toxic alk. of Opium & capsules of *Papaver* spp., esp. *P. rhoeas*; red color said due to accompanying substances adhering to the crystals of the alk.

Rhoeo spathacea (Sw.) Stearn (R. discolor Hance; Tradescantia d. L. Hér.) (monotypic) (Commelin.) (sFla [US], Mex, CA, WI): oyster plant; cordobán (Cu); hb. u. as expect. in pulm. dis.; astr. hemostat. in hemorrhages, menorrhagia; dye plant; ornamental hb. (—> *Tradescantia*).

Rhoicissus capensis Planch. (Vit.) (sAf): wild grape; w. vine; monkey rope; rt. u. as anthelm. (calf) (Zulus); frt. in conserves; split vine u. as rope & in basket making - **R. cuneifolia** Planch. (sAf): wild grape; decoc. of decorticated rt. u. as enema for painful menstruation; to facilitate birth; for sterility & impotence - **R. digitata** Gilg & Brandt (Af): ophthalmic remedy (Zimbabwe); frt. to make jam; split vine as rope - **R. erythroides** Planch. (Af): monkey rope; u. in epilepsy; rt. decoc. nerve stim. & for GC; leprosy (Tanz); purple frt. edible - **R. revoilii** Planch. (Tanz): tuberous rt. u. for swellings; st. sap appl. to cuts, wounds, sores & burns; has anesthetizing properties; rts. mixed with milk fed to calf of cow producing insufficient milk.

Rhopalostylis sapida H. Wendl. & Drude (Areca s. Forst.) (Palm.) (nNZ & Chatham Is): nika(u) palm; a hardy tree; source of a palm cabbage (central sprout) as delicious food; st. pith sl. lax, u. to relax pelvic muscles (eaten by woman at birthing); sap drunk at time of labor; lvs. u. for building native huts (Maoris); u. in diarrh., dys.; only palm sp. native to NZ.

rhubarb: *Rheum* spp.; specif. (in medicine, pharmacy): rhiz. & rts. of *R. palmatum, R. officinale* & other Chin spp. of *R.*; also appl. to *Rumex alpinus* - **Austrian r.**: rhapontic r. - **black-hearted r.**: inferior r. with dark areas due to decay - **bog r.**:

Petasites hybridus - **Brazilian r.**: 1) *Euphorbia cyparissias* (u. as r. substit.) 2) *Trimezia juncifolia, T. cathartica* (u. as r. subst.) - **Canton r.**: a var. of Chin r. fr. Tibet; flats & rounds - **Chinese r.**: Rheum (NF); chiefly *R. palmatum*, the most important medicinal r. - **common round r.**: one of best known types of Chin r.; transverse segments of entire rhiz. or rt. - **Crimea r.**: Krimea r. - **Danish r: Dutch r.**: r. formerly imported by the seAs comml. concerns of these countries - **East Indian r.**: Chin r. - **English r.: European r.**: rhapontic r. - **false r.**: *Thalictrum flavum* - **r. flats**: large pieces of rhiz. or rts. cut in two lengthwise (Chin r.) - **French r.**: 1) rhapontic (chiefly) 2) *R. palmatum*, etc. (at times) - **garden r.: German r.**: rhapontic r. - **high dried r.**: Chin r. dr. while strung on cords, often with holes clear through segment (incl. Se-Tschouen flats and common rounds); has strong empyreumatic odor - **hill r.: Himalaya(n) r.: Indian r.**: Rheum Indicum (L.): usually *R. emodi* (also *R. moorcroftianum, R. spiciforme, R. webbianum*) 2) (Calif [US]) *Peltiphyllum peltatum* - **Krimea r.**: rhapontic r. - **monk's r.**: *Rumex alpinus & R. patientia* - **mountain r.**: *Rumex alpinus* - **officinal r.**: *Rheum palmatum*, etc. - **poison r.**: bog r. - **poorman's r.**: *Thalictrum dioicum* - **rhapontic r.**: *R. rhaponticum* - **r. root**: rhiz. & rts. of *R.* spp. - **r. rounds**: common round r. - **Russian r.**: *R. undulatum* or possibly *R. palmatum* - **Se-Tschouen r.**: artif. dr. Chin r.; as flats - **Shanghai r.**: Chin r. fr. Yangtse-kiang reg. - **Shensi r.**: *R. palmatum* prod. fr. Kukunor dist. near Tibet; considered finest type of Chin r.; in both flats & rounds - **stick r.**: kind of English r.; rather cylindrical, c. 3 cm X 13 cm; more or less irreg. on surface as though unequally shrunk in drying - **Turkey r.**: 1) *Rumex scutatus* 2) **Turkish r.**: Chin r. re-exported through Turk ports - **waved-leaved r.**: *R. undulatum* - **white r.**: *Asclepias tuberosa* - **wild r.**: *Rumex hymenosepalus*; *Begonia obliqua*.

rhubarbe du Népaul (Fr): Nepal rhubarb - **r. de Perse**: Indian r.

Rhus L. (Anacardi.) (subtr. & warm temp. zones worldwide): sumac(h)s; ca. 300 spp. of shrubs, vines, & trees; bk. & lvs. u. as tan; now often with toxic spp. split off into *Toxicodendron* (former sect.); bk. & lvs. u. as ton. - **R. aromatica** Ait. (Schmaltzia crenata Greene) (e&cNA): fragrant (or lemon) sumac; polecat bush; rt. bk. c. VO, 22% tannins, gallic acid, resin; u. as diur. (VD), in kidney & bladder dis.; astr. in diarrh. dys. (Cortex Rhois aromaticae) (radicis); formerly in diabetes (?!); var *serotina* Rehd. (cUS): bk. u. for colds (Comanches) - **R. chinensis** Mill. = *R. javanica* - **R. ciliata** Licht. (Af): kareebos; rt. u. for tanning (Thlaping) - **R. copallina** L. (e&cUS): dwarf (or shining) sumac; flameleaf s.; wing-rib s.; frts. added to water to improve taste; lvs. smoked with tobacco (Am Indians) (Calif [US]); lvs. & bk. c. tannin; u. as tan ("American sumac"); lvs. to perfume tobacco; frts. acidulous, astr. sometimes u. in beverages; rts. in syph., red dye; fresh juice for warts; sd. oil for hemorrhoids - **R. coriaria** L. (Med reg): (Syrian) sumac; tanner's s.; lvs. c. gallotannins; u. as tanning subst. ("gambuzzo") and as yellow dye (for cordovan leathers); many comm. vars. recognized; lvs. u. to perfume tobacco - **R. cotinus** L. = *Cotinus coggygria* - **R. dentata** Thunb. (sAf): Rhodesian currant; frt. eaten raw or mixed with milk - **R. divaricata** (Greene) McNair (non E. & Z. of sAf) = *Toxico-dendron radicans* subsp. *divaricatum*; rt. decoc. u. in colds & flu; mixt. with *Scabiosa columbaria & Cussonia paniculata* as colic remedy; wd. u. to make strong knobkerries (club weapons of sAf natives) - **R. diversiloba** T. & G.: *Toxicodendron diversilobum* - **R. erosa** Thunb. (sAf): bessingbos (Du); remedy for diarrh. (human & cattle); snuff plant of Sotho; branches u. for brooms & hedges - **R. glabra** L. (e&cNA): smooth/common/Pennsylvania/white/upland sumac(h); u. astr. acidulous refrig. in beverages ("Indian lemonade" made fr. frt. by steeping husks in hot water, straining, & sweetening the aqueous ext.); as tan; bk. & lvs. c. tannin, u. astr.; "Am sumac galls" (gall apples) c. 60% tannin; u. as tan; rts. —> dye; herbage when bruised —> milky juice exudate; u. for warts; frt. decoc. mixed with glycerin & water, u. as gargle (537) - **R. gueinzii** Sond. (sAf): bastard karee; inf. of decorticated rt. u. in bilharziasis (checks hematuria); lf. decoc. for relief of biliousness (sTswana) - **R. integrifolia** B. & H. f. (sCalif, Baja Calif [Mex]): lemonade bush; l. berry; coast sumac; sour berry; frt. u. to make sour beverage; shrub or small tree cult. for ornam. foliage - **R. javanica** L. (R. semialata Murr.; R. chinensis Mill.; R. japonica Buch. Ham.) (eAs, esp. Chin, Jap; Haw Is): Chin sumac; C. nutgall tree; mai-kok-kyin (Shan); lvs. & petioles bear galls (Chin, Japanese, or sumac galls) by sting & eggs of an aphis; c. 78-80% tannin, starch; u. astr. (mostly in veter. med.); lf. decoc. appl. to skin rash (2448); in tanning & dyeing; in mfr. tannin & gallic acid; frt. u. in colic (Burma) - **R. kirkii** Oliv. (Af): u. for menstrual difficulties - **R. laevigata** L. (Af): bk. u. as tan; acidulous frt. eaten by children - **R. laurina** Nutt. = *Malosma l.* - **R. longipes** Engl. (tropAf): rts. u. in abdomin. pain, indiges. (Moz); vapor fr. boiled rt. inhaled for grippe; rt. in male

& female sterility; rt. cold water inf. for difficult labor (Moz) (2445) - **R. margaretae** C. E. Moss (Af): rt. u. as remedy in GC. - **R. metopium** L. = *Metopium toxiferum* - **R. natalensis** Bernh. (s&cAf): u. as tenicide; lv. & rt. for GC; rt. for habitual abortion and repeated stillbirths; bark (rich in tannins) u. poultice for sores (Tanz); lf. juice in dermatoses, boils; frt. in stomach dis.; for flu; as vulner. - **R. perniciosa** HBK = *Pseudosmodingium perniciosum* - **R. pontanini** Maxim. (China): u. much like *R. javanica* - **R. radicans** L. = *Toxicodendron r.* - **R. semialata** Murr. = *R. javanica* - **R. succedanea** L. = *Toxicodendron succedaneum* - **R. sylvestris** S. & Z. = *Toxicodendron sylvestre* - **R. taitensis** Guill. (PI to Polynesia): lvs. u. med. on Tonga (86); green lvs. burned & smoke inhaled for deafness (Guadalcanal); wd. much u. also - **R. terbinthifolia** Schlecht. & Cham. (swMex to CR): agrillo; u. in gastric cancer (with *Gnaphalium viscosum*); decoc. u. to bathe external tumors; in steam baths for syph. and childbirth (Zapotec Indians) - **R. toxicodendron** L. = *Toxicodendron toxicarium* - **R. trilobata** (Nutt.) Gray (Ore, Calif [US], wMex): skunk b(r)ush (sumac) (SC unpleasant odor); lemita; frt. edible; in beverages; exts. active against nasopharynx carcinoma (gallic acid chief cytotoxic agt. (Utah [US]); lf. tea for coughs, colds - **R. tripartita** Grande (Eg-Mor): lac sumac; inf. of frts. & lvs. u. for gastrointest. ailments - **R. typhina** L. (R. hirta Sudworth) (eNA): staghorn (or velvet) sumac; bucks' horn; lvs. u. similarly to those of *R. copallina* & *R. glabra* as tan ("American sumac"); bk. astr. u. for hemorrhoids; rt. bk. for internal dis. (Menominees); in sore throat (Micmacs); frts. astr. acidulant; in mfr. of vinegar & beverage like lemonade; lvs. & frts. said to be cancer inhibitory; juice u. for warts; dark wd. for cabinet making; important wildlife food (esp. as slow to decay) - **R. undulata** Jacq. (SAf): garra; lf. decoc. u. for postpartum troubles (Hottentots) - **R. venenata** DC = *Toxicodendron vernix*; term "**rhus venenata**" also u. in medical circles with reference to poison ivy (&c.) dermatitis - **R. verniciflua** Stokes (R. vernicifera DC) (Chin, Jap, Ind): urushi (Jap); (Japanese) lacquer or varnish tree; bk. source of a varnish resin (Lacca Japonica; Japanese or Chin lacquer); u. in fine lacquers & varnishes; gives smooth surface resistant to heat and solvents; sds. yield pl. wax similar to Jap. tallow; dr. sap u. for intest. worms, esp. ascarids; young lvs. edible; u. as anthelm.; sds. hemostat.; latex may cause allergic reactions (SC vernix, L. varnish) - **R. vernix** L. = *Toxicodendron v.* - **R. viminalis** Vahl (sAf): karreewood; milk inf. of lvs. u. as enema for abdom. upsets (children); wd. almost indestructible & u. for fence posts & hut building - **R. virens** Lindh. (Tex, Mex): evergreen sumac; live s.; lvs. mixed with tobacco smoked by Am Indians & Mex; good for teeth/gums.

rhus dermatitis: r. poisoning: r. venenata: dermatitis venenata: irritation with erythema & urticaria, sometimes intense following contact with poison ivy (&c.); various treatments incl. tannins, Tinct. $FeCl_3$ (local); antigens (parenteral or oral).

Rhynchelyt(h)rum repens C. E. Hubbard (Gram) (w&sAf): Natal grass; an annual pasture grass now introd. into warm Am; cult. for dry bouquets - **R. roseum** Stapf & Hubb. (Tricholaena rosea Nees) (sAf; natd. tropAm.; cult. sFla [US] & along Gulf coast): Natal grass, umuru (Burundi); chopped lvs. & inflorescences u. in making of plaster appl. strongly with pressure to disarticulated joint (527); meadow grass for grazing & hay.

Rhynchosia minima DC (Leg) (eAs, natd. sFla [US], WI, Mex to trop&subtr SA): pissabed, frijolillo (pequeño); lvs. c. flavones; decoc. u. as eyewash for ophthalmia; strong decoc. u. in VD, abort.; rt. u. in mal. (Hond, Nic); lvs. u. as diur., cardioactive; sds. in pod u. as coffee (roasted) (but bitter & said toxic) - **R. phaseoloides** DC (Dolichos p. Ktze) (tropAm): pulquitas; sds. u. as beads; toxic & supposed to produce a kind of insanity; with fat in ointments for various uses (Mex) - **R. pyramidalis** (Lam) Urban (WI, CA, Mex): pega palo (Domin); virility vine; ojo de cangrejo; pretty sds. u. as beads in making bracelets & collars; with curare-like action (alk.-traces); general ton. action in fatigue (old men) (528); reputed aphrodisiac & androgenic (unproven) (2516); poisonous - **R. reniformis** (Pursh) DC (R. similicifolia [Walt.] Wood) (seUS): dollar weed; small pl. u. as diaph. & for kidney calculi - **R. volubilis** Lour. (Chin, Jap): sd. decoc. u. for lumbar & abdom. pains; headache; in anuria, enteritis; drug mentioned in ancient Chin pharmacopeia (ca. 3000 BC) (529): var. *leiocarpa* Chien.

Rhynchospora cyperoides (Sw.) Mart. (Cyper.) (CA, WI): sds. boiled & u. for fever (Domin Rep).

rhyolite: mineral mined near Rhyolite, Nev (US); u. for cat litter; igneous rock similar to granite (quartz, feldspar, mica).

Rhynchostylis retusa Bl. (Orchid.) (Ind): pl. emoll.

Rhyssopterys (Ryssopterys) **timoriensis** Juss. (Malphigi.) (nAustral, PI, Taiwan+): lvs. of vine appl. for internal pain (PI).

rhytidoma: rhytidome: outer scaling bk., inner periderm produced independently in epidermis or cortex or phloem; occ. in plates; spontaneous or adventitous cork; v. borke.

Rhytidophyllum tomentosum Mart. (Gesneri) (Jam): touch-me-heart; search-my-heart; lvs. brewed as tea (bad tasting); u. by country folk for colds (530).

Ribes L. (Grossulari./Saxifrag.) (temp. regions of world): the currants & gooseberries; frt. small, sour; u. in jellies, preserves, jams, pies, juices, &c.; most frts. are good sources of vit. C; cult. for frt. in US; some spp. are u. as brown dyes - **R. americanum** Mill. (e&cNA): (American) black currant (bush); golden-flowered c.; frt. u. much like that of *R. nigrum* & *R. rubrum*; a good source of vit. C - **R. aureum** Pursh (wNA): golden currant; Missouri (or buffalo) c.; berries u. in pies, jellies, with characteristic flavor; compounded in pemmican with buffalo meat, tallow, &c. (Am Indians) - **R. binominatum** Heller (Ore, Calif [US]): "Siskiyou gooseberry"; a trailing bush; frt. eaten - **R. cereum** Dougl. (Calif-BC): squaw (or wax) currant; berries edible; u. for stomach pains (Hopi, Ariz [US]) (1746) - **R. divaricatum** Dougl. (wBC to cCalif): common (or straggly) gooseberry; bk. or rt. decoc. appl. to eyes for soreness & weakness of sight (Bella Coola Indians, BC) - **R. grossularia** L. = *R. uva-crispa* - **R. hirtellum** Mx. (NA, incl. Alta [Cana]): black currant; wild gooseberry; frt. shiny & bristly; u. for "winter cough" - **R. hudsonianum** J. Richards. (Alas, Minn [US], BC [Cana]): (northern) black currant; u. in tea (unsweetened) for colds (BC ca. 1920); frt. scarcely edible - **R. lacustre** Poir. (Alas-Penn-Calif [US], Labr): (black) swamp currant; s. gooseberry; frt. edible; sts. u. to make pleasant tea u. in colds & diarrh. (cent. BC Indians) - **R. laxiflorum** Pursh (Alas to nCalif [US]; eAs): (trailing black) currant; western b. c.; rt. decoc. u. locally for inflamed eyes (Bella Corda Indians, BC) - **R. lobbii** A. Gray (BC-Calif): gummy gooseberry; frt. edible - **R. nigrum** L. (Euras, natd. NA, Austral; often cult): black currant; lvs. u. as diur. diaphor. esp. in edema, gout, rheum. strangury, diarrh. colics; constit. of many Eu household tea mixts.; frt. small, black, with charact. bed-bug-like odor; pleasantly sour; c. vit. C, A, B factors, rutin; u. in jellies, jams, pies, juice popular in GB; u. for stomach pains; cough; colds (quinsy berry); in mfr. liqueurs, vitamin prepns.; black currant paste of comm. (GB); syr., puree; liqueur de cassis (frt. juice, sugar, alc.); buds. u. as arom.; black currant juice (sugo del Ribes nero) u. for acute bronchitis (It); alternate host in pine blister rust - **R. orientale** Desf. (nInd, Pak, Nepal): berries u. as purg. (Lahul) - **R. reclinatum** L. = *R. uva-crispa* - **R. oxyacanthoides** L. (Cana, nUS: moist woodlands): smooth or wild gooseberry; dark red berries edible when ripe (Alta); in jams & pies - **R. petraeum** Wulf. (Euras, nAf - mts.): rock (red) currant; frt. sour but pleasant tasting; u. to prep. cooling bev., frt. wine, tarts, jellies, preserves - **R. rubrum** L. (R. vulgare Lam.) (Euras widely cult.): (northern) red currant; frt. c. large amounts citric, acetic, succinic acids, smaller amts. oxalic & malic acids; invert sugar (1727); u. acidulous arom. esculent, refrigerant food; in prepn. medl. syrups; in marmalades, juices, berry wines; thirst-quenching drink in fever; fresh juice said virucidal - **R. sanguineum** Pursh (wNA): red flower(ing) currant; bloody c.; fresh berries eaten; ornam. shrub (fls. red, in most other spp. green or cream colored) - **R. sativum** Syme = *R. rubrum* - **R. setosum** Lindl. (Alta-Ont [Cana], Mich-Neb [US]): wild gooseberry; frt. first yellow to green; frt. edible - **R. uva-crispa** L. (R. grossularia L.; R. reclinatum L.) (nEuras, nAf, natd. NA): (English) (or European) gooseberry; cat-berries; berry (tree); frt. c. citric, malic, tartaric, acetic, kinic acids; sugars (6-8%), pectins; proteins; very popular frt. esp. in Britain; in 18th cent. London, there were gardens where people went to partake of gooseberries (1728); frts. green, yellow, red; often of large size (-2.5 cm diam), very delicious; eaten fresh, cooked in pies, sauces, syrups, in wines; widely cult. UK (var. *sativus* DC) - **R. velutinum** Greene (Ore, Calif [US]): plateau gooseberry; frt. edible - **R. viscosissimum** Pursh (nwNA): sticky currant; berries edible.

ribes (It): currant - **r. nero**: black c.

rib-grass: *Plantago lancifolia, P. lanceolata, P. major, P. media.*

riboflavin: vit. B_2 or G; present in all living cells; rich sources incl. yeast, malt, milk, eggs, kidney, liver; now prepd. by isolation fr. fermentation of bacteria or mostly by synth.; u. in pellagra, cheilosis, vit. B_2 deficiency states; in persisting entercolitis as after antibiotic & sulfonamide therapy; skin dis., esp. psoriasis; esp. important to health of skin; u. for scaly lesions around eyes; in X-ray injury of liver, &c.; usage —> yellowish urine; in flour enrichment; is photosensitive (preserve in light-resistant container); daily min requirement 2 mg. adm. orally or IM.

ribonuclease: ribosenuclease: RNAse: enzyme (phosphatase) which depolymerizes ribonucleic acid; occurs in pancreas, liver, &c., & in higher plants.

ribonucleic acid: RNA: a nucleotide polymer important in genetics; u. in anemias, &c.

ribose: pentose, simple sugar; $C_5H_{10}O_5$; occ. in RNA, &c.

ribosome: tiny body of cell contg. RNA or DNA; believed by some to represent what were once bacterial symbionts of the cell.

rib-wort: *Plantago lanceolata, P. cordata, P. media, P. virginica.*

rica-rica (Arg): *Acantholippia deserticola.*

rice: *Oryza sativa* (husks, FO) - **black r.**: wild r. - **r. bran**: 1) husk & outer bran layers of kernel & part of germ of grain 2) same minus husk; r. polishings (**r. b. oil**: r. oil - **r. b. wax**: hard wax sepd. fr. r. b. oil fr. unhusked r.; u. in cosmetics like carnauba wax [2209]) - **brown r.**: r. hulled but not polished; still has bran layer & most of germ - **coated r.**: polished r. which is overlaid with a thin mixt. of corn syrup & talc —> pearly luster - **converted r.**: processed r. - **r. flour**: Oryzae Farina: fine flour obt. by grinding & sifting rice grains; nearly pure starch; readily distinguished by micro examn. - **hungry r.**: *Digitaria exilis* - **r. hulls: r. husks**: outer floral envelopes around r. grain; u. by Chin. as detergent (scrub backs with husks in bag) - **Indian r.: Manchurian water r.**: wild r. - **r. meal**: 1) mixt. of rice bran & polishings 2) ground dehusked grain of r. - **r. oil**: FO obt. by solvent extn. of r. bran (15% yield); very stable culinary oil (salad oil) (cf. peanut oil); blown o. u. in technology, mostly in textile ind. (softening agt. & lubricant for cotton, rayon, wool) (in cooking, advantage in deep frying: does not retain flavors, hence can be reused); the oil-free bran can be u. in cereal & breads - **r. paper**: fine paper prepd. not fr. rice but fr. the pith of *Tetrapanax papyrifera (qv.)* - **polished r.**: white r. - **r. polishings**: Perpolitiones Oryzae (L.): containing inner (*not* outer) bran layers plus outer layers of r. kernel with some of germ (pericarp, spermoderm, embryo, aleurone layer, starchy endosperm in part); u. as source of ext. u. to present natural vit. B_1 (thiamine) in human & cattle foods (mostly u. in PI & seAs); c. 15% FO & source of r. p. oil (Jap Phar.) = r. oil - **processed r.**: white r. parboiled, steamed, & dr. before milling; retains much of original vit. cont. - **puffed r.**: breakfast cereal prepd. by heating rice under pressure then releasing same suddenly whereby steam in cells of rice expands & grain explodes - **red r.** (Miss [US]): *Oryza rufipogon* - **r. root(s)**: fr. *Fritillaria* spp. (Calif [US]) - **Tuscarora r.: water r.**: wild r. - **r. wax**: r. bran w. *(qv.)* - **white r.**: polished r.; prepd. by removal of husks & outer layers of grain; mostly endosperm of grain; since the polishings are so rich in unsatd. fatty acids & vit., &c., this represents a vit. poor & otherwise deficient nutr.; its consumption as the main article of diet in the Orient has often resulted in beri-beri - **wild r.** (**, Canadian**): *Zizania aquatica* & 2 other Z. spp. - **yellow r.**: grains of r. seasoned with saffron, &c.

rich weed: 1) *Cimicifuga racemosa* 2) *Collinsonia canadensis.*

Richardia africana Kunth = *Zantedeschia aethiopica* - **R.** (*Richardsonia*) **brasilensis** Gomez (Rubi) (Br, seUs, tropAf): rt. emet., source of a false ipecac rt., Radix Ipecacuanhae amylaceae (R. I. undulatae) ("undulated ipecacuanha"); "white starchy ipecac"; c. traces of an alk. (emetine?); u. antipyr. (Br); adulter. of ipecac (poaya alba; p. branca; ipecac alba; i. farinosa) - **R. scabra** L. (R. indica) (seUS, Mex, tropAm): undulated (or white) or false ipecac; Mexican clover (Ga [US]); (Florida) pus(s)ley (fr. water parsley); Florida purslane; rt. may c. emetine; u. diaph. emet.; adulter. of true ipecac; hb. u. in salve for sores (folk med., Miss [US]); good food for hogs; pl. a common pernicious ann. weed, growing best in wetter places (*Richardia* Kunth [Ar.] = *Zantedeschia*).

Richardsonia Kunth = *Richardia* L.

Richeria grandis Vahl (Euphorbi) (WI): bois bandé; bk. decoc. u. as aphrod.; timber u. for boards (Domin, Caribs) (2328).

ricin: very toxic protein mixture (albumins) (toxalbumin) fr. *Ricinus communis* (castor bean); represents ca. 3% of sd.; one of most toxic substances known, exceeded only by botulism and tetanus toxins; m.l.d. 0.001 microgram/g of body wt. (mice); 20 mg. IV may be lethal (man); hemagglutinin & proteolytic effects; castor oil is flash-steamed to coagulate proteins (incl. r.), which are then filtered out; has striking anti-cancer properties in combination with monoclonal antibiotic cmpds.; being u. experimentally in skin & breast cancers; u. to treat rheum., arthritis, diabetes, etc. (combines as conjugate with antibodies); in studies of cell surface properties; must be handled with greatest care (1856) - **Ricini Semen** (L.): castor bean - **ricinine**: xall. toxic alk. of prec.; insecticide props. - **ricino** (It): *Ricinus c.*

ricinoleic acid: characteristic unsatd. OH-fatty acid of castor oil & ergot oil, where it occ. as glyceride; u. to prep. dry cleaning soaps, u. contraceptive jellies; in finishing textiles; Na ricinoleate u. for varicose veins.

Ricinus communis L. (monotypic) (Euphorbi) (orig. trop Af, cult. SA, CA, Ind, USSR; natd. near habitations): castor bean tree; c. oil plant; palma Christi; fast-growing shrub to small tree; sds. (Ricinus, L.) c. ricin, ricinine (alk.), source of castor oil, valuable FO (yield 43-60-70%) c. ricinoleic acid, linoleic acid, stearic acid, mostly as glycerides; u. in acute (not chronic) constipation; in diarrh. enteritis (531); u. in hypnotic overdose (1731); toxic (2-6 sds. lethal); exter. u. in skin dis. (itch, tinea; for rectal itching); rubbed daily on skin for skin cancer; in hair oils (1730), pomades; most extensively u. in tech. fields as lubricants (fine machinery, airplanes, esp. jet planes), mfr. soaps, sulfated hydrogenated castor oil (SHCO); in making Turkey Red oil (u. in dye ind.); in textile ind. (fillers), plastics; rust preventative; illuminant; in auto enamels (2215); sd. press cake poisonous, has been u. as rodenticide; fertilizer, fuel; source of lipase; heated lvs. u. as compress for intest. pains & to massage spleen (Domin, Caribs); pl. toxic to livestock; insecticidal props.; pl. stalk in room to repel mosquitoes; cult. ornam. (attractive fls.) (532); in SpAm cult. as shade tree; large scale cultn. in trop. India (largest prodn. of oil, Chin, Br, US) - **R. zanzibariensis** Nicols = *R. communis* var.

rickets: rachitis; severe bone defects due to deficiency of vit. D & UV radn. (daylight), affecting Ca and phosphate metabolism.

Rickettsiae: very small rod-shped pleomorphic microorganisms living intracellularly & producing some common diseases, ex. typhus.

Rieselkaemme (Ge): field comb for plucking fls.; incl. Kamillenriesler, for picking camomile fls.

Rifampicin™: **Rifampin**: semisynth. antibiotic; u. antibact., effective tuberculostatic; u. perorally.

rifamycins: gp. of antibiotics obt. fr. *Streptomyces mediterranei* Margolith & Beretta; u. as antibacterials & antivirals. (Rifampin: semisynth. deriv. of r.) (E).

rifatto (It): woody residue (marc) left after extn. of juice fr. licorice rt.; u. principally in paper mfr.; as fuel; to adult. pd. licorice (1447).

rijs (Du) (pron. rice): rice - **rijsta(a)fel** (Neth, Java): curry; lit. rice table.

rimu: *Dacrydium cupressinum.*

rind (of fruit): cortex; outer part of frt.; tech. represents the epicarp; ex. peel of lemon & other citrus frts.

Rinde (Ge): bark.

Ringelblume (Ge): marigold, *Tagetes & Calendula* spp. - **sumpf R.**: *Caltha palustris.*

Ringer's solution: similar to but not identical with Locke-Ringer's soln. c. NaCl 8.6; KCl 0.30; $CaCl_2$ 0.33; aq. dest. to 1000.; u. parenterally in dehydration, acidosis, alkalosis, &c. - **lactate(d) R. s.**: similar to prec. but superior in some kinds of acidosis, since c. sodium lactate which helps buffer action of blood.

Rinodina orcina (Ach.) Mass. (Variolaria o. Ach.) (Buelli./Pertusari. - Lichenes) (US, Eu): thallus source of a violet dye; u. to dye woolen goods (Fr).

Rinorea castanaefolia (Spreng.) O. Ktze. (Viol.) (SA): lvs. consumed as veget. by Negroes (Br) - **R. elliptica** (Oliv.) O. Ktze (Tanz): rt. decoc. u. for threatened abortion; rt. chewed & juice swallowed for snakebite - **R. floribunda** Merr. (seAs): rts. boiled & appl. to allay fever - **R. ilicifolia** (Oliv.) O. Ktze. (trop eAf): rt. pounded & soaked in water; the inf. is drunk for cough - **R. physiphora** O. Ktze. (Br): lobolobo; bk. u. as astr. & febrif.; mucilag. lvs. cooked & eaten by Negroes (near Rio de Janeiro) (41).

rinse: strained liquid prepn. u. as shampoo to dye hair (ex. henna lf. decoc.).

ris (Russ): **riso** (It-m.): rice.

risdon: *Eucalyptus risdoni* (Tasm).

"rising" (seUS slang): boil.

risotto (It): rice dish flavored with wine in which r. fritters are dipped in FO, then cooked in water, colored with saffron, then meat added.

Ristocetin: Spontin (Abbott); antibiotic complex prod. by *Nocardia lurida* Grundy & al.; obt. fr. soil from Garden of the Gods (Colo [US]); u. in resistant staph. inf., endocarditis (selected cases); reagent to det. platelet aggregation.

ristra de chile (NM [US]): string of red peppers (Capsicum) hung up outside of house.

rita: ritha: *Sapindus laurifolius*; *S. mukorossii.*

Ritchiea reflexa Gilg & Benedict (Capparid.) (trop wAf): pl. u. to treat Guinea worm; rts. for earache; cult. ornam. evergreen vine (ex. Fla [US]).

Rivea corymbosa (L) Hallier fil. = *Turbina c.*

Rivina humilis L. (1) (Phytolacc.) (seUS, WI, Mex to Arg, Pe—> pantrop): rouge plant (hb. or subshrub with red berries); bloodberry; pl. decoc. u. in diarrh. (Virgin Is); decoc. in jaundice (nCaicos); frt. juice u. as red dye (Tonga, Guat).

riz sauvage (Qu): wild rice.

rizine: rice partially converted with superheated steam.

Rizinus (Ge): Ricinus, castor bean (531).

rizoma (Por): rhizome.

roach: 1) butt of cannabis cigarette 2) cockroach.

road oil: heavy residue fr. petroleum oil mixed with soap soln.; u. in oiling dirt & gravel roads.

rob: inspissated juice of elderberries, juniper berries, carrots, etc.

robin: toxic protein of *Robinia pseudoacacia*; phytotoxin - **r. runaway** (NC): *Glechoma hederacea.*

Robinia pseudoacacia L. (Leg) (e&cUS, Mex; natd. BC-Calif; cult. Eu, sAf): black (false) locust (tree); (yellow) locust (tree); (false) acacia; bk.

(twigs/rt.) c. robin (toxalbumin), phasin (toxalbumin), pyrocatechol tannins, resin; u. ton. emet. in hyperacidity (homeop.), gastric ulcer, obstipation; lvs. u. arom., insecticide, antibacterial; fls. (as Acacia Fls.) c. robinin (kaempferol glycoside) u. arom. spice, for prodn. of arom. waters, sherbets (Eu); lvs. u. in making persimmon beer (Ga [US]); source of nerol (u. perfum., org synth.); in vermin control; earlier yellow dye (silk, paper); sds. c. robin (toxic) but cooked & eaten (Am Indians); wd. valuable timber (2668); var. *decaisneana* Carr.; with light rose fls.; ornam. tree.

Robinsonella discolor Rose & Baker (Malv.) (Mex.): u. med.; ornam. shrub.

roble (Sp): oak (tree), typically *Quercus robur* (533) - **r. raspada, madera**: Limousin wood - **r. sabána** (Vera Cruz, Mex): *Tabebuia pentaphylla* - **r. sabañero** (Cu): *Tecoma lepidophylla.*

robust: v. coffee.

rocambole (**, wild**) (& variant spellings): *Allium scorodoprasum.*

Roc(c)ella DC (Roc[c]ell. - Ascolich. - Lichenes): Canary weed; orchal(l); orchilla (Sp); lichen growing close to sea in drooping masses fr. branches of trees & shrubs; source of cudbear (persio), litmus, &c.; prod. in various parts of world; ex. Ind, Baja Calif (Mex), whence formerly exported to Engl in large amts. for mfr. fast dyes (534); c. antibiotics - **R. fuciformis** (L) DC (Indones, SriL, Moz, wMed reg.): Angola (or Mauritius) weed; thallus is one of chief sources of archil (orchill); lichen preps. u. to treat eczema of head & to cleanse hair (Ind) (42) - **R. fucoides** Vain. (R. phycopsis Darb.) (Eu-Atl coasts): herbe de Magador; thallus c. roccellic acid, antibiotic, active against TB m. o.; source of blue dye u. to color British broadcloth; alc. tinct. u. in thermometers - **R. linearis** Wain. (wAf): formerly exp. as "Orcela weeds" & "Canary rock moss" to Eng & Ge for use in dye indy. - **R. montagnei** Bel. (Ind, Moz, Madag): one of sources of cudbear (orseille); c. orcin, lecanoric acid; roccellic acid - **R. portentosa** (Mint.) Darb. (Ch, Pe): exts. inhibit growth of tobacco mosaic virus - **R. tinctoria** DC (Azores, Canary Is, Cape Verde, Iberian & Fr coasts): (dyers') Canary weed; orchilla (weed); orseille (lichen); a source of archil (cudbear) & litmus; u. as drug (Madras); in red, purple, orange dyes (silk, wool), with or without mordant; coloring matter for syrups & elixirs; reagent in lab.; indicator; suggested as alternate to amaranth in pharm. prepns. (NF) (2202); prod. Med reg., Holland; GB, &c.

Rochelle salt(s): sodium potassium tartrate (SC disc. by Seignette, a LaRochelle [Fr] apothecary, 1672).

rocio del sol (Sp): *Drosera rotundifolia.*

rock brake: *Pellaea atropurpurea* - **r. candy**: (crystals): "sugre candie"; sucrose in large xals. prepd. by melting granulated sugar, then letting crystallize; u. for cough, colds, hoarseness, respir. dis.; often added to rye whiskey ("rock & rye") for throat irritation ("phlegm"); culinary uses, confection (known since ca. AD 200) (1861) - **r. c. syrup**: sucrose in aq. soln. left behind after crystn. of r. candy; u. sweetener - **r. crystal**: pure natural crystalline form of SiO_2 - **r. oil**: petroleum - **r. salt**: natural crystall. form of NaCl; u. for cattle - **r. vine**: *Mitchella repens* - **rock**: v. under parsley; rose; salt; tripe.

rockberry: Uva Ursi - **rock-cork**: asbestos.

rocket: 1) *Eruca sativa* 2) (US addicts): cannabis cigarettes 3) *Hesperis matronalis* - **blue r.**: *Aconitum napellus* - **crambling r.**: *Sisymbrium officinale* - **garden r.**: rocket (3) - **r. larkspur**: var. under l. - **r. salad**: r. (l.) - **sweet r.**: *Hesperis matronalis.*

rockfoil: *Saxifraga* - **rockrose family**: Cistaceae - **rockweed**: *Ascophyllum nodosum* - **r. family**: Fucaceae - **rock-wood**: asbestos.

"rocks" (wUS-slang): "crack"; cocaine base u. by addicts.

Rocky Mountain grape (**root**): *Mahonia aquifolium* - **R. M. tea**: herbal veg. laxative c. senna, sass., sambuc. fl., fennel, Hepatica, Tarax., Podoph - **R. M. spotted fever**: American s. f.: acute febrile dis. caused by *Rickettsia rickettsii*, transmitted by ticks to man; may be prevented with **R. M. s. f. vaccine**, prepd. as sterile suspension of microorganisms cultured in embryonated egg.

roco (US: mid-1800s): **roco** (Sp): **rocu** (Sp): annatto.

Rodentia: gnawing mammals mostly of small size; strong chisel-like incisor teeth; ex. rabbit; beaver.

Rodgersia aesculifolia Batal. (Saxifrag.) (Chin): crushed rt. made into plaster to use as antisep. for malignant sores.

Rodriguezia secunda Kunth (Orchid.) (cBr): reputed contraceptive; shavings of rt. u. as hot water inf. to drink & also rub over woman's body.

Rodung (Ge): uprooting; grubbing; digging up; clearing.

roe (**, fish**): egg masses or testes & sperm of fishes; much u. as food.

Roemeria caudata Sw. (Papaver.) (Br): lvs. c. VO; u. med.

Rogeria adenophylla J. Gay (Pedali.) (trop wAf): antipyr., diaph.; for dys.; part of compn. used as remedies for fatigue (Senegal).

Rogge (Ge): rye.
rogue: of plants, a variation or mutant with no known causes.
rohan bark: *Soymida febrifuga.*
Rohdea japonica Roth (Lili.) (eAs): rhiz. u. as heart ton., diur.; mixed with saliva, dressing for sores.
Rohrkassie (Ge): *Cassia fistula* - **Rohrzucker**: cane sugar, sucrose.
roid (Gaelic): *Myrica gale.*
rokambolle (Eg): *Allium sativum* v. *ophioscorodon.*
Rolandra fruticosa (L) Ktze. (monotypic) (Comp) (CA, WI): téte negresse (Domin); lf. inf. u. in amenorrh. (Trin); for sores (Domin); to improve woman's ability to make tapestry (Cuna Indians-Panama); rt. inf. drunk for stomachache, GC (Domin, Caribs) (2328).
Rollinia A. St. Hil (Annon.) (tropAm): shrubs & trees, some with edible frt.; one sp. u. in prep. hair tonics (Co) to aid hair growth - **R. emarginata** Schlecht. (Arg, Br): era hichu; frt. arom. & edible with pleasant flavor - **R. exalbida** Mart. (Br): sds. u. as astr.; c. ca 40% FO (Annonaceae oil) - **R. salicifolia** Schlecht. (Par, Br): bk. u. astr.
roly-poly: *Salsola kali*, as tumbleweed.
romã (Por): pomegranate.
romaine (lettuce): cos l.
Romaji: tranliteration into Roman alphabet letters of Japanese, Okinawan, &c.
Roman meal: coarsely grd. meal of wheat & rye with some flaxseed; u. dietary (536) - **Roman**: v. under chamomile; wormwood.
romero (Sp): rosemary.
Romische Muenze (Ge): spearmint.
romperropa (Cu): *Tecoma lepidophylla.*
Romulea bulbocodium Sebast. & Mauri. (Irid.) (Med reg.): Arab shepherds dig bulbs & eat (Algeria) (1733) - **R. rosea** Eckl. (sAf): half-ripe frts. eaten by children as "frutang"; ingestion of pl. by stock may prod. stomach impaction, sometimes death.
ronce (It Fr f.): blackberry, bramble; *Rubus* spp. - **r. noir** (Fr): **r. sauvage**: shrubby wild blackberries of Eu formerly & still at times lumped together under the collective name of *Rubus fruticosus.*
roob: rob.
roo-bar (Chin): fenugreek.
rooibos tea: fr. *Aspalathus linearis.*
Roosenhout (Du): rosewood; *Amyris balsamifera.*
root: 1) the descending axis of a pl. devoid of lvs. & with branches in no regular order as found on the st.; ex. belladonna rt., althea rt. 2) (layman): any underground part(s) of pl. (except bulb, &c.) - **r. beer**: popular beverage similar to old-fashioned "sarsaparilla" drink; fermented inf. of (variously) sassafras (formerly), sarsaparilla, dandelion, black birch, spruce (spruce beer one variety), wild cherry bk., wintergreen (lvs.), sweet flag, anise, spikenard, cinnamon, cloves, allspice, lemon, vanilla, ginger, nutmeg, anise, coriander, peppermint, yeast, hops, caramel, maple syrup, molasses, honey, sugar, tar, alc., water (1795) - **r. booster**: comb. of rts. of ginger, ginseng, sassafras, sarsaparilla (to aid absorption); supposed aphrod. properties - **r. doctor** (seUS): herb "doctor"; often Negro - **tuberous r.**: a swollen rt. serving for food storage; it shows no anomalous structure; ex. Jalap, Aconite - **rootstock**: rhizome.
roots, starchy (food classifn.): cassava, sweet potato; potato; yam; coco-yam; &c.
rope: 1) strong heavy cord of twisted hemp, manila, or other fibers; also appl. to various kinds of lianas 2) (US slang): Cannabis.
roqueta (Arg): rocket salad.
Rorip(p)a armoracia Hit. = *Armoracia rusticana* - **R. austriaca** Bess. (Nasturtium austriacum Crantz) (Cruc.) (seEu, swAs): lvs. & to a certain extent the rt. c. cardioactive glycoside similar to strophanthin - **R. nasturtium** Beck = *R. nasturtium-aquatium* (L) Hayek = *Nasturtium officinale* - **R. palustris** Bess. (R. islandica Borbas; Radicula palustre Moench) (nHemisphere, incl. Jap; seUS): marsh (water)cress; u. anti-scorbutic; hb. u. ton. stim. diur.; in salads.
Rosa acicularis Lindl. (Ros.) (Siber, Far East): frts. u. in rheum.; astr. - **R. X alba** L. (e&cEu, wAs): white rose; fls. u. for prep. rose oil, incidental to use of pl. for marking boundaries between rose fields in Bulg. (white fls. in contrast with red) - **R. arkansana** Porter (cPlains States, US): prairie rose; wild r.; rt. u. in wash for inflamed eyes (Omaha Indians, eNeb [US]) - **R. banksiae** f. Ait. (Chin): Banks' rose; u. rt. bk. as astr., for tanning - **R. bracteata** J. C. Wendl. (As, esp. sChin; natd. seUS): Macartney rose; Chickasaw r.; introd. ca. 1870 fr. Eng, now a pest in La, Miss, Ala (US); evergreen shrub - 10 m; cult. ornam. - **R. californica** Cham. & Schlecht. (Ore [US] - nBaja Calif [Mex]): ripe frts. eaten raw or stewed (after frost); popular in SpAm as "macuatas" - **R. canina** L. (Euras, nAf): (wild) dog rose; hip r.; wild brier; hep tree; ripe frt. c. frt. acids, VO, large amts. vits. C, A, & P, pectin, coloring matter; ripe frts. (Cynosbati Fructus) u. to prep. frt. juice, marmalades; dog rose pulp (London [Eng] Phar.) (ca. 1800-); lvs. u. as tea substit.; u. astr. anthelm. in confec.; incomplete ripe frt. u. esp. as source of natural vit. C (syrups) (r. hip syr.) (Eu); sds. (as Semen Cynosbati) u. astr., in herbal teas (ingredient); rose fungus, grown on pl. transmitted through sting of *Cynips rosae*; u. astr. (tannin), in

urin. dis.; cult. ornam. (1734) - **R. centifolia** L. (wAs now widely cult., esp. in Eu): cabbage (or pale or great province) rose; fresh fls. (petals) c. 0.2% VO, tannin ("rosatannic acid"), malic acid, quercetin; coloring matter; u. astr. in diarrh. phthisis, arom. astr. in mouth, eye, & gargle waters (for mouth odor); to improve odor & appearance of tea mixts., snuff, &c.; perfume ind.; in prepn. stronger rose water, r. oil (in perfumes); dry petals u. in sachets (light red in color ["Rosa pallida"] in contrast with dark red of *R. gallica*); rose buds (fr. Mor, Fr) - **R. cinnamomea** L. (R. majalis Herrm.) (eEu, USSR): cinnamon rose; ripe & dr. frts. (as of some other *R.* spp.) c. vit. C, tocopherol, quercetin & its glycosides, popular in cEu (Hagebutten) as mild astr. in enteritis, diur. for stones; cough remedy, VO, FO (u. in skin cosmetics) - **R. dahurica** Pall. (Baikal region, Amur basin): frts. u. in rheum. - **R. damascena** Mill. (origin unknown, possibly wAs; AsMinor [535] 2208A]): Damask r.; fresh fls. an important (or chief) source of r. oil (otto of r.); u. arom. perfume, in cosmetics (chief prodn. Fr, Bulg, Turk [c. 1952-] USSR, Morocco) - **R. eglanteria** L. (R. rubiginosa L.) (Euras, natd. NA): eglantine; sweet-briar; lvs. have odor of ripe apples when bruised; fls. u. in colic & diarrh. (Iran); in confectionery; ornam.; fls. symbol of revolution (Fr ca. 1900) - **R. fedtschenkoana** Regel (Pamir, cAs): frts. u. as source of vit. C - **R. gallica** L. (Euras, nAf; introd & natd. US): French r.; red r.; dr. petals (dark red color) c. VO, rosa-tannic acid; coloring principle (indicator props.); u. astr. ton. vehicle, mostly as Mel, FE, or Inf.; fresh petals u. prep. r. oil; u. perfume, cosmet., natural vit. C source; aq. prep. of petals u. in hay fever; (chief prodn. Fr, Bulg) - **R. gymnocarpa** Nutt. (wNA): wood rose; frt. edible; famine food (BC [Cana] Indians) - **R. indica** L. (R. chinensis Jacq.) (Chin, cult. Ind): Bengal rose; (blush) Chin r.; kat gulab (Bengali); frts. appl. to wounds, sprains, injuries, ulcers (Chin) - **R. laevigata** Mx (Chin, Jap, natd. seUS): Cherokee rose; frt. u. in dys., "spermatorrhea"; plant a pest in sUS - **R. mollis** Sm. (Euras): frt. u. as in *R. canina* - **R. moschata** Herrm. (Med reg. to nInd): musk rose; "mosqueta" (Ch); "coral" (Ch); astr. anthelm. refriger.; to "clear the kidneys & bladder"; sts. & lvs. with some antibiotic and antitumor activity; ornam. shrub - **R. multiflora** Thunb. ex Murr. (Jap): Japan (or bramble) rose; rose galls produced by sting of gall wasp; c. much tannin, u. in insomnia (folk med.) in homeop. med. for urinary retention; ornam. shrub - **R. nutkana** Presl. (wCana): Nootka rose; petals & frt. edible; lvs. chewed & appl. to bee sting for relief - **R. pendulina** L. (Eu): wild rose; frt. with high vit. C content (-2.6%); u. to prep. juices, concentrates, conserves - **R. pomifera** Herrm. (Euras): lvs. u. as tea substit. (Deutcher Tee); frts. (rose hips) made into preserves & sauces; beverages; rose wine & r. honey known to ancient Romans; popular in Bavaria - **R. roxburghii** Tratt. f. *normalis* Rehd. & Wils. (China): frts. u. in native med. for dyspepsia - **R. rubiginosa** L. = *R. eglanteria* - **R. rugosa** Thunb. (eAs, nChin, Jap-Kamschatka, Sakhalin): rugosa rose; frts. (haws) large & with high vit. C content; u. as food (Ainu); popular in cultn. for frts. & ornam. (hedges) & as root-stock - **R. sempervirens** L. (Med reg.): evergreen rose; cult for floral VO - **R. taiwanensis** Nak. (eAs): petals u. as astr. - **R. virginiana** P. Mill (e&cNA s to Ala): low pasture rose; cult. ornam.

rosa de Castilla (Mex): *Rosa gallica* - **r. de Cruz** (Mex): - **r. de Jamaica** (WI): *Hibiscus sabdariffa* - **r. de montaña**: *Brownea grandiceps* Jacq. - **r. morada** (Mex): *Tabebuia pallida* - **r. de pasión** (Sp): *Passiflora caerulea*

Rosaceae: Rose Fam.; sometimes split into several families: Malaceae, Amygdalaceae, Pomaceae, &c.; hbs. shrubs, & trees; mostly of temp. regions around the globe; one of largest fam. of pls. (ca. 3,100 spp.); lvs. alternate; fls. regular, much alike & mostly recognizable as to fam. at glance; 5-merous (often double-petaled); stamens numerous; frt. variable; 1) drupe (ex. plum) 2) comp. frt. made up of drupels (ex. blackberry) 3) collection of achenes on dry or succulent receptacle (ex. strawberry) 4) hip (ex. rose) 5) pome (ex. apple) 6) dry follicles (ex. bridal wreath) & others,. ex. cherry; mountain ash; meadow sweet.

rosadillo (Mex): *Swietenia* spp. - **rosal Castel-lano** (Sp): *Rosa gallica* - **rosas mosquetas pétalas** (Sp): musk rose petals.

Roscoea procera Wall. (R. purpurea Sm.) (Zingiber.) (nInd, wHimalay, Nep): fleshy rt. u. as ton., in seminal debility; in veter. med.; c. alks. & steroids.

rose: *Rosa* spp. - **African r.**: *Papaver rhoeas* - **althea r.**: *Althaea rosea* - **attar of r.**: r. oil - **r. bay**: *Nerium oleander* - **r. buds**: fr. **cabbage r.**: *R. centifolia* - **Cherokee r.**: *R. laevigata* - **China r.** (Jam): *R. indica* - **Chinese r.**: *Hibiscus rosa-sinensis* - **r. compound, oil of**: arbitrary name for cocaine alk. in oil u. by some hospitals to mask identity fr. patient (danger!) - **Damask r.**: *R. damascena* - **r. de Provins** (Fr): *R. gallica* & its v. *provincialis* - **dog r.**: *R. canina* - **essence of r.**: r. oil - **r. family**: Rosaceae - **French r.**: *R. gallica* - **r. gentian**: *Sabbatia angularis* -

r. geranium: *Pelargonium odoratissimum* - **green r.**: *R. chinensis* v. *viridiflora*; petals replaced by narrow green lvs. (cAs) - **Guelder r.**: *Viburnum opulus* v. *roseum* (cult. form) (**wild G.r.**: *V. opulus*) - **r. hips**: r. frt. fr. *R. canina*, &c. - **holly r.**: rock r. (1) - **hundred-leaved r.**: cabbage r. - **Japan r.**: *R. rugosa; Camellia japonica* - **r. jar**: potpourri: sachet contg. rose petals, cloves, allspice, mace, nutmeg, cinnamon, orris, &c. - **r. leaves**: r. petals - **Macartney r.**: *R. bracteata* - **r. malloes: r. melloes**: liquid storax; corruption of rasamala (*qv.*) - **r. mallow**: *Hibiscus rosa-siensis* - **Mexicali r.**: *Selaginella* sp. - **moss r.**: *R. centifolia* v. *muscosa* (&c.) - **musk r.**: *R. moschata* - **r. oil: otto of r.**: VO steam-dist. fr. fresh fls. of *R. damascena, R. gallica, R. alba, & R. centifolia* (chief prodn. USSR, Bulg, Fr) - **pale r. (leaves)**: cabbage r. - **r. paprika**: *Capsicum annuum* var.; v. Rosenpaprika - **Persian r.**: damask r. - **r. petals**: fr. French r. - **pink r.**: r. gentian - **provence r.: provins r.: red r. (leaves)**: French r. - **rock r.**: 1) *Cistus* spp. 2) *Helianthemum canadense* - **r. rouge** (Fr): French r. - **sage r.**: 1) *Turnera ulmifolia* 2) *Cistus* spp. (obsol.) - **r. of Sharon**: 1) *Hibiscus syriacus* (usually) 2) *Hypericum calycinum* (**yellow-flowered r. of S.**) 3) *Narcissus tazetta* 4) (of Bible): *Colchicum autumnale* (prob.) 5) *Lilium candidum* - **r. tree**: 1) *Rhododendron* spp. 2) *Rosa* (certain spp.) - **r. tremière (noire)** (Fr): hollyhocks - **r. water, stronger**: the aq. distillate remaining after the separated VO has been removed in distn. of rose petals for r. oil; u. to perfume ointments, cosmetics to prepare by diln. **r. water** (u. to correct aromas; in toiletries, often with glycerol for skin treatment; warts (folk med.) - **r. w. ointment**: cold cream; first prep. by Galen (ca. 180 AD); sometimes now prep. with liquid petrolatum in place of the sweet almond or persic oil, such prods. tending to rancidify less - **wil(d) r.** (Ala [US] slang): *Rosa eglanteria* - **r. willow**: *Cornus sericea*.

rose-apple: *Eugenia jambos* - **roseau à balais** (Fr): *Phragmites maxima* - **rose-bay**: *Rhododendron* - **rosella** (Guat): **roselle** (Fr Eng): *Hibiscus sabdariffa* & its v. *altissima*.

rosemary: 1) *Rosmarinus officinalis* (fls. VO) (also **garden r.**) 2) (Fla) *Santolina chamaecyparissus* - **marsh r.**: *Limonium carolinianum* - **sage r.**: old name for *Cistus* spp. (WI) (43) - **wild r.**: *Croton linearis*.

Rosenpaprika (Ge): Hungarian paprika fr. which stalks, stems, placentae, &c., have been removed & residue then pd.; has fine red color & savory flavor.

roseta (SpAm): 1) roasted or fried corn grains 2) *Tribulus terrestris* 3) spiny frt. of any hb. - **rosewood**: 1) rt. wd. of *Convolvulus scoparius, C. floridus* (VO dist. therefore called **true original r. oil**) 2) Chinese r. 3) *Amyris balsamifera* 4) *Aniba rosaeodora* - **Brazilian r.**: 1) *Physocalymma scaber-rimum* 2) *Machaerium villosum* & other *M.* spp. 3) *Dalbergia nigra* - **Chinese r.**: *Dalbergia latifolia* - **false Brazilian r.**: *Jacaranda* spp. - **Guiana r.**: *Amyris balsamifera* (VO).

rosha (grass) oil: fr. *Cymbopogon martinii* - **roshé oil: rosé o.**: VO of rose geranium.

rosilla (Mex): 1) *Bidens pilosa* var. *radiata* 2) *Commelina pallida*, &c. 3) **r. de puebla**: *Helenium mexicanum*.

rosin: pine resin residual mostly fr. steam distn. of crude gum turpentine ("dip") fr. *Pinus palustris, P. caribaea*, etc. (or fr. Euras spp. as *P. pinaster, P. longifolia*); c. abietic acid, resene, traces of VO; u. mfr. of plasters, ointments, adhesive tapes, soaps, &c.; exter. as dusting pd. on sores; int. in enteritis, catarrhs, &c. (Eu); to mfr. polymers (satd. cyclic terpene hydroperoxides) u. to make synth. rubber polyester resins, &c.; prodn. of r. oil, r. spirit, &c.; tech. in varnishes, sealing wax, paper mfr., insulation, &c. (grades are based on color; the lightest are best) - **black r.: fiddler's r.**: v. black resin - **gum r.**: that prep. fr. crude dip - **liquid r.**: tall oil - **r. oil: r. spirit**: oily liquid obt. by dry distn. of rosin at atmospheric pressure; obt. in 6-8% yields; u. exter. in skin dis.; mostly in technology (solvent for S, &c.; in varnishes, lacquers, &c.; in axle greases, &c.; in mfr. carbon black (u. in printing); turpentine oil substit.; several grades - **r. weed**: *Silphium laciniatum* - **white r.**: light colored var. of r. - **wood r.**: r. obt. by steam distn. of pine stumps, waste lumber, &c. as distinct fr. gum r. - **yellow pine r.**: r.

rosinol: rosin oil.

Rosmarienbaum (Ge) (rosemary tree): probably an imaginary tree; encountered in a dream (539).

rosmarin (Fr, Gr, Roum): **rosmarino** (It): foll.

Rosmarinus officinalis L. (Lab) (Med reg, widely cult.): rosemary; lvs. (Rosmarinus; Folia Rosmarini [L]); u. as spice (with strong antioxidant props.); choleretic, stom. spasmolyt. carm. arom. (esp. for bad mouth odors) stim. nervine; emmen.; in med. teas (Ec); abort. (misuse); sed. in climacteric; for rheum. (Co); in liqueurs (Benedictine, &c.); insecticide; moth repellant; fls. (Flores Rosmarini) u. flavor. arom. diaph. in baths; VO prepd. fr. flg. tops or leafy twigs; u. to perfume pharmaceuticals, in hair lotions (540); moth repellant (sewn into cloth bags & placed in drawer or closet) (chief prodn. rosemary oil: Sp, Fr, Mor) (rosemary exports fr. Valencia); rosemary from Sp and Por very similar; sometimes obtained from wild-growing plants; Eng & Fr production is mostly fr. cultivation.

Rosoil™: single distd. tall oil (541).

rossed: descriptive of bks. with outer corky layers removed, leaving inner & middle bk.; ex. rossed sassafras; wild cherry bk. - **rosser** (logging term): man who trims rough bk. fr. log to get ready for hauling or skidding to sawmill.

Rosskastanie (Ge): horse chestnut.

Rosskuemmel (Ge): name appl. to at least 10 spp. (in 8 genera) of Umbelliferae; incl. *Anthriscus sylvestris*; *Peucedanum palustre*; *Oenanthe aquatica*; *Laser trilobum*, &c.

rot: 1) (Ge): red (also roth) 2) (Sw): root.

rotang: *Calamus r.*

rote Gurke (Ge): *Carex arenaria* - **r. perle** (SpanAm): *Begonia semperflorens*.

rotenoids: gp. of compds. structurally related to rotenone; found in various taxa of *Derris*, *Tephrosia*, &c.

rotenone: complex org. comp. representing active insecticidal constit. of many pl. spp., incl. *Derris elliptica* & other *D.* spp., *Pachyrrhizus* spp., *Tephrosia* spp., etc.; u. insecticide, for moth-proofing, in sprays & pds. (for int. u.) as parasiticide (veter. med.) (in demodectic mange); grubicide; antiprotozoal; u. as dust & wettable powders; compd. has rel. low toxicity for warm blooded animals - **r. roots**: probably Derris or cube roots.

rotes Sandelholz (Ge): red saunders.

Rothmannia capensis Thunb. (sAf) (Rubi): candlewood; juice of heated frt. appl. to wounds & burns; rt. powd. appl. locally to rheum. & leprosy; wd. u. for engraving, tool handles, &c. - **R. globosa** Keay (Af): pd. rt. rubbed into incisions in skin over affected parts in leprosy; frt. edible - **R. whitfieldii** Dandy (Af): juice of unripe frt. appl. to new & old wounds & eczema.

Rotkraut (Ge): red cabbage.

roto: *Scopolia japonica*.

Rottboellia exaltata L. f. (Gram) (tropAs, now pantrop incl. sFla [US], WI): lvs. u. anodyne (PI); diur. (Indochin); fodder (hay); forage; lvs. u. to make mats; hairs of leaf sheaths irritant to skin when pl. handled.

rottenstone (powdered): 1) an earthy mineral representing a friable impure argillaceous (clayey or siliceous) stone derived fr. limestone after the Ca compds. have been leached out by weathering (action of water over a long period of time), hence rendered soft & friable; commonly u. to polish softer metals, &c. as abrasive; *cf.* tripoli 2) any siliceous material, esp. infusorial earth obt. by natural decompn. of chert; u. for polishing, as abrasive; filter, rubber filler, &c.; incl. 3) pumice.

rotten wood: wd. rendered spongy & granular by insect & fungus attack; u. as toilet pd. for babies (eCana Indians).

Rottlera Willd. = *Mallotus*, *Trewia*, &c. - **R. schimperi** Hochst. & Steud. = *Croton macrostachys* - **R. tinctoria** Roxb.: *Mallotus philippinensis* with **rottlerin** as active princ. - **Rottlerae Glandulae**: Kamala.

rottlerin: complex chromane deriv.; chief active princ. & most toxic compd. of *Mallotus philippinensis*.

Rotula aquatica Lour. (1) (Ehreti/Boragin.) (eAs, Br): st. decoc. u. diaph. & diur.; for piles, bladder stone, &c. (Ind).

Rotulae (L.): lozenges; small medicated discs.

Roucheria griffithiana Planch. (Lin) (eAs, EI): akar biji (Mal): hb. u. in prepn. dart poison of Sakais ("Bhoi") (c. brucine).

roucou (Fr Ec, &c.): *Bixa orellana*.

rouge: 1) red ferric oxide; u. as cosmet. coloring for cheeks, for polishing, &c. (fr. rouge, Fr, red) 2) dye fr. safflower - **r. d'assiette** (Fr) (plate red): purified carthamin - **r. végétale** (Fr): carthamin.

rough bent (grass) (US): *Agrostis hiemalis* - **rough**: v. sunflower - **round mandrake: r. ransom**: *Gladiolus communis* - **r. quantities**: large amounts, as in wholesale purchases.

R(o)umelia: Rumelien (Ge): former Turkish territory incl. sBulgaria; Macedonia; Albania; Thrace; one of chief prodn. areas of rose oil, beeswax, sugar beet, sunflower sd., cotton, &c.

round-leaved: v. under dogwood; plantain.

Roupala mollis Pittier (Prote.) (Ve): mapurite; decoc. of leafy branch tips u. in grippe; lf. decoc. for intermittent fevers; bk. decoc. in brandy to regulate menstr. - **R. montana** Aubl. (Trin, nSA): Trinidad roupala; bk. inf. u. as nerve stim. (Trin).

Roupel(l)ina boivinii Pichon = *Strophanthus b.*

Rourea glabra HBK (Connar.) (tropAm): palo de chilillo (Mex); rt. u. to poison coyotes (mata cachorro, Bo); sds. u. for skin dis.; decoc. u. as gargle in catarrh; rts. c. long resistant fibers u. for cordage & formerly in Orient to punish slaves (in place of leather) - **R. mimosoides** Planch. (Indones): rt. decoc. u. for colic - **R. minor** Leenh. (R. erecta Merr.) (pantrop): cangoura (Sal); palo santo (Par); rt. decoc. drunk for fever & postpartum; emet.; toxic to carnivora (u. as dog poison) but not herbivora; sd. ext. toxic but —> inert on drying; wd. u. in airplane mfr. & restricted (*cf.* Lignum Vitae) (542) - **R. obliquaefoliolata** Gilg = *Roureopsis o.* - **R. oblongifolia** Hook. & Arn. =

R. glabra - **R. volubilis** Merr. (R. heterophylla Planch.) (PI): frt. u. to destroy dogs; similar uses to *R. minor*.

Roureopsis obliquifoliolata (Gilg) Schellenb. (Connar.) (tropAf): bk. & rts. u. by natives for elephantiasis; bk. u. as rub in rheum. (Congo).

Rovamycin: spiramycin (Clichy, Fr).

rovere (It): oak.

rovo (It): blackberry (bush), bramble.

rowan tree: *Sorbus* spp.

royal jelly: bee j.: gelée royale (Fr): Propolis (L.); Queen Bee Jelly; subst. secreted fr. head (cephalic or pharyngeal) glands of worker honey b. (*Apis mellifera*) & deposited in cells as nutrient for b. larvae; furnishes sole food for queen b. (essential for devt.), altho small amts. are also fed worker larvae; whitish yellow thick liquid mass, with pH 2.5-4; c. protein, FO, sugars, resins, wax, balsam, choline, cholinesterase, biopterin, abundance of vits.; esp. rich in pantothenic acid; antibiotic (ca. 0.25% activity of penicillin) (1858); first mention in med. by Scribonius Largus (*fl.* AD 40); now again becoming popular in med. (Eu); u. geriatrics (claimed aid to longevity), in sexual impotence, arteriosclerosis, hypercholesterolemia, cosmetics (to improve complexion) (1858), falling of hair, insomnia; for burns (543); occas. u. as arom. & in incenses; made up in creams; excessive claims - **r. palm: r. poinciana (tree)**: *Delonix regia*.

Royena decidua Burch. (Eben.) (sAf): blue bush; rt. decoc. u. purg. & in diarrh., for menstrual diffic.; frt. ("poison peach") sometimes eaten; sd. coffee substit.; cause of animal poisoning - **R. fischeri** Guerke (Tanz): rt. rubbed on teeth to whiten; remedy in stomach dis. of children; pounded rt. made into poultice with butter & appl. to scabies; boiled rts. mixed with white flour & adm. for bilharziasis (Tanz) - **R. lucida** L. (sAf): bow tree; enema in menstrual pain; frt. stones as coffee substit. - **R. villosa** (sAf): vesicant; counterirritant; purg.; frt. inedible (—> *Diospyros*).

Roystonea elata (Bartram) Harper (Palm.) (sFla [US]): Florida royal palm; ornam. tree; some consider conspecific with *R. regia* - **R. oleracea** (Jacq.) Cook (WI esp. Trin, neSA): Caribee (or Brazilian) royal palm; palmiste; tree -35 m; lf. buds eaten as veget. (WI); much planted as ornam. - **R. regia** (Kunth) O. F. Cook (Cu, natd. Fla): (Cuba) royal palm; palma de monte (Cu); (tree -25 m); frts. ("palmiche" [Cu]); u. as food (Am Indians), hog fodder; rich in FO ("Pará butter"); lf. buds eaten as veget. (WI); trunks u. in construction; as wharf piles; many other uses; ornam. tree (Fla).

rozella: *Hibiscus sabdariffa*.

RU 486 (Fr): abortion "pill" (abbreviation); mfd. by RousselUlaf (Fr); abortient; said useful in Cushing's dis., &c.

rubbed: reduced to pd. by rubbing, as in mortar; referred to drugs.

rubber (SC u. to rub out pencil marks on paper): Elastica (L.): caoutchouc; a colloidal elastic solid substance of much complexity; c. chiefly polyisoprene & hydrocarbon polymers with m. wt. ca. 300,000; obt. by processing of the latex of many different pl. spp.; also now obt. by synthetic processes (for listing of rubber plants, refer to Appendix G; for serials, see App. C-1) (544) - **r. adhesive plaster**: r. plaster - **African r.**: obt. fr. *Landolphia florida* & other *L.* spp.; also fr. *Funtumia elastica* - **Assam r.**: India r. - **Bahiá r.**: prod. by *Hancornia speciosa* Gomez (Br) - **biscuit r.**: crude r. coagulated in shallow vessels & removed as a block - **r. bleeders**: workmen on plantation who cut into the rubber tree to secure latex - **Borneo r.**: fr. *Willughbia coriacea*, *Urceola* spp. - **Caucho r.**: fr. *Castilla ulei* & other *C.* spp. of Amazon region - **Ceará r. (plant)**: *Manihot glaziovii* - **Central American r. (tree)**: Panama r. - **chemical r.**: synthetic r. - **crepe r.**: crude r. after passing through corrugated rollers of a machine - **creped r.**: raw sheet unvulcanized r. not chemically treated - **crude r.**: that derived fr. pl. materials - **dry r.**: prod. made fr. natural liquid latex as soon as possible after collecting (coagulating with formaldehyde or acetic acid) - **East Indies r.**: plantation-produced r. fr. seAs - **guayule r.** fr. *Parthenium argentatum* - **hard r.**: rubber vulcanized (usually by heating with 30% sulfur and accelerator [catalyst]) - **India r.: Indian r.**: caoutchouc fr. r. plant (*Ficus indica*) - **I. r. tree**, term also applied to *Hevea guianensis* - **iré r.: Lagos silk r.**: *Funtumia elastica* - **Landolphia r.**: fr. *L.* spp. - **latex r.**: that fr. r. tree, *Hevea* spp. - **Mangabeira r.**: fr. *Hancornia* spp. - **natural r.**: prod. obtained fr. r. trees; essential in many areas (ex. airplane tires); imported - **Panama r.**: *Castilla elastica* - **Pará r.**: fr. *Hevea brasiliensis* & other *H.* spp. (SA) - **r. plant**: *Ficus elastica*; the first pl. u. for rubber mfr.; ornam. pl. - **plantation r.**: rubber fr. Pará r. trees (cult. in Af, Mal, Cey, Burma, &c.) - **r. plaster**: plaster with an India r. base; c. 33% India rubber, Burgundy pitch, olibanum, 33% orris rt., & other filler - **South American r.: r. tree, Pará**: Pará r. - **vulcanized r.**: r. treated with sulfur chemically or by mechanical means (pounding, heating) - **wild r.**: obt. fr. Pará r. trees growing wild in Br.

rubber-vine (, **Cryptostegia**): *Cryptostegia* spp.

rubella: German (or 3 day) measles; acute viral contagious dis. of children & young adults, with fever, rash, &c.; active immunity is acquired by use of **r. virus vaccine (live)** prepd. on duck embryo tissues; often combns. of r. v. with measles and/or mumps vaccines used.

ruberythric acid: primaveroside of alizarin; occ. in *Rubia tinctorum*; hydrolyzes to release alizarin (+ glucose + xylose).

Rubi Fructus (L.): blackberry frt.; u. to prep. syrup & wine - **R. Idaei Fructus**: raspberry frt.

Rubia chinensis Regel & Maack (Rubi.) (Chin): extr. u. to cure lumbago & other bodily pain - **R. cordifolia** L. (eAs tropAf): Indian madder; m. of Bengal; rhiz. & rt. u. in rheum. jaundice, menstr. pain; to heal wounds (Chin); in native med. against leprosy, antispastic; dye (Ind); rt. decoc. for certain urin. dis (PI) - **R. peregrina** L. (R. lucida L) (Med reg., Near East): wild (or Levant) madder; u. for urinary calculi; rhiz. sometimes u. for prepn. madder (red dye); cult. in As Minor - **R. petiolaris** DC (sAf): rhiz. u. in kidney dis., dys.; rheum; pd. rt. u. as dusting powder - **R. tinctoria** L. (R. tinctorum L.) (sEuras, nAf): (common or red) madder; robbia (It); Krapp (Ge); rhiz. c. glycosides, alizarin, rubiadin; purpur-oxanthin; erythrozyme; dyestuff; earlier u. as a red dye; now largely replaced by synth. coal tar dyestuffs but natural dye is superior (with glowing red color); for Dutch cheese, &c.; u. in ancient times as med.; u. still in Eu as diur. ton., aper. in arthritis, astr. for bed sores; in kidney dis., esp. for urinary stones (several other *R.* spp. with similar props.).

Rubiaceae: Madder Fam.; hbs., shrubs, & trees; widely distr. in temp. & trop. zones; lvs. opposite, usually entire; fls. perfect; frt. capsule, berry, nutlet; pls. often rich in alks.; ex. *Cinchona*, *Coffea*, *Cephalanthus*, *Cephaëlis*; ca. 10,400 spp.

Rubigo Ferri (L): red iron rust (FeO).

rubreserine: red compound derived fr. eserine by oxidation (exposure to air).

rubric: purity r. (USP): assay standard requirement of off. compendium monograph.

Rubus L. (Ros.) (mostly in nHemisph): brambles; ca. 300 spp.; shrubs; numerous segregates & apomictic forms of *R. fruticosus* recognized; with biennial canes; incl. blackberries, raspberries, dewberries, loganberries, &c.; specif. (L) blackberry rt. bk., widely u. in diarrh. & dys. (folk med.) - **R. aboriginum** Rydb. (R. almus Bailey) (ne&cTex [US]): Mayes dewberry; frt. edible - **R. allegheniensis** Porter (R. nigrobaccus Bailey) (e&cNA): common (or sow-teat) blackberry; high-bush b.; Allegheny b.; one source of b. bark; u. astr. ton. in diarrh. (1859) - **R. caesius** L. (Euras, esp. eEu, natd. NA): (European) dewberry; lvs. c. 6% tannin, org. acids, inositol; u. astr. in gargles, for menorrhagia, diarrh. skin dis.; diaph.; substit. for tea; frt. edible; rt. u. folk med. - **R. canadensis** L. (e&cNA): low (or smooth) blackberry; dewberry; "bilberry"; rt. bk. u. as astr. in diarrh. chronic dys. & cholera infantum; popular with US physicians (19th cent.) - **R. chamaemorus** L. (Euras; arctic & subarctic areas circumpolar; natd. Cana, neUS): cloudberry; baked-apple berry; plaquebière (Qu [Cana]); mountain raspberry; rt. u. in cough, TB, fever (Maritime Indians); frt. u. in mfr. of a beer, jellies, conserves, &c.; berries u. as edible frt. (Scandia), for scurvy & edema; lvs. in bladder dis. - **R. cochinchinensis** Tratt. (eAs): inf. of dr. frts. u. as aperient at childbirth - **R. coreanus** Miq. (eAs): "mayberry" (548); dr. frts. inf. u. as ton. & to normalize renal functions; ornam. shrub - **R. cuneifolius** Pursh (eUS): sand blackberry; "bilberry"; u. like *R. allegheniensis*; off. in NF XIV - **R. discolor** Weihe & Nees (R. procerus P. J. Muell.) (c&eEu, cult. & occasionally natd. e&cUS-Calif): Himalaya(n) berry; H. blackberry; a very large thorny shrub with widely extended arching canes & bearing very large, sweet succulent frt. for which widely cult. - **R. flagellaris** Willd. (Cana & nUS): (American) dewberry; spherical, edible berry; "blackberry rose" rt. bk. (c. 29% tannin); u. as for *R. cuneifolius* (549) - **R. fruticosus** L.: old collective name now little u. (except as a pharm. title) for the European wild shrubby blackberry or bramble spp.; frt. edible; lvs. rt. bk. u. med., yellow & black dyes; incl. *R. caesius* L., *R. ursinus* Cham. & Schlecht., *R. ulmifolius* Schott, & other *R.* spp. of subgen. *Eubatus*, sect. *Morifera* (545) - **R. glaucus** Benth. (Ec, Co, CR, &c.): mora de castilla; delicious frt. - **R. hawaiiensis** A. Gray (HawIs): Hawaiian raspberry; akala; large frt. u. as food; ashes u. for scaly scalp; heartburn; vomiting - **R. hispidus** L. (seUS): (swamp) dewberry; ground berry; ripe frt. u. in dys., unripe frt. u. for urinary calculi; rt. u. as astr. in intest. irritation - **R. idaeus** L. (R. framboesianus Lam.) (Euras, natd. NA much & widely cult): (European red) raspberry; wild r.; dr. ripe frt. c. citric, malic, & other frt. acids, color, aromatics; u. flavor, vit. source, nutr., in teas for colds (Eu), liqueurs; fresh frt. expressed for r. juice, u. in making r. syr. now popularly u. as flavor for med. as well as in foods; lvs. u astr. in diarrh. & intest. dis.; as uter. tonic & relaxant (mares); in menstr. difficulties (inf.) (Jewish people, New York City, NY [US], &c.) (2203); lvs. c. tannins, gallic & ellagic acids, succinic & lactic acids; vit. C, flavonoids (546); subsp. *sachalinen-*

sis Focke (var. strigosus Maxim.) (R. s. Maxim.): wild red raspberry; American (or common) r. r.; u. as prec.; forma *albus* Fern. with amber-white frt.; "white blackberries" (547) - **R. inedulis** Rolfe (trop eAf): frts. inedible since so sour - **R. laciniatus** Willd. (Eu, natd. eUS): cut-leaved blackberry; frt. edible; comm. trade (Eu) - **R. lasiococcus** Gray (wNA): dwarf bramble; creeping raspberry; frt. edible; cooling & astr. - **R. leucodermis** Dougl. (wNA): black cap; (white stem) raspberry; frt. eaten fresh & dr. (winter use); also sprouts & young lvs. eaten (Am Indians) - **R. loganobaccus** (L.) Bailey (R. ursinus Cham. & Schlecht. var. 1. Bailey) (Pac coast US & Cana esp. Calif, Ore, much cult. [Pac states]): loganberry; fresh frt. esculent, sweet, sour; flavor; in preserves, &c.; canned; dr.; l. juice; syr. (2229); incl. boysenberry = *cv. Boysea*; (some believe loganberry is a hybrid between *R. idaeus* [or its subsp. *sachalinensis*] & *R. ursinus*; others think distinct sp.) - **R. ludwigii** E. & Z. (sAf): (wild) bramble; pd. rt. taken for stomachache (Zulu); rt. u. as remedy for colds & indig. - **R. mirus** Bailey (cult. Fla [US]): marvel dewberry; frt. edible - **R. moluccanus** L. (seAs): rt. decoc. in dys. & urin. trouble; to treat mild stroke; heated lvs. appl. to relieve abdom. pain - **R. nigrobaccus** Bailey = *R. allegheniensis* - **R. niveus** Thunb. (Ind): makhdim; rt. decoc. u. in GC & renal dis. (with *Rubia cordifolia*) (124); frt. eaten by natives - **R. occidentalis** L. (e&cNA): black (cap) raspberry; popular for delicious frt., u. in jellies, pies, tarts, &c.; mild lax. dr. in diarrh. (also rt.); frt. juice u. to mark meats in market - **R. palmatus** Thunb. = *R. coreanus* - **R. parviflorus** Nutt. (wNA): (western) thimbleberry; frt. edible; ornam. shrub - **R. phoenicolasius** Maxim. (eAs, natd. ne&cUS): (Jap) wineberry; wine raspberry; frt. small, edible, acid (somewhat insipid), cherry red (classed as a raspberry); sold in US. markets; common in Jap; cult. ornam. & for frt. - **R. pinnatus** (sAf): (South African) blackberry; rt. decoc. drunk for bronchial dis., in chronic diarrh.; twig tips u. by Europeans in sAf for abdomin. cramps & rt. for convuls., acidity - **R. procerus** P. J. Muell. = *R. discolor* - **R. pubescens** Raf. (R. triflorus Rich.) (Cana, nUS): catherinettes (Qu [Cana]); dwarf raspberry; dewberry; rts. u. in diarrh.; for irregular menstrn. (Maritime Indians) - **R. rigidus** Sm. (sAf): bramble; rt. decoc. for diarrh. dys. & to facilitate childbirth - **R. roribaccus** Rydb. (seCana, neUS: Ky-WVa): Baker (&c.) dewberry; frt. edible - **R. rosaefolius** Smith (pantrop): framboea (Br); frt. popularly eaten (Br, sAf); rt. u. in med. (Br Phar.); cult. ornam. in sUS - **R. spectabilis** Pursh (wNA, as BC): salmon berry; erect shrub; frt. (salmon to scarlet colored) edible raw or mixed with seal oil (Alaska Indians); peeled shoots eaten (Vancouver, BC); favorite food (some wAm Indians); fine brittle spines on st. —> rash on skin - **R. strigosus** Mx. = *R. idaeus* subsp. *sachalinensis* - **R. tokkura** Sieb. = *R. coreanus* Miq. (548) - **R. trivialis** Mx. (seUS): southern dewberry; trailing blackberry; frt. eaten raw & u. in jams, preserves, &c.; u. like *R. hispidus*; rt. chewed for diarrh.; popular home beverage (SC); c. much tannin; inf. of herbage u. for gastric dis. (Seminole Indians) - **R. ulmifolius** L. Schott (incl. var *inermis* Focke) (R. i. Pourvet) (Euras, nAf): comm. exploited frt.; lvs. u. like those of *R. caesius* - **R. ursinus** Cham. & Schlecht. (Or [US] to wMex): blackberry; California dewberry; edible frt.; var *loganobaccus* Bailey = *R. l.* - **R. villosus** Ait. = *R. flagellaris* - **R. vitifolius** Cham & Schlecht. (Calif [US]): Pacific blackberry; fine flavored edible frt. eaten fresh or dr.; rt. inf. u. to check diarrh. (2441).

ruby wood: red saunders.

Ruchgras (Ge): *Anthoxanthum odoratum*.

ruda (Sp): 1) rue 2) *Zanthoxylum elephantiasis* - **r. del campo**: *Thamnosma texanum* - **r. carbana**: *Galega officinalis*.

Rudbeckia hirta L. (Comp.) (eNA, nMex): black-eyed Susan; coneflower; "yellow daisy"; "nigger-head"; rt. u. folk med. (Kan [US]); homeop. med.; honey pl.; ornam. ann. hb.; var. *angustifolia* Perdue (R. divergens T. V. Moore) (SCar-Fla [US]): cold inf. of pl. u. for headache & fever (Seminole Indians) (also related taxa) (Rudbeckia named after Olof Rudbeck, scientist friend of Linnaeus) - **R. laciniata** L. (eUS & Cana): brown-eyed Susan; thimble-weed; cone sunflower; perenn.; decoc. u. as stim. diur. ton. vulner.; formerly in kidney dis. (Eclectics); var. *hortensia* Bailey, cult. double form; Golden-glow - **R. purpurea** L.= *Echinacea p.*

rudd (nEng): red earth occ. in large slabs; u. to mark sheep (cognate of ruddy).

ruderal: 1) (Por): *Portulaca oleracea* 2) (adj.) found growing in waste places.

Rudgea viburnoides Benth. (Palicourea densiflora Mart.) (Rubi) (Br): coto; bk. u. astr. antirheum. depur.; lf. tea u. as "blood purifier."

rue (, common): orig. "herb of rue & grace" (rue = remorse, sorrow since toxic but healing) = *Ruta*. - **early meadow r.**: *Thalictrum dioicum* - **r. family**: Rutaceae - **garden r.**: *Ruta graveolens* (hb.; oil) - **meadow r.**: *Thalictrum flavum* - **wild r.**: *Peganum harmala*.

Ruebe (Ge): rape; turnip; beet - **gelbe R.**: carrot - **rote R.**: garden beet.

Ruebse (Ge): *Brassica campestris* & var. *oleifera*.

Rueckenmark (Ge): spinal medulla - **Rueckenstrang** (Ge): spinal cord.

Ruellia caroliniensis (Walt.) Steud. (subsp. ciliosa [Pursh] R. W. Long [R. c. Pursh]) (Acanth.) (seUS): East Tennessee pinkroot; wild petunia; Kentucky (or false) p.; u. adulter. of Spigelia (true pinkroot); rhiz. & rts. (Ruellia, L.); fls. of *Ruellia axillaris* are not terminal as in Spigelia; plant of R. is hairy; var. *heteromorpha* (Fern.) R. W. Long (dry pineland & woods, Fla [US]): hot decoc. of whole pl. u. for stomach dis. (Seminole Indians) - **R. tuberosa** L. (trop Am, incl. Tex [US], WI, s to SA): "saltapetrico" (Cu); lvs. u. antipyr. in chronic bronchitis (decoc.); tuberous rt. u. as purg. (Mart); in heart dis. (CaymanIs) ("heart bush"); Mayans (sMex) u. in chest colds (550).

Rufochromomycin: antineoplastic subst. fr. *Streptomyces flocculus* Waksman & Henrici & *S. rufochromogenes* Dubost & al.; u. in advanced carcinoma; antibacterial.

rufur pill: rufus pill: Pil. Aloes & Myrrh.

Ruhr (Ge): dysentery; "bloody flux".

Ruhrwurzel (Ge): ipecac rt.

ruibarbo (Sp, Por): rhubarb.

rum: spiritous liquor mfd. fr. sugar cane products (molasses fermented & prod. dist.; u. as sed.; off. in Swiss Ph., Fr Ph. ("rhum") - **bay r.**: v. under *Pimenta acris* (551).

rumarama: *Drimys axillaris*.

Rumex L. (Polygon.) (cosmop., mostly nHemisph): fiddle docks; sorrels; usually coarse hbs., many spp. coarse weeds; thick yellow rhiz. of several spp. u. —> black dyes; many with med. usage; herbage of some spp. eaten as greens - **R. abyssinicus** Jacq. (Ethiopia): Spanish rhubarb dock; mokmoko; rhiz. u. as purg. for tapeworm; cough rem (Tanz); rhiz. u. to give reddish color to butter; cult. as potherb - **R. acetosa** L. (Euras, natd. Cana, neUS): garden/field/common sorrel; hb. c. vit. C, K oxalate, oxalic acid (552); quercitrin; u. for roundworms, spring tonic, in sorrel soup (an herbal soup); diur. stom. in bronchial dis., cough; exter. in skin dis.; (fresh) rt. u. in skin dis., esophagitis; diur. (inf.); frts. c. 5% tannin, vit. C; u. in diarrh., skin dis.; as red dye - **R. acetosella** L. (Euras, natd. NA): sheep sorrel; u. like *R. acetosa*; only in folk med.; lvs. as veget.; eaten fresh (children) (Qu [Cana]); claimed of value in cancer (canker?) (Ala [US] farmers, using juice) - **R. alpinus** L. (Euras-mts.): monk's rhubarb; mountain r.; rhiz. u. as lax.; replacement for rhubarb rt.; adulter. of gentian rt.; veget. & salad pl. - **R. aquaticus** L. (Euras, Arctic NA): water dock; w. sorrel; rhiz. (Radix Britannicae) c. chrysophanic acid, nepodin; u. astr. for herpetic eruptions; diarrh. dys.; young lvs. as veget. - **R. confertus** Willd. (As, Russ): rts. c. emodin, nepodin, neposide, &c.; prepns. hypotensive, bacteriostatic, purg. - **R. conglomeratus** Murr. (Euras, nAf, natd. US): one source of Radix Lapathi (also fr. *R. obtusifolius*); u. lax. astr. in diarrh.; sds. u. against mal. - **R. cordatus** Desf. (sAf): dock; u. as tan. - **R. crispus** L. (Euras, natd. NA, Mex, Ch): yellow/curled/curly/narrow/sour dock; garden patience; rhiz. & rts. c. emodin & nepodin & glycosides; u. in skin dis., astr. in diarrh. ton. lax. spasmolytic; earlier in scurvy; much like rhubarb; frt. u. in checking diarrh. esp. of children; inf. in cramps; prod. Fr - **R. dentatus** L. (Ind): rts. c. anthraquinone derivs.; u. astr. in cutaneous dis. - **R. hydrolapathum** Huds. (Eu): (great) water dock; bloodwort; u. as popular ton. astr. (cEu); rhubarb substit. - **R. hymenosepalus** Torr. (Tex, Calif [US], Mex): canaigre; wild rhubarb; w. pie plant; tanner's (or red) dock; tubers (*cf. Dahlia*) u. astr. in sore throat & sore lips (Pima Indians), in diarrh. (Pawnees); chiefly in prepn. tanning ext. (ca. 30% tannin); chewed for decaying teeth & pyorrhea (Calif & NM Indians); wrongly claimed ton. ("ginseng"); petioles u. like rhubarb for pies (205) - **R. japonicus** Houtt. (eAs): gishi-gishi; u. in Chin med. for swellings, hemostat., antifungal in ringworm; lax. - **R. lanceolatus** Thunb. (sAf): smaller dock; rhiz. mildly lax.; decoc. u. to treat sterility, in tapeworm; hot decoc. of pl. u. to wash wounds & bruises - **R. madarensis** Lowe (tropAf): lf. juice with water u. as eye drops in ophthalmia; lf. antipyr. & diur. (Por) - **R. maritimus** L. (R. wallichii Meissn.) (Euras, natd. neUS): golden dock; u. as lax. - **R. nepalensis** Spreng. (Ind, Nepal, Chin, sAf): lf. decoc. u. in bilharziasis; rt. lax., substit. for rhubarb rt., astr.; antifungal, hemostat. (Chin) - **R. obtusifolius** L. (Euras, nAf; natd. sAf, NA): bitter/European/broad-leaved dock; u. like *R. crispus*; rt. (Radix Lapathi acuti) u. to treat skin dis. & nettle sting; lvs. eaten as potherb ("greens") (Af) (mostly prod Fr) - **R. pallidus** Bigelow (Eu, eCana, neUS): similar uses to other spp. - **R. patientia** L. (seEu, wAs, natd. ne&cUS): monk's rhubarb; patience (dock); herb p.; spinach dock; rt. substit. for rhubarb; u. lax.; hemostat. & antifungal (Chin); cult. for early greens - **R. pulcher** L. (sePacif & US, Mex, WI, natd. Eu): fiddle dock; u. med (Mex Phar.) - **R. salicifolius** Weinm. var. *mexicanus* (Meisn.) C. L. Hitchc. (R. m. Meisn.) (Mex; NM, PacNA, BC-

Calif): lengua de vaca; u. in the native medicine of Mex; willow dock; boiled pl. u. as poultice for swellings; rt. decoc. u. to improve blood, enema for severe constipation; sds. edible - **R. sanguineus** L. (Euras to nIran): bloody (veined) dock; rt. c. anthraquinone derivs.; u. med (Fr Phar.) - **R. scutatus** L. (Euras.): Schild-Ampfer; French sorrel; perenn. hb. (Herba Acetosa romana) u. med. formerly; cult. as veget., good for soups, salads, &c. (tart) - **R. triangulivalvis** Rect. f. (R. mexicanus Meissn.) (Cana, US, Mex): white (or Mex) dock; lengua de vaca (Mex); rts. u. in liver dis., intest. dis., jaundice, irreg. menstr. (Houma Indians, La [US]); lvs. fried in fat & appl. to inflamm. (Mex) - **R. usambarensis** Dam. (trop eAf): lf. juice u. for liver & stom. dis., skin dis.; chewed to relieve thirst - **R. venosus** Pursh (wNA): veiny dock; winged d.; lf. & petiole edible - **R. verticillatus** L. (e&cNA): swamp (or water) dock; u. to cleanse foul ulcers (Am Indians); lf. decoc. formerly as bath to prevent smallpox - **R. vesicarius** L. (nAf, As, esp. Indo-Pak): u. antisept. in dys.; for snake bite & scorpion sting; lvs. boiled & eaten (Bedouin, Saudi Arabia).

Rumohra adiantiformis Ching (Polystichum a. J. Sm.) (Davailli./Pterido.) (sHemisph trop & subtr): rt. (Radix Calagualae) u. in chronic resp. dis., asthma, whooping cough; cult. ornam.

rumu (Peru): cassava.

Rungia pectinata Nees (R. parviflora Nees) (Acanth.) (eAs): decoc. u. to clear excretory systems & in liver dis.; rt. antipyr. (Ind) - **R. repens** Nees (Ind): dr. pd. pl. u. for fevers & cough; vermifuge; appl. to scalp with castor oil for tenia capitis.

runner = stolon.

running snake spores: Lycopodium.

Runzelblume (Ge): **Runzelwurzel**: Aletris.

Ruprechtia hamanii Blake (Polygon.) (nSA): mazamorra (Ve); sds. of tree are eaten.

rupturewort: *Herniaria glabra.*

rusa: *Cymbopogon martinii* (VO).

Rusby, Henry Hurd (1855-1940) (US): pl. explorer; pharmacognosy prof.; dean of pharm. school (CUCP), curator, NY [US] Bot. Gard.; introduced cocillana, miré, caapi, pichi (553).

rusbyine: the alk. found in Cocillana.

Rusci Oleum (L.): rect. birch tar oil.

ruscogenin: sapogenin fr. *Ruscus aculeatus*; u. to treat hemorrhoids (1955-).

Ruscus aculeatus L. (Lili) (sEuras nAf; cult. Fla): butcher's broom; knee (or box) holly; sweet broom root; Mediterranean lily; rt. (as Radix Rusci [or Brusci]) c. saponins, ruscogenin; sucrose; u. to treat venous & capillary insufficiency in varicose veins, thrombophlebitis, hemorrhoids, indolent ulcer of leg (ulcus cruris), venous congestion, in folk med as diur. diaph.; tender shoots edible (*cf.* asparagus); u. aper.; in prostatic tumors; dr. pl. sprays u. for decoration - **R. hypoglossum** L. (sEu AsMin, nAf): double tongue; lvs. formerly u. as "Laurus alexandrina" (c. ruscogenin, sucrose) as emmen. in urin. dis. difficult birth; spermicidal; rt. in bladder pain; sds. coffee substit.; for hemorrhoids; cult. ornam. - **R. hypophyllum** L. (sEuras): broad-leaved butcher's broom; "asparago italiano" (Br, where cult. ornam.); lvs. c. 4.6% crude saponin, with tigogenin; u. like prec. - **R. hyrcanus** G. Woron. (Russ.): saponins of rhiz. & rts. may hinder arteriosclerosis & lower cholesterol level of blood; u. in steroid synthesis.

rush: 1) *Juncus* spp. 2) some *Scirpus* spp. 3) *Equisetum hyemale* - **American r.**: *Juncus* spp. - **r. family: rushes** = Juncaceae - **Guinea r.**: *Cyperus articulatus* - **Japanese r.**: *Juncus effusus* - **leaf r.**: *Cyperus* spp. - **mountain r.** (wUS): *Ephedra trifurca, E. antisyphilitica* - **scouring r.**: r. (2) - **spike r.**: *Eleocharis* spp. - **sweet r.**: Calamus.

rusot (Ind): aq. extr. of *Berberis lycium, B. asiatica*; dark brown sticky mass (*cf.* opium)) rich in berberine; sold in native bazaars.

Russian flies: Cantharis - **R. oil**: purified liquid petrolatum (heavy), purified for internal use - **Russian**: v. under ergot; cantharis; licorice; spearmint.

Russoel (Ge): birch tar oil.

Russula emetica (Fr) S. F. Gray (Russul. - Basidomyc. - Fungi): (Euras, Austral, NA): pungent russula; "sickener"; acrid agaric; thallus (as Agaricus Emeticus) c. muscarine, choline, dihydroergosterol; toxic; emetic; u. in homeop. (Eu) like *Amanita muscaria* - **R. foetens** Fr. (temp zone): fetid (or stinking) russula; frt. bodies sl. poisonous but eaten boiled & salted; in Russ served with sour cream & vodka.

rusts: 1) fungi of Order Uredinales (Class Basidiomycetes) which infect many living pl. spp., incl. a number of drug pls.; frequently seen on lf. drugs, ex. *Malva* lf. 2) pl. disease caused by these organisms or with rust-like symptoms (ex. infection by *Puccinia* spp.) - **rustweed**: *Polygonum aviculare*.

rut of hearts: *Lycoperdon cervinum* (hearts: corruption of harts).

Ruta chalepensis L. (R. bracteosa DC) (Rut.) (seEu, nAf+): fringed rue; Syrian (or Aleppo) r. Bible; u. similarly to *R. graveolens* & *R. montana*; c. VO, chalepensin (furocoumarin), elemol (sesquiterpene), inf. u. by natives of Guat (Arura) & Mex for measles, scarlet fever, headache, cardiac states; in skin dis. (Saudi Arabia); much u. in Mor.; off. in Tunis Phar.; cult. Ch (554) - **R. graveolens** L. (sEu; cult. & natd. in eUS, sAf): (common or garden) rue; bitter herb; countryman's treacle; rue hb. (Ruta [L., It, Rom, Mex]); c. VO, rutin, bergapten, NBP, tannin; inf. u. emmen. in amenorrh.; abort., ton, sed. in earache, carm. in colic; antisep.; in fainting attacks (Guat); epilepsy (Haiti); claimed most bitter pl.; flavor (not recommended); source of VO (prod. Sp Fr); alexiter. (US in colonial days) (72) (ref. 630) - **R. montana** L. (Med reg.): mountain rue; hb. u. as emmen. abort.; like *R. graveolens* but stronger in action.

ruta capraria (It): goat's rue.

rutabaga: *Brassica napobrassica.*

Rutaceae: Rue Fam.; mostly woody pls. (shrubs), many spp. are small trees; most spp. trop. & subtrop., few in temp. zones, of both hemisph.; fls. perfect, usually 3-5 sepals & petals, many stamens; styles fused to appear single; frt. capsule, samara-like, or berry-like, with many chambers (hesperidium); ex. *Citrus* (most important gen. economically), *Coleonema.*

Rutamycin: Oligomycin D.

Rutidea parviflora DC (Rut) (w tropAf): lf. juice instilled into eye for conjunctivitis.

rutin: quercetin rutinoside; rutoside; this flavonol glycoside of quercetin occ. in many pl. spp., incl. buckwheat, *Ruta* (rue), tobacco, forsythia, hydrangea; some eucalyptus spp.; pansy; found commonly in Rut., Polygon., & other fams.; most prepd. fr. buckwheat & *Sophora japonica* (-22%); hydrolyzes to quercetin + rutinose (simple sugar); u. to decrease capillary fragility, in pulmon. hemorrhage, diabetic retinitis, vascular purpura, cataract (*sic*), epistaxis, vulner. (wound healing); rheum.; lymphedema (Austral, NZ, 1984); grippe (1735); said to improve physical endurance (555).

Ryania speciosa Vahl (Flacourti) (tropAm WI [Trin] Guiana, Ve, Br): bois l'agli; all parts (esp. wd.) c. insecticidal princ. ryanodine & cyanodine; ryanine (alk.); u. insecticide for (European) corn borer, soybean caterpiller, &c.; advantages: low toxicity for warm-blooded animals in diluted form; is both a stomach and contact poison for insects; fish poison; var. *tomentosa* Monachino (Patrisia parviflora A DC).

ryanodine: Ryanex (Merck); Ryanicide; insecticidal alk. fr. *Ryania speciosa* Vahl (rts. & sts.); u. in combating apple codling moth, sugarcane borer, European corn borer, &c. (in dusts & sprays).

rye: *Secale cereale*; term also appl. to grain, flour, & whiskey made therefr. - **Bohemian r.**: rye & wheat flour mixed - **ergot of rye**: spurred r. - **Italian r.**: *Lolium multiflorum* - **Robbins r.**: *Polytrichum juniperinum* - **r. smut: spurred r.**: ergot; *Claviceps purpurea* - **wild r.**: different grasses; incl. spp. of *Elymus* (ex. *E. robustus*, *E. virginicus*, *E. canadensis*, &c.).

Ry-Krisp™: rye hard tack.

Rytigynia schumannii Robyns (Rubi) (trop eAf): lf. u. for rheum. & pleurisy - **R. canthioides** Robyns (trop wAf): lf. u. to wash ulcers of feet.

Ryzamine B™: vit. B complex prepd. fr. rice polishings.

Ss

Saat (Ge): seed(s) - **Saatgut** (Ge): seed as a product; young crop - **Saatweizen** (Ge): cult. wheat.

Saba senegalensis Pichon (Apocyn.) (wtropAf): latex fr. st. mixed with water & drunk for pulmon. dis. (even TB); green frts. effective diur.; crushed lvs. place on scalded places; hemostat.

sabadilla (seed): **sabadille** (Fr): cevadilla, *Schoenocaulon officinale* - **s. dusts**: pd. c. 10-20% s. with diluents, such as sulfur, talc, slaked lime; u. pediculicide; insecticide.

Sabal Adans. (Palm.) (warm Am, incl. WI): 40 spp. of low to tall palms; also specif. (L) frt. of *Serenoa repens* (saw palmetto) - **S. causiarum** (Cook) Beccari (Inodes c. Cook; I. glauca Dammer) (PR, Domin Rep, Anegada [Virg Is]): yarey (de Cuba); Puerto Rico hat palm; palma de sombrero; lvs. u. to make hats (Cu), baskets, mats, &c. - **S. etonia** Swingle (c&sFla [US]): scrub palmetto; s. cabbage; apical bud boiled & eaten like cabbage **- S. mauritiiformis** (Karsten) Griseb. & H. Wendland (S. allenii Bailey) (sMex to n coastal SA:): palma de guagara; carat; soso (Cuna Indians, Pan); young lf. sheaths u. for female dis. (Pan) - **S. mexicana** Mart. (S. texana Becc.) (Tex [US], cMex, Guat): cabbage tree; Mexican (or Texas) palmetto; ext. sed. to nervous system, digestive stim.; appetizer; formerly resp. & sexual remedy; for tumors in arm (Mayas); lvs. u. in thatching roofs - **S. minor** Pers. (seUS to Okla): dwarf (or bayou) palmetto; (little) blue stem; rt. juice u. for sore eyes, rt. decoc. for hypertension, kidney dis. (Houma Indians, La [US]); the palm most frequently browsed by cattle; ornam. shrub/tree - **S. palmetto** (Walt.) Lodd. (seUS coastal area [NCar-Fla], w to St. Andrew's Bay, Fla; WI): cabbage palm(etto); c. tree; thatch palm; swamp cabbage; tree -25 m; apical bud edible, u. in salads or boiled & eaten ("cabbage"), this terminal bud steeped in gin & drunk for fish poisonings; frt. ("bayas negras") u. as food (Am Indians) made into syr. & wine; trunks u. for building (SC) & piling; sheaths of young lvs. for brushes; lvs. for thatching; leaf blades ornam. for religious ceremonies; gum sometimes u. as tragacanth substit. (poor); rt. astr., source of tan; cult. ornam. (Calif, Fla Indians) - **S. pumos** Burret (Mex): palma (real); lvs. u. to make hats (sombreros) (Cu) - **S. serrulata** Nutt. = *Serenoa repens* - **S. yapa** Wright (Inodes japa Standl.) (sMex, WI, CA): bayleaf palm; botan (Guat); extr. u. as nerve sed.; appetizer; for respir. dis.

Sabatia (Sabbatia) angularis (L.) Pursh (Gentian.) (e&cNA): rose pink; square-stem(med) rose gentian; American (red) centaury; bitter bloom; dr. hb. intensely bitter; u. bitter ton.; c. "erythrocentaurin" (NBP), gentiopicrin - **S. bartramii** Wilbur (Fla [US]): marsh pink; pl. decoc. drunk hot or cold; claimed of specific value in indigest. - **S. brevifolia** Raf. (S. elliotti Steudel) (seUS-pine savannas): quinine flower; q. herb; hb. c. gentiopicrin; u. antimal., bitter ton; substit. for quinine - **S. campestris** Nutt. (cUS): (American) centaury; u. similarly to other spp.; hb. decoc. boiled till dark green for use as "chill tonic"; long u. by Mex & Indians of Tex (US); pl. flowers several times/yr. (Tex).

Sabdariffa Kostel. = *Hibiscus.*

Sabia gracilis Wernham (Sabi) (eAs-Af, Cameroons): lf. inf. u. as expect. (Chin, Indochin) - **S. japonica** Maxim. (seAs): lvs. & branchlets inf. u. in rheum.; sds. pectoral & antipyr. - **S. olacifolia** Stapf. (seAs): alc. inf. u. for dislocations, sprains, lumbago.

Sabicea vogelii Benth. (Rubi) (Sierra Leone): u. by natives in GC.

sabicu: *Lysiloma latisiliqua.*

sabila (WI, Carib reg): aloe.

Sabin: Sabina (L.): 1) **Sabinae Cacumina**: savin, *Juniperus sabina*, tops (oil) utilized 2) *Sabina* Mill. = *Juniperus* - **s. macho** (Española): *Cyrilla racemiflora* - **sabino (tree)** (Mex): *Taxodium distichum* - **s. macho** (NM [US]): (lit. male savin) *Juniperus communis*; *J. deppeana* - **sabinol**: terpene alc., chief component of VO of Sabina.

sabonete (Br): frt. - **saboneteiro**: tree of *Sapindus saponaria* & other *S.* spp.

sabra (Hebrew): prickly pear frt. (native of Israel); also name for native-born Israelites.

sabugueirinho (do campo) (Br): *Borreria centranthoides* Cham. & Schlecht.

sabugueiro (do Brasil) (Br): *Sambucus australis.*

sac fungi: Ascomycetes (Ascomycotina).

sacaca (Br): *Croton cajucara;* (nBr); u. in sorcery.

sacate: zacate.

saccelli (It): wafers; cachets.

saccharase: inverting enzyme wh. hydrolyzes sugars contg. b-fructose residue, esp. sucrose (—> glucose + fructose); raffinose; gentianose.

Sacchari Faex (L.): theriaca; molasses - **saccharine exudation**: a sugary or sweet-tasting outpouring, as fr. a bk. incision; ex. Manna - **saccharole** (Fr): prep. made with sugar in large excess; some are liquid, ex. honeys; some soft, ex. electuaries or preserves; some dry, ex. saccharures: some solid, ex. pastilles & tablets; syrups are not classed here.

Saccharomyces cerevisiae Meyen (Saccharomycet. - Blastomycetes - Deuteromyc - Fungi): brewer's yeast; Faex (L.); a cultivated sp. with many individual strains; c. nucleic acids; mostly u. in brewing (for alc. productn.) (Cerevisia [L.] beer); & baking to raise dough (CO_2 prod. utilized) (living & humid cells attach. to starchy absorbent base) (2532); u. in prep. medicinal yeast (**S. Siccum**, *qv.*); large doses int. with honey & water for rhus poisoning; antibiotic prepn.; mild lax. - **S. kefir** Beijerinck: chief component of kefir grains (together with mixture of various other yeasts & bacteria); u. to prod. keffir (*qv.*); chief prodn. Bulg, USSR - **S. sake** Yabe: one of yeasts involved in prepn. of sake (Japanese rice wine); acts to saccharify rice starch - **S. Siccum** (L.): dr. yeast; cells dead, hence of no value for brewing or breadmaking (2350); c. 40-50% proteins; 25% carbohydrates; FO, ergosterol; vit. B complex, enzymes, mineral salts (rich in chromium & selenium); u. mild lax.; for furunculosis, exanthemata, acne, & other skin dis.; nutr. in hypovitaminosis B, G, &c.; in pellagra, neuritis; ingredient in multivitamin prepns.; substit. for meat extr.; source of ergosterol (for synth. vit. D); raw material for synth. of Cortisone, &c.; reduces cholesterol level in blood; improves glucose tolerence; binding agt. for tablet masses; pill excipient; no known allergies to yeast; with antibiotic properties; potential food source of future; possible fodder.

Saccharomycetaceae: Yeast Fam.; unicellular organisms which grow out to form buds & simple chains before fragmenting (asexual reproduction); also reproduce by formation of ascospores; generally act on certain sugars to prod. alc. & CO_2; *Saccharomyces*, the genus of industrial importance.

saccharose (Eng, Du): sucrose.

Saccharum L. 1) (Gram.) (OW, trop): tall perenn. hbs.; chief member, the sugar cane 2) (L.) sugar, specif. sucrose 3) (comml.) invert sugar - **S. Acer(i)num** (L.): maple sugar - **S. Album**: sucrose - **S. arundinaceum** Retz. (Gram) (eAs, Indo-Pak to Chin): decoc. u. int. for boils (Mal); rt. for skin dis. (Mal); lf. sheaths source of munj fiber; lf. blades u. for thatch, paper making; fl. stalks u. to make chairs, stools, &c. - **S. bengalense** Retz. (S. ciliare Anders.; S. munja Roxb.; S. sara Roxb.) (Pak, incl. Gilgit; nwInd, Afgh): pen reed grass; munj g.; abundant wild in Punjab, covering large land areas; large grass -4 m; st. furnishes munj fiber; food & med.; young grass fodder for livestock; lvs. u. as thatch & to make paper; st. u. in urin. dis., refrig. aphrod.; smoke of burning rt. u. for burns, scalds - **S. Canadense**: maple sugar - **S. Candidum**: **S. Cand(isat)us** (L.): rock candy (*qv.*) - **S. fuscum** Roxb. (Ind): sts. u. to make writing pens (Ind village schools) - **S. Lactis** (L.): lactose - **S. Malti**: maltose - **S. officinarum** L. (indig. Melanesia, tropAs; much cult. trop &subtr countries [OW, NW]): (common) sugar cane; noble c.; a giant grass (-6 m); a cultigen; a chief source of sucrose (-26% of juice fr. st.), cane syrup; molasses; rum, alcohol (u. in int. combustion engines); one of most efficient pls. for converting sunlight energy into carbohydrate (1800); defoliated st. preferred for syrup prodn.; st. also source of sugar wax; u. like carnauba wax (foot powder prepns., furniture polishes; insulation material, in steroid hormone synthesis) (1799); FO u. in cosmetics; rt. u. as emoll. stim. diur. (folk med. s&eAf); pl. vulner. & in lung dis.; cane chewed as sweetmeat; pl. u. as parent in breeding comm. sugar cane hybrids; fibrous st.

residue (bagasse) after expressing sugar juice u. to make paper, cardboard, plastics, chemicals, pitch, charcoal & as fuel in sugar mills; chief prodn. WI (Cu, Jam, Haiti); PI; Mex; Br; Haw Is; Austral; Java; Ind; Braz is chief sugar cane & cane sugar producer (556) (v. under sugar; sucrose) (serials-Appendix C-1) - **S. Purificatum**: pure sucrose - **S. Rubrum** (L): crude sucrose; raw sugar - **S. Saturni** (L, Sp Phar.): lead acetate - **S. sinense** Roxb. (nInd): Chinese (or India) cane; earlier u. as source of sugar & syr. (ca. 1000 BC); cane chewed - **S. spicatum** sensu Lour. = *Imperata cylindrica* P. Beauv - **S. spontaneum** L. (OW trop; esp. As, natd. PR [WI]); wild cane; "thatch grass"; kash(a) (Sanscr); khus; kans; Kahn grass; lvs. appl. to abscesses & scabies (Indones); rt. decoc. diur. (PI) in vesical calculi, for burning sensation; pl. lax. aphrod.; lf. u. as emmen.; fiber u. to make ropes; lvs. for brooms; st. (culm) u as arrow shafts (PNG); fodder for buffalo; young shoots boiled with rice (Java) & eaten; pl. u. as lax. aphrodis.; in Ind in phthisis, scrapings fr. young shoots in water for inflamed eyes (topically); parent in sugar cane breeding - **S. Tostum** (L): **S. Ustum** (L): caramel - **S. Uvae** (L): **S. Uvarum** (L): grape sugar.

saccharure (Fr): dry saccharolé: medicated sugar; ex. Abstractae (USP, 1880), a type made by extr. pl. drug with 95% alc., evaporating extr. with added lactose; prepn. is 200% strength.

Sac(c)oglottis ceratocarpa Ducke (Houmari.) (Co): smoke of burning bk. inhaled for recurrent cough (TB?) (Makuara Indians).

sachet: sweet bag; u. for perfuming clothes, chests, &c.; c. verbena, rose geranium, &c.

saclac: lactose (contraction of Saccharum Lactis).

sadao (Siamese): *Melia azadirachta.*

saddle leaf: lf. of s. tree - **s. (l.) plant**: *Sarracenia purpurea* - **s. tree**: *Liriodendron tulipifera.*

Sadebaum (Ge): *Juniperus sabina.*

sadleroside: glycoside of 3-epi-corotoxigenin + boivinose; occ. in *Strophanthus boivinii.*

safflower, American: dyer's s.: Egyptian s.: *Carthamus tinctorius* (sd. oil) - **false s.**: *Onopordon* sp. close to *O. sibthorpianum.*

Saff-o, Safflo, Safflor, Saffola™: trade names for safflower oil.

saffraanwortel (Dutch): turmeric.

saffron: *Crocus sativus* - **African s.: Alicante s.: American (false) s.: bastard s.: dyer's s.: Egyptian s.: false s.**: safflower, *Carthamus* - **French s.**: - **Indian s.**: *Curcuma longa* - **meadow s.**: 1) *Colchicum autumnale* (corm & sd.) 2) *Cochlearia officinalis* - **Mexican s.**: 1) safflower 2) *Buddleja marrubiifolia* - **Peruvian s.**: American s. (u. in foods [vermicelli], varnishes to color hides, skins, &c.) - **s. root**: turmeric - **Sicilian s.**: *Crocus longiflorus Rafin* - **Spanish s.: true s.: yellow s.**: s.

Saflor (Ge): safflower.

saframi (Finn): **Safran** (Fr, Ge, Nor, Dan, Roum, Urdu): saffron - **s. de terre:** s. d'Inde (Fr): **Safranwurz** (Ge): turmeric - **safran des fleuristes** (Fr): *Crocus sativus* - **sáfrány** (Hung): saffron - **safrol: safrole**: allyl-methylene-dioxybenzene; const. of many VO, ex. sassafras, nutmeg, &c.; u. perfum., mfr. of heliotropin (also u. perfum.), denaturing soaps; formerly as anod.

Saft (Ge): juice.

sag (Urdu): any pot herb.

Sagapenum: gum-resin fr. *Ferula persica, F. szowitziana.*

sage (leaves): *Salvia officinalis* - **s. of Bethlehem**: *Mentha spicata* - **blue s.** 1) *S. azurea* 2) (US slang): Cannabis - **s. brush**: term appl. to a number of different kinds of pls.; esp. *Artemisia frigida, A. tridentata, A. arbuscula* & other *A.* spp. (**black s. b.** (wUS): 1) *Artemisia nova* 2) *A. arbuscula* (claimed antimalar) 3) *A. tridentata*, &c. - **carpet s.**: *Artemisia frigida* - **s. cheese**: green c.: s. lvs. are added to American c. (or cheddar), giving a special flavor & mottling color - **clammy s.: clary (s.)**: *S. sclarea* - **coast s.**: *Artemisia californica* - **cultivated s.**: *S. officinalis* cult. in gardens (Ge, It, US) - **Cyprus s.**: *Salvia cypria, S. triloba*; similar to Greek s., but with more pleasant aroma, also than of foll. - **Dalmatian s.**: *S. officinalis* cult. in Yugoslav; said best grade - **s. of the desert** (Ariz [US]): saguaro cactus - **s. oil**: VO of *Salvia officinalis* - **domestic s.: garden s.**: *S. officinalis* - **dwarf s. brush**: *Artemisia bigelovii* - **fringed s. b.**: *A. frigida* - **Greek s.**: *S. triloba* - **green s.**: garden s. - **hoary s.**: *Artemisia vulgaris* - **Indian s.** (US): *Eupatorium perfoliatum* - **Jerusalem s.**: *Phlomis* - **meadow s.**: *S. officinalis, S. pratensis* - **mountain s.** (Colo [US]): 1) *Artemisia frigida* 2) *Teucrium scorodonia* 3) term often very loosely appl. to many different pl. spp. with no real significance, ex. **wild m. s.** (Calif [US]): garden s. - **muscatel s.**: sclary s. - **s. orange** [Va (US)]: incorrect form of "Osage orange" - **prairie s.**: hoary s. - **purple s. b.**: a figment of the imagination; the s. b. of Utah [US] is a greenish gray; *v.* purple s.) - **purple s.**: 1) *Atriplex canescens* 2) (Calif) *Salvia leucophylla, S. reflexa, S. dorrii* 3) (Tex [US]) *Leucophyllum texanum* 4) *Artemisia tridentata* var. *vaseyana* & other *A.* spp. (2432) - **purple-topped s.**: *S. horminum* - **red s.**: 1) *Lantana camara* - 2) **scarlet s.**: *S. splendens* - **sclary s.**: *S. sclarea* - **Spanish s.**: *Salvia lavandulifolia,*

S. bicolor, S. triloba - **Syrian s.**: *Salvia triloba* - **Texas s.**: *S. coccinea* - **thistle s.**: *Salvia carduacea* - **true s.**: garden s. - **white-leaved s.**: hoary s. - **wild s.**: 1) (Euras) *S. verbenaca* 2) (wUS) s. brush 3) (US-local) Indian s. 4) red s. 5) (US) *S. columbariae*; *S. lyrata*, &c. 6) *S. officinalis* growing wild on Dalmatian & Adriatic islands 7) *Eupatorium perfoliatum* - **wood s.**: *Teucrium canadense, T. scorodonia.*

sagebrush: *v.* sage brush.

Sageretia theezans Brongn. (Rhamn.) (seAs): decoc. of frt. & branchlets said effective as wash for acne & freckles.

Sagina saginoides (L.) Dalle Torre (Caryophyll.) (Chin): pl. decoc. taken with sugar to stop hemorrhage; pl. pulped & appl. to abscesses.

Sagittaria L. (Alismat.) (mostly Am): arrowheads; swamp potatoes; tubers of some spp. starchy, edible (roasted); eaten by Am Indians & Chin; lvs. also u. med. - **S. cuneata** Sheldon (NA, mostly n&w): wapato; arum-leaved arrowhead; corms eaten as remedy for indigest. (Ojibwa Indians) - **S. intermedia** M. Micheli (WI): "flechera" (Cu); u. in same manner as **S. lancifolia** L. (seUS, Mex, WI s to Peru): (lance-leaved) arrowhead; crushed lvs. poulticed on sciatica, chronic itch, snakebite; u. for shock after alligator bite (Seminole Indians) - **S. latifolia** Willd. (wNA, Mex): wapato (Am Indians); tule potato; (common) arrowhead; duck potato; tuberous rts. boiled or baked like true potato (Am Indian); relished by BC Indians (Cana) (559); eaten by Chin in Calif; rt. decoc. for indigest. (Chippewa); corms pounded & u. as poultice on wounds & sores (Potawatomi) - **S. sagittifolia** L. (S. major Scop.) (Euras): (old world) arrowhead; lvs. & tubers popular med. in eAs (Chin); tuber decoc. u. in eye dis., GC; bruised lvs. appl. to sores, insect & snake bites; & in itching; cult. in aquaria & ponds (Chin, Jap) for edible corms (eaten boiled); corms & lvs. astr. vulnerary; var. *obtusa* Wieg. (S. variabilis Engelm.) (cUS): rts. coll. in spring boiled & eaten (taste like salsify) (560).

sagittal section: vertical section of object in anterio-posterior direction which divides body into 2 lateral halves.

sago: palm starch materials obtained chiefly fr. st. of sago palm, *Metroxylon sagu*; but also fr. taxa of *Cycas* & *Zamia*; *Caryota urens, Phoenix sylvestris, Arenga pinnata, Borassus flabellifer, Mauritia vinifera, Brahea serrulata, Encephalartos caffer*, &c. - **Brazilian s.**: fr. sweet potato - **s. flour**: fr. *Cycas* spp. - **French (indigene) s.: German s.**: prep. fr. wheat or Irish potato starch - **s. japonicus** (L.): **Japanese s.**: fr. *Cycas revoluta* - **pearl s.**: s. in shining hard grains prep. by quick drying - **Portland s.**: fr. Arum - **s. starch: true s.**: obt. fr. medulla of trunk of sago palm, *Metroxylon sagu* - **white s.**: Japanese s.

Sagotia racemosa Baill. (Euphorbi) (nSA, Guianas): arataciú (Br); fragrant rts. scraped in water u. to scent clothing & perfume baths; sold in markets.

sagou blanc (Fr) = tapioca.

sagu (Mex): sago.

saguaro (Mex): *Cereus giganteus.*

Sagus rumphii Perr.: *Metroxylon laeve.*

sahlab: salep.

sahuaro (cactus): saguaro.

saias de mulher (Br): *Datura fastuosa.*

Saigonzimmt (Ge): Saigon cassia.

saim: seam; lard (dial & obs. form).

sainfoin: *Onobrychis viciaefolia.*

Saint(e): St. - **St. Ann's bark**: red cinchona - **St. Andrew's cross**: *Hypericum* (Ascyrum) *hypericoides* - **St. Anthony's rape**: *Ranunculus bulbosus* - **St. Bennet's herb**: *Conium* - **St. Catherine's prunes**: *Prunus domestica* - **St. George's herb**: *Valeriana officinalis* - **St. Germain tea**: a kind of laxative species; c. senna, manna, elder fls. fennel, anise & potassium bitartrate - **St. Ignatia bean** (or **poison nut**): sd. of *Strychnos ignatia* - **St. John Long's Liniment**: acetic turpentine linim. (NF VIII) - **St. John's blood**: *Hypericum perforatum* v. *angustifolium* - (**S. J. bread**: *Ceratonia siliqua* - **S. J. wort** (Eu): S. J. blood - **S. J. w., marsh**: *Elodea* spp.) - **St. Lucie cherry**: *Prunus mahaleb* - **St. Mary's thistle**: *Silybum marianum* - **St. Peter's root: *Succisa pratensis*** - **St. Peter's wort**: *Ascyrum* spp. - **St. Victor's Balsam**: benzoin tr. comp.

Saintpaulia ionantha Wendl. (Gesneri) (tropAf): African violet; usambara v.; popular ornam. house plant. (20 spp. & var.)

saké (Jap): rice wine.

sakar (Russ): sugar.

Sal: 1) (L, Sp, Por): salt 2) pungency, etc. 3) (Ind) teak tree & many other pl. entities 4) (Ind) *Shorea robusta* - **S. Absinthii** (L.): potassium carbonate (lit. salt of wormwood, SC wormwood ash c. mostly this salt) - **S. Aeratus**: $KHCO_3$ - **S. Amarum**: $MgSO_4$ - **S. Ammoniac**: NH_4Cl - **S. Ammoniacum Volatile**: ammonium carbonate - **S. Anglicum**: S. Amarum - **S. Argentii**: $AgNO_3$ - **S. Benjamin: S. Benzoin**: benzoic acid - **S. Carolinum**: Carlsbad salt - **S. dammar**: resin fr. *Shorea robusta* - **S. de Duobus** (early name): K_2SO_4 - **S. de Saturno Purificada** (Sp): lead acetate, purified - **S. Diureticus** (L): potassium acetate - **S. Enixum** (early): $KHSO_4$ -

S. Glauber: Glauber's salts, Na_2SO_4 - **S. Kissingen**: K. salts; usually artificial mixt. of $MgSO_4$, KCl, & NaCl; u. laxative - **S. Lactis**: milk sugar, lactose - **S. Limonis**: salts of lemon; oxalates mixt. u. for cleaning - **S. Marinum**: sea salt, NaCl - **S. Nitri**: KNO_3 - **s. plant**: s. tree - **S. Polychrestum**: S. Glauber - **S. Polychristicum** (Seignette): **S. Polyseignette**: NaK Tartrate - **S. Prunella: S. Prunellae: S. Prunelle**: fused KNO_3, sometimes with sulfur - **S. Saturni**: lead acetate - **S. Sedativum Hombergi**: boric acid - **S. Seignette** (continental Eu): S. Polyseignette - **s. soda**: Na_2CO_3 - **S. Tartar(i)**: S. Absinthii - **s. tree**: *Shorea robusta* - **S. Vichy**: natural or artificial Vichy salts; latter of varying compn.; may c. $MgSO_4$, K_2CO_3, K_2SO_4, NaCl, $NaHCO_3$, sod. phosphate - **S. Volatile**: ammonium carbonate.

sala plant: *Shorea robusta* - **salab** (Per): salep - **salabreda: salabreida**: a grade "D" type of Senegal Acacia, obt. fr. *A. albida*; designated as "gomme friable" (friable gum); also rendered sadra-beida, sadra-brada - **salad tree** (sUS): *Cercis canadensis* - **salap**: salep - **salarata** (NM [US], Sp): baking soda - **Salatdistel** (Ge): *Triptilion spinosum.*

Salacia brachypoda (Miers) Peyritsch (Celastr.) (Br): abacatel; frt. u. as food (561); sds. u. as digestive - **S. fluminensis** Peyr. (Br): sds. (Semen Salaciae) u. in domestic med.; sweet frt. flesh eaten - **S. krausii** Harv. (Af): "castanha mineira"; u. in teniasis & epilepsy (children) - **S. prinoides** DC (Ind, seAs, PI, Austral): rt. decoc. u. in amenorrh. & dysmenorrh.; abort.; formerly in diabetes - **S. senegalensis** DC (S. angustifolia Ell.) (tropAf): rt. bk. pd. u. as vulnerary, strewed on infected wounds; rt. decoc. in mal. & GC; antidiarrheal.

Salamandra salamandra L. (Salamandridae - Urodela - Amphibia) (c&sEu): (European fire) salamander; a tailed lizard-like amphibian; adults are terrestrial; c. samandarine (chief alk.) +; u. in Eu homeop. in epilepsy; to raise b.p.; analeptic; loc. anesth.; poison glands have secretion with activity like that of convulsants picrotoxin and strychnine.

salal (pron sal-al): *Gaultheria shallon.*

Salala (Oman; Dhofar province): ancient distribn. point for frankincense.

salap: salep.

Salat (Ge): salad; lettuce - **S. distel** (Ge) (lit. salad thistle): *Triptilion spinosum.*

Salbe (Ge-f.): salve, ointment - **Salbei (blaetter)** (Ge): sage (lvs.).

salep: source of s. root (tuber) - **s. des indes occidentales** (Fr): *Maranta* arrowroot - **oriental s.: s. root: s. tuber: Salep Tubera** (L): tubers of *Orchis latifolia, O. maculata, O. muscula, O. morio* & other *O.* spp. (also *Eulophia* spp.); *Dactylorhiza* spp. c. mucilage with mannan, polysaccharide made up of D-mannose units (1801); u. med. & food - **royal s.**: bulbs of *Allium macleanii.*

Salicaceae: Willow Fam.; shrubs & trees of temp. to cold regions of both OW & NW; lvs. simple, alternate, deciduous, entire; fls. in catkins; pl. diecious, entire; frt. capsule; genera: *Salix, Populus.*

salicaire (Fr): *Lythrum salicaria.*

salicin: glycoside occurring in willow & poplar bks.; amt. varies as to season, &c.; u. analg. bitter ton. (dogs); reagent; is hydrolyzed by ß-glucosidase (or emulsin) —> saligenin + ß-glucose; uter. relaxant (claim); topical use in arthritis (1985) (1738).

Salicornia brachiata Roxb. (Chenopodi.) (Ind): ashes of pl. considered useful for mange, itch, as emmen. & abort. - **S. europaea** L. (S. herbacea L.) (Euras, Af, natd. NA): (marsh) samphire; chicken claws; pigeonfoot; glasswort; found in alkaline soils, esp. at margins of pools; st. u. as condim. in pickles (pickles with vinegar); ash (barilla) formerly u. in mfr. glass & soap; hb u. as diur. antiscorbutic - **S. fruticosa** L. = *Sarcocornia f.* - **S. virginica** L. (e. & Pacif. US, WI, sMex, OW trop): (salt marshes, sea margin) pickleweed; American (or woody) glasswort; pl. decoc. u. as cold remedy, in pertussis (SC); young pls. eaten raw or cooked.

salicylic acid: occurs as methyl ester in wintergreen, sweet birch, *Polygala* spp., &c.; free & combined in fls. of *Filipendula ulmaria*; most prep. by synth.; u. preserv. reagent; in mfr. salicylates, incl. aspirin; in med. as keratolytic, antirheum.; to remove corns, warts, &c.; in veter. med (E) - **s. a. plaster**: u. as corn remover (in typical "corn plaster") based on keratolytic props.

saligenin: saligenol: salicyl alcohol: o-OH-benzyl alc.; aglycone formed in hydrolysis of salicin (with emulsin as catalyst); u. in acute rheum.

salinigrin: picein: phenolic glucoside found in various pls., incl. *Salix nigra* (hence the name); hydrolyzed —> m-OH-benzaldehyde.

saliva: liquid secrn. of oral region, chief component of sputum (latter may c. pus, blood, &c., besides mucus); c. mucins (polysaccharides); enzymes; acts as solvent, antibacterial, digestant (for stomach, &c.), moistener, lubricant for bolus, coagulant (2501); & other properties; sometimes u. in folk med.

Salix L. (Salic.) (nTemp zone): willows; ca. 450 spp. trees, shrubs, or subshrubs; widely distr.; generally found growing on margins of streams; next to birch, the most cold-hardy gp. of trees; spp. u. as windbreaks, to check soil erosion; bk. source of salicin, yellow dyes; wd. source of willow charcoal; also bk. (Cortex Salicis [L]) of *Salix alba* &c. - **S. acmophylla** Boiss. (Pak-Baluchistan): bada; bk. decoc. u. as antipyr. (Ind med.) - **S. alba** L. (Euras, natd. e&cNA): white (or common) willow; bk. c. tannin (4%), salicin (9.5%); lvs. of this & other spp. c. salicin; bk. u. simple bitter, ton. astr. antipyr. antiperiod. antirheum. in "ague" (= chills of mal.); most popular willow bk. in med. (USP 1820-80) (that of *S. fragilis* also u. a good bit); lvs. u. as fumitory with or without tobacco; as tea adulter.; as antipyr.; inner bk. u. as emergency food; bk. in tanning; wd. to make charcoal; twigs u. to mfr. baskets; wd. (hard) to make wooden shoes (Nether); honey pl.; ornam. shade shrub/tree; var. *caerulea* Smith (Eng): bat willow; SC white wd. u. to make cricket bats - **S. amygdaloides** Anderss. (eNA): peach-leaf willow; inf. of crushed lvs. u. inter. & exter. in mal., &c. (Mex); quinine substit.; wd. a fair fuel - **S. babylonica** L. (Af, As, natd. NA): weeping willow; catkin & young twigs decoc. u. for "slow" fever; appl. exter. to sores; lf. inf. (willow tea) (bush tea) for rheum. (sAf); bk. of twigs & rt. u. int. for rheum. fever, GC (Chin); tree widely cult. as ornam. (esp. in cemeteries, symbol of mourning) - **S. berlandieriana** A DC (sTex [US], neMex): Mexican ash; Texas fresno; lvs. crushed for inf. u. for mal. & other fevers (Mex) - **S. bonplandiana** HBK. (NM, Ariz [US], Mex to Guat): bk. u. in med. as antipyr. in fevers - **S. caprea** L. (Euras, natd. eUS): goat willow; pussy w.; round-leaved w.; sallow; wd. light, u. for vineyard supports, mfr. charcoal, fl. VO (attar) u. in perfumes (Kashmir); cult. ornam. - **S. caroliniana** Michx. (eNA w to Okla [US]; Cu): (coastal plain) willow; inner bk. decoc. u. for severe colds, fever, diarrh. (2451) - **S. cinerea** L. (Euras, nAf, natd. eCana, eUS): gray (florists') willow; pussy w.; bk. c. 0.1-1% salicin; 1-2% triandrin; 1-2% picein; tannin; vimalin; pipecolinic acid; lvs. c. salicin, &c., u. in fever, rheum. - **S. cordata** Mx. (NA & Arctic reg.): fresh bk. u. for bruises & skin eruptions (Thompson Indians); to stim. appetite; in colds (1739) - **S. daphnoides** Vill. (nEuras): young shoots & catkins eaten fresh or in seal oil (Eskimos); cambium eaten; twigs u. as floss; u. in rheum.; ornam. shrub - **S. discolor** Muhl. (e&cNA): (large) pussy willow (SC woolly appearance of floral gps.); tree 4-6 m; hardy; bk. ("black willow bark") c. salinigrin (not salicin); u. astr. bitter ton.; antispas. deterg. antiperiod; chewing tobacco substit.; buds as "antiluetic"; fl. gps. u. as ornam.; bk. inf. u. to stop bleeding (Potawatomi Indians); ornam. shrub or small tree - **S. exigua** Nutt. (NA): jarita (NM); sandbar willow; lvs. chewed to harden gums, in pyorrhea (Am Indians); lvs. & bk. u. antipyr. - **S. fragilis** L. (Euras, natd. e&cNA): crack (or brittle) willow; bk. u. as main source of salicin (8%), fragilin; u. like *S. alba*; bitter ton. in coughs, grip, bladder dis.; wd. u. to make charcoal for gunpd.; tree ornam. & for shade - **S. humilis** Marsh. (e&cNA): small pussy (or upland) willow; rt. tea u. as enema in "flux" (Meskwaki Indians) - **S. lasiandra** Benth. (wNA): red willow; inner bk. u. as source of a flour (Am Indians); suspension of charred wd. taken for diarrhea (Bella Coola Indians, BC) - **S. lasiolepis** Benth. (wUS, wMex): arroyo willow; white w.; ahuejote; bk. decoc. antipyr. u. before quinine available; diaph. (rt. bk. preferably); exter. for itch; lf. inf. u. to check diarrh.; inner bk. u. as substit. for chewing tobacco (2441) - **S. lucida** Muhl. (eCana & neUS w to SDak): shining willow; squaw bush; bk. u. in hemorrhaging, asthma (Maritime Indians) (1939) - **S. mucronata** Thunb. (S. capensis Thunb.) (sAf): native/river/wild willow; lf. decoc. u. for muscular rheum.; in fever; branch inf. u. as appetizer; bk. antipyr. - **S. nigra** Marsh. (e&cNA): black willow; bk. rich in tannins; good source of salicin (1882), salinigrin; picein (piceoside); u. for fevers; "sexual ton. & sed."; in gout, asthma; also u. much like *S. discolor*; wd. soft, u. for excelsior, artificial limbs; tree planted as windbreak; "buds" (blooms) - **S. pentandra** L. (Eu, natd. Cana + US): bay (leaved) willow; bk. u. medicinally like other *S.* spp.; ornam. for public grounds - **S. purpurea** L. (Euras, nAf, natd. ne&cNA): (purple) osier; basket willow; p. w.; bk. c. populin, salicin (perhaps richest, with 4-7%) (2005); bk. u. in mfr. salicin; bitter, antipyr.; twigs u. to make baskets (bk. for salicin mfr. by-product); introd. into US for basket making; shrub u. to control erosion - **S. X rubens** Schrank (Euras, occ. in NA): Bedford willow; hybrid pl.; bk. u. astr. ton. antipyr. - **S. rubra** Huds. (hybrid of *S. purpurea* X *S viminalis*) (Eu, cult. in NA, &c.); branches u. to make baskets; bk. u. med; cult. for centuries - **S. sitchensis** Sanson (Pac NA, also neAs): Sitka (or silky) willow; bk. decoc. u. ton.; supple twigs u. to make baskets, &c. - **S. subserrata** Willd. (S. safsaf Forsk.; S. aegyptiaca L.) (n&e tropAf): Jordan willow; African w.; bk. c. 0.6% salicin; lf. u. lax.; twigs u. for baskets; lvs. for black dye - **S. tetrasperma** Roxb. (Ind, Pak): Indian willow (tree); safsaf rumy (Arab); bk. c. 0.1% salicin (91); u. med. as antipyr.; wd. u. for fancy & cabinet work, pencils; to make charcoal for gunpowder; lvs. u. in bladder stone - **S. triandra** L. (Euras to Jap): almond-leaved willow (Mandelweidl [Ge]); French w.; bk. c. salicin (<0.1%), salireposide, triandrin; rutin; twigs u. to make baskets - **S. viminalis** L. (Euras, natd. eCana, neUS+): (common) osier; o. willow; basket w.; bk. c. 8-12% tannin, salicin (<0.1%), salicortin; lf. ext. antimicrobial (562); shrub cult. for use in making baskets, chairs, &c. (Eng) - **S. woodii** Seem. (sAf): woods willow; bk. u. in prepns. for burns & for urin. trouble in male (Lesotho).

sallet (seUS): salad (greens), *Phytolacca.*

sally (bushes) (trees) (Ire): *Salix* spp.

Salmea scandens (L.) DC (Comp.) (sMex to Par; WI): baigua; bejuco de chile; vine u. in nephritis, when rt. chewed, sensation is numbed for tongue & lips, u. in toothache; fish poison (harmless to humans) (Totanac Indians, sMex).

Salmiakgeist (Ge): ammonia spirits.

salmine: protamine in sperm of salmon; u. like protamines, antidote to heparin, &c.

Salmo (Salmonidae - Pisces): the salmon; important bony fishes of Atl & Pac Oc; valuable food fishes; source of protamine (also genus *Oncorhynchus*); source of salmon oil (FO by-prod. of s. canning industry; drying oil u. paints, soaps); rich in vit. D; has been u. in infantile rickets (2006); Alaskan Chinook salmon oil u. to lower cholesterol in blood & b. p. (omega) - **salmon berry**: *Rubus spectabilis* (SC frt. color of salmon flesh).

Salmonella enteritidis Castellani & Chalmers (**S. paratyphi**) (serotype A) Castellani & Chalmers (Enterobacteriaceae): strains A & B cause of a mild paratyphoid fever in man; strain C causes enteric fever (As, Af) - **S. e. schottmuelleri** (Winslow & al.) Bergey & al. (S. paratyphi serotype B): organism which causes paratyphoid fever (some strains affect lower animals) (nUS, nEu) - **S. e. hirschfeldii** (S. paratyphi serotype C); causes paratyphoid fever in humans (As, Af, neEu) (enteric fever); all these organisms u. in prepn. typhoid & parathyphoid vaccines - **S. typhi** Warren & Scott: the bacterial organism (bacillus) causing typhoid fever.

saloop: sassafras tea (usually as beverage) with sugar & milk; sold in coffee houses, Fleet St., London (in the past!); claimed useful for skin dis., rheum., dypsomania; sometimes prepd. fr. cuckoo fl. (*Cardamine* sp.), salep, &c.

salopian: 1) saloop 2) Shropshire (Eng).

Salpichroa diffusa Miers (Solan.) (Pe, Ec, Co): pepinillo; frts. u. in helados (ice cream) - **S. origanifolia** (Lam.) Thell. (S. rhomboidea Miers) (Arg, Ur, natd. Sp, Fr It): cock's eggs; uvilla (del campo) (Ur); c. alk. tannins; berry edible —> symptoms of intoxication (loquacity, &c.); frt. u. in preserves; cult. US as ground cover, as bee pl.; vicious weed (Fla [US]).

salsa: 1) (Br): *Petroselinum sativum*, parsley 2) (Sp) hot sauce (tomatoes, jalapeño; onion; coriander, red pepper, &c.).

salsapariglia (It): **salsaparrilla** (Por, Br): Salseparrilha: sarsaparilla.

salsepareille (Fr): 1) sarsaparilla 2) (Qu [Cana]): *Aralia nudicaulis* - **s. de Virginie**: bamboo brier, *Smilax rotundifolia.*

salsifi: salsify: *Tragopogon porrifolius* - **black s.**: **Spanish s.:** *Scorzonera hispanica.*

Salsola aphylla L. f. (Chenopodi.) (sAf): pl. ash u. in soap making; pl. useful as fodder - **S. foetida** Del. (Pak, wInd): pl. u. as vermifuge; ash appl. to itch; pl. camel food - **S. kali** L. (Euras, now widely distr. in nTemp Zone): (prickly) (common) saltwort; Russian thistle; barilla plant; kali; glasswort (SC u. in glass making formerly); pl. u. as anthelm. (Indones); emmen. stim. diur. cath.; young lvs. sometimes eaten; barilla (pl. ash) formerly u. as source of salt & proposed as source of Na oxalate; very polymorphic (varies greatly with age); stock poison; var *tenuifolia* G.F.W. Meyer (S. pestifer A. Nels.) (cAs, wNA): Russian thistle; (Russian) tumble weed; bad weed over much of US but young plants foraged in very dry years in semiarid areas; rt. eaten as starvation food during "the great depression" (ca. 1930) (Cana, US); good soil holder; st. base weak so that pl. separates in windy weather & rolls across plains, hence "tumbleweed," the chief one of several; important cause of hay fever - **S. richteri** Kar. (USSR): hb. c. salsolidine, salsoline (alk.) (563); u. as agt. in hypertension - **S. soda** L. (sEu n to Hung): pl. formerly u. to obtain soda - **S. tamariscifolia** Lag. (sSp plains): source of Spanish wormseed; u. anthelm.

salt: prod. of acid & base (usu. metal); specif. table s. - **common s.:** occ. widely in nature, esp. in sea water; obt. by mining rock salt; or by evapn.; brine obt. fr. the sea, salt lakes, salt wells; forms cubic white (colorless) crystals; u. as preservative in foods (ex. fish); curing of hides; flavoring foods (2714); mfr. Cl_2, Na, & compds., &c. - **sea s.**: produced by evapn. of sea water; c. NaCl (78%), Mg, Ca, S, K, Fe, F, Cu, Mo, P, I, Mn, Zn, Co, & traces of other elements; some claim advantages over common table salt, such as formation of fewer dental caries (2412) - **table s.**: common s., NaCl, sodium chloride.

salt bush: *Atriplex* spp., also appl. to at least 4 other genera - **s. rock moss**: *Chondrus* - **rock s.**: native NaCl fr. mined deposits; occurs in form of large crystals - **s. Seignette**: Rochelle salt - **s. (of) wormwood**: K_2CO_3.

saltfish: cod (&c.) salted but *not* dr. (*cf.* klipfish).

"salts": usually **Epsom salts** ($MgSO_4$) (*toxic!*) - **s. of lemon: s. of sorrel**: potassium acid oxalate - **smelling s.:** English s.; perfumed ammonium carbonate, with such aromatica as violet, anise, rose, bergamot oils u. to revive persons fr. a faint (formerly popular) - **s. of tartar**: potassium bitartrate; earlier K_2CO_3.

saltwort: *Batis* & *Glaux* spp. - **prickly s.**: *Salsola kali.*

salutaridine: morphinan-dienone alk. (precursor of morphine); occ. in *Croton* & *Papaver* spp., incl. *Croton balsamifera* Jacq., which also c. norsinoacutine.

salva brava (Por): *Salvia verbenaca* - **salvador** (SpAm): *Physalis angulata.*

Salvadora oleoides Dcne. (Salvador.) (Pak, Afgh, arid areas): pil (Punjabi); wanh; sds. rich in FO u. as stim. in painful rheum. & after childbirth (564); fresh lvs. & bk. with odor of cress; lvs. & bk. c. trimethylamine, alk.; frt. aphrod.; lvs. purg. - **S. persica** L. (tropAs, Af): toothbrush tree; salt bush; rt. bk. acrid, vesicant, u. in ankylostomiasis; lvs. eaten, mustard flavor; u. diur.; sds. c. 43% FO, kikuel oil; smaller sts. & twig u. as toothbrush (Af); said by some to be the mustard tree of the Bible.

Salvia aethiopis L. (Lab) (eMed reg., nAf, Ukraine [USSR], nIran; natd. sOre, Wash, nCalif [US]): African sage; Mediterranean s.: rt. u. for hemoptysis & pulmon. dis.; hb. u. med. (Ukraine); serious weed in wUS (wOre, nCalif) - **S. africana** L. (S. africana-coerulea L.) (sAf): wild African (or purple) sage; hb. c. VO u. like *S. officinalis*, as in coughs, colds, & chest dis.; pertussis; twig inf. (with Epsom salts, lemon) u. in abdom. dis., as diarrh. colic, flatulence, indigest. - **S. apiana** Jeps. = *S. polystachya* - **S. argentea** L. (Med reg.): silver sage; white s.; u. as tea; cult. ornam - **S. ballotaeflora** Benth. (Tex [US]): shrubby blue sage; u. exter. & inter. as tonic, arom, &c. - **S. bridgesii** Britton (Pe, Bo): "matico"; highly esteemed by natives for med. values - **S. carduacea** Benth. (Calif, Ariz [US], Baja Calif [Mex]): thistle sage; sds. ground after roasting & made into flour; u. as mush; also in cold drinks (Calif Indians) - **S. chamdaeagnea** Berg (sAf): aromatic sage; lf. inf. u. in cough, pertussis, colds, bronchitis, in diarrh. (Europeans in sAf) - **S. chia** Fernald (Mex): chia; mucilag. sds. u. to make refreshing drink; sds. c. unsatd. FO, u. like linseed oil; in foods, paints (ancient Mex) - **S. coccinea** Burch. (seUS, Mex, tropAm, natd. sAf): Texas (or South American) sage; fls. deep scarlet (scarlet s.); many variations known; c. salvianin (monardein) (anthocyanin, red dye); cult. ornam. - **S. columbariae** Benth. (swUS): chia (this name appl. to some other *S.* spp.); Calif c.; wild sage; sds. u. to make flour, important food of wIndians (Paiute [2471]); sds. rich in FO (drying), mucil. u. to make poultices, cooling mucilag. beverage (Am Indians); in VD - **S. cypria** Ung. & Kotschy = *S. fruticosa* - **S. divinorum** Epling & Játiva-M. (sMex): yerba (hojas) de María (Pastora); believed cultigen; u. as hallucinogenic agt. by Mex Indians (Mazatecs) (chewed or as inf.) (for divining, as in diagnosis of disease) - **S. dorrii** (Kell.) Abrams subsp. **carnosa** Abrams (S. carnosa Dougl.) (wUS-Rocky Mt area): purple sage; hb. c. VO, carnosol (with bactericidal activity); sds. parched, grd., & baked into bread; hb. u. in colds, headache, gastralgia, in eye washes (Am Indians) (var. *incana* [A. Gray] J. L. Strachan) - **S. fruticosa** Miller (Sicily, Cyprus to Syr.): Cyprus sage; chahomilia (Cyprus); VO like that of *S. lavandulifolia*; c. 75% eucalyptol; u. med. much like *S. officinalis*; for sophistication of same; u. in beverages (locally) - **S. fulgens** Cav. (Mex): Mexican (red) sage; cardinal s.; cult. as ornam. - **S. glutinosa** L. (Euras) yellow sage; y. sclary; u. med. & cult. by the Ancients (Elelisphakos); young plants u. in omelets - **S. hispanica** L. (Am trop; CA, Mex, SA): (wild) chia; chia blanco; frts. u. much as for *S. chia* & *S. urticifolia*; sold in open pkgs.; soaked in water to form mucilagin. beverage "chia"; also u. in eye dis.; in liver dis. (Salv); sd. chief source of chia sd. oil (FO); v. chia - **S. horminum** L. = *S. viridis* - **S. hyptoides** Mart. & Gall (Latin Am): hierba de reuma; hb. u. as poultice for tumors, &c. (Salv) - **S. lanata** Roxb. (Pak): good substit. for *S. moorcroftiana* - **S. lavandulifolia** Vahl (S. officinalis L. subsp. l. Gams) (Sp, wFr, esp. Andalusia): Spanish sage; hb. u. mostly as flavor, to some extent in med.; adulter. or replacement for *S. officinalis*; source of VO, u. cosmet. adulter. of spike oil - **S. lavanduloides** HBK. (sMex, CA): "verbena"; herb inf. u. in mal. (Guat) - **S. leucophylla** Greene (Calif: Orange Co-Kern Co [US]): wild California sage; purple s.; VO c. camphor; has been marketed; rt. ingred. of certain med. prepns. (UP, Ind); cult ornam. - **S. lyrata** L. (e&cUS): cancer weed; (lyre-leaved) sage; wild s.; lvs. & juice u. for warts, cankers, &c.; rts. made into salve for sores (Catawbas) - **S. macrophylla** Benth. (Ec): inf. u. for kidneys & liver - **S. mellifera** Greene (Calif

[US]): black (or blue) sage; hb. c. VO c. 40% camphor (for prodn. c.?); fls. furnish superior honey; similarly *S. somonensis* Greene; cult. - **S. miltiorrhiza** Bunge (Chin): rt. u. as circ. stim., in rheum. for inflamm. of bladder & kidneys - **S. moorcroftiana** Wall. (Ind): rts. for colds & coughs; sds. in hemorrhoids, dys. & colic; lvs. u. for guinea worm & itch (poultice) - **S. nemorosa** L (As, sSiber, AsMin, seEu): Hainsalbei (Ge) (wood sage); lvs. yield 0.04% VO u. in USSR (-1926-); u. like *S. officinalis* & adulter thereof - **S. occidentalis** Sw. (Fla [US], Mex, WI, CA): banderilla (Cu); ina osi (Cuna Indians, Pan); mozote de gallina (Salv); "catnip" (Bah); "wild mint" (Trin); decoc./ inf. of hb./lvs. u. in backache, gastric dis. dys. diarrh.; worms (sMex, Trin) - **S. officinalis** L. (Med reg., esp. Dalmatia, Albania, It, Sp, Turk): (garden) sage (subspp. *major, minor*); green s.; common s.; lvs. c. 0.5-2.5% VO, with camphor, thujone, cineole, borneol; tannin; NBP (picrosalvin); ursolic acid; u. flavor, condim. carm.; strongly antioxidant; to inhibit perspiration; for convalescents; astr. in bronchitis; in hair dressings; mouth washes; as roach repellant; in spices, esp. for pork sausage, poultry stuffing, soups, sauces, cheeses; in liqueurs; teas; for prodn. VO; rt. u. in folk med. (Eu, ex. It); fls. & sds. also u. - **S. palaefolia** HBK (Co): "mastranto"; u. as cardiac tonic & in arteriosclerosis; to destroy insects on animals - **S. polystachya** Ort. (S. apiana Jeps.) (sCalif [US], Mex): salvarial; white sage; chia seed; c. mucilage; u. to clear eye of foreign particles; lf. decoc. for coughs & colds; lvs. locally for stiff neck - **S. pratensis** L. (sEu, natd. neUS): wild sage; meadow s.; lvs. u. as for *S. officinalis* but less popular since lower in VO content; mucilagin. frts. u. for eye dis. - **S. procurrens** (Arg): hiedra terrestre; hb. u. ton. (5% inf.) - **S. reflexa** Hornem. (US: Rocky Mt & plains areas; Mex): Rocky Mountain (or purple) sage; lanceleaf s.; hb. u. in rheum. mal. fever, ton astr. in gastritis - **S. repens** Burch. (sAf): rt. (c. VO) u. in diarrh. indigest., also adm. to cattle; lf. added to bath for sores; smoke u. to disinfect hut & drive out insects; mixed with tobacco - **S. rugosa** Dryand. (sAf): inf. u. as tea; lotion for sores - **S. runcinata** L. f. (sAf): pl. burned to smoke out (disinfect) hut after sickness & to drive out bugs; pl. decoc. for urticaria (Europeans); mixed with tobacco - **S. scabra** L. f. (sAf): lf. u. to medicate newborn babies; purg. - **S. sclarea** L. (Med reg., sEu, nAf, Iran): (s)clary (or muscatel) sage; common clary; c. wort; hb. c. 0.2-0.5% VO with 1-linalool as chief component (gives strong odor of lavender due to linalyl acetate); sclareol (diterpene), NBP; pectin (565); u. flavor, condim. for indiges., flatulence, in toilet waters; mouthwashes, lavender waters, soaps, perfumes (eau de Cologne); wines (vermouth, muscatel), liqueurs; added to beers (hop substitute); anhidrotic; adulter. of Digitalis; stock poison (Austral) & serious weed pest (prodn. Ge, Fr, Sp, Dalmat) - **S. serotina** L. (Fla [US], WI, tropAm): "catnip"; catnit (Bah); little woman; lf. decoc. for gastralgia, indig., colds, &c. (Bah); stim. (Cu); vermifuge - **S. shannonii** Donn.-Smith (Mex, CA): monte amargo; pl. decoc. for mal. (Salv) - **S. sisymbrifolia** Skan. (sAf): milk decoc. for sore throat; water decoc. lotion for sores & swellings - **S. splendens** Sellow ex Roemer & Schultes (SA, esp. Br rain forests; natd. US[?]): scarlet sage; many forms known in cult.; cult. ornam. (blooms nearly entire yr. round) - **S. stachyoides** Benth. = *S. mellifera* - **S. tiliaefolia** Vahl (Mex, CA): píojillo; hb. u. to kill head lice (Guat) (44); sds. u. to make beverage (sometimes mixed with barley water) - **S. triangularis** Burch. (sAf): decoc. for liver sickness in cattle; in mixts. for sterility (female) - **S. triloba** L.f. (Sp, Gr, AsMin, Syr, Crete): Greek (or Syrian) sage; u. similarly to *S. officinalis*; hb. u. as tea; source of Greek sage oil; "(mountain) apple oil"; & of some Spanish sage oil (VO); adulter. of true sage (566); VO made up of cineole chiefly; cult. sFla (US) - **S. urticifolia** L. (se&eUS; Mex): nettle (leaved) sage; one source of chia seed - **S. verbenaca** L. (Med reg., seEu, wAs): wild clary; vervain (sage); "clear eye"; frt. u. in treatment of ophthalmia; once valued by Celts; one of hbs. u. by ancient Romans under name "Verbenaca"; symbol of friendship - **S. virgata** Jacq. (Med reg., sEu, nwAs): yilancik (Turk); hb. c. VO (mostly terpenes & sesquiterpenes); u. med. - **S. viridis** L. em. Battandier & Trabut (Med reg. to Iran in dry places): bluebeard; Roman sage; purple-topped clary; VO u. to flavor beers & wines; good honey plant; source of Herba Hormini (fr. Gr. hormao, stim. to venery); hb. u. as stim. tea; sd. as aphrod.; & in eye dis.; cult. chiefly in cEc; may be progenitor of *S. italica.*

salvia: 1) (Sp, It, Swed, Roum): *Salvia officinalis* & other *S.* spp. 2) (SpAm) name appl. to many different genera & spp. incl. *Hyptis, Buddleia, Croton, Lippia, Phlomis, Pluchea,* &c. - **s. betónica** (Mex): *Lippia geminata* - **s. brava** (Por): *Salvia verbenaca* - **s. de prado** (Sp): *S. pratensis* - **s. real**: 1) (Sp) *Salvia officinalis* 2) (Mex) *Buddleia perfoliata* - **s. salvaje**: s. brava - **s. santa**: 1) (Guat) *Buddleja americana*, &c. 2) (SA) *Lippia dulcis.*

salwood: wd. of sal, *Shorea robusta.*

Samadera indica Gaertn. (Simaroub.) (Ind, SriL, Indones): "samadara" (bark tree); niepa bark tree; kathai (Burma); bk. rt. frt. u. antipyr., bitter ton. stom.; bk. & frt. in rheum.; bk. inf. remedy for constipation; inf. of very bitter sds. u. as antipyr.; lvs. macerated & mixed with coconut oil u. to kill hair lice (Solomon Is); bruised lvs. appl. in erysipelas; lf. inf. insectiside; var. *lucida* Blatter (S. l. Wall.) u. similarly (—> *Quassia*).

saman tree (SA): *Albizia* sp.; *Pithecellobium saman.*

samaun tree: *Pangium edule.*

samara: "key"; 2-winged frt. of *Acer* spp.

sambuco (It): following.

Sambucus: 1) (Caprifoli/Sambuc): elders; hbs., shrubs, or small trees; widely distr. 2) (specif.): elder flowers fr. *S. nigra*, &c. - **S. australis** Cham. & Schlecht. (Br, Ch): lvs. fls. u. diaph.; bk. purg.; exter. as vulner. - **S. c(a)erulea** Raf. (S. glauca Nutt.) (wNA, esp. Calif-dry places): blue elder(berry); blueberry elder; shrub with dark blue or black berries; frt. edible, u. diur. (nwIndians); eaten fresh or dr. (Indians); in wine, jellies, pies; pl. reputedly toxic to livestock, probably not (205); wd. u. for bows (wIndians) - **S. canadensis** L. (e&sUS: Fla-Tex; Cana: NS): American elder; (common) e.; fls. c. rutin, VO, mucil. gum, resin, tannin, u. diaph. diur. stim.; tea in colds (Ala [US]); bk. u. as diur.; purg. frt. lax. u. in foods (pies; jellies), to make elderberry wine, rt. decoc. cure-all (2451); sts. to prepare whistles - **S. ebulus** L. (Euras, nAf): dwarf elder; danewort; frt. c. VO valeric acid, HCN glycoside (traces), tannin; u. mild lax., diur, diaph, in prepn. Succus Ebuli (dr. juice); in prepn. elderberry wines & to color wines; as purple dyes (leather, yarns); in prepn. sambucyanin (u. as indicator); rt. c. VO & HCN glycosides (traces); tannin; saponin (—> diur.); u. diaph. lax. mild diur.; hb. (Folia Ebuli) u. diur. diaph. weakly lax. (now little u.) - **S. intermedia** Carr. (Trin, Bah): white elder; sureau; lvs. & fls. inf. u. in colds, coughs, flu, fevers (Trin); as bath for chicken pox, rashes, prickly heat (567) - **S. javanica** Reinw. (seAs): pl. u as analg. antirheum.; diur. purg. decoc. as bath for skin dis., painful swellings; pl. poisonous - **S. mexicana** Presl (seUS, Mex, CA, nSA): sauco (blanco); Mexican elder; frt. eaten fresh & dr.; frt./fl. decoc. u. in mixt. for asthma or chronic cough; chest colds; antipyr. in measles (Co); liver dis.; mal. syph.; pneumonia; lvs. bound on forehead to relieve headache; bk. purg. (Co); cult. as shade tree & as medl. pl.; sometimes grows to large size (-10 m); var. *bipinnata* (Co): bk. u. in catarrhs, grip; fls. diaph. diur. - **S. nigra** L. (Euras, nAf): European elder (berry) (bush); common e.; bour tree; bk. cath. diur. diaph.; fls. diur. diaph. in rheum., depur., lax.; gargle; in teas, poultices, as inhalation; for hemorrhoids; lvs. u. lax. diur. diaph.; frt. c. tannin, color, sugar, citric acid (1802); u. in colds, rheum., gout, for prodn. extr. & juice; as purple, blue, green, black dye; lvs. shoots, bk., &c. c. sambunigrin (HCN glycoside) (toxic); u. lax. diur. diaph.; rt. u. diur. in edema; emetocath.; st. pith u. absorptive for sectioning pl. materials; frt. u. in prepn. elderberry wine & non-alc. beverages, desserts; to color Port wine (Por) (exp. fr. It, Hung, Pol, USSR) - **S. racemosa** L. (Euras to Jap): European red elder; scarlet-berried e.; bk. & rt. bk. u. as diur. diaph. lax.; frt. u. in folk med. as decoc. for fever, retention of urine, colds; purg. for cattle (Chin); juice diaph.; in marmalades; sd. FO induces vomiting & purg.; var. *arborescens* Gray (S. callicarpa Greene) (wNA: Alas-Calif): (Pacific) red elderberry; redberry elder; berries of shrub eaten by Indians (esp. in winter) (steamed & stored under ground); nwIndians u. in med.; var. *pubens* Koehne (S. p. Mx.) (NA: Alas-Newf-Ga-Ore): American red elder; red-berried (or stinking) e.; cult. ornam. shrub - **S. sieboldiana** Graebner (eAs): lvs. diur.; fls. diaph. lf. decoc. for nephritis; rt. bk. decoc. for heartburn (Jap) - **S. simpsonii** Rehd. (Fla to La [US], CA, WI): (southern) elder (berry); sureau; lvs. as decoc. u. to treat colds & fever (Domin, Caribs) (2328) - **S. vulgaris** Neck. = *S. nigra* - **S. wightiana** Wall. ex Wight & Arn. (Kashmir, nPak & Ind): small frts. u. as lax.; cooked & eaten - **S. williamsii** Hance (Chin, Kor): decoc. u. for curing spasm (Chin); in treating fractures & injuries; lvs. boiled as food (Kor).

sambunigrin: cyanogenetic glycoside found in bk. & lvs. of *Sambucus nigra.*

Sämerei (Ge): sds. or grains of various kinds.

Sammtrose (Ge): *Rosa gallica.*

samohú (Arg): *Chorisia speciosa.*

Samolus valerandi L. (S. floribundus HBK) (Primul.) (SpAm, SA, As): hb. antiscorbutic; young lvs. sometimes u. in salads or as spinach; emergency food plant.

samp: edible portion of *Zea mays* grain (kernel minus hull & germ), a coarse maize hominy.

sampfen wood: fr. *Caesalpinia sappan.*

samphire (SC Herbe de *Sainte Pierre*): 1) *Crithmum maritimum* (sometimes pickled) 2) *Salicornia europaea* (**marsh s.**), *Sarcocornia* spp.

sample: a small portion of a drug or other material representative of the entire lot or batch being studied; purchases are often made on a basis of samples - **sampling**: samples of crude drugs are selected by hand or obt. with a special tool called a sampler; this consists of a hollow drill with sharp knife edge or sawtooth edge, which is drilled into bales, &c. by means of a standard brace.

Sampson root (Sampson's): 1) *Echinacea angustifolia* 2) S. snakeroot; *Gentiana saponaria, G. villosa, G. catesbaei* 3) *Psoralea psoralioides* & var. *eglandulosa, Dalea pedunculata* 4) (US voodoo): *Asarum canadense* - **black S. root**: *Echinacea purpurea.*

Samuela carnerosana Trel. (Agav.) (swNA): c. samogenin (steroidal sapogenin); fls. & frts. eaten; lf. fiber ("palma fiber") u. for twine, brushes, &c.

sana (Arab): senna.

San Agustin (Mex): *Stenotaphrum secundatum.*

Sanamycin™ = Cactinomycin (Bayer).

Sanchezia pennellii Leonard (Acanth.) (Co): lvs. u. as hemostatic - **S. thinophila** Leonard (Co): fls. decoc. u. as hemostatic (bleeding fr. initiation rites with hair pulling) (2299).

sanchi: *Panax noto-ginseng.*

sand: particles of quartz; silicon dioxide (silica); SiO_2; occ. in nature in enormous amts., also as gravel (grains 2 mm or more); other forms incl. flint; chert; agate; amethyst; onyx; bloodstone; chalcedony; jasper, &c.; hard, water-insol; u. in mfr. glass; as abrasive; cleaner; in grinding; as filtering agt.; in sand baths; anti-foaming agt; anti-caking agt; another form is diatomaceus earth (*qv.*); chronic inhalation —> lung fibrosis (silicosis); the gemstones of this gp. u. in jewelry & as art objects; in antiquity, for many medl. purpos.; sandstone, a sedimentary rock important in bldg.; many classes of sand: filter; sea; in medl./pharm. uses, purified with acid & calcined.

sand acid: HF - **s. box tree**: *Hura crepitans* - **s. bur**: 1) *Tribulus terrestris* 2) (US) *Cenchrus* spp. (also **s. spur**).

sandal wood (also as **sandalwood**): *Santalum album* (wd. VO) - **African s. w.**: *Pterocarpus santalinus* - **Australian s.w.**: *Santalum lanceolatum*; *Eucarya* spp. - **East Indian s.w.**: from Indones - **false s. w.**: *Caesalpinia sappan* - **s. w. family**: Santalaceae - **S. w. Island**: former name of Sumba (Indones) (SC large exports of s. w.) - **red s.w.**: Santalum Rubrum - **West African s. w.**: *Santalum spicatum* - **West Indian s. w.**: *Amyris balsamifera* - **yellow s. w.**: s.w.

sandarac (gum): **Sandaraca** (L.): **sandarach**: resin fr. *Callitris quadrivalvis* - **Australian s.**: similar prod. fr. *Callitris* spp. native to Austral - **German s.: Sandaraca Germanica** (L.): exudn. fr. *Juniperus communis.*

Sanddorn (Ge): *Hippophäe rhamnoides*, sea buckthorn.

Sandelholz (Ge): sandalwood - **sanderswood**: *Pterocarpus santalinus*, red saunders.

sandflower (Man): *Anemone patens* - **sandia** (Sp): watermelon - **Sandriedgras** (Ge): *Carex arenaria* - **sand-spurry**: *Spergularia salina.*

Sandoricum koetjape Merr. (S. indicum Cav.) (Meli) (Ind, seAs): sentul; rt. u. carm. stom. astr., in diarrh. dys.; general ton. (Mal); pounded bk. appl. in ringworm; fresh lvs. appl. to skin as diaph.; decoc. in baths appl. in ringworm; fresh lvs. appl. to skin as diaph.; decoc. in baths antipyr. (PI); cult. for edible frt. (Fla [US]).

sandwort: *Plantago arenaria* +.

sang: 1) (Fr-m.): blood 2 (US slang) ginseng, as in **s. diggers** (ginseng collectors) - **sanga wood** (sAf): u. as subst. for cork & rubber stoppers (compressed, then soaked in water) - **sang(-de)-dragon** (Qu [Cana]): Sanguinaria - **sangre de drago** (Sp): dragon's blood; appl. to rt. (Mex) of *Jatropha spathulata* & several other pl. spp. incl. *Pterocarpus draco* - **sangsue** (Fr): leech - **sangu** (Ec): sweet potato.

sanguijuela (Sp): leech.

sanguinaire (Fr): Sanguinaria.

Sanguinaria canadensis L. (1) (Papaver.) (eNA fr Cana to Mex): blood (or red) root; Indian paint; (red) puccoon; tetterwort; "turmeric"; rhiz. c. 4-7% alk., incl. sanguinarine (chief alk., [-6.5%] found also in other taxa of Papaveraceae), chelerythrine, protopine, coptisine, sanguirubine, &c. (1887); u. stim. emet. expect. sialagogue; in chronic bronchitis, bad colds; emmen. abort. (Abenakis Indians); sternutat., ton., diaphor., digest. disturb., aphrod., in chronic eczema, skin cancer (folk med.); in mouthwashes recently for preventing dental plaque (tartar); bleeding gums & periodontitis; does not stain teeth; antifungal (1904); for fever in livestock (US Negroes); tonic for mules; Am Indians u. as emet. anointing on swellings, aching joints, &c.; in rattler bite; war paint for painting hair, face, shoulders, clothes (hence name "Indian paint"); fls. u. in tea mixt. (Eu); proper time for collection of drug = flg. season (568); chemosurgical treatment of skin cancer consists in appln. to the area of thick black paste (bloodroot ext. + satd. $ZnCl_2$ soln. + SbS_3) (569); cult. ornam.

sanguinaria (CR): *Rhoeo discolor.*

sanguinarine: pseudochelerythrine: isoquinoline alk. obt. fr. rhiz. of *Sanguinaria canadensis* & other Papaver. (*Chelidonium, Elscholtzia, Glaucium, Bocconia, Stylophorum* (also *Fumaria* [*Dicentra*]); assay method (570); free base colorless, quaternary salts reddish; fluorescent (in org. solvents); u. as mitogenic agt.; to produce glaucoma (lab. animals); to produce polyploidy in plant genetics (1905); antifungal.

sanguinary: *Achillea millefolium*; *Capsella bursa-pastoris*, &c.

Sanguis (L): blood, actually of ox; u. in form of dr. blood formerly as "restorative" - **S. Draconis (Gummi)** (L): dragon's blood (resin); fr. *Daemonorops draco*.

Sanguisorba canadensis L. (Poterium canadense Gray) (Ros) (eNA): Cana burnet; American great b.; "tobacco leaf" (Newf [Cana]); pl. u. as astr. in hemorrhage, ton. - **S. hakusanensis** Mak. (or *S. stipulata* Raf.) (Jap): rt. decoc. u. for diarrh., to stop vomiting & abdom. pain.; boiled lvs. eaten - **S. media** L. = *Sanguisorba canadensis* - **S. minor** Scop. (Poterium sanguisorba auct.) (Euras): (salad) (garden) burnet; hb. c. tannin, flavone, u. astr. in diarrh., hemostat.; condim.; in soups, salads, to prep. Sauce Ravigotte (+ chives + chervil); lvs. with a fresh cucumber-like odor; in wines & brandies - **S. officinalis** L. (Euras, natd. US): burnet (blood wort); hb. u. like prec. sp. for catarrhs, hemorrhages, dys. diarrh. in folk med.; freshly pressed juice ingested for TB; lvs. as spring veget. & salad - **S. polygama** Nylander (Poterium p. Waldst. & Kit.) (Fin, nwRussia): u. med. much like other *S.* spp. (astr.): FO in paints.

Sanguisuga medicinalis Savigny: **S. officinalis** Sav. = *Hirudo m.*: (medicinal) leech (sanguisuga, It) (Eu); v. under leech.

sanicle: 1) any sp. of *Sanicula* 2) any of several other pl. spp. usually with qualifying term, ex. American s. - **American s.**: *Heuchera americana* - **black s. (root)**: *Sanicula marilandica* - **European s. (herb)**: *S. europaea* (also rt.) - **Indian s.** (Am): **white s.**: *Eupatorium rugosum* (rt.) **wood s.**: *S. europea* - **Yorkshire s.**: *Pinguicula vulgaris*.

Sanicula bipinnata H. & A. (Umb.) (Calif [US]): poison sanicle; p. snakeroot; pl. has fetid odor; some believe poisonous to cattle (205); young lvs. edible (boiled or leached); boiled lvs. u. as poult. for snakebite (Butt Co., Calif) (571) - **S. bipinnatifida** Dougl. (Calif): purple sanicle; rt. decoc. "cure all"; lf. inf. poultice for snakebite (571) - **S. canadensis** L. (e&cUS): black snake (or pool) root; lvs. & rts. u. in med. like *S. europaea* - **S. crassicaulis** Poepp. (Calif): Pacific sanicle; gambelweed; lvs. u. as poultice for wounds, rattlesnake bites (571) - **S. europaea** L. (Euras, Af): (wood) sanicle; s. wort; hb. c. saponins, resin, tannin-bitter subst., allantoin; u. stom. stypt.; in necrotizing stomatitis, tumors, eczema, furuncle, chronic catarrhs, vulner.; rt. u. similarly as astr. in hemorrhages, catarrhs.; vulner. - **S. gregaria** Bickn. (e&cNA): clustered snakeroot; u. by Meskwaki Indians as astr. to stop nosebleed (inhale fumes of burning pl.) (572) - **S. marilandica** L. (Cana, US): black snakeroot; (b.) sanicle; American sanicle; a tall smooth hb.; u. astr. antispas. antiperiod - **S. tuberosa** Torr. (Calif [US], Lower Calif [Mex]): turkey pea; tuberous rts. edible.

Sankt Peterswurzel (Ge): rt. of *Succisa pratensis*.

san-lo: cheap grade of Opium prep. fr. residues.

Sansevieria Thunb. (also Sanseverinia Petagna.) (Agav.) (Af, As): bow-string hemp; mother-in-law's tongue; snake plant (US); perenn. hb. much cult. as houseplant - **S. abyssinica** N.E. Br. (trop eAf): lvs. source of fiber made into kilts for women - **S. aethiopica** Thunb. (trop eAf): rt. u. as rem. in diarrh. hemorrhoids, abdom. pain (by chewing & swallowing saliva); lf. fiber u. to make strong cord - **S. hyacinthoides** Druce (S. guineensis Willd.; S. thyrsiflora Thunb.; Cordyline guineensis Britt.) (sAf; natd. WI, CA, tropAm, sFla [US]): African bowstring hemp; sweet snakeplant; rhiz. decoc. to expel intest. (round & tape) worms; rts. chewed for hemorrhoids; lf. juice in earache; vulner., for dental caries; lvs. furnish strong fiber; ornam.; exptl. cult. for fiber (Fla) - **S. kirkii** Bak. (eAf): pd. rt. appl. to slow healing wound on foot (Tanz); rhiz. purg. with weak digitalis action - **S. senegambica** Bak. (S. guineensis A. Chev.) (w tropAf, Senegal): African bowstring hemp; rt. crown has been u. in otitis; fiber u. for fish lines, &c.; lf. pulp or juice appl. to ulcers, smallpox sores, &c.; lf. juice u. locally to eye & ear dis.; rt. u. ton. stim.; rts. & lvs. u. parturient, abort. - **S. spicata** Haw. = *S. hyacinthoides* - **S. trifasciata** Prain (prob. of w tropAf origin; cult. Ind, Mal): warmed lf. juice dropped in ear for earache; lf. decoc. as hot fomentation on itch (Mal); u. much like *S. senegambica* - **S. zeylanica** Willd. (Ind, seAs): Ceylon bowstring hemp; much u. in native med.; rt. u. in card. dis., purg. ton. in Hansen's dis., fever, rheum., cough, GC, TB; juice in snakebite (573); source of valuable textile fiber (Ceylon bowstring hemp) u. for sails & in paper mfr.; lvs. c. 10% crude saponin.

Santa Claus's whiskers: *Nectandra cinnamomoides* - **Santa Maria** (Por): *Allamanda cathartica* - **santal citrin** (Fr): *Santalum album*. - **S. rouge** (Fr.): red sandal (wood), Santalum Rubrum.

Santalaceae: Sandalwood Fam.; hbs., shrubs, & trees; lvs. entire; frts. drupes or nut-like; many spp. are parasites on rts. of other pls.

Santali Lignum (L): wd. of *Santalum album* - **santalin**: red xall. coloring matter of Santalum Rubrum.

Santalina Baill. = *Enterospermum* Hiern (Rubi.).

santalol: sesquiterpene alc. found in VO of *Santalum album* wd.

Santalum album L. (Santal.) (Ind, SriL, to Mal, PI): (white) sandalwood, Indian s.; wd. has up to 6% VO with 90% santalol, santene, santalene, santalic acid; wd. u. for prodn. VO (Indonesia); wd. u. in incenses; in funeral pyres of wealthy or outstanding persons (Ind); VO u. as diaph. in skin dis. urin. tract inflammn. GC leukorrhea; expect. disinf.; stom. (Chin); in perfumes, soaps, &c.; lvs. c. proline, l-allo-proline, antisep., lf. wax (most prodn. of wd. in Timor [Indonesia]; danger of extirpation of sp.); many other *S.* spp. have fragrant wds. (576) - **S. austro-caledonicum** Vieillard (New Caled, New Hebr): source of New Caled sandalwood (emmanjo wd.) with over 90% santalol in VO; valuable wd. yellow, scented - **S. cunninghamii** Hook. f. (NZ): source of New Zealand sandalwood - **S. ellipticum** Gaud. (Hawaii): Hawaiian sandalwood; wd. was the first profitable export fr. HawIs (shipped to Chin, where u. for furniture, incense, &c.; Chin called Haw Is the "Sandalwood Islands"), lack of proper conservation depleted the forests and killed the trade - **S. freycinetianum** Gaud. (Oceanica): was exploited for Tahiti sandalwood (Haw Is) - **S. lanceolatum** R. Br. (trop wAustral): plum bush; p. wood; Australian sandalwood; Queensland s.; source of Australian s. oil, u. much like that of *S. album*; c. lanceol (sesquiterpenoid); wd. yellow, dense, easy to polish; lf. & bk. decoc. u. as purg. (Austral aborigines) (2263) - **S. marchionense** Skottb. (Pac Is): fragrant wd. u. in ceremonies of the natives; chips of wd. made into leis worn about forehead at feasts; source of scented wood oil u. to anoint the body & for embalming - **S. obtusifolium** R. Br. (Austral): wd. decoc. drunk for constip. & for aches & pains (Austral aborigines) (2263) - **S. Rubrum** (L): red saunders; fr. *Pterocarpus santalinus* - **S. spicatum** DC = *Eucarya spicata* R. Br - **S. yasi** Seem. (Pac Is incl. Fiji): Fiji sandalwood tree; ahi (Tonga); wd. u. for cabinet work but mainly source of VO; Fiji sandalwood oil u. in perfumes (86).

santene: monoterpene (cyclic) component of many VO, incl. *Santalum album* (name), *Mentha rotundifolia, Abies balsamea*, &c.

Santolina chamaecyparissus L. (S. incana Lam.) (Comp.) (sEu, wMed reg): lavender cotton; ground cypress; holy herb; evergreen shrub with lvs. pinnately toothed, whitish tomentose; fl. heads (as Flores Abrotani montani [femini]) u. as vermifuge, also in ringworm; hb. u. as abort. (Styria, Austria) (574); for intest. worms, spasmolytic, stom.; moth repellant; adulter. of Matricaria & rosemary (actually replacement rather than adulter. of latter); cult. orn. sUS: Va, Tenn & southw. - **S. rosmarinifolia** L. (sEu, nAf): santoline (Fr); u. as vermifuge; to replace rosemary (575) as flavor.

Santonica (L): levant wormseed; unexpanded fl. heads of *Artemisia cina*; much fr. Russia (577) - **Barbary s.**: *Artemisia ramosa & A. herba-alba* - **Indian s.**: *Artemisia maritima*, &c., source of Indian santonin (578) - **santonin**: inner anhydride or lactone of santoninic acid; chief active princ. of Santonica; isolated fr. *Artemisia maritima* L. & some other *A.* spp.; mostly fr. Pak, Turkestan, sRussia; after World War II, Russia; shipped s to the west as reverse lend-lease; tot. synthesis (in 1954); u. as anthelm. (round worms [1909]); has also been u. in pertussis (1742) - **santoninless s.**: a product shipped fr. Iran - **s. substitutes**: 1) *Quisqualis* sds. 2) many other spp. of *Artemisia* 3) *Alsidium helminthochorton.*

Sanvitalia abertii Gray (Comp.) (Calif, Ariz, Tex [US], nwMex): Abert's sanvitalia; yerba fria; fl. heads green yellow, chewed —> salivary flow (sialagog) and pleasant cool sensation in mouth (refrigerant) u. by Navajo Indians; lf. decoc. in stomach trouble; pl. not abundant - **S. procumbens** Lam. (Mex): ojo de gallo; vaquita; lf. decoc. u. in indiges. dys. (Yuc); for resp. dis.; in diarrh. (Guerrero); cult. ornam.

São João, raiz de (Br): *Berberis laurina*, rt.

Sap Green: v. green, sap.

Sap Red: pigment formed by interaction of humic acid; prepd. formerly from swamp water (Fla [US]) &c. (579).

sapayo (CR): cushaw.

sapé (Br): *Imperata exaltata.*

sa-pe-pa (Co): 1) *Chondrodendron toxicoferum* 2) *Abuta splendida.*

sapin (Fr): fir, *Abies* spp - **s. de Douglas**: *Pseudotsuga menziesii* - **s. noir**: *Picea mariana*, &c. - **sapind**: *Sapindus* spp.

Sapindaceae: Soapberry Fam.; vines, shrubs, & trees of trop.; lvs. usually alternate & compd.; frts. of many types, often berry- or drupe-like; many spp. rich in saponins & u. as detergents; ex. foll. (581).

Sapindus drummondii H. & A. (*S. saponaria L.* var. *d.* L. Bens.) (Sapind.) (c&swUS, nMex): western (or Drummond's) soapberry; very similar to *S. saponaria*; frt. coats u. for washing clothes (Okla, [US], Mex); with high content of saponin (c. 30%); u. as soap substit.; may produce dermatitis; formerly u. in prepn. beer (foaming); sds. c. 22-24% FO (non-drying, not easily rancidified) (580); u. in soap mfr. (*Sapindus* <sapo [L]: soap + indicus [L]: Indian) - **S. laurifolius** Vahl = *S. tri-*

foliatus - **S. marginatus** Willd. (s&swUS, nMex): western soapberry (tree); jaboncillo; pulp of berries u. as soap substit.; & to make beer foam (formerly); wd. useful for baskets, &c. - **S. mukorossi** Gaertn. (S. utilis Trabut) (Ind, Burma, Chin, Jap): soapnut tree; ritha; rito; rts. c. 11% (38%!) saponin, u. deterg.; fls. u. in conjunctivitis, various eye dis.; frts. with high saponin yield u. in epilepsy (Burma); lotion to remove freckles & tan fr. skin; for cleaning clothing, polishing silver & jewels; sds. in necklaces; bk. decoc. expect. - **S. oahuensis** Hillebr. (Haw Is): sds. u. as cath. (natives) - **S. rarak** DC (seAs, Java): soapberry; frt. u. much like foll. as soap substit. in scabies, as insecticide; fish poison; frt. c. 14% rarik-saponin - **S. saponaria** L. (sFla [US], WI, SA [1886]): (southern) soapberry (tree); soaptree; "Indian soapnut" (Fla); saboneteiro (Mex, WI, CA to Br); jaboncillo (Mex); bk. rt. entire frt. c. saponins (frt. 8-20%); frt. ("soap nut"; "s. berry") u. as expect.; in chlorosis, leucorrh., colics, meteorism; & urin. dis.; rt. u. as astr. ton.; fish poison (Am Indians); sds. as deterg. (delicate fabrics) & to make saponin; FO of sd. kernel (soap tree oil) u. in soap mfr.; sds. as rosary beads (SA); lf. inf. u. for snakebite, stings of rays (Ve); frt. in prepn. hair washes, &c.; bk. decoc. in dys. (2328); rt. & lvs. also u. med. (1906) - **S. trifoliatus** L. (S. laurifolius Vahl; S. emarginatus Vahl) (Pak, Ind, SriL): soapnut tree; soaptree; poongum oil plant; Indian filbert; frt. flesh c. 4-5% saponin; sds. (54% of frts.) kernels c. 30-45% FO (soap tree fat; Oleum Sapindi) (1907); rt. (bk.) u. as expect. in rheum. sciatica; sd. ton. in skin dis. & snakebite; sds. u. as pessaries to stim. uterus in amenorrh. & parturition (Ind); frt. & sds. contraceptive (anti-implantation); cult. for frts, ornam, (Bengal, WI, nSA) - **S. utilis** Trabut (*S. mukorossi* var. u.) (tropAs, Alg planted Fla, PR [US]); frts. c. -38% saponin; sds. c. 60% saponin (1908); best yields in gen.; u. similarly to *S. saponaria* (582); cult. ornam - **S. vitiensis** (A. Gray) Seem. (Fiji): aq. inf. of leaf drunk for lung trouble; lvs. squashed in water & drunk for sore stomach (Degener).

sapinette (Fr): spruce, *Picea* spp.

Sapium P. Browne (Euphorbi) (trop & subtr of world): ca. 140 spp.; shrubs & trees with milky juice; Am spp. source of natural rubber - **S. biloculare** (S. Wats.) Pax (S. salicifolium Torr.; Sebastiania bilocularis S. Wats.) (swAriz [US]-Baja Calif & Sonora [Mex]): yerba de (la) fleche; sap u. to make arrow poison; as fish poison; u. to poison enemies; said lax; sds. infested with larva —> Mexican jumping bean (*qv.*) - **S. bussei** Pax (Tanzania): u. as ankylostomicide - **S. ellipticum** Pax (cAf): lf. u. for sore eyes, maggoty wounds; bk. decoc. in stomatitis; boiled rt. as fomentation over enlarged spleen (babies) - **S. grahami** Prain (tropAf): u. med. (Guinea), lvs. rts. in leprosy (Nig); pd. lf. ext. u. to facilitate extn. of Guinea worm; bk. & lf. ext. u. to mark red & orange signs on face - **S. indicum** Willd (eAs): lvs. to relieve fever (local); inf. for GC; in compn. as poultice for yaws; frt. as fish poison - **S. lanceolatum** Huber (Br): murupita; u. in eczema, herpes with pain (shingles); curubas - **S. laurocerasus** Desf. (S. laurifolius Griseb.) (WI): tabaiba; wd. u. for boxes, crates, paper pulp, plywood - **S. salicifolium** HBK (swUS, Mex, cult. Jap): yerba de la fleche (arrow hb.); rts. bk. lvs. u. as mild lax. (*cf.* senna); rt. bk. & FO u. lax. purg. diur. (Chin) - **S. sebiferum** Roxb. (Ind, Chin, natd. seUS): Chin tallow tree; candle tree; rt. bk & FO u. lax. purg. diur. (Chin); frt. mesocarp source of (Chin) vegetable tallow; u. in soaps, admixed with other waxes, as illuminant; "stillingia oil" is expr. fr. sd. kernels; u. food, drying oil, for hair care & in skin dis. (frt. wall fat) - **S. sylvaticum** Torr. = *Stillingia sylvatica* - **S. verum** Hemsl. (sCo, Ec): one of chief sources of white or virgin rubber (caucho blanco).

Sapo (L): soap; glycerides of Na, K, &c.; known since ca. 2500 BC; u. as deterg. cleanser - **S. (Medicatus)** (L).: a good quality of neutral white Castile soap - **Sapocresol**: prod. similar to Cresol - **sapodilla (fruit)**: fr. *Sapota achras* - **s. gum**: gum chicle - **sapogenin**: the aglycone of a sapotoxin - **sapon wood**: Brazil wood.

Saponaria levantica (Caryophyll.) (Med reg.): Levant (or Russian) soaproot; u. deterg. - **S. officinalis** L. (Euras, natd. NA): soapwort; bouncing Bet; root (Radix Saponariae [rubrae]) c. 2.5-5% saponin (highly variable yields have been given, 2.5-34%); hb. & lvs. also c. saponin; saponins of strongly acting type, called "sapotoxins" in older lit.; saporubin; saponoside; u. as expect. in bronchitis, coughs, &c.; diaph. purg. lvs. u. emet. in rheum.; rt. u. diur. appl. to contusions, hemorrhoids, tumors, &c.; formerly added to toothpowders; techn. in scouring waters, cleaning agts.; widely distr. as weed - **S. vaccaria** L. = *Vaccaria pyramidata* Medik.

saponins: gp. of complex (non-nitrogenous) pl. principles, often colloidal pl. glycosides (saponosides) (steroidal or triterpenoid) (steranes with spiro-ketal side chains); widely distr. among pl. spp. & in pl. organs; foam with water (when shaken); act as fish poisons; toxic to other animal organisms on entering blood stream (destroy RBC; act to hemolyze [lake] blood cells); emulsify FO

& resins; ex. diosgenin (<Dioscoreaceae); many uses incl. the foll.; foam builder in fire extinguishers; detergents in textile indy.; washing agt. for fine fabrics (silks & woolens); emulsifier for fats & oils; in ceramics (—> lighter wt. ware); in facial creams, &c.; specif. the saponin obt. fr. *Quillaja saponaria* or *Saponaria* spp. (1911).

Sapota Mill. = *Archras, Manilkara.* - **S. achras** P. Mill. = *Manilkara zapota* (L) van Royen.

Sapotaceae: Star Apple Fam.; shrubs & trees of the trops; numerous lacticiferous sacs in the cortex & medulla of st. & in lvs.; ex. *Palaquium*; *Pouteria*; *Manilkara*; lvs. alter. & evergreen; fls. regular, axillary; frts. berries.

sapote (Sp): 1) *Pouteria sapota* 2) (Pe) *Capparis angulata* or *C. scabrida* 3) (Pe) *Matisia cordata* 4) (PI) *Diospyros ebenaster+* 5) *Salicornia europaea* (also **mammee s.**) - **red. s.**: *Pouteria sapota* - **white s.**: *Casimiroa edulis.*

sapotoxin: a class of highly active (hence toxic) saponins.

sappan (wood): *Caesalpinia s.*; *C. echinata.*

sapucaea (beans): **sapucaia (beans)** (Br): *Lecythis* spp.

sapucainha (Br): *Carpotroche brasiliensis* - **sapuyo** (Par): pumpkin.

sapwood: alburnum, outer lighter wd. of trunk & branches with still living cells as distinct fr. heart wd.; ex. Quassia.

sara (Pe): maize.

Saraca asoca De Wilde (Leg) (Ind): cult. ornam. shrub (Ind) - **S. declinata** (Jack) Miq. (Burma, Thail, Malay arch.): u. as tonic to give strength (2448); decoc. of young lvs. u. in fever; trunk bk. considered by some to be anti-maleficient (against evil; magic) (Camb); lvs. eaten - **S. indica** L. (Indo-Pak; eBengal, c&eHimalay, Kumaon, sInd): asok(a); sok (Thai); a sacred tree of the Hindus, being planted near temples for worshipping & fls. u. at temple offerings; bk. u. astr. ton. esp. in uter. dis., incl. menorrh.; cult. for beautiful fls. (NB: do not confuse *Saraca* & *Saracha*).

Saracha contorta R. & P. (Solan.) (tropAm, esp. Pe): ahuaimantu; fls. admin. in childbirth (as inf.) (folk med.); frt. edible; lvs. u. as poultice for pain - **S. dentata** R. & P. (Pe): lvs. with lard u. as poultice for tumors - **S. jaltomata** Schlecht. = *Jaltomata edulis* Schlecht. - **S. procumbens** R. & P. (Jaltomata p. J. L. Gentry) (Pe): lvs. said emoll. analgesic, cleansing & healing.

sarachs: male fern (Abu Mansur; Ibn el Baitar).

saragatona rubia, semilla de (Sp): blond psyllium sd. - **saranae** (Thail): *Mentha arvensis* - **sarapia (cristalizada)** (Ve, Co): tonka (with xals.).

Saraka™: lax. prepn. once claimed obt. fr. *Saraca indica*; actually made up of bassorin (fr. Karaya) & Frangula ext. (1912).

Sarapin™: prepn. of aq. distillate of *Sarracenia purpurea* u. for relief of low back pain, paresthesias, neuromuscular pains, neuralgia, &c. (by local injection) (583).

Sarcandra glabra Nak. (Chloranthus glaber Mak.) (1) (Chloranth.) (seAs): st. & lvs. u. locally for contusions, bone fractures (Chin); lvs. & rts. in rheum (Chin); lf. decoc. drunk as astr. anti-emet. (Chin); lvs. appl. for internal pain, to prevent boils; rts. in cough med. (PI).

sarcine: Sarkin (Ge): hypoxanthine; alpha-sarcine; widely distr. in pl. & animal kingdom (ex. potato juice, tea, hops); often synthesized; precursor of uric acid (high levels in gout & some neoplastic dis.); believed active compd. (as arabinoside [metabolite]), in animal transfer factor (u. to give immunity in viral & bact. dis., malignancies); as arabinoside may be useful in viral eye dis. (herpetic), zoster (shingles).

Sarcobatus vermiculatus Torr. (1) (Chenopodi) (wPlains & Rocky Mts [US], adj. Mex, semi-arid areas): (Mexican) greasewood; chico; sds. sometimes eaten; ash of pls. formerly u. in diarrh., rectal bleeding (Nev Indians); forage for cattle (but sometimes poisonous) (soluble oxalates); wd. & sts. c. FO; u. fuel.

sarcocarp: the fleshy portion of a frt., such as of a drupe.

Sarcocaulon burmanni Sweet (Gerani.) (sAf): Bushman's candle; gifdoorn; u. in diarrh.; pd. rt. u. as poultice; other *S.* spp. u. med.; cult. as succulent.

Sarcocephalus Afzel. = *Nauclea.*

Sarcocolla Boehm. = *Penaea* L.

sarcocolla: gum-like prod. long known & formerly u. in med. (Middle Ages); possibly obt. fr. *Penaea fucata* (*qv.*) or *Astragalus sarcocolla* (*qv.*) or *A. mucronata* (Iran).

Sarcocornia fruticosa (L.) A. J. Scott (Arthrocnemum f. (L.) Moq.) (Chenopodi) (nAf, sEu): glasswort; samphire; hb. u. like *Salicornia europaea*; diur. antiscorb. source of soda.

Sarcodes sanguinea Torr. (monotypic) (Monotrop.) (Calif, Ore [US]): snow plant; pl. inf. u. to help build blood (in pneumonia); inf. of dr. pl. u. as wash on ulcers, cankers, sore mouth, itch, toothache; sts. eaten after cooking (*cf.* asparagus) (Calif Indians); rare pl. (parasitic on fungi).

Sarcodina: Rhizopods, one class of the Protozoa in which members possess pseudopodia; ex. *Amoeba.*

sarcolactic acid: l-lactic a.; occ. in blood & muscle fluids of animals (in small amts.), increasing after vigorous work; u. for muscle fatigue, aches, cramps, exhaustion; in tabs. with Arnica (Homeop. med).

Sarcolobus banksii Roem & Schult. (S. narcoticus Miq.: S. spanoghei Miq.) (Asclepiad.) (seAs): gonggom; rt. & inner bk. rasped to a powd. u. to poison predatory animals (tigers, wild dogs, &c.) (Java); c. "sarcolobia," resinous subst. - **S. globosus** Wall. (seAs): lvs. made into paste with *Aleurites* nuts & u. to rub rheumatic joints & for dengue fever; sds. u. to poison varmints (antidote: fresh coconut juice); frt. peels made into a pleasant candy; peel also eaten as dessert or side dish.

Sarcomycin: Sarkomycin (Takeda): antibiotic prod. by *Streptomyces erythrochromogenes* Waksman & Henrici of soil (Jap); anti-tumor active, prolonging life span of cancer-injected mice.

Sarcopoterium spinosum (L.) Spach (Poterium s. L.) (monotypic) (Ros.); (eMed reg., It, seEu, nAf, AsMin, Syr+); prickly shrubby burnet; thorny b.; bellan (Syr); very common undershrub; hb. & rt. bk. rich in tannins, insulin-like subst.; u. astr. in catarrhs., antidiabetic (reduces sugar in blood & urine) (2511); pl. grown as hedge (Syr); u. as fuel, to prevent soil erosion; may have furnished crown of thorns (Jesus Christ on cross); ornam.

sarcosine: methylglycine, an org. base of the animal body, incl. sea urchins; u. in synth. of antienzyme agts. (toothpastes); occ. in actinomyces cultures, acid hydrolysis of peanut proteins, &c.

Sarcostemma acidum Voigt (S. brevistigma W. & A.) (Asclepiad) (seAs): soma (Sanskrit); sap u. as refreshing beverage, to prod. alc. drink; insecticide(!) - **S. brunonianum** W. & A.: (sInd, SriL): u. like *S. acidum*; dr. sts. emetic - **S. intermedium** Dcne. (Ind): pl. bitter, cooling; eaten by rhinoceroses; u. similar to *S. acidum*; major component of soma - **S. stocksii** Hook. f. (Ind, Pak): soma (Kanarese); u. like *S. acidum*; major component of soma - **S. viminale** R. Br. (e&sAf): caustic bush; c. vine; u. as emet. in heartburn; latex u. for veld sores, varicose ulcers, septic cuts; allergenic; milk substit. (Ind); pl. eaten; latex fish poison; sometimes considered to be "soma" (intoxicant).

Sarcostigma kleinii W. & A. (Icacin.) (Kashmir, Ind): sd. —> oil u. in treating rheum.

sardine: name given several small marine fishes, including *Sardinia pilchardus* (Fr, It, Po, Sp), juvenile herring, &c; commercially canned (Fam. Clapeidae).

sargam (Eg): *Lens esculenta.*

Sargasso (or **Sargassum) weed**: *Sargassum* spp.

Sargassum C. Agardh (Sargass - Fucales - Phaeophyc - Algae) (all oceans): gulf weed; some 260 spp. of brown algae, supported on ocean surface by air-bladders in thallus; tremendous numbers in the Atl Oc w of Af make the "Sargasso Sea" lying betw. Azores & WI (585); some spp. are useful as foods (As) (*S. pyriforme* Ag., *S. acanthocarpa*); several are the source of alginic acid & alginates; others are u. as fertilizers, stock foods, & in tanning - **S. confusum** Agardh (eAs coasts): thallus c. sargalin (hypotensive phenol), hydrolyzes to anthelm. agt. - **S. cuneifolium** (Haw): grape weed; u. as food (HawIs) - **S. echinocarpum** J. Ag. (Haw): sea weed; limu kala; thallus eaten fresh or after fermentn.; often combined with fish heads & salt - **S. enerve** Ag. (wPac Oc): hondawara; chief sp. u. for food by fishermen in Jap; added to soy sauces; in soups; to prep. alginic acid/alginates; as decoration on New Year's Day (Jap) - **S. fusiforme** Setch. (coasts of eAs): u. to treat goiter, glandular dis. (Chin); as veget. & in soups (Amoy, Chin); dr. thallus to prep. tea (cooling) (Chin); raw material in mfr. alginic acid & alginates, iodine, potash (Jap) - **S. linifolium** (Turn.) Ag. (Ind Oc, Med Sea+): u. in urin. dis.; esp. gravel (calculi) (Ind); antiscorbutic (Martinique); c. iodine; u. in prepn. alginic acid & alginates - **S. natans** (L) J. Meyen (Atl Oc coasts): Sargasso weed; thallus c. FO with high vit. A & D content; sarganin (antibiotic); u. med. (lithiasis, renal colic; goiter) (SA); important source of alginates - **S. polyceratium** Montagne (Caribbean Sea coasts): alga has antibiotic activity (1287) - **S. thunbergii** O. Ktze. (coasts of sChin, Kor, Jap): c. pantothenic acid, biotin, inositol; antibiotic to TB organism; u. med. (Chin) - **S. vulgare** Ag. (coasts of Atl & Pac Oc, Carib Sea): protein fractions show lipolytic, anticoag. & hypoglycemic activity; u. in folk med. of various countries as anthelm; many other *S.* spp. such as *S. tortile* Ag. & *S. fluitans* Borgesen are utilized in mfr. chemicals, foods, &c.

sargenoside: glycoside of *Strophanthus sarmentosus*; made up of sarmentogenin + D-digitalose + glucose.

Sargentodoxa cuneata Rehd. & Wils. (monotypic) (Sargentodox.) (Chin+): st. u. for ascarids; stomach ulcers; rt. & st. u. in rheum.; pl. insecticide.

sarghana (Arab): orpine, *Telephium* spp.

sarhamnoloside: glycoside of sarmentologin & L-rhamnose.

sariette (Fr-f.): 1) *Serratula tinctoria* 2) *Satureja hortensis.*

sarkomycin (Jap): antibiotic prod. of *Streptomyces* sp. similar to *S. erythrochromogenes*; anti-cancer & weakly bacteriostat. (586).

sarmentocymarin: complex glycoside hydrolyzing to sarmentogenin + sarmentose.

sarmentogenin: aglycone (steroidal sapogenin) of sarmentocymarin, rhodexin A, & several other glycosides; obt. fr. *Strophanthus sarmentosus.*

sarmentoloside: glycoside of sarmentologenin + deoxytalose; occ. in *Strophanthus sarmentosus.*

sarmentose: a sugar, methyl ether of deoxyhexomethylose; derived fr. sarmentocymarin, glycoside of *Strophanthus sarmentosus.*

sarmentosides: glycosides of sarmentosigenin + talomethylose; occ. in *Strophanthus sarmentosus*, &c.

sarmiento (Sp): branch or shoot of the grape vine.

sarmutoside: glycoside of sarmutogenin + D-sarmentose; occ. in *Strophanthus* spp.

sarnovide: glycoside of sarmentogenin + D-digitalose; occ. in *Strophanthus sarmentosus.*

Sarothamnus Wimm.: *Cytisus.*

Sarracenia L. (Sarraceni.) (Atl NA): trumpets; Indian cup (587); pitcher plants; insectivorous pl.; (traps roaches, &c.); sometimes odorous (588); found in bogs, wetlands; named after Dr. Michel Sarrasin (589); of current medical interest; many hybrids - **S. flava** L. (Fla-Ala-Va [US]: bogs): (golden) trumpets; huntsman's horn; trumpet plant (or leaf); trumpet pitcher plant; bonnet lily (Ga [US]); rhiz. u. astr. dem. in dyspep., pyrosis, diarrh.; analg; food ("gator tator"); hb. (as tinct.); ton. stomach. in gastralgia; rhiz. & tops claimed to have antitumor activity; u. in folk med (in moonshine exts.; Okefenokee Swamp) - **S. minor** Walt. (S. variolaris Mx.) (seUS-savannas): hooded pitcher plant; (spotted) side-saddle plant; smallpox plant; dr. hb. formerly u. for smallpox (doctrine of signatures); rt. decoc. u. for skin rash, eruptions (2451) - **S. purpurea** L. (eCana, US fr. Newf [Cana] to Fla; peat bogs): Indian cup; I. jug (Newf, where prov. flower); common (or northern) pitcher plant; rhiz. c. alks. "sarracenine," serapine (anesthetic princ.); u. stim. ton lax., in dys.; diur. diaph.; in smallpox (tinct.); for relief of pain (neuralgia, sciatica, peripheral neuritis, alcoholic neuropathy, tabes dorsalis, cancers) by local and spinal injection (590) - **S. rubra** Walt. (seUS): sweet pitcher plant; cult. ornam.

Sarraceniaceae Dum. (Dicot.) (NA, SA): sarracenia family; 3 genera & ca. 20 spp., chiefly *Sarracenia* (8 spp.).

sarrasin (Fr): buckwheat.

sarrapia (Br): sarapia - **sarriette** (Fr): 1) *Satureja hortensis*, &c. 2) *Lepidium* (pepperwort), &c.

sarsaparilla: 1) rt. of several *Smilax* spp. (off.: *S. aristolochiaefolia*; *S. regelii*; & *S. febrifuga* & undetermined sp.) (USP XIV) 2) beverage formerly popular; c. VO of anise, orange, sassafras, also methyl salicylate, cologne spts., caramel, sugar, water; formerly popular medicinal, as Ayer's S. (1885); Hood's s. (C. I. Hood & Co.) (-1884-) - **American s.**: *Aralia nudicaulis* (NCar [US]); *A. racemosa* - **brown s.**: *S. regelii* - **Central American s.: Costa Rica s.**: undetermined *S.* sp. - **Ecuadorian s.**: *S. febrifuga* - **false s.**: American s. - **gray s.**: *S. aristolochiaefolia* - **Honduras s.**: brown s. - **Indian s.** *Hemidesmus indicus* - **Jamaica s.: Lima s.**: undetermined *S.* sp. - **Mexican s.**: gray s. - **Pará s.**: Brazilian s.; fr. *Smilax papyracea* - **red s.**: Jamaica s. - **small s.**: American s. - **Tampico s.**: gray s. - **Texas s.**: yellow s. - **Vera Cruz s.**: gray s. - **Virginia s.: white s.: wild s.**: false s., *S. glauea* - **yellow s.**: *Menispermum canadense*; sometimes *Aralia nudicaulis.*

sarsa(sa)pogenin: the sapogenin of **sarsaponin** (a saponin glycoside occurring in sarsaparilla); u. in mfr. of the pregnane series of compds. incl. cortisone; u. in psoriasis.

sarsepareille(Fr): sarsaparilla.

sarson, yellow (wPak): *Brassica trilocularis*; sd. prod. FO

sarveroside: glycoside of sarverogenin + D-sarmentose; occ. in *Strophanthus gerrardii.*

"sarvice" (Tenn [US]): serviceberry.

Sarzae Radix (L): Sarsaparilla.

Sasa albo-marginata Mak. & Shib. (Gram.) (Chin): u. as hemostat. - **S. kurilensis** Mak. & Shib. (Jap): charred lf. & st. u. as antidote in fish poisoning (Ainu) - **S. veitchii** (Carr.) Rehd. (Chin, Jap): bamboo grass; c. bamfoline (polysaccharide fraction) with antitumor properties (Ehrlich's carcinoma, sarcoma).

sasafrás (Sp): 1) (Hisp.): *Ocotea* spp., incl *O. cuneata* Urb. 2) *Sassafras albidum..*

Saskatoon (service) (berry) (Indian name): *Amelanchier alnifolia* & possibly related *A.* spp.

"sass" (US street name): capsule with petrol. ether extr. of sassafras bk. or rt.; u. psychotropic (-1975-) (*cf.* LSD).

sassafac (Ala-Ga [US] slang): Sassafras (591).

Sassafras: 1) genus; *Sassafras* Nees & Eberm. (foll. paragraph) 2) (misnomer): *Ocotea* spp. - **Australian s.**: *Atherosperma moschatum* - **black s.**: 1) Brisbane s. 2) camphor tree - **Brazilian s. oil**: fr. *Ocotea pretiosa* - **Brisbane s.**: *Cinnamomum oliveri* - **s. medulla** (L.): s. pith - **s. nuts**: Pichurim beans - **Santa Catarina s.**: Brazilian s. - **s. soap**: homemade soap stirred with a s. stick or aromatized otherwise with s. - **swamp s.**: *Magno-*

lia glauca - **Tasmanian s.**: Australian s. - **white s.**: a form of *Sassafras albidum* with rt. bk. whitish, & with smooth lvs., buds, & young twigs; said to be of inferior med. value.

Sassafras albidum (Nutt.) Nees (S. officinale Nees & Eberm. & var. *albidum* Blake; S. variifolium O. Ktze.) (Laur.) [eUS & Cana w to eKan [US]; eMex): (common or American) sassafras; sassafraz (Br): mitten (or sassafras) tree; ague tree; cinnamon wood; "satisfac" (sUS folk, facetious); inner rt. bk. (Sassafras, L.): c. 8% VO with c. 80% safrole, 6% tannin, sassafrid (red pigment), resin, &c.; u. diur. diaph. arom.; prepn. beverage (sassafras tea; also prepd. with rt.); ice tea; formerly in prepn. root beer, &c. (such use of safrole & derivs. now forbidden in US [1960-] because of liver cancerogenic potential); stim. alter. spring ton. (s. tea or saloop formerly regarded as remedy for intoxication); source of VO; wd. (rt.) c. 1-2% VO; u. diur. depur. in skin dis. gout, rheum.; syph. (592); pith (S. Medulla) c. mucil. VO; u. dem. in eye lotions (rept. of such use by NY State Indians ca. 1750, Peter Kalm) (1913); VO formerly prepd. Tenn, Ken, SCar, Ark [US] (1914); u. flavor in beverages, candies, chewing gum; arom. in skin lotions, soaps, insecticides, parasiticides, in skin prepns. for itch; lice; insect bites; ant repellant; arom. in library paste, mucilage of postage stamps; disinf.; lvs. comm. article, u. in prepn. gumbo filet; trunk bk. also comm. article formerly; shipload of *S.* & fine wd. sent to Eng fr. Jamestown, Va (US) 22 June, 1607 (593) (name fr. Salsafras [Sp] fr. Am Indian word]) - **S. tzumu** Hemsl. (Chin, Indochin): white inner bk. said anthelm. parasiticide for scabies, head lice; eye inflamm. (decoc.); antipyr. for nausea & vomiting (Chin); wd. & rts. u. as diaph. in rheum. (Indochin).

sassafrassy: sassafras.

sassafraz brasileiro (Br): **s. do Brasil**: Brazilian sassafras - **sassifraga** (It): Sassafras; s. bark.

sassaparilla (US slang): **sassy**: Sarsaparilla.

sassy: 1) (w&cAf): *Erythrophleum guineense* (also saucy; sasswood; sassy tree) 2) (sUS): sassafras (sassy tree).

satin leaf: *Chrysophyllum cainito.*

satin leaves & s. pods: *Lunaria annua.*

"satisfac" (Ala [US], perhaps facetious): sassafras (*cf.* sassafac).

Satsuma: S. orange: emerald tangerine, *Citrus unshiu* (C. reticulata*)*.

Satureia (L.): **Satureja** L. (Lab) (temp & warm regions): savory; ca. 200 spp. of arom. hbs. & subshrubs; many u. for flavor & in med. - **S. acinos** (L) Scheele (Thymus a. L.) (Euras, natd. eNA): spring savory; mother-of-thyme; perenn. hb. u. arom. carm. - **S. brownei** Briq. = *Micromeria b.* - **S. calamintha** (L.) Scheele = *Calamintha nepeta* Savi and subsp. *glandulosus* P. W. Ball) - **S. chamissonis** Briq. (Micromeria c. Greene) (wNA): yerba buena; lvs. of vine c. VO, palmitic & other carboxy acids, resin, with micromeritol & micromerol (alc.) (2333); lvs. u. carm. diaph. antipyr.; vine (st.) depur., in colic; arom. - **S. douglasii** (Benth.) Briq. (Micromeria d. Benth.) (wUS): yerba buena; arom. hb. u. by w. coast Indians in beverage (Oregon tea) (594) - **S. eugenioides** (Griseb.) Loes. (Arg, Co): muña-muña; hb. c. VO u. stim. aphrod., digestant; diaph. - **S. hortensis** L. (Euras): (summer) savory; hb. (Herba Saturejae, Satureja [L]) c. 0.3-2% VO with 30% carvacrol, u. flavor, condim. carm. stim antispasm. expect. diaph. emmen.; popularly u. (Ge) in diarrh. of all kinds, anthelm.; herbal baths; cooking spice, herbal ext., in "mixed herbs"; mfr. of sausages; in perfume indy. - **S. macrostema** (Benth.) Briq. (Mex): té de monte; inf. u. in stomach & other digestive dis. (mostly women, children); natives u. with alc. drinks - **S. montana** L (Euras, Af; cult. cEu, Arg): (winter) savory; hb. u. flavor, kitchen spice; source of VO (Fr); u. with bean dishes & as salt substit. - **S. nepeta** Scheele = *Calamintha n.* - **S. thymbra** L. (eMed reg, also Sp): Canadian savory; strongly branched subshrub; c. VO with pinene, cymol, dipentene, bornyl acet., hb. u. as spice, stim., & disinfectant - **S. viminea** L. (Micromeria obovata Benth.) (WI): "all heal"; shrub -4 m; pl. boiled with ginger & decoc. u. for colic (Jam) - **S. vulgaris** K. Maly = *Clinopodium vulgare* L.

Saturnus (Saturni) (L.): lead.

Satyria warscewiezii Klotzsch (Eric.) (CA, SA): palo de miel (CR): epiphytic shrub with showy red fls.; small purple frt. edible & u. to make jellies & preserves; u. med (CR) (595).

Satyrium nepalense D. Don (Orchid.) (Ind, Nep): tubers ("saleb misri") u. in ton. prepns.; substit. for *Roscoea procera.*

sauce (Sp): *Salix* spp., esp. *S. pentandra* - **s. blanco** (Sp): white willow, *Salix alba* - **s. hediondo** (Arg): *Fagara coco* - **saúco** (Sp): elderberry, *Sambucus nigra* & other S. spp - **s. amarillo** (Cu): *Tecoma stans.*

Sauerbeere(n) (Ge): 1) *Berberis vulgaris* (frt.) 2) *Ribes* spp. (frt.) (esp. *R. petraeum* Wulf.) - **Sauerdorn (, gemeiner)** (Ge): *Berberis vulgaris* - **Sauerkirsche** (Ge.): sour cherry, *Prunus cerasus.*

Sauerkraut (Ge, Eng): pickled or salted cabbage, prepd. by fermentation with yeast; the juice a mild lax. & important source of vit. C in the winter.

saufre végétale (Fr): Lycopodium.
sauge (Fr, It): sage - **s. sclarée**: sclary sage.
Säugetier (Ge): mammal.
saule (Fr-m.): willow, *Salix* spp.
saunders, red: *Pterocarpus santalinus* - **white s.: yellow s.**: *Santalum album*.
Sauohr (Ge): *Plantago major* (lit. sow ear).
Saurauia aspera Turczaninov (Actinidi. [Dilleni]) (sMex): edible frt. decoc. u. as emoll. (much mucilage); u. in coughs, colds, & throat irritation (1816) - **S conferta** Warb. (Solomon Is): sts. roasted & placed in ears of deaf person; lvs. placed in dog's mouth before hunting opossum to make hearing more acute - **S. glabrifolia** Merr. (PI): young lvs. appl. to treat headache & for pain - **S. napaulensis** DC (Indochin, Ind, Nep): bk. u. to poultice areas with splinters (helps draw out); frt. u. to adulter. honey (Vietnam) - **S. rubicunda** (A. Gray) Seem. (Fiji): lf. juice squeezed fr. lf. into eye for eye trouble - **S. schumanniana** Diels (SolomonIs): frt. & buds heated & appl. to sores on legs.
Sauromatum venosum (Ait.) Kunth (incl. *S. guttatum* Schott) (Ar) (tropAs, Af): voodoo lily; v. bulb; inflor. "disgustingly fetid" (carrion odor attracts flies); tubers u, exter. as stim. poultice (Ind.): condim., food (after processing); ornam. hb.
Sauropus androgynus Merr. (Euphorbi) (seAs): rts. u. in fever; lf. juice u. to treat sore eyes - **S. changianus** S. Y. Hu (Chin): pl. u. med.; has antibiotic properties - **S. quadriangulus** Muell-Arg. (Ind): dr. lvs. smoked for tonsillitis - **S. spatulaefolius** Beille (eAs): rt. decoc. for sore throat.
Saururus cernuus L. (Saurur.) (e&cNA): (common) lizard('s) tail; water dragon; swamp lily; rt. ("black sarsaparilla") u. in pleuritis; emoll., discutient; homeop.; c. manassantin A, claimed to have tranquilizing effects (Koppaka Rao) (cult. orn.) - **S. chinensis** Baill. (eAs): underground st. & fls. diur. & lax. in edema; crushed lvs. to cleanse abscesses & heal suppurating flesh; in mal.; insecticide; parasiticide (Chin) - **S. lucidus** Don. (SA): rt. u. as pungent arom., ext. for sores.
sausage tree: *Kigelia pinnata*.
Saussurea amara DC (Comp.) (Euras): formerly u. as antisyphilitic - **S. arenaria** Max (nEuras): rt. u. as ginseng substit - **S. candicans** C. B. Clarke (Indo-Pak-mts.): sds. u. as carm. (horses); remedy for horse bites - **S. hypoleuca** Spreng. (s&eAs): lvs. purg. & antisyphil. (Indochin) - **S. lappa** C. B. Clarke (S. costus Lipsch.) (Kashmir [cult]): kut(h) (Punjabi, Kashmir); ooplate (Bombay); rhiz./rts. c. 4-7% VO ("costus oil") with myrcene, caryophyllene, humulene; "saussurine" (alk.) (now believed artifact formed in isolation process); represents "costus root" of the ancients; u. diur. emmen. anthelm., ton. in coughs, bronchial asthma; fever, dyspep.; rheum.; aphrod.; "universal antidote"; for abdomin. tumors (Ind); against cancer (folk med. of Chin); in prepn. of nerve ton.; clothes moth repell. (Ind); fumitory (Chin); VO highly bactericidal (1915), fixative in perfume indy. (VO); u. arom. perfum. in Chin for incense; fraction of VO hypotensive (bibliog. [628]) - **S. pulchella** Fisch. (Siber): c. sesquiterpene lactone, saurin (antimicrobial); u. as antipyr. styptic, in rheum. (folk med.) - **S. transcaucasica** D. Sosn. (Azerbaijan): pl. F. E. & total alk. prepns. u. as hypotensive.
sauterelle (Fr-f.): locust; grasshopper.
Sauvegesia erecta L. (Ochn.) (Mex to SA, WI): decoc. of crushed hb. taken orally for stomachache (777); —> Creole tea.
savadilla: *Schoenocaulon officinale*.
savage mariolain(e): *Origanum vulgare*.
savakim: savakin: *v.* suak.
savanna(h): flat grasslands of tropics & subtr without trees - **s. flower**: *Urechites suberecta*.
Savannah: important seaport of Ga (US), chief naval stores market of world (turpentine oil, rosin) (596).
savin: *Juniperus sabina* - **horse s. berries**: *Juniperus communis*, frt. - **Indian s. (tree)**: *Caesalpinia vesicaria* - **s. tops**: young green twig tips of savin.
savon (Fr-m.): soap - **savory (leaves)**: - **summer s.**: *Satureja hortensis* - **winter s.**: *S. montana*.
savory, alpine: *Calamintha alpina*.
Savoy: 1) **S. cabbage**, *Brassica oleracea* v. *capitata* 2) *Paradisea liliastrum* (also **savoye**).
saw grass (or **sawgrass**) (sUS): *Cladium jamaicensis, C. effusum* & other *C.* spp. or other sedges/grasses with teeth on lf. edge (as sedges/grasses in Everglades of Fla [US]) - **s. palmetto**: *Serenoa repens* - **s. wort** (or **sawwort**): *Serratula tinctoria*.
sawdust: wood meal obt. fr. sawmills; purified s. reagent; purified by treatment with 1% NaOH, 1% HCl, thoroughly washed, & dr.; u. as adsorbent; sometimes admixed in breads (rarely).
Saxifraga L. (Saxifrag.) (nTemp Zone, Arctic, mts. worldwide in both OW & NW): saxifrages; rockfoil; some 400 spp. of ann. to perenn. hbs.; rts. of several Am spp. u. med.; rts. sometimes mistaken for Rheum (Pak) (<saxum, stone +frango, to break; as thought to dissolve "stone" concretions [urin.]) - **S. bronchialis** L. (Sib; Alas [US]): u. in pleuritis, spaghitis (quinsy) - **S. ciliata** Royle (Ind, Pak): zakham-i-hayat; rt. appl. to boils & in ophthalmia; ton. in fevers, dys. diarrh. coughs;

antiscorb. (Azad Kashmir, Pak); (pl. noted in Kaghan Valley; Gilgit, Pak) - **S. cotyledon** L. (sEu): pyramidal saxifrage; lady's cushion; u. as tea substit.; cult. ornam. - **S. dactylites** L. (Euras, nAf, natd. NA): "whitlow grass"; hb. u. in icterus, "glandular hardenings" (Ge) - **S. erosa** Pursh = *S. miranthidifolia* - **S. granulata** L. (Eu, nAf): (meadow) saxifrage; (white) stone-break; hb. u. as lithontryptic (for stone) in Middle Ages & in current folk med.; effect uncertain; in chest dis. (tubers, sd. also u.) - **S. ligulata** Bell. = *Bergenia l.* - **S. micranthidifolia** Steud. (eUS s to Ga): lettuce saxifrage; mountain l.; "deer tongue"; hb. u. for making salads (mt. people, sPa [US]) - **S. peltata** Torr. = *Peltiphyllum p.* Engl. - **S. stolonifera** Meerb. (S. sarmentosa L. f.) (Chin, Jap): beefsteak (saxifrage); strawberry geranium; sailor plant; pl. juice dropped into ear for earache; to stop cough, pertussis; exter. for burns, frostbite (Jap folk med.); cult. ornam. - **S. umbrosa** L. (wEu): London('s) pride; chickens; pl. u. diur.; cult. ornam. - **S. virginiensis** Mx. (e&cNA): early (or rock) saxifrage; supposed to destroy calculi formed in kidney.

Saxifragaceae: Saxifrage Fam.: a large gp. mostly of small hbs. with a few shrubs & trees; distr. through temp. latitudes; fls. usually bisexual, regular, with 4-5 sepals, 4-5 petals, 4-10 stamens, 2-5 carpels; frt. usually a capsule or berry; ex. currants, gooseberry (*Ribes* spp.).

saxifrage, dotted: *Saxifraga nelsoniana* D. Don (S. punctata L.) - **s. (root)**: **small (burnet) s.**: *Pimpinella saxifraga* - **swamp s.**: *S. pensylvanica* L.

saxifrax: sassafras (in common use by sUS Negro).

saxitoxin: mussel (or clam) poison; paralytic shellfish poison; red pigment abundantly produced by dinoflagellates (*Gonyaulax* spp.) harbored in clams & mussels; with neurotoxic effect; often found in "red tides" (esp. in Gulf of Mexico); compd. may improve efficiency of local anesthetic similar to tetrodotoxin.

Scabiosa L. (Dipsac.) (Euras, Af): scabious; pincushion flower; herbs.; ca. 90 spp., many cult. orn. - **S. atropurpurea** L. (S. maritima L.) (sEu): sweet scabious; hb. & fls. u. for scabies (itch), hence name - **S. columbaria** L. (Euras, nAf): (wild) scabious; rice flower; hb. c. scabioside (glycoside); rt. decoc. for colic; rt. chewed for heartburn; dr. pd. pl. as dusting powder for infants; lf./fl./rt. u. for scabies, herpetic eruption, other skin dis.; for condylomata & carcinomata - **S. japonica** Miq. (Jap) sap of fresh lvs. u. to treat gastric dis.; antiemetic - **S. stellata** L. (sEu, wMed reg.): star scabious; u. for itch - **S. succisa** L. *Succisa pratensis* - **S. transvaalensis** S. Moore (sAf): rt. decoc. u. as lotion for sore eyes.

scabious: 1) *Succisa pratensis* 2) *Erigeron philadelphicus* 3) *Scabiosa* spp. - **sweet s.**: *Erigeron heterophyllus* - **wood s.**: s. (1) - **scabish**: *Oenothera biennis*.

scabrin: insecticidal principle fr. *Heliopsis helianthoides* subsp. *scabra*.

Scaevola floribunda A. Gray (Goodeni) (Fiji): u. as stomach med.; hair dye - **S. frutescens** Krause (Goodeni.) = *S. taccada* - **S. oppositifolia** Roxb. (NG): juice of shoots sprayed on areas of eye and ear inflamn. - **S. sericea** Vahl (S. frutescens Krause) (Oceania): frt. juice squeezed into coconut meat & eaten as "sea" medicine; white fls. (& frt.) u. as wash for inflamed eyes (Caroline Is); lvs. crushed & given with coconut to dogs to make scent keener (Solomon Is) - **S. spinescens** R. Br. (Austral): prickly fanflower; rt. inf. drunk for GI(T) pain (Austral aborig.) (2263) - **S. taccada** (Gaertn.) Roxb. (S. koenigii Vahl; S. sericea Forst.) (seAs, Pac area incl. Austral, Polynes): shrub or small tree; bk. u. in ringworm & "leprosy" (Tonga [86]); rt. u. with *Guettarda speciosa* in VD (Gilbert Is [103]); frts. esculent, juice u. as eye wash; lvs. emet. aphrod. for irreg. menses; fls. u. in conjuntivitis; charred pith with betel, &c. in diarrh. (Indones); cold water inf. in stomach trouble (Mad); lf. decoc. appl. to sores (NG); hard wd. useful.

scag (slang): Heroin.

scale, glandular: epidermal gland of multicellular type & more or less globose shape; ex. gl. scales (or trichomes) of peppermint lf. - **s. insects**: small structurally degenerate insects (Homoptera) which have the appearance of scales on the bark; incl. some serious pests, such as the San Jose scale.

scallion: *Allium* spp., esp. shallot & leek.

Scammonia (L): **Scammoniae Radix** (L): (Levant) scammony rt. fr. *Convolvulus scammonia*.

scammonin: ether-sol. fraction of jalap resin, turpeth resin, &c. - **Scammonium** (L.): scammony gum-resin obt. by incision of the living rt. of *Convolvulus scammonia* (not by solvent extraction) - ether-soluble part of fraction of scammony gum-resin, jalap resin, turpeth resin, &c.; represents oligosaccharide acid complex, which hydrolyzes to jalapinolic acid, rhamnose, fucose, glucose, &c.

scammony (root): *Ipomoea orizabensis* - **Aleppo s.**: 1) *Convolulus scammonia* rt. 2) term now also appl. to Mexican s. - **s. de France** (Fr): *Cynanchum acutum* - **Levant s.**: Aleppo s. - **Mexican s. (root)**: *Ipomoea orizabensis* -

Montpellier s.: gum-resin of *Marsdenia erecta* - **s. resin**: galenical resin obt. mostly fr. Mexican s.; small amts. fr. Levant s. - **s. de Smyrna**: *Oxystelma esculenta* - **spurious s.**: s. de France - **wild s.**: *Ipomoea pandurata*.

Scandia: Scandinavia.

Scandix cerefolium L. = *Anthriscus c.* - **S. grandiflora** L. (Umb.) (seEu, wAs): shepherd's needle; young pls. eaten as salad (parts of Gr) - **S. pecten-veneris** L. (Euras, natd. Pac NA): Venus' comb; lady's comb; young st. tops (shoots) occas. u. as salad hb.

Scaphium macropodum Beumée (S. affine Pierre) (Sterculi) (seAs): sds. soaked in water several hrs. & adm. with sugar, has lax. effect & u. in diarrh. dys.; given in fever, sunstroke, &c.; very popular in Far East.

scarlet fever antitoxin: s. f. streptococcus a.: sterile aqueous soln. of antitoxic subst. fr. blood serum of healthy animals immunized against toxin of the spp. of *Streptococcus* believed causative of scarlet fever; formerly u. in prophylaxis and treatment of s. fever (scarlatina); now mostly superseded by antibiotics (incl. s. f. anti-scarlet fever globulin).

scat: fecal masses of mammals.

scatula: flat rectangular box u. to dispense pds. & pills (fr. M. L. for parallelipiped).

sceau d'or (Fr): Hydrastis.

Sceletium tortuosum N. E. Br. (Aizo.) (sAf): kanna; Africans use as a narcotic; c. mesembrine (*cf.* cocaine); lf. decoc. u. as sed., intoxicant, to increase body strength; lf. chewed for toothache; mydriatic; some addiction to prepn. of above-grd. parts called "kougoed."

Sceptrocnide macrostachya Maxim. (Laportea m. Ohwi) (monotypic) (Urtic.) (Jap): miyama-irakusa; herbage u. occas. as veget. (Akita, Jap).

Schachtelhalm (Ge): *Equisetum* sp.

Schaf (Ge-m.): sheep - **Schafgarbe** (Ge): milfoil, *Achillea millefolium*.

Schale(n), bittere (Ge): bitter orange peel(s) - **süsse S.**: sweet o. p.

Scharbockskraut (Ge): *Cochlearia* - **Scharfhahnenfuss** (Ge): *Ranunculus acris*.

Scharlachwurm (Ge): cochineal.

Schattenmorellen (Ge): type of sour cherry (SC does better in shaded areas; ex. n. side of bldgs., hence Nordkirsche [north cherry]).

Scheelea butyracea Karsten (Palm.) (Co, Br; cult. Domin Rep): New Granada oil palm; palma dulce; frt. source of FO, Cuesco fat; u. cooking purposes (but sl. rancid) & u. to make soap - **S. costaricensis** Burret (CR): corozo; a tall palm; source of FO - **S. martiana** Burret (Attalea excelsa Mart.) (Equator: Br, esp. Bahia): Scheelea palm; urucuryсiro; coqueiro urucuri; kernel source of vegetable FO (ouricuri fat or oil); valuable in mfr. soaps & artif. lard (*cf.* copra oil); also ouricury wax (wax fr. underside of lvs. similar to syagrus wax in indl. use) - **S. preussii** Burret (Guat, Mex): corozo palm; potential source of FO; other *S.* spp. similarly used, ex. **S. zonensis** Bailey (Pan).

Scheele's green: S. mineral: a copper arsenic comp., cupric arsenite; u. as pigment, insecticide.

Schefflera J. R. & G. Forst. (Arali) (trop & subtr): over 200 spp. of shrubs, vines, trees (-25 m); popular ornam. & honey plants (*Brassaia*) - **S. aromatica** Harms (Indones esp Java): u. as side dish (lablab) of rice table (Sunda) - **S. capitata:** c. spermicidal saponins - **S. cumingii** Harms (PI): said useful in treating stomach dis. - **S. delavayi** (Franch.) Harms (Chin): lf. & bk. decoc. u. as sudorific & to relieve mechanical injury - **S. digitata** Forst. (NZ): "seven leaf"; wd. soft, u. to "make" fire (Maoris) - **S. elliptifoliola** Merr. (PI): decoc. u. after childbirth by mothers - **S. insularum** Harms (PI): juice of fresh lvs. u. as purg. - **S. octophylla** (Lour.) Harms (Jap, Chin, Indochin, eAs): fresh branches of tree u. as wash for itch; lvs. u. for flu (Chin); wine inf. in rheum.; bk. & lvs. diur. diaph. - **S. odorata** Merr. & Rolfe (PI): bk. u. as bechic; bk. scrapings appl. hot to aching limbs; resin vulner.; lvs. decoc. antiscorbutic - **S. piperoidea** Elm. (PI): ton. after childbirth & to treat stomach dis. - **S. trifoliata** Merr. (PI): lvs. with wine given as emmen.; to treat tympanites of children - **S. venulosa** Harms (eAf); c. saponins; to treat tympanites of children - **S. vitiensis** (A. Gray) Seem. (Fiji): aq. inf. of leaf drunk for lung trouble - **S. volkensii** Harms (eAf): gum u. as remedy for coughs, colds, lung dis.; ornam.; with 5-7 leaflets.

Scheinquirl (Ge): verticillaster.

Schellenbaum (Ge): *Thevetia peruviana* (name also appl. to *Cerbera* spp; *Hippomane mancinella*).

Schepti: tapeworm remedy made fr. frt. of *Phytolacca dodecandra* (Ethiopia).

schi: shea.

Schick test (control): dil. inactivated diphtheria toxin u. for detn. of immunity to the disease; u. intradermally.

Schierling (Ge): Conium - **Schierlingstanne**: *Tsuga canadensis*.

Schilddrüse (Ge): thyroid gland - **Schildkraut** (Ge): Scutellaria.

Schima wallichii Korth (1) (The.) (seAs): sap of bk. very corrosive; subsp. *noronhae* Bloemb. u. for otitis; in mixts. for nausea, &c.; bk. to stun fish. anthelm., rubefac. (Ind); wd. used in construct.

Schinopsis balansae Engl. (Anacardi.) (Arg, Par): (willow leaf) red quebracho; wd. c. 25% tannin; u. to produce tannins; ext. of Lignum Quebracho colorado (Santa Fe quebracho) u. to treat burns; as tan for high-grade leathers (2190) - **S. lorentzii** Engl. (S. quebracho-colorado F.A. Barkley & Meyer) (Arg, Par, Ur): (Lorentz) red quebracho; logs imported fr. Argentina (-1955-) for use in prepn. tannin & in tanning hides; source of valuable red timber (said hardest, heaviest, most durable wd. known [2353]); fuel wd.; bk. also u. for prepn. tannins; wd. (c. 16-20% tannin) exts. c. 63-4% tannins (solid "crystal") (1916); soft pastes less tannin (45%) (*cf.* kino); plant produces a dermatitis venenata in susceptible individ. (allergic reaction).

Schinus L. (Anacardi.) (Mex-Arg): resinous diecious trees; some 30 spp. - **S. areira** L. (nArg, widespread): aguaribay; lf. c. rutin, quercitrin, isoquercitrin; u. antisept. (extern.); resin purg.; bk. & lvs. appl. exter. for edema of feet - **S. dependens** DC (wSA, Ch, Br+): Peru peppertree; shrub u. in rheum. (Ch); bk. c. balsaminic resin; u. in gout, syph. as vulner.; frt. in bladder dis.; sds. (huingan) u. as stom. diur. in bladder dis.; in hysteria, in prepn. beverage (chicha) - **S. latifolius** Engl. (Ch): Chilean pepper tree; p. shrub; molle; lilén; u. as ton. in rheum., bronchitis, antispasm.; frts. in prepn. alc. bever. - **S. molle** L. (Mex through Andes to Ur; cult. & natd. Calif [US]): (California) pepper tree (597); Australian p. t.; Peruvian mastic tree; "pico" (Mex); small pink frt. ("molle seed") c. 3-5% VO with phellandrene, limonene, carvacrol; early claims of piperine incorrect; frt. u. to make a beverage (with atole); purg. (veterin.); substit. for & adulter. of black pepper & cubeb; much u. in local med. (1917); contraceptive (Ur); bk. u. exter. for wounds & sores; "resin" (oleogumresin) exudate (bk.) ("American mastic") u. for eye dis. & chewed to harden gums (Pe); purg. & in gout; rheum.; lvs. u. exter. like bk., int. as diur.; shoots u. to clean teeth, strengthen gums, give pleasant odor & taste in mouth; cult. shade tree - **S. polygamus** Carr. (Br, Ch): molle; incienso; boroco; trementina; frts. u. in prepn. chicha (beverage), as stom. diur. in hysteria, in cases of bladder dis.; balsamic resin fr. bk. u. against gout; syph., wounds, & in myiasis prod. by pricking house flies (Rio Grande do Sul, Br) - **S. terebinthifolius** Raddi (Br, Par): (American or Brazilian) pepper tree; Mexican p.; frt. c. schinol, terebinthone; u. (also lvs.) in baths for wounds & sores; inf. diaph.; balsam fr. lvs. & frt. u. in tumors; condim.; bk. u. as stim. ton. astr., exter. in rheum., gout; exhalations of pl. thought by some injurious; cult. orn., natd. (Fla [US]) - **S. weinmanniaefolius** Engl. (Br. Par): bk. stim, ton, astr.; u. exter. in gout, rheum.; gargle & to wash wounds (Par).

Schinzent (Ge): ginseng.

Schisandra (Schizandra) **chinensis** Baill. (Schisandr. - Magnoli.) (Chin, neJap, nKor, Amur): Chin lemonade tree; limonnika (Russ); chosen-gomishi (Jap); small climbing shrub; frt. u. as ton. (Chin) (85), astr. sed. for cold symptoms (cough+); strengthening agt.; inf. of frt. & vine u. antipyr. & against seasickness (Ainu); sds. u. as stim. in fatigue, shortness of breath, muscul. pains (598); lvs. u. as stim. in depressions, after hard exertions; condiment.

Schistostemon macrophyllum Cuatrecasas (Houmiri.) (Br): arom. frts. eaten for female sterility; lf. & bk. decoc. u. for heavy colds, bronchitis, constip. TB.

schiu: shiu.

Schizachyrium scoparium Nash (Gram) (NA): broom sedge ("b. sage"); broom (beard grass); (little) bluestem; bunchgrass; wiregrass; rhiz. & rts. in folk med. as alter. ton. diaph. in erysipelas; st. ashes u. for syph. sores & int. for other VD (Am Indians); stems u. to sweep yards (seUS).

Schizobasopsis Macbride = *Bowiea* Hook. f. & Haw. (Lili).

Schizocarphus nervosus F. v.d. Merwe (Lili) (sAf): bulb u. as dys. remedy - **S. rigidifolius** F. v. d. Merwe (Scilla rigidifolia Kunth) (sAf): wild squill; bulb decoc. u. for nervousness (children), rheum. fever; toxic plant (—> *Scilla*).

schizogenous oil gland: cavity formed by separation of cells within a tissue & later formation of a secreting epithelium lining of same; ex. eucalyptus.

Schizoglossum atropurpureum E. Mey. (Asclepiad) (Af): rt. edible - **S. shirense** N. E. Br. (eAf): uzari (Nyamwezi tribe); African root; rt. of tree u. as stom. aphrod. in dys. & diarrh. (also entire pl. u.) (*cf. Gomphocarpus*) (c. uzarin, glycoside) (1919).

Schizomycetes: Bacteria (lit. fission fungi).

Schizonepeta tenuifolia Briq. (Nepeta japonica Maxim.) (Lab) (eAs): flg. pl. u. in fever, sore throat, blood in feces (Chin); decoc. of whole pl. plus brown sugar u. for cough.

Schizophyta: acellular or thalloid plants without differentiated tissues; made up of 2 classes: *Schizomycetes* (Bacteria) & *Cyanophyceae* (sometimes also viruses); at times called the procaryotes (all other organisms Eucaryotes).

Schizostachyum dielsianum Merr. (Gram) (PI): baboos; rhiz. decoc. u. as refreshing drink; young shoots u. to reduce opacity of the cornea -

S. dumetorum Munro (Chin). tang chuk; u. as drug; rhiz. sold in drug stores - **S. glaucifolium** Munr. (Samoa): Samoan bamboo; lf. ash mixed with "samoan oil" for burns - **S. lumampao** Merrill (Bambusa l. Blanco) (PI): "glebuk sao"; u. for puerperium.

Schizozygia coffaeoides Baill. (Apocyn.) (monotypic) (Tanz): lvs. boiled & patient sits in steam for painful eyes.

Schkuhria pinnata O. Ktze. ex Thellung (Comp.) (sAf): khaki bush; pd. lf. swallowed with water for mal., influenza, colds; cattle browsing causes tainted milk; to kill fleas (wetting floors); st. & sd. inf. u. in mal. (Bo).

Schlafmuetzchen (Ge): *Escholtzia cailifornica.*

Schlange (Ge-f.): snake, serpent - **Schlangenwurzel:** snakeroot; Serpentaria.

Schlechtendalia sinensis (Aphides-Hemiptera-Insecta): the aphis sp. which attacks *Rhus javanica* & gives rise thereby to Chin & Jap galls.

Schlehe (Ge-f.): sloe.

Schleichera oleosa Oken (S. trijuga Willd.) (monotypic) (Sapind.) (Ind, trop seAs to subtrAustral): kussum; pongro (Camb); gyt (Burma); Ceylon oak; large tree (-25 m); acidulous frt. edible & marketed (khussambi nuts); source of FO (kusum oil) (ca. 70% of kernel); edible, in mfr. soap; promotes hair growth; base of Macassar (hair) oil (599); u. with *Sapindus* frt. as hair wash; bk. astr. antimal. (inf.); in tanning; inf. u. as dressing for furuncles (immature) & adenitis; wd. strong, tough, but warps, splits, cracks during drying; u. for wheels, axles, source of 2nd class charcoal; lvs. eaten with rice; host tree for insects producing the best type of lac (Mirzapore lac) fr. wh. a gum-resin is obtained; sds. insecticidal.

Schlempe (Ge): Vinasse (Ge): residue fr. grape fermentation u. as fertilizer or potash source.

Schlerocaryum wallichianum (Terminalia horrida Steud.) (Santal): c. spermicidal saponin.

Schlohdorn (Ge): *Prunus spinosa.*

Schmaltzia Steud. = **Schmalzia** Desv. = *Rhus.*

Schmalz (Yiddish): chicken fat; u. in cooking.

Schmetterling (Ge-m.): butterfly.

Schmidelia L.: *Allophyllus.*

Schminke (Ge): rouge.

Schneeballrinde (Ge) (snowball [tree] bark): *Viburnum opulus* - **Schneeglöckchen** (Ge): *Galanthus* spp.

Schnittbohnen (Ge): string beans, *Phaseolus vulgaris.*

Schoenobiblus peruvianus Standl. (Thymelae.) (Col+): crushed frts. & rt. u. in prepn. of curare of Kofan Indians.

Schoenocaulon jaliscense Greenman (Lili) (Mex): cebadilla; rts. u. to kill maggots in wounds (powd. or decoc. as wash) - **S. officinale** Gray (Mex, CA, Ve, Pe): sabadilla; Indian caustic barley; sds. c. 1-5% mixed sterol alk. called collectively "veratrine"; incl. cevadine, veratridine, sabadine, sabadilline, &c.; u. as parasiticide esp. for lice (head) (decoc. tinct., or vinegar prepn.); insecticide (flies); in cattle wash powders; in homeop. for coughs, migraine, &c. (toxic!); in neuralgia (salves).

Schoenoplectus articulatus Palla (Scirpus a. L.) (Cyper.) (Med reg., eAs-wet places): pl. u. as purg.; rhiz. eaten (—> *Scirpus*).

Schoepfia schreberi J. F. Gmelin (Olac.) (sMex, WI, CA, nSA): shrub or tree; bk. decoc. u. to bathe boils & pimples (Totonac, sMex).

Schokolade (Ge-f.): chocolate.

Schöllkraut (Ge): *Chelidonium majus.*

Schönteten's reaction: wet test for aloes (green fluorescence with borax).

Schopftintling (Ge): *Coprinus comatus.*

Schorf (Ge): scab. - **Schotendotter** (Ge): *Erysimum* spp. - **Schotenpfeffer** (Ge): Capsicum.

Schotia afra Bodin (S. speciosa Jacq.) (Leg) (sAf): boerboon (Du) (boer bean); frt. & sd. edible; bk. astr. (16% tannin) u. med. & as tan; hard durable wd. - **S. brachypetala** Sond. (sAf): African walnut; Boer bean; bk. decoc. u. for heartburn; rt. c. tannin, u. in diarrh. & heartburn; sd. edible after roasting.

Schradera marginata Standl. (Rubi) (Co): quera; young shoots of this epiphyte chewed by some Indians to preserve teeth (teeth blackened).

Schrankia nuttallii Standl. (*Mimosa quadrivalvis* L. var. n. [DC] Barneby) (Leg) (s&cUS): cat claw; sensitive brier; shame weed; Adam weed; ointment prepn. for fungus infections, incl. athlete's foot (claimed effective).

Schraubenbaum (Ge): screw palm. (2577).

Schrebera swietenioides Roxb. (Ole.) (Ind): moha (Hindi); ghanta (Sanskrit); rt. u. in Hansen's dis.; frt. paste appl. to boils, cuts, wounds (124).

Schreckstoff (Ge): repellant.

Schumanniophyton magnificum Harms (Rubi) (wAf): stim. aphrod. (but harmful); stim. for hunting dogs; fish poison.

Schwamm (Ge-m.): 1) fungus 2) sponge 3) excrescence - **Schwanz** (Ge-m.): 1) tail 2) penis (slang) - **schwarze Schlangenwurzel** (Ge): Cimicifuga - **schwarzer Pfeffer** (Ge): black pepper - **s. Senf.**: b. mustard - **Schwarznuss** (Ge): *Juglans nigra* - **Schwarzpappel** (Ge): *Populus nigra* - **Schwarzwurz** (Ge): Symphytum.

Schwein (Ge-m.): hog - **Schweinesschmalz** (Ge): lard - **Schweinfurth's green**: copper acetoarsenite; u. insecticide, &c.

Schweinfurthia sphaerocarpa A. Br. (Scrophulari) (Pak): frt. & powd. lvs. u. in typhoid fever; pd. pl. snuffed up nose for epistaxis.

Schweitzer's reagent: Cuoxam; cupric oxide dissolved in 20% or stronger ammonia water; u. to dissolve cotton, silk, linen; wool is insol.

Schweizertee (Swiss tea): Thea Helvetiae; mixt. of hbs.; hyssop; thyme; melissa; wormwood; ground ivy (Glechoma); coltsfoot; yarrow; &c.

Schwenkia americana L. (Solan.) (WI, continental tropAm; natd. wAf): tabaco cimarrón (Cu); ann. hb.; pl. decoc. u. in rheum. pain & swellings (Nig); as cough med. (Ghana); in chest complaints (Angola); c. cardiac glycoside; rt. decoc. lax. of choice for infants only a few days old; to strengthen children; antidote (Ivory Coast).

Schwertlilie (Ge): *Iris* - **blasse S.**: *I. pallida* - **stinkende S.**: *I. pseudacorus*.

Sciadopitys verticillata Sieb. & Zucc. (monotypic) (Taxodi) (sJap): umbrella pine; u. fir; parasol p.; koyamaki (Jap); sd. FO (Oleum Sciadopytis) u. as lamp oil; in dye and varnish indy.; wd. useful, resistant to moisture; ornam. planted near temples; bk. made into oakum.

Sciadotenia toxifera Krukoff & Smith (Menisperm.) (Ec, Co): u. to prep. curare (arrow poison) (Canelos Indians, Ec) (formerly Witoto Indians, Col) (2301).

Scilla (L. It): **Scillae Bulbus** (L.): squill - **Scilla autumnalis** L. (Lili) (Euras, nAf): winter hyacinth; (autumn-flowering) squill; u. much like off. squill - **S. cooperi** Hook. f. (sAf): wild squill; bulb with digitaloid action but less active than off. squill; u. as soothing med. to pregnant women; adm. to cow to ensure a succession of calves of same sex(!) - **S. hyacinthiana** (Roth) Macb. (S. indica Baker) (Ind.): south Indian squill; similar constits. & therap. uses as *Urginea maritima* & u. as substitute therefor; expect. cardiotonic, diur.; exports to Ge - **S. indica** Roxb. = *Urginea i.* - **S. lanceaefolia** Bak. (sAf): wild squill; bulbs c. hemolytic sapogenin, alk.; u. expec. diur.; local stim., in some skin dis. - **S. maritima** L. = *Urginea m.* - **S. nonscripta** Hoffmgg. & Link = *Endymion n.* - **S. rigidifolia** Kunth = *Schizocarphus r.* - **S. sinensis** (sChin): pounded bulbs appl. to abscesses, mastitis.

scilladienolides: glycosides found in some pls. & in the skin & skin glands of some toads; cardioactive compds.; ex. bufalin, bofogenin, &c. - **scillarenin**: aglycone of proscillaridin A.; with cardiotonic action of squill - **scillarens**: active glycosides of white squill (A: xall.; B: amorphous) - **scillaridin**: the aglycone formed on hydrolysis of scillaren A. - **scille (, maritime)** (Fr): squill.

Scincus officinalis Laub. (Scincidae - Ord. Squamata - Reptilia) (Eg, Nubia, Arab-deserts [Sahara]): skinc; skink; a desert lizard (-21 cm); all parts u. (dr.) int. as stim. & aphrod. (73).

Scirpus L. (Cyper.) (worldwide; usu. wet places): bulrush; tule; fibrous parts u. in basket making; sds. sometimes u. as food (Am Indians); tubers of Mex sp. u. as hallucinogen (1739) - **S. articulatus** L. = *Schoenoplectus a.* - **S. atrovirens** Willd. (Mex): u. as hallucinogen - **S. cernuus** Vahl (widely distr., incl. Pac NA & SA): low club-rush; pl. ground up with a locust to make med. for sick children (sSotho, Af) - **S. grossus** L.f. (S. kysoor Roxb.) (tropAs): rt. astr. u. to check diarrh. & vomiting (Hindu med.); lax., diur. for fevers; GC - **S. lacustris** L. (S. acutus Muhl.; S. validus Vahl) (NA incl. WI): American great (or soft stem) bulrush; (true) b.; sts. u. to make chair seats, matting, &c.; rhiz. & young shoots u. as food (Ariz [US] Indians); rhiz. u. astr. diur.; grown to purify water (Ge, Neth); ornam. - **S. maritimus** L. (nEuras; adventive in Am): salt marsh club-rush; tubers said galact. emmen. antipyr.; to prevent polycythemia, amenorrh (Kor); cooked with Chinese wine as digestive (Taiwan); rhiz. sometimes found as adulter. of *Carex arenaria* rhiz. - **S. microcarpus** Presl. (S. rubrotinctus Fern.) (e&cNA): bulrush; hb. u. for sore throat; rhiz. u. for abscesses (Maritime Indians) (1739) - **S. supinus** L. (trop wAf): pl. pounded in water & u. as scorpion sting antidote (nNig).

Scleranthus annuus L. (Caryophyll.) (Euras, nAf, natd. e&cNA): gravel chickweed; (annual) knawel; a common low-spreading ann. weedy pl. with inconspicuous greenish fls.; formerly u. in cancer - **S. perennis** L. (Euras): perennial knawel; Polish cochineal (since bears an insect with dye properties); source of med. Herba Scleranthi perennis.

sclereid: sclerified cell: stone cell; a sclerenchymatous cell of various shapes, often approx. equidiametric, with thick, lignified cell walls bearing numerous prominent pits - **sclerenchyma**: tissue composed of sclereids & fibers (in phloem & xylem); furnishes hardness & support (ex. nut shells) - **sclerophyllous** (adj.): with thick, leathery lvs. - **sclerotic cell**: sclereid - **sclerotium**: hardened mass of mycelium (resting stage of fungal life cycle); ex. ergot.

sclererythrin: reddish pigment in ergot; u. as a color test for ergot.

Scleria laevis Retz. (Cyper.) (eAs): sds. chewed in the betel quid for cough (Mal) - **S. lithosperma** Sw. (seAs, Oceania): yebey (Caroline Is); rt. decoc. u. as post-partum protective med. (Mal); antipyr. (nwSolomon Is); for cough (Caroline Is) - **S. pergracilis** Kunth (seAs): hb. u. as antipyr. & to treat hoof and mouth dis. (Indones).

Sclerocactus polyancistrus Br. & R. (Cact.) (swUS deserts): (Mohave) bisnaga cactus; frt. u. to make candy; st. c. water (emergency source).

Sclerocarpus divaricatus Benth & Hook. (Comp.) (Salv): "calacote"; u. for fever.

Sclerocarya birrea (A. Rich.) Hochst. (Anacardi.) (trop wAf e to Tanz): "birr" (Wolof, Senegal); pl. u. as antivenom, for rabies(?!) (Sen); in combns. for syph. leprosy; lvs. c. tannins & flavonoids; decoc. or macerate u. in hypoglycemia p.o.; pos. effects with low toxicity; rts. in bilharziasis, scabies (in baths) (Tanz); toothache (Sen); bk. decoc. purg., u. as paste for inflamm. (local); frts. eaten as euphoriant (Senegal); bk. c. catechol tannins, u. in dys. (Nig); subsp. **caffra** Kokuaro (S. caffra Sond.) (sAf, Madag): cider tree; marula (Lobedu tribe); frt. edible, sought by baboons & elephants; sometimes elephants consume fermented frt. & become inebriated; u. to make beer & spirits; some believe hallucinogenic; bk. c. 4% tannin, u. in diarrh. dys. bilious vomiting, mal.; gum exudn. c. tannin u. with soot to make inks; frts. insecticidal, source of maroola nuts; var. *oblongifoliolata* Engl. (eAf): bk. decoc. u. to wash wounds; lf. juice in GC; sds. source of FO - **S. schweinfurthiana** Schinz (Af): produces sweet-sour juicy plum-like frts. with nutty-tasting seeds also eaten; frts. u. to make beer-like beverage (Ovambos) (Namibia).

scleroproteins: albuminoids: proteins found in harder parts of animal bodies (such as horns), serving to protect & support; ex. keratin, fibroin (found in Arthropoda); collagens (of connective tissues).

Scleropyrum wallichianum Arn. (Pyrularia wallichiana DC) (Santal.) (seAs incl. Thail): c. spermicidal saponin; frts. edible; bk. of st. u. in edema & icterus; rt. bk. u. for buboes; fl. & frt. for brain dis.

Scobitost™: roasted sawdust u. as dusting powd. for wounds (sometimes with iodoform) (-1915-).

scoke: *Phytolacca americana.*

Scolopendrium vulgare Newm. = *Asplenium s.*

Scolophyllum ilicifolium Yamazaki (1) (Scrophulari.) (Thail, Vietnam, Camb): crushed pl. appl. to wounds or sores (Camb).

Scolymus hispanicus L. (Comp.) (Med reg., sEu): Spanish oyster plant; golden thistle; tap rts. boiled & eaten; salsify formerly more popular & often cult.

scopa (It): *Calluna vulgaris.*

Scoparia dulcis L. (Scrophulari) (tropAm, now pantrop): hierba de pajarito (Mex); sweet broom; teeth brush (Virgin Is); c. alk. (602); tops u. for local appln. gums of teething babies (decoc.) (Virgin Is); in dys. (Curaçao); dig. dis. (Camb); entire pl. u. in GC & female dis. (Ve); antipyr in mal (seAs); purg. in colics (Pan); branches placed in drinking wells to give water "cool taste"(!) (WI); branches u. in mfr. brooms, &c. (wiwirisisibi) (Suriname); hb. u. for bruises (Mex); for healing wounds (Ec); dr. hb. to "purify the blood" (Guyana); lvs. u. for tea for griping of baby (Jam); lf. tea u. in bath & as cooling drink (DominRep); inf. u. for contusions (Bo); u. with broom to kill fleas, hen lice, &c. (Salv, Par); weed.

Scoparius (L): broom tops, *Cytisus s.*

Scopola (L): **s. root**: *Scopolia carniolica* (also **scopolia root**) - **Scopola** Jacq. = *Scopolia* Jacq. corr. Link.

scopolamine: hyoscine (formerly chiefly u. in US, Eng; newer term preferred to avoid confusion with hyoscyamine) (1920): characteristic alk. of *Scopolia* spp., some *Datura* spp., &c.; u. (salt as HBr) anticholinergic, analg. (608); in daytime sedation; considered generally best drug for motion sickness (603) (topical appln.); liquid but forms cryst. monohydrate; often u. as **s. aminoxide** (HBr).

scopolamine "stable": soln. of scopolamine rendered stable by addn. of sugar alcohols, such as mannitol, dulcitol, &c. (marketed -1915-) - **scopoletin**: methyl-aesculetin; fluorescent comp. occurring in Belladonna, Scammony, Ipomoea, & Gelsemium roots.

Scopolia carniolica Jacq. (S. atropoides Bercht. & Presl; Hyoscyamus scopolia L.) (Solan) (s&eEu, esp. Russ & Balk): nightshade-leaved henbane; European scopol(i)a; "Russian belladonna"; rhiz. c. 0.3-0.8% total alk., much L-hyoscyamine (0.22-0.4%), less atropine (0.03%) & scopolamine (traces); lvs. with less tot. alk. but proportionately more scopolamine (45-66% of total alk.); similar uses to Belladonna rt. but rather more narcotic (analg.; comm. source of alks.); morphol. differences fr. belladonna (604); u. in paralysis agitans; in folk med. for rheum., gout - **S. japonica** Maxim. (Jap, Chin): Japanese scopol(i)a; J. belladonna; J. henbane; rhiz. (Japanese belladonna root) u. similarly to *S. carniolica*; expressed juice of rhiz. u. in Far East for neuralgia, asthma; ointment in hemorrhoids; substit. for belladonna - **S. lurida** Dun. (S. anomala Airy-

Shaw) (Himalay-Nepal, Sikkim): with high alk. cont., hence u. to prepare alk.; to dilate pupils (tinct. of lvs.); otherwise like *S. carniolica* - **S. mutica** Dun. = *Hyoscyamus muticus* - **S. parviflora** Nak. (eAs): rts. u. in eye lotions; anthelm.; to prevent catching common cold (Kor) - **S. tangutica** Maxim. (Anisodus tanguticus Maxim.) (wChin, Tibet): c. 2.6% tot. alk. (rts.), 2.9% (shoots), of which 0.4% L-hyoscyamine, 0.26% scopolamine; cuskhygrine (rts.), anisodamine; u. for prodn. of alks.; in med.

Scordium (L): *Teucrium s.*, hb.

Scorpio (L): genus of African scorpions, utilized in Zodiac sign - **S. europaeus** L. = *Euscorpius italicus.*

scorpions: members of order Scorpiones; occ. in warmer areas; considerable usage in folk med.; the cause of frequent envenomenations & poisonings, sometimes fatal (antivenin u. to treat) in trop &subtr Am & Af (ex. Arg); more than 600 spp. known; v. under *Euscorpius*, *Scorpio*.

scorza (It): bark.

Scorzonera (L, It): *Scorzonera* L. spp. (Comp.); winter asparagus; snake weed or garden viper grass (fr. Sp *escorzonera*, viper rt.) - **S. hispanica** L. (Euras, c&sEu to cAs): viper's grass; Spanish salsify; rt. c. glycosides, histidine, crude fiber, starch; mucil., sugar; u. nutr. condim. stom. diaph. diur. antipyr.; to stim. appetite; in pulmon. dis., colds, hypochondria; fresh pl. veget.; like salsify & dandelion in prop.; dr. as coffee substit. - **S. humilis** L. (Eu. Caucasus): humble viper's grass; Kleine Schwarzwurzel (Ge) (small black root); rt. u. as preceding; fl. heads u. to adulter. Arnica fl.; young lvs. as veget. (spinach) - **S. kirghisorum** (Pamiro-Alai, USSR): considered valuable rubber plant (605) - **S. tau-saghyz** Lipsch. & Bosse (cAs., USSR): tausaghyz; rts. u. as source of natural rubber (latex) (30% of dr. rts.): so-called agrarcaoutchouc (farm rubber), u. in World War II along with latex of *Taraxacum koksaghyz* (&c.).

Scotch broom = *Sarothamnus scoparius* - **S. thistle:** *Onopordum acanthium*; *Cirsigum vulgare*.

scouring rush: *Equisetum hyemale* ("scouring brush" appar. only a curious error or typo).

scrambled eggs: *Corydalis.*

scrap: crude gum turpentine congealed on the tree trunk (*cf.* Gum Thus of commerce).

scraper: knife; a tool u. in turpentining; forms new channels secreting oleoresin.

scrapfish: discarded fragments of fresh fish, collected & fed to minks, &c.; also whole fish of turbot, whiting, &c.

screw-palm: screw pine: *Pandanus* spp. found on trop seashores; frt. flesh often edible. - **screwpod**: *Prosopis juliflora.*

scrofula: scrophula: TB infection of cervical (neck) lymph glands; appar. commoner in past days, since frequently mentioned then as indication - **s. plant**: *Scrophularia marilandica, S. nodosa* - **scrofulaire** (Fr): *Scrophularia* spp.

Scrophularia aquatica L. (S. umbrosa Dumor.; S. auriculata L) (Scrophulari) (Euras fr. Brit Isles to nAf): water betony; poor man's salve; rhiz. u. like that of *S. nodosa* - **S. buergeriana** Miq. (eAs): rt. antipyr. ton. somewhat cardioton.; in mal., tonsillitis - **S. koraiensis** Naika (Kor): rts. u. in Korean folk med. for fever, typhus; mal; c. scropheanoside, harpagoside (iridoid glycosides), aucubine - **S. marilandica** L. (e&cNA): carpenter's square; (Maryland) figwort; M. figroot; hb. u. emmen. alter. vulner. rt. inf. u. in insomnia, restlessness - **S. ningpoensis** Hemsl. (Chin): bitter saltish rt. u. for dry mouth, sore throat, boils - **S. nodosa** L. (Euras, natd. eNA): (great) figwort; common f.; throatwort; rhiz. c. aucubin, saponins, bitter subst. alk.; decoc. of rhiz. & rts. u. in skin dis. scrofula, erysipelas; hb. c. saponins, diosmin, tannin; u. as diur. in eczema; similar to rhiz. - **S. oldhami** Oliv. = *S. buergeriana.*

Scrophulariaceae: Figwort Fam.; c. 4,500 spp. of hbs. shrubs, & trees; cosmop.; fls. usually with irreg. often lipped corolla; stamens 5, pistil bicarpellate; frt. capsule; pls. bitter, often narcotic or toxic; ex. *Digitalis, Chelone, Mimulus.*

scullcap: *Scutellaria lateriflora* & other *S.* spp. - **European s.**: *S. galericulata* - **hairy s.**: *Scutellaria elliptica* - **heart-leaved s.**: *Scutellaria ovata* - **hyssop s.**: *Scutellaria integrifolia* - **mad dog s.**: *S. lateriflora* - **marsh s.**: European s. - **narrow leaved s.**: *S. integrifolia* - **slender s.**: marsh s. - **southern s.**: heart-leaved s. - **western s.**: 1) *S. incana.* 2) (US misnomer): *Teucrium* spp. ("germander").

scumpia (Russ.): *Cotinus coggygria.*

Scuppernong: *Vitis rotundifolia*, a wild grape, with amber green frts.

Scurrula gracilifolia Danser (Loranth.) (seAs): lf. decoc. u. in lumbago, to make teeth firm & hair to grow; frts. to brighten vision - **S. liquid-mabaricola** Danser (Taiwan): branches chief ingred. in remedy for menorrhagia, menstr. pain, postpartum ills - **S. parasitica** L. (Chin, seAs): whole pl. (shrub/parasitic on trees) (Liquidambar, &c.) u. as kidney, to quiet pregnant uterus; shoot decoc. for pain esp. of legs (2448) & many other uses.

scurvy grass: s. weed (, **common**): *Cochlearia officinalis.*

Scutellaria altissima L. (Lab) (Eu): scullcap; hohes Helmkraut; helmet flower; u. similarly to *S. galericulata* - **S. baicalensis** Georgi (S. macrantha Fisch.) (Russ, Chin, Jap, Kor): large-flowered skullcap; Baical skullcap; rt. (Jap drug "wogon") c. wogonin, baicalin, scutellarin (glycosides); VO; u. astr. in diarrh.; acts to normalize blood pressure; in shingles - **S. discolor** Colebr. (As, Af): u. to treat pain in loins (Sudan) - **S. elliptica** Muhl. (S. pilosa Mx) (e&cUS): hairy skullcap; decoc. of hb. u. to promote menstruation, in diarrh, (Cherokee Indians) - **S. galericulata** L. (circumboreal: nEuras, Cana, nUS-Calif): common (or marsh) scullcap; European s.; hb. u. ton. spasmolytic, in some areas against mal.; in CNS dis., as Parkinsonism, tremors; as lax. (Delaware Indians); as heart med. (Ojibway Indians); antipyr.; rabies(!) (Cana); in muscle spasms, in menstrual cramps; insomnia: adulter. of *S. lateriflora* - **S. incana** Biehler (S. canescens Nutt.) (e&cUS, eCana): western skullcap (comm.); hb. u. as replacement for & adulter. of *S. lateriflora*; hb. u. in folk med. - **S. indica** L. (eAs): decoc. ingested for injuries; poultice appl. to skin for fungous dis.; carm. - **S. integrifolia** L. (e&cUS): hyssop scullcap; hb. u. much like *S. lateriflora* - **S. japonica** Burm. = *Plectranthus japonicus* - **S. lateriflora** L. (US & Cana): (side-flowering) skullcap; mad-dog s.; mad (dog) weed; woodwort; hb. c. scutellarin, VO, FO, tannin, resin; u. bitter ton. antipyr. antispasm. diur. nervine; formerly reputed in rabies(!), chorea, neuralgias, epilepsy: this is one of the most adulterated drugs - **S. luzonica** Rolfe (PI): decoc. for stomach pains - **S. macrantha** Fisch. = *S. baicalensis* - **S. ovata** Hill (S. cordifolia Muhl.) (s&cUS): southern skullcap; heart-leaved s.; hb. u. bitter; adulter. of *S. lateriflora* - **S. parvula** Mx. (s&cUS): small skullcap; hb. u. as antidiarrheal in treatment of "flux" (Fox Indians) - **S. versicolor** Nutt. = *S. ovata*.

scutellarin: glucuronide of scutallarein (flavonoid); fr. lvs. of *Scutellaria altissima*, *Centaurea scabiosa*, &c.

scutellum: shield-shaped organ attaching embryo to endosperm; considered by some to be the grass cotyledon (Gram-cereal grains).

Scutia buxifolia Reiss. (Rhamn.) (Br): "coronicha"; stem bk. c. glycosides of anthranols & anthrones; ext. with digitaloid acitivity; diur.; in folk med. u. in cardiovasc. dis. - **S. martynia** Kurz (As, Af): cat thorn; lf. in ointment appl. to hasten birth; frt. edible, appar. astr.

scyllitol: cocositol; polyhydroxy alc. of *Cocos nucifera* (lvs.), *Cornus florida* (lvs.), *Helinus ovatus, Calycanthus* spp., dogfish; similar to inositol.

Scythian lamb: *Cibotium barometz*.

Scytosiphon lomentaria (Lyngb.) Link. (Scytosiphon - Phaeophyc - Algae): (Atl & Pac Oc): "sugara" (Jap); u. as stock feed with high food values (horses); in soups (Jap).

sea: v. also under heath; onion; side grape; thrift (root); wrack.

sea ash: *Zanthoxylum clavaherculis* - **s. bladder**: *Nereocystis* spp. - **s. coal**: mineral c. shipped by sea - **s. cole**: s. kale - **s. cucumber:** taxa of Holothuroidea (Echinodermata) found in seas worldside; elongated sausage-shaped animals attached to substrates of ocean floor; collected for food (Galápagos Is, SA+) & u. as aphrod.; extensive collections threaten extermination in some areas - **s. devil**: octopus - **s. horse:** *Hippocampus* spp. (*H. coronatus*, &c.) (Hippocampedae - Solenoichthyes [Ord.] - Pisces) (widely distr. in seas): bony fish with body in vertical position & head at right angles; prehensile tail u. to grasp eelgrass, &c; -15 cm long; u. as ton. stim. in Chin med. - **s. kail: s. kale**: *Crambe maritima* & other *C.* spp. - **s. lettuce**: *Ulva lactuca; Fucus vesiculosus* - **s. otter**: *Enhydria lutris* L. (Pac Oc): flesh u. as meat; intest. u. as food for sick; "specif." in scurvy (1921) - **s. rockets**: *Cakile* spp. - **s. tangle tent**: dr. cylindrical piece of kelp stipe; u. as bougie to dilate orifices of body, esp. os uteri - **s. urchins:** echinoids strongly armed with heavy spines; more or less globular Echinodermata (Echinoidea) of marine habitats, usu. near shore; gonads (Foe), &c. valued as food by Japanese+; some spp. harmful to coral, devour bivalves, &c.; important gen. *Strongylocentrotus* (*S. purpuratus*, wNA), & other *S.* spp.; *S. droehbuchiensis* (worldwide); at times overfished.

seablite: *Suaeda maritima* & other *S.* spp.

seal: marine mammal with streamlined body & carnivorous habits - **gray s. oil**: marine type FO obt. fr. blubber of *Halichoerus grypus* (Phocidae) (Atl Oc); of drying type, highly unsatd. (iodine no 162.2); u. to make soap & as lubricant.

sealing wax: fusible plastic combn. of resins & shellac & color u. to seal documents, &c.; formerly widely u. in correspondence.

seam: (hog's) lard.

seaweed: 1) any pl. growing in sea water 2) (specif.): Algae (name "sea plant" sometimes now preferred because of economic value).

seawhip: Gorgonian coral.

sebacic acid: ipomic a.: $(COOH)_2$ $(CH_2)_8$: fatty acid occ. in *Exogonium purga* (Ipomoea jalapa) (resin), olive & castor oils, yeasts; &c.; for comm. prodn. obt. fr. castor oil by heating with alkali; ethyl sebacate (ester) also u.; raw material in synth. resins, plasticizers, synth. fibers, &c.; emoll. in prepns. to protect damaged skin; diisopropyl sebacate as skin moisturizer in lotions for dry skin.

Sebaea crassulaefolia Cham. & Schl. (Gentian.) (sAf): inf. remedy in snakebite & for "stitch" (in side).

Sebastiania Spreng. (Euphorbi) (trop worldwide): some Mex spp. (esp. *S. bilocularis*) prod. frt. (infested with larva) called "Mexican jumping beans" (v. under "j") - **S. bilocularis** Wats. (Mex, swUS): hierba de la flecha; tree said poisonous; smoke fr. burning wd. & even sleeping several hrs. in shade of tree lvs. said to prod. sore eyes; latex u. as arrow & fish poison (27) - **S. chamaelea** Muell. Arg. (Ind, now in As, Af, Austral): pl. juice in wine u. as astr. in diarrh., ton., in syph.; appl. to head in vertigo (Ind) - **S. palmeri** Riley (Mex): latex u. as arrow poison; frt. furnish jumping beans - **S. pavoniana** Muell. Arg. (Mex): palo de la fleche; shrub; frt. forms (Mexican) jumping beans (due to butterfly larva); latex u. to poison arrows.

sebestens = *Cordia myxa* & *C. sebestena* (& other *C.* spp) (also **sebesten plums** & **s. fruits**).

sebo (Sp): tallow; suet - **s. ovino**: mutton t.: suet - **s. vacuno**: beef tallow.

Sebum (L.): Sevum: 1) suet; tallow 2) (physiol.): oily prod. (**s. cutaneum**) secr. by cutaneous sebaceous glands; serves to lubricate skin & hair (1922) - **S. Ovile** (Eu Phar.): suet.

Secale cereale L. (Gram) (Euras, now cultd. worldwide): (common) rye; chief cereral grain of nEu (Ge, USSR, Scand); very closely related to wheat (Triticum); u. as food (flour, for "black bread") (eEu) fodder; mfr. whiskeys (chief use in US, gins, beers); bread often prepd. with caraway; in mfr. rye starch (Amylum Secalis); u. nutr. for powds. pills, tablets, &c.; in technology (1923); source of good straw - **S. Cornutum** (L): S. Luxurians (L) (1491): ergot of rye.

Secamone afzelii K. Schum. (Asclepiad.) (trop wAf): milky latex u. as galactagog (in form of macerated lf); in drinks & massages for the heart; lf. decoc. purg. - **S. emetica** R. Br. (Ind): rt. u. as emet. - **S. gerrardii** Harv. (sAf): rt. u. as remedy in spinal dis. (toxic) - **S. micranthes** Dcne. (eAs+): lf. decoc. u. as vulnerary (Hainan); rt. source of beverage "Laroe" (Java) - **S. parvifolia** (Oliv.) Bullock (Tanz): rts. eaten with honey for hoarse cough; pd. lvs. with water as wash for scabies; rt. decoc. for abdom. pain, stomach dis., & snakebite.

sèche (Fr): cuttle bone.

Sechium edule Swartz (Sicyos edulis Jacq.) (monotypic) (Cucurbit.) (Mex, WI, CA): chayote (squash); "chote" (Cu); chocho; vegetable pear; choke plant; vine bears large frt. cult. & eaten as delicious veget.; frts. (green or whitish) made into sweetmeat (CA); rhiz. also eaten; rich in starch (25%) u. as starch source; rhiz. u. as diur. purg. in pulmon. dis.; pulp for cataplasms; sd. u. to prep. emulsion for appl. as dem. in inflamm.; eaten (fried).

Seckelblume (Ge): *Ceanothus americanus*.

seco-: indicates ring cleavage compd.

secretin(e): gastrointestinal endocrine (polypeptide) secreted by mucous membrane of duodenum & jejunum wh. stimulates secretion of water & bicarbonate by the pancreas & liver; amino acid sequence of porcine & bovine secretin (polypeptide) identical; u. as diagnostic aid for pancr. & gallbladder function; injectable form u. (1905-).

secretory tissues: those pl. tissues which prod. VO, resins, gums, enzymes, water, milky juice (latex), tannin, etc.; incl. secretion sacs (or cavities), secretory canals (or ducts), secretion cells, latex cells, laticiferous ducts, etc.

Sectio: section: (bot.): sub-genus.

Securidaca longipedunculata Fres. (Polygal.) (tropAf): violet (or fiber) tree; wild wisteria; rt. c. saponins, incl. one lethal to molluscs; methyl salicylate; rt. u. antirheum. for colds; purg.; boiled with flour & swallowed for GC (irritation); bk. u. as ton. anthelm., abort., in prepn. arrow poison; lvs. u. purg. (Ethiopia), locally for wounds & inflamm., in snakebite; pd. lvs. u. as snuff in ordeal brew (Angola) (one of several ingred.); twigs source of buaze fiber, u. for nets, &c. (*cf.* flax); rt. u. as intravaginal poison (often lethal!) (suicide) (Tanz) (606).

Securinega suffruticosa Rehd. (S. ramiflora Muell. Arg.) (Euphorbi) (eAs): Asiatic securinega: lvs. of shrub u. for remedy for bruises (Taiwan); u. in Russia for prepn. of securinine, convulsant, nerve stim., u. in place of strychnine in therapy (inhibits GABA) (2354) - **S. virosa** Pax & K. Hoffm. (S. microcarpa Muell. Arg.; Fluggea v. Baill.) (OW trop, Ind, Chin to Austral): lf. decoc. u. as lax. detergent, vulner. for boils, wounds (Chin), ext. has antibiotic props.; powd. charcoal fr. burned wd. u. as cicatrizant (PI); rt. decoc. u. to treat VD, diseased teeth & gums; in bilharziasis (Tanz); lvs. in VD; gum fr. sts. u. mucil.; bk. c 9% tannins, u. as fish poison; inf. of young lvs. u. in pain (int.) (2263).

sedano (It): celery.

sedge: spp. of Fam. Cyperaceae, esp. of genus *Carex* - **broom s.**: v. broom (**pointed broom s.** (Miss

[US]): *Carex scoparia*) - **cotton s.**: *Eriophorum* - **sand s. (root)**: pointed broom s. - **sword s.**: *Lepidosperma* - **white s.**: *Carex canescens* - **white top s.**: white tip rush: *Dichromena latifolia* ("white tops").

Sedum acre L. (Crassul.) (Euras, natd. US, Cana): mossy (or biting) stonecrop; wallpepper; love-entangle; hb. u. as hypotensive in high blood pressure sedamine); locally skin irritant (inflamm. vesication), mydriatic; in skin cancer; purg. emet.; fresh flg. pl. u. by Eu Homeop. for bleeding hemorrhoids (1924) - **S. album** L. (Euras, natd. eUS): (wild) prick madam; (tall) white stonecrop; lvs. & sts. u. in salads; boiled as greens - **S. alpestre** Vill. (S. repens Schleich.) (cEu-mts.): Steinkorn (Ge); u. med. similar to *S. acre* - **S. dendroideum** Moç. & Sessé ex DC (Mex-on cliffs): "siempreviva"; pl. juice astr. u. to harden gums, in hemorrhoids, chilblains, dys.; appl. to forehead to stop nosebleed - **S. fimbriatum** (Turcz.) Franch. (Chin+): u. as hemostat. in intest. bleeding, dys.; exter. as poultice for boils, piles, & wounds - **S. praealtum** DC (Mex, Guat): Santa Polonia; lf. juice dropped in eye to alleviate irritation; for cataracts (causes to disappear); for oral inflammns. - **S. purpureum** Schultes (Eu): garden orpine; live-forever; Aaron's rod; cult. in rock gardens; lvs. u. for boils & carbuncles (1739) - **S. rosea** Scop. (Rhodiola r. L.) (Arctic, Euras, Cana, nUS on rocks): rose root (Rosea radix of early apothecaries); rt. u. in headache & scurvy; in the "troubles of old age" (gold root) (607); folk med. (Sib, Vladivostok) nutr. in salads & kitchen veget. - **S. rupestre** L. (Eu, natd. eUS): (sharp-pointed) stonecrop; thrift; trick madam; prickmadam; hb. u. as cath. emetic - **S. telephium** L. (Euras, esp. Russ; natd. eNA): live long; l. forever; (garden) orpine; similar c. as *S. acre*; u. for bleeding hemorrhoids (folk med.); juice heals wounds; ornam. hb. (rock garden): formerly u. as cicatrizant; astr. in dys. diarrh. hemoptysis; diur.

seed: the fertilized & ripened ovule containing an embryo; with one or two sd. coats and with or without endosperm (albumen); ex flaxseed; physostigma; diagnostic tissues (microsc.) include: sclerenchymatous layer of testa; epidermis of testa, sometimes with char. trichomes; perisperm & endosperm; parenchyma cells often with characteristic contents, ex. aleurone, FO, starch, etc.; "seed" as popularly u. often represents the fruit (ex. anise "seed") - **s. lac:** fr. *Croton lacciferum* - **s. vessel:** fruit - **"white s."**: *Plantago ovata.*

Seegras (Ge): *Zostera marina.*

Seegurken (Ge): **Seewalzen:** sea cucumber.

seepage: 1) the act or process of leaking, oozing, or slowly percolating through porous materials, such as soil 2) the liquid which has so leaked, etc.

seer (Ind, Pak): unit of wt., one kg, but varies.

sehav (seEu. Jews): sorrel u. in soups.

seiba (Sp): soft wood.

Seidelbastrinde (Ge): mezereon bk.

Seidenpflanzengewaechse (Ge): Asclepiadaceae.

Seifenrinde (Ge): soap bark, Quillaja.

seigle (Fr): rye - **S. ergoté** (Fr): ergot.

Seignette('s) salt: Rochelle salt, sodium potassium tartrate.

seille (Fr): 1) *Hedysarum coronarium* 2) rye.

sel (Fr-m.): salt - **s. de Seignette**: Seignette's s.

Selachoidei: sub-order of Sharks; source of shark liver oil.

Selaginella Beauv. (Selaginell - Lycopsida - Pteridoph.): spikemoss; some 700 spp. of moss-like ferns; mostly trop, a few temp.; one sp. u. in female dis. (Cuna, Indians; Pan) - **S. flabellata** Spring (Solomon Is): fronds u. in feverish headache; rt. to control menstruation - **S. helferi** Wall. (Thail): dr. pd. lvs. appl. to burns; also u. with *Pteridium* sp. to treat snake bite (2448) - **S. involvens** Spring (eAs): u. in cough, gravel, rectal prolapse, anti-hemorrhagic - **S. lepidophylla** Spring (CA, Mex, seUS): rock spikemoss; doradilla; resurrection plant (sTex [US]) u. as diur. (in some parts of Mex) (Mex Phar.); in dyspepsia, biliousness; sold as curiosity - **S. rupestris** Spring (sAf): pd. rhiz. made into ointment & rubbed into venereal sores; smoked with *Lycopodium clavatum* for headache - **S. tamarisciana** Spring (eAs): u. therap. like *S. involvens.*

selajit (Pak, Ind): fossilized material of unknown origin (possibly excreta of monkeys, etc.) highly prized in Ayurvedic & Unani med. in wAs; generally occurs in Himalaya Mts in nearly inaccessible places; deserves study.

Selenicereus grandiflorus Britt. & Rose = *Cereus g.* Mill.

Selenipedium chica Reichb. f. (Orchid.) (CA, SA): vanilla chica; formerly u. in place of vanilla.

selenite: native calcium sulfate; u. to make plaster of Paris; as polishing agt.; for glazing windows (formerly).

Selenium: Selenum (Ge): Se; highly toxic non-metallic element; present in soils; essential element but apparently adequate in normal diets (thus rich in garlic); u. in seborrhea - **s. sulfide**: anticancer claims made (2455); stim. immune reactions in allergies to cold; said useful in TB, pneumonia; much u. in technical areas. related to vit. E (in muscle health activity); claimed to oppose the aging process.

self-heal: *Prunella vulgaris.*

selguacha (Sp; Calif [US]): *Datura inoxia.*

Selinum benthami S. Wats. (Umb.) = *Cnidium officinale* - **S. monnieri** L. = *Cnidium m.* - **S. officinale** Vest sec. Roth. (Mak.) = Peucedanum officinale = *Cnidium officinale* - **S. officinale** Mak. (Kor, Jap): rhiz. u. for headache, lameness, female dis., common cold (Kor); decoc. in headache (Jap); VO of rhiz. in anemia - **S. palustre L.** = *Peucedanum p.* - **S. vaginatum** C. B. Clarke (Ind): rt. c. selinidin, vaginol, selinon; u. as sed. analg. & hypotensive agt. (folk med.).

Sellerie(laub) (Ge): celery (foliage).

Selloa glutinosa Spreng. = *Gymnosperma g.*

Selzer (water): natural (or artificial mineral water charged with CO_2 (effervescent; sparkling); u. in mixed drinks, &c.

selva (Sp): forest, woods.

Semecarpus anacardium L. f. (Anacardi.) (Ind): (cashew) marking nut; Oriental cashew nut (tree); marking (or ink) tree; frt. pericarp c. anacardol, urushiol (bhilawanol), cardol & semecarpol (phenols), anacardic acid, tannins, gallic acid, chuchu(n)arine (alk.); sd. c. 49% FO (kernel); u. as wooden floor dressing to prevent termite ("white ant") attack; u. as aphrod.; vermifuge, in Hansen's disease; after heating, as food oil; tech. in varnishes, inks; frts. (Fructus Anacardii orientalis) ("Dhobie's nut") chewed in Arab countries; (brain) stim., hallucinogenic; "Cardol(e)um pruriens" (or "vesicans") is a solid ext. ("balsam") made with alc.-ether fr. frt.; shell still sometimes u. as vesicant, discutient, for warts, corns, &c., insecticide; shell liquid antioxidant (u. to stabilize Pyrethrum exts., &c.); produces great reduction in viscosity of mineral oils to which added; becomes black on exposure; juice u. with slacked lime (lime water) to make indelible marking ink for printing on cotton cloth; green frts. u. as bird lime & as tan.; tree (-10 m) cult. in trop. (Chin, EI, Austral, Br [Recife]) - **S. decipiens** Merr. & Perry (Guadalcanal): bk. macerated & appl. to tinea & other skin eruptions - **S. travancoricus** Bedd. (Ind): u. similarly to *S. anacardium.*

Semen, Seminis (L.): seed. v. also Faba(e); Fructus - **S. Abelmoschi:** of *Hibiscus abelmoschus* - **S. Amomi**: pimenta frt. - **S. Anisi stellati**: star anise (frt.) - **S. Badian(ae)**: *Illicium verum* (frt.) - **S. Bonduc:** *Caesalpinia bonducella*, sds. - **S. Cali:** *Mucuna urens*, sds. - **S. Canariense**: *Phalaris canariensis*, sd. - **S. Cardui**: of *Cnicus benedictus* - **S. Cataphysis**: **S. Cinocephali**: of *Plantago psyllium* - **S. Cinae: S. Contra**: Levant wormseed, *Artemisia cina* (etc.), fl. heads - **S. Cocculi Indici**: of *Anamirta cocculus* - **S. Cydoniae**; quince sd. - **S. Cynae:** S. Cinae - **S. Cynomii**: *Plantago psyllium, P. indica*, &c. - **S. Cynosbati:** *Rosa canina* sd. - **S. Foenugraeci:** of *Trigonella foenumgraecum* - **S. Guaranae:** of *Paullinia cupana* - **S. Guizotiae** (oleiferae): Niger sd., *Guizotia abyssinica* - **S. Ispaghulae**: of *Plantago ovata* - **S. Mariae**: of *Silybum marianum* - **S. Melanthii: S. Nigellae**: *Nigella sativa*, sd. - **S. Papaveris**: maw seeds; *Papaver somniferum*, sds. - **S. Paradisi**: of *Aframomum melagueta* - **S. Pedicularis:** stavesacre sd., *Delphinium staphisagria* - **S. Plantaginis:** of *Plantago asiatica* - **S. Psyllii**: 1) of Plantago psyllium &c. 2) **S. (Plantaginis) Pulicariae**: of *Plantago arenaria* - **S. Ramtillae**: S. Guizotiae - **S. Sabadillae**: of *Schoenocaulon officinale* - **S. Sanctum**: *Santonica* - **S. Sinapis**: of *Brassica nigra* - **S. Staphisagria:** of *Delphinium staphisagria* - **S. Stramonii**: of *Datura stramonium* - **S. Strychni**: of *Strychonos nux-vomica.*

semence(s) (Fr): seed(s) - **semencina** (Por): **semenzina** (It): Santonica.

semente (Por): **semenza** (It): seed.

semi-fossil: recent fossil: referring to resins, etc., dug out of the ground representing exudns. fr. trees (etc.) now decayed away but of spp. still extant.

semilla (Sp): seed - **s. amapola**: poppy sd. - **s. de tartago** (Sp, Por): *Euphorbia lathyrus.*

Semina (L): seeds, pl. of Semen (L) - S. Strychni (L.): Nux Vomica - **S. Tonco:** tonka beans.

semolina (It): the purified "middlings" (gluten) of high quality (hard) wheats (*Triticum turgidum*) (durum) u. to prep. spaghetti, macaroni, vermicelli, etc. (pasta) (fr. simila [L.] = finest wheat flour).

Sempervivum tectorum L. (Crassul.) (cEu, eAf): (common or root) house leek; "welcome home husband however drunk you be"(!!); hb (or lvs.), preferably fresh, u. antispas. in dysmenorrh., amenorrh., astr. styptic, deterg. in wounds, burns, refrig. in tumors (folk med.).

sen (Sp fr. Arabic): sena (Sp, It, Arab) (pron. seen-a) (derived fr. Abyssinia area of growth): Senna - **sennar**: var. of Acacia - **séné d'Alexandrie** (Fr): Alexandrian Senna.

senape (It-m.): mustard.

séné (Fr): Senna.

Senebiera DC = *Coronopus.*

Senecio L. (Comp.) (cosmop): groundsel; some 2,000 spp. (largest gen. of fam.) incl. herbs, vines, shrubs & trees; hb. of some spp. formerly u. for feeding to caged birds; alks. much studied; genus split into several genera; some spp. cult. ornam. - **S. abyssinicus** Sch. Bip. (tropAf): lf. appl. to wounds & sores to aid healing (cicitrizant), in eye dis.; in yaws & syph. - **S. anonymus** Wood (seUS): hb. said to have antitiumor properties -

S. **asperulus** DC (sAf): hb. decoc. appl. as lotion to aching legs & feet; for colds (horses) - **S. aureus** L. (e&cNA): (American golden) ragwort (or senecio); female regulator; squaw weed; unkum; golden groundsel; rt. (Senecio, L.) (as "life root") c. senecionine, &c.; u. emmen. in menstr. disturbances; diur. (NF IV, V); vulner. (Am Indians) in hemorrhages; fresh flg. hb. u. (Eu Homeo) - **S. bellidifolia** HBK (Mex) "calamapati"; lvs. u. for bladder & kidney dis.; pl. ground with olive oil & appl. as poultice to boils, tumors, inflammn.; pl. decoc. u. as cure for ulcers & sores; depur. in syph. - **S. brachypodus** DC (Af): rt. u. in syph.; inf. for colds & other resp. dis. - **S. brasiliensis** Less. (Br): flor das almas; tasneirinha; maria mole; cravo do campo; catiao; cardo morto; entire pl. (hb.) used in galenical prepns. (heart dis.?) (609); lvs. c. senecionine, jacobine, brasilinecine: toxic pl.; in Braz Phar. - **S. bupleuroides** DC (sAf): ragwort; ingredient in remedy for chest dis.; may be stock poison - **S. campestris** DC (S. glabellus DC) (nEuras): fl. heads u. as diur. hematic (Kor); parasiticide in skin dis. & in clothes (Chin); weed in Chin - **S. cancellatus** DC (Hispaniola): senecon; pl. appl. to sore eyes, hemorrhoids, swollen breasts of nursing mothers; hb. decoc. purg. - **S. canicidus** Moç. & Sessé (Mex): yerba de la puebla; c. senecic (senecioic) acid (toxic): claimed of value in epilepsy (reduces attacks); dog poison. (Mex Phar.) - **S. cervariaefolius** Hemsl. (Mex): rhiz. c. glycoside (resembling digitalis type), oleoresin, tannin; emet. cath.; exter. anodyne: toxic - **S. cineraria** DC (Cineraria maritima L.) (Med reg., cult. NA): dusty miller; silver grounded; pl. u. deterg. vulner.; source of Succus Cinerariae Maritimae; u. in collyria for eye dis., esp. cataract prevention & remedy (Ve Phar.); in migraine; emmen. - **S. concolor** DC (sAf): lf. poultice appl. to cuts, wounds, swellings, also hot decoc. - **S. coronatus** Harv. (sAf): pl. decoc. u. as emet. cath.; toxic - **S. cruentus** DC (Canary Is): florists' cineraria; a common house pl. - **S. discolor** DC (Jam); West Indian elm; shrub lvs. u. in decoc. for colds, fever, gastric dis. - **S. dregeanus** DC var. *discoideus* Harv. (sAf): rt. decoc. u. as emet. in chest colds & in "madness" - **S. erraticus** Wimm. & Grab. (Euras, nAf): hb. c. alk.; stock poison of importance - **S. erubescens** Ait. (sAf): rt. decoc. u. in nausea, rheumatic fever; lf. mixed with tobacco - **S. filifolius** Harv. (sAf+): groundsel; yerba del caballo; (hb.) u. in rheum. - **S. fraudulentus** Phillips & Smith (sAf): u. for palpitations, phthisis; in cough & difficult breathing - **S. fuchsii** Gmel. (cEu): hb. u. as tea for diabetes, styptic in uter. hemorrhage (Ge); said may cause liver cancer - **S. glabellus** Poir. (se&cUS, nMex): butterweed; yellow top; lvs. u. as astr. ton. vulner.; weed - **S. gracilis** Pursh = *S. aureus* - **S. grayanus** Hemsley (S. decomposita) = *Cacalia decomposita* Gray - **S. haitiensis** Krug & Urb. (Haiti): senecon; u. like *S. cancellatus* - **S. hieracifolius** L. = *Erechtites h.* - **S. hubertia** Pers. (Hubertia ambavilla; S. a.) (Madag, Bourbon Is): senecon; ambaville blanche; lvs. u. diur.; in rheum.; vulner. expect. - **S. ilicifolius** Thunb. (sAf): guanobos; toxic, causing human seneciosis, "breading poisoning" (pl. a weed in wheat fields): c. retrorsine, senecionine: similar to *S. retrorsus* - **S. imparipinnatus** Klatt (Tex [US]): groundsel (Tex): toxic to livestock - **S. integerrimus** Nutt. (wCana, nwUS): ragwort; butterweed; hb. c. integerrimine, senecionine; livestock poison - **S. isatideus** DC (Af): poisonous ragwort; hb. c. isatidine, retrorsine; toxic to stock animals; causes horse staggers (dunsiekte) - **S. jabobaea** L. (Euras, now cosmop, incl NA): (common) ragwort; stinging Willie (US); (tansy) ragwort; ragweed; staggerwort; St. James wort; hb. c. senecionine; senecine (amorph. alk.); jacobine; jacoline; VO, FO; analg. in dysmenorrh., cramps, colic; hemostatic; *cf.* ergot; yellow dye; stock poison (causes cirrhosis, liver dis., in cattle): plant pest in Austral, being controlled by insects - **S. johnstonii** Oliv. (eAf): tree form; found on volcanoes - **S. latifolius** DC (sAf): Dan's cabbage; lf. in paste appl. to burns & wounds; in rituals; c. senecifoline; poisonous with props. similar to those of *S. jacobaea* - **S. lobatus** Pers. = *S. glabellus* - **S. longiflorus** Sch. Bip. (sAf): pl. u. to make snuff - **S. lyrata** L. (sAf): rt. decoc. u. in colic; inhalation for colds - **S. maranguensis** O. Hoffm. (sAf): remedy for syph. GC yaws - **S. orbicularis** Sond. (sAf): rt. decoc. taken for shivering fits during fever; lf. appl. locally to bubo - **S. praecox** (Cav.) DC (sMex): palo loco; lf. decoc. u. exter. for "rheumatism" & wounds; hb. off. in several Mex Phar. - **S. oxyriaefolius** DC (Af): pd. rt. made into cake & fed to overcome female sterility - **S. platyphyllus** DC (Caucasian reg., USSR): hb. c. platyphylline, seneciphylline (rt.): said antispasm. hypotensive - **S. quinquelobus** DC (sAf): u. in flu (orally, enema); sd. colic remedy - **S. retrosus** DC (Zambia; sAf): ragwort; Dan's cabbage; c. alk. (retrorsine; isatidine; scleratine); u. as cicatrizant in skin dis.; toxic to cattle (& humans?); a chief cause of seneciosis (610) - **S. rhomboideus** Harv. (sAf): remedy for female sterility - **S. rhyncholaenus** DC (sAf): smoke fr. burning lf. inhaled for colds; rt. decoc. in colics - **S. riddellii** T. & G. (cUS):

causes "walking dis." in horses (Neb [US]); u. rheum. arthritis (Navaho) - **S. ruwenzoriensis** S. Moore (Kenya, Congo): hb. u. to reduce inflammn. in the limbs; important cattle poison - **S. sagittatus** Sch. Bip. = (*Cacalia hastata* L.) (Russia, China): u. as folk med. (Perm); u. as vulner., anodyne, u. to relieve hemorrhages (epistaxis, hemoptysis, menorrhagia), swellings, inflamed eyes (Chin, Tibet) - **S. scandens** Buch-Ham. (eAs): climbing groundsel; German ivy; pl. decoc. u. in hemorrhoids & eye dis.; exter. for dermatitis caused by heat, infantile skin dis. (Chin) - **S. serratuloides** DC (sAf): lf. decoc. drunk for skin eruptions; rt. appl. to burns & sores to promote healing; cicatrizant: lvs. u. antiluetic - **S. serratus** Sond. (sAf): decoc. u. to wash persons with swollen limbs or internal tumors - **S. smallii** Britt. (eUS, esp. se): Small's squawweed; hb. u. in TB (Catawba Indians) - **S. sonchifolius** Moench = *Emilia sonchifolia* - **S. speciosus** Willd. (Af): lf./st. decoc. u. for pleurisy & other chest pains; lf. in edema - **S. stuhlmannii** Klatt (eAf): lvs. u. as ulcer dressing (where aloe juice has failed); lf. appl. as healing appln. to wounds - **S. subscandens** Hochst. = *Crassocephalum s.* - **S. tolucanus** DC (Mex): hb. c. toxisenicine, VO; u. in Mex folk med.; off. in Mex Phar. - **S. tomentosus** Mx. (eUS): yaw leaf; supposed useful in treating yaws (Md, Del [US]) - **S. vernus** L. (S. salignus DC) (Mex): jarilla; jaralillo; decoc. appl. as wash for fever; passed over eyelids to treat ophthalmia; in baths for rheum. - **S. vulgaris** L. (Euras, nAf, natd. Alas, Cana, nUS): (common) groundsel; grinsel; hb. an ancient drug, c. senecionine, retrorsine, seneciphylline; senecine (amorph.) u. emmen. in menstr. irreg., dysmenorrh., anthelm. hemostat.; discutient in hypertrophic gum changes with bleeding; astr. in cholera; in colics; vulner.; in icterus, sciatica; weed - **S. vulneraria** DC (Mex): hb. u. as decoc. appl. as lotion for wound healing.

Senecionine: alk. found in several *Senecio* spp.; hypod. inj. prod. hepatic necrosis, kidney fatty degeneration (83).

Senega (snake) root: *Polygala senega* (rt.) - **northern S.**: chiefly fr. Minn (US) & Man (Cana) - **southern S.**: mostly fr. seUS - **white s.**: Polygala boykinii.

Senegal gum: var. of Acacia, specif. of wAf origin - **S. root:** *Tinospora bakis.*

Senegalia Raf. = *Acacia.*

Senegambia: region of trop wAf around the Senegal & Gambia Rivers; now mostly in Senegal & Mali.

senegin: saponin found in *Polygala senega, P. alba, P. amara,* etc.

Seneka (snake) root: *Polygala senega.*

senevol (Por): **senevolico** (Sp): mustard oil; thiocyanates.

Senf(samen) (Ge): mustard (sd.).

seng (US): corrupted form of ginseng.

senisa (Tex [US]): *Leucophyllum texanum.*

Senna: 1) (L.): off. Senna leaf fr. *Cassia angustifolia, C. acutifolia.* (prod. Ind, Eg, Chin, Sudan) 2) (US): *C. marilandica* & other native *C.* spp. - **Alexandria(n) s.**: *C. acutifolia* - **American s.:** *C. marilandica* - **S. Amomi:** pimenta frt. - **Arabian s.**: Tinnevelly s. - **avaram (or avaran) s.**: *C. auriculata* - **bladder s.:** *Colutea arborescens* - **Bombay s.**: similar to Tinnevelly s., but less carefully prep. & an inferior prod. - **Brazilian s.**: *Cassia angustifolia* - **Central American s.**: *C. occidentalis, C. reticulata* - **coffee s.:** *C. occidentalis* - **Congo s.**: *C. angustifolia* - **dog s.**: *C. obovata; S. italica*; u. adult. of true s. - **East India s.**: Bombay s. - **s. fruit:** s. pods - **golden shower s.**: *Cassia fistula* - **s. husks:** *C. acutifolia* - **India s.**: 1) TV s. 2) Senna Indica, fr. *C. auriculata* (worthless prod.) - **Italian s.:** *Cassia italica* - **s. & manna**: a mixture of these drugs commonly sold as lax. (US) for home use - **Maryland s.**: wild s. - **Mecca S.: Mocha S.**: lvs. of wild growing pls. of *Cassia angustifolia* or *C. italica* (Arabia), shipped fr. Por, Sudan & Bombay - **Nubia s.**: Alexandria s. - **Palthé S.**: fr. *Cassia auriculata*; adultern. of Tinnevelley S. - **s. pod(s)**: frt. of *C. acutifolia* & *C. angustifolia*; chief prodn. Ind, Eg - **s. stalks**: composed of rachis, petiole, stems (twigs) of senna; constitute adult - **Syrian S.**: Italian S. - **Tinnevelly s.: TV s.**: *C. angustifolia* - **wild s.**: leaflets of *Cassia marilandica* & *C. hebecarpa* - **w. s. of Europe**: *Globularia alypum.*

sennaar: sennari gum: var. of Acacia.

Sennae Fructus (L.): **Sennesbalge** (Ge): senna pods - **Sennesblätter** (Ge): senna lvs. - **sennosides**: (A & B): glycosides of Senna lf. (isolated by Stoll et al.); believed chief lax. princ.; u. in tablets.

sensitiva (Sp): **sensitive brier: s. pea** (US): **s. plant** (US): 1) *Mimosa pudica* 2) **s. vine**: *Schrankia* spp. - **wild s. plant.**: *Cassia nictitans* - **s. tree**: *Albizzia julibrissin.*

sepal: segment of the calyx, usually green in color; separated or fused together.

Sephadex™: modified dextran; u. in chromatography & ion exchange electrophoresis; numerous vars. available.

Sepia officinalis L. (& other *S.* spp.) (Sepiidae - Decapoda Cephalopoda - Mollusca) (OW seas): cuttle fish; food (arms roasted in FO); a calcareous shell ("cuttlefish bone," Sepia "sea foam") (part of the internal skeleton) enclosed in a sac of the mantle c. $CaCO_3$ (80-85%), &c.; u. gathered on beach; cuttlebone u. as antacid, in tooth pds., for Ca addn. to diet of caged birds; polishing agt. for jewelers; marking clothing (tailor); in uter. dis., pregnancy (circulat. problems); sepiá (pigment) in inks (Chin); u. in "surma" (eyelid cosmetic of India) & in eye dis., in Unani med. as antacid, carm. for headache, insomnia, etc.

septentrional(is): (L.): north(ern).

septfoil: *Potentilla tormentilla.*

Sequoia sempervirens (Lamb.) Endl. (1) (Taxodi.) (Calif, sOre [US]): (California) redwood ; considered tallest known tree (-105 m) (2007); diam. -10.5 m; bk. c. tannin, u. astr. antisep.; very valuable lumber (612); state tree of Calif.

Sequoiadendron gigantea (Lindl.) Buchh. (1) (Sequoia g. [Lindl.] Dcne.) (Taxodi) (Calif [US]): giant sequoia; big tree (of California); mammoth tree; giant redwood; one of largest & oldest known trees; planted as ornam. (613); wd. —> shingles.

sereni textil: *Hibiscus sabdariffa v. altissima.*

Serenoa repens (Bartr.) Small (S. serrulata (Mx.) Nicholson) (monotypic) (Palm) (seUS, coastal plain fr. SCar to La; natd. CA, SA): saw palmetto; shrub p.; "berries" (really drupes) (Serenoa, L.; Sabal) c. VO, FO (with free fatty acids, esters) (sd. c. FO, with different compn.), cycloartenol, phytosterol, aliph. fatty acids; tannin, resin, invert sugar (28%), mannitol; liposoluble complex marketed; u. diur. sed. anti-inflamm. in cystitis (1st u. by Am Indians); in functional difficulties arising fr. prostatic adenoma (benign p. hypertrophy [u. now banned by FDA]); prostatits; aphrod. in impotency, in lung dis.; to aromatize cognacs (1926); starvation food (Am Indians); lf. inf. u. in dys. & stomach pains; sts. u. as tan; honey pl.; frt. prodn. chiefly in eFla (US) region of Cape Canaveral (1925); better product when collected within sight of ocean; Permixon™ (active principles, 1982) (614) (615); u. as anti-androgenic, anti-edematous; st. soft tissues u. as cork substit. (WWII); lvs. yield wax (but of no comml. value).

Seriania Schumacher = Serjania Mill.

Serica (L.): silk.

"sericea" (Lespedeza): Lespedeza cuneata G. Don (L. sericea Benth.) (= name formerly u.) (sUS).

sericin: silk glue, substance u. by silkworm in cementing two silk fibers to form one thread.

Sericocarpus bifoliatus (Walt.) Porter (Arnica b. [Walt.] Ahles) (Comp.) (seUS): white-topped aster; lf. inf. u. as diaph. mild lax., for fever, headache, sea sickness (1927) (—> *Aster*).

Sericocoma avolans Fenzl. (Amaranth.) (Af): u. to treat rheum. (Moz).

Sericographis mohintli Nees = *Justicia m.* Hemsl. = *J. spicigera.*

sericoside: glycoside of sericic acid (a triterpene acid) fr. rts. of *Terminalia sericea.*

sericulture: prodn. of raw natural silk fr. silkworms; mostly now in As; formerly on small scale in Carolinas (US).

Sericum (L.): Serica.

serine: amino acid (non essential) found in silk (sericine, silk fibroin) & other protein materials; inhibits uptake of salts by plant u. in reagents - **s. proteinases:** trypsin, plasmin, thrombin.

seringueira (Br): *Hevea brasiliensis.*

Seriola dorsalis (Carangidae - Pisces) (Pac Oc off US): Calif yellow-tail; mackerel-like carnivorous fish; u. as food; canned sometimes as tuna; liver oil (FO fr. Austral yellow-tail) rich in vits. A & D (36 & 110 X min. assays of codliver oil (resp.).

Serissa japonica Thunb. (S. foetida Comm.) (Rubi) (eAs): sts. & lvs. u. to treat cancer, carbuncles (Chin); rt. anthelm (Indochin).

Serjania curassavica Radlk. = *Paullinia pinnata* - **S. inebrians** Radlk. (Sapind.) (CR, Hond): barbasco; vine crushed & u. as fish poison; honey fr. pl. may be toxic - **S. mexicana** Willd. (Mex to Co, Ve, Jam): barbasco; lambeador; pl. decoc. u. in rheum. & syph.; sts. chewed for toothache (Guat); pieces of vine u. to stupefy fish; sts. as rope; substit. - **S. subdentata** Juss. (S. paniculata Hitch.) (WI): 3-finger (Bah); 5-finger; soapberry; st. decoc. u. in VD, to enrich blood ("blood building," Bah); vine decoc. u. as bath water for rash; sts. u. as fence material in corrals (Cu) - **S. triquetra** Radlk. (Mex. CA): carretilla; u. med. (Mex); fish poison.

sero-bacterin: a suspension of killed bacterial cells which have been sensitized by treatment with an immune serum specific for the bacterial sp.; u. to develop active immunity.

Seromycin™: cycloserine (Lilly).

seron (Fr, Sp): **seroon: serron**: bale or box (sometimes small) covered or bound with hide, u. to ship various crude drugs fr. foreign parts.

serotonin: enteramine: 5-hydroxy-tryptamine; an indole vasoactive amine widely distr. in body (blood), *Urtica dioica*, bananas, tomatoes; serves as neurotransmitter; at times occurs in pathologically large amts. in body; increases efficency of nerve transmission in the brain (1928); imbalance in obsessive/compulsive disorder (OCD); antidiur.; sed. protective against X-radiation.

serpão (Br): wild thyme; appl. to various *Thymus* spp. (ex. *T. serpyllum*, & its v.).

Serpasil™: first marketed prod. of reserpine (Ciba).

serpent: snake (Ophidia), esp. when of large size.

Serpentaria (L.): **serpentary:** *Aristolochia serpentaria; A. reticulata* - **English** s.: *Polygonum bistorta.*

serpine: reserpine.

serpol (Sp): Thymus serpyllum.

Serpyllum (L): *Thymus s.*

Serratia marcescens Bizio (monotypic) (Enterobacteri) (widely distr.): gm.- facultatively anerobic bacillus; some strains with deep red pigment; considered harmless saprophyte until quite recently; now known to be pathogenic, at times causes septicemia, pulmon. dis., pneumonia, endocarditis, may cause hospital epidemics & deaths (Calif) (occ. where immunity very low); c. polyribosomes; u. to treat primary brain malignancies (an orphan drug); u in biol. warfare expts. (ca. 1970).

Serratula L. (Comp.) (Euras): sawwort; genus of hbs; formerly included *Liatris, Carphephorus, Vernonia+* spp. which are now separated into these genera - S. **cerinthaefolia** Sibth. & Sm. (Comp.) (AsMinor): source of behen root; u. med. in Near East as a nervine, aphrod. resuscitant - **S. chinensis** S. Moore (Chin): u. as substit. for *Cimicifuga heracleifolia* Kom. - **S. scordium** Lour. (Vernonia squarrosa R. Br.) (Chin, Ind, Nepal): u. as diaph. emmen. antisep. - **S. tinctoria** L. (Euras, mostly cEu): (common) saw-wort; dyer's s.; hb. sts. & rt. c. serratulan, on drying converted into serratulin, yellow dye; rich in ecdisteroid (-0.7%), trideca-tetrain-diene: hb. & rt. u. exter. for sores, hemorrhoids, broken bones (vulner.); formerly important yellow dye for wool, cotton, silk, linen (Dutch pink or yellow madder) (Hung).

serum (L.): 1) the watery portion of liquids; specif. whey, the serum of milk (also **serum lactis**) 2) the aqueous medium of animal blood as separated fr. the blood clot (wh. cont. fibrin & corpuscles); must be distinguished fr. blood plasma, the liquid portion of unclotted blood; term "blood serum" often u. to refer to **immune b. s.** (with specific immune bodies) - **anti-A blood grouping s.**: derived fr. high titer serums of individual humans with or without stimulation by inj. of group specific red cells, etc., or similarly fr. lower animals; anti-A serum agglutinates human red cells which c. A. agglutinogens (gps. A, AB, sub-gps. A1, A2, A3, A1B, A2B) - **anti-B serum** agglutinates human red cells contg. B agglutinogens (gps. B, AB, A1B, A2B); these serums are u. to type blood as prelude to blood transfusion so as to avoid serious complications - **S. Antimeningococcicum** (L.): antimeningococcic s.; obt. fr. blood of animal immunized with cultures of several types of *Neisseria meningitidis* Murray (meningococci) prevalent in the US; u. in cerebrospinal meningitis where there are epidemics; vaccine now preferred - **S. Antipneumococcicum** (L.): antipneumococcic serum, type specific; obt. fr. blood of animal immunized with cultures of *Streptococcus pneumoniae* Chester (pneumococcus), one of types for which a serum has been prep.; u. in early pneumonias; now displaced by antibiotics - **anti-Rh typing s.**: derived fr. blood of persons who have developed specific Rh antibodies; u. as diagnostic means to determine if individual is Rh positive or negative; specialized serums are u. to determine the exact Rh type of person by an agglutination reaction; Rh typing is necessitated in blood transfusions, diagnosis of cause of erythroblastosis foetalis, medicolegal investigations, etc. - **anti-snake bite s., North American**: Antivenin; serum prep. by immunizing animals against venom of snakes of Crotalus gp. - **S. Humanum Normale** (L.): normal s. (v. under normal) - **S. Immune Morbillosi Humanum** (L.): human measles immune serum; sterile s. obt. fr. blood of healthy human who has survived measles infection - **S. I. Scarlatinae Humanum** (L.): **human scarlet fever i. s.:** sterile s. obt. fr. blood of healthy human surviving attack of scarlet fever; antibiotics now being used - **meningitis s.**: antimeningococcic s. - **s. therapy**: use of blood fr. immune animals to treat dis.

service berries: *Amelanchier* spp.

servilleta (PR): towel gourd, *Luffa* - **s. sanitaria** (Sp): sanitary napkin.

sesame (seed): Oriental s.: *Sesamum.*

sesamin: lignan fr. sds. of *Sesamum indicum* (616); bk. of Fagara spp., & c.; synergistic to pyrethrum insecticides & fungicides; may be toxic.

sesamolin: lignan related to sesamin; occ. in sesame sd. oil; powerful synergist of pyrethrum; a 1:1 mixt. of pyrethrum & sesamolin has 31x the activity as insecticide as p. alone.

Sesamum alatum Thonn. (Pedali) (tropAf): decoc. u. to treat sterility (Senegal); sd. as foodstuff (wAf) - **S. angolense** Welw. (hot Af): mucilaginous inf. u. in diarrh.; in skin dis. pounded rts. put on bleeding sores (Tanz): lf. + rt. decoc. u. in measles (eye drops); rt. decoc. in coughs; at times eaten as veget. - **S. angustifolium** Engl. (Af): rt. & st. inf. in diarrh.; appl. to wounds & burns; esp. to protect against tsetse fly (sticky inf. traps fly so it cannot bite); in coughs & pertussis; sts. as soap substit. (Tanz); food - **S. calycinum** Welw. (sAf): lf. decoc. u. to bathe eyes, as nasal douche; formerly to treat smallpox - **S. capense** Burm. f. (sAf): "wild foxglove"; lf. decoc. u. in mal.; fresh lvs. chewed like tobacco - **S. indicum** DC (S. orientale L.) (prob. native to India; widely cult. in trop, esp. Ind [1929], Chin, Indochin, At, WI, NA [Fla, SCar, Tex. Calif]; CA, SA): (Oriental) sesame; benne (oil plant); til (oil plant); ann. hb.; sd. food, (u. esp. to prepare pastry & halva) (important in African cooking along with spices [capsicum, ginger, &c.]); galactagog (Yuc); exter. in plasters; in pediculicides; to prod. FO (sesame oil; benne, gingely, teel oil); c. sesamin, sesamolin (ca. 1%) (synergistic with pyrethrum fl.); u. nutr. emoll. substit. for olive oil; lax., in cosmetics (ex. Teel), iodized oils; in mfr. margarine (*cf.* safflower oil); soaps; pharmaceuticals; injectables; medium & solvent in IM inj. (1930); in penicillin therapy (with beeswax); in prepn. sesamolin; lvs. u. aphrod., in mal. cough; emoll., chewed like tobacco - **S. radiatum** Schum. (trop eAf): black beni-seed; foodstuff; u. similarly to prec.

Sesban Adans. = **Sesbana** = **Sesbania** Adans.

Sesbania aegyptiaca Pers. (Leg) (Ind, Pak, tropAf): (Egyptian) sesban; janter (Bengal); fls. & unripe frts. edible; sds. (c. 6% FO, oleanolic acid & another triterpenic sapogenin); lvs. fls. rt. u. in Muslim med. by hakims (Pak) for snakebite & scorpion sting; sds. u. for hemorrhage & catarrhs.; in splenomegaly (Eg); cult. in Burma for lvs. u. in maturative poultices; found in ancient Egyptian tomb garlands (fr. time of Ahmes I) - **S. bispinosa** Steud. ex Fawc. & Rendle (tropAs, Af): zamarke (Nig); sds. mixed with flour appl. to skin dis. (incl. ringworm); int. with milk for Guinea worm (wAf): pl. vulner.; fiber u. to make ropes; source of a gum - **S. drummondii** Cory (Daubentonia d. Rydb.) (Fla to Tex [US], Mex): rattlebox; coffeebean; rattle bush; senna-bean; pl. c. saponins; stock poison; sds. as coffee substit. but toxic, esp. to children (2580) - **S. grandiflora** Poir. (Agati g. Desv.) (Ind, seAs, Austral, NZ, natd. WI, often cult.): corkwood tree; gallitio blanco (Cu); bitter bk. ton. & antipyr.; lvs. diur. lax.; tender pods, fls. & lvs. eaten (Indochin) fried & in curries (pot herb) (SriL); salads; lvs. resolvent (in cataplasms); sds. emmen.; fls. & rts. also u. in mal.; soft wd. u. to make charcoal; stoppers (spongy like cork) fibers u. for cords - **S. javanica** Miq. (seAs): fls. edible & u. with sugar in salt pork; pl. u. to fertilize soil - **S. longifolia** DC (Daubentonia l. DC) (tropAm+): "coffee bean"; sds. poisonous to poultry - **S. pachycarpa** DC = S. *bispinosa* (617) - **S. paludosa** Prain (trop): med. usage (Camb; Mex): said to have similar potential as source of gum, &c., as *S. bispinosa* (617) - **S. pubescens** DC (trop, esp. Af): asrati (Ghana); sds. edible after roasting; u. in stim. & invigorating beverage like coffee (Senegal); pl. u. as fish poison (Ghana) - **S. punicea** Benth. (Daubentonia p. [Cav.] DC) (tropAm [SA], now seUS): purple sesban; purple (or Brazil) rattlebox; flame plant (or bush) (Ga [US]): pl. toxic (esp. sds.) to livestock & humans (said due to saponins present) (618) - **S. sesban** Merr. (tropAs, Af): Egyptian sesban; pd. lvs. in water as vermifuge (Indones); sds. in bronchitis, excessive menstr. hemorrhage; pd. sd. with flour appl. to skin for itch (Ind); fresh lf. juice u. anthelm; in fever & guinea worm (wAf); lvs. u. as poultice for absn. of hydrocele (Ind); food, fodder - **S. vesicaria** (Jacq.) Ell. (Glottidium vesicarium Harper) (seUS): coffee bean (weed); bagpod; bladderpod; sds. c. FO toxic principle; however, apparently edible after cooking; stock poison.

seseli (herb): *Laserpitium siler*; *L. latifolium.*

Seseli diffusum Santapau & Wagh. (S. indicum Wight & Arn.) (Umb) (India plains): meadow saxifrage; frt. c. FO, VO with limonene, seselin, bergapten (kinds of coumarin), selinene; u. carm. condim. stim. stomach. anthelm. (round worms); in cattle med. (619) - **S. libanotis** W. D. J. Koch (Euras, nAf): mountain meadow saxifrage; m. spignel; rt. & rhiz. (Seseleos Radix) (incl. var. *daucifolia* Franch. & Sav.) u. carm. (Chin), to regulate GI functions (Indochin); in Jap med.; arom.; tranquilizer (potentiates barbiturates); fresh lvs. u. as condim. & stim. (Ainu) - **S. sibiricum** Benth. (S. libanotis subsp. s. Thell.) (nAs): hb. u. to treat mental dis.; rts. locally in blending beverages; med. for livestock - **S. tortuosum** L. (Cnidium dubium Thell.) (seEu): Marseilles seseli; French hartwort; frt. u. stim.

sesquiterpene: hydrocarbon, usually $C_{15}H_{24}$, theoretically composed of 3 monoterpene units, generally "head to tail"; frequent component of many VO, as of *Cedrus atlantica, Peucedanum ostruthium*; ex. cadinene, caryophyllene, santonin.

Sesuvium portulacastrum L. (Aizo./Sesuvi.) (pantropics): sea(side) purslane; hemostatic; u. for wounds; one of best exter. antidotes for stings of toxic fishes, using prolonged baths with decoc. of entire pl. (Senegal); cult. in eAs as vegetable & offered on markets.

seta de prado (Sp): *Agaricus campestris.*

Setae Dolichi (L.): **S. Mucunae: S. Stizolobii:** Mucuna, cowage hairs, fr. *Mucuna pruriens.*

Setaria glauca (L) Beauv. (S. lutescens Stuntz) (Gram) (tropAf): pearl (or cat-tail) millet; foxtail grass (Alta); yellow f.; pigeon grass; source of crude drug (Jap); cultigen; cult. for cattle food; common serious weed (Cana, US+) - **S. italica** (L.) Beauv. (Euras, cult. & natd. in many parts of OW, NW, incl. US): fox-tail (or Italian) millet; German (or Hungarian) m.; considered ton. diur.

emoll. astr. (diarrh., nosebleed, claw wounds of bears, tigers) (Chin); decoc. of crushed grains u. in fever (Indones); human food (grain), animal fodder (hay); bird seed; cult. since pre-history, esp. in Med reg.; many var., ex. var. *nigrofructa* Bailey (Hungarian grass) - **S. palmifolia** Koenig Stapf (OW, trop, esp. Ind): palm grass; lf. decoc. u. for cough & to "purify" the blood (Chin); young sprouts eaten with rice by women in confinement (Indones); lf ash in water appl. locally for skin dis. (1529); cult. ornam. (sUS) - **S. pumila** (Poiret) Roem. & J. A. Schultes (S. glauca auct. pl.; S. lutescens f. pumila [Poir.] Soo) (sEu, Af, natd Austral+) pearl (or cat-tail) millets, foxtail grass (Alta), yellow foxtail; y, bristle grass; source of crude drug (Jap); cult. for cattle food; u. like other *S.* spp.; common serious weed (Cana, US, Ge, Saudi Arabia) - **S. sphacelata** Stapf & C. E. Hubbard (sAf): u. to treat open wounds (Uganda); inflorescence to make brooms; for hay, silage, pasture - **S. sulcata** Bertoli Raddi (sAf): crushed & mixed with water as poultice for bruises (Zulu) - **S. verticillata** (L.) P. Beauv. (Af, incl. Sahara): burr (or ginger) grass; bristly foxtail; grains sometimes u. as food; to make alc. beverage; sts. to weave hats; pl ash cooking salt (Uganda) - **S. viridis** (L.) P. Beauv. (Euras, natd. NA [cosmop.]): green bristlegrass; g. foxtail; puss grass; panicle decoc. for diarrh. (Chin); st. decoc. for eyewash (Indochin); common serious weed grass; lf. decoc. u. in cough (Chin): dr. lvs. & sts. rubbed on carbuncles, boils, psoriasis, ringworm; sds. u. for eye dis. (Chin); whole pl. mixed with water & appl. to bruises (Ind).

setolone (It.): horse-tail, *Equisetum* spp.

setting: planting out, as mint runners.

setwall: Valerian.

Seuce (Ge): infect. dis.; epidemic dis.

sève du pin (Fr): fresh sap fr. *Pinus pinaster.*

seven sisters: 1) (WVa [US]): queen of the meadow (of Ohio?); *Filipendula ulmaria* 2) (Eu): *Euphorbia helioscopia, E. peplus* 3) (in Lit.): spp. of climbing rose, exp. *Rosa multiflora* (also **s. s. rose**) - **seven-barks**: 1) wild hydragea, *Hydrangea aborescens* 2) *Physocarpus* spp.

Sevenbaum (Ge): savin.

Severinia buxifolia (Poir) Ten. (S. monophylla Tanaka) (Rut.) (eAs): rt. decoc. for mal. & in resp. illness (Taiwan); lvs. & branches in resp. dis. (Indochin); tree u. for grafting citrus.

Sevin™: Arylam: contact insecticide with low toxicity; vet. u. in exter. parasites (1957-).

Sevum Praeparatum (L.): suet, prepared (NF).

sezernieren (Ge): to secrete.

shadblow (plant): shad(bush): *Amelanchier* spp., esp. *A. canadensis.*

shaddock: *Citrus maxima.*

shafsufiam (Pers.): spearmint.

shakar (Urdu): crude granulated sugar (cognate of Saccharum).

shake: rough shingle u. to cover barns & other rustic bldgs. (*cf.* shingle).

shale: fine grained layered rock of sedimentary origin; made up one-third each of quartz, clay, and miscellanea, latter including iron salts, org. matter, &c.; some with admixed petroleum (**oil s.**) - **S. oil (crude)** (fr. Colo [US]): sepd. by destructive distn. of fossilized org. matter (kerogens) in shale; u. in psoriasis, eczema, neurodermatitis; said to give marked improvement (45).

shallots: *Allium ascalonicum.*

shallu: *Sorghum bicolor* subsp. *bicolor.*

shaman: medicine man among the Am Indians, & al.

shame vine (seUS): a sensitive pl., *Schrankia* spp. - **s. weed:** *Abrus precatorius.* (SC lvs. curl before storm).

shame-briar: *Schrankia uncinata* - **shameface** (NC): *Geranium maculatum.*

shamer (Jam): *Mimosa pudica.*

shamrock: term appl. variously in different countries to *Trifolium dubium* (generally recognized as Truc S.), *T. repens* (hort. v.) (white clover), *Oxalis acetosella, Medicago lupulina* (black medic), etc. (SC seamrog, Gaelic, name given many pl. spp. bearing lvs. with 3 leaflets) - **Irish s.: true s.**: *T. dubium* (much argument over identity) (lf. u. by. St. Patrick to illustrate doctrine of the Trinity) - **water s.:** *Menyanthes trifoliata, Trillium erectum* - **white s.:** *T. repens.*

shank dog (NY [US] slang): shame weed.

sharifa (Hind): *Annona squamosa* (frt.).

shark liver oil: FO fr. livers of various spp. of sharks, esp. of sub-order Selachoidei; rich in vit. D & esp. A, sometimes with very large unitage of latter.

sharks: active & voracious elasmobranch fishes (Pisces - Elasmobranchii - subord. Squali), typically of warm sea waters; several products as foods (meat), &c., are obtained fr. them; ex. soupfin s., blue s., dogfish s., mackerel s., hammerhead s., nurse s., sand s., basking s. (-10 m long), white s. (-12 m long), whale s. (-18 m long); skin u. as leather (shagreeen), also as abrasive in med., liver & liver oil (vits. A [esp. high], D); artificial skin in burn cases, &c.; cartilage recently u. to inhibit tumor (cancer) growth (apparently on basis that sharks do not have cancer); admin. in capsules; sold at health food stores; meat good protein source; dark m. u. in sausages; hog food; shite m. marketed (Panama, &c.); hide u. in hard leathers (shoes; attache cases; hand bags; bill folds; sandpaper substit.; blood (immune value); fins u. to make f. soup (Chin); cartilage food supplement for cattle; in scleroderma, piles; teeth —> abrasive pd.

sharp sow: *Sonchus oleraceus* .

shattering: dropping or scattering of lvs., etc., esp. in the field.

shave grass: *Equisetum* spp.

shawnee haw: shonny h.

shea butter: FO fr. sd. of *Butyrospermum parkii, Vitellaria paradoxa,* **s. b. tree.**

sheep: *Ovis aries* & other *O.* spp. (Bovidae), ruminant mammals, source of many econ. prods., incl. wool, meat (mutton, lamb), wool fat, &c.

sheep: v. also under laurel; sorrel.

sheep berry bark: fr. *Viburnum lentago* - **s. pills** (seUS): **s. shadney** (Fla, Ala [US] Negro): sheep droppings u. folk med. to make a tea, sweetened & fed very young babies - **s. weed:** *Lithospermum arvense.*

shell: 1) outer cover encasing such organisms, as shellfish; (sea s. oft. in early Brit Phar.) 2) peel, outer covering of frt, nut, &c. - **s. flowers:** *Moluccella laevis, Alpinina zerumbet*; native (NA) s. fl. include various spp. of Iridaceae, as *Nemastylis geminiflora* Nutt., *Alophia drummondii* R. C. Foster, &c. (1319).

shellac: (gum) lac; (Gummi) Lacca (L); a colored resinous secretion on the bodies of insects, esp. *Tachardia lacca* Kerr and other *T.* spp. (Coccidae - Homoptera) (seAs, EI); these scale insects subsist on the bk. of various tree spp., such as *Ficus religiosa, F. indica; Schleichera oleosa,* &c.; at the twig tips the fertilized females hold fast, sucking the pl. juices & secreting a resinous exudation in which they become immersed; u. in varnishes, for coating tablets, in cosmetics; - prod. Ind (chief); Thail, Bangladesh, &c. (1824) - **Arizona s.: California s.:** lac exudn. on *Larrea tridentata* (*qv.*) - **earth s.:** *Xanthorrhoea* spp. - **Japanese s.:** prepd. fr. latex of *Rhus vernicifera*; sometimes acts as allergen - **orange s.:** one of chief comm. vars., in thin pieces, brownish yellow.

shelled: with pericarp or husk removed.

shell-fish: food classifn. as distinct fr. finfish; incl. univalves (ex. snails) and bivalves (ex. oyster), crustaceans (crabs, lobsters, shrimp), &c.

Shen Nung (2891-2637, BC): "father" of Chinese pharmacognosy; produced earliest Chinese pharmacopeia (620).

Shepherdia argentea (Pursh) Nutt. (Lepargyr(a)ea a. Greene) (Elaeagn.) (wCana, wUS): (silver) buffalo berry; bullberries (Mont [US]); soap berry pl.; beef suet tree; frt. scarlet, sour but edible; eaten fresh & in the winter preserved & dr.; eaten with beef (Am Indians); u. in female puberty rites, in jams, jellies, wines (Plains Indians); ornam. shrub; hedges - **S. canadensis** Nutt. (L. c. Greene) (Cana, esp. BC, nUS): soapberry; (western or Canadian) buffalo berry; soopolallie (Chinook jargon); cranberry (misnomer); shrub a calcicole (prefers chalky soils); typically bushes found near streams; orange-red berries picked after first frost; frt. whipped —> heavy froth ("Indian ice cream") considered delicacy; thought to counteract mosquito stings; frt. supposed worm remedy; branches in hair ton. & dye; rt. as parturient (74); decoc. of branches drunk after childbirth as contraceptive (with saskatoons & *Cornus stolonifera*); Eskimos store frt. for winter use.

shepherd's blanket: *Verbascum thapsus* - **s. purse:** *Capsella bursa-pastoris.*

sherbet: (fr. Arabic): 1) flavored sweetened beverage 2) flavored sweetened ices, eaten like ice cream 3) (Eng): kind of Seidlitz pds. put up in pairs, formerly popular as effervescent mild lax.

shiah zira: siah z.

shialkata (Ind): FO of *Argemone mexicana.*

shield fern: *Polystichum aculeatum, P. setiferum* - **s. f. root:** Aspidium.

shikimi (Jap): *Illicium religiosum,* star anise.

shikimic acid: arom. carboxy acid, occ. in star anise, gymnosperms, & many other pls.; plays important part as precursor in biosynthesis of alkaloids flavonoids, &c.

shikonin (+): (+): alkannin; alkanna red; fr. *Alkanna tinctoria* rt.; also prod. by cell tissue culture (2355); u. as natural dye in foods & cosmetics; with astr. properties.

shillelagh (Gaelic): sapling (as of oak or blackthorn) u. as cudgel (SC Irish town famed for its oaks) (pron. shilaylee).

shim (Ind): *Dolichos lablab.*

shingles: thin pieces of wood u. to cover (by overlap) a roof, &c., to keep out rain, cold, &c.; in US red cedar is u.; *cf.* shakes.

shin-leaf: *Pyrola* spp., esp. *P. elliptica.*

shinnery: dense growth of low trees, esp. of the shin oak, *Quercus incana,* &c.

shirayuki (Jap): sodium glutamate.

Shiraz gum, insoluble: exudn. prod. of almond tree & of several spp. of jungle trees, such as *Albizia lebbek* & *A. odoratissima.*

shirwin sticks: muggles cigarettes (Cannabis) laced with AcHo (Dallas, Tx [US]).

shisham (Urdu, Indo-Pak): *Dalbergia sissoo* (wd.).

shittam (2691) **bark:** Cascara Sagrada - **Shittem tree** —> **Shittem wood** (Tex [US]): *Bumelia lanuginosa, B. lycioides* - **shittim wood:**

1) (Bible): *Acacia seyal* or possibly *A. arabica* (621); the ark was made of this (622) (also Shittah w.) 2) *Halesia carolina*+ (eUS) (2657).

shiu (Jap) **oil: shiusho** (Jap): VO fr. the s(c)hiu camphor tree, an inadequately known sp. or spp. of *Cinnamomum,* possibly *C. camphora* v. *glaucescens*; c. linalol; u. perfum. & soap indy.; possible substit. for bois de rose (prodn. Formosa, Jap).

shoe-black plant: shoe-flower: *Hibiscus rosa-sinensis* - **shoe-make** (usu. midwest, Ala, Miss [US]; slang): sumach (shumach): *Rhus glabra,* &c. - **shoe-maker's heel:** *Chenopodium Bonus-henricus.*

shonny: v. under haw.

shoo-fly: (US): 1) *Hibiscus trionum* 2) *Nicandra physalodes* (usu.) 3) (sUS): *Baptisia tinctoria.*

shoot (US, slang): inject (drug).

Shorea (Roxb.) (Dipterocarp.) (seAs, Oceania): 357 spp. of resiniferous trees; source of (green) damar (gum) (EI) (*cf. Hopea*), others furnish FO; chief timber trees of As - **S. aptera** Burck (Borneo): sds. c. 40-(50)-60% FO, Borneo tallow, tenkawang fat; u. cooking fat, illuminant; in Eu to make soaps, candles; replacement for cocoa butter; produced by several other pl. sds. - **S. cambodiana** Pierre (Indochin): heart wd. u. in compd. decoc. as vermifuge; young sts. with *Barringtonia acutangula* sts. u. in decoc. as antifebrile - **S. crassifolia** Ridle (Malay): source of Resina Dammar u. med., for lighting purposes - **S. gysbertiana** Burck (Borneo): sds. source of solid fat, teglan fat, enkabank fat, with sweet buttery taste u. to treat oral aphthae; kernel FO ("shorea fat") u. as substit. for cocoa butter - **S. harmandii** Pierre (Indochin): astr. bk. u. to treat dys. - **S. hypochra** Hance (Mal, Borneo): damar temak; sds. (illipe kernels) c. 50% FO, Borneo tallow, u. food & for lamps, &c.; gum damar (resin) also produced - **S. macrophylla** (DeVriese) Ashton (Borneo): prod. illipe or engkabang seeds, source of FO (Borneo tallow) u. like cocoa butter (in chocolate) & coconut o. but bitter - **S. robusta** Gaertn. f. (nInd): sal (plant); salwa; source of sal bk. (c. 32% tannin) u. as tan. & dye (typewriter ribbons); sa(u)l resin or s. damar (arom. resin) also called "dhoma" or "dhuna" gum, u. incense, in carbon papers; to caulk boats; distd. to give "chua oil" (u. in attars & incense, in skin dis.); frt. (with sugar) u. purg. (124) & in GC; lvs. & bk. u. in tea for carcinoma (sBengal); wd. fine lumber (like teak) - **S. stenoptera** Burck (Mal, Borneo+); sds. source of FO (illipe oil) u. as emoll., salve on wounds; cocoa butter substit. (also *S. macrophylla* Ashton) - **S. stylosa** Foxw. (S. ciliata Foxw.) (PI): kusmus; forest tree is source of a valuable lumber wd. - **S. talura** Roxb. (S. robusta Roth non Gaertn. f.) (n&cInd): sa(u)l: source of sal resin ("lal dhuna"), technically u. resin similar to damar; valuable wd. - **S. thorelii** Pierre ex Laness. (Thail): crushed bk. u. for itch caused by urticaceous plants; with *Dicranopteris* - **S. tumbuggaia** Roxb. (sInd): guggilamu (Tel.); wd. of tree u. for door & window frames, &c.; trunk yields resin (Congo copal; C. dust) u. locally as incense; exter. stim. in varnishes, &c. - **S. vulgaris** Pierre (Indochin): bk. renders sows sterile; source of important resin - **S. wiesneri** Schiffn. (Mal, EI+): source of the true Dam(m)ar = (Batavian d.) resins (d. gum); Resina Dammar (Jap Phar., 1907) c. dammaroresenes, dammarolic acid, &c.; formed as gummy exudn. fr. wounds in trunk; similar grades come fr. other *S.* spp.; u. in med. plasters, shellacs, dyes, oil painting, varnishes, &c.

short: v. under buchu.

shortening: edible fats u. in baking to make texture of bread & pastry "short" or crumbly; formerly lard u., now often hydrogenated FO, esp. cottonseed oil; to give shore "soft" crumb.

shot, Indian: *Canna indica.*

shotgun (prescription, &c.): a complicated mixt. of remedies given in the hope that one or more of the ingredients would be effective or that a combn. would be more effective.

shrimp: marine decapod crustacean, mostly of small size (but at times up to 15 cm long); most are edible; the most valuable single catch fr. the sea (623); *Crangon* & other genera u. med. as anti-inflamm., aphrod. &c. (Unani med.) - **brine s.:** *Artemia salina* (*qv.*).

shrooms (US slang): psilocybin-contg. mushroom.

shrub: perenn. woody pls. a few feet in height, with branches originating close to the soil line; ex elderberry (1746).

shuck (US slang): husk of corn ear; also verb, to remove same; also, generally, frt. shell or husk.

shukkar (powder) (Ind): crude brown sugar fr. sugar cane.

shumac: sumac.

si: v. also "sy."

siah zira: *Bunium aitchisonii.*

sialic acids: group of amino sugars widely distr. in animal kingdom; component of "bifidus factors" of human milk & of gangliosides of brain/nerve tissues.

Siam: kingdom of seAs now officially designated Thailand (pron. tie-land).

Siam v. under Benzoin.

Siamese vine: *Butea superba* (rts. or tubers) & butea seeds.

sickle pod: *Cassia obtusifolia.*

Sicana odorifera Naud. (Cucurbita odorata Vell.) (1) (Cucurbit.) (WI, sMex to Br): cassabanana; c(u)rua; frt. u. for sore throat (raw flesh, juice, or decoc.); sd. inf. purg. anthelm. antipyr.; frt. peel & sds. inf. u. in uterine complaints, int. hemorrhoids (Br): frt. in preserves (Pan); to scent linens.

Sickingia rubra K. Schum. (Rubi) (Br): source of cantagallo bk. c. harman, harmine, red dye; tannins; u. antipyr.; wd. useful - **S. tinctoria** (HBK) Schum. (Pe): bk. ext. u. to treat ulcers & rheum. - **S. viridiflora** K. Schum. (Br): similar to prec.; u. antipyr. (casca de Arariba branca): wd. source of dye - **S. williamsii** Standl. (Pe): u. similarly to *S. tinctoria* (—> *Simira*).

Sicydium monospermum Cogn. (Cucurbit.) (Br) (monotypic): sds. c. some 30% FO, sds. u. as purg. vermifug, emet.; FO as drastic.

Sicyos angulatus L. (Cucurbit.) (eNA): star (or bur) cucumber; cancer herb; rts. c. starch(?); sds. with saponin (curcurbitacin B); rts. & sds. u. bitter, diur. (folk med.); in cancer.

Sida L. (Malv.) (warm regions, esp. Am): Indian mallow; bast of some spp. furnishes a strong fiber: hb. rich in mucilage; ann. or perenn. hbs. - **S. abutilon** L. = *Abutilon theophrasti* - **S. acuta** Burm. f. (S. carpinifolia L. f.) (seUS, Mex, pantrop): hornbeam-leaved sida; escobilla (Pan); kwala (Cuna Indians); lvs. u. in skin dis., as anthelm. abort (Af); expect., in cough (Guat) kidney dis. (Indian); rt. bitter, stom. ton.; sd. c. 7% FO; hb. u. med. (Cuna Indians, Pan); st. fibers u. to make cordage; pl. eaten somewhat by stock; common weed (Pan, Ind+) - **S. ciliaris** L. (Tex [US], Mex, s to Par, WI): yerba (de) pinda; decoc. of pl. for bad chest colds; decoc. u. in baths & to rub on head for common cold; int. in kidney dis. (Curaçao) - **S. cordifolia** L. (tropAm, Fla, sTex, pantrop): heartleaf sida; malva blanca (Pan); borla (Sanscr.); lei ilima (Haw [US]); fls. orange-yellow; rts. u. astr. ton. aphrod. (also sds.), refrig.; decoc. (with ginger) in mal.; cough, inflammn. (Par); rt. for dis. of nervous syst. & in "autointoxication" (Ind); sds. in GC; rt. arom. bitter, dem. diur. (Thail); fiber source; pl. juice u. rheum. (Ind); u. similarly to *S. rhombifolia* - **S. fallax** Walp. (Pac Is): ilima (Haw); large shrub with yellow fls.; fls. u. to make leis (hence called "ilima-lei"); garlands u. as symbols of welcome or farewell; territorial fl. of Hawaii, also fl. of Oahu; fls. fed to infants; as lax.; in combns popularly u. for asthma, debility (624); uter. dis. - **S. floribunda** HBK = *Sidastrum paniculatum* - **S. glutinosa** Commers. (tropAm, fr. Mex to nSA; WI; introd. into Af, seAs): chivatera (Ve); pl. u. emoll. lax. in intest. dis., to dispel flatulence; lf. decoc. for ophthalmia - **S. linifolia** Cav. (tropAm, now pantrop): given in pregnancy (rt. cortex) to facilitate sepn. of placenta (Lagos, Nig, wAf) (62) - **S. ovalis** Kostel. (Pe): rts. u. like Althea - **S. paniculata** L. = *Sidastrum paniculatum* - **S. rhombifolia** L. (S. rhomboidea Roxb.; S. retusa L.) (pantrop & subtr, incl. Af, Ind, PI, seUS): Cuba jute; rhomboid ilima (Haw); brown jute sida (Canary Is); tea plant; t. weed; iron weed; wire weed (Ala [US]); common sida (Austral); mes(e)be; sd. c. 17% FO; rt. c. alk.; lvs. u. as beverage tea (faux thé; cha inglez); inf. u. as dem. expect. in rheum. pulm. TB, antipyr; exter. for snakebite; lf juice in dys.; entire pl. incl. rts. mashed & soaked in water to form mucilaginous liquid u. to treat bladder & urethral dis.; rts. u. stom. dys., intest. dis., hemorrhages; fever; to incr. male potency (2451); rts. eaten raw by indig.; decoc. drunk for diarrh (Austral aborig.) (2263); antiperiod.; for infant diarrh. (CR); inner bk. prod. strong fiber, "Queensland hemp" (c. 83% cellulose); in clinic u. in various forms of TB (625); lvs. to dress ulcers (Mal); pl. shows heliotropism - **S. sabdariffa** L. = *Hibiscus s.* - **S. spinosa** L. (S. alba L.) (WI, tropAm, natd. eUS, Haw, sAs): prickly sida; false (or prickly or spiny) mallow; pl. decoc. for dys. (Mex); to bathe children's sores (Barbad); rt. u. stom. (Ind); u. to make paper (Geo. W. Carver); has been u. as diur. dem. vulner; a widely distributed weed - **S. tomentosus** Mx. (eUS); yaw leaf; supposed useful in treating yaws - **S. triloba** Cav. (Mex): malvavisco; fresh pl. decoc. u. in dys.; rt. decoc. u. to reduce inflamm. of ulcers; Mex Phar. - **S. urens** L. (WI, eMex, to Peru; also occ. in wAf, Indones): wind bush; malva montes; pl. decoc. u. for griping in infants - **S. veronicaefolia** Lam. (WI, SA, OWtrop, incl. Ind): bariara; pl. u. in dys., as "blood purifier" (124); for cuts; in diarrh. of pregnancy; fls. (with sugar) for burning sensation in micturition; pl. decoc. in GC (PI).

Sidastrum paniculatum Fryxell (ined.) (Sida floribunda HBK) (Malv) (tropAm, incl. WI): hb. u. vermifuge (near Lima, Peru) (626).

Sideritis hirsuta L. (Lab) (seEu, esp. Balkans, Med reg.): (German) ironwort; Berufskraut (Ge): hb. c. VO tannin (2.7%), NBP; u. astr. arom. vulner. antipyr. in fevers, in hysteria, for menstr. disorders, stitch (unilateral pain in side), ton.; in arom. baths (Eu) - **S. montana** L. (Eu, Med reg. to Afghan.): Gliedkraut (Ge); u. med., &c. - **S. perfoliata** L. (Turk): c. flavonoid glycosides, phenylpropanoid glycoside; u. diur. digestive; carm; in GI(T) dis.; anti-inflamm. - **S. roeseri** Boiss. & Heldr. (cEu); u. as component in various Ge herbals (breast teas) - **S. romana** L. (Eu, Med reg.): thé de campagna; hb. u. similarly to *S. hirsuta* - **S. scardica** Griseb. (Eu, Balkans, esp. Bulgaria): hb. u. as drug in Bulg & Yugoslav (627), as tea substit. (Pueringer tea) - **S. theezans** Boiss. & Heldr. (Gr): wasteway tea (La [US]) (SC u. in wasting dis. [progressive emaciation & weakness]); lvs. & flg. tops u. as arom. tea (Greek tea) (tea substit.); sold in markets. (2727).

sideritis herb: (US): consists sometimes of hb. of *Sideritis* spp. such as *S. linearifolia,* but more often of *Stachys recta* L. & related *S.* spp.; *S. pungens* noted in New Orleans (La [US]) markets.

Sideroxylon dulcificium A DC (Sapot.) (tropAf): miraculous berry (tree); fresh ripe frts. impart sweet taste to bitter, acid, & acrid flavors (ex. quinine, lime juice) if eaten immediately afterwards; u. to sweeten palm wine - **S. glabrescens** Miq. (Mal archipelago): bangka; latex u. as chewing gum base - **S. inerme** L. (sAf): milkweed; white milkwood; bk. astr.; inf. emet.; frt. edible; bk. decoc. given to stock for gall-sickness; wd. favorite lumber - **S. laurifolium** Benth. & Hook. (Austral): sweet bark; coondoo; bk. sweet, astr.; ext. c. glycyrrhizin; suggested for lozenges for throat dis. - **S. mastichodendron** Jacq. = *Mastichodendron foetidissimum* - **S. obovatum** Gaertn. f. (Af, Austral): ironwood; bk. u. as ton.; wd. (bois d'acouma) - **S. resiniferum** Ducke (Br): source of a balata rosada - **S. tomentosum** Roxb. = *Xantolis tomentosa.*

side-saddle flower, purple: *Sarracenia purpurea* - **red. s. f.**: *S. rubra* - **yellow s. f.:** *S. flava* - **s. plant:** *S. minor* (etc.).

siempreviva (Sp): *Sempervivum tectorum.*

sienna: argillaceous mineral c. iron oxides, etc., u. as pigment, esp. in oil colors - **burnt s.:** after calcination of **raw s.**, crude prod., a brownish ocher (fr. Lt, Cyprus).

Sierra Leone: independent state of wAf (Commonwealth of Nations); with variable climate; v. under copal tree; pepper.

sieve tube: elongated cell occurring in the phloem which conducts cell sap upwards in the spring & downward in the summer & fall; SC the sieve plates or pitted areas in wall; always accompanied by companion tube.

siftings: fine powd. of drug removed by use of sieves; represents pieces of st., sds., sand, &c. (ex. Senna).

Sigesbeckia (Siegesbeckia) glabrescens Mak. (Comp.) (eAs): frt. u. to correct minor deafness & blindness (Kor): above-grd. parts for rheum. pain, numbness of limbs (Chin) - **S. orientalis** L. (OW pantrop): Indian weed; guérit vite (lit. heals quickly; Bourbon); herbe St. Paul; common St. Paul's wort; hb. u. in ringworm & other skin dis. (Hainan); entire pl. in arthritis, neuralgic pains, gout (Chin), diarrh. dys., VD (124); styptic (PI); anthelm. - **S. pubescens** Mak. (Chin): hb. u. for sores, in rheum., weak back, boils, tumors, similarly to *S. glabrescens.*

Sigmamycin™ (Pfizer): combn. of Oleandomycin & tetracycline (synergistic) formerly marketed (1956-).

silage: fodder (either green or ripened) changed into succulent winter feed by limited fermentative process, using a closed chamber, the silo; this process gives agreeable flavor due to formation of sugars, acetic acid, &c., & prevents spoiling (u. for maize, clover, millet, etc.); means also to ensile.

silajit: silajat(u); (Indo-Pak): variously referred to as asphalt, mineral resin, &c.; found as exudation fr. rocks, &c. in mt. areas (Himal), appearing during hot summer months; suggested of pl. origin or even fr. animals (apes' feces!!); c. FO proteins (amino acids), gum, minerals, &c., hence believed of vegetable origin; u. locally as antisep. & systemically, sometimes considered almost as heal-all, thus in nerve dis., diabetes, &c.

Silbermaenteli (Swiss Ge): *Alchemilla alpina* & *A. hoppenana* (*qv.*).

sild: young herring canned as sardine (Norway) (other than brisling).

Silene alba Krause (S. latifolia Poiret; Melandrium album [Mill.] Garcke) (Caryophyll.) (Euras; natd. NA): white cockle; w. campion; rt. sometimes called Radix Saponariae albae; u. as substit. for saponaria; weed - **S. apetala** Willd. (Euras): juice u. in eye dis. (Sp); pl. u. as emoll. (Sp) - **S. aprica** Turcz. (eAs): pl. u. as remedy in anuria; breast cancer (Chin) - **S. burchelii** Ott. (sAf): u. to treat scrofulosis; in tonic bath after serious illness & in exhaustion - **S. capensis** Ott. (sAf): catch fly; u. in many dis., esp. fevers & delirium; no antibiotic activity - **S. dioica** (L) Clairville (Lychnis d. L.) (Euras, natd. Cana, nUS): red campion; rt. c. saponin; rt. & lvs. u. as detergent; ornam. hb. - **S. gallica** L. (orig. Med reg.; now cosmop.): French catchfly; gunpowder weed; pl. u. as substit. for *S. vulgaris*; believed stock poison (Af, Austral) - **S. macrosolen** Steud. (eAf, esp. Ethiop): rts. (Radix Ogkert or Sarsari) u. in tapeworm - **S. menziesii** Hook. (BC [Cana], Wash, Ore [US]): bladder campion; catchfly; in inf. u. as eyedrops for cataract (BC Indians) - **S. nutans** L. (Euras, nAf): u. similarly to *S. vulgaris* & *Saponaria officinalis* - **S. rupestris** L. (Euras-mts.): catch fly; pl. tolerates Cu in soil, hence serves as indicator pl. to early mining ventures (incl. prehistoric) - **S. virginica** L. (eNA): fire pink; c. saponin; pl. u. vermif., nervine - **S. vulgaris** Garcke (S. cucubalus Wibel; S. inflata Sm.) (Euras, nAf; natd. temp. NA): bladder campion; rattle box; hb. (Herba Silenis inflata) u. in folk med. as tea for bladder pain & chronic cystitis: rt. (Radix Behen albi) u. like Saponaria; fls. give off scent at night.

Siler divaricatum Benth. & Hook f. (Ledebouriella divaricata Hiroe) (monotypic) (Umb.) (Euras, Sib, Chin): rt. (Fang Feng) c. no alk. or tannin; u. diaph. antipyr. expect. in colds, flu, headache, backache; lvs. diaph.; fls. in circul. disturb.; sds. to treat obstinate colds (Chin, Jap) - **S. trilobum** Crantz = *Laser trilobum* (—> *Laserpitium*).

"Silex": quarts SiO_2.

silica: silicon dioxide, SiO_2; occ. naturally as opal, amethyst, flint, quartz, sand, amorphous (in walls of diatoms, in spicules of sponges, &c.); crystalline s. found as sand; insol. in all acids except HF; colloidal s. as thickener, suspending agt., stabilizer, lubricant, &c. in tabs - **s. gel:** silicic acid (precipitated silica): H_2SiO_3 (approx.): occ. in nature as opal; u. as adsorbent in chromatography; absorbent of vapors, incl. atm. water, desiccant (thus to keep photo films dry in trop. humid climates); insecticide, suspending, disintegrating, & anticaking agt.

siliceous earth, purified: Terra Silicea Purificata (L): Kieselgu(h)r; fossil flour; f. earth; siliceous earth; infusorial e.; diatomaceous e.; consists of the frustules & fragments of diatoms (class Bacillarieae - Algae), found as fossil formations in many parts of the world (largest deposits at Lompoc, Calif [US] & in Austral); this deposit is calcined, purified by boiling with dil. HCl, washing, and heating; u. as filtering aid (syrups), absorbent, in dermatological pastes; clarifier; polishing agt.; filler; insulator; esp. good as dispersing agt. for insecticidal dusts & liquids (1286); esp. for roaches.

silicicolin: anthricin: desoxypodophyllotoxin; a lignan isolated fr. lvs. of *Juniperus silicicola;* also *Anthriscus sylvestris, Podophyllum peltatum* & other pls.; said anti-tumor agt.

silic(u)le: silique - **Siliqua Dulcis** (L.): frt. of *Ceratonia siliqua,* St. John's bread - **silique:** long slender pod-like frt. of Fam. Cruc.; it splits down the 2 sides to set free 1 or 2 rows of sds.; ex mustard.

silk: Serica (L); fibrous extrusion prod. of the s. worm; largely prod. in Jap & sAs (Chin, Ind, Russia); c. proteins (sericin, fibroin, &c.); silk very useful & valuable fiber u. in mfr. of some pharmaceuticals; fabrics, sutures, surgical ligatures, etc.; largely replaced in Am by cheaper synth. fibers, such as nylon, rayon, etc. - **s. chrysalid oil:** chrysalis oil; special FO u. in Industry.

silk cotton: vegetable s. - **s. floss:** kapok - **s. f. tree:** *Chorisia speciosa* (cult) - **s. grass:** various members of Gram., Amaryllid., Bromeli., esp. *Yucca* spp., *Stipa comata, Oryzopsis hymenoides, Karatas plumieri* - **s. moth:** s. worm - **s. plant** (US): *Plantago rugelii* - **s. seed aster:** *Aster bifoliatus* - **s. tassel** (**, Fremont**): *Garrya fremontii* - **s. tree:** *Albizzia julibrissin* - **vegetable s.:** *Ceiba pentandra* (kapok); *Bombax ceiba*; *Calotropis gigantea*; &c.

silk-cotton tree: *Ceiba pentandra*; *Cochlospermum religiosum* - **Silksamen** (Ge): **silk-seed:** *Petroselinum crispum,* frt. - **silk-vine:** *Periploca graeca* - **silkweed:** *Asclepias syriaca* & other *A.* spp. - **silkworm:** the larva (or caterpillar) of *Bombyx mori* L. & other spp. of *B., Antheraea,* & of other genera & spp. (Ord. Lepidoptera - Insecta); the silk is secreted & wrapped around itself by the larva as it enters the pupal (or chrysalis) stage of the life history; *Bombyx* spp. live on *Morus* (mulberry) leaves; some silkworms of other genera feed on *Quercus, Ailanthus, Ricinus,* and other tree lvs.; larvae ("worms") and wing scales u. in homeopath. med.; chrysalis or "silk pod" u. as styptic, ton., astr., in leukorrh., chronic diarrh., &c.; s. chrysalis oil (pupa oil), special FO u. in indy.; refined & u. in paints, soaps (indy.); hair tonics, as food; worm & cocoon both u. in Unani med. (cardioton., hypotensive, hypocholesterolemic); cocoon ash u. in surma.

Silocel™ a type of kieselgur u. for heat insulation.

Silphium compositum Mx. (Comp.) (eUS): rosinweed; strong stim., u. for "weakly females" (Cherokees); rt. boiled with mullen for swellings - **S. laciniatum:** L. (*S. gummiferum* Ell.) (US: Oh to Ala [2752] & Tex): rosin weed; compass w. (or plant); lvs. turned vertically to avoid midday sun, lower lvs. arranged facing north & south (hence common name); u. in homeopathy (esp. fresh pl.); rt. diaph. stim. in debility, in chest dis., cough, painful chest, gum of pl. u. as styptic, antispasm; tall hb. (-3.5 m); pl. c. resin, with terpenes; hb. u. emet.; antipyr. in asthma; diur. in cystitis; formerly cult. in US - **S. perfoliatum** L. (e&cCana & US): cup (plant); ragged cup; hb. u. as ton. diaph. u. as restorative for deafness; diur. in cystitis, dys; u. homeop. med.; rts. u. as poultice for bleeding wounds, as decoc. in backache, amenorrh., hemoptysis (Chippewas); in smoke treatments for neuralgia, rheum. (Missouri R. Indians); rt. decoc. as emetic (Winnebago) - **S. pinnatifidum** Ell. (S. terebinthinaceum Jacq. var. *pinnatifidum* Gray) (Oh to Ala, Ga [prairies] [US]): prairie dock; resin with agreeable fragrance & bitterish taste chewed formerly by Indians & traders to cleanse the teeth or mouth & sweeten their breath (638) - **S. terebinthaceum** Jacq. (eNA-prairies): prairie (bur)dock; turpentine shrub; u. for asthma.

Silphium (L): kind of asafetida, u. in Hippocratic med.; prob. fr. *Thapsia garganica* or *Ferula* sp.; very popular in ancient med. as diaph. contraceptive.

silver: v. also under poplar.

silver bell tree: *Halesia* - **s. berry:** *Elaeagnus commutata* - **s. dollar**(s) (**flowers**) (US): *Lunaria annua* - **s. feather:** *Potentilla anserina* - **s. leaf:** 1) s. feather 2) *Impatiens capensis* 3) *Stillingia sylvatica* 4) *Argyrolobium* (Leg.) 5) several other pls. - **s. manso**: *Cayaponia globosa* - **s. rod:** 1) *Eupatorium serotinum* (SC white inflor. contrasts with adjoining golden rod) 2) *Solidago bicolor* - **s. shilling** (BC): s. dollar - **s. tree:** 1) (Ga) *Halesia carolina* 2) *Picea* spp. 3) *Melaleuca leucodendron* - **s. weed:** 1) *Potentilla anserina* (usual appln.) 2) *Spiraea tomentosa* 3) *Impatiens capensis.*

silvering (eUS): *Baccharis halimifolia.*

Silvia Allem. = *Endiandra; Mezilaurus.*

silvic acid = silvinic acid = sylvic acid.

silvichemicals: naval stores indy. chemicals.

silviculture: cultn. of forest trees; forestry: adj. silvic(al).

Silybum marianum (L.) Gaertn. (Carduus marianus L.) (Comp.) (Med reg., wAs, natd. NA, SA [pampas], sAustral): St. Mary's/holy/milk/lady's/blessed/white/variegated thistle; chardon blanc (Fr); c. argente (Fr); frts. ("Kenguil seeds") c. 16-28% FO; silymarin (a chromanone-flavone [1749]); u. as bitter ton., choleretic [for icterus, liver & gallbladder calculi]; for liver dis. (hepatitis, cirrhosis, &c.) in splenopathy; in varicose veins; appl. to wounds & sores (antibiotic); lvs. & rts. cooked, eaten in former times; ext. powder reduces injury to liver by alc. & certain drugs; silymarin = mixt. of flavanlignans (incl. silybin, &c.) histamine, flavonoids, &c.; u. skin dis., esp. psoriasis as source of FO; silybin (silymarin) (u. in liver dysfunction); adulter. of *Cnicus benedictus*; roasted frt. coffee substit.; "sd. (press) cake" u. as fodder; in comm., white, gray, & black "seeds" are recognized; decoc. hb. & rt. u. in mal. dropsy, &c.; juice for carcinoma of breast & nose (640); hb. stock poison, thought due to nitrate (nitrite) content.

silymarin group: mixt. of principles extd. fr. the frt. of *Silybum marianum,* made up chiefly of 3 isomers: silybin (chief component), silidianin, & silicristin; represent flavanolignans; antihepatotoxic agt. u. to treat liver dysfunctioning such as hepatitis, chronic icterus, gallstones, cirrhosis, cholecystopathy, &c.

Simaba cedron Planch. (Quassia c. Baill.) (Simaroub.) (CR to Co, Br): cedron (tree); vela de muerto (Ve, CA); patisleno (Ve); cedron seed (Semen Cedronis; S. Simabae [c.]); (rattlesnake's bean) u. by natives for fever & bitter ton. (very bitter); in snakebite & dog bite (rabid); sed., antispasm. in dyspeps.; sds. sold for snakebite & carried by travelers (641); vermif.: pd. cotyledons mixed in brandy & rubbed into wound (Ve, Co), also taken intern.; sd. c. FO, starch, glycosides with cedrin (valvidin), cryst. bitter princ.; bk. u. as "cure" for mal. (642); pl. decoc. u. against insects attacking herbarium specimens; widely cult. in Latin America - **S. ferruginea** St. Hil. (Br): bk. (as calunga or celunga bk.) antipyr. & stim. - **S. guaianensis** Aubl. (Guianas, nBr): bk. (Cortex Arubae) u. as bitter agt.; rt. bk. occurs in comm. as "true simaruba bark" - **S. salubris** Engl. (sBr): bk. as Calunga bark u. as for *S. ferruginea* - **S. valdivia** Planch. (Br, Co): u. against snakebite & rabies (Br) (*Simaba* Aubl. —> *Quassia).*

Simarouba Aubl. (conserved name) (Simaruba) (Simaroub.) (trop&subtr Am): trees & shrubs, bitter tasting; closely related to *Quassia* (to which often referred) & *Simaba* (643), now often referred to *Quassia* - **S. amara** Aubl. (Quassia s. L. f.) (nSA, esp. Guianas, Br, Bo, Pe, CA, WI): Orinoco (simaruba) bark; simar(o)uba bk.; c. 20-27% tannins; VO, NBP, simarubin, simarolide; u. as bitter ton., antipyr. astr. in diarrh. dys.; in tanning; abort. diur.; in mal. (644); oil sds. - **S. glauca** DC (Quassia g. L.f.) (sFla [US], sMex, WI, CA, maybe nSA): paradise tree; bitter wood; aceituno (negrito); simarouba bark (Jamaica b.) u. in mal. (inf., CR) (also bitter sap); frt. purg. emet. but also edible - **S. officinalis** DC = *S. amara* - **S. versicolor** St. Hil. (Br, Bo): paraiba; (hookworm); parahyba (Br); bk. & lf. decoc. gastric ton. astr. in dys. antiluet. anthelm. u.; rts. similarly, also antipyr.; pd. bk., u. exter. as insecticide, parasiticide (—> *Quassia*).

Simaroubaceae: Quassia (or Ailanthus) Fam.; mostly shrubs & trees of trops; lvs. in spiral order & pinnate (mostly); frt. usually drupe-like; members mostly bitter; many spp. u med.; ex *Samadera, Ailanthus, Quassia, Brucea, Simaba, Castela, Picrasma, Picramnia.*

Simaruba: Simarubaceae: incorrect spellings of original *Simarouba* Aubl. - **Simaruba** (Br): s. bark: *Simarouba glauca* (&c.).

Simira Aublet (Arariba Mart.; Sickingia Willd. (Rubi) (tropAm): vario taxa u. for lumber; dyes;

med. as febrifuge (Sickingia) (*Arariba rubra* source of alk. aribine).

'simmon (Ala [US] slang): persimmon - **'s. beer**: made fr. frt. of persimmon.

Simmondsia chinensis (Link) C. Schneid. (S. californica Nutt.; Buxus chinensis Link) (monotypic) (Simmondsi./Bux.) (sCalif [US], nwMex, Arg): jojoba (bush); jojove; goat-nut; frt. u. partur. vulner.; sds. edible, u. as coffee substit.; source of a liquid wax (c. 48%); much like & u. in place of sperm oil, in automatic transmissions (clings to metals); c. C_{20}-C_{22} straight chain fatty acids & esters; u. like olive oil in cosmetics (hair ton., facial creams, shampoos, conditioners, sun screens, moisturizing creams & ointments, in polishes, soaps); oil is rel. indigestible, good for those wishing to reduce wt. (645); cult. in arid lands (Calif, Ariz [US], Mex) for prodn. wax.

Simonilla: Simonillo (Sal): 1) *Conyza filaganoides* 2) *Calea sacatechici* 3) *Baccharis amara.*

simple: 1) a medicinal pl. or pl. drug, SC originally believed that each drug entity had a single specific therapeutic application 2) drug or medicine composed of a single ingredient - **s. elixir:** a pleasantly arom. vehicle & base for more complex elixirs; c. orange spt. comp., syrup, alc. 22%, water (< simplus [L.]) - **simpler:** collector of simples; drug collector - **simpler's joy:** (sUS): *Verbena hastata, V. officinalis.*

sim-sim: *Sesamum indicum.*

simulo (Por): *Capparis coriacea.*

sinalbin: complex crystall. glucoside found in white (yellow) mustard sd., *Sinapis alba;* hydrolyzes to form acrinyl isothiocyanate (only slightly volatile but pungent), acid sinapine sulfate (s. hydrogen sulfate) (alk.) & d-glucose (dextrose).

Sinapis alba L. (Brassica a. Rabenh.; B. hirta Moench) (Cruc.) (Euras, cult. & natd. NA): white (or yellow) mustard; ann. hb. mostly occurring as weed; also cult. for salad & sd. (yellow mustard seed), ripe sds. c. sinalbin (*qv.*), sinapine, FO (20-30%); mucil. proteins (25%), starch; u condim. flavor, emet.; skin irritant (rubefacient, vesicant); milder than black m. (odorless), hence better for int. use & in baths (1750); u. to make prepd. mustard (spice) along with black m. (mixed); important because retains sharpness since non-volatile; for prepn. FO of mustard; entire sds. u. in pickling cucumbers; to dry out bottles; prod. in Cana - **S. arvensis** L. (Brassica kaber L. C. Wheeler; B. arvensis Rabenhorst) (Euras, nAf, natd. NA): charlock; "California rape"; serious weed pest; sds. u. diur., source of FO with 4% free oleic acid; sds. used in inferior mustards; substit. for & adulter of black & white mustards; importantly u. in USSR - **S. dissecta** Lag. (Med reg.): sds. ("gardal mustard"), source of FO - **S. nigra** L. = *Brassica n.*+ (sometimes in Mex appl. to *Potentilla aurea*).

sinapism: blistering agt. u. as poultices, plasters, &c.; ex *Thapsia* (SC originally Sinapis [mustard] so u.).

sindoc: sintoc.

Sindora cochinchinensis Baill. (Leg) (Indochin): bois de luxe; bk. inf. u. with other materials in dys. diarrh bronchitis; aril chewed with betel - **S. inermis** Merr. (PI): trunk source of kayu-galu oil, u. in skin dis., perfumes; as lamp oil - **S. sumatrana** Miq. (seAs): prickly pod with cooling props. u. in combns. for profuse uter. bleeding - **S. supa** Merr. (PI): wd. FO (supa oil) u. in skin dis., as illuminant, in varnishes, paints, transparent paper, to adulter. other FO; wd. useful - **S. tonkinensis** A. Chevalier (Vietnam, Camb): go dau; bk. u. to tenderize & color fish fillets; wd. in construction (not attacked by insects).

sinfito (Sp): *Symphytum officinale.*

Singapore gum: 1) S. damar 2) S. manila (prod. in Borneo, shipped through Sing) - **singe** (Fr-m.): monkey; ape.

sinicuiche: *Heimia salicifolia.*

sinigrin: glucoside occurring in black mustard sd. & horseradish rt.; hydrolyzes to allyl isothiocyanate, potassium hydrogen sulfate, & dextrose - **sinigrinase:** enzyme active in hydrolysis of sinigrin, &c. = thioglucosidase = myrosinase - **sinistrin:** carbohydrate closely related to inulin; obt. fr. *Urginea maritima,* etc. (1931-A).

sinine: alk. fr. *Fraxinus chinensis* (*qv.*); u. antimal.

Sinningia speciosa Hiern (Gesneri.) (Br): "gloxinia" (of cultn.); ornam.; highly variable.

Sinomenium acutum Rehd. & Wils. (Cocculus diversifolius Miq. non DC) (monotypic) (Menisperm.) (eAs): rhiz. or rt. st. c. alk. stepharine, sinomenine, acutumine, sinoacutine, &c.; rt. decoc. diur. & to treat neuralgia, arthritis, stiff shoulder; st. u. for rheum. swollen joints, itching skin (Kor, Chin, Jap); inj. of sinomenine gives muscle flexibility, numbness (Jap Phar. VII [rhiz. or rt. or st.]; Kor Phar. II [rhiz]).

sinoside: glycoside of sinogenin + L-oleandrose; occ. in *Strophanthus divaricatus* - **sinostroside:** glycoside of sinogenin + L-diginose.

sintoc: Sinctozimmtbaum (Ge)**:** *Cinnamomum javanicum* - **s. bark: sintok bk.: Cortex S.:** fr. *Cinnamomum sintok* (s.: a Malay name).

sin valor (Sp): of no value; appl. to samples (muestras).

Siomycins (A, B, C, D): antibiotics of cyclic depsipeptide type obt. fr. *Streptomyces sioyaensis*

Nishimura & al. (Jap) (1969-); active against gm.+ bacteria.

Siparuna apiosyce (Mart.) A DC (Monimi.) (Br): wild coffee tree; lvs. (Limoeiro bravo, Br Phar.) c. VO; u. in med. (Br) - **S. cujabana** DC (Br): wild lemon tree; bk. u. diaph. abort. (Br); VO c. safrole - **S. guianensis** Aubl. (S. panamensis A DC) (Br, Bo, WI, CA): capitu; lf. decoc. appl. to treat mange, skin dis. (Br); int. for stomach dis. & to dispel flatulence (Pe, Br); lf. inf. for rheum. (Ve, Co); lf. decoc. diaph. in fever (Br); pl. (shrub) decoc. u. as sitz bath for expectant mothers, fever (Suriname) - **S. nicaraguensis** Hemsl. (sMex to Pan): wild coffee; kez; chuche; lf. decoc. for colds, rheum. (Pan); in catarrh. flu (Guat) - **S. pauciflora** A. DC (CR to Pe): limoncillo; pasmo; pl. decoc. drunk or u. in baths for fever, chills, grippe, weakness: u. by Cuna Indians (Pan) in fever - **S. petiolaris** DC (Ve): u. in arom. cataplasms, &c. - **S. thea** DC (Chin): lvs. u. to aromatize tea.

Siphocampylus foliosus Griseb. (Campanul.) (Arg): wd. c. siphocampiline, lvs. foliosine (alks.) - **S. odontosepalus** Vatke (Ve): tuckura; pl. decoc. u. as bath for drowsiness; antidote to poison; several *S.* spp. u. as rubber source (latex).

Siphonostegia chinensis Benth. (Scrophulari.) (Chin): hb. c. cardiac stim.; u. to allay bleeding fr. wounds.

sipo de Chumbo (Sp): *Cuscuta umbellata.*

Sippe (Ge): taxon.

sirih leaf (Borneo, etc.): *Piper betle.*

siris (Pak): *Albizzia procera, A. lebbeck.*

sirop gomme (Fr): syrup of acacia.

sirup: v. syrup.

sisal: fiber fr. *Agave sisalana*; u. binder (or harvester) twine, etc. (SC *Sisal,* exporting port of Yucatan, Mex) - **false s. hemp:** fr. *A. decipiens.*

sisam (Pak): sissoo.

sisana (NM [US], Sp): tumbleweed.

sisiak: *Daniella ensifolia.*

Sisomicin: antibiotic prod. by *Micromonospora inyoensis*; somewhat like Gentamicin; made up of 2 linked saccharides: garosamine & sisosamine; broad spectrum antibacterial activity; u. in bacterial pneumonias; related to Netilmicin.

Sison ammi L. = *Trachyspermum ammi* - **S. amomum** L. (Umb) (Eu, AsMin): (field) honewort; stone parsley; frt. u. arom. carm. diur. (646): condim.

sissai: sissoo (Indo-Pak): *Dalbergia sissoo.*

Sistrurus catenatus Raf. (Crotalidae - Reptilia) (US, Mex, esp. cMex): massasauga; poisonous snake (-1.3 m); venom toxins c. hemolysins & hemorrhagins; u. as hemostatic.

Sisymbrium alliaria Scop. = *Alliaria petiolata* - **S. altissimum** L. (Cruc.) (cAs to nBalkans; natd. all of Eu & NA): tumble mustard; Hungarian hedge m.; lvs. & fls. astr. & antiscorb. (Ind); sds. c. 20% semi-drying FO with 35% linolenic acid; oil u. technically - **S. irio** L. (Euras; natd. eUS): London rocket; rock gentle; hb. u. as stim. antipyr.; in indurations of breast & scrotum; expect. in asthma (also sds. [Pak]); lf. inf. throat & chest dis. (Sp) - **S. loeseli** L. (Euras, natd. nUS): winter rocket; lvs & fls. given in scurvy & scrofula (Ind); sprouts u. as veget. - **S. officinale** (L) Scop. (Euras, nwAf, natd. NA, Jap): hedge (or bank) mustard; hb. (Herba Erysimii) c. glucosinolates & cardenolides (sds.); u. in folk med. as antiscorbut., in laryngitis, hoarseness, coughs, asthma, pulmonitis, lithontryp. in jaundice, kidney stones, &c. - **S. orientale** L. (Med reg. to wHimalay): sds. c. sinigrin, sinapine, &c.; u. as stim. (Ind, Iran); for cancer - **S. sophia** L. = *Descurainia s.*

Sisyrinchium acre Mann (Irid.) (Haw Is [US]): yellow-flowered hb.; juice formerly u. to dye the skin with bluish designs & for painless tattooing - **S. angustifolium** P. Mill. (S. gramineum Curtis) (e&cNA, natd. in Chin): blue-eyed grass; pl. decoc. with brown sugar u. as remedy for hydrophobia (Chin)(!) - **S. arenicola** Bickn. (eCana & US): blue-eyed grass; fever grass (seUS); "blue-grass"; "tea" fr. rt. u. as lax., "spring tonic," antipyr.; other *S.* spp. u. similarly (seUS) - **S. bellum** S. Wats. (Calif [US], Lower Calif [Mex]): (California) blue-eyed grass; rts. u. as purg., lf. inf. antipyr. (Maidu Indians, Calif) - **S. schaffneri** S. Wats. (Chihuahua, Mex): rts. u. to cure dis. of teeth - **S. tinctorium** HBK (Mex to Ve, Pe, Bo): espadilla (de loma): pl. decoc. u. popularly as purg., liver tonic, for grippe.

Sitodrepa panicea L. (Ptinidae - Coleoptera): the drug-store beetle; worst insect pest of stores of crude drugs, etc.

sitosterol, beta-: sterol occurring in germ oils of wheat, rye, corn, cottonseed; tall oil, soy beans, calabar beans, &c.; reduces absn. of cholesterol, hence u. as anticholesteremic in hyperlipoproteinemia, coronary atherosclerosis (1932); to treat prostatic adenoma (overgrowth); in synthesis of hormones.

Sium L. (Umb.) (cosmop.): water-parsnips; glabrous perenn. hbs. found in wet places (647) - **S. angustifolium** L. = *S. erectum* Huds. = *Berula erecta* - **S. graecum** Lour. = *Oenanthe javanica* - **S. latifolium** L. (Eu, Af): (great) water parsnip; European (or common) w. p.; frt. c. VO with perillaldehye; u. purg. diur. antiscorb.; toxic; rt. u. diur., to adulter. valerian rt.; frt. u. to adulter. phellandrium frt. (mixed) - **S. ninsi** L. = **S. sisarum** L. (Euras): skirret; rt. (Radix Sisari) u. as expect., to reduce salivation; tuberous rt. as veget. (chiefly in salads, broths [like oyster plant]); arom. as coffee substit.; the Chin & Jap pl. sometimes recognized as *S. sisarum* var. *ninsi* Pres., often cult., called ninzi (or ninjin); rt. is u. to adulterate ginseng rt. - **S. suave** Walt. (NA [Cana & US]: BC to cCalif to NS to Fla): (hemlock) water parsnip; fls. poisonous; young lvs. & sts. edible cooked; rts./tubers edible raw or cooked (Calif Indians) (571) - **S. thunbergii** DC (Berula t. Wolff): (sAf): water parsnip; rt. held in mouth or chewed for toothache (Eu colonists); stock poison.

six naught six (606): Salvarsan; arsphenamine (SC represented sample No. 606 of Ehrlich's experimental series); its use in syph. now largely abandoned.

size: sizing: any gelatious soln. u. to glaze paper, stiffen fabric, coat wd. before painting, &c.; ex. starch, gums.

skag (US slang): cigarette (or stub or butt) (incl. cannabis).

"skeety" (sUS, slang): mesquite (tree).

skeleton weed (Ida [US]): *Chondrilla juncea.*

Skelly-Solve™: petroleum naphtha solvents, better than petrol. ether (1930).

skimmetin: umbelliferone.

Skimmi Adans. = *Illicium.*

Skimmia japonica Thunb. (Rut.) (eAs, esp. Jap, Taiwan): Japanese skimmia; Miyamashikima (Jap); lvs. c. seseline, skimmianine, eduline; VO with skimmene; wd. of twigs c. skimmin; triterpenes, skimmon (taraxeron), skimmiol (taraxerol); pl. has cardioactivity; u. to aromatize tea; cult. ornam. shrub; poisonous; st. & lvs. ton. restorative; carm. (Chin); var. *repens* Ohwi (S. r. Nakai) (Jap): lvs. u. med. - **S. laureola** Decne. (Indo-Pak, wHimal): neera (Kashmir); nehar; hb. c. VO with terpene hydrocarbons, esters, esp. linalyl acet.; u. perfum.; lvs. u. as spine & in folk med. (smallpox formerly & in vet. med.); in prodn. linalyl acetate; frt. abort (Ind); poisonous.

skimmianine: beta-fagarine: toxic xall. alk fr. *Skimmia japonica, Fagara coco, Illicium religiosum* (frts.).

skimmin: OH-coumarin glycoside releasing skimmetin (umbelliferone) on hydrolysis; occ. in *Daphne mezereum, Ferula* spp., *Skimmia laureola*, camellia fls.

skimmiol: taraxerol.

skimmion: taraxerone: a triterpene isolated fr. *Alnus glutinosa* & other *A.* spp.

skin, artificial: prepd. fr. soln. of fibroblasts fr. infant foreskin & collagen extd. fr. human placenta or fr. skin of cows & pigs ("Testskin"); u. in place of animal skin for testing cosmetics, &c.; may be useful in skin grafts.

skipper (eUS): maggot infesting (cured) meats.

skirret: *Sium sisarum.*

skullcap: scullcap.

skunk: mammal member of *Mephitis* (esp. *M. mephitis* L.) or other genera (Mustelidae: Carnivora) (NA); most active at night; offensive smelling secretion sprayed from 2 perianal glands (1753); c. mercaptans & dicrotyl sulfide; u. in homeop. for cough, neurasthenia, insomnia - **s. bush:** *Garrya fremontii* - **s. cabbage:** 1) (eastern) *Symplocarpus foetidus* (toxic) 2) (western) *Lysichton americanum* (non-toxic) (yellow s. c.) (2656) - **s. grease**: fat u. in folk med. like any other animal fat - **s. weed:** 1) *Croton texensis* 2) (Miss [US]): stramonium 3) *Cleome* spp.

slack (Scot): *Porphyra umbilicalis* var. *laciniata;* also the sauce prepd. fr. thallus.

slash (NA): kind of muddy swamp but with bottom (648) - **s. pine:** appl. to *Pinus elliottii* (mostly), *P. echinata,* & *P. taeda* (best appl. to latter since it tolerates swampy areas, *cf.* term loblolly pine).

slaw: kind of salad using fresh cabbage = **cole slaw.**

Slevogtia Reichb. = *Enicostema.*

slipper flower: *Pedilanthus tithymaloides.*

slippery: v. under elm.

Sloanea domingensis Urb. (Elaeocarp) (Haiti): bois coq; lf. decoc. u. in stomachache, colic, migraine - **S. rhodantha** Caparon (Mad): entire dugout canoes made fr. trunk of tree; wd. u. in construction; rice vans made using buttresses; other *S.* spp. employed for their wood.

sloe: 1) *Prunus spinosa* 2) (sUS): *P. umbellata* 3) (s.US): *Viburnum prunifolium* - **s. berries: sloes**: frt. of *Prunus spinosa*, u. to prepare **s. gin**.

sloke (Ireland): slack.

slop worms (Ga [US] Negro): earthworms raised in yard.

Slow Reacting Substances: S-R-S; SRS-A: s.r.s. of anaphylaxis; subst. causing slow contraction of some smooth muscle tissues; *cf.* leukotrienes.

slurry: thin aqueous suspension of small solid particles like thin mud; appl. to many pharmaceutical intermediate products; fr. slur (Old Eng); mud, esp. watery mud of thin consistency (*cf.* slur: stain, spot, as if caused by mud) (ex. kaolin, cement [mixts. in water]).

smack (US, Cana, slang): Heroin.

small: v. under bibernel, burnet, fennel flower, hemlock, houseleek, periwinkle, pimpinella, saxifrage, balsam (seaside), Solomon's seal, spikenard, &c.

smallpox: variola: acute viral often fatal infectious disease, highly dangerous, and formerly the cause of many deaths; eradication brought about by use of s. vaccine (which is no longer needed as there are no cases now left in the entire world) - **s. plant**: *Sarracenia minor* - **s. vaccine**: suspension of the vesicles of vaccinia (or cowpox) obt. fr. healthy vaccinated cow calves; must be prepared aseptically; u. as prophylactic against smallpox.

smartweed: *Polygonum hydropiper, P. punctatum.*

Smeathmannia laevigata Soland. (Passiflor.) (Senegal): lf. inf. u. as drink or in baths for severe headaches, febrile lameness; decoc. of branches to treat injuries & ocular dis.

smeech: strong smoky smell.

smellage: *Levisticum officinale* - **smelling salts**: ammonium carbonate with arom. ammonia spts.; u. to revive fainting person; resp. stim.

Smilacaceae: a fam. sometimes separated out of Lil. to incl. *Smilax, Nemexia*, etc.

Smilacina racemosa Desf. (Convallaria r. L.) (Lili) (Cana & cUS): small (or false) Solomon's seal; Solomon's plumes; S.'s zigzag; false spikenard; rhiz. u. astr. ton. expect. (Am Indian); berries sometimes eaten - **S. stellata** Desf. (Cana, nUS to sCalif): false Solomon's seal; rt. inf. u. to regulate menstr.; lf. inf. contraceptive (Nev [US] Indians).

Smilax L. (Smilac./Lili.) (worldwide trop&subtr): greenbriers; ca. 400 spp., mostly climbing shrubs; some US spp. have large rts. u. to prod. "conti chatee" (Am Indians), a red flour; many spp. u. in folk med. (649); frt. of some seUS. spp. added to sweet gum to soften; pls. popular for Christmas decorations (seUS) - **S. angolensis** Welw. (wAf): rt. c. saponin u. in GC & rheum. - **S. aristolochiaefolia** Mill. (Mex, Guat): Mexican (or Tampico) sarsaparilla; rts. c. parillin, sarsaparilloside; sarsasaponin; sitosterol glycoside (sitosterolin); orig. u. as alter. in syph., later as "blood purifier" ("depurative"), flavor; adjuvant in dermatoses, esp. psoriasis, in gout, chronic rheum.; in mfr. root beer; to prep. contee (Am Indian food) - **S. aspera** L. (Med reg. to Ethiopia & Ind): rough bindweed; tuberous rts. c. steroidal sapogenins (no parillin), much tannin; u. as nerve & general ton., substit. for Indian sarsa - **S. auriculata** Walt. (seUS-WI): wild bamboo; tuber fecula u. to prep. red coontie (jelly) u. as food by Am Indians - **S. bona-nox** L. (e&cUS, eMex): bull brier; bamboo (or saw) b.; Chin brier; a common prickly vine; rt. used in tea ("Chin ale") formerly u. in Carolinas for fevers, GC (650); fecula u. as food (Am Indians) - **S. bracteata** Presl (PI): rhiz. decoc. emmen.; uter. ton.; for postpartum hemorrhage - **S. campestris** Griseb. (Arg, Par): rt. u. diur. diaph.; lvs. u. in rheum., VD (Arg) - **S. china** L. (S. japonica A. Gray) (Chin, Kor, Jap, Indochin+): Chin root; (green) brier root; dr. tuberous rt. of climber c. tannins, gum, resin, smilax saponins, coloring matter, starch; rt. u. (as Radix Chinae ponderosae) in gout, syph., frambesia; antidote in Hg poisoning; aphrod.; shipped to Eu from 1535 - **S. corbularia** Kunth (seAs): tuberous rt. u. in ven. dis. (Chin, Thai) - **S. cordifolia** H. & B. (sMex): cocolmecan (Mex); cozolmecatl (Mex); u. in dropsy - **S. coriacea** Spreng. (WI): bell-apple; frt. said edible; veget. parts of pl. u. as abort. vermifuge (Virg Is) - **S. domingensis** Willd. (Mex, WI, CA): raiz de Chin; zarza; tuberous rhiz. u. as ton. & in VD (esp. WI) - **S. excelsa** L. (Gr, Bulg, Turk, Iran): Anatolian sarsaparilla; c. sapogenin similar to sarsasapogenin: parillin (651) - **S. febrifuga** Kunth (S. purhampuy Ruiz) (Pe, Ec): Ecuador(ian) sarsaparilla; source of an off. Sarsaparilla like *S. regelii* - **S. ferox** Wall. (Ind): water ext. of ground rt. u. for gastric dis. - **S. glabra** Roxb. (Ind, sChin): u. similarly to *S. china* - **S. glauca** Walt. (e&cUS, Mex): bramble (or saw) brier; wild sarsaparilla; (2009); boiled with *Sabal* sp., beaten rts. for stomach trouble (Ala [US] Indians) - **S. glycyphylla** Sm. (Austral): c. glycyphyllin (phloretin rhamnoside); lf. decoc. u. as ton. antiscorb., in cough & chest dis. - **S. guianensis** Vitman (Lesser Antilles): cold water rt. inf. u. in GC (Domin, Caribs: 2328) - **S. havanensis** Jacq. (sFla & Keys [US]; WI): China brier; chaney-vine; rt. decoc. for pain in testes & kidneys; aphrod. to promote "courage" (potency); in VD - **S. herbacea** L. (S. peduncularis Muhl.) (e&cNA): carrion flower (fr. bad odor); Jacob's ladder; u. in prepns. sometimes with *Jeffersonia diphylla*; frt. eaten - **S. hispida** Muhl. (eCana, e&cUS): hell-fetter; hispid (or bristly) greenbrier; bristly sarsaparilla; bristly cat-brier; a common spiny vine; u. similarly to other Am *S.* spp.; u. as "bad magic" (Ojibwa Indians) (2008) - **S. japecanga** Griseb. (Br): dr. rts. u. as depurat. in syph. gout, rheum., skin dis., & as antipyr.; Br Phar. - **S. kerberi** Apt. (Mex): source of medicinal Mexican Sarasaparilla - **S. kraussiana** Meissn. (Tanz): wild sarsaparilla; lf. or rt. decoc. u. in cough; as diur. in VD; eyes exposed to steam fr. hot decoc. for ophthalmia; famine food (British soliders in Ashanti wars) - **S. lanceaefolia** Roxb. (Ind, sChin): rt. u. like *S. china*; reputed antisyphilitic, antirheum. - **S. laurifolia** L. (eUS, esp. se; WI): laurel-leaved green brier; chaney brier (Bah); blaspheme-vine; "bamboo" (brier or vine); tuberous rt. decoc. u. in VD to "improve the blood" (Bahama Out Isl); to incr. potency (2451);

vigorous pl. with evergreen lvs., very strongly armed; rt. starchy food of Seminole (Am Indians) (bread & mush); young shoots eaten like asparagus; frts. eaten by birds - **S. medica** Schlecht. & Cham. (Mex): Mexican (or Veracruz) sarsaparilla; an important official Sarsaparilla; props. & uses similar to those of *S. aristolochiaefolia* - **S. megacarpa** A & C DC (Burma, Assam, seAs): rts. u. in local med. for cough and for women after delivery (Laos) - **S. mexicana** Griseb. (Mex, CA to Ve): coce(e)h; zarzaparilla; tuberous rts., often of very large size, u. in syph. skin dis.; rheum.; popular with brujos (Ec) - **S. moranensis** Mart. & Gal. (sMex): "mecapatli"; "palo de la vida"; rt. decoc. u. for kidney dis. - **S. myrtillus** A DC (Thai, Assam): lf. & rt. ext. u. med. - **S. nipponica** Miq. (S. herbacea var. n. Maxim.) (Jap, Taiw, Kor): tachi-shiode (Jap); fresh lf. u. for boils & eye dis. (Ainu); in skin dis., back pain, wounds, &c.; young lvs. boiled & eaten as veget. (Jap) - **S. officinalis** HBK (CR, Hond, Co): red sarsaparilla; one source of off. Sarsaparilla rt.; u. similarly to *S. medica*; cult. Jam - **S. ornata** Hook. fil.: *S. regelii* - **S. ornata** Lem.: *S. aristolochiaefolia* - **S. ovalifolia** Roxb. (Ind, Mal, Indochin): u. like *S. china*; pd. rt. in fever, VD (124); dys. rheum. - **S. papyracea** Poir. (sMex-Guiana; cBr): Chin smilax; salsaparilha de Maranhao; variously cited as source of Pará, cult. Jam, & Ec Sarsaparillas; u. in VD (Haiti); cult. Jam, Haiti - **S. parvifolia** Sessé & Moç. (India [Assam]): lf. & rt. exts. u. in med. - **S. pseudochina** L. (S. tamnifolia Mx.) (Far East): u. like *S. china*; rt. eaten at times - **S. pumila** Walt. (seUS, incl. sAla): wild sarsaparilla; s. vine; rt. decoc. u. in rheum. (folk med.) (2247) - **S. regelii** Killip & Morton (S. ornata Hook. f.) (CA, Co; cult. Jam): Honduras (or Jamaica or brown) sarsaparilla; similar c. & u. to *S. aristolochiaefolia*; diur. diaph. in skin & GU dis., rheum. formerly in syph.; in many herbal teas, cataplasms, gargles, &c. - **S. rotundifolia** L. (e&cNA): common greenbrier; "bamboo"; catbrier; horsebrier; bullbrier; fecula fr. rhiz. u. as food by swIndians (soup, bread); rhiz. u. to make bowls of brier root pipes (1860-) - **S. saluberrima** Gilg (S. utilis Hemsl.) (Hond, Salv, Guat, Nic; cult. Jam): may be source of cult. Jam sarsa. - **S. siamensis** T. Koyama (eAs): lf. juice u. for ear dis. (pus) (124) - **S. smallii** Morong (S. lanceolata L.) (seUS): wild smilax; china brier; bamboo vine; jack(son) brier; evergreen lvs. ("evergreen") much u. in comm. winter decorations (652); lvs. forage; tubers u. as food (Am Indians), alter. aphrod. - **S. tamnoides** L. (se incl. Tex & cUS): china root; hellfetter; bristly greenbrier; u. like *S. china*, as in fever (Am Indian) - **S. vitiensis** A DC (Fiji): u. med. - **S. zeylanica** L. (S. leucophylla Bl.) (eAs): rts. u. in VD, rheum. pains in lower extremities; in bloodless dys. (Ind); substit. for *S. china*: rt. decoc. for irreg. menses; tops u. as veget. (PI).

smilax, southern: *Smilax bona-nox* - **s. vine**: *Bignonia capreolata* - **wild s.**: *Smilax smallii.*

Smithia conferta Sm. (Leg) (Ind+): pl. u. as lax. in rheum., ulcers, female sterility; to reduce effects of old age (wrinkles, &c.!!) - **S. sensitiva** Ait. (Ind, Andamans): pl. in lotion form. u. for headache.

"smoke" (US slang): diluted alc. esp. denatured or wood alc.; a dangerous & disagreeable beverage - **s. bush: s. plant:** *Cotinus coggygria; C. obovatus* (western s. tree) - **s. tree**: 1) s. plant 2) *Psorothamnus spinosus* (SC st. sections at times smoked like cigar) = **s. weed** (swUS).

smooth: v. under alder; sumac.

smut corn: corn s.

snagrel: *Aristolochia serpentaria.*

snail (Ord. Pulmonata - Class Gastropoda - Ph. Mollusca): member of the univalve mollusk animals; common garden snails (*Helix hortensis, H. nemoralis, H. pomatia*) are common articles of diet (cooked) (Eu, esp. Fr & Swit [escargot]); *H. pomatia* L. (Eu, natd. NA): c. digestive enzymes; u. in folk med. for cough - **cone s.:** c. shells: *Conus* spp. (Conidae - Gastropoda - Mollusca) (trop. oceans, esp. on coral reefs): venom has analg. props. u. in cancer, severe pains.

snake: legless reptile of suborder Serpentes (Ophidia) (ord. Squamata); of 3 types 1) constrictor (ex. python) 2) venomous (ex. cobra) 3) fanged (ex. king snake) (2356): venom of some taxa u. in med. (ex. cobra, rattlesnake [v. *Crotalus*; venoms]); skin u. in leather mfr.; some spp. eaten - **snake berry**: *Duchesnea indica* - **s. grass**: v. under grass - **s. head**: *Chelone glabra* - **s. milk**: *Euphorbia corollata* - **s. oil**: body oil of serpents (sometimes rattlesnake) claimed of medicinal value; term generally applied to nostrums sold by itinerant showmen; connotation of no value since sometimes or even generally not a product of snakes; sometimes mixt. of mustard oil & benzine; liquid of variable & indeterminate nature; maybe term was derived from "Seneca Oil" - **s. plant** (NC): *Prunella vulgaris* - **s. root**: v. snakeroot - **snakebite root**: Sanguinaria - **snakegourd**: *Trichosanthes anguina* - **snakeroot** (also as **snake root**): term alone generally refers to 1) Serpentaria (46) 2) *Asarum canadense* 3) Cimicfuga 4) *Senecio aureus* 5) *Actaea* 6) *Eupatorium urticaefolium* 7) *Liatris* spp. - **black s.r.**: 1) Cimicifuga 2) *Sanicula marilandica* - **Brazil s. r.**: *Chiococca alba*, rt. - **button s.r.**: 1) *Eryngium aquaticum* 2) (seUS): *Liatris spicata, L. scariosa* (SC button-shaped tuberous rt.) - **Canada s.r.**: *Asarum canadense* (VO) - **corn s.r.**: *Liatris spicata* - **heart s.r.**: Canada s. - **Kansas s.r.**: *Rudbeckia hirta.* - **large button s.r.**: *Liatris squarrosa*- **large s.r.**: *Liatris scariosa* - **prairie s.r.**: *Liatris pycnostachya* - **rattle s.r.**: Cimicifuga - **Sampson('s) s.r.**: v. under "Sampson" - **Seneca s.r.**: **senega s.r.**: **seneka s.r.**: *Polygala senega* - **Sungree s.r.**: Virginia s. - **Texas s.r.**: *Aristolochia reticulata.* - **Viginia s.r.**: *Aristolochia serpentaria* - **white s.r.**: *Eupatorium urticaefolium.*

snake's head: *Fritillaria meleagris.*

snakeweed: 1) *Scorzonera* spp. 2) Serpentaria 3) *Xanthocephalum* 4) *Prunella vulgaris* - **snakewood**: *Cecropia*; *Colubrina* - **West Indian s.**: *Colubrina ferruginosa* - **snakewort**: *Polygonum bistorta.*

snap-bean (Ala [US]): *Phaseolus vulgaris* - **snapdragon**: common name appl. to pls. of many genera, ex. *Antirrhinum* (usually), *Aquilegia, Digitalis, Fumaria, Impatiens, Linaria, Ruellia, Asarina,* &c. - **(wild) s.**: *Linaria vulgaris* - **snapper shark** (Austral): fish whose liver FO is u. as source of natural vit. A.

sneezeweed (US): 1) *Eupatorium capillifolium* (confusion?) 2) *Helenium tenuifolium, H. autumnale,* & other *H.* spp. 3) *Achillea ptarmica* - **orange s.**: **tall s.**: **western s.**: *Dugaldia hoopesii.*

sneezewort: *Helenium autumnale.*

sneezing powder: *Mucuna pruriens* trichomes.

sniff(l)er (US slang): cocaine addict.

snoose (US slang): **snooze**: snuff.

snort (US slang): snuff (cocaine).

snow ball: *Symphoricarpos albus*, etc. - **s. b. tree**: 1) *Viburnum opulus* v. *roseum* (cult. form) (also as **viburnum s. b.**) 2) *Chiococca alba* (also as **s. berry tree**); *C. brachiata* - **s. in summer**: *Arabis alpina; Cerastium tomentosum* - **s. berry**: s. ball - **s. berry, creeping**: *Chiogenes hispidula* - **s. boys** (Indiana [US]): *Hepatica* spp. - **s. brush**: *Ceanothus* spp. - **s. drop**: *Galanthus nivalis* - **s. d. tree**: *Chionanthus; Halesia* - **s. flake**: *Leucojum* - **s. lepopard:** *Panthera uncia* (Felidae) (Ladakh): bones u. as aphrod. by natives of Asia - **s. on the mountain(s)**: *Euphorbia marginata* - **s. peas**: *Pisum sativum* with edible pods - **s. plant**: 1) *Cerastium tomentosum, C. fontanum* 2) *Sarcodes sanguinea* Torr.

snuff: "dip"; a form of pd. tobacco formerly u. by sniffing up the nostrils in order to promote a vigorous sneeze (hence the name); still a pharmaceutical form in Eu where u. as sternutatory (1754) since ca. 1870 in US u. by chewing & sucking; u. less than formerly & steadily declining - **hallucinogenic s.**: v. under *Virola* - **mint s.**: u. to substitute for snuff based on tobacco - **s. stick: s. swab** (Ala [US]): wooden stick u. to rub snuff into gums.

soap: a salt of one or usually more higher fatty acids (mostly stearic, oleic, & palmitic) made by a process of saponification (releasing glycerin as a by-product) with a metal, esp. an alkali metal; sol. in water & alc.; hard (Na) soap less sol. than soft (K) soap; u. deterg. solvent, dem. emoll. irritant, lax.; ex. sodium oleate; soaps long known (2266) - **black s.**: green s. - **Castile s.**: hard s., usually made fr. NaOH & olive oil but any bland FO may be used (1885) - **curd s.**: Sapo Animalis: made fr. animal oils or fats with NaOH; insol. in alc.; mostly u. in household - **disinfectant s.**: s. made up with some disinfectant such as phenol (5-10%), naphthol (5%), $HgCl_2$ (O.1%), other Hg comps., cresol, etc. - **floating s.**: made with air incorporated to render density less than water's - **green s.**: medicinal soft soap - **g. s. tincture**: soft soap liniment - **hard s.**: Sapo Durus; NaOH soap; made up in bars or as pd. - **hard water s.**: marine s. - **s. liniment**: camphor & soft soap - **liquid s.**: 1) s. in aqueous soln. with color & perfume, for use in public dispensers (soaps made FO except coconut & palm kernel) 2) ammonia liniment; lime liniment - **marine s.**: s. which forms lather with salt water; made with coconut or palm kernel oils or with resin & NaOH - **Marseilles s.**: Castile s. made with olive oil & soda - **medicated s.**: s. incorporated with such therapeutic agts. as pine tar (5%), sulfur (10%), salicylic acid (5%), etc. - **medicinal soft s.**: KOH soap, in which the glycerin is not removed fr. the end prod.; occurs as unctuous mass; u. in skin dis. - **mottled s.**: red s. - **oil s.**: soft s. made with KOH & veget. FO (not coconut o.) or fatty acids, producing a bland, semi-solid, neutral jelly-like prod. u. to wash fine (smooth) surfaces (ex. enamelled & varnished surfaces) - **potash s.**: a soft soap - **red s.**: s. with $FeSO_4$ added & allowed to oxidize so as to form reddish gray brown colors in the bar - **salt water s.**: marine s. - **soda s.**: a hard soap - **soft s.**: medicated s. - **s. s. liniment**: comp. of medicinal s. s. 65%, lavender oil, alc.; u. deterg. before surgical operations, in skin dis. - **solid s. liniment**: soft s. l. made solid with sodium stearate; also with added camphor, thyme & rosemary oils, ammonia water, alc. - **s. substitutes** (Calif [US]): 1) *Chlorogalum* spp. 2) *Cucurbita foetidissimum* 3) *Ceanothus integerrimus* - **superfatted s.**: s. with excess FO, u. medicinally to leave skin softer in texture; generally made with olive oil or lanolin; c. 3-5% excess fat - **tar s.**: s. with pine tar incorporated; u. in shampoos - **transparent s.**: made so with alc., glycerin, sucrose, etc.

soap-bark: 1) *Quillaja saponaria* (usually) 2) *Pithecolobium bigeminum* (also as **soap bark**) - **soap berries**: *Sapindus saponaria; S. marginatus* - **soapberry family**: Sapindaceae - **soap-bloom** (wUS): *Ceanothus* spp. - **soap-bulb:** *Chlorogalum pomeridianum* - **soap-bush**: *Ceanothus integerrimus* - **soap-plant, Indian**: 1) soap-bulb 2) *Yucca elata* & other *Y.* spp. - **soap-root**: 1) *Saponaria officinalis* 2) *Gypsophila struthium* 3) soap-bulb - **Levant s.**: *Acanthophyllum* spp. - **Russian s.**: soap-bulb - **white s.**: *Lychnis dioica* - **soap-stone**: a form of talc or French chalk in lumps - **soap-tree**: 1) Quillaja 2) *Sapindus saponaria; S. trifoliata* 3) *Gymnocladus chinensis* - **soap-weed**: 1) *Yucca glauca, Y. elata* 2) *Chlorogalum* spp. - (**small s.-w.**: *Yucca glauca*).

soapwort: *Saponaria* and *S. officinalis*.

sobreiro (Por): cork oak.

Socotra: **Sokotra: Soqotra**: island of Indian Oc off Somalia; source of **Socotrine aloes**: island now part of Rep of Yemen.

Socratea exorrhiza Wendland (Iriartea e. Mart.) (Palm.) (Br): palmito (amargo) exudate fr. prop rts. u. as admixture with curare (Mayongong Indians) (653); palm heart u. med. (CR).

soda: Na_2O; NaOH; Na_2CO_3; $NaHCO_3$: must determine precise meaning by context - **s. mint tablets**: $NaHCO_3$ tabs. flavored with peppermint oil & sugar - **s. water**: water charged under pressure with CO_2; releasing pressure gives effervescence; u. for beverages.

sodium: Natrium (L, Ge): Na; alkali metal; in combined forms abundant in earth's crust & ocean waters; occ. in body tissues, mostly as NaCl (ca. 0.5% of blood); many Na salts u. in med.-pharm.: **s. bicarbonate** (E); **s. biphosphate**; **s. chloride** (v. below); **s. iodide**; **s. nitrite** (E); **s. phosphate**; **s. salicylate**; **s. sulfate** (E); **s. thiosulfate** (E); & others (v. below).

sodium alginate: algin; Kelgin™: made by extrn. of kelps, such as *Laminaria claustonii* & other *L.* spp., with dil. soda soln.; u. as a replacement for acacia, agar, &c.; adjuvant (to potentiate antibody response to human gamma globulin in rabbit) (&c.) - **S. arabinate:** v. *Acacia arabica* - **s. ascorbate**: injectable vit. C: Na comp. of vit. C; u. in sterile aq. soln. for parenteral inj., being preferable for this to ascorbic acid itself - **s. borate**: borax (technically = the decahydrate): u. as detergent & antiseptic, astr. to mucosa; in acne, red mange (dogs); muscicide; to kill "skippers" (maggots) in meat; to destroy roaches & ants; contraceptive to rats; numerous industrial uses; toxic (use with care!) - **s. caseinate**: Na salt of casein made by dissolving casein in NaOH & evapng.; white tasteless odorless water-sol. powd. u. in paints, foods, textile mfg., &c. - **s. cellulose phosphate**: u. in adsorpt. hypercalcinuria (excess Ca in urine); approved 1982 - **s. chloride**: (common) salt; NaCl; occ. naturally as rock salt, halite (mineral), sea salt, in salt lakes & springs (source of "evaporated s."); essential nutrient; u. orally as emet. & lax. (hypertonic solns.); stim. thirst (to prevent calculi); stomach electrolyte replenisher; in diarrh. of infants; IV inj. to incr. blood vol. in hypovolemia; to combat dehydration (fr. diarrh., &c.) (E); locally to irrigate wounds (anti-inflamm.); rectal douche: numerous tech. uses such as to make freezing mixts., food preservn. & seasoning; in dyes & soaps; cake s. ("Blusalt") u. for farm animals (with Fe, Cu, Zn, Mn, Co, I) - **s. citrate**: trisodium citrate (dihydrate): u. as anticoagulant for collected blood (E); systemic alkalizer; diur. expect. diaph.; sequestering agt. (trace metals); emulsifier; acidulant; tech. in photography, &c. - **s. fluoride**: NaF: prepd. by fusing NaOH with cryolite (mineral made up of $3NaF.AlF_3$; occ. in Greenland & Ural Mts); u. as prophylactic for dental caries (measured amts. put into drinking water of many cities) (fluoridation); pesticide, esp. as insecticide for roaches & ants; anthelm. (veter. med.) (E) - **s. fluoroacetate**: (Comp.) 1080; CH_2FCOOH; extremely poisonous water-soluble compd.; u. as rodenticide, mostly to kill rats; formed by *Dichapetalum cymosum* (& other *D.* spp.) being formed enzymatically from inorg. F; plants favor areas where high F content in ground water - **s. gentisate**: salt of gentisic acid, found in gentian; made synth.; u. antirheum. in acute rheum. fever; analg.; said to decrease toxic symptoms of large doses of salicylates - **s. glutamate**: monosodium glutamate; prep. fr. wheat gluten, etc.; u. to give chicken flavor to soups, sauces, etc.; in hepatic coma - **s. hydnocarpate**: Na salt of fraction of fatty acids of chaulmoogra oil; u. in leprosy - **s. lactate**: much u. in techn.; sterile aq. soln. (1.9%) u. parenterally in acidosis alone or combined with Ringer's soln.; also to alkalinize urine, as in admn. of sulfonamides (E) - **s. lauryl sulfate**: a mixture of the alkyl sulfates, chiefly s. l. sulfate, u. as anion wetting agt. (ex Dreft), in soapless shampoos (ex. Drene), liquid dentifrices, brushless shave cream - **s. morrhuate**: s. salts of fatty acids of codliver oil; 5% soln. u. as sclerosing agt. for varicose veins; may add 5% alc. as preservative -

s. nitrate: $NaNO_3$; most obtained by recrystallization fr. Chile saltpeter (crude $NaNO_3$); mostly u. as fertilizer; represents remains of excreta (droppings) of guano birds accumulating in the arid areas near the coast of Chile; with many technical uses (2754) - **s. pectate**: prep. fr. *Agave sisalana* & other *A.* spp.; u. to replace gums (acacia, tragacanth), agar, etc. - **s. polypectate**: prod. fr. citrus rinds; u. for colloidal dispersions of metal salts, emuls. stabilizer, culture media, dietary - **s. psylliate**: soap representing Na salt of mixed liquid fatty acids obt. fr. kernels of sd. of *Plantago ovata* & other *P.* spp.; u. as sclerosing agt. in varicose veins (ex. hemorrhoids) - **s. sulforicinate**: mixt. of Na salts of sulfonated fatty acids of castor oil - **s. tallowate**: in soap mfr., prod. formed by interaction of Na salt with mixed fatty acids of animal fats - **s. valproate**: v. valproic acid.

Soenda Islands: Sunda I.

Soframycin™: Neomycin B.

soft chess (US): *Bromus mollis.*

soga de muerto (Pe): *Banisteriopsis caapi*

soil: a mixture of mineral matter (obt. by rock weathering) with humus (fr. the decay of pl. & animal tissues); soils vary greatly in particle size, water retention, air capacity, richness in chief elements (N, P, K, Ca,·etc.) and in trace elements (B, Cu, etc.); c. silicates of Al, Fe, Ca, Mg; free SiO_2; a study of the soil is essential to the proper cultivation of drug plants.

Soja (L): Soya: **Soja hispida** Moench: **S. max** (L) Piper = *Glycine max*, soy bean.

solacauline: glyco-alkaloid of *Solanum acaule.*

solamargine: glyco-alkaloid of *Solanum marginatum, S. nigrum* (frt.).

Solanaceae: Nightshade Fam.; usually hbs., sometimes shrubs & trees; ca. 2,600 spp.; mostly growing in trop.; economically important; lvs. alternate, fls. often showy; calyx & corolla each 5-cleft; frt. capsule or berry; many members are rich in alks.; fam. thus includes many poisonous spp. & numerous drug pls.; ex. *Datura, Physalis.*

solanain: proteolytic enzyme of *Solanum elaeagnifolium.*

solancarpine: solanine-S: solasonine (654); alk. occurs in *Solanum sodomeum* & some other *S.* spp.

Solandra Sw. (Swartzia J. F. Gmelin) (Solan.) (Mex, tropAm): chalice vine; some spp. u. as emoll. for wounds, tumors - **S. decipiens** (Br): lvs. u. as adulter. of Maranham pilocarpus; one source of "sacred" hallucinogen (by inhalation of smoke fr. burning dr. lvs.) (Br Indians); lvs. emollient to tumors - **S. grandiflora** Sw. (S. nitida Zucc.) (Mex, WI, tropAm): "chamico bejuco"; bk. toxic (alks); cult ornam. vine: u. med. (Mex) - **S. guerrerensis** Mart. (sMex): hallucinogen; *cf.* peyote; frt. & rts. considered more potent than lvs.; c. tropine alks.; admin. per ano; other *S.* spp. similarly used.

Solani Fructus (S. Radix) (L.): *Solanum carolinense*, berries (rt.).

solanidine: aglycone of solanine, chaconine, solacauline, & other glycosides (spp. of *Solanum, Veratrum* &c.).

solanine: steroidal glycoalkaloid of *Solanum* spp., as *S. tuberosum, S. dulcamara,* &c.; *Lycopersicon* spp.; mostly obt. in practice fr. seed potatoes (is bitter princ. of same); formerly u. in asthma, epilepsy (1933) - **solanine-S**: solasonine.

solano (It): **s. negro** (Sp): *Solanum nigrum.*

Solanum L. (Solan.) (worldwide, trop & subtr): nightshades; hbs., shrubs, vines, & trees of diverse types: some 1,800 spp., comprising many important economic pls., such as potato, tomato, egg plant, bittersweet, &c.; a number of toxic pls. also such as *S. carolinense* (frt.) - **S. acanthifolium** Drege (sAf): bitter (or devil's) apple; frt. juice appl. to "sandworms" & ringworm - **S. acaule** Phil. (Pe): a wild potato; c. solacauline; u. in breeding program for cult. potato - **S. aculeastrum** Dun. (Nigeria+): apple of Sodom; ash of frt. rubbed into scarifications over knees for arthritis there; in ringworm of cattle & horses - **S. aculeatissimum** Jacq. (Af): frt. & rt. u. for colds; in dysmenorrhea - **S. adoense** Hochst. (eAf): frt. u. as wound dressing - **S. aethiopicum** L. (trop wAf): young shoots & frts. u. as sed. in intractable vomiting (Nigeria); carm., colics (eSudan); lvs. u. as veget.; cult. - **S. albicaule** Kotschy (tropAf to wInd): pl. decoc. u. for ulcers - **S. albidum** Dun. (Br, Pe): "jurubeba branca" (Br); inf./decoc. u. for kidney, liver, & splenic dis. (sold in market stalls) (Br); lf. u. in cataplasms for cancers, &c. (Pe) - **S. americanum** Mill. (S. nigrum L. var. *a.* O. E. Schulz) (sFla, WI, Mex to Pe, Arg): inkberry; mora; gooma-bush; lf. poultice analg. for erysipelas, neuralgia, rheum. (Pe); lf. inf. sed. in asthma (PR, Trin); lf. juice smeared on shingles (Trin); fl. decoc. pectoral (Salv) - **S. anomalum** Thonn. (Nigeria): frts. u. lax. digest. - **S. argillicolum** Dun. (Ve, Neth, WI): patia di zumbi; rt. decoc. in diabetes (Aruba); lf. decoc. formerly in syph. - **S. auriculatum** Ait. = *S. mauritianum* - **S. aviculare** Forst. f. (Austral, NZ): kangaroo apple; gunyang (seAustral): source of solasodine (steroid alk.); frt. c. 0.8-3.5% steroids but edible when ripe; closely related to *S. laciniatum* Ait. - **S. bahamense** L. (sFla [US], WI): canker berry (Bah); frt. decoc. for weak back, bed-wetting, thrush (children) (Bah); for missed menses -

S. bojeri Dun. (Af): lf. appl. in ulcers; antisyph. - **S. campylacanthum** Hochst. (trop eAf): rt. & frt. in hot milk for fever (Masai); frt. for dehairing hides - **S. capense** L. (sAf): nightshade; frt. pulp rubbed on ringworms, warts; rt. u. in milk for dys., cough, in dysuria - **S. capsicoides** All. (S. aculeatissimum auct. non Jacq.) (orig. eBr, spreading till pantrop): cockroach berry; c. poison; pounded pl. mixed with lime juice & appl. to ringworm (Jam) - **S. carolinense** L. (sCana, e&cUS): horse nettle; ball n.; bull n.; frt. (horse nettle berries) are richest part of pl. in alk. (solanine, solanidine, solasonine, solamargine); u. sed. antispasm. in epilepsy, tetanus(!), (male) aphrod. (CA); diur.; rt. u. anod. sed. diur.; chewed for toothaches (655) - **S. chacoense** Bitt. (Arg): a wild potato; c. chaconines (steroid alkaloid glycosides yielding solanidine on hydrolysis); u. in genetic work; cult. - **S. ciliatum** Lam. = *S. capsicoides* - **S. cornutum** Lam. (S. angustifolium Mill.) (cMex to Hond): ayohuistle (Mex); pl. u. sed. narcotic, resolvent - **S. crispum** R. & P. (Ch, Pe): natri; pl. inf. u. antipyr. ton., in headache caused by exposure to sunshine; in intestinal washes for fever (321); c. natrine, solanine, chaconine(?); one of drugs called "paolo natri" (SA); ornam. shrub or small tree - **S. cumingii** Dun. (PI): sds. sed. for toothache; lvs. as poultice to mitigate pain - **S. cymosum** (wSA): kusmailla; u. sudorific, lax. - **S. demissum** Lindl. (Mex): Mexican wild potato; c. demissine (glycoside of demissidine); tubers edible but remainder of pl. toxic - **S. diffusum** R. & P. (Ec, Pe): cold water inf. u. for stomachache & diarrh. (777) - **S. diplacanthum** Dammer (Tanz): u. as remedy for miscarriages - **S. diversiflorum** F. Muell. (North Territory Austral): green frts. rich in solasodine; some claim outer parts of frt. eaten - **S. dulcamara** L. (Euras, nAf; introd. & natd. into NA): (European) bittersweet; (woody) nightshade; fls. pale violet or blue or blue white; scarlet berry (crimson or bright red, not orange); sts. (Dulcamara, L.) c. steroid alks., soladulcidine, solamarines, solamargine, &c.; in chronic skin dis. (esp. where itching irrit.), eczemas; chronic bronchitis, asthma, dyscrasias; anod. in gout, arthritis, rheum.; emetic, narcotic, diur. (twigs & rts. also u.); lvs. u. diaph. (esp. Homeop.); prod. Ge, Fr, It; distinctions from American bittersweet (*Celastrus scandens*): color of frt.; leaf nature - **S. elaeagnifolium** Cav. (S. hindsianum Benth.) (sUS, Mex s to Arg, Ch): white horse (or bull) nettle; silverleaf nightshade; whiteweed; trompillo (Tex [US]); buena mujer; "mariola"; c. alks., solanain (protease); u. sternutatory (sold in Mex-Am herbal stores, swUS); frt. u. as rennet substit. to curdle milk (Navaho Indians) in making cheese; for tooth dis.; poison ivy; mixed with brain tissue to tan buckskins (Kiowa Indians); appl. topically for tonsillitis; pd. insufflated into bronchial area for catarrh., &c.; frt. diaph.; in rheum.; toxic to livestock - **S. erianthum** D. Don (S. verbascifolium auct. non L.) (sTex, Fla [US], Mex, WI, s to Pe & Ur, natd. tropAs): guardolobo; guastomate (Mex); potato tree; rt. decoc. drunk to reduce fever; heated lvs. appl. to forehead for headache; &c. (frts. to sores & aches [Mex]) - **S. esculentum** Dun. = *S. melongena* - **S. esculentum** Mill. = *Lycopersicum e.* - **S. excisirhombeum** Bitt. (Pe, Ch): japichina; u. toothache remedy - **S. ferox** L. (S. indicum L.) (Ind & trops): rt. decoc. u. for pains over body; in syph. (Mal); in diabetes (656); expect; frts. u. astr. carm.; frt. sour-relish in curries - **S. ficifolium** Ortega = *S. torvum* - **S. galeatum** André (Austral, SA): naranjilla (Co); frt. edible - **S. gayanum** (Remy) Phil. f. (Ch): natri; reputed febrifuge; source of paolo natri; *cf. S. crispum* - **S. giganteum** Jacq. (sAf, Ind) large shrub or tree (-7 m); lf. u. as dressing for foul ulcers, also ointment with fresh juice & lf.; berry u. to curdle milk; u. in throat abscess - **S. glaucum** Dun. (Ur, Br): duraznillo blanco; hb. u. in enema (decoc.) for painful colics, antipyr.; in indiges. diarrh.; rt. u. as analg. stim. - **S. gracile** Link = *S. nigrescens* - **S. gracilipes** Dcne. (Pak, wInd): halun (Punj); frt. & lf. juice appl. to otitis - **S. grandiflorum** R. & P. (Mex+; Ind): fruto do lobo; c. grandiflorine; scopolamine; frt. delicious, peach-like; said hallucinogenic; grd. frt. u. in poultice for inflammns. (Br) - **S. hartwegi** Benth. (Mex to CR; Martinique): lava-plato; lf. decoc. u. for infect. of ears, nose, & sinuses - **S. heterophyllum** Lam. (Pe, Ve, FrGuiana): ayoc mullaca (Mexia); sds. caustic, u. on skin spots - **S. hirtum** Vahl (Yuc to Co, Ve, Trin): huevo (de gato); rt. decoc. u. as diur. in edema; for bleeding hemorrhoids (Ve) - **S. hispidum** Pers. (sMex to Pe; Ind): campucas(s)a; sosa; lvs. c. spirostane saponins; lvs. crushed & appl. as poultices to wounds, bruises (1743) - **S. incanum** L. (S. coagulans Forsk.) (s&cAf, Ind): bitter apple; frt. juice u. for dandruff (Europeans); rt. for abdom. pains; lf. & rt. u. in cough, colic, GC; frt. juice c. rennet-like enzyme, u. to curdle milk & prep. cheese; frt. inf. to remove external benign tumors; in carcinomata; frt. juice & rt. u. to prep. arrow poisons: crushed frts. to treat epilepsy; some believe causes esophageal cancer: sometimes regarded as *S. melongena* var. - **S. incertum** Dun. = *S. nigrum* - **S. indicum** L. = *S.*

ferox L. - **S. insidiosum** Mart. (Br): jumbeba do Rio; rts. u. as stom. - **S. jacquinii** Willd. = *S. xanthocarpum* - **S. jamaicense** Mill. (sMex to Pe; WI): tomate del diablo; pl. u. to treat leprosy (Pan) - **S. juciri** Mart. (Br): jequirioba; lvs. u. sed. (as inf.) - **S. khasianum** Clarke (Ind): u. med.; source of raw materials u. in steroid syntheses. - **S. kioniotrichum** Bitter (wAmaz basin, Ec): bk. of tree as aq. prepn. u. as purg. (777) - **S. laciniatum** Ait. (Austral, NZ): kangaroo apple; richest source of solasodine (1-2%), related to & u. in prepn. of diosgenin (for steroid synthesis); frt. eaten when "dead" ripe; cult. in seEu, USSR) - **S. luteum** P. Mill. (S. villosum P. Mill.) (Eu, Med reg., AsMin): c. solasonine, solamargine, solavillin (glycoside); u. in homeop. med. (fresh flowering pl.) (1935) - **S. lycopersicum** L. = *Lycopersicon esculentum* - **S. mammosum** L. (native to WI, natd. Mex, CA to Pe; introd. OW): bachelor's pear; nipple nightshade; chich(it)a (Guat); tintoma (Pe); turkey berry; udder plant; titty plant (Fla [US]); frt. decoc. in colds, asthma; frts. c. steroid alks. solanidine, solasodine (u. in steroid mfr.); poisonous; u. to kill roaches, mice, rats; sds. u. for colds (Salv, Hond); lf. decoc. u. in kidney & bladder dis. (CR) - **S. mauritianum** Scop. (S. auriculatum Ait.) (SA, natd. pantrop): frt. c. solasonine, solamargine; frt. decoc. u. as insecticide; sd. ext. exter. in rheum.; lvs. u. as tobacco substit. (Br); lf. ext. in hemorrhoids - **S. melongena** L. (S. esculentum Dun.) (sAs, now pantrop, esp. Af; widely cult.): (garden) eggplant; Guinea squash; aubergine (Fr); mad apple; berenjena (Sp); garden egg; frt. eaten as veget. (thoroughly cooked!), in curries (Ind); frt. & lvs. with decholesterinizing action (reduces cholesterol in blood by some 50%), u. cholagog, diur. (101); rt. u. as stim. in asthma (Ind); chilblains; rt. peel u. to blacken teeth (high class Chin women); lvs. u. for hemorrhoids (PI); in toothache, in skin dis.; hb. adm. for intoxication (Ur); sds. stim. - **S. merkeri** Dammer. (trop eAf): frt. u. as soap (Tanz); rt. u. for cough, swelling of gums, hookworms; lf. as styptic; vars.: black, purple, white (fewer seeds) - **S. muricatum** Ait. (Pe; cult. SA, CA, for frts): pepino (morado); p. dulce; frt. edible, melon-like, eaten raw; in jams & preserves; u. in med. (Co); said good for kidneys - **S. nigrescens** Mart. & Gal. (S. gracile Otto/Link) (CA, Mex, Fla [US]): black nightshade; "deadly n." (misnomer); toxic pl. but lvs. u. as potherb (Mex); frt. edible; u. in fevers - **S. nigrum** L. (cosmop. in temp. & torrid zones) (now generally considered a complex or gp. of spp. to incl. *S. americanum, S. nodiflorum, S. pseudogracile* Heiser): black (or garden) nightshade; houndsberry; "bilberry"; hb. c. small amts. mydriatic alks. incl. solanines, chaconines, solasonine; sed. for stomach & bladder spasms; weak narcotic (670), vulner.; adulter. of belladonna lf.; pot herb (care!); decoc. for erysipelas (Mex): rt. u. expect., frt. & lvs. lax. (Ind); frts. u. diur. in edema (Ind): in eye dis. (UP, Ind); for asthma (swAf); appl. to suppurating wounds; frt. edible (some cult. vars. have delicious frts. called "wonder berry," "sun b.," &c.) (but ripe berries may be sl. toxic); u. in pies, &c.; considered undesirable but non-toxic contaminant of canned peas, where occasionally occur; berries not harmful when fed to cats (657) - **S. nodiflorum** Jacq. (Austral, now pantrop.): glossy (or black) nightshade; lf. u. as poultice or dressing on ulcers, abscesses, furuncles, carbuncles, swollen glands (cAf); u. in resp. dis., skin eruptions (Haw), fever (Guad), colic (Pe) - **S. nudum** HBK (S. parcebarbatum Bitter) (sMex to Pan): sauco (CA); chillo; pl. u. antipyr. in colds, fever (Pan); in kidney dis. (Guat): for liver dis. (Pan, Choco Indians) - **S. ochraeoferrugineum** Fern. (S. diversifolium Schlecht.) (Mex to Pan): salvadora; pl. u. to treat leprosy (Pan) - **S. opacum** A. Br. & Bouché (Austral): green berry nightshade; "blackberried n." (658); frt. & lvs. c. solanine, possibly also mydriatic alk. (658); hb. u. med. (sAf); stock poison (Austral) (may be confused with *S. nigrum*) - **S. panduraeforme** Drege (tropAf): rt. decoc. u. in hemorrhoids; pd. rt. for aphthae (children); smoked for cough; frt. juice for sore eyes; ointment fr. burnt pl. rubbed on skin of legs for rheum. - **S. paniculatum** L. (tropBr): jurubeba; rt. c. jurubine (steroidal alk.) with jurubidine (aglycone) (not present in lvs. & frt.); rt. lf. & frt. u. as stom., in gallstone colics, lax., diur. ton.; for liver, bladder, & uter. dis.; anti-tumor - **S. pectinatum** Dun. (Ec): edible frt. pulp rubbed into hair to give luster & control head lice; acidic frts. eaten to prevent vomiting fr. scorpion stings (659) - **S. pharmacum** Klotzsch. (eAf): lvs. c. solanine, &c.; appl. locally to skin lesions - **S. phyllanthum** Cav. (Ch, Pe): papa cimarrona; appl. to wounds - **S. probolospermum** Bitter (Pe): shopta lvs. inf. u. in flatulence - **S. procumbens** Lour. (S. hainanense Hance) (Hainan, Chin): rt. u. in GC, in wine intoxication - **S. pseudocapsicum** L. (Eu, tropAm, Af): Jerusalem cherry (SC frt. resembles same but is actually a berry not a drupe); (Madeira) winter c.; Natal c.; c. solanocapsine; solacapsine, solacapine; pl. anodyne, in visceral pains; feeble narcotic; stock poison; frt. has been fatal to children; popular cult. ornam. - **S. pseudolulo** Heiser (Co): lulo comun; frt. deliciously edible - **S. pubescens** Roxb. = *S.*

verbascifolium - **S. pulverulentum** Pers. (Pe, Bo) lvs. & fls. c. VO; nunuma (Pe); frt. u. antipyr. emet. - **S. quitoense** Lam. (Co, Ec, Pe [much cult.]): naranjilla; Quito orange; lulo; frt. edible, said much like orange; u. for prepn. frt. pulp juice with 5-11% sugars: u. in drinks & sherbets: frt. added to maté tea - **S. radicans** Hill (Pe): kusmaillu; u. as purg. (enema) in folk med. - **S. rechingeri** Witasek (Solomon Is): bk. macerated & heated & appl. to sore legs - **S. repandum** G. Forst. (Oceania): sou sou (Fiji): berries eaten with yam or in soups & stews; fresh juice as drink; med. (Fiji; New Caled); much cult. - **S. rostratum** Dunal (Mont, Neb, Wyo [US] to Mex): buffalo bur(r); sandbur; duraznillo; fl. decoc. u. in cough & pertussis; supposed stock poison; similar to *S. cornutum* - **S. rubetorum** Dun. (sAf): fumes fr. heating frt. pulp inhaled for toothache - **S. sanitwongsei** Craib (Thail): frt. c. "solanine," u. in diabetes (Bangkok) (659) - **S. scabridum** Dun. (Ve): lf. inf. u. as bath in mal. (Kunana, nVe, Ind) - **S. seaforthianum** Andr. (Br): ornam. shrub - **S. sessiliflorum** Dun. (Ec, cult.): frts. eaten or juice sucked for thirst & in beverages; rubbed into hair to make it shine; pulp imbibed for scorpion sting or spider bite to prevent vomiting; frts. rubbed into insect stings to relieve pain (659), pd. sds. u. to relieve irritation fr. chewing coca lvs. (Co) - **S. sisymbrifolium** Lam. (tropAm, now cosmop. trop): prickly tomato; frts. eaten in times of food scarcity (Ind); rt. u. for kidney dis. & rheum. (Arg) - **S. sodomeum** L. (Af, eMed reg.): apple of Sodom; bitter a.; Dead Sea a.; lvs. & frt. u. as diur. in cystitis, edema; cough (sAf); rt. bk. u. in impotency & barrenness; human & stock poison - **S. somniculentum** Kunze ex. Schlecht. = **S. somniferum** Ktze. = *Atropa belladonna* - **S. straminifolium** Jacq. (Co+): frt. (in vinegar) u. as condim. - **S. supinum** Dun. (sAf): pl. c. solanine; lf. & frt. u. as diur. expect. in cough; frt. u. to curdle milk - **S. surattense** Burm. f. (Ind, pantrop): rigni; frt. juice u. for ear dis. (124); in fever, cough - **S. tomatillo** Remy (Ch): natri; hb. u. similarly to *S. crispum*, as antipyr., also hypotensive; c. tomatillidine - **S. tomentosum** L. (sAf): u. for sore throat; in syph. (Zulus) - **S. topiro** HBK (Co, Ve, Pe): tupiro; cocona; frt. esculent, delicious, juice u. as beverage; pl. u. insecticide (head lice) (660) - **S. torvum** Sw. (CA, WI, now pantrop): devil's fig (Austral); plate-bush; gully-bean; c. solasodine, jurubine; pd. rt. u.in stomachache (with honey) (124); in combns. for GC (Domin, Caribs); in mumps (Guat); oil expressed fr. frt. u. in arthritis (Cu); immature frts. u. as soap (Co; PR); frts. eaten (usually unripe) in curries (Ind); suspected stock poison (Austral); rt. decoc. u. as antidote for poisons; u. after birth to reduce blood loss (PI) - **S. trilobatum** L. (Ind): rt. u. bitter ton.; fls. & frts. u. as expect. in cough - **S. tripartitum** Dun. (trop): cuti cuti (Bo); c. solapalmitine, solapalmitenine; shows significant tumor inhibition activity; u. to "cleanse the blood" (Bo) - **S. tuberosum** L. (Andes reg. SA; now widely cult.) (1934): (common or "Irish") potato; Kartoffel (Ge); pomme de terre (Fr): tubers furnish highly nutritious food in reducing diets (1744); u. in mfr. p. flour (663); p. starch (Amylum Solani), alc.; the most popular starch on Eu continent & with many uses; "seed potatoes" (tubers for propagation; chiefly fr. Cana) source of solanidine (u. in steroid synthesis); bread made fr. mixt. of p. flour with rye (or wheat) flour (serves as humectant), in mfr. sausages, confect.; very important rt. crop in temp. zone & cult. at high elevations in trop&subtr regions (ex. India); u. mfr. dehydrated & canned potatoes, p. chips (=Saratoga c.) (thin sliced potatoes fried in cottonseed oil or lard); p. "flakes" (dehydrated mashed potatoes); p. syrup & p. sugar much u. in Eu; juice of tubers (esp. of red ones) u. in stomach dis. as spasmolytic & antacid; concd. juice & in glycerin soln. u. analg. in gout, rheum.; green tubers & new sprouts c. solanidine, &c., & are toxic: lf. decoc. u. for chronic bronchitis (Yuc): raw potato wrapped in cloth & appl. to inflammed eye for relief (Burm), also for hot water burns: sliced raw potato placed on burns (Jewish folk med.) (1755); fl. decoc. for chest complaints: cure for scurvy (661) (662): diseased potatoes u. in homeop. med. (Solanum tuberosum aegrotans) - **S. uporo** Dun. (Polynesia, incl. Tahiti, Rarotonga): polo tonga (Tonga); cannibal apple; lvs. u. in med. (Tonga; 86); cult. for frt., Fiji - **S. verbascifolium** L. (pantrop.): potato (or tobacco) tree (Austral); mullen nightshade (sUS); u. to treat dys., intest. pain (Taiwan); decoc. of lvs. in vertigo (Mal); frt. roasted to give poultice for ulcers, boils, &c. (Mex); rt. decoc. for leucorrh. (Indones); in diarrh. dys. (PI); lf. inf. as rinse for sores in mouth; lvs. rubbed on skin (Solomon Is); berries edible - **S. villosum** P. Mil. = *S. luteum* subsp. *v.* - **S. vulgatum** Willd. = *S. nigrum* - **S. wendlandii** Hook. f. (CR, cult. US, Ind): Costa Rican nightshade; ornam. climber u. for arbors & porches - **S. xanthocarpum** Schrad. & Wendl. (Ind, SriL, seAs): katali (Hind); c. solasonine, "solanine," solasodine, solamargine(?), sitosterol; important in Hindu med. (664); in folk med.; rt. u. as expect. in cough, asthma (also fls. & st.); frt. edible.

solasonine: glyco-alk. found in *Solanum luteum, S. carolinense, S. sodomeum*, &c.

Soldanella alpina L. (Primul.) (cEu): alpine soldanella; a. gravel-bind (or snowbell); hb. c. saponins, cyclamin, quercetin, kaempferol; u. as gentle purg. (folk med.).

soldier's cap: *Dicentra cucullaria* - **s. shrub**: Matico.

soleliver oil (Jap): liver FO fr. *Solea solea* (Soleidae), the sole, & other flat-fishes; u. like codliver oil.

Solena heterophylla Lour. = *Melothria h.*

Solenostemma argel Hayne (S. oleifolium Nectoux) (monotypic) (Asclepiad.) (Upper Eg, Nubian Desert [Sudan], Arab): arg(h)el; lvs. & fls. of shrub u. as purg., antipyr.; added to Alexandrian senna lvs. as corrective or adulter.; pd. lvs. u. to heal sores of camels: lf. decoc. u. for colic, neuralgia & sciatica (Lib); c. kaempferol, argeloside; cynanchin.

Solenostemon ocymoides Schum. & Thonn. (Lab.) (w tropAf): tubers u. for craw-craw; pl. juice heated & appl. to yaws; lf. juice u. as stom. carm. in colic (Ghana); lvs. c. VO; u. anticonvuls. in infantile convulsions; in colic; lvs. & tubers u. as veget., sometimes cult.

Solidago L.: 1) (Comp.) (mostly NA): ca. 140 spp. of perenn. hbs. with mostly brilliant yellow masses of fl. heads, hence common name "goldenrod"; some spp. may cause pollinosis: state flower of Ala [US]; some spp. give rich yield of rubber latex. (1936); some spp. u. in arthritis (Ala (SC solidus [L.]; whole, firm; referring to reputed qualities as vulnerary; i.e., to make tissues firm); 2) (L) (specif): *S. odora* - **S. californica** Nutt. (swOre to sCalif [US]): California goldenrod; lotion of boiled lvs. & sts. u. to heal cuts & sores of man & domestic animals (665) - **S. canadensis** L. (e&sCana, most of US): Canadian (or high) goldenrod; yellow weed: mariquilla (Mex); hb. c. VO with cyclocolorenone, caryophyllene, borneol, cadinene, terpene hydrocarbons (pinene, phellandrene, dipentene) (666); quercitrin, rutin; u. carm., diur. stim. (Mex-Am herbal stores); sds. eaten by Am Indians; source of Canadian goldenrod oil; adulter of *S. virgaurea* hb. - **S. chilensis** Meyen (Ch): hb. u. as astr. in dys. & intest. ulcers, colicky pains - **S. gigantea** Ait. (NA, widely distr.): giant goldenrod; hb. c. VO saponins; u. as gargle tea in pharyngitis, tonsillitis, stomatitis - **S. graminifolia** (L) Salisb. = *Euthamia g.* (L) Nutt. - **S. leavenworthii** T. & G. (seUS; eNCar-Fla): claimed Am pl. richest in caoutchouc (T. A. Edison); potential use as domestic source of rubber in case of emergency: hybrid with *S. gigantea* (wild in Everglades) produced very large plant with high rubber content (667) - **S. mexicana** Berl. (Mex): u. exter. on wounds & sores - **S. microcephala** Bush = *Euthamia tenuifolia* - **S. microglossa** DC (Br, Arg): hb. c. alk.; lvs. & inflor. u. as vulner. & astr. - **S. nemoralis** Ait. (e&cNA): compass goldenrod (Man [US]) (668); gray g.; dyersweed g.; rt. inf. u. by Houma Indians to cure "yellow jaundice" (doctrine of signat.): pl. source of oil of g.; hb. c. VO with pinenes & phenols; u. folk med. - **S. odora** Ait. (e&cUS, sCana): sweet(-scented) goldenrod; lvs. & tops with odor of fennel or anise (not sassafras); source of goldenrod oil (VO, with estragol, borneol, bornyl acetate, limonene); hb. off USP 1820-82; hb. u. carm. stim. diur. diaph.; sometimes to make beverage tea (as Blue Mountain Tea) (669) with flavor similar to that of other hb. teas like catnip - **S. paniculata** DC = *S. mexicana* - **S. petradoria** Blake = *Petradoria pumila* - **S. rigida** L. (e&cNA): stiff goldenrod; hard-leaved g.; hb. c. VO with terpenes, prob. borneol & another terp. alc.; hb.. u. in GU dis. (Chippewa Indians); fls. in lotion for bee stings (Meskwakis); astr. stypt. - **S. rugosa** Ait. (e&cNA swamps): tall hairy goldenrod; hb. c. VO ("goldenrod oil") with pinenes & limonene; u. folk med. in fever (Miss. Negroes) - **S. sempervirens** L. var. *mexicana* Fern. (S. mexicana L.; S. stricta Ait.): (seUS, Mex, WI): seaside goldenrod; "pluma de oro" (Cu); narrow-leaved g.; rts. c. dihydromatricaria ester; u. for wounds & warts: specimens fr. Fla Everglades rich in caoutchouc (T. A. Edison) - **S. speciosa** Nutt. var. *rigidiuscula* T. & G. (S. r. Porter) (e&cNA): showywand goldenrod; u. in pulmon. dis. (Chippewa Indians) - **S. ulmifolia** Muhl. (e&cUS): elm-leaved goldenrod; smoke fr. burning hb. u. to revive unconscious persons (Meskwakis) - **S. virga-aurea** L. = **S. virgaurea** L. (Euras, nAf): (European) goldenrod; woundwort; hb. c. catechol tannins, VO, saponin, alk.-like subst. (sds.); hb. (as Herba [Solidaginis] Virgaureae or H. Consolidae Saraceniae) u. as diur. in kidney stones, nephritis, urinary retention; astr. in diarrh. enteritis; as mouthwash for loose teeth, antipyr. in rheum. arthritis; fresh flg. tops u. in Eu homeop.; replacement for Fol. Orthosiphonis; cult. It, Austria.

solidism: doctrine of 18th cent. medicine widely held that considered all dis. due to alteration of solid parts of body (bones, muscles, arteries, nerves, &c.), which were endowed with "vital" properties.

Sollsia = *Mammillaria.*

Sollwert (Ge): theoretical (nominal) value.

Solomon's seal: *Polygonatum officinale, P multiflorum*, & other *P*. spp. (**clustered S. s.**: *Convallaria racemosa* - **false S. s.**: *Smilacina* spp. - **small S. s.**: *Smilacina racemosa*).

Solorina crocea (L.) Ach. (Peltigeraceae-Lichenes) (North & Alpine regions): **saffron solorina**: u. red, purple, orange dyes.

Solutio Tanogeni (L.) (Hung): epinephrine soln. 1:1000.

solution, disclosing (dental): I, KI, ZnI_2, glycerin, water; appl. to tooth surface by dentist to reveal film & plaque; useful in cleaning teeth.

soma: 1) (Ind): plant or mixt. of plants u. to bring long life, health, & psychic experiences 2) *Sarcostemma viminale*, latex regarded as supernatural remedy (Vedic med.), thus to confer eternal youth(!) 3) *Amanita muscaria* sporophores 4) Ephedra 5) the non-reproductive tissues of animal body.

Somali Republic: **Somalia**: country of neAf occupying the "Horn" of Af - v. under "myrrh."

somalin: cardiac glycoside fr. rt. of *Adenium somalense*; similar to strophanthin in pharmacolog. properties; made up of digitoxigenin & D-cymarose.

somario (Mex): sweet gum (tree).

somatic: 1) body wall of animal 2) chromosome no. of body cells, 2X that of gametes.

somatostatin: natural polypeptide hormone; secreted by hypothalamus; acts to inhibit release of growth hormone; produces hypoglycemia & hypothyroidism; sometimes u. to stop hemorrhage.

somatotropin: somatotropic (or growth) hormone: STH: GH; species-specific growth hormone of anterior pituitary gland; Human growth hormone (HGH) formerly obt. fr. anterior pituitary; now synthesized by recombinant DNA techniques: u. to treat short stature associated with Turner's Syndrome - **bovine s.** (BST): b. growth hormone (BGA); Posilk™; u. to increase milk prodn. in cows.

Sommerteufelsauge (Ge): Adonis.

Sonchus arvensis L. (Comp.) (Euras, natd. Cana, nUS): field/tree/corn sow-thistle; milk t.; swine t.; gutweed; latex c. taraxasterol, inositol; hb. c. lactucopicrin, FO; lvs. appl. to swellings (Indones.) in virus dis. (Java); u. like *S. paluster* (Eu); to suppress tumors; given in jaundice (Ind); diur. in kidney dis. - **S. asper** Hill. (Euras, nAf, natd. NA, SA, &c.): spiny (leaved) sow-thistle; sharp-fringed s.; pounded lvs. appl. to wounds or boils (Ind); warts; u. similarly to Taraxacum (Eu) - **S. brachyotus** DC (Chin, seAs): pl. c. choline, caoutchouc, mannitol, lactucerol; entire pl. u. to promote circulation, for leukorrh., dysmenorrh. (Indones) - **S. oleraceus** L. (Euras, nAf, natd. NA, SA): (common) sow-thistle; milky (or tufty) thistle; lvs. bitter, c. flavonoids, taraxasterols, caoutchouc, vit. C; u. lax. galactagog., diur. ton. antipyr.; rt. anthelm.; inspissated brown latex u. as hydragog., with belladonna & aromatics in dropsy; anti-opium remedy by Chin formerly (671); u. as pot vegetab. (Lapland); eaten in time of want; fresh lvs. for stomach dis. (Pe); juice (latex) u. for warts & some types of carcinoma - **S. paluster** L. (S. palustris L.) (Euras): hb. c. lactucopicrin; rt. c. inulin; u. like Taraxacum, sometimes as vegetab. - **S. schweinfurthii** Oliv. & Hiern. (Af): lf. decoc. to treat habitual abortion; for chicken pox; remedy for sore eyes.

Sonde (Ge): peroral probe u. to force feed exptl. animals.

son-lo: san-lo.

Sonnenthau (Ge): sundew, *Drosera* spp.

Sonneratia caseolaris (L) Engl. (S. acida L. f) (Sonnerati. / Lythr.) (seAs to Austral, eAf); lampa tree; common in mangrove swamps; of which they are an important component: bk, with 9-15% tannin, u. as tan; frt. & lf. u. as antipyr. & in aphthous stomatitis; frt. edible, with cheese-like flavor; to treat swelling & sprains; young frts. to flavor chutneys; fermented frt. juice as styptic (Ind); air rts. (pneumatophores) u. as cork substit.; lvs. with NaCl as cataplasm for burns & sl. wounds - **S. griffithii** Kurz (Mal): pounded rt. appl. to ringworm - **S. ovata** Backer (Mal): bogen; frts. consumed as a syrup (roodjak) & given to sick persons to stim. appetite.

Sonora gum: **Sonora Gummi** (L): 1) mezquite gum fr. *Prosopis juliflora, P. glandulosa*, & other *P.* spp.; similar in appearance to poorer or aver. grades of acacia; largely sol. in water; only 3-14% remains insol. (1937); 2) (less commonly): *Larrea tridentata*, exudn. (really misnomer; prod. should be called American gum lac.).

sont (ancient Eg.): *Acacia arabica*; gum u. as adhesive by tomb robbers.

sontol (CR): *Cymbopogon citratus*.

sooji: similar to rawa but coarser; pasta.

soopolallie (BC Indians): *Shepherdia canadensis*.

soorma: (Urdu): kohl or color for eye lids & e. lashes; c. graphite, etc.; said healthful to eye; name also appl. to metal applicator (Indo-Pak).

soos (Arg): licorice rt.

soot: Fuligo (Ligni) (L): finely divided carbon particles fr. incomplete combustion of resinous wds. of *Picea* & *Pinus* spp.; seen in smoke; obtained as a rule by ppt. of solid on cool surface; u. styptic in folk med.; in dyeing hair; paints for dermographic pencils; in printers & India inks; *cf.* lamp black.

soothing elixir: paregoric, opium tr. camphorated - **s. herb**: dill.

sophistication: 1) adulteration 2) substitution 3) addn. of something to improve quality of a drug, &c.

Sophora alopecuroides L. (Leg.) (cAs, Caucasus, wSiber): c. sophocarpine, sophoramine (vulner.), aloperine; drug u. in folk med., seeds strongly hallucinogenic; insecticide - **S. angustifolia** S. & Z. = **S. flavescens** Ait. (neAs): ku-shen (Chin, NZ); shrubby sophora; rt. c. matrine, methyl cystisine, anagyrine; hydroxymatrine; many uses in Chin med., ex. bk. u. antipyr. in colds, & sore throat; claimed curative in amebic & bacillary dys.; in breast cancer; alk. occ. in Dysentol (Jap proprietary) (2217) - **S. griffithii** Stocks (Keyserlingia g. Boiss.) (Pak - Gohar Reserve For.; Afgh): pl. juice put into sore eyes; rt. decoc, appl. warm to head for headache; pd. sd. in FO for head lice - **S. japonica** L. (Chin, Kor, Jap.): (Chin or Japanese) pagoda tree; Chin scholar tree; huai-chiao (Chin); "wai fa"; "Chin yellow berries" (Sophora flowers) (actually unopened fl. buds) u. hemostyptic in brain hemorrhages (85), as epistaxis (also sap); hypertension; fl. buds source of rutin (17-22%); also fls. (May) (-30%) & lvs. (5%); formerly for dyeing mandarin clothing yellow, occasionally for easter egg dyeing: insecticidal; vermifuge; ext. fr. lvs. & frt. u. to adulterate opium: fls. u. in brewing beer (Ge) - **S. lupinoides** L. = *Thermopsis lanceolata* R. Br - **S. microphylla** Ait. (Ch, NZ): kowhai (Maori); a stately tree; c. cytisine, anagryine; bk. u. stim. purg., internal pain; chronic rheum. gout; bruises, skin dis., internal blood clots; bk. when boiled yields guaiac resin-like subst.; pods u. in making inks; very hard wd. u. to make rollers, cogwheels, gears, in bridge bldg. - **S. pachycarpa** Schrenk (or C.A, Mey.) (e&cAs, Siber): sds. c. (+)-sparteine (pachycarpine), pachycarpidine, sophoramine; u. in arterial inflamm. & muscle disorders; pl. not grazed as so bitter - **S. secundiflora** (Ortega) Lag. (S. speciosa Benth.) (sNM, Tex [US], Mex): bean; Texas mt. laurel; frijilito; poison beans (sds.) (colorins) c. cytisine (sophorine) u. in small amts. (mixed with mescal) as intoxicant (Mex & Tex Indians) (delirium, excitement, followed by deep sleep); very toxic, single sd. may cause death in human: wd. source of yellow dye - **S. subprostrata** Chun & Chen (Chin, Manchuria): rt. c. pterocarpine; flavonoids; rt. u. as remedy in sore throat, cough, icterus (Chin) - **S. tomentosa** L.(pantrop&subtr, esp. seAs): sea-coast laburnum (Austral); sds. & rt. bk. u. in cholera, colic, dys.; rt. as purg. (Mal); sds. mixed with opium to alleviate unpleasant by-effects (Camb); beans u. to produce intoxication (Chin); sd. oil for painful joints; juice u. as fish poison (eAf); sd. insecticidal; lf. emet. & purg.; lvs. powd. & mixed with coconut oil & tied around limb with broken bone(s) (Fiji) - **S. wightii** Bak. (eAs, Ind): kusin (or kusan) root; c. bitter alk.; u. in Jap med.

sophorin: glycoside found in *Sophora* spp.; identical with rutin - **sophorine**: alk. fr. *Sophora secundiflora*; poisonous; later identified as cytisine.

Sopubia cana Harv. (Scrophulari.) (sAf): decoc. with *Salvia runcinata* u. for dysmenorrh. & threatened miscarriage - **S. delphinifolia** G. Don (Ind): dudhali (Bombay); pl. juice astr., appl. to feet to heal sores caused by exposure to water - **S. scabra** G. Don (sAf): rt. u. for gangrenous rectitis, body sores; inner rt. bk. for typhoid & other fevers.

sorb: sorbier (Fr): *Sorbus* spp. - **sorbin**: sorbose: monosaccharide occurring in frt. of *Sorbus aucuparia*; mostly now prep. by bacterial fermentation of sorbitol; u. mfr. vit. C.

Sorbaria sorbifolia A. Br. (Ros.) (wInd, Pak, natd. eUS, Cana): false spir(a)ea; pl. c. HCN; cult. woody shrub.

sorbic acid: propenyl acrylic acid: $CH_3CH=CHCH=CH.COOH$; isolated fr. frt. of *Sorbus aucuparia*, hence the name; u. as fungistatic agt. in preservation of foods (esp. cheese), &c.; SC active against yeasts & bacteria; u. in mycotic vaginitis.

sorbier des oiseaux (Fr): *Sorbus aucuparia*.

sorbit: sorbitol: sorbol: an alcohol, $C_6H_{14}O_6$, found in ripe frt. of cherries, plums, *Sorbus* spp., & other Rose Fam. members; prep. by fermentation; u. as "natural" sweetening agt. (esp. for diabetics), advantage: reduced tooth decay; less plaque (sugar-free gums); diur., lax. dehydrating agt.; plasticizer; humectant; in anti-freeze mixt. & many technical fields (ex. in synthesis of vit. C); ingredients in cough syrups, vitamin & iron tonics, antacid suspensions, polio virus vaccine, antibiotics inj., ointments, creams, lotions, &c.

sorbose, 1-: sorbin: sorbinol: a ketose sugar; occ. in *Sorbus* frt.; formed by oxidn. by sorbose bacteria (*Acetobacter suboxydans* Kondo); in *Passiflora edulis* frt. (with pectin); prod. from sorbitol by action of various m. o., as *Glucobacter* DeLay subsp. *suboxydans* DeLay & Frateur.: u. as roborant in intoxications, pancreopathies, hepatopathies; starting material in synth. (mfr.) of vit. C.

Sorbus L. (Ros) (nTemp Zone): mountain ash; rowan; some 90 spp.; gen sometimes absorbed into *Pyrus*; thus by some incl. in subgenera *Sorbus* & *Aronia* of *Pyrus* - **S. americana** Marsh.

(neNA): American mountain ash; roundwood; dogberry; c. flavonoids (*cf. S. aucuparia*); robins eating frt. become intoxicated; bk. u. in homeop. med. as astr. ton. antisep.; cinchona substit.; astr. frt. u. in folk med.; ornam. shrub or small tree (1938) - **S. aria** (L) Crantz (Eu): white beam; (white) b. tree; parched frts. eaten for coughs & catarrh.; frts. edible, astr. u.in diarrh.; u. to make vinegar, brandy, frt. sauce; lvs. u. expect.; wd. u. like that of *S. aucuparia*; cult. ornam. - **S. aucuparia** L. (Euras; cult. & natd. NA): (European) mountain ash; rowan (tree); witch-(hazel) wood; shrew ash (2267); frt. (rowan berries) c. malic (mostly), citric, succinic, tartaric acids; sorbic acids (obt. fr. parasorbic acid [lactone]) (672)); sorbitol (in ripe berries), sorbicortol I & II (1756); frt. u. as mild lax. (FE), ext. u. choleretic in obstructive jaundice, gall bladder & bile duct dis., hepatitis (Sorparin, McNeil); uter. stim. (emmen.); diur. (reputedly in diabetes); antibiotic (*sic*); frt. edible as preserves, jellies, compotes; bird food; coffee substit.; component in some Ge brandies, Russian vodkas; fls. mild lax.; lvs. & fls. as tea (at times as adulter.) or in tea mixts.; bk. u. as tan; tree often associated in magic rites (673) - **S. cydonia** L. = *Cydonia oblonga* - **S. domestica** L. (Pyrus sorbus Gaertn.) (Euras, esp. sEu; nAf): service tree; sorb apple tree; frt. c. malic acid, sorbitol, &c.; frt. eaten after frost; u. in wines; wd. valuable, very heavy (s. g. 0.79) - **S. sitchensis** M. Roemer (Alas to cCalif to Mont [US]): Pacific (or Sitka) mountain ash; frt. edible when ripe or after cooking: inf. of branches u. for enuresis of children (BC [Cana] Indians); bk. decoc. u. for rheum. & colds (S Carrier Indians); frt. eaten raw as purg. - **S. torminalis** (L) Crantz (Euras): checker tree; service t.; frt (Elsbeere) edible, u. to make brandy, wine (Aliziergeist, made in Alsace); vinegar.

sorgho (Fr): **s. à sucre** (Fr): sorgho sucre; sorgo (Fr, Sp, Eng): *Sorghum bicolor*.

Sorghum bicolor (L) Moench (S. vulgare Pers.; Holcus sorg(h)um L.) (Gram) (As, Af, many vars. cult. in Eu & Am): sorghum; (common) Guinea corn; Tunis grass; cereal grain grown in many areas too dry for other cereals; pl. occas. toxic (HCN); weed; chief types: 1) sweet (sorgho) (formerly called *S. saccharatum* Moench); c. sucrose; u. in prepn. sorghum syrups (popular in seUS), molasses, & sugar (sorghum cane crushed & boiled down) 2) grain sorghum (formerly *S. caffrorum* Beauv.): a) kaf(f)ir (corn) (Ethiopia to Natal; introd. US 1875); Hegari (prob. a form) (674); u. forage (pasture) in semi-arid regions; fodder; much u. for mfr. beer (sAf); other grain sorghums incl. b) durra (formerly *S. durra* Stapf), Egyptian corn, u. for poultry feed, also for human food c) milo (formerly *S. subglabrescens* Schweinf. & Aschers.), much cult. swUS; grain u. to feed cattle & poultry (green sts. may be toxic); sorghum sd. u. as substit. for grain in mfr. whisky (Confed. Army, ca. 1864) d) shallu (formerly *S. roxburghii* Stapf) (Af, Ind): grain crop e) feterita (formerly *S. caudatum* Stapf) (cAf): grain (Sudan); these & other grain millets u. as food like maize (meal in mush, pudding, pancakes, may be mixed with flours) (1939) f) Chin sorghum, Chin (sugar) cane; kaoliang (formerly *S. nervosum* Bess.) (eAs): source of grain, sugar, fodder; in alc. prodn.; straw u. to cover roofs; widely u. for many purposes (Chin); subsp. *drummondii* de Wet & Harlan (S. d. Steud.) (Guinea, natd. at scattered locations in La, Miss, Ala [US]): chicken corn; a grain corn, u. for fodder 3) broom sorghum (formerly *S. vulgare* var. *technicum* Jav.) (cult. worldwide): broomcorn; panicle rays of inflorescence u. for whisk brooms, brushes; (after removing frt.): frt. or grain ("seed") u. as diur. lithontryptic (675) - **S. halepense** Pers. (Euras, natd. NA, SA, esp. sUS): Johnson('s) grass (SC introd. into Ala [US] by Col. Wm. Johnson) (676); c. dhurrin (cyanogenetic glucoside, releasing HCN); pl. u. as forage, fodder; sometimes poisonous on acct. of HCN; rt. u. as substit. for sarsaparilla ("blood purifier"); sd. ton. diur., human food (Ind); bad weed; important allergenic sp. (seUS) - **S. sudanense** (Piper) Stapf (Sudan; cult. & natd. US as Tex): Sudan grass; u. for pasture & hay; sometimes cited as *S. bicolor* subsp. *drummondii* de Wet & Harlan (S. d. Steud.) - **S. virgatum** (Hack.) Stapf = *S. bicolor*.

Sorindeia juglandifolia Planch. (Anacardi.) (trop wAf): lf. decoc. u. as lax. for bilious complaints, gargle for mouth sores - **S. madagascariensis** Thou. ex. DC (Tanz): rts. eaten for stomachache; frt. edible, with delicious flavor.

soro (Por): serum.

Sorparin™: ext. of *Sorbus aucuparia* frt. (tablets) formerly u. in liver dis., esp. where deficient prothrombin secretion, icterus, &c.

sorrel: *Rumex* spp., esp. subgenus *Acetosella* - **common English s.**: *R. acetosa* - **Guinea s.**: **Jamaica s.**: *Hibiscus sabdariffa* - **ladies' s.**: *Oxalis corniculata* - **mountain s.**: *Oxalis acetosella* - **rose s.**: Jamaica s. - **sheep s.**: 1) mountain s. 2) (seUS): *Rumex acetosella* - **s. tree leaves**: fr. *Oxydendrum arboreum* - **wood s.**: 1) *Oxalis acetosella* 2) *Xanthoxalis* spp.

sorry plant: (eCana): *Arisaema triphyllum*, Indian turnip.

Sortenregisterstelle (Ge): registration office for pl. sorts (vars.).

sorts: grade of drug composed of odds & ends, irregular pieces, & generally inferior materials.

sorva (gum): **s. pequena** (Br): **sorveira** (Br): **sorvinha** (Br): **sorwa** (Br): a chicle subst. fr. *Couma utilis, C. amara, C. macrocarpa.*

sotol: 1) *Dasylirion texanum* & other *D.* spp. 2) alc. spts. (beverage) prepd. fr. maguey (*Furcraea* & *Agave* spp.), & fr. sotol (*Dasylirion* spp.).

souari nuts (, **common**): sds. of *Caryocar* spp.

souche (Fr): rhizome.

souci: (Fr-f.): marigold.

Soudan: Sudan.

Soulamea amara Lam. (Simaroub.) (Mal, PapNG+): bitter king; lvs. pounded & u. as hot fomentations (Solomon Is); frt. u. as antidote for bite of sea snakes (önykken [Dutch]) (squashed & rubbed on affected part) (Caroline Is); rt. & frt. u. as lax. in cholera; colic, pleurisy, cough; frt. antipyr.; rt. emetic; juice fr. heated lvs. appl. exter. to destroy head lice.

soul-food: diet typical of southern US blacks, such as collard greens, black-eyed peas, corn bread, fish, hog parts (ex. chitterlings) &c. - **s. plants**: **s. vine**: several spp. of lianas of SA jungle u. to produce hallucinogenics (ex. *Banisteriopsis caapi*).

sounds, fish: swimming bladders of many fishes, incl. sturgeons (*Acipenser*), carp (*Cyprinus*), & weak-fish, *Otolithias regalis* (Cana).

soup concentrates: **soup mixes**: often contg. herbs.

sour dirt: clay u. by blacks (Ala, Miss [US]) for GI ailments, having acid taste - **s. dough starter:** yeast in water - **s. grass (pie)**: *Digitaria insularis; Rumex acetosa, Bothriochloa pertusa, Carex* spp., *Juncus* sp., *Oxalis acetosella, O. corniculata, O. stricta*, & c. (1940) - **s. gum**: *Nyssa sylvatica* - **s. salt**: citric acid, tartaric acid (u. in Jewish cooking) - **s. sop**: *Annona muricata* - **s. wood: sourwood**: *Oxydendrum arboreum.*

sour: v. also under cherries.

sourberry: 1) *Vaccinium oxycoccos* 2) *Berberis vulgaris.* (v. also under Sauerbeere).

Souroubea crassipetala de Roon (Marcgravi.) (Co): lvs. of vine u. as astr. tea in mouthwash for sores in mouth - **S. guianensis** var. *corallina* Wittm. (Co): lf. inf. u. as anxiolytic in elderly - **s. guian.** var. *cylindrica* Wittm. u. similarly.

sous (Arg): licorice rt.

southern: v. under blackhaw; gentian; prickly ash; senega - **s. root**: *Carlina acaulis* - **s. wood**: *Artemisia abrotanum* - **Tartarian s. w.**: *Artemisia cina.*

sow berry: sowberry: 1) *Berberis vulgaris* 2) *Vaccinium oxycoccos* - **s. bread**: *Cyclamen purpurascens* - **s. fennel**: v. under f.

sowa: *v.* soyah.

Soxhlet (extraction apparatus): continuous extn. distn. apparatus in which solvent is recycled over & over again; u. extn. of drug, &c. with ether, &c. (anal.).

soy: soya (Sp): Soja: soybean (fr. Shoyu [Jap]?).

soy bean: **soybean**: *Glycine max*, source of FO, &c.; other products of soy include: phytic acid & Ca phytate (= inositol); tyrosine; high value protein (HVP); and fermentation nutrients - **green s. b.**: mung(h). bean (Chin) - **s. b. cake**: press cake fr. soy sd. - **s. sauce**: consists of s. protein hydrolyzed; with meaty flavor; u. Chin, Jap+.

soyah (Hind): *Anethum sowa*, Indian (or Japanese) dill.

Soymida febrifuga Juss. (Meli) (s&cInd, SriL, Mal): rohan(a) (Hind); rt. bk. c. 13-14% tannin, NBP, pectin, gums, sitosterol; u. antipyr. astr. in diarrh. ton.; in gastric dis. (124); as tan; clear gum obt. in large pieces, with good adhesive properties; valuable comm. wd. (East Indian mahogany).

spagyric medicines: earlier applied to medicinals of alchemists; now designation for compound medicinals having a mineral (inorg.) base (Paracelsus' ideas).

Spanische Hopfen (Ge): **Spanish hops**: *Origanum vulgare* var *creticum.*

Spanischer Flieder (Ge): lilac - **Spanischen Fliegen** (Ge): Cantharis.

Spanish: v. also under anise; bugloss; carnations; digitalis; ergot; flies; hops; larkspur; moss; paprika; pepper; psyllium; saffron; sage; white.

Spanish bayonet: *Yucca* spp., esp. *Y. aloifolia* - **S. broom**: *Spartium junceum* - **S. brown**: highly impure dark ferric oxide; u. prime paint, colors mortars - **S. dagger**: *Yucca* spp., esp. *Y. gloriosa* & *Y. aloifolia* - **S. earth**: influsorial e. - **S. flies**: Cantharis fr. Spain - **S. needles**: *Bidens bipinnata.*

Sparganium L. (Spargani.) (worldwide in cooler areas): bur-reeds; rhiz. rich in starch, u. as food (Am Indians) - **S. stoloniferum** Buch.-Ham. (nChin): rhiz in small pieces u. to treat int. hemorrhages; emmen., abort. sed.

Spargel (Ge): asparagus.

sparkleberry: farkleberry: *Vaccinium arboreum.*

sparrowgrass: **sparrygrass** (regional, US): asparagus.

Sparamycin A= Tubercidin.

spartase: mixt. of K & Mg aspartates (v. aspartic acid).

sparteine: liquid alk. of pryidine group (lupanine sub-group); occ. in broom tops (*Cytisus scopar-*

ius), *Sophora pachycarpa, Thermopsis lanceolata, Anagyris foetida, Lupinus* spp.; and other plant spp.; u. in cardiac arrhythmias (*cf.* quinidine); to raise blood pressure in hypotonia; oxytocic, to strengthen labor pains (where uterine inertia); as pre-medication before operations; u. mostly in form of **s. sulfate**.

Spartina Schreb. (Gram) (Am, Eu, Af, saline coastal marshes): cord or marsh grass; suggested as good source of protein in form of hay or food ext. - **S. cynosuroides** Roth (NA): cord grass; yields coarse hay; st. & lvs. u. to make binder-twine, paper, for roofing - **S. maritima** Fern. (S. stricta Roth) (NA, SA, Eu): spurt grass; valuable to fix sands & marshy soils; u. for this in Neth - **S. patens** Muhl. (NA): salt meadow grass; yields hay ("salt hay") u. mostly for packing - **S. pectinata** Link (S. michauxiana Hitchc.) (nUS): freshwater (or prairie) cord grass; slough g. (wUS); tall marsh g.; u. as hay; ornam. hortic. (cult.) - **S. schreberi** Gmel. (NA): prairie grass; type sp. - **S. townsendii** H. & J. Groves (sEng, Fr): u. as hay.

Spartium junceum L. (monotypic) (Leg.) (Med reg., natd. SA, Ec): (Spanish) broom; weaver's b.; rush b.; fls. c. cytisine, anagyrine, sparteine (similar to *Cytisus scoparius* but with more alk. & more toxic [cytisine]); adulter. of same; fls. u. as lax. (also sds.), emet. diur.; for kidney stones (Gr); rt. inf. abort. (Ec Indians); dr. fls. smoked for asthma (Ec); fls. source of VO u. perfumery; yellow dye; bk. source of fiber u. in cardboard, & conveyor belts; sts. u. in basketry; pl. toxic - **S. scoparium** L. = *Cytisus scoparius*.

Spartothamnella juncea (Walp.) Briq. (Dicrastylid./Verben.) (Austral): decoc. u. for lung dis., cough, postpartum fever (2263).

spasm ("spazzum") **root** (eCana): *Caulophyllum thalictroides*.

Spathiphyllum cochlearispathum Engl. (S. cannaefolium [Dryand.] Schott) (Ar.) (CA): lvs. u. for swellings & warts - **S. candicans** Poepp. & Endl. (S. cannaefolium [Dryand] Schott) (tropAm, Ve, Guian): lvs. with vanilla odor u. to perfume tobacco (Pe Indians) (1941) - **S. cleavelandii** (CR+): peace lily; p. plant; white flag; popular ornam.; cult. in malls, &c.; in cut flower trade - **S. phryniifolium** Schott (tropAm): similar.

Spathodea campanulata Beauv. (Bignoni.) (trop wAf, now pantrop): (African) tulip tree; tulipa africana; fountain t.; flame of the forest; bk. u. as dressing for skin dis.; bruised lvs. & fls. laid on wounds (Sen); bk. decoc. in constipation & GI dis. (Ghana); fls. appl. to heal ulcers (Indochin): u. in witchcraft ("baton de sorcier," Ghana); sds. edible; cult. tropAm & OW.

Spathoglottis eburnea Gagnep. (Orchid) (Camb): tubers u. as food - **S. lobbii** Reichb. f. (seAs): hot decoc. u. as fomentation for rheum., also small amt. drunk (Mal); vulner. (Indochin) - **S. plicata** Bl. (Mal, Indones): lvs. as hot poultice u. for rheum. arthritic pains.

Spatholobus ferrugineus Benth. (Leg) (Indones): sap drunk for colic and after childbirth; pounded st./lf. decoc. u. in menstrual dis.; st. inf. for cough - **S. littoralis** Hassk. (seAs): sap drunk as stim, also as footwash - **S. parviflorus** Ktze. (Thai): st. decoc. drunk for "upset" stomach "poor circulation" (folk med.) (2448) - **S. roxburghii** Benth. (seAs): maula; lvs. med. (Burma); bk. decoc. remedy in edema, worms, bowel dis. (Ind); rt. pd. in VD; source of Bengal or Balasa kino, u. in place of Malabar kino; rts. c. rotenone, u. insecticide (Ind).

Spathyema foetidus Raf. = *Symplocarpus f.*

spatlum: *Lewisia rediviva*.

spearmint: 1) *Mentha spicata* (*M. viridis*) 2) *M. cardiaca* 3) *M. longifolia* (misnomer) - **common (wild) s.**: *M. spicata* - **Scotch s.**: *M. cardiaca* - **true s.**: *M. spicata*.

spearwort: *Ranunculus lingua; R. flammula*; &c.

speciation: evolutionary process by which new spp. are formed.

species (sing. & pl.) (**speciei** or **specii**) (L.) (genitive): 1) a kind, species 2) spices, drugs, ingredients, etc. 3) a tea - **s. ad infusum pectorale** (L): S. pectorales - **s. aromaticae** (L.): aromatic tea; off. in various Eu pharmacopeias with differing formulas: often incl. sage, marjoram, peppermint, lavender, thyme, wild t., cloves, cubeb, hyssop, origanum, rosemary, orange lf., absinthium, etc. - **s. laxantes** (L.): **s. laxativae**: laxative tea; off. in many phar. with varying formulas: sometimes incl. senna, senna resin, linden fls., fennel, potassium tartrate, sodium tartrate, tartaric acid, anise, licorice, elder fls., etc. - **s. pectorales** (L.): **pectoral species**; breast tea; u. as exp.; off. in many phar. with such various ingredients as mallow fls., red poppy fls., mullen fls., star anise, pearl barley, marshmallow rt., licorice rt., althea lvs., life everlasting fls., coltsfoot lvs., veronica hb., elder fls., orris rt., anise, fennel, maiden hair fern hb., thyme, hyssop hb., etc.

speck: spike lavender.

Spectinomycin (US): an antibiotic (aminocyclitol type) fr. cultures of *Streptomyces spectabilis* Mason & al.; broad-spectrum type, u. (as HCl) esp. in GC (urethral, cervical) (IM injections); also in veter. med. (enteritis of dogs; coccidiosis); 1961-; *cf.* tetracycline (1757) (E).

spectrometer: spectroscope provided with calibrated scale or with photoelectric photometer to measure intensities of radiation.

spectrophotometry: a type of spectrometry in which wavelength of radiant energy is measured with a photometer (visible spectrum, UV, X-rays).

"speedball" (US, slang): injectable mixt. of cocaine & morphine (or Heroin) is used; more recently, of Heroin & amphetamine; a "recreational drug."

speedwell: 1) *Veronica officinalis* 2) *Peucedanum oreoselinum* (rt.) - **field s.**: *V. arvensis.*

Speichel (Ge): spittle; sputum.

Speisezwiebel (Ge): onion

speke: speck.

spell (seUS Negro): conjure.

spelt: *Triticum spelta.*

Spergula arvensis L. (Caryophyll.) (Eu, natd. NA, SA): (corn) spurrey; hb. c. saponin; u. as diur. (Co); sds, u. for bread with wheat in famine times (Scand); FO u. pulmon. TB; fodder; weed.

Spergularia marginata Kittel (Caryophyll.) (Med reg.; Af oases): rt. c. saponins, sapogenins; u. as replacement for Senega root - **S. marina** (L.) Griseb. (Arenaria m. L.) (Euras, nAf, Am, NZ) (saline places): sandwort. hb. u. med. as antiphlogistic - **S. rubra** J. & C. Presl (Euras, nAf, natd. NA): (red) sand-spurrey; hb. c. arom. resin, saponins; said useful in cystitis, lithiasis, ureteral colic, dysuria; folk med. in Malta; starvation food.

sperm oil: waxy FO fr. cranial cavity of sperm whale, *Physeter catodon*; formerly u. to lubricate high-grade machinery, automatic transmissions, &c.; sperm whale oil now disallowed in commerce of US (&c.) as measure of conservation; being replaced by jojoba oil, crambe o., limnanthes o., &c.

Spermacoce centranthoides Kunt. (Borreria c. Cham. & Schlecht.) (Rubi) (Br, Arg): saburgueirinho do campo (Br); buttonweed; gauycuru (Arg); u. abort.; u. like ipecac in many galenicals; rt. c. phenols, tannins. alk.; off Bras. Ph. I. - **S. hispida** L. = *S. flexuosa* Lour. = *Borreria articularis* F. N. Williams - **S. stachydea** DC (Borreria s. [DC] Hutch. & Dalz.) (trop wAf): u. diur. - **S. tenuior** L. (Fla [US], WI, tropAm): button weed; rt. decoc. emet. (Cu); for intest. dis.; in elephantiasis - **S. tetraquetra** A. Rich. (Fla [US], WI, sMex to Hond): pond-bush; pl. deoc. u. in skin dis. (Yuc); cold remedy (Bah) - **S. verticillata** L. (Borreria v. Mey.) (WI, tropAm, tropAf): herba de garro (Cu); vassourinha de botão (Br); wild margaret (Trin); rt. u. in leprosy (exter.), int. in amebic dys., schistosomiasis (Nigeria); u. diur., in GC (Sen); hb. decoc. u. in fevers, colds, diabetes (Trin); u. med. in neBr (esp. rural areas); u. in place of tobacco in cigar stores.

spermaceti = Cetaceum (L.) (*qv.*) (fr. sperma [sperm] & cetus [whale]) (Old Lat.) (ketos) (2750).

Spermatophyta: Spermatophytes: Embryophyta; the division (phylum) of flowering seed-bearing pls., regarded as the highest division of the pl. kingdom; ca. 300 fams & 300,000 spp. recognized; 2 large subdivisions: Gymnospermae (Coniferopsida) (ca. 72 fams.) and Angiospermae (flowering plants, Anthophyta) (with 2 large classes, Monocotyledonae [ca. 64 fams.] and Dicotyledonae [ca. 313 fams.]); the great majority of crude drugs originate fr. this division.

spermine: gerontine; polyamine occ. (as phosphate) in semen, pancreas, yeast, & almost all tissues; derived fr. spermidine; first observed in 1678 (van Leeuwenhoek): u. in biochem. research.

Spermoedia clavus Fr. = *Claviceps purpurea* (sclerotial stage).

Spermolepis divaricata (Walt.) Raf. ex. Ser. (Apium divaricatum B. & H.) (Umb.) (c&eUS, w to Neb): fool's parsley; forked scaleseed; rough-fruited spermolepis; pl. poisonous, nauseant - **S. gummifera** Brongn. & Griseb. (Arillastrum gummiferum); similar.

Sphacelia: conidial state of *Claviceps* spp. (ergot) (fr. sphakelos, Gr, smut, blight, gangrene, disease) (fr. sphakelo, Gr, to kill).

sphacel(in)ic acid: resinous acidic product of ergot; ergotin(e) of Wiggers; thought to prod. gangrene.

Sphaeralcea angustifolia: G. Don (Malv.) (wUS, nMex): narrow-leaved globe (or desert) mallow; sts. chewed by Hopi Indians; u. med. (Mex Phar.) - **S. bonariensis** (Cav.) Griseb. (Arg): malvavisco; tipichata; lf. u. as expect. digestive; rt. u. exter. as emoll.; lf. & fl. u. as antiphlogistic (exter.) - **S. coccinea** (Nutt) Rydb. (Man [Cana], Tex-Ore [US]): scarlet globemallow; pl. chewed & paste appl. to flesh wounds & to prevent burns ("the medicine of the heyoka") (Dakota Indians): forage - **S. fendleri** A. Gray (wTex [US], nMex): scarlet globemallow; yerba de la negrita; u. antitussive, expect., lax.; to "loosen" tumors (fresh lf. decoc.); to promote hair growth (shampoo with lvs. & fls.); sold in Mex-Am health stores (swUS, nMex).

Sphaeranthus africanus L. (Comp.) (MascareneIs; seAs): lf. sap u. as gargle for pharyngitis (Indochin); stom., anthelm. (PI) - **S. cochinchinensis** Lour. = *S. africanus* - **S. cyathuloides** O. Hoffm. (Tanz): cold inf. u. to wash children with skin dis. - **S. indicus** L. (seAs, Austral): East Indian globe thistle; pl. decoc. u. as ton. (Burma); said to be a fish poison; sds. fried in sesame oil u.

as aphrod.; anthelm. (Ind); diur. (Java) - **S. peguensis** Kurz ex C. B. Clarke (seAs): lvs. in paste u. as styptic; also as hot fomentation - **S. senegalensis** DC (tropAf, Chin); chenuet; ground lvs. in plaster for maturing large furuncles; macerated hb. u. topically to combat fatigue; in prepns. for rheum (Sen) made into pillows - **S. suaveolens** DC (OWtrop): u. to scent hair (Ethiopia): lvs. & capitula inf. as antipyr. (Burundi) (527); pl. u. as contraceptive (Moz) (679).

Sphaerophysa salsula (Pall.) DC (Leg) (cAs on salt steppes): hb. c. sphaerophysine (proto-alk., active princ.) (Russ Phar.) u. as hypotensive in hypertony; in uterine atony.

Sphaeropteris: Bernh. (Cyathe.) (tropAs, Pacif): large tree ferns: sap collected, left stand to thicken u. as anesth. in toothache (Ec Indians) (678) - **S. aterrima** Schott. (CA): lanose material around apical meristem of this tree-fern is chopped up, moistened, & placed on skin to reduce swelling of bruises (2444).

Sphagnum L. (Sphagn. - Bryophyta) (worldwide, bogs): bog (or peat) moss; the source of peat, widely harvested (1942); u. as fuel, fertilizer (680), litter, soil builder, to pack plants; absorbent in surgical dressings, bandages, sanitary napkins (said antiseptic or bacteriostatic (I_2); better absn. than cotton); in mfr. peat tar, liquid fuels (USSR), pitch; many spp. are represented, such as **S. cuspidatum** Ehrh.; **S. cymbifolium** Ehrh.; **S. imbricatum** Warnst.; **S. magellanicum** Brid.; **S. palustre** L.; **S. papillosum** Warnst., &c.

Sphenocentrum jollyanum Pierre (Menisperm.) (trop wAf): rts. c. bitter & acid subst., u. as chewsticks, in constipation, stom.; render things eaten to taste sweet; frts. (with *Piper guineense*) for cough (Ghana).

Sphenodesme involucrata (Presl) B. L. Robinson (Verben.) (Ind, Mal, sChin, Borneo): yan duk; vine or tree; u. locally in med. (Thai) - **S. pentandra** Jack (seAs): rt. and/or lf. decoc. rubbed into rheumatic parts (Mal) - **S. triflora** Wight (se.As): said antipyr. but also "stupefying".

Sphenomeris chusana Copeland (Stenoloma chusanum Ching) (Dennstaedti-) (Chin, PapNG): st. (rhiz?) of fern chewed for toothache (Chin): u. as dem. appl. to burns; antipyr. in dys.; said to promote hair growth & preserve black color(!) (ashes in sesame oil u. similarly).

Sphenostylis erecta (Bak. f.) Hutch. (Leg) trop wAf): rts. crushed & macerated in water to furnish a beverage u. against dys.; aphrod.; wds. edible (Cong) - **S. marginata** E. Mey. (tropAf): umukarakara (Burundi); u. as vermifuge (20-30 frts. swallowed while fasting without drinking; adult dose) (Burundi) - **S. schweinfurthii** Harms (tropAf): "yam bean"; woody pl. cult. for edible sds. & frts. - **S. stenocarpa** Harms (tropAf): herbaceous vine cult. by natives for sds. (food).

spherocrystals: spherites: microscopic globose bodies found in tissues of pls. of many fams., esp. Labiatae, Campanulaceae, Oleaceae; consist of hesperidin, diosmin, or other flavonoids: also appl. to ppt. of inulin in alc. & c.; sometimes Ca oxalate (radiating mass of needles) (after pretreatment): possibly starch grains may be regarded as s.; insol. in water, dil. acids, even hot chloral soln., sol. in alkalies; supposed function to protect cells against excessive insolation.

sphingomyelins: diamino-phosphatides; occ. in nerve tissues (myelin sheaths), milk, &c.; consist of sphingosine (an amino alc.) + choline + fatty acid + H_3PO_4.

Spica Celtica (L.): *Valeriana celtica* - **S. Indica**: *Nardostachys jatamansi*.

spice: condiment: arom. pl. material u. to season food, &c.; carminative; some 35 important spices include pepper, caraway, coriander, fennel, ginger, marjoram, mustard, cloves, sage, &c.; often with high antioxidant properties (ex: rosemary) (2443) - **s. bag**: cloth bag filled with spices (chamomile, wormwood, &c.) heated & appl. to jaw for toothache (19th c) - **s. berry**: 1) *Gaultheria procumbens*; *G. hispidula* 2) (Ala [US]): frt. of s. bush - **s. bush**: *Lindera benzoin*; *Calycanthus floridus*: or almost any arom. shrub, such as wintergreen - **S. Islands:** Moluccas; part of Indonesia (Maluku); sPac Oc lying between Celebes & PapNG, s of the Philippines; formerly chief prodn. area for cloves, nutmegs (Banda), pepper, &c. - **s. money**: formerly u. in Baltic - **s. parisienne:** mixt. thyme, coriander, mace, &c. - **s. poultice**: mixt. cinnamon, clove, nutmeg, pimenta; appl. over epigastrium to aid flatulence; over abdomen for colic - **s. wood**: s. bush.

spick: spicknarde, French: spike lavender - **spicknel**: *Meum athamnticum*.

spider: member of order Araneida (Araneae) of Div. Arachnida; with 8 legs (instead of 6 as in insects) & cephalothroax + abdomen; some spp. venomous; some 50,000 spp. worldwide - **black widow s.**: *Latrodectus* (*qv.*) - **brown recluse s.**: **b. s.**; *Loxosceles reclusa*; small (ca. 3 cm across, incl. legs); found around house in storage areas; violin marking on back, hence "fiddle-back spider"; bite gives local pain usually but may prod. systemic effects & occasionally death: venom collected - **s. web**: cobweb: web woven by members of order Arachnida; sometimes of exceeding regularity of pattern; hallucinogens have been assayed on spiders using web form as criterion of effect; web u. in folk medicine, as styptic (in nose bleed); spider silk u. in med. ("dragline silk").

spider flower: *Cleome* spp. - **s. plant**: *Saxifraga stolonifera; Chlorophytum* spp., esp. *C. comosum* var. *vittatum*; *Anthericum* spp.; u. to reduce pollution in the home (CO_2; NO_2; &c.).

spiderwort: 1) *Tradescantia* spp. esp. *T. virginiana* (SC formerly u. for s. bite) 2) *Paradisea liliastrum.*

Spigelia L. (Spigeli./Logani.) (warm Am): perenn. or ann. hbs.; often with brilliant fls.; named after Adrian van Spiegel (L. Spigelius) late 1500s; said to have directed prepn. of first herbarium - **S. anthelmia** L. (sFla [US], Mex to Br, WI, natd. seAs, Af.): (Demerara) pink root; p. weed; worm grass; "horse poison"; said excellent vermifuge, esp. for tapeworm; purg. (Choco Indians); rts. u. to execute criminals (Cuna Indians); lvs. repel flies & roaches; cattle poison: toxic, esp. when fresh; rt. u. sed.; hb. c. spigeline, tannin; u. in cardiac dis. (Eu; homeop. & "biochemists") - **S. flemmingiana** Cham. & Schlecht. (Br): hb. considered anthelm. & diaph. - **S. glabrata** Mart. (tropSA, esp. Br): u. similarly to *S. anthelmia*; anthelm. diaph. antipyr. "excitant" - **S. humboldtiana** Cham. & Schlecht. (sMex to Pan; Co to Par): lombricera (blanca); rhiz. decoc. u. to expel tapeworm; general vermifuge (Guat); pl. decoc. stim. tonic - **S. longiflora** Mart. & Gal. (sMex: Tabasco State): espigelia; hierba del burro(?); lf. decoc. u. to expel intest. worms - **S. marilandica** L. (e&cUS, esp. se): (Maryland) pink root; Indian pink; (M.) worm grass; w. root; dr. rhiz. & rts. coll. in fall (**Spigelia**, L.) u. as anthelm. for roundworm; often adulter. with *Phlox carolina*: toxic (1943) - **S. pedunculata** HBK (Co): rt. u. as vermif. - **S. splendens** Wendl. ex Hook. (CR): rhiz. u. as vermifuge.

spigélie (du Maryland) (Fr): Spigelia.

spignet: *Aralia racemosa.*

spike disease: serious pathology of *Santalum album*, occcasioning much economic loss - **s. lavender**: *Lavandula latifolia* (VO).

spikenard: 1) *Aralia nudicaulis, A. spinosa* (usually) 2) *Valeriana celtica* 3) also *Nardostachys jatamansi* - **American s.**: *Aralia racemosa* - **Celtic s.:** *Valeriana celtica* - **Chinese s.**: *Nardostachys jatamansi* - **small s.**: *Aralia nudicaulis* - **s. tree**: *Aralia spinosa* - **true s.**: Chinese s. - **wild s.**: *Smilacina racemosa.*

Spilanthes acmella Murr. (Comp.) (EI, now pantrop): Indian rupture-wort; alphabet plant; hb. (Abckraut); u. for toothache, in scurvy, bladder & kidney dis., kidney stones; fish poison (Burma); in suppressed menstruation; in pulmon. paralysis; hb. & st. sharp tasting; sometimes cult. (683) - **S. alba** L'Hér. (Mex to Pe, Ve): hierba de sapo; formerly u. in scurvy, dental caries; much like *S. oleracea*: to relieve toothache (Guat, Salv, Ec [777]) - **S. americana** Hieron. (S. oppositifolia D'Arcy) (se&cUS, s to Pe, Arg): calabaza; lvs. eaten for liver trouble; in sore throat; eaten as salad or cooked as greens - **S. mauritiana** DC (Af, As): Brazil (or Pará) cress; fl. heads & rts. u. in tooth dis., lf. juice in pyorrhea; mouthwash; hb. as antipyr., in articular rheum., snakebite; kidney stones (Ind): frt. insecticide, fish poison - **S. oleracea** L. (also Jacq.) (S. acmella Murr. var. o. L.) (tropAm, esp. Br, WI, pantrop): Pará cress; cabbage spilanthes; Paraguay Roux; hb. u. for mouth & dental washes, for gum dis.; inter. in bladder dis. & prostate tumor; gout. rheum; antiscorb.; as standardized ext. u. for tendon sheath inflammn., small wounds, insect bites; fls. u. dentalgia, diur. in bladder dis., in prep of tinct. (684) - **S. uliginosa** Sw. (WI): boton d'or; crushed fls. & buds scrubbed on teeth (FrWI) - **S. urens** Jacq. (Par): u. mostly in dental dis.; antiscorb.

Spilanthus= *Spilanthes.*

Spillbaumrinde (Ge): Euonymus.

Spina Cervina (Baccae, etc) (L.): buckthorn (berries, etc.).

spinach: term applied to plants of 11 or more genera but especially to *Spinacia oleracea* - **Ceylon s.: Indian s.: Malabar s.**: *Basella rubra* - **mountain s.**: 1) *Atriplex hortensis* 2) *Chenopodium album* - **New Zealand s.**: *Tetragonia tetragonioides* - **perennial s.:** *Chenopodium bonus-henricus.*

Spinacia oleracea L. (Chenopodi) (Euras, prob. origin. in wAs): (garden) spinach; prickly seed(ed) s.; espinaca (Sp); widely cult. for greens; pl. rich in vits., iron (3 mg%), &c. (1944); lvs. u. nutr. mild ton. lax., muscle relaxant; to "regulate" digestion & bile flow; formerly u. in anemia (Fe content; however lower Fe than water cress, in cooking part is lost, & another part is not resorbed); also high oxalic acid count (360 mg%) ppts. Ca salts esp. from milk, hence not advised in nutrition of children (685); juice u. to color candy, &c., green; indust. source of chlorophyll: antitumor (folk med).

spinasterol: sterol fr. FO of spinach lvs., *Polygala senega,* rt. sds. & lvs of alfalfa, &c.

spindelfoermig (Ge): fusiform, tapering towards each end of axis.

spindle-tree: *Euonymus atropurpureus, E. europaeus* - **American s.**: *E. atropurpureus* - **s. family**: Celastraceae.

spines: thorns; sharp indurated protuberances fr. pl. surfaces considered to be modified lvs., stipules, branches, &c.; with protective function (ex. *Liquidambar styraciflua* frt.) sometimes hazardous (ex. *Prunus spinosa*) (686).

Spinifex: (nAustral): name appl. to several genera of grasses (*Triodia*+); u. as cattle forage.

Spinne (Ge): spider - **Spinn(en)gewebe** (Ge): **Spinnennetz**: spider web, cobweb.

spinulosin: antibiotic isolated fr. *Penicillium spinulosum* Thom.

Spiraea L. (Ros.) (nTemp Zone): spirea; meadow sweet; shrubs with alternate simple lvs. & white or rose-colored fls. in panicles or corymbs; VO of lvs. & rhiz. c. salicaldehyde ("spirige Saeure" [Ge], hence aspirin) - **S. alba** Du Roi (e&cNA): queen of the meadow; rt. u. as ton.; lvs. astr. diur. (Am Indians); lvs. as substit. for tea. (363); for "bloody flux" (Meskwakis.) - **S. aruncus** L. = *Aruncus silvester* - **S. betulifolia** Pallas (BC [Cana]): flat-topped spiraea; pre-flowering pl. u. as decoc. for menstrual pains; flg. plant u.in teas for poor kidneys, rupture, colds, abdom. pains (447) - **S. douglassii** W. J. Hooker (BC to nCalif [US]): Douglas' spiraea; hardhack; boys smoked wild spirea lvs. wrapped in newspaper (Vanc [Cana]) or in clay pipes. - **S. filipendula** L. = *Filipendula vulgaris* Moench. - **S. hypericifolia** L. (Eu, Siber): may wreath; fls. u. astr.; cult. ornam. shrub - **S. japonica** L. f. (Jap, Kor, Chin to wAs, natd. US.): c. spireine & perhaps other alks.; ornam. shrub - **S. latifolia** Borkh. (eCana, neUS): common meadow-sweet; thé du Cana (Que [Cana]); u. astr. (folk med.); cold water inf. u. to prevent nausea & vomiting; as beverage tea (Abanakis); ton. attractive ornam. shrub with creamy white fragrant fls. & handsome smooth lvs. - **S. prunifolia** S. & Z. (Chin, Jap, Kor): bridal wreath; u. antipyr.; cult. ornam. shrub - **simpliciflora** Nak. (cult. Chin, Formosa, Kor): rt. u. as antipyr. antimal. & emetic (Kor) - **S. salicifolia** L. (Euras): (willow-leaved) meadow sweet; u. med. (Ind); ornam. shrub - **S. sorbifolia** L. = *Sorbaria s.* - **S. stipulata** Muehl. = *Porteranthus stipulatus* - **S. tomentosa** L. (eCana, neUS): steeple-bush; hardhack; silverleaf; pink meadow sweet (or spirea); fls. deep rose; lvs. branchlets, frts. mostly hairy; pl. c. tannins, esp. rich in rts.; rt., bk. & lvs. astr. u. in diarrh. "cholera infantum" (summer diarrh.), dys., hemorrhage; as ton. - **S. ulmaria** L. = *Filipendula u.*

Spiramycin(um): Rovamycin (& other trade names); antibiotic substance of erythromycin type; obt. fr. cultures of *Streptomyces ambofaciens* Pinnert-Sindico (fr. soils of nFr); isol. by Cosar et al (1952), Pinnert-Sindico & al. (1954-); antibacterial active against gm.+ m. o.; ex. *Bacillus subtilis*, & esp. *Streptococci*; also *Klebsiella pneumoniae* (gm.-); *Neisseria*; in bronchopneumonia, &c.; u. orally; similar spectrum to erythromycin. (1, 11, 111).

Spiranthera odoratissima St. Hil. (Rut.) (Br, Ve): "manaca"; u. for blood disorders (Br).

Spiranthes Rich. (Orchid.) (cosmop): ladies' tresses; pearl-twist; rts. of some spp. u. as "blood purifier" (decoc., Seminole Indians, Fla [US]) - **S. autumnalis** L. C. Rich. (S. spiralis Chevall.) (Euras, nAf): (Our) Lady's traces; tubers u. as aphrod.; in homeop. - **S. diuretica** Lindl. (Ch): hb. u. as strong diur. (in parts of Ch)- **S. sinensis** Ames (S. australis auct. non Lindl.) (Ind, eAs, Austral): nejibana (Jap): made into broth (cooking with pork) u. as ton. for weakness; in kidney dis. (Chin).

spirea: *Spiraea* spp. - **pink s.**: *S. japonica.*

spirein: glycoside fr. *Filipendula ulmaria, F. kamtschatica* (Spiraea k.), &c.; hydrolyzes (with emulsin) to salicaldehyde + D-glucose; may be the same as helicin.

spirit: any distilled liquid, generally alcoholic (v. also spirits) (1535) - **cologne s.**: ethyl alc. - **colonial s.: Columbian s.**: refined methanol - **s. gum**: alc. soln. of sandarac & damar; u. as adhesive (ex. to attach mustache to face of actor, etc.) - **methylated s.** (GB): denatured ethanol - **s. of nitre**: crude nitric acid (do not confuse with **sweet s. of nitre**: ethyl nitrite spt.) - **s. of salt(s)**: crude HCl, muriatic acid - **s. (of) turpentine**: turpentine oil obt. by distn. of oleoresin fr. *Pinus palustris, P. taeda, P. caribaea, P. pinaster*, etc. - **perfumed s.**: Cologne water - **white s.**: mixt. of petroleum hydrocarbons with b. pt. 150°-200°; u. as solvent in paints, &c.

spirits: spirit (*qv.* also) - **s. of ants**: formic acid s. - **s. of glonoin**: glycerly trinitrate s. - **s. of hartshorn**: ammonia s. - **s. of mindererus**: ammonium acetate soln. (SC Raymond Minderer) - **mineral s.**: petroleum fraction with high flash point (ca. 170-458° F); u. as turpentine oil substit. - **neutral s.**: "raw" alcohol - **s. of myrcia**: bay rum - **s. of nitroglycerin:** glycerly trinitrate s. - **s. of sea salt**: HCl - **s. of turpentine**: VO of *Pinus* spp. - **s. of wine**: ethyl alc. - **s. sal volatile** (Eng): aromatic ammonia s. - **wood s.**: methanol (**green w. s.: standard w.s.**: purified methanol).

Spiritus (L): 1) spirit 2) alcohol - **S. Frumenti**: whiskey - **S. Inflammabilis**: whiskey, alcohol, &c. - **S. Juniperi**: generally gin, although might be j. oil - **S. Lethalis**: CO_2 - **S. Odoratus**: cologne s. - **S. Rectificatus**: **S. Vini Rectificatus** (SVR): 90% ethanol - **S. Sacchari**: rum - **S. Veneris**: acetic acid - **S. Vini Gallici**: brandy - **S.**

Vini Rectificati: ethyl alc. - **S.V. Vitis**: brandy - **S. Vitrioli**: dil. sulfuric acid.

spirochaetes: higher bacteria with spiral configuration; some spp. produce certain diseases, notably syphilis.

Spirodela polyrrhiza (L.) Schleid. (Lemna p. L.) (Lemn.-Monocots.) (temp&trop regions, incl. NA): water flaxseed; (greater) duckweed; pl. floats free in the water; pl. grown on waste water (dairy farms) & fed to cattle & pigs (US); dr. pl. u. as diur. diaph., to aid circulation (Chin).

spiroketal: double ring compd. with one C atom in common & with an O gp. as lactone in each ring.

Spirolobium cambodianum H. Baill. (Apocyn.) (seAs); alc. macerate of rts. u. as lotion to rub on swollen limbs; rt. aq. inf. u. as antipyr. (Indochin).

Spiroplasma (Spiroplasmat. - Bacteria): close to *Mycoplasma*; extremely small celled organism shows pleomorphy, with spherical to helical shapes; incl. the pleuropneumonia-like organisms; several spp. incl. *S. citri*; may be pathogens.

Spirostachys africanus Sond. (Euphorbi) (tropAf): African (or Cape) sandalwood; tambootie; bk. c. oleoresin, u. as purg. & emet. (Moz); latex caustic, irritant, dangerous to eyes; heart wd. resistant to termites, borers, rats, &c.

spirostanol glycosides: 6-ring steroidal compds.; ex. parillin.

Spirulina Turpin (Oscillatori. - Cyanochloronta - Algae) (worldwide-fresh or brackish water): filamentous blue-green algae made up of coiled (helical) trichomes; often found in mineral & thermal springs; also on surface of shallow fresh water ponds (2753); sometimes form algal "mats"; u. as nutr. in form of algal cakes (chiefly *S. maxima* & *S. platensis* [Nordst.] Geitl.) marketed in Chad Rep, Af & elsewhere; dr. material being furnished as dark green pd. or tabs. (health food stores) for use in dieting (to reduce wt.) & as "rejuvenator"; actually a protein prod. (60-70%) with essential & non-essential amino acids; chlorophyll; FO with cholesterol, sitosterol; vitamins A, B complex, E; minerals: some obt. fr. Lake Texcoco (near Mex City) (*S. maxima*); value as "diet pill" in doubt; may produce hypercholesterolemia.

Spitzbeeren (Ge): frts. of *Berberis vulgaris*.

Spitzklette (Ge): *Xanthium strumarium*.

Spitzwegerich (Ge): *Plantago lanceolata*.

spleen: splen (L.): large ductless (endocrine) gland of the upper part (left side) of abdomen; functions to break down the RBC, releasing hemoglobins; source of RBC early in life (called "graveyard of RBC); protects against radiation sickness.

spleenwort: *Asplenium* spp. - **black s.**: *A. adiantum-nigrum* - **common s.**: *A. trichomanes* - **fern s.**: *Comptonia peregrina* - **maidenhair s.**: common s.

splenins: substances wh. occur in spleens of mammals; **s. A** is anti-inflammatory, decreases capillary permeability; u. in rheumatic fever; **s. B** incr. capillary permeability & bleeding time.

spliff (Jam - Rastafarians): Cannabis.

split-rock: *Heuchera americana*.

splitters: those taxonomists who identify many variants as formal named taxa (esp. species), split genera, &c.

spodium: crude bone charcoal (ivory black).

spodogram: a diagram, photo, &c., u. to demonstrate distrn. of ash (spodis, Gr), esp. of woody parts of plant to show structure, localization of minerals, &c.

spogel seed: *Plantago ovata*.

Spondianthus preussii Engl. (Euphorbi.) (trop wAf): very toxic pl., much u. in criminal poisoning (Nig): bk. u. as rat poison (Cameroon) - **S. ugandensis** Hutch. (wAf): cold inf. of leafy shoots u. to kill trespassing goats (etc.) without leaving post-mortem signs (Zaire); useful wd.

Spondias dulcis Soland. ex. Parkins. (S. cytherea Sonn.) (Anacardi.) (South Seas & pantrop; cult in some places, as Fla [US]): Otaheite apple (SocietyIs); ambarella; bk. astr. u. with *Terminalia* bk. in diarrh. (Indochin); bk. & lvs. inf. u. as panacea (Guadalcanal, Solomon Is): frt. edible ("Jew apple"), in jellies, preserves, marmalades, in syrup with agar; aper., antisep.; in "biliousness"; bk. in toothache; lf. as flavor (Fiji) - **S. lutea** L. = **S. mombin** L. (WI, Mex, to Br; now pantrop): hog plum; jam p.; hoba (Sp); jobo hembra (Cu); joboban (Domin Rep); frt. edible; eaten fresh; said galactagog; lvs. frt. u. cath. antipyr.; lf. inf. for intern. pains (Domin Rep); in yaws (wAf); bk. c. tannin, resins, decoc. u. as expect. in severe cough, tenifuge, vulner. in wounds (wAf); lvs. for colds, GC; lvs. or sds. u. for sore throat (Domin, Caribs) - **S. pinnata** Kurz (S. mangifera Willd.) (s&seAs): frt. (mango plums; amra); edible; u. in dyspep. (Burma); bk. & gummy exudn. u. in mal. (Indochin), bk. astr. u. in dys. (Mal); bk. decoc. u. in asthma (PI) - **S. purpurea** L. (WI, cont. tropAm; pantrop) Spanish plum; golden apple; ciruela (amarilla); acidulous frt. edible (fresh, boiled, dr. preserved); frt. juice fermented to give wine; tree planted to give a "living fence" (Cu); give excellent support for growing orchids; wd. pulp in paper mfr.; frt. u. for heat rash, boils, pustules, hives (Yuc) (79) - **S. tuberosa** Arruda (neBr): hog plum; imbu; delicious edible frt.; u. to make dessert, imbuzada (frt. boiled in milk);

source of beverage - **S. venulosa** Mart. (Br): source of caju gum; bk. u. med.

sponge: most primitive invertebrate muticellular animal (phylum Porifera); mostly marine organisms attached to a support; the skeletons are widely u. for cleansing purposes (as in the bath); local appln. of medication, ex. vaginitis, athlete's foot; contraceptive; to absorb menses ("natural sea sponge" recommended), &c.; some small, others weigh hundreds of kg; rich in iodine (-14%) (*cf.* algae [-1.5%]); wool and velvet sponges furnish bath (or toilet) s.; chief prodn. Bahamas, Cuba (1992-), Tarpon Springs, Fla (US); fungal dis. produced great havoc in indy. (1938-1945); some sponges c. antibiotics, thus *Microciona prolifera* (Demospongiae) is active against *Pseudomonas pyocanea* (1883) - **beauty s.**: *Luffa* - **burnt s.**: c. iodine & is u. by Homeop. as alt. in thyroid deficiency states, as simple goiter, obesity - **s. clipping**: trimmings in prep. of s. for the market; u. in drug trade (in mfr. burnt s.); to stuff large sponges, etc. (2696) - **finger s.**: *Axinella Polycapella* (Fla) - **gelatin s., absorbable**: sterile water insol. s. of gelatin base; u. as hemostat. by local appln. with gentle pressure; also in diagnosis of cancer by laying over body area suspected of being cancerous to trap cells - **glove s.**: *Spongia cheiris* (Fla) - **s. gourd**: *Luffa* spp. - **Maltese s.**: *Cynomorium coccineum* - **Mediterranean s.**: *Euspongia officinalis* var. *s. melon,* esp. *L. cylindrica* - **roast s.**: burnt s. - **(sheep's) wool s.**: *Hippiospongia lachnes*; chief type produced in Fla; good bath sponge - **s. tent**: small plug of compressed s. u. to dilate os uteri (abort.) - **vegetable s.**: *Luffa* spp. - **Zimocca s.**: fine quality of silk s. fr. eMed Sea; fr. *Spongia zimocca* O. Schmidt; best quality fr. Cyreniaca (Libya).

Spongia fluviatilis (L): *Spongilla lacustris* - **S. officinalis** L. (Euspongia o. L.) (Spongidae - Porifera) (Med Sea; WI; PI): sponge (*qv.*); c. diiodotyrosine; bromine; u. mostly as burnt s. in hypothyroidism, simple goiter; widely u. for washing body, in household, as pillow material, for suppressing noise, lininge helmets, &c., in antiquity (1900 BC) then in disuse until ca. AD 1830; dr. s. skeleton absorbs 30x own wt. water (synthetic sponges only 4x): fine-textured sorts (silk, velvet) popular; vaginal contraceptive (2010) - **S. somnifera** (somnifacient sponge): sponge saturated with inf. of opium, lettuce juice, hyoscyamus juice, mulberry juice, hemlock (*Cicuta*), mandragora, in wine or water, dried, and moistened as needed; u. inserted in nose for anesthesia; Salerno, ca. AD 1100 - **S. Tosta = S. Usta** (L): burnt sponge fr. *Spongia officinalis*, &c.

Spongilla lacustris L. (Spongia l. L.) (Spongillidae - Porifera) (still & flowing fresh waters Euras, NA, SA): river sponge; badiaga; c. no iodine, little Br, 30-40% silicates, zoosterols, toxin (protein?); u. in folk med. for glandular swellings, scrofula; in homeop.; in rhinitis, simple goiter (struma); Grave's dis.; is u. exter. to massage body in rheum., gout (Russ, Rom); activity depends on presence in sponge of algae (formic acid, acting with the silica needles) (also used in ointments) (687): also of interest here are *S. fragilis* Leydy; *Ephydatia fluviatilis* L. (harder); *E. muelleri* Liebk.

Spongipellis spumeus: (Sow. ex. Fr.) Pat. (Polyporus s. Sow. ex. Fr) (Polypor - Fungi) (Eu): c. bacteriostatic principles.

spoolwood: *Betula papyrifera.*

spoonflowers: *Xanthosoma* spp. - **spoonwood**: *Kalmia latifolia.*

spore: asexual reproductive cell, usually with a highly resistant cell wall; ex. Lycopodium - **sporeling**: young sporophyte.

Sporobolus buckleyi Vasey (Gram) (Tex [US], eMex): crawly grass; dropseed; furnishes palatable forage - **S. capensis** Kunth (eAs): appl. locally to wounds & snakebite - **S. poiretii** (Roem. & Schult.) Hitchc. (trop, As, introd. to eUS & SA): smutgrass; good forage grass - **S. pyramidatus** Hitchc. (cUS, WI, Mex to Pe): brak grass; match g.; decoc. of dr. grass u. as diur. in urin. irritation; in severe fever (Neth, WI); lax. (Aruba) - **S. virginicus** Kunth. (seUS, WI, Mex to Ch): crab grass; (seashore) rush g.; dr. grass decoc. for urin. irritation, kidney dis.; as gargle - **S. wrightii** Munro (sUS, cMex): sacaton; valuable forage & hay source in (semi) arid areas; good winter range.

sporophore: a branch or portion of a thallus which bears one or more spores; ex. Agaricus (fruiting body).

sport: a mutation.

"spot": on spot: commodity sold for immediate delivery; many drug goods are thus sold by brokers for quick turnover at favorable prices - **s. flower**: Spigelia.

spotted: v. under alder; carduus; cowbane; medic; plantain; wintergreen.

spray: medicated vapor formed by atomizer.

"spreading factor": hyaluronidase.

spreading: v. under dogbane.

Sprekelia formosissima Herb. (monotypic) (Amaryllid.) (sMex-CA); jacobaea lily; bella donna l.; bulbs c. tazettine, haemanthamine, haemanthidine; acts as emet., cardiac poison; u. as cataplasm in inflammatory swellings (Antilles); fls. antispasmodic; cult. ornam.

Spreu (Ge): chaff.

spring: v. under adonis.

spring-beauty: *Claytonia virginica.*

spring(s): natural flow of water fr. beneath the earth's surface; the water of some regarded as of med. value (mineral s.); ex. Bethesda s., s. of Jerusalem (Israel) long visited by Jews for healing purposes; Brunnen (Ge): Spring b.: fountain; some s. became famous as vacation resorts; ex. Epsom (Eng); Saratoga S., N, (US).; Leukerbad hot springs in Leukerbad valley (Gemmi Pass, Switz) are rich in $CaSO_4$; baths found beneficial by Goethe (AD 1797).

Spross (Ge): shoot; branch; axis of pl.; stalk; sprout; generally means above-ground axis or stem - **S. Pilze** (Ge): yeasts; Saccharomycetales (Endomycetales).

sprouts: shoots fr. germinated sds.; often very rich in vit. C, ex. Mung bean s.; wheat s. claimed anti-cancerogenic.

spruce: *Picea* spp. - **s. beer**: prepd. fr. *P. mariana* & other *P.* spp. - **black s.: bog s.**: *P. mariana* - **cat s.**: white s. - **common s.**: *P. abies* - **s. extract (, liquid)**: tanning prep. fr. bk. of *Picea abies*; c. 27% tannin - **flowering s.**: *Euphorbia corollata* - **s. gum**: resin exudn. fr. *Picea mariana, P. rubens*, or fr. *Abies balsamea*; u. masticatory (in chewing gums) - **hemlock s.**: *Tsuga* - **Norway s.:** *Picea abies* - **s. oil**: VO obt. fr. 1) needles & young twigs of *Tsuga canadensis* 2) needles & twig tips of *Picea mariana* - **red s.:** 1) *Picea rubens* 2) *P. mariana* - **skunk s.** (eNA: lumbermen): *Picea glauca* - **s. tops:** fr. *Picea abies* - **white s.**: *Picea glauca* & other *P.* spp.

spud (slang): Irish potato.

spunk: punk; touchwood; tinder - **s. (wood)**: 1) *Polyporus igniarius*; less commonly *Fomitopsis officinalis; Fomes fomentarius* 2) woody materials taking fire readily; tinder.

spurge: 1) *Euphornbia* 2) *Pachysandra* - **caper s.**: *E. lathyrus* - **s. family**: Euphorbiaceae - **s. flax (bark)**: *Daphne mezereum* **flowering s.: large f. s.**: *E. corollata* - **ipecac(uanha) s.**: *E. ipecacuanha* - **s. laurel**: *Daphne laureola*, etc. - **pillbearing s.**: *Euphorbia pilulifera* - **tramp's s.**: flowering s.

spurred rye: ergot.

spurry: 1) *Spergula arvensis* 2) *Marchantia polymorpha* - **sand s.**: *Spergularia* spp.

squalane: prepd. fr. squalene by hydrogenation; u. in prepn. of embrocations in cosmetics, suppostories; in chromatography; fine lubricant in watch & chronometer oils; perfume fixative.

squalene: spinacene; perhydrosqualene; unsaturated paraffin hydrocarbon widely distr. in animal & some pl. organisms; representing intermediate prod. in biol. cholesterol synthesis in liver; found in shark liver oil, olive oil, wheat germ oil, yeasts; colorless, odorless, almost tasteless compd.; u. to determine content of olive oil in mixts. with other bland FO since there is much more in olive oil than in these others; obt. chiefly as by-product in molecular distn. of vit. A fr. fish liver (esp. shark l.) oils; skin lubricant u. extensively in cosmetics; inhibits hair growth; fungistatic to some dermatophytes (SC *Squalus*, genus of dogfishes).

Squalus (Squalidae - Pisces): (Atl, Med, Pac Oc): dogfish; grayfish; smaller type of sharks; occ. in migrating shoals - **S. acanthias** (L) (mostly Atl Oc, Med Sea): spring (or spiny or picked) dogfish; grayfish; "rock salmon"; roussette; "blue dog"; spurdog; chien de mer (Fr Gaspé) spined dogfish; marketed fresh, frozen, smoked, semi-preserved in jelly; canned ("grey-fish," World War II); liver processed for l. oil (c. vit. A, D); skin u. to make a leather; fuel (Iceland); FO of liver as paint base (Iceland) - **S. suckleyi** Girard (Pac Oc): Pacific coast dogfish; P. shark; profitable source of liver oil (during World War II) (avg. 6,000 USP units vit. A/gm); meat u. by Japanese; since dogfish eat herring and small salmon, considered a nuisance in US & sugg. that Jap fishing for this fish be encouraged (689).

Squamata: Order of Class Reptilia; to it belong the snakes, lizards, & chameleons.

square (seUS): ovoid fl. bud of cotton; consists of 3 protective bracts grouped together around young fl. - **s. stalk**: 1) *Scrophularia marilandica* 2) *Monarda didyma.*

squash: 1) (US): *Cucurbita maxima* of edible type (Am Indian name); chief pie s. of US 2) (Brit): beverage frt. juice esp. lemon or orange - **s. flowers**: fr. *C.* spp.; staminate fls. eaten after frying (US) - **Hubbard s.**: *C. maxima* v. - **pear s.**: *Sechium edule* - **spaghetti s.:** var. of *C. pepo* - **summer s.:** *Cucurbita maxima, C. pepo & C. moschata* cultivars (non-keeping) - **winter s.**: similar to summer s. but keeping.

squaw-bush (US folk name): *Viburnum opulus* - **s. - mint**: *Hedeoma pulegioides* - **s. - root**: 1) *Cimicifuga* 2) *Caulophyllum* 3) *Conopholis americana* 4) *Apocynum cannabinum* (Choctaw Indians, Ala [US]) 5) many other pl. spp. - **s. vine**: *Mitchella repens* - **s. weed**: 1) *Senecio aureus, S. obovatus* 2) (Miss-Negro [US]): *Stylosanthes biflora* 3) *Aster puniceus*; *Cimicifuga racemosa*; *Eupatorium rugosum.*

Squibb's diarrhea mixture: Opium & Chloroform Mixt. Comp.
squid: calmar (Fr): pelagic marine cephalopod; carnivorous molluscs; great variations in size (2 cm-18 m long), with 8 arms having suction disks; popular sea food; caught off e coast of NA (esp. Va [US]); related to octopus & cuttlefish (with which often confused); u. in med. & physiol. res. since has single large nerve (nerve physiol. & pharmacol.).
squill (root): Radix Scillae - **autumn s.**: *Scilla autumnalis*.
Squillae, Radix Scillae: fr. *Urginea (Drimia) maritima* & *U. indica* - **Indian s.**: *U. indica* - **Louisiana s.**: *Crinum americanum* - **male s.**: red s. - **Mediterranean s.**: *U. maritima*, white var. - **red s.**: **Spanish s.**: *U. maritima*, red color form; u. raticide - **white s.**: s.
squirrel corn (root): 1) Corydalis 2) *Aletris farinosa* - **s. nuts** (US slang): acorns.
squirting cucumber: v. cucumber.
Sri Lanka (new name): Ceylon.
SSS: TM of prepn. made in Atlanta, GA (US); may stand for Sarsaparilla; stillingia, sodium iodide.
St. - See Saint(e).
stable: (adj.): permanent or relatively so.
Stachelmohn (Ge): *Argemone mexicana*.
stachydrine: alk. found widely distributed among plants, incl. *Stachys* spp.) (to ca. 0.2%); *Citrus* lvs.; *Chrysanthemum* spp., *Galeopsis grandiflora, Betonica officinalis,* alfalfa, &c.; in animal tissues; dimethyl deriv. of proline, an amino acid, hence may be considered more amino acid than alk.; u. as systolic depressant.
stachyose: tetrasaccharide of tubers of *Stachys floridana* & other *S.* spp. & found in many other pl. spp.; $C_{22}H_{42}O_{21}.4H_2O$; the commonest of the tetrasaccharides; hydrolyzes to galactose (2) + glucose + fructose.
Stachys L. (Lab.) (n&s trop&subtr): betony; woundwort; mostly strong smelling weeds of wide distrn.; some 320 spp. (< stakys (Gr) stachys, spike, since all have spicate inflorescenses) - **S. aethiopica** L. (sAf): u. in snakebite; for "black quarter" (cattle dis.) by smoke inhaln. - **S. alopecuros** (L.) Benth. (sEu, incl. Balkans): yellow betony; formerly u. in phthisis, abdom. dis., & bladder disturbances - **S. alpinus** L. (*S. alpina* L.) (s&cEu, As mts.): betony; alpine stachys; hb. u. to adulter. *S. officinalis* & *Veronica officinalis* - **S. annua** L. (S. annuus L.) (Eu, natd. neUS): hedge nettle; sorceress herb(?); hb. c. flavones; u. in chronic catarrhs; fls. in insomnia (folk med.), female dis., to adulterate fls. of *Lamium album* - **S. arvensis** L. (wEu, advent. neUS, worldwide): field nettle; woundwort (US); hb. u. as antidiarrheal (folk med.); toxic to cattle (Austral) - **S. aspera** Mx. (S. hyssopifolia Mx. var. *ambigua Gray*) (seUS, eAs): (rough) hedge-nettle; clownheal; sts. & lvs. carm., astr., deodorizing; for colds (with hot spirits) (Chin) - **S. baicalensis** Fisch. (Siber, Chin): pl. arom. u. med. (Chin, Russia) - **S. betonica** Benth. = *S. officinalis* - **S. bullata** Benth. (S. californica Benth.) (s& wCalif [US]): California hedge nettle; mastransia (Sp); hairy perenn. hb. 4-8 dm high, with wrinkled lvs., small purple fls., odorous; hb. (st. & lf.) inf. u. as wash for sores & wounds, esp. saddle galls; sts. & lvs. soaked in hot water & u. as poultice - **S. byzantina** K. Koch (S. lanata Jacq.) (Caucasus; natd. eUS+): woolly hedge-nettle; lamb's ears; whole plant & tot. alkaloid prepns. show hypotensive activity (2390) - **S. drummondii** Benth. (sMex): samana; ann. hb. u. in baths for the chills of fever (2342) - **S. floridana** Shuttlew. (S. sieboldii Miq.; S. affinis Bunge; S. tuberifera Naud.) (Chin, Jap [cult.], natd. seUS): Chinese (or Japanese) artichoke; knotroot; crosne (du Japon [Fr]); tubers c. stachydrine, stachyose (1758), &c.; tubers ("Japan tubers"; chorogi) edible (salted or pickled), with taste reminiscent of cauliflower; also food for hogs, &c.; tuber steeped in wine u. for flu, colds (China) - **S. germanica** L. (seEu: Gr+): downy woundwort; hb. u. in menstrual disturbances; abdom. dis.; fever - **S. mexicana** Benth. (S. ciliata Dougl.) (BC [Cana] to Ore [US]): great field nettle; hedge n.; tubers & rhiz. eaten; pl. u. as poultice for cuts, bruises (571) - **S. officinalis** (L.) Trev. (S. betonica Benth.; Betonica officinalis L.) (Euras, natd. US): (wood) betony; common b.; woundwort; hb. (lvs.) c. betonicine, stachydrine, tannin (15%), NBP; old popular Eu drug; stom. carm. nervine ton. sed. aperient, pyrosis, in snuffs for treating colds (Eu); lvs. thrust into nostrils to give sneezing, astr. in diarrh.; rt. emet.; pl. long u. in magic (amulets), esp. cEu & Balkans - **S. palustris** L. (*S. paluster* L.) (nTemp reg., orig. Eu): hedge nettle; marsh betony; (marsh) woundwort; clown's all-heal; hb. u. as vulner. antispasm. emmen. in menstr. disturb., antipyr.; to adulter. *S. officinalis* hb.; tubers edible - **S. recta** L. (S. rectus L.) (s&cEu, wAs): upright hedge nettle; decoc. of perenn hb. u to wash children to protect fr. dis. (Russ); u. for catarrh., fever, epilepsy, hysteria (Grozny, USSR); u. med. (Sp): to aid circulation - **S. rugosa** Ait. (sAf): hb. inf. u. ton. galactagog; inf. in local treatment of ophthalmia - **S. sieboldii** Miq. = *S. floridana* - **S. sylvatica** L. (S. silvaticus L.) (Euras, Himal.): hedge woundwort; h. nettle; pl. u. vulner. expect. emmen. diur. ton. in

diarrh. & colics; yellow dye - **S. tuberifera** Naud. = *S. floridana.*

stachys, field: *Stachys arvensis.*

Stachytarpheta cayennensis (Rich.) Vahl (Verben.) (WI, Yuc, to Pe, Arg): wild broom (Trin); pastor; lf. inf./decoc. of branches (bitter) in fever, incl. mal. (Pe-Ve); in dys., vermifuge - **S. dichotoma** Vahl (sMex to Br): gervão roxo (Br); lvs. u. med. (Br); in prepn. galenicals; sometimes as beverage tea - **S. fruticosa** (Millsp.) B. L. Robinson (WI): blue flower (Bah); Bahama vervain; lf. decoc. bath for prickly heat; int. as lax. & for worms - **S. indica** Vahl (OW trop): Brazilian tea; (East Indian false) vervain; lvs. u. in cardiac complaints, in fever; topically for sores, vulner.; juice as purg. emet.; lf. inf. as anthelm.; rt. decoc. in GC, anthelm. emmen., as abort. - **S. jamaicensis** (L) Vahl (sFla [US] WI, Mex to Arg: natd. OW trop): Jamaica vervain; mes; "Brazilian tea," lvs. u. emmen. purg. anthelm.; deodor. in arom. baths (PR); lf. inf./decoc. in fever, pneumonia; juice of whole pl. u. as ton. emet. expect. diaph. anthelm. in eczema & vitiligo (Trin); emmen. stim. purg., &c. (Mad) (689); lvs. u. to adulter. tea (Austria+) - **S. laevis** Mold. (sBr): pl. u. as stom. (690) - **S. mutabilis** Vahl (Mal, now pantrop): tea u. to aid menstr.

Stachyurus himalaicus Hook. f. & Thoms (Stachyur.) (eAs): pith of young shoots u. in diabetes, leukorrh.; pith soaked in wine u. as diur. (Chin).

Stacte: 1) balsamic exudn. of *Liquidambar* spp. 2) pure form of myrrh (sometimes so used) 3) exudn. of *Styrax officinalis* (691).

Staehelina L. (Comp.): some *Liatris* taxa have formerly been placed in this genus; also in *"Anonymos"* & *Serratula.*

staff tree (, climbing): *Celastrum scandens* - **s. vine**: *Solanum dulcamara.*

stag (US slang, soldiers, Vietnam): Heroin

stag-bush: *Viburnum prunifolium* - **stagger bush**: *Lyonia mariana* - **stagger weed**: 1) *Stachys arvensis* 2) *Lamium amplexicaule* - **staggered** (adj.): arranged alternately, as the rivets in a boiler.

staghorn: v. under sumac.

stag's horn: Lycopodium.

stalk: lengthened support to any organ, as **leaf s.** (petiole), flower s. (peduncle), etc.; preferably not u. with meaning of stem.

stamen: the pollen producing (or male) sexual organ of higher fls. (microsporophyte); each consists of anther & filament; adjectives include **staminate** (fl. without female organs); **staminiferous.**

staminode: false stamen bearing no pollen.

stand oil: litho oil: linseed (&c.) oil thickened by heat; u. in lithography.

standard deviation: square root of arithmetic mean of squares of deviations of various items fr. the arithmetic mean of the whole: s. d. = O.

standing cypress: *Gilia rubra.*

Standort (Ge): stand, station, place of growth, habitat, &c.

Stanleya Nutt. (Cruc.) (wNA): some spp. u. as potherbs (Am Indians); some taxa are toxic, some accumulate Se. - **S. pinnata** (Pursh) Britt. var. *bipinnata* (Greene) Rollins: accumulates much Se. (-3%).

stannous fluoride: v. under tin.

stans (L): erect.

Stapelia L. (Asclepiad.) (trop & sAf): carrion flowers; c. plant (SC bad stench of fls.); some 100 spp. found in semi-arid areas (named after J. B. van Stapel, Dutch physician) - **S. dummeri** neBr. (tropAf): carrion flower; st. c. purgative anthraquinone compds.; juice u. in otalgia - **S. flavirostris** neBr. (sAf): hot inf. u. to bathe limbs for numbness (sSotho) - **S. gigantea** neBr. (trop & sAf): toad plant; huge fls. (-46 cm diam.); sd. c. cardioactive glycosides; st. u. sed. in hysteria, emet. purg. - **S. hirsuta** L. (sAf): juice c. aloe-like bitter principles - **S. reflexa** Haw. (EI): hb. u. as ton. & antipyr. - **S. semota** neBr. (trop eAf): kawala (Kamba); juice of entire pl. u. to treat wounds & ulcers locally (pl. pounded in water juice, squeezed into wound & wound washed with liquid (499) - **S. variegata** L. (Arabia & sAf): toad flower; hb. u. as ton. antipyr.

staphidine: ext. of dry "currants" (grapes) prod. in districts of Gr; u. as sugar substit.; furnished as is or bleached.

Staphisagria (L.): stavesacre, *Delphinium s.*

Staphylea trifolia L. (Staphyle) (eNA): (American) bladder nut; cult. ornam. shrub.

staple: length of cotton fiber: short s., under 25 mm, medium s., 25-30 mm, long s., 30-40 mm; Sea Island cotton runs to 54.5 mm staple, & hence is regarded as a superior fiber; under 19 mm (3/4 in) called "no staple" because cannot be spun.

star anise: *v.* under anise - **s. apple family**: Sapotaceae - **s. bloom** (US.): Spigelia - **s. fruit**: 1) s. apple; *Chrysophyllum cainito* 2) *Damasonium stellatum* Thuill. (Eu) 3) *Averrhoa carambola* - **s. grass**: *Chamaeliriuim luteum*; *Aletris farinosa* (usually); *Helonias* - **Japanese s. a.**: *Illicium anisatum* - **s. of Bethlehem**: *Ornithogalum umbellatum+* - **s. of the earth**: *Plantago coronopus*; also spp. of *Geum, Silene, Coronopus* - **s. root**: *Aletris farinosa, Chamaelirium luteum* - **s. wort**: v. under starwort.

starch: Amylum (L.): a polysaccharide carbohydrate widely distr. among pls.; believed one of early prods. of photosynthesis fr. water & CO_2; in body hydrolyzed first to dextrins, ultimately to sugars; occurs in granules in pl. tissues, occasionally quite distinctive for the sp. & often valuable in diagnosis, in conjunction with other characters; tasteless white powd., lyophilic; forms hydrosol in boiling water; give char. blue-black color with I_2; u. for prickly heat, pruritis, intertrigo (handful as needed in hip bath); u. in pharm. as thickening agt. for regulation of viscosity, gel formn., compressed tablet disintegration; starch eating noted in sAla [US] (Negroes) (*cf.* clay eating) (2384); the following are the most important starches: arrowroot (chiefly fr. *Maranta arundinacea*) (*v.* arrowroot); arum (fr. *Arum maculatum* & other *A.* spp.); barley (fr. *Hordeum vulgare*); canna (fr. *Canna edulis*); cassava (fr. *Manihot esculenta*); maize (corn) (fr. *Zea mays*) (now imported fr. Eu into US!!); dioscorea (fr. *Dioscorea alata,* &c.); manioca = cassava; oat (fr. *Avena sativa*); potato (fr. *Solanum tuberosum*); Queensland (*v. Q.* arrowroot); rice (fr. *Oryza sativa*); rye (fr. *Secale cereale*); sago (fr. *Metroxylon rumphii*, etc.; *Cycas* spp.); tacca (fr. *Tacca leontopetaloides*); taro (fr. *Colocasia esculenta*); wheat (fr. *Triticum aestivum*); the principal starches u. in pharm./med. are maize, wheat, & potato (2694).

starch blockers: **s. channel b.:** substances (proteins) wh. interfere with the regular hydrolysis of starch to sugars (inhibit animal amylase); & hence reduce food absorption & the tendency to obesity to a certain extent; types: 1) *Phaseolus vularis* var. Great Northern (phaseolamines, glycoproteins); *Vigna sinensis*, &c. 2) oligosarccharides fr. Actinomycetes 3) wheat (1981-) - **s. glycerite**: mixt. of 10% starch in 20% water & 70% glycerin plus 0.2% benzoic acid (preserv.); u. as emoll. & pill (tab.) excipient - **s. gum**: dextrin.- **s. index**: % of starch granules larger than a certain diameter; of diagnostic use (ex. Cartagene ipecac, 15-212%; Rio i., 7-10% - **s. iodide paper**: paper impregnated with KI + starch; u. to test for oxidg. agts. —> free I_2 —> blue - **s. mucilage**: prepn. of 2.5% starch in water made by boiling - **pregelatinized s.**: similar with ruptured starch grains; in dried form - **soluble s.**: maize or potato s. treated with dil. HCl to destroy gelatinizing properties; u. as indicators; enzymes assays - **s. sponge:** absorbable surgical/dental dressing (poultice) u. to control bleeding on burns, &c.; food value - **s. sugar**: glucose.

starfish: sea star; common cross-fish; *Asterias* & other genera (Asteriidae - Asteroidea - Echinodermata): incl. *Asterias rubens* L., *A. vulgaris*, &c. (central areas of Atl. coasts of NA); with 5 arms fr. central disk; predator on oysters & other shell fish; u. in homeopath. med.; in uter. dis. (Hippocrates); in swellings & glandular hardening; in mammary carcinoma (2751).

Stärke (Ge): starch - **S. Gummi** (Ge): dextrin.

starwort: *Chamaelirium luteum* (also **drooping s.**); also spp. of *Aster, Aletris, Stellaria, Callitriche*, &c.

stathmokinetic: producing a state of arrested mitosis (due to mitotic poison) (ex. colchicine; oncovine) - **stathmokinetics**: genetics.

Stathmosetelma pedunculatum (Dcne.) K. Schum. (Asclepiad.) (Tanz): rt. purg., may be toxic; as male aphrod. (rt. cooked in flour & eaten).

Statice L.: transferred to *Limonium* & *Armeria* - **S. limonium** L. = *Limonium vulgare.*

Staude (Ge): perennial hb., shrub.

Staudtia capitata Warb. (S. stipata Warb.) (Myristic) (w tropAf): sd. c. 30-50% solid fat; u. as parasiticide in skin conditions; bait for traps; wd. of fine quality; u. to make paddles, lead pencils, &c. - **S. kamerunensis** Warb.) (wAf): sd. source of solid FO, stau(d)tia butter edible (Cameroon).

Stauntonia hexaphylla Decne. (Lardizabal.) (eAs): st./rt. decoc. u. as diur. (China, Jap); bk. u. in blenorrhea & to regulate menses (Indochin); frt. edible; cult. orn.

stavesacre (seed): *Delphinium staphisagria.*

"steam doctor" (US 19th cent.): Thomsonian practitioner.

steapsin: pancreatic lipase.

stearic acid: saturated fatty acid; $CH_3(CH_2)_{16}$ COOH (actually in nature mixed with a little palmitic acid); solid to ca. 55° C; found (as glyceride) in some pl. FO & in many animal FO, incl. tallow (of cattle) (fr. stearo [Gr], tallow); u. in ointments, suppos., &c. (to raise m. p.); to coat enteric pills; in candles, modeling compds., &c.

stearin: tristearin: glyceryl tristearate (usually mixed with small proportion of glyceryl palmitate); solid white or slightly yellowish greasy substance; occ. in many pl. and animal FO; u. pharm. mfg., textile sizes; to make candles (formerly); in "solid" alc. (canned heat-Sterno) cosmetics (creams, lipsticks, &c.); in flame-jellies, u. in warfare - **beef s.**: solid residue fr. tallow after removing tallow (or oleo) oil; u. as lard substit. - **stearoptene:** oleoptene: solid portion of VO (ex. camphor) (oxygenated) as contrasted with the liquid portion (eleoptene) - **stearyl alcohol**: pentadecyl alc.; pentadecanol; cocculus FO, also in spermaceti; $C_{17}H_{35}CH_2OH$; occurs in pl. tissues; ex. in ragweed pollen; in commerce, prod. is a mixt. of solid alcohols consisting chiefly of s. a. per se; u. mfr. cosmetics, emulsions, anti-foaming agts., lubricant, chem. mfr., etc. (*cf.* cetyl alc.).

steatite: soapstone: French chalk: talc.

Stechapfel (Ge): Stramonium.

Stechpalme (Ge): holly

steeple bush (root): *Spiraea tomentosa.*

Steganotaenia araliacea Hochst. (Umb) (tropAf): popgun tree; Transvaal carrot tree; rt. chewed for sore throat; bk. chewed for asthma: u. for swellings due to allergy.

stegma: (pl. **stegmata**): elongated tabular pl. cells nearly filled with silica (SiO_2) fragments.

Stegnosperma halimifolium Benth. (S. scandens [Lunan] Standl.) (Phytolacc.) (Mex, WI, CA, nSA): ojo de zanate (Sinaloa); fls. year around; lvs. mixed with those of *Bursera microphylla* and boiled —> shampoo appl. for headache; rts. u. as soap substit.; rt. bk. u. as blue dye; reputed cure for rabies - **S. watsonii** (Mex): chapa color; remedy for rabies(?!).

Stegobium paniceum (L.): *Sitodrepa panicea.*

Steineichenholz (Ge): limousin bark.

Steinpilz (Ge): *Boletus* (edible mushroom).

Stelechocarpus burhakol Hook. f. & Thoms. (Annon.) (Mal, Indones): keppel tree; frt. edible; was a great favorite of sultan's court ladies (Java); cult.

Stellaria cuspidata Willd. (Caryophyll.) (Ch) quilloiquilloi; u. to treat hemorrhoids, anal fissures; refrigerant - **S. dichotoma** L. var. *lanceolata* Bge. (eAs): rt. u. in mal. fever; kala azar (children) (inner Mongolia, Kansu) - **S. holostea** L. (Alsine h. Britt.) (nEuras, Med, natd. neUS): snake grass; star g.; greater stitchwort; Easter bells; u. similarly to foll. - **S. media** (L) Cirillo Alsine m. L) (Euras, natd. NA, As): (common) chickweed; small c.; starwort; chickenwort; hb. u. exter. in skin dis. (eczema), in poultices for ulcers, hemorrhoids; in TB, dem. refrigerant, in ophthalmia; in medl. baths; as potherb & in salads; frequent weed in back yards - **S. palustris** Retz. (Eu, natd. Qu [Cana], Mich [US], &c.); stitchwort; hb. u. astr. bitter ton. - **S. saxatilis** Buch.-Ham. (eAs): decoc. in broth u. for cough & hemorrhage; as tinct. in wine for rheum. (Chin); as poultice fr rheum. pain (Indones) - **S. uliginosa** Murr. = *Minuartia stricta* Hiern.

stellasterol: sterol obt. fr. starfish, *Asterias rubens.*

Stelle (Ge): bureau; office.

Stellera L. = *Wikstroemia* + *Dendrostellera* (Thymelae.).

stem: that part of the pl. axis (usually ascending) which bears lvs. or lf. modifications; ex. Dulcamara, Thuja (leafy twigs); the chief microscop. characters are: cork; fibro-vascular tissues; medullary rays; parenchyma, often with starch, young stems also show epidermis (often with trichomes), cortex, etc.

Stemmadenia donnell-smithii Woodson (Apocyn.) (sMex to Pan): chapon(a); cachito tree; bk. decoc. u. in rheum (Salv); latex of sd. pod u. for ophthalmia, toothache (chewed), masticatory & skin troubles.

Stemodia maritima L. (Scrophulari) (WI, nSA): hierba de iguana (Cu); pond crab (Bah); poor man's strength (Bah); pl. inf. u. in gastralgia, pain, dropsy, &c.; lf. decoc. u. for bath as tonic; lvs. brewed for discomfort fr. eating (Bah); dr. lvs. smoked like tobacco (Bah) - **S. viscosa** Roxb. (Ind): dr. pl. inf. c. mucilage, u. dem.

Stemona (Stemone) collinsae Craib (Stemon.) (close to Lili) (seAs): insecticide; to make emulsions for insecticides (Thail); tuber inf./decoc. drunk to destroy ectoparasites (fleas, lice); stomach pains (Thail) (2448) - **S. japonica** Miq. (Chin): tuber inf. u. in TB (pulmonary); cough, skin dis.; parasiticide (hair lice); vermifuge; planted for med. use (Jap) - **S. sessilifolia** Franch. & Sav. (eAs): u. similarly to prec. in some areas; erect sp.; substit. for Chinese stemona rt. (foll.) - **S. tuberosa** Lour. (sChin, Siam, Burma, Indochin, Jap, Mal, Ind): po-pu; peh-pu; pai-pu (Chin); "wild asparagus"; tuberous rts. (Chin stemona rt) of this liana yield alks. stemonine, isostemonine, &c.; u. as vermifuge (bach-bo); drug cut into strips, decoc. u. (dose 7-10 g); rts. u. cooked with pig's feet in rheum. (Chin); reported useful as insecticide (lice, &c.) (692); u. as dem. in dis. of air-passages, cough, carm., young st. potherb.

Stenandrium dulce Nees var. *floridana* Gray (Acanth.) (Fla [US]): "rattlesnake flower"; hot decoc. of pl. drunk for snakebite (Seminole Indians).

Stenocereus thurberi (Engelm.) Buxb. (Cereus t. Engelm.) (nwMex): organ pipe (cactus); pitaya dulce; slabs u. as analg. (appl. described under *Pachycereus pringlei*); frt. a food delicacy (Suri Indians) (—> *Lemaireocereus*).

Stenochlaena palustris Bedd. (Blechn.-Pterid.) (eAs): juice u. to treat skin dis. (Thail); young lvs. eaten as lablab with rice & in soup (Java).

Stenocline incana Baker (Comp.) (Madag): u. as stom. ton. astr. aphrod.

Stenolobium stans Seem. = *Tecoma s.*

Stenoloma Fée = *Sphenomeris.*

Stenophyllus Rafin. = *Bulbostylis.*

Stenotaphrum dimidiatum (L.) Brongn. (S. glabrum Trin.) (Gram) (Am & Af, trop sAs, natd. Br): (American) buffalo grass; horse g.; rhiz.

(Radix Graminis [Bi]) decoc. u. as diur. diaph. (Br) - **S. helferi** Munro (seAs): decoc. u. at parturition to arrest hemorrhage & as "protective med." - **S. secundatum** (Walt.) O. Ktze. (seUS, subtr&trop Am, Af): St. Augustine grass; (seaside) quick grass; rt. u. diur. (Br, Madag); tinct. of rt. u. for urinary gravel, stomach "acidity," wind, gallstones (sAf): cult. as lawn grass (coarse) in coastal cities; u. to bind sandy soils; var. *variegatum* Hort. u. as basket plant.

Stephania abyssinica Walp. (Menisperm) (tropAf): fresh lvs. mild purg.; st. juice mixed with milk as emet.; rts. for roundworms; rt. decoc. for sores (Zulus); lf. juice as snakebite antidote - **S. brachyandra** Diels (Chin): Central Chin stephania; rt. stock edible; u. med. (Chin) (2098) - **S. capitata** Spreng. (Java, nBorneo, Mal): lvs. c. dicentrine, crebanine, phanostenine; u. as expect. in asthma, fever; substit. for *Cyclea barbata* Miers - **S. cepharantha** Hayata (Jap, Taiwan); tubers c. bis-benzyl oxyquinoline alks. cepharant(h)ine, cepharanoline, tetrandrine (ixtacanthine gp. of alk.); cycleanine (isochondrodendrine gp.); cepharanthine active against TB (Jap) & leprosy organisms (693); effective in pertussis; also antipyr. analg., anti-inflamm. - **S. dielsiana** (Chin): u. as analg. hypnotic - **S. elegans** Hk. f. (nInd): rajpatha; rts. u. to treat anasarca (generalized edema) (2256) - **S. glabra** Miers = *S. rotunda* Lour - **S. gracilenta** Miers (nInd): rts. u. as substit. for those of *S. elegans* & *S. hernandifolia*: rt. c. magnoflorine, papaverine, sinoacutine, isosinoacutine (2257) - **S. hernandiifolia** Walp. (S. discolor Spreng.) (trop OW, as Ind, Indones, Austral): rts. c. picrotoxin, isotrilobine (trilobine gp. of alk.); u. in fever, diarrh., dyspep., kidney dis.; rheumatoid arthritis (Chin); fish poison - **S. japonica** Miers (Ind to sChin, esp. Taiwan, Jap, seAs): st. & tubers c. stephanine, metaphanine (hasubanan deriv.), "metastephanine" (694): u. in mal. & as ton. for persons in poor health (Jap); rts. (powd. in drink) in stomachache (Chin); hb. u. for itch (PI) - **S. rotunda** Lour. (S. glabra Miers) (Ind, Indochin, Burma): tubers c. stepharine, palmatine, &c.; u. as ton. (Vietnam); in TB, asthma, diarrh. dys. fever; sed. hypotensive - **S. salomona** Diels (Solomon Is): lvs. rubbed on skin for pain - **S. sasakii** Hayata (Jap): tubers source of cepharant(h)ine: also c. steponine (quaternary base) - **S. tetrandra** S. Moore (Chin, Taiwan): rts. c. tetrandrine, sed. analg. antiphlogistic, strongly tuberculostatic; rt./lf. decoc. u. for neuralgia, acute arthritis, stomach ulcers. (2748)

Stephanopodium peruvianum Poepp. & Endlich. (Dichapetal.) (Co): u. antipyr.

Stephanotis floribunda (R. Brown) Brongn. (Asclepiad.) (Mad): Madagascar jasmine, u. as ornam. (fls. u. in wedding corsages, &c.); climbing shrub for greenhouse or outdoors.

Stephanskoerner (Ge): Staphisagria.

stepmother: pansy.

steppe (Russ): semi-arid grassy plains, typical of Siberia, usu. plural.

Sterameal: Sterilized animal meal: prod. made by steam digestion of mammalian bodies (after removal of hide); u. fertilizer.

sterane: 1) earlier designation for cyclopentanoperhydrophenanthrene; the nucleus of the steroids, &c. 2) (TM) (cap.): prednisolone.

Stercolepis gigas Ayres: California jewfish; a percomorph fish, the liver of which is u. for prodn. FO; rich in vits. A & D.

Sterculia alata Roxb. (seAs) (Sterculi): sds. narcotic, u. to replace opium; sds. eaten by Mois (Indochin) despite tingling & drowsiness caused - **S. apetala** (Jacq.) Karst. (sMex to Pe, Ve, Par; cult. Fla [US]): Panama (nut); kupu, kuppa (Cuna Indians, Pan); castana (Mex); fl. decoc. expect. in asthma; bk. inf. stom. (PR); mucil. exudn. u. emoll.; lvs. u. exter. for inflammn. (Hond); bk. decoc. in mal. (Guat); frts. edible; sds. eaten raw or cooked, FO extd. - **S. appendiculata** K. Schum. (trop eAf): lf. & esp. petiole u. as purg. for relief of abdom. pains; rt. bk. for hookworm, bk. added to baths to strengthen children with mal. - **S. balanghas** L. (eAs): frt. juice u. as antidiarrh.; tree produces a gum - **S. chicha** A. St. Hil. (Guianas, Br): sds. (castanha de Maranhao = Maranhao seeds) edible, source of FO u. as lubricant (watches) & in paints - **S. cinerea** A. Rich. (Tanz, Kenya, Uganda): tree source of Tartar Gum, a clear yellowish gum like tragacanth, u. like karaya & to adulter. acacia - **S. cordata** Bl. (S. javanica R. Br.) (Indones): kalong(an); sds. sold in markets as "pranadjiwa"; bitter sds. u. in chest complaints - **S. ferruginea** R. Br. (PI, endem.): bk. decoc. u. in puerperium - **S. foetida** L. (OW, tropAf to Austral): stink tree; "wild almond"; letkok (Burma); lvs. of large tree u. aperient, diur. (Vietnam); bk. diaph. in rheum. edema (PI); abortif.; frt. astr. (Burma), frt. rind u. to treat GC, edema; sd. ("Java olives") u. as source of FO edible if roasted, u. for soaps, &c.; u. int. for scabies & other skin dis.; tree yields gum with good swelling capacity & strong adhesive power of karaya type; fls. with vile smell (hence names) - **S. graciliflora** Perk. (PI, endem.): lvs. appl. in paralysis - **S. hypochra** Pierre (Indochin): trom; st. source of greenish yellow gum, chewed as ton. by Vietnamese - **S. lepidostellata** Mildbr. (Indones+): sd. decoc. drunk for colds (Solomon Is) - **S. lychnophora** Hance = *Scaphium macropodum* Beumée - **S. mbosya** Engl. (trop eAf): mohosia (Gogo); bk. decoc. u. for indigestion - **S. plantifolia** L. = *Firmiana simplex* Wight - **S. polyphylla** R. Br. (Moluccas): lvs. u. in cataplasms for dislocations; sds. source of FO - **S. populifolia** Roxb. (Hildegardia p. (Roxb.) Schott & Endl.) (Indochin): sds. diur. - **S. pruriens** K. Schumann (Co): sd. FO u. to treat mange ("sarua") (Takanos Indians) - **S. quinqueloba** K. Schum. (Af): lvs. & rt. u. in mal.; bk. in strengthening baths for children after malarial attacks - **S. rhynchocarpa** K. Schum. (cAf): rt. & bk. decoc. u. in mal.; bk. in gastric dis.; bk. fibers (phloem fibers) u. for rope making - **S. roseiflora** Ducke (Co): st. decoc. u. as strong diur. - **S. rupestris** Benth. (Austral): tree trunk produces a gum - **S. scaphigera** Wall. (Scaphium scaphigerum Guibourt & Planch.) (Ind, seAs): gum produced fr. sds. by maceration of frt. pericarp; this colloidal matter is u. for diarrh & dys. (Chin) and as food; bk. u. as diur. and antipyr. in rheum. - **S. tomentosa** Guill. & Perr. (S. setigera Dell.) (trop wAf): karaya gum tree; bk. u. in dys., enteritis (tannin), anthelm. for roundworms; gum u. as emoll. lax., in dyeing cloth (Eu) - **S. tragacantha** Lindl. (wEquator, Af): tragacanth (gum) tree; one source of sterculia gum (*qv.*) (African [gum] tragacanth); young lvs. u. as potherb; young shoots u. as vermif. (also inner fibrous bk.); frt. & sds. boiled & u. for digestive dis. (Sierra Leone); macerated bk. or lvs. appl. as poultice to whitlows, boils, &c.; twigs u. antiluet.; wd. ashes rich in K, u. in soap making; wd. u. to make fences - **S. urceolata** Sm. (Moluccas, Sunda): bk. u. as emmen.; rt. exter. in head pains - **S. urens** Roxb. (cInd, SriL): kuteera gum tree; shaw (Burma); a source of sterculia gum (Indian tragacanth); sd. c. 24% FO; lvs. u. in cattle dis. - **S. villosa** Roxb. (Ind, Pak): udal; a source of sterculia gum; bk. fiber u. for ropes, elephant harness, bags, &c.

sterculia gum: Sterculia Gummi (L.): karaya gum; kadaya gum; Indian g.; gum kute(e)ra; Indian tragacanth; a relatively recent prod.; gummy exudn. fr. trunk & branches of *Sterculia urens*, *S. villosa*, *S. tragacantha*, & other *S.* spp. & fr. *Cochlospermum gossypium* & other *C.* spp.; rich in bassorin; u. dem. in resp. dis., as cough; emulsifying agt., mech. lax. (due to great swelling), hair fixative (wave sets) & other comestics; denture powder (Fasteeth +); toothpaste; in baking indy.; prep. artificial whipped cream; a poor adhesive; in candies (gum drops), some diabetic foods; in ice cream powders; tablets adjuvant; thickening & filling agt. (food indy.); to heal excoriated or irritant skin as in ostomies; for coating of adhesive surfaces of colostomy bags; cigar binding; in printing, textile, & cosmetic indy.; adulter. of acacia; frequent hypersensitivity reported (695) (1946).

Sterculiaceae: Sterculia or Cola Family; usually shrubs, lianes, or trees of tropics & subtr.; some 80 genera & 1,500 spp.; lvs. simple; frt. a follicle, nut, or capsule; fls. Malvaceae-like; ex. kola, baobah, cacao.

Stercus Diaboli (L): asafetida (lit. devil's dung).

stereome: the supporting tissues of a fibrovascular bundle; includes collenchyma & sclerenchyma.

Stereospermum chelonoides DC (Bignoni.) (Ind, Burma, Indochin, seAs): thakutpo (Burma) rts. lvs. fls. u. as antipyr.; fls. aphrod.; very hard wd. u. for furniture; wd. fibers so tough destroy set of saw - **S. ephorioides** (Madag): secretes a water-sol. gum somewhat like tragacanth - **S. fimbriatum** A.P. DC (Mal): juice of lvs. u. for earache; crushed lvs. & lime appl. to scabies; rt. decoc. given as protective med. after delivery (Mal); lvs. & rts. u. in local popular med. - **S. kunthianum** Engl. (tropAf): wild jacaranda; golombi (nNig); pod u. as cough remedy (Tanz) (also bk. & rts.); lf. inf. u. to wash wounds, ulcers; rt. decoc. for VD (syph.) & diarrh. dys.; young bk. chewed by girls to stain lips; rt. u. in "rana" (dis. with hematuria, &c.) (with other rts.); rts.in poultice for syph. sores (Tanz); bk. &c. much u. in witchcraft - **S. personatum** D. Chatterjee (S. tetragonum DC; S. colais (Dilwyn) Mabb.) (Ind, SriL, Burma): trumpetflower; yellow snaketree; padri; rt. decoc. in asthma, cough, excessive thirst; lf. decoc. in chron. dysp.; lf. juice in FO for ear & tooth dis., rheum.; rt. & lf. decoc. antipyr.; rt. bk, & fls. with sugar for refreshing drink; wd. for tea chests, &c. - **S. suaveolens** DC (Ind; Nepal): pad (Hind); padal (Punjabi); rt. bk. c. NBP, u. diur.

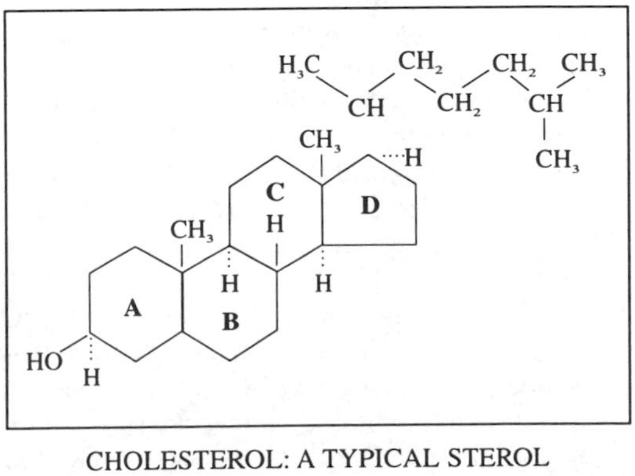

CHOLESTEROL: A TYPICAL STEROL

ton. refrig. esp. in comb. with other drugs (Ayurvedic med.); fls. to check hiccup (with honey); aphrod. (as confection).

Stern (Ge): star - **Sternanis** (Ge): star anise - **S. dolde** (Ge): **grosse S.**: *Astrantia major* - **Sternwurzel**: *Aletris farinosa.*

Sternbergia fischeriana Rupr. (Amaryllid.) (eMed reg.): bulbs c. alk. fraction (with lycorine, hippeastrine, galanthamine, &c.) with emet., expect. antipyr. & cytotoxic props. - **S. lutea** (L) Spreng. (Med reg, sEu, wAs): yellow amaryllis; common sternbergia; bulbs. c. lycorine, tazettine; u. drastic(?!); cult. ornam. hb. (*cf. Colchicum*).

Sternwurzel: *Aletris farinosa.*

steroid: generic term for a large series of compds. with same structural basis as sterols; tetracycles with 17-C phenanthrene nucleus; classified as sterols, saponins, cardiotonic agents, bile acids, carcinogens, pro-vitamins, sex hormones, &c.; term as currently u. refers to sterols, adrenocorticol hormones, & androgenic hormones - **s. hormones**: incl. testosterone, androsterone, estrone, estradiol, estriol, progesterone, corticosterone, desoxycorticosterone, etc.

sterols: class of complex cyclic alc. (lipid) of solid type readily distinguished fr. FO by the fact that they are non-saponifiable; related structurally to the steroids (phenanthrene derivs); important in living organism; occ. in plants (phytosterols) & animals (zoosterols); generally run 1-3% of total lipids; typical ex. cholesterol (commonest sterol) (v. formula) (rich in lanolin); zymosterol (in yeasts); ergosterol (in ergot & yeasts); spinasterol (in spinach & senega); bile acids (taurocholic, glycocholic, cholic); estrogens, androgens, adrenal cortical hormones, digitaloid aglycones, some-saponins - **marine s.**: those of sea animals; ex. chaliasterol.

sterrick root (hysteric root) (US. slang): *Cypripedium pubescens.*

Steudnera virosa Prain (Ar.) (Ind): bish kachu (Bengali): u. med.; toxic.

Stevia Cav. (Comp.) (NW trop&subtrop): 229 spp. of herbs & shrubs; many with folk med. uses (697) - **S. cardiatica** Perkins (Bo): said u. to treat heart dis. - **S. eupat(o)ria** (Spreng.) Willd. (Mex): yerba del borrego; u. diur. animal **- S. glandulosa** Hook. & Arnott (Mex): u. to treat fevers - **S. lucida** Lag. (Mex to Ve): chilca; chirca; hierba dela arana; resionous lvs. & fls. u. as decoc. for bath for rheum.; antiphlogistic - **S. oligocephala** DC (tropAm): sweet but less so than *S. rebaudiana* - **S. pilosa** Lag. (S. confera DC) (Mex): u. antipyr. in mal., cath. diur. - **S. puberula** Hook. (Pe): lima-lima; stom. med.; u. as tea substit. - **S. rebaudiana** Bertoni (Eupatorium rebaudianum Bert.) (Par); yerba dulce; caa-ehe; all parts of pl. esp. lvs. intensely sweet due to glycoside, stevioside (rebaudin, steviol, [e]stevin) which is extremely sweet (some 300x sweeter than sucrose); hb. u. as sweetening agt. for matte, &c. (powd. & mixed with food) (sweetest of all *S.* spp. studied) (1340) (BC [Cana]); oral contraceptive (2391); in diabetes; to regulate blood pressure - **S. rhombifolia** HBK (Guat+): tea u. for dysmenorrh. - **S. salicifolia** Cav. (sNM [US], Mex): hierba de la mula; aq. decoc. (or alc. inf.) u. as rub for rheum., arthritis; for contusions; fish poison (Mex) - **S. serrrata** Cav. (Guat): anis silverstre; u. as cough remedy - **S. trifida** Lag. (sMex): pastilleja; perenn. hb. u. to aid digestion (2342); rt. & fl. inf. u. in dys.

stick (US slang): Cannabis cigarette.

sticker (US slang): thorn, bramble.

stick lac: pieces of branches & twigs with shellac still attached; fr. various pls. such as *Aleurites laccifea* - **sticks** (US slang): cannabis - **stick seed**: 1) *Bidens frondosa* 2) *Lappula* spp. - **sticktight**: *Bidens bipinnata+.*

sticklewort: *Agrimonia eupatoria.*

stickweed: 1) *Verbesina alternifolia* 2) *Agrimonia gryposepala* & other *A.* spp. (**stickworrt**) 3) *Ambrosia artemisiifolia* var. *elatior* 4) (eCana) *Galium* spp. 5) *Aster lateriflorus* var. *hirsuticaulis* 6) *Lappula virginiana* 7) *Hackelia floribunda.*

Sticta crocata (L) Ach. (Pseudocyphellaria c. Vain.) (Stict - Ascolichenes - Lichen.): (worldwide Temp. zone): gold edge lichen; yellow sticta; u. brown dye for woollens - **S. pulmonaria** (L) Ach. = *Lobaria pulmonaria* (L) Hoffm - **S. verrucosa** Fink = *S. scrobiculata* Ach. = *Lobaria s.* DC.

Stiefmuetterchen (Ge): *Viola tricolor.*

Stiel (Ge): st.; stalk; peduncle.

stiff cock (US slang): *Diospyros crassinervis.*

stigma: portion of the pistil which receives the pollen; often incl. in drug with the style (structure connecting stigma to ovary); ex. Crocus.

Stigmaphyllon A. Juss. = **Stigmatophyllum periplocifolium** A. Juss. (Malpighi) (WI): bejuco San Pedro (Cu); soldier vine (Bah); decoc. u. for dandruff; bk. u. in back pain, &c. (Bah) - **S. sagraeanum** A. Juss (WI): nature vein; bk. inf. u. to treat GC.

stigmasterol: steroid occ. in many pls.; originally found in Physostigma sd. FO; constn. (698); starting material u. in synthesis of cortisone, &c.

Stigmata Maydis (L): **stigmates de maïs** (Fr): corn silk.

still: retort & condenser u. in sepn. VO, &c.; condensed fr. "distillery."

Stillingia Garden ex L. (Euphorbi): (mostly Am) some 40 spp. - **S. acutifolia** Benth. (Mex): pavil; milky juice u. to treat sores, ulcers, spots on skin (locally): toxic - **S. sebifera** (L) Mx. = *Sapium sebiferum* - **S. sylvatica** Garden ex. L. (S. salicifolia Raf.) (seUS): queen's root; q. delight; yaw r.; silver leaf; "Indian flea root"; stillingia rt. & rhiz. c. VO (3-4%); glycoside, alks. FO, resin with r. acids; mostly in spindle-shaped pieces (699); u. lax. emet. diur. expect. in rheum. liver injury, skin dis.; after childbirth (Creek Indians); formerly in syph. & scrofula; poisonous (700); dr. pl. said to repel fleas - **S. texana** I. M. Johnston (cTex [US]): Queen's delight; latex u. for ringworm - **S. treculiana** I. M. Johnston (Tex): trecul stillingia; stinkweed; pl. c. HCN glycosides; stock poison.

stimulin: 1) subst. in blood serum wh. stimulates phagocytic activity 2) Nikethamide (synth., stimulant).

Stincus marinus (L): *Scincus officinalis*.

sting ray: *Dasyatis* (Dasyatidae) & other genera of Batoidea (Chondrichthyes - Pisces): long tail of this fish provided with saw-edged spines wh. produce serious wounds that heal slowly because of venom; venom slows heart & may produce heart stoppage (also called stingaree).

stinger (US slang): French brandy with white mint.

stinging plant: *Urtica dioica+*, *Laportea crenulata* & *L. stimulans*, *Loasa* spp., some spp of *Euphorbium* & *Hydrophyllaceae*.

stink bush: *Illicium floridanum* - **s. fruit**: durian - **s. leaf** (Ga): *Ailanthus altissima* - **s. pot:** 1) *Phallus impudicus* (mushroom) 2) musk turtle (US) (*cf.* s. gem [Fla]); *Sternotherus odoratus* (eNA), with strong odor 3) *Amorphophallus* spp. 4) *Smilax herbacea*.

Stinkasant (Ge): Asafetida.

stinking balm: Hedeoma - **s. goosefoot**: *Chenopodium vulvaria* - **s. l'épinet** (Trin): *Zanthoxylum* spp. - **s. nightshade**: *Hyoscyamus* spp. - **s. weed**: *Chenopodium anthelminticum* - **stinkweed**: 1) *Datura stramonium* 2) *Stillingia treculeana*.

Stipa L. (Gram.) (temp & trop regions of world): feather/spear/needle grasses; porcupine g. (some spp.); a bunch grass; some 310 spp.; an important gp. of forage pls.; some toxic - **S. avenacea** L. (Piptochaetium avenaceum (L.) Parodi) (eNA): black oat grass; black seed needle grass; valuable forage pl.; cult. ornam - **S. capillata** L. (Ind, wHimal): needle grass; forage, fodder; fed to milch mares; their milk makes a superior kumyss - **S. comata** Trin. & Rupr. (wNA): needle & thread; cult. for forage (cattle); injurious to sheep, fiber penetrating skin & working inwards - **S. ichu** (Ruiz & Pavon) Kunth (Mex s to Arg): ichu; u. fodder - **S. inebrians** Hance (Mongol): a poisonous grass similar in effects to *S. robusta* - **S. leptostachya** Griseb. (Arg): similar effects to *S. robusta*; c. HCN - **S. pekinensis** Hance var. *planifolia* (S. extremiorientalis Hara) (China, eSiber.): spike u. in med. - **S. pennata** L. (Euras): (soft) feather grass; forage; tufted perenn. grass cult. for ornam. - **S. robusta** (Vasey) Scribn. (S. vaseyi Scribn.) (cUS, Rocky Mt. states; nMex): both fresh & dr. grass hypnotic; sleepy grass; porcupine g.; popote (NM [US]); c. diacetone alc.; both fresh & dr. grass has narcotic effect on stock, esp. horses & to lesser extent sheep; stupefaction (symptoms sleepiness, confusion, weak & irreg. pulse & resp.); rts. u. in kidney dis. (natives' folk med.); fibers u. to make brooms, brushes (Am Indians) - **S. sibirica** Lam. (Ind-wHiamal) (Chin, Siber, nInd): poison grass of Kashmir; pl. c. HCN, sometimes toxic; appl. to bad sores & wounds (Kashmir); (similarly, *S. capensis* Thunb. [S. tortilis Desf.], nwInd); causes intoxication & often death to domesticated animals (horses+) - **S. spartea** Trin. (wCana, nwUS): porcupine grass; forage pl.; however irritant to cattle, sheep, dogs, &c. (sharp-pointed frt. penetrate the skin, producing mechanical injury, sometimes death) (701) - **S. tenacissima** L. (Sp, nAf, natd. NA): esparto (grass); (h)alfa (grass); fiber & straw u. in mfr. paper (esp. fine book papers) (Sp, Por, Mor, Alg); source of fiber (or esparto) wax (similar to candelilla wax); fiber also u. to make sails, mats, rope; cult. ornam., &c. - **S. viridula** Trin. (wNA): var. *green Stipa* disease-resistant; green needle grass; sleepy grass(?); good for forage (grazing) grass; reports of stock poisoning (520).

stipe: stalk of 1) fern frond, analogous to petiole of phanerogams; esp. the base (ex. Aspidium) 2) pistil 3) mushroom (or other fungus).

Stipites Cerasorum (L): frt. stalks of *Prunus cerasus* - **S. Dulcamarae** (L): Dulcamara (stems collected after lf. fall) - **S. Laminariae** (L): laminaria stalks (v. *Laminaria*).

Stirabout (Irish, later, Eng): porridge made by stirring oatmeal, &c. into boiling water or milk.

stirps: race, permanent var., or sp. of organism (unofficial category).

stitchwort (also **great[er] s.**): *Stellaria holostea*.

Stixis scortechinii (King) Jacobs (Capparid.) (Indones): rt. juice u. to treat sore eyes - **S. suaveolens** (Roxb.) Pierre (Indochin): lvs. u. to treat eye dis.

Stizolobium R. Br. = *Mucuna* - **S. deeringianum** Bort. = *M. deeringiana* (Bort.) Merr. = *M. pruriens* var. *utilis* - **S. pruritum** Piper = *M. pruriens* (L) DC.

stock: *Mathiola* spp.

stockfisch: (Ger, Swed, Nor): fish dr. hard on poles without salt in open air; esp. cod, haddock, hake, ling, torsk.

Stoebe cinerea Thunb. (Comp.) (sAf): bobo-bos; arom. pl. u. in heart trouble; pl. favorite bedding materials for camping trip.

Stoechados Flores (L): inflorescense of *Helichrysum arenarium* (also **F. S.**) - **S. F. arabicae (purpureae)**: fls. of *Lavandula stoechas.*

stogie tree: (wVa [US]): *Catalpa.*

Stokesia laevis (Hill) E. Greene (monotypic) (Comp.) (seUS): Stokes' aster; source of epoxy fatty acid oils; cult. perenn. ornam. hb.

stolon: runner; horizontal st. growing above the earth's surface; serves for propagation; usually have longer internodes than rhiz; ex. Bermuda grass.

stoma (pl. **stomata**): air pore in lf. & young st. which is closed by the change in shape of two guard cells which surround it; serves to admit air to the interior tissues of the leaf to permit exit of gases & water vapor fr. the lf. interior; of diagnostic use in pd. drug indentification.

stomach, desiccated hog's: powered s.; the dr. & defatted s. wall of the hog, *Sus scrofa*; c. factors which incr. number of red blood cells in the blood of persons suffering fr. pernicious anemia; adm. orally in anemias, etc. - **s. drops**: Tinctura Amara, bitter tincture (NF V) - **s. lining (hog)**: defatted & desiccated prod.; u. as source of pepsin, mucin, intrinsic factor of hog's stomach - **stomachic tincture**: stomach drops - **Stomachus Pulveratus** (L): desiccated stomach.

stomatal index: percentage of stomata in epidermal cells (incl. stomata); formula: S.I. = (S X 100)/(S + E), where S = no. of s. & E = no. of epidermal cells; cannot use no. s. per sq. mm since individual variation in cell size; best carried out with camera lucida; repeat for 5-10 fields (400-500 cells); specify lower or upper epidermis (702): index is a constant for pl. sp. - **s. number**: average number of s. per sq. mm of epidermis; recorded as range (min-aver.-max.) for each surface of lf. & ratio of lower to upper counts.

stomate: stoma.

stone: v. under mints; nuts.

stone cell: slereid - **s. crop ditch**: *Penthorum sedoides* - **s. fruits**: frts. of *Prunus* spp. (large drupes) - **s. leek**: *Allium fistulosum* - **s. linden tree**: *Phillyrea latifolia* v. *media*, bk. & lvs. - **s. oil**: orig. crude petroleum; later, an artifical prod. - **s. plant**: *Lithops*; some are "windowed" plants (with translucent upper surface) - **s. root**: *Collinsonia canadensis.*

stone-crop: *Sedum* - **ditch s.**: *Penthorum sedoides* - **gold moss s.**: *S. acre* - **Virginia s.**: ditch s. - **stone-fern**: *Asplenium ruta-muraria.*

stool: stand; a rhiz. or stump wh. produces shoots or suckers.

stopfen (Ge): constipate.

storax: Styrax; balsam fr. *Liquidambar styraciflua* & *L. orientalis*; term (as distinct fr. styrax) sometimes u. to indicate rel. solid prod.; also *Styrax officinalis* - **American s.**: balsam prod. fr. *L. styraciflua* - **black s.**: made by Gr monks on Symi Isl. (chiefly); a mixt. of olibanum & liquid s.; u. fumigant & incense in Gr Cath. churches - **s. calamita**: solid prod., generally made by mixing sawdust or sweet gum tree bk. pd. or residue fr. cinchona preps. with liquid s.; u. incense in Gr Cath. churches (SC formerly packed in reeds = calami) - **factitious s.**: artificial s.: prepd. by mixing gum thus, rosin, castor oil, & perhaps glycerin & pine tar oil, &c. - **levant s.**: balsam fr. *Liquidambar orientalis*; mouse-gray semiliquid product; sometimes preferred to American s. - **liquid s.**: s., more usually of the levant type - **s. officinalis** (L): s. calamita - **s. oil**: VO of s.; much more (75) in American than in levant (1%) - **s. purifacatus** (L): **purified s.**; water cont. 15% as against 30% for the crude - **solid s.**: 1) s. usually of Am type 2) exudn. fr. *Styrax officinalis* - **Turkish s.**: levants.

storesin: the hard resin fr. styrax, made up of storaresin, cinnamic acid ester of storesinol.

"stories": levels of elevation in a forest; ex. upper story.

stork: *Ciconia ciconia* (Ciconiidae - Aves) (Eu, nwAf): common European s.; white s.; large migratory wading bird; commonly builds nests on chimneys or house tops (Austria; Switz, &c.); considered semidomesticated (nesting on houses gave rise to association with delivery of baby); flesh u as stom.; in sexual & nerv. dis. in Unani. med., also blood & excreta in skin dis.; bile in eye dis. (night blindness); eggs.

storksbill: *Erodium circutarium*+ - **strongly scented s.:** *Pelargonium graveolens.*

Stoughton's elixir: compound gentian tincture.

stramen: straminis (L.): straw - **s. avenae**: oat s.

stramoine (Fr): Stramonium - **s. d'est**: *Datura metel.* - **stramonio** (It.): **Stramonium** (L): 1) *Datura s.* & its var. (form) *tatula* (**purple s.**) 2) (French Bourbon) (Réunion): *D. metel* & its var. *alba* - **green-stem stramonium** (med.): *D. stramonium* - **red-stem s.** (med.): purple s.: *D. stramonium* var. *tatula* - **stramony** (US. slang).

strangler (fig): *Ficus aurea.*

Stratiotes aloides L. (monotypic) (Hydrocharit.) (Eu): water soldier; in summer pl. floats up to surface of water & bears fls. then sinks in the fall; axillary shoots produce young plants which sink to bottom & remain over winter: rt. & frt. serve as nutrients (manure); cult. orn.

Strauch (Ge-m.): shrub, bush.

straw: (dr.) st. of cereal grain or pulse, esp. after threshing & drying (usually matured st.) (*cf.* hay, usu. immature); ex. wheaten straw - **s. flower:** Strohblumen (Ge): *Helichrysum bracteatum* Willd., *H. arenarium* Moench, &c.- **s. process**: hay p.; method of prodn. of pl. principles by growing for a relatively short period, then cutting down & extracting the pl. with vol. solvents; process appl. to Cinchona; camphor (US) Opium for morphine (Hung - von Kabay process; Hung, Yugoslav; Austral); sometimes poppy capsules after sd. removal (1947).

strawberry: *Fragaria* spp. (SC pl. covered with straw to keep soil moist, berries clean, & to hasten ripening); esp. *F. vesca*; or because growing so close to ground as tho strewn there; or because frt. spitted on straws & thus sold (long ago) - **s. bush**: *Euonymus americanus, E. obovatus, E. atropurpureus*, &c. - **meadow s.**: **scarlet s.:** *Fragaria ananassa* - **s. tree**: 1) *Arbutus unedo* (also **European s. t.**) 2) *Euonymus atropurpureus* - **Virginia s: wild s.**: 1) *Fragaria virginiana*, &c. 2) *Potentilla simplex* & related *P.* spp.

Streblus asper Lour. (Mor.) (tropAs, Ind, Mal): grey wood of Seroet; rt. bk. c. ca. 30 cardenolides; incl. asperoside; "streblid" (toxic bitter princ.) (703): lvs. c. milk-clotting enzyme u. to clabber milk (704); bk. u. in paper mfr. (the paper tree of Siam) (Khoi paper) (white ants will not eat); bk. u. for dys. hemorrhoids, skin dis. (folk med., Thail); lvs. for kidney pain, galact.; seeds as ton. to stim. appetite; rts. u. in lumbago, rt. could be u. as heart ton.; supernatural pl. (Ind).

Strelitzia augusta Thunb. (Mus./Strelitzi) (sAf): bird of Paradise tree; cult. as ornam. worldwide - **S. reginae** Banks ex Dryander (sAf): bird of Paradise flower; the commonest *S.* sp. in cultn. as ornam. for showy fls. & banana-like lvs.; sd. u. to hasten milk souring (Cape of Good Hope).

Streptocaulon baumii Decne. (Asclepiad.) (PI): latex & lvs. u. as vulner.; for pinworms - **S. juventas** (Lour.) Merr. (Indochin): rt./tuber u. as "rejuvenator" esp. for the aged; for bloody stools; in bitters & as a "moderator"; rt. u. as ginseng replacement.

Streptococcus lactis (Lister) Loehnis = *Lactococcus lactis* Loehnis et al. (Streptococc.) (Bacteria) - **S. pyogenes** Rosenbach (S. erysipelatos (*sic*) Rosenbach): m. o. causative of numerous infections; incl. scarlet fever; the toxin formerly u. to det. if patient has immunity or not & as prophylactic; antitoxin formerly u. in treating scarlet fever (now better controlled with antibiotics); source of nisin (antimicrobial u. as preservative in dairy foods, &c.) - **S. thermophilus** Orla-Jensen: occ. as starter in Swiss cheese & yogurt; also involved is *Lactobacillus bulgaricus* Rogosa & Hansen.

streptodornase: thymonuclease; enzyme found in filtrates fr. cultures of some hemolytic streptococci; occ. in many sds. (ex. *Zea*), intest. mucosa, pancreas; promotes hydrolysis of desoxyribonucleoproteins & desoxyribonucleic acid; u. with streptokinase to remove infected tissues & dead tissues; u. before skin grafts, in serious wounds, burns, osteomyelitis, sinusitis, emphysema.

Streptogyna crinata P. Beauv. (Gram) (trop wAf): pith u. for dental caries & for dental dis.; scabrid infloresc. u. to catch mice by stuffing into holes.

streptokinase: streptococcal fibrinolysin; enzyme (activator) produced by hemolytic streptococci; serves to activate profibrinolysin (plasminogen) plasmin (a blood enzyme which digest fibrin), thus removing clots fr. blood stream (1948); esp. for pulmon. embolisms; injected into artery in heart to remove clot, a use (Thrombolysis) approved by FDA (for coronary artery thrombosis (1982); disadv.: temporarily suppresses clotting power of blood; of urokinase and acylated plasminogen s. complex (under plasminogen).

streptolin: a form of streptothricin.

Streptolirion volubile Edgew. (1) (Lili) (Chin): crushed pl. parts appl. to burns & scalds.

Streptoludigin: antibact. antibiotic fr. *Streptomyces lydicus* DeBoer & al. (1957-).

Streptomyces Waksman & Henrici (Streptomycet. - Actinomycetales - Schizophyta [Bacteria]) (previously classified under Ascomycetes [Fungi]); microorganisms of special importance in prodn. of antibiotics; thus, spp. prod. streptomycin, streptothricin, chloromycetin, terramycin, aureomycin, &c.; mostly soil forms; organisms remain attached in chains, not breaking into rods (distinction of fam. fr. Actinomycetaceae); mostly saprophytic, a few parasitic; some 160 spp. incl.: **S. albidoflavus** Waksman & Henrici: one source of vit B_{12}; **S. albus** Waksman & Henrici: source of alborixin & other antibiotics; **S. antibioticus** (Waksman & Woodruff) Waksman & Henrici; source of actinomycins, oleandomycin; &c.; **S. aureofaciens** Duggar: source of chlortetracycline (CTC) (=Aureomycin), first broad spectrum antibiotic; active against bacteria, viruses, rickettsiae, & protozoa (1948-) - **S. colombiensis** Prudham & al.: producer of vit. B_{12} - **S. erythreus** Waksman & Henric: source of Erythromycin - **S. fradiae** Waksman & Henrici: source of Neomycin; organism important in microbiological transformations of steroids, &c.; **S. griseocarneus** Benedict & al. (source of hydroxystreptomycin); **S. griseus** Waksman & Henrici: source of man prods. incl. streptomycin, streptocin, grisenin, actidione, viractin, cyanocobalamin, animal protein factor, viridogrisein, griseoviridin, rhodomycetin, amicetin, candicidin, &c.; **S. humidus** Nakazawa & Shibata (fr. soil, Jap): source of hydroxystreptomycin; **S. kanamyceticus** Okamvami & Umezawa: source of kanamycins; **S. lavendulae:** source of streptothricins (incl. pleocidin), lavandulin, ehrlichin, & other antibiotics - **S. lincolnensis** Mason & al.: source of Lincomycin; **S. natalensis** Struyk et al. (sAf): source of pimaricin (natamycin) (u. as antimycotic for epidermophytosis & candidamycoses, also antiviral) - **S. olivaceus** Waksman: one source of vit. B_{12}; source of granaticin & synergistins (antibiotics); **S. orchidaceus**: source of Cycloserine (Seromycin): u. in TB; (also fr. *S. lavenduae*); **S. reticuli** Waksman & Henrici: one source of hydroxystreptomcycin ("reticulin") (not to be confused with the protein called reticulin) & rotaventin (antibiotic); **S. rimosus** Waksman: source of oxytetracycline (Terramycin) (1951-); **S. roseochromogenus** Waksman & Henrici: source of antibiotic & vit. B_{12}; **S. venezuelae** Ehrlich & al: source of chloramphenicol (Chloromycetin) (note: these species are selections from among many other *S.* species).

Streptomycin: a gp. of antibiotic principles (or any one) obt. fr. *Streptomyces griseus*; u. in control of Gm.- dis.; of service in TB, esp. when u. with PAS or isoniazid (2011); u. as salt (HCl, phosphate, double salt [$CaCl_2$], sulfate); adm. IM SC IV orally (for intest. inf.) (E) (1950); ototoxic - **S.A.:** streptomycin per se - **S.B.:** mannoside s. - **S.C.**: dihydro s.

streptonigrin: bruneomycin: rufochromomycin; antitumor antibiotic fr. *Streptomyces flocculus* Waksman & Henrici; u. antineoplastic.

streptonivicin = Novobiocin.

Streptopus amplexifolius (L) DC (Lili) (Euras, NA, mostly w): liver-berry; (clasping-leaved) twisted stalk; berries edible; ornam. perenn. hb.; frt. & st. inf. u. for sickness in general (Montagnais Indians) - **S. roseus** Mx. var. *perspectus* Fassett (Cana, nUS): sessile-leaved twisted stalk; fl. inf. u. diaph. (Montagnais Indians); rt. inf. as poultice for stye (Chippewa); for cough (Ojibwe; Potawatomi Indians).

Streptosiphon hirsutus Mildbr. (1) (Acanth.) (eAf): crushed rt. appl. to wounds.

streptothricins: mixt.of antibiotics prod. by *Streptomyces lavendulae* (Actinomyces l.) (2097); u. similarly to streptomycin; formerly m. o. classed as *Streptothrix* sp., hence name; active against Gm. neg.-bact., discarded delayed toxicity (Waksman 1942) (1949).

streptovaricin: complex mixt. of several macrolide antibiotics produced by *Streptomyces spectabilis* Mason & al. chief one: **s. C**: u. antimicrobial, antiviral; also u. as feed supplement (Dalacin Upjohn).

streptovirudin: complex nucleoside prod. by *Streptomyces griseoflavus* Waksman & Henrici subsp. *thuringiensis* Thrum & al. inhibitory to gm.+ bacteria, mycobacteria, &c.

streptozocin: **streptozotocin**: nitro-sugar isolated fr. fermentation broth of *Streptomyces achromogenes* Okami & al. (ca. 1959-); antineoplastic (cytotoxic) properties; diabetogenic (rats); antiapudoma drug; may be itself a carcinogen.

stress (disease): variation fr. homeostasis of individual caused by stimulus/stimuli, such as injuries, heat, cold, &c.; failure to adjust may cause disease, disability, death.

stretchberry: *Smilax bona-nox.*

Streublumen (Ge) (lit. strewn fls): **Streumuster**: (lit. spread design or pattern) design of an individual fl. or fl. bouquet irregularly distributed over the surface in porcelain & faience painting; similarly in weaving & needlework (18th cent.) - **Streupulver** (Ge): Lycopodium.

strewing herb: pls. scattered over an area to release perfume (VO) when stepped on by crowd of people; formerly popular in GB (ex. peppermint).

Striga asiatica Ktze. (S. lutea Lour.) (Scrophulari) (eAs): witchweed; a root parasite; decoc. stom. vermifuge; in poultices - **S. gesnerioides** Vatke (S. orobanchioides Benth.) (Ind, tropAf): u. in diabetes (48) - **S. hermontheca** Benth. (S. senegalensis Benth.) (tropAf): hemiparasitic pl.; lf. inf. u. in thrush (e tropAf); u. in leprosy (esp. rts.); for dermatoses (local appl. with friction) (Seneg.) - **S. pubiflora** Klotzsch (Tanz): lf. inf. u. as gargle for persistent cough; rt. decoc. in stomach dis. (also rts. eaten).

strigol, D- L-: furanone deriv., isolated fr. exudates of cotton (*Gossypium hirsutum*) rt., acts to stim. sd. germination of *Striga asiatica*; u. in microbiol. synthesis (2589).

strigose: surface beset with short stiff hairs; *cf.* villous (with long soft hairs), canescent (hoary).

string beans: v. beans.

stringybark, peppermint: *Eucalyptus piperita.*

Strobilanthes auriculatus Nees (Acanth.) (Ind, Burma): pounded lvs. rubbed on body for intermittent fevers - **S. callosus** Nees (Ind): bk. emoll. u. as fomentation in tenesmus; extern. in parotitis; fls. vulner. - **S. crispus** Bl. (Indones): dr. lf. inf. u. in diabetes, as diur., lax.; for kidney, bladder, & GB dis.; replacement for *Orthosiphon stamineus* lvs. - **S. flaccidifolius** Nees (Baphicacanthus cusia Bremek.) (tropAs, esp. Java, Sumatra): rt./rhiz. decoc. u. as antipyr., to counteract opium poisoning (Chin); pounded lvs. in cough (Mal); as astr.; popular ornam. (lvs.); source of blue dye (Assam indigo).

Strobilanthopsis linifolia Milne-Redhead (Acanth.) (Af): rt. decoc. in GC.

strobile: a cone; scaly multiple frt.; ex. infloresences of hops; Lycopodium - **Strobili** (L) (pl. of Strobilus): strobiles.

Strobili Humuli (L): **S. Lupuli** (L): Humulus.

stroboside: glycoside of corotoxigenin + D-boivinose; occ. in *Strophanthus boivinii.*

Stroh (Ge): straw.

Strombosia grandifolia Hook. f. (Olac.) (w tropAf): u. like foll. - **S. pustulata** Oliv. (w tropAf): bruised bk. & rts. incorporated in palm oil & salve smeared on body for shriveled skin, withered hands.

Strongylodon lucidus Seem. (Leg) (Guadalcanal, Solomons): heated lvs. rubbed on boils - **S. macrobotrys** A. Gray (PI): jade vine; large cult. ornam. vine.

strophanthin; Strophanthinum (L): a bitter glycoside obt. fr. *Strophanthus kombé & S. hispidus*; u. card. ton. diur.

strophanthol: glucocymarol; cardiotonic glycoside (705).

Strophanthus amboensis Engl. & Pax (Apocyn.) (swAf): sds. c. several glycosides, incl. intermedioside, kwangoside; u. med. **- S. boivinii** Baill. (Mad): all parts toxic; sd. c. 17+ glycosides, incl. madagascoside, 17-ß-H-madagascoside, zettoside, boistroside, & others (cardiac glycosides) (sarmentogenin gp.); u. med.; decoc. u. to poison dogs & varmints - **S. bracteatus** Franch. = *S. preussii* - **S. caudatus** Kurz (Burma, Mal): sd. c. caudoside, divaricoside; latex u. as arrow poison, fish poison; var. *marckii* Planch. (S. cumingii A. DC; S. letei Merr.) (PI): lanot (Iloko); latex & rt./st./bk. u. as arrow poison; c. h-strophanthin, cymarin (706) - **S. courmontii** Sacl. (trop eAf, Malawi): sds. c. sarveroside, panstroside; u. arrow poison; rt. decoc. antirheum. aphrod.; commonest adulter. of off. Strophanthus sd. - **S. divaricatus** Hook. & Arn. (s. & eAs): sds. u. in artheriosclerosis (China); c. divaricoside, caudoside (cardioactive glycosides) - **S. eminii** Asch. & Pax. (s. & eAf): sds. c. cymarin, emicymarin, cymarol; rts. u. as emet. abortif. (Tanz); rts. & sds. u. arrow poison; adulter. of off. Strophanthus; sd. hair ("vegetable down") u. for padding - **S. gerrardi** Stapf (sAf): sds. c. sarveroside, panstroside; cardioactive glycosides - **S. glaber** Cornu = *S. gratus* - **S. gracilis** K. Schum. & Pax (wtrop Af): sds. c. many glycosides, incl. ouabain (g-strophanthin), sarmentoside, strogoside; ext. of trunk wd. & sds. u. as lance poison for elephant hunts (wAf); sds. u. carioton. in cases of digitalis sensitivity; lvs. u. antipyr.- **S. grandiflorus** Gilg = *S. petersianus* - **S. gratus** Baill. & Franch. (w tropAf): sds. c. many glycosides, incl. ouabain (g-strophanthin), sarmentoside, strogoside; ext. of trunk wd. sds. u. as lance poison for elephant hunts (wAf): sds. u. cardioton. in cases of digitalis sensitivity; lvs. u. antipyr. - **S. hispidus** DC (trop wAf): "Transvaal" strophanthus; source of brown s. sd.; c. h-strophanthin; cymasrin; cymarol; u. as for others - **S. mirabilis** Gilg (trop eAf - Kenya): c. cymarin, cymarol; u. similar to other *S.* spp. - **S. nicholsonii** Holmes (trop eAf): c. & u. similar to those of *S. mirabilis*; sd. serves as adulter. of off. strophanthus (*S. kombé*) - **S. petersianus** Klotzsch (trop e & sAf): sds. c. carmentocymarin, samentogenin; formerly known as "S. lanuginosus" in commerce; arrow poison - **S. preussii** Engl. & Pax (S. bracteatus Franch.) (trop Af): dietwa (Ghana); sds. c. 46 heterosides (glycosides) incl. periplocin, alliperiplocymarin; sap appl. to wounds & sores; arrow poison; juice u. as coagulant for cauotchouc (latex of rubber tree, *Funtuma elastica*); young lf. eaten as veget(!) - **S. sarmentosus** (DC, trop wAf): tio(k)h (Malo); sds. c. carmentogenin, sarverogenin, &c.; arrow poison; heart activity weak; as raw material in mfr. of cortisone & other adrenocortical hormones (objective of expeditions); adulterant of off. s. sd. (*S. hispidus*); cult. ornam. (Fla [US], &c.) - **S. schuchardtii** Pax. (Angola): sds. c. sarveroside, leptoside, panstroside, &c. (707) - **S. speciosus** Reber (sAf): sds. c. christyoside (corotoxigenin glycoside) (with digitaloid action), strospeside; u. as arrow poison; in homicides; roasted rt. u. in snakebite (stock & humans) - **S. thollonii** Franch. (708) (trop cAf): sd. similar to *S. sarmentosus*; c. sarmentosides, tholloside, lokundjoside; bipindoside, zenkoside; sd. u. arrow poison - **S. vanderijstii** Staner (Zaire): sds. c. many glycosides, incl. digistroside, odorosides; similar to others in usage - **S. verrucosus** Stapf (trop eAf): sds. c. cardiac glycosides; u. as arrow poison - **S. wallichii** A. DC (trop As): sds. c. cardiotoxic glycosides -

S. welwitschii Schum. (trop Af). sds. c. intermedioside, panstroside but often lacking in glycosides - **S. wightianus** Wall. (Ind): sds. c. divaricoside, caudoside; similar to *S. sarmentosus.*

strophanthus seed: generally appl. to *Strophanthus hispidus* and *S. kombe* - **brown s. s.:** *S. hispidus* - **green s.:** *S. kombé*

Stropharia cubensis Earle (Psilocybe c. [Earle] Sing.) (Strophari. - Basidiomy. - Fungi): (WI as Cuba; Mex, Fla): a dung mushroom; found growing on horse dung - **S. mexicana**: *Psilocybe m.*

strophiole: caruncle *(qv.).*

strospeside: glycoside consisting of aglycone gitoxigenin + glycone D-digitalose; occ. in *Digitalis* spp. as *D. grandiflora, Nerium oleander, Strophanthus* spp. as *S. boivinii.*

Struchium sparganophora Ktze. (1) (Comp.) (sMex to trop SA, WI, natd. wtrop Af) yerba de faja; lf. decoc. for "fits" & convulsions (Suriname); in poultice for sprains & contusions (WI).

struma: goiter.

Strumpfia maritima Jacq. (1) (Rubi) (sFla, WI, sMex): romero falso; weak decoc. for colds & to expel kidney stones (Curaçao); abort.

Struthanthus haenkeanus (Presl) Standl. (Loranth.) (Mex): toje; decoc. of herbage u. to wash wounds, bites, & stings - **S. marginatus** Blume (Loranthus m. Desb.) (tropSA): semi-parasite on twigs of woody pls. (*cf.* our mistletoe); lvs. off. as Herva de Passarinho (Br Phar.) u. to prep. galenicals - **S. syringifolius** Marp. (Guyana, Ve, Br): lf. inf. u. for dentalgia (Guyana); formerly u. to produce rubber; weed pest in SA coffee plantations.

Struthiola L. (Thymelae.) (s&tropAf): bk. of spp. u. like Mezereum - **S. thomsonii** Oliv. (trop eAf): raw rts. chewed for cough; rt. inf. drunk with milk for stomachache.

struvite: crystals of $MgNH_4PO_4$ complex found in canned shellfish & often mistaken for ground glass but harmless.

Strychni Semen (L): Nux Vomica.

Strychnina (L): **strychnine:** chief alkaloid of Nux Vomica; found in severa; other *Strychnos* spp.; extremely bitter; synthesized by Woodward (1954); u. spinal stim., u. for alcoholism (1951); aphrod.; poison for rodents, wolves, &c.

Strychnos L. (Logani) (trop. both OW & NW): ca. 220 spp. of shrubs & trees opp. entire lvs.; frt. a berry, sometimes with hard shell; bk. ext. of SA. u. in prepn. curare, an arrow poison of SA, Ind; alks. u. in modern med. (1953) - **S. aculeata** Sol. (w&cAf; rain forests): sds. c. brucine (traces) (questioned), strychnofendlerine, &c.; frts. c. saponin; no strychnine; u. as fish poison, soap substit. - **S. brachiata** R. & P. (Co, Pe, Bo): rt. of vine u. in making arrow poison (Kofan Indians, Col) - **S. castelnaeana** Wedd. (Pe, Br): Amazon poison nut; gure (Tecuan Indians); urari-uva; bk. chief source of curare (Oregonas Indians, Pe) & other Indians of upper Amazon; c. mavacurine &c. - **S. cinnamomifolia** Thwaites (Ind): sd. c. strychnine, brucine, hydroxymethoxy-strychnine; rt. decoc. u. in rheum., fever, epilepsy, elephantiasis - **S. cocculoides** Bak. (s&eAf): Kaffir (or wild) orange; sd. c. brucine, chlorogenic acid; pulp of frt. edible; rt. chewings to treat eczema, &c. - **S. cogens** Benth. (Guyana, Br, Ve, Co): arimaru (Macusi Indians); this bush rope ca. 35 m long found on "terra firme"; one of major spp. u. in prepn. curare (Tecuna & Macusi Indians) (709) - **S. colubrina** L. (Ind, SriL): snakewood; sds. bk. & wd. c. strychinine & brucine; wd. (Lignum colubrinum) = snake wood, u. against snakebite; mal.; fresh lvs. in paste for sores; rt. in diarrh. (may be conspecific with *S. minor*) - **S. crevauxiana** Baill. = S. crevauxii Planch. = S. curare Baill. = *S. guianensis* - **S. densiflora** Baill. (cAf incl. Nigeria): bk. rts. sds. c. small amt. strychnine; u. as ordeal poison - **S. depauperata** Baill. (SA): u. in prepn. curare (arrow poison) - **S. dysophylla** Benth. (sAf): poisonous klapper; may c. alk.; frt. pulp u. to stave off hunger; in dys. - **S. erichsonii** Rich. Schomb. (Guy, Suriname, Br, Ve, Co): occ. as giant "bush rope" sometimes over 36 m long; bk. u. as aphrod.; considered most potent sp. for curare mfr. (Maku, Am Indians) - **S. fendleri** (Ve.): bk. u. as stim. in mal.; said source of strychnine; actually c. diaboline, strychnofendlerine, but no strychnine - **S. gautheriana** Pierre (S. malaccensis Benth.) (seAs, esp. Indochin): c. strychnine & brucine; "Malacca poison nut" u. in rheum. syph. leprosy, rabies(!), &c.; in prepn. arrow poison (hoang-nan; hwang-nao) - **S. gerrardii** N. E. Br. (trop eAf): bk. c. alk. u. in colic (Moz); ripe frt. pulp edible, refreshing (citric acid) - **S. gubleri** Planch. (SA): a source of curare - **S. guianensis** Mart. (S. crevauxiana Baill.; S. crevauxii Planch.; S. curare Benth./Baill.) (Guianas, Ve, Co, Br+): bejuco (Pe); b. de mavacure (Ve); st. & rt. bk. u. in prepn. curare - **S. henningsii** Gilg (trop e&sAf): st. bk. c. 22 known alks., incl. diaboline, holstiine, henningsamine; bk. u. in stom. dis., colics, as purg. anthelm.; rt. in rheum. fever - **S. icaja** Balli. (trop w&cAf): (i)kaja; akazga; bundou; rt. c. icajine, strychnine; rt. u. as fish poison; antimal.; small doses as intoxicant, diur.; to obtain curare (710); much u. as ordeal poison, arrow poison (woorari p.; quarari) - **S. ignatii** Berg (S. ovalifolia Wall.) (PI, Borneo, Java; cult. in Ind, Chin, sVietnam, Mal): (Saint)

Ignatius bean; Ignatia bean; dr. ripe sd. of climbing shrub; c. 1.5-2.4% strychnine, brucine, caffeic acid, chlorogenic acid (caffeetannic acid), FO, wax, starch; u. similarly to Nux Vomica as spinal stim. (resp., circ.), convulsant, bitter ton.; poison; in prodn. of strychnine & brucine: frt. eaten by monkeys ("monkey apples"), &c.; rt. bk. u. to prep. Javanese & Mal blowpipe arrow poisons (1952) (SC Ignatius [de] Loyola, founder of Jesuit Society) - **S. innocua** Del. (sAf): Kaffir orange; monkey apple; rts. taken with food for ankylostomiasis (Tanz); frt. pulp eaten, u. in marmalades - **S. javariensis** Krukoff (Br, Co): secondary source of curare; bk. chewed to relieve toothache (Tikuna Indians) - **S. jobertiana** Baill. (nSA, basin of Amazon; Co.): quilio huasco (Ec Indians): rt. preferred source of curare; st. & rt. bk. also so u. - **S. lucens** (Tanz): sts. in water u. for ankylostomiasis - **S. lucida** R. Br. (S. ligustrina Bl.) (Indones): sds. c. strychnine, brucine; wd. u. antipyr. wd. scrapings in weak arrack u. as ton. to stim. appetite, anthelm. - **S. madagascariensis** Poir. (sAf, Madag): rt. decoc. purg., for yaws; frt. edible, frt. pulp eaten - **S. malaccensis** Benth. - *S. gautheriana* - **S. melinoniana** Baill. (Guianas); white evil-doer (Guyana); doberdoca (Surinam); c. melinonines (quarternary indole alks.); vine u. in indig. med. & as aphrod.; also source of curare - **S. minor** Dennst. (S. multiflora Benth.) (PI): bk. decoc. u. emmen., throat troubles; bitter rts chewed by older women for uter. prolapse - **S. mitscherlichii** Rich. Schomb. (Amazon basin, SA): "white evil-doer" (Guyana); bk. ingredient in curare; bk. u. as animal poison (Col) - **S. nigricans** (Br, Ve, Pe): unha de cigana; curare ingredient - **S. nux-blanda** A. W. Hill in Kew Bull. 1917: 189 (sInd, Burma, Thail): sds. c. no alk.; u. adulter. of Nux Vomica (1747); frt. fish poison (Burma) - **S. nux-vomica** L. (tropInd, SriL, Mal, nAustral; cult., wAf): nux vomica; poison nut; Quaker buttons; (common) vomit nut; mostly obt. fr. wild-growing trees (common in Madras, Tennessarim, Ind.); sds. c. 2.5-3% alks (strychnine, brucine, vomicine, novacine) (1954), chlorogenic acid (igasuric a.), loganic acid, FO sterols, frt. pulp c. loganin (glycoside) (not in sd.); sds. u. bitter ton., spinal stim. to incr. muscle tone, in exhaustion, weakness; stim. activity of respir. & circulatory centers; in paralyses, impotency, chronic alcoholism (711); collapse; antidote to misuse of med. and beverages; poisonous; formerly u. to kill varmints (wolves, Ge), parasites; comm. source of strychnine, brucine - **S. pachycarpa** Ducke (Br): potential curare source - **S. panamensis** Seem. (sMex, Hond): "guaco"; u. as remedy for pain (Hond) - **S. panurensis** Sprague & Sandwith (wAmazon reg.): frt. strongest in alks.; valued ingredient of curare - **S. peckii** B. L. Robinson (CA, Co, Ve, Br): tie-tie (Belize); bk. secondary ingredient in curares; rt. u. in curare by Kofan Indians+; frt. pulp edible - **S. pedunculata** Benth. (Guyana, Ve): yakki (as ingred. in curare of Macusi Indians) (Guyana) - **S. potatorum** L. f. (literally of drinkers) (tropAf, As, esp. Ind, Burma): clearing nut(tree); sds. are placed in water for drinking to clear or clarify it by ppt. of impurities; sds. c. diaboline; lf. c. angustine; sds. u. in Ayurvedic med. as components of many complex remedies, esp. in ophthalmology; for corneal ulcers; adulter. of Nux Vomica; furnishes useful timber - **S. pseudo-quina** St. Hil. (Br, Par): quina do campo (Br); common bushy pl. source of a spongy yellow & very bitter bk., reputed antipyr. in mal.; u. like Cinchona; stom. strong purg. (49) - **S. pungens** Soler. (trop e&sAf): monkey apple; wild orange; frt. pulp acidulous (citric acid), edible; no alk. in sd.; crushed lf. with water appl. to sore eyes; unripe frt. emet. in snakebit (Nyamwezi) - **S. rheedii** Clarke (S. aenea A. W. Hill) (Ind): sds. c. brucine; u. like Nux Vomica: said source of colubrina wood; v. *S. colubrina* - **S. roborans** Hill (Thail): wd. u. in Thai med. - **S. solimoesana** Krukoff (wBr): ira (Jamamadi Indians); bk. c. calebassine, curarine, u. as chief ingred. in arrow & blowgun dart poisons (mixed with *S. toxifera*); curare - **S. spinosa** Lam. (trop&sAf, Madag): monkey orange; klapper; Natal (or Kaffir) orange; a small tree; frt. pulp edible, refreshing, sour (fr. citric acid) sds. lvs. c. akagerine, kribine, sds. c. only trace of alk.; green frt./rt. u. as antipyr. & in snakebite; rts. cooked in porridge for cough, asthma (Tanz); lf. decoc. analg.; var. *lokua* E. A. Bruce: u. in combns. to treat insanity (Tanz) (712) - **S. stuhlmannii** Gilg (tropAf): mupalapande; rts. u. admixed with other pls. in insanity (Tanz) (712) - **S. subcordata** Spruce (Br, Co): iraryrana; u. in prepn. of curare by Cauichanas; c. caracurine, mavacurine - **S. tietue** Lesch. = *S. ignatii* - **S. toxifera** Rob. Schomb. (Pan, Guian, Ve, Ec, Br): devil-doer (Guyana); urari; curare poison nut; st. bk. c. many alks, incl. caracurines, flurocurine, toxiferine derivs.; st. bk. chief source of "Brit Guiana" (or calabash) curare (or woorari) u. as arrow poison; also u. med. - **S. triclisioides** Bak. (trop wAf, As?): bowihi (nNigeria); frt. edible but somewhat nauseant; u. asemet. for children; sds. weak in alk. but poisonous - **S. trinervis** Mart. (S. triplinervia Mart.) (Br.): fava de S. Ignacia; cruzeiro; stated to be source of curare: sd. c. strychnine, brucine (ca.

1.8%) - **S. usambarensis** Gilg (tropAf): "kinyarwanda"; lf./rt./bk. c. 31 or more alks., incl. usambarensine, several dimeric alks. with curarizing props. (no O) (C-curarine, C-calebassine, afrocurarine, dihydrotoxiferine), u. to make poison arrows; frt. toxic - **S. vacacoua** Balli. (Mad): sds. c. bakanosin (glycoside) - **S. wallichiana** Stelld. ex. DC - *S. cinnamomifolia.*

Stryphnodendron barbatimao (barbatimam) Mart. (Leg.) (indig. to Minas Gerais, São Paulo, now thruout Br): barbatimão; avaremotemo; bk. of this lge. tree c. 18-27-38-46% tannin; red dye; bk. (cortex adstringens [brasiliensis]; écorce de virginité) u. astr. antisep. styptic, in diarrh., uter. hemorrhage; inf. u. in hemoptyses, for GC; ext. as gargle in pharyngeal sores; exter. for washing & flushing (prostitutes) (1955); in comm. tanning (called Casca Virginidade); wd. as lumber; lvs. (c. 6.7% tannin) u. as ton - **S. coriaceum** Benth. (neBr): barbatimão de nordeste; frts. eaten in dry periods by cattle, giving toxic effects, photosensitizing.

stucco: artifical marble made with $CaSO_4$, sand & pd. marble held together with gum arabic.

Stupa (BPC): jute tow; short fibers fr. *Corchorus* spp. in rolls; u. as absorbent - **stupe:** cloth or tow dipped in liquid (medicament or even warm water) & appl. to sore, wound, etc.; ex. **turpentine s.** (fr. Gr. stype, tow; *cf.* styptic) (not cognate with stoup: vessel, cup, etc. esp., for holy water).

Sturmhut (Ge): aconite. (lit. rain hat).

style: part of pistil which connects stigma & ovary; in drugs often found with stigma attached; ex. Zea (corn silk) - **s. appendages: s. arms:** divisions of the style, usually corresp. to the no. of carpels.

Stylochiton (Stylochaeton) natalensis Schott (Ar.) (sAf): rt. decoc. u. for chest dis., earache - **S. warneckii** Engl. (w tropAf): pl. u. as famine food; yellow dye.

Stylophorum diphyllum (Mx) Nutt. (Papaver.) (ne&cUS. not at all common wild, sometimes cult.): wood (or celandine) poppy; yellow (or horn) p.; rt. (Stylophori Radix) c. several alks., incl. chelidonine (as salt of chelidonic acid), stylopine, coptisine, sanguinarine, protopine, berberine, diphylline; narcotic, diur. lithontriptic, vulner; rt. ("large golden seal") u. as substit. for Hydrastis.

stylopodium: thickened glandular basil portion of style (ex. in Umbelliferae).

Stylopyga orientalis: *Blatta o.* (roach).

Stylosanthes biflora (L) BSP (S. elatior Sw.) (Leg) (eUS): pencil flower; p. plant; afterbirth weed; squaw weed (Miss [US] Negro); u. as uter. sed.; in cramps, emmen. (decoc.) - **S. erecta** P. Beauv. (S. guineensis Schum. & Thonn.) (trop wAf): lf. decoc. u. for cough; rt. decoc. invigorating ton. - **S. guayanensis** Sw. (Guat, Hond, Guianas): yerba del campo; pl. inf. u. for "pains in the side"; cult. as fodder, "stylo" (Austral) - **S. hamata** Taub. (sFla [US], WI, sMex. to Co, Ve): donkey weed; Caribbean stylo; hb. decoc. u. for colds, kidney dis. - **S. mucronata** Willd. (w tropAf, Ind; natd. Queensland, Austral): wild lucerne (Queensl); hb. inf. for cough (or smoked with tobacco for same); in infant diarrh., rt. male aphrod. (Tanz): fodder - **S. procumbens** Willd. (Br): u. as purg. diur. - **S. viscosa** Sw. (WI, Mex to Br, Pe; natd. w tropAf) "comino sabanero (Cu); poor man's friend; hb. decoc. (or inf. in rum) u. for colds, fever, headache, kidney dis., nervous dis., aphrod. (Af).

Styphelia malayana J. J. Sm. (Epiacrid.) (Mal): lf./rt/decoc. u. for body & stomach aches; lvs. appl. exter. as paste - **A. tameiameiae** (Cham.) Muell. (Haw Is [US]): pukiawe; bk. u. to make red & brown dyes. (713).

styptic cotton: c. saturated with $FeCl_3$ soln. - **s. weed:** *Cassia occidentalis.*

Stypticin™: cotarnine chloride.

Styracaceae: Benzoin Fam.; shrubs or small trees native to tropAm mostly, but sometimes found in other trop.; lvs. simple alternate; frt. berry, fleshy drupe, or with dry pericarp.; fls. regular in gps.; many taxa bear a balsam in secretory receptacles of bk. and wd.

styracin: cinnamyl cinnamate; found in storax, Sumatra benzoin, etc.

Styrax: (Styrac.) (worldwide trop. & subtr.); woody pls. with white fls. in racemes (1956) (L.) (Eng.): balsam fr. *Liquidambar orientalis & L. styraciflua*; sometimes u. (as distinct fr. "Storax") to indicate a rel. liquid prod.; arom. resin fr. *S. officinalis* (solid s.); Benzoin (sometimes so u. in Ge ref. books) - **S. americana** Lam. (seUS, n to Va & sIll): mock orange; spring o.; (American) snowbell; fls. arom.; u. as expect. arom. vulner. for chest dis. - **S. argenteus** Presl (Mex, CA): estoraque (Guat.); resino (CR, Nic); gum exudn. burned as incense in churches; bk. u. in Salv as fish poison - **S. aureus** Mart. = *Pamphilia aurea* - **S. benzoides** Craib. (Thai, Indochin [sVietnam]); Siam snowbell; along with *S. tonkinensis* the source of Siam benzoin; formerly u. only by natives (714) - **S. benzoin** Dryand, (Indones, exp. Java, Sumatra, Sunda Is; Mal): Benjamin tree; chief source of the off. Sumatra Benzoin (other sources: *S. paralleloneurus, S. sumatranus*); this balsam c. resin with benzoresinol cinnamate, triterpenoid acids; sumaresinolic acid & esters; VO with styracin (cinnamyl cinnamate), styrol, benzaldehyde, vanillin, phenyl propyl cinnamate; u. antisep. stim. expect. in cough; vulner.; fat preservative; in shellacs, textile printing; in inhalations & incenses; in perfumes; cosmetics ("virgin milk") (for skin fissures); sorts: Palembang, Padang, Penang (1970) - **S. camporum** Pohl (Br): st. source of pleasantly odorous resin (estoraque) u. as incense in churches (715) - **S. ferruginea** Nes & Mart. (Br, Bo): resin u. as for *S. camporum:* called "benjui" (benzoin) or "estoraque" (storax) - **S. formosana** Matsum. var. *hayataiana* (Perkins) Li (S. suberifolius Hook. & Arn. var. *hayataianus* [Perkins] Mori) (China): Hayata's snowbell; stated as source of Siam benzoin - **S. grandifolia** Ait. (seUS): "storax"; shrub/small tree; ornam. - **S. japonica** Sieb. & Zucc. (Jap, China, Kor, PI): ego-no-ki; snowbell; wd. useful; u. for umbrella handles; sds. source of FO (ego oil); ornam. shrub/small tree (-10m) (1968) - **S. martii** Seub. (Br): canella poca; bk. u. stim. antipyr - **S. obassia** S. & Z. (Chin, Jap, Korea): sd. kernels c. 18% FO. (1972) - **S. officinalis** L. (orig. Levant, Gr, later cult. It, Fr): storax tree; the storax (balsam) of the ancients; "solid storax" is prod. fr. bk.; do not confuse with the storax (styrax) fr. *Liquidambar* spp.; sometimes seen in commerce but no longer important; u. expect. perfumes; incenses (Jew's incense; prescribed by Moses) - **S. paralleloneurus** Perkins (seAs): one of several sources of Sumatra benzoin - **S. pearcei** Perk. var. *bolivianus* Perk. (Bo): source of a benzoin-like resin (Bolivian benzoin; estoraque) c. benzoic (little) & cinnamic (more) acids; u. as incense (1969) - **S. pohlii** A DC (sBr, Par): produces a resin like benzoin - **S. reticulata** Mart. (Br, Bo): as for *S. ferruginea* - **S. subdenticulata** Miq. (Sumatra): said source of Penang benzoin - **S. sumatranus** Smith (seAs, Indones): one of sources of Sumatra benzoin (*cf. S. benzoin*) - **S. tessmannii** Perkins (Co): balsam fr. tree trunk u. for painful dental caries, packing the resin (softened with heat) into the cavity (Tikuna Indians) (2299) - **S. tonkinensis** (Pierre) Craib (Thail, Vietnam+): Tonkin snowbell; one of chief sources of Siam benzoin (1967); many similar constits. as under *S. benzoin*; incl. phenyl prophyl cinnamate, resinotannol cinnamate, siaresinolic acid (siaresinol), benzaldehyde; coniferyl cinnamate, c. benzoate (chief constit.); free balsamic acids but less cinnamic acid than in Sumatra b.; triterpenoid acids (1974) (siaresinolic acids = siaresinol = 19-OH-oleanolic acid); vanillin; best sort in grains (*in graniis*); u. mostly in cosmetics, esp. for skin blemishes; fixative in perfumes; to preserve FO (esp. lard) (preferred to Sumatra b.) (1975); mfr. many toilet articles; int. as espect. exter. antisep. & disinfect. for itch; chafed nipples; wash for freckles; binding agt. for tabs; Tinct. B. u. to produce honey flavor; wd. u. in making wooden shoes, matches; other uses similar to those of *S. benzoin* - **S. warscewiczii** Perk. (CR): quitirrici; bk. inf. drunk for gastritis & other GI(T) complains; bk. decoc. appl. exter. to wound & ulcers.

styrax, American: s. fr. *Liquidambar styraciflua* - **s. balsam**: fr. *L. orientalis* - **s. calamita** (L): **s. calamitus**: storax calamita - **crude s.: s. crudus** (L): prod. fr. *L. orientalis* as supplied by collectors - **s. depuratus** (L): Storax Purificatus - **S. family:** Styracaceae - **Levant s.: liquid s.: S. Liquida** (L): prod. fr. *L. orientalis* - **S. Rubra** (L): storax bark - **solid s.:** *s. solidus* (L): resinous prod. fr. *S. officinalis.*

styrene: styrol: phenylethylene, hydrocarbon found in styrax, pine tar, benzoin; made synth.; u. in plastics & synth. rubber & resin mfr.

styrone: cinnamyl alc.

sua (Ec.): black dye prep. fr. *Genipa americana.*

Suaeda fruticosa Forsk. (S. verà Forssk. ex J. Gmelin) (Chenopodi) (nTemp. Zone incl. Med reg.): alkali seep-weed; ink bush; lf poultice appl. for ophthalmaia, sores; lf. int. emet.; wooly excrescence of branch tips appl. with empyreumatic oil to sores on camels' backs: camel fodder - **S. monoica** Forsk. (Ind, Pak): pl. incorp. into ointment for wounds - **S. physophora** Pall. (Russ): anthelm (folk med.); in heart dis. - **S. suffrutescens** Wats. (swUS): desert seep-weed; source of black dye (Am Indians).

suak(im), gum: gum suakin: v. of Acacia (gum), similar to Gum Talha.

suakwa: towel gourd, *Luffa* spp.

subcontinent: Indo-Pakistan.

Suber (L): cork: corky part of bk.; phelloger - **S. Quercinum** (L): cork fr. *Quercus suber.* (cork oak).

suberin: a waxy subst. made up largely of condensations of esters of polymerized fatty acids, esp. phloionic & phellonic acids; impregnates walls of cork cells, etc., & serves to waterproof same - **suberised:** impregnation or deposit of suberin in tissues.

subex: portion of st. with bracts or scales.

sublimation: process of vaporizing a solid & condensing it again to a solid without melting; ex. camphor.

sub-shrub: one some 0.6-1.0 m high.

subspecies: the primary or chief subdivision of a sp.; based on morphological differences, as result of geographical, ecologic, & other factors of environment.

substance, P: SP: neurotransmitter; an undecapeptide of the protein gp. called tachykinins (neuropeptides); disc. by von Euler & Gaddum (1931); occ. in brain & intestines of vertebrates; isolated fr. cattle brain; acts to contract extravascular smooth muscles; as vasodilator, depressant —> analgesia & hyperalgesia; sialagog; related to locomotion.

substances, primary plant: those rel. simple compds. implicated in the basic metabolism of the plant - **Secondary p. s.:** more elaborate compds. produced in the plant, frequently with med. or other uses; ex. alkaloids, VO, glycosides, tannins.

substitution, drug: replacement of drug by another similar material; may be illegal (adulteration) or legal (replacement of similar drug during wartime emergency, &c.).

substrate: substratum: 1) material upon which an enzyme reacts 2) chem., biol., or bacteriological medium 3) the substances (as soil) upon which a pl. grows.

subtilin: peptide antibiotic; (isol. 1944); prod. by *Bacillus subtilis* Cohn; occ. worldwide in atmosphere; u. formerly in amebic dys.; food preserv.

succata: citron of the grocers.

succedaneum: following, taking the place of, hence substitute.

succinic acid: Acidum Succinicum (L): Bernsteinsaeure (Ge); $(CH_2COOH)_2$; a dicarboxy acid found in many pls. & pl. prods. (resins); u. in asthma; u. in dye, perfume, lacquer mfr.; photography; in prepn. of salt substit. (succinates) (SC. 1st discov. in Succinum by direct distn. [Agricola, 1546]).

Succinimycin: antibiotic derived fr. *Streptomyces olivochromogenes* Wakeman & Henrici (1963).

succinite: Succinum (L): (gum) amber; Baltic a.; fossil resin fr. oleoresins of extinct pines (*Pinites succinifera* Goepp [*Pinus succinifera* Conwentz]) & many other plants (2496); dists. —> amber oil (Oleum Succini), u. like turpentine oil in liniments; formerly as spasmolytic in asthma; amber u. to make jewelry, tobacco pipe mouthpieces, amber colophony (u. for shellacs): pl. & animal inclusions often informative of fossil organisms; Electrum (L), amber, when rubbed generates static electricity (origin of name), off. in earliest Br Phar.

Succisa arvensis L. (Dipsac.) = *Knautia a.* - **S. pratensis** L. (Scabiosa succisa L.) (Dipsac.) (Euras, nAf, natd. locally eCana, neUS): meadow s.; devil's bit (scabious); St. Peter's root; blue buttons; hb. (as Herba Morsus diaboli) c. tannin; u. in diarrh. edema, exter in skin injuries & lesions; in tumors; inflammn., contusions, sores, ulcers, abscesses, wet ("weeping") eczemas, erysipelas, scabies; rt. u. astr. in diarrh.; as gargle in throat dis.; oral ulcers; blue dye; adulter of valerian; lvs. as tea substit. (early); in leukorhea; (douches); orally in stomachache, vertigo, toothache, paronychia (whitlow); fresh lvs. in salads (spring); various folk med. uses.

succory: *Cichorium intybus*:; lvs. & rts. u. - **garden s.:** *C. endivia* - **wild s.:** *Sabatia angularis.*

succotash (Eng fr. Am Indian): popular dish of cracked maize (75%), green (Lima) beans (25%), & pork boiled together.

succulent: desert plant with fleshy appearance due to water storage in st. & lvs.; seen in many *Crassula, Opuntia,* & *Euphorbia* taxa.

Succus Alterans (L): Alterative juice: fresh juices prepd. fr. *Phytolacca americana*, *Stillingia sylvatica*; *Smilax "sarsaparilla," "Lappa minor," Zanthoxyolum carolinianum*; originally prepd. by W. McDade of Montgomery, Ala (US); later marketed by Eli Lilly & Co.; u. in blood dis. incl. syph. (2713) - **S. Cerasi:** cherry j. - **S. Glycyrrhizae:** licorice juice (*qv.*) - **S. Limonis:** lemon juice; diluted with water & sweetened forms pure lemonade, u. as beverage, vit. C source, flavoring agt. in foods; formerly popular in preventing colds (1957), arthritis (-1954-), keeping "regular," &c. (USP, 1860-1900) - **S. Liquiritiae** = S. Glycyrrhizae. - **S. Pomorum:** fresh apple j.; sweet cider - **S. Rubi idaei:** raspberry j. - **S. Thebaicus:** Opium - **S. Viridis:** sap green, dr. juice of *Rhamnus cathartica* (pigment).

Suchtmittel (Ge): addictive agent.

sucker: sucoir (Fr. m): new shoot arising from the rts. of a pl.

sucrase: saccharase: invertase: enzyme wh. aids in hydrolysis of sucrose.

sucre (Fr-m.): 1) sugar 2) sucrose.

sucrochemistry: chem. of sucrose & its derivatives.

sucrose: Sucrosum (L): cane sugar; beet sugar; Saccharum (L); saccharose; a disaccharide sugar commonly found in pls., ex. sugar beet, carrot, sugar maple, *Madhuca indica*, in fact all chlorophyll-bearing pls.; first u. medicinally; u. sweetener, nutr., in candies, rock candy (*qv.*), &c.; appl. directly to varicose & other ulcers, wounds, chronic bed sores (decubitus ulcers) (appl. after sterilization of wound, &c.) promotes healing (716); preservative; dem., antioxidant; excipient for tabs.; tablet coating; in fermentaton industries (2013); incorporation of vit. A (for serials re. sucrose v. App. C-1) - **s. polyester** (SPE): nonabsorbable lipid ("no-fat fat"); synth. fat substit.

u. in diets for wt. loss (P & G, Cinc.) & to reduce blood serum cholesterol (1987-) ("Olestra").

sucupira (Br): *Bowdichia nitida, B. virgilioides*

Sudafed™ (BW): pseudoephedrine.

Sudan (, The): republic of neAf, s of Egypt; traversed by the Nile River; chief source of Acacia (mostly fr. Kordofan province); also produces senna, *Hyoscyamus muticus,* and minor pl. drugs, as *Hibiscus sabdariffa, Albizzia anthelmintica, Adansonia digitata,* &c. - **S. grass:** v. under grass.

suelda consuelda (CR): *Commelina communis*: (Ve): *C. virginica; C. erecta* (Mex); also spp. of *Callisia, Anredern,* &c.

Suessmost (Ge): unfermented preserved frt. juices, preserved by pasteurization & bacterial filtration, CO_2 atmosphere; addn. preservatives; u. as beverages; mild lax., vitamin source.

Suesswurzel (Ge.): licorice.

suet: mutton s.: prepared s.: purified abdominal fat of sheep, *Ovis aries,* prep. by melting & straining; c. oleic, stearic, palmitic, linoleic, & myristic acids as glycerly esters; u. as base for ointments & cerates; has firmer consistency than lard on account of more stearin; benzoinated s. is made by a mixture of 3% benzoin.

sue-pow: 1) sponge u. in cleaning & cooling an opium pipe 2) the opium extr. fr. this sponge; often re-used.

sugar: 1) class of soluble carbohydrates with sweet taste 2) (specif): sucrose (named derived fr. Sarkara, Sanscr) - **s. acid:** oxalic acid - **African s. cane:** *Sorghum bicolor* - **s. alcohol:** ethanol prep. fr. sugar cane molasses (esp. in Cu & PR) - **bar s.:** finely granulated s. u. in mixed drinks - **barley s.:** sucrose melted slowly & then left to solidify - **beet s.**: sucrose fr. *Beta vulgaris,* sugar beet - **brown s.:** currently, pure sucrose + molasses (light b. with 12% m.; dark b. with 13% m.); formerly unrefined s. preferred over white s.; u. to stop bleeding (folk med.) - **s. bush:** 1) grove of s. maple trees 2) *Protea* spp. - **cane s.:** sucrose fr. sugar cane, *Saccharum officinarum* - **cariogenic s.:** one wh. favors formation of dental cavities when u. in foods, incl. sucrose, fructose, lactose - **carob s.:** s. extd. fr. frts. of *Ceratonia silique* - **Chinese s. cane:** *Sorghum bicolor* (S. vulgare, v. saccharatum) - **coarse powdered s.:** u. by bakers: to make tablets - **s. color(ing):** caramel - **common s.:** sucrose - **compressible s.**: sucrose 95-8% with starch, malto-dextrin, or invert sugar added (NF); u. in making compr. tabs - **confectioner's s.**: finely pd. sucrose + 3% starch (to prevent caking) (3x sugar) - **corn s.:** glucose - **cube s.:** sucrose in cubes for table use in beverages - **cut loaf s.:** loaf s. - **date (tree) s.:** palm s. - **Demerara s.:** raw

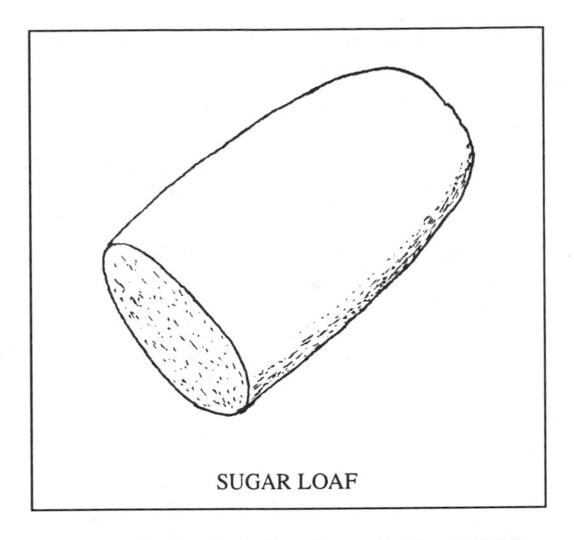

SUGAR LOAF

cane s. in lumbs (<D., Guyana) (Cu+) (717) - **fruit s.:** levulose - **granulated s.:** sucrose in large crystals - **grape s.:** dextrose (glucose) - **s. grove:** gp. of s. maple trees tapped for prodn. of maple s. - **s. gum (tree):** 1) *Eucalyptus cladocalyn, E. corynocalyx* 2) (US) loosely appl. to sugar maple, sweetgum, &c. - **honey s.:** dextrose - **horse s.:** *Symplocos tinctoria* - **invert s.:** mixt. of equal pts. of glucose & levulose; prepd. by hydrolysis of sucrose; occ. in honey, frts., &c.; does not crystallize - **"Kleenraw s.":** pure sucrose + 5% molasses - **liquid s.:** solns. of sugars; 2 types ("medium invert" = mixt. of sucrose & invert s.; sucrose) - **s. loaf:** conical cake of sucrose shaped somewhat as in adjoining figure - **malt s.:** maltose - **maple s.:** sucrose fr. *Acer saccharum, A. rubrum, A. platanoides* - **milk s.: s. of milk:** lactose fr. *Bos taurus* milk (whey) - **s. nuts:** peanuts with added sugar - **palm s.:** sugar fr. sap of one of several palm spp - **s. plum:** 1) *Amelanchier canadensis 2) Uapaca guineensis* 3) kind of candy in small balls; bon bon - **s. polyester:** sucrose p. (*qv.*) - **powdered s.:** confectioners s. - **refined s.;** purified sucrose in usual large crystals - **s. sand:** 1) crude lactose 2) prod. separating & left as residue on evaporaton of sugar maple juice; mostly malic acid - **s. spheres** (NF XVI): 62-91% sucrose; base chiefly starch but may have added color; these small pellets are u. in capsules for deposition in layers & coating (to limit absorption rate of medicinal) - **starch s.**: glucose - **s. substitutes:** sorghum; honey; maple sugar; boiled down watermelon juice; boiled down corn cobs: saccharin; dulcin &c. - **"sweeter s.":** levolose, some 1-1/2 times sweeter than sucrose - **s. tree:** (Oh valley [US], esp. WVa to NCar): *Acer saccharum* -

turbinado s.: Demarara s. - **wood s.:** xylose - **yellow s.:** sucrose with a small amt. of molasses.

sugarberry: *Celtis* spp.

sugo (It): juice.

Suidae: Hog Fam.; thick-skinned hornless omnivorous mammals which do not chew the cut; ungulates; ex. wart-hog

suif de mouton purifié (Fr): prep. suet.

Suillus bellini Watl. (Boletus b. Inz.) (Bolet. - Boletales - Hymenomyc. - Basidiomycet - Fungi) (Eu+): edible mushroom, not well known (1958); homeop. med.

suint: s. de laine (Fr.): impure (or crude) wool fat, representing dr. perspiration of sheep deposited on wool; u. antisep.

sukkar (Heb.): sukker (Nor, Dan): soukar (Arab): sucrose, sugar.

sulfite pulp: prep., by digesting wood chips in a soln. of SO_2 & $Ca(HSO_3)_2$: this solvent removes the lignin & leaves fairly pure cellulose; source of wood cellulose (c. wadding); component of newsprint, book papers, etc.

sulfonated bitumen: Ichthyol; ammonium ichthyolsulfonae obt. by treatment of dist. prod. fr. mineral deposits (a peculiar bituminous shale) found in the Tyrol (Austria); this thick brown subst. has been much u. as emoll. stim. antisep., etc., in skin dis., rheum., etc., in ointments, dermatological soaps; cosmetics; employed both in regular & in veter. med.

sulfur: suphur: S.; a volcanic material element; u. fumigant, insecticide, in mfr. s. compds., etc - **S. ballinum** (L.): S. Vivum - **s. dioxide:** sulfurous (acid) anhydride; SO_2, colorless acrid non-flammable intensly irritant gas; produced in combustion of sulfur, some coals, &c.; available in cylinders (compressed); u. as preservative (frts., veg., &c.); u. as disinfectant, fungicide, bactericide, fumigant for insects, &c.; bleaching agt. (glucose, sucrose, &c.); antioxidant to prevent darkening of frt. &c.; atm. pollutant; chief cause of acid rain - **s. flowers:** sublimed s. - **S. Lotum** (L): washed s. - **S. Nigrum:** S. Vicum - **precipitated s.:** finely pd. S. obtained by pptn. fr. Ca sulfide with acid - **roll s.:** sublimed s. which has been melted & cast as a cylinder - **sublimed s.:** obt. by a process sublimation; as fine xall. pd., slowly sol. in CS_2 - **vegetable s.:** Lycopodium - **S. Vivum:** crude subst. left behind in retort after re-sublimation of crude s. - **volcanic s.:** native s.; often found in vicinity of volcanoes - **washed s.:** sublimed s. washed with dil. ammonia water, then dist. water, strained, & dr..

sulfure (Fr): **sulfuret:** sulfide - **sulfuric ether:** ethyl ether, $(C_2H_5)_2O$ - **sulfur-weed:** *Peucedanum officinale.*

sulla (Malta): *Hedysarum coronarium.*

Sulphas Americanus Australis. (L): quinine sulfate.

sulphur: previous common spelling of sulfur, still u. in Britain.

Sultan: v. under coffee.

sumac(h): sumack: usually *Rhus* spp., sometimes specif. *R. copallina* or *R. glabra* (rt. bark comml. item) - **American s.:** 1) divi-divi, *Caesalpinia coriaria* 2) red s. - **s. berries:** red s. frt. - **black s.;** *R. copallina* - **Chinese s.:** *R. chinensis, Ailanthus altissima* - **climbing s.**: *Toxicodendron toxicarium* - **dwarf s.:** black s. = **s. family:** Anacardiaceae - **flame leaf s.:** black s. - **hairy s.:** *R. typhina* - **Pennsylvania s.** red. s. - **poison s.:** *Toxicodendron vernix* - **red s.: scarlet s.: sleek s.:** *Rhus glabra* - **staghorn s.:** hairy s. - **swamp s.:** 1) (usually) poison s. 2) climbing s. - **sweet s.:** 1) *Rhus aromatica* 2) *Schmaltzia* spp. - **tanner's s.:** *Coriaria myritifolia* - **upland s.:** red s. - **white s.:** red s., but may sometimes refer to poison s.

sumaresinol(ic acid): 6-OH-oleanolic acid; occ. in Sumatra & Siam benzoin.

Sumatra: large island of Indonesia v. under benzoin.

Sumbaviopsis albicans (Bl.) J. J. Sm. (monotypic) (Euphorbi.) (seAs): st. lvs. u. to treat fever (Laos, Indochin).

sumbul (root): Sumbulus moschatus (L.) (Homeop.): *Ferula sumbul.*

summer: v. under adonis: savory.

summer farewell: *Kuhnistera pinnata* - **s. sweet:** *Clethra alnifolia.*

Summitas (plur. **Summitates**) (L): top(s), consisting of sts., lvs., usually with fls. and/or frts.; ex. henbane - **S. Fabianae:** S. Pichi - **S. Juniperi**: of *Juniperus communis* - **S. Pichi:** of *Fabiana imbricata* - **S. Sabinae:** of *Juniperus sabina.*

Sumpfgarbe (Ge): *Achillea ptarmica.*

sumpweed: *Iva annua* var. *macrocarpa* Jackson.

sunburst: (seUS): *Gaillardia* spp.

"sun-chokes" (sUS slang): Jerusalem artichokes (fr. sunflower artichoke).

Sun cholera mixture: Mixt. of Opium & Rhubarb Comp. (NF V) (SC first published in *N. Y. Sun*).

sun drops (US): various taxa of Onagraceae, specific. *Oenothera fruticosa.*

Sunda Islands: Soenda Is; a chain of large islands in the southern part of Indonesia (formerly Netherlands, EI), incl. Burneo, Sumatra, Java, Timor; Celebes, important prodn. area for many drug plants, spices, rubber, &c.

sundew: *Drosera* spp. - **English s.:** *Drosera anglica* - **s. family:** Droseraceae - **long (leaved) s.:** *D.*

longifolia - **round-leaved s.:** *D. rotundifolia*: usual source of s. herb. (drug).

sunflower: 1) *Helianthemum nummularium* 2) **common sunflower:** *Helianthus annuus*+ - **s. family:** Compositae (Asteraceae) - **false s.:** *Helenium autumnale* - **garden s.:** 1) common s. (also as **common g. s.**) 2) *Helianthemum nummularium* - **Indian s.:** *Carthamus tinctorius* (sd.) - **little s.:** garden s. (2) - **Mexican s., showy:** *Tithonia speciosa*+ - **rough s.: wild s.:** *Helianthus divaricatus.*

sungazer (US slang): sunflower.

sunlight: sunshine: exposure to s. in moderation is u. to reduce daytime fatigue; excessive exposure may result in **sunstroke** (marked by high temp.; convulsions, coma); chronic overexposure results in some individuals in skin cancer.

Sunn-hemp: *Crotalaria juncea.*

superoxide dismutase: SOD: enzyme of body cells wh. opposes excessive oxidn.; by preventing free radicals fr. developing & scavenging H_2O_2 in tissues; said to reverse aging process in osteoarthritis, &c.

superphosphate: fertilizer prod. c. 18-20% available P_2O_5, made by treating phosphate rock with sulfuric acid (**normal** or **ordinary**); or c. 43-48% available P_2O_5 by treating p. rock with phosphoric acid (**triple**).

Suppe (Ge): soup.

suppressor cells: leukocytes of the T-cell lymphocyte group, which suppress the immune reaction.

suprarenal (gland), desiccated: Suprarenalum (L): adrenal gland; the dr. partially defatted & pd. suprarenal gl. of animals u. as food by man; c. small amts. of epinephrine, corticosterone, etc.; sometimes u. in treatment of Addison's dis., altho potency very low or null; made up of medulla (secreting epinephrine, &c.) and cortex (many steroidal hormones).

suprarenalin™: epinephrine.

suprasterols: terminal irradiation products of ergosterol beyond calciferol; with no antirachitic properties.

sureau (Fr-m.): *Sambucus* spp., esp. *S. nigra.*

sure-death:. *Amanita phalloides.*

surgeon's: v. under agaric.

Surgibone: sterile bone & cartilage of young calves & calf embryos; processed to reduce antigenicity; u. in reconstructive surgery.

surgical dressing: prods. composed of such fibrous materials as cotton, wool, sphagnum moss, etc., often in the form of fabrics; designed to protect a skin lesion, to apply disinfectant or vulner. substances, to absorb secretions, etc. - **S. gut:** cat gut (fr. sheep intestine); u. as suturing material of the absorbable type - **s. silk:** fibrous prod. of the silk worm, *Bombyx mori* (etc.) - **s. suture: s. thread:** u. for surgical sewing, etc.

Suriana maritima L. (monotypic) (Suriani) (worldwide coasts): bay cedar; fls. u. for bloody vomiting & diarrh. (powd. & mixed with coconut water & milk) (Caroline Is); lf. branches decoc. u. as bath for arthritis (Cu); bk. & lf. decoc. u. to clean old sores, lvs. eaten with "atole" for rectal hemorrhage (Yuc, Mex) (2463).

Surinam(e): formerly Dutch Guiana; self-governing territory of the Netherlands; v. under quassia.

surinamine: andirine: N-methyl-tyrosine; amino acid, at times classed as alkaloidal amine; occ. in *Andira retusa*, rhatany bk., & other Leg., & other fams.

surma (Indo-Pak): **surmah**: paste appl. to eye lids (& eye?) ("eye collyrium") c. native antimony trisulfide (Sb_2S_3) (stibnite) mixed with several oils; said to have a cooling effect; not hazardous to health as used; (al) kohl (Arab) (2717).

suron (Fr-m.): ceron; a kind of bag or bale.

Sus scrofa L. v. **scrofa** (v. domesticus Gray) (Suidae - Artiodactyla - Mammalia) (Eu): the domesticated hog; the source of pork meat prods. (largest usage of all meats) (roasts, chops, bacon); dog food; lard, pd. stomach, pepsin, pancreatin, liver ext., endocrine glands (insulin, heparin, prolactin), etc., hog bristles (fr. Chin, Sib); u. for brushes (come in wooden boxes; malodorous); heart valves u. in card. surgery; to replace human skin after burns, other injuries; pig skin u. as leather (footballs); brain u. to aid memory (Chin med.) (2012; muzzled hogs u. to hunt truffles (underground mushrooms).

sushi (Jap): popular cold sea food meal contg. fish (eel, octopus, squid, shrimp, tuna) (sometimes living), seaweed, & vinegared rice.

suesse Mandeln (Ge): sweet almonds.

Süssholz (Ge): licorice.

Sutera atropurpurea Hiern (Scrophulari) (sAf): Cape saffron; has been comm. source of saffron (substit.) (sAf); fl. u. as antispasm., anodyne, stim.; lf. antispasm.; flavor and coloring agt. - **S. burkeana** Hiern (sAf): lf. inf. u. as local appl. in conjunctivitis (Zimbabwe) - **S. filicaulis** (sAf): lf. cold water inf. u. for chest dis. - **S. floribunda** O. Ktze. (sAf): rt. decoc. u. for chest colds in children - **S. hispida** Druce (sAf): u. like prec. - **S. pinnatifida** O. Ktze. (Af): lf. juice appl. locally to sore eyes (sheep); forage for stock.

suterberry: *Zanthoxylum americanum* (bk. & berries).

Sutherlandia frutescens R. Br. var. *incana* E. Mey. (Leg): (sAf): cancer bush; inf. u. int. for cough, chest colds, douche for uter. prolapse; gastric dis.; cancers; popular among both natives & Eu settlers - **S. humilis** Phill. & Dyer (sAf): kankerbossie (Afrikaans); u. int. for cancer & for flu; bk. & lf. u. similarly - **S. microphylla** Burch. (sAf): lf. inf./decoc. u. int. for gastrointest. dis., cancer, uter. troubles, liver dis., rheum. & many other uses (129).

Sutilaina: Sutilains: crude proteolytic enzyme fr. *Bacillus subtilis*: u. exter. for débridement of dirty wounds & sores (esp. deep burns), in ointments & liniments (USP XVI) (Flint); tested on casein.

sutures: fine cords u. in surgery for closing wounds, operational areas, incision, etc.; of two types: **absorbable** (ex. catgut) which are not removed; **non-absorable** (ex. silk, wire), which are generally removed after healing is complete.

swale: marshy ground, a hollow, &c.

swallowroot, orange: *Asclepias tuberosa* - **gigantic s.:** *Calotropis gigantea* - **white s.** *Cynanchum vincetoxicum.*

swamp: low, flat, wetland area with a few trees & shrubs & fresh water; seasonally flooded bottomland - **s. beggar's tick:** *Bidens tripartitus* - **s. elder (root):** may be s. alder is intended - **s. hairstrong:** *Peucedanum palustre* - **s. root:** 1) *Xanthorrhiza simplicissima* 2) *Lysichton americanum* 3) *Eupatorium purpureum* 4) (US slang): illict alc. liquors usually of inferior quality - **s. sow fennel**: *Peucedanum palustre* - **s. weed:** *Selliera radicans.*

swamp: v. under aster, cabbage, dogwood, elder, helebore, laurel, milkweed, sassafras, sumach, valerian.

swampseed (US slang): rice.

Swartzia J. F. Gmel.: *Solandra* - **S. brachyrhachis** Harms (Leg) (Br): sds. crushed & mixed with food to kill intest. parasites (718) - **S. cabrerae** Cowan (Co): lvs. mixed with farina & eaten for use as vermifuge (718) - **S. conferta** Spruce (Co): st. decoc. u. for diarrh. (Tukano Indians) (718) - **S. decipiens** Holmes (SA): lvs. u. to adulter. Pilocarpus - **S. madagascariensis** Desv. (warm Af): snake bean; widely u. in med., bk. for diarrh; frt. emet.; lf. for schistosoma-carrying snail; pericarp & sd. insecticide (esp. for termites); sd. pod inf. for gastric dis.; rt. raw or as inf. sex stim.; in TB (Tanz) (712); frt. juice for middle ear suppuration; bark abort.; fish poison - **S. microcarpa** Spruce (Co): frt. inf. for diarrh. - **S. nitida** (Zucc.) Standl. (sMex): copa de oro; water contained in calyn before flower opens appl. as collyrium in ophthalmia (2342); widely cult. as ornam - **S. schomburgkii** Benth. (Co): inf. of young pods effective as vermifuge - **S. simplex** Sprengel var. *grandiflora* Cowan (Co): c. alk.; crushed boiled lvs. rubbed on abdomen for "liver inflammn" (718); the following *Swartzia* spp. are u. as fish poisons: *S. auriculata* Poeppig (Co); *S. fistuloides* Harms (Gabon, Af); *S. pendula* Spruce (Co) (pounded bk. u. by Witoto Indians); *S. schultesii* Cowan (Co); *S. sericea* Vogel (Co).

sweat: perspiration (refined term); sudor (L); liquid prod. by eccrine sweat glands of skin; salty, usually acid (pH 4.5-7.5); human s. formerly u. in med. (listed in older Br Phar.).

sweat plant (eCana): **sweating p.**: *Eupatorium perfoliatum* - **s. root:** *Polemonium reptans.*

sweet, abdominal (or **stomach**): pancreas - **s. bracken:** *Myrrhis odorata* - **s. brake:** Aspidium - **s. breads:** glands of mammals (esp. calf) (thymus [immature animal], brain [Ind]), pancreas, testes, &c.) u. for food - **s. cane:** Calamus - **s. cicely**: *Osmorhiza longistyia* - **s. fern:** s. bracken - **s. gale (herb)**: *Myrica gale; M. asplenifolia* - **s. leaf:** *Symplocos racemosa; S. tinctoria* - **s. meadow:** *Spiraea salicifolia* - **s. pepperbush:** *Clethra* spp. - **s. pod:** *Ceratonia siliqua* - **s. root:** 1) *Osmorhiza longistylis* 2) Calamus - **s. scented female fern:** s. bracken - **s. shrub** (seUS): *Calycanthus floridus* & other *C.* spp. - **s slumber root:** Sanguinaria - **s. sop:** *Annona squamosa* - **s. sub**: (Ala [US] slang): sweet shrub - **s. suckle** (US, Negro): honeysuckle - **s. tree:** *Acar nigrum* - **s. weed:** Althaea - **s. William, wild**: *Phlox divaricata* - **s. wood:** licorice, Cascarilla.

sweet: v. also under almond, balm, balsam, basil, bark, bay, birch, brier, bush, cassaba, cherries, chocolate, clover, cinnamon, cumin, elder, elm, fennel, flag, gum, haw, laurel, life everlasting, marjoram, myrtle, oak, oil, orange (peel), pellitory, potato, rush, scabious, scented bay, scented goldenrod, scented life everlasting, scented virgin's bower, scented water lily, smelling trefoil, sumach, vernal grass, wine, woodruff.

sweet(e)ning (seUS): sugars - **long s.:** molasses - **short s.:** sugar - **sweetheart:** *Talinum triangulare* - **sweethearts** (Austral): *Galium aparine.*

Sweetia elegans Benth. (Leg) (Br): perobinha campestre (Braz Phar.); perobinha; rt. bk. u. like cinchona; also in dysmenorrh., asthma, epilepsy; wd. u. in carpentry, joinery; to make charcoal - **S. fruticosa** Spreng. (Br): sucupira; wd. u. as timber - **S. panamensis** Benth. (WI, sMex, to Co, Ve): Honduras (or bitter) bark; quina silvestre; Billly Webb (sweetia); huesillo (Mex): off. as Cascara Amarga (NF, V, VI); bk. c. sweetine (picramnine), decoc. u. in coughs & colds; mal.; syph.; antidiabetic (Pen): diur., in skin dis. (scrofula); simple bitter (1142).

sweetmeat: any dainty food with sweetener (esp. sugars).

Swertia alata Royle (Gentian) (nwHimalay, Pak, Ind): hb. u. as adulter. of *S. chirata*; less bitter - **S. angustifolia** Buch. Ham. (Himalayas, Pak, Ind): similar to *S. alata* - **S. caroliniensis** Ktze. - *Frasera c.* - **S. chinensis** Franch. (Chin, Kor, Jap): hb. (as Herba Swertia, Jap Phar.) u;. as bitter ton. astr. anthelm.; u. in Jap since ancient times - **S. chirata** Buch.-Ham. (S. chirayita Lyons) (nIndo-Pak, Nepal); chiretta (plant); chirayata (Hind.) East Indian balmony; hb. u. as simple bitter. stom., ton.; antiperiod. in mal.; anthelm.; in yellow fever: hop substit.; rts. dr. boiled & made into soup with NaCl u. as remedy in fevers & flu (Sikkim); to aid digestion; cult. - **S. cordata** Clarke (Pak): hb. c. ursolic acid, mangiferin, &c.; u. as substit. for *S. chirata* - **S. japonica** Makino (Jap, Kor, Chin): Japanese chirata; semburi; hb. u. as bitter stom.; decoc. in diarrh. dys.; off. in Jap Phar. (To-Yaku); u. rather like *S. chinensis*: c. swertiamarin acet. (bitter princ.) u. as hair ton. - **S. javanica** Bl. (Java, Sumatra): hb. u. as stom. - **S. lutea** Vest. = *Gentiana l.* - **S. paniculata** Wall. (Ind): u. as substit. for *S. chirata* - **S. perennis** L. (temp. Euras): hb. c. gentiopicrin, swertiamarin; u. as bitter - **S. pulchella** (Sw.) Druce = *Centaurium pulchellum* - **S. punicea** Hemsl. (Chin): hb. eaten cooked with port as ton. for deafness - **S. purpurascens** Wall. (temp. Pak, Ind): u. as substit. for *S. chirata.*

Swietenia Jacq. (Meli) (tropAm): mahogany (trees); large trees with dark red wd. furnishng the valuable mahogany of commerce; finest cabinet wd.; u. mostly as a veneer wd.; in some spp., gum exudes fr. truck - **S. humilis** Zucc. (Mex, Guat): cobano; rosadillo; sds. toxic, produce emesis & catharsis; in nephritis; inf. u. for chest pains; bk. u. bitter ton., astr. antipyr.; minor source of mahogany wd. (Mexican m.) - **S. macrophylla** G. King (sMex to Pe, Br, natd. seAs); Honduras mahogany; aguan; c. swietenolide (acatate) (NBP) (719); bk. decoc. u. as antipyr. in chest dis., nervousness; sds. as hypotensive (Mal); sds. in ointment to restore hair(?); important lumber tree valuable for wd. ("Honduras mahogany") with many uses (furniture, cabinet work, pianos); cult. - **S. mahagoni (mahogani)** Jacq. (WI, CA, SA, s to Pe; cult. OWtrop; Calif, Fla [US]): (Barbados or West Indian or Cuba) mahogany u. for furniture (1959); rosadillo; bk. decoc. u. for catarrh., tetanus(!), aphrod. abort. febrifuge (Jam); as dye; sd. c. purg. FO; chief source of mahogany wd.; bk. prod. a catechu-like gummy ext. marketed in Ind.

swift: birds of fam. Micropodidae; fly at high velocity hence name; nests are made of materials glued together with mucilaginous salivary secretion; the cleansed nest of oriental spp. (great caves of Indonesia, Thail, &c.); u. to make "bird's nest soup," a great delicacy of Chin (prepd. with pork & other meats) (2288).

swine: members of fam. Suidae (ungulates), incl. hogs, boars, pigs, &c. (2357) - **s. bread:** *Cyclamen purpurascens* - **s. family:** Suidae.

Swinglea glutinosa (Blanco) Merr. (Rut) (PI): u. locally in med. for skin dis.; cult. ornam.

Swiss: v. under chard.

switch cane: *Arundinaria tecta* (sUS) - **s. grass**: *Panicum virgatum.*

swizzlestick: *Quararibea fieldii.*(SC twig twirled between hands in glass to make "swizzle" (mixed drinks; chocolate).

sword-fish: *Ziphias gladius* (Xiphiidae, Pisces): large marine fish with extension of upper jaw; important game & food fish: formerly one source of percomorph liver oil; flesh a luxury food.

Syagrus amara (Jacq.) Mart. (Rhyticocos amara Becc.) (Palm) (WI): overtop palm; apical bud ("cabbage") formerly used as source of wine; frt. u. to make fermented drink; kernels edible; also source of FO mixed with grd. sds. of *Bixa orellana* to make body paint; young lvs. u. as roof thatch (Domin, Caribs) (2328) - **S. capitata** Glassman (Butia c. Beccari; Cocos odorata Barb. Rodr.; C. leiospatha Barb. Rodr, s. l.) (nom. nudum) (sBr, Ur): butia (do campo) (Br); cabecudo (Br); Brazil(ian) b. palm; SA jelly p. (SC u. to prep. jellies); u. much like *S. coronata* (1748) - **S. comosa** Mart. (Br): babão; "palmito," palm heart (apical bud) bitter & u. as stom.; frt. pulp diur.; sd. endosperm roasted & u. in diarrh.; frt. pulp u. to prep. refreshing beverage (720) - **S. coronata** (C. Mart.) Beccari (Cocos c. Mart.) (Br): SA jelly palm; Brazil(ian) butia palm; (o)uricury; urucury; butia (do campo) (Br): licury (palm); sd. kernel c. 66% FO, much u. for cooking fat & in mfr. of soaps; in prepn. of a jelly; lf. furnishes a wax (ouropaido wax) much like carnauba wax; u. for protective coatings; illumination; weaving thread lubrication; thickening of ointments (154) - **S. leptospatha** Burret (sBr): butia do campo; frt. c. FO, vit. A; u. to prepare a liqueur - **S. macrocarpa** Barb. Rodr. (Br): Maria rosa; sd. kernel c. 25-28% FO u. as food (cooking) oil - **S. oleracea** (Mart.) Beccari (Br, Par): guahiro; "palmito" eaten & u. as ton. carm. stom. in hysteria; frt. pulp edible; endosperm of sd. c. FO u. in cooking (720) - **S. picrophyllla** Barb. Rodr. = *S. oleracea* - **S. romanzoffiana** Glassman (Arecastrum romanzoffianum Becc.) (Br, Par, neArg, Ur): queen palm; pindo; "palmito" edible altho bitter & eaten by poor people (also frt. & shoots [cogollos]); furnishes sago; frt. with gummy fibrous pulp & sweet flavor (1012), makes good forage; frt. as syrup or wine supposed useful for chest dis.; ornam. tree - **S. schizophylla** Glassman (Br): ariciryroba; aricury (palm); juice of unripe frt. u. for ophthalmia: cult. in Ind - **S. yatay** Glassman (Arg, Ur): yatay; fat very nutritious; u. in folk med.

sycamine fruit: *Morus nigra,* black mulberry.

sycamore: 1) US: *Platanus occidentalis* 2) (Eng): **s. maple:** *Acer pseudoplatanus* 3) (Med reg.): *Ficus sycomorus* - **false s.:** *Platanus occidentalis.*

sycomore: sycamore (721), *Ficus sycomorus.*

syconium: syconus: a multiple frt. consisting of succulent torus on the interior of which is a large cavity cont. the akene-like individual frts.; ex. fig (really an inside-out multiple frt.).

Sydenham's laudanum: wine of Opium.

Sylphium: plants u. medic. by the ancients; represents a *Ferula* sp. now extinct; u. as antipyr. & sudorific.

sylvestrene: monoterpene of *PInus sylvestris* VO.

sylvic acid = abietic a.

sylvie (Fr): *Anemone nemorosa.*

symbiont: symbiote: organism involved in a symbiosis (beneficial assn. of organisms).

sympathetic: relating to the **sympathetic nervous system**, one of the two autonomic n. systems, the other being the parasympathetic.

sympathin: mixt. epinephrine and norepinephrine (sepd. fr. the spinal cord) or one or the other of these, usu. latter; term now in disuse; sometimes appl. to sympathomimetic prods. of epinephrine.

sympatholytic: antagonistic to the activity prod. by stimn. of the sympathetic nervous system; ex. ergotamine gp. of drugs.

sympathomimetic: relating to drugs wh. produce effects simulating thos. prod. by stimn. of the sympathetic nervous system; ex. epinephrine, ephedrine.

sympatric: 2 or more plants or animals living in same area or overlapping areas.

Symphonia fasciculata Baill. (Guttif./Clusi.) (Mad): sds. rich in FO, u. as food & as cosmetic agt. - **S. globulifera** L. f. (WI, CA, SA, s to Br; natd. tropAf): cerillo (CR); hog gum tree; "boar wood" (Jam); trees tapped by Am Indians for wax-like gum-resin, called hog gum, mani canani, or anani, u. as purg., also as surrogate for Caranna, u. as ton. & vulner. healing balsam, also illuminant (722); appl. to dislocations, broken bones (Ve); resin mixed with beeswax & charcoal u. as cement for arrowheads & as wd. glue; substit. for cobblers wax (Br); as matrix; u. for torches & plastics in dancing masks (Co); for calking boats; int. as drastic; hog or doctor's gum is actually a resin; gamboge substit. in comm.; frt. eaten (hog plum) for flatus (Hond): bk. inf. u. to bathe ulcers (Guyana); ashes of bk. appl. to wounds (Makuna Indians, Amazon) (906); frt. edible; wd. hard & heavy, useful; waika chew stick (Belize) - **S. microphylla** R. E. Schultes (Co): main; resin u. much like preceding - **S. utilissima** R. E. Schultes (Co): u.; as prec. esp. for calking dugout canoes.

Symphorema polyandrum Wight (Verben.) (nwInd): tapa-tonda; fls. eaten for leukorrh.

Symphoricarpos albus (L) S. F. Blake (Caprifoli.) (Cana & nUS): snowberry; St. Peter's wort; c. saponins, tannins; rt. u. as mild skin irritant, int. to increase defensive vigor of body, ton. antiluent.; in hyperemesis gravidarum; ornam. shrub: berries toxic: not eaten by birds or rodents - **S. longiflorus** A. Gray (wUS: Calif, Ore, e to Utah & wTex): (desert) snowberry; waxberry; pl. decoc. u. for indig. or stomach pains (Nev [US] Indians) - **S. occidentalis** Hook. (cent. US & Cana): wolf berry; rt. u. as astr. ton. shrub 0.3-1.1 m high, growing on knolls of badger holes commonly; lvs. to 1 dm; buckbrush; dull greenish white berries eaten by mammals & birds (*S. albus* fruit larger, white) - **S. orbiculatus** Moench (c&eUS, Mex): Indian currant; coralberry; red waxberry; berries red to purplish red; lf. inf. u. for ophthalmia, weak inflamed eyes (also inner bk. decoc.) (Am Indians); rt. & st. u. as diur. antipyr. rt. u. as for *S. occidentalis*: ornam. shrub - **S. racemosus** Mx. = **S. rivularis** Suksdorf = *S. albus* - **S. rotundifolius** A. Gray (Rocky Mt. [US]): round-leaf snowberry; ornam., shrub - **S. vulgaris** Mx. = *S. orbiculatus.*

Symphytum officinale L. (Boragin.) (Euras, natd. eCana, US): (common) comfrey; blackwort; rhiz. & rts. (as Radix Consolidae) of perenn. pl. prod. e&cEu, esp. Hung); c. allantoin, at times pyrrolizidine alks. (hepatotoxic) as lasiocarpine, retronecine, consolodine, symphytocynoglossine, hence some advise against int. use (2399); tannins; mucil.; aspartic acid; u. exter. as vulner. for healing periosteal dis., fractured bones, arthritis, thrombophlebitis; muscular rheum., for slowly healing wounds; mild astr. u. decoc. in mouthwashes in stomatitis, pharyngitis, periodontitis; styptic; u. int. as dem. in rhinitis; gastric ulcer; pectoral (esp. in Eu as black rt. honey for children); hb. c. allantoin, tannin, consolidine; appl. as lf. poultice in skin cancers (1960); lvs. u. in salads & like spinach; adult. mixed with chicory of dandelion coffee - **S. orientale** L. (Turk): c. allantoin, symphytine, has anti-tumor activity (cell culture 9 KB) - **S. tuberosum** L. (sEu): c. allantoin in all organs; fructans, starch; u. similarly to *S. officinale.*

Symplocarpus foetidus (L.) Nutt. (Ictodes foetidus Bigel.) (monotypic) (Ar.). (e&cNA, neAs): skunk (or swamp) cabbage; skunk weed; fetid hellebore; hb. u. as antispasm., for cancer, snakebite; fresh hb. & rt. u. in Eu Homeop.; tubers & sds. u. as spasmolyt. in cough, asthma, edema; lotion fr. rts. u. for "prairie itch" (scabies) (Iowa [US]) (723); to reduce swellings (Abenakisi Indians, eCana); "chou plant" (1961); rt. adulter. of Veratrum Viride: young lvs. & lf. stalks boiled & eaten like spinach (Am Indians); bears fond of eating rts.; cult. ornam.

Symplocos alstonia L'Hér. (Symploc.) (Co) lvs. u. as tea substit. - **S. loha** Buch.-Hamilton (Symploc.) (SriL): "wal bombu"; bk. decoc. u. by exter. appln.; lvs. in dyeing - **S. odoratissima** Choisy (Java): seriwan; lvs. u. like bk. of *S. racemosa*; for sprue (also inner bk.) - **S. paniculata** Wall. (S. crataegoides Buch.-Hamilton) (Ind, Chin, Jap): dankyat (Burma); bk. u. as ton. & in ophthalmia - **S. penangiana** King & Gamble (Mal): lf. juice drunk for stomachache - **S. perakensis** King & Gamble (Mal): bk. u. as vermifuge - **S. poilanei** Guill. (Chin): boiled lvs. u. to treat mange - **S. racemosa** Roxb. (nInd, sChina, Indochin): lodh (or lode) tree; sweet leaf; bk. once thought a form of Cinchona, hence called variously "China (nova) brasiliensis," "China nova," &c.; also called lodebark, lotur (or autour) bk. (724); Cortex S. r.; with bitter subst. (autourin ["californin"]), alks. harman (loturine), loturidine; u. bitter subst. (cinchona substit.) in febrile dis.; intest. dis., such as dys., ophthalmia, astr., uter. ton. in menorrhagia, &c., appl. as vulner. for furuncles, in gargles to firm up the gums, lotions for ulcers, &c. (Ayurvedic med.); gainal suppurations, to prevent premature birth & abort.; lvs. & bark as dye - **S. spicata** Roxb. (e&sAs): lod(h)ra; lodh; bk. c. saponins, much Al (as in other *S.* spp.), red coloring matter; u. much like *S. racemosa* - **S. thaeiformis** Guerke (CA, Co): Sante Fe tea; lvs. u. as tea surrogate - **S. tinctoria** L'Hér. (seUS, n to Del); (common) sweetleaf; horse sugar; yellow wood; lvs. sweet, chewed; eaten gladly by stock; bitter rts. u. as ton. stom. digestive; in beverage tea (SC); for asthma (Gullah Negroes SC); bk. lvs. frts. —> yellow dye; wd. u. in turnery.

Synadenium cupulare L. C. Wheeler (Euphorbi) (sAf): latex appl. to painful dental caries; exter. as embrocation; dr. lf. chewed for asthma; irritant poison - **S. glaucescens** Pax (Tanz): lvs. purg.; latex to treat boils - **S. grantii** Hook f. (trop eAf): African milk bush; "cathedral cactus"(?); remedy in Hansen's dis.; dr. lf. appl. to cuts in skin, to relieve pain of backache, &c.; rt. in earache (ear drops); latex very irritant, u. to provoke malicious injury of cattle - **S. volkensii** Pax (Tanz): latex styptic; rt. in mal.

synaptase = emulsin.

syndets: contracted form of "synthetic detergents" - **syndoc:** sintoc.

synecology: study of relations between gps. of organisms & the environment.

Synedrella nodiflora (L) Gaertn. (Comp.) (sFla [US], WI, now pantrop): Saigon (Guam); nodeweed; lvs. u. to cleanse teeth; boiled lvs. u. as lax. (Ghana); young lvs. to prep. lablab (Java; side dish of rice table); lvs. to poultice sores & placed on head for headache (Mal); lvs. in ointment u. as embrocation in rheum. (Indones): pl. juice put into ears for earache (Mal).

synergistic: medicinal substances which together achieve more qualitatively or quantitatively than is expected fr. their additive dosage; ex. pyrethrum & sesamin.

Synergistin: antibiotic from *Streptomyces* strains; active against Gm.+ bacteria; types: A (nautral), B (acid) (Pfizer) (1955-); several other names appl.

Syngonanthus Ruhl. (Eriocaul.) (tropAm & Af): ca. 200 spp. of monecious perenn. bog herbs - **S. elegans** (Bong.) Ruhl. (Br, esp. Minas Ger): "sempervivum," "sempreviva" (trade) (SC fls. durable after drying, resembling strawflowers; dyed in many colors, widely u. as ornam. in artif. decorations) ("flor de Diamantina") (SC abundant in D. dist. of M G); large amts. shipped to u. in US - **S. glandulosus** Gleason (Ve, Guyana): "guanak" (Ve); entire pl. boiled to make decoc. u. in toothache (725).

Synnematin B: Penicillin N: antibiotic fr. growth products of *Cephalosporium* sp., *Tilachlidium* sp., &c. (1951-); useful against typhoid fever m. o., &c. (1962).

synonym: 1) scientific name previously appl. to a pl. or animal, which is now superseded 2) common or vernacular names for pls., animals, etc. 3) (general) word having the same or nearly the same meaning as another word.

Synpetalae: sub-class of Dicotyledons, in which both calyn & corolla of fls. are present in fls. & the segments in each are more or less united along their adjacent margins; ex. *Datura;* morning-glory; Mentha.

Synsepalum dulcificsum Daniell (Sapot.) (trop wAf, fr. Ghana to Zaire; cult PR, Jam, sFla [US]): miracle (or miraculous) fruit; miraculous berry; agbayun; pulp of fresh berry has lingering sweetness effects, makes salt, sour foods (ex. lemons) taste sweet; also reduces appetite; u. for centuries in Af; c. glycoprotein, miraculin (Miralin™) (726); u. to sweeten sour palm wines & beers, breads, &c.; also to dye foods red (anthoyan) - **S. glycodorum** Wernham (tropAf): similar properties to prec.

synthetic: not occurring as or derived fr. natural products; made artificially (ex. camphor); v. under camphor; oleovitamin D; rubber - **s. detergents:** syndets; cleansing materials differing fr. soaps & saponins; mfrd. in chem. laboratories often fr. natural fats; ex. lauryl sulfate; Dreft™.

Synt(h)omycin: chloramphenicol.

syntype: one of 2 or more specimens cited by author when no holotype was cited or when any one of 2 or more specimens was designated simultaneouly as types.

Syrian: v. also under alkanet; tragacanth.

Syrian gum: flake tragacanth; Carmania - **S. herb mastich:** *Teucrium marum.*

Syringa vulgaris L. (Ole), (seEu, AsMinor): (common) lilac; syring; c. syringin, glycoside (syringenin, aglycone); fragrant fls. c. glycosides, VO (with terpene alcs., &c.); frt. (capsule) & bk. u. as antipyr., esp. for chronic mal. (fresh lvs. also); astr. ton.; very popular ornam. flg. shrub; many cv. & hybrids.

syrup (<Arabic): aq. prepn. of sugar (85 gm%) (w/v) of thick viscous consistency; u. as sweetener, flavor - **carob s.:** Syrupus Ceratoniae (L): prepd. fr. frt. of *Ceratonia silique*; at one time u. to preserve tamarinds, &c. - **S. cusinier** (sirop c.) (Fr): sarsaparilla syr. comp. - **golden s.:** refined molasses - **high fructose s.:** HFS: simple sucrose s. with addl. invert. sugar (glucose + fructose); sweeter - **high fructose corn s.:** HFCS: c. up to 90% fructose; prepd. by enzymic process, incl. invertase; very sweet; u. to sweeten soft drinks, &c., in place of sucrose or invert sugar; for sweetening frts. - **simple s.: sugar s.:** prepd. by evapgn. sucrose soln. to a thick syrup - **sugar cane s.:** prepd. by evapn. down the expressed juice of s. c. - **s. of violets:** orig. produced fr. petals of *Viola* spp., u. as indicator (18th cent.), displaced by litmus; later this syr. prepd. fr. iris tinct. & colored with violet dye - **s. rhubarb & potassium compound** (NF 1890, 1900): Neutralizing Cordial (728); u. as lax.

Syrupus (L.): syrup - **S. Balsamicus** (L.): tolu syrup - **S. Communis:** molasses - **S. Diacodii:** syrup of poppy heads; formerly appl. to carob syrup ("Diacodium") (729) - **S. Empyreumaticus: S. Fuscus: S. Hollandicus:** molasses - **S. Rhei et Potassii Compositus:** Syrup Rhubarb & Potassium Compound.

Syzygium aromaticum (L.) Merr. & Perry (Eugenia caryophyllata Thunb.; E. caryophyllus Bullock & Harrison) (Myrt.) (native to Molucc, PI; now cult. far & wide in trop insular climates, ex. Mad, SriL, Malaya, Zanzibar, Réunion, Congo, Br) (1966): (aromatic) clove tree; fl. buds dr. in sun are the cloves of commerce; broken & defective cloves source of VO (clovebud oil) (16-21%) with eugenol (85-92% of VO); cloves u. as cond. stim. flavor, stomach, antispasm. corrective, arom. aphrod. antiem.; contracept. (2358); antisep. carm. in bronchitis, &c. (0.7-1.5 pd.); astr.; (2720); VO. exter. loc. anesth. analg. in dentalgia; excellent germicide; rubefac., in dental preps. (tooth pastes, mouth washes), to perfume soaps, in prepn. liqueurs; in cigaretes (kretek) (sAs) (1963); lure in beetle traps; to mask alc. in breath (cloves chewed); in pomanders; frt. (mother of cloves) appetitive (seAs); u. much like cloves but with lower VO content; clove stalks (Festucae Caryophyllorum) (= frt. stalks & young twigs) u. to produce "clove stem oil" (1964) & eugenol; rt. also so u.; clove leaf oil (fr. lvs. & twigs) in Mad, Indones, (730): also u. as eugenol source: fls. & bk. u. in Chinese med. - **S. cordatum** Hochst. (s&eAf): water (berry) tree; waterwood; frt. acidulous, edible; lf. inf. u. for stomach dis. diarrh.; purg. emet. in TB; pd. bk. u. as fish poison; wd. in construction - **S. corynathum** (F. Muell.) L.A.S. Johnson (Australasia): "sour cherry" (frt. edible but sour) (NSW-Queensland [Austral]): lvs. & buds u. for lung trouble (Fiji) - **S. corynocarpum** Muell. (Tonga, Fiji, Samoa): fragrant myrtle; heahea (Tonga); seasea (Fiji); lvs. u. int. for fever (Samoa); topically for inflamm. (Sam); bk. u. med. (Ton); frt. eaten (86) - **S. cumini** Skeels (S. jambolanum DC.; Eugenia c. Druce; E. jambolana Lam.) (tropAs-Ind, Mal, sChin, Indones, Austral): jambol(an)(plum); jambul (seed); jamboo; Indian allspice; Java plum; frt. c. VO, FO, resin; edible, formerly u. in diabetes (wine); ton.; to strengthen teeth; against annelids (incl. leeches); wine vinegar u. as carm.; stom.; sds. c. VO. resin, FO, jambosine (alk.); starch (40%); jambolin (glycoside); bk. with 19% tannins (gallic & ellagic types); u. astr. in dys. diarrh.; in asthma; as tan & dye: frt. popular among the poor people of Ind (*cf.* olive); juice u. as refreshing beverage - **S. fibrosum** Hartley & Perry (Queens [Austral], PapNG.): sd. u. as abort. - **S. flavescens** Merr. & Perry (PapNG+): boiled lvs. edible; u. as cure for stomach pains (New Britain) - **S. gerrardi** Hochst. (s&eAf.): (forest) waterwood; bk. inf. drunk for lung TB (Zulu); said to ease pain & cough; frt. edible - **S. guineense** DC (trop c&wAf): water-pear; w. berries; frt. u. in dys. diarrh. (also bk. decoc.): frt. & bk. cond., in spicing brandy; wd. for railroad ties sleepers) - **S. jambos** (L) Alst. (Eugenia j. L.) (tropAs; widely cult. in trop.): rose apple; Malabar plum tree; plum rose; thabye (Burma); lvs. & bk. c. alk., jambosine; bk. with tannin, resin; lvs. decoc. u. as ton. diur. expect. (Guat); for sore eyes (Burma); jamosa rt. u. for epilepsy (Cu); roasted pd. sds. u. in diabetes (Nic); wd. u. to mfr. first class charcoal - **S. malaccense** (L.) Merr. & Perry (Eugenia malaccensis L.) (seAs, Ind, Mal, Polynesia): Malay (rose) apple; frt. esculent; u. with sds. to make refreshing beverage; bk. u. to make mouthwash for thrush (Indones); lvs. diur. u. in baths & lotions; bk. & lvs. u. med. (Tonga) (86) - **S. neurocalys** Gray (Fiji): frt. oil u. as lotion - **S. operculatum** Gamble = *Cleistocalyx operculatus* Merr. & Perry - **S. owariense** Benth. = *S. guineense* - **S. polyanthum** Walp. (Burma to Java): bk. lvs. & rt. u. as poultice or itch; lf. & bk. inf. for diarrh (Indones) - **S. rowlandii** Sprague (trop wAf): sagbei (Sierra Leone); pounded bk. mixed with clay & spices & rubbed on body for rheum. & other pains; bk. u. for dye.

Tt

tabac (Fr-m.): tobacco; snuff - **t. de montagne: t. des Vosges:** *Arnica montana* (SC pd. lvs. cause sneezing, like snuff) - **t. du yab** (Qu [Cana]): *Verbascum thapsus* - **t. suisse** (Fr) Arnica: **tabacca: tabacco** (It): **tabaco** (Sp, Por) (1759): **Tabacum** (L.): tobacco.

tabacheer = tabaschir.

tabachin (Mex): *Delonix regia.*

tabajillo (Mex): *Hedeoma piperita.*

Tabak (Ge, Dutch, Russ, Pol, Rom): **tabako** (Jap): tobacco.

tabaquillo (Mex): *Hedeoma piperita*; *Lippia umbellata*; *Calamintha macrostema*; *Wigandia urens* var. *caracasana*; & other pl. spp., esp. of Labiatae.

tabas(c)hir (Sanscr): **tabasheer (tabashir)** (Pers.): **tabashira** (Arg): **tabaskir**: 1) common bamboo plant 2) bamboo manna, silica, concretions inside st./rt. cavity, esp. of *Bambusa arundinacea*; u. in asthma.

Tabasco (pepper): **T. chillies**: *Capsicum frutescens* var. *conoides* Bailey; frts. -2 cm; Mex type; a pungent var. u. in med. & for culinary purposes (**Tabasco sauce**) (La [US]) (1760).

Tabebuia avellanedae Lorentz ex. Griseb. (Bignoni) (Br, Arg, Par+): pau d'arco; lapacho (colorado); red lepacho; ipe roxo (sBr): inner bk. c. lapachol u. to treat cancer (taheebo tea) (Indians); wd. u. in cabinet making (durable -400 yrs.) (similarly u. in cancer is bk. of *T. altissima* [lapacho morado]); u. mal Chin.) - **T. bahamensis** (Northrop) Britt. (Bah): "big man" (Bah); decoc. u. for anuria pain; bk. inf. for "building up" health (Bah) - **T. caraiba** (Mart.) Bur. (Arg): lapacho amarillo; st. bk. u. against cancer - **T. cassinoides** DC (Br): cacheta; tamanqueira; c. cytotoxic quinones; anticancer properties; wd. u. in toys, chests, pencils, &c. - **T. donnell-smithii** Rose (Mex, CA): gold tree; cortez (blanco); wd. u. for veneering, cabinet work ("white mahogany") - **T. guayacan** (Seem.) Hemsl. (CA [Nic, Pan]; Mex; Ec, Ve, Co): guayacán; wd. hard & durable; u. to make furniture; ornam. tree - **T. heterophylla** Britt. (T. pentaphylla auct.) (Mex, WI, CA to Ve): roble (blanco) (Oaxaca, Cu, PR); bk. u. diur. antipyr. alexiteric for snakebite (also frt. & lvs.); antidote for manchineel; decorticated frt. u. as diur.; frt. peel hypn.; subsp. **leucoxylon** (T. leucoxyla subsp. pallida Stehlé) (T. p. Miers): lvs. u. to make tea; lvs. & bk. boiled as cold cure (St. Kitts, WI); bk. u. as plaster on corns; lvs. as poultice (Domin) on sores; important timber tree; wd. u. for planking, frames, veneer, &c.; cult. ornam. - **T. impetiginosa** (C. Mart. ex DC) Standl. (Pe): "impetiginosa"; u. in impetigo; cancer (Campas) (2578) - **T. insignis** (Miq.) Sandwith var. *monophylla* Sandwith (SA-Br-Amazon area): bk. inf., u. for stomach ulcers - **T. ipé** (Mart.) Standl. = Tecoma i. - **T. neochrysantha** A. Gentry (SA: Amazonia): bk. decoc. u. for mal., chronic anemia, & ulcer pains (Tikuna Indians) (2579) - **T. obscura** (Bur. & K. Schum.) Sandwith (Br-Amazon): tahuari; dr. fls. u. in food for rheum. & irreg. menstr.; cancer - **T. obtusifolia** Bur. (Br): similar to *T. cassenoides* & u. in the same way - **T. ochracea** (Cham.) Standl. (tropSA): u. in mal. (2578) - **T. rosea** DC (T. pentaphylla Hemsl.) (cMex to CA to Ve, Ec): may bush; m. flower; roble blanco (=white oak); (pink) trumpet tree; maculiz; bk. decoc. home remedy for fevers, colds, mal. headache (Co); cancer (Mex); adm to dogs to prevent rabies(!?) (Guat); lf & bk decoc. as antipyr. (Mex); bk. rich in tannins; wd. useful; ornam. tree (-30 m) - **T. serratifolia** (Vahl) Nichols. (WI-Co): roble (Cu); yellow poui; washiba; c. quinones; u. in cancer (Co); large tree furnishing valuable timber.

Tabernaemontana L. (Taberna Miers) (Apocyn.) (tropics): some 100 spp. of trees & shrubs; many c. alks. (732); many of the original species have been transferred to some 26 other genera; thus f. i. *Hazunta* - **T. alba** Mill. (mainland tropAm): lvs. antipyr. purg. anthelm.; bk. ton. antipyr.; latex to destroy warts; lf. smeared with lard, heated, & appl. to swellings & chilblains, for pustules (sMex) (79) - **T. amblyblasta** Blake (Guat, Hond): cojón; juice mixed with chicle to chew (local use); also appl. to draw larvae fr. flesh; wd. (comm. lumber) - **T. amygdalifolia** Jacq. (CA, WI, Co, Ve): cojón (macho); latex purg., appl. carefully to blepharitis granulosa (Mex); latex for warts, lvs. as cataplasm for tumors, wounds (Co) (2512); bk. decoc. for fevers, syph. (PR, CA) - **T. australis** Muell. Arg. (Arg, Par): latex u. for healing warts - **T. bufalina** Lour. (Ervatamia b. Pichon) (Chin, Indochin): pl. u. as emoll. lax.; frt. latex u. for extn. of thorns fr. flesh; emoll.; bitter rts. for gastric dis., chewed for sore throat; crushed lvs. appl. to boils - **T. citrifolia** L. (WI): lechoso (Cu); pegojo (Cu): lvs. u. antipyr. (Antilles); purg. (Mart); acrid latex fr. twigs u. for toothache (Antilles): appl. to remove warts; hemostatic; for herpes simplex (Cu); remedy for swellings & chilblains; lf. smeared with lard, seared in hot coals, & appl. to pustules (Totonac; sMex) - **T. coronaria** (L) Stapf = *T. divaricata*. - **T. crassa** Benth. (trop e, c &wAf): latex u. for ringworm, to coagulate rubber (Sierra Leone); disinf. to wounds, for leprosy, to calm insane (Ivory Coast); for abscesses, boils, carbuncles (Réunion): rt. & bk. decoc. for coryza, sinusitis (cAf Rep, Congo) - **T. crispa** Roxb. (Ind): bk. c. alk., u. astr. in dys. (inf.), also rt. - **T. dichotoma** Roxb. (Ervatamia d. Burkill) (Ind, SriL): "Eve's apple"; lvs. & bk c. alk. perivine, apparicine, &c., FO; purg.; to treat ulcers, fistulae (SriL); latex appl. to boils, wounds; rts. chewed for toothache - **T. dinhensis** Pitard (sVietnam): rt. inf. u. for colic & indigest. - **T. divaricata** (L) Stapf (T. coronaria; Ervatamia c. Stapf) (Ind, Oceania, now pantrop): crepe jasmine; widely & much u.; pounded lvs in drink with sugar for coughs; anticonvulsive; rts for eye dis. (Mal); in jungle fever (Vietnam); latex cooling, appl. to sore eyes, in toothache (SriL); wd. u. refrigerant, as incense, in perfumes (Ind) - **T. donnell-smithii** Rose = *Stemmadenia d.* Woods - **T. eglandulosa** Stapf (trop cAf): rt. u. against snakebite - **T. grandiflora** Jacq. = *Stemmadenia g.* - **T. heyneana** Wall. (Ervatamia h. Cooke) (Ind): pl. u. much like *T. divaricata*; bk. u. for fevers - **T. hystrix** Steud. (Br, Par, Arg): bk. u. as bitter; u. exter. for slipped disk, in hernias - **T. laevis** Vell. + = *Geissospermum laeve* - **T. pachysiphon** Stapf = *Conopharyngia p.* - **T. pandacaqui** Poir. (T. laurifolia Blanco) (PI): latex appl. as vulner.; lf. decoc. in dys.; lvs. placed as cataplasm on abdomen to induce menstr. & hasten delivery; bk rts. decoc. u. for GI dis. - **T. parviflora** Heyne = T. salicifolia Wall. = *Hunteria zeylanica* - **T. penduliflora** K. Schuman (tropAf): lvs. u. as uter. sed. (Nigeria) - **T. sphaerocarpa** Bl. (Ind, seAs): lf. & bk. decoc. rubbed on skin for fever; lvs. appl. to sprained ankles; frt. u. for toothache - **T. sralensis** Pierre (Kampuchea): frts. u. to treat skin eruptions; rt. may be added to betel quid; rts. for snakebite - **T. utilis** Arn. = *Lacmellea u.*

tabernaemontanain: proteolytic enzyme fr. *Stemmadenia* (*Tabernaemontana*) *grandiflora* - **tabernaemontanine**: indole alk. occurs in *Tabernaemontana sphaerocarpa* & several other *T.* spp.

Tabernanthe iboga Baill. (Apocyn.) (trop wAf, incl. Congo): iboga; bocca; aboua; lvs. bk. wd. & esp. rt. c. indole alks.. incl. ibogaine, ibogaline (most potent); rt. u. nerve & muscle stim. antipyr. aphrod.; hallucinogenic (ibogo); in small doses for overcoming fatigue & hunger; hypotensive; alk. source; bk. & lf. u. as stim. (1761) - **tabernanthine**: alk. fr. preced., also fr, *Tabernaemontana* & *Stemmadenia* spp.; u. analges., serotonin antagonist.

tabersonine: indole alk. of Apocyn.; occ. in *Amsonia tabernaemontana*; also found in *Tabernaemontana* and *Stemmadenia* spp.; sds. of *Voacanga africana* said good source; u. in alk. synthesis.

tabla bark: flat bark: Cinchona bk. dr. under weights to flatten it out to facilitate handling & shipping.

Tabriz gum: type of tragacanth or may be Hercules Gum (sTabriz [Iran]).

tacamahac(a): 1) oleoresinous gums fr. spp. of *Calophyllum*, *Protium*, *Bursera*, *Canarium*, etc. 2) *Populus balsamifera* (*P. tacamahacca* (& its "resin": existence as a comml. prod. mythical, & the name a misnomer) 3) resin fr. *Fagara octandra* - **tacamahac orientale**: fr. *Calophyllum inophyllum*.

Tacazzea apiculata Oliv. (Asclepiad.) (sAf): bk in water or milk u. as tonic (Zulus).

Tacca aspera Roxb. (Tacc.) (seAs, Bangla Desh): tubers u. in hemorrhagic diathesis, skin dis., leprosy; ton. - **T. chantrieri** André (Mal, Burma,

Thail; cult. Taiwan): devil flower; "Jew's beard"; pl. u, as cure-all by natives - **T. cristata** Jack (Mal): tubers pulped to make poultice for appln. to rash caused by "hairy" caterpillars - **T. leontopetaloides** (L) O. Ktze (T. involucrata Schum. & Thonn.; T. pinnatifida Forst.) (tropAf, As, Australasia, Oceania): pia plant; large starch-rich tubers cult. for mfr. South Sea (East Indies or Tahiti) arrowroot (comml. prod.); tubers u. to make bread flour; stiffening agt. in laundry; in poi; as size in mfr. tapa (bark cloth); lvs. to make hats; roasted rhiz. (after boiling) eaten, considered of value for dys. & diarrh.; crushed & mixed with sugar & water u. for heart dis.; impotence - **T. palmata** Bl. (Mal, Sunda Is, PI): tangeling (Java); tubers u. as ton. stom. in gastric dis., dentalgia; impotency, falling hair; appl to snake- & centipede bites (natives-Indones) - **T. paxiana** Limpricht f. (Mal, Chin, Indochin): said u. as cure-all.

taccada (pith): *Scaevola frutescens, S. sericea.*

Tachardia lacca = *Laccifera lacca.*

Tachia guianensis Aubl. (Gentian.) (Guianas, Col, Br): caferana: juciarara; bk. & wd. c. "caferanine"; tachinin (NBP); rt. very bitter u. (as Radix Quassiae paraënsis) like gentian, swertia; as antipyr. prophyl. for mal., &c. (home med.); "sore stomach"; pd. lvs. added to coca to improve flavor (Witoto Indians) (2299).

Tachiadenus carinatus (Desrousseaux) Grisebach) (Gentian.) (Madag., e.coast): u. as aperitive, ton., in dis. of urin. organs, digestive, respir., & circul. apparatus (2513); u. in making beer (bitters).

Tachigali schultesiana Dwyer (Leg.) (Co): fl. decoc. u. as "cure" for TB, as gargle (hot daily for several wks.) (Taiwano Indians).

tachysterol: product of one stage in irradiation between ergosterol & precalciferol.

tackiness: stickiness to touch; shabby (general use).

taco (Sp): fried tortilla stuffed with cheese, chicken, roast pork, &c. (Mex).

tacoyana (Mex): tree which prod. Sonora gum, prob. *Prosopis* spp.

Tadehagi godefroyanium (Ktze) H. Ohashi (Leg) (Indochin, Thail): lf. & st. decoc. u. as mouthwash for dental pains - **T. triquetrum** (L.) Ohashi (Leg) (seAs): rt. decoc. u. to treat. mal.; also for diarrh. (Thail) (2448); subsp. **rodgeri** (Schindler) H. Ohashi: lf. inf. u. as beverage to replace tea; u. to kill mites; in pickling brine for fish.

taffia (WI, La [US]): **tafia**: rum-like drink from sugar cane.

tag: v. under alder.

Tagetes L. (Comp.) (Am trop&subtrop): marigolds; some 50 spp.; many are popular ornam. - **T. erecta** L (Mex, CA, WI, SA): big marigold; African m. (misnomer; once thought native to Af); copete; fl. heads c. glycosides quercetagitrin & tagetiin (quercetagetin + glucose) (733); terthienyl (ascaricidal props.); fl. decoc. u. for chest colds, emmen. VD (Cu); yellow dye; pl. decoc u. for flatulence, colic, constip. indigestion (Mex); also emmen. & vermifuge (Br); cult. ornam. hb. - **T. filifolia** Lag. (T. pusilla HBK) (Mex to Bo, Arg): (anis-) anis; flor de muerto; anisillo (de monte); decoc. u. as carm. diur. digestive, &c.; carried by druggists; frt. purg. anthelm.; fl. heads u. for prepn. of helenien (xanthophyll dipalmitate) (u. with vit. A as adaptogen); to treat malignant tumors (Cu) - **T. lucida** Cav. (T. florida Lw.) (Mex, CA; cult. US, Eu): periquilo; yerbanis; (yerba de) Santa Maria (Mex); sweet-scented marigold; flg. tops u. as emmen. antispasm.; dr. pl. as antiperiod. in mal.; hallucinogenic; hb. decoc. for colds, fevers, upset stomach; tea as hot beverage; lvs. & fl. heads u. to perfume bath water for babies; spice in food prepn., thus in soups in place of tarragon; for smoking out mosquitoes; ornam. hb. - **T. micrantha** Cav. (swUS, cMex): little marigold; bitter ball; hb. inf. u. for colds, fevers, summer complaint, stomach troubles (Navaho); tranquilizer, to improve circulation; to dye wool yellow - **T. minuta** L. (sBr, Arg, Ch, seUS, Mex, natd. Af): wild (or Mexican) marigold; lf. u. diur. diaph. anthelm.; lvs. crushed & appl. to swellings (Br); VO u. as anthelm. (Eu); in perfumery; wd. killer; as larvicide - **T. multifida** DC (Mex, CA): inf. of anise-scented hb. u. as diur. (home med.) - **T. patula** L. (Mex, cult. Euras): French (or spreading) marigold; fl. heads c. quercetagitrin, helenien; decoc. u. as carm. (PI); dye; stim. emmen.; flg. hb. as anthelm.; sds. & rts. u. as purg. (Am Indians).

tagged: labeled with radioactive element.

tagua (nut tree) (Co, Par): *Phytelephas* spp.

Tahini (near East): sesame butter (canned) (pasty).

Tahiti: largest island of Fr Polynesia (sPac Oc); exports copra. vanilla; v. under arrowroot.

Taiá = taioba (Por, Br): *Colocasia antiquorum*; also (Br) rhiz. of other Araceae u. similarly, esp. *Xanthosoma nigrum*; X. *mafaffa.*

taiga (Russ): "black forest"; virgin forest (mostly coniferous) zone of nHemisph (esp. Siber) lying between Euras tundra & steppes & hardwood forests further s.

taioba (Br): **toia** (more correct): 1) taro 2) *Xanthosoma nigrum.*

Taiuiá: *Cayaponia.*

tai-voy (Chin): fennel.

Taiwania cryptomerioides Hayata (monotypic) (Taxodi.) (Taiwan-central mt. ranges): a large tree (-60 m); ornam.

Taka-diastase™: Koji; enzyme fr. *Aspergillus oryzae*, with amylase, glycerophosphatase, phenol sulfatase, pyrophosphatase III; &c.; u. as digestant like diastase; in prepn. sake (Jap); to convert maize starch to sugar in mfr. whiskey.

take-all weed (Austral): *Lithospermum arvense*.

talatisine: alk. of *Aconitum talassicum*.

Talauma dubia DC (Magnoli.) (Br) a kind of tulip tree; bk. u. in therapy (also some other *T.* spp); fl. source of VO u. in perfumery - **T. mexicana** (DC). G. Don (sMex, CA): flor de corazón; cocte; bk decoc. u. as cardioton. like digitalis; antipyr.; to relieve mal.; decoc. of petals u. for various card. dis. (heart attack palpitations), ton. antiperiod. in epilepsy; to flavor chocolate; fls. u. to perfume houses (—> *Magnolia*).

talba, gum: talha.

talc: Talcum (L.): French chalk; finely pd. native hydrous magnesium silicate, mined in US (Ga), Chin, USSR, Fr, &c., u. as dusting pd. alone or in combination, usually perfumed; deodorizer; excipient in tabs. pills, & suppos.; as pigment, filler (in tabs., &c.); carrier for insecticides, fungicides; in mfr. crayons; lubricant; absorbent in glove & shoe pds.; filtering medium, insulator, cosmetic pd.; appl. directly to heart thru incision —> stimn. (sl. irritant) (1983); however, may produce granuloma; carcinogenic (734) (chief prodn. It, Fr, Eng, US); **Spanish t.**: prod. in Sp but appar. u. there & not exported - **talca**: talha - **Talcum**: talc.

Talg (Ge, Nor, Dan): tallow; grease; fat; FO.

talh(a), gum: Ghezireh gum (type of Acacia); fr. *A. seyal*, etc.

tali: 1) (Senegal-Mandé): *Erythrophloeum guineense* (bk.) 2) (Senegal-Soci): *Detarium senegalense* 3) (Senegal-Peul): *Spondias mombin* 4) *Tephrosia vogelii* Hook. f.

Talinum aurantiacum Engelm. (Portulac.) (Tex, NM, Ariz [US], nMex): orange fameflower; tuberous rts. u. in fever, ague (Lipan Indians) - **T. caffrum** E. & Z (sAf): rt. inf. u. for nervousness, gastric pain; lf. inf. u. for stomach upsets (children) - **T. crispatulatum** Dinter (Moz): pl. juice u. as emoll. for cuts & wounds; tuber anesthetic & narcotic - **T. cuneifolium** (Vahl) Willd. (Af, Ind): aby; u. as aphrod. (Tanz); cult. as a spinach - **T. paniculatum** (Jacq.) Gaertn. (Portulaca p. Jacq.) (735) (native to WI; now sUS, Mex to Ve; cult. As): panicled fame flower; pink baby-breath; tuberous rt. u. as aphrod. (Indones); pl. emoll. (Indochin); antiscorbutic; substit for *Portulaca oleracea* (Indochin); cult for med. in Hanoi; lvs. u. like spinach - **T. triangulare** (Jacq.) Willd. (T. fruticosum [L.] Juss.) (Fla [US], WI, CA, SA, eAs): bokkolille (Surinam); verdolaga (Cu); (pink) purslane; fame flower; sweetheart; eyes plant (SC usage in Bah); u. as ton. for general weakness (Chin); possible ginseng substit.; for inflamm. & to reduce swellings; lvs. boiled & u. as cooling bath to relieve eye strain (Bah); u. as potherb like spinach (Cu; some use in Fla [US]); as salad, popular food in PI, somewhat like okra (mucilaginous) (purslane flavor), rich in vitamins; often cult. (SriL; Ghana; Fla).

Talisia olivaeformis Radlk. (Sapind.) (Ve): cotoperiz (Ve); frts. edible.

talka gum: talki g.: talha g.

tall oil: FO with oleic, linoleic, & rosin acids; obt. fr. coniferous wd. pulp indy. (fr. **tall** Swed., pine) (pron. to rhyme with "pal"); mostly obt. fr. the sulfate or kraft pulp process, (2367); t. o. fatty acids important in indy.

tallo (Sp): shoot; sprout, st. which bears lvs.

tallol: tall oil.

tallow: heavy fat obt. by rendering various parts of bodies of cattle ("**beef t.**") and sheep ("**mutton t.**"); c. various glycerides of fatty acids ("drippings"); the finest kinds are the suets or internal fats; **beef t.** is more liquid than mutton t.; c. mostly stearin with smaller amts. of palmitin & olein; u. to make soap, prod. stearic acid, as lubricant, etc.; **mutton t.** is harder but rancidifies more readily - **t. bayberry**: *Myrica cerifera* - **Japan t.**: wax, Japanese - **J. t. tree**: *Rhus succedanea* - **Marfura t.**: *Trichilia emetica* - **t. oil**: pale yellow liquid FO expressed fr. tallow; s. g. 0.943-6; u. as lubricant, in leather dressings, soap mfr., &c. - **t. seed oil**: FO obt. fr. the sd. of *Sapium sebiferum* (Chin) - **t. shrub**: *Myrica cerifera*+ - **t. tree** (**, Chinese**): *Sapium sebiferum* - **vegetable t.** (**, Chinese**): prod. of prec.

talluelo (Sp): caulicle.

talo (Por): 1) st. 2) petiole 3) thallus 4) taro (also **tallo**) (Br).

Talwin (Pentazocine): synthetic analg. replacing morphine; once claimed to have low addictive potential; now often sought by addicts.

tamale (Mex CA): **tamala**: **tamalli** (Mex): a ground meat-crushed Indian corn (meal) dish spiced with capsicum (& garlic), often with rice and pumpkin flour & enriched with a FO; usually wrapped in corn husks or banana lvs & cooked in oven.

tama(n)quare (Br): *Caraipa psidifolia*, *C. densifolia*, & other *C.* spp. (also tamacoari).

tamarac (Fr): **tamarac(k) (larch)**: 1) *Larix laricina*, *L. lyallii*, *L. occidentalis*, &c. (bk.) 2) *Pinus contorta* (bk.) 3) *Picea rubens*.

tamarin: **tamarind(s)**: *Tamarindus indica* - **black t.**: **East Indian t.**: var. of tamarind sometimes considered superior; with pulp more acid in taste and darker (almost black) than the West Indian. t. - **wild t. (wood)**: *Lysiloma bahamensis* - **Tamarinde** (Ge): **tamarindo** (Sp, Por, It): tamarind; second name also the beverage prep. fr. t., popular with It & PR people (street vendors, New York City, NY [US]); purified pulp with 20% sugar added (Arg Ph.).

Tamarindus indica L. (1) (monotypic) (Leg.) (tropAf; cult. Ind, EI, WI): tamarind tree; Indian dates; frt. pulp c. 7% tartaric acid, also malic, succinic, citric, & other acids; 30% invert sugar; has the highest tot. % of organic acids & sugars of all frts.; fresh ripe frt. u. as food; frt. packed in molasses, sugar, or brine; u. mild lax., refriger. in fevers, etc.; in beverages & soda water syr.; flavor in Worcestershire sauce; chewing tobacco; preservative; t. pulp u. in foods, mfr. pectin, K bitartrate, alc.; in gargle for sore throat (Cey); in prep. confection of senna, t. paste (u. in cooking), t. pulp, t. wine, t. sd. u. in Ind (2675), t. kernel powder c. t. polyose, polysaccharide; u. as binding agt. in tabs., suspending agt. for pds., emulsifying agt.; substit. for sizing starches in textile (cotton & jute) indy. (Ind) (in place of corn, sago, & wheat flours) (736); lvs. & fls. u. as mordant in dyeing; lvs. poultice for boils (Ind) (SC tamr-i-Hindi [Arab]: date of India); wd. useful.

tamarinier (Fr): tamarind tree.

tamarisk: foll.; occasionally confused with tamarack.

Tamarix L. (Tamaric.) (Euras): tamarisk (tree); 54 spp., thrive in alkaline soils, typical habitat, banks of Jordan & Indus Rivers; natd. (spreading fr. cultn.) in Calif (US) are *T. chinensis* Lour. & *T. parviflora* DC (garden tamarisk) (ex. Cache Creek [near Sacramento, Calif]) - **T. aphylla** Karst. (nAf, swAs; natd. Ariz, NM [US]): Athell (tamarisk); desert pine; salt cedar; wd. u. in syph. & herpes simplex (nAf); pl. produces tacaont galls; c. tannin, inf. u. in stomach ache, as tan. (also bk.); bk. mixed with oils & kamala as aphrod.; wd. u. to make plows & other equipment; fuel (arom.); tree or shrub planted for shade & as wind-break - **T. articulata** Vahl = prec. - **T. chinensis** Lour. (Chin, Indochin, natd. wUS, nMex.): salt cedar; leafy tips of young branches u. as diur. diaph. to treat alc. poisoning; ton.; in syph. & scaly skin dis.; juice u. formerly for smallpox (Chin) (-1936-); exter. as wash for skin allergies; galls nearly 50% tannin, astr.; cult. swEng; sts. u. to make lobster pots (Cornwall); often confused with *T. gallica* (T. anglica Webb) - **T. dioica** Roxb. (Ind, Pak): jhau (Hind); galls with ca. 40% tannin u. astr. int. in dys. diarrh.; manna lax. ex pect. detergent; wd. u. for Persian water wheels - **T. gallica** L. (wEu to Indo-Pak & beyond): French tamarisk or tamarix (not tamarack); Indian tamarix; salt cedar; "jhau"; bundu (Wolof); fls. earlier u. in jaundice; lf. & bk. in leukorrhea, hemoptysis; frt. in sterility; bk. u. simple bitter ton. (NM [US]); astr. in ulcers; in tanning & dyeing; similarly the galls (40-45% tannin); macerated lvs. u. for rhinitis; lf. branch decoc. in conjunctivitis (washes) (Seneg); branches for pipe stems; twig tips as hops substit.; st. & lvs. sumac adulter.; manna u. as lax. expect. deterg.; shrub or tree found near rivers & backwaters (1762); in desert places, roots found to 100 ft. depth to reach water - **T. mannifera** Ehrbg. (T. gallica var. m. Ehrbg.) (Eg, Iran, Afgh, sArab): manna tamarisk; m. tree; puncture of twigs by *Coccus manniparus* Ehr. prod. exudn. called "manna of the Bedouins," Jew's m., m. of (Mt.) Sinai; a sacchar. substit.; may be the manna of the Bible; c. sucrose, levulose, dextrin; u. as confection, med. (Orient); tree also prod. galls - **T. senegalensis** DC = **T. gallica.**

tambour (Fr, New Orleans, La [US]): piece of cloth for covering over sieve for protection when sieving such irritants as mustard, cantharis, &c.

tame root: yerba mansa, *Anemopsis californica.*

Tampico: seaport of Tamaulipas State, neMex - **T. bristles: T. fibers**: fibers generally fr. *Agave lecheguilla*; u. for polishing wheels, circular power brushes, & other types of brushes; term also appl. to other fibers exp. through Tampico (Tamaulipas State [Mex]) - **T. jalap**: *Ipomoea simulans* (rt.).

tampon: vaginal plug; plug appl. to other orifices of the body.

tamr (Arab): date - **tamr i hindi** (Arab): tamarind.

Tamus communis L. (Dioscore.) (Euras, incl. GB, Med reg., nAf): black bryony (or b. bindweed); rhiz. c. sterols incl. campesterol, ß-sitosterol, stigmasterol; u. for arthritic pains, rheum., contusions, inflamm. neuralg. skin sores; vulner. (esp. by Homeopaths); in small doses as purg. diur. emmen.; values confirmed (737); shoots eaten like asparagus (Gr).

tan: tanning agent; subst. appl. to skins & hides to preserve, form leather, & make resistant to decay, infestation, etc.; types: natural (ex. acacia, *Larix, Byrsonima, Tsuga*) & artificial (ex. chromium potassium sulfate).

Tanacetum balsamita L. = *Chrysanthemum b.* L. (*Balsamita major* Desf.) - **T. cinerariifolium** (Trev.) Schultz-Bip. = *Chrysanthemum c.* - **T. fruticulosum** Ledeb. (Comp.) (Ind, Pak): hb. u. ton.

anthelm. - **T. gracile** Hook. f. & Th. (Pak, Ind): u. as antipyr. - **T. parthenium** Sch.-Bip. (Chrysanthemum p. (L.) Bernh.) (sEuras such as Fr [waste places]; natd. US, SA): wild chrysanthemum; feverfew; "Santa María" (Ec); perenn. hb. c. VO, NBP; u. as bitter ton. antipyr. emmen. anthelm. carm.; in "trop. anemias" (Co); lf. decoc. u. to bathe nervous children (Ec); fl. heads in tea for intermittent fever, Parkinsonism, female hysteria, migraine, menstr. irreg.; boiled with rice in diarrh.; vermif.; renewed interest recently for migraine (effective), &c.; formerly adult. of Anthemis; insecticide (but much weaker than regular pyrethr. fls.); cult. (Eu) largely as ornam. - **T. umbelliferum** Boiss. (Iran, Afgh, Pak, sUSSR): sweet pellitory; umbel tansy; rts. c. VO alk. flavonoids; u. ton. aphrod. deobstruent; in rheumatoid arthritis; snakebite (12); formerly imported into Ind (738) - **T. vulgare** L. (Chrysanthemum v. Bernh.) (Euras, natd. NA, in waste places): (common) tansy; double t.; (golden)-buttons; bitter b.; parsley fern; char. by its flat-topped corymbs of small bright yellow disc-shaped fl. heads (*cf.* golden-rod) with strong unpleasant odor & by its dark green dissected fern-like lvs.; dr lvs. & flg. tops c. bitter arom, narcotic VO with thujone ("tanacetone"), umbellulone, L-camphor, L-borneol, isothujone, &c.; hb. c. resins, tannins, malic acid, &c.; tea u. in folk med. as anthelm. (mostly vet.) vulner. arom. bitter, ton, stom., purg. flavor; emmen. weakly abort. (2677); diur.; overdose of VO may be lethal; customary not to sell except on Rx (except 10% in VO for mosquito bites); hb. repellant for ants, so planted around houses; hb. u. in neuralgia, migraine; gout; exter. in rheum.; widely cult.; prod. Va (US), Ge.

Tanaecium nocturnum Bureau & K. Schumann (Bignoni) (Br, Co): u. aphrod.; mind-altering drug; lvs. u. as ritual snuff ("koribo") with psychoactive properties (Paumari Indians); sed.; inf. of bk. u. similarly (739); fresh lvs. c. HCN.

tanaka (Jap): *Digenea simplex.*

tanbark: any bk. (esp. of oak) useful for tanning purposes; the exhausted bk. (marc) is u. on ground for circus rings, arenas, etc., to reduce dust - **t. oak**: 1) *Lithocarpus densiflora* (Calif [US]) 2) (eUS): *Quercus prinos* - **t. tree**: *Tsuga canadensis.*

tanchagem menor (Por): *Plantago lanceolata* - **t. major: tansagem:** *P. major.*

tanekaha bark: fr. *Phyllocladus trichomanoides.*

tang (Eng, Nor, Dan): seaweeds - **black t.**: **sea t.**: large coarse seaweeds, esp. *Fucus* spp., as *F. vesiculosus* - **Tang Kohle** (Ge): charcoal of *Fucus* spp.

tangelo (fr. tange(rine) + (pome)lo): hybrid between *Citrus reticulata* var. *deliciosa* (tangerine) & *C. paradisi* (pomelo; grapefruit); frt. seedless with pleasant flavor.

tangerine: *Citrus reticulata* Blanco var. *deliciosa* (var. *tangarin*) (orange of Tangiers).

Tanghinia venenifera Poir. = *Cerbera venenifera* - *Tanghinia* Thouars taxa are now generally transferred to *Cerbera.*

tang(k)awang: *Hopea aspera* DeVriese (Java) - source of tangkawang fat fr. sds.; u. in ointments & supposit.

tangkuei (Chin): *Angelica anomala* v. *chinensis* - **tangle**: *Laminaria* spp. - **tanguin**: *Cerbera venenifera.*

tania: **tanier** (PR): blue tania: *Xanthosoma* spp., *Colocasia esculenta*, & other edible Araceae.

tanisette (Qu [Cana]): *Gaultheria* (*Chiogenes*) *hispidula.*

tank: parsnip.

tankage: dr. animal residues, a by-product of slaughter houses; usually freed fr. FO, gelatin; u. as fertilizer & feeding stuff.

tannase: esterase type of enzyme; acts to hydrolyze tannins - **Tanne(nbaum)** (Ge): fir tree, *Abies* spp. - **tanner's**: v. under sumac - **tannia**: *Xanthosoma sagittaefolium* - **tannic acid**: tannin; gallotannic acid; found in bk. wd. lvs. frts. of many pl. spp.; mfrd. fr. nutgalls of AsMin, Jap, & Chin (on *Quercus* & *Rhus* spp.); occ. as whitish to pale yellowish amorphous powd. or shining masses; readily sol. in water, alc., insol. in ether; complex & variable in compn.; condensed tannins (flavonol derivs.) & hydrolyzable t. (sugar esters with trihydroxybenzenecarboxylic acids); u. med. as astr. in burns, ulcers, sore throat, diarrhea, etc. (although rather toxic), & carcinogenic; exter. in rhus poisoning; in tanning hides; to provide exptl. hepatic cirrhosis in rats by administr. soln. PO for a prolonged period (740); in making X-rays of intestine (1862); in clarification of wines & beers; in mfr. inks; dyeing; for prepn. tannin; as reagent; in photography; rubber indy.; etc. - **tannier**: tania.

Tannigen™: acetyltannic acid.

tannin: 1) class name for complex phenolic org. compds., fr. plants; many of which are glycosides; char. by strong astringency, producing green-black or blue-black colors or pptes. with iron salts; ppte. proteins (hence u. in leather tanning), alks. etc.; mordant in dyeing; classified into phlobatannins; hydrolyzable (or non-phlobatannins) (subdivisions ellagitannins, gallotannins, etc.); pseudotannins 2) (specif.) tannic acid; tannic acid (*qv.*) is one of the gallotannins (2510) - **t. de mimosa** (Fr): catechutannic acid; occurs in many spp., ex. *Laguncularia racemosa, Eucalyptus smithii, Acacia catechu* (hence the name).

tanning: process of changing natural raw hide into leather by the action of tannin, Cr salts, &c.

tansy: common t. (herb): *Tanacetum vulgare* (also u. fls. lvs. sds. VO) - **double t.**: *T. vulgare* v. *crispum* - **wild t.**: *Potentilla anserina.*

tanzy: tansy.

tapa: (Fiji; Marshall Is): inner bk. of *Broussonetia papyrifera* (paper mulberry); u. to make **t. cloth**, chief clothing material in Polynesia; also made fr. *Pipturus albidus*, breadfruit, *Ficus* sp., &c.

tapate (CR): Stramonium.

tape: plaster often in roll consisting of fabric (usually cotton with or without rayon) coated on one side with self-adhesive plaster mass such as natural rubber, with filler, such as zinc oxide; u. to hold together, support & protect body parts, as in tendinitis, wounds, &c.

tapenade: capers - **taperos** (sFr): frt. of *Capparis spinosa.*

tapioca (<Am Indian): Mandioca, Manioc(a); fr. *Manihot esculenta*; food delicacy made by heating cassava meal to partly convert the starch; u. to make puddings, as thickener in cooking, (soups, &c.) supplementary feeding of milch cows, hogs, &c.; especially good as food for convalescents & children; useful as adhesive - **t. dextrin**: d. made fr. t.; u. as warp size in textile indy. - **factitious t.**: an artificial t. made fr. potato starch by forming into lumps by heat & moisture treatment - **flake t.**: masses of t. formed when the moistened starch is raked on hot plates & dried - **granulated t.** (& pellets): forms u. for puddings; useful for adhesives, etc. - **t. flour**: u. in adhesive pastes, emulsifiers, component of candies, fine pastries, thickener in soups, puddings; as a starch, in hog feeds, etc. - **pearl t.**: prep. in globules & flakes; form eaten as dessert (in puddings, etc.) in US.

Tapiphyllum velutinum Robyns (Rubi) (tropAf): cold inf. of rt. u. for worms & stomach pains.

Tapiri(r)a edulis T.S. Brandeg. (Cryptocarpa e. Standl.) (Anacardi) (wMex): ciruela; frt. pleasantly edible, but sometimes bitter (perhaps unripe) - **T. guianense** Aubl. (Ve): palo de sabo; balsamic secretion u. in med.

tappuakh (Hebrew): apple.

Tapura amazonica P. & E. (Dichapetal.) (eEc): frts. eaten - **T. guianensis** Aubl. (Ve): palo de sebo; balsam (exudate) u. as medl.; fish poison - **T. peruviana** Krause (Co+): calentura chiricaspi (Co); lf. inf. u. in fevers (741); timber wood.

tar: a viscid thick dark-colored semi-liquid of organic comp.; obt. by destr. distn. of organic substances & bituminous minerals; ex. pine tar, juniper t., asphalt; where term u. alone, refers to pine t. (2392) - **American t.:** prod. obt. by heating pine wood (US); pix liquida - **Archangel t.**: former comml. form of wood t. - **Barbadoes t.**: thick black petroleum prod. obt. fr. B.; asphaltite; asphaltic bitumen; manjak - **beechwood t.**: Pix Fagi (L): wood tar obt. by dry distn. of wd. of *Fagus* spp., such as *F. sylvatica* - **coal t.**: v. under coal - **common t.:** wood t. fr. *Pinus sylvestris,* &c.- **t. heel**: cognomen appl. to NCar (US); said based on pine tar anecdote of Am Civil War - **t. oil, rectified** (or **redistilled**): VO distd. fr. pine tar - **pine t.:** Pix Pini (L): obt. by destructive distn. of wd. of *Pinus palustris* & other P. spp. & of *Abies* spp. (US) or of *P. sylvestris* & *Larix sibirica* & other spp. of each genus (Eu); c. benzene, toluene, xylene, phenol., creosol, turpentine oil, resins, guaiacol, phlorol, styrene, naphthalene, retene, pyrocatechol, etc.; u. antisep. stim. in chronic skin dis. (where scale formation & thickening); irritant exp. in subacute bronchitis & pulmon. dis.; very important technical usages; pine t. is dist. to prod. rectified pine tar oil; the residue is Pix Navalis, black pitch; somewhat like pine t.; in ointments, cerates, & plasters - **Stockholm t.**: dry distn. wood t. fr. stumps & rts. of *Pinus sylvestris,* & other Eu conifers; u. in shipbuilding; term now obsolete - **t. water: t. tea:** cold water inf. fr. p. tar (fr. *Pinus rigida,* etc.) at one time very popular in US & Eu. for lung dis., cancer(!?), apoplexy, throat dis., &c.; later displaced by Chinese tea (2527) - **t. weed**: 1) *Grindelia squarrosa* 2) *Hemizonia virgata* 3) *Madia* spp. (**t. weed, California blue:** *Trichostema lanceolatum*) - **wood t.**: prod. obt. fr. dry distn. of various wds., usually of beech or pine fam. - **w. (tar) creosote**: obt. by distn. of wd. tar; u. as preservat. antiseptic - **w. t. pitch:** pine p.: the residue after distn. the VO fr. pine t.

Tara spinosa Britt. & Rose: **tara** (Cu): *Caesalpinia spinosa* Ktze, a source of tannin.

taragon: tarragon.

Taraktogenos Hassk. = *Hydnocarpus* - **T. blumei** Hassk. = *H. kurzii.*

taramira (Urdu): **taramiri**: *Eruca sativa.*

tarantula: name appl. to several genera & spp. of large spiders, incl. *Avicularia avicularia* L. (Aviculari.) (tropCA, WI, SA): bird spider; sometimes carried in bunches of bananas; bite harmless (no worse than bee sting); u. (folk med.) in gastrointest. dis.; *Eurypelma spinicrus* Latr. (Theraphosidae): Cuban t. found in swUS (2772).

tarassaco, radice di (It): **Taraxaci Radix** (L): Taraxacum - **taraxacin**: xall. bitter princ. of Taraxacum; possibly a mixture.

Taraxacum Weber (Comp.) (temp zones worldwide): dandelions; some 70 spp. of perenn. hbs. - **T. densleonis** Desr. = *T. officinale* - **T. erythropodium** Kitag. (Chin): pugongying; rts. diur. bitter ton. galactagog - **T. heterolepis** Nak. & Koidz. (Chin): u. like prec. - **T. hibernum** Steven (T. megalorrhizon Handel-Mazzetti) (USSR): krim-saghyz; krim-sagis; source of a natural rubber; cult. in eEu - **T. koksaghyz** Rodin (T. bicorne Dahlstedt) (cAs, USSR, Turkestan): kok-saghyz; Russian dandelion; latex u. to produce kok-sagis rubber (natural r.) (cult. in eEu) - **T. laevigatum** DC (T. erythrospermum Andr.) (Euras, now widely distr. in nCana, US): (red-seed) dandelion; hb. c. apigenin & luteolin glycosides; vits. A, C, D; u. as bitter, diur. (in bladder dis.), cholagog in gall bladder dis., in liver dis. (Kneipp); depur. in chronic eczema; fresh young lvs. as salad; greens raw or boiled; tap rt. (Taraxacum, L.) c. taraxac(er)in (NBP), taraxerol, inulin (much more in fall), resin, choline, sugars, VO (traces) (743); dandelion root u. as bitter ton. hepatic stim. in cholecystopathies, diur. mild lax. in hemorrhoids; in rheum. (gypsies); roasted rt. u. as coffee substit.; d. tea; d. wine (prepd. fr. fls.); purple dye; pl. a frequent cause of pollinosis; weed - **T. megalorhizon** (Forssk.) Hand.-Mazz. (wAs): krim-saghyz; u. much like *T. koksaghyz* - **T. mongolicum** Hand.-Mazz. (Chin): pu-kung-ying; pl. u. as bitter, diur., in leukorrhea; rubber source - **T. officinale** Wigg. (T. sinense DC; Leontodon t. L.) (Euras, nAf, weed in many parts of world, esp. more temp.; thus does poorly in Fla [US]): (common) dandelion; bitterwort; "Irish daisy"; cankerwort; puff ball; pissenlit (Fr) (diur.); c. & u. much like those of *T. laevigatum*; many forms recognized; all pl. parts u. in folk med. of many countries for warts & cancers, ex. mammary carcinoma; fl. in prepn. d. wine; fl. heads starvation food (Cana, 1930s); rts. coffee substit. - **T. sinense** L. = *T. officinale* - **T. sinicum** Kitag. (Chin): u. like *T. mongolicum* - **T. tau-saghyz** = *Scorzonera t.* Lipsch. & Bosse.

taraxerol: friedoolean-enol; steroid compd. found in *Taraxacum officinale, Tilia cordata, Phragmites communis, Bridelia micrantha*, &c.

taray (Mex): *Caesalpinia bonducella*

Tarchonanthus camphoratus L. (Comp.) (sAf): camphor wood; African fleabane; Hottentot tobacco; fresh lvs. chewed by Muslims as medl.; smoked by Hottentots; lf. inf. u. for abdom. pains, diaph. in asthma; wd. u. to make musical insts., fancy joinery; wd. with odor of camphor.

tare: *Vicia sativa+*.

Tarenna asiatica Gaertn. (Rubi) (Ind): kura (Madras State); lvs. u. in skin dis. - **T. attenuata** Pitard (Indochin): crushed lvs appl. to sores in children - **T. grandifolia** Ridl. (Mal): fls. astr., chewed with betel for cough - **T. incerta** Kod. & Val.: probably = *T. attenuata* - **T. littoralis** (Hiern) Brem. (eAf coasts): frts. of shrub edible - **T. sambucina** Dur. ex Drake (Micronesia to Tuamotu Is): manonu (Tonga); inner bk. inf. u. in "mumu tuaula" (cellulitis with septicemia) (Am, Samoa); in med. in Tonga (86); in combns. for asthma (Am, Samoa); bk. macerated in water, strained, & drunk for sour stomach and as tonic after sickness (Fiji) - **T. spinosa** = *Chomelia s.* Jacq. (Mex, nSA): "malachuita"; "limoncillo"; edible frt. u. in fevers, fls. very fragrant - **T. stellulata** Ridl. (Mal): lf. decoc. u. for bowel dis. - **T. stenantha** Merr. (PI, endem.): lvs. appl. to external pain.

tari: *Caesalpinia digyna, C. spinosa.*

tariric acid: 6, 7-stearolic acid; fatty acid isomeric with linoleic acid; derived fr. sd. FO (cariri fat) of *Picramnia tariri* & other *P.* spp.; u. in mf. of Iodostearine.

taro: (Pac Oc Is): *Colocasia esculenta.*

tarragon: *Artemisia dracunculus* - **tarro** 1) (Salv): gourd 2) (Br): taro.

tártago: 1) (Sp): spurge 2) (Arg): castor bean 3): *Jatropha curcas* 4) (Por): *Euphorbia lathyrus.*

tartar: 1) argols; K acid tartrate, found as deposit (crust) pptd., in wine casks by increasing alc. content of the wine; **cream of tartar** is formed by purifying t. 2) deposit on teeth, chiefly Ca phosphates - **tartar emetic**: antimony & potassium tartrate - **t. sauce**: c. capers, olives, cucumbers, pickles, mayonnaise.

tartarated (Brit): tartrated (US) (SC tartarus (L.) soda tartarata, etc. - **Tartarian lamb**: *Cibotium barometz* - **Tartarus** (L.): crude potassium bitartrate; argols, red - **T. Natronatus**: **T. Potassico-Sodicus: T. Sodico-Potassicus: tartrate de potasse et de soude** (Fr): potassium sodium tartrate - **tartrated antimony**: tartar emetic - **t. soda**: sodium potassium tartrate.

Tartarian: v. also under moss, southernwood.

tartaric acid: COOH $(CHOH)_2$COOH; natural org. carboxy acid widely found in plants; occurs in 4 stereoisomeric forms; d-tartaric acid from argol is u. as acidulant in effervescent medl. salts, baking powd., &c.; buffering agt.; obt. chiefly from argol. - **Tartuffeln** (Ge): potatoes.

tartufo(lo) (It): 1) truffle 2) *Helianthus tuberosus.*

tarwar (Hind): *Cassia auriculata* (sd.)

tarweed: *Madia* spp.; *Hemizonia* spp.; & several other plants.

Tasmanian blue eucalyptus: *Eucalyptus globulus.*

tasneirinha (Br): *Senecio vulgaris.*
tassel (US): male inflorescence of maize (&c.) - **t. flower**: 1) *Amaranthus caudatus* 2) *Emilia sagittata.*
tasso (It): yew tree - **tassobarbasso** (It): mullen.
tatajuba (Br): **tatauba** (Br): *Chlorophora tinctoria.*
tators (sUS): (informal); potatoes.
taurine: 2-aminoethanesulfonic acid; amino ethonic acid; occ. in red meats, fish, human milk, lung, muscle, heart muscle, bile (with cholic acids) (source) & many other animal tissues; beets; higher titer in left ventric. muscle in congestive heart failure; u. prophylactically in stroke, hypertension; epilepsy; antiarrhythmic; SID syndrome; addns. to vit. combns. (Jap); component of taurocholic acid (2327).
taurocholic acid: one of 3 bile acids found in human bile; formed by fusion of cholic acid with taurine (formed fr. cysteine); u. as choleretic.
taurus (L): bull; male of *Bos taurus*, &c.
Tausendgueldenkraut (Ge): *Centaurium erythraea.*
Tauschia bicolor Constance & Bye (Umb) (Chihuahua, Mex): masiawari; lvs. u. as potherb (quelite) - **T. tarahumara** Const. & Bye (Chihuahua, Mex): "huve"; rhiz, u. ground up & in fat or FO to rub on body for rheum.; piece of rhiz. placed in cavity for toothache; lvs. as edible greens.
Taverniera abyssinica Rich. (Leg) (neAf): c. flavonoids; rts. u. in native med. (Ethiopia) as antipyr. & analg. (confirmed by expts.) - **T. cuneifolia** Arn. (Leg): lvs. appl. to ulcers as poultice.
taxa: pl. of taxon.
Taxodium ascendens Brongn. (Taxodi.) (Va to Fla to La [US]): pond cypress; upland c.; less common & smaller tree than *T. distichum*; ornam. tree - **T. distichum** L. C. Rich. (seUS n to Del; along Miss Riv n to Mo, sIndiana; esp. in swamps ["cypress s."]); bald (or deciduous) cypress; southern c.; sabino (Sp, Mex); s. tree; one of the loveliest conifer spp. (-45 m); resin fr. cones u. as diur. carm. vulner.; emmen. (sTex [US], nMex); wd. much u. for packing cases, greenhouse constrn.; water tanks; a valuable lumber for house building (long lasting); air roots ("knees") are aerating organs, popularly sold as novelties to tourists, much u. for wd. carvings - **T. mucronatum** Tenore (sTex [US], Mex, Guat): Mexican cypress; Montezuma c.; trees often attain great age & size; much u. in Mex folk med. (ahuehuete): bk. ext. & VO distd. fr. wd. chips u. astr. in diarrh. bronchitis, &c.; as emmen. (744).
taxol: paclitaxol: active principle in *Taxus brevifolia* bk.; u. in ovarian & breast cancer (not curative but prolongs life of patient); recommended for use by the FDA.
taxon (pl. **taxa**) (New Greek): any entity or unit in pl. & animal classifn., at any level; represents organisms of similar characteristics; important in classifn.; ex. genus, class.
taxonomy, botanical: systematic botany, or the science of classification of pls.; in early days, mostly based on macroscopic characters, now also on microscopy, chemistry, genetics, &c.
Taxus baccata L. (Tax.) (Euras, nAf): common/English/American yew, if (Fr); lvs. c. taxine (toxic alk.); myricyl alc.; lf. decoc. occasionally u. in folk med. as anthelm., emmen. (poisonous); branches u. to make long bows (famous Eng weapon); lvs. bk. sds. toxic but not frt. pulp; wd. u. in veter. med. in struma of horses - **T. brevifolia** Nutt. (Pacif NA, Alaska-cCalif [US]): Pacific (or western) yew; California y.; wd. scrapings in vaseline u. for sun-burn (BC [Cana] Indians); st. bk. c. taxol (active agt. with potent antileukemic & tumor-inhibiting props., esp. in ovarian cancer); branch wd. strong u. to make bows, paddles, spear handles (Am Indians) - **T. canadensis** Marsh. (eCana, neUS w to Ia): American yew; ground hemlock; lvs. u. as hypotensive, in teas (tisanes) in rheum. (Qu [Cana] Indians: Chippewas) - **T. cuspidata** Sieb. & Zucc. (eAs): Japanese yew; lvs. u. as antidiabet. abort. - **T. wallichiana** Zucc. (c&seAs, Afghan, Chin): Himalayan yew; "thuner"; lvs. & sts. u. to prep. beverage drunk as tea for coughs & colds; frt. eaten.
tayberry: frt. of *Rubus idaeus* X *R.* sp., with delicate flavor.
Taylor's Elixir of Yellow Cinchona: El. Cinchonae Flavae (ca.1860).
tayote (PR): chayote.
tayuyá (Br): 1) depuratives in general, but esp. *Cayaponia* (*Trianosperma*) spp. as *C. pendulina* 2) anthelm. in general, but esp. *Apodanthera smilacifolia.*
té (It): **té** (Sp): **te** (Nor): tea - **t. de España**: *Chenopodium ambrosioides* - **té de milpa** (Mex): *Bidens pilosa, B. tetragona* - **té del pais** (Mex): **té del pan** (Mex): *Lippia alba* - **t. limón**: lemongrass lvs.
tea: 1) *Thea sinensis*, or beverage prep. fr. its lvs. & buds 2) a medicinal prep. consisting of a mixture of various hbs. or the inf. prep. therefr. 3) (nUS: addicts' slang): Cannabis - **Abyssinian t.**: **African t.**: khat t. - **alfalfa t.**: hb. u. in beverage sometimes with ground peppermint added - **Algerian t.**: *Paronychia* sp. - **Arab t.**: fr. *Cistus albidus* - **Arabian t.**: khat t. - **Assam t.**: *Thea assamica* (lvs. sds. fls.) - **t. berry**: *Gaultheria*

procumbens - **beverage t.**: tea (1) - **black t.**: the commonest form of Chinese t. u. in US; prep. fr. lvs. of *Thea sinensis* by drying, rolling, then fermenting in dark warm place; have dark color & charact. flavor - **Blue mountain t.**: *Solidago odora* - **Bohemian t.**: *Lithospermum officinale* - **Bourbon t.**: fr. *Angraecum fragrans* (of Bourbon Isl); tea substit. formerly quite popular; said not to occasion insomnia, nervousness, etc. - **breast t.**: Species Pectorales - **Brigham Young t.**: teamster's t. - **bush t.**: inf. of lvs. of *Ledum groenlandicum*; *Cyclopia buxifolia* P. Kies & 9 or more other *C.* spp. (sAf); in addn. 12 other pl. taxa are u. in sAf for the very popular bush teas (129) - **cambric t.**: beverage of weak tea with milk & sugar - **Canadian t.**: t. berry - **cat t.**: khat t. - **Chinese t.**: *Thea sinensis* - **cold t.** (Jam): coconut milk boiled down with *Cecropia peltata* lvs. & let stand for several hrs.; u. to cure colds (51) - **common t.**: Chinese t. - **Continental t.**: Labrador t. - **Croatian t.**: Bohemian t. - **crystal t.**: *Ledum palustre* - **desert t.**: *Ephedra californica* - **Faham t.:** Bourbon t. - **t. family**: Theaceae (Ternstroemiaceae) - **Formosa (oolong) t.**: form of Chinese t. - **Georgian t.**: Gruzinian t., Russian prod. from Grusinien area of Georgian Republic (USSR) - **green t.**: 1) a form of ordinary t. fr. *Thea sinensis*, made by drying lvs. until soft, then rolling, then again drying in sun or in hot air; green color is retained, flavor is quite different fr. that of black t. 2) (nPak): popular beverage of unknown sp. origin - **Hudson Bay t.**: Labrador t. - **Indian t.**: 1) inf. of *Ledum groenlandicum,* &c. (or pl. itself) 2) *Gaultheria hispidula* - **James t.**: Labrador t. - **Java t.**: *Orthosiphon stamineus* - **Jersey t.**: New J. t. - **Jerusalem t.**: *Chenopodium ambrosioides* (v. *anthelminticum*) - **Jesuit's t.**: 1) maté 2) Mexican t. 3) *Psoralea glandulosa* - **kafir t.**: khat t. - **kelp t.**: fr. mixt. of kelps (seaweeds); rich in iodine - **kettle t.**: cambric t. - **khat t.**: lvs of *Catha edulis*, or beverage prep. therefr. - **Labrador t.**: *Ledum groenlandicum* & *L. decumbens* - **lemon t.**: *Cymbopogon citratus* - **liberty t.** (USA, Revolutionary War Period): popular beverage fr. various plants, incl. bee balm - **marsh t.**: crystal t. - **Mexican t.**: 1) *Chenopodium ambrosioides* 2) *Psoralea glandulosa* 3) *Ephedra* spp. (Mex) 4) *Capraria* spp. 5) *Croton monanthogynus* Mx. (1863) 6) *Hedeoma drummondii* Benth. (1864) - **Mormon t.**: *Ephedra antisyphilitica*, *E. viridis*, &c. - **mountain t.**: 1) *Gaultheria procumbens* 2) New Jersey t. - **narrow-leaved Labrador t.**: crystal t. - **Navaho t.**: 1) *Thelesperma megapotamicum* 2) *Heuchera bracteata* 3) **Nevada t.**: *Ephedra viridis*, &c. - **New Jersey t.:** *Ceanothus americanus* - **t. oil**: t. seed oil - **oolong t.**: common t. prepd. by partial fermentation - **orchid t.**: Bourbon t. - **Oregon t.**: *Satureja chamissonis* - **Oswego t.**: *Monarda didyma* - **Paraguay t**: maté, *Ilex paraguariensis* - **patriot t.** (USA, 1770s): inf. of many plant spp., the favorite: anise goldenrod - **pectoral t.**: Species Pectorales - **Philippine t.**: *Ehretia microphylla* - **pigeon t.**: *Cajanus cajan* - **t. plant**: *Lycium halmifolium*, etc. - **red t.** (leaves): *Malus hupehensis* - **Saint Bartholomew's t.**: maté - **Salvador t.**: Canadian t. - **t. seed oil**: FO fr. sds. of *Camellia sasanqua*, *Thea sinensis* & v. - **Siberian t.**: *Bergenia crassifolia*; perhaps also appl. to *Epilobium angustifolium* - **squaw t.** (wUS): for Ephedra spp. - **Sudanese t.**: *Hibiscus sabdariffa* - **sun t.**: beverage t. prepd. by infusing in sun rather than with hot water - **swamp t.**: *Ledum palustre* - **teamster's t.**: Brigham Young t.: wUS *Epedra* spp., incl. *E. antisyphilitica*, *E. nevadensis* - **violet t.**: inf. of lf of *Viola odorata* (sAf), u. in indig. & gastric upsets - **whorehouse t.**: teamster's t. - **willow t.**: inf. of *Salix babylonica* lf. commonly u. in rheum. (sAf).

tea-fungus: symbiosis of yeasts (such as *Schizosaccharomycodes ludwigii*) & bacteria (such as *Acetobacter xylinum* Yamada) (745) (v. under Kombucha).

teak (, **Burma**): *Tectona grandis*, source of teakwood.

tearal: *Eupatorium perfoliatum.*

teasel (, **common**): *Dipsacus sylvestris* - **fuller's t.**: *D. fullonum*.

tea-tree: **Australian:** *Leptospermum laevigatum+* - **black t.**: **brigalow t.**: *Melaleuca preissiana* - **broad-leaved t.**: *Melaleuca quinquenervia*, *M. viridiflora* - **coast t:** Australian t. - **lemon-scented t.:** *L. flavescens* v. *citratum*, - **narrow-leaf** (-**leaved**) **t.**: *Melaleuca linariifolia* - **t. oil**: VO fr. tea (or ti) tree: *Melaleuca* & *Leptospermum* spp.

tea-weed: *Sida* spp.

Tecata (Sp-Cal): Heroin.

Teclea amaniensis Engl. (Rut) (sAf): bk. bitter (c. hesperidin) u. as headache remedy (Swahili) - **T. grandifolia** Engl. (tropAf, incl. Ghana): bk. bitter c. evoxanthine; chewed for cough; twigs u. as chew-sticks; some antitumor activity - **T. nobilis** Delile (tropAf): bk. c. hesperidin, u. in GC; rt. as anthelm.; lf. & st. decoc. as diaph. (Tanz) - **T. simplicifolia** Verdoorn (tropAf): lf. u. to treat pneumonia (Masai), decoc. in milk for lung dis. (Zigula); rt. dococ. for sharp stomach pain (Tanz) - **T. sudanica** A. Chev. (Mal): c. tecleine, flindersiamine; lvs. u. as replacement for those of *Combretum micranthum* in commerce; hypotensive - **T. unifoliata** Baill. (tropAf): rt. decoc. u. in GC &

as ton. (mixed with cow blood & milk) (Masai); wd. u. to make clubs, pestles, walking sticks - **T. utilis** Engl. (Af): wd. u. to make knife handles, ax-shafts, drums - **T. verdoorniana** Exell (Nigeria): lf. juice u. in conjunctivitis.

Tecoma araliacea DC (Bignoni.) (Br): wd. c. lapachol; lapacho wd. comm. timber - **T. chrysotricha** Mart. (Br): source of valuable wd. (ipé tabaco wd., ipé do campo) - **T. grandiflora** Lois. = *Campsis g.* - **T. impetiginosa** Mart. (Br): black ipé; bk. u. exter. as astr. in GC, leucorrh. joint inflammns; exanthemata; wd. valuable - **T. ipé** Mart. (Br, Par, Ur/Arg): lepacho negro; source of lapacho wood, comm. timber, u. in skin dis.; heart wd. c. lapachol (antineoplastic), chrysophanic acid; trunk bk. astr. u. against cancer - **T. lapacho** K. Schum. (Br, Arg): source of comml. wd. lapacho wd. - **T. leucoxylon** Mart. (WI, Guianas, Br): bk. u. as tan (Guian); wd. (green heart wd; green ebony wd.) c. lapachol & skin irritant resin; u. as comm. timber. (*cf. Diospyros chloroxylon*) - **T. mollis** HBK = *T. stans* var. *velutina* DC - **T. ochracea** Cham. (Br): brown ipé; bk. & lvs. u. as astr. in syph., flatulence; in gargles & eye waters; wd. (brown ipé) u. in herpetic eruptions; as comm. timber - **T. radicans** Juss. = *Campsis r.* - **T. stans** (L) Kunth (Stenolobium s. HBK) (Fla to Tex, NM [US]; WI, Mex, CA, SA): yellow trumpet-flower; y. elder; hardy y. trumpet; tronador(a); retama; palo del arco (Mex); lvs. c. tecomine, tecostanine; u. diur. for colic, ton. to stim. appetite; Herba Tronadorae in diabetes; rt. decoc. u. diur. ton. antiluet. vermif.; in mfr. beer (like hops); lvs. (with *Cuscuta*) as antipyr. (Virgin Is); fl. decoc. for stomach ache (2342); sd. kernels source of FO, garapa oil (Br) (20% yield), semi-drying; cult. ornam. shrub; var. *velutina* DC (T. mollis HBK) (LatinAm): retamo; lvs. & sts. inf. drunk for TB (Pe) - **T. undulata** G. Don = *Tecomella u.*

Tecomaria capensis Spach (Bignoni) (e&sAf): Cape (or Kaffir) honeysuckle; C. trumpet-flower; pd bk. u. in pneumonia, high fever, to relieve pain; lf. decoc. for enteritis - **T. nyassae** Baill. (tropAf): rt. decoc. in combns. for cough; wd. u. to make hut poles & doors; cult. ornam.

tecomate (Salv): **tecomatillo** (Salv): calabash tree, *Crescentia cujete, C. alata.*

Tecomella undulata Seem. (monotypic) (Bignoni) (swAs, Arabia): rahura (Pak); bk of young branches u. in syph. (Ind, Pak); wd. u. to make wooden combs (Pak).

Tectaria irregularis Copel. (Aspleni./Aspidi.) (Indochin): woolly scales u. as styptic like those of *Cibotium*; pd. rhiz. & rts. u. as poultice for itch; lf. inf. for diarrhea - **T. variolosa** (Wall.) C. Chr. (Indochin): inf. of young boiled fronds u. to treat colic, stomachache.

Tectona grandis L. f. (Verben.) (Ind, SriL, Mal, Indones; cult. Af+): teak; common t.; huge trees; wd. (bois de tech) c. tectol, lapachol (antineoplastic); pd. wd. u. to treat skin dis. & as vermif., in mal.; teakwood is very hard & highly resistant to termites; chiefly in ship construction, making furniture; lvs. in cholera, aphthous stomatitis ("thrush"); diur.; in dyeing silk & basketwork yellow; bk. u. in dermatitis (Thail), as tan.

tectoquinone: methyl anthraquinone, crystall. compd. found in teak wd. & in tar prod. therefrom (resin).

tee (Ge) (formerly Thee): tea.

teel (Ind): *Sesamum indicum* (FO).

Teep (fr. Tee + Pulver [Ge] = tea + powder): prepn. of fresh plant made by trituration with sugar, &c. & dr. without heat & powd. (Ge Homeop.).

Teer (Ge): tar, esp. pine t. - **Teeröl**: rect. pine. t. oil.

Teesdalia nudicaulis (L). R. Br. (Cruc.) (Eu, Med, natd. NA): shepherd's cress; young pl. u. as vegetable.

t'ef: **teff (grass)** (Amharic): *Eragrostis tef.*

tegmen: 1) inner sd. coat developed fr. the secundine of ovule 2) glume of grass fl. - **tegmenta**: pl of **tegmentum**: 1) outer covering of scales of a leaf bud 2) spermoderm (sd. coat, when single) 3) indusium of a fern.

teichoic acids: phosphoric acid polymeric esters with glycerol or ribitol; occ. in some bacterial cell walls (often 40-60% of content) & cytoplasm; obt. fr. *Lactobacillus casei,* &c.

Teijsmanniodendron Koorders (Verben.) (seAs, Indones): 14 spp.; at least one sp. u. with coconut oil to form ointment u. for hemorrhoids.

teinture (Fr): tincture.

tejido (Sp): tissue; membrane.

tejocote (Mex): *Crataegus mexicana.*

teleology: the doctrine of purposiveness in natural phenomena, as opposed to the doctrine of mechanism, which latter now prevails.

Teleostei: the bony fishes; a sub-class (largest) of Pisces; ex. salmon, walleyed pike, carp, catfish, trout, herring, eel, cod, mullet, mackerel (formerly named *Teleostomi*).

telepathine: harmine.

Telfairia occidentalis Hook. f. (Cucurbit.) (wAf): (common) oyster nut tree; young shoots & lvs. u. as pot-herb; sds. cooked & eaten like beans; sd. kernels c. 48% non-drying FO (**telfairia oil**) u. in cooking, soap mfr.; frt. shell u. for utensils; dr. st. tissue u. like loofah sponge - **T. pedata** Hook. f. (tropAf): Zanzibar oil vine; o. plant; oyster nut;

koeme; frt. pulp bitter & gives headache after ingestion; source of jiconga (jekungu) nuts (kernel edible after roasting) c. 50% bitter tasting FO; u. as food (unpleasant odor), to make soap, candles; for stomach dis., rheum.; lf. bitter ton.; rt. ext. for hydrocele.

Teliostachya alopecuroidea (Vahl) Nees (Lepidagathis a. R. Br) (Acanth.) (tropAm, incl. CA, nSA, WI): lvs. u. to make soothing tea for nervous children (Domin, Caribs) (2328) - **T. lanceolata** Nees (+ var. *crispa* Nees) (Co-Pe): hb. u. for stomachache (777); u. as hallucinogen & added to ayahuasca; cult.

Telitoxicum peruvianum Mold. (Menisperm.) (Co-Pe): bk. of treelet is chief ingred. of Barasana curare. (also **T. minutiflorum** [Diels] Mold. so u.).

Tellerrot (Ge) (plate red): purified carthamin; process finished on plate.

Tellicherry bark: *Holarrhena antidysenterica* (bk.) - **T. peppers**: superior kind of black pepper (*Piper nigrum*) of Indian origin with full flavor, much u. in curry flavoring.

Telomycin: polypeptide antibiotic fr. *Streptomyces* sp. from Fla [US] soil (1957).

telondella (Mex): *Jatropha spatulata*.

Teloschistes candelarius (L) Fink (Xanthoria candelaria Th Fr) (Buelli. - Gymnocarpeae - Lichenes) (worldwide): ljus mansa lichen (US = on trees): thallus u. as brown dye - **T. flavicans** (Swartz) Norm. (Borrera f. Sw.) (both hemisph): yellow borrera; borrera lichen; c. anthraguinones; u. as brown dye; test plant for air pollution (grows only in pure air).

Telosma procumbens Merr. (Asclepiad.) (PI): latok; lf. decoc/inf. u. to cleanse wounds, ulcers, scabies; lvs. as cataplasm appl to forehead in headache; young frts. cooked & eaten.

temoé lawak = temu lawak.

témoin (Fr): control.

tempate (CR): *Jatropha curcas*.

template: pattern or gage; term u. in genetics for means of replicating.

templin oil: 1) VO fr. cones of *Pinus mugo* & some other *P.* spp. 2) VO fr. green cones of *Abies alba* (usually).

temu lawak (Java): *Curcuma longa*; sometimes *C. xanthorrhiza*.

tendu (leaves) (Ind): bidi (l.).

Tenerif(f)e: largest of the Canary Is w. of nAf; v. under Coccus.

tengah bark: fr. *Ceriops* spp.

teniposide: podophyllotoxin deriv. u. as antineoplastic.

Tennecetin: Natamycin (Massengill) (SC researched at Univ. Tenn [US]) (1890).

teocinte (SpAm): **teosinte**: *Euchlaena mexicana*; believed closely related to, possibly ancestor of, maize (Indian corn) (*Zea mays* subsp. *mexicana*).

tentoxin: cyclic tetrapeptide occ. in *Alternaria alternata* (Fr) Keissler (A. tenuis); induces chlorosis & effective as natural herbicide.

tepal: perianth segment, like either sepal or petal, where not clearly differentiated; ex. tulip.

teparí: **tepar(r)y**: **t. bean**: 1) *Phaseolus acutifolius* var. *latifolius*; beans popular with (frt).

tepejilote (Mex): *Chamaedorea elatior*.

Tephrosia Pers. (*Cracca L.*) (Leg.) (worldwide): goat's rue; devil's shoestrings; sand pea; etc; hbs. or shrubs; c. 400 spp.; many spp. c. rotenone, deguelin; u. fish poison, insecticide (746) - **T. aequilata** Bak. (tropAf): u. to relieve abdom. pains - **T. apollinea** (Delile) Link (Eg. Arabia): silver-leaved senna; lvs. u. like senna; adult. of off. senna; source of a low grade indigo - **T. bracteolata** Guill. & Peyr. (tropAf): rts. u. in syphilis (pregnant female); fodder for horses - **T. candida** DC (Af, Ind, natd. WI): boga medeloa (Ind); rt. bk. & sds. c. tephrosin & rotenone (-0.5%;) bk. & lf.; u. fish poison, insecticide; cover crop between tea bushes (on t. plantations) - **T. capensis** Pers. (sAf): rt. inf. u. as emet., in biliousness; cooked rt. in card. palpitations; arrow & fish poison - **T. cathartica** (Sessé & Moç.) Urb. (Galega c. Sessé & Moç.) (Mex, WI, esp. PR, Lesser Antilles): sen (amargo) (Domin); indigo marron (Guad); u. as cath. (PR: hoja-sen) - **T. cinerea** (L.) Pers. (Antilles, Guian, Br): West India (WI) fish poison; barbasco (blanco); añil (cenizo); pl. c. rotenone; u. fish poison; insecticide; lf. decoc. u. for fever, severe colds, cardioton - **T. densiflora** Hook. f. (tropAf): fish poison bean; sd. c. tephrosin (0.3% in sd.); HCN; rt. u. arrow poison; lf. & pod as fish poison; cult as cover crop - **T. diffusa** Harv. (sAf): rt. decoc. rubbed well into hair for head lice; insecticide; fish poison - **T. grandiflora** Pers. (sAf): rt. decoc. parasiticide; fish poison - **T. hispidula** Pers. (T. elegans Nutt.) (seUS, n to Va): hoary pea; rt. chewed for cough (Choctaw Indians) - **T. kraussiana** Meissn. (Af): rt. inf. u. for night coughs; may c. saponin - **T. latidens** Standl. = T. virginiana - **T. lupinifolia** DC (sAf): rt. decoc. taken by mouth to induce abort.; for suicide, pounded rt. inserted into vagina —> death in ca. 12-24 hrs. - **T. macropoda** Harv. (e&sAf: Natal): fish bean (or poison); (i)lozane (Zulu); c. rotenone (0.3-0.4% rt); toxicarol; deguelin; rt. (& twigs) u. as piscicide, insecticide, molluscicide; in wash for fleas on dog; rt. u. in typhoid fever, anthelm. (cattle) - **T. nitens** Benth. (tropAm): lvs. of shrub u. to stupefy fish (Br) - **T. noctiflora** Bak. (Tanz): rt.

chewed with water for cough; emetic; rt. decoc. aphrod.; in GC - **T. obcordata** Bak. (Sen): vulner., antisep. to cicatrize wounds (pd. rts., leafy sts.); rts. & lvs. in impotence - **T. petrosa** Blatter & Halb. (Ind): boiled lvs. eaten for syph. - **T. piscatoria** Pers. = **T. purpurea** Pers. (T. leptostachya DC) (e tropAf, as Tanz; natd. As, Austral, Oceania): ash vetch; wild indigo (Ind); bastard i.; s(h)arpankha; rts. c. rotenone (7%), tephrosin, isotephrosin, deguelin; "cubé rt." u. as diur., ton. emm. (Indochin), for abortion (Nepal); spasmolytic, bitter astr. stom. in liver dis., tumors, for coughs; purg. (Ind); rt. ext. to stupefy fish (Polynesia); toxic to stock (Austral); insecticide; lvs. c. 2.5% rutin, u. anthelm. ton. (Ind): common weed (Ind) (747) - **T. rosea** F. Muell. (Austral): Flinders River poison; a fish poison; considered a stock poison - **T. seminole** Shinners (sFla [US]): devil's shoestring; u. as bitter astr., to stop epistaxis (pl. decoc.) (Seminole Indians) - **T. senna** H. & B. (tropAm): sen (Co); lvs. u. as purg.; substit. for senna - **T. sinapou** A. Chev. (Co, Ec): barbasco; shrub (cult. in gardens, Ec): u. as fish poison - **T. tinctoria** Pers. (Ind, SriL): lvs. u. as adulter. of senna; source of inferior indigo - **T. toxicaria** (Sw.) Pers. (WI, Mex, CA, nSA, incl. Guyana, natd. Af, Oceania): barbasco; timbo (these not exclusive names); fish bean (Af); rt. c. deguelin, sumatrol, toxicarol; possibly rotenone; u. insecticide; fish poison; arrow poison; purg.; in scabies; blue dye - **T. villosa** Pers. (Ind): lf. juice u. in edema; in diabetes; var. *incana* Roxb. (Af): c. saponins; lf. & rt. u. anthelm. - **T. virginiana** (L) Pers. (T. latidens Standl.) (sCana, e&cUS, Mex): Devil's shoestring(s); catgut (plant): hoary pea; turkey p.; rts. woody parts, & sds. c. rotenone, dehydrorotenone, deguelin, tephrosin; u. insecticide, vermif., lax.; bladder "trouble" (Am Indians); as "nerve medicine," in diabetes (folk rem.); insecticide - **T. vogelii** Hook. f. (tropAf, incl. Angola, Tanz, Sen): fish bean (Zambia); f. poison; tuba (& many other names in Af); lvs. c. rotenone, deguelin, tephrosin; u. as fish, & arrow poison; insecticide (also twigs, rts. & sds.); rt. in dental pains, purg. rheum (native Af med.); bk. & lf. inf. anthelm.; lvs. abort.; in yaws (decoc.).

tephrosin: hydroxy-deguelin; a chromone deriv.; fr. lvs. of *Tephrosia vogelii*, derris, cubé rt., &c.; toxic to cold-blooded animals; u. in insect poisons.

tepopote (swUS, Mex): *Ephedra antisyphilitica*.

tepozan (Mex): Buddleja americana.

tequila (Mex): strong alc. beverage made fr. *Agave* spp.

Teramnus labialis Spreng. (Leg) (WI, CA to Ve, natd. OW trop): horse (or rabbit) vine; pl. decoc for stomachache (Ve Indians); frt. stom. antipyr. in rheum. (Ind).

teratogen: product wh. causes mutations; *cf.* mutagen.

teratology: science of monsters & monstrosities, esp. fetuses.

terebene: mixture of terpenes, mostly dipentene (dl-limonene) & terpinene obt. by the action of cold sulfuric acid on turpentine oil; u. exp. & inhalant in bronchitis.

terebi(ni)c acid: a monocarboxy ring acid prod. by oxidn. (HNO_3) of turpentine oil, hence the name.

terebinth tree = *Pistacia terebinthus*.

Terebinthina (L.): crude gum turpentine oleo-resin as taken fr. the pine tree - **T. Abietina: T. Abietis:** oleoresin fr. *Abies* yreligiosay - **T. Argentoratensis:** Strasburg (or Alsace) turpentine; fr. *Abies alba*; no longer common in commerce - **T. Canadensis**: oleoresin fr. *Abies balsamea* - **T. Cocta**: pine resin obt. by distn. or heating t. oleoresin - **T. Communis**: American (or European) crude gum turpentine - **T. Gallica: T. Laricina: T. Laricis: T. Veneta:** Venice turpentine fr. *Larix decidua*.

térébinthine commune (Fr): crude gum turpentine oleoresin. fr. *Pinus* spp.

Terebinthus Mill. = *Pistacia*; *Metopium*; *Bursera* - **T. brownei** Jacq. = *Metopium b.* - **T. lentiscus** Moench = *Pistacia* l.

Terfezia (Tul.) Tul. (Terfezi. - Pezizal es - Fungi) (nAf, Alg, arid regions): truffle; subterr. frtg. bodies (ascocarps) edible ("Kames"); symbionts with Cistaceae spp.; u. to treat skin & eye dis. (antibiotics with Gm.- bact. inhibition; folk med.

teri: tari - **teriaga vêneta em po** (Por): pd. Venetian theriac; Pulvis theriac; *v.* Venetian theriac.

Terminalia arjuna W. & A. (Combret.) (Ind, Mal): "arjuna myrobalan"; kahu(a); bk. decoc. u. cardiac stim., diur., astr. for open wounds; poison antidote; ash for scorpion sting; wd. economically useful, for carts & agr. implements; myrobalans prod. in Bangladesh (-1950-) - **T. avicennioides** Guill. & Perr. (tropAf): trunk source of a spongy reddish material burned as incense (ouolo [mogo]); rt. decoc. u. in ascites, edema; rt. bk. u. exter. for slow healing wounds (Sen) - **T. baumannii** Engl. & Diels (sAf): rt. decoc. u. in diarrh.; pl. u. to make black dye - **T. bellerica** (*bellirica*) Roxb. (Pak, Ind, SriL, Burma, Mal): bedda nut tree; harra (Nep); frt. (belleric myrobalan[s]) c. 17% tannins (more in unripe frt. than ripe); astr. in diarrh. dys. eczema, chronic sores, in throat dis.; ton., sl. lax., anthelm.; in abdom. tumors (ulcers of portal vein system); carcinomata; in scorpion sting; greatly reputed in Ayurvedic med.;

in tanning indy; frt. edible; peel & lvs. u. as tan., to dye black; tree source of a dark gum (trunk smooth surface, without cracks) - **T. bialata** Steudel (Burma, Andamans): bk. decoc. u. in dys. & as postpartum ton.; wd. valuable, of 2 kinds (white, gray) - **T. calamansanai** Rolfe (PI): bk. astr. u. lithontryptic; u. int. & exter. - **T. catappa** L. (tropAs, Austral, Pac Is, pantropic; widely cult., ex. eAf): Indian (or tropical) almond (almendro); myrobalan; sd. (as "country almond") edible, u. in carcinoma (Mex); c. 50-60% FO (catappa oil; talisai oil [PI]) with almond odor; chiefly olein, palmitin, linolein; u. in cooking, substit. for almond oil; med.; in cosmetics; bk. rich in tannin, u. in diarrh. dys. gastric dis.; exter. in skin dis.; sap of young lvs. u. in Hansen's dis. (PI), scabies; lvs. u. purg., vermifuge (2453); frt. edible, popular, somewhat purg.; frt. juice u. as dye & to stain teeth black - **T. chebula** (Gaertn.) Retz. (tropAs, widely cult.): (black) myrobalan; harra (Ind); dr. frt. known as True Myrobalans (formerly BP); unripe c. 20-45% tannin (ellagitannins); u. astr. exter. in wounds, ton. lax. anthelm.; dentifrice; stim. appetite (with milk); cardiotonic; with high repute in Ayurved med., very important in med. of Middle Ages; u. tan, nutgall substit. (WW II), in dye indy.; prepn. fountain pen ink (Pak); wd. at times u. for smoking (748) - **T. citrina** Roxb. (Ind to PI): bingas (PI); frts. (yellow myrobalans) u. astr. in obstinate diarrh., thrush; much like *T. chebula* - **T. comintana** Merr. = *T. citrina* - **T. coriacea** W. & A. (tropAs, esp. Ind): bk. u. as moderately potent cardioton. - **T. edulis** Blco. = *T. bellerica* - **T. laxiflora** Engl. (w tropAf): lvs. (deprived of juice) & rts. u. in dys. - **T. macroptera** Guill. & Perr. (tropAf): bk. lf. decoc. diur. lax.; lvs. u. for ringworm & other skin dis.; galled frts. in dys. - **T. microcarpa** Decne. (T. edulis Blco.) (Java, PI): calumpit (PI); fleshy frt. edible, acid, sl. astr., for preserves; in acute diarrh.; in eye lotions, skin lotions (eczema, wet herpes); lvs. & sds. u. in diabetes; bk. tannin-rich, u. astr. tan. - **T. moluccana** Lam. = *T. catappa* - **T. nigrovenulosa** Pierre (S. Vietnam, Camb): chieu lieu; bk. inf./decoc. u. for diarrh. - **T. paniculata** Roth (swInd): kindal (Bomb); bk. diur. carioton. - **T. pierrei** Gagnep. (Indochin): frts u. as astr. & ton. - **T. sericea** Burch. (tropAf): Assegai (or yellow) wood; silver terminalia; rt. decoc. bitter, u. to stop purging, in dys., hydrocele; stomach dis.; pd. bk. taken with mealie meal for diabetes; bk. ext. u. as yellow dye; wd. in furniture - **T. splendida** Engl. &.Diels (tropAf): pd. bk. u. as snuff (women, Sudan) - **T. superba** Engl. & Diels (trop. wAf): shingle wood; Indian almond; rts. c. resin, u. as lax.; bk. decoc. u. for dys. diarrh. - **T. tomentosa** W. & A. (T. alata Heyne) (Ind, SriL): an (Hindi, Punj.): bk. as paste for boils, bk. decoc. in diarrh. (Burma), dys.; frts. eaten to reduce wt. in edema (124); cardioton., diur.; in ulcers; trunk of large tree prod. tears of gum with strongly adhesive props., eaten by natives; lvs. bear "tasar" silkworm; bk. u. in tanning.

termites: "white ants"; insects representing many thousands of spp. of order Isoptera; chief importance lies in the great destruction they cause esp. in the tropics (ex. telegraph poles; papers); genera, incl. *Termes* & *Termopsis*; sometimes live in massive strong structures called termitaria, up to 9 m tall, and made by cementing bits of earth, vegetation, &c., together; these animals are frequently eaten & very nutritious; queens are eaten as aphrod. (Ind); antimicrobial; for growth stimn. ("termitin"); made into ointment, with ox hump fat, the insects are u. for hemorrhoids (Unani med.).

Ternstroemia gymnanthera Beddome (T. japonica Thunb.) (The.) (Jap, Chin, South Kor, Ind): mokkoku (Jap); lvs. u. to allay mal. (mt. people, Taiwan); bk. & rt. u. in dys. (Jap); wd. u. in building ships (saponins protect wd. against termites) - **T. pringlei** Standl. (Mex): flor de tilia; pl. with cyanide odor; decoc. with cognac u. in pneumonia, cough, throat infections, hemoptysis - **T. robinsonii** Merr. (Indones, NG): crushed bk. u. to stupefy fish; kill head lice; wd. u. to make tool handles - **T. tepezapote** Schlecht. & Chem. (Taonabo t. Szysz.) (sMex, CA): river craboo (Belize); bk. decoc. u. in snakebite (Guat); fl. decoc. u. to bathe rheum. parts of body.

Ternstroemiaceae: Theaceae (*qv.*).

terpene: unsaturated hydrocarbon, typically cyclic, with the general formula $(C_5H_8)_n$; four types are: hemiterpenes (n = 1), **true terpenes** (n = 2), sesquiterpenes (n = 3), & diterpenes (n = 4); triterpenes, polyterpenes; these & their derivatives, phenols, alcs., ketones, aldehydes, esters, etc., represent the chief fragrant components of most VO; ex. menthol, pinene, menthone - **terpeneless oils**: kind of VO (esp. citrus oils) fr. which the terpene hydrocarbons have been removed by various methods, principally fractional distillation under reduced pressure, volatile solvents extn. etc.; ex. t. o. of lemon, t. o. of orange (removal of limonene, etc., leaving larger proportion of citral, citronellal, etc.) (2018).

terpenoid: all isoprene derivs. regardless of functional gps.; incl. such terpene-like compds. as oxygenated terpenes (ex. t. alcs.), sesquiterpenoids (ex. zingiberene), diterpenoids (ex. abietic, palustric. & pimaric acids), triterpenoids (ex. sterols); some alks. (ex. nupharidine), tetraterpenoids (ex. carotenoids), rubber, &c. (term terpene restricted to hydrocarbons) (1866).

Terpentin (Ge): turpentine - **Venetianischer T.**: Venice t.

terpilenol = terpineol.

terpinene: monoterpene hydrocarbon found in VO of various pls.: 3 isomers known: a- (in cardamom, juniper, VO); ß- (*Pittosporum tenuifolium*-VO); ß- is also made by synthesis; g- (VO of coriander, *Mosla japonica*).

terpineol(um): p-menthen-ol; a liquid terpene alc. with lilac fl. odor; occ. usually only in small quantities in many VO; found in 3 forms: d-a- (in lovage; petitgrain oil-*Backhousia angustifolia*); 1-a- (*Pinus palustris*, VO); dl-a- (*Melaleuca leucadendron, Leptospermum flavescens*); now chief source by synthesis; u. perfuming agt., in cosmetics, soaps; to denature FO for soap mfr.; corrective for iodoform odor.

terpin hydrate: product made by oxidizing the terpenes (esp. pinene) of turpentine oil; u. stim. expect. (more accurately terpinol hydrate).

terpinol(um): mixt. of terpenes, esp. terpineol, dipentene, terpinine, &c., found in wood tissues; u. in resp. inflammn.

terpinolene: monocyclic terpene hydrocarbon, $C_{10}H_{16}$; oily colorless liquid; found in VO of *Coriandrum sativum*, manila elemi, &c.

Terra (L.): earth - **T. Alba**: 1) kaolin clays 2) gypsum; Ca sulfate dihydrate (usual meaning) 3) (less commonly) burnt alum; magnesia; $BaSO_4$; fuller's earth - **T. Blanca**: whiting; crude $CaCO_3$ - **T. Cariosa**: rotten stone - **t. firma** (L): **t. firme** (Br): elevated unflooded terrain, solid ground, as in contrast to boggy or swampy terrain (Amazon Basin) - **T. Fullonica**: **T. Fullonum**: Fuller's earth - **T. Japonica**: cutch; ext. of *Acacia catechu* - **T. Merita** (L.): *Curcuma longa* - **T. Rossa**: red soil (lit.); (Greater Antilles): characteristic of bauxite soils - **T. Sigillata (alba, rubra)**: (white & red) bole - **T. Silicea Purificata**: purified siliceous earth.

Terramycin™: antibiotic prod. by *Streptomyces rimosus* Waksman; wide spectrum type, active on bacteria, viruses, Rickettsiae, etc.; appl. in pneumonia, septic sore throat, septicemia, peritonitis, GC, trachoma, amebiasis, actinomycosis, etc.; powerful growth stim. for farm animals.

terrapin (US): fresh water turtles, with meat having market value for food, &c.

terrarium: closed box with glass sides & top for growing pls. & animals indoors; is modified Wardian case (Nathaniel B. Ward, ca. 1872) u. for transporting live pls. across ocean.

terre à porcelaine (Fr): fuller's earth.

terre blanche (Fr): terra blanca.

terreic acid: antibiotic metabolite compd. produced by *Aspergillus terreus* Thom; quite stable in acid medium, decomposes in alkaline; a monobenzoquinone (1942-).

terthienly, α: compd. from *Tagetes* spp. (NA) u. as ascaricide.

teshuino (NM [US], Sp): beverage made fr. sprouting maize.

tesota: *Olneya t.*

Tessaria integrifolia R. & P. (1) (Comp.) (T. legitima DC) (sCA; c&sSA): bobo (Br, Pe, Arg): alisa (Arg); pajara bobo (Pe); lvs. & rts. u. as diur. & in hepatitis (Pe); wd. light, easily worked, u. for small indoor articles (furniture, &c.); in paper mfr. (749); bk. tan. - **T. sericea** (Nutt.) Shinners (Pluchea s. [Nutt.] Cav.) (swUS, Mex): sts. inf. u. for sore eyes (Mex Indians).

Tessmannia moesiekei Harms (Leg) (Af: Congo): source of Congo (or West African) Copal.

test, Borntraeger: reaction with emodin; dissolve aloe (&c.) in boiling water, cool, filter, shake filtrate with ether, and shake sepd. ether fraction with ammonia water (red = +) - **Schmidt's t.**: 1) for sugars: Pb acetate added & pptd. with ammonia water; on heating, white ppt. remains unchanged if sucrose or lactose present but orange color if glucose present 2) for glucose: ammonium molybdate gives white ppt. - **Schoenteten's t.**: to soln. of aloe (&c.) add small amt. borax, heat, pour few drops into test tube of water (green = +).

Testa Ovi (L.): egg shell; u. antacid.

Testiculi Cervi (L.): sheep testes; u. as aphr., etc. (folk med.).

testis (plur. testes): testicle; orchis (Gr); the primary male sexual organ (gonad); paired organs contained in scrotum & suspended externally to body cavity; occas. article of food ("mountain oysters"); secretes testosterone & other male hormones; formerly dr. testis was administered by physicians p. o. but active princlples digested, hence worthless; u. in Unani med. (bull, tiger, fowl) to incr. sexual potency, as aphrod., promote hair growth, &c. (2741).

testoide (Fr) (adj.): androgenic effect of steroids.

testosterone: the primary male sex hormone; secreted by the interstitial cells of the testis; obtained from bull testes or derived by synth. fr. cholesterol, &c.; u. to promote secondary sex characters; to treat male hypogonadism, cryptorchidism, eunuchoidism, impotency (of transient value); lack of sexual libido in both sexes, etc. (2499); mostly u. in form of esters; however, t. is administered as pellet by implantation technic (E) - **methyl t**:

ester admin. orally or sublingually, mostly in maintenance therapy - **t. patch:** furnishes absorption of. t. through skin; approved by FDA (Testoderm) - **t. propionate: Testosteroni Propionas** (L.): ester sol. in alc., FO, etc. much u. by IM inj., **t. cypionate** and **t. enanthate** are also in use - **Testoviron**™: methyl testosterone propionate.

Testudinaria Salisb. = *Dioscorea* - **T. elephantipes** Lindl. - *D. e.* Engl.

tetanus: lockjaw; infectious dis. with contraction & spasm of the voluntary muscles of body (spasmodic closing of jaw), including trismus (lockjaw), opisthotonos (arching of back), &c.; death - **t. antitoxin** (aq. soln. of the antitoxic substances prepd. fr. blood serum or plasma of healthy horses immunized against the toxin): is u. as prophylactic (passive immunization) where indicated (E) - **t. & gas gangrene a.**: obt. fr. mammals immunized against toxins of *Clostridium tetani* (organism of t.), *Cl. perfringens*, & *Cl. septicum* (last two are organisms of gas gangrene); u. as prophylact. where either or both dis. are prevalent - **t. toxoid** (systemic) and **adsorbed t. toxin** (adsorbed on $AlPO_4$ &c) u. similarly but produce an active immunization (E); t. toxoid u. in flocculation tests; combn. of adsorbed t. toxoid with diphtheria toxoid also in use - **T. immune globulin** (immunoglobulin) (also called **human t. antitoxin [human]**) (prepd. fr. blood plasma of humans who have been immunized with t. toxoid): u. in prophylaxis & treatment of tetanus (E) - **t. "vaccine"**: t. toxoid.

Tetracarpidium conophorum Hutch. & J. M. Dalz. (1) (Euphorbi) (trop wAf): lvs. appl. locally for headache (Nig); lvs. frt., & sds. eaten.

Tetracera alnifolia Willd. (Dilleni) (w tropAf): pl. juice dropped into eye as collyrium for cataracts; combns. in snakebite; lvs. astr., u. as fish poison; paste fr. leafy twigs appl. to headache, backache, rheum.; alc. ext. u. as antipyr.; lvs. in aphrodis. prepn. - **T. indica** Merrill (T. assa DC) (Mal, Indones): pounded pl. appl. to scabies; sap in cough; lvs. & st. tips in poultice for festering fingers; rough lvs. u. as sandpaper - **T. loureiroi** Craib (Mal): pl. u. for scabies - **T. macrophylla** Wall. (Mal): pulped lvs. & rts. appl. to foul syphilitic ulcers - **T. potatoria** Afz. (trop wAf): water tree (Sierra Leone): pd. lf. or lf. decoc. given for coughs or held in mouth for toothache; lf./rt./st. decoc. u. in VD - **T. sarmentosa** Vahl. subsp. *asiatica* Hoogl. (seAs): decoc. antipyr. ton. (Indochin) - **T. scandens** Merr. (T. volubilis Merr.) (tropAm, seAs): chaparro tietie; sap u. in cough (Indones); sds. & lvs. u. as diaph. antipyr. diur. in syph.; one of the "water vines," the large st. of which when cut supplies sap clear as water; a source during dry months of year - **T. sessiliflora** Triana & Planch. (Mex, CA, Co): bejuco de agua (Mex); rough lvs. u. to polish wd.; sts. as rope substit.

Tetraclinis articulata (Vahl) Mast. (Callitris quadrivalvis Vent.) (monotypic) (Cupress.) (sSp, nAf): arar tree; juniper gum tree; bk. & branches source of sandarac gum (resin) (g. juniper; pounce); u. loc. stim. in gout, rheum., "risings," &c. (1884); temporary dental fillings; for incense; techn. in shellacs & varnishes (retouching s. in photography; tablet s.); glass & porcelain cements; lf. decoc. abort.; crushed lvs. as cataplasm; for migraine, neck pains, insolation; tar fr. old wd. u. in dermal dis. (750); valuable wd. u. to build houses; make cabinets; chief prodn. Mor; export port: Essaouira (Mogador).

Tetracycline: generic name of antibiotic, with many proprietary names, such as Achromycin, Panmycin; produced by *Streptomyces viridifaciens* Pridham & al., &c., also made by removal of one Cl atom fr. molecule of Chlortetracycline (Aureomycin™); u. much like latter, with fewer by-effects; anti-amebic, antibacterial, antirickettsial; HCl salt popular form; one of few agents effective against *Chlamydia trachomatis*, cause of trachoma & STD, lymphogranuloma venereum, non-gonococcal urethritis; Lyme's dis. (early); essential in treating some forms of mal. (E) (1501).

Tetradymia canescens DC (Comp.) (wUS): inf./decoc. of pl. u. as physic.; decoc. in VD (Navaho Indians [US]); toxic to sheep (photosensitization).

tetraenoic: fatty acid with 4 double bonds (ex. arachidonic acid).

tetra-ethyl-thiuram-disulfide: u. to prevent mold growth in sds.

Tetragastris balsamifera Ktze. (Hedwigia b. Sw.) (Burser.) (WI, esp. Hispaniola): abey (hembra); gommier (encens); pigwood; bk. decoc. u. as antipyr.; resin (resina cachibu; "Balsam Hedwigia") u. as substit. for Peru Balsam, in chest complaints, dyspep.; placed in hollow at base of neck for hoarseness; sometimes confused with *Bursera gummifera* - **T. panamensis** O. Ktze. (Hedwigia p. Engler) (CA, incl. Pan, s to Pe): bois cochon; frt. edible - **T. stevensonii** Standl. (CA): incienso del monte; copal-pom (Guat); elemi (resin); resin u. in popular med. to cure wounds; incense in religious rites.

Tetraglochin alatum (Gillies ex Hook. et Arn.) O.K. (Ros.) (Chile, Pe, Arg): "horizonte"; lvs. & twigs. u. as diur. (Ch); c. triterpenes &/or sterols, flavonoids.

Tetragonia tetragonioides O. Ktze. (T. expansa Murr.) (Aizo.) (Austral, NZ, eAs + natd. Calif, Ore [US]): New Zeland spinach; hb. c. much saponin, alks.(?), d- pinitol; boiled & eaten as spinach; dr. pl. decoc. drunk for stomach cancer; fresh pl. eaten for scurvy (Br); entire pl. u. to treat stomach ulcers (Jap); will grow in sea water; (cult. dry warm areas, Calif+).

Tetragonotheca helianthoides L. (Comp.) (seUS): pine-land (or Florida) ginseng; said u. as folk med. (751).

tetrahydrocannabinols: THC: most active constits. of Cannabis (marijuana); dibenzo pyranol derivs.; several isomers are known: Δ^9-THC is the major active constit.; synthetic THC is a mixt. (racemate) and much less active; u. tranquilizer, hallucinogenic; inebriant; anti-emetic in patients with cancer chemotherapy; hypertensive glaucoma; must be u. with care (legal restraints).

tetraiodothyronine: thyroxin (qv.)

Tetrameles nudiflora R. Br. (Datisc.) (1) (seAs): kapong; bk. u. lax., antibilious; in combns. for rheum. (Indochin).

Tetramerium hispidum Nees (Acanth.) (sMex to Pan): olotillo; fl./lf. decoc. diur.; pl. decoc. at childbirth to aid expulsion of lochia (Yuc Indians).

Tetranthera Jacq. = *Litsea.*

Tetrapanax papyrifer (Hook.) K. Koch (Aralia papyrifera Hook.; Fatsia p. Benth. & Hook.) (monotypic) (Arali) (sChin, Taiwan, cult. Fla+ [US]): (Oriental) rice-paper plant; r. p. tree (of Formosa); white pith u. as diur. sed. in diabetes, fevers, cough, for surgical dressings (Chin; exported to Mal for this use); source of Chin "rice-paper" u. to make artificial flowers, Bible paper; ornam. shrub.

Tetrapleura tetraptera (Schum. & Thonn.) Taub. (Leg) (Sen-Tanz-Angola-Nig-Cameroon): fish poison; bk. & frt. u. antipyr. (Gabon), emet. ton. in VD, leprosy, rheum. in schistosomiasis (wAf); frt. pulp in soups - **T. thonningii** Benth. (trop wAf): pod u. in "black soup" of natives of Calabar region (sNig); wd. u. to make doors, benches (some regard this sp. as identical with *T. tetraptera*).

Tetrapteris (Tetrapterys) methystica R. E. Schultes (Malpighi) (Co, Br): bk. u. to make "caapi" (strong narcotic beverage) by Makú Indians of Ira-Igarapé (Rio Negro), &c. - **T. mucronata** Cavanilles (Br): u. like prec. - **T. silvatica** Cuatrecas. (Co): ash fr. burned lvs. mixed with oil or grease & appl. to fungal patches on skin (Makunas) - **T. styloptera** Juss. (Co): rasped bk. u. to prep. bitter drink u. as antypyr.; lf. ashes mixed with oil appl. to itching infections of skin (fungal?) (Makunas).

tetrasaccharides: oligosaccharides with 4 sugar units; ex. stachyose.

Tetrastigma harmandii Planch. (T. strumarum Gag.) (Vit.) (Indochin, PI): sap fr. crushed lvs. drunk for headache & fever; decoc. diur. (PI) - **T. lanceolarium** Planch. (Vitis l. Wall.) (tropAs, Austral): lf. poultice appl. to boils; pl. juice u. for coughs; frt. eaten raw or cooked - **T. loheri** Gagnep. (PI, endem.): st. decoc. for dys.; young lvs. u. for condiment - **T. serrulatum** Planch. (Ind): aerial parts ext. has anticancer activity; frt. edible.

Tetrastylidium engleri Schwacke (Olac.) (sBr): c. saponins with cough-moderating action; sds. yield ca. 15% bitter-tasting FO, tatu oil of comm. - **T. grandifolium** Sleumer (SA): sds. c. 16% FO, tatu oil u. in technology.

Tetrazygia bicolor Cogn. (Melastomat.) (sFla [US], Bah, Cu): cordobancillo; lf. decoc. u. to reduce night sweats (Bah).

Tetrodo(n)toxin: TTX; fugu poison; potent neurotoxin found in ovaries & liver of many spp. of Tetraodontidae, such as globe fish (*Spheroides rubripes*), Japanese puffer fish (*Spheroides* spp.), ocean sunfish (*Mola mola*), &c.; ingestion as food may be lethal; u. as reagent in physiolog. research (blocking agt.) (1963-) (1763); u. in zombie creation (Haiti) (2509); with unique chem. structure; acts as paralyzant (LD 1 mg); u. as analg. in neuralgia, arthritis, &c. (Jap).

Tetrorchidium didymostemon Pax & K. Offm. (Euphorbi.) (trop wAf): bk. soaked in water or rum & u. as purge; to treat swellings.

tetrose: simple sugar with 4 C atoms; ex. apiose.

tetterwort: *Chelidonium majus.*

Teucrium L. (Lab) (cosmopol., esp. Med reg.): germander; ca. 100 spp. of hbs. & shrubs (752) - **T. africanum** Thunb. (sAf): padda foot; ton. in snakebite; diaph., antipyr.; for hemorrhoids; lf. decoc. for sore thorat - **T. botrys** L. (sw&cEu, natd. eCana, US): "oak of Jerusalem"; hb. c. isoquercitrin, diosmin, choline; u. like *T. scordium* - **T. canadense** L. (Cana, US, WI, widely distr.): American germander; wood sage; u. as antipyr. - **T. capense** Thunb. (sAf): gallsick bush; pl. soaked in brandy & u. as stom. ton. (old Dutch people, sAf); inf. drunk for hemorrhoids; decoc. in flu, fever; carm.; in diabetes; exter. appl to whitlow - **T. chamaedrys** L. (c&sEu, wAs, nAf): (common) germander; German chamaedrys; hb. (Herba Chamaedryos) c. VO (with caryophyllene), tannin, scutellarin; hb. tea u. ton. stom. diur. vulner. expect.; exported fr. Austria - **T. chamaepitys** L. = *Ajuga c.* - **T. creticum** L. (sEu, AsMin, Eg, but not Crete!): Cretan germander; hb. u. as arom. & in homeop. - **T. cubense** Jacq. (Cu+, Am fr. Tex &

swUS to Arg): small coast germander; Jeremiah bush (Bah); agrimonia (Cu); hb. c. clerosterol, triacontane, eugarzasadon; u. in stomachache, as amebicidal; in Bah decoc. u. as bath for itching - **T. flavum** L. (Med reg.): yellow germander; teucrio amarillo (Sp): source of Herba Teucrii (flavi) - **T. fruticans** L. (wMed reg.): true germander; hb. c. marrubiin, u. as vulner. (Herba Teucrii veri) - **T. incanum** Aitch. & Hemsl. (sAf): lf. u. for sore throat, as ton. in snakebite - **T. iva** L. = *Ajuga i.* - **T. marum** L. (wMed reg.): cat thyme (SC attracts cats); marum germander; lvs. & flg. tops (or above-grd. parts) = Herba Mari (veri); hb. c. marrubiin, tannins; u. as expect. in pharyngitis, sternutatory (in snuffs), for chronic rhinitis, emmen. antispasm in cholecystitis, arom. diur. - **T. montanum** L. (sEu, AsMin): mountain germander; source of both Herba Mari veri & H. Polii montani - **T. polium** L. (Med reg., AsMin, Saudi Arab, Eg [nwDesert]): mountain germander; ja'-dah (Arab); for fevers (Bedouins); to treat fungal dis., antiseptic, diaph. ton.; lvs. in stomachache, as tisane (nIsrael) (753); diur. ton. in hemorrhoids; sometimes as adulter. of or replacement for *T. marum*; var. *aureum* Schreb. u. similarly - **T. riparium** Hochst. (sAf): strong inf. as emet. in snakebite; decoc. as ton. - **T. scordium** L. (Euras, Siber): water germander; wood garlic; English treacle; polymountain; hb. = Herba Scordii vulgaris, c. VO NBP tannin; u. in upper resp. tract congestion; incl. chronic bronchitis, TB; diaph. anthelm. ton. vulner.; for hemorrhoids; pd. lvs. advertised as "cancer cure", Chicago (754) - **T. scorodonia** L. (wEu, Med reg.; natd. eNA): germander (or wood) sage; w. germander; garlic s.; Herba Scorodoniae u. as vulner. (exter.); for chronic bronchitis, nose & throat inflammn., in obesity; in skin dis., as gargle - **T. stocksianum** Boiss. (Pak): hb. u. as antipyr. & for cardiac pain - **T. viscidum** Bl. (T. stoloniferum Roxb.) (eAs): hb. u. as diaph., antidote for poisons; to stop intest bleeding (Chin); stim. ton. antipyr. (Indochin); pl. inf. u. as wash for feet itching fr. wading in dirty water (Chin).

Teufelsauge (Ge-f.) (devil's eye): Adonis.

Teufelsdreck (Ge) (devil's feces): Asafetida.

Teufelskralle (Ge-f.): (devil's claw): *Phyteuma hemisphaericum.*

Texas tea (US slang): Cannabis.

Texas: v. also under bluebonnet; krameria; nutgalls; sarsaparilla; snakeroot.

Textularia: gp. of Foraminifera (Rhizopoda-Protozoa) found in chalks; shells composed of series of chambers arranged in spiral.

thalamus: 1) receptacle or torus; subex 2) gray matter located on each side of 3d ventricle (cavity) of the brain (<Gr = inner chamber).

Thalassia testudina Koenig (Hydrocharit. - Monocot.) (Atl coasts of Fla [US], WI, SA): (Caribbean) turtle grass; "sea weed"; lvs. coll. for fertilizer; grows in sea water as deep as 30 m.

Thalia dealbata Roscoe (Marant.) (seUS, n&w to Mo): thalie; powdered thalia; ornam. perenn. hb. - **T. geniculata** L. (T. welwitschii Ridl.) (tropAm & wAf): lvs. & rts. u. u. in combns. for snakebite; lvs. u. as packing material; sts. u. for basketwork, &c.

Thalictrum L. (Ranuncul.): meadow rue; hbs. found worldwide; ca. 90 spp.; c. alks.; several cult. ornamental spp., esp. *T. delavayi* Franch. (Chin) - **T. anemonoides** Mx. (Ranuncul.) = *Anemonella thalictroides* Spach. - **T. aquilegifolium** L. (cooler nHemisph, esp. Eu, Jap): meadow-rue; karamatsu-so (Jap); perenn. hb.; rt. inf. u. in intest. "catarrh.," bitter ton., locally for small wounds (Ainu); in Jap as antisept., lf. as local anesth. - **T. dasycarpum** Fisch. & Lall. (Cana w to Alta; nUS, w to Ariz): purple meadow-rue; rts. c. thalicarpine & thalidasine with hypotensive & crytostatic activity; rts. u. for fever (Obijwa Indians); lvs. & sds. u. for cramps (Potawatomi Indians) - **T. delavayi** Franch. (wChin); u. in local med.; popular cult. ornam. - **T. dioicum** L. (e&cCana, US): early meadow-rue; e. mountain rue; quicksilver weed; poor man's rhubarb; c. berberine & other alks.; rt. u. purg.; bruised lvs. & sts. appl. wounds & sores (Indians); ornam. perenn. hb. - **T. fendleri** Engelm. ex Gray (wUS): (Fendler's) meadow-rue; ruda de la sierra; rt. inf. u. in GC (Navajo Indians); rt. decoc. for colds; sores (Indians); emet.; digitalis-like activity reported - **T. flavum** L. (Euras): yellow meadow-rue; rhiz. c. berberine, thalisopyrine, thalicarpine; rt. u. as diur. purg., in jaundice, epilepsy; as yellow dye - **T. foetidum** L. (eEu, cAs mts.): c. fetidine, thalfetidine (alks.); hb. tinct. u. in hypertonia (Russ) - **T. foliolosum** DC (Ind, Pak- Himal): lohangi; perenn. hb.; rts. c. isoquinoline alks., incl. berberine, jatrorrhizine, palmatine, magnoflorine; rt. ext. u. as ton. antipyr. diur. cath. in flatulence, jaundice, dyspepsia, &c.; collyrium to improve eyesight (ophthalmia) (whole pl. u. for eye dis. [124]) - **T. hernandezii** Tausch. (Mex to SA): alboquillo de campo; costicapatli; cozticapatli; rt. c. alk., yellow coloring matter, resin; rt. diur. purg., decoc. u. in rheum. (Arg) - **T. kuhistanicum** Ovcz. & Koczk. (USSR): c. thalmine, thalmidine, &c.; u. in folk med. for dis. of digestive tract, mal. TB, kidney dis., hemorrhage, malignant tumors, &c. (2428) -

T. lucidum L. (T. angustifolium L.) (se&eEu): sds. c. HCN; pl. u. as diur. (Ukraine folk med.) - **T. macrocarpum** Gren. (Eu-Pyrenees): (big fruit) meadow-rue; rt. c. thalictrine, macrocarpine; u. in folk med. - **T. mexicanum** DC (Mex): rt. u. as diur. & exter. in eye dis. - **T. minus** L. (Euras, e&sAf); lvs. c. thalmine, orientin; rts. c. thalictamine, magnoflorine & many other alks.; lf. inf. or rt. decoc. u. in fever (sAf natives & in Lahul [nInd]) bitter addn. to beer; var. *hypoleucum* Miq. (var. elatum auct. japon.): c. "elatrine," adiantifoline; dr. pl. u. as home remedy, stom. & in diarrh. (Jap) (755) - **T. occidentale** A. Gray (wCana, US): frt. inf. u. for chest pains; smudge to repel insects; inf. of frts. as hair & body scent; inf. in hiccups (Blackfoot Indians, BC [Cana]); dr. frt. decoc. for colds, chills, fevers (Flathead Indians, BC); rt. chewed for loosening phelgm (Gitksan Indians, BC); rt. poultice for open wounds (Thompson Indians, BC) (756) - **T. polycarpum** (Torr.) S. Wats. (Calif [US]): meadow-rue; juice fr. crushed lvs. & sts. u. to wash head for headache (Indians, Mendocino Co., Calif [US]) - **T. revolutum** DC (e&cNA, Mex): purple, skunk, or wax-leaved meadow-rue; rt. c. thalicarpine; hypotensive, antispasm. - **T. rubellum** L. (Manchuria): rhiz & rtx. u. as ton. in leucorrh., amenorrh. - **T. rugosum** Ait. (sEu, nAf): c. many alks. incl. berberine, thalidasine; rt. ext. hypotensive, inhibits tumor growth; antibiotic - **T. simplex** L. (Euras.): c. many alk. incl. thalictrine, thalsimine; thalisamine; exts. analg. sed. hypotensive, spasmolytic - **T. thalictroides** (L) Eames & Boivin (Anemonella t. Spach) (e&cNA): rue anemone; (anemone) meadow-rue; wind flower; rt. u. purg. diur.; tubers edible; cult. as ornam.- **T. thunbergii** DC = *T. minus* var. *hypoleucum* - **T. tuberiferum** Maxim. (T. filamentosum Maxim. var. tenerum Ohwi) (Jap, Kor, Manchur): parts above-ground u. as kidney ton.

thalleioquin test: color test (ammonia, Br_2) positive for quinine & quinidine.

thallium (T1): metal occurring in earth's crust - **t. sulfate**: $T1_2SO_4$; u. as rat & ant poison; reagent (poison!); formerly u. for night sweats (TB).

Thallophyta: Thallophytes: the ribbon (or strap) plants; the lowest in evolutionary status of the four phyla (or divisions) of the P1. Kingdom; the pl. body in the gp. consists of the **thallus**, rel. simple & undifferentiated as compared with the higher pls. (no distinct st., lvs., rts., etc.); subdivided into Algae (sea weeds), Fungi (molds, toadstools, etc.), & Lichens; "AFL."

Thamnosma montana Torr. & Frem. (Rut) (swUS, nwMex): turpentine broom; cordocillo; prepns. of shrub u. as general ton., to "purify" the blood (Ariz [US]); decoc. in GC (Pima, Apache Indians).

Thamnolia sp. (*T. vermicularis* [Sw.] Ach.?) (Usne. - Discolichenes) (cosmop.): cachile de venado (Ve); decoc. u. for stomachache.

thao (Chin): agar.

Thapsia garganica L. (Umb.) (Med reg.): deadly carrot; death c.; smooth thapsia; thapsie; rt. bk. irritn. cath., by alc. extn. produces poisonous irritant gum resin, Resina Thapsiae (Laser [Cyrenaicum]); u. as skin irrit. in plasters; by malingerers to simulate erysipelas, &c. (*cf.* croton oil); rt. u. for pulm. inflammn., exter. in skin dis. & rheum; to dye hair(?): lf. emet. antidiarrh.; var. *silphium* (Viv.) DC (T. s. Viv.) (nAf-ripolitania, Libya to Mor): this is not the silphion (Sylphium Cyrenaicum) of the ancients (Hippocrates et al.) as earlier thought; prod. of rt. bk. = gum-resin (Resina Thapsiae; thapsia resin), more acrid & irritant than that of *T. garganica* (757) - **T. villosa** L. (sEu, Sp, Alg): yields a resin ("false turpeth"; "faux turpith") weaker than that of prec.; rt. ext. u. like prec.

Thaspium trifoliatum (L) Gray (T. aureum Nutt.) (Umb.) (e&cUS): (purple) meadow-parsnip; round heart (plant); (golden) alexanders; hb. u. as vulner. antiluet.; in uter. dis., chorea; in homeop.; rt. also u. med.

Thaumatococcus daniellii Benth. (monotypic) (Marant.) (trop wAf): miraculous fruit (or m. berry); Katemfe (Sudan); soft aril around sds. in frt.; c. thaumatin, intensely sweet, protein (1600x as sweet s sucrose) (10x sweeter than Aspartame); rendering food (breads), beverage (tea, &c.) much sweeter; u. to improve acidity of palm wine, &c.; sweetness due to polypeptides - **"thaumatine"** (1600 x as sweet as sucrose): breaks down when heated (cooking).

THC: tetrahydrocannabinol (*qv.*) (active principle of Cannabis).

thé (Fr): tea - **t. de Blankenheim**: prep. based on *Galeopsis ochroleuca*; was popular in Ge in consumption - **t. d'Europe: t. de la Gréce**: *Salvia officinalis* - **T. de Saint-Germain: t. de santé: Species Purgativae**.

Thea (L): Chinese tea - **Thea** L. (The.) = *Camellia* - **T. sinensis** L. = *Camellia s.* (L) Ktze.

Theaceae (Ternstroemiaceae; Camelliaceae): Tea Fam.; shrubs & trees; lvs. evergreen; fls. regular, perfect; pentamerous; frt. capsule, berry, semidrupe; ex. *Camellia* (*Thea*).

Thebaicum (L.): Opium (old term).

thebaine: one of the chief Opium alks.; constitutes c. 0.2-0.5% (of Ind Opium) (in some Opiums to 1.5%); acts as stim. rather than narcotic.

thecophore: false pedicel of frt. formed by continuation of pericarp (frt. wall); ex. cubeb "stalk."

Thecostele poilanei Gagnep. (Orchid.) (Vietnam): decoc. combined with arsenic & rice u. to kill rats.

Thedford's Black Draught: proprietary senna prepn. mfd. until recently in Montgomery, Ala (US).

Thee (Ge): tea.

theelin: estrone; follicular hormone; keto-hydroxyestrin - **di-hydro-t.**: estradiol; dihydrofollicular hormone; di-OH-estrin.

theelol: estriol; follicular hormone hydrate; tri-OH-estrin.

theine: charact. alk. of Thea, tea; identical with caffeine.

Thelepogon elegans Roth (monotypic) (Gram.) (tropAf & As): dr. hb. fed to horses as bitter ton., to counteract effect of too long green feeding.

Thelesperma longipes Gray (Comp.) (swUS, nMex): wild tea; longstalk greenthread; hierba de San Nicolas; diur.; u. as beverage tea (Am Indians) (Apache; Pueblo [ex. Isleta, NM]); much like beverage tea; nerve stim.; said beneficial to teeth; u. for sore gums, toothache - **T. megapotamicum** (Spreng.) Ktze. (T. gracile [Torr.] Gray) (swUS, Mex, sSA): "Navajo tea"; cota (SpAm); antipyr. diur. (758); lf. & fl. decoc. as beverage (NM & Ariz [US] Indians); fls. furnish yellow dye, other pl. parts brown dye - **T. subnudum** A. Gray (Colo, NM, Ariz, Utah [US]): green thread; diur.; lvs. sts. fls. u. to dye wool orange (Navahos).

Theligonum cynocrambe L. (Theligon.) (Med reg.): young shoots formerly u. as lax.; vegetable food.

Thelypteris dentata E. St. John (Aspleni./Thelypterid./Polypodi.) (Tanz, Am): fresh lvs. crushed & appl. to bacter. & fungal inf. of body parts - **T. hexagonoptera** Weatherby (Dryopteris h. (Mx.) C. Christens.) (e&cUS & Cana) broad beech-fern; reported to have hypotensive activity; c. no phloroglucides or filicin (1889) - **T. kunthii** (Desv.) Mort. (T. normalis Moxley) (seUS, WI, Mex, nSA): (maiden) fern (Bah); lalla; pd. dr. pl. sprinkled on open cuts & dermatitis produced by *Metopium toxiferum* (Bah).

Themeda arguens Hack. (Gram.) (Indones): u. as poultice on hips for lumbago - **T. gigantea** Duthie (Indones): young shoots eaten for fever - **T. triandra** Forssk. (Chin to Austral, Af): red (oat) grass; kangaroo g. (Austral): rts. diur. antipyr. stom., in GC (Chin); good grazing & hay grass; HCN poisoning sometimes reported; grains a famine food (natives).

Theobroma albiflorum DeWild. (Sterculi.) (Co): cacao montaras; c. cimarron; sds. sometimes mixed with those of *T. cacao*; frts. u. to make jam or sauce - **T. angustifolium** Moç & Sess. (or DC) (sMex, CA): cacao (de) mico; c. silvestre; monkey c.; cushta (Salv); source of cacao de Soconusco (famous chocolate); cult. Mex - **T. bicolor** Humb. & Bonpl. (sMex, CA, nSA): (cacao) pataste; bacao (Co); sd. u. like that of *T. cacao*; sometimes admixed with true cocoa bean to which similar but inferior; prod. known in comm. as werbra cacao (Hond, Nic), wariba; tiger c.; cult. in LatinAm; rt. u. in jams - **T. cacao** L. (tropAm, chiefly sMex, CA): cacao tree; (common) cocoa t.; chocolate (nut) tree; cacaoyer; source of cocoa bean (cacao (beans), Theobromatis Semina, L.): kernels expressed for FO, cocoa butter (Oleum Theobromatis or O. Cacao) (v. cacao butter), u. in prepn. cosmetics as emoll. (esp. for sunburn), dem., suppositories (adv.: m. pt. at near body temp., does not become rancid), in confectionery; appl. to burns, wounds, cracked lips; sd. also c. alks. theobromine, caffeine; proteins, cacao red, starch, cellulose; occurring only in fermented beans are vol. acids (ex. acetic) 0.2%; VO (0.001%) (costs more than rose oil) with linalool, linalyl acetate, geraniol, amyl acetate, a. propionate, a. butyrate; propylacetate, (759); sds. decoc. u. as beverage (breakfast cocoa, chocolate) popular in Am long before coming of white man; these stim. diur.; in "lumbago"; sds. u. in mfr. cocoa butter, theobromine; cocoa shells (Cortex Cacao) is by-prod. of mfr. chocolate & cocoa; c. alks., FO, crude protein, cacao red; u. diur., arom.; prodn. theobromine, caffeine; coffee & tea substit.; cacao shells u. as adulter. cocoa; fertilizer; compressed into briquettes (for heating purposes); complementary fodder for milch cows on pasturage (with fermented potatoes, other foods) (760); small abortive pods constit. of shampoo u. to promote hair growth (Caribs of Domin) (2328); cult. worldwide in trop, esp. Af (Ghana chief source), Mal, SA; chief vars.: Criollo; (Ec [Guayaquil], Ve [Caracas], Java, Nic); forastero (Trinitario [Trinidad], Amazonas [Pará], Br [Bahía], Cundeamor [Java], Caribbean [Cuba], Lagos [Nigeria]) (v. also under cacao, cocoa) (for Serials see App. C-1) (1764) - **T. glaucum** Karst. (Co): sds. source of cocoa of good quality - **T. martiana** Dietr. (T. sylvestris Aubl.) (nBr): sds. u. to make beverage, sometimes mixed with those of *T. cacao* - **T. subincanum** Mart. (T. obovata Klotzsch) (Br): sd. furnishes an inferior cocoa; bk. ashes u. as narcotic snuff in combn. with Virola resin, sometimes with tobacco (SA Indians); frt. edible.

theobromine: charact. white odorless alk. of cocoa & chocolate; also found in kola, tea; prepd. fr. cocoa shells by distn. & other processes (761); acts as card. stim. diur. vasodilator; several salts u. such as **T. Calcium Salicylate** (Theocalcin™); mixt. or double salt of Ca t. and Ca salicylate; u. diur. coronary vasodilator, card. muscle stim.; less irritant to stomach than **t. and sodium salicylate** (Diuretin) separately; formerly claimed hypnotic! (2414).

theocin: v. theophylline.

Theophrastus (ca. 370-287, BC) of Lesbos, Gr; regarded as the first writer on botany & materia medica in Eu; a pupil of Aristotle; he described many important crude drugs; ex. opium, ergot, cinnamon, etc.

theophylline: dimethyl-xanthine; both anhydrous & monohydrate u.; alk. occurs in tea, also synth.; u diur, in cardiac dis., edema, in asthma; but now being replaced by corticosteroids; occasionally severe adverse reactions - **t. ethylenediamine**: Aminophyllin: u. diur. card. stim. - **t. (sodium) and sodium acetate** (Int. Ph.): Theocin Soluble™; u. as diur. smooth muscle relaxant - **t. sodium glycinate**: equilibrium mixt. of th. sodium & aminoacetic acid; u. to treat bronchospasm & coronary artery dis.

Theragra spp. (Gadidae - Pisces) (marine waters): pollack; one sp. source of a caviar substit. (Jap) - **T. chalcogrammus** (nPac): Alaska pollack; liver oil like that of cod.

therapeutics: therapy: the application of drugs in clinical medicine; a field of med. dealing with the use of remedies in treating dis.; the healing art (term first u. in 1541).

Theraphosa & other genera (Avecularidae [Theraphosidae] - Arachnida): bird spiders; "tarantula"; large spiders (-10 cm); similar to *Avicularia avicularia* L. (CA, SA); venom u. in GI complaints (homeopathic); venom collected.

therapic acid: unsatd. fatty acid, apparently arachidonic a.

theriac: Theriaca (L.): **Theriace** (L.): (lit. treacle or molasses): popular form of med. in 17th cent.; many formulas were developed; one by Sir Walter Raleigh had 40 components (sds. hbs. bks. & wds.), typical of this type of prep.; believed to act as antidote against poisons in general (sometimes only for use against animal poisons, ex. snake venom); soothing draught; later, treacle; derived fr. theriacus (*qv.*). - **T. Andromachi**: an arom. astr. spiced confection c. opium, & originally more than 70 other ingredients - **T. Pulvis** (L.): **Venetian t.**: v. under "V."

Theriackwurzel (Ge): 1) Angelica rt. 2) Valeriana.

Theriacus (L.): antidote for animal poisons, esp. snake venom.

Theridion curassavicum (Araneida - Arachnida - Arthropoda) (CA): orange spider; black spider of Curaçao; close to *Latrodectus mactans* (black widow spider); u. in homeop. for migraine, hysteria, climacteric dis.

Thermopsis caroliniana M.A. Curtis (Leg) (seUS: NCar-Ga): golden peas; Aaron's rod; c.quinolizidine alk.; rts. nodulated; cult. as bright perenn. ornam. in gardens - **T. fabacea** (Pall.) DC (eAs): sds. c. cytisine, methyl cytisine; pods u. for dis. of mouth, throat, & teeth (Chin) - **T. lanceolata** R. Br. (T. lupinoides Link) (cAs, Siber, Russ, Kamchatka, Mong, Chin): hb. (Herba Thermopsidis [lanceolatae]); termopsie (Russ); choc chin-sek; c. thermopsine, thermopsidine, pachycarpine (sparteine), cytisine, methyl cytisine, flavonoids (formononetin); strong expect. (Ukraine) & emet. u. for cough & as ipecac substit. (Russ), blood pressure-reducing; sds. c. cytisine (with strong lobeline activity) & (+)-sparteine (source); poisonous; intense research (USSR, 1932-4) on Herba Thermop. (USSR Ph. VIII); cult. - **T. pinetorum** Greene (swUS: NM, Ariz, Utah, Colo): golden pea; decoc. u. as cough med. (Navaho) - **T. rhombifolia Nutt.** (c&n Great Plains [US]): golden pea; false lupin(e), yellow bean, golden banner; dr. fls. u. in smoke treatment for rheum. to reduce pain & swelling: pl. toxic to cattle; sds. toxic to children.

Thesium angulosum DC (Santal.) (sAf): remedy in heartburn - **T. chinense** Turcz. (Kor, Chin, Jap): st. & lvs. expect.; entire pl. u. to treat diarrh. ulcers & abscesses on head - **T. humifusum** DC (wEu): bastard toadflax; c. alks.; u. in folk med. - **T. hystrix** A. W. St. Hill (sAf): st. decoc. u. as expect. in cough & pulm. TB; in kidney & bladder dis.; abort. - **T. minkwitzianum** B. Fedtsch. (eTurkestan, Armenia): rt. parasite; c. thesine (alk.-hypotensive); u. in folk med. - **T. utile** A. W. St. Hill (sAf): st. chewed for relief of gastric dis.

Thespesia danis Oliv. (Malv.) (trop eAf): rt. u. for stomach pains, swollen stomach; frt. edible (Tanz) - **T. lampas** Dalz. & A. Gibs. (wInd-penins.): ban-kapas (Hind, Beng) rt. & frt. ext u. in acute GC (with sugar & water). syph.; pd. bk. mixed with *Hydnocarpus wightiana* FO in leprosy, appl. to ulcers; fiber - **T. populnea** Soland. (OW trop, esp. coastal forests of Ind): Portia (or bendy) tree; false gamboge tree; frt. juice c. antibiotic, u. to treat herpes, also bk. decoc. in cholera, dys., hemorrhoids; as lotion (women, Mariana ls); lvs. emoll.; bk. & rts. in gall bladder dis., asthma; colds (Solomon ls): skin dis. (as psoriasis) (decoc); bk. decoc. in thrush (Fiji); fls. & rts. c. gossypol; ripe sds (Portia sds.) with 19%

FO; fls. u. as yellow dye; ornam. small tree; considered sacred in some places.

Thevetia ahouai A DC (Apocyn.) (Guian, Br, Co): c. cardioactive glycosides, incl. thevefolin; u. emet. cath. antipyr.; to cure ulcers - **T. cuneifolia** A.DC = **T. ovata** A DC (Mex, Guat): regalgar; cascabel; c. cardioactive glycosides; sds. said toxic, carried in pockets to prevent hemorrhoids (Mex Indians) - **T. peruviana** Schumann (T. neriifolia Juss.) (sFla & Keys [US], Mex, WI, CA, SA): lucky nut (WI+); trumpet flower (Fla); yellow oleander; cabalonga (Cu); st. bk., c. theviridoside, peruvoside; u. bitter, antipyr. purg.; lvs. c. cardenolides of thevetin type; u. purg., in card. dis.; lvs. & rts. u. earlier for tumors (Indians, Mex); entire pl., esp. sds. toxic c. thevetins A and B; neriifoline, &c.; u. in card. insufficiency (digitaloid cardioton.), potent insecticide; in prepn. peruvoside: juice soaked in cotton appl. to cavity for toothache (Yuc) - **T. plumeriaefolia** Benth. (sMex to Salv): chilindron (blanco); inf. of latex & sds. u. as antipyr. in rheum.; sd. inf. u. as cardioton. - **T. yccotli** DC (T. thevetioides K. Schum.) (sMex, SA): joyota; yoyote; &c.; sds. c. 40% FO (non-drying), thevetins, thevelose; u. as substit. for digitalis; in hemorrhoids, alexiteric (rattlesnake bite); lvs or juice u. for cutaneous dis., deafness, ulcers (Aztecs); toxic; latex to allay pain of scorpion sting, toothache (2342).

thevetigenin: aglycone of thevetins, mixt. of glycosides thevetin A & t.B. (thevetin B) (cerberoside); hydrolyzes to release digitoxigenin, L-thevetose ($C_5H_{13}O_4CHO$), & gentiobiose; u. cardiotonic, intermediate in steroid synthesis to prod. digitoxigenin, &c.

thevetin A: glycoside of cannogenin + L-thevetose + gentiobiose (2 glucose).

thiamine: Vitamin B_1; aneurine; a heterocyclic compd. made up of pyrimidine & thiazole moieties; occ. in pls. & animals, esp. cereal grains, bran, egg (yolk), yeast, liver, milk, &c.; now mostly prepd. by synth.; the anti-neuritic factor u. in beri beri, neurasthenia, polyneuritis (in lower animals); various neuritides; in alcoholism; in deficiency states with fatigue, anorexia, GI dis., irritability, tachycardia, &c.; generally u. as **t. hydrochloride**; **t. mononitrate**; &c. - **t. pyrophosphate hydrochloride** = cocarboxylase.

thimbleberry: *Rubus odoratus* (*western t.*: *R. parviflorus); R. occidentalis.*

thimbleweed: Rudbeckia laciniata.

Thiocol™: potassium guaiacol sulfonate.

thiocolchicine: **thiocolchicoside**; **thiocolquicoside**; Coltrax (Br): Coltramyl; Coltromy; alk. fr. *Colchicum autumnale* (1954-) (762); acts as central ganglioplegic agt.; u. as long-acting relaxant to voluntary muscle in myositis, tetanus, &c. (but no curariform effects noted); antihistamine in allergy, gout (*cf.* colchicoside, 3-deme-colchicine glucoside).

thioctic acid: lipoic acid; dithiolanyl valeric acid; growth factor found in natural sources (animals, pls., ex. liver, yeasts); with growth stim. props. (bacteria, protozoa) u. to treat liver dis. (chronic hepatitis; alc. fatty liver), *Amanita* poisoning; detoxicant in heavy metal poisoning (2248).

thiodiamine: ext. fr. bk of *Crataeva roxburghii*; said useful in cholera.

thioglucosidase: enzyme wh. catalyzes hydolysis of mustard oil glucosides, sinigrin, sinalbin, glucocheirolin; occ. in Cruc. in admixt. with myrosulfatase.

thiolutin: antibiotic fr. various strains of *Streptomyces albus* Waksman & Henrici.

Thiostrepton: Bryamycin: polypeptide antibiotic (with S) furnished by *Streptomyces azureus* Kelly et a. (NM [US] soil) (1955-); antibacterial against certain bacteria resistant to penicillin, &c.

thistle: term mostly appl. to spiny hbs. of Fam. Comp. esp. of genera *Cirsium, Cnicus, Carduus, Onopordon, Echinops, Silybum, Carlina, Centaurea*, etc. - **binniguy t.** (Austral): yellow t. - **black t.**: common t. - **blessed t.**: 1) *Cnicus benedictus* 2) *Silybum marianum* 3) (sometimes) *Centaurium erythraea* - **blue t.**: *Echium vulgare* - **boar t.**: 1) Canada t. 2) *Cirsium vulgare* - **bull t.**: plume t. - **California t.**: **Canada t.**: **Canadian t.**: *Cirsium arvense* - **Carline t.**: *Carlina* spp. - **common t.**: *Cirsium vulgare* - **corn t.**: Canada t. - **cotton t.**: *Onopordum acanthicum*; oat t. (ex. of vernacular name no longer u.) - **crab t.**: *Onopordium acanthium* - **creeping t.**: **cursed t.**: Canada t. - **distaff t.** (Calif [US]): wild spp. of *Carthamus,* esp. *C. lanatus* & *C. boeticus* - **t. family**: Comp./Carduaceae - **fragrant t.**: *Cirsium pumilum* - **globe t.**: *Echinops* - **green t.**: common t. - **holy t.**: *Silybum marianum* - **horse t.**: corn t. - **Jamaica t.**: *Argemone mexicana* (Mexican t. [800]; also appl. to *A. albiflora* & *Cnicus conspicuus* Hemsl.) - **Mary t.**: blessed t. - **milk t.**: 1) *Sonchus oleraceus* 2) *Lactuca scariola* 3) *Carduus marianus* - **musk t.**: 1) crab t. 2) *Carduus nutans* - **oat t.**: *Cirsium arvense* - **pasture t.**: fragrant t. - **perennial t.**: corn t. **perennial sow t.**: *Sonchus arvensis* - **plume**(**d**) **t.**: *Cirsium* spp., esp. *C. pumilum* & *C. vulgare* - **poppy t., prairie**: **white p. t.**: white t. (**poppy t. yellow:** yellow t.) - **Russian t.**: *Salsola kali, S. pestifer* - **St. Benedict's t.** (Ont [Cana], etc.): *Cnicus benedictus* - **St. Mary**('**s**) **t.**: Mary t. - **Scotch t.**: **Scottish t.**: *Cirsium vulgare, Carduus nutans* -

sow t.: *Sonchus* spp., esp. *S. oleraceus* (SC lf. edges cut simulating teats of sow) - **spear t.** (, **common**): common t. - **spiny t.** (, **large-flowered**): *Carduus nutans* f. *spinosissima* - **star t.**: *Centaurea, calcitrapa* - **swamp t.**: *Cirsium muticum* - **variegated t.**: Mary t. - **white t.**: *Argemone alba* - **yellow (flowering) t.** (, **Jamaica**): *Argemone mexicana.*

Thix: fine purified talcum, mineral clay consisting primarily of Al silicate; thickening agt. for cosmetics, emulsion dyes, &c.

thixotropic gel: one with the prop. of becoming mobile when shaken vigorously & then of reverting to a typical stiff gel when left stand again; u. in some dispensing forms.

Thladiantha calcarata C. B. Clarke (Cucurbit.) (Chin & adj. areas): fresh rts. rubbed on sores & swellings; cooked frt. antipyr., u. for urinary incontinence, sds. for cough - **T. dubia** Bge. (Chin): red hail stone; pinyin; rt. cholag. in jaund., galact. in agalactia, diur. for urin. dis. deafness, &c.; sds. ton. astr. - **T. grosvenori** (Swingle) C. Jeffrey (Chin): c. a very sweet glycoside; u. med. - **T. nudiflora** Hemsl. (Chin): pulped rt. appl. as antisep to knife wounds; exter. for headache (Taiwan).

Thlaspi arvense L. (Cruc.) (Euras, natd. Cana, US): stink-weed; fanweed; Frenchweed; (field) penny-cress; Mithridate mustard; frt. a flat, deeply notched at top, broad-winged silicle, 1- 1.8 cm long; sd. c. FO u. like rape sd. oil; edible, salad oil, lamp oil; in paints & varnishes; sds. u. diur. in rheum.; sds. u. as ton. (Chin); in eye dis. - **T. bursa-pastoris** L. = *Capsella b.*

tholloside: glycoside of sarmentosigenin A + rhamnose; occ. in *Stroph. thollonii.*

Thomson, Samuel (1769-1843) (Mass [US]): founder of a formerly popular school of medical thought, Thomsonianism (or Herbalism), which held to the use of remedies only of pl. origin; in its early phases, chief stress was placed on Lobelia.

Thomsonia nepalensis Wall. (monotypic) (Ar.) (Ind, Nepal, Sikkim): pl. acrid, toxic (—> *Apomorphophallus*).

Thomsonian Number Six: Tr. Capsicum & Myrrh (NF VII) - **Thomson's Fever Syrup**: a typical formula of S. Thomson (*qv.*); Syr. Rhubarb 4 fl. 0oz, Tr. Valer. 2 fl. oz, Cinnamon Oil 20 drops, piperine 10 gr, $NaHCO_3$ 20 gr.

thorn: 1) hardened & pointed tip of a stem branch; ex. honey locust 2) *Crataegus* spp. ex. *C. intricata, C. oxyacantha* - **black t. (blossoms)**: 1) sloe, *Prunus spinosa* 2) certain *Crataegus* spp. - **t. bush**: *Crataegus* spp. (**African t. b.**: *Acacia* spp. of Af) - **Egyptian t.**: *Acacia senegal*, tree - **Jerusalem t.**: *Parkinsonia aculeata* - **May t.**: t. bush - **red t.**: *Crataegus oxyacantha* & other *C.* spp. - **t. tree:** 1) *Balanites aegyptiaca* 2) *Gleditsia triacanthos* 3) *Crataegus* spp. 4) other plants, incl. *Acacia* spp.; ex *A. paradoxa, A. herpophylla, Mespilus germanicus* (*Ulex europaeus*) - **tree t.**: *Crataegus viridis* - **white t. (blossoms)**: *Crataegus oxyacantha.*

thornapple: 1) *Datura stramonium* (usually) 2) *Argemone mexicana* - **blue-flowered t.**: *D. stramonium* var. *tatula* - **common t.**: *D. stramonium* - **hairy t.**: *D. metel* - **horny t.: purple (flowered) t.**: blue-flowered t. - **recurved t.**: hairy t.

thorough-wax: thorow-wax: 1) (US) *Bupleurum rotundifolium* 2) *Eupatorium perfoliatum.*

thoroughwort: 1) *Eupatorium perfoliatum* 2) certain other spp. of *Eupatorium* (SC lvs. in opposite pairs appear as though single lf. passed through by st.).

Thottea dependens Klotzs. - **T. grandiflora** Rottb. (Aristolochi.) (Mal): for former, pounded lvs. appl. as poultice for skin dis.; bk rubefacient; for latter, rt. u. as ton. in fever & "ague"; given to women after confinement.

Thouinia discolor Griseb. (Sapind.) (WI): three-finger; naked wood; lf. decoc. drunk for "weakness" (i.e. impotence); lvs. dried & pd. & sprinkled on sores (Bah) - **T. paucidentata** Radlk. (sMex, Guat, Belize): canchunup (Maya); bk. boiled in salt water & decoc. u. as expec. for cough; sap smeared on chiclero ulcer with lvs. as poultice; bee tree.

thousand leaf: *Achillea millefolium.*

threonic acid: apparently L-threonine, an essential amino acid; CH_3.CHOH.$CHNH_2$. COOH; occ. in egg, milk, gelatin, & other protein materials; u. nutr.

thrift: *Armeria* spp. - **sea (side) t. (root)**: 1) *A. maritima* (v. *elongata*) 2) *Limonium carolinianum* & other *L.* spp.

Thrinax radiata Lodd. ex Schultes & Schultes (T. wendlandiana Becc.) (Palm.) (sFla & Keys [US], sMex, WI, nSA): sea thatch; t. palm; frts. edible, u. (ext.) as ton. nutr. diur.; in TB, bronchitis; to develop mammary glands & "vitalize" other sex organs.

throat root: *Geum rivale.*

thrombin: protein enzyme-like substance prep. fr. bovine prothrombin through interaction with thromboplastin & calcium; appl. in sterile form locally to stop capillary bleeding; acts by forming fibrin fr. fibrinogen; also human t.

thrombocytes: blood platelets; important part of blood coagulation mechanism.

thromboxanes: prostaglandin endoperoxide derivs. that stimulate platelet aggregation; occ. in prostate, platelets, leucocytes, & other tissues; released when tissues injured; conducive to occlusive vascular disease; inhibited by aspirin.

Thryallis glauca Ktze. (Malpighi,) (Mex, CA): corpionchi (Taxco); ramo de oro; lf. decoc. emoll. u. to heal wounds; shrub sometimes cult.

Thuarea involuta R. Br. (Gram) (Indochin coasts): inf. of roasted pl. taken for headache.

Thuja L. (Cupress.) (As, eNA): arbor vitae; l'arbre de vie (tree of life); (SC bk. u. in scurvy [-1535-], thought "miraculous" cure); evergreen trees & shrubs; 6 spp. - **T. occidentalis** L. (Cana; neUS, w to Minn): (American) (eastern) arbor vitae; (northern) white cedar; arbre de vie; "cedre" (Qu [Cana]); leafy young twigs (Thuja, L.) c. VO (cedar leaf oil) with a-thujone, pinene, borneol, fenchone, sabinene; twigs u. exter. (pd. to make cataplasm) in rheum. & gout; local irrit. hemostat., in swellings, warts, & condylomata; for prodn. VO; formerly fresh lvs. & twigs u. as diur. diaph. expect. vermif. emmen., uter. stim. (most consider too toxic); bk. u. in folk med. (Eu); VO u. perfume, lavender oil replacement (WW II); insect repellant (sometimes combined with citronella oil); VO of wd. ("white cedar oil") & cones also u.; toxic; abort. (763); lvs. & twigs u. to make "fir pillows" (nNew Eng [US]) - **T. orientalis** L. (Biota o. Endl.) (Iran to Chin, Jap, wSiber): oriental arbor vitae; inclined cypress; a traditional Chin med.; lf. decoc. or juice hemostat., in duodenal & gastric ulcers, GC, colds, rheum. parasitic skin dis.; resin mixed with pine rosin u. as resolvent for tumors; pd. bk in lard appl. burns. scalds; decoc. of twigs or branches for dys. & cough; frts. & pd. rts. also u. in Chin med. - **T. plicata** Donn ex D. Don (wNA, Alas-nwCalif [US]): western (or Pacific) red cedar; (giant) arbor vitae; (British Columbian) cedar; decoc. of boughs u. as hair wash to eliminate dandruff & improve scalp; bath for arthritic & rheumatic limbs; for this also drunk diluted (danger since toxic); weak soln. as sweathouse ton (BC [Cana] Indians); wd cause of red cedar asthma (BC) (esp. in lumber mills); inhibits fungal growth: wd. u. for sashes, interior finish, "cedar" chests, fences; inner st. bk. u. to make rope, blankets, fish nets, utensils, as water bailers, mats, clothing, (nwIndians hats & dress); rts. u. in basket weaving; cambium u. as food (fresh, cooked, dr.) (w.Indians) (Fr Ph. VII).

thujene: dicyclic terpene fr. VO of *Boswellia serrata, Thuja occidentalis.*

thujone: a terpene ketone; occurs in many VO; ex. VO of thuja, tansy, absinthe.

Thujopsis dolabrata Sieb. & Zucc. (1) (Cupress.) (Jap): hiba tree; wd. & lvs. c. thujaplicins, these & other components of VO antibiotic: wd. u. for cabinet work, cooperage, &c.

Thujyl alcohol: tanacetyl alc.: dicyclic terpene alc. fr. VO of *Artemisia absinthium, Thuja wareana,* &c.

Thunbergia alata Bojer ex Sims (Acanth.) (natd. widely in trop): "black-eyed Susan"; u. medl. (Fiji+); ornam. per. twining plants - **T. atriplicifolia** E. Mey. (sAf): Natal primrose: lf. or unripe frt. u. to make hairwash (Zulus+) - **T. chrysops** Hook. (trop wAf): clock vine; lvs. rubbed in hands & juice squeezed out to heal cuts - **T. cynanchifolia** Benth. (Nig): lvs. u. for colic, coughs - **T. grandiflora** Roxb. (seAs): Bengal clock vine; lvs. u. in stomach trouble (Chin); lf. decoc for this, also local appln. of lvs. to abdomen (Mal) - **T. hossei** C. B. Clarke (seAs): hb. u. to make poultice to apply to cut or broken bone to reduce swelling & speed healing (Thail) (2448) - **T. lancifolia** T. Anders. (e&cAf): rt burned & mixed in porridge to be eaten as contraceptive; aq ext. of rt. drunk for bilharziasis; rt. crushed with lvs. & inhaled for constipation; rt. inf. u. for pains (stitch) (Moz); cicatrizant; macerated lvs. appl. to burns (Zaire) - **T. laurifolia** Lindl. (seAs): fls. u. in eye dis. (Burma); lvs. in excessive menses, ulcers (China); juice dropped in ear for deafness (Mal).

thunder god vine: *Tripterygium forrestii* - **t. plant**: *Sempervivum tectorum* - **thunderwood**: *Rhus vernix.*

Thunnus thynnus (Thunnidae - Ord. Acanthopteri-Pisces) (wnAtl Oc s to nSA): **T. salinus** (Fla [US]): western north Atlantic bluefin tuna (fish); common tunafish; horse mackerel; albacore; FO of liver is percomorph liver oil (Oleum Percomorphum); of about the same potency in vits. A & D as halibut liver oil, & u. in the same way; *Neothunnus macropteris* (Pacific US) u. similarly.

thur, gum: v. of Acacia.

Thuricide™: Dipel™: live spores of *Bacillus thuringiensis* Berliner; insecticide for larvae of Lepidoptera (butterflies, moths, such as cabbage butterfly); non-toxic to warm blooded animals.

thus: 1) one grade of frankincense, Olibanum 2) *Picea excelsa*, oleoresin 3) gum t. (*qv.*) - **t. Americanum:** gum t. - **t. commune**: resin fr. *Juniperus oxycedrus* - **t. gum**: v. under gum - **t. judeorum** (L., Jews' t.): balsam fr. *Liquidambar orientalis* - **t. libycum**: *Dorema ammoniacum*, gum resin - **t. masculum: t. orientale**: oleoresin fr. *Boswellia* spp., frankincense - **t. vulgare:** 1) Burgundy resin, oleoresin fr. *Picea excelsa* 2) oleoresin fr. *Pinus sylvestris* (collected by ants).

Thuya L. – *Thuja* L; also Fr. for *Thuja occidentalis.*

thylakoids: intracellular flattened membrane sacs characteristic of the Cyanophyta.

Thym (Ge): thyme.

Thymbra spicata L. (Lab) (Eu, Med. reg.): "thymon" of Theophrastus; flg. tops u. med. (Span, Ph. III, IV); cult. ornam. shrub. (the ancients applied name "Thymbra" to various Labiatae members).

thyme: *Thymus vulgaris* - **basil t.** - 1) creeping t. 2) mountain t. 3) *Calamintha nepeta* subsp. *glandulosa* - **t. camphor**: thymol - **common t.: garden t.**: *T. vulgaris* - **creeping t.**: wild t. - **lemon t.:** *T. citriodorus* - **mother of t.:** wild t. - **mountain t.**: *T. acinos* - **red t. oil**: VO of *T. vulgaris* - **white t. oil**: 1) (genuine) redistd. red t. oil 2) (comm.) mixt. of fractions of red t. oil, oils of pine, rosemary, eucalyptus, origanum, &c. - **wild t.**: = *T. serpyllum.*

Thymelaea Endl. (Thymelae.) (Med reg., temp. As): bk. of some spp. u. like Mezereum - **T. tartonraira** (L) All. (Med reg.-rocky shores): lvs. u. in Sardinia officially as emet. & purg.

Thymelaeaceae: Mezereon Fam. (mostly of sHemisp); shrubs or small trees, lvs. evergreen or deciduous; leathery; fls. perfect, small; frt. a nut, berry, or drupe; ex. *Daphne* spp., *Lasiosiphon, Gnidia.*

Thymian (Ge): common thyme.

thymidine: combn. of thymine & ribose, found in cell nucleus; obt. fr. herring sperm; found in peas, rye, wheat (germ); important for formation of blood cells; u. in synth. of azidothymidine, u. experimentally to treat AIDS.

thymine: a methylpyrimidine base, contd. in nucleic acids; a component of deoxyribonucleic acid (DNA).

thymol: a terpene phenol fr. VO of *Thymus vulgaris* & *Monarda punctata*, among other pl. spp.; acts to prevent mold, mildew growth; parasiticide; disinfec.; antisep.; occ. in white crystals; popular (ca. 1935) as dental desensitizer, in ethanol/ether soln., appl. topically (Hartman's solution) - **t. iodide**: important antisep. formerly very popular.

thymonucleic acid = deoxyribonucleic acid (DNA).

thymosins: series of hormones fr. calf thymus gland; polypeptides u. to incr. immunity of indivdual by aiding production of **thymus-dependent lymphocytes** (T cells); may have some control over aging process, cancer, AIDS, &c. (sometimes u. with aspirin); fraction 5 esp. potent as immunopotentiating agt.

Thymus algeriensis Boiss. & Reut. (Lab) (nAf, Tun to Mor): djertil (Arab); lvs. & flg. branches condim. stom. diaph. antispasm. (esp. in pertussis), circ. stim., aphrod. - **T. broussonettii** Boiss (nAf): zeitra; lf. & flg. branches u. as inf. for coryza, rheum. arthritis; pain; as gargle in sore throat; decoc. in jaundice, as galact. vermif. emmen. diur. antisep. for intestine; as plaster appl. to abdomen for dig. dis.; spice - **T. capitatus** Hoffm. & Link (Coridothymus c. Rchb.) (Med reg., esp. Sp): Spanish origanum; S. hops; headed thyme; a spice plant; hb. source of red brown VO (Spanish or Andalusian) thyme (or origanum) oil (inferior to true thyme oil); c. phenols (40%-65%), esp. carvacrol, but little or no thymol; d-pinene; hb. u. int. for cough, some skin dis. (nAf); diur. (2539); formerly cult. in Cyprus - **T. ciliatus** Benth. (nAf): lf. c. VO with thymol; u. in snuffs for otalgia; in folk med. of natives - **T.** X **citriodorus** Schweiger & Koerte (cult. sFr & It), a hybrid (*T. pulegioides* X *vulgaris*): lemon thyme; VO c. citral; source of important part of comm. thyme herb - **T. cunila** E. H. L. Krause = *Satureja hortensis* - **T. herba-barona** Loisel (Sard, Cors): baron's herb thyme; VO rich in phenols, esp. carvacrol: u. flavor - **T. hirtus** Willd. (Sp): tomillo limonero; VO with high content of citral, limonene, & terpene alcs.; source of "thyme lemon oil" - **T. integer** Griseb. (Cyprus): thymari; u. in folk med. (2740) - **T. mastichina** L. (Med reg., esp. Sp): (herb) mastich; mastick thyme; VO with 64-67% cineole; hb. & VO occur in drug commerce - **T. odoratissimus** MB (sRuss, AsMin): source of a VO with 8% phenols - **T. piperella** L. (Calamintha p. Reichb.) (Sp, Valencia): VO c. thymol, carvacrol, cymol; u. carm. - **T. praecox** Opiz & subsp. *britannicus* (Ronn.) Holub. (n&wEu): wild thyme; VO (serpolet oil) u. in med.; u. to flavor soups, meat, &c. - **T. pulegioides** L. (T. serpyllum L. subsp. chamaedrys Fries) (cw&nEu): pillolet (Fr); lemon (or bank) thyme; produces an important part of the serpyllum herb on the market; c. specially valuable VO with lemon scent - **T. schimperi** Ronn. (Ethiopia): hb. u. as condim.; inf. antispasm. in gastralgia; vermif.; in colds, coughs, otitis - **T. serpyllum** L. (T. glabratus Hoffmseg & Link) (Euras, nAf, natd. eNA): mother of thyme; creeping (or wild) t.; satar (Arab); very common small hb.; hb. (as Herba Serpylli) c. up to 1% VO with thymol, cymol; carvacrol; pinene; tannins (5-8%), serpyllin (NBP), luteolin glucoside; u. spice (condim.), stom., ton. (in Haiti, &c.); expect. antispasm. in pertussis, asthma, & other resp. dis.; sds. anthelm, abort.; in "heart conditions"; exter. in herb baths, &c. (rheum. itch, neuralgias); in various teas (Species); homeop. med (Serpyllum); formerly for prodn. of thymol: as ground cover (gardeners)

- **T. vulgaris** L. (c&sEu, wAs; cult. Af, As, NA): (common) thyme; garden t.; satar rasmi (Arab); hb. c. 0.4-2.5-4% VO with much thymol, carvacrol, cymol; tannins (10%), saponin, resin, NBP; lvs. exported fr Fr, Sp; u. as expect. popular in pulmon. & resp. dis. (as dry coughs, whooping cough, bronchitis, laryngitis, asthma); stom. in anorexia, dyspep., chronic gastritis; ton. flavor, condim (meats, esp. sausages); baked in bread, u. in sandwiches, supposed stim. to student in taking exams. (Lebanon); carm. aperitive; diaph. (Cu); prodn, VO ("red t. oil") & thymol (chief prodn. Sp, Fr); narrow & broad lf. forms - **T. zygis** L. (Sp, Por, sFr, nAf): Spanish thyme; linear-leaved thyme; source of VO (with *T. vulgaris*); this VO rich in thymol, carvacrol, cineole; var. *gracilis* Boiss. (Sp): tomillo rojo; source of Spanish thyme oil.

thymus (**gland**): endocrine organ located in anterior chest area, well developed in infant, gradually deteriorates with age; secretes thymosin; related to natural immunological defenses; said directly related to aging.

Thyridolepis mitchelliana (Nees) S. T. Blake (Gram) (Austral.): mulga grass; grows in arid places; good fodder.

thyrocalcitonin = Calcitonin.

thyroglobulin: protein of thyroid gland wh. cont. org. bound iodine; it is the reservoir source of thyroxin & triiodothyronine.

thyroid (**gland**): **Thyroideum** (**Siccum**) (L.): dr. gland deprived of fat & connective tissues; obt. fr. animals u. as food by man; usually administered in hypothyroidism (cretinism, myxedema, some types of obesity); in hypercholesterolemia; in form of CT; gland c. thyroglobulin; given orally (tabs).

thyrotrop(h)in: TSH; Thytropar™; thyroid-stimulating hormone of the anterior pituitary lobe - **thyrotropin-releasing hormone**: TRH: a tripeptide neurohormone; u. to treat neuromuscular dis., esp. amyotrophic lateral sclerosis ("Lou Gehrig's disease").

thyroxin(**e**): T_4: tetraiodothyronine; an amino acid of the thyroid gland; rich in iodine; levothyroxine (the L-form) is the active physiol. form; isolated fr. the gland or mfrd. synthetically; represents circulating form wh. is transformed into active "sensitized" form in peripheral tissues (764) (E); generally u. as Na salt; directly responsible for heart rate.

Thyrsanthera suborbicularis Pierre (monotypic) (Euphorbi.) (Indochin): rt. inf. u. in mal.; ingred. in detergent solns. u. by women postpartum.

Thysanocarpus elegans Fisch. & Mey. (Cruc.) (Calif, Ariz, NM [US], nMex): lace-pod; decoc. of whole pl. u. to relieve stomachache; sds. u. in pinole mixts. (meals) (Indians of Medocino Co., Calif).

Thysanolaena maxima O. Ktze. (monotypic) (Gram) (Indochin, PI): tiger grass; lvs. of perenn. grass u. for pin worms; rts. u. to alleviate swollen feet; cult. ornam. (sFla, Calif [US]).

Thysselinum Hoffm. = *Peucedanum* L.

ti palm: *Cordyline terminalis* - **t. tree**: tea t.: 1) *Cyrilla* spp. 2) *Cliftonia monophylla* 3) *Melaleuca* (esp. *M. leucodendron*) 4) *Leptospermum* spp. 5) *Cordyline* spp. esp. *C. fruticosa* 6) *Lyonia mariana* (SC) (pron. like "tie" in (1) and (2), like "tea" in others) - **Australian t. t.**: *Leptospermum pubescens* - **scrub t. t.**: *L. attenuatum* - **white t. t.**: *Cyrilla* spp.

tianquispepetla (Mex): *Alternanthera sessilis*.

Tiarella cordifolia L. (Saxifrag.) (eCana, neUS): coolwort; (false) mitrewort; foamflower; hb./lvs. (Tiarellae Folium, L.) inf. u. diur. ton., in urin. lithiasis, formerly in strangury (urin. suppression); rts. u. diur.; to "loosen phlegm in chest."

TIBA: 2,3,5-triiodo-benzoic acid; u. to accelerate flowering of petunia, zinnia, pineapple, &c.

Tibabat: **Tibb**(**i**): Muslim science of medicines (practiced in Near East); related to Unani; less formal & more along the lines of folk medicine.

tibey (**blanco**): *Hippobroma longiflora* - **t. tupa**: *Lobelia portoricensis*.

Tibouchina aspera Aubl. (Melastomat.) (Br): lvs. & fls. as sed. pectoral - **T. holocericea** Baill. (Br): lf. decoc.u. for sore throat - **T. longifolia** Baill. (sMex to Pe, Bo): mullaca; talchinal (Salv); lf. decoc. held in mouth for toothache (Salv); appl. to eye dis.

tickletongue (**tree**) (La [US]); prickly ash.

tickweed: *Hedeoma pulegioides*.

Tieghemella africana Pierre (Sapot.) (tropAf, esp. Gabon, Congo): sds. source of fat, like Ndjawe butter, u. med. - **T. heckelii** Pierre ex A. Chev. (Dumoria h. Chev.) (trop wAf): fine wood (makore wood) u. for cabinet work, autos (formerly); sds. source of FO, dumori butter u. as food; u. med. (Guinea).

tienchi (**ginseng**): rt. of *Panax notoginseng*.

Tierkohle (Ge) : animal charcoal.

ti-es (Fla [US]): *Lucuma nervosa* (*Poutaria*).

tiger: *Panthera tigris* L. (*Felis t.*) (Felidae-Carnivora-Mammalia) (As, incl. Siber): many parts of this large cat (-230 kg); are u. in Unani med. (ex. flesh for gastric & uter. colics); FO in rheum, sex. debility (exter); testes (aphrod. & cardioton.); excreta (u. like Antabuse to curb alcohol intake); t.

bone (Os Tigris [L]); hu gu (Mandarin Chinese) (bone of penis) u. in Chin. med. in wines for arthritis; analg. & cardioton.; tongue in asthma; bile to improve eyesight; milk to improve skin texture; excreta (acts as Antabuse) - **t. flower**: *Tigridia pavonia* - **t. nut** (Brit): *Cyperus esculentus*, sd. - **t. snake** (Austral): *Notechis scutatus*; toxic serpent; venom strongly hemostatic; dr. venom u. locally in hemophilia, blood dyscrasias, &c.; antibiotic (Staphyloccus).

Tiger Balm: Baume du Tigre (Hydro-Metaux) (Société Pharma-Metal): famous pharm. prepn. of Singapore; ointment c. menthol, camphor, cajuput oil, &c.; *cf.* Vick's Vapo-Rub (1765).

tiger-bone: Os Tigris (L); hu gu (Mandarine Chinese); bone of penis; u. in med. wines for analg. antiinflamm. in arthritis (Chin med.).

tighteye: titi.

tiglic acid: C_4H_7COOH; unsaturated fatty acid, found widely spread in plants, ex. croton oil, *Anthemis, Geranium* spp.

Tiglium (L.): *Croton t.*, sd. - **tiglio, flori di** (It): *Tilia platyphyllos.*

tigloidine: tiglic acid ester of pseudotropine; isolated fr. *Duboisia myoporoides, Datura inoxia*; now of synth. origin; u. anticholinergic, CNS depressant, in Parkinsonism, paraplegia; HBr salt u. in med. as atropine substit. (2368).

tigonin: only saponin in *Digitalis lanata* lvs.; aglycone (or genin) = tigogenin.

Tigridia pavonia (L. f.) DC (Irid.) (sMex, CA): tiger flower; bulb formerly u. by aborigines as food, antipyr., to incr. fertility; ornam.

tikhar: starch of *Curcuma* spp. (many variant spellings).

tikitiki (Jap): rice polishings.

tikor: *Curcuma angustifolia.*

Ti-kwang (Honan, China-cult.): pl. sp. related to comfrey; rt. u. antipyr.

til: teel (Indo-Pak): *Sesamum indicum.*

Tilapia (Cichlidae - Pisces) (tropics): cichlid fishes of fresh water - **T. aurea** (Steindachner) (T. nilotica L.) (Af): small fish (-18 cm); surface feeder, eats larvae of mosquito, hence sometimes u. in mal. control; to reduce growth of noxious water plants; excellent food; introd. into seUS, PI+; cult. in salt water (Jap).

Tilia L. (Tili) (nTemp Zone): linden (tree); basswood; plane tree (hort. name); source of lime fls. (esp. *T. platyphyllos)*, popular med. esp. in Ge (marketed with & without lvs.); bk. u. in folk med. (eEu); bee trees (fls. very fragrant) (some Am spp. have toxic honey); trees sometimes infested with aphids (prod. honey dew); wd. u. to make charcoal; shade & ornam. trees - **T. americana** L. (T. glabra Vent.) (e&cNA): American linden or lime; whitewood; silver l.; one of several good sources of linden fls.; boiled bk. appl. to burns (forest Am Indians); pl. fibers fr. inner bk. u. as sutures (Am Indians); rt. u. for worms (Marit. Indians) in teas for epilepsy, headache, spasmodic cough; frts. & fls. ground to edible paste with chocolate flavor; bk. u. for suppurating wounds (Marit Indians); to make basswood splints to support limbs (hardens on drying) (Am Indians); wd. considered best one for carving (2369) - **T. argentea** Desf. = *T. tomentosa* - **T. caroliniana** P. Mill. (T. floridana Small) (seUS, n. to sIll; nMex): (Florida) basswood (or linden); Carolina (or southern) b.; tilio; one of silver lindens; distinguished by unpleasant odor of infl.; fl. inf. u. for cough (antispasm.) - **T. cordata** Mill. (cEu): bast (tree); small-leaved lime; winter l.; stone l. (also appl. to *T. platyphyllos, T. europaea*); fls. c. VO with geraniol & its acetate, eugenol, linalool, &c.; bk. c. taraxerol (tiliadin) (triterpenoid); vanillin; frt. c. 58% FO (linden frt. [sd.] oil) (like olive o.); fls. u. diaph., antipyr., in rheum. as diur. stom. sed.; tea substit.; bk. as vulner. (Russ); in liver & gallbladder dis.; wd. u. to make charcoal with strong adsorption; nectar narcotic to bees; wd. meal as fodder - **T. X europaea** L. (*T. cordata* Mill. X *T. platyphyllos* Scop.) (cEu): Dutch (or European) linden; c. & u. like those of *T. platyphyllos*; much cult. - **T. heterophylla** Vent. (e&cUS): white basswood; inner bk. u. for making ropes (Civil War); wd. useful lumber; bee tree - **T. X juranyiana** Simk. (*T. cordata* X *T. tomentosa*) (Eu): silver linden; fls. u. like those of *T. platyphyllos* - **T. mesembrinos** Merr. (Indochin): fls. u. in inf. like *T. europaea*, &c. - **T. mexicana** Benth. (Schlecht.?) (sMex): tila (Sp); fls. (flor de tila) u. in inf. as hypnotic, tranquilizer, popular in flatulence; intest. colic; congested hemorrhoids - **T. microphylla** Vent. = *T. cordata* - **T. neglecta** Spach = *T. americana* - **T. parvifolia** Ehrh. = *T. cordata* - **T. petiolaris** DC (Euras, natd. US): (pendent) silver linden; weeping s. l.; *cf. T. americana* - **T. platyphyllos** Scop. (T. grandifolia Ehrh.) (cEu, natd. US): large-leaved /summer/big-leaf linden; female lime; classed as one of the stone lindens; fls. c. VO with geranyl acetate, phenyl ethanol, farnesol, linalool; flavonoid glycosides, mucilage, tannin, sugars; popularly u. in Eu med. (767), as diaph. antispasm. diur. stom. sed. dem. analg. in colds, cramps, colics, influenza, diarrh. rheum.; in mouthwashes, gargles, baths, cosmetics (linden water); one of most widely u. tisanes in Fr; fl. VO u. in perfumes: tuberous rts. u. med. (Fr Ph.); inner bk. (Fr Ph. IX;

1980) u. in dyskinesia of biliary excretory ducts; cholecystectomy, gallstones, liver & biliary dis. - **T. pubescens** Ait. = *T. caroliniana* - **T. sylvestris** Desf. = *T. cordata* - **T. tomentosa** Moench. (T. alba W. & K.) (s&cEu, As): (European) silver lime tree; a silver l.; very arom. fls. u. as substit. for those of *T. platyphyllos*; to aromatize champagnes, &c. (prodn. Yugoslavia, It) - **T.** X **vulgaris** Hayne: close to *T.* X *europaea* (*T. cordata* X *T. platyphyllos*) (Eu. natd. US): u. much like parent spp.

Tiliaceae: Linden Fam.; hbs. but mostly shrubs & trees; lvs. simple; fls. regular, commonly with 5 sepals & 5 petals; frt. capsule, drupe, or berry; ex. *Corchorus, Grewia*.

Tiliacora chrysobotrya Welw. (Menisperm.) (Angola): u. med. (767) - **T. funifera** Engl. ex Diels (trop wAf): rt. bk. c. funiferine, tiliacorine; u. in gastritis, hernias, menstr. difficult.; for tanning - **T. racemosa** Colebr. (T. acuminata Miers) (Indones, Java, Ind): rt. bk. c. tiliacorine, corine, neuromuscular blocking like curare; comminuted rt. in water drunk for snakebite (Nep); bk. & lvs. c. cardiac poison.

tiliadin: taraxerol: steroid found in *Tilia cordata, Taraxacum officinale, Alnus glutinosa* & other plants.

Tillandsia aeranthos L. B. Smith (Bromeli) (sBr, Par, Arg, Uru): cravo-do-mato; pl. decoc. u. int. as diur. in GC; ornam. (1822) - **T. benthamiana** (Mex): pl. cooked & taken for anemia, kidney dis. - **T. recurvata** L. (sTex, Fla [US], WI, Mex to Ch, Arg): ball moss; old man's beard; 2- leaved wild pine; grows on oaks, &c.; lf. decoc. u. emmen. in leucorrh., gall bladder dis. (Cur) pd. lvs. in butter appl. to hemorrhoids (Arg.): ornam. - **T. usneoides** L. (Dendropogon u. Raf.) (seUS n to Va; tropAm s to Arg): Spanish (or Louisiana or Florida) moss; hair m; old man's beard; an epiphyte of lichen-like appearance, hanging on branches of trees, (esp. of live oaks), telephone wires, &c.; c. 47% cellulose, galactans, flavonol type heteroside with antibiotic properties (active against gm.+ bacteria); a wax, estrogenic subst. (*sic*), cycloartenol; pl. retted 6 wks, furnishes fiber ("black moss," moss fiber) u. in upholstering trade like horse-hair ("vegetable horse-hair") to stuff mattresses, pillows, furniture, auto seats, &c. (gathered pls. kept moist until fleshy exterior parts are rotted off [retted], well dr.); sterilized fibers u. as absorbent like cotton, with higher absorptive capacity, for dressing wounds; fresh pl. fodder; starvation emergency food (boiled); dr. pl. tinder, threads; packing materials; in folk med. as emoll. for inflamm. swellings, rheum, hernias; pd. in ointment or direct appl. for hemorrhoids; a forest by-product of the south (769); wax u. as substit. for carnauba wax - **T. xiphoides** Ker (Arg-Bo): fls. u. to prepare med. for chest dis. (rural med.).

tiller: sprout esp. fr. rt. or lower lf. axils.

tilleul (Fr-m.): **tilo** (Sp.): linden, *Tilia* spp.

timber: woody growth sufficiently large for mfr. useful articles.

timbó: 1) (Br) *Tephrosia brevipes, T. cinerea, T. nitens, T. rufescens, T. toxicaria* 2) (Br) *Paullinia pinnata* & other spp. 3) (Arg): *Pithecellobium multiflorum* (also as **t. blanco**) 4) *Indigofera lespedezioides* - **t. boticário**: *Lonchocarpus peckolti, L. nicou* - **t. di cipo**: *Serjania curassavica* +.

timo (It): 1) thyme 2) thymus gland.

Timonius timon (Spreng.) Merr. (T. sericeus K. Sch.) (Rubi.) (Indones, PapNG, Austral): rt. chewed on sea voyages for sea sickness; in rheum. fever; in fever (PapNG); inner bk. decoc. drunk for colds, flu, fever (2263).

timothy: *Phleum pratense*.

timu (Ec-Jivaro Indians): *Lonchocarpus nicou*, barbasco.

tin: Stannum (L): Sn: a metallic element present in such minerals as cassiterite (stannic oxide; u. in polishes, incl. fingernail polishes); several salts u., such as - **t. difluoride**: stannous fluoride: SnF_2; u. as ingredient in toothpastes to prevent caries.

Tinantia fugax Schneider (Commelinaceae) (Mex, SA): hbs. like *Tradescantia*; fresh lvs. c. cyanogenetic glycosides; u. hemostatic.

Tinctura (L): tincture - **T. Balsamica** (Fr Codex. 1925): Tr. Benzoin Comp. - **T. Hiera**: wine of aloes - **T. Japonica**: Tr. Catech - **T. Maeconii**: Opium Tr. - **T. Melampodii**: tincture of *Helleborus niger* - **T. Stomachica**: Cardamom Tr. Comp. - **T. Thebaica**: Opium Tr.

tincture: hydroalcoholic (or alc.) soln. of crude drug or chemical substance usually with a strength of 10% (w/w) prepd. by various means - **Huxham's t.**: Tr. Cinchona Comp. with saffron - **t. of benzoin compound**: popular prepn. of world's pharmacopeias; compn. varies in various compendia; typical formula (Croat Sloven Pharm) (pts. by wt.): Tr. Benzoin 100, Tr. Aloe 15, Balsam Peru 4, alc. 40; Of Fr Codex 1925: Benzoin 60 Balsam Tolu 60, angelica rt. 10, hypericum hb. 20, 10 each of aloe, myrrh, olibanum, alc. (80%) 720.

tinder: t. wood: spunk w.: fr. *Fomes igniarius, Fomitopsis officinalis*.

tinea: ringworm.

Tinea pellionella (L) (Tineidae-Lepidoptera-Insecta): case-bearing moth; a moth, the larva of which attacks many crude drugs, ex. orris, almonds, capsicum; clothes - **T. zeae**: corn meal moth; larva commonly attacks crude drugs. as well as woolen clothes, upholstery, &c.

tinguaciba (Br): *Zanthoxylum rhoifolium.*

Tinnea antiscorbutica Welw. (Lab) (tropAf): inf. of shoot & lf. u. as antiscorb. (Angola).

Tinnevelly: Tirunelveli: town of Madras, sInd; name applied to type of Senna.

Tinomiscium philippinense Miers (Menisperm.) (PI): milky sap u. in combns. as eye wash; frt. c. picrotoxin, u. as fish poison - **T. tonkinense** Gagnep. (Indochin, Mal): latex u. for tooth decay (with care); decoc. fr. boiled rts. u. as poultice in rheum.

Tinospora bakis Miers (Cocculus b. Rich.) (Menisperm.) (trop wAf, Seneg, Nig, Ind): bakis; Senegal root; rt. c. colombin (NBP) 3%, palmatine, not sangoline or pelosine; u. bitter ton. diur. antipyr. emmen. cholagog in biliousness; rheum. (770); in yellow fever (Seneg) - **T. capillipes** Gagnep. (sChin): rts. u. to treat sore throat, laryngitis, bitter - **T. cordifolia** (Willd.) Miers (Cocculus cordifolius DC) (seAs, as Burma): heart-leaved moonseed; giloe (Hindi); a climbing shrub; sts. with bk. (gu[i]-lancha bark) c. tinosporine, berberine, sal gilo (alk. prepn.) picroretin; giloin (bitter glucoside); giloin (non-glucosidal bitter) u. bitter ton. antipyr. antiperiod. in mal. (claimed more effective than quinine), aphrod.; dietetic in diarrh. & chronic dys.; starch of rt. & st. similar to arrowroot; drug u. exter. as alixiter - **T. crispa** Hook f. & Thumb. (T. rumphii Boerl.; T. tuberculata Beum.) (Indones, esp. Java, Mal, PI): makubuhay; gulancha (Ind); twigs u. as antipyr. ton. in debility, diur.; to prep. arrow poisons (Mal); c. picroretin (NBP) palmatine, berberine (traces) (claimed) (771); inf. of scraped vine u. as ton. in mal., quinine substit.; to flavor cocktails & cordials; sts. ("Stipites Menispermi") u. in fevers, stomachache, jaundice: very popular in folk med. of As - **T. glabra** Merr. (Indones): pounded lf. poultice appl. to abdomen for constip. - **T. malabarica** (Lam.) Miers (seInd, Chin): gaduchi (Ayurvedic med.); pl. ton.; fresh lvs. & sts. u. in chron. rheum. (Chin); st. u. to prod. starch (Gaduchi satva); sometimes cult. - **T. negrotica** Diels (PI): cut st. boiled in coconut oil rubbed on swollen rheum. joints - **T. reticulata** Miers (PI): burnt lvs. u. for pinworm; juice of fresh sts. in mal.; decoc. ton. parasiticide, cardiac; makes pigs drowsy; macerated lvs. & sts. vulner. - **T. sagittata** Gagnep. (w&cChin): u. similarly to *T. capillipes;* exter. for boils - **T. sinensis** Merr. (T. tomentosa Miers) (Indoch): ext. of green st. drunk for rheum.; ton.

Tiquilia canescens A. Richardson (Coldenia c. DC) (Boragin.) (swUS, Mex): yerba del pobre; guachichile; huichichile; lf. decoc. u. to bathe wounds (Mex), in sweat baths; to prevent going to sleep after a meal - **T. greggii** A. Richardson (NM, Tex [US], Mex): yerba de la cachuca; plume coldenia; u. to treat GC - **T. latior** (Johnston) A. Richardson (Coldenia hispidissima A. Richards var. l. Johnston) (Ariz, Utah [US]): supais; rt. decoc. drunk for stomach dis. (Arg) - **T. paronychioides** A. Richardson (wSA: Ec to Ch): flor de arena; hb. u. in GC (1767).

tiquisque (CR): *Xanthosoma* sp.

tisana: **tisane** (Fr, La [US]): herbal tea, cooling draught prepd. by inf., sometimes medl.; sometimes decoc. of pearl barley, &c. or of some hb.; anciently u. in many combns. (Hippocratic med.) (mints, matricaria, tilia, anise, marrubium, elder fls., sage, lemon balm, agrimony, &c., lvs. fls. sts. with honey) (fr. ptisana, L. = barley) (2589).

tissa: *Spergularia rubra.*

tissue: complexes of cellular and non-cellular materials of organism - **t. factor**: thromboplastin - **t. plasminogen activator** (TPA): "clot buster"; occ. in bacteria, mammalian tissues, esp. in vascular walls; inj. to dissolve blood clots in situ in coronary arteries, &c. (u. to treat mcyocardial infarction [heart attack]) embolism; stroke; converts plasminogen to plasmin wh. dissolves fibrin; *cf.* urokinase: said twice as effective a thrombolytic as streptokinase: does not provoke hemorrhages; prod. by genetic engineering (inserting gene into *E. coli*); costs ca. $2,000/dose (Genentech); streptokinase considered by some just as good & much cheaper.

tiswin (wNA Indians): liquor.

titanium: Ti: metallic element of transition series; abundant in the earth's crust, where found as rutile, ilmenite, &c.; u. in alloys to give light-weight metals; as surgical aid to fix fractures - **t. dioxide**: TiO_2; important white pigment in paints, artifical teeth, dusting & face powders, &c. - **t. sulfate**: $TiOSO_4$; u. in dermatology.

titer: 1) level or concn. of dissolved subst. in soln. (as detd. by titration) 2) solidification temp. of fatty acids of FO.

Tithonia diversifolia A. Gray (Comp.) (swUS; Mex, CA, WI; natd. tropAs, Af): Mexican sunflower; jalacate; rt. decoc. for stomachache (small children, Tanz), fl. heads u. like arnica; common weed (sAs): often cult. ornam. - **T. rotundifolia** Blake (T. speciosa Hook.) (Mex, CA, WI): orange (or Mexican) sunflower; "titonia" (Cu); acate (Salv); lvs of ann. hb. u. in mal.; hemostatic (decoc.); antipyr. (Mayas); cult. ornam. - **T. scaber-**

rima Benth. (Mex, CA): mirasol (Salv); lvs. u. as bath for colds, fever (Salv); antipyr. - **T. tubaeformis** Cass. (nMex): polocote; weed in abandoned fields; u. as diur.

Tithymalopsis Klotz. & Garcke = *Euphorbia.*

titi (pron. tie-tie) (nFla, sAla, sGa [US]): 1) *Cyrilla racemiflora* (also **black t., red t., white t.**) 2) *Cliftonia monophylla* (sometimes).

tlachichinoa (Mex): *Tournefortia hartwegiana.*

tlacopetl (Mex): *Aristolochia tequilana*, &c.

toad: member of order Anura (Div. Amphibia): many of fam. Bufonidae, & gen. *Bufo* (incl. *Bufo bufo* L.) (Euras) - *Bufo bufo* L. (Bufonidae) (Euras, Af): (graue) Kroete (Ge); u. since antiquity; still popular in Chin, Jap; u. in edema (significant use); c. digitalis-like glycosides (many found) (bufogenin, &c.); u. in epilepsy, inflamm. skin dis. (erysipelas, herpes, furuncles, &c.); u. in Unani med. (*Bufo melanostictus*); in typhoid fever (Ga [US]) (folk med.) (2722) *B. marinus* (CA, SA): giant toad; -22 cm long (head & body) - **t. skin:** c. bufotenin, &c. - **t. venom** obt. by expression of secretions of parotid glands of living toads; long u. in Chin med.; to stim. formation of corticosteroids in rheum. arthritis & inf. dis.; in epilepsy; c. digitaloid principles such as the bufogenins (bufotalin, bufotalinin, bufalin) (hence u. like digitalis in heart dis.), bufotoxins, organic bases (bufotenin), cinobufotalin, &c. (2739).

toad: v. also under lily.

toadfish: *Opsanus* spp. (Pisces); venom claimed to act like insulin; also hypotensive compds. present.

toadflax: *Linaria*, esp. *L. vulgaris* - **bastard t.**: 1) (US) *Comandra* spp. 2) (GB): *Thesium alpinum, T. humifusum* - **blue t.**: *Linaria canadensis* - **common t.**: **yellow t.**: *Linaria vulgaris.*

toasted: roast(ed).

tobacco: *Nicotiana tabacum, N. rustica,* other *N.* spp.; the dr. lf. prep. for smoking, chewing, sniffing, etc. (leaf tobacco), in form of cigarettes, cigars, smoking & chewing tobaccos, snuff, etc. (772) (for Serials see App. C-1) - **Australian t.**: *Duboisia* spp. (apparently) - **Indian t.**: 1) *Lobelia inflata* (SC NA Indians smoked) 2) *Antennaria* spp. 3) Cornus amomum - **t. juice**: 1) (US slang) sputum of a t. chewer. 2) t. liquor - **t. leaf**: prohibited entry into Comoro Is - **t. liquor**: a fluidext. of t. u. as insecticide - **t. mosiac virus:** TMV: the best known & most studied virus; causes t. m. disease; is combn. of protein with RNA; composed of many units of polypeptides, with 158 amino acids in known sequence forming each polypetide subunit; valuable in the study of relationship of chemical structure to biol. activity - **mountain t.**: *Arnica montana* - **poison t.**: henbane, Hyoscyamus - **rabbit t.** (seUS): *Gnaphalium obtusifolium* - **t. sauce**: praiss (fr. Fr, presser, to press): juice squeezed fr. t. leaf; prohibited entry into Comoro Is - **t. sickness, green**: nicotine poisoning often seen in t. producing areas of Ky (US), &c.; prod. emesis, exhaustion, semi-narcosis, weakness, collapse, death; esp. bad when t. collected in the rain - **smokeless t.**: chewing tobaccos, snuff (ex. Skoal), &c.; c. nitrosamines; considered carcinogenic & hazardous to health - **t. water**: t. liquor - **wild t.**: *N. rustica, Lobelia inflata.*

tobaccoism: tobaccosis: state of chronic ill health due to excessive use of tobacco.

tobaso (NM [US], Sp): *Sorghum halepense.*

tobie (seUS Negro): **toby**: hand or jack.

Tobramycin: broad spectrum antibiotic of aminoglycosidic gp.; made up of a nebramycin complex fr. *Streptomyces tenebrarius* Higgens & Kastner (*cf.* gentamycin); bactericide for *Pseudomonas aeruginosa, Citrobacter* spp. &c.; ca. 1967-); incl. Nebcin (Lilly) - **t. sulfate** u. by injection.

tocopherol, alpha: Vitamin E (*qv.*) - **beta-t.**: found along with alpha-t. & gamma-t. in natural prods.; less active therap. than (but similarly to) vit. E - **gamma-t.**: see prec.; less active than alpha.-t. - **delta-t.**: common segment of vit. B complex as found in wheat germ oil, soy bean, & other FO; most active antioxidant of gp. but much less active than vit. E; u. as food preserv.

Tocosamine™: sparteine sulfate for IM use (oxytocic).

Tocoyena longiflora Aubl. (Rubi.) (Br): rt. c. emetine, cephaeline; u. as emet.; ipecac substit.; toxic.

Toddalia asiatica Lam. (T. aculeata Pers.) (monotypic) (Rut.) (tropAs, Af, Mad): wild orange tree; cockspur o.; kaki todali (Malabar); rt. (bk.) (as Lopez root) c. toddaline (glycosidal alk.); berberine(?); skimmianine; u. as arom. ton. stim. antipyr.; weak inf. for constitutional weakness; formerly more popular; rt. & bk. u. considerably in Af in cholera, diarrh., rheum.; yellow dyestuff; rt. bk. c. diosmin (773) - **T. lanceolata** Lam. = *Vepris l.*

toddy: 1) drawn sap of any one of several spp. of palms (ex. *Caryota urens*, **t.palm**), esp. when fermented; ex. of palmyra, coconut, wild date (Ind); *Arenga saccharifera; Raphia vinifera; Acrocomia vinifera, Nypa* spp. 2) synth. alc. beverage - **t. tree**: *Mammea americana.*

toe sack: (sUS): gunny s.; burlap bag.

tofa: soybean curd, a popular nutritious food (rich in protein) (*cf.* cottage cheese); cakes in Oriental cuisines; said to inhibit carcinogens, as nitrosamines (2758).

toffee (Ge) (Eng): hard chewable candy made fr. brown sugar, caramel, butter, &c.; orig. in Lancashire, Eng; (*cf.* taffy).
tofu (Jap) (teo-fu): tofa.
toia: (Br): taioba.
Tokio purple: *Lithospermum erythrorhizon.*
tola: wt. u. in Ind = 11.7 g. (ca.) - **t. seeds**: *Tylostemon mannii* sds.
tolerance: 1) the ability of an animal to resist the potent or poisonous effects of a drug, generally acquired by the continued use of the substance 2) specified max. or min. allowance for error in an assay requirement (in measuring, weighing, etc.); also variation in standard weights, etc. 3) upper & lower limiting % allowed for impurity or foreign subst. (ex. H_2O, ash, &c.).
tolguache (Mex.): Stramonium.
Tollkirsche (Ge) (fool's cherry): belladonna (pl. or frt.)
Tollkraut (Ge) (fool's herb): belladonna pl.
Tolmiea menziesii (Pursh) T. & G. (monotypic) (Saxifrag.) (PacNA fr. Alas to Calif [US]): thousand mothers; youth-on-age; lvs. chewed fresh for leg & body swellings; sprouts eaten raw in spring (BC [Cana] Indians) (1768); cult. orn.
toloache (Mex): Stramonium & other *Datura* spp.
toloman (Fr): topinambour.
tolu: t. balsam: a balsam obt. fr. *Myroxylon balsamum* - **fluid t.**: a concentrate of t. balsam (with strength 1:7) ("**Liquor Tolu**") - **t. resin**: Resina Tolutana (L.): t.
toluachi (Mex): *Datura* spp. as *D. meteloides.*
Toluifera balsamum L. = *Myroxylon balsamum* Harms - **T. pereirae** Baill. = *Myroxylon p.* Klotzsch.
Tomate: (Ge, Sp, Por): tomato, *Lycopersicon esculentum* - **t. frances** (Sp): *Cyphomandra betacea.* - **tomatidine**: aglycone of tomatine, occ. in rts. & lvs. of Rutgers tomato (USDA 1948); starting with t., simplest procedure (& cheapest source) for prepn. of cortisone, testosterone, progesterone; steroid secondary amine.
tomatillo: 1) (Sp) *Physalis philadelphicus (P. ixocarpa)* ("green tomato"), *P. angulata* 2) (NM [US], Sp): *Solanum elaeagnifolium* 3) *Lycium* (some spp.).
tomatin(e): cryst. glycosidal alk. fr. wild tomato lvs.; tomatidine linked to 4 sugar mols. (glucose, xylose, galactose); antibiotic, antifungal.
tomato: *Lycopersicum esculentum* - **t. catsup**: sauce made with tomatoes, black pepper, pimenta, ginger, mace, NaCl, coccus (sometimes also vinegar, cloves, capsicum) - **cherry t.**: tomato with small frts. the size of cherries, *L. esculentas* var. *cerasiforme* - **Mexican t.**: *Physalis ixocarpa* - **plum t.**: with small type of frt. - **privy t.**: cherry t. growing near outdoor toilets - **t. puree**: a paste-like prepn. of t. pulp prepd. fr. strained t. - **Roma t.:** rather pearshaped frt.; preferred for canning - **strawberry t.:** *Physalis alkekengi* - **tree t.**: *Cyphomandra betacea* - **wild t.**: *Solanum* spp., esp. *S. triflorum* (wUS) & *S. carolinense.*
tomillo (Sp): *Thymus vulgaris* or *T. zygis* (growing in Sp) 2) (Mex): Mexican thyme.
"tommy-toes" (US slang): cherry tomato.
tomography: sectional radiography, the art of making X-rays of plane sections of solid bodies.
ton, metric: tonne: 1000 kg.
tonca: tonco (bean): **Tonco Semina** (L.): tonka.
tonga: 1) mixture of rhiz. of *Epipremnum pinnatum* & bk. of *Premna taitensis* in eq. parts; or beverage fr. sds. u. antirheum. alt., in neuralgia (Fiji Is) 2) (Pe, Bo): *Datura sanguinea.*
Tongaline™: Elixir of Tonga & Salicylates.
tongue: v. under adder('s).
tonic cup: quassia cup.
tonka (bean): **Dutch t.**: sd. of *Dipteryx odorata* - **English t.**: *D. oppositifolia,* sd. - **tonkin** (Chin): strong bamboo u. for fishing poles, &c.
Tonkin(g): former name of area around Hanoi [nVietnam].
tonne: metric ton = 1000 kg, ca. 2204.6 lbs.
ton-pang-chong (root): *Rhinacanthus rasatus.*
tonoplast: protoplasmic membrane of living cell surrounding & separating vacuole fr. cytoplasm = vacuole wall.
Tonquin bean: tonka b. - **T. musk**: term appl. to finest type of m.; appar. derived fr. areas near Tibet (Shong-kin or Tong-ting-ho) (775) where collected.
toon (Ind, Pak): *Toona sureni, T. ciliata.*
Toona M. Roem. (Meli.) (OW trop.): toon; in part assigned to *Cedrela* but latter is NW, *Toona* OW - **T. calantas** Merr. & Rolfe (T. ciliata Merr.) (PI): Indian red wood; bk. decoc. astr., antisep. u. to cleanse wounds; fl. decoc. antispasm; bark edible; wd. valuable mahogany - **T. sinensis** M. Roem. (Cedrela s. Juss.) (Chin to Mal): bk. decoc. u. in measles; pd. rt. diur.; tender lvs. eaten in spring as carm.; wd. durable - **T. sureni** Merr. (T. febrifuga Roem.; Cedrela toona Roxb.) (Ind to seAs, Austral): cedrela tree; tuni (Ind); fls. u. as emmen. (Ind); for red & yellow dyeing; edible bk. u. as febrifuge (NSW); in diarrh.; wd. red, excellent for musical instruments, &c.
toot (US slang): cocaine.
"toothache": *Hyoscyamus* spp., sds. - **t. tree**: 1) *Aralia spinosa* (bk.) 2) *Zanthoxylum americanum, Z. clava-herculis* - **"toothbresh"** (Ala [US], etc): **toothbrush**: snuff stick - **t. plant** (WI, Br): *Goua-*

nia domingensis; twigs u. to make t. - **tooth-pick plant: tooth-picks**: *Ammi visnaga.*

toothwort: *Dentaria* spp.

topee tamboo (Trin): **topinamb(o)ur (de Cayenne)**: 1) *Calathea allouia* 2) *Helianthus tuberosus.*

Topitracin: bacitracin (for local or topical use).

toponymics: toponymy: study of place names: 1) geographic 2) anatomic.

top(s): above-ground part of pl. with edible rts. (ex. beet tops).

toquilla (Ec): palm fiber fr. *Carludovica palmata.*

tor (Hindi): *Cajanus cajan*, a kind of pulse - **torch flower: t. lily**: *Kniphofia uvaria* - **torchwood** (WI): **mountain (fragrant) t.:** *Amyris balsamifera* - **torcnon** (Haiti): *Luffa.*

Tordylium apulum L. (Umbell.) (eMed reg., incl. Yugo): lvs. of this perenn. hb. eaten as veget. in Gr - **T. officinale** L (sEu): frt. (Frutos de tordilio) u. med., off. in Sp Ph. II-IV.

Torenia asiatica L. (Scrophulari) (eAs): lvs. u. to treat GC (Indochin) - **T. concolor** Lindl. (Ind, Chin): decoc. gargled for mouth sores (Taiwan) - **T. polygonoides** Benth. (seAs): whole lvs. appl. as poultice to lower abdomen in edema (Mal).

tori (, Indian): *Hedera helix*; tolu balsam (fr. *Myroxylon toluifera*).

toria (Urdu) (Pak): oil sd.

Torilis arvensis (Huds.) Link (Umb.) (seUS): hedge parsley; pl. yields some 10% mannitol; has been u. as sugar substit. - **T. japonica** DC (T. anthriscus Gmel.) (Euras, nAf; natd. NA+): hogweed; hemlock chervil; c. VO with cadinene; frts. u. as ascaricide, expect. (Kor), antipyr. (Jap).

Tormentilla L. = *Potentilla.*

tormentil(la): *Potentilla erecta* (P.t.), rt.

tornasol (Sp): litmus.

tornillo (Mex): *Prosopis pubescens, P. glandulosa.*

torn-weed: *Liatris (Lacinaria).*

toro (NZ): *Persoonia t.*

toronja (Sp): 1) grapefruit 2) (Mex): *Solanum refractum* - **toronjil** (Sp): 1) *Melissa officinalis* 2) (Guat, Yuc): *Mentha citrata* 3) (Mex): *Lippia alba* 4) (LatinAm): *Thymus vulgaris; Salvia microphylla* - **toronjo** (Sp): 1) grapefruit tree 2) (Co): *Solanum galeatum.*

torr: unit of pressure, the pressure of l mm Hg at 0° C & standard gravity (SC Torricelli, inventor of barometer).

torrefaction: process of roasting organic substances; the constituents are modified but not charred; ex. **torrefied coffee**.

Torresea cearensis Allem. = *Amburana c.* A. C. Smith.

Torreya californica Torr. (Tumion californicum Greene) (Tax.) (Calif [US]): California nutmeg; juicy frts. u. like nutmegs.; nuts buried in mud for several mos. then boiled or roasted & eaten (inner part of sd. edible, outer part astr); rt. strands u. in basket-making (571); lvs. c. VO - **T. grandis** Fort. (Chin): Chinese torreya; sds. rich in FO, u. as anthelm.; nerve poison; roasted sds. eaten - **T. nucifera** (L) Sieb. & Zucc. (Taxus n. L.) (Jap): kaya; Japanese torreya; frts. c. estrogens; u. as lax. anthelm. expect.; sd. edible, vermif. (Chin); cold expressed FO fr. sd. (Kaya oil) u. in cooking; hot pressed FO u. as lamp oil & in dye and shellac industries; wd. u. for boxes, water pails, &c. - **T. taxifolia** Arnott (Tumion taxifolium Greene) (nwFla [US], esp. along Appalachicola River); Florida torreya; stinking cedar; gopher tree; durable wd. u. for fence posts (polecat wood) (776).

Torrington's Drops: Benzoin Tr. Comp.

torta (Sp, Mex): **Torte** (Ge): round cake or bread of corn meal that is u. in Mex dressed with chile.

tortilla (Sp, Mex): Indian corn meal pancake, unleavened (Sp, omelet) - **t. de mesa** (Sp): corn bread.

tortoise (US): land turtle with hind limbs stump-shaped - **t. oil**: Schildkroetenoel (Ge); FO obt. fr. giant sea turtles (Mex) by extn. of muscles & sex organs; u. for skin care as such or as emuls. or cream.

Torula Pers.: false yeast; pseudo y.; "wild y." (Pasteur); *Candida utilis* Lodder & Kreger-van Rij (Torulopsis u. Lodder) (Blastomycetes) ("food yeast") is cult. on waste prod. such as sulfite waste liquor (paper indy.) (1868); prod. u. as foodstuff for man & lower animals; tasteless (hence preferred to regular yeast which is bitter); good vit. source, esp. for B complex.

torus: receptacle (of fl. head)

tosilado (Sp.): **tossilagem** (Por): **tossilagina** (It): *Tussilago farfara,* coltsfoot.

tostada (Sp): slice of toasted bread (fr. **tostus, a, um** [L]: toasted, roasted).

Totaquina (L.) (USP XII, XIII, NF IX, X): **Totaquine**: a mixture of xallizable Cinchona alks. which c. 7-12% quinine (anhydrous) & 70-80% total anhydrous xallizable alks., the balance being composed of lactose, starch, sucrose, & other diluents; represents ordinarily a readily available & cheap form of antimalar. therapy, still much u. abroad (as As) (68) (1275).

Totopon: Opium total alkaloid prepn.

totumo (Mex): *Crescentia cujete.*

touch-me-not: 1) (sUS): *Impatiens pallida+* 2) (US): *Schrankia* spp. - **touchwood** (fungus): *Fomes fomentarius, Fomitopsis officinalis.*

Tournefortia angustiflora R. & P. (Boragin.) (Co to Pe): vine st. cut into sections & steeped in water overnight, then drunk as purg. before use of *Banisteriopsis* hallucinogen (777) - **T. argentea** L. f. (tropAs): velvet-leaf; lvs. occas. smoked in place of tobacco (Seychelles natives); lf. decoc. for poisoning fr. fish (skipjack, &c.) wh. has eaten toxic seaweeds (Marshall Is) - **T. bicolor** Sw. (tropAm): mirette; lvs. mixed with FO u. as poultice for boils & carbuncles (Domin, Caribs) (2328) - **T. densiflora** Mart. & Gal. (Mex, Guat): hierba rasposa; pl. decoc. u. in intest. dis; remedy for wounds & pimples (Oaxaca) - **T. glabra** L. (sMex to Pe, WI): limoncillo; lvs. poulticed to feet as remedy for flu; pl. decoc. u. exter. for rheum. (Veracruz); root decoc. as diur. (Guat) - **T. gnaphalodes** R. Br. (sFla [US], WI, sMex; CA, Ve): temporana (PR); decoc. of leafy twigs u. as antipyr., emmen.; abort. - **T. hartwegiana** Steud. (T. capitata Mart. & Gal.) (Mex): hierba del zapo; decoc. int. & exter. for snakebite u. med. (Mex Ph.) - **T. hirsutissima** L. (sFla, SpAm): horse bath; hog hook; crushed lvs. appl. to skin to remove chiggers ("niguas") (WI); lf. decoc. as mouth rinse for stomatitis & to heal oral ulcers; rt. decoc. diur. - **T. mexicana** Vatke (sMex): rt. & lvs. u. for inflammn., GC; in scabies, oral ulcers (2342) - **T. montana** Lour. (seAs): lvs. u. as antipyr. (mt. people); rts. in TB with hemoptysis; rheum. (Indochin) - **T. petiolaris** DC (sMex to Salv): pie de guaca; rt. decoc. popular diur. (Guat) - **T. poliochros** Spreng. (sFla, WI): sara-wine (Bah); pl. decoc. u. to bathe sores (Bah) - **T. sarmentosa** Lam. (Chin to Austral): rts u. to hasten delivery of after-birth; appl. to suppurating sores; lvs. u. to destroy larvae in ulcers in cattle (PI) - **T. volubilis** L. (sFla, sTex; WI, sMex to Ve): soldier vine; s. wine; vine decoc. u. to bathe painful muscles, drunk as male aphrod., to relieve menstrual cramps; pl. decoc. drunk as ton. (Aruba).

tournesol (Fr): 1) litmus 2) *Helianthus* spp.

tourteau (Fr): oil cake.

tous-les-mois (starch): 1) Canna starch 2) (Fr): marigold.

toute épice (Fr): allspice.

tow (pron. to rhyme with "dough"): 1) short coarse fibers separated fr. the longer ones in processing (in cotton, linters); incl. hemp, jute, & flax 2) (Eng pharm.): jute fiber of good aver. quality (Stupa, BPC) - **t. sack** (sUS): gunny sack; a general utility container on the farm, etc.; commonly u. for potatoes & other rt. vegetables.

towel-gourd: *Luffa* spp. - **Singkwa t.**: *L. acutangula* - **Suakwa t.**: *L. cylindrica.*

toxic: poisonous; sometimes appl. to person very ill (poisoned with disease) (toxon, Gr bow, since often used to shoot poisoned arrows; *cf. Taxus,* yew [branches u. to make bows]).

toxicarol: tephrosin.

toxicodendrin: **toxicodendrol**: older name for urushiol.

Toxicodendron Mill. (Anacardi) (As, Am): gen. split out of *Rhus*; toxic plants (1769) - **T. diversilobum** Greene (Rhus d. Torr. & Gray) (BC [Cana], Wash [US] to nwMex): (Pacific) poison oak; (western) p. o.; dermal poison pl., lf., st., or rt. juice appl. to warts (after cutting off top) & ringworm; tr. of fresh lvs. u. in eczema, skin dis. (*cf. T. radicans)*; rt. coll. in dormant season, cut & dr., decoc. drunk for immunity to "rhus dermatitis"; juice u. as black dye; st. u. in basketry (571) - **T. globosum** Pax. & Hoffm. (T. capense Thunb. = *Hyaenanche globosa* Lamb. & Vahl) (sAf): hyena bane; h. poison; frt. u. as bitter ton.; cadavers are treated with sds. to kill hyenas, jackals, &c.; arrow poison (Bushmen) - **T. quercifolium** Greene = *T. toxicarium* Gillis - **T. radicans** (L) Ktze. (Rhus r. L.) (eCana, US, Mex): (common) poison ivy; p. oak; fresh lvs., &c., can produce dermal irritation, even systemic symptoms (best remedy suggested: bathe with very hot water [50-55°C]); dr. lvs. u. as antipyr., diaph, diur, stim.; in prepn. poison ivy & p. oak antigens; rt. bk. u. in Eu homeop.; nine subspp., incl. **divaricatum** (Greene) Gillis - **T. succedaneum** (L) Ktze. (R. succedanea L.) (Jap, Chin, Mal, nInd): (Japanese) wax tree; J. vegetable w. t.; frt. wall source of Japan(ese) wax (sumac w.; Cera Japonica), u. in ointment & pomade prepn., plasters; for floor & furniture polish, candles; latex basis for natural shellac; gall apples u. as ton. expect. in clysters for abdom. tumors (Ind) - **T. sylvestre** Ktze. (Rhus sylvestris Sieb. & Zucc.) (Jap, Chin, Kor): source of Jap shellac & plant wax like *T. succedaneum* - **T. toxicarium** (Salisb.) Gillis (T. toxicondendron (L) Britt.; Rhus t. L.) (e&cUS): poison oak; lvs. (&c.) c. urushiol (the "poisonous" principle), acts as skin irritant, vesicant; u. much like *T. radicans* - **T. vernicifluum** (Stokes) Barkl. (Rhus verniciflua Stokes; R. vernicifera DC) (Chin, Ind; cult. Jap): urushi (Jap); Chinese lacquer (tree); (Japanese) lacquer (or varnish) tree: source of a varnish resin (Lacca Japonica; Japanese or Chinese lacquer); u. in fine lacquers & varnishes; gives smooth surface resistance to heat & solvents; sds. yield pl. wax similar to Jap tallow; dr. sap u. for intest. worms, esp. ascarids; young lvs. edible; u. as anthelm.; sds. hemostat.; latex may cause allergic reactions (SC vernix, L., varnish) - **T. vernix** (L.) Ktze.

(Rhus v. L.) (e&cUS, eCana, swamps): poison (or swamp) sumach; p. elder; thunderwood (Ala, Ga [US]); shrub or small tree (-7 m); very potent dermal irritant & vesicant, even the exhalations of pl. may prod. swelling & inflammn. of skin; u. like *T. radicans*; fresh bk. & lvs. (eq. pts.) ("Rhus venenata") u. in Homeop. med. (Ge) in rheum. & neuralgias (—> *Rhus*).

Toxicodendrum Thunb. = *Hyaenanche.*

Toxicophlaea Harv. = *Acokanthera.*

Toxicoscordion Rydb. = *Zigadenus.*

toxiferine: alk. fr. calabash curare (778) with dehydro-t., most potent alks. present (2392).

toxin: an extremely soluble toxic prod. of bacterial cells, &c., wh. in the case of an exotoxin diffuses out of the living cell; u. to confer active immunity; ex. scarlet fever t. - **toxin-antitoxin**: a mixt. of t. & a. with the former in slight excess; u. to prod. active immunity; ex. diphtheria t. a. - **Toxinum Diphthericum Detoxicatum** (L.): d. toxoid. - **T. Scarlatinae Streptococcicum**: Scarlet Fever S. Toxin.

toxinology: study of toxins of all kinds.

toxisterol: terminal prod. of excessive irradiation of ergosterol; is one step beyond calciferol; toxic product.

Toxocarpus villosus Decne. (Asclepiad.) (Ind, SriL): whole pl. u. to treat sprue.

toxoid: toxin in which the toxic properties have been destroyed or neutralized; u. to develop active immumity; ex. diphth. & tetanus t. - **adsorbed t.** = "**purified t:**" toxoid combined with $Al(OH)_3$ or Al phosphate; produces higher levels of immunity because of persistence in the body, but action somewhat delayed.

Toxoidum Diphthericum (L.): diphtheria toxoid - **T. Tetanicum:** tetanus toxoid.

Toxosiphon macropodus (K. Krause) Kallunki (Rut.) (Br, Bo, + Ec): u. in curare (Ec).

toyon: *Heteromeles arbutifolia.*

TPA: tissue plasminogen activator.

trace elements: chemical e. found in very small amts. in pl. or animal tissues but considered essential for life; ex. B, Zn.

trachea: tubular cell of the xylem which conveys water; term now generally replaced by "vessel" (or "duct") - **tracheid**: a fusiform cell wh. like a trachea conducts water but in addition is of value as a strengthening & supporting cell; tracheids constitute typically the chief bulk of secondary xylem in coniferous wds. (Gymnosperms) (walls bear bordered pits); also common in lf. lamina & petiole tissues usually; types incl. pitted, annular, spiral, scalariform, reticulate.

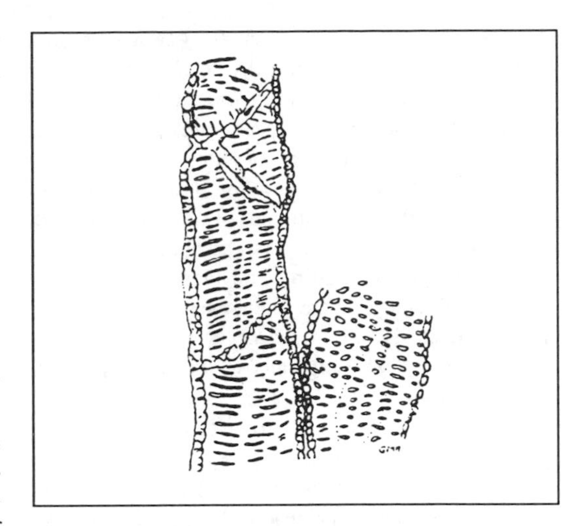

PITTED-RETICULATE TRACHEIDS:
ROOT, *LIATRIS TENUFOLIA*

Trachelospermum asiaticum Nak. (Apocyn.) (eAs): twigs u. to treat rheum. laryngitis, ulcerations (Kor); st. lf. inf. roborant (Taiwan) - **T. jasminoides** Lem. (Chin): pl. decoc, with malted rice u. in rheum.; emmen. anodyne; in sore throat & arthritis; for lumbar & knee pains; ton. for the aged.

tracheophytes: vascular plants (ferns & seed plants) (SC trachea = vessel).

Trachycarpus fortunei (Hook. f.) H. A. Wendl. (T. excelus H. A. Wendl.) (Palm.) (scChin, nBurma): windmill (or chusan) palm; sds. u. as hemostatic to control capillary bleeding (uterus, bowel, resp. tract), in hypertension; diarrh.; rts. uter. stim. (birth control); petiole in hypertension; all parts of tree utilized; thus, lf. base fibers (Chin coir) u. for ropes, mats, &c.; lvs. for hats; one of hardiest palms, widely cult. ornam, (2452).

Trachylobium verrucosum (T. hornemannianum Hayne T. Oliv.; T. gaertnerianum Hayne) (monotypic) (Leg) (eAf, Mad, Maurit): source of Zanzibar (or anime) copal; linde c.; & perhaps of Mad copal; hard transparent resin a semi-fossil prod.; most valuable copal for varnishes; u. for wound dressing (eAf); u. med. (Mad); sometimes put under *Hymenaea* - **T. mossambicensis** Klotz. = *T. verrucosum.*

Trachypogon spicatus O. Ktze. (T. plumosus Nees) (Gram) (China, s&eAf): grey tussock grass; rt. c. antibiotic; lf. u. as tea substit.

Trachyspermum ammi (L) Sprague ex Turrill (T. copticum [L.] Link; Carum copticum [L.] Benth. & Hook.; Ptychotis ajowan DC; Ammi copticum [L.]) (Umb.) (Eg, Crete; swAs, Ind): ammi (officinal) (Fr); amum; aj(o)wan; frt. ("ajowan seed"/ "ajava seed") c.VO with thymol, carvacrol, &c.; u. as spice in foods (Ind), carm. in indigest., dyspep. arom. stim. ton. expect. antispasm. in colic; once important in prodn. of natural thymol (sepn. on cooling, leaving the "oil of ajowan") - **T. roxburghianum** (DC) Sprague (Carum r. Benth. & Hook. f.) (Ind, seAs): persil glauque (Fr); frts. u. carm. stom. stim. in hiccup; bladder pain; spice for food; cult. in many Indian gardens.

Tradescantia L. (Commelin.) (Am): spiderworts; perenn. hbs. with regular fls.; for med. uses v. *T. ohiensis* - **T. discolor** L'Hér. = *Rhoeo spathacea* - **T. diuretica** Mart. = **T. elongata** G. F. W. Mey. = *Tripogandra elongata* - **T. erecta** Cav./Jacq. = *Tinantia fugax* Scheidw. - **T. fluminensis** Vell. Conc. (Br, Par, Ur): wandering Jew (name also appl. to *Zebrina pendula*); cult. basket plant - **T. hirsuta** HBK (CA, Co, Br): trapoer ava (Br) pl. ext. u. in GC & leucorrhea; rts. in baths for rheum., in urin. retention - **T. multiflora** Sw. (Co, Ve): suelda (con suelda); pl. decoc. u. as bath for joint pain; pl. as poultice for hernias, burns, dislocations, rheum.; pl. believed antidiabetic - **T. ohiensis** Raf. (eUS, n to Minn): Job's tears, widow's tears; kiss-me-by-the-gate; formerly lf. inf. in wine for spider bite (hence name), scorpion sting, &c.; astr. diur. (Am Indians), kidney & stom. dis., in "cancers" (Indians); ornam. - **T. roseolens** Small (T. longifolia Small) (swFla [US]): spiderwort; pl. u. ton. in mal. (other *T.* spp. of area may be so used) - **T. virginiana** L. (e&cUS): (common) spiderwort; flower of a day; rt. u. dem.; young shoots & lvs. sometimes u. as salad & potherb - **T. zanonia** (L) Sw. (Br): fresh or dr. lvs. u. to stanch the flow of blood fr. surface wounds.

traditional medicine: a more or less rational style of medicine with its own specific theoretical basis and systematic order; not to be confused with folk medicine.

tragacanth (gum): 1) gummy exudn. of *Astragalus gummifer* & other *A.* spp. of As 2) hog gum; fr. *Symphonia globulifera* - **common t.**: t. in sorts - **flake t.**: best grade of t.; in ribbonlike colorless pieces - **hog t.**: Karaya (Sterculia) gum - **t. in sorts**: common t., a low grade of t., consisting of irreg. tears, etc. - **Indian t.**: hog t. - **leaf gum t.**: **Persian t.**: flake t. - **Russian t.**: that prod. in Transcaucasia & Turkmenistan - **Syrian t.**: flake t. - **vermiform t.: vermicelli t.**: tragacantha vermicularis: type made up of small long worm-shaped pieces, sometimes attributed to *Astragalus cyllenaeus* Boiss. & Heldr (Gr) - **Tragacantha** (L.): tragacanth - **T. Indica**: Sterculia gum - **tragacanthin**: the component of gum tragacanth wh. in water forms a hydrosol; the other component, bassorin, present in larger amt., swells to form a hydrogel.

Tragasol: gum fr. sd. of *Ceratonia siliqua.*

Tragia involucrata L. (Euphorbi) (peninsular Ind, Pak): chural (Assam); parasanta (Hind); u. in rabies(!), scabies (Assam); rt. u. diaph. alter. - **T. meyeriana** Muell. Arg. (sAf): rt. decoc. drunk for bladder region pain, taken only in evening - **T. nepetifolia** Cav. (swUS, Mex): hoobox; popox; contact with trichomes produces pain, inflammn., swelling; u. as counteririt., in rheum., and as remedy in sores & cancer - **T. rupestris** Sond. (sAf): lf. rubbed on forehead for headache, produces stinging sensation. - **T. volubilis** L. (Mex to Pe; WI; trop wAf): creeping (or twining) cowitch; vine nettle; u. as analg. diur. (Virg Is); diaph. (Mex); in leprosy (Ve) - **T. yucatanensis** Millsp. (Yuc, Belize, Guat, Hond): ortiguilla; rt. u. in gastralgia, GC; lf u. counterirrit. (Yuc Indians).

Tragopogon porrifolius L. (Comp.) (Euras, Med reg., Austral; natd. NA; cult. sAf): (purple) salsify; (purple) goat's beard; vegetable oyster; o. plant (SC flavor of rts.); c. natural rubber (0.1%); rt. u. expect. nutr. aperitif; as relish in salads; rhiz. as vegetable (boiled [*cf.* asparagus] [fried, stewed] [esp. cult.]); roasted & powd. as substit. for coffee; latex fr. rts. chewed by BC [Cana] Indians - **T. pratensis** L. (Euras, natd. US, Cana): (yellow) goat's beard; (meadow) salsify; (Johnny) go-to-bed-at-noon (SC fl. heads closed at midday); rt. u. diur. nutr.; rt. decoc. u. for heart burn, anorexia, liver dis.

trailing arbutus: *Epigaea repens* - **t. mahonia:** *Berberis.*

traiteur (Fr La [US]) (lit. treater): healer who uses folk medicines, &c.

Trametes serialis Fr. (Polypor. - Fungi) (NA, Eu+): found on dead wood; c. bacteriostatic antibiotics.

tranquilizers, plant: pl. growth inhibitors; ex. maleic hydrazide.

transaminase: a transferase wh. is concerned with an NH_2 gp.

transferase: enzyme wh. catalyzes the transfer of radicals (such as phosphoryl, methyl, &c.) fr. one compd. to another; ex. phosphorylase.

transfusions (blood): adminstration of blood or b. component in therapy - **autologous blood t.**: use of one's own blood collected beforehand & stored until needed; recommended to use to avoid infection with AIDS, hepatitis, &c., since comm. blood is occasionally contaminated.

translocation: movement in pl. sap of various compds.

transplantation: grafting or transfer of organs fr. one organism to another, usually of the same or related sp.

Trapa L. (Trap./close to Onagr.) (OW warm): water chestnut; w. caltrops; horn-nut; 15 spp.; much u. in As foods; ex. in subgum, chow mein (Chin) (available canned) - **T. bicornis** Osbeck (T. bispinosa Roxb.; T. cochinchinensis Lour.) (Chin, Ind, SriL; cult. Chin, Jap, Kor): leng; ling nut; frt. with sds. (Singhara nuts) roasted & eaten as staple food (Kashmir, China, &c). u. antipyr. & in sunstroke (Chin); fls. & frt. shells u. in diarrh.; frt. roasted for fever & headache (Indoch); u. as ton. & to treat stomach dis., ulcers (2448); frt. 2-horned (v. fig.); sometimes sold as "Chinese lily tubers" (pedlars); cult. in Mexicali area & Chihuahua ("Buffalo nut") in Mex; u. for adornment; sometimes *T. bispinosa* considered a separate sp. - **T. maximowiczii** S. Korsch. var. *tonkinensis* Gagn. (Indochin): sds. u. antipyr.; inf. of rind drunk as ton. for debility & fever; sds. in diarrh. - **T. natans** L. (T. quadrispinosa Roxb.) (Med reg., Euras, cAf; natd. NA, Austral): water caltrop; w. (chest) nut; horn n.; frt. blackish brown, 4-horned; sds. c. starch 52%, crude protein 8-10%, FO rich in Fe; edible (fresh, roasted, parboiled & baked); formerly in cataplasms, ferruginous ton.; for diarrh., as coffee substit.; frts. u. to make rosaries: cult. in sChin, Mex.

trapoeraba (Br): *Tradescantia diuretica*, hb.

Trasylol™: aprotinin (*qv.*).

Traubenzucker (Ge): glucose (<Traube: grape cluster).

traumatic acid: long-chained fatty acid ($C_{10}H_{18}(COOH)_2$), isolated fr. green bean pods; a wound hormone of plants.

Traumaticin(e): gutta percha soln. in $CHCl_3$; u. for closure of small wounds, dentistry; sometimes combined with active medication.

Trautvetteria caroliniensis (Walt.) Vail (Ranuncul.) (NA, eAs): false bugbane; rts. u. to poultice boils (to mature) (Indians at Bella Coola, BC [Cana]) (779).

Travase™: sutilains in ointment base.

traveller's joy: *Clematis viorna, C. vitalba*, &c. - **t. palm: t. tree**: *Ravenala madagascariensis*.

treacle: orig. theriac (antidote); in Brit prescriptions, generally means refined molasses or syrup; US: molasses.

tread softly: **treadsaft**: **treadsalve** (seUS): *Cnidoscolus stimulosus*.

trebol (Sp): clover - **t. acuatico**: *Menyanthes trifoliata* - **t. blanco**: white c.

Treculia africana Dcne. ex Trécul (Mor) (w tropAf): African breadfruit (tree); okwa (eNigeria); sds. are ground into a meal u. in soups, beverages, &c.; wd. marketed as "African boxwood"; bk. u. as expect. in cough med., lax. (Ghana); in leprosy; sap appl. on cotton to cause carious tooth to drop out.

tree of heaven (NA): **heaven t.**: t. of paradise (Cana): *Ailanthus altissima*; also names appl. to other trees - **t. of life**: date palm (Bible); *Thuja*, &c.

tree: v. also primrose, tomato.

tree: woody plant with single main st. & generally quite large size (arbor, L.).

trèfle (Fr): *Trifolium* spp. - **t. aquatique** (Fr): **t. d'eau**: *Menyanthes trifoliata*.

trefoil: *Lotus* spp. - **bird's foot t.**: *L. corniculatus* - **Bokhara t.**: *Melilotus alba* - **marsh t.**: *Menyanthes trifoliata* - **sweet smelling t.**: *Eupatorium cannabinum* - **tick t.**: *Desmodium* spp. - **yellow t.**: *Melilotus officinalis, Medicago lupulina*.

trehalose: mycose; a sugar; one of chief natural pl. disaccharides; consists of 2 glucose mols. combined; occ. in fungi, yeasts, lichens; also in trehala manna; as a storage or reservoir carbohydrate: u. in bacterial media, &c.

Trema amboinensis F. - Vill. = *T. orientalis* var. *a.* Lauterb. - **T. cannabina** Lour. (Ulm.) (Solomon Is): lvs. heated with lime appl. to boils; lvs. eaten for gastric dis.; for pain in body - **T. guineensis** Fic. (tropAf): pigeon wood; Rhodesian elm; "charcoal tree"; lvs. u. ton. cough remedy; vermifuge (Togo); tea fr. roasted wd. u. in dys. (Lagos, Nig); bk. in hookworm & roundworm; wd. u. in construction (may = *T. orientalis*) - **T. lamarckiana** Blume (sFla [US], WI): bit-root (Bah); pain-in-back; lvs. u. in combns. for strengthening back & blood; to restore menstr. flow; bk. chewed to prevent colds (Bah); rts. in combns. u. to incr. female fertility - **T. micrantha** Blume (T. floridana Britt.) (s peninsula Fla & Keys; WI; Mex, CA, to Pe & Arg): Jamaican nettle tree; Bastard Bay cedar; capulin (macho); bk. & lf. decoc. u. to bathe skin eruptions; lf. poultice for stiff neck or limbs (CR); timbers u. for rafters & strips of bk. for tying house rafters (Ec) - **T. orientalis** Bl. (Far East, incl. Oceania): rt. decoc. ingested for sore tongue, diarrh., hematuria (Mal); macerate of soft wd. u. as poultice for swellings (PI); pl. u. in epilepsy (Ind); rts. int. for diarrh. skin rash, hypertension, lax. (Samoa); lvs. u. med. (Tonga) (86); var *amboinensis* Lauterb. (seChin, Austral, Polynesia): mangele (Tonga); lvs. u. med. (Tonga) (86).

tremble (FrCana): *Populus tremuloides*.

Tremella fuciformis Berk, (Tremell. - Hymenomycetes - Basidiomyc. - Fungi): (worldwide [sub]trop): silver ear; frtg. bodies u. in pulmon. dis., TB, gastric ulcers (bleeding) (Chin); thought to have antibiotics; food (Chin), decoc. as tonic, sweet u. by elderly.

trementina (Sp, It): turpentine - **t. de abeto** (Mex): Terebinthina Abietina.

trepang: bèche de mer: sea cucumber: sea slug; seveal holothurians (Echinodermata); coll. on coral reefs, sPac; u. in foods, as gelatinous soups (Chin, PI), invalid dietaries (Austral); dr. eviscerated animal on market.

Treponema pallidum Schaudinn (Spirochaet. - Spirochaetales - Bacteria): causative organism of syphilis (humans); u. to prepare antigens for serological tests; formerly m. o. classed under Protozoa - **T. pertenue** Castellani & Chalmers): causative organism of yaws (frambesia; parangi).

Trester (Ge-m. pl.): grounds, husks, residues.

tretinoin: retinoic acid; vit. A acid; u. topically in acne, psoriasis, &c.; keratolytic in basal cell carcinoma; 13-cis-form u. in severe cystic acne; to prevent neoplastic dis. (exptl.). (natural form is all trans-).

Trevesia cheirantha Ridl. (Arali.) (Mal): pounded lvs. u. as poultice for fractured bones, skin dis.; rub for rheum. - **T. sundaica** Miq. (Indones): one component in "protective med."

Trevoa trinervia Miers ex Hook. (Rhamn.) (Chile): trevo; bk. c. trevoagenine A & B; u. vulner. esp for cuts & burns.

Trewia nudiflora L. (Euphorbi.) (Ind-hot parts): bhillaura (Hind); lvs. u. as refreshing ton.; rt. decoc. for flatulence; locally for gout & rheum.; wd. u. to make carved images, tea chests, &c.

TRH: thyrotropin releasing hormone.

Triaca (L, It): Electuarium Theriacae; theriac (*qv.*) a solid prepn. of great complexity; because of its "shotgun" composition, formerly presumed useful as general remedy in dis.; the most famous of the gp. was Venetian theriac (*qv.*) - **T. Venetiana** (It): **triaga veneta em po** (Por): Venetian theriac (*qv.*).

Triacetyloleandomycin: v. Troleandomycin.

triacontanol: wax alc.; along with other members of series, claimed to improve stamina & physical endurance (780).

Triadenum virginicum (L.) Raf. (Hypericum v. L.) (Guttif./Clusi.) (eNA): marsh St. John's wort; High John the Conqueror; rt. ("Big John root") believed most potent med. of forest, carried in pocket as charm (US, sNegro) (781); lf. tea u. for chest dis.; fls. for colics; in ointment for wounds, sores, swellings, &c. (782) - **T. walteri** (Gmel.) Gleason (T. petiolatum [Walt.] Britt.; Hypericum p. Walt.) (se&cUS): names & uses similar to prec.

Trianosperma Mart. = *Cayaponia*.

Trianthema decandra L. (Aizo.) (Ind): rt. u. aper. in asthma; orchitis; lf. juice dropped into nostrils for unilateral headache - **T. pentandra** L. (sAs, trop Af): horse purslane; pl. u. astr. in abdom. dis., abort., hb. in headache; exter. in plasters; weed in many areas of OW - **T. portulacastrum** L. (T. monogyna L.) (sFla [US], WI, Mex, tropAm, OWtrop.): horse purslane; pigweed (Ariz [US]) verdolaga (blanca); hb. c. trianthemine, saponin; lvs. & st. u. in diarrh. & paralysis of cattle; source of Punernava (Ind Ph.); u. diur, &c. (*cf. Boerhaavia diffusa*): in edema, anemia; cardiac dis.; as ton. ("rejuvenant"); rt. decoc. abort. (PI); eaten raw as salad (NM [US]), served as greens (Ariz).

Triasyn B™: combn. vitamins B_1, B_2, niacinamide (Lannett) (capsules & tabs.).

Triatoma spp. (Reduviidae - Hemiptera - Insecta) (CA, SA): kissing bug; Mexican bed bug; cone-nosed bugs; carries & transmits Chagas' disease.

tribe: sub-family in plant & animal classification.

Tribroma bicolor O. F. Cook (monotypic) (Sterculi) (CA, SA cult.): patashte; sd. u. like cocoa; is sometimes confused with *Theobroma bicolor* (—> *Theobroma*).

Tribulus alatus Del. (Zygophyll.) (Pak, Ind, tropAf): frts. u. as in case of *T. terrestris* - **T. cistoides** L. (seUS, WI, CA to Arg, natd. Austral, Oceania, tropAf): turkey blossoms; (large yellow) caltrop; abrojo (rojo) (Mex, Cu); puncture weeds (Fla [US]); pls. often fragrant; lvs. & rt. u. diur. kidney dis.; decoc. for rheum. ton.; juice + honey appl. to oral ulcers; decoc. of lvs. & frts. u. to bathe & poultice rheum. parts: may be poisonous to cattle (783) - **T. terrestris** L. (T. lanuginosus L.) (Ind, Pak, SriL, Med reg.; natd. tropAf): (land) (or small) caltrop(s); puncture vine; chota; gokhru (Bengali); gockroo; gogru; frts. ("seeds") (burra gokroo) spherical, 4-spined; of 5 carpels; all pl. parts u., esp. rt. & frt.; lvs. c. saponins (u. in hormone synthesis); food, ton. galactagog, diur. muc. memb. inflammn.; sds. pounded & given in honey as ton. to improve skin luster, remove dryness & wrinkles; in icterus (sInd); frt. (as Fructus Tribuli Ikshugandha, F. T. lanuginosi) u. as diur. aphrod. in impotency, analg. in painful micturition, urinary stone; aper.; ton.; to "brighten eyesight" (Jap) (1770); bad weed worldwide.

Tricalysia nyassae Hiern. (Rubi) (Tanz): rt. decoc. u. in mal.; rt. inf. for diarrh., rt. cooked in porridge for bilharziasis; poultice on sores - **T. ovalifolia** Hiern var. *acutifolia* (Tanz): fresh rts. chewed or u. as inf. for snakebite; lvs. u. as poultice for headache - **T. sphaerocarpa** Gamble (Ind): roasted sds. u. like coffee (with similar odor & taste) (similarly *T. coffeoides* Hutch. & J.M. Dalz) (trop wAf).

Trichachne californica Chase (Gram) (swUS, Mex, SA): cottontop; fair forage grass - **T. insularis** Nees (Digitaria i. [L] Mez) (sUS, Mex, WI to Arg): sourgrass; pl. decoc. in cystitis, nephritis; rt. diur.; panicles u. to decorate churches & homes (genus often transferred to *Digitaria*).

Trichanthera gigantea Nees (Acanth.) (CR to Pe, Surinam): palo de agua; yatago (Ve); decoc. of leafy branch tips u. as antipyr. diur.; lf. decoc. drunk for rheum. hepatitis, diur. in kidney dis. (Co); branch decoc. u. for skin defects in cattle; juice u. to treat windgall in horses (Ve).

Trichilia arborea DC (Meli) (Yucatan, CR): chobenche; bk. decoc. antipyr.; lf. juice rubbed on limbs for convulsions - **T. capitata** Klotzsch (tropAf): rts. u. for snakebite (Moz) - **T. cipo** DC (Ve): bk. decoc. u. as "cure" for mal. & other fevers (Kuripako Indians, Rio Guainia) (2299) - **T. dregeana** Sonder (sAf): mutuhu; bk. macerate u. as enema for kidney dis. (Venda) (784) - **T. havanensis** Jacq. (sMex, CA, WI, nSA): bastard lime; cot; quina sylvestre (Guat) limoncillo (de montana); bk. decoc. in mal., bladder dis.; lf. decoc. in hot baths for rheum., skin dis., int. for urin. dis. - **T. heudelotti** Planch. (Nig): bk. for sores, ulcers (local), purg.; lf. decoc. for heart trouble - **T. hieronymi** Griseb. (Arg, Par): bk. u. as tan. - **T. hirta** L. (sMex to Br, WI): broom wood; wild mahogany; rt. strong purg.; lvs. ton. steeped in wine & appl. to tumors & swellings: lf. inf. u. in asthma, bronchitis (Cu); sd. FO appl. as hair dressing supposedly to stim. growth: wd. useful, to make canoes, oars, &c. - **T. micrantha** Benth. (Co): smoke of burning lvs. u. to treat lung dis. (Barasana Indians) (2299) - **T. oblonga** DC (Br): pimenteira; carrapeta; lf. inf. u. in pleurisy - **T. pleeana** DC (Co): tree bk. astr., u. as antipyr. (inf.) (Taiwano Indians) (2299) - **T. prieureana** A. Juss. (trop wAf): rt. u. to treat GC; antipyr. purg. - **T. roka** Chiov. (T. emetica Vahl) (trop&sAf, Saudi Arab): Cape (or Natal) mahogany; red ash; bk./lf. inf. u. as enema in sore back, lumbago; rt. antipyr. purg.; bk. decoc. in indig. dys.; vermifuge (784); lf. in dys. frt. appl. locally for eczema; sd. emet., to relieve itch; sd. c. FO ("mafura" or "mawah") tallow u. in cosmetics, soap mfr., in rheum.; frt. flesh source of FO u as culinary oil - **T. septentrionalis** DC (Co): lf. inf. u. as antipyr. (Ingano Indians) (2299) - **T. singularis** DC (Co): lf. inf. u. as antipyr. (Tikuna Indians) (2299) - **T. subcordata** Guerke (eAf): sd. source of mafureira fat u. in soap & candle mfr. - **T. trifolia** L. (sMex, CA, Curaçao; nCo, Ve): cerezo (macho): lf. decoc. to stop dys.; promotes labor & delivery in childbirth; abort.; rt. decoc. male aphrod.

trichinella extract: aq. ext. of larvae of *Trichinella spiralis* (nematode) u. in diagnosis of **trichiniasis**.

trichoblast: base of trichome.

Trichocereus chiloensis (Colla) Br. & R. (T. quisco Remy; Cereus q. Gay) (Cact.) (Chile):frt. —> medl. tea u. in rectal inflamm., proctitis, acute diarrh., fever; also as beverage; frt. edible - **T. pachanoi** Br. & R. (wSA): huachuma (Pe); achuma (Bo); giganton (Ec); San Pedro; c. mescaline; u. by curanderos to diagnose dis. (divination); hallucinogenic: widely cult. in area (785).

Trichocline argentea Griseb. (Brachyclados stuckerti Spegazzini) (Comp.) (Arg): hb. & rts. falsely called "Punaria ascochingae," "Punaria tea"; u. as smoking material & also int. as aq. ext. in asthma similarly to stramonium lvs.; antirheum. - **T. incana** Cass. (Arg): yerba calmante; "arnica" (misnomer); entire pl. prepn. u. as stomach. digest. fl. heads u. in compress for contusions & wounds; ashes of entire pl. u. as dentifrice (52) - **T. reptans** (Wedd.) Rob. (Arg., Chaco region): contrayerba; coro (Arg); rhiz. smoked for "stomachache"; rt. pd. & mixed with tobacco to smoke as narcotic (other *T.* spp. u. similarly); also reported u. in chicha as narcotic.

Trichocoronis rivularis Gray (Comp.) (sTex [US], Mex, CA): water pl.; u. for mouthwash (Mex).

Trichoderma viride Fr. (Hypocreaceae - Hyphomycetes - Fungi): found in soils; c. trichodermin, antifungal antibiotic; antineoplastic; converts cellulose to glucose; u. in *Candida albicans* infections in man.

Trichodesma angustifolium Harv. (Boragin.) (swAf; As): fls. u. antipyr., pectoral; rts. u. as vulner. (swAf) - **T. zeylanicum** R. Br. (tropAs, Af, Austral): camel bush; similar u. as prec.; emoll. diur. (Ind): sd. meal u. as food; decoc. appl. to sores (Austral aborig.) (2263): pl ash mixed with wild salt & eaten for dry cough (Tanz).

Tricholaena repens Hitchc. (T. rosea Nees) (Gram) = *Rhynchelytrum roseum*.

Tricholepis glaberrima DC (Ind) (Comp.): u. in skin dis., leucoderma; as nerve ton., aphrod.

trichome: hair-like outgrowth of epidermal cell of pl., hence sometimes called pl. hair; occurs on lvs. sts., fls., frt. etc.; two general classes: glandular & non-glandular (or mechanical); some make lvs. velvety to touch; others glutinous, etc.; ex. lupulin; *Mucuna pruriens* - **glandular t.** producing secretions fr. a unicellular or multicellular head supported by stalk of non-glandular cells (see Fig.) - **stellate t.**: fascicled hair; looked down on, it appears star-shaped; ex. *Sida*.

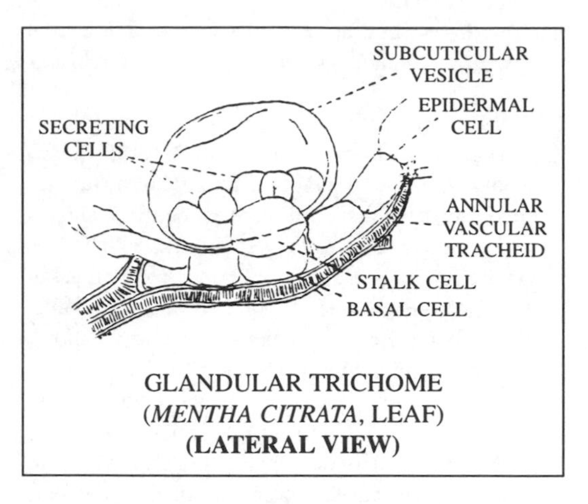

GLANDULAR TRICHOME
(*MENTHA CITRATA*, LEAF)
(LATERAL VIEW)

Trichomycin: antibiotic fr. *Streptoverticillium hachijoense* (Hoysoya et al.) Locci et al (Jap): antifungal, antiprotozoan.

trichophytin: class of protein substances extd. fr. mycelia of fungi, incl. *Trichophyton epidermophyton* & other spp.; *Microsporum lanosum* & other spp.; u. as antigen for dermal tests to determine nature of infection, also for treatment.

Trichosanthes amara L. (Cucurbit.) (WI): frt. u. as lax like colocynth; rat poison; sds. with FO which may produce a tetanic convulsion - **T. anguina** L. (Ind, Chin, seAs): (edible) snake gourd; Chinese cucumber; frt. u. purg. anthelm.; inner pulp u. for syr. for cough; edible after boiling & draining off the bitter water; sds. refrig., for liver tumors; prob. = *T. cucumerina* - **T. bracteata** Voigt (Ind): makal (Bengal): frt. smoked for asthma; hydragog cath.; rt. u. in pulmon. dis. (pneumonia) of cattle; frts. & rts. very bitter; frt. pulp mixed with coconut oil & u. locally in ear dis., hemicrania (Ind); saponin, NBP, "trichosanthin" (Protochlorophylla), green coloring matter different fr. chlorophyll (786) - **T. cordata** Roxb. (Ind): rt. u. ton. for enlarged spleen & liver; dr. fls. stim. - **T. cucumerina** L. (T. cucumeroides Maxim.) (Jap, Chin, seAs, Austral): bitter gourd; (wild) snake g. (SC long & narrow frt.); janglichichonda (Hind); pl. esp. rt. u. ton.; antipyr.; bitter frt. u. as bitter ton. lax.; decoc. of lvs./sts. u. in skin dis., emmen.; sds. antifebrile, anthelm.; in stomach dis; lf. juice emet.; in poisonings & convalescence fr. fevers (patola, Sanscr.) (Ayurvedic med.): young frt. may be eaten (as veget.); ornam. - **T. dioica** Roxb. (Ind): pointed gourd: rt. hydragog, antipyr.; frt. in spermatorrhea; fresh juice of unripe frt. u. as lax. - **T. homophylla** Hay. (Taiwan): rt. u. for stomachache - **T. kirilowii** Maxim. (Chin, Kor): Chinese melon; sds. u. in popular med. as astr. ton.; in mammary carcinoma; tuberous rts. u. as expect. to treat hemorrhoids & skin dryness; c. trichosanthin; starch fr. rhiz. spread on ulcers, wounds, chicken pox, &c.; var. *japonica* Kitam. (T. j. Regel) (Jap): u. similarly - **T. ladam** Miq. (Ind, Indones): yellow nearly odorless fat fr. sds. (kadam oil; k. fat) u. in native med. - **T. multiloba** Miq. (Chin): u. similarly to *T. kirilowii* - **T. nervifolia** L. (Ind): pl. u. bitter ton. antipyr.; rt. purg. - **T. palmata** Roxb. = **T. pubera** Bl. = *T. bracteata*. - **T. quinquangulata** A. Gray (PI): mature sds. powd., boiled in coconut oil, & appl. to itches; pd. sds. in wine int. for stomachache - **T. tricuspidata** Lour. (seAs): lvs. u. to poultice boils; lf. sap drunk for diarrh. (Indones) - **T. trifolia** (L.) Merr. (Indonesia): lvs. u. to treat "paralysis" & edema - **T. uniflora** Hao (Chin): gua lou zi; u. much like sds. of *T. kirilowii* for abnormal pulm. and intest. functions, fever, cough, constipation, & as expect. - **T. villosa** Bl. (Indones): sap drunk for looseness of bowels; crushed lvs. appl. to body in fever - **T. wallichiana** Wight (Ind, seAs): crushed frts. appl. to head for headache; bitter ripe crushed frt. mixed with opium & porcupine bile u. as poison - **T. wawraei** Cogn. (Mal): pounded lvs. appl. to abdomen to treat ague.

trichosanthin: Tra Hua Fen (THF) (Chin): polypeptide obt. fr. *Trichosanthes* spp; u. for abortion (Chin) (-1986-); inhibits HIV replication; may destroy HIV in macrophages; purified ext. (c. agent [or compound] Q) in use.

Trichostema (Gronov.) L. (Lab) (NA, mostly Calif): blue curls; spp. u. as source of honey - **T. dichotomum** L. (ne&seUS): bastard pennyroyal; found in mint farms; u. as fish poison - **T. lanatum** Benth. (sw to wc Calif [US]): (woolly) bluecurls; romero; vinegar weed; camphor w.; a shrub; lf./fl. decoc. u.for colds, fever, general debility, as bath in smallpox (earlier); lvs. chewed & put in cavity of aching tooth; lvs. to stupefy fish (Am Indians); cult. ornam. - **T. lanceolatum** Benth. (Ore to nCalif [US]): vinegar weed; turpentine (or camphor) w.; blue curl(s); (California blue) tar weed; "blue flower" (bee keepers); lf. inf. u. to wash head in feverish headaches, wash in typhoid fever (Am Indians); lvs. crushed into bed rolls to repel fleas (pioneer cowboys); fish poison (stupefier); considered best fall honey of Calif; u. in prepn. medicines (205) - **T. oblongum** Benth. (Wash, Ida to Calif [US]): mountain blue-curls; mashed pls. u. as fish "stunner" (Am Indians).

Trichostigma octandrum H. Walt. (Phytolacc.) (sFla [US], WI, sMex to Arg): hoop vine (or withe); basket wythe; lf. decoc. taken for suffocation or choking (Haiti); lvs. appl. to wounds (Co); cult. ornam.

trichotecines: Nivalenol: mycotoxin produced by *Fusarium tricinctum* (Corda) Sacc., *F. sporotrichioides* Sherbakoff, *F. poae* Park, *F. nivale*, *Trichothecium* spp., &c. (Fungi Imperfecti); causes root rot in wheat, oats, &c.; hemorrhagenic agt., causes toxic aleukia (= leukopenia) in farm animals & man; supposedly u. as chemical warfare agt. in seAs.

Trichothecin: antibiotic (antifungal) produced by *Trichothecium roseum* (Pers.) Link (Fungi Imperfecti) (1948-).

trick (US Negro): amulet or charm (v. hand).

tricken bag (seUS Negro): jack.

Triclisia dictyophylla Diels (T. gilletii (DeWild.) Staner (Menisperm.) (wAf, esp. Ghana)): lvs. & twigs of this liana (woody vine) c. isoquinoline alks., triclisine, tricliseine; exts. u. for anemia, edema of legs; in diarrh., pyorrhea; as arrow & lance poison; claimed component stronger antimal. than quinine - **T. patens** Oliv. (trop wAf): sts. cut & placed in palm wine to make more intoxicating (Sierra Leone) - **T. saccleuxii** Diels (trop eAf): fresh rt. decoc. drunk for intest. worms; raw rt. in VD; rt. chewed for kidney dis. (786); remedy in sore throat.

tricolpate: pollen grains possessing 3 furrows or expansion folds; ex. belladonna.

Tricoryne platyptera (Lili) (Austral): crushed lvs. appl. bound to wounds, cuts, sores; apply fresh lvs. daily for 2-3 days (Austral aborig.) (2263).

tri-o-cresyl phosphate: TCP: tritolyl phosphate; synth. additive u. in some gasolines to reduce preignition & spark plug "miss" (lead scavenger in gasoline); in plastics as plasticizer; notorious use as diluent in F.E. Ginger ("jake") (corruption of Jamaica [ginger]); causing "jake paralysis" (polyneuritis) (ca. 1936).

tridace: tridaceo (Br): **Tridaceum** (L.): **tridacio** (Sp): Ext. Lactucae sativae; Lactucarium; dr. juice obt. by contusion of st. of *Lactuca sativa*.

Tridax procumbens L. (Comp) (tropAm, esp. Mex, WI; natd. Ind+): romerillo (Cu); hierba del toro (Salv); dhusera (Nepal); fl. heads chewed for coughs & colds; bronchial catarrh; pl. made into a paste & appl. to boils (787); lvs. u. parasiticide (insect larvae) (placed on meat [CA]); pantrop weed.

Triebe (Ge): shoot - **T. spitze** (Ge): shoot tips.

trienoic: fatty acid with 3 double bonds (ex. linolenic acid).

Trientalis borealis Raf. (T. americana Pursh) (Primul.) (Cana & nUS): starflower; s.-anemone; chickweed-wintergreen; inf. u. in any sickness but esp. TB (Qu [Cana] Indians) - **T. europaea** L. (Euras [n]; Jap): rhiz. c. saponins, u. as emet. vulner.; cult. ornam.

trifolianol: a phytosterolin (= plant sterol glycoside) found in *Trifolium incarnatum*.

trifoliati (It) = truffles, actually *Boletus edulis* Bull. (porcini) (funghi t.).

trifoliin: isoquercitrin: flavone glucoside occ. in *Trifolium repens, Aesculus hippocastanum* (fls.); *Tropaeolum majus*, &c.

Trifolium L. (Leg) (worldwide temp. & subtr. except seAs, Austral): clovers; trefoil; some 238 spp.; ann., bienn., & peren. hbs. important as forage, honey, & cover crops; fls. u. in mfr. clover wine (vin de trèfle) (Qu [Cana]); one of chief sources of honey in eCana & US - **T. alexandrinum** L. (Med reg., nAf, Eg, Syria): Egyptian clover; berseem (clover); cult. for fodder & as green manure; sds. c. FO; trypsin & chymotrypsin inhibitors; no folk medl. uses known - **T. amabile** HBK (Mex to SA): Aztec clover; hb. mixed with maize grain u. as food (chucan) eaten by Pe Indians - **T. ambiguum** Bieb. (AsMin, seEu): kura clover; forage; valuable for honey prodn. - **T. aquaticum** (L.) = *Menyanthes trifoliata* - **T. arvense** L. (Euras, nAf, natd. eCana): rabbit-foot (clover); hare's food trefoil; hb. & fls. u. for gout (folk med.); in dys. diarr. expect. in bronchitis; no good as fodder (sts. hard & bitter) - **T. aureum** Pollich (T. agrarium L.) (Euras, nAf; natd. neUS+): (big) hop (or yellow) clover; u. to prep. clover omelet, &c. - **T. dubium** Sibth. (Eu, natd. NA): the true Irish shamrock (supposedly; argument over sp. hence the sp. name); coll. in County Cork (Ire), & shipped packed with NaCl to keep fresh; lf. natl. emblem of Ireland; lf. with 4 leaflets considered extra lucky (4 leaf clover); forage crop - **T. fibrinum** Tab. = *T. aquaticum* - **T. fragiferum** L. (Euras, natd. nUS): strawberry (headed) clover; fodder crop; green manure; honey pl.; pl. salt tolerant - **T. globosum** L. (Chin): u. as remedy in eye illnesses (Taiwan) - **T. hirtum** All. (Med reg., AsMin, NA, natd. eUS): rose clover; furnishes pasturage - **T. hybridum** L. (Euras, nAf, natd. Cana, nUS): alsike (clover); (misnamed since not a hybrid); Swedish c.; bastard c.; u. as forage & hay: shows antibiotic activity against TB organism - **T. incarnatum** L. (Eu, nAf, natd. US): crimson clover; Italian c.; ann./bienn. hb. cult. as cover crop, green manure, chiefly in orchards; pasture, hay crop; ornam.; common on roadsides (seUS) - **T. lappaceum** L. (Med reg. AsMin): lappa clover; forage crop with special merits - **T. medium** L. (Euras, natd. NA):

zigzag clover; u. for hay, pasture, & erosion control; in med. (Ukraine) - **T. montanum** L. (Euras): mountain trefoil; u. in gout. rheum. (Ukraine +) (788), often along with *T. repens* - **T. pannonicum** L. (c&sEu, wAs): Hungarian clover; cult. for fodder; rel. drought-resistant; u. med. (Ukraine) - **T. pratense** L. (Euras, nAf, natd. NA+): red clover; purple c.; brood c.; clover tops (or blossoms) (Trifolium, L.; Flores Trifolii rubri, L.), c. VO (with furfurol), caffeic acid, cumaric acid; salicylic acid; flavones; tannin; chief source of clover honey; fls. sed. in whooping cough; colds; hb. fls. & sd. as expect. diur., in gout; exter. for inflammn. burns, sores; dr. fls. added to flour for breads (extender); fl. heads u. in prep. of fraudulent cancer remedy (789): detergent; chief prodn. Ge, USSR; important fodder, rotational crop; cult. ornam. ("shamrock!"); sometimes c. HCN - **T. procumbens** L. (Euras, nAf, natd. US+): low hop-clover; u. similarly to *T. pratense* - **T. repens** L. (Euras, nAf, natd. NA): white clover; Dutch c.; fls. (as Flores Trifolii albi) u. in rheum. gout; detergent; inf. for pertussis; locally for sores & ulcers; mixed in flour (starvation food): young pls. u. as veget.; fodder; pl. c. estrogen; adult. with long peduncles - **T. strepens** Crantz = *T. aureum* - **T. subterraneum** L. (Med reg., Eu, wAs, nAf, natd. Austral): subterranean fodder clover; subclover; u. for pasturage, silage fodder (hay); in erosion control; has estrogenic effect (77); antifertility props. due to isoflavonoids present - **T. wormskioldii** Lehm. (Pacif NA): marsh clover; this is one of several *T.* spp. of which the rhiz., lvs. & fls. are eaten raw (Indians, Mendocino Co., Calif [US]).

Triglochin maritimum L. (T. maritima) (Juncagin.) (nTemp zone, NA fr. Alas to nMex): seaside arrow-grass; lvs. c. HCN (toxic); sds. sold as birdseed in Paris markets.

triglycerides: FO compds. in which each of the 3 OH gps of glycerol is occupied by a fatty acid (hence "neutral fats"); commonly found in fats of foods & in body fat; believed more significant than cholesterol in devt. of atherosclerosis & in heart attacks. (2372).

trigo (Sp, Por): wheat - **t. candeal** (Ch): white wheat, *Triticum aestivum* - **T. sarraceno** (Cu, PR): buckwheat.

Trigonella caerulea Ser. (Leg) (eMed reg., AsMin natd Am+): blue melilot; sweet trefoil; in folk med. as vulner., for tumors, edema, hepatic cirrhosis (Ind); eye dis.; spice, addn. to herb cheeses; moth repellant; to adulter. melilot hb. (*Melilotus officinalis*); often as ornam. - **T. corniculata** L. (T. elatior Sibth.) (Med reg., As; cult. nInd): bitter frt. u. as astr. styptic, appl. to swellings & bruises; sds. in back pain; in childbirth, sds. & lvs. as spice & vegetable (kasuri methi, Punjab, Kashmir) - **T. foenum-graecum** L. (sEuras: Med reg. [native], Pak, Ind, Israel, Syr, Chin, nAf [Ethiop]; cult. Calif [US] [WW II]): (common) fenugreek; mathi; mathee; methi (Ind, Pak); hilba (Eg); ann. hb.; sd. (Foenum Graecum, L., lit. Greek hay) c. mucilage (20-30%) (mannogalactans); proteins (20-28%); FO; VO with typical odor; trigonelline; NBP; resin; steroidal sapogenins diosgenin, gitogenin, &c.; no starch (1771); u. dem. emoll. flavor (in maple syrup flavors) (790); long eaten like lentils; as poultices, enemata, gargles (for sore throat), plasters; for hemorrhoids, catarrhs; esp. in veterinary med. (condition powders) nutr. to fatten stock); similarly u. for women (Tunisia); galactagog (Eg) (in teas); carm. ton. aphrod.; dye; lvs. u. as greens (as veget. or pot herb), in curries (Ind, Pak); entire pl. u. dr. as flavor (*cf.* mint); potential source of hormone precursors (diosgenin; botogenin [0.8-2.2%]) claimed contraceptive; (cult. Ethiopia); green manure (Calif [US]); insecticide; chief prodn. Ind, Mor, Calif; cult. for millenia (Assyria) - **T. occulta** Delile (Ind): u. like prec.; sds. in dys. - **T. polycerata** L. (Ind): u. like prec. esp. in diarrh. - **T. uncata** Boiss. (Iran, Afgh): similar but narcotic.

trigonelline: nicotine acid N-methyl-betaine; nitrogenous base (sometimes referred to as "inactive" alk.); found in *Trigonella foenum-graecum*, & in many other sds. (Cannabis, Strophanthus, *Pisum sativum*, oats, coffee, &c.) & in animals (ex. sea urchin).

Trigonia paniculata Warm. (Trigoni., Dicots) (Br): lvs. (both fresh & dr.) much u. to prep. beverage tea.

Trigonostemon howii Merr. & Chun (Euphorbi) (seAs): lf. decoc. u. as wash for scabies (Indochin) - **T. longifolius** Baill. (Mal): juice rubbed on stings of bees & jellyfish.

triiodthyronine: T_3 =liothyronine.

Trilisa Cass. (close to *Carphephorus* Cass.) (Comp.) - **T. odoratissima** (Willd.) Cass. (Carphephorus o. (J. F. Gmelin) Herb.); Liatris o. (Walt.) Willd.) (seUS n to NCar): deer('s) tongue; dog/hound's t.; "Carolina vanilla" (Eng); vanilla plant; wild v.; evergreen perenn. hb., rel. conspicuous in winter; fresh lf. with rather disagreeable odor but on drying a day or two develops pleasant aroma, due to coumarin formed; c. VO, FO, resin, coumarin, wax, sugar, mucilage, "liatrol" (with persistent but not pleasant odor); lf. ("Liatris" [L.]) u. arom. ton. stim. diaph. diur., formerly in pertussis, "nervous states," bruises, sprains, &c.; rt. u. diur. diaph. (Homeop.); lf. chiefly u. to aromatize snuffs, cigars, cigarettes, & chewing tobacco; in perfume indy.; substit. for tonka beans; to imitate vanilla ext.; moth proofing; pl. said injurious to foraging stock (1137); rt. u. as diur. diaph. (Homeop.).

trillin: glucoside fr. *Trillium erectum*; fenugreek sd.; made up of diosgenin + glucose.

Trillium L. (Lili.; in some floras Trilli.) (NA, As): trillium; wakerobin; birthroot; wood lily; perenn. hb.; lvs. in whorls of 3 (hence name; tres (3) + ium (characteristic of) - **T. cernuum** L. (eNA w to Wisc [US]): nodding trillium; (n.) beth root; birth root; drooping t.; rt. & berries u. in epilepsy; rt. astr. ton. antisep. emmen. expect. emet. diaph.; for tumors (by Am Indians) (791); ornam. hb. - **T. chloropetalum** (Torr.) Howell (wWash to sCalif [US]): giant wake-robin; white trillium; rhiz. decoc. drunk for deep sleep & sickness in general; bruised lvs. & crushed rhiz. u. as poultice for boils (Indians, Mendocino Co. [Calif]) - **T. erectum** L. (T. pendulum Willd.; T. purpureum Kin) (eNA w to Mo [US]): stinking Benjamin; purple trillium; squawroot; (American) beth root; wakerobin; rhiz. & rt. (Trillii Radix, L) c. saponins trillin & trillarin; chlorogenin, nologenin; tannin, resin; u. ton. emet. expect. astr. hemostatic; antispasm.; uter. stim.; when chewed incr. salivary flow; fls. with fetid odor; hb. u. in poultices for tumors & ulcers; in leucorrh., diarrh.; for infant maladies (Abenakisi Indians); popular med. among Am Indians & early settlers: hb. u. in homeop. med. - **T. grandiflorum** (Mx.) Salisb. (eNA w to Minn [US]): white lilies; snow trillium; wakerobin; large-flowered (white) trillium; rt. inf. u. for sore nipples (Potawatomi Indians); grated rt. as poultice for eye swelling; decoc. for cramps (Menominee Indians): lvs. u. as greens - **T. ovatum** Pursh (BC [Cana], wPacif US): (western) wake-robin; w. trillium; pl. eaten boiled as greens: ornam. - **T. petiolatum** Pursh (BC to Ore [US]): petioled wake-robin; round-leaved trillium; rts. steeped in hot water & "tea" drunk as appetizer (BC Indians) (447) - **T. sessile** L. (e half of US): toadshade; (sessile-flowered) wake-robin; rts. emet. astr. ton. antisep. emmen. expect. diaph. (782) - **T. undulatum** Willd. (T. erythrocarpum Mx.) (e half of NA): painted trillium (or beth-root); wild pepper; rts. ton. emmen. expect. diaph., &c. (792).

trilopo (It): *Hamamelis virginiana.*

trimester: a 3-month period; one of 3 subdivisions in term of pregnancy.

Trimeria grandifolia Warb. (Flacourti.) (sAf): wild mulberry; lf. inf. u. with other pls. for abdom. difficulties in general; wd. u. to make yoke-keys.

trimethyl (1,3,7)-xanthine: = caffeine.

trimethylamine: $N(CH_3)_3$; occ.as degradation prod. from plant & animal substances; in conjugated form in animal tissue, esp. of fish; present in hb. of *Chenopodium vulvaria*; u. as insect attractant; in prepn. quaternary compds.

tri-methyl-colchinic acid: u. like colchicine; claimed less toxic & more active.

Trimezia cathartica (Mart.) Benth. (Irid.) (Br): ruibarbo do Brasil; u. as lax. (Co); substit. for rhubarb rt. - **T. juncifolia** (Klath) Benth. (Br): similar to prec. - **T. lurida** Salisb. (tropAm): u. lax. (Co) - **T. martinicensis** Herb. (WI, sMex to Br; natd. sFla [US]): dragon's blood (Trin); walking iris; wild scallion; bulb decoc. u. for chills (Guat); in flu (Trin); as diur. in oliguria & emmen. (Trin).

trincomali (wood): *Berrya cordifolia.*

Trinia glauca (L.) Dum. (Umb.) (c&sEu): honewort; rt. u. purg.

trinomial: bot. (or zool.) name made up of gen. sp. & var. names.

Triolena pluvialis Wurdack (Melastomat.) (Co, Pe): ground pl. boiled in water, & placed in mouth 5 mins. for toothache (777).

trioses: sugar with 3 carbons; ex. glycerose (mixt. glyceric aldehyde & its ketone).

Triopteris jamaicensis L. (Malpighi) (WI): cough vine; cough remedy (Bah).

Triosteum perfoliatum L. (Caprifoli.) (e&cUS): (common) horse gentian; (Dr.) Tinker's weed; wild coffee; fever root; f. wort; bastard ipecac; rhiz. & rt. (USP 1820-70) u. antipyr. purg. antirheum. emet. diaph. ton.; adulter. of Senega; fresh rt. u. in homeop. med.; lvs. diaph.; hard sds. dr. & roasted u. as coffee substit. formerly; in mal. (Am Indians) - **T. sinuatum** Maxim. (Indochin): rt. inf. u. to treat abscesses of breast.

tripa(s) de Judas (guts of Judas) (Mex): *Cissus sicyoides* L. (*C. tiliacea* HBK).

tripe: inner lining of stomach of beef (less frequently of veal, mutton, pork); u. as food - **t. de roche** (Qu [Cana]): **rock t.**: 1) *Polypodium virginianum* 2) *Gyrophora* (*qv.*) *erosa, G. proboscoidea: Umbilicaria* (Lichenes): u. med.

Tripetalum cymosum Schumann (monotypic) (Gutt.) (NG): frt. u. to stain teeth black.

triphala (Ind): famous Ayurvedic med.; mixt. of frts. of *Terminalia bellerica, T. chebula, & Phyllanthus emblica*; u. as lax.

Triphasia trifolia P. Wils. (T. aurantiola Lour.) (Rut) (seAs, esp. Mal; natd. widely, incl. sTex [US], Mex): limeberry; limoncillo (Mex); lvs. u. in resp. dis., in several native herb tonics; syr., mucilag. frts. u. as expect.; frt edible, raw or in jams; c. VO.

Triplaris americana L. (Polygon.) (CA): hormigo (CR); tabaco(n) (de monte); satteuala (Cuna Indians, Pan): long John; pano negro (Hond); pl. u. med. (Hond); pl. infested with aggressive ants (CR).

triple dye jelly: tragacanth base with combination of 3 dyes: Gentian Violet, Brilliant Green; Flavine; formerly u. for burns, scalds, antisep., etc. (keep fr. eyes) - **t. rose water**: stronger rose w.

triplex pills: Pilulae Triplex; Pills of Aloes, Mercury, & Podophyllin (NF V).

Tripogandra elongata Woods. (Tradescantia diuretica Mart.; T. elongata G.F.W. Mey.) (Commelin.) (SA): hb. (as Herba Tradescantiae) u. in homeop. med., galenical prepns. (Br) of hb. & of whole pl. as diur.

tripoli: rotten stone (*qv.*) (SC found orig. in Tripoli [nAf]).

tripolite (earth): diatomaceous earth (not tripoli); (Tenn, Miss [US]).

Tripsacum dactyloides L. (Gram.) (c&seUS, WI): (eastern) gama grass; near relative of *Zea*; u. forage, fodder - **T. laxum** Nash (CA): gama grass; a forage grass.

Tripterygium regelii Sprague & Takeda (Celastr.) (Chin): bk. c. pristimerine; u. as insecticide - **T. wilfordii** Hook. f. (T. forrestii Loes.) (sChin, Jap): thunder god vine; pd. rt. bk. u. as insecticide (793); c. alks. wilfordine; wilforine, wilforgine, wilfortrine; tripterine, ext. ("T2") effective in rheumatoid arthritis, lupus (Chin); ext. interferes with spermatozoa maturation, hence serves as male contraceptive (*cf.* gossypol) (*T. hypoglaucum* Hutch. similar in use).

Triptilion spinosum R. & P. (Comp.) (Ch): salatdistel (794); clonqui (Ch); rt. u. as diur. emoll. in liver & stomach dis.; u. exter. as cataplasms in popular med.

trisaccharide: complex sugar contg. 3 simple sugars (monoses); ex. raffinose; melezitose; gentianose.

Trisetum flavescens (L.) Pal. (Med. reg., Eu): forage grass; prod. vit. D in UV irradiation (*cf.* animals).

trisodium citrate: sodium c.

trisporic acid: sex hormone of Mucorales (Fungi) wh. elicits formation of zygospores.

Tristemma coronatum Benth. (Melastomat.) (Ghana): lvs. u. in enema prepns for lumbago, &c. - **T. littorale** Benth. (Liberia): lvs. rubbed in water for mouthwash for infants with stomatitis.

Tristerix tetrandrus (R. & P.) Mart. (Loranth.) (Ch): quintral del álamo; u. to reduce blood cholesterol (—> *Macrosolen; Trigonastemon*).

Tritaxis gaudichaudii Baill. (Euphorbi) (Indochin): lf. decoc. u. to treat dental caries; also sores of animals.

Triteleiopsis palmeri Hoover (monotypic) (Lili) (wUS, Mex): Indian children fond of eating fresh bulbs.

triticale: fertile hybrid of wheat X rye (Triticum X Secale).

triticin: a polyfructosan (polylevulosan) (inulin-like carbohydrate) fr. rhiz. of *Agropyron repens* (3.5-5-7.9%); may be identical with phlein. (not found in *Triticum* spp., wheats).

Triticum L. (Gramineae-tribe Hordeae [Triticeae]) (Euras): tall ann. or perenn. grasses with close terminal spikes (also L. title of doggrass rt. [rhiz.] of *Agropyron repens*) - **T. aestivum** L. (T. durum Desf.; T. sativum Lam.; T. spelta L.; T. vulgare Vill.) (Euras; orig. in Trascaucasia; cultigen; cult. worldwide): (common) wheat; bread w.; many vars. recognized; chief types: hard & soft; semihard; red & white; spring & winter; bearded & non- bearded; dwarf; spelt (hardy; for bread; subsp. *spelta* (L.) Thell. (T.s.L): (ordinary) spelt (grain) was the earliest cult. very similar to regular w.); mostly widely produced pl. food in the world (2415); grain (frt.) source of w. flour (795), starch; w. germ (2416); bran; gluten (flour) (aleuronat[e]) & w. germ oil, w. syrup; vodka; semolina; (u. med. ancient Gr); w. straw. u. in mfr. hats, mats, packing, in stables, &c. - **T. hybernum** L. = *T. aestivum* - **T. monococcum** L. (Eu): einkorn (grain); both wild & cult.; one-grained wheat; small spelt; supposed to be the most primitive wheat; cult. for fodder, also (rarely) for human consumption - **T. repens** L. = *Agropyron r.* - **T. turgidum** L. (T. durum Desf.; T. dicoccum Schrank); (wAs): emmer (wheat); e. (grain); English w.; turgid w.; rivet w.; 2-grained w.; incl. durum (T. durum Desf.), hard, durum, flint, or macaroni w.; source of semolinas; u. mostly for pastas, such as macaroni, spaghetti, vermicelli, &c.; convar. *polonicum* (L.) MacKey (T. pol. L.) (eEu): Polish w.; a hard grain.; subsp. **dicoccoides** (Koern.) Thell. (T. d. Koerner): wild emmer; believed the primitive type of w.

Tritonia crocata (L) Ker (Irid.) (sAf): u. as surrogate for Spanish saffron.

trituration: rubbing substance in mortar with pestle.

Triumfetta heudelotii Planch. (Tili.) (Sen): decoc. of various parts u. as antitussive - **T. hirsuta** Sprague & Hutch. (sAf): rt. inf. (cold) u. for dys.; emet. - **T. lappula** L. (sMex to Pe, Bo; WI; natd. tropAf): bur(weed); crown of thorns; mozote (colorado); pega-pega; decoc. of leafy sts. u. as diur. in diarrh., for colds (CR); for "hang overs"; lvs. & fls. crushed in water & appl. to itching wounds (Pe); twigs & lvs. u. to clarify sugar solns; starch source - **T. palmatiloba** Dunkley (sAf): aq. ext.

(warm) instilled into ear for otitis & earache - **T. pilosa** Roth. var. *tomentosa* Szysz. (sAf): macerate of lvs. u. in eye dis., aphrod.; to wash the hair (Zulus) - **T. procumbens** Forster f. (Micronesia+): te kiaon (Gilbert Is); u. for poultices, as soap; inf. or decoc. drunk at childbirth (Gilbert Is) - **T. rhomboidea** Jacquin (T. bartramia L.) (pantrop): bishakhopra (Nepal); rt. hot inf. taken to hasten onset of labor & facilitate birth (Zulus); in impotency, sterility; rt. & bk. u. for diarrh. hemorrhages GC (Ind); rt. & lf. emoll. (PI); pl. made into paste appl. to boils & pimples (727); in skin dis., abscesses (sChin); pl. source of soft glossy fiber - **T. semitriloba** Jacq. (sFla [US], sMex to Pe & Par; WI): bur(r) bush; cadillo (altea); rt. decoc. in VD, liver dis.; lf. & rt. decoc. for diarrh. dys.; bk. fiber to make rope, cloth.

trivalent: tervalent: having a valency of 3, combining power with other elements, &c.

trivial name: species n.

Trixis angustifolia DC (Comp.) (Mex): hierba del aire; pl. u. in rheum. - **T. californica** Kell. (wTex [US] to nwMex): American trixis; rts of this small shrub u. as tea by woman before delivery to hasten childbirth; lvs. smoked like tobacco (Seri Indians) - **T. divaricata** Spreng. (Br, Co): u. in amenorrhea - **T. glandulifera** Benth. & Hook. (Arg): yerba del ciervo; hb. u. diur. diaph. carm.; in dis. of kidney & bladder; cystitis; contraceptive(ur); for measles (796) - **T. inula** Crantz (T. radialis Ktze.; T. frutescens P. Br.) (Tex [US] to Ch & Ve): falsa arnica; a. del monte; hb. decoc. u. in VD, exter. in rheum (Ve); in diabetes (Mex); decoc. on wounds & ulcers (Cu); decoc. in milk for poultices for tumors & swellings; contracept. - **T. longifolia** D. Don (Mex): lobo buase; fl. heads placed in ear for earache - **T. pringlei** Rob. & Greenm. (Mex): yerba de la vibora; decoc. u. for snakebite.

Trochomeria vitifolia Hook. f. (Cucurbit.) (Angola): mbumba ia xole; fleshly rt. u. in croup; also eaten (797).

Trochus (Trochidae - Gastropoda - Mollusca): large gen. of marine snails; source of trochus shell (Austral); pearl shell; top shell; some spp. u. to make pearl buttons.

Troleandomycin: triacetyloleandomycin (obsolete name): semi-synth. macrolide antibiotic; prepd. fr. oleandomycin; antibacterial; u. in asthma, &c.

Trollius asiaticus L. (Ranuncul.) (eRuss to Arctic As): kintien (Chin); hb. c. saponins; u. in scurvy - **T. chinensis** Bunge (Chin, neAs): kintien (Chin): u. like prec. - **T. europaeus** L. (nEuras): mountain globe flower; fls. formerly u. as antiscorb. - **T. laxus** Salisb. (eUS): globe flower; pl. acrid.

trombeiteira (Br): *Datura arborea* - **trompatila:** *Bouvardia.*

trompeta (Mex): *Datura arborea.*

trompetil(l)a (Mex): 1) *Datura stramonium* (usually) *D. arborea* 2) *Bouvardia ternifolia* 3) *Tecoma stans* 4) *Jacobinia spicigera* 5) 4 or more other pl. spp. - **trompillo**: 1) (Ariz [US]): *Solanum elaeagnifolium* 2) (Bo): cocillana bk. 3) (Salv): *Lafoensia* 4) (Mex): *Cordia boissieri* (& other pl. supp.).

trona: urao: Na_2CO_3. $NaHCO_3$. $2H_2O$; alkal. mineral deposited fr. salt springs, salt lakes, &c.; formerly u. in mfr. soda (anagram of natron).

tronadora 1) (Mex): *Sebastiania pavoniana* 2) (SpAm): Cannabis 3) (SpAm): *Tecoma stans.* - **t. retama**: *Tecoma mollis.*

tropacocaine: alk. fr. Pe & Java Coca lvs.

Tropaeolum majus L. (Tropaeol.) (Pe, Co; natd. Fla, Calif [US]): (common) nasturtium; garden n.; large Indian cress; hb. u. in urin. inf., cystitis, nephritis; in bronchitis, scurvy, eye dis.; meteorism; lvs. eaten in salads; young lvs. show antibiotic activity, attributed to benzyl mustard oil; fl. buds & young green sds. u. in pickles (pickling cucumbers) for pungency ("Indian cress"); pl. juice u. in itch; arom. frts. as flavor; ornam. for salads; widely cult. garden ornam. - **T. minus** L. (Pe): Indian cress; u. like prec. - **T. peltophorum** Benth. (SA-Andes): c. antibiotics; lf. buds & young frt. u. as capers substit. - **T. pentaphyllum** Lam. (Br, Arg): u. for scurvy; cult. in gardens - **T. peregrinum** L. (T. canariense) (Pe, Ec): canary bird flowers; pls. & fls. u. as antiscorb.; garden ornam. - **T. tuberosum** Ruiz et Pavon (wSA, Pe): anyu; cultigen cult. for edible tubers; these rich in starch & vit. C & u. as food by SA Indian women (not men); c. methoxybenzyl glucosinolate, reduces testosterone level in blood; u. as emmen. antiaphrod. insecticide, diur. nematocide (798).

tropane: bicyclic compd., $C_8H_{15}N$; considered base of tropane alks., ex. atropine.

Trophis racemosa Urban (T. americana L.) (Mor) (Mex to Co; WI): ramoon tree; ramón (de bestias) (Cu); branches with lvs. fodder for domestic animals; lvs. u. as galactagog for nursing mothers; bk. c. tannin u. astr.; wd. u. in rural construction.

tropic acid: phenolic carboxy acid; acid component of ester alks. of mydriatic gp.; ex. atropine, hyoscyamine, scopolamine.

Tropidia curculigoides Lindl. (Orchid.) (Mal): rt. decoc. ingested for diarrh.

tropine: nitrogenous base found in various mydriatic alks.; thus, atropine is the tropic acid ester of α-tropine.

tropism: movement towards (+) or away from (-) a stimulus (light, gravity, &c.) of some part of a pl. or animal (*cf.* taxis where entire organism moves).

trubbes (archaic): truffles: *Tuber* spp.

true love (seUS): *Trillium erectum.*

true: v. also under bittersweet; cinnamon; coto (bark); cramp bark; mace; saffron; sage; unicorn root.

trueno (Mex): *Ligustrum vulgare.*

truffles: subterranean fungi (*Tuber* spp) (also other genera in Pezizales & Elaphomycetales.) popularly eaten as luxury; considered aphrod. (black & white); dug up by trained dogs, hogs; u. in prepn. of paté de foie gras, garnish, stuffings, &c.; cult. in truffle "orchards" or "farms" (Fr).

trumpet climber: t. creeper (seUS): *Campsis radicans* - **t. flower:** *Thevetia peruviana* - **t. plant**: *Sarracenia flava* - **t. tree**: *Cecropia* spp. - **t. vine** (seUS): t. creeper - **t. weed**: 1) *Lactuca canadensis, L. sagittifolia* 2) *Eupatorium purpureum* - **trumpets**: *Sarracenia flava.*

Trutta trutta L. (Salmonidae-Pisces): trout; testes & sperm u. as source of protamine.

truxillic acid: cocaic acid: dicarboxylic acid; found in coca lvs.

truxilline, a- and **b-**: alks. of coca lf.; truxillic acid esters of methyl ecgonine (cocamine & isococamine may be mixts.).

Truxillo: Trujillo: city of Pe; v. under *Erythroxylon*

Trypeta arnicivora Löw. = *T. arnicae L.*. (Trypetidae - Diptera - Insecta): arnica fly; commonly found as infestant of *A.* fl. heads; formerly thought to confer therap. properties (& toxicity) on Arnica: related to fruit fly.

trypsin: peptonizing enzyme of pancreatic juice; formed in small intestine by action of enterokinase (a peptidase) on trypsinogen; acts on proteins to furnish amino acids; pure enzyme (**crystallized t.**) u. to dissolve necrotic tissue (ex. in inoperable cancers) & pus (in tubercular empyema, &c. [aerosol]) - **t. inhibitor**: substance wh. inhibits various proteases (incl. t.); anti-trypsin: found in body tissues (incl. pancreas), plants (esp. soy); u. to treat fibrinolysis (1446).

trypsinogen: precursor of trypsin, wh. latter is formed by activation with enterokinase; secreted by glands of pancreas.

Tryptar™: trypsin (Armour).

tryptase: fibrinolysin.

tryptophan, L: indole amino-propionic acid; an essential amino acid (not synthetized by human body); many natural sources incl. casein; prod. by tryptic digestion of caseinogen; natural precursor of serotonin (5-OH-tryptamine); important in food as growth factor; u. to treat depression, often with pyridoxine (coenzyme in metabolic reactions involving t.) & to induce sleep (reduces sleep latency) ("non addictive hypnotic") (799); in doses of 0.5-1 gm; sed. in manic-depressive psychosis; relief of chronic pain (PMS); to aid memory; supposed to protentiate antidepressant effects of MAOI; may prod. eosinophilia myalgia (danger!); serious reactions (paralysis) reported (1989-) (removed fr. market US '89) (contamination?); prod. Japan (2511).

tsano (Switz, Fr): *Quercus robur*: *Q. petraea* - **t. oil** (isano; sano) (wAf): FO fr. sd. kernels of *Ongokea gore.*

tsanuga (Ec): brew of tobacco lvs. u. by witch doctors (produces visions).

tsao (Chin): jujube.

Tschirch, Alexander (1856-1939): outstanding Swiss pharmacognosist; author of *Handbuch der Pharmakognosie* & *Die Harze und ihre Harzebehaelter.*

tschongott tree: *Semecarpus venosus.*

tsetse (fly) (pron. tset-see): *Glossina* spp. (Diptera), a fly of wh. one sp. transmits African sleeping sickness in man, another sp. causes n(a)gana, a dis. of cattle, horses, goats (cAf).

tsiology: dissertation on teas (fr. tisane).

Tsoongia axilliflora Merr. (Vit.) (Indochin): lf. & bk. decoc. u. in scabies.

tsoubaki oil: camellia seed oil: Chinese (or Japanese) rose oil: fr. *Camellia japonica*; FO u. in cosmetics.

Tsuga canadensis Carr. (T. americana Farw.; Pinus c. L.) (Pin.) (eNA s to Ala [US]): (eastern) hemlock; common h.; Cana h.; h. spruce; twig tips source of VO ("hemlock oil"; "spruce oil") c. borneol & b. acetate, limonene, myrcene; u. analg. ton. diur. rubef. cath.; bk. c. tannin, u. int. astr. in diarrh. night sweats, &c., exter. in leucorrh. (F. E.) (801); diaph. stim. (2721); in tanning (chief source of tannin in America); source of hemlock pitch (domestic Burgundy p.); u. irrit. vesic.; young twigs u. in scurvy; trunk yields Canada balsam (also fr. *Abies* spp., esp. *A. balsamea)*; lvs. emmen. abort.; u. fresh in homeop. med.; fresh lf. decoc. u. as tisane or beverage for rheum; prod. Cana - **T. heterophylla** (Raf.) Sargent (Pac coast fr. Alas to nwCalif [US]): western (or Pacific) hemlock (-65 m); u. like *T. canadensis* for prepn. lf. needle VO & hemlock bk.; lf. inf. u., as tea (wIndians); inner bk. as food (Alas [US] Indians) cakes as bread substit. (Bella Bella Indians, BC [Cana]); wd. u. in many ways, for boxes, crates, sashes, pulpwood, &c.; bk. u. as tan. - **T. mertensiana** Carr. (Pac area fr. Alas & Mont to cCalif [US]): mountain (or black) hemlock; lf. inf. u. as tea (NA Indians); bk. u. astr., tan. - **T. sieboldii** Carr. (Jap): tsuga (Jap); toga (Jap); bk. u. as tan; wd. useful.

tsunomata (Jap): *Chondrus ocellatus*

tuba: 1) (PI): various fish poisons, incl. *Jatropha curcas*; *cocculus indicus*; *Dalbergia* spp.; etc. 2) (PI): fermented sap of various palms tapped at crown of plant, esp. coconut & nipa; also dist. liquor fr. same - **t. root:** derris r.: rts. of *Derris elliptica, D. malacensis, D. trifoliata.*

Tubarine™: tubocurarine Cl (BW).

tube à essai (Fr-m.): test-tube.

tuber: a short, solid, underground st. or st. segment, borne usually at the end of slender creeping branches; ex. *Corydalis*; potato - **Tubera Ari** (L): T. Aronis. - **T. Aristolochiae cavium** fr. *Corydalis cava* - **T. Aronis** (L.): tuber of *Arum maculatum* - **T. Chinae**: fr. *Smilax china* - **T. Jalapae**: fr. *Exogonium purga* - **T. Salep**: tuberous rts. of *Orchis mascula*, etc. - **tubercle**: a small nodular growth on roots of Leg. Fam. spp. enclosing nitrogen fixation bacteria.

Tuber (Tuber. - Tuberales - Ascomycot. - Fungi): truffle; gastronomically important food materials; incl. several important spp. as follows - **T. aestivum** Vitt. (Eu - Med. reg.) (said best in GB) (common) truffle; summer t.; truffe blanche (Fr); frt. bodies a popular food for centuries (in wine, butter, oil, salads, soups; baked; ingred. of some liver sausages; collected under trees, using trained hogs & dogs (muzzled) who dig for them; also **T. brumale** Vitt. (winter t.); *T. borchi* - **T. melanosporum** Vitt. (sEu, NA): (French or Perigord t.); truffe de France; edible frtg. bodies of fine flavor, coll. summer & fall; sold in markets - **T. uncinatum** Chatin (Burgundy t.).

tubercidin: antibiotic produced in medium of growth of *Streptomyces tubercidicus* Nakamura; purine; has anti-tubercular, antifungal, antineoplastic props.; 1957-.

tubercle: a small nodular growth on roots of Leg. Fam. spp. enclosing nitrogen fixation bacteria.; also a tuberous rt.

tuberculin: a sterile soln. of the metabolic prods. of an ext. of the TB bacillus; u. diagnosis & therapy of TB - **purified protein derivative of t.**: PPD; t. in a liquid medium free fr. proteins; u. in diagnosis of TB (E); - **Tuberculinum Pristinum** (L.): old tuberculin(um); OT; concentrated t.; t. Koch; crude t.; a sterile soln. of the prods. of growth of the tubercle bacillus with c. 50% glycerin; u. in diagnosis & treatment of TB; u. intracutaneously & SC.

tuberin: a protein fr. the potato, *Solanum tuberosum.*

Tuberkulosentee (Ge) (tuberculosis tea): *Polygonum aviculare.*

tuberose: *Polianthes tuberosa.*

Tubiflora Gmel. = *Elytraria* Rich. (Acanth.).

Tubocurarine chloride: quaternary base alk. obt. fr. *Chondodendron tomentosum* & related *C.* spp.; reduces tone of skeletal muscle to point of paralysis; u. (as inj.) as adjunct to surgical anesthesia to increase (or induce) striated (voluntary) muscle relaxation; u. in spastic dis., esp. s. paralysis; to soften convulsions in shock (or convulsion) therapy of insanity; in diagnosis of myasthenia gravis, etc.

tubulin: kidney hormone; u. in nephritic hypertension.

tuckahoe (Algonquin Indians): various tuberous root foods of aborigine, incl. Indian turnip, *Orontium aquaticum, Peltandra virginica*, the subterranean fungus called "Indian loaf," "Indian bread" (*Poria cocos* Wolf) (Polypor.) (resting stage of large edible fungus) found in gps., esp. in aspen groves (Sask); at times mistaken for pemmican; u. by Am Indians for poultices & to treat rheum.; called "ground medicine" (53).

tucom: tucum bravo: t. do Amazonas: t. kernels: t. nuts: tucuma piranga: *Astrocaryum vulgare (A. aculeatum);* source of FO fr. frt.

tuffahh (Arab): apple.

tukamerian (Bombay): **tukmaria** (Ind): *Ocimum basilicum.*

tukham langa (Punjabi): *Lallemantia royleana.*

Tulbaghia acutiloba Harv. (Lili) (sAf): lotion fr. cooked pl. u. to wash incisions made in skull of child in cases of depressed fontanelle: young pl. eaten as veget. - **T. alliacea** L.f. (sAf): wild garlic; rt. u. as purg. (small doses); in bath for rheum. & paralysis; decoc. antipyr.; anthelm. - **T. cepacea** L. f. (sAf): bulb u. for pulmon. TB & as anthelm. - **T. violacea** Harv. (sAf): wild garlic; u. as remedy in pulmon. TB; cold water ext. of pl. postive against *Mycobacterium tuberculosis.*

tule (NM [US], Sp): large bulrushes, *Scirpus* spp.; term also appl. to underground parts of *Sagittaria* spp. (**t. potato**); *Typha* spp. (in SpAm also appl. to certain spp. of *Cyperus,* etc.).

tulip: **tulipe** (Fr): *Tulipa* spp. - **blue t.**: *Anemone patens* - **t. poplar**: t. tree - **Mariposa t.**: *Calochortus* spp. - **t. tree**: *Liriodendron tulipifera* (bk.).

Tulipa edulis Bak. (Lili.) (eAs): edible tulip; bulb u. for abscesses, boils, buboes, snake & insect bite; goiter (Chin); source of starch; sprout held in mouth for oral dis. - **T. gesneriana** L. (c&wAs, now widely cult.): common (garden) (or late) tulip; c. tulipine (aconitine-like alk.); tuliposides (antibiotic; fungostatic) bulbs u. exter. as emoll., in bronchial inflammn., sprains, contusions; although classed as poison, bulbs of some vars. u. to make bread during food shortages (not very palatable) (1944-45) (Neth); during 1600s, bulbs traded at fantastic prices (801-A); very popular garden ornam. (perenn. hb.) - **T. sylvestris** L. (eMed reg., natd. Par): wild tulip; u. similarly to prec. - **T. turkestanica** Rgl. (Uzbekistan): c. tulipine, flavonoids; u. in heart dis. rheum. (folk med.).

Tully's powder: morphine pd. comp.

Tulpe (Ge-f.): tulip.

tulsi (Hind) *Ocimum sanctum.*

tumbavaqueros (Mex): *Ipomoea stans.*

tumbleweed: name appl. (wUS) to ripened bunches of various pl. spp. wh. detach at st. base and roll in the wind, scattering their seeds; usually appl. to *Amaranthus graecizans* & *A. albus* (tumbling pigweed); but also to *Salsola kali* (Russian thistle); *Cycloloma atriplicifolium* (winged pigweed); *Corispermum nitidum*+ (bugseed); *Sisymbrium altissimum* (tumbling mustard); *Digitaria cognatum* (witchgrass); &c.

tumeric: common misspelling of turmeric.

Tumion Rafin. = *Torreya* Arn.

tun (Sanscrit) (tunn = tunna): *Cedrela toona.*

tun: large cask.

tuna: 1) a percomorph fish, *Thunnus thynnus* (Atl Oc); *Neothunnus maeropterus* (US, Pac); popular food packed in cottonseed & other FO or in water 2) edible frts. of **t. cactus**, *Opuntia* spp. (some 10 spp. SC in Calif [US]) - **t. real**: *Opuntia cochinellifera.*

tundra (Russian): wild treeless area with stunted shrubs, mosses, & lichens stretching as a broad belt across Eurasia in a general e-w direction; with permanently frozen subsoil.

tung: *Aleurites fordii* & other *A.* spp. (FO).

tunica: surface covering layer(s) of bud & shoot apex enclosing the corpus.

tunicamycin: nucleoside antibiotics produced by *Streptomyces lysosuperificius.*

tunicin: cellulose (no pentoses) found in sea squirts (Tunicata) (attached primitive chordates), &c.

tunu gum: 1) inferior form of gutta percha fr. *Castilla fallax* 2) may be intended for tuna gum fr. *Opuntia* spp.

tupakka (Finn): tobacco.

tupelo (gum): *Nyssa aquatica, N. sylvatica.* - **t. berries**: frt. of *Nyssa ogeche.*

tur (Indo-Pak): tor.

turacin: common pigment fr. tarakoo (African bird); c. Cu.

Turbina cormybosa (L) Raf. (Ipomoea c. (L.) Roth; I. sidaefolia Choisy; Rivea c. Hallier) (Convolvul.) (sUS, Mex, WI, Middle Am, nSA): morning glory; ololiuqui; sds. c. ergot (ergoline) alks., ergometrine, lysergic (acid) amide (ergine), clavine series (penniclavine, chanoclavine); corymbosine; u. as hallucinogen by sMex Indians; widely cult. & u. formerly by teenage youth; chewing hard sds. harmful to teeth; sds. u. to prep. fermented agave beverage (piule) with hallucinogenic props. (483); honey plant (Leon): antispasm.; contraceptive; to stim. labor in childbirth (SpAm).

turbinado (sugar): crude sugar refined only by centrifuging; sold in health food stores, &c.; v. under sugar.

turbot: name u. for several Eu flat-fishes incl. flounders & halibuts; do not confuse with burbot (round fish member of cod fam.).

tureve (Bo): *Rhamnidium elaeocarpum.*

turic gum: var. of acacia.

turio (Bo): **turion** (Fr): **Turione** (Ge): bud esp. winter b.

Turionis Juniperi (L): bud of *Juniperus communis* - **T. Pini**: **Turiones Pini:** pine shoots (buds, branch tips, tops) of *Pinus* spp., esp. *P. sylvestris.*

turkey: *Meleagris gallopavo* (Meleagrididae - Galliformes - Aves) (eUS s to Mex, but now widely distr. & domesticated): American t.; meat a delicacy, esp. with truffles; popular Xmas chief meat course; meat of wild bird more flavorsome than that of domestic - **t. berry**: 1) *Symphoricarpos orbiculatus* 2) *Callicarpa americana* 3) (Ala+ [US]): *Mitchella repens* - **t. corn: t. pea**: *Dicentra canadensis* - **t. red**: 1) madder 2) red Fe_2O_3 - **t. red oil**: sulfated castor oil.

Turkey: v. also under paprika; rhubarb.

Turkish delight: **T. paste**: candy of gum type in cubes dusted with. pd. sugar; "good for the stomach"; originated in Turkey.

Turk's cap (Fla [US]): *Malvaviscus grandiflorus* HBK; less commonly *M. arboreus* Cav. - **T. c. lily**: *Lilium martagon; L. michauxii.*

turma (Co): potato.

turmeric: *Curcuma longa* - **Aleppy t.**: the most esteemed comm. var. of *Curcuma longa* (Ind; Chin) - **common t.**: *Curcuma longa*, rt. - **Indian t.**: Hydrastis - **long t.**: form of t. with slender elongated rhiz. - **Madras t.**: important comm. var. of *C. longa* - **red t.**: Sanguinaria - **T. root:** 1) common t. 2) Hydrastis (misnomer) - **round t.**: a form of t. with ovate or pyriform pieces.

Turnera diffusa Willd. (T. aphrodisiaca Ward.; T. diffusa var. a. Urb.) (Turner.) (sTex [US], Mex, WI, CA, SA): damiana; lvs. c. VO tannin, NBP, resin, u. mild ton., aphrod. (usu. with strychnine, P., &c.); stim. diur. nerve ton. mild purg.; expect. (Jam.); "general cure" (Mex [US]); astr. (Br);

decoc. for children with nocturnal incontinence (bed wetting) (Bah); in colds, VD; in kidney, menstr. & pregnancy dis.; hb. inf. u. as tea substit.; lvs. u. as flavor for liqueurs (ex. panzanita) (with FDA approval) - **T. humifusa** Endl. = *T. microphylla* Desv. = *T. diffusa* - **T. subulata** J. E. Sm. (tropAm; natd. Mal): rts. u. to poultice boils - **T. ulmifolia** L. (T. angustifolia Mill.; T. mollis HBK) (sUS, WI, cMex to Pe, Arg): amaranto; catbush (Bah); yellow flower; "dash along"; margarita de los campos (Nic); u. ton., for indigest. bronchitis (Mex); lf. inf. u. for liver & kidney dis. (Grand Cayman Is); for insomnia; as general beverage; weed.

Turneraceae: Turnera Fam.; hbs. & shrubs of trop & subtrop; lvs. alternate, pinnately veined; 10 genera represented, as *Erblichia, Turnera* (with 110 spp.).

turnesol: (Sp): sunflower.

turnip: *Brassica rapa* - **angel's t.** (Am Negro): *Apocynum androsaemifolium* - **deer t.** (Fla [US]) *Liatris* spp. - **Hungarian t.:** 1) *B. oleracea* v. *gongylodes* 2) kohlrabi - **Indian t.**: *Arisaema triphyllum* and its subsp. *stewardsonii* - **Swedish t.**: *Brassica napus*. -**wild t.**: 1) B. campestris 2) Arisaema triphyllum.

turnsole: 1) *Lecanora tartarea* 2) *Heliotropium* spp.

turpentine: 1) the crude oleoresin fr. *Pinus palustris, P. pinaster*, & other *P.* & Pin. Fam. spp. 2) (in commerce, etc.): spirits (VO) of turpentine prepd. by distillation of prec. - **Alsace t.: Alsatian t.**: t. fr. *Abies alba* (*A. pectinata*) - **American t.: common A. t.** : fr. *P. palustris, P. caribaea, P. taeda* - **Austrian t.**: fr. *Pinus nigra* - **Bordeaux t.**: oleoresin fr. *P. pinaster,* & other *P.* spp. of swFr - **Canada t.**: C. balsam: t. fr. *Abies balsamea* - **Chian t.: Chios t.**: t. fr. *Pistacia terebinthus* - **crude t.**: Terebinthina (NF) - **Cyprus t.**: Chian t. - **dammara t.**: fr. *Shorea* sp - **dombeya t.**: fr. *Dombeya excelsa* (Sterculi.) or fr. *Araucaria araucana* - **European t. (, common)**: fr. *P. sylvestris* - **gum t.**: 1) American t. or Bordeaux t. 2) turpentine oil dist. fr. the gummy exudate of the tree (& not fr. stumps, wd., etc.) - **Hungarian t.**: t. fr. *Pinus mugo*: possibly Venice t. - **Jura t.**: oleoresin fr. *Picea abies*. - **larch t.**: oleoresin fr. *Larix decidua* - **lump t.**: crude oleoresin of seUS pines - **t. oil**: VO dist. fr. oleoresin turpentine or fr. wd. or isolated in process of making paper by sulfate process. c. α- and ß-pinenes, dipentene, terpinolene, &c.; u. solvent, disinfect. vermif. (**ozonized t. o.**: u. as antidote in phosphorus poisoning; in guaiac test for occult blood) - **t. (oil) substitute**: Terebinthina artificiale (L): various colorless org. liquids, esp. petroleum distillates u. in non-med. areas such as in shellacks, floor polishes, shoe creams, &c., the odor may be masked with essential oils (802) - **spirits of t. (, gum): t. oil:** fr. oleoresing (as distinct fr. other sources, such as wd.) - **Strasbourg t.: Strasburgh t.: Strassburg t.**: Alsatian t. - **stump t**: wood t. (oil) - **t. sunflower**: *Silphium terebinthaceum* - **Venetian t.**: **Venice t.**: larch t. - **Virginian t.:** t. - **t. weed**: *Silphium laciniatum* - **white t.** : crude t. - **wood t.**: "pine oil"; VO obt. fr. wd. (mostly stumps) of *Pinus palustris* or other *P.* spp. by one of 3 methods: 1) steam distn. 2) destructive distn. 3) fr. wd. pulp mfrd. by sulfate process (by-prod.); distinguish fr. gum t. oil made by distn. of oleoresin; u. in synth. of terpin hydrate, terpineol, &c.; as solvent, disinfectant, deodorant, antifoaming agt.; flotation of Pb, Zn, Cu ores; in textile scouring; additive to pharmaceuticals (flavor, perfume), insecticides, emulsions, polishes, bath oils, &c.; c. chiefly isomeric tertiary & sec. cyclic terpene alcs. ex. borneol, terpineol.

turpeth (root): Turpethum (L): *Ipomoea turpethum.*

Turpinia formosana Nakai (T. arguta sensu auct. Taiwan) (Staphyle.) (China): u. for boils.

"turps": turpentine oil, tech. grade, as u. in painting, etc.

Turraea floribunda Hochst. (T. heterophylla Sond.) (Meli) (sAf): rt.u. as emet. in rheum. edema, heart dis. (Zulus); bk. u. emet. purg. (Tanz) - **T. heterophylla** Smith (trop wAf): lvs. u. as bitters, in rum for impotency (Ghana); wood harder than mahogany - **T. nilotica** Kotschy & Peyr. (tropAf): "litakombwa" (Tanz); rt. boiled in water u. for toothache (Tanz); dr. lf. considered toxic - **T. obtusifolia** Hochst. (sAf): lf & bk. u. for gastrointest. compl., drastic - **T. robusta** Guerke (Af): lf. u. in diarrh. wd. u. to make spoons, &c. - **T. stolzii** Harms (Tanz); lf. inf. drunk for fever, &c., rt cooked with maize eaten as cure for ulcer - **T. villosa** Benn. (Ind): rt. appl. to fistulas; u. int. for leprosy.

turtle: reptile of Ord. Chelonia (Testudinata); worldwide; some spp. are u. for edible meat (soup) (ex. loggerhead t.); green t.; hawk's bill t. (Hond.); snapping t.; soft-shelled t. (one of best); edible eggs, food luxury, (ex. snapping t.); "tortoise" shell (u. for ornament); t. oil; leather (fr. flippers) - **t. bloom**: *Chelone glabra*. (SC corolla resembles head of turtle) - **t. head (, white)**: *Chelone glabra* - **t. oil:** FO obt. by extn. of muscles & genital organs of giant sea turtles; c. much linoleic acid; cholesterol, squalene; tocopherols; u. in cosmet., some consider beneficial to skin: as vehicle for VO, stimulins, tissue exts., iodized compds., phosphorated compds.; in esthetic prepns (803) (Schildkroetenoel, Ge) - **sea t., green:** *Chelonia mydas* (trop. oceans): giant turtles (ca. 900 kg); legs like giant flippers; flesh & eggs sought for foods; tea fr. dr. penis drunk to aid conception; dr. intest. decoc. u. as remedy for motion sickness (sea; car) (Seri Indians, nwMex); eggs valued for restoring virility (Mex.).

tus: (gum) thus.

tusca (Arg): *Acacia micrantha.*

Tusch(e) (Ge): India ink.

Tussacia Rchb. = *Chrysothemis* Dcne.

Tussilago farfara L. (Comp.) (Euras, nAf; natd. eCana, neUS): (common) coltsfoot; clay weed; **tussilago leaves** c. mucil. VO resin; heads u. dem. exp. in cough, bronchitis, asthma, diarrh., in many teas; when young carcinogenic; lvs. smoked for asthma; fls. u. dem., in gargles; rt. u. as folk med. agt. (chief prodn. It) (suggested replacement: Althea lvs.) (1778).

tutin: glycoside fr. *Coriaria thymifolia;* convulsant action like that of picrotoxin.

tutti fruti (It, Am) (lit. all fruits): combn. of various dr. frts. (as pineapple, cherries, maraschino cherries, &c.) appl. to ice cream, pies, confections, &c.

tutun (Turk): tobacco.

tuya (Mex): *Thuja occidentalis.*

twilight sleep: combn. of analgesia-sedation-amnesia induced by a mixt. of morphine (or Demerol [meperidine] or barbiturate [such as Na Amytal]) with scopolamine; at times u. as pre-operative sed. induction in childbirth, but not always recommended (may cause asphyxia of child [esp. in premature cases]) & prolonged labor.

twin berry: *Mitchella repens* - **twin-flower**: *Linnaea borealis* - **twin-leaf**: *Jeffersonia diphylla* - **pointed t.**: *Zygophyllum apiculatum.*

twitch grass: *Agropyron* (*Elymus*) *repens.*

Tylophora brevipes F.-Vill. (Asclepiad.) (PI): rt. c. alks., decoc. u. as emet. emmen. carm.; diaph. expect. - **T. brownii** Hay. (Chin): juice fr. st. u. on wounds (Taiwan) - **T. cissoides** Bl. (Indones): lvs. appl. to treat pain of abdom.; rts. & lvs. u. for thrush - **T. conspicua** N. E. Br. (w tropAf): lvs. heated with peppers appl. to yaws; pounded lvs. appl. to ulcers & wounds - **T. crebriflora** S. T. Blake (Austral): c. tylocrenrine, active against leukemia & various tumors - **T. fasciculata** Ham. (Ind): c. tylophorine; lvs. u. to promote granulation of open wounds; rt. & lvs. emet.; rt. pressed juice u. with milk as ton.; lvs. & rts. as rat eradicant - **T. hirsuta** (Wall.) W. & A. (Ind-Himal): pl. c. alks.; with anti-leukemic, anti-cancer, and antiamebic properties - **T. indica** (Burm. f.)Merr. (T. asthmatica Wight & Arn.) (Ind, Mal): country ipecacuanha; Indian ipecac(uanha); rt. & lvs. u. emet. expect. in dys.; rt. favored as replacement for ipecac; lvs. (tylophora leaves) chewed in asthma (small doses admin. once daily for 6 days or more give relief of asthmatic symptoms; unique remedy) (804); native physicians use for hydrophobia(!!) - **T. laevigata** Decne. (Mauritius Is): coffee climber; rt. u. like ipecac ("false ipecac") - **T. ovata** Dcne. (eAs): u. as ton. astr. styptic, in scrofula (Chin) - **T. perrottetiana** Dcne. (PI): lvs. u. as vulner. - **T. polyantha** Volk. (Caroline Is): whole pl crushed, mixed with cold water, & drunk for general debility, sleepiness, tiredness - **T. tenuis** Blume (Ind-Mal): lvs. u. to treat scabies (Mal); in urticaria, asthma, arsenic poisoning.

tylose: tylosis (plur. tyloses): intrusive growth fr. 1 cell into cavity of a second cell, producing bladder-like extensions of wd. parenchyma or wd. ray cells into the bodies of adjacent vessels or tracheids; resultant blocking up of vessel cavity during change of sapwood to heartwood - **Tylose™**: methyl or carboxy-methyl cellulose.

Tylosin: macrolide antibiotic fr. *Streptomyces fradiae* Waksman & Henrici strain (Thail soils); u. antibacterial in vibrionic scour (pigs) & resp. inf. in lower animals.

Tylostemon Engl. = *Beilschmiedia* Nees.

Tylostoma (Tulostoma) brumale Pers. (Tylostomataceae - Gasteromycetes - Basidiomyc. - Fungi) (Eu, NA): spores appl. to cord of newborn infant to prevent inflammn. and remedy for suppuration (Pima Indians).

Tyn(n)anthus elegans Miers (Bignoni.) (Br): cipo-cravo; sts. u. for correction of odor & taste (in iodine prepns); ton. stim. stom. carm. in galenic prepns. (Br); st. lf. rt. u. aphrod. (guanabara) - **T. fasciculatus** Miers (Br): u. same as prec.

type specimen: that herbarium specimen fr. which a pl. sp. is first described; in the absence of such, descriptions may be substantiated by means of photographs, sketches, paintings, etc.

Typha angustata Bory & Chaub. (Typh.) (Med reg., Euras, nAf, Jap): lesser cat-tail; rt. stock u. as astr. diur., in insanity & epilepsy (Ind); pollen u. as diur. styptic; for menstr. diffic.; in dropsy (cooked with pork) (Chin); lvs. u. to make mats, baskets, &c. (Pak) - **T. angustifolia** L. (Euras, nAf, NA, SA): cat-tail (flag); (lesser) reed mace; rhiz. u. in dys. GC sores, like *T. latifolia*; with antitumor activity; food of aborigines (all parts, esp. rhiz.), &c. (starch); lvs. u. to make mats, baskets (Pak); frt. spike fibers ("cat-tail down"; "cat-tail wool") u. for stuffing upholstery, mattresses, paper mfg., for insulation, &c. (prod. Bengal); as moxa (Ind);

pollen as lycopodium substit.; food prepn (1772); pls. as fuel, to cover roofs, for ropes, substit. for raffia fiber - **T. australis** Schum. & Thonn. (c tropAf): cat-tail; roseau (Fr); rhiz. famine food; central parts of immature spike a food delicacy (Hausa) - **T. capensis** Rohrb. (sAf, eAs, PI): rt. u. in diarrh. VD diur. in urethritis; to aid expulsion of after birth; woolly inflorescence u. as vulner. mechanical hemostatic (PI); stamens as astr. styptic (Chin); sd. as appln. to burns; rhiz. edible - **T. davidiana** Hand.-Mazz. (As): u. similarly to other *T.* spp. - **T. domingensis** Pers. (tropAm, US): source of tabua fibers, u. as substit. for kapok (Br) - **T. elephantina** Roxb. (Ind, Pak.): elephant(s) grass; pitz (Kashmir); erect perenn. marsh grass, 2-5 m tall, lvs. 3-angled; white-margined; rhiz. u. astr. diur. in measles, dys. GC; down of ripe frt. u. as appl. to wounds & ulcers (*cf.* absorbent cotton, kapok); pollen eaten; sts. u. to make ropes, baskets, mats, studied for possible textile use (Pak) - **T. latifolia** L. (worldwide in temp & trop zones; incl. NA, Euras, Af, Oceania): (common) cat-tail; bullrush; u. similarly to *T. angustifolia*; sold in Mex Am herbal stores (aguapa), u. as diur. & in insect bite; fls. u. to treat burns, sores (Am Indians); stamens & pollen u. as astr. & styptic (Chin); floss of female spike u. to dress wounds & in serious burns; lf. decoc. to stop uterine hemorrh. & diarrh.; rhiz. stim. astr. antipyr.; rt. u. to dress scalds (spread as paste) (Omaha Indians); sometimes toxic to cattle (Fla [US]); fiber (plush) from frtg. spike u. like kapok in life preservers, cushions, pads, &c.; spikes dipped in FO, &c., & u. as torch (colonial US); ornam.; to purify polluted water - **T. orientalis** Presl. (T. japonica Miq.) (Orient: Chin, Jap, PI, Ussuri): carbonized pollen u. as styptic; raw pollen for abdomin. pain, dysmenorrh., amenorrh; young shoots pickled, cooked, & eaten (Chin); frtg. heads soaked in oil & u. as torches; rhiz. ton. cooling, diur. in mastitis.

typhoid combined vaccine: t. & paratyphoid vaccine, with t. & paratyphoid A & B bacilli represented - **t. prophylactic: t. vaccine**: sterile suspension of killed organisms of *Salmonella typhosa* White, organism of typhoid fever; u. for active immunization.

Typhonium divaricatum Dcne. (Ar.) (Chin, Jap, Ind, Mal, PI): tubers rubefac., u. in diarrh.; food in times of scarcity - **T. giganteum** Engl. (Chin) rhiz. u. for strokes, tentanus(!!), localized headache; insecticidal - **T. roxburghii** Schott (Chin): u. exter. for skin eruptions (very irritant) - **T. trilobatum** Schott (Ind): fresh tuber acrid u. as strong stim., poultice for scirrhous tumors; tubers u. for hemorrhoids; eaten with bananas for gastric dis.; cooked tubers edible; lvs. heated to soften u. as poultice for boils. (Thail).

Typhonodorum lindleyanum Schott (T. madagascariensis Engl.) (monotypic): (Ar) (eAf coast, Mad, Mascarenes): sds. & tubers edible; u. as famine food.

typhus vaccine: a sterile suspension of killed rickettsial organisms of a strain of epidemic t. rickettsiae (*Rickettsia prowazeki* da Rocha-Lima) selected for antigenic efficiency; cultured on chick embryos; u. as prophylactic to t. fever.

tyramine: sympathomimetic amine (*cf.* epinephrine) prod. by decarboxylation of tyrosine (*cf.* histamine); occurs in ergot (toxic constit.), mistletoe, ripe cheese, putrified animal tissues (one of "ptomaines"); toxic with pressor effect; destroyed by enzyme tyraminase (monoamino oxidase); u. in circulatory collapse & as ecbolic.

Tyrian purple: natural dyestuff fr. *Murex* spp. (molluscs of e. Med Sea); popular in antiquity but very expensive; dibromo-indigo.

tyrocidin: one of components of antibiotic, tyrothricin.

Tyroglyphus (Acarus) **dimidiatus** (Tyroglyphidae) (Acaridae) (Arachnida): cheese mite; found in unpackaged cheese, drugs, etc.; other mites (acarids) which attack crude plant materials include the following: **T. farinae** *(Acarus siro):* flour (or meal) mite; common in cereal prods., oil sds., etc.; **T. longior** (on fermenting honey); **T. siculus** (attack Cantharis); **T. siro** Gerv.: cheese mite (attacks mustard meal, tobacco, &c).

tyrosamine: tyramine.

tyrosinase: enzyme wh. catalyzes oxidn. of tyrosine to give melanin pigments (cause of darkening of cut surfaces of frts., potatoes) & certain phenols (ex. catechol); occ. in many bacteria, mushrooms, higher plants, crustacea, butterflies, &c.

tyrosine: an amino acid with acid phenolic gp.; p-OH-phenyl-amine propionic acid; first found in cheese (Liebig, 1846); occ. widely in pls. & animals.

tyrothricin: an antibiotic mixture of princ. extracted fr. a culture of soil bacteria, principally Bacillus brevis Migula; first isolated by Dubos; consists of mixture of gramicidin (20%) & tyrocidine (50%); u. to best effect against Gm.+ organisms, also Gm.- cocci; can be u. only topically & not parenterally; ineffective orally; u. in form of soln., spray, troches, or lozenges (USP. XI).

tyty: titi .

Uu

Uapaca bojeri Baill. (Euphorbi.) (Mad): "tapia" (not "tapa"); bk. u. in diarrh.; frt. edible; wd. u. as fuel; tree is host to several Lepidoptera u. to furnish silk & as food (pupae) - **U. kirkiana** Muell. Arg. (tropAf): wild loquat; pl. remedy in amebiasis; frt. tasty (*cf.* pear) & much eaten; u. to prep. a sweet beer; famine food; wd. useful timber.

Uapacaceae (Dicot Fam. sometimes recognized as distinct fr. Euphorbi. (Mad & tropAf): trees or shrubs; monotypic with *Uapaca.*

ube (Mal): **ubi**: white yam.

ubiquinones: coenzymes Q; gp. of widely distributed FO-sol. substances, sometimes classed as vitamins; benzoquinone ($C_6H_4[CO]_2$) derivs.; occ. in most aerobic organisms from bacteria & pl. to animals; concerned with oxidative phosphorylation; cardiovascular agt. (ubique, L., everywhere), ind. CoQ_{10} (2628).

ubiquitins: large moleculear (76 amino acids) polypeptides with proteolytic function; present in all eukaryotes (incl. humans, insects, plants); with many functions: protein degradation; heat shock; immunity reactions; &c. (SC ubique = everywhere).

uchu (Pe): Capsicum.

ucuuba fat (Br): bicuiba.

udo (Jap): *Aralia cordata.*

Ueberpruefung (Ge): examination testing; detn.; control; preventing.

UFA: unesterified fatty acid(s): lack suggested as cause of duodenal ulcer, &c.

ugali (eAf): maize meal mixt.; a staple food.

Uganda: republic of eAf; exports cotton, coffee, tea, &c.; v. under aloes.

ugli fruit: product of a *Citrus* hybrid (grapefruit X tangerine [mandarin]) wh. looks like a grapefruit but has a wrinkled yellow skin & better flavor (Jamaica); cult. Calif, Fla (US) (<ugly, because of appearance).

uintjes (sAf): nut grass, *Cyperus rotundus*

ule (Sp): true rubber

Ulex europaeus L. (Leg.) (Eu, natd. wNA [Pacific coast]): (common) gorse; (common) furze; whin; c. cytisine (formerly called ulexine) (808); fls. c. ulexoside; u. yellow dye kindling (Ireland) - **U.** gallii Planch. (wEu): (dwarf) goarse; said hybrid of *U. europaeus*; uterin. stim.

ulexine: **ulexoside**: v. prec.; u. in edema.

ulje (Croat): oil.

ulluco (Bo): **ullucu** (Pe): **Ullucus tuberosus** Caldas. (*Melloca tuberosa* Lindl.) (monotypic) (Baselaceae) (Co, Bo, Pe, Ch): tuberous rts. edible; c. over 33% starch in dry wt., sugars, proteins.

Ulmaceae Mirbel: Elm Fam.; trees of temp&trop zones; alternate simple lvs., frt. a samara; ex. *Ulmus, Celtis, Zelkova.*

Ulmenrinde (Ge): elm bark; generally that fr. *Ulmus procera* intended.

Ulmus L. (Ulm.) (nTemp Zone); the elms; 18 spp. of deciduous shrubs & trees, with alternate uniequilateral dentate lvs.; inner bk. source of fibers; pollen causes hay fever in some; cult. ornam. & shade trees - **U. alata** Mx. (se&cUS): winged (or cork) elm; wahoo (elm); tree usually small but sometimes to 20 m ("giant wahoo"); bk. u. as cataplasm for inflammns.; inner bk. u. to fasten cotton bale covers; wd. u. for tool handles (does not split easily) - **U. americana** L. (e&cNA, e of Rockies): American (or white) elm; bk. u. as poultice for gunshot wounds (US, Rev. War); bk. decoc. u. in cough & colds (Mohican Indians); in pulm. hemorrhage (Penobscot Indians); tree subject to dis. (Dutch elm disease) (806); wd. hard & heavy, u. in flooring, boat & ship building; ornam. tree (- 40 m) (variable forms) - **U. campestris** L.: this sp. has been split between *U. glabra & U. procera* - **U. crassifolia** Nutt. (sUS, nMex): cedar/basket/red elm; wd. u. for furniture & wheel hubs (Mex) - **U. fulva** Mx. = *U. rubra* - **U. glabra** Huds. (U. scabra Mill.) (nEuras) Scotch/wych/mountain elm; large tree (- 40 m) with large & abundant frt.; sd. rich in FO of edible quality; ornam. street tree - **U. japonica** Sarg. (neChin, Jap, Sakhalin): Japanese elm; bk. u. diur. purg., for nerve dis. (Kor); wd. u. in turnery - **U. laevis** Pall. (U. effusa Willd.) (se, c&eEu): European white elm; spreading branched elm; inner bk. u. as mucilagin. astr.; antidiarrh. in folk med.; chronic coughs (homeop.); sd. rich in FO of cdible qual. - **U. macrocarpa** Hance (Chin): sds. c. NBP, mucilage; u. as vermif., antipyr., exter. in skin dis. - **U. minor** P. Mill. (U. carpinifolia Gled.) (Euras, nAf): inner bk, u. as for *U. glabra*; u. as adjuvant in formulas for antirheumatic agts. - **U. montana** Stokes (n&cEu [mts.]; wAs; Jap): mountain (& wych) elm; inner bk. u. much like that of *U. laevis*; small pieces of pl. u. as charm against witches; dairy maids used to place tiny bits in a churn so witches could not prevent milk fr. being turned into butter; wd. useful in many ways, ex. turnery - **U. parvifolia** Jacq. (U. chinensis Pers.) (Chin, Kor, Jap): Chinese elm; bk. u. antipyr. expect., diur. soporific; lvs. fls. & frt. u. as diur. dem. antiphlog. (Yupei, Chin); cult. in wUS as shade & ornam. tree (-25 m) & windbreak (805); in nUS, lvs. become red to purple, in sUS, remain green - **U. procera** Salisb. (e&cEu; natd. neUS): English elm; field e.; bk. c. mucil. tannin; u. dem. emoll. diur. astr. in diarrh. in surgery (fresh inner bk. off. in Ge Homeop. Pharm.); cult. ornam. tree (- 55 m): may produce dermatitis; frts. u. in Morocco folklore med.; devastated by Dutch e. dis. (1970s-) - **U. rubra** Muhl. (U. fulva Mx.) (e&cNA): slippery/red/gray elm; the dr. inner bk. (Ulmus [Fulvae Cortex], L.) very mucilagin. & arom., c. mucil., arom. princ. (like fenugreeek in odor); u. dem. emoll. poultice (69); expect. in sore throat, bronchitis; chewed to relieve thirst; for intest. dis. (1785); early settlers boiled bear fat with bk. to prevent rancidity; claimed useful in rhus poisoning (807); slivers of bk. have been u. to pierce "bag of waters" in uterus & initiate abortion; white colored bark sometimes claimed superior to red in medic. - **U. thomasii** Sarg. (eCana, neUS w to SD & Okla): rock (or cork) elm; wd. very strong, u. for heavy agricult. implements, chairs, &c.; ornam.

ultracentrifuge: lab. implements developing centrifuge forces of more than 100,000x gravity.

ultramarine (**blue**): natural mineral pigment (in lapis lazuli, &c.); u. as bluing in laundry, food color; formerly u. to whiten sucrose (*cf.* bluing clothes).

ultraviolet light: UVL: light of short wave lengths obt. fr. mercury vapor lamps, etc.; u. in fluorescent analysis of crude drugs; UV radiation is called "actinic radiation" (SC chem. biological effects).

ultra-virus: (filterable) v.

Ulva L (Ulvaceae-Chlorophyceae-Algae): sea lettuce; grows in shallow sea water; green fronds eaten; u. for iodine content in med. (one of 4 genera of Algae of Linnaeus) - **U. fasciata** Delile (eAs coastal waters, Haw Is [US]): limu papahapapa (Haw): thallus u. as food, esp. in Hawaii; there boiled with squid - **U. lactuca** L. (cosmop. in oceans. incl. Eu) : sea lettuce; (lettuce) laver; u. in salads & as garnish; as vegetab. & in soups; u. for gout (eAs); fodder for hogs+; to fertilize the soil; popular in health foods ("to fill in wrinkles of face"[!!]): has been much studied - **U. latissima** L. (Pac Oc esp., Ch & Co coasts): "luche" (Ch) u. for human food, often along with *Porphyra kunthiana*: pl. fed to pigs & ducks; cult. in Singapore area - **U. nematoida** Kuetz. (Pac Oc): found on coasts of Ch, Pe (Juan Fernandez); u. as food in New Caled - **U. penniformis** Mart. (Pac Oc): u. med. in parts of Siberia - **U. pertusa** Kjellm. (Pac Oc): lacy sea lettuce; u. in Chin med. to reduce high body temps.; eaten in Amoy reg. (Chin) (with oysters, noodles, or meat).

umba opopanax: opopanax (gum) fr. *Opopanax chironium.*

umbel: 1) flat-topped inflorescence characteristic of fam. Umbelliferae members 2) (US): *Cypripedium pubescens.*

Umbelliferae: Parsley (or Carrot) Fam.; mostly hbs., few shrubs, chiefly of nTemp. Zone; one of largest pl. families (3,100); members show striking similarities in habit; fls. very small, have 5-merous calyx & corolla, 2-chambered inferior ovary; umbel type of inflorescence bear frts. wh. in some spp. are poisonous (ex. poison hemlock, cowbane); others edible (ex. carrot, celery, parsnip, parsley); some are important condim. (ex. caraway, anise, asafetida, coriander).

umbelliferone: OH-coumarin; the aglycone of skimmin; occurs in resins of many spp. of Fam. Umb.

(ex. sumbul, galbanum); also by synth.; u. in sun lotions & creams; prod in dry distn. of asafetida.

Umbellularia californica (H & A) Nutt. (Laur.) (Calif, sOre [US]): Calif (bay) laurel (or olive); pepperwood; Oregon myrtle; spice tree; bay tree; lvs. of this large tree u. as condim. in soups; substit. for bay lvs; as stim. anod. (in stomachache); decoc of fresh lvs. u. as disinfect. wash & appl. to scalp for headache; in rheum (exter.); lvs. repel fleas & lice; source of VO (u. to exterminate fleas) c. umbellulone (bicyclic terpene ketone, u. as fungicide, bactericide); frt. eaten by Calif Indians (formerly); sds. roasted & eaten (Am Indians); rts. u. to prep. beverage; wd. ("myrtlewood") u. in sOre & nCalif (US) for fine cabinet work, jewelry, utensils, &c.; wd. very heavy, strong, with beautiful grain, multicolored. (2650).

umber: an earthy iron-rich mineral; u. pigment (**raw** & **burnt** types).

umbilical cord (human): functions to provide blood components (RBC, WBC, plaquettes [act to prevent hemorrhages]); used in medullary aplasia, leukemia, constitutional anomalies (ex. syndrome of Fanconi); function similar to that of bone marrow.

Umbilicaria esculenta (Miyoshi) Minks (Gyrophora e. Miyoshi) (Umbellicari. - Discolichenes - Lichenes): (Jap): rock mushroom; eaten as delicacy in Jap (ex. in soy soups); c. NBP; anti-tumor activity against bone cancer (inhibition & regression of cancer) (1773) - **U. havaasii** Llano (U. fuliginosa Havaas) (Denmark): blistered lichen; u. as red brown dye - **U. pustulata** (L) Hoffm. (nHemisph, esp. Ge, Scand): blistered umbilicaria; source of red, purple, & brown dyes (wEu) - **U. vellea** (L) Ach. (Gyrophora vellerea Nyl.) (Eu): fleecy umbilicaria; u. as purple dye.

Umbilicus erectus DC (Cotyledon umbilicus L.) (Crassul.) (seEu, AsMin): lvs/hb. u. as mild diur. - **U. rupestris** Dandy (U. pendulinus DC; Cotyledon r. Salisb.; C. umbilicus var. *tuberosus L.*) (Med reg., AsMin to Mor): (nodding) navelwort; kidneywort; (small) pennywort; lvs. (as Folia Cotylendonis) u. as mild diur.; cooling agt.; ton. fresh lvs. in homeop. med.

umbill root: *Cypridedium flavum.*

umbrella plant: 1) *Podophyllum peltatum* 2) *Peltiphyllum peltatum* 3) *Cyperus involueratus* 4) *Schefflera aromatica* - **u. thorn**: *Acacia planifrons* (Ind) - **u. tree** (US) 1) *Magnolia virgininana* (bk.) 2) *M. tripetala* 3) *Melia azedarach* v. *umbraculiformis* (umbrella china[berry] tree) (809) 4) *Schefflera actinophylla* (SC lvs. palmately compd.).

umburana seeds (Br): *Torresea cearensis.*

umiri balsam: **umbery b.**: prod. secreted in the bk. & wd. of *Humiria balsamifera* (etc.) (Br).

una de gato (CR): *Martynia jussieui* (*M. fragrans*).

unab: *Ziziphus mauritiana.*

Unani medicine: ancient medical practice of the Indian subcontinent representing the Greco-Arabian system; utilizing some 2,000 crude drugs including 200 of animal orgin; many drugs are from the ancient materia medica of Gr; since Unani is practiced by Muslims, Unani names are in Urdu; there are many more animal drugs than in Ayurveda (*cf.* Ayurveda; Tibbi).

Uncaria africana G. Don (Rubi.) (cAf-thickets in wooded hills): rt. inf. & lf. juice u. for hardened abscesses; bk. chewed for cough - **U. elliptica** R.Br. (U. dasyoneura Korth) (Mal): c. rutin (-20%) - **U. ferrea** DC (Indones): lf. decoc. u. to cleanse wounds & ulcers; rt. inf. u. for enteritis - **U. gambir** (Hunter) Roxb. (Ourouparia g. Baill.; Nauclea g. Hunt.) (Mal, SriL, Indones, esp. Borneo): Bengal gambir (plant); gambier (plant); source of pale catechu (or gambir) fr. aq. ext. of tree; c. catechin (up to 75%), catechu-tannic acid (22-45-50%), catechu red, quercetin, rutin; u. astr. much like cutch; in diarrh.; addn. to betel chew; as dyestuff & tan.; prod. in Borneo, where cult - **U. guianensis** Gmel. (Guiana): hb. decoc. u. in dys. - **U. rhynchophylla** Miq. (Jap, Chin): st. c. rhynchophylline; st. u. for childhood dis. esp. convulsions (Jap); in dizziness - **U. sclerophylla** Roxb. (Mal, Indochin+): lvs. u. exter. as paste for sores; bk u. for dyeing yarns - **U. sessilifructus** Roxb. (Chin): twigs u. as sedative for children; in fever & irritability; also in abdom. pain & faintness - **U. sinensis** (Oliv.) Havil. (Chin): u. similarly to *U. rhynchophylla* - **U. tonkinensis** Havil. (U. sessilifructus Drake) (Indochin): inf. u. as "depurative"; bitter bk. to replace areca nuts in chew; considered antipyr.; for headache & lumbago.

Uncinia hamata Urban (Cyper.) (sMex to Arg, WI): bird-catching sedge; pl. u. as remedy in fever (Ven Indians).

uncomocomo: *Dryopteris athamantica.*

uncum root: fr. *Senecio aureus.*

undecylenic acid: CH_2: $CH(CH_2)_8COOH$; occurs in perspiration; u. in fungal skin infections, psoriasis; usually with zinc undecylenate in ointments.

undi (Hind): *Calophyllum inophyllum* .

undulated: with wave-like appearance; v. under ipecacuanha.

Ungernia minor N. (Lili/Amaryllid) (cAs): c. ungminorine (alk. with hypotensive activity) - **U. servertzovii** (Regel) A. Fedtsch. (Kazakstan, Uzbekistan, Armenia): pl. c. alks. ungerine, hippeastrine, unsevine (with hypotensive & spasmolytic activity; u. in folk med. for wounds & sores - **U. trisphaera** Vved. (cAs): c. lycorine, ungeremine, tazettine, pancratine, &c.; u. med. - **U. victoris** Vved. (cAs) c. galanthamine (alk.), hypotensive; cult. orn. tree.

Ungnadia speciosa Endl. (monotypic) (Sapind.) (sNM, Tex [US], nMex mts.): Mexican buckeye; Spanish b.; monilla (nuts); shrub or tree with thin light gray bk. & alternate pinnate lvs.; frt. a 3-lobed capsule with black lustrous sds.; sds. c. FO 50% (edible); HCN-glycoside; u. emet.; toxic; u. like marbles by children; frt. edible; honey pl.; ornam. (said formerly large exports to Ge).

unguent: Unguentum (L.): ointment - **U. Basilicum**: c. purified black pitch 14, rosin 14, yellow wax 14, olive oil 58 - **U. Fuscum**: brown o. (NF V) - **U. Glycerini**: starch glycerite. - **U. Matris**: U. Fuscum - **U. Nervinum**: 1) o. with dog fat (17th cent.) 2) comp. rosin o. - **U. Pediculi: U. Pediculosis**: stronger mercurial o. - **U. Refrigerans:** rose water o.: cold cream - **U. Tetrapharmacum**: U. Basilicum.

"Unguentine catus": *Aloe* spp.

Unguiculata: gp. of mammals with claws or nails; incl. Insectivora, Edentata, Primates, &c. (animal cohort).

Ungulata: ungulates, mammals with hooves; ex. zebras.

unha de vaca: 1) (Br): *Bauhinia forficata* 2) (nBr): uni: 2564; all *B.* spp.

unicate: a single specimen; solitary; growing alone.

unicorn: a mythical, one-horned animal; formerly supposed u. in med. (prob the narwhal tusk was actually employed) (2563) - **false u.** (root): *Chamaelirium luteum* (*Helonias dioica*) - **u. horn: u. plant**: **(true) u. root:** *Aletris farinosa.*

Uniola L. (Gram) (Am): sea oats; 4 spp.; a pasture grass - **U. paniculata** L. (seUS, WI, eMex, sea coasts): sea oats; dr. pl. decorative & previously widely collected; since the pl. serves to anchor the sand dunes & prevent erosion, it is classed as a protected sp. (Ga, Ala [US]) & cutting them is now a misdemeanor - **U. palmeri** Vasey (Mex): grains u. as food by Cocopa Indians.

unit: 1) a standard in weight & measure 2) (specif.): 500 cc (of blood or plasma for parenteral administration) - **Florey u.**: Oxford u - **international u.**: a definite wt. of standard subst. decided upon by the Permanent Commission on Biological Standardization (UNO) (abbrev. IU) - **Oslo u.** (for vit. D): 1.66 Oslo u. = 1 int. u. - **Oxford u.**: minim. amt. of penicillin which dissolved in 50 cc of meat broth completely inhibits growth of test strain of *Micrococcus aureus.*

United States Pharmacopoeia: USP: non-governmental compilation of drugs and medicines bearing the legal enforceable standards of strength and quality of medicinal agents u. in the US; a companion compendium of standards is the National Formulary (NF): formerly published at 10 yr. intervals, now at 5 (USP [I] [1820] - XXII [1990]).

Unonopsis veneficiorum (Mart.) R. E. Fries (Annon.) (Co, Ec): "choon" (Puinave); bk. & rt. of tree u. in arrow poisons (curare) (Co); contracep.; timber for house rafters (Ec).

unstectla (NA Indians): Spigelia.

upas tree: *Antiaris toxicaria.*

upland: higher regions or inland; v. under sumach.

uplate (Ind): uplet: ooplate: Costus, *Saussurea lappa.*

Uragoga Baill.: *Cephaelis; Psychotria* - **U. granatensis** Baill.: *Cephaelis acuminata.*

urao (Sp): hydrated Na_2CO_3.

urari: curare; fr. *Strychnos toxifera,* etc.

Uraria cochinchinensis Schindler (Leg) (Indochin): u. in local med. as "depurative" (Vietnam) - **U. constricta** T. C. Chen (Vietnam): inf. of boiled lvs. u. to combat cough - **U. cordifolia** Wall. (Ind, seAs): lf. inf. u. to kill larvae of fish pickling brine - **U. crinita** Desv. (s&seAs): rt. decoc. for diarrh. (Mal); for colic & fever in children (Camb); vermif. (Laos); crushed lvs. appl. for lice; cover crop, green manure; remedy for resp. oppression (Vietnam) - **U. lagopodioides** (L.) Desv. (eAs): odorbau (Nepalese); pl. paste heated in mustard oil & appl. to treat boils (727); ingred. in wash to treat hemorrhoids (Chin): lf. & rt. decoc. for dys. & resp. dis. (Vietnam) - **U. picta** Desv. (tropAf, As): entire pounded pl. u. for GC (Af); for snakebites (sInd) - **U. rufescens** Schindler (Ind, seAs): u. in local med. for certain skin dis. (Vietnam) - **U. vespertilionis** Bukhuisen f. (Ind, sChin, seAs): fl. inf. u. before meals to combat "forest fever."

Urceola Roxb. (Apocyn.) (seAs): caoutchouc vine; several spp. yield Borneo rubber (esp. in Malayan area) - **U. elastica** Roxb. (Borneo, Sumatra): tree trunk furnishes rubber; frt. edible - **U. esculenta** Benth. ex Hook. f. (Burma): Rangoon caoutchouc; caoutchouc vine; source of Penang and Borneo caoutchouc; frt. edible.

Urceolaria Hook. = *Lecanora*; **Urceolaria** Ach. = *Diploschistes* - **U. scruposa** (Schreb.) Ach. = *Diploschistes s.* - **urd** (Urdu): *Phaseolus mungo.*

urea: carbamide: $CO(NH_2)_2$; found in urine & other animal prods., results fr. protein catabolism; in large amts. in some fungi, in smaller amts. in higher pls.; first org. compd. synthesized (Woehler, 1828) (ob alcohol); much u. in chem. mfr. (barbiturates); in med. to aid healing processes of infected wounds & ulcers; skin cancer (1774); orally as diur. & to test kidney function; to incr. effectiveness of sulfonamides; u. hypertonic solns. (**sterile u.**) to reduce intracranial & intraoc. pressures (IV); in migraine; antisep.; in ammoniated dentifrices; in animal feeding to supplement low protein diets (esp. with molasses); penetrating moisturizer; fertilizer (2365).

urease: enzyme wh. accelerates change of urea into NH_3 & CO_2; present in jack beans (richest source); soy beans; many invertebrates; many bacteria (ex. *Micrococcus ureae*, wh. occurs in stale urine).

Urechites karwinskii Muell. Arg. (Apocyn.) (Mex, Guat, Salv): loroco; rts. c. loroquine as chief alk.; rts. toxic & u. as animal poison; fls. u. as spice; fls., buds, & younger tender shoots cooked with rice & eaten - **U. lutea** (L) Britt. (Pentalinon luteum) (s peninsula Fla, Keys [US], WI): wild allamanda; clavelitos (Cu); yellow nightshade(?); pl. toxic; cattle sometimes poisoned (circul. difficulties); sts. & lvs. u. to kill ants; dr. pd. lvs. u. to destroy wild dogs (Cu); u. like foll. in folk med. - **U. suberecta** Muell. Arg. (WI as DominRep, SA): yellow nightshade; Jamaica n.; Dominican viper tail; savannah flower; hb. lvs. c. glycosides "urechitin" & urechitoxin (cardiac poisons); u. in heart dis., (cardioton.), edema; latex very poisonous, u. to make curare poison (arrows); in fever, colic, as purg. (sometimes this sp. is considered conspecific with *U. lutea*).

Urecolaria scruposa (Schreb.) Ach. = *Diploschistes s.*

Urena lobata L. (Malv.) (cosmopol. in trop. as weed): Florida jute; "malva"; guaxima (roxa); cadillo (blanco); Indian mallow; pata perro (Ve); lvs. u. emoll. dem. vulner. in GC hemostat. contraceptive; for liver dis. (Ve); fls. u. as expect. & pectoral in cough (Trin, Br Indians); whole pl. decoc. u. in cough; bk. expressed juice u. in dys. (Caribs, Domin); sds. c. FO urease; 3% inf. sds. potent vermifuge (Br); st. furnishes a textile fiber, carrapicho, u. as jute replacement (claimed stronger than jute) (810) (cult.); lvs. appl. as poultice to ulcers (Pl): u. in Tonga med. (86) - **U. rigida** Wall. (Laos): kème onh (Laotian); pl. commonly u. in traditional med. (Laos) - **U. sinuata** L. (U. lobata L. var. s. L.) (cosmpol. tropFla [US], WI, SA): bur mallow; duck foot; wild cotton; u. similarly to *U. lobata* (Domin); lf. inf. in colds, cough, TB, flu (Trin) in dyspepsia (Trin) in syph.; rt. & lvs. as emoll. expect.; adulter. of patchouli lvs.; frt. edible.

Urera Gaud. (Urtic.) (warm Am & Af): pls. provided with powerful stinging hairs; one sp. ("take") is an important medicinal among the Cuna Indians (Pan) as counterirrit. in rheum. & backache - **U. baccifera** (L) Wedd. (sMex, WI, CA, SA): chichica(s)te; pl. u. in fainting & for all kinds of pain (myalgia, arthritis, pulled muscles, snakebite; stingray sting, fire ant sting); edema. int. hemorrhages (Haiti); rt. antihemorrhagic (Ec); lf. inf. in erysipelas (Ec); rt. decoc. GC (PR); lf. brushed against legs of children as punishment (Ec) (777); fiber u. to make ropes & twine; frt. edible - **U. cameroonensis** Wedd. (tropAf): obondji; cult. & lvs. u. as veget - **U. caracasana** (Jacq.) Griseb. (sMex, WI, CA, tropSA): ortiga (blanca); pl. u. in syph. (Mex Indians); u. to treat muscular pain & discipline children (Ec) (777); rts. boiled in honey & u. in gastralgia & as vermifuge (Yuc); lvs. rubbed over body in scorpion sting (Pan); lf. inf. in orange juice u. in infantile diarrh.; to cure effects of poison ivy (Mex) - **U. laciniata** (Goudot) Wedd. (CR to Ve & Pe): u. in myalgia & to discipline children (Ec) (777) - **U. obovata** Benth. (trop wAf): lf. inf. in dys.; lvs. & frts. boiled with pork for children's dis.; bast u. to make fishsnares.

urethan(e): $NH_2COOC_2H_5$; this amine is synthesized but also occurs naturally in port wine, &c.; u. as antineoplastic but partly superseded; as anesthetic for small animals; actually may be carcinogenic.

Urginea altissima Bak. (Lili.) (sAf): slangkop; toxic to livestock; fleshy bulb scales heated & appl. to gouty limbs & rheumatic swellings; pd. bulb. u. int. for bronchitis, asthma, hoarseness, flu; diur. dem.; substit. for squill - **U. burkei** Bak. (sAf): slangkop; stock poison; bulb rubbed over body for paralysis; aq. prepn. of bulb abort. & for circul. dis. (816); rat poison - **U. coromandeliana** Hook. f. (seInd, Coromandel Coast): bulb u. as substit. for squill - **U. delagoensis** Bak. (sAf): u. med. by Africans; to preserve animal hides against destruction by dogs - **U. indica** Kunth (Ind & Pak [Sind]; Af [Sen]): Indian squill; u. like & substit. for squill fr. *U. maritima* but inferior (811); weed - **U. maritima** (L) Bak. (U. scilla Steinh.) (Med reg., coasts of Eu, nAf, wAs): (Mediterranean) squill; white (or Greek) s.; sea onion; bulb c. glucoscillaren A; with emulsin hydrolyzes to scillaren A (chief glucoside); scillaren B; scillarenase (transforms Scillaren A to proscillaridin A); polyfructosans (sinistrins); bulb u. since the days of the ancient Egyptians (812); u. cardioton. in heart insufficiency (as in aging heart, mitral stenosis, cardiac edema); diur. expect. in congestive bronchitis, in bronch. & pulm. dis.; crushed bulb appl. to hemorrhoids (nIsr); pl. decoc. u. in fever (Ind) (124); red squill is a color var. with red bulb scales, c. scilliroside, scillaren A, "scillitin" (bitter subst.) (813); has similar cardiac activity as white var. but also high specific toxicity for rats; u. as rodent poison (esp. for rats); prod. Sp, Por, It, Mor - **U. rubella** Bak. (sAf): bulb decoc. drunk for colic (Basutoland); cardioton. props.; toxic to rats (—> *Drimia*).

uric acid: org. acid of purine gp.; colorless crystall. solid; chief end-product of avian & reptilian excretion; occ. also in sds. of some pls. such as *Melilotus officinalis*; obt. mostly fr. excreta of large snakes (zoos); u. diur.; in chem. mfr. (caffeine synthesis); reagent: Na & K salts (urates) deposited in joints in gout (Harnsaeure, Ge).

uricase: enzyme (dehydrogenase); found in liver, spleen, & kidney of most lower mammals (not in primates); converts uric acid to allantoin (in presence of O_2); u. in detn. of uric acid.

uricury: ouricury.

urine: chief liquid excretion of mammals; formed by processing of blood by kidneys; c. 25% urea, salts, org. acids, pigments, hormones, water; much u. in med. folklore in all parts of world; ex. Miss (US); for fungus inf. (Ind) - **cow u.** (general well-being); Yuc (in bath water for colds; int. in childbirth to expel placenta; as wash in fever; in ear for otalgia; in poultices, etc.) (6) - **human u.**: claimed to have medl. values as general ton. (Ind, Chin) (2366); exter. for snakebite (ancient Israel); urine of other animals also u. (ex. elephant, sheep, rabbit) (1668).

uriza (Br): *Pogostemon heyneanus*.

urobilin: class name of series of bile pigments wh. occur in feces & urine; represent dipyrrylmethane compds., forming pyrrolenone or pyrrolidone rings; u. as color reagent for Cu, Hg, Zn; precursor = urobilinogen.

Urocystis Rabenh. (Ustilagin. - Ustilaginales - Basidiomycot. - Fungi): smuts which are pathogenic on various plants, esp. grasses; ca. 60 spp. - **U. occulta** (Wallr.) Rabenh.: causes leaf smut of rye; med.

Urofollitropin (systemic) (USP XXII): folliculotropic hormone.

urogastrone: polypeptide hormone isolated fr. urine; appears identical to epidermal growth factor (EGF), hence now designated EGF-urogastrone; thus both inhibit gastric acid secrn. (like enterogastrone) & stimulate epithelial cell proliferation.

urokinase: proteolytic enzyme in blood & urine (1869) wh. activates plasminogen to form plasmin, wh. dissolves blood clots; u. in thrombolysis (to dissolve blood clots) in lungs (pulmonary embolism); also by direct inj. into coronary artery to remove clot & restore circ. in heart (in heart attack); active per ora.

uronic acid: a unitary component of gums & mucilages, in combination with various monosaccharides; 3 chief compds. are: 1) glucuronic (or glycuronic) acid: occurs in Acacia (& other pl. & animal prods.) 2) galacturonic acid: found in mucil. of elm & linseed 3) mannuronic acid: found in algae, as pectin-like subst.

uroporphyrins: several metal-free porphyrins (large colored heterocyclic compds.) found in normal urine & feces.

Urospatha antisylleptica R. E. Schultes (Ar) (Co): unripe dr. spadices (inflorescences) u. orally as contraceptive (natives) (Barasana Indians) - **U. caudata** Schott (Br): juice of rhiz. u. in skin dis. - **U. somnolenta** R. E. Schultes (Co): pl. ashes u. as poultice for ulcers & infected cuts.

Urospermum picroides (L) F. W. Schmidt (Comp.) (sEu, Med reg.): hb. c. 2 flavonoids (1775); u. exter. in inflammns., int. for stomach pain.

Urostigma Gasp. = *Ficus* L.

Ursinia abrotanifolia Spreng. (Comp.) (sAf): lf. c. VO; u. for cough - **U. tenuiloba** DC (sAf): hot milk decoc. of rt. drunk as cough remedy (Zulus).

ursode(s)oxycholic acid: ursodiol; found in bear bile in large amt. (hence name ursus, L., bear) (source) also human; cholagog; u. to eliminate gallstones (cholesterol type) (popular in Jap); u. perorally.

ursolic acid: urson(e); prunol; complex crystall. OH-cyclic triterpene acid; found in waxes of cuticle of many frts. (ex. apple [source]), prune, grape, cherry, cranberry, pear, uva ursi; also lvs. (of *Arctostaphylos uva-ursi*); u. as emulgent (Ca & Na salts) in pharmaceuticals & foods.

Ursus L. (Ursidae - Mammalia) (worldwide): the bears; omnivorous - **U. americanus** Pallas (Euarctos a.) (OW; NA:Alas-Fla): (American) black bear; cinnamon & brown vars.; gallbladder u. med. (Chin); considered very valuable u. as tranquilizer & sed. in heart dis., epilepsy, &c. ($200/g.); meat & fur utilized; body fat utilized in folk med.; blood u. in osteoporosis - **U. arctos** L. (U. horribilis: U. arctos var. horribilis): (northern NA): grizzly bear; big (or Alaskan) brown b.; fur, hide, & flesh (meat) utilized: fat formerly sought & used (*cf. Euarctos*) - **U. maritimus** Phipps (Thalarctos maritimus) (arctic & circumpolar reg. of NA, Euras): polar bear; fur & meat used.

Urtica ballotaefolia Wedd. (Urtic.) (Ec): ortiga; caballo china; rts. component of med. much favored in pneumonia (fide W. Camp) - **U. cannabina** L. (Euras) (temp Siberia, Iran): hemp (leaved) nettle; fiber prod. fr. sts. - **U. dioica** L. (Euras cosmop; lacking in c&sAf): (great) stinging nettle;

(common) n.; big sting nettle; Swedish hemp; stinging hairs c. acetylcholine, histamine (also in lvs.); serotonin, HCOOH, acetic acid; lvs. c. chlorophyll (rich), tannins, vitamins (B complex, C), silicic acid; KNO_3; hb. u. as diur. ton. "depurative" ("blood purification"), hemostat.; in gout, rheum. diarrh., uter. & other hemorrhages, pleurisy, icterus, anemia; expect. in chronic bronchitis, in cough & shortness of breath; for hemorrhoids, skin dis.; sexual stim.; in early prostatitis; exter. in hair washes, &c.; for mfr. chlorophyll, yellow & green dyes (camouflage-WW II); fibers u. in Middle Ages to make fabrics (nettle cloth); young lvs. eaten as veget. (1777); u. in folk med. (1776); rt. c. tannin; u. astr. in diarrh. & int. dis; hair washes; to stim. hair growth ("nettle hair washes"); sds. u. by Eu homeop. in rheum., hair & skin dis., (washes); var. *angustifolia* Schlecht. (U. breweri S. Wats.) (wNA): fiber u. in fishnets; var. *gracilis* Taylor & MacBryde (U. gracilis Ait.) (Cana & nUS): common American nettle; close to var. *procera* Wedd. (dr. lvs. u. for nosebleed [Abenakis]) (814).; u. similarly to subsp. *dioica*: likewise var. *holoserica* C. L. Hitchc. (U. h. Nutt.) (wUS, esp. Calif, nwMex): common (or hoary) nettle; big stinging n.; fresh branches u. to strike body in arthritis, rheum.; lf. decoc. in colds; rt. decoc. in baths for rheum. limbs; u. in sweat baths for grippe & pneumonia (Calif, Span, & Ind people); var. *lyallii* C. L. Hitchc. (U. l. S. Wats.) (Alas to Ore [US]): Lyall's nettle; u. in rheum. (folk); bast u. as source of fiber (Indians of wWash [US]); u. for fishlines, &c. - **U. ferox** Forst. (NZ): tree nettle; bast inf. u. to treat eczema & VD - **U. lobulata** E. Mey. (sAf): u. in indigenous med. for snakebite & ophthalmia - **U. massaica** Mildbr. (tropAf): lvs. eaten boiled by natives - **U. membranacea** Poiret (Med reg., nAf): hb. (Herba Urticae, [Por]) u. as emmen. & aphrod. (also sd.) - **U. mexicana** Liebm. (Mex): rt. u. as diaph. (815) - **U. nivea** L. = *Boehmeria n.* (L) Gaud. - **U. pilulifera** L. (Med reg., sAs): Roman (or burning) nettle; hb. u. like & substit. for *U. dioica* hb. (Sp & elsewhere); sd. earlier known as Semen Urticae romanae; appl. exter. to wounds (nIsr) - **U. spat(h)ulata** Smith (SA): rupá (chico); ortiga chica (Arg); lvs. u. int. as diur. in hematuria, dys. skin dis.; stinging hairs much more toxic than European spp. - **U. thunbergiana** Sieb. & Zucc. (Jap): ira-kusa; young lvs. & sts. eaten as veget.; older sts. source of fiber - **U. urens** L. (Euras, natd. NA, now cosmop.): dwarf/burning/dog nettle; u. similarly to *U. dioica*; pl./lvs. inf. u. as compress for burns; snuff for epistaxis; excessive menstr.; rubef. (Cu) to irritate limbs with paralysis, rheum., &c.; in impotence; aphrod. (Af), as food relish (Af); in fish poisons; bk. inf. as galactagog (sEu; Af): hemostat. in uter. & int. hemorrhages; burns; local anesth. (nIsr); sds. diur; rt. stim.

Urticaceae Juss. (worldwide trop. & temp.): nettle fam.; some 52 genera and 1,050 spp.; some with stinging hairs; certain authors include it in Ulmaceae & at times in Moraceae.

urtication: med. treatment by application of nettles to irritate skin surfaces, as counterirrit., etc.

Urtinktur (Ge) (lit. original tincture): 10% alc. tincture fr.: 1) fresh plant juices 2) dried fls. or pl. parts 3) soln. or dilution of same.

urucú(m) (Br): **urucuuba**: *Bixa orellana.*

urushiol: collective name for a series of toxic phenolic compds. found in mixt. as an oily subst. in several *Toxicodendron* spp.; formerly called toxicodendrol; believed chief cause of dermatitis in rhus poisoning (fr. *Toxicodendron radicans, T. Toxicarium, T. vernicifluum*, &c.); represent catechol derivs. with single long side-chains; very strong allergen; sometimes u. as poison ivy desensitizer.

Urwald (Ge): (primeval) forest.

Usnea Dillen (Usne. - Discolichenes - Lichenes) (cosmop.): old man's moss; (old man's) beard lichens; pendulous or erect lichen mostly found growing on trees or shrubs; some spp. u. as hair pd. (Ind); pls. dr. & stored in boxes for use as kind of towel (BC [Cana] Indians) - **U. barbata** (L.) Wigg. (Eu, NA): bearded (or drooping) usnea; tree hair; c. usnaric & barbatolic acids; u. antibiotic (Homeop.); dem. expect.; as red, orange, purple dyes; occurs on bk. of trunks & branches of trees & on dead wd. - **U. dasypoga** (Ach.) Roehl. (c&nEuras): c. usnic acid; antibiotic, u. in med. by Java natives - **U. diffracta** Wain. (cosmopol.): c. usnic acid; u. as anti-inflamm. drug (817) - **U. florida** Wigg. (Eu, NA, cosmop.): flowering (or Florida) usnea; occ. on bk. of trunks & branches of mostly broad-leaved trees & on dead wd.; described by Dioscorides as "Bryon"; c. usnaric & lobaric acids; u. to prep. antibiotics (esp. for TB), (u. in combn. with streptomycin, PABA, &c.); in pertussis; as bitter ton.; to prep. usnic acid - **U. hirta** (L) Wigg. (Euras): mostly on bk. of pines; c. usnic & thamnolic acids; u. in blenorrhagia, hemorrhage; diarrh. dys.; dyspepsia: with antibiotic props. - **U. lobata** Hoffm. (cosmop.): earlier u. as ton. bitter, astr.; in pertussis, hemorrhoids - **U. longissima** Ach. (Euras, nNA): maidenhair moss; old man's beard; c. barbatinic acid; protocetraric acid; u. as expect., in night sweats & for local treatment of sores (Chin); absorbent for wound dressing, baby diapers, sanitary napkins, & to wipe salmon (not washed in water since shrinks skin) (BC Indians) - **U. philippina** Wainio (PI): pl. decoc. (red) u. as stom. (140) - **U. plicata** (L) Wigg. (Lichen plicatus L.) (nUS s to Gulf of Mex, Calif [US], Eu): pleated (or plaited) usnea; grows on coniferous trees & old wd.; u. bitter ton. & in whooping cough; astr. stypt. vulner.; u. like *U. florida*; source of green & yellow dye, for woolens (Eu).

Usnea (L): **Usnea Cranii humani** (Muscus cranii): "moss" on human skull of executed person left hanging for long periods; popular med. in medieval & renaissance periods in Eu; u. in hemorrhages, epilepsy, &c.; of uncertain identity but proposed to represent lichens *Parmelia saxatilis* (L) Ach. or *P. omphalodes* (L) Ach.; or possibly *Usnea* spp. or even mosses.

usnic acid: **usninic a.**: occ. in many lichen spp., esp. *Cladonia & Usnea;* antibacter. antibiotic u. in inflamm. skin & mucosa dis., as TB; good results in skin dis. (818).

Uspulun™: hydroxy-mercuric-chlorophenols; u. since 1914 as fungicide, disinfect. for sds.; c. in Semesan: also u. in rubber tapping.

Usteria guineensis Willd. (Logani) (monotypic) (trop w&sAf): frt. decoc. given to infants with cough (Lib); inf. of branches for infants with fever (Sen).

Ustilago maydis (DC) Corda (U. zeae Unger) (Ustilagin. - Basidioiomyc. - Fungi) (sNA, Eu): corn smut; c. ergot; maize smut; forms parasitic growth on *Zea mays*; c. ustilagic acid (antibiotic subst.); "sclerot(in)ic acid"; trimethylamine; ustilaginine (older lit.); u. in dysmenorrh., menorrhagia, hemorrhage fr. uterine fibroid tumor; alopecia, seborrhea, nail growth disturbances; as an obstetric aid, ecbolic; hemostat. like ergot (mostly in Eu); in NA, mostly by Negroes (who also use for abort. [in whisky]); sometimes fatal to stock (cattle+); consumed as food & considered delicacy (cuiteche) (Mex).

Utricularia aurea Lour. (U. flexuosa Vahl) (Lentibulari.) (s&seAs): crushed pls. u. for epidermal irritations fr. summer heat & for eruptions of pustules (children) (Indochin); in urin. dis. (Nepal) - **U. bifida** L. (eAs): u. as antipyr. & for swellings; in urin. dis. (Ind) - **U. uliginosa** Vahl (eAs): u. like *U. aurea* - **U. vulgaris** L. (Euras, now natd. NA): bladderwort; insectivorous pl.; pl. u. as diur.

uva (Sp, Por, It): grape - **Uva Passa** (pl. **Uvae Passae** [L]): raisin(s) - **U. P. Minor(es):** currant(s) - **u. d'orsa** (It): **u. orsina:** Uva Ursi - **Uvae Ursi Folia** (L): lvs. fr. Arctostaphylos uva-ursi.

Uvaria acuminata Oliv. (Annon.) (Tanz): lf. decoc. in cough; rt. decoc. for dysmenorrhea, dys., chest dis., snakebite; frt. edible, u. for stomachache - **U. afzelii** Scott Elliott (w tropAf): frt. edible; lvs. u.in fevers, cystitis, nephritis, as purg. (Lib) - **U. caffra** E. Mey. (sAf): rt. decoc. drunk for cough (Venda) - **U. calamistrata** Hance (nVietnam): branches with lvs. sent to Chin for use like yeasts in fermentation - **U. catocarpa** Bak. (Mad): senasena; sds. stim.; u. in pharm. prepns - **U. chamae** P. Beauv. (trop wAf): rt. ("finger root") u. as purg. antipyr.; rt. bk. u. in dys. & in resp. dis. (Ghana); prep. u. as eye wash; st. bk. c. uvaritin & iso-u. (benzyl flavanones) (cytotoxic); lf. juice u. for wounds, sores, lf. inf. for injuries, for ophthalmia; frt. as ingred. for children's fevers - **U. dulcis** Dunal (Ind, seAs): rt. bk. astr. stim. alter - **U. kirkii** Hook. f. (Tanz): fresh lvs. crushed & juice drunk for mental dis. - **U. leptocladon** Oliv. (trop eAf): rt. decoc. in combn. u. in epilepsy of children; as diur., in GC, dys., for sunstroke, tonsillitis; frt. edible (Tanz) - **U. narum** Bl. (Ind): rt. and oil fr. rts. u. in various dis.; arom. - **U. purpurea** Bl. (U. grandiflora Wall.) (Mal+): cooked lvs. eaten for flatulence; lf. & rt. decoc. u. for stomachache & as postpartum protective med.; lvs. as poultice for skin dis, &c.; sds. in cough; freshly cut rts. with odor of eau de cologne: lvs. arom. eaten (Chin); frts. edible - **U. rufa** Bl. (tropAs): alc. tinct. of rts. u. as ecbolic (PI); frts. edible & very popular.

Uvularia grandiflora Sm. (Lili.) (eCana, e&cUS): large-flowered bellwort; merry-bells; hb. u. as ton. dem. for snakebite; in ringworm - **U. perfoliata** L. (eNA): bellwort; wood daffodil; hb. u. like prec.; ornam. hb. - **U. sessilifolia** L. (eNA): bellwort; little merrybells; straw-lily; young shoots substit. for asparagus; rts. edible when cooked; emergency food plant; rt. inf. u. for backache, in ointment for sore muscles & tendons (Potawatomi Indians); pl. u. as ton. dem. nervine, hepatic (whites).

uyama (Co): squash.

uzara (Af): *Xysmalobium undulatum* R. Br.; name was formerly applied to *Schizoglossum shirense* N.E. Br., certain *Asclepias* spp., *Gomphocarpus fruticosus & G. physocarpus* and to *Dicoma anomala* Sond. (misnomer).

uzarigenin: aglycone of **uzarin** = cardioactive digitaloid glycoside made up of uzarigenin + D-glucose (2 mols.); occ. in *Xysmalobium undulatum.*

uzaron: rt. ext. (with alc.) of *Xysmalobium undulatum*; represents crude tot. glycosides formerly u. in diarrh.; in many medl. specialties (Eu).

Vv

vacca (L, Sp.): cow.

Vaccaria pyramidata Medik. (V. segetalis Garcke; V. vulgaris Host.; Saponaria vaccaria L.) (monotypic) (Caryophyll.) (nEuras, natd. NA): cow herb; c. cockle; popular forage for cattle; pl. u. as galactagog for cattle; frt. (sd.) u. diur; in ointment for eczema, parapsoriasis; styptic & anodyne for cuts; appl. to boils, scabies (Chin); rt. c. alk. saponin, "lactosin" (polysaccharide), rich in starch; sd. (frts.) c. glycoside, toxic to cattle (sAf where pl. is a weed); source of starch (1779).

vaccenic acid: trans-11-oleic acid; occ. in butter fat & other FO; less rancidity tendency than oleic acid; u. in cosmetics, foods, &c.; promotes growth in rats.

Vaccina Typhosa et Paratyphosa (L.): typhoid & paratyphoid combined vaccines.

vaccine: a suspension of living, dead, or retarded organisms wh. serves as an antigen for conferring active immunity; ex. smallpox v., rabies v. - **anti-smallpox v.**: smallpox v. - **cholera v.**: sterile suspension of killed strain(s) of *Vibrio cholerae;* u. to confer active immunization (short-lived; must repeat in 6 mos. or less) - **enteric v.**: typhoid v. - **Jennerian v**: smallpox v. (SC Jenner first prep. prod.) - **meningococcal polysaccharide v.**: antigen prepd. fr. capsule polysaccharides of meningococci; u. during epidemics, &c. - **mixed enteric v.**: typhoid & paratyphoid vaccines - **plague v.**: a sterile suspension of killed p. bacilli (*Pasteurella pestis)* of a strain with high antigenic activity; u. parenterally for active immunization of bubonic plague (or "black death") - **pneumococcus v: pneumococcal polysaccharide v.**: a multivalent v. with capsular polysaccharides of *Streptococcus pneumoniae* (23 types), chief causes of pneumococcal dis. in US; u. to immunize in nursing homes, &c. - **rabies v.**: v. rabies - **smallpox v.**: v. smallpox - **typhoid v.**: v. typhoid - **t. & paratyphoid vaccines**: typhoid combined vaccine: containing suspensions of killed organisms of typhoid & of paratyphoid A & B; u. for active immunization against the respective infections - **typhus v.: epidemic t. v.**: sterile suspension of killed rickettsial organisms of strain(s) of epidemic type selected for high antigenic efficiency; obt. fr. cultures on egg yolk sac; u. for immunization against t. fever - **v. virus (glycerinated)**: smallpox v. - **yellow fever v.**: living culture of attenuated strain of yellow fever virus selected for high antigenic efficiency & safety.

Vaccinia (L): cow pox.

vaccinia immune (human globulin): sterile soln. of globulins derived fr. blood of human donors immunized with vaccinia virus (smallpox vaccine); it can be used both as immunizing prophylactic and in therapy of smallpox.

Vacciniaceae S. F. Gray: Heath Fam.; mostly shrubs or small trees; fls. 4- or 5-parted; calyx attached to wholly or partly inferior ovary (chief distinction fr. Ericac. Fam. wh. usu. has superior ovary); petals united; 8-10 stamens; several chambered ovary develops into true berry; frts. rich in vitamins; ex. blueberry, huckleberry, cranberry; most floras maintain this as integral part of Fam. Ericac. rather than separate.

Vaccinium angustifolium Ait. (incl. var. *laevifolium* House) (Eric./Vaccini) (NA, Arctic to neUS, s to Va): low (bush) blueberry; "low sweet"; "early sweet"; sugarberry; important comm. source of blueberries on market; frt. esculent, antiscorb. diur.; lvs. & bk. astr. - **V. arboreum** Marsh. (Batodendron a. Nutt.) (seUS, n to Mo): farkleberry; sparkleberry; tree huckleberry (shrub to small tree -9 m); "huckleberry trees"; winterberry (Clay County, Ala [US]); said the most widely distrib. sp.; lvs. astr. u. in diarrh.; rt. decoc. u. as astr. in diarrh. intestinal dis.; diur. (rts. & berries as tinct. in gin); bk./rts. decoc. u. as gargle & wash for sore throat, oral abscesses, stomatitis, cough, &c.; in psoriasis; cont. no arbutin (819); black frt. eaten in pies (for "huckleberry pies") jellies, cookies, wines, &c.; wd. u. to make tool handles; canes (820) - **V. arctostaphylos** L. (Bulg, Turkey in Europe; AsMin, wCaucasus): Caucasian tea; dr. & fermented lvs. u. as substit. for Chin tea; called Broussa or Kutai tea - **V. caespitosum** Mx. (NA, esp. Ore [US]): dwarf blueberry; frt. edible - **V. corymbosum** L. (eNA: NS [Cana] to SCar [US]): (black) high bush blueberry; lvs. c. neomyrtillin, producing hypoglycemia; of possible value orally in diabetes (824); food - **V. crassifolium** Andr. (seUS): creeping blueberry; frt. edible - **V. erythrocarpum** Mx. (Hugeria erythrocarpa Small) (seUS: WVa-NCar): (southern) mountain cranberry; dingleberry; bearberry; frts. variable in flavor; sought after in making jellies - **V. hirsutum** Buckl. (NCar, Ga, Tenn): hairy (or bear) huckleberry; frt. edible - **V. macrocarpon** Ait. (e&cNA-bogs): large (or American) cranberry; frt. c. oleanolic & ursolic acids; u. same as & generally preferred to *V. oxycoccos*; widely cult. for food since 1840 (NJ, Mass, Wisc [US]) (828); consumed cooked or as jelly (eaten with chicken); cranberry juice u. as popular beverage; claimed of value as urin. antisep. in chronic pyelonephritis (to lower urinary pH); cystitis, & other UTI; to deodorize urine (821): source of ursolic acid & derivs (methyl; calcium) u. as emulsifying agt. in pharmaceuticals, foods (822) - **V. membranaceum** D. Douglas (Ore [US] to BC [Cana]): black blueberry; big whortleberry; frt. edible, important crop of Klamath Indians - **V. myrtilloides** Mx. (V. canadense Kalm) (Cana, nUS s to Va): sourtop (or velvet leaf) blueberry; "Fraser Valley" b. (BC); Canada b.; "sour top"; large berries esculent; sold in stores & canned - **V. myrtillus** L. (nEuras; Balkans-mts.): bilberry; (European) whortleberry; blaeberry; huckleberry; lvs. weakly antidiabetic (neomytrillin); u. in mild diabetes (folk med.) (824); bladder weakness; vomiting; cough; exter. in eye & mucosal inflammn., skin dis. (cranberry poultice in erysipelas) (823); lvs. (as Folia Myrtilli) c. quercetin & its glycosides; 6-20% tannins; myrtillin-a (glucoside of delphinidin); anthocyanoside complex fr. fresh frts. marketed; no arbutin; u. astr. in cystitis, bronchitis, diarrh., hemorrhage; as beverage tea; frt. (as Fructus Myrtilli) c. myrtillin-a, 1.5% arbutin (ericolin) (none in unripe berries), 5-12% tannin, fruit acids, sugars, pectin; u. as attractive food (dr. bilberries imported fr. Russ [825]); in blueberry wine & spt.; in preserves, muffins, pies, &c.; juice to color red wines; u. astr. in diarrh., dys. pinworms; exter. in eczema, gum dis.; in degenerative retinal conditions; antiscorbut. (Eu); in prepn. Extr. Myrtilli; in prepn. blue, purple, brown dyes; entire bush as tan. - **V. ovalifolium** J. E. Smith (Pac nwUS, Alas, Cana): oval-leaved bilberry/blueberry; mathers (Newf [Cana]); frt. u. as food (fresh & preserved) (rel. sweeter) - **V. ovatum** Pursh (Pac coast, NA): evergreen/winter/Calif. huckleberry; winter blueberry; shrub 2-3 m high; frt. very sweet, u. in preserves, pies, jams, jellies, &c.; in houshold decorations (Calif, Ore [US]) (evergreen lvs.) - **V. oxycoccos** L. (nEuras, natd. NA): (small/bog) cranberry; frt. c. vacciniin, keto acids; eaten after cooking with sugar; considered inferior to *V. macrocarpon*; or eaten raw (USSR); antiscorb. frt. juice delicious u. as cooling agt., to replace lemon juice in punches; lvs. c. arbutin, cannivorine (alk.); lvs. u. to prep. beverage tea; cranberry juice ("cocktail") considered of value in urinary infections; var. *intermedium* Gray (Oxycoccus palustris Pers.): marshwort - **V. pallidum** Ait. (V. alto-montanum Ashe) (seUS n to Maine): (blue) huckleberry; (low bush) blueberry; delicious frt. eaten & canned; u. in pies, muffins, tarts, assorted pastries - **V. parvifolium** Smith (Pac US, fr. AleutianIs to Calif): red bilberry; r. huckleberry; berries red, esculent - **V. scoparium** Leigberg (wNA fr. Alas to Calif, e to Utah [US]): grouseberry; grouse huckleberry; dwarf red whortleberry; frt. light red, sweet & tasty - **V. stamineum** L. (eUS s to Fla, sCana): deerberry; squaw huckleberry; "gooseberry"; frt. variable in flavor, sometimes palatable; also u. as red dye; ornam. shrub - **V. uliginosum** L. (nEuras, natd. nNA): alpine blueberry; bog bilberry; frt. (as Fructus Uliginosi) c. benzoic acid, free & esters; u. (brandy ext.) in gastritis, enteritis, cystitis; frt. juice u. in colic, &c.; popular food (nwIndians); lvs. c. hyperoside, no arbutin, ursolic acid; formerly u. like berries in med. & as adulter. of uva ursi; subsp. *occidentale* Hultén (V. occ. Gray) (BC [Cana] to Calif, e to Nev [US]): western blueberry (or huckleberry); western bog h.; swamp h.; frt. edible, raw or cooked - **V. vitis-idaea** L. subsp. *minus* Hultén (Cana, nUS) (typical var. in Eu, esp. Sweden): mountain (or rock) cranberry; cowberry; (red) whortleberry; ling(on) berry; lingen; low creeping shrub; lvs. c. 4-9% arbutin, pyrogallol tannin, hydroquinone; u. like & as replacement for uva ursi, in lithiasis, kidney dis., gout, rheum; for diarrh.; in diabetes (value); formerly for colds & chronic cough (folk med.); fls. u. in rheum. dis.; frt. red & rather bitter, c. aliphatic & arom. acids, sugars, pectin, vit. C; dr. frt. u. as astr. in diarrh., hemorrhages; in cooling drinks, as seasoning for meats; as food in preserves popular in Ge & Sweden; jellies; prepn. of pleasant brandy; entire pl. u. as tan.

Vaccinum Febris Flavae (L.): yellow fever vaccine - **V. Pestis**: plague v. - **V. Rabies**: rabies v. - **V. Typhosum**: typhus v. **V.Variolae**: smallpox v.

vacuna (Sp): vaccine virus; or vaccine (generally) - **vacuno** (Sp): pertaining to cows - **vacunos** (Arg): cattle in general.

vagem (Por, Br): leguminous pl.; green beans.

Vagnera Adans.: *Smilacina.*

Vahlia capensis Thunb. (Bistella c.) (Vahli./Saxifrag.) (sAf): u. as remedy for sore eyes (with *Wormskioldia longipedunculata*).

vaid (Indo-Pak): medical man (& pharmacist) who follows the Ayurvedic system of medicine.

vainilla (Sp): Vanilla.

valepotriate: collective name for series of lipophilic monoterpene derivs. with iridoid basic skeleton fr. valerian; u. as tranquilizer, sed.; popularity increased demand for valerian rt. in recent past.

valeranone: jatamansone: satd. bicyclic sesquiterpenoid ketone, occ. in VO of *Nardostachys jatamansi* & *Valeriana officinalis*; u. sed. antiulcer.

valerian: *Valeriana* spp., usually *V. officinalis* - **African v.**: 1) *V. officinalis* v. (*V. capensis*) 2) *Fedia cornucopiae* - **American v.**: ladies' slipper rt., *Cypripedium pubescens* & other *C.* spp. - **Belgian v.**: *V. officinalis*, named for country of chief prodn. - **cat's v.: common v.**: *V. officinalis* - **Eastern v.**: Japanese v. - **English v.: European v.: great wild v.**: common v. - **Greek v.**: *Polemonium caeruleum, P. reptans* (also **American G. v.**) - **Indian v.**: *V. wallichii* - **Japanese v.:** *V. faurioi* - **large v.**: *V. phu* - **officinal v.**: common v. (& Indian v.) - **v. oil**: VO dist. fr. *V. officinalis* rt.; u. sed., tobacco flavor, etc. (chief prodn. Fr, Eng, Ge, Neth) - **v. root**: fr. common v. - **swamp v.**: *V. dioica.*

Valeriana L. (Valerian.) (Euras, Af, Am): valerians; ca. 250 spp. of perenn. hbs. with strong smelling rts; some Am spp. u. as food (Am Indian) (name <valeo, valere [L] to be strong; or Publius A. L. Valerianus, Roman emperor) - **V. adscendens** Turcz. (Pe): hornamo morado; u. as strong purg. (827) - **V. angustifolia** Tausch. = *V. fauriei* Briq. - **V. capensis** Thunb. (sAf): Cape (or African) valerian; u. much like *V. officinalis* in epilepsy, hysteria & nervous conditons; also in typhoid fever, as vermif., diaph.; VO u. as antispasm. - **V. celtica** L. (Nardus c. auct.) (Eu-Alps): (Celtic) nard; "golden n."; Spica celtica; Celtic spikenard; echter Speik (Ge); rhiz. c. 0.4-1.5-1.8% VO with odor more like that of patchouli than of common v. (829); c. valtrate; fls. with coumarin-like aroma; rhiz. u. as nerve "tonic" (popular med.), in dys.; to "strengthen" the heart; mostly now in cosmetics, to perfume soap ("Speikseife" [Ge] = spikenard soap) (therap. values claimed); for incenses; replacement for *Nardostachys jatamansi*; to aromatize baths (It) - **V. ceratophylla** HBK (Mex): rhiz. offic. in Mex Phar. ("raiz del oso"); rhiz. boiled, liquid turns red, take 1 cup at night for nervousness & sleeplessness (Ciudad Juarez, Mex) - **V. dageletiana** Nak. (Kor): u. for neuralgia, leprosy - **V. dioica** L. (Euras swamps): (small) marsh valerian; (bastard) v.; rhiz. (as Radix Valerianae palustris) u. like *V. officinalis* for wh. sometimes u. as adulter: weaker action (830) - **V. edulis** Nutt. (BC [Cana], Wash, Ore [US], nwMex): tobacco root; rhiz. c. starch, & are cooked & eaten (Klamath Indians + [Or]) - **V. fauriei** Briquet (eAs, Jap, Chin, Kor); Japanese valerian; kesso root; rhiz. c. 3-7-8% VO (produced) with bornyl acetate & isovalerate, kessyl alcohol & acetate, &c. (with unpleasant flavor; once thought toxic); valepotriate; tannin; u. similarly to *V. officin.*; said being u. in US 1918-9 (831) - **V. flaccidissima** Maxim. (Jap, Kor): u. as sed. (eAs) - **V. hardwickii** Wall. (Ind-Himalaya-Pak; seAs): "nikan"; rhiz. u. antispasm. carm. stim. diaph. aphrod. (Chin, Jap); in hysteria, epilepsy, chorea, shell shock, neurosis (Nepal): suggested as replacement for *V. officinalis* (54); u. similarly to foll. - **V. jatamansi** Jones (V. wallichii DC) (Ind, Pak-Himal): Indian valerian; Indische Narde (Ge); to be distinguished fr. *Nardostachys jatamansi* (entirely different VO); "spike nard"; rhiz. (Rhiz. Valer. Indicae) similar to that of common v. but with less VO (0.35-1%); & has been allowed as substit.; said to have more agreeable taste than *V. offic.*; VO with isovaleric, acetic, capronic acids & esters; also present linarin isovalerate (flavonoid), valepotriate; u. in urin. dis. & incense (Kumaon Hills); recommended as substit. for imported *V. officin* (54) (off. in US & Br Phar.) (2252) - **V. latifolia** = *V. officinalis* var. l. Miq. = *V. fauriei*) - **V. leschenaultii** DC (V. brunoniana W. &A.) (sInd: Nilgiris & other mts.): Indian valerian; recommended as substit. for European v. (832) - **V. mexicana** DC (Mex): Mexican v.; rhiz. c. only traces of VO; valtrate (5%); u.

like *V. offic.* as substit. but generally considered inferior - **V. microphylla** HBK (Ec): rt. of this pungent aromatic pl. u. as heart stim. & in nervousness - **V. officinalis** L. (Euras): (common) valerian; (great) wild v.; officinal v.; dr. rhiz. & rts. c. 0.5-1.7% VO with borneol & bornyl isovalerate & other esters; isovaleric & valerenic acids; valepotriate (collective term); alkaloids: valerine, valerianine, chatinine; valerian tannin; a-pyrryl-methyl ketone (833): good quality rts. should be light to brownish yellow, with strong odor of borneol esters & none of isovaleric acid (834): very attractive to cats; rhiz. (& VO) u. sed. antispasm. in hysteria, nervous unrest, hypochondriasis; emotional states; insomnia; carm. in colic; in asthma, stomach-intestinal dis.; uter. sed. in menstrual irregul. (esp. VO) (1781); for VO prodn (esp. Neth); component of many prepns. (tinct. ext., &c.), esp. proprietary items; popular in herbal teas; hb. u. as condim. (baker's hartshorn [Eng]); VO to flavor tobacco (835); prod. Belg (unavailable during WW II, then replaced with Ind v., Ge, Fr, Eng) - **V. paniculata** R. & P. (Guat to Pe, Arg): valeriana (del pais); rt. decoc. antispasm., sed. for nervous dis.; rt. inf. u. for weak, nervous children (Arg); fls. u. med. (Par); lf. decoc. with *Cnicus benedictus* & latex of lettuce to lower blood pressure (Guat) - **V. pavonii** Poepp. & Endl. (nVe to Pe): kiniyapa; hb. decoc, u. as bath for fever (Ve Indians) - **V. phu** L. (Siber, Ural Mts, Cauc): (great) garden valerian; (large garden) v.; rhiz. (as Radix [or Rhizoma] Valerianae majoris) u. as substit. for & adulter. of *V. officin.* - **V. procera** HBK (Mex): hierba del gato; richer in VO (3%) than *V. officin.*; u. similarly to "foreign" valerian; recommended for female nervousness (835); in Mex Phar. - **V. pyrolaefolia** Dcne. (Ind-high mts.): recommended as replacement for Eu valerian; inferior to *V. jatamansi* - **V. salicariifolia** Vahl (Ur): indigenous med. pl., u. like ofc. val. (1053) - **V. sambucifolia** Mikan f. (nGe, Scand, USSR): similar to *V. officin.* but somewhat inferior; has replaced latter in Swed (836); cult in Pol, Hung - **V. scandens** L. (Fla [US] Mex, WI, CA, SA): starchy rts. eaten - **V. scorpioides** DC (Mex to CR): u. like *V. paniculata* - **V. sitchensis** Bong. (Alas to Ore e to Ida [US]): mountain heliotrope (or valerian); northern v.; st. pounded up with hummingbird's heart to make strong love potion (BC Indians) (837) - **V. sorbifoila** HBK (Mex to nSA): u. like *V. paniculata* in Guat - **V. sylvatica** Banks = *V. dioica* subsp. *s.* F.G. Mey. (830) - **V. toluccana** DC (Mex, CA): hierba del gato (Mex); low level of VO; u. antispasm. in folk med.; in liver dis., substit. for *V. officin.* in Mex Phar. - **V. tuberosa** L. (sEu, nAf): sembel (Tun); u. as sed.; u. by ancients (Nardos oreia of Galen) - **V. wolgensis** Kasakevicz (USSR): rts. u. as sed.; c. 0.53% VO - **V. wallichii** DC = *V. jatamansi.*

Valerianaceae: Valerian Fam.; hbs & shrubs of wide distribution (mostly nHemisph); lvs. opposite; fls. small, bisexual or unisexual; genera: *Valeriana, Valerianella, Centranthus.*

Valerianella Mill. (Valerian.) (wEu to cAs): corn (or field) salad; lamb's lettuce; some 50 spp. of ann. or bienn. hbs.; popular salad & pot herb pls.; peddled in Paris by vegt. vendors - **V. adscendens** (Valeriana a. Turcz.) (Pe): hornamo morado; strong purg; considered a "magic plant" & cult. by the San Pedros cult in Pe (Indians) (1782) - **V. locusta** (L) Betcke (V. olitoria Poll.) (Euras, nAf; natd. eUS); lamb's lettuce; corn salad; mâche (Fr); lvs. u. as salad, sometimes cooked (rarely); alimentary (Tun); u. antiscorb. (Tun); cult. veget. (Eu, US).

valerianic acid: valeric acid: several forms known, the chief being **normal v. a.** & isovaleric a.; latter occurs in many pl. & animal spp., ex. *Viburnum opulus* bk. & frt., perspiration (esp. of foot), dolphin blubber, etc.; u. sed., parasiticide.; intermediate in perfume mfr.; not sed.

valine, L-: amino iso-valeric acid; occ. in *Lupinus* seedlings, *Phaseolus aureus*, &c.; essential amino acid, not synthesized in human body; obt. by hydrolysis of fish proteins; u. nutrient. in treating amyotrophic lateral sclerosis.

Valinomycin: antibiotic with large cyclic peptide structure prod. by *Streptomyces fulvissimus* Waksman & Henrici; u. nematocide; insecticide.

Vallaris solanacea (Roth) O. Ktze. (V. heynei Spreng.) (Apocyn.) (Ind, seAs): nabu-nwe (Burma); juice appl. to sores (Burma); pl. c. cardiotonic glycosides; bk. u. as antipyr. (Indochin).

Vallesia glabra (Cav.) Link (V. cymbifolia Orteg.) (Apocyn.) (Fla Keys [US], WI, Mex, CA, tropSA): otatave (Mex); juice of shrub (or small tree) u. for ophthalmia (Mex); frt. edible sometimes eaten (838) - **V. hypoglauca** Ernst (Ve): source of amargosa bark(?) (839).

Vallisneria americana Mx. (Hydrocharit.) (NA): eelgrass; tape grass; aquatic herb; a favorite food for ducks - **V. spiralis** L. (OW & NW trop): decoc. u. in leucorrh. (Chin); when chewed with sesame or made into tea said to incr. appetite: stomach; lvs. u. as salad (Vietnam): u. in aquaria.

valonea: valonia (cups): acorn cups; fr. Turkish oak, *Quercus aegilops* (Gr [Crete]) & *Q. macrolepis*; source of v. extract; u. mfr. ellagic acid; for tanning fine leathers (such as Mor) (usually combn.

formulas) - **v. beards**: scales of cup(ule) sold as tirnac or trillo; sometimes tannin-free.

valproate: salt of valproic acid, usually Na (or Mg).

valproic acid: 2-propyl valeric acid; anticonvuls. in epilepsy (petit mal, grand mal, myoclonic); sed.; listed as essential drug (WHO) (840) but with some risks; Na salt u. mostly (E).

valtrate: one of the 3 chief components of valepotriate; complex isovaleryl compd. fr. *Valeriana* spp.; derivs. are didrovaltrate & acevaltrate (Valmane Kili Chem., BRD).

valuation of drugs: based on various methods: morphological, anatomical, organoleptic, physical, chemical, biological (physiological), pharmacological, microbiological, &c.; instrumentation now plays a large role.

Valvanol™ (Ge): chlorinated phenols in alc. soap soln.; u. in 0.5-10% dilns. as disinf. (prods. for human & for veter. usage).

Valvoline™: petrolatum.

vanaspati (Ind, Pak): **v. ghee: vegetable ghee**, in which cottonseed, peanut, & other FO replaces butter; now often margarine, with hydrogenated vegetable FO; chiefly u. as cooking fat.

Vancocin™: (Lilly): **vancomycin**: antimicrobial antibiotic fr. *Streptomyces orientalis* (soils fr. India, Indonesia); disc. 1955; a glucopeptide; effective against resistant staphylococcus infections (esp. s. endocarditis, s. enteritis & enterocolitis), osteomyelitis, pneumonia, septicemia, &c. (orally, parenterally); quite toxic & hence should be u. with great care; claimed to control plaque formation on teeth, reducing dental decay, gingivitis (gum inflammn.) & bone loss (trials on animals infected with *Streptococcus mutans* (forms plaque) (841); formerly thought effective against spirochetes (Vincent's angina).

Vanda roxburghii R. Br. (Orchid.) (tropAs, esp. Ind) rt. u. in rheum.; lvs. pounded to paste & appl. to body in fever; lf. juice put into aural meatus for otitis media - **V. spathulata** Spreng. (Ind): fls. given in TB, asthma, as powd.

vandal root: *Valeriana officinalis*.

Vandellia crustacea Benth. (Lindernia c. F. Muell.) (Scrophulari.) (Br): bitter blain; douradinha do campo (Br); hb. u. as diur. - **V. diffusa** L. (L. d. Wetts.) (tropAm): bitter blain; hb. c. NBP; u. as emet. antiperiod. antipyr. diur. - **V. pyxidaria** Maxim. (Ind+): pl. u. to treat GC (as "ghrita") (—> *Lindernia*).

Van Diemen's Land: Tasmania (former appellation [ca. 1850]).

vaniglia (It): Vanilla.

Vangueria acutiloba Robyns (Rubi) (Tanz): rt. bk. decoc. drunk against hepatitis; rt. decoc. taken as vermif.; bk. inf. in mal. - **V. apiculata** K. Schum. (c&cAf): rt. anthelm. for roundworm; frt. edible . - **V. canthioides** (wAf): lf. u. topically in yaws - **V. emirensis** (Madag): rt. u. astr. for wounds, eye dis. - **V. infausta** Burch. (sAf, Moz): wild medlar; rt. decoc. u. for menstr. diffic. esp. dysmenorrh. (Moz), expect. for cough; anthelm. for roundworms; in dental pain (Moz); rt. inf. as aphrod. (Venda) - **V. madagascariensis** J. F. Gmel. (V. edulis Vahl) (tropAf, Mad): voavanga; vanguerin; frts. esculent when overripe (*cf.* medlar): widely cult. - **V. neglecta** K. Schum. = *Rytigynia schumannii* - **V. spinosa** Roxb. (Ind): allu (Hind); atu (Bombay State); frt. refrig. cholagog. to stim. flow of phlegm & bile; lvs. u. in diphth.; frt. eaten & often sold in bazars - **V. tomentosa** Hochst. (e&sAf): small white medlar; rt. c. triterpenoid acids (vanguerolic & tomentosolic acids); vanguerin (saponin); rt. decoc. u. as anthelm., esp. for ascaris; rt. in fermented drink u. to treat GC (Tanz); lf. was appl. to eruptions of smallpox; poultice fr. boiled lvs. u. for sores on lips (Tanz); frt. edible - **V. venosa** Hochst. (trop wAf): frt. edible.

Vanilla Mill.: 1) (Orchid.) (OW, NW): vines of trop & subtrop regions 2) (L.): frt. of *V. planifolia* & *V. tahitensis* 3) (vernac.): v. leaf - **V. fragrans** (Salisb.) Ames (Orchid.) = *V. planifolia* - **V. griffithii** Reichb. f. (Indones, Mal): fls. pulped in water are laid on body when violent fever (Mal) lf. juice rubbed into hair to promote growth - **V. guayanensis** Splitgerb. (Guianas, Br): pods ("vanillons") with heliotrope odor, u. flavor; *cf. V. pompona* - **V. mexicana** Mill. (V. inodora Schiede; V. aromatica Sw.) (Mex, WI, nSA) scentless vanilla; vanille batarde; rt. decoc. u. for syph.; appl. exter. on ulcers, chancres, cancers; pl. juice vermifuge (Cu): formerly u. to perfume tobacco - **V. palmarum** Lindl. (Braz): frt. u. like that of *V. planifolia* - **V. phaeantha** Reichenb. f. (sFla [US], WI): leafy vanilla; pods sometimes sold as vanilla - **V. planifolia** Jackson (V. aromatica Willd. (*partim*)) (sFla, Mex, WI, SA): vanilla (beans); Mexican v.; v. plant; original home: seMex; cured full grown unripe frt. (sun process in Mex, hot water process in Bourbon Is): frts. c. 1.5-3% vanillin, vanillic acid, vanillyl alc., sugars (15% glucose & fructose, 35% sucrose), resins, mucilage; u. as flavor (fine aroma), esp. for confectionery, condim. originally in foods & beverages, now also in med. as flavor, corrective, a pharmaceutical necessity; partly now replaced by synth. vanillin; very important in flavoring chocolate; as aphrod. sed. in hysteria; menstr. diffic.; in complex perfumes for the perfume & cosmetic indy.;

trade prod. incl "crystallized" (showing vanillin crystals on pod) & " non-c." (prod. in sMex & many trop islands. esp. Mad, Réunion, Java, WI [Cu]; SriL, &c. [SC vainilla, Sp., small pod]) (2234) - **V. pompona** Schiede (V. grandiflora Lindl.) (Mex, WI [Lesser Antilles, esp. Guadeloupe], CA, tropSA): frt. called Pompona vanilla (Pompona Bova; Vanille bouffie); West Indian vanilla; cult. on Guad & Mart; similar to vanillons - **V. tahitensis** Moore (Tahiti): Tahiti vanilla; much like *T. planifolia* in properties & uses; may possibly represent a perenn. vine form of *V. pompona* (842) - **V. wrightii** Reichenb. f. (V. palmarum Griseb.) (WI, Guianas, Br): prod. vanilla beans of low value.

vanilla bean: a popular misnomer for v. fruit (capsule) - **Carolina v.**: 1) *Trilisa odoratissima* 2) *Achlys triphylla* - **v. extract**: alc. prepn. of v. frt - **v. plant: v. leaf**: Carolina v. (l) - **v. powder**: mixt. of sugar (+ starch) (+$CaCO_3$) & vanilla (or vanillin); u. to flavor ice cream - **v. sugar**: v. powder - **v. trilisa**: v. leaf.

Vanilla varieties: Bourbon: prod. on Is of Bourbon; chiefly exp. to Fr - **Mauritius v.**: fr. the Seychelles; most is sent to Eng - **Mexican v.:** considered the most superior v.; chiefly u. in US & Mex - **Pompona v.**: frt. of *V. pompona* - **Réunion v.**: Bourbon v. - **v. splits**: more mature frts. which have been cut up - **Tahiti v.**: that cult. on Tah & the Haw Is (US); *V. tahitensis, V. planifolia, V. p.* var. *angusta*.

vanillin an arom. aldehyde, vanillic acid aldehyde, occuring naturally in many prods., such as vanilla, Balsam of Peru, storax, lupine sds., etc.; u. flavor, perfume, in cosmetics, to fortify vanilla ext., etc.; most u. is now synth. fr. eugenol, guaiacol, lignin (by-prod. of wood pulp indy.); formerly considered aphrod. nerve stim, appetite stim. (844) - **sodium v.**: much u. as flavor, perfume.

vanillons (Fr): frt. of wild specimens of *Vanilla planifolia*; u. as arom. in tobacco indy.

Vantanea parviflora Lam. (Houmiri.) (Ve): bk. rasped into fermenting chicha (prepd. fr. *Manihot esculenta*) to improve flavor & potency (1783).

Van Tiegham cell: u. for hanging drop prepns. in microscopy.

Van Van oil (New Or, La [US] voodoo): lemongrass oil, u. to cleanse doorsteps (of "evil spirits").

vara (Sp): rod - **v. de coche** (Domin): *Cryptostegia* - **v. de San José**. (Sp) (lit. St. Joseph's stick): 1) (LatAm): *Althaea rosea* 2) (Pan): *Agave panamana* Standl. 3) *Cassia emarginata* (seMex) 4) *Polianthes tuberosa* (Mex) 5) *Leonotis nepetifolia* (PR).

varec (Eng, Fr): **varech** (FrQu): **varek: varette**: 1) marine algae, esp. spp. of *Fucus, Laminaria, Rhodymenia*, etc. (sometimes specif. *Fucus*) Normandy 2) (usually): the ash of kelps, a crude soda, with iodine, etc.; a source of iodine.

varianose: polysaccharide synthesized fr. glucose by *Penicillium varians* Smith.

varicella: chickenpox (a virus dis.) - **v. zoster immune globulin**: specific immune prod. derived fr. plasma of human donors with high levels of varicella antibodies; u. to prevent or modify the dis.

Varidase™: purified mixt. of streptokinase & streptodornase (enzymes prod. by hemolytic streptococci); u. to remove clotted exudates fr. wound surfaces, thus hastening the healing process (1888); in bacterial endocarditis (2235).

varietas (L.): **variety**: in bot. there is considerable variation in meaning; orig. intended as a subdivision of a sub-species; now often intended to denote trivial differences, thus, at times to indicate races, horticultural forms, & even only abnormalities.

Variolaria Pers. = *Pertusaria* DC (usually as subgen.); &c. - **V. oreina** Ach. = *Rhizocarpon o.*

varix: dilated & tortuous vein; hence varicose veins.

varnish tree, Chinese: *Aleurites fordii; Firmiana platanifolia* - **Japanese v. t.**: *Rhus verniciflua; Firmiana platanifolia.*

Varthemia iphionoides Boiss. & Bl. (Comp.) (Near East): lvs. as tisane for headache & indigest.; for body aches; fresh lvs. appl. to wounds (Israel).

vasaka (Sancr, Beng): *Adhatoda vasica.*

vascular bundle: term now generally u. in preference to fibrovascular b. (*qv.*) since function is chiefly that of conduction.

Vasculose = crude lignin.

Vaselin(e): petrolatum; as trademark (Chesebrough who 1st isolated, 1871) in US; mostly generic abroad; prob. fr. Wasser (Ge) + elaion (Gr) (water + oil); since supposed origin: H_2O + methylcarbides.

vasicine: alk. from *Adhatoda vasica*; oxytocic.

vasoconstrictine: epinephrine - **vasopressin**: pitressin.

vassourinha (Br): *Scoparia dulcis.*

vastago (Sp) bud; shoot.

Vatairea guianensis Aubl. (Leg.) (Guiana, Br): bk. decoc. u. to dress ulcers; sap employed med. for ringworm; wd. u. for decorative work, furniture.

Vataireopsis speciosa Ducke (Leg) (Br, esp. Amaz reg.); bk. source of araroba powder (po da Bahia) u. in med.; wd. for carpentery; jare; u. by Indians to "cure" leprosy; said extremely poisonous to man; fish poison (—> *Vatairea*).

Vateria copallifera (Retz.) Alston (V. acuminata Heyne; Elaeocarpus copalliferus Retz.) (Dipterocarp.) (sInd, SriL): source of one of the best dammars; u. in fine varnishes; wds. formerly u. to make tea boxes; frt. edible - **V. indica** L. (sInd): source of exudate called "white" dammar (or "piney resin") (or Indian copal), u. to make varnishes (in turpentine oil soln.) ("piney varnish"); photographer's varnish ($CHCl_3$ soln.); as ton. carm. expect. in chronic bronchitis; formerly in cholera, vomiting; frts. as tan.; sds. source of FO ("piney tallow"; Malabar t.); appl. in rheum.; u. as cooking fat, in making soap & candles & torches (mixed with coconut oil) - **V. macrocarpa** Gupta (Kerala, Ind): vellapiney; wd. used much like that of *V. indica*.

Vatica L. (Dipterocarp.) (Ind, seAs, incl. Mal): vatique (Fr); several spp. yield resins (dammars), timbers, &c. thus: **V. chinensis** L. (V. roxburghiana Bl.) (swInd, SriL): njato (resin); **V. lancaefolia** Bl.: VO u. to flavor chewing tobacco; **V. papuana** Dyer (PapNG, Indones): dammar hiroe; cambium tissue appl. to cuts & wounds (2261); **V. rassak** Bl. (IndoMal, Borneo): source of "rose dammar" & VO.

Vaupesia cataractarum R. E. Schultes (monotypic) (Euphorbi) (wBr, Co): ma-ha-wa-soo; sds. rich in cyanophoric glycosides; FO; thin watery red latex borne at times; large sds. collected fr. ground; boiled & eaten (natives) (toxic before boiling).

Vavaea amicorum Benth. (Meli) (Oceania): "false sandalwood" (Fiji); bk. u. for internal pains (PI).

Veegum™: complex colloidal Al Mg silicate; prepd. fr. natural mineral prod.; u. emulgent, thickening, binding & suspending agt.; to incr. viscosity; stabilizer; in pharm. & cosmetic prepns.

vegetable: 1) a plant (member of p. kingdom) 2) such plants that are u. as food by man or lower animals; or any part of such a plant (st. rt. lf. fl. frt., &c) - **v. antimony**: *Eupatorium perfoliatum* - **v. calomel**: Podophyllum - **v. fibrin**: gluten f.: a brown material extr. fr. wheat gluten - **v. gold:** saffron - **v. mercury**: 1) Podophyllum 2) *Calotropis gigantea* 3) Manaca 4) *Jussiaea repens* - **v. pear** (La [US]): *Sechium edule* - **v. powder**: Lycopodium. - **v. salt:** potassium tartrate - **v. sponge**: *Luffa* spp. - **v. sulfur**: v. powder - **v. wadding**: *Gymnema sylvestre*.

vegetable: v. also under agar; albumen; gelatin; ivory; musk; pepsin; rennet; silk; sponge; wax.

vegetal: 1) vegetable: relating to plants 2) either pl. or animal functions.

vegi-lecithin: soybean l.

Veilchen (Ge-n.): violets - **Veilchenwurzel** (Ge): orris rt. (SC "violet rt." as has odor of violets).

vein islet: segments of the area of a leaf surface (ventral or dorsal) separated by terminal divisions of the conductive strands of vascular bundles; these areas are either completely or nearly surrounded by the bundle divisions - **v. i. number** : the number of v. i. per sq mm of leaf surface; a constant for the sp. for an average part of the lf. & for a fully grown lf.

vela de dios (NM [US], Sp) (candle of God): *Yucca* spp.

velamen: multilayered outer coating of aerial rts. of orchids, aroids, &c.; cells are dead & empty, give silvery appearance.

"vellarin": amaroid (NBP); once supposed active principle of Herba Hydrocotylis (also in rts.) (*Centella asiatica*).

vellón (Sp): sheep fleece.

Vellozia equisetoides Bak. (Vellozi.) (e tropAf): tree lily; u. to treat asthma; small sts. u. as toothbrushes; section of bole as scrubbing brush (Tanz).

velvet leaf: 1) *Abutilon theophrasti* 2) *Chondodendron tomentosum* - **v. plant**: 1) *Verbascum thapsus* 2) *Gynura aurantiaca* - **v. seed:** *Guettardia* spp. - **v. tree:** *Phellodendron amurense*.

velvet : v. also under bean; dock.

venadero (CR): *Cissampelos pareira*.

venalis, venale (L.): comml; technical (grade).

Venetian red: Fe_2O_3. - **V. theriac**: the most famous of the theriacs (popular in the days of polypharmacy); prep. in various of the large cities of Eu, esp. Paris, with appropriate ceremonies; made as a pd. which could be converted into a paste by adding wine.

Venice: v. under turpentine.

Venice sumac: *Cotinus coggygria*.

Venidium arctotoides Less. (Comp.) (sAf): lf. juice in epilepsy, indigest. & locally appl. to wounds.

venins, snake: s. venoms: specifically zootoxins; exotoxins prod. by snakes; can sometimes be counteracted by anti-venins; venins are also produced by scorpions, spiders, toads, etc.; venoms are very complex mixts., c. neurotoxins, hemagglutinins, hemotoxins, leukotoxins, endotheliotoxins, &c.; have been variously u. as hemostatics & to reduce blood clots (enzyme pholase acts directly to dissolve clots at remote places of circulatory system (preference for copperhead v.) - **toad v.**: t. poison: secretion of skin glands of various toad spp.; c. bufalin, bufotalin, bufotoxin (cardioactive steroids) in rich amts.

venkel (Dutch): fennel.

venom: poison; specifically zootoxins, toxic subst. secreted by animals (snake, insect, millipede, &c.); chiefly u. is snake v. (venins).

venomology: science of animal toxins (venoms).

Venostasin™: popular pharm, prepns. of wGermany (BRD); c. *Aesculus hippocastanum* with vit. B_1; u. for venous congestion.

Ventilago calyculata Tul. (Rhamn.) (Ind, Nep): juice of bk. & young shoots appl. to body for pain fr. malarial fevers - **V. denticulata** Willd. (Ind): pitti; bk. & dr. frt. u. as paste for itches & burns; poultice for headache (124) - **V. dichotoma** Merr. (PI): pd. bk. mixed with oil appl. to certain skin dis. - **V. madraspatana** Gaertn. (Ind, SriL, Burma, Java): pitti (Hindu); rt. bk. u. int. for atonic dyspep., debility, mild fever, exter. in skin dis., itch; as purplish to black dye for wool, silk, cotton - **V. viminalis** Hook. (Austral): supplejack(s); ashes of pl. mixed with native tobacco (*Nicotiana* sp.?) to make stronger chewing tobacco; external uses of rts. & bk.: rheum.; toothache; swellings, cuts, sores, & to restore hair in baldness (aborig. of Austral) (2263).

ventura: Jamaica dogwood.

Venus: goddess & symbol of love & of coitus (adj. venereal); copper - **V. finger**: *Cynoglossum officinale* - **V. hair**: *Adiantum capillus-veneris* - **V. looking glass**: *Legousia hybrida* - **V. mercenaria** (Veneridae-Pelecypoda-Mollusca) (Atl coast): round (or little neck) clam; quahog; cherrystone c. (when small); popular food; shells u. as wampum by Am Indians.

Vepris bilocularis Engler (Rut.) (Ind): wd. boiled in FO; prepn. u. in eye & ear dis., rheum, asthma; rt. decoc. in biliousness (Nep) - **V. lanceolata** G. Don. (Toddalia l. Lam.) (trop e&sAf): (white) ironwood; c. alks. pd. rt. u. for flu, colic, emet.; frt. ("African cubebs") u. as substit. for & adulter. of Cubebs; twigs. boiled & patient sits over pot covered with blanket for fever; wd. tough, u. for hammer & pick handles - **V. undulata** (Thunb.) Verdoorn & C. A. Smith (sAf): muumano; pd. lvs. u. exter. on incisions in forehead for headache; rt. macerate u. for infertility; rt. decoc. u. in menorrhagia.

ver (Fr-m.): worm; maggot; mite; moth; etc. - **v. à soie**: silk worm - **verac**: kelp; *cf.* varec - **vératre vert** (Fr): *Veratrum viride.*

vera (Ve): *Bulnesia arborea.*

Verain's balsam: comp. benzoin tincture.

Veratrina (L): **veratrine (mixture)**: mixed alks. fr. sds. of *Schoenocaulon officinale* (sabadilla) *not* fr. *Veratrum* spp.; c. cevadine (-75%) (chief insecticidal princ.); veratridine (-25%); smaller amts. of sabadine, sabadinine (cevine), sabadilline (cevadilline), veracevine; crystalline veratrine = pure cevadine; very toxic; should not be u. internally (but in past was u. in 2 mg dose in myasthenia gravis, MD; exter. in rheum. neuralgias; for lice (both stomach & contact poison to insects); in veter. med. as emet. (esp. hogs); ruminatoric.

veratroidine: in older lit. referred to as amorphous "veratrine" fr. *Veratrum viride*; now believed secondary decompn. prod.

veratrum, American: false v.: *Veratrum viride.*

Veratrum L. (Lili.) (nTemp Zone): false hellebores; perenn. hbs. (c. 30 spp.) with short, thick rhiz.; many spp. cont. alks. of steroid type with modified rings C & D; ex. jervine; some spp. stock poisons - **V. album** L. (V. lobelianum Bernh.) (Euras): (European) white hellebore; (white) false h.; w. veratrum; c. ester alks. (protoveratrines A & B; germerine); alkamines (jervine, pseudojervine, rubijervine, veratramine) (845): rhiz. & rts. u. med. in manner of *V. viride*; as adulter. of same; parasiticide, in vermin control, esp. against lice (veter. med.); formerly in MD, locally in neuralgia, skin dis., myasthenia; in snuffs; insecticide; source of hypotensive alks. - **V. californicum** Dur. (Pac, Cana, US [Wash to Calif, e to Utah] nwMex): (Calif) false hellebore; C. h.; white h.; western h.; "wild Indian corn"; c. veratramine, cycloposine (teratogenic); rt. decoc. u. as contracept. (Nevada Indians); rhiz. & rts. ashed & ash adm. in blood dis., esp. syph.; scraped rt. may be u. as snuff for colds (BC [Cana] Indians); int. use has killed stock & humans; ornam. (846) - **V. eschscholtzii** Gray = *V. viride* - **V. fimbriatum** Gray (Calif): fringed false hellebore; c. jervine, germitrine, neogermitrine, &c.; has hypotensive props. - **V. frigidum** Schlecht. & Cham. (Stenanthium f.) (Mex): cintul; u. med. (Mex) (200) - **V. formosanum** O. Loes. (Taiwan): rhiz. u. to treat polio, paralysis; lf. for scabies & ringworm; toxic pl. - **V. mackii** Regel var. *japonicum* Shimizu (V. japonicum O. Loes.) (Jap): rhiz. bactericidal, u. in skin dis. - **V. nigrum** L. (seEu, nAs): black false helleborine; rhiz. c. jervine, rubijervine; u. for abdomin. illnesses, skin parasites; u. in dressings for skin dis., boils; not to be confused with *Helleborus niger*; u. as substit. for *V. album* - **V. viride** Ait. (Alas s to Ore [US], e to Atl coast): (green) false hellebore; American f. h.; white (or green) h.; itchweed; Indian poke; rhiz. & rts. c. jervine, pseudojervine, rubijervine, veratramine, protoveratridine, germerine, germidine (hydrolyzes to germine), veratrosine, germitrine, protoverine, & other alks. (847); u. card. depressant in essential hypertension, puerperal eclampsia (hypertensive toxemia of pregnancy) chiefly (848); also in auricular fibrillation: acts partly by reflex vasodilation, improving circulation; has been bioassayed by *Daphnia magna* method (of Viehoever; Craw) (849); resp. sed.; emetic; in toothache (Am Indians); insecticide; poisonous; must not be confused with black hellebore (*Helleborus niger*) or white h. (*H. viridis*); much coll. in Appalachian mts. region (US) (2243).

Verbasci (Flavi) Flores (L.): **verbasco (grande)** (It): mullen (fls.), *Verbascum thapsus*, etc.

Verbascum austriacum Schrad. (V. orientale Neilr.) (Scrophulari.) (seEu): fls. u. to adulter. agrimony herb (*Agrimonia eupatoria*) - **V. blattaria** L. (Med reg. to cAs; natd. NA): moth mulle(i)n; u. med. like *V. thapsus*; against moths; said repellant to roaches - **V. erianthum** Benth. (Pak, Afgh): fl. decoc. u. for eye dis. (Afgh) - **V. litigiosum** Sampaio (V. crassifolium Hoffmg. & Link): source of Flores Verbasci (Por Ph.) - **V. lychnitis** L. (Euras, nwAf; natd. eNA): white mulle(i)n; fls. u. to destroy mice (*sic*) - **V. nigrum** L. (cEu to Siber): black (rooted) mulle(i)n; dark m.; sd. as fish poision; fls. u. to adulter. Verbascum Flowers & hb. u. to adulter. agrimony herb - **V. phlomoides** L. (Eu, nAf; natd. e&cUS): woolly (or clasping) mulle(i)n; fls. c. acid (& neutral) saponins, sapogenin, NBP, mucil. FO, VO (traces), hesperidin, xanthophyll (coloring matter); Verbasum Flowers long u. as mild expect. dem. & diaph. in colds, resp. tract inf., asthma, catarrhs; diur. antirheum. (for rheum. pains), in diarrh.; component of pectoral species & of many other teas; antiphlogistic; antiviral values demonstrated; esp. popular in Hung; lvs. also u. like fls., together in asthmatic bronchitis; fresh fls. pressed on warts will remove (*sic*) (2236); claimed that fresh pls. in cellar will drive away rats & mice - **V. pulverulentum** Vill. (w, c&sEu): u. for hemorrhoids, in tapeworm, & to stupefy fish - **V. sinaiticum** Benth. (eAf, swAs to Turkestan): source of ternacha root; u. med. - **V. sinuatum** L. (Med reg. to Iran): lvs. c. verbascoside; u. in eye dis.; frts as fish poison - **V. speciosum** Schrad. (eEu [USSR+] natd. Ore [US]): showy mulle(i)n; lvs. & fls. u. in Russ med.; off. in Rom Phar. - **V. thapsiforme** Schrad. (Eu, AsMin, Mor): wool mulle(i)n; c. & u. much as foll. - **V. thapsus** L. (Euras, natd. NA [widely distr.], SA, Jap, NZ): common (or flannel) mulle(i)n; hag taper; flannel leaf; f. plant; long taper; (velvet) mullen; candlewick; gordolobo; verbasco; dr. lvs. (Verbasci Folia; mullen leaves) c. saponin, mucil. resin, NBP; decoc. u. emoll. dem. in coughs, colds, catarrhs.; antipyr. (2451); in diarrh. dys.; treatment for some types of baldness (value?) (850); in poultices; foot soaks for sprains, swollen feet; smoked in cigarettes for asthma (Ozark mountaineers) (1785); to line mocassins (Am Indians); lf. inf. ingested for rheum. (Ala [US]) (2313); fls. c. mucil. saponins; FO, VO (traces) thapsiac acid; u. dem. inf. in catarrhs.; expect. (in breast tea); to adulter. Digitalis; locally in otalgia; fresh fls. u. for warts (2236); sd. c. mucil. saponoside, 30-35% FO (oil of mullen) (851); u. fish poison; rts. c. verbascose; verbasterol, aucubin; pieces of rt. on string u. as collar for teething baby (Abenakis Indians); most prodn. Fr, Ge, Hung (2765) - **V. virgatum** Stokes (Med reg. n to Cornwall [Eng], nwAf; natd. NA, Austral): purple-stamen mulle(i)n; Aaron's rod; u. by Am Indians as antiscorb. (852).

Verbena L. 1) (Verben.) (mostly Am): verbenas; vervains; ann. or (mostly) perenn. hbs. or sub-shrubs; often cult. ornam; odoriferous; much u. in tisanes (Fr) 2) (specif. L.): dr. aerial portions of *V. hastata* 3) (Sp): *Aloysia triphylla* 4) (Madrid) a kind of festival before a celebrated saint's day; SC this plant ("holy herb") u. to adorn altars in offering sacrifices, &c.; later this name given to all green branches of sacred trees, &c., u. for religious purposes (may be fr. verveine [Fr]) (fr. Celtic fer [take away] + faen [stone]: since u. in bladder stones) - **V. ambrosifolia** Rydb. (Glandularia bipinnatifida Nutt.) (cUS [Tex, Ariz, Colo, SD]; Mex): ragweed verbena; western pink v.; moradilla (NM [US]); pd. hb. mixed with pitch gives plastic appl. for backache & skin dis.; bath with leaves u. in rheum. or neuritis ("nerves") (Tex); also tea int. - **V. bonariensis** L. (SA, now CA, US, Euras, Af, Oceania): South American vervain; "(Buenos Ayres) verbena"; yerba de los hechiceros; lvs. u. as decoc. for contusions (Mad), periodic fevers, chronic catarrhs.; lvs. & young sts. u. for certain liver & gastric dis. (Sp & SpAm); c. alk. - **V. bracteata** Lag. & Rodr. (NA, incl. Mex): hb. u. to stop hemorrhage (Navaho Indians); for scrofula - **V. bracteosa** L. = *V. bracteata* - **V. brasiliensis** Vell. (SA, esp. Br; WI, CA, natd. US, NZ): u. in fever, anemia, edema, pleurisy (Sotho people, Br) - **V. callicarpiaefolia** Kunth = *Lippia c.* - **V. canadensis** (L) Britt. (Glandularia c. [L] Nutt.) (e&cUS): rose vervain; u. as "stomach medicine" (Omaha Indians) - **V. capitata** Forsk. = *Phyla nodiflora* - **V. carolina** L. (sw US to CA): Mexican vervain; hb. as tea for headache (Sonora) (853); mal. (Mex) - **V. ciliata** Benth. (Glandularia bipinnatifida Nutt.) (Tex [US], Arg, Mex, CA): Mex vervain; "verbena" (Sp); hb. u. antipyr. - **V. crithmifolia** Gill. & Hook. (Ec): té de burro; u. in blenorrhagia (855) - **V. demissa** Moldenke (Ec): "verbena echada"; hb. u. antipyr. (854) - **V. elegans** HBK (swUS, Arg, Mex): showy verbena; lvs./hb. u. as tea in gastric dis. (Chihuahua); var *asperata* Perry (Tex, nMex): hb. decoc. u. in gastric dis. & catarrhs. (Chihuahua) - **V. glabrata**

HBK (Ec, Co.): "verbena"; hb. u. medl. (Firmin); to treat fever, gastric colic, as purg. in liver & kidney dis.; vaginitis (Co) (856) - **V. gracilescens** (Cham.) Herter (Br, Bo, Par, Ur, Arg): found in anthesis every month of the year - **V. hastata** L. (sCana, US): (American) blue vervain; common v.; b. verbena; simpler's joy; ague weed; hb. c. verbenalin, hastatoside; aerial parts u. diaph. expect. antipyr. astr.; lvs. u. in gastralgia (Teton Dakota Indians); lf. decoc. as beverage tea (Omahas); leaf tea u. in epistaxis (Mich [US] Indians) emet. (US): local appln. (Fla [US]); lvs. rolled & smoked like cigar in asthma (sMich-1930-) ton. vulner.; emmen. in diabetes; small sds. u. for pinole (Calif Indians) u. for colds, edema, icterus, sed. (Am Indians); roasted sds. ground to make bread meal; rt. u. as expect. rt. boiled, u. for toothache (Mex) (V. pinnatifida Ph.) - **V. hispida** R. & P. (Ec, Pe, Br, Par, Ch, Arg): u. as substit. for *V. officin* - **V. X hybrida** Voss (US, LatAm): garden v.; pl. decoc. u. in stomachache (Mex) - **V. indica** L. = *Stachytarpheta i.* - **V. jamaicensis** L. = *Stachytarpheta j.* - **V. laciniata** (L.) Briq. (Ec, Pe, Bo, Ur, Ch): inf. u. to aid menstruation: roasted lvs. to replace tea - **V. litoralis** HBK (Mex, CA, SA, natd. US [Calif]; introd. Oceania, &c.): false vervain; Cayenne v.; coast verbena; "verbena" (Par); pl. juice u. in skin dis. (scrofulous), pl. inf. for hypertension (Guat); hb. as febrifuge, bitter, in fever, grippe, catarrhs, typhoid fever, mal. vermif. (hookworm) (857); in sprains & broken bones, contusions (topically) (Bo); beverage tea (Ve, Ec); for skin & sex organ dis. (Easter Is); contraceptive (Ur); purg. (777); in cough (LatinAm); inf. for icterus & neuralgia (Arg): shoots in potions as diaph. in "catarrhs" (2435); as "general remedy" (Par); incl. var. *caracasana* Briq. - **V. macdougalii** Heller (wUS: NM, Wyo): "vervena"; "dormilon"; tea u. as diur. (NM, Tex); green lvs. mashed & put on gums for toothache - **V. maritima** Small (Fla-pinelands): tea fr. fls.; u. antidote for water mocassin bite (Seminole Indians) - **V. menthaefolia** Benth. (wUS, Mex, CA): "vervena"; lvs. mashed in water & drunk for chills & fever (Guat); pl. inf. in stom. dis., to stop vomiting (Sonora [Mex]) - **V. minutiflora** Briq. (Br, Arg): u. to treat gastric & dig. ailments (Br) - **V. montevidensis** Spreng. (sUS, CA, Br, Arg, Par, Ur): "anil"; u. popular med. (Br, Arg) - **V. nodiflora** L. = *Phyla n.* - **V. officinalis** L. (Euras, Af, natd. eUS): (European) vervain; common v.; hb. (as Herba Verbenae) c. glycosides (verbenalin), hastatoside, NBP, tannin, mucil. VO with citral, limonene, geraniol (2216); u. as bitter, stim. ton., astr. galact. emmen. diur. expect. in pertussis; chronic bronchitis; diaph. antispasm. styptic; anthelm. vulner. in wounds, ulcers, dog bites (Hainan Is, Jap); in dropsy (Indochin); int. tumors (Mex); in eczema: asthma; kidney dis.; considered a panacea by Hippocrates; in eye dis. (Doctrine of Signatures, bright-eyed corolla); rt. ext. u. for liver & gb. dis.; lvs. u. as tea substit. (tisanes); pernicious weed (US) - **V. origenes** R. A. Phil. (nwArg, Ch): rica rica; u. in relaxing arom. baths; cult. in Arg - **V. pary-cary** Fr. Allem. (Br): u. for snakebite (Am Indians) - **V. peruviana** (L.) Britt. (Pe, Br+): u. for eye pains; abdom. pains - **V. pinetorum** Mold. (Arg, Mex): pl. decoc. u. in fever - **V. rigida** Spreng. (V. venosa Gill. & Hook.) (seUS, Mex, WI, CA, SA, natd. sAf): rt. decoc. u. in heartburn, colic (sAf) - **V. sororia** D. Don = **V. spuria** L. = *V. officinalis* - **V. stricta** Vent. (c&eUS, eCana); hoary vervain; u. in rheum.; rt. & st. boiled in vinegar for 12 hrs. then decoc. rubbed on afflicted parts (US, Negro) - **V. tenuisecta** Briq. (seUS, Mex, WI, nSA [Br], Ur, Eu, nAf): u. as diur. lax. (Ur) - **V. triphylla** L'Hér. = *Aloysia t.* - **V. urticifolia** L. (eNA, Mex): white (or nettle-leaved) vervain; hb. u. like that of *V. officinalis*; in urinary calculi; rt. decoc in milk & water u. in rhus venenata; ton. antipyr. for skin eruptions - **V. variabilis** Moldenke (Pe): rts. u. against fleas (Am Indians) - **V. wrightii** A. Gray (Glandularia w. Umber) (swUS, Mex): desert verbena; Wright verbena; irritant medicine of Navajo Am Indians (858).

verbena, common: *Verbena officinalis* - **Indian v. oil**: *Cymbopogon citratus; C. flexuosus* - **lemon (scented) v.**: *Lippia citriodora* - **moss v.** : *Verbena laciniata* - **sand v.**: *Abronia spp.* - **wild v.**: common v.

Verbenaceae J. St. Hil.: Vervain Fam.; hbs., shrubs, & trees (rarely) of trop., subtrop. & temp. zones; fls. irregular, more or less 2-lipped, white, pink, or blue; calyx tubular; corolla gamopetalous, tubular, with 4- to 5-fld limb; spp. often c. VO, a few glycosides, etc.; technical prods. incl. teak wood & several crude drugs.

verbenalin: bitter glycoside of *Verbena officinalis* & *V. hastata* (richest in fls.).

verbenol, d-: dicyclic terpene alc. found in VO of incense (*Boswellia carteri*).

verbenone, d-: dicyclic terpene ketone found in VO of *Aloysia triphylla*.

Verbesina acmella L. (Comp.) = *Spilanthes a.* - **V. alata** L. (WI, Guianas): inflammation bush; camomille rouge; fl. head inf. taken as stom. & antispasm. strong brew emet.; for muscular strains (Virgin Is) - **V. alba** L. = *Eclipta prostrata* L. - **V. capitaneja** Nees (Mex): capitaneja (vernac. name); hb. u. as vulner. to wash venereal ulcers - **V. columbiana** B. L. Robinson (Ve): decoc. u. as bath for blindness (Am Indians) - **V. crocata** Less. (Mex): lvs. crushed to prep. cooling beverage, antipyr.; for wounds - **V. encelioides** (Cav.) B. & H. (swUS to SA): flor de santa; mirasol; crownbeard; anil del muerto; c. alk.; u. diaph., diur.; for boils & skin dis. (Hopi Indians); lf. u. exter. for tumors, warts; decoc. int. for cancers (Arg): sold in health food stores; subsp. *exauriculata* J. R. Coleman (Arg): u. for stom. trouble (Navaho); for boils, skin dis.; spider bites (Zuñi) - **V. gigantea** Jacq. (Mex, CA, WI): capita negra; rt. decoc. held in mouth for toothache; u. stom. antispasm. (Haiti) - **V. helianthoides** Mx. (Actinomeris h. Nutt; Pterophyton h. E. J. Alexander) (s&cUS, SA): gravel-weed crownbeard; sunflower c.; diabetes weed (Ga [US]) (ca. 1880); gravelweed; pl. u. diur. in edema, cystitis (folk med.) - **V. scabriuscula** Blake (Guat): saqi mank; pl. decoc. u. for refreshing & healing baths - **V. tonduzii** Greenm. (CR): rt. decoc. u. to expedite delivery in cows - **V. virginica** L. (s&cUS): tickweed; frost-weed; (white) crownbeard; rt. decoc. diaph. depur.; pounded rts. soaked in water u. for fever (Choctaw Indians).

verdigris: mixed basic cupric acetates; c. subacetate; u. med. as dusting pd. & in ointments as pigment; in wood preservn.

verdolaga: *Portulaca oleracea.*

Verdorbenheit (Ge): rancidity of FO.

verduras (Sp, Por): greens; vegetables.

vered(e)lung (Ge): improvement; budding; grafting - **Veredlungsanlage:** sd. prodn. plot.

verek: *Acacia senegal.*

vergamota (Br): *Mentha gentilis.*

verger (Fr-m.): orchard.

vergerettes du Canada (Fr): *Erigeron canadensis.*

Vergissmeinnicht (Ge-m.): forget-me-not.

Veriloid™: hypotensive ester alk. fraction of *Veratrum viride*, extd. by a special process; u. in hypotensive medication; orally & parenterally.

vermil(l)ion: Vermillon (Ge): red pigment prepd. fr. cinnabar; native form of red mercuric sulfide; u. in (artists') paints, rubber, sealing wax; formerly u. in skin dis., syph.; antibacterial.

vermouth: wormwood-flavored wine.

vermur (Sp, Am): absinthe.

vermiculite: collective name for series of yellowish to brown porous mica-rich minerals (Colo, Wyo, Mont [US], Transvaal); montmorillonoids; "expanded" forms absorb much water & remain free flowing; u. as filler, insulation, packing material; for mineral bedding; pl. propagation.

vernalization: procedure for inducing pl. to flower & fruit earlier; carried out by chilling bulbs or sds. (esp. grains) in darkness after bringing to early stage of germination, then storing at low temps (0-5°C) for 35-45 days; shortens transition from vegetative to reproductive phase.

vernis (Fr-m.): **Vernix** (L): varnish; glaze.

Vernonia Schreb. (Comp.) (worldwide): ironweeds; some 1,100 spp. of perenn. hbs. with leafy sts. - **V. ambigua** Kotschy & Peyr. (e tropAf): rt. u. as expect. (natives) - **V. amygdalina** Del. (trop wAf): bitter leaf; tree vernonia; rt. bk. inf. drunk for schistosomiasis; st. bk. bitter ton. in fevers & for intest. upsets; twigs & rt. u. for chewsticks (cult. for this purpose); lvs. antipyr. ("native quinine"); lf. decoc. for cough (Ghana): twigs chewed as stom. ton., appetizer; lvs. rubbed on body for itch, ringworm, &c.; lvs. good protein source & u. as vegetab. - **V. angola** O. Hoffm. (sAf): juice appl. as styptic to wounds; pl. purg. for cattle; lvs. u. in thoracic pain; fish poison (Konde) - **V. angustifolia** Mx. (seUS): ironweed; pl. c. sesquiterpene lactones glaucolide-ß and baldvernin; rts. u. as arom. bitter ton. alter. aperient; in female dis. - **V. anthelmintica** (L.) Willd. (sAs, [Ind, SriL, Afghan], seAs): (kinka oil) ironweed; purple fleabane; vernonia; "sds." (achenes) u. as anthelm. for ascarids and oxyurids (pinworms) (often combined with other substances); for leucoderma & other skin dis.; leprosy (Ayurvedic med.); abort.; lf. in colic, edema; exter. in gout, rheum; in combns. for mal.; natives said to u. to counteract snake venom (value?) (859) - **V. arborea** (Buch.-Ham.) (seAs): cabbage tree; inf. u. stim. digestive; rt. inf. for fever (Mal) (var. *javanica* C.B. Clarke); wd. u. in mfr. matches - **V. aschenborniana** Schauer. (Mex, CA): suquinay; hb. u. in gastric dis. (Hond, Guat) (44) - **V. bahamensis** Griseb. (Bah): white sage; lf. decoc. drunk for jaundice (Inagua Is) - **V. brachycalyx** O. Hoffm. (Tanz): rts. chewed for stomachache, swollen gums, as purg,; lf. inf. u. in mal. - **V. brachylaenoides** S. Moore (cAf): u. med. (Zambia) - **V. brevipetiolata** Sch. Bip. (SA): u. med. (Br) (860) - **V. canescens** HBK (sMex to Pan & nSA): caratillo; lf. inf. u. to stanch hemorrhage & reduce inflammn.; fl. inf. as enema for severe constipation - **V. cinerea** Less. (tropAf, As): rt. & frt. anthelm. (for roundworm); fls. in conjunctivitis; entire pl. as expect. antipyr.; in skin dis., for baths, as wound bandaging; expressed juice in hemorrhoids - **V. cistifolia** O. Hoffm. (Af): rt. u. in coughs & catarrhs. by natives - **V. colorata** Drake (trop wAf): bitter tree; b. leaf; rt. or twig u. as chewstick (tooth cleaner); chewed as gastric ton. (also lvs.) lvs. stim. antipyr.

astr. in tonsilitis, stomatitis; rt. as emet. expect.: in GC; exter. in skin dis.; rt. bk. bitter, u. astr. antipyr. in scabies - **V. conferta** Benth. (w tropAf): cabbage tree; lf. & bk. decoc. u. in colics, lax.; pd. bk. u. as vermif. (children) (Sierra Leone); antidiarrh.; ash fr. burned sts. u. to make soap ("soap tree") - **V. corymbosa** Less. (sAf): macerated lf. in epilepsy; rt. decoc. in dys. & as anthelm.; for menstr. irregularity & as abort.- **V. cuneata** Less. (Oceania): rt. sap drunk for fever (Solomon Is) - **V. cymosa** Bl. (Indones): fresh pl. u. to treat stomachache (Java); tea fr. pl u. similarly (Sumatra) - **V. deppeana** Less. (sMex to CR): suquinay (hembra); palo de asma (Salv); pl. decoc. u. for stomachache. asthma (Salv) - **V. divaricata** Sw. (Jam, Grand Cayman): old man bitter bush; pl. decoc. drunk to relieve upset stomach - **V. divergens** Benth. (Ind): bara pithol; rt. decoc. u. in stomachache, intest. dis.; lf. as paste appl. for headache (124) - **V. elaeagnifolia** DC (Indochin): wd. decoc. (in combn.) u. for bronchitis; smoke fr. burning wd. u. to treat nasal ulcers - **V. exsertiflora** Baker (tropAf): lvs. u. with other pls. for insanity (712) - **V. galamensis** (Cass.) Less. (Tanz): lf. decoc. drunk for stomachache; source of epoxy oils - **V. glaberrima** Welw. (cAf): lvs. u. against psoriasis; lvs. accumulate Cu - **V. glabra** Vatke (Af): lvs. & rt. u. as diur. in GC, amebic dys.; herb juice u. as vulner. - **V. grantii** Oliv. (Af): rt. & lf. juice u. in GC & bilharziasis; rt. decoc. as bath for skin dis. (children) (Tanz) - **V. gratiosa** Hance (Taiwan): young shoots pounded & appl. to swellings - **V. guineensis** Benth. (trop wAf): rt. chewed raw as aphrod.; rt. decoc. emet.; pounded lvs. appl. to face to relieve toothache (Guinea) - **V. hildebrandtii** Vatke (trop eAf): lvs. u. in diarrh.; arrow poison; cardiac activity - **V. hirsuta** Sch. Bip. (sAf): bitter in digest. difficulties, in colics; antipyr. - **V. holstii** O. Hoffm. (Tanz): rt. u. for relief of abdom. pain - **V. hymenolepis** A. Rich. (Ethiopia): lvs. c. vernolepin (an elemanolide dilactone), vernomenin; vernolepin inhibits carcinosarcoma in rats - **V. iodocalyx** O. Hoffm. (sAf): very bitter decoc. u. in epilepsy; lf. in abdomin. troubles; rt. to facilitate childbirth - **V. lasiopus** O. Hoffm. (Tanz, Rwanda): lf. decoc. for stomach ache, female sterility; rt. decoc. as male sex stim. - **V. leiocarpa** DC (sMex to Nic): hoja blanca; pl. decoc. u. for asthma (Salv) - **V. macrocyanus** O. Hoffm. (Angola): u. antispasm. in colic; fish poison - **V. menthaefolia** Less. (Cu): rompezaraguey (legitimo); decoc. of leafy branches u. diur. & in severe diarrh.; rubbed on as liniment in rheum.; cult. for med. use by Cuban immigrants in Fla - **V. missurica** Raf. (cUS): Missouri (or Drummond's) ironweed; aq. decoc. u. for dandruff; fls. source of purple dye - **V. natalensis** Sch. Bip. (Af): u. as remedy in mal. & other fevers; expect. in cough; inf. u. as lotion for hemorrhoids; as enema for pain in loins - **V. nigritiana** Oliv. & Hiern (trop wAf, tropAm): Nigerian ironweed; jubu jamba (Gambia); batiator (Sen); source of batiator rt., c. vallarin, vernonin (with weak digitalis action); rts. u. purg., in hemorrhoids; emetic, antidysenteric in amebiasis; (*cf.* ipecac); antipyr. - **V. noveboracensis** Mx. (eUS): ironweed; flat tops; purple fls. fragrant or slightly so (1788); hb. u. as adulter. of *Eupatorium maculatum* or *E. purpureum* (1335) - **V. oligocepahalus** Sch. Bip. (sAf): silver-leaved vernonia; variously u. in colics, dys. rheum. diabetes, as stom.; lf. as purg. - **V. parviflora** Vahl (Am+): hb. u. for dys. diarrh. (NZ) - **V. patens** HBK (sMex to Pan & nSA): tuete (blanco); crushed lvs. put into nostrils for epistaxis; (CR); decoc. u. as vermif. - **V. patula** Merr. (V. chinensis Less.) (seAs): u. ton. & antidiarrh. (Indochin) rt. antipyr. (Mal) - **V. pauciflora** Less. (tropAf): lvs. cooked in tea or porridge u. in chest or lung pain (Tanz); crushed fls. u. as scent by hunters (Nigeria) - **V. pectoralis** Baker (Mad): lf. decoc. as ton. - **V. pinifolia** Less. (sAf): u. to wipe dirt from a child (Lesotho) - **V. pumila** Kotschy & Peyr. (trop wAf): rts. eaten raw or boiled with cereal as stom. & in GC (natives) - **V. rheedis** Kostl. (Ind): lf. u. as arom. diaph. stom. - **V. roxburghii** Less. (Ind): u. similarly to *V. cinerea* - **V. scabra** Pers. (Br, Ve): palotal; pl. decoc. u. for eye dis., as bath; vermif. arom. ton. stom. - **V. schiedeana** Less. (Mex, CA): aroma (Mex); u. as rem. in colic; ton. antiphlogistic (Guat); to treat sores of domestic animals - **V. scorpioides** Pers. (sMex to Br, Par; Trin): wild patchouli; rt. decoc. aphrod. (Trin); to treat mange (Br) - **V. senegalensis** Less. = *V. colorata* - **V. shirensis** Oliv. & Hiern (V. woodii O. Hoffm.) (sAf): wild cotton; u. for stomachache; lf. inf. in chronic cough, flus, colds, febrile states; antiparasitic for hair parasites; burned wood rubbed into scarifications for rheum. - **V. smithiana** Less. (Tanz): pl. ash appl. with oil to syphilitic lesions & also to white spots on skin - **V. solanifolia** Benth. (Chin): ashes of frtg. branch mixed with sesame oil as dressing for abscesses - **V. stuhlmannii** O. Hoffm. (Af - savannas): rt. & lf. u. in lung inflammn. & purulent abscesses - **V. subuligera** O. Hoffm. (Tanz): u. as galact. (humans, cattle) - **V. teres** Wall. (Ind, seAs): pl. u. for ulcers, wounds, dislocations, (Vietnam); fl. heads ascaricidal. - **V. triflosculosa** HBK (sMex to Pan): barreto; lvs. mashed & appl. to sprains (poultice); resinous sap to cauterize wounds (CR) - **V. volkameriaefolia** DC (seAs): rt. decoc. taken to aid urination (Thail) (2448) - **V. zanzibarensis** Less. (Tanz): rt. decoc. drunk for stomachache, chest dis.; st. decoc. u. for spleen & kidney dis.; to relieve strangulated hernia; emet.

verodoxin: glycoside found in *Digitalis purpurea*; made up of gitaloxigenin + D-digitoxose.

Verona: city of nIt between Venice & Milan; trade center; v. under Orris.

Veronica agrestis L. (Scrophulari) (Euras, nAf; natd. US, WI): field (or garden) speedwell; pl. decoc. emmen., entire pl. boiled with ginger oil to make liniment for elephantiasis (Haiti); decoc. for hemorrhage; broth (with pork) as ton. (Chin) - **V. americana** Schwein. (NA, incl. Mex; neAs): (American) brooklime; speedwell; pl. diur. antiscorb. emmen. antipyr.; young sts. & lvs. edible, rich in vit. C - **V. anagallis-aquatica** L. (V. anagallis L.) (Euras, nAf, NA, SA): water speedwell; brook pimpernel; u. as digestive aid, to prevent beri beri (Kor); antipyr. gargle for throat dis. (Chin); decoc. for piles (Taiwan); adulter. of *V. beccabunga* hb. - **V. arvensis** L. (Euras, natd. NA): corn speedwell; u. like *V. beccabunga* - **V. beccabunga** L. (Euras, nAf; natd. eNA): (European) brooklime; speedwell wort; hb. c. aucubin, NBP, tannin; hb. u. in folk med. as diur. lax., to stim. appetite, depur.; in skin dis., bladder stones, bleeding gums; found in many Eu tea mixts. - **V. chamaedrys** L. (Euras, nAf, natd. NA): germander speedwell; u. much like *V. officinalis* - **V. hederifolia** L. (Euras; natd. US): ivy-leaved speedwell; u. much like *V. beccabunga* - **V. officinalis** L. (Euras, natd. NA [some consider indig. to NA]): (common) speedwell; male s.; gypsyweed; hb. c. aucubin (glycoside), NBP, tannins, saponins, alk.(?), VO (traces), mucil. org. acids, resins; u. as expect. in bronchial dis. (tea), bronchial asthma, pulmon. TB, chronic catarrhs.; diur. in cystitis, diaph.; alter., for wounds, gout, rheum., scrofula, urinary calculi; the pruritus of aging: in severe ulcerative colitis (862) - **V. parviflora** Vahl (Am+): bird's eye; hb. u. for dys. diarrh. (NZ) - **V. peregrina** L. (NA, Alas to Fla; CA, SA; now Euras, Austral): purslane speedwell; neckweed; u. in folk med. for scrofula, &c. - **V. salicifolia** Forst. (NZ): shrub to small tree; koromiko; hb. (Herba Salicifolia) u. by Maoris in diarrh. & dys. - **V. scutellata** L. (Iceland; Euras, natd. NA): marsh speedwell; hb. u. as anitscorb.; in abdom. congestion; gastric dis. diarrh. dys. (Maoris) (1870); in salads - **V. spicata** L. (Euras, natd. eNA): (upright) spiked speedwell; hb. c. D-mannitol (1.7%), luteolin glycoside; u. like *V. officinalis* - **V. triphyllos** L. (Euras, nAf, natd. US): variable-leaved speedwell; u. in jaundice (folk med.) - **V. virginica** L. = *Veronicastrum virginicum.*

veronica: **véronique** (Fr): 1) *Veronica* spp., usually *V. officinalis* 2) (Br): *Dalbergia subcymosa.*

Veronicastrum virginicum (V. sibiricum) (L) Pennell (Veronica sibirica L.) (Scrophulari) (Sib, Kor, nChin, Jap, neNA): tall veronica; Culver's root (or physic); "Leptandra"; rhiz. & rts. c. "leptandrin" (glycosidal amaroid); resins, tannin, NBP FO sitosterol; u. similarly to *Veronica anagallis-aquatica*; st. & lf. decoc. drunk as lax. & in rheum.; u. cholagog cath. emet. hepatic stim. in hepatitis, diaph. ton.; hemorrhoids (Rhizoma Leptandrae virginiae).

verosterol: a phytosterol isolated fr. rhubarb rt.

verst (Russ): unit of length, ca. 1 km (ca. 2/3 mi).

Vertavis™: a biologically standardized pd. Veratrum Viride; marketed in tabs.

Vertebrata: vertebrates: animals with a vertebral column or backbone; constituting nearly all of the Chordata; ex. bear, cobra.

verticil: verticillium: whorl.

verticillaster: apparent whorl; inflorescence made up of two opposite gps. of short-stalked fls. in axils of lvs., so arranged as to look like an actual whorl; ex. *Mentha, Salvia,* & other Lab. spp.

verticillate: whorled.

vervain: *Verbena* spp., esp. *V. hastata* or *V. officinalis* - **American v.: blue v. (herb)**: *V. hastata* - **common (European) v.: European v.**: *Verbena officinalis* - **v. family** : Verbenaceae - **native v.: official v.**: *V. officinalis* - **v. tea**: inf. prepd. fr. *Verbena hastata*; u. as spring ton. - **vervaine** (Fr-f.): *Lippia citriodora.* - **verveine officinale** (Fr-f.): *Verbena officinalis.*

Vespa carbro L. (Vespidae - Hymenoptera - Insecta): hornet; venom c. histamine, serotonin, acetylcholine, hemolyzing poison, free amino acids; in occasional cases, may cause anaphylaxis, sometimes fatal; u. like bee venom in polyarthritis, &c.; homeop.

vessel: the fundamental tissue element in xylem area for conduction of water & dissolved salts through Angiosperm sts. & rts.; actually a vertical series of cylindrical cells in xylem fuse & become altered into a single continuous tube, often of great length; carries water upwards fr. rt. to lvs., etc.; the cell wall shows characteristic sculpturing (annular & spiral in earlier cells, scalariform, pitted, & reticulate in later cells); "duct" & "trachea" are equivalent terms, the latter now going out of use.

vessie (Fr-f.): bladder.

Vestia foelida Hoffman (V. lyciodes Willd.) (1) (Solan.) (Ch, Pe, Arg, Br) (monotypic): chuplin;

wd. inf. u. as enema for reducing fever; lf. decoc. u. in baths as antirheum.; c. fraxetin; 1-acetyl-3-carbomethoxy-B-carboline (first alk. of this type found in fam.) (863): ton. stom. anthelm.; rabies(?!); exter. in gangrene.

vet (Dutch): fat, fatty (**vetolie**: fatty oil).

vetch: *Vicia* spp., esp. *V. sativa* - **common v.**: *V. sativa* - **hairy v.**: *V. villosa* - **Hungarian v.**: *Vicia pannonica* - **kidney v.**: *Anthyllis vulneraria* - **Russian v.** : **Siberian v.**: **villous v.**: *Vicia villosa* - **vetchling**: *Lathyrus sativus.*

vetiver(t) (root): **vetiver** (Fr): **Vetiveria** (L): rt. of *Vetiverria zizanioides.*

Vetiveria nigritana Stapf. (Gram.) (Sen): rhiz. u. in infant diarrh.; in prepn. of perfumed waters; to attract fishes - **V. zizanioides** Nash ex Small (sAs; widely cult. & natd. in LatinAm): vetyver; khus-(khus); tropical sweet grass; cockroach g.; khas; rt. c. 1-3.3% VO (vetiver oil) with vetivenolenes, vetivenones, &c.; very arom., u. in perfumes as arom. to flavor canned asparagus, &c.; inf. as carm. stom. stim. ton. diaph. antipyr.; analg. in headache, neuralgia; appl. exter. in rheum.; to scent clothing in wardrobes; in primitive type of air conditioning (khas ki tatti) (window screens of rt. wetted with arom. waters & fanned) (Ind, Pak) (1789): repels roaches & moths; for abdom. tumors (Ind); abort. (Maurit); VO in soaps, cosmetics (chief prodn. Ind, Mal, Java, Reunion Is, La [US]); rt. u. in baskets, brushes (174).

vetolie (Dutch): fatty oil.

ve-tsin (Chin): sodium glutamate.

viborana (CR): *Asclepias curassavica.*

Vibramycin™: doxycycline; broad spectrum antibiotic derived from oxytetracycline (obt. fr. *Streptomyces rimosus*); bacteriostatic; u. in GC (acute), syph., *Chlamydia trachomatis* inf., resp. inf., &c.; u. as HCl hydrate salt ("hyclate").

Vibrio cholerae Pacini (V. comma [Blanchard]) (Spirill. - Schizomycetes) (widely dispersed): cause of cholera, serious, often fatal, dis.; killed organisms u. to prep. vaccine, active immunizing agt.

viburnin: resinous bitter princ. fr. *Viburnum prunifolium, V. opulus* - **viburno** (Sp): *Viburnum prunifolium* - **v. comun**: *V. lantana.*

Viburnum acerifolium L. (Caprifoli.) (eCana & US s to Ga): maple-leaved viburnum; dockmackie; arrow-wood; inner bk. u. for tea drunk for cramps, colic (Menomini Indians); formerly appl. exter. to tumors - **V. cassinoides** L. (e&cNA): witherod; wild-raisin; sweet viburnum; nanny-berry; frt. edible; lvs. u. as tea ("false Paraguay tea"); bk. occurs as adulter. of Viburnum Prunifolium (3572); mildly sed. to uterus; berries eaten like bluets (frt. of *Vaccinium angustifolium, V. myrtilloides*) - **V. dentatum** L. (e&cUS): (southern) arrow-wood (bush); bk. u. diur, adulter. of Vib. Prunifolium; sd. oil is a non drying FO - **V. dilatatum** Thunb. (Chin, Jap): lvs. & shoots as anthelm. & corrective; decoc. u. to clean maggoty sores (Chin); frt. anthelm. (children) & for tapeworm in lower animals - **V. edule** Raf. (V. pauciflorum T. & G.) (Alas, Cana & nUS): mooseberry; (viburnum) squashberry; pimbina; high (bush) cranberry; very sour frt. eaten fresh or preserved for winter use (Eskimos); highly valued for jam & jellies; bk. of this ornamental shrub chewed (or u. as decoc.) in cough & pertussis; inner bk. inf. u. as purg. & in dys. - **V. foetidum** Wall. (Ind & seAs): lf. juice u. int. for menorrhagia & postpartum hemorrhage fever (Burma) - **V. inopinatum** Craib. (seAs): decoc. fr. upper parts drunk for stomachache (2448) - **V. lantana** L. (Euras, natd. eNA): wayfaring tree; twistwood; bk. u. diur. deterg. for ulcers; lvs. & frt. astr. irrit.; inf. u. as mouthwash & gargle; u. yellow dye: lvs. & bk. c. valeric acid glycoside; frt. c. pectin, resin, phytosterol; lvs. fls. & frt. c. catechol tannins; ursolic acid in bk. lvs. fls. - **V. lantanoides** Mx. (V. alnifolium Marsh.) (eNA-shady places): hobble bush; witch-hobble; tanglelegs; bk. has uterine sed. properties; u. aper. diur.; occurs as adulter. of Viburnum Prunifolium st. bk.; marketed as (Southern) Cramp Bark - **V. lentago** L. (e half NA): sheepberry; sweet viburnum; wild raisin; nannyberry; inner bk. u. as diur. tea (Ojibwas), aperient; occ. as adulter. of Vib. Prunifolium; cooked frt. edible - **V. nudum** L. (e&cUS, mostly se) possum/shonny/swamp haw; smooth witherod; bk. u. as substit. for Vib. Prunifolium - **V. odoratissimum** Ker-Gawl. (V. chinense Zeyh.) (Chin): u. in homeop. med.; fls. u. to aromatize tea - **V. opulus** L. (Euras, nAf, natd. US): European cranberry bush; (high) cranberry; Guelder rose; wild snowball; bk. (cramp bark) u. as uter. sed., antispasm. for cramps, menstr. pain, in threatened abort.; aper. expect. alter. ton. in hysteria; in homeop. med.; for many years completely replaced by the similar (but inactive) bk. of *Acer spicatum* (864); lvs. & frt. c. "viburnin," NBP; var *roseum* L. (cult.): snowball (tree); s. bush; cult. bush or small tree with large sterile fls. - **V. plicatum** Thunb. (V. tomentosum Thunb.) (Jap, Chin): Japanese snowball; cult. as ornam. bushy shrub - **V. prunifolium** L. (eUS w to Kans): (black) haw; sweet h.; stagbush; sheepberry; nanny b.; sweet viburnum; dr. st. twig, or rt. bk. (latter specified in NF, Cortex Vib. prunif.; "Viburnum" (BP+); c. "viburnin" (resinous bitter subs.) (older lit.);

salicin, saligenin; other glycosides; VO, valeric & other acids; amentoflavone, scopoletin; arbutin (traces); aesculetin; bk. u. uter. sed. (2373); antispasm. in dysmenorrhea, & other menstr. diffic.; in menorrh., threatened abortion, after-pains; hysteria & other states of pregnancy; habitual (or threatened) abortion (some consider of little value); astr. ton. diur.; antiasthmatic; bk. often adulterated (2417); fresh frt. (blue black drupes) u. in Eu homeop. med.; eaten (Am Indian females) - **V. rufidulum** Raf. (V. prunifolium var. *ferrugineum* T & G.; *V. ferrugineum* Small) (seUS): southern black haw; props. & u. identical with those of prec. - **V. trilobum** Marsh. (V. oxycoccus Pursh; V. opulus var. americanum Ait.) (sCana, nUS): (highbush) cranberry; pimbina; American c. bush; supposed source of (true) cramp bark; u. uter. antispasm. sed.

viburnum: **black haw v.**: *Viburnum prunifolium* - **maple-leaved v.**: *V. acerifolium* - **pear-leaved v.**: **sweet v.**: **tree v.**: *V. lentago*.

vicenin: flavonoid glycoside found in *Vicia, Passiflora*, &c.

Vicia L.:(Leg.) (worldwide): vetches; tares; cult. for forage, fodder, edible sds.; ornam. - **V. americana** Muhl. (NA, esp. wNA, BC [Cana] to sCalif [US]): American vetch; perenn. hb. with purple to violet corollas; young pl. rts. whole pods, black sds. boiled or baked & eaten; crushed lvs. u. in bath to relieve muscle soreness (Calif Indians) - **V. articulata** Hornem. (Euras, esp. Med reg.): one-flowered vetch; cult. as cattle fodder, green manure - **V. benghalensis** L. (sEu, nAf, natd. seUS; cult. Pac Coast [US]): purple vetch; cult. as hay crop, green manure; no data on folk med. use - **V. caroliniana** Walt. (eNA): wood vetch; u. med. (Cherokee Indians) - **V. cracca** L. (Euras to Jap, natd. Cana, nUS): tufted vetch; Gerard v.; Cana pea; u. as galactagog cooked with pork & mixed with wine; cattle fodder; green manure; as test substance for blood group detn. (phytohemagglutinins); cult. orn. scrambler - **V. ervilia** Willd. (sEu, Med reg.): (true) bitter vetch; ervil; officinal tare; sds. u. nutr. (soups), astr. formerly in diarrh. urin. diffic.; forage, fodder; cult. - **V. faba** L. (V. vulgaris Moench; Faba v. Moench) (swAs, nAf, cultigen, widely cult. & natd. in many places, ex. Eu, Cana, Calif [US], Chin): broad/field/pigeon/tick/Windsor bean; horse b. (var. *equina* Pers.); haba (Sp); fava b.; Saubohne (Ge); the large sd. is the chief bean of history, known since the earliest times; sds. c. vicine, convicine (pyrimidine glucosides), phytin; levodopa, sugars, citric acid, enzymes (α-carboxylase, amylases), many essential amino acids; u. human food & animal forage & fodder; sds. sometimes toxic when ingested raw, may result in lathyrism & favism; (inhalation of pollen may give untoward symptoms, believed allergenic); toasted beans (habas tostadas) sold on st. corners like peanuts (Bo); cult. for waterfowl (BC [Cana]); bean meal much u. in med. in antiquity & Middle Ages, thus for warts, cancer (exter. & int.); cough, urinary dis. VD; currently, seedlings source of L-dopa (865); roasted as coffee substit.; lvs. u. as emoll. discut. - **V. hirsuta** S. F. Gray (Euras, nAf, natd. sCana, US): common tare; "bindweed"; hairy t.; sds. u. as food after alkali treatment & boiling (water thrown away [HCN]); similar in flavor to lentils - **V. ludoviciana** Nutt. subsp. *leavenworthii* Lassetter & Gunn (V. l. T. & G.) (cUS): Leavenworth vetch;; spring forage pl.; green manure - **V. pannonica** Crantz (seEu, swAs): Hungarian vetch; cult. for pasturage, hay, & green manure - **V. sativa** L. (Euras, natd. eNA & worldwide): common/spring/wild vetch; tare; sds. u. earlier for measles, tumors; have antidiabetic activity; adulter. of pd. pepper; deterg. astr.; for food; an important fodder pl. - **V. unijuga** A. Br. (Siber, Chin, Kor, Jap.): two-leaved vetch; cult. as forage crop. - **V. villosa** Roth (V. dasycarpa Ten.) (Euras, nAf, natd. NA): hairy/winter/Russian vetch; u. winter legume; soil builder (turned under); for grazing, hay; sd. nutr.; silage (866).

Vicks Vapo-Rub™: "Vick's salve" (Vick Chem. Co.): very popular domestic remedy (US); ointment with menthol, camphor, oils of eucalyptus (or eucalyptol), thyme (or thymol), nutmeg, turpentine, cedar leaf; petrolatum; u. as inunction for chest, nasal passage, &c., in case of colds, &c.

Vicoa indica DC (Comp.) (Ind): bondre; pl. paste & pd. u. in mastitis & to stop bleeding (Uttar Pradesh, Ind) (—> *Pentanema*).

Victoria amazonica (Poeppig) Sowerby (Nymphae.) (Br [Amazona], Guyana): giant (or royal) water lily ("Victoria regia"); enormous (-2 m across) lvs. floating on water (support child); fls. aromatic, fugitive, -5 dm across; lvs. u. to soften ulcers & infected wounds; roasted sds. edible.

Vicugna vicugna Molina: **vicuña** (SA): *Lama vicugna*.

vid (Sp): grape; *Vitis vinifera*.

Vidarabine: adenine arabinoside; ara-A; purine nucleoside; prod. by *Streptomyces antibioticus* strain and *Cryptotethia crypta* (Caribbean sponge) (spongo-adenosine); u. antiviral antibiotic in herpes simplex (eye ulcers), herpes zoster (shingles), herpes encephalitis; in keratitis, keratoconjunctivitis.

Vigna aconitifolia (Jacq.) Maréchal (Phaseolus aconitifolius Jacq.) (Leg.) (As, esp. Ind): moth (or mat) bean; sds. (dal) u. as food, for starch prepn.; pods cooked as veget.; pls. for forage, erosion control - **V. ambacensis** Wilw. (c tropAf): lvs. dr. & smoked to control cough (Congo) - **V. angularis** Ohwi & Ohashi (As widely cult., esp. in Chin, Jap): adzuki bean; in Chin med. sds. considered strongest "yang," u. in kidney dis., constipat., boils, difficult labor, &c.; important food in Orient: sds. boiled & eaten; sprouted sds. as veget.; bean meal & paste u. in confections: forage crop - **V. aureus** (Roxb.) Hepper = *V. radiata* - **V. cylindrica** Skeels = *V. unguiculata* - **V. desmodioides** Wilczek (trop eAf, incl. Zaire): sometimes confused with *V. racemosa* Hutch. & S. Dalziel - **V. haumaniana** Wilczek var. *pedunculata* Wilczek (Congo): rts. u. as purg. - **V. luteola** Benth. (V. repens Ktze.) (sMex to Pe; WI, natd. Af): frijol de monte; pl. decoc. in coughs & colds (VI, B [Cana]); pasturage - **V. marina** Merr. (V. lutea Gray) (pantrop.): lvs. heated over fire & appl. while hot to sore (Karkar Is, PapNG) (867); for circumcision; eye dis. (Samoa); lvs. u. med. (Tonga) (86) - **V. multiflora** Hook. f. (tropAf): rts. u. as vermif. by natives - **V. mungo** (L) Hepper (Phaseolus m. L.) (prob. EI, Ind, Pak, a cultigen): black gram; urd; small, black-seeded bean; the chief pulse of Ind, rainy season crop: popular in trop. as food, in curries, as flour to make bread, &c.; sds. u. as astr. u. in diets for fevers & to strengthen the eyes; sds. poulticed to abscesses as suppurative; for tumors (Ind) - **V. radiata** (L.) Wilczek (Phaseolus aureus Roxb.; P. radiatus L.) (seAs, now OW, NW trop): mung bean; green (or golden) gram; "green soy bean" (health food stores); closely related to *V. mungo*; sds. green, brown, or mottled; the small beans are u. for sprouting (Chin, US): u. as veget. in Oriental dishes (as bean sprouts in chop suey, &c.); sds. made into "bean threads" for use in prepn. soups. &c.; both sds. & green pods popular food in As.; u. in edema, diarrh. diur.; sprouts in threatened abortion; commonly cult. in trop. (868); var. *sublobata* Verdcourt (Phaseolus trinervius Wight & Arn.) (Ind, Mal): wild mung - **V. trilobata** Verdcourt (Phaseolus trilobatus Schreb.; P. trilobus Air.) (Ind, Chin): ambi bean; lf. ton. antibilious; ext. in eye dis.; sds. edible; rt. rich in starch, u. in Chin ("koh") as emet. diaph. antiphlog.; entire pl. sed. hypnotic (Ind) - **V. umbellata** Ohwi & Ohashi (Phaseolus calcaratus Roxb.) (temp&tropAs): rice/Mambi/Oriental bean; sds. food; young pods, lvs. & sprouts also as veget.; forage, cover crop, green manure; in med. (off. Chin Phar.) - **V. unguiculata** (L.) Walp. subsp. **u.** (V. sinensis Savi) (center of origin: Ethiopia; now widely distr. OW trop): cow/Crowder/black-eyed/southern pea; bannette (Fr); cow bean; inf. of whole pl. u. emet; flour of dr. sds. softened by boiling in water u. as paste as emoll. poultice for stubborn boils (Cu, Af); frts. u. in biliousness, liver dis. & intoxication (Camb); flg. tops u. in incipient leprosy; sds. popular as food (esp. seUS) (869) (but not popular in LatinAm); pods cooked & eaten like string beans ("French beans"); lvs. young shoots & rts. also u. as foods; scorched sds. as coffee substit.; fodder; cover crop; peduncles give strong fiber; widely cult. in trop. & subtr. since earliest times; u. smother crop, green manure, forage, stock feed; although called "pea" really a type of bean; subsp. *cylindrica* (L) Verdcourt (V. c. Skeels; V. catjang Walp.) (Ind, SriL; widely cult.): catjang; sds. c. starch, proteins, FO, globulins "vignin" & "phaseolin" (870); inhibitors to trypsin & chymotrypsin; vitamins; pods & sds. food; forage; hay crop; subsps. *sesquipedalis* (L) Verdcourt (V. s. Fruwirth) (Ind, Af, widely cult.): yard-long pea (or bean); asparagus bean; with long limp pod (871); lvs. boiled with rice for headache; pulped lvs. with alum u. to stop flow of maternal milk; lf. juice u. in med. (Mal, Indones): immature pods u. as veget. like snap beans; ripe dry beans, young lvs. & sprouting sd. also u. as food; forage. - **V. vexillata** (L) A. Rich. (Austral+): wild cow pea; rts. chewed & eaten for constip. (Austral aborigines) (2263).

vigne (Fr-f.): vine, specif. grape(vine) - **v. sauvage** (lit. wild vine): *Vitis vinifera* (wild form); *Chondrodendron tomentosum; Citrullus colocynthis; Clematis vitalba; Solanum dulcamara.*

Viguiera Kunth (Comp.) (warm Am): some 160 spp. - **V. buddleiaeformis** B. &. H. (Mex): "yellow flower of Toluca"; c. alk. viguetenine. desacetyl-v., budleine A & B; with anticancer activity - **V. dentata** Spreng. (cMex to Hond, WI): jarilla; pl. decoc. u. as cough remedy; honey plant; bundles of dr. sts. lighted u. as torches - **V. excelsa** B. & H. (sMex): liga; rt. (raiz del manso) reputed vulner. u. to cicatrize wounds & cleanse ulcers; decoc. u. in dys - **V. sphaericephala** (DC) Hemsl. (sMex): vara blanca cabezona; lf. decoc. u. in gastric dis.

vilca (Latin Am): huil(l)ca: *Anadenanthera colubrina.*

vilca bark: Vilcae Cortex (L.): *Piptadenia macrocarpa.*

Villaresia conghona Miers (monotypic) (Icacin.) (nSA): gongonha (Br); lvs. u. as a sort of maté (beverage) (Br) (—> *Citronella*).

Villarsia ovata Vent. (Gentian./Menyanth.) (sAf): this water pl. u. since early Colonial days as bitter, antiscorb. antipyr.; in ointment for hemorrhoids.

Villebrunea frutescens Bl. (Urtic.) (seAs): pl. inf. u. in respir. affect. (Indochin) - **V. rubescens** Bl. (Indones): st. sap u. for dysuria; wash for inflamed eyes - **V. trinervis** Wedd. (PI): juice fr. st. of vine drunk as remedy for common cold or bad cough.

villikinin: subst. extd. fr. intest. mucosa which stim. intestinal villi.

villous: bearing long soft hairs.

vin (Fr-m.): wine - **v. de bibittes** (Qu [Cana]): wine made with granules representing an association of yeasts & molds seeded in the frt. juice (**v. d'abeilles**) - **v. de bettes** (Qu): a wine made fr. beet juice, raisins, sugar, etc. - **v. d'espagne**: sherry - **Vin Mariani** (Marian's wine): prepn. of fresh coca lvs. in wine popular ca. 1880-ca. 1910); many claims of benefit; off. in Fr Phar., Codex 1884; outlawed like other coca products in US - **v. ordinaire** (Fr): common table or dinner wine; the wine drunk in Fr fr. day to day.

vina (PR): roselle.

vinagrillo (Sp): 1) weak vinegar 2) *Oxalis* spp. (SpAm)

vinagreira: fiber of *Hibiscus sabdariffa.*

vinasse: residual liquid left behind in retort after distn. of fermented beetroot molasses grapes, &c. u. as source of K_2CO_3 & fertilizer.

vinblastine: alk. fr. *Catharanthus roseus*, antineoplastic, mostly in Hodgkin's dis. (E); mostly as sulfate hydrate.

Vinca erecta Rgl. & Schmalh. (Apocyn.) (cAs; wEu): periwinkle; some 30 alks. have been isolated fr. rts. incl. vincanine, vinanidine, reserpinine; u. hypotensive, analeptic. - **V. herbacea** Waldst. & Kit. (Euras, natd. Mass [US]): a trailing hardy pl. with firm lvs. (872); lvs. c. alks. (more in rts.), incl. herbaine, herbaceine, lochnerinine, hervine, herbaline, &c.; hb. decoc. u. against tumors (873) - **V. madagascariensis** = *Catharanthus roseus* - **V. major** L. (Eu: Med reg., sRuss, esp. Transcaucasia; Ind): (greater) periwinkle; large p.; big leaf p.; hb. (Herba Vincae [pervincae] majoris) c. reserpine, reserpinine (pubescine), vincine, vincamine, carapanaubine (vinine), & ca. 40 others; vincoside (protocatechuic acid gluco-rhamnoside) (874); tannin; u. mostly in homeop. & popular med. as astr. bitter ton. vulner. hemostat. in hemoptysis; antigalact.; abort., in diabetes (Natal, Af); more recently similarly to foll. sp. - **V. minor** L. (Eu, esp. It, Austr; AsMin, Ind) (common/lesser/small)/periwinkle; evergreen; (running) myrtle; lvs. c. many alks., incl. vincamine (reduces hypertension), vincine, eburnamenine, vincanorine, minovine: ursolic acid, ornol, β-sitosterol, triacontane; lvs. off. in Fr Codex (Pervenche officinale); & in Ge & US Homeop. Phar.; earlier u. in folk med. for tumors, menorrhagia & other hemorrhages; astr. bitter ton. (small doses) diur.; much like *V. major*: vulner. (in Vulnerary Species); in large doses purg. & diaph. - **V. pervinca** = *Catharanthus roseus* - **V. pubescens** Urv. (Transcaucasia) = *V. major* - **V. pusilla** Murr. = *Catharanthus pusillus* - **V. rosea** L. = *Catharanthus roseus.*

vincamine: chief alk. of *Vinca minor & V. major*; u. as vasodilator to reduce blood pressure, esp. in cerebral sclerosis (cerebrovascular dis.); found in "**v. fraction**" along with vincine, vincinine, & vincaminine.

Vincetoxicum Wolf (Asclepiad.): many spp. now trasferred to *Cynanchum, Gonolobus, Matelea* - **V. amplexicaule** Sieb. & Zucc. = **V. atratum** Morr. & Dcne. = *Cynanchum a* - **V. edule** (Hemsl.) Standl. (Gonolobus e. Hemsl.) (sMex, CA): gueto; cuayote (CR); large young frts. cooked & eaten as veget. - **V. hirudinaria** Medik. = **V. officinale** Moench = *Cynanchum vincetoxicum* (vincetoxico, Sp; Vincetoxicum, L.) - **V. edule** (Hemsl.) Standl. (Gonolobus e. Hemsl.) (sMex, CA): gueto; cuayote (CR); large young frts. cooked & eaten as veget. - **V. hirudinaria** Medik. = *Cynanchum vincetoxicum* .

vincristine: leurocristine; Oncovin™; alk. isolated fr. *Catharanthus roseus* u. as antineoplastic, in acute lymphocytic leukemia (877) (esp. useful for children); Hodgkin's dis., lymphosarcoma; reticular cell sarcoma; rhabdomyosarcoma; neuroblastoma; Wilms' tumor (876); u. mostly as sulfate (E).

vine: 1) twineing herbaceous pl. 2) (specif.) grape; g. vine (Brit) - **v. family**: Vitaceae - **pipe v.**: *Aristolochia sipho* - **poison v.**: *Rhus toxicodendron* - **porch v.**: Kudzu - **smoke v.**: *Adlumia* spp. - **squaw v.**: *Mitchella repens* - **wild v.**: *Vitis labrusca*, etc.

vinegar: acetum (L); an acetous liquid generally made by fermentation with *Acetobacter aceti* (Pasteur) Beijerinck (Mycoderma aceti Pasteur), *A. pasteurianum* Beijerinck, *Bacterium curvum* Henneberg, &c., of alc. liquid (incl. wine, cider, dil. alc. [spirit], beer, barley malt fermentation, &c.); c. dil. acetic acid (c. 4-10%), its esters, acetone, &c.; u. flavor, seasoning, in pickles, salads, sauces, &c.; many folk med. uses (2735) - **v. acid**: acetic acid - **aromatic v.: balsamic v.:** combination of VO of bergamot, cassia, clove, lavender, lemon, with tinct. tolu, tinct. benzoin; & alc.; (BPC 1934) - **beer v.: British v.**: prepd. fr. beer or fermented wort - **cider v.**: prepd. by continued fermentation of hard (apple) cider; also fr. pear juice, palm juice, &c. - **glucose v.**: fr. corn syrup - **grain v.**: fr. (grain) alc. - **grape v.**: fr. grape juice (2 types, red & white) - **malt v.**: beer v. - **v. of the four thieves**: an herbal vinegar u. by thieves during the medieval Great Plague who by covering their faces with cloths soaked in the vinegar supposedly protected them fr. infection & permitted them to plunder houses, people, &c., unrestrained by any fear of catching the infection (1894); many formulas have been proposed - **radical v.**: glacial acetic acid - **spirit v.**: that prepd. fr. alc. prods. of rye, corn, barley - **sugar v.**: prepd. fr. alc. products of sugar, syrup, molasses. honey, &c. - **toilet v.**: aromatic v. - **v. tree** (Qu [Cana]): *Rhus typhina* (SC frt. u. to make v.) - **v. weed**: *Trichostema lanceolatum* - **white v.**: spirit v. - **wine v.**: grape v.

vinettier (Fr): *Berberis.*

Vinland: early Scand name for NA given by Norsemen ca. 1000 AD (SC wild grapes so common).

vino (It-m., Sp-m.): **vinho** (Por-m.): wine.

Vinum (L): wine - **V. Adustum** (L): alc. - **V. Ardens** (L): brandy - **V. Album** (L): white wine - **V. Crematum**: (L): brandy - **V. Gaduol** (L): - **V. Jecoris** (L): wine of codliver oil - **V. Portense** (L): port wine - **V. Rubrum** (L): red wine - **V. Xerense: V. Xericum** (L): sherry.

vinyl: radical, -CH=CH_2

Viola L. (Viol.) (cosmop, mostly nTemp Zone; Andes): violets; some 400 spp. of mostly perenn. hbs. or subshrubs (rarely); in warm climates, may bloom year around (Tex [US]) (878) (see under violet for common names) - **V. adunca** Sm. (Alas, Calif, nUS; Cana): western dog (or hooked spur) violet; rts. & lvs. eaten during labor (Makah Indians, Wash [US]); whole pl. decoc. for stomach pain (Calif Indians) - **V. betonicaefolia** Sm. (V. partrinii DC) (sChin, Indo-Pak, Indochin, Austral): lvs. sts. fls. u. to make dressing for wounds & sores, also inf. as purg. & for coughs & colds & to make a refreshing beverage (tisane) u. during the hot season (in mixt. with other pl. materials, incl. *Berberis asiatica, Lactuca indica*, &c.) - **V. biflora** L. (Ind, Pak): lvs. emoll. lax.; fls. emollient, diaph. antispasm. pectoral; rt. emet. - **V. canadensis** L. (e&cNA): Cana violet; tall white v.; rt. decoc. u. by Ojibwas for pains in bladder region - **V. canescens** Wall. (Ind): "bansa"; fl. decoc. u. to relieve throat pain in colds (Uttar Pradesh) - **V. canina** L. (Euras fr. Med reg. to Jap): dog('s) violet; rt. c. violine, saponins; u. as expect. in coughs & colds - **V. chinensis** G. Don (Chin): rhiz. decoc. u. as sed. in diarrh.; pounded rhiz. appl. to abscesses, swellings, & scrofula - **V. cinerea** Boiss. (Pak, Ind): pl. u. like *V. odorata* - **V. conspersa** Reich. (eCana, eUS, s to Ala): American dog violet; whole pl. u. for tea for treatment of heart trouble (Ojibwas, Flambeau) - **V. decumbens** L. f. (sAf): wild violet; u. as poultice for painful varicose swellings in leg - **V. diffusa** Ging. (Taiwan): decoc. u. as antipyr. carm. - **V. distans** Wall. (Indochin): u. like *V. diffusa* - **V. fimbriatula** Sm. (e&cUS s to Ga): fever (or fringed) violet; rich in mucilage; u. as substit. for slippery elm. bk. - **V. floribunda** Jord. = *V. odorata* - **V. glabella** Nutt. (Pacific coast fr. Alas [Aleutians] to Calif): stream violet; rt. decoc. purg. - **V. hirta** L. (Euras): hairy violet; u. in med. Ukraine, USSR - **V. hondoensis** W. Becker & Boiss. (Jap): pl. in wash u. exter. for abscesses (or pounded & laid on as poultice) - **V. japonica** Langsd. (Jap, sKor, Kyurus): water macerate of pl. appl. to inflamed eyes - **V. klossii** Ridl. (PapNG): pl. juice to allay toothache - **V. maderensis** Lowe (swEu): fls. off. in Por Phar. - **V. mirabilis** L. (Euras to Jap): broad-leaved violet; wonder v.; rhiz. purg.; pl. decoc. u. for heart dis., palpitation, shortness of breath (Ukr); c. salicylic acid - **V. obliqua** Hill (V. cucullata Ait.) (e NA): common (or marsh) blue violet; u. med. by Utah Indians [US]; hb. & fls. u. to make soup; ornam. perenn. hb. - **V. ocellata** T. & G. (Calif, Ore [US]): two-eyed violet; heart's ease; pinto pansy; cult. ornam. - **V. odorata** L. (Euras, natd. Af, NA, SA): sweet (-scented) violet; garden v.; English v.; viola (suavis); rt. (Viola, L.) c. violine (or viola-emetine), saponins, violutoside (gaultherin) (glycoside yielding methyl salicylate, glucose, arabinose) NBP; u. expect., emet., purg. replacement for ipecac & senega; in skin inflammn., ulcers, wounds; claimed to remove cancerous growths in stomach & intest. (Mex, Arg); fls. c. VO (with ionone derivs.); malic acid; glycosides (violaquercitrin, violarutin); wax, mucil.; u. as expect. emet. nervine, sed., mild cath. emoll. antisep. in candidiasis (dis. of skin+); for prepn. v. syrup (admin. to children for bronchial catarrhs., &c.); hb. u. as diaph. expect., antisept., in lithiasis (also rts. & sds.); sd. c. saponin, myrosin, salicylic acid; u. diur purg. emet.; VO fr. fls. very expensive, u. in some perfumes; state pl. of Fla (US) - **V. palmata** L. (eNA): palmate (or early blue) violet; pl. rich in mucil. u. to thicken soups (US Negro, "wild okra") - **V. pedata** L. (e, c&sUS): bird's foot (or pansy) violet; pl. rts. c. mucil. u. as expect. lax. emet. - **V. pinnata** L. (Chin): bruised entire pl. appl. to foul ulcers & cancers; anodyne, cleansing, made into pills to treat cancerous ulcers; often u. as pot-herb; thought to purify blood

V. pubescens Ait. (neNA s to Tenn, w to Neb): downy yellow violet; rt. decoc. u. for sore throat (Ojibwa Indians); var. *eriocarpa* Russell (V. pensylvanica Mx.) (seUS): smooth yellow violet; u. similarly - **V. serpens** Wall. (V. repens Buch.-Ham.) (nInd, common at Dehra Dun): pl. antipyr. diaph. fls. emoll. dem. in "biliousness," pulm. dis., in syr. for infant dis.; rhiz. rt. emet. (Nepal) - **V. sylvestris** Lam. (Ind, Kashmir): pl. u. for resp. dis.; sts. lvs. fls. bruised & appl. to wounds & sores - **V. symei** Baker = **V. syrtica** (Eu, Russ): c. violaquercetin, salicylic acid; u. cath., in skin dis., "to purify the blood" - **V. tricolor** L. (Euras, widely natd. US+): (wild) pansy; hearts-ease; Johnny jump-up; cat's face; viola del pensiero (It); pansy hb. coll. during flowering; hb. (Herba Jaceae) c. saponins, viola-quercitrin (rutin) (lvs. 0.13%, fls. 23%); violanin (in petals-33%) an anthocyanin, (delphinidin triglycoside, a blue chromoglycoside); tannin, sugar, mucil.; u. expect. diur. diaph. purg. in urin. dis. (care!); exter. in skin dis. hemorrhoids; int. in gout, arthritis, as stom.; gastric & bladder dis.; depur. (folk med.) (1790); u. esp. in pediatrics; fls. (petals) u. in skin dis., chronic bronchitis, catarrhs., gargle; rt. inf. u. in dys. of child (Ind) (879); subsp. *arvensis* (Murray) Gaud. (V. a. Murr.) (Euras cosmop.): wild pansy; with pale yellow petals; said less useful in therapy; a source of rutin; var. *maxima* (var. hortensis DC): (cultivated) pansy; ornam. - **V. triloba** Schwein. (eUS w to Tex) u. similarly to *V. palmata*; cooked as greens, considered lax. (SC [US]); bruised lvs. emoll.; rts. emet. purg. (2431).

Violaceae Batsch: violet fam.; some 23 genera & 830 spp. of ann. or perenn. hbs. or shrubs, worldwide; chief genus *Viola*; others *Hybanthus, Leonia.*

violanin: anthocyanin glycoside, made up of delphinidin combined with glucose, rhamnose, & p-hydroxy-cinnamic acid; occ. in petals of *Viola tricolor*, &c.; a chromoglycoside: blue dyestuff; emetic.

violaquercitrin: violarutin: rutin (named fr. discovery in *Viola tricolor* & other *V.* spp.).

violaxanthin: carotenoid pigment occ. in *Viola tricolor; Iris pseudacorus,* alfalfa, & many other plants.

violet: *Viola* spp., esp. *V. odorata*; fls. candied (or "crystallized") & u. to garnish salads - **African v.**: *Saintpaulia ionantha* - **bird's foot v.**: *Viola pedata* - **bland v.**: *Viola macloskeyi* Lloyd subsp. *pallens* M. S. Baker (V. blanda Willd.) - **blue v.** (flower): *V. obliqua* - **Canada v.**: *V. canadensis* - **canker v.**: *Viola rostrata* Pursh - **common v.**: *V. odorata* - **Confederate v.** (Ala [US]): *Viola priceana* Pollard - **dog tooth v.**: *Erythronium* spp. - **English v.:** common v. - **v. family** Violaceae - **flame v.**: *Episcia cupreata* & *E. reptans* - **v. flowers**: common v. - **garden v.**: pansy v. - **v. herb**: common v. - **lance-leaved v.**: *Viola lanceolata* L. - **long-spurred v.**: canker v. - **pansy v.**: *Viola tricolor* var. *maxima* - **v. root: sweet v.**: fr. common v. - **Trinity v.: wild v.**: pansy v. - **yellow v.**: *Viola tricolor* (wild); *V. pubescens; Cheiranthus cheiri.*

violeta (Sp, Por): 1) *Viola odorata* 2) *Anoda hastata* - **violrot** (Swed): lit. violet rt.; Iris rhiz.

Viomycin: antibiotic subst. prod. by *Streptomyces floridae* Bartz & al., *S. vinaceus* Jones, & other *S.* spp. (1951-); suppressive action on TB organism (tuberculostatic) (as sulfate).

Viorna Reichb. = *Clematis.*

Viosterol: Calciferol: ergosterol which has been irradiated with UV light; vit. D_2.

viper: a snake, *Vipera berus* (Viperidae) (Euras): common European v.; the bite is poisonous, but rarely lethal; venom is u. (Ind) to check hemorrhage in vital tissues (as eye) - **v. grass: v. root**: *Scorzonera hipanica.*

Vipera Laur. (Viperidae - Ophidia [Serpentes] - Reptilia): (true) vipers; limited to Euras & Af (not pit vipers); venomous - **V. ammodytes** L. (Eu, esp. Austria, Balkans): the most dangerous venomous snake of Eu; males to 90 cm long; venom u. in neuralgias, rheum. pain caused by tumors, styptic - **V. aspis** L. (V. redii) (s&cEu, esp. Fr, It, Elba): asp; male grows to 75 cm length; venom u. in rheum., thrombosis, styptic, septic fever - **V. berus** L. (c&eEu; As [Siber]): common (European) viper; adder; peliade; venom u. in asthma, in ointment for vasomotor rhinitis, in rheum., pelvic pain (ischias); by homeop. (fresh venom in triturations) in phlebitis, varicose ulcers; to check hemorrhage in vital tissues (as eye) (Ind.); dr. viper formerly used; body fat (grease) also u. med. - **V. lebetina** L. (Eu): venom c. proteases, phosphoesterases, &c.; u. similarly to other viper venoms - **V. russelli** Shaw (Ind, seAs): Russell's Viper; brightly colored serpent; venom u. in menorrhagia, Werlhof's disease (idiopathic thrombocytopenic purpura), hemophilia, &c.; to remove blood clots from brain.

Viperidae: viper fam. (Ord. Squamata, sub-ord. Serpentes - Cl. Reptilia): incl. *Vipera* (true vipers), *Bitis* (*qv.*), *Cerastes*, but not pit vipers (Crotalidae) (rattlesnakes, &c.).

Vipers: 1) any venomous snake 2) members of serpent fams. Viperidae and Crotalidae 3) (specif.): *Vipera* spp., & sometimes *V. berus* - **v.'s bugloss**: *Echium vulgare* - **Malaysian pit v.**: *Calloselasma rhodostoma* (Boie) (Viperidae) (Mal, Thail, Vietnam, Laos, Indones [Java]): venom u. to remove blood clots in brain.

Viractin: antibiotic subst. fr. ferment. mother liquors of *Streptomyces griseus*; u. antiviral in upper respir. tract dis. (colds; Asian flu).

Virectaria major Verdc. (Rubi) (Burundi): crushed rts. & fresh lvs. u. as vulner. by appln. to wounds; inf. of lvs. & fls. u. for stomach pain.

Vireya Bl. = *Rhododendron.*

virgin: as u. in pharmacy, first prodn. or extra pure quality - **v. dip**: first grade of crude gum turpentine coll. fr. pine tree - **v. (olive) oil**: that fr. first cold expression of barely ripe, hand-picked, best grade olives; c. chlorophyll, giving greenish coloration - **Virginia(n) creeper**: *Parthenocissus quinquefolia* - **V. dip**: virgin d. - **Virginia**: v. also under snake root, stonecrop - **Virginische Schlangenwurzel** (Ge): Serpentaria - **virgin's bower**: *Clematis virginiana* (also **common v. b.**); *C. recta* (also **upright v. b.**); *C. flammula* (also **sweet-scented v. b.**) - **v. milk**: a mixt. of benzoin tr. with water or rose w. (rendered opalescent).

Virginiamycin: virgimycin: antibiotic mixt. obt. fr. organism close to *Streptomyces virginiae* Grundy & al; (A_1, M_1 fractions); u. antibacterial, antifungal, food additive.

viroid: smallest known agts. of infectious diseases; represents small fragment of RNA without protein shell (as in virus); causes some dis. formerly thought viral in origin (ex. hop stunt); tomato bunchy top; spindle tuber of potato (880).

Virola bicuhyba (Schott) Warb. (V. oleifera [Schott] A. C. Smith) (Myristica b. Schott) (Myristic.) (Br): bicuiba; sds. ("oil nuts") rich in arom. concrete FO (bicuhyba or ucuhyba fat; sangue (de b.); c. fatty acid esters, incl. isopropyl myristate; u. by local inunction for gout, rheum., ischias, hemorrhoids; in prepn. of myristic acid & glyceryl trimyristate; aril sl. arom.; sd. u. in colic, as stom. (in large doses toxic); bk. astr. in diarrh.; dr. latex as plaster, substit. for copaiba - **V. calophylla** Warb. (Co, Ec): inner bk. squeezed to prod. red resinous subst. appl. to fungous inf., scabies, mite infestations (similarly for *V. peruviana*) (Ec Indians) & to treat skin sores; processed & snuffed as hallucinogen. - **V. calophylloidea** Mgf. (Co): u. similarly to prec. (Co Indians) (region of Vaupés River); sd & bk u. in form of intoxicating snuffs (epena) (Waika Indians); c. tryptamine; death prod. in overdoses - **V. elongata** Warb. (Colombian Vaupés): resin fr. inner bk. ingested directly by Maku Indians for hallucinogenic purposes: like others of the gp., c. mono- and di-methyl tryptamines & derivs. (881); resin u. to prep. hallucinogenic snuff (epena); this considered most important source; u. often with additives, esp. *Justicia pectoralis* var. *stenophylla*; resin also u. to treat erysipelas, bk. inf. to wash wounds (Br) - **V. guatemalensis** (Hemsl.) Warb. (Guat to Pan): sd. called African oil nut; kernel rich in FO - **V. otoba** (H & B) Warb. (Myristica o. H & B) = *Dialyanthera o.* - **V. peruviana** Warb. (Co): bk. c. tryptamines, u. hallucinogen. - **V. sebifera** Aubl. (WI, Guian, Br) red acuuba; sds. c. odorless virola fat (or v. tallow); u. in mfr. candles, soap, ointments; for cataplasms in rheum.; bk. c. N,N-dimethyl-tryptamine; u. as hallucinogen ("cuajo negro") by Waica Indians - **V. surinamensis** Warb. (nBr, Guian, WI): white ucuuba; cuajo; sds. ("oilnuts") c. 72% FO virola (or oilnut) fat; with much myristin & free fatty acids - **V. venezuelensis** Warb. (Ve): cuajo; sd. c. FO (virola fat), VO; u. principally in rheum.

virology: science of viruses & of viral diseases.

virracas (Pe): *Arracacia xanthorhiza.*

Virus Vaccinicum (L.): smallpox vaccine.

virus: extremely small agt. (10-250 nm) (Prokaryotae) causative of infections of pls. & animals; some consider living organisms, others complex chem. substances with large, self-reproducing molecules, made up of sheath of protein encasing core of nucleic acids; depends on living cells which it penetrates for reproduction; some attack bacteria, called (bacterio)phages; some cause cancers - **plant v.**: one wh. can replicate only within living pl. cell.

visammin: khellin.

viscera (singular, viscus): internal organs of an animal; in case of fish, includes all organs except liver, stomach, kidney, milt (male reproductive organs with sperm), roe (ovary with eggs); one source of vitamin oils.

vischio (It): **visco**: *Viscum album.*

viscid (pron. vis-kid): viscous - **viscosity** (pron. vis-kos-ity), prop. of being **viscous** (pron. vis-kus); sticky, glutinous, resistant to flow.

viscogen: calcium saccharate u. in whipping cream.

Viscum album L. (Visc.) (Eu, nAs, nwAf): (European) mistletoe; visgo (Por); vischio (It); gui (Fr); Mistel (Ge); semi-parasite on many tree spp., esp. *oaks* (preferred), apple trees, birch, beech; entire pl. (branches) (Herba Visci [albi], Viscum) c. lupeol, tyramine, acetylcholine (883); histamine, toxic proteins (mistletoe lectins) (viscotoxins) (882); u. as hypotensive, in arteriosclerosis, essential hypertension (884); arthroses; dermatoses; prophylaxis & treatment of (cancerous) tumors; claimed effective in lung cancer (studies; 1994); in epilepsy (since Roman times) (2274); berries u. for bird lime, a viscous mass placed on poletops to entrap birds, frt. (berry) eaten by birds, but sd. with layer of mucilage (viscin) around sd. is scraped off with bill on tree branch where sd. adheres & germinates; gen. name fr. "viscid" (frt. flesh); prodn. in sGe, Tyrol, nIt, Hung, Balkans) - **V. articulatum** Burm. f. (tropAs, Austral): sts. much jointed, hence u. for aching limbs with fever (doctrine of signatures); cooked with pork, broth given for TB in children; ton. for general weakness (Chin); decoc. in bronchitis (Indochin); in neuralgia (Taiwan) lvs. u. as adhesive for fractures, bruises (Sikkimi) - **V. capense** L. (sAf): (Cape) mistletoe; Mistel; rubbed on warts to make disappear (Eu, in sAf); formerly st. astr. in epilepsy, St. Vitus' dance; asthma, bronchitis; in nose bleed (frts.) - **V. coloratum** Nakai = *V. album* var. *c. Ohwi* - **V. monoicum** Roxb. (Ind+, Burma): dr. lvs. u. as substit. for strychnine & brucine (transfer of alks. fr. host[?]) - **V. orientale** Willd. (As & Austral, trop, esp. Ind, sChin): banda (Hind); lvs. pd. made into paste appl. to ear dis. & for pustules; as substit. for Nux Vomica (Ind) - **V. pauciflorum** L.f. (sAf): u. as astr. - **V. quercinum** auct. ex Reichb. = *Loranthus europaeus*.

Vismia brasiliensis Choisy (V. laccifera Mart.) (Gutt.) (Br): bloodwood; source of drastic gum-resin like gamboge; u. ("lacre") to treat skin dis.; caulking for boats (Ve) (2430) also inhalant for bronchitis - **V. cayennensis** Pers. (n&wSA): sroinani pina; swinani; u. fish poison (Surinam); for "lota" (skin dis.) (bk.; expressed juice): st. source of a yellow gum-resin like gamboge, drastic - **V. ferruginea** HBK (Belize to Ve, Co): sangre de perro; orange latex u. to treat white skin spots (San Blas Indians); skin sores; to dye fabrics - **V. guianensis** Choisy (Pan, Guianas to Co, Ve, Br): ach(i)otillo; latex u. as drastic (Co); appl. locally to skin dis. (CR); bk. ton. & antipyretic (Br): food coloring - **V. leonensis** Hook. f. (trop wAf): bk (with yellow resin) pounded & incorporated into ointment u. for craw-craw; inf. of young lvs. u. in gleet, as purg.; for throat inflammn. (Sierra Leone) - **V. mexicana** Schlecht. (sMex to Hond): camparaguey; decoc. of latex u. as mouthwash & for toothache; fresh lvs. appl. to forehead for headache.

visnaga = bisnaga (*qv.*).

visnagin: furanochromone; xall. compd. found in *Ammi visnaga*; breaks down to prod. khellin & khellinin; acts as antispas. (2238).

Visnea mocanera L. f. (The.) (monotypic) (Canary Is, Madeira): frt. mócan (mocanes) eaten, furnishes syrup (charcherquem) u. for hemorrhages; rt. astr.

Vitaceae: (Vitidaceae); vine (or grape) fam.; woody vines or shrubs, mostly trop/subtr; with alternate palmately lobed or compd. lvs.; fls. perfect, small, greenish, stamens & petals opp. & equal; frt. a berry with large, hard sds.; chief gen. *Vitis*; others *Parthenocissus; Ampelopsis, Cissus.*

vital dye: v. stain: intravital s.: one injected into living animal body to stain certain cells or tissues selectively; rel. non-toxic.

vitamer: gp. of chem.-related subst. wh. specifically relieves vitamin deficiency in pl. or animal; ex. D. vitamers.

vitamin: substance essential to the maintenance of normal health in vertebrates; occurs in rel. small amt. in most foods of natural origin; believed to serve principally as catalyst in many metabolic processes; when absent or present in subminimum amts., vitamins bring on deficiency dis. which in aggravated form may result in death (1791); accessory food factor; component of a normal diet; other proposed designations (2566) - **vit. A:** retinol & r. esters; vit. A_2 & vit. A_3 distinguished fr. vit. A_1 (= "vit. A")(neovit. A is an isomer also found naturally); essential for normal state of mucosal epithelium (incl. resistance to infection), visual acuteness, etc. (E); occurs in codliver & other similar fish oils (**fat-soluble v. A;** oleovitamin A), in milk fat, carotene & many cereals (as oats, wheat germ, etc.); neovitamin A constitutes ca. 35% of both natural & synth. vit. A; the mammalian liver converts pro-vit. A (carotenoids) into carotene & vit. A (ß-carotene; vit. A deficiency —> night blindness, optical dis.; death (2419); u. in vaginal inf., warts (claim), for nyctalopia (night blindness), xerophthalmosis, hyperkeratosis, tooth & epithelial tissue growth; in chronic rhinopharyngitis sicca; ozena (2418); common infections (885); total synth. vit. now available (886); some consider anticancerogenic (also C & E); incr. immume properties of blood (1792) - **vitamin A acid**: retinoic acid; natural prod.; u. in keratosis, to treat severe acne - **vit. B**: now called vit. B complex; sometimes u. to designate vit. B_1 - **vit. B complex**: a gp. of water-soluble vitamins occurring often in association, as in liver, yeast, rice polishings, etc.; incl. B_1, B_2, B_6, nicotinic acid, nicotinamide, pantothenic acid, biotin, folic acid, adenylic acid, inositol, para-amino-benzoic acid, & perhaps others - **vit. B_c**: folic acid - **vit. B_p:** choline - **vit. B_T;** L-carnitine: factor required for growth & survival of *Tenebrio molitor* (mealworm) - **vit. B_x:** para-amino-benzoic acid - **vit. B_1:** thiamine (hydrochloride) (*qv.*) (formerly known as **vit. B**) - **vit.B_2**: riboflavin (*qv.*): vit. G -

vit. B_3: an undetermined factor said required for growth maintenance of pigeon; variously appl. to niacin(amide), pantothenic acid & vit. E - **vit. B_4**: undetermined factor said required to prevent paralysis of rat & chick - **vit. B_5**: undetermined factor said required for growth maintenance of pigeon; variously apppl. to niacin(amide), pantothenic acid, & vit. E. - **vit. B_6**: pyridoxine (*qv.*) - **vit. B_q**: pl. growth regulator (depressant); dominozide; has been synth. - **vit. B_8**: adenylic acid (occurs in muscle & yeast; necessary to muscular energy) - **vit. B_{10}; vit.B_{11}**: former designations for folic acid type compds. - **vit. B_{12}**: cyanocobalamin; accessory food factor; component of a normal diet; subst. obt. fr. liver or through growth of soil bacteria; c. cobalt; believed to be anti-anemia factor present in liver; hematopoietic; reduces precancerous lung cells in lung of smoker (with folic acid); u. in pernicious anemia, retarded growth in children, etc. (E) (as hydroxocobalamin) (vit. B_{12a}) in neuropathy, trigeminal neuralgia (IM, often gave remarkable relief); in pain of herpes zoster (shingles) (inj.) (2394) - **B_{13}**: related to orotic acid - **vit. B_{14}**: xall. compd. fr. human urine; has inhibiting effect on growth of some neoplastic tissues - **vit. B_{15}**: pangamic acid; found in apricot kernels; no definite comp. established; some consider an imaginary vitamin; supposed to give energy (2565) (2523) - **vit B_{17}**: amygdalin: D-mandelonitrile-ß-D-glucosido 6-ß-D-glucoside; "laetrile"; "nitriloside"; cyanogenic glycoside fr. ground apricot pits (& sds. of other Rosaceae); recommended for cancer (E. T. Krebs, 1920-); proofs lacking; danger of cyanide poisoning - **vit. C**: ascorbic acid (*qv.*) - **vit. D**: the antirachitic vitamin (or factor) (some class as hormone); u. in infectious dis., colds, heat prostration; exter. for pain (sciatica, gout, &c.), night leg cramps; codliver oil rubbed in; can be overdosed (danger!); the following fractions now recognized: **vit. D_1**: impure calciferol (with lumisterol) (obsolete term); synthetic form of vit.; **vit. D_2**: calciferol (*qv.*); **vit. D_3**: activated (or irradiated) 7-dehydrocholesterol; "chick vit D" (SC usefully employed in treatment of poultry where vit. D_2 is almost useless agent; u. to treat P & Ca metabolic dis. such as rickets, nutritional defects, osteomalacia, spasmophilia, to aid callus formation in fractures, &c.; **vit D_4**: 22:23-dihydrovitamin D_2; made by irradiating 22:23-dihydroergosterol; **vit D_5** irradiated 7-dehydrositosterol - **vit. E:** tocopherol; reproductive factor; anti-sterility vit.; anti-dystrophic vit.; term generally considered equiv. to a-tocopherol or its acetate (greatest biol. activity); β-tocopherol, less potent, may also be present; some of richest sources are FO of wheat germ, corn, sunflower seed, alfalfa, lettuce, olive, oats; synthesized 1938; some consider vit. of natural origin superior; FO of animals very poor vit. source; however, occurs in rich amts. in placenta; u. in sterility, habitual & threatened abortion, premenstrual syndrome (also vit. B_6); muscular dystrophy, myositis, sclerosis of legs, peripheral neuropathy, nocturnal leg cramps ("restless legs"); intermittent claudication; chronic torticollis (wry neck); card. & hepatic dis.; u. in skin care (increases skin elasticity)); vasomotor disturbances of menopause; appl. directly to plantar & other types of warts; antioxidant in FO; mostly u. as esters (acetate: succinate) (2375); said to prevent heart attack (prophyolactic) in many cases - **vit F** (term obsolete): certain essential unsaturated fatty acids, esp. linoleic, linolenic, arachidonic, needed by small animals (some state for humans also); u. in allergies, eczema - **vit G**: vit B_2: riboflavin (*qv.*) - **vit H**: biotin (*qv.*); coenzyme R; antiseborrheic vit. - **vit. H^1**: paraamino-benzoic acid (*qv.*) -**vit K**: coagulation factor; antihemorrhagic vit.; a series is known, of which vit. K_1 and K_2 (naphthoquinones) are of natural origin; others are synthetic (naphthalene derivs.); vit. K rarely deficient - **vit. K_1**: phylloquinone; antihemorrhagic v.; obt. fr. green pls. or by synth.; u. in prothrombin deficiency; to reverse anti-coagulating action of Dicumarol (etc.); u. orally or IV (SC Koagulations vitamin, Ge, coagulating vit.) - **vit. K_2:** antihemorrhagic vit., found in microrganisms, soil cultures; decayed fish meal; etc.; u. like v. K_1 but rather less active - **vit. K_3**: Menadione (*qv.*): methyl naphthoquinone; menaphthone; has vit. K activity - **vit. L**: lactation factors (water-soluble); occ. in liver, yeast; not yet thoroughly studied - **vit L_1**: o-aminobenzoic acid (fr. beef liver) - **vit. L_2**: an adenine deriv. (fr. yeast) - **vit. M**: folic acid (*qv.*) (SC lack prod. macrocytic anemia in *Macacus* spp. (macaque; monkeys) - **vit. P (complex):** term sometimes appled to the permeability factor; often associated in the pl. kingdom, esp. in Citrus frts., with vit. C; claimed to maintain normal capillary resistance & elasticity; widely distributed series of compds. (bioflavonoids) with vit. P activity, incl. citrin, rutin, eriodictyol, hesperidin chalcone (hesperidoside); quercitrin, esculin (esculoside), etc.; use of term discouraged, since vit. nature inadequately demonstrated - **vit. PP**: pellagra preventive factor; P. P. factor: niacin(amide) (*qv.*) - **vit. R: vit. S**: "chick development factors"; biotin(?) - **vit. T:** termitin; tegotin; **vit. T Goetsch**; termite vit.; "platelet factor"; complex of growth-stimulating factors; orig. obt. fr. termites; also found in roaches, fungi; extd. fr. yeast, *Candida (Torula) utilis*; said to play role in control of growing & regenerating of tissues; said increases spermatozoa count & quality (2237) - **vit. U**: cabagin-U; anti-ulcer factor; Stokstad-Manning factor; obt. fr. cabbage lvs. & other green veget.; supposed essential to chick growth; u. to treat peptic ulcer (U for ulcer) (anti-ulcer) - **vit. V**: "chick development factor"; biotin(?); para-aminobenzoic acid - **vit. W**: "Elvehjem factor" - **vit. X**: vit. P or biotin(?); term formerly in use.

vitamin, anti-anemic: folic acid - **antiberiberi v.:** vit. B complex - **anticheilosis v.:** vit. B_2 - **anti-eczematic v.:** vit. F - **anti-hemorrhagic v.**: vit. P - **anti-pernicious anemia v.:** vit. B_{12} - **anti-pneumonia v.**: vit. J - **anti-sterility v.** : vit. E - **anti-xerophthalmic V**: vit. A.

vitaminology: the chemistry & application of vitamins.

vite edera (It): ivy (vine) - **vitebranca** (It): bryony.

Vitellaria paradoxa Gaertn. f. (Butyrospermum parkii Kotschy) (Sapot) (tropAf in upper Guinea, upper Nile reg): shea tree; shea butter (u. culinary fat) is obt. fr. oil-rich sds.; fat u. as base of Kiuma Ointment™; u. in arthritis; also nut press cake prod.; bk. u. in leprosy & dys. (Nigeria).

Vitellus Ovi (L): egg yolk; source of vitellin = lecithin.

Vitex L. (Verben.) (worldwide trop & temp): chaste trees; some 250 spp. of trees & shrubs; lvs. opposite - **V. agnus-castus** L. (Euras, Med reg.): hemp-tree; chaste lamb; monks' pepper (tree); chaste tree; pl. with strong arom. odor; frt. popular in homeop. med.; with 4-celled stone; sds. arom. u. stim. carm. in depressions, impotence (aphrod.); in eye dis.; gastritis; rheum (Trans Jordan); pepper substit. (2240); supposed to reduce sex desire (by medieval monks) (Moenchspfeffer); bk. stim. ton.; lvs. spice (Af) (2239) (3576) - **V. altissima** L. f. var. *alata* Mold. (sAs, Ind): ahay; bk. u. in rheum. swellings; wd. yields yellow dye - **V. amboniensis** Guerke (trop & sAf): mtalali (Swahili); antidote for snakebite (Swahili) - **V. cofassus** Reinw. (eMal, wPolynesia): "father"; wd. u. to make bowls, drums, &c. - **V. compressa** Turcz. (WI, SA): white fiddlewood; frt. edible; wd. durable.; good tree for reforesting - **V. congolensis** DeWild & T. Dur. (tropAf, esp. Congo, Cameroon, Liberia): aanion; bk. inf. u. as enema; frt. flesh u. for eye dis.; wd. u. to make tomtoms - **V. cuneata** Schum. = *V. doniana* Sweet - **V. cymosa** Bert. ex Spreng. (SA): taruma (do alagado); bk. u. as expect. & antidysent. (Bo); frt. edible; wd. u. for making yokes for draft animals - **V. divaricata** Sw. (WI, Ve, Guian): (black) fiddlewood; decoc. of branches u. for GC; burnt pd. bk. u. appl. to ulcers (Domin, Caribs); ripe frt. edible; hard wd. valuable, for scantlings, cabinet work - **V. doniana** Sweet (tropAf, Cu): West African (or black) plum; ink tree; African oak (or teak); edi (Nigeria); frt. edible, u. in anemias, avitaminoses A & "B"; with sugar & crushed Randia sd. furnishes black cosmetic, lasts 3-4 days on skin; lvs. as tea substit. (beverage); rt. u. in GC; tree sometimes cult. for frt. - **V. djumaensis** DeWild. (Zaire): kanfulu; frt. as purg. (natives); lf. decoc. (hot) u. as wash by women & to free themselves of lice - **V. excelsa** Mold. (Amazonas, Br): taruma; very large tree (-60 m); wd. like teak, u. for ship building - **V. flavens** HBK (cSA): taruman (tuira); wd. u. for house building (Ec) - **V. gaumeri** Greenm. (sMex, CA): blue blossom (or flower); tough wd. much u. for boats, &c. - **V. gigantea** HBK (Ec): moconto; wd. u. for decking for bridges & ships - **V. glabrata** R. Br. (sAs, Austral+): bk. & rt. astr. med.; frt. edible; wd. u. for cartwheels; fuel - **V. grandifolia** Guerke (trop wAf): fo(ti); lf. decoc. u. as ink (Sierra Leone); rum made fr. plum-like frts. (pulp edible); in sweetmeats; wd. termite-proof - **V. leucoxylon** L. f. (Ind): "jamela"; lf. decoc. u. both for anemic & fevered patients - **V. longisepala** King & Gamble (Mal): flowery leban; kahua (coffee); timber u.; frts. may be u. as coffee substit. - **V. lucens** T. Kirk (V. littoralis A. Cunn.) (NZ): New Zealand teak (tree) (or oak); puriri (tree); lf. decoc. in bath for sprains, backache; inf. u. in ulcers, esp. under ear, sore throat; wd. hard & heavy, u. for underwater purposes - **V. madiensis** Oliv. (tropAf): filu (tando); lf. decoc. diaph.; bk. inf. in diarrh.; frt. edible; var. *milanjiensis* Pieper (c tropAf); wd. u. for mortars to pound manioc - **V. megapotamica** (Spreng.) Mold. (sBr, Ur, Par, Arg): chaste tree; taruma (mirim); taruman (pardo) (Br); cinco folhas (Br); grows well in marshy lands, even where standing water; lvs. (with VO: c. pinene, cineole, camphene) u. in decoc. to remove urinary stones (diur.); as depur. (in proprietaries); taruma bk. u. in syphilitic conditions; wd. useful; drupes edible (but sd. kernel c. intoxic. alks.); cult. as shade tree - **V. micrantha** Guerke (trop wAf): feve(i); lvs. appl. to cure craw-craw; wd. u. as lumber in construction - **V. mollis** HBK (wMex): coyotomate (black coyote); lf. decoc. u. for gastralgia, bronchial dis. (99); cough; stim.; lvs. & frts. astr. in gastralgia; diarrh.; black frts. eaten raw or cooked, leave indelible spot; bk. u. as antipyr.; odor & taste much like those of black tea - **V. mombassae** Vatke (s&cAf): frt. edible; pl. decoc.

to strengthen & aromatize tobacco; to incr. penile size - **V. negundo** L. (eAf, Pak, Ind e to Chin, PI+): Indian privet; 5-leaved chaste tree; negundo (chaste tree); lvs. c. VO (with ionone, sabinene); aucubin; nishindine (alk.); arom. lvs. u. ton, antipyr. vermif. (children), insect repellant; antiparasit.; juice in eye pain (UP) & for "fetid discharges"; smoked like tobacco for catarrhs. & headache; smoke u. to fumigate huts; stuffed into pillows for headache & nasal congestion; in baths for after childbirth; to clean maggoty ulcers; decoc. in cow urine appl. to sprains & swellings (UP); macerated lvs. appl. to forehead for headache, in toothache, rheum. (Ind); fls. as card. ton., astr. in diarrh.; fls. & sds. exter. as poultice for ulcers; frt. nerve ton., emmen. cath. vermif.; rt. ton. antipyr. expect. diur. in edema. analg. (Ind); in (cobra) snakebite (also lvs.); rt. scrapings mixed with betel mixt. & chewed to give sterility; many other uses recorded; may represent "Soma" (887); var. *intermedia* (P'ei) Mold. (trop worldwide, often cult.): u. med. (Chin); var. *heterophylla* Rehd. (nChin; cult.): (cutleaf) chaste tree; frt. u. as carm.; wd. in construction - **V. orinocensis** HBK var. *multiflora* Huber (SA): taruma (frondosa); u. for fenceposts (Ve) - **V. paniculata** Lam. = *V. negundo* - **V. parviflora** A. L. Juss. (PI): roble de filipinas; molave(- batu); bk. decoc. u. for obesity & edema - **V. patula** E. A. Bruce (sAf): mucoro; lax esp. for domestic animals. - **V. payos** (Lour.) Merr. (seAf): purro; u. to treat asthma, cough (natives); FO u. lax.; frts. eaten with relish; FO u. lax. esp. veterin.; cult. Br - **V. penduncularis** Wall. (sAs, Ind, Pak, Burma, Thailand, Indochin): "tinnok"; lvs. c. vitexin, alks. (traces); inf./decoc. u. in mal. (black water fever) ("cure") (Assam), eaten as veget., in ophthalmia; ripe frts. eaten (888) - var *roxburghiana* C. B. Clarke (Ind, Bangla Desh, Burma): "bhadu" (Bengal); "goda"; bk. u. exter. for chest pains; wd. u. for posts & beams - **V. phaeotricha** Mildbr. (wAf): "kparseh"; bk. & lvs. u. in lotion to wash ulcers - **V. pinnata** L.f. (tropAs, Ind, Mal to PI): pinnate chastetree; tree or shrub; lvs. u. in fever, poultices for wounds; lf. juice u. in anorexia; lvs. (& bk.) believed to activate growth of beard; bk. decoc. appl. to wounds (Sunda); decoc. in gastralgia (Mal) often in childbirth; rt. inf. in colic of infants (Kampuchea); frts. ext. u. to prevent convulsions; sds. vermif.; wd. useful in many ways, ex. to make wooden bells - **V. polygama** Cham. (Br): "maria preta"; drupes edible - **V. pooara** Corbishley (sAf): "stinkbessie"; dark purple frt. popularly eaten by natives; with vit. C - **V. pseudolea** Rusby (Bo, Pe): wild olive; aceituno del monte; tree; bk. medl. in rheum; sweet frt. edible (*cf.* olives) (by children) - **V. pubescens** Vahl = *V. pinnata* - **V. pyramidata** B. L. Robin. (Mex): "capulin"; tree; frt. edible, decoc. u. in diarrh.; st. lf. decoc. u. as expect. in chest dis.; in irreg. menstr.; wd. in construction - **V. quinata** (Lour.) F. N. Will. (seAs): various-leaved chaste-tree; "ashoi"; bk. of shrub or tree u. arom. ton. to stim. appetite (nVietnam); wd. u. for pillars, &c.; frt. u. in neuralgia (Indochin) - **V. rapinoides** Guillaum. (New Hebrides): wd. u. to build native homes - **V. rehmanni** Guerke (sAf): "mokwele"; lf. c. aucubin, ecdysterone, u. as inf. for "stomachache" (enema); in hysteria (Zulu): useful wd. - **V. rivularis** Guerke (tropAf): frt. edible - **V. simplicifolia** Oliv. (tropAf): "plum tree"; decoc. u. in snakebite (Ivory Coast); bk. u. as lotion to treat skin dis., toothache; twigs for "tooth sticks" and "chew sticks" (Nigeria) - **V. strickeri** Vatke & Hildebr. (trop cAf): "mugombo" (Tanz) lf. decoc. u. to rinse mouth for swollen gums; lf. juice u. orally for snakebite; rt. decoc. u. in stomachache - **V. taruma** Cham. = *V. megapotamica* - **V. thyrsiflora** J. G. Baker (w tropAf): ngunge-di; lvs. boiled & u. to alleviate pain of ulcers; myrmecop hilous - **V. triflora** Vahl (Ve, Br, Pe, Bo+): mama de cachoira; m. cachorra; frt. u. emmen. diur.; lvs. in cystitis & urethritis (Br); rt. ton & febrifuge - **V. trifolia** L. (V. trifoliata L.; V. lagundi Farnsworth) (tropAf, As to Austral): Indian wild pepper (or privet); hamago (Jap); "hand of Mary"; lvs. c. VO (with pinene, camphene, cineole, methanol, &c.), vitricine (alk.); lvs. strongly resolvent (discutient) for tumors, u. in splenomegaly; lf. decoc. u. in acute rheum., intermittent fever; frt. vermif., emmen. in headache; sds. sed., analg.; rt. ton. antipyr. expect. contraceptive; fresh lvs. placed on wounds; cult. med. pl. (Thail); var. *bicolor* (Willd.) Mold. (eAf, tropAs to Oceania): 2-colored chaste-tree; lala(tea); agulundi (Pacific Is); lvs. (juice) u. in fever (Samoa); as poultice for abscesses; toothache, rheum, &c.; pl. juice to repel mosqutoes (Ryukus); lvs. rubbed on body for fever; in med. (PI); var. *heterophylla* (Mak.) Mold. (OW trop): vone lang (Hainan Is); u. antipyr. (Sumatra); in med. (Mauritius); var. *simplicifolia* Cham. (V. rotundifolia L.) (OW trop, Jap, PI+): beach vitex; abisa (Ind); "so pa"; lvs. u. as decoc. as sed. (women in childbed); pulm. TB (Hainan Is); lvs. st. wd. u. as bath for backache, swollen feet, rheum (Jap); to combat soil erosion: to make sweets (dark jelly) (Thail); to treat glandular tumors (Chin); var. *subtrisecta* (Ktze.) Mold. (Ind, Java, PI, &c.): salt bush; "blue vitex"; bk. & rts. u. as antipyr. (Thail); branches burned

& smoke directed to aching joints (Sumatra); u. med. (Hainan); var. *variegata* Mold. (cult. Fla [US]): grown as hedge (sFla); often produces allergy (asthma) - **V. tripinnata** (Lour.) Merr. (Vietnam): lf. sometimes u. with betel as masticatory; var. *pubescens* Mold. (Laos, near Vientiane): pl. u. med. during childbirth; var. *variegata* Mold. (Haw, Fla [US]): lvs. put in irrigation water to protect rice pls. fr. pests; dr. pd. lvs. u. to repel insects in grains & clothes - **V. umbrosa** Sw. (Jam): (yellow) fiddlewood; wd. u. for boards & framing purposes - **V. umbrosa** G. Don = *V. doniana* - **V. vestita** Wall. (s&seAs): flowering (or black) leban; pl. u. for abortion (Indones); rt. u. for snakebite (Thail) - **V. welwitschii** Guerke (c tropAf): nungu; bk. u. in prepn. of enemata; wd. u. for pestles; frt. edible.

viticchio (It): *Convolvulus arvensis.*

Vitis L. (Vit.) (temp areas worldwide, on all continents): the vine or grape; some 80 spp.; the vines (climbing pls. with tendrils) are much cult. - **V. acida** Morales = *Cissus trifoliata* L. - **V. adnata** Wall. = *Cissus adnata* Roxb. - **V. aestivalis** Mx. (e&cUS): summer (or pigeon) grape; berries edible; u. for cough (Choctaw Indians) (852); sometimes cult.; esp. in Eu where u. as stocks since *Phylloxera*-resistant - **V. californica** Benth. (Ore, Calif [US]): Calif wild grape; furnishes edible grapes; lf. inf. as poultice for wounds, in snakebite; lf. juice for "loose" bowels; lvs. chewed to allay thirst; vine sts. for rope, in basketry (Calif Indians) - **V. caribaea** DC (CA, WI): wild grape; uva silvestre; parra cimarrona; sts. u. as diur.; as ton. (in combns.) (Cu); frts. refreshing - **V. discolor** Dalz. = *Cissus sicyoides* - **V. hederacea** Ehrh. = *Parthenocissus quinquefolia* - **V. hildebrandtii** (trop eAf): rt. c. tannin, u. in snakebite - **V. labrusca** L. (neUS): fox grape; wild vines; one of best Am grapes (said to have "foxy" flavor char. of Am wild grapes); eaten raw, as grape juice, syr., jelly, jam, wines, as flavor in candies; sds. source of FO; sd. cake food for livestock; vars. Concord, Catawba, Niagara, Delaware; sometimes called **V. labruscana** Bailey; lvs. u. in rheum. (folk med.) (2377) - **V. lanceolaria** Wall. (seAs, Indones): ojod; pl. juice u. in cough; lvs. u. like sorrel by natives - **V. pallida** W. & A. = *Cissus p.* (*C. populnea* Guill. & Perr.) - **V. pentagona** Diels & Gilg (eAs): nho; rts. u. for cough, diur.; in bronchitis & GC (Kampuchea): frts. edible but sour; source of inferior wine but good vinegar - **V. polystachya** Wall. = *Ampelocissus p.* - **V. quadrangularis** Wall. = *Cissus q.* - **V. repens** W. & A. = *Cissus r.* - **V. rotundifolia** Mx. (s, e&cUS w to Okla; Mex): muscadine, scuppernong (with amber green frt.), bullace grape; (southern) fox g.; wild growing vine; frt. deliciously esculent but with thick skin; u. mfr. of grape juice, wine; pl. resistant to phylloxera (insects), hence u. as grafting stock for *V. vinifera* - **V. sessilifolia** Baker (Br): cipo de maõ de sapo (frog-hand vine); rt. u. as alexiteric in snakebite ("certain remedy") (hence pl. called "infallivel" [infallible]); tuberous rhiz. c. coumarin (0.033%) of fresh); lf. decoc. u. in cystitis (int.), in baths for rheum., wash for ophthalmia, as cataplasm in furunculosis, &c. (2242) - **V. stefaniana** (eAf): rt. u. in mal. - **V. sicyoides** Morales = *Cissus s.* - **V. thunbergii** Sieb. & Zucc. var. *adstricta* Gagnep. (Taiwan): decoc. of lf. sap u. as wash for wounds & swelling - **V. tiliifolia** H. & B. (cMex to Co & Ve; WI): parra (bronca); water-wise; st. sap u. diur., for VD, lf. decoc. antipyr.; crushed lvs. appl. to erysipelas - **V. trifolia** L. = *Cayratia t.* - **V. vinifera** L. (c&sEu; wAs; now cult. worldwide in temp regions): (common or English) grape; (g.)vine; common (or wine) v.; European g.; fresh lvs. u. as astr. stypt., for lithiasis, in herbal remedies, (fresh mild decoc. of lvs. of red grapes u. in venous stasis) (edema, "heavy legs," stasis ulcers); hemorrhoids; cutaneous capillary fragility (petechiae, ecchymoses) (2420); tea mixt. & ext. in epicurean cooking for stuffed vine leaf (grape lf. roll) (dalmados); grape lvs. coll. in May & u. for winter pilaus (Armenia); juice exuding after clipping tendrils u. exter. for rashes, int. in intest. bleeding, esp. in dys.; for rheum. (2765); in eye waters; tendrils (Pampini Vitis) occas. still u. as exter. prepn. in folk med.; unripe frt. juice (verjus, agresta) u. in obesity (Syr. of Verjus) (omphacium [Ancients]), as acidulant & diur.; juice of sts. u. as lithontriptic, for ophthalmia; ripe frt. u. nutr. (in pies, cakes, biscuits, puddings, &c.), flavor, refriger., in mfr. sugar ("Korinthin") & staphidine (*qv.*) (Gr); in mfr. grape jelly, g. juice (mostly however fr. Am grapes) (u. to overcome fatigue), g. vinegar, wine (*qv.*) (u. sed. in gout, &c.), brandy (cognac); dr. frt. are raisins (Rosinen [Ge] [2421]) & "currants" (*qv.*); of several types: Muscatels (sun dr.); Valencias (dipped in KOH, lavender oil, rosemary oil, olive oil); Sultanas (small seedless sour, fr. Levant, esp. Turkey), Thompson (very sweet seedless fr. Calif); syr. of raisins with honey consistency; mfr. of wine (or cognac or raisin) oil, u. arom. in brandies; sds. expressed to give grape sd. oil (FO) u. as culinary, in techn. trades (linseed oil substit.); charcoal (fr. press residues) is "Frankfort black"; proanthocyanidin oligomers fr. sds. marketed; fresh grapes delicious dessert frt.; raisins u. as lax., diur, nutrient (esp.

for hair), stom., in hoarseness; in baked goods, &c.; vars. of table grapes include Malaga, Mission, Emperor, Sultanina, &c. (889); a popular folk remedy for arthritis proposes soaking white raisins in gin (brandy?) for several days, then 9(!) raisins are consumed (but not the liquor!) (1995) - **V. vulpina** L. (V. cordifolia Lam.) (e, s&cUS): winter/frost/chicken grape; river bank g.; wild (fox) g.; frt. edible, acidulant; nutr.; refriger.; antiscorb.; sometimes cult. as cover.

vitoriera (Co): pumpkin.

Vitrellae (Br): therapeutic substances inhaled fr. a crushed glass capsule; ex. amyl nitrite.

vitriol: sulfuric acid - **blue v.**: cupric sulfate - **elixir of v.**: aromatic H_2SO_4 - **green v.**: crude ferrous sulfate - **oil of v.**: conc. sulfuric acid (SC oily consistency) - **red v.**: various iron minerals, such as a ferric-ferrous sulfate (Swed); red iron oxide - **Roman v.**: **Salzburg v.** : cupric sulfate (tech.) - **white v.**: zinc sulfate.

Vittaria graminifolia Kaulfuss (V. filifolia Fée) (Vittari./Adiant.): fronds of the epiphytic fern are boiled & decoc. drunk for colic (2444).

Viverra civetta Schreb. (Viverridae - Carnivora) (nAf: Ethiopia+): African civet (cat); civet is an unctuous secretion fr. a pouch lying between anus & genitalia of both sexes; c. civetone, VO, resin; u. perfume fixative; cosmetic (Arabs); in cosmetics; civet is shipped in buffalo or ox horns covered with chamois, cloth & sealed with wax - **V. zibetha** L. (Iran, Ind to sChin to Mal): Asiatic civet (cat); c. & u. similar to prec.; form. u. for cramps; among several other spp., one found in Indones.

Viverridae: Civet Fam.; small carnivorous mammals related to the cats, dogs, & hyenas; useful spp. for extermination of vermin; ex. civet (cat).

Viviania esculenta (2146) (Viviani./Gerani) (Ind, Br): u. lax., in chronic liver hypertrophy; hemorrhoids; diabetes - **V. marifolia** Cav. (Ch): té de burro (colorado); oreganillo; u. stom.

viviparous (in bot.): propagating by bearing young plants instead of sds. (ex. life plant *Kalanchoë pinnata*); or bearing sds. which germinate before being shed.

viznaga: barrel cactus, *Ferocactus or Echinocactus* spp. (pitaya).

Voacanga africana Stapf (Apocyn.) (wAf: Sierra Leone, Nigeria, Togo): bk. & sds. c. alks., tabersonine, voacamine, vobtusine, voacangine, voacorine, &c.; with cardioactive, hypotensive props. (890); latex u, to adulterate rubber; as temp. filling for carious teeth; rt. fiber to make mats: decoc. of leafy branches u. as panacea, fortifying; to reduce CNS activity in aged; rt. decoc. for puerperium (Sen); st. & rt. bk. as prophylactic for infant dis. (Nigeria) - **V. megacarpa** Merr. (PI): u. in drinks for int. pains.; rt. decoc. u. for "rigid abdomen"; source of chicle - **V. schweinfurthii** Stapf (Af): alk. rich; u. in heart dis., rheum. - **V. thouarsii** R. & S. (Af): latex u. to make bird lime (glue) (also var. *obtusa* Pichon); sd. FO u. to prepare pomades (salves); sds. c. tabersonine (comml source), voacamine; u. to treat tumors, cerebral vascular deficiencies (hypotensive) (Mad).

Voandzeia subterranea DC/Thou. (monotypic) (Leg.) (tropAf, Madag): Bambara groundnut; Congo goober; earth pea; growth rather similar to peanut vine (altho with different mechanism), with frts. buried in soil; sds. rich in starch with low protein & FO cont., of high food value; eaten boiled, roasted, & fried; pl. believed a cultigen; lvs. u. for abscesses & envenomation (Senegal) (—> *Vignà*).

Vochysia bifalcata Warming (Vochysi.) (Br): gum u. as agglutinating agt. for compressed tabs. similarly to Acacia (891) - **V. columbiensis** Marcano-Berti: tree bk. u. in arrow poisons; supposed kind of curare - **V. ferruginea** Mart.(Nic to Br, Amazonian area): red yemeri; mayo colorado; lf. & bk. u. int. to heal ulcerating wounds; esp. of nasal & buccal areas; lf. & bk. u. antipyr. (2716); component of a kind of curare - **V. guatemalensis** J. D. Smith (sMex to Pan): white yemeri; bribri; chancho blanco; wd. furnishes good lumber. (2716) - **V. laxiflora** Stafleu (Co): tree lvs. boiled with Coca for cases when urinary stoppage is painful; bk. for same; dr. pd. bk. rubbed into sores; smoke inhaled for asthma, &c. - **V. lomatophylla** Standl. (Co, Ec, Pe): sacha-alfaro; lvs. & bk. of tree u. as abort.; perhaps u. as contraceptive (Campa tribe, Pe) (892) - **V. obscura** Warming & Mart. & **V. saccata** Stafleu (both of Amazonian Br): u. by the native Indians, the first as a contraceptive, the latter as a "powerful" febrifuge (bk. decoc.).

Vogel (Ge-m.): bird - **Vogelskirsche**: *Prunus avium.*

volatile oil: essential o.; substance of ordorous nature wh. disappears in the course of a few hrs. when exposed to the atmosphere; obt. fr. pl. materials (lvs. fls. frt. bk., etc.) by steam or direct distn., by expression (citrus), by extrn. with volatile solvents, FO, etc.; generally made up of mixtures of terpenes & their derivatives; highly characteristic of some pl. fam., ex. Laurac., Myristic., Umb., Lab.

Volkameria L. = *Clerodendrum.*

volutin: nucleoprotein; found in yeasts, some algae, &c.

Volvariella volvacea (Bull. ex. Fr.) Sing. (Plute. - [Amanit.] - Agaricales - Basidiomyc. - Fungi)

(As, NA+). Chinese (or straw) mushroom; popular food (Chin), in Chin med. (nutr. ton.) (2213).
vomiquier (Fr): Nux Vomica pl.
vomit wort: *Lobelia inflata.*
voodoo: Vodoun (means God): **voodooism**: religious superstitious practices based on sorcery & fetishes wh. originated in various parts of Af & is now also practiced in Haiti & La+ (sUS) (also "hoodoo"); in their practices, a number of pl. materials are used (2487; 2488); associated with conjuration, superstition-based remedies, & witchcraft; formerly it involved elements of snake worship, bestiality, human sacrifices, &c. - **v. bull** (US Negro): *Arum cornutum.*
Vorderasien (Ge): Near East; wAs, incl. Syria, Iraq, Iran, &c. - **V. indien** (Ge): India proper (now Indo-Pak) (*cf.* Hinterindien).
vorgestreckt (Ge): exserted (spoken of pistils &/or stamens projecting beyond corolla).
Vorlage (Ge): receiver (still).
Vorstufe (Ge): precursor.
Vouacapoua Aubl. (*Andira* Juss.): **V. americana** (*Andira inermis*): *A. excelsa.*
Vouapa Aubl: = *Macrolobium* Schreb.
vrille (Fr): tendril
vulcanite: a hard rubber made by vulcanization (heating or otherwise treating rubber with c. 30% its wt. of sulfur).
vulnéraire (Fr): **vulnerary**: *Anthyllis vulneraria.*
vulvaire (Fr): *Chenopodium vulvaria.*
vuss: furze.

waahoo = wahoo.

wabayo: *Acokanthera* spp.

Wacholder (Ge): *Juniperus communis.*

Wachs (Ge): wax; beeswax - **Wachsmyrtel**: *Myrica* spp. - **Wachsbusch** (Ge): **Wachs-gagel**: *Myrica cerifera.*

Wade's drops: Benzoin Tr. Comp.

wafer ash: *Ptelea trifoliata* (bk.).

Wagatea indica Dalz. (monotypic) (Leg) (swInd): wagati; bk. u. as appln. for skin dis.; rt. u. in pneumonia —> *Moullava*).

Wahlenbergia marginata DC (W. gracilis DC) (Campanul.) (seAs): rt. u. for pulm. dis. (Indochin); pl. in skin eruptions (Indones) - **W. undulata** DC (sAf): decoc. as eye lotion (Zulus); for intest. ulceration (children) (sSotho); pl. as spinach.

wahoo: 1) *Euonymus atropurpureus* (bark) 2) *Ulmus alata.*

wai fa (China): *Sophora japonica*

wake-robin: 1) *Trillium* spp. (**ill-scented w. r.**: *T. erectum*) 2) *Arisaema triphyllum* 3) *Arum maculatum.*

waken beggar: pd. white hellebore.

Wal (Ge-m.): whale.

Waldeinsamkeit (Ge): *Lysimachia nemorosa.*

Walderbeere (Ge-f.): wild strawberries, *Fragaria* spp.

Waldgeissbart (Ge): *Aruncus sylvester.*

Waldmeister (Ge): *Asperula odorata.*

Waldwolle (Ge): wood wool.

wall: v. also under moss; pellitory - **w. rue**: *Adiantum ruta-muraria* (**w. spleenwort**).

Wallacea insignis Spruce (Ochn.) (Amaz, Br): c. rotenone-like compds.; u. as insecticide.

Wallenia laurifolia (Jacq.) Sw. (Myrsin.) (Jam, Cu, Española): "camagua"; wd. u. in rural carpentery - **W. yunquense** (Urban) Mez (WI): pl. shows cytostatic activity (rats); c. myrsin saponin, stigmasterol.

wallflower: *Cheiranthus cheiri* - **western w.**: *Erysimum asperum.*

Wallichia disticha T, Anders (Palm.) (nInd): berries irritant to skin.

Wallis, T. E. (1876-ca. 1972) D. Sc.: Reader in Pharmacognosy, Univ. of London; an outstanding Brit pharmacognosist; active in developing methods of microanalysis.

walnoot (Du): **Walnuss** (Ge): walnut.

Walnussrinde, graue (Ge): white walnut bk., fr. *Juglans cinerea.*

walnut: *Juglans* spp., esp. *J. regia* or *J. nigra*; source of w. oil, FO fr. sd. kernel - **American w.: black w.**: *J. nigra*: hulls (or shells or peel or shucks); lvs.; bk. - **English w.: European w.**: *J. regia*: bk.; hulls; lvs. - **w. family**: Juglandaceae - **gray-barked w.**: white w. - **Otaheite w.**: *Aleurites mollucana* - **shell bark w.:** white w. - **w. shell flour**: commodity of commerce obt. fr. J. spp. - **true w.**: English w. - **unripe w.**: understood as obt. fr. prec. - **white w.:** 1) *J. cinerea* (bk.) 2) *Carya ovata.*

Walrat (Ge): **walschot** (Dutch): spermaceti.

Walsura elata Pierre (Meli) (Indochin): bk. c. tannins, decoc. u. in combns. in diarrh. & dys. - **W. multijuga** King (Mal, Sumatra, Banka Is, PI): boiled bk. held in mouth for mumps; scraped bk. eaten for poisoning (PI) - **W. piscidia** Roxb. (seAs, esp. Ind): bk. u. to stupefy fish ("fish poison"); c. saponins, tannin.

Waltheria americana L. (Sterculi.) = *W. indica* var. *americana* - **W. doura-dinha** St. Hil. (Br): "douradinha" (lit. golden) lf. of flg. pl. u. as expect. antisyph. vulner.; diur. diaph. digitalis substit. - **W. glomerata** Presl (Mex to Pan): "palo del soldado" (soldier wood); u. for wounds (pd. strewn on wounds): to stanch bleeding - **W. indica** L. var. *americana* (L.) R. Br. (trop worldwide, incl. Fla [US], WI, SA): Florida waltheria; "malva blanca" (Cu); bretonica (Mex); bretonica macho (Ve); hb. u. as sudorific; refrigerant; antipyr. (CR); in rheum.; antisyph. (PI); in colds (Togo); in thrush ("ea") (Haw); emoll. (Antilles); rt. purg.; abort. (Ghana); lf. & fl. mucilag.; fiber pl.; rt. decoc. for pain in urinating; lvs. as poultice for suppuration; lvs. in erysipelas; rts. in leprosy (Tanz).

wampee: **wampi** (fruit): *Clausena lansium.*

wandering Jew (US): 1) *Tradescantia fluminensis* 2) *Zebrina pendula.*

wandflower: *Dierama pulcherrum* (sAf).

wanga (seUS, WI; Negro): amulet (v. hand).

Wanze (Ge-f.): bug - **Wanzenkraut** (Ge): name appl. to several medic. pls., ex. *Euphorbia peplus; Cimicifuga racemosa* (bugwort; bugbane); male fern; &c.

wap(p)ato: *Sagittaria latifolia, S. cuneata.*

war: waras (dye) (tropAs): 1) the hairs fr. frts. of *Flemingia grahamiana* 2) Kamala.

Warburg's apparatus: W. manometer: a very efficent apparatus for determining the O_2 consumption & CO_2 prodn. of incubated tissues; u. in metabolism studies - **W. pills: W. tincture**: these W. prepns. c. quinine sulfate, aloe, rhubarb & several arom. condim. & stim. drugs; u. as antiperiodics formerly.

Warburgia salutaris (Bertol. f.) Chiov. (W. ugandensis Sprague) (Canell.) (OW trop): bk. macerate drunk for cough; decoc. for sore throat; backache; pd. appl. to skin sores; fever tree; bk. u. in mal. purg. expect. (eAf).

Warfarin: anticoagulant compd. (Na & K salts); u. as rodenticide, esp. for rats; orig. obt. fr. spoiled sweet clover; u. as important anticoagulant (Coumadin+); small doses daily to reduce risk of stroke & systemic embolism; now by synthesis; named for **W**isc. **A**lumni **R**es. **F**ound.

warp: the threads in a fabric which run lengthwise & are crossed at right angles by the weft (or woof).

wars: warus: war.

wart taker (seUS slang): voodoo practitioner.

wartwort: *Euphorbia helioscopia.*

was (Dutch): wax.

wasabi (Jap): foll.

Wasabia japonica (Miq.) Matsum. (Eutrema wasabi (Sieb.) Maxim.) (Cruc.) (Jap, Sakhalin): "wasabi" (Jap); green horse radish; rhiz. appl. to neuralgic pain; antisep.; eaten as diur., juice drunk as antidote for fish poison (Jap); rts. twigs, & petioles u. as spice esp. for raw fish (*cf.* horseradish); cult. in wet places.

wash rag (gourd): *Luffa aegyptiaca*; sponge g.

Washingtonia filifera (Linden) Wendland (Palm.) (swUS, esp. Calif [US], nwMex): California fan palm; cannon p.; frts. & sds. eaten fresh or dr. (Am Indians); young apical bud roasted & eaten; lf. fibers u. for baskets.

Wasicky, Richard (1884-1970): outstanding pharmacognosist; refugee from Austria to Br; Univ. São Paulo (893).

Wasserholder (Ge): *Viburnum opulus* - **Wasserklee**: *Menyanthes* - **Wasserreis**: *Zizania aquatica* - **Wasserschierling**: *Cicuta virosa* - **Wasserschwertlilie**: *Iris pseudacorus.*

wata (Jap): cotton.

water: H_2O; Aqua (L); Wasser (Ge); eau (Fr); agua (Sp, Por); naturally occurring liquid solvent & medium; natural w. never entirely pure since cont. dissolved salts (river, rain, spring, well, tap); called "the universal solvent"; essential to existence; of primary importance as nutr.; solvent in many pharm. prepns.; w. is sometimes u. therapeutically, int. & externally (hydrotherapy); ex. holding hands in hot water (62°C) 30 mins./day said useful in curing warts; very hot w. said excellent remedy for poison ivy dermatitis & insect sting (using wash cloth); w. "diet" for reducing wt.; cleansing agt.; childbirth process sometimes carried out in deep bath (such as Jacuzzi) (redn. of pain) (*sic*); various types: tap (or hydrant) water; distilled w. has been u. in mixing alc. drinks, considered to have superior taste to tap w.; in past centuries, drinking of w. as sole beverage often considered unhealthful or harmful; redistilled w. (double distilled w.); pyrogenfree w. (w. for injection); demineralized w., &c.; ice (frozen w.) u. for cooling, preservation, ingestion for leg cramps; boiling w. for cooking foods; v. under "ice" - **bacteriostatic w.**: u. for injections - **bottled w.**: u. by many health faddists (Perrier™, &c.); only advantage over tap w. is rel. purity and better flavor - **w. cress (herb)**: *Nasturtium* spp.- **w. cup**: *Sarracenia purpurea* - **deionized w.**: **demineralized w.**: water with minerals removed by adsorption without use of heat; has many of the uses of distilled w. - **distilled w.**: obt. as distillate fr. boiling of tap water; mandatory in many areas of med., such as parenteral solns. ("water for injection") (E); u. in emergency (fallout) shelters; sometimes u. in mixed drinks - **European w. cress**: *Nasturtium officinale* - **w. glass**: sodium silicate; u. for preserving eggs - **gripe w.** (Ind): dill w. - **heavy w.:** D_20; deuterium oxide; occ. naturally in water of all sources; prepd. by various methods from water; u. to study chem. reactions (mechanism; rate); inhibits pl. & animal growth; u. in nuclear reactors - **iced w.:** appln. of or immersion in of lesser burns is one of the best therapies; acts promptly to relieve pain & reduce redness & swelling - **w. leaf**: *Hydrophyllum* spp.; *Brasenia schreberi* - **w. l. family**: Hydrophyllaceae - **w. lettuce**: Pistia stratiotes - **w. lily**: water-lily (*qv.*); water hyacinth (sAla [US]) - **w. maudlin**: *Eupatorium cannabinum* - **w. nuts**: *Trapa natans* - **w. pore**: w. stoma - **purified w.**: that obt. by distn., ion-exchange treatment, reverse osmosis, &c. (not for parenteral use) - **pyrogen-free w.:** w. for injection; distd. w. rendered free of fever-producing proteins (such as bacteria & their metabolites) - **sea w.**: considered by some to be of therap. value since trace elements are present; sold as concentrate sometimes (894) - **w. star-wort**: *Callitriche verna* - **sterile w.**: prepd. by distn. or reverse osmosis; u. for inhalations, injections, irrigations - **w. stoma**: hydathode - **w. witching**: detecting water underground by means of holding branch of tree or shrub (witchhazel; willow [Alta]; &c.); indicated by spontaneous movement of branch (claimed).

water: v. also under avens; bugle; chestnuts; dock; dropwort; eryngo; fennel (seed); figwort; germander; hemlock; horehound; hyacinth; mint; parsnip; pepper; plantain; radish; reed; shamrock; weed.

waterbouts: *Caltha palustris* - **water-cress**: *Nasturtium officinale* & various other *Nasturtium* spp. (European w. c.); *N. officinale* (**marsh w. c.: w. c.**) - **water-feather**: *Myriophyllum proserpinacoides+* - **water-lily: sweet(-scented) w.**: *Nymphaea odorata+* - **white w. (, European)**: *N. alba* - **white w., large**: sweet w. - **yellow w.**: *N. mexicana; Nelumbo lutea* - **watermelon (, wild)**: *Citrullus lanatus* (frt., sds.) - **water-milfoil**: water-feather - **water-nymph**: *Nymphaea odorata* - **water-soluble extractive**: an assay determination u. for some crude drugs where no effective constituent assay is available; u. for Aloe, Gentian, etc. - **w.-s gum**: term appl. in commerce to such gums as Acacia (which diffuses in water to form hydrosols) & Tragacanth (which diffuses to form chiefly hydrogels); the purpose is to distinguish fr. alcohol-soluble gums, such as "Gum" Myrrh, "Gum" Thus, "Gum" Camphor, etc., & fr. prods. sometimes called "gums" because of their physical nature & orgin (as "Gum" Opium).

wattle: **Australian w.**: *Acacia decurrens+* - **w. gum**: fr. Wattle Bark, bk. of *Acacia* spp. of Austral, eAf (Kenya), &c.; ext. u. in tanning.

Wau (Ge): *Reseda luteola.*

wauhoo: wahoo.

wax: class of subst. composed essentially of mixtures of esters of monohydric solid straight-chain alcs. with higher fatty acids; loosely and incorrectly applied to solid hydrocarbons, ex. paraffin w., &c.; sometimes c. free fatty acids & alcs. & even solid hydrocarbons; with thick cohesive plastic consistency; greatly diverse in origin: some waxes of mineral origin, others fr. pls., some fr. insects, & some fr. mammals; of considerable economic value; u. in lipsticks & other cosmetics; ex. beeswax, carnauba w., candelilla w., &c. - **bayberry w.**: *Myrica cerifera*, etc. - **bee's w.: beeswax**: prod. of the honey bee, *Apis mellifera* (**white**: Cera Alba; **yellow**: Cera Flava; bleached w.) - **w. berry**: 1) bayberry 2) *Symphoricarpos albus* - **bleached w.**: w. fr. the honeycomb of the bee, *Apis mellifera*, whitened by exposure to sun & moisture or artificially with SO_2 fumes - **Brazil w.**: carnauba w. (*qv.*) - **Brazilian w.**: Carnauba; ouricuri; cauassu wax; candelilla; sugar cane w.; *Coccoloba cerifera* wax (comml.?); sisal w. (comml.?) - **Bute Inlet w.**: large masses of wax found in water at Bute Inlet, BC [Cana] believed prod. of pollen of *Pinus contorta* (1369) - **candle w.** 1) paraffin w. 2) spermaceti (formerly) 3) crude stearin (formerly) - **carnauba w.**: fr. *Copernicia prunifera* - **China w. : Chinese w.**: Cera Chinensis (L.): fr. *Fraxinus chinensis* - **w. dolls**: *Fumaria officinalis* - **earth w.**: lignite w. - **green w.:** myrtle w. - **Japan(ese) w.**: 1) Cera Japonica fr. *Rhus succedanea* (*qv.*) 2) *Ligustrum* - **lignite w.**: v. under lignite - **microcrystalline w.**: obt. fr. petroleum (still bottom fraction) - **mineral w.: w. montan**: lignite w. - **myrtle w.**: bayberry w. - **ocotilla w.**: fr. *Fouquieria splendens* - **w. plant, one-flowered**: *Monotropa uniflora* - **red w. berry**: *Symphoricarpos orbiculatus* (apparently; frt. red to purplish red) - **salt. w.**: u. p.o. for allergy, asthma, hypovolemia - **sealing w.**: v. under *s.* - **Spanish grass w.**: esparto w. - **sperm w.**: prod. fr. *Simmondsia californica* - **spermaceti w.:** Cetaceum (*qv.*) - **sumac(h) w.**: Japan w. - **w. vegetable**: *Ceroxylon andicola* - **vegetable w.**: 1) fr. prec. 2) Japan w. - **white w.**: Cera Alba: bleached w. - **wool w.**: refined form of wool fat - **w. work:** celastrus scandens - **yellow w.**: Cera Flava, fr. Apis mellifera.

waxweed: *Cuphea* spp.

waxwood: *Myrica cerifera.*

waxwork: *Celastrus scandens.*

way bread: *Plantago major.*
wayfarer's tree: wayfaring t.: *Viburnum lantana.*
waythorn (berries): *Rhamnus cathartica.*
waywort: *Anagallis arvensis.*
weakfish: *Cynoscion regalis* & related spp. & genera (Otolithidae): food fishes & source of fish sounds.
Wedelia asperrima Benth. (Comp.) (Austral): sunflower daisy; c. wedeloside (amino-sugar) wh. inhibits growth of tumors in rats prod. by aflatoxin B_1; pl. kills sheep (nwQueensland [Austral]) - **W. bahamensis** (Britt.) O. L. Schulz. (Bah, Turks Is): wild marigold (Bah); lf. decoc. as bath water for itching skin; int. (with sugar) in colic & flatulence; dr. lvs. smoked in pipes (make one "high") - **W. biflora** DC (Wollastonia b. (L) DC; Adenostemma biflorum Less.) (seAs, Oceania): pounded lvs. u. as poultice or vulner.; antispasm; juice as snuff appl. to wounds, festers (895); pl. decoc. antipyr. in mal.; decoc. of fresh rt. diur. emmen; lf./rt. decoc. in TB - **W. calendulacea** Less. = **W. chinensis** Merr. (eAs, Austral): lvs. ton. alt. in coughs, headache, skin dis., baldness (Mal); u. liver tonic; has antihepatotoxic activity; pl. inf. u. to neutralize effects of impure water (Indochin) - **W. gracilis** L. (WI): consumption weed; pl. decoc. u. for colds & chest complaints, antipyr. - **W. natalensis** Sond. (sAf): lf. & st. inf. u. as emet.; rt. inf. u. in diarrh. & dys.; antipyr.; bruised lf. steeped in water u. as vulner. for wounds (Zulu) - **W. oblonga** Hutch. (eAf): fls. u. as yellow brown dye to stain pipe stems - **W. strigulosa** B. & A. (Indones): sts. u. as antipyr. - **W. trilobata** A. S. Hitchc. (OW trop, natd. sFla [US], sMex, WI, CA, Co): Spanish verveine (Bah); water weed; lf. & fl. head inf. u. in amenorrhea, dys. (Trin); pl. decoc. antipyr. (Jam) & for colds (Jam): pounded lvs. u. as poultice on sores (Domin, Caribs); pl. inf. in GC, liver dis. (Pe); pl. decoc. u. as ton. (Belize, Creoles); in stomach & kidney dis (Guat); fl.-head inf. u. as carm. & nerve ton. (PR) - **W. wallichii** Less. (Ind): pl. as vulner. to wounds.
weed: 1) a plant growing out of its place, whose virtues (i.e., value & uses) have not yet been discovered (Emerson); usu. very difficult to eradicate; many weeds are u. in med., ex. dandelion; *Rumex*; Oregon grape, &c. 2) (US slang): tobacco or cigar 3) (wUS. slang): Cannabis or c. cigarette - **blue w.**: *Echium vulgare* - **divine w.** (US slang): 1) tobacco 2) Coca - **w. grass**: *Lophiola aurea* Ker-Gawl. (Haemodor.); *Conostylis americana* - **poverty w.**: term appl. to many pls., ex. spp. of *Antennaria,* esp. *A. plantaginifolia*; *Anaphalis margaritacea*; *Chrysanthemum*; *Gnaphalium*, esp. *G. obtusifolium; Iva axillaris* - **rich w.**: one growing in rich soil; ex. *Collinsonia canadensis; Cimicifuga racemosa; Eupatorium rugosum; Ambrosia trifida* - **stick w.** v. stickweed - **tumble w.**: v. tumble-weed - **water w.** : *Elodea canadensis* - **white w.**: 1) *Lepidium draba* 2) *Lithospermum arvense* 3) *Chrysanthemum leucanthemum* - **yellow w.**: *Reseda luteola.*
weedhead (US slang): user of Cannabis.
weedicide: herbicide for weeds.
weeping Polly (US slang): paper grass - **w. spruce**: *Tsuga canadensis.*
weft: the fibers of a fabric which extend at right angles to the fibers running lengthwise known as the warp.
Wegekresse (Ge): **Wegekreuze**: *Berteroa incana.*
Wegerich (Ge): *Plantago* - **grosser (breiter) W.**: *P. major* - **Indischer w.**: *P. arenaria* - **mittlerer W.**: *P. media* - **niederliegender W.**: *P. arenaria* - **spitz(en) W.**: *P. lanceolata.*
Weichselbaum (Ge): **Weichselkirschbaum**: **Weichselkirsche**: 1) *Prunus mahaleb* 2) *P. cerasus.*
Weide (Ge): willow.
Weigela Thunb. (Caprifoli.) (eAs): ca. 12 spp. of showy shrubs; frt. inedible; grown as ornam.; close to *Diervilla* - **W. coraeensis** Thunb. (Indochin): rt. depurative - **W. floribunda** C. A. Mey. (Diervilla f.) (Chin, &c.): lf. decoc. u. as wash for "virulent" sores (Chin); bath for skin dis. esp. scabies (Indochin) - **W. florida** A DC (nChin, Kor): very popular cult. ornam. shrub.
Weihrauch (Ge): frankincense; incense.
Weinessig (Ge): vinegar.
Weingeist (Ge): alc.
Weinmannia bifida Poepp. (Cunoniaceae) (Pe): huichullu; u. in med. by Pe Indians - **W. blumei** Planch. (seAs): bk. considered medl. (Java) - **W. humilis** Engler (Br): grami-munha-miuda; lvs. u. for wounds, bruises - **W. pinnata** L. (W. glabra L.f.) (tropAm, WI): lorito (CR.); bk. astr., c. 10% tannin; has been u. as Cinchona adult.; tan; gum exudn. occurs - **W. paulliniifolia** Pohl ex. Seringe (sBr): gramimunha; bk. rich in tannins, mostly u. in treating wounds; wd. fr. trunk useful - **W. trichosperma** Cav. (Ch): palo santo; pl. c. tannins; u. as astr. & vulner. (Araucanos Indians).
Weinraute (Ge): common rue.
Weinstein (Ge): tartar (lit. wine stone).
Weintraub (Ge): cluster of grapes.
weisser Leim (Ge): gelatin - **w. Senf(same)**: white mustard (seed) - **w. Zimt**: Canella (alba) - **weisses Wachs** (Ge): white beeswax.
Weiszdorn (Ge): hawthorn, *Crataegus,* esp. *C oxyacantha.*
Weizen (Ge.): wheat.
weld: *Reseda luteola.*

Welschkorn (Ge): maize.

Wendlandia formosana Cowan (Rubi) (Taiwan): lvs. usefully appl. to swellings - **W. populifolia** = *Cocculus carolinus.*

Wendtia calycina Griseb. (Ledocarp./Gerani) (Arg): té del burro; té andino; hb. u. as carm. in indigest. dyspep., gastric spasms, gastric acidity, hepatoton., &c.

Werneria dactylophila Sch. Bip. (Comp.) (Arg): fofosa; pupusa; inflor. u. for indig,; stomachache; enteritis; decoc. is effective but with unpleasant taste.

wermoutört (Swed): **Wermut** (Ge): vermouth hb. (or pl.), Absinthium - **Wermutwein** (Ge): vermouth.

Wesson Oil™: food grade of cottonseed oil; named after Wesson, Miss [US], where first prepd. (ca. 1880).

West Indian snake wood bark 1) *Colubrina ferruginosa* 2) *Cecropia peltata* - **W. I. tamarind**: var. of t. fr. WI prep. by separating epicarp of frt., then packing pulp in barrels with hot syrup.

West Indies: archipelago (large gp. of islands) in the Caribbean Sea incl. greater Antilles, Lesser A (incl. Netherlands A), Bahs; rich in crude drugs, ex. sarsaparilla, ginger, tamarinds, quassia, orange peels, &c. (*cf.* Antilles).

Weymouth pine: *Pinus strobus* (SC planted extensively at Longleat, Wiltshire, England, by Lord Weymouth [ca. 1715]).

whale: any large member of Cetacea (not smaller porpoises & dolphins [Delphinidae]); incl. right or baleen w. (Balaenidae), sperm w. (*Physeter* of Physeteridae) - **blue w.:** *Balaenoptera* (Sibbaldus) *musculus* (Balaenopteridae - Mysticae - Cetacea): Blauwal (Ge); sulfur bottom w. (one of the whalebone or toothless whales); up to 33 m long (the largest animal which ever lived); widely distr. in the oceans; a source of Cataceum - **right w.**: *Eubalaena* spp. & *Balaena* spp. (SC the correct whale for production of whale-bone, whale oil, &c.) (esp. *Eubalanaea glacialis*) (considered endangered animals) - **North Atlantic r. w.**: *E. glacialis* - **Pacific r. w.**: *E. japonica* (Lacépède).

whalebone: baleen; horny subst. formed fr. upper jaw of some whales, u. to strain out sea water & retain food (plankton or small organisms) in mouth and then swallowed; u. to make combs, brushes, corset stays, &c., formerly popular.

whale-puke (US slang): ambergris.

wheat: cereal grass, *Triticum aestivum*; source of **w. bread, w. grain (w. berry)**; **w. flour; w. starch**; &c. - **Egyptian w.**: *Sorghum bicolor* - **false w.** (US slang): *Agropyron repens*, couch grass - **w. germ oil**: prepd. by expression or solvent extraction of w. germ; rich in linoleic, oleic, and linolenic acids; satd. fatty acids; unsaponifiable matter (tocopherols, or vit. E); sitosterol; dihydrositosterol, phospholipids, &c. - **Inca w.**: *Amaranthus caudatus* - **Turkey w.** : *Zea mays.*

wheaten (adj.): made of wheat.

whey: the serum of milk, separated after curdling; rich in lactose, vits, minerals; u. to make glucose and galactose mixt. (sweeter) using lactase; substrate in prodn. single-cell protein.

whig plant: Anthemis.

whin: *Ulex europaeus* & other *U.* spp.

whiskey: whisky: alc. spirits dist. fr. mash prep. fr. barley, rye, maize, potatoes, wheat, etc.; u. stim. ton. appetizer, sed.; misused as intoxicant - **moonshine w.**: that prod. illegally often under primitive conditions, & in hidden or remote places.

white: v. also under acacia; agaric; ash; avens; balsam; bay; birch; bole; bryony; cedar; chestnut oak; chocolate; cinnamon; clay; clover (heads); cohosh; dogwood; elaterium; elm; galls; gentian (root); gram; hellebore; horehound; Indian hemp; ipecacuanha; lily; lin; mallow; mint; mustard (seed); oak; pepper; pine; pond-lily; poplar; poppy; rhubarb; rosin; sandalwood; sanicle; sapota; saunders; senega; soap root; soap wort; squill; sumac; swallow-wort; turpentine; virtriol; walnut; water-lily; wax; weed; willow; wine.

white fraxinelle : *Dictamnus albus* - **w. hearts**: *Dicentra cucullaria* - **w. leaf**: 1) *Spiraea tomentosa* 2) tragacanth - **w. liquor** (sUS Negro dial.): alc. - **w. murdah** (Eng): *Terminalia arjuna* - **w. of (the) egg**: albumen fr. ovum of *Gallus domesticus*, domestic fowl - **w. root**: *Asclepias tuberosa; Polygonatum multiflorum, P. officinale*; & some other plants - **Spanish w.**: S. whiting - **w. thorn**: *Crataegus* spp., esp. *C. monogyna* - **w. top**: *Eucalyptus smithii* - **w. top weed**: *Erigeron annuus* - **w. wood (bark)**: 1) *Liriodendron tulipifera* 2) *Drypetes crocea* 3) *Tilia americana* - **whiting**: prep. $CaCO_3$, dr. in the open air or in a kiln, ground, levigated, & dr. again; u. in whitewash, mfr. putty, cleaning silverplate, etc. **(Spanish w.**: 1) w. 2) bismuth subnitrate u. as pigment).

whiteman's foot (Am Indians): *Plantago major* (SC lf. shape & appearing after whites settled Indian territory).

whitewort: *Polygonatum* spp.

Whitfieldia elongata Willd. & Dur. (Acanth.) (trop eAf): rts. chewed for snakebite - **W. longifolia** T. Anders (tropAf): lf. juice u. for black dye; sts. u. for spindles: sd. as love charm.

whortleberry: *Vaccinium myrtillus* (lvs.); *V. vitis-idaea* (frt.) (**red w.**) - **blue w.**: *Gaylus-sacia frondosa* - **w. family**: Vacciniaceae.

wickiup (wNA-Am Indians): willow teepee or hut u. for sweat baths - **wickopy: wickup herb**: *Epilobium angustifolium; E. palustre* - **wicky**: *Kalmia hirsuta, K. angustifolia* var. *carolina*.

wicopy (nUS): *Dirca palustris*.

Widerthorn (Ge): *Drosera rotundifolia*.

Wiegemesser (Ge): chopping knife; cleaver.

Wienertraenkchen (Ge): infusion of senna (lit. Viennese drink).

Wiesensafran (Ge): Colchicum.

Wigandia caracasana Kunth (Hydrophyll.) (sMex, CA to Co, Ve): (hoja de) San Pablo; hb. u. in syph. (Mex); lf. decoc. in rheum (CR) - **W. kunthii** Choisy (Mex, CA): ortiga (grande); hairs sting like a nettle; lvs. u. as antisyph., in metrorrhagia (Mex); lf. decoc. in rheum (CR).

Wikstroemia Endl. (Thymelae.) (temp As, Austral, Oceania): some 80 spp. of shrubs to trees; bk. of some spp. u. like Mezereum in seAs - **W. androsaemifolia** Dcne. (W. candolleana Meissn.) (Chin, Mal, Indones): paste of lvs. u. as poultice to heal wounds of water buffalo; wd. u. to make incense sticks; concrete arom. resin present - **W. chamaedaphne** Meissn. (Chin): fl. buds u. as diur.; in baldness, lumbago - **W. indica** C. A. Mey. (W. viridiflora Meissn.) (Ind, seAs, PI, Austral): small-leaf selago; rt. u. as antipyr. lax. antisyph.; pl. u. in syph., arthritis, & some cancers (Chin); bk. emet. for poisoning (Vietnam); to destroy false membrane of diphtheria; bk. diur.; purg. vesicant, in schistosomiasis; bk. fiber u. to make ropes, paper (Jap); c. wikstromin (flavonoid glycoside), wikstromol (lignan); fish poison - **W. japonica** Miq. = *W. trichotoma* - **W. lanceolata** Merr. (PI): rt. decoc. u. in amebic dys. - **W. ovata** C. A. Mey. (PI): lvs. emet. cath.; bk. filings in wine as ton., mal.; quite toxic - **W. pollanei** Léandri (Vietnam): rt. reported purg.; furnishes bk. u. to make cords - **W. ridleyi** Gamble (Mal): lvs. eaten as aper.; bk. to poultice boils & as fish poison - **W. rotundifolia** C. A. Mey. (Tonga): lala vao; rt. ext. u. as purg (86) - **W. sandwicensis** Meisn. (Haw Is [US]): akia; bast fibers of bk. make good fiber for rope & tapa; beaten branches, lvs. & frt. u. to stupefy fish in fish ponds; wd. useful (896); rt. u. as tonic & for cough - **W. trichotoma** (Thunb.) Mak. (Jap, Kor): u. to treat typhoid fever, cough; bk. u. to make strong paper (bark notes) (Jap).

Wilcoxia striata Britt. & Rose = *Cereus striatus*.

wild: v. also under: angelica; basil; bergamot; black cherry; briar; cardamom; carrot; celandine; chamomile; cherry; cinnamon; coffee; cotton; cucumber; fennel, geranium; ginger; guelder rose; horehound; hyssop; indigo; ipecac; jalap; laurel; lemon; lettuce; licorice; mace; marjoram; mint; mustard; nutmeg; oats; passion flower; patience root; pepper; pine; pink; plum; potato; radish; rosemary; sarsaparilla; scammony; senna; succory; sunflower; tamarind; tansy; thyme; tobacco; turnip; wormseed; yam; zerumbet.

wild alum root: a. root - **w. china**: *Sapindus marginatus* - **w. rice:** *Zizania aguatica*.

Wildkirsche (Ge): *Prunus serotina*.

wilfordine: complex cryst. alk. occ. in *Tripterygium wilfordii*; related alks. found with it are wilforine, wilforgine, wilfortrine, &c.; ester of a C_{15} polyalcohol with acetic acid & ring carboxylic acid; u. insecticide.

Willardia Rose = *Lonchocarpus*.

willow: term usually appl. to *Salix* spp. - **w. apple**: w. gall - **ball w.**: button w. - **w. bark**: usually fr. *S. fragilis* - **basket w.**: *S. viminalis* - **bat w.** (Eng): *S. alba* var. *caerulea* - **black w. (root)**: *S. nigra* (bk.); *S. discolor* - **button w.**: *Cephalanthus occidentalis*. - **crane w.**: button w. - **desert w.**: *Chilopsis linearis* - **diamond w.** (Mont [US]): *S. eriocephala* - **dwarf w.**: *S. humilis* - **European (white) w.**: *S. alba* - **false w.**: *Baccharis angustifolia* - **w. family**: Salicaceae - **w. gall**: any of various galls on *Salix* spp., but esp. the **pine-cone w. g.**, common on *S. exigua, S. discolor*, &c. - **heart-leaved w.**: *S. rigida* - **w. oak**: gp. of *Quercus* spp. with long narrow lvs. & smooth or undulated margins; ex. *Q. imbricaria* - **w. poplar (buds)**: *Populus* spp. yielding medl. buds - **purple (osier) w.**: *Salix purpurea* - **pussy w.**: *S. discolor* (pistillate catkins on branches u. for decoration) - **red w.**: *Salix lasiandra* - **w. rose**: w. gall - **Salicon w.**: European w. - **silver w.**: *Elaeagnus* spp. - **sweet w.**: *Myrica gale* - **weeping w.**: *S. babylonica, S. caprea, & S. purpurea* (in last two spp. as vars. *pendula*) - **white w.** (NA): *S. alba*; *S. lasiolepis* (bk.).

willow-herb, (giant): bay w.: *Epilobium angustifolium* - **purple w.**: *Lythrum salicaria* - **yellow w.**: *Lysimachia vulgaris*.

Willughbeia cochinchinensis Pierre (Apocyn.) (seAs): rt. u. int. for icterus, heartburn, dys.; exter. on yaws; latex smeared on yaws & sores (Mal), dressing for wounds; source of an inferior rubber (Indochin) - **W. tenuiflora** Dyer (Indones): dango; latex source of inferior kind of rubber; u. in skin dis. by natives.

wind flower: *Anemone* spp. (Pulsatilla); also spp. of *Thalictrum*, &c. - **w. root**: *Asclepias tuberosa* - **windowed plants**: Mesembryanthemaum: spp. of

Fam. Aizoaceae which have spaces in the lf. tips for admission of light.

wine: 1) fermented grape juice (*Vitis* spp.) with alc. content up to about 15% (natural w.) using yeast; often u. medicinally (897) 2) pharm. prep. of med. substance in wine (usually white w.) - **dry w.**: sour w. with very little sugar left - **w. ether**: ethyl pelargonate - **fortified w.**: with added alc. (usually brandy); ex. port, sherry - **w. lees**: sediment which collects in bottom of w. vats; c. 20-35% potassium bitartrate & up to 20% calcium tartrate - **w. of beef**: beef ext. in alc., syr., & sherry; u. ton. - **w. of beef & iron**: elixir carnis et ferri (L.): like prec. with iron & ammonium citrate - **oil of w.: w. oil**: ethyl oenanthate; cognac essence; e. pelargonate; or mixt. of e. caprate, caprylate, butyrate, amyl caprate, etc.; prepd. by distn. of press refuse fr. wine making with little w. + H_2SO_4; u. flavor in mfr. artifical cognac & brandy flavor; *cf.* ethereal oil (**heavy oil of w.**: prod. of dist. of definite proportions of alc. or ether [or both] with sulfuric acid; c. diethyl sulfate, ethylene polymers, etc.; mixed with equal vol. of ether constitutes ethereal oil) - **port w.**: port; sweet strong (fortified) wine orig. fr. Oporto, Port; u. to aid recovery in convalescence - **red w.**: prep. fr. grapes with reddish skins, the color being conferred in expressing the juice fr. the whole grape; said to reduce serum cholesterol; dry r. w. valuable for stim. appetite - **resinated w.**: retsina - **sherry w.**: fortified dry to sweet amber-colored w. orig. fr. Spain - **sour w.**: acidic; dry w. - **Spanish w.**: sherry - **sparkling w.**: effervescent w. in which CO_2 is prod. by fermentation within the bottle; ex. champagne - **spirits of w.**: alc. - **still w.**: one lacking effervescence; ex. Mosel w. - **w. stone**: cream of tartar, potassium bitartrate, which xallizes on walls of wine storage tanks - **sweet w.**: with considerable original sugar left unfermented by addn. of alc.; ex. angelica - **white w.**: made fr. "white" grapes or by skinning colored grapes before expression of juice.

wineberry: 1) *Rubus phoenicolasius* (Chin, Jap) (**Japanese w.**) 2) *Aristotelia racemosa* (NZ tree; frt. u. to make wine).

wing scale: *Atriplex canescens*

wino (US slang): person addicted to the use of wine; this often taken with spirits.

winter: v. also under asparagus; cherry; clover; grass; green; savory.

winter berry (, common): *Ilex verticillata* (also **Virginia w. b.**); *I. glabra* - **w. bloom**: *Hamamelis virginiana* - **w. chilled: w. strained**: process of leaving FO (etc.) stand outside in cold weather (in tanks, drums, etc.), then straining to remove solidified matter, such as stearin - **w. sweet**: 1) *Acokanthera spectabilis* 2) *Origanum vulgare*.

Wintera (L.): **Winteranus Cortex: winter's bark**: fr. *Drimys winteri*.

Winterana canella L.: *Canella winterana* (*Winterana* —> *Canella*; *Drimys*) - **wintergreen**: 1) *Gaultheria procumbens* (lf. VO) 2) *Pyrola* spp. 3) contraction of w. oil, methyl salicylate - **artificial w.**: synthetic methyl salicylate - **bitter w.**: *Chimaphila umbellata* - **w. extract: w. flavoring: w. oil**: methyl salicylate, either synth. or obt. fr. *Gaultheria procumbens* or fr. *Betula lenta* - **"w. plant"** (swGa [US]): *Polygala* spp. - **spotted w.**: *Chimaphila maculata*.

winterize: prevent freezing by low temps. such as by adding liquid oils to fat, or alc. to water.

Wirbelsaeule (Ge): spine.

wiregrass (seUS): *Aristida stricta* - **w. counties** (Ala [US]): Mobile, Baldwin, Escambia, Geneva, Henry, & Covington - **w. region** (or section): se corner of Ala (9 counties [around Dothan]), swGa [US], and nwFla [US]; mostly flat pine woods; SC characteristic grassy covering (esp. weedy *Cynodon dactylon*) ("wiregrass country").

wireweed: 1) *Polygonum aviculare* 2) (seUS): *Aster lateriflorus* 3) (Ala [US]): *Sida* spp.

Wisteria Spreng. = *Wisteria* Nutt. (Kraunhia Rafin.)

Wisteria Nutt. (Leg.) (As, NA): grape flower vine; 6 spp. of woody vines; some spp. toxic (bk.) (wistarin, glycoside; toxic resin); cult. ornam. (SC Caspar Wistar, Prof. Anat., Pa [US]) - **W. floribunda** DC (Jap): fuji, (Japanese) wisteria; leaflets u. as substit. for & to adulter. tea; popular ornam. vine - **W. sinensis** (Sims) Sweet (W. chinensis DC; Glycine s. Sims) (Chin): Chin wisteria; C. kidney bean tree; bk. & rt. c. wistarin (toxic glycoside) & a toxic resin; u. for gastric cancer (Jap); sds. diur.; lvs. as tea substit.; bast fibers; u. for threads & cloth (Jap); ornam. pl. (1797).

witchgrass: *Agropyron repens* (*Elymus*).

witch-hazel: *Hamamelis virginiana*: bk. & lvs. - **w. family**: Hamamelidaceae - **w. water**: Aqua Hamamelidis (L.): distilled w. extract; obt. by water-macerating recently cut but partly dr. dormant twigs of w. for one day, then distilling until distillate represents 85% of orig. wt. of twigs; then adding 15% (V/V) of alc.; or preferably by simultaneous distn. of alc. & water over pl. materials.

witchweed: *Striga* spp., esp. *S. lutea*.

witchwood: *Sorbus aucuparia*.

Withania coagulans Dunal (Solan.) (Indo-Pak, Afghan): vegetable rennet; shapianga (Pushto); punirband (Indo-Pak) (means "the cheesemaker"); frt. c. withanin (casein coagulating rennet-like enzyme); u. to make cheese (substit. for rennet) also u. in native med. as emet. diur. antipyr. antisep. - **W. somnifera** Dunal (incl. var. *flexuosa* (L) Dunal (W. f. [L] Hassk.) (Med reg.; Ind, Pak, sAf): winter cherry; asgand, aswagandha (Urdu, Hind); rt. c. narcotic alks., anaferine, tropine, withanine (898); rt. popular aphrodisiac; ton; in TB, sed. in senile debility; rheum.; abort (Pak); local appln. to furnuncles, sores; lf. c. withanolides; u. as emmen.; in urin. inf., antipyr. in nervousness; in cheese prepn.; frt. c. lab-enzyme as in *W. coagulans*, u. to coagulate milk; rts. important Ayurvedic drug (629).

withe rod (**bark**) : *Viburnum cassinoides.*

Witheringia solanacea L'Hérit. (Solan.) (sMex to CA, WI, Pe, sBo): lf. inf. u. for pain (Pan) ("viata dolor"; lf. inf. u. as hypnotic [Pe]).

witloof (Brussels): **w. chicory**: var of *Cichorium intybus* in wh. lvs. are blanched & u. raw as salad; popular veget. (developed in Belgium).

Wittiocactus (Wittia) panamensis Britt. & Rose (Cact.) (Pan, Co, Ve): tubopaica; pl. u. as rem. for sunstroke & eye ailments.

woad: *Isatis tinctoria* - **woadwaxen, common**: *Genista tinctoria.*

wogonin: methoxyflavone fr. rt. & hb. of *Scutellaria baicalensis*; SC, wogon, Jap for rt.

wolds: weld.

Wolf (Ge): 1) *Nardus stricta* L. 2) *Canis* spp. (wild) - **W. oel** (Ge): 1) Oleum Rusci 2) Oleum Hyoscyami, FO of sds. of *Hyoscyamus niger* (Wolfsdisteloel).

Wolfbeerenoel (Ge): Ol. viride coctum: olive oil with chlorophyll.

wolf-berry: *Symphoricarpos albus; S. occidentalis* (wUS, Cana) - **Wolferlei** (Ge): Arnica.

wolf flowers (Tex [US]): *Lupinus subcarnosus* - **w. peaches** (old name): tomatoes.

wolf's bane: wolfsbane: 1) *Aconitum napellus* 2) *Arnica montana* (**winter w.**) 3) *Eranthis hyemalis* - **w. claw(s)**: *Lycopodium clava-tum.*

Wollastonia DC = *Wedelia.*

Wollfett (Ge): wool fat - **Wollkraut**: mullen.

wolveboon: *Toxicodendron capense.*

wolvet (Dutch): wool fat.

woman's tongue (Jam): mother-in-law tongue.

wonder berry: cult. v. of *Solanum nigrum*; edible frt. u. in pies, etc.

wonga wonga vine: *Tecoma australis; Pandorea pandorana.*

wood: xylem; all that hard fibrous portion of sts., branches, & rts. of a perennial exogenous pl. (trees & shrubs) axis lying outside of the pith and which occurs inside the interfascicular cambium layer & is produced therefrom; incl. vessels, wood fibers, wood parenchyma, tracheids; med. wds. incl. Quassia, Santalum Rubrum; also found in some endogenous sts. (ex. palm tree); the high value of wood appears in the many prods. of pine wh. incl. the lumber, w. pulp, cellulose & its esters, turpentine oil, rosin, tall oil, soap, sawdust (in explosives), wall board, charcoal, tar, tar oil, methyl alc., acetone, acetic acid, plywood, cellophane, viscose, textiles, explosives, paper, etc.; wood distillation products are derived fr. *Acer, Fagus, Quercus, Betula*, &c., spp. & include flammable gases (u. to heat retort), "pyroligneous acid" (liquid mixt.), wood tar, wood charcoal (residue in retort), &c. 2) (hort.): branches or branch wd.; in frt. tree, leaf-bearing as against the more valuable frt.-bearing branches - **acid w.**: chemical w. - **w. alcohol**: methanol - **w. bind**: **w. bine: woodbine**: 1) *Parthenocissus quinquefolia* 2) *Lonicera* spp. - **chemical w.**: that u. for distn. in chem. mfr. - **compression w.**: the heavier & hardier wd. prod. in the lower part of tree trunk when st. is bent down as sapling from its vertical position; this wd. is higher in lignin (ex. normal pecan wd, c. 26% l., compr. wd. 38%) - **"everlasting w.:** 1) cypress ("wood eternal") 2) sassafras wd. - **fat w.**: v. under "lighter wd." - **w. fiber**: w. wool (2) - **w. louse** (plant w. lice): one of numerous spp. of small terrestrial crustacea of genera *Armadillidium* (esp. *A. vulgare*); *Oniscus*; *Cubaris*; *Porcellio*, &c.; *Armadillidium pictum* Brandt (Armadillidiidae - Isopoda [pill bugs; sow bugs; wood lice]) (close to the millipedes) (Eu, America); common terrestrial pillbug; in damp places; able to roll into a ball; still carried in some pharmacies (Ge) under name "millipedes (levanticae)" (L); u. as diur. in edema, asthma; bronchial catarrh.; mostly now uses restricted to homeopaths & folk medicine - **w. naphtha**: methanol - **Nicaragua w.**: 1) *Caesalpinia echinata*; one of red woods u. for dyeing 2) *Haematoxylon brasiletto* - **w. pulp**: disintegrated wd. for use principally in paper mfr.; chief types are: mechanical; chemical (sulfite; soda; Kraft or sulfate) - **w. row**: *Marchantia polymorpha* - **w. vinegar**: pyroligneous acid - **w. wool: fir w.**: 1) (Eu): fibers prep. fr. needles of *Pinus sylvestris* (etc.) by steaming or by boiling in dil. alkali soln.; u. in mfr. of several med. prods. as by-prods. in Ge, esp. as surgical dressing; also in mfr. clothing, bagging, stuffing of pillows, mattresses, etc. (fir wool ext.) 2) (US): excelsior; fine wd. shavings in long strips, mostly fr. *Populus* &*Tilia* wds.; mostly u. in packing.

wood: v. also under charcoal, creosote; tar; betong; scabious; sorrel.

Woodfordia fruticosa (L) Kurz (W. floribunda Salisb.) (Lythr.) (seAs): dadki (Ind); dhyaro (Nep); bk. decoc. appl. to relieve sprains & pains (Nep); lf. & fl. decoc. u. ton. in pregnancy; dys. liver dis. (124); fls. astr. in hemorrhoids (Nep) (727); fls. lvs. & frt. u. as diur. for dysuria & hematuria; in rheum (Indones).

woodroof (, sweet[-scented]): **woodruff**: *Galiam odoratum* - **woodskin**: bark, peeling of tree sts. u. to make canoes, &c.

Woodwardia cochinchinensis Ching (Polypodi./ Blechn.) (Indochin): chain fern; rhiz. decoc. u. for dis. of vertebral column, lameness, trembling of old age - **W. japonica** (L.f.) J. E. Sm. (Jap, Chin, Kor, Indochin): mild ton.; insecticide - **W. radicans** (L) Sm. (Indochin): u. similarly to *W. cochinchinensis*.

woody (nighshade): *Solanum dulcamara*.

woof: weft.

wool: 1) animal hair, esp. when matted & crinkled 2) (specif.) sheep's hair; v. sheep's w. - **absorbent (cotton) w.** (Brit): absorbent cotton - **w. fat (, refined)**: anhydrous lanolin; purified anhydrous fat-like substance fr. sheep's wool (*Ovis aries*); actually a wax, sol. in alc.; c. cholesterol, isocholesterol, as esters of higher fatty acids; u. inunction, vehicle; base for ointments, creams, cosmet.; prep. hydrous lanolin - **fir w.: pine w.**: v. wood w. - **w. grease**: lanolin - **hydrous w. fat**: lanolin - **sheep's w.**: prep. fr. fleece of sheep (*Ovis aries*) by thorough cleaning & washing in soap & water; u. sometimes in surgical dressing - **w. tree**: 1) *Ceiba pentandra* & other *C.* spp.; source of kapok 2) balsa tree - **vegetable w.**: 1) *Ceiba pentandra* (kapok) 2) *Girardinia palmata* (fiber) - **wood w.**: v. under wood.

woollen (adj.): **woolly**: of or relating to wool - **woolly pyrol**: *Phaseolus mungo*.

woorali: woorara: woorari: curare; fr. *Strychnos toxifera* (etc.).

Worcestershire sauce: condiment prepd. by one of several formulas, incl. soy, tamarinds, garlic, asafoetida, shallots, onions, chile, cloves, vinegar, lime juice, anchovies, pickled herring, &c. (SC first prepd. in Worcester, Eng).

worm bark, Surinam: West Indian w. b.: 1) *Andira inermis* 2) *A. retusa* - **w. candies**: santonin tablets - **w. grass**: *Spigelia* - **w. moss**: *Alsidium helminthochorton* - **worm-root (, American)** (US): *Spigelia marilandica*.

Wormian bones: *Ossa Suturalia* (L): small irregular bones in sutures of human skull; formerly u. med. & off. in Br Phar.

wormseed: 1) *Chenopodium ambrosioides* & its v. *anthelminticum* (in US) 2) *Artemisia cina* (etc.) (in Brit) - **American w.**: w. (1) - **European w.: Levant(ic) w.**: Santonica (L.): *Artemisia cina, A. pauciflora,* & some other *A.* spp. - **w. oil**: fr. **wild w.**: American w.

Wormskioldia longipedunculata Mast. (sAf): (Turner.): Rhodesian pimpernel; u. as remedy for sore eyes; as earache remedy; analgesic.

wormwood: 1) *Artemisia absinthium* (VO) 2) *Ambrosia elatior* - **Alpine w.**: génépi des Alpes - **common w.**: *Artemisia absinthium* - **gentile w.**: *Artemisia valesiaca* All. - **Indian w.**: *Artemisia dracunculus* - **Levant w.**: *Artemisia cina* - **linear-leaved w.**: Indian w. - **Pontic w.**: *A. pontica* - **Roman w.**: 1) Pontic w. 2) *Ambrosia elatior* 3) *Corydalis* sp. - **wild w.**: *Artemisia vulgaris*.

wort (pron. wurt) (Middle Eng; orig. meant root; med. or food; pl. or any part): 1) herb; root; usually as a suffix 2) (specif.): *Vaccinium myrtillus* 3) (brewing): inf. of malt or some other cereal or cereals; earliest stage of beer mfr. - **bell w.**: *Uvularia perfoliata* - **blood w. (, striped)**: *Hieracium venosum* - **cat's w.**: *Nepeta cataria* - **cross w.**: *Eupatorium perfoliatum* - **fever w.**: *Eupatorium perfoliatum* - **fig w.**: *Scrophularia* spp., esp. *S. aquatica, S. nodosa* - **frost w.**: *Helianthemum canadense* - **marsh w.**: 1) *Vaccinium oxycoccus*. var. *intermedium*; *Apium nodiflorum* - **master w.**: *Heracleum lanatum* (etc.) - **mug w.**: *Artemisia* spp. esp. *A. vulgaris, A. ludoviciana* - **rag w.**: *Senecio* spp. - **St. Peter's w.**: *Ascyrum hypericoides* (etc.) - **St. Johns w.**: *Chelidonium majus; Hypericum*, esp. *H. perforatum* - **sneeze-w.**: *Helenium autumnale* - **soap w.**: 1) *Saponaria* spp. esp. *S. officinalis* 2) *Gypsophila paniculata* (**white s.**: *Lychnis dioica)* - **soldier's w.**: *Achillea millefolium* - **spider-w.**: 1) *Tradescantia* spp., esp. *T. virginana* 2) *Paradisea lilia-strum* - **star-w.**: 1) *Chamaelirium* 2) *Aster* 3) *Aletris farinosa* (rt.) (also **mealy s.**) (**dropping s.:** *Chamaelirium*) - **tetter w.**: *Chelidonium majus* - **thorough w.**: v. under t. - **vomit w.**: *Lobelia inflata* - **yellow w.**: *Chlora perfoliata* - **youth w.**: 1) *Drosera rotundifolia* 2) *Heracleum lanatum* - **wortcunning** (Middle Eng): a knowledge of hbs. & skill in their usage; herbalism - **wortle-berries**: *Vaccinium myrtillus*.

woundwort: corn w.: 1) *Stachys palustris* (SC soft lvs. u. for centuries as bandages) 2) *Laserpitium latifolium* (**downy w.**) - **soldier's w.**: *Achillea millefolium*.

wrack (, sea): 1) *Fucus vesiculosus* & other *F.* spp. 2) *Posidonia oceanica* 3) *Zostera marina,* &c.

wrenbush: decorated holly bush (St. Stephen's Day) (SC wrens hung fr. it).

Wrightia antidysenterica R. Br. (W. ceylanica R. Br.) (Apocyn.) (Ind, SriL): bk. (Conessi Bark) & sds u. in dys. & as anthelm. & antipyr. - **W. natalensis** Stapf (sAf): saddle pod (Swaziland); pd. root (+ bk) u. in beer as aphrod. (Venda) - **W. pubescens** R. Br. (Indones+): latex u. to treat dys.; lf. decoc. antipyr. (Hainan) - **W. tinctoria** R. Br. (Ind, SriL, Burma): bk. c. lupeol, pseudoindican; lvs. source of indigo-like dye; bk. & sds. u. as ton. aphrod., in dys., skin dis.; bk. substit. for kurchi bark; latex source of caoutchouc, as styptic (Rajasthan, Ind); wd. useful - **W. tomentosa** Roem. & Schult. (W. arborea [Dennst.] Mabb.) (Ind, Indoch+): bk. u. in dys. renal dis., menstr. disturb., in baths for gout; sds. antipyr. u. as styptic (Nep); c. yellow dye; wd. u. in carvings, etc.

wrightine: alk. of *Wrightia antidysenterica* sds.; same as conessine; occ. in some other plants.

wu chu yu (Chin): *Evodia rutaecarpa.*

Wuerzburg (Bavaria, Germany): ancient city on the Main River; one of chief wholesale drug market centers of Eu; earlier called Herbopolis (= city of herbs); SC. formerly important for cultn. of herbs; here was published Pharmacopoea Herbipolitana (1796); center of wine industry (dating fr. 6th cent.) (2723).

Wunderbaum (Ge) (wonder tree): *Ricinus communis* - **Wundklee** (Ge): *Anthyllis* spp.

Wurmfarn (Ge): Aspidium - **Wurmkraut** (Ge): *Filipendula ulmaria* - **Amerikanische W.**: Am wormseed - **Wurmsame(n)** (Ge): Santonica.

wurrus: wurs: *Flemingia congesta* (frt. hairs).

Wurstkraut (Ge): Majorana.

wurt (Middle Eng): wort.

wurus: wurrus.

Wurz (Ge): 1) suffix in compd. words, means hb.; wort; ex. Korallenwurz, coral-wort: *Corallorhiza* 2) Wurzel, root.

Wurzel (Ge): root - **W. haube**: rt. cap - **W. huelle**: velamen - **W. stamm**: **W. stock**: rhizome (lit. rt.-st.) - **W. tasche**: velamen.

Würze (Ge): spice.

Würzigkeit (Ge): spiciness.

Wyethia amplexicaulis Nutt. (Comp.) (BC [Cana] to Nev [US]): northern mule-ears; smooth dwarf sunflower; rt. after heating & fermentation eaten (Am Indians) - **W. helenioides** (DC) Nutt. (Calif [US]): gray mule-ears; rts. bitter & arom., u. as domestic remedy in colds, coughs, asthma, throat dis. flu, &c.; said formerly prescribed by many Calif physicians, esp. in country towns; found in some drug stores (899) ("turpentine root") - **W. mollis** A. Gray (sOre, nCalif, Nev [US]): mountain mule-ears; woolly dwarf sunflower; rt. u. in cataplasms appl. to swellings (3583); rt. decoc. physic; weak decoc. of rt. in VD, TB, "blood" ton.; sds., cypse/as & young shoots edible.

wynne grass: *Melinis minutiflora.*

xanthaline: papaveraldine: xall. benzyl isoquinoline alkaloid of Opium Poppy, possibly formed during processes of extraction.

xanthan gums: polysaccharide mixts. prod. by fermentation with *Xanthomonas campestris* (Pammel) Dowson (Bacteria); c. incorporated glucosyl, mannosyl, and glucosyluronic acids with acetyl & pyruvic acid acetal; u. as emulsifier & stabilizer in foods, pharmaceuticals, cosmetics.

xanthatin: antibiotic fr. *Xanthium pennsylvanicum, X. americanum* - **xanthine**: purine comp. obt. fr. animal organs, yeast, potatoes, etc.; also synth.; u. as reagent - **x. oxidase**: enzyme found in milk & many animal tissues; oxidizes many purines.

xanthic: referred to all fls. with yellow, sometimes also to all colors except blue.

Xanthium L. (Comp.) (cosmop.): cockleburs; frts. as inf. u. in folk med. for rheum. arthritis (900), exter. as styptic; lvs. u. to adulter. Stramonium: hay fever plants - **X. americanum** Walt. = *X. strumarium* var. *glabratum* - **X. canadense** Mill. = *X. strumarium* var. *c.* - **X. catharticum** HBK (Pe, Ch): yerba de Alonso; u. in liver dis.; "depurative" **X. chinense** P. Mill. (Austral, natd. WI, sMex): rt. decoc. as diur. in kidney dis., to expel gravel & stones; lf. decoc. u. for scrofula, ringworm, pimples, facial blemishes; cancer (Cubans cult. plant in Miami for this use) = *X. strumarium* var. *glabratum* - **X. echinatum** Murr. = *X. italicum* Moretti - **X. pensylvanicum** Wallr .= *X. strumarium* var. *canadense* - **X. pungens** Wall. (As, Austral): sd. FO u. in foods, also as paint base (semidrying); pl. poisonous to stock; causes dermatitis (itching, redness) in humans - **X. riparium** Itzingson & Bertsch) (X. album H. Scholz) (nGe): sd. FO as food, in soap mfr., in prepn. of high mol. alcs., alc. sulfonates; in cosmetics; press cake as fodder - **X. speciosum** Kearney = *X. strumarium* var. *canadense*. - **X. spinosum** L. (orig. SA, now natd. warm temp & subtrop reg., incl. US, Med reg.): clotweed; spiny clotbur (or burweed); Spanish nettle (Calif [US]); S. thistle (Calif); hb & lvs. (Xanthium Herb) u. diur. diaph. in diarrh. dys., antiperiod. in mal.; in diabetes (folk med.); in hepatic dis. (Arg); formerly in rabies (1798); burs u. for diarrh. (NM [US]); lvs. for snakebite; rt. u. diur emoll. in liver & stomach dis. (Ch): burs injurious to horses & cattle; sds. c. hydroquinone, often toxic to livestock (esp. young pigs) (sUS) - **X. strumarium** L. (Euras; natd. eUS): (common) cocklebur; sea (or small) burdock; source of Herba Lappae minoris (lvs.) & sds.; hb. u. as diur. diaph. sed., in mal. for sore throat (inf. with honey [Fr]) adulter. of Stramonium; frt. u. as dem. hemostyptic; sds. yield a bland FO u. as lamp oil, to make varnishes; pl. juice u. for goiter, cancer, rt. u. in cancer & tumors, venomous bites & stings: rt. decoc. given in maté to alcoholic to prod. repugnance to drinking (Andes reg.).; meat washed in rt. inf. keeps longer (Arg): sds. burned & fumes inhaled for piles (Rajasthan, Ind); var. *canadense* (P. Mill.) T. & G. (X. commune Britt.; X. echinatum Murr.; X. italicum Moretti; X. pensylvanicum Wallr.; X. saccharatum Wallr.; X. speciosum Kearney) (e, c&sNA): sea burdock; burweed; cocklebur; c. xantholides (xanthinin, xanthatin); pl. u. antiluetic; mashed fresh lvs. put on screw-worm sores (cattle); bur decoc. in diarrh.; pl. poisonous & a bad weed - var. *glabratum* (DC) Cronq. (X. americanum

Walt.; X. chinense P. Mill.; X. macrocarpum DC var. m. DC; X. orientale L.) (e&cNA, incl. Mex): clotbur; c. xanthatin.

Xanthoceras sorbifolia Bge. (Sapind.) (nChin): wd. & branches off. (Chin); sds. edible; ornam. flg. shrub/tree.

Xanthocillin: antibiotic complex isolated fr. strains of *Penicillium notatum*; antibacterial; u. in feed supplements: ca 1956-; "Brevicid."

Xanthomonas campestris (Pammel) Dowson (Bacteria): microorganism u. to produce xanthan gums fr, carbohydrates; u. in pharm. prepns.; NF XIV.

xanthone: a flavanone deriv.; u. ovicide, larvicide.

xanthophyll: lutein; yellow coloring matter accompanying chlorophyll in most green pl. parts; a carotenoid alc.; usually isolated fr. algae, nettles, yellow petals of many pl. spp.

Xanthophyllum excelsum Miq. (Xanthophyll./Polygal.) (Indones): hb. useful in colic - **X. flavescens** Roxb. (Ind): rts. c. lupeol, oleanolic acid; hard wd. u. for construction, temporary fencing - **X. glaucum** Wall. (Indochin): lvs. u. like hops in making rice beer - **X. lanceolatum** J. J. Sm. (Indones): FO expressed fr. sds. u. to treat thrush, as food; mfr. soap & candles (suir or sioerna oil).

xanthoproteic reaction: yellow coloration with formation of **xanthoprotein** fr. tyrosine- or tryptophan-contg. proteins when heated with strong HNO_3; "nitrodyes."

xanthopterin: pigment occ. widely in insect & other animals; (crab: urine); first found in butterfly wings (900A); converted by yeast enzymes into folic acid; u. in anemias.

xanthopuccine = canadine.

xanthopurpurin: glucoside of *Rubia tinctorum*; di-OH-anthraquinone + glucose.

Xanthoria candelaria (L.) Kickx. (Teloschistes candelarius (L) Fink) (Teloschist. - Lecanorales - Lichenes) (worldwide in temp & subarct): thallus source of yellow dye u. to dye woollens (Sweden) - **X. parietina** (L.) Th.Fr. (Teloschistes parietinus [L] Norm.); (Parmelia p. [L.] Ach.) (worldwide): (common yellow) wall lichen; u. as quinine substit.; c. chrysopicrine; bitter ton., in scrofula, phthisis (formerly); brown dye.

Xanthorhiza simplicissima Marsh. (X. apiifolia L'Hérit.) (monotypic) (Ranuncul.) (eNA): (southern) yellow root; rts. & rhiz. c. berberine (1.2-1.3%), jatrorrhizine, magnoflorine, oxyacanthine (901); sometimes u. as source of berberine but low yield; more often u. as bitter ton. in mouth cankers, sores, inflamm. & soreness of gums, esp. fr. the use of dentures (Miss [US]); popular in folk med. (country people), decoc. in diabetes & childbirth (Ala [US] Negro), sore throat, eyes, &c. (902); rts. source of yellow dye.; adultern. of Hydrastis. (variants of spelling: Zanthorrhiza, Zanthorhiza, Xanthorrhiza).

Xanthorrhoea Sm. (Xanthorrhoe./Lili.) (Austral): source of yacca gum, also called (red) gum accroides; grass tree resin, Botany Bay gum; black boy g.; (gum) acaroid; c. erythroresino-tannol, VO with styrol, free & esterified cinnamic acid, flavonoids, &c.; resins u. now mostly in technol., chiefly in shellacs, lacquers, & varnishes; floor stains, pyrotechnics; as rosin substit. in paper mfr., &c.; formerly in med. as alc. tinct. (6%) for phthisis & chronic catarrh: chief grades: coarse & fine pd. - **X. arborea** R. Br.: black boy; yacca; grass tree; source of xanthorrhoea resin (red acaroid resin; r. gum accroides); u. as shellac substit.: tender young lf. bases & extremities of young shoots u. as food by Aborigines (Austral) - **X. australis** R. Br. (NSW, Victoria, Tasmania): Australian grass tree; source of red acaroid resin (fr. lf. bases); c. resin acid esters, such as p-coumaric acid ester of erythroresinotannol; formerly u. as antispasm. in catarrhal dis., carm., mostly now in technol. (grass tree gum) - **X. hastilis** R. Br. (X. resinosa Pers.; Aloe hastile) (e&seAustral): (spearleaf) grass tree; lf. bases & sts. important source of yellow acaroid resin; c. p-coumaric acid; u. mostly in varnish (shellac) business; in paper mfr.; perfumes (Eu); formerly, alc. tr. u. for phthisis & chronic catarrhs, diarrh., stomachache, night sweats; sds. source of FO with unsatd. fatty acids - **X. quadrangulata** F. v. Muell. (sAustral): source of red acaroid resin; formerly u. by aborigines as cement to fix stone axe heads to hafted handles.

Xanthosoma Schott (Ar.) (warm Am): spoon flowers; malanga; some 50 spp. of large hbs.; rhiz. & lvs. of some spp. u. as food (*cf.* taro) - **X. atrovirens** K. Koch & Bouché (WI, nSA): darkleaf malange; tubers sometimes eaten; lf. inf. u. for chills (Domin, Caribs) - **X. edule** Schott (Surinam): tubers edible, with 62% starch - **X. jacquinii** Schott (WI, Ve): yautia palma; edible; fed to hogs (WI) - **X. lindenii** (André) Engler (Co): Indian kale; cult. in glasshouses as ornam. - **X. mafaffa** Schott (sBr): yautia; mafaffa; rascadera; cult. (Br) - **X. nigrum** (Vell.) Conc. Mansf. (X. violaceum Schott) (SA, esp. Br): primrose malanga; Indian kale; tubers (starch 62%) & lvs. eaten - **X. sagittifolium** (L) Schott (WI, SA): (yellow) yautia; tan(n)ier; tannias; eddoes; malanga blanca; "coco" (Jam); tubers (tuberous rhiz.) esculent after boiling; somewhat like taro in food quality; for this purpose cult.

widely in trop.; source of starch; lvs. sometimes boiled & eaten as spinach.

Xanthoxalis Small = *Oxalis* - **X. corniculata** (L) Small = *Oxalis c.*

Xanthoxyli Fructus (L.): prickly ash berries; obt. fr. *Zanthoxylum americanum & Z. clava-herculis.*

Xanthoxylum Miller = *Zanthoxylum.*

Xantolis cambodiana (Pierre) van Royen (Sideroxylon c. Pierre) (Sapot.) (Indochin, Thail): gratings of rt. u. in quotidian fever (Mal); rt. lf. & wd. decoc. galactagog - **X. tomentosa** (Roxb.) Rafin. (Sideroxylon tomentosum Roxb.) (Ind-Indochin): frt. pulp u. in cholera; black sds. rubbed in vinegar & given to lick as astr.; hard wd. u. for mallet heads, &c.; pl. yields a gutta percha; frt. edible & is conserved.

xenechiche (Mex): *Heimia salicifolia.*

xenobiotic: relating to symbionts living together.

xenon: Xe; noble gas element; occ. in air & in gases of thermal springs; obt. fr. liquid air; u. in anesthesia (exptl.); diagnostic aid (Xe 133); in electric lamp bulbs; series of nine natural & artificial isotopes known.

Xenopus (Pipidae-Algossa-Anura-Amphibia) (sAf): South African Toad; African Clawed "Frog"; living animal u. in thyroid assay; pregnancy tests.

Xeranthemum annuum L. (Comp.) (sEu, Med reg., wAs): common immortelle; pink (purple, annual) everlasting; sds. c. FO; ornam. hb. with fl. heads white, purple, violet, reddish; u. med (Por Phar. III, IV) - **X. cylindraceum** Sibth. & Sm. (sEu): widely cult. as ornam.

Xereswein (Ge.): sherry wine (SC *Jerez*, Sp city in production area).

Xerochlamys pilosa Bak. (Sarcolaen.) (Mad): hatskikana; rt. bk. u. to flavor alc. liquors.

Xeroderris Roberty = *Ostryocarpus.*

Xeromphis nilotica (Stapf) Keay (Rubi) (Tanz): frts. chewed as male aphrod.; boiled rts. eaten with porridge for syph. - **X. obovata** (Hochst.) Keay (Tanzania): "mupongolo"; rts. u. in mixt. for epilepsy, insanity (712); rt. bk. decoc. in asthma, bilharzia; rts. for snakebite; rts. in abdom. pains; rts. mixed with meat & cooked as aphrod. (903) - **X. uliginosa** Maheshwari (Ind): pesar; unripe frts. u. in dys. (124), diarrh. (genus close to *Randia*) (—> *Catunaregam*).

Xerophyllum tenax (Pursh) Nutt. (Lili) (wNA): (common)bear grass; Indian basket g.; western turkey-beard; pine lily; rts. roasted or boiled for food; strong lvs. u. to make waterproof baskets, hats, &c. (Am Indians); ornam.

Xerophyta retenervis Bak. (Vellozi.) (SA Indians): st. decoc. drunk to prevent uterine lesions following childbirth.

xerophthalmia: state of acute vit. A deficiency; characterized by drying of conjunctiva, etc. - **xerophthol**: Vit. A - **xerophyte**: desert or arid land pl.; ex. most Cactaceae; sage brushes - **xerosis**: dryness of skin or eyes.

Xerospermum glabratum Radlk. (Sapind.) (Ind): frt. sweet, edible - Several *X.* spp. are u. in Mal med.

X-factor: the Fe or Hb factor of blood necessary to grow *Hemophilus influenzae*, &c.

Ximenia americana L. (Olac.) (trops, esp. Af, Br, SA, sFla [US]): tallow nut; Spanish/monkey/mountain/seaside/sour/wild lime/olive/plum; hog p. (Mex); yellow p. (Austral); tigrito (Ve, Br); tallow wood (Fla) plum; sd. husks rich in FO with santalic acid, &c.; u. as food oil, in soap & cosmetic mfr. (Br); butter (ghee) substit. (Ind); sd., kernel purg.; wd. & rts. with acetylene fatty acids as glycerides; u. exter. in sores, menorrhag. (Tanz); for trichophytosis, as astr.; rt. as antipyr. & for diarrh. (in calves) (sAf): lvs. c. sambunigrin, tannin, resin, pinitol, flavonoid glycoside, u. for coughs, fever, wounds, toothache, emet. & as eye lotion; fls. fragrant; frt. of tree edible like cherry; lf. rich in HCN, toxic to stock (Austral): wd. brown, arom., valuable; sandalwood substit.; u. in turnery & to make toys - var. *microphylla* Welw. (Af): frt. edible; ripe frt. brewed to make beer (Transvaal) - **X. caffra** Sond. (sAf): Kaffer/Natal/sour/wild plum; lvs. c. tannins & cyanogens; u. as anti-inflammatory & in eye lotions; rt. in fever, diarrh. syph., hookworm dis.; frt. edible.; sd. c. FO u. as cosmetic - var. *natalensis* Sond. (sAf): large sour plum; lf. decoc. for hookworm.

Xiphidium caeruleum Aubl. (Haemodor.) (sMex, WI, CA to Br): soskia (Cuna Indians, Pan); "walk fast" (Trin); palma (del norte); u. for female dis.; rubbed on feet & knees of infants to prevent cracks in the skin.

Xiphiidae: Swordfish Fam.; characterized by the long simple dorsal fin & esp. by the beak or prolongation of the upper jaw for which the animal is named.

Xiphius gladius (Xiphiidae): common swordfish; liver FO classified as a percomorph liver oil; u. in vit. A & D therapy.

xuxu (Br): chuchu.

xylan: a pentosan (pentoside) (complex polysaccharide) made up of units of xylose; occurs in wood, (wheat) straw, tamarinds, &c.

Xylaria obovata Berk. (Xylari. - Sphaeriales - Pyrenomyc. - Ascomycet. - Fungi) (eAs): pd. sporocarp with coconut oil u. on wounds caused by burns (Mal).

xylem: wood portion of st. & rt.; made up of vessels, tracheids (Gymnosperms), wood parenchyma, wood rays, wood fibers; on account of its heavy lignification & strength, the important component of lumber u. in construction, furniture mfr., etc.; v. under wood.

xylene: xylol: dimethyl-benzene: $C_6H_4(CH_3)_2$; obt. fr. distn. of coal & wood tars; a mixture of 3 isomers, chiefly consisting of m-**xylene**; u. solvent, esp. for Canada balsam, u. in microscopy as clearing agent & solvent.

Xylia evansii Hutch. (X. dolabriformis formis Benth.) (Leg.) (trop wAf, Burma): wd. of large tree valuable; "one of best timbers of world" for making railroad ties ("sleepers"); twigs u. as chewsticks; lvs. beaten in water prod. foam; - **X. xylocarpa** (Roxb.) Theob. (seAs, Burma, Mal): bk. decoc. u. in worms, leprosy, diarrh., GC, ulcers; sds. FO u. in rheum., leprosy, hemorrhoids.; wd. valuable. & much used for ties (sleepers); var. *kerrii* I. Nielsen (Burma, Indochin): bk & frts u. in hemoptysis (Camb).

Xylinabaria minutiflora Pierre (Apocyn.) (Indochin): bk u. in dys.

xylitol: a pentitol (sugar alc.); occ. in small amt. in edible mushrooms; frts. & vegets. (yellow plum 0.94%), in metabolic cycle; prepd. by hydrogenation of xylose (fr. wd., &c.); u. as sucrose substit., sweetener for diabetics, in chewing gums; in diet to prevent formation of dental plaque & caries (ADA); has pleasant cooling taste, more refreshing than sucrose; some objections to use (904).

Xylocarpus benadirensis Mattei (Meli.) (eAf): a mangrove (also other *X.* spp.); fluid fr. crushed unripe frt. drunk as aphrod. (Tanz); bk. u. as tan. (Moz) - **X. granatum** Koenig (seAs, Oceania, eAf): pl. astr.: rt. u. in cholera; bk. in dys.; frt. & sd. in diarrh.; sd. kernels ton. & for colic: bk. decoc. u. to prep. "vaikahi," u. during pregnancy (Tonga); lf. inf. (or lf. chewed) for thrush (Fiji) - **X. moluccensis** M. Roem. (OW trop, esp. eAf, Ind, SriL, Mal, Molucc): Molucca crabwood; tree classed as a mangrove; bk. rich in tannins (30%); astr. u. as tan, u. in dyes (Guam); sds. source of carapa oil or fat; rt. decoc. u. for stomach pains (Tanz); tree (mangrove) yields resin; valuable wd. (u. in veneers, &c.).

Xylopia aethiopica (Dunal) A. Rich. (Annon) (w&c tropAf, esp. Congo & Gabon): Guinea/Ethiopian/African/Senegal/Negro pepper; kani (fing) (Bambara); frt. c. 2% VO with cineole, cuminal; odor like cinnamon + pepper + pimenta + nutmeg; frt. u. condim. (like Malegueta pepper); good for mixed spices, esp. when ground; replacement for red chillies; u. in Af to add pungency to snuff; expect.; emmen. for skin eruptions; now shown to have value in mal.; sds. u. as ascarifuge; rt. ex pect.; for centuries imported into Eu (905): for cough (mixed with rhiz. of *Cyperus articulatus*) (Nigeria) - **X. aromatica** Mart. (WI, Guianas, Co): malagueta (brava) (Cu); frt. Guinea pepper; stim. digestive, carm.; bk. bitter, considered good febrifuge (Cu); weak tea fr. lvs. u. as diur. to reduce edema of legs; sds. carm. (Br Indians, 1935) (906) - **X. brasiliensis** (Mart.) Spreng. (Br): pimenta de macaco; frt. u. as replacement for pepper & cubebs; in GC - **X. discreta** (L. f.) Sprague & Hutchins. (Af): frt. & lvs. c. cineole; u. similarly to other spp. - **X. frutescens** Aubl. (CA, Guianas, Br.): frt. spice, u. to make "asthma coffee" ("Asthma koffie"), and as false cubebs; lvs. or sds. decoc. u. for abdom. pain (CR) (2444); bk. fibers u. to make ropes - **X. grandiflora** St. Hil. (tropAm, esp. Cu, Pan, Br): fls. u. ton. carm (Br); VO with peppery flavor - **X. odoratissima** Welw. ex Oliv. (Af): small bitterwood; arom. rt. taken for female sterility - **X. parviflora** Benth. (tropAf = Congo; Sen, sAf): poivre de Sédhiou (Fr); tall bitterwood; aq. decoc. of frts. & lvs. (eq. pts) u. as beverage for its expect. & anti-cough properties: frt. vermif. (Angola); rt. decoc. as vaginal douche for menorrhagia; rt. inf. drunk for dysmenorrh.; rt. maceration drunk to "incr. size of penis" (Venda); for male sterility; wd. termite-resistant (2445) - **X. polycarpa** Oliver (Coelocline p. A. DC) (trop wAf): bk. c. berberine: latex c. resins; u. to dress ulcers - **X. quintassii** Engl. & Diels (w tropAf): ghay-dee (Lib); soft inner bk. made into mouthwashes; beaten bk. rubbed on hands with knotty swellings; frt. u. med.; excellent wd.; stringy inner bk. u. for ropes - **X. sericea** St. Hil. (Br): frts. u. as condim. - **X. vallotii** Hutch. & J. M. Dalz. (w tropAf): bk. & rts. u. med.; frt. u. as spice as "poivres de Sédhiou" (Fr) - **X. villosa** Chipp (w tropAf): (attié) bello (Ivory Coast); sds. pounded & u. as poultice for boils; wd. termite-proof & u. for house posts, &c.

xylose: wood sugar; a pentose sugar found as glycosides in woods, straw, hulls, etc.; obt. fr. corn cobs, peanut husks, etc.; u. as diabetic sweetener, in dyeing, etc.

xylosidoglucose: disaccharide consisting of xylose + glucose; ex. primaverose.

Xylosma flexuosa Ktze. (X. flexuosum Hemsl.) (Flacourti) (sMex to CR): sweetwood; bk. decoc. given in TB (Mex); remedy in heart ailments (CR): fragrant wd. valued for fuel - **X. longifolium** Clos (Pak, Ind): katari; young lf. ext. resembles opium in action & is u. with it (Assam); wd.

u. for fences - **X. racemosa** Miq. (Chin): rt. u. in difficult parturition.

Xymenia = *Ximenia.*

Xymalos monospora (Harv.) Warb. (Monimi.) (trop & sAf): lemonwood; pd. bk. u. in colic; berries edible; wd. u. for ornam. furniture, beehives.

Xyris Gronov. ex L. (Xyridaceae) (trop & subtrop): yellow-eyed grass; rush-like pls. with yellow (less often white) corollas - **X. ambigua** Beyl. (seUS): lvs. u. for colds & pulm. dis. (inf. rubbed on chest) (Seminole Indians, Fla [US]) - **X. caroliniana** Walt. (e&sUS): Carolina yellow-eyed grass; lvs. & rts. u. in skin dis. - **X. complanata** R. Br. (Vietnam): yellow-eyed grass; bruised lvs. & juice mixed with brandy & u. for itch, ringworm, leprosy - **X. indica** L. (Ind, seAs, Austral+): pl. u. as "cure" for ringworm, itch, leprosy; lvs. u. in local pharm. (Camb) - **X. elliottii** Chapman (seUS): yellow-eyed grass; pl. decoc. rubbed on patient's breast for colds & pulmon. dis. (Seminole Indians) - **X. jupicai** L. C. Rich. (X. communis Kunth) (tropAm): rt. u. emet. cath.; in eczema, leprosy - **X. melanocephala** Miq. (Indones): sts. made into coarse mats u. by natives - **X. pauciflora** Willd. (Ind): mundari; hb. u. to cure insomnia (Mundas); bulbs eaten.

Xysmalobium dilatatum Weimarck (Asclepiad.) (sAf): rt. c. cardenolides, pregnane derivs., dialtosides; u. in folk med. (sAf) - **X. heudelotianum** Dcne. (w tropAf): tuberous rts. ("yakhop") food (Sen); stom. bitter ton.; eaten as med. for stomach troubles (nNigeria) - **X. stellatum** (sAf): latex c. glycosides; appl. to ulcers; said to kill maggots - **X. undulatum** R. Br. (sAf): milk bush; wild cotton; uzara; rt. c. uzarin (glucoside), uzaroside; sd. c. frugoside; has digitaloid heart action; u. against diarrh. dys. spastic dysmenorrhea; enuresis; "uzara" was previously thought to come from spp. of *Asclepias, Dicoma, Schizoglossum,* &c.

(Note: See entries also under "I" because often interchangeable)

yacca gum (**, South Australian**): red gum accroides (or acaroid resin) fr. *Xanthorrhoea* spp. (fine, coarse pds.); sol. in alc.

yachole (NM [US], Sp): raspberry.

yack(h)a gum: yacca g.

yagalache (Mex): caspi (Co) (*qv.*).

yagé (**del monte**): **yajé**: narcotic of SA Indians prepd. fr. *Banisteriopsis caapi* & *B. inebrians*; caapi.

yageine: harmine.

yagrumomacho (Ve): *Oreopanax capitatum.*

yak: *Quercus robur.*

yam: 1) *Dioscorea* spp. (true yam) (with orange & reddish flesh) 2) often referred (incorrectly) (US) to sweet potato, *Ipomoea batatas* (with whitish flesh) (ex. "yellow yams") - **Asiatic y**: white y. - **Atlantic y**: *D. villosa* - **y. bean**: *Pachyrrhizus erosus* - **Chinese y**.: *D. sinensis; D. esculenta* - **coco y.**: *Colocasia esculentum* var. *antiquorum* - **y. family**: Dioscoreaceae - **Mexican y.: tropical y.**: *D. composita* - **true y.**: *Dioscorea* spp. - **uve y.: white (negro) y.: white y., common**: *D. alata* - **wild y. (root)**: Atlantic y.

yamagua (Cu): *Guarea guara.*

yam-bean: 1) *Pachyrrhizus erosus*, sds. 2) *Dolichos.*

yankee weed (seUS): 1) *Helenium tenuifolium* 2) *Eupatorium capillifolium* (SC perhaps as disliked weed).

yapon maté: lvs. of *Villaresia congonha.*

yara yara: comml. name for Nerolin, subst. for neroli oil - **yaragua** (Co, PR): *Melinis minutiflora.*

yarb (dial.): herb.

"yarbin it" (NCar [US] rural): collecting medicinal plants.

yareta: llareta: *Laretia acaulis* (source of gum-resin) (Umb.): perennial low shrubs, typical of sCh; rt.

yarey: various palms of the WI, incl. *Sabal causiarum* Beccari (much u. for hat manufact. in Cu & PR); *Copernicia molineti* Léon; *Sabal pumos* Burret; *Hemithriax compacta* (Griseb. & Wendl.) Hook. f. (Cu), &c. - **y. de Cuba: y. cubano:** *Copernicia baileyana* Léon.

yarmor: wood turpentine.

yarn: 1) spun wool, flax, &c. 2) fibers twisted giving strands of rope.

yarrow: *Achillea millefolium* (fls. lvs.) - **noble y.**: *A. nobilis* (fls. lvs.).

yatanine: alk. fr. sd. of yat-tan-tzu, u. in amebic dys. (Chin).

yatchmen (Russ): barley.

yate: *Eucalyptus cornuta.*

yat-tan-tzu (Chin): *Brucea javanica.*

yaupon: *Ilex vomitoria; I. cassine.*

yautia: 1) *Colocasia esculenta* 2) *Xanthosoma sagittaefolium.*

yaw root: *Stillingia sylvatica* - **y. weed**: *Morinda umbellata.*

yccotli: *Thevetia y.*

yeast: *Saccharomyces* spp. (Saccharomycetaceae) - **baker's y.:** *Saccharomyces cerevisiae*; SC u. to raise dough in making bread - **bottom y.**: form which thrives at rel. low temperatures at or near the bottom of the container; u. in mfr. lager beers - **brewer's y.:** *Saccharomyces cerevisiae* (fr. *cerevisia*, L., beer), with its many vars.; u. in brewing & baking industries; is the off. medl. y.; the viable y. is often put up as **y. cakes: compressed y.** (u. in baking) - **crude y:** the original yeast product containing living yeast cells, active enzymes, &c., wh. can be u. in making bread, wine, &c.; distinguish fr. prepared y. which is sterile - **dried y.: dry y.**; the off. non-fermenting y. u. for internal medication, sometimes in tablets; rich in vit. B fractions, good quality proteins, & minerals - **y. family**: Saccharomycetaceae - **top y.**: var. of brewer's y. which grows at or near the surface of the vat & does best at summer temperatures, causing much CO_2 formation & the froth which is characteristic of many brews.

yebraj (Ind): belladonna rt.

yedra trepadora (Sp): English ivy, *Hedera helix*.

yellow bark: Cascara Sagrada - **y. bell(s)**: 1) (eUS as nGa): *Erythronium americanum* 2) (Pac US): *Fritillaria pudica* 3) (Calif [US]): *Emmenanthe penduliflora* 4) *Allamanda cathartica* var. - **y. berries**: *Rhamnus infectoria* - **y. enzyme** (of Warburg) = **y. ferment** = flavoprotein = yellow oxidative ferment; yellow pigment; actually a phosphorylate, neoflavin phosphate; flavin e.: one of several dehydrogenases functioning in tissue respiration - **y. fever vaccine**: prepd. fr. living cultures in embryonated chick egg of attenuated strain of **y. f. virus**, with high antigenic activity & safety - **y. gum**: 1) Sierra Leone copal 2) *Eucalyptus paludosa*, tree - **y. jacket**: kind of wasp, *Vespa diabolica* (fam. Vespidae) & other Hymenoptera spp. with painful sting; forms nest hanging fr. trees or shrubs or in the ground; abdomen with yellow-black pattern - **y. pigment**: y. enzyme - **y. pileweed**: *Ranunculus acris* - **y. rain**: clouds of feces of oriental bees, contg. much pollen; interpreted by seAs natives as poison chemical gas - **y. rattle**: *Rhinanthus* spp. - **y. root**: 1) Hydrastis 2) Curcuma 3) *Coptis trifolia* 4) *Xanthorhiza simplicissima* 5) *Jeffersonia diphylla* 6) *Celastrus scandens* - **y. top(s)**: 1) *Solidago* spp. 2) *Senecio jacobaea* 3) (Ala [US]): *Helenium amarum* - **y. weed**: *Solidago* spp. - **y. wood** (bark): 1) *Cladrastis* spp. 2) *Cotinus obovatus* 3) *Zanthoxylum americanum* (**y. w. bark, berries**; (prickly y. w.: *Zanthoxylum clavaherculis*; **United States y. w.**: *Cladrastis kentukea*) - **yellowtail, Australian**: *Seriola* spp.; source of liver FO (**California y.**: *S. dorsalis*).

yellow: v. also under adder's tongue; bark; basilicon; beeswax; broom; centaury; cinchona; corydal; dextrin; dock; flowered rhododendron; gentian; immortelle; jasmine; jessamine; ladies bedstaw; loosestrife; mustard; parilla; pileweed; pond lily; puccoon; resin; rooted water dock (v. under dock); rosin (pine); saunders; wax; willow herb.

Yemen: sw area of Arabian peninsula; the most fertile portion; produces millet, wheat, barley, grapes, Mocha coffee, khat, myrrh (**Y. m**).

yen-chee (Chin): smoking opium, esp. residue left in pipe bowl after being smoked.

yengi (flower): *Sophora japonica*.

yen-hock (Chin): - **yen-hok(e)**: hooked needle u. to hold opium pellets while cooking - **yen-hu-so** (Chin): *Corydalis ambigua*.

yen-ji (Jap): **yenju:** yengi.

yen-she(e) gow (Chin): tool for cleaning residue fr. opium pipes for reuse - **yen-shee:** yen-chee.

yerb (US dialect): variant of herb; frequently u. in combns. such as **y. doctor**.

yerba (Sp): 1) herb (v. also hierba, more correct form); term now often considered outmoded or even obsolete 2) maté (other entries under "hierba") - **y. amarga** (Sp): *Castela* - **y. Bahia** (Pe): *Paspalum notatum* - **y. blanca** (Mex): *Boerhaavia erecta* - **y. buena**: 1) *Mentha* spp. (mint, spearmint, peppermint, etc.): *M. aquatica* (Arg); *M. piperita* (Mex); *M. citrata* (Guat); *Mentha rotundifolia* (Arg) & other *M.* spp.; *Phyla scaberrima* (Mex) 2) *Micromeria chamissonis M. douglassii* - **y. b. colocha** (Guat): **y. b. colonia** (Yuc): **y. b. de comer** (Yuc): **y. b. menta** (Guat): *M. citrata* - **y. Carmen** (Sp): Phytolacca - **y. de cucaracha**: *Haplophyton cimicidum* - **y. de la golondrina** (Bo) (lit. swallow herb): *Verbena gracilescens* - **y. de la niña** (Cu): *Euphorbia prostrata* - **y. de (la) vibora (vivora)** (NM [US], Mex) (snake weed): 1) *Gutierrezia sarothrae* 2) *Aplopappus spinulosus* 3) *Brogniartia intermedia* 4) *Daucus pusillus* 5) *Zornia diphylla*, &c. - **y. de la virgen:** term appl. to various common plants, ex. *Tiquilia canescens* (Mex); *Polygala asperuloides* (Co); *Loeselia mexicana*, &c. - **y. del burro** (Arg): *Cassia bicapsularis; C. occidentalis* - **y. del manso**: y. mansa - **y. hedionda** (SpAm): *Cassia occidentalis* - **y. Luisa**: *Lippia citriodora* - **y. mala** (NM [US], Sp): tumble-weed - **y. mansa** (means "mild herb") (Mex, NM): *Anemopsis californica* - **y. maté**: *Ilex paraguariensis* - **y. melado** (PR): yaragua - **y. mora** (CR) *Solanum nigrum* - **y. reuma** : *Frankenia grandifolia* var. *campestris* - **y. sagrada**: *Lantana brasiliensis* - **y. santa** : 1) (US) *Eriodictyon californicum* & other *E.* spp. 2) (Mex) *Piper sanctum* 3) (Mex, CR): *Pluchea odorata* - **y. virgen**: (Par): mate?

yerbabuena (Mex): *Mentha piperita* - **yerbal (yerbales**, pl.) (sSA, esp. Arg): plantation of *Ilex paraguariensis* trees - **yerbatero**: 1) (Arg): grower of or dealer in yerba maté 2) (Mex, etc.) person practicing herbal med.

yercum fiber: bark fiber of *Calotropis gigantea*.

yerra (Ind): *Gossypium nadan.*

Yersinia pestis (Lehmann & Neumann) van Loghem (Pasteurella p. Bergey & al.) (Enterobacteri.) (cosmopol.): the bacillus wh. causes bubonic plague (the "black death" of medieval times) & pneumonic plague (plague pneumonia).

yew: *Taxus baccata* (bk., berries, lvs., wd.) - **common y.**: **English y.**: *Taxus baccata.*

yezgo (Sp): *Sambucus ebulus.*

yick-chee (Chin): *Ziziphus jujuba*

yin-yang: male & female principles in balance to give good health; some Chin medicine is based on this principle: thus, one medicine of yang type is anteater scales(!).

ylang-ylang (oil): *Canaga odorata.*

yoco (Co, Pe): *Paullinia yoco* (stim. drug).

yoghoort (Turk): **yoghurt: yogurt**: non-alc. semisolid prod. made by evaporating & fermenting milk of cows, goats, or other ruminants; orig. fr. Balkans, Turkey, Armenia, Persia; milk usually boiled before fermentation, sometimes being evapd. to half vol.; prod. eaten with dates or bread or in candy bars; also u. med. for many purposes; excellent nutrient; in diets for reducing wt.; differs fr. kumiss: frozen in place of ice cream (fewer calories), reduces blood cholesterol levels; easily digested; good Ca source (best dietary source); u. (p. o.) for vaginal yeast inf.; much higher % of lactic acid than Kumiss; very soft curd; little or no alc.

Yohimba: Yohimbe (or **yohimbéhé**) (Congo): *Corynanthe johimbe* (bk.) (BPC) - **yohimbine,** α_1: 1) other names are rauwolscine and coryonantheidine; xall. alk. of yohimbe bk. fr. *Rauwolfia canescens* & c.; salts u. in treating forms of sexual impotence (esp. to promote erection); arteriosclerosis, angina pectoris, &c.; other names are rauwolscine and coryonantheidine.

yolk (, egg): Ovi Vitellus (L.): the most nutritive part of the egg of the hen, *Gallus domesticus*; c. 30% FO, vitellin, inorganic salts, coloring matter, cholesterol, water; u. nutr. emulsifying agt.

yoloxochitl (Mex): *Talauma mexicana.*

yoncopin: *Nelumbo lutea* (a water lily).

York gum: *Eucalyptus loxophleba*, tree.

yoshinokuzu (Jap): starch fr. Kudzu tubers; popular in Jap where said superior to corn starch, etc., in cooking, etc.

youngberry: sweet reddish berry; deriv. fr. loganberry, *Rubus loganobaccus*: possibly a hybrid of *R. loganobaccus* X *R. flagellaris* (dewberry); u. as popular food.

Youngia japonica DC (Crepis j. Benth.) (Comp.) (eAs, Polynesia, Austral): u. as antipyr. & in snakebite; decoc. is bechic & wash for boils (Taiwan); shoots u. as veget. in Chin.

Youngken, Heber W(ilkinson), Sr., (1885-1963): late Research Professor, Massachusetts Coll. of Pharmacy, Boston, Mass [US]; outstanding teacher & research worker in pharmacognosy; special interests: *Viburnum, Aconitum, Myrica, Plantago, Capsicum, Rheum.*

youth root: *Drosera* spp.

Ysop (Ge): hyssop.

yuca (root) (CA, SA): mandioca, *Manihot esculenta.*

Yucca L. (Agav.) (trop&subtr NA): Spanish bayonet (plant); S. dagger; soap weed; lf. fibers u. to make brushes, baskets, roofing, textiles, in "balsam wool" for insulation of houses (1980); lvs. u. as source of sugary component of cattle fodders; sts. roasted & eaten (Calif [US] Indians); wd. u. for splints, &c.; rhiz. & rt. u. as saponin source, soap substit. (2379), detergent; frt. of some spp. edible; usually cooked (*cf.* apple); state flower of NM [US] (907); some spp. live for centuries (1999) - **Y. aloifoiia** L. (WI, Mex; natd. sFla to NCar [US], Ind+): Spanish bayonet (or dagger); aloe yucca; frts. edible, may be purg. (called "banana"); fleshy rts. edible (Hispaniola); fls. fried & eaten; rhiz. soap substit. (Mex); inf. of young lvs. diur.; lf. fiber of this (& other *Y.* spp.); u. by pioneers to make rope & to supply strings for hanging cured meats, tying up vines, &c; lvs. c. tigogenin, sarsasapogenin, gitogenin, &c.; aloifoline (active against Lewis lung tumor); ornam. - **Y. australis** (Engelm.) Trel. (Mex): palma (grande); lf. & st. fiber ("ixtle de palma") u. for cordage, brushes, mats, &c., & dipped in pitch for torches in mines; fls. cooked & eaten; frts. edible; ("datiles"); young lvs. & sts. fermented & distd. to give alc.; spongy interior of trunk u. to make pads ("sudaderos") for pack animals; hollowed trunks as beehives - **Y. baccata** Torr. (swUS): blue/banana/datil yucca; palmilla; large frts. eaten raw, dr., roasted, or as meal; sds. similarly; frts in prepn. of fermented beverage; rts. (amole) u. deterg. & lax.; lf. fibers useful for rope, twine, burlap; fresh fl. buds eaten; lf. juice c. 0.8% sarsasapogenin; claimed juice aids penetration of water into soil & aids soil flocculation - **Y. brevifolia** Engelm. (Y. arborescens Trel.) (sCalif, nAriz, sUtah [US]): Joshua tree; yucca t.; striking feature of Mojave Desert; trees -12 m high; frts & sds. eaten by Am Indians; frt. with about same food value as dr. orange peel; wd. u. to make boxes, &c., as packing materials - **Y. campestris** McKelvey (Tex [US]): plains yucca; yucca chigeuta; st. decoc. u. as vermif. & vermicide; inf. of pollen (not petals) u. as emet. (Tex Indians) - **Y.**

elata Engelm. (swUS, nMex): palmilla (NM; Chihuahua); soapweed; soap-tree yucca; chopped sts. an emergency fodder for stock; lvs. emergency forage; veter. lax.; rts. (amole) deterg. in mfr. shampoo soaps; lf. fibers u. in cordage, woven into mats & cloth; as jute substit. (1871); in making paper; rts. u. to wash clothes; rt. ext. u. as foaming agt. in beverages; fl. stalks & lower st. eaten (Am Indians) - **Y. elephantipes** Regel (Y. guatemalensis Bak.) (sMex to CA [Pan]; introd. swUS, sFla, &c.): itabo; izote; palmito; fls. popular food (wUS, CA, Mex): in salads or fried; considered stom. ton. & for kidney dis.; decoc. as diur. esp. in albuminuria; heart of st. boiled & decoc. taken for kidney dis. (908); fls. u. in salads, cordials, &c.; lf. fiber u. for twine & textiles - **Y. filamentosa** L. (eUS n to NJ): silkgrass; spoonleaf-yucca; Adam's needles; bear-grass; rt-stocks yield 6% saponins (1976) with several genins; u. for cleansing; lvs. seared over fire & u. as twine to hang hams & shoulders in smokehouse (2378); rts u. in GC, rheum. antiluet. (homeop.) (US & Ind cult.) - **Y. flaccida** Haw. (Y. smalliana Fern.) (Ala to NCar [US]): weak-leaf yucca; cult. ornam - **Y. glauca** Nutt. ex Fraser (cUS, SD to NM & adj. Mex): (small) soapweed; palmilla ancha; Adam's thread & needle; the common yucca of n&eNM; frt. shells c. many sapogenins, incl. neotigogenin, hecogenin, gitogenin, manogenin, sarsasapogenin; pounded amole rts. also c. saponins, u. soap substit., esp. to wash hair; hair lotion (Navaho); to curdle milk (like rennet); in GC; in the sed. wine of the Penitentes (NM); lvs. to make stable brooms; frt. eaten by some Am Indians; chopped lvs. livestock feed; macerated lvs. source of good fiber; st. pith u. as punk for starting fires with fire drill - **Y. gloriosa** L. (NCar to Ga, natd. nInd, seAs): Spanish dagger; mound lily; lamparas de dies (Lord's candle); lvs. c. 5.7% crude saponins, incl. tigogenin (u. as starting material for steroid synthesis), chlorogenin, gitogenin, &c.; rhiz. u. in mfr. of Costa Rica arrowroot; as deterg.; pl. u. in rheum, edema, ulcers, asthma; in steam bath for "swelling" (Thail) (2448); frt. purg. (Ind); ornam. - **Y. grandiflora** Gentry (Mex): datil; ripe frt. sweet & edible (better roasted than raw) - **Y. macrocarpa** Coville = *Y. torreyi* Shafer. - **Y. madrensis** Gentry (Sierra Madre Mts, Mex.): "soco"; rts. u. as kind of soap; tender green frts. eaten (Warihic Indians); lvs. c. sapogenin; rts. c. similogenin - **Y. rupicola** Scheele (wTex, adj. Mex): twisted-leaf (or Texas) yucca; u. med. (Mex) - **Y. schidigera** Roez. (Y. mohavensis Sarg.) (sCalif to Arg, Mex): Mojave yucca; lf. fibers u. during WW II; *cf. Y. baccata*; lf. mesophyll rich in sarsasapogenin (2.9%) & markogenin; frt. eaten: source of a toddy - **Y. schottii** Engelm. (sAriz, nwMex): pl. c. saponins, incl. glycosides of yuccagenin & kammogenin; these show anti-inflamm. activity against carrageenan-induced edemas; cult. ornam. - **Y. torreyi** Shafer (Y. macrocarpa Coville) (wTex, sAriz, nMex): large-seed yucca; palma (NM); p. criolla, p. angosta (Chihuahua); lvs. u. to make baskets (sNM); sts. u. formerly as food by natives; fl. inf. u. in cough - **Y. whipplei** Torr. (sCalif, nMex): Our Lord's candle; inflor. eaten by Am Indians (roasted; flavor of asparagus); grd. sds. as porridge; lf. fiber nearly as strong as henequen; reputedly very fast grower (909).

yucca gum: yacca g. - **y. leaf powder**: obt. fr. various *Y.* spp.; c. sarsasapogenin; precursor of cortisone.

yuquilla (CR): *Maranta indica.*

yute: jute: *Corchorus capsularis.*

Zz

zabila (WI, etc.) sabila, aloe.
zacachichic (Mex): *Conyza filaginoides.*
zacate (Mex): 1) generic designation for low grassy pl. spp. which cover the fields & serve for pasturage; also appl. to some Cyperaceae which resemble grasses 2) (NM [US], Sp): grass; hay 3) *Nolina longifolia* - **z. de limon** (CR): *Cymbopogon citratus* - **z. de olor** (CR): vetiver. - **z. gordura** (Salv): *Melinis fminutilora* - **zacatán**: tall wild grasses 1) (Mex) *Epicampes macroura* 2) (Mex): *Muehlenbergia distichophylla* 3) (PR): *Andropogon ischaemum; Bothriochloa bicornis,* etc. 4) (Mex): *Sporobolus wrightii* 5) various other plants.
zacatechichi (Mex): *Calea zacatechichi.*
Zafarán = Saffron.
zafra (Arg): sugar cane harvest - **zafrán** (Sp): saffron.
zaftig (Ge): saftig (Ge): juicy, succulent.
zahina (Sp): *Sorghum vulgare* - **zahirmohra kahatai** (Urdu): type of diamond u. in traditional med.
Zahn (Ge-m.): tooth - **Zahnwehrinde** (Ge): prickly ash bk., Zanthoxylum.
zajino (CA, Col): *Caesalpinia eriostachys.*
Zakatón: zacatón.
Zakynthos: Zante.
Zalacca Bl. = *Salacca.*
Zaluzania augusta (Comp.) (Mex): alta misa; u. med. (Mex).
zaman: *Pithecellobium saman*, rain tree.
Zambesia: large area of seAf located in Mozambique lying in the drainage area of the Zambesi River.
zambo (Ec): pumpkin - **zamboa** (Ch): citron.
Zambuk™: popular ointment (Cana) containing boric acid, chlorophyll, eucalyptus & sassafras oils, oil of swallows (infused oil of elder flowers), camphor, petrolatum (vaseline); u. as inunction (chest, &c.) in case of colds, &c., & cold sores, for cuts & bruises, sore feet, chapped hands & chilblains, &c.
Zamia L. (Zami. - Gymnosperm.) (trop&subtr Am): coonti(e); comptie; wild sago; small primitive fern-like shrubs or small trees, with tuberous rhizomes; often planted for ornam. (SC conti hateka, Seminole for white [bread] root or plant) - **Z. angustifolia** Jacq. (WI, natd. sFla [US]): guayara (Cu); lvs. c. flavone compds., amentoflavone, bilobetin, sequoiaflavone, ginkgetin, &c. - **Z. floridana** DC = *Z. integrifolia* - **Z. furfuracea** L. f. (sMex, CA): camotilla; poisonous to livestock; rhiz. u. to kill noxious animals & in criminal poisoning.; ext. of tubers slightly toxic to Am cockroaches; source of starch (natives) (910) - **Z. integrifolia** Ait. (Fla, WI): Florida arrowroot; co(o)ntie (fern); underground st. early source of flour of the Fla aborigines & Seminoles (trunk & rhiz. pith); later as source of Florida arrowroot, a kind of starch; staple food of Seminole Indians ("Seminole bread") & early settlers; lvs. u. as decoration - **Z. latifoliolata** Prenleloup (WI, CA): starch fr. st. u. as food, in laundry; pl. juice toxic - **Z. lindenii** Regel (Ec, Co): sds. u. as food by Choco Indians (Co) - **Z. pumila** L. (Z. umbrosa Small) (Fla, Fla Keys [US]; WI, Bah): Florida arrowroot; coontie; pigmy zamia; commonly found on Indian kitchen middens & aboriginal village sites of neFla; source of Bahama arrowroot - **Z. skinneri** Warsc. (CR, Pan): goma; elkia (Cuna

Indians, Pan); sds. eaten in bread; sticky sap of trunk appl. in snakebite (Bribri Indians, CR) - **Z. spiralis** R. Br. = *Macrozamia s.* (Austral).

Zamiaceae (Cycadales - Gymnospermae) (warm areas of globe): with 8 genera (ex. *Zamia, Bowenia, Ceratozamia, Dioun,* &c.) & ca. 100 spp.

Zamioculcas loddigesii Schott (1) (Ar.) (Tanz): rt. u. as local appl. to ulceration; poultice appl. to inflamed penis ("mshipa") - **Z. zamiifolia** (Lodd.) Engl. (Z. lodigesii Schott.): u. similarly.

zanahoria (Sp): carrot.

Zanha africana (Radlkof.) Engl. (Sapind.) (Tanz): rt. bk. with vaseline appl. to skin with fungus inf.; rts. pounded & rubbed on aching legs; decoc. to ease childbirth, in prostatitis; rt. bk. emet.; in insanity; rt. mixed with other pls. as purg. & emet.; for epilepsy (712) - **Z. golungensis** Hiern (trop wAf): frt. edible; wd. u. to make furniture.

Zanonia indica L. (Z. macrocarpa Bl.) (monotypic) (Curcurbit.) (Indomal): bandolier fruit tree; chirpoti (Hind); lvs. in baths & linim. to reduce inflammn.; frts. (c. FO) & lvs. u. for asthma & cough; edible but cath.

Zanosar™: streptozocin Upjohn.

Zante: one of Ionian Islands off Peloponnesus peninsula of Gr where corinthes are raised for export.

Zantedeschia aethiopica (L) Spreng. (Richardia africana Kunth) (Ar.) (sAf): arum/Egyptian/pig/trumpet lily; calla (lily); l. of the Nile; white arum; Ethiopian (or African) lily; common calla; whole leaf appl. as poultice to sores, boils, wounds, insect bites, burns, gout, rheum.; supposed value in preventing evapn. & stim. granulation; rt. inf. drunk for infertility (due to uter. lesions) & snakebite (Venda); rt. u. as vesicant; young lvs. & petioles boiled or roasted & eaten; cult. as starch source (Af) & pig feed; ornam. (calla fls. u. to decorate funerals, altars, shrines, &c.). - **Z. hastata** Engl. (sAf): decoc. u. to prevent repeated miscarriages (Zulu).

Zanthorhiza L'Hérit. = *Xanthorhiza* Marshall.

Zanthoxylum L. (*Xanthoxylum*; *Fagara*) (Rut.) (NA, eAs): ca. 250 spp. of shrubs & trees; some spp. furnish good lumber - **Z. acanthopodium** DC (nInd, Himal): timur (Hind); dr. frts. spice, eaten in stomach dis., esp. flatulence; or as tea for same (Sikkim); frt. ext. in mixt. with water drunk for cholera (Assam); sds. diaph. antipyr. in prepn. tooth powd.; frts. distd. yield VO, wartara oil; rich in linalool, u. in perfum.; pl. insecticidal against house flies; fish poison - **Z. ailanthoides** Sieb. & Zucc. (Chin): lf. inf. for chills & flu; frts. for indig., sunstroke, stim. - **Z. americanum** Mill. (eCana, e&cUS): northern (or common) prickly ash; toothache tree; clavalier (Qu); dr. bk. c. resins, VO, tannins, alk., berberine: xanthoxaletin (xanthoxyloin); bk. u. stom. stim. diur. diaph. ton. sialagog, antirheum., spasmolytic, in pneumonia; chewed for dentalgia (hence called "toothache bark") (911); frt. c. VO with citral, &c.; frt. of this northern sp. mostly u. med.; u. stim. diaph. ton. carm. stom., irrit., in sore throat (Chippewas) - **Z. armatum** DC (Z. alatum Roxb.) (Chin, Jap, Ind): chin jill (Chin) (generally given as Chin Chiao); timru (Ind); frt. c. VO with antibacterial props (*cf.* eucalyptus oil); u. antisep. deodor., in catarrhs.; frts. anthelm. stim. anodyne, in dental dis. (to check decay); rt. decoc. in digest. dis., frt. condim. (Chin, Ind); bk. lvs. frt. u. as fish poison; pl. insecticidal: twigs u. to brush teeth; lvs. u. as carm. emmen. anthelm.; sts. to make walking sticks - **Z. avicennae** DC (PI, seChin): kangai; very bitter bk. u. as ton. astr.; st. decoc. stom. ton., counter-poison for snakebite - **Z. budrunga** Wall. (Z. rhetsa DC) (Ind, Pak, Burma, Himal reg.): bazna tree; frt. ("false cubebs") u. as arom. stom. astr. in dyspeps. rheum.; rt. bk. in kidney dis.; frt. furnishes FO - **Z. bungeanum** Max.: **Z. bungei** Planeh. (Chin): Szechuan pepper; chen shiao; frt. u. in Chin med. as analgetic, for chills, diarrh. dys. vermif, in ascariasis; for skin sores; spice; cult. - **Z. capense** Harv. (Fagara capensis Thunb.) (sAf): fever tree; small knobwood; frt. u. as stom. in flatulence; lf. int. for intestin. parasites; bk. locally & int. for snakebite - **Z. caribaeum** Lam. (sMex, CA, WI, Co, Ec, Ve): (bastard) prickly yellow; scorpion tree; bk. & branchlets chewed for toothache (Hond, Yuc); lf. decoc. appl. to scorpion bites & rheum. pains - **Z. carolinianum** Lam. = **Z. clava-herculis** L. (seUS to Ark): (southern) prickly ash; sea ash; pepperwood; Hercules' club; tickle tongue; sting tongue; wild orange; bk. & frt. of small tree with similar uses & const. as *Z. americanum*; bk. lvs. & frt. u. stim. sialagog; in toothache; also insecticide (912): bk. made up in whiskey & u. as ton. in rheum., &c. (Fla [US]) (2254); bk. of this southern sp. mostly u. med. - **Z. coco** Gill = *Fagara c.* - **Z. coriaceum** A. Rich. (sFla, WI): Biscayne prickly ash; heartless club; cold inf. of rt. u. to stim. appetite; singed rt. inf. u. in diarrh.; in combns. for GC; bk. decoc. in rheum & syph. (Cu); lvs. infused in gin for stomachache (Inagua) - **Z. culantrillo** H. et B. = *Z. hyemale* - **Z. cuspidatum** Champ. (seAs): arom. sds. diaph. antipyr.; emmen.; lvs. u. in rabies (Ind) - **Z. davyi** Waterm. (Fagara d. Verdoorn) (sAf): knobwood; u. emet. ton. in severe coughs; exter. in snakebite - **Z. fagara** Sarg. (Z. pterotis HBK) (seTex, sFla, lower Fla Keys [US]; WI, nMex to Co, Br): lime (or southern) prickly ash; wild l.; colima; satinwood; yellow wood; yellow sandalwood (Virgin Is) (comm. timber); bk. decoc. diaph., stim.; wd. & bk. decoc. in syph. (Mex); bk. of trunk & rts. u. as ton. (in form of ext.) (Virgin Is); lvs. sialagog; bk. source of yellow dye; lf. inf. drunk for rheum. & snakebite (Pe); bk. & lvs. u. stom., in skin dis., shooting pains (growing p.) - **Z. flavum** (tropAm): wd. (as Jamaican or WI satin wood) u. in cabinet-making, &c. - **Z. fraxineum** Willd. = *Z. americanum* - **Z. hastile** (sphalm.) Steud. = **Z. hostile** Wall. = *Z. acanthopodium*; *Z. armatum* - **Z. hyemale** St. Hil. (wSA): white fls. u. in rheum.; lvs. u. in snakebite (with brandy) & in cuts - **Z. limonella** Alston = *Z. budrunga* - **Z. macrophyllum** Oliv. = *Fagara macrophylla* - **Z. martinicense** (L) DC (WI, incl. PR): prickly yellow: y. Hercules; espino (rubial); yellow (or white) prickle; lf. decoc. general ton. for colds; appl. to ulcers & sores; wd. tinct. to alleviate tightness of chest (difficult respir.); bk. inf. diaph. diur. antipyr. in rheum. syph. (Cu); for aching teeth (wd. chewed); formerly much more popular in med.; wd. u. for many utilitarian purposes - **Z. microcarpum** Griseb. (sMex to Br; WI): ikor (Cuna Indians, Pan): alligator-toothed prickly yellow; bk. u. as analg. & stim.; wd. u. for semi-fine furniture: dr. spines u. med. - **Z. myriacanthum** Wall. (Ind): smoke fr. burning sds. inhaled for treating ulcerated noses; wd. u. to make tea-boxes - **Z. naranjillo** Griseb. (Br): laranja brava (Br); lvs. c. VO with linalool, phellandrene; lvs. u. diaph., diur., deobstruent; exter. for lesions - **Z. nitidum** (Roxb.) DC (Z. hamiltonianum Wall.) (Ind-Himalay reg.): rt. u. in stomach & toothache; boils; insecticide, fish poison; frt. arom. stim. stom., condim.; in beverages; sds. yield 3.5 -5% VO with pleasant odor, formerly called "evodia oil" & u. to deodorize iodoform. (913) - **Z. olitorium** Engl. = *Fagara olitoria* - **Z. ovalifolium** Wight (Ind): frt. & bk. similar to those of other Ind spp.; wd. u. in inlay & cabinet work - **Z. oxyphyllum** Edgew. (Ind-Himalay.): frt. u. as condim. in chillies; bk. stim., stom. digestive; in colic; diaph. - **Z. panamense** P. Wils. (Pan, CR): a(r)cabu; dr. detached spines u. (pd. or decoc.) to treat sores (Cuna Indians, Pan); base of spine sometimes engraved & u. to stamp designs on paper or cloth - **Z. piperitum** (L) DC (Korea, Manchur, nChin, Jap): sansho; (Japanese) pepper; sds. & young lvs. u. as spice; sap of young lvs. appl. to insect stings & cat bites; bk. stom. & antipyr.; frt. u. int. for sties; vermif.; frt. pericarp diaph., antispasm., anthelm.; in lung & bronchial dis.; aphrod.; peeled sds. diur. for edema & pleurisy - **Z. pterota** HBK = *Z. fagara* - **Z. rigidifolium** (Herzog) Waterm. (cAf): important medicinal pl.; rts. u. as chewing sticks; c. chelerythrine & other alks. (2438) - **Z. schinifolium** Sieb. & Zucc. (Fagara schinifolia Engl.) (Chin, Jap, Kor, Manchur): sds. c. VO with estragole, bergapten; u. vermif.; pericap to treat asthma, cough; for painful & swollen breast (Chin); pd. lf. appl. to contusions - **Z. senegalense** DC = *Fagara xanthoxyloides* - **Z. simulans** Hance (Chin, seAs): hua-chiao; pericarp & lvs. u. as stim. carm. diaph., stom. (formerly for roundworms); red. frt. u. as spice (Szechuan pepper) (poivre de Chine); c. VO, sansho oil - **Z. spinifex** DC (WI; Ve): bois flambeaux; bk. "essence" given to combat infections (Fr, WI) - **Z. tachuelo** Little (Ec): bk. decoc. u. as wash for painful legs; also ingested - **Z. zanthoxyloides** = *Fagara z.*

zanthoxylum bark: z. berries: fr. *Z. americanum* or *Z. clava-herculis*.

Zanzibar: island off the eAf coast; formerly independent sultanate, with flag displaying image of clove; now with Pemba a part of Tanzania (republic); v. under aloes; pepper.

zapallo (Sp): squash.

zapatero (Sp): *Casearia praecox*.

zapel(l)o (Sp): papaw.

zapota: **zapote (gum)**: chicle; v. under *Manilkara zapota* - **z. blanco** (Mex): *Casimiroa edulis* - **mamey colorado z.** (Mex): *Calocarpum sapota; Manilkara zapota.*

zapotillo (Sp): *Diospyros texana*

zaragatona (Sp): psyllium sd. - **z. española**: **z. francesca**: **z. negra**: *Plantago psyllium* - **z. rubia**: *Plantago ovata.*

zarsaparilla, Vera Cruz (Sp): *Carex arenaria.*

zarza (CR): **zarzamora** (Mex): *Rubus fruticosus* - **zarzaparilla** (LatAm): sarsaparilla, various *Smilax* spp.

Zataria multiflora Boiss. (monotypic) (Lab.) (Afghan, Iran, Pak): hb. u. as arom. stom. stim. diaph.; broken lvs. (with thyme odor) u. to adulter. safflower (914).

Zauber(ei) (Ge): magic; witchcraft.

Zauberhasel (Ge): witch-hazel.

Zaukentabak (Ge): chewing tobacco.

Zaunruebe (Ge): bryony.

Zea L (Gram.) (tropAm): maize; Indian corn; **Zea** (L.): (specif.): corn silk, dr. or fresh stigmas & styles of pistillate fls. of *Z. mays*; the spike ("ear") is an aggregate fruit with the grains (ovaries) arranged in rows (much variance: 8-20) on the central dry "cob"; u. diur. in card. ailments (Haiti) - **Z. mays** L. (NA; prob. domesticated in CA, developed in Mex & nSA; now widely cult. OW, NW, trop to temp. zones): maize; (Indian) corn;

"Indian wheat" (formerly); mealies (sAf); ann. grass (1978); a cultigen cult. for many centuries in the warmer Americas; thought derived fr. *Z. mays* subsp. *mexicana* (Schrader) Iltis (teosinte) (sMex, CA); ranks third largest prodn. among plant food crops of world; grain is source of corn meal, c. starch (Amylum ([Maydis] Zeae) (917) (the most available & cheapest of all starches in US); hominy; dextrose; dextrins; grits; corn steep liquor; coffee substit. (parched grain; Confederate days); corn oil (FO); cereals (corn flakes, &c.); source of glucose (corn syrup), sorbitol; corn liquor, moonshine (2423); alc.; proteins (esp. zein); green corn given for loose bowels (horses); corn silk ext. dissolves carbonate (not oxalate) type of kidney stone (915); u. as fumitory (Pe Indians; children; in Hollywood films during cig. shortage); corn cobs as fodder, fuel, tobacco pipes, in mfr. furfurans, &c.; tiny ears (30 mm long) pickled in vinegar as delicacy ("épis de maïs," Fr); lvs. u. as fodder, in silage, &c.; sts. in paper mfr.; v. under "corn"; var. *indentata* Bailey: dent corn; the chief maize of US agriculture; pl. usually tall (-7.3 m); var. *indurata* Bailey (Arg): flint maize; flint corn; with hard grains; good protein source; considered aphrod. (Domin, Caribs); var. *tunicata* St. Hil.: pod corn; believed primitive type; var. *everta* Bailey: pop corn; small kernel, hard, when heated explodes turning kernel inside out & incr. volume up to 35 X; var. *rugosa* Bonaf. (Z. saccharata Sturt.): sweet (or sugar) corn; (NA); popularly eaten as "corn on the cob"; var. *amylacea-saccharata* Bonaf. (Pe, Mex): starchy sweet corn.; var. *amylacea* Sturt. (SA): soft maize (916); var *ceritina* Kulesh. (Am, eAs): waxy maize (1979).

zeaxanthin: caroteinoid pigment occurring as yellow leaflet crystals in yellow corn frts. (obt. fr. yellow corn grits); one of most widespread carotenoid alcs.

zebra plant: 1) *Calathea zebrina* (usually); also appl. to *Cryptanthus zonatus* and *Aphelandra squarrosa* var. *louisae*.

Zebrina pendula Schnizl. (Tradescantia z. Boiss.) (Commelin.) (Fla [US], Mex, WI, CA): wandering Jew (in part); inch plant; moradilla (sMex); cucaracha (Cu); lvs. rubbed with sugar & water u. per ora for diarrh. (79); lvs. said to scare away roaches; u. to cover keys (islands); ornam. (—> *Tradescantia*).

Zeder (Ge): cedar, *Cedrus* spp.

Zedoaria (L.): **zedoary** (**rt.**): *Curcuma zedoaria*.

zegoma oil: fr. *Perilla frutescens*.

zehe Knoblauch (Ge): clove of garlic - **z. Ingwer** (Ge): race of ginger.

Zehneria scabra (L. f.) Sond. (Cucurbit.) (trop eAf): lf. mild lax. (Moz); in GC, intest. worms (enema), griping pain during childbirth, wounds (Rwanda); rt. decoc. emet. cath. diur (Moz).

Zeidane™: DDT.

zein: alc.-soluble protein found in maize grain (2.5-10%) (dry basis); extd. fr. gluten meal; of industrial importance; u. as coating & granulating agt. in tablet mfr. (film coating); in making plastics, paper coatings, adhesives, solid color printing; edible coating for foods (918).

zeitig (Ge): mature; seasoned; early.

Zeitlosenknollen (Ge): Colchicum corm.

zeitoside: glycoside of uzarigenin + D-boivinose; occ. in *Strophanthus boivinii*.

Zelkova serrata (Thunb.) Makino (Ulm.) (Jap, Kor, Chin): Japanese zelkova; wd. useful for making ships, furniture, &c.; cult. ornam. tree; bonsai.

Zellstoff (Ge): 1) cellulose 2) protoplasm.

zengo: 1) (Jap): *Angelica decursiva* & var. 2) (Korea): *Anthriscus nemorosa*.

zenia (Ala [US] slang): *Zinnia*.

Zentner (Ge): 50 kg; metric hundredweight = 110.23 lbs avg.

zenze(n)ro (It): ginger, *Zingiber*.

zeolite: natural hydrated Ca & Mg silicates; u. to soften hard water by base exchange principle (Ca in water replaces Na in zeolite); mineral is purified by reversal of process; synth. z. also mfrd.

Zephyranthes Herb. (Atamasco Rafin.) (Amaryllid.) (warm Am): zephyr (or flame) lily; fire l.; inf. of bulbs of sFla (US) spp. (such as *Z. atamasco* Herb., *Z. simpsonii* Chapman, *Z. treatiae* S. Wats.) u. in dentalgia (81) - **Z. atamasco** (L.) Herbert (seUS n to Va): atamasco (or Easter) lily; zepher l.; bulbs eaten in times of food scarcity by Creek Indians - **Z. candida** Herb. (sAf): star of Bethlehem; lf. u. as remedy in diabetes; pl. cult. & escape (NC); bulb edible after roasting (Br) (919) - **Z. grandiflora** Herb. (Z. carinata P. Wils.) (Z. rosea Lindl.) (Mex, WI, CA, natd. China): bulb decoc. u. to relieve fever; bulb pounded to paste u. as poultice for abscesses; brujita rosada (Cu); lily grass (Bah); paste of pulped bulbs appl. as cooling & purifying agt. for diseased areas (Chin); subterranean parts heated & boiled to give decoc. u. for colds, coughs, TB (Bah).

zerno (Russ): corn.

zerumbet (root): wild z.: *Zingiber zerumbet*.

zerumbone: sesquiterpene ketone; occ. in *Zingiber zerumbet*.

zest: zeste (Fr-m.): 1) section of orange &/or other *Citrus* frt.; also citrus outer peel (or flavedo, u. as flavor as in mixed drinks) 2) partition membrane (nuts).

Zeuxine strateumatica (L) Schlechter (Adenostylis s. [L] Ames) (Orchid.) (warm As, Ind, seAs, PI): tubers u. locally as salep (Ind); as sed. ton. in neuralgia.

Zexmenia frutescens (Mill.) Blake (Bidens fruticosa L.) (Comp.) (Mex, Guat to Pan): zactah (Yuc); pl. c. lasidol angelate (sesquiterpenoid); acts as ant repellant; pl. u. for infantile fevers (Guat); wd. ashes u. to keep fingers smooth (spinning women, Salv) - **Z. podocephala** Gray (sTex [US], Mex): "pionya"; decoc. of tuberous rts. u. for stomach ailments (Mex) - **Z. pringlei** Greenm. (sMex): "arnica"; decoc. of dr. or fresh pl. u. to wash sores (Mex).

Zeyhera DC = **Zeyheria** Mart. (Bignoni.) (Br): u. in Br against syph. - **Z. tuberculosa** D. Don (Br, Ch): sd. kernels u. as ton.; sd. FO, "zeyhera oil" is non-drying; exts. show anti-tumor activity.

Zeylonzimmt (Ge): Ceylon cinnamon.

Zibeth (Ge): **Zibethum (Verum)** (L.): civet, fr. *Viverra zibetha.*

Zieria Sm. (Rut.) (eAustral): several spp. c. cyanogenetic glycosides & toxic to livestock - **Z. smithii** Andr. (seAustral): lanolin(e) bush: chief const. VO with safrole & methyl eugenol; u. to relieve headaches (aborigines). some spp. c. berberine-like alk.; stock poison.

Zieridium gracile Baill. (Rut.) (New Caled): u. med.

Ziest (Ge): *Stachys* spp. (herb).

Zigadenus Mx. (**Zygadenus** Endl.) (Lili.) (NA, CA, cooler As): death camas(s); several spp. in the Rocky Mt area; fls. yellow, cream colored, greenish; poisonous pl. - **Z. glaucus** Nutt. (Z. elegans subsp. g. Hultón) (Anticlea e. Rydb.) (NA, Alas, Ariz/Mo, Minn [US]; natd. USSR)[?]): mountain death camas; alkali grass; u. emet.; toxic pl. (stock); white camash; glaucous zygadene - **Z. mexicanus** Hemsl. (Mex): u. med. (Mex) - **Z. nuttallii** Gray (Toxicoscordion n. Rydb.) (cUS, Kans, Tex): death (or poison) camas(s); u. emet.; stock poison (920) - **Z. venenosus** S. Wats. (wNA, BC to Mont to Calif [US], Siberia[?]): (meadow) death camas(s); white c.; death quamash; hog('s) potato; hb. c. colchicine-like alks., zygadenine + hypotensive esters of this & germine (mostly in above-ground parts); u. emet. & to reduce blood pressure; bulbs u. exter. for boils, bruises, rheum. (cataplasms) (nwAm Indians); formerly as arrow poison ("poison onion") (Okanag. Indians) (447): poisonous to humans, livestock, bees, &c. (may be confused with edible true camas).

zimbro (Por): juniper.

Zimm(et) (Ge): **Zimt** (Ge): cinnamon.

zinc: Zn; transition metal element; abundant in earth's crust as smithsonite, z. blende, &c.; essential metal (trace element) in nutrition; however high Zn/Cu ratio conducive to coronary heart dis.; in insulin Zn salts; in numerous alloys (ex. brass); many salts u. in pharm./med.; ex. **z. acetate** (astr.); **z. bacitracin**; **z. caprylate**; **z. chloride** (astr. antisep. in ulcers; desensitizing teeth); **z. citrate** (toothpastes); **z. oxide** (protective, astr. in ointments & pastes); **z. propionate** (antifungal); **z. stearate** (protective in dusting powders, ointments; in mfr. tablets [lubricant]); **z. sulfate** (astr. [ophthalmic solns.]); **z. undecylenate** (antifungal) - **z. gelatin**: mixt. of ZnO, gelatin, glycerin, & purified water, furnishing a kind of jelly; melted & appl. externally as a protective ("boot") & to support varicosities, &c., of lower limbs (USP) - **z. gluconate**: u. for acne, prostatic enlargement; as deodorant; in virus infections; wound healing - **z. insulin protaminate**: p. z. insulin (*qv.*) - **z. oil**: Oleum Zinci (L.): ZnO & olive oil in equal pts. rubbed together - **z. phosphide**: Zn_3P_2; u. rat & field mouse poison; aphrod.

zingerone: phenolic ketone compd. formed fr. gingerol by treatment with $Ba(OH)_2$; is pungent & sweet odored; u. in flavors.

Zingiber Boehm. (Zingiber.) (trop eAs, nAustral): gingers; perenn. hbs. with arom, props.; some 100 spp.; also specif. (L) for ginger rhiz., drug, & food; fr. *Z. officinale* (word fr. Sanscrit, sringavera) - **Z. amaricans** Bl. (Java): int. for stomach pains, leg cramps, puerperal inf., ton.; exter. as rubefac., irrit. in fever & numb feet (Mal); rhiz. juice to stim. appetite - **Z. aromaticum** Vahl (Indones): stim. for mucosa of stom. & intestines; exter. to relieve pain - **Z. chrysostachys** Ridol. (Mal): rhiz. decoc. for fever; rhiz. pounded & u. as poultice - **Z. griffithii** Baker (Mal): bruised pl. u. as poutice; rubbed over body in enteric fever; in combns. in asthma - **Z. mioga** (Thunb.) Rosc. (Jap); Japanese ginger; myoga g.; rhiz. has bergamot-like odor & taste; c. zingerone, shogaol; a comml. prod. but considered inferior to regular ginger; generally peeled & limed (with $CaSO_4$ or $CaCO_3$); types distinguished as "hands" & "fingers"; cult. in Jap. for edible shoots, fls. & frts. - **Z. officinale** Roscoe (seAs. or Moluccas; now widely cult. OW & NW trop): (common) ginger; cultigen; rhiz. partly (Af; Cochin; Calicut) or entirely (Jamaica ginger) decorticated; rhiz. c. pungent VO with gingerol (chief pungency) & zingerone (minor pungent princ.), zingiberol (odor), zingiberene; resin; starch, FO, sugars; u. as carm. stim. stom. digestive, aperitive (to stim. appetite); expect., odor & taste corrective; astr.; in stom. cramps, dyspepsia, gastritis (subacid), in diarrh. & other digest. dis.; menstrual pain (1977); beverage flavor (esp. in carbonated beverages, as ginger beer) (2255); condim. in pickles, conserves, &c.; in ginger ale, in cakes, confectionary, &c.; wines, brandy; pd. rt. (1.5 gm) u. as antinauseant in motion sickness (*mal de mer*) (said superior to Dramamine); added to purg. to prevent griping (ex. in Beecham's Pills); small amt. dr. ginger added to dry ingredients of bread aids yeast in raising dough; in trop. eAf for headache, rheum. in cough prepns., galactagog; in prepn. of oleoresin (substandard prod. not suitable for this); fresh (green) & u. (dr. pd. or crystalized); cult. in Jam, Ind, trop wAf (921); Chin, Austral (922); small prodn. Indones; Mex (923), Cu (raw); other WI, &c. - **Z. purpureum** Roscoe (Z. cassumar Roscoe; Z. cassumunar Roxb.) (Indochin): cassumun(i)ar (or Bengal) root (tubers); Rhizoma Cassumunar; cassum(un)ar ginger; rhiz. u. like ginger, as condim. in med. (u. for diarrh. & colic by natives); vermifuge for children (Mal); not as pungent as true ginger; u. as ginger substit. (very poor, however, as it looks, smells, tastes differently [like camphor]); as adulter. of whole rhiz.; cult. - **Z. zerumbet** (L) Sm. (tropAs; widely cult. in trop&subtr [ex. Fla (US), Br, Martin]): zerumbet (or wild) ginger; zerumbet; product: Rhizoma Zerumbet; u. stim. ton. arom.; appetizer, in gastralgia (Ind); like true ginger (*Z. officinale*) but less arom.; as adulter. of whole ginger; in combns. for fever & childbirth (Indones); vermif., after childbirth, in poultices; also in magic med. (Mal); in infertility (inf. of bk. scrapings drunk) (Tonga) (923A); rhiz. & rts. u. for infant's cough & in gastric dis. (Fiji); mostly u. exter. not int.; in arrow poisons(?): also ornam.

Zingiberaceae Lindl. Ginger Fam.; perenn. hbs. native to tropics; chiefly in eHemisph; ca. 1,300 spp.; with tuberous rhiz. or fleshy rts.; lvs. lanceolate; c. VO often arom. also pungent princ.; characteristic starch grains, with beak near pt. of origin. Fam.; of great economic importance; ex. ginger, cardamom, turmeric; EI. arrowroot; grains of Paradise.

zingiberene: a sesquiterpene, chief component of VO of ginger.

zingiberone: zing(h)erone; arom. ketone; one of chief components of ginger root oil.

Zinnia L. (Comp.) (Am): zinnia; "youth & old age"; cult. ornam.; Z. spp. as "china paya" u. in gastric dis. (Pe); fls. both single & double - **Z. elegans** Jacq. (Z. violacea Cav.) (Mex): old maid's pink; "youth & old age"; widely cult. ornam. hb. thus by Ec Indians (zinia) - **Z. grandiflora** Nutt. (Mex; Tex[US]): plains zinnia; cinco llagas (Chihuahua): decoc. drunk as astr. in diarrh.; diur. in sexual dis. (Navahos) - **Z. multiflora** L. = *Z. pauciflora* L. - **Z. peruviana** (L.) L. (sTex [US], Mex): redstar zinnia; u. in eye dis. (Mex).

Zinnkraut (Ge): *Equisetum hyemale.*

ziola (Pol): species; tea.

Zitrone (Ge-f.): lemon; citron.

Zitronengrasoel (Ge): lemongrass oil.

Zitwerblueten (Ge): **Zitwersamen**: Cina; *Artemisia cina* & related spp. - **Zitwerwurzel** (Ge): 1) Zedoaria 2) (much less commonly): ginger; aconite; Calamus.

Zizania aquatica L. (Gram.) (e&c NA cult. eAs): wild (or Canadian or Indian) rice; water oats; minomin (Am Indians); formerly staple food of Am Indians, still u. by Chippewas; now popular in sophisticated restaurants; u. as side-dish like rice; also in hash, sauces, sea foods, &c.; as gruel with wild licorice tea for "cholera morbus" (severe enteritis) (Am Indians); game-bird attractant; actually not related to true rice (similar only in thriving in water); most coll. in Minn. & Wisc. (924) - **Z. latifolia** Stapf (Z. caduciflora Hand.-Mazz.) (Taiwan); shoots prescribed for fever, as diur. & to relieve thirst; lvs. ton.; rhiz. ash mixed with egg white u. as ointment for burns; cult. & the solid culms & young shoots u. as vegetab. - **Z. texana** Hitchc. (cTex [US]): Texas wild rice; u. similarly to *Z. aquatica*.

Zizia aurea Koch (Thaspium aureum var. *apterum* A. Gray) (Umb.) (e&cNA): golden Alexander(s); wild parsley; hb. (c. arom. VO) u. as diur. (Eu homeopath.); rt. u. as antipyr. (Meskwaki Indians).

zizim (Mex-Yuc): *Artemisia vulgaris.*

Ziziphora Mill. (Zizyphora Dum.) (Lab) (Med reg. to cAs): field basils; some 25 spp. of ann. hbs. - **Z. capitata** L. (seEu): cult. ornam. - **Z. clinopodioides** M. Bieb. (Iran, Baluch, Turkestan): maurai (Pushtu); dr. pl. decoc. u. for typhus fever; lf. inf. drunk in hot weather (928); hb. c. VO with pulegone 47-58-70% - **Z. tenuior** L. (Pak [NWFP, Baluch, USSR]): hb. u. stim. carm. expect.; sds. antipyr., dys. (pd. & mixed with buttermilk); hb. c. VO with 78% pulegone; flavor for yogurt.

Ziziphus Mill. (*Zizyphus* Adans.) (Rhamn.) (tropAm, Af, As, Austral): jujube; deciduous or evergreen shrubs or trees; some cult. ornam. or frt. trees; some spp. u. as fish poisons - **Z. abyssinica** Hochst. (cAf): catch-thorn; hot lf. u. as fomentation to chest in pneumonia; rt. decoc. abort.; for toothache, sores, eye wash; to prevent elephantiasis; frt. u. to make potent alc. beverage (kachaso); wd. u. to make wagons - **Z. cambodiana** Pierre

(Laos): mak mai; u. med. in Laos - **Z. chloroxylon** (L) Oliv. (Jam): cogwood; wd. hard & tough - **Z. elegans** Wall. (Mal): rt. bk. decoc. u. to treat fever - **Z. helvola** Sond. (sAf): rt. decoc. taken for menstrual troubles - **Z. joazeiro** C. Mart. (Br): joa'; lvs. u. in dyspeps. & digest. disturb.; macerated bk. u. in urin. dis., GC, emet. antipyr.; decoc. of chips in syr. u. for cough; wd. & rts. c. saponins; lvs. & frts. as fodder; frts. sometimes consumed by natives - **Z. jujuba** Mill. (Z. vulgaris Lam.) (Med reg., temp., e&wAs): (common or Chin) jujube (tree); French j.; "date tree"; jujuba tree; lotus; Chin date; azufayfo (Sp); frt. (Zizyphus (L) (Sp & Por Phar.) (Fructus Jujubae); orange-red (drupes) c. galactose, fructose, glucose, mucil., tannin, tartaric acid; u. dem. sialagog, nutrient, mild lax.; expect. in bronchial dis., chest colds, coughs, ton.; as pastes, pastilles, syr.; frts. esculent, raw, dr. or cooked; in pickles as glacé frt.; said most popular frt. (eaten by 400 million Chinese); frt. superficially resembles date when fresh, prune when dr.; paste & lozenges made (2676); bk. u. for diarrh. (Burma), in fever (Chin); lvs. u. as plasters, as food for the Tussa silkworm; source of shellac; cult. As Min, Eu (lt. Fr, Sp) Fla, Calif (US); var. *inermis* Rehder: anti-allergic action of value - **Z. lotus** (L) Lam. (Med reg.): lotus (tree); African l.; frt. u. much like that of *Z. jujuba*; believed the lotus fruit, food of the "lotus eaters" of the Odyssey; planted as spiny hedge - **Z. mauritiana** Lam. (Z. jujuba Lam. non Mill.) (s tropAs; cult. Af, hot Am): (Indian) jujube; coolie plum; dunk(s); "bayar" (or "ber") (Ind); frt. edible ("Chinese dates"), eaten raw unripe or ripe, cooked, preserved; decoc. of frt. (or lvs. & rts.) given for bilious attack (Haiti); lvs. & bk. u. for tanning (Ind); u. for colic; pounded lvs. appl. as dressing to wounds (Af); lf. juice drunk for GC (Tanz): frt. u. to stupefy fish (Ind): 50 g. pd. gall given twice a day for dys. (727) - **Z. mistol** Gris. (Arg+, Andes): mistol (Arg): Argentine jujube; frt. edible (as "table food"), u. to make a syrup, to prep. bolanchao (Arg dish) & chicha (Arg); ext. as antispasmodic - **Z. mucronata** Willd. (tropAf): buffalo (or Cape) thorn; shiny leaf; "wait a bit"; rt. macerate or decoc. for menorrhagia, infertility; in dys.; lumbago; GC; urinary incontinence; general body pains (Venda); frt. scarely edible but roasted for coffee substit.; hot inf. of bk. for cough; lf. as poultice to boils & other septic skin swellings; silk worms of tree prod. silk of some interest; sds. u. in rosaries - **Z. nummularia** W. & A. (Pak, Ind): lvs. appl. to scabies & boils - **Z. obtusifolia** Gray (Condalia o. Weberb.) (swTex [US] nMex): graythorn; lotebush (Tex); (white) crucillo; abrojo; st. & rt. decoc. appl. to sores (horses); bk. soap substit. (Tamaulipas) - **Z. oenoplia** Mill. (Ind, Indochin): bk. chewed & spewed over body of newborn baby for fever (Indochin): rt. decoc. to heal fresh wounds (Ind); frt. in stomachache - **Z. rugosa** Lam. (Ind, Pak): bk. pd. u. in diarrh. (mixed with *Terminalia bellerica*) (124) & menorrhagia; fls. with betel leaf & lime u. for latter - **Z. sativa** Gaertn. = Z. *jujuba* Mill. - **Z. spina-christi** (L) Willd. (eAf to Middle East; natd. NWI) Christ's thorn; nubk tree (Israel); ap(p)eldam (NWI); frt. consumed by Arabs (tastes like dr. apple), may represent lotus of the Ancients; foliage eaten by camels; lf. decoc. u. as cold remedy (Curaçao); believed hypotensive by some: u. to make bread in Sudan; believed to have furnished crown of thorns of Jesus on cross - **Z. vulgaris** Lam. = Z. *jujuba* Mill. - **Z. zeyheri** Sond. (sAf): rt. decoc. (or tinct.) u. in diarrh. dys.; rt. inf. blood ton. purg.; lf. u. to dress swellings; frt. sweet & palatable, u. for porridge & as coffee substit.

zloto glow (Pol): asphodel (lit. gold head) - **z. kwiat**: *Chrysanthemum segetum* - **zloty deszcz** (Pol): laburnum.

zoapatle (Mex): *Montanoa tomentosa*

zoethout (Dutch): licorice.

Zoisia Aschers. & Graebn. = *Zoysia* Willd.

Zoll (Ge-m.): inch.

Zonitis (Meloideae - Coleoptera - Insecta): one of the blister beetle genera.

zontol (SpAm): sontol

zoonosis (pl. **zoonoses**): disease of lower animals transmissible to man; ex. brucellosis, anthrax, Q-fever, salmonellosis.

zoopherin: animal protein factor (APF) = vitamin B_{12}.

zoospore: motile spore found in Fungi; asexually produced; "swarm spore."

zoosterin: zoosterol; sterol occurring usually in animals; ex. cholesterol; v. sterol.

zoosterols: zoosterins: collective name for sterols found in animals; both free & as esters; ex. cholesterol; coprosterol; SC zoon (Gr): animal, living organism; zoe (Gr): life.

zootic acid: HCN.

zopilocauve (Salv): Jamaica dogwood.

Zornia bracteata (Walt.) J. F. Gmel. (Leg.-Hedysareae) (seUS, Mex, sAf): bracted zornia; five-finger grass (but with 4 leaflets per lf.); cinquefoil; pl. u. lax. diur. - **Z. diphylla** Pers. (Z. reticulata Sm.) (Mex to Pe, Ur; OW trop in places): raiz de vibora; trencilla; pl. decoc. u. in diarrh. & dys. (CR); diur. (Co); antipyr (Mex); rt. for induction of sleep in children (Ind): emergency stock food in dry weather (wAf) - **Z.**

glochidiata DC (trop e&wAf): lvs. cooked & u. as veget.; u. in kwashiorkor. (786) - **Z. gibbosa** Spanog. var. *cantoniensis* Ohashi (trop worldwide, as Taiwan): u. locally as med. herb (Taiwan) - **Z. latifolia** Sm. (Br): u. as hallucinogenic in place of Cannabis. - **Z. pratensis** Milne-Redhead (cAf): remedy for scurf (Burundi) - **Z. setosa** Bak. f. (trop eAf): hb. u. for stomachache (sometimes in combns.); fresh rt. chewed & juice swallowed for same (786).

Zosima orientalis Hoffm. (Umb.) (Pak, Ind): pl. u. for cough & bowel dis.

Zostera marina L. (Zoster. - Monocot.) (worldwide fresh waters): eel grass; grass-wrack; sea w.; s. ware; "sea weed" (appl. to dr. lvs. though not an alga); grass weed; alva marina; u. for packing material (for glassware, &c.), pillows, bedding for horses, &c.; mfd. into street paving material (*sic*); lvs. u. to treat goiter, edema, women's diseases (Chin); balls formed of pl. called "pilae marinae" (sea balls or spheres); pl. important food of some water birds (ex. geese).

Zoysia japonica Steud. (Z. pungens auct. jap.) (Gram.) (eAs): Korean lawn grass; cult. perenn. grass for turf (US, mostly in warm areas) - **Z. matrella** (L) Merr. (Z. pungens Willd.) (seAs, Austral, NZ): Manila grass; Japanese lawn grass; birodo-shiba; u. as good lawn turf grass (seUS); does well in Fla; remains green during winter months - **Z. tenuifolia** Willd. (eAs, Chin, Jap): Korean velvet grass; korai-shiba; cult. for turf (Calif, Fla [US]).

zubrooka (Russ): *Anthoxanthum odoratum*.

zucca (pl. **zucce**) (It): **zucca indiana; z.** (**melon**); pumpkin; also gourds of various spp. esp. *Cucurbita pepo* (*cf.* zucchina) (**zuca** [It]: *Cucurbita maxima*).

zucchero (It): sugar - **z. di latte** (It): lactose - **zucchetta** (pl. **zucchette**) (It scient. term): *Cucurbita pepo* (v. *italica* Tod.?) - **zucchina** (pron. zukeena) (diminutive of *zucca*) (It- common market name): zucchini - **zuchette** (Fr): *Cucumis sativus* var. - **zucchini** (pl. of **zucchino**): **zuccino**, zuccini, (It): small zucca; zucchini squash; *Cucurbita pepo* L. var. *medullosa* Alef. (cultivar zucchini), race of vegetable marrows with pl. dwarfed & tendrils aborted; or one at the young stage; kind of very long squash (-1.3 m long, 60kg) with smooth thin rind; eaten fried or boiled & seasoned with oil, often with white fls. still attached (popular It).

Zucker (Ge-m.): sugar - **Z. futterruebe** (Ge): mangel - **Z. ruebe** (Ge): sugar beet - **Z. rohr** (Ge): sugar cane - **Z. saeure** (Ge): oxalic acid.

Zuelania guidonia Britt. & Millsp. (Z. roussoviac Pittier; Laetia g. Sw.; sometimes referred to *Casearia*) (monotypic) (Flacourti) (sMex to Pan; WI): water (or cuffey) wood; "guaguasí," "guagasí;" one of the most famous medl. spp. of Cu (938); lvs. & bk. bitter, appl. to wounds & chronic ulcers as detergent; lf. decoc. u. as bath in fevers (Yuc); antimal., in syph.; frts. roasted & eaten (Yuc); resin (white, arom.) flowing fr. natural or artificially made incision in trunk of tree/shrub is u. as diur. antiluet. (with bk.) (Cu).

Zukker (Dan.): sugar.

zumaque (Sp): sumach, incl. poison ivy.

zumpkin: hybrid of zucchini squash & pumpkin.

Zunder (Ge): tinder, punk; *Fomes fomentarius* ("German tinder") (also **Z. schwamm**).

Zwergschwein (Ge): guinea pig (lit. dwarf pig).

Zwetsche (Ge): **Zwetschge**: (damson) plum, common p.; *Prunus domestica*; prune.

Zwieback (Ge, Eng): kind of biscuit or rusk baked in a loaf & cut & toasted (Brussels biscuit).

Zwiebel (Ge): onion; bulb.

Zwitterbluete (Ge): Santonica.

Zygadenus Endl. = **Zigadenus** Mx. (latter preferable spelling).

Zygia P. Br. (Leg) (tropAs, Austral, Am): some spp. referred to *Pithecolobium, Albizia,* &c. - **Z. brevifolia** Br. = *Pithecellobium brevifolium* (*P. pallens*).

Zygopetalon (**Zygopetalum**) Hook. (Orchid.) (warm Am): one unidentified sp. u. to alleviate cardiac pain (decoc. of pseudobulbs); "matum-biaya" (937).

Zygophyllaceae: caltrop family; hbs., shrubs, & trees of trop & subtr; lvs. opposite, compd., stipulate; fls. regular, pentamerous; frt. usually a capsule; ex. *Guajacum; Larrea*.

Zygophyllum coccineum L. (Zygophyll.) (nAf, Near East to Pak): bean caper; frt. c. chinovasic acid, zygophyllin (NBP), u. as decoc. or inf. in hypertension, rheum. gout, coughs, asthma, colic, diur.; sds. said anthelm. (Pak) - **Z. cornutum** Coss. (Tun): bou griba; u. as hypoglycemic - **Z. fabago** L. (nAf, sRuss, AsMin, Mex): (Syrian) bean caper; fl. buds steeped in vinegar and eaten like capers (u. since olden times); lf. u. as anthelm. & antiluetic (nAf) - **Z. gaetulum** Emberger & Maire (Mor): berraya (Arab); dr. pd. lvs. hemostatic, u. as plaster on mature boils & abscesses; to treat eczema; lf. inf. as antisep. for babies (750) - **Z. morgsana** L. (sAf): pl. u. for cramps, paralyses, shock injury; fresh pl. toxic gives diarrh. - **Z. simplex** L. (nAf, sAs): carmal; hb. u. for removal of acanthokeratodermia (horny spots on hands or feet) - **Z. tridentatum** DC = *Larrea tridentata* - **Z. waterlotii** Maire (w tropAf, as Senegal): lf. (Folia Zygophylli, L.) u. in inflammns. attributed to blood poisoning.

zymase: endocellular enzyme of yeast; by fermentation converting dextrose & levulose to alc. & CO_2.

zymine: pancreatin.

zymochemistry: chem. of fermentation (<zyme = emzyme).

zymogen: inactive form of an enzyme; activated to become an enzyme by contact with a kinase; ex. pepsinogen; prorennin.

zymohexase: aldolase: enzyme wh. as catalyst splits fructose diphosphate into dihydroxy-acetone & phosphoglyceric aldehyde or condenses these last two into former; occ. universally in cells; best sources: muscle; yeast.

zymosan: anticomplementary factor; protein-carbohydrate complexes found in yeast cell wall; acts to enhance specific defenses of organism through activation of the properdin system.

zymosterol: sterol occurring in yeasts.

zymurgy: art & science of brewing, wine-making, & distilling-among the chief comml. applns. of fermentation science.

Zypresse (Ge): *Cupressus* spp.

APPENDICES

APPENDIX A: BIBLIOGRAPHY

REFERENCES AND NOTES CITED BY NUMBER IN TEXT

1. *Dorvault's L'Officine,* ed. 18. (Vigot-Paris) (1948).
2. Muir, John. *Thousand Mile Walk to the Gulf.*
3. *Proc. A PhA.* 21:620; 1874.
4. *Pharm. Formul.*, 682.
5. *Handbook of Commercial Information on India.* pp. 438-441.
6. Steggerda, J. "Yucatan plants," *BAE Bull.* 136:189-226 (1943).
7. Lawrence, T. E. *Oriental Assembly* (1939).
8. Darlington. *Agricultural Botany* (1847).
9. Bentham. *Flora Austral.* 1:230; 1863-1878.
10. Woodworth, R. H. "Economic plants of St. John, US Virgin Islands," *Bot. Mus. Leaflets.* (Harvard Univ.), 11:29-56 (1943).
11. Standley, Paul C. "Trees and shrubs of Mexico," *Contr. U.S. Nat. Herbar.* 23 (1920-1926).
12. Meyer, T. "Arboles indigenas de frutos comestibles....," *Lilloa.* 3:233-242; 1938.

12A. Taiyab & Khan. *Hamdard.* 32:61-3: 1989.

13. *Publns. Mississippi Hist. Soc.* 4:350.
14. *U.S. Nat. Museum Rept.* 1896:1025-56.
15. Anchel, Marjorie. *JBC.* 177:169 (1949).
16. *Drugg. Circ.*16:128 (1872).
17. Teit, J. A. "Ethnobotany of the Thompson Indians of B.C.," *BAE Ann. Rept.* 45:481 (1930).
18. Gilmore, M. R. "Use of plants by the Indians of the Missouri River region," *BAE Ann. Rept.* 33:71 (1919).
19. *Ber. d. d. pharm. Ges.* 14:205.
20. Zoernig, H. *Arzneidrogen.* I:192 (c. 1909).

20A. *Chem. Gaz.* 10:375-7 (1852).

21. *Amer. Drugg.* Sept. 1937:146.
22. McIlwaine. *Memphis Down in Dixie.* p. 58 (1948).
23. *Contr. U.S. Nat. Herbar.* 12:100 (1908).
24. Partridge, E. *Dictionary of Underworld (Slang)*; 1950.
25. Rusby, Henry H. *Jungle Memories* (1933).
26. Ernst. *J. Bot.* 3:277 (1865).
27. Goldman, E. A. *Contr. US Natl. Herbar.* 16:333; ca. 1920.
28. Couch, James F. *JACS.* 61:3327 (1939)
29. Barker-Webb & Berthelot. *Canary Islands.*
30. Ducke, A. *Bol. Tec.* 4 (1945).
31. Schultes, R. E. *Bot. Mus. Leaflets.* 15:40 (etc.) (1951).
32. Instituto Medico Nacional. *Datos para la materia medica mexicana* (México D. F.) (1894).
33. *FDA. NJ.* 15954 (1950).
34. Freise, F. W. "Plantas medicinaes Brasileiras," *Boletim de Agricultura* (São Paulo). 34a:252 (1933).
35. *JAMA.* 127:1129 (1945) (editorial); *Bull, Johns Hopkins Hosp.* 76:93.
36. Youngken, Heber W. *Textbook of Pharmacognosy*, ed. 6:418 (1950).
37. Matthews, A. W. *JA PhA.* 28:1030.
38. LaWall, C. H. *Four Thousand Years of Pharmacy.* p. 312; 1927.
39. *Nature.* 146:277 (1940) (retrorsine).
40. Giboin, Lucien. *Epitomé de botanique et de matière médicale de l'Inde* (Pondichéry, India) (1949).
41. Blake, S. F.. *Contr. U.S. Natl. Herbar.* 20:501 (1924).
42. Biswas, Kalipada. *J. Royal Asiatic Soc. of Bengal, Sci.* 13:75 (1947).
43. Miller, William. *Dictionary of the English Names of Plants* (1884) (see 619A).
44. Blake, S. F.. *Contr. U.S. Natl. Herbar.* 24:98 (1922).
45. *J. Investigative Dermat.* 15:163 (1950).
46. *Dr. Circ.* 38:183 (1894).
47. Lumholtz, Carl. *Unknown Mexico*, II:257 (1902).
48. Chopra, R. N. *Indigenous Drugs of India* (Calcutta) (1933).
49. Peckolt, Theodor. "Volksbenennungen der brasilianischen Pflanzen u. Produkte derselben in brasilianischer (portugiesischer) u. von

Tupisprache adoptirten Namen," *Pharm. Sci. Series Monograph* No. 15. Milwaukee, Wisc. 1907 - Lacombe, Fco. Henrique Luiz. "Quina do campo," *Anais da Fac. de Odont. e Farm. de Univ. de Minas Ger*. 4: 266-277; 1939.

50. Pittier, H. F. *Contr. U.S. Natl. Herbar*. 13:456; 1909.
51. *Am. Drugg*. Sept. 1937, p. 142.
52. *Anales de la Universidad Montevideo*, 1936:201.
53. *Stories of New Jersey*. Bull. 12 (1938/9).
54. Bal, S. N. & Prasad, S.: *Indian J. Pharm*. 5:30-36; 1943.
55. Walker: *Contr. U.S. Natl. Herbar*. 28:594; 1941.
56. *Washington Post*. III-18-51.
57. *Ariz. Agr. Exp. Sta. Tech. Bull.*
58. Curtin, L.S.M.: *Healing Herbs of the Upper Rio Grande* (1947).
59. Hocking, G. M.: *Collier's Encyclopedia*; NYC, article on "Glycosides," 1950.
60. *Idem*. article on "Alkaloids."
61. *JAMA*, 114:1625 (1940).
62. Dalziel, J. M. *Useful Plants of West Tropical Africa* (1948).
63. Mautner, H. G., *et al*.: *J A Ph A, Sci. Ed*., 42:294 (1953).
64. Rousseau. *Mémoire* 2, *Jardin Bot*. Montréal (1946).
65. Williams, L. *Exploracio Botanicas en la Guayana Venezolana*, I (1942).
66. Farwell, Oliver A. *J A Ph A*, 7:854; 1918.
67. *Arch. Intern. de Pharmacodynamie et de Thér*., 88:253 (1951).
68. Hocking, G. M. "Totaquine," *Med. Digest* (Bombay), 12:61-75 (1944).
69. Hocking, G. M. *Ozarks Pharmacist* (Clarksville, Ark), 2: No. 1, pp. 4, 12; 1950.
70. Hocking, G. M. *Ind. J. Phar*., 7: 89-92 (1945) (Rheum).
71. Hocking, G. M. *El Farmaceutico* (NYC), 19:20 *et passim* (Oct. 1943).
72. Barton, *Trans. Amer. Philos. Soc*., 3:100-115 (1793).
73. Hoppe, Heinz A. *Drogenkunde*, Ed. 4 (1945).
74. Hocking, G. M. *Rocky Mountain Drugg*., Nov. 1949 (Ethnobotanical Notes).
75. Hocking, G. M. *Ibid*., Aug. 1950 (*Populus*).
76. Hocking, G. M. *Ibid*., May 1951 (Folk remedies for arthritis).
77. *Austral. J. Agr. Res*. 1951:457.
78. Hocking, G. M. *Rocky Mountain Drugg*., Oct. 1951 (Southern superstitions).
79. "The Tajin Tononac," *Smithsonian Inst. of Social Anthropology, Publn*. 13 (1952).
80. Hooper, D.: *Phar. J*., (4) 35:519 (1912).
81. Small, John K.: Manual of Flora of the SE US (NYC) (1933).
82. Hocking, G. M. "Pharmacognosy, chemistry, and therapeutics of *Lacinaria (Liatris) spicata* and *L. tenuifolia*," MS thesis, U. Fla 1932 (Portion published as: *J A Ph A*. 25:15-18 [1936]).
83. Wakim et al. *J. Pharm. Exp. Ther*. 5:1946 - Brown, H. C. *Lancet*. 1923: I: 382.

83A. Brown, H. C. *Lancet*. 1923. 3:382.

84. *Rept. Provincial Museum* (B.C.) 1931 (B9):9; Hardy, Eric. *Nature* 134:54-5, 327; 1934.
85. Bristol, M. L. *Bot. Museum Leaflets* 19:191; 1961 - Cult. comm. St. Petersburg (Fla [US]) (ca. 1940).

85A. Yao-Wu Shou-Tse (*Pharmac. Handbook*) (Peiping; 1957).

86. Yuncker, T. G. Plants of Tonga; *Bishop Museum Bull*. 220; 1959. Now often referred to *G. repens* (L) I.M. Johnston.
87. Vacca, D. D. & R. A. Walsh: *J A Ph A S. E*. 43:24-6; 1954.
88. Anderson, E. & L. Harris: *J A Ph A S E* 41:529; 1952.
89. Pratt & Yuzuriha: *J A Ph A S E* 48:69+; 1959.
90. Svoboda, G. H. et. al. *J A Ph A S E* 48:659-666; 1959.
91. Fahmy. L. R. & Abdel-Latif. *J A Ph A S E*. 37:276-283; 1948.
92. Steyermark, J. et al. "Contr. to Flora of Venezuela," *Fieldiana, Bot*. 28, #4; 1957.
93. Moldenke, H. N. *Lilloa* 10:365-9; 1944.
94. Bachstez & Aragon. *J A Ph A S E* 34:170; 1945.
95. Jahn & Leutschaft. *Arzneimittel-Forschung Nov. 1953:561*.
96. Tokin, B. P: *Phytonzide* - Berlin (DDR) 1956.
97. Cuatrecasas, J. "Revision of Humiriaceae" (monograph), *Contr. U.S. Nat. Herb*. 35: Part 2; 1961.
98. Cruz & West. *Philipp. J. Sci*. 46:131-139; 1931.
99. Moldenke, H. N. *Phytologia* 5:465+ (1959); 15:265 (1967); 46:42 (1980).
100. Holmes, E. M. *Chem & Drugg*. 94:60+ (1921) - *Ph. J*. 106:17 (1921).
101. Roffo, A. H. *Bol. Inst. Med. Exper*. (Bs.As.) 20:515-84; 1943.
102. Standley, Paul C. *Flora of Guatemala*; 1937/38.
103. Moul. "Onotoa Atoll, Gilbert Is.," *Atoll Res. Bull*. 57; 1957.
104. Tang-Shui-Liu. Abies*: Monograph of the genus*. Forestry Dept., Taiwan Univ., Taipei; 1971.
105. U. as an LSD replacement: Schultes, R. E. and Hofmann, A. *Bot. and Chem. of Hallucinogens*, p. 254; 1980. Pl. also cult. as ornam. in US. (fls. foliage): "(baby) Hawaiian woodrose" (*Argyreia nervosa*).
106. Beath, O. A. *J A Ph A* 15: 265-269; 1926.

107. Kaurrote. *Farm. Mezeotiseski Zhur.* 1916; per *Ph. J* . 1916:328; *Y. Bk. Br. Ph. C* . 1916:249 (Cult. in Bessarabia & Crimea).
108. Ritchie, David. *Chem & Drugg.* 6/23/45:647; *Ph. J.* Apr. 7, 1945.
109. Kupchan, S. M. et al. *J. Med. Chem.* 7:803; 1964.
110. Bergen, Fanny D. *Popular American Plant Names*; 1893+.
111. Vogel, V. J. *American Indian Medicine*; 1970.
112. Altschul, S. v. Reis. "The genus *Anadenanthera* in Amerindian culture"; Botanical Museum; Harvard Univ.; 1972.
113. Schoepf, D. *Materia Medica Americana*; 1787.
114. *Current Sci.* 1959:54; *J A Ph A* V; 1927.
115. Moldenke, H. N. *Phytologia* 7:190-; 1960.
115A. Bisset, N. G. *J. Ethnopharm.* 1:325-841 (1979); 4:247-336 (1981).
116. *Nat. Geog. Mag.* 1956.
117. 1) Morton, J. *Atlas of Medicinal Plants of Middle America.* 1981:844 2) *Nature Mag.* Jan. 1958:41-3.
118. *Tribuna Farm.* 1952:84.
119. Tamm: *Planta Med.* 1962:134-
120. Townsend, Wyeth. *Expedition to the Pacific.*
121. Moldenke, H. N. *Phytologia* 7:388; 1961.
122. *Anal. Chem.* 21:502; 1949.
123. *Science* 173:1190; 1971.
124. Paul, S. R. *Quart. J. Crude Drug Research* 15:79-97; 1977 - *Phytologia.* 55:436, 438; 1984.
125. Kupchan, S. M. 1972: re *Datisia.*
126. Manresa, M. *Philippine Agric.* 13:113; 1924.
127. Elmore, F. E. *Ethnobotany of the Navajo*; 1944.
128. Kupchan, S. M. et al. *Phytochemistry* 16:1834; 1977.
129. Watt, J. M. & J., Breyer-Brandwijk. M. G. *Medicinal and Poisonous Plants of South and East Africa*, ed. 2. 1962.
130. Kokwaro, J. G. *Medicinal Plants of East Africa*; 1976.
131. Boorsman. *Pharm. Centralhalle* 43:155; ca. 1902.
132. According to R. G. Wasson in "Soma — divine mushroom of immortality": (ca. 1967) based on the Rig-Veda (ca. 1500, B.C.). (RGW's book reviews the several plants proposed as representing the Soma of ancient Asia [31 or more plant species or groups]).
133. *J. Endocrinol.* 1955:94.
134. Wasicky. 1957 - Pridham. *Cornell Ext. Bull.* 231. 65 pp. 1932 (many refs.) (*G.* spp. [Brasil] lvs. 6.38% saponins; bulbs 8.68%).
135. *Pharmazie* 22:208; 1967.
136. Fryxell, P. A. - *U S D A, T. B.* 1491 (114 pp.); 1976.
137. Ayensu, E. S. *Medicinal plants of the West Indies*; 1981.
138. *J. Chem. Soc.* (London); 1962.
139. Adam, J. G. *La Pharmacopée Sénégalaise Traditionelle*; 1974.
140. Quisumbing, E. *Medicinal Plants of the Philippines*; 1951.
141. Glover, T. *Philos. Trans. Royal Soc.* XI:628; 1676.
142. Uphof, J. C. T. *Dictionary of Economic Plants*, ed. 2; 1968.
143. Levring., T. et al. *Marine Algae*; 1969.
144. Singh & Siwatibau: *Medicinal Plants in Fiji*; WHO; 1977.
145. Chopra, R. N. et al. *Glossary of Indian medicinal plants*; 1956.
146. Snyder, H. R. et al.; 1954.
147. Pharmacopoea Gallica IX: Supplem.; 1980/ Hirschhorn. H. H. *Journal of the Herbalist of Erdendorf,*" p. 13; 1980.
148. Githens. *Univ. Penn. African Handbook* 6; 1949.
149. Seton, Ernest Thompson. *Lives of Game Animals* (4 v.); 1929.
150. Meyer, E. *Heilerde, Anwendung and Wirkung*; 1957.
151. *Ala. Coll. State Rept.* 1918:32-4. - MacDonald, M. B., Glaser, A. Ph. J. (London) 4 Oct. 1930; Tenn. Agr. Exp. Sta. 26; 1929 - *J. Dairy Sci*: 16:401; 1933. (*CA* 24:2508).
152. Taylor, Mary Susan. "Syllabus for the medicinal and edible native plants of Butte County, Cal."; 1977.
153. Karrer, W. *Chem Zentralblatt* 1936, II:2727; *Helv. Chim. Acta* 1943:1353.
154. Darmer, G. *Sanddorn als Wild u. Kulturpflanze* (book); 1952 - *Pharmazie* 8:112; 1952 - Prep. by Diata, Ingolstadt - Danau. (concentrate u. in med.).
155. Mabberley, D. J. *The Plant-book: A Portable Dictionary of the Higher Plants.* Cambridge Univ. Press: Cambridge, NY, 1987-1989.
156. Chopra, R. N. & al. *Glossary of Indian Med. Plants*; 1956.
157. Jarvis, D. C. *Folk Medicine*; 1958. - Venturi, Alb. "De mellis origine et usu diss. historico-medico," 40 pp. (1765).
158. Train, et al. *Med. Uses of Plants, Nevada*; 1957 - Research of Univ. Utah (-1949-).
159. Gr. Brit. Med. Act., Part III: - Ph J. 4 Feb., 1978.
160. Nemnich. *Allgemein Polyglotten Lexikon*; 1793-5.
161. Sulit, M. D. *Makiling Echo* 13:5-24; 1934.
162. Altschul, S. van Reis: *Drugs & Foods from Little Known Plants*; 1973.
163. *C. R.* (Paris) 157:727; 1913 (Greshoff - distrn. of HCN in pl. kingdom).
164. Hoppe, H.A. *Drogenkunde*, ed. 8, v. 1; 1975.

165. Danos, B. *Gesellschaft f. Arzneiforschung*, Helsinki; 1972.
166. Rakoto - Ratsimamanga - *Eléments de Pharmacopée Malagasy*; 1969.
167. Erichsen-Brown, C. *Use of Plants for the Past 500 Years*; 1979.
168. Morton, J. *Quart. J. Crude Drg. Research* 15:152-192; 1977.
169. *Japanese Drug Directory* (Tokyo). 1973.
170. Patino, V. M. *Econ. Bot.* 22:4; 1968 - Schultes, R. E. *Etnologiska Studien.* 32:115-38; 1972.
171. See 142.
172. Cheo, T. Y. *Bot. Bull. Acad. Sinica* 1:298-308; 1947.
173. UP. Toronto 29 Aug. 1934.
174. (Khus). *Jl. Ind. & Trade.* 9:872-3; 1959 - Louisiana *Vetiver* Farms, Pleasant Ridge, Hammond, CA.
175. Chandra & Pandey. *Int. J. Crude Drug Res.* 21:25; 1983.
176. *Hagers Handbuch d. Pharm. Praxis*, ed. 4. 5:237; 1976.
177. *Flora Medicinal de Colombia*; Inst. Cienc. Natur., Univ. Nac., Bogota, 1974-5.
178. Youngken, H. W. Jr. *Psychedelic Review*, 1964.
179. Duke, James A. *Isthmian Ethnobotanical Dictionary*. 1972.
180. Brazier, J. A. *Chem. Gazette.* 10:375-7; 1852.
181. Rasmussen, R. A. & Jones, C. A. *Phytologia.* 12:15-19; 1973.
182. Cheo, T. Y. *Bot. Bull. Acad. Sin.* 3:137; 1949.
183. *Memoir, Instituto Butantan.* 9 (1935):301-18.
184. *Int. J. Crude Drug Res.* 25:25; 1983. Jasmine oil is obtained by enfleurage; very expensive ($8500/lb.).
185. *Flora of Taiwan* v.4: 138; 1978.
186. Standley, P. C. *Trees & Shrubs of Mexico.* 1926.
187. Schultes, R. E. *Bot. Mus. Leaflets* 15:39; 1951.
188. Burlage, H. M. *Index of Plants of Texas.* 1968.
189. Wyman, L. C. *Navajo Ind. Med. Ethnobotany.* 1941.
190. Jacobs, M. L. & Burlage, H. M. *Index of Plants of North Carolina.* 1958.
191. Watt, G. *A Dictionary of the Economic Products of India*, v. 4: 555-6; 1890.
192. Hooker, J. D. *Flora of British India.* 1872-97.
193. Hartwell, J. L. et al. *J Am. Chem. Soc.* Aug. Sept. 1952.
194. See 156.
195. *Pharmacopoeia of the People's Republic of China* (China Ministry of Public Health) ed. 11; Peking; 1977.
196. Landa, L. Montoya, R. *Lista de maderas y plantas medicinales hondurianas* (Univ. Nac. Auton, Hond); 1977.
197. Barileskaia, V. *Plantes Méd. de Guinée.* 1969.
198. Rageau, J. *Plantes Méd. de la Nouvelle-Calédonie, Noumea.* 1957.
199. Singh, A. & Siwatibau, S. *Medicinal Plants in Fiji.*WHO; 1977.
200. Diaz, J. L. *Indice y Sinonimia de las plantas medicinales de México.* 1976.
201. Cahalane, V. H. "Biol. Survey of Katmai Natl. Monument," *Smithsonian Misc. Coll.* v. 138, #5, 1959.
202. Roti-Michelozzi, G. *Adumbratio Florae Aethiopicae*; Webbia 1954-8.
203. Rusby, H. H. *J A Ph A.* 13:101; 1924; Youngken, H. W. *Ibid.* 14:195; 1925.
204. *Papers . . Michigan Acad. Sci.* 47:160-2; 1961.
205. Twisselmann. "Flora of the Temblor Range," *Wasmann J. Biol.* 1956.
206. Eldridge, Jona. "Bush medicine in the Exumas & Long Island, Bahamas: A field study," *Econ. Bot.* 29:307-332; 1975.
207. Duke, James A. "Ethnobotanical observations on the Cuna Indians," *Econ. Bot.* 29:278-93; 1975.
208. Wong, W. "Some folk medicinal plants from Trinidad," *Econ. Bot.* 30:103-42; 1976.
209. Bandoni, A. L. *et al.* "Survey of Argentine medicinal plants: folklore & phytochemical screening, II" *Econ. Bot.* 30:161-85; 1976 (I: *Lloydia.* 35:69-80; 1972)
210. Mitsuhasi, H. "Medicinal plants of the Ainu," *Econ. Bot.* 30:209-17; 1976.
211. Johnson, E. J. & T. J. "Economic plants in a rural Nigerian market," *Econ. Bot.* 30:375-81; 1976.
212. See 129.
213. Uphof, J. C. T. *Dictionary of Economic Plants*, ed. 2. (J. Cramer, Lehre; 1968).
214. Morton, J. "Folk and medicinal plants of Central America," *Quart J. Crude Drug Res.* 15:165-92; 1977.
215. Waksman, Selma A. *The Actinomycetes*, 2 v. 1961 (Williams & Wilkins); W. believed these organisms belonged to the Bacteria not Fungi.
216. Moldenke, H. N. *Phytologia.* 6:262-352, 448; 1958.
217. *Idem.* 7:7-; 1959
218. *Idem.* 32:74; 1975.
219. *Mfg. Chem.* Aug. 1956:319-20.
220. Wirth, E. & Regl. *Amer. J. Bot.* 1937:68.
221. *Taxon.* 6:89-90 (Schultes); 1957.
222. *Atlanta* (Ga) *Constitution.* 19-XII-73. p. 12.
223. Coulter, J. M. *Contr. U.S. Nat. Herb.* 3:3907; 1895.
224. Kupchan, S. M. *J. Am. Chem. Soc.* 88:3074; 1966.
225. Perry, Lilly M. *Medicinal Plants of East & Southeast Asia* (MIT Press, Cambridge, Mass). 1980.
226. Salle, A. J. et al. *Arch. Biochem. & Biophysics* 32:21-3; 195 - Hiner, L. D. & Merrell, K. J. *J. A. Ph. A.* 31:209-12; 1942.

227. Diaz. *Indice & sinonimia de las plantas medicinales de México*. 1976.
228. Goethe, C. M. *Elfin Forest*. p. 22
229. *Contrib. U. S. Nat. Herb*. 10:34-126; 1906.
230. Gunther, Erna: *Ethnobotany of Western Washington*, ed. 2. 1973.
231. Berger, F. *Synonyma-Lexikon*; Wien:1954/5 - Pardes & Semsen. Washington, DC; 1938.
232. Ayensu, E. S. *Medicinal Plants of West Africa*. 1978.
233. Gildemeister, E. & Hoffman, Fr. *The Volatile Oils*, ed. 3; 909.
234. *Amer. Pharmacy*, N.S. 19(13):40; 1979.
235. See 117 (1).
236. Dalziel, J. M. *The Useful Plants of West Tropical Africa*. 1948.
237. Robbins, Harrington, Freire-Marreco. *Ethnobotany of the Tewa Indians*. 1916.
238. Ferreira. *Trees & Shrubs of South Africa*, 2. 1952.
239. Letter received from Prof. Davidson (U.B.C.) ca. 1930.
240. Knebels. *Apoth. Ztg*. 1892:112.
241. *Hagers Handbuch d. Pharm. Praxis*, ed. 4. 4:254-6; 1973.
242. *Merck Index IX*. 1976; product declared a fraud by FDA (1971).
243. Modarski, M. *Arch. Imm. Terap. Doswiad*. 4: 1956.
244. See 213.
245. *Rept. Govt. Chemist, Tanganyika*; 1955.
246. Palmer in Standley, *Trees & Shrubs of Mexico*. 1924.
247. See 232.
248. See 198.
249. Dallimore, W. & Jackson, A. B. *Handbook of Coniferae & Ginkgoaceae*. 1967.
250. Turner, Nancy J. *et al*. "Ethnobotany of the Okanagan-Colville Indians of BC & Wash", *BC Prov. Museum. 21*; 1980.
251. Pacheco, P. & al. *Quimica de las Plantas Chilenas Usados en Medicina Popular*, I. Univ. Concepcion; 1977.
252. *Pharmacopoeia of Eastern Medicine*, Karachi; 1969.
253. Istituto Italo Latino Americana (Roma): *Simposio Internazionale sulla Medicina Indigena e Popolare dell America Latina*; 1977.
254. Dragendorff, Georg. *Die Heilpflazen der verschiedenen Voelker u. Zeiten* .... 1898 (1967).
255. Kerharo, J. & Adam, J. G. *La Pharmacopée Sénégalaise Traditionnelle*; Vigot Frères, Paris; 1974.
256. Nakanisi, S. *J. Pharm. Soc. Japan*. 59:730; 1939.
257. See 129.
258. Hart, J. A. "Ethnobotany of....Western Mont.," *Bot. Mus. Leaflets*. 27:261-307; 1979.
259. *Time*. 12 Mar., 1956: p. 37.
260. Kroll. *Water over the Dam*; p. 228.
261. Rollins, R. C. & Shaw E. A. *The Genus* Lesquerella *(Cruciferae) in North America*; p. 279; 1973.
262. See 227.
263. Kupchan, S. M. et al. *J. Org. Chem*. 38:1853; 1973.
264. Ford, R. I. in Hand, W. D. *American Folk Medicine*. 143-7; 1976.
265. Train, P. *et al*. *Med. Uses of Plants by Indian tribes of Nevada*. 1957.
266. Chopra, R. N. *et al*. *Supplement to Glossary of Indian Medicinal Plants*. 1969.
267. Berman, p. 413.
268. Heyne, K. *De Nuttige Planten van Indonesië*; 1950.
269. Moldenke, H. N. *Phytologia*. 13:358 (1966); 39:39-100 (1975).
270. See 167.
271. Plunkett, E. R. *Endocrinology*. 49:1-7; 1951 - *L. ruderale* c. "lithospermic acid" with antigonadotropic activity.
272. *J. Arnold Arboretum*. 23:123; 1978.
273. *Deutsches Homöopathisches Arzneibuch*. 1934.
274. Morton, J. *Quart J. Crude Drug Res*. 15:165; 1977.
275. Persinos, I. *Planta Medica*. 15:361-; 1967.
276. Alami, R. et al. *Medical Plants in Kuwait*. 1974.
277. *New Eng. J. Med*. 262:263; 1960.
278. Vogel, V. J. See 111
279. Vaughan. *Strange Malady* (allergy).
280. Komarov, V. L. *Flora of the USSR* (-1964-).
281. See 130.
282. See 253 (Venezuela).
283. Lamba, S. S. *Econ. Bot*. 24(2): 1970.
284. See 242.
285. Dickerson, C. W. *J A Ph A S E*. 48:243-5; 1939. Stem -13 cm. high with 2-3 cordate lvs.; berries dull red, spotted (fall).
286. Schultes, R. E. *Bot. Museum Leaflets* (Harvard Univ.). 1953.
287. Parker, A. C. *36th Ann. Archeology Repts*. 1928:9-26 (Ontario).
288. See 227.
289. Schmid, R. *Naturw. Rundschau* 23(1):1970.
290. Pennington, C. W. *Tarhumar of Mexico*, p. 118; 1963.
291. See 254.
292. Oliver, B. *Medicinal Plants in Nigeria*. 1960.
293. See 199.
294. Kupchan, S. M. *Chem, Eng. News*. 3 June, 1974.
295. University of Saskatchewan (Saskatoon) research, Aug. 1968 (*Canad. Res. & Devt*. 1(4):31-2, 1968).
296. Howard. *Lloydia*. 6:142; 1943.

297. Porcher, F. P. *Resources of the Southern Fields & Forests*; 1863 (reprint 1970).
298. Miller, W. *Dictionary of English Names of Plants*. 1884.
299. British Pharmaceutical Codex; 1973.
300. See 198.
301. See 266.
302. *Calif. Geol. Survey, Bot.* 1:78; 1889
303. Darby, J. *Botany of the southern States*, II:355-7; 1856.
304. Bohlmann, F. & al. *Phytochemistry.* 18:1228; 1979.
305. Morton, Julia: *Atlas of Medicinal Plants of Middle America*. 1981.
306. Partington. *Ann Sci.* 11:1-26; 1955.
307. Boelcke, O. (1946).
308. Burkill, I. H. *Dictionary of the Economic Products of the Malay Peninsula*. 1935.
309. Moldenke, H. N. *Phytologia.* 17:47; 1965.
310. Moldenke, H. N. *Ibid.* 13:345; 1966.
311. Kuhn, O.; 1955.
312. Gunther. See 230.
313. Sargent. *Plantae Wilsonianae*, I:381-3; 1911-3.
314. Shenstone, 1955.
315. Koelz, W. N. *Quart J. Crude Drug Res.* 17:1056; 1979.
316. Dominguez, X. A. *Planta Med.* 15:401; 1967.
317. Lindley: *Bot. Reg.* 24, Misc. No. 141:1838.
318. Roig y Mesa, J. T. *Plantas Medicinales de Cuba*. 1945 - *J. Wash Acad.* 1919:435.
319. Stopp, K. "The medicinal plants ... in New Guinea," *Econ. Bot.* 17:16; 1962.
320. *The State Pharmacopoeia of the Union of Soviet Socialist Republics*, ed. 10; Moscow; 1968.
321. See 251.
322. Fournet. *Plantes Médicinales Congol.* 1972.
323. See 255.
324. See 140.
325. See 140.
326. Chin, T. Q. *Chinese J. Physiol.* 9:267-74; 1935.
327. Chamberlin, R. V. *Memoirs, Am. Anthropol. Assn.* 2:331-; 1911.
328. Lauritsen, L. P. *Flora og Fauna.* 1928:18-20.
329. See 129. Kittens & young cats dislike catnip but gradually become attracted to it.
330. Castañeda, M. (Mex) *Science.* 1943-4.
331. Liberali, C. H. Chem. study; Lab. Clinico, Rio de Janeiro; 1932.
332. See 159.
333. *Pharmacien de France.* 1983: 670-2.
334. *Siam Dept. Public Health.* 1922:63, 64; *Obs. Naucl. Ind.* 19; 1839 - Siamese: Kratom.
335. Schaller, R. *Nat. Hist. Bull., Siam Soc.* 14:56; 1947 (inf. drunk).
336. See 250.
337. *Lloydia.* 33:111; 1970.
338. Shelley, W. B. *Science.* 122:469; ca. 1955.
339. Stevenson, R. L. *Island Nights Entertainment.* 1892.
340. von Loesecke, H. W. *Bananas*, ed. 2. 1950.
341. See 198.
342. Sherry. *Int. J. Crude Drug Res.* (*Nepeta cataria*).
343. Rusby, H. H. *National Health.* 8:116; 1921.
344. *Ohio State Hist. Quart.* 46:81-102; 1937.
345. Anon. "El Nopal. Inst. Nac. Invest. Forest., México," *Publ. Esp.* no. 34; 1981.
346. *Ber. d. d. chem. Ges.* 51:1608. Removed from GRAS list. 1968: banned in animal fats; 1971 (developed cysts & kidney damage in rats).
347. See 130.
348. *Amer. J. Pharm.* 1883:631.
349. Bushnell, D. I. "Choctaw....," *Bur. Am. Ethn. Bull.* 48; 1909.
350. "The Galax Gatherers" (painting) by Edward D. Guerrant; 1910.
351. Tanret, G. *Compt. Rend.* (Par) 194:914-6; 1099-1101; 1932 (neriin gives nerigenin + glucose).
352. de Toledo. "São Paulo," 1950; Lowman & Kelly - *J A Ph A.* 1945.
353. Fairchild, H. G. *The World Was My Window*, p. 271; 1938.
354. (*Olea*). Rikli, M. *Lebensbedingungen u. Vegetationsverhaeltnisse der Mittel Meerlaender und atlantischen Inseln* (map): 88; 1912.
355. Genesis 8:11.
356. Herout. *Plant Med.* Suppl. 16:90; 1968.
357. Moldenke, H. N. *Phytologia.* 12:25; 1965.
358. Stern, F. C. *Paeonia, a Study of the Genus* (Quaritch, London). 1946.
359. Schneider, Albert, 1912, p. 128.
360. Roys, R. L. *Ethnobotany of the Maya.* 1931.
361. Martinez, M. *Plantas Medicinales de México*, ed. 2. 1939.
362. Soine, T. O. *J A Ph A S E.* June, 1956.
363. See 167.
364. Benson, L. *Cacti of Arizona.* Univ. Ariz. Press, Tucson; 1969.
365. Legrand, R. & J. *Thérapie* (Paris). 6:103-7; 1951.
366. *Chem & Drugg.* 138:395; 17 Oct. 1942.
367. Butenandt, A. & Martin. *Ann.* 495; 198-210; 1932.
368. Burton, H. H., Bishop, Cal. from A. Schneider, 1912 (ref. 59).
369. *J. Econ. Entomol.* 36:937-8; 1943.
370. Roempp, H. *Chemie-Lexikon*, ed. 6. 1966.
371. *U.S. Dept. Agr.—F. B.* 990 (1918; 1923).
372. *Pharm. Ztg.* 1884:59 (Maisch).
373. *Natural History.* 68:269-75; 1959.
374. per R. M. Harper (Ala.) (conversation).
375. Lehmann, B. L. *Deutsch. Ap.-Ztg.* 111 (No. 40): 1971.

376. *JAMA*. 1912:1607.
377. Iwakawa. *Arch. Expt. Path. Pharmakol.* 67:119; 1912.
378. Harris, A. S. + Ruth. R. H. *Bibliog. US Dept. Agr. For. Serv. Res. Papers PNW*, 105; 1970.
379. Rousseau, J. *Mémoires du Jardin Botanique de Montréal*, 2: Partie, 19; 946.
380. Roys, R. L. *Ethnobotany of the Maya*. 1931 (1976).
381. See 274.
382. Verbal communication; Alexander City, Ala.
383. Ford, Tennessee Ernie: "This is my story, this is my song...," 1963.
384. Clark, T. D. *Pills, Petticoats, & Plows — The Southern Country Store*. 1944.
385. Chandler, R. F. *et al.* (Maritime Indians). *J. Ethnopharm.* 1:49-68; 1979.
386. See 379.
387. Hepburn, J. S. *Amer. J. Pharm.* 96:804-9; 1924.
388. Munro: *The Middle Ages*, ed. 2; 307 - *Z.f.u.d Lebensmittel.* 85:426-36; 1943 (substitutes).
389. Degener, O. *Ferns & Flowering Plants of Hawaii National Park*. 1930.
390. *Romeo & Juliet* (Shakespeare). I, ii: 52-4 (Romeo tells Benvolio plantain lf. is excellent for "broken shin").
391. Clevenger, R. E. *Drug Markets* 29:236; Sept. 1931 - Stephenson, White (USDC). *Ibid.* 29:455; Nov. 1931 (trade) - Fuchs L. ca. 76; 3069; 1932 (histology).
392. Little, John & Johnstone, D. B. (Agr. Expt. Sta., Burlington, Vt), Amer. Chem. Soc. 1950.
393. Environment S. W., Aug. 1965, p. 4.
394. Pickering, D. S. Somerset, PA. ca. 1944.
395. Paschkis, Heinrich. *Folia Patchouli des Handels* (chemistry) - *Zts. f. allgem. Oesterr. Ap. Verein* 17:415-20; 1778.
396. Krochmal, Arnold. Possibly another Honduran *Dryopteris* sp. (1967).
397. Faull, Joesph Horace. *Trans. Royal Canadian Inst.* 11:185-209; 1916-7 - Lowe, J. L. *State U. Coll. For. Syracuse U. Tech. Pub.* 80 (1957).
398. Per R. M. Harper, Tuscaloosa, Ala (personal communication)
399. Scott, E. & Maurice, James. *The White Poppy*. 1969.
400. Cantoria, Magdalena: *Quart. J. Crude Drug Res.* 14:97-; 1976.
401. *Bol. Acad. Nac. Cienc. Cordoba* 19:379-. 1913.
402. *Oesterreich. Aertze-Ztg*. 3(30):13; 1933.
403. Dorsey, J. L. *Ann. Int. Med.* 10:628-31; 1936.
404. Liddell & Scott's *Greek-English Lexicon* (abridged) (XVIII; 1880).
405. Warren, L. E. *J A Ph A.* 22:294-7; 1933.
406. Molfino, J. F. *Campo y Suelo Arg.* 32:45; 1948. Hb. c. menthone, pulegone, limonene.
407. Vestal, P. A. *Papers Peabody Mus. 40; 1952.*
408. Wyman, L. C. *Univ. New Mex. Bull.* 366; 1941.
409. Dalziel. See 236.
410. See 111.
411. Hartwell, J. W. *Lloydia*. 31:71; 1968.
412. La Llave & Lex. - Nov. Veg. Desc. fasc. i: 11.
413. See 209.
414. *Ohio Sta. Hist. Quarterly* (Columbus) 45:265-72; 1936.
415. Miller, E. R. "Studies of W. D. Salmon," *J. Biol. Chem* 44:481-; 1920.
416. See 231.
417. Mr. Parry, Merck. Montréal.
418. Rellan, D. R. *J. Assn. Physicians of India* 14:629; 1966/Weil, Andrew T. *Econ. Bot.* 1965 (unpleasant side effects using grd. prod).
419. Breton, J. F. M. Univ. Bordeaux theses, 1925-6, 2.
420. Chikamatsu, H. *Chem. Abst.* 59:7571; 1963.
421. Rusby, Henry Hurd. *Drugg. Bull.* 4:212-6; 1890.
422. Lima, D. P. *Rev. Reunidas.* 18:18 (Aug. 1943) (ind.).
423. *Chopra's Indigenous Drugs of India*, ed. 2. 1956.
424. "Smoking & Health" Rept. (US) Surgeon General (USD HEW); 1979 - Wynder, E. L. et al. *Penna. Med. J.* 57:1073; 1954 (cause of lung cancer). Chief hazards of cig. smoking: pulmonary cancer & other cancers; bronchial dis. (as emphysema), cardiovasc. dis.; gastric ulcer.
425. "Nicotinic acid cures pellagra but a beefsteak prevents it" (Henry E. Sigerist).
426. Lloyd Brothers (Cincinnati, OH). *Bull.*
427. Isaiah 41:14.
428. See 145.
429. *Therapeutic Problems of Today*, 2:13-33; 1943 (TEVA, Middle East Pharm. & Chem. Works Co. Ltd.).
430. Guenther, Ernst. *Amer. Perf. E. O. R.* 45 (11):41-3; 1943.
431. Trelease, W. *Trans. St. Louis Acad. Sci.* 7:493; ca. 1897.
432. (*Opuntia spp.*). *Gardeners Chronicle* 1885. *O. cochin.* & other spp. produce a dermatitis-like scabies at times.
433. Fairchild, David: *The World Grows Round My Door*; pp. 286-8; 1947.
434. Basu, N.K. *J A Ph A S E.* Sept. 1956.
435. *Nature Mag.* Jan. 1935:33-4 ("save the souvenirs").
436. See 179.
437. Plantae Colombianae, IV. - Univ. Nac. Bogota (Col), *Facult. de Agron. Rev.* 5:59-79; 1943.
438. Branigan, G. V. (T.D. Ungerer & Co.) *Drug Cosm. Ind.* May, 1944.
439. *Nature.* 171:343; 1953/*Lancet.* 1942, I:732-3, 534-6, 600.

440. Koch, R. A. & Haines & Hollingsworth. *JAMA*. 129:491; 1945/Jacoby & Ollswang, A. H. *Amer. J. Syph*. 34:60-1; 1950.
441. See 230.
442. Yamovsky, E. *Food Plants of the NA Indians*; 1936.
443. *Prevention*. Apr. 1974:189.
444. *Lynn Index* 6:264; 1969.
445. Jardin, C. *Agents....contre la lèpre*. 1962
445A. Schultes, R. E. "De plantis toxicariis e mundo novo tropicale commentationes XXIII," *Bot. Mus. Leafl.* 26:191; 1978.
446. Silverstone. "N. H. Museum Los Angeles, Calif," *Sci. Bull.* 77; 1976.
447. See 250.
448. Murrill, Wm. A. *Amer. Bot.* 36:1-3; 1930 - Maisch. *Amer. J. Pharm.* 53:81; 1881 ("pinckneyin" NBP) - Naudin. *Ibid.* 57:161; 1885.
449. Sanders, Bob. Radio station WAUD, Auburn, Ala.
450. Perry, L. M. See 225. The best celery said from Ravenna Italy (Rabelais. *Work*, Book 4, Ch. 7 [1592]).
451. *Naval Stores Rev.* Jan. 1967 (Lettine & Doorenbos) (*P. taeda*). This fast growing tree is of lumber size in 25-30 yrs., furnishing the chief commercial lumber of the seUS.
452. *Lloydia.* 33:355-60; 1970.
453. Per Carl G. Hartman, Ortho Research Foundation.
454. Nera. *J A Ph A S E.* 38:34; 1949.
455. Cooke, M.C. *Pharm. J.* 1(3):56-; 1870.
456. *Platanus hybrida* is about the most common tree in Paris; ex. along embankments, boulevards, park enclosures, &c. (GMH).
457. Sorgdrager.
458. Liogier, A. *Flora de Cuba*, IV: 123; 1963 (1974) - *Diccionario Botanico de Nombres Vulgares de la Española*, p. 63; 1974.
459. Friese, F. W. *Boletin de Agricultura.* 34:257; 1933.
460. "Poke Salad Annie" (song by Elvis Presley) ("poke salad in a poke sack").
461. Hoerhammer, L.; 1955.
462. Wohl & Goodhart. *Modern Nutrition.*
463. Caswell - Massey, NYC; 1970 - < pomum de ambra (med. <) = apple or ball of amber.
464. *Populus: a bibliography* (world literature, 1964-74); South For. Exp. Sta. 1976.
465. Goldman. *US Nat. Herb.* v. 16 (3 [#14]);1916.
466. Hampson, M. A. "The laverbread industry in south Wales & the laverweed," *Fisheries Investig.* Ser. II, v. XXI, #7; 8 pp.; 1957.
467. Andrews, Roy C. *Ends of the Earth.* 70-3; 1929.
468. Standley, P. C. *Field Mus.* 18:429; 1937.
469. Stitt, E. R. *Mil. Surgeon,* June, 1940.
470. Lopez-Palacios, *S. Flora de Venezuela*. Verbenaceas; 1977.
471. See 361.
472. Mayer, T. *Lilloa.* 1938.
473. Williams, L. *Exploraciones Botan. en la Guayana Venezolana.* 1942.
474. Wasicky, R. *Pharm. Post.* 1913.
475. Nichols, P. F. & Reed, H. M. *Hilgardia.* 6 (#16): (Apr. 1932) - "Nutritive value of California prunes" (pamphlet).
476. Reported from Jasper, Ala (US). Chilton Co. famous for its peaches: "...a promise of peach hickories" (Hurston, Z. N. *Dust Tracks in a Road.* 1942).
477. Harvey, A. G. *Douglas of the Fir*; Harvard Univ. Press, Cambridge, Mass.; 1947.
478. See 472.
479. McKenny, M. *The Savory Wild Mushroom*, ed. 2. 1971.
480. See 199.
481. Ford, K. C. *Las Yerbas de la Gente: A Study of Hispanic American Medicinal Plants.* 1975.
482. Carvalho, J. C. M. *Ceres.* 5:166; 1944.
483. Schultes, R. E. & Hofmann, A. *Botany & Chemistry of Hallucinogens*, ed. 2. 1980.
484. See 142.
485. See 130.
486. Ernest Thompson Seton.
487. A treatise on Pulsatilla. *Lloyd Bros. Drug Treatises* #28; 1913.
488. Siddappi, G. S. *Indian Farming.* 4:196-8; 1943.
489. Grayson, D. *Adventures of David Grayson*: "The Friendly Road" (Chap. 5). 1913 (Ray S. Baker).
490. Moldenke, H. N. *Lilloa.* 10:365; 1944.
491. Bur. Amer. Ethn. 136:189-; 1943. Rts. found of no value in diabetes (Squibb Inst. Med. Res.); long black spines thought poisonous (Yuc).
492. *Calif. Geol. Surv. Bot.* 3:342. Sirop de capillaire well-known as cough medicine (Calif before 1912) (A. Schneider).
493. *Contrib. US Nat. Herb.* 23:140; 1920.
494. Sigal, A. E. *Klin. Med. S.S.S.R.* 33:24-8 (No. 11); 1955 (Russ).
495. Ogden, P. M. "Tung Oil: Florida's Infant Indy.," *Econ. Geog.* 5:348.
496. Haddad, D. Y. *J. Egypt. Med. Assn.* 20:117-24; 1937.
497. *Malayan Agr. J.* 17:47, 268; 1929.
498. Blaine, G. *Nursing Times.* 47:584-6; 1951 (in surgical practice).
499. See 130.
500. Miller, E. R. "Chemical investigations of some spp. of genus *Pycnanthemum*" (Ph.D. diss. Univ. Minn.); 1917 - *Univ. Calif Publ. in Bot.* v. 20, #3 (systematics); 1960(?).
501. Correll, W. G. *Pharm. Rev.* 14:32; 1896 - Alden. *Ibid.* 16:414; 1898 - Miller, E. R. *Pharm. Expt. Sta., Univ. Wisc., Circ.* 2.

502. *Merck Index IX*; 1976.
503. Buckley+. *J A Ph A.* 188:#4; 1955.
504. Oliveri & Denaro. *Am. Drugg*. July 1935:104.
505. Mohr. *Pharm. Rundschau.* 1883:136
506. Kern, E. E. *Sovetskiye Subtropiki.* 1931 (#1) (8), p. 68 - Watrow, R. C. & Barnes, H. V. *Bibliographic Bull. USDA,* Bul. 17.
506A. Ryan, Victor H. (Res. Dir.): "The cork oak: some geographic & economic aspects," Crown Cork & Seal Co. Inc.; 1948 - Kern. E. E. Trudy po prikladnoi botanike, genetiki, i selektsii 20: 359-412; ca. 1941 (cork o. in the Crimea & Caucasus).
507. Roig, J. T. *Plantas Medicinales de Cuba.* 1945.
508. Shampoos: B. Pat 545, 405 (ca. 1942).
509. *U.S. Publ. Health Rept.* v. 62 (#1) (1.3.47).
510. *Crown Colonist.* 14 (150):304; 1944.
511. *Pharm. J.* 5/19/45: p. 253.
512. *Hagers Handbuch der Pharmazeutischen Praxis* ..., ed. 4; 1967-1980.
513. Kipling, Rudyard. *Eye of Allah.*
514. Pomini, L. *Erbistoria Italiana*; ca. 1960 (A. Goris).
515. Schuler. 1956 (Bantu) - Rindl & Groenewoud. Trans. R. Soc S.A.F. 1932.
516. Canescine - Ciba (ca. 1955).
517. Davis, E. W. & Yost, J. A. *Bot. Museum Leaflets* (Harvard Univ.). 29:159-; 1983.
518. *Griffith's Dispensatory.* 1848.
519. Greshoff, M. Distribution of HCN in vegetable kingdom; Phytochemical investigations at Kew.
520. *Hagers Handbuch d. Pharm. Praxis*, ed. 6, vol. VIB:94; 1978.
521. Varga *et al. Herba Hungarica.* 1982:139 - Ozarowski, A. *Farm Polska.* 30:557; 1974.
522. Chung, T. H. "Economic plants wild-grown in Korea," *10th Pac. Sci. Congress*; 1961.
523. See 292.
524. Survey. *Herba Polonica.* 29:65-81; 1983.
525. Lavietes, David B. (Brooklyn, NY). US Pat. 2,367,360; 1945 (seasoning for 7 weeks). "Burrs" (bases) sometime so large that 2 men were required to haul them.
526. *Rhodora:* name of a US journal of botany (*cf.* "The Rhodora," poem by R. W. Emerson).
527. See 274.
528. Soba, J. G., Sanchez, E. E., Damrau, F. (ca. 1960).
529. Shen Nung Pent'sao King (Shen Nung period, ca. 3200-2737, B.C.).
530. Procter, George. Inst. Jamaica, Kingston, Jamaica.
531. "Hast du mal im Darm Verdruss" (trouble) "Nimm ein Loeffel Rizinus" (German apothecaries). The first record (in modern times) of the use of castor oil was in 1764 in St. Kitts (WI), when it was reported (Peter Canvane) being u. by the Am. Indians for VD. It was also u. for bilious complaints. The word "castor" was first used in French WI for *Vitex agnus-castus*, later for *Ricinus c.*
532. *The Garden.* 56:315; 1899 - Daugherty. *USDA YBk.* 1904:287 - *Mo. Bot. Gard Bull.* 17; 130-1; 1929.
533. *cf.* El Paso de los Robles (=Paso Robles, Calif).
534. Goldman, E. A. *Contr. U.S. Natl. Herb.* 16:312; 1916.
535. Anon. *Perf. Ess. Oil Rec.* 11:210-; 1930 (*CA* 14:2675).
536. Term used in Canada; same as "Dr. Jackson meal" (US) (named after Robert G. Jackson, M.D.).
537. *U.S. Dept. Agr. Bull.* 706; 14 pp., 1944.
538. *Brazilian Bull. (NYC) v. 3; Jan. 1, 1946.*
539. Herder, J. G. *Deutsche Volkslieder.*
540. Leclerc, H. *J. Prat.* (Paris). 65:293-6; 1951 - Burkhart, Harriet. *Nature Mag.,* Mar. 1959: 149-51 - Vernazza, N. & Nadali, P. *Res. Inst. for Investigation of Mea Plank (Belgrade).* 1:1-9; 1981 (VO)
541. WVa Pulp & Paper Co.; mfr.
542. 1) Pharmacodynamic study of seeds (*CA* 1940) 2) American Dyewoods, Inc.
543. *Look* (magazine). 19-X-54 issue.
544. Anon. "A guide to the literature on rubber," U.S. Dept. Agr. 1931.
545. Davis, H. A. & al. *Castanea.* 32:20-37 (1967); 33:50-76, 206-41 (1968); 34:154-79; 235-56, 276-94 (1970+).
546. Franzer, Hartweg, & C. Stern. *Z. physiol. Chem.* 121:195-220; 1922.
547. Welsh. *Bull. Miss. Hist. Soc.* 4:350 (Pioneer) (ft. white with bluish tinge).
548. Luther Burbank, Calif.
549. Sneed, C. S. "Med. properties," *South. Med. & Surg.* July (Augusta, GA) n.s. 11:723-81; 1955.
550. Barkley, F. A. "*Ruellia* in Texas," *Amer. Midl. Nature.* 42:1-8; 1949.
551. Morse, Willard H. "Bay rum, its derivation, botany, chemistry, & adulteration," *Pharm. J.* 118:105-6; 1927.
552. Purdie, I. A. *Pharm. J.* 118:105-6; 1927.
553. *World Who's Who in Science*, ed. 1; Marquis Co.; 1968 (p. 1458).
554. Baghlaf, A. U. & al. *Herba Hungar.* 22:39-; 1983.
555. Rutin & related flavonoids; Mack Publishing Co.; ca. 1965 ($7.50); USDA publ. 3/8/43.
556. Artschwager, E. *US Dept. Agr. Handbook.* 122; 1958. The story of sugar (Am Sugar Co.) - *Cane Sugar Handbook.* Molasses was once being used to surface roads (Ind).
557. Goldman. *Contr. U.S. Nat. Herb.* 18:320; 1916.

558. Best farm land in India (2 X 10^6 acres) being ruined by this grass with its many deep roots; smothers out other pls. (Camp, J. L. *Far Eastern Advertiser*. 51).
559. *Rept. Provincial Mus. B.C.* 1942:HH29.
560. *27th Ann. Rept. Bur. Amer. Ethn.* 1905-6 (Omaha tribe).
561. Liberali, C. H. *Lab. Clinico* (Rio de Janeiro) 1935 (comp. treatment values).
562. *Wealth of India, Raw Materials*, v. 9:175; 1973.
563. Orekhoff, A. & Proskurnina, N. *Bull. Acad. Sci. USSR*. 6:960-; 1936.
564. *Chem. Ztg*. 20:#49; 1896.
565. *Khim. Prir. Soedin*. 1979:222.
566. Kritikos, P. G. & Steinegger, E. *Pharm. Acta Helv.* 26:48; 1951.
567. Ayensu, E. S. *Medicinal Plants of the West Indies*. 1981.
568. Farwell, O. A. *Amer. J. Pharm*. 87:970-8; 1915.
569. Mohs, Frederick E. (MD). *Roche Med. Image & Commentary*, Mar. 1970: p. 7
570. Blome, W. H. *Pr A Ph A*. 51:284-6; 1893.
571. See 152.
572. See 167.
573. Scheindlin & Dodge. *Am. J. Pharm*. 119:1947.
574. Hockauf. *Osterr. Ztsch*. 41:82; c. 1902 (*Pr A Ph A*. 51:765).
575. Lemordant, D. & al. "Plantes utiles et toxiques de Tunisie," *Fitoterapia*. 1977:191-214.
576. Harrison, Tom. *Savage Civilization*. Knopf; 1937 (much re *Santalum album* trade, esp. pp. 133-47). Sandalwood said most valuable of all woods.
577. O'Connor, P. J. *Ph. J.*, pp. 118-59; 1927.
578. Chopra, R. N. & Chandler. *Ind. Med. Gaz*. 59, #11; 1924.
579. Beal, Jane H. Comml. devt. in WW I (ca. 1914-8); later prod. more cheaply in Germany.
580. Dermer, O. C. & Crews, L. T. *J. Amer. Chem. Soc*. 61:2697-8; 1937.
581. Radlkofer, L.A.T. *Das Pflanzenreich*.
582. *Rev. Horticole*. 67:196; 1895.
583. Marketed by The Chex Co. (div. Day & Frick, Philadelphia, Pa. [US]) formerly (High Chem. Co., Phila. - 1965-81).
584. Train, P. & al. *Medicinal Uses of Plants by Indian Tribes of Nevada*. 1952.
585. *cf.* "The Ancient Mariner"; Columbus was said encouraged to continue his westward voyage on observing *Sargassum*
586. Umezawa; 1954 - (Bristol).
587. It is legendary that a plant grows where an Indian dies.
588. In warm weather, so many putrefying insects occur that "I have known their putrescence to render the swamp where the plant abounds highly offensive" (Torrey, *Flora of ... New York*. 1943, p. 42). However others have claimed an attractive odor (adhesive substance).
589. (1659-1736). Born in France, practiced med. in Canada, Physician to the King; studied natural history & medicine; wrote "Traité sur le pleurésie" (unpubl.; only in MS).
590. Hepburn. *Amer. J. Pharm*. 1928 - Anesthetic Div., NYU Coll. Med.; Merck - Bates, W. *et. al.* (Grad. Hosp., Philadelphia). Blanden (claimed Indians u. as anesth.).
591 Word list for Georgia. *Dialect Notes*. 5 (ca. 1918). Many variants.
592. Alc. ext. of sassafras rt. bk. diluted with water to ppt. safrole; this can be used safely as flavor.
593. Fisher, John. *Old Virginia*.
594. Gray, A. *Bot. U.S. Exp*. (1838/42) i. 206, p.15; 1854 - Engler & Prantl IV. III. A300; c. 1912.
595. Klotzsch: *Linnaea*. 24:22; 1951.
596. 1881:133, 139 bbl. of spirits of turpentine; 564,026 bbls. of rosin; aggregate value ca. \$4 x 10^6; second only to cotton in value (Pres. Bd. of Trade; 1883).
597. Lindsay, F. G. *Proc. Calif. Pharm. Soc*. 1884.
598. Shass, E. Iu. *Feldsher i Akusherka*. #4:56-8; 1951.
599. *Agric. Ledger*. 1905: #1.
600. Fairchild, D. *The World Was My Garden*; 1956.
601. Said "the most characteristic" flower of the Alberta plains.
602. (*Scoparia dulcis*). New alkaloid. Vaquero, Jose M. Rodrigues (Tucuman, Arg). *Baillon's Hist. d. Plantes*. 9:398; 1867-95 - *Bull. Herb. Boissier* II.:8, 4. In Curaçao, S.d.; not common & is very similar to a common plant called "Willing Sally."
603. *Drug News Weekly*. 25-IV-66; p. 14 - *Lancet*. 1944, I:127, 804 (air sickness).
604. Kraemer, Henry. *Amer. J. Pharm*. 80:459-464; 1908 - *Bot. Mag. 28:ii26*.
605. Afanasiev, C. S. *C.R. (Doklady) Acad. Sci. URSS*. 18:479-82; 1938.
606. Hedberg, I. & al. *J. Ethnopharm*. 9:237-260; 1983.
607. *Pharmacien de France*. 1972:871.
608. *Presse Méd*. 59:859; 1951 (Scopos) - Luboldt, Walter. *Diss. Marburg*. 1895:58 pp. (just 1985, II) (scopolamine & its cleavage prods.).
609. Machado, Adelino. *Lab. Clinico Silva Araujo*. No. 30 (Rio de Janeiro); ca. 1940.
610. *Farming in South Africa*. 16:69-70; 1941.
611. Zazilbash, N. A. *Pharm. J*. 163:108-9; 1949. Rts. & underground parts of *Polygala* (esp. fresh) have methyl salicylate odor.
612. The word *"Sequoia"* contains all vowels.
613. Bibliog. Univ. Calif. (Berkley). "*Sequoia gigantea* (big trees of California)," 474 pp. (typed) ms (1937) (WPA 165-3-6305). Anno-

tated; incl. books, repts., pamphlets, serials, newspapers.
614. Manufactured by Pierre Fabres (France).
615. *Pharm. Archives.* 1899:101-16.
616. Tocher, J. F. isolated & named (*Berichte*; 1893).
617. Natl. Acad. Science (US). *Tropical Legumes; Resources for the Future.* 1979.
618. Ellis, M. D. *Dangerous Plants, Snakes, Arthropods, & Marine Life.* 1975.
619. Watt. *Dict. of Econ. Prod. of India* VI, Part 2: p. 545, ca. 1890 - *Wealth of India, Raw Materials.* IX:302; 1972.
619A. Brandel, Robt. & Dester, H. (Cincinnati Coll. Med.).
620. Schramm, G. *Naturw. Rdsch.* 1958:64-5 (Int'l Pharmaz. Gesellschaft Kongress; 1957).
621. Deuteronomy 10:3.
622. Exodus 25:10.
623. In a recent year, catch of 214 million lbs. worth $73 million (US).
624. Akana, A. *Hawaiian Herbs of Medical Value*; Tuttle; 1972.
625. Instituto Mesbe (Berlin); Dieseldorff, E. P. "Las plantas medicinales del departmento de Alta Verapaz; Guatemala, CA," ref. in *Morton's Atlas.*
626. *Pharm. Ztg.* thru *Pharm. Record.* 8:14; 1888.
627. *Pharm. Monatsh. Beibl. Pharm. Post.* 10:5-6; 1938.
628. *Medicinal & Aromatic Plants Abstracts.* 5:501-2; 1982.
629. *Ibid.* 5:273-282; 1983.
630. *Ibid.* 5:167-175; 1983.
631. *Ibid.* 5:85-91; 1983.
632. *Ibid.* 4:435-40; 1982.
633. *Ibid.* 3:75-9; 1981.
634. *Ibid.* 4:171-3; 1982.
635. *Ibid.* 4:351-9; 1982. Also Chopra. *Indig. Drugs of Ind.*, p. 376-8; 1958. Occ. Himal to Travencore; much u. in India meds. (Caius. J. F. *Ind. Med. Res. Mem.* #6; 1927).
636. *Ibid.* 4:257-264; 1982.
637. *Ibid.* 4:83-94; 1982.
638. *Silphium* was described for Montgomery or Lowndes Co. (Alabama) by Wm. Bartram but not recently found there by R. M Harper (Ala prairies).
639. Perm region, USSR.
640. Hartwell. *Lloydia.* 31:71; 1968.
641. Engler Prantl. *Die natuerlichen Pflanzenfamilien* IV. 2. 215; ca. 1892.
642. Maranhao. *Campo de Bra.* Esperanca, Brazil.
643. *Bull Torrey Bot. Club.* 71:226-234; 1944. - *Ap. Ztg.* 49:144-7; 1934.
644. *Offiz. Oesterr. Bericht, Weltanstellung* V - 1867.
645. Per Soviet botanists; Knoepfler, N. B. & Vix, HLE. *J. Agr. Food Chem.* 6:118; 1958 - Duisberg. *J. Animal Sci.* 1952.
646. *Science.* 1952. *Sison amomum* (frt) occurs in Fr. Ph. I (1818) and Dan. Ph. I (1772).
647. *Sium* (cium) (in Shakespeare).
648. Hall, James. *Legends of the West.* 1833.
649. Coker, W. C. "The Woody Smilaxes of the US," *J. Elisha Mitchell Sci. Soc.* 60:#1; 1944
650. Vogel, V. J. *American Indian Medicine*, p. 54, F. N. 48.
651. *Folia Pharm.* (Istanbul) II (12):1953 (15 pp.).
652. Much collected around Evergreen, Ala (US), since ca. 1890 (R. M. Harper), *Birm. News*, 13 Nov. 1963.
653. *J. Ethnopharmacology.* 10:176; 1984.
654. Briggs, Lindsay H. *J. Amer. Chem. Soc.* 59: 2467-8; 1937.
655. (*Solanum car.*). Reported from Marshall County, in Ala/Thrush., M.C. *Amer. J. Pharm.* 69:84-9; 1897. When mowed plants have very unusual odor.
656. Kleiner, Israel S. *Science.* 79:273; 1934.
657. McClosky & Eisenberg (US FDA). *JAOAC.* 27:301-9; 1944
658. Webb, L .J. *Guide to the Medicinal and Poisonous Plants of Queensland.* 1948.
659. Smith, H. M. See entry in P. Investigated also by F. N. Carr (London). Frt. exposed for sale in markets; probably other spp. *Solanum coagulans, S. melongena, S. trilobatum./Sci.* 23 Dec. 1927.
660. Sarmiento; 1958.
661. Dana, C. A. Two years before the mast (N.Y.); 1840.
662. Burton, W. G. *The Potato.* 1948. Research at ERRL (Pa) (USDA).
663. Potato farina (flour) being u. in Cornwall (County) (England) in 1794 (Fraser, R. General views of the County of Cornwall; London) (S.E.) 1794 (advantage: could be kept an indefinite period).
664. Dymock, W. *Pharmacographia Indica*; 3 vols. & appendix; 1880-3 (reprinted by Hamdard, Karachi, Pakistan); 1972.
665. See 142.
666. Wehmer, C. *Die Pflanzenstoffe*, ed. 1. 1911.
667. Halgrim. *Sunrise.* Dec. 1932: 79 (photo).
668. Seton, E. T. "Trails of an Artist-Naturalist" (poem in praise of flowers).
669. Meehan, Thomas. *Am. J. Pharm.* 51:377; 1879 (*PrA Ph A.* 28:40) (Berks Co., Pa [US]).
670. *Solanum nigrum.* Young shoots c. glutinous subst. ("chansillo") which chewed gave nerve sedation & sleep.
671. San Francisco & Portland, Ore. *Pac. Pharm.* 2:113; 1908.
672. Sole constituent of Vogelbeeroel obtained by steam distn. of acidified juice of ripe frt. of SA.

673. Rhys Hibbert Lectures. *Sorbus* believed magic; ex. in "The laidley worm of Spindleston Heugh" in *English Fairy Tales* (Joseph Jacobs).
674. Georgeson. *Kaffir Corn* (book).
675. Hartley. *Broom Corn* - Anon. *Broom Corn & Brooms*; Orange Judd Co., 1912. Now crop produced in It., Mor.
676. *Garden. Mo.* 1882: 59:122-3. (Ark [US]).
677. Instituto Interamericano de Ciencias Agricolas. *Doc. & Inf. Agr.* #54; 1977 (Bibliography with annotations) - Matagrin, A. *Le Soja & les Industries du Soja* ca. 1939. Other products of soy include phytic acid & Ca phytate; inositol; tyrosine; high value protein (HVP); and fermentation nutrients.
678. Davis & Yost. *Bot. Mus. Leaflets.* 29:159-; 1983.
679. Amico, A. "Med. plants of s. Zambesia (Mozambique)," *Fitoterapia.* 48:101-139; 1977.
680. Johnson, S. W. *Peat & Its Uses as Fertilizer & Fuel*; NYC. 168 pp; 1866. Wilson, B.D., & al - Cornell U. *AES Memoir.* 88; 1936 (genesis & compr.).
681. The story of wooden nutmegs of Conn (US) is only a legend. It is not known who originated it. Immortalized by T. C. Haliburton, *The Clockmaker* ("Sam Slick"). It may have originated from the sale of wormy nutmegs as worthless as the wooden (Fuller: *Tinkers & Genius*; p. 64; 1955).
682. Eigsti, O. J. & Tenney, B. *Colchicine*. Univ. Okla. Press; 1942 - Eigsti. O. J. *Colchicine in Agriculture....* 470 pp.; 1955.
683. Wagner, Philip. *Beitraege zur Kenntnis der neueren Droguen - Plumbago ceylanica, Capraria biflora, Spilanthes acmella.* 1897.
684. *Pharm J.* 2:(3)512-; 1972.
685. Knueckel, F. *Med. Welt.* 1960:1502.
686. Re spines: Rose, S. C. *Proc. Fla. State Hort. Soc.* 90:122-3; 1978. Lists 31 genera or families (ex. Palmae - petioles).
687. Schindler, H. & Frank, H. *Tiere in Pharmazie u. Medizin.* 1961.
688. *Fishery News International*, June, 1969, p. 84.
689. *Flore de Madagascar Fam.* 174; 1956.
690. Moldenke, H. N. *Phytologia.* 2:369; 1947.
691. Exodus 30:34.
692. *Chin. Med. J.* 54:151-8; 1938 - *Bull. Econ. Indochine.* 40:539-42; 1937 - *J A Ph A.* 29:391+; 1940.
693. Hocking, G. M. "Stephania," *Acta Phytotherapeutica.* 3, #6:6-8; 1956.
694. Kariyone, T. & Kimura, Y. *Japanese-Chinese Medicinal Plants*, ed. 2. 1949 (through Perry, *Med. Plants*).
695. Klecker, Ernst. *Arch. Exp. Path. Pharmakol.* 161:596; ca. 1930.
696. Boakye-Yeadom, K. *Quart. J. Crude Drug Res.* 15:201-2; 1977.
697. Soejarto, D. D. & al. *Bot. Mus. Leafl.* 29:1-25; 1983.
698. Fernholz, E. *Ann.* 507:123-38; 1933 (*C. A.* 28:168).
699. Holm, Theodor. *Merck's Rep.* 29:36-8; 1911 (*Med. Plants of NA* #49).
700. 400 sheep lost from eating *Stillingia sylvatica;* said very poisonous also to goats (L. A. Scott, Shamrock, Fla [US]; 1930); also Ocala, Fla.
701. E. T. Seton. In *Trail of an Artist-Naturalist*, pp. 226-8 discussed method of action; see more complete discussion in *Knowledge* 1-IX-1888 (Miller Cansty).
702. Wallis, T. E. *Analytical Microscopy*, ed. 3:180; 1965.
703. Visser, H. C. *Nederl. Tijd. v. Pharm.* 8:204-8; 1896.
704. *J. Sci. Ram.* (Imp. Inst. Sci., Bangalore); *J. Ind. Inst. Sci.* 35A:215-222; 1953.
705. Ravina, A. & Chamanes, H. & Garas, F. *Presse Méd.* 59:682-3; 1951.
706. Wells, Albert H. *Philipp. J. Sci.* 26:9-19; 1925 (chem. & pharmacolog. studies).
707. Foppiano, R., Salmon, M. R. & Bywater, W. G. *J. Amer. Chem. Soc.* 74:4537; 1952.
708. In Morot: *J. de Bot.* 7:299; 1893.
709. Schultes, R. E. *Bot Mus. Leafl.* 29:360; 1983.
710. Heckel & Schlagdenhauffen. *J. de Pharm.* 1882.
711. Tolvinski, K. K. *Pharm. Record* 7; 1887.
712. Mathias, M. E. *Taxon.* 31:488-94; 1982.
713. Lamb, S. H. *Native Trees & Shrubs of the Hawaiian Islands.* 1981.
714. Kerr. *Pharm. J. & Trans.* Dec. 21, 1912:977 (*Y B A Ph A.* 1:172).
715. *Schweizer. Wcschr. f. Chem. u. Pharm.* 1:237; 1912.
716. Barnes, J. W. 1966.
717. Formerly sold in Vancouver, BC (Cana); San Francisco (etc.) stores.
718. Schultes. R. E. *J. Ethnopharm.* 1:80; 1979 (*Swartzia*).
719. Asima, Chatterjee: *Proc. Ind. Sci. Cong.* 1955:135.
720. Plotkin, M. J. & al. *J. Ethnopharm.* 10:157; 1984.
721. Luke 19:14.
722. Pittier. *Contr. U.S. Natl. Herb.* 13:456; 1912 - Billington; 1895.
723. Petersen, W. J. *The Story of Iowa,* 2: 758; 1952
724. *Real-Enzyklopaedie.* 8: 326.
725. *fide* Steyermark (Venezuela).
726. Mirialin Co., Hudson, Mass (US) is said to have cult. 1 million shrubs for marketing of Miralin (dose 1-2 mg.); failed when could not get FDA clearance. - Cultivated by Bob Snow

(Lakeland, Fla) - I M M Employee (*Pop. Mech.* June 1966:107).

727. Manandhar, N. P. *Int. J. Crude Drug Res.* ca 1985.
728. Originally an eclectic remedy based on Elixir Rhei Alkalinus (Mistura Rhei Alkalin NF IV-V).
729. Schneider, W. *Lexicon zur Arzneimittel Geschichte* V/I:
730. British Standards Institute has published standards for this VO.
731. *Science.* 17 July, 1942:59-60.
732. Van Beek, T. A. et al. *J. Ethnopharmacol.* 10 (1):1-156; 1984 (Monograph on *Tabernaemontana* L.).
733. Mahai, H. S. *J. Indian Chem. Soc.* 15 (2):87-8; 1938.
734. No longer u. in operating rooms (formerly talc appl. to abdominal cavity as lubricant); should not be used in vicinity of vulva (danger of uterine or ovarian cancer).
735. A record of 21 synonyms for this sp. will be found in *Ann. Mo. Bot. Garden.* 48:86-8; 1961.
736. *J. Ind. & Trade*, Feb. 1953:250.
737. Capusso, F. *et al. J. Ethnopharm.* 8:327-9; 1983.
738. CSIR. *Wealth of India* 10:126; 1976.
739. Schultes, R. E. *J. Ethnopharm.* 11:17-32; 1984.
740. Koltay, M. & Korpassy, B. *Archivio de Vecchi per l'Anat. Patol. e la Med. Clin.* 17:307-28; 1951.
741. Davis, E. W. & Yost, J. A. *Bot Mus. Leafl.* 29:159-; 1983.
742. Greshoff, Maurits. *Rept. Colonial Museum, Haarlem* (Netherlands); 1906.
743. Power, F. B. et al. *J. Chem. Soc.* 101:2411-29; 1912.
744. Meyer, G. G. & al. *Folk Medicine and Herbal Healing.* 1981.
745. *Sydowia.* 11: 380-8; 1957. *Zbl. Bakter.* 180(1):401; 1961.
746. Pinto, Ferreira. *O timbo.* Lab. Clinico, Rio de Janeiro; 1935.
747. Dymock. *Pharm. J.* (3)8; 1878. Some consider *T. piscatoria* a separate valid sp.
748. Extensive bibliography. *Med. & Arom. Plants Abstracts.* 5:337-46; 1983 - Magene, P. *Les Badamiers (Terminalia).* 1-112; 1914.
749. Correa. *Diccionario das plantas uteis do Brasil* = I:311; 1926.
750. Boulos, L. *Medl. Plants of No. Africa.* Reference Publications; 1983. Wood = Thyine (or sweet) wood (Revelations 18:12) (Duke, J. A. *Medl. Plants of Bible.* 1983).
751. Chiefland, Fla (US). *Fide* the late Erdman West (botanist; Univ. Fla).
752. McClintock, Elizabeth. "*Teucrium* of New World" (monograph), *Brittonia.* 5:491-510; 1946 (classifn. based partly on trichomes).
753. Stoopel fraud. *JAMA.* 113:1751-2; 1939 - Fraud order issued by Post Office Dept. (Gehes Codex).
754. *Cassia corymbosa* Lam. (nSA, natd. NA): u. for fertility control; often cultd.
755. Nakagima, T. *J. Pharm. Soc. Jap.* 65B:422-4; 1945.
756. Turner, N. J. *J. Ethnopharm.* 11:181-41; 1984.
757. Anon. *Pharm. J.* (3) 4:598; 1874.
758. See 744.
759. Jack, Bainbridge, & Davies. *J. Chem. Soc.* 1908; 1912 (also hexyl butyrate & h. propionate).
760. Haring, F. & Kubitz, M. *Z. Tierernaehr.* 10:211-23; 1955.
761. U.S. Pat. 2,422,874; 1,851,872; 1,833,597.
762. *Hagers Handbuch d. pharm. Praxis* 4:237-8; 1973.
763. *Ztsch. des allgem. Oesterr. Apoth.-Verein* 31: 128, 153 (is *Thuja* an abortive?) - *Lloyd Bros. Treatise - Thuja occidentalis.*
764. Lachaze, A. & al. *C. R. Soc. Biol.* 146:50-2; 1951.
765. Sometime imported by Colgate Palmolive Peet Co. (soap manufacturer).
766. Flos Tiliae. *Pharm. Acta Helv.* 8:70-; 1933.
767. *Not. Farm. Porto.* 5:193; 1939.
768. A very popular German tree; *cf.* Unter den Linden, a principal street of Berlin.
769. Corfield, G. S. *J. Geog.* 42:308-17; 1943 - Tschirch, A. *Handbuch der Pharmakognosie* II; 252, 256; 1917.
770. See 292.
771. Marañon, Joaquin. *Philippine J. Sci.* 33:357-61; 1927.
772. Etymology of the word "tobacco." Ernst, A. (Caracas) - *American Anthrop.* 2:133-41; 1889.
773. Dey, B. B. & al. *Arch. Pharm.* 271:477; 1933 - Gaignault, J. C. & al. *Fitoterapia.* 55:37-57; 1984.
774. Brown, N. F. *Gardeners' Chronicle.* 51:180, 259: 1882.
775. Tschirch, A. *Handbuch der Pharmakognosie.* II:1159; 1917.
776. Some believed wd. u. in construction of Noah's ark. Forest in Liberty County, Fla (US) (e bank of Apalachicola River near Bristol).
777. Vickers, W. T. & al. "Useful plants of the Siona Indians of eEcuador," *Fieldiana, Bot.* N.S. #15; 1984.
778. Schmid & Karrer. *Helv. Chim. Acta.* 30:1162; 1947. Structure: *Ibid.* 1959:394.
779. Turner, Nancy. *J. Ethnopharm.* 11:190; 1984.
780. Also tetracosanol, octacosanol, hexacosanol - "Prometol" (Lee Vitamin Co.) - U.S. Pat. 3,031,376; 1962.

781. Winslow, D. J. in *Keystone Folklore Quarterly*. 14:59; 1969.
782. See 167 (p. 381).
783. Bailey & Gordon. *Plants Rep. Pois.* 1887:7.
784. Arnold, H. J. & al. *J. Ethnopharm.* 11:35-74; 1984.
785. Ostolaza, C. *Cact. Succ. J. (US).* 56:102-4; 1984 - Davis, T. W. *Bot. Museum Leafl.* 29:367-86; 1983 - Dobkin, M. *Econ Bot.* 22:151; 1968.
786. Tschirch, A. *Schweiz. Wochenschr. Chem. Pharm.* ca. 1892 (*Pharm. Centralh.* 1892:499 - *Proc. A Ph A.* 1893:624).
787. Manandhar, N. P. *Int. J. Crude Drug Res.* 1985.
788. *J. de Pharm. de Belgique.* 5(#5):1933.
789. Kokwaro, Ju. *Medicinal Plants of East Africa.* 1976/A. Schneider, Calif (*Pr. A Ph A.* 29:222; 1912).
790. Artificial maple syrup (Crescent Manufacturing Co., Seattle).
791. Schoepf, J. D. *Materia Medica Americana.* 1787.
792. Jacobs & Burlage. *Index of Plants of NCar.* 1951.
793. Tien Hsi Cheng. *J. Econ. Entom.* 38:491; 1945 (melon leaf beetle).
794. Ulrich, W. *Internat. Woerterbuch der Pflanzennamen.* 1875.
795. Bailey, Clyde H. *Chemistry of Wheat Flour.* Reinhold. 1945 c.
796. Saggese, D. *Yerbas Medicinales Argentinas*, ed. 9; ca 1950.
797. Gossweiler, J. *Noms Indigenas de Plantas de Angola.* 1953.
798. Johns, T. et al. *J. Ethnopharm.* 5:149-161; 1982.
799. Hartmann, E. *Arch. Gen. Psychiatr.* 31:394-7; 1974 ("natural hypnotic"). Studied at Sleep-Wake Disorders Unit, Montefiore Hosp. NYC.
800. Schulz, Ellen D. *Texas Wild Flowers*, p. 105; 1928.
801. Marketed as Pinus Canadensis (Kennedy's Light & K. Dark) for use as "mucous astringens" (Abican & Darpin; Rio Chemical Co.).

801A. *Cf.* story of Dutch workman who ate very valuable tulip bulbs for his lunch in place of onions.

802. *Pharm. Ztg.* 1930, #100.
803. Aval, N. *Rev. Fr. Corps Gras.* 3:105-7; 1956.
804. Shivpuri. *Ann. Allergy.* 30:407-12; 1972 - *J. Allergy.* 43:145, 1969.
805. One of the few trees which survive the aridity of NM (US) &c.
806. *Ulmus amer.* 400,000 elm trees with Dutch elm disease in Denver, Colo; to remove would cost $500 million ($1250/tree) & would raise city temp. 7° in summer & lower it 7° in winter. (IR film photos for early detection).
807. Persons (Montgomery, Ala) claimed best treatment he had; W. S. Merrell marketed poultice as proprietary.
808. Gerrard - ulexine; *J. de Pharm. et de Chim.* 14:334, 469; 1886 (*Pharm. Record.* 7:227; 1887).
809. *Bulletin van het Koloniaal Museum te Haarlem*; 1905.
810. *Gard. Monthly.* 1882: 8.
811. Chopra & Mukerjee. *Ind. Med. Gaz.* 12:666; 1874 (ca.) - *Ind. J. Med. Res.* 1936:666. Sometimes claimed equivalent (BP).
812. Mueller, Rudolf. *Beitraege zur Geschichte der Offizinellen Drogen.*
813. Barthels (Bulbus Scillae) - "Ueber die Meerzwiebel," diss. Hamburg 1926.
814. Hermann, F. J. *Amer. Midl. Natur.* 35:773-8; 1946.
815. *Hagers Handbuch d. Pharm. Praxis,* ed. 4. v. 6C:363; 1979.
816. Cont. transvaalin (=scillaren A); cardiac activity studied - Baetze, S. *Naunyn-Schmiedebergs Arch. Exp. Path. Pharmak.* 218:375-84; 1952 - George, Ernest. *J. Soc. of Chem. Ind.* 8:14-27; 1925.
817. Otsaka. *C. A.* 77:574; 1972.
818. Koenigsbauer, H. "Ueber die Behandlung bakterieller Hautkrankheiten mit Usninsaeure," *Hautarzt.* 6:501-4; 1955 (usnic acid).
819. Ramstad; 1954.
820. Conecuh County, Alabama.
821. The late popular radio star Arthur Godfrey claimed cranberry juice was a good tonic; drank a pint a day - Re lowering urine pH: Bodel, PT et al. (1958). Used to prevent urin. tract inf. (1 pint/day) (*Vaccinium macrocarpa*); cranberry juice c. anti-stickness subst.; said to prevent retention of urinary calculi, thus reducing risk of urinary calculi (kidney stones) formation.
822. National Cranberry Association, Hanson, Mass (US) - Claimed of use as emulsifying agt. &c. (Nealy, Walter A., Res. Lab., Cranberry Canners, Inc.) (*Spice Mill.* Dec. 1943).
823. *Medical Brief.* 1877:155-6.
824. Edgar, N. K. (New Era Pharm. Lab., Tenafly, NJ) - *DCI.* 35:479-80; 1934. (with photos).
825. Imported from USSR by Amtorg Trading Corp. (15 - IV - 46).
826. Epoxide with 3 OH gps., esterified with isovaleric acid, acetic acid, &c. (Depakena Abbott; Valmane (Kali-Chemie, BRD).
827. *Bot. Museum Leafl.* 29:377; 1983.
828. 1959 crop contaminated with weed killer (aminotriazole) & destroyed (250,000 lbs.) (UPI 27-XI-59).
829. Haensel, N. *Geschaeftsbericht*, Apr.-Sept. 1929 (*C. A.* 5:1156) (*V. celtica* - constants, &c.) - Schwartz, Hans. *Seifensieder-Ztg.* 68:61; 1941).

830. Schaffhauser, Heinz: *Arch. Expt. Path. Pharmakol.* 201:391-6; 1943 - subsp. *sylvatica* (Cana nUS-sw to wet hillsides of Sierras [Calif]): strong scented; rts. like those of *V. off.*; Indians u. rts. as food.
831. Kraemer. *Mich. Acad. Sci.* 1919:189.
832. *Rept. Proc. Central Indig. Drugs Comm.* (India) 7:326; & elsewhere.
833. Now believed this compd. is produced on processing (Clonga, E. *C. R.* (Paris) 200 (#9): 780; 1935.
834. Strazewicz, W. J. *Wiadomosci Farmac.* 63:147-9; 1936 (*C. A.* 31:8115).
835. Guenther, Ernest S. *Drug Cosm. Ind.* 53:160-1, 222-4; 1943 (*BA.* 18:5354).
835A. Cabrera, L. G. "Plantas curativas de México," *México*, ed. 1: 115-9; 1943 (1945).
836. Söderberg, Iva. *C. A.* 12:1234; 1924.
837. *U.S. Dept. Agr. DB.* 545:49-50 - Turner, N. & al. *B.C. Prov. Mus. No. 21*; 1980.
838. *C. A.* 42:4648 (chemistry); 1954.
839. *Jour. Bot.* 8:375; 1870.
840. WHO. *T. Rept. Ser.* 685:1983.
841. Englander. *Archives of Oral Biol.* 1971-2.
842. Zeven, A. C. & Zhukovsky, P. M. *Dictionary of Cultivated Plants & Their Centres of Diversity*. Centre for Agr. Publishing & Documentation, Wageningen, Netherlands; 1975.
843. Bouriquet, Gilbert: "Le vanillier & la vanille," *Enc. Biol.* 46:748 pp.; 1954 (Sect. Technique d'Agr. Tropicale, Div. de Défense des Cultures, Nogent-sur-Marne [Seine], Fr) - *Ann. Rept. Puerto Rico Agr. Expt. Sta* (USDA) 1937:16-26.
844. Wolff, L. *Amer. J. Pharm.* 1879:584-5 (once thought toxic).
845. Poethke, W. *Südd. Ap. Ztg.* 88:370; 1948; *cf. Ibid.* 87:22; 1947.
846. Kesselring, Wilhelm. *Gartenfl.* 75:58-60; 1926.
847. Jacobs, Walter A. & Lyman, C. Craig. XXV *J. Biol. Chem.* 155:565-; 1944 - 160:555-65; 1945 - Holm T. *Merck's Rep.* 24:109-11; 1915.
848. Bryant & Fleming. 1940 - Rhoads, E. E. & al. *Am. J. Obst. Gyn.* 61:914; 1951.
849. Craw, R. O. & Treloar, A. E. - *J A Ph A.* 40:345-8; 1951 (assay depending on survival time of Daphnia; use of Craw units). - Christensen, B. V. & McLean, A. P. *J A Ph A.* 25:414-7; 1936; 28:74-82; 1939.
850. Tincture reported helpful in hair prepns. (per Harry Taub); ca. Mar. 1945 - Newspaper article by Geo. Jean Nathan - *JAMA* editorial claimed worthless as baldness cure; ca. June, 1937.
851. Oil of mullen recommended for USP (*A Ph A.* 1901).
852. Campbell, T. N. *J. Wash. Acad. Sci.* 41:285-90; 1951 (med. plants of Choctaw, Chickasaw, & Creek Indians in early 19th cent.) (*Verbascum virgatum*).
853. Pennington.
854. Moldenke, H. N. *Phytologia.* 4:183; 1953.
855. Hunziker.
856. Killip & Smith.
857. Moldenke, H. N. *Phytologia.* 10:56-79; 1984.
858. Wyman & Harris; 1941.
859. Majumdar, D. N. *Indian July Pharm.* 5:61-4; 1943.
860. *Contrib. Sci. Los Ang. Museum* #63.
861. Mathias, M. E. *Taxon.* 31:488-494; 1982.
862. Feher, M. *Inter. J. Crude Drug Res.* 1986.
863. Simposio Internazionale - 348-9; 1977.
864. Youngken, H. W. *J A Ph A.* 21:444; 1932.
865. Hornbeam extract gave dramatic relief in Parkinsonism (Natl. Parkinsonian Institute, Miami, Fla [US]).
866. *Miss. Agr. Exp. Sta. Bull.* 408.
867. *Int. J. Crude Drug Res.* 22:116; 1984 (Holdsworth).
868. Morse, W. J. *U.S. Dept. Agr. BPI.* ca. 1942 - Cult. in Kurram Agency, NWFP, Pakistan, where u. in burns.
869. Eaten customarily at New Year's with hog jowl (deep South).
870. Osborne & Campbell. *J. Amer. Chem. Soc.* 1897:494.
871. Specimen in Ala. 1.1 m. long bean (1 yd., 6 ft.) (ca. 1970); sometimes longer.
872. Said very common in Curitiba (Paraná, Brazil) in gardens & along railings, &c.
873. Hartwell. *Lloydia.* 30:71-; 1967.
874. Hegnauer, R. *Chemotaxonomie der Pflanzen*; Birkhaeuser Verlag, Basel; 6 vols.; 1962-73.
875. Janot, M. & LeMen, J. *C. R. Paris.* 238:2550-2; 1954. (lvs. may c. serpinine) - Orechoff, A. *& al.* 1934 - Plat, M. *& al. Bull. Soc. Chim. Fra.* 1965:2497.
876. *JAMA.* 30 Oct. 1967.
877. James G. Armstrong *& al.* (Lilly Lab. for Clinical Res.) (AACR, Apr. 1962).
878. A conserve of violets was u. as flavor in pharmacopeial prepns. in the XVII-XVIII cent. ("grateful flavor); made by heating fls. with sugar.
879. Kraemer, Henry. *Diss. Marburg*, 4, 67 pp. 5 pls.; 1897 (Netolitzky p. 97) (most exhaustive treatment of anatomy of mucilage & gland cells).
880. Diener, Theodore M. 1971.
881. McKenna, D. J. & al. *J. Ethnopharm.* 12:179-211; 1984.
882. Winterfeld, K. "Ueb. die Wirkstoffe der Mistel," *Arch. Pharm.* 280:23-26; 1942.
883. Jarisch, A. "Zur Frage der Azetylkolins in der Mistel," *Arch. Pharm.* 280:241-2; 1942.
884. Kochmann, M. *Arch. Exp. Path. Pharm.* 161:553-; 1931 (pharm'col) - Colbatch, J.

Dissertation concerning M. (41 pp; London; 1725).

885. Lack of vitamin A leads to blindness for 20,000 infants/yr. in the world (1975) (WHO).

886. Chauchard, D. & al. *C. R. Soc. Biol.* 145: 1823-4; 1951 (natural & synth. vit. A - comparison of chronaxic effects, i.e., time needed to excite tissue with electric current).

887. Hartwich. *Die Genussmittel S.* 126; 1911.

888. Bougerede, L. J. *Indian For.* 11:595-600; 1935.

889. Enormous grape vine stocks (0.6-1.5 m. diam.) noted in Kurram Valley, Pakistan (G.M.H).

890. *J. Linn. Soc.* 30:87; 1894 (Lilly research).

891. Wasicky, R. thesis, "Faculd. Farm. y Bioquim.," Univ. São Paulo, 79 pp.; 1979.

892. Altschul, S. von Reis. "Drugs & foods from little known plants," Notes in Harvard Univ. *Herbaria*; Harvard Univ. Press; 1973.

893. *World Who's Who in Science*; 1968.

894. Some NCar (US) pharmacies (-1961-).

895. Lf. sap dropped on tropical sore or scabies or new cuts (PapNG) (*Int. J. Crude Drug Res.* 22:113; 1984); pl. u. to treat wounds (Tonga).

896. Lamb, S. H. *Native Trees & Shrubs of the Hawaiian Islands*; 1981.

897. Wine ". . . the most healthful & hygienic of beverages" (Louis Pasteur) - Ellwanger, G.H. "Medit. on gout ... its cure through ... wine" NYC - Dodd Mead. Lucia, Salvatore P. *Wine as Food & Medicine*. Blakiston; 1954.

898. Majundar, D. N. *J. Ind. Inst. Sci.* 16A:29-39; 1933) - Kopaczewski, W. *Thérapie.* 3:98-103; 1948 (Phys. & Biol. char. of frt. juice).

899. Schneider, Albert; 1912.

900. Tea fr. cockleburs gave relief in rheum. & forms of arthritis (Dr. R. E. Gardner, Johns Hopkins Univ., School of Hygiene & Public Health; ca. 1945).

900A. Schoepf & Becker. *Ann.* 507:266; 1933; 524: 49-55, 124-6; 1936 (*CA* 30:8166).

901. *Chem. News* 1862: #204 - *Pharm. Rundschau* 1886:35 - *Amer. J. Pharm.* 1886:161.

902. Collected in swamps & river banks, thus near Waverly, Ala, Benson Springs, Fla; commer. coll. in Tenn.

903. *J. Ethnopharm.* 11:168; 1984.

904. Used in Orbit chewing gum in place of sucrose; however, bladder stones & tumors may develop; hence withdrawn (*Sugar y Azucar*) 1978 (4):136-42.

905. Griebel, C. (Berlin): *Zts. f. Unter. der Lebensmittel.* 85(5):426-36; May, 1943 - Ascherson. *Bot. Ztg.* 1876:34, 321.

906. *Bot, Museum Leafl.* 29(3):83- ; 1971.

907. Hood, M. P. *New Mexico.* 21 (#12):13, 36-7; Dec. 1943 ("Yucca's useful").

908. See 274.

909. Albrecht, Gustav. *Sci. Monthly.* Oct. 1952 (Sci. Fiction Mag. 1953) (*Yucca whipplei* var.) (facetious); also *Science* 1952. (fls. supposed to shoot up in few seconds) (photos).

910. Clevenger, Joseph F. *J A Ph A.* 10:837-49; 1921 (report on the zamia starch industrial situation in Fla).

911. Dohme. *Drugg. Circ.* 41(3):64; 1897 (pharmacog. & histology).

912. La Forge, F. B. & Haller, L. J. U.S. Pat. 2,328,726; 7 Sept. 1943; essential active ingredient of insecticide ext. of *Z. clava-herculis.*

913. *Pharm. Ztg.* 1888 (*Pharm. Record* 8; 1888), with ref. to *Pharm. Centralhalle*; 1887.

914. *Pharm. J.* (per *New Remedies* May, 1883: 128).

915. Dzhamalieva; 1954/Eden, W. G. & al. - *J. Econ. Ento.* ca. 1970 (compn. corn silk).

916. Burtt-Davy, Joseph. *Maize, Its History, Handling, & Uses*, 871 pp. (sAf).

917. Akbar, M. Abdel (Fac. Agr. Cairo Univ. Gizah, Eg). *Pak J. Sci.* 18:112; 1966 (sepn. of starch fr. maize).

918. Manufr. Colorcan, Inc., West Point, Penn, 1968-

919. Peckolt.

920. Rydberg, P. A. *Bull. Torr. Bot. Club.* 1903: 272.

921. *Chem. & Drugg.* 31-VII-1942 (prodn. of ginger in Sierra Leone under govt. control, 1942-) *Crop Sci.* i: 91-3.

922. Bartlett, R. A. *Agr. Gaz. NSW.* 33:818; 1922 (*Bot. Abst.* 12:2463) (ginger culture: yield 2-4 tons/acre; planting, mfg, harvesting).

923. Grown in Mexicali area (Mex): 12,000# sent to US (1942) (Chinese g.?). Ramon de la Saigra. *Histoire de l'Ile de Cuba, Botanique.* 1843, 83-.

923A. *J. Ethnopharm.* 12:328; 1984 (Tonga).

924. Jenks, A. E. The wild rice gatherers of the Upper Lakes - Exhibits in Milwaukee Public Museum; Amer. Museum Nat. His. - Janness, Diamond. *Canad. Geog. J.* 2:477-82; 1931 (wild rice—photos & text, manner of harvesting by Ind. detailed) - Capen, Ruth G. & LeClerc, J. A. *J. Agr. Res.* 77 (#3):1948. (chem. compn.). - Thayer, T. *An American Girl* (1933).

925. Ebbinger, John C. *Mem. N.Y. Bot. Gard.* 10:279-304; 1964 (taxonomy of *Luzula* subgen. *Pterodes*).

926. Lettuce leaf cigarette marketed in Israel, called "Long Life" (c. 60-70% less tar than tobacco) (19-I-76).

927. Sigueira, Rubens de & Pechnik & Cruz. *Tribuna Farm.* 21:1-5; 1953 (chem. & pharmacol. of mate) - Hennings, C. R. *Ber. d. d. Ph. G.* 30:22-6; 1920: p. 438.

928. Lam. Illustr. i. 63 (figure) - Bieberstein: Fl. Taur. Caucas. i: 17.

929. *Acer* frt. a samara - Maple lf. emblem of Canada; u. in BC on auto license plates.
930. *Acer negundo* L. - C. tumor inhibiting saponins with aglycones acerotin & acerocin - Fresh bk. listed in *Schwabe: Pharm. Homeopath.* ed. 2, 1929 - Box elder County, Utah (SC profuse growth of these trees near Brigham City). Also B. E. (village, U.) now renamed Brig City.
931. *A. saccharum* - Called sugar tree (Ohio valley, esp. wVa-NCar) - Sap c. maple sugar 1.5-3 kg./tree/ann. - St. Johnsbury, VT, center of m. sugar (syr.) industry - Wd. lumber for flooring, furniture.
932. Lamson-Scribner, T. *Nature Mag.* Sept. 1934: 125 (good photos & desc.). Pl. placed in slow fires to furnish mosquito repellant smoke (NA Indians) (*Ford Times* XI/60: p. 62).
933. *Adonis* named after handsome youth, beloved of Aphrodite (Venus); he was slain by a wild boar & the flower sprang up in his place.
934. Basu, U. P. (Calcutta). *J. & Proc. Instit. of Chemists* (India). Dec. 1945 (Indian agar fr. *Gracilaria confervoides*). FWS Separate No. 263 (feeding studies).
935. *Chem. & Drugg.* 30-IX-44; p. 351. (agar fr. Australia).
936. Bauer, K. H. & Rajdhan, T. C. *Pharm. Zentralh.* 72:146-152; 1931 (native to Kashmir).
937. (*Zygopetalum*). Gines, H. et al. "Florula de la cuenca del Rio Negro" in *La region de Perija...Soc. Cienc. Nat. La Salle*, Maracaibo, Ve.
938. (*Zuelania guidonia*). Leon, H. & Alain, H. *Flora de Cuba* iii, 329-30; ca. 1960 (1974).
939. Amer. Chem. Soc. 1977 - Kresse, H. (Texas, A&M), 1976.
940. *JACS.* 1920 - Beattie, W. R. *U.S. Dept. Agr. FB.* 232 (culture & uses) - Arthur D. Little. *Industrial Bull*: 1959.
941. (VO). Kotlarenko, M. P. *Sovetskiye Subtropiki* No. 1 (8) p. 75 (1931).
942. Sukev, F. N. Thesis, Univ. Wash.
943. Kurram Valley, Pakistan.
944. In large field of Palatine Hill, Rome, continuous loud cracking noise of pods opening in hot sun noted (Sept. 1951) (GMH). Many illustrations of decorative use in *Enciclopedia Universal Ilustrada* (vol. 1).
945. von Humboldt, F. H. A. *Reise.*
946. Has been called "the most deadly [higher] plant in the world."
947. "Flore du Cambodge, du Laos, et du Viet-Nam," *Scrophulari.* v. 21:1985.
948. *Current Science.* 12:209; July 1943 (active principle).
949. *Apoth.-Ztg.* 1888:250.
950. Couch. *Perf. Rec.* 13:172; 1920 - *Amer. J. Pharm.* 1922 (VO 0.184-0.316% variously described as having peppermint & thyme leaf odor).
951. *Current Science.* 1954:53 (Ind).
952. Sodium pectate prod. comm. fr. *Agave sisalana* in Eng, the sisal flesh being treated to remove waxes and sol. glycosides (new technics); marketed prod. nearly white pd.; sisal prod. fr. B.E.A. plantations formerly.
953. Heyne, K. *Die nuttige planten van Nederlandsch Indie*, ed. 2, v.1, p. 422; 1927.
954. The tree of *A Tree Grows in Brooklyn* (novel). One of the few trees that survives this city environment; banned from Atlanta (Ga [US]) city streets because tree has a strong stench.
955. Hoeffler, Paul L. *Africa Speaks*, p. 266; 1931.
956. Stacey, B. F. "Medicinal plants of the Indians," *Proc. A Ph A.* 21:619; 1873.
957. Viehoever, Arno. *J. Assoc. Off. Agr. Chem.* 4:153-4; 1920 (coll'n.).
958. Kinabrew, R. G. (Univ. Miss.). *Tung Oil in Mississippi.*
959. Alfacene Co., Newcastle, Neb (FDA 1956).
960. Maass, H. *Alginsaeure und Alginate* (XI + 465 pp.); 1959.
961. Mordarski, M. *Arch. Immun. Terap. Doswied.* 4: 1956 (antibacter.) - Kovaler, A. A. *Veterinarija SSSR.* 12:27-30; 1955 (phytoncides u. to treat *Trichomonas* inf. in cattle).
962. *Chem. & Drugg.* 30/IX/44: p. 351. Sweet onions (Vidalia, GA) can be eaten like apple; Onion Festival there. Chinese onions preferred because less rapid deterioration.
963. *Sci.* 29/XI/57 (in Ca) (value sometimes attributed to Se (antioxidant) - Rae, J. *Mfg. Chem.* (GB) June, 1951 - Tokin, B. P. (Univ. Tomsk, USSR). *Die Phytonzide*; 1956 - *Sci.* (NS) 99: #2563, 12 (11 Feb. 1944) (phytonicides act as bactericides, protozoicides, zymocides, ovicides) - Gilroy, Calif = "Garlic capital of the world."
963A. Ginsburg, B. A. *Stomatologija* (USSR) 1951, #2: p. 56 (1-2 g. ext. injected daily submucosally for 20-25 days beside other special local treatment).
964. Komis, A. *Revue de Pathologie Comparative et d'Hygiène Générale.* 51:294-8; 1951 (A.s. apoxtaxin as antagonist to allergic state). Also to stings of bees, &c. - *Rev. Méd. Suisse Rom.* 49:205-6; 1929. *Cf.* to ramps.
965. *Fide* John A. Beutler, Ph.D. - *Cf.* Roch - *Med. Rev.* 25:153; 1922 (Tinct. in resp. affections); Keeran Tunney (Irish) prepd. garlic soup for Tallulah Bankhead which "cured her cold" overnight (with potatoes, onions, butter, milk) (*Tallulah: Darling of the Gods* [1973]).
965A. Crewe, J. E. *Minn. Med.* 22:538; 1939 (burns treated with ointment c. *Aloe barbadensis* pd., mineral oil, white soft petrolatum - Uhl & Rowe. Research at U. Tenn (1968) to study "healing substance" in aloe (MJ $24,000).
966. Tonn, A. *Pharm. Monatshefte.* 12:200; 1931 (alc. ext. of bark used).

967. Tschirch, A. *Handbuch der Pharmakognosie* II:363-6; 1912.
968. Boymond, P. (Geneva). *J. Suisse de Pharm.* 82:229-33; 245-9; 1944 ("Principes toxiques de l'Amanite phalloide").
969. *Fide* R. M. Harper.
970. *Phytologia.* 4:455; 1953 (H. N. Moldenke).
971. Henry, John P. and A. L. Herring. "Bot. survey of Memphis, Tenn & surrounding territory," *J. Allergy.* 1:163; 1930 (photos of giant ragweed).
972. Holmes, Edward M. *Pharm. J.* (IV). 60:328-9; 1925.
973. Balls, E. K. *Early Use of California Plants.* 1965.
974. "Nothing can be more beautiful on earth" (John Gibson; 1838).
975. Sidi, E. *et al. Bull. Soc. Franç. de Dermat. et de Syphil.* 1951:490-2.
976. *Proc. Am. Pharm. Assn.* 34:427 - Quimby, Maynard, W. *Econ. Bot.* 7:89-92; 1953.
977. Brett, Charles H. *J. Agr. Res.* 73, No. 3; 1946. - *Cf.* Featherly, H. I. Okla. Agr. Mech. Coll. Stillwater, Okla.; 1942.
978. Rose, J. N. *Contr. US Nat. Herb.* 8:4; 1903.
979. Wehmer, C. *Die Pflanzenstoffe,* ed. 1:394; 1911.
980. Bocek. *Econ. Bot.* 38:340-; 1984.
981. Altschul, Siri von Reis. *Drugs & Foods from Little Known Plants.* 1973.
982. Matthew 4: 28-9.
983. *J. Ind. & Trade.* 4:1133-6; 1954 (cashew indy.) - Sebastinek, K. M. *Proc. Indian Acad. Sci.,* Sect. B. 1955:239-48.
984. Johnson, Maxwell O. *The Pineapple* (306 pp.) Pacific Press, Honolulu, Hawaii; 1935 - Bromelin (pineapple enzyme c. in Albromol [Hart Co.]).
985. Braun. *Univ. Basel Diss*; 1933.
986. *Encyclopedia van Nederlandsch-Indië*; 1895-1905.
987. *Kew Bulletin*; 1886.
988. Holden, Margaret, Seegal, Beatrice Carrier, & Baer, Harold. *Proc. Soc. Exp. Biol. & Med.* 66:54-60; 1947.
989. Angostura bark u. to prep."Angostura Bitters"; orig. made at Angostura or Ciudad Bolivar (Ve); later (1906) also mfd. at Port-of-Spain, Trin.; u. antimalarial, &c. (WI).
990. Klobb et. al. *Bull. Soc. Chim.* (4) 7:940; 1910 - *C. R.* 138:763; 1904 - *Bull. Soc. Chim.* (3) 27:1229; 1902.
991. Rae, John. *Mfg. Chemist*, Feb. 1945:50-1.
992. Leclerc, H. *J. de médecine et de chirurgie pratiques.* 121:142-6; 1950 - *Flora URSS* (Komarov, V. L.). II:56-7; 1934.
993. Apioline (Etablissements Rigaud) - Brasil Marca 144 347; 1947 (Re toxicity). Barni, M. *Minerva Medico-legale.* 72:5-7; 1952.
994. (Bee sting therapy). Guyton. *J. Soc. Entom.* 40:469-; 1947 - Thompson. *Lancet.* 19 Aug. 1933 - Bee honey: "healing for mankind" (Koran XVI: 69).
995. "East west" common name in NCar (near Statesville), since rts. grow e&w supposedly (per Edw. J. Alexander, *NY Bot. Gard.*).
996. Woodson, Robt. C. *Ann. Mo. Bot. Garden* 17 (histology, &c.): *cf.* Youngken, H. W. *J A Ph A.* ca. 1940 (assay in digitalis units, SBP).
997. "Martin's Cancer Powder" was mixt. with $As_2 0_3$.
998. Chapman - 1870.
999. Twisselmann. *Flora of the Temblor Range* (Calif) - *Wasmann J. Biol.* 1956.
1000. *Agric. Ledger.* 1904, #1.
1001. National Peanut Festival, Dothan, Ala; held annually.
1002. Rusby, Henry Hurd. *J NY Bot. Gard.* 2:94-123; 1901 (nature & uses). Kinds of peanuts: 1) jumbo (eaten, usually salted) Virginian 2) oil source (med.) (ex. Dixie Runner) 3) Spanish (in candy) (salted).
1003. *J. Ethnopharmacology. 6:351; 1982.*
1004. Name has caused confusion with bark of *Zanthoxylum clava-herculis*; sometimes name "prickly alder bark" has been seen (Donzelot).
1005. *A. araucana* grows well in Eng (Kew +) with climate similar to that of sCh; name "monkey puzzle" first appl. by man in Co. Cornwall; "alma chesnut" (*Nature Mag.*, Nov. 1934: 227-9; 235).
1006. Lawn & shrubbery tree in Eng (Knight & Step. *The Living Plant.* p. 218) (good figure) - Adapted to countries with light rainfall; "tender" in Ga & Carolinas); dies in Fla (hot wet weather) (Fairchild. *Exploring for Plants*, p. 20).
1007. Hawksworth, Frank G. & Wien, A. "Dwarf mistletoes," *USDA, Agr. Handbook*, 401; 1972.
1008. *(Arctium lappa).* Rich in polysaccharides (J. C. Krantz, Jr. [1930]).
1009. Fauconnet, L. "Contribution à l'étude chimique des bardanes," *Pharm. Acta. Helv.* 1943:419-25. Tea formula: 10% ea. of b. frt. & Triticum.
1010. Wieslander, A. E. & al. "Notes on the genus," *Madroño.* 5:38-47; 1939 (Calif & wNev [US]).
1011. Safford. *Contr. US Nat. Herb.* 9:187-; 1905 (much on chewing of areca) - Lloyd, J. U. "*Areca catechu* L." (8-p. repr. fr. *West. Drugg.* 5/97).
1012. Meyer, T. *Lilloa.* 1938.
1013. Sarkar, S. N. *Current Sci.* X/1941:405 (sensitive test for detection) - J. L. Stewart (1894).
1014. Georgi, C. D. V. *Malayan Agr. J.* IX/29. *Ibid.* II/31 (variations in amt. of ether ext. of tuba

root [*D. malaccensis*]; active princ. & methods of extn.).
1015. *NM AES Bull.* 308; 1943 (prodn. frt.) - *Nature Mag.* VI/34 (illus.).
1016. *J. de Pharm. de Belg.* 5: #3; 1933.
1017. Bergen, Fanny D. *Popular American Plant Names.* 1893. Some believed that *Daucus pusillus* was found in areas where rattlers occurred (Slosson).
1018. Moore, R. H. *P. R. Exp. Sta. Circ.* 24. 17 pp.; 1943 (culture of *Derris elliptica* in Puerto Rico).
1019. *Anais da Fac. de Farm e Odont.* 8:85-6; 1951.
1020. *J A Ph A S.E.* 16:1955.
1021. Stern, K. R. *Brittonia.* 13:1-57; 1961.
1022. Lindsey.
1023. Davis, E. W. *Bot. Museum Leafl.* 29 (4):378; 1983.
1024. *Malayan Agr. J.* IX/28 (periodic harvesting of tuba rt., *Derris elliptica*) - *Ibid.* IX/30 (derris or tuba root).
1025. *J. Pharmacol.* 57:196; 1942.
1026. Greenish, H. G. *Pharm. J.* 105:474-5; 1920 (Uzara: what is it?).
1027. Rizzini, C. T. & Paolo, Occhioni. *Rodriguesia* 1957: 1+ (pictures of CaOx raphide crystals).
1028. Taylor, Wm. Randolph. *The Marine Algae of Florida.* 1928.
1029. Schultes, R. E. "Ethnopharmacological notes," *Bot. Museum Leafl.* 26:230; 1978.
1030. *Gardeners Monthly.* 1881:280; 1882:6, 79-80 (cultn. of teasel in NY [US]).
1031. Manresa, Miguel. *J. Dept. Agr. So. Af.* 7:296; 1923 (drug of dubious value).
1032. Gupta, J. C. & al. *Indian J. Med. Res.* 39:255-60; 1951. Glucoside isolated in pure form; low toxicity; lowered carotid BP; mild stim. to organs innervated with autonomous cholinergic nn. (syn. Marsdenia v. T. Cooke).
1033. *Pharm. Praxis.* 1911(1):1.
1034. *Trans. Proc. Royal Soc. So. Austral.* 37:113-24; 1933.
1035. Mitchell, W. *Ph. J.* 144:137; 1940 (new alkaloids). Koehler 3:26+.
1036. Schultes, R. E. *Bot. Mus. Leafl.* 15:77; 1951.
1037. *USDA Leaflet* 129. 1937 (prodn. & prepn.). Rasapen (propr.) u. urin. antisep.
1038. White, J. C. *Dermatitis venenata....* p. 216; 1887.
1039. Hocking, G. M. *Bull N. F. Committee.* 13:41-8; 1945.
1040. Hardin, J. W. *Bull. Torr. Bot. Club.* 100:178-84; 1973.
1041. Sometimes genus wrongly rendered as "Atrabotrys."
1042. *Cf.* title of popular novel *Riders of the Purple Sage* by Zane Gray.
1043. Nath, M. C. *Z. Physiol. Chem.* 247:9-13; ca. 1937 - *Ibid.* 249:71-81; ca. 1938 - Nath, M. C. & Gupta. *Ind. J. Med. Res.* 27:121-9; 1939 (artostenone).
1044. *Biol. Central Am. Bot.* iii:354-.
1045. *Parfum. Mod.* 14:122-3; 1921 (Jahandiez) (biblio.).
1046. "Summary of the literature on milkweeds...& their utilization," *USDA Bibl. Bull.* 41 pp.; 1943.
1047. Sparrow, F. K. *J. Agr. Res.* 73, No. 3, 1943 (*Asclepias syriaca*, types of pods found in Michigan) - Barrell & Miller. *Science.* 1940.
1048. Kral, R. *Brittonia.* 12:233-278; 1960 (monograph) - Fairchild. *Explor. for Plants*, p. 20 (thrives & fruits at Kew, England, in hot summers).
1049. (Asparagine). Abderhalden, E. *Biochemische Handlexikon,* v. 4:597; 1911 - Stieger. *Z. Physiol. Chem.* 86:245; 1914 - *J. Biol. Chem.* 145:45; 1942.
1050. Rothlin, E. "Tesis," *Bs.As.* 1918 ("Contribución al estudio de A.").
1051. *Atoll Research Bull.* 262:6; 1963.
1052. (*Asplenium*). *Drugg. Circ.* 16:128; 1872 - *Cf. Gray's Bot.*
1053. Gonzalez, Matias & Lombardo, Atilio. *Indigenous Med. Plants of Uruguay* - *Rev. Farm.* (*Bs. As*). 88:297-309; 1946.
1054. *Cf.* "The Crazyweed" (poem): "He must have eaten crazy weed before he went to bed."
1055. General Foods Corp. imported 220,000 lbs. of t. kernels at one time in 1944; tucum oil also imported (1944).
1056. Murimuri oil imported ca. 1942 (Durkee Famous Foods) - "The Absinthe Drinkers" (painting by Degas in Louvre, Paris).
1057. Greenish, Henry G. & Pearson, Constance E. *Pharm. J.* 106:2-3; 1921 - *Idem. Year Book of Pharm. & Trans. Brit. Pharm. Conf.* 1922/23: 646-50.
1058. Schoepf, J. D. *Materia Medica Americana*, p. 112; 1787.
1059. van Laren, A. J. *Pharm. J.* 111:631-2; 1923 (reprinted fr. *Pharm. Week.* 61:1057-61; 1924) (cultn.).
1060. Maplethorpe, Cyril W. *Pharm. J.* 113:106; 1924 (found no appreciable amt. santonin) (*A. gallica, A. maritima*).
1061. Holmes, E. M. *Pharm. J.* 124: Jan. 11, 1930 (mode of growth) - Simonsen, John J. *Ibid.* 111:632-4; 1923.
1062. Qazilbash, N. *J. Univ. Peshawar.* 1:120-131; 1933 (comml utilizn. of *Art.* grown in nwKashmir).
1063. FDA prosecution; case (1954); "salt bush tea."
1064. Hall, H. M. & Clements, F. E. Carnegie Inst.
1065. "Diaplex" (propr.) *FDA, DD NJ.* 642; 1943.
1065A. Hand, C. J. E. & Tyler, C. *J. Sci. Food Agr.* 6: 743-754; 1955. Found some favorable effect from feeding moderate amts. of this seaweed meal to hens (wt., egg prodn.; water & miner-

al metabolism, &c.). Higher rates (20%) were unfavorable.

1066. King, Harold & Ware. *J. Chem. Soc.* 1941: 331-7 (action attributed to high alk. content of rt. [10.8%]) [hyoscyamine 0.008%]).

1067. Case at Haleyville, Ala, 9 Aug. 1962.

1068. *Nature.* 1971.

1069. Acts as phagorepellant for locusts & grasshoppers - *J A Ph A.* 21:242; 1932 (ext. of lvs. repels swarming locusts & grasshoppers since taste is repellant; u. for garden crops) - Natural pesticides from the neem tree (*A. indica*) & other trop. plants - *Proc. 2nd Int. Neem Conf.*, Rauisch - Holzhausen, FRG; 1985.

1070. Calcutta Chemical Co., Calcutta, Ind - Margo Soap: toilet soap with neem oil.

1071. Kremers, E. *J A C S.* 45:717-23; 1923 (azulene fr. milfoil).

1072. Arreguine, Victor (*Bs. As.*, Arg 1918) (*Rev. Inst. Bact. B. A.* 3:389-96) - Moreira, E. A. *Trib. Farm.* 34:27-30; 1966.

1073. Renewed interest manifest (1975).

1074. Pavan. *Anais d. Fac. Farm. e Odont.* 1952.

1075. Sayre, L. E. *Drugg. Circ.* 41(2):32; 1897.

1076. Organ built of bamboo ca. 1825, Las Pinas, PI; still in use (1975) - USDA Expt. Sta. for bamboo, near Savannah, Ga; Fla; Chico, Calif - Bamboo is relished by gorillas, panda (their only food), &c.

1077. Hammerman, A. F. *Trudy po prikladnoi Bot. Genetike, Selektsii.* 22:165-208; 1929 (yajé, narcotic of the Indians) - Schultes, R. E. & Hofmann, A. *Hallucinogens*, ed. 2, pp. 163+; 1980.

1078. Macht, David I. *J A Ph A* 16:1056-9; 1927 (with James A. Black) (Pharmacolog. note) - This sp. common in Fla (Citrus, Lake, Marion, Sumter Cos, esp. between Ocala & Inverness); in dry fields & woods; gathered in fall, cut, & dried.

1079. Wallis, T. E. & Dewar, T. *Quart. J. Pharm. & Pharmacol.* 1933: 19 (compar. study of anatomy of leaves of *Barosma* spp.).

1080. *Proc. Indian Sci. Congress.* 1937.

1081. *J A Ph A.* 31:193; 1942 - *JAMA.* 127:992; 1945 ("obvious advantages of m.c. over nat. gums)."

1082. *Anal. Fac. Quim. Cienc.* (Univ. Chile). 29/30: 110; 1977/78.

1083. Michigan white bean u. in prepn. celebrated & trad. congressional bean soup (Washington, DC).

1084. *The Thompson Begonia Guide*, ed. 2, 4 vols., 1976-78 (E. J. Thompson, Southampton, NY). There is a large collection of literature on *Begonia* at NY Bot. Garden.

1085. *Miss. State Geol. Surv. Bull.* 57; 1943 (esp. pp. 90-106) (Monroe Co.).

1086. Sternberg, T. H. and Taylor, P. R. *Urologic & Cutaneous Review.* 55:348-352; 1951.

1087. "Genus *Berberis* in Asia," *J. Bot.* 1942.

1088. Short, G. R. A. *Trans. Brit. Pharm. Conf. Yearbook Pharm.* 1925/26:365-84; 1926 (comparative study of structure of stems of *B. arist.*, &c.).

1089. Arata. *Répert. de Pharm.* 1892: 45.

1090. Cromwell, B. T. *Biochem. J.* 27:860-72; 1933. Berberine content increases from year to year because not used; seems product of protein metabolism; 30-yr.-old plant whole rts. bark 11.0%, wood 2.80%, rts. at soil surface 11.17%; not very much variation through year.

1091. Dekker. *Pharm. Weekbl.* 43:942; 1906.

1092. Costa, Oswaldo de Almeida & Silva. *Rev. soc. brasil. quim.* 4:199-201; 1933 (analysis) - Stellfeld, Carlos. *Trib. Farm.* 2:142-3; 1934 (anal.).

1093. Mason. *Chem. News.* 121:61; 1920.

1094. Schulz, E. R. *J A Ph A.* 15:33-9; 1926.

1095. Watt, G. *Commercial Products of India* (1908).

1096. Khasanov, T. Kh., Malikov, v. M., & Rakhmankulov, U. *Khim. Prir. Soedin. SSSR.* 1979: 226-7.

1097. Mangold leaf c. carbohydrates (Campbell, A U. *J. Agr. Sci.* 4:248-57; ca. 1911).

1098. *Year Book Amer. Pharm. Assn.* 7:438.

1099. Norite-A, adsorptive material; u. to decolorize sugar juices, &c. (American Norit Co., Jacksonville, Fla).

1100. The late E. E. Scherff (Chicago Nat. Hist. Museum, Chicago, Ill) was a specialist in gen.

1101. Siddiqui; 1945.

1102. Blaeberry, a community in BC (Canada).

1103. Lewalle, J. & Rodegem, F. M. *Quart. J. Crude Drug Res.* 8:1257-70; 1968 (med. plants of Burundi).

1104. Holdsworth, D. K. *So. Pac. Commission Tech. Papers* #175; 1977 (med. plants of Papua New Guinea).

1105. Dewey, Lyster H. "Ramie, a fiber-yielding plant," US Agr. Dept. May, 1929.

1106. Rakshit, Jitendra Nath. *Analyst.* 48:169; 1923. (drug plants u. in dropsy).

1107. Ryamond-Hamet. *C. R. Soc. Biol.* 121:431-4; 1936.

1108. *Hagers Handbuch d. Pharm. Praxis*, ed. 4. 6A: 555; 1977.

1109. E. G. Lucas, Michigan State University, E Lansing; 1950.

1110. Elisofen, Eliot: "The Nile," 1964: fig. 91 - An old Tamil song records 801 uses of the tree - Zoete, Beryl de. *Dance & Drama in Bali* (Faber & Faber, Ltd., London). pp. 2-3 ("the metallic leaves of lontar palms. . . .").

1111. One of the three gifts of the Wise Men (NT) - Midbary, S. *Hassedeh* (Tel-Aviv). 23:318-9; 1943 (in Hebrew) (the Olibanum & the Myrtus in the life of our forefathers; both prods. important in life & ceremonies of the Jews).

1112. "Mesquite grass" (*Bouteloua* spp., esp. *B. rigidiseta* Hitchc.); *cf.* Henry, O. O. " Double-eyed deceiver."
1113. Bennison, E. C. *Pharm. J.* 120; 318-9; 1928.
1114. Van Arman, C. G. *Proc. Soc. Exp. Biol. Med.* 79:356-9; 359-63; 1952 (origin).
1115. McPherson. *J. Lipid Res.* 23:221; 1982.
1116. Worth, Miklos. *Ideal Health Books* (NY) 48 pp. ("prolongation of youth," history, recipes).
1117. Scott, Walter. *Old Mortality*, p. 88 ("charger of broth thickened with oatmeal and colewort") ("Colewort" also appl. to *B. napus* and *Crambe* spp.) - *B. oleracea* var. *acephala*, giant cabbage kales of Eng up to 2.8 m high; sts. u. for rafters; made into walking sticks (Jersey) (adv. cows cannot eat lvs.).
1118. *J. Physiol.* 94:249; 1938.
1119. Such as near Ada, Ohio (US).
1120. Liu, Shao-Dwang et al. *J. Pharm. Soc. Jap.* 63:579; 1943; *C. A.* 45:1731; *C. A.* 36:4198.
1121. Lockwood, T. E. *J. Ethnopharm.* 1:147-64; 1979.
1122. Monachino. *Phytologia.* 4:342-7; 1953.
1123. *Bryonia alba*: Hager. *Handbuch d. pharm. Praxis* ed. 4, 3:523; 1972 - Appl. to black eyes to remove discoloration (care !) by black "eye" artists.
1124. Davis, E. W. & Yost, J. A. *Bot. Mus. Leafl.* 29: 203; 1983.
1125. Hotchkiss, Neil. "Bulrushes & bulrushlike plants of eNA," *Fish & Wildlife Serv. Circ.* 221:21 pp.; 1965.
1126. Bennet, A. W. *On the Medicinal Products of the Indian Simarubeae & Burseraceae* (London) octavo; 1875.
1127. Read, B. E. & Schmidt, Carl. *Proc. Soc. Exp. Biol. Med.* 20:395-6; 1923 (pharmacol. of tang kuei).
1128. Schindler, H. *Süddeutsch Apoth. Ztg.* 88:427-30; 1948.
1129. *Merck's Index* X:1308; 1984.
1130. Only fuel available free in Baltistan is this plant.
1131. Simon de León (Thesis) found 26% tannin (AOAC leather method) in *Caesalpinia* from Guerrero while that from Veracruz had traces only.
1131A. Hydrocotyles Folia (Ve Ph+): supposedly disc. by Chang-Li-Un (1677-1733) ca. 1700; called Fu-(Yi)-Tieng (the life plant); recommended for palpitations; nervous unrest, &c.
1132. Neff, E. A. *J. Trop. Med. Hyg.* IX/29; Marsden, Lt. Col. E. *Ser. Sci. & Ind. Res.*, NZ; *Mfg. Chem.* 38:219; 1945.
1133. "La Dame aux Camelias" by A. Dumas was one of the first stories about prostitutes; it is basd on Dumas' own experiences with Marie Duplessis (Marguerite Gautier of the novel), who wore a bouquet of white camellias for 25 days of the month, then red for 5 days, an obvious symbol of menstruation (*cf.* Hugo [Marion Delorme]; Augier [L'Aventurière], and La Traviata [1853]).
1134. Freise, F. W. *Plantas Medicinales Brasileiras.* São Paulo. 468-9; 1934.
1135. Test for cannabis: heated over wire gauze produces characteristic odor.
1136. Negrete, M. J. & Yankelevich, G. *Rev. del Inst. de Salubridad y Enfermedad. Trop.* 17:65-8; 1958 (local anesthesia prod. by capsicin); . . . capsaicin < capsicin. as local anesthetic appl. directly to tooth; by release & depletion of substance P in dental pulp produces analgesia or hyperanalgesia; u. ca. 1850.
1137. Mireya, D. Correa A. & Wilbur, R. L. *J. Elisha Mitch. Sci. Soc.* 1969 (revision of genus *Carphephorus*).
1138. Re saguaro cactus: Shreve, F. & Wiggins, I. L. *Vegetation and Flora of the Sonora Desert*, v. 1: 147-9; 1964 - *Arizona Highways*, Dec. 1948.
1139. Kuhlmann, G. *Memorias do Instituto Oswaldo Cruz.* 21, pt. 2, 389; 1931 - Rolfs, P. H. & C. R. *A cultura da Sapucainha* - Secretaria da Agricultura, Estado de Minas Gerais; Serie Agricola, Num. 6 - *Drug Markets* 29:350-2; 1951.
1140. Baring Gould. *Historic Oddities*, pp. 151, 153. True story of Pastor Pieter Nielsen, obscure priest of the Roager Heath, Denmark, who became Bishop of Rebe (Denm.) ca. 1465. He was fond of caraway & said, "It is good for the stomach & it clears the brain" (imaginary conversation). He could not offer distinguished guests beer so offered them a mug of dill-water made from "carraway." "It is given to infants to enable them to retain their milk. It is good for adults to make them recollect their promises."
1141. Cooke, Mordecai Cubitt. *Pharm. J.* s. 3, 1: 1007-8; 1871.
1142. Thompson. *Amer. J. Pharm.* 56:331; ca. 1884 (Cascara Amarga).
1143. LePrince. *C. R.* 115:286. - 129:60 (per Tschirch Handbuch d. Pharm.).
1144. Conrad, H. C. *Agr. in the Americas*, IV/44 (broken into bits; preserved teeth into ripe old age).
1145. Schwartz, Louis et al. *Indust. Med.* 14:500- ; 1945 (skin hazards in mfr. & use of cashew nut shell liquid-formaldehyde resins).
1146. *U.S. Nat. Mus. Rept.* 1896:1025-56 (lamp of Eskimos).
1147. Comm. ext. of bk. added to brine (2: 1000) u. in prepn. of freshly prepd. salted fish; retarded devt. of rancidity (*C. A.* Apr. 10; 1953).
1148. (*Castela*). Holmes. *C. & D.* 1921:40 - *J. Trop. Med.* 1923:145 - *Brit. Med. J.* 1923 (1):57.
1149. Sedative Tablets (Brisland), with pure pd.
1150. *Catalpa.* Early yellowing of lvs. (Sept).

1151. Parry, Elsie A. (NYC) (explorer) - "Khat, blessing or curse?" *Nature Mag.* Apr. 1959: 208-10 + 219. (chewing of lvs. a form of escapism in Yemen) (5 figs.) - *Arch. Pharm.* Jan. 1870 - *Nat. Geog. Mag.* 1917:173.

1152. Taylor, Gart C. *Amer. J. Pharm.* 99:214-32; 1927 (résumé of invest. of chem. pharmacol. & clinical use) - U.S. Pat. 2, 254, 051 (Flint Eaton) (Ceanothyn; u. for hemorrhage) (*Amer. Drug.* June 1929: 50) - Wirth, E. H. *Amer. J. Pharm.* 98:503-14; 1926.

1153. Hatch, Wm. Collins. *Mass. Med. J.* 18:248-9; 1898.

1154. Roig y Mesa, Juan Tomas. *Plantas Medicinales* (Cuba); 1945.

1154A. Gorrie, R. Maclagan (IFS). *Ind. Forest. Rec.* 17: pt. 4, 1-140 + 6 pls.; 1933 ("The Sutlej deodar - its ecol. & timber prodn."). Sale of wd. not permitted in Pak; however, wd. was being smuggled in openly on camel back (in 1951) fr. Afghanistan. - Note novel of Kipling, *Under the Deodars* (re Simla & nInd where tree grows).

1155. Zderkiewicz, T. *Roczniki Nauk Rolniczyck* 75, ser. A, "Plants," No. 3; ca. 1957 (biol. of Centaury in connection with cultn.).

1156. Taylor, E. B. *Plant World.* 9:39, 43; 1906.

1157. *Bull. Imp. Inst.* 17:296-7; 1919.

1158. Found in cemeteries & yards of seAla (US); range w to Miss, n to SCar; much destroyed by fire; VO said of interest.

1159. San Giorgio (Anna): *Catalogo poliglotto delle Piante.* 329; 1870 (Cerea erba).

1160. Gallizioli, Filippi. *Elementi Botanico-Agrarj,* v. 3, No. 654; 1810 (Firenze): "Cactier à grandes fleurs"; serpent; frt. substance has an acid agreeable (piacevole) flavor.

1161. Reece, Richard. *A Treatise on the Antiphthisical Properties of the Lichen Islandicus.* 96 pp.; 1805 - Berg, Theodor. *Zur Kenntniss der in der* Cetraria islandica *Ach. vorkommenden lichenins u. jodblaeuende Stoffes.* Dorpat. 40 pp.; 1872 - Guesdon, H. *Le lichen d'Islande....* Paris. 63 pp.; 1901 (acids) - Mercier, Justin. *Contribution à L'étude Pharmacologique des Cestrées.* Paris, 134 pp.; 1913 - Salkowski, E. *Z. Physiol. Chem.* 110:155-66; 1920 (carbohydrates).

1162. Wellendorf, M. *Nature.* 198:1086-7; 1063 (compn.); 1966.

1163. Greene. *Amer. J. Pharm.* 1878:250 - Moser, John. *Ibid.* 89:291-6; 1917 (gives diffs. of H. & Aletris; notes misleading synonyms of 2; errors of descrip.; histology).

1164. *Amer. J. Pub. Health.* 13:543-6; 1923 - However, negative opinion of Castillo, Roberto Levi (Ecuad) - *Sci.* 100:266; 1944 (*Chara fragilis* Desvaux found of no pronounced value).

1165. Guerra, R. Y. R. *et al. Archivos de Pediatria del Uruguay.* 22:367-8; 1951 (dermatosis in course of C. therapy).

1166. Naito, K. *Acta Med. Biol.* 2:695-705; 1954 (more marked on cocci than bacilli; morph. eff. noted on *Pseudomonas aeruginosa, P. fluorescens, & E. coli).*

1167. Jensen, A. L. *The White House*; 1958. (Mrs. Taft was instrumental in bringing Japanese cherry trees to Wash, DC; sent by Mayor of Tokyo; also much cult. at Macon, Ga (Macon Cherry Blossom Festival, annual event).

1168. Backwoods people around Charleston, WVa use lvs. in tea for kidney trouble (*Chimaphila*).

1169. Dilling, Walter J. *J. Pharm Exper. Ther.* 26:397-411; 1926 (constits. & therap. pharmacol. of *Mondia w.*).

1170. Bovet, Daniel. *Etat Actuel du Problème du Curare - De Claude Bernard à L'anesthésiologie Moderne.* 48 pp.; 1960.

1171. Lacombe, Fco. Henrique Luiz. *Anais da Facul. de Odontologia e Farmacia, Univ. Min. Gerais.* 3:551-62; 1939 (contr. to study of Paineira).

1172. Zuco, F. M. *Rend. d. Acad. d. Lincei (Roma).* 1890:571-5 - *Amer. J. Pharm.* Nov. 1890:579 - Karrer, W. *Konst. u. Vorkommen d. org. Pflanzenstoffe,* ed. 2, 1976.

1173. Andrews, H. S. *J. Kentucky Sta. Med. Assn.* 49:279-80; 1951 (carob flour) - Frt. eaten for heartburn (Geiger, 1830) - Malaspina, M. *Minerva Medica.* 4:264-72; 1952 (c. flour in diet of infant).

1174. Hall, H. M. & Clements: Carnegie Inst.

1175. *Household Words* 6: #138; pp 208-10; 1853 - Lamson-Scribner, F. *Nature Mag.* XI-34:207.

1176. Schmitt, Alfred: *Bot. Arch.* 40:516-59; 1940 (lactucopicrin, NBP).

1177. Parrish, Edward. *Eclectic Med. J.* 20:197-200; 1861.

1178. Bowie, McCool, Miss.

1179. Bartholomew, E. T. & Sinclair, W. B. *The Lemon Fruit; Its Compn....*" Univ. Calif Press; 1951 ($4.50) - Painting of lemon tree was masterpiece of Sir Frederic Leighton; 1856.

1180. Gathering of citrons reproduced in book by A. L. Baldry (*Leighton*) (s.d.) (Old Damascus) - *J. Agr. Res.* 53, #11; 1937 (chem. & ferm. studies on citron).

1181. (*Citrus*). Hume, H. Harold. *Citrus Fruits,* ed. 2; 1957 - *The Citrus Industry,* v. 1 (1967), v. 2 (1968). Univ Calif - *The International Citrus Crops Bibliography* (1974-83) (Biol. Abstr.) - Lord, Earl L. "A taxonomic revision of genus *Citrus"* (esp. cultivated spp.) MS thesis, Univ. Fla; 1929. Citrus VO have high terpene hydrocarbon content & on standng develop terebinthinate flavor & odor; hence terpeneless oils preferred. Citrus frt. u. to make syrups, wines, & vinegars (orange v. [with 5.5% acetic acid]) (lemon v.), marmelades, &c.

1182. Saelhof, Clarence D. (Chi). *Amer. J. Dig. Dis.,* Aug-Sept. 1955. (CVP) u. to treat hemorrhagic cystitis, trigonitis; effects comparable to

Gantrisin™, Orange oil (& other citrus oils) claimed insecticidal (Univ Calif research, 1983).

1183. Wester, P. J. "Food Plants of the Philippines," Bur. Agr. *Bull.* 39:94; 1925 - Termed a lemon-like citrus frt.; "Kusaie."

1184. Tang, T. H. & Chao, Y. S. *Pharm. Archives.* 11 (4):60-4; 1940.

1185. Cult. as garden crop (Chiengmai Prov., Thailand) (paktamlung [Thai]; u. in foods).

1186. Ukers, W. H. *All about Coffee.* 1935 - Jacob. "Coffee: The Epic of a Commodity," s.d. ($3.50) - "Coffee making," *Holiday* (Mag.) summer, 1949 - *Sci. Mo.* 70:185-90; 1950; coffee houses once very popular in England; among others, Lloyd's Coffee House (London), a meeting place for businessmen, became later Lloyd's of London, famous insurance agency.

1187. Bengis, R. O. *J. Biol. Chem.* 97:99-113; ca. 1938 (kahweol isolated fr. non-sap. matter of coffee oil).

1188. *Pharm. Zentralh.* 81:1-5;1940 (c. charcoal useful on surface wounds; esp. suppurating & foul smelling; higher mineral cont.) - Koenigsfeld. Carbo - *Klin. Wchschr.* 18:609; 1939 (0.85-0.87% caffeine).

1189. Sprecher von Bernegg, Andreas. *Tropische u. subtrop. Weltwirtschaftspflanzen.* III: 214-256; 1934 (Enke) (extensive review) - Heckel, Edouard. *Les Kolas africains* (1893).

1190. *Rept. U.S. Commissioner of Patents—Agr.* 1854:383, 384. (tanyah keeps longer and is as sweet as potato) - Hood, S. C. "The Dasheen," *U.S. Dept. Agr. Farm. Bull.* 1396 - Adlung. "Tropenpflanzer, Ztsch. f. tropische Landwirtschaft," Berlin 16:547-, 609-, 662-; 1912 - 22:191-201 (most important foods) - Taro suggested for use as meal (flour, farina, fecula) with better flavor than irish potato meal, &c. - Grown in Mexicali area, Mex, 1942. Poi was basic food of native Hawaiians in old days (good physique, longevity) (Degener, *Plants of Haw. Natl. Park*, 79-85; 1930; 1945).

1191. *Not. Judg. FDA (DD)* 4675; 1956 (crude natural unrefined seepage petroleum oil).

1192. Engler, A. & Diels, L. in A. Engler, "Monograph. Afrikan. Pflanzenfam. u. Gattungen III. Combretaceae," *Combretum.* 116 pp.; 1899.

1193. Peyer, W. & Liebisch, W. *Pharm. Zentralh.* 70:197-; 1929 (authors give bot. pharmacog. & micro descr. of plant, also chem. anal.; u. in lues, as worm remedy).

1194. *Amer. J. Pharm.* 1898:321 (u. in wounds, hemorrhages, ear & upper air passages).

1195. Was used in propr. "C & C" (*Commelina caroliniana* or Child & Chick) (Daytona Beach, Fla [US]).

1196. Ulrici; 1905.

1197. Vale, Francisco Pereira. *Anais da Faculdade de Odontologia e Farmacia da Univ. de Minas Gerais.* 4:238-43; 1939 (A Cicuta - Socrates) - *Chem. Drugg.* 12/8/45, p. 621 (Antony Stoerck, Vienna, overenthusiastic for C.).

1198. Heckel & Schlagdenhauffen. *Ph. J.* 1896:243 (seri-gbeli).

1198A. Hunter, Matthew V. *Chem. Drugg.* 139:304; 3/20/43 (insecticidal properties not due to its alkaloids).

1199. Hirschfeld, E. "Studium ueber die Heilpflanzen," *Kyklos, Jahrb. d. Inst. f. Geschichte der Medizin*, v. 2:145; 1929 - Eicholtz. *Dtsch. med. Wchschr.* ca. 1936 - Brunn v. Geschichtliches ueber das Maiglöckchen - Shubov, M. I. *Terapevticheskii Arkhiv.* 23:59-62; 1951 (new glycosidal prepn. in the treatment of circulatory insufficiency) - Convallariae Flos has been/is off. in 8 world pharmacopeias, incl. Helv.V, Gall. III-IV, Pol II, USSR IV-VI.

1200. *Trans. 7th Ann. Meeting, Iowa Sta. Med. Soc.* (1956) (u. in calculi of the bladder with pain on urination; given prepn. copaiba & benzoic acid, stones excreted *per via naturalis).*

1201. Specialists on *Copaifera*: J. Dwyer (Loudonsville, NY); J. Hutchison, former keeper, Museums of the Royal Botanic Gardens, Kew, Eng (revising gen.) - Macbride. *Flora of Peru*, III:120- ; 1936/38.

1202. Caranday was said superior to ouricury wax (Markley, 1953) (Guarani).

1203. McIsaac, R. *Trans. Med. Phys. Soc. Calcutta* 3:1827 (*Goetting. gel. Anzeigen.* 1829, II:999) (description of a root called Misimee Teeta used medicinally in Assam) - Hooper. *Pharm. J.* 34(4):482; 1912.

1204. Nozoe, T. *et al. Bull. Chem. Soc. Japan.* 1955: 594-6.

1205. Ohlsson, Erik. *Svensk Kem. Tids.* 44:121-9; 1932 (favorable results in TB of cavy) - Steinmann, B. *Schweiz. Med. Wchschr.* 81:633-7; 1951 (combined treatment of tuberculous meningitis with streptomycin, PAS, & chaulmoogra oil).

1206. Desprez. *Caducée* (Paris). 3:75-; 1903 - Valenti. *Arch. di Farmacolog. Sper* (Roma) 24:23, 33, 65; 1917 - Valenti. *Gior. di. Clin. Med. Parma.* 2:161-5; 1921 (pharmacotherapeutic study of a new iodized fatty substance: iodated chaulmoogra oil) - *J. Ind. Eng. Chem.* 19: 939-42; 1927 - Roos, Ann. *Man of Molokai.* 1943.

1207. *Rept. of Govt. Chemist, Khartoum* (Williams). 1933.

1208. *S. Af. Med. J.* 25:284-6 - Joyce & Curry. *Cannabis, Botany & Chemistry.* 1970 (J. & A. Churchill, London, $11.40).

1209. O'Shaughnessy, Sir Wm. Brooke (1804-89). *On the Preparations of the Indian Hemp*, ca. 1840 (NYPL has copy) (effects & uses) - Robinson, Victor. *An Essay on Hasheesh, Historical & Experimental*, ed. 1 (1912)/ed. 2

(1930) (Dingwall Rock, NY) - Bromberg, Walter. *Amer. J. Psychiatry.* Sept. 1934 - Typical large seizure: 50 tons of marijuana seized in Fla brought fr. Ve (24 Dec. 1973).

1210. (*Cap. frut.*) Hort. var. ("pepper plants") at St. Augustine, Fla, Chas. F. Hopkins, Prods. Inc., (1931): "datil" - *Chemurgic Digest.* 30/IV/42 - *NM Agr. Exp. Sta. Bull.* 306; 1943 (contents of vits. A & C) - Gilpin, Laura. *"The Rio Grande, River of Destiny.* 1949 - Chili peppers strung on cords in long loops (ristras) (picture) hung fr. vigas & festooned over roofs. Both Span. Americans & Indians subsist largely on beans & chile.

1211. Hart, Ralph (H. Products Co., NY). *Ind. Eng. Chem.* 22:329-331; 1930 (Anal. Ed.). (includes info. on gum, with many names).

1212. Plowright, T. R. *J. Pediatrics*, St. Louis. 38:16-21; 1951 (Carob flour [Arobon] in controlled series of infant diarrheas) - *Med. Welt.* 20:747-9; 1951 (Carob, a new dietetic: II. c. meal in clinical use for inspissation [antidiarrheal]).

1213. (Chelidonium alks.). Gadamer, J. *Arch. Pharm.* 262:249-77; 452-500; 1924 - Osada, Shoji (Univ. Marburg). *Ibid.* 262:501 (conversion of alks. of aporphine series into alks of chelidonine series) - Bersch, H. W. *Ibid.* 272:129; 1934 (constitution).

1214. Gadamer, J. *Ap. Ztg.* 39:1569; 1924.

1215. *Ap. Ztg.* 52:1038; 1937 - *Die Deutsche Heilpflanze.* 5(#7):98-104; 1939 - Bandelin, F. J. *J A Ph A.* Oct. 1956.

1216. Rape (seed) (or colza) is an ancient crop in As; currently it is a big crop of Canada. As a food oil (chief use), it was said to cause myocardial lesions in rats (due to large % of erucic acid). A strain has been developed low in erucic acid (in BRD erucic acid limited to max. 5% in oils for food use) wh. represents a better food product. (*Chem. Eng. News.* 50(39):6; 1972.)

1217. VO fr. Univ. Fla. Med. Pl. Garden low in ascaridole (27-8%) (ca. 1930).

1218. Rusby, H. H. *Jungle Memories*, p. 340; 1933. Quinoa (pron. keen-wa) (cereal mill, Bo).

1219. *Chem & Drugg.* 5/15/42: p. 166.

1220. Rentz, Edward. *Skand. Arch. Physiol.* 56:36-117; 1929 (pharmaco-dynamics: vasodilator).

1221. Engelmann, G. *Amer. J. Sci. & Arts,* 43:333-45; 1942 (monograph) - Yuncker, T. G. *Torr. Bot. Club. Mem.* 18:113-331; 1932 (gen. *Cuscuta*) - Gaertner, E. E.: B.S. A. thesis for Ont Agr. Coll. (unpublished) (1944).

1222. *Chem. Eng. News* 37:37-8; 1959 (Sept. 21).

1223. Van Rossem, Walter J. U.S. Pat. 2,265,118.

1224. Reedman, E. J. (Summerside, Quebec) & Buckley, L. *Canad J. Res.* D21:348-57; 1943 (Gelose as agar substit.).

1225. MacFarlane & Bell. *Proc. & Trans. Nov. Scotia Inst. Sci.* 18:124; 1933 - *Drug Cosm. Ind.* 35:495, 567-; 1934 - *Chem. Drugg.* 1945. p. 587, 659 (soln. of chondrus with saccharin u. during Occupation as sugar substit. in Jersey Is; Chondrus thriving indy., selling at 5/-/lb.) - *Fish Boat* 7 (Apr):26-7; 1962.

1226. *J. Pharm. Chim.* 6:58-62; 1932 (M. Fauré) (diffs. nat. & artif. camphors).

1227. Juan T. Roig y Mesa (Santiago de las Vegas) letter (1945): *Cassia grandis* rather common in 3 eastern provinces of Cuba in forests; popular in Orient; frt. sold in markets (dealers) at Santiago & Guantanamo.

1228. Anchel, Marjorie (NYBG). *J. Biol Chem.* 177: 169-73; 1949 (antibiotic identified as rhein).

1229. Pollard, C. L. *B. Torr. B.C.* 21:208-222; 1894 (gen. *Cassia* in NA) - *Ibid.* 22:513-6; 1895 (so. *Cassia* taxa) - Slama, Frank J. *J A Ph A.* 23:550-9; 1934.

1230. Cult. in Ind (Hallacary Estate, Coonoor) (1944); also at Dehra Dun, Saharanpur, Nilgiris, Mysore, Calcutta (Chopra, R. N., Chopra, I. C., Handa, K. L., and L. D. Kapur. *Indigenous Drugs of India*).

1231. Cortone™ (Merck): Annotated bibliography (several vols.) (Merck. USA).

1232. Laverne Block Co., Bay Minette, Ala (Eiland) mfr. dogwood spindles (-1956-).

1233. Diels, L., Samuelson, Hannig, & Winkler. *Die Pflanzenareale.*

1234. Hesser, & al. *J. Neuro-psychiatry.* 1941 (describes techniques u. to immobilize cats for neurophysiologic studies).

1235. Robertson, W. A. *Chem. Drugg.* 18 (Mar. 1944):316-7 (cultn. minor prods.).

1236. Tot. extracts of C. bark incl. Quinetum; Cinchona Febrifuge (CF), & Totaquine; *cf.* Cinchona Alkaloids, Merck c. 1930.

1237. Graham, J. D. P. *Brit. Med. J.* (2) 951; 1939 (*NY Daily News*, art. by Dr. Irving S. Cutter) (use in hypertension) - Metzger, P. *Med. Welt.* 20:814-6; 1951 (clinical experiences in cardiac treatment with Crataegutt & the combn. with digitalis) - Bohm, K. *Arzneimittelforsch.* 6:38-41; 1956 (results of C. research. II. Animal expt. findings with total exts. & isolated pure substances) - Beringer, G. M. *Amer. J. Pharm.* 76:283-4; 1904 (speaks of plants as only of European origin).

1238. Baechler, L. *Diss. Basel.* 1927:114 pp. (Monographie der Mehlbeeren; chemische Untersuchungen ueber die Fruechte).

1239. Babcock Ernest Brown. *Univ. Cal. Publns. in Bot.* v. 21, 22; 1945 (I: tax., phylogeny, distrn., evol.; II: Systematic treatment).

1240. Chapin, Robert M. *U.S. Dept. Agr. Bull.* 1208: 1924 (chem. & phys. methods for control of saponified cresol solns.).

1241. England, J. W. *Amer. J. Pharm.* 58:82-; 1888 - Nichols, A. B. *Ibid.* 92:391-3; 1920 (lists forms of $CaCO_3$ u. in phy. & med.).

1242. Kronfeld, M. *Geschichte des Saffrans.* 1892 - DeMori, Alessandro. "The aquila saffron," *III. Int. Congress of the Int. Fed. for the Promotion of Prodn. & Utilization of Med., Arom., & Allied Plants.* 1929, pp. 501-23 - Koelner, P. *Safranzunft* (=Saffron guild) *zu Basel.* 1935.

1243. *Crotalaria* alks., consult: Carmack (Pa); Rockef. Inst. Med. Res. (NY); Adams (Ill.); Noller (Stanford Univ.).
1244. Neal, W. M., Rusoff, L. L., & Ahmann, C. F. *J. Amer. Chem. Soc.* 57:2560-1; 1935 (isolation & properties).
1245. Cash & Dilling. *J. Pharmacol. Exp. Ther.* 5:235; 1913
1246. Baillon. *Bull. Soc. Lin. Paris.* II:865; 1890.
1247. *Amer. Dr. Circ.* 1:121; 1857 - VO c. 3%+ (anal. of J.B. Dancy, Miss): "malambo" name used by Colombian Ind.
1248. Payson, E. B. *Ann. Mo. Bot. Gard.* 14:211-358; 1927 (monograph of the sect. *Oreocarya of Cryptantha*).
1249. Kraemer, H. *Amer. J. Pharm.* 1924:197 - Wehmer. *Pflanzenstoffe*, ed 2, pp. 1203-5 - *College J.* (N Y C P). 3:45; 1858 - *Worcester J. Med.* 10:305; 1855 (under new remedies: pumpkin sd. rem. for tape worm).
1250. (*Cucurbita pepo).* Pumpkin, "largest known" grown in Conn [US] 1985 (wt. 515 lbs); specimen 755 lbs. 12 ft. circumference (Canada, 1989) - Scheibe, A. *Der Oelkurbis* (*Cuc. pepo*). 1977 - In Russia, pumpkins u. only as hog food; considered unfit for humans!
1251. Zucca melon (BC [Cana]). Grown at Osoyoos, BC (4' long). Pulp cut into cubes, colored red & green, and candied for Xmas cakes.
1252. *Science.* 3 May 1968:545-7 (importance of chemotaxonomy using FA to show phylogenetic position).
1253. Pulzker & Jungkunz. *Pharm. Acta Helv.* 13: 29; 1938 - Gilan quality, Teheran, Iran ('37)
1254. *Cymb. pend.* (Dunal) Baill. (*Adansonia.* 8: 268; 1868) was mentioned centuries ago for use in med. ("Mataboni"; "Guanabano" [Belize]).
1255. Cult. São Paulo (Br).
1256. Nesterenko, P. *Sovetskiye Subtropiki* (Sov. Subtrop.) 1931, 2(9):134 - *J. Simla Nat. Soc.* 1:55; 1885 - "Sabalin" (Burma) - Meyer Bros. Drugg. 11:90-; 1890.
1257. Billington, Horace W. L. *Chem. Drugg.* 47:120; 1895. (decoc. u. in wAf) - *Misc. Inform. Kew* 1906, No. 8 (Stapf).
1258. *JAMA.* 119:917; 1942 (11 July).
1259. Gathercoal, E. N. *J A Ph A.* 8:26-32; 1919 (couch grass vs. Bermuda grass) - Farmacopeia Portuguesa IV (1946).
1260. Note work of Dr. Florencio Bustinza ca. 1927 (rich in enzymes).
1261. L. G. Holm (Univ. Wisc.) writing book on nutgrasses (ca. 1973) "world's worst weed" (*CEN* 29-V-73) - Asenjo, C. *J A Ph A.* 30/31: (1941-2).
1262. Chevalier, J. *C. R.* (Paris) 150:1068-; 1910 (variation in strength of sparteine acc. to stage of vegetation).
1263. *C. canariensis* is a shrub to 1.8 m. high with fragrant yellow fls. in head-like clusters.
1264. Dangschat, Gerda & Fischer, Hermann O. L. *Naturwissenschaften.* 27:756-7; 1939 - Kern, W. & Fricke, W. (Inst. f. angewandte Pharmazie, Techn. Hochschule, Braunschweig) *Pharm. Zentralh.* 80:349-52; 1939 (obtained higher yields [aver. 0.4%] than Kubler [0.3-0.5%]).
1265. *Arch. d. Pharm.* 1955
1266. *Ind. Eng. Chem.* 23:413; 1931 - *C. nuc.* cult. at Orissa, also in trop. Ind & Pak, esp. near coast; valuable tree in Bangladesh.
1267. Unna, K. R., *et al.* N.Y. Acad Sci. v. 54 (Art. 3): *Curare & Anti-curare Agents* (sponsored by Roche). 240 pp.; ca. 1950 - Curare concentrate was marketed ca. 1947 by Transandino Co. (biol. assayed). While calabash curarc is 10-100 X as strong as tube c., the latter in the form of tubocurarine is chiefly u. in therapy (Awe, Walther: *Südd. Ap. Ztg.* 88:330-; 1948) (Much work by Wieland) - "Curare" of the physician/surgeon is usually succinyl choline chloride (a synthetic).
1268. First samples of curare sent in 1595 to Eng by Sir Walter Raleigh ("urari").
1269. Michaelis, A. A. (ca. 1894): *Der Kaffee als Genuss- u. Heilmittel nach seine bot. chem. diaetet. u. mediz. Eigenschaften* - Eichler, O. *Kaffee u. Koffein*, V + 160 pp., 24 figs. in text; 1938 (RM 8.70). Use of coffee said associated with higher pancreatic cancer rates.
1270. Largely cult. on lower mt. slopes of sInd (Cochin) at 1000-5000 ft.
1271. *Chem. Drugg.* 52:549; 2 Apr. 1898; cult. in Darjeeling Distr. (1936); Nilgiris; Burma.
1272. Dixon, W. E. - *Brit. Med. J.* 1920, ii: 113-21; 1920 (In expts. found Cinchona Febrifuge (CF) more efficacious than quinine in benign tertian mal., due to quinidine & cinchonidine; amorph. alks. no good in mal.) (Acton, HW).
1273. Quinetum (Chinetum) c. 50-70% cinchonidine with quinine, cinchonine, & amorph. bases - "In the doses recommended by the Commission (1 g. per day per 70 kg. body wt.), the Amsterdam quinetum (80% crystallizable alks. incl. 15% quinine) is quite as efficacious as quinine" *(League of Nations Health Orgn. Malaria Commission; C.H. Malaria*/161 p. 3 (25 Apr. 1931) - Q. Sulphate - price ca. 1.5 X that of Quinetum (*Merck Index* 1907) - Name Quinetum origin. DeVrij ca. 1872.
1274. Quinidine: reputedly greater antimalarial activity than quinine and u. specifically to treat chloroquine-resistant cases of malaria (*Plasmodium falciparum*); but not generally u. because of circulatory effects.

1275. Totaquine: Chick, Oliver in Allen's *Comm. Org. Analysis*. 7:1929 (ed. 5) states that commerical quinine may cont. cupreine since prepd. fr. *Remijia pedunculata* bark (formerly comm. used) - Totaquine was manfd. by a number of foreign firms, incl. Roche (1929), B & W (London), Howards (London) (Type II), and a firm in Colombia (SA) (ca. 1941, said to produce 500-600 kg./mo.) & some natl. govts. (incl. Govt. of Madras, State of Turin, Italy).
1276. *League of Nat. CH/Mal./159* (alk. cont. bk of *Cinch. succ.* & *C. ledger.*).
1277. *League of Nat., Malaria Commission, Sect. of Trop. Med. Rept.* 6/16/32 - Taylor, Norman. "Cinchona," *Fortune*. 9:76-86; 1934 (Jan.) - BPI (USDA) Library - Cinchona - references, ca. 1944 (copy at NYBG Libr.) - Duran-Reynals, (Mrs.) M. L. *The Fever Bark Tree*. Doubleday-Doran, NY; 1946. $2.75 - Formerly, cinchona bk. powder u. in malaria (dose: 20-60 gr. 2 to 3 X/day (antiperiodic) (10-15 gr. tonic).
1278. Brazil, Oswaldo Vital, Seba, R. A., & Campos, J. S. *Bol. Inst. Vital Brazil*. 1944:268- (curarizing props. of methylated alks.).
1279. *Schimmel Rep*. Okt. 1906:25; Okt. 1909:51 - *Chem. Drugg*. 74:81; 1909 (yield from fresh peel —> 0.15% VO).
1280. Wolfson, Wm. Q. *et al.* (Univ. Mich) *J. Mich. State Med. Soc.* 49:1058; 1950. (colchicine was u. with ACTH [initial] in acute gouty arthritis & found very effective; used c. in doses sufficient to prod. diarrh.) - Eigsti, O. J. & Dustin, P. *Colchicine in Agr. Med....* (Colchicine Research Foundation, Inc., Normal, Ill, later Goshen, Indiana) (Ia. Sta. Coll. Press, 470 pp.; 1955) - Eigsti, A. J. *Amer. J. Pharm.* 1957. "Induced Polyploidy" with bibliography - *Cf.* No. 682. Vogl. Anat. Atlas t. 33) (C. sds. c. 1.20% alk. [SB]).
1281. Grown in Illyria, Macedonia, Afghan, Ind (Himal); exp. fr. last 2 (Schamballit).
1282. Some say occ. naturally with colchicine. *Monatsh. Chem.* 9:870; 1888 - Oberlin. *C. R. 43:1199; 1956* - Huebler. *Arch. Pharm.* 171: 193; 1865; Gohar & al. *J. Pharm. Pharmacol.* 3:415; 1951 - Walz (older lit.) - Zeisel thought it formed during extn. by decompn.
1283. (Curcuma) Dieterle, H. & Kaiser, Ph. *Arch. Pharm.* 270:413-8; 1932 (constits. of rhiz.). Gruber, A. *Deut. Med. Wchschr.* 53:1299-1301; 1927 - Peyer, W. & Huenerbein, E. *Ap. Ztg.* 47:112-5; 1932. (Commercial brands of Emoe Lawak [turmeric] examined; histol.) *Ibid.* 49:112-8: 1932 - *J. de Pharm. de Belgique* 5(3); 1933 - *Pharm. Acta Helvet.* 8:72-; 1933 - Turmeric is used in boiler compds. ("Era Formulary"). Sole supplier: India (exports to Sri Lanka) (1953) (see also 1997).
1284. *Bericht von Schimmel & Co.* Apr. 1893 (figure in color) - General Custer dressed his hair with cassia oil.
1285. *Chem. Drugg.* 52:549; Apr. 2, 1898 (1565).
1286. Calvert, Robert. *Amer. Chem. Soc. Monograph* #52; ca. 1941 (251 pp.) (Reinhold) - *Miss. State Geol. Survey, Bull.* 51:24 (Montgomery Co.) - Various uses of diatomite: 1) in salt dispenser tops to prevent salt caking & flow stoppage 2) diatomaceous filters in filter presses 3) heat insulation (Superex Blocks) 4) building blocks 5) furnace-lining bricks 6) important component in asphalt battery box 7) silver polishes (Gorham). Some 100,000 tons/year u. in US.
1287. Pilocarpine also has been u. for reabsorption of cataract - Geiger. "Pharmacog. of Pilocarpus leaves" (diss., 1897-).
1288. Lloyds' Brothers was first to market prepn. (1886-); said to have effect on 0_2-carrying power of blood serum; useful for reabsorption of cataract (Wm. A.Lurie, M.D., New Orl. [US], 1940).
1289. Greshoff, *Bull. Kew* 9; 1909 (comml. use?) - Mueller-Dietz et al. *Arzneipflanzen in der Sowjetunion* (Berlin). 1960-8.
1290. Dyerberg. *Lancet.* 1978-80.
1291. Viehoever, Arno & Sung, Le Kya. "Common & oriental cardamoms," *J A Ph A.* 26:871-87; 1937.
1292. Card. cult. abundantly in rich moist forests of hilly tracts of Kanara, Mysore, Coorg. Travancore, Cochin, Bombay, Madura, &c., at 2500-5000 ft.; "Ceylon" card. packed in cases.
1293. Baranov, A. L. *J. Ethnopharm.* 6:346; 1982.
1294. (*E. sonch*) Small red fl. heads (*cf.* small carnation); corolla 7-8 mm. long; blooms all year. Natd. WI introduct. to sFla (peninsular Fla & the Keys).
1295. The synthesis was carried out by B.W. Co. at Dartford, Eng in 8 stages.
1296. Brandegee: *Zoe.* 1:83; 1896 (Baja Calif).
1297. Chopra, Chopra, & Ghosh. *Indian J. Med. Res.* 28:469-74 (saponin) - Chakravarti, D. *& al. Proc. Ind. Sci. Cong.* 1955:128 (sds. c. entagenic acid, new triterpenoid).
1298. Found in Cana & eUS s to Ala, Ga; occurs in railroad cuts & embankments, stream banks, old fields (chiefly mts. & piedmont), n to Greenl, Newfld, Alas - *Dom Canada Dept. Agr. Circ.* 74, N.S.
1299. Ergosterol most important pro-vitamin D; ca. 1% yield fr. dry mycelium of *Penicillium* spp.; irradiated gives vit. D_4 (antirachitic vit.) also lumisterol & tachysterol; 2-5% yield fr. yeasts (*Sci.* 13: Oct 1944, p. 333).
1300. Acree, Jacobson, Haller. *J. Org. Chem.* 10: 236, 445; 1945 - c. affinin (amide).
1301. Tschirch, Alexander. *Handbuch der Pharmakognosie*. 3:839-40; 1925 - Mager, A. "Ueber die Zusammensetzung des Eriodictyolglykosides," *Hoppe Seylers Zts. f. Pharm. Chem.* 274:109-15; 1942.

1302. Thornber, J. J. *Ariz. Agr. Expt. Sta. Bull.* 52:27-58; 1906 - *Idem. Plant World.* 10:208; 1907 (sp. is drought resistant).
1303. This annual herb is raised in Austria; weed in Austral (Carne, W. M. *J. Dept. Agr. W. Austral.* II. 4:390; 1927) - *Trans. Calif Agr. Soc.* 1873: 487-501.
1304. Lucius Junius Moderatus Columella in his *De Re Rustica* (60 A.D.) wrote re *Eruca sativa* "Excitet ut venere tardos Eruca maritos" (Eruca excites carnal desire in late marriages) (transl. in Nisard. *Les Agronomes Latins*, Paris; 1877).
1305. Jacobs, Joseph. *Proc. Amer. Pharm. Assn.* 46:205; 1898.
1306. Hiller, K. *et al. Pharmazie.* 30:105-9; 1975 ("Constits. of some Saniculoideae"; 22: "Isolation of new sapogenin esters fr. *E. gig.").*
1307. Dalma, G. *Ann. Chim. Applicata* 25:569-71; 1935 - Schlittler, E. *Helv. chim. Acta.* 24:63-76, 319-32; 1941 - Ruzicka, L., Dalma, G. & Scott, W. E. *Ibid.* 23:753-64; 1940 (cassidine, cryst. alk. fr. *E. guineense*).
1308. Ruzicka, L. & Dalma, G. *Helv. Chim. Acta.* 22:1516-23; 1939 (cassaine hydrolyzed to cassainic acid; both have digitaloid activity).
1309. Peckolt, Th. *Volksbenennungen des brasil. Pflanzen*; 1907.
1310. Hegnauer, R. *J. Ethnopharm.* 1981:279-93, esp. 286-8.
1311. Martindale Wm. *Coca and Cocaine.* 1884 (ed. 2, 1897) (76 pp.) (another ed. printed in blue ink on green paper: *Coca, Cocaine & Its Salts* [69 pp., H.C. Lewis, London; 1886]) - Mortimer, W. G. *Peru, History of Coca, The Divine Plant of the Incas.* xxxi + 576 pp.; 1901, J .H. Vail, NY - Glasenfeld, B. *Deutsche Med. Wchschr.* 46:185; 1930 (increase in cocaine addiction in Berlin) - Countair-Suffel & Giroux. *Bull. Acad. de Méd.* Paris. 88:65-8; 1922 (means to reduce c. traffic) - Niessen, Jan. *Aerztl. Mitt.* 28:595-8; 1927 (physicians as c. "bootleggers") - Rosenfeld, L. M., *Deut. Med. Wchschr.* 54:998-1001; 1928 (psychic disorders resulting fr. morphine & c. habit) - Merzbecher. *Muench. Med. Wchschr.* 76:2016-9; 1929 (coca chewing among SA inhabitants) - Jacoby, C. *Arch. f. Exp. Path. U. Pharmakol.* 159:495-515; 1921 (significance for peripheral actions in explaining chewing of coca lvs. by Indians) - Muntch, N. *Guy's Hosp. Gaz.* 46:425-9; 1932 - Mellini, F. *Archivio de Farmacol. Sper. & Scienze Affini.* 79:21-4; 1949 (histol. str. of c. lvs.) - *Sci. Mo.* Feb. 1950 - Arthand, Claude & Hebert-Stevens, F. *The Andes, Roof of America* (chap. on coca, the green gold of the Andes). 1956 - Ashley. *Cocaine.* St. Martins Press, NY; 1975 - A. Weil, M.D. (lecture) stated that c. leaf is u. pwd. & carbonized, placed in mouth & kept until all absorbed to give stim. Coca Cola™ generally believed superior when whole c. leaf was used in prod. - Coca & coca paste exported fr. Peru - Shortage in supply of legal cocaine (for Brompton's Mixt. &c.) reported (*A Pharmacy.* 25 Oct. 1978).
1312. *Foreign Commerce Weekly.* 7 Aug. 1944, pp. 8-9.
1313. *Merck Index* X. 1848; 1984.
1314. Red gum imp. in boxes, 56# net wt. fr. Australia (ca. 1944)
1315. Tomaszewska, E. & M. *Arboretum Kornickie.* 2; 1957 (g. p. in lvs.).
1316. Melvin Calvin, Nobel prize winner.
1317. Neidhart, Guenther. *Die Pharmazie*, ca. 1947.
1318. Melton, William E. & L. E. Sayre. *J. Amer. Pharm. Assn.* 14:308-17; 1925 (dr. flg. tops - analysis) - Rue u. as substit. (per Madaus).
1319. Howard, T. M. *Plant Life.* 14:114-20; 1958.
1320. Specimen of *Eucalyptus regn.* reported in 1872 as 435 ft. (135 m) high to a point where it had broken off in the fall. The total plant was estimated at ca. 500 ft. (150 m). The "Baron Tree" (1868) was reported at 464 ft. (142 m). Currently the highest trees recognized (*Sequoia sempervirens*) are 385 ft. (117 m) high.
1321. *Evernia prunastri* (Lichen Prunastri [L]): Burger, Alfons M. *Riechstoff-Industrie u. Kosmetik.* 11:41-4; 1936 (review incl. uses) - Finnemore. *Essential Oils.* pp. 1-2.
1322. Chen, A. Ling & Chen, K. J. *J. Amer. Pharm. Assn.* 22:716-; 1933 (constituents of Wu chu yu (*Evodia rutaecarpa*) .
1323. *E. purg.* was cult. successfully in cinchona plantations in Nilgiris (Ind); the pl. needs abundant rainfall.
1324. Jackson, John P. *Pharm. J.* 1876:681.
1325. Trimble. *Amer. J. Pharm.* 1890:73 (also found in *E. cannabinum*).
1326. *Ber. d. d. Pharm. Ges.* 14:205; 1905.
1327. *Zeitsch. des Oesterr. Apoth. Ver.* 36:144; ca. 1894. This sp. is natd. in many parts of Ind; thus, common in Bombay gardens.
1328. Blouch, C. H. *Amer. J. Pharm.* Mar. 1890.
1329. Rothrock, J. T. Notes on econ. bot. of the western US, from *Rept. on the bot. coll. made in Nev., U. etc.*, fr. *Rept. upon US Geogr. Survey VI. Bot.*; *US Geol. Surv. Wheeler Rept. Bot. US* (*New Rem* 8:233; 1879).
1330. *Pharm. J.* 3(7):849; 1874
1331. Warren, L. E. *J Amer. Pharm. Assn.* 7:510-2; 1918.
1332. Eberle, E. G. *Proc. Amer. Pharm. Assn.* 53:304; 1905 (Med. Pl. Tex).
1333. Maisch, J. M. *Amer. J. Pharm.* 99:154; 1887 - Heckel, E. *C. R.* 152:1825-7; 1911.
1334. Wayne, E. S. *Pr. A Ph A.* 15:400; 1867; *Pharm J.* 2(10):27; 1868-9 (Fla plants).
1335. Farwell, Oliver Atkins. *J. Amer. Pharm. Assn.* 13:121; 1924 (adulteration).

1336. Lloyd, J. U. *Pharm. Rev.* 23:30; 1905 (confusion of *E. perfol.* & *E. macul.* thru ignorance in substitution of former).
1337. "I myself have been troubled with the ague but a dose of boneset will set me up" (Quaker letter, 1831) (*Nation* 133:433) - *Amer. J. Pharm.* 1851:206 - Jackson, John R. *Pharm. J.* 33(6): 462-; 1875 ("very best of indigenous antiperiodics as a substit. for quinine") - Latin, Geo. *Ibid.* 52:392; 1880 (chem.) - Dana, *Ibid.* Mar. 1887 - Franz, F. W. *Ibid.* 60:77; 1888 - Diggins, Frank M. *Ibid.* 60:121; 1888 - Herbert & Trimble. *Ibid.* 1890:74 - Kaercher, Henry F. *Ibid.* 64:510; 1892 (anal rhiz. & rts.) - Mundy. *Ecl. Med. J.* 65:572; 1905 (indications) - Fyfe, John Wm. *Ibid.* 66:321; 1906 - Farwell, O.A. *Merck's Rept.* 16:220; 1907 (carelessness in coll. & handling) - Holm, Th. *Ibid.* 17:326-8; 1908 - Bigney, A. *J. Proc. Indiana Acad. Sci.* 25:163-4; 1909 - Gane, E. H. *Drug Topics.* 228:1910 (thinks *E.* should be excluded fr. USP) - Howes, P. E. *J. Therap. & Diet.* 5:203; 1911 (indic. in pneumonia) - Rabe, R. P. *Hahnemann Mo.* 46:399; 1911 (use in intermittent fevers; vesical irritation in females) - Henkel, Alice. *Spatula.* 19:82; 1912/3 - Grosh, Daniel M. *Merck's Rep.* 22:170; 1913 -X-Rayser II. *Chem. Drugg.* 87:467; 1915 - Johnson, Wm. M. *Spatula.* 22:295-6; 1916 (pioneer med.) - Anon. *N A R D J.* 23:504-5; 1916 (cultn. & harv.) - Editor, *Ecl. Med. J.* 78:97-8; 1918 - Ewe. *Proc. Penn. Pharm. Assn.* 1920:110 (st. cont.) - Clute, Willard N. *Amer. Botanist.* 36: 11-16; 1930 ("The bonesets") - Gruber, *Amer. Nat.* 25:8; 1900 (ca.) - Mansfield's "Histology," p. 82; ca. 1930 - Heeve, Wm. L. *Natl. Ecl. Med. Assn. Quart.* 5:131 (claimed that *E.* produces great cell proliferation). In US, used to heal, unite (doctrine of signatures) ("consolidante"; herbe à sonder).
1338. Maisch, Johm M. *Amer. J. Pharm.* 63:321-; 1891 ("sovereign cure for rheumatism").
1339. Shaw. *Amer. J. Pharm.* 1892.
1340. (*Stevia r.*) *Pharm. Centralhalle.* 50:435; 1909 - Aranda, J. B. *Rev. Agricult. Asuncion.* July 1941:24-9 - *Mo. Bull. Agr. Sci & Pract.* July 1939: 271T-273 - U. by Lorillard Tobacco Co. to sweeten snuff, chewing tobacco.
1341. Whitehall, A. S. (Attica, Ind). *New England Bot Med. & Surg. J.* (n.s.) 5:122-5; 1851 (Howe [1845] discovered cause of milk sickness) - Sackett, W. G. *J. Inf. Dis.* 24:231-59; 1919 - Bukey, F. S. *J. Amer. Pharm. Assn.* 14:595-9; 1925 (partial anal. of frt.).
1342. Diehl, C. Lewis. *Proc. 24th Ann. Meeting of A. Ph. A.* 24:141, 734; 1876 (*Rept. on Progress of Pharmacy*).
1343. Cortland, G. F. *& al. J. Amer. Pharm. Assn.* 20: 448-54; 1931. Also work at U.N.C.
1344. Holm, Th. "Med Plants of No. Am," #39 *Merck's Rep.* 19:126-8; 1910.
1345. *Lancet*, thru *Pharm. Record.* 7:18; 1887 (Dr. John Reed, Port Germain, sAustralia).
1346. Holm, Th. *Merck's Rep.* 18:115-8; 1909.
1347. Harris, Loyd. *J. Amer. Pharm. Assn.* 20:1281-6; 1931 (seed).
1348. Univ. Calif Dept. Bot. correspondent; per Albert Schneider.
1349. Goldman. *U.S. Nat. Herb.* 16:342.
1350. Irwin claimed *C. prostrata* to be rattlesnake bite remedy; exptl. detn. using slut (bitch) ("yerba de la niña") (*Dr. Circ.* 5:107; 1861).
1351. Bekhtin(e), P. V. *Bolnitchenaia Gazeta Botkina.* No. 19; 1891 (Dr. L. V. Popoff's Clinic, St. Petersburg, Russia).
1352. Ephedra is used for tanning goat & lamb skins (Pak). 22 million kg/yr. prod. in Baluchistan for mfr. ephedrine - *Pakistan Today & Tommorrow.* 1951: "ephedra is the "financial mainstay of the Forest Dept." (Baluchistan). It is also a productive occupation for the tribal peoples. 1950-1, 26,000 maunds was exported (Rs. 550,000).
1353. Miura. *Berlin. Klin.Wochschr.* ca. 1888; per *Pharm. Record.* 8:15; 1888 (recomm. as mydriatic) - Kelly, J. W. claimed ephedrine good for itching.
1354. Grozdov, B. V. *Bot Zhurn. SSSR.* 21:93-5; 1936 - Bolomaz, V. A. *Trudy Briansk. Lesokhoz. Inst.* 4:473-99; 1940 (as tech. plant).
1355. "Tsentral'nyi nauchno-issledovatel'skii Institut Lesnogo khoziaistva (Leningrad)," *Bulletin.* 179 pp.; 1938 - Lisitsyn, D. I. *Biokhimia* 8: 182-7; 1943 (gutta p. fr. bks. of rts. & trunks of *E. verrucosa & E. velutina*) - Bosse, G. K. "Probleme maksimal'noi guttonosnoste bereskletov," *Akad. Nauk S S R Inst. Lesa Trud.* 55-78; 1947.
1356. One lime (or ration of 1. juice) was given each day to British navy personnel, as a scurvy preventive (and to ward off colds); hence British sailors called "limeys." Custom since time of Capts. Cook, Bligh, Vancouver.
1357. Fla physician advertised grapefruit as "cure" for flu in 1932; stopped by AMA. The sour-bitter flavor of this frt. is much esteemed by some, esp. with sugar added.
1358. Developed at Winter Park, Fla, 1917.
1359. Mason, David J. *J. Amer. Pharm. Assn.* 27:42-7; 1938 (review of literature from several aspects of flavoring agt. & vehicle).
1360. *(Fortunella japonica* Swingle*)*.
1361. Other vernacular names: fingered citron; f. lemon; Buddha's hand.
1362. Thus, fresh peel is used in making Calvert Gin (advt.).
1363. Other common names: golden (or orange) apple - Florida oranges appear to have more juice than Calif and are larger.
1364. Danesi e Boschi. *Staz. sperim. agrar. Ital.* 28:699; 1895 - Koenig-Boemer, *Nahrungsmit-*

telchemie, 4A. 849; 1903 - Anon. *Riv. Ital. Essen. Prof.* 9:214; 1927.

1365. *USDA BAE. Bull.* 136:222.

1366. Pittier. *Contr. U.S. Nat. Herb.* 20:102-105; 1918 (palo de vaca).

1367. Beille, Lucien. *Bot Pharm.* ii, 599; 1909 - Kanehira. *Famous Trees*, p. 100; 1917 - Bailey, F.M. *Compn. Cult Queensland Plants.* 89; 1909-13.

1368. B.s. the wood of the tree c. 4-8% semi-solid "guaiac wood" oil with guaiol (alc.), having little odor value, but esters with same; u. in fine soaps, talcs, facial powders. Wd. u. to make humanoid figures (Arg Indians). Tree from Oran & Gran Chaco provinces.

1369. Williams, M. Y. *Trans. Royal Soc. Canada* (3) 51: sect. 4, 13-7; 1957.

1370. "Mountain pink"; u. as urinary antiseptic.

1371. Schneider, Hoya S. III. *Congresso Sul-Americana de Quimica* (ca. 1937).

1372. Pickled frt. eaten (Af) but hazardous (may c. cucurbitacin); Kaffir, food (or bitter) melon; sd. eaten (Af, As), hypotensive; frt. source for food & water (Kalahari Desert); sd. FO u. as illuminant.

1373. (Colocynthin). Walz. *Neues Jahrb. Pharm.* 9: 16, 25; 1858 - Naylor, Chappel. *Pharm. J.* 25:117; 1907 - Power, F. B. & Moore, C. W. *J. Chem. Soc.* 97:99; 1910.

1374. See #1285. - Recommended in ozena (rhinitis sicca) (0.1 g./30 ml. olive oil with 30,000 I.U. vit. A.) (2-3 wks. treatment using 3 x per day).

1375. Tomaszewski, E. & M. *Arboretum Kornickie.* 2; 1957 (chem. compn. of frts. of edible *Cornus mas* & *C. officinalis*).

1376. Barksdale, I.S. *So. Med. & Surg. J.* 89:27-32; 1927 - *Amer. J. Physiol.* 105:53; 1933; 111:68; 1939 - *Pop. Mech.* 1930: p. 265 - U.S. Pat. 1,626,321 (1927) (aq. ext.).

1377. Comin, J. & Deulofen. *An. As. Quim. Arg.* 1955: 83-97 (*Fagara coco* grows from Cordoba to Bolivia; cheaper to collect in nArg or in Bo [per Enrique Moisset de Espanes, *Fac. Cien. Med.*, Cordoba, Arg; 1946]).

1378. Deulofeu, Venancio & al. *Sci.* n.s. 102:69-71; 1945.

1379. "Of all the garden herbs, none is of greater virtue than Sage" (Thomas Cogan, 1596).

1380. Beech nuts are u. as a food by humans & swine (to fatten in fall before slaughtering) (Peterson, W. J. *Deutsch. Lebensmittel Rundschau.* 1943) (compn.); however it is often difficult to find the nuts. Extensive study showed many years have crop failures; in good years, 30% are eaten by squirrels, etc., 20% show incomplete nut formation, some claim the nuts ripen only in the nUS (Gysd, L. W. *J. Wildlife Mgmt.* 35:516-9; 1971).

1381. Asafetida, compn. Subrahmanyan, V. L. Sastry, L. V. L. & Srinivasan, M. *J. Sci. Ind. Res. India.* 13A:382-6; 1954 - *J. Appl. Chem.* (need of further study for use in heart dis.; specifications needed for allowable types) - Chem. & phys. detns. of gum & VO - Clevenger, Joe F. *J A Ph. A. S. E.* 21:668; 1932 - Sastry, L.V. L. *J.S.I.R. India.* 14:585-590; 1955 (assessment of quality; chem. anal. not adequate, organoleptic tests should be retained) - In treatment of irritable colon: Rahlfs, V. W. & Moessinger, P. *Deutsch. Med. Wchschr.* 104(4):140-4; 1979 - As uter. sed., reduces ut. irritation in threatened abortion - Warman. *Der Frauenarzt.* Aug. 1895 - As hair tonic (with naphthalene & kerosene, Sheldon, E. N. U.S. Pat. 841,057 (8 Jan 1907) (claimed to promote hair growth) - Asafetida packed in soft hides (1944-) - *Arch. Pharm.* 1931:269-91 (microsublimation).

1382. Holmes, Edward M. *Pharm. J.* 115:633; 1925 (Sumbul root of commerce) - Bauer, M. U. Breslau diss. 1915.

1383. Standley. *Contr. U.S. Nat. Herb.* 20:1-35; 1917 - *Tex. AES Bull.* 208 (1917) - *Tex. AES Bull.* 466; 1932 (F. culture Tex) - Traub & Frap. *Proc. Am. Soc. Hort. Sci.* 1928:306-310 - Ploss, E. *Acta Phytotherap.* 16:61-66; 1969 - Dawson, W. R. *Amer. Drugg.* Oct. 1926:19-21 (in ancient materia medica).

1384. Ficus latex extd. with org. solvents such as acetone, $CHCl_3$, &c. & resultant material dr. (U.S. Pat. 1,616,291: 1 Feb. 1927) - Fernan-Nuñez M. *F. laurifolia* most efficient vermifuge against *Oxyuris & Trichocephalus* - *JAMA.* 88:903-5; 1927 - Caldwell, Fred C. & Elfreda, L. *Amer. J. Trop. Med.* 9:471-82; 1929 (fresh latex kept cool & in dark bottle, active after 1 year).

1385. Robbins, B. H. & Lamson, P. D. *J. Biol. Chem.* 106:725-8; 1934 - BHR. *Ibid.* 87:251; 1930; *Proc. Soc. Expt. Biol. Med.* 32:892-, 894-; 1934/5; Fairchild Co., *Handbook of Digestive Ferments.*

1386. Asenjo, Conrado F. *Puerto Rico J. Publ. Health Trop. Med.* 15:141-7; 1939 (*Ficus pumila*) (rev. of lit. on latex of *F.* spp.).

1387. *B. Torr. Bot. Club* 85:71-2; 1953 (tree popular in the sUS; frts. interesting (15 cm. across); large lvs. Trees survived several winters outside, Brooklyn Bot. Garden (Geo. Kalmbacker). Excessive coffee use leads to restlessness, sleep disturbances, cardiac palpitation, GI tract irritation, diarrhea; even to mild delirium; increased gastric acidity. Some medicos advise coffee use be limited to 2 cups/day.

1388. FO fr. sds. of several genera u. in leprosy: *Carpotroche*, *Hydnocarpus*, *Mayna*; *Oncoba*; *Taraktogenos*; *Lindackeria*; &c. - Many centuries ago, natives of Africa, Asia and South America discovered medl. properties of these FO & u. them in treating leprosy, skin dis., wounds, &c.

1389. Florifundia (Guan, Guerr Oaxaca, [Mex]) also incl. *Datura sanguinea, D. suaveolens* (*floripondia*).
1390. Folic acid levels are often low in serum of elderly patients who show depression (forgetfullness, apathy, psychoses, hallucinations). Sometimes treated with f. a. supplement, as Folvite (Lederle); chief danger of use in multivitamin products is danger of masking symptoms of pernicious anemia
1391. Overholts, L. O. *Polyporaceae of the U.S., Alaska & Canada*. U. Mich. Press; 1953, 1967.
1392. Henry, Augustine. *Pharm. J.* 112:439-40; 1924 (*F. officinalis* known as "garikun" in the Indian trade).
1393. *Fouquieria splendens* rts. u. in baths for fatigue (decoc.) (Am Indians) also appl. to painful swellings (pd.). Abbott, Helen C. D. S. (n.d.) (Ocotillo wax).
1394. "Rheumatism Herb" grows in alkali or salt lands; is 8-30 cm. high; and is collected May to Aug. 1.
1395. *Fraxinus excelsior* (called in error "mountain ash") is supposedly a good malaria remedy (20-30 g. rts/day in 500 ml. boiling H2O; let cool) (*C & D.* 10 June 1944; p. 630).
1396. LaWall & Forman. *J A Ph A.* 6:22-; 1927 (3-sided flakes).
1397. Hanbury. *Scientific Papers* (*Fraxinus chinensis* harbors *Coccus Pe-la* Westw.); 1876.
1398. Fructose useful in mild diabetes since needs no insulin for utilization. Advantages: smaller amts. needed to sweeten, hence less u. & fewer calories; less cariogenic than sucrose or glucose (25% less); reduces symptoms of hypoglycemia; does not prod. hypertriglyceridemia; increases rate of alc. metabolization, hence reduces hangover, ameliorates alc. intoxication, since forms glycogen rapidly in liver (rationale for use of tomato juice).
1399. *C & D.* 138:413; 24 Oct. 1942 (Eire govt. paid $3/ton; 100 tons awaiting shipment at Burtonport, Co. Donegal).
1400. Fumaric acid prod. by fermentative process (Chas Pfizer & Co); u. as replacement for tartaric and citric acids in beverages, baking powd., in dyeing, making synth. resins, &c.; antioxidant (often as salts).
1401. Term "funori" appl. to both alga & product; related to agar; "funoran" term now u. for compd. type present (*cf.* agarose) - Tressler, D. K. *Marine Products of Commerce.* NY 1923 - Chapman. *Seaweed & Their Uses.* 1980.
1402. Peters, F. N. Jr. & Brownlee, H. J. *Furfural & Other Furan Compounds* (Reinhold); 1943.
1403. Peterson, John E. (Univ Mo) (1958) (assoc. with wood staining).
1404. *Merck Jahresbericht.* 9:76, 133; 1895 - Carrière in expts. found aq. ext. of galega (doses up to 5 g./day) greatly incr. milk secretion; recommended for women nursing babies, improving milk quality, quieting babies, & improving health (Schulz, Hugo. *Vorlesungen ü. Wirkung u. Anwendung der deutschen Arzneipflanzen.* 1921) - Lactogenic properties first discov. by Gillet-Damitte (1873); milk secrn. in cows increased 35-50% after 24 hrs.; aid in bust devt. (Inverni, Carlo Boccaccio. *Piante Medicinali e Loro Estratti in Terapia.* Bologna; 1933) - Berger, Geo. & White, F. D. *Bioch. J.* 17:836-8; 1923 (sds. c. luteolin and its glycoside, galuteolin (hydroxy flavones).
1405. Tanret, Georges. "Galega, recherches chimiques et physiologiques sur la graine" (Thèse, Univ Paris, 1917) - Barger, G. & White, F. D. *Biochm. J.* 17:827-35; 1923 (constitution; meth. isol.; color reaction. & c.) - Aronin, David. "Galégine et diabète" (Thèse, Univ Paris, 1929, v. 2 #210; 26 pp.).
1406. Elmore, F. H. *Univ. NM Bull.* 1944.
1407. Laboulbene. "Theorie sur la production des diverses galles végétables," *C. R.* 114:720; 1892 - Fockeu. *Contribution à l'Histoire des Galles.* Lille; 1889 - Boehner, Konrad. *Geschichte der Cecidologie.* 2 v. ca. 1933.
1408. Jamieson, G. S. *Oil & Soap.* 20:202-3; 1943.
1409. *Univ. Calif. Agr. Exp. Sta. Rept.* 1895:1896-7.
1410. Chau. *Chinese Physiological J.* 1931 (alkaloids).
1411. Hoyer, O. & Wasicky, R. *Pharm. Post.* 51:145; ca. 1918 (*Gentiana asclepiadea* L. as substit. for *G. lutea*).
1412. Anon. *Gard. Chron.* III, 90:389; 1931 (*Gentiana lutea).*
1413. Gajewski, W. *Polish Bot. Soc., Mongraphiae Botanicae* v. 4: 415 pp/1957 (cytogenetic study of 36 spp. & 120 interspecific hybrids).
1414. Rathnasabapathy, V. *et al.* (Madras Med. Coll.) *Ind. J. Vet. Sci. & Anim. Husbandry.* 21:95-106; 1951.
1415. Laing, R. M. & Gourlay, H. W. *Trans. N.Z. Institute.* 62: parts I, II (pt. I. narrower non-foliose forms; pt. II: foliose forms).
1416. Sprecher, A. *Le Gingko biloba L.* (Génève, Suisse); 1907 - *Gard. Mo.* 1882:122 (frt. u. to allay fatigue, Jap, Chin) - Selma, Ala, has oldest tree in Ala (planted 1882 ca.); Gingko Festival, Monroe, Mich; frts. stink c. gingkolic acid, gingkol, bilobol (toxic substances); lf. ext. u. in arteriosclerosis, vascular injuries in diabetes, intermittent claudication; Raynaud's dis.; resp. dis. like common cold (Volkner, J. H.; 1967); rhinitis, bronchitis. Frt. u. in asthma, TB, GC as anthelm. (Chin); sds. expect. sed. & in skin dis.
1417. *Gleditschia triacanthos*: large tree seen near Camp Carr (NJ [US]) with trunk ca. 4 ft. diam.
1418. Okamura, Kentaro. *Descriptions of Japanese Algae.* 560-4; 1936.
1419. *Anthropos.* 22:288; 1927.

1420. *J. Chem. Soc.* 1915: 546- (*G. superba*) (Methonica Radix).
1421. Zimmerman & Burgemeister. *N.Y. State J. Med.* 50:693-7; 1950 (effect of glutamic acid on borderline & high grade defective intelligences) - Weiss, Dess A. *Nervous Child.* 9:28; 1951 (improves speech in retarded children, improves sleeping habits, regulates digestion somewhat; aids in amnesia [not over 9 g./day in children; sed. effect noted] [regimen; give 4 weeks , interrupted 1 wk., then 4 wks. & so on]). This excitatory amino acid is now being thought as a possible factor in development of Alzheimer's and Huntington's diseases (research - Univ. Fla Sch. Phy. 1987).
1422. ex. Entona.
1423. Leffingwell, Georgia & Lesser, Milton A. *Food Packer.* Apr. 1945: 64 - Glycerin substitutes: *Ph.J.* 2 Oct. 1943, 122 - Use of 2,3-butylene glycol + 10-20% H_2O (not for int. use) (*C. A.* 37:2134); *APC* 276,705: direct red. of sugar by electrolysis of sorbitol, mannitol, & other poly-OH alcs. (glyc. substit.).
1424. Smythe. *Amer. J. Pharm.* 1890: p. 121 - Dragendorff. *Heilpflanzen der versch. Völker u. Zeiten.* p. 667; 1898.
1425. *J. Biol. Chem.* 56:513; 1920 (ca.).
1426. Nilkantum, S.V. *J. Sci. Technology, India.* 2: 39-51; 1936 (prepn. of many salts).
1427. *Bol. Agricultura y Ganaderia.* III: 95-100; 1904 (comm. phases of prodn. of "Goma Brea Argentina").
1428. Williamson, H. B. *Victorian Nat.* 47:110-2; 1930 (taxonomic notes) - Common name "clover tree" - *Agr. Gaz. NSW.* 39:885; 1928 - *J. Proc. Royal Soc. NSW.* 62:369 (1878); 369-78 (1934) (ca.).
1429. Called "Chaw stick" (BWI); pl. decoc. u. to harden gums (Jam); formerly dr. st. exported to US & Eu to make tooth powder; bee plant.
1430. Gourds were u. to make "gourdians" (musical instruments): mfd. near Russellville, Ark (1945-).
1431. *Merck Rept.* 1919, 310-2 (Nov.) (Holm, Th.) (herbage balsamic viscid).
1432. Fischer, J. L. *Med. Herald* (Louisville, Ky). Mar. 1888; *Pharm. Record.* 8: 387 (c. "robustic acid"[?], grindelin [NBP]) - Power, F. B. & Tutin, F. *Proc. A Ph A.* pp. 192-200 (1905); pp. 337-344 (1907).
1433. *G. humilis* H. & A (G. cuneifolia Nutt.): u. in pulmon. troubles by Indians & settlers of Calif; as wash for poison oak (A. Schneider, 1912).
1434. *Guaicacum off.* - Saponin-like subst. (*Pharm, Centralhalle.* 33:685; 1892); Paetzold, Ernst. "Resin & wd of *G. offic.* and of palobalsam," diss. Strassburg 1901 (believed active princ. of resin, wd. & bk. = saponin body) - *C & D.* 14 July, 1945: p. 31 - *U S D Comm. Spec. Circ.* 234:190 - Method of prepn: logs scarified & suspended horizontally in air on appropriate suppports; heat applied, resin melts & is collected in calabash cups. Sometimes obt. fr. incisions in bk. of tree.
1435. Guarana: Gagno, Nelson. *Rev. Inst. Adolfo Lutz,* II. #1; 1951; *Rev. Flora Med.* XIV. 5, 213 (1947); *Rev. Brasileira de Farmacia.* 29:211-2; 1948.
1436. Ministerio de Agricultura, Rep. de Cuba, 1942.
1437. Imported into Eu since ca. 1850; oil press cake u. as fodder.
1438. The first chewing gum in U.S. was spruce gum (Maine, 1848) ("Trunk spruce" & other brands).
1439. *Food Manuf.* Oct. 1944:356-66 (effects on gustation, sense of taste). *Acta Phytother.* 9:134-6; 1962 - Occ. Deccan peninsula; Concan to Travancore, Banda - *Nature.* Apr. 1887.
1440. "Tripe de roche" - St. Adrien d' Irlande, Côte Mejantic, Richardville, Québec (convent) (letter).
1441. Also called Holland Balsam, Dutch Oil, D. Drops (Tully drops); u. in eczema; in France mixed with fetid animal fats; sometimes the name was appl. to Juniper Tar (per *Merck Index* X). Originated by de Koning Tilly, on Achterstraat at Haarlem (Neth) ca. 1720.
1442. *C & D.* 1:140; 1880 (Hachettine).
1443. Todd, A. R. & Lyons, J. A. *Mfg. Chem.* 16:319; ca. 1962; Orth & Riedl: *Ann.* 663:83; 1963 (research).
1444. Sometimes called *Caragana argentea* (incorrect). Found on salt steppes of Transcaucasus, Turkestan. Plant with large violet fls. & pinnate lvs.
1445. *International Soybean Bibliography* (BA) (1983). Soy bean "milk" is u. when baby cannot tolerate cow's milk (colic) since hypoallergenic (Liquid Sobee); as whipping agt. (whipped cream); stabilizer in beer & other beverages - *The Soybean Industry in the US: A Selected List of References on the Economic Aspects of the Industry.* 1920-31 - Soy one of chief exports of US ($5.4 billion in 1984).
1446. Soybean antitrypsin (trypsin inhibitor) u. to treat fibrinolysis (Almquist, H. G. & al. *Arch. Biochem. Biophysics.* 35:352-4; 1952).
1447. "Rifatto" u. in mfr. of cellulose for paper - *Staz. Sper. Agrar.* 53:393-4; 1920.
1448. Wild growing plants found in Iraq west of Baghdad in desert areas; dug up by hired coolies (Forbes & Campbell, NY Company) - Seen in the Euphrates River area near Ran El Ain (T. E. Lawrence. *Oriental Assembly.* 1939) - Lemonade or sherbet sellers & liquorice water sellers noted in Eg (Butcher, E. L. *Things Seen in Egypt.* p. 84; 1923) - Mitlacher, W. - cult. in Moravia - *Pharm. Praxis.* 1911. 289-.

1449. (Books, pamphlets). Brown, Harry Bates & Ware, J. O. *Cotton: History, Species, Varieties, Morphology, Breeding, Culture, Diseases, Marketing, & Uses*. ed. 3:566 pp.; 1958 (McGraw-Hill, NY) - Dickson, Harris. *Story of King Cotton*. 309 pp.; 1939, Funk & Wagnall, NY - Brooks. Story of Cotton. 1948 (Rand McNally & Co.) - Vale, Edmund. The World of Cotton. ca. 1951 (Robert Hale) - Tomkins, D. A. *Cotton and Cotton Oil* - Wilkinson, F. *The Story of the Cotton Plant* (Orange Judd.) - Tisdale, H. B. "The cotton industry in Alabama before the [confederate] War," API B.S. thesis 1911 - Dowdell, L. and Tucker, W. H. "The growth and devt. of the cotton industries of Ala" (with questionnaires from all cotton textile mfrs.) AP B.S. thesis, 1919 - Ward, Kyle (Jr). *Cotton-Chem. & Chem Technology*. 1955. Fryxell, Penn - *The Natural History of the Cotton Tribe*. ca. 1980. Texas A&M Univ. Press - Bennett, H. W. *Histology of Cotton Seed & Fiber*. 1932 - Novels: Faulkner, John. *Dollar Cotton*. 1942; Harcourt Brace, NY; Mark Twain. *Life on the Mississippi* (chap. 30, 33, 39).
1450. Fl. Ext. of Cotton Root Bark (USP 1870-90; NF III-VII; DAB V).
1451. Cotton seed sometimes claimed galactogenic or galactagog.
1452. (Gossypol). Dowell & Menaul. *J. Agr. Res.* 1923; cotton meal varies in amt. of poison contained; may be destroyed by autoclaving then air drying - Anon. *J. Agr. Res.* 52(1):1936 (effect of $CaCO_3$ & $NaHCO_3$ on toxicity) - Harms, W. S. & Holley, K. T. *Proc. Soc. Exp. Biol. Med.* 77:297-9; 1951 (hypoprothrombinemia induced by g.).
1453. Mitlacher, Wilhelm. *Pharmakognosie; Lehrbuch f. Aspiranten der Pharmazie* IV. 1909 (Carl Fromme).
1454. "Pond's Extract" developed by Theron T. Pond (1846) - *Chem. & Drugg*. 12 Jan. 1946, p. 34 (First British distilled witch hazel now being mfrd. at Torquay by Mr. E. Quaint, MPS, who devised new process of extn. competitive to US prod.) - Ruemele, T. *Perf. Ess. Oil Rec.* 41: 323-5, 237; 1950 (lvs. c. tannins, FO, VO; preps: soln.; tinct.; formulas given) - Ext. (soln) u. for venous dis. as varices, ulcus cruris, phlebitis. For itching, &c. (Tucks).
1455. Holm, Th. *Merck's Rep.* 17:115-7; 1908 (Med pl. of NA, 15).
1456. (*Hed. coron.*). Garland flower; butterfly lily (common names). Recommended as source of paper (stems ?).
1457. (*Hel. tub.*). Do not boil too long, this may make tough; scrub tubers, do not peel; let steam in their own juices; the water in which boiled gelatinizes; dig up in fall or winter.
1458. *Hel. ann.* achene is second in importance as source of FO (soy - #1) Prod. chiefly USSR, Arg - U. in food snacks, (roasted in soy oil, seasoned with sea salt), bird seed, in ensilage. "Russian peanuts" - *BMJ* 115:72; 1947 (rich protein source (-50%) - *The Sunflower for Food, Fodder, & Fertility* by E. F. Hurt (London, 1946).
1459. *Heimia salicifolia*: C. alks. sinicutchine, lythrine, lyfoline; u. in bronchitis; tea in dys. (toxic). The tea is said to bring back memories of the distant past.
1460. Silva, M. Rocha E. *Histamine*. 1955 (Thomas). Found in many ointments, myalgia, lumbago, &c.; ex. Imadyl Inunction, Imahist.
1461. Spores of *Clostridium botulinum* may occur in honey & produce infant botulism caused by toxin. Recommended that honey not be given children 0-1 yr.
1462. Croydon, Surrey. *Materia Medica*. ed. 5. 1832; gargle for sore throat: HC1 f3j; Barley water f. oz. viij, Mel Ros. f3ii.
1463. *J. Ethnopharm.* 1: 89-94; 1979.
1464. *H. kurzii* could not survive Fla climate; being cult. by Zonal Exp. Sta. of Aromatic & Drug Plants of Abkhasia, Transcaucasus, Soukhoum (USSR) ca. 1935 (Dir. Nosarov).
1465. Schwartz, J. R. *Odontoiatria e protesi dentaria* 19:278-84; 1952 (use of hydrocolloids (alginates) for taking impressions for construction of inlays).
1466. Using *Asc. hyper.*, man passed 22 stones; no need to operate (Ketchum, Okla, 15 May 1944) ("wonderful herb" for kidney stones).
1467. *Bull. Pharm.* Sept. 1897:401 (*Pr A Ph A.* 46:855); Rudolf, Norman F. (*Abrus precatorius*).
1468. *Nature Mag.* Jan. 1935, p. 14. (Cashew nut).
1469. Coreil, F. *Etude Toxicologique de la Coque du Levant et de la Picrotoxine*. 127 pp.; 1930 (Gaston Doin & Cie. Paris) - Reportedly picrotoxin added to beer (US; ca. 1899).
1470. *Merck Index X*: #848; 1983.
1471. Jaretzky, R. & Breitweiser, K. *Deut. Ap. Ztg.* 58:37-9; 1943 (*Euonymus eur.* - Perianth, bk, lvs. sd. oil of doubtful value as lax.).
1472. Budagyan, F. *C. A.* 31:2698 (compn. of beech nut meal); *B. A.* 1947:4386 (comp. beech nut lvs.).
1473. Shepard, C. C. *Bull. World Health Org.* 38: 135-; 1968 - BCG vaccine against leprosy.
1474. Mechling, W. H. *Anthropologica.* 8:239-63; 1959.
1475. *Amer. J. Med. Sci.* Nov. 1942 - (*Capsella b.p.*) - 3 saponins found (2 = acid) (non hemolyzing). Sd. (pl.?) u. effectively to combat dis. in rabbits (Calif); effective in aiding clotting of blood in hemophilacs & to incr. clot resistance.
1476. Hammarsten's C., highly purified casein ("T.B. Physiol. Chem." ed. 7 (Hammarsten) v. 4:619; 1911).
1477. Wood found in commerce but soft.

1478. O.T. Tobit (Apocrypha).
1479. Yo. H. *Jap. J. Med. Sci. IV. Pharmacol.* 12: *89-*90; 1940 (effect of emetine HCl & carpaine HCl on dys. amebas [culture]; card. stim.).
1480. Claimed perfect quinine substit.; discovered by Dr. T. J. Naingoolan, Medan, Sumatra; probably mythical.
1481. Scott, Sir Walter. *St. Ronan's Well*, p. 243: "...wi' his poor wizened houghs as blue as a blawort" (*Cent. cyanus).*
1482. (*E. pil.*). *J A Ph A.* 42:607; 1953 (antispasmodic princ.) - Marsset, A. M. *Contributions à l'Étude Botanique, Physiologique, et Thérapeutique de l'*Euphorbia pilulifera (Le Mans, Fr, 1884) - Holmes, E. M. *Pharm. J.* 110:162-3; 1923 - Euphorbia Syrup Comp. (formerly mfrd. by Sharp & Dohme.)
1483. Valenzuela, P. *et al. J. Philipp. Pharm. Assn.* 1947.
1484. *Cirsium vulgare* Ten. (Carduus lanceolatus L.; Cnicus 1. Willd.) (do not confuse *Cirsium vulgare* Airy-Shaw of sAf) - Fish, Jessie G. *NJ Expt. Sta. Bull.* 113:1919 - Potter, H. C. *Nature* 107:120-1; 1921 (source of industrial alc.).
1485. Tang, T. H. *et al. Pharm. Archives* 11(4):1940 (exts. of bituminous coals showed estrogenic activity [Shantung Province, China]).
1486. *Colchicum luteum* Bak. one of the Hermodactyli; at times confused with *Merendera aitchisonii* Hook. f.
1487. Chief source Pak; most work done by General Mills ca. 1959; Wallerstein, Nichols (worked on starches prepd, in food ind.) (Plant Sci. Sem. 1957).
1488. *U.S. Bur. Stand. Circ. Lett.* LC. 500; 4 June, 1937 (levulose fr. artichokes).
1489. (Ergot). Garriques, A. *Les Plantes en Médecine: Le Seigle et l'Ergot*" (Doin, Paris) (1921). 254 pp. - Corner on ergot supplies - See *Drug Markets*, 6 Sept. 1927 - Fiero, George W. "Ergot—A Pharm. Chemical Study," 793 pp.; Univ. Wisc. Ph.D. diss., 1932 - *NY Herald Trib.* 21 Oct. 1941 (pigeons u. on very small scale to separate grain fr. ergot (Staten Is, warehouse roofs) - Youngken, H. W. Jr. *Econ. Bot.* 1(#4):1947 - Barger, Geo. *Ergot & Ergotism.* 1921 - *Contr. U.S. Nat. Herb.* 7(#3):299; 1903 - Brady, L. *Lloydia.* 25:1; 1962 - Arends, G. *Pharm. Ztg*. 70:566; 1925 (testing e. prepns.) - US Patents: No. 1,645, 096 (mfr. of prepn.); 2,056,360 (McCrea, A. to PD Co., process for artif. e.); 2,156,242 (25 Apr. 1939; Kharasch to Eli Lilly); 2,255,124 (9 Sept 41; extn. of alk. with EtoNa to improve process; Moore, Marjorie B. to Abbott Labs.); 2,390,575 (Dyas, Clair S. to Lederle; stabilized soln. of e. alk-salts; 12 Nov. 1945) - Thompson patents to Lilly for ergotamine, ergonovine.
1490. Ergotamine tartrate - u. to overcome physical effects of "shell shock" (nervousness, trembling, gastric upset). All US ships in WW II to be supplied with the drug for "convoy fatigue." (Dr. Daniel Blain, Dept. Med. Chair., US War Shipping Admin., US Off of War Inf.) - *Mfg. Chem.* 15:259; 1944.
1491. Ray, John. *Catalogus Plantarum Angliae*, p. 269 ("Secale luxurians"); 1670.
1492. Pesco, C. *Oleaginosas da Amazonia.* 1941.
1493. Black, Otis F. & Kelley, J. W. *Amer. J. Pharm.* 99:748-51; 1927.
1494. Boulos, L. *Medicinal Plants of North Africa.* Reference Publications, 1983.
1495. *Pak. J. For.* July:257-60;1951.
1496. Igolen, G. *Parfums Fra.* 14:266-70, 300-4; 1936.
1497. Power, F. B. & Tutin. *J. Amer. Chem. Soc.* 28:1176-; 1906.
1498. Muehlemann & Kasermann. *Pharm. Acta Helv.* 17:154-76; 1942.
1499. *Dyera costulata* for bubble gum - During World War II, replaced by synthet. substit.
1500. Sillert, F. W. (gutta Katrani, G. Soh.). *Hist. Boliv. Co.* (Miss) p. 118; 1948.
1501. WHO. *Tech. Rept. Ser.* 711:1984 (Advances in chemotherapy).
1502. *J. Amer. Pharm. Assn.* Mar. 1952 (defn.).
1503. Catgut u. also for fishing tackle; tennis & other rackets. J. A. Blue, 1955.
1504. Faust, E. C. & Thomen, L. F. *Proc. Soc. Exper. Biol. Med.* 47:485-7; 1941 (Ficin, proteolytic enzyme, fr. *Ficus glabrata* - parasiticide). Latex u. exter. in hemorrhage; in beer, cheese, & meat indy. large exports fr. Iquitos, ca. 1945.
1505. Risi, A. *Arch. Int. Pharmacodynamie Thérap.* 61:428-46; 1939. Ravina, A. *Presse Méd.* 42:1302; 1934.
1506. Term "gum tree" first u. by Collins; 1862.
1507. This plant is one of the mangroves; sds. not permitted by Int. Code.
1508. Dandiya, P. C. & Vorhora, S. B. *Research & Development of Indigenous Drugs.* New Delhi (Ind) (1989).
1509. *Caltha palustris* lvs. edible (spring), boiled with lard —> good dish; generally listed as poisonous pl. with helleborine; to avoid poison, boil herb 1 hr. & discard water (Abenakis Indians).
1510. Beneke - *Mycologia.* 55:287; 1969.
1511. Camphene occ. in Ceylon cironella oil. turpentine oil, lavender oil, & c. and is synthesized. "The hissing of the camphene lamps" (in theater at Victoria, BC [*Colonist.* 23 May, 1862]).
1512. Cannabis was cultivated for hemp in eTenn when revolution cut off the Russian supply (ca. 1920). Cultn. in Am goes back to 1611 (Jamestown, Va) (2303).

1513. Hot cherry peppers were u. in prepns. to treat hemorrhoids (Pouvert, Edwin K. U.S. Pat. 3,781,424 (Virginia Beach, Va). The peppers were steeped in olive or peanut oil for 14-28 da., the ext. decanted off, the solid residue being u. orally or locally as paste, emuls., oily suspen.

1514. Hickory bk. syr. (Mo [US]): scaly bk. peeled, washed in cold H_2O covered with same in kettle, boiled till color of maple syrup is lost, strained, brown sugar added, & boiled till of desired thickness (*Farm J.* Dec. 45).

1515. During WWII, nearly all senna came fr. Ind. - often badly insect infested & dirty (stones, &c.).

1516. Senna glycosides are very popular, in constipation - von Thus, Lippmann, R. *Deutsche medizinische Wochenschrift.* 76:835-6; 1951, using Pursennid (1-4 dragees/day). Others u. have been Glysennid (Sandoz) (sennosides A,B) (12 mg. tabs); Senokot (Purdue Frederick) (to reduce high & excessive straining at stool, thus protecting vascular system) (Walpern, Alfred: Stuyvesant Clinic, NYC, 1960) - Marri, R. *Settimana Medica.* 37:591-2; 1949).

1517. Beaver castors were placed in alc. & u. for rheum. & sore throat, abortif. & aphrod. (Gaspian med., Québec).

1518. *Catharanthus roseus* (Lochnera rosea Rchb.; Vinca pervinca Chula): decoc. of new fls. u. in malar. (folk med., sGuat coast) & for fevers - *Revista agricola* (Guat). 20(7/9):34; 1943 - White, Cyril T. *Queensl. Agr. J.* 23:143-4; 1925; 28:355; 1927.

1519. *Carica p.* frt. & sd. with significant antibact. activity; u. for gastric trouble; frts eaten like melon (yellow frt.); planted with red-frt. melon (PR) - Max Wallerstein patented process of clarifying beers during processing; later he established 1st papaya plantation in sFla.

1520. "Hands": prepared amulets esp. wrapped in red flannel (WI, esp. Haiti; seUS); sold in herb shops (mojo, trick, wanga).

1521. *Cedrela odorata* L. (C. mexicana M. J. Roem.): said potential source of camphor in Tex (AP, Austin, Tex, 25 Jan 1945).

1522. Kapok underwear - life-saving (Sweden; 1952).

1523. *Centranthus ruber* - *D. Apth Ztg.* 124:1213-6; 1984.

1524. Baumert & Halpern. *Arch. d. Pharm.* 231: 644-8; 1893.

1525. Chile fr. SpAm - C. con carne (SpAm, US), chile with meat (& sometimes beans); very popular dish, esp. swUS.

1526. Chinese shavings formerly u. in beauty shops for finger wave settng; in place of karaya & quince sd. Immersed in water 1 oz./gal. viscid mucil.; comb dipped in this drop 2-6 in. long.

1527. Choline not believed an essential vitamin: u. as c. chloride or one of various c. esters. Value questioned by some (dosage 5-10 gm/da.).

1528. Buhach Insect Powder (B. Prod. & Mfg. Co., Rochester, NY) outstanding (1956-).

1529. Manandhar, N. P. *Medicinal plants of Nepal* ed. 3. (Kathmandu). 1982, p. 122; *Elaeocarpus sphaericus* and *E. tuberculatus* are the object of religious veneration in India. The putamina (stones of fruits) are worn as amulets supposed to repel venomous snakes, &c. When placed between copper plates, some degree of movement of putamina is noted.

1529A. Bodner, C. C. and Gereau, R. E. (Botnoc ethnobotany). *Econ. Bot.* 42:307-369; 1988.

1530. *Bot. Bull. Acad. Sin.* NS. 26:213-20; 1985.

1531. Cassava (tapioca) bread made fr. mixed c. & wheat flour; popular in some places (ex. Indonesia) - *Agric. in the Americas.* 3(5):93-; 1943 (manioc said cheapest source of starch). Areas of high consumption of c. (Af) associated with low levels of diabetes & obesity in population (Thorburn, A. W. 1987).

1532. Tidestrom, I. & Kittell. *Flora of Ariz & NM* (1941) - Ayensu. *Med. Plants of W. Africa.* 1978.

1533. Sts. of some spp. roasted & eaten (Am Indians); sds. eaten whole with frt. are sometimes recovered fr. stool, washed, ground, & u. again as food.

1534. Provides chief summer food of grizzly & black bears (per E. T. Seton).

1535. The first spirits were prepd. by distn. of wine.

1536. Clay eating observed in black children (clay banks, Troy, Ala) (ca. 1965) & elsewhere - *Cf. Lancet.* II:614; 1978.

1537. Blake, S. F. *Contr. U.S. Natl. Herb.* 24:14; 1936.

1538. *Cnicus benedictus* - Cult. in Tenn by Chattanooga Med. Co. (formerly); R M Harper collected near Ellaville, Ga; now procured fr. Calif (ca. 1960) - "It was the blessed thistle that built the Patton Hotel at Chattanooga" (Samuel C. Hood) (letter). "Get you some of this distilled carduus benedictus and lay it to your heart. It is the only thing for a qualm." (Shakespeare, *Much Ado About Nothing* III, iv).

1539. *Cnidoscolus stim.* common in c&nFla, in sandy soils; uncommon in sFla - Sds. u. as food; sds. c. mostly FO & protein, much histidine; much unsatd. fatty acids (*J. Agr. Res.* 26:259-60; 1924 [Menaul, Paul]) - U. in psoriasis (dry eczema or dermatitis), each case needing 4-8 oz. for cure (BDH 1939).

1540. (*C. gossyp.*). Robinson, H. H. *J C S.* 89:1496-1505; 1906 (gum comp). Ketra gum (kutira, ketirah, &c.) (sometimes called karaya); u. by Arabs & Pers materia medica writers for Tragacanth.

1541. This small palm (1-5 m high) is esp. common in Rio Grande do Sul State (Br) - Bondar, G. *Campo* (Rio de J). 12:52; Mar. 1941) (cera de ouricuri em po).
1542. Sometimes applied to *Cassia ligustrina, C. marilandica*
1543. Ridley, H. N. *Agric. Bull. Straits & Fed. Malay States.* 6:45-7; 1907 - *Just Jahres Bericht* III: 638; 1907.
1544. Studied by Ashram Bhati (*J. Ind. Chem. Soc.* 1950) - Chopra, *et al. Indigenous Drugs of India.* 285-7; 1958.
1545. *Phytologia.* 54(7):681; 1984.
1546. Condurango - Origin of name: cundar, cuntur (Pe): eagle, condor. plus ango (angu ?), vine. Because it was believed the condor u. the lvs. of *Gonolobus* as antidote for snake venom. Bk. entered use in therapy in 1871 - Luchsinger, Fr. (Basel). *Beitraege zur Kenntnis der Inhaltsstoffe der Condurangorinde* (1924).
1547. Jute products: canvas, tarpaulin also made fr. jute. Jute mills blt. in ePak after partition; prevously all handled by Calcutta factories.
1548. Hume, H. H. *Bull Torr. Bot. Cl.* 65:79-87; 1938 (monograph).
1549. U.S. Pat. 3,173,792 (16 Mar. 1965; Mustakas, G. C. & Kirk, L. D. - method of obt. detoxified meal fr. sds. contg. both isothiocyanate & thiooxazolidone).
1550. Nestler. *Pharm. Zentralhalle.* 64:148; 1923 (adult. Crocus).
1551. Rattlesnake oil good for stiff joints, rubbed into skin (Am Indians) - *J. Wash. Acad. Sci.* 41:229; 1951 - U. in rheum. (seUS) - Lanman. *Adventures,* p. 157 (1856) - R. venom is collected (Ross Allen Reptile Inst. Fla) (high value: $980-1600/oz.): u. to prep. antitoxin. Flesh eaten (US) (of poultry meat); sold for c. $6.00/lb.
1552. "Swallowed a pumpkin seed" euphemism for becoming pregnant.
1553. Cydonium (also flax & psyllium sds.) - US Pat. 2,278,464 (Albert Musher to Food Mfg. Corp.) (*C. A.* 36:P4972) (prepn. for food use, *cf.* puffed rice).
1554. Dal or pulses u. in Ind, incl. *Cicer arietinus, Dolichos biflorus, D. lablab, Glycine hispida, Lathyrus sativus, Lens esculenta, Phaseolus mungo, P. radiatus, P. trilobus, P. vulgaris, Pisum arvense, P. sativum, Vigna catiang.*
1555. Gerlach, Geo. H. (Vick Chem. Co) *Econ. Bot.* 2(4):436-54; 1948.
1556. Sorgdrager. *Engele Indische Genees Kruiden (Desmodium triquetrum).*
1557. *Dichondra sericea*: oreja de gato; o. de ratón
1558. Ruda, K. *Acta Fac. Pharm. Brun. et Bratislav.* 1:33-49; 1958 (*Pharm. Zentralh.* 99(4); 1960).
1559. Death fr. digoxin overdose (Cardiac unit, Toronto Hospital for Children) (ca. 1982) (21 babies died).
1560. Unripe persimmons consumed may produce bezoar in intest.; may be confused with cancer & may need surgical removal (NIH?) (*AP.* 22: Nov. 1961). Frt. fed to hogs (seUS).
1561. Tonka. *Boletim do Mus. Nac. do Rio de J.* 1:127-35; 1924 (Roguette-Pinto, E.) (physiol. action) - Sobrinho, Penido. *Rev. Flora Med.* 3:531-7: 1937 (case recorded of pulm. TB lesions u. oil of T. by int. admin. & inj.).
1562. In some 80% of cases, gallstones made up of cholesterol.
1563. *Cicer arietinum.* Santa Barbara Co. (Calif) grew nearly 200 mill. lbs. in 1942.
1564. Cinnamaldehyde stopped feeding of boll weevils (1982) (antifeedant).
1565. Harlot says to man: "I have perfumed my bed with myrrh, aloes, and cinnamon; come let us take our fill of love until the morning; let us solace ourselves with love" (Proverbs 7:17-18).
1566. *C. pareira* L. - "*Sina* well compounded" (Dacca) (Kipling, Rudyard: "Kim."); 1893.
1567. *Citrus taiwanica* - Said will prosper in the Atlanta, Ga (US) area.
1568. Efficiency Products, Inc., Somerville, NJ
1569. Codliver oil - Fr. livers rotted in barrel; u. to make soap esp. to wash floor of fishing buildings.
1570. Bocek, B. R. *Econ. Bot.* 38:240-56; 1984
1571. Marchand. *Amer. J. Pharm.* 52:573; 1880 (incl. list of spp. furnishing agar [Jap isinglass]).
1572. Licorice mass packed in flat wooden cases (52 lbs. net wt) (M.E. Martin Co., NYC [US])
1573. Webb, L. J. *Guide to Med. & Poisonous Plants of Queensland*; 1948.
1574. Ext. of calendula u. to make toilet cream for babies (cult. [Burpee]) - *Sat. Eve. Post.* 3-12-38).
1575. Candelilla wax smuggled into US fr. Mex (ca. 1970).
1576. *The Carreta* by B. Traven (novel); 1970.
1577. Assam Tea - seeds - *Biochem. Zts.* 19:310-67; 198 (1909) (saponin) - *J A Ph A.* 1:226-7; 1912 (tea sd. oil).
1578. See serials in App. C-1 - Ukers, W. H. *All about Tea* (2 vol.) (1935).
1579. Sterling, C.M. - *Amer. J. Pharm.* 80:361-8; 1908 (histology)
1580. "With juice of cursed hebenon in a vial, and in the porches of my ears did pour the leperous distillment; whose effect holds such an enmity with blood of man...." (Shakespeare, *Hamlet,* I, V: 61-5 ("hebenon" in folios, "hebona" in quartos; probably henbane but possibly either yew or juice of ebony) (Glossary of Herley edition).
1581. Liu, J. C. *Lingnaam Agr. Rev.* (YBAPhA 16:170; 1929) (cult. of henbane in Chin) - Newcomb, E. L. *Amer. J. Pharm.* 86:531-42;

1914 - 87:1-10; 1915 - Lundstrom. *Farm. Revy.* 1918: No. 35-9 (Swedish henbane) - Dafert, Otto & Siegmund, Otto *Heil—U. Gewuerz-Pflanzen.* 14:76,98; 1932 (fertilizn. expts. with stramonium & henbane) - Henbane seed was off. USP, 1830-70.

1582. Ibogaine mentioned in French movie, *More* (1969). Said to obliterate drug addiction (1993) ("miracle") (psychedelic controlled). Crude drug u. by natives in stalking game, when hunter remains motionless for up to 2 days while retaining mental alertness!
1583. *Amer. J. Pharm.* 101:551-; 622-; 687-; 1929 (Sze Yee Chen: Phytochemical study of *Illicium religiosum* Siebold [Mang Tsao]) - Hin-Chueng Hoh. *Sunyatsenia* (Canton, China) 4, 33(#3/4):272-89 (The star anise tree in Kwangsi; much on prodn.).
1584. Dr. Bedford Shelmore (Coll. Med. Dallas, Tex) found of no value.
1585. Studied at Univ. Florida, Gainesville, Fla
1586. Bay, J. C. *Trans. Acad. Sci. St. Louis.* 6:151-9; 1892-4) (materials for a monograph on inulin; long bibliography) - Colin, H. *C. R.* (Paris) 166:224, 305-7; 1918 (genesis of inulin in plants; transformation of i. in the tuber of Topinambour during the rest period) - Grafe, V. & Vouk, V. *Biochem. Zts.* 43:424-433; 1912 (inulin metabolim in chicory) - Colin, H. *J. Fabre Sucr.* 61(16); 1920 (I. prodn. in *Dahlia, Hel., Cichor.*—highest cont. in all in fall). *C.A.* 32:5121; 1932 (effect of I. on meristem in Comp. rts.).
1587. Hansen, K. Fr. W. *Ber. d. d. chem. Ges.* 64 - 67 (1931) (bitter subst. in inula rt.).
1588. Steiner, R. P. *Folk Medicine, the Art and the Science.* Amer. Chem. Soc., Washington, DC. 125-; 1986 (*Allium sat.*).
1589. Mosig, A. *Ap. Ztg.* 62:11-3; 1950 ("Die Topinambur als Arzneipflanze") - Cremer, H. D. *et al. Zts. f. Hyg. u. Infektionskrankheit.* 132: 384-93; 1951 (*H. tub.* in human nutrition, significance. II. Possibility of bacterial decompn. of inulin in human intestine) - Wettstein, W. Von. *Der Zuechter* 10:9-14; 1938 (cultn. of *H. tuberosus* [topinambur]).
1590. Moser, John. *Amer. J. Pharm.* 89:291-6; 1917 (Pharmacog. of Helonias).
1591. Small, J. K. *Torreya.* 1:107-8; 1901 (2 spp. of *Chamaelirium*, incl. *C. obovale* sp. nov.) - Holm, Th. *Merck's Rep.* 23:268-9; 1914 (*Med. Plants of North America* series).
1592. Stout, A. B. *The Wild Species & Garden Clones, Both Old & New, of the Genus* Hemerocallis. Macmillan, NY (US); 1934.
1593. *JAMA.* 207:1807; 1969 - A stroma-free hemoglobin soln. developed by a team of investigators at Michael Reese Hosp., Chicago (US), showed promise as a blood substitute. No toxic reactions were found in animal experiments.
1594. Close, Geo. C. *Ann. Rept. CPCNY.* 1876:56-7. - Hepatica was temp. popular for treatment in flu epidemics in 1830 in NYC. Many syrups & balsams appeared on market.
1595. Scottish herring oil: Herring Industry Board (rules for prodn. & marketing); 1945.
1596. *Hibiscus sabdariffa* - Holmes, E. M., *Ph. J.* 1904 (fleshy calyces u. as refrigerant, &c.; u. by bakers, confectioners, &c., in Chicago, 15 Aug. 1931) - Frt. & dry seed c. pectin, color. - Cult. in San Salvador, 1942. *U.S. Dept. Agr. Leaflet 139.*
1597. Gerber, Donald A. (Brooklyn, NY) - Large doses of 1-histidine in rheumatoid arthritis (-1971-).
1598. Genus named after Thos. Hoy, Eng gardener; 1599.
1599. Essig. *College Entomology*. 214-5; 1940.
1600. Humulus c. 2-30 mg. estrone/100 gm. hops - Koch, W. & Heim, G. *Muench. Med. Wchschr.* 1953: 845.
1601. *Mfg. Chem.* Mar. 1944: 109 - Freake, A. *Observations on the* Humulus lupulus *of Linnaeus* (London, ed. 2, 84 pp. s.d.). Subtitle: "With an account of its use in gout & other diseases with cases and communications" - Myrick, Herbert. "The Hop" (Orange Judd Co.) - "Resins in hops, analysis," *C. A.* 1932: 1383; 5377-8 (Pepit, P.) - Griesedick, A. *The Flagstaff Story*. 1951 - *Herb Growers Mag.* 1953:118-22 - *J. Inst. Brewing.* 61:472-87 1955 (Compar. studies of methods of hop analysis. Estimation of humulone & other hop resin constituents [4 methods]) - Hops as tea (bad tasting) for acidosis & nervousness proposed (capsules) = solid extr. of hops.
1602. Daniel, K. *Hippokrates.* 1949:826 - *Merck's Jahresbericht.* 1950:164 - Hypericat™: c. 3.5 mg. hypericin & 5 mg. cholorophyll/cc.; recommended for dystonia, migraine, exogenous & endogenous depressions.
1603. *JCSIR* (Austral). 11(#4); 1938 - *Hypericum perforatum*, control with insects; *cf. Chrysolinum* spp. (since 1929).
1604. Antibiotic properties of *Hypericum* (*NY Times,* Aug. 23, 1956) (Mich State Univ.).
1605. Britt, Lewis C. (Univ. Wash Ph.D., 1937) - Common forage plant of eOre; u. emetic (Piutes, Shoshones, Nev [US]); dr. lvs. pd. stirred in water & drunk for stomach disorders (Navaho).
1606. Karayone, T. & Kimura, Y. *Jap, Chinese Med. Plants*. ed. 2; 1949.
1607. Jamieson, Geo. S. (USDA). *Veg. Fats & Oils*, ed. 2. p. 39; 1943 (Shinning & Hillier).
1608. Stanford, E. E. *Nature Mag.* 1958:80 (bitter, only sl. sweet).
1609. *Herba Hungarica.* 22:39-42; 1983 (VO compn.).
1610. *Juniperus scopulorum* - Common around Missoula, Mont (VO, *cf.* savin). Specific areas:

Cieniuga Cañon (Sandia Mts); Malpais (s of Grants, NM) (Ornamental tree).

1611. Kallikrein action (capillary permeability) - Decourt, R. & Ducrot. *C. R. Séances de la Soc. de Biol.* Paris 145:353-6; 1951.

1612. Open mines in Cornwall near St. Austell.

1613. Willis, J. H. *Handbook to plants in Victoria*, II. Dict. 1972.

1614. Nishida, Hashima. *J. Dept. Agricult., Kyushu Imp. Univ.* 2: 277; 1930. - Kleins. *Handbuch* ... II: 51 (process).

1615. Grazer, Fred. *Pharm. J.* 3(16):128-9; 1885 (Sonora gum).

1616. *Bot. Museum Leafl.* 23(#2):61-92, 1971 (*Latua*).

1617. *Bollettino, Societa Med. - Chir. Della Provincia di Varese.* 1951:138-9 (veg. lecithin mostly from soybeans as by-prod. of prepn. s. oil).

1618. *Herba Hungar.* 22:37- ; 1983.

1619. Meyer, G. G. & al. *Folk Medicine* (1981) - Ford, K. C. *Anthrop. Papers* (Univ. Mich) #60; 1975.

1620. Mors & Rizzini. *Useful Plants of Brazil*, pp. 65-6; 1966.

1620A. *Nect. p.-m.* c. 2% VO with safrole, terpenes, cineole, eugenol; 30% FO with much glyceryl laurate; much starch. Some tannin; resin.

1621. Albers, C. C. *Am. J. Pharm. Educ.* 1966:255-60 ("Introd. to simple lipids").

1622. Kafuku: *J. Chem. Ind. Japan* (*C. A.* 10:2386; 1916) (VO fr. *Liquidambar formosana* [Styrax]).

1623. Morgenbesser, Herbert. *Tobacco* (NY) 141(25):26; 1955 (desharpening ingredient for tobacco).

1624. Inner bk. of sweetgum tree chewed by children (near Ecofino, nFla) (lasted all day with sweet taste, like ball gum) - Wd. of *L. styraciflua* u. as bait for beaver poisons (strychnine) since beavers esp. fond of this wd. (Ala) - Bk. of this tree u. in sUS in diarrh., esp. in children, dys. (histology of seedling st., lf. petiole, epidermis, midrib) (Holm, Theo. *Merck Rep.* 17:31-4; 1908).

1625. Jane, P. H. thesis, Calif. Coll. Ph. 1902 - *Contr. US Nat. Herb.* 1(#3)343-; 1878 (*Lithocarpus densiflora* Rehd., Calif tan bark oak).

1626. *Lophophora* - Prentiss, D. W. & Morgan, F. P. "Mescal in therapy," *Ther. Gaz.* 20:4; 1891 (antispasm. sed., hallucinogen; tinct. prepd. after removal of hairy center; brought relief in cramps & griping; stopped pain in colic & headache, reduced nervousness, melancholia; dose pd. 0.5-1 gm.) - Cairns, Huntington. *Atl. Month.* 144:638-45; Nov. 1929 - *Denver Post* Apr. 30, 1950 (use by Indians of nUS) - Huxley, Aldous. *The Doors of Perception.* 1954 (0.4 gm. mescaline with half glass water: sensations described; thinks may be superior to cocaine, alc., barbiturates, &c.; harmless because non-addictive) (*cf. Sat. Review of Lit.* 6/II/54 p 14).

1627. Dalziel, J. M. *The Useful Plants of West Tropical Africa.* London; 1948.

1628. Tomato juice u. as bath to combat skunk "perfume" for person sprayed; also appl. to relieve poison ivy; to "stimulate love" (hence "love apples). Raw frt. rubbed on boils (folk med) - T. plant sensitive to 1 part gas in 200,000 of air (lvs. turned black); more sensitive than canaries (Crocker, Wm., Boyce Thompson P. R.) (*P. Mech.* ca. 1928, p. 247) - Grafting of t. plants on *Datura stramonium* to prod. more cold-resistant t. attempted, but frt. gave "food poisoning" (Hawkins Co., Tenn ,21 Apr. 1964 [AP]).

1629. Tanaka, T. *Fol. Pharmacol. Jap* (Brev.). 28: 74-5; 1940 ("Effects of emetine & lycorine on *in vitro* cult. of fibroblasts").

1630. Ling, T. M. & Buckman, J. *Lysergic Acid (LSD 25) & Ritalin in the Treatment of Neurosis*; ca. 1965 (rept. of several cases treated by authors; pharmacol., contraindications, &c.).

1631. Sobrinko, L., Meniucci, A. & Slywitch, Przemyls Warsis. *Anais da Fac. de Odont e Farm. de Univ. de Minas Gerais.* 4:247-52; 1939 ("Analysis of *Lithraea molleoides*" [tannins]).

1632. Ude, W. & Heeger, E. F. *Pharm. Zentralhalle* 82:193-8; 1941 (General review of subject with 14 refs.) (Cantharis).

1633. Ward, Kingdon. *Plant Hunting on Edge of World.* p. 293; ca. 1940

1634. Wiesner. *Rohstoffe des Pflanzenreichs*, ed. 4; I. 758-9; 1927/28.

1635. Cornforth, J. W. & Henry, A. J. (Wellcome Chem. Lab., Khartoum) *J. Chem. Soc. (Lond)* 1952:597-601 (Hydroxy-stachyine: cis- & trans- 3-, (ca. 10% by wt. of frt. husks [dried]).

1636. *Amer. Drugg.* Sept. 1937:54-5, 78 (suggests that maggot therapy may be more effective than allantoin).

1637. Greathouse, Glenn A. & Rigler, N. E. *Plant Physiol.* 15:563-4; 1940 (rts. c 2.2-2.5% berberine, 0.4% berbamine) - Anon. *Am. J. Bot.* 25:743-8; 1938.

1638. *Malus sylvestris* var. *paradisiaca* Bailey, the paradise apple - very small tree; much u. for dwarfing - ERRL (USDA) res. on M. s.

1639. *J. Ethnopharmacol.* 1:30; 1979.

1640. Cassava (Eng, It): Casabe (Sp): Cazabe (Sp): Cassabe (Por & Catal); Kassava (Ge);<cazabi (Haiti, yuca bread) - Fécula de cazabe (Sp) = tapioca (Brazilian arrowroot).

1641. *Bull. Imp. Inst.* 40:257; 1942 - *Mfg. Chem.* 14: 31; 1943 (Nigerian cassava starch to replace that from NEI; for use of dextrin mfrs). *Manihot utilissima* ("garri").

1642. *Man. esc.* - Difficult to remove all HCN, some converted into SCN (thiocyanate), this traps iodine & changes thyroid metabolism (Hypothyroidism); esp. bad where I intake is low and inadequate (goiter) in such regions, cassava is goitrogenic - McGaughey, C. A. *Brit Vet. J.* (London) 107:27-80; 1951 (HCN poisoning in nutrias caused by cassava) - The effect of "cassava factor" in the diet producing HCN toxicity in some places (cassava one of chief food sources of tropics); prepn. of c., pictures - H P C Melville, Arawah women, Wapisiana tribe, Brit Guiana - USDA, F.B. 167; 1903.

1643. *Bot. Museum Leafl.* 29:205; 1983 (*Mansoa standleyi).*
1644. Condurango: Luchsiner. "Beitraege zur Kenntnis des Inhaltsstoffe der Kondurangorinde," Basel, inaug. diss.; 1924 (could not confirm Kubler's finding of conduritol in *C.* bark; believed different kinds of such in commerce) - Freise. *Pharm. Zentralbl.* 75:423; 1934 - Kern, W. & Momsen, B. *Arch. Pharm.* - Kern, W., Fricke & Steger, H. *Arch. Pharm.* 276:463-; 1938 - 278:145-56; 1940 (confirms Kubler's findings on constitution) - Cond. bark coll. in Loji region, Ec.
1645. *Matricaria* - fls. u. in hair care to incr. blond reflection of hair ("shampooing chamomile") - Cont. bisabolol, useful in peptic ulcers; c. allantoin (?) promoting tissue granulation, aids wound healing.
1646. *Econ. Bot.* 1955:647.
1647. Brooker, S. G. & Cooper, R. C. *New Zealand Medicinal Plants.* 1961.
1648. *J. de Pharm. von Elsass Lotheringen.* 31/32: 79; 1904 (*Jahresbericht d. Pharm.* 1904:78).
1649. *Metopium: Mid. Drugg. & Pharm. Rev.* 44: 149, 218, 288 (bibliog.) - *Pharm. J.* Oct., Nov. 1909; *Amer. J. Pharm.* 1910:499-506 - Gum in irreg. translucent or transparent pieces imcompletely sol. in water.
1650. Gisvold, O. *J A Ph A.* 33:270; 1944.
1651. Hanzlik. *J A Ph A.* 18:375-; 1929 (1803 generally considered actual date of discovery of morphine by Sertuerner) - M. is synergistic with papaverine (ca.1958).
1652. Abreu, B. E. & Emerson, G. A. "Skin absorption of insulin with *Mucuna pruriens," Univ. Cal. Publns. in Pharmacol.* Vol. 1, No. 3; 49-54; 1938 (use not recommended) - Sollmann, T. & Pilcher, J. D. *J. Pharmacol.* 9:309-340; 1917.
1653. Sanyal, A. K. & al. *J. Pharm. Pharmacol.* 15:283; 1963 (value of ripe bananas in peptic ulcer believed due to serotonin; prevents histamine & induced gastric hyperacidity & acute peptic ulceration; dr. pulp 1 gm./kg. body weight/day as effective as antacid therapy; *cf.* Sanyal (*Ind. J. Pharm.* 13:358; 1951. Sinber et al. - *Ind. J. Med. Res.* 49:681; 1961) - Anon. *Nutritive & Therapeutic Values of the Banana* (A digest of the sci. lit.). Res. Dept. United Frt. Co., 1936. 143 pp. (chem. phcog. phcol. [nutr. expts], medical [therap.]. 292 refs with abstr.) - Green skins chewed & swallowed for bowel disorders (Austral. aborigines) - *Drug Topics* 16Jl. 1945 - Slocum's Banana Wafers. - Simmonds. *Evolution of the Banana*; John Wiley & Sons, Inc., NYC; ca. 1950.
1654. *Pharm. Record.* 8:115-; 1888 (Epitome of Org. Mat. Med. [series] - figures & descrip. of prepn. of Balsam Peru).
1655. Stout, F. M. & al. (OSU) found antibiotic in Pacific hake fish: preserved hamburger meat kept fresh several weeks at room temp.
1656. Konzett, Vienna, 1955 (narcotine to control cough in cats (mech. stim. cough) d. 0.5-1 mg./kg.
1657. *Beaver* (HBC). Spring, 1956.
1658. Moldenke, Harold N. *Phytologia.* 46:176; 1980.
1659. Seil, Harvey A. *J A Ph A.* 11:904-6; 1922 (compn. of *Nect. c.)* - Gstirner, F. *Heil—U. Gew. Pfl.* 15:41-; 1933/4 (coto & paracoto bk).
1660. Nectar of fl. nectary always cont. large amt. sucrose (Bonnier: gum, dextrin, mannite, small amts. of N & P compds., H_2O 60-85%) (Haberlandt, 508).
1661. Chah, N.C. *I U S J.* Aug. 1970:60.
1662. *J. Chem. Soc. Lond.* 1962.
1663. Courtney, Roger. *African Argosy.*1953 (p. 28) (tobacco smoked mixed with various herbs & cow dung) (cAf [Kenya]).
1664. Nicotine in tabs. (2 mg.) to help overcome smoking habit ("Nicorette") 1984. Addiction to nicotine is comparable to that of Heroin (acc. to medical authorities). N. gum & patches are also u. in place of cigarettes, since far the far greater risk from smoking lies in the smoke, not in the nictoine.
1665. Voerman, G. L. *Pharm. Weekblad.* 22:617; 1887 - Klimont, J. *Pharm. Post.* 51:561-2; 1916 (Ca) - Sd. oil also u. as quick drying oil in paints, mfr. soaps, &c.
1666. E. Pulmer - noted - before 1906.
1667. *J. Ethnopharmacol.* 1:34; 1979 (sometimes fermented with maize added).
1668. Vohora, S. B. & Khan, M. S. Y. *Animal Origin Drugs Used in Unani Medicine.* 1978. New Delhi, Ind.
1669. Meyer, J. & al. *Rev. Immunol. Therap. Antimicrob.* 19:414-6; 1955 (pantothenic acid plays important role in formation of antibodies).
1670. Dr. C. P. Li *et al.* (Nat. Inst. Health).
1671. Awe, Walther. *Arch. d. Pharm.* 274:442; 1936 (constits. of poppy petals, rhoeadine, rheagenine.). A formula formerly popular in English hospitals was as follows: Rx

    Syr. Rhoeados 3j
    Liq. (soln.) Morph. HCl 3s
    Aq. Chloroform. ad ℥ss
    M. s.a.

1672. Saenger, Werner. *Der Mohn* (Berlin; ca. 1949) - Scott, James Maurice. *The White Poppy.* Funk & Wagnall, NYC; 205 pp.; 1969 (title of book fr. white p. latex when fresh) - Heeger, E. F. & Poethke, W. *Pap. Somn.* 1947 (book) (cult. chem. uses) - Poppy sds. u. in bagels (pastry) c. small amts. of morphine (Fla prison parolees not allowed to eat bagels with poppy sds.) - Kapoor, L. D. *Opium Poppy: Botany, Chemistry, & Pharmacology* (Haworth Press. Binghamton, NY; $39.95 hardbound).

1673. Modarski, M. *Arch Imm. i Terapii Doswiadczahne* v. 4; 1956.

1674. *Trudy po prikladnoi Bot. Genet. Selektsii.* 22:209-76; ca. 1930 (Nicolaieff, V. "Morph. & classif. of guayule plant").

1675. Hurston, Zora Neal. *Mules & Men*; 1935. App. 4 - For GC, good handful of may pop rts., 1 pt. ribbon cane syrup, 1/2 plug Brown's Mule tobacco cut up, KI (50 cents worth); take mixt. 3 X/day as ton.

1676. *Aerztlich. Praxis.* 7:18; 1953 (Pototschnig, H.) (prepn., Peloidin.) - Prof. Giuseppe Cerie, (Univ Torino) 1954 - report before 3 Int. Kongress f. Moorforschung in Meran (Moor = peat).

1677. Weiser, Harry B. (ed). *Colloid Symposium Monographs,* v IV; 1926 (pectin jellies) - Keys, Ancel: claimed 15 gm. pectin/day lowered blood cholesterol (= 2 apples) (1961) - Daily doses of pectin being u. to reduce blood serum cholesterol (ca. 1976); also consumption of oranges & grapefruit is recommended.

1678. *Veroeff. a.d. Gebiete der Medizinalverwaltung* 30:235-; 1929 - Rozenfeld, A.D. *Farm. Zhurn.* 1930:183-6; *C. A.* 25:2811 (harmine prepd fr. root in 3% yield; all reactions common to banisterine & harmine) - Kruger, J. *Versuche* ... 1931 (*Berlin, Theses* [Vet.] 1931) - Gunn, J. A. (Oxford) *Heffters Handbuch d. Experim. Pharmakol.-Ergaenzungswerk.* 5:184-96; 1937 (mentions other alks. of *P. harmala,* incl. harman, tetrahydronorharman, harmol, ethylharmol, tetrahydroharmine, vasicine, &c. - Several of these stim. cerebral cortex of mammals, to prod. hallucinations, tremor, clonic convulsions) - U. in Parkinsonism.

1679. Tambovskaya, Vira. *Visnik Prikladnoi Botanika.* 1930 (*Ztsch. f. angew. Chemie Not.*) (Kharkow) 1 (#3/4):147-50; 1930/ (daily variation in amt. of VO in *Pelargonium roseum* Willd.).

1680. Griebel, C. *Z. Lebensmittel-Untersuch.* 102:432-5; 1955 (microscopy of Avocado).

1681. Schindler, H. *Arzneimittelforsch.* 7:747; 1958 (comprehensive acct. of biochem. phcol.).

1682. Viehoever, Arno. *Thai Science Bull.* (Bangkok) 2 (#1):1-99; 1940 (edible & poisonous beans of the lima type *Phaseolus lunatus* L.) (cyanogenetic glycosides, &c.).

1683. Portugal not hot enough for quality dates; trees planted ca. 1758 still bearing ca. 1926 (Fairchild, D.) - Surgeon Gen. of US Navy examined hundreds of dates bacteriologicallly & discovered no microorganisms, even when not clean; the sugar seems to preserve them (Fairchild, D. *Exploring for Plants*, p. 85; 1931) - Date growing in US (*US Dept. Agr. Leaflet* 170; 1939) - Difficulty in getting planting stock for Calif, finally Frenchman got materials for W. Swingle, because of recovery of his health in Calif - Ancient Egypt, columns with capital representing date palm fronds - "Palmiform" or "dactyliform" - Usurped by Rameses II (Louvre, Paris).

1684. *Phragmites maxima* - Mucil. secrn. of stem may be "garum" of late Grecian physicians; on this plant in Am at times a manna-like secretion of sugar appears - Panicles u. for brooms, Radix Arundinaris (Ned. Ph.) - In *Woman of the River* (movie) (1957) Sophia Loren shown cutting cane (reed) at mouth of River Po, Italy (very lush growth); u. for shelter.

1685. *Ann. farm. Quim. São Paulo.* 23:19-27; 1984.

1686. *Bur. Amer. Ethn. 27 Ann Rept.* 488, 584-5; 1905-6 (Omaha tribe).

1687. Kovacs, O. *C. A.* 56:1544; 1962 - *C. A.* 55:5995; 1961 (re sperm whale oil) (utilization & compn.).

1688. Wills, Cecil M. *The Case of the Calabar Bean* (Hodder & Stoughton Ltd., London). 250 pp.; 1939.

1689. *J. Iowa Sta. Med. Soc.* 36:109-110; 1946 (editorial) (value of physostigmine in arthritis) - *JAMA.* Feb. 2, 1946 (uses of p.) HT (BW); Endo Prod. Inc. "Esertropin."

1690. Prorok, C. *JAMA.* June 9, 1978:2419 (physostigmine as memory restorative; by inhibition of cholinesterase, leaving Ac choline active for longer period, enhances cholinergic neuronal activity). *Sci.* 1978; 705: 1039-41 (1 mg. doses aided memory in older people, also doses of 0.25-0.5 mg.).

1691. Martinenghi, C. *Bollettino Soc. Med.—Chir. della Provincia di Varese* 1950: 107-9 (case of phytobezoar, with ulcer of the great curve of stomach).

1692. Skinner, W. A. & Parkhurst, R. M. (Chem. SRI) ca. 1964 - Aklilu Lemma, Inst. Past., Haile Sellasie I Univ. Addis Ababa, Ethiopia.

1693. Berry juice of Phyt. u. in "Ann Phillips Reducing Berries" (nostrum).

1694. Frankforter, G. B. *Amer. J. Pharm.* 69:134-7; 1897 (chem. study; believes rt. does not rapidly deteriorate on standing; much K in ash; destr. distn.) - *Merck's Rep.* 1907:312-4 - Phyt. berries u. in prepn. of "white lightning" & "third rail liquor" (NCar).

1695. Ishii (Jap). *C. A.* 6:1172; 1912 (wild yam tubers with mucilage) (glucoproteins) - *Tabulae Biol.* III: 1415 - Molisch & Wiesner...[difficultly sol. in H0H, insol. in alc.]).

1696. *Samuel Pepys' Diary*, v. 5. Swallowed piece of Cyprus turpentine size of hazel nut AM & PM for pus in urine (*Diary.* 31 Dec. 1664): also Cassia bk.

1697. *Plantage major* - Fresh lvs. put on swollen foot or rheumatic part after soaking in water 20 mins., previously; u. in plasters for pain (Abenakis).

1698. Miquel. *Flora Ind. Batavia.* 1859.

1699. Rageau, J. *Plantes Médicin. de Nouvelle Calédonie* (ORSTOM); 1973.

1700. Carslaw, R. Workman & Neill, J. *Brit J. Dermatology.* 75:280; 1963 (podophyllin in plantar warts). Podophyllin (isolated 1975) - Deriv. u. as antineoplastic agt., Etoposide; 1983 (testicular tumor) - Kremers, E. *Phytochemical Terminology.*

1701. Hazelton, L. W. (podophyllin IV inj. incr. prodn. of hepatic bile in anesthetized dogs) (*J. Pharmacol. Exp. Ther.* 72:26-; 1941) - Tomskey, G. C. & Vickery, G. W. & Getzoff, Paul L. - *J. Urol.* 48:401-6; 1942 (P. in granuloma inguinale). - Tsarvev, M. V. - *Farmatsiya* 6(#6):16-22; 1943 (chem.) - Haber, H. *Brit. J. Venereal Dis.* 1945: 63-4 (v.21) (5% pod. in tannic acid to treat venereal warts) - Shelton, A. L. (letter 5 July 1945) (use of P. in condyloma acuminata of penis) - MacGregor, J. V. *Brit. Med. J.* Apr. 28, 1945, I:593- (in soft warts) - King, Lester S. & Sullivan, Maurice - *Sci.* 104:244-5; 1946 (P. & colchicine in treating condylomata acuminata) - Reisch, Walter J. *& al.* - *Amer. J. Ob. Gyn.* 1947 (June) - P. in papillomas) (urethral) (8% p. in ointment, excellent results) - Epstein, E. - *J. Invest. Dermatology* 17:7; 1951 (P. in acanthosis nigricans) - Duthie, D. A. & McCallum, D. I. *Brit. Med. J.* 1951, JL 28; 216-8 (P. in plantar warts with elastoplast & p.) - Firstater, M. *Revista med. de Cordoba.* 40:70-3; 1952 (treatment of vesical papillomas) - Loran, M. R. & al. *Cancer Res.* 12:279-80; 1952 (tumor depressant effect of the non-dialyzable buffer-soluble fraction of P.) - Hartwell, J. L. & Schrecker, A.W. "The Chem. of Podophyllum," *Fortschritte der Chemie org. Naturstoffe* v. 15; 1958 - Podophyllum derivs. u. in lung cancer (antineoplastic): 1) Etoposide (VP 16) 2) teniposide (VM 26) (semisynth. deriv. of podophyllotoxin) (recommended by FDA Oncology Advisory Committee).

1702. *Lake State Forest Expt. Sta* (St. Paul, Minn) *Paper* #49 (silvical char. of quaking aspen).

1703. *Garden Bull. SS.* 6:115; 1929 (prepns. in Mal medicine).

1704. *J. Ethnobiol.* 1 (1):135, 64; 1981.

1705. *The Century Encyc.* 35: 678 (Propolis [Gr & Eng]) (< pro, before + polis city); u. to prevent cold air draughts in hive & to strengthen the cells; said collected sometimes from viscid buds of various trees (odor of storax).

1706. Zwinger, Hegner. *Analysis Fructuum Amygdalorum* (Basel; 1703) (monog.) - *Calif. Agr. Expt. Ext. Serv. Circ.* 103; 1937 (almond culture in Calif).

1707. Pumice u. in ancient Rome to remove facial hair in lieu of shaving.

1708. Stylomycin Lederle (fr. *Streptomyces albo-niger;* structure elucidated by Coy Waller et al[Lederle]). Useful in trypanosome infections by interference with nucleoside synthesis.

1709. Also said that robins become "drunk" on berries of *Pyracantha.*

1710. Cos, Alpha. "The cultivation of ginseng," *Amer. Agricult.* 54:121-2; 1894 - Kains, Maurice G. *Ginseng, Its Cultivation, Harvesting, Marketing.* 53 pp.; 1897; 1910 - Lin Yutang. *My Country & My People* (1935) says (p. 135), "We have discovered the magic tonic & building qualities of ginseng for which I am willing to give personal testimony as to its being the most enduring and most energy-giving tonic known to mankind, distinguished by the slowness & gentleness of its action." - "Ginseng" (pamphlet, 64 pp., Brooklyn Bot. Gard. Aug. 1938 (WPA 46-9=7-3-69) - Shalen, Mrs. Harry *Fur. Fish, Game* 76(#12):38; Dec. 1943 (also No. 11: 3809) ("Ginseng, the big five prong"). - Harris, Jennie E. *Nat. Hist. Mag.* 1948; *Mag. Digest J.* 1950, 75-7 ("Give me some ginseng—I need strength.") selling for as high as $400/oz.; US exports ca. $1 X 10^6 worth/yr. - *Atl. Constit. Mag.* 26 Oct. 1975 (8, 10, 13, 14) (Woodhead, Henry - re Jake Plott ["Stalking the wild ginseng"]).

1711. Carre, P. T. *Wine Rev.* Jan. 1944: 15, 16, 19, 21, 23 (location of cork trees in Calif) - *Chemurgic Digest.* 28 Feb. 1942.

1712. Hardy. *Return of the Native* (Eng, ca. 1850 - red ocher as pigment; dealer in this was called a reddleman).

1713. Tsarev, M. V. *Farmatsiya.* (#3):27-33; 1938. (*C. A.* 34:1129).

1714. Gunther, Ernest. *DCI.* 55:538, 9, 620; 1944 - Produced in Santa Catarina State, Br (s of São Paulo); u. like Jap artificial sassafras oil (fraction of camphor oil).

1715. *Bot. Museum Leafl.* 22:241-; 1969.

1716. Olive oil: believed to reduce blood cholesterol; color light yellow to deep green; best It. oil fr. Tuscany; best grades: virgin oil & extra-virgin oil (1st extn.; low smoking pt. [< 300°F]); u. on salads (sauteed). Olive oil is considered by many not only as a nutrient but also to prevent coronary heart disease and increase life expectancy because of the particular balance of the fatty acids (as glycerides) in the oil. The ideal ratio of these is said to be saturated/ monounsaturated/polysaturated: 1:1:1 (in this case for instance, palmiticoleic and linoleic/ linolenic). It is now claimed that polyunsaturated fats in excess may lead to accelerated aging, carcinogenesis, liver dis., & atherosclerosis. It is no wonder that olive oil has been considered sacred in many religions (Jewish, Christian, &c.) and that the olive is a symbol of peace, fertility, strength, and purification. Ol. olivae solidifies at ca. 0-2°C (approx. temp. at which water freezes).

1717. Hosie, Sir A. *On the Trail of the Opium Poppy*, 2 vols. - Holt, E. *The Opium Wars in China*. 1965 (Putnam).

1718. Piccoli, Leonard J., Sperapolice, M. A., and Hecht, Morris. *A Pharmacologic Study of Devils's Club Root* (ca. 1940) - *J. Ethnopharmacol.* 7:313-20; 1983.

1719. *Z. Physiol. Chem.* 211:5-15 (1932); *J. Pharm. Chim.* 5:530-4; Müller, Klara. *Ueber die Bedeutung d. Saponin fuer die system, Stellung d. Araliaceen* (Berlin: 61 pp.; 1932).

1720. Orange fl. water u. in Ramos Phiz (New Orleans, La).

1721. *J. Ethnopharmacol.* 1:411-6; 1979).

1722. *JAMA*. 163:425; 1959 (reserpine for alcoholism; 71% improved or well; in placebo gp. only 21% better or well) - *cf.* Lereboullet, J. *Journées Thérap.* 1958.

1723. Clark, R. H. & Gillie, K. B. *Amer. J. Pharm.* 96:1924 (wd. & bk. assays) - *J. de Pharm. de Belg.* 5:#8; 1933 - Wiley, Leonard. *Nature Mag.* 36:Feb. 73-5, 1980.

1724. R. palm. has deep blood red flattened fronds; palmately divided into oblong wedge-shaped lobes.

1725. *Remijia* (Cuprea) - Planchon - *Pharm. Ztg.* #96:823 (1884); #105:899 (*Jahresber d. Fortschr. Pharm.* (NF 18/19:225-6; 1883/4) (bot. study of gen. with most detail on the 2 spp. which furnish "China cuprea") - Hodgkin, John. *Ph. J.* (3) XIV. 217 (ca. 1884) (China bicolorata or C. Tecamez may originate fr. a *Remijia* sp. [False cinchona barks]) - Genuine Mexican Cuprea contd. 4% alks. of which 1% NaOH-sol. (Paul Elderfield analysis ca. 1942) - Flueckiger, F. A. *Vorwerke Neues Jahrb. f. Pharmacie u. verwandte Faecher.* 36: 291-302; 1871 - Weidlein, E. R. (Mellon Inst.) *Sci.* 1 May, 1942, p. 459 ("...no beneficial action of quinine which cannot be accomplished equally well by hydroxy-ethyl apocupreine, a drug which is greatly superior to quinine in pneumonia, in infections of the eye & throat, & in certain asthmatic conditions.") - Hesse, O. *Arch d. Pharm.* 240:652-5; 1902 (This bk. today is scarcely used at all in mfr. of quinine since the largest imports of good cult. cinchona bks. fr. Java, Ceylon, Brit, India, SA makes this superfluous. Also these bks. have a high content in quinine.).

1726. Zupko, A. G. *& al.* Reserpine prolonged survival time of rats receiving tetanus & botulism exotoxins (1959) - Smith,, T H F & Rossi, G. V. Reserpine significantly lowered serum cholesterol levels in rats using chronic administration (1959).

1727. Franzen, H. & Helwert, F. *Z. Physiol. Chem.* 124:65-74; 1923 (Chem. const. of green plants XXII, Currants).

1728. Scott, W. S. *Green Retreats. A History of Vauxhall Gardens.* 1956.

1729. *Ribes* - taxonomy - Wyckoff (USDA, PBPI, Spokane, Wash).

1730. Castorol, a perfumed castor oil, made by Calcutta Chemical Co.

1731. Castor oil u. in large doses as antidote for overdose of sleeping pills, such as methaqualone - Local names: "taking oil'; "drawed oil" (Porter).

1732. Killip, T. M. Wassim. *Rollinia cordifolia* Szyszyl. & other *R.* spp.

1733. Fairchild, David. *Exploring for Plants.* 82-3.

1734. Sayings: "He who loves the rose should put up with the thorns" (Old Turkish proverb) - *cf.* "No rose is without thorns."

1735. Biskind, M. S. & Martin, W. C. *Am. J. Dig. Dis.* 21:177; 1954 (combn. eq. parts rutin & vit. C in grippe, flu with good results [D: 200 mg. tid]) - Kugelmass, I. N. *JAMA.* 115:519; 1940 (in vascular purpura) - *Blood Disorders in Children.* Oxford U. Press; 1940.

1736. Samuelson, Gunnar. *Svensk Botanisk Tidskrift* 30:697-721; 1936 (Chinese spp. of *Rheum*) (16 spp., incl. 5 new ones) - Staeglich, H. *Suedd. Ap. Ztg.* 88:332-3; 1948 (*Rheum palmatum & R. undulatum*, possibilities of med use).

1737. Koelz, Walter N. "Persian Diary I," *Asa Gray Bull.* N.S. 3:159.

1738. Aspircreme™ - Scientific Publishers, Jodhpur, India; 1986 (Compilation fr. *J. of the Bombay Nat. Hist. Soc.* [1935-1944]).

1739. Jungle rose pods.

1740. Tarahumara Indians - *J. Ethnopharm.* 1:23-48; 1979.

1741. Chen, K. K. et al. *J. Pharmacol.* 75:69, 79, 83; 1942 (toxic effects of *Senecio* alks.) - Rusch, Harold P. (1956) (*Senecio* alks. in diet of sAf natives may be factor in common devt. of hepatic cancers).

1742. Oelssner, W. *Pharmazie* 6:515-20; 582-92; 1951 (*Artemisia cina*) (Word "Cina" < Semen [L. seed]) (fl. heads originally considered to be seeds). Santonin < Santonion, designation u. by Dioscorides < city of Santonium (today's Saintonge) where *Artemisia cina* grew (between Pyrenees & Garonne). S. pills said u. as currency in China.

1743. *Solanum hispidum* Pers. - Berries furnish yellow dye, cosmetic; lvs. roasted & u. to expel worms.

1744. *Am. Potato J.* 1960:82-3 (expts. at Douglass College with potatoes in reducing diet showed wt. reduction on such diet).

1745. (*Euthamia tenuif.*). 3-8 dm. high; with woody base, branched; fl. heads yellow, tiny, in many terminal clusters.

1746. Vestal, P. A. *Bot. Museum Leafl.* 8:153-168; 1940.

1747. *St. n-b - C & D.* 18 Mar. 1944:316-7 (studied during World War I).

1748. Souza, Amaro Henrique de (Rio de J.). *Syagrus (Cocos) leiospatha* Barb. Rodr. (per Glassman, *nomen nudum*) - Butia do campo. Rio Grande do Sul; frt. c. FO with vit. A; u. in liqueurs. Plant studied probably *S. capitatus* (Mart.) Glassman.

1749. Silymarin popular in Eu - "Legalon" Madaus (etc.).

1750. Mustard bath prepn: 4 tablespoonfuls mixed in water & added to comfortably warm bath; gives incr. O consumption & CO_2 output; 25% improvement of circulation; gives relaxation & euphoria. First industrial nurse was employed in a mustard factory (Norwich, Eng 1878).

1751. *Sunset New Western Book*, p. 239.

1752. *Sium* possibly referred to by Shakespeare in *Macbeth* where he mentions use of rhubarb & cyme in "scouring" (*cf.* vet. use) the English out of Scotland (later edns., senna & cumin).

1753. After skunk spray "wears off," the animal sprayed becomes more playful (inebriant ?) (Marlin Perkins, in *Wild Kingdom* [TV show]).

1754. "Dipping for snuff" is a common practice in seUS; formerly ostensibly u. as dentifrice; appl. to teeth with soft pine stick.

1755. Raw potatoes cut up in water u. to relieve bad burns (applied all day) (*fide* Dr. Leon Wilken) (Gretna, La, formerly).

1756. Denoff, C. G. & Zellner, J. *Monatsh.* 59:207-13; 1932 (*Sorbus aucuparia* bark c. ceryl alc. or similar; phlobaphenes; tannins; invert sugars; choline; amygdalin [traces]).

1757. Spectinomycin (Trobicin Upjohn) - Isolated by Mason & al. (1961) - Tricycle, aminocyclitol type of AB. Has been u. in proctitis, pharyngitis - Said specific in acute GC (1 dose cured 90% of cases with no cross-resistance to penicillin; mostly u. in US Army; once claimed to cure GC in a single dose).

1758. Bourdon, D. *& al. Bull. Soc. Sci. Bretagne.* 29: 7-16; 1954 (stachyose 60-80% of dr. wt. of tuber).

1759. Origin of word: tabaco (Carib Ind. of Haiti, referring to smoking pipe not the product).

1760. Tabasco sauce prepd. fr. red pepper (seed) from Tabasco, México; made since ca. 1870.

1761. Tabernanthe under investigation by Lilly ca. 1957.

1762. *Tamarix gallica* - Other uses claimed: frt. & lvs. u. as astr. in dys. diarrh., leucoderma; splenic dis.; eye dis. - Steam fr. cooked lvs. u. in piles, ulcers, wounds; c. polyphenols, sterols.

1763. Liver & gonads of Jap globe fish; formerly most toxic venom known. Several Gulf of Mexico fish carry this poison; believed used to make zombies in Haiti.

1764. Bywaters, W. W. *Modern Methods of Cocoa & Chocolate Manufacture.* Blakiston, 1930. (Also J. & A. Churchill, London) - Chatt, Eileen M. *Cocoa: Econ. Crops,* vol. 3. Interscience Publ. NYC - Sulc, Delimar. *Farm. Glasnik.* 3:49-55 (Eng summary) (better results from rinsing with hot water than from roasting Cacao prod.) - Achcenich, R. Cacao hulls c. vit. D; after ferment. & sunlight exposure to give vit. D; irradn. with u.v.1., increased (ca. Mar. 1937) - Mejia F. B. & Caviria M. E. *Bol. Agr. Soc. Antioq. Agr. Colombia.* 1937:1042-62 (cacao in Tumaco region) - Kaden, O. F. *Tropenpfl.* 41:14-20; 1938 (proposal for controlled cult. of cacao trees) - Urquhart, D. H. *Cocoa* (Longmans Green) 230 pp. 1, 1955 - Van Hall, C. J. J. *Cocoa.* London; 1914 ("the standard work") - Chocolate will kill dogs in rel. small doses (theobromine acts as CNS & cardiac muscle stimn.) - "The Chocolate Girl" painting u. by Baker Chocolate Co., Dorchester, Mass - painted ca. 1743 by Jean-Etienne Listard.

1765. Tiger Balm - Compn: menthol 8; camphor 24; Ol. Menth. 15.9; ol. Caryophy. 0.5; procaine 0.10; ol. cajuput 0.9; solid paraffin 18.4; petrolatum 18.4.

1766. Spanish Moss - Research at U.F., CP (ca. 1953-5): wax (extn. method devd. by Robt. Bennett) (believed of value for car polishes, plastic reinforcements, inks, cosmetics, foods, & animal feeds [Lauretta Fox & S. F. Feurt]).

1767. Richardson, Q. T. *Rhodora.* 79:467-572; 1977.

1768. Turner, Nancy J. & al. *Ethnobotany of the Nitinaht Indians of Vancouver Island.* 1983.

1769. Harlow, Wm. W. *Poison Ivy & Poison Sumac.* NY Sta. Coll. Forestry, Syracuse Univ. 1945 - *U.S. Dept. Agr. F.B.* 1166 - *Field Museum Leaflet* 12 - "Rhus dermatitis" - Univ. Chi. Press, 1923 - McNair listed 250 remedies used or proposed.

1770. Tribulus < tribolos (Gr) = caltrop: war weapon u. against enemy cavalry; strewn on ground caltrops always have point turned up (< tribulo - afflict, oppress) (*cf.* tribulation) - Nerinji fruit u. as diur. in fever, spermatorrhea.

1771. Fenugreek - Salgues, R. *Bull. Sci. Pharmacol.* 46:77; 1939 - *Nature.* 151:195-6; 1943 (saponins).

1772. Curtin, L. S. M. *By the Prophet of the Earth.* 1949 (1984).

1773. Richardson, D. *The Vanishing Lichens.* 1974.

1774. *Lancet.* 26:115-8 (Jan. 1974) (Danopolous, Evangelos [Athens]) (inj. 10% soln. of urea in 0.9% saline around lesions of skin cancer; 73% benefit [112 patients]; later addl. treatment of traumatized surface with pd. urea [91% benefit]). Esp. good for around eyes, &c.

u. also in migraine; 20 grains repeated (*BMJ*, 1943, II, 201).

1775. Metwally, A. M., Saleh, M. R. I., and Amer, M. A. *Planta Med.* 23:94-8; 1973 (isoln. 2 flavonoids fr. *Urospermum picroides*).

1776. *Urtica dioica* - U. in sntiquity & still in folk med. for diabetes, anemia, epistaxis, eczema, wounds; diur. - In Nepal, long known in Ayurvedic med. for weakness, fainting, (patient lashed with fresh pl. bundles; footbath with NaC1 soln. u. for weak & tired legs).

1777. Nettle lvs. u. to make nettle porridge (*Pepys Diary*, 1661) - Excellent fodder: higher nutrient content than any grass hay incl. alfalfa; also good chicken feed.

1778. *Tussilago farfara* (< tussus, cough (cf. pertussis) + agere = to drive away). Name "coltsfoot" orig. recorded by Huloel (1552) - *Coltsfoot Rock* (candy) mfrd. formerly by Necco (New England Candy Co).

1779. *Cereal Chem.* 21:430; 1944 (cow cockle starch prepd. by method of Dimler; protein dissolved in 0.03% NaOH soln.; grains spherical, small, uniform in size; fiber sepd. by screening thru 170 mesh screen.

1780. Schultes, R. E. *Bot. Museum Leafl.* 9:194-5; 1941.

1781. Valerian - Other values sometimes claimed: sed. (day-time), analg. hypnotic, in epilepsy, aphrod., for GIT dis.; to furnish drug store odor (?); admin. in capsules - Kionka, H. "Die Wirkung d. Baldrians," *Arch. Inter. Pharmacodynamie.* 13:215-44; 1904/5.

1782. *Bot. Museum Leafl.* 29:377; 1983.

1783. Schultes, R. E. *J. Ethnopharmacol.* 1:89-94; 1979.

1784. Other constits. vit. C, resin, pectin, ash 15.6% (*J. Pharm.* 28(3):47; 1855) - *Gehe & Co. Gesch. -Ber.* 1887, Apr.

1785. Randolph, Vance. *J. Amer. Folklore.* 40:78-93; 1927 (folk beliefs in the Ozark Mountains).

1786. *Chem. Ztg.* 32:605, 618, 647; ca. 1913 (distrn. odoriferous princ.).

1787. Roure-Bertrand. *Schimmels Rep.* 4/07:126-.

1788. Gruber. *Amer. Nat.* 25: 8 (also vol. 23, 24).

1789. For centuries u. for air conditioning (cooling); air both cooled & scented; *cf.* use in NM of excelsior within chicken-wire cage plus pie plate perforated above) + fan.

1790. *Viola tricolor*: considered as heart-strengthening for racing pigeon u. as syr. of pl. in drinking water; also for humans as syr.

1791. Vitamins were so named by Casimir Funk (1884-1967) (Swiss, later Eng) (< vita, life + amine, as first thought amines [like thiamine]).

1792. Lecoq, R. *C. R. Soc. Biol.* 145:1817-9; 1951 (axerophthol acetate & palmitate, IV inj. effect on plasma alkaline reserve of rabbit).

1793. *Phytologia.* 8:85; 1940.

1794. Vinegar of the Four Thieves: concerns story of robbers who u. a vinegar prepd. fr. rue, sage, crisp mint (= spearmint), wormwood, & lavender with wine vinegar poured over; this was placed in hot ashes in the earth & left for 4 days, then strained into bottles, & camphor added. As prophylactic for contagious dis., rinsed out mouth with prepn., drew it into nose, washed & rinsed several times a day, or took a few spoonfuls per ora several times a day, to prevent taking the disease, usu. the plague.

1795. Root beer was first formulated by Chas. Elmer Hires (Jefferson Med. College, Penna.) in 1866; first marketed in 1869. Some formulas contained licorice, dandelion, dock rt., *Tsuga canadensis* tips, *Ptelea trifoliata* sds., *Lindera benzoin* bk., bay lvs., egg white, &c.

1796. Bandelin & Tuschoff. *J. Am. Pharm. Assn. PPE.* 14:106, 126; 1953.

1797. Ottow. *Proc. A Ph A.* 35:170; 1887.

1798. Doliber. *Alumni Assn., Mass. Coll. Phy.* 7 Feb. 1878 (*Xanthium spinosum* u. as remedy [F1.Ext] in hydrophobia; but no cure on record since its introduction despite strong odor & taste ["Probably will prove useful for something."]).

1799. Balch, Royal T. "Wax & fatty byproducts fr. sugarcane," Sugar Res. Found. (NYC), *Technology Rept. Ser.* No. 3 (ca. 1948).

1800. Barnes, A. C. *The Sugar Cane.* ca. 1960 (Leonard Hall) - Porter, G. R. *The Nature & Properties of Sugar Cane.*

1801. Salep noted in bazar in Quettta, Pakistan (1951) - "The best salep of Cabul" (Kipling, R. "Kim").

1802. Scheele. *Crells Ann.* 2:291-; 1785 - Enz - *Vierteljschr. f. Pharm.* 8:311; 1859 (chem. of frt.).

1803. Masterman, Nancy K. & Chapell, V. N. *Cornell Bull. for Homemakers* #730; 1947 (Frozen e.).

1804. *Heil-u. Gewuerz-pflanzen.* 14:112; 1932.

1805. Dykeman, Wilma - "The French Broad" (river):
If she has flu, your daughter Pansy,
Give her quickly a dose of tansy.
If her fever runs quite high,
There's Joe Pye weed for you to try
If she has a cough that's deep,
Give white pine syrup, it's very cheap.

1806. Internal combustion fuel.

1807. Nigrelli, Ross F. *et. al. Zoologica.* 1959.

1808. Bushmen claimed e. trees over 500 ft. high; never officially verified (Victoria, NSW & Tasmania). One specimen in Ceram Is, Moluccas, said to be 428 ft. high (British Admiralty Pilot) (per Dunbabin, Thomas: "Nature's skyscrapers") - Euc. trees considered a fire hazard in Calif, wd. so rich in VO, very combustible; threat of forest fires (ex. San Francisco Bay

Hills) - 2 million trees over 2,700 acres killed by freeze (-150 ft. high).

1809. Eucalyptus frts. u. as flea repellants; tied around dog's neck, stringing cord thru like a neck band or collar; claimed to work well (Hippies; U. Cal. Berkeley, ca. 1970).

1810. Cantharidal test u. for diagnosis of genital carcinoma; by detn. of reactive conditions (H. Fischer: *Krebsarzt* [Wien] 6:138-47; 1951).

1811. Karow, Edward Otto. "The production of citric acid in submerged cultures" (Rutgers Univ. Ph.D. diss. 1942).

1812. Oldham, F. K. *et al. Essentials of Pharmacology.*

1813. Bergamot oil chief const. of eau de cologne, u. in almost every perfume combn.; cost $17-19/lb. (1966). Synth. made by Glidden at $5./lb. prepd. fr. α-pinene (fr. turpent. oil).

1814. "Sour-sweet orange" is sometimes applied to some of the milder forms of the bitter o., *Citrus aurantium.*

1815. Sherlock Holmes: "Quick Watson, the needle!" Sniffing of c. may produce erosion of nasal septum.

1816. Schultes, R. E. "Plantae Mexicanae, VI," *Bot. Museum Leafl.* 8:189-199; 1940.

1817. Chauvin & Lenormand (Fra) (1957) (bee bread).

1818. Mode of prepn. of egg may be partly cause of varying data on cholesterolemia; thus, raw egg had 433 mg %, soft boiled, 524 mg., fried in butter, 660 mg. (Pollak: *J. Amer. Geriatrics Soc.* 6:614; 1958).

1819. Wilken, L. & Nasir, S.S.

1820. Nigroids (licorice pellets) (in tins) (Ferris & Co. Ltd. Bristol, Eng) (c. licorice ext., menthol); popular remedy for the throat.

1821. Koop, C. E. & al. (Sch. Med. Un. Pa) *Surgery* 15:839-58; 1944) (gelatin as plasma substitute) - Gelatin dissolved in orange juice is taken to improve nails, skin, hair (some say prevents baldness !); u. by some theatrical stars, unproven value.

1822. Reitz, R. "Bromeliaceas," *Flora Ilus. Catarinense.* 1983.

1823. Mildner, T: *Dragoco Rept.* 26:156-70; 1979 (scents u. to prevent body [incl. mouth] odor; mouth stink in scurvy noted).

1824. Scheiber, J. *Lacke und Ihre Rohstoffe.* Leipzig; 1926.

1825. Turner. *Herbal* (1551).

1826. Clark, E. W. *Amer. Perfumer Cosmetics.* 77:89; 1960 (lanolin).

1827. No. Dak. Sta. Univ. Agr. Exp. Sta. linseed expts. 1968; several mfrs. interested.

1828. Liatrin discov. in 1973 by Kupchan *et al*; cytotoxic.

1829. Harris, Geo. W. "Mrs. Yardley's Quilting" (chewing poplar bark as substitute for tobacco chewing.)

1830. Lupulon antimicrobial; restricts acid prodn. in *Lactobacillus bulgaricus* Rogosa & Hansen.

1831. Lysozyme disc. 1922 by Fleming in nasal secretions; actually first antibiotic studied by F.; also found in other secretions & tissues (saliva); Florey *et al.* found it to have little activity *in vivo.*

1832. Parfentjev, Ivan A. (Yale Univ. Med. School) (malucidin [lit. destroying evil]).

1833. Byrum, W. R. & Dale, L. B. Jr. *J A Ph A SE.* Apr. 1956:134-6 (sapote gum prod. in Pe u. there as size in textiles, felt hats, glue, component of water paints; useful emulsifying & suspending agt.; mixt. of water-sol. polyuronides [90-94%]).

1834. Red maple SC red buds of spring, acc. to some; others suggest red lvs. of spring & fall; red blossoms.

1835. *News Week.* Aug. 27, 1962:55 (cajuput oil said most popular medicine in VietNam).

1836. Peppermint formerly u. in "gouty" & colicky pains of stomach & bowels (*Creme de menthe*); in "low" headache; all disorders arising fr. flatus (ca. 1790).

1837. Meyer, Erich (Sonneborn Sons, Inc.). *White Mineral Oil & Petrolatum* (135 pp.); 1950. Many uses noted, ex. in air filters, apple washing, cattle sprays, corn earworm insecticides (esp. with pyrethrum).

1838. Grenada's flag has nutmeg in dead center embraced by horizontal blue, gold, green bars.

1839. Bouquet. *Invent. Pl. Med. Tox. Cong. Brazil* 32:1967.

1840. Baumbach. *Romance of Daffodils.* Greenwich Books, Publisher, NY.

1841. Uphof stated that he found *Neviusia alabamensis* in Mo. (1 specimen only, & sent to Eu).

1842. Niacin in 1 gm. dose reduced cholesterol by 37% but may exacerbate ulcers & diabetes; use ASA 1 tab/day; much cheaper than cholestyramine.

1843. von Euler, J. "Tobacco alkaloids" (Wenher-Gren Symposium; Pergamon Press; 1964 - Inferior tobacco (makhorka) lvs. u. as source of citric acid (USSR, 1934).

1844. Novobiocin with unusual biosynth. origin (amino acids, acetate, carbohydrates incl. glucose); action much like penicillin, mostly Gm+ bacteria; useful to alternate for pen. when resistant Staphylococci arise.

1845. "Flore du Madagascar" (fam. 87) (1982).

1846. Petrova, 1911

1847. Wilde & Howard. *J. Phcol. & Exp. Ther.* 130: 232-7; 1960 (ouabain acts directly on kidney by inhibiting tubular resorption of Na; diur. action comparable to that of aldosterone inhibitors) - Sammarteno, V. *Arch. Farmacol. Oper.* 79:1-9; 1949 (direct contact).

1848. Moore, H. E. Jr. *An Annotated Checklist of Cultivated Palms* (1963-4) (Internat. Palm Society, Miami, Fla 33143).

1849. Chestnut, V. K. *J. Ass. Off. Agr. Chem.* 3:387-97; 1920 (samples of papain prepd. by Fairchild, H., H. Rusby & J. E. Higgins) - Abernethy & Kientz: *J. Chem. Educ.* 1959: 582-4 - Babin, R. M. *Diss. Bordeaux.* 1953 - Burger, A. *Riechstoffe u. Aromen.* 1955:207-10 - Bergman, Max & Fruton, Jos. S. *Sci.* 86:496-7; 1937 (nature of papain activity) (papain is marketed in tins packed in cases).

1850. DL-phenylalanine involved in reducing degradation of endorphins in body, hence prolongation of analgesic effect of same; proven value in arthritis, low back pain, fibrositis, whiplash.

1851. Cori, C. F. *J. Phcol.* 23:99-106; 1974 (in phlorizin poisoning, insulin gives fall in blood sugars in rabbits).

1852. Renouard fils, A. *J. de Pharm. et de Chim.* 9(5):484-5; 1884 ("Produits naturels du dattier") (*Annales Agronom.* 9:551; 1883). Over 4 million trees cult. in Sahara alone; fibers extd. fr. net-like tissue at base of stem, called "life" (Eg) or ghimbusu (Af) u. to make cords, &c.; sd. kernels ground in water & fed to camels; another prod. date oil.

1853. ß-pinene is present in rel. smaller amts. with α-pinene; obt. by isomerization process; u. to prep. polyterpene resins (prodn. 25 million lbs./year (Arizona Chem. Co., Internat. Paper, Co., Panama City, Fla).

1854. Juice of frt. good bird food.

1855. Guava frt. rich source of vit. C (2.5-3% in dried; cf. dr. rose hips) - Guava paste (crème de guayaba) (Fla) (Cu): c. guava pulp, g. juice, citric acid, sugar.

1856. Ricin being u. in cancer therapy (with monoclonals) (exptl.).

1857. Blum, M. S., Novak, A., and Taber, S. (Baton Rouge, La; 7 Sept. 1959).

1858. *Bull. of Endemic Dis.* Baghdad 1:118-22; 1955 (Akrawi, F.) - *Boll. Chim. Farm.* 94:95-8; 1956 (Vole, S.) - Rembold, Heinz. *Vitamins & Hormones.* 23:359-; 1965.

1859. Dr. blackberries permitted to be sold only for medl. usage (SBP Co).

1860. Stacker, T. J. *Pac Drug Rev.* 39(#6)52; 1927 (Will cascara be farmed?) (Dealer, W. J. Lake, & Co., Seattle).

1861. Rock candy considered better for phlegm than "white sugar" (A.D. 1584). Used for centuries to alleviate throat irritation.

1862. Tannic acid has been u. in enemas; to sharpen X-ray pictures (added to Cyclodrast, &c.); however liver damage reported after int. use (1964).

1863. Schulz, Ellen D. *Texas Wild Flowers* (1928), p. 188.

1864. Whitehouse, Eula. *Texas Flowers in Natural Colors,* p. 121; 1936.

1865. Terpeneless lemon & orange oils with 40-50% citral (instead of some 4%); naturally much more expensive but with several advantages: 1) much more stable (longer life) 2) no terebinthinate odor developed 3) 16 X stronger flavor, hence need smaller amts. 4) more sol. in H_2O

1866. Examples of terpenoids: vitamin D; some hormones.

1867. Also smoked by children: coffee; rabbit tobacco; grape vine (like cigar), &c. (observed near Econfina, Bay Co, Fla).

1868. *Torula* yeasts cult. with molasses in Jam, Wisc (US); source of food protein.

1869. Secreted by kidney nephron; prod. is obtained fr. human urine; prod. popular in Eu.

1870. *New York Times.* 5 Sept. 1942.

1871. *Popular Mechanics.* 69:512; 1938 (In exptl. farm, 30 tons of yucca/acre obtained worth $30./ton; land worth $2 per acre [or less]). Source of y. fiber, machinery developed to process - Meehan, Thos. *Gard. Mo.* 24:345; 1882 (early settlers used fibers of y. lvs. to hang up meat to dry or smoke; useful in florist packing houses to tie up plants, also in vineyards to tie up grapes) - Sayman's Vegetable Wonder Soap (St. Louis) was advertised to cont. saponins of yucca.

1872. Strittmatter, D. A. *J A Ph A S E.* 44(7):411-4; 1955 (canna gum).

1873. *Amer. Pharmacy* NS. 26(3):11-12; 1986.

1874. Ferenczy, L., Gracza, L., & Jakoby, I. *Naturwissen.* ca. I/1959 (antibacterial subst. fr. hemp seed resin) - Abel, E. L., *The Scientific Study of Marihuana* (Nelson & Hall, Chicago) (1976). However, it is said that cannabis use reduces gastric acid level & reduces protection to cholera bacteria, wh. are destroyed by acid. In US, cannabis was available fr. Univ. Mississippi drug garden on prescription.

1875. Cetaceum also obt. fr. blue, fin, & bottlenose whales; also dolphin & shark(!?) oils.

1876. Suda *& al. Chem. Pharm. Bull.* 17:2377; 1969.

1877. Other genera of Micrococcaceae: *Staphylococcus & Planococcus - Micrococcus* spp. occ. in soil, water, skin; mostly saprophytic or commensal (colonization of surface of healthy tissues without attacking tissues); very close to *Staphylococcus* which is mostly now reserved for pathogens; however, some may occur in diseases but are passive or of low grade virulence; ex. *M. tetragenus* Haupt (Gaffkya tetragena Trevisan) found in phthisis, pus of wounds, &c. - *M. pyogenes = Staphyloccus albus & S. aureus.*

1878. Developed in Gt Brit (Upjohn).

1879. Eymard. *J. Pharm. Chem.* 1890:243 (berries of *Phytolacca* & elder are considerably u. to color wines & liqueurs (pls. cult. in corner of vineyard) (Sp, Port). Berries thought purg. (dose of 60 gm. gave only 1 bowel movement; 80 gm. gave cath. action). Color brightened with acids; alkalies produced bottle green - André, G. *C. R.* 144:276-8; 1907 (compn. ext.).

1880. Sources of naval stores: 1) living tree (gum resin) 2) 1st growth stumps 3) pulp & paper indy.

1881. Cook, Richard H. & Goodrich, Forest J. *J. Amer. Pharm. Assn.* 26:1252-5; 1937 (Pericarp & sd. of frt. of *Sambucus callicarpa* Greene [S. glauca Nutt?] studied separately (extractive, VO, FO, ash, &c.); frt. non-toxic to white rats & dogs.

1882. Jowett & Potter. *Ph. J.* 15(4):157-; 1902 (salicin content in *Salix* spp. varies acc. to time of year) - *cf. J. Chim. Méd.* 7:17 - Braconnet. *Ann. Chem.* 44:308, 513 - Herberger. *Pharm. Zentralbl.* 1836:428 - *Jahrb. Prak. Pharm.* 1838, I:157.

1883. Nigrelli, Ross F., Jakowska, Sophie, & Calventi, Idelia. *Zoologica.* 3 Dec. 1960 (exts. fr. *Microciona prolifera,* Atlantic sponge, antibacterial.

1884. Sandarac u. by Arabs int. for diarrh. & hemorrhoids; more recently it is being added to plasters (Tschirch, A. *Die Harze.* 3. A. 536-60; 1936).

1885. Powd. castile soap u. to polish playing cards to make them slip easily (joke concern, Asbury Park, NJ).

1886. *Sapindus saponaria* - Largest trees in Fla found on the 10,000 Islands fr. Caxambas Bay to Cape Sable region & on Key Largo (s of Miami & in Everglades). Trees - 15 m. high; fls. May-June; frts. July, Aug.

1887. Kozniewski. *Anzeig. Acad. Wiss. Krakau* 1910: A. 235 (CC 1910, II, 1932) (sepn. of sanguinarine, chelerythrine, & protopine) (also Noller, CR).

1888. Used following vaginal surgery (ex. hysterectomy) to eliminate foul smelling discharges (ReMine, W.H. & Murphy. S. R.; 1959).

1889. Coward, R. H. & Harris, L. E. *J A Ph A S E.* 45:325-6; 1956 (*Dryopteris hexagonoptera* Mx.).

1890. Holton, F. *Antibiotics & Chemotherapy.* 1959 (Tennecetin, work for Massengill).

1891. Chloromycetin said to have been responsible for 43% of total sales of Parke Davis & Co. in 1960; in 1963, 31% tot. sales ($59 million) - Injectable form taken off market in 1969, some talk of removing C. fr. market - At least 43 trade names in use in 1970.

1892. *Apoth. Ztg.* 47:351; 1932 (m. oil).

1893. Cavallito, C. J. & Bailey, J. H. (WCC) - *Sci.* NS 100:390; 1944 (several antibiotics of higher plants inactivated by cysteine & some of its esters [not by methionine, alanine, serine]).

1894. Resolution to have syrup of ipecac available over counter approved by House of Delegates, AMA (1985). Some opposed to O/C status because of misuse by young women with anorexia nervosa & bulimia (danger of accumulation of emetine in body). Ipecac considered safest & most effective emetic.

1895. *C. tom.* very handsome & with showy fls.; common in sandy & semiswampy soils on coastal plain (Hond, Bel, in poor quality forests) (Ranson. *Ph. J.* 18:787; 1887).

1896. Citroen (Neth), lemon (*cf.* Citroen automobile).

1897. *Citrus sinensis* - In citrous juices, VO fr. peel said to add special tang to taste; recommend use of squeezer to extract this by expressing entire frt. (as with Hamilton Beach Extractor).

1898. Cola: not everyone is enamored of the cola flavor of so many beverages; note popularity of "Uncola," Sprite, &c.

1899. Laqueur, E. *et al. Deut. Med. Wchschr.* 54: 265-7; 1928 - Fellner, O. O. *Ibid.* 54: 922; 1928. also p. 923 (estrogenic subst. prod. hyperplasia of secretory part of mammary gland & of nipple in castrated or normal male g. pig or dog; in castrated undevd. female —> milk prodn.).

1900. *Malus* - "Delicious" Apples biggest seller in groceries (as A & P) but inferior to many others (as "Winesap") at same price. One grocer called the Delicious (red) the "sorriest eatin' apple there is." Favored by grocers since long lasting with little deterioration or bruising.

1901. Triticum is being marketed in recent times.

1902. Blum, M. S. *& al. Science* ca. July 1958 (active against *Strep. pyogenes, Micrococcus pyogenes, E. coli, L. casei,* & some fungi).

1903. Schultes, R. E. *J. Ethnopharm.* 1:211-39; 1979.

1904. Greathouse, Glenn A. *Pl. Physiol.* 14:377-80; 1939 (*Sanguinaria canadensis* resistant to root rot caused by *Phymatotrichum omnivorum;* sanguinarine prevented growth of m.o. at 2.5 ppm).

1905. Little, Thomas M. *Sci.* (2486) 188-9; 21 Aug. 1942 (tetraploidy in *Antirrhinum majus* induced by sang. HC1) (use in pl. genetics by Cambridge Labs., Wellesley, Mass).

1906. Freise, F. W. *Bol. de Agric.* 1933:450.

1907. Kasao-Menou. *J S C I.* 29:1431; 1910 - Peckolt & Dymock (see Ubbelohde, Oele u. Fette, II: 504; 1920) (fat compn. of *Sap. trifoliata:* oleic acid 61.5%; n-eicosanic 21.9; stearic 8.5; palmitic 5.6; lignoceric 25; unsapon. 1.2; latter with 30% phytosterol).

1908. Trabut, 1896. *Winterstein & Blau. Fieger* (see Czapek, *Bioch.* 2.A.III: 534) - Jacobs. *J. Biol. Chem.* 63:621; 1925; 64: 379; 1926.

1909. Massagetov, P. S. *Arch Pharm.* 270:392-5; 1932 (estimation of santonin in pl. material).

1910. Davenport, C. J. *China Med. J.* 32:133; 1918 (santonin seeds: shih-chun-tzü).

1911. Nord, Gustav. U.S. Pat. 2,301,787 (10 Nov. 1942) (recovery of saponin fr. cactus juice: diatomaceous earth filter aid added to expressed plant juice, this agitated & filtered, filtered juice heated to 85°-110° with sepn. of ppt.; cooled rapidly to room temp; again filtered —> liquid rich in saponin - Kofler, Ludwig. *Die Saponine*. 1927.

1912. Saraka (Union Pharmaceutical Co. Ind., Montclair, NJ).

1913. Sassafras pith (USP 1830-1900; NF 1910-20) u. for diseased eyes (NY State natives); young stems cut in half & medulla scraped out, put in water, & left stand; liquid u. to wash eyes; Indians preserved pith (Peter Kalm. *Travels in North America* II. 616; 1770 (1937). The gum (mucilage) is water-soluble & said to be the ony one not pptd. fr. aq. soln. by alc.

1914. Brooks, C. M. *History of Sassafras* - S. oil plant blt. at Yellville (Marion Co.) Ark, 1945; also Hickory Valley, Tenn.

1915. Chopra, R. N. & Premankur, De. *Indian Med. Gaz.* 59:540-5; 1924 (many uses listed); VO = 1.0% (Schimmel & Co. 1892; Semmler & Feldstein; 0.86% - Ghosh & Chatterjee).

1916. Lewton, Fredrick L. *Amer. J. Pharm.* 71:22-3; 1899 (several types of quebracho: blanco, colorado, moreno, prieto, negro, &c.; white & red discussed; large works for processing ext. at Hamburg, Frankfurt, &c.).

1917. Cremonini, A. *Ann. Chim. Applic.* 18:361-5; 1928 (describes frt.) - *Ibid.* 24(20):309-12; 1930 (adulterant of *Piper nigrum*) - Wessel, A. G. "The peppertree of California," Calif CP thesis, 1904.

1918. Barkley, Fred A. *Brittonia.* 1944 (monograph with 27 spp.).

1919. Linsker. *Med Klinik.* 1914:937 - Ochsenius - *Muench. Med. Wchsch.* 115:1720 - Hennig. *Arch. Pharm.* 255:382 (introd. ca. 1918) (Liquor U. Merc, 2%).

1920. Dumez, A. G. *Amer. J. Pharm.* 86:329-49; 1914 (history of names hyoscine & scopolamine; suggests USP should use latter name only; physiol. action, tests, m. pt. dispute; rotatory power, &c.).

1921. Large industry centered around sea otter (loutre) (marine) in 18th cent.; single animal would furnish 40-50 lbs. of meat. Animal now nearly extinct in BC waters, few left in Siberia & off Alaska (*Rept. Prov. Mus. of Nat. Hist.* [BC] 1928:12-4).

1922. Wheatley, V. R. *SPC.* Feb. 1956:181-4 (compn. of human sebum given, showing 5% squalene [possibly u. in sterol synthesis]).

1923. *Beidermanns Zntr. f. Agriculturchemie.* 42: 683; 1913 (chem. compn. of rye & its milled products; distrn. of substn. in grain) - Kern, Klar, Ludwig. *Biochem. Zeit.* 1927:186 (ketonealdehyde mutase in wheat & rye as well as in soy beans).

1924. Gyr. Franz. "Pharmakognostische Untersuchungen ueber einige als Arzneidrogen verwendete *Sedum*-Arten," ETH, Zuerich, Ph.D. diss 1940.

1925. Saw palmetto formerly collected in large amts. near Cape Canaveral, Fla by R. C. Burns ("Palmetto King"); estimated 200,000 lbs./yr. used. Berry of coastal region preferred as skin thicker & richer in oil, "syrups in flesh are heavier & oilier, sweeter"; "honey" or syr. secreted at the ripening stage accumulates outside frt., more profusely; stands more mistreatment in drying process to still give a good grade dry berry. When dr., heavier oil content in pit or between pit & outer shell, esp. after drying several months; better retention of volatile principles. Some claimed Canaveral region berries best; however, examn. of prod. fr. Ormond to Ft. Lauderdale showed much the same (RCB, 1928). Plant observed inland as far as 4 mi. south of Compass Lake, Fla ca. 35 mi. fr. Gulf of Mexico (GMH).

1926. Leung, A. Y. *Chinese Herbal Remedies*; Universe Books, NYC; 1984.

1927. *S. bifoliatus* - Found in low flat woods or near swamps (Welaka, Fla). Dr. lvs. with slight odor. Said of exceptional value in fever using 4 oz. ground lf. in 1 pt. water, steeped (2 doses); diaphoresis produced, lets patient rest; mild laxative; only slightly bitter taste; no griping or nausea; said to break "any" fever in 30 mins. —> euphoria; also u. for headache.

1928. Serotonin is isolated fr. beef serum, molluscs, coelenterates; has been synthesized (1951); discov. 1948-; known as "neurohumor" (CNS); "motor hormone" (GIT); "tissue hormone" (repairs connective tissue) - Ref: *Serotonin in Health & Disease*, 5 vols. 1978. (v. 3 = CNS).

1929. *Sesamum ind.* called gingelly is cult. in Ind (Andhra, Madras, Hyderabad, Rajasthan, Maharashtra, Vindhya Pradesh); prodn. area (1952-3) 5,860,000 acres; 460,000 tons sd. produced.

1930. Sesame - *Conn. Agr. Expt. Sta. Ann. Rept.* 1903:190 (anatomy, esp. microscopic) - Refining oil to use as a parenteral vehicle (Lakeside Labs proj. U Neb CP 1946-7) - Sievers, A. F. *Chemurgic Dig.* 1946 (expts. with s.).

1931. Skelly Solve - designations: A = n-pentane bp 28-30°; B. = petrol. naphtha esp. n-hexane bp 60-80°; C = essentially heptane, b p 90-100° D,E,F,G,H, L.; S = petrol. solvent; boiling range 309-385°

1931A. (Sinistrin). Schmiedeberg. *Z. physiol. Chem.* 4: 112; 1879 - *J. Agric. Chem.* 1879:130 - v. Riedemeister - *Diss. Dorpat*, 46; 1880.

1932. Joyner, C. Jr. *& al. Am. J. Med. Sci.* 230:636-47; 1955 (Sitosterol given 10 g./day for 4 wks. —> redn. of serum cholesterol [no incidents noted]; also noted in electrophoresis, redn. of lipid zone ß without modification of lipid fraction α).

1933. (Solanine). Potatoes most extensively grown in Maine, Ida, NY, Minn, Mich (summer); Fla, Tex (winter) - Solanine extd. fr. "Solanum du

Pérou" (several *S.* spp.) - Ramdohr, F. & Neger, F. W. *Pharm. Zentralh.* 39:521-3; 1898 - Ilano, Josef. *Bull. Int. Acad. Polonaise Sci., Classe Méd.* 1936:399-412; *Ber. ges. Physiol. exp. Pharmakol.* 99:674 (solanine is less toxic than solanine-I for intact animals); equally effective —> hemolysis & necrosis; neither incr. body temp. (Ber. 69: 811-8 1937)- may be regarded as on borderline between steroids & alkaloids.

1934. Potatoes ingested furnish vitamin C in good amts. The potato is basic in many countries to the meat course of the meal, thus in Ireland & Germany; it has been popular in Eu since its introduction in 1584. The dietetic properties are close to those of rice.

1935. Guth, P. *J. A. Ph. A.* 27:217; 1938 (*S. Vill.*, phytochem. & phcol) - McClosky & Eisenberg (USDA). *J. Ag. Off. AC.* 27:308-9; 1944 (berries not harmful to cat when fed them).

1936. *J. Agr. Res.*; 47:149-152; 1933 (rubber cont. of several goldenrod spp.) - Goriunova, A. G. & Bosse, G. G. *Sovetsk. Subtrop.* (Moskva) 1938 (2):95-9; 1938 (golden rod at the rubber indy. farm) - Henry Ford & Harvey Firestone financed efforts of botanists in finding rubber-yielding U.S. plants; Edison tested 15,000+ plant taxa for rubber & planned to produce such rubber at Ford's plant near Savannah, Ga.

1937. Anderson, Ernest, Sands, Lila, & Sturgis, Nelson. *Amer. J. Pharm.* 97:589-92; 1925 (pl. gums of swUS - cholla, mesquite).

1938. US Pat. No. 1,749,812 (process for pulverizing mountain berries) (al.).

1939. Farwell, B. "Burton" (Richard F. B.):1963 (p. 106: dinner of holcus cakes soaked in sour milk with thick coating of red pepper (Harar, 1954-5) (p. 104: gate of holcus stalks).

1940. "Sour grass" also appl. to *Panicum fasciculatum* Swartz, *Triachne insularis* Nees, *Paspalum conjugatum* Berg, *Xerophyllum setifolium* Mx. (s. g. pie).

1941. Bunting, Geo. S. "*Spathyphyllum,*" *Mem. NY Bot. Gard.* 10:1-54; 1960

1942. *Pop. Mech.* 37:92-3 (Jan. 1922) (powd. peat u. as fuel for heating furnaces after being thoroughly dried & pulverized (with consistency of a fine flour; delivered in large air-tight containers in trucks to heating plant) - Project to ext. humic acids (u. as soil conditioners & fertilizers) (U. Minn. 1954) (large prodn. of peat on Lulu Is, BC).

1943. Spigelia - Stockberger, W. W. *USDA, BPI, Bull.* No. 100:41-4 - *Essays & Observations, Physical & Literary.* 1:386-9; 1754 (Edinburgh) ("Of the anthelmintic virtues of the root of the Indian pink," being part of a letter from Dr. John Lining, physician at Charlestown in South Carolina, to Dr. Robert Whytt, prof. of medicine in the Univ. of Edinburgh - Fée, A. L. A. *Cours d' Histoire Nat. Pharm.* tome 2: 390-2; 1928 (roots very slender, fibrous, having a slight marked odor, a bitter & astr. taste, & a very marked external resemblance to the Virginia Serpentaria; color brown) - Analysis in *Jap. Ph.* IX, 197; 1976 - S. used considerably in pharmacies of nFrance; u. for a long time in NA - Dr. Garden first to give details on the properties of this pl. It is called "unstectla" by the natives; Dr. Barton called it febrifuge - *Ph.J. & Tr.* 1887:91 - commonly u. in Eng as vermifuge (Hare, A. H. *Practitioner.* July 1886/p. 61); in large doses depresses cardiac action & resp. and —> muscular power loss - Anon. *Am. Bot.* 45:120-1; 1939 (seldom seen transplanted in cult.) - Fechner. *Pfl. Analyse.* 27:113; 1828 - Stabler. *Ph. Rundschau.* 1887:731 - *Farmers' Bull.* (USDA) #1999.

1944. Schwietzer, C. H. *Med. Klinik.* 29:1271; 1960 (spinach & blood sausage); Fe content 3-4 mg. % in fresh lvs.; only 10% resorbed by human; to absorb 10 mg. Fe, human must eat 3 kilos.

1945. Decline in sponge prodn. in Cuba due to prevalence of a dis. of sponge beds; app. ran natural course & disappeared in latter part of 1941 (US Dept. Comm.) (*C&D.* 9-5-42) - Sponge clippings prod. by Amer. Sponge & Chamois Co. Inc. (NYC), bleached, u. to stuff other sponges, &c. Formerly paid men to cart away.

1946. Figley, Karl D. (MD) (Toledo, Oh) *JAMA.* 114:747-8; 1940 (many cases of sensitivity to Karaya (or Indian) gum). *Ibid.* p. 1284 (Alvarez, Walter C., letter: reported on recent case of patient who "suddenly began to have 3 severe attacks of migraine in a week. When it was found that these attacks of headache started when she began to use a gummy laxative, this use was stopped & that was the end of the headache.") Other references: Bullen S. S. *J. Allergy.* 5:484; 1934 - Feinberg, S. M. *JAMA.* 105:515 - Ivy, A. C. & Isaacs, Bertha L. *Amer. J. Dig. Dis.* 5:315; 1938 - Solis-Cohen. *Pharmacotherapy* - Pringsheim. *Chem. of Monosaccharides.* 1933 - Porges, M. *Med. J. & Rec.* 128:89; 1923 - Parsons, M. *Ibid.* 129:70; 1924.

1947. *Amer. J. Pharm.* 106:351-2; 1934 - *OPD.* Aug. 6, 1934: p. 45.

1948. Johnson, Alan J. *& al.* Amer. Coll. Surgeons, 44th Ann. Meeting; 1958 - Villavicencio, J. Leonel & Warren, R. - *Ibid.*

1949. *Proc. Soc. Exp. Biol. & Med.* 49:207-10; 1942 (bacteriostatic & bactericidal; obt. fr. soil actinomycete [formerly *Streptothrix*]).

1950. Waksman, S. A. *Streptomycin* (Williams & Wilkins, Baltimore); 1950.

1951. Tolvinski, K. K. *Pharm. Rec.* 7:1887 (man ca. 34 yrs. old; hard drinker 2-3X/mo.; tried to control with chloral, bromides, opium, &c. Desire, nausea, vomiting, &c. remained, tried strychnine nitrate 1 mg. 3X/da. Patient better next day, soon resumed work; continued treatments for 6 wks. 1mg. twice a day; 4 mos. later no inclination to drink.

1952. *S. ignatius*; Clementi, Hector. Univ. Nac. Bs. Aires, Arg. Thesis #200; 1942.
1053. Dop, Paul. *C. R.* 150:1256-; 1910 (*Str.* spp of eAs).
1954. Nux Vomica, bot, & chem. - *Ph. J.* 15(3):1-7; 1884/5 - Morrison, Robert & Bliss, A. Richard, Jr. "Ratios of stry. & brucine & rel. potency," Lloyd Bros. Treatise.
1955. Correia, M. Pio. *Diccionario das plantas uteis do Brasil*, I:268-9; 1926 ("...que faria outrora procurada pilas prostitutas e outras mulheres de ma vida.") (Por) - Rodrigues, J. Barbosa. *Hortus Fluminensis.* 154-5; 1894.
1956. Engler & Prantl. *Styrax.* 4:172-80; Nachtrag. 281 - 172; 1890.
1957. Advertising slogans for Sunkist Lemons: "Keep regular this healthful way"; "When you take cold, take lemons."
1958. Chiappella, A. R. *Ztsch. Untersuch, Nahr—U. Genussmittel.* 13:384-9; 1907 (descriptive & analytical data on a little known edible mushroom).
1959. Mahogany, Jamaica wood, 1671; some sold as part of a cargo in London, 1703; use for furniture well established by ca. 1730; the hard Spanish var. favored; ex. of use cited (Boswell, J. *Life of S. Johnson.* 4:485).
1960. Rowson, J. M. *J. Roy. Micr. Soc. GB.* 75:119-128; 1955 (anatomy of comfrey rt. & its detection in admixt. with chicory in dandelion coffee) - Macalister, C. J. *The* Symphytum Officinale *and Its Contained Allantoin* (John Babe, Sons & Danielson, Ltd., London; 1936).
1961. *Symplocarpos foetidus.* Pres. Truman as a boy worked in a Mo (US) drug store where he thought one of the jars contained "icy toad's feet"; he learned in 1956 that this was derived from "Ic. todes foet" (*Ictodes foetidus*, former bot. name for skunk cabbage) (AP).
1962. *JAMA.* 157:989; 1955 - *J A Ph A. P P* ed. May 1955:274.
1963. Clove cigarettes u. in Ind & Indonesia; c. 70% tobacco, 30% cloves; eugenol is inhaled; dangerous respiratory irritant.
1964. *Mfg. Chem.* 15:226; 1944 (Clove stem oil, Zanzibar; since vanillin is obt. from guaiacol & sulfite waste liquors, which are cheaper, this material needs other uses) (Res. Colonial Prod. Research Council, GB).
1965. Bandes, J. *et al. Gastrenterology.* 18:391-9; 1951 (clinical response of patients with peptic ulcer to topical mucigogue [eugenol]).
1966. Tidbury, G. E. *The Clove Tree* (ca. 1950) - Williams, W. B. *Coffee & Tea Ind.* 84(4):42; 1961 (cloves in world commerce) - Cloves cult. in Cochin (1936) & elsewhere in sInd, mostly imported into Ind - *Nova Briataka* (Yugoslav newspaper) claimed clove tree growing in Yugoslavia - Clove Growers Assn. (photos).
1967. Holmes, E. M. *Ph. J.* 1916 (*Y Bk A Ph A* 5:183) (*Styrax tonkinense*, source of Siam benzoin).
1968. Asahina & Momoga. *Arch. Pharm.* 252:56; 1914 (*Styrax japonica*, character of the saponin).
1969. Hartwich & Wichmann. *Schw. Wcschr. Chem. Pharm.* 1:237; 1912 (*YB A Ph A.* 1:174; 1912) (*Styrax pearcei*, source of an East Bolivian incense resin known as "estoraque" or "benjui").
1970. Aschoff, Hermann. *Proc. Am. Pharm. Assn.* 31:126 (1862) (*S. benzoin*); 12:106 (1864) - Holmes, E. M. *Ph. J. & Trans.* 13 Dec. 1890: 518-9; *Proc. A Ph A.* 39:400; 1891 (*S. Benz.* source of Palembang benzoin) - Rosenthaler. *Arch. d. Pharm.* (*C. A.* 19:2261) (coloration of benzoin—true & spurious).
1971. Battier, Albert. "L'amelioration du collapsus pulmonaire dans le pneumothorax artificiel par l'injection sous-cutanée de cinnamate de benzyl cholestérine," Univ. Paris, thesis. 1931 (61 pp).
1972. Asahina, Y. *Yr. Bk. A Ph A.* 56:207; 1908 (*Styrax obassis* S & Z, constituents; sd. kernels c. 18% FO).
1973. Britt, L. C. "Benzyl benzoate—correlation of clinical & exptl. findings," U. Wash. thesis ca. 1930 - Hassan, M MA & al. *Analytical Profiles of Drug Substances*, v. 10:55-74; 1976.
1974. Winterstein, Alfred & Egli, Robt. *Z. physiol. Chem.* 202:207-17; 1931 (resin acids are strong fish poisons; 0.5% soln. lethal to goldfish in 30-50 mins.) - *C. A.* 26:732.
1975. Riley, Donald Edwin. "Study of Siam benzoin & its preservative action on lard" (U. Fla. M.S.P. thesis; 1933).
1976. Chernoff, L. H., Viehoever, Arno, & Johns., Carl O. *J. Biol. Chem.* 28:437-43; 1917 (*Y. filamentosa* saponin; pl. obt. fr. vicinity of Baltimore, Md) (Schultz, V. W. *Arbeit. Ph. Inst. Dorpat.* 1896:110; reported sap. fr. *Y. filamentosa*, evidently an older name).
1977. Ginger tea: ASA 1 tab., ginger FE, f3j, hot water 1 teacupful (EH).
1978. Maize sometimes called a giant grass; specimens - 7.4 m. high (nr. Washington, Iowa; ca. 1939 [Don Radda]).
1979. Sargent, F. L. *Corn Plants, Their Uses & Ways of Life* (Houghton Mifflin & Co.) (ca. 1902) (popular & very interesting acct.) - Wallace, Henry A. & Brown, Wm. L. *Corn & Its Early Fathers* (Mich. State Univ. Press, E. Lansing, Mich; 134 pp.; 1956) (a history of *Z. mays* incl. the devt. of hybrid corn: a very valuable background for corn research) - Misc. uses: boiled corn as fish bait (Lake St. Basil, Macedonia, Greece) (*Aleia*, Sept. 1955) - Corn silk to make handrolled cigs. in Hollywood (tobacco outer 1/3, rest c. s. since in 90% of scenes, only a few puffs taken (during cig. famine, Columbia Studios; *Tobacco.* 5-24-45 p. 4) - Lye

(or big) hominy - Corncob syrup (Ohio): After shelling 24 ears, boil rapidly 20 mins. in suff. water to cover; add 3-4 cups of sugar; half teaspoon cream of tartar; boil to syrupy consistence (sometimes brown sugar added) - For sore mouth, wash out mouth with decoc. of blue corn (New England's Rarities; p. 101) - For warts: get grains of yellow hard seed corn off cob, pick wart with sharp end, let bleed on grain, & feed corn to a black hen, wart gone by morning (Tenn) (Tennessee Ernie Ford: "This is my story, this is my song" [1963]).

1980. Jain, S. K. *Glimpses of Indian Ethnobotany.* Oxford & IBH Publ. Co.; 1981.

1981. Actidione = cycloheximide; by-prod. in mfr. of streptomycin; u. as raticide (rat poison); u. against fungi & yeasts (not bacteria); seed treatment (wheat smut); various fungous pl. diseases; good repellant for food pkgs. but toxicity & high cost make it impractical (Welch, 1954); anticarcinogen (research at UB, $40,000 grant, 1960); fungicide u. to control pine blister rust; Acti-Dione (Upjohn).

1982. Smith, 1912.

1983. Molisch, H. S. *Ber. Wien. Acad. Wiss. Nat. Math. Klass.* 1911:87; 143 (coumarin).

1984. van de Koppel, C. & van Duuren, A. J. *Indische Mercuur.* 62:221; 1940 (*Ber. afdeel. Handelsmuseum Koninkl. Ver. Koloniaal Inst.* 149, 1940) (comparison of Surinam tonka beans with those from elsewhere).

1985. Amphomycin - Bristol ca. Oct. 1953. Active against Gm.+ bacteria, African trypanosomiasis (sleeping sickness) in mice.

1986. Brown, F. *& al. Nature.* 1956:188-.

1987. Fish, M. S. (Lab. Chem. Nat. Products, NIH) - *Trans. Soc. Biol. Psych.* XI:25-6; 1956 (plant hallucinogenic snuffs).

1988. *Pharm. Zentralh.* 72:721; 1931-32 - Trease & Evans. *Pharmacognosy*, ed. 10; 1972.

1989. *Chem. Industry.* July 1946 (WCC studying prods. fr *Asarum canadense,* found as active an antibiotic as penicillin [claim]).

1990. MacDonald in *Antibiotics* (Gottlieb) (Springer-Verlag NY, 1967).

1991. Aveeno™ (Copper Labs.): colloidal oatmeal; u. in dermatoses.

1992. Morton, Julia. *Folk Remedies of the Low Country*; 1974.

1993. Alazopeptin (Amer. Cyanamide) - Trials on mice.

1994. Appln. of alc. to nose = new "cure" for hay fever (Cleveland, Oh) (AP dispatch, 8-13-34).

1995. van Royen, P. *Blumea.* 10(#7):432-666; 1960 (monograph).

1996. (Re Sassafras). Old Mississippi ballad "Tocowa" (Arkansas?) (Sulphur Springs): "I got so thin on sassafras tea I could hide behind a straw/Indeed I *was* a different man when I left Tocowa." (Hudson, Arthur P. *A Patch of Mississippi Balladry* (1927); text from S. R. Hughston, Esq., Courtland, Ala, typewritten).

1997. Meyer, Th. M. & Koolhaas, D. R. "Temu Lawak and Kumis Kuching," Dept. of Economic Affairs, No. 10, Comml. Intelligence Service, Bureau of Commerce, Batavia - cJava n.d. (ca. 1937), 31 pp.

1998. Domermuth, C. H. & Rittenhouse, J. G. *Mycoplasmataceae,* a bibliography & index 1852-1970 - Va Polytechnic Inst. *Res. Div. Bull.* 6:136 pp.; 1971.

1999. Gentry, H. J. *The Agave Family in Sonora.* USDA, A R S, *Agr. Handbook* 399: p. 169; 1972 - McKelvey, "S. Yuccas of swUS., Part 1," *Mo. Bot. Gard. Rept.* 13:18 (16+ spp. in Texas).

2000. Neal, J. M., Sato, P. T., Howald, W. N., & McLaughlin, J. L. *Sci.* 176:1131-3; 1972.

2001. *Roempp Chemie Lexikon*, ed. 8, 1:456; 1981 (1 part in 60 million is bitter).

2002. Moldenke, H. N. *Phytologia.* 57:467; 1985.

2003. Bachstez, M. & al. *Materiae Veget.* 1:375-7; 1954.

2004. *Eczacukuj Vyktebu* (Turkey). 4:199-206; 1962.

2005. (*Salix purpurea*) (c. salipurpuroside; isosalipurpuroside 1-3%) - *C. R.* 196:816; 1933 - 192:1478; 1931 - *Bull. Acad. Med.* 106(3):353; 1931 - *Bull. Soc. Chim. Biol.* 13:588, 814; 1931.

2006. Salmon oil, Alaskan (body FO) (Chinook salmon) u. to lower cholesterol & BP.

2007. *Sequoia* - Redwood - "Founder's Tree" 364 ft. high; supposed tallest tree; few hundred ft. east of Dyerville bridge (Dyerville, Calif) on Redwood Highway - Another redwood reported 368.7 ft. - "General Sherman Tree": 272.4 ft.; girth 101-1/2 ft. (Dunbabin, Thos. *Nature's Skyscrapers.* 267 pp.) (Found. Amer. Resource Mgt.); 1957.

2008. Gilmore, M. R. Fritz, Emanuel. Annotated bibliography, *Papers, Mich. Acad. Sci.* 17: 119+; 1933.

2009. Caponetti, J. D. & Quimby, M. W. *J A Ph A.* Oct. 1956: 691-6 (comp. anatomy of various *S.* spp.).

2010. Sponge u. as vaginal contraceptive - left in place 30 hrs. after s.i. Misuse in some cases may lead to toxic shock syndrome.

2011. Classed as first line antituberculosis agt., useful in acute cases; given IM.

2012. *Fide* David Hand

2013. Lippmann, E. V. *Geschichte des Zuckers,* Berlin; 1925 (572 pp.).

2014. Miller, G. N. "The genus *Fraxinus* in North America north of Mexico," *Cornell Univ. Agr. Exp. Sta. Memoir* 335; 64 pp.; 1955.

2015. *Cf.* use of this or similar substance in Russia to detect informers (1985).

2016. *Cf.* "bruisie" (or "brusey"), mashed Irish potatoes with much Irish butter added (Co.

Cork, Ireland) (Woodfort) (<bruise: crush [?]) (GMH).

2017. "The primrose pale is Nature's meek & modest child" (Balfour). Ariel's song:

Where the bee sucks, there suck I,
In a cowslip's bell I lie,
There I couch when owls do fly
On the bat's back I do fly
After summer merrily
Merrily, merrily, shall I live now
Under the blossom that hangs on the bough.
(Shakespeare, *The Tempest*)

2018. Usually applied to citrus oils: distn. &c. leaves aldehyde (as citral, &c.); (ex. lemon oil 40-50% citral as prepd.) Advantages: 1) better keeping qualities 2) smaller amts. needed in formulations. Disadv.: expensive.

2019. Sometimes considered as synonym for *K. integiora* (resp.) *K. tomentosa* Bak.

2020. Found in *Rauwolfia serpentina*, *R. tetraphylla*, *R. vomitoria.*

2021. *Arch. der Pharm.* 315:477; 1982 (Rauwald, H. W.).

2022. Hisaw, F. L. & Zarrow, M. X. *Vitamins & Hormones* (NY). 8:151-78; 1950 - (Walczak, S. R. Garden City, Mich ca. 1960).

2023. Cansdale. *Animals and Man*, p. 146.

2024. (Mg). *Amer. J. Cardiology.* 57:486; 1986.

2025. Molasses: *Revista de Chimica Industrial* (Rio de Janeiro) 9:#95; Mar. 1940 (utilization in the prepn. of extracts) - Surprising accident in Boston in 1919, when a molasses tank burst, killing 21 persons; could smell m. for years in area (*Smithsonian.* Nov. 1983).

2026. Miller, E. A. (AES, AMC, Texas) (*C. ensiformis).*

2027. Barton, B. S. *Phil. Mag.* 22:97-103; 1779, 204-11; 1792-3 (a botanical description of the *Podophyllum dyphyllum*, now called *Jeffersonia virginica*) [*Jeff. diphylla*]).

2028. (*Joannesia princeps*). *Brazil: Min. da Agr. Bol.* 31(8):1-6; 1942 (cultn.).

2029. *Sci.* 100:10; 1944 (wild walnuts c. 1.5-3.7% vit. C [wt. of nut]).

2030. Bauer, R. (work at Innsbruck, Pharm. Inst.) juglone actually occurs only in fresh lvs. or up to 3 days old; bk still much u. in Ge.

2031. Walnut shells ground & mixed with Al_20_3 u. to clear jets of corrosion, dirt, & grease without overhaul (Tinker Air Base, Oklahoma City).

2032. Knight, Albert E. & Step, E. *The Living Plant.* pp. 353-5; 1905 (jumping beans jump away fr. tree where growing; found in morass, near Alamos, Mex).

2033. Hough, Romeyn B. *Handbook of the Trees*, ed. 3; 1924 (Osage orange trees - trunk 3 ft/diam.) (Large trees noted at home of Wm. Faulkner, author, in Oxford, Miss [GMH]) - *Alberta Med. Bull.* Jan. 1944: p. 18 (Res. at Coll. Phys. & Surg., of Alta., Calgary: gave no increase in pulse rate like most other card. stim.) - Clopton, John R. *Chemurg. Dig.* 1954 - *J. Am. Oil Chemists Soc.* 26:470-2; 1945 - *Ibid.* 30:156-9; 1953 (antioxidant use) - Wolfrom et al. *J A C S.* 64:406-18; 1946 (pigment).

2034. It is said the the idea for barbed wire came from observation of the spines of the leaf axils of Osage Orange; these are so formidable as to obstruct passage; wd. u. for fence posts ("one hole outlasts ten fence posts" [facetious]); formerly planted as living fences; sleepers (on r.r. tracks), war clubs, &c.; lvs. have been u. to feed silk worms.

2035. *Magnolia acuminata* - thousands in Clarke Co., Ala in lime hills (*Clarke Co. Democrat* 1959). Spotty distrn. La-Ga, Fla.

2036. Wehmer. *Pflanzenstoffe*, ed. 2, v. II: 1296 (VO also fr. branch bk. & young fructifications with different physical props) (Tommasi) - *Magnolia grandiflora* blooms Apr.-Jun., sometimes few blossoms until Oct., fl. lasting 3-4 days. Frt. light pinkish when ripe, bright coral red sd. (Montgomery, Ala).

2037. Sweet bay (*M. virginiana* or *Persea borbonia* (unlikely) early use (ca. 1850) at Sebastian, Fla (between Daytona & Palm Beach) wd. u. as stake for spitting meat (slices of pork, venison & bear placed alternately) (old timer) (using wd. fr. tree in fl., then sweetish, otherwise "bitter as quinine").

2038. Chevalier, A. *Rev. Bot. Appl.* 18:204, 593; 1938 (cult. majorana).

2039. *Jl. Ind. Trade.* 1954.

2040. Emmel, M. W. *U. Fla. Agr. Exp. Sta. Press Bull.* 587 ("known remedy") - NF VIII: fowl dose: 0.5 gm. - Smith, Robert M. *Ark. AES Bull.* 431; 1943 (vermifuge treatment with K. & egg production) - *USDA BAI, Bull. 12* (poultry tapeworm).

2041. Hunt, Mabel Leigh. *Better Known as Johnny Appleseed* (book) (1950) - Johnny Appleseed (born John Chapman [1774-1845]) is described as "quite an old man...intilligent (sic)...but slightly demented...full of pleasant stories...." Last visit to Ohio 1842-3; died at Ft. Wayne, Indiana. He carried the works of Swedenborg with him.

2042. Mustard, Margaret J. *Sci.* 104:230-1; 1947; (vit. C content of *Malpighia* frts. & jellies).

2043. Petricic, J. *Farm. Glasnik.* 4:69-84; 1948 (detn. of viscosity of exts. of some drugs of *Malva*; using Hoeppler's viscosimeter successfully).

2044. *Mammea americana* - Frt. round & fleshy; 8-16 cm. diam., 2-4 large sds. (LeComte); of no industrial value (Pesce); decoc. of husk & pericarp antiparasitic.

2045. Mandragora root - Wisselingh. *Pharm. Weekbl.* v. 58:1921 (anatomy) - *Apoth. Ztg.* 48:#10;

1933 - Alique Tomico, Angel - *Medicamenta* (Madrid) 13:175-6; 1950 (2 cases of intoxicn. by root). Also see Lucian (Greece).

2046. *Mangifera indica* - *USDA Div. Pomology Bull.* 1; Bailey. "Cyclop. Am. Hort." 978 - *Pr. A Ph A.* 25:218; 1878 (cult. in Calif, Fla).

2047. Haupt, Paul (ca. 1922) - Manna mixed with honey-like exudn. fr. punctured bk. of tamarisk; mixt. baked.

2048. Conzatti, C. *Generos de la Veg. Méx.* p 410.

2049. Arrowroot cooked —> dessert: 2 drams added to 1 oz. milk or water, poured over this a half pint of boiling milk or water, stirred briskly, 1 dram sugar added; 1 dram. brandy or cream may also be added; useful for invalids.

2050. Hertel. *Amer J. Pharm.* 1890:273-; 327-. (Morrison) - Rusby, H. H. *NY J. Pharm.* 9:93-4; 1892 (marvelous new drug; Valparaiso, Chile, 1885-supposed great discovery; MD not anxious to reveal, wanted recompense, finally revealed it was only horehound!) - Gordin. *J A C S.* 30:265-; 1908 - Poser, F. B. *JCS.* 105:2280-91; 1914 (const. fls. of *Mat. chamomilla*) - Beguin, Ch. *Pharm. Acta Helv.* 7: 332; 1932 (glucides of fls.) - Ruemele, T. - *Manuf. Chem.* 22:276-; 1951 (C. in cosmetics) - Casetti, A. & Venanson, M. T. *Boll. Chim. Farm.* 94:#4; 1955 (La camomilla).

2051. Freise. *Plantas Med. Brasil* (VO).

2052. Alfalfa - Holmes. *Ph. J.* 1882 (adult. for Belladonna rt. on Engl Market) - Law suit of propr. 8-9 Oct. 1934 (WPB expert witn.) - *Bull. Pharm.* 1894:139 (cult. SA) - King, Carroll & Ball, C. D. Jr. *JACS.* 61:2910-1; 1939 (alf. sd. FO 11.8%; of this 4% = sterols, in part spinasterol or very similar) (Hager, 1976 *Handb.* spinasterols (3), glucoside; toxic saponins, sojasapogenol A,B,C) - Musher, S. US Pat. 2,198,214 (stabilizing alf. to retain carotene during drying & storage by use of a sugar & phosphate in small amts) (1940) - Alfalfa County, Okla so called since ca. 1905; since very large prodn. & distrn. of alfalfa.

2053. *The American Eagle* (Estero, Fla, 14 Dec. 1929 - cajeput tree becoming pest in Fla; Fla Forest Service/Lebeau, P. Traité de pharmacie chimique, t. 3, 2147 (1946).

2054. Hudson & Sherwood. *J A C S.* 42:116-25; 1920 (method of prepn. melezitose fr. crystd. honey formed due to pine honeydew).

2055. Rogosa. "Visnik sil's'ko Gospodans'koi," *Naiky ta Dosvidnoi Spravy* (Kharkov). 6:81; 1929 (*J. Agron. Sc. & Agr. Res.*).

2056. Braun, H. *Fortschr. d. Therap.* 18:80-; 1942 - Melissa one of oldest medl. plants.

2057. Bokori, J. *& al. Acta Veter. Acad. Sci. Hung.* 5:431-46; 1955 (poisoning of horses, 2 cases deaths; active princ. not isolated).

2058. Musk plant taken fr. Vancouver Is(?) to England ca. 1833 by Douglas (1798-1834) - Search for very rare pl. on w. coast of BC ca. 1930 by Kew Gardens - Ward, M. E. - *Rhodora* 6:227-8; 1904 (Mass) - Saunders, C. F. *J. Bot.* 61:176-; 1923 (scentless) - Borza, A. *Bulet. Grad. Bot. Univ. Cluj.* 13:52-3; 1934 (*M. mosch. & M. guttatus* in Romanian flora) - Grant, A. L. *Ann. Mo. Bot. Gard.* 11:99-388 (monograph of genus) - Holmes, E. M. - *C & D.* 100:258-9; 1924 (scented var. said almost non-existent in BC today) - *Idem. Gard. Chron.* III, 75:78-9; 1924 - Hill, Arthur W. *Gard. Chr.* III. 88:259; 1930 - Redgrove, H. Stanley. *Ibid.* III. 89:202-3; 1931 - Stubbs, Frederick J. *Ibid.* III. 89:116; 1931 - Chevalier, Charles. *Rev. Hort.* 106:32-4; 1934.

2059. Bhajuratna Manisharsha Joti, Kalimpong (1944) (Tibet musk).

2060. Tonkin (or Tonquin) musk was named after 1) Shong-kin (Tibetan border city doing trade in musk) 2) Tong-ting-ho (upper course of Yangtse Kiang River at Chinese-Tibetan border. In the 18th cent., all the musk came through Tonkin (however Tschirch believed name came from country of origin).

2061. *M. gummosus* was reported not found around Ensenada abundantly but 350 mi. s. around Santa Rosalia & across Gulf of California around Guaymas, Sonora. The other sp. in gen., *M. eruca*, also of Lower California.

2062. Lode, G. *Fettchem. Umschau* 42:205-7; 1935 (comp. FO); u. as moisturizer, emollient, in sun creams; in hair care ("conditioner").

2063. Mineral waters were drunk by Pres. Eisenhower & many members of Congress (*Readers' Dig.* 1960).

2064. Mitella < mitra (L) (diminutive), cap or mitre; SC ripe seed vessel with 2 pointed lobes resembled small mitres.

2065. *M. vert.* - Plant grows in hills of coco palm & marshy zones of Cordoba, also Tucuman (nArg) & sBo.

2066. *Drug Markets,* Mar. 1980.

2067. Wallace Brothers - Drug dealers, NCar.

2068. Research by Dr. Gilberto Rivera, Trop. Sch. Med., Univ. Puerto Rico (ca. 1942).

2069. Rivera, Gilberto. *Amer. J. Pharm.* 113:281-97; 1941; 114:72-87; 1942 (prelim. chem. & phl. stud. on "Cundeamor" [*Momordica charantia*]).

2070. Chamberlaine, Wm. *A Practical Treatise—Dolichos*, 121 pp.; 1812.

2071. May be conspecific with *F. xanthoxyloides.*

2072. Gasster, Marvin (MD). *Geriatrics.* 1964. (banana extremely useful for geriatric & obese patients; give satiety & satisfaction, flavorsome, high bulk of low caloric food).

2073. Ryan, G. M. *J. Bombay Nat. Hist. Soc.* 15:586-93; 1904 (wild plantain (*Musa superba*) eaten as frt. & veget., also sheathing petiole & root stock); no real st. above ground; actually what appears to be a stem is a spurious herbaceous shoot made up of closely packed convolute lf.

sheaths, with a tuberous rt. stock at base; lvs. u. as dinner plates (Bombay). Frt. eaten during small pox attack, thought "specific" in rabies ("Bassein Taluka"). Lf. fiber u. for floor matting & for tying parcels; cult. as ornam.

2074. *Musa japonica* = hardiest of bananas; grown at Messrs. Veitch's nurseries at Exeter, Cornwall (GB) but not expected to ripen frt.

2075. Parry, Wm. H. *Metal Industry.* 18:89; 162; 1920.

2076. Chevalier, Auguste. *Rev. Bot. Appl.* 2:633-6; 1922 (Asiatic spp. with edible fruits) - Youngken, H. W. *J A Ph A.* 12:484-8; 1923 (bark of *Myrica cerifera*) - Bayberry candles for Xmas sold by Will & Baumer Candle Co. (2 lengths) - In the comic strip "Pogo" by Walt Kelly, "bayberry tart" was facetious; however, some Asiatic *M.* spp. have edible frt. (*cf.* citation) - M. Dubreuil produced 6,000 lbs. of myrtleberry wax in a single year; this was the only source of illumination in Louisiana at that time (ca. 1750) (Asbury-French Quarter, p. 33).

2077. *Harpers Weekly.* 1:173-4; 1857 ("A nutmeg plantation") (pictorial history of nutmeg; figs. of prepn. &c.; quite detailed acct. of prepn. &c. in Penang).

2078. Smith. *JCS.* 73; Perkin, *JCS.* 81 (Myriticolorin).

2079. Myrtle wax harder than beeswax, tastless, with balsamic odor (like rosemary): England imports (imported) prod. from sAf & u. in mfr. phonograph discs.

2080. Police, Gesnaldo. *Boll. Pesca, Piscicolt. e Idrobiol.* 6:298-369; 1930 (nutritive value and pathol. action of lamellibranch molluscs [mussels, oysters, &c.]) (exposure to long drought & sun prod. "ptomaines"; living in stagnant water —> mussel toxin) - Mueller, Hellmut (Hooper Found. Med. Res., Calif). *J. Phcol.* 53:67-89; 1935 (mussel poison) - Gibbard, J. & al. *Cana. P. H. Jl.* Apr. 1939:193-7 - Bendien, Wm. M. & Sommer, H. *Proc. Soc. Exp. Biol. Med.* 48:715-7; 1941 (paralytic shellfish poison, purification).

2081. Wakeman, Nellie. "The Monardas, a Phytochemical study," *Bull. Univ/Wisc.* No. 448 (Sci. ser.); Aug. 1911 - Christensen, B. V. & Justice, R. S. *J A Ph A.* 26:11-; 387-; 466-; 469-; 1937 (I-IV) (study of several *M.* spp.) - Fellowship, Univ. Wisconsin, phytochemistry esp. of the Monardas (Fritzsche Bros.) - McClintock, Elizabeth & Epling, C. *Univ. Cal. Publns. in Bot.* 20(2):147-94; 1943 (reviews of the genus *Monarda*).

2082. *Monarda didyma.* Linné. *Mat. Med.* 1772, ed. 2 - Schoepf, J. D. *Materia Medica Americana.* p. 7; 1787 (*Lloyd Libr. Bull.* 6, Reprint ser. 3, 7 - Virey, J. J. *J. Pharm. et Chem.* 1815:89 - *Repertorium f. die Pharmacie.* 8:234; 1824 - (Thee von Oswego) - *Pharm, J. Mar.* 1857 - Stearns, F. *Proc. A Ph A.* 7:271; 1858 (med. plants of Mich).

2083. *Monarda fistulosa*: Register of *Arch. d. Pharm.* 1822-51 - Melzner, E. H. & Kremers, eds. *Amer. J. Pharm.* 68:539 - Miller, E. R. *Pharm. Exp. Sta. Univ. Wisc. Circ.* 4:1918 - Rang, K. H. (anatom. study, U. Wisc) - Harwood, Arthur A. *J. Am. Phar Assn.* 20:1268-72; 1931 (inorg. const. of several parts).

2084. *Monards punctata*: Hood, S. C. *US Dept. Agr. Bull.* 372;1916 (comm. prod. thymol) - *Schimmels Rep.* 1917:100; 1918: 87 - Collens. *West Ind. Bull.* 17:50; 1926 (thymol cont.) - Fonda, Lyman D. Univ. Fla. M.S. thesis, 1927 - Expts. at U Fla drug garden gave yields of 16-17 lb. VO/a. (price $3.50 to as high as $10./lb.) - Shinners, Lloyd H. - *Field & Lab* 21: 89-92; 1953 (nomenclature of vars.). - "A man living in Brookhaven, sMiss, called the 'Indian Doctor' uses large amounts of horsemint in preparation of liniments (and) sells them for the treatment of rheumatism and muscular soreness & numerous other ailments...used externally only.... This (man) paid the children in that area a penny for each pound of horsemint they collected" (Chas. H. Young).

2085. Watercress (*Nasturtium officin.*) "capital of the world" is Huntsville, Ala (but producers closing down, July 1969) (Oviedo [Fla] has also been called by this designation).

2086. Young girl smelling lotus flowers—dramatic fragment of decoration of a limestone Memphis mastaba style of 5th Dynasty, Egypt) (Louvre).

2087. *U.S. Dept. Agr. Rept.* 100 - *US Senate Document* 190 - McConnaughey, E. "Sea vegetation" (*Naturegraph*, 1985), pp. 221-2 (effective use of mucilage in 2nd degree burns - *cf. Aloe vera*).

2088. *Pharm. J.* 4/11/1978.

2089. *West. Drugg.* Jan. 1895:21-.

2090. Kalm, Peter. *Travels in North America*, v. 2; 1870 (*Acer rubrum*).

2091. Adeps Hominis. *Pharmacopoeia Universalis.* 32, 67, 1835-.

2091A. Several other *Agropyron* spp. are similarly u., as *A. dasystachyum* Scribn. & Sm. (thickspike wheatgrass); *A. desertorum* Schultes (A. cristatum Gaertn.) (crested w.) (fr. Russia, extensively planted in the Great Plains, Alta south); *A. elongatum* Beauv. (tall w.); & others.

2091B. Blood (serum) albumin - advantages: more stable, no risk of hepatitis or AIDS, long life span, no cross matching needed - Technetium aggregated albumin u. for X-ray photos of brain, lungs, thyroid, &c. (half life of Tc99 = 66 hrs.), preferred over radio-iodinated (half life 192 hrs.).

2092. Allium - in history - *Janus.* 23:167-; 1918.

2093. Marshmallow paste formerly (-1932-) cont. althea root; prod. called Hippolyte, handled by Charles, & Co. (grocery) bought up by Gristede (Hip-O-Lite Marshmallow Creme [pd. topping & meringue pd.], St. Louis, Mo).
2094. *Amorphophallus* (Araceae) - Chawla, H. M. & Chibber, S. S. *Indian J. Pharm.* 38:109-110; 1976 (some extractives fr. same: betulinic acid; stigmasterol, ß-sitosterol & its palmitate, lupeol, triacontane, glucose, galactose (considerable), rhamnose, xylose.
2095. Dent, Dr. John Yerbury (ed. *Brit. J. Addiction*) uses apomorphine HC1 for dypsomania (alcoholism). Based on aversion technic u. in c Eu, administers alc. beverages with apomorphine; causes vomiting (room not cleaned; disgust) (tabs. administered with increases hourly until vomiting; repeated 3-4 days with mild diet; sometimes with vit. B_1 over 4-8 wks.) - *Cf.* Keely Cure for alcoholism.
2096. Biol. Survey of Texas; 1905 - Bailey, Vernon. *NA Flora* #25 (granddaddy gray beard). Frts. of shrub or tree are pickled; summer food of bears.
2097. *Streptomyces lavendulae* is the source of some 20 other antibiotics.
2098. Pellacani, Paoli. *Arch. Exp. Path. Pharm.* 16: 440-51; 1882-3 (active constit.) - Feng, Y. X. *& al. Acta Pharm. Sinica.* 18:849-51; 1984 (morph., histology & chem. of this & 17 other medl. *S.* spp.).
2099. *Ocimum bas.* - Rept. of the Govt. Chemist for the year 1933: p. 16: air dr. wt. yield, lvs. & small sts. (excl. woody parts) 0.93%; lvs. only 1.59%.
2100. Charabot, Eug. *L'*Ocimum canum *Source Naturelle de Camphre Droit* - Blaque, G. & J. Maheu. *Etude Botanique de l'*Ocimum canum, *Plante à Camphre* - Ungerer & Co. (NY): pamphlets.
2101. Hurley, John. *The Tree, the Olive, the Oil*; ca. 1920 - Bitting, K. G. *The Olive* (Glass Containers Assn of Amer. Chicago) ca. 1920 - *USDA FB.* 1249:33 pp. (rev. 1939) - Soroa y Pinedo, Jose Maria de. *El Aceite de Oliva* (Editor. Dosset, Madrid; 1949) - On the return of the dove to Noah on the arc, the olive branch it held was significant. "An olive, the token of succeeding plenty and the badge of peace, the two greatest of earthly blessings and which usually go hand in hand together" (Thomas Swift ca. 1700, bro. of Jonathan S.) - The Franciscans at the Garden of Gethsemane near Jerusalem claim that the olive trees there are over 3,000 years old.
2102. Tanret, G. *Schweiz. Ap. Ztg.* 71:1933 - *Bull. Soc. Chim. Biol.* 14:724-44; 1932.
2103. *Olearia arg.* - *Chem. Drugg.* 91:39.
2104. The chief mission cactus and the source of best vars. of edible tunas, as well as one of the most commonly cult., is the var. formerly known as *Opuntia megacantha*, commonly as La Tuna (or Rancheria) prickly pear. Cultivated in sCalif, it escaped and is now common in Lower Calif. Called "spineless cactus" but is difficult to see why since this sp. as cult. in Jam & Haw Is has well-developed spines up to 3.5 cm. long.
2105. *Origanum vulgare* vars. Durango, Monterey.
2106. Austin, A. "Rice Book," *La. Agr. Bull.* 61 - Williams, R. R. *J. Am. Dietetic Assn.* 28:209-22; 1952 (rice in Asiatic diets) - Some products of rice on market incl. rice bran, r. hulls, rough r., uncoated & coated rice, polished rice. R. u. in prepn. of oral electrolyte solns. u. for diarrhea (p.o.) (Oryza). Long grain rice: claimed easier to cook, with grains separate. Medium grain r. when cooked, becomes rather sticky; less preferable for eating. Short grain r. u. in mfr.; not generally sold in retail stores.
2107. Several thousand acres of wild sweet anise were located near Chicago (1945).
2108. *Amer. J. Pharm.* 94:429; 1922.
2109. Williams, R. J. *Enzymologia.* 9:387-94; 1941 (review of pantothenic acid) - RJW. *J. Am. Chem. Soc.* 60:2719; 1938 - *Fette u. Seifen.* 48:211-2; 1941.
2110. Pronounced in US mess-keet´ (Americanized).
2111. *Brit. J. Exp. Path.* 19:53-65; 1938 (mesquite gum cross-reaction with *Pneumococcus* antiserum, type II) - This gum is similar to acacia & is u. as substit. for same; potential source of arabinose & glucuronic acid.
2112. The lvs. & bk. of this shrub or small tree c. prulaurasin hydrolyzing to racemic mandelonitrile & glucose; drupe 10-13 mm. long.
2113. Bk. u. to make cherry water. (astr.); trunk source of cherry gum, u. as adulter. tragacanth (Kordofan); wd. hard & strong u. for furnit. walking sticks; handsome fls., tree often cult. as ornam. (Burma) *(Prunus cerasoides).*
2114. Madden, T. (MD) (Dublin). *Philosoph. Trans. Royal Soc.* B (London). 37: 84-100. "An Account of two women being poisoned by the Simple Distilled Water of Laurel-leaves, and of several experiments upon dogs; by which it appears that this Laurel-Water is one of the most DANGEROUS POISONS hitherto known."
2115. Flores Amygdali. Almond blossoms, Fl. Amygdalarum - mentioned in "Tabulae Maestro Salerno"; ca. 1130 (*Prunus dulcis*).
2116. *Prunus institia* L. - wHimalayas: Jarhwal to Kashmir (frt. cult.).
2117. Cases of infection of wounds from sloe thorns were reported (Staph.; tetanus; allergic reactions; tumor formation). *Lancet.* 1960:310, 433. These were answered by v. Mauch. (*Med. Monatsspiegel* 1960(7): 160), who claimed the infections were due not to the thorns but to the practice of the shrike (bird) of hanging his booty on the thorns to ripen & then to con-

sume it. The rotted insect is source of infection. Toxic subst. are not known in the plant.

2118. *Amer. J. Pharm.* 1875:53 - Wild cherries, dried (*Prunus serotina*) were formerly handled by S. B. Penick & Co. (fr. NCar, Ore, west US)

2119. Prune powder has the laxative props. of prunes plus pleasant flavor; finely milled fibers act as colloid, absorbing water & providing soft bulk & demul. action.

2120. Cherry gum u. as thickening agt. for cloth printing (fr. *Prunus cerasus, P. avium*, &c.) (Diamalt Akt. Ges., Munich, DRP 707847 (ca. 1938-41). *Bibl. Sci. Ind. Repts.* 3:981; 1946.

2121. *Business Week* 7/XI/42 (cork prodn. fr. Douglas fir); Wash state with potential ann. output of 100,000 tons; same fr. Ore; high grade cork in 4 grades; mfg. processes: grinding, drying, screening.

2122. Berger, F. *Pharm. Zentralh.* 77:389-91; 1936 (*Pterid. aquil.* u. to adulterate herbs of male fern, maidenhair fern, polypody; adultern. easily detected fr. spores).

2123. Patch, Edgar L. *J A Ph A.* 2:1103, 683; 1913 (often large amts. water found in red saunders [7-44%] [ash 1.2%]; for unknown reason, shippers wet down before shipping; reasonable H_2O cont. 5-7%) - Wd. u. in cabinet making.

2124. *Puer. lob. - Hawaii Bull.* 60:52; 1929 - *USDA, FB* 1125:45-7; 1940 - *Y Bk Agr.* 1908:249-51; 1937:1006 - *Dept. Circ.* 89:5-7.

2125. Kudzu starch u. in cooking & technology; said superior to corn starch; keeps dry, not easily wetted as is corn s.; 13 oz. pkg., quality "Hime" (ca. 1944) (Japan).

2126. *J. Bact.* Aug. 1946 (Dr. Seegal) (Carlson (USNR), Klamath Falls - Cogeshall; WRU).

2127. *Chem. Drugg.* 1943, II:226; 1944, Aug. 19: 203 (Lipari pumice stone).

2128. Lloyd, J. U. *West. Drugg.* 19:202-5; 1897 (balaustion flowers) - Waring, *Pharm. of India*; 1968 (rind u. in dys.) - Hodgson, Robert W. *Cal. Agr. Exp. Sta. Bull.* 276:161-92; 1917 - P. fl. u. as popular motive in Turkish embroidery designs (ca. 1670) (*Ciba Review.* #102:3659, 3680; ca. 1960) - (*Cf.* Assyrian bas reliefs 885-860, B.C.) (Metropolitan Museum).

2129. Becker, J. *Ztsch. Pflanzenzuecht.* 6:215-21; 1918 (inheritance of certain fl. characters in *Papaver rhoeas*) - Wiedmann, Friedrich. *Ueb. Bestandteile der Bluethen v.* Papaver rhoeas *zur chem. Charakteristik der Familie der Papaveraceen - Druck d. Akad. Buchdruckerei v. F. Straub.* 33 pp.; 1901 - Schmid, Leopold. *Ueb. die Konst. d. Farbstoffes des Klatschmohns* (Papaver rhoeas) - *Sitzungsber. Akad. Wiss. Wien - Nature. Cl. Abt.* II.B. 139: 1049-60; 1930 - 141:205-14; 1932.

2130. *Western Drugg.* 27:774-; 1905 (Opium poppy formerly grown by Shakers; ca. 1870, good quantity of "Vermont Opium" on market; 1 man sold $3,000. worth in a year; 1 man in Calif had over 100 a., grown for seed. One Confederate physician (Dr. Bartholomew Egan) reported that native white poppies gave an opium equal to regular comm. grades (ca. 1864); Porcher expresses a similar view. Opium was grown in Mass (late 1800s).

2131. *Paris quadrifolia*: stamens show long narrow connective prolonged into a point; lvs. in whorls of 4 on flg. stalk.

2132. van Dersal, Wm. R. "Native woody plants of the US," USDA. Misc. publ., 303; 1938.

2133. Typical desert plant of region fr. Toreon, NM to Ft. Stockton, Tex; latex fr. bk. emergency rubber source, u. in some auto tires.

2134. Ruggy, G. H. & Smith, J. *J A Ph A.* 29:207-8; 1946 (chem. studies on physiol. active subst. in *Passiflora inc.*) - *Ibid.* 245-9 (phcol. study of active princ.).

2135. Bullington, S. D. "Passion flower in epilepsy, & other neuroses," *Nashville J. Med. & Surg.* 1897 - Phares, D. L. *Richmond Med. J.* 1867 (said remedy for tetanus !!).

2136. Pope, W. T. *Haw. AES Bull.* 74:1-22; 1935 (*Passiflora edulis*) - Poore, H. D. *Fruit Prod. J.* 14:264-8; 285; 1935 (*P. edulis*) - Jewell, W. R. - *J. Dept. Agr. Victoria* 31:609; 1933 - "Yellow juice u. in Hawaiian Punch" (R. J. Reynolds) - Only P. sp. grown comm. in US (Hawaii).

2137. Vian(n)a, Artur Lourenco: "Anais Fac. Odontol. Farm.," *Univ. Minas Gerais.* 3:357-93; 1938/9 (contrib. to study (pharmacognostic) of *Pass. quadrangularis*) (thesis).

2138. *Cooking the Indian Way* (Spring Books, London)(1962) (prepn. peel mangoes & boil frt. in a little water to give pulp; strain & sweeten; chill & serve mixed with cream or milk) (same amt. milk but half the cream).

2139. Plasmids: small closed-loop mol. of DNA found outside chromosome; can be transferred fr. one strain to another & u. in introducing new genetic information; u. by recombinant DNA technics.

2140. Rudd, Velva E. *Contr. US Nat. Herb.* 32(2):1956.

2141. Rudd, V. E. *Ibid.* 32(1);279-384; 1966 (monograph).

2142. Rosengarten, F. Jr. "A neglected Mayan galactagogue, Ixbut" (*Euphorbia lancifolia*) *Bot. Mus. Leafl.* 26:277-309; 1978.

2143. Rudd, V .E. *Ibid.* 32(6):1968.

2144. Rudd, V. E. *Ibid.* 32(3):1968.

2145. Rodin, R. J. *Mo. Bot. Gard. Monographs in Syst. Bot.* 9:1985.

2146. Sp. cited in "Index Kewensis" as a pharmaceutical name published in Wehmer's *Pflanzenstoffe* (1911).

2147. Prentiss, Daniel Webster (MD) (1843-99) (papers) (386-7): Remarkable change in the color of the hair fr. light blonde to black in a patient while under treatment with pilocarpine (1881)

- Change in color of the hair fr. white hair of old age to black with jaborandi (1889) - Change of hair color fr. internal use of pil. (1889) (*J. Amer. Med. Sci.*?) - *Experientia.* 16:373-; 1960.

2148. *Trans. Med. & Phys. Soc.* (Ind) 4:118; 1841 (anise root, sorif-ki-jir) placed at night in earthen pot in moonlight; in morning ground & mixed with sugar & water —> draught with cooling props - Found in bazars of Hyderabad, Khyrpur, Shikarpur (India).

2149. Michlin, Jacob. "Recherches pharmaco-anatomiques sur *Pimpinella* spp.," (Univ. Lausanne, thèse, 1926) - In pottery fr. very ancient remains found anise, peas, barley, &c., on Sanborin & Therasia (Greek archipelago isl.) (studied by Christomanos, Univ. Athens, before 1870).

2150. Shyreu, E. W. *J. Pharmacol. Exp. Ther.* 1935: 1-12 (drugs fr. *Plantago* gen.) (per Tessene (1967/8, Wisc).

2151. Pollen, prices per gram (1958): catalpa, $6.00; black locust, $4.00; *Magnolia grandiflora,* $4.00; Bermuda grass, $3.50; red clover, $5.00.

2152. *Piptostegia* variously equated to *Operculina, Convolvulus, Ipomoea, & Merremia.*

2153. *Piscidia erythrina. A Treatise on Hydrastis & Piscidia* (1928) (Lloyd Bros.) - Acc. to some, curare-like alks. present (Dragendorff-*Heilpflanzen,* 329; 1898 [1967]).

2154. Akacic, B. *Farm. Glasnik.* 3:153-8; 1947 *(Pistacia tereb.,* localization of tannins in various parts of lf.; 2 tannin types: catechin, gallotannin. Quant. methods (skin pd., modif. Wolf-Gawalowsky method) gave same results; lf. c. 23% tannins. Qual. similar to sumacs. Tree found along Adriatic sea coasts in large numbers.

2155. Farwell, O. A. *Bull. Torrey Bot. Club.* 42:247-58; 1919 (Mich spp. of *Polygonatum).*

2156. Rao, N. V. S. & Seshadri, T. R. *Curr. Sci.* 10:413; 1941 (effect of storage on quality of pongamia oil) (*P. pinnata (P. glabra*) [Indian]).

2157. Wurtz, A. *Bull Fran. Pisciculture.* 28:41-52; 1955 (action of poplar lvs. in small fish culture ponds) - Anon. "Populus; a bibliography of world liter. 1964-74," *USDA For. Serv. Res. Pap.* SO-124; 1976.

2158. Origin of name *Populus:* (NL) 1) pamp - pap - to swell, as in papula (pimple), papilla (nipple), pampinus, young vine shoot or lf.) 2) palpulus (palpitare to tremble); paipalllo (shake, vibrate) 3) arbor opuli (tree of the people (u. to decorate Roman public places; ex. Piazza Poppolo, Roma) - US expression: "poplar tall" - "Right as an aspen leaf she 'gan to shake" (Chaucer, *Troilus & Cressida* III; 1200).

2159. *Hardwood Record* 26 (Aug. 25, 1908) pp. 18-9 (fls. appear long before the lvs. in Man.; also NC where prod. comml.; easy to distinguish fr. lf buds) - Rouleau, Ernest. *Rhodora* 48:103-; 1946 (*Populus balsamifera* of L. not a *nomen ambiguum).*

2160. *Populus deltoides* subsp. *monilifera* Ait. discovered by Dr. W. M. Alter (England, Ark) ca. 1880 to be powerful "antiperiodic," almost a specific cure for "swamp fever or malarial hematuria" trunk or rt. bk. used (*Merck's Rep.* 1904).

2161. Madaus, G. *Lehrbuch der Biologischen Heilmittel,* ed 2. III: 2209; 1979 (since antiquity, *Populus nigra* [also *P. alba*] u. in med. Galen called it "kerkis"; bk. u. as tenifuge (China). Lvs. c. 19% tannin [*ca.* 32:3189]).

2162. *Schweiz. Apoth. Ztg.* 71:101; 1933 (*Potentilla*).

2163. Wolfred, Morris. Univ. Wash. diss. 1944: 60 pp ("*Potentilla anserina & P. argentea*: a phytochem. study"). A concentrated prepn. of *Potentilla anserina* called Meliomen (Asta A. G. Chem. Fabrik Brackwede [Westfalen] Germany) is recommended in the treatment of spastic dysmenorrhea; sometimes obtainable with 9% added Hydrastis ext. - A good picture of Herba Anserinae appears in *Thoms Handbuch prakt. u. wiss. Pharmazie Bd.* 5:1029.

2164. *Ap.-Ztg.* 48: #24; 1933 (Tr. Primulae) - Source of much facial eczema in Ge (ca. 1930); campaign to exterminate plant).

2165. *Prenanthes altissima* described in Fernald, M. L. *Gray's Manual of Botany,* ed. 8:1561; 1950. - Grows in woodland thickets, Newfoundland to cGa.

2166. Youngken, Heber W. Jr. Ph.D. diss. Univ. Minn. 1942 (*Potentilla anserina* L. & *P. argentea* L.) - *P. anserina* occurs around the Great Lakes; in Wisc, is "quite abundant on sandy beaches along Lake Michigan, particularly in Sheboygan, Manitowoc, & Door Counties"; sometimes a weed along railroads inland; Milwaukee County.

2167. Bourke, Norma & Dorman, K. W. *Univ. Fla. Eng. Expt. Sta. Bull.* 10: 1946 (spray gun u. for pine tree gum flow stimulation).

2168. Muc. fr *Psyllium* hulls u. to treat ulcers (Kober, P. A. to G. D. Searle & Co., US Pat. 2,132,484 (1939) - Extn. pectins & mucil. fr. sds. (Fre. Pat. 694,460; 1929; Olivier, Henri) (extd. with water in a closed vessel at temp above 100° C.) - US Pat. 2,115,491-2 (1938) (fatty acid prepns. fr. P. sd. oil are sol. salts u. by inj.).

2169. *US Dept. Comm. Spec. Circ.* 328; 1930 (prodn. of *Psyllium* sd. in Fr) - Howes, F. N. *Bull. Misc. Inf. Kew.* 1931 (cultn. of psyllium sd.) - Montague, Dr. Joseph F. Psyllium *Seed: The Latest Laxative* (Montefiore Hosp. for Intest. Ailments, NYC) 170 pp.; 1932 - Bottini, O. *Ann. Ist. Super. Agrar. Portici* 9:75-82; 1937-38 (mucil. may at times be u. in place of acacia, esp. in dye printing; FO fr. mucil.-free sd. may be u. to prep. varnishes [mixed with linseed oil]; residue u. for feed). Psyllium mucilage is prepd. by boiling the P. sds. for some time in the proper amt. of water; the mucilage separated is u. in dressings for various textiles (silk; muslin; lace).

2170. Maisch, J. M. *Amer. J. Pharm.* 61:449-53; 1989 (origin of false senega rt.)

2171. Maisch, J. M. *Ibid.* 64:177-82; 1992 (*Polygala alba*).

2172. Preuss, Paul. *Expedition.* 1899/1900, p. 100 - Nicaragua - "blue-flower ipecac" rts. good emet. without toxic by-effects.

2173. Wherry, Edgard T. *J. Wash. Acad. Sci.* 17:181-4; 1927 (presence of free Me sal in some Amer. *Polygala* spp.) - *P. polygama* inflorescence with strong odor of Me Sal (specimen gathered in Fla summer, 1931).

2174. Fernald, Merritt Lyndon. *Rhodora.* 4:133-4; 1902 (Seneca snakeroot in Maine) - Hood, S. C. *Vt. Sta. Rept.* 1907:371 - Holm, Theodor - *Merck's Rep.* 16:155-7 (Med. Pl. of NA).

2175. Podophyllum - Warren, L. E. *J A O A C.* 10:272; 1927 - *Ibid.* 13:117; 1930 - Warren, L. E. *J A Offic, Agr. Chem.* 18:555-7; 1935 - (resin assay) - Haworth, Robert D., Richardson, Thos., & Sheldrick, Geo. *J C S.* 1935: 1576-81 ("Natural phenolic resins") - Husa, W. J. & Fehder, P. *J A Ph A.* 26:1246-7; 1937 (extn.) - Viehocver, Arno & Mack, Harry. *J A Ph A.* 27:632-43; 1938 (biochem., chief active const. podophyllotoxin) - Uhl, A. H. *J A Ph A.* 27:595-6; 1938 (prepn. of resin of P.). - Husa, W. J. & Lee, D. W. *J A Ph A.* 28:593-7; 1939 (extn.).

2176. Jeffreys, T. *Remarks on the Efficacy of Matico as a Styptic and Astringent.* London, 1845 ed. 44 pp.

2177. *Piper longum* well-known as drug of India since time immemorial; cult. in Bengal, Madras (limited), Nadia, Bekrampur (attempts).

2178. Garabed Bishirgian (1885-1943) sponsored a pool which almost captured the white pepper market in London in 1935; failed because he did not know that black p. can be converted into white p. by stripping.

2179. *JAMA.* 115:77; 1940 (letter). Claimed by patient that *Polygonum aviculare* is of value in diabetes (old European treatment).

2180. Trapeznikov, A. *Sem. Med.* 1894:270 (*Polygonum aviculare* - Decoc. very active in diarrh.) (1-4 gm/da).

2181. Kobert, R. "Ueb. Kieselsaeurehaltige Heilmittel insonderheit bei Tuberkulose" (Rostock) (*Historische Studien aus der Pharmakol. Inst.*, Univ. Dorpat, Halle, 1889/96).

2182. Parry. *Essential Oils*, ed. 4, 269-70; ca. 1930 - Werner, H. W. *J A Ph A.* 20:445-8; 1931 (Based on MS thesis, Univ. Fla. 1929) (*P. muticum*).

2183. Juetner, R. & Siedler, P. *Pharm. Zentralh.* 50: 1431-5; 1912 - Costa, D. *Giorn. Chem. Ind. Appl* 4: 91; 1921 (analysis) - De Mori Alessandro. *III. Int. Congr. of Int. Feder. for the Promotion and Utilization of Med., Arom. & Allied Plants*, pp. 328-48; 1929.

2184. Abel, Jos. *Amer. J. Pharm.* 8:520; 1860 (Persian insect powder, lately introd. into market; Dr. Koch found originated in Caucasus reg. Cult. US with success; much adulter. with chamomile, fleabane, &c.) - McLaughlin Gormley, King, Minneapolis, MN. process for extn. of Pyr. Fla (ca. 1942) - Gnadinger, an officer in this firm, wrote several volumes on Pyrethrum Flowers.

2185. Holmes, J. S. *Common Forest Trees of No. Carolina,* ed. 4, 1929 - Little, E. L. Jr. *USDA For. Serv.* "Checklist of U.S. Trees"; *Agr. Handbook* No. 541; 1979.

2186. *Textile Colorist.* 50:845-6; 1928; *Ibid.* 23:709(*Prosopis jul.* as source of dyes & tans.) - In Mex, u. to dye khaki colors; wd. bk. & pod c. tannins.

2187. *USDA Bull.* 1194:1-19; 1973 (mesquite bean kernels rich in protein, sds. c. pentosans & galactans) - Fraps, G. S. *Tex. Agr. Exp. Sta. Bull.* 315:1924 (digestibility, productive values, &c.; pods rich in sucrose).

2188. Anon. *J. de Pharm. d. Anvers.* 1862 - *Anthemis cotula* (Maruta c. DC) valuable surrogate for insect powder, u. for bedbugs, fleas, flies, &c. (ants not affected) (troublesome weed in some parts of US).

2189. Historic tree specimen of *Quercus laurifolia* at Montgomery, Ala (capitol grounds) fr. Va battlefield.

2190. Quebracho. *USDA For. Circ.* 42; 1906 - *USDA, Dept. Bull.* 1409:20 - Kerr, Geo. A. "Q. forests of SA," *Commodities of Comm. Series* No. 9 - Wilson, J. A. & Merrill, H. B. *Analysis of Leather & Materials Used in Making It.* McGraw-Hill; 1931 - Gramm, H. "Die Gerbstoff u. Gerbmittel" (1933), vol. XII of *Chemie in Einzeldarstellungen*, ed. 7 (486 pp.) - Freudenberg, K. "Tannin, cellulose, lignin" (also ed. 2. of *Chemie der natuerl. Gerbstoffe*), pp. 63-5; 1933.

2191. *Brass World.* 8:271 (1912) (In tumbling zinc or aluminum goods, get high polish with soap bark; cannot use soap as with brass or steel; ex. Zn eyelets; 2 photos).

2192. Burini Guletti, A. *Biochim. e terapia spec.* 18: 341-7; 1931 (*Ranunculus* - probable presence of saponins & distrn.).

2193. Once thought juice of *Ranunculus sceleratus* ingested brought death, the man laughing as he died (Apuleius); prob. confusion with toxic member of fam. Ranunculaceae which produces rectus (pseudo-laughter) due to fatal spasm - grin remains on face of dead.

2194. First isolated by Winckler (1835).

2195. *Pharm. Taschenbuch*, ed. 6 (Kaiser) p. 794 gives details for Universal Reagent: chloral 45.0, Fe alum 4.0; heat in 10 ml. dist. water, filter. add I_2 - 0.4 g. diss. in EtOH (96%) 40 ml., then anilin sulfate 1.0 in dist. water 15 ml., diss. with heat. Add 30 ml. glycerin & 0.1 g. Sudan III. Let stand 24 hrs. filter, keep in brown bottles - In tests, red = cutin, suberin, wax, FO, VO, resin, &c.; yellow = lignin; colorless or red; violet = cellulose; colorless = gum, mucil.; blue = starch; yellow brown = aleurone; black blue = tannin; crystall. sepn. = alk.
2196. In red tide, water discolored; variously called yellow, brown, "rusty," gun metal color, dark green, red, purple, "like oil streaks," milky, murky, turbid, &c. Consistency almost syrupy, sometime stringy; ruins (darkens) paint.
2197. Largest stand in US said to be in Elliott State Forest, Coos. Co., Ore, where cut is limited. Largest collector & processor of the bk. was Henry Collison (Port Orchard, Wash) (*USP Convention* 1940, *Abstr. Proc.* p. 19).
2198. Named after Reinecke (*Ann.* 126:114, 117; 1863).
2199. *Renealmia exaltata* Roscoe - Monandr. Pl. Scitamin. t. 65; 1828 (plate). Tree found mostly in lowland forests.
2200. Steele, James G. "The buckthorns of California," *Pacific Record of Med. & Pharm.* (SF); *Pharm. Record* 7:41-2; 1887 (reprint of article); *Apoth. (Ztg)* 1887:277 - *Pharm. Centralh.* 1887:25 (taste bitter & nauseous; odor weak but characteristic).
2201. Colored figures in Migahid, A. M. *Flora of Saudi Arabia*, vol. 1:402-403; 1978 (ed. 2).
2202. *Roccella* produces good color in Alkaline Arom. Soln.; when amaranth used, haze developed; hence C. superior; quality of C. inferior to pre-war (Cusick, PD & Co).
2203. Raspberry leaf, aqueous ext. produced smooth muscle stim., esp. of uterus; anticholesteremic; spasmolytic (Beckett, A. H. & al. *Chelsea*, 1954) - *Cf.* Burn, ca. 1939-40 - Inf. u. for menstrual difficulties (1 oz. lvs./20 oz. boiling water; dose 10-15 fl. oz. per ora). - Sci. studies supported by Smith Dorsey, Univ. Neb. -1947.
2204. Koepfli. *J A C S.* 1932.
2205. R. Serp. *Indian J. Med. Res.* 32:183; 1944 - *Ibid.* 29:763; 1941 (R. N. Chopra & al.).
2206. Genus named after Leonhard Rauwolf; wrote *Travels into the Eastern Countries*, viz., Syria, Palestine; publ. 1738.
2207. *Rauwolfia tetraphylla*: Deger, Erwin: *Rev. Agricltura* (Inst. Nac Quim.—Agr. de Guatemala). 1938 (chem. study; found practical yield of 6.5% alkaloids; not toxic in small doses) - Pacheco Herrarte, Mariano. *El Imparcial.* July 17, 1936 (Guatemala) - Floriani. *Rev. Farm.* (Bs. As.) Sept. 1929 - Magdalena Orero, Maria. *Tesis* (Univ. Mex. 1939) - Mookerjee, Asima. *J. Indian Chem. Soc.* 18:33; 1941 (rauvolscine).
2208. Pega palo = *Rauwolfia pyramidalis* & other plants (Domin Rep) (exporting plans, 1957).
2208A. Damask rose growers in Turkey planning to expand prodn. (6 Jan. 1952). Bulgaria & USSR supplying Am market (34,000 oz. @ $25./oz [1951])> Attar of roses ("gulyag" in Turk) comes from s. provinces (vilayets) of Isparta & Burdur. Women pick fls. while dew is still on them & immediately distd.
2209. *J. Amer. Chem. Soc.* Nov. 1954 (simultaneous recovery of wax & oil fr. rice bran by filtration extraction).
2210. Dey, Kanai Lal. *The Indigenous Drugs of India...* (Calcutta, 1896), p. 267: Himalayan rhubarb fr. *R. emodi*, *R. moorcroftianum,* R. webbianum, *R. spiciforme*). 1,000 maunds (400,000 lbs) exported ann fr. Kangra distt. - Ghouse, M. & Katti, M CT. *J. Indian Inst. Sci.* 16A, Pt. 1 (109); 1933 - Bal & Prasad proposed *R. emodi* be u. in place of *R. palmatium.*
2211. Viehoever, A. *J A O A C.* 16:527-31; 1933 - 20: 561-4; 1937 - *Amer. J. Pharm.* 109:561 (rept. on rhubarb & rhaponticum) - Planchon, Louis. *Précis* (*Rh. rhapont.*) - von Ledebour, C. F. *Flora Rossica.* 1842-53 - *R. rhapont.* off. in Fr Codex, Ven Ph., Mex Ph. (also *R. emodi*).
2212. Kennedy, H. A. *De Rhabarbaro.* Edinburgh; 1754 - Pereira, J. *Elements of Materia Medica.* v. 2, pt. 1:492; 1839-40 - B. & R. *Materia Med.* p. 574 - Hosseus, C. C. *Arch, d. Pharm.* 240:419-24; 1911 (*R. palmatum* as source of good off. rhubarb) - *Ap. Ztg.* 31: Nos. 96-8; 1916 - (*Rheum*). NF XI minimum standard was NLT 30% dil. alc. sol. extractive - *Cf.* US Treasury Standards: 40% water-sol. extractive) - Rhubarb & Calomel Tablets & capsules, sometime popular (Ga).
2213. Chang, S.-T. *The Chinese Mushroom—Morphology, Cytology, &c.* Chinese Univ. of Hong Kong; 1972.
2214. Dannenfeldt, K. H. *Iris.* 73:382-397; 1982.
2215. *Ricinus*: Ubbelohde & Svanoe. *Seifensieder-Ztg.* 46:681; 1919 (hardened castor oils) - Fiero, Geo. W. "Tasteless castor oil obtained by solvent extraction," Pharm. D. thesis, Univ. Buffalo, 280 pp.; 1931 - *Chemurgic Digest* 1:59; 30 Apr. 1942.
2216. Makboul. *Fitoterapia.* 58:50-51; 1986.
2217. Fukushima & al. *C. A.* 76:158, 326; 1972 - *PB* 95813 (Library of Cong., Washington, DC) (1951).
2218. Lopez-Palacios, S. *Rev. Fac. Farm., Univ. Los Andes* 23:24; 1983.
2219. Schindler. *Brazil. Med. Pl.* 1884.
2220. *Agave cantala* Roxb. - Maguey; manila mague; may be source of Cantella gum; close to *A. angustifolia* Haw. (both in Group *Rigidae*).

2221. Tinozzi, C. C. *Dermatologia.* 3:103-5; 1952 (therapy of herpes zoster with cocarboxylases; also u. in nephropathies, multiple sclerosis, &c.).
2222. *Chem. & Drugg.* 119:393 (history of Cologne water).
2223. Sayre, Chas. B. *Univ. Ill. Agr. Expt. Sta. Bull.* 309 (expts. in culture & forcing of witloof chicory ["French endive"]).
2224. Animal expts. were negative for cycloserine in TB but clinically found effective; tests at NY Med. Coll. on 750 patients; a reevaluation of many antibiotics required (*C & EN.* 16 May, 1955).
2225. Power, F. B. & Chestnut, V. K. *J. Agr. Res.* 26: 69-75 (chem. exam. of chufa, tubers of *C. esculentus* showed 28.9% FO, 12% starch, gum, no caffeine); parched & ground for coffee substit. in Germany
2226. *Econ. Bot.* 1983, p. 118.
2227. *Cancer.* 5:405; 1952 (Layton, L. L. - chondroitin sulfate synth. by tissues of normal dba mice; such synth. decreased with age of animal).
2228. *Hirudo medicinalis* is claimed an endanged species. It is used much more than for simple blood letting; thus, u. to encourage graft adoption; to study nervous system, &c. Extolled in medicine by Andre Soubiran (MD). The healing oath (1954). Leeches sold for bait (Indian Lake, Logan Co. Oh [US]; ca. 1938; 20¢/doz.).
2228A. *Econ. Bot.* 36:178; 1982 (Nagaland, NE Indians).
2229. Originated in garden of Judge Logan, Calif (1881-).
2230. Balsam fr. trunk claimed u. in urin. dis. (Havard, V. *Proc. U.S. Nat. Mus.* 8:449-533; 1885).
2231. *J. Allergy.* 17:178-; 1946 (pollen counts).
2232. (*Rhinoceros*). Collected (horns) by Wakamba & Walungulu tribesmen of Kenya; kill rhino. with poisoned arrows - Fischer, Helen. *Field & Stream.* VII/54:44-7 (Royal Tsavo National Park).
2232A. Dutschewska & Kuzmarov. *J. Nat. Prod.* 45: 295-310; 1982.
2233. Decaris, L. J. *Ann. Pharm. Franc.* 11:44-6; 1953 (*C. A.* 47:7029) (prod. by cultures of *S. lavendulae*).
2234. Hefliger. *Beitraege z. Anatomie der Vanilla-Arten.* - Kelby, Harry G. "Vanilla beans," *Drug. Cosm. Ind.* 34:455-6; 1934 (cover & text pictures) - *The Story of Vanilla.* Livermore & Knight Co., Providence, RI - Vanilla is the most popular ice cream flavor (US).
2235. Varidase may help endocarditis by reducing valvular growths - Parker, B. M. *J. Lab. Clin. Med.* 1958/9.
2236. Roemer, H. "Pharm. Post" (in *Therap. Gazette* 15:718; 1891). Fresh fls. of *Verbascum phlomoides & V. thapsus* always u. with success against warts: pressed on wart with fingers, finally rubbed smartly. Should be repeated to ensure results.
2237. Urgell, J. M. (Barcelona, Spain) (before June, 1957).
2238. Samaan, Karam. *Quart. J. Pharm. Pharmacol.* 6:174; 1933 (visnagin).
2239. *Vitex agnus-castus* - Hanea in *Anatomy of Melancholy* (Burton); u. by L. Sterne in *Tristram Shandy,* v. VI, ch. 37; 1760-67 (apparently S.'s mistranslation of agnos [Gr] (supposed powerful in preserving chastity).
2240. Griebel, C. *Zeitschrift f. Unters. der Lebensmittel.* 85:511-5; 1943 (Moenchspfeffer als Pfefferersatz: *Vitex agnus-castus).*
2241. *Pharm. Era.* 1919: Greek "staphos" (currant or grape): extract condensed to almost solid form by heat-vacuum process.
2242. Peckoldt, Theod. (Rio de J.) *Ztsch. d. Allgem. Oesterr. Ap. Ver.* 31:830-3; 1893) (*Vitis sessilifolia* Baker).
2243. Viehoever, A. *J. Amer. Pharm. Assn.* 10:581-93; 1921 (domestic & imported veratrum) - Gilgut & Youngken, H. W. exptl. growth on peat moss & ordinary soils (-1950-) - Veriloid (Riker Phl. Co.) c. alkavervir (alk. ext. of hypotensive principles).
2244. Ergots found on wUS grasses: *Elymus* sp., commonly *E. condensatus* Presl (giant wild rye); *Agropyron* spp., commonly *A. spicatum* Scribn. & Smith; *Koeleria aristata* Pers. (June grass); *Phalaris arundinacea* L. (reed canary grass).
2245. Schunck de Goldfiem, Jean. *Revue de Path. Comp. et d'Hyg. Générale.* 46:131-; 1946 (Lucernes u. in Algerian phytomedicine: *Medicago sativa, M. lupulina, L. & M. "arborescens"* (*M. arborea* L. ?) - u. as hydragog, cholagog, to give wt. & height increase; internal hemostatic, to "balance" blood, in sick convalescent chldren; u. as salad).
2246. Amusing comments appear in *Pepys's Diary* (IV: 246) - after drinking Epsom waters, people walked into nearby open doors to defecate, with separate male & female places (ca. 1670).
2247. Informant, Mr. Gunn, Auburn, Ala.
2248. Used in NY cancer clinic; potentiates action of vit. E; Lipoicin, &c. (trade marks).
2249. *Lancet.* 4 Nov. 1972: Insulin & glucose & K. administ. in severe heart failure with promising results.
2250. Oliver-Bever. *Quart. J. Crude Drug Res.* 6: 861; 1966.
2251. Zechmeister, L. *Fortschr. Chemie Org. Naturstoffe.* Bd. 6:10. 13 (1950-56) - Blout, Elkan Rogers (MD) - structure of yohimbine - Harvard U. 1943-.
2252. (Indian valerian). Kobert R. *Historische Studien aus dem Pharmakol.* Inst. d.k. Univ., Dorpat, 1889 - Lindenberg. *Untersuch. Pharm.*, Inst. Dorpat, 1887 - *Pharm. z.f. Russland* 1886:523 - Bullock, K. *Pharm. J.* 115:122-5;

ca 1925 (oleoresin ca 0.78%) (*cf.* genuine root with 2-3%; no alk. but large amt. free acid, mainly valeric).

2253. Cocx, M. M. A. *Pharm. Weekblad.* 56:735-55; 1915 (mostly bot. re *Valeriana officinalis*, some chem) - Hahmann, C. *Ap. Ztg.* 43:11-3; 1928 (adulteration of valerian).
2254. Prickly Ash Bitters (Sherman's) (also S. compound P.A.B.); popular patent med. in Ala 1880-1900; c. buchu, emodin; principally an alc. beverage in prohibition States (*cf.* Peruna, Wine of Cardui, &c.).
2255. Ginger beer (Idris), made in Eng orig. 1880; c. aq. ext. ginger, acacia, capsicum ext.
2256. Lal, V. K. Ph.D. thesis, Banaras Univ., p. 261; 1980.
2257. Mohan, T., Ph.D. thesis. Banaras Univ. (Ind); 1986.
2258. Uniyal, M. R. *Nagarun.* 9:276; 1966.
2259. Tallest tree (Douglas fir) at Vancouver, BC before fire of 1886 was 328 ft. high and 14 ft. in diam. at thickest place; stood near Georgia St.
2260. Marker, Russell E. & Lopez, Josefina. "Steroidal sapogenins," series #160-171: *J A C S.* 69:2373-5 through p. 2404; 1947 (studies made at Lab. of Botanica Mex, SA [Texoco, Mex]; study of at least 4 genera of plants).
2261. Humm, H. J. *Fla. State Univ. Stud.* (contr. to Sci. 1-6); 1950 (ca.).
2262. Coville, F. V. "Plants used by Klamath Indians of Oregon," *Contr. US Nat. Herb.* 5:87-108; 1897 - Holdsworth. *Medicinal Plants of Papua NG.* 1977.
2263. (*Eucalyptus microtheca* & other *E.* spp.). Turnbull, R. F. *Econ. Bot.* 4:99-131; 1950.
2264. Blake, S. F. "Eastern Guatemala and Honduras," *Contr. US Nat. Herbarium.* 24 (4); 1922 - *Int. J. Crude Drug Res.* 24:19-24; 1986.
2265. A comprehensive account of chimo is found in the following paper: Kamen-Kaye, Dorothy: Chimo: an unusual form of tobacco in Venezuela - *Botanical Museum Leaflets, Harvard Univ.* 23(1):1-53; 1971. By irritation of vessel wall, this agent induces tissue growth with obliteration of vein; prod. is usually combined with local anesthetic & antimicrobial agt.
2266. Soap ("sope" in Bible: "For though thou wash thee with much sope, yet thine iniquity is marked" (Jeremiah 2:22 (KJV) - "He is like fuller's sope" (Mal. 3:2 (KJV) - If we went back to soap in place of detergent, it would reduce pollution (per *Commoner*, Dec. 1971) (soap is of natural origin hence easily degraded by bacteria).
2267. Black, W. G. *Folk Medicine.* 1883, London, p. 67 (Frazer. *Golden Bough*, i:83; ii:168-; 1890).
2268. Value due to methylene-dioxy-phenyl group in sesamin (Budowski, *J. Amer. Oil Chem. Soc.* 41:280, 104; 1964 - Haller, H. L. *et al.* 1942; Gertler, S. I.).
2269. *Scolopendra subspinipes* (centipede): size 16 cm long fully developed; *Bol. Inst. Vital Brasil.* v. 4:#27, p. 5 (ca. 1944). Dioscorides Pedacius. *De Materia Medica* (A.D. first century).
2270. Moldenke, H. N. *Phytologia.* 60:483-96; 1986.
2271. Standley, Paul: *Contr. US Nat. Herbarium.* 23:895+; 1924.
2272. Balzac, Fausta. *Annali della R. Accademia di Agricoltura di Torino.* 58:19 Dec. 1916 ("Le Artemisie de vermouths e de Genepis" [incl. analytical key to several spp.]).
2273. Peanut milk has been in use since ca. 1943 (India, Af); value lies in use to replace cow's milk for infants with allergy to same. P. butter has been used since ca. 1890 (St. Louis, Mo - prescribed by physicians for invalids) (use of peanut in foods 33%, non foods 67%). Research ca. 1970 indicated that peanut oil in monkeys produced more arterial hardening than butter fat; some have considered this oil therefore inferior to other unsatd. fats in diet.
2274. *Pharmazie.* 40:97-104; 1985 (192 refs.) (mystic plant of the Druids of ancient Britain).
2275. Chaney, W. R. *Econ. Bot.* 32:118-25; 1978. Cedar of Lebanon is a symbol of modern Lebanon; used on stamps, flag, &c.
2276. Fong, Harry & al. *J. Pharm. Sci.* 61:1818; 1972.
2277. Karg. *Riechst. Arom. Kosmet.* 28:45-150; 1978 (CocaCola & related beverages) (per Roempp).
2278. Boswell, Jas. 18 Apr. 1778 - Using the word "chocolade"
2279. Coconut milk is u. to preserve sperm for up to 14 days at room temp. (u. in artificial insemination in Af, SA, Ind) (-1973-).
2280. Dalla Torre, L. & al. *Minerva Medica.* 43: 853-61; 1952 (action of natural & decaffeinized coffee on cardiovascular apparatus).
2281. Capper & Moser. *Amer. J. Med. Sci.* 239: 1055, 261-71 1960 (careless admin. of corticosteroids may —> adrenal insufficiency, manifested under stress or intercurrent disease; months or even yrs. after treatment) - Hazards in use of c. more susceptibility to infection, diabetes, osteoporosis; healing of wounds slowed; suppression of adrenal glands. Of most value in collagen diseases (arthritis, &c.) - Use in therapy of mental diseases (with vit. C) (D'Angelo, C. *Lavoro neuropsichiatrico.* 10:88-97; 1952).
2282. Senn, H. A. *Rhodora.* 41:317-67; 1939 (NA spp. of *Crotalaria*).
2283. Outstanding alcoholics include Jack London, Dylan Thomas, Sinclair Lewis, O'Henry, John Barrymore, Thomas Wolfe, and many others.
2284. The name of Al-Kohl (the kohl) was u. in medieval Latin & refers to any powder obt. by sublimation (or distn.): meaning quintessence, such as spirits of wine (= alcohol).

2285. Plant fiber detd. by non-hydrolysis by human alimentary enzymes; dietary f. incl. pectins (decrease as plant ages) —> celluloses, hemicelluloses, (incr. with age of plant) —> lignins (accumulate); also incl. some storage polysaccharides, ex. galactomannans: not incl. in crude fiber - It is claimed that fibers (such as brans or whole grains) reduce colon & rectal cancer, appendicitis, gallstones, varicose veins, phlebitis, diabetes(?), coronary heart dis.(?), hernias (because of less strain), &c.
2286. "Nemo me impune lacessit" (no one assaults me with impunity). Expression applied to Scotch thistle (barefoot Danish soldier carrying sword).
2287. *Bot. Marina.* 6:143-6; 1964.
2288. *Smithsonian.* 14(6):66-75; 1983.
2289. The word "assassin" is believed to have been derived from "hashish" becasue of the nefarious use of this drug by one Hassan. In his castle at Alamut (in present nIran), he conducted a kind of professional "Murder for Hire," in which the inmates participated in fanatical religious killings, presumably under the influence of Cannabis. This activity was carried out over the period approx. 1090-1250 (some now think Cannabis may not have been involved) (Iyer. *Smithsonian.* 17(7):145-62; 1986).
2290. Radin, J. B. *Med. Times.* July 1961 - Brooks, L. *Dis. Colon & Rectum.* 1(5): Sept. 1953 - Sometimes combined with Na carboxymethyl cellulose as in Bacid™.
2291. *Riedels Ber.* 1912:49 (Lobelia, ash 5.10.3%; AIA - 1.6%) - Steinegger *et al. Pharm Acta Helv.* 27:113-27; 1952.
2292. Giant bamboo has extraordinary growth rate (-0.5 in./hr); after 42 days stops growing & never re-starts. Plant blossoms & dies at age of 120 yrs (*Phyllostachys* spp.).
2293. Legare, Cecil. *E. Mercks Ber.* 1901 (use of *Ephedra nevadensis).*
2294. Moldenke, H. N. *Phytologia.* 61:82-115; 1986. Burkill (1966) says: "An untidy seashore shrub, with somewhat soft greenish-brown or black fruit (from which it gets the first part of its names terong jambul and limau lelang), found from Bombay to the Pacific; in the (Malay) Peninsula it is round the coasts. The Malays apparently make no use of it, but to the eastward it seems to be held medicinal in Celebes, in the Philippine Islands, and in the Pacific, in Guam."
2295. *Bot. Museum Leafl.* 30:274-5+; 1986.
2296. Codfish is the emblem of Massachusetts, with its effigy hung in many places. About 3 centuries ago, in the early days of colonial Mass, codfish from the sea saved many from starvation; later, preserved in salt, it was the first export & chief revenue for some time.
2297. *Pharmacien de France.* 1986:601.
2298. Ayensu, E. *Medicinal Plants of West Africa.* 1978.
2299. *Bot. Museum Leaflets.* 30:255-85; 1986.
2300. *Ibid.* 247-54; 1986.
2301. *Ibid.* 14:117; 1950.
2302. *Ibid.* 30:239-46; 1986.
2303. Hopkins, Jas. F. *A History of the Hemp Industry in Kentucky.* Univ. Ky Press; 1951 - First crop recorded for Ky in 1775 but grown elsewhere in America much earlier (Jamestown, Va, 1611; recommended by Spanish as early as 1545) - *Bot. Mus. Leaflets.* 13:293; 1949.
2304. Shibata & al. *Z. Krebsforsch.* 71:107; 1968.
2305. In the annual rattlesnake hunts (Tex, Ala+), when enormous numbers of snakes are killed, the venom, meat, rattles, &c., are utilized (AP Big Spring, Tex; 24 Mar. 1964; tells of hunt at which 427 lbs. rattlers were caught).
2306. Stanwell-Fletcher, T. C. *Driftwood Valley* (BC). 1947. p. 170.
2307. Carr, D. J. (ed.). *Sydney Parkinson.* p. 77; 1983.
2308. Tauber, E. *Minerva Pediatrica.* 4:232; 1952 (tincture used).
2309. Let *Helenium* stand in homemade whisky for 2 wks., strained, & u. as liniment (seUS folk remedy).
2310. Types of honey: comb; extracted; strained.
2311. (*Hydrastis canadensis*). Mohr found in Ala (Cullman & Chilton Co.). Also R. M. Harper found in DeKalb Co. (1956); said found in Miss & La; scarce in sUS areas.
2312. Talaat. *Brit. Med. J.* 1960, II:944.
2313. Hurston, Zora Neal. *Mules & Men.* 1935 - Sweet gum bark & mullen cooked with lard —> ointment; appl. in fistula.
2314. At a dose of one 50 mg. tab. 3 times a day.
2315. *Nic. bigelovii* var. *quadrivalvis* coll. in sOre by David Douglas, 1828. Said extinct in BC, with only 2 herbarium specimens left (Vancouver Museum; 1986). Cult. by Haida & Tlingit Indians (nwAm) (this native tobacco orig. chewed with burnt clam shells).
2316. *Mfg. Chem.* June, 1943:167-70 (compilation of iodine compds., uses).
2317. Jalapa, Mexico, was a receiving depot for used autos from US in early 1900s, Hence term applied as "Jalapa" corrupted to "jalopy" to old cars, later to dilapidated vehicles.
2318. Apple aroma & flavor due to mixt. of free acids (acetic, HCOOH, propionic, &c.), acetate esters (incl. isobutanol), amines, acetaldehydes, HCHO, acetone, &c., having odor of fresh ripe apples (Roempp) - Apple pie is said the most popular in US dessert.
2319. Mango juice u. to prevent stings any time during season (wIndia, Rajputana) (per Sood).
2320. Rhinitis sicca, also call ozena and atrophic rhinitis, has been treated with instillation of 5 drops olive oil contg. menthol (0.33%) plus

vitamin A (30,000 I.U.) 3 times/day; treatment for 2-3 wks.

2321. The coral snake injects venom by chewing on the body area; the snake is red, black, yellow-banded. The antibody is prepd. fr. blood of healthy horse immunized against the venom.

2322. Kissam, Samuel. "Anthelmintic quality of the *Phaseolus zuratensis, siliqua hirsuta,* or cow-itch," diss. 1771 - Calvino, E. M. *Rend. Atti Accad. Lincei.* 31:166-72; 1922 - deCalvino. *Rev. Cubana Medica.* 33:265-78; 1922.

2323. (Cali nuts) *Pharm. Ztg. Russland.* 1887:122+ - *Jahrb. f....Pharmakogn.* NF 22:34; 1887.

2324. Soejarto, D. D. & Farnsworth, N. R. "Correct name for Siberian ginseng," *Bot. Mus. Leafl.* 26:339-343; 1978.

2325. Klapholz, R. & Zellner, J. *Monatsh.* 47:179-83; 1926 (chem. of *Oenothera biennis*: phytosterols [lvs. & fls], ceryl alc., phlobaphene, tannins, invertase).

2326. Opium said more analgetic than equiv. amt. morphine (Macris & al.) (1958).

2327. Huxtable, R. & Barbeau, Andre (ed.). *Taurine & Neurological Disorders* (1978) (468 pp.) (high concn. in heart tissues, being chief amino acid of heart; action multiple [inotropic, antiarrhythmic], ubiquitous in amimal tissues, almost absent fr. plants; important to CNS, retina, animal behavior, useful in "partial" epilepsy, against gallstones, mold, &c.).

2328. (*Sapindus sap*). Hodge, W. H. & Taylor, D. Ethnobot. of the island Caribs of Dominica - *Webbia.* 12:513-644; 1957-). Said to be 500,000 trees in Fla- valuable yield *(Am. For.).*

2329. Per Mrs. Mattie Lou Bridges née Willis (Ga/Fla).

2330. Taylor, R. J. & Boss, T. R. "Biosystematics of *Quercus garryana* in Relation to its Distribution in the State of Washington," *N. W. Sci.* 49: 49-56; 1975.

2331. Feyertag, E. & Zellner, I. *Monatsh.* 47:601-9; 1927 (chem. of *Rh. hirsutum*).

2332. Linoleic acid —> arachidonic acid —> prostaglandins acts to prevent gastic ulcers). Unsatd. vegetable oils in small amts. recommended for this effect.

2333. Power, F. B. & Salway. *J A C S.* 30:251; 1908 (chem. examn.).

2334. Tai Ginseng, "the Asiatic wonder drug" said to keep one young & to give fresh power (Dr. Poehlmann & Co. Wuerzburg, BRD, advt. *DAZ* 1955).

2335. *D. Apt. Ztg.* 1951:91-3 (Ebring & Corcilius); *Parfuemerie & Kosmetik.* 1956:578 (Huet).

2336. Boulos, L. *Med. Plants of N. Africa.* 1983.

2337. Hoch, J. *Amer. J. Pharm.* 1934:166-70 (Legend & history of *Passiflora*) - Crucifixion suggested by cross-arrangements; lobed lvs. around fl. are hands of Christ's persecutors; stalked pistil with 3-knobbed style represents nails of crucifixion; sepals & petals may be the 10 apostles (12 minus Peter who denounced Christ & Judas who betrayed him).

2338. Oliver, Daniel. *Hook. Icon. Pl.* 24:Pl. 2331; 1894 (bot. descript. & plate of *Pilocarpus microphyllus*).

2339. Kupchan & Lacadie. *J. Org. Chem.* 40:654-6; 1975.

2340. Fine specimen seen in "Italian House" (greenhouse) at Duke Gardens, NJ (1985).

2341. Wiley, L. *Wild Harvest* (1966). pp. 55-62.

2342. Rivera, Irene. *An. Inst. Biol. Mex.* 14:37-67; 1943.

2343. (*Convolvulus scammonia*). Said u. by Cleopatra as contraceptive.

2344. Besides its use as aftershave lotion and hair tonic, bay rum was popular among the depressed city dwellers of our big cities during the days of prohibition as an intoxicant (close to Bay Spts. Comp.) - *West Ind. Bull.* 12; 513-20; 1912 (Fishlock, W. C. (Virgin Is) (bay rum industries of St. Thomas & St. John).

2345. Members of Pinaceae often associated with death; *cf.* yews at cemetery. Evergreens are symbolic of eternity (since evergreen).

"Come away, come away, death,
And in sad cypress let me be laid;
Fly away, fly away, death;
I am slain by a fair cruel maid,
My shroud of white, stuck all with yew."
(Shakespeare)

2346. Pine monoculture in contrast with the natural forest is not fitted to support the complex life of the latter. Wild life conservationists are opposed to clear cutting.

2347. Cole, Wm. D. *True.* 40 (Dec.):54-7, 119-22; 1959 ("The vegetable that drives men mad") (Tells of Gordon Wasson's experience in sMex [Oaxaca]; clairvoyance claimed developed) (Discovery by Conquistadores).

2348. *J. Gerontology.* 15, #1; 1960 (Ranke) (vit. B_6) - Suggested that pyridoxine deficiency may result in sub-par intelligence in children.

2349. Beng, J. T. *Ueber Indonesische Volksheilunde an Hand d. Pharm. Indica* (1684).

2350. Dried yeast cont. few or no living y. cells (BPC requirement) & has no fermenting power. Living y. cells detract fr. the vitamin value since they consume the vitamins.

2351. Singer, R. *Agaricales—Modern Taxonomy.* ed. 2; 1962.

2352. Fishman, M. L. "Chemistry & Functions of Pectins," *ACS Symposium Ser.* 310; 1986 (Gelatinization of pectins increased synergetically by addn. of alginates).

2353. Quebracho wood: Durland, W. D. *J. For.* 21: 600-3; 1923 - *Geograph. Rev.* 14:227-41; 1924 - di Lullo, O. *Tesis Sobre el Paaj, un Nueva Dermatitis Venenata* (1930) - Hilbert, F. L. *Hide & Leather.* 93:Nos. 4, 7, 11; 1937.

2354. *Arnoldia.* 41:188; 1981; photo & desc. of *Securinega suffruticosa.*
2355. Staba, E. J. *J. Nat. Prod.* 48:203; 1985.
2356. Minton, S. A. (ed.). *Snake Venoms and Envenomation*, ix + 188 pp. 1970 (Dekker).
2357. Swine (general term): hog (boar [or sow]) <— pig (young) <— shoat (shote) (very young).
2358. Female holds one clove in mouth to prevent conception(!) (ca. A.D. 1000 - Arabs of Af).
2359. Tétényi & al. *Herba Hungarica.* 25:27-42; 1986.
2360. Anon. *The Winged Bean, a High-Protein Crop for the Tropics,* ed. 2. Natl. Acad. Press (Nat. Acad. Sci.) ix + 47 pp., many figs.; 1982.
2361. Bourne, L. B. *Brit. J. Industr. Med.* 13:55-8; 1956 (dermatitis from mansonia wood).
2362. (Nematodes). Wilmore, R. P. *Sat. Eve. Post* 228: Oct. 15, p. 36; 1955 (some nematodes in Af up to 4 ft. long; several nematode toxicants have been used: MeBr, EDB, DD Mixt., etc.).
2363. (Nucleases). *Les Nucléases: l' Application à l'Étude des Acides Nucléiques.* 281 pp.; 1964 (much also on nucleic acids).
2364. Power, F. B. & Salway, A. H. *Trans. Chem. Soc.* (Lond). 105:767-78; 1914.
2365. (Urea). US Dept. Agr. Circ. 679; 1943 (prodn. & fertilizer uses).
2366. (Urine). A former prime minister of India (M. Desai) told of drinking his own urine as a health measure (TV program, ca. 1965).
2367. Sandermann, W. *Naturharze, Terpentinoel, Talloel—Chemie U. Technologie.* Springer-Verlag, 1960, 491 pp.; detailed data on several resins, volatile oils, & FO of natural origin; with comprehensive bibliography.
2368. (Tigloidine). Trautner, E. M. & Noack, C. H. *Med. J. Australia.* 1:751-4; 1951.
2369. Thayer, Maj. Ralph E. (Burlington, Vt) (said excellent woods for wood carving are basswood [best]; also good: mahogany, red cedar, clear pine, some hardwoods) (1965).
2370. (Tisanes). *Le Moniteur.* 29 Aug. 1959 (consumption in Fr; 650 tons per yr. = 260 m. cups = 6 cups/person/yr.) (5 hbs. most used: tilia, mentha, camomile, verbena, orange flowers).
2371. *Ancistrocladus extensus* Wallich (seAs, wMal): rt u. in dys.; young lvs. in flavoring (Thail).
2372. *CEN.* Oct. 27 1975, p. 3 (letter from G. A. Halaby, Ph.D., Stokeley-Van Camp, Indiana) (medium chain triglycerides [MCT] claimed to lower tot. cholesterol in blood by 50%—more than long chain fatty acids of conventional fats) (MCT=C^6- C^{10} fatty acid glycerides) (20 yrs. research by Kounitz, Greenberg, Shah, et al).
2373. (*Viturnum prunifolium+*). Schamelhout, A. *Ann. Pharm.* (Louvain). 3:113; 1897 - Munch, James C. "Pharmacognosy, chem., & pharmacol. of *Viburnums,*" *Pharm. Arch.* Parts 4-8; ca. 1940 - Evans, W. E., Krantz, J. C. & al. *J. Phcol.* 25:147, 174-7; 1942 (active pterine princ. in *V. p.*) (Evans. *Fed. Proc.* 1942, I: 151) - Grote, I. W. & Woods. *J Am Pharm Assn.* 36:191; 1947 - Sloane, A. B. & Latven, A. R., & Munch, J. C. *J. Am. Pharm. Assn.* 37:132; 1948 - 38:452-9; 1949 - Youngken, H. W. *Ibid.* 17:330-5; 1928 - Collected mostly in Ky, Tenn, & NCar (US) (2 grades).
2374. de Groot, A. P. *Nature.* 183:1191; 1959 (dietary CLO effect on serum cholesterol level of rats).
2375. (Vit. E). Vitamin Foundation, AH/API, 1955 (Antagonism by CLO, TCP (tricresyl phosphate), CCl_4; treated by giving larger doses of E).
2376. Hodges, W. H. "Goethe's Palm," *Principes* 26:196; 1982 - (In the Botanical Garden of Padua (Italy) there is present "an obviously frail chastetree (*Vitex agnus-castus*), planted in 1550."
2377. (*Vitis* sp.). *NY St. Agr. Exp. Sta. Tech. Bull.* 96. 121, 146, 181, 1921 (chemical studies of grape pigments & of g. juice).
2378. Editor (Thos. Meehan). *Gardeners Mo.* 24: 345; 1882 (*Yucca* lvs., uses by early settlers).
2379. Sayman's Vegetable Wonder Soap™ (Sayman Products Co., St. Louis) contained Na salt of veg. FO with water & perfume; claimed in advertising to contain saponins of yucca (*FDA* 24513; 1926).
2380. Humm, H. J. *Fla. State Univ. Stud.* (contrib. to Sci.) 1-6; ca. 1950.
2381. (*Jacaranda caroba*) de Barros, F. Paes. *Memorias d. Instituto Butantan.* 16:357-9; 1942.
2382. (Oak Moss). Glenk, Robt. *J A Ph A.* 13:1028; 1925-31; 1924. *Evernia prunastri* - abundant on live & water oaks of sUS.
2383. Tomatoes—altho technically the part eaten is a fruit, it is treated in commerce as a vegetable.
2384. Eating of (Argo) starch noted among cotton pickers near Dothan, Ala; in one instance, woman purchased in case lots (informant: Larry Dyess, Hartford, Ala, ca. 1965).
2385. Morton, Julia F. *Econ. Bot.* 37:164-73; 1983.
2386. Falsone, G. & Crea, A. E. G. *Liebigs Ann. Chem.* 1979:1116-21 (3 new 4-desoxyphorbol esters fr. *Euphorbia biglandulosa* Desf.).
2387. Karlsen, J. & al. *Planta Med.* 17:281-93; 1969 (fennel cult. extensively in se&neBulg - *Foeniculum Dulce* (sweet var.) mostly as cremocarp, larger & heavier frt.; ann. - *F. vulgare* (bitter) mostly as mericarps; longer & broader; annual.
2388. (*Fraxinus nigra*). Winnebago Indians near Wisc. Dells u. black ash (or swamp a.) sts. cut into 6-8 ft. lengths & pounded until strips of bk. come off, this cut into proper width & made into baskets (letter fr. Chief).
2389. Ash tree is said to possess magic powers of healing the sick; babies were said to be cured

if they were passed through a cleft made in the trunk. Man & animals were cured (*sic*) from sickness by touching the ash twigs gathered from the tree where a shrew mouse had been buried (hence "shrew ash") (*cf.* J. G. Frazer. *Golden Bough*. 1890) (other trees also so used, as oak).

2390. Aliev, R. K. & Damirov, I. A. *Pharmazie*. 21: 457-9; 1966 (drugs u. in Azerbaijan).

2391. Planas & Kúc. *Sci.* 162:1007; 1968 (exptl. studies on rats support contracepn. activity).

2392. Tar was u. to impregnate cloth to make it waterproof; called "tarpaulin." This material was u. to make the hats of British sailors, hence "tars."

2393. FO rich in oleic acid are said to be protective against heart attack.

2394. (Vit. B_{12}). *Neurol.* 2:131; 1952 (Felder, Hoff).

2395. Lentz, D. L. "Jicaque of Honduras," *Econ. Bot.* 40:710-9; 1986 - Rts. u. with *Euphorbia hirta* as vag. astr. & to relieve anal itching (lvs. mashed & inserted).

2396. Galanti (1954) re *Geranium*.

2397. Powdered mummy said sold by a NYC pharmacy for $40./oz. (-1972-).

2398. Iceberg ice is u. in making vodka, cosmetics, mixed drinks, &c. In Alaska, costs $7.00 per bag (1989).

2399. Hirono, I. & al. *J. Nat. Cancer Inst.* 61:865-70; 1978 (lvs. & rts. of *Symphytum officinale* fed to rats produced hepatocellular ademomas, occasionally hemangioendothelial sarcoma of liver).

2400. Graham, S. & al. *J. Nat. Cancer Inst.* 61: 709-14; 1978 - Amt. of beef or other meats eaten not related to cancer incidence. However, frequent ingestion of vegetables, esp. cabbage, Brussels sprouts, & broccoli, reduced risk of colonic or rectal cancer, esp. colonic. (sauerkraut, coleslaw) (Mormons have low colon cancer rate but are big beef eaters).

2401. Kritchevsky, D. (Wistar Inst. Anat. & Biol.) found peanut oil prod. atherosclerosis more than corn oil - *Atherosclerosis.* 14:53; 1971.

2402. Lechner-Knecht, Sigrid. *Pharm. Ztg.* 128: 1874-6; 1983 ("Beitrag zur ayurvedischen Arzneikunde"; description of the oldest medical system in the world).

2403. Nosek, J. J. *Ces. Lek. Ces.* 90:679-681; 1951 (treatment of arthritis with vit. C & corticotropic hormones).

2404. Tragacanth allergy. *J. Allergy.* 14:203 (1943) (Gelfand); 18:214; (1947) - (Brown, E. B.); 26:311 (1949) (Gelfand).

2405. Quercitrin has been found in *Thuja occidentalis* lvs., *Solidago canadensis*, *Litsea glauca* (spring lvs.), *Leucaena flava* lvs., tea lvs., *Polygonum sachilenense* lvs., *Euphorbia longana* lvs., *Pieris japonica* lvs., *Prunus tomentosa* fresh lvs., *Houttuynia* sp. lvs., &c.

2406. *Chemurgic Digest.* 15/IV/42 pp 52-3 (Acorns as possible raw material for FO [*Quecus alba* & red oaks]).

2407. McKamey, L. *Secret of the Orient: Dwarf* Rhapis excelsa. Rhapis Gardens, Gregory, Tex, 54 pp.; 1983 (discusses *R. excelsa* Henry, a dwarf form developed by Jap; also other taxa incl. cultivars.).

2408. *Rheum* - Lf. laminas dangerous; may be lethal (oxalic acid); however u. as antiscorbutic & petioles widely u. in preserves & pies.

2409. Hoch, J. H. *A Survey of Cardiac Glycosides and Genins.* Univ. So. Carolina Press, Columbia, 99 pp.; 1961 ($3.50) (compilation of 3 tables + extensive bibliography; I: sources; II: hydrolytic products of glycosides; III: animal assay doses of natural & semi-synth. compds.).

2410. Williams, M. Y. & Pillsbury, E. W. *Canad. Geog. J.* 56:184-201; 1958 (general article re Gulf Islands of BC [Straits of Georgia between Vancouver & Vancouver Island]).

2411. Denison, E. "*Poncirus trifoliata*—A 'divided' Tree," *Amer. Hort. Mag.* 42:116-20; 1963.

2412. Sea salt beneficial to health; c. 70% NaC1, 30% of some 37 or more other elements (*C & EN.* 30 Apr. 1956, p. 2107).

2413. Joseph, M. *Amer. J. Surg.* 87:905; 1954 - *Q. S. Digest. J.* 54:38-9 (*Capsella bursa-pastoris*; Koagamin [Chatham]).

2414. Theobromine as hypnotic (*Merck's Rep.* Oct. 1904, from *Pharm. J.* No. 3434:260) (insomnia in cardiac dis.).

2415. *Triticum aestivum* furnishes white flour & bread of spongy kind, hence is of superior quality, although other wheats are just as or more nutritious.

2416. Wheat germ rich in vit. B_1, B_6; u. in cereals, breads, salads; FO rich in vit. E.

2417. Farwell, A. O. (*Merck's Rep.* 17:34; 1908) (adulteration & substitn. of crude drugs).

2418. Strandbyard, Ebba (Denmark). *Arch. Otolaryngology.* 59:485; 1954) (vit. A in ozena).

2419. Vit. A deficiency very common in third world countries, resulting commonly in xerophthalmia & blindness and in death (lowered immunity).

2420. *Pharmacien de France.* 1988 (4):191.

2421. The term "Rosinen" for raisins resulted in a curious error in Tschirch's *Handbuch d. Pharmakognosie* (ed. 1, vol. 2(1):39; 1912) in which a figure (photo) shows enormous numbers of casks of pine rosin (not raisins !) stored at the Port of Savannah, Georgia.

2422. Kaufman, Z. *Ges. Med.* 55:1; 1927 (aq. & alc. exts. of bean husks lowered blood sugar of fasting rabbits & diminished hyperglucemia following glucose injections).

2423. "White mule" (moonshine) made in Fla from maize chicken feed + sugar + water; let stand in barrels only 3 days, then distd. giving 120 proof spirits.

2424. *J. Ethnopharmacol.* 2:389-393; 1980.

2425. Hartwell, J. L. *Plants Used Against Cancer.* Quaterman Publns.; 1982.

2426. Hoffer. *J. Clin. Exper. Psychopath.* 18:131; 1957 (vit. C).

2427. Oguntimein, B. O. & Erhun, W. O. *Nigerian J. Pharm. Sci.* 1:25-30; 1985.

2428. Kurbanov, M. M. & al. *Rastit. Resur.* 20:125; 1984.

2429. Rask, M. R. *J. Neurol.* 1(5):1-19; 1980 - 9:12-18; 1988 - *Orthoped. Med. Surg.* 6:295-302, 211-8; 1985.

2430. Clark, L. *The Rivers Ran East* (1953), p. 286: *Cecropia peltata* (cetico) - Latex relieved pain & stopped infection, per author's experience (Peruvian doctors had given up).

2431. (*Ilex glabra*). Morton, Julia F. *Folk Remedies of the Low Country.* 1974 (SC). In Miss, Negroes prepared broth fr. rts. u. in mal. & rheum: lvs. u. to "loosen phlegm" in throat: lvs. u. as lax. (ch. 12) (Billy N. McAdoo, Univ. M. [1950]).

2432. The plant of the title of Zane Grey's *Riders of the Purple Sage* is definitely sagebrush (*Artemisia* sp.) and not a *Salvia* sp. The name may have existed only in the imagination of the novelist!

2433. Burt, Joseph B. "Phytochem. study of *Monarda pectinata,*" *J A Ph A.* 25:602-8; 1936 (herb); 682-7; 1936 (VO) (detns. of extractive, &c., of fl. lf. st., rt.; VO of pl. grown in Neb, with 0.49% yield of VO; 78% made up of phenols incl. thymol & carvacrol.

2434. *Brugmannsia (Datura) candida* juice said to make person into a zombie (unfaithful husbands et al.) (2430).

2435. Ruiz-Teran, L. & Lopez-Palacios, S. *Rev. Fac. Farm.* No. 25:1-95; 1985 (1987).

2436. Actually one can write very well by simply dipping an old (rusty) steel pen into the fresh juice gall (Chesnut, V. K. *Contr. US Nat. Herb.* 7:1902).

2437. Many publications on *Liatris* by W. Herz and R. P. Shamma *et al.* (FSU) *J. Org. Chem.* 41: 1248; 1976 & others.

2438. Adesina, S. K. & al. - *Pharmazie.* 41: 747; 1988.

2439. Jardine. *Amer. J. Pharm.* 55:576; 1883.

2440. *Prunus leiocarpa* - Collected near Kerman, Sheraz (Iran) & sold in bazaars (1906); Persian (or cherry) gum; u. in foods (djedki-i-ardjin); ketirah (gum)-i-ardjen. Exported. Formerly much on market fr. Persia.

2441. Chestnut, V. K. "Plants used by the Indians of Mendocino Co., Calif," *Contr. US Nat. Herb.* 7(#3); 1902.

2442. FAO. *Forestry.* 14:326 pp.; 1960 (Rome) (*Pinus radiata* D. Don).

2443. (Spices). Important in commerce of antiquity; camel caravans crossed Asia; Phoenicia; Italy; Gt. Brit., Portugal, Netherlands, Far East.

2444. Hazlett, D. L. *Econ. Bot.* 40:337-52; 1980.

2445. Jansen, P. C. M. & Mendes, O. *Plantas Medicinais Seu Uso Tradicional en Moçambique*; 1983; rts. u. as cataplasm appl. exter. to sprains (Camb); antipyr. (Laos); rt. inf. u. in kidney dis. (Vietnam); lvs. u. as tea for stomachache.

2446. According to Boswell, Dr. Samuel Johnson always carried in his pocket orange rind, which he nibbled occasionally for its medicnl. value.

2447. Coral was one of the most used drugs in antiquity - Int. as diur., urin. antisep. ton., in impotence, in Unani medicine; better known are its uses in jewelry & ornaments.

2448. Anderson, E. F. "Ethnobotany .... N. Thailand," *Econ. Bot.* 4:38-53; 1986.

2449. Joyal, E. "Ethnobot. notes for Ecuador," *Econ. Bot.* 41:163-87; 1987.

2450. Bureau of Chemistry, U.S. Dept. Agr. (1925).

2451. Morton, Julia F. *Folk Remedies of the Low Country.* 1974.

2452. Essig, F. B. *Econ. Bot.* 41:411-7; 1987.

2453. Morton, Julia F. *Econ. Bot.* 39:101-12; 1985.

2454. Raie, M. Y., Ahmed, D. "The fatty acid composition of *Leucaena,*" *Pak. J. For.* 32:76-80; 1982.

2455. Low selenium intake believed associated with higher rates of cancer (esp. GI & prostate); reduction in incidence greater when used along with vits. A & E.

2456. (Heroin). Many bizarre uses are reported in the older literature, for ex. to cure morphine addiction; to treat spermatorrhea (nocturnal pollution), &c.

2457. Fournet, A. *Plantes Méd. Congolaises.* CRSTOM. 1979,

2458. Formerly piturine was sometimes regarded as a mixture of hyoscyamine, scopolamine, atropine, &c.

2459. (Asafetida). Martinez, D. *Pharmazie.* 43:720-722; 1988.

2460. In the Mediterranean area, heart attack rates are significantly lower, due it is said to the Mediterranean diet, which consists chiefly of whole white flour (pasta), olive oil, red wine, and fish foods.

2461. Kapadia, G. J., Subba Rao, G., & Morton, J. F. *Carcinogens & Mutagens in the Environment,* vol. 3: 3-12; 1983.

2462. Cottonseed oil said to occasion low fertility rate (China) because of gossypol.

2463. Morton, Julia F. *Atlas of Medicinal Plants of Middle America.* 1981.

2464. *Rhus emodi* - origin = Emodus (Emodon), name given to Himalayas by Ptolemy.

2465. Poppea, wife of Nero, perfumed her breath by chewing small balls of myrrh (*sic*).

2466. Estimate that 500 million tons/yr. of methane are formed on our planet.
2467. Regular consumption of Cassava associated with little or no diabetes, suggested cause = slow digestion & absorption (Teuscher. *Lancet.* Apr. 4, 1987, p. 765).
2468. Tomato juice, dangers: rich in K, Na. acetic acid, 5-OH-tryptophan: may be incompatible with MAO inhibitors for patients with cardiac failure, ulcers, dyspepsia, &c. (*New Eng. J. Med.* 284:1105; 1971).
2469. Baser, K. H. C. & *al.* "Herbal drugs in Turkey," *Studia Culturae Islamicae.* No. 27:1986.
2470. Kvist, L. P.+ "Lowland Eucador," *Opera Botanica.* 92:83-107; 1987.
2471. *Salvia columbariae* - Seeds said good nutritive: one tablespoonful per day said adequate emergency diet!!
2472. It has been stated that 100 mg. + calcium per day will give 22% lower risk of hypertension; while 300+ mg. magnesium per day results in 23% lower risk of hypertension.
2473. (*Liatris*) Herz, W. & Sharman, R. P. *J. Org. Chem.* 40:392; 1975; *Ibid.* 41:1248-53; 1976.
2474. Choline claimed to prevent fatty & cirrhotic liver, anemia, edema, &c.; lipotropic factor (prevents formation of FO hepatitis in liver) (Alexander, H. D., & Engel, R. W. [ca. 1952] - API).
2475. Kunkel, G. *Plants for Human Consumption* (1984) (Koeltz).
2476. Queen Elizabeth Park, Vancouver, BC, has giant gunnera plants with huge lvs. 10 ft. high & 6-7 ft. broad; rhubarb leaf shaped; this is largest species in genus.
2477. Named after Franklin S. Earle (1856-1929), botanist, prof. at Alabama Polytechnic Institute (Auburn, Ala) (1896-1901), who collected the plant s. of Auburn ca. 1896 (sandy plain).
2478. *Myrica inodora* Bartram - Collected on Upper Tensaw River near Hall's Creek, Ala. Bartram found the pl. lacking in fragrance & u. by inhabitants for "production of wax for candles, for which purpose it answers equally well with bees-wax or preferably as it is harder & more lasting in burning."
2479. *Malus sylvestris*: "An apple a day keeps the doctor away" (School of Salerno, It, ca. A.D. 900); value proven in clinical expts.
2480. See ref. (2475).
2481. Riddell, J. Leonard. *A Synopsis of the Flora of the Western States.* 1835.
2482. Galeano, G. G. & Skov, F. *Principes.* 33:108-112; 1989.
2483. Felger, R. S.+ *People of the Desert & Sea.* 1985.
2484. Lutomski, J. & al. (*Panax viet.*). *Herba Pol.* 34:151-; 1988.
2485. Marquin, G. & al. *Pharmazie.* 1985.
2486. Plowman, T. *Brittonia.* 34:442-57; 1982 (*Erythroxylum*).
2487. Davis, W. *Passage of Darkness.* UNC, Chapel Hill, NCar; 1988.
2488. Ivora, M. D. & al. *Pharmazie.* 1990 (re *Centaurea*).
2488A. Tallant, R. *Voodoo in New Orleans.* Pelican Publ., Gretna, La (1946; 1983) (book on voodoo).
2489. Villiers, J. E. *Flore de Gabon*, No 31 (Mus. Nation. d'Hist. Natur., Paris).
2490. Lopez-Palacios. *Rev. Fac. Farm. Univ. de los Andes* 1985 - Roys, R. L. *The Ethnobotany of the Maya.* Philadelphia, Penn; 1931.
2491. *Apoth. Ztg.* 1919:28 (*Helianthus*).
2492. Fats are now highly suspect of producing harmful effects on the circulation, &c. Minimum fats are found in such meats as chicken, veal, lamb, fish, lean beef, &c.
2493. Thus in Bolivia, the chief export is coca and cocaine paste (1989) exceeding all other exports combined.
2494. Davis, W. (see No. 2487) (re zombification).
2495. Colonies of bees are rented to facilitate pollination of blueberries, cranberries, and other crops.
2496. Succinum (amber) is known to have been formed in geological times from resins of various plant taxa, incl. *Liquidambar*, *Taxodium*, Cupressaceae, Leguminosae, &c. (215 taxa known up to 1989).
2497. Morton, Julia F. *& al. Econ. Bot.* 41:221-; 1987 (*Argania*).
2498. gamma-linoleic acid = cis 6,9,12-octadecatrienic acid; commonest source = *Oenothera* spp. (evening primrose).
2499. Parenteral administration of testosterone enanthate produces a reduction of sperm count, resulting in temporary sterility; proposed as a means of male contraception.
2500. *Acalypha* sp. pelito negro (Ve); ash u. as lye; & in prepn. of chimo (chewing tobacco prepn.)
2501. Cats (and other pets) have been u. considerably in hospitals, nursing homes, etc. as mood-changing agts. for senile persons, autistic children, etc.
2501A. Saliva, coagulative action in tonsillectomy - Giannoni, E. & Bernicchi, L. *Atti Acad. Med. Chir. di Perugia* N.S. v.1:49-51; 1949/50.
2502. "Flore du Congo Belge et du Ruanda-Urundi," *Spermatophyta.* v. 6:1954.
2503. Tupelo honey prod. in Fla (Wewahitchka Lakes, Calhoun Co.). Barges with hives moved along the channel (trees at edge of or in water). This honey does not solidify. It is popular with Appalachian coal miners, who take it to work & consume it for quick energy (Tenn+).
2504. Dominguez (1877).

2505. *Lancet.* 1967, I. 458 (Epilators & depilators).
2506. Polykrates, G. "Wawanauterei u. Pokimapueteri," Nat. Mus., Denmark, *Ethnographic Ser.* 13 (1969).
2507. Hansen's disease (leprosy) - In 1963, 207 patients left in H. D. settlement at Leprosarium - Kalaupapa, Molokai, Hawaii. There is also H. D. Sanatorium in Louisiana - Wilbert, J. *Bot. Museum Leafl.* 24:275-335; 1976.
2508. Isaiah 38:21 - Bouquet, J. "Mat med. indigène de l'Afrique du Nord," *Bull. Soc. Pharmacol.* 28:22-36; 73-84; 1921.
2509. Davis, Wade. *Passage of Darkness* (UNC Press) (1988).
2510. Tannin 7% ext. u. for cold sores, fever blisters, &c.
2511. Quisenbery, T. M. & Gjerstad, G. *Quart. J. Crude Drug Res.* 7:957-64; 1967.
2512. Dorr, L. "Review of NA *Callirhoë,*" *Mem. NYBG.* 56 (1990).
2513. Chicken gizzard; considered a luxury; "craved" by character in *Laddie* (novel by G. Stratton Porter).
2514. Glycyrrhiza - Large amts. of rt. said found in Tutankhamen's tomb.
2515. Cattle raised in parts of India solely for sale of fecal masses (u. as fertilizer & as fuel) (said to be 79 drug products fr. cattle in USP XVII).
2516. Mastic is u. in m. (serum) test for syphilis, using cerebrospinal fluid (CSF) to produce ppt. in colloidal m. soln.
2517. Per Barkley.
2518. *Brassica napus* - Colza oil - 14% linolenic acid, 10% α-linoleic; proper amts. of n-6 fatty acid (5-8% tot. calories); when this proportion is raised to 15-20%, causes incr. in plaques, favoring thrombosis; consumption of n-3 FA in moderate amts. (0.5-1% of tot. calories) said to prevent thrombosis.
2519. Brassica - Broccoli - Good for soups, pasta, &c. (Pres. Geo. Bush [US] averse to this vegetable) - Cauliflower - Eaten raw & in soups.
2520. Champagne - Best not to store in refrigerator; best chilled with ice bath as needed (in container with coarse salt).
2521. Jade - u. in lusters, hairdressing, bangles; symbol of protection; occ. often in huge pieces (boulders weighing up to 36 tons).
2522. Chocolate-coated ants sold & eaten US (as fad) formerly.
2523. Angostura Aromatic Bitters™: popular bitter & flavor for foods, mixed drinks, &c.; bitter principle said to be gentian (not Angustura bark) with other herbs; orig. in Venezuala (Angustura; now Ciudad Bolivar) but now made in Trinidad; a secret formula but said to c. cinnamon, clove, mace, nutmeg, orange & lemon peels, prunes incl. crushed pits.
2524. (Blow fly maggots). Pavillard, Robin, & Wright, Eric. *Nature.* 1957 (abstr. in *Med. News* 16/XII/57, p. 3).
2525. Ca = good sources: turnip & mustard greens; kale; dandelion greens; cress - Inferior sources (since oxalic acid present); spinach; chard, beet greens.
2526. (Cerationia) Advantages: 1) non-allergenic 2) no xanthine derivs.
2527. Tar water "to cheer but not inebriate" (Bishop Berkeley, *Siris* [1744]) ("Further thoughts on tarwater" [1752]) - Later replaced in slogan by tea.
2528. Other names applied to Pharmacognosy (1st u. 1815) are: Materia medica (ca AD. 70) (later split into pharmacognosy & pharmacology); pharmacographia (1830), pharmacography (1900); pharmacomathy (1882); pharmacop(a)edia (1901); organic materia medica (1884). Subfields: pharmacoergasy (cultivation & collection); pharmacoemporia (comm. handling, export & import, treatment of drug by importers); pharmakodiakosmie (comm. sorts; packaging); pharmacobotany; pharmacozoology; pharmacochemistry (now phytochemistry); &c.
2529. Concha (L.): term also u. to mean mussel (or m. shell), shellfish producing purple dye (Murex) (*qv.*), pearl oyster, &c.
2530. (Coriander). Long history of use as spice (fresh lvs. &c.) since ca. 5000 B.C.
2531. Yams (*Dioscorea* spp. of seAs and other tropical areas) are not often seen in the US. They are more fully distinguished as Chinese yams or C. potatoes. What are sold as yams in the US are usually sweet potatoes (*Ipomoea* spp.)
2532. Granular yeast in an envelope or yeast compressed into cakes is generally seen.
2533. Probably from *Allium ursinum* L., leek, of nEu (rams or ramson[s]).
2534. Seeds notorious for producing intest. flatus (high carbohydrate content).
2535. Helianthus (Jerusalem artichoke) considered unfit for human food, better for swine, even after addn. of raisins, ginger, dates, &c.
2536. Pomo d'oro (It) (gold apple) SC early tomatoes were yellow in color.
2537. Word "rubber" first u. by J. Priestley.
2538. Matzoh prod. mostly by B. Manischewitz Co. (Jersey City, NJ).
2539. (Thyme). *A Dietary of Helth* (sic) by Andrew Boorde ca. 1530.
2540. (Colchicine) *Hosp. Pharmacy* 26:251; 1991.
2541. Hager. *Pharm. Praxis.* 3:866; 1972.
2542. (Glutathione). Major thiole (-SH) compd. of cell; isolated fr. yeast; reducing agt. u. in redox systems; deficiency produces cataracts (*Sci.* Aug. 1986, p. 553).
2543. (Glechoma hederacea). U. to replace regular tea (Jonas Hanway, London).

2544. Guano palm (palma de guano): *Coccothrinax argentea*; FO fr. sds. u. to revive the sense of smell.
2545. Cypress noted along Via Appia (Rome) (GMH); VO of shoots & lvs. u. as insect repellent, esp. for moths; wd. for building.
2546. *Bauhinia forticata* Link u. in proprietary "Bauintrato" (prod. of Lab. Clinico Silva Araujo [L.C.S.A.], Rio de Janeiro).
2547. Phytolacca - Mode of use in folk med. fill glass bottle nearly full with fully ripe berries; pour in QS. Brandy to fill vessel, cork. Shake occasionally over 2 wks. for extn. of juice. For chronic rheum., use 1-3 wine glasses once in 24 hrs. on empty stomach (if too strong, dilute with water) (Plough Boy, 1822).
2548. Erythronium - Source of starch (Amylum Erythronii), Jap Ph. I-IV. u. in med. (*cf.* Moeller, *Mikroskopie*. 138:1905).
2549. Kerharo, J. & Bouquet, A. *Bull. Soc. de Pathologie Exotique*. 43:56-65; 1950.
2550. Cannabis - 265 plants seized valued at $600,000. (news, Apr. 1990): with value of $2642 per plant. As with most hallucinogens & illegal addictive drugs, control has greatly inflated value.
2551. Alpha-linolenic acid has 3 double bonds, gamma-linolenic acid (GLA) with 3 double bonds shifted closer to COOH gp. GLA is richest in evening primrose oil, also FO of black currant & some fungi.
2552. *Liquidambar s.* (Mex): Genth: *Liebigs Ann.* 46:124; ca. 1910.
2553. Belladonna frt. are esp. toxic to children; deaths fr. ingestion frequent. However, frts. eaten by various mammals and birds seem innocuous (Tschirch. *Handbuch d. Pharm.* ed. 1—Bd III: 281; 1923).
2554. *The Merck Index* XI, 1989.
2555. *Flore de Madagascar et des Comores. Fam.* 168 (Klackenberg) (1990).
2556. *Rhynchosia pyramidalis* (pega palo) - Farnsworth, N. R. & al. *J. Pharm. Sci.* 56:967-70; 1967.
2557. In India, tulsi (*Ocimum sanctum*) grown in house "only to please the gods and his womenfolk" (Kipling, *From Sea to Sea*).
2558. Ants: red harvest or agricultural worker ants collected to supply "ant farms" (Utah & nearby states, US) (50,000/ann.) *Pogonomyrmex* spp. (Formicidae).
2559. Inks made with lvs. of *Pterolobium stellatum* remained on written documents under water for 24 hrs. without any change whatever (Schimper—label on herbarium specimen, 1854).
2560. Fructose can be managed in diabetic ketoacidosis, hence useful in early management of this complication. F. & NaCl inj. NF XI-XIII u. as fluid nutrient & electrolyte replenisher (F. metabolized without insulin).
2561. *Oryza*: economically used species are *O. barthii* A. Chev. *O. brevliguluta* A. Chev. & Roerich) (tropAf & Ind); *O. brachyantha* Chev. & Roehrich (tropAf & Ind); and *O. granulata* Nees (seAs). Annual rice prodn. for 1952: China: 51,825,000 tons; India: 40 million tons; US: 2,433,000 tons
2562. *Allium sat.*: Garlic Capital of the World: Gilroy, Calif, with its Garlic Festival in Aug. - Allium is u. for colds as ton. & in heavy metal poisoning.
2563. Unicorn (Einhorn, Ge; licorne, Fr) was u. in medieval medicine (Catelan. *Hist. de la Nature de la Licorne* [1624]).
2564. "uni" (Jap): eggs & roe of sea urchin u. in suchi, &c.
2565. Vitamin B_{15}: calcium pangamate; "Dedyl"; dimethyl glycine; DMG; occ. in vit. B complex fr. rice bran, blood, &c.; variable mixt. of Ca & Na gluconate; N,N-dimethyl-glycine; glycine; diisopropylamine dichloroacetate (DIPA), &c.; u. as hypotensive in cardiovasc. dis.; rheum. dis.; diabetes; emphysema; fatigue; &c. (claims unsupported); representa amino deriv. of glucuronic acid; not proven as vitamin.
2566. Other supposed vitamins: vit. A aldehyde (vit. A; retinal) (vit. A_2 = retinol); vit. B_4: adenine (in agranulocytosis); vit. C_2: esculoside ("antihemorrhagic"); vit. I: vit. B_7; mesoinositol (in steatorrhea); vit. J: choline; vit. L: folic acid (?); vit. N: thioctic acid (lipoic a.); growth factor; vit. O: carnitine (antidystrophic) - The discovery of vitamins is variously attributed to primitive peoples (ex. Chinese: vit. B_1; Am Indians: vit. C) and to such scientists as Manille Ide (Univ. Louvain) - The name "vitamin" originated with Casimir Funk (1912) (vita: life + amine, since some v. are amines) (such as Thiamine).
2567. Liatris may have come fr. liazo (Gr) = to come forth.
2568. Calcite is a mineral form of $CaCO_3$; it is a valuable optical mineral, partly imported, also found in mountains near Sante Fé, NM (US); it was used in place of glass in windows of houses of Am Indians in Ácoma, NM; still remains (GMH).
2569. (*Kochia scoparius*). Other reported uses incl. frts. & sds. u. as diur. ton. for impotence, GC, kidney dis,; shoots & lvs. u. in digestive disturbances.
2570. Calcitonin - Salmon injection u. in Paget's dis.; ca. 50 x more potent than human c.
2571. Antivenin: see *Venoms*, Publ. 44, AAAS; 1956, p. 381 (Buckley, E.).
2572. Products chiefly u. in US (Rx) (1992): digoxin; digitoxin; deslanatoside; ouabain.
2573. Cholesterol.
2574. High John the Conqueror is regarded by some African Americans as an African spir-

it who in the olden days gave hope to the slaves in the US. Eventually he is thought to have taken up residence in the roots of a plant wh. symbolically was dressed with perfume & kept on the person or in a secret place in the homes (Hurston, Zora. *Amer. Mercury.* 57:450-8; 1957).

2575. The first "synthetic" solid fat shortening for cooking was Crisco (Proctor & Gamble, Cincinnati, Ohio).

2576. Madstones were supposed to act by a process of absorption of the toxic liquid (*Tile & Till.* 25: 41-43; 1939).

2577. A recent monograph on *Mimosa* (Barneby. *Mem. NYBG.* 65; 1991) classifies *Schrankia nuttallii* as *Mimosa quadrivalvis* L. var. *nuttallii* (DC) Beard. Out of 15 other vars. of this sp., var. *angustata* (T. & G.) Barneby (S. a.var. bradycarpa Chapman) occurs in the seUS (Va-Fla.-Ala-Tex).

2578. Gentry, A. H. *Ann. Mo. Bot. Gard.* 79:53-61; 1992.

2579. Schultes, R. E. and Raffauf, R. F. *The Healing Forest.* 1990.

2580. *Sesbania* (Daubentonia): Nuessle, Noel N. "Coffee bean weed poisoning of cattle on Payne's Prairie, Fla" (U. Fla MS. Pharm. thesis, 1955).

2581. Isano (or tsano) oil. *Lipids.* 12:669-75; 1977 (w. African tree).

2582. *Amer. J. Chinese Med.* 16 (1988).

2583. (Musk deer): >54,000 m. d. were killed in 1967 for their musk. Hence considerable concern that the animal will be exterminated.

2584. In recent yrs. there have been many (mostly perennial) spp. transferred from this genus to other genera (such as *Tanacetum, Leucanthemum,* &c.). However, this is often controversial and will take some time to settle down.

2585. Orycha = operculum of *Strombus gigas* & other spp. (Strombidae - Heterogastropoda - Mollusca); when burnt this gives a penetrating aroma; u. in perfumery of ancient Hebrews (ca. 1,000 B.C.) (with *Sticta*, Galbanum, &c.). In Unani med., shell u. as antacid (calx) in colic; pd. in eye dis.

2586. Schroeder. *Circulation.* 35:570; 1976 - *J. Chronic Dis.* 23:123; 1970 (Cr. in human heart muscle 0.01 mg./100 g.).

2587. Cheese (Caseus) (L.): therapeutic use discussed in Serapion's List (1525): "The grete Herball whiche giveth perfect knowledge and understanding of all maner of herbes and there gracyous virtues" (sic).

2588. Generic names used: *Acourtia* (for NA spp.); *Peziza* (SA spp.) (Ref. *Phytologia.* 27:238; 1973).

2589. (Tisanes). Usu. prepd. in earthenware or heat-resistant glass vessel, using 1 dram per cup; let infuse 15 mins. (some 30 mins.); sometimes mixed with black or green tea.

2590. (Cannabis). Paper has been made from hemp; the US Declaration of Independence (original document) is said written on such.

2591. Cinnamon water (prepd. fr. c. oil) (USP 1820-1975) has been u. by hop-pickers to wash their hands after picking hops (to prevent fungal infections) (*JAMA.* 176:752; 1961).

2592. Other genera with species named "cedar" include: *Calocedrus; Toona; Guazuma; Soymida; Libocedrus; Weddringtonia; Cryptomeria; Pinus; Acacia*(!!); *Melea; Chukrasia.*

2593. *cf.* Wuerzburg (historic city of Bavaria, Ge), important spice trade center in the past; its name means "spice stronghold."

2594. Alcorn, J. D. *Huastec Maya Ethnobotany* (Univ. Texas) (1984).

2595. *Callirhoë* (genus) name of hot springs near Machaeus (e of Dead Sea); Herod bathed here (springs now called Zerka Ma'in).

2596. *Fusarium* mycotoxin claimed to be the "yellow rain" said u. as a Chinese warfare agent (poison gas) in the Orient (proved to be false; actually was bee feces).

2597. Iron when in excess is now considered to be of potential danger in heart dis. and associated with heart attacks.

2598. Soy: there is some evidence that ingestion of soy may reduce risk of prostate cancer.

2599. Charles J. Sih, Professor, School of Pharmacy, Univ. Wisconsin, Madison.

2599A. Common names of Cannabis prepns.: gunja (resinous prod.); charras (chars) (resin). majun (majoon) (prepns. of C. in chocolate or other sweetmeats).

2600. *Catha edulis* (khat) said used over longer time than coffee. Leaves are u. in asthma, chest dis., debility, diabetes, flu, lethargy, stomach dis. This drug produces logorrhea (excessive volubility, talkativeness) (action—*cf.* methamphetamine).

2601. Cantharidin u. to remove warts: prepn. in acetone & collodion appl. with glass rod & covered with air-tight dressing; gives a blister in 24 hrs.; in 7-10 days, tear off dead skin & wart.

2602. Orellana, S. L. *Indian Medicine in Highland Guatemala* (1987).

2603. *Achillea millefolium*: fls. white or pinkish or yellowish. Yellow fls. (bright) in *A. holosericea* Sibth. & Sm.; *A. chrysocoma* Friv. Less bright color in *A. clypeolata* Sibth. & Sm., *A. coarctata* Poir. (per Polunin).

2604. *Albizia saman* (Jacq.) F. Muell. = *Pithecollobium saman* (tree planted by Mark Twain ca. 1866 in Hawaii (big island) near Kona; still alive in 1990).

2605. Saggese, D. *Yerba Medicinal Argentina*, ed. 5; 1949.

2606. Hammerschmidt, D. E.

2607. Turner, Nancy. *Econ Bot.* 31:461-470; 1977.

2608. "Ye pay tithe of mint and anise and cummin" (Matt. 23:23).
2609. Some specialists now place *Tectona* in the Labiatae (Lamiaceae), also *Clerodendron* & other genera.
2610. Silphium was highly valued by the ancient Greeks (who introduced it) & Romans, being considered sometimes more valuable than an equal wt. of gold. The plant source may have been *Thapsia garganica* L. (Med. area), but this is disputed by some. It was collected in Cyrenaica (= Libya) & eventually so intensively that the pl. was eradicated by the 1st cent. A.D. One of the most celebrated drugs of antiquity, it is now almost unknown. The resinous subst. fr. the st. & rts. was u. as Thapsia Resina in homeopathy; also the fresh pl. The rt. ("false turpeth root") was earlier u. int. in pulmonitis. Rt. bk. c. thapsic acid, wax, &c.; it acts as purg. & emet.; on the skin it produces itching —> vesication.
2611. Petroleum was first prod. for indy. use in Pennsylvania (Titusville and nearby).
2612. *Dryopteris filix-mas:* use of oleoresin dangerous because while it paralyzes the tape worm it may also paralyze the voluntary muscle of the patient. It is now mainly replaced by quinacrine.
2613. (Epinephrine bitartrate). Ophthalmic ointment u. to reduce conjunctival congestion & control hemorrhage & to reduce intraocular pressure in simple glaucoma; sl. mydriatic.
2614. *Eucalyptus* sp. cult. in Ala & Ga (US) said u. to furnish a new fabric fiber.
2615. *Fritillaria imperialis* L. proposed to cultivate for starch prodn.
2616. Horseshoe crab isolate ("T22") the subject of Japanese research for treatment of AIDS infection.
2617. *cis-* and *trans-*fatty acids: in the hardening of fats to make margarine, the treatment procedure converts the *cis*-oleic acid (f.e.) (m. pt. 13° C) to the *trans*-form (called elaidic acid (m. pt. 51.5°C). This is supposed by some to have a bearing on heart dis., hence some prefer to use the "soft" margarine rather than the "hard."
2618. Infusion: u. in prepn. of many beverages besides tea; ex. on (excursion) liners such as "Ile de France" & "Normandie" - Where infusions of camomile, linden fls., verbena, & mint are/were served at the ends of meals.
2619. Liebig's Extract of Beef. First appeared in 1865 as "Extract of meat"; promoted early as a tonic for weakness, & for digestive dis.; later was mostly u. as flavor in foods & as soup base.
2620. *Juglans regia*: in As bk. u. to clean & color teeth (Pak); as lipstick substit.; wd. u. for carving.
2621. Tsumura & Co., Japan (-1990-).
2622. Kaolin mixt. with pectin popular for diarrhea, &c.; ex. Kaopectate (Upjohn).
2623. Kombologion (combologion) = beads fingered by Greeks. Called the "rosary of the Orient"; rosary, string of beads.
2624. Krameriaceae was formerly placed with or at least close to the Leguminosae. It is now excluded & regarded as close to the Polygalaceae & is placed under Order Polygales rather than Rosales as formerly.
2625. "Naval stores" is also interpreted by some to include masts, planking, & live oak ribs & keels of a ship.
2626. Prostaglandins (incl. thromboxane [T], prostacyclin). These 2 are in balance to reduce risk of thrombosis; where endothelium is damaged, T —> platelet aggregations which may in some cases —> atheromatous plaque.
2627. Quail: *Lophortyx californica*, valley (or California) quail found in Calif, Ore (US), BC (Cana); distinguished by a plume on head.
2628. CoQ_{10}™: sometimes claimed useful in incr. cell energy (esp. in heart cells), antioxidant, & in hypertension, to stim. immune system, to slow aging process; in cancer, obesity, &c. (10 mg. dose).
2629. Balm (or Balsam) of Gilead (named after district of Gilead (now in W. Jordan) (near Mt. Gilead & e. of Jordan River). Long u. for healing & cosmetic properties. The monks in nearby Jericho (Israel) said to prod. it from frts. of *Balanites aegyptiaca* (zafgim) (mastic?).
2630. Alcoholic drinks (1-3 per day) said to reduce chances of coronary dis. (News item; 3-V-93). - Alcohol when consumed with barbiturates greatly increases the lethal effect of same.
2631. Bee sting - Said to stim. immune system & NS in multiple sclerosis (sic); palliates arthritis.
2632. *Laser* Borkh. (s&c Eu, wAs): 4 spp. (in olden times, *L. foetida* u. as source of an asafetida).
2633. (*Lumbricus terrestris*). Earthworms (Lumbrici, L.) are named in ancient medical writings, thus listed in Aliphita (Saliternian MS) (ca. 1387) & in the list of Serapion (1100-1300). The worm was contained in Unguentum Armarium (ca. 1100), an ointment u. to anoint the weapon wh. made a wound, resulting in healing (sic) of the same(!). Dioscorides (2: 61) mentions an ointment made with goose grease or prepd. by cooking earthworms in oil. A physician, C. P. Paullini (Muenster) (d. 1712) wrote a work on "Lumbrici terrestri" (1703). Earthworms were official in Brit Phar. I. Another earthworm species (identified as *Pheretimas sperguillus* [F. Perrier] [*P. posthuma*]) of As was u. in Chin med. (entire animal) as antipyr. antispas. for wheezing (in many combinations) and in Unani med. as a dr. pulverized prod. in snakebite, TB, nerve dis., &c. Earthworms (*Lumbricus terrestris*) importantly u. as bait (wrigglers, red worms) for fishes;

also other taxa of soil inhabiting oligochetes u.; sold in bait shops. Cult. in worm farms for the fishing trade.

2634. *Melia azederach* (China tree) berries are said to make birds (esp. robins) drunk (as evidenced by the bird's behavior when loaded with berries) (berries fall all through the winter). Mules & cows also eat (CWB) (Hartmen, Carl. "A list of trees & shrubs occurring in the vicinity of Huntsville, Texas," *Trans. Tex. Acad. Sci.* 2(2):66-90; 1913). When u. to "worm" stock, the animal is fed all it will eat. - 2-5% ext. of dr. lvs. is effective in repelling acridia (locusts, grasshoppers, crickets) (*Arch. Inst. Pasteur, Algérie.* 15:427-32; 1937). At least one death is reported (3 yr. old child) fr. eating berries (sometimes claimed to c. alks.), also rabbits after inj. (Carratala. *Rev. Assoc. Med. Argentina.* 53:338-40; 1935) - This tree is regularly found growing beside each house in rural areas of the south.

2635. Methyl salicylate, natural (sweet birch oil) is said to be much less toxic than the synthetic oil; the same is true it is said of the salts (salicylates) (synth. said depressant & irritant to kidneys).

2636. Belém (Pará, Brazil) is called "the city of mango trees" because these line the streets there.

2637. Muava bark; c, muawin (heart poison) (Jacobsohn: *Diss. Univ. Dorpat*, 1892).

2638. (*Acanthus*). In Rome, capitals of pillars decorated with acanthus lvs. (Museo de las Termes). Capitals in Temple of Marti Vengador; Temple of Jupiter Stator (Corinthian).

2639. Old Calif mining sites often marked by *Ailanthus* trees planted originally by homesick Chin. cooks et al. (*Time.* 11-V-41, p 34) - *Ailanthus* honey (unpleasant odor) prod. in London.

2640. (*Allium*). Festival of the Ramps held annually at Cosby, Tenn (US). In WVa (US) there is a Society for Preservation of Allium Tricoccum (SPAT).

2641. (*Ananas*). Anthelmintic values disproven (Hernandez Morales & Asenjo - *P. R. J. Publ. Health & Trop. Med.* 1942:119).

2642. Sup Vatna root cont. estrogenic subst. prob.a glycoside. EtOH ext. fr. c. 0.002 gm. dr. pd. prod. estrus in mice.

2643. (*Dahlia*). Fresh tubers c. 27% sugars & 10.33% inulin (dr. tubers 61.95% inulin). Ciba advertised Phytine Tablets (= Phytin). Inulin considered as chief source of fructose (E.S. Haber); however prod. by hydrolysis of sucrose now considered better method of prodn.

2644. (Clove bark). Ref. Schwarzk, S.A. *Lehrb. d. Colonial-u. Spezerei-Waarenkunde* (1854).

2645. *Erianthus giganteus* (Walt.) Muhl. (E. saccharoides Mx.): sugarcane plumegrass (s&eUS (NY-Fla,Cu); specimen coll. at Auburn, Ala (US) 3.6 m. high (*Erianthus* is now often transferred to *Saccharum*).

2646. (*Haematoxylon*). In his early career before turning buccaneer, Wm. Dampier (also famous later as an explorer) was in the logwood industry in Campeachy Bay. Investment wiped out by a hurricane which swept the Caribbean in June, 1678.

2647. Kusso tablets are u. regularly by natives of Ethiopia to control worms. Kassa (took title of King of Kings & Theodore) ca. 1850 began his career by trading in Kusso (Ludwig, Emil. *The Nile.* 1957).

2648. (*Hypericum*). "Grandma's Salve" c. Cera flava; tallow; turpentine oil; Johnswort ext. (decoc. of lvs. & small sts.) (Hartford, Ala [US]); given for skin infections (hands, &c.); pruritus (itching) esp. of shingles. Hypericum sd. said toxic to any small animal with white hair(!) - Claims of value in diabetes are probably false.

2649. (*Helianthus tuberosus*). In an exhastive study made in England during a food crisis (World War I; 1917), to determine which pl. would furnish max. food amts. in case of a blockade, it was found that *H. tuberosus* would best fill the bill, yielding the most food (sugar) per acre of all plants studied. Tuber c. 10-12 & fructose ($100/# during that period), furnishing 20-22 tons of tubers/acre in GB - Other uses of plant: prodn. acetone, BuOH fuel (heat, power) - *H. tuberosus* raised in Penn (US) by Dutch.

2650. (*Umbellularia californica*). Cult. commer. in Coos Co. (Bridge Myrtle Point.), Ore (US) (200 acres) (VO prod.) - VO c. eugenol, 1-pinene, cineole, safrole, &c. - Trees 300 yrs. old & 45 m. high (Ref. U. Calif. *AES Rept.* 178 - *ACJ.* 4:206- [1882]; *JCS.* 85:6291 [1904]).

2651. (*Ungnadia*). Schaedler. *Pharm. Ztg.* 1889 - Fette Oele - 2.A.564; Cheel & Penfold - *JCSI Trans.* 38:76; 1919.

2652. (*Crataegus aestivalis*). May Haw(thorn). The Mayhaw Festival is an annual event in Colquitt, Ga (US) - Mayhaw Belt = triangular area bounded by the Flint & Chatahootchee Rivers (Ga) / Mayhaw, Ga,/ Mayhaw, Miss (US) - Fruit of this sp. like a small apple, size of cranberry.

2653. Addn. of lecithin (soy) (5% +) to soap gives better cleansing by incr. emulsifying & lathering power (Lederer. *Seifensieder Ztg.* 60:919 [1933]; *cf. J.Soc. Ch. Ind. Japan.* 32:595 [1934]).

2654. Bitter root plant = *Lewisia* Pursh, esp. *L. rediviva* (State Flower of Montana (US) - *cf.* Bitter Root Valley; B. R. Mountains (border Mont & Ida) - Plant is a beautiful garden ornam.

2655. (*Liriodendron tulipifera*). Lvs. u. in poultices & ointments for local appln. in inflammns., bruises, rheum. headache, &c.

2656. (Skunk cabbage). A man had a pig & fed it s. c. (green vegetable). When the pig was slaughtered, it was found unfit to eat on acct. of the horrible smell of s. c. Name s. c. wrongly appl. to *Veratrum* sp. (false hellebore), per Hultén & H. St. John (*Svensk. Bot. Tidskrift.* 28(2); 1934; 25:457-64; 1931).

2657. (Kharasch +). Cont. stadin (steroid saponin) 3,000 X as sweet as sucrose; one of sweetest subst. known.

2658. *Viburnum brachycera*: specimen claimed as oldest flowering plant—over 10,000 yrs. old. Single plant widely proliferating (Moldenke. *Wild Flower.* 33: 4-8; 1957).

2659. Nasir, S. S. & Wilken, L. O. Jr. *J. Phar Sci.* 53:794-; 1966.

2660. Geo. Washington Carver listed 41 products available fr. the sweet potato (*G. W. Carver Museum Bull.* 31;39).

2661. Gentian was some 6 times as bitter as Calumba (informal expt., GMH).

2662. Ref: Hess & F. Merck. *Ber. Chem. Ges.* 52: 1976; 1919. - Lloyd, H. A. & al. *JACS.* 50: 1506; 1958.

2663. (*Pastinaca*). Inflammn. & vesication caused in some people by handling parsnip pl. (flowers). Wild parsnip (cult. pl. run wild) believed poisonous, but Pammel gives extended list of experimenters who ate pl. without ill effects of any kind. Often mistaken for cowbane (*Oxypolis* Raf. spp.) wh. is very poisonous - Parsnip has best flavor after being frosted.

2664. *Paulownia* named after Anna Paulowna (1795-1865), Princess of the Netherlands - Soft pubescence is prominent feature of lvs & fls. of *P. tomentosa.*

2665. Mezquite - *Prosopis* spp. - Pods u. to prepare a kind of syrup (put up in beer bottles) & u. as a "fortificante" (aphrodisiac) (Peru) (per Higbee).

2666. (Rhodium oil). Applied to fish bait - Said to incr. catch: mixed with mash or a drop can be put on a worm or similar bait.

2667. *Rhodymenia* - Some 10 spp.; sold for 65¢ per lb. in Washington Market, NYC, 1-10-46; also smaller pkgs. (*Minn. Bot. Studies* 2:205-); die Rosen-rotalge.

2668. Black locust tree ca. 350 yrs. old (1980s) in the Jardin des Plantes, Paris (from Robin's garden [hence generic name *Robinia*]).

2669. *Senecio platyphyllus* - Herb & rt. c. platyphylline; studied 1955-7, found antispasmodic without unpleasant side effects (*Am. Review of Soviet Med.* Dec. 1943: 155-7) (review by David I. Macht) - *Klinicheskya Meditsina.* 31: 55-68; 1943.

2670. Sarsaparilla in psoriasis-Deneke - *Deutsch. med. Wchschr.* 1936, #9 - Thurmon, F. M. *New Engl. J. Med.* 227:128-32; 1942.

2671. *Smilacina racemosa* (S. amplexicaulis Nutt. ex Baker) (misnomer since no resemblance to a Smilax). A commoner plant than *Polygonatum* (Solomon's Seal) (smaller plant with small & narrower lvs. altho with much the same general appearance). Berries red-dotted, globose, (in other spp. berries are red or blackish). Rootstocks creeping; lack the characteristic scars ("seals").

2672. *Arundinaria gigantea* - Canebrake sections of central prairies belt of Ala (lime sect.); swamp cane largely destroyed by high water (1916) some still growing in Sumter Co. Belmont (1948); plant may be killed by cold.

2673. Blundell Peat Co. Ltd., Lulu Island (Vancouver, BC) - Cutting peat into bricks ca. 14" X 10" X 6". Trenches 3 ft. to 4 ft. deep; piled up in stacks ca. 7 ft. high; said that ca. 50% of area of Lulu Is is covered by peat bogs; charcoal fr. Sphagnum said to be u. to polish microscope lenses (c. 1970).

2674. *Stillingia* - Name from Stillingfleet, Engl. botanist; Linné named in 1756; u. by early settlers - seUS; 1828, Young; "nettle potato root." Reputedly u. in the proprietary SSS (Stillingia, swamp sumac, sod. iodide).

2675. Tamarind sd. u. extensively in India for many purposes, such as 1) FO (no VO) ca. 20% by wt.: this veget. oil u. in foods, paints, as fuel, is a drying oil like linseed 2) hunger food, u. in parts of India during famines (outer skin removed by roasting & soaking, then boiled or fried or dried & ground into flour 3) outer coating u. as drug; astr., poultice for boils; int. in diarrh. & chronic dys. 4) fed to pigs (good for fattening) 5) sd. rubbed up with water & u. as "specific" for Delhi boil 6) sd. pd. & boiled to form paste & mixed with thin glue to prod. a very strong wood cement; also u. as size for blankets in India 7) sd. gum like locust gum 8) raw sd. chewed like betel - "Tamarind Fuzz" = beverage served cold to children (Jamaica drug stores). In México, a confection is made which is a sweetmeat considered a luxury - sold for 20-25¢/lb. (ca. 1950) (*cf.* Turkish confection).

2676. Gum drops u. as med. (drug stores) for cough (early 1800s). Later jujube paste lozenges sold as sweetmeat in d .s. "Gum troches" (US Disp. 1833) c. acacia, starch, sugar, rose water (*cf.* Edinburgh Pharm. 1756 [Trochischi bechici alba (sugar, trag.pd. orris rt., rose water]) - US Disp. V (1843). Jujube paste (acacia, sugar, decoc. of frt. of *Zizyphus vulgaris*).

2677. Popular compns. are u. with pennyroyal and rue oils, apiol (green, parsley); quinine sulfate.

2678. *Tilia europaea* cultivar *pallida* is the large-leaved linden tree of Unter den Linden (Berlin) (*T. europaea* may be conspecific with *T. vulgaris*). The best loved tree of Germany.

2679. *Trifolium arvense* (fresh plant) listed in Schwabes Polyglotta (Pharmacopoeia Homeopathica) ed. 2, 1929 (per Tschirch).

2680. *Trifolium erectum* var. *erectum*: 4-parted: 4 perianth segments, 8 stamens, 4 carpels

("Quadrillium") (Patillo, J. Dan, *Atlanta Constitut. Sunday Magazine.* 8-26-73, p. 5).

2681. Ipomoea - The Hakims (physicians in Arabia, Iran, Pakistan, &c.) use "turbeth" in med., often only the inner substance of the rt. "And I remarked that turpeth is sold in London without the inner substance having the appearance as if perforated" (Honigberger, John Martin. *35 Years in the East* [1852] vol. 2: p. 360) ("t. pipes").

2682. *Lathyrus* is derived from a Gr word (referring to a plant mentioned in Theophrastus). However, Lathyrus also was applied to aking of Egypt (Ptolemy VIII) ruled 117-107, B.C., son of Cleopatra, who had banished him to Cyprus.

2683. (*Lunaria annua*). Other common names are: money plant; St. Peter's penny; honesty (because the flower does not hide [its seeds]).

2684. Carbon filaments in the old incandescent electric light bulbs were found to last for very long periods of time. For instance, Byer's Opera House (now Palace Theater) Ft. Worth, Texas, had a light bulb lasting 1908-1969-? (see UPI dispatch 26 May 1969).

2685. Sometimes other plant species have been confused with & described as cannabinus: *Datisca cannabina; Hibiscus cannabinus; Urtica cannabinua* - The high valuation placed on cannabis may be illustrated by the following case of police seizure in Opelika (Ala), when 200 marijuana plants were seized & evaluated at $350,000 (ence with an av. value of $1250./plant) (*Lee Co. Eagle*) (ca. 1990).

2686. "The hickory told me manifold
Fair tales of shade, the popular [poplar] tall
Wrought me her shadowy self to hold."
(Song of the Chattahoochee, Lanier).

2687. (*Canavalia*). Jack Bean grows sporadically in Venezuela; not cult.; ca. 1924, a coffee planter of Caracas had large fields of the sp. & u. as part of a fertilizer scheme for coffee tree growing. Sds. u. pd. as flour - *C. ensiformis* c. antibiotic, canavalin, isolated 1944 by D. L. Farley; active against Gm.+ & Gm.- MO & in lobar pneumonia.

2688. (Essential fatty acids). Function: 1) precursors of prostaglandins 2) important in synthesis of biomembranes 3) prod. lower plasma cholesterol levels. Hazards in use; hydroperoxides formed —> free radicals (Ref. *Am. J. Clin. Nutrition.* 23:86; 1970). Margarines c. significant amts. of trans. fatty acids formed from the *cis*- f. a. in the process of hardening; believed to be hazardous (heart dis.).

2689. (*Castela texana*). "Common on gravelly bluffs of the lower Rio Grande from Eagle Pass downward." (*Botany of Western Texas.* 1891-4).

2690. (*Castalia; Nymphaea*). Cult. in Kenilworth Aquatic Gardens, DC (now operated under the Natl. Parks Service) ca. 1946 - Both genera featured.

2691. The use of this word (shittam & variants) has given rise to much confusion in some places. Thus, f. i.; *Rhamnus purshiana* has been stated as occurring in the Holy Land, a view based on name occurring in Bible.

2692. (Tragacanth scrub). T. shrubs form a characteristic vegetative society of the eastern Mediterranean Sea region & the adjoining Persian zone, the most singular form of thorn bush of the steppes.

2693. "Stachys...," *Phytochem.* 10:2537-8; 158; 1971.

2694. (Starch). References: Tschirch. *Handbuch der Pharmakognosie.* Bd. 2 - *Sprecher v. Bernegg.* 1:280-1.

2695. (Strophanthus). Schlotterbeck & Watkins. *Ber. Chem.* 1962:7 - *Arch. Pharm.* 1890:441 - 1893:136 - *Phytochem.* 7:41; 1968 - 9:69; 1969.

2696. (Sponge). U. in tech. areas (ceramics, lithography, painting, decorating, pottery industries, &c.). Many other types of sponges: (gulf) grass; yellow, &c., are in use.

2697. Redonnet, T. Alday. *Farmacognosia.* 12:85-125; 1959.

2698. Kombologion beads also made fr. amber; fingered by Greeks; this is the rosary string of the orient.

2699. (*Juniperus macropoda*). Girth 2.-2.2 m. considerable numbers oted near Ziarat (Baluchistan - Pak) (GMH); extensive forests in state of Baluchistan.

2700. (*Juniperus osteosperma*). *C. A.* 20:22: 146; 24: 579.

2701. Gold sold at Ophir (in modern Yemen) in Biblical times (I. Kings 9: 128).

2702. (*Lallemantia royleana*) sds. - A. Ghose. *Indian Agr. Chem. & Drugs.* 1897:258/*Ztsch. Oesterr. Ap.-Verein.* 1893:26.

2703. (Adonis). Term "pheasant's eye" also appl. to *Narcissus poeticus*, *Dianthus plumarius* L. var. *annulatus*, and *Primula* spp.

2704. (*Prunus armeniaca*). Chem.: Rabak, USDA, *BPI Bull.* 133:1908.

2705. (*Randia echinocarpa*). Bye, R. *et al. Anal. Biol. Univ. Mex. Ser. Bot.* 62:87-106; 1966.

2706. (*Cyclamen*). Aroma extremely attractive: "I'd lay down my life for that scent" (*Dragoco.* 1/86: p.14).

2707. (*Chaenomeles speciosa*). Frt. can be used to help jell elderberries, &c. & to make "japonica apple jelly," also wine jelly. Directions: cut up whole frt when ripe, remove any fls., cover with water, and cook in pressure cooker 10-15 mins., cool & release pressure, juice strained or squeezed fr. frt. pulp.; add 1 1/4 cups sugar per cup juice. Boil 1-2 mins.

2708. It is obvious that bot. nomenclatural rules will forbid using the same bot. names for 2 different organisms. Both names, however, seem to be well established!

2709. For iron assimilation, various administration forms are used (*Schw. Wchsch.* 73:1176; 1943 - "Absorption." *Amer. J. Dig. Dis.* 11:244; 1944 - "Administration Fe," Hazelton. *J. Ph-col. E. T.* 83:158-: 1945.

2710. The specific name is too confusing since these spp. have been found in at least 6 states besides Utah, viz., Mont, Ore, NM, Ariz, Tex, Calif. Some spp. cont. HCN (glycoside) (Greshoff).

2711. Bittersweet weed has been u. as a symbol of hatred, ex. in a demonstration against Laval ("traitor") at Thaïs Cemetery close to Paris (during or after WWII) - Bittersweet branch was thrust into earth at his temporary tomb, along with spitting & stamping on the ground.

2712. (strawberry). Probably orig. strayberry since spread by runners; or may be fr. straw-like stems of pl. - The strawberry is not a true berry but actually an enlarged fleshy torus.

2713. (S.A.). *Propaganda for Reform, Proprietary Medicines* (Council, AMA), ed. 5, pp. 1102-3; ca. 1920.

2714. (sodium chloride; salt). "Can that which is unsavory be eaten without salt? or is there any taste in the white of an egg?" (Job 6: 6).

2715. (Aerosporus). *Bacilllus aerosporus* Greer: isolated from city water of Chicago, Ill (*J. Inf. Dis.* 42:508 [1948]).

2716. (*Vochysia*). Flores, E. M. *Arboles y semillas del Neotropica.* 2(2):29-52; 1993.

2717. (Surma). Considered harmless; however, sometimes galena (native lead sulfide) has been used & this is feared possibly hazardous. ([al] kohl [Arab]).

2718. The origin of the word "Nyssa" is not well established (prob. name of nymph of antiquity). For details see Eyde. *Rhodora.* 61:209- ; 1959.

2719. (*Cochlearia*). Bateman's Spirit of Scurvy Grass (1680) (John Hooper sold [& mfd.?]).

2720. (Clove). Resemblance of bud to phallus (Doctrine of Signatures).

2721. (*Tsuga canadensis*). J. H. Kennedy's Extract of Pinus Canadensis highly extolled in its day. Recommended by Dr. James Morin Sims (b. in Ala, 1813).

2722. (toad). U. in folk med. - Toad split open & tied around neck of patient.

2723. The modern University of Wuerzburg has an outstanding pharmacy school with a very large drug garden. Several famous scientists have been associated with the Univ. Virchow; V. Koelliker; Ernst v. Bergmann (surgery).

2724. (*Juniperus c.* wood). Bergen. *Pharm. Monatshefte.* 12:242-244; 1931.

2725. Juniper Oil, Russian: <*Juniperus excelsa.* VO showed pos. rotation (7-8°); Italian/Hungary VO showed neg. rotation, otherwise seemed normal (*Schimmels Ber.* ca. 1903-4).

2726 Pasta fr. Senikuba: known in Italy since at least 1284 A.D. These types also include (elbow) macaroni; lasagna; ravioli; noodles & others (some 150 varieties). Advantages: 1) cheap 2) non-fattening.

2727. (*Suderutus theezans*). Patient bathed in hot decoc. of hb. then rubbed with cod liver oil.

2728. Pearls formerly official in Phar Germ II, Brit Phar. I, & many early pharmacopeias. Classes: 1) Margarita orientalis (EI pearls): largest & finest 2) M. occidentalis (WI. pearls) medium size. (*Margaritifera* spp); *Pinctada* spp. (NA) 3) Marg. textilis (pd. p.) small p., u. in med. (formerly); *Ostrea edulis* (Pteriidae) (warm seas), pearl oysters; Meleagiina (off eAs coasts); source of most comm. pearls. Pearls also from many oysters & clams (ex. *Mya*); some fr. fresh waters.

2729. *Phoradendron* (American mistletoe) and *Viscum* (European mistletoe) rather similar in morphology. Frt. of *Viscum* white or yellowish & viscid while *Phoradendron* frt. is white & glabrous, also smaller (4-6 mm) than in *Viscum* (6-10 mm). Stem of *Phor.* smaller (2-4 dm. high) than in *Viscum* (-1 m). Pharmacol. action different: *Viscum* lowers BP; *Phor.* raises BP.

2730. (Heroin). As analg. dose: 5-10 mg. p.o., antitussive dose: 1.5-6 mg p.o. Formerly, physician dispensing of Heroin permitted (*cf.* J. Barksdale: "Doctor Bill"; 1970) (popular Cocillana syr. incorporated in the 1920s-'30s era) ["best cough syrup"] [*cf.* Heroin & Tolu]). Intense euphoria (orgasmic sensations) led to high rates of addiction; in US; opposed by declaration of H. as controlled substance; later the mfr., importation, & use of H. strictly prohibited in the US.

2731. *Elaeocarpus* alkaloids: *Austral. J. Chem.* 22: 301-6; 775-782.

2732. Jorge, L. I. F. et al. *Rev. Farm. Bioquim* X (Univ S. Paulo). 29:69-72; 1993.

2733. Schultes, R. E. & Raffauf, R. F. *The Healing Forest* (medicinal & toxic plants of N.W. Amazonia); 1990.

2734. Fatty acids: 1) satd. 2) unsatd. C_nH_{2n-1} (2n - 3; 2n - 5). trans- (produced in hydrogenation).

2735. Vinegar has been used to treat sore throat, as disinfectant, for dandruff, age spots, to improve complexion, in hiccups, &c. A combination of apple cider v. & honey was u. by a celebrated Vermont (US) physician, D. C. Jarvis (*Folk Medicine.* 1958).

2736. Miller, A. G. & al. *Plants of Dhofar.* 1988.

2737. (*Ceratonia siliqua).* Cult. ornam. & shade tree; around Gulf of Mexico coast but in this area does not produce frt. (rots in moist rainy climate); successful cult. in Calif, sAriz, swTex (US); sds. u. in rosaries; poultry feeds; pl. a good soil binder.

2737A. (Silphium) (Silphion [Gr]) (giant fennel plant). Fr. areas near Cyrene (nAf) (Libya) u. as early term contraceptive (500 B.C.); collected to extinction, cultivated elsewhere but not successfully (hence, very expensive).

2738. (cucumber). "As cool as a cucumber" (popular expression), indicating value in cooling & refrigerating salads, &c. (in warm weather); also u. in sour pickles; said to prevent seasickness.

2739. Licking toad—> a "high" since glands of head of toad prod. hallucinogens.

2740. (*Thymus integer.*). Ref. Bellomaria et al. *Pharmazie.* 49:684-; 1994 (chem).

2741. (testes). Testicular ext. was u. by oral ingestion (as early as 1890s) for senility, "insanity," tabes dorsalis, mal., TB, cancer, &c. (!!!); such usage worthless.

2742. (Bitumen). "Jellied petroleum" early prod. in the Dead Sea where it floated to surface & was collected by the Nabateans (centered at Petra) & sold to Egyptian dealers for use in the mummification process (B.C.); also u. in those times to caulk boats, attach points to spears & arrows; as lax., in liniments &c.; later u. by Romans in int. med. (ref. *Aramco World*, May 1994, 26-31).

2743. Walrus baculum ca. 18 in. long; sketched at Pioneer Museum, Tilamook, Ore (US) - Photo of racoon baculum (Auburn Univ. *Bull.* 402; 1970).

2744. Larkspur lotion u. for head, body, & pubic lice, esp. the latter, popularly called "crabs." Formula: ground larkspur 100, in acetic acid (50), glycerin (50), alc. (100), water sufficient to make 1000 (parts by vol).

2745. (Queensland arrowroot). Very large (largest?) starch grains; can be seen as tiny dots to the naked eye; readily digested; u. by invalids & children; said better than West Indian arrowroot (Barbados a. very similar).

2746. (*Asclepias tuberosa* & other *A.* spp). Floss u. 1) to replace kapok in life jackets (6 X as buoyant as cork) 2) in linings of flying suits to retain warmth (as warm as wool). Formerly cult. in Upper Michigan (US).

2747. A paper on New World *Aster* by Nesom (*Phytologia.* 77:141-297; 1994) recommends numerous important changes in nomenclature viz., transfers varially from *Aster* to *Symphyotrichum*; *Eucephalus*; *Oclemena*; & *Chloracantha.*

2748. The toxic Chinese plant, *Stephania tetrandra* (chronic renal damage) has been confused with *Aristolochia heterophylla* Hemsl. (fang chi). Sales of both are prohibited in France.

2749. Canna starch (West Indian arrowroot) has been claimed the largest starch grains (*cf.* 2745); they have a glistening appearance *en masse.* With boiling water, it forms a cloudy but very tenacious jelly & represents a food readily digested & suitable for invalids & infants.

2750. *Pentaspadon motley* Hook. f. (Anacardi) (seAs & islands): st. furnishes FO u. in treating ringworm & other skin dis.

2751. Schindler, H & Frank. H. H. *Tiere in Pharmazie u. Medizin* (1961).

2752. (*Silphium laciniatum*). Was described from Montgomery or Lowndes County (Ala) (prairies) by Wm. Bartram (ca. 1760) but was not found there in recent years by R. M. Harper.

2753. *Spirulina* (Oscillatori. - Cyanophyc. - Algae): widely distr. in Lake Natron (Af) (reputedly most saline lake in world); furnishes food for flamingos & source of red pigment of their feathers.

2754. Sodium nitrate is derived fr. caliche (& costra) deposits of nChile (Atacama Desert near coast) & Peru (guano islands) (Chinca Is); at times deposits were up to 80 m. thick (now largely exhausted); guano birds furnishing this manure are large sea birds, living in huge colonies.

2755. Petroleum (Oil): Seneca Oil (name may be derived fr. "Snake Oil") (or vice versa). Early u. for rheum., sores, later as illuminant. Am Indians also u. to calk canoes, as war paint.

2756. (*Pleurotus o.*). Ref: *Pharmazie.* 50:241-2; 1995.

2757. (Tiger). Value est. at $10,000 (food, med., hide) (ca. 1995).

2758. (Jap): tofu; dioufou (Chin): "the cow of China"; 2nd largest import fr. China into the US; u. for 5,000 yrs. in China.

2759. "We saw greate store of the spane (=spawn) of whale whereof they make spermaceti" (Dallan, 1600).

2759A. With linseed & soy oils. there is danger of spontaneous combustion on rags; also paints contg. same— (care!!).

2760. (*Populus tremuloides*). Said most widely distributed tree in NA.

2761. (*Hebe salicifolia*). Ref. Brocker, S. G. & al. *New Zealand Medicinal Plants.* 1981

2762. (Holy) incense of Bible: c. stacte: onycha (onyx); frankincense ("incense"): galbanum - Also u. in incenses: sweet bark ; myrrh; cassia; spikenard; costus.

2763. (Eel): (Fluss)aal: body slender & snake-like; remarkable life history of Am eel (*Anguilla bostoniensis*): breeds in Sargasso Sea (wAtl Oc.), migrates to NA fresh water rivers; later both sexes migrate to the Sargasso Sea, where they spawn, & then die. The European eel has a similar life history (Schindler). Eel "serum" is toxic but u. sometimes in med. for certain heart disturbances, rheum., endocarditis, nephritis; to reduce tendency to uremia & oliguria.

2764. (*Phyla scaberrima*). VO with verbena scent; Enrico Caruso u. verbena scent liberally as a personal perfume.

2765. (*Verbascum thapsus*). The "staked plain" (llano estacado) of the western plains (US) was so named for the numerous stalks of *V. t.* growing in such great numbers.

2766. Baking potash was a bicarbonate prepd. fr. wd. ashes, popularly used in early frontier days. On the w. coast where saltwort (*Salsola kali*) grew, this was burned to ash, leached with water, and evapd. to dryness, & burned to whiteness, equal wt. water added, and left stand for 1 wk. or more, & evapd. to dryness. This was a mixt. of Na and K bicarbonates. The same process was appl. to hardwood ashes, giving $KHCO_3$, baking potash.

2767. (Propadrine). Ref: *De re medica* (Lilly), ed. 3 (1951): p. 430.

2768. (Vitis). White raisins were put into gin, the gin poured off after a while, & the grapes eaten (for arthritis) (current folk medicine).

2769. (Pigeon). Worldwide there are some 300 spp. of pigeon, incl. spp. of *Columba, Ectopistes*, &c.). The young are fed on "pigeon's milk" (crop m.), representing regurgitated food of the parent. The living bird is u. in the assay of some vitamins, &c.

2769A.Buckland. *Curiosities of Natural History*. 1857.

2770. (*Balanites aegyptiaca*). The Koran says that the bitter frt of this pl. is the food of the "damned" (in Hell).

2771. *Phoenix canariensis* lvs. are u. by floating (with plants & similar attachments) (kannizzati) on the sea surface in order to attract fishes (esp. dolphin & lampuki) which are then easily netted (Med area).

2772. (Tarantula). Venom of this spider was thought to produce a form of neurosis so that the bitten person was stimulated to practice a fast dance (tarantella) (Rossini).

2773. Cottonwood, *Populus* spp. esp. *P. deltoides* & *P. trichocarpa,* but also *P. angustifolia, P balsamifera, P. fremontii, & P. heterophylla.* As a child in Edmonton Alta (Cana), GMH noted cotton-like masses abundantly distributed on the ground under the c. trees in early spring.

2774. *Phyllostachys bambusioides*: flowering may or may not produce death of the entire grove but reduces the number of cultures for many years. In smaller plants, death is a certainty. The flowering cycle is unknown, varying from 1-60 yrs. or more; in some bamboo species, this may be as much as 150 years. Heavy fertilization of soil (5 X normal) seems to prevent flowering.

2775. The mandarin orange is smaller than the tangerine & has a different flavor (more sour).

2776. Saffron is very expensive because of the very high labor costs: the 3 pistil lobes picked from each fl. are so light that it takes ca. 225,000 stigmas to make a pound of prod. (Sp). The flowering period is only two weeks; the product must be carefully dried. S. is used in many dishes: risotto (It); seafood stews; Scand sweet breads; saffron cake (Cornwall, Eng); &c.

2777. (*Chimaphila umbellata*). U. in a process to "make human organs transparent" (1939); said valued by Am Indians in edema & urin. dis.; rts. & sts. sometimes u. in beverage tea.

2778. The dr. yellow fls. are u. as food flavor (Chin, Jap).

2779. (*Canella winterana*). False winter's bark; lvs. & bk. arom.; rossed (=scraped); dr. bk. u. as antipyr. (decoc., Br); flavor; alc. macerate as rub for rheum. pains; abort. (Cu); bk. c. VO (with eugenol, cineole), resin (no tannin); lf decoc. in bath for rheum. (Out Is); bk. u. in prepn. Hiera Picra (with aloe) (sp. named for John Winter; "canella" from fluted character of quills).

2780. *Adenanthera* L.: bead trees; genus related to *Mimosa*; sds. reminiscent of those of *Abrus.*

2781. *Capsicum annuum*: var. *annuum* cultivar 'Ancho': (chili) poblano; green capsicum popular for centuries in Mexico; cult. mostly in Mex, less so in US (NM, Tex, Calif); eaten green mostly; sometimes dr. (sometimes eaten like peanuts with beer) (SC ancho = broad; large; poblano is appl. to the fresh green pepper).

2782. *Carapa guianensis* Aubl.: royal mahogany; caobilla; bk. bitter, u. in dys.; ulcers; eczema; in tanning; lvs antipyr. vermif.; sds. furnish FO u. for nutrition, candles; for skin disease; insecticide; as fuel.

2783. (*Cecropia*). Trees celebrated for myrmecophily, in which the ants often display a defensive posture.

2784. (*Cedrus*). Another sp. of the genus is *C. brevifolia* (Hook. f.) Dode (Cupressus cedrus) (Cyprus).

2785. Chili Serranus (Mex; NM, Tex [US]): green hot peppers said more popular in many areas; said originated fr. Serranas (mt. ridge of Mex); u. fresh or dr.; many types.

2786. Some taxonomists consider this gen. monotypic, others believe it includes hundreds of species

2787. (*Hippomane mancinella*). Found in Oriente province, Cuba; "poison tree"; latex —> blindness (sic); frt. toxic —> diarrh. & "inaito" (dis.); ingestion poisons animal flesh; fls. astr.

2788. *Hovenia dulcis* Thunb. (Rhamn.) (China, Kor, Jap): Chinese (or Japanese) raisin (tree): frt. pedicels fleshy & sweet: u. med. for hangover; valuable timber.

2789. *Stegnosperma halimifolium* Benth. (S. scandens (Lunan) Standl.) (Phytolacc.) (Mex, WI CA, nSA): lvs mixed with those of *Bursera microphylla* and boiled —> shampoo appl. for headache; rts. serve as soap substit.; r. bk. u. as blue dye.

2790. *Psygmorchis pusilla* (L.) Dodson & Dressler (Orchid) (CA): dikik; pl. (an epiphyte) cooked & eaten for colic.

2791. antitrypsin, alpha: AAT; a plasma protein (globulin) originating in liver (also prod. by yeasts) wh. acts to inhibit some proteolytic enzymes (incl. trypsin), which may attack liver tissue; hence u. in emphysema (where AAT-deficient).

2792. ([High] John the Conquerer). The term was sometimes (probably erroneously) referred to subterranean parts of *Hypericum* sp., *Potentilla erecta, Geranium maculatum.*

2793. *Gaultheria antipoda* Forst. f. (NZ) (native): snowberry; papapa; lf. decoc. u. as healing agt.) (appl. to wounds): to heal cuts (decoc. on poultice); lf. inf. u. in asthma.

2794. *Haplopappus fremontii* (Gray) Greene (in Colo [US]): reputed kidney dis. cure.

2795. *Hedyotis scandens* Roxb. (Ind+): Kimprong (Himal): c. epi-glutinose, sitosterol; st. & lf.ext u. for arthritic pain (Bangladesh); rt. u. for jaundice, dys.

2796. *Iostephane heterophylla* (Cav.) Benth. (Comp.) (sMex): (hierba del) manso: tuberous rts. u. to incr. female fertility (popular); relieves pain; sold by druggists.

2797. *Lobelia tupa* L (Ch, Pe): hb. psychodelic. perhaps hallucinogenic; sed. to treat toothache; smell of plant may —> emesis.

2798. *Pimelodendron amboinicum* Hassk. (Euphorbi) (Moluccas): bk purg.; sds. edible; milky latex u. as varnish.

APPENDIX B

GENERAL REFERENCE WORKS ON PHARMACOGNOSY AND ECONOMIC BOTANY

(Most important references indicated with an asterisk [*])

Agard. *Medical Greek and Latin at a Glance* (1937).

*Ainsworth, G. C. and G. R. Bisby. *Dictionary of the Fungi* (Kew, Surrey) (ed.7) (1983).

Anon. *Commercial Nomenclature* (3 v.) (Washington, DC) (1897).

Arends, G. *Synonymen-Lexikon* (Leipzig) (1891).

Backer, C. A. *Dictionary of Scientific Names of the Wild and Cultivated Plants of the Netherlands and Netherlands East Indies* (Groningen) (c. 1936).

Bailey. L. H. *Standard Cyclopaedia of Horticulture* (NYC) (1914-1917); *Manual of Cultivated Plants* (NYC) ed. 2 (1949).

*Bedevian, A. K. *Illustrated Polyglottic Dictionary of Plant Names, English, Fr, Ge, It, Turk, Armen, Arab, L.* (Cairo) (1936).

*Berger, F. *Synonyma-Lexikon der Heil- und Nutzpflanzen.* Wien; 1954-5.

Berger, F. *Handbuch der Drogenkunde: Erkennung, Wertbestimmung, u. Anwendung.* 7 volumes, incl. Register. 1967.

Brady, George S. *Materials Handbook* (NYC) (ed. 7, 1951).

Britten and Holland. *Dictionary of English Plant Names* (1886).

*Boulos, L. *Medicinal Plants of North Africa* (Reference Publications, Inc.). 1983.

Burkill, I. H. *A Dictionary of the Economic Products of the Malay Peninsula* (2 v.) (London) (1935).

Celebioglu, Sarim. *Dictionnaire de Botanique en 7 Langues* (Istanbul) (in press, 1955) (4 v.).

Chaplet. A. *Dictionnaire des Produits Chimiques Commerciaux et des Drogues Industrielles* (Paris) (ed. 3, 1947).

*Chopra, R. N. & al. *Glossary of Indian Medicinal Plants.* 1956.

*Chopra, R. N. *Indigenous Drugs of India.* 1933.

Clute, Willard N. *American Plant Names* (1940).

Correa, M. P. *Diccionario das Plantas Uteis do Brasil* (2 v.) (1916-1931) (monographs in alphabetical sequence; work remains incomplete).

Dalziel, J. M. *The Useful Plants of West Tropical Africa.* 1948.

Dayton, W. A. et al. *Range Plant Handbook* (Washington, DC) (1937).

Dorvault, F. *La Oficina de Farmacia* (ed. 2, Sp) (transl. fr. 17th Fr ed.) (Madrid) (1930).

*Dragendorff, G. *Die Heilpflanzen der Verschiedenen Voelker und Zeiten.* 1898 (1967) (Werner-Fritsch, Muenchen).

*Duke, J. A. *Ethnobot. Dictionary.*

Duke J. A. & Ayensu, E. S. *Medicinal Plants of China.* 2 vol.; 1985.

Falck, A. *Die Offizinellen Drogen U. Ihre Ersatzstoffe* (1928).

Foster, Frank P. *An Illustrated Encyclopaedic Medical Dictionary . . . Latin, English, French, and German Languages* (12 v.) (NYC) (1880-1893).

Fournier. *Le Livre des Plantes Médicinales et Vénéneuses de France* (3 v.) (Paris) (1947).

Gardner, Wm. *Chemical Synonyms and Trade Names* (London) (ed. 5, 1948).

Gatin, C.-L. *Dictionnaire Aide-Mémoire de Botanique* (Paris) (1924).

*Gerth van Wijk, H. *A Dictionary of Plant Names* (2 v.). (Possibly the most complete listing ever attempted, but unfortunately it is overloaded with book names and lacks many important genuine folk names.) The Hague (1911) (reprint 1971—Asher. Amsterdam).

Githens, T. S. *Drug Plants of Africa.* University Museum (Penn); *African Handbooks.* No. 8 (1949).

Graa, Albert. *Manual of International Pharmacy* (Hoboken, NJ) (1911).

Graa, Albert. *Polyglot Pharmaceutical Lexicon* (NYC) (1904) (reprinted in part in Basle, 1924).

Gray's Manual of Botany (Fernald, M. L.) (NYC) (ed. 8, 1950).

Gunther, E. *Ethnobotany of Western Washington.* 1973 (1945).

**Hagers Handbuch der Pharmazeutischen Praxis,* ed. 4; vol. I-VIII (bound in 11). 1980.

Hedges, H. T. *A Polyglot Index of All the Principal Articles in the Materia Medica in Latin, English, French, German, Swedish, Norwegian, and Danish* (Chicago) (1884).
Hedrick, U. P. *Sturtevant's Notes on Edible Plants* (Albany, NY) (1919).
Hegnauer, R. *Chemotaxonomie der Pflanzen* (6 v.) (1973).
Hellbusch, E. *Fachwörterbuch f. den Chemikalienhandel* (Berlin) (1921).
Heraud, A. *Dictionnaire des Plantes Médicinales* (Paris) (ed. 7, 1949).
Hill, Albert F. *Economic Botany* (NYC) (ed. 1, 1937) (ed. 2, 1951).
Hobbs, C. E. *Botanical Handbook of Most of the Crude Vegetable Drugs* (1876).
Hooker and Jackson. *Index Kewensis (with Supplements)* (Oxford, England) (1895-date).
*Hoppe, H. A. *Europaeische Drogen*. 2 v. (1948-1951).
Howes, F. N. *Vegetable Gums and Resins* (Waltham) (1949).
Huysse, A. C. *Alphabetische Lijst van de Verschiedende Benamingen Onger Geneesmiddelen* (Leiden) (1943).
Ibañez, Manuel. *Diccionario de Sinonimos Quimicos, Farmacéuticos, y Botanicos Usada en Farmácia* (Puebla) (1942).
Jackson, B. D. *Glossary of Botanic Terms* (London) (ed. 4, 1928) (+ supplement).
Jaeger, Edmund C. *A Source-Book of Biological Names and Terms* (Springfield, Ill) (ed. 1, 1949).
Jain, S. K. & DeFillipps, R. A. *Medicinal Plants of India*. 2 vol.; 1991.
Karel, L. and E. S. Roach: *A Dictionary of Antibiosis* (NYC) (1951).
*Karrer, W. *Konstitution und Vorkommen der Organischen Pflanzenstoffe* (excl. Alkaloide) (ed. 2, 1976). Also *Ergaenzungsband I*, 1977.
*Kartesz J. T. *A Synonymized Checklist of the Vascular Flora of the US, Canada, and Greenland* (ed. 2, 2 v.) (1994) (Timber Press, Portland, Ore).
Klemperer, G. and E. Rost. *Handbuch der Allgemeinen U. Speziellen Arzneiverordnungslehre* (ed. 15, 1929).
Koenig, J. K. and Frerich. *Warenlexikon* (Brunswick) (ed. 12, 1911).
Kokwaro, J. O. *East African Plants*.
Kunkel, G. *Plants for Human Consumption*. 1984.
Laurent, E. *Lexicum Medicum Polyglottum* (Paris) (1902).
Lerch, J. Zd. *et al. Synonyma, Latinska, Ceska, Polská i Nemeeká* [Latin, Czech, Polish, and German Synonyms] (Prague) (1890).
Littré, E. and C. Robin. *Dictionnaire de Médecine, de Chirurgie, de Pharmacie* (Paris) (ed. 13, 1873).
Lyons, A. B. *Plant Names, Scientific and Popular* (Detroit) (ed. 1, 1900; ed. 2, 1907).
Macmillan, H. F. *Tropical Planting and Gardening* (London) (ed. 5, 1949).
*Mabberley, D. J. *The Plant-Book, a Portable Dictionary of the Higher Plants*. Cambridge Univ. Press, 1987. 1989.
Madaus, G. *Lehrbuch der Biologischen Heilmittel*. (3 v.) (1938) (1979 reprint).
Mantell, C. L. *The Water-Soluble Gums* (NYC) (1947).
Martindale's Extra Pharmacopoeia (ed. 30) (London) (1993).
Martinez, M. *Las Plantas Medicinales de México*. ed. 2; 1939.
Mayer, A. W. *Chemisches Fachwoerterbuch* (2 v.).
Mayne, R. G. *An Expository Lexicon of the Terms . . . in Medical and General Science* (London) (1860) ca. XII (1996).
**The Merck Index: An Encyclopedia of Chemicals, Drugs, and Biologicals* (ed. XI) (Rahway, NDak) 2335 pp. (1989).
Meyer, Hans. *Buch der Holznamen* (Hannover) (1933).
*Morton, J. F. *Atlas of Medicinal Plants of Middle America: Bahamas to Yucatan*. c. 1981.
Morton, J. F. *Folk Remedies of the Low Country*. 1974.
Nemnich, P. A. *Allgemeines Polyglottenlexicon der Naturgeschichte* (Hamburg) (1793-8).
Nickell, J. M. *Botanical Ready Reference* (Chicago) (1880; 1911).
Northrop, S. A. *Glossary of Scientific Names* (c. 1949).
Ohwi, J. *Flora of Japan* (1965) (Smithsonian).
Parham, J. W. *Plants of the Fiji Islands*. 1964.
*Penso, G. *Index Plantarum Medicinalium Totius Mundi Eorumque Synonymorum*. D. E. M. F. Milano; 1984.
*Perry, L. M. *Medicinal Plants of East and Southeast Asia: Attributed Properties and Uses*. MIT Press (1980).
Peyer, A. *Pars Pro Toto: Brevarium Medicum Internationale* (1950) (Abbreviations).
Pollock, Allan. *A Botanical Index to All the Medicinal Plants* (NYC) (1872).
Quisumbinq, E. *Medicinal Plants of the Philippines*. 1951.
Ramstad, E. *Modern Pharmacognosy*. 1959.
Rangel, M. *Diccionario de Sinonimos Quimicofarmaceuticos* (1950).
Record, S. and R. W. Hess. *Timbers of the New World* (New Haven) (1943).
Rehder, Alfred. *Manual of Cultivated Trees and Shrubs Hardy in NA*. ed. 2 (NYC) (1940).
Rolland, Eugène. *Flore Populaire* (11 v.) (Paris) (1896-1914). (Perhaps the most complete listing of vernacular names ever published; but work never was completed.)
Rosenberg, Hugo. *Pharmacompendium* (Sp transl., Buenos Aires) (1925).
Rousseau, C. *Poliglota (Farmacia) Vademecum de Internacia Farmacio* (Paris) (1911) (in Esperanto, with 9 other languages).

Rudolphy, John. *Pharmaceutical Directory of All the Crude Dugs Now in General Use* (1866, 1877, 1910, etc.) (title varies).

Schindler, H. & Frank H. *Tiere in Pharmazie und Medizin.* 1961.

Schlomann, A. *Hoyer-Kreuter Technological Dictionary* (ed. 6, 1932) (in 3 v.) (Berlin).

Smith, John. *Dictionary of Popular Names of the Plants ... in ... Economy* (London) (1882).

Standley, P. C. "Flora of the Panama Canal Zone," *Contr. U.S. Nat. Herbar.* (v. 27) (1928).

*Standley, P. C. *Trees & Shrubs of Mexico* (2 v.) (-1926).

Stanley, T. D. & Ross, E. M. *Flora of Southeastern Queensland* (Australia) (3 v.) (1989).

Stapf. *Index Londinensis* (an extensive listing of illustrations of ferns and flowering plants) (Oxford) (1929-1941).

Stedman's Medical Dictionary, ed. 17 (Baltimore) (1949).

Steinmetz, E. F. *Codex Vegetabilis* (Amsterdam) (1947).

Sudworth, G. B. *Checklist of the Forest Trees of the US*, USDA. Misc. Circ. 72 (1927).

Sweringen, H. V. *Pharmaceutical Lexicon* (Philadelphia) (1873, 1882).

Tavera, T. H. Pardo de. *Medicinal Plants of the Philippines* (Philadelphia) (1901).

Taylor, Norman. *Encyclopedia of Gardening* (ed. 2, 1948).

Trease, G. E. & W. C. Evans. *Pharmacognosy* (Baltimore) (ed. 10, 1972).

*Tschirch, Alexander. *Handbuch der Pharmakognosie* (Leipzig). ed. 1, 2 (1908-*c.* 1940).

Tyler, V. E., Brady, L. R. & Robbers, J. E. *Pharmacognosy* (ed. 8, 1981).

Ulrich, Dr. William. *Internationales Wörterbuch der Pfanzennamen* (ed. 2, 1875).

*Uphof. J. C. T. *Dictionary of Economic Plants* (ed. 2, 1968).

Vanstone, J. H. *Dictionary of the World's Commercial Products* (London) (1930).

Various authors. *Flore de la Nouvelle-Calédonie et Dependances.* 1967-date.

*Watt, George. *Dictionary of the Economic Products of India* (10 v.) (1889-1896) (In course of revision under title *The Wealth of India*, New Delhi; ed. 1, [11 vol.] + 2 suppl.) (-1976) ed., 2 v. 1-2- (1985-8).

Watt, S. M. & Breyer, Branwijk, M. G. *The Medicinal & Poisonous Plants of Southern & Eastern Africa.* 1962.

Webster's New International Dictionary (Springfield, Mass) (1939).

Wehmer, C. *Die Pfanzenstoffe*, ed. 2 (2 v.) and *Ergaenzungsband* (1929-1935).

*Willis, J. C. *Dictionary of the Flowering Plants and Ferns* (Cambridge). ed. 6 (1931, 1951).

Wittstein, G. C. *Etymologisch-Botanisches Handwörterbuch* (1852) (Carl Junge-Ansbach). Later republ. as *Handwörterbuch der Pharmakognosie der Pflanzen* (1882).

*World Health Organization (WHO) (Geneva). *The Selection of Essential Drugs* (209). 1979.

Youngken, H. W. *Text Book of Pharmacognosy* (ed. 6, 1950).

*Zeven, A. C. & Zhukowsky, P. M. *Dictionary of Cultivated Plants and Their Centers of Diversity.* (1975).

Zimmerman, C. T. and I. Lavine. *Handbook of Material Trade Names* (Dover, NH) (1953).

APPENDIX C

LIST OF IMPORTANT SERIALS IN THE FIELD OF PHARMACOGNOSY AND RELATED AREAS

(including both continuing and discontinued titles)

It is extremely difficult to draw up a list of serials in the field of pharmacognosy that shall be neither too restricted on the one hand, nor too broadly extended on the other. This comes from the fact that pharmacognosy cuts across so many fields of divergent interest, where nevertheless it seems that all clamor for attention. Even within the field of pharmacognosy proper, there is much variety of specialization: some pharmacognosists are particularly interested in morphology, others in histology and microscopy, some again specialize in phytochemistry, others genetics; individuals sometimes go into cultivation and commerce, and so on.

To make a list of reasonable brevity, it was decided *not* to include here serials primarily devoted to the following fields, unless there is known to be a considerable content of pharmacognosy proper in the particular organ: general science, pharmacy, pharmaceutical chemistry, biology, botany, zoology, medicine, endocrinology.

The data presented are as follows: **Title** (in bold face), with *Subtitle* in italics; Publishing or sponsoring organization or firm; Place of publication; Dates published (range from first year to last; if known to be still current: "to date"); Miscellaneous information (contents, sponsoring groups, mergings, etc.). Unless specially noted otherwise, the interval between publication dates of a serial is understood to be monthly, and one year between volumes. *Sequential titles* are in italics.

The following catalog includes a number of the more important titles included in two lists of the compiler published earlier[1]. There are in addition many new titles not found in these lists. However, the majority of the 592 titles (or slightly fewer since some are duplicated by cross-reference) appearing in the second list cited are not repeated here.

[1] 1) Periodicals pertaining to pharmacognosy. *Amer. J. Phar. Educ.*, April, 1943. 2) Serials pertaining to pharmacognosy and phamacology. Check lists (S. B. Penick & Company, 50 Church St., New York 7, NY, 1944).

Only the titles of serials which make a significant contribution in one way or another to the field are listed here.

Abbreviations Used in the Listing

Am	American
Ar	Archiv
A. R.	Annual Report(s)
Ber	Bericht(e)
Bul	Bulletin(s)
C. R.	Compte(s) Rendu(s)
J	Journal
Jb	Jahrbuch (Jahrbuecher)
Jber	Jahresbericht(e)
Mitt	Mitt(h)eilungen
Phar	Pharmaceutical
Proc	Proceedings
Rep	Report(s)
S. B.	Sitzungsbericht(e)
Tr.	Transactions
Wschr	Wochenschrift
Z	Zeitschrift
Zbl	Zentralblatt
Ztg	Zeitung

Acta Phytochimica (Iwata Institute of Plant Biochemistry). Tokyo (1922).

Agricoltura Coloniale (Istituto Agricolo Coloniale Italiano), Firenze (1907-1944). Cont. as *Rivista de Agr. Subt. (qv.).*

Agricultural Ledger, The. Calcutta (1892-1912). Very fertile source of data on drug plants and products of India, each number being a monograph on some drug or other economic plant. Coll. indices: 1892-9; 1900-5; 1906-1911/12.

Am J. of Pharmacy (& *The Sciences Supporting Public Health*) (Philadelphia College of Pharmacy and Science). Philadelphia, PA (1825-date).

Am. Perfumer and Essential Oil Review. NYC (1906-date). Title varies: *Am. Perfumer, Am. Perfumes, Cosmetics, Toilet Preparations.*

Angewandte Botanik; Z. fuer Erforschung der Nutzpflanzen (Vereinigung f. Ang. Bot.). Berlin. (N.S. 1919-date).

Annales de la Société Belge de Microscopie. Bruxelles (1874-1907) (*Bulletin; Mémoires*).

Annales des Falsifications. Paris (1908-date); cont. of *Rev. Intern. des Falsif.*

Annales des Sciences Naturelles, Série **Botanique.** Paris (1834-date).

Annales du Musée Colonial de Marseille (1907-date).

A. R. on Essential Oils, Synthetic Perfumes, etc. Published by Schimmel and Company, Aktiengesellschaft, Miltitz bei Leipzig (1873-date) (also Fr and Ge editions). Title varies; sometimes semi-annual.

Antibiotics (Washington Institute of Medicine). Washington, DC (1951-date). (Ed. Henry Welch.)

Anzeiger für Schädlingskunde . . . Berlin (1925-date).

Apotheker-Ztg. (Deutsch. Apothekerverein). Berlin (1886-1945; 1950-1953) then fused with *Pharm. Zeitung.* Other titles: *Deutsche Ap.-Ztg. Standesztg. Deutscher Apotheker.*

Ar. der Pharmazie (u. Berichte der deutschen pharmazeutischen Gesellschaft). Berlin (1821-date).

Atti del (Primo) Congresso Nazionale sugli Antibiotici. Pavia (1949-date).

Australasian J. of Pharmacy, The. Melbourne (1885-date).

Badger Pharmacist, The (Univ. of Wisconsin). Madison (1930-date).

Bericht von Schimmel & Co. A. G. (usw.), Miltitz bei Lepzig (*cf. A. R. on Essential Oils. . .*) (index in this Ge, not in Eng edn.).

Bericht von Variochem VVB. Schimmel, Miltitz bei Leipzig, ueber Aeth. Oele, Riechstoffe usw. (-1948-); similar to prec. item but with more complete bibliographical portion.

Ber. ueber die pharmakognostische Literatur aller Laender (Deutsche Pharmazeutische Gesellschaft). Berlin (1896-1908/09). In *Justs Bot. Jber.*; also separately. Abstracts of the literature.

Bibliographical Contributions from the Lloyd Library (Lloyd Brothers, Inc.). Cincinnati (1911-18).

Bibliography of Agriculture (US Dept. of Agriculture Library), Washington, DC (1942-date). Cont. of *Plant Science Literature* (indexed listing).

Bibliography of Forestry and Forest Products (FAO, UNO). Washington, DC, Roma. Ann. (1948-date).

Biologia, a Monthly News-letter; supplement to *Chronica Botanica,* Waltham, MA (1947-date).

Biological Abstracts. Philadelphia (1927-date). This and *Chémical Abstracts* are the most important abstracting organs. Ann. indices only (BA BIOSIS).

Bollettino Chimico-Farmaceutico, Il (Societa farmaceutica). Milano (1861-date).

Botanische Litteratur der Zelle. Jber. ueber die Fortschritte der Anatomie u. Physiologie (1872-1907?).

Botanisches Centralblatt (International Assn. of Botanists). Jena (1800-1945). Abstract journal; earlier vols. cont. only lists of titles without abstracts. Coll. index for v. 1-60.

Botany-Current Literature (US Dept. Agr.). Washington, DC; cont. as *Plant Science Literature.* An index listing of the literature.

British & Colonial Druggist, The. London (1884-1915); cont as foll.

Brit. & Col. Pharmacist. London (1916-date).

British Chemical (and Physiological) Abstracts. London. Important abstract organ (1926-date). Cont. as *British Abstracts.* Sections: A: Pure Chem.; B: Applied Chem.; C: Analysis and Apparatus.

Bul. of Miscellaneous Information. Royal Botanic Gardens, Kew, London (1896-1954). General Index for 1919-1928. Pharmaco-botanically important. Often alluded to erroneously as "Bulletin of (Royal) Kew Gardens," "Kew Bulletin," etc.

Bul. of the Lloyd Library and Museum of Botany, Pharmacy, and Materia Medica. Cincinnati, (1934-date): Series: Reproduction, Botany, Mycocological, Entomological, Pharmacy, Historical, etc.

Bul. of the National Formulary Committee. Washington, DC (1930-1951). Cont. as *Drug Standards.*

Bul. of the Shanghai Science Institute, Suajgua (-1929-1937). Incl. pharmacog.

Caesar und Loretz in Halle a/S.: Preis-verzeichnis. . . Special-handlung f. veget. Drogen im ganzen u. bearbeiteten Zustande, Halle (1897-1906).

Cellule, La. J. of original investigations in cytology, biology, and general histology, Louvain (1884-date). Coll. index for v. 1-40.

Chemical Abstracts (Am. Chem. Soc.). Columbus (1906-date). Perhaps the most complete abstract organ published today. Coll. indices published.

Chemicals and Drugs (US Dept. Commerce). Later cont. as **Chemicals, Drugs, and Pharmaceuticals** (irr.). Washington, DC (1917-date).

Chemisches Zbl. Berlin. Very important abstract organ. (1830-date).

Chemist and Druggist. London (Eng) (1859-date).

Chemurgic Digest, The (Natl. Farm Chemurgic Council). Columbus, NYC (1942-date). Semi-mo., later mo.

Chemurgic Papers (Natl. Farm Chemurgic Council), Columbus, NYC (-1941-date) (mimeo.) (serially numbered). Series of topical papers.

Chronica Botanica. Waltham, MA (1935-1941). Earlier primarily a new organ in botany, then each volume was a series of monographs on botanical subjects.

Ciba Symposia (Ciba). Summit, NJ (1939-fall, 1951); actually series of monographs.

Circulars—pub. by many agricultural experiment stations of the various states, USDA, and of foreign countries.

Colonial Plant and Animal Products (Colonial Office). London (1950-1956). Supersedes *Bul. of the Imperial Institute* (1903-1948) —> *Tropical Science.*

Commercial Fisheries Abstracts. (Fish &) Wildlife Service, US Dept. Interior), Washington, DC (1948-date).

Commercial Fisheries Review. (Fish &) Wildlife Service, USDI), Washington, DC (1939-date). In this and prec. considerable data on products of whale, shark, cod, seaweeds, etc.

C. R. du . . . Congrès International de Pharmacie. 1) Braunschweig, 1854 2) Paris, 1867 3) Wien, 1869 4) St. Petersburg, 1874 5) London, 1881 6) Bruxelles, 1885 7) Chicago, 1893 8) Bruxelles, 1897 9) Paris, 1900 10) Bruxelles, 1919 11) The Hague, 1913 (considerable variation in titles).

Congrès International des Plantes Médicinales et plantes à essences, Rapports. 1) Wien, 1927 2) Budapest, 1928 3) Venezia, Padua, Vicinaza, 1929 4) Paris, 1931 5) Bruxelles, 1925 6) Prag, 1936. Versions also in Ge, It, etc.

Current List of Medical Literature (Armed Forces Medical Library, formerly Army Medical Library). Washington, DC (1932-date). An indexed list.

Deutsche Apotheker-Ztg. Cont. of Apoth.-Ztg. (*qv.).* In 1943 became a monthly, incorporating Süddeut. Ap.-Ztg., Wiener Pharm. Wochenschrift, Die Krankenhaus Apotheke. Discont. II-15-45.

Deutsche Heilpflanze, Die (Reichsverband der Deutschen Heil- u. Gewürzpflanzenanbauer). Stollberg i/Sa. (1934-c. 1949). The partial extract of this published as a supplement (Beilage) to the *Deutsche Apotheker-Ztg.* has frequently been confused with the original full work.

Deutsche Parfümerie-Ztg., Die. Leipzig; Heidelberg (1914-1943). Reappeared as Parfümerie u. Kosmetik (1949-94).

Digest of Comments on the Pharmacopoeis of the United States and on the National Formulary (Hygienic Laboratory Bul. of the US Public Health Service). Washington, DC (1905-1922). Useful abstracting of data on official compendia, products, prepns., processes, etc. Annual. Articles arranged alphabet. by OLT.

Drug and Cosmetic Industry, The. Pittsfield, MA and NYC (1932-date) (cont. of *Weekly Drug Markets* [1914-1916], *Drug and Chemical Mts.* [1916-1926], and *Drug Mts. [*1926-1932]).

Drug Standards (Committee on National Formulary). Washington, DC (1951-1960); cont. of *Bul. of the National Formulary Committee.*

Drug Topics. NYC (1883-date) (monthly, then weekly, now semi-mo.). The chief pharmacy news organ in US; mostly distributed through wholesalers to drug retailers.

Drug Trade News. NYC (1925-1972) (monthly, then semi-mo.). News organ. of drug mfg. indy.

Drug Treatises (Lloyd Bros., Inc.). Cincinnati (Nos. 1-25, ca. 1904-1911).

Economic Geography. Worcester, MA (etc.) (1925-date).

Excerpta Botanica Sectio B (Stuttgart) (1959-1996).

Farmacista Italiano, I (Sindicato Nazionale Fascista dei Farmacisti). Roma (ser. II, anno I, 1933-1936); cont. of *Scienza del Farm.*

Farmacognosia. Anales del Instituto José Celestino Mutis, Madrid (1942-1967 ca.); semi-ann.

Fitoterapia. Rivista di Studi ed Applicazioni delle Piante Medicinali (Dott. Inverni e Della Beffa S.p.A.). Milano (1929-date); tri-mestral.

Folia Pharmaceutica (Malik Zafir, Cagaloglu Yokusa). Basak ap Istanbul (Turkey) (1949-69).

Food. . ., with which is included *The Canning and Food Trades J.,* London (Eng) (1932-date).

Food, Drug, Cosmetic Reports (also *Supplements*). *Drug and Cosmetic Edition.* Washington, DC (1939-date); mimeo.

Forestry Abstracts. Quarterly. A part of this is issued and bound separately under the title *Forest Products and Utilization* (constitutes parts 3, 7, 8). Oxford (1939-1954).

Fortschritte der Botanik *unter Zusammenarbeit mit mehreren Fachgenossen.* Berlin (1931 (1932)-1938 (1939)) *et seqq.* 1954; ann.; covers morphology, taxonomy, physiology, ecology, etc. (older title: 1879-1886).

Fortschritte der Chemie organischer Naturstoffe. Berlin, later Wien (1938-1979); ann.

Galenica—Bul.: *Bul. Galenica* Berr. 1933-date.

Geographical societies, clubs, etc.—Bul., J., and other serial publns., ex. *National Geographic Mag.* (1889-date); *J. of the Manchester Geographical Society* (1885-date).

Handelsber. von Gehe and Co. Dresden (1886-1903-); sometimes publ. in *Pharm. Centralhalle.*

Heil-und Gewürz-Pflanzen (Deutsche Hortus-Gesellschaft). München (1917-1939).

Herb Grower Magazine. Falls Village, CT (1947-date).

Herb J. New Rochelle, NY+ (1936-39).

Herba (Hungary) —> *Herba Hungarica,* Budapest (1962-date).

Herba Polonica. Poznan, 1962-date.

Herbarist (Herb Soc. of America). Boston (1935-date).

Hippokrates. Stuttgart (form. Leipzig) (1928-date); semi-mo.; attempts a synthesis of various schools of medicine, incl. allopathy, homeopathy, psychotherapy, *Naturheilkunde, Konstitutionstherapie,* and esp. the use of herbal remedies.

Index Medicus; cont. as *Quarterly Cumulative Index Medicus.*

Indian J. of Pharmacy, The (Dept. of Pharmaceutics, Benares Hindu Univ.). Benares (1939-date).

Insect Control Committee, Abstract Bul. (Coordination Center, NRC). Washington, DC (-1941-date).

International Perfumer, The. *A review of progress and research.* East Molesey, Surrey (1950-1974); quarterly; cont. as *Cos. World News.*

Jb. Dr. Madaus—Die Arzneipflanzen in der modernen Heilkunde. Radebeul/Dresden.

Jber. des Instituts für angewandte Botanik. Hamburg (1912-1938-).

J. de Pharmacie et de Chimie. Paris (1809-date).

J. of Allergy. St. Louis (1930-date).

J. of Applied Microscopy (and Laboratory Methods) (Bausch and Lomb). Rochester (1898-1903).

J. of Economic Entomology. Menasha (1908-date).

J. Natural Products (v. Lloydia).

J. of Practical Pharmacy, The (1950-date); nearly entirely in Japanese.

J. of Technical Methods and Bul. of the International Assn. of Medical Museums. Toronto (1925-1939-?).

J. of the Am. Phar. Association, The. Washington, DC (1912-date); since 1940 publ. in separate parts as *Scientific Edn.* and *Practical Phar. Edn.*; with later changes.

J. of the Bombay Society of Natural History. Bombay (1892-date).

J. of the Chinese Phar. Assn. Shanghai (1939-1940-); partly in Occidental languages.

J. of the Pharmaceutical Association of Siam (Thailand), The. Bangkok (1st ser. 1930-; 2nd ser. 1936-; 3d ser. 1947-1958?); in Eng and Siamese.

J. of the Shanghai Science Institute, The. Shanghai (*c.* 1933-1945?); Sect. III incl. Biology, with Pharmacognosy, etc.

Just's Botanischer Jber. Berlin, Leipzig (1873-1939); v. 1-10 as *Botanischer Jber. Hrsg. v. Just;* authors' index from 1873; subject index beginning in 1927. Very thorough abstract coverage, altho always several years behind.

"Kew Bul.": *v. Bul. of Miscellaneous Information.*

Kleinwelt. Muenchen (1909-1915), then merged to form *Mikrokosmos.*

Lingnan Science J. (Lingnan Univ.). Canton (1922-1951); formerly *Lingnaam Agricultural Review.* Incl. much on Chinese crude drugs.

Lloydia (Lloyd Library). Cincinnati (1938-1978); quarterly. *J. Natural Products* (1979-date).

Manufacturing Chemist and Pharmaceutical and Fine Chemical Trade J. London (1930-1963).

Manufacturing Perfumer, London (1936-1939 when merged to form prec.).

Materiae Vegetabiles (organ. of The International Commission for Plant Raw Materials), Den Haag (1952-1957). Very important journal on prime materials —> *Qualitiàs Planatra & Materiae Vegetabiles* (1958-87).

Médecine Végétale. Bruxelles (-1901-1911).

Merck's Annual Report *of recent advances in pharmaceutical chemistry and therapeutics* (E. Merck). Darmstadt (1887-date); ann.; many issues were publ. only in Ge under the title **E. Merck's Jber.** Coll. index. for v. 1-25.

Mikrochemie (*vereinigt mit Mikrochimica Acta).* Wien (1923-1938).

Mikrochimica Acta. Wien (1937-1938); then merged with *Mikrochemie.* (1938-53) (-1994)

Mikrokosmos (Deutsche Mikrologische Gesellschaft). Stuttgart, Leipzig (1907-date); combined with *Kleinwelt* and Z. *fuer angewandte Mikroskopie u. klinische Chemie* (1896-1910).

Mycological Notes and **Mycological Writings** (serial publications of the Lloyd Library). Cincinnati.

New Commercial Plants (T. Christy and Co.). London (1878-1885); cont. as foll.

New Commercial Plants and Drugs (1886-1889-?); compiled by Thomas Christy, FLS, etc.; each illustrated number with index. An editorial note indicates an intention to continue this serial by issuance of monthly leaflets of 4 to 8 pp. each for eventual collation into ann. volumes.

Oil, Paint, and Drug Reporter. NYC (1871-1971); weekly; mostly news and comml. information on raw materials; most complete comml. coverage for crude drugs; price data cont. as Chem. Mark. Rep.

Oele, Fette, Wachse. Wien; name changed to **Oele, Fette, Wachse, Seife, Kosmetik.** Wien (1935-1938), then merged with **Fette und Seifen** (1894-date).

Pacific Pharmacist, The. San Francisco (1907-1919).

Pallas, Waltham (-1947); "A review of progress in international relations and cooperation in science"; spec. Ann. number of *Chronica Botanica.*

Parfümerie und Kosmetik. Heidelberg (1949-date).

Parfumerie Moderne, La. Paris (1908-date).

Parfums de France. Grasse (1923-1938).

Peking Natural History Bul. Peiping (1926-1934-date); quarterly.

Pharmaceutical Archives. Milwaukee (1898-1903); Madison (1936-*c.* 1945).

Pharmaceutical Review. Milwaukee (1896-1908); cont. of *Pharmazeutische Rundschau* (1883-1895).

Pharmaceutisch Weekblad voor Nederland. Amsterdam (1864-date). Mostly in Dutch language.

Pharmazeutische Ztg. Bunzlau, later Berlin (1855-1938; resumed 1947-date).

Pharmazeutische Ztg. (Verband deutscher ang. Apotheker). Reichenberg (Czecho-Slovakia) (1919-1937-1954); weekly or biweekly (apparently not available in US).

Pharmazeutische Zentralhalle fuer Deutschland. Dresden (1860-date).

Phamazie, Die. Berlin, 1946-date. Also **Beihefte.**

Philippine J. of Science, The (Bureau of Science). Manila (1906-date). Sec. *B,* Trop. Med.; Sect. *C,* Botany (E. D. Merrill, ed.) (sectioning ended in 1918).

Plant Science Liteature (U.S. Dept. Agr.). Washington, DC; (1934-1942); cont. of *Botany-Current Literature;* cont. as *Bibliography of Agriculture.* Indexed listing.

Planta Medica. Stuttgart (1953-date).

Pratical Microscopy. NYC (1934-1936).

Proceedings of the Central Indigenous Drugs Committee. India (-1901-1916-).

Profumi Italica. San Remo (1923-1927).

Quarterly Cumulative Index Medicus (AMA). Chicago (1927-date). Indexed listing.

Rep. of the Indigenous Drugs Committee of Madras (-1924-).

Rep. of the Phar. Experiment Station (Univ. of Wisconsin). Madison (1914-1918).

Rep. of the Proc. of the . . . International Phar. Congress. Eng version of title, *C. R. du . . . Congrés (q.v.);* there is also a Ge version.

Revista da Flora Medicinal. Rio de Janeiro (1929-date).

Revista Farmacéutica. Buenos Aires (1859-date).

Revue de Botanique Appliquée et d'Agriculture Tropicale (Laboratoire d'Agronomie Coloniale). Paris (1921-date).

Revue de Phytothérapie (École Française de Phytothérapie). Neuilly-sur-Seine, Paris (1937-1938-).

Revue des Fermentations et des Industries Alimentaires. Bruxelles (1925-date).

Revue des Marques de la Parfumerie et de la Savonnerie, La. Paris (1923-1939).

Revue Internationale des Falsifications. Paris (1887-1908); cont. as *Annales des Falsifications....*

Rivista de Agricoltura Subtropicale e Tropicale. Firenze (1945-date).

Roure-Bertrand Fils, Scientific and Industrial Bul. Evreux) (-1909-1921); also Fr and Ge versions exist.

Royal Asiatic Society, various branches—J., Proc., Tr., as of Bengal (Calcutta), Great Britain and Ireland, Malayan, North Chinese, Ceylon, Bombay, Korea, Straits, etc.

Schweizerische Apotheker-Ztg. Zuerich (1863-date); articles in Fr and Ge mostly, some in It.

"Schimmel's Reports": common designation for *A. R. on Essential Oils. . .; Semi-Ann Rep. on Essential Oils. . . (qv.).*

Science (Am. Assn. for the Advancement of Science). Lancaster (1881-date).

Scientia Pharmaceutica. Wien (1930-date); supplement to *Pharm. Monatshefte.*

Scienza del Farmaco, La. Roma; 1933-5; cont. as *Farmacista Italiano.*

Seifensieder-Ztg. u. Revue ueber die Harz-, Fett- u. Oelindustrie. Augsburg (1874-date). During World War II, united with *Allgemeine Oel- u. Fett-Ztg.* and with *Z. Mineraloele.*

Semi-Annual Rep. on Essential Oils and Synthetic Perfumes, Schimmel and Co., A. G. Miltitz bei Leipzig (from 1873-1885, as *Ann. Rep.),* then *Semi-ann.* until 1915, since then *Ann.*; Ge and Fr edns.; coll. index every 5 yrs. since 1900.

Soap, Perfumery, and Cosmetics. London (1928-date). "S.P.C."; title varies.

Sovetskie Subtropiki *(Soviet Subtropics).* Sukkum (1936-?); cont. of *Subtropiki.*

Soviet Plant Industry Record. Moskva (1937-?).

Special Crops. Skaneateles, NY (1890-1936); was devoted mostly to interests of growers of ginseng, hydrastis, and other drug plants.

Subtropiki. Sukkum (1929-1930); cont. as *Sovetskie Subtropiki.*

Süddeutsche Apotheker-Ztg. Stuttgart (1860-1861) (1950); merged with *Apotheker-Ztg. c.* 1950. Beilage (Supplement) was also published; title changes.

Sunyatsenia. *J. of the Botanical Institute, Sun Yatsen Univ.,* Canton (1930-date). Good source for econ. botany.

Svensk Farmaceutisk Tidskrift. Stockholm (1897-date).

Taxonomic Index, The (Am. Soc. Plant Taxonomists), Lancaster (1938-date) (early vols. mimeo., currently printed & part of Brittonia).

Théses, École Supérieure de Pharmacie de Montpellier; Faculté de Pharmacie del'Université de Paris; etc.

Transactions of the Am. Microscopial Society. Menasha (etc.) (1880-date).

Tribuna Farmacéutica. Curitiba, Brasil (1932-date), Tropical Science, London (1959-date).

Turtox News (General Biological Supply House, Inc.). Chicago (1923-date).

United States Department of Agriculture: A. R.; Yearbook; Circulars; Farmers' Bul.; Leaflets; Miscellaneous Publns.; Technical Bul.; etc. (see "Index to Publns. of the USDA," 1901-1925; 1926-1930; 1931-1935; 1936-1940).

United States Department of Commerce: many serial publications of interest; ex. *Domestic Commerce,* Washington, DC; 3 issues per mo.

University of Wisconsin—Bul., General Series; Bul., Science Series; Circulars of Phar. Experiment Station (some are in *Bul. of General Series);* **(Laboratory) Guides;** etc.

Visnik Prikladnoi Botaniki (*Journal for Applied Botany*). Charkiv. (1930).

Vitamin und Hormone. Leipzig (1940-date).

World Trade Notes on Chemicals and Allied Products (US Dept. of Commerce). Washington, DC (1928-date); weekly.

Z. des allgemeinen Oesterreichischen Apotheker-Vereins. Wien (1847-1921) (First 16 vols. as *Oesterreichische Z. fuer Pharmazie*); merged with *Pharmazeutische Post* (1868-); latter merged (1938) with *Wiener Pharmazeutische Wochenschrift* (1868-1943).

Z. fuer das Landwirtschaftliche Versuchswesen in Deutsch-Oesterreich. Wien (1898-1925).

Z. fuer wissenschaftliche Mikroskopie und fuer Mikroskopische Technik. Leipzig (1884-1936-date).

APPENDIX C-1

LIST OF PERIODICALS DEVOTED TO CERTAIN ECONOMIC PLANTS OR SPECIAL PLANT PRODUCTS

(including both continuing and discontinued titles)

Note: Some of the following are essentially commercial organs.

I. SUGAR

Sugar Reference Book and Directory (Sugar, NYC) (1931-1944-).

American Sugar Industry and Beet Sugar Gazette. Later title changed to **Sugar-Azucar**. NYC (1919-date). Cont. as *Int. Sugar J.*

International Sugar J. Manchester, London, Eng, 1899-date.

(Louisiana) Planter and Sugar Manufacturer. Later, title changed to **Sugar Manufacturer**, New Orleans (US) (1915-date).

Sugar Bul., The (New Orleans [1922-date]).

Sugar Cane, The. Manchester, Eng, 1869-98. Cont. as *Int. Sugar J.*

Sugar J. New Orleans & Crowley, La (1938-date).

Gazeta Cukrownicza (Sugar Gazette). Warszawa (1893-1937-date) (re. sugar and s. beets).

Weekly Statistical Sugar Trade J. NYC (discont.).

Sugar Beet J., **The** (Farmers & Mfrs. Beet Sugar Assn.). Saginaw, Mich (1936-date).

Bul. du Trust Sucrier. Moskva (-1937-).

Mededeelingen—Proefstation voor de Java Suikerindustrie.

New Agriculture. New Falls, Ida, San Francisco (sugar beets only) (1917-date).

Beet Sugar Gazette. Chicago (discont.).

III. COTTON

Am Cotton Planter. Montgomery, Ala (US) (1853-61?) (discont.).

Algodão (*Brazil Cotton Review).* Rio de Janeiro (1934-).

Algodón y sus Industrias, El (Fed. Int. Algonera Delegacion Española). Barcelona (1932-1935-?).

Cotton & Cotton Oil News —> Cotton & Cotton Oil —> Cotton Gin & Oil Mill Press (-date).

Cotton Oil Press, The (1917-1935). —>Cotton.

III: RUBBER, NATURAL

Caoutchouc et la Gutta Percha, Le. Paris.

Archief voor de Rubbercultuur in Nederlandsch-Indië. Buitenzorg. (Bogar), 1917-41; 1948-.

India Rubber J., London —> *Rubber J & Internat. Planter.*

India Rubber Review, Akron, Ohio (discont.) —> *Tire & TBA Rev.*

India Rubber World. NYC (1889-date) (now **Rubber World**).

Rubber Age, NYC (1917-date).

Bul. du Syndicat des Planteurs de Caoutchouc de l'Indochine (1909-date).

A. R. Rubber Research Institute of Malay. Kuala Lumpur.

IV. COFFEE

Café de El Salvador, El (*Revista de la Asociación Cafetalera de El Salvador*). San Salvador (1930-date).

DNC: Revista do Depto. Nacional do Café. Brazil (1924-1943-).

Mysore Coffee Experiment Station: Bul., Circulars, etc.; Bangalore (India).

Revista Cafetalera de Guatemala (1945-54-?).

V. TEA

Archief voor de Theecultuur in Nederlandsch-Indië (Batavia) (1927-date) (continuation of **Thee**).

Tea and Coffee Trade J. NYC (1901-1984; 1985-date).

Verslagen, Gouvernements . . . (*v.* under cinchona below).

VI. CHOCOLATE

Archief voor de Cacao en Andere Kleine Cultures in Nederland Indië. Semerang (1927-1928).

Research (British Assn. of Research for Cocoa, Chocolate, etc.). London SW1 (1920-); irreg.

Abhandlungen ueber Kakao u Schokolade (Gordian). Hamburg (1919-1924-).

Bul. de l'Office International des Fabricants du Chocolat et du Cacao. Schaerbeck (Belgique).

Z. fuer Kakao, Zucker u. deren Erzeugnisse, Gordian, Hamburg (1895-1934-date?).

VII. CITRUS

California Citrograph. Riverside; Los Angeles (1915-date).
Citrus. Messina, Italy. (1913-).
Citrus. Tampa (Fla) (1938-1941); cont. as *Citrus & Veg. Crops* (1942-date).
Citrus Industry, The. Bartow, Fla (1915-date).
Citrus Leaves. Redlands, Los Angeles, Calif (1921-57).
Papers (of the) Citrus Experiment Station, Calif. Univ Calif, Berkeley.

VIII. TOBACCO

Tabaksplant, De (Ver. van Tabaksfabrikanten in Nederland). Kuilenburg (1873-date).
Tobacco. NYC (1886-date); weekly.
Tobacco Leaf, The. NYC (1865-date); weekly.
United States Tobacco J. NYC (1874-date).

IX OLIVES

Aceite de Oliva de España, El (Feder. Exp. Ol. Ol. Español.). Madrid (1929-1934-).
Fig and Olive J. Los Angeles; cont. of *Olive J.* (discont.) 1916-22 —>*Orchard & Farm.*
Importer (Olive Oil Assn.) (cont. as foll.) (1928-34).
Olive Oil. NYC (discont.).
L'**Olivicoltore.** Roma (1923-1940-) (cont. as **Olivicultura** (1946-date).
Proc. of the California State Convention of Olive Growers. Sacramento (1987-93).
Olivos. Madrid (1927-1934-date).

X. CINCHONA

Cinchona—Archief voor de Kina-cultuur *onder Redactie van het Proefstation voor Kina.* Bandoeng. Java (various sub-titles). (1924-date).
Jber. der Zimmerschen Chininfabrik in Frankfurt a/M.
Verslagen, Gouvernements Kina en **Thee Onderneming te Tjinjiroean (Pengalengan), Dept. van Landbouw** [*Rep. Government Cinchona and Tea Plantation at Tjinjiroean [Pengalengan] Agricultural Dept.].* Batavia.

XI. MISCELLANEOUS (Titles are self-explanatory)

Am Tung News. Poplarville, Picayabe, Miss (1949-1951) (discont.).
Tung World. Gulfport &c. (1946-date).
Spice Mill, The. NYC (1878-date) (covers spices, vanilla, cloves, tea, coffee).
Am. Bee J. Hamilton, Ill, Philadelphia (1861-62; 1866-date).
Naval Stores Review. Savannah; New Orleans (1890-1954; weekly).
Pine Institute of America: Technical Bul.; Abstracts, Chemical Section (1927-1931).
Soybean Digest, The. Hudson, Ia. (1940-date).
Wines and Vines. San Francisco (1919-date); cont. of *California Grape Grower.*
Wine Review, The (1933-1934). In 1934, merged with *Wine News* (discont.), merged with *Wines & Vines* (1950).
Bul. (of the) Mysore Sandal Spike Investigation Committee. Bangalore. ("Spike" is a serious disease of *Santalum album.*)
Ginseng Garden, Ginseng J. and Golden Seal Bull, The. Arrowsmith, Ill (1912-19?), Scranton, Pa (US) (discont.). (1902-66)?)
Rice J. New Orleans (1897-date).
Olivicultura (1946-date).

APPENDIX D

TERMS DESCRIBING PROPERTIES AND THERAPEUTIC USES OF DRUGS, PESTICIDES, SOME PATHOLOGIES (DISEASES, DISORDERS, SYMPTOMS), &C.

Therapeutics (< thérapeutique (Fr) < therapeutica (L): **Therapy:** the branch of medicine concerned with the use of remedial measures for diseases or disorders of the body; the art of healing.

abirritant: diminishing or relieving irritation; soothing.
abluent: detergent.
aborticide: expels & kills embryo (ex. lead salts).
abortient: abortifacient: abortive: produces expulsion of fetus in premature childbirth by causing uterine contractions (ex. $KMnO_4$).
absorbefacient: promoting absorption.
absorbent: promotes the taking up outside the body of fluids by sucking and/or other mechanical processes (ex. kaolin, starch, cellulose, chalk dressing).
abstergent: abstersive: cleansing.
acaricide: kills acari (mites), esp. the sp. causing itch (ex. sulfur).
accelerator: 1) agt. wh. speeds up function or process 2) catalyst.
acidifier (urinary): producing acid reaction of urine after oral administration (systemic or local); ex. NH_4C1.
acro-narcotic: combines acrid & narcotic (general nerve sedative) properties.
activator: renders another subst. active.
adaptogen: natural non-toxic product wh. increases non-specific resistance to adverse influences, such as stress; reducing stress or increasing body's ability to overcome stress; effective against some bacteria, viruses, & toxins; normalizes blood pressure & endocrine balance (ex. *Centella asiatica*); sometimes now equated with similar earlier term, alterative.
adjuvant: used to aid in the action (or administration) of another drug (ex. jalap to anthelmintics).
adrenergic: sympathomimetic.
adrenolytic: inhibits action of adrenergic nerves; (adrenal cortex inhibitor).
adsorbent: in solid form takes up a fluid on its surface in the molecular state and not by capillary attraction (which is absorption) (ex. activated charcoal).
adstringent: astringent.
aerosol: liquid with medication wh. is atomized into a fine spray for inhalation.
agglutinant: adhesive subst. wh. holds body parts together during healing.
ague: old term for mal.
alexipharmic: general antidote to poisons, esp. venoms, &c.; protective agt.
alexiteric = alexitive: same as prec.
alkali(ne): alkal(in)izer: neutralizes systemic acidity (*cf.* antacid).
alpha-beta-blockers: cause blockade of both alpha- and beta-adrenergic receptors.
alpha-blockers: agts. wh. combine with compd. to cause blockade of alpha-adrenergic receptors; ex. phentolamine.
alterative: presumed to alter (i.e., modify or affect) the course of a disease for the better; often by stimulating nutrition & repair & promoting the expulsion of waste matter; thus to improve morbid conditions by modifying metabolism, inducing a gradual return to normal function; usually given over long periods in convalescence & in treating chronic diseases (*cf.* discutient, resolvent); ex. organic (or plant): sarsaparilla, codliver oil & other vitamin materials; inorganic (or mineral): mercurials, As_2O_3, sulfur, iodides.
amara (L): bitters (NBP).
amethystic agents: drugs wh. reduce effects of alcohol; promotes "sobering up"; ex. ephedrine; aminophyllin; DOPA.
am(o)ebicide: kills amebae, esp. *Entamoeba histolytica* (ex. ipecac).
anabolic: stimulates constructive metabolism.
an(a)esthetic: numbs or paralyzes or destroys feeling or sensation temporarily, esp. of pain; types: a) general: affects whole system by bringing unconsciousness & (general) anesthesia (ex. $CHCl_3$, N_2O, Et_2O, EtBr) b) local: acts on specific parts of system (paralysis of sensory nerves of local area) (ex. cocaine, extreme cold [through ether spray]) c) spinal: actually form of local a.

induced by injection into spinal subarachnoid space (ex. Nupercaine) d) others: rectal & intravenous: used to induce general anesthesia; preoperative; refrigeration; etc.

analeptic: stimulates brain activity (medulla, midbrain) (ex. picrotoxin), especially to restore consciousness, &c.

analgesic: relieves or reduces perception of or sensitivity to pain by quieting (or depressing) nerve sensory centers, or by reducing conductivity of nerve fibers (without sensibly affecting other parts of body); types: a) general or systemic (ex. morphine, codeine, acetanilid, aspirin) b) local (ex. aconite; methyl salicylate; liniments).

analgetic: **analgic:** making insensitive to pain.

anaphodisiac: allays (lessens) or depresses (reduces) sexual desire (appetite) or ability (function, activity); acts by lessening excitability or depressing genital centers; ex.: cocaine, camphor, tobacco, belladonna, henbane, stramonium, bromides, opium, vegetable diet, purgation.

anaphoretic: stops or reduces perspiration.

anasarca: massive edema.

anastaltic: high astringent; stypic.

androgen: male hormone.

angina: sore throat, &c.

angiotonic: increases vascular tension.

an(h)idrotic: anhydrotic: checks (or prevents) or lessens (or reduces) sweating (or perspiration) (the opposite of diaphoretic); ex.: ergot, opium, belladonna, hyoscyamus, stramonium, zink salts, talcum, acids locally; local cold (*cf.* antihydrotic).

anodyne: v. analgesic (allays pain).

anoretic: reduces appetite (ex. amphetamine).

anorexiant: anorexic (agent): resulting in absence or diminution of appetite; ex. ephedrine.

anovulatory: anovular: prevents ovulation, hence contraceptive.

antagonist: opposes directly in one or all of its physiological actions; ex. morphine to atropine, muscarine to atropine, strychnine to barbiturates, and the reverse.

antalgesic: antalgic: analgesic.

antalkaline: checks over-alkalinity by acidification, etc.

anterotic: decreases sexual desire (ex. bromides): *cf.* anaphrodisiac.

an(t)gonorrheic: v. antigonorrheic.

anthelmintic: provides remedy or cure for worms, usually in the alimentary tract; which destroys (vermicide) or expels (vermifuge) worms inhabiting the intestinal canal; also *cf.* tenifuge, tenicide, etc.

an(t)hysteric: v. antihysteric.

ant(i)acid: tends to check or neutralize hyperacid conditions, in various parts of body, particularly stomach or urine: a) direct or local acion on stomach (ex. $CaCO_3$,$MgCO_3$, lime water, $NaHCO_3$) b) indirect or systemic action on urine (ex. sol. acetate, Na_2HPO_4).

antiadiposity: reduces body wt.

antiallergent: antiallergic; prevents or reduces allerice (hypersensitivity) reaction.

antiamebic: opposes amebae (a. dysentery).

antianemic: to remedy anemia.

antiaphrodisiac: v. anaphrodisiac.

antiarrhythmic: prevents cardiac irregularity of beat.

antiarthritic: provides a remedy against arthritis, gout, etc. (ex. cortisone).

antiasthmatic: palliative for asthma; ex. cubebs, lobelia, tripelennamine.

antibacterial: destroys or prevents growth of bacteria (bacteriostatic) by physical or chemical means (ex. clove oil).

antibechic: remedy for cough (ex. honey).

antibilious: remedy for "biliousness" (minor hepatic disorder).

antibiotics: substances or compds. prod. by Fungi or other organisms capable of inhibiting in small doses the life processes of microorganisms; conceived as basically a competitive force between fungi and bacteria; however, besides antibacterials, they include antimycotics, antivirals, antirickettsials, antiparasitics, &c.; they play a very important role in modern medicine & have saved countless lives since com. introduction. Chem. compn. is very varied: polypeptides; aminoglycosides, macrolides, oligosaccharides, nucleosides, sesquiterpense, polyenes, &c.

antiblastomycosal: opposes yeast-like organisms causing dis.

antiblenorrhagic: used against gonorrhea (ex. silver salts).

antibronchospasmodic: to reduce bronchial spasms.

anticatalyst: reduces or retards catalytic action.

anticholagog(ue): lessens secretion & decreases flow of bile from liver (ex. Sb, As, P prepns.).

anticholeric: to treat cholera.

anticholester(ol)emic: lowers serum lipids, esp. cholesterol.

anticholinergic: blocks parasympathetic nervous system; opposes cholinergic activity = parasympatholytic (ex. atropine).

anticoagulant: reduces or prevents blood coagulation; ex heparin; dicumarol.

anticonceptive: contraceptive.

anticonvulsant: anticonvulsive: checks convulsions.

antidepressant: stimulates mood of depressed person.

antidiarrhetic: tends to prevent diarrhea (ex. astringents, as tannin, nut-gall, gambir, chalk).

antidiuretic: prevents or reduces passage of urine, other than by reducing liquids perorally, sometimes by increasing perspiration.

antidoloritic: eases pain & inflammation.

antidontalgic: something u. for toothache.

antidote: counteracts poison (by chemical, physiological, or physical means); ex., starch for iodine, tannic acid for digitalis (gives insol. tannate).

antidysenteric: prevents or abates dysentery (ex. emetine).

antieczematic: remedies eczema.

antiemetic: allays or prevents tendency to be nauseated & to vomit (ex. bromides, cocaine, codeine, atrophine).
antienzymatic: retards activity of enzyme(s).
antifebrile: antipyretic.
antifeedant: pl. subst. which inhibits the normal feeding of insect.
antiferment: prevents fermentation.
antifilarial: against nematodes.
antifoam agent: surface active subst. u. to reduce surface tension of liquids (ex. soy oil).
antifungal: opposing fungus growth & reproduction.
antigalactagog(ue): antigalactic: decreases secretion & checks flow of milk in nursing mothers (ex. belladonna [Tr.], henbane, stramonium, atropine).
antihemolytic: prevents blood hemolysis (laking).
antihemorrhagic: prevents bleeding (ex. protamine sulfate).
antihidrotic: antihydrotic: v. anhidrotic.
antihistamine: antihistaminic: prevents or reduces physiological effects of histamine in a more or less direct fashion & not by simply counteracting the manifest effects of histamine on the body (ex. prophenpyridamine).
antihypercholester(ol)emic: agt. wh. reduces excessive levels of cholesterol in blood serum.
antihypertensive: reduces high blood pressure.
antihypnotic: opposes sleep.
antihysteric: relieves hysteria (nervous excitement in women & children chiefly) (ex. bromides, valerian).
anti-infective: v. antibiotic.
anti-inflammatory: opposes inflammation.
antileishmanial: opposes kalaz azar, trypsomiasis.
antilethargic: reduces drowsiness.
antileukemic: opposes leukemia.
antilithic: prevents formation or aids in solution of calculi (stones of various ducts of body, esp. urinary & to lesser extent biliary concretions) (ex. water, Li & other metallic salts).
antiluetic: used in treating syphilis (lues) (ex. penicillin, NaI, KI).
antilyssic: remedy for rabies (ex. rabies vaccine).
antilytic: inhibits lysis in destruction of cells.
antimalarial: remedy against malaria (prevents or cures) (ex. cinchona).
antimetabolite: mechanism combined with enzyme to prevent a metabolic process; often as immunosuppresant & antineoplastic (ex. aminopterin thioguanine, methotrexate, purinethol) (6-mercaptan purine).
antimethemoglubinemia: opposes methemoglobinemia.
antimetrorrhagic: reduces profuse uterine bleeding.
antimutagen: opposes mutagenic effects of some substances.
antimycotic: against fungal microorganisms, esp. the filamentous fungi (ex. undecylenic acid).
antinauseant: reduces or abolishes nausea (ex. phenobarbital; papain; $NaHCO_3$).
antineoplastic: opposes malignancies.
antinephritic: counteracts kidney inflammn.
antineuralgic: relieves neuralgic pain (ex. cocaine).
antineuritic: moderates nerve inflammn.
antiobesity: antiobesogenic: counteracts overweight state.
antiparasitic: removes or destroys animal & vegetable parasites (chiefly of the skin, like itch, ringworm, headlice) (ex. mercurials, I_2, larkspur, sulfur, sulfides, phenol—applied as lotions or ointments).
antiperiodic checks symptoms of periodic febrile disease; sometimes specif. antimalarial (ex. quinine in malaria).
antiperistaltic: reduces intestinal activity.
antiperspirant: anhidrotic.
antiphlogistic: reduces or allays inflammation (swelling, redness, heat), esp. in bronchial area; term becoming obsolete (ex. cold locally, aconite, clay poultice).
antipodagric: used in gout (ex. colchicum).
antipolycythemic: reduces number of red blood cells.
antiprotozoan: counteracts Protozoa (esp. pathogenic).
antipruritic: used to relieve itching (ex. phenol).
antipsoric: antipsoriatic: active against psoriasis.
antipsychotic: major tranquilizer.
antipyretic: reduces febrile temperatures of body (ex. quinine, antipyrine, aconite, camphor, acetanilid).
antipyrotic: to treat burns & "heartburn."
antirachitic: used in treating rickets (ex. vitamin D).
antirheumatic: relieves or prevents muscular rheumatism, incl. lumbago, gonorrheal r., &c. (ex. salicylates).
antirickettsial: agt. against rickettsial diseases.
antiscabetic: anitscabious: against the itch (mites).
antischistosomal: agents against flukes, trematodes, &c.
antiscorbutic: agent in scurvy (scorbutus, L.) (ex. citrus fruits, lime juice).
antiseborrheic: against dermal accumulations of fatty subst.
antiseptic: inhibits (or retards) or prevents the growth & reproduction or arrests the development of bacteria or other microorganisms causative of infection or other deleterious processes (ex. alcohol, $KMnO_4$); types: digestive, intestinal, ophthalmic, oral, genital, urinary, etc.
antiserum: that contg. antibody/antibodies.
antisialagog(ue): checks, diminishes, or prevents salivation (flow of saliva or buccal mucus) (ex. atropine, opium, tannic acid).
antisialic: checking salivation.
antisoporific: reduces sleepiness.
antispasmodic: allays, relaxes, or controls spasms, esp. of bronchi (ex. narcotics, valerian, bromides, $CHCl_3$, ether, tobacco).
antispastic: antispasmodic to skeletal muscles.
antisquamic: reduces or prevents scaliness of the skin.
antistrumatic: against goiter.

antisudorific: v. *anhidrotic.*
antisyphilitic: remedies syphilis (caused by protozoan, *Treponema pallidum*) (ex. mercurials, penicillin, bismuth compds., arsphenamines).
antitensive: hypotensive.
antitetanic: tends to relax tetanic muscle contractions (ex. curare).
antithermic: cools.
antithyroid: inhibits or depresses thyroid activity (ex. thiouracil).
antitoxic (serum): absorbs toxins or poisons generally.
antitreponemal: active against *Treponema.*
antitrichomonal: active against *Trichomonas.*
antitrypanosomal: active against trypanosomes.
antitubercular: antituberculotic: used for TB (ex. streptomycin).
antitussive: cough remedy (ex. cocillana); relieves cough in some way, esp. by interference.
antivenereal: 1) antisyphilitic 2) anaphrodisiac.
antivenin: antivenom.
antivinous: discouraging the use of wine or other alc. beverages; also reducing effects of alc. intoxication; reducing the desire for alcohol.
antivirotic: harmful to viruses & hence is used to treat virus infections (ex. oxytetracycline).
antixerophthalmic: eliminating xerophthalmia (use of vit. A).
antizymotic: prevents or arrests ordinary fermentative processes (ex. benzoid acid, creosote, antiseptics, excess of fermentative end-products).
aperient: mild laxative, causing no irritation (ex. manna, sulfur).
aperiodic: having no definite period.
aperitive: 1) stimulates appetite (*cf.* apertitif Fr) ex. bitters, Calumba 2) aperient.
aphrodisiac: Korean ginseng; rhinoceros horn(?); anise, gotu kola; licorice, mandrake, yohimbine, sarsaparilla, donqual; stimulates sex desire (appetite) or power (ability) (ex. cantharis, nux vomica, cannabis, blood; meat diet).
aromatic: with appetite-stimulating properties (spicy odor or/and taste) (ex. cinnamon, ginger, cardamom).
aromatic bitter: combines aromatic & bitter properties (ex. wild cherry).
arthropod repellant: a. toxicant: insecticide.
ascaricide: destroys round worms (*Ascaris* spp.) (ex. santonin).
ascites: dropsy: abnormal collection of serous fluid in peritoneal cavity.
astringent: contracts tissues & checks secretions (as perspiration), allays inflammation, acts as styptic (stops capillary blood flow), stimulates epithelial growth locally, hardens tissues to form crust on surface, hence acts as antiseptic & protective (ex. tannic acid, alum, silver nitrate, ferric chloride, lead acetate, witch-hazel).
ataractic: ataraxic: tranquilizer.
bacillicide: fatal to bacilli.
bacteriolytic: destroys by dissolving bacteria.
bactericide: kills bacteria (ex. phenol).
bacteriostatic: arrests or prevents growth and reproduction of bacteria & other microorganisms (ex. sulfonamides).
balsamic: healing or soothing (ex. benzoin).
bechic: remedies cough (ex. white pine syrup).
beta blockers: beta adrenergic blocking agents combined with & producing blockade of beta-adrenergic receptors; u. in hypertension; angina pectoris (ex. propanolol).
biliousness: combn. of malaise, constipation, headache, bloating, &c. (believed [probably in error] due to disorder of bile flow).
bioflavonoid: group of plant compds. believed to strengthen blood vessel walls; sometimes called vitamin P (ex. rutin).
blennorhagia: blennorrhea: discharge fr. a mucous surface, esp. appl. to gonorrhea.
bitter(s): subst. wh. stim. taste buds at back of tongue, specifically, & to give better taste, & stim. or improve appetite; types: 1) simple (or plain): ex. nux vomica, gentian, quassia 2) aromatic (ex. yarrow, absinthe) 3) mucilaginous (ex. colombo, cetraria) 4) astringent (ex. condurango).
brash: heartburn in which some acid (or tasteless water) is regurgitated into mouth.
bronchodilator: expands air passages in lungs; ex. methoxyphenamine.
calcemic: increasing Ca titer of blood (use of Ca therapy).
calcium blocking agents: calcium channel blockers: formerly calcium antagonists; reduces intake of Ca into cells, reducing coronary vessel resistance & hypertension.
calefacient: calorifacient: produces sensation of warmth in part to which applied (ex. capsicum).
calmative: soothes or sedates the nerves (ect.) (ex. valerian).
cancerogenic: carcinogenic: agent producing cancers.
cardiac depressant: c. depressor: c. sedative: reduces (lessens, depresses, decreases) frequency or force or both of heart action (beat) (frequently when heart is over-stimulated), thus lowering blood pressure (ex. aconite, Veratrum viride, HCN).
cardiac stimulant: c. tonic: cariotonic: increases (or stimulates) frequency (rate) &/or force of heart action & raises blood pressure (ex. alcohol [small doses], camphor, et_2O, strychnine, heat); "cardiac tonic" is sometimes differentiated from "c. stimulant" as acting directly on the cardiac muscle, strengthening & slowing contractions (ex. digitalis, adonis, caffeine [sometimes]).
carminative: fr. carmino (L) to cleanse, as in carding of wool; promotes expulsion from or reduces formation of gas in stomach & intestines, hence prevents griping & colic (ex. ginger, $NaHCO_3$, belladonna, pepper, asafetida, mustard, aromatic vol. oils).
cascado (Moluccas): a skin dis.
catalytic: producing catalysis, i.e., speeding reaction or process.

catastaltic: restraining or inhibiting any process.
cathartic: serves as a fairly powerful purgative to aid in & hasten emptying of intestines (ex. castor oil).
caustic: cauterant: cauterizer: destroys tissue to which applied (ex. KOH, $ZnCl_2$ & other Zn salts, CaO, NaOH, mercury salts, mineral acids, high heat).
cell proliferant: causes reproduction of cells of one type.
central nervous system (CNS) **depressant**: reduces activity of CNS.
cephalic: remedies conditions of the head, esp. headache; term now mostly in disuse.
cerebral depressant: c. depressor: c. sedative: decreases functional activity of higher nerve centers of brain (ex. alcohol in large doses, narcotics, anesthetics, opium, hypnotics, bromides) - **c. stimulant**: increases functional activity of higher nerve centers of brain (incl. cerebellum) (ex. alc. in small doses, caffeine, tea, ammonia).
chalybeate: contains iron & used as a tonic in anemia, chlorosis, etc.; ex. spring waters (as at Roseburg, Or [US]).
chemotherapeutic agent: u. in treatment of bacterial, fungal, & viral infections; parasitic disorders, & neoplasms.
chemotic: causes fluid —> soft jelly-like solid; specif. producing edema of eye conjunctiva (swelling around cornea).
cholagog(ue) (cathartic): promotes or stimulates flow or removal of bile causing purgation of green & liquid stools (ex. aloe, calomel, podophyllum, colocynth).
cholecystographic agent: u. in producing X-rays of gallbladder, &c.
choleretic: increases bile secretion or formation by liver (ex. bile salts).
cholinergic: gives effects similar to those of acetycholine.
ciliary: relating to eyelashes (cilia, L.); ex. c. excitant; c. humectant.
cicitrizant: aids healing wounds, ulcers, &c., by formation of cicatrix (=scar).
circulatory stimulant: increases the functional efficiency of the blood movement.
coccidiostat agent: u. to combat coccidiosis (infection by coccidia [Protozoa]) in lower animals (u. in poultry feeds, &c.).
condiment: spice; sauce or relish for food, which creates appetite & aids digestion (ex. cloves); combines props. of flavor & carminative.
condition (powder): conditioner: kind of tonic intended to stim. appetite of stock (bring into better condition), esp. horses; kept in boxes where animal has free access; sometimes considered not advantageous (fr. conditus [L], seasoned, savory) (*cf.* conspergo [L], to sprinkle, strew all, bestrew over, "pepper" [now rare], bespatter, scatter).
conditioner: condition hair against sun & wind.
conjunctivitis: inflammation of the lining of the eyelids and adjacent eyeball.
conspergative: conspergent: dusting powder either for body or for pharmaceutical preparations, such as pills, suppositories, etc. (ex. lycopodium).
constipant: produces retarded movements of the bowels; constipates (ex. opium).
constringent: astringent (older term).
constructive: a tonic that aids in reconstruction of bodily tissues (ex. multivitamin preps.).
contraceptive: interferes with the process of conception following sexual intercourse (ex. lemon juice soln.).
convulsant: producing convulsions (violent involuntary contraction of several muscles) (clonic, tonic); ex. strychnine.
copragog(ue): cathartic.
cordial: stimulating heart action (fr. cor, cordis [L], heart); also stomachic.
coronary dilator: dilating or widening the coronary arteries; ex. nitroglycerin.
correctant: corrective: corrigent: modifies side effects, esp. harsh actions such as griping or purging; thus, it improves the therapeutic properties of a medicinal agent; at times, term used to mean agent which restores health.
corroborant: strengthening.
corrosive: destroys tissue (ex. KOH).
cosmetic: improves condition & appearance of the skin (ex. talc) or otherwise improves appearance of person.
costive: agt. wh. reduces intestinal activity; constipating.
cough depressant: reduces cough; ex. methadone.
counter-irritant: irritates in one place or part in order to modify or improve conditions or check processes (such as congestion) at some other remote point or part (ex. capsicum, iodine, cantharis, mustard, volatile oils).
craw craw: onchocerciasis, dis. of wAf.
curaremimetic: drug wh. has similar action to that of curare; paralyzant.
curarisant: action of curare (&c.).
cyanosis: bluish discoloration of skin resulting from deficient oxidation of blood.
cycloplegic: paralyzes the ciliary muscle within the eye-ball & thus removes the power of accommodation (ex. amphetamine).
cynanche: severe sore throat & choking, as seen in diphtheria.
cystic: acts in some manner on the urinary bladder or gall bladder.
débridement: removal of macerated, lacerated, infected, or contaminated tissues.
decongestant: decongestive: reduce congestion.
Delhi belly: diarrhea (*E. coli* infection).
deliriant: delirifacient: excites cerebral functions so as to disorder or confuse the mentality (ex. cannabis, alc., N_2O, belladonna) (all deliriants are narcotics, but not all narcotics are deliriants).
demulcent: soothes, softens, relaxes, & protects mucous surfaces to which applied (ex. acacia, licorice, glycerin, bland FO, egg white).
dental obtundent: relieving tooth pain.

dentifrice: cleanses the teeth and gums or aids in that process (ex. myrrh tincture, chalk, borax).
deobstruent: removes obstructions in bowel; aperient (ex. cascara).
deodorant: deodorizer: removes, masks, absorbs, adsorbs, or destroys foul odors (ex. HCHO, Cl_2, CaO, H_2O_2, charcoal).
deoxidizer: removes oxygen from a chemical combination.
depilatory: kills growth of, or removes hair (ex. BaS, X-rays, T1 Acet.).
depletive: removes fluids or solids; causes emptying or evacuation.
depressant: diminishes functional activities of any cell, tissue, organ, system or organism (ex. aconitine).
depurant: depurative: 1) removes impurities or waste materials, specific. cleanses foul sores (ex. H_2O_2) 2) "purifies" the blood, as a "blood tonic" (ex. sassafras).
derivant: derivative: withdraws blood from the seat of a disease to some other part of body, usually to relieve congestion (ex. capsicum, cupping) (*cf.* counter-irritant).
dermatotherapy: applied in the treatment of the skin & its diseases, in burns, etc. (ex. benzoin, ichthyol).
desensitizer: deprives of sensitivity or sensation.
desiccant: desiccative: desiccatory: dries up moist surfaces (ex. ZnO, Zn Stearate, boric acid).
detergent: detersive: cleanser; often a surfactant; ex. guillaja.
detoxicant: reduces or removes a poison or poisonous property of a substance.
devalvant: depilatory; agt. u. to remove hair.
diagnostic (re)agent: used in the determination of a disease or complaint (ex. Barfoed's reagent).
diaphoretic: causes sweating or produces increased perspiration (ex. alc., aconite, pilocarpine, ammonium acet., pimpinella).
diapyetic: promotes suppuration.
diathermy: heals local tissues by the aid of high frequency currents.
dietary supplement: dietetic: addition of certain foods or food elements or drinks to diet to improve health; ex. banana.
digestant: digestive: aids the process or speed of digestion, often by increasing the efficiency of breakdown & absorption of food in the stomach & intestine (ex. pepsin, pancreatin, & other enzymes; acids).
diluent: 1) reduces the strength of a drug preparation, thus often rendering less irritant (ex. lactose) 2) sometimes specific., increases fluidity of blood or other body liquid (ex. glucose soln.).
discutient: causes a tumor, exudate, or other pathological formation to disappear; resolves congestions; reduces swellings (*cf.* alterative).
disinfectant: destroys infection & putrefaction by destroying pathogenic, fermentative, & otherwise deleterious bacteria, molds, etc. (ex. $KMnO_4$, alc., heat, mercuric bichloride, comp. cresol soln., sod. benzoate in catsup).
diuretic: induces or stimulates flow of urine, hence increases urinary secretion & output (ex. buchu, juniper, gin, beer, trifolium, water perorally, hot baths).
dolorific: causes pain.
dope: 1) depressant (ex. Opium) 2) stimulant for race horses (ex. strychnine).
drastic (cathartic): acts with much irritation as a powerful purgative (ex. gamboge).
dusting powder: absorbs moisture, thus lubricating surfaces, preventing friction, & reducing infection (ex. talcum powder).
dyspnea: difficult or labored breathing.
ecbolic: increases uterine contractions & thus 1) aids in or hastens expulsion or delivery of child during birth 2) produces abortion (ex. ergot, savin, strong purgatives).
eccritic: eccoprotic: promotes excretion (mild purgative).
electrolyte replenisher: parenteral replacement of ionized substances (such as NaC1).
eliminant: evacuant.
embrocation: represents the external application of a liniment.
enema: liquid injected into rectum —> evacuation.
emetic: causes or produces vomiting (ex. mustard, NaCl, $ZnSO_4$, tartar emetic, apomorphine).
emmenagog(ue): promotes, aids (stimulates, strengthens), increases, re-establishes (restores), or produces normal (or healthy) menstrual function (flow of menses) (ex. hydrastine, ergot, quinine, cotton root bark, savin, & other ecbolics in small doses).
emollient: soothes, softens, relaxes, & protects the skin (external tissues) (ex. petrolatum, cocoa butter, lard, glycerin, olive oil, poultices).
emulgent: 1) extracts from the blood, as urinary components by the glomeruli of kidney 2) stimulates urinary or bile flow.
emulsifier: emulsifying agent: emulsive: used in preparing an emulsion (ex. acacia).
endermic: administered through the skin by rubbing (ex. testosterone propionate).
energizer: psychic stim. such as amphetamines.
enteric: acts in some way on the intestines.
enuretic: causes urinary incontinence, specif. bed-wetting; reducing bed-wetting.
envenomation: effects of bites or stings of snakes, insects, &c.
enzymatic: causes enzymatic fermentation (ex. papain).
epilatorium: epilatory: removes hair.
epispastic: blistering agent (ex. iodine) (*cf.* vesicant, pustulant).
errhine: increases nasal secretions or discharge, resulting in sneezing (ex. snuff, ginger, pd. soap, euphorbia, soap bark, pepper, pyrethrum, many powdered drugs) (*cf.* sternutatory).
escharotic: powerful caustic, destroying tissue when in contact & producing eschars (scars or dry

crusts of dead tissue, which form a protective coating); used to remove warts, etc. (ex. KOH, NHO_3, $AgNO_3$, phenol. alum, $HgCl_2$) (*cf.* caustic).
esculent: edible; fit for food (ex. *Pastinaca sativa*).
estrogen(ic): estrogenic hormone: female hormone.
eupeptic: promotes normal digestion.
euphoriant: euphoric: produces an artificial state of happiness or well-being (ex. morphine, amphetamine).
evacuant: promotes excretion, esp. of feces, i.e., acts to empty the gastro-enteric canal (ex. frangula).
excipient: increases physical properties conducive to prepn. of prescription or formula.
excision: surgical removal of body structures or growth (ex. hemorrhoidectomy).
excitant: stimulates vital activity in any part of organism (ex. whisky).
expectorant: controls cough by increasing (promoting) or decreasing (discouraging) bronchial secretions, hence used in bronchitis, asthma, etc.; types: 1) stimulating e. (ex. ipecac, apomorphine, lobelia, NH_4Cl) 2) soothing (or sedative) e. (ex. glycyrrhiza, belladonna tr.) (more antitussive) & facilitates expectoration & discharge of mucus 3) anodyne (ex. codeine).
febrifuge: v. antipyretic.
fecal softener: mild form of laxative.
feticide: destruction of fetus.
fibrillolytic: destruction or dissolution of fibrillae (or fine fibers) of muscle, nerve, &c.
fibrinolytic: destruction (dissolution) of fibrin (blood clot).
fixative: agent wh. fixes animal or plant tissues in desired condition.
flavorants, food: flavoring agent: flavors u. in food industry; aids administration by adding a pleasant taste to a medicine, etc. (ex. vanilla).
fodder: prepared food for livestock (ex. clover hay).
f(o)eticide: destroys the embryo *in utero*.
food: provides nourishment & sustenance to the body (ex. milk, olive oil, cocoa butter).
forage: food growing in the field for domestic (or stock) animals (ex. Dallis grass).
framb(o)esia: yaws.
fumigant: fumigating agent: used to destroy infection by exposure to fumes or smoke (ex. sulfur).
fumitory: somethine smoked; boys often smoke pl. materials (1867).
fungicide: destroys fungi (ex. $CuSO_4$).
fungistat(ic): inhibits the growth of fungi.
galactagog(ue) galactopoietic: stimulates (increases) secretion and/or flow of milk (ex. pilocarpus, liquids).
galactophyge: galactophygous: decreases secretion and/or flow of milk (ex. turpentine oil for cattle).
ganglionic blocking agent: interruption of nerve impulses at autonomic ganglia.
gastric acidifier: increases acid content of stomach - **g. sedative:** reduces gastric irritation, thus allaying nausea & vomiting (ex. lime water, calomel) - **g. stimulant:** increases gastric function, thus acting as a tonic (ex. wine).
general anesthetic: produces loss of consciousness (with insensibility to pain & abolition of reflexes) (ex. ether, $CHCl_3$, C_2H_5Cl).
germicide: kills germs (pathogenic microorganisms) (ex. thymol).
gestagen: progestational hormone; progestin.
glucogenic: producing glucose.
gonadotropin: stim. growth & activity of gonad.
gonococcide: kills gonorrhea (GC) organisms.
growth retardant: discourages or diminishes increase of stature.
gyniatric: agent u. in menstrual dis. & other related conditions of female.
haem-: v. *hem-*.
hallucinogen(ic): psychomimetic; produces hallucinations in humans; delir(ifac)ient; incl. mescaline; LSD; harmine; cannabis; adrenochrome; ibogaine.
Hansen's disease: leprosy (2507).
helicide: kills snails, slugs, &c.
helminthagog(ue): v. vermifuge.
hemagog(ue): promotes menstrual or hemorrhoidal bloody discharge.
hemat(in)ic: improves quality (condition) of blood, esp. by increasing number of erythrocytes (ex. Fe, Mg, liver).
hematopoietic: v. hemopoietic.
hemocoagulant: produces coagulation of blood.
hemolytic: is destructive to the blood cells (ex. benzene).
hem(o)poietic: has to do with blood formation (ex. iron).
hemorrhoidal agent: to treat varicose veins of rectal area (piles); ex. tannin, phenol, epinephrine.
hemostatic: arrests (stops) internal bleeding (hemorrhage) (ex. F. E. ergot, adrenalin, stypticin, gelatin, witch-hazel, dil. sulfuric acid) (*cf.* styptic).
hepatic: acts on liver (*cf.* two following def.) - **h. depressant:** decreases or depresses (or reduces) liver function - **h. stimulant:** increases or stimulates liver function (ex. aloe, rheum, arsenic).
herbicide: to kill herbaceous weeds.
herpes simplex: acute virus dis. with water blisters on lips & nose; treated with **herpetic**.
humectant: keeping moist.
hydragog(ue) (cathartic): produces copious watery discharges in stool (ex. jalap, colocynth).
hydrochloretic: increases secretion of relatively thin bile with high water content (ex. dehydrocholic acid).
hydrotherapy: hydriatic (or water) treatments for disease, of which there are said to be over 200 kinds (ex. hot, cold applications, vapor [sprays], Swedish & colon massage, salt glows, etc.).
hydrotic: v. diaphoretic.
hypercalcemic: producing levels of Ca in blood higher than normal.
hypercholester(ol)emic: producing high levels of cholesterol in blood.
hypertensive: raising the level of the blood pressure (agt. = hypertensor).

hypnagog(ue): hypnic: hypric: induces (produces, causes) or promotes normal sleep (resembling the natural) without lessening consciousness to pain (hence, individual in pain may be wakeful) (ex. barbital, luminal, allonal, trional, sulfonal, chloral hydrate, paraldehyde).

hypodermic: injected thru the skin; types: subcutaneous; intradermal (intracutaneous); intravenous; &c.

hypodermoclysis: introduction of liquids into the subcutaneous tissues (mostly normal saline).

hypoglycemic: agt. wh. lowers levels of glucose in blood.

hypokinetic: abnormally decreasing motor activity of muscles of body.

hypopigmenting agent: inhibits melanin formation.

hyposensitive: reduces sensitivity.

hypostyptic: moderately or mildly styptic (stopping blood).

hypotensive: reducing blood pressure (agt. hypotensor).

incisive: is employed in the teething of babies (ex. orris).

icterus: jaundice.

inebriant: providing intoxication.

inflammation: condition manifested by redness, swelling, heat, & pain.

inhalant: subst. to be inhaled into lungs (ex. epinephrine; arom. spirits of ammonia).

inhibitor: suppresses action.

insecticide: kills insects & related organisms (ex. pyrethrum flowers).

insectifuge: drives away or repels insects & related organisms (ex. citronella oil).

incisive: is employed in the teething of babies (ex. orris).

intestinal astringent: has a puckering effect on intestinal mucosa & is used to combat diarrhea, etc. (ex. rhatany root).

intoxicant: 1) excites or stupifies (ex. alc., mescal) 2) poisons (ex. lead salts).

intrathecal: earlier expression for intraspinal.

inunction: applied by rubbing or anointing (ex. mercury).

irritant: causes local stimulation (*cf.* counter-irritant).

jaundice: condition with more or less pigmentation of skin in certain areas.

keratitis: inflammn. of cornea of eye.

keratolytic (agent): produces keratolysis (softening & peeling of horny layer of skin).

lachrymator: lacrimator: causes tears by irritant action (ex. tear gas, benzene bromide).

lactagog(ue): v. galactagog.

lactifuge: checks milk secretion.

laxative: mild purgative, which acts without causing pain or violence (ex. liquid petrolatum).

lenient: lenitive: 1) allays irritation & eases pain 2) demulcent 3) mild purgative.

leprostatic: inhibits growth of the m.o. *Mycobacterium leprae.*

leucotaxic: having power of attracting leukocyters to form pustules.

levigating agent: one wh. makes smooth by reducing grit; producing impalpable powder.

lipotropic: prevents or reduces accumulation in excess of fat in liver or other places.

lissive (Fr): makes skin, &c., smooth (fr. lissos, Gr).

lithagog(ue): lithic: litholytic: lithontriptic: lithontrypic: lithotripic: lithotryptic: said to aid in removal or expulsion of calculi or stones from the bladder, kidney, gall bladder, bile ducts, etc. (surgical means are most effective).

local analgesic: reduces pain, itch, &c., in superficial area.

local an(a)esthetic: destroys pain & sensation on local application (ex. cocaine) - **l. sedative:** relieves pain or discomfort by local application - **l. stimulant:** increases functions of repair, etc., by causing topical irritation.

loosener: (US slang): cathartic.

lubricant: agent wh. makes slippery smooth; ex. tragacanth mucilage.

lupus erythematosus (LE): skin dis. often with lesions & frequently with systemic involvement; formation of butterfly pattern on cheeks & nose (cuataneous LE).

luteolytic: producing degeneration of the corpus luteum, resulting in menstruation or abortion.

lymphagog(ue): agent promoting formation of lymph.

lymphogranuloma (inguinale): lympopathia venereum = venereal lymphogranuloma: sexually transmitted dis. (sometimes spoken of as the 4th or 5th VD).

lytic: brings about gradual subsidence in symptoms of disease (ex. antitoxins).

malodorant: subst. producing an unpleasant or repulsive odor.

marasmus: tissue wasting due to starvation, &c.; seen in infants <1 yr. old; similar to kwashiorkor.

masticatory: improves local conditions in mouth, incl. saliva increase, by chewing of some material (ex. betel).

metabolic: has a bearing on the constructive (anabolism) & destructive (catabolism) processes within the body.

metabolite: v. antimetabolite.

meteorism: abdominal bloating due to gas either inside or outside of GIT.

Mexicali revenge: Montezuma's r.: diarrhea.

migraine: unilateral headache (hemicrania), often associated with nausea & other symptoms.

milk modifier: additions to milk of cows, &c., in order to make it more similar to human milk.

mineralizer: increases or alters the amount of minerals or inorganic substances in the body (ex. mineral spring water).

miotic: v. *myotic.*

miticide: destroys mites (tiny arthropods, often parasitic on humans & other animals).

motor depressant: m. paralyzant: paralyzes motor part of nervous system (brain & spinal cord) &

depresses its functions, esp. by lessening activity of cord (ex. opium, alc., strychnine [large doses], tobacco) - **m. exitant: m. stimulant:** increases motor activity of central nervous sysem (ex. nux vomica, strychnine [small doses], digitalis).
mucigog(ue): stimulates formation & flow of mucus.
mucolytic: dissolves or destroys mucus.
myasthenia gravis: serious weakness of voluntary muscles; progressive autoimmune dis. (Aristotle Onassis died of this).
mydriatic: dilates eye pupil (ex. belladonna, homatropine, henbane, atropine).
myelographic: relating to X-rays of spinal cord after injection of contrast medium.
myotic: diminishes the size of or contracts eye pupil (ex. pilocarpine, physostigmine, nicotine, arecoline).
narcotic: tends to paralyze nervous system, producing (with proper dosage) sleep, stupor (complete insensibility) & death (ex. opium, cannabis).
nasoconstrictor: contracts the mucous membranes of the nasal passages, making respiration easier & more comfortable (ex. ephedrine).
nauseant: causes sensation of sickness at stomach, wh. may or may not proceed to emesis (ex. warm salt water).
neonatal: relating to the newly born; u. to cover first 4 weeks of life.
neoplastic suppressant: reducing or preventing growth of neoplasms (new tissue or tumor).
nephritic: used to cure or relieve kidney diseases, as bladder stone (ex. methenamine).
nephrotropic agent: specially affecting kidney tissues.
nerve sedative: allays nervous excitement (quiets the nerves) (ex. bromides, aconite, belladonna, valerian, hyoscyamus), having a calming, depressant, & quelling effect on the CNS - **n. stimulant:** increases nervous excitement & irritability (stimulates the nerves) (ex. caffeine, nux vomica) - **n. tonic:** v. foll.
nervine: term in former use for accelerating, exciting, stimulating, or toning nerve activity (ex. barbituates).
nervous (or restless) legs: nocturnal myoclonus: (therapy: empirin compound with codeine).
neuritic: neurotic: acts on central nervous system (ex. valerian, NaBr).
neuroleptic: drug u. to improve mental health (incl. tranquilizers, antipsychotics).
neurotransmitter: chemical substance produced by nerve cell wh. enhances or inhibits responses in receptors or neighboring cells (ex. acetylcholine).
non-specific therapy: protein t.: when a foreign protein is administered parenterally in the treatment of disease (ex. typhoid vaccine).
nootropic: affects neurons favorably.
nutrient: nutriment: nutritive: medicinal food (v. food).
obstipant: corrects obstipation (stubborn constipation).
obstruent: causing obstruction or blocking.
obtundent: reducing sensitivity or pain (ex. demulcent).
odont(alg)ic: relieves or reduces severity of toothache (ex. $CHCl_3$, clove oil).
onchocerciasis: infection by *Onchocerca* spp. (nematodes).
oncology: science of neoplastic growth in individuals (esp. cancers).
ophthalmic (agent): relating to eye treatment.
opiate: promotes sleep.
opioid: substance with opiate properties but not derived fr. Opium.
orthopedic: relates to the correction of physical deformities.
oxytocic: increases expulsive power of uterus & aids in childbirth by stimulation of uterine contractions (ex. pituitrin, ergot) (*cf.* ecbolic).
palliative: relieves but does not cure (ex. acetylsalicylic acid).
panacea claimed to cure all or many diseases (ex. ginseng).
paralyzant: paralyzer: causes loss of function of any type.
parasiticide: destroys parasites (animal or plant organisms wh. live off other organisms) (ex. sulfur, kerosene).
parasympatholytic: annuls effects of parasympathetic nervous system activity.
parasympath(ic)mimetic: produces effects similar to those fr. stimulation of parasympathetic nervous system; i.e., simulates action of parasympathetic nervous system; ex. pilocarpine, eserine.
parenteral: introduced into the body by some means other than the gastrointestinal route; generally refers to hypodermic injections.
parturient: parturifacient: aids in childbirth by inducing or accelerating labor (ex. ergonovine).
pectoral: used for benefit of diseases of chest & esp. of lungs; often specif. for expectorant (ex. *Prunus virginiana* in cough).
pediculicide: destructive of lice (parasitic arthropods).
perfume: imparts fragrance (ex. rose oil).
perlèche: contagious dis. characterized by inflammn. & scales at corner of mouth (esp. in children).
pesticide: u. against dis. & infestants of plants/animals.
peristaltic: relating to regular contraction of intestines.
phagorepellant: discourages feeding or eating.
phantastic(a): psychomimetic: producing fantasies (phantasia).
phthisiology: science of tuberculosis (TB).
phygogalactic: stopping the secretion of milk.
physic: a strong cathartic of a strength intermediate between a purgative & a hydragog; ex senna (Phyisican, one who uses physics in practice; indicating great importance formerly).
phytocide: plant killer; includes herbicides, arboricides, &c.
phytopharmaceutic: relating to effect of drugs on plants; ex. gibberellic acid.

placebo: has no therapeutic action but is designed to please or satisfy the patient (ex. lactose tablets, dough pill).
pneumococcide: destructive of *Streptococcus pneumoniae.*
poison(ous): when taken into body in relatively small amounts is deleterious to health or is even lethal (ex. prussic acid).
poikoilothermia: variation of body temperature, varying with environmental temp.
poikiluria: abnormally frequent urination, as in diabetes.
polycythemia: increase in red blood cells.
polycythemic suppressant: reduces number of red blood cells.
pre-anesthetic: narcotic administered prior to a general anesthetic.
procoagulant: subst. which is a precursor to blood coagulation; ex. prothrombin, thrombokinase, thrombin.
progestational: having effects like that of progestin.
progestin: progestagen = gestagen; type of female endocrine protective to fetus; tends to protect uterus from conception.
prophylactic: prevents disease, etc.; to layman, often connotes sexual p. (ex. HgCl, $KMnO_4$, small pox vaccine).
protective: prophylactic against disease & injury, often by mechanically shielding surface areas to wh. applied from external influences (as light, air, friction, water, etc.); hence, often used to cover wounds, burns, etc. (ex. adhesive plaster, collodion, ointments).
proteolytic (enzyme): breaks up proteins by hydrolysis.
prothrombinogenic: promotes formation of thrombin.
protozoacide: destroys Protozoa (incl. amebae).
prurient: causes pruritus (itching).
psychedelic: altering the mental state, supposedly revealing emotions, &c., wh. are normally repressed.
psychic energizer: psychostimulant: activates the mind.
psychodysleptic: produces delusional state of mind.
psychotherapy: psychotherapeutics: treatment of mental disorders, mostly using psychological methods.
psychotica: the field covering all angles of drugs with hallucinogenic effects.
psycho(to)mimetic: produces changes in thought, perceptions, or moods without causing major disturbances of autonomic nervous system or other serious disabilities (ex. mescaline); other names are psychotogen; psychedelic; "consciousness expanding."
psychotropic: general term for agents acting on mental processes incl. analgesics, euphoriants, sedatives, tranquilizers, hypnotics, inebriants, stimulants, & psychotomimetics.
ptalagog(ue): v. sialagog.
ptarmic: errhine.
puerperium: period of confinement after childbirth, to allow recovery.
pulmonary: used to treat lung & chest disorders - **p. sedative:** reduces coughing by relieving irritation.
pungent: has a sharp & acrid taste (ex. horsemint).
purgament: purgative: purge: causes evacuation of the bowels; term "purgative" used both as generic & specific; when generic, classified into the following types (in approx. increasing order of strength): 1) aperient 2) eccoprotic 3) laxative 4) purgative (in specific sense) (ex. senna) 5) cholagog 6) cathartic 7) hydragog 8) drastic.
pustulant: produces pustules (pus-containing lesions) usually for purposes of counter-irritation (ex. croton oil, $AgNO_3$).
pyretic: pyretogen: pyrogenic: producing fever or associated with fever.
pyrosis: heartburn; water pang; (water) brash; often related to gastric esophageal reflex (GER).
Q fever: Queensland f.; **query f.**: an acute self-limited infection found in Australia.
radiological: relating to radioactive substances, X-ray, &c.
radiopaque medium: remains unpenetrated by X-rays, hence provides contrast in taking X-ray photographs; X-ray contrast medium.
raticide: poisonous to rats (ex. warfarin, red squill).
reducing agent: 1) used to reduce obesity (ex. thyroid, phytolacca fruit) 2) prevents oxidation (ex. ascorbic acid).
reflex stimulant: causes increased activity of the vital functions by counter-irritation, inducing a reflex action (ex. strong acid).
refrigerant: refrigerative: produces bodily cooling effect, which relieves fever, restlessness, thirst (ex. vegetable or fruit acids, dil. mineral acids, ammonium acetate soln.).
relaxant: reduces tension or strain of tissue, organ, system, etc. (also used specifically for laxative).
renal depressant: decreases or suspends flow of urine by reducing kidney action (ex. ergot, morphine).
repellant: repellent: drives away, esp. vermin, as rodents (ex. peppermint oil for mice).
resistogen(ic): agt. which reduces stress.
resolvent: causes solution of tissue or exudate, thus allaying inflammation, dispersing abnormal (morbid) swelling, etc.; *cf.* alterative.
respiratory depressant: r. sedative: decreases force or slows rate of respiration (ex. morphine) - **r. stimulant:** increases or accelerates force (depth) and/or frequency of respiration (ex. strychnine, atropine, smelling salts [NH_3]).
restorative: resumptive: renews strength & vigor (ex. tonics, aromatic ammonia spirits).
revulsant: revulsent: revulsive: acts as a derivative (draws blood from some other part) or as a counter-irritant, when applied locally (ex. cantharis plaster).
rheum: watery nasal discharges as seen in colds (*cf.* **rhume** [Fr]: cold).

roborant: tonic; strengthening agt.; giving strength & health.
rodenticide: destroys rats, mice, and other rodents (mammals with gnawing incisor teeth) (ex. red squill).
roentgenographic: X-ray.
rubefacient: causes reddening of skin (due to dilation of blood vessels) (ex. camphor, turpentine, analgesic balm, I_2, capsicum, mustard) when applied locally.
ruminant: bringing up food fr. stomach to chew, as done by cattle & other ruminants (Ruminantia) generally.
ruminatoric: aiding digestion in rumen (first & largest part of the 4-part stomach in ruminants).
saccharine: used for sweetening purposes (ex. sucrose).
saline (cathartic): causes bowel evacuation by osmotic or salt action on the intestinal walls (ex. certain salts, such as Epsom salts).
salivant: producing saliva.
saponaceous: has the action or properties of a soap (ex. quillaja).
sapremia: contamination of circulating blood by absn. of products of putrefaction.
scabicide: kills the itch mite (*Sarcoptes scabei*) wh. causes scabies.
Schreckstoff (Ge): subst. producing shock, such as histamine injection.
sclerosing agent: causes hardening or inward growth from the walls of veins; u. in treatment of varicose veins & hemorrhoids (ex. sodium morrhuate).
seasoning: spice.
secretory depressant: causes reduction of body secretions.
sedative: 1) soothes (or allays) irritability, chiefly in hysteria & other nervous diseases (ex. narcotics, allonal, $CHCl_3$) 2) lessens excitability or activity of a system or of an organ; quiets or calms hyperactivity (respiratory, gastric, etc.).
sialagog(ue): increases salivation by stimulating secretion & flow of saliva (ex. mercurials, ginger, tobacco, horseradish, iodine compds., pilocarpine).
siccative: dries up.
simple purgative: causes active purgation without inflammation or depression.
sinapism(us): blistering agent: vesicant (ex. mustard, cantharides).
skeletal muscle relaxant: reduces tone of muscles supporting skeletal frame.
solvent: a liquid or other substance wh. holds one or more substances in solution (ex. acetone).
somnifacient: somniferic: somniferous: somnific: soporific: v. hypnotic.
sorbefacient: produces or aids (or promotes) or facilitates absorption of exudates (ex. iodides).
spasmodic: concerning the sudden involuntary contraction of muscle or set of muscles.
specific: has direct curative or prophylactic influence on certain individual diseases (ex. diphtheria antitoxin).
spermicidal: relating to spermaticides (spermicides), u. to kill spermatozoa, as in some contraceptive methods.
spice: condiment; term somewhat more restricted than "condiment"; incl. cinnamon, coriander, &c.
spinal stimulant: v. reflex stimulant.
spirocheticide: destroys spirochetes (type of spiral bacteria formerly assigned to the Protozoa).
spray: inhalant.
staphylocide: destroys staphylococcus organisms (ex. sulfonamides).
sternu(at)ory: v. errhine.
stimulant: increases or augments normal functional activity of specific portions of body (reflex, local); as intestinal, cardiac, respiratory, etc.
stimulins: subst. in blood wh. were believed to stimulate phagocytes & aid in destruction of m.o.
stiptic: styptic: checks hemorrhage or bleeding externally by local astringent action (ex. alum, ferric subsulfate, tannin, cold).
stomachal: stomachic: stimulates or increases appetite & promotes gastric digestion by increasing secretion of gastric juices (ex. gentian, nux vomic, cinchona, aromatic VO, bitters).
sudorific: v. diaphoretic.
suppurant: forms & produces pus.
surface active agent: surfactant: u. to reduce surface tension & as wetting agent; ex. aerosol O. T.; saponin.
surrogate: u. as substitute.
sweetener: produces a sugar-like taste; used to increase palatability of medicines, etc. (ex. maltose).
sympathicolytic: sympatholytic: sympathoparalytic: inhibits sympathetic nerve cell transmission.
sympath(ic)omimetic: producing effects similar to those produced by stimulating the sympathetic nervous system; or simulating stimulation of the sympathetic nervous system (through the end organs).
synanche: cyanche: severe sore throat.
syndrome: symptoms in aggregate or complex of symptoms of a disease.
synergist: aids the action of another drug in one way or another (ex. bromides & chloral for hypnotic use; calomel & jalap for purgative use).
t(a)eniacide: t(a)enicide: kills tapeworm (*Taenia* spp.) (ex pelletierine).
t(a)eniafuge: t(a)enifuge: expels tapeworms (ex. pomegranate, aspidium, pepo).
thickener: thickening agent: u. in pharmaceutical formulations to make a liquid less thin; ex. colloidal clay.
thrombolytic: to remove thrombi (clots) in blood vessels such as those of heart as in coronary thrombosis (ex. streptokinase; urokinase).
thyroid inhibitor: reduces activity of thyroid gland; ex. thiouracil.
tolerance: the ability of enduring the influence of a drug or poison, particularly when acquired by continued use of or exposure to the substance.

tone: normal activity, strength, functionability, sensitivity, &c., of mucsle (&c.) as seen in health.
tonic: improves or increases general bodily tone & vitality, restoring strength & energy (besides this general use, there are also specific tonics, such as blood, nerve, etc.) (ex. elixir or iron, quinine, & strychnine; cinchona; gentian; iron; arsenic; nux vomica).
topic(al): an external local application of remedial substance.
tranquilizer: having a quieting effect.
trichomonacide: destructive to *Trichomonas* & related Protozoa (ex. suramin sodium).
tuberculostatic: inhibiting tuberculosis (TB) organism or infection; ex. streptomycin.
turista (Mex slang): diarrhea seen among tourists.
trypano(somi)cide: destroys trypanosomes (Protozoa) (ex. suramin sodium).
uricolytic: relates to the hydrolysis of uric acid into urea.
uricosuric: promoting urinary excretion of uric acid.
urinary acidifier: promotes a relatively acid urine (ex. ammonium chloride) - **u. alkal(in)izer** promotes relatively alkaline urine (ex. potassium citrate) - **u. antiseptic:** stops growth of bacteria, &c., in urinary tract.
urographic agent: in X-ray photography of urinary tract, contrast medium.
urticaria: conditon where skin is characterized by development of weals (or risings) wh. result in burning and itching.
uterine sedative: reduces hypermotility of uterus, hence prevents or tends to prevent miscarriage (ex. viburnum prunifolium) - **u. stimulant**: stimulates contraction of uterus on menstruation.
vascular constrictor: causes diminution or constriction of lumen of small blood vessels (ex. epinephrine, digitalis) - **v. dilator:** causes enlargement (or dilation) of lumen of small blood vessels (ex. alc., ether [peripheral cutaneous], opium [cutaneous]).
vasoconstrictor: v. vascular constrictor.
vasodilator: v. vascular dilator.
vasomotor depressant: lowers arterial pressure, resulting in vasodilatation (ex. nitroglycerin) - **v. stimulant:** increases arterial pressure, resulting in vasoconstriction (ex. epinerphrine).
vasopressor: stimulates muscular tissue of arteries & capillaries.
vehicle: serves as a medium in administration of medicines, usually without therapeutic action of importance (ex. syrup).
venereal: 1) aphrodisiac 2) remedies sexual diseases (ex. penicillin).
vermicide: kills (or destroys) (intestinal) worms (ex. thymol, pepo, CCl_4).
vermifuge: expels or aids in expulsion of worms often by paralysis (ex. arecoline, scammony, castor oil, quassia, chenopodium, santonin, strong purgatives).
vertigo: sensation due to lack of equilibrium, indicating inner ear disturbance.
vesicant: vesicatory: irritates skin sufficiently to cause watery blisters or vesicles to form (ex. capsicum, cantharis, mustard, stronger ammonia water, croton oil).
viru(li)cide (Fr): destroys viruses.
vitiligo: leukoderma (a progressive depigmentation whitening of the skin).
vomitant: vomitive: v. emetic.
vulnerary: healing agent for wounds (ex. Peru balsam, benzoin).
wetting agent: one which causes spreading of a liquid on a surface, mostly by reduction of surface tension of liquid.
zomotherapy: involves treatment of disease by administration of raw meat diet, muscle plasma, meat juice, etc.
Zugmittel (Ge): vesicant.
zymic: zymolytic: zymotic: has fermentative properties.

APPENDIX E

DIAGRAMS OF TYPES OF INFLORESCENCES AND OF FLOWERS

Inflorescence (flower groupings)—chief types:

Raceme: an elongated flower group formed by numerous flowers growing on stalks (pedicels) along the sides of a common stalk (the peduncle), with well-developed internodes; ex. *Linaria repens* (simple raceme) (Fig. A).

Panicle: a compound or branched raceme; ex. *Yucca filamentosa, Agave chrysantha, Avena sativa* (oats) (Fig. B).

Spike: a raceme with sessile flowers; term sometimes extended to racemes with very short pedicels; ex. *Verbena hastata* (Fig. C).

Varieties of spike: **Catkin** (or **Ament**); ex. *Salix* (willow)
Spadix; ex. Calla lily
Strobile (or **Cone**); ex *Pinus* (pine)

Corymb: a flat-topped raceme, in which the upper flowers have short, the lower flowers progressively longer, pedicels; ex. *Crataegus* (may-haw) (Fig. D).

Umbel: a flower cluster in which the pedicels all seem to originate from a common point at the top of the peduncle and are surrounded by a circle of bracts (simple umbel [Fig. E-1]) ex. *Daucus carota* (carrot); cherry. A *compound umbel* represents a grouping of simple umbels (Fig. E-2).

Cyme: a flower group with the oldest flowers at the top or center; thus, similar to a corymb or umbel except that the central flowers are the first to develop; ex. *Saponaria* (Fig. F).

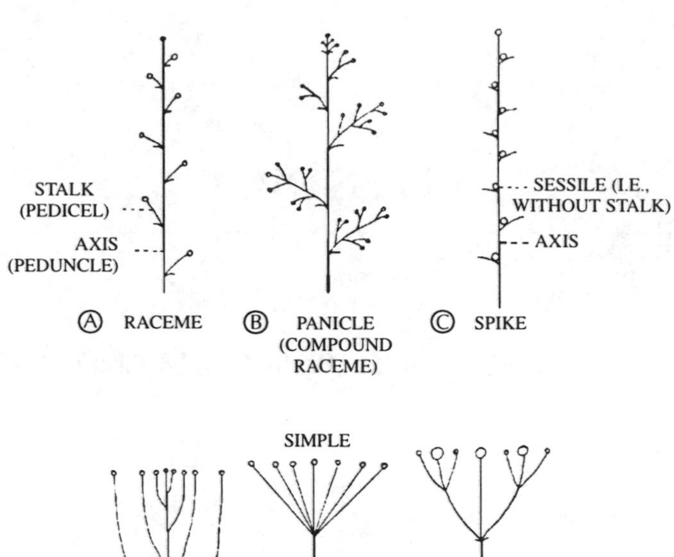

TYPES OF INFLORESCENCE

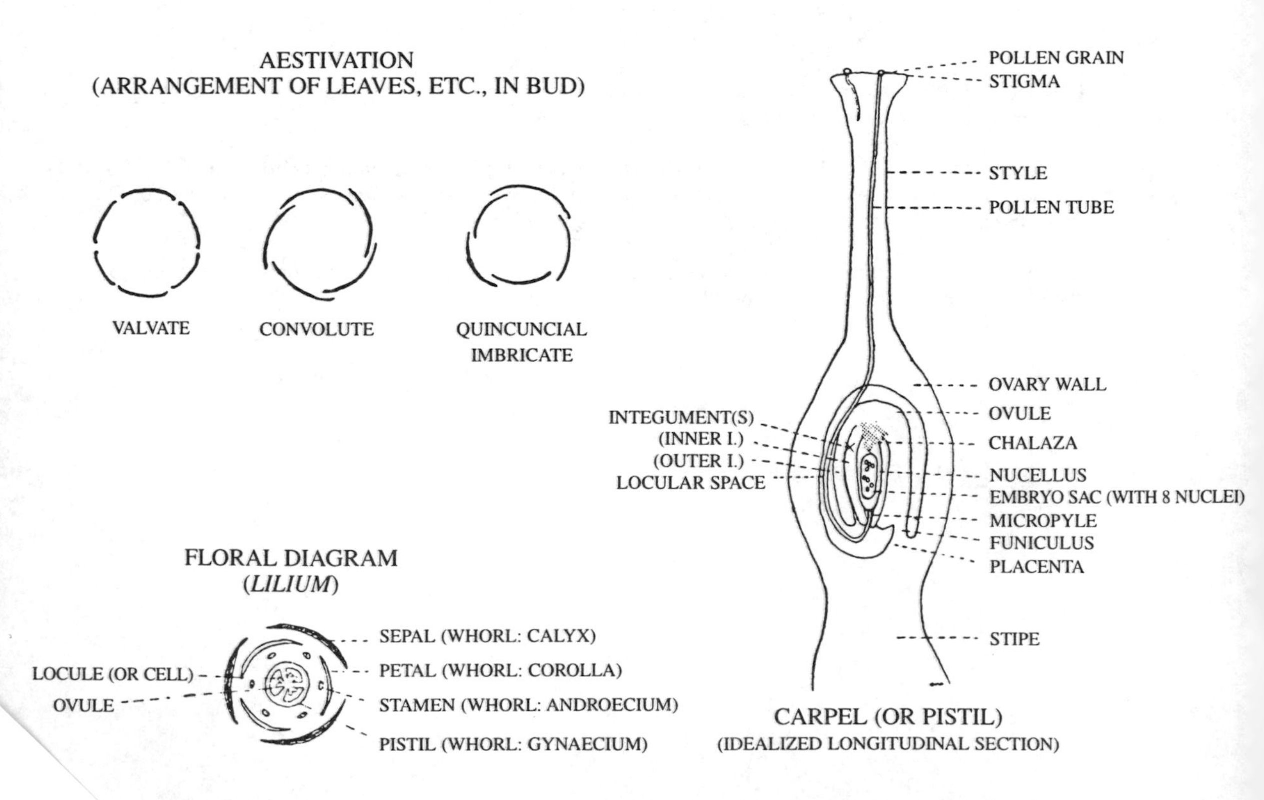

CARPEL (OR PISTIL)
(IDEALIZED LONGITUDINAL SECTION)

APPENDIX F

SCHEME OF CLASSIFICATION FOR PLANTS AND ANIMALS WITH EXAMPLES OF SOME OF THE GROUPINGS IN ONE PLANT AND ONE ANIMAL SPECIES*

Italicized bracketed terms are Latin equivalents*

GROUPINGS	EXAMPLES	
KINGDOM (*Regnum Vegetabile)*	Plantae (Plants)	Animalia (Animals)
Phylum (Division) (*Divisio)*	Spermatophyta	Chordata
SUB-PHYLUM (*Sub-Divisio)*	Angiospermae	Vertebrata
CLASS (*Classis)*	Dicotyledoneae	Mammalia
SUB-CLASS (*Sub-Classis)*	Polypetalae	Theria
ORDER (*Ordo)*	Rosales	Ungulata
SUB-ORDER (*Sub-Ordo)*	Rosineae	Artiodactyla*
FAMILY (*Familia*)	Rosaceae	Bovidae†
SUB-FAMILY (*Sub-Familia*)	Rosoideae	Bovinae
TRIBE (*Tribus)*	Potentilleae	Bovini
SUB-TRIBE *(Sub-Tribus*)	Rubinae	
GENUS (*Genus*)	*Rubus*	*Bos*
SUB-GENUS (*Sub-Genus*)	Ideobatus	
SECTION (*Sectio*)	Ideanthi	
SUB-SECTION (*Sub-Sectio*)	Idaei	
(SERIES—sometimes used)	Eu-idaei	
(SUB-SERIES—sometimes used)		
SPECIES (Species)	*idaeus*	*taurus*
Author of species	Linné	Linné
SUB-SPECIES (*Sub-Species*)		
VARIETY (*Varietas)*	*strigosus* (Mx.) Maxim.	
SUB-VARIETY (*Sub-Varietas*)		
FORM (*Forma*)	*albus* (Bailey) Fernald	
BIOLOGICAL FORM (*Forma Biologica*): RACE		
INDIVIDUAL (*Individuum*)		

*Infra-Order: Pecora
† Super-Family: Bovoidea

Conventional designation of plant and animal species:
1) Plant: *Rubus idaeus* L. v. *strigosus* (Mx.) Maxim. f. *albus* (Bailey) Fern.
2) Animal: *Bos taurus* L.

* Both English and Latin group names as used in the International Code of Botanical Nomenclature.

APPENDIX G

PLANTS YIELDING NATURAL RUBBER

Apocynaceae (Dogbane Fam.): several genera & spp.
Apocynum cannabinum (Apocyn.)
Artocarpus elastica, A. integrifolia (Mor.)
Asclepias stellifera (sAf), *A. subulata, A. syriaca* (Asclepiad.)
Brosimum alicastrum (Mor.)
Carpodinus hirsuta (Apocyn.) → Glu Rubber
C. lanceolatus & other spp. → African Rubber
Castilla costaricana, C. elastica (Mor.) → Central American (Panama; Castilla; Carthagena; Guayaquil) Rubber
Cecropia peltata (Mor) → Caucho Rubber
Cryptostegia grandiflora, C. madagascariensis (Asclepiad.) → Palay Rubber
Cynanchum ovalifolium (Asclepiad.) → Penang Rubber
Dyera costulata Hooker (*Alstonia c.* Miquel*)* → Jelutong or Dead Borneo
Euphorbia intisy (Euphorbi.) (Madagascar)
Euphorbiaceae (Milkweed Fam.): several genera & spp.
Ficus elastica (Mor.) → (East) India (or Assam) Rubber
F. hispida
F. vogelii → form of African Rubber
Funtumia elastica (Apocyn.) → Lagos Silk Rubber (form of African R.)
Guayule (shrub): *v. Parthenium*
Hancornia speciosa (Apocyn.) → Bahîa or Mangaheira Rubber
Hevea brasiliensis (Euphorbi.) → Pará Rubber
H. guianensis → Hevea Rubber (also *H. Benthamiana, H. discolor, H. pauciflora*)
Kickxia africana (Apocyn.) → Kickxia Rubber (form of African Rubber)
K. elastica & other spp. → Silk (or Lagos) Rubber
Landolphia florida, L. heudelotii (wAf), *L. kirkii* (eAf), *L. madagascariensis* (Madagascar Rubber), *L. owariensis* (wAf) → Landolphia Rubber
L. thollonii → Root Rubber
Leuconotis anceps, L. eugenifolia, L. gigantea, L. griffithii, L. subavenis, & other spp. (Apocyn.)
Manihot glaziovii (Euphorbi.) → Ceará or Pernambuco Rubber
Micrandra minor, M. siphonoides (Euphorbi.) → Pará Rubber
Mimusops balata (Manilkara bidentata) (Sapot.) → Balata (Rubber)
Moraceae (Mulberry Fam.): many genera & spp.
Ochrosia spp. (Apocyn.)
Palaquium gutta (Sapot.) → Gutta Percha
Parthenium argentatum (Comp.) → Guayule Rubber
Periploca canescens (Asclepiad.) → Black Rubber
P. gracea
Sapium verum & other spp. (Euphorbi.) → Rubbers of Columbia & Bolivia
Scorzonera kirghisorum (Comp.) → Russian Rubber
Siphocampylus caoutchouc, S. giganteus, &c. (campanul.)
Solidago leavenworthii, S. stricta (Comp.) → Goldenrod Rubber
Urceola spp., such as *U. elastica, U. esculenta* (Apocyn.) → Penang (or Borneo) Rubber
Willughbeia edulis, W. firma (Apocyn.) → Getah Borneo (or Chittagong) Rubber